The Elements

Name	Symbol	Atomic Number	Atomic Mass*
Actinium	Ac	89	(227)
Aluminum	Al	13	26.98
Americium	Am	95	(243)
Antimony	Sb	51	121.8
Argon	Ar	18	39.95
Arsenic	As	33	74.92
Astatine	At	85	(210)
Barium	Ba	56	137.3
Berkelium	Bk	97	(247)
Beryllium	Be	4	9.012
Bismuth	Bi	83	209.0
Bohrium	Bh	107	(267)
Boron	B	5	10.81
Bromine	Br	35	79.90
Cadmium	Cd	48	112.4
Calcium	Ca	20	40.08
Californium	Cf	98	(249)
Carbon	C	6	12.01
Cerium	Ce	58	140.1
Cesium	Cs	55	132.9
Chlorine	Cl	17	35.45
Chromium	Cr	24	52.00
Cobalt	Co	27	58.93
Copernicium	Cn	112	(285)
Copper	Cu	29	63.55
Curium	Cm	96	(247)
Darmstadtium	Ds	110	(281)
Dubnium	Db	105	(262)
Dysprosium	Dy	66	162.5
Einsteinium	Es	99	(254)
Erbium	Er	68	167.3
Europium	Eu	63	152.0
Fermium	Fm	100	(253)
Flerovium	Fl	114	(289)
Fluorine	F	9	19.00
Francium	Fr	87	(223)
Gadolinium	Gd	64	157.3
Gallium	Ga	31	69.72
Germanium	Ge	32	72.61
Gold	Au	79	197.0
Hafnium	Hf	72	178.5
Hassium	Hs	108	(277)
Helium	He	2	4.003
Holmium	Ho	67	164.9
Hydrogen	H	1	1.008
Indium	In	49	114.8
Iodine	I	53	126.9
Iridium	Ir	77	192.2
Iron	Fe	26	55.85
Krypton	Kr	36	83.80
Lanthanum	La	57	138.9
Lawrencium	Lr	103	(257)
Lead	Pb	82	207.2
Lithium	Li	3	6.941
Livermorium	Lv	116	(292)
Lutetium	Lu	71	175.0
Magnesium	Mg	12	24.31
Manganese	Mn	25	54.94
Meitnerium	Mt	109	(268)
Mendelevium	Md	101	(256)
Mercury	Hg	80	200.6
Molybdenum	Mo	42	95.94
Neodymium	Nd	60	144.2
Neon	Ne	10	20.18
Neptunium	Np	93	(244)
Nickel	Ni	28	58.70
Niobium	Nb	41	92.91
Nitrogen	N	7	14.01
Nobelium	No	102	(253)
Osmium	Os	76	190.2
Oxygen	O	8	16.00
Palladium	Pd	46	106.4
Phosphorus	P	15	30.97
Platinum	Pt	78	195.1
Plutonium	Pu	94	(242)
Polonium	Po	84	(209)
Potassium	K	19	39.10
Praseodymium	Pr	59	140.9
Promethium	Pm	61	(145)
Protactinium	Pa	91	(231)
Radium	Ra	88	(226)
Radon	Rn	86	(222)
Rhenium	Re	75	186.2
Rhodium	Rh	45	102.9
Roentgenium	Rg	111	(272)
Rubidium	Rb	37	85.47
Ruthenium	Ru	44	101.1
Rutherfordium	Rf	104	(263)
Samarium	Sm	62	150.4
Scandium	Sc	21	44.96
Seaborgium	Sg	106	(266)
Selenium	Se	34	78.96
Silicon	Si	14	28.09
Silver	Ag	47	107.9
Sodium	Na	11	22.99
Strontium	Sr	38	87.62
Sulfur	S	16	32.07
Tantalum	Ta	73	180.9
Technetium	Tc	43	(98)
Tellurium	Te	52	127.6
Terbium	Tb	65	158.9
Thallium	Tl	[illegible]	204.4
Thorium	Th		232.0
Thulium	Tm		168.9
Tin	Sn		118.7
Titanium	Ti		47.88
Tungsten	W		183.9
Uranium	U		238.0
Vanadium	V		50.94
Xenon	Xe		131.3
Ytterbium	Yb		173.0
Yttrium	Y		88.91
Zinc	Zn		65.41
Zirconium	Zr		91.22
			(284)
		115	(288)
		117	
		118	(294)

*All atomic masses are given to four significant figures. Values in parentheses represent the mass number of the most stable isotope.

**The names and symbols for elements 113, 115, 117, and 118 are pending verification by IUPAC.

SECOND CANADIAN EDITION

CHEMISTRY

The Molecular Nature of Matter and Change

Martin S. Silberberg

Patricia Amateis
Virginia Tech

Sophie Lavieri
Simon Fraser University

Rashmi Venkateswaran
University of Ottawa

Mc Graw Hill Education

CHEMISTRY: THE MOLECULAR NATURE OF MATTER AND CHANGE
SECOND CANADIAN EDITION

ISBN-13: 978-1-25-908711-0
ISBN-10: 1-25-908711-5

2 3 4 5 6 7 8 9 TCP 22 21 20 19 18

Printed and bound in Canada.

Director of Product Management: *Rhondda McNabb*
Product Manager: *Kevin O'Hearn*
Executive Marketing Manager: *Joy Armitage Taylor*
Product Developer: *Erin Catto*
Senior Product Team Associate: *Stephanie Giles*
Supervising Editor: *Jessica Barnoski*
Photo/Permissions Editor: *Monika Schurmann*
Copy Editor: *Julia Cochrane*
Plant Production Coordinator: *Scott Morrison*
Manufacturing Production Coordinator: *Emily Hickey*
Cover Design: *Jodie Bernard*
Cover Image: *Dejan Patic/Getty Images*
Interior Design: *Jodie Bernard*
Page Layout: *Aptara®, Inc.*
Printer: *Transcontinental Printing Group*

To my dearest: my husband, my kids, my friends, my students.

Sophie Lavieri

I would like to dedicate this book to my family, namely, my daughter, Nikhila; my son, Amaresh; my husband, Hari; my parents, Mrs. Indira and Dr. A Venkateswaran; and my parents-in-law, Mrs. Padmavathy and the late Mr. V. Parameswaran, with love.

Rashmi Venkateswaran

ABOUT THE AUTHORS

Martin S. Silberberg received a B.S. in Chemistry from the City University of New York and a Ph.D. in Chemistry from the University of Oklahoma. He then accepted a position as research associate in analytical biochemistry at the Albert Einstein College of Medicine in New York City, where he developed methods to study neurotransmitter metabolism in Parkinson's disease and other neurological disorders. Following six years in neurochemical research, Dr. Silberberg joined the faculty of Bard College at Simon's Rock, a liberal arts college known for its excellence in teaching small classes of highly motivated students. As head of the Natural Sciences Major and Director of Premedical Studies, he taught courses in general chemistry, organic chemistry, biochemistry, and liberal-arts chemistry. The small class size and close student contact afforded him insights into how students learn chemistry, where they have difficulties, and what strategies can help them succeed. Dr. Silberberg decided to apply these insights in a broader context and established a textbook writing, editing, and consulting company. Before writing his own texts, he worked as a consulting and development editor on chemistry, biochemistry, and physics texts for several major college publishers. He resides with his wife Ruth in the Pioneer Valley near Amherst, Massachusetts, where he enjoys the rich cultural and academic life of the area and relaxes by cooking, gardening, and singing.

Patricia G. Amateis graduated with a B.S. in Chemistry Education from Concord University in West Virginia and a Ph.D. in Analytical Chemistry from Virginia Tech. She has been on the faculty of the Chemistry Department at Virginia Tech for 28 years, teaching General Chemistry and Analytical Chemistry. For the past 13 years, she has served as Director of General Chemistry, responsible for the oversight of both the lecture and lab portions of the large General Chemistry program. She has taught thousands of students during her career and has been awarded the University Sporn Award for Introductory Teaching, the Alumni Teaching Award, and the William E. Wine Award for a history of university teaching excellence. She and her husband live in Blacksburg, Virginia, and are the parents of three adult children. In her free time, she enjoys biking, hiking, competing in the occasional sprint triathlon, and playing the double second in Panjammers, Blacksburg's steel drum band.

Sophie Lavieri studied in Venezuela and received a B.Sc. in Chemical Engineering from the Universidad Metropolitana (UM), a Pharmacy degree from the Universidad Central de Venezuela (UCV), an M.Sc. in Biology from the Instituto Venezolano de Investigaciones Científicas (IVIC), and a Ph.D. in Medicinal Chemistry from UCV. She worked for one year in industry doing research and coordinating a quality control laboratory and polymer production. She then accepted a position as Assistant Professor at UM, where she developed audiovisual resources for the general chemistry laboratories. After six years, she joined the Faculty of Pharmacy at UCV, where she did research in molecular modelling for compounds with possible anti-tumour and/or anti-HIV activity. During this time, she was the National Coordinator of Advanced Training in Chemistry for the Venezuelan Chemistry Teachers Association. In 1997, Dr. Lavieri joined the Department of Chemistry and the McKnight Brain Institute at the University of Florida as a Visiting Professor, where she synthesized and tested novel compounds as potential therapeutics against Alzheimer's disease. In 2001, she came to Simon Fraser University (SFU), and she is now a Senior Lecturer, teaching and developing teaching aids for first- and second-year undergraduate students in general and organic chemistry. She maintains close contact with the students and emphasizes interactive lessons. These protocols have afforded her insight into how to optimize the student learning process and engender more successful teaching outcomes. For this work Dr. Lavieri won Excellence in Teaching Awards at both the Faculty of Science and the university levels. Dr. Lavieri developed the EC4U (Experimental Chemistry for Us) program, which involved visiting hundreds of classrooms all over Canada, to bring science alive for K–12 students. She also founded the SFU Science in Action (SIA) outreach program, which offers a day of science immersion to thousands of grade school children each year. For this work and her lifetime of outreach, she won the 2009 YWCA Woman of Distinction Award in Education, Training, and Development and the 2014 Waldo Briño Inspirational Latin Award, and was named one of Canada's 10 Most Influential Hispanic Canadians in 2014. Dr. Lavieri decided to use her extensive teaching experience to contribute to the writing of laboratory manuals and textbooks for first-year undergraduate students. When not teaching, Dr. Lavieri enjoys travelling and singing with two different choirs that have performed live all over British Columbia and have been featured on both radio and television. She resides with her husband, Luis, in North Vancouver, British Columbia, from where she enjoys keeping in touch with her two grown children, Mariel and Luis Felipe.

Rashmi Venkateswaran received her B.Sc. (Hon) in Chemistry from Carleton University in Ottawa and her Ph.D. in Chemistry from the University of Ottawa. She taught at a progressive magnet school in Charleston, South Carolina, and also taught chemistry and physics at Limestone College in South Carolina before returning to Canada, where she was invited to take a position as Senior Instructor and Undergraduate Laboratory Coordinator at the University of Ottawa. In this position, she was encouraged to develop new teaching methods and use emerging technology to improve student learning at the first-year level in large class sections of 420 students. She also redesigned and renewed the first-year laboratory component of the course. Currently, she is focusing on applying novel pedagogical approaches to the teaching of smaller classes for students in first year with little or no previous chemistry background. The smaller groups of 100 to 150 students have allowed her to work more closely with the students and teach in a more personal and collaborative environment. In 2012, Dr. Venkateswaran was nominated by students for the 2012 Capital Educators' Award. Prior to writing this text, Dr. Venkateswaran has been a reviewer of many textbooks and journals and has been active in publishing chemistry articles for the general public. She has also written chemistry content for other textbooks. Dr. Venkateswaran is a strong proponent of outreach programs in science for the general public, with a view to reviving interest in science and mathematics in general and chemistry in particular. She performs shows for the public and at schools called "The Magic of Chemistry" in which she uses chemical demonstrations to illustrate real-life chemistry in a way that encourages interest in both youth and adults. Dr. Venkateswaran also has a B.A. and an M.A. in South Indian Classical Music (Vocal), which she obtained in India after completing her B.Sc. and before beginning her Ph.D. She is a professional performing artist and has given concerts across North America and in India. She relaxes by singing and cooking with her husband, Hari; her son, Amaresh; and her daughter, Nikhila.

BRIEF CONTENTS

DETAILED CONTENTS

CHAPTER 1 Keys to the Study of Chemistry 2

CHAPTER 2 The Components of Matter 36

CHAPTER 3 Stoichiometry and Chemical Equations 84

CHAPTER 4 Gases and the Kinetic-Molecular Theory 148

CHAPTER 5 Thermochemistry: Energy Flow and Chemical Change 198

CHAPTER 9 The Shapes of Molecules 354

CHAPTER 10 Theories of Covalent Bonding 374

CHAPTER 11 Intermolecular Forces: Liquids, Solids, and Phase Changes 402

CHAPTER 12 The Properties of Mixtures: Solutions and Colloids 458

CHAPTER 13 Periodic Patterns in the Main-Group Elements 504

CHAPTER 14 Kinetics: Rates and Mechanisms of Chemical Reactions 554

CHAPTER 15 Equilibrium: The Extent of Chemical Reactions 610

CHAPTER 16 Acid-Base Equilibria 658

CHAPTER 19 Electrochemistry: Chemical Change and Electrical Work 806

CHAPTER 20 Organic Compounds and the Atomic Properties of Carbon 872

CHAPTER 21 Organic Reaction Mechanisms 932

CHAPTER 22 Special Topics in Organic Chemistry 986

CHAPTER 23 The Elements in Nature and Industry 1030

List of Sample Problems

(Molecular-scene problems are shown in colour.)

Chapter 1

Chapter 2

Chapter 3

Chapter 4

Chapter 5

Chapter 6

Chapter 7

Chapter 8

Chapter 9

Chapter 10

Chapter 11

Chapter 12

Chapter 14

Chapter 15

Chapter 16

Chapter 17

Chapter 18

Chapter 19

Chapter 20

Chapter 21

Chapter 24

Chapter 25

PREFACE

At the core of natural science, chemistry is so crucial to an understanding of medicine, health sciences, molecular biology, genetics, pharmacology, ecology, atmospheric science, engineering, nuclear studies, materials science, politics, law, and many other fields that it has become a central requirement for an increasing number of academic majors. Furthermore, chemical principles are at the core of many key societal issues, including climate change, energy options, materials recycling, diet and nutrition, and medicine and disease. Clearly, the study of chemistry plays an essential role in our world.

What Sets This Book Apart

For seven editions, *Chemistry: The Molecular Nature of Matter and Change* has been recognized for setting the standard among general chemistry textbooks. The second Canadian edition continues to maintain that unparalleled reputation of keeping pace with the evolution of student learning while adapting for the needs of Canadian undergraduate students and instructors. The text still contains the most accurate macroscopic-to-molecular illustrations, consistent step-by-step worked problems, and an extensive collection of end-of-chapter problems, with a wide range of difficulties and applications targeting student interests in engineering, medicine, materials science, and environmental studies. Three hallmarks make this text a market leader.

Visualizing Chemical Models—Macroscopic to Molecular

Chemistry deals with observable changes caused by unobservable atomic-scale events, requiring an understanding of a size gap of mind-boggling proportions. One of the text's goals coincides with that of so many instructors: to help students visualize chemical events on the molecular scale. Thus, concepts are explained first at the macroscopic level and then from a molecular point of view, with the text's ground-breaking illustrations always placed next to the discussion to bring the point home for today's visually oriented students.

Thinking Logically to Solve Problems

The problem-solving approach, based on the four-step method widely accepted by experts in chemical education, is introduced in Chapter 1 and employed consistently throughout the text. It encourages students to plan a logical approach to a problem, and only then proceed to solve it. Each sample problem includes a check, which fosters the habit of taking a breath to assess the reasonableness and magnitude of the answer. Finally, for practice and reinforcement, each sample problem is followed immediately by a similar follow-up problem, for which an abbreviated solution, not merely a numerical answer, is given at the end of the chapter. In this edition, solving problems and visualizing models have been married through an increased number of molecular-scene problems in both worked examples and homework sets.

Applying Ideas to the Real World

An understanding of modern chemistry influences attitudes about climate change and health care, while also explaining the spring in a running shoe and the display of a laptop screen. Today's students may enter one of the emerging chemistry-related hybrid fields—biomaterials science or planetary geochemistry, for example—and their text should point out the relevance of chemical concepts to such career directions. The *Chemical Connections* and *Tools of the Laboratory* boxed essays (which include problems to reinforce learning), the fewer but more pedagogical margin notes, and the many applications woven into the chapter content are up-to-date, student-friendly features that complement the text content.

Refining the Standard: An Evolving Learning System

With this second Canadian edition, the authors, together with key members of the editorial, sales, and marketing teams, consulted extensively with student and faculty users to deliver a text optimized for the needs of Canadian courses in general chemistry. From chapter reviews, focus groups, and one-on-one interviews with instructors, we were pleased to learn that everyone still loved the pioneering, accurate molecular art; the stepwise problem-solving approach; the abundant mix of qualitative, basic quantitative, and applied end-of-chapter problems; and the thorough, student-friendly coverage of mainstream topics. We were also gratified to see the positive market response to the key content changes and updates made to adapt this text for the needs of students and instructors in Canada.

While retaining the time-tested strengths of seven U.S. editions, we have moved toward perfecting the content and learning approach for Canadians in the following ways:

Expanded Material on Organic Chemistry. Due to the increasing interest of students in the fields of health sciences and medicine, and renewed interest in the applications of chemistry to the fields of biochemistry, biomedical science, and biopharmaceuticals, this edition features expanded content in the field of organic chemistry. Chapter 20 focuses on nomenclature and the different functional groups, while

Chapter 21 studies the major reaction types that occur in this field. Chapter 22 focuses on polymers and organic macromolecules as well as techniques of analysis in organic chemistry. The three chapters together provide a high level and in-depth examination of organic content. The glossary has also been expanded to include the additional terminology in these chapters.

Adherence to IUPAC Recommendations. The entire text is IUPAC consistent.

Consistent Use of Metric and SI Units. All reference to units other than metric and SI units has been removed from the book to ensure consistency with IUPAC recommendations.

Updated and Action-Oriented Learning Objectives. The learning objectives provided at the end of each chapter use action-oriented verbs to allow students to understand the learning objective and to assist instructors in clearly identifying the learning objectives and skills required in each chapter. New in this edition, rather than a summary of the key points of each chapter at the beginning, students are given a list of goals they should be able to achieve by the end of the chapter.

Chapter Openers and Photos. The chapter openers highlight chemistry that occurs in our daily lives, posing questions relevant to students and then providing answers in the chapter. The opener is then followed by a bulleted list of the key points that will be taught in the chapter, expressed as action-oriented learning goals.

Enhanced Canadian Content. This edition of the book highlights Canadian research and researchers in the appropriate chapters. Many chapters contain vignettes about Canadians who have contributed significantly to chemical history. There are numerous references to Canadian geographical, environmental, and health issues as well as common Canadian icons. In a few cases, the authors have directly contacted Canadian researchers and asked them to provide first-hand information regarding their own work (the work has then been directly included in the text). The second edition contains even more references to Canadian research and researchers.

Consistent Emphasis on Quantity Rather Than Unit. The authors have consistently guided students to understand the difference between measured quantities and the units used to measure these quantities by referring to them separately. In sample, worked, follow-up, and end-of-chapter problems, students are asked to find quantities rather than units (e.g., "What volume of HCl(*aq*)...?" rather than "How many millilitres of HCl(*aq*)...?").

Updated Chemistry. The most recent periodic table has been included, with the recently IUPAC-approved names for elements 114 and 116, as well as the recent discovery of element 117. The chapter on materials chemistry includes references to new and innovative materials, and the chapter on electrochemistry shows new applications of solar and fuel cells.

Referencing. In an effort to teach by example, the authors have included references where applicable to new material. As we expect students to reference cited work, the authors have also done so.

End-of-Chapter Problems. Where possible, additional end-of-chapter problems of the integrated and comprehensive type have been added to help students see connections between material that is connected but placed in different chapters.

Content Changes to Individual Chapters

Chapter 1:
- Redefines states of matter in terms of energy
- Introduces the plasma and Bose-Einstein condensate states
- Contains new figures

Chapter 2:
- Contains a historical update on the Milliken experiment, which ties back to the scientific method
- Introduces the Higgs boson in the context of other subatomic particles
- Emphasizes the different ways of visualizing molecules
- Contains a new chart on naming acids of oxoanions
- Introduces extraction under separation techniques

Chapter 3:
- Emphasizes limiting reactants and the effect on reactants and products as well
- Introduces symbols wherever possible to encourage students to identify quantities using the standard symbols for those quantities
- Clarifies terms such as *concentration* and *precipitate*

Chapter 4:
- Express atmospheric pressure with more consistency using SI units
- Explores methods other than Van der Waal's for dealing with non-ideal gases
- Clarifies how to move from proportion to equality in an empirical relationship
- Clarifies terms in the root mean square expression
- Includes more forward referencing

Chapter 5:
- References to gas mileage changed to fuel efficiency
- Mentions microcalorimetry and ice calorimetry
- Clarifies open, closed, and isolated systems
- Includes new problems incorporating the heat capacity of the calorimeter
- Clearly specifies enthalpies of neutralization and dissolution in context
- Clarifies the reference standard in enthalpy

Chapter 6:
- Features new sidebars on the scientists instrumental in developing the quantum model of the atom
- Clarifies key terms in the chapter

Chapter 7:
- Arrows in orbital diagrams representing electrons drawn with a single head to be consistent with the arrows representing electron movement in organic chapters
- Clarifies the concept of diamagnetism

Chapter 8:
- Notes that a 100% ionic bond doesn't exist
- Refers to the restricted use of the octet rule
- Relates the strength of the bond to the relative reactivity of compounds, giving an example with explosives
- Adds information about dative bonds
- Includes S═N—A for Lewis diagrams
- Changed Figure 8.11, presenting a single large crystal, instead of loose salt
- Added an example of Lewis structures for more complex molecules
- Another sample problem about predicting relative lattice energy from ionic properties added
- Changed to a fish hook arrow for *one*-electron movements and a double hook arrow for *two*-electron movements
- Corrected the representations of N_2 in Sample Problem 8.8b
- Indicates that bond energies work because H is a state function

Chapter 9:
- Highlights some unique coordination, such as the two coordinate C in a carbine
- Corrected Figure 9.9
- Removed references for write-ups
- Includes the "1" when there is one lone pair in the VSEPR notation (AX_mE_n)
- Keeps the direction of the dipole moment from the negative to the positive charge (the "head" of the arrow should be at the electron-deficient region), and eliminated the cross from the arrows' tails
- Changed the backgrounds in Figure 9.10

Chapter 10:
- Addresses the fact that the orbitals in methane are not all equivalent and are not sp^3 orbitals, which can only be explained via molecular orbital (MO) theory
- Addresses the fact that MOs deal with covalent bonds; therefore the bond order (Equation 10.1) only deals with the covalent portion of the bond
- Highlights the hybridization of external atoms
- Changed the perspective of the lobes in Figure 10.11
- Corrected Figures 10.22 and 10.27
- Writes formulas using words and symbols when appropriate
- Changed atomic orbitals to the same colour

Chapter 11:
- Added a picture of ice-fishing to show surface ice of lakes
- Added a Chemical Connections to Nanotechnology in Canada with examples
- Changed Figure 11.1
- Corrected Figure 11.5
- Changed Dr. Burk's photo (Figure 11.12)
- Includes "Qualitative" in "Quantitative aspects of Phase Changes"
- Improved Figure 11.14A

Chapter 12:
- Changed the example of the Nile Delta (Figure 12.24) to a Canadian example (the difference in the water colour when the Fraser River (in British Columbia) enters the ocean)
- Added biological examples and pictures of osmosis
- Removed the word "ionic" from the title of Table 12.3
- Added photos of William Henry and François-Marie Raoult

Chapter 13:
- Updated the caption in Figure 13.16
- Mentions carbenes in the organic section
- Added Dr. Michael D. Fryzuk's work on the activation of N_2
- Added an image of F_2
- Changed the photochemical smog picture
- Referenced in the outline that the main-group elements are also referred to as representative elements
- Added the uses of S
- References Chapter 19 when mentioning that alkali metals are powerful reducing agent
- References Chapter 24 when mentioning phosphate as a nonrenewable resource

Chapter 14:
- Includes information about and contributions made by Henry Eyring
- Includes additional referencing to kinetics of radioactive decay (first-order)
- Includes updated information about pharmacokinetics
- Expands and clarifies the section on pseudo-first order kinetics
- Clarifies the different rate laws (instantaneous, integrated, etc.)
- Replaces partial derivatives with simple derivatives for ease of student understanding
- Includes derivation of the Michaelis-Menten equation and information on both contributors

Chapter 15:
- Clarifies how the reaction is affected when Q is greater than, less than, or equal to K
- K always used to identify the thermodynamic equilibrium constant, with other K values being specifically tagged
- Includes examples where students have to solve a quadratic to arrive at the solution

Chapter 16:
- Emphasizes the importance of the contribution of H^+ ions by water in the presence of a very weak acid
- Reaction tables now consistently called ICE tables for easy student recognition
- Clarifies relative acid/base strengths
- pH scale adjusted to show that it is possible to have pH values greater than 14 and less than 0

Chapter 17:
- Includes additional clarification on proper titration techniques

Chapter 18:
- Includes additional photos to demonstrate spontaneous processes
- Clarifies visual representation of the microstates

Chapter 19:
- Includes additional examples where potential is used to determine *K* values
- Includes additional mnemonic devices to help students remember terminology
- Includes updated information on new work in battery research

Chapter 20:
- Addresses both ways of expressing degrees of unsaturation (DU and IHD)
- Added examples showing functional groups
- Encourages the use of models to visualize 2-D to 3-D
- Addresses the use of oxygen as an example of heteroatom in Figure 20.2
- Changed units in Figure 20.9 to kJ
- Ensures students know that the skeleton is not the final step when drawing an organic molecule
- Added that the 1,3-diaxial interactions in Figure 20.15 are really gauche butane interactions
- Corrected the name of TNT
- Improved the perspective in 2,2-dimethylpropane

Chapter 21:
- Refers to previous chapters when analyzing kinetics and thermodynamics variables
- Clarifies the nucleophilic attack on a carbocation, and the product formation by hydrogenation
- Corrected some mechanisms
- Added a still photo of an ether explosion in progress

Chapter 22:
- Deleted the reference to Dr. Vy M. Dong and added Dr. Tomas Hudlicky's work in the Chemical Connections to Canadian Research in Organic Chemistry
- Corrected Figure 22.35

Chapter 23:
- Highlights the fact that the world's supply of readily available ore containing phosphorus has already peaked
- Clarifies that phosphate is actually removed by surface complexation

Chapter 24:
- Includes information on transition metals in biological systems
- Relativistic effects added to explain lanthanide contraction
- Updated information on application of MO theory in explaining metal ligand complexes with diagrams

Chapter 25:
- Highlights that General Fusion in Burnaby, British Columbia, is working on developing a fusion generator
- Deleted the list of participating universities in TRIUMF

Flexibility in Topic and Chapter Presentation

The second Canadian edition of *Chemistry: The Molecular Nature of Matter and Change* has been revised to keep the mainstream topic sequence working optimally for teacher and student. But every course is unique, so flexibility has been built in: many section and subsection breaks allow topics to be rearranged with minimal loss of continuity. Likewise, several chapters can be taught in different orders. For instance, gases (Chapter 4) can be covered in sequence to explore the mathematical modelling of physical behaviour or, with no loss of continuity, just before liquids and solids (Chapter 11), to show the effects of intermolecular forces on the three common states of matter. In fact, feedback has indicated that many instructors move chapters, sections, and topics around, for example, covering descriptive chemistry (Chapter 13) and organic chemistry (Chapters 20–22) in the more traditional placement at the end of the course. Because the topic sequence is so flexible, any instructor can feel comfortable making such changes to suit his or her course.

Keeping Pace with the Evolution of Student Learning

The goal of Connect is to usher in a new era of meaningful online learning that balances the conceptual and quantitative aspects of this most vital discipline. Within Connect, each problem retains Silberberg's problem-solving methodology and is enhanced with specific hints, as well as feedback for common incorrect answers—all authored and accuracy-checked by teams of dedicated chemistry professors with many years of classroom experience.

In addition to the specific hints and feedback, many problems offer students the opportunity to engage in chemical drawing that can be assessed directly within their homework. To provide this important pedagogical experience, Connect has partnered with PerkinElmer's ChemDraw, widely considered the "gold standard" of scientific drawing programs and *the* cornerstone application for drawing and annotating molecules, reactions, and pathways. This collaboration of Connect and ChemDraw features an easy-to-use, intuitive, and comprehensive course management and homework system with professional-grade drawing capabilities.

The Silberberg Learning System

Many pedagogical tools are woven throughout the chapters to guide students on their learning journey.

Chapter Openers

Each chapter introduces familiar applications of chemistry in daily life, and a series of photos depicting common applications, related to the main topic of the chapter. This is followed by a bulleted list of main topics and a chapter outline, which shows the sequence of topics and subtopics in the chapter. In the sidebar panel, *Concepts and Skills to Review* lists key material from earlier chapters that students should understand before starting to read the current chapter.

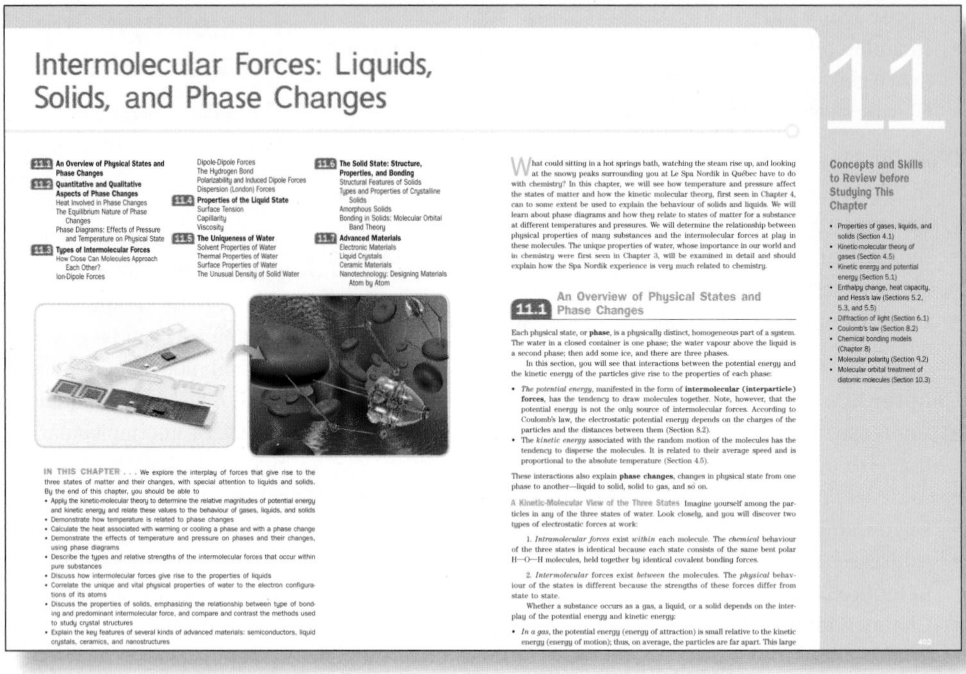

Problem Solving

A worked-out *Sample Problem* appears whenever an important new concept or skill is introduced. The problem-solving step helps students think through chemistry problems logically and systematically. The universally accepted four-step approach of plan, solve, check, and practise is used consistently for every sample problem in the text. The steps are as follows:

- **Plan** analyzes the problem so that students can use what is known to find what is unknown. This step develops the habit of thinking through the solution before performing calculations. Most quantitative problems are accompanied in the margin by a *Road Map*, a block diagram that is specific to the problem and leads students visually through the planned steps.
- **Solution** presents the calculation steps *in the same order* as they appear in the plan and in the road map.
- **Check** fosters the habit of going over one's work with a rough calculation to make sure the answer is both chemically and mathematically reasonable—a great way to avoid careless errors. Where appropriate, after the check, a comment may be added if the solution involves the making of an assumption or if the solution is valid under a particular set of conditions.
- **Follow-Up Problem** presents a similar problem to provide immediate practice, with an abbreviated multistep solution appearing at the end of the chapter. Where appropriate in the first several chapters, students are asked to draw their own road map to solve the follow-up problem.

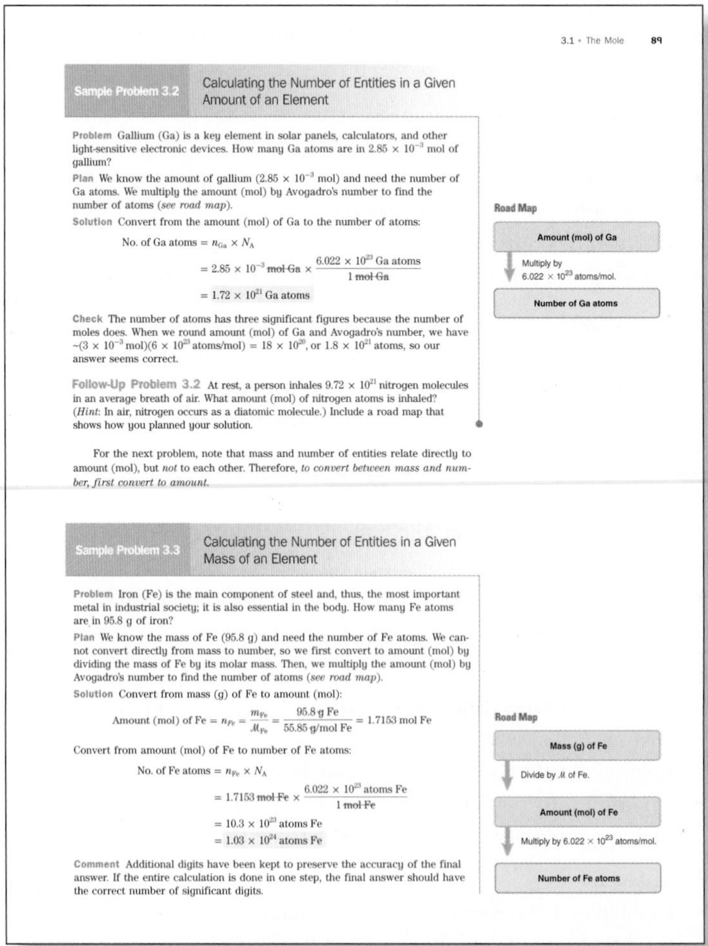

Three-Level Illustrations

As the art that set the standard for chemistry textbooks, these illustrations connect the macroscopic and molecular levels of reality with the symbolic level in the form of a chemical equation.

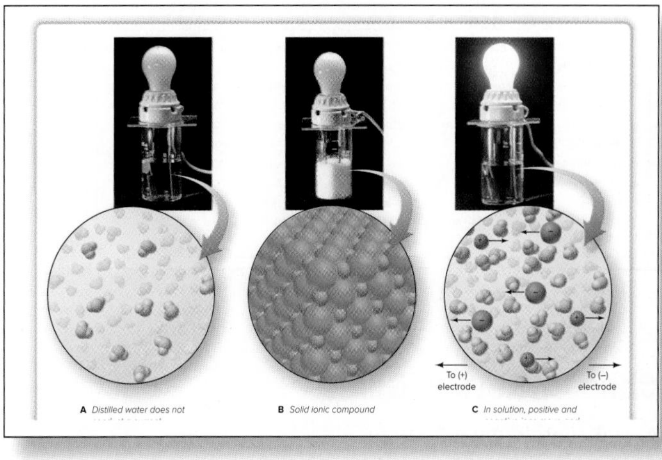

Constructing a Conversion Factor A **conversion factor** is a *ratio that is used to express a quantity in different units.* Suppose that we want to know the distance of the 150-km car trip in metres. To convert kilometres to metres, we use *equivalent quantities*:

$$1 \text{ km} = 1000 \text{ m}$$

From this equation, we can construct two conversion factors. Dividing both sides by 1000 m gives one conversion factor (shown in blue):

$$\frac{1 \text{ km}}{1000 \text{ m}} = \frac{1000 \text{ m}}{1000 \text{ m}} = 1$$

Dividing both sides by 1 km gives the other conversion factor (the inverse):

$$\frac{1 \text{ km}}{1 \text{ km}} = \frac{1000 \text{ m}}{1 \text{ km}} = 1$$

Since the numerator and denominator of a conversion factor are equal, multiplying a quantity by a conversion factor is the same as multiplying by 1. Thus, *even though the number and unit change, the size of the quantity remains the same.*

To convert the distance from kilometres to metres, we choose the conversion factor with kilometres in the denominator, because it cancels kilometres and gives the answer in metres:

$$\text{Distance (m)} = 150 \cancel{\text{km}} \times \frac{1000 \text{ m}}{1 \cancel{\text{km}}} = 150\,000 \text{ m}$$

$$\text{km} \Rightarrow \text{m}$$

Optimized Presentation

Text paragraphs are concise, with presentation of content optimized through the use of subheads, numbered paragraphs, and lists. Main ideas are delineated, resulting in a more student-friendly study format, and key calculations are inset for clarity and comprehension.

Annotated Figures

Modern, explanatory figures describe chemical processes through instructional labelling and realistic, three-dimensional art.

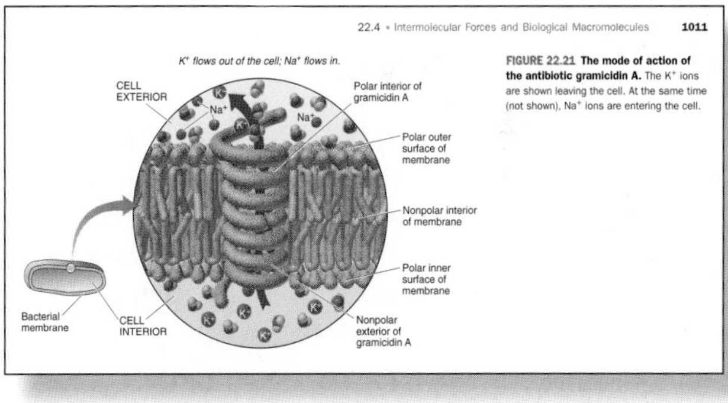

Applications

Chemical Connections essays show the interdisciplinary nature of chemistry by applying chemical principles directly to related scientific fields, including physiology, geology, biochemistry, engineering, and environmental science. *Tools of the Laboratory* essays describe the key instruments and techniques that chemists use in modern practice to obtain the data that underlie their theories. Both essay features now include several problems to enhance learning and relevance.

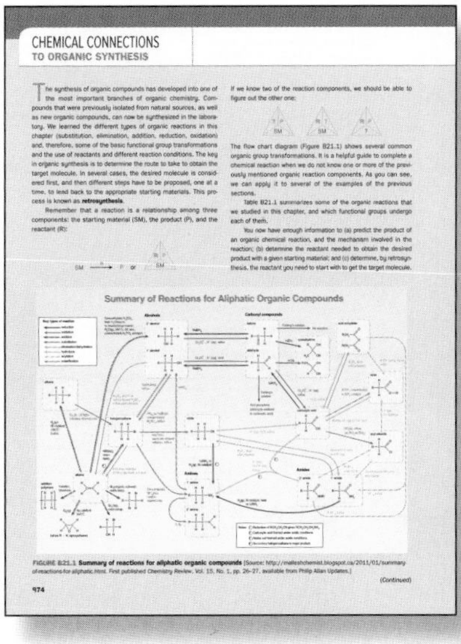

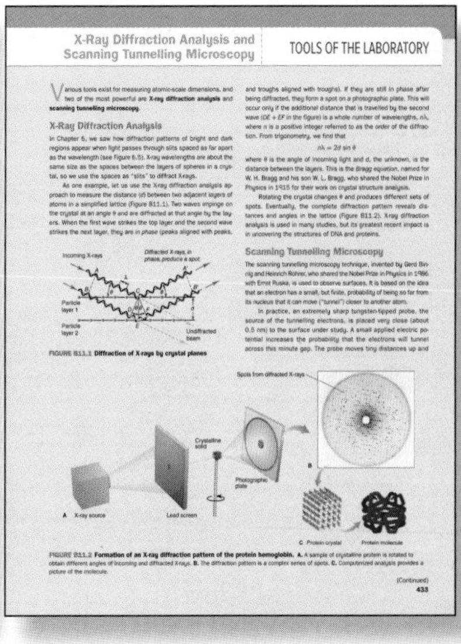

SUMMARY OF SECTION 5.6

- Standard states are a set of specific conditions used to determine thermodynamic variables for all substances.
- When 1 mol of a compound forms from its elements, with all the substances in their standard states, the enthalpy change is the standard enthalpy of formation, $\Delta_f H°$.
- Hess's law allows us to picture a reaction as the decomposition of the reactants to their elements, followed by the formation of the products from their elements.
- We can use tabulated $\Delta_f H°$ values to find $\Delta_r H°$, or we can use known $\Delta_r H°$ and $\Delta_f H°$ values to find an unknown $\Delta_f H°$.
- Because of major concerns about climate change, chemists are developing energy alternatives, including coal and biomass conversion, hydrogen fuel, and non-combustible energy sources.

Section Summaries

Concise, bulleted summary lists conclude each chapter section, restating the major ideas just covered.

Chapter Review Guide

A rich catalogue of study aids ends each chapter to help students review its content.

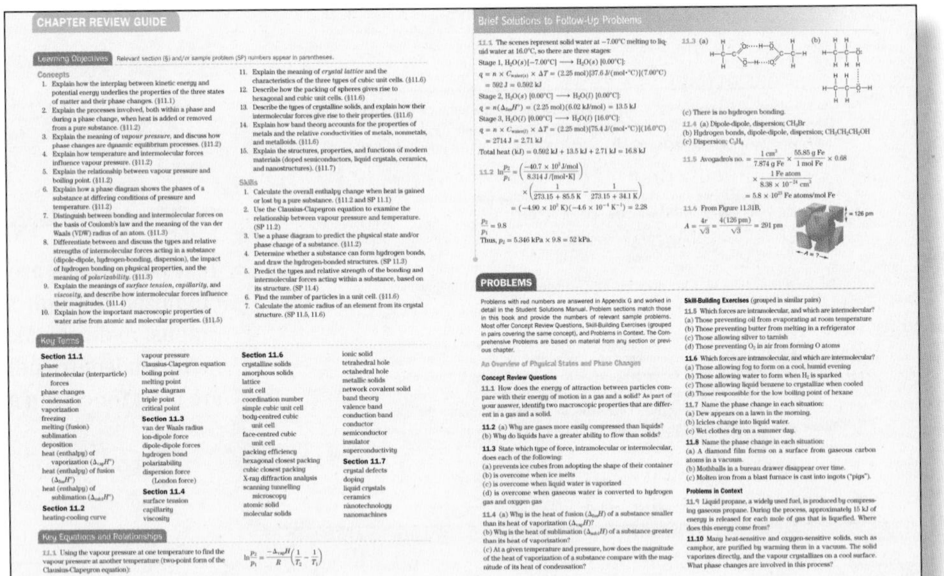

- **Learning Objectives**, referenced to section and/or sample problems, are action oriented and focus on key concepts and skills developed in the chapter.
- **Key Terms** are boldfaced and defined within the chapter and listed here by section, as well as being defined again in the *Glossary*.
- **Key Equations and Relationships** from within the chapter are numbered and listed here.
- **Brief Solutions to Follow-Up Problems** are provided to double the number of worked sample problems. These provide multistep calculations at the ends of the chapters, rather than just a numerical answer at the back of the book. Road maps are supplied for those follow-up problems that ask students to prepare one in planning their solution.

End-of-Chapter Problems

An exceptionally large number of qualitative, quantitative, and molecular-scene homework problems end each chapter. Three types of problems are keyed by chapter section, with comprehensive problems following:

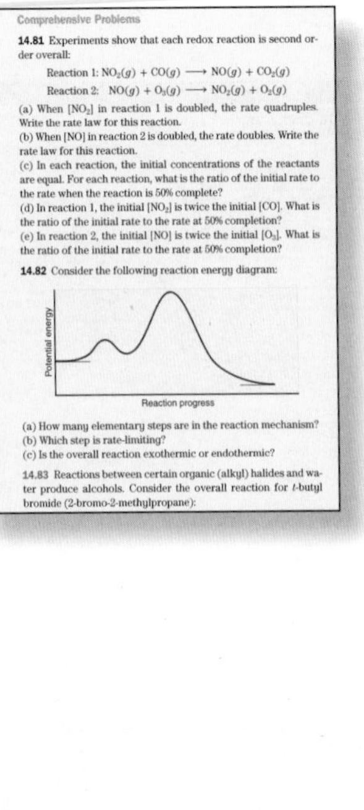

- **Concept Review Questions** test students' qualitative understanding of key ideas.
- **Skill-Building Exercises** are usually grouped in pairs that cover a similar idea, with one of each pair answered in the back of the book. These exercises begin with simple questions and increase in difficulty, gradually eliminating students' need for multistep directions.
- **Problems in Context** apply the skills learned in the skill-building exercises to interesting scenarios, including examples from industry, medicine, and the environment.
- **Comprehensive Problems**, most based on realistic applications, are more challenging and rely on concepts and skills from any section of the current chapter or from previous chapters.

MARKET-LEADING TECHNOLOGY

McGraw Hill Education connect®

Learn without Limits

McGraw-Hill Connect® is an award-winning digital teaching and learning platform that gives students the means to better connect with their coursework, with their instructors, and with the important concepts that they will need to know for success now and in the future. With Connect, instructors can take advantage of McGraw-Hill's trusted content to seamlessly deliver assignments, quizzes, and tests online. McGraw-Hill Connect is a learning platform that continuously adapts to each student, delivering precisely what they need, when they need it, so class time is more engaging and effective. Connect makes teaching and learning personal, easy, and proven.

Connect Key Features

SmartBook®

As the first and only adaptive reading experience, SmartBook is changing the way students read and learn. SmartBook creates a personalized reading experience by highlighting the most important concepts a student needs to learn at that moment. As a student engages with SmartBook, the reading experience continuously adapts by highlighting content based on what each student knows and doesn't know. This ensures that he or she is focused on the content needed to close specific knowledge gaps, while it simultaneously promotes long-term learning.

Connect Insight®

Connect Insight is Connect's new one-of-a-kind visual analytics dashboard—now available for instructors—that provides at-a-glance information regarding student performance, which is immediately actionable. By presenting assignment, assessment, and topical performance results together with a time metric that is easily visible for aggregate or individual results, Connect Insight gives the instructor the ability to take a just-in-time approach to teaching and learning that was never before available. Connect Insight presents data that helps instructors improve class performance in a way that is efficient and effective.

Simple Assignment Management

With Connect, creating assignments is easier than ever, so instructors can spend more time teaching and less time managing.

- Assign SmartBook learning modules.
- Edit existing questions and create your own questions.
- Draw from a variety of text-specific questions, resources, and test bank material to assign online.
- Streamline lesson planning, student progress reporting, and assignment grading to make classroom management more efficient than ever.

Smart Grading

When it comes to studying, time is precious. Connect helps students learn more efficiently by providing feedback and practice material when they need it, where they need it.

- Automatically score assignments, giving students immediate feedback on their work and comparisons with correct answers.
- Access and review each response; manually change grades or leave comments for students to review.
- Track individual student performance—by question, by assignment, or in relation to the class overall—with detailed grade reports.
- Reinforce classroom concepts with practice tests and instant quizzes.
- Integrate grade reports easily with Learning Management Systems, including Blackboard, D2L, and Moodle.

Instructor Library

The Connect Instructor Library is a repository for additional resources to improve student engagement in and out of the class. It provides all the critical resources instructors need to build their course.

- Access instructor resources.
- View assignments and resources created for past sections.
- Post your own resources for students to use.

ACKNOWLEDGEMENTS

This part of any book is always the most fun to write. It gives us the opportunity to thank all the people who had a hand in ensuring the book could be successfully completed but whose names will not appear on the cover.

I, Rashmi, would like to start by acknowledging the immense contribution made by my family. Life is always a challenge, more so for those having a full-time job and children, but the workload imposed by writing went above and beyond the normal. My 18-year-old son, Amaresh, was a helpful sounding board when I was not sure if what I was trying to express was being said clearly. My 8-year-old daughter, Nikhila, was always there to remind me that I had a family too! It was a lot of extra work for my husband, Hari, and without the cooperation of my family, there is no way this writing would ever have been completed. My parents were always there to help when we needed the extra hands, and, during the summer, my mother-in-law ably substituted for the extra hands without any hesitation at all. I would also like to thank my dear friend and colleague, Mr. Robert Nadon, for always being there to help me and my family and for teaching us all the value of a smile.

I, Sophie, would like to say an ENORMOUS thank you to Luis, my husband, for his love and support, to my son Luis Felipe, for the diagrams, to my daughter-in-law Rachel, for reading the chapters (it was wonderful to have someone who did not know anything about chemistry explaining the different concepts after reading the chapters!), and to my daughter Mariel and my friends for their unconditional support during the writing process.

We would like to thank the reviewers who offered their insight, ideas, and feedback. Reviewers for the second Canadian edition were the following individuals:

John Carran, *Queen's University*
John Paul Canal, *Simon Fraser University*
Philip J. Dutton, *University of Windsor*
Michael Hempstead, *York University*
Robert Hilts, *Grant MacEwan University*
Felix Lee, *University of Western Ontario*
Andrew McWilliams, *Ryerson University*
Vadoud Niri, *Brock University*
Scott Smith, *Wilfrid Laurier University*

We would like to thank Mr. Clement Kazakoff, Mr. Alexander (Sander) Mommers, Dr. Wendy Pell, Dr. Kevin J. Smith, Dr. Erika Plettner, Dr. David Vocadlo, Dr. Karyn Ho, Dr. Tomas Hudlicky, Dr. Soledade Pedras, Dr. Vladimir Kitaev, Dr. Howard Trottier, Ms. Jean Bennett, Dr. Travis Fridgen, Dr. Darrin Richeson, and Dr. Bianca J. van Lierop for their contributions of advice, content assistance, and chapter revision, without which many improvements to the text would have been impossible.

In today's educational milieu, no book can stand alone, and this text is no different. It is accompanied by a stunning array of complementary pedagogical tools, all designed to assist the instructor and the student. Each of these individual components was reviewed for consistency with the ideas expressed in the book, and we would like to thank Patrick Crewdson for his work on the Computerized Test Bank, the instructor PowerPoint Presentations, Connect, and LearnSmart, for performing the technical check, and overall for his helpful and constructive feedback.

Our thanks to our McGraw-Hill Education team: James Booty—Senior Product Manager, Erin Catto—Product Developer, Jessica Barnoski—Supervising Editor, and Julia Cochrane—Copy Editor. We are sure there were numerous other people who played key roles behind the scenes, and our heartfelt thanks go to them as well.

I, Rashmi, cannot complete this part of the book without acknowledging one other person, and that is my wonderful co-author Dr. Sophie Lavieri. She has been a joy to work with and it is truly my pleasure and privilege to have been able to work with her. I am proud to be able to call her my friend and I genuinely hope our friendship will continue to grow with the second edition of this book.

I, Sophie, want to say that it was awesome to have the opportunity to work together, and I look forward to doing so in many more projects in the future. I do not believe in coincidences. Thanks again, Rashmi, for being YOU and for being there, not only for the book, but for me!

Thank you to all of you who choose to read this book. We have tried to make chemistry come alive by making it relevant to our world and our lives today. The first Canadian edition was infused with our enthusiasm and love for chemistry, and our sincere hope is that, with the second edition, our passion for chemistry will have seeped through to you! Enjoy!

SECOND CANADIAN EDITION

CHEMISTRY

The Molecular Nature of Matter and Change

Keys to the Study of Chemistry

IN THIS CHAPTER . . . We discuss some central ideas about matter and energy, the process of science, units of measurement, and how scientists handle data. By the end of this chapter, you should be able to

- Define the fundamental concepts of matter and energy and the changes they undergo
- Discuss the origins of chemistry, including some major missteps
- Explain how scientists build models to study nature
- Determine different problem-solving methods in chemistry, including unit conversion and using modern unit systems for mass, length, volume, density, and temperature
- Use the concept of uncertainty in data collection to differentiate between accuracy and precision

1

Concepts and Skills to Review before Studying This Chapter

- Using exponential (scientific) notation (Appendix A)

Why is chemistry potentially the most essential, exciting, and engaging course you may ever take? In today's world, with its rapid advances, it is almost impossible to survive without a basic knowledge of chemistry and the tremendous impact it has on our day-to-day lives. Chemistry is essential to our daily hygiene, health, and happiness (imagine life without soap, pain relievers, or electricity); elementary to our transportation, telecommunications, and technology (imagine life without cars, cell phones, or computers); and vital to our sustenance, supplies, and society (imagine life without food, plastics, and energy). Chemists apply the fundamental principles of chemistry in a myriad of ways to improve all of the above in ways that are less wasteful and more efficient and that epitomize the principles of "reduce, reuse, and recycle." In this book, we hope to show you the wonders of the world of chemistry, all of which stem from the impossibly small but unimaginably powerful atom.

1.1 Some Fundamental Definitions

A good way to begin our exploration of chemistry is to define chemistry and a few central concepts. **Chemistry** is *the study of matter and its properties, the changes that matter undergoes, and the energy that is associated with those changes.*

The Properties of Matter

Matter is the "stuff" of the universe: air, glass, planets, students—*anything that has mass and volume.* (In Section 1.5, we discuss the meanings of *mass* and *volume* in terms of how they are measured.) Chemists want to know the **composition** of matter, *the types and amounts of simpler substances that make up matter.* A *substance* is a type of matter that has a defined, fixed composition.

We learn about matter by observing its **properties**, *the characteristics that give each substance its unique identity.* To identify a person, we might observe height, weight, hair and eye colour, fingerprints, and DNA patterns until we arrive at a unique identification. To identify a substance, we observe two types of properties, physical and chemical, which are closely related to two types of change that matter undergoes:

- **Physical properties** are characteristics that a substance shows *by itself, without changing into or interacting with another substance.* These properties include melting point, electrical conductivity, and density. A **physical change** occurs when a substance *alters its physical properties,* ***not*** *its composition.* For example, when ice melts, there are many changes in physical properties, such as hardness, density, and ability to flow. But the composition of the sample does *not* change: the sample is still water. The photograph in Figure 1.1A shows what a physical change looks like in everyday life. The "blow-up" circles depict a magnified view of the particles making up the sample. The particles lie in a repeating pattern in the icicle, whereas they are jumbled in the droplet. However, *the particles are the same* in both forms of water.

  Physical change (same substance before and after):

  water (solid form) ⟶ water (liquid form)

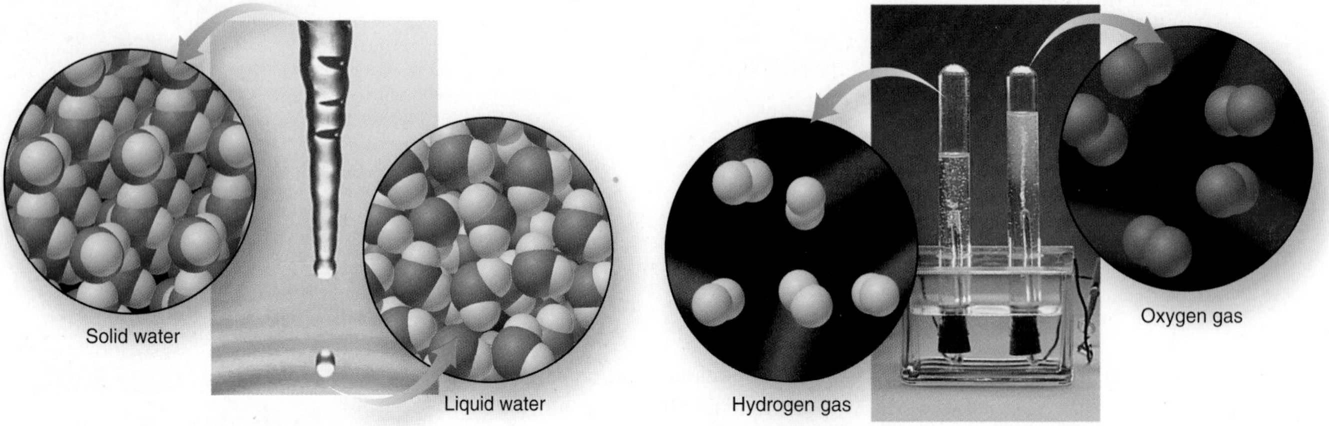

A Physical change:
Solid state of water becomes liquid state. Particles before and after remain the same, which means composition did ***not*** *change.*

B Chemical change:
Electric current decomposes water into different substances (hydrogen and oxygen). Particles before and after are different, which means composition ***did*** *change.*

FIGURE 1.1 The distinction between a physical change and a chemical change

- **Chemical properties** are characteristics that a substance shows *as it changes into or interacts with another substance (or substances)*. Chemical properties include flammability, corrosiveness, and reactivity with acids. A **chemical change (chemical reaction)**, occurs when *a substance (or substances) is (are) converted into a different substance (or substances)*. Figure 1.1B shows the chemical change (reaction) that occurs when you pass an electric current through water: the water decomposes (breaks down) into two other substances, hydrogen and oxygen, which bubble into the tubes. The composition *has* changed: the final sample is no longer water.

  Chemical change (different substances before and after):

$$\text{water} \xrightarrow{\text{electric current}} \text{hydrogen} + \text{oxygen}$$

Let us work through a sample problem that uses atomic-scale diagrams to distinguish between a physical change and a chemical change.

Sample Problem 1.1 Visualizing Change on the Atomic Scale

Problem The diagrams below represent an atomic-scale view of a sample of matter, A, undergoing two different changes, left to B and right to C:

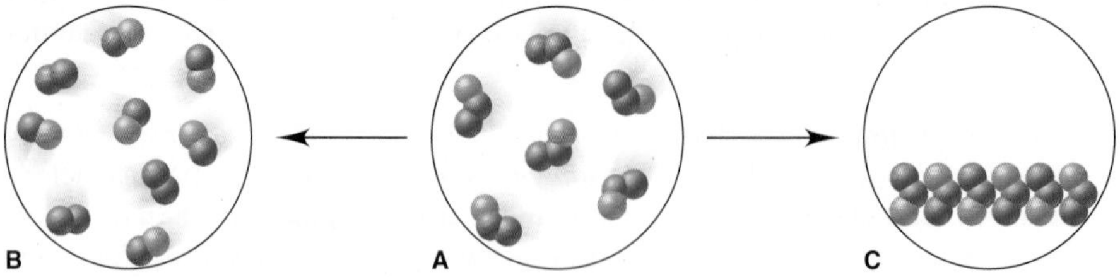

Decide whether each diagram shows a physical change or a chemical change.

Plan Given diagrams of the changes, we have to determine whether each diagram represents a physical change or a chemical change. The number and colours of the little spheres that make up each particle tell its "composition." A diagram with particles of the *same* composition but in a different arrangement depicts a *physical* change, whereas a diagram with particles of a *different* composition depicts a *chemical* change.

Solution In A, each particle consists of one blue sphere and two red spheres. The particles in A change into two types of particles in B, one made of red and blue spheres and the other made of two red spheres; therefore, they have undergone a chemical change to form different particles. The particles in C are the same as those in A, but they are closer together and arranged differently; therefore, they have undergone a physical change.

Follow-Up Problem 1.1 Is the following change chemical or physical? (Compare your answer with the one in Brief Solutions to Follow-Up Problems at the end of the chapter.)

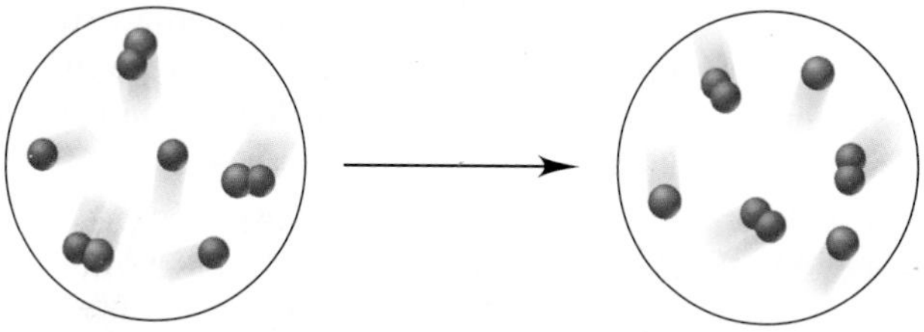

Like water, hydrogen, oxygen, or any other real substance, copper is identified by *its own set* of physical and chemical properties (Table 1.1).

The States of Matter

Matter occurs commonly in *four physical forms* called **states**: solid, liquid, gas, and plasma, although a fifth state of matter, the Bose-Einstein condensate (BEC), has been found to exist under extremely low temperature conditions. We will define the states and see how temperature can change them. In this text, however, we will mostly deal with the three common states: solid, liquid, and gas.

Defining the States On the macroscopic scale, each state of matter can be defined in many different ways. The easiest to understand is perhaps in terms of kinetic energy:

- A **solid** contains particles with low kinetic energy; this results in strong intermolecular forces between the atoms or molecules, causing the solid to have a fixed shape and a fixed volume that does not conform to the shape of the container.

TABLE 1.1 Some Characteristic Properties of Copper

Physical Properties		Chemical Properties	
Easily shaped into sheets (malleable) and wires (ductile)	 	Slowly forms a blue-green carbonate in moist air	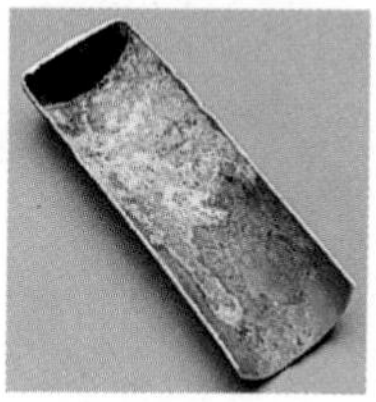
Can be melted and mixed with zinc to form brass		Reacts with nitric or sulfuric acid	
Density = 8.95 g/cm^3 Melting point = 1083°C Boiling point = 2570°C		Slowly forms a deep-blue solution in aqueous ammonia	

Solids are *not* defined by rigidity or hardness: solid iron is rigid and hard, but solid lead is flexible and solid wax is soft.

- A **liquid** contains particles with greater kinetic energy than the particles in a solid; there are weaker intermolecular forces between the atoms or molecules, allowing the liquid to have a varying shape that conforms to the shape of the container, but its volume is fixed; depending on the gravitational force, a liquid may or may not have *an upper surface.*
- A **gas** contains particles with much greater kinetic energy than the particles in a liquid. There are very weak intermolecular forces between the atoms or molecules, allowing the gas to spread indefinitely if left unconfined; if placed in a container that is enclosed, gases will conform to the shape of the container and fill it completely. Gases do *not* have a fixed volume.
- A **plasma** contains particles with very high kinetic energy. It behaves very much as a gas does in that it fills the entire container and does not have a fixed volume (it is *fluid*, like a liquid or a gas). It is different from a gas in that it is composed of a mixture of neutral atoms, electrons, and ions although it has an overall charge of zero. Unlike a gas (which is neutral), a plasma is significantly affected by electrical and magnetic fields. Although plasma is not a state commonly encountered on Earth except in a laboratory environment, it is believed to be one of the most commonly encountered forms of matter in the universe.
- The **Bose-Einstein condensate (BEC)** was first formed in 1995 when a rubidium sample was cooled to a temperature close to −273°C, or absolute zero. At this low temperature, the kinetic energy of the particles in the sample was so low that the atoms, rather than existing individually, began "clumping" together to form something akin to a "superatom." The BEC exhibits extremely interesting properties, including superfluidity, and allows scientists to study quantum mechanics in an experimental setting.

On the atomic scale, each state is defined by the relative positions of its particles (Figure 1.2, *circles at bottom*):

- In a *solid*, the particles lie next to each other in a regular, three-dimensional *array*.

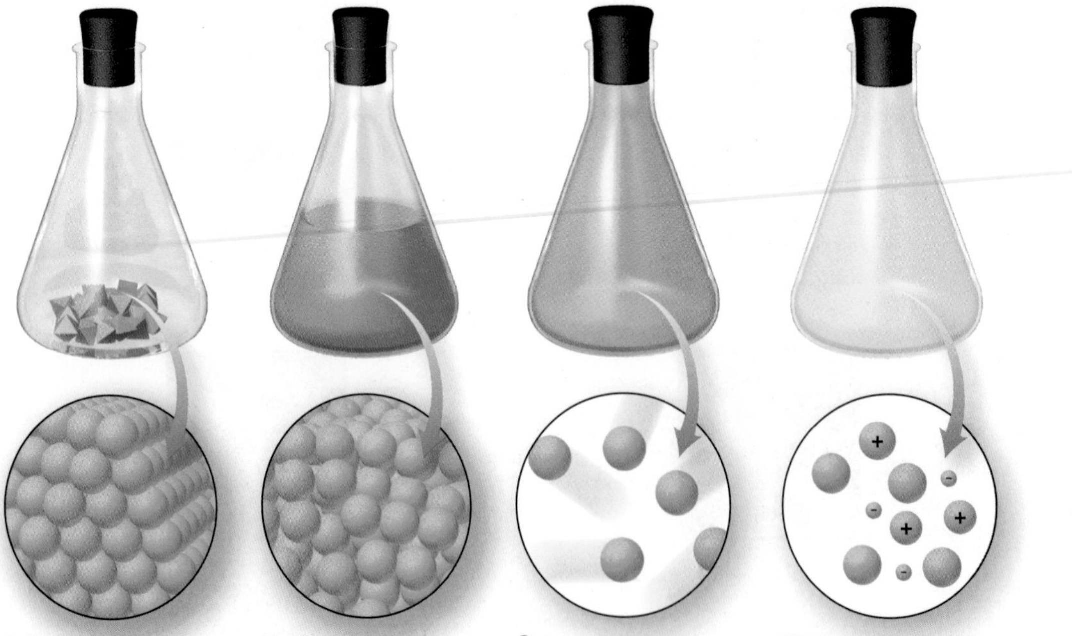

Solid
Particles are close together and organized.

Liquid
Particles are close together but disorganized.

Gas
Particles are far apart and disorganized.

Plasma
Particles, consisting of neutral atoms, ions, and electrons, are far apart and disorganized in the absence of an electric or magnetic field.

FIGURE 1.2 The physical states of matter

- In a *liquid*, the particles also lie close together but move randomly around each other.
- In a *gas*, the particles have large distances between them and move randomly throughout the container.
- In a *plasma*, the particles have large distances between them and move randomly throughout the container in the absence of an electric or magnetic field. At times, the plasma may exhibit a coherent and collective property due to the generation of a magnetic or electric field within the plasma itself.

Temperature and Changes of State Depending on the temperature and pressure of the surroundings, many substances can exist in each of the three common physical states and also undergo changes in state. For example, as the temperature increases, solid water melts to liquid water, which boils to gaseous water (also called *water vapour*). Similarly, as the temperature drops, water vapour condenses to liquid water, which, with further cooling, freezes to ice. In a steel plant, solid iron melts to liquid (molten) iron and then cools to solid iron again. Plasma states are generally formed when a substance is exposed to a huge electric potential or heated to extremely high temperatures.

The main point is that *a physical change caused by heating can generally be reversed by cooling.* This is *not* generally true for a chemical change. For example, heating iron in moist air causes a chemical reaction that yields the brown, crumbly substance known as rust. Cooling does not reverse this change; rather, another chemical change (or series of changes) is required.

The following sample problem provides practice in distinguishing some familiar examples of physical and chemical changes.

Sample Problem 1.2 Distinguishing between Physical and Chemical Changes

Problem Decide whether each process is primarily a physical change or primarily a chemical change, and explain briefly:

(a) Frost forms as the temperature drops on a humid winter night.

(b) A cornstalk grows from a seed that is watered and fertilized.

(c) A match ignites to form ash and a mixture of gases.

(d) Perspiration evaporates when you relax after jogging.

(e) A silver fork tarnishes slowly in the air.

Plan The basic question we ask to decide whether a change is chemical or physical is, "Does the substance change composition or just change form?"

Solution **(a)** Frost forming is a physical change: the drop in temperature changes water vapour (gaseous water) in humid air to ice crystals (solid water).

(b) A seed growing involves a chemical change: water; substances from the air, fertilizer, and soil; and energy from sunlight cause complex changes in the composition of the seed.

(c) A match burning is a chemical change: the combustible substances in the match head are converted into other substances.

(d) Perspiration evaporating is a physical change: the water in sweat changes its form, from liquid to gas, but not its composition.

(e) Tarnishing is a chemical change: silver changes to silver sulfide by reacting with sulfur-containing substances in the air.

Follow-Up Problem 1.2 Decide whether each of the following processes is primarily a physical change or a chemical change, and explain briefly. (See Brief Solutions to Follow-Up Problems at the end of the chapter.)

(a) Purple iodine vapour appears when solid iodine is warmed.

(b) Gasoline fumes are ignited by a spark in an automobile engine's cylinder.

(c) A scab forms over an open cut.

The Central Theme in Chemistry

Understanding the properties of a substance and the changes it undergoes leads to the central theme in chemistry: *macroscopic-scale* properties and behaviour that we can see are the results of *atomic-scale* properties and behaviour that we cannot see. The distinction between chemical change and physical change is defined by composition, which we study macroscopically. But composition ultimately depends on the makeup of substances at the atomic scale. Similarly, macroscopic properties of substances in all of the states arise from the atomic-scale behaviour of their particles. Picturing a chemical event on the molecular scale, even one as common as the flame of a laboratory burner (*see margin*), helps to clarify what is taking place. What is happening when water boils or copper melts? What events occur in the invisible world of minute particles that cause a seed to grow, a neon light to glow, or a nail to rust? Throughout the text, we return to this central idea:

*We study **observable** changes in matter to understand their **unobservable** causes.*

Methane and oxygen form carbon dioxide and water in the flame of a laboratory burner. (Carbon is *black*, oxygen is *red*, and hydrogen is *blue*.)

The Importance of Energy in the Study of Matter

Physical and chemical changes are accompanied by energy changes. **Energy** is often defined as *the ability to do work.* Essentially, all work involves moving something. Work is done when your arm lifts a book, when a car's engine moves the wheels, or when a falling rock moves the ground as it lands. The object doing the work (arm, engine, or rock) transfers some of the energy it possesses to the object on which the work is done (book, wheels, or ground).

The total energy that an object possesses is the sum of its potential energy and its kinetic energy.

- **Potential energy** is *the energy due to the **position** of the object relative to other objects.*
- **Kinetic energy** is *the energy due to the **motion** of the object.*

Let us examine four systems that illustrate the relationship between these two forms of energy: a weight raised above the ground, two balls attached by a spring, two electrically charged particles, and a fuel and its waste products. The following two concepts are central to all of these systems:

1. *When energy is converted from one form to another, it is conserved, not destroyed.*
2. *Situations of lower energy are more stable and, therefore, favoured over situations of higher energy (less stable).*

The four systems are described below:

- *A weight raised above the ground* (Figure 1.3A). The energy you exert to lift a weight against gravity increases the weight's potential energy (energy due to its position). When you drop the weight, this additional potential energy is converted to kinetic energy (energy due to motion). Since the system with the weight elevated is higher in potential energy and *less stable*, the weight will fall when released to result in a system that is lower in potential energy and *more stable.*
- *Two balls attached by a spring* (Figure 1.3B). When you pull the balls apart, the energy you exert to stretch the relaxed spring increases the system's potential energy. This change in potential energy is converted to kinetic

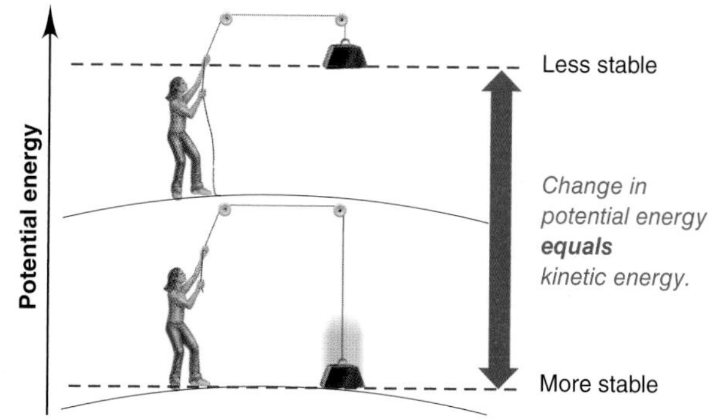

A **A gravitational system.** *Potential energy is gained when a weight is lifted. It is converted to kinetic energy as the weight falls.*

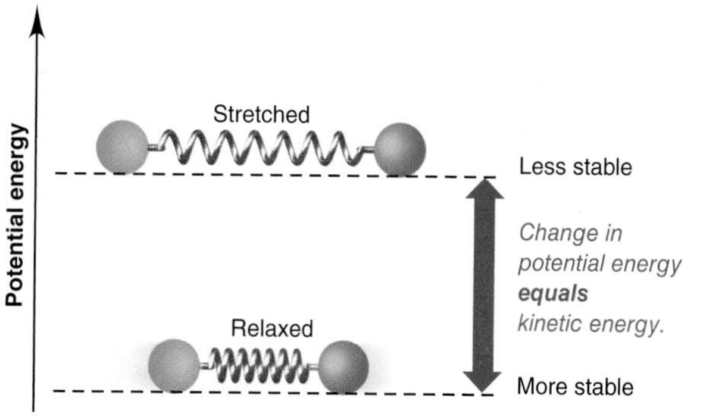

B **A system of two balls attached by a spring.** *Potential energy is gained when the spring is stretched. It is converted to the kinetic energy of the moving balls as the spring relaxes.*

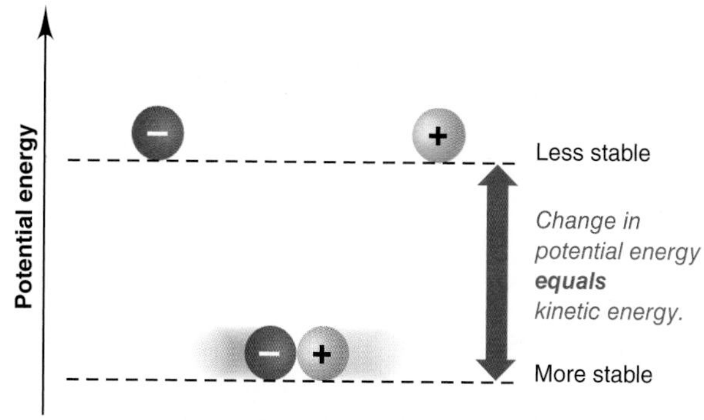

C **A system of oppositely charged particles.** *Potential energy is gained when the charges are separated. It is converted to kinetic energy as the attraction pulls the charges together.*

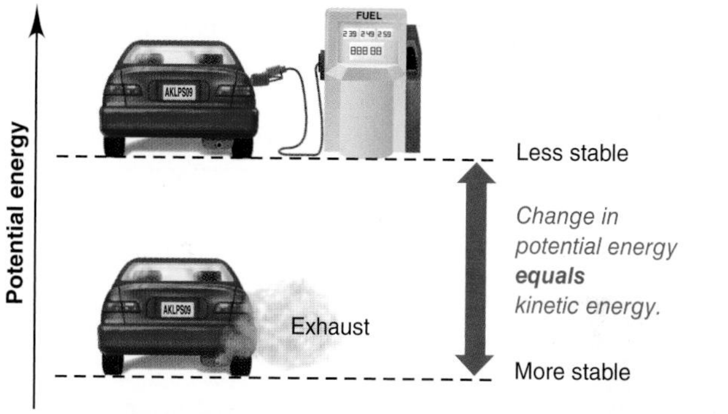

D **A system of fuel and exhaust.** *A fuel is higher in chemical potential energy than the exhaust. As the fuel burns, some of its potential energy is converted to the kinetic energy of the moving car.*

FIGURE 1.3 Potential energy is converted to kinetic energy. The dashed horizontal lines indicate the potential energy of each system before and after the change.

energy when you release the balls. The system of the balls and spring is less stable (has more potential energy) when the spring is stretched than when the spring is relaxed.

- *Two electrically charged particles* (Figure 1.3C). Due to interactions known as *electrostatic forces, opposite charges attract each other, and like charges repel each other.* When energy is exerted to move a positive particle away from a negative particle, the potential energy of the system increases. This increase is converted to kinetic energy when the particles are pulled together by the electrostatic attraction. Similarly, when energy is used to move two positive (or two negative) particles together, their potential energy increases and is converted to kinetic energy when they are pushed apart by the electrostatic repulsion. Charged particles move naturally to a more-stable situation (lower energy).
- *A fuel and its waste products* (Figure 1.3D). Matter is composed of positively and negatively charged particles. *The chemical potential energy of a substance results not only from the relative positions of its particles but also from the attractions and repulsions among them.* Some substances have more potential energy than others. For example, gasoline and oxygen have more chemical potential energy than the exhaust gases they form. This difference is converted into kinetic energy, which moves the car, heats its interior, makes the lights shine, and so on. Similarly, the difference in potential energy between the food and the air we take in and the wastes we excrete enables us to move, grow, keep warm, study chemistry, and so on.

SUMMARY OF SECTION 1.1

- Chemists study the composition and properties of matter and how they change.
- Each substance has a unique set of *physical* properties (attributes of the substance itself) and *chemical* properties (attributes of the substance as it interacts with or changes to other substances). Changes in matter can be *physical* (different form of the same substance) or *chemical* (different substance).
- Matter exists in five physical states: solid, liquid, gas, plasma, and the Bose-Einstein condensate (BEC). The behaviour of each state is due to the arrangement of the particles.
- A physical change caused by heating may be reversed by cooling. A chemical change caused by heating can be reversed only by other chemical changes.
- Macroscopic-scale changes result from atomic-scale changes.
- Changes in matter are accompanied by changes in energy.
- An object's potential energy is due to its position; an object's kinetic energy is due to its motion. Energy that is used to lift a weight, stretch a spring, or separate opposite charges increases the system's potential energy, which is converted to kinetic energy as the system returns to its original condition. Energy changes form but is conserved.
- Chemical potential energy arises from the positions and interactions of the particles in a substance. When a higher-energy (less-stable) substance is converted into a lower-energy (more-stable) substance, some potential energy is converted into kinetic energy.

1.2 Chemical Arts and the Origins of Modern Chemistry

This brief overview of early breakthroughs, and a few false directions, describes how the modern science of chemistry arose and progressed.

Prechemical Traditions

Chemistry originated in a prescientific past that incorporated three overlapping traditions: alchemy, medicine, and technology:

1. *The alchemical tradition. Alchemy* was an occult study of nature that began in the 1st century C.E. and dominated thinking for over 1500 years. Originally influenced by the Greek idea that matter strives for "perfection," alchemists later became obsessed with converting "baser" metals, such as lead, into "purer" metals, such as gold. The alchemists' names for substances and their belief that matter could be magically altered were very difficult to change over the centuries. Their legacy to chemistry was in their technical methods. They invented distillation, percolation, and extraction, and they devised apparatus still used routinely today (Figure 1.4). But perhaps even more important was their encouragement of observation and experimentation, which replaced the Greek approach of explaining nature solely through reason.

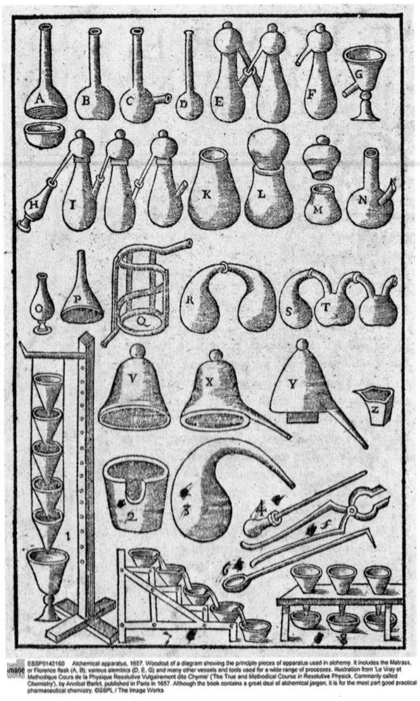

FIGURE 1.4 **Alchemical apparatus**

2. *The medical tradition.* Alchemists also influenced medical practice in medieval Europe. Ever since the 13th century, distillates and extracts of roots, herbs, and other plant matter have been used as sources of medicines. The alchemist and physician Paracelsus (1493–1541) considered the body to be a chemical system and illness to be an imbalance that could be restored by treatment with drugs. Although many early prescriptions were useless, later prescriptions had increasing success. Thus began the alliance between medicine and chemistry that thrives today.

3. *The technological tradition.* For thousands of years, pottery making, dyeing, and especially metallurgy contributed greatly to people's experience with materials. During the Middle Ages and the Renaissance, artisans published books that described how to purify, assay, and coin silver and gold; how to use balances, furnaces, and crucibles; and how to make glass and gunpowder. Some books introduced quantitative measurement, which was lacking in alchemical writings. Many creations from those times are still considered marvels today, throughout the world. Nevertheless, the skilled artisans showed little interest in *why* a substance changes or *how to predict* its behaviour.

The Phlogiston Fiasco and the Impact of Lavoisier

Chemical investigation in the modern sense—inquiry into the causes of changes in matter—began in the late 17th century. At that time, most scientists explained **combustion**, the *process of burning*, with the *phlogiston theory*. This theory proposed that combustible materials contain *phlogiston*, an undetectable substance that is released when the material burns. Highly combustible materials, such as charcoal, were thought to contain a lot of phlogiston, and slightly combustible materials, such as metals, only a little.

But inconsistencies continuously arose.

Phlogiston critics: Why is air needed for combustion, and why does charcoal stop burning in a closed vessel?
Phlogiston supporters: Air "attracts" phlogiston out of the charcoal, and burning stops when the air in the vessel is "saturated" with phlogiston.

Critics also noted that when a metal burns, it forms its *calx* (a residual substance), which weighs more than the metal. This led them to ask the following question:

Phlogiston critics: How can the *loss* of phlogiston cause a *gain* in mass?
Phlogiston supporters: Phlogiston has negative mass.

As ridiculous as these responses seem now, it is important to remember that, even today, scientists may dismiss conflicting evidence rather than abandon an accepted idea.

The conflict over phlogiston was resolved when the young French chemist Antoine Lavoisier (1743–1794) performed several experiments:

1. He gently heated mercury in a measured volume of air at a certain temperature in a closed container, yielding mercury calx and four-fifths of the original air. The contents of the closed container before and after the reaction *had the same total mass.*
2. When he placed a burning candle in the remaining air, the flame was extinguished.
3. He strongly heated mercury calx in a closed container at a higher temperature than what he used in his first experiment. The mercury calx decomposed into two products—mercury and a gas—whose *total mass equalled the starting mass of the calx.*

Lavoisier named the gas *oxygen* and the metal calxes *metal oxides*. His explanation of his results made the phlogiston theory irrelevant:

- Oxygen, a normal component of air, combines with a substance when it burns.
- In a closed container, a combustible substance stops burning when it has combined with all the available oxygen.
- A metal calx (metal oxide) weighs more than the metal because its mass includes the mass of the oxygen.

Lavoisier's new theory triumphed because it relied on *quantitative, reproducible measurements*, not on strange properties of undetectable substances. Because this approach is at the heart of science, many propose that the *science* of chemistry began with Lavoisier.

SUMMARY OF SECTION 1.2

- Alchemy, medicine, and technology placed little emphasis on objective experimentation, focusing instead on mystical explanations or practical experiences, but these traditions contributed some apparatus and methods that are still important today.
- Lavoisier overthrew the phlogiston theory by showing quantitatively that oxygen, a component of air, is required for combustion and combines with a burning substance.

1.3 The Scientific Approach: Developing a Model

Unlike our prehistoric ancestors, who survived through *trial and error*—gradually learning which types of stone were hard enough to shape others, which plants were edible and which were poisonous—we employ the *quantitative theories* of chemistry to understand materials, make better use of them, and create new ones:

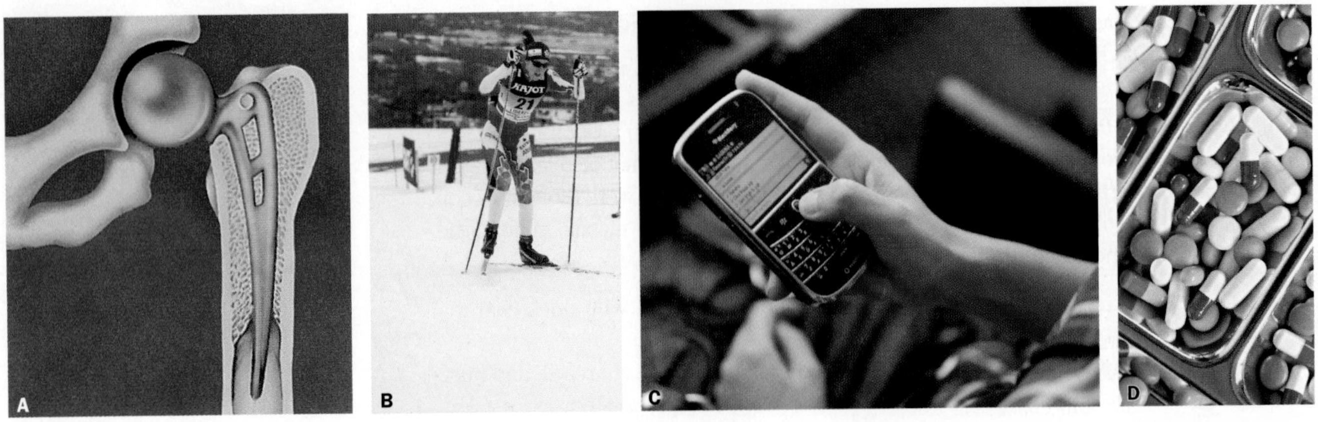

FIGURE 1.5 Modern materials in a variety of applications. **A.** High-tension polymers in synthetic hip joints. **B.** Specialized polymers in clothing and sports gear. **C.** Liquid crystals and semiconductors in electronic devices. **D.** Medicinal agents in pills.

specialized drugs, advanced composites, synthetic polymers, and countless other materials (Figure 1.5).

To understand nature, scientists use an approach called the **scientific method.** It is not a step-by-step checklist but rather a process involving creative propositions and tests aimed at objective, verifiable discoveries. There is no single procedure, and luck often plays a key role in discovery. In general terms, the scientific approach includes the following parts (Figure 1.6):

- *Observations.* These are the facts that our ideas must explain. **Observations** can be divided into two types: **qualitative** and **quantitative**. Qualitative observations are used *to describe things we can see, hear, smell, or feel in an experiment.* (*Never* taste chemicals in the laboratory!) Quantitative observations are *numerical pieces of information obtained in the laboratory,* and they are often called **data**. While the most useful observations made in the laboratory are usually quantitative (because they can be analyzed to reveal trends), the importance of qualitative data should not be underestimated. When *the same observation is made by many investigators in different situations with no clear exceptions,* it is summarized, often in mathematical terms, as a **natural law**. The observation that mass remains constant during a chemical change—made by Lavoisier and numerous experimenters since—is known as the law of mass conservation (Chapter 2).
- *Hypothesis.* Whether derived from observation or from a "spark of intuition," a **hypothesis** is *a proposal made to explain an observation.* A sound hypothesis may be intuitive, or it may be derived from a set of observations. It need not be correct, or the *only* correct explanation, but it must be *testable by experiment.* Indeed, a hypothesis is often the reason for performing an experiment: if the results do not support it, the hypothesis must be revised or discarded. Hypotheses can be altered, but experimental results cannot. A hypothesis can include a mathematical relationship between the variables in an experiment.
- *Experiment. A set of procedural steps that tests a hypothesis,* an **experiment** often leads to a revised hypothesis and new experiments to test it. An experiment

FIGURE 1.6 The scientific approach to understanding nature. Hypotheses and models are mental pictures that are revised to match observations and experimental results, *not* the other way around.

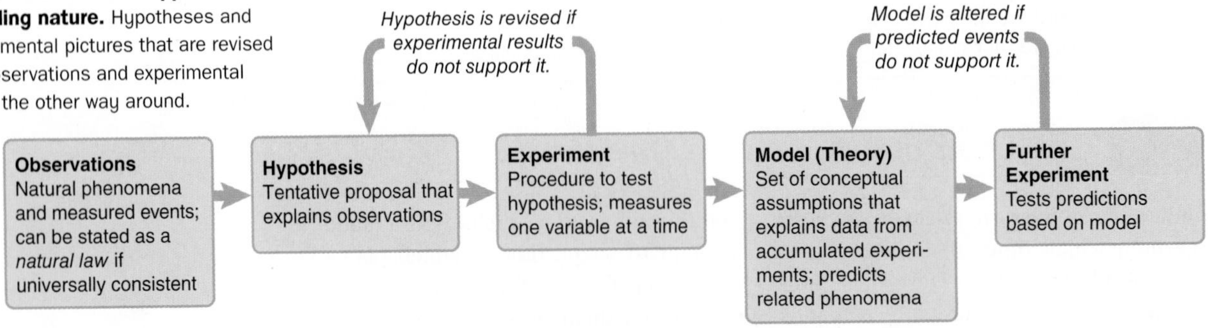

typically contains at least two **variables**, *quantities that can have more than one value.* A well-designed experiment is controlled. A **controlled experiment** *measures the effect of one variable on another while keeping all the other variables constant.* Experimental results must be *reproducible* by others. Both skill and creativity play a part in experimental design. If the hypothesis is valid, the data produced in an experiment should show that the mathematical relationship between the variables is consistent. If the mathematical relationship between variables cannot be shown using the experimental data, the hypothesis must be revised. Many empirically derived relationships, such as the gas laws, have been termed "laws" because the data consistently support the mathematical relationship between the variables.

- *Model.* Formulating conceptual **models (theories)**, based on *experiments* that test *hypotheses* about *observations* distinguishes scientific thinking from speculation. As hypotheses are revised according to experimental results, a model emerges to explain how a phenomenon occurs. A model is a *simplified,* not an exact, representation of some aspect of nature that we use to *predict* related phenomena. Ongoing experimentation refines the model to account for new facts.

Lavoisier's overthrow of the phlogiston theory demonstrates the scientific method of thinking. *Observations* of burning and smelting led to the *hypothesis* that combustion involved the loss of phlogiston. *Experiments* showing that air is required for burning and that a metal gains mass during combustion led Lavoisier to propose a new *hypothesis,* which he tested repeatedly with quantitative *experiments.* Accumulating evidence supported his developing *model (theory)* that combustion involves combination with a component of air (oxygen). Innumerable *predictions* based on his theory have supported its validity, and Lavoisier himself extended his theory to account for animal respiration and metabolism.

SUMMARY OF SECTION 1.3

- The scientific method is a process designed to explain and predict phenomena.
- Observations lead to hypotheses about how or why a phenomenon occurs. When repeated with no exceptions, observations may be expressed as a natural law using a mathematical relationship between variables.
- Hypotheses are tested by controlled experiments and revised when necessary.
- If reproducible data support a hypothesis, a model (theory) can be developed to explain the observed phenomenon. A good model predicts related phenomena but must be refined whenever conflicting data appear.

1.4 Chemical Problem Solving

In many ways, learning chemistry is learning how to solve chemistry problems, not only those in exams or homework, but the more complex ones in life and society. The sample problems in this book are designed to strengthen your skills for solving problems. In this section, we describe the problem-solving approach. Most problems include calculations, so let us first discuss how to handle measured quantities.

Units and Conversion Factors in Calculations

All measured quantities consist of a number *and* a unit: a person's height is "178 cm," not "178." Ratios of quantities have ratios of units, such as km/h. (We discuss some important units in Section 1.5.) To minimize errors, make it a habit to *include units in all calculations.*

The arithmetic operations used with quantities are the same as those used with pure numbers; that is, units can be multiplied, divided, and cancelled:

- A carpet measuring 3 m (metres) by 4 m has the following area:

$$\text{Area} = 3\text{ m} \times 4\text{ m} = (3 \times 4)(\text{m} \times \text{m}) = 12\text{ m}^2$$

- A car travelling 350 km (kilometres) in 7 h (hours) has this speed:

$$\text{Speed} = \frac{350\ \text{km}}{7\ \text{h}} = \frac{50\ \text{km}}{1\ \text{h}}\ (\text{often written } 50\ \text{km/h},\ 50\ \frac{\text{km}}{\text{h}},\ \text{or } 50\ \text{km}\cdot\text{h}^{-1})$$

- In 3 h, the car travels this distance:

$$d = 3\ \text{h} \times \frac{50\ \text{km}}{1\ \text{h}} = 150\ \text{km}$$

Constructing a Conversion Factor A **conversion factor** is a *ratio that is used to express a quantity in different units.* Suppose that we want to know the distance of the 150 km car trip in metres. To convert kilometres to metres, we use *equivalent quantities*:

$$1\ \text{km} = 1000\ \text{m}$$

From this equation, we can construct two conversion factors. Dividing both sides by 1000 m gives one conversion factor (shown in blue):

$$\frac{1\ \text{km}}{1000\ \text{m}} = \frac{1000\ \text{m}}{1000\ \text{m}} = 1$$

Dividing both sides by 1 km gives the other conversion factor (the inverse):

$$\frac{1\ \text{km}}{1\ \text{km}} = \frac{1000\ \text{m}}{1\ \text{km}} = 1$$

Since the numerator and denominator of a conversion factor are equal, multiplying a quantity by a conversion factor is the same as multiplying by 1. Thus, *even though the number and unit change, the size of the quantity remains the same.*

To convert the distance from kilometres to metres, we choose the conversion factor with kilometres in the denominator, because it cancels kilometres and gives the answer in metres:

$$\text{Distance (m)} = 150\ \cancel{\text{km}} \times \frac{1000\ \text{m}}{1\ \cancel{\text{km}}} = 150\ 000\ \text{m}$$

$$\text{km} \Rightarrow \text{m}$$

Choosing the Correct Conversion Factor It is easier to convert if you first decide whether the answer expressed in the new units should be a larger or a smaller number. In the previous case, we know that a metre is *smaller* than a kilometre, so the distance in metres should be a *larger* number (150 000) than the distance in kilometres (150). The conversion factor has the larger number (1000) in the numerator, so it gives a larger number in the answer.

Most important, the *conversion factor you choose must cancel all units except those you want in the answer.* Therefore, set the unit you are converting *from* (beginning unit) in the *opposite position in the conversion factor* (numerator or denominator) so that it cancels and you are left with the unit you are converting *to* (final unit):

$$\cancel{\text{beginning unit}} \times \frac{\text{final unit}}{\cancel{\text{beginning unit}}} = \text{final unit} \quad \text{as in} \quad \cancel{\text{km}} \times \frac{\text{m}}{\cancel{\text{km}}} = \text{m}$$

Or, in cases that involve units raised to a power:

$$(\cancel{\text{beginning unit}} \times \cancel{\text{beginning unit}}) \times \frac{\text{final unit}^2}{\cancel{\text{beginning unit}^2}} = \text{final unit}^2 \quad \text{as in} \quad (\cancel{\text{m}} \times \cancel{\text{m}}) \times \frac{\text{km}^2}{\cancel{\text{m}^2}} = \text{km}^2$$

Or, in cases that involve a ratio of units:

$$\frac{\cancel{\text{beginning unit}}}{\text{final unit}_1} \times \frac{\text{final unit}_2}{\cancel{\text{beginning unit}}} = \frac{\text{final unit}_2}{\text{final unit}_1} \quad \text{as in} \quad \frac{\cancel{\text{km}}}{\text{h}} \times \frac{\text{m}}{\cancel{\text{km}}} = \frac{\text{m}}{\text{h}}$$

Conversion factors can also be used to convert from one set of units to another (certain metric units to SI units, for example) or between systems of units (imperial to metric, for example). *The use of conversion factors in calculations* is often referred to as the factor-label method or **dimensional analysis** (because units represent physical dimensions). We use dimensional analysis throughout this book. We also use metric and SI units extensively. For more practice using non-metric and non-SI units, please go to Connect.

A Systematic Approach to Solving Problems

The approach used in this book to solve problems emphasizes reasoning, not memorizing, and is based on a simple idea: plan how to solve the problem *before* you try to solve it, check your answer, and then practise with a similar follow-up problem. In general, the sample problems consist of several parts:

1. *Problem.* This part states all the information you need to solve a problem, usually framed in some interesting context.
2. *Plan.* This part helps you *think* about the solution *before* juggling numbers and pressing calculator buttons. There is often more than one way to solve a problem, and the given plan is one possibility. The plan will
   - Clarify the known and unknown: What information do you have, and what are you trying to find?
   - Suggest the steps from known to unknown: What ideas, conversions, or equations are needed?
   - Present a road map (especially in the early chapters) or flow diagram of the plan: The road map has a box for each intermediate result and an arrow showing the step (conversion factor or operation) used to get to the next box.
3. *Solution.* This part shows the calculation steps in the same order as in the plan (and the road map).
4. *Check.* This part helps you check that your final answer makes sense: Are the units correct? Did the change occur in the expected direction? Is it reasonable chemically? To avoid a large math error, we also often do a rough calculation and see if we get an answer in the same ballpark as the actual answer. Here is a typical ballpark calculation from everyday life: You are at a music store and buy three CDs at $7.97 each. With 13% sales tax, the bill comes to $27.02. In your mind, you know that $7.97 is about $8, and 3 times $8 is $24; with the sales tax, the cost should be a bit more. So, your quick mental calculation *is* in the same ballpark as the actual cost.
5. *Comment.* This part appears occasionally to provide an application, an alternative approach, a common mistake to avoid, or an overview.
6. *Follow-Up Problem.* This part presents a similar problem that requires you to apply concepts and/or methods used in solving the sample problem.

Of course, you cannot learn to solve chemistry problems, any more than you can learn to swim, by reading about it, so here are a few suggestions:

- Follow along in the sample problem with pencil, paper, and calculator.
- Try the follow-up problem as soon as you finish the sample problem. A feature called *Brief Solutions to Follow-Up Problems* appears at the end of each chapter, allowing you to compare your solution steps and answer.
- Read the sample problem and text again if you have trouble.
- Try the practice quizzes available on Connect. For each chapter, two interactive quizzes provide conceptual and problem-solving practice and offer feedback for review.
- The end-of-chapter problems review and extend the concepts and skills in the chapter, so try as many as you can. (Answers are given in Appendix G for problems with a red number.)

Let us apply this systematic approach in a unit-conversion problem.

Sample Problem 1.3 Converting Units of Length

Problem To wire your stereo equipment, you need 325 cm (centimetres) of speaker wire, which sells for $2.78/m (metre). How much does the wire cost?

Plan We know the length of wire in centimetres (325 cm) and the price in dollars per metre ($2.78/m). We can find the unknown cost of the wire by converting the length from centimetres to metres. The price gives us the equivalent quantities

Road Map

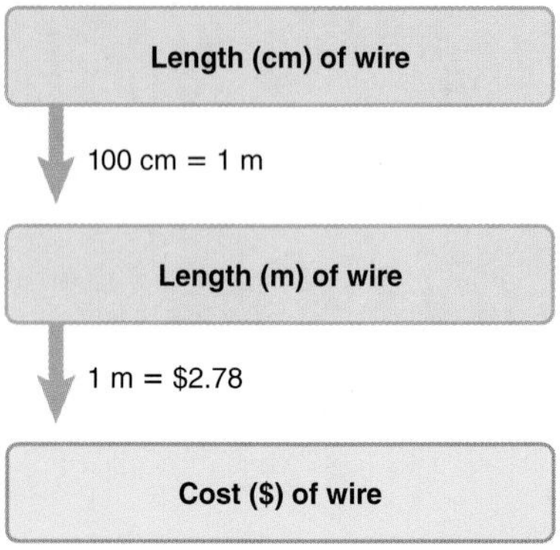

(1 m = \$2.78) to convert metres of wire to cost in dollars. The road map starts with the known and moves through the calculation steps to the unknown.

Solution Convert the known length from centimetres to metres. The equivalent quantities alongside the road map arrow are needed to construct the conversion factor. We choose 1 m/100 cm, rather than the inverse, because it gives an answer in metres:

$$\text{Length (m)} = \text{length (cm)} \times \text{conversion factor} = 325\ \cancel{\text{cm}} \times \frac{1\ \text{m}}{100\ \cancel{\text{cm}}} = 3.25\ \text{m}$$

Convert the length in metres to cost in dollars:

$$\text{Cost (\$)} = \text{length (m)} \times \text{conversion factor} = 3.25\ \cancel{\text{m}} \times \frac{\$2.78}{1\ \cancel{\text{m}}} = \$9.04$$

See Rules for Rounding in Section 1.6 for instructions on how to round with multiplication.

Check The units are correct for each step. The conversion factors make sense in terms of the relative unit sizes: the number of metres is *smaller* than the number of centimetres (a metre is *larger* than a centimetre). The total cost seems reasonable: a little more than 3 m of wire at a little under \$3/m should cost about \$9.

Comment **1.** We could also have strung the two steps together:

$$\text{Cost (\$)} = 325\ \cancel{\text{cm}} \times \frac{1\ \cancel{\text{m}}}{100\ \cancel{\text{cm}}} \times \frac{\$2.78}{1\ \cancel{\text{m}}} = \$9.04$$

2. There are usually alternative sequences in unit-conversion problems. Here, for example, we would get the same answer if we first converted the cost of wire from \$/m to \$/cm and kept the wire length in centimetres. Try it yourself.

Follow-Up Problem 1.3 A doll manufacturer needs 31.5 cm^2 of fabric to make one doll's dress. Its Dutch supplier sends the fabric in bolts of exactly 200 m^2. How many dresses can be stitched with two bolts of fabric (1 m = 100 cm)? Draw a road map to show how you plan the solution.

SUMMARY OF SECTION 1.4

- A measured quantity consists of a number and a unit.
- A conversion factor is a ratio of equivalent quantities (and, thus, equal to 1) that is used to express a quantity in different units.
- The problem-solving approach used in this book has four parts: (1) plan the steps to the solution, which often includes a flow diagram (road map) of the steps, (2) perform the calculations according to the plan, (3) check to see if the answer makes sense, and (4) practise with a similar problem and compare your solution with the one at the end of the chapter.

1.5 Measurement in Scientific Study

Measurement has a rich history, characterized by the search for *exact, invariable standards*. Measuring for trade, building, and surveying began thousands of years ago, but, for most of this time, it was based on standards that could vary: a yard was the distance from the king's nose to the tip of his outstretched arm, and an acre was the area tilled in one day by a man with a pair of oxen. Our current system of measurement began in 1790 in France, when a committee, of which Lavoisier was a member, developed the *metric system*. Then, in 1960, another committee in France revised the metric system and established the universally accepted **SI units** (from the French **S**ystème **I**nternational d'Unités).

General Features of SI Units

The SI system is based on seven **base units (fundamental units)**, each identified with a physical quantity (Table 1.2). All other units are **derived units**, combinations of the seven base units. For example, the derived unit for speed, metre per second (m/s), is the base unit for length (m) divided by the base unit for time (s). (Derived units that are a ratio of base units can be used as conversion factors.) For quantities much smaller or larger than the base unit, we use decimal prefixes and exponential (scientific) notation (Table 1.3). (If you need a review of exponential notation, see Appendix A.) Because the prefixes are based on powers of 10, SI units are easier to use in calculations than imperial units.

Some Important SI Units in Chemistry

Here, we discuss units for length, volume, mass, density, temperature, and time. Other units are presented in later chapters.

Length The SI base unit of length is the **metre (m)**. In the metric system, it was originally defined as 1/10 000 000 of the distance from the equator to the North Pole, and later as the distance between two fine lines engraved on a corrosion-resistant metal bar. More recently, the first exact, unchanging standard was adopted: 1 650 763.73 wavelengths of orange-red light from electrically excited krypton atoms. The current standard is exact and invariant: 1 m is the distance that light travels in a vacuum in 1/299 792 458 of a second.

Biological cells are often measured in micrometres (1 μm = 10^{-6} m). On the atomic scale, nanometres (10^{-9} m) and picometres (10^{-12} m) are used. Many proteins have diameters of about 2 nm; atoms have diameters of about 200 pm (0.2 nm). An older unit still in use is the angstrom (1 Å = 10^{-10} m = 0.1 nm = 100 pm).

Volume Any sample of matter has a certain **volume (*V*)**, which is the amount of space it occupies. The SI unit of volume is the **cubic metre (m^3)**. In chemistry, we often use non-SI units: the **litre (L)** and the **millilitre (mL)** (note the uppercase L).

TABLE 1.2 SI Base Units

Physical Quantity (Dimension)	Unit Name	Unit Abbreviation
Mass	kilogram	kg
Length	metre	m
Time	second	s
Temperature	kelvin	K
Electric current	ampere	A
Amount of substance	mole	mol
Luminous intensity	candela	Cd

TABLE 1.3 Common Decimal Prefixes Used with SI Units

Prefix*	Prefix Symbol	Word	Conventional Notation	Exponential Notation
tera	T	Trillion	1 000 000 000 000	1×10^{12}
giga	G	Billion	1 000 000 000	1×10^{9}
mega	M	Million	1 000 000	1×10^{6}
kilo	k	Thousand	1 000	1×10^{3}
hecto	H	Hundred	100	1×10^{2}
deka	da	Ten	10	1×10^{1}
—	—	One	1	1×10^{0}
deci	d	Tenth	0.1	1×10^{-1}
centi	c	Hundredth	0.01	1×10^{-2}
milli	m	Thousandth	0.001	1×10^{-3}
micro	μ	Millionth	0.000 001	1×10^{-6}
nano	n	Billionth	0.000 000 001	1×10^{-9}
pico	p	Trillionth	0.000 000 000 001	1×10^{-12}
femto	f	Quadrillionth	0.000 000 000 000 001	1×10^{-15}

*The prefixes most frequently used by chemists appear in bold type.

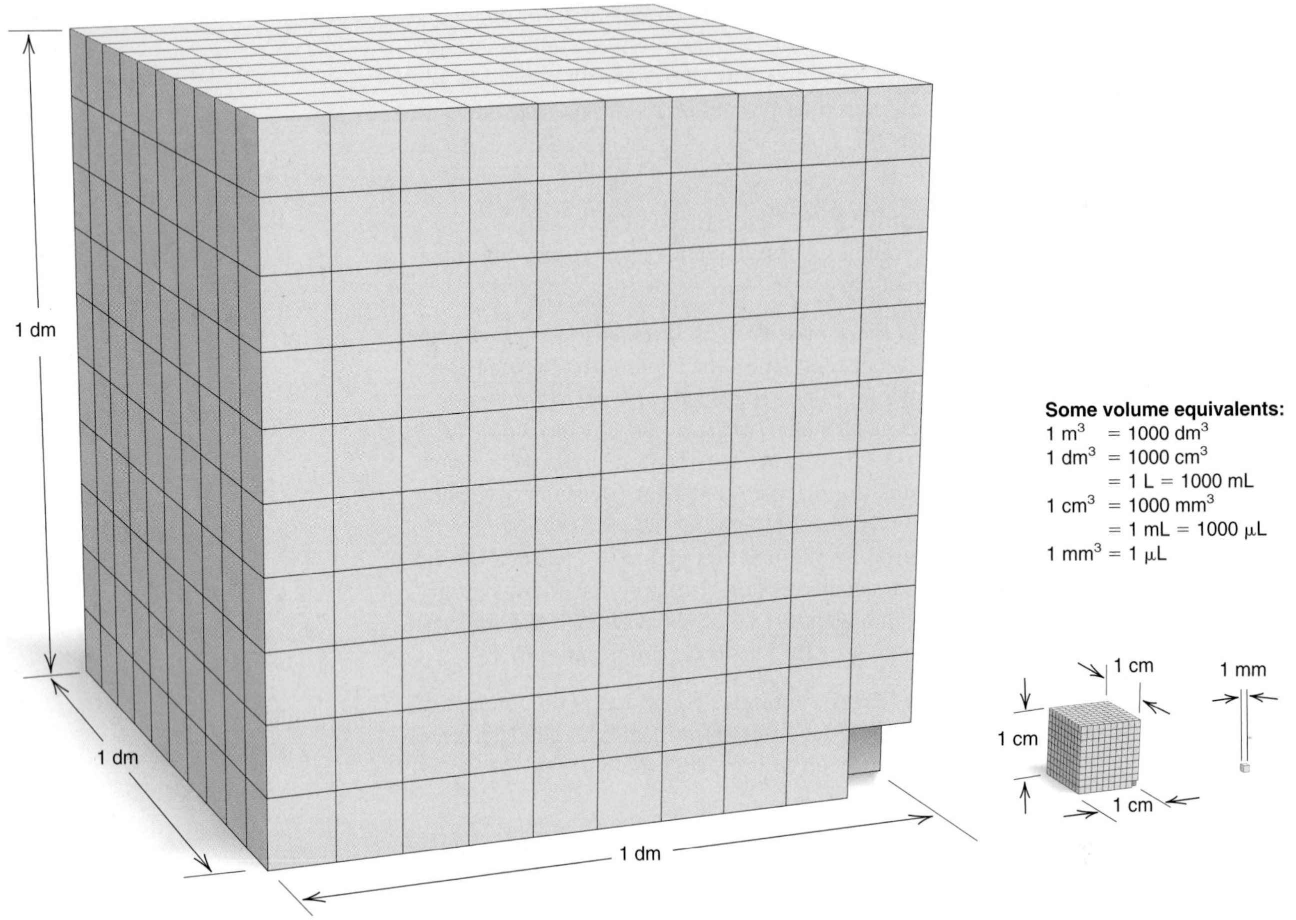

FIGURE 1.7 Some volume relationships in SI: from cubic decimetre (dm^3) to cubic centimetre (cm^3) to cubic millimetre (mm^3)

Physicians and other medical practitioners measure body fluids in cubic decimetres (dm^3), which are equivalent to litres:

$$1 \text{ L} = 1 \text{ dm}^3 = 10^{-3} \text{ m}^3$$

1 mL, or $\frac{1}{1000}$ of a litre, is equivalent to one cubic centimetre (cm^3):

$$1 \text{ mL} = 1 \text{ cm}^3 = 10^{-3} \text{ dm}^3 = 10^{-3} \text{ L} = 10^{-6} \text{ m}^3$$

Figure 1.7 is a life-size depiction of the two 1000-fold decreases in volume from 1 dm^3 to 1 cm^3 and then to 1 mm^3. The edge of a 1 m^3 cube would be about 2.5 times the width of this book opened flat.

Figure 1.8A shows some laboratory glassware for working with volumes. Erlenmeyer flasks and beakers are used to contain liquids. Graduated cylinders, pipettes, and

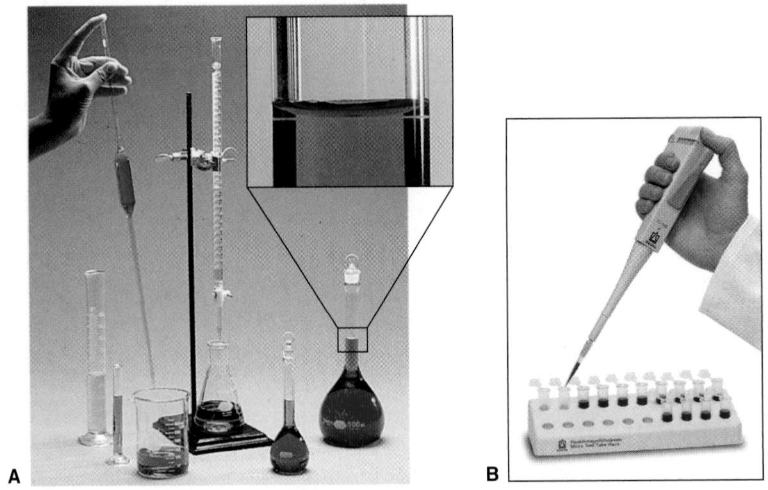

FIGURE 1.8 Common laboratory volumetric glassware. A. From left to right are two graduated cylinders, a pipette being emptied into a beaker, a burette delivering liquid to an Erlenmeyer flask, and two volumetric flasks. **Inset.** In contact with the glass neck of the flask, the liquid forms a concave meniscus (curved surface). **B.** An automatic pipette delivers a given volume of liquid to each test tube.

burettes are used to measure and transfer liquids. Volumetric flasks and pipettes have a fixed volume, which is indicated by a mark on the neck. Solutions are prepared quantitatively in volumetric flasks, and specific amounts are put into cylinders, pipettes, or burettes to be transferred to beakers or flasks for further steps. In Figure 1.8B, an automatic pipette transfers liquid accurately *and* quickly.

Sample Problem 1.4 Converting Units of Volume

Problem The volume of an irregularly shaped solid can be determined from the volume of water it displaces. A graduated cylinder contains 19.9 mL of water. When a small piece of galena, an ore of lead, is added, it sinks and the volume increases to 24.5 mL. What is the volume of the piece of galena in cm^3 and in L?

Plan We have to find the volume of the galena from the change in volume of the cylinder contents. The volume of galena in mL is the difference between the volumes before (19.9 mL) and after (24.5 mL) adding it. Since mL and cm^3 represent identical volumes, the volume in mL equals the volume in cm^3. We then use equivalent quantities ($1\ mL = 10^{-3}\ L$) to convert mL to L. The road map shows these steps.

Road Map

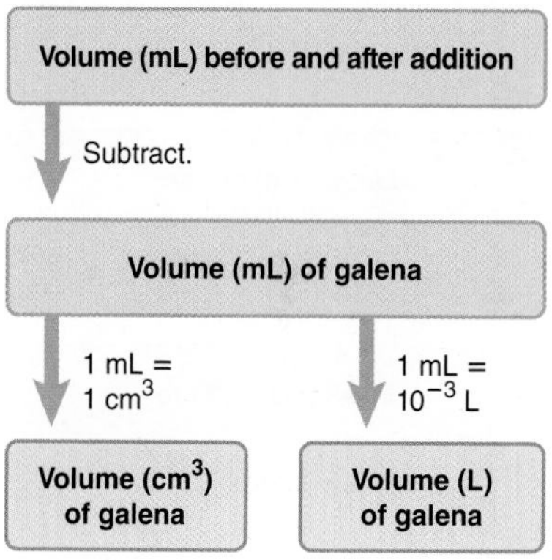

Solution Find the volume of galena:

$$\text{Volume (mL)} = \text{volume after} - \text{volume before} = V_f - V_i$$
$$= 24.5\ \text{mL} - 19.9\ \text{mL} = 4.6\ \text{mL}$$

Convert the volume from mL to cm^3:

$$\text{Volume } (\text{cm}^3) = 4.6\ \cancel{\text{mL}} \times \frac{1\ \text{cm}^3}{1\ \cancel{\text{mL}}} = 4.6\ \text{cm}^3$$

Convert the volume from mL to L:

$$\text{Volume (L)} = 4.6\ \cancel{\text{mL}} \times \frac{10^{-3}\ \text{L}}{1\ \cancel{\text{mL}}} = 4.6 \times 10^{-3}\ \text{L}$$

Check The units and magnitudes of the answers seem correct, and it makes sense that the volume in mL would have a number 1000 times the same volume in L.

Follow-Up Problem 1.4 Within a cell, proteins are synthesized on particles called ribosomes. Assuming ribosomes are spherical, what is the volume (in dm^3 and μL) of a ribosome whose average diameter is 21.4 nm? Draw a road map to show how you planned the solution.

Mass *The quantity of matter that an object contains* is its **mass**. The SI unit of mass is the **kilogram (kg)**, the only base unit whose standard is an object—a platinum-iridium cylinder kept in France—and the only one whose name has a prefix.*

The terms *mass* and *weight* have distinct meanings:

- *Mass is constant* because an object's quantity of matter cannot change.
- **Weight** *is variable* because it depends on the local gravitational field that is acting on an object.

Because the strength of the gravitational field varies with altitude, you (and other objects) weigh slightly less on a high mountain than you do at sea level.

Does this mean that weighing a sample on laboratory balances in Queen's County, Prince Edward Island (sea level), and on Mount Logan, Yukon Territory (about 6.0 km above sea level), gives different results? No, because these balances measure mass, not weight. (We actually "mass" an object when we weigh it on a balance, but we do not use this term.) Mechanical balances compare the object's mass with masses built into the balance, so the local gravitational field pulls on them

*The names of the other base units are used as the root words. For units of mass, however, we attach prefixes to the word "gram," as in "microgram" and "kilogram"; thus, we say "milligram," never "microkilogram."

equally. Electronic (analytical) balances generate an electric field that counteracts the local field, and the current needed to restore the pan to zero is converted to the equivalent mass and displayed.

Sample Problem 1.5 Converting Units of Mass

Problem Many computer communications are carried by optical fibres in cables. If one strand of optical fibre weighs 540 mg/m, what is the mass (kg) of a cable made of six strands of optical fibre, each long enough to link St. John's, Newfoundland and Labrador, and Vancouver, British Columbia (7.55×10^3 km)?

Plan We have to find the mass of the cable (kg) from the given mass per length of fibre (540 mg/m), number of fibres per cable (6), and length of the cable (7.55×10^3 km). Let us first find the mass of one fibre and then the mass of the cable. As shown in the road map, we convert the length of one fibre from km to m and then find its mass (mg) by converting m to mg. Then we multiply the mass of the fibre by 6 to get the mass of the cable, and finally convert mg to kg.

Road Map

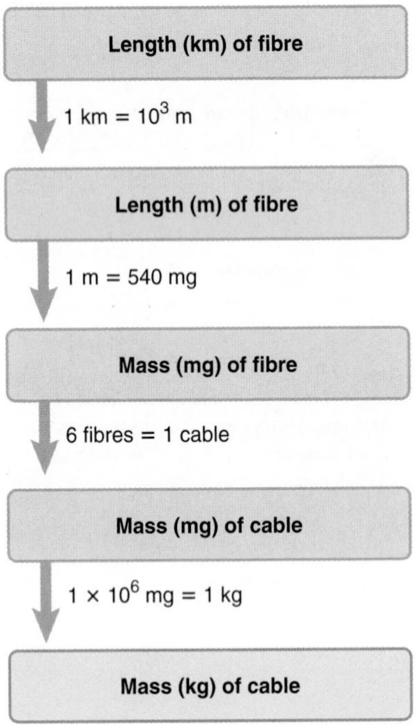

Solution Convert the fibre length from km to m:

$$\text{Length (m) of fibre} = 7.55 \times 10^3\ \cancel{\text{km}}\left(\frac{10^3\ \text{m}}{1\ \cancel{\text{km}}}\right) = 7.55 \times 10^6\ \text{m}$$

Convert the length of one fibre to mass (mg):

$$\text{Mass (mg) of fibre} = 7.55 \times 10^6\ \cancel{\text{m}}\left(\frac{540\ \text{mg}}{1\ \cancel{\text{m}}}\right) = 4.0\underline{8} \times 10^9\ \text{mg}$$

Find the mass of the cable (mg):

$$\text{Mass (mg) of cable} = \frac{4.0\underline{8} \times 10^9\ \text{mg}}{1\ \cancel{\text{fibre}}} \times \frac{6\ \cancel{\text{fibres}}}{1\ \text{cable}} = 2.4\underline{5} \times 10^{10}\ \text{mg/cable}$$

Convert the mass of the cable from mg to kg:

$$\text{Mass (kg) of cable} = \frac{2.4\underline{5} \times 10^{10}\ \cancel{\text{mg}}}{1\ \text{cable}} \times \frac{1\ \text{kg}}{10^6\ \cancel{\text{mg}}} = 2.4 \times 10^4\ \text{kg/cable}$$

Check The units are correct. Let us think through the relative sizes of the answers to see if the answers make sense. The number of m should be 10^3 times the number of km. If 1 m of fibre weighs about 0.5 g, about 10^7 m should weigh about 5×10^6 g. The mass of the cable should be six times as much, or about 30×10^6 g. Since 1 kg is 1000 g, the number of kg should be 10^{-3} times the number of g.

Follow-Up Problem 1.5 An intravenous nutrient solution is delivered to a hospital patient at a rate of 1.5 drops per second. If a drop of solution weighs 65 mg, on average, what mass (kg) is delivered in 8.0 h? Draw a road map to show how you planned the solution.

Figure 1.9 shows the ranges of some common lengths, volumes, and masses.

Density The **density (*d*)** of an object is its mass divided by its volume:

$$\text{Density} = \frac{\text{mass}}{\text{volume}} \tag{1.1}$$

We isolate each of these variables by treating density as a conversion factor:

$$\text{Mass} = \text{volume} \times \text{density} = \cancel{\text{volume}} \times \frac{\text{mass}}{\cancel{\text{volume}}}$$

or

$$\text{Volume} = \text{mass} \times \frac{1}{\text{density}} = \cancel{\text{mass}} \times \frac{\text{volume}}{\cancel{\text{mass}}}$$

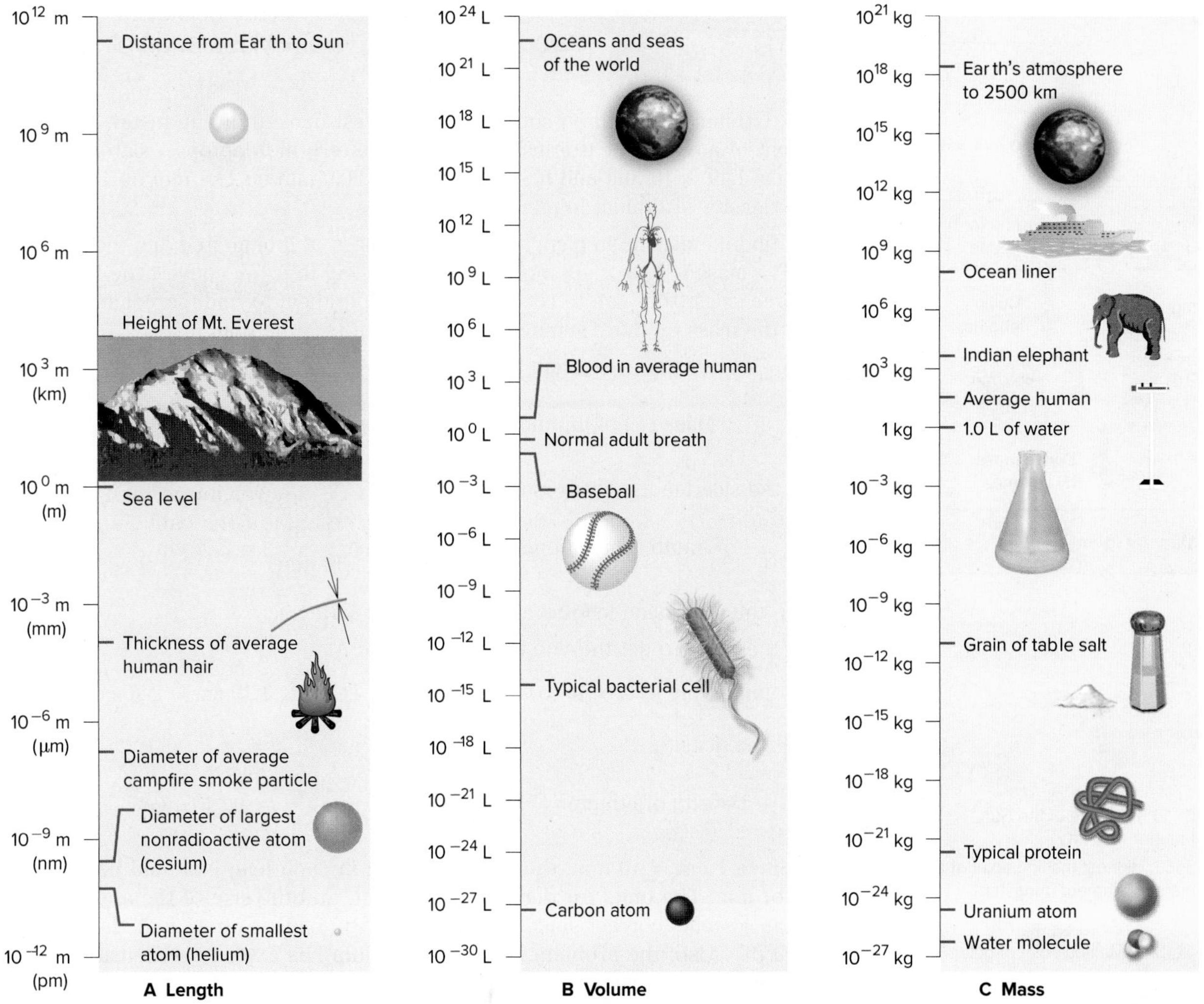

FIGURE 1.9 Some interesting quantities of length (A), volume (B), and mass (C). The vertical scales are exponential.

Because volume can change with temperature, so can density. At a given temperature and pressure, however, the *density of a substance is a characteristic physical property and, thus, has a specific value.*

The SI unit of density is kilogram per cubic metre (kg/m^3), but, in chemistry, we commonly use units of g/L (g/dm^3) or g/mL (g/cm^3) to measure densities (Table 1.4). The densities of gases are much lower than the densities of liquids or solids (see Figure 1.2).

TABLE 1.4 Densities of Some Common Substances*

Substance	Physical State	Density (g/cm^3)
Hydrogen	Gas	0.000 089 9
Oxygen	Gas	0.001 33
Grain alcohol	Liquid	0.789
Water	Liquid	0.998
Table salt	Solid	2.16
Aluminum	Solid	2.70
Lead	Solid	11.3
Gold	Solid	19.3

*At room temperature (20°C) and normal atmospheric pressure (1 bar).

Road Map

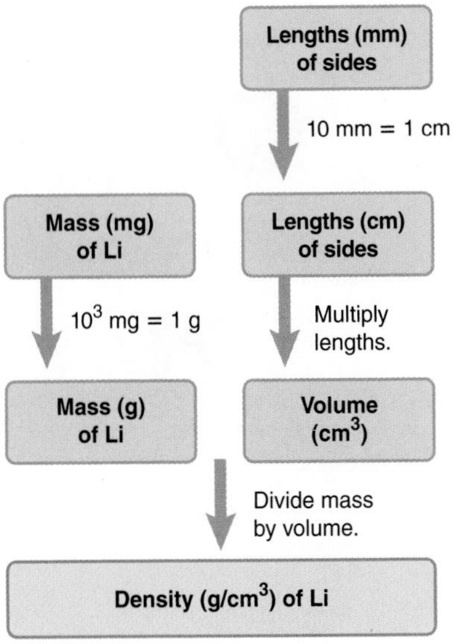

Sample Problem 1.6 Calculating Density from Mass and Volume

Problem Lithium, a soft, grey solid with the lowest density of any metal, is a key component of advanced batteries, such as the battery in a laptop. A slab of lithium weighs 1.49×10^3 mg and has sides that are 20.9 mm by 11.1 mm by 11.9 mm. Find the density of lithium in g/cm^3.

Plan To find the density in g/cm^3, we need the mass of lithium in g and the volume in cm^3. The mass is 1.49×10^3 mg, so we convert mg to g. We convert the lengths of the three sides from mm to cm and then multiply them to find the volume in cm^3. Dividing the mass by the volume gives the density (see the road map).

Solution Convert the mass from mg to g:

$$\text{Mass (g) of lithium} = 1.49 \times 10^3 \cancel{\text{mg}}\left(\frac{10^{-3}\text{ g}}{1\ \cancel{\text{mg}}}\right) = 1.49\text{ g}$$

Convert the side lengths from mm to cm:

$$\text{Length (cm) of one side} = 20.9\ \cancel{\text{mm}}\left(\frac{1\text{ cm}}{10\ \cancel{\text{mm}}}\right) = 2.09\text{ cm}$$

Similarly, the other side lengths are 1.11 cm and 1.19 cm.

Multiply the sides to get the volume:

$$\text{Volume (cm}^3) = l \times w \times h = 2.09\text{ cm} \times 1.11\text{ cm} \times 1.19\text{ cm} = 2.76\text{ cm}^3$$

Calculate the density:

$$\text{Density of lithium} = \frac{\text{mass}}{\text{volume}} = \frac{m}{V} = \frac{1.49\text{ g}}{2.76\text{ cm}^3} = 0.540\text{ g/cm}^3$$

Check Since 1 cm = 10 mm, the number of cm in each length should be $\frac{1}{10}$ the number of mm. The units for density are correct, and the size of the answer (~0.5 g/cm^3) seems correct since the number of g (1.49) is about half the number of cm^3 (2.76). Also, the problem states that lithium has a very low density, so this answer makes sense.

Follow-Up Problem 1.6 The piece of galena in Sample Problem 1.4 has a volume of 4.6 cm^3. If the density of galena is 7.5 g/cm^3, what is the mass (in kilograms) of this piece of galena? Draw a road map to show how you planned the solution.

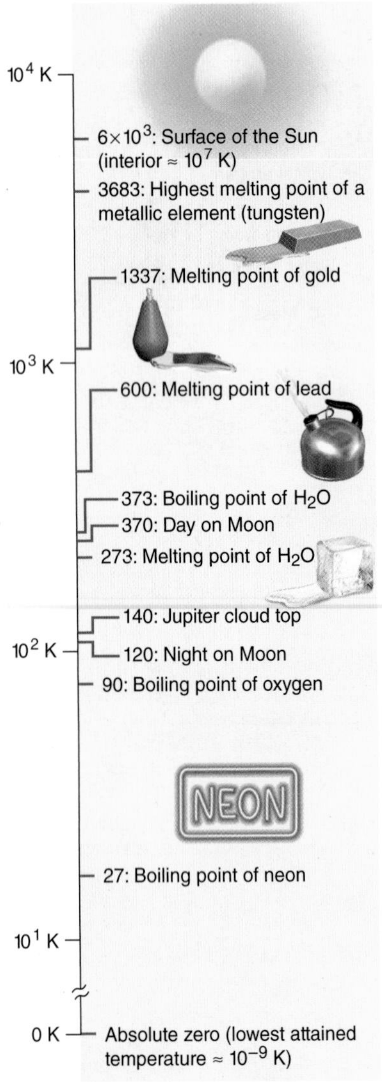

FIGURE 1.10 Some interesting temperatures

Temperature There is a noteworthy distinction between temperature and heat:

- **Temperature (T)** is *a measure of how hot or cold one object is relative to another object.*
- **Heat** is *the energy that flows from an object with a higher temperature to an object with a lower temperature.* When you hold an ice cube, it feels like the "cold" flows into your hand, but, actually, heat flows from your hand to the ice.

In the laboratory, we measure temperature with a **thermometer**, *a narrow tube containing a fluid that expands when heated.* When the thermometer is immersed in a substance hotter than itself, heat flows from the substance through the glass into the fluid, which expands and rises in the tube. If a substance is colder than the thermometer, heat flows to the substance from the fluid, which contracts and falls within the tube.

We will consider two temperature scales: the Celsius (°C, formerly called centigrade) scale and the Kelvin (K) scale. The *SI base unit of temperature* is the **kelvin** (**K**, with no degree sign, °). Figure 1.10 shows some interesting temperatures in the Kelvin scale, which is preferred in scientific work (although the Celsius scale is still used frequently). In the United States, the Fahrenheit scale is still used in many situations, such as weather reporting and body temperature.

The two scales differ in the temperature of the zero point. Figure 1.11 shows the freezing and boiling points of water in both scales. The **Celsius scale** sets the

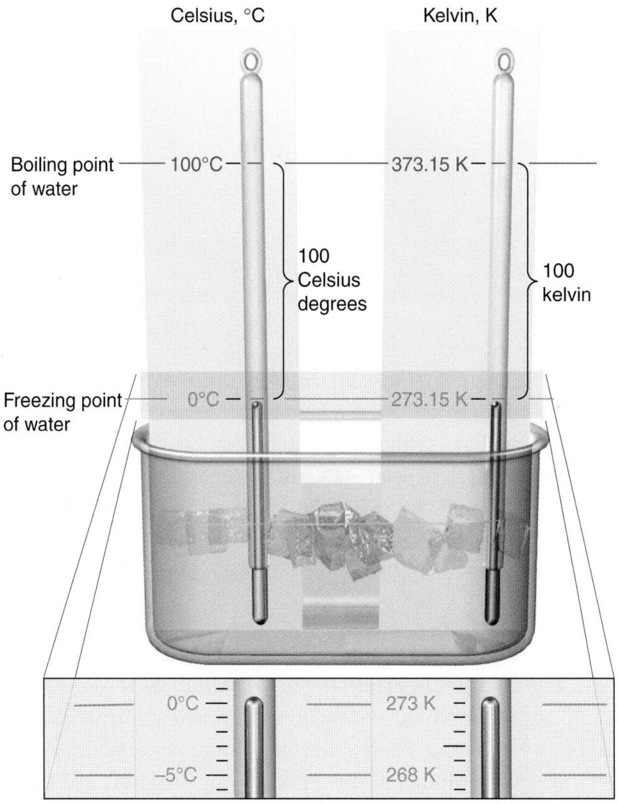

FIGURE 1.11 Freezing and boiling points of water in the Celsius and Kelvin (absolute) scales. At the bottom of the figure, a portion of each thermometer scale is expanded to show the sizes of the units.

freezing point of water at 0°C and the boiling point (at normal atmospheric pressure) at 100°C. The **Kelvin (absolute) scale** uses the *same intervals* as the Celsius scale—$\frac{1}{100}$ of the difference between the freezing and boiling points of water—but it has a *different zero point*; that is, 0 K, or *absolute zero*, equals −273.15°C. Thus, in the Kelvin scale, *all temperatures are positive*; for example, water freezes at +273.15 K (0°C) and boils at +373.15 K (100°C).

We convert between the Celsius and Kelvin scales by remembering the different zero points: 0°C = 273.15 K, so

$$T\text{ (in K)} = T\text{ (in °C)} + 273.15 \quad (1.2)$$

And, therefore,

$$T\text{ (in °C)} = T\text{ (in K)} - 273.15 \quad (1.3)$$

Table 1.5 compares the two temperature scales.

TABLE 1.5 The Two Temperature Scales

Scale	Unit	Size of Interval (Relative to K)	Freezing Point of H_2O	Boiling Point of H_2O	*T* at Absolute Zero	Conversion
Kelvin (absolute)	kelvin (K)	—	273.15 K	373.15 K	0 K	To °C (Equation 1.3)
Celsius	Celsius degree (°C)	1	0°C	100°C	−273.15°C	To K (Equation 1.2)

Sample Problem 1.7 Converting Units of Temperature

Problem A child has a body temperature of 311.9 K, and normal body temperature is 37.0°C. What is the child's temperature in degrees Celsius (°C)? Does the child have a fever?

Plan To see if the child has a fever, we convert from K to °C (Equation 1.3) and compare our result with 37.0°C.

Solution Convert the temperature from K to °C:

$$T\text{ (in °C)} = T\text{ (in K)} - 273.15 = 311.9 - 273.15 = 38.8\text{°C}$$

Yes. The child has a fever. See Rules for Rounding in Section 1.6.

Check From everyday experience, you know that 38.8°C is a reasonable temperature for someone with a fever.

Follow-Up Problem 1.7 Mercury melts at −39°C, lower than any other pure metal. What is its melting point in kelvin (K)?

Time The SI base unit of time is the **second (s)**, which is now based on an atomic standard. The best pendulum clock is accurate to within 3 s per year, and the best quartz clock is 1000 times as accurate. The most recent version of the atomic clock is over 6000 times as accurate as that—within 1 s in 20 million years! Rather than the oscillations of a pendulum, the atomic clock measures the oscillations of microwave radiation absorbed by gaseous cesium atoms cooled to around 10^{-6} K: 1 s is defined as 9 192 631 770 of these oscillations. Chemists now use lasers to measure the speed of extremely fast reactions that occur in a few picoseconds (10^{-12} s) or femtoseconds (10^{-15} s).

Extensive and Intensive Properties Some variables are *dependent on the amount of substance that is present*; these variables are called **extensive properties**. On the other hand, **intensive properties** are *independent of the amount of substance*. Mass and volume, for example, are extensive properties, but density is an intensive property. Thus, a litre of water has 1000 times the mass of a millilitre of water, but it also has 1000 times the volume, so the density, the *ratio* of mass to volume, is the same for both samples.

Another important example concerns heat, an extensive property, and temperature, an intensive property: a vat of boiling water is capable of transferring more heat (that is, it contains more energy) than a cup of boiling water, but both samples are the same temperature.

SUMMARY OF SECTION 1.5

- The SI unit system consists of seven base units and numerous derived units.
- Exponential notation and prefixes based on powers of 10 are used to express very small and very large numbers.
- The SI base unit of length is the metre (m); on the atomic scale, the nanometre (nm) and picometre (pm) are commonly used.
- Volume (V) units are derived from length units. The most important volume units are the cubic metre (m^3) and the litre (L).
- The *mass* of an object—the quantity of matter in it—is constant. The SI unit of mass is the kilogram (kg). The *weight* of an object varies with the gravitational field.
- Density (d) is a characteristic physical property of a substance and is the ratio of its mass to its volume.
- Temperature (T) is a measure of the relative hotness of an object. Heat is energy that flows from an object at higher T to an object at lower T.
- Temperature scales differ in the size of the degree unit and/or the zero point. For scientific uses, temperature is measured in kelvin (K) or degrees Celsius (°C).
- Extensive properties, such as mass, volume, and energy, depend on the amount of a substance. Intensive properties, such as density and temperature, do not.

1.6 Uncertainty in Measurement: Significant Figures

All measuring devices—balances, pipettes, thermometers, and so on—are made to limited specifications, and we use our imperfect senses and skills to read them. Therefore, we can *never* measure a quantity exactly; put another way, every measurement includes some **uncertainty**. The device we choose depends on how much

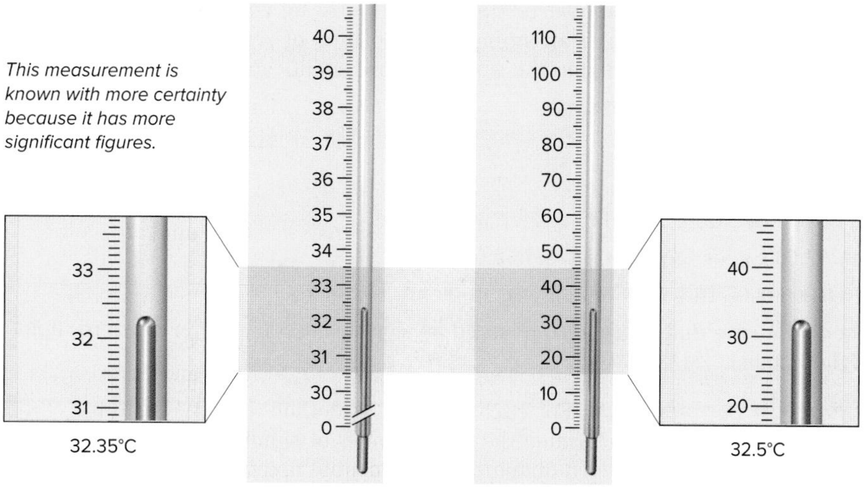
This measurement is known with more certainty because it has more significant figures.
32.35°C
32.5°C

FIGURE 1.12 The number of significant figures in a measurement. The thermometer on the left is graduated in increments of 0.1°C; the thermometer on the right is graduated in increments of 1°C.

uncertainty is acceptable. When you buy potatoes, a supermarket scale that measures in 0.1 kg increments is acceptable; it tells you that the mass is, for example, 2.0 ± 0.05 kg. The "± 0.05 kg" term expresses the uncertainty: the potatoes weigh between 1.95 kg and 2.05 kg. Needing more certainty than this to weigh a substance, a chemist uses a balance that measures in 0.001 kg increments and finds that the substance weighs 2.036 ± 0.0005 kg, or between 2.0355 and 2.0365 kg. The greater number of digits in this measurement means that we know the mass of the substance with *more certainty* than we know the mass of the potatoes.

We *always estimate the rightmost digit* of a measurement. The uncertainty can be expressed with the ± sign, but generally we drop the sign and *assume an uncertainty of one unit in the rightmost digit.* The digits we record, both the certain and the uncertain ones, are called **significant figures**. There are four significant figures in 2.036 kg and two in 2.0 kg. *The greater the number of significant figures, the greater is the certainty of a measurement.* Figure 1.12 shows this point for two thermometers.

Determining Which Digits Are Significant

When you take a measurement or use a measurement in a calculation, you must know the number of digits or figures that are significant. *All the digits are significant, except zeros used only to position the decimal point.* To determine which digits are significant in a measurement with a decimal point, you can use the following procedure:

1. Make sure that the measurement has a decimal point.
2. Start at the left, and move right until you reach the first nonzero digit.
3. Count this digit and every digit to its right as significant.

A complication can arise when zeros end a number:

- If there *is* a decimal point and the zeros lie either after or before it, they *are* significant: 1.1300 g has five significant digits, or figures, and 6500. has four.
- If there is *no* decimal point, we assume that the zeros are *not* significant, unless exponential notation clarifies the quantity: 5300 L is *assumed* to have two significant figures, but 5.300×10^3 L has four, 5.30×10^3 L has three, and 5.3×10^3 L has two.
- A terminal decimal point indicates that zeros are significant: 500 mL has one significant figure, but 500. mL has three (as do 5.00×10^2 mL and 0.500 L).

Sample Problem 1.8 Determining the Number of Significant Figures

Problem For each quantity, underline the zeros that are significant figures (sf) and determine the total number of significant figures. For parts (d) to (f), express each quantity in exponential notation first.

(a) 0.0030 L **(b)** 0.1044 g **(c)** 53 069 mL
(d) 0.000 047 15 m **(e)** 57 600. s **(f)** 0.000 000 716 0 cm^3

Plan We determine the number of significant figures by counting digits, as just presented, paying particular attention to the position of zeros in relation to the decimal point, and underlining the zeros that are significant.

Solution **(a)** 0.0030 L has 2 sf.

(b) 0.1044 g has 4 sf.

(c) 53 069 mL has 5 sf.

(d) 0.000 047 15 m, or 4.715×10^{-5} m, has 4 sf.

(e) 57 600. s, or 5.7600×10^{4} s, has 5 sf.

(f) 0.000 000 716 0 cm^3, or 7.160×10^{-7} cm^3, has 4 sf.

Check Be sure that every zero counted as significant comes after nonzero digit(s) in the number.

Follow-Up Problem 1.8 For each quantity, underline the zeros that are significant figures and determine the total number of significant figures (sf). For parts (b) to (f), express each quantity in exponential notation first.

(a) 31.070 mg **(b)** 0.06060 g **(c)** 850.°C

(d) 200.0 mL **(e)** 0.0000039 m **(f)** 0.000401 L

Significant Figures: Calculations and Rounding

Measuring several quantities typically results in data with differing numbers of significant figures. In a calculation, we keep track of the number in each quantity so that we do not have more significant figures (more certainty) in the answer than in the data. If we do have too many significant figures, we must **round** the answer.

The general rule for rounding is that *the least-certain measurement sets the limit on certainty for the entire calculation and determines the number of significant figures in the final answer.* Suppose that you want to find the density of a new ceramic. You measure the mass of a piece of the ceramic on a precise laboratory balance and obtain 3.8056 g. Then you measure the volume as 2.5 mL by displacement of water in a graduated cylinder. The mass has five significant figures, but the volume has only two. Should you report the density as 3.8056 g/2.5 mL = 1.5222 g/mL or as 1.5 g/mL? The answer with five significant figures implies more certainty than the answer with two. But you did not measure the volume to five significant figures, so you cannot possibly know the density with this much certainty. Therefore, you report 1.5 g/mL, the answer with two significant figures.

Rules for Arithmetic Operations There are three rules for arithmetic operations:

1. *For multiplication and division,* the answer contains the same number of significant figures as there are in the measurement with the *fewest significant figures.* Suppose that you want to find the volume of a sheet of a new graphite composite. The length (9.2 cm) and width (6.8 cm) are obtained with a ruler, and the thickness (0.3744 cm) is obtained with a set of calipers. The calculation is

$$\text{Volume (cm}^3) = 9.2\text{ cm} \times 6.8\text{ cm} \times 0.3744\text{ cm} = 23.422464\text{ cm}^3 = 23\text{ cm}^3$$

Even though your calculator shows 23.422464 cm^3, you report 23 cm^3, the answer with two significant figures, the same as in the measurements with the lower number of significant figures. After all, if the length and width have two significant figures, you cannot possibly know the volume with more certainty.

2. *For addition and subtraction,* the answer has the same number of decimal places as there are in the measurement with the *fewest decimal places.* Suppose that you want the total volume after adding water to a protein solution. You have 83.5 mL of solution in a graduated cylinder, and add 23.28 mL of water from a burette. The calculation is shown in the margin. Here the calculator shows 106.78 mL, but you report the volume as 106.8 mL, because the measurement with fewer decimal places (83.5 mL) has one decimal place.

$$\begin{array}{r} 83.5\text{ mL} \\ +\ 23.28\text{ mL} \\ \hline 106.78\text{ mL} \end{array}$$

Answer: Volume = 106.8 mL

3. *For logarithms,* the answer has the same number of significant figures *after the decimal place* as there are in the number whose logarithm was taken. This rule

can be applied to both natural logarithms (ln) and logarithms with base 10 (log). Suppose that you are given a strong acid with a concentration of 0.0105 mol/L. The acid has a concentration given to three significant figures. (The zeros before the 1 are not significant. If you are not sure whether the zeros in a number are significant, convert the number to exponential notation.) To find the pH of the acid, take the negative base 10 logarithm of the acid concentration (Chapter 16):

$$\text{pH} = -\log[\text{H}^+] = -\log\,(0.0\mathbf{105}) = 1.\mathbf{979}$$

Thus, when we take the logarithm of a number with x significant figures, the logarithm should have x digits *after* the decimal.

The same rule applies in reverse when taking inverse logarithms. If you have a strong acid with a pH of 2.42 and you want to know the concentration of the acid, you take 10 to the negative power of the pH. The pH has two significant figures *after* the decimal, and so you determine the concentration of the acid to two significant figures.

$$[\text{H}^+] = 10^{-\text{pH}} = 10^{-2.42} = 3.8 \times 10^{-3}\ \text{mol/L}$$

Rules for Rounding You usually need to round the final answer to the proper number of significant figures or decimal places. Notice that we removed the extra digits when calculating the volume of the graphite composite above, but we removed the extra digit and increased the last digit by 1 when calculating the total volume of the protein solution. The general rule for rounding is that *the least-certain measurement sets the limit on the certainty of the final answer.* Here are detailed rules for rounding:

1. If the digit removed is *more than 5*, the preceding number increases by 1. For example, 5.379 rounds to 5.38 if you need three significant figures and to 5.4 if you need two significant figures.
2. If the digit removed is *less than 5*, the preceding number remains the same. For example, 0.2413 rounds to 0.241 if you need three significant figures and to 0.24 if you need two significant figures.
3. If the digit removed is *5*, the preceding number increases by 1 if it is odd and remains the same if it is even. For example, 17.75 rounds to 17.8, but 17.65 rounds to 17.6. If the 5 is followed only by zeros, rule 3 is followed; if the 5 is followed by non-zeros, rule 1 is followed: 17.6500 rounds to 17.6, but 17.6513 rounds to 17.7.
4. *Always carry one or two additional significant figures through a multistep calculation and round* ***only*** *the final answer.* Do not be concerned if you string together a calculation to check a sample or follow-up problem and find that your answer differs in the last decimal place from the answer in the book. To show you the correct number of significant figures in text calculations, *we will indicate the correct number of significant figures in regular type and underline any extra digits that are carried in intermediate steps*; the extra digits will be carried only to show you that the calculation gives the correct answer.

Note that you should never read a number directly from a calculator, since calculators give answers with too many figures. You must correctly round the displayed result.

Significant Figures in the Lab The measuring device you choose determines the number of significant figures you can obtain. Suppose that an experiment requires a solution made by dissolving a solid in a liquid. You weigh the solid on an analytical balance and obtain a mass with five significant figures. It would make sense to measure the liquid with a burette or a pipette, since both of these devices measure volumes to more significant figures than a graduated cylinder. If you do choose to use a graduated cylinder, you will have to round more digits, and some certainty in the mass value will be wasted (Figure 1.13). With experience, you will be able to choose a measuring device based on the number of significant figures you need in the final answer.

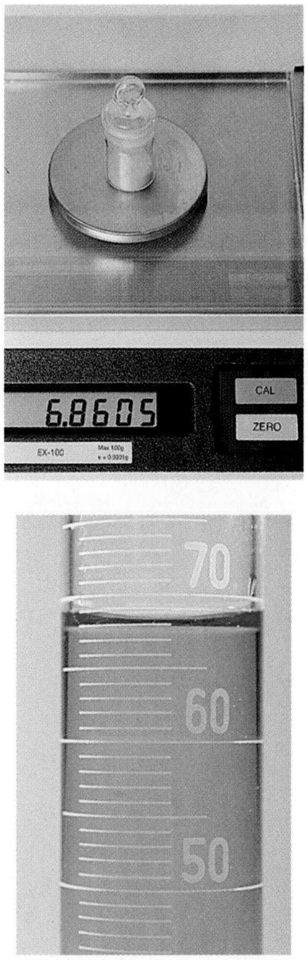

FIGURE 1.13 Significant figures and measuring devices. The mass measurement (6.8605 g) has more significant figures than the volume measurement (68.0 mL).

Exact Numbers An **exact number** *has no uncertainty associated with it.* Some exact numbers are part of a unit conversion: by definition, there are exactly 60 min in 1 h, 1000 μg in 1 mg, and 760 mm Hg in 1 atm. Other exact numbers result from actually counting items: there are exactly 3 coins in my hand, 26 letters in the English alphabet, and so on. Therefore, unlike a measured quantity, *exact numbers do not limit the number of significant figures in a calculation.*

Sample Problem 1.9 Significant Figures and Rounding

Problem Perform the following calculations, and round the answers to the correct number of significant figures:

(a) $$\frac{16.3521\ \text{cm}^2 - 1.448\ \text{cm}^2}{7.085\ \text{cm}}$$ **(b)** $$\frac{(4.80 \times 10^4\ \text{mg})\left(\frac{1\ \text{g}}{1000\ \text{mg}}\right)}{11.55\ \text{cm}^3}$$

Plan We use the rules just presented in the text: **(a)** We subtract before we divide. **(b)** We note that the unit conversion involves an exact number.

Solution (a) $$\frac{16.3521\ \text{cm}^2 - 1.448\ \text{cm}^2}{7.085\ \text{cm}} = \frac{14.904\ \text{cm}^2}{7.085\ \text{cm}} = 2.104\ \text{cm}$$

(b) $$\frac{(4.80 \times 10^4\ \text{mg})\left(\frac{1\ \text{g}}{1000\ \text{mg}}\right)}{11.55\ \text{cm}^3} = \frac{48.0\ \text{g}}{11.55\ \text{cm}^3} = 4.16\ \text{g/cm}^3$$

Check Note that we lose a decimal place in the numerator of the answer to (a), and we retain 3 sf in the answer to (b) because there are 3 sf in 4.80. Rounding to the nearest whole number is always a good way to check:

(a) $\frac{16 - 1}{7} \approx 2$; **(b)** $\frac{\left(\frac{5 \times 10^4}{1 \times 10^3}\right)}{12} \approx 4.$

Follow-Up Problem 1.9 Perform the following calculation, and round the answer to the correct number of significant figures: $$\frac{25.65\ \text{mL} + 37.4\ \text{mL}}{73.55\ \text{s}\left(\frac{1\ \text{min}}{60\ \text{s}}\right)}$$

Precision, Accuracy, and Instrument Calibration

We may use the words *precision* and *accuracy* interchangeably in everyday speech, but they have distinct meanings for scientific measurements. **Precision**, or *reproducibility*, refers to *how close the measurements in a series are to each other*, and **accuracy** refers to *how close each measurement is to the actual value.* These terms are related to two widespread types of error:

1. **Systematic error** *produces values that are* ***either*** *all higher or all lower than the actual value.* This type of error is part of the experimental system, and it is often caused by a faulty device or by a consistent mistake in taking a reading.
2. **Random error**, in the absence of systematic error, *produces values that are higher* ***and*** *lower than the actual value.* Random error *always* occurs, but its size depends on the measurer's skill and the instrument's precision.

Precise measurements have low random error, that is, small deviations from the average. *Accurate measurements have low systematic error; although efforts should be made to keep random errors low, it is not uncommon to have high random error and* ***still*** *obtain an accurate average result.*

Suppose that each of four students measures 25.0 mL of water in a graduated cylinder of known mass and then determines the mass of the water *plus* the cylinder on a balance. If the density of water is 1.00 g/mL at the temperature of the experiment, the *actual* mass of 25.0 mL of water is 25.0 g. Each student performs the operation four times, subtracts the mass of the empty cylinder, and obtains one of four graphs (Figure 1.14). In graphs A and B, random error is large; that is, precision is low. Note, however, that in graph B there is also a systematic error (all the values are high), whereas in graph A the average of the values is close to the actual value. In graphs C and D, random error is small; that is, precision is high (the masses are reproducible). In graph D, however, the accuracy is high as well (all the values are close to 25.0 g), whereas in graph C the accuracy is low (there is a systematic error).

Systematic error can be taken into account through **calibration,** *comparing the measuring device with a known standard.* The systematic error in graph C, for example, might be caused by a poorly manufactured cylinder that reads "25.0" when it actually contains about 27 mL. If this cylinder had been calibrated, the student could have adjusted all the volumes measured with it. The student should also have calibrated the balance with standardized masses.

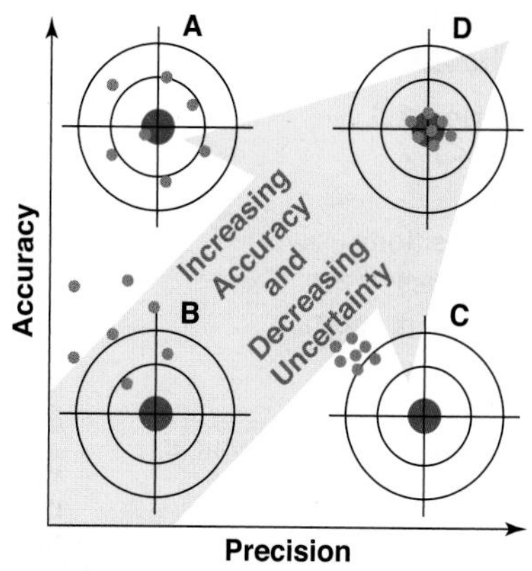

FIGURE 1.14 Precision and accuracy in a laboratory calibration. A. Low precision (large random error) but average value close to actual. **B.** Low accuracy and low precision. **C.** High precision but low accuracy (systematic error). **D.** High precision (low random error) and high accuracy.

SUMMARY OF SECTION 1.6

- The final digit of a measurement is always estimated. Thus, all measurements have some uncertainty, which is expressed by the number of significant figures.
- The certainty of a calculated result depends on the certainty of the data, so the answer has as many significant figures as in the least-certain measurement.
- Excess digits are rounded in the final answer with a set of rules.
- The choice of laboratory device depends on the certainty needed.
- Exact numbers have as many significant figures as the calculation requires.
- Precision refers to how close values are to each other. Accuracy refers to how close values are to the actual value.
- Systematic errors give values that are either all higher or all lower than the actual value. Random errors give some values that are higher and some that are lower than the actual value.
- Precise measurements have low random error. Accurate measurements have low systematic error and low random error.
- A systematic error is often caused by faulty equipment and can be compensated for by calibration.

CHAPTER REVIEW GUIDE

Learning Objectives

Relevant section (§) and/or sample problem (SP) numbers appear in parentheses.

Concepts

1. Distinguish between physical and chemical properties and changes. (§1.1; SPs 1.1, 1.2)
2. Characterize the states of matter. (§1.1)
3. Demonstrate the nature of potential energy and kinetic energy, and show their interconversion. (§1.1)
4. Approach a phenomenon scientifically and differentiate between observation, hypothesis, experiment, and model. (§1.3)
5. Identify the common units of length, volume, mass, and temperature, as well as their numerical prefixes. (§1.5)
6. Characterize the distinctions between mass and weight, heat and temperature, and intensive and extensive properties. (§1.5)
7. Recognize uncertainty in measurements, and correctly apply the use of significant figures and rounding. (§1.6)
8. Differentiate between accuracy and precision, and between systematic and random errors. (§1.6)

Skills

1. Create conversion factors in calculations, and apply a systematic approach of plan, solution, check, and follow-up for solving problems. (§1.4; SPs 1.3–1.5)
2. Calculate density from mass and volume. (SP 1.6)
3. Convert between temperatures in degrees Celsius and kelvin. (SP 1.7)
4. Determine the number of significant figures (SP 1.8), and round to the correct number of digits. (SP 1.9)

Key Terms

Section 1.1
chemistry
matter
composition
properties
physical properties
physical change
chemical properties
chemical change (chemical reaction)
states
solid
liquid
gas
plasma
Bose-Einstein condensate (BEC)
energy
potential energy
kinetic energy

Section 1.2
combustion

Section 1.3
scientific method
observations
qualitative
quantitative
data
natural law
hypothesis
experiment
variables
controlled experiment
models (theories)

Section 1.4
conversion factor
dimensional analysis

Section 1.5
SI unit
base units (fundamental units)
derived units
metre (m)
volume (V)
cubic metre (m^3)
litre (L)
millilitre (mL)
mass
kilogram (kg)
weight
density (d)
temperature (T)
heat
thermometer
kelvin (K)
Celsius scale
Kelvin (absolute) scale
second (s)
extensive properties
intensive properties

Section 1.6
uncertainty
significant figures
round
exact number
precision
accuracy
systematic error
random error
calibration

Key Equations and Relationships

1.1 Calculating density from mass and volume:

$$\text{Density} = \frac{\text{mass}}{\text{volume}};\ d = \frac{m}{V}$$

1.2 Converting temperature from °C to K:

$$T\ (\text{in K}) = T\ (\text{in °C}) + 273.15$$

1.3 Converting temperature from K to °C:

$$T\ (\text{in °C}) = T\ (\text{in K}) - 273.15$$

Brief Solutions to Follow-Up Problems

1.1 Chemical. The red and blue and separate red particles on the left become paired red and separate blue particles on the right.

1.2 (a) Physical. Solid iodine changes to gaseous iodine.

(b) Chemical. Gasoline burns in air to form different substances.

(c) Chemical. In contact with air, substances in torn skin and blood react to form different substances.

1.3 Number of dresses

$$= 2\ \text{bolts} \times \frac{200\ \text{m}^2}{1\ \text{bolt}} \times \frac{(100)^2\ (\text{cm})^2}{1\ \text{m}^2} \times \frac{1\ \text{dress}}{31.5\ \text{cm}^2}$$

$$= 127\ 000\ \text{dresses}$$

See Road Map 1.3.

1.4 Radius of ribosome (dm) $= \frac{21.4\ \text{nm}}{2} \times \frac{1\ \text{dm}}{10^8\ \text{nm}}$

$= 1.07 \times 10^{-7}$ dm

Volume of ribosome (dm^3) $= \frac{4}{3}\pi r^3 = \frac{4}{3}(\pi)(1.07 \times 10^{-7}\ \text{dm})^3$

$= 5.13 \times 10^{-21}\ \text{dm}^3$

Volume of ribosome (μL) $= (5.13 \times 10^{-21}\ \text{dm}^3)\left(\frac{1\ \text{L}}{1\ \text{dm}^3}\right)\left(\frac{10^6\ \mu\text{L}}{1\ \text{L}}\right)$

$= 5.13 \times 10^{-15}\ \mu\text{L}$

See Road Map 1.4.

Road Map 1.3

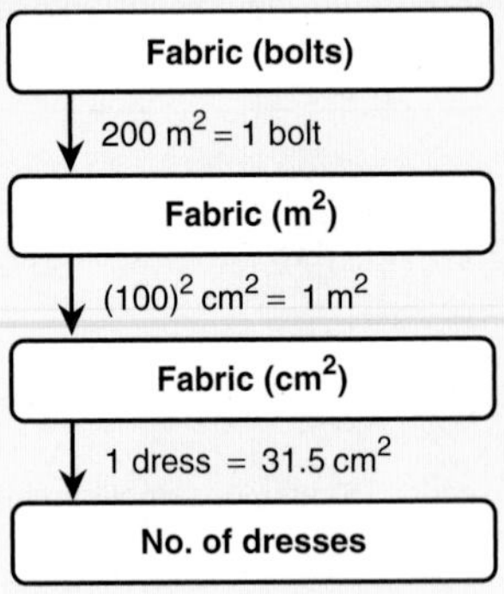

Road Map 1.4

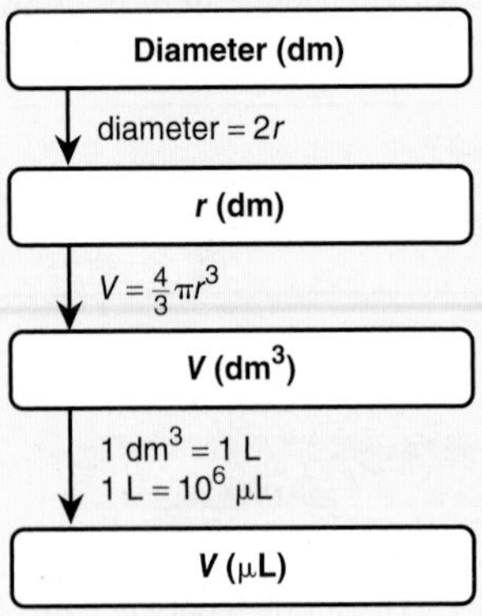

1.5 Mass (kg) of solution

$$= 8.0\ \text{h} \times \frac{60\ \text{min}}{1\ \text{h}} \times \frac{60\ \text{s}}{1\ \text{min}} \times \frac{1.5\ \text{drops}}{1\ \text{s}} \times \frac{65\ \text{mg}}{1\ \text{drop}} \times \frac{1\ \text{g}}{10^3\ \text{mg}} \times \frac{1\ \text{kg}}{10^3\ \text{g}}$$

$= 2.8$ kg

See Road Map 1.5 below.

1.6 Mass (kg) of piece of galena $= 4.6\ \cancel{\text{cm}^3} \times \dfrac{7.5\ \cancel{\text{g}}}{1\ \cancel{\text{cm}^3}} \times \dfrac{1\ \text{kg}}{10^3\ \cancel{\text{g}}}$

$= 0.034\ \text{kg}$

See Road Map 1.6.

1.7 T (in K) $= -39 + 273 = 234$ K

See Rules for Rounding in Section 1.6 for how to round with addition.

1.8 (a) $31.\underline{070}$ mg, 5 sf

(b) $0.06\underline{0}\ \underline{60}$ g or 6.060×10^{-2} g, 4 sf

(c) $85\underline{0}.$°C or 8.50×10^2 °C, 3 sf

(d) $2.\underline{000} \times 10^2$ mL, 4 sf (e) 3.9×10^{-6} m, 2 sf

(f) $4.\underline{0}1 \times 10^{-4}$ L, 3 sf

1.9 $$\frac{25.65\ \text{mL} + 37.4\ \text{mL}}{73.55\ \text{s}\left(\dfrac{1\ \text{min}}{60\ \text{s}}\right)} = \frac{63.0\ \text{mL}}{1.22\underline{6}\ \text{min}} = 51.4\ \text{mL/min}$$

Road Map 1.5

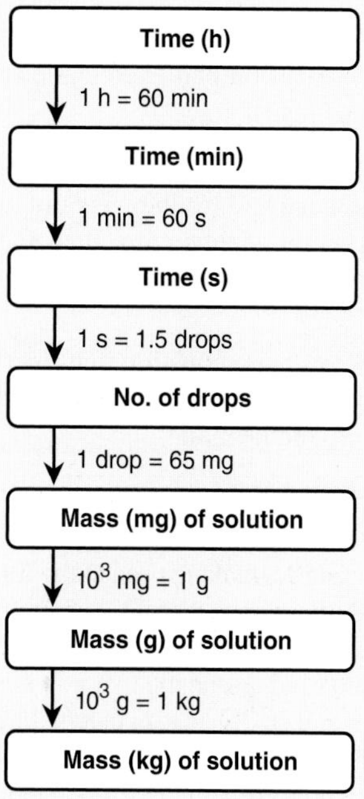

Road Map 1.6

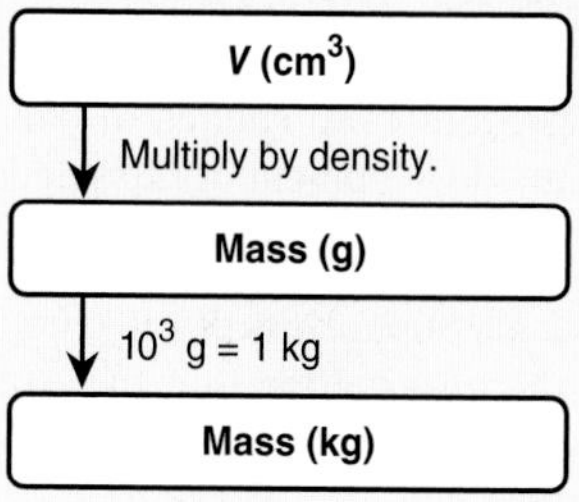

PROBLEMS

For help with using scientific (exponential) notation, see Appendix A.

Problems with **red** numbers are answered in Appendix G and worked in detail in the Student Solutions Manual. Problem sections match those in this book and provide the numbers of relevant sample problems. Most sections offer Concept Review Questions, Skill-Building Exercises (grouped in pairs covering the same concept), and Problems in Context. The Comprehensive Problems are based on material from any section.

Some Fundamental Definitions

(Sample Problems 1.1 and 1.2)

Concept Review Question

1.1 Diagrams A to D represent atomic-scale views of different samples of substances:

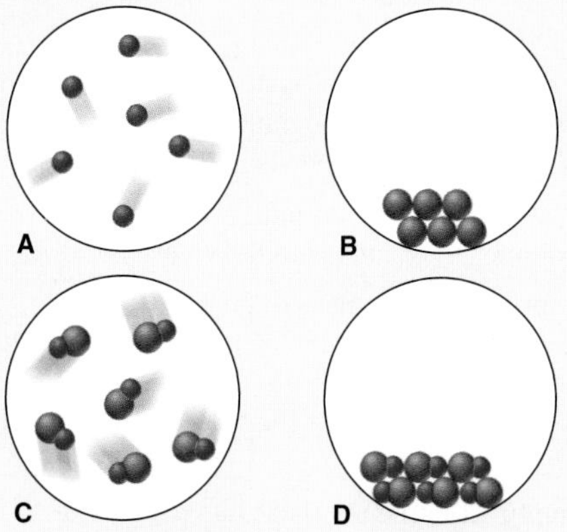

(a) Under one set of conditions, the substances in A and B mix, and the result is depicted in C. Does this represent a chemical change or a physical change?
(b) Under a second set of conditions, the same substances mix, and the result is depicted in D. Does this represent a chemical change or a physical change?
(c) Under a third set of conditions, the sample depicted in C changes to that in D. Does this represent a chemical change or a physical change?
(d) After the change in part (c) has occurred, does the sample have different chemical properties? Does it have different physical properties?

Skill-Building Exercises (grouped in similar pairs)

1.2 Describe solids, liquids, and gases in terms of how they fill a container. Use your descriptions to identify the physical state (at room temperature) of (a) helium in a toy balloon; (b) mercury in a thermometer; (c) soup in a bowl.

1.3 Use your descriptions from Problem 1.2 to identify the physical state (at room temperature) of (a) the air in your room; (b) tablets in a bottle of vitamins; (c) sugar in a packet.

1.4 Define *physical property* and *chemical property*. Identify each type of property in each statement:
(a) Yellow-green chlorine gas attacks silvery sodium metal to form white crystals of sodium chloride (table salt).
(b) A magnet separates a mixture of black iron shavings and white sand.

1.5 Define *physical change* and *chemical change*. State which type of change occurs in each statement:
(a) Passing an electric current through molten magnesium chloride yields molten magnesium and gaseous chlorine.
(b) The iron in discarded automobiles slowly forms reddish-brown, crumbly rust.

1.6 Which is a chemical change? Explain your reasoning.
(a) Boiling canned soup (b) Toasting a slice of bread
(c) Chopping a log (d) Burning a log

1.7 Which changes can be reversed by changing the temperature?
(a) Dew condensing on a leaf
(b) An egg turning hard when it is boiled
(c) Ice cream melting
(d) A spoonful of batter cooking on a hot griddle

1.8 For each pair, which has higher potential energy?
(a) The fuel in your car or the gaseous products in its exhaust
(b) Wood in a fire or the ashes after the wood burns

1.9 For each pair, which has higher kinetic energy?
(a) A sled resting at the top of a hill or a sled sliding down the hill
(b) Water above a dam or water falling over the dam

Chemical Arts and the Origins of Modern Chemistry

Concept Review Questions

1.10 The alchemical, medical, and technological traditions were precursors to chemistry. State a contribution that each made to the development of the science of chemistry.

1.11 How did the phlogiston theory explain combustion?

1.12 Supporters of the phlogiston theory had trouble explaining the observation that the calx of a metal weighs more than the metal itself. Why was this observation important? How did the phlogistonists respond?

1.13 Lavoisier developed a new theory of combustion that overturned the phlogiston theory. What measurements were central to his theory, and what key discovery did he make?

The Scientific Approach: Developing a Model

Concept Review Questions

1.14 How are the key elements of scientific thinking used in the following scenario? While making toast, you notice that it fails to pop out of the toaster. Thinking that the spring mechanism is stuck, you check and notice that the bread is untoasted. Assuming that you forgot to plug in the toaster, you check and find that it *is* plugged in. When you take the toaster into the dining room and plug it into a different outlet, you find that the toaster works. Returning to the kitchen, you turn on the switch for the overhead light and nothing happens.

1.15 Why is a quantitative observation more useful than a non-quantitative observation? Which of the following observations is (are) quantitative?
(a) The Sun rises in the east.
(b) A person weighs one-sixth as much on the Moon as on Earth.
(c) Ice floats on water.
(d) A hand pump cannot draw water from a well that is more than 10.5 m deep.

1.16 Describe the essential features of a well-designed experiment.

1.17 Describe the essential features of a scientific model.

Chemical Problem Solving

(Sample Problem 1.3)

Concept Review Question

1.18 When you convert kilometres to centimetres, how do you decide which part of the conversion factor should be in the numerator and which should be in the denominator?

Skill-Building Exercises (grouped in similar pairs)

1.19 Write the conversion factor(s) for
(a) cm^2 to m^2 (b) km^2 to cm^2 (c) km/h to m/s (d) kg/m^3 to g/cm^3

1.20 Write the conversion factor(s) for
(a) cm/min to mm/s (b) m^3 to cm^3
(c) m/s^2 to km/h^2 (d) mL/s to L/min

Measurement in Scientific Study

(Sample Problems 1.4 to 1.7)

Concept Review Questions

1.21 Describe the difference between intensive and extensive properties. Which of the following properties are intensive: (a) mass; (b) density; (c) volume; (d) melting point?

1.22 Explain the difference between mass and weight. Why is your weight on the Moon one-sixth your weight on Earth?

1.23 For each of the following, state whether the density of the object increases, decreases, or remains the same:
(a) A sample of chlorine gas is compressed.
(b) A lead weight is carried up a high mountain.
(c) A sample of water is frozen.
(d) An iron bar is cooled.
(e) A diamond is submerged in water.

1.24 Explain the difference between heat and temperature. Does 1 L of water at 65 K have more, less, or the same quantity of energy as 1 L of water at 65°C?

1.25 In a set of ratios, such as $\frac{d}{V}$, conversion between units is not necessary as long as *both* quantities are expressed using the same base units (for example, kg and kg/L, or mg and mg/mL). However, if temperature is one of the quantities and it is given in °C but should be in K, this rule does not apply. Explain why.

Skill-Building Exercises (grouped in similar pairs)

1.26 The average radius of a molecule of lysozyme, an enzyme in tears, is 1430 pm. What is its radius in nanometres (nm)?

1.27 The radius of a barium atom is 2.22×10^{-10} m. What is its radius in nanometres? In picometres?

1.28 What is the length, in nanometres, of a 100. m soccer field?

1.29 The centre on a basketball team is 0.001 96 km tall. How tall is the player in millimetres (mm)?

1.30 A small hole in the wing of a spacecraft requires a 20.7 cm^2 patch.
(a) What is the area of the patch in square kilometres (km^2)?
(b) If the patching material costs NASA \$3.25/mm^2, what is the cost of the patch?

1.31 The area of a telescope lens is 7903 mm^2.
(a) What is the area in square metres (m^2)?
(b) If it takes a technician 45 s to polish 135 mm^2, how long does it take her to polish the entire lens?

1.32 The mass of a paper clip is commonly described as being 1 g. How many paper clips are required to have a mass of 3.56 kg?

1.33 There is a mass of 2.36×10^{21} g of oxygen in the atmosphere. How many tonnes (t) of oxygen are present (1 t = 1000 kg)?

1.34 The average density of Earth is 5.52 g/cm^3. What is its density in (a) kg/m^3; (b) mg/mm^3?

1.35 The speed of light in a vacuum is 2.998×10^8 m/s. What is its speed in (a) km/h; (b) cm/min?

1.36 The volume of a certain bacterial cell is 2.56 μm^3.
(a) What is its volume in cubic millimetres (mm^3)?
(b) What is the volume of 10^5 cells in litres (L)?

1.37 (a) What volume (m^3) of milk is in 946.4 mL?
(b) What volume (L) of milk is in 835 mm^3?

1.38 An empty vial has a mass of 55.32 g.
(a) If the mass of the vial is 185.56 g when filled with liquid mercury (d = 13.53 g/cm^3), what is its volume?
(b) What would be the mass of the vial if it were filled with water (d = 0.997 g/cm^3 at 25°C)?

1.39 An empty Erlenmeyer flask has a mass of 241.3 g. When filled with water (d = 1.00 g/cm^3), the flask and its contents have a mass of 489.1 g.
(a) What is the volume of the flask? (b) What is the mass of the flask when it is filled with chloroform (d = 1.48 g/cm^3)?

1.40 A small cube of aluminum measures 15.6 mm on a side and has a mass of 10.25 g. What is the density of aluminum in g/cm^3?

1.41 A steel ball-bearing with a circumference of 32.5 mm has a mass of 4.20 g. What is the density of the steel in g/cm^3? (V of a sphere = $\frac{4}{3}\pi r^3$; C of a circle = $2\pi r$.)

1.42 Perform the following conversions:
(a) 18°C (a pleasant spring day) to K

(b) −164°C (the boiling point of methane, the main component of natural gas) to K
(c) 0 K (absolute zero, theoretically the coldest possible temperature) to °C

1.43 Perform the following conversions:
(a) 37°C (the body temperature of many birds) to K
(b) 3410°C (the melting point of tungsten, the highest of any metallic element) to K
(c) 6.1×10^3 K (the surface temperature of the Sun) to °C

Problems in Context

1.44 A 25.0 g sample of each of three unknown metals (A, B, and C) is added to 25.0 mL of water in a graduated cylinder. The final volumes of the three metals are depicted in the circles below. Given their densities, identify the metal in each cylinder: zinc (7.14 g/mL), iron (7.87 g/mL), nickel (8.91 g/mL).

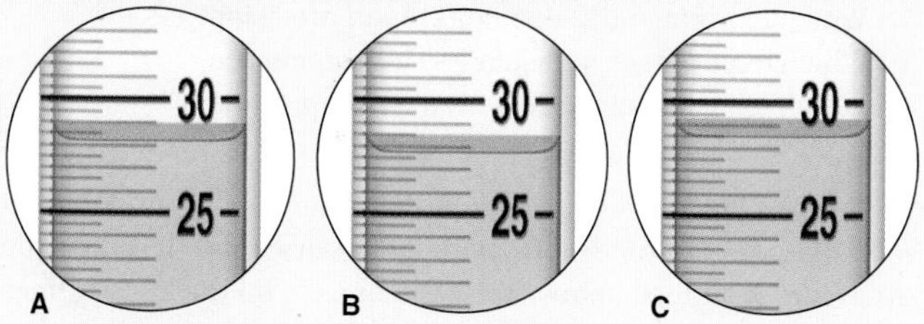

1.45 The distance between two adjacent peaks on a wave is called the *wavelength*.
(a) The wavelength of a beam of ultraviolet light is 247 nm (nanometres). What is its wavelength in metres?
(b) The wavelength of a beam of red light is 6760 pm (picometres). What is its wavelength in nanometres?

1.46 Each of the beakers depicted below contains two liquids that do not dissolve in each other. Three of the liquids are designated A, B, and C, and water is designated W.

(a) Which of the liquids is (are) more dense than water, and which is (are) less dense?
(b) If the densities of W, C, and A are 1.0 g/mL, 0.88 g/mL, and 1.4 g/mL, respectively, which of the following densities is possible for liquid B: 0.79 g/mL, 0.86 g/mL, 0.94 g/mL, or 1.2 g/mL?

1.47 A cylindrical tube, 9.5 cm high and 0.85 cm in diameter, is used to collect blood samples. What volume (dm^3) of blood can it hold? (V of a cylinder $= \pi r^2 h$.)

1.48 Copper can be drawn into thin wires. What length of 34-gauge wire, with a diameter of 0.1601 mm, can be produced from the copper in 5.01 kg of covellite, an ore of copper that is 66% copper by mass? (*Hint*: Treat the wire as a cylinder: V of a cylinder $= \pi r^2 h$; d of copper $= 8.95$ g/cm^3.)

Uncertainty in Measurement: Significant Figures
(Sample Problems 1.8 and 1.9)

Concept Review Questions

1.49 What is an exact number? How are exact numbers treated differently from other numbers in a calculation?

1.50 Which procedure(s) decrease(s) the random error of a measurement: (1) taking the average of more measurements; (2) calibrating the instrument; (3) taking fewer measurements? Explain.

1.51 A newspaper reported that the attendance at Slippery Rock's home football game was 16 532.
(a) How many significant figures does this number contain?
(b) Was the actual number of people counted?
(c) After Slippery Rock's next home game, the newspaper reported an attendance of 15 000. If you assume that this number contains two significant figures, how many people could actually have been at the game?

Skill-Building Exercises (grouped in similar pairs)

1.52 Underline the significant zeros in each number:
(a) 0.41 (b) 0.041 (c) 0.0410 (d) 4.0100×10^4

1.53 Underline the significant zeros in each number:
(a) 5.08 (b) 508 (c) 5.080×10^3 (d) 0.050 80

1.54 Round each number to the indicated number of significant figures (sf):
(a) 0.000 355 4 to 2 sf (b) 35.8348 to 4 sf (c) 22.4555 to 3 sf

1.55 Round each number to the indicated number of significant figures (sf):
(a) 231.554 to 4 sf (b) 0.008 45 to 2 sf (c) 144 000 to 2 sf

1.56 Round each number in the following calculation to one less significant figure, and find the answer:

$$\frac{19 \times 155 \times 8.3}{3.2 \times 2.9 \times 4.7}$$

1.57 Round each number in the following calculation to one less significant figure, and find the answer:

$$\frac{10.8 \times 6.18 \times 2.381}{24.3 \times 1.8 \times 19.5}$$

1.58 Carry out each calculation, making sure that your answer has the correct number of significant figures:
(a) $\dfrac{2.795 \text{ m} \times 310 \text{ m}}{6.48 \text{ m}}$ (b) $V = \frac{4}{3}\pi r^3$, where $r = 17.282$ mm
(c) 1.110 cm + 17.3 cm + 108.2 cm + 316 cm

1.59 Carry out each calculation, making sure that your answer has the correct number of significant figures:
(a) $\dfrac{2.420 \text{ g} + 15.6 \text{ g}}{48 \text{ g}}$ (b) $\dfrac{7.87 \text{ mL}}{161 \text{ mL} - 8.44 \text{ mL}}$
(c) $V = \pi r^2 h$, where $r = 6.23$ cm and $h = 4.630$ cm

1.60 Write each number in exponential notation:
(a) 131 000.0 (b) 0.000 47 (c) 210 006 (d) 2160.5

1.61 Write each number in exponential notation:
(a) 282.0 (b) 0.0380 (c) 4270.8 (d) 58 200.9

1.62 Write each number in standard notation. Use a terminal decimal point when needed:
(a) 5.55×10^3 (b) 1.0070×10^4
(c) 8.85×10^{-7} (d) 3.004×10^{-3}

1.63 Write each number in standard notation. Use a terminal decimal point when needed:
(a) 6.500×10^3 (b) 3.46×10^{-5}
(c) 7.5×10^2 (d) 1.8856×10^2

1.64 Convert each number to correct exponential notation:
(a) 802.5×10^2 (b) 1009.8×10^{-6} (c) 0.077×10^{-9}

1.65 Convert each number to correct exponential notation:
(a) 14.3×10^1 (b) 851×10^{-2} (c) 7500×10^{-3}

1.66 Carry out each calculation, paying special attention to significant figures, rounding, and units. (J = joule, the SI unit of energy; mol = mole, the SI unit for the amount of a substance.)

(a) $\dfrac{(6.626 \times 10^{-34}\ \text{J·s})(2.9979 \times 10^{8}\ \text{m/s})}{489 \times 10^{-9}\ \text{m}}$

(b) $\dfrac{(6.022 \times 10^{23}\ \text{molecules/mol})(1.23 \times 10^{2}\ \text{g})}{46.07\ \text{g/mol}}$

(c) $(6.022 \times 10^{23}\ \text{atoms/mol})(2.18 \times 10^{-18}\ \text{J/atom})\left(\dfrac{1}{2^2} - \dfrac{1}{3^2}\right)$, where the numbers 2 and 3 in the last term are exact

1.67 Carry out each calculation, paying special attention to significant figures, rounding, and units:

(a) $\dfrac{4.32 \times 10^{7}\ \text{g}}{\frac{4}{3}(3.1416)(1.95 \times 10^{2}\ \text{cm})^3}$ (The term $\frac{4}{3}$ is exact.)

(b) $\dfrac{(1.84 \times 10^{2}\ \text{g})(44.7\ \text{m/s})^2}{2}$ (The term 2 is exact.)

(c) $\dfrac{(1.07 \times 10^{-4}\ \text{mol/L})^2(3.8 \times 10^{-3}\ \text{mol/L})}{(8.35 \times 10^{-5}\ \text{mol/L})^2(1.48 \times 10^{-2}\ \text{mol/L})^3}$

1.68 Which of the following statements include exact numbers?
(a) The Horseshoe Falls at Niagara are 50.9 m high.
(b) There are 8 known planets in the Solar System.
(c) There are 402 students in the classroom.
(d) There are 1000 mm in 1 m.

1.69 Which statements include exact numbers?
(a) The speed of light in a vacuum is a physical constant; to six significant figures, it is $2.997\,92 \times 10^{8}$ m/s.
(b) The density of mercury at 25°C is 13.53 g/mL.
(c) There are 3600 s in 1 h.
(d) Canada has 10 provinces and 3 territories.

Problems in Context

1.70 How long is the metal strip shown below? Be sure to answer with the correct number of significant figures.

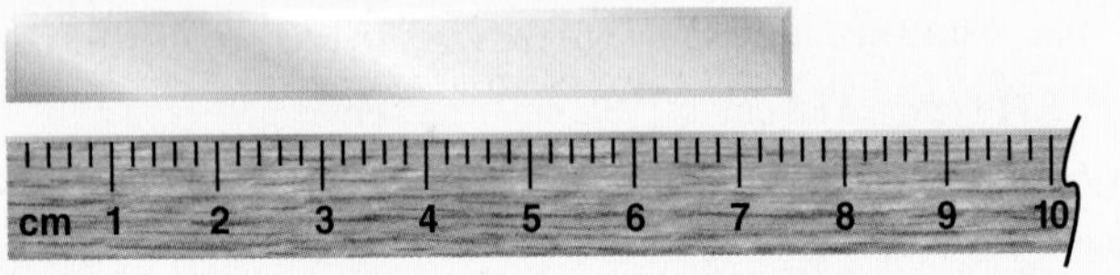

1.71 These organic solvents are used to clean compact discs:

Solvent	Density (g/mL) at 20°C
Chloroform	1.492
Diethyl ether	0.714
Ethanol	0.789
Isopropanol	0.785
Toluene	0.867

(a) If a 15.00 mL sample of a CD cleaner has a mass of 11.775 g at 20°C, which solvent does the sample most likely contain?
(b) The chemist analyzing the cleaner calibrates the equipment and finds that the pipette is accurate to ±0.02 mL, and the balance is accurate to ±0.003 g. Is this equipment precise enough to distinguish between ethanol and isopropanol?

1.72 A laboratory instructor gives a sample of amino-acid powder to each of four students (I, II, III, and IV), and the students take the mass of the samples. The true value is 8.72 g. Their results for three trials are

I: 8.72 g, 8.74 g, 8.70 g   II: 8.56 g, 8.77 g, 8.83 g
III: 8.50 g, 8.48 g, 8.51 g   IV: 8.41 g, 8.72 g, 8.55 g

(a) Calculate the average mass from each set of data, and state which set is the most accurate.
(b) Precision is a measure of the average of the deviations of each piece of data from the average value. Which set of data is the most precise? Is this set also the most accurate?
(c) Which set of data is both the most accurate and the most precise?
(d) Which set of data is both the least accurate and the least precise?

1.73 The following dartboards illustrate experiments in which the types of errors often seen in measurements occur. The bull's-eye represents the actual value, and the darts represent the data.

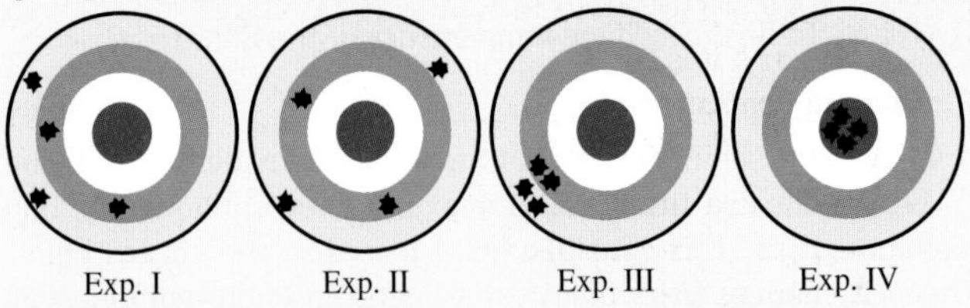

(a) Which experiments yield the same average result?
(b) Which experiment(s) display(s) high precision?
(c) Which experiment(s) display(s) high accuracy?
(d) Which experiment(s) show(s) a systematic error?

Comprehensive Problems

1.74 Two blank potential energy diagrams are given. Beneath each diagram are objects to place in the diagram. Based on the situations described below, draw the objects on the dashed lines to indicate higher or lower potential energy, and label each situation as more or less stable:

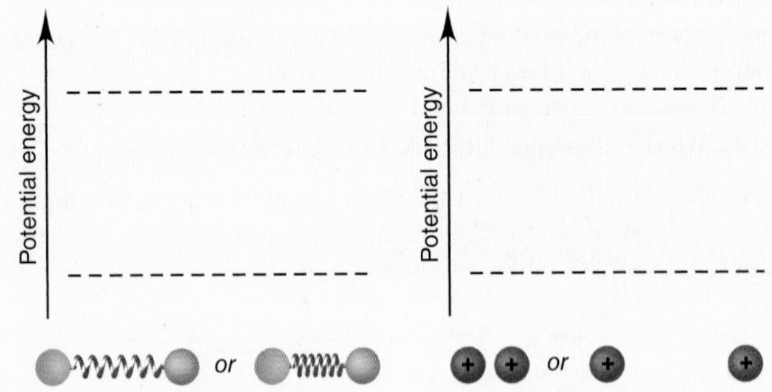

A Two balls attached to a relaxed *or* a compressed spring   **B** Two positive charges near *or* apart from each other

(a) Two balls attached to a relaxed *or* a compressed spring
(b) Two positive charges near *or* apart from each other

1.75 The diagrams below illustrate two different mixtures. When mixture A, at 273 K, is heated to 473 K, mixture B results.

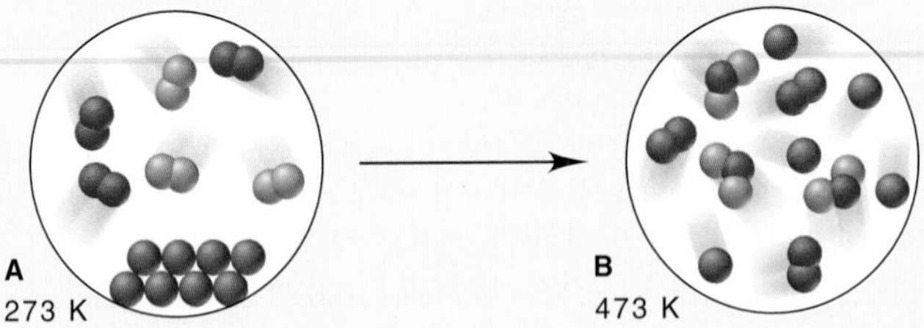

(a) How many different chemical changes occur?
(b) How many different physical changes occur?

1.76 Bromine is used to prepare the pesticide methyl bromide, as well as flame retardants for plastic electronic housings. It is recovered from seawater, underground brines, and the Dead Sea. The average concentrations of bromine in seawater (d = 1.024 g/mL) and the Dead Sea (d = 1.22 g/mL) are 0.065 g/L and 0.50 g/L, respectively. What is the mass ratio (the proportion of the masses) of bromine in the Dead Sea to bromine in seawater?

1.77 An Olympic-sized pool is 50.0 m long and 25.0 m wide.
(a) What volume (m^3) of water (d = 1.0 g/mL) is needed to fill the pool to an average depth of 146 cm?
(b) What is the mass (kg) of water in the pool?

1.78 At room temperature (20°C) and normal atmospheric pressure, the density of air is 1.189 g/L. An object will float in air if its density is less than the density of air. In a buoyancy experiment with a new plastic, a chemist creates a rigid, thin-walled ball that weighs 0.12 g and has a volume of 560 cm^3.
(a) Will the ball float if it is evacuated?
(b) Will it float if it is filled with carbon dioxide (d = 1.830 g/L)?
(c) Will it float if it is filled with hydrogen (d = 0.0899 g/L)?
(d) Will it float if it is filled with oxygen (d = 1.330 g/L)?
(e) Will it float if it is filled with nitrogen (d = 1.165 g/L)?
(f) For any situation in which the ball will float, how much weight must be added to make it sink?

1.79 Asbestos is a fibrous silicate mineral with remarkably high tensile strength. It is considered extremely hazardous to health, however, because airborne asbestos particles can cause lung cancer. Grunerite, a type of asbestos, has a tensile strength of 3.5×10^2 kg/mm^2. (Thus, a strand of grunerite with a 1 mm^2 cross-sectional area can hold up to 3.5×10^2 kg.) The tensile strengths of aluminum and Steel No. 5137 are 1.8×10^3 kg/cm^2 and 3.5×10^3 kg/cm^2, respectively. Calculate the cross-sectional areas (mm^2) of wires of aluminum and of Steel No. 5137 that have the same tensile strength as a fibre of grunerite with a cross-sectional area of 1.0 μm^2.

1.80 Earth's oceans have an average depth of 3800 m, a total surface area of 3.63×10^8 km^2, and an average concentration of dissolved gold of 5.8×10^{-9} g/L. (a) What mass (g) of gold is in the oceans? (b) What volume (m^3) of gold is in the oceans? (c) Assuming that the price of gold is $1611.46 CAD/oz t (troy ounce), what is the value of gold in the oceans? (1 oz t = 31.1 g; d of gold = 19.3 g/cm^3.)

1.81 Brass is an alloy of copper and zinc. Varying the mass percents (amounts present by mass) of the two metals produces brasses with different properties. A brass called *yellow zinc* has high ductility and strength and is 34% to 37% zinc by mass.
(a) Find the mass range (g) of copper in 185 g of yellow zinc.
(b) What is the mass range (g) of zinc in a sample of yellow zinc that contains 46.5 g of copper?

1.82 Liquid nitrogen is obtained from liquefied air and is used industrially to prepare frozen foods. It boils at 77.36 K.
(a) What is this temperature in °C?
(b) At the boiling point of liquid nitrogen, the density of the liquid is 809 g/L and the density of the gas is 4.566 g/L. What volume (L) of liquid nitrogen is produced when 895.0 L of nitrogen gas is liquefied at 77.36 K?

1.83 A jogger runs at an average speed of 9.4 km/h. (a) How fast is she running in m/s? (b) What distance (km) does she run in 98 min? (c) If she starts a run at 11:15 a.m., what time is it after she covers 14.5 km?

1.84 Diagrams A and B depict changes in matter at the atomic scale:

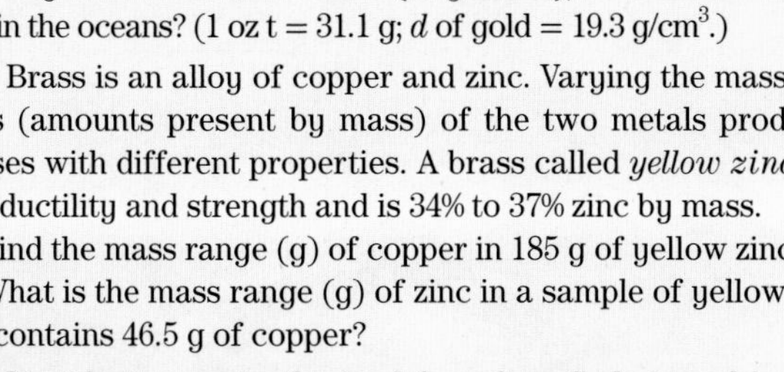
A

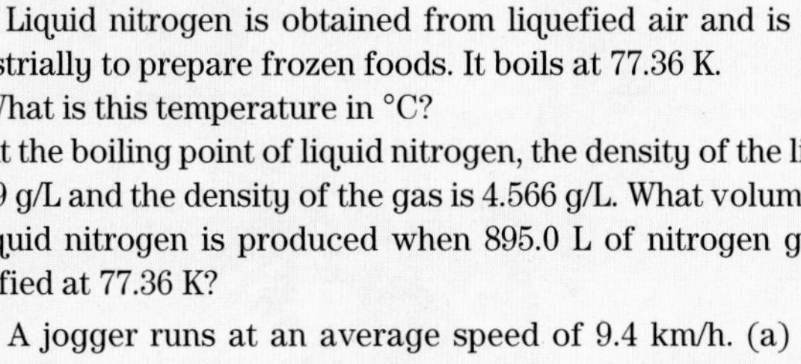
B

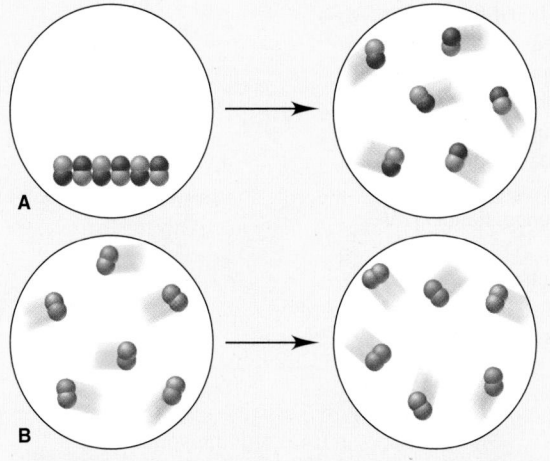

(a) Which of these diagrams show(s) a physical change?
(b) Which show(s) a chemical change?
(c) Which of the changes result(s) in different physical properties?
(d) Which result(s) in different chemical properties?
(e) Which result(s) in a change in state?

1.85 If a temperature scale were based on the freezing point (5.5°C) and boiling point (80.1°C) of benzene and the temperature difference between these points was divided into 50 units (called °X), what would be the freezing and boiling points of water in °X? (See Figure 1.11.)

1.86 Earth's surface area is 5.10×10^8 km^2; its crust has a mean thickness of 35 km and a mean density of 2.8 g/cm^3. The two most abundant elements in Earth's crust are oxygen (4.55×10^5 g/t, where t stands for tonne; 1 t = 1000 kg) and silicon (2.72×10^5 g/t). The two rarest nonradioactive elements in Earth's crust are ruthenium and rhodium, each with an abundance of 1×10^{-4} g/t. What is the total mass of each of these elements in Earth's crust?

1.87 A sheet of zinc is used to manufacture washers. The sheet is 24.0 cm long, 12.0 cm wide, and 0.5 cm thick. Each washer has an outer diameter, d, of 1.0 cm and an inner diameter, d_i, of 0.5 cm. How many washers can be made from the sheet in the diagram, assuming that the washers touch at a point on the edge? What mass *and* what volume of zinc remain after the washers have been cut? The density of zinc is 7.049 g/cm^3. (*Hint*: The centre of the washers must be included in the amount of zinc that remains.)

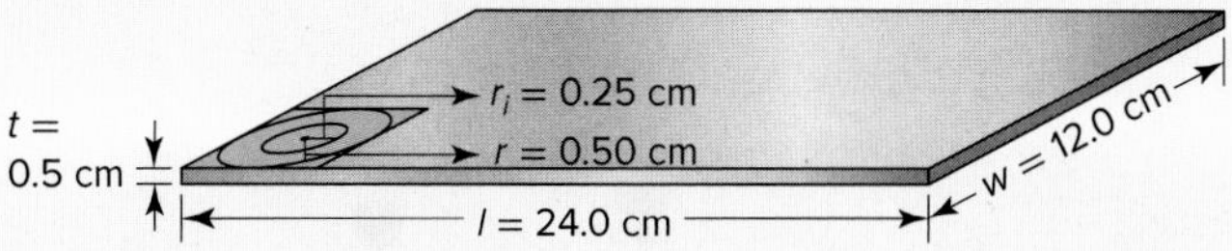

1.88 A glass tube with an internal diameter of 10.3 mm and a length of 25 cm is filled with mercury (density 13 534 kg/m^3). (a) What mass of mercury is in the tube? (b) What length of tube, with an inner diameter of 8.4 mm, would be required to hold the volume of ethanol (with density 789.00 kg/m^3) with the same mass as in part (a)?

1.89 In the Indiana Jones movie *Raiders of the Lost Ark*, Indy is shown removing a golden idol from a booby-trapped surface and replacing it with a similar size bag of sand.
(a) Explain, using scientific terms, whether this is scientifically reasonable or not.
(b) If the idol he removed was pure gold (density 19.3 g/cm^3) and had a volume of 53.5 mL, what was its mass?
(c) What volume of sand (density 1600 kg/m^3) would he have needed to keep from triggering the trap?
(d) Is this consistent with your answer to (a)?

The Components of Matter

IN THIS CHAPTER . . . We examine the properties and composition of matter on the macroscopic and atomic scales. By the end of this chapter, you should be able to

- Relate the three types of matter—elements, compounds, and mixtures—to the simple chemical entities that they comprise—atoms, ions, and molecules; more-complex entities, such as network covalent solids and macromolecules, will be discussed later
- Correlate the defining properties of the types of matter to laws discovered in the 18th century concerning the masses of substances that react with each other
- Describe in detail the 19th-century atomic model that was proposed to explain these laws
- Explain how certain 20th-century experiments led to our current understanding of atomic structure and atomic mass
- Explain how the elements are organized in the periodic table and introduce the two major ways in which elements combine
- Derive the name and formula of a compound and calculate its mass
- Depict molecules using models used by chemists
- Classify mixtures and explain how to separate them
- Differentiate between the components of matter

2

"Could anything at first sight seem more impractical than a body which is so small that its mass is an insignificant fraction of the mass of an atom of hydrogen?"

— J. J. Thomson.

Atoms are the building blocks of chemistry; everything is made of atoms; atoms are the tiniest particles that cannot be divided. These are facts about the atom that are learned from childhood and that are taken for granted with little thought as to how we know this or when such thoughts became prevalent. For those who were at the forefront of the effort to understand matter and its nature, it was a very different story. Conflicting ideas of what matter was, what differentiated one type of matter from another, and how far matter could be broken down until it became unbreakable caused huge divides among scientists. From Democritus, who first coined the term *atom* from the Greek word "atomos" meaning "that which cannot be cut," to Dalton, who proposed the first concrete framework for the properties of an atom, there was a lapse in time of over 2000 years. After much of the scientific community was finally convinced that atoms were real, it is unsurprising that the realization, first vocalized by J. J. Thomson, that atoms were in fact composed of even smaller particles came as a shock! The current model of the atom has a number of subatomic particles; in addition to the proton, neutron, and electron, atoms are known to be composed of gluons, muons, leptons, quarks, fermions, and bosons. How did scientists arrive at the current model of the atom?

This chapter briefly describes some of the important steps in the development of the model of the atom. The current model, the quantum model, is able to account for known data about the atom and was able to predict information about recently discovered elements, thus making it a reasonable model. The recent observation of the Higgs boson (2012) is one addition to our current body of scientific knowledge that supports the quantum model. This does not mean that at some time in the future, a better model may not be developed. Scientists constantly incorporate new data into existing models and revise hypotheses to improve knowledge.

Once we know how atoms of various elements differ from one another, we need to be able to express the multiple ways in which they are capable of combining. This chapter will explain the basics of chemical nomenclature and briefly discuss the periodic table as well.

Concepts and Skills to Review before Studying This Chapter

- Physical and chemical changes (Section 1.1)
- States of matter (Section 1.1)
- Attraction and repulsion between charged particles (Section 1.1)
- Meaning of a scientific model (Section 1.3)
- SI units and conversion factors (Section 1.5)
- Significant figures in calculations (Section 1.6)

2.1 Elements, Compounds, and Mixtures: An Atomic Overview

Matter can be classified into three types based on its composition: elements, compounds, and mixtures. Elements and compounds are the two kinds of substances: a **substance** is *matter whose composition is fixed.* Mixtures are not substances because they have a variable composition.

1. *Elements.* An **element** is *the simplest type of matter with unique physical and chemical properties. It consists of only one kind of atom* and, therefore, cannot be broken down into a simpler type of matter by any physical or chemical methods. Each element has a name, such as silicon, oxygen, or copper. A sample of silicon contains only silicon atoms. The *macroscopic* properties of a piece of silicon, such as colour, density, and combustibility, are different from those of a piece of copper because the *submicroscopic* properties of silicon atoms are different from those of copper atoms. *Each element is unique because the properties of its atoms are unique.*

In nature, most elements exist as populations of atoms, either separated or in contact with each other, depending on their physical state. Figure 2.1A shows atoms of an element in its gaseous state. Several elements occur in molecular form: a **molecule** is *an independent structure of two or more atoms bound together* (Figure 2.1B). Oxygen, for example, occurs in air as *diatomic* (two-atom) molecules.

2. *Compounds.* A **compound** consists of *two or more different elements that are bonded chemically* (Figure 2.1C). That is, the elements in a compound are not just mixed together: their atoms have joined in a chemical reaction. Many compounds, such as ammonia, water, and carbon dioxide, consist of molecules. Other compounds, such as sodium sulfate (which we will discuss shortly) and silicon dioxide, do not. No matter what the compound, however, one defining feature is that *the elements are present in fixed parts by mass* (fixed mass ratio). This occurs because *each unit of the compound consists of a fixed number of atoms of each element.* For example, consider a sample of ammonia. It is 14 parts nitrogen by mass and 3 parts hydrogen by mass *because* 1 nitrogen atom has 14 times the mass of 1 hydrogen atom, and each ammonia molecule consists of 1 nitrogen atom and 3 hydrogen atoms:

Ammonia gas is 14 parts N by mass and 3 parts H by mass.
1 N atom has 14 times the mass of 1 H atom.
Each ammonia molecule consists of 1 N atom and 3 H atoms.

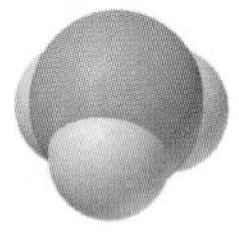

Another defining feature of a compound is that *its properties are different from the properties of its component elements.* Table 2.1 shows a striking example: soft, silvery sodium metal and yellow-green, poisonous chlorine gas are very different from the compound they form—white, crystalline sodium chloride, or common table salt!

Unlike an element, a compound *can* be broken down into simpler substances—its component elements. For example, an electric current breaks down molten sodium chloride into metallic sodium and chlorine gas. By definition, this breakdown is a *chemical change*, not a physical change.

3. *Mixtures.* A **mixture** consists of *two or more substances (elements and/or compounds) that are physically intermingled.* Because a mixture is *not* a substance, in contrast to a compound, *the components of a mixture* **can** *vary in their parts*

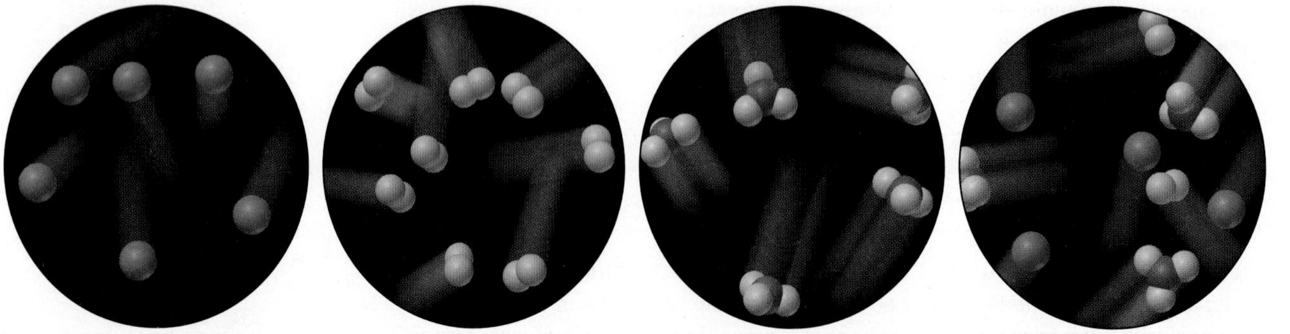

A Atoms of an element **B Molecules of an element** **C Molecules of a compound** **D Mixture of two elements and a compound**

FIGURE 2.1 Elements, compounds, and mixtures on the atomic scale. The samples depicted here are gases, but the three types of matter also occur as liquids and solids. Most of the atoms and molecules are ionized when forming plasma.

TABLE 2.1 Some Properties of Sodium, Chlorine, and Sodium Chloride

Property	Sodium	+	Chlorine	→	Sodium Chloride	
Melting point	97.8°C	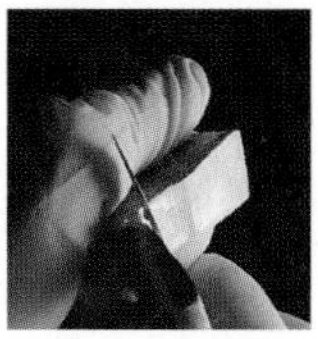	−101°C		801°C	
Boiling point	881.4°C		−34°C		1413°C	
Colour	Silvery solid		Yellow-green gas		Colourless (white) solid	
Density	0.97 g/cm^3		0.0032 g/cm^3		2.16 g/cm^3	
Behaviour in water	Reacts		Dissolves slightly		Dissolves freely	

by mass. A mixture of the compounds sodium chloride and water, for example, can have many different parts by mass of salt to water. On the atomic scale, a mixture consists of the individual units that make up its component elements and/or compounds (Figure 2.1D). Thus, *a mixture retains many of the properties of its components.* Salt water, for instance, is colourless like water and tastes salty like sodium chloride.

Unlike compounds, mixtures can be separated into their components by *physical changes*; chemical changes are not needed. For example, the water in salt water can be boiled off, a physical process that leaves behind solid sodium chloride. The following sample problem will help to differentiate these types of matter.

The elements, compounds, and mixtures shown in Figure 2.1 are depicted using specifically coloured atoms. At the end of Section 2.8 in this chapter, a diagram is provided that shows which atoms are assigned which specific colour. The atoms of a particular element are depicted using the colour scheme provided to make it easier to consistently identify a particular element or molecule from the visual depiction. This colour scheme for atoms has been universally adopted in the field of chemistry.

Sample Problem 2.1 Distinguishing between Elements, Compounds, and Mixtures at the Atomic Scale

Problem The diagrams below represent atomic-scale views of three samples of matter:

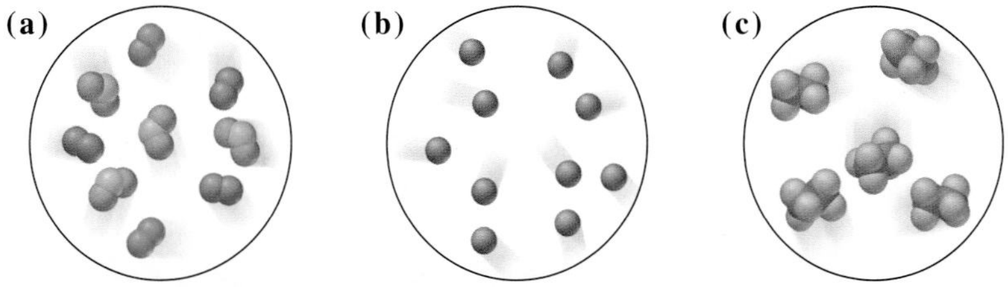

Describe each sample as an element, a compound, or a mixture.

Plan We have to determine the type of matter by examining the component particles. If a sample contains only one type of particle, it is either an element or a compound; if it contains more than one type, it is a mixture. Particles of an element have only one kind of atom (one colour), and particles of a compound have two or more kinds of atoms.

Solution **(a)** Mixture: There are three different types of particles. Two types contain only one kind of atom, either green or purple, so they are elements. The third type contains two red atoms for every one yellow atom, so it is a compound.

(b) Element: The sample consists of only blue atoms.

(c) Compound: The sample consists of molecules, each with two black atoms and six blue atoms.

Follow-Up Problem 2.1 Describe the following reaction in terms of elements, compounds, and mixtures. (See Brief Solutions to Follow-Up Problems at the end of the chapter.)

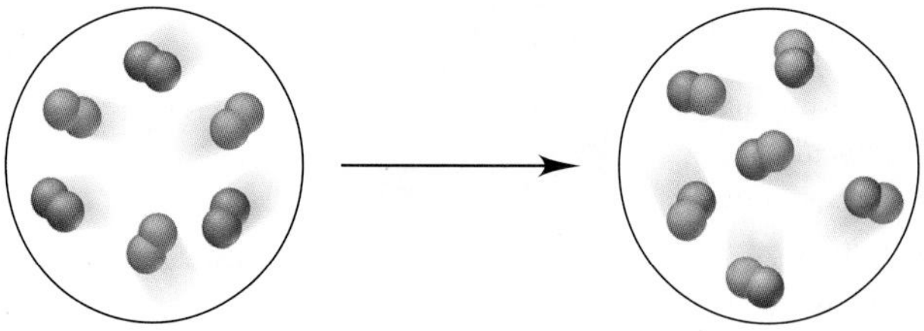

SUMMARY OF SECTION 2.1

- All matter exists as elements, compounds, or mixtures.
- Elements and compounds are substances, which are types of matter with a fixed composition.
- An element consists of only one type of atom and occurs as a collection of individual atoms or molecules.
- A compound contains two or more elements that are chemically combined and exhibits different properties from its component elements. The elements occur in fixed parts by mass because each unit of the compound has a fixed number of each type of atom. Only a chemical change can break down a compound into its elements.
- A mixture consists of two or more substances mixed together, not chemically combined. The components retain their individual properties, can be present in any proportion, and can be separated by physical changes.

2.2 The Observations That Led to an Atomic View of Matter

Any model of the composition of matter had to explain the *law of mass conservation* and the *law of definite (or constant) composition.* As you will see, an atomic theory developed by John Dalton in the early 19th century explained these mass laws and another now known as the *law of multiple proportions.*

Mass Conservation

The most fundamental chemical observation of the 18th century was the **law of mass conservation**: *the total mass of substances does not change during a chemical reaction.* The *number* of substances may change and, by definition, their properties must, but the *total amount* of matter remains constant. (Lavoisier had first stated this law on the basis of his combustion experiments.) Figure 2.2 illustrates mass conservation because the lead nitrate and sodium chromate solutions (*left*) have the same mass as the solid lead chromate in sodium nitrate solution (*right*) that forms when they react.

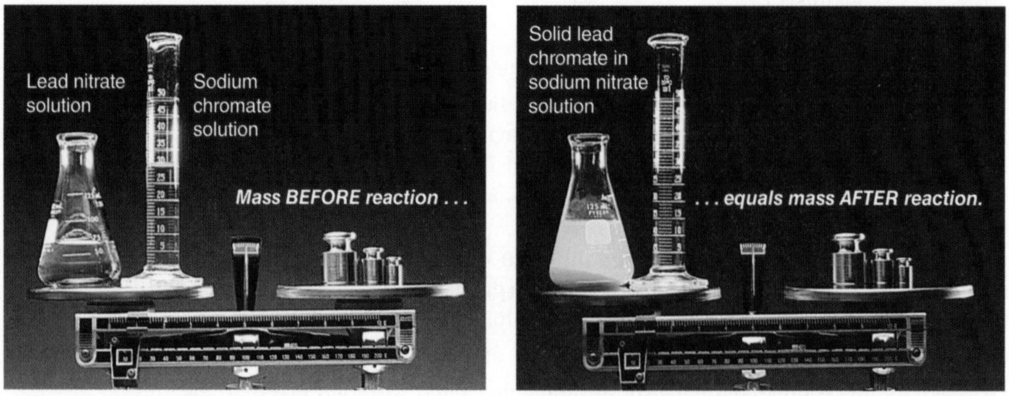

FIGURE 2.2 The law of mass conservation

Even in a complex biochemical change that involves many reactions, such as the metabolism of the sugar glucose, mass is conserved:

$$180 \text{ g glucose} + 192 \text{ g oxygen gas} \longrightarrow 264 \text{ g carbon dioxide} + 108 \text{ g water}$$
$$372 \text{ g material before} \longrightarrow 372 \text{ g material after}$$

Mass conservation means that, based on all chemical experience, *matter cannot be created or destroyed.*

To be precise, we now know, based on the work of Albert Einstein (1879–1955), that the mass before and after a reaction is not *exactly* the same. Some mass is converted to energy, or vice versa, but the difference is too small to measure, even with the best balance. For example, when 100 g of carbon burns, the carbon dioxide that is formed weighs 0.000 000 036 g (3.6×10^{-8} g) less than the carbon and oxygen that reacted. Because the energy changes of *chemical* reactions are so small, for all practical purposes, mass *is* conserved. Later in this book, you will see that energy changes in *nuclear* reactions are so large that mass changes are easy to measure.

Definite Composition

The sodium chloride in your salt shaker is the same substance whether it comes from a salt mine, a salt flat, or any other source. This fact is expressed in the **law of definite (or constant) composition**, which states that *no matter what its source, a particular compound is composed of the same elements in the same parts (fractions) by mass.* The **fraction by mass (mass fraction)** is *the part of the compound's mass that each element contributes.* It is obtained by dividing the mass of each element by the mass of the compound. The **percent by mass (mass percent, mass %)** is *the fraction by mass expressed as a percentage* (i.e., multiplied by 100).

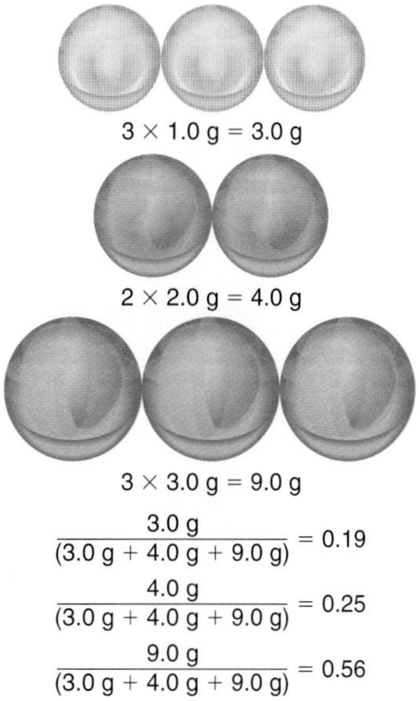

For an everyday example, consider a box that contains three types of marbles: yellow marbles weigh 1.0 g each, purple weigh 2.0 g each, and red weigh 3.0 g each. Each type makes up a fraction of the total mass of marbles, 16.0 g. The *mass fraction* of yellow marbles is their number times their mass divided by the total mass: $\frac{3 \times 1.0 \text{ g}}{16.0 \text{ g}} = 0.19$. The *mass percent* (parts per 100 parts) of yellow marbles is $0.19 \times 100 = 19\%$ by mass. The purple marbles have a mass fraction of 0.25 and are 25% by mass of the total, and the red marbles have a mass fraction of 0.56 and are 56% by mass of the total.

Similarly, in a compound, each element has a *fixed* mass fraction (and mass percent). For example, calcium carbonate, the major compound in seashells, marble, and coral, is composed of three elements: calcium, carbon, and oxygen. The following results are obtained from a mass analysis of 20.0 g of calcium carbonate:

Analysis by Mass (grams/20.0 g)	**Mass Fraction (parts/1.00 part)**	**Percent by Mass (parts/100 parts)**
8.0 g calcium	0.40 calcium	40.% calcium
2.4 g carbon	0.12 carbon	12% carbon
9.6 g oxygen	0.48 oxygen	48% oxygen
20.0 g	1.00 part by mass	100.% by mass

The mass of each element depends on the mass of the sample—that is, more than 20.0 g of compound would contain more than 8.0 g of calcium—but *the mass fraction is fixed no matter what the size of the sample is.* The sum of the mass fractions (or mass percents) equals 1.00 part (or 100%) by mass. The law of definite composition tells us that pure samples of calcium carbonate, no matter where they come from, always contain 40.% calcium, 12% carbon, and 48% oxygen by mass (Figure 2.3).

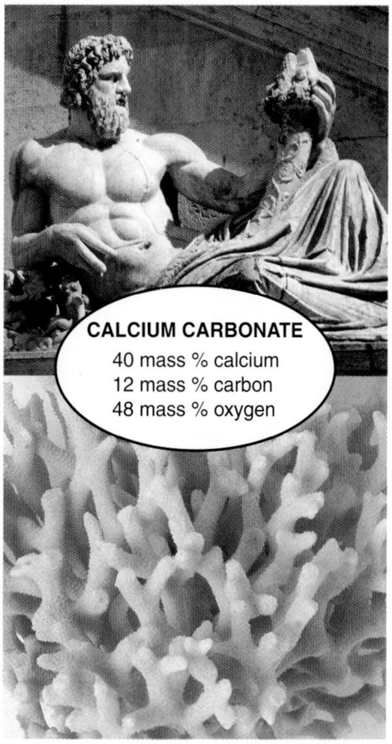

FIGURE 2.3 The law of definite composition. Calcium carbonate occurs in many forms (such as marble, *top*, and coral, *bottom*), but the mass percents of its elements are always the same.

Because a given element always constitutes the same mass fraction of a given compound, we can use this mass fraction to find the actual mass of the element in any sample of the compound:

$$\text{Mass of element} = \text{mass of compound} \times \frac{\text{part by mass of element}}{\text{one part by mass of compound}}$$

Or, more simply, we can skip the need to find the mass fraction first and use the results of mass analysis directly:

$$\text{Mass of element in sample} = \text{mass of compound in sample} \times \frac{\text{mass of element in compound}}{\text{mass of compound}} \quad (2.1)$$

Sample Problem 2.2 Calculating the Mass of an Element in a Compound

Canada is the second-largest producer worldwide of uranium (roughly 18% of total output) and was just recently removed from first place by Kazakhstan, which now produces roughly 33% of the world output of uranium. For more information on uranium, see Chapter 25. Pitchblende is the most important compound of uranium. Mass analysis of an 84.2 g sample shows that it contains 71.4 g of uranium, with oxygen the only other element. What mass of uranium is in 102 kg of pitchblende?

The MacArthur River Uranium Mine in northern Saskatchewan is the world's largest uranium mine.

Road Map

Mass (kg) of pitchblende

↓ Multiply by mass ratio of uranium to pitchblende from analysis.

Mass (kg) of uranium

↓ 1 kg = 1000 g

Mass (g) of uranium

Plan We have to find the mass of uranium in a known mass (102 kg) of pitchblende, given the mass of uranium (71.4 g) in a different mass of pitchblende (84.2 g). The mass ratio of uranium to pitchblende is the same for any sample of pitchblende. Therefore, using Equation 2.1, we multiply the mass (kg) of the pitchblende sample by the ratio of uranium to pitchblende from the mass analysis. This gives the mass (kg) of uranium, and we convert kilograms to grams.

Solution Find the mass (kg) of uranium in 102 kg of pitchblende:

$$\text{Mass (kg) of uranium} = \text{mass (kg) of pitchblende} \times \frac{\text{mass (g) of uranium in pitchblende}}{\text{mass (g) of pitchblende}}$$

$$= 102\text{ kg } \cancel{\text{pitchblende}} \times \frac{71.4\text{ }\cancel{\text{g}}\text{ uranium}}{84.2\text{ }\cancel{\text{g pitchblende}}} = 86.5\text{ kg uranium}$$

Convert the mass of uranium from kg to g:

$$\text{Mass (kg) of uranium} = 86.5\text{ }\cancel{\text{kg}}\text{ uranium} \times \frac{1000\text{ g}}{1\text{ }\cancel{\text{kg}}} = 8.65 \times 10^4\text{ g uranium}$$

Check The analysis showed that most of the mass of pitchblende is due to uranium, so the large mass of uranium makes sense. Rounding off to check the math gives

$$\sim 100 \text{ kg pitchblende} \times \frac{72}{84} = 100 \text{ kg pitchblende} \times \frac{6}{7} \approx 86 \text{ kg uranium}$$

Follow-Up Problem 2.2 What mass (t) of oxygen is present in a sample of pitchblende that contains 2.3 t of uranium? (*Hint:* 1 t = 1000 kg; remember that oxygen is the only other element in pitchblende.)

Multiple Proportions

An observation that applies when two elements form more than one compound is called the **law of multiple proportions**: *if elements A and B react to form two compounds, the different masses of B that combine with a fixed mass of A can be expressed as a ratio of small whole numbers.* Consider two compounds that carbon and oxygen form; let us call them I and II. These compounds have very different properties: the density of carbon oxide I is 1.25 g/L, whereas the density of carbon oxide II is 1.98 g/L; carbon oxide I is poisonous and flammable, but carbon oxide II is not. Mass analysis shows that

- Carbon oxide I is 57.1 mass % oxygen and 42.9 mass % carbon
- Carbon oxide II is 72.7 mass % oxygen and 27.3 mass % carbon

To see the phenomenon of multiple proportions, we use the mass percents of oxygen and carbon to find their masses in a given mass, say 100 g, of each compound. Then we divide the mass of oxygen by the mass of carbon in each compound to obtain the mass of oxygen that combines with a fixed mass of carbon:

	Carbon Oxide I	Carbon Oxide II
g oxygen/100 g compound	57.1	72.7
g carbon/100 g compound	42.9	27.3
g oxygen/g carbon	$\frac{57.1}{42.9} = 1.33$	$\frac{72.7}{27.3} = 2.66$

If we then divide the grams of oxygen per gram of carbon in II by that in I, we obtain a ratio of small whole numbers:

$$\frac{2.66 \text{ g oxygen/g carbon in II}}{1.33 \text{ g oxygen/g carbon in I}} = \frac{2}{1}$$

The law of multiple proportions tells us that, in two compounds of the same elements, the mass fraction of one element relative to the other element changes in *increments based on ratios of small whole numbers.* In the carbon oxide example, the ratio is 2/1—for a given mass of carbon, carbon oxide II contains *2 times* as much oxygen as carbon oxide I, not 1.583 times, 1.716 times, or any other intermediate amount. In the next section, we will explain the mass laws on the atomic scale.

SUMMARY OF SECTION 2.2

- The law of mass conservation states that the total mass remains constant during a chemical reaction.
- The law of definite composition states that all samples of a given compound have the same elements present in the same parts by mass.
- The law of multiple proportions states that, in different compounds of the same elements, the masses of one element that combine with a fixed mass of the other element can be expressed as a ratio of small whole numbers.

2.3 Dalton's Atomic Theory

With 200 years of hindsight, it may be easy to see how the mass laws could be explained by an atomic model: matter existing in indestructible units, each with a particular mass. However, the mass laws were a major breakthrough in 1808 when

John Dalton (1766–1844) presented his atomic theory of matter in *A New System of Chemical Philosophy.*

Despite having no formal education, Dalton began teaching science at age 12 and then studied colour-blindness, an affliction still known as *daltonism.* At 21, he started recording daily weather data, continuing this project for the rest of his life. His results on humidity and dew point led to a key discovery about gases (Section 4.4) and eventually to his atomic theory.

Postulates of the Atomic Theory

Dalton expressed his theory in a series of postulates. Like most great thinkers, he integrated the ideas of others into his own. As we go through the postulates, presented here in modern terms, let us see which were original and which came from others.

1. All matter consists of **atoms**, *tiny indivisible particles of an element that cannot be created or destroyed.* (This derives from the "eternal, indestructible atoms" proposed by Democritus more than 2000 years earlier and reflects mass conservation as stated by Lavoisier.)
2. Atoms of one element *cannot* be converted into atoms of another element. In chemical reactions, the atoms of the original substances recombine to form different substances. (This rejects the alchemical belief in the magical transmutation of elements.)
3. Atoms of an element are identical in mass and other properties and are different from atoms of any other element. (This contains Dalton's major new ideas: *unique mass and properties* for the atoms of a given element.)
4. Compounds result from the chemical combination of a specific ratio of atoms of different elements. (This follows directly from the law of definite composition.)

How the Atomic Theory Explains the Mass Laws

Let us see how Dalton's postulates explain the mass laws:

- *Mass conservation.* Atoms cannot be created or destroyed (postulate 1) or converted into other types of atoms (postulate 2). Therefore, a chemical reaction, in which atoms are combined differently, cannot possibly result in a mass change.
- *Definite composition.* A compound is a combination of a *specific* ratio of different atoms (postulate 4), each of which has a particular mass (postulate 3). Thus, each element in a compound constitutes a fixed fraction of the total mass.
- *Multiple proportions.* Atoms of an element have the same mass (postulate 3) and are indivisible (postulate 1). The masses of element B that combine with a fixed mass of element A give a small whole-number ratio because different numbers of B atoms combine with each A atom in different compounds.

The *simplest* arrangement consistent with the mass data for carbon oxides I and II in our earlier example is one atom of oxygen combining with one atom of carbon in compound I (carbon monoxide) and two atoms of oxygen combining with one atom of carbon in compound II (carbon dioxide):

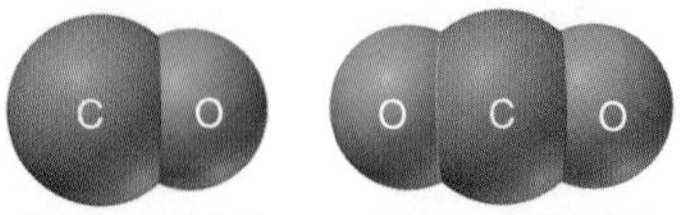

Carbon oxide I (carbon monoxide)   Carbon oxide II (carbon dioxide)

Let us work through a sample problem that reviews the mass laws.

Sample Problem 2.3 Visualizing the Mass Laws

Problem The diagrams below represent an atomic-scale view of a chemical reaction:

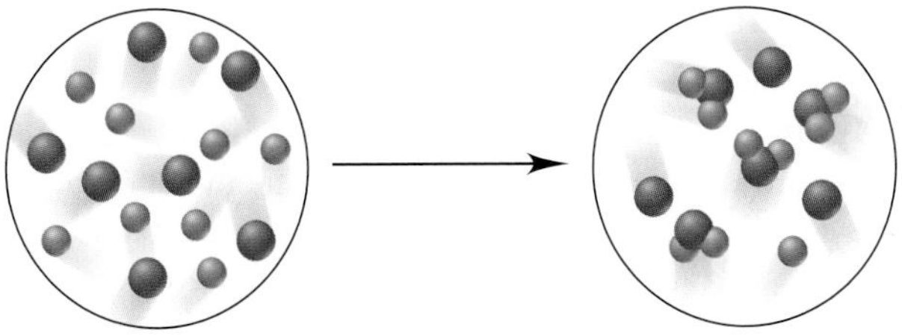

Which of the mass laws—mass conservation, definite composition, or multiple proportions—is (are) illustrated?

Plan We note the numbers, colours, and combinations of atoms (spheres) in the diagrams to see which mass laws are illustrated. If the numbers of each atom are the same before and after the reaction, the total mass did not change (mass conservation). If a compound that always has the same atom ratio forms, the elements are present in fixed parts by mass (definite composition). When the same elements form different compounds and the ratio of the atoms of one element that combine with one atom of the other element is a small whole number, the ratio of their masses is a small whole number as well (multiple proportions).

Solution There are seven purple and nine green atoms in each circle, so mass is conserved. The compound formed has one purple atom and two green atoms, so it has definite composition. Only one compound forms, so the law of multiple proportions is not illustrated.

Follow-Up Problem 2.3 Which sample(s) best display(s) the fact that compounds of bromine (*orange*) and fluorine (*yellow*) exhibit the law of multiple proportions? Explain.

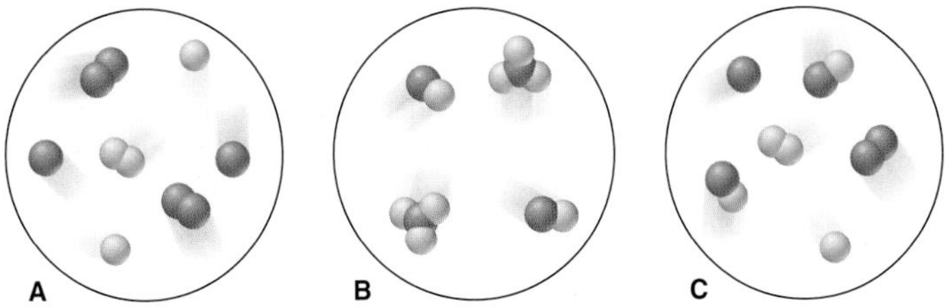

SUMMARY OF SECTION 2.3

- Dalton's atomic theory explained the mass laws by proposing that all matter consists of indivisible, unchangeable atoms of fixed, unique mass.
- Mass is conserved during a reaction because the atoms retain their identities but are combined differently.
- Each compound has a fixed mass fraction of each of its elements because it is composed of a fixed number of each type of atom.
- Different compounds of the same elements exhibit multiple proportions because each compound consists of whole atoms.

2.4 The Observations That Led to the Nuclear Atom Model

Dalton's model established that masses of reacting elements could be explained in terms of atoms. However, it did not establish why atoms bond as they do. Why, for example, do two, and not three, hydrogen atoms bond with one oxygen atom in a water molecule?

Moreover, Dalton's solid model of the atom did not predict the existence of subatomic charged particles, which were observed in later experiments that led to the discovery of *electrons* and the atomic *nucleus*. Let us examine some of these experiments and the more-complex atomic model that emerged from them.

Discovery of the Electron and Its Properties

For many years, scientists knew that matter and electric charge were related. When amber is rubbed with fur, or glass is rubbed with silk, positive and negative charges form—the same charges that make your hair crackle and cling to your comb on a dry day. Scientists also knew that an electric current could decompose certain compounds into their elements. But they did not know what a current was made of.

Cathode Rays To discover the nature of an electric current, some investigators tried passing current through nearly evacuated glass tubes fitted with metal electrodes. When the electric power source was turned on, a "ray" could be seen striking the phosphor-coated end of the tube and emitting a glowing spot of light. The rays were called **cathode rays** because *they originated at the negative electrode (cathode) and moved to the positive electrode (anode).*

Figure 2.4 shows some properties of cathode rays based on these observations. The main conclusion was that *cathode rays consist of negatively charged particles found in all matter.* The rays appear when these particles collide with the few remaining gas molecules in the evacuated tube. Cathode ray particles were later named *electrons.* There are many familiar cases of the effects of charged particles colliding with gases or hitting a phosphor-coated screen:

- In a neon sign, electrons collide with the gas particles in the tube, causing them to give off light.
- In older televisions and computer monitors, the cathode ray passes back and forth over the coated screen, creating a pattern that we see as a picture.

Mass and Charge of the Electron Two classic experiments and their conclusions revealed the mass and charge of the electron:

1. *Mass/charge ratio.* In 1897, the British physicist J. J. Thomson (1856–1940) measured the ratio of the mass of a cathode ray particle to its charge. By comparing this ratio with the mass/charge ratio for the lightest charged particle in solution, Thomson estimated that the cathode ray particle weighed less than $\frac{1}{1000}$ as much as hydrogen, the lightest atom! He was shocked because this implied that, contrary to Dalton's atomic theory, *atoms contain even smaller particles.* Fellow scientists reacted with disbelief to Thomson's conclusion, thinking he was joking.

2. *Charge.* In 1909, the American physicist Robert Millikan (1868–1953) measured the *charge* of the electron. He did so by observing the movement of oil droplets

FIGURE 2.4 Observations that established the properties of cathode rays

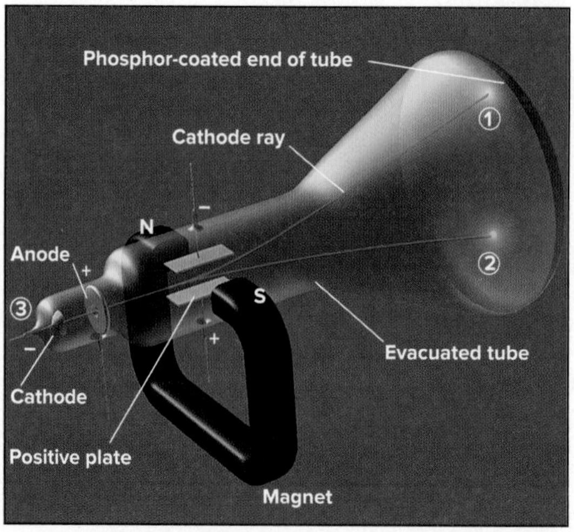

OBSERVATION	CONCLUSION
1. Ray bends in magnetic field.	Consists of charged particles
2. Ray bends toward positive plate in electric field.	Consists of negative particles
3. Ray is identical for any cathode.	Particles found in all matter

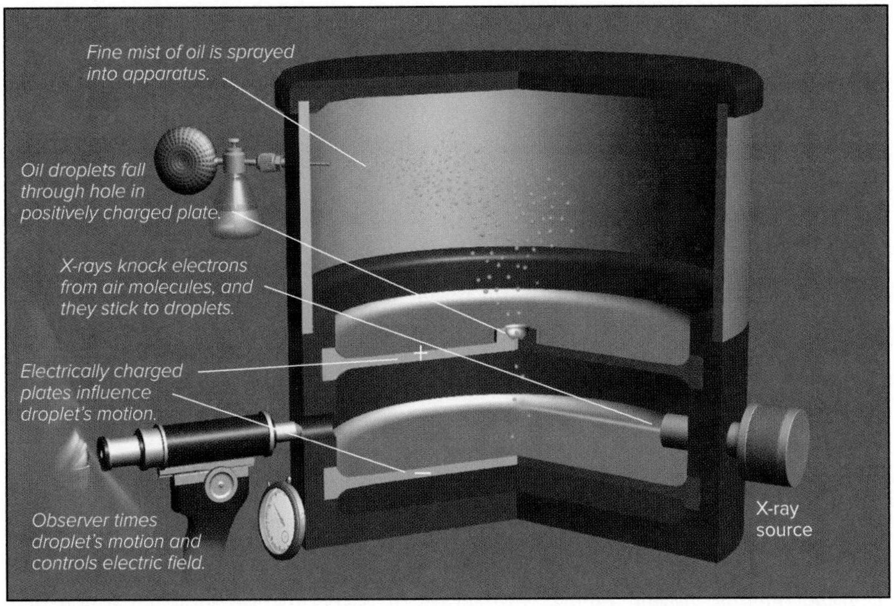
Fine mist of oil is sprayed into apparatus.
Oil droplets fall through hole in positively charged plate.
X-rays knock electrons from air molecules, and they stick to droplets.
Electrically charged plates influence droplet's motion.
Observer times droplet's motion and controls electric field.
X-ray source
+
−

FIGURE 2.5 Millikan's oil-drop experiment for measuring the charge of an electron. The total charge on an oil droplet is some whole-number multiple of the charge of the electron.

in an apparatus that contained electrically charged plates and an X-ray source (Figure 2.5). X-rays knocked electrons from gas molecules in the air within the apparatus, and the electrons stuck to an oil droplet falling through a hole in a positively charged plate. With the electric field off, Millikan measured the mass of the droplet from its rate of fall. Then, by adjusting the field's strength, he made the droplet hang suspended in the air and, thus, measured its total charge.

After many tries, Millikan found that the total charge of the various droplets was always some *whole-number multiple of a minimum charge.* If different oil droplets picked up different numbers of electrons, he reasoned that this minimum charge must be the charge of the electron itself. Remarkably, the value that he calculated over a century ago was within 1% of the modern value of the electron's charge, $-1.602\ 176\ 565\ (35) \times 10^{-19}$ C (C stands for *coulomb*, the SI unit of charge).

What was most interesting about Millikan's research was the fact that it led to one of the great scientific controversies of the time. A Viennese researcher by the name of Felix Ehrenhaft (1879–1952) was studying a similar problem. Millikan used a fixed set of guiding assumptions from which he never wavered while performing his experiments, leading him to conclude that he had determined the charge on the electron. Although Ehrenhaft performed similar experiments, he used a very different set of assumptions, which led him to quite different conclusions. Ehrenhaft was convinced that what Millikan had measured was the fractional charges on subelectrons. The Millikan-Ehrenhaft controversy lasted from 1910 to 1923 and was the subject of heated discussion involving such famous scientists as Einstein, Planck, Born, and Schrödinger. The entire situation once more came into the spotlight in 1978 when Gerald Holton (1922–) found two of Millikan's notebooks from which it appeared that Millikan had used only part of his data to arrive at his conclusions. The issue of scientific fraud and falsification of data tainted Millikan's work for some time. It is interesting to note that Millikan always maintained a clear sense of belief in his work, and it was finally shown that in fact, the data he had excluded had not met the strict set of guiding assumptions that the data he published had met.

As students of a largely experimental science, chemists can learn some very interesting lessons from the Millikan-Ehrenhaft controversy and the subsequent issues that arose from it. It is essential to maintain very clear notes when performing experiments in the lab. While it is critical to explain data that are published, it is sometimes equally if not more critical to explain data that are excluded. It is also important to persevere in achieving a goal and to have conviction in our work.

3. *Conclusion: calculating the mass of an electron.* The electron's mass/charge ratio (from work by Thomson and others) and the value for the electron's charge can be used to find the electron's mass, which is *extremely* small:

$$\text{Mass of electron} = \frac{\text{mass}}{\cancel{\text{charge}}} \times \cancel{\text{charge}} = \left(-5.686 \times 10^{-12}\ \frac{\text{kg}}{\cancel{\text{C}}}\right)(-1.602 \times 10^{-19}\ \cancel{\text{C}})$$
$$= 9.109 \times 10^{-31}\ \text{kg} = 9.109 \times 10^{-28}\ \text{g}$$

Discovery of the Atomic Nucleus

The presence of electrons in all matter posed some major questions about the structure of atoms. Matter is electrically neutral, so atoms must also be neutral. However, if atoms contain negatively charged electrons, what positive charges balance them? If an electron has such a tiny mass, what accounts for an atom's much larger mass? To address these issues, Thomson proposed his "plum pudding" model—a spherical atom composed of diffuse, positively charged matter with electrons embedded like "raisins in a plum pudding."

In 1910, New Zealand-born physicist Ernest Rutherford (1871–1937) tested this model and obtained an unexpected result (Figure 2.6):

1. *Experimental design.* Figure 2.6B shows the experimental setup, in which tiny, dense, positively charged alpha (α) particles emitted from radium are aimed at gold foil. A circular zinc-sulfide screen registers the deflection (scattering) of the α particles by emitting light flashes when the particles strike it.
2. *Expected results.* With Thomson's model in mind (Figure 2.6A), Rutherford expected only minor, if any, deflections of the α particles because they should act like bullets and go right through the gold atoms. After all, an electron should not deflect an α particle any more than a table-tennis ball would deflect a baseball.
3. *Actual results.* Initial results were consistent with this idea, but then the unexpected happened (Figure 2.6C). A few α particles were deflected, and 1 in 20 000 was deflected by more than 90° (it went "backwards").
4. *Rutherford's conclusion.* Rutherford concluded that these few α particles were being repelled by something small, dense, and positive within the gold atoms. Calculations based on the mass, charge, and velocity of the α particles and the proportion of these large-angle deflections showed that
   - an atom is mostly space occupied by electrons
   - *in the centre is a tiny region*, which Rutherford called the **nucleus**, *that contains all the positive charge and essentially all the mass of the atom*

FIGURE 2.6 Rutherford's α-particle scattering experiment and discovery of the atomic nucleus

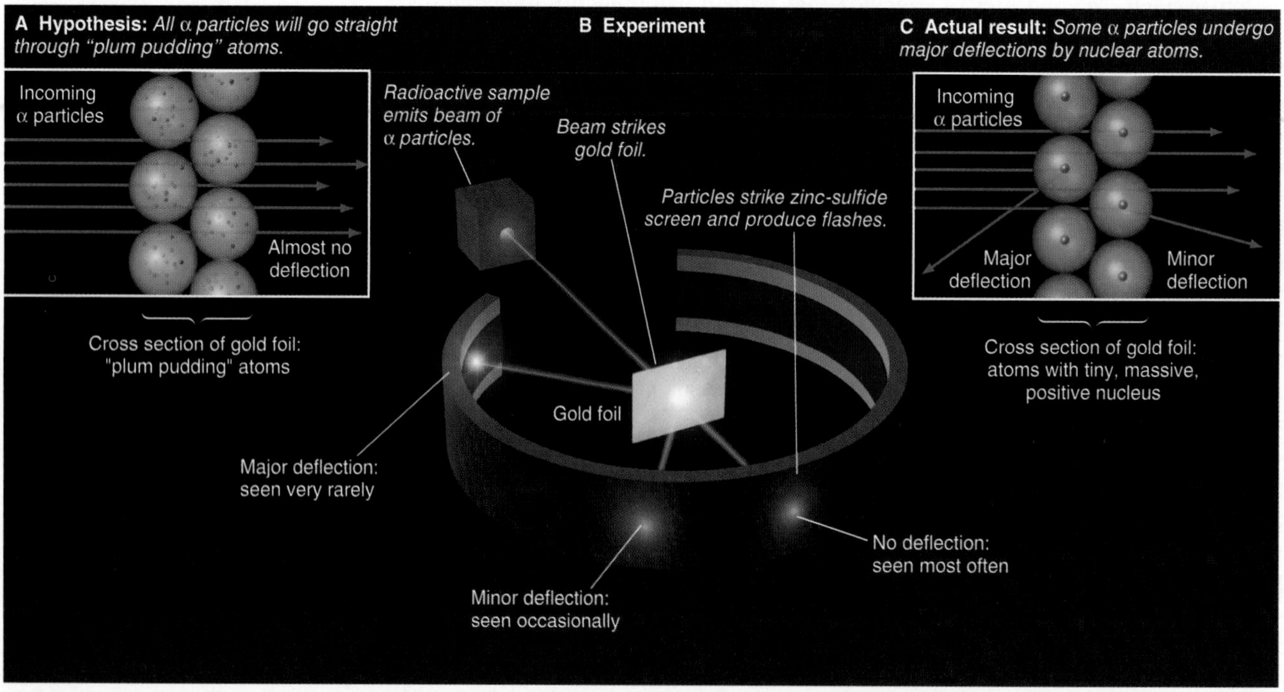
A Hypothesis: All α particles will go straight through "plum pudding" atoms.
Incoming α particles
Almost no deflection
Cross section of gold foil: "plum pudding" atoms
B Experiment
Radioactive sample emits beam of α particles.
Beam strikes gold foil.
Particles strike zinc-sulfide screen and produce flashes.
Gold foil
Major deflection: seen very rarely
Minor deflection: seen occasionally
No deflection: seen most often
C Actual result: Some α particles undergo major deflections by nuclear atoms.
Incoming α particles
Major deflection
Minor deflection
Cross section of gold foil: atoms with tiny, massive, positive nucleus

Rutherford proposed that positive particles lay within the nucleus and called them *protons*.

Rutherford's model explained the charged nature of matter, but it could not account for the atom's entire mass. After more than 20 years, in 1932, James Chadwick discovered the *neutron*, an uncharged dense particle that also resides in the nucleus.

Ernest Rutherford (1871–1937) made enormous contributions to the fields of chemistry and physics. Born in New Zealand, Rutherford was Macdonald Professor of Experimental Physics from 1898 to 1907 at McGill University in Montréal. He received a Nobel Prize in 1908 for "his investigations into the disintegration of the elements and the chemistry of radioactive substances," work he did while at McGill. He is also the only Nobel Prize winner in Science to have performed his most famous work *after* obtaining the Nobel Prize! His astonishment at the results obtained in his famous gold foil experiment is clear in his words, "I remember two or three days later Geiger [one of his co-workers] coming to me in great excitement and saying, 'We have been able to get some of the α particles coming backwards. . . .' It was quite the most incredible event that has ever happened to me in my life. It was almost as incredible as if you fired a 15-inch shell at a piece of tissue paper and it came back and hit you."

SUMMARY OF SECTION 2.4

- Several major discoveries at the beginning of the 20th century resolved questions about Dalton's model and led to our current model of atomic structure.
- Cathode rays were shown to consist of negative particles (electrons) that exist in all matter. J. J. Thomson measured their mass/charge ratio and concluded that they are much smaller and lighter than atoms.
- Robert Millikan determined the charge of the electron, which he combined with other data to calculate the mass of the electron.
- Ernest Rutherford proposed that atoms consist of a tiny, massive, positive nucleus surrounded by electrons.

2.5 The Atomic Theory Today

Dalton's model of an indivisible particle has given way to our current model of an atom with an elaborate internal architecture of subatomic particles. In this section, we examine that model and see how Dalton's theory stands up today.

Structure of the Atom

An *atom* is an electrically neutral, spherical entity composed of a positively charged central nucleus surrounded by one or more negatively charged electrons (Figure 2.7). The electrons move rapidly within the available volume, held there by the attraction of the nucleus. An atom's diameter ($\sim 1 \times 10^{-10}$ m) is about 20 000 times the diameter of its nucleus ($\sim 5 \times 10^{-15}$ m). The nucleus contributes 99.97% of the atom's mass, occupies only about one-quadrillionth of its volume, and is incredibly dense—about 10^{14} g/mL!

An atomic nucleus consists of protons and neutrons (the only exception is the simplest hydrogen nucleus, which is a single proton). The **proton (p^+)** has a *positive charge*, and the **neutron (n^0)** has *no charge*; thus, the positive charge of the nucleus results from its protons. The *magnitudes* of the charges possessed by a proton and by an **electron (e^-)** are equal, but the *signs* of the charges are opposite. *An atom is neutral because the number of protons in the nucleus equals the number of electrons surrounding the nucleus.* Some properties of these three subatomic particles are listed in Table 2.2.

Atomic Number, Mass Number, and Atomic Symbol

The **atomic number (Z)** of an element equals *the number of protons in the nucleus of each of its atoms. All atoms of an element have the same atomic*

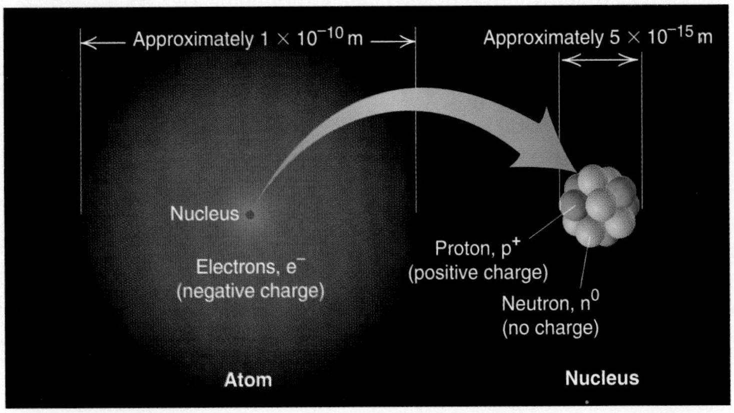

FIGURE 2.7 General features of the atom

TABLE 2.2 Properties of the Three Key Subatomic Particles

	Charge		Mass		
Name (Symbol)	**Relative**	**Absolute (C)***	**Relative (u)†**	**Absolute (g)**	**Location in Atom**
Proton (p^+)	1+	$+1.602\ 18 \times 10^{-19}$	1.007 27	$1.672\ 62 \times 10^{-24}$	Nucleus
Neutron (n^0)	0	0	1.008 66	$1.674\ 93 \times 10^{-24}$	Nucleus
Electron (e^-)	1−	$-1.602\ 18 \times 10^{-19}$	0.000 548 58	$9.109\ 39 \times 10^{-28}$	Outside nucleus

*The coulomb (C) is the SI unit of charge.
†The unified atomic mass unit (u) equals $1.660\ 54 \times 10^{-24}$ g; it is discussed later in this section.

number, and the atomic number of each element is different from that of any other element. All carbon atoms ($Z = 6$) have 6 protons, all oxygen atoms ($Z = 8$) have 8 protons, and all uranium atoms ($Z = 92$) have 92 protons. There are currently 118 known elements, of which 90 occur in nature and 28 have been synthesized by nuclear scientists.

The **mass number (A)** is *the total number of protons and neutrons in the nucleus of an atom.* Each proton and each neutron contributes one unit to the mass number. Thus, a carbon atom with 6 protons and 6 neutrons in its nucleus has a mass number of 12, and a uranium atom with 92 protons and 146 neutrons in its nucleus has a mass number of 238.

The **atomic symbol** (or *element symbol*) of an element *is based on its English, Latin, or Greek name,* such as C for carbon, S for sulfur, and Na for sodium (from the Latin *natrium*). Often written with the symbol are the atomic number (Z) as a left *sub*script and the mass number (A) as a left *super*script; for example, ${}^{A}_{Z}X$ represents element X. Since the mass number is the sum of the protons and neutrons, the number of neutrons (N) equals the mass number minus the atomic number:

$$\text{Number of neutrons} = \text{mass number} - \text{atomic number, or } N = A - Z \qquad (2.2)$$

Thus, a chlorine atom, represented by ${}^{35}_{17}Cl$, has $A = 35$, $Z = 17$, and $N = 35 - 17 = 18$. Because each element has its own atomic number, we also know the atomic number given the symbol. For example, instead of writing ${}^{12}_{6}C$ for carbon with mass number 12, we can write ${}^{12}C$ (spoken "carbon twelve"), with $Z = 6$ understood. Another way to name this atom is carbon-12.

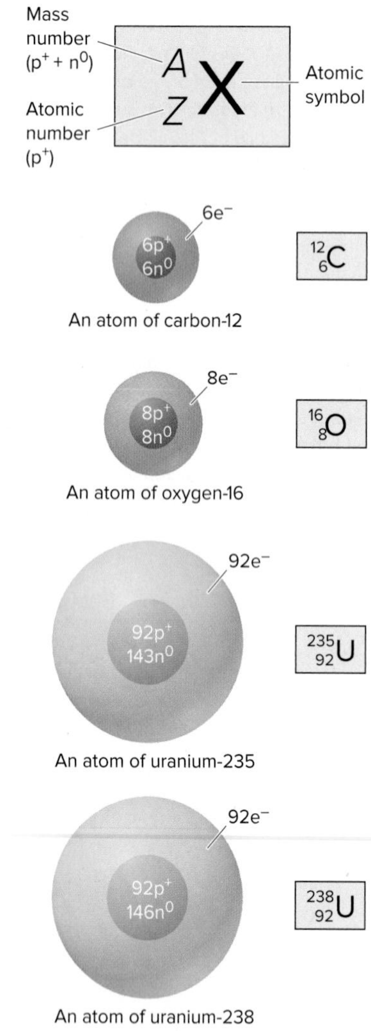

FIGURE 2.8 The ${}^{A}_{Z}X$ notations and spherical representations for four atoms. The nuclei are not drawn to scale.

Isotopes

All atoms of an element have the same atomic number but not necessarily the same mass number. **Isotopes** of an element are *atoms that have different numbers of neutrons and therefore different mass numbers.* For example, all carbon atoms ($Z = 6$) have 6 protons and 6 electrons, but only 98.89% of naturally occurring carbon atoms have 6 neutrons ($A = 12$). A small percentage (1.11%) have 7 neutrons ($A = 13$), and even fewer (less than 0.01%) have 8 ($A = 14$). *A natural sample of carbon has the three natural isotopes* ${}^{12}C$, ${}^{13}C$, and ${}^{14}C$ *in the relative proportions* 98.89%, 1.11%, and less than 0.01%. Five other carbon isotopes—${}^{9}C$, ${}^{10}C$, ${}^{11}C$, ${}^{15}C$, and ${}^{16}C$—have been created in the laboratory. Figure 2.8 depicts the atomic number, mass number, and symbol for four atoms, two of which are isotopes of the element uranium.

The chemical properties of an element are primarily determined by the number of electrons, so *all isotopes of an element have nearly identical chemical behaviour,* even though they have different masses.

Sample Problem 2.4 Determining the Number of Subatomic Particles in the Isotopes of an Element

Problem Silicon (Si) is a major component of semiconductor chips. It has three naturally occurring isotopes: ${}^{28}Si$, ${}^{29}Si$, and ${}^{30}Si$. Determine the numbers of protons, neutrons, and electrons in each silicon isotope.

Plan The mass number (A; left superscript) of each of the three isotopes is given. Recall that the mass number is the sum of the protons and neutrons. From the list of elements on the inside front cover of this book, we find the atomic number (Z, number of protons), which equals the number of electrons. We obtain the number of neutrons by subtracting Z from A (Equation 2.2).

Solution From the list of elements, the atomic number of silicon is 14. Therefore,

$$^{28}\text{Si has } 14p^+, 14e^-, \text{ and } 14n^0 \ (28 - 14).$$
$$^{29}\text{Si has } 14p^+, 14e^-, \text{ and } 15n^0 \ (29 - 14).$$
$$^{30}\text{Si has } 14p^+, 14e^-, \text{ and } 16n^0 \ (30 - 14).$$

Follow-Up Problem 2.4 How many protons, neutrons, and electrons are in **(a)** $^{11}_{5}Q$; **(b)** $^{41}_{20}R$; **(c)** $^{131}_{53}X$? What elements do Q, R, and X represent?

Atomic Masses of the Elements

The mass of an atom is measured *relative* to the mass of an atomic standard. The modern standard is the carbon-12 atom, whose mass is defined as *exactly* 12 atomic mass units. Thus, the **unified atomic mass unit (u)** is *$\frac{1}{12}$ the mass of a carbon-12 atom.* Based on this standard, the ^{1}H atom has a mass of 1.008 u; in other words, a ^{12}C atom has almost 12 times the mass of a ^{1}H atom. We will continue to use the term *unified atomic mass unit* in this book, although the unit name **dalton (Da)** is also acceptable; thus, one ^{12}C atom has a mass of 12 Da, or 12 u. The unified atomic mass unit is a unit of relative mass, but it has an absolute mass of $1.660\,54 \times 10^{-24}$ g.

The isotopic makeup of an element is determined by **mass spectrometry**, *a method for measuring the relative masses and abundances of atomic-scale particles very precisely.* (Mass spectrometry is discussed in the Tools of the Laboratory section that follows.) For example, using a mass spectrometer, we measure the mass ratio of ^{28}Si to ^{12}C as

$$\frac{\text{Mass of } ^{28}\text{Si atom}}{\text{Mass of } ^{12}\text{C standard}} = 2.331\,411$$

From this mass ratio, we find the **isotopic mass** of the ^{28}Si atom, *the relative mass of this silicon isotope*:

$$\text{Isotopic mass of } ^{28}\text{Si} = \text{measured mass ratio} \times \text{mass of } ^{12}\text{C}$$
$$= 2.331\,411 \times 12\text{ u} = 27.976\,93\text{ u}$$

Along with the isotopic mass, the mass spectrometer gives the relative abundance as a percentage (or fraction) of each isotope in a sample of the element. For example, the relative abundance of ^{28}Si is 92.23% (or 0.9223).

From such data, we can obtain the **atomic mass** of an element, *the* ***average*** *of the masses of its naturally occurring isotopes weighted according to their abundances.* Each naturally occurring isotope of an element contributes a certain portion to the atomic mass. For example, multiplying the isotopic mass of ^{28}Si by its fractional abundance gives the portion of the atomic mass of Si that is contributed by ^{28}Si:

$$\text{Portion of Si atomic mass from } ^{28}\text{Si} = 27.976\,93\text{ u} \times 0.9223 = 25.8031\text{ u}$$
(retaining two additional significant figures)

Similar calculations give the portions contributed by ^{29}Si ($28.976\,495\text{ u} \times 0.0467 = 1.3532\text{ u}$) and by ^{30}Si ($29.973\,770\text{ u} \times 0.0310 = 0.9292\text{ u}$). Adding the three portions together (rounding to two decimal places at the end) gives the atomic mass of silicon:

$$\text{Atomic mass of Si} = 25.8031\text{ u} + 1.3532\text{ u} + 0.9292\text{ u}$$
$$= 28.0855\text{ u} = 28.09\text{ u}$$

The atomic mass is an average value; thus, while no individual silicon atom has a mass of 28.09 u, we consider a sample of silicon in the laboratory to consist of atoms with this average mass.

Mass Spectrometry

Mass spectrometry is a powerful technique for measuring the mass and abundance of charged particles (their mass/charge ratio, m/z). The first mass spectrometer was constructed in 1912 by J. J. Thomson (Figure B2.1), who noted that ionized atoms or molecules with positive charge in a vacuum, exposed to parallel electrostatic and magnetic fields, arrived at the detector in parabolic arcs. No two ions arrived at the same spot unless they had the same m/z ratio. The fact that Thomson observed different arcs for the same ionized atoms led to the idea that different atoms of the same element could have different masses. This, in turn, led to the realization of the existence of isotopes.

Francis Aston received the Nobel Prize in Chemistry in 1922 "for his discovery, by means of his mass spectrograph, of isotopes, in a large number of nonradioactive elements, and for his enunciation of the whole-number rule." One of the first major applications of the mass spectrometer was in the Manhattan project, where it was used not only to separate radioactive isotopes but also to collect them after separation. Mass spectrometry received major post-war interest from the petrochemical industry, in an effort to know more precisely what products were being produced during cracking processes.

FIGURE B2.1 J.J. Thomson, the father of mass spectrometry

How a Mass Spectrometer Works

A: Could be Gas Chromatograph (GC) or Liquid Chromatograph (LC/HPLC)

B: Could be ICP (Inductively Coupled Plasma) → completely destroys molecule; only atoms remain
EI (Electron Impact) → medium-high fragmentation
MALDI (Matrix Assisted Laser Desorption Ionization) → low-medium fragmentation
FAB (Fast Atom Bombardment) → little fragmentation
ESI (Electrospray Ionization) → little fragmentation

C/E: Could be Quadrupole → mass range to 1000 m/z; low resolution
Ion Trap → mass range to 1000 m/z; low resolution
TOF (Time of Flight) → mass range to 10 000 m/z; medium-high resolution
Electromagnet → mass range to 5000 m/z; high resolution
FTICR (Fourier Transform Ion Cyclotron Resonance) → mass range 5 to 10 000 m/z; extremely high resolution

D: Collision with inert gas or another method of enhancing internal energy of ion of interest

F: Could be MCP (Multi Channel Plate)
EM (Electron Multiplier)
Scintillator + PMT (Photon Multiplier Tube)

The chemical separator (if necessary) is used to separate the original mixture into the component molecules to be analyzed. The ion source then ionizes the sample. The ion source is chosen based on the type of sample to be analyzed and the extent of fragmentation desired. The method of ionization determines whether the ions have a little or a lot of internal energy remaining. If a molecule has very little internal energy after ionization, it will stay as it is and give the molecular ion. However, if there is sufficient internal energy, bonds of the molecular ion can and will be broken, leading to fragment ions. These fragment ions contain information about the structure of the molecular ion. The tunable mass filter then allows the detection of a range of all the ions generated inside the ion source by the m/z ratio. Different mass filters vary in their mass range and resolution (precision). The detector records either the charge induced or the current produced when an ion passes by or hits the surface.

If, for example, we had a mixture of CO (27.9949 u), N_2 (28.0061 u), and C_2H_4 (28.0313 u), it would appear that we have three molecules, all with a nominal m/z of 28, which would give a single peak using a low-resolution mass filter such as a quadrupole. If we were to use a high-resolution mass filter, such as a magnet or ICR, the three ions in our example would actually give three distinct peaks. A mass measured with sufficient accuracy can thus serve as proof of structure. However, even a low-resolution mass filter would allow us to distinguish the three molecules in our example based on their fragmentation patterns. (For example, fragmentation of CO would not give a peak at m/z 14; similarly, fragmentation of N_2 would not give a peak at m/z 12 or m/z 16.) Thus, mass spectrometry is helpful for determining the structure of molecules based on exact mass and/or fragmentation patterns (Figure B2.2).

Every molecule has a characteristic fragmentation pattern, which leads to the idea of molecular fingerprints. These fragmentation fingerprints have been collected into databases and are available to identify unknown substances. Major applications include blood analysis and reference for analysis in crime labs, quality-control labs, and environmental labs.

If an instrument only has one mass analyzer, the observed fragmentation pattern is that of *everything* generated in the ion source. If there is a second mass analyzer, however, we could use the first analyzer to select one particular ion from all those generated in the ion source and use the second analyzer to measure the

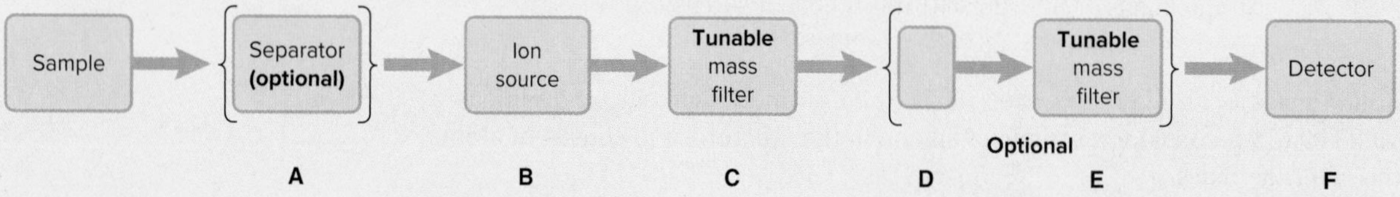

fragmentation pattern of only the *ion* selected by the first analyzer. In order to do this, we need to put extra energy into the selected ion to make it fragment. This can be done in various ways; collision with an inert gas is commonly used. These secondary mass spectra/MS-MS spectra can be very helpful when the fragmentation pattern obtained from the ion source is ambiguous because of the presence of contaminants. MS-MS performs a function for ions that is similar to the function of the chemical separator for molecules.

Most commonly used instruments have a GC or LC chemical separator. An EI/Quad (such as an HP benchtop GC-MS) is used for routine analysis, both qualitative and quantitative, of small organics. A (GC) EI/Magnetic instrument is used for accurate mass measurements and ppm-level environmental analysis, while an (LC) ESI/TOF or MALDI-TOF is more suitable for the study of large molecules, such as peptides. Some of the most significant mass spectrometers used in the industry are designed, and in some cases built, in Woodbridge, Ontario. These include the SCIEX mass spectrometers and some PerkinElmer ICP mass spectrometers (SCIEX: sciex.com; PerkinElmer NexION ICP-MS: www.perkinelmer.ca/en-ca/Catalog/Product/ID/NexION350X).

Mass spectrometry is now used to measure the mass of virtually any atom, molecule, or molecular fragment. The technique is being used to study catalyst surfaces, forensic materials, fuel mixtures, medicinal agents, radiocarbon dating, and many other samples, especially proteins. In fact, John B. Fenn and Koichi Tanaka shared part of the 2002 Nobel Prize in Chemistry for developing methods to study proteins by mass spectrometry.

Problems

B2.1 Chlorine has two naturally occurring isotopes, ^{35}Cl (abundance 76%) and ^{37}Cl (abundance 24%), and it occurs as diatomic (two-atom) molecules. In a mass spectrum, peaks are seen for the molecule and for the separated atoms.
(a) How many peaks are in the mass spectrum?
(b) What are the m/z values of the heaviest particle and the lightest particle?

B2.2 When a sample of pure carbon is analyzed by mass spectrometry, peaks X, Y, and Z are obtained (random order). Peak Y is taller than peaks X and Z, and peak Z is taller than peak X. What is the m/z value of the isotope that is responsible for peak Z?

A

B

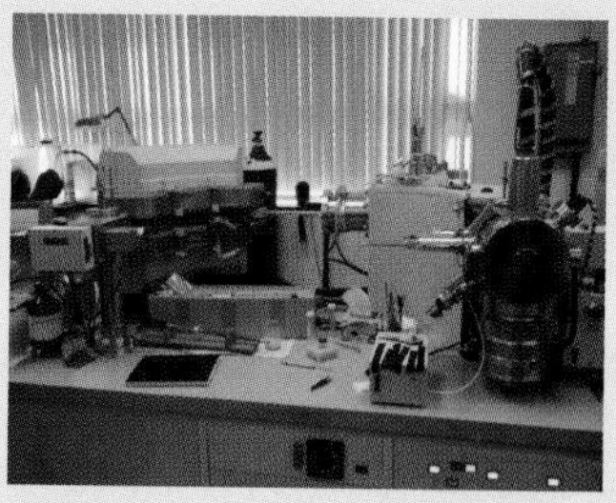
C

D

E

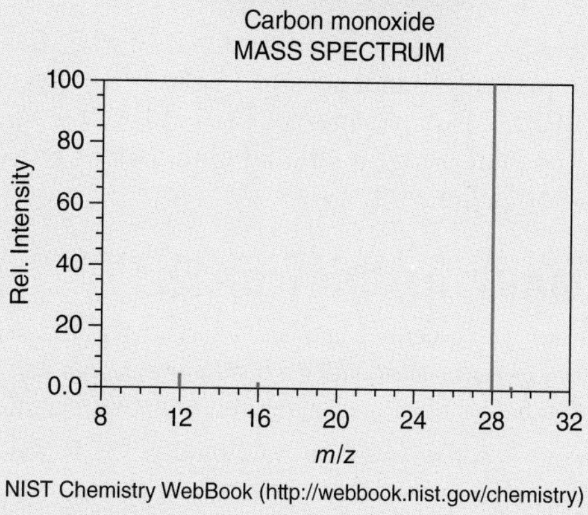

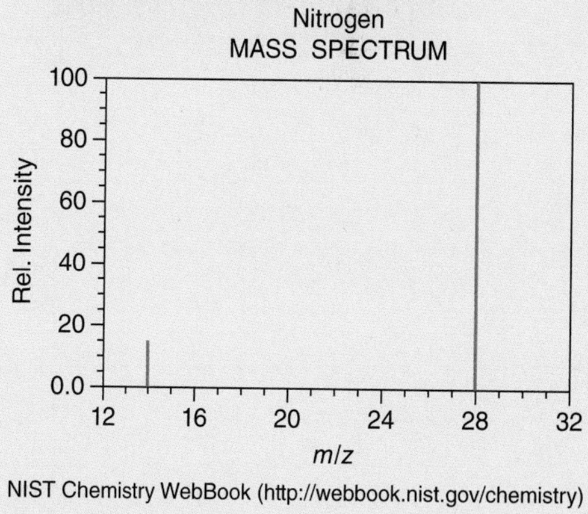

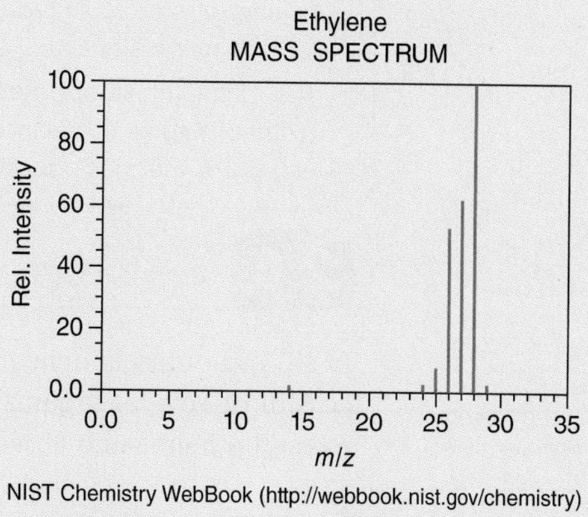

F

FIGURE B2.2 Mass spectrometers and spectra. A. The size of the sample quadrupoles relative to the size of a magnet. **B.** TOF tube. **C.** Magnetic Sector-Double Focusing KRATOS CONCEPT 1S. **D.** Benchtop HP GCMSD unit. **E.** Benchtop Varian 500 ITMS. **F.** Mass spectra of CO, N_2, and C_2H_4.

Sample Problem 2.5 Calculating the Atomic Mass of an Element

Problem Silver (Ag; $Z = 47$) has 46 known isotopes, but only two—^{107}Ag and ^{109}Ag—occur naturally. Given the following data, calculate the atomic mass of Ag:

Isotope	Mass (u)	Abundance (%)
^{107}Ag	106.905 09	51.84
^{109}Ag	108.904 76	48.16

Road Map

Mass (g) of each isotope

↓ Multiply by fractional abundance of each isotope.

Portion of atomic mass from each isotope

↓ Add isotopic portions.

Atomic mass

Plan From the mass and abundance of the two Ag isotopes, we have to find the atomic mass of Ag (the weighted average of the isotopic masses). We divide each percent abundance by 100 to get the fractional abundance and then multiply this value by each isotopic mass to find the portion of the atomic mass contributed by each isotope. The sum of the isotopic portions is the atomic mass.

Solution Find the fractional abundances:

$$\text{Fractional abundance of } ^{107}\text{Ag} = \frac{51.84}{100} = 0.5184$$

Similarly, fractional abundance of ^{109}Ag = 0.4816.

Find the portion of the atomic mass from each isotope:

$$\begin{aligned}\text{Portion of atomic mass from } ^{107}\text{Ag} &= \text{isotopic mass} \times \text{fractional abundance}\\ &= 106.905\,09 \text{ u} \times 0.5184 = 55.42 \text{ u}\\ \text{Portion of atomic mass from } ^{109}\text{Ag} &= 108.904\,76 \text{ u} \times 0.4816 = 52.45 \text{ u}\end{aligned}$$

Find the atomic mass of silver:

$$\text{Atomic mass of Ag} = 55.42 \text{ u} + 52.45 \text{ u} = 107.87 \text{ u}$$

Check The individual portions seem right: $\sim$100 u × 0.50 = 50 u. The portions should be almost the same because the two isotopic abundances are almost the same. We rounded each portion to four significant figures because that is the number of significant figures in the abundance values. This is the correct atomic mass (to two decimal places); in the list of elements (inside front cover), the atomic mass is rounded to 107.9 u.

Follow-Up Problem 2.5 Boron (B; $Z = 5$) has two naturally occurring isotopes. Find the percent abundances of ^{10}B and ^{11}B, given the following data: atomic mass of B = 10.81 u, isotopic mass of ^{10}B = 10.0129 u, and isotopic mass of ^{11}B = 11.0093 u. (*Hint*: The sum of the fractional abundances is 1. If x = abundance of ^{10}B, then $1 - x$ = abundance of ^{11}B.)

SUMMARY OF SECTION 2.5

- An atom has a central nucleus, which contains positively charged protons and uncharged neutrons and is surrounded by negatively charged electrons. An atom is neutral because the number of electrons equals the number of protons.
- An atom is represented by the notation ^A_ZX, in which Z is the atomic number (number of protons), A is the mass number (sum of protons and neutrons), and X is the atomic symbol.
- An element occurs naturally as a mixture of isotopes, which are atoms with the same number of protons but different numbers of neutrons. Each isotope has a mass relative to the ^{12}C mass standard.
- The atomic mass of an element is the average of its isotopic masses weighted according to their natural abundances. The atomic mass is determined using modern instruments, especially the mass spectrometer.

2.6 Elements: A First Look at the Periodic Table

In 1871, the Russian chemist Dmitri Mendeleev (1836–1907) published the most successful of several organizing schemes as a table of the elements. He listed the elements by increasing atomic mass and arranged them so that elements with similar chemical properties fell in the same column. The modern **periodic table of the elements,** based on Mendeleev's version (but arranged by *atomic number,* not mass), is one of the great classifying schemes in science and an indispensable tool to chemists—and chemistry students.

Organization of the Periodic Table One common version of the modern periodic table appears in Figure 2.9 (and inside the front cover). It is formatted as follows:

1. Each element has a box that contains its atomic number, atomic symbol, and atomic mass. (The mass in parentheses is the mass number of the most stable isotope of the element.) The boxes lie, from left to right, in order of *increasing atomic number* (number of protons in the nucleus).
2. The boxes are arranged into a grid of **periods** (*horizontal rows*) and **groups** (*vertical columns*). Each period has a number from 1 to 7. Each group has a number from 1 to 18.
3. Groups 1, 2, 13, 14, 15, 16, 17, and 18 (two on the left and six on the right) contain the *main-group elements.* Groups 3 to 12 contain the *transition elements.* Two horizontal series of *inner transition elements,* the lanthanides and the actinides, fit *between* the elements in group 3 and group 4 and are placed below the main body of the table.

FIGURE 2.9 The modern periodic table. As of June 2012, elements 114 and 116 have been officially recognized. Elements 113, 115, 117, and 118 are pending verification by IUPAC.

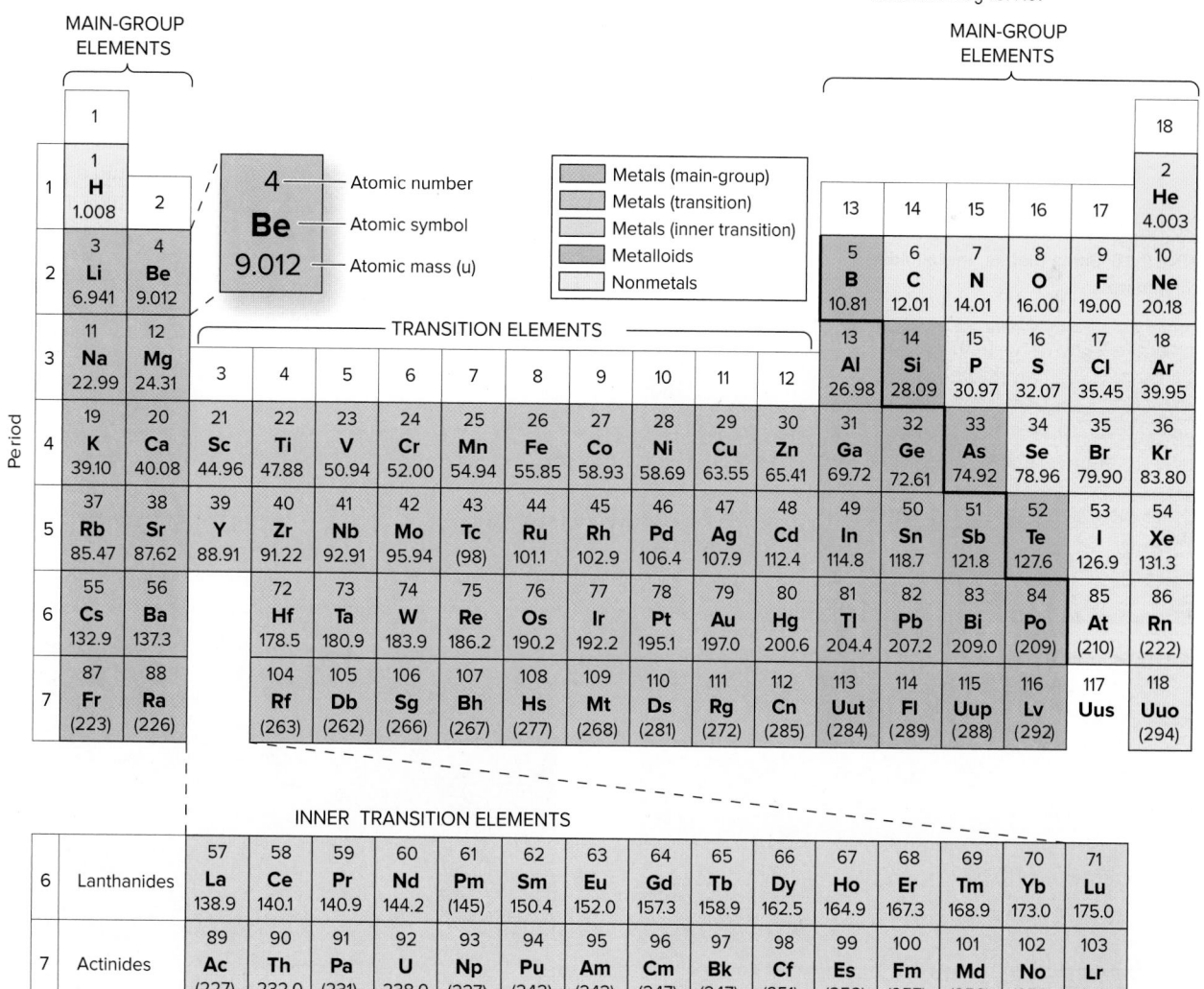

Key: 4 — Atomic number; Be — Atomic symbol; 9.012 — Atomic mass (u)

Legend: Metals (main-group); Metals (transition); Metals (inner transition); Metalloids; Nonmetals

MAIN-GROUP ELEMENTS (groups 1–2) | TRANSITION ELEMENTS (groups 3–12) | MAIN-GROUP ELEMENTS (groups 13–18)

Period	1	2	3	4	5	6	7	8	9	10	11	12	13	14	15	16	17	18
1	1 H 1.008																	2 He 4.003
2	3 Li 6.941	4 Be 9.012											5 B 10.81	6 C 12.01	7 N 14.01	8 O 16.00	9 F 19.00	10 Ne 20.18
3	11 Na 22.99	12 Mg 24.31											13 Al 26.98	14 Si 28.09	15 P 30.97	16 S 32.07	17 Cl 35.45	18 Ar 39.95
4	19 K 39.10	20 Ca 40.08	21 Sc 44.96	22 Ti 47.88	23 V 50.94	24 Cr 52.00	25 Mn 54.94	26 Fe 55.85	27 Co 58.93	28 Ni 58.69	29 Cu 63.55	30 Zn 65.41	31 Ga 69.72	32 Ge 72.61	33 As 74.92	34 Se 78.96	35 Br 79.90	36 Kr 83.80
5	37 Rb 85.47	38 Sr 87.62	39 Y 88.91	40 Zr 91.22	41 Nb 92.91	42 Mo 95.94	43 Tc (98)	44 Ru 101.1	45 Rh 102.9	46 Pd 106.4	47 Ag 107.9	48 Cd 112.4	49 In 114.8	50 Sn 118.7	51 Sb 121.8	52 Te 127.6	53 I 126.9	54 Xe 131.3
6	55 Cs 132.9	56 Ba 137.3		72 Hf 178.5	73 Ta 180.9	74 W 183.9	75 Re 186.2	76 Os 190.2	77 Ir 192.2	78 Pt 195.1	79 Au 197.0	80 Hg 200.6	81 Tl 204.4	82 Pb 207.2	83 Bi 209.0	84 Po (209)	85 At (210)	86 Rn (222)
7	87 Fr (223)	88 Ra (226)		104 Rf (263)	105 Db (262)	106 Sg (266)	107 Bh (267)	108 Hs (277)	109 Mt (268)	110 Ds (281)	111 Rg (272)	112 Cn (285)	113 Uut (284)	114 Fl (289)	115 Uup (288)	116 Lv (292)	117 Uus	118 Uuo (294)

INNER TRANSITION ELEMENTS

Period	Series														
6	Lanthanides	57 La 138.9	58 Ce 140.1	59 Pr 140.9	60 Nd 144.2	61 Pm (145)	62 Sm 150.4	63 Eu 152.0	64 Gd 157.3	65 Tb 158.9	66 Dy 162.5	67 Ho 164.9	68 Er 167.3	69 Tm 168.9	70 Yb 173.0	71 Lu 175.0
7	Actinides	89 Ac (227)	90 Th 232.0	91 Pa (231)	92 U 238.0	93 Np (237)	94 Pu (242)	95 Am (243)	96 Cm (247)	97 Bk (247)	98 Cf (251)	99 Es (252)	100 Fm (257)	101 Md (258)	102 No (259)	103 Lr (260)

Classifying the Elements One of the clearest ways to classify the elements is as metals, nonmetals, and metalloids. The "staircase" line that runs from the top of group 13 to the bottom of group 16 is a dividing line:

- The **metals** (three shades of *blue* in Figure 2.9) lie in the large lower-left portion of the table. About three-quarters of the elements are metals, including many main-group elements and all the transition and inner transition elements. They are generally shiny solids at room temperature (mercury is the only liquid) and conduct heat and electricity well. They can be tooled into sheets (are malleable) and wires (are ductile).
- The **nonmetals** (*yellow*) lie in the small upper-right portion of the table. They are generally gases or dull, brittle solids at room temperature (bromine is the only liquid) and conduct heat and electricity poorly.
- The **metalloids (semimetals)** (*green*), which lie along the staircase line, have properties between those of metals and nonmetals.

Figure 2.10 shows examples of these three classes of elements.

Keep in mind the following two major points:

1. In general, elements in a group have *similar* chemical properties and elements in a period have *different* chemical properties.
2. Despite this classification of three types of elements, in reality, there is a gradation in properties from left to right and top to bottom.

It is important to learn some of the group (family) names. Group 1, except for hydrogen, consists of the *alkali metals*, and group 2 consists of the *alkaline earth metals*. The elements in both of these groups are highly reactive. The *halogens*, group 17, are highly reactive nonmetals, whereas the *noble gases*, group 18, are relatively unreactive nonmetals. Other main groups (13 to 16) are often named for the first element in the group; for example, group 16 is the *oxygen family*.

Two of the major branches of chemistry have traditionally been defined by the elements that are studied. *Organic chemistry* focuses on the compounds of carbon, specifically those that contain hydrogen and often oxygen, nitrogen, and a few other elements. This branch is concerned with fuels, drugs, dyes, and the like. *Inorganic*

FIGURE 2.10 Some metals, metalloids, and nonmetals

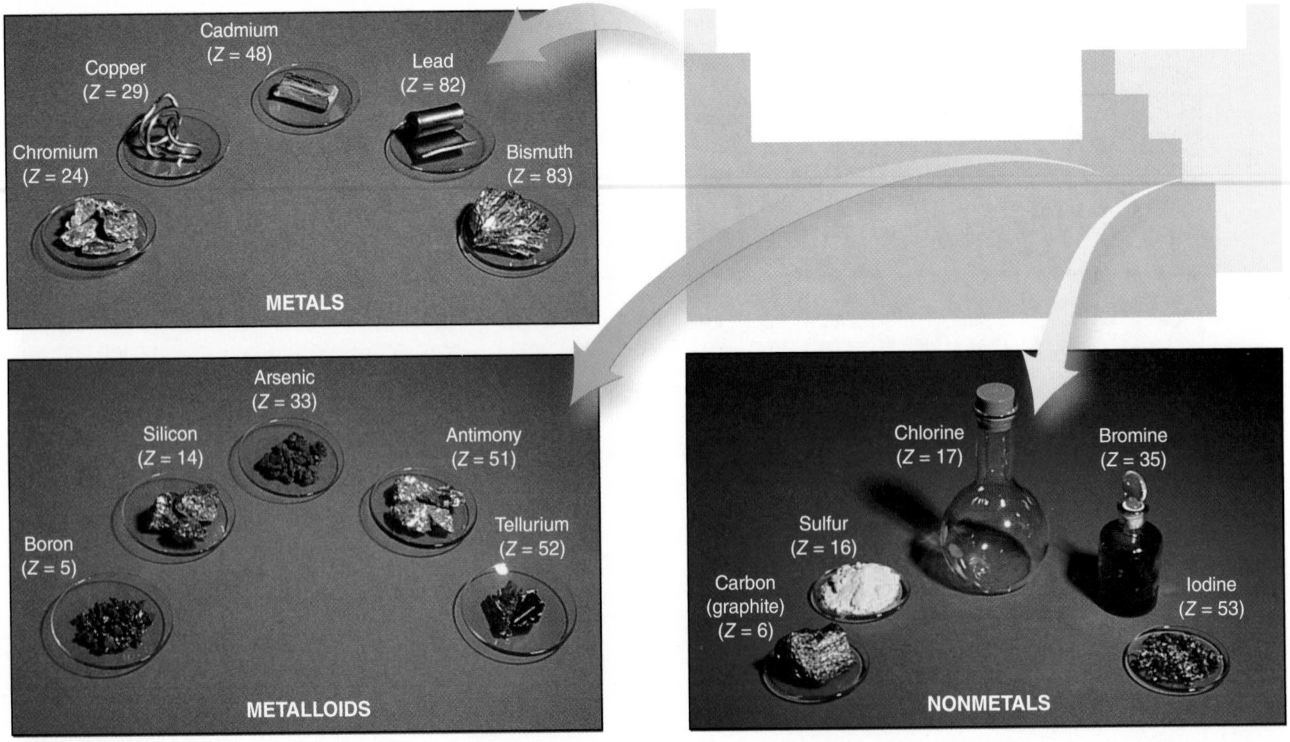

chemistry, on the other hand, focuses on the compounds of all the other elements and is concerned with catalysts, electronic materials, metal alloys, mineral salts, and the like. With the explosive growth in biomedical and materials sciences, the line between these branches has all but disappeared.

SUMMARY OF SECTION 2.6

- In the periodic table, the elements are arranged by atomic number into horizontal periods and vertical groups.
- Nonmetals appear in the upper-right portion of the table, metalloids lie along a staircase line, and metals fill the rest of the table.
- Elements within a group have similar behaviour, whereas elements within a period have dissimilar behaviour.

2.7 Compounds: An Introduction to Bonding

Only a few elements occur unbound, or as single atoms, in nature. The noble gases—helium (He), neon (Ne), argon (Ar), krypton (Kr), xenon (Xe), and radon (Rn)—occur in air as separate atoms. In addition to occurring in compounds, oxygen (O), nitrogen (N), and sulfur (S) occur in their most common elemental form as the molecules O_2, N_2, and S_8, and carbon (C) occurs in vast, nearly pure deposits of coal. Certain metals—copper (Cu), silver (Ag), gold (Au), and platinum (Pt)—are sometimes found uncombined. Aside from these few exceptions, *the overwhelming majority of elements occur in compounds combined with other elements.*

Elements combine in two general ways, and both ways involve *the electrons of the atoms of interacting elements*:

1. *Transferring electrons* from one element to another to form **ionic compounds**
2. *Sharing electrons* between atoms of different elements to form **covalent compounds**

These processes generate **chemical bonds**, *the forces that hold the atoms together in a compound.* This section introduces compound formation, which we will discuss in much more detail in later chapters.

The Formation of Ionic Compounds

Ionic compounds are composed of **ions**, *charged particles that form when an atom (or small group of atoms) gains or loses one or more electrons.* The simplest type of ionic compound is a **binary ionic compound**, *a compound composed of two elements.* It typically forms *when a metal reacts with a nonmetal*:

- Each metal atom *loses* one or more electrons and becomes a **cation**, *a positively charged ion.*
- Each nonmetal atom *gains* one or more electrons and becomes an **anion**, *a negatively charged ion.*

Binary ionic compounds can form in many ways: one way occurs when the metal atoms *transfer electrons* to the nonmetal atoms. These compounds may also be formed as a result of acid/base or precipitation reactions. The resulting large numbers of cations and anions attract each other and form an ionic compound. *A cation or anion derived from a single atom* is called a **monatomic ion**; we will discuss polyatomic ions, those derived from a small group of atoms, later.

The Case of Sodium Chloride *All binary ionic compounds are solid arrays of oppositely charged ions.* The formation of the binary ionic compound sodium chloride (common table salt) from its elements is shown in Figure 2.11. In the electron transfer, a sodium atom *loses* one electron and forms a sodium cation, Na^+. (The charge on the ion is written as a *right superscript.*) A chlorine atom *gains* the electron and becomes a chloride anion, Cl^-. (The name change when the nonmetal atom becomes an anion is discussed in the next section.) The oppositely charged ions (Na^+ and Cl^-) attract each other, and the similarly charged ions (Na^+ and Na^+, or Cl^- and Cl^-) repel each other. The resulting solid aggregation is a regular array of alternating Na^+ and Cl^- ions that extends in all three dimensions. Even the tiniest visible grain of table salt contains an enormous number of sodium and chloride ions.

Coulomb's Law The strength of the ionic bonding depends, to a great extent, on the net strength of these attractions and repulsions and can be described by *Coulomb's law: the energy of attraction (or repulsion) between two particles is directly proportional to the product of the charges and inversely proportional to the distance between them:*

$$\text{Energy} \propto \frac{\text{charge 1} \times \text{charge 2}}{\text{distance}}$$

This can be summarized as shown in Figure 2.12 and expressed as follows:

- Ions with higher charges attract (or repel) each other more strongly than ions with lower charges.
- Smaller ions attract (or repel) each other more strongly than larger ions, because their charges are closer together.

Predicting the Number of Electrons Lost or Gained *Ionic compounds are neutral* because they contain equal numbers of positive and negative *charges.* Thus, there are equal numbers of Na^+ and Cl^- ions in sodium chloride, because both ions are singly charged. However, there are two Na^+ ions for each oxide ion, O^{2-}, in sodium oxide because two 1+ ions balance one 2− ion.

Can we predict the number of electrons that a given atom will lose or gain when it forms an ion? For elements in groups 1, 2, and 13 to 18, we usually find that metal atoms lose electrons and nonmetal atoms gain electrons to *form ions with the same number of electrons as in an atom of the nearest noble gas* (group 18). Noble gases have a stability that is related to their number (and arrangement) of electrons. Thus, a sodium atom ($11e^-$) can attain the stability of a neon atom ($10e^-$), the nearest noble gas, by losing one electron. Similarly, a chlorine atom ($17e^-$) can attain the stability of an argon atom ($18e^-$), its nearest noble gas, by gaining one electron. Thus, in general, elements located near a noble gas form monatomic ions as follows:

- *Metals lose electrons:* Elements in group 1 lose one electron, elements in group 2 lose two electrons, and aluminum in group 13 loses three electrons.
- *Nonmetals gain electrons:* Elements in group 17 gain one electron, oxygen and sulfur in group 16 gain two electrons, and nitrogen in group 15 gains three electrons.

FIGURE 2.11 The formation of an ionic compound. A. The two elements as seen in a laboratory. **B.** The elements on the atomic scale. **C.** The electron transfer from Na atom to Cl atom forms Na^+ and Cl^- ions. **D.** Countless Na^+ and Cl^- ions attract each other and form a regular three-dimensional array. **E.** Crystalline NaCl occurs naturally as the mineral halite.

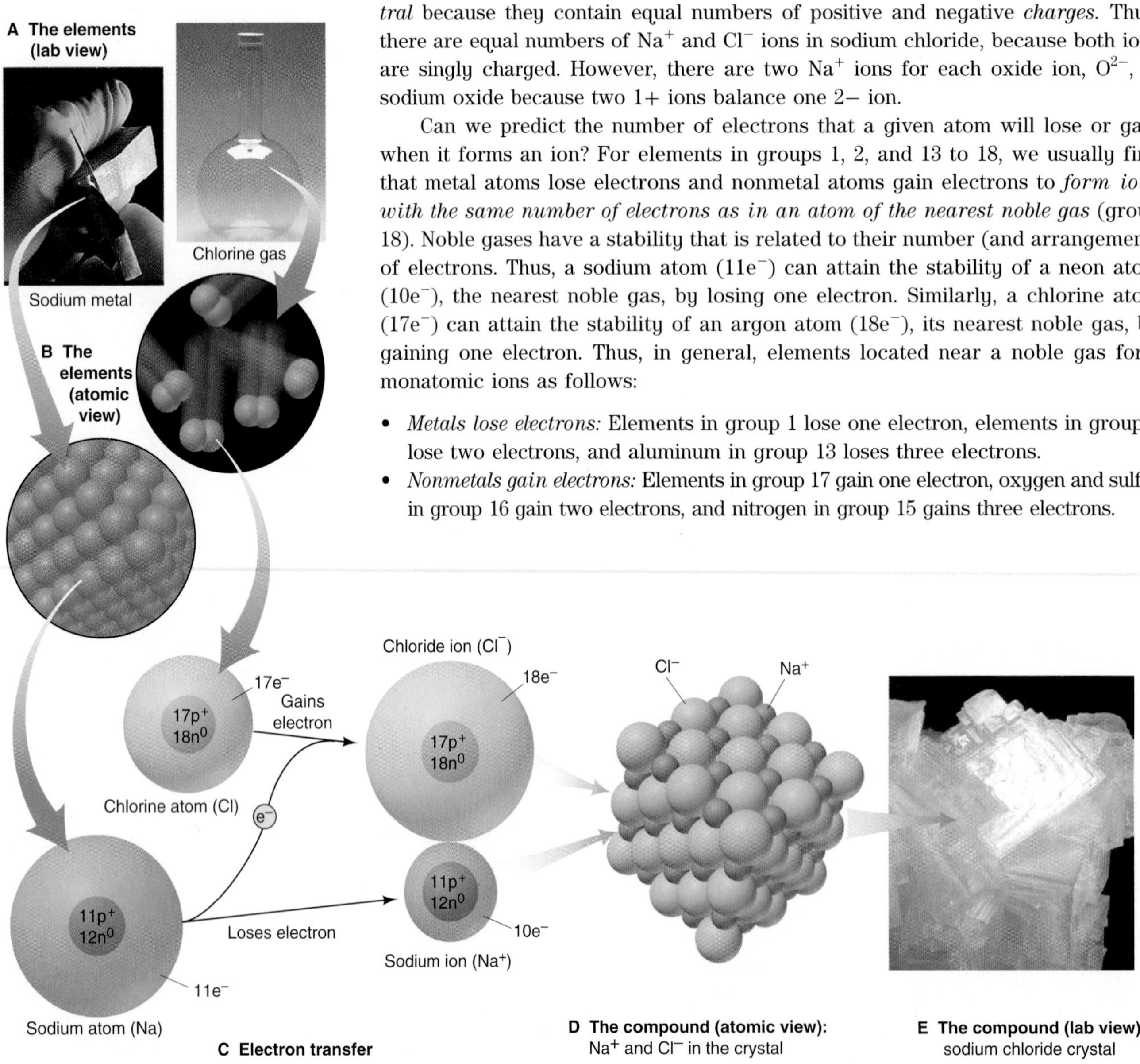

In the periodic table in Figure 2.9, the elements in group 17 appear to be "closer" to the noble gases than the elements in group 1. In truth, both groups are only one electron away from the number of electrons in the nearest noble gas. Figure 2.13 shows a periodic table of monatomic ions that is cut and rejoined as a cylinder. Note that fluorine (F; $Z = 9$) has one electron *less* than the noble gas neon (Ne; $Z = 10$), and sodium (Na; $Z = 11$) has one electron *more*; thus, fluorine and sodium form the F^- and Na^+ ions. Similarly, oxygen (O; $Z = 8$) gains two electrons and magnesium (Mg; $Z = 12$) loses two electrons to form the O^{2-} and Mg^{2+} ions and attain the same number of electrons as neon. In Figure 2.13, notice that species in a row have the same number of electrons.

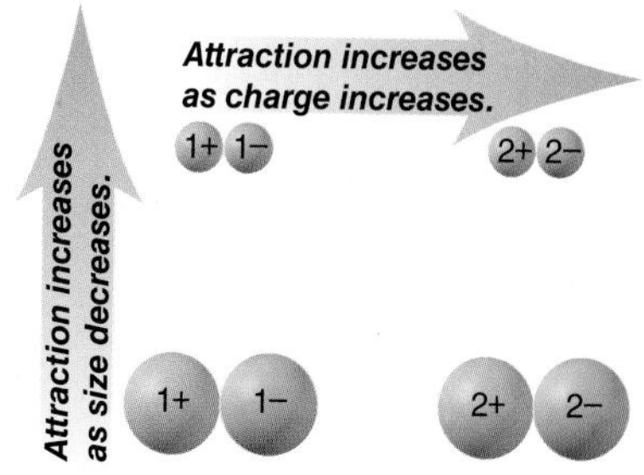

FIGURE 2.12 Factors that influence the strength of ionic bonding

Sample Problem 2.6 Predicting the Ion That an Element Forms

Problem What monatomic ions does each element form?

(a) Iodine ($Z = 53$) **(b)** Calcium ($Z = 20$) **(c)** Aluminum ($Z = 13$)

Plan We use the given value of Z to find the element in the periodic table and see where its group lies relative to the noble gases. Elements in groups 1, 2, and 3 *lose* electrons to attain the same number of electrons as the nearest noble gas and become positive ions; those in groups 15, 16, and 17 *gain* electrons and become negative ions.

Solution (a) I^- Iodine ($_{53}I$) is in group 17, the halogens. Like any member of this group, it gains 1 electron to attain the same number of electrons as the nearest group 18 member, in this case, $_{54}Xe$.

(b) Ca^{2+} Calcium ($_{20}Ca$) is in group 2, the alkaline earth metals. Like any group 2 member, it loses 2 electrons to attain the same number as the nearest noble gas, $_{18}Ar$.

(c) Al^{3+} Aluminum ($_{13}Al$) is a metal in the boron family (group 13) and, thus, loses 3 electrons to attain the same number as its nearest noble gas, $_{10}Ne$.

Follow-Up Problem 2.6 What monatomic ion does each element form?

(a) $_{16}S$ **(b)** $_{37}Rb$ **(c)** $_{56}Ba$

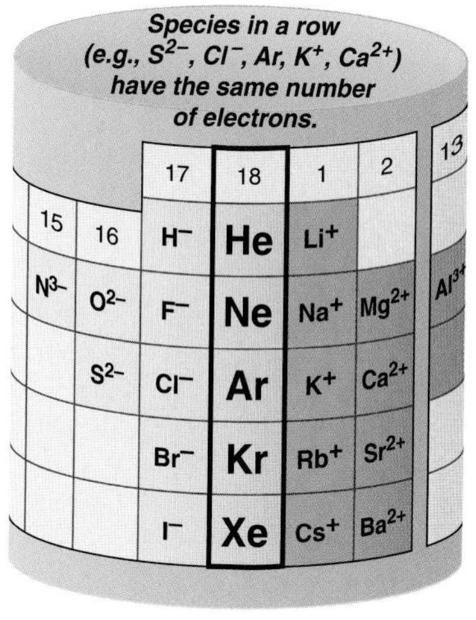

FIGURE 2.13 The relationship between the ion formed and the nearest noble gas

The Formation of Covalent Compounds

Covalent compounds form when elements, usually nonmetals, share electrons. The simplest case of electron sharing occurs not in a compound but between two hydrogen atoms (H; $Z = 1$). Imagine two separated H atoms approaching each other (Figure 2.14). As they get closer, the nucleus of each atom attracts the electron of the other atom more and more strongly. As the separated atoms begin to interpenetrate each other, repulsions between the nuclei and between the electrons begin to increase. At some optimum distance between the nuclei, the two atoms form a **covalent bond**, *a pair of electrons mutually attracted by the two nuclei.* The result is a hydrogen molecule, in which each electron no longer "belongs" to a particular H atom: the two electrons are *shared* by the two nuclei. A sample of hydrogen gas consists of these diatomic molecules (H_2)—pairs of atoms that are chemically bound and behave as an independent unit—*not* separate H atoms. Figure 2.15 shows other nonmetals that exist as molecules at room temperature.

Atoms of different elements share electrons to form the molecules of a covalent compound. A sample of hydrogen fluoride, for example, consists of molecules in which one H atom forms a covalent bond with one F atom; water consists of molecules in which one O atom forms covalent bonds with two H atoms:

Hydrogen fluoride, HF

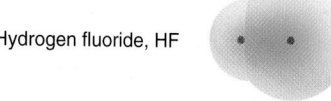

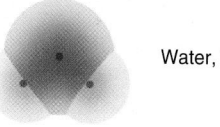

Water, H_2O

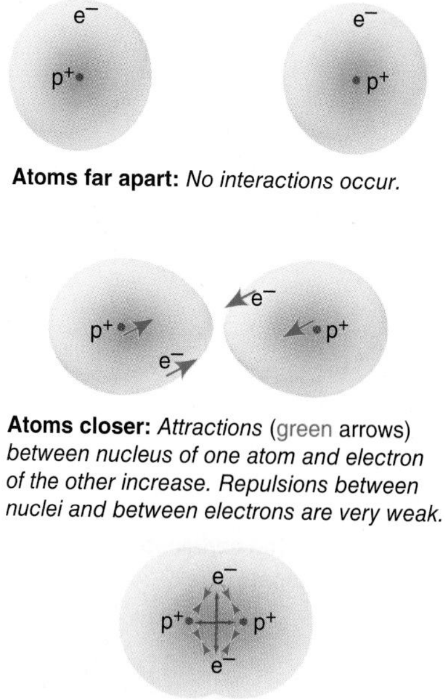

Atoms far apart: *No interactions occur.*

Atoms closer: *Attractions (green arrows) between nucleus of one atom and electron of the other increase. Repulsions between nuclei and between electrons are very weak.*

Optimum distance: *H_2 molecule forms because attractions (green arrows) balance repulsions (red arrows).*

FIGURE 2.14 Formation of a covalent bond between two H atoms

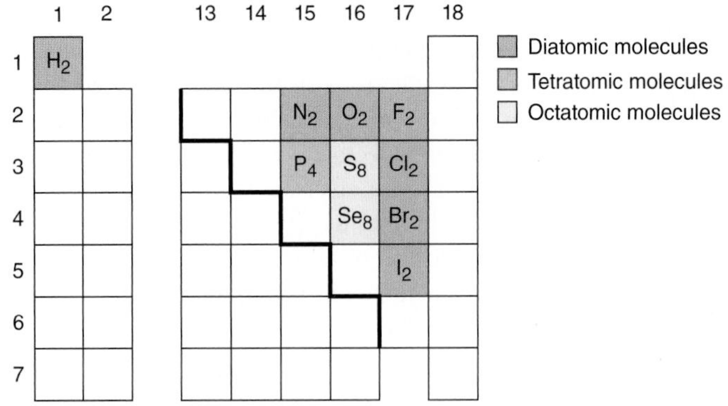

FIGURE 2.15 Elements that occur as molecules

Distinguishing the Entities in Covalent and Ionic Substances There is a key distinction between the chemical entities in covalent substances and in ionic substances. *Most covalent substances consist of molecules.* A cup of water, for example, consists of individual water molecules lying near each other. In contrast, under ordinary conditions, *there are no molecules in an ionic compound.* A piece of sodium chloride, for example, is a continuous array, in three dimensions, of oppositely charged sodium and chloride ions, *not* a collection of individual sodium chloride "molecules."

Another key distinction between covalent and ionic substances concerns the nature of the particles attracting each other. Covalent bonding involves the mutual attraction between two (positively charged) nuclei and the two (negatively charged) electrons that reside between them. Ionic bonding involves the mutual attraction between positive and negative ions.

Polyatomic Ions: Covalent Bonds within Ions Many ionic compounds contain **polyatomic ions**, which *consist of two or more atoms bonded* ***covalently*** *and have a net positive or negative charge.* For example, Figure 2.16 shows that a crystalline form of calcium carbonate (*left*) occurs, on the atomic scale, as an array of polyatomic carbonate anions and monatomic calcium cations (*centre*). The carbonate ion (*right*) consists of a carbon atom covalently bonded to three oxygen atoms, and two additional electrons give the ion its 2− charge. In many reactions, the polyatomic ion stays together as a unit.

SUMMARY OF SECTION 2.7

- Although a few elements occur uncombined in nature, the great majority exist as compounds.
- Ionic compounds form when a metal *transfers* electrons to a nonmetal, and the resulting positive and negative ions attract each other to form a three-dimensional array. In many cases, metal atoms lose and nonmetal atoms gain enough electrons to attain the same number of electrons as in atoms of the nearest noble gas.
- Covalent compounds form when elements, usually nonmetals, *share* electrons. Each covalent bond is an electron pair mutually attracted by two atomic nuclei.
- Monatomic ions are derived from single atoms. Polyatomic ions consist of two or more covalently bonded atoms that have a net positive or negative charge due to a deficit or excess of electrons.

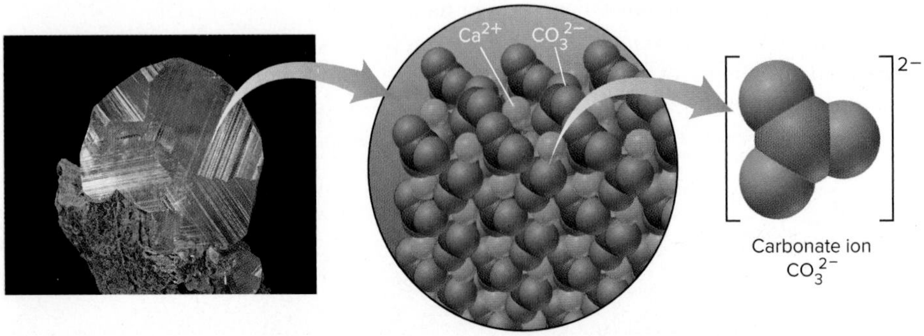

FIGURE 2.16 The carbonate ion in calcium carbonate

2.8 Formula, Name, and Mass of a Compound

In a **chemical formula**, *element symbols and often numerical subscripts show the type and number of each atom in the smallest unit of the substance.* In this section, you will learn how to write the name and formula of ionic compounds and simple covalent compounds, how to calculate the mass of a compound from its formula, and how to visualize molecules with three-dimensional models. To make learning the name and formula of a compound easier, we will rely on various rules. So, be prepared for a bit of memorization and a lot of practice!

Binary Ionic Compounds

Let us begin with two general rules:

- For *all* ionic compounds, *the name and formula give the positive ion (cation) first and the negative ion (anion) second.*
- For all *binary* ionic compounds, *the name of the cation is the name of the metal, and the name of the anion has the suffix -ide added to the root of the name of the nonmetal.*

For example, the anion formed from brom*ine* is named brom*ide* (brom + ide). Therefore, the compound formed from the metal calcium and the nonmetal bromine is named *calcium bromide.*

In general, if the metal of a binary ionic compound is a main-group element (groups 1 and 2), it usually forms a single type of ion; if it is a transition element (groups 3 to 12), it often forms more than one type of ion. We will discuss each case in turn. Some of the more common monatomic ions are presented in Table 2.3.

Compounds of Elements That Form One Ion The periodic table presents some key points about the formula of main-group monatomic ions:

- Monatomic ions of main-group elements have the same ionic charge; the alkali metals—Li, Na, K, Rb, Cs, and Fr—form ions with a 1+ charge; the halogens—F, Cl, Br, and I—form ions with a 1− charge; and so on.
- For cations, the ion charge equals the group number: Na is in group 1 and forms Na^+; Ba is in group 2 and forms Ba^{2+}. (Exceptions in Figure 2.17 are Sn^{2+} and Pb^{2+}.)
- For anions, the ion charge equals the group number minus 18; for example, S is in group 16 ($16 - 18 = -2$) and, thus, forms S^{2-}.

Because an ionic compound consists of an array of ions rather than separate molecules, its formula represents the **formula unit**, *the* ***relative*** *numbers of cations*

TABLE 2.3 Common Monatomic Ions*

Charge	Formula	Name
Cations		
1+	H^+	Hydrogen
	Li^+	**Lithium**
	Na^+	**Sodium**
	K^+	**Potassium**
	Cs^+	Cesium
	Ag^+	**Silver**
2+	**Mg^{2+}**	**Magnesium**
	Ca^{2+}	**Calcium**
	Sr^{2+}	Strontium
	Ba^{2+}	**Barium**
	Zn^{2+}	**Zinc**
	Cd^{2+}	Cadmium
3+	Al^{3+}	Aluminum
Anions		
1−	H^-	Hydride
	F^-	**Fluoride**
	Cl^-	**Chloride**
	Br^-	**Bromide**
	I^-	**Iodide**
2−	**O^{2-}**	**Oxide**
	S^{2-}	**Sulfide**
3−	N^{3-}	Nitride

*The ions are listed by charge; those in boldface are the most common.

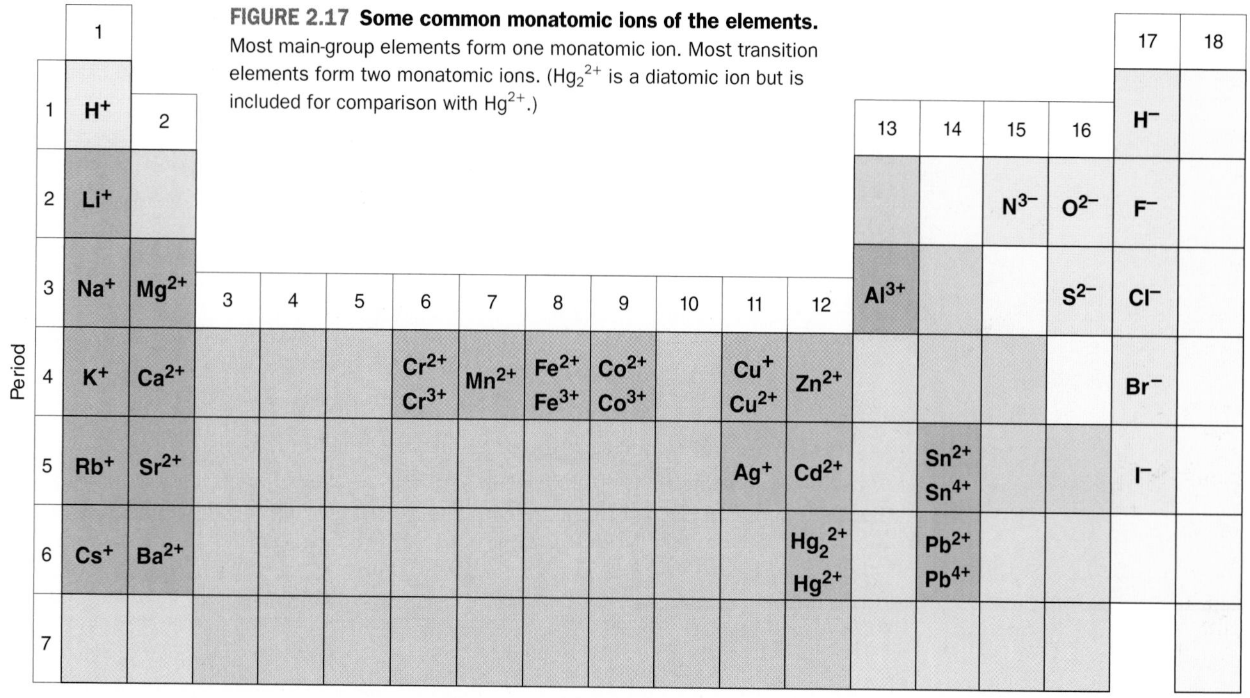

FIGURE 2.17 Some common monatomic ions of the elements. Most main-group elements form one monatomic ion. Most transition elements form two monatomic ions. (Hg_2^{2+} is a diatomic ion but is included for comparison with Hg^{2+}.)

and anions in the compound. The compound has zero net charge, so the positive charges of the cations balance the negative charges of the anions. For example, calcium bromide is composed of Ca^{2+} ions and Br^- ions, so two Br^- balance each Ca^{2+}. The formula is $CaBr_2$, not Ca_2Br. For help in writing the formula of this and other compounds, remember the following rules:

- The subscript refers to the element *preceding* it. The *subscript 1 is understood* from the presence of the element symbol alone (that is, we do not write Ca_1Br_2).
- The charge (without the sign) of one ion becomes the subscript of the other:

$$Ca^{2+} \quad Br^{1-} \qquad \text{gives} \qquad Ca_1Br_2 \qquad \text{or} \qquad CaBr_2$$

- The subscripts need to be reduced to the smallest whole numbers that retain the ratio of ions. Thus, for example, for the Ca^{2+} and O^{2-} ions in calcium oxide, we get Ca_2O_2, which we reduce to the formula CaO.*

The following two sample problems apply the rules we just discussed. In Sample Problem 2.7, we name the compound from its elements. In Sample Problem 2.8, we find the formula.

Sample Problem 2.7 Naming Binary Ionic Compounds

Problem Name the ionic compound formed from each pair of elements:

(a) Magnesium and nitrogen

(b) Iodine and cadmium

(c) Strontium and fluorine

(d) Sulfur and cesium

Plan The key to naming a binary ionic compound is to recognize which element is the metal and which is the nonmetal. When in doubt, check the periodic table. We place the cation name first, add the suffix *-ide* to the nonmetal root, and place the anion name last.

Solution **(a)** Magnesium is the metal; *nitr-* is the nonmetal root: magnesium nitride

(b) Cadmium is the metal; *iod-* is the nonmetal root: cadmium iodide

(c) Strontium is the metal; *fluor-* is the nonmetal root: strontium fluoride (Note the spelling is fluoride, not fl*ou*ride.)

(d) Cesium is the metal; *sulf-* is the nonmetal root: cesium sulfide

Follow-Up Problem 2.7 For the following ionic compounds, give the name of each element and its periodic table group number:

(a) Zinc oxide

(b) Silver bromide

(c) Lithium chloride

(d) Aluminum sulfide

Sample Problem 2.8 Determining the Formula of a Binary Ionic Compound

Problem Write the formula for the compounds named in Sample Problem 2.7.

Plan We write a formula by finding the smallest number of each ion that gives the neutral compound. This number appears as a *right subscript* to the element symbol.

*Compounds of the mercury(I) ion, such as Hg_2Cl_2, and peroxides of the alkali metals, such as Na_2O_2, are the only two common exceptions; in fact, reducing the subscripts for these compounds would give the incorrect formula for each, namely HgCl and NaO.

Solution

(a) Mg^{2+} and N^{3-}; three Mg^{2+} ions (6+) balance two N^{3-} ions (6−): Mg_3N_2

(b) Cd^{2+} and I^-; one Cd^{2+} ion (2+) balances two I^- ions (2−): CdI_2

(c) Sr^{2+} and F^-; one Sr^{2+} ion (2+) balances two F^- ions (2−): SrF_2

(d) Cs^+ and S^{2-}; two Cs^+ ions (2+) balance one S^{2-} ion (2−): Cs_2S

Comment **1.** The subscript 1 is understood and so not written; thus, in (b), we do *not* write Cd_1I_2.
2. Ion charges do *not* appear in the formula for a compound; thus, in (c), we do *not* write $Sr^{2+}F_2^-$.

Follow-Up Problem 2.8 Write the formula for the compounds named in Follow-Up Problem 2.7.

Compounds with Metals That Form More Than One Ion As noted earlier, many metals, particularly the transition elements (groups 3 to 12), can form more than one ion. Table 2.4 lists some examples; see Figure 2.17 for their placement in the periodic table. Names of compounds containing these elements include a *roman numeral within parentheses* immediately after the metal ion's name to indicate its ionic charge. For example, iron can form Fe^{2+} and Fe^{3+} ions. The two compounds that iron forms with chlorine are $FeCl_2$, named iron(II) chloride (spoken "iron two chloride"), and $FeCl_3$, named iron(III) chloride.

We are focusing here on systematic names, but some common (trivial) names are still used. In the common names of some metal ions, the Latin root of the metal is followed by either of two suffixes (see Table 2.4):

- The suffix *-ous* for the ion with the lower charge
- The suffix *-ic* for the ion with the higher charge

Thus, iron(II) chloride is also called ferr*ous* chloride and iron(III) chloride is called ferr*ic* chloride. (Memory aid: There is an ***o*** in *-**o**us* and *l**o**wer*, and an ***i*** in *-**i**c* and *h**i**gher.*)

TABLE 2.4 Some Metals That Form More Than One Monatomic Ion*

Element	Ion Formula	Systematic Name	Common (Trivial) Name
Chromium	Cr^{2+}	Chromium(II)	Chromous
	Cr^{3+}	**Chromium(III)**	Chromic
Cobalt	Co^{2+}	Cobalt(II)	
	Co^{3+}	Cobalt(III)	
Copper	**Cu^+**	**Copper(I)**	Cuprous
	Cu^{2+}	**Copper(II)**	Cupric
Iron	**Fe^{2+}**	**Iron(II)**	Ferrous
	Fe^{3+}	**Iron(III)**	Ferric
Lead	**Pb^{2+}**	**Lead(II)**	
	Pb^{4+}	Lead(IV)	
Mercury	Hg_2^{2+}**	Mercury(I)	Mercurous
	Hg^{2+}	**Mercury(II)**	Mercuric
Tin	**Sn^{2+}**	**Tin(II)**	Stannous
	Sn^{4+}	Tin(IV)	Stannic

*The ions are listed alphabetically by metal name; those in boldface are the most common.
**This ion has been included in both this table and the table for polyatomic ions (Table 2.5). It is a single element that has been included in this table for reasons of completeness for the charge on the mercury ion and in Table 2.5 because there is more than one atom in the ion.

TABLE 2.5 Common Polyatomic Ions*	
Formula	**Name**
Cations	
NH_4^+	**Ammonium**
H_3O^+	**Hydronium**
Hg_2^{2+}	**Mercury(I)**
Anions	
CH_3COO^- (or **$C_2H_3O_2^-$**)	**Ethanoate (or Acetate)**
CN^-	Cyanide
OH^-	**Hydroxide**
ClO^-	Hypochlorite
ClO_2^-	Chlorite
ClO_3^-	**Chlorate**
ClO_4^-	**Perchlorate**
NO_2^-	Nitrite
NO_3^-	**Nitrate**
MnO_4^-	**Permanganate**
CO_3^{2-}	**Carbonate**
HCO_3^-	**Hydrogen carbonate** (or **bicarbonate**)
CrO_4^{2-}	Chromate
$Cr_2O_7^{2-}$	**Dichromate**
O_2^{2-}	Peroxide
PO_4^{3-}	**Phosphate**
HPO_4^{2-}	Hydrogen phosphate
$H_2PO_4^-$	Dihydrogen phosphate
SO_3^{2-}	Sulfite
SO_4^{2-}	**Sulfate**
HSO_4^-	Hydrogen sulfate (or bisulfate)

*Boldface ions are the most common.

No. of O atoms (increasing ↑)	Prefix	Root	Suffix
	per	*root*	ate
		root	ate
		root	ite
	hypo	*root*	ite

FIGURE 2.18 Naming oxoanions. Prefixes and suffixes indicate the number of oxygen (O) atoms in the anion.

Sample Problem 2.9 Determining the Name and Formula of Ionic Compounds of Metals That Form More Than One Ion

Problem Give the systematic name for the given formula or the formula for the given name of each compound: **(a)** Tin(II) fluoride **(b)** CrI_3 **(c)** Ferric oxide **(d)** CoS

Solution **(a)** Tin(II) ion is Sn^{2+}; fluoride is F^-. Two F^- ions balance one Sn^{2+} ion: tin(II) fluoride is SnF_2. (The common name is stannous fluoride.)

(b) The anion is I^-, iodide, and the formula shows three I^-. Therefore, the cation must be Cr^{3+}, chromium(III) ion: CrI_3 is chromium(III) iodide. (The common name is chromic iodide.)

(c) *Ferric* is the common name for iron(III) ion, Fe^{3+}; oxide ion is O^{2-}. To balance the charges, the formula is Fe_2O_3. (The systematic name is iron(III) oxide.)

(d) The anion is sulfide, S^{2-}, which requires that the cation be Co^{2+}. The name is cobalt(II) sulfide.

Follow-Up Problem 2.9 Give the systematic name for the formula or the formula for the given name of each compound:
(a) Lead(IV) oxide **(b)** Cu_2S **(c)** $FeBr_2$ **(d)** Mercuric chloride

Compounds That Contain Polyatomic Ions

Many ionic compounds contain polyatomic ions. Table 2.5 shows some common polyatomic ions. Remember that *a polyatomic ion stays together as a charged unit.* For example, the formula for potassium nitrate is KNO_3: each K^+ balances one NO_3^-. The formula for sodium carbonate is Na_2CO_3: two Na^+ balance one CO_3^{2-}. *When two or more of the same polyatomic ion are present in a formula unit, this ion appears in parentheses with the subscript written outside.* For example, calcium nitrate contains one Ca^{2+} and two NO_3^- ions and has the formula $Ca(NO_3)_2$. Parentheses and a subscript are *only* used if *more than one* of a given polyatomic ion is present; thus, sodium nitrate is $NaNO_3$, *not* $Na(NO_3)$.

Families of Oxoanions As Table 2.5 shows, most polyatomic ions are **oxoanions** (or *oxyanions*), *ions in which an element, usually a nonmetal, is bonded to one or more oxygen atoms.* There are several families of two or four oxoanions that differ only in the number of oxygen atoms. The following simple naming conventions are used with these ions.

For families with two oxoanions:

- The ion with *more* O atoms takes the nonmetal root and the suffix *-ate*.
- The ion with *fewer* O atoms takes the nonmetal root and the suffix *-ite*.

For example, SO_4^{2-} is the sulf*ate* ion, and SO_3^{2-} is the sulf*ite* ion; similarly, NO_3^- is nitr*ate*, and NO_2^- is nitr*ite*.

For families with four oxoanions (a halogen bonded to O) (Figure 2.18):

- The ion with the *most* O atoms has the prefix *per-*, the nonmetal root, and the suffix *-ate*.
- The ion with *one less* O atom has just the root and the suffix *-ate*.
- The ion with *two fewer* O atoms has just the root and the suffix *-ite*.
- The ion with the *least (three fewer)* O atoms has the prefix *hypo-*, the root, and the suffix *-ite*.

For example, for the four chlorine oxoanions, ClO_4^- is *per*chlor*ate*, ClO_3^- is chlor*ate*, ClO_2^- is chlor*ite*, and ClO^- is *hypo*chlor*ite*.

Hydrated Ionic Compounds Ionic compounds called **hydrates** *have a specific number of water molecules in each formula unit*, which is shown after a centred dot in the formula and noted in the name by a Greek numerical prefix before the word *hydrate*. Table 2.6 shows these prefixes. For example, Epsom salt has seven water molecules in each formula unit: its formula is $MgSO_4 \cdot 7H_2O$, and its name is magnesium sulfate *hepta*hydrate. Similarly, the mineral gypsum has the formula $CaSO_4 \cdot 2H_2O$ and the name calcium sulfate *di*hydrate. The water molecules, referred to as "waters of hydration," are part of the hydrate's structure. Heating can remove some or all of them, leading to a different substance. For example, when heated strongly, blue copper(II) sulfate pentahydrate ($CuSO_4 \cdot 5H_2O$) is converted to white copper(II) sulfate ($CuSO_4$).

TABLE 2.6 Numerical Prefixes for Hydrates and Binary Covalent Compounds

Number	Prefix
1	mono-
2	di-
3	tri-
4	tetra-
5	penta-
6	hexa-
7	hepta-
8	octa-
9	nona-
10	deca-

Sample Problem 2.10 Determining the Name and Formula for an Ionic Compound That Contains Polyatomic Ions

Problem Give the systematic name for the given formula or the formula for the given name of each compound:

(a) $Fe(ClO_4)_2$ **(b)** Sodium sulfite **(c)** $Ba(OH)_2 \cdot 8H_2O$

Solution (a) ClO_4^- is perchlorate, which has a 1− charge, so the cation must be Fe^{2+}. The name is iron(II) perchlorate. (The common name is ferrous perchlorate.)

(b) Sodium is Na^+; sulfite is SO_3^{2-}, and two Na^+ ions balance one SO_3^{2-} ion. The formula is Na_2SO_3.

(c) Ba^{2+} is barium; OH^- is hydroxide. There are eight (*octa-*) water molecules in each formula unit. The name is barium hydroxide octahydrate.

Follow-Up Problem 2.10 Give the systematic name for the formula or the formula for the given name of each compound:

(a) Cupric nitrate trihydrate **(b)** Zinc hydroxide **(c)** LiCN

Sample Problem 2.11 Recognizing the Incorrect Name and Formula of an Ionic Compound

Problem Explain what is wrong with the name or formula at the end of each statement, and correct it:

(a) $Ba(C_2H_3O_2)_2$ is called barium diacetate.

(b) Sodium sulfide has the formula $(Na)_2SO_3$.

(c) Iron(II) sulfate has the formula $Fe_2(SO_4)_3$.

(d) Cesium carbonate has the formula $Cs_2(CO_3)$.

Solution (a) The charge of the Ba^{2+} ion *must* be balanced by *two* $C_2H_3O_2^-$ ions, so the prefix *di-* is unnecessary. For ionic compounds, we do not indicate the number of ions with numerical prefixes. The correct name is barium acetate.

(b) Two mistakes occur here. The sodium ion is monatomic, so it does *not* require parentheses. The sulfide ion is S^{2-}, *not* SO_3^{2-} (which is sulfite). The correct formula is Na_2S.

(c) The roman numeral refers to the charge of the ion, *not* the number of ions in the formula. Fe^{2+} is the cation, so it requires one SO_4^{2-} to balance its charge. The correct formula is $FeSO_4$. [$Fe_2(SO_4)_3$ is the formula for iron(III) sulfate.]

(d) Parentheses are *not* required when only one polyatomic ion of a kind is present. The correct formula is Cs_2CO_3.

Follow-Up Problem 2.11 State why the formula or name at the end of each statement is incorrect, and correct it:

(a) Ammonium phosphate is $(NH_3)_4PO_4$.

(b) Aluminum hydroxide is $AlOH_3$.

(c) $Mg(HCO_3)_2$ is manganese(II) carbonate.

(d) $Cr(NO_3)_3$ is chromic(III) nitride.

(e) $Ca(NO_2)_2$ is cadmium nitrate.

Prefix	Root	Suffix
per	*root*	ic acid
	root	ic acid
	root	ous acid
hypo	*root*	ous acid

No. of O atoms (increasing upward)

FIGURE 2.19 Naming oxoacids. Prefixes and suffixes indicate the number of oxygen (O) atoms in the anion.

Acid Names from Anion Names

Acids are an important group of hydrogen-containing compounds that have been used in chemical reactions since before alchemical times. In the laboratory, acids are typically used in water to form aqueous solutions. When naming an acid and writing its formula, we consider acids as anions that are connected to the number of hydrogen ions (H^+) needed for charge neutrality. The two common types of acids are binary acids and oxoacids:

1. *Binary acid* solutions form when certain gaseous compounds dissolve in water. For example, when gaseous hydrogen chloride (HCl) dissolves in water, it forms *hydrochloric acid*:

   prefix *hydro-* + nonmetal *root* + suffix *-ic* + separate word *acid*
   hydro + chlor + ic + acid

   This naming pattern holds for many compounds in which hydrogen combines with an anion that has an *-ide* suffix.

2. *Oxoacid* names are similar to the names of the oxoanions, except for two suffix changes (see Figure 2.19):
   - The *-ate* in the anion becomes *-ic* in the acid.
   - The *-ite* in the anion becomes *-ous* in the acid.

   The oxoanion prefixes *hypo-* and *per-* are retained. Thus,

   BrO_4^- is *per*brom*ate*, and $HBrO_4$ is *per*brom*ic* acid.
   IO_2^- is iod*ite*, and HIO_2 is iod*ous* acid.

   (Memory aid: There is an ***o*** in *-**o**us* and *lower*, and an ***i*** in *-**i**c* and *higher*.)

Sample Problem 2.12 Determining the Name and Formula for Anions and Acids

Problem Name each anion, and give the name and formula for the acid derived from it:
(a) Br^- **(b)** IO_3^- **(c)** CN^- **(d)** SO_4^{2-} **(e)** NO_2^-

Solution (a) The anion is bromide; the acid is hydrobromic acid, HBr.
(b) The anion is iodate; the acid is iodic acid, HIO_3.
(c) The anion is cyanide; the acid is hydrocyanic acid, HCN.
(d) The anion is sulfate; the acid is sulfuric acid, H_2SO_4. (In this case, the suffix is added to the element name *sulfur*, not to the root *sulf-*.)
(e) The anion is nitrite; the acid is nitrous acid, HNO_2.

Comment We must add *two* H^+ ions to the sulfate ion to obtain sulfuric acid because SO_4^{2-} has a 2− charge.

Follow-Up Problem 2.12 Write the formula for the name or the name for the formula for each acid:
(a) Chloric acid **(b)** HF **(c)** Acetic acid **(d)** Sulfurous acid **(e)** HBrO

Binary Covalent Compounds

Binary covalent compounds are typically formed by the combination of two nonmetals. Some—such as ammonia (NH_3), acetic acid (CH_3COOH), and water (H_2O)—are so familiar that we use their common names, but most are named systematically:

- The element with the lower group number in the periodic table comes first in the name. The element with the higher group number comes second and is named with its root and the suffix *-ide*. For example, nitrogen (group 15) and fluorine (group 17) form a compound that has three fluorine atoms for every nitrogen atom. The name of the compound is nitrogen trifluoride, and the formula is NF_3.

(*Exception*: When the compound contains oxygen and any of the halogens chlorine, bromine, and iodine, the halogen is named first.)

- If both elements are in the same group, the element with the higher period number is named first. Thus, the group 16 elements sulfur (period 3) and oxygen (period 2) form sulfur dioxide, SO_2.
- Covalent compounds use Greek numerical prefixes (see Table 2.6) to indicate the number of atoms of each element. The first element in the name has a prefix *only* when more than one atom of it is present; the second element *usually* has a prefix. When the second element name begins with a vowel, we usually drop the vowel attached to the prefix. For example, we say dinitrogen tetroxide, not dinitrogen tetraoxide.

Sample Problem 2.13 Determining the Name and Formula of a Binary Covalent Compound

Problem (a) What is the formula for carbon disulfide?

(b) What is the name of PCl_5?

(c) Each molecule in a compound consists of two N atoms and four O atoms. Give the name and formula for the compound.

Solution (a) The prefix *di-* means "two." The formula is CS_2.

(b) P is the symbol for phosphorus; there are five chlorine atoms, which is indicated by the prefix *penta-*. The name is phosphorus pentachloride.

(c) Nitrogen (N) comes first in the name (lower group number). The compound is dinitrogen tetroxide, N_2O_4.

Follow-Up Problem 2.13 Give the name or formula for each compound:
(a) SO_3 **(b)** SiO_2 **(c)** Dinitrogen monoxide **(d)** Selenium hexafluoride

Sample Problem 2.14 Recognizing the Incorrect Name and Formula of a Binary Covalent Compound

Problem Explain what is wrong with the name or formula at the end of each statement, and correct it: **(a)** SF_4 is monosulfur pentafluoride. **(b)** Dichlorine heptoxide is Cl_2O_6. **(c)** N_2O_3 is dinitrotrioxide.

Solution (a) There are two mistakes. *Mono-* is not needed if there is only one atom of the first element, and the prefix for four is *tetra-*, not *penta-*. The correct name is sulfur tetrafluoride.

(b) The prefix *hepta-* indicates seven, not six. The correct formula is Cl_2O_7.

(c) The full name of the first element is needed, and a space separates the two element names. The correct name is dinitrogen trioxide.

Follow-Up Problem 2.14 Explain what is wrong with the name or formula at the end of each statement, and correct it: **(a)** S_2Cl_2 is disulfurous dichloride. **(b)** Nitrogen monoxide is N_2O. **(c)** $BrCl_3$ is trichlorine bromide.

The Simplest Organic Compounds: Straight-Chain Alkanes

Organic compounds typically have complex structures that consist of chains, branches, and/or rings of carbon atoms bonded to hydrogen atoms and often to atoms of oxygen, nitrogen, and a few other elements. At this point, we will lay the groundwork for naming organic compounds by focusing on the simplest ones. Rules for naming more-complex organic compounds are detailed in Chapter 20.

TABLE 2.7 The First 10 Straight-Chain Alkanes

Name (Formula)	Model
Methane (CH_4)	
Ethane (C_2H_6)	
Propane (C_3H_8)	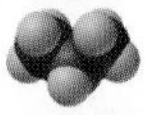
Butane (C_4H_{10})	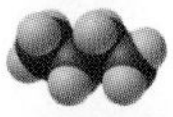
Pentane (C_5H_{12})	
Hexane (C_6H_{14})	
Heptane (C_7H_{16})	
Octane (C_8H_{18})	
Nonane (C_9H_{20})	
Decane ($C_{10}H_{22}$)	

Hydrocarbons, the simplest type of organic compound, contain *only* carbon and hydrogen. *Alkanes* are the simplest type of hydrocarbon; many function as important fuels, such as methane, propane, butane, and the mixture that makes up gasoline. The simplest alkanes to name are the *straight-chain alkanes* because the carbon chains have no branches. Alkanes are named with a *root*, based on the number of C atoms in the chain, followed by the suffix *-ane*. Table 2.7 gives the name, molecular formula, and space-filling model (discussed shortly) of the first 10 straight-chain alkanes. Note that the roots of the four smallest alkanes are new, but the roots of the larger alkanes are the same as the Greek prefixes shown in Table 2.6.

Masses from a Chemical Formula

In Section 2.5, we calculated the atomic mass of an element. Using the periodic table and the formula for a compound, we can calculate the **molecular mass** of a formula unit of a compound as *the sum of the atomic masses*:

$$\text{Mass of formula unit} = \text{sum of atomic masses} \qquad (2.3)$$

The mass of a water molecule (using atomic masses to four significant figures from the periodic table) is

$$\begin{aligned}\text{Molecular mass of } H_2O &= (2 \times \text{atomic mass of H}) + (1 \times \text{atomic mass of O})\\ &= (2 \times 1.008\text{ u}) + 16.00\text{ u} = 18.02\text{ u}\end{aligned}$$

Ionic compounds do not consist of molecules, so *the mass of a formula unit* is termed the **formula mass**, not the *molecular mass*. To calculate the formula mass of a compound with a polyatomic ion, *the number of atoms of each element inside the parentheses is multiplied by the subscript outside the parentheses.* For barium nitrate, $Ba(NO_3)_2$,

$$\begin{aligned}&\text{Formula mass}\\ &= (1 \times \text{atomic mass of Ba}) + (2 \times \text{atomic mass of N}) + (6 \times \text{atomic mass of O})\\ &= 137.3\text{ u} + (2 \times 14.01\text{ u}) + (6 \times 16.00\text{ u}) = 261.3\text{ u}\end{aligned}$$

We can use atomic masses, not ionic masses, because electron loss equals electron gain, so electron mass is balanced. In the next two sample problems, the name or molecular depiction is used to find a compound's molecular or formula mass.

Sample Problem 2.15 Calculating the Molecular Mass of a Compound

Problem Using the periodic table, calculate the molecular (or formula) mass of
(a) Tetraphosphorus trisulfide **(b)** Ammonium nitrate

Plan We first write the formula. Then we multiply the number of atoms (or ions) of each element by its atomic mass (from the periodic table) and find the sum.

Solution (a) The formula is P_4S_3.

$$\begin{aligned}\text{Molecular mass} &= (4 \times \text{atomic mass of P}) + (3 \times \text{atomic mass of S})\\ &= (4 \times 30.97\text{ u}) + (3 \times 32.07\text{ u}) = 220.09\text{ u}\end{aligned}$$

(b) The formula is NH_4NO_3. We count the total number of N atoms even though they belong to different ions:

$$\begin{aligned}&\text{Formula mass}\\ &= (2 \times \text{atomic mass of N}) + (4 \times \text{atomic mass of H}) + (3 \times \text{atomic mass of O})\\ &= (2 \times 14.01\text{ u}) + (4 \times 1.008\text{ u}) + (3 \times 16.00\text{ u}) = 80.05\text{ u}\end{aligned}$$

Check You can often find large errors by rounding atomic masses to the nearest 5 and adding:

(a) $(4 \times 30) + (3 \times 30) = 210 \approx 220.09$. The sum has two decimal places because the atomic masses have two decimal places.

(b) $(2 \times 15) + 4 + (3 \times 15) = 79 \approx 80.05$

Follow-Up Problem 2.15 Find the molecular (or formula) mass of each compound:

(a) Hydrogen peroxide **(b)** Cesium chloride **(c)** Sulfuric acid **(d)** Potassium sulfate

Sample Problem 2.16 Using Molecular Depictions to Determine Formula, Name, and Mass

Problem Each diagram represents a binary compound. Determine the formula, name, and molecular (or formula) mass of each compound.

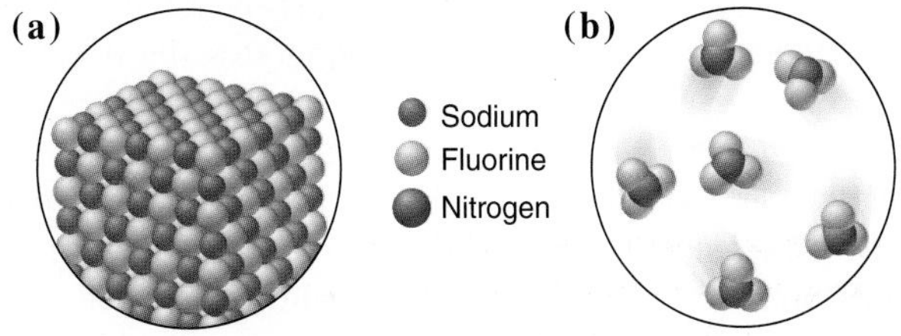

Plan Each compound contains only two elements, so, to find the formula, we find the simplest whole-number ratio of one atom to the other. From the formula, we determine the name and the molecular (or formula) mass.

Solution (a) There is one brown sphere (sodium) for each green sphere (fluorine), so the formula is NaF. A metal and nonmetal form an ionic compound, in which the metal is named first: sodium fluoride:

$$\text{Formula mass} = (1 \times \text{atomic mass of Na}) + (1 \times \text{atomic mass of F})$$
$$= 22.99 \text{ u} + 19.00 \text{ u} = 41.99 \text{ u}$$

(b) There are three green spheres (fluorine) for each blue sphere (nitrogen), so the formula is NF_3. Two nonmetals form a covalent compound. Nitrogen has a lower group number, so it is named first: nitrogen trifluoride:

$$\text{Molecular mass} = (1 \times \text{atomic mass of N}) + (3 \times \text{atomic mass of F})$$
$$= 14.01 \text{ u} + (3 \times 19.00 \text{ u}) = 71.01 \text{ u}$$

Check (a) For binary ionic compounds, we predict ionic charges from the periodic table (see Figure 2.13). Na forms a 1+ ion, and F forms a 1− ion, so the charges balance with one Na^+ per F^-. Also, ionic compounds are solids, consistent with the diagram. **(b)** Covalent compounds often occur as individual molecules, as shown in the diagram.

Rounding gives $25 + 20 = 45$ in (a) and $15 + (3 \times 20) = 75$ in (b), so there are no large errors.

Follow-Up Problem 2.16 Each diagram below represents a binary compound. Determine the name, formula, and molecular (or formula) mass of the compound.

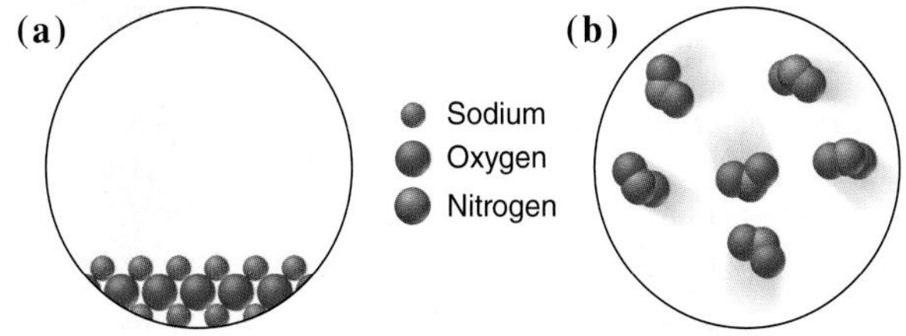

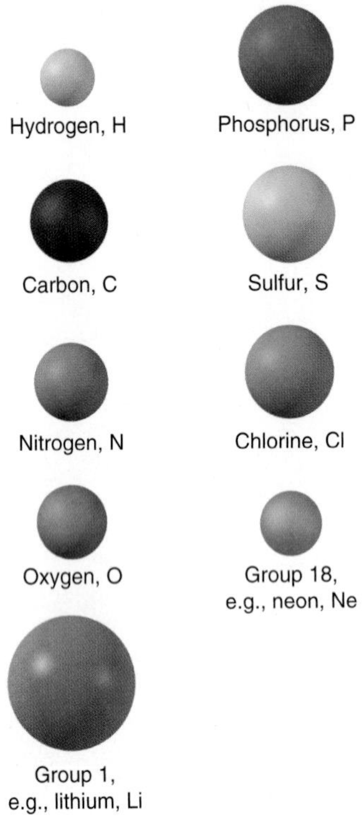

Representing Molecules with a Formula and a Model

To represent objects that are too small to see, chemists often use a formula and a model, of which there can be different types. Each conveys different information, as shown for water below:

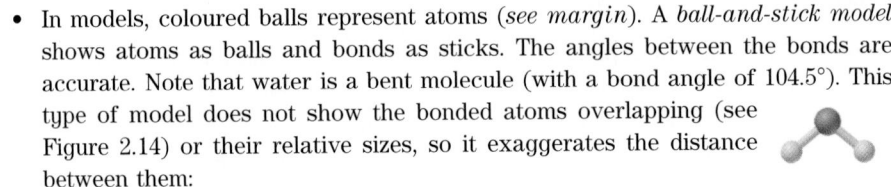

- A **molecular formula** *uses element symbols and often numerical subscripts to give the* ***actual*** *number of atoms of each element in a molecule of the compound.* (Recall that, for ionic compounds, the *formula unit* gives the *relative* number of each type of ion.) The molecular formula for water is H_2O: there are two H atoms and one O atom in each molecule: **H_2O**
- A **structural formula** *shows the relative placement and connections of the atoms in a molecule.* It uses symbols for the atoms *and* either a pair of dots (*electron-dot formula*) or a line (*bond-line formula*) to show the bonds between the atoms. In water, each H atom is bonded to the O atom, but not to the other H atom: **H:O:H** **H–O–H**
- In models, coloured balls represent atoms (*see margin*). A *ball-and-stick model* shows atoms as balls and bonds as sticks. The angles between the bonds are accurate. Note that water is a bent molecule (with a bond angle of 104.5°). This type of model does not show the bonded atoms overlapping (see Figure 2.14) or their relative sizes, so it exaggerates the distance between them: 
- A *space-filling model* is an accurately scaled-up image of the molecule, so it shows the relative sizes of the atoms, the relative distances between the nuclei (centres of the spheres), and the angles between the bonds. However, the bonds are not shown, and it can be difficult to see each atom in a complex molecule: 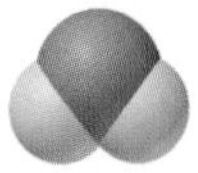

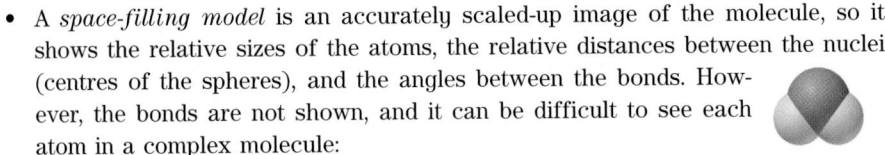

Every molecule is minute, but the range of molecular sizes, and thus molecular masses, is enormous. Table 2.8 shows some diatomic and small polyatomic molecules, as well as two extremely large molecules, called *macromolecules*, deoxyribonucleic acid (DNA) and nylon.

SUMMARY OF SECTION 2.8

- An ionic compound is named with the cation first and the anion second. If a metal can form more than one ion, the charge is shown with a roman numeral.
- The name of an oxoanion has a suffix, and sometimes a prefix, attached to the root of the element name to indicate the number of oxygen atoms.
- The name of a hydrate has a numerical prefix indicating the number of associated water molecules.
- Acid names are based on anion names.
- In the name of a binary covalent compound, the first word is the element farther left or lower down in the periodic table, and a prefix shows the number of each atom.
- The molecular (or formula) mass of a compound is the sum of the atomic masses.
- A chemical formula gives the number of atoms (molecular) or the arrangement of atoms (structural) of one unit of the compound.
- Molecular models convey information about bond angles (ball-and-stick) and relative atomic sizes and distances between atoms (space-filling).

2.9 Mixtures: Classification and Separation

In the natural world, *matter usually occurs as mixtures.* A sample of clean air, for example, consists of many elements and compounds physically mixed together, including O_2, N_2, CO_2, the noble gases (group 18), and water vapour (H_2O). The oceans are complex mixtures of dissolved ions and covalent substances, including

TABLE 2.8 Representing Molecules

Name	Molecular Formula (Molecular Mass, in u)	Bond-Line Formula	Ball-and-Stick Model	Space-Filling Model
Carbon monoxide	CO (28.01)	$C\equiv O$		
Nitrogen dioxide	NO_2 (46.01)	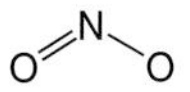		
Butane	C_4H_{10} (58.12)	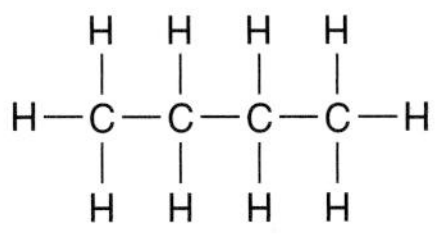	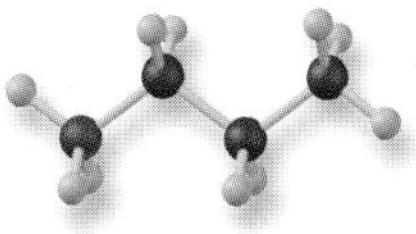	
Aspirin (acetylsalicylic acid)	$C_9H_8O_4$ (180.15)	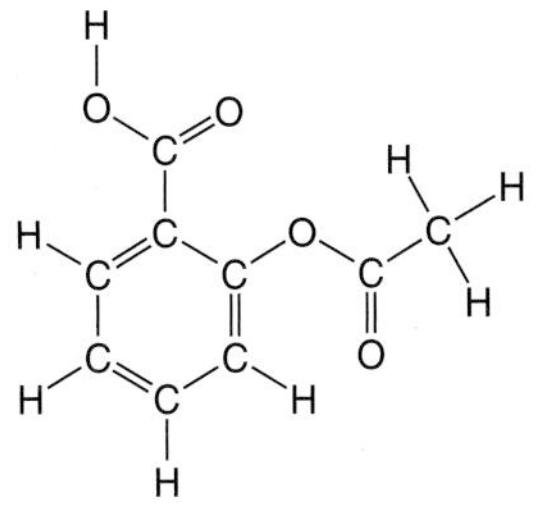	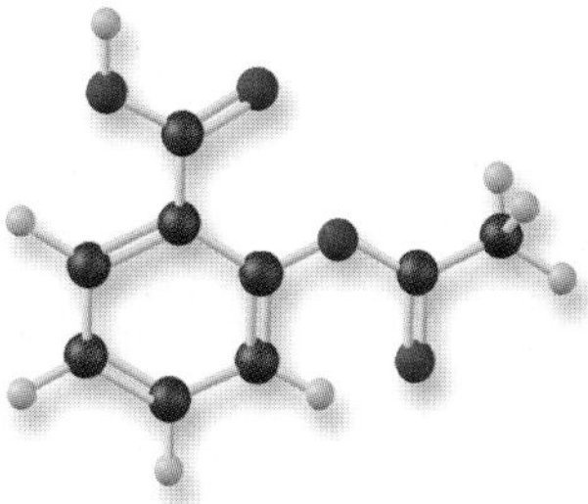	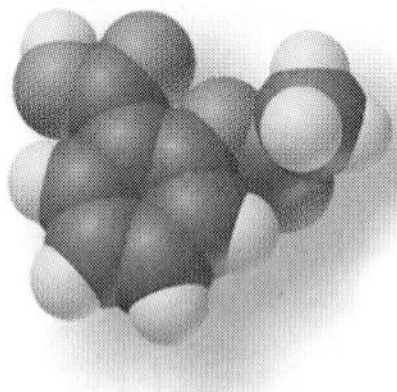

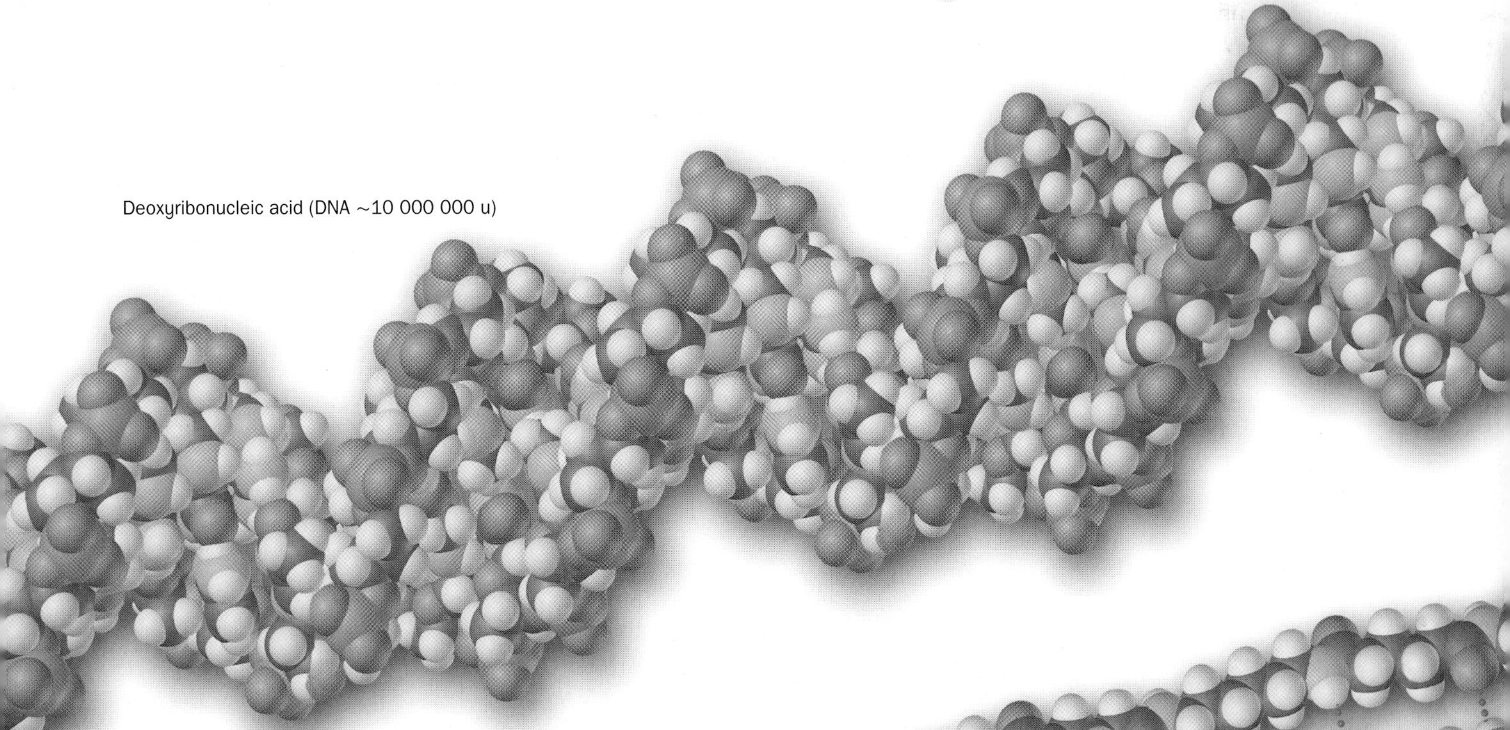

Deoxyribonucleic acid (DNA ~10 000 000 u)

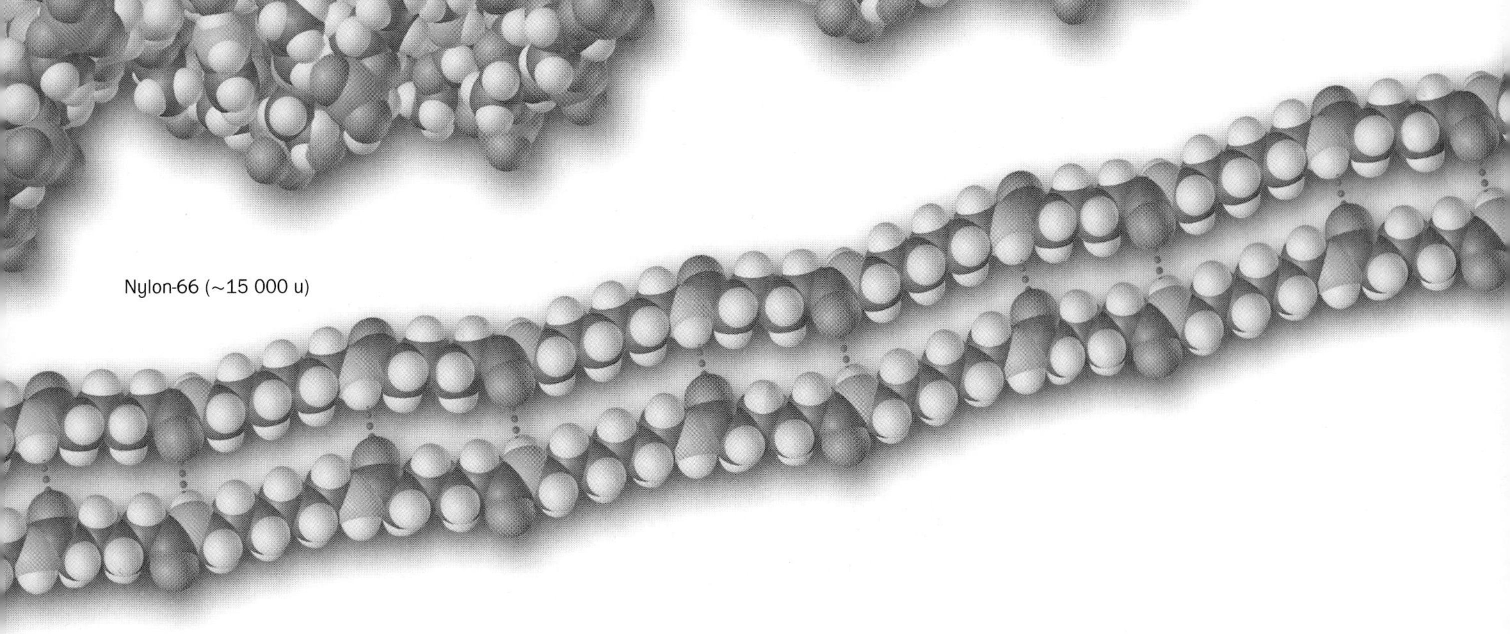

Nylon-66 (~15 000 u)

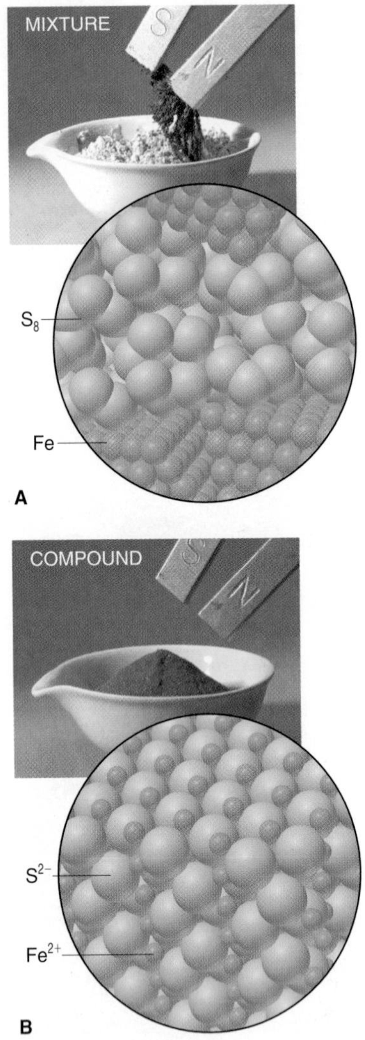

FIGURE 2.20 The distinction between mixtures and compounds. **A.** A *mixture* of iron and sulfur consists of the two elements. **B.** The *compound* iron(II) sulfide consists of an array of Fe^{2+} and S^{2-} ions.

Na^+, Mg^{2+}, Cl^-, SO_4^{2-}, O_2, CO_2, and of course H_2O. Rocks and soils are mixtures of numerous compounds, including calcium carbonate ($CaCO_3$), silicon dioxide (SiO_2), aluminum oxide (Al_2O_3), and iron(III) oxide (Fe_2O_3). Living things contain thousands of substances: carbohydrates, lipids, proteins, nucleic acids, and many simpler ionic and covalent compounds.

There are two broad classes of mixtures:

- A **heterogeneous mixture** *has one or more visible boundaries between the components.* Thus, its composition is *not* uniform, but rather varies from one region to another. Many rocks are heterogeneous, having individual grains of different minerals. In some heterogeneous mixtures, such as milk and blood, the boundaries can be seen only with a microscope.
- A **homogeneous mixture (solution)** *has no visible boundaries because the components are individual atoms, ions, or molecules.* Thus, its composition *is* uniform. A mixture of sugar dissolved in water is homogeneous, for example, because the sugar molecules and water molecules are uniformly intermingled on the molecular level. We have no way to tell visually whether a sample of matter is a substance (element or compound) or a homogeneous mixture.

Although we usually think of solutions as liquid, they can exist in all the common physical states. For example, air is a gaseous solution of mostly oxygen and nitrogen molecules, and wax is a solid solution of several fatty substances. *Solutions in water*, called **aqueous solutions**, are especially important in the chemistry lab and constitute a major portion of the environment *and* a major portion of all organisms.

Recall that mixtures differ from compounds in three major ways:

1. The proportions of the components can vary.
2. The individual properties of the components are observable.
3. The components can be separated by physical means.

The difference between a mixture and a compound is well illustrated using iron and sulfur as components (Figure 2.20). Any proportion of iron metal filings and powdered sulfur forms a mixture. The components can be separated with a magnet because iron metal is magnetic. If we heat the container strongly, however, the components form the compound iron(II) sulfide (FeS). The magnet can no longer remove the iron because it exists as Fe^{2+} ions chemically bound to S^{2-} ions.

Chemists have devised many techniques for separating a mixture into its components. Some of the common techniques are described in the next Tools of the Laboratory section.

An Overview of the Components of Matter

Understanding matter at both the observable scale and the atomic scale is the essence of chemistry. Figure 2.21 is a visual overview of many key terms and ideas in this chapter.

SUMMARY OF SECTION 2.9

- Heterogeneous mixtures have visible boundaries between the components.
- Homogeneous mixtures (solutions) have no visible boundaries because mixing occurs at the molecular level. They can occur in any physical state.
- Components of mixtures (unlike components of compounds) can have variable proportions, can be separated physically, and retain their properties.
- Separation methods are based on differences in physical properties and include filtration (particle size), crystallization (solubility), distillation (volatility), and chromatography (solubility).

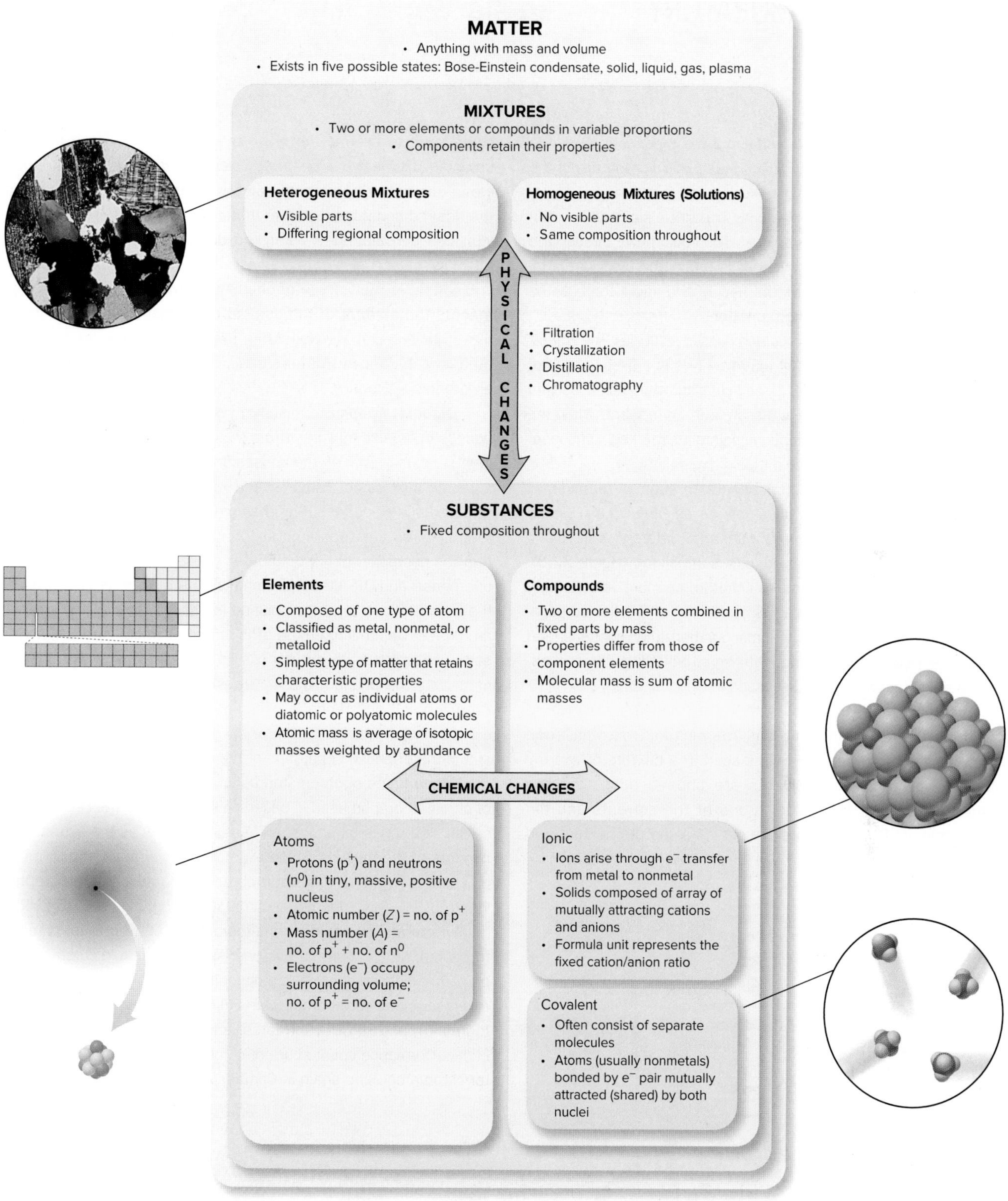
MATTER
• Anything with mass and volume
• Exists in five possible states: Bose-Einstein condensate, solid, liquid, gas, plasma
MIXTURES
• Two or more elements or compounds in variable proportions
• Components retain their properties
Heterogeneous Mixtures
• Visible parts
• Differing regional composition
Homogeneous Mixtures (Solutions)
• No visible parts
• Same composition throughout
PHYSICAL CHANGES
• Filtration
• Crystallization
• Distillation
• Chromatography
SUBSTANCES
• Fixed composition throughout
Elements
• Composed of one type of atom
• Classified as metal, nonmetal, or metalloid
• Simplest type of matter that retains characteristic properties
• May occur as individual atoms or diatomic or polyatomic molecules
• Atomic mass is average of isotopic masses weighted by abundance
Compounds
• Two or more elements combined in fixed parts by mass
• Properties differ from those of component elements
• Molecular mass is sum of atomic masses
CHEMICAL CHANGES
Atoms
• Protons (p^+) and neutrons (n^0) in tiny, massive, positive nucleus
• Atomic number (Z) = no. of p^+
• Mass number (A) = no. of p^+ + no. of n^0
• Electrons (e^-) occupy surrounding volume; no. of p^+ = no. of e^-
Ionic
• Ions arise through e^- transfer from metal to nonmetal
• Solids composed of array of mutually attracting cations and anions
• Formula unit represents the fixed cation/anion ratio
Covalent
• Often consist of separate molecules
• Atoms (usually nonmetals) bonded by e^- pair mutually attracted (shared) by both nuclei

FIGURE 2.21 The classification of matter from a chemical point of view

Basic Separation Techniques

Some of the most challenging laboratory techniques involve separating mixtures and purifying the components. All of the techniques described here depend on the *physical properties* of the substances in the mixture; no chemical changes occur.

Filtration is based on *differences in particle size* and is often used to separate a solid from a liquid. The liquid flows through the tiny holes in filter paper, and the solid is retained. In vacuum filtration, reduced pressure speeds the flow of the liquid through the filter. Filtration is used to purify tap water.

Crystallization is based on *differences in solubility*. The *solubility* of a substance is the amount that dissolves in a fixed volume of solvent at a given temperature. Since solubility often increases with temperature, the impure solid is dissolved in hot solvent. When the solution cools, the purified compound crystallizes. A key component of computer chips is purified by a type of crystallization.

Distillation separates components through *differences in* ***volatility****, the tendency of a substance to become a gas. Simple* distillation separates components with *large* differences in volatility, such as water from dissolved ionic compounds (Figure B2.3). As the mixture boils, the vapour is richer in the more volatile component, in this case, water, which is condensed and collected separately. *Fractional* distillation uses many vaporization-condensation steps to separate components with small volatility differences, such as those in petroleum (discussed in Chapter 12).

Extraction is used to separate the components of a mixture based on *differences* in **solubility**. The mixture from which one component is to be separated is placed in a mixture of two immiscible solvents. The solvents are chosen such that the desired component is soluble in one of the solvents and the other (undesired) components are not. Shaking the mixture and solvents in a *separation* or *extraction funnel* allows the desired component to be transferred to the solvent in which it is more soluble. At times, the desired component may be slightly soluble in one solvent and more soluble in the other. The amount of the desired component that is transferred to the preferential solvent can be calculated using the *distribution coefficient*, $K_D = (m_1/V_1)/(m_2/V_2)$, where m_1 represents the amount of the desired component in the volume of one of the solvents, V_1, and m_2 represents the amount of the desired component in the volume of the other solvent, V_2. Generally, extraction is used to separate organic substances from inorganic substances using water and an organic solvent.

Chromatography, originally used to separate coloured compounds in organic mixtures (from the Greek *chromos*, meaning "colour"), is now one of the most commonly used techniques to separate mixtures of chemical and/or biological compounds, both coloured and colourless. Chromatography is based on the principle that mixtures are separated based on the differences in the way that the components are distributed between two phases: the mobile phase and the stationary phase. The mixture is dissolved in the mobile phase, and the solution is passed through the stationary phase. Depending on the affinity that each component has for each phase, and the differences in the polarities of the components and the phases, the movement of the components varies. Even a small difference in affinities becomes magnified as the components travel the length of the stationary phase. The net result is the effective separation of the components of the mixture.

Many types of chromatography are available today. The type of chromatography used is based on the nature of the components (solid, liquid, gas, aqueous, and so on), as well as the actual components of the mixture. Some examples of chromatographic methods are GC (gas chromatography); LC (liquid chromatography; Figure B2.4); GLC (gas liquid chromatography; Figure B2.5); HPLC (high-performance liquid chromatography); TLC (thin-layer chromatography); affinity chromatography; supercritical fluid chromatography; ion-exchange chromatography; size-exclusion chromatography; reversed-phase chromatography; FPLC (fast protein liquid chromatography); and HPCCC, the latest and best-performing version of high-performance countercurrent chromatography. You will encounter chromatography again in Chapter 22.

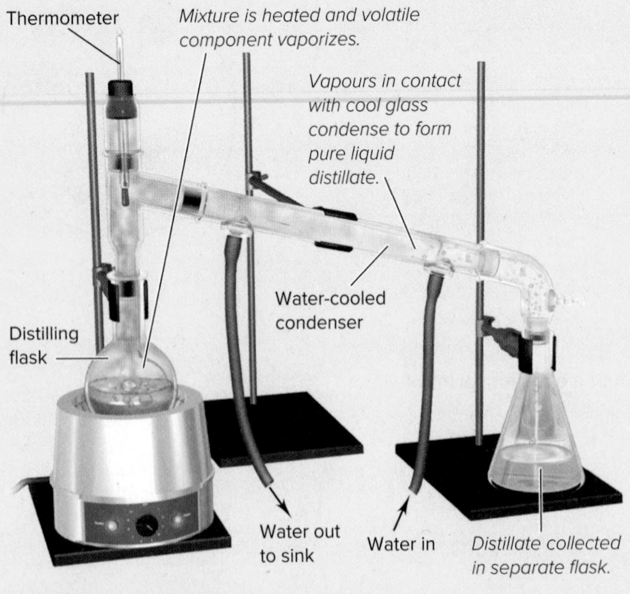
Thermometer
Mixture is heated and volatile component vaporizes.
Vapours in contact with cool glass condense to form pure liquid distillate.
Water-cooled condenser
Distilling flask
Water out to sink
Water in
Distillate collected in separate flask.

FIGURE B2.3 Distillation

FIGURE B2.4 Laboratory and large-scale column chromatography

Problem

B2.3 Name the technique(s) and briefly describe the steps for separating each of the following mixtures into pure components:
(a) Table salt and pepper
(b) Drinking water contaminated with soot
(c) Crushed ice and crushed glass
(d) Table sugar dissolved in ethanol
(e) Two pigments (chlorophyll *a* and chlorophyll *b*) from spinach leaves

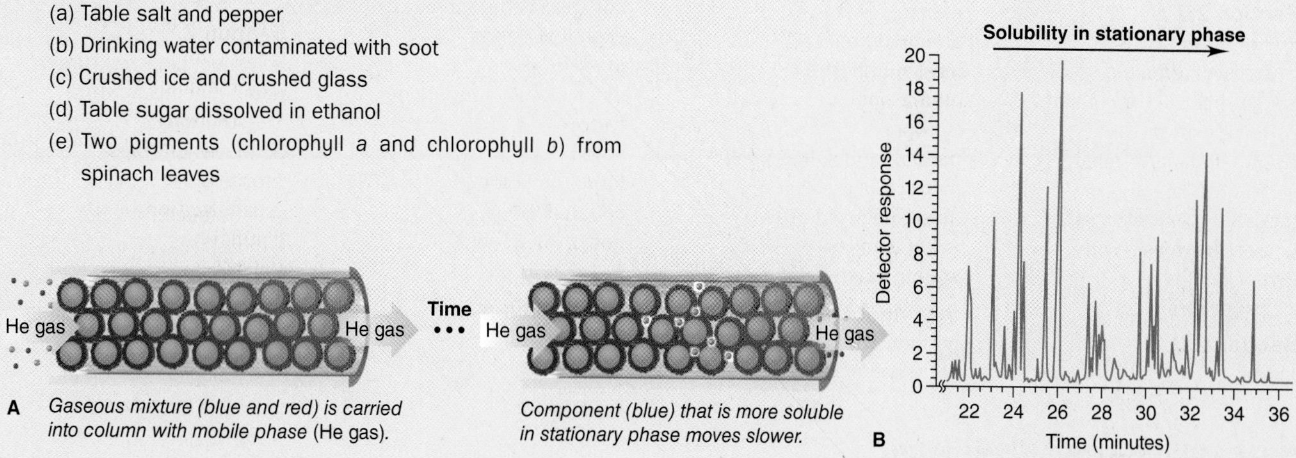

FIGURE B2.5 Principle of gas-liquid chromatography (GLC). A. The stationary phase is shown as a viscous liquid (*grey circles*) coating the solid beads (*yellow*) of an inert packing. **B.** Typical spectrum obtained using GLC.

CHAPTER REVIEW GUIDE

Learning Objectives

Relevant section (§) and/or sample problem (SP) numbers appear in parentheses.

Concepts

1. Define the characteristics of the three types of matter—element, compound, and mixture—on the macroscopic and atomic scales. (§2.1)
2. Discuss the significance of the three mass laws—mass conservation, definite composition, and multiple proportions—and identify their key characteristics. (§2.2)
3. Summarize the postulates of Dalton's atomic theory and how it explains the mass laws. (§2.3)
4. Compare and contrast the major contributions to our understanding of atomic structure of experiments by Thomson, Millikan, and Rutherford. (§2.4)
5. Characterize the structure of the atom, the main features of the subatomic particles, and the importance of isotopes. (§2.5)
6. Explain the format of the periodic table, and identify the general location and characteristics of metals, metalloids, and nonmetals. (§2.6)
7. Compare and contrast the essential features of ionic and covalent compounds and the distinction between them. (§2.7)
8. Name different types of compounds (ionic, molecular, acids, simple organic) (§2.8)
9. Categorize the types of mixtures and their properties. (§2.9)

Skills

1. Distinguish between elements, compounds, and mixtures on the atomic scale. (SP 2.1)
2. Apply the idea of the mass ratio of element to compound to find the mass of an element in a compound. (SP 2.2)
3. Visualize the mass laws. (SP 2.3)
4. Express the subatomic makeup of an isotope using atomic notation. (SP 2.4)
5. Calculate an atomic mass from isotopic composition. (SP 2.5)
6. Predict the monatomic ion formed from a main-group element. (SP 2.6)
7. Name and write the formula for an ionic compound formed from the ions in Tables 2.3 to 2.5. (SPs 2.7–2.12, 2.16)
8. Name and write the formula for a binary covalent compound. (SPs 2.13, 2.14, 2.16)
9. Calculate the molecular or formula mass of a compound. (SP 2.15)

Key Terms

Section 2.1
substance
element
molecule
compound
mixture

Section 2.2
law of mass conservation
law of definite (or constant) composition
fraction by mass (mass fraction)
percent by mass (mass percent, mass %)
law of multiple proportions

Section 2.3
atoms

Section 2.4
cathode rays
nucleus

Section 2.5
proton (p^+)
neutron (n^0)
electron (e^-)
atomic number (Z)
mass number (A)
atomic symbol
isotopes
unified atomic mass unit (u)
dalton (Da)
mass spectrometry
isotopic mass
atomic mass

Section 2.6
periodic table of the elements
periods
groups
metals
nonmetals
metalloids (semimetals)

Section 2.7
ionic compounds
covalent compounds
chemical bonds
ions
binary ionic compound
cation
anion
monatomic ion
covalent bond
polyatomic ions

Section 2.8
chemical formula
formula unit
oxoanions
hydrates
binary covalent compounds
molecular mass
formula mass
molecular formula
structural formula

Section 2.9
heterogeneous mixture
homogeneous mixture (solution)
aqueous solutions
filtration
crystallization
distillation
volatility
extraction
solubility
chromatography

Key Equations and Relationships

2.1 Finding the mass of an element in a given mass of compound:

Mass of element in sample

$$= \text{mass of compound in sample} \times \frac{\text{mass of element in compound}}{\text{mass of compound}}$$

2.2 Calculating the number of neutrons in an atom:

$$\text{Number of neutrons} = \text{mass number} - \text{atomic number}$$

or

$$N = A - Z$$

2.3 Determining the mass of a formula unit of a compound:

$$\text{Mass of formula unit} = \text{sum of atomic masses}$$

Brief Solutions to Follow-Up Problems

2.1 There are two types of particles reacting (*left circle*), one with two blue atoms and the other with two orange atoms; the depiction shows a mixture of two elements. In the product (*right circle*), all the particles have one blue atom and one orange atom; this is a compound.

2.2 Mass (t) of pitchblende

$$= 2.3\ \text{t}\ \cancel{\text{uranium}} \times \frac{84.2\ \cancel{\text{g}}\ \text{pitchblende}}{71.4\ \cancel{\text{g uranium}}} = 2.7\underline{1}\ \text{t pitchblende}$$

Mass (t) of oxygen

$$= 2.71\ \text{t pitchblende} - 2.3\ \text{t uranium} = 0.41\ \text{t oxygen}$$

2.3 Sample B best shows the law of multiple proportions. Two bromine-fluorine compounds appear. In one compound, there are three fluorine atoms for each bromine; in the other, there is one fluorine atom for each bromine. Therefore, in the two compounds, the ratio of fluorines combining with one bromine is 3/1.

2.4 (a) $5p^+$, $6n^0$, $5e^-$; Q = B
(b) $20p^+$, $21n^0$, $20e^-$; R = Ca
(c) $53p^+$, $78n^0$, $53e^-$; X = I

2.5 $10.0129x + 11.0093(1 - x) = 10.81$

$$0.9964x = 0.1993$$

$$x = 0.2000 \text{ and } 1 - x = 0.8000$$

% abundance of $^{10}B = 20.00\%$
% abundance of $^{11}B = 80.00\%$

2.6 (a) S^{2-}; (b) Rb^+; (c) Ba^{2+}

2.7 (a) Zinc (group 12) and oxygen (group 16)
(b) Silver (group 11) and bromine (group 17)
(c) Lithium (group 1) and chlorine (group 17)
(d) Aluminum (group 13) and sulfur (group 16)

2.8 (a) ZnO; (b) AgBr; (c) LiCl; (d) Al_2S_3

2.9 (a) PbO_2; (b) copper(I) sulfide (cuprous sulfide); (c) iron(II) bromide (ferrous bromide); (d) $HgCl_2$

2.10 (a) $Cu(NO_3)_2 \cdot 3H_2O$; (b) $Zn(OH)_2$; (c) lithium cyanide

2.11 (a) $(NH_4)_3PO_4$; ammonium is NH_4^+ and phosphate is PO_4^{3-}.
(b) $Al(OH)_3$; parentheses are needed around the polyatomic ion OH^-.
(c) Magnesium hydrogen carbonate (or magnesium bicarbonate); Mg^{2+} is magnesium and can have only a 2+ charge, so it does not need (II); HCO_3^- is hydrogen carbonate (or bicarbonate).
(d) Chromium(III) nitrate; the *-ic* ending is not used with roman numerals; NO_3^- is nitr*ate*.
(e) Calcium nitrite; Ca^{2+} is calcium and NO_2^- is nitr*ite*.

2.12 (a) $HClO_3$; (b) hydrofluoric acid; (c) CH_3COOH (or $HC_2H_3O_2$); (d) H_2SO_3; (e) hypobromous acid

2.13 (a) Sulfur trioxide; (b) silicon dioxide; (c) N_2O; (d) SeF_6

2.14 (a) Disulfur dichloride; the *-ous* suffix is not used.
(b) NO; the name indicates one nitrogen.
(c) Bromine trichloride; Br is in a higher period in group 17, so it is named first.

2.15 (a) H_2O_2, 34.02 u; (b) CsCl, 168.35 u; (c) H_2SO_4, 98.09 u; (d) K_2SO_4, 174.27 u

2.16 (a) Na_2O; this is an ionic compound, so the name is sodium oxide.

Formula mass
$= (2 \times \text{atomic mass of Na}) + (1 \times \text{atomic mass of O})$
$= (2 \times 22.99\ \text{u}) + 16.00\ \text{u} = 61.98\ \text{u}$

(b) NO_2; this is a covalent compound, and N has the lower group number, so the name is nitrogen dioxide.

Molecular mass
$= (1 \times \text{atomic mass of N}) + (2 \times \text{atomic mass of O})$
$= 14.01\ \text{u} + (2 \times 16.00\ \text{u}) = 46.01\ \text{u}$

PROBLEMS

Problems with **red** numbers are answered in Appendix G and worked in detail in the Student Solutions Manual. Problem sections match those in this book and provide the numbers of relevant sample problems. Most offer Concept Review Questions, Skill-Building Exercises (grouped in pairs covering the same concept), and Problems in Context. The Comprehensive Problems are based on material from any section or previous chapter.

Elements, Compounds, and Mixtures: An Atomic Overview
(Sample Problem 2.1)

Concept Review Questions

2.1 What is the key difference between an element and a compound?

2.2 List two differences between a compound and a mixture.

2.3 Which are pure substances? Explain.
(a) Calcium chloride, used to melt ice on roads, consists of two elements, calcium and chlorine, in a fixed mass ratio.
(b) Sulfur consists of sulfur atoms combined into octatomic molecules.
(c) Baking powder, a leavening agent, contains 26% to 30% sodium hydrogen carbonate and 30% to 35% calcium dihydrogen phosphate by mass.
(d) Cytosine, a component of DNA, consists of H, C, N, and O atoms bonded in a specific arrangement.

2.4 Classify each substance in Problem 2.3 as an element, a compound, or a mixture, and explain your answers.

2.5 Explain the following statement: The smallest particles unique to an element may be atoms or molecules.

2.6 Explain the following statement: The smallest particles unique to a compound cannot be atoms.

2.7 Can the relative amounts of the components of a mixture vary? Can the relative amounts of the components of a compound vary? Explain.

Problems in Context

2.8 The tap water found in many areas of Canada leaves white deposits when it evaporates. Is this tap water a mixture or a compound? Explain.

2.9 The following diagrams represent mixtures. Describe each diagram in terms of the number(s) of elements and/or compounds present.

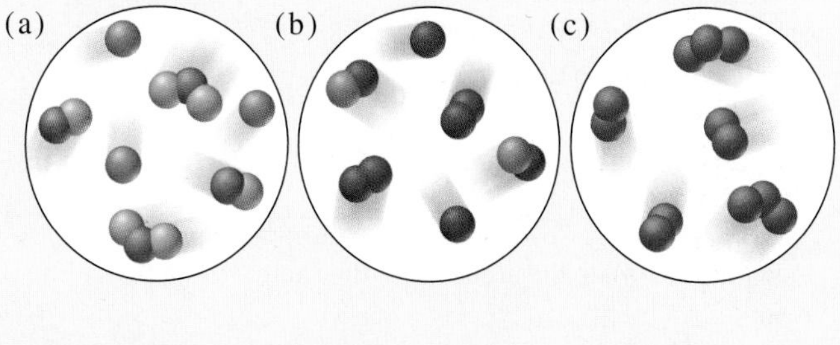

2.10 Samples of illicit "street" drugs often contain an inactive component, such as ascorbic acid (vitamin C). After obtaining a sample of cocaine, government chemists calculate the mass of vitamin C per gram of sample, and use it to track the cocaine's distribution. For example, if different samples of cocaine, obtained on the streets of New York, Vancouver, and Paris, all contain 0.6384 g of vitamin C per gram of sample, they very likely came from a common source. Do these street samples consist of a compound, an element, or a mixture? Explain.

The Observations That Led to an Atomic View of Matter
(Sample Problem 2.2)

Concept Review Questions

2.11 Why was it necessary for separation techniques and methods of chemical analysis to be developed before the laws of definite composition and multiple proportions could be formulated?

2.12 To which classes of matter—element, compound, and/or mixture—do the following apply: (a) law of mass conservation; (b) law of definite composition; (c) law of multiple proportions?

2.13 In our modern view of matter and energy, is the law of mass conservation still relevant to chemical reactions? Explain.

2.14 Identify the mass law that each of the following observations demonstrates, and explain your reasoning:
(a) A sample of potassium chloride from Chile contains the same percent by mass of potassium as a sample from Poland.
(b) A glass bulb contains magnesium and oxygen before use and magnesium oxide afterward, but its mass does not change.
(c) Arsenic and oxygen form one compound that is 65.2 mass % arsenic and another that is 75.8 mass % arsenic.

2.15 Which of the following diagrams illustrate(s) the fact that compounds of chlorine (*green*) and oxygen (*red*) exhibit the law of multiple proportions? Name the compounds.

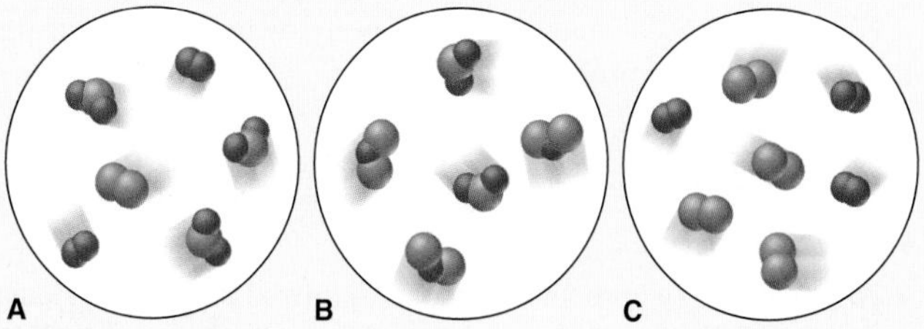

2.16 (a) Does the percent by mass of each element in a compound depend on the amount of compound? Explain.
(b) Does the mass of each element in a compound depend on the amount of compound? Explain.

2.17 Does the percent by mass of each element in a compound depend on the amount of the element that was used to make the compound? Explain.

Skill-Building Exercises (grouped in similar pairs)

2.18 State the mass law(s) demonstrated by the following experimental results, and explain your reasoning:

Experiment 1: A student heats 1.00 g of a blue compound and obtains 0.64 g of a white compound and 0.36 g of a colourless gas.
Experiment 2: A second student heats 3.25 g of the same blue compound and obtains 2.08 g of a white compound and 1.17 g of a colourless gas.

2.19 State the mass law(s) demonstrated by the following experimental results, and explain your reasoning:

Experiment 1: A student heats 1.27 g of copper and 3.50 g of iodine to produce 3.81 g of a white compound; 0.96 g of iodine remains.
Experiment 2: A second student heats 2.55 g of copper and 3.50 g of iodine to form 5.25 g of a white compound; 0.80 g of copper remains.

2.20 Fluorite, a mineral of calcium, is a compound of the metal with fluorine. Analysis shows that a 2.76 g sample of fluorite contains 1.42 g of calcium. Calculate (a) the mass of fluorine in the sample; (b) the mass fractions of calcium and fluorine in fluorite; (c) the mass percents of calcium and fluorine in fluorite.

2.21 Galena, a mineral of lead, is a compound of the metal with sulfur. Analysis shows that a 2.34 g sample of galena contains 2.03 g of lead. Calculate (a) the mass of sulfur in the sample; (b) the mass fractions of lead and sulfur in galena; (c) the mass percents of lead and sulfur in galena.

2.22 Magnesium oxide (MgO) forms when the metal burns in air.
(a) If 1.25 g of MgO contains 0.754 g of Mg, what is the mass ratio of magnesium to oxide?
(b) What mass of Mg is in 534 g of MgO?

2.23 Zinc sulfide (ZnS) occurs in the zinc blende crystal structure.
(a) If 2.54 g of ZnS contains 1.70 g of Zn, what is the mass ratio of zinc to sulfide?
(b) What mass, in kilograms, of Zn is in 3.82 kg of ZnS?

2.24 A compound of copper and sulfur contains 88.39 g of metal and 44.61 g of nonmetal. What mass of copper is in 5264 kg of the compound? What mass of sulfur?

2.25 A compound of iodine and cesium contains 63.94 g of metal and 61.06 g of nonmetal. What mass of cesium is in 38.77 g of the compound? What mass of iodine?

2.26 Show, with calculations, how the following data illustrate the law of multiple proportions:

Compound 1: 47.5 mass % sulfur and 52.5 mass % chlorine
Compound 2: 31.1 mass % sulfur and 68.9 mass % chlorine

2.27 Show, with calculations, how the following data illustrate the law of multiple proportions:

Compound 1: 77.6 mass % xenon and 22.4 mass % fluorine
Compound 2: 63.3 mass % xenon and 36.7 mass % fluorine

Problems in Context

2.28 Dolomite is a carbonate of magnesium and calcium. Analysis shows that 7.81 g of dolomite contains 1.70 g of calcium. Calculate the mass percent of calcium in dolomite. On the basis of the mass percent of calcium, and neglecting all other factors, which is the richer source of calcium, dolomite or fluorite (see Problem 2.20)?

2.29 The mass percent of sulfur in a sample of coal is a key factor in the environmental impact of the coal. The sulfur combines with oxygen when the coal is burned, and the oxide can then be incorporated into acid rain. Which of the following coals would have the smallest environmental impact?

	Mass (g) of Sample	Mass (g) of Sulfur in Sample
Coal A	378	11.3
Coal B	495	19.0
Coal C	675	20.6

Dalton's Atomic Theory
(Sample Problem 2.3)

Concept Review Questions

2.30 Which of Dalton's postulates about atoms are inconsistent with later observations? Do these inconsistencies mean that Dalton was wrong? Is Dalton's model still useful? Explain.

2.31 Use Dalton's theory to explain why potassium nitrate from India or Italy has the same mass percents of K, N, and O.

The Observations That Led to the Nuclear Atom Model

Concept Review Questions

2.32 Thomson was able to determine the mass/charge ratio of the electron, but not its mass. How did Millikan's experiment allow the electron's mass to be determined?

2.33 The following charges on individual oil droplets were obtained during an experiment similar to Millikan's: -3.204×10^{-19} C; -4.806×10^{-19} C; -8.010×10^{-19} C; -1.442×10^{-18} C. Determine a charge for the electron (in C, coulomb), and explain your answer.

2.34 Describe Thomson's model of the atom. How might it account for the production of cathode rays?

2.35 When Rutherford's co-workers bombarded gold foil with α particles, they obtained results that overturned the existing (Thomson) model of the atom. Explain.

The Atomic Theory Today
(Sample Problems 2.4 and 2.5)

Concept Review Questions

2.36 Define *atomic number* and *mass number*. Which can vary without changing the identity of the element?

2.37 Choose the correct answer. The difference between the mass number of an isotope and its atomic number is (a) directly related to the identity of the element; (b) the number of electrons; (c) the number of neutrons; (d) the number of isotopes.

2.38 Even though several elements have only one naturally occurring isotope and all atomic nuclei have whole numbers of protons and neutrons, no atomic mass is a whole number. Use data from Table 2.2 to explain this fact.

Skill-Building Exercises (grouped in similar pairs)

2.39 Argon has three naturally occurring isotopes: ^{36}Ar, ^{38}Ar, and ^{40}Ar. What is the mass number of each isotope? How many protons, neutrons, and electrons are present in each?

2.40 Chlorine has two naturally occurring isotopes: ^{35}Cl and ^{37}Cl. What is the mass number of each isotope? How many protons, neutrons, and electrons are present in each?

2.41 Do both atoms in each pair have the same number of protons? Neutrons? Electrons?
(a) $^{16}_{8}$O and $^{17}_{8}$O (b) $^{40}_{18}$Ar and $^{41}_{19}$K (c) $^{60}_{27}$Co and $^{60}_{28}$Ni
In which pair(s) do the atoms have the same Z value? N value? A value?

2.42 Do both atoms in each pair have the same number of protons? Neutrons? Electrons?
(a) $^{3}_{1}$H and $^{3}_{2}$He (b) $^{14}_{6}$C and $^{15}_{7}$N (c) $^{19}_{9}$F and $^{18}_{9}$F
In which pair(s) do the atoms have the same Z value? N value? A value?

2.43 Write the ${}^{A}_{Z}X$ notation for each atomic depiction:

(a) (b) (c)

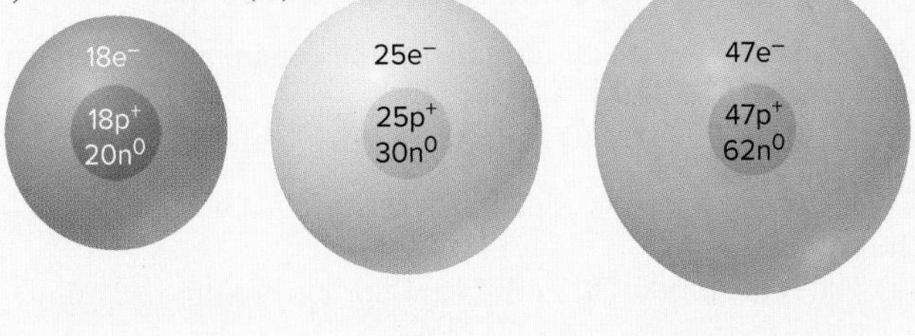

2.44 Write the ${}^{A}_{Z}X$ notation for each atomic depiction:

(a) (b) (c)

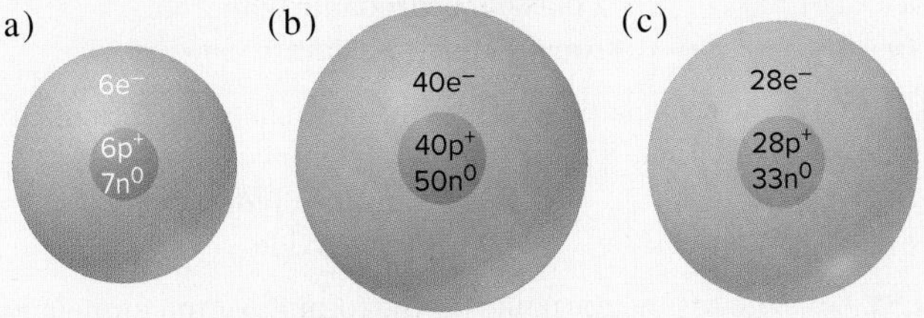

2.45 Draw atomic depictions similar to those in Problem 2.43 for the following atoms: (a) ${}^{48}_{22}Ti$; (b) ${}^{79}_{34}Se$; (c) ${}^{11}_{5}B$.

2.46 Draw atomic depictions similar to those in Problem 2.43 for these atoms: (a) ${}^{207}_{82}Pb$; (b) ${}^{9}_{4}Be$; (c) ${}^{75}_{33}As$.

2.47 Gallium has two naturally occurring isotopes: ${}^{69}Ga$ (isotopic mass = 68.9256 u, abundance = 60.11%) and ${}^{71}Ga$ (isotopic mass = 70.9247 u, abundance = 39.89%). Calculate the atomic mass of gallium.

2.48 Magnesium has three naturally occurring isotopes: ${}^{24}Mg$ (isotopic mass = 23.9850 u, abundance = 78.99%), ${}^{25}Mg$ (isotopic mass = 24.9858 u, abundance = 10.00%), and ${}^{26}Mg$ (isotopic mass = 25.9826 u, abundance = 11.01%). Calculate the atomic mass of magnesium.

2.49 Chlorine has two naturally occurring isotopes: ${}^{35}Cl$ (isotopic mass = 34.9689 u) and ${}^{37}Cl$ (isotopic mass = 36.9659 u). If chlorine has an atomic mass of 35.4527 u, what is the percent abundance of each isotope?

2.50 Copper has two naturally occurring isotopes: ${}^{63}Cu$ (isotopic mass = 62.9396 u) and ${}^{65}Cu$ (isotopic mass = 64.9278 u). If copper has an atomic mass of 63.546 u, what is the percent abundance of each isotope?

Elements: A First Look at the Periodic Table

Concept Review Questions

2.51 How can iodine (Z = 53) have a higher atomic number and yet a lower atomic mass than tellurium (Z = 52)?

2.52 Correct each statement:
(a) In the modern periodic table, the elements are arranged in order of increasing atomic mass.
(b) Elements in a period have similar chemical properties.
(c) Elements can be classified as either metalloids or nonmetals.

2.53 What class of elements lies along the "staircase" line in the periodic table? How do the properties of these elements compare with the properties of metals and nonmetals?

2.54 What are some characteristic properties of elements to the left of the elements along the "staircase"? To the right of the elements along the "staircase"?

2.55 All of the elements in groups 1 and 17 are quite reactive. What is a major difference between them?

Skill-Building Exercises (grouped in similar pairs)

2.56 Give the name, atomic symbol, and group number of the element with each Z value, and classify it as a metal, metalloid, or nonmetal:
(a) Z = 32 (b) Z = 15 (c) Z = 2 (d) Z = 3 (e) Z = 42

2.57 Give the name, atomic symbol, and group number of the element with each Z value, and classify it as a metal, metalloid, or nonmetal:
(a) Z = 33 (b) Z = 20 (c) Z = 35 (d) Z = 19 (e) Z = 13

2.58 Fill in the blanks:
(a) The symbol and atomic number of the heaviest alkaline earth metal are ______ and ______.
(b) The symbol and atomic number of the lightest metalloid in group 14 are ______ and ______.
(c) Group 11 consists of the *coinage metals*. The symbol and atomic mass of the coinage metal whose atoms have the fewest electrons are ______ and ______.
(d) The symbol and atomic mass of the halogen in period 4 are ______ and ______.

2.59 Fill in the blanks:
(a) The symbol and atomic number of the heaviest nonradioactive noble gas are ______ and ______.
(b) The symbol and group number of the period 5 transition element whose atoms have the fewest protons are ______ and ______.
(c) The symbol and atomic number of the first group 16 element displaying a metallic nature are ______ and ______.
(d) The symbol and number of protons of the period 4 alkali metal atom are ______ and ______.

Compounds: An Introduction to Bonding

(Sample Problem 2.6)

Concept Review Questions

2.60 Describe the type and nature of the bonding that occurs between reactive metals and nonmetals.

2.61 Describe the type and nature of the bonding that often occurs between two nonmetals.

2.62 How can ionic compounds be neutral if they consist of positive and negative ions?

2.63 Given that the ions in LiF and the ions in MgO are similar sizes, which compound has stronger ionic bonding? Use Coulomb's law in your explanation.

2.64 Are molecules present in a sample of BaF_2? Explain.

2.65 Are ions present in a sample of P_4O_6? Explain.

2.66 The monatomic ions of groups 1 and 17 are all singly charged. In what major way do they differ? Why?

2.67 Describe the formation of solid magnesium chloride ($MgCl_2$) from large numbers of magnesium and chlorine atoms.

2.68 Describe the formation of solid potassium sulfide (K_2S) from large numbers of potassium and sulfur atoms.

2.69 Does potassium nitrate (KNO_3) incorporate ionic bonding, covalent bonding, or both? Explain.

Skill-Building Exercises (grouped in similar pairs)

2.70 What monatomic ions do potassium (Z = 19) and iodine (Z = 53) form?

2.71 What monatomic ions do barium (Z = 56) and selenium (Z = 34) form?

2.72 For each ionic depiction, give the name of the parent atom, its mass number, and its group and period numbers:

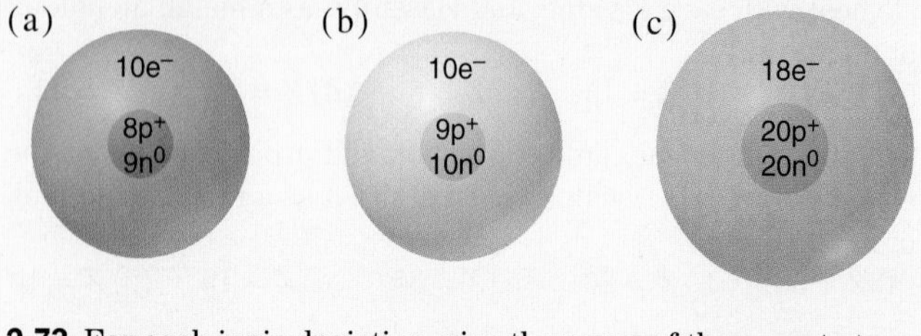

2.73 For each ionic depiction, give the name of the parent atom, its mass number, and its group and period numbers:

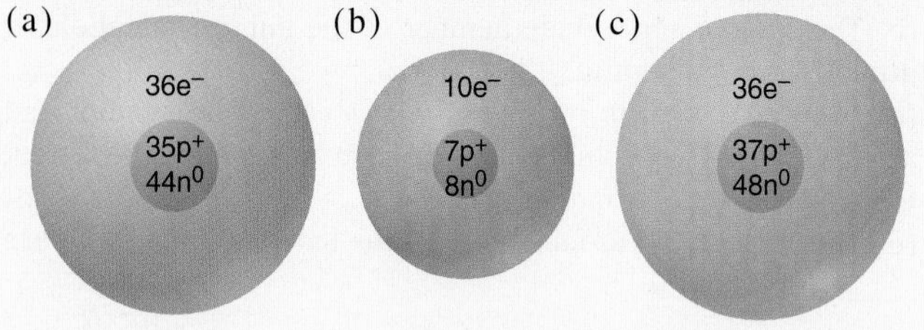

2.74 An ionic compound forms when lithium ($Z = 3$) reacts with oxygen ($Z = 8$). If a sample of the compound contains 8.4×10^{21} lithium ions, how many oxide ions does it contain?

2.75 An ionic compound forms when calcium ($Z = 20$) reacts with iodine ($Z = 53$). If a sample of the compound contains 7.4×10^{21} calcium ions, how many iodide ions does it contain?

2.76 The radii of the sodium and potassium ions are 102 pm and 138 pm, respectively. Which compound has stronger ionic attractions, sodium chloride or potassium chloride?

2.77 The radii of the lithium and magnesium ions are 76 pm and 72 pm, respectively. Which compound has stronger ionic attractions, lithium oxide or magnesium oxide?

Formula, Name, and Mass of a Compound

(Sample Problems 2.7 to 2.16)

Concept Review Questions

2.78 What information about the relative numbers of ions and the mass percents of the elements is in the formula MgF_2?

2.79 How is a structural formula similar to a molecular formula? How is it different?

2.80 Consider a mixture of 10 billion O_2 molecules and 10 billion H_2 molecules. In what way is this mixture similar to a sample that contains 10 billion hydrogen peroxide (H_2O_2) molecules? In what way is it different?

2.81 For what type(s) of compound(s) do we use roman numerals in the names?

2.82 For what type(s) of compound(s) do we use Greek numerical prefixes in the names?

2.83 For what type of compound are we unable to write a molecular formula?

Skill-Building Exercises (grouped in similar pairs)

2.84 Give the name and formula for the compound formed from each pair of elements:
(a) Sodium and nitrogen (b) Oxygen and strontium
(c) Aluminum and chlorine

2.85 Give the name and formula for the compound formed from each pair of elements:
(a) Cesium and bromine (b) Sulfur and barium
(c) Calcium and fluorine

2.86 Give the name and formula for the compound formed from each pair of elements:
(a) ${}_{12}L$ and ${}_{9}M$ (b) ${}_{30}L$ and ${}_{16}M$ (c) ${}_{17}L$ and ${}_{38}M$

2.87 Give the name and formula for the compound formed from each pair of elements:
(a) ${}_{37}Q$ and ${}_{35}R$ (b) ${}_{8}Q$ and ${}_{13}R$ (c) ${}_{20}Q$ and ${}_{53}R$

2.88 Give the systematic name for the formula or the formula for the name:
(a) Tin(IV) chloride (b) $FeBr_3$ (c) Cuprous bromide (d) Mn_2O_3

2.89 Give the systematic name for the formula or the formula for the name:
(a) Na_2HPO_4 (b) Potassium carbonate dihydrate
(c) $NaNO_2$ (d) Ammonium perchlorate

2.90 Give the systematic name for the formula or the formula for the name:
(a) CoO (b) Mercury(I) chloride
(c) $Pb(C_2H_3O_2)_2 \cdot 3H_2O$ (d) Chromic oxide

2.91 Give the systematic name for the formula or the formula for the name:
(a) $Sn(SO_3)_2$ (b) Potassium dichromate
(c) $FeCO_3$ (d) Copper(II) nitrate

2.92 Correct each incorrect formula:
(a) Barium oxide is BaO_2.
(b) Iron(II) nitrate is $Fe(NO_3)_3$.
(c) Magnesium sulfide is $MnSO_3$.

2.93 Correct each name:
(a) CuI is cobalt(II) iodide.
(b) $Fe(HSO_4)_3$ is iron(II) sulfate.
(c) $MgCr_2O_7$ is magnesium dichromium heptaoxide.

2.94 Give the name and formula for the acid derived from each anion:
(a) Hydrogen sulfate (b) IO_3^- (c) Cyanide (d) HS^-

2.95 Give the name and formula for the acid derived from each anion:
(a) Perchlorate (b) NO_3^- (c) Bromite (d) F^-

2.96 Many chemical names are similar at first glance. Give the formula for the different species in each set:
(a) Ammonium ion and ammonia
(b) Magnesium sulfide, magnesium sulfite, and magnesium sulfate
(c) Hydrochloric acid, chloric acid, and chlorous acid
(d) Cuprous bromide and cupric bromide

2.97 Give the formula for the different compounds in each set:
(a) Lead(II) oxide and lead(IV) oxide
(b) Lithium nitride, lithium nitrite, and lithium nitrate
(c) Strontium hydride and strontium hydroxide
(d) Magnesium oxide and manganese(II) oxide

2.98 Give the name and formula for the compound whose molecules consist of two sulfur atoms and four fluorine atoms.

2.99 Give the name and formula for the compound whose molecules consist of two chlorine atoms and one oxygen atom.

2.100 Correct the name to match the formula for each compound:
(a) Calcium(II) dichloride, $CaCl_2$ (b) Copper(II) oxide, Cu_2O
(c) Stannous tetrafluoride, SnF_4 (d) Hydrogen chloride acid, HCl

2.101 Correct the formula to match the name of each compound:
(a) Iron(III) oxide, Fe_3O_4 (b) Chloric acid, HCl
(c) Mercuric oxide, Hg_2O (d) Dichlorine heptaoxide, Cl_2O_6

2.102 Give the number of atoms of the specified element in a formula unit of each compound, and calculate the molecular or formula mass:
(a) Oxygen in aluminum sulfate, $Al_2(SO_4)_3$
(b) Hydrogen in ammonium hydrogen phosphate, $(NH_4)_2HPO_4$
(c) Oxygen in the mineral azurite, $Cu_3(OH)_2(CO_3)_2$

2.103 Give the number of atoms of the specified element in a formula unit of each compound, and calculate the molecular or formula mass:
(a) Hydrogen in ammonium benzoate, $C_6H_5COONH_4$
(b) Nitrogen in hydrazinium sulfate, $N_2H_6SO_4$
(c) Oxygen in the mineral leadhillite, $Pb_4SO_4(CO_3)_2(OH)_2$

2.104 Write the formula for each compound, and determine its molecular or formula mass:
(a) Ammonium sulfate (b) Sodium dihydrogen phosphate
(c) Potassium bicarbonate

2.105 Write the formula for each compound, and determine its molecular or formula mass:
(a) Sodium dichromate (b) Ammonium perchlorate
(c) Magnesium nitrite trihydrate

2.106 Calculate the molecular or formula mass of each compound:
(a) Dinitrogen pentoxide (b) Lead(II) nitrate
(c) Calcium peroxide

2.107 Calculate the molecular or formula mass of each compound:
(a) Iron(II) acetate tetrahydrate (b) Sulfur tetrachloride
(c) Potassium permanganate

2.108 Give the formula, name, and molecular mass of each molecule:

(a)
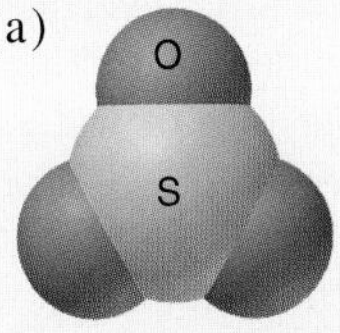

(b)
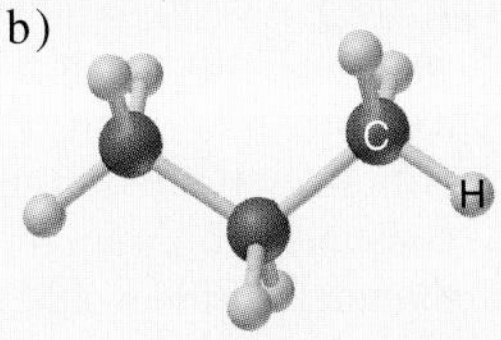

2.109 Give the formula, name, and molecular mass of each molecule:

(a)
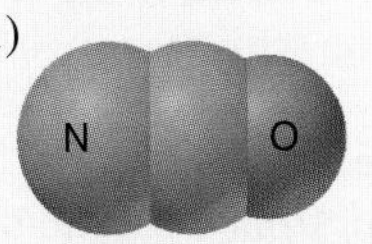

(b)
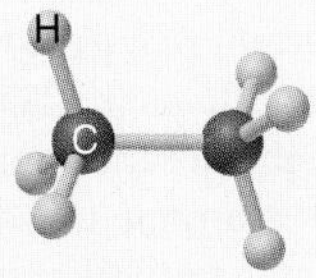

Problems in Context

2.110 Before the use of systematic names, many compounds had common names. Give the systematic name for each compound:
(a) Blue vitriol, $CuSO_4 \cdot 5H_2O$ (b) Slaked lime, $Ca(OH)_2$
(c) Oil of vitriol, H_2SO_4 (d) Washing soda, Na_2CO_3
(e) Muriatic acid, HCl (f) Epsom salt, $MgSO_4 \cdot 7H_2O$
(g) Chalk, $CaCO_3$ (h) Dry ice, CO_2
(i) Baking soda, $NaHCO_3$ (j) Lye, NaOH

2.111 Each circle contains a representation of a binary compound. Determine the name, formula, and molecular (formula) mass of the compound:

(a) (b)
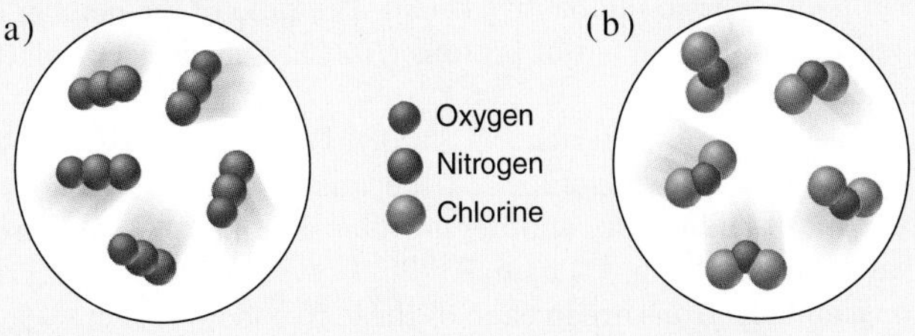

Mixtures: Classification and Separation

Concept Review Questions

2.112 In what main way is separating the components of a mixture different from separating the components of a compound?

2.113 What is the difference between a homogeneous mixture and a heterogeneous mixture?

2.114 Is a solution a homogeneous or heterogeneous mixture? Give an example of an aqueous solution.

Skill-Building Exercises (grouped in similar pairs)

2.115 Classify each of the following as a compound, a homogeneous mixture, or a heterogeneous mixture:
(a) Distilled water (b) Gasoline (c) Beach sand
(d) Wine (e) Air

2.116 Classify each of the following as a compound, a homogeneous mixture, or a heterogeneous mixture:
(a) Orange juice (b) Vegetable soup (c) Cement
(d) Calcium sulfate (e) Tea

Problems in Context

2.117 Which separation method is operating in each procedure? (a) Pouring a mixture of cooked pasta and boiling water into a colander; (b) Removing coloured impurities from raw sugar to make refined sugar.

2.118 A quality-control laboratory analyzes a product mixture using gas-liquid chromatography. The separation of components is more than adequate, but the process takes too long. Suggest two ways, other than changing the stationary phase, to shorten the analysis time.

Comprehensive Problems

2.119 Helium is the lightest noble gas and the second-most abundant element (after hydrogen) in the universe.
(a) The radius of a helium atom is 3.1×10^{-11} m; the radius of its nucleus is 2.5×10^{-15} m. What fraction of its spherical atomic volume is occupied by its nucleus? (V of a sphere $= \frac{4}{3}\pi r^3$.)
(b) The mass of a helium-4 atom is $6.646\,48 \times 10^{-24}$ g, and each of its two electrons has a mass of $9.109\,39 \times 10^{-28}$ g. What fraction of this atom's mass is contributed by its nucleus?

2.120 From the following ions (with their radii in pm), choose the pair that forms the strongest ionic bond and the pair that forms the weakest ionic bond:

Ion:	Mg^{2+}	K^+	Rb^+	Ba^{2+}	Cl^-	O^{2-}	I^-
Radius:	72	138	152	135	181	140	220

2.121 Give the molecular mass of each compound depicted below, and provide a correct name for any that are named incorrectly:

(a)
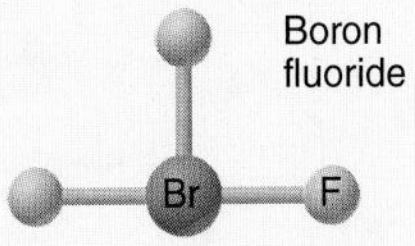

(b)

(c)
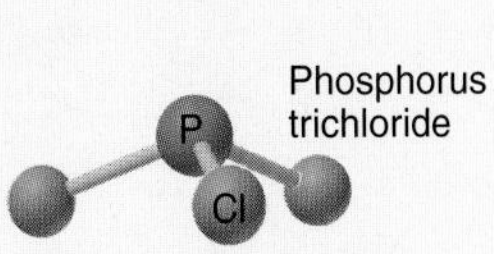

(d)
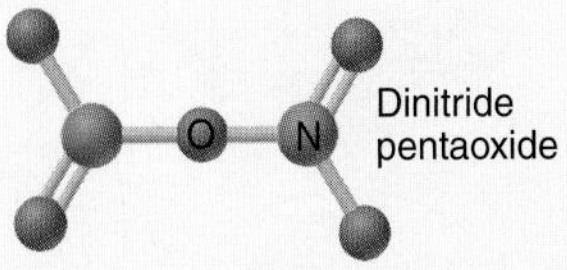

2.122 Polyatomic ions are named by patterns that apply to elements in a given group. Using the periodic table and Table 2.5, give the name of each ion:
(a) SeO_4^{2-} (b) AsO_4^{3-} (c) BrO_2^- (d) $HSeO_4^-$ (e) TeO_3^{2-}

2.123 Ammonium dihydrogen phosphate, formed from the reaction of phosphoric acid with ammonia, is used as a crop fertilizer as well as a component of some fire extinguishers.
(a) What are the mass percents of N and P in the compound?
(b) What mass of ammonia is contained in 100. g of the compound?

2.124 Nitrogen forms more oxides than any other element. The percents by mass of N in three different nitrogen oxides are (I) 46.69%; (II) 36.85%; and (III) 25.94%. For each oxide, determine (a) the simplest whole-number ratio of N to O and (b) the mass of oxygen, in grams per 1.00 g of nitrogen.

2.125 The number of atoms in 1 dm^3 of aluminum is nearly the same as the number of atoms in 1 dm^3 of lead, but the densities of these metals are very different (see Table 1.4). Explain.

2.126 You are working in the laboratory preparing sodium chloride. Consider the following results for three preparations of the compound:

Case 1: 39.34 g Na + 60.66 g $Cl_2 \rightarrow$ 100.00 g NaCl
Case 2: 39.34 g Na + 70.00 g $Cl_2 \rightarrow$ 100.00 g NaCl + 9.34 g Cl_2
Case 3: 50.00 g Na + 50.00 g $Cl_2 \rightarrow$ 82.43 g NaCl + 17.57 g Na

Explain these results in terms of the laws of mass conservation and definite composition.

2.127 Diagrams A to I depict various types of matter on the atomic scale. Choose the correct diagram(s) for each description:
(a) A mixture that fills its container
(b) A substance that cannot be broken down into simpler substances
(c) An element that has a very high resistance to flow
(d) A homogeneous mixture
(e) An element that conforms to the walls of its container and displays an upper surface
(f) A gas that consists of diatomic particles
(g) A gas that can be broken down into simpler substances
(h) A substance that has a 2/1 ratio of its component atoms
(i) Matter that can be separated into its component substances by physical means
(j) A heterogeneous mixture
(k) Matter that obeys the law of definite composition

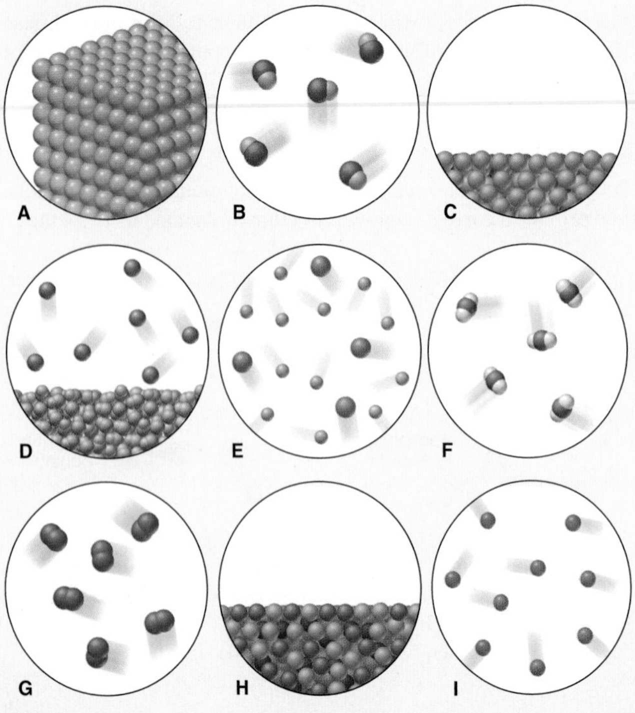

2.128 The seven most abundant ions in seawater make up more than 99% by mass of the dissolved compounds. They have the following abundances, in units of mg ion/kg seawater: chloride 18 980; sodium 10 560; sulfate 2650; magnesium 1270; calcium 400; potassium 380; hydrogen carbonate 140.
(a) What is the mass percent of each ion in seawater?
(b) What percent of the total mass of ions is sodium ion?
(c) How does the total mass percent of alkaline earth metal ions compare with the total mass percent of alkali metal ions?
(d) Which make up the larger mass fraction of dissolved components, anions or cations?

2.129 The diagram below represents a mixture of two monatomic gases undergoing a reaction when heated. Which mass law(s) is (are) illustrated by this reaction?

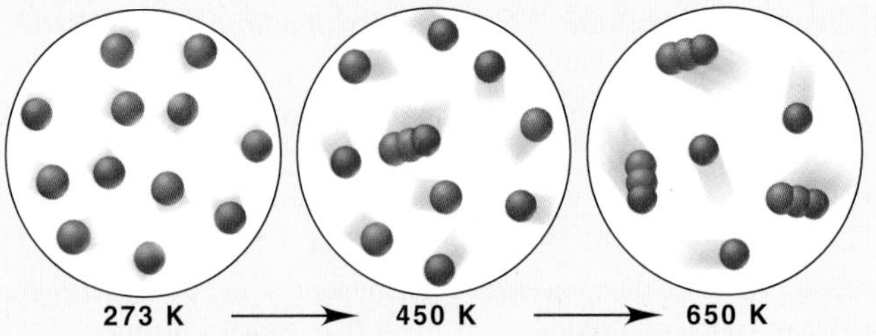

2.130 Succinic acid (*right*) is an important metabolite in biological energy production. Give the molecular formula, molecular mass, and mass percent of each element in succinic acid.

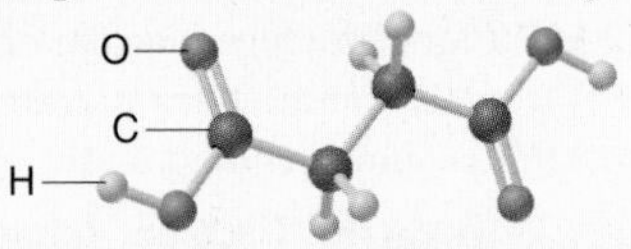

2.131 Fluoride ion is poisonous in relatively low amounts: 0.2 g of F^- per 70 kg of body weight can cause death. Nevertheless, in order to prevent tooth decay, F^- ions are added to drinking water at a concentration of 1 mg of F^- ions per litre of water. What volume (L) of fluoridated drinking water would a 70 kg person have to consume in one day to reach this toxic level? What mass (kg) of sodium fluoride would be needed to treat an 8.50×10^7 L reservoir?

2.132 Antimony has many uses, for example, in infrared devices and as part of an alloy in lead storage batteries. The element has two naturally occurring isotopes, one with mass 120.904 u and the other with mass 122.904 u.
(a) Write the notation for each isotope.
(b) Use the atomic mass of antimony from the periodic table to calculate the natural abundance of each isotope.

2.133 Dinitrogen monoxide (N_2O; nitrous oxide) is a greenhouse gas that enters the atmosphere principally from the breakdown of natural fertilizer. Some studies have shown that the isotope ratios of ^{15}N to ^{14}N and of ^{18}O to ^{16}O in N_2O depend on the source, which can thus be determined by measuring the relative abundance of molecular masses in a sample of N_2O.
(a) What different molecular masses are possible for N_2O?
(b) The percent abundance of ^{14}N is 99.6%, and that of ^{16}O is 99.8%. Which molecular mass of N_2O is least common, and which is most common?

2.134 Nuclei differ in their stability, and some are so unstable that they undergo radioactive decay. The ratio of the number of neutrons to the number of protons (*N/Z*) in a nucleus correlates with its stability.
(a) Calculate the *N/Z* ratio for (i) ^{144}Sm; (ii) ^{56}Fe; (iii) ^{20}Ne; and (iv) ^{107}Ag.
(b) The radioactive isotope ^{238}U decays in a series of nuclear reactions that includes another uranium isotope, ^{234}U, and three lead isotopes: ^{214}Pb, ^{210}Pb, and ^{206}Pb. How many neutrons, protons, and electrons are in each of these five isotopes?

2.135 Use the box colour(s) in the periodic table below to identify the element(s) described by each of the following:

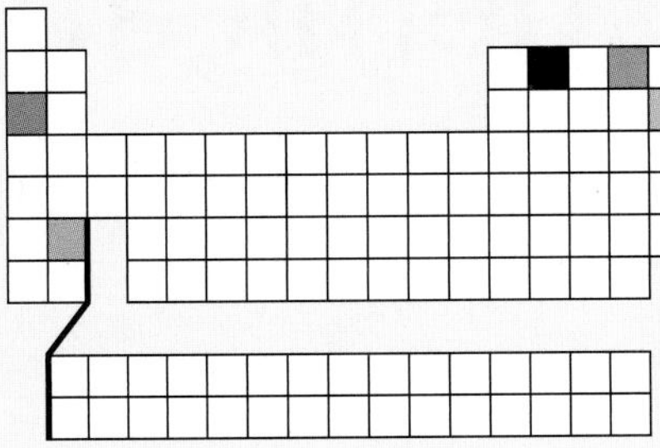

(a) Four elements that are nonmetals
(b) Two elements that are metals
(c) Three elements that are gases at room temperature
(d) Three elements that are solids at room temperature
(e) One pair of elements likely to form a covalent compound
(f) Another pair of elements likely to form a covalent compound
(g) One pair of elements likely to form an ionic compound with formula MX
(h) Another pair of elements likely to form an ionic compound with formula MX
(i) Two elements likely to form an ionic compound with formula M_2X
(j) Two elements likely to form an ionic compound with formula MX_2
(k) An element that forms no compounds
(l) A pair of elements whose compounds exhibit the law of multiple proportions

2.136 The two isotopes of potassium with significant abundance in nature are ^{39}K (isotopic mass 38.9637 u, 93.258%) and ^{41}K (isotopic mass 40.9618 u, 6.730%). Fluorine has only one naturally occurring isotope, ^{19}F (isotopic mass 18.9984 u). Calculate the formula mass of potassium fluoride.

2.137 Boron trifluoride is used as a catalyst in the synthesis of organic compounds. When this compound is analyzed by mass spectrometry (see the first Tools of the Laboratory in this chapter), several different 1+ ions form, including ions representing the whole molecule as well as molecular fragments formed by the loss of one, two, and three F atoms. Given that boron has two naturally occurring isotopes, ^{10}B and ^{11}B, and fluorine has one, ^{19}F, calculate the masses of all the possible 1+ ions.

2.138 Nitrogen monoxide (NO) is a bioactive molecule in blood. Low NO concentrations cause respiratory distress and the formation of blood clots. Doctors prescribe nitroglycerin, $C_3H_5N_3O_9$, and isoamyl nitrate, $(CH_3)_2CHCH_2CH_2ONO_2$, to increase NO. If each compound releases one molecule of NO per atom of N it contains, calculate the mass percent of NO in each.

2.139 TNT (trinitrotoluene; *right*) is used as an explosive in construction. Calculate the mass of each element in 1.00 kg of TNT.

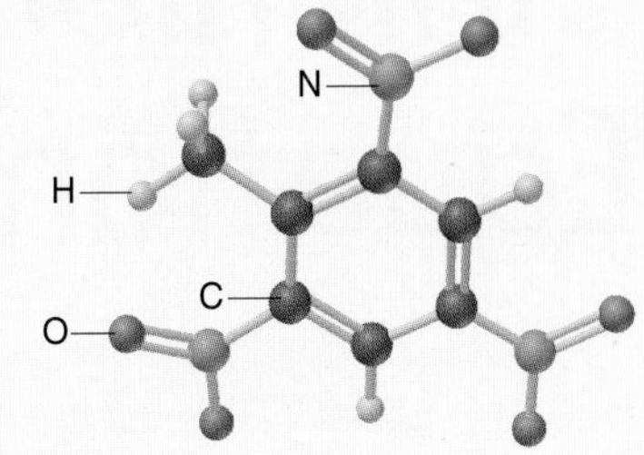

2.140 The anticancer drug Platinol (cisplatin), $Pt(NH_3)_2Cl_2$, reacts with a cancer cell's DNA and interferes with its growth.
(a) What is the mass percent of platinum (Pt) in Platinol?
(b) If Pt costs $32/g, what mass of Platinol can be made for $1.00 million? (Assume that the cost of Pt determines the cost of the drug.)

2.141 In the periodic table below, give the name, symbol, atomic number, atomic mass, period number, and group number of (a) the *building-block elements (red)*, which occur in nearly every biological molecule, and (b) the *macronutrients (green)*, which are either essential ions in cell fluids or part of many biomolecules.

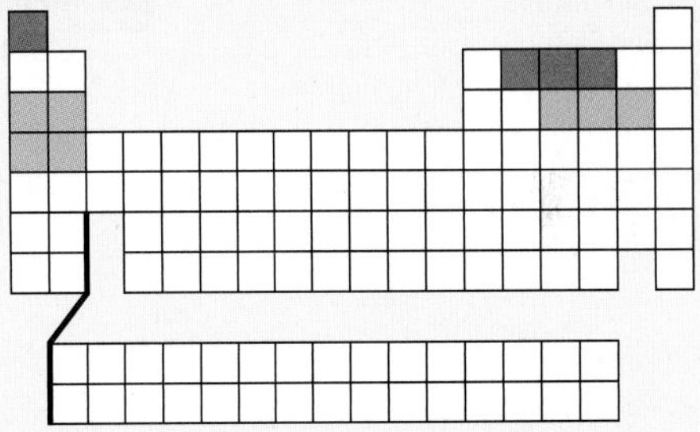

2.142 The following block diagram classifies the components of matter on the macroscopic scale. Identify blocks (a) to (d).

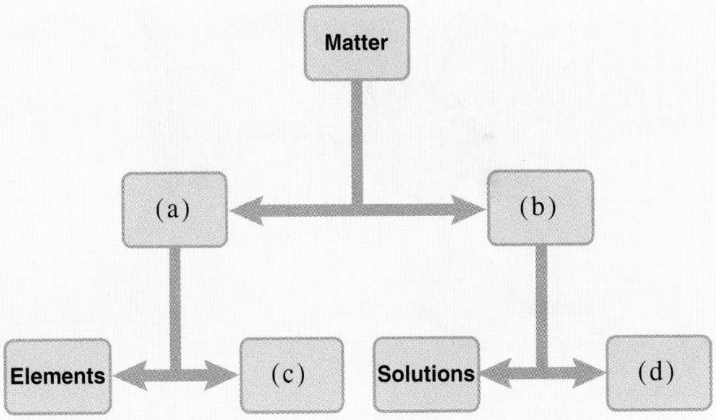

2.143 Which of the following steps in an overall process involve(s) a physical change? Which involve(s) a chemical change?

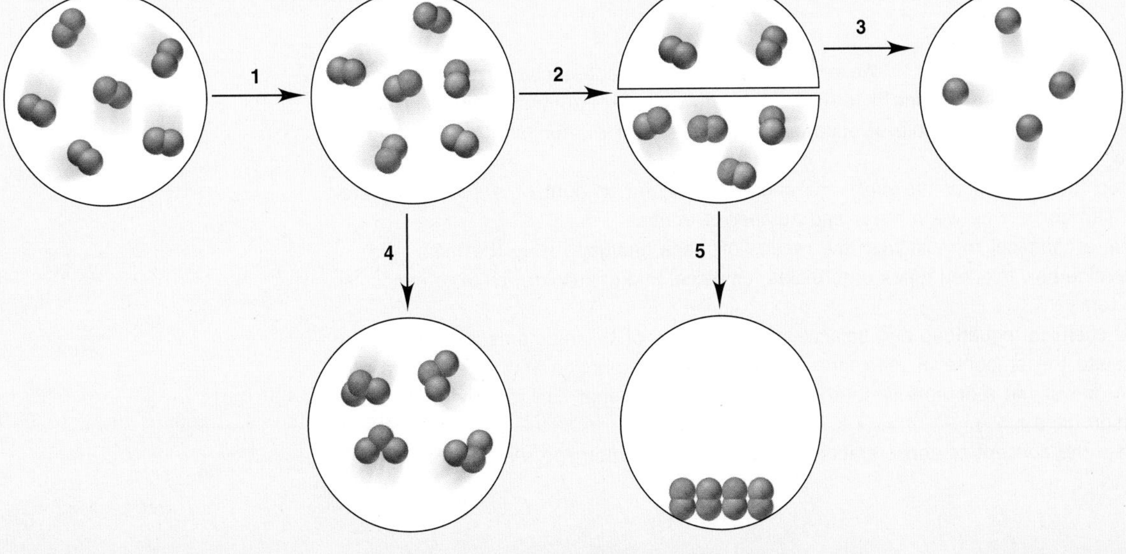

Stoichiometry and Chemical Equations

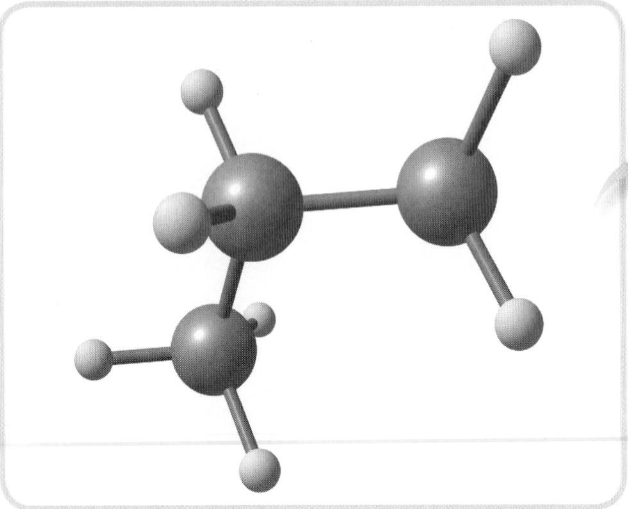

IN THIS CHAPTER . . . We relate the mass of a substance to the number of chemical entities constituting it (atoms, ions, molecules, or formula units) and apply this relationship to a formula or equations. By the end of this chapter, you should be able to

- Explain the concept of the *mole*, the chemist's unit for amount of a substance, and use it to convert between mass and number of entities
- Derive a chemical formula from the results of mass analysis using the mole
- Determine how two key types of formulas (empirical and molecular) relate to molecular structures
- Write chemical equations and balance them in terms of the reactants and products
- Calculate the amounts of reactants and products in a reaction and determine if and when one of the reactants limits the amount of product that can form and, thus, the reaction yield
- Explain the concept of concentration, and apply stoichiometry to reactions in solution

3

Concepts and Skills to Review before Studying This Chapter

- Atomic mass (Section 2.5)
- Name and formula for compounds (Section 2.8)
- Molecular (or formula) mass (Section 2.8)
- Molecular and structural formula, and ball-and-stick and space-filling models (Section 2.8)

Two slices of bread, two tablespoons of peanut butter, and one teaspoon of raspberry jam makes the perfect peanut butter and jam sandwich. It would be very hard to make this perfect sandwich with only one slice of bread and impossible to make it without the jam! Chemical reactions are no different. In order for a chemical reaction to work perfectly, we need to have all the components of the reaction present and in just the right amounts. If we do not have one of the components or we do not have enough of one of the components, it will affect whether the reaction can take place and whether or not we can get the amount of product we want. While we know instinctively what we need to make a sandwich and what to do if we want more than one sandwich, it takes practice to transfer this same idea, called **stoichiometry** (from the Greek words *stoicheion* meaning element and *metron* meaning measure), to chemistry and chemical reactions.

This chapter introduces the idea of measurement in chemistry, describing the various units we use to determine the stoichiometry that relates one reactant to another or the amount of a reactant to the amount of product. We will also learn how to express a reaction using a chemical equation that shows the stoichiometric relation between the reactants and products. We will determine what happens if we have a reaction in which not all the reactants are present in their stoichiometric amounts and how this affects the reaction yield. Finally, we will learn about the properties of water, the solvent we will encounter most frequently in this book.

3.1 The Mole

In daily life, we often measure things by counting or by mass: we measure a mass of flour or rice, but we count eggs or doughnuts. Then we use counting units (a dozen doughnuts) or mass units (a kilogram of sugar) to express the amount. Similarly, daily life in a laboratory involves measuring substances. We want to know the numbers of chemical entities—atoms, ions, molecules, or formula units—that react with each other. But how can we possibly count or weigh such minute objects? As you will see, chemists have devised a unit, called the *mole*, to *count chemical entities using their mass.*

Defining the Mole

The **mole (mol)** is the SI unit for *amount of substance.* It is defined as *the amount of a substance that contains the same number of entities as the number of atoms in 12 g of carbon-12.* This number, called **Avogadro's number** (symbol N_A, and named in honour of the 19th-century Italian physicist Amedeo Avogadro), is enormous:

$$\text{One mole (1 mol) contains } 6.022\,141\,29 \times 10^{23} \text{ entities} \qquad (3.1)$$

For most calculations, it is sufficient to use a value of 6.022×10^{23} for Avogadro's number. A counting unit, such as *dozen,* tells you the number of objects but not their mass; a mass unit, such as *kilogram,* tells you the mass of the objects but not their number. The mole tells you both—the *number* of objects in a given *mass* of substance:

1 mol of carbon-12 contains 6.022×10^{23} carbon-12 atoms *and* has a mass of 12 g

What does it mean that the mole unit allows you to count entities using the mass of the sample? Suppose that you have a sample of carbon-12 and want to know the number of atoms in the sample. You find that the sample has a mass of 6 g, so it is exactly 0.5 mol of carbon-12 and contains 3.011×10^{23} atoms:

6 g of carbon-12 is 0.5 mol of carbon-12 and contains 3.011×10^{23} atoms

16
S
32.07

Knowing the amount (mol), the mass (g), and the number of entities becomes very important when we mix different substances to run a reaction. The central relationship between masses on the atomic scale and on the macroscopic scale is the same for elements and compounds:

- *Elements.* The average mass in *unified atomic mass units (u)* of one atom of an element is the *same numerically* as the average mass in *grams (g)* of one mole of atoms of the element. In fact, the choice of the value for Avogadro's number was intended to make the numeric value of the mass of *one atom of* ^{12}C (in units of u) equal to the mass of *one mole of* ^{12}C *atoms* (in units of g). Recall, from Chapter 2, that each atom of an element is considered to have the *average atomic mass* given in the periodic table (*see margin*). Thus,

1 atom of S has an average mass of 32.07 u and 1 mol (6.022×10^{23} atoms) of S has an average mass of 32.07 g.

1 atom of Fe has an average mass of 55.85 u and 1 mol (6.022×10^{23} atoms) of Fe has an average mass of 55.85 g.

Note, also, that since atomic masses are relative, the ratio of average atomic masses of Fe and S equals the ratio of average molar masses of Fe and S, or,

$$\frac{\text{average mass of 1 atom of Fe (u)}}{\text{average mass of 1 atom of S (u)}} = \frac{55.85}{32.07} = \frac{\text{average mass of 1 mol of Fe (mol)}}{\text{average mass of 1 mol of S (mol)}}$$

- *Compounds.* The average mass in *unified atomic mass units (u)* of one molecule (or formula unit) of a compound is the *same numerically* as the average mass in *grams (g)* of one mole of the compound. Thus, for example,

1 molecule of H_2O has an average mass of 18.02 u and 1 mol (6.022×10^{23} molecules) of H_2O has an average mass of 18.02 g.

1 formula unit of NaCl has an average mass of 58.44 u and 1 mol (6.022×10^{23} formula units) of NaCl has an average mass of 58.44 g.

Here, too, because masses are relative, we can say,

$$\frac{\text{average mass of 1 molecule of } H_2O \text{ (u)}}{\text{average mass of 1 formula unit of NaCl (u)}} = \frac{18.02}{58.44} = \frac{\text{average mass of 1 mol of } H_2O \text{ (g)}}{\text{average mass of 1 mol of NaCl (g)}}$$

There are two key points to remember about the importance of the mole unit:

- The *mole* lets us relate the *number* of entities to the *mass* of a sample of those entities.
- The mole maintains the *same numerical relationship* between mass on the atomic scale (unified atomic mass units, u) and mass on the macroscopic scale (grams, g).

In everyday terms, a grocer *does not* know that there are one dozen eggs from their mass or that there is 1 kg of beans from their count, because eggs and beans do not have fixed masses. However, by taking a mass of 54.938 045 g (1 mol) of manganese, a chemist *does* know that there are 6.022×10^{23} manganese atoms, because all manganese atoms have an average atomic mass of 54.938 045 u. Figure 3.1 shows 1 mol of some familiar elements and compounds.

FIGURE 3.1 One mole (6.022×10^{23} entities) of some familiar substances. From left to right: 1 mol of copper (63.55 g), liquid H_2O (18.02 g), sodium chloride (table salt, 58.44 g), sucrose (table sugar, 342.3 g), and aluminum (26.98 g).

Determining Molar Mass

The **molar mass ($\mathcal{M}$)** of a substance is the mass per mole of its entities (atoms, molecules, or formula units) and has units of grams per mole (g/mol). In reality, because the atomic mass of each element listed in the periodic table represents the weighted average of the isotopic masses for this element, the molar mass we calculate is also a weighted average molar mass. For the sake of convenience, we will simply refer to it as the molar mass. The periodic table is indispensable for calculating molar mass:

1. *Elements.* To find the molar mass, look up the atomic mass and note whether the element is monatomic or molecular.

- *Monatomic elements.* The molar mass is the periodic-table value in grams per mole.* For example, the molar mass of neon is 20.18 g/mol, and the molar mass of gold is 197.0 g/mol.
- *Molecular elements.* You must know the formula to determine the molar mass (see Figure 2.15). For example, in air, oxygen exists most commonly as diatomic molecules, so the molar mass of O_2 is twice the molar mass of O:

$$\text{Molar mass } (\mathcal{M}) \text{ of } O_2 = 2 \times \mathcal{M} \text{ of O} = 2 \times 16.00 \text{ g/mol} = 32.00 \text{ g/mol}$$

Sulfur exists most commonly as octatomic molecules, S_8:

$$\mathcal{M} \text{ of } S_8 = 8 \times \mathcal{M} \text{ of S} = 8 \times 32.07 \text{ g/mol} = 256.6 \text{ g/mol}$$

2. *Compounds. The molar mass is the sum of the molar masses of the atoms in the formula.* Thus, from the formula for sulfur dioxide, SO_2, we know that 1 mol of SO_2 molecules contains 1 mol of S atoms and 2 mol of O atoms:

$$\mathcal{M} \text{ of } SO_2 = \mathcal{M} \text{ of S} + 2 \times (\mathcal{M} \text{ of O}) = 32.07 \text{ g/mol} + (2 \times 16.00 \text{ g/mol}) = 64.07 \text{ g/mol}$$

Similarly, for ionic compounds, such as potassium sulfide, we determine $\mathcal{M}$ using the formula unit for potassium sulfide (K_2S), which tells us that each formula unit contains 2 mol of K atoms and 1 mol of S atoms:

$$\mathcal{M} \text{ of } K_2S = 2 \times (\mathcal{M} \text{ of K}) + \mathcal{M} \text{ of S} = (2 \times 39.10 \text{ g/mol}) + 32.07 \text{ g/mol} = 110.27 \text{ g/mol}$$

Thus, *the subscripts in a formula refer to both individual atoms (or ions) and moles of atoms (or ions).* Table 3.1 summarizes these ideas for glucose, $C_6H_{12}O_6$ (*see margin*), the essential sugar in energy metabolism.

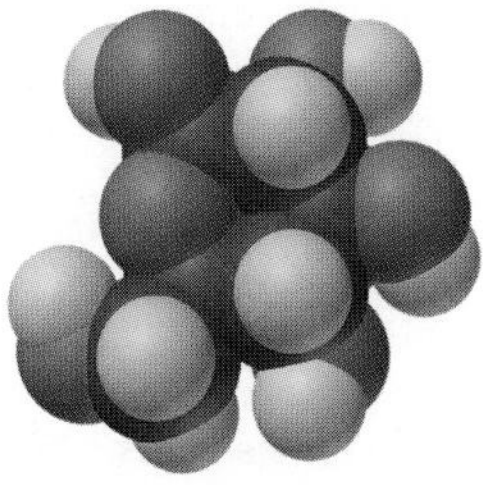
Glucose

Converting between Amount, Mass, and Number of Chemical Entities

One of the most common skills in the lab—and on exams—is converting between amount (mol), mass (g), and number of entities of a substance.

1. *Converting between amount and mass.* If you know the amount of a substance, you can find its mass, and vice versa. The molar mass ($\mathcal{M}$), which expresses the equivalence between 1 mol of a substance and its mass in grams, is the conversion factor.
   - *From amount (mol) to mass (g),* multiply by the molar mass:

$$\text{Mass (g)} = \text{amount (mol)} \times \text{molar mass} \left(\frac{\text{g}}{\text{mol}}\right)$$

$$m \text{ (g)} = n \text{ (mol)} \times \mathcal{M}\left(\frac{\text{g}}{\text{mol}}\right) \quad (3.2)$$

TABLE 3.1 Information Contained in the Chemical Formula for Glucose, $C_6H_{12}O_6$ ($\mathcal{M}$ = 180.16 g/mol)

	Carbon (C)	Hydrogen (H)	Oxygen (O)
Atoms/molecule of compound	6 atoms	12 atoms	6 atoms
Moles of atoms/mole of compound	6 mol of atoms	12 mol of atoms	6 mol of atoms
Atoms/mole of compound	$6(6.022 \times 10^{23})$ atoms	$12(6.022 \times 10^{23})$ atoms	$6(6.022 \times 10^{23})$ atoms
Mass/molecule of compound	6(12.01 u) = 72.06 u	12(1.008 u) = 12.10 u	6(16.00 u) = 96.00 u
Mass/mole of compound	72.06 g	12.10 g	96.00 g

*The mass value in the periodic table has no units because it is a *relative* atomic mass, given by the atomic mass (in u) divided by 1 u ($\frac{1}{12}$ mass of one ^{12}C atom in u):

$$\text{Relative atomic mass} = \frac{\text{atomic mass (u)}}{\frac{1}{12} \text{ mass of } ^{12}\text{C (u)}}$$

Therefore, you use the same number for the atomic mass and for the molar mass.

- *From mass (g) to amount (mol)*, divide by the molar mass $\left(\text{multiply by } \frac{1}{\mathcal{M}}\right)$:

$$\text{Amount (mol)} = \text{mass (g)} \times \frac{1 \text{ (mol)}}{\text{molar mass (g)}}$$

$$n \text{ (mol)} = \frac{m \text{ (g)}}{\mathcal{M} \text{ (g/mol)}} \qquad (3.3)$$

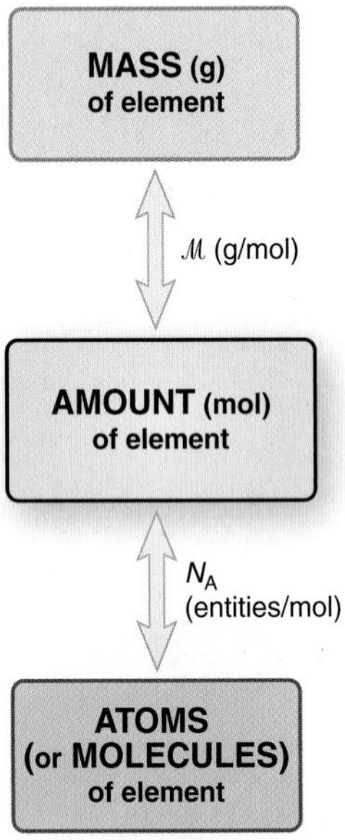

FIGURE 3.2 Mass-mole-number relationships for elements. To find the number of atoms (or molecules) in a given mass, or vice versa, first convert to amount (mol). For molecular elements, Avogadro's number gives *molecules* per mole.

2. *Converting between amount and number.* Similarly, if you know the amount (mol), you can find the number of entities, and vice versa. Avogadro's number, which expresses the equivalence between 1 mol of a substance and the number of entities it contains, is the conversion factor.
   - *From amount (mol) to number of entities*, multiply by Avogadro's number:

$$\text{No. of entities} = \text{amount (mol)} \times \frac{6.022 \times 10^{23} \text{ entities}}{1 \text{ mol}}$$

$$\text{No. of entities} = n \text{ (mol)} \times N_A \left(\frac{\text{no. of entities}}{\text{mol}}\right) \qquad (3.4)$$

   - *From number of entities to amount (mol)*, divide by Avogadro's number:

$$\text{Amount (mol)} = \text{no. of entities} \times \frac{1 \text{ mol}}{6.022 \times 10^{23} \text{ entities}}$$

$$n \text{ (mol)} = \frac{\text{no. of entities}}{N_A \left(\frac{\text{no. of entities}}{\text{mol}}\right)} \qquad (3.5)$$

Amount-Mass-Number Conversions Involving Elements We begin with amount-mass-number relationships of elements. As Figure 3.2 shows, *begin by converting mass or number of entities to amount (mol).*

Let us work through a series of sample problems that show these conversions for both elements and compounds.

Sample Problem 3.1 Calculating the Mass of a Given Amount of an Element

Problem The Royal Canadian Mint uses silver (Ag) to make special coins, such as the Maple Leaf Raindrop and the Maple Leaf Forever. What mass of Ag is in 0.0342 mol of silver?

Plan We know the amount of Ag (0.0342 mol) and have to find the mass (g). To convert units of *moles* of Ag to *grams* of Ag, we multiply by the *molar mass* of Ag, which we find in the periodic table (*see road map*).

Road Map

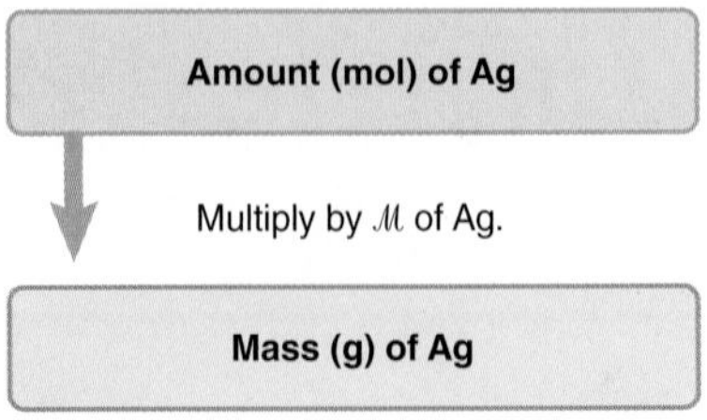

Solution Convert from amount (mol) of Ag to mass (g):

$$\text{Mass (g) of Ag} = m_{Ag} = n_{Ag} \times \mathcal{M}_{Ag} = 0.0342 \cancel{\text{mol Ag}} \times \frac{107.9 \text{ g Ag}}{1 \cancel{\text{mol Ag}}} = 3.69 \text{ g Ag}$$

Check We rounded the mass to three significant figures because the amount (mol) has three significant figures. The units are correct. About 0.03 mol × 100 g/mol gives 3 g; the small mass makes sense because 0.0342 is a small fraction of a mole.

Follow-Up Problem 3.1 Graphite is the crystalline form of carbon that is used in "lead" pencils. What amount of carbon (mol) is in 315 mg of graphite? Include a road map that shows how you planned your solution. (See Brief Solutions to Follow-Up Problems at the end of the chapter.)

Sample Problem 3.2 Calculating the Number of Entities in a Given Amount of an Element

Problem Gallium (Ga) is a key element in solar panels, calculators, and other light-sensitive electronic devices. How many Ga atoms are in 2.85×10^{-3} mol of gallium?

Plan We know the amount of gallium (2.85×10^{-3} mol) and need the number of Ga atoms. We multiply the amount (mol) by Avogadro's number to find the number of atoms (*see road map*).

Road Map

Amount (mol) of Ga

↓ Multiply by 6.022×10^{23} atoms/mol.

Number of Ga atoms

Solution Convert from the amount (mol) of Ga to the number of atoms:

$$\text{No. of Ga atoms} = n_{\text{Ga}} \times N_{\text{A}}$$

$$= 2.85 \times 10^{-3} \text{ \sout{mol Ga}} \times \frac{6.022 \times 10^{23} \text{ Ga atoms}}{1 \text{ \sout{mol Ga}}}$$

$$= 1.72 \times 10^{21} \text{ Ga atoms}$$

Check The number of atoms has three significant figures because the number of moles does. When we round amount (mol) of Ga and Avogadro's number, we have ~(3×10^{-3} mol)(6×10^{23} atoms/mol) = 18×10^{20}, or 1.8×10^{21} atoms, so our answer seems correct.

Follow-Up Problem 3.2 At rest, a person inhales 9.72×10^{21} nitrogen molecules in an average breath of air. What amount (mol) of nitrogen atoms is inhaled? (*Hint*: In air, nitrogen occurs as a diatomic molecule.) Include a road map that shows how you planned your solution.

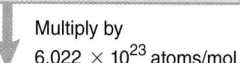

For the next problem, note that mass and number of entities relate directly to amount (mol), but *not* to each other. Therefore, *to convert between mass and number, first convert to amount.*

Sample Problem 3.3 Calculating the Number of Entities in a Given Mass of an Element

Problem Iron (Fe) is the main component of steel and, thus, the most important metal in industrial society; it is also essential in the body. How many Fe atoms are in 95.8 g of iron?

Plan We know the mass of Fe (95.8 g) and need the number of Fe atoms. We cannot convert directly from mass to number, so we first convert to amount (mol) by dividing the mass of Fe by its molar mass. Then, we multiply the amount (mol) by Avogadro's number to find the number of atoms (*see road map*).

Road Map

Mass (g) of Fe

↓ Divide by $\mathcal{M}$ of Fe.

Amount (mol) of Fe

↓ Multiply by 6.022×10^{23} atoms/mol.

Number of Fe atoms

Solution Convert from mass (g) of Fe to amount (mol):

$$\text{Amount (mol) of Fe} = n_{Fe} = \frac{m_{\text{Fe}}}{\mathcal{M}_{\text{Fe}}} = \frac{95.8 \text{ \sout{g} Fe}}{55.85 \text{ \sout{g}/mol Fe}} = 1.7153 \text{ mol Fe}$$

Convert from amount (mol) of Fe to number of Fe atoms:

$$\text{No. of Fe atoms} = n_{\text{Fe}} \times N_{\text{A}}$$

$$= 1.7153 \text{ \sout{mol Fe}} \times \frac{6.022 \times 10^{23} \text{ atoms Fe}}{1 \text{ \sout{mol Fe}}}$$

$$= 10.3 \times 10^{23} \text{ atoms Fe}$$

$$= 1.03 \times 10^{24} \text{ atoms Fe}$$

Comment Additional digits have been kept to preserve the accuracy of the final answer. If the entire calculation is done in one step, the final answer should have the correct number of significant digits.

Check Rounding the mass and the molar mass of Fe, we have $\sim\frac{100\ \text{g}}{60\ \text{g/mol}} = 1.7$ mol. Therefore, the number of atoms should be a bit less than twice Avogadro's number: $< 2(6 \times 10^{23}) = 1.2 \times 10^{24}$, so the answer seems correct.

Follow-Up Problem 3.3 Manganese (Mn) is a transition element essential for the growth of bones. What is the mass (g) of 3.22×10^{20} Mn atoms, the number found in 1 kg of bone? Include a road map that shows how you planned your solution.

Amount-Mass-Number Conversions Involving Compounds Only one new step is needed to solve amount-mass-number problems involving compounds: we need the chemical formula to find the molar mass and the amount of each element in the compound. The relationships are shown in Figure 3.3, and Sample Problems 3.4 and 3.5 apply these relationships to compounds with simple and more complicated formulas, respectively.

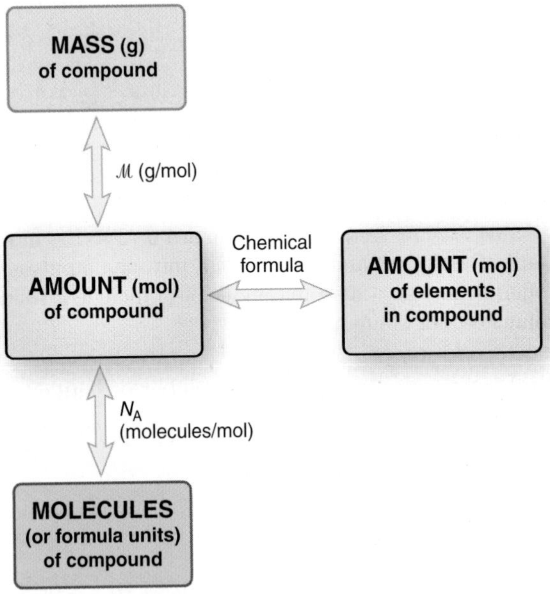

FIGURE 3.3 Amount-mass-number relationships for compounds. Use the chemical formula to find the amount (mol) of each element in a compound.

Sample Problem 3.4 Calculating the Number of Chemical Entities in a Given Mass of a Compound I

Problem Nitrogen dioxide is a component of urban smog that forms from gases in car exhaust. How many molecules are in 8.92 g of nitrogen dioxide?

Plan We know the mass of the compound (8.92 g) and need to find the number of molecules. As you just saw in Sample Problem 3.3, to convert mass to number of entities, we have to find the amount (mol). To do so, we divide the mass by the molar mass ($\mathcal{M}$), which we calculate from the molecular formula (see Sample Problem 2.15). Once we have the amount (mol), we multiply by Avogadro's number to find the number of molecules (*see road map*).

Road Map

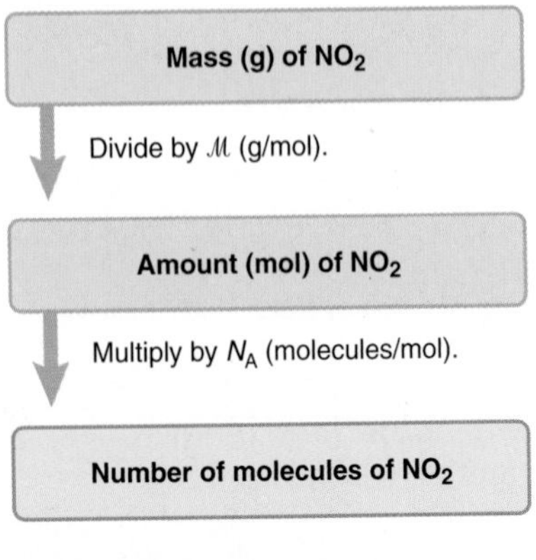

Solution The formula is NO_2. Calculate the molar mass:

$$\begin{aligned}\mathcal{M} \text{ of } NO_2 &= (1 \times \mathcal{M} \text{ of N}) + (2 \times \mathcal{M} \text{ of O})\\ &= 14.01\ \text{g/mol} + (2 \times 16.00\ \text{g/mol})\\ &= 46.01\ \text{g/mol}\end{aligned}$$

Convert from mass (g) of NO_2 to amount (mol):

$$\begin{aligned}\text{Amount (mol) of } NO_2 = n_{NO_2} &= \frac{m_{NO_2}}{\mathcal{M}_{NO_2}}\\ &= \frac{8.92\ \cancel{\text{g}}\ NO_2}{46.01\ \cancel{\text{g}}/\text{mol}\ NO_2}\\ &= 0.1939\ \text{mol}\ NO_2\end{aligned}$$

Convert from amount (mol) of NO_2 to number of molecules:

$$\text{No. of molecules} = n_{NO_2} \times N_A$$
$$= 0.1939\ \cancel{\text{mol } NO_2} \times \frac{6.022 \times 10^{23}\ NO_2\ \text{molecules}}{1\ \cancel{\text{mol } NO_2}}$$
$$= 1.17 \times 10^{23}\ NO_2\ \text{molecules}$$

Check Rounding, we get (~0.2 mol)(6×10^{23}) = 1.2×10^{23}, so the answer seems correct.

Follow-Up Problem 3.4 Fluoridation of drinking water to prevent tooth decay is carried out by adding sodium fluoride to water. What is the mass (g) of sodium fluoride in a litre of water that contains 1.19×10^{19} formula units of the compound? Include a road map that shows how you planned your solution.

Sample Problem 3.5 Calculating the Number of Chemical Entities in a Given Mass of a Compound II

Problem Ammonium carbonate is a white solid that decomposes with warming. It has many uses, for example, as a component in baking powder, fire extinguishers, and smelling salts.

(a) How many formula units are in 41.6 g of ammonium carbonate?

(b) How many O atoms are in this sample?

Plan (a) We know the mass of the compound (41.6 g) and need to find the number of formula units. As in Sample Problem 3.4, we find the amount (mol) and then multiply by Avogadro's number to find the number of formula units. (The *road map* shows the steps for part (a).)

(b) To find the number of O atoms, we multiply the number of formula units by the number of O atoms in one formula unit.

Road Map

Mass (g) of $(NH_4)_2CO_3$

↓ Divide by $\mathcal{M}$ (g/mol).

Amount (mol) of $(NH_4)_2CO_3$

↓ Multiply by 6.022×10^{23} formula units/mol.

Number of formula units of $(NH_4)_2CO_3$

Solution (a) The formula is $(NH_4)_2CO_3$ (see Table 2.5). Calculate the molar mass:

$$\mathcal{M} = (2 \times \mathcal{M}\text{ of N}) + (8 \times \mathcal{M}\text{ of H}) + (1 \times \mathcal{M}\text{ of C}) + (3 \times \mathcal{M}\text{ of O})$$
$$= (2 \times 14.01\text{ g/mol N}) + (8 \times 1.008\text{ g/mol H}) + 12.01\text{ g/mol C}$$
$$+ (3 \times 16.00\text{ g/mol O})$$
$$= 96.094\text{ g/mol }(NH_4)_2\,CO_3$$

Convert from mass (g) to amount (mol):

$$\text{Amount (mol) of }(NH_4)_2\,CO_3 = n_{(NH_4)_2CO_3}$$
$$= \frac{m_{(NH_4)_2CO_3}}{\mathcal{M}_{(NH_4)_2CO_3}}$$
$$= 41.6\ \cancel{\text{g }(NH_4)_2\,CO_3} \times \frac{1\text{ mol }(NH_4)_2\,CO_3}{96.094\ \cancel{\text{g }(NH_4)_2\,CO_3}}$$
$$= 0.4329\text{ mol }(NH_4)_2\,CO_3$$

Convert from amount (mol) to formula units:

Formula units of $(NH_4)_2\,CO_3$

$$= n_{(NH_4)_2CO_3} \times N_A$$
$$= 0.4329\ \cancel{\text{mol }(NH_4)_2\,CO_3} \times \frac{6.022 \times 10^{23}\text{ formula units }(NH_4)_2\,CO_3}{1\ \cancel{\text{mol }(NH_4)_2\,CO_3}}$$
$$= 2.61 \times 10^{23}\text{ formula units }(NH_4)_2\,CO_3$$

(b) Find the number of O atoms:

$$\text{No. of O atoms} = 2.61 \times 10^{23}\ \cancel{\text{formula units }(NH_4)_2\,CO_3} \times \frac{3\text{ O atoms}}{1\ \cancel{\text{formula unit }(NH_4)_2\,CO_3}}$$
$$= 7.83 \times 10^{23}\text{ O atoms}$$

Check In (a), the units are correct. Since the mass is less than half the molar mass ($\frac{\sim 42}{96} < 0.5$), the number of formula units should be less than half Avogadro's number ($\frac{\sim 2.6 \times 10^{23}}{6.0 \times 10^{23}} < 0.5$).

Comment A *common mistake* is to forget the subscript 2 outside the parentheses in $(NH_4)_2CO_3$, which would give a much lower molar mass.

Follow-Up Problem 3.5 Tetraphosphorus decoxide reacts with water to form phosphoric acid, a major industrial acid. In the laboratory, the oxide is a drying agent.

(a) What is the mass (g) of 4.65×10^{22} molecules of tetraphosphorus decoxide?

(b) How many P atoms are present in this sample?

The Importance of Mass Percent

For many purposes, it is important to know how much of an element is present in a given amount of compound. A biochemist may want the ionic composition of a mineral nutrient; an atmospheric chemist may be studying the amount of carbon in a fuel; a materials scientist may need the metalloid composition of a semiconductor. In this section, we find the composition of a compound in terms of mass percent and use it to find the mass of each element in the compound.

Determining Mass Percent from a Chemical Formula Each element contributes a fraction of a compound's mass, and this fraction multiplied by 100 gives the element's mass percent. Finding the mass percent is similar on the molecular and molar scales:

- *For a molecule (or formula unit) of a compound*, use the molecular (or formula) mass and chemical formula to find the mass percent of any element X in the compound:

$$\text{Mass \% of element X} = \frac{\text{atoms of X in formula} \times \text{atomic mass of X (u)}}{\text{molecular (or formula) mass of compound (u)}} \times 100$$

- *For a mole of a compound*, use the molar mass and formula to find the mass percent of each element on a mole basis:

$$\text{Mass \% of element X} = \frac{\text{moles of X in formula} \times \text{molar mass of X (g/mol)}}{\text{mass (g) of 1 mol of compound}} \times 100 = \frac{n_{\text{X in formula}} \times \mathcal{M}_{\text{X}}}{\mathcal{M}_{\text{compound}}} \quad (3.6)$$

As always, the individual mass percents add up to 100% (within rounding). In Sample Problem 3.6, we determine the mass percent of each element in a compound.

Road Map

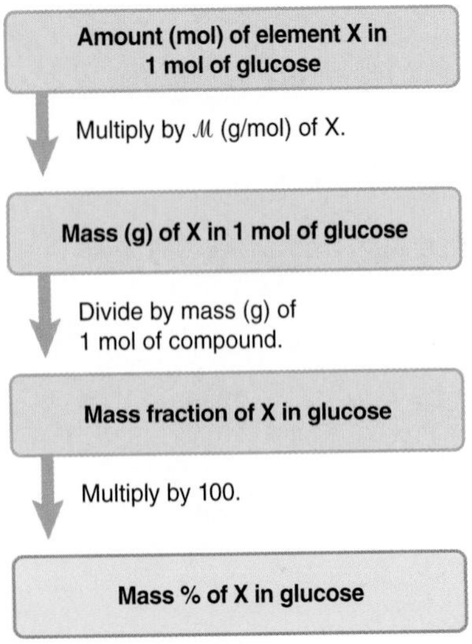

Sample Problem 3.6 Calculating the Mass Percent of Each Element in a Compound from the Formula

Problem In mammals, lactose (milk sugar) is metabolized to glucose ($C_6H_{12}O_6$), the key nutrient for generating chemical potential energy. What is the mass percent of each element in glucose?

Plan We know the relative amounts (mol) of the elements from the formula (6 C, 12 H, 6 O), and we have to find the mass % of each element. We multiply the amount of each element by its molar mass to find its mass. Dividing each mass by the mass of 1 mol of glucose gives the mass fraction of each element, and multiplying by 100 gives the mass %. The calculation steps for any element (X) are shown in the road map.

Solution Convert amount (mol) of C to mass (g):

We have 6 mol of C in 1 mol of glucose, so

$$\text{Mass (g) of C} = m = n \times \mathcal{M} = 6 \cancel{\text{mol C}} \times 12.01 \frac{\text{g}}{\cancel{\text{mol C}}} = 72.06 \text{ g C}$$

Calculate the mass of 1 mol of glucose ($C_6H_{12}O_6$):

$$\mathcal{M} = (6 \times \mathcal{M} \text{ of C}) + (12 \times \mathcal{M} \text{ of H}) + (6 \times \mathcal{M} \text{ of O})$$
$$= (6 \times 12.01 \text{ g/mol C}) + (12 \times 1.008 \text{ g/mol H}) + (6 \times 16.00 \text{ g/mol O})$$
$$= 180.16 \text{ g/mol } C_6H_{12}O_6$$

Find the mass fraction of C in glucose:

$$\text{Mass fraction of C} = \frac{\text{total mass of C}}{\text{mass of 1 mol glucose}} = \frac{72.06 \text{ g C}}{180.16 \text{ g glucose}} = 0.4000$$

Change to mass %:

$$\text{Mass \% of C} = \text{mass fraction of C} \times 100 = 0.4000 \times 100$$
$$= 40.00 \text{ mass \% C}$$

Combine the steps for each of the other elements in glucose:

$$\text{Mass \% of H} = \frac{\text{amount (mol) H} \times \mathcal{M} \text{ of H}}{\text{mass (g) of 1 mol glucose}} \times 100 = \frac{12 \cancel{\text{mol H}} \times \dfrac{1.008 \text{ g H}}{1 \cancel{\text{mol H}}}}{180.16 \text{ g glucose}} \times 100$$
$$= 6.714 \text{ mass \% H}$$

$$\text{Mass \% of O} = \frac{\text{amount (mol) O} \times \mathcal{M} \text{ of O}}{\text{mass (g) of 1 mol glucose}} \times 100 = \frac{6 \cancel{\text{mol O}} \times \dfrac{16.00 \text{ g O}}{1 \cancel{\text{mol O}}}}{180.16 \text{ g glucose}} \times 100$$
$$= 53.29 \text{ mass \% O}$$

Check The answers make sense. Even though there are equal numbers of moles of O and C in the compound, the mass % of O is greater than the mass % of C because the molar mass of O is greater than the molar mass of C. The mass % of H is small because the molar mass of H is small. The sum of the mass percents is 100.00%.

Comment From here on, you should be able to determine the molar mass of a compound, so this calculation will no longer be shown.

Follow-Up Problem 3.6 Agronomists base the effectiveness of fertilizers on their nitrogen content. Ammonium nitrate is a common fertilizer. Calculate the mass percent of N in ammonium nitrate.

Determining the Mass of an Element from Its Mass Percent Sample Problem 3.6 shows that *an element always constitutes the same fraction of the mass of a given compound.* We obtained this fraction by dividing the total mass of the element in a formula unit by the mass of 1 mol of the compound (see Equation 3.6). We can use this fraction directly to find the mass of the element in any mass of the compound:

$$\text{Mass of element} = \text{mass of compound} \times \frac{\text{mass of element in 1 mol of compound}}{\text{mass of 1 mol of compound}} \quad (3.7)$$

For example, to find the mass of oxygen in 15.5 g of nitrogen dioxide, we have

$$\text{Mass (g) of O} = 15.5 \text{ g } NO_2 \times \frac{2 \text{ mol} \times \mathcal{M} \text{ of O (g/mol)}}{\text{mass (g) of 1 mol } NO_2}$$
$$= 15.5 \cancel{\text{g } NO_2} \times \frac{32.00 \text{ g O}}{46.01 \cancel{\text{g } NO_2}} = 10.8 \text{ g O}$$

Sample Problem 3.7 Calculating the Mass of an Element in a Compound

Problem Use the information in Sample Problem 3.6 to determine the mass (g) of carbon in 16.55 g of glucose.

Plan To find the mass of C in the sample of glucose, we multiply the mass of the sample by the mass of 6 mol of C divided by the mass of 1 mol of glucose.

Solution Find the mass of C in a given mass of glucose:

$$\text{Mass (g) of C} = \text{mass (g) of glucose} \times \frac{6\ \text{mol C} \times \mathcal{M}\ \text{of C (g/mol)}}{\text{mass (g) of 1 mol glucose}}$$

$$= 16.55\ \cancel{\text{g glucose}} \times \frac{72.06\ \text{g C}}{180.16\ \cancel{\text{g glucose}}} = 6.620\ \text{g C}$$

Check Rounding shows that the answer is in the ballpark: 16 g times less than 0.5 parts by mass should be less than 8 g.

Follow-Up Problem 3.7 Use the information in Follow-Up Problem 3.6 to find the mass (g) of N in 35.8 kg of ammonium nitrate.

SUMMARY OF SECTION 3.1

- A mole of substance is the amount that contains Avogadro's number (N_A, 6.022×10^{23}) of chemical entities (atoms, molecules, or formula units).
- The mass (g) of 1 mol of an entity has the same numerical value as the mass (in atomic mass units, u) of the individual entity. Thus, the mole allows us to count entities by weighing them.
- Using the molar mass ($\mathcal{M}$, g/mol) of an element (or compound) and Avogadro's number as conversion factors, we can convert among amount (mol), mass (g), and number of entities.
- The mass fraction of element X in a compound is used to find the mass of X in a given amount of the compound.

3.2 Determining the Formula for an Unknown Compound

In Sample Problems 3.6 and 3.7, we used the formula for a compound to find the mass percent (or mass fraction) of each element in the compound *and* the mass of each element in any size of sample of the compound. In this section, we do the reverse: we use the masses of elements in a compound to find the formula. Then, we look briefly at the relationship between molecular formula and molecular structure.

Let us compare three common types of formulas using hydrogen peroxide as an example:

- The **empirical formula** is derived from mass analysis. It shows the *lowest* whole number of moles and, thus, the *relative* number of atoms of each element in the compound. For example, in hydrogen peroxide, there is 1 part by mass of hydrogen for every 16 parts by mass of oxygen. Because the atomic mass of hydrogen is 1.008 u and the atomic mass of oxygen is 16.00 u, there is one H atom for every O atom. Thus, the empirical formula is HO.

Recall the following from Section 2.8:

- The *molecular formula* shows the *actual* number of atoms of each element in a molecule: the molecular formula for hydrogen peroxide is H_2O_2, which is twice the empirical formula.
- The *structural formula* shows the relative *placement and connections of atoms* in the molecule: the structural formula for hydrogen peroxide is H—O—O—H.

Let us see how to determine empirical and molecular formulas.

Empirical Formula

A chemist studying an unknown compound goes through a three-step process to find the empirical formula:

1. Determine the mass (g) of each component element.
2. Convert each mass (g) to amount (mol), and write a preliminary formula.
3. Convert the amounts (mol) mathematically to whole-number (integer) subscripts.

To accomplish the conversion, follow these steps:

- Divide each subscript by the smallest subscript.
- If necessary, multiply through by the *smallest integer* that turns all subscripts into integers.

Sample Problem 3.8 demonstrates this process.

Sample Problem 3.8 Determining an Empirical Formula from Amounts of Elements

Problem A sample of an unknown compound contains 0.21 mol of zinc, 0.14 mol of phosphorus, and 0.56 mol of oxygen. What is the empirical formula?

Plan We are given the amount (mol) of each element as a fraction. We use these fractional amounts directly in a preliminary formula as subscripts of the element symbols. Then we convert the fractions to whole numbers.

Solution Use the fractions to write a preliminary formula, with the symbols Zn for zinc, P for phosphorus, and O for oxygen:

$$Zn_{0.21}P_{0.14}O_{0.56}$$

Convert the fractions to whole numbers:

1. Divide each subscript by the smallest subscript, which is 0.14:

$$Zn_{\frac{0.21}{0.14}}P_{\frac{0.14}{0.14}}O_{\frac{0.56}{0.14}} \longrightarrow Zn_{1.5}P_{1.0}O_{4.0}$$

2. Multiply through by the *smallest integer* that turns all subscripts into integers. We multiply by 2 to make 1.5 (the subscript for Zn) into an integer:

$$Zn_{(1.5\times2)}P_{(1.0\times2)}O_{(4.0\times2)} \longrightarrow Zn_{3.0}P_{2.0}O_{8.0} \quad \text{or} \quad Zn_3P_2O_8$$

Road Map

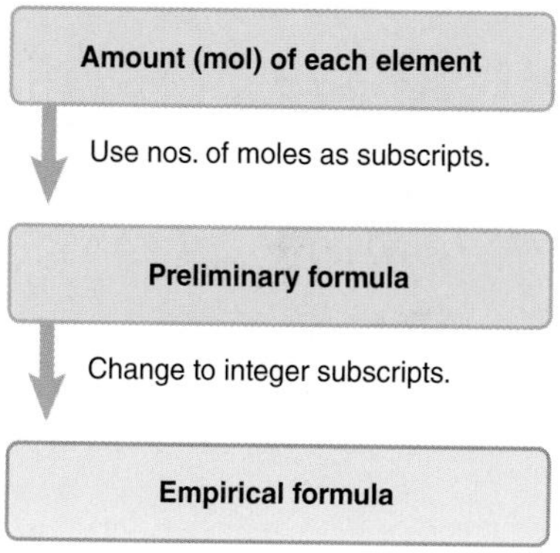

Check The integer subscripts must be the smallest integers with the same ratio as the original fractional numbers of moles: 3/2/8 is *the same ratio* as 0.21/0.14/0.56.

Comment A more conventional way to write this formula is $Zn_3(PO_4)_2$. The compound is zinc phosphate, formerly used widely as a dental cement.

Follow-Up Problem 3.8 A sample of a white solid contains 0.170 mol of boron and 0.255 mol of hydrogen. What is the empirical formula?

Sample Problems 3.9 to 3.11 show how other types of compositional data are used to determine chemical formulas.

Sample Problem 3.9 Determining an Empirical Formula from Masses of Elements

Problem Analysis of a sample of an ionic compound yields 2.82 g of Na, 4.35 g of Cl, and 7.83 g of O. What are the empirical formula and name of the compound?

Plan This problem is similar to Sample Problem 3.8, except that we are given element *masses* that we must convert into integer subscripts. We first divide each mass by the element's molar mass to find the amount (mol). Then we construct a preliminary formula and convert the amounts (mol) to integers.

Solution Find the amount (mol) of each element:

$$\text{Amount (mol) of Na} = n = \frac{m}{\mathcal{M}} = 2.82\ \cancel{\text{g Na}} \times \frac{1\ \text{mol Na}}{22.99\ \cancel{\text{g Na}}} = 0.123\ \text{mol Na}$$

$$\text{Amount (mol) of Cl} = n = \frac{m}{\mathcal{M}} = 4.35\ \cancel{\text{g Cl}} \times \frac{1\ \text{mol Cl}}{35.45\ \cancel{\text{g Cl}}} = 0.123\ \text{mol Cl}$$

$$\text{Amount (mol) of O} = n = \frac{m}{\mathcal{M}} = 7.83\ \cancel{\text{g O}} \times \frac{1\ \text{mol O}}{16.00\ \cancel{\text{g O}}} = 0.489\ \text{mol O}$$

Construct a preliminary formula:

$$Na_{0.123}Cl_{0.123}O_{0.489}$$

Convert to integer subscripts (divide all the subscripts by the smallest subscript):

$$Na_{\frac{0.123}{0.123}}Cl_{\frac{0.123}{0.123}}O_{\frac{0.489}{0.123}} \longrightarrow Na_{1.00}Cl_{1.00}O_{3.98} \approx Na_1Cl_1O_4 \text{ or } NaClO_4$$

The empirical formula is $NaClO_4$; the name of the compound is sodium perchlorate.

Check The numbers of moles seem correct because the masses of Na and Cl are slightly more than 0.1 of their molar masses. The mass of O is greatest and its molar mass is smallest, so it should have the greatest number of moles. The ratio of subscripts, 1/1/4, is the same as the ratio of moles, 0.123/0.123/0.489 (within rounding).

Follow-Up Problem 3.9 An unknown metal M reacts with sulfur to form a compound with the formula M_2S_3. If 3.12 g of M reacts with 2.88 g of S, what are the names of M and M_2S_3? [*Hint*: Determine the amount (mol) of S, and use the formula to find the amount (mol) of M.]

Molecular Formula

If we know the molar mass of a compound, we can use the empirical formula to obtain the molecular formula, which uses as subscripts the *actual* numbers of moles of each element in 1 mol of compound. For some compounds—such as water (H_2O), ammonia (NH_3), and methane (CH_4)—the empirical and molecular formulas are identical, but for many other compounds, the molecular formula is a *whole-number multiple* of the empirical formula. As you saw, hydrogen peroxide has the empirical formula HO. Dividing the molar mass of hydrogen peroxide (34.02 g/mol) by the empirical formula mass of HO (17.01 g/mol) gives the whole-number multiple:

$$\text{Whole-number multiple} = \frac{\text{molar mass (g/mol)}}{\text{empirical formula mass(g/mol)}} = \frac{34.02\ \cancel{\text{g/mol}}}{17.01\ \cancel{\text{g/mol}}} = 2.000 = 2$$

Multiplying the empirical formula subscripts by 2 gives the molecular formula:

$$H_{(1\times2)}O_{(1\times2)} \text{ gives } H_2O_2$$

Instead of giving compositional data as the masses of the elements, analytical laboratories provide mass percents. To find the molecular formula, we first determine the empirical formula and then the whole-number multiple:

1. Assume 100.0 g of compound to express each mass percent directly as mass (g).
2. Convert each mass (g) to amount (mol).
3. Derive the empirical formula.
4. Divide the molar mass of the compound by the empirical formula mass to find the whole-number multiple.
5. Multiply each subscript in the empirical formula by the whole-number multiple.

Sample Problem 3.10 Determining a Molecular Formula from Elemental Analysis and Molar Mass

Problem During excessive physical activity, lactic acid ($\mathcal{M}$ = 90.08 g/mol) forms in muscle tissue and causes muscle soreness. Elemental analysis shows that this compound contains 40.0 mass % C, 6.71 mass % H, and 53.3 mass % O.

(a) Determine the empirical formula for lactic acid.

(b) Determine the molecular formula.

(a) First we need to determine the empirical formula.

Plan We know the mass percent of each element and must convert the mass percent to an integer subscript. The mass of the sample of lactic acid is not given, but the mass percents are the same for any sample of lactic acid. Therefore, we

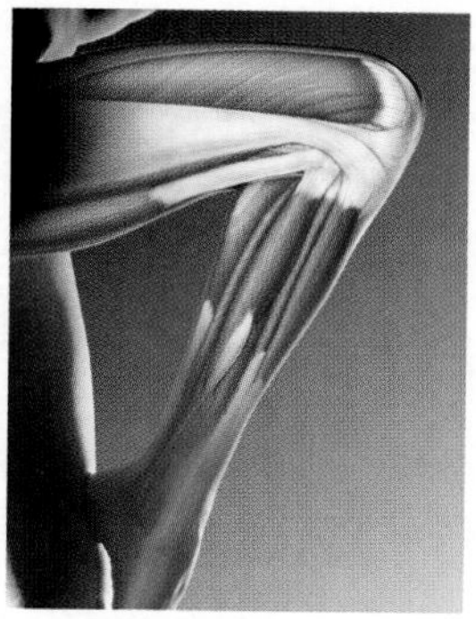

assume that there is 100.0 g of lactic acid and express each mass % as a number of grams. Then we construct the empirical formula, as in Sample Problem 3.9.

Solution Express mass % as mass (g) by assuming 100.0 g of lactic acid:

$$\text{Mass (g) of C} = \frac{40.0 \text{ parts C by mass}}{100 \text{ parts by mass}} \times 100.0 \text{ g} = 40.0 \text{ g C}$$

Similarly, we have 6.71 g of H and 53.3 g of O.

Convert from mass (g) of each element to amount (mol):

$$\text{Amount (mol) of C} = \text{mass of C} \times \frac{1}{\mathcal{M} \text{ of C}} = 40.0 \text{ }\cancel{\text{g C}} \times \frac{1 \text{ mol C}}{12.01 \text{ }\cancel{\text{g C}}} = 3.33 \text{ mol C}$$

Similarly, we have 6.66 mol of H and 3.33 mol of O.

Construct the preliminary formula:

$$C_{3.33}H_{6.66}O_{3.33}$$

Convert to integer subscripts:

$$C_{\frac{3.33}{3.33}}H_{\frac{6.66}{3.33}}O_{\frac{3.33}{3.33}} \longrightarrow C_{1.00}H_{2.00}O_{1.00} = C_1H_2O_1$$

The empirical formula is CH_2O.

Check The numbers of moles seem correct: the masses of C and O are each slightly more than 3 times their molar masses (for example, for C, $\frac{40 \text{ g}}{12 \text{ g/mol}} > 3$ mol), and the mass of H is over 6 times its molar mass of 1.

(b) Now we need to determine the molecular formula.

Plan The molecular formula subscripts are whole-number multiples of the empirical formula subscripts. To find this whole-number multiple, we divide the given molar mass (90.08 g/mol) by the empirical formula mass, which we find from the sum of the elements' molar masses. Then we multiply each subscript in the empirical formula by the multiple.

Solution The empirical formula mass is 30.03 g/mol. Find the whole-number multiple:

$$\text{Whole-number multiple} = \frac{\mathcal{M} \text{ of lactic acid}}{\mathcal{M} \text{ of empirical formula}} = \frac{90.08 \text{ }\cancel{\text{g/mol}}}{30.03 \text{ }\cancel{\text{g/mol}}} = 3.000 = 3$$

Determine the molecular formula:

$$C_{(1\times3)}H_{(2\times3)}O_{(1\times3)} = C_3H_6O_3$$

Check The calculated molecular formula has the same ratio of moles of elements (3/6/3) as the empirical formula (1/2/1) and corresponds to the given molar mass:

$$\begin{aligned}\mathcal{M} \text{ of lactic acid} &= (3 \times \mathcal{M} \text{ of C}) + (6 \times \mathcal{M} \text{ of H}) + (3 \times \mathcal{M} \text{ of O})\\ &= (3 \times 12.01 \text{ g/mol}) + (6 \times 1.008 \text{ g/mol}) + (3 \times 16.00 \text{ g/mol})\\ &= 90.08 \text{ g/mol}\end{aligned}$$

Follow-Up Problem 3.10 One of the most widespread environmental carcinogens (cancer-causing agents) is benzo[*a*]pyrene ($\mathcal{M}$ = 252.30 g/mol). It is found in coal dust, cigarette smoke, and even charcoal-grilled meat. Analysis of this hydrocarbon shows 95.21 mass % C and 4.79 mass % H. What is the molecular formula for benzo[*a*]pyrene?

Combustion Analysis of Organic Compounds Still another type of compositional data is obtained through **combustion analysis**, which is used to measure the amounts of carbon and hydrogen in a combustible organic compound. The unknown compound is burned in an excess of pure O_2, and the H_2O and CO_2 that form are absorbed in separate containers (Figure 3.4). By weighing the absorbers before and after combustion, we can find the masses of CO_2 and H_2O and then use these masses to find the masses of C and H in the compound; from our results, we can find the

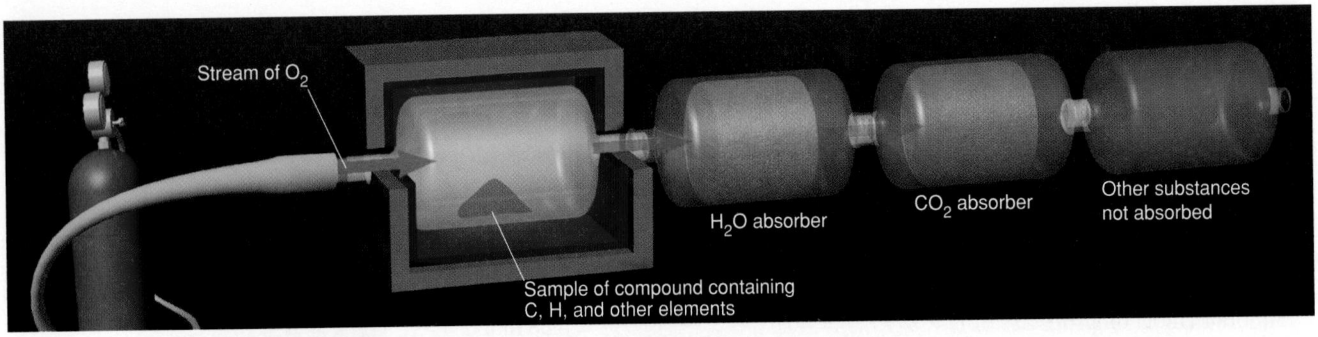

FIGURE 3.4 Combustion apparatus for determining formulas for organic compounds. A sample of an organic compound is burned in a stream of O_2. The resulting H_2O is absorbed by $Mg(ClO_4)_2$, and the CO_2 is absorbed by NaOH on asbestos.

empirical formula. Many organic compounds also contain oxygen, nitrogen, or a halogen. As long as the third element does not interfere with the absorption of H_2O and CO_2, we can calculate its mass by subtracting the masses of C and H from the original mass of the compound.

Sample Problem 3.11 Determining a Molecular Formula from Combustion Analysis

Problem Vitamin C ($\mathcal{M}$ = 176.12 g/mol) is a compound of C, H, and O that is found in many natural sources, especially citrus fruits. When a 1.000 g sample of vitamin C is burned in a combustion apparatus, the following data are obtained:

$$\text{Mass of } CO_2 \text{ absorber after combustion} = 85.35 \text{ g}$$
$$\text{Mass of } CO_2 \text{ absorber before combustion} = 83.85 \text{ g}$$
$$\text{Mass of } H_2O \text{ absorber after combustion} = 37.96 \text{ g}$$
$$\text{Mass of } H_2O \text{ absorber before combustion} = 37.55 \text{ g}$$

What is the molecular formula for vitamin C?

Plan We find the masses of CO_2 and H_2O by subtracting the masses of the absorbers before and after the combustion. From the mass of CO_2, we use Equation 3.7 to find the mass of C. Similarly, we find the mass of H from the mass of H_2O. The mass of vitamin C (1.000 g) minus the sum of the masses of C and H gives the mass of O, the third element present. Then we proceed as in Sample Problem 3.10: we calculate the amount (mol) of each element using its molar mass, construct the empirical formula, determine the whole-number multiple from the given molar mass, and construct the molecular formula.

Solution Find the masses of the combustion products:

$$\text{Mass (g) of } CO_2 = \text{mass of } CO_2 \text{ absorber after} - \text{mass before}$$
$$= 85.35 \text{ g} - 83.85 \text{ g} = 1.50 \text{ g } CO_2$$
$$\text{Mass (g) of } H_2O = \text{mass of } H_2O \text{ absorber after} - \text{mass before}$$
$$= 37.96 \text{ g} - 37.55 \text{ g} = 0.41 \text{ g } H_2O$$

Calculate the masses (g) of C and H using Equation 3.7:

$$\text{Mass of element} = \text{mass of compound} \times \frac{\text{mass of element in 1 mol of compound}}{\text{mass of 1 mol of compound}}$$

$$\text{Mass (g) of C} = \text{mass of } CO_2 \times \frac{1 \text{ mol C} \times \mathcal{M} \text{ of C}}{\text{mass of 1 mol } CO_2} = 1.50 \cancel{\text{g } CO_2} \times \frac{12.01 \text{ g C}}{44.01 \cancel{\text{g } CO_2}}$$
$$= 0.409 \text{ g C}$$

$$\text{Mass (g) of H} = \text{mass of } H_2O \times \frac{2 \text{ mol H} \times \mathcal{M} \text{ of H}}{\text{mass of 1 mol } H_2O} = 0.41 \cancel{\text{g } H_2O} \times \frac{2.016 \text{ g H}}{18.02 \cancel{\text{g } H_2O}}$$
$$= 0.046 \text{ g H}$$

Calculate the mass (g) of O:

$$\text{Mass (g) of O} = \text{mass of vitamin C sample} - (\text{mass of C} + \text{mass of H})$$
$$= 1.000\text{ g} - (0.409\text{ g} + 0.046\text{ g}) = 0.545\text{ g O}$$

To find the amounts (mol) of the elements, we divide the mass (g) of each element by its molar mass. We get 0.0341 mol of C, 0.046 mol of H, and 0.0341 mol of O.

Construct the preliminary formula:

$$C_{0.0341}H_{0.046}O_{0.0341}$$

To determine the empirical formula, we divide through by the smallest subscript:

$$C_{\frac{0.0341}{0.0341}}H_{\frac{0.046}{0.0341}}O_{\frac{0.0341}{0.0341}} = C_{1.00}H_{1.3}O_{1.00}$$

We find that 3 is the smallest integer that makes all the subscripts into integers:

$$C_{(1.00\times3)}H_{(1.3\times3)}O_{(1.00\times3)} = C_{3.00}H_{3.9}O_{3.00} \approx C_3H_4O_3$$

Determine the molecular formula:

$$\text{Whole-number multiple} = \frac{\mathcal{M}\text{ of vitamin C}}{\mathcal{M}\text{ of empirical formula}} = \frac{176.12\ \cancel{\text{g/mol}}}{88.06\ \cancel{\text{g/mol}}} = 2.000 = 2$$

$$C_{(3\times2)}H_{(4\times2)}O_{(3\times2)} = C_6H_8O_6$$

Check The masses of the elements seem correct. Carbon makes up slightly more than 0.25 of the mass of CO_2 ($\frac{12\text{ g}}{44\text{ g}} > 0.25$), as do the masses in the problem ($\frac{0.409\text{ g}}{1.50\text{ g}} > 0.25$). Hydrogen makes up slightly more than 0.10 of the mass of H_2O ($\frac{2\text{ g}}{18\text{ g}} > 0.10$), as do the masses in the problem ($\frac{0.046\text{ g}}{0.41\text{ g}} > 0.10$). The molecular formula has the same ratio of subscripts (6/8/6) as the empirical formula (3/4/3) and the preliminary formula (0.0341/0.046/0.0341), and it gives the known molar mass:

$$(6 \times \mathcal{M}\text{ of C}) + (8 \times \mathcal{M}\text{ of H}) + (6 \times \mathcal{M}\text{ of O}) = \mathcal{M}\text{ of vitamin C}$$
$$(6 \times 12.01\text{ g/mol}) + (8 \times 1.008\text{ g/mol}) + (6 \times 16.00\text{ g/mol}) = 176.12\text{ g/mol}$$

Comment The subscript we calculated for H was 3.9, which we rounded to 4. If we had strung the calculation steps together, however, we would have obtained 4.0:

$$\text{Subscript of H} = 0.41\text{ g }H_2O \times \frac{2.016\text{ g H}}{18.02\text{ g }H_2O} \times \frac{1\text{ mol H}}{1.008\text{ g H}} \times \frac{1}{0.0341\text{ mol}} \times 3 = 4.0$$

Follow-Up Problem 3.11 A dry-cleaning solvent ($\mathcal{M}$ = 146.99 g/mol) that contains C, H, and Cl is suspected to be a cancer-causing agent. When a 0.250 g sample was studied by combustion analysis, 0.451 g of CO_2 and 0.0617 g of H_2O were formed. Find the molecular formula for the solvent.

Chemical Formulas, Molecular Structures, and Isomers

A formula represents a real, three-dimensional object. The structural formula makes this point, with its relative placement of atoms, but do empirical and molecular formulas contain structural information?

Different Compounds with the Same Empirical Formula The empirical formula tells us nothing about molecular structure because it is based solely on mass analysis. In fact, different compounds can have the *same* empirical formula. NO_2 and N_2O_4 are inorganic examples, and there are numerous organic examples. For instance, many compounds have the empirical formula CH_2 (the general formula is C_nH_{2n}, where n is an integer greater than or equal to 2): ethylene (C_2H_4) and propylene (C_3H_6) are the starting materials for two common plastics. Table 3.2 shows some biological compounds with the same empirical formula.

TABLE 3.2 Some Compounds with Empirical Formula CH_2O (Composition by Mass: 40.0% C, 6.71% H, 53.3% O)

Name	Molecular Formula	Whole-Number Multiple	$\mathcal{M}$ (g/mol)	Use or Function
Methanal or formaldehyde	CH_2O	1	30.03	Disinfectant; biological preservative
Ethanoic or acetic acid	$C_2H_4O_2$	2	60.05	Acetate polymers; vinegar (5% solution)
2-hydroxypropanoic or lactic acid	$C_3H_6O_3$	3	90.08	Causes milk to sour; forms in muscles during exercise
2,3,4-trihydroxybutanal or erythrose	$C_4H_8O_4$	4	120.10	Forms during sugar metabolism
Ribose	$C_5H_{10}O_5$	5	150.13	Component of many nucleic acids and vitamin B_2
Glucose	$C_6H_{12}O_6$	6	180.16	Major nutrient for energy in cells

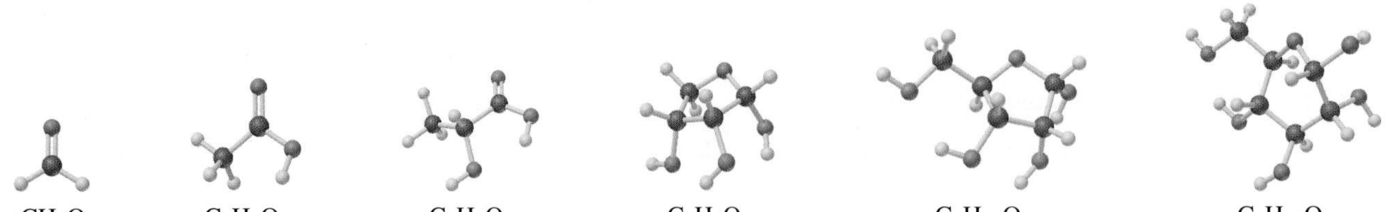

CH_2O $C_2H_4O_2$ $C_3H_6O_3$ $C_4H_8O_4$ $C_5H_{10}O_5$ $C_6H_{12}O_6$

Isomers: Different Compounds with the Same Molecular Formula A molecular formula also tells us nothing about structure. Different compounds can have the *same* molecular formula because their atoms can bond in different arrangements to give more than one *structural formula*. **Isomers** are *compounds with the same molecular formula, and thus the same molar mass, but with different properties. Constitutional*, or *structural, isomers* occur when the atoms link together in different arrangements. Table 3.3 shows two pairs of examples. The left pair, butane and 2-methylpropane, share the molecular formula C_4H_{10}. One has a four-C chain and the other has a one-C branch off a three-C chain. Both are small alkanes, so their properties are similar, but not identical. The two compounds with the molecular formula C_2H_6O have very different properties; indeed, they are different classes of organic compound: one is an alcohol and the other is an ether. We will study these compounds in more detail in Chapter 20.

TABLE 3.3 Two Pairs of Constitutional Isomers

	C_4H_{10}		C_2H_6O	
Property	Butane	2-Methylpropane	Ethanol	Methoxymethane or Dimethyl Ether
$\mathcal{M}$ (g/mol)	58.12	58.12	46.07	46.07
Boiling point	−0.5°C	−11.6°C	78.5°C	−25°C
Density (at 20°C)	0.579 g/mL (gas)	0.549 g/mL (gas)	0.789 g/mL (liquid)	0.001 95 g/mL (gas)
Structural formula	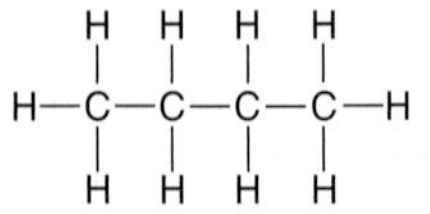	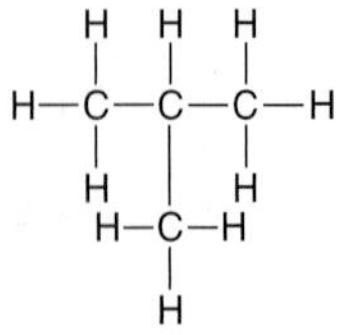	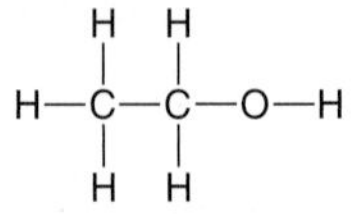	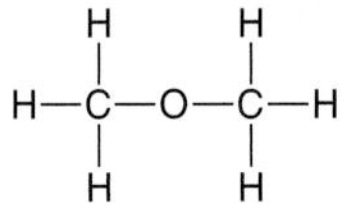
Space-filling model	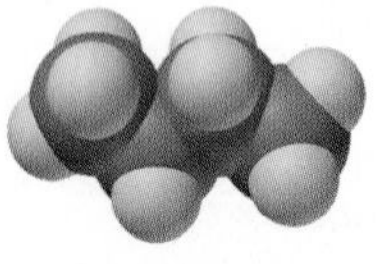		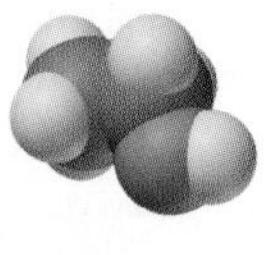	

As the number and kinds of atoms increase, the number of constitutional isomers—that is, the number of structural formulas that can be written for a given molecular formula—also increases: C_2H_6O has two structural formulas (Table 3.3), C_3H_8O has three, and $C_4H_{10}O$ has seven. Imagine how many structural formulas there are for $C_{16}H_{19}N_3O_4S$! Of all the possible isomers, only one is the antibiotic ampicillin (Figure 3.5). We will discuss these and other types of isomerism in more detail in Chapter 20.

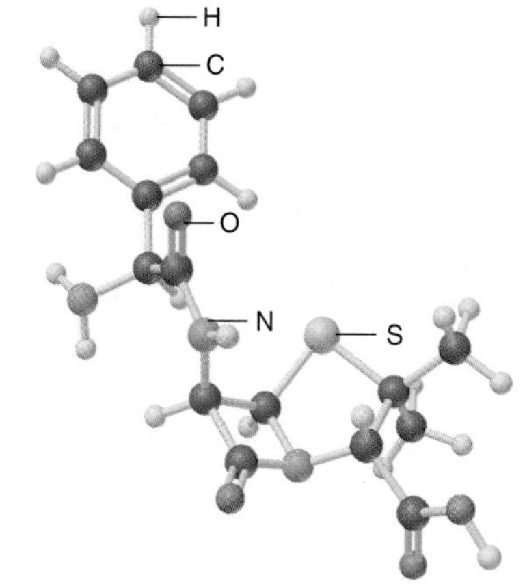

FIGURE 3.5 The antibiotic ampicillin

SUMMARY OF SECTION 3.2

- From the masses of the elements in a compound, the relative numbers of moles can be found, giving the empirical formula.
- If the molar mass of a compound is known, the molecular formula, which gives the actual numbers of moles of the elements, can also be determined.
- Combustion analysis provides data about the masses of carbon and hydrogen in an organic compound. These masses can be used to obtain the formula for the compound.
- Atoms can bond in different arrangements (structural formulas). Two or more compounds with the same molecular formula are constitutional isomers.

3.3 Writing and Balancing Chemical Equations

Thinking in terms of amounts, rather than masses, allows us to view reactions as large populations of interacting particles rather than grams of material. For example, for the formation of HF from H_2 and F_2, if we take the masses of the substances, we find the following:

Macroscopic level (g): 2.016 g of H_2 and 38.00 g of F_2 react to form 40.02 g of HF.

This information tells us little except that mass is conserved. However, if we convert these masses (g) to amounts (mol), we get more information:

Macroscopic level (mol): 1 mol of H_2 and 1 mol of F_2 react to form 2 mol of HF.

This information reveals that an enormous number of H_2 molecules react with just as many F_2 molecules to form twice as many HF molecules. Dividing by Avogadro's number shows the reaction between individual molecules:

Molecular level: 1 molecule of H_2 and 1 molecule of F_2 react to form 2 molecules of HF.

Thus, *the macroscopic (molar) change corresponds to the submicroscopic (molecular) change* (Figure 3.6). This information forms the essence of a **chemical equation**, *a statement that uses formulas to express the identities and quantities of substances in a chemical or physical change.*

Steps for Balancing an Equation To present a chemical change quantitatively, the equation must be *balanced: the same number of each type of atom must appear on both sides.* As an example, here is a description of a chemical change that occurs

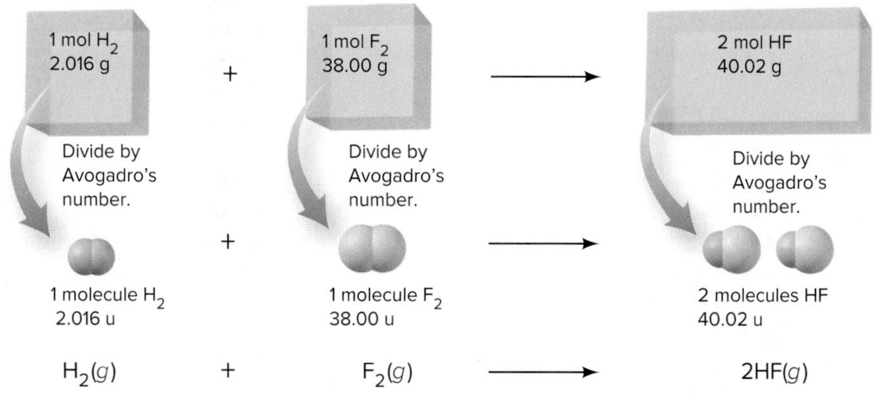

FIGURE 3.6 The formation of HF on the macroscopic and molecular levels

in many fireworks and in a common lecture demonstration: a magnesium strip burns in oxygen gas to yield powdery magnesium oxide. (Light and heat are also produced, but we are concerned only with substances here.) Converting this description into a balanced equation involves the following steps:

1. *Translating the statement.* We first translate the chemical statement into a "skeleton" equation: the substances present *before* the change, called **reactants**, are placed to the left of a yield arrow, which points to the substances produced *during* the change, called **products**:

reactants	yield	product
$\_\_Mg + \_\_O_2$	$\longrightarrow$	$\_\_MgO$
magnesium and oxygen	yield	magnesium oxide

   At the beginning of the balancing process, we put a blank *in front of* each formula to remind us that we have to account for its atoms.

2. *Balancing the atoms.* By shifting our attention back and forth, we *match the numbers of each type of atom on the left and the right of the yield arrow.* In each blank, we place a **balancing (stoichiometric) coefficient**, a numerical multiplier of *all the atoms* in the formula that follows it. In general, balancing is easiest when we do the following:
   - Start with the most complex substance, the one with the largest number of different types of atoms.
   - End with the least complex substance, such as an element by itself.

   In this example, MgO is the most complex, so we place a coefficient 1 in that blank:

$$\_\_Mg + \_\_O_2 \longrightarrow \underline{1}MgO$$

   To balance the Mg in MgO, we place a 1 in front of Mg on the left:

$$\underline{1}Mg + \_\_O_2 \longrightarrow \underline{1}MgO$$

   The O atom in MgO must be balanced by one O atom on the left. One-half an O_2 molecule provides one O atom:

$$\underline{1}Mg + \underline{\tfrac{1}{2}}O_2 \longrightarrow \underline{1}MgO$$

   In terms of numbers of each type of atom, the equation is balanced.

3. *Adjusting the coefficients.* There are several conventions about the final coefficients:
   - In most cases, *the smallest whole-number coefficients are preferred.* In this example, one-half of an O_2 molecule cannot exist, so we multiply the equation by 2:

$$2Mg + 1O_2 \longrightarrow 2MgO$$

   - We used the coefficient 1 to remind us to balance each substance. However, a coefficient of 1 is implied by the presence of the formula, so we do not write it:

$$2Mg + O_2 \longrightarrow 2MgO$$

   (This convention is similar to not writing a subscript 1 in a formula.)

4. *Checking.* After balancing and adjusting the coefficients, always check that the equation is balanced:

$$\text{reactants (2 Mg, 2 O)} \longrightarrow \text{products (2 Mg, 2 O)}$$

g	for gas
l	for liquid
s	for solid
aq	for aqueous solution

5. *Specifying the states of matter.* The final equation also indicates the physical state of each substance or whether it is dissolved in water. The abbreviations used for the states are shown in the margin. From the original statement, we know that the Mg "strip" is solid, O_2 is a gas, and "powdery" MgO is also solid. The balanced equation, therefore, is

$$2Mg(s) + O_2(g) \longrightarrow 2MgO(s)$$

As you saw in Figure 3.6, *balancing coefficients refers to both individual chemical entities and moles of entities*. Thus,

2 atoms of Mg and 1 molecule of O_2 yield 2 formula units of MgO.
2 moles of Mg and 1 mole of O_2 yield 2 moles of MgO.

Figure 3.7 depicts this reaction on three levels:

- *Macroscopic level (photos)*, as it appears in the laboratory
- *Molecular level (blow-up circles)*, as chemists imagine it (with darker-coloured atoms representing the stoichiometry)
- *Symbolic level*, in the form of the balanced chemical equation

Keep in mind several key points about the balancing process:

- A coefficient operates on *all* the atoms in the formula that follows it:

  2MgO *means* $2 \times (\text{MgO})$, *or* 2 Mg atoms + 2 O atoms
  $2\text{Ca(NO}_3)_2$ *means* $2 \times [\text{Ca(NO}_3)_2]$, *or* 2 Ca atoms + 4 N atoms + 12 O atoms

- Chemical formulas *cannot* be altered. In step 2 of the example, we *cannot* balance the O atoms by changing MgO to MgO_2 because MgO_2 is a different compound.
- Other reactants or products *cannot* be added. Thus, we *cannot* balance the O atoms by changing the reactant from O_2 molecules to O atoms or by adding an O atom to the products. The description of the reaction mentions oxygen gas, which consists of O_2 molecules, *not* separate O atoms.
- A balanced equation remains balanced only if you multiply all the coefficients by the same number. For example,

$$4\text{Mg}(s) + 2\text{O}_2(g) \longrightarrow 4\text{MgO}(s)$$

  is also balanced because the coefficients have just been multiplied by 2. However, *by convention*, we use the *smallest* whole-number coefficients to balance an equation.

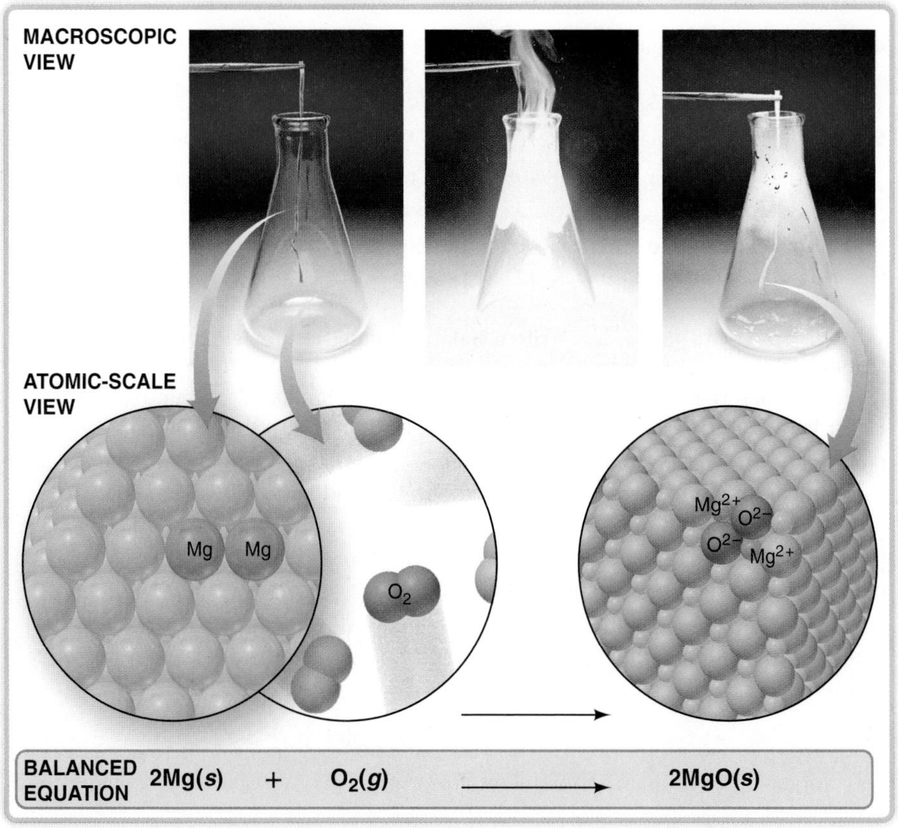

FIGURE 3.7 A three-level view of the reaction between magnesium and oxygen

Sample Problem 3.12 Balancing Chemical Equations

Problem Within the cylinders of a car's engine, the hydrocarbon octane (C_8H_{18}), one of many components of gasoline, mixes with oxygen from the air and burns to form carbon dioxide and water vapour. Write a balanced equation for this reaction.

Solution **1.** *Translate* the statement into a skeleton equation (with blanks for the coefficients). Octane and oxygen are the reactants; "oxygen from the air" implies molecular oxygen, O_2. Carbon dioxide and water vapour are the products:

$$\_\_C_8H_{18} + \_\_O_2 \longrightarrow \_\_CO_2 + \_\_H_2O$$

2. *Balance the atoms.* Start with the most complex substance, C_8H_{18}, and balance O_2 last:

$$\underline{1}\ C_8H_{18} + \_\_O_2 \rightarrow \_\_CO_2 + \_\_H_2O$$

The C atoms in C_8H_{18} end up in CO_2. Each CO_2 contains one C atom, so 8 molecules of CO_2 are needed to balance the 8 C atoms in each C_8H_{18}:

$$\underline{1}\ C_8H_{18} + \_\_O_2 \rightarrow \underline{8}\ CO_2 + \_\_H_2O$$

The H atoms in C_8H_{18} end up in H_2O. The 18 H atoms in C_8H_{18} require the coefficient 9 in front of H_2O:

$$\underline{1}\ C_8H_{18} + \_\_O_2 \rightarrow \underline{8}\ CO_2 + \underline{9}\ H_2O$$

There are 25 atoms of O on the right (16 in $8CO_2$ plus 9 in $9H_2O$), so we place the coefficient $\frac{25}{2}$ in front of O_2:

$$\underline{1}\ C_8H_{18} + \frac{25}{2}O_2 \rightarrow \underline{8}\ CO_2 + \underline{9}\ H_2O$$

3. *Adjust the coefficients.* Multiply the equation by 2 to obtain whole numbers:

$$2C_8H_{18} + 25O_2 \rightarrow 16CO_2 + 18H_2O$$

4. *Check* that the equation is balanced:

$$\text{Reactants (16 C, 36 H, 50 O)} \longrightarrow \text{Products (16 C, 36 H, 50 O)}$$

5. *Specify* states of matter. C_8H_{18} is liquid; O_2, CO_2, and H_2O vapour are gases:

$$2C_8H_{18}(l) + 25O_2(g) \longrightarrow 16CO_2(g) + 18H_2O(g)$$

Comment This is an example of a combustion reaction. *Any* compound containing C and H that burns in an excess of air produces CO_2 and H_2O.

Follow-Up Problem 3.12 Write a balanced equation for each of the following reactions:

(a) A characteristic reaction of group 1 elements: chunks of sodium react violently with water to form hydrogen gas and sodium hydroxide solution.

(b) The destruction of marble statuary by acid rain: aqueous nitric acid reacts with calcium carbonate to form carbon dioxide, water, and aqueous calcium nitrate.

(c) Halogen compounds exchanging bonding partners: phosphorus trifluoride is prepared by the reaction of phosphorus trichloride and hydrogen fluoride; hydrogen chloride is the other product. The reaction involves gases only.

(d) Explosive decomposition of dynamite: liquid nitroglycerine ($C_3H_5N_3O_9$) explodes to produce a mixture of gases—carbon dioxide, water vapour, nitrogen, and oxygen.

Visualizing a Reaction with a Molecular Scene A great way to focus on the rearrangement of atoms from reactants to products is to visualize an equation as a molecular scene. Here is a representation of the combustion of octane we just balanced:

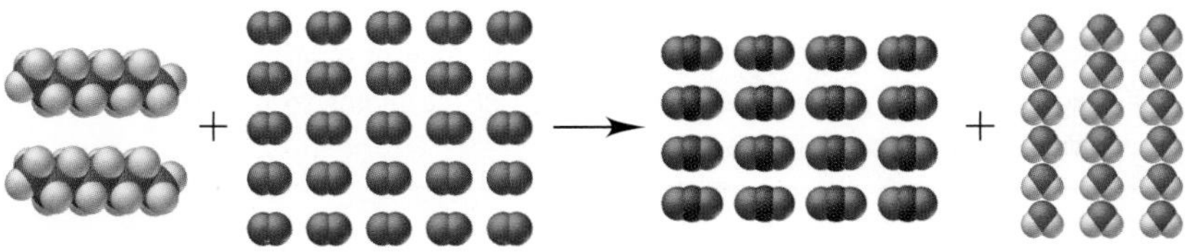

$2C_8H_{18}(l)$ + $25O_2(g)$ $\longrightarrow$ $16CO_2(g)$ + $18H_2O(g)$

Now let us work through a sample problem to do the reverse—derive a balanced equation from a molecular scene.

Sample Problem 3.13 Balancing an Equation from a Molecular Scene

Problem The following molecular scenes depict an important reaction in nitrogen chemistry (nitrogen is blue; oxygen is red):

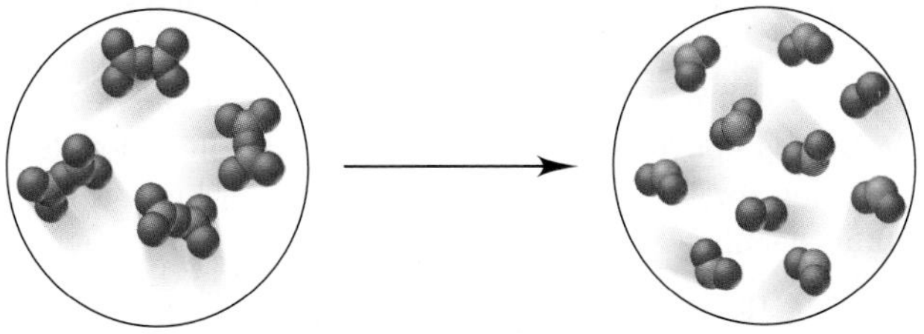

Write a balanced equation for this reaction.

Plan To write a balanced equation, we first have to determine the formulas for the molecules and obtain coefficients by counting the number of each molecule. Then, we arrange this information in the correct equation format, using the smallest whole-number coefficients and including the states of matter.

Solution The reactant circle shows only one type of molecule. It has two N and five O atoms, so the formula is N_2O_5; there are four of these molecules. The product circle shows two different molecules, one with one N and two O atoms, and the other with two O atoms; there are eight NO_2 and two O_2. Thus, we have the following equation:

$$4N_2O_5 \longrightarrow 8NO_2 + 2O_2$$

Writing the balanced equation with the smallest whole-number coefficients and all the substances as gases, we have

$$2N_2O_5(g) \longrightarrow 4NO_2(g) + O_2(g)$$

Check The equation is balanced: reactant (4 N, 10 O) $\longrightarrow$ products (4 N, 8 + 2 = 10 O)

Follow-Up Problem 3.13 Write a balanced equation for the important atmospheric reaction depicted below (carbon is *grey*; oxygen is *red*):

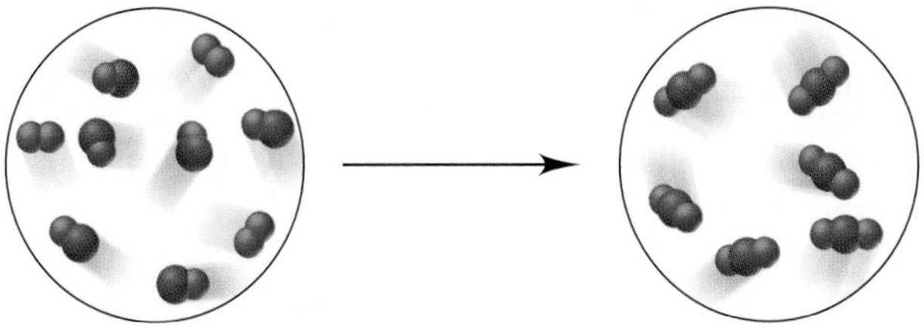

SUMMARY OF SECTION 3.3

- A chemical equation has reactant formulas on the left of a yield arrow and product formulas on the right.
- A balanced equation has the same number of each type of atom on both sides.
- Balancing coefficients are integer multipliers for *all* the atoms in a formula and apply to the individual entities or to moles of entities.

3.4 Calculating Quantities of Reactant and Product

A balanced equation is essential for all calculations involving chemical change: *if you know the amount (mol) of one substance, the balanced equation tells you the amount (mol) of the other substances.*

Stoichiometrically Equivalent Molar Ratios from a Balanced Equation

In a balanced equation, *the amounts (mol) of substances are stoichiometrically equivalent to each other,* which means that a specific amount of one substance is formed from, produces, or reacts with a specific amount of the other. The quantitative relationships are expressed as *stoichiometrically equivalent molar ratios,* and we use these ratios as conversion factors to calculate the amounts. For example, consider the equation for the combustion of propane, a hydrocarbon fuel used in cooking and water heating:

$$C_3H_8(g) + 5O_2(g) \longrightarrow 3CO_2(g) + 4H_2O(g)$$

If we view the reaction quantitatively in terms of C_3H_8, we see that

1 mol of C_3H_8 reacts with 5 mol of O_2
1 mol of C_3H_8 produces 3 mol of CO_2
1 mol of C_3H_8 produces 4 mol of H_2O

Therefore, in this reaction,

1 mol of C_3H_8 is stoichiometrically equivalent to 5 mol of O_2
1 mol of C_3H_8 is stoichiometrically equivalent to 3 mol of CO_2
1 mol of C_3H_8 is stoichiometrically equivalent to 4 mol of H_2O

We chose to look at C_3H_8, but any two of the substances are stoichiometrically equivalent to each other. Thus,

3 mol of CO_2 is stoichiometrically equivalent to 4 mol of H_2O
5 mol of O_2 is stoichiometrically equivalent to 3 mol of CO_2

and so on. A balanced equation contains a wealth of quantitative information relating individual chemical entities, amounts (mol) of substances, and masses of substances, as shown in Table 3.4.

Consider a typical problem that shows how stoichiometric equivalence is used to create conversion factors: in the combustion of propane, what amount of O_2 is consumed when 10.0 mol of H_2O is produced? To solve this problem, we have to find the molar ratio between O_2 and H_2O. From the balanced equation, we see that, for every 5 mol of O_2 consumed, 4 mol of H_2O is formed:

5 mol of O_2 is stoichiometrically equivalent to 4 mol of H_2O

As with any equivalent quantities, we can construct two conversion factors, depending on the quantity we want to find:

$$\frac{5 \text{ mol } O_2}{4 \text{ mol } H_2O} \text{ or } \frac{4 \text{ mol } H_2O}{5 \text{ mol } O_2}$$

TABLE 3.4 Information Contained in a Balanced Equation

Viewed in Terms of	Reactants $C_3H_8(g)$ + $5O_2(g)$	⟶	Products $3CO_2(g)$ + $4H_2O(g)$
Molecules	1 molecule C_3H_8 + 5 molecules O_2	⟶	3 molecules CO_2 + 4 molecules H_2O
Mass (u)	44.09 u C_3H_8 + 160.00 u O_2	⟶	132.03 u CO_2 + 72.06 u H_2O
	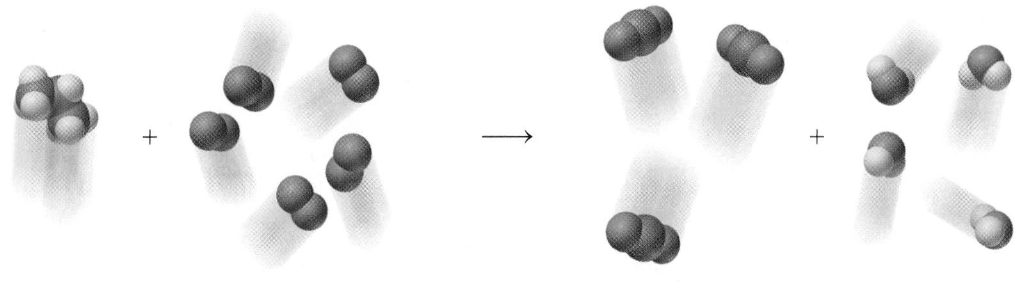		
Amount (mol)	1 mol C_3H_8 + 5 mol O_2	⟶	3 mol CO_2 + 4 mol H_2O
Mass (g)	44.09 g C_3H_8 + 160.00 g O_2	⟶	132.03 g CO_2 + 72.06 g H_2O
Total mass (g)	204.09 g	⟶	204.09 g

Since we want to find the amount (mol) of O_2 and we know the amount (mol) of H_2O, we choose "$\frac{5 \text{ mol } O_2}{4 \text{ mol } H_2O}$" to cancel "mol H_2O":

$$\text{Amount (mol) of } O_2 \text{ consumed} = 10.0\ \cancel{\text{mol } H_2O} \times \frac{5 \text{ mol } O_2}{4\ \cancel{\text{mol } H_2O}} = 12.5 \text{ mol } O_2$$

mol H_2O ⟹ (molar ratio as conversion factor) mol O_2

*You **cannot** solve this type of problem without the balanced equation.* Here is an approach for solving *any* stoichiometry problem that involves a reaction:

1. Write the balanced equation.
2. When necessary, convert the known mass (or number of entities) of one substance to the amount (mol) using its molar mass (or Avogadro's number).
3. Use the molar ratio to calculate the unknown amount (mol) of the other substance.
4. When necessary, convert the amount (mol) of the other substance to the desired mass (or number of entities) using its molar mass (or Avogadro's number).

Figure 3.8 summarizes the possible relationships among quantities of substances in a reaction, and Sample Problems 3.14 to 3.16 apply three of these relationships in the first chemical step of converting copper ore to copper metal.

Sample Problem 3.14 Calculating Quantities of Reactants and Products: Amount (mol) to Amount (mol)

Problem Production of the Canadian penny (one cent coin) was discontinued in May 2012. Until then, in a lifetime, the average Canadian used about 680 kg of copper in coins, plumbing, and wiring. Copper is obtained from sulfide ores, such as copper(I) sulfide (chalcocite), in a multistep process. After the ore is ground, it is "roasted" (heated strongly with oxygen gas) to form powdered copper(I) oxide and gaseous sulfur dioxide. What amount (mol) of oxygen is required to roast 10.0 mol of copper(I) sulfide?

Road Map

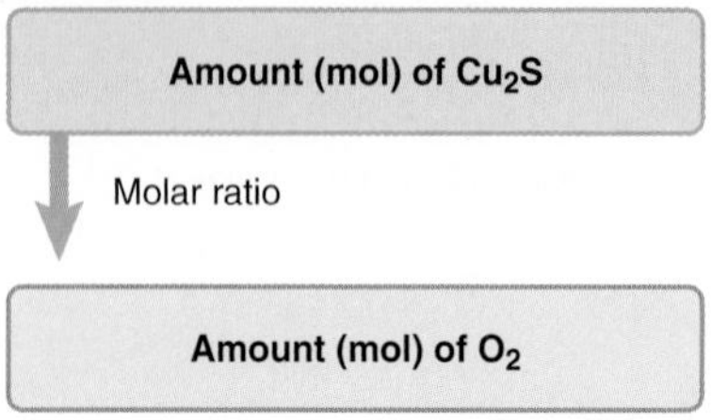

Plan We *always* write the balanced equation first. The formulas for the reactants are Cu_2S and O_2, and the formulas for the products are Cu_2O and SO_2, so we have

$$2Cu_2S(s) + 3O_2(g) \longrightarrow 2Cu_2O(s) + 2SO_2(g)$$

We know the amount of Cu_2S (10.0 mol) and must find the amount (mol) of O_2 that is needed to roast it. The balanced equation shows that 3 mol of O_2 is needed for 2 mol of Cu_2S, so the conversion factor for finding the amount (mol) of O_2 is $\frac{3 \text{ mol } O_2}{2 \text{ mol } Cu_2S}$ (*see road map*).

Solution Calculate the amount of O_2:

$$\text{Amount (mol) of } O_2 = 10.0 \cancel{\text{mol } Cu_2S} \times \frac{3 \text{ mol } O_2}{2 \cancel{\text{mol } Cu_2S}} = 15.0 \text{ mol } O_2$$

Check The units are correct, and the answer is reasonable because this molar ratio of O_2 to Cu_2S (15/10) is identical to the ratio in the balanced equation (3/2).

Comment A *common mistake* is to invert the conversion factor; that calculation would be

$$\text{Amount (mol) of } O_2 = 10.0 \text{ mol } Cu_2S \times \frac{2 \text{ mol } Cu_2S}{3 \text{ mol } O_2} = \frac{6.67 \text{ mol}^2 \; Cu_2S}{1 \text{ mol } O_2}$$

The strange units would alert you that an error was made when setting up the conversion factor. Also, this answer, 6.67, is *less* than 10.0, whereas the equation shows that there should be *more* moles of O_2 (3 mol) than moles of Cu_2S (2 mol). Be sure to think through the calculation when setting up the conversion factor and cancelling units.

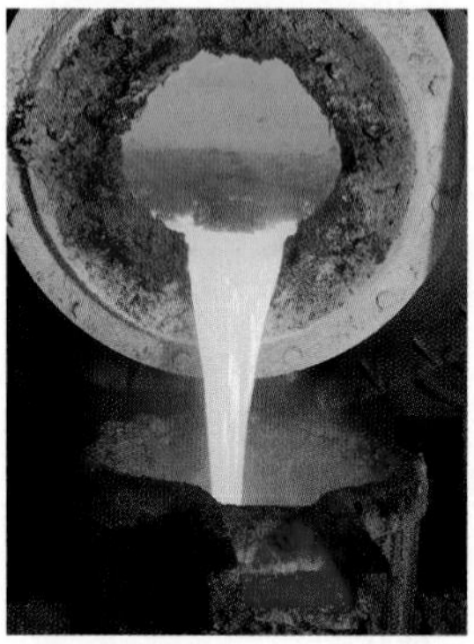

Follow-Up Problem 3.14 Thermite is a mixture of iron(III) oxide and aluminum powders that was once used to weld railroad tracks. It undergoes a spectacular reaction to yield solid aluminum oxide and molten iron. What amounts (mol) of iron(III) oxide and aluminum powders are needed to form 3.60×10^3 mol of iron? Include a road map that shows how you planned your solution.

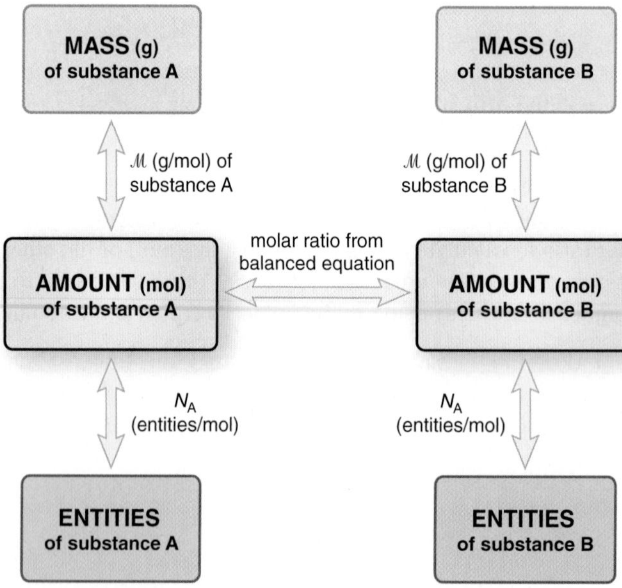

FIGURE 3.8 Summary of amount-mass-number relationships in a chemical equation. Start at any box (known) and move to any other box (unknown) by using the conversion factor on the arrow. As always, convert to amount (mol) first.

Sample Problem 3.15 Calculating Quantities of Reactants and Products: Amount (mol) to Mass (g)

Problem During the roasting process, what mass (g) of sulfur dioxide forms when 10.0 mol of copper(I) sulfide reacts?

Plan We can use the balanced equation in Sample Problem 3.14. Here we are given the amount of reactant (10.0 mol of Cu_2S) and need the mass (g) of product (SO_2) that forms. We find the amount (mol) of SO_2 using the molar ratio (2 mol SO_2/2 mol Cu_2S) and then multiply by its molar mass (64.07 g/mol) to find the mass (g) of SO_2 (*see road map*).

Road Map

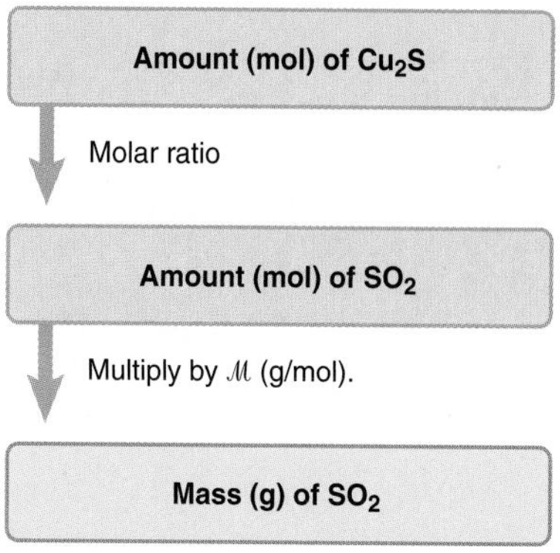

Solution Combining the two conversion steps into one calculation, we have

$$\text{Mass } (g) \text{ of } SO_2 = 10.0\ \cancel{\text{mol } Cu_2S} \times \frac{2\ \cancel{\text{mol } SO_2}}{2\ \cancel{\text{mol } Cu_2S}} \times \frac{64.07\ \text{g } SO_2}{1\ \cancel{\text{mol } SO_2}} = 641\ \text{g } SO_2$$

Check The answer makes sense, since the molar ratio shows that 10.0 mol of SO_2 is formed and each mole weighs about 64 g. We rounded to three significant figures.

Follow-Up Problem 3.15 In the thermite reaction, what amount (mol) of iron forms when 1.85×10^{25} formula units of aluminum oxide reacts? Write a road map to show how you planned your solution.

Sample Problem 3.16 Calculating Quantities of Reactants and Products: Mass to Mass

Problem During the roasting of chalcocite, what mass (kg) of oxygen is required to form 2.86 kg of copper(I) oxide?

Plan In this problem, we know the mass of the product, Cu_2O (2.86 kg), and we need the mass (kg) of O_2 that reacts to form it. Therefore, we must convert from mass of product to amount of product to amount of reactant to mass of reactant. We convert the mass of Cu_2O from kilograms to grams and then use the molar mass to convert mass (g) to amount (mol). Then, we use the molar ratio (3 mol O_2/2 mol Cu_2O) to find the amount (mol) of O_2 required. Finally, we convert the amount of O_2 to mass in grams and then kilograms (*see road map*).

Road Map

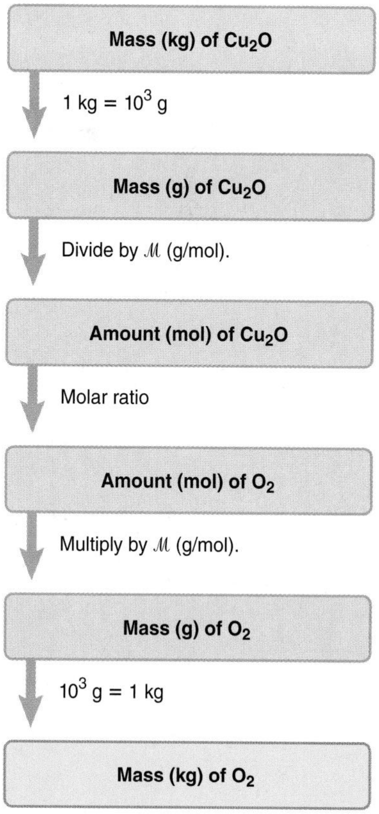

Solution Convert from m (kg) of Cu_2O to amount (mol) of Cu_2O by combining the mass unit conversion with the mass-to-amount conversion:

$$\text{Amount (mol) of } Cu_2O = 2.86\ \cancel{\text{kg } Cu_2O} \times \frac{10^3\ \cancel{g}}{1\ \cancel{\text{kg}}} \times \frac{1\ \text{mol } Cu_2O}{143.10\ \cancel{\text{g } Cu_2O}} = 20.0\ \text{mol } Cu_2O$$

Convert from amount (mol) of Cu_2O to amount (mol) of O_2:

$$\text{Amount (mol) of } O_2 = 20.0\ \cancel{\text{mol } Cu_2O} \times \frac{3\ \text{mol } O_2}{2\ \cancel{\text{mol } Cu_2O}} = 30.0\ \text{mol } O_2$$

Convert from amount (mol) of O_2 to m (kg) of O_2 by combining the amount-to-mass conversion with the mass unit conversion:

$$\text{Mass (kg) of } O_2 = 30.0\ \cancel{\text{mol } O_2} \times \frac{32.00\ \cancel{g}\ O_2}{1\ \cancel{\text{mol } O_2}} \times \frac{1\ \text{kg}}{10^3\ \cancel{g}} = 0.960\ \text{kg } O_2$$

Check The units are correct. Rounding to check the math in the final step, for example, we get ~30 mol × 30 g/mol × $\frac{1\ \text{kg}}{10^3\ \text{g}}$ = 0.90 kg. The answer seems reasonable: even though the amount (mol) of O_2 is greater than the amount (mol) of Cu_2O, the mass of O_2 is less than the mass of Cu_2O because $\mathcal{M}$ of O_2 is less than $\mathcal{M}$ of Cu_2O.

Comment The three related sample problems (3.14 to 3.16) highlight the main focus when solving stoichiometry problems: *convert the information given into amount (mol)*. Then you can use the appropriate molar ratio and any other conversion factors to complete the solution.

Follow-Up Problem 3.16 During the thermite reaction, how many atoms of aluminum react for every 1.00 g of aluminum oxide that forms? Include a road map that shows how you planned your solution.

Reactions That Occur in a Sequence

In many situations, a product of one reaction becomes a reactant for the next in a sequence of reactions. For stoichiometric purposes, when the same (common) substance forms in one reaction and reacts in the next, we eliminate this substance in an **overall (net) equation**. The following steps are used to write an overall equation:

1. Write the sequence of balanced equations.
2. Adjust the equations arithmetically to cancel the common substance.
3. Add the adjusted equations together to obtain the overall balanced equation.

Sample Problem 3.17 shows these steps by continuing the copper recovery process we started in Sample Problem 3.14.

Sample Problem 3.17 Writing an Overall Equation for a Reaction Sequence

Problem Roasting is the first step in extracting copper from chalcocite. In the next step, copper(I) oxide reacts with powdered carbon to yield copper metal and carbon monoxide gas. Write a balanced overall equation for the two-step sequence.

Plan To obtain the overall equation, we write the individual equations in sequence, adjust the coefficients to cancel the common substance(s), and then add the equations together. In this problem, only Cu_2O appears as a product in one equation and a reactant in the other, so it is the common substance.

Solution Write the individual balanced equations:

$$2Cu_2S(s) + 3O_2(g) \longrightarrow 2Cu_2O(s) + 2SO_2(g) \quad \text{(Equation 1; see Sample Problem 3.14)}$$
$$Cu_2O(s) + C(s) \longrightarrow 2Cu(s) + CO(g) \quad \text{(Equation 2)}$$

Now we need to adjust the coefficients. Since 2 mol of Cu_2O forms in equation 1 but 1 mol of Cu_2O reacts in equation 2, we double *all* the coefficients in equation 2 to use up the Cu_2O:

$$2Cu_2S(s) + 3O_2(g) \longrightarrow 2Cu_2O(s) + 2SO_2(g) \quad \text{(Equation 1)}$$
$$2Cu_2O(s) + 2C(s) \longrightarrow 4Cu(s) + 2CO(g) \quad \text{(Equation 2, doubled)}$$

Finally, we need to add the two equations and cancel the common substance. We keep the reactants of both equations on the left and the products of both equations on the right:

$$2Cu_2S(s) + 3O_2(g) + \cancel{2Cu_2O(s)} + 2C(s) \longrightarrow \cancel{2Cu_2O(s)} + 2SO_2(g) + 4Cu(s) + 2CO(g)$$

or

$$2Cu_2S(s) + 3O_2(g) + 2C(s) \longrightarrow 2SO_2(g) + 4Cu(s) + 2CO(g)$$

Check The equation is balanced:

$$\text{reactants (4 Cu, 2 S, 6 O, 2 C)} \longrightarrow \text{products (4 Cu, 2 S, 6 O, 2 C)}$$

Comment 1. Even though Cu_2O *does* participate in the chemical change, it is not involved in the reaction stoichiometry. An overall equation *may not* show which substances actually react; for example, C(*s*) and $Cu_2S(s)$ do not interact directly in this reaction sequence, even though both are shown as reactants.
2. The SO_2 that forms in the copper recovery contributes to acid rain, so chemists have devised microbial and electrochemical methods to extract metals without roasting sulfide ores. Such methods are examples of *green chemistry*; we will discuss another one of these methods later in this chapter.
3. The reactions in copper recovery were used here to show how to obtain an overall equation. The actual extraction of copper is more complex, as you will see in Chapter 23.

Follow-Up Problem 3.17 The SO_2 that forms in copper recovery reacts in air with oxygen and forms a gas called sulfur trioxide. Sulfur trioxide, in turn, reacts with water to form a sulfuric acid solution that falls in rain. Write a balanced overall equation for this process.

Reaction Sequences in Organisms Multistep reaction sequences, called *metabolic pathways*, occur throughout biological systems. (We will discuss them again in Chapter 15.) For example, in most cells, the chemical energy in glucose is released through a sequence of about 30 individual reactions. The product of each reaction is the reactant of the next, so that all the common substances cancel. The overall equation is

$$C_6H_{12}O_6(aq) + 6O_2(g) \longrightarrow 6CO_2(g) + 6H_2O(l)$$

We eat food that contains glucose, inhale O_2, and excrete CO_2 and H_2O. In our cells, these reactants and products are many steps apart: O_2 never reacts *directly* with glucose, and CO_2 and H_2O are formed at various, often distant, steps along the sequence of reactions. Even so, the molar ratios in the overall equation are the same as if the glucose burned in a combustion chamber filled with O_2 and formed CO_2 and H_2O directly (Figure 3.9).

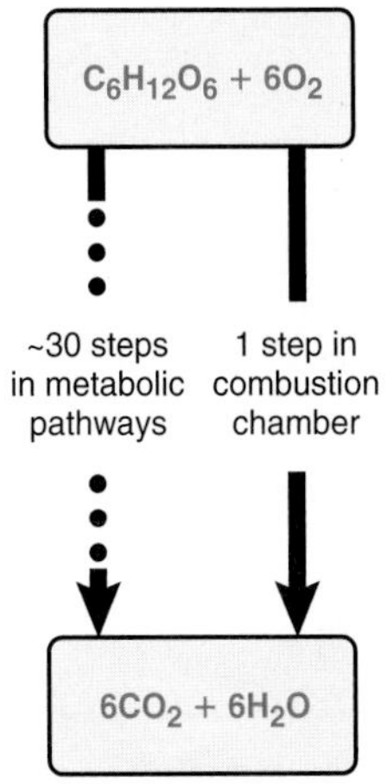

FIGURE 3.9 An overall equation equals the sum of the individual steps

Reactions That Involve a Limiting Reactant

In problems up to now, the amount of *one* reactant was given, and we assumed that there was enough of the other reactants to react with it completely. For example, suppose that we want the amount (mol) of SO_2 that forms when 5.2 mol of Cu_2S reacts with O_2:

$$2Cu_2S(s) + 3O_2(g) \longrightarrow 2Cu_2O(s) + 2SO_2(g) \quad \text{(Equation 1; see Sample Problem 3.14)}$$

We assume that the 5.2 mol of Cu_2S reacts with as much O_2 as needed. Because all the Cu_2S reacts, its initial amount of 5.2 mol determines, or *limits*, the amount of SO_2 that can form, no matter how much more O_2 is present. In this situation, we call Cu_2S the **limiting reactant** (or *limiting reagent*).

Suppose, however, that you know the amounts of both Cu_2S *and* O_2, and you need to find out how much SO_2 forms. You first have to determine whether Cu_2S *or* O_2 is the limiting reactant—that is, which one is completely used up—because it limits how much SO_2 can form. The reactant that is *not* limiting is present *in excess*, which means the amount that does not react is left over. ■ To determine which reactant is the limiting reactant, we use the molar ratios in the balanced equation to perform a series of calculations to see *which reactant forms less product.*

■ **Limiting "Reactants" in Everyday Life** Situations that involve limiting "reactants" arise in business all the time. The manager of a car-assembly plant must order more tires if there are 1500 car bodies and only 4000 tires, and a clothes manufacturer must cut more sleeves if there are 320 sleeves for 170 shirt bodies. You have probably faced such situations in daily life as well. A muffin recipe calls for 2 cups of flour and 1 cup of sugar, but you have 3 cups of flour and only $\frac{3}{4}$ cup of sugar. Clearly, the flour is in excess, and the sugar limits the number of muffins you can make. Suppose that you are in charge of making cheeseburgers for a picnic, and you have 10 buns, 12 meat patties, and 15 slices of cheese. Here, the number of buns limits how many cheeseburgers you can make. What if there are 26 students and only 23 microscopes in a cell biology lab? You will find that situations involving limiting "reactants" are almost limitless.

Determining the Limiting Reactant Let us clarify these ideas in a much more appetizing situation. Suppose that you have a job making ice cream sundaes. Each sundae requires two scoops (170 mL) of ice cream, one cherry, and 50 mL of syrup:

$$\text{2 scoops (170 mL)} + \text{1 cherry} + \text{50 mL syrup} \longrightarrow \text{1 sundae}$$

A mob of 25 ravenous children enter, and each one wants a sundae with vanilla ice cream and chocolate syrup and a cherry on top. You have 4.0 L of vanilla ice cream (at 85 mL per scoop), 30 cherries, and 1 L of syrup. Can you feed all of them? A series of calculations based on the balanced equation shows the number of sundaes you can make from each ingredient:

$$\text{Ice cream: No. of sundaes} = 4.0\ \cancel{\text{L}} \times \frac{1000\ \cancel{\text{mL}}}{1\ \cancel{\text{L}}} \times \frac{1\ \cancel{\text{scoop}}}{85\ \cancel{\text{mL}}} \times \frac{1\ \text{sundae}}{2\ \cancel{\text{scoops}}} = 23.5\ \text{sundaes}$$

$$\text{Cherries: No. of sundaes} = 30\ \cancel{\text{cherries}} \times \frac{1\ \text{sundae}}{1\ \cancel{\text{cherry}}} = 30\ \text{sundaes}$$

$$\text{Syrup: No. of sundaes} = 1000\ \cancel{\text{mL syrup}} \times \frac{1\ \text{sundae}}{50\ \cancel{\text{mL syrup}}} = 20\ \text{sundaes}$$

Of the reactants (ice cream, cherry, and syrup), the syrup forms the *least* product (sundaes), so it is the limiting "reactant." When all the syrup has been used up, some ice cream and cherries are "unreacted" so they are in excess:

$$\text{4.0 L (47 scoops)} + \text{30 cherries} + \text{1 L syrup} \longrightarrow$$
$$\text{20 sundaes} + \text{600 mL (7 scoops)} + \text{10 cherries}$$

Figure 3.10 shows a similar example with different initial (starting) quantities.

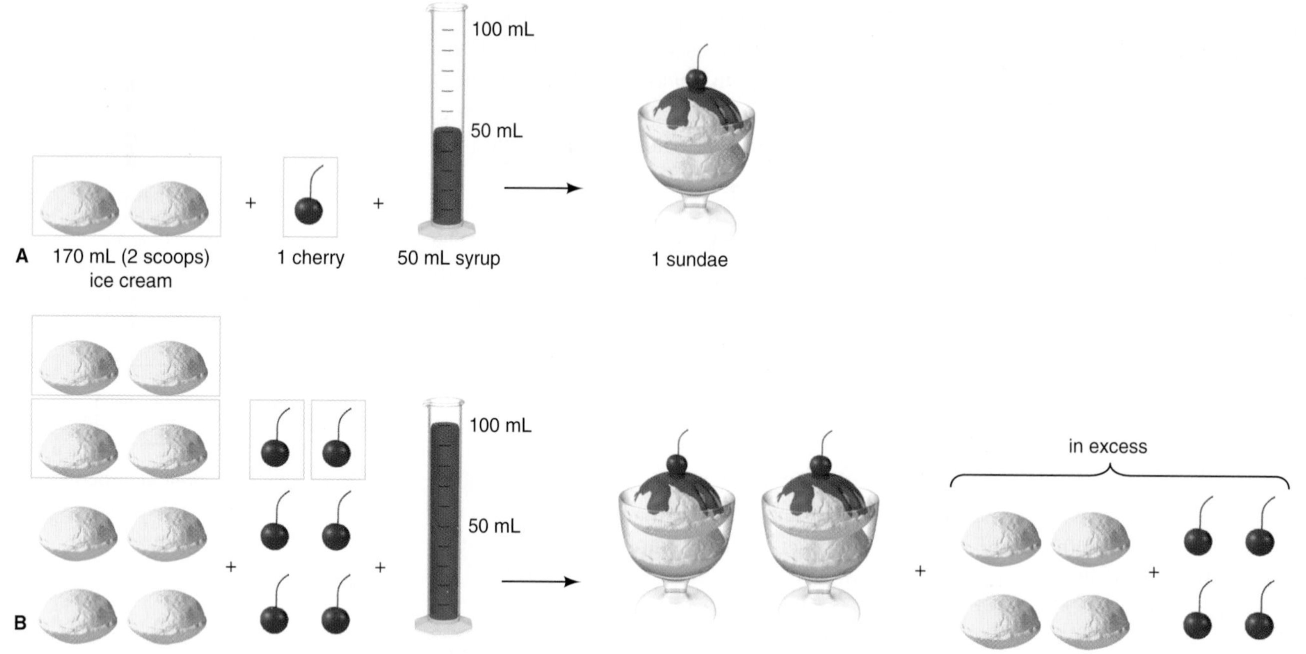

FIGURE 3.10 **An ice cream sundae analogy for limiting reactants. A.** The "reactants" form the "product" (sundae). **B.** Only two sundaes can be made before all the syrup is used up, so syrup is the limiting "reactant"; four scoops of ice cream and four cherries are "in excess."

Using Reaction Tables in Limiting-Reactant Problems A good way to keep track of the quantities in a limiting-reactant problem is to use a *reaction table.* The balanced equation appears at the top, as the column headings. The table shows the following information:

- *Initial* quantities of reactants and products *before* the reaction
- *Change* in the quantities of reactants and products *during* the reaction
- *Final* quantities of reactants and products remaining *after* the reaction

For example, for the ice cream sundae "reaction," the reaction table would look like this:

Quantity	170 mL (2 scoops)	+	1 cherry	+	50 mL syrup	⟶	1 sundae
Initial	4000 mL (47 scoops)		30 cherries		1000 mL syrup		0 sundaes
Change	−3400 mL (40 scoops)		−20 cherries		−1000 mL syrup		+20 sundaes
Final	600 mL (7 scoops)		10 cherries		0 mL syrup		20 sundaes

The body of the table shows the following important information:

- In the *Initial* line, "product" has not yet formed, so the entry is "0 sundaes."
- In the *Change* line, since the reactants (ice cream, cherries, and syrup) are used during the reaction, their quantities decrease, so the changes in their quantities have a *negative* sign. At the same time, the quantity of product (sundaes) increases, so the change in its quantity has a *positive* sign.
- For the *Final* line, we *add* the *Change* and *Initial* lines. Notice that some reactants (ice cream and cherries) are in excess, while the limiting reactant (syrup) is used up.

Solving Limiting-Reactant Problems In a limiting-reactant problem, *the quantities of two (or more) reactants are given, and we first determine which reactant is limiting.* To do this, just as we did with the ice cream sundaes, we use the balanced equation to solve a series of calculations to see how much product forms from the given quantity of each reactant: the limiting reactant is the reactant that yields the *least* quantity of product.

The following problems examine these ideas from several angles. In Sample Problem 3.18, we solve the problem by looking at a molecular scene; in Sample Problem 3.19, we start with the amounts (mol) of two reactants; and in Sample Problem 3.20, we start with the masses of two reactants.

Sample Problem 3.18 Using Molecular Depictions in a Limiting-Reactant Problem

Problem Nuclear engineers use chlorine trifluoride to prepare uranium fuel for power plants. The compound is formed as a gas by the reaction of elemental chlorine and fluorine. The circle in the margin shows a representative portion of the reaction mixture before the reaction starts (chlorine is *green*; fluorine is *yellow*).

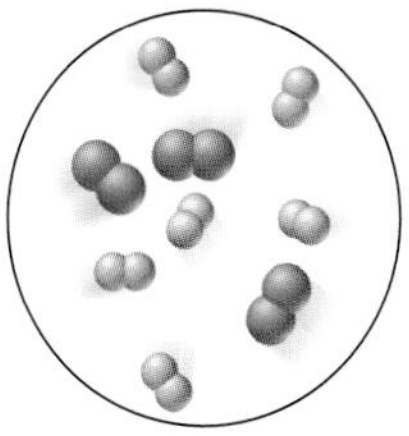

(a) Find the limiting reactant.

(b) Write a reaction table for the process.

(c) Draw a representative portion of the mixture after the reaction is complete. (*Hint*: The ClF_3 molecule has Cl bonded to three individual F atoms.)

Plan (a) We have to find the limiting reactant. The first step is to write the balanced equation, so we need the formulas and states of matter. From the name, chlorine trifluoride, we know that the product consists of one Cl atom bonded to three F atoms, or ClF_3. Elemental chlorine and fluorine are the diatomic molecules Cl_2 and F_2, and all three substances are gases. To find the limiting reactant, we find the number of molecules of product that would form from the numbers of molecules of each reactant: whichever forms less product is the limiting reactant. **(b)** We use these numbers of molecules to write a reaction table. **(c)** We use the numbers in the *Final* line of the table to draw the scene.

Solution (a) The balanced equation is

$$Cl_2(g) + 3F_2(g) \longrightarrow 2ClF_3(g)$$

$$\text{For } Cl_2\text{: Molecules of } ClF_3 = 3\ \cancel{\text{molecules of } Cl_2} \times \frac{2 \text{ molecules of } ClF_3}{1\ \cancel{\text{molecules of } Cl_2}}$$

$$= 6 \text{ molecules of } ClF_3$$

$$\text{For } F_2\text{: Molecules of } ClF_3 = 6\ \cancel{\text{molecules of } F_2} \times \frac{2 \text{ molecules of } ClF_3}{3\ \cancel{\text{molecules of } F_2}}$$

$$= \tfrac{12}{3} \text{ molecules of } ClF_2 = 4 \text{ molecules of } ClF_3$$

Because it forms less product, F_2 is the limiting reactant.

(b) Since F_2 is the limiting reactant, all of it (6 molecules) is used in the *Change* line of the reaction table:

Molecules	$Cl_2(g)$	+	$3F_2(g)$	$\longrightarrow$	$2ClF_3(g)$
Initial	3		6		0
Change	−2		−6		+4
Final	1		0		4

(c) The representative portion of the final reaction mixture (*see margin*) includes 1 molecule of Cl_2 (the reactant in excess) and 4 molecules of the product, ClF_3.

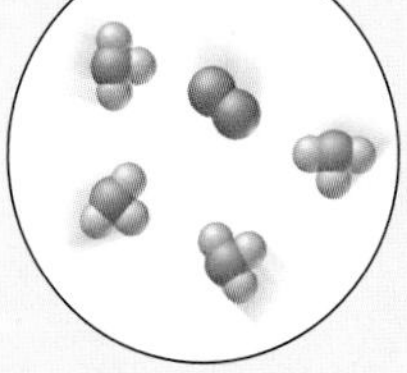

Check The equation is balanced: reactants (2 Cl, 6 F) $\longrightarrow$ products (2 Cl, 6 F). As shown in the circles, the numbers of each type of atom before and after the reaction are equal. Let us think through our choice of the limiting reactant. From the equation, one Cl_2 needs three F_2 to form two ClF_3. Therefore, the three Cl_2 molecules in the circle, depicting the reactants, need nine (3 × 3) F_2. But there are only six F_2, so there is not enough F_2 to react with the available Cl_2, or, put another way, there is too much Cl_2 to react with the available F_2. From either point of view, F_2 is the limiting reactant.

Follow-Up Problem 3.18 B_2 (B is *red*) reacts with AB as shown below:

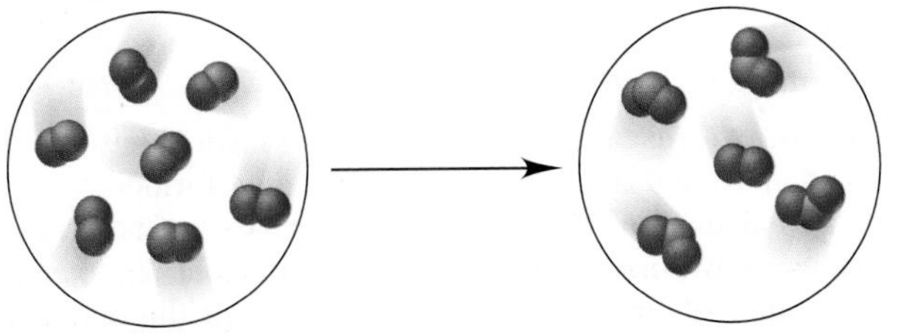

Write a balanced equation for the reaction, and determine the limiting reactant.

Sample Problem 3.19 Calculating Quantities in a Limiting-Reactant Problem: Amount to Amount

Problem In another preparation of ClF_3 (see Sample Problem 3.18), 0.750 mol of Cl_2 reacts with 3.00 mol of F_2. **(a)** Find the limiting reactant. **(b)** Write a reaction table.

Plan **(a)** We find the limiting reactant by calculating the amount (mol) of ClF_3 formed from the amount (mol) of each reactant. The reactant that forms fewer moles of ClF_3 is limiting. **(b)** We enter those values into the reaction table.

Solution **(a)** We need to determine the limiting reactant.

Find the amount (mol) of ClF_3 from the amount (mol) of Cl_2:

$$\text{Amount (mol) of ClF}_3 = 0.750\ \cancel{\text{mol Cl}_2} \times \frac{2\ \text{mol ClF}_3}{1\ \cancel{\text{mol Cl}_2}} = 1.50\ \text{mol ClF}_3$$

Find the amount (mol) of ClF_3 from the amount (mol) of F_2:

$$\text{Amount (mol) of ClF}_3 = 3.00\ \cancel{\text{mol F}_2} \times \frac{2\ \text{mol ClF}_3}{3\ \cancel{\text{mol F}_2}} = 2.00\ \text{mol ClF}_3$$

Therefore, Cl_2 is limiting because it forms fewer moles of ClF_3.

(b) Write the reaction table, with Cl_2 limiting:

Amount (mol)	$Cl_2(g)$	+	$3F_2(g)$	$\longrightarrow$	$2ClF_3(g)$
Initial	0.750		3.00		0
Change	−0.750		−2.25		+1.50
Final	0		0.75		1.50

Check Let us check that Cl_2 is the limiting reactant by assuming, for the moment, that F_2 is limiting. If that were true, all 3.00 mol of F_2 would react to form 2.00 mol of ClF_3. However, based on the balanced equation, obtaining 2.00 mol of ClF_3 would require 1.00 mol of Cl_2, and only 0.750 mol of Cl_2 is present. Thus, Cl_2 must be the limiting reactant.

Comment A major idea to note from Sample Problems 3.18 and 3.19 is that the relative quantities of reactants *do not* determine which reactant is limiting, but rather the quantity of product formed, which is based on the *molar ratio in the balanced equation*. In both problems, there is more F_2 than Cl_2. However,

- Sample Problem 3.18 has an F_2/Cl_2 ratio of 6/3, or 2/1, which is less than the required molar ratio of 3/1, so F_2 is limiting and Cl_2 is in excess
- Sample Problem 3.19 has an F_2/Cl_2 ratio of 3.00/0.750, which is greater than the required molar ratio of 3/1, so Cl_2 is limiting and F_2 is in excess

Follow-Up Problem 3.19 For the reaction in Follow-Up Problem 3.18, what amount (mol) of product forms from 1.5 mol of each reactant?

Sample Problem 3.20 Calculating Quantities in a Limiting-Reactant Problem: Mass to Mass

Problem A fuel mixture that was used in the early days of rocketry consisted of two liquids, hydrazine (N_2H_4) and dinitrogen tetroxide (N_2O_4), which ignite on contact to form nitrogen gas and water vapour.

(a) What mass (g) of nitrogen gas forms when 1.00×10^2 g of N_2H_4 and 2.00×10^2 g of N_2O_4 are mixed?

(b) Write a reaction table for this process.

Plan The amounts of two reactants are given, which means that this is a limiting-reactant problem. **(a)** To determine the mass of product formed, we must find the limiting reactant by calculating which of the given masses of reactant forms *less* nitrogen gas. As always, we first write the balanced equation. We convert the mass (g) of each reactant to amount (mol) using the molar mass of the reactant and then use

the molar ratio from the balanced equation to find the amount (mol) of N_2 that each reactant forms. Next, we convert the lower amount of N_2 to mass (*see road map*).
(b) We use the values based on the limiting reactant for the reaction table.

Solution (a) Write the balanced equation:

$$2N_2H_4(l) + N_2O_4(l) \longrightarrow 3N_2(g) + 4H_2O(g)$$

Next, we need to find the amount (mol) of N_2 from the amount (mol) of each reactant.

For N_2H_4:

$$\text{Amount (mol) of } N_2H_4 = n = \frac{m}{\mathcal{M}} = 1.00 \times 10^2\ \cancel{g\ N_2H_4} \times \frac{1\ mol\ N_2H_4}{32.05\ \cancel{g\ N_2H_4}} = 3.12\ mol\ N_2H_4$$

$$\text{Amount (mol) of } N_2 = 3.12\ \cancel{mol\ N_2H_4} \times \frac{3\ mol\ N_2}{2\ \cancel{mol\ N_2H_4}} = 4.68\ mol\ N_2$$

For N_2O_4: $$\text{Amount (mol) of } N_2O_4 = 2.00 \times 10^2\ \cancel{g\ N_2O_4} \times \frac{1\ mol\ N_2O_4}{92.02\ \cancel{g\ N_2O_4}} = 2.17\ mol\ N_2O_4$$

$$\text{Amount (mol) of } N_2 = 2.17\ \cancel{mol\ N_2O_4} \times \frac{3\ mol\ N_2}{1\ \cancel{mol\ N_2O_4}} = 6.51\ mol\ N_2$$

Thus, N_2H_4 is the limiting reactant because it yields less N_2.

Convert from amount (mol) of N_2 to mass (g):

$$\text{Mass (g) of } N_2 = m = n \times \mathcal{M} = 4.68\ \cancel{mol\ N_2} \times \frac{28.02\ g\ N_2}{1\ \cancel{mol\ N_2}} = 131\ g\ N_2$$

(b) With N_2H_4 as the limiting reactant, the reaction table looks like this:

Amount (mol)	$2N_2H_4(g)$	+	$N_2O_4(g)$	$\longrightarrow$	$3N_2(g)$	+	$4H_2O(g)$
Initial	3.12		2.17		0		0
Change	−3.12		−1.56		+4.68		+6.24
Final	0		0.61		4.68		6.24

Check The mass of N_2O_4 is more than the mass of N_2H_4, but the amount (mol) of N_2O_4 is less because its $\mathcal{M}$ is much higher. Rounding for N_2H_4: $100\ g\ N_2H_4 \times \frac{1\ mol}{32\ g} \approx 3$ mol; ~3 mol $\times \frac{3}{2} \approx 4.5$ mol N_2; ~4.5 mol × 30 g/mol ≈ 135 g N_2.

Comment 1. Keep in mind this *common mistake* when solving limiting-reactant problems: The limiting reactant is not the *reactant* that is present in smaller quantity [whether by mass or amount (mol)]. Rather, it is the reactant that forms less *product* [whether by mass or amount (mol)].

2. An *alternative approach* to finding the limiting reactant compares "How much is needed?" with "How much is given?" That is, based on the balanced equation, follow these steps:

- Find the amount (mol) of each reactant that is needed to react with the other reactant.
- Compare this *needed* amount with the *given* amount in the problem statement. There will be *more* than enough of one reactant (excess) and *less* than enough of the other (limiting).

For example, the balanced equation for this problem shows that 2 mol of N_2H_4 reacts with 1 mol of N_2O_4. The amount (mol) of N_2O_4 that is needed to react with the given 3.12 mol of N_2H_4 is

$$\text{Amount (mol) of } N_2O_4 \text{ needed} = 3.12\ \cancel{mol\ N_2H_4} \times \frac{1\ mol\ N_2O_4}{2\ \cancel{mol\ N_2H_4}} = 1.56\ mol\ N_2O_4$$

The amount of N_2H_4 that is needed to react with the given 2.17 mol of N_2O_4 is

$$\text{Amount (mol) of } N_2H_4 \text{ needed} = 2.17\ \cancel{mol\ N_2O_4} \times \frac{2\ mol\ N_2H_4}{1\ \cancel{mol\ N_2O_4}} = 4.34\ mol\ N_2H_4$$

We are given 2.17 mol of N_2O_4, which is *more* than the 1.56 mol of N_2O_4 needed, and we are given 3.12 mol of N_2H_4, which is *less* than the 4.34 mol of N_2H_4 needed. Therefore, N_2H_4 is limiting, and N_2O_4 is in excess.

Follow-Up Problem 3.20 What mass (g) of solid aluminum sulfide can be prepared by the reaction of 10.0 g of aluminum and 15.0 g of sulfur? What mass (g) of the other reactant is in excess?

Road Map

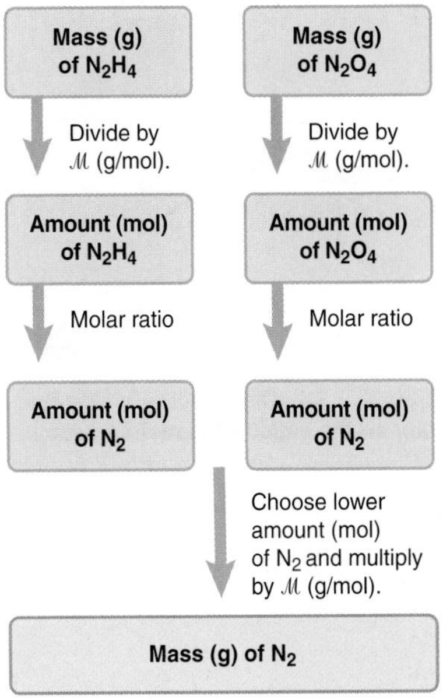

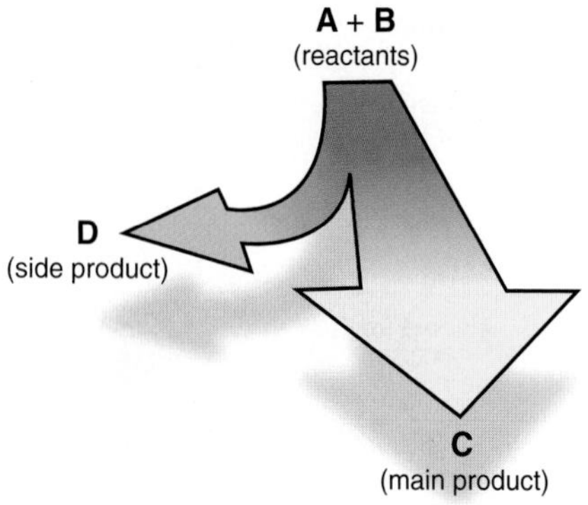

FIGURE 3.11 The effect of a side reaction on the yield of the main product

Theoretical, Actual, and Percent Reaction Yields

Up until now, we have assumed that 100% of the limiting reactant becomes product, that ideal methods exist for isolating the product, and that we have perfect lab techniques to collect all the product. In theory, this may happen. In reality, however, it does not, and chemists recognize three types of reaction yield:

1. *Theoretical yield.* The *amount of product calculated from the molar ratio in the balanced equation* is the **theoretical yield**. However, there are several reasons the theoretical yield is *never* obtained:
   - Reactant mixtures often proceed through **side reactions** that form different products (Figure 3.11). In the rocket fuel reaction in Sample Problem 3.20, for example, the reactants might form some NO in the following side reaction:

   $$N_2H_4(l) + 2N_2O_4(l) \longrightarrow 6NO(g) + 2H_2O(g)$$

   This reaction decreases the amounts of reactants available for N_2 production.
   - Even more important, many reactions seem to stop before they are complete, so some limiting reactant is unused. (We will see why in Chapter 15.)
   - Physical losses occur in every step of a separation (see Tools of the Laboratory, Section 2.9). Some solid clings to filter paper, some distillate evaporates, and so on. With careful techniques, we can minimize, but never eliminate, such losses.
2. *Actual yield.* Given these reasons for obtaining less than the theoretical yield, the amount of product that is actually obtained is the **actual yield**. Theoretical and actual yields are expressed in units of amount (mol) or mass (g).
3. *Percent yield.* The **percent yield (% yield)** is the actual yield expressed as a percentage of the theoretical yield:

$$\% \text{ Yield} = \frac{\text{actual yield}}{\text{theoretical yield}} \times 100 \qquad (3.8)$$

By definition, the actual yield is less than the theoretical yield, so the percent yield is *always* less than 100%.

Sample Problem 3.21 Calculating Percent Yield

Road Map

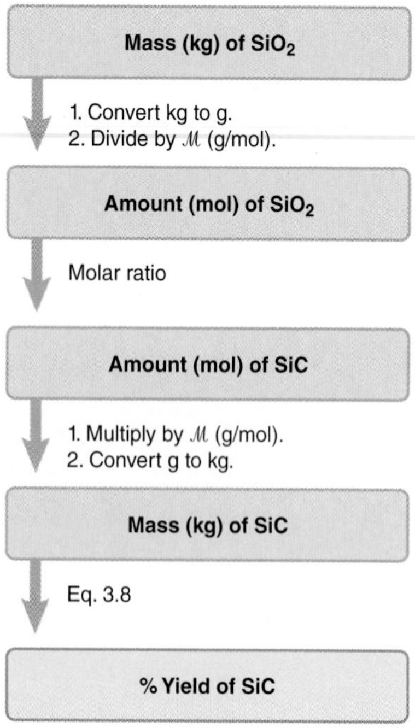

Problem Silicon carbide (SiC) is an important ceramic material that is made by reacting sand (silicon dioxide, SiO_2) with powdered carbon at a high temperature. Carbon monoxide is also formed. When 100.0 kg of sand is processed, 51.4 kg of SiC is recovered. What is the percent yield of SiC from this process?

Plan We are given the actual yield of SiC (51.4 kg), so we need the theoretical yield to calculate the percent yield. After writing the balanced equation, we convert the given mass of SiO_2 (100.0 kg) to amount (mol). We use the molar ratio to find the amount of SiC formed and convert it to mass (kg) to obtain the theoretical yield. Then, we use Equation 3.8 to find the percent yield (*see road map*).

Solution Write the balanced equation:

$$SiO_2(s) + 3C(s) \longrightarrow SiC(s) + 2CO(g)$$

Convert from mass (kg) of SiO_2 to amount (mol):

$$\text{Amount (mol) of } SiO_2 = n = \frac{m}{\mathcal{M}} = 100.0\ \cancel{\text{kg } SiO_2} \times \frac{1000\ \cancel{\text{g}}}{1\ \cancel{\text{kg}}} \times \frac{1 \text{ mol } SiO_2}{60.09\ \cancel{\text{g } SiO_2}}$$
$$= 1664 \text{ mol } SiO_2$$

The molar ratio is 1 mol SiC/1 mol SiO_2. Convert from amount (mol) of SiO_2 to amount (mol) of SiC:

$$\text{Amount (mol) of } SiO_2 = \text{amount (mol) of SiC} = 1664 \text{ mol SiC}$$

Convert from amount (mol) of SiC to mass (kg):

$$\text{Mass (kg) of SiC} = 1664\ \cancel{\text{mol SiC}} \times \frac{40.10\ \cancel{\text{g SiC}}}{1\ \cancel{\text{mol SiC}}} \times \frac{1 \text{ kg}}{1000 \text{ g}} = 66.73 \text{ kg SiC}$$

Calculate the percent yield:

$$\% \text{ Yield of SiC} = \frac{\text{actual yield}}{\text{theoretical yield}} \times 100 = \frac{51.4 \text{ kg SiC}}{66.73 \text{ kg SiC}} \times 100 = 77.0\%$$

Check Rounding shows that the mass of SiC seems correct: ~1500 mol × 40 g/mol × $\frac{1 \text{ kg}}{1000 \text{ g}}$ = 60 kg. The molar ratio of SiC/SiO_2 is 1/1, and $\mathcal{M}$ of SiC is about two-thirds ($\sim\frac{40}{60}$) of $\mathcal{M}$ of SiO_2, so 100 kg of SiO_2 should form about 66 kg of SiC.

Follow-Up Problem 3.21 Marble (calcium carbonate) reacts with hydrochloric acid solution to form calcium chloride solution, water, and carbon dioxide. Find the percent yield of carbon dioxide if 3.65 g is collected when 10.0 g of marble reacts.

Yields in Multistep Syntheses In the multistep synthesis of a complex compound, the overall yield can be surprisingly low, even if the yield of each step is high. For example, suppose that a six-step synthesis has a 90.0% yield for each step. To find the overall percent yield, *express the yield of each step as a decimal, multiply the yields together, and then convert back to percent.* The overall recovery is only slightly more than 50%:

$$\text{Overall \% yield} = (0.900 \times 0.900 \times 0.900 \times 0.900 \times 0.900 \times 0.900 \times 100) = 53.1\%$$

Multistep sequences are common in laboratory syntheses of medicines, dyes, pesticides, and many other organic compounds. For example, the antidepressant sertraline is prepared from a simple starting compound in six steps, with yields of 80%, 80%, 50%, 100%, 48%, and 30%, respectively. Therefore, the overall percent yield is only 4.6% (Figure 3.12). Because a typical synthesis begins with large amounts of inexpensive, simple reactants and ends with small amounts of expensive, complex products, the overall yield greatly influences the commercial potential of a product.

Starting with 100 g of 1,2-dichlorobenzene,

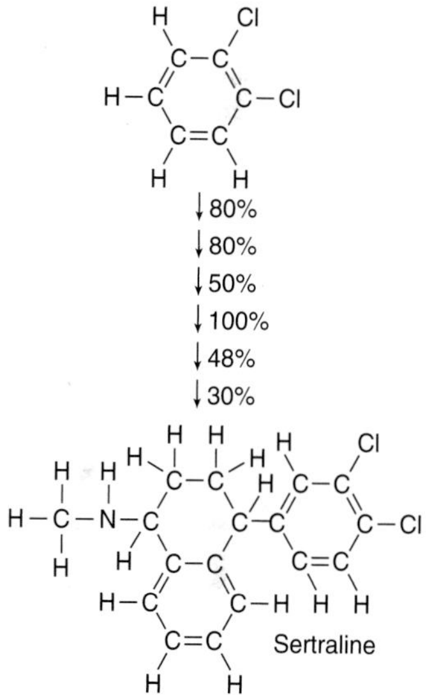

the yield of sertraline is only 4.6 g.

FIGURE 3.12 Low overall yield for a multistep biomedical synthesis

Atom Economy: A Green Chemistry Perspective on Yield In the relatively new field of **green chemistry**, academic, industrial, and government chemists develop methods that reduce or prevent the release of harmful substances into the environment and reduce the amounts of energy resources used.

One way that green chemists evaluate a synthetic route is by focusing on its *atom economy*, the proportion of reactant atoms that end up in the desired product. The efficiency of a synthesis is quantified in terms of the *percent atom economy*:

$$\% \text{ Atom economy} = \frac{\text{amount (mol) product} \times \text{molar mass of desired product}}{\text{sum of the molar masses of all products}} \times 100$$

Consider two synthetic routes—one starting with benzene (C_6H_6), the other with butane (C_4H_{10})—for the production of maleic anhydride ($C_4H_2O_3$), a key substance in the manufacture of polymers, dyes, medicines, pesticides, and other products.

Route 1: $2C_6H_6(l) + 9O_2(g) \rightarrow \cdots \rightarrow 2C_4H_2O_3(l) + 4H_2O(l) + 4CO_2(g)$

Route 2: $2C_4H_{10}(g) + 7O_2(g) \rightarrow \cdots \rightarrow 2C_4H_2O_3(l) + 8H_2O(l)$

Let us compare the efficiency of these routes in terms of percent atom economy.

Route 1:

$$\% \text{ Atom economy} = \frac{2 \times \mathcal{M} \text{ of } C_4H_2O_3}{(2 \times \mathcal{M} \text{ of } C_4H_2O_3) + (4 \times \mathcal{M} \text{ of } H_2O) + (4 \times \mathcal{M} \text{ of } CO_2)} \times 100$$

$$= \frac{2 \times 98.06 \text{ g}}{(2 \times 98.06 \text{ g}) + (4 \times 18.02 \text{ g}) + (4 \times 44.01 \text{ g})} \times 100$$

$$= 44.15\%$$

Route 2:

$$\% \text{ Atom economy} = \frac{2 \times \mathcal{M} \text{ of } C_4H_2O_3}{(2 \times \mathcal{M} \text{ of } C_4H_2O_3) + (8 \times \mathcal{M} \text{ of } H_2O)} \times 100$$

$$= \frac{2 \times 98.06 \text{ g}}{(2 \times 98.06 \text{ g}) + (8 \times 18.02 \text{ g})} \times 100$$

$$= 57.63\%$$

From the perspective of atom economy, route 2 is preferable because a larger percentage of reactant atoms ends up in the desired product. It is also "greener" than route 1 because it avoids the use of the toxic reactant benzene and does not produce CO_2, a gas that contributes to global warming. However, keep in mind that, in reality, the yield must also be taken into consideration. A process with a slightly lower atom economy may be preferable if its yield is higher than a process with a higher atom economy.

SUMMARY OF SECTION 3.4

- The substances in a balanced equation are related to each other by stoichiometrically equivalent molar ratios, which are used as conversion factors to find the amount (mol) of one substance given the amount of another.
- In limiting-reactant problems, the quantities of two (or more) reactants are given, and the limiting reactant is the reactant that forms the lower quantity of product. Reaction tables show the initial and final quantities of all the reactants and products, as well as the changes in these quantities.
- In practice, side reactions, incomplete reactions, and physical losses result in an actual yield of product that is less than the theoretical yield (the quantity based on the molar ratio from the balanced equation), giving a percent yield that is less than 100%. In multistep reaction sequences, the overall yield is found by multiplying the yields for each step.
- Atom economy, or the proportion of reactant atoms found in the product, is one criterion for choosing a "greener" reaction process.

3.5 Fundamentals of Solution Stoichiometry

In popular media, you may see a chemist in a lab coat, surrounded by glassware, mixing coloured solutions that froth and release billowing fumes. Most reactions in solution are not this dramatic, and good techniques require safer mixing procedures. However, it is true that aqueous solution chemistry is central to laboratory chemistry. Liquid solutions are easier to store than gases and easier to mix than solids, and the amounts of substances in solution can be measured precisely. Many environmental reactions and almost all biochemical reactions occur in solution.

We know the amount of a pure substance by converting its mass into the number of moles. For a dissolved substance, we need the *concentration*—the amount in moles per volume of solution—to find the volume that contains a given amount in moles. In this section, we first discuss the most common way to express concentration, and that is in mol/L (Chapter 12 discusses other concentration units). Then, we see how to prepare a solution of a specific concentration and how to use solutions in stoichiometric calculations.

Expressing Concentration in Terms of Moles per Litre

A solution consists of a smaller quantity of one substance, the **solute**, dissolved in a larger quantity of another, the **solvent**. When it dissolves, the solute's chemical entities become evenly dispersed throughout the solvent. The **concentration, c**, of a solution is often expressed as *the quantity of solute dissolved in a given quantity of solution.*

Concentration is an *intensive* property (like density or temperature; see Section 1.5) and, thus, is independent of the volume of a solution: a 50 L tank of a solution has the *same concentration* (solute quantity/solution quantity) as a 50 mL beaker of the solution. We can express the concentration of a solution in units of *moles of solute per litre of solution*:

$$\text{Concentration (mol/L)} = \frac{\text{amount of solute (mol)}}{\text{volume of solution (L)}}$$

$$c = \frac{n}{V} \qquad (3.9)$$

The unit mol/L is often given the symbol M (not to be confused with the molar mass, $\mathcal{M}$). A solution that has a concentration of 3 mol/L can also be called a "three molar (3 M) solution." Concentration expressed in mol/L is often called **molarity, *M*.**

Sample Problem 3.22 Calculating the Concentration of a Solution

Problem Glycine has the simplest structure of the 20 amino acids that make up proteins. What is the concentration (mol/L) of a solution that contains 0.715 mol of glycine in 495 mL?

Plan The concentration is the amount (mol) of solute in each litre of solution. We divide the amount (mol) (0.715 mol) by the volume (495 mL) and then convert the volume to litres to find the concentration in mol/L (*see road map*).

Road Map

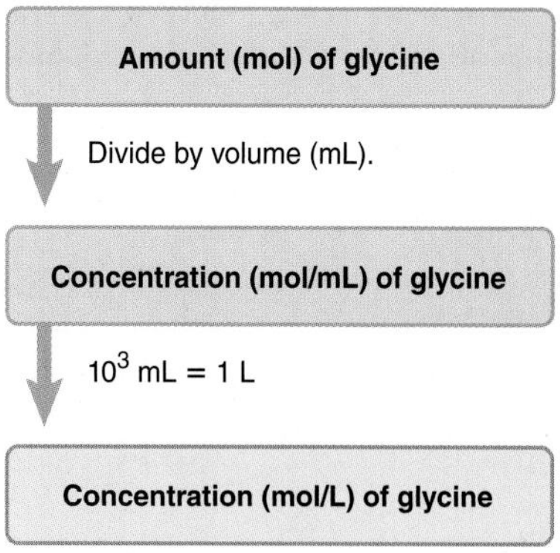

Solution Concentration (mol/L) = [Glycine] $= \frac{n}{V} = \frac{0.715 \text{ mol glycine}}{495 \cancel{\text{mL soln}}} \times \frac{1000 \cancel{\text{mL}}}{1 \text{ L}}$

$$= 1.44 \text{ mol/L glycine}$$

Check A quick look at the math shows about 0.7 mol of glycine in about 0.5 L of solution, so the concentration should be about 1.4 mol/L.

Comment We use the square brackets around a species to denote the term *the concentration of.* In the follow-up problem, we will solve for the concentration of KI, or [KI].

Follow-Up Problem 3.22 What amount (mol) of KI is in 84 mL of 0.50 mol/L KI? Include a road map that shows how you planned your solution.

Amount-Mass-Number Conversions Involving Solutions

Like many intensive properties, concentration (mol/L) can be used as a *conversion factor* between volume (L) of solution and amount (mol) of solute. From the concentration (mol/L), we can find the mass or the number of entities of solute (Figure 3.13), as you will see in Sample Problem 3.23.

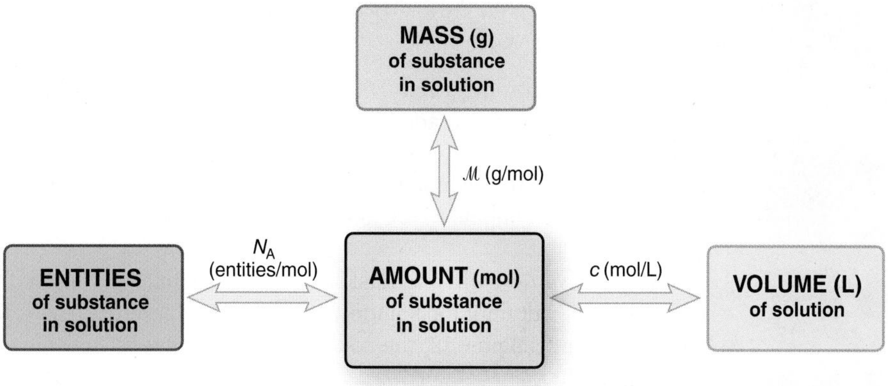

FIGURE 3.13 Summary of amount-mass-number relationships in solution. The amount (mol) of a substance in solution is related to the volume (L) of the solution through the concentration (c; mol/L). As always, convert the given quantity to amount (mol) first.

Sample Problem 3.23 Calculating the Mass of Solute in a Given Volume of Solution

Problem Biochemists often study reactions in solutions that contain phosphate ions. These solutions are commonly found in cells. What mass of solute is in 1.75 L of 0.460 mol/L sodium hydrogen phosphate?

Plan We need the mass (g) of solute, so we multiply the known volume of the solution (1.75 L) by the known concentration (0.460 mol/L) to find the amount (mol) of solute. Then we convert the amount to mass (g) using the solute's molar mass (*see road map*).

Road Map

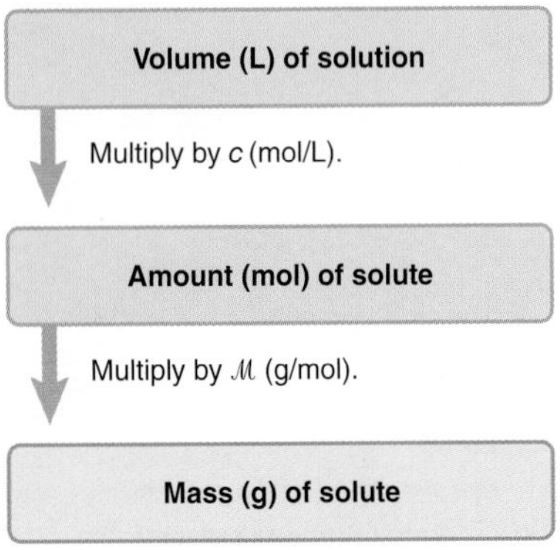

Solution Calculate the amount (mol) of solute in solution:

$$\text{Amount (mol) of } Na_2HPO_4 = n = V \times c = 1.75\ \cancel{\text{L soln}} \times \frac{0.460 \text{ mol } Na_2HPO_4}{1\ \cancel{\text{L soln}}} = 0.805 \text{ mol } Na_2HPO_4$$

Convert from amount (mol) of solute to mass (g):

$$\text{Mass (g) } Na_2HPO_4 = m = n \times \mathcal{M} = 0.805\ \cancel{\text{mol } Na_2HPO_4} \times \frac{141.96 \text{ g } Na_2HPO_4}{1\ \cancel{\text{mol } Na_2HPO_4}} = 114 \text{ g } Na_2HPO_4$$

Check The answer seems to be correct: ~1.8 L of 0.5 mol/L solution contains 0.9 mol, and 150 g/mol × 0.9 mol = 135 g, which is close to 114 g of solute.

Follow-Up Problem 3.23 In biochemistry laboratories, solutions of sucrose (table sugar, $C_{12}H_{22}O_{11}$) are used in high-speed centrifuges to separate the parts of a biological cell. What volume (L) of 3.30 mol/L sucrose contains 135 g of solute? Include a road map that shows how you planned your solution.

Preparing and Diluting Molar Solutions

Notice that the volume term in the denominator of the concentration expression in Equation 3.9 is the *solution* volume, *not* the *solvent* volume. This means that you *cannot* dissolve 1 mol of solute in 1 L of solvent to make a 1 mol/L solution. Because the solute volume adds to the solvent volume, the total volume (solute + solvent) is *more* than 1 L, so the concentration is *less* than 1 mol/L.

Preparing a Solution Correctly preparing a solution of a solid solute requires four steps. These steps can be illustrated by the preparation of 0.500 L of 0.350 mol/L nickel(II) nitrate hexahydrate [$Ni(NO_3)_2 \cdot 6H_2O$]:

1. *Determine the mass of the solid.* Calculate the mass of the solid needed by converting from volume (L) to amount (mol) and then to mass (g):

$$\text{Mass (g) of solute} = m = V \times c \times \mathcal{M} = 0.500\ \cancel{\text{L soln}} \times \frac{0.350\ \cancel{\text{mol } Ni(NO_3)_2 \cdot 6H_2O}}{1\ \cancel{\text{L soln}}} \times \frac{290.82 \text{ g } Ni(NO_3)_2 \cdot 6H_2O}{1\ \cancel{\text{mol } Ni(NO_3)_2 \cdot 6H_2O}} = 50.9 \text{ g } Ni(NO_3)_2 \cdot 6H_2O$$

2. *Transfer the solid.* We need 0.500 L of solution, so we choose a 500 mL volumetric flask, add enough distilled water to dissolve the solute fully (usually about half of the final volume, or 250 mL of distilled water in this example), and transfer the solute. Wash down any solid that is clinging to the neck using a wash bottle containing the solvent.
3. *Dissolve the solid.* Swirl the flask until all the solute is dissolved. If necessary, wait until the solution is at room temperature. (As we will discuss in Chapter 12, the solution process may be accompanied by heating or cooling.)
4. *Add solvent to the final volume.* Add distilled water to bring the solution volume to the line on the neck of the flask. This process must be carried out precisely and often requires practice to achieve perfectly. Ideally, the final amount of solvent should be added dropwise using a pipette. Once the required amount of solvent has been added, cover the flask, and mix thoroughly again.

Figure 3.14 shows the last three steps; as explained in step 4, it is best to use a pipette to add solvent dropwise.

FIGURE 3.14 Laboratory preparation of solutions

Diluting a Solution A concentrated solution is converted to a dilute solution by adding solvent, which means that the volume of the solution increases, but the amount (mol) of solute stays the same. As a result, the dilute solution contains *fewer solute particles per unit*

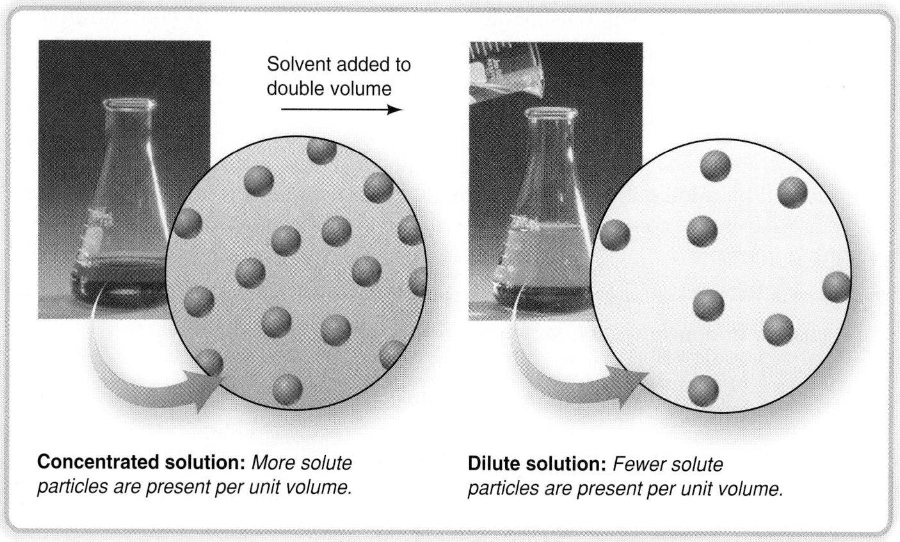
Solvent added to
double volume
Concentrated solution: *More solute particles are present per unit volume.*
Dilute solution: *Fewer solute particles are present per unit volume.*

FIGURE 3.15 Converting a concentrated solution to a dilute solution

volume and, thus, has a lower concentration than the concentrated solution (Figure 3.15). If you need several different dilute solutions, prepare a concentrated solution (*stock solution*) that you can store and dilute as needed.

Sample Problem 3.24 Preparing a Dilute Solution from a Concentrated Solution

Problem Isotonic saline is 0.15 mol/L aqueous NaCl. It simulates the total concentration of ions in many cellular fluids, and its uses range from cleaning contact lenses to washing red blood cells. How would you prepare 0.80 L of isotonic saline from a 6.0 mol/L stock solution?

Plan To dilute a concentrated solution (soln), we add only solvent, so the *amount (mol) of solute is the same in both solutions*. We know the volume (0.80 L) and concentration (0.15 mol/L) of the dilute (dil) NaCl solution needed, so we find the amount (mol) of NaCl that it contains. Then we find the volume (L) of concentrated (conc; 6.0 mol/L) NaCl solution that contains the same amount (mol). We add solvent *up to* the final volume (*see road map*).

Road Map

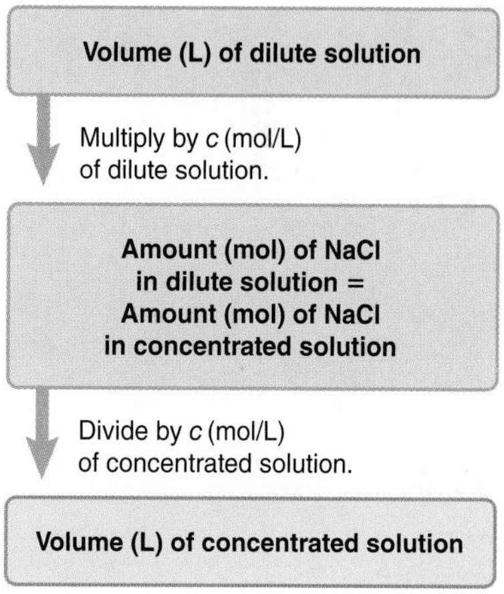
Volume (L) of dilute solution
Multiply by *c* (mol/L) of dilute solution.
Amount (mol) of NaCl in dilute solution = Amount (mol) of NaCl in concentrated solution
Divide by *c* (mol/L) of concentrated solution.
Volume (L) of concentrated solution

Solution Find the amount (mol) of solute in the dilute solution:

$$\text{Amount (mol) of NaCl in dil soln} = n = V \times c = 0.80\ \cancel{\text{L soln}} \times \frac{0.15\ \text{mol NaCl}}{1\ \cancel{\text{L soln}}}$$
$$= 0.12\ \text{mol NaCl}$$

Next, we need to find the amount (mol) of solute in the concentrated solution. Because we add only solvent to dilute the solution,

$$\text{Amount (mol) of NaCl in dil soln} = \text{amount (mol) of NaCl in conc soln}$$
$$= 0.12\ \text{mol NaCl}$$

Find the volume (L) of concentrated solution that contains 0.12 mol of NaCl:

$$\text{Volume (L) of conc NaCl soln} = V = \frac{n}{c} = 0.12\ \cancel{\text{mol NaCl}} \times \frac{1\ \text{L soln}}{6.0\ \cancel{\text{mol NaCl}}}$$
$$= 0.020\ \text{L soln}$$

To prepare 0.80 L of dilute solution, place 0.020 L of 6.0 mol/L NaCl in a 1.0 L graduated cylinder, add distilled water (~780 mL) to the 0.80 L mark, and mix thoroughly.

Check The answer seems reasonable because a small volume of concentrated solution is used to prepare a large volume of dilute solution. Also, the ratio of volumes (0.020 L/0.80 L) is the same as the ratio of concentrations [0.15 mol/L/(6.0 mol/L)].

Follow-Up Problem 3.24 A chemical engineer dilutes a stock solution of sulfuric acid by adding 25.0 m^3 of 7.50 mol/L acid to enough water to make 500. m^3. What is the concentration of sulfuric acid in the diluted solution in g/mL?

Solving Dilution Problems To solve dilution problems and other problems involving a change in concentration, apply the following relationship:

$$c_{dil} \times V_{dil} = \text{amount (mol)} = c_{conc} \times V_{conc} \qquad (3.10)$$

In this relationship, c and V are the concentration and volume of the *dil*ute (subscript "dil") and *conc*entrated (subscript "conc") solutions. Using the values in Sample Problem 3.24, for example, and solving Equation 3.10 for V_{conc} gives

$$V_{conc} = \frac{c_{dil} \times V_{dil}}{c_{conc}} = \frac{(0.15\ \cancel{\text{mol/L}})(0.80\ \text{L})}{(6.0\ \cancel{\text{mol/L}})} = 0.020\ \text{L}$$

Notice that Sample Problem 3.24 had the same calculation broken into two parts:

$$V_{conc} = 0.80\ \cancel{\text{L}} \times \frac{0.15\ \cancel{\text{mol NaCl}}}{1\ \cancel{\text{L}}} \times \frac{1\ \text{L}}{6.0\ \cancel{\text{mol NaCl}}}$$
$$= 0.020\ \text{L}$$

In the next sample problem, we use a variation of this relationship, with molecular scenes showing numbers of particles, to visualize changes in concentration.

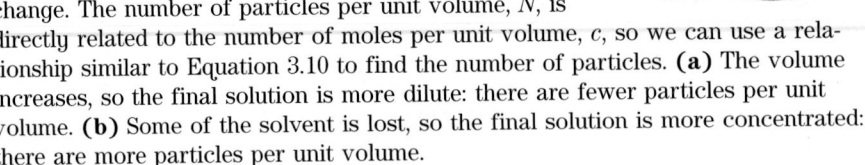

Sample Problem 3.25 Visualizing Changes in Concentration

Problem The circle shown at right represents a unit volume of a solution. Draw a circle representing a unit volume of the solution after each of these changes:

(a) For every 1 mL of solution, 1 mL of solvent is added.

(b) One-third of the solvent is boiled off.

Plan Given the starting solution, we have to find the number of solute particles in a unit volume after each change. The number of particles per unit volume, N, is directly related to the number of moles per unit volume, c, so we can use a relationship similar to Equation 3.10 to find the number of particles. **(a)** The volume increases, so the final solution is more dilute: there are fewer particles per unit volume. **(b)** Some of the solvent is lost, so the final solution is more concentrated: there are more particles per unit volume.

Solution (a) Find the number of particles in the dilute solution, N_{dil}:

$$N_{dil} \times V_{di} = N_{conc} \times V_{conc}$$

Thus,

$$N_{dil} = N_{conc} \times \frac{V_{conc}}{V_{dil}} = 8\ \text{particles} \times \frac{1\ \cancel{\text{mL}}}{2\ \cancel{\text{mL}}} = 4\ \text{particles (\textit{see margin})}$$

Solution (a)

(b) Find the number of particles in the concentrated solution, N_{conc}:

$$N_{dil} \times V_{dil} = N_{conc} \times V_{conc}$$

Thus,

$$N_{conc} = N_{dil} \times \frac{V_{dil}}{V_{conc}} = 8\ \text{particles} \times \frac{1\ \cancel{\text{mL}}}{\frac{2}{3}\ \cancel{\text{mL}}} = 12\ \text{particles (\textit{see margin})}$$

Solution (b)

Check In (a), the volume is doubled (from 1 mL to 2 mL), so the number of particles should be halved: $\frac{1}{2}$ of 8 is 4. In (b), the volume is $\frac{2}{3}$ of the original, so the number of particles should be $\frac{3}{2}$ of the original: $\frac{3}{2}$ of 8 is 12.

Comment In (b), we assumed that only solvent boils off. This is true with nonvolatile solutes, such as ionic compounds, but in Chapter 12, we will encounter solutions in which both the solvent *and* the solute are volatile.

Follow-Up Problem 3.25 The circle labelled A represents a unit volume of a solution. Explain the changes that must be made to A to obtain the unit volumes in B and C.

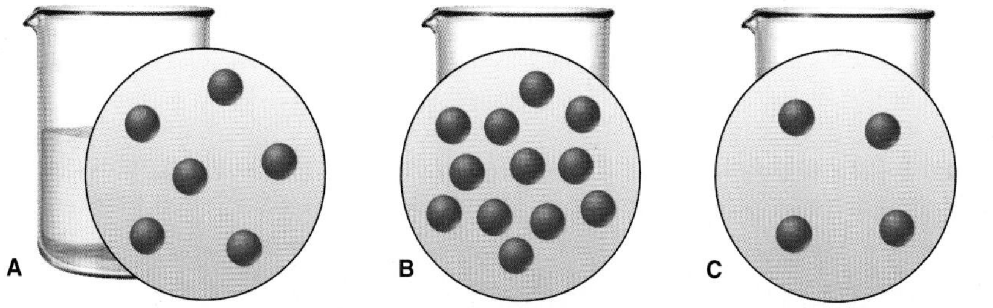

Stoichiometry of Reactions in Solution

Solving stoichiometry problems for reactions in solution requires the additional step of converting the volume of the reactant or product in solution to the amount (mol):

1. Balance the equation.
2. Find the amount (mol) of one substance from the volume and concentration.
3. Relate the amount to the stoichiometrically equivalent amount of another substance.
4. Convert to the desired units.

Sample Problem 3.26 Calculating Quantities of Reactants and Products for a Reaction in Solution

Problem Specialized cells in the stomach release HCl to aid digestion. If they release too much, the excess can be neutralized with an antacid. A common antacid contains magnesium hydroxide, which reacts with the acid to form water and magnesium chloride solution. Suppose that you are a government chemist who is testing commercial antacids. You use 0.10 mol/L HCl to simulate the acid concentration in the stomach. What volume (L) of "stomach acid" reacts with a tablet that contains 0.10 g of magnesium hydroxide?

Plan We are given the mass (0.10 g) of magnesium hydroxide, $Mg(OH)_2$, that reacts with the acid. We also know the concentration of the acid ([HCl] = 0.10 mol/L). We must find the volume of the acid. After writing the balanced equation, we convert the mass (g) of $Mg(OH)_2$ to amount (mol) and use the molar ratio to find the amount (mol) of HCl that reacts with it. Then we use [HCl] to find the volume (L) that contains this amount (*see road map*).

Road Map

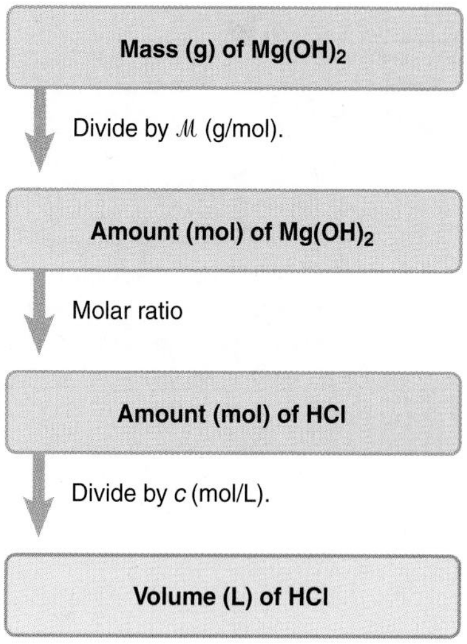

Solution Write the balanced equation:

$$Mg(OH)_2(s) + 2HCl(aq) \longrightarrow MgCl_2(aq) + 2H_2O(l)$$

Convert from mass (g) of $Mg(OH)_2$ to amount (mol):

$$\text{Amount (mol) of } Mg(OH)_2 = n = \frac{m}{\mathcal{M}} = 0.10 \cancel{\text{g Mg(OH)}_2} \times \frac{1 \text{ mol Mg(OH)}_2}{58.33 \cancel{\text{g Mg(OH)}_2}}$$
$$= 1.7 \times 10^{-3} \text{ mol Mg(OH)}_2$$

Convert from amount (mol) of $Mg(OH)_2$ to amount (mol) of HCl:

$$\text{Amount (mol) of HCl} = 1.7 \times 10^{-3} \cancel{\text{mol Mg(OH)}_2} \times \frac{2 \text{ mol HCl}}{1 \cancel{\text{mol Mg(OH)}_2}}$$
$$= 3.4 \times 10^{-3} \text{ mol HCl}$$

Convert from amount (mol) of HCl to volume (L):

$$\text{Volume (L) of HCl} = V = \frac{n}{c} = 3.4 \times 10^{-3}\ \text{mol HCl} \times \frac{1\ \text{L}}{0.10\ \text{mol HCl}} = 3.4 \times 10^{-2}\ \text{L}$$

Check The size of the answer seems reasonable: a small volume of dilute acid (0.034 L of 0.10 mol/L) reacts with a small amount of antacid (0.0017 mol).

Comment In Chapter 4, you will see that this equation is an oversimplification, because HCl and $MgCl_2$ exist in solution as separated ions.

Follow-Up Problem 3.26 Another active ingredient in some antacids is aluminum hydroxide. Which is more effective at neutralizing stomach acid: magnesium hydroxide or aluminum hydroxide? (*Hint*: "Effectiveness" refers to the amount of acid that reacts with a given mass of antacid. You already know the effectiveness of 0.10 g of $Mg(OH)_2$.)

Except for the additional step of finding amounts (mol) in solution, limiting-reactant problems for reactions in solution are handled just like other such problems.

Sample Problem 3.27 Solving Limiting-Reactant Problems for Reactions in Solution

Road Map

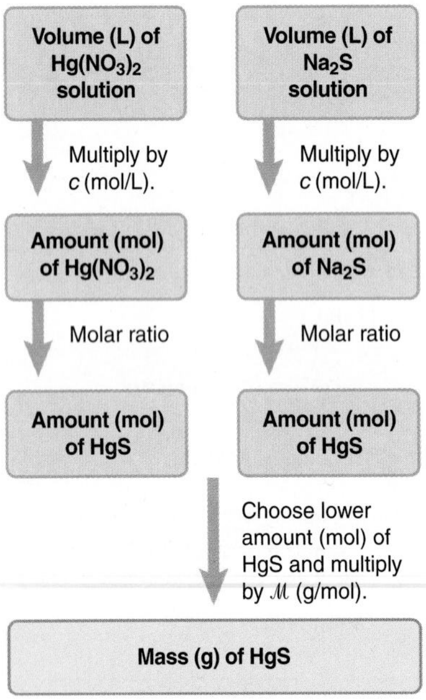

Problem Mercury and its compounds have many uses, from fillings for teeth (as a mixture containing silver, copper, and tin) in the past to the current production of chlorine. Because of their toxicity, however, soluble mercury compounds, such as mercury(II) nitrate, must be removed from industrial wastewater. One removal method reacts the wastewater with a sodium sulfide solution to produce solid mercury(II) sulfide and sodium nitrate solution. In a laboratory simulation, 0.050 L of 0.010 mol/L mercury(II) nitrate reacts with 0.020 L of 0.10 mol/L sodium sulfide. **(a)** What mass of mercury(II) sulfide forms? **(b)** Write a reaction table for this process.

Plan This is a limiting-reactant problem because *the quantities of two reactants are given*. After balancing the equation, we determine the limiting reactant. From the molarity (concentration in mol/L) and volume of each solution, we calculate the amount (mol) of each reactant. Then we use the molar ratio to find the amount of product (HgS) that each reactant forms. The limiting reactant forms a smaller amount (mol) of HgS, which we convert to mass (g) of HgS using its molar mass (*see road map*). We use the amount of HgS formed from the limiting reactant in the reaction table.

Solution (a) Write the balanced equation:

$$Hg(NO_3)_2(aq) + Na_2S(aq) \longrightarrow HgS(s) + 2NaNO_3(aq)$$

Find the amount (mol) of HgS formed from $Hg(NO_3)_2$ (by combining the steps):

$$\text{Amount (mol) of HgS} = 0.050\ \text{L soln} \times \frac{0.010\ \text{mol Hg(NO}_3)_2}{1\ \text{L soln}} \times \frac{1\ \text{mol HgS}}{1\ \text{mol Hg(NO}_3)_2}$$
$$= 5.0 \times 10^{-4}\ \text{mol HgS}$$

Find the amount (mol) of HgS from Na_2S (by combining the steps):

$$\text{Amount (mol) of HgS} = 0.020\ \text{L soln} \times \frac{0.10\ \text{mol Na}_2\text{S}}{1\ \text{L soln}} \times \frac{1\ \text{mol HgS}}{1\ \text{mol Na}_2\text{S}}$$
$$= 2.0 \times 10^{-3}\ \text{mol HgS}$$

$Hg(NO_3)_2$ is the limiting reactant because it forms fewer moles of HgS. Convert the amount (mol) of HgS formed from $Hg(NO_3)_2$ to mass (g):

$$\text{Mass (g) of HgS} = m = n \times M = 5.0 \times 10^{-4}\ \text{mol HgS} \times \frac{232.7\ \text{g HgS}}{1\ \text{mol HgS}} = 0.12\ \text{g HgS}$$

(b) With $Hg(NO_3)_2$ as the limiting reactant, the reaction table looks like this:

Amount (mol)	$Hg(NO_3)_2(aq)$	+	$Na_2S(aq)$	$\longrightarrow$	HgS(s)	+	$2NaNO_3(aq)$
Initial	5.0×10^{-4}		2.0×10^{-3}		0		0
Change	-5.0×10^{-4}		-5.0×10^{-4}		$+5.0 \times 10^{-4}$		$+1.0 \times 10^{-3}$
Final	0		1.5×10^{-3}		5.0×10^{-4}		1.0×10^{-3}

A large excess of Na_2S remains after the reaction. Note that the amount of $NaNO_3$ formed is twice the amount of $Hg(NO_3)_2$ consumed, as the balanced equation shows.

Check As a check on our choice of the limiting reactant, let us use the alternative method described at the end of Sample Problem 3.20 (see Comment).

Find the amount (mol) of reactants given:

$$\text{Amount (mol) of Hg(NO}_3)_2 = n = V \times c = 0.050 \text{ L soln} \times \frac{0.010 \text{ mol Hg(NO}_3)_2}{1 \text{ L soln}}$$
$$= 5.0 \times 10^{-4} \text{ mol Hg(NO}_3)_2$$

$$\text{Amount (mol) of Na}_2\text{S} = n = V \times c = 0.020 \text{ L soln} \times \frac{0.10 \text{ mol Na}_2\text{S}}{1 \text{ L soln}}$$
$$= 2.0 \times 10^{-3} \text{ mol Na}_2\text{S}$$

The molar ratio of the reactants is 1 $Hg(NO_3)_2$/1 Na_2S. Therefore, $Hg(NO_3)_2$ is limiting because there is less of it than we would need to react with all the available Na_2S.

Follow-Up Problem 3.27 Despite the toxicity of lead, many mass-produced toys are still found to contain a significant proportion of this metal.

(a) When 268 mL of 1.50 mol/L lead(II) acetate reacts with 130. mL of 3.40 mol/L sodium chloride, what mass (g) of solid lead(II) chloride can form? (Sodium acetate solution also forms.)

(b) Using the abbreviation "OAc" for the acetate ion, write a reaction table for the process.

Figure 3.16 provides a visual review of all the stoichiometric relationships we have discussed in this chapter.

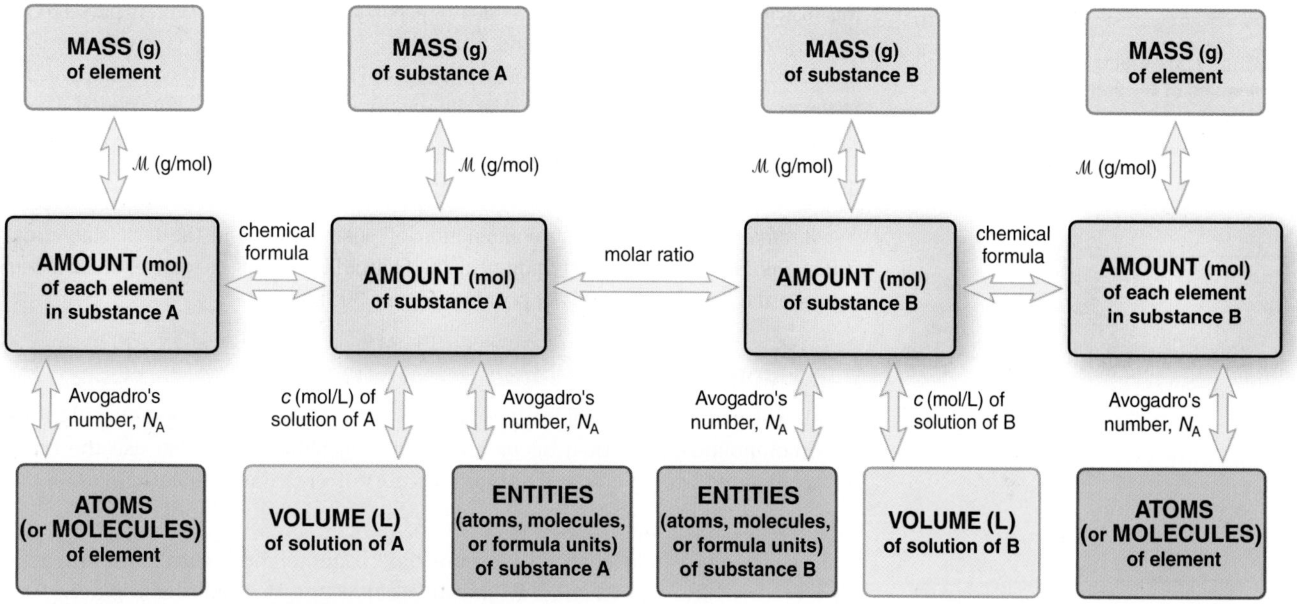

FIGURE 3.16 **An overview of amount-mass-number stoichiometric relationships**

SUMMARY OF SECTION 3.5

- When reactions occur in solution, the amounts of reactants and products are given in terms of concentration and volume.
- Concentration is often expressed as the number of moles of solute dissolved in 1 L of solution. A concentrated solution is converted to a dilute solution by adding solvent.
- By using mol/L as a conversion factor, we can apply the principles of stoichiometry to reactions in solution.

3.6 The Role of Water as a Solvent

For any reaction in solution, the solvent plays a key role that depends on its chemical nature. Some solvents passively disperse the substances into individual molecules. Water is much more active, interacting strongly with the substances and sometimes even reacting with them. In this section, we look at how the water molecule interacts with both ionic and covalent solutes. Why do certain compounds dissolve while others do not appear to do so? Why do we say that some compounds dissolve while we say that others dissociate, and what is the difference? This section is a brief introduction to the topic, since you need to understand what happens when certain types of chemical compounds are placed in water. You will learn more about the topic in Chapter 11.

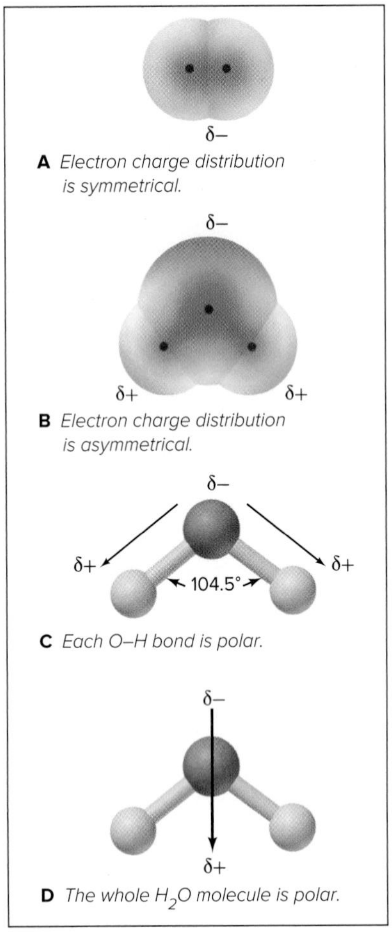

δ–
A Electron charge distribution is symmetrical.
δ–
δ+
δ+
B Electron charge distribution is asymmetrical.
δ–
δ+
δ+
104.5°
C Each O–H bond is polar.
δ–
δ+
D The whole H_2O molecule is polar.

FIGURE 3.17 Electron distribution in molecules of H_2 and H_2O

The Polar Nature of Water

On the atomic scale, the great solvent power of water arises from its *uneven distribution of electron charge* and its *bent molecular shape*, which create a *polar molecule*:

1. *Uneven charge distribution.* Recall, from Section 2.7, that the electrons in a covalent bond are shared between the atoms. In a bond between identical atoms—as in H_2, Cl_2, and O_2—the sharing is equal and the electron charge is distributed evenly between the two nuclei (see the symmetrical shading in the space-filling model in Figure 3.17A). In covalent bonds between different atoms, the sharing is uneven because one atom attracts the electron pair more strongly than the other atom does.

   For example, in each O—H bond of water, the shared electrons are closer to the O atom because an O atom attracts electrons more strongly than an H atom does. (We will discuss the reasons in Chapter 9.) This uneven charge distribution creates a polar bond, a bond with partially charged "poles." In Figure 3.17B, the asymmetrical shading shows this distribution, and the δ symbol indicates a partial charge. The O end is partially negative, represented by red shading and δ–, and the H end is partially positive, represented by blue shading and δ+. In the ball-and-stick model in Figure 3.17C, the polar arrow points to the positive pole, and the tail marks the negative pole, by IUPAC convention.

2. *Bent molecular shape.* The sequence of the H—O—H atoms in water is not linear: the water molecule is bent with a bond angle of 104.5° (Figure 3.17C).

3. *Molecular polarity.* The combination of polar bonds and the bent shape makes water a **polar molecule**: the region near the O atom is partially negative, and the region between the H atoms is partially positive (Figure 3.17D).

Ionic Compounds in Water

In this subsection, we consider two closely related aspects of aqueous solutions of ionic compounds: how they occur and how they behave. We also use the formula for a compound to calculate the amount (mol) of each ion in solution.

How Ionic Compounds Dissolve: Replacement of Charge Attractions In an ionic solid, oppositely charged ions are held together by electrostatic attractions (see Figure 1.3C and Section 2.7). Water separates the ions by *replacing these attractions with stronger attractions between several water molecules and each*

FIGURE 3.18 An ionic compound dissolving in water. The inset shows the polar arrow and partial charges (not shown in the rest of the scene) of each water molecule.

ion. Picture a granule of a soluble ionic compound in water: the negative ends of some water molecules are attracted to the cations, and the positive ends of other water molecules are attracted to the anions (Figure 3.18). Dissolution occurs because the *attractions between each type of ion and several water molecules outweigh the attractions between the ions*. Gradually, all the ions separate (dissociate), become **solvated** (*surrounded closely by solvent molecules*), and then move randomly in the solution.

For an ionic compound that does not dissolve significantly in water, the attraction between the ions is greater than the attraction between the ions and water. Actually, these sparingly soluble substances *do* dissolve to a very small extent, usually several orders of magnitude less than so-called soluble substances. For example, NaCl (a "soluble" compound) has over 4×10^4 times the solubility of AgCl (a "sparingly soluble" compound):

$$\text{Solubility of NaCl in } H_2O \text{ at } 20°C = 365 \text{ g/L}$$

$$\text{Solubility of AgCl in } H_2O \text{ at } 20°C = 0.009 \text{ g/L}$$

In Chapter 12, you will see that dissolving involves more than a contest between the relative attractions of ions for each other or for water. It occurs because it frees ions from the solid to disperse randomly through the solution.

How Ionic Solutions Behave: Electrolytes and Electrical Conductivity When an ionic compound dissolves, the solution's *electrical conductivity*, the flow of electric current, increases dramatically. When electrodes are immersed in distilled water (Figure 3.19A) or pushed into an ionic solid (Figure 3.19B), no current flows, as shown by the unlit bulb. However, in an aqueous solution of the compound, a large current flows, as shown by the lit bulb (Figure 3.19C). Current flow implies the *movement of charged particles*: when the ionic compound dissolves, the separate solvated ions move toward the electrode of opposite charge. A substance that conducts a current when dissolved in water is an **electrolyte**. Soluble ionic compounds are *strong* electrolytes because they dissociate completely and conduct a large current.

Calculating the Amount (mol) of Ions in Solution *The formula for the soluble ionic compound tells the amount (mol) of the component ions in solution.* For example, the equation for dissolving KBr in water to form solvated ions is

$$KBr(s) \xrightarrow{H_2O} K^+(aq) + Br^-(aq)$$

("H_2O" above the arrow means that water is the solvent, not a reactant.) Note that 1 mol of KBr dissociates into 2 mol of ions—1 mol of K^+ and 1 mol of Br^-. Sample Problems 3.28 and 3.29 apply these ideas, first with molecular scenes and then in calculations.

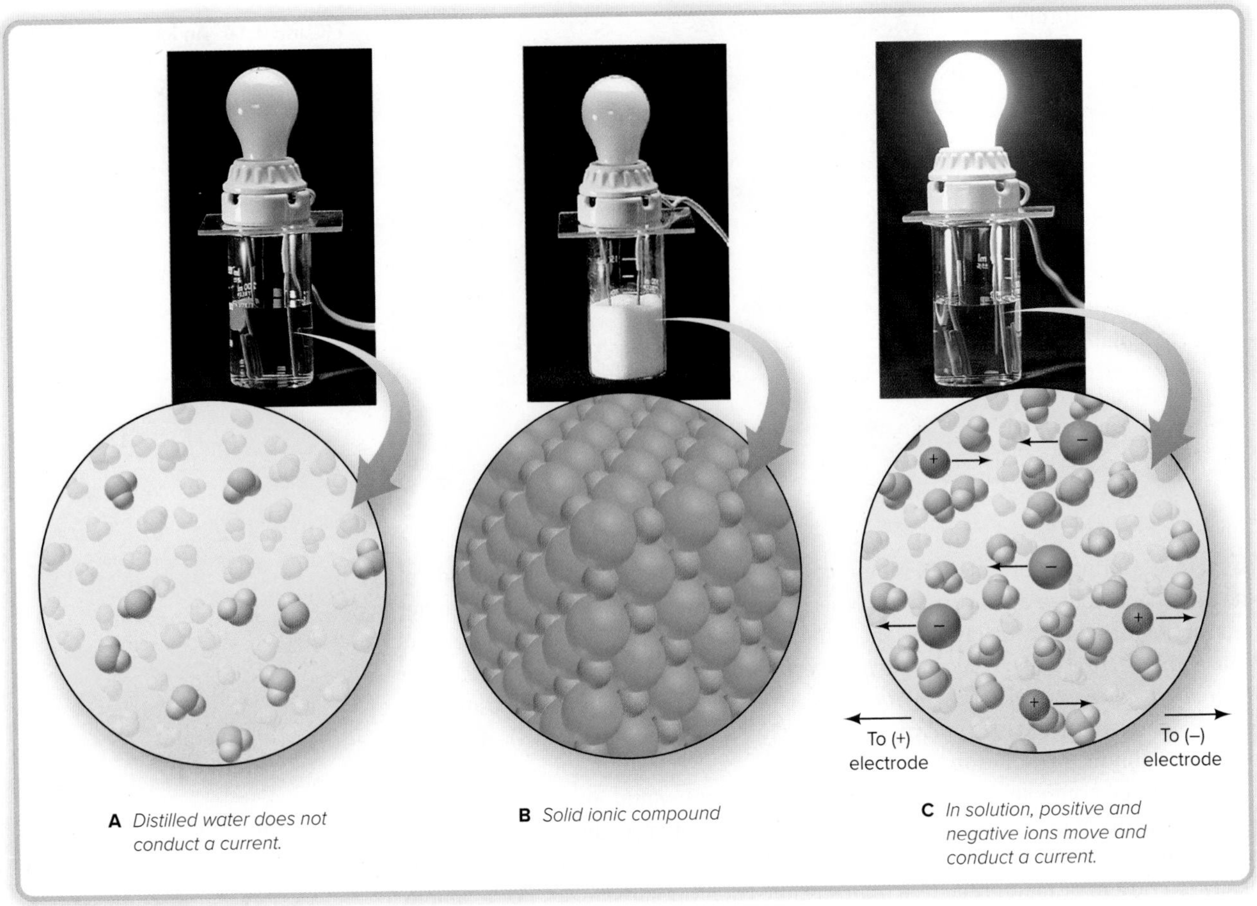
To (+)
electrode
To (–)
electrode
A Distilled water does not conduct a current.
B Solid ionic compound
C In solution, positive and negative ions move and conduct a current.

FIGURE 3.19 The electrical conductivity of ionic solutions

Sample Problem 3.28 Using Molecular Scenes to Depict an Ionic Compound in Aqueous Solution

Problem The beakers contain aqueous solutions of the strong electrolyte potassium sulfate.

(a) Which beaker best represents the compound in solution (water molecules are not shown)?

(b) If each particle represents 0.1 mol, what is the total number of particles in solution?

A
B
C
D

Plan (a) We determine the formula and write an equation for 1 mol of compound dissociating into ions. Potassium sulfate is a strong electrolyte, so it dissociates completely, but, in general, *polyatomic ions remain intact in solution.*

(b) We count the number of separate particles and then multiply by 0.1 mol and by Avogadro's number.

Solution **(a)** The formula is K_2SO_4, so the equation is

$$K_2SO_4(s) \xrightarrow{H_2O} 2K^+(aq) + SO_4^{2-}(aq)$$

There are two separate 1+ particles for every 2− particle, so beaker C is best.

(b) There are 9 particles, so the total amount (mol) of particles is 0.9 mol, and we have

$$\text{No. of particles} = 0.9\ \cancel{\text{mol}} \times \frac{6.022 \times 10^{23}\ \text{particles}}{1\ \cancel{\text{mol}}} = 5.420 \times 10^{23}\ \text{particles}$$

Check Rounding to check the math in (b) gives $0.9 \times 6 = 5.4$, so the answer seems correct. The number of particles is an exact number since we actually counted them. Thus, the answer can have as many significant figures as in Avogadro's number.

Follow-Up Problem 3.28 **(a)** Which strong electrolyte is dissolved in the water in the beaker in the margin: LiBr, Cs_2CO_3, or $BaCl_2$? (The water molecules are not shown.)

(b) If each particle represents 0.05 mol, what mass (g) of compound was dissolved?

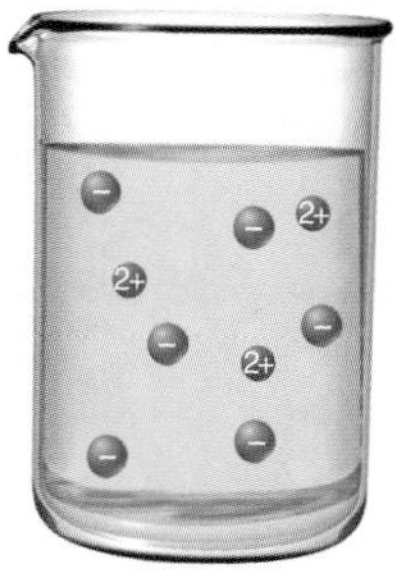

Sample Problem 3.29 Determining Amount (mol) of Ions in Solution

Problem What amount (mol) of each ion is in each solution?

(a) 5.0 mol of ammonium sulfate dissolved in water

(b) 78.5 g of cesium bromide dissolved in water

(c) 7.42×10^{22} formula units of copper(II) nitrate dissolved in water

(d) 35 mL of 0.84 mol/L zinc chloride

Plan We write an equation that shows 1 mol of compound dissociating into ions. **(a)** We multiply the amount (mol) of ions by 5.0. **(b)** We first convert m, mass (g), to n, amount (mol). **(c)** We first convert formula units to n, amount (mol). **(d)** We first convert V, volume, and c, concentration, to n, amount (mol).

Solution **(a)** $(NH_4)_2SO_4(s) \xrightarrow{H_2O} 2NH_4^+(aq) + SO_4^{2-}(aq)$

Calculate the amount (mol) of NH_4^+ ions:

$$\text{Amount (mol) of } NH_4^+ = 5.0\ \cancel{\text{mol }(NH_4)_2SO_4} \times \frac{2\ \text{mol } NH_4^+}{1\ \cancel{\text{mol }(NH_4)_2SO_4}} = 10.\ \text{mol } NH_4^+$$

The formula shows 1 mol of SO_4^{2-} per mole of $(NH_4)_2SO_4$, so 5.0 mol of SO_4^{2-} is also present.

(b) $CsBr(s) \xrightarrow{H_2O} Cs^+(aq) + Br^-(aq)$

Convert from mass (g) to amount (mol):

$$\text{Amount (mol) of CsBr} = 78.5\ \cancel{\text{g CsBr}} \times \frac{1\ \text{mol CsBr}}{212.8\ \cancel{\text{g CsBr}}} = 0.369\ \text{mol CsBr}$$

Thus, 0.369 mol of Cs^+ and 0.369 mol of Br^- are present.

(c) $Cu(NO_3)_2(s) \xrightarrow{H_2O} Cu^{2+}(aq) + 2NO_3^-(aq)$

Convert from formula units to amount (mol):

$$\text{Amount (mol) of } Cu(NO_3)_2 = 7.42 \times 10^{22}\ \cancel{\text{formula units } Cu(NO_3)_2} \times \frac{1\ \text{mol } Cu(NO_3)_2}{6.022 \times 10^{23}\ \cancel{\text{formula units } Cu(NO_3)_2}} = 0.123\ \text{mol } Cu(NO_3)_2$$

$$\text{Amount (mol) of } NO_3^- = 0.123\ \cancel{\text{mol } Cu(NO_3)_2} \times \frac{2\ \text{mol } NO_3^-}{1\ \cancel{\text{mol } Cu(NO_3)_2}} = 0.246\ \text{mol } NO_3^-$$

Therefore, 0.123 mol of Cu^{2+} is also present.

(d) $ZnCl_2(aq) \longrightarrow Zn^{2+}(aq) + 2Cl^-(aq)$

Convert from volume (mL) and concentration (mol/L) to amount (mol):

$$\text{Amount (mol) of } ZnCl_2 = 35\ \cancel{mL} \times \frac{1\ \cancel{L}}{10^3\ \cancel{mL}} \times \frac{0.84\ mol\ ZnCl_2}{1\ \cancel{L}} = 2.9 \times 10^{-2}\ mol\ ZnCl_2$$

$$\text{Amount (mol) of } Cl^- = 2.9 \times 10^{-2}\ \cancel{mol\ ZnCl_2} \times \frac{2\ mol\ Cl^-}{1\ \cancel{mol\ ZnCl_2}} = 5.8 \times 10^{-2}\ mol\ Cl^-$$

Therefore, 2.9×10^{-2} mol of Zn^{2+} is also present.

Check Round off to check the math and see if the relative numbers of moles of ions are consistent with the formula. For instance, in (a), 10 mol NH_4^+/5.0 mol SO_4^{2-} = 2 NH_4^+/1 SO_4^{2-}, or $(NH_4)_2SO_4$. In (d), 0.029 mol Zn^{2+}/0.058 mol Cl^- = 1 Zn^{2+}/2 Cl^-, or $ZnCl_2$.

Follow-Up Problem 3.29 What amount (mol) of each ion is in each solution?

(a) 2 mol of potassium perchlorate dissolved in water

(b) 354 g of magnesium acetate dissolved in water

(c) 1.88×10^{24} formula units of ammonium chromate dissolved in water

(d) 1.32 L of 0.55 mol/L sodium bisulfate

Covalent Compounds in Water

Water also dissolves many covalent (molecular) compounds. Table sugar (sucrose, $C_{12}H_{22}O_{11}$), beverage (grain) alcohol (ethanol, CH_3CH_2OH), and automobile antifreeze (ethylene glycol, $HOCH_2CH_2OH$) are some familiar examples. All contain their own polar bonds, which interact with the bonds of water. However, most soluble covalent substances *do not* separate into ions, but remain intact molecules. For example,

$$HOCH_2CH_2OH(l) \xrightarrow{H_2O} HOCH_2CH_2OH(aq)$$

As a result, their aqueous solutions do not conduct an electric current, and these substances are called **nonelectrolytes**. (As you will see shortly, there is a small group of H-containing molecules that act as acids in aqueous solution and *do* dissociate into ions.) Many other covalent substances, such as benzene (C_6H_6) and octane (C_8H_{18}), do not contain polar bonds, and these substances do not dissolve appreciably in water.

SUMMARY OF SECTION 3.6

- Because of polar bonds and a bent shape, the water molecule is polar, and water dissolves many ionic and covalent compounds.
- When an ionic compound dissolves, the attraction between each ion and water replaces the attraction between ions. Soluble ionic compounds are electrolytes because the ions are free to move and, thus, the solution conducts electricity.
- The formula for a soluble ionic compound shows the amount (mol) of each ion in solution, per mole of compound dissolved.
- Water dissolves many covalent substances with polar bonds. These compounds are nonelectrolytes because the molecules remain intact and, thus, the solution does not conduct electricity.

3.7 Writing Equations for Aqueous Ionic Reactions

Chemists use three types of equations to represent aqueous ionic reactions. Let us examine a reaction to see what each type of equation shows. When solutions of silver nitrate and sodium chromate are mixed, brick-red, solid silver chromate (Ag_2CrO_4) forms. Figure 3.20 depicts the reaction at the macroscopic level (*photos*),

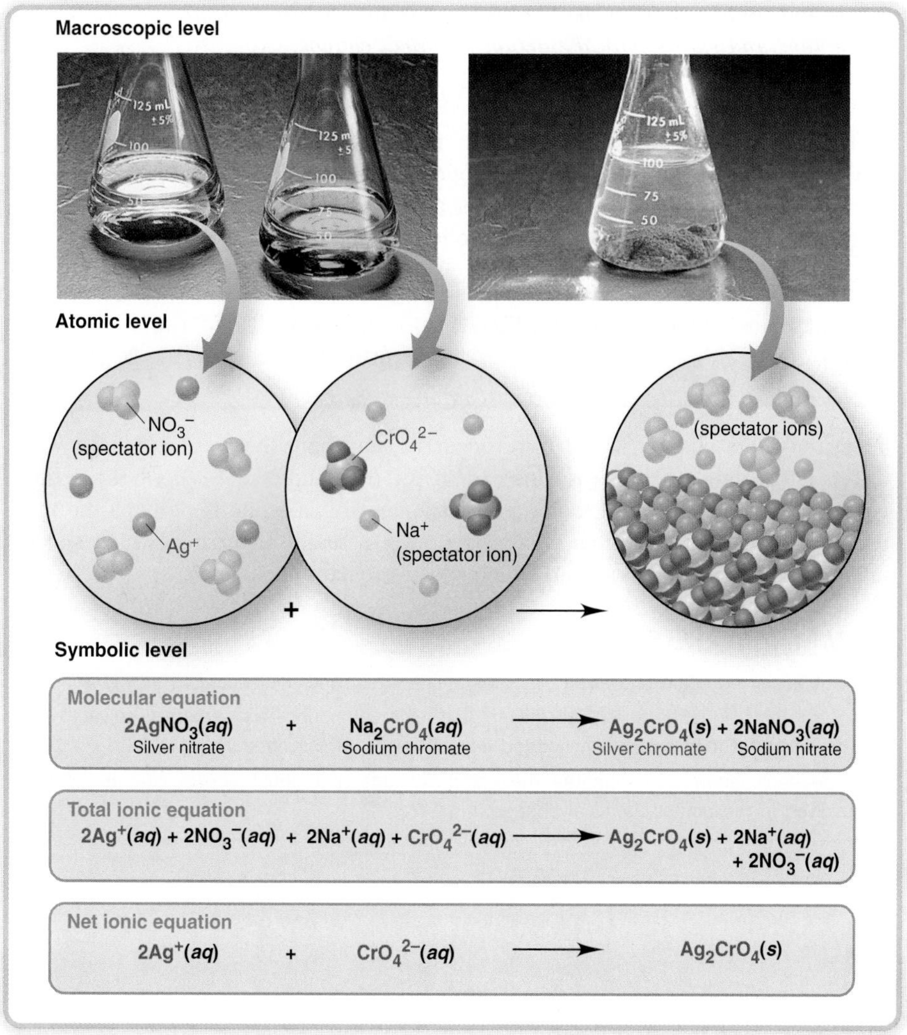

FIGURE 3.20 An aqueous ionic reaction and the three types of equations

the atomic level (*blow-up circles*), and the symbolic level with the three types of equations (reacting ions are in *red* type):

- The **molecular equation** (*top*) reveals the least about the species that are actually in solution because *it shows all the reactants and products as if they were intact, undissociated compounds*. Only the designation for solid, (*s*), tells us that a change has occurred:

$$2AgNO_3(aq) + Na_2CrO_4(aq) \longrightarrow Ag_2CrO_4(s) + 2NaNO_3(aq)$$

- The **total ionic equation** (*middle*) is much more accurate because *it shows all the soluble ionic substances dissociated into ions*. The $Ag_2CrO_4(s)$ stands out as the only undissociated substance:

$$2Ag^+(aq) + 2NO_3^-(aq) + 2Na^+(aq) + CrO_4^{2-}(aq) \longrightarrow Ag_2CrO_4(s) + 2Na^+(aq) + 2NO_3^-(aq)$$

The charges also balance: four positive and four negative for a net zero charge on the left side, and two positive and two negative for a net zero charge on the right side.

Notice that $Na^+(aq)$ and $NO_3^-(aq)$ appear unchanged on both sides of the equation. These are called **spectator ions** (shown with pale colours in the atomic-level scenes). They are not involved in the actual chemical change but are present only as part of the reactants; that is, we cannot add an Ag^+ ion without also adding an anion, such as the NO_3^- ion.

- The **net ionic equation** (*bottom*) is very useful because *it eliminates the spectator ions and shows only the actual chemical change*:

$$2Ag^{+}(aq) + CrO_4^{2-}(aq) \longrightarrow Ag_2CrO_4(s)$$

The formation of solid silver chromate from silver ions and chromate ions *is* the only change. As another example, suppose that we had mixed solutions of potassium chromate, $K_2CrO_4(aq)$, and silver acetate, $AgC_2H_3O_2(aq)$, instead of sodium chromate and silver nitrate. The three ionic equations are given below:

Molecular: $2AgC_2H_3O_2(aq) + K_2CrO_4(aq) \longrightarrow Ag_2CrO_4(s) + 2KC_2H_3O_2(aq)$

Total ionic: $2Ag^{+}(aq) + 2C_2H_3O_2^{-}(aq) + 2K^{+}(aq) + CrO_4^{2-}(aq) \longrightarrow Ag_2CrO_4(s) + 2K^{+}(aq) + 2C_2H_3O_2(aq)$

Net ionic: $2Ag^{+}(aq) + CrO_4^{2-}(aq) \longrightarrow Ag_2CrO_4(s)$

Thus, the same change would have occurred, and only the spectator ions would differ: $K^{+}(aq)$ and $C_2H_3O_2^{-}(aq)$ instead of $Na^{+}(aq)$ and $NO_3^{-}(aq)$. Of these three equations—molecular, total ionic, and net ionic—the last one is perhaps the most useful as it clearly shows us what reaction is actually occurring, which species undergo a chemical change, and which species remain unchanged.

SUMMARY OF SECTION 3.7

- A molecular equation shows all the substances intact and undissociated into ions.
- A total ionic equation shows all the soluble ionic compounds as separate, solvated ions. The spectator ions appear unchanged on both sides of the equation.
- A net ionic equation eliminates the spectator ions and, thus, shows only the species involved in the actual chemical change.

CHAPTER REVIEW GUIDE

Learning Objectives

Relevant section (§) and/or sample problem (SP) numbers appear in parentheses.

Concepts

1. Define *mole unit.* (§3.1)
2. Explain the relationship between the mass of a chemical entity (atomic mass units) and the mass (g) of 1 mol of the entities. (§3.1)
3. Describe and differentiate between the relations among amount of substance (mol), mass (g), and number of chemical entities. (§3.1)
4. Convert between amount-mass-number information in a chemical formula. (§3.1)
5. Describe the difference between the empirical and molecular formulas for a compound. (§3.2)
6. Explain how more than one substance can have the same empirical formula and the same molecular formula (isomers). (§3.2)
7. Explain the importance of balancing equations for the quantitative study of chemical reactions. (§3.3)
8. Convert between amount-mass-number information in a balanced equation. (§3.4)
9. Determine the relation between amounts of reactants and amounts of products. (§3.4)
10. Explain why one reactant limits the amount of product. (§3.4)
11. Discuss the causes of yields that are lower than expected and the distinction between theoretical and actual yields. (§3.4)
12. Explain the meaning of *concentration.* (§3.5)
13. Describe the effect of dilution on the concentration of a solute. (§3.5)
14. Describe how reactions in solution differ from reactions of pure reactants. (§3.5)
15. Explain why water is a polar molecule and how it dissociates ionic compounds into ions. (§3.6)
16. Compare and contrast the differing nature of the species that are present when ionic and covalent compounds dissolve in water. (§3.6)
17. Compare and contrast the distinctions among strong electrolytes, weak electrolytes, and nonelectrolytes. (§3.6)
18. Compare and contrast the three types of equations that specify the species and the chemical change in an aqueous ionic reaction. (§3.7)

Skills

1. Calculate the molar mass of any substance. (§3.1; SPs 3.5, 3.6)
2. Convert between amount of substance (mol), mass (g), and number of chemical entities. (SPs 3.1–3.5)
3. Use mass percent to find the mass of an element in a given mass of a compound. (SPs 3.6, 3.7)
4. Determine empirical and molecular formulas for a compound from the mass percents and molar masses of the elements. (SPs 3.8–3.10)

5. Determine a molecular formula from combustion analysis. (SP 3.11)
6. Convert a chemical statement or a molecular depiction into a balanced equation. (SPs 3.12, 3.13)
7. Using stoichiometrically equivalent molar ratios, convert between amounts of reactants and products in reactions of pure and dissolved substances. (SPs 3.14–3.16, 3.26)
8. Write an overall equation from a series of equations. (SP 3.17)
9. Solve limiting-reactant problems for reactions of pure and dissolved substances. (SPs 3.18–3.20, 3.27)
10. Calculate percent yield. (SP 3.21)
11. Calculate the concentration and mass of a solute in solution. (SPs 3.22, 3.23)
12. Prepare a dilute solution from a concentrated solution. (SP 3.24)
13. Using molecular depictions, describe changes in concentration. (SP 3.25)
14. Using molecular scenes, depict a soluble ionic compound in water. (SP 3.28)
15. Using the formula for a compound, determine the amount (mol) of moles of ions in solution. (SP 3.29)

Key Terms

stoichiometry

Section 3.1
mole (mol)
Avogadro's number, N_A
molar mass ($\mathcal{M}$)

Section 3.2
empirical formula
combustion analysis
isomers

Section 3.3
chemical equation
reactants
products
balancing (stoichiometric) coefficient

Section 3.4
overall (net) equation
limiting reactant
theoretical yield
side reactions
actual yield
percent yield (% yield)
green chemistry

Section 3.5
solute
solvent
concentration, c
molarity, M

Section 3.6
polar molecule
solvated
electrolyte
nonelectrolytes

Section 3.7
molecular equation
total ionic equation
spectator ions
net ionic equation

Key Equations and Relationships

3.1 Number of entities in one mole:

1 mol contains 6.022 141 29 × 10^{23} entities

3.2 Converting amount (mol) to mass (g) using $\mathcal{M}$:

$$\text{Mass (g)} = \text{amount (mol)} \times \text{molar mass}\left(\frac{\text{g}}{\text{mol}}\right)$$

$$m\ (\text{g}) = n\ (\text{mol}) \times \mathcal{M}\left(\frac{\text{g}}{\text{mol}}\right)$$

3.3 Converting mass (g) to amount (mol) using $\frac{1}{\mathcal{M}}$:

$$\text{Amount (mol)} = \text{mass (g)} \times \frac{1\ (\text{mol})}{\text{molar mass (g)}}$$

$$n\ (\text{mol}) = \frac{m\ (\text{g})}{\mathcal{M}\ (\text{g/mol})}$$

3.4 Converting amount (mol) to number of entities:

$$\text{No.of entities} = \text{amount (mol)} \times \frac{6.022 \times 10^{23}\ \text{entities}}{1\ \text{mol}}$$

$$\text{No. of entities} = n\ (\text{mol}) \times N_A\left(\frac{\text{no. of entities}}{\text{mol}}\right)$$

3.5 Converting number of entities to amount (mol):

$$\text{Amount (mol)} = \text{no.of entities} \times \frac{1\ \text{mol}}{6.022 \times 10^{23}\ \text{entities}}$$

$$n = \frac{\text{no. of entities}}{N_A\left(\frac{\text{no. of entities}}{\text{mol}}\right)}$$

3.6 Calculating mass percent:

Mass % of element X

$$= \frac{\text{amount (mol) of X in formula} \times \text{molar mass of X (g/mol)}}{\text{mass (g) of 1 mol of compound}} \times 100$$

$$= \frac{n_{X_{\text{in formula}}} \times \mathcal{M}_X}{\mathcal{M}_{\text{compound}}}$$

3.7 Finding the mass of an element in any mass of compound:

$$\text{Mass of element} = \text{mass of compound} \times \frac{\text{mass of element in 1 mol of compound}}{\text{mass of 1 mol of compound}}$$

3.8 Calculating percent yield:

$$\%\ \text{Yield} = \frac{\text{actual yield}}{\text{theoretical yield}} \times 100$$

3.9 Defining concentration:

$$\text{Concentration (mol/L)} = \frac{\text{amount of solute (mol)}}{\text{volume of solution (L)}}$$

$$c = \frac{n}{V}$$

3.10 Diluting a concentrated solution:

$$c_{\text{dil}} \times V_{\text{dil}} = \text{amount (mol)} = c_{\text{conc}} \times V_{\text{conc}}$$

$$c_{\text{dil}} \times V_{\text{dil}} = n = c_{\text{conc}} \times V_{\text{conc}}$$

Brief Solutions to Follow-Up Problems

3.1
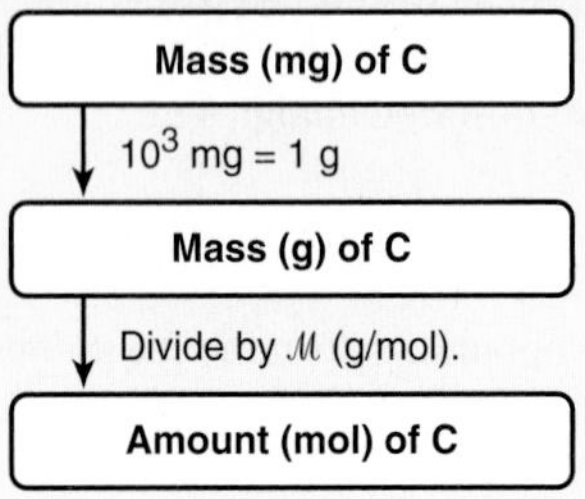

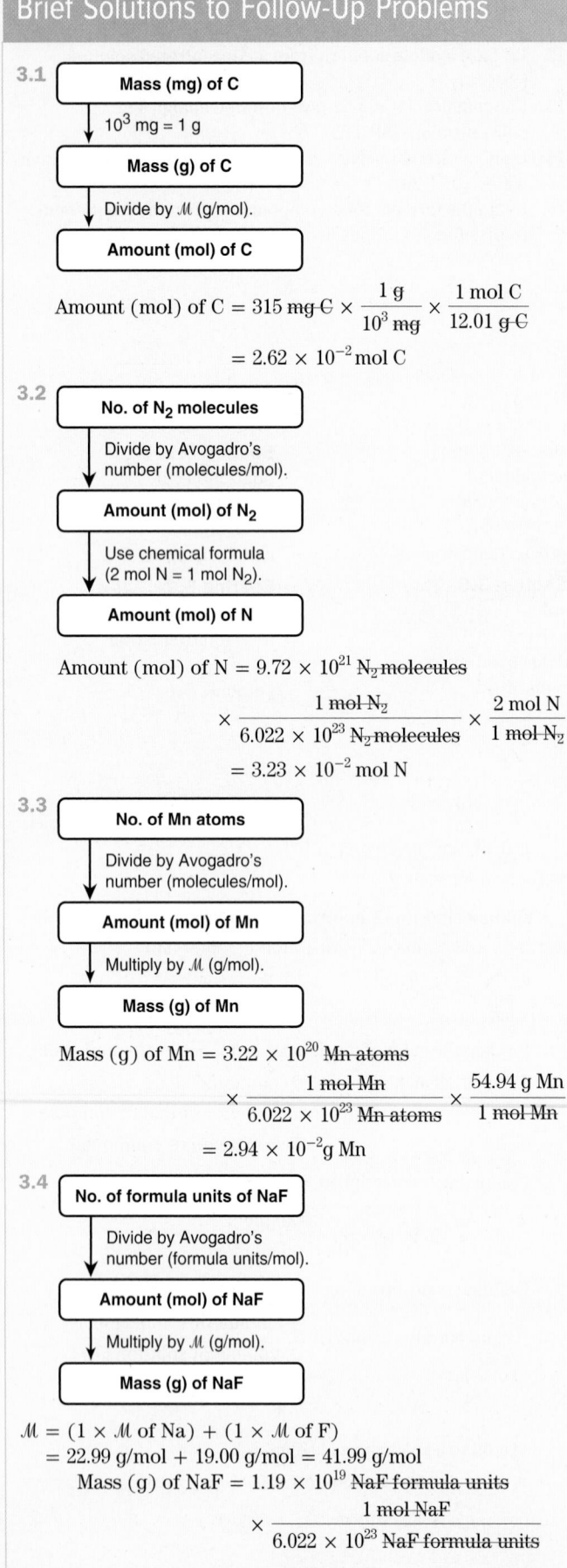

$$\text{Amount (mol) of C} = 315\ \text{mg C} \times \frac{1\ \text{g}}{10^3\ \text{mg}} \times \frac{1\ \text{mol C}}{12.01\ \text{g C}}$$
$$= 2.62 \times 10^{-2}\ \text{mol C}$$

3.2
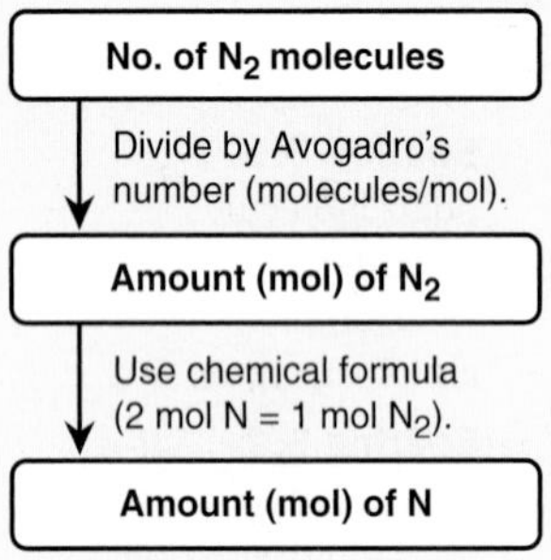

$$\text{Amount (mol) of N} = 9.72 \times 10^{21}\ \text{N}_2\ \text{molecules} \times \frac{1\ \text{mol N}_2}{6.022 \times 10^{23}\ \text{N}_2\ \text{molecules}} \times \frac{2\ \text{mol N}}{1\ \text{mol N}_2}$$
$$= 3.23 \times 10^{-2}\ \text{mol N}$$

3.3
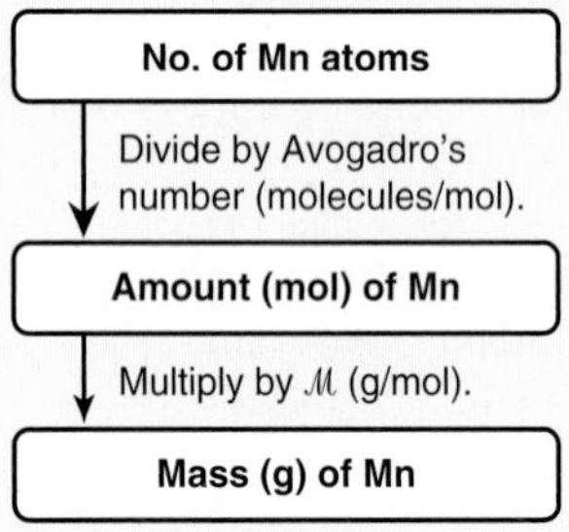

$$\text{Mass (g) of Mn} = 3.22 \times 10^{20}\ \text{Mn atoms} \times \frac{1\ \text{mol Mn}}{6.022 \times 10^{23}\ \text{Mn atoms}} \times \frac{54.94\ \text{g Mn}}{1\ \text{mol Mn}}$$
$$= 2.94 \times 10^{-2}\text{g Mn}$$

3.4
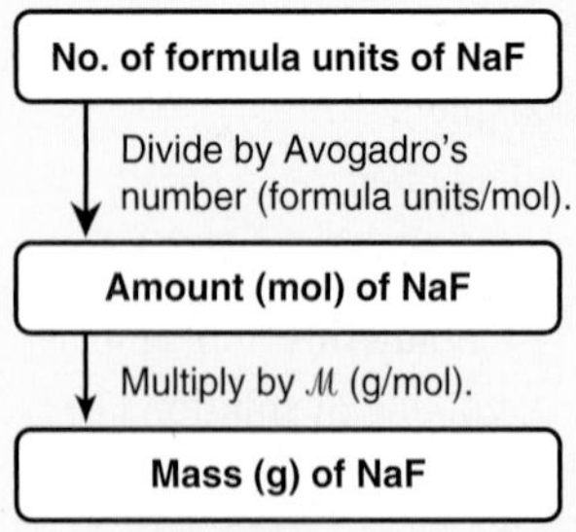

$$\mathcal{M} = (1 \times \mathcal{M}\ \text{of Na}) + (1 \times \mathcal{M}\ \text{of F})$$
$$= 22.99\ \text{g/mol} + 19.00\ \text{g/mol} = 41.99\ \text{g/mol}$$
$$\text{Mass (g) of NaF} = 1.19 \times 10^{19}\ \text{NaF formula units} \times \frac{1\ \text{mol NaF}}{6.022 \times 10^{23}\ \text{NaF formula units}}$$
$$\times \frac{41.99\ \text{g NaF}}{1\ \text{mol NaF}} = 8.30 \times 10^{-4}\ \text{NaF}$$

3.5 (a) Mass (g) of P_4O_{10}

$$= 4.65 \times 10^{22}\ \text{molecules P}_4\text{O}_{10} \times \frac{1\ \text{mol P}_4\text{O}_{10}}{6.022 \times 10^{23}\ \text{molecules P}_4\text{O}_{10}} \times \frac{283.88\ \text{g P}_4\text{O}_{10}}{1\ \text{mol P}_4\text{O}_{10}}$$
$$= 21.9\ \text{g P}_4\text{O}_{10}$$

(b) $$\text{No. of P atoms} = 4.65 \times 10^{22}\ \text{molecules P}_4\text{O}_{10} \times \frac{4\ \text{atoms P}}{1\ \text{molecule P}_4\text{O}_{10}}$$
$$= 1.86 \times 10^{23}\ \text{P atoms}$$

3.6 $$\text{Mass \% of N} = \frac{2\ \text{mol N} \times \frac{14.01\ \text{g N}}{1\ \text{mol N}}}{80.05\ \text{g NH}_4\text{NO}_3} \times 100$$
$$= 35.00\ \text{mass \% N}$$

3.7 $$\text{Mass (g) of N} = 35.8\ \text{kg NH}_4\text{NO}_3 \times \frac{10^3\ \text{g}}{1\ \text{kg}} \times \frac{0.3500\ \text{g N}}{1\ \text{g NH}_4\text{NO}_3}$$
$$= 1.25 \times 10^4\ \text{g N}$$

3.8 Preliminary formula: $B_{0.170}O_{0.255}$

Divide by smaller subscript: $B_{\frac{0.170}{0.170}}O_{\frac{0.255}{0.170}} = B_{1.00}O_{1.50}$

Multiply by 2: $B_{2\times1.00}O_{2\times1.50} = B_{2.00}O_{3.00} = B_2O_3$

3.9 $$\text{Amount (mol) of S} = 2.88\ \text{g S} \times \frac{1\ \text{mol S}}{32.07\ \text{g S}} = 0.0898\ \text{mol S}$$

$$\text{Amount (mol) of M} = 0.0898\ \text{mol S} \times \frac{2\ \text{mol M}}{3\ \text{mol S}} = 0.0599\ \text{mol M}$$

$$\text{Molar mass of M} = \frac{3.12\ \text{g M}}{0.0599\ \text{mol M}} = 52.1\ \text{g/mol}$$

M is chromium, and M_2S_3 is chromium(III) sulfide.

3.10 Assuming 100.00 g of compound, we have 95.21 g of C and 4.79 g of H:

$$\text{Amount (mol) of C} = 95.21\ \text{g C} \times \frac{1\ \text{mol C}}{12.01\ \text{g C}} = 7.928\ \text{mol C}$$

Similarly, there is 4.75 mol H.
Preliminary formula: $C_{7.928}H_{4.75} \approx C_{1.67}H_{1.00}$
Empirical formula: C_5H_3

$$\text{Whole-number multiple} = \frac{252.30\ \text{g/mol}}{63.07\ \text{g/mol}} = 4$$

Molecular formula: $C_{20}H_{12}$

3.11 $$\text{Mass (g) of C} = 0.451\ \text{g CO}_2 \times \frac{12.01\ \text{g C}}{44.01\ \text{g CO}_2}$$
$$= 0.123\ \text{g C}$$

Similarly, there is 0.006 90 g H.
Mass (g) of Cl = 0.250 g − (0.123 g + 0.006 90 g) = 0.120 g Cl
Amount (mol) of elements: 0.0102 mol C; 0.006 85 mol H; 0.003 39 mol Cl
Empirical formula: C_3H_2Cl
Whole-number multiple = 2
Molecular formula: $C_6H_4Cl_2$

3.12 (a) $2Na(s) + 2H_2O(l) \longrightarrow H_2(g) + 2NaOH(aq)$

(b) $2HNO_3(aq) + CaCO_3(s) \longrightarrow CO_2(g) + H_2O(l) + Ca(NO_3)_2(aq)$

(c) $PCl_3(g) + 3HF(g) \longrightarrow PF_3(g) + 3HCl(g)$

(d) $4C_3H_5N_3O_9(l) \longrightarrow 12CO_2(g) + 10H_2O(g) + 6N_2(g) + O_2(g)$

3.13 From the depiction, we have

$$6CO + 3O_2 \longrightarrow 6CO_2$$

or

$$2CO(g) + O_2(g) \longrightarrow 2CO_2(g)$$

3.14

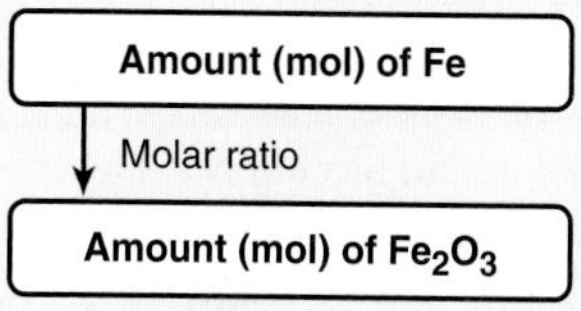

$$Fe_2O_3(s) + 2Al(s) \longrightarrow 2Al_2O_3(s) + 2Fe(l)$$

$$\text{Amount (mol) of } Fe_2O_3 = 3.60 \times 10^3 \text{ mol Fe} \times \frac{1 \text{ mol } Fe_2O_3}{2 \text{ mol Fe}} = 1.80 \times 10^3 \text{mol } Fe_2O_3$$

$$\text{Amount (mol) of Al}(s) \text{ needed} = n_{Al} = 2\, n_{Fe_2O_3} = 2(1.80 \times 10^3 \text{ mol}) = 3.60 \times 10^3 \text{ mol}$$

3.15

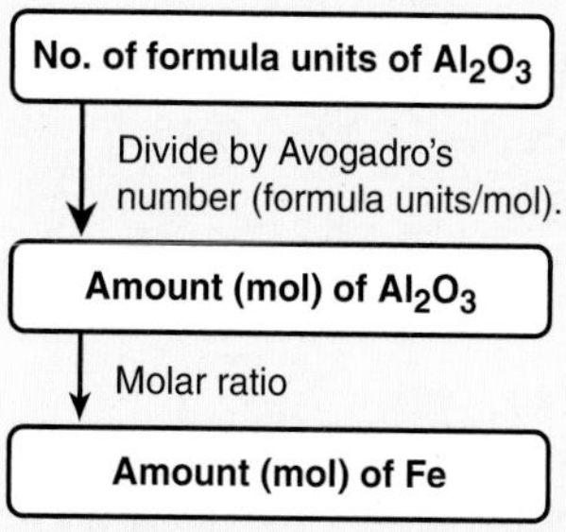

$$\text{Amount (mol) of Fe} = 1.85 \times 10^{25} \text{ } Al_2O_3 \text{ f.u.} \times \frac{1 \text{ mol } Al_2O_3}{6.022 \times 10^{23} \text{ } Al_2O_3 \text{ f.u.}} \times \frac{2 \text{ mol Fe}}{1 \text{ mol } Al_2O_3} = 61.4 \text{ mol Fe}$$

3.16

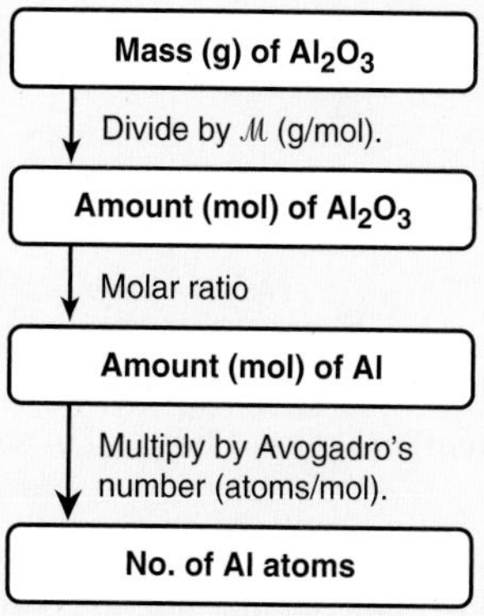

$$\text{No. of Al atoms} = 1.00 \text{ g } Al_2O_3 \times \frac{1 \text{ mol } Al_2O_3}{101.96 \text{ g } Al_2O_3} \times \frac{2 \text{ mol Al}}{1 \text{ mol } Al_2O_3} \times \frac{6.022 \times 10^{23} \text{ Al atoms}}{1 \text{ mol Al}} = 1.18 \times 10^{22} \text{ Al atoms}$$

3.17

$$\begin{array}{l} 2SO_2(g) + O_2(g) \longrightarrow 2SO_3(g) \\ 2SO_3(g) + 2H_2O(l) \longrightarrow 2H_2SO_4(aq) \\ \hline 2SO_2(g) + O_2(g) + 2H_2O(l) \longrightarrow 2H_2SO_4(aq) \end{array}$$

3.18 $4AB + 2B_2 \longrightarrow 4AB_2$, or $2AB(g) + B_2(g) \longrightarrow 2AB_2(g)$

$$\text{For AB: Molecules of } AB_2 = 4AB \times \frac{2AB_2}{2AB} = 4AB_2$$

$$\text{For } B_2\text{: Molecules of } AB_2 = 3B_2 \times \frac{2AB_2}{1 B_2} = 6AB_2$$

Thus, AB_2 is the limiting reactant; one B_2 molecule is in excess.

3.19

$$\text{Amount (mol) of } AB_2 = 1.5 \text{ mol AB} \times \frac{2 \text{ mol } AB_2}{2 \text{ mol AB}} = 1.5 \text{ mol } AB_2$$

$$\text{Amount (mol) of } AB_2 = 1.5 \text{ mol } B_2 \times \frac{2 \text{ mol } AB_2}{1 \text{ mol } B_2} = 3.0 \text{ mol } AB_2$$

Therefore, 1.5 mol of AB_2 can form.

3.20 $2Al(s) + 3S(s) \longrightarrow Al_2S_3(s)$

Mass (g) of Al_2S_3 formed from 10.0 g of Al

$$= 10.0 \text{ g Al} \times \frac{1 \text{ mol Al}}{26.98 \text{ g Al}} \times \frac{1 \text{ mol } Al_2S_3}{2 \text{ mol Al}} \times \frac{150.17 \text{ g } Al_2S_3}{1 \text{ mol } Al_2S_3} = 27.8 \text{ g } Al_2S_3$$

Similarly, mass (g) of Al_2S_3 formed from 15.0 g of S = 23.4 g Al_2S_3. Thus, S is the limiting reactant, and 23.4 g of Al_2S_3 forms.

Mass (g) of Al in excess

$$= \text{total mass of Al} - \text{mass of Al used}$$
$$= 10.0 \text{ g Al} - \left(15.0 \text{ g S} \times \frac{1 \text{ mol S}}{32.07 \text{ g S}} \times \frac{2 \text{ mol Al}}{3 \text{ mol S}} \times \frac{26.98 \text{ g Al}}{1 \text{ mol Al}}\right)$$
$$= 1.59 \text{ g Al}$$

(We would obtain the same answer if sulfur were shown more correctly as S_8.)

3.21 $CaCO_3(s) + 2HCl(aq) \longrightarrow CaCl_2(aq) + H_2O(l) + CO_2(g)$

$$\text{Theoretical yield (g) of } CO_2 = 10.0 \text{ g } CaCO_3 \times \frac{1 \text{ mol } CaCO_3}{100.09 \text{ g } CaCO_3} \times \frac{1 \text{ mol } CO_2}{1 \text{ mol } CaCO_3} \times \frac{44.01 \text{ g } CO_2}{1 \text{ mol } CO_2} = 4.40 \text{ g } CO_2$$

$$\% \text{ Yield} = \frac{3.65 \text{ g } CO_2}{4.40 \text{ g } CO_2} \times 100 = 83.0\%$$

3.22

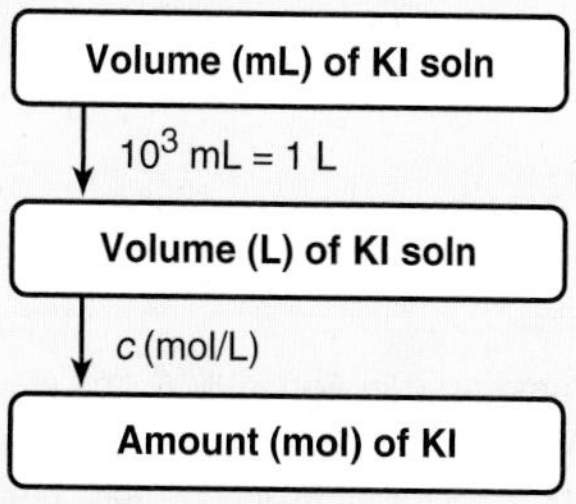

$$\text{Amount (mol) of KI} = 84 \text{ mL soln} \times \frac{1 \text{ L}}{10^3 \text{ mL}} \times \frac{0.50 \text{ mol KI}}{1 \text{ L soln}} = 0.042 \text{ mol KI}$$

3.23

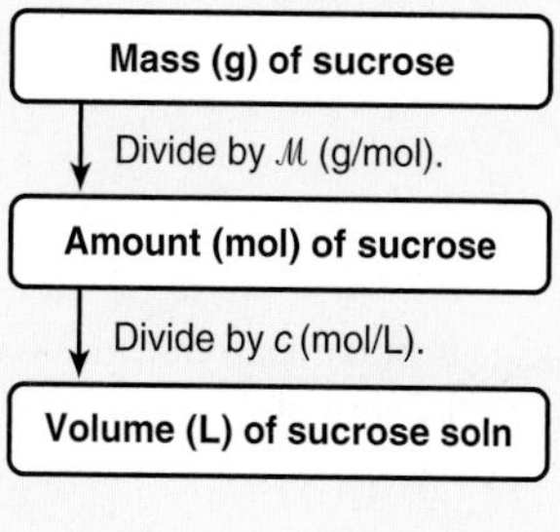

Brief Solutions to Follow-Up Problems (continued)

Volume (L) of sucrose soln

$$= 135\ \text{g sucrose} \times \frac{1\ \text{mol sucrose}}{342.30\ \text{g sucrose}} \times \frac{1\ \text{L soln}}{3.30\ \text{mol sucrose}}$$

$$= 0.120\ \text{L soln}$$

3.24 $$c_{\text{dil}}\ \text{of}\ H_2SO_4 = \frac{7.50\ \text{mol/L} \times 25.0\ \text{m}^3}{500.\ \text{m}^3} = 0.375\ \text{mol/L}\ H_2SO_4$$

Mass (g) of H_2SO_4/mL soln

$$= \frac{0.375\ \text{mol}\ H_2SO_4}{1\ \text{L soln}} \times \frac{1\ \text{L}}{10^3\ \text{mL}} \times \frac{98.09\ \text{g}\ H_2SO_4}{1\ \text{mol}\ H_2SO_4}$$

$$= 3.68 \times 10^{-2}\ \text{g}\ H_2SO_4/\text{mL soln}$$

3.25 To obtain B, the total volume of solution A was reduced by half:

$$V_{\text{conc}} = V_{\text{dil}} \times \frac{N_{\text{dil}}}{N_{\text{conc}}} = 1.0\ \text{mL} \times \frac{6\ \text{particles}}{12\ \text{particles}} = 0.50\ \text{mL}$$

To obtain solution C, $\frac{1}{2}$ of a volume of solvent was added for every volume of A:

$$V_{\text{dil}} = V_{\text{conc}} \times \frac{N_{\text{conc}}}{N_{\text{dil}}} = 1.0\ \text{mL} \times \frac{6\ \text{particles}}{4\ \text{particles}} = 1.5\ \text{mL}$$

3.26 $Al(OH)_3(s) + 3HCl(aq) \longrightarrow AlCl_3(aq) + 3H_2O(l)$

Volume (L) of HCl neutralized

$$= 0.10\ \text{g}\ Al(OH)_3 \times \frac{1\ \text{mol}\ Al(OH)_3}{78.00\ \text{g}\ Al(OH)_3} \times \frac{3\ \text{mol HCl}}{1\ \text{mol}\ Al(OH)_3} \times \frac{1\ \text{L soln}}{0.10\ \text{mol HCl}}$$

$$= 3.8 \times 10^{-2}\ \text{L soln}$$

Therefore, $Al(OH)_3$ is more effective than $Mg(OH)_2$.

3.27 (a) For $Pb(C_2H_3O_2)_2$:

$$\text{Amount (mol) of}\ Pb(C_2H_3O_2)_2 = 0.268\ \text{L} \times \frac{1.50\ \text{mol}\ Pb(C_2H_3O_2)_2}{1\ \text{L}} = 0.402\ \text{mol}\ Pb(C_2H_3O_2)_2$$

$$\text{Amount (mol) of}\ PbCl_2 = 0.402\ \text{mol}\ Pb(C_2H_3O_2)_2 \times \frac{1\ \text{mol}\ PbCl_2}{1\ \text{mol}\ Pb(C_2H_3O_2)_2} = 0.402\ \text{mol}\ PbCl_2$$

$$\text{For NaCl: Amount (mol of NaCl)} = \frac{3.40\ \text{mol NaCl}}{1\ \text{L}} \times 0.130\ \text{L} = 0.442\ \text{mol NaCl}$$

$$\text{Amount (mol) of}\ PbCl_2 = 0.442\ \text{mol NaCl} \times \frac{1\ \text{mol}\ PbCl_2}{2\ \text{mol NaCl}} = 0.221\ \text{mol}\ PbCl_2$$

Thus, NaCl is the limiting reactant.

$$\text{Mass (g) of}\ PbCl_2 = 0.221\ \text{mol}\ PbCl_2 \times \frac{278.1\ \text{g}\ PbCl_2}{1\ \text{mol}\ PbCl_2} = 61.5\ \text{g}\ PbCl_2$$

(b) Using "Ac" for the acetate ion and having NaCl as the limiting reactant:

Amount (mol)	$Pb(Ac)_2$	+ 2NaCl	⟶ $PbCl_2$	+ 2NaAc
Initial	0.402	0.442	0	0
Change	−0.221	−0.442	+0.221	+0.442
Final	0.181	0	0.221	0.442

3.28 (a) The compound is $BaCl_2$.
(b) Mass (g) of $BaCl_2$

$$= 9\ \text{particles} \times \frac{0.05\ \text{mol particles}}{1\ \text{particle}} \times \frac{1\ \text{mol}\ BaCl_2}{3\ \text{mol particles}} \times \frac{208.2\ \text{g}\ BaCl_2}{1\ \text{mol}\ BaCl_2} = 31.2\ \text{g}\ BaCl_2$$

3.29 (a) $KClO_4(s) \xrightarrow{H_2O} K^+(aq) + ClO_4^-(aq)$; 2 mol of K^+ and 2 mol of ClO_4^-
(b) $Mg(C_2H_3O_2)_2(s) \xrightarrow{H_2O} Mg^{2+}(aq) + 2C_2H_3O_2^-(aq)$; 2.49 mol of Mg^{2+} and 4.97 mol of $C_2H_3O_2^-$
(c) $(NH_4)_2CrO_4(s) \xrightarrow{H_2O} 2NH_4^+(aq) + CrO_4^{2-}(aq)$; 6.24 mol of NH_4^+ and 3.12 mol of CrO_4^{2-}
(d) $NaHSO_4(s) \xrightarrow{H_2O} Na^+(aq) + HSO_4^-(aq)$; 0.73 mol of Na^+ and 0.73 mol of HSO_4^-

PROBLEMS

Problems with **red** numbers are answered in Appendix G and worked in detail in the Student Solutions Manual. Problem sections match those in this book and provide the numbers of relevant sample problems. Most offer Concept Review Questions, Skill-Building Exercises (grouped in pairs covering the same concept), and Problems in Context. The Comprehensive Problems are based on material from any section or previous chapter.

The Mole

(Sample Problems 3.1 to 3.7)

Concept Review Questions

3.1 The atomic mass of Cl is 35.45 u, and the atomic mass of Al is 26.98 u. What are the masses (g) of 3 mol of Al atoms and 2 mol of Cl atoms?

3.2 (a) What is the amount (mol) of C atoms in 1 mol of sucrose ($C_{12}H_{22}O_{11}$)?
(b) How many C atoms are in 2 mol of sucrose?

3.3 Why might the expression "1 mol of chlorine" be confusing? What change would remove any uncertainty? For what other elements might a similar confusion exist? Why?

3.4 How is the molecular mass of a compound the same as the molar mass, and how is it different?

3.5 What advantage is there to using a counting unit (mol) for the amount of substance, rather than a mass unit?

3.6 You need to calculate the number of P_4 molecules that can form from 2.5 g of $Ca_3(PO_4)_2$. Draw a road map and write a plan, without doing any calculations.

3.7 Each balance weighs the indicated numbers of atoms of two elements:

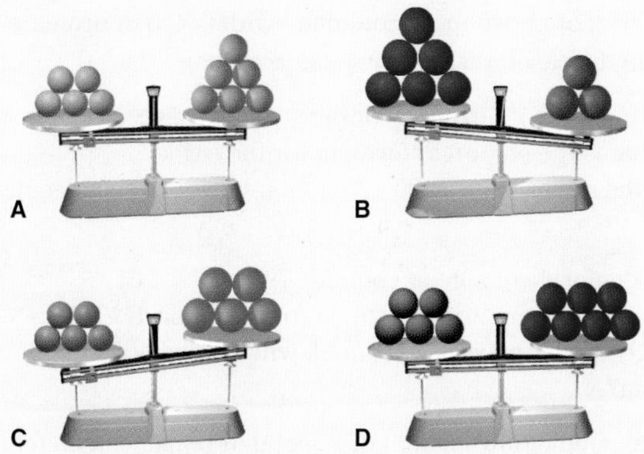

For each balance, determine which element, *left*, *right*, or *neither*, (a) has the higher molar mass; (b) has more atoms per gram; (c) has fewer atoms per gram; (d) has more atoms per mole.

Skill-Building Exercises (grouped in similar pairs)

3.8 Calculate the molar mass of each of the following:
(a) $Sr(OH)_2$ (b) N_2O_3 (c) $NaClO_3$ (d) Cr_2O_3

3.9 Calculate the molar mass of each of the following:
(a) $(NH_4)_3PO_4$ (b) CH_2Cl_2 (c) $CuSO_4{\cdot}5H_2O$ (d) BrF_3

3.10 Calculate the molar mass of each of the following:
(a) SnO (b) BaF_2 (c) $Al_2(SO_4)_3$ (d) $MnCl_2$

3.11 Calculate the molar mass of each of the following:
(a) N_2O_4 (b) C_4H_9OH (c) $MgSO_4{\cdot}7H_2O$ (d) $Ca(C_2H_3O_2)_2$

3.12 Calculate each quantity:
(a) Mass (g) of 0.68 mol of $KMnO_4$
(b) Amount (mol) of O atoms in 8.18 g of $Ba(NO_3)_2$
(c) Number of O atoms in 7.3×10^{-3} g of $CaSO_4{\cdot}2H_2O$

3.13 Calculate each quantity:
(a) Mass (kg) of 4.6×10^{21} molecules of NO_2
(b) Amount (mol) of Cl atoms in 0.0615 g of $C_2H_4Cl_2$
(c) Number of H^- ions in 5.82 g of SrH_2

3.14 Calculate each quantity:
(a) Mass (g) of 6.44×10^{-2} mol of $MnSO_4$
(b) Amount (mol) of compound in 15.8 kg of $Fe(ClO_4)_3$
(c) Number of N atoms in 92.6 mg of NH_4NO_2

3.15 Calculate each quantity:
(a) Total number of ions in 38.1 g of SrF_2
(b) Mass (kg) of 3.58 mol of $CuCl_2{\cdot}2H_2O$
(c) Mass (mg) of 2.88×10^{22} formula units of $Bi(NO_3)_3{\cdot}5H_2O$

3.16 Calculate each quantity:
(a) Mass (g) of 8.35 mol of copper(I) carbonate
(b) Mass (g) of 4.04×10^{20} molecules of dinitrogen pentoxide
(c) Amount (mol) and number of formula units in 78.9 g of sodium perchlorate
(d) Number of sodium ions, perchlorate ions, chlorine atoms, and oxygen atoms in the mass of the compound in part (c)

3.17 Calculate each quantity:
(a) Mass (g) of 8.42 mol of chromium(III) sulfate decahydrate
(b) Mass (g) of 1.83×10^{24} molecules of dichlorine heptoxide
(c) Amount (mol) and number of formula units in 6.2 g of lithium sulfate
(d) Number of lithium ions, sulfate ions, sulfur atoms, and oxygen atoms in the mass of the compound in part (c)

3.18 Calculate each of the following:
(a) Mass percent of H in ammonium bicarbonate
(b) Mass percent of O in sodium dihydrogen phosphate heptahydrate

3.19 Calculate each of the following:
(a) Mass percent of I in strontium periodate
(b) Mass percent of Mn in potassium permanganate

3.20 Calculate each of the following:
(a) Mass fraction of C in cesium acetate
(b) Mass fraction of O in uranyl sulfate trihydrate (the uranyl ion is UO_2^{2+})

3.21 Calculate each of the following:
(a) Mass fraction of Cl in calcium chlorate
(b) Mass fraction of N in dinitrogen trioxide

Problems in Context

3.22 Oxygen is required for the metabolic combustion of foods. Calculate the number of atoms in 38.0 g of oxygen gas, the amount absorbed from the lungs at rest in about 15 min.

3.23 Cisplatin (*below*), or Platinol, is used in the treatment of certain cancers. Calculate (a) the amount (mol) of compound in 285.3 g of cisplatin; (b) the number of hydrogen atoms in 0.98 mol of cisplatin.

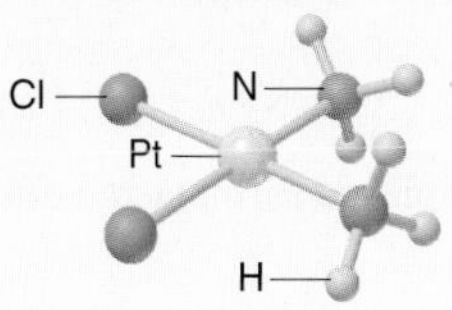

3.24 Allyl sulfide (*below*) gives garlic its characteristic odour. Calculate (a) the mass (g) of 2.63 mol of allyl sulfide; (b) the number of carbon atoms in 35.7 g of allyl sulfide.

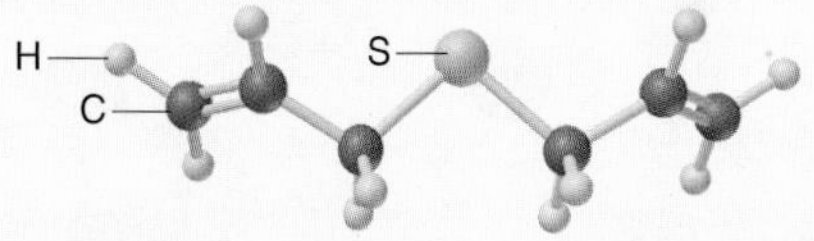

3.25 Iron reacts slowly with oxygen and water to form a compound that is commonly called rust ($Fe_2O_3{\cdot}4H_2O$). For 45.2 kg of rust, calculate (a) the amount (mol) of compound; (b) the amount (mol) of Fe_2O_3; (c) the mass (g) of Fe.

3.26 Propane is widely used in liquid form as a fuel for barbecue grills and camp stoves. For 85.5 g of propane, calculate (a) the amount (mol) of compound; (b) the mass (g) of carbon.

3.27 The effectiveness of a nitrogen fertilizer is determined mainly by its mass % N. Rank the following fertilizers, most effective first: potassium nitrate; ammonium nitrate; ammonium sulfate; urea, $CO(NH_2)_2$.

3.28 The mineral galena is composed of lead(II) sulfide and has an average density of 7.46 g/cm^3.
(a) What amount (mol) of lead(II) sulfide is in 1.00 m^3 of galena?
(b) How many lead atoms are in 1.00 dm^3 of galena?

3.29 Hemoglobin, a protein in red blood cells, carries O_2 from the lungs to the body's cells. Iron (as ferrous ion, Fe^{2+}) makes up 0.33 mass % of hemoglobin. If the molar mass of hemoglobin is 6.8×10^4 g/mol, how many Fe^{2+} ions are in one molecule?

Determining the Formula for an Unknown Compound
(Sample Problems 3.8 to 3.11)

Concept Review Questions

3.30 What is the difference between an empirical formula and a molecular formula? Can they ever be the same?

3.31 List three ways that compositional data may be given in a problem that involves finding an empirical formula.

3.32 Which of the following sets of information allow(s) you to obtain the molecular formula for a covalent compound? For each set that allows this, explain how you would proceed. (Include a road map and a plan for the solution).
(a) Amount (mol) of each type of atom in a given sample of the compound
(b) Mass percent of each element and the total number of atoms in a molecule of the compound
(c) Mass percent of each element and the number of atoms of one element in a molecule of the compound
(d) Empirical formula and mass percent of each element
(e) Structural formula

3.33 Is $MgCl_2$ an empirical formula or a molecular formula for magnesium chloride? Explain.

Skill-Building Exercises (grouped in similar pairs)

3.34 What are the empirical formula and empirical formula mass for each compound?
(a) C_2H_4 (b) $C_2H_6O_2$ (c) N_2O_5 (d) $Ba_3(PO_4)_2$ (e) Te_4I_{16}

3.35 What are the empirical formula and empirical formula mass for each compound?
(a) C_4H_8 (b) $C_3H_6O_3$ (c) P_4O_{10} (d) $Ga_2(SO_4)_3$ (e) Al_2Br_6

3.36 Give the name, empirical formula, and molar mass of the compound depicted in Figure P3.36.

3.37 Give the name, empirical formula, and molar mass of the compound depicted in Figure P3.37.

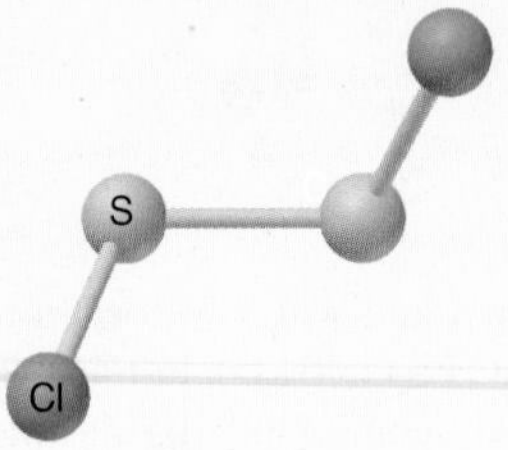

Figure P3.36

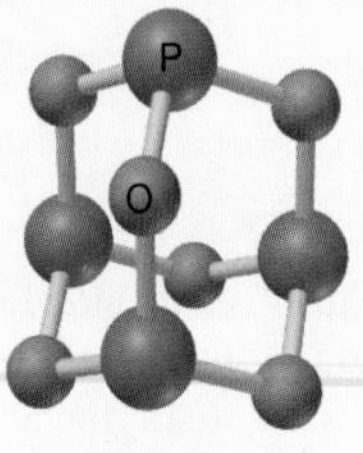

Figure P3.37

3.38 What is the molecular formula for each compound?
(a) Empirical formula CH_2 ($\mathcal{M}$ = 42.08 g/mol)
(b) Empirical formula NH_2 ($\mathcal{M}$ = 32.05 g/mol)
(c) Empirical formula NO_2 ($\mathcal{M}$ = 92.02 g/mol)
(d) Empirical formula CHN ($\mathcal{M}$ = 135.14 g/mol)

3.39 What is the molecular formula for each compound?
(a) Empirical formula CH ($\mathcal{M}$ = 78.11 g/mol)
(b) Empirical formula $C_3H_6O_2$ ($\mathcal{M}$ = 74.08 g/mol)
(c) Empirical formula HgCl ($\mathcal{M}$ = 472.1 g/mol)
(d) Empirical formula $C_7H_4O_2$ ($\mathcal{M}$ = 240.20 g/mol)

3.40 Find the empirical formula for each compound:
(a) 0.063 mol of chlorine atoms combined with 0.22 mol of oxygen atoms
(b) 2.45 g of silicon combined with 12.4 g of chlorine
(c) 27.3 mass % carbon and 72.7 mass % oxygen

3.41 Find the empirical formula for each compound:
(a) 0.039 mol of iron atoms combined with 0.052 mol of oxygen atoms
(b) 0.903 g of phosphorus combined with 6.99 g of bromine
(c) A hydrocarbon with 79.9 mass % carbon

3.42 An oxide of nitrogen contains 30.45 mass % N.
(a) What is the empirical formula for the oxide?
(b) If the molar mass is 90 ± 5 g/mol, what is the molecular formula?

3.43 A chloride of silicon contains 79.1 mass % Cl.
(a) What is the empirical formula for the chloride?
(b) If the molar mass is 269 g/mol, what is the molecular formula?

3.44 A sample of 0.600 mol of a metal M reacts completely with excess fluorine to form 46.8 g of MF_2.
(a) What amount (mol) of F is in the sample of MF_2 that forms?
(b) What mass (g) of M is in this sample of MF_2?
(c) What element is represented by the symbol M?

3.45 A 0.370 mol sample of a metal oxide (M_2O_3) weighs 55.4 g.
(a) What amount (mol) of O is in the sample?
(b) What mass (g) of M is in the sample?
(c) What element is represented by the symbol M?

Problems in Context

3.46 Nicotine is a poisonous, addictive compound found in tobacco. A sample of nicotine contains 6.16 mmol of C, 8.56 mmol of H, and 1.23 mmol of N [1 mmol (1 millimole) = 10^{-3} mol]. What is the empirical formula for nicotine?

3.47 Cortisol ($\mathcal{M}$ = 362.47 g/mol) is a steroid hormone involved in protein synthesis. Medically, it has a major use in reducing inflammation from rheumatoid arthritis. Cortisol is 69.6% C, 8.34% H, and 22.1% O by mass. What is its molecular formula?

3.48 Acetaminophen (*below*) is a popular non-Aspirin pain reliever. What is the mass percent of each element in acetaminophen?

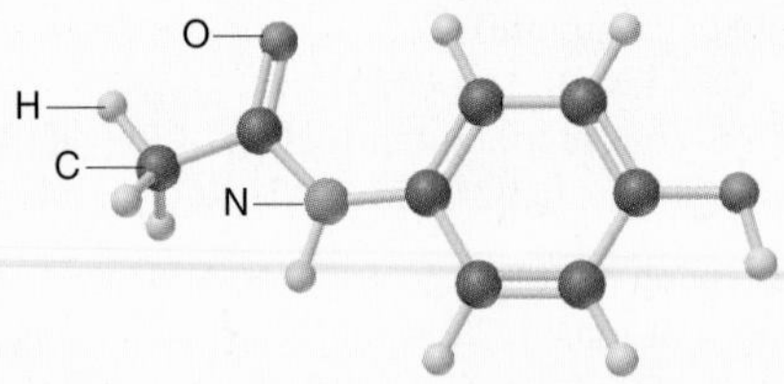

3.49 Menthol ($\mathcal{M}$ = 156.3 g/mol), the strong-smelling substance in many cough drops, is a compound of carbon, hydrogen, and oxygen. When 0.1595 g of menthol was burned in a combustion apparatus, 0.449 g of CO_2 and 0.184 g of H_2O formed. What is menthol's molecular formula?

Writing and Balancing Chemical Equations
(Sample Problems 3.12 and 3.13)

Concept Review Questions

3.50 What three types of information does a balanced chemical equation provide?

3.51 How does a balanced chemical equation apply the law of mass conservation?

3.52 Three students are balancing the following equation:

$$Al + Cl_2 \longrightarrow AlCl_3$$

In the process,

- Student I writes $Al + Cl_2 \longrightarrow AlCl_2$
- Student II writes $Al + Cl_2 + Cl \longrightarrow AlCl_3$
- Student III writes $2Al + 3Cl_2 \longrightarrow 2AlCl_3$

Is the approach of student I valid? student II? student III? Explain.

3.53 The scenes below represent a chemical reaction between elements A (*red*) and B (*green*):

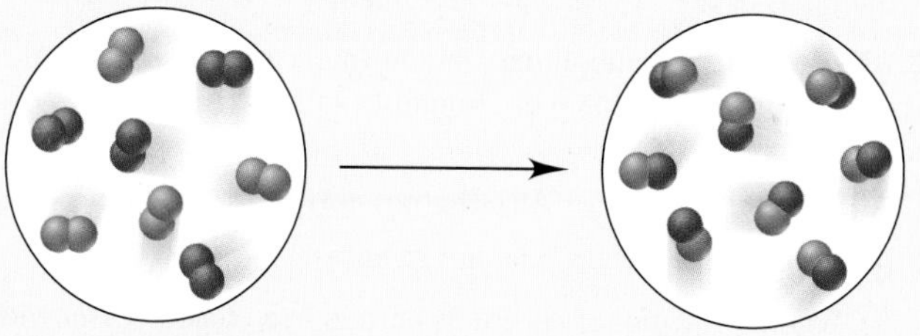

Which of the following best represents the balanced equation for the reaction?

(a) $2A + 2B \longrightarrow A_2B_2$ (b) $A_2 + B_2 \longrightarrow 2AB$
(c) $B_2 + 2AB \longrightarrow 2B_2 + A_2$ (d) $4A_2 + 4B_2 \longrightarrow 8AB$

Skill-Building Exercises (grouped in similar pairs)

3.54 Write a balanced equation for each skeleton equation by inserting the correct coefficients in the blanks:

(a) $\_\_Cu(s) + S_8(s) \longrightarrow \_\_Cu_2S(s)$
(b) $\_\_P_4O_{10}(s) + \_\_H_2O(l) \longrightarrow \_\_H_3PO_4(l)$
(c) $\_\_B_2O_3(s) + \_\_NaOH(aq) \longrightarrow \_\_Na_3BO_3(aq) + \_\_H_2O(l)$
(d) $\_\_CH_3NH_2(g) + \_\_O_2(g) \longrightarrow \_\_CO_2(g) + H_2O(g) + \_\_N_2(g)$

3.55 Write a balanced equation for each skeleton equation by inserting the correct coefficients in the blanks:

(a) $\_\_Cu(NO_3)_2(aq) + \_\_KOH(aq) \longrightarrow \_\_Cu(OH)_2(s) + \_\_KNO_3(aq)$
(b) $\_\_BCl_3(g) + \_\_H_2O(l) \longrightarrow \_\_H_3BO_3(s) + \_\_HCl(g)$
(c) $\_\_CaSiO_3(s) + \_\_HF(g) \longrightarrow \_\_SiF_4(g) + \_\_CaF_2(s) + \_\_H_2O(l)$
(d) $\_\_(CN)_2(g) + \_\_H_2O(l) \longrightarrow H_2C_2O_4(aq) + \_\_NH_3(g)$

3.56 Write a balanced equation for each skeleton equation by inserting the correct coefficients in the blanks:

(a) $\_\_SO_2(g) + \_\_O_2(g) \longrightarrow \_\_SO_3(g)$
(b) $\_\_Sc_2O_3(s) + \_\_H_2O(l) \longrightarrow \_\_Sc(OH)_3(s)$
(c) $\_\_H_3PO_4(aq) + \_\_NaOH(aq) \longrightarrow \_\_Na_2HPO_4(aq) + \_\_H_2O(l)$
(d) $\_\_C_6H_{10}O_5(s) + \_\_O_2(g) \longrightarrow \_\_CO_2(g) + \_\_H_2O(g)$

3.57 Write a balanced equation for each skeleton equation by inserting the correct coefficients in the blanks:

(a) $\_\_As_4S_6(s) + \_\_O_2(g) \longrightarrow \_\_As_4O_6(s) + \_\_SO_2(g)$
(b) $\_\_Ca_3(PO_4)_2(s) + \_\_SiO_2(s) + \_\_C(s) \longrightarrow \_\_P_4(g) + \_\_CaSiO_3(l) + \_\_CO(g)$
(c) $\_\_Fe(s) + \_\_H_2O(g) \longrightarrow \_\_Fe_3O_4(s) + \_\_H_2(g)$
(d) $\_\_S_2Cl_2(l) + \_\_NH_3(g) \longrightarrow \_\_S_4N_4(s) + \_\_S_8(s) + \_\_NH_4Cl(s)$

3.58 Convert the following descriptions into balanced equations:
(a) When gallium metal is heated in oxygen gas, it melts and forms solid gallium(III) oxide.
(b) Liquid hexane burns in oxygen gas to form carbon dioxide gas and water vapour.
(c) When solutions of calcium chloride and sodium phosphate are mixed, solid calcium phosphate forms and sodium chloride remains in solution.

3.59 Convert the following descriptions into balanced equations:
(a) When lead(II) nitrate solution is added to potassium iodide solution, solid lead(II) iodide forms and potassium nitrate solution remains.
(b) Liquid disilicon hexachloride reacts with water to form solid silicon dioxide, hydrogen chloride gas, and hydrogen gas.
(c) When nitrogen dioxide is bubbled into water, a solution of nitric acid forms and gaseous nitrogen monoxide is released.

Problem in Context

3.60 Loss of atmospheric ozone has led to an ozone "hole" over Antarctica. The loss occurs, in part, through three consecutive steps:
(1) Chlorine atoms react with ozone (O_3) to form chlorine monoxide and molecular oxygen.
(2) Chlorine monoxide forms ClOOCl.
(3) ClOOCl absorbs sunlight and breaks into chlorine atoms and molecular oxygen.
(a) Write a balanced equation for each step in the sequence.
(b) Write an overall balanced equation for the sequence.

Calculating Quantities of Reactant and Product

(Sample Problems 3.14 to 3.21)

Concept Review Questions

3.61 What does the term *stoichiometrically equivalent molar ratio* mean, and how is it applied to solving problems?

3.62 The circle below represents a mixture of A_2 and B_2 before they react to form AB_3.

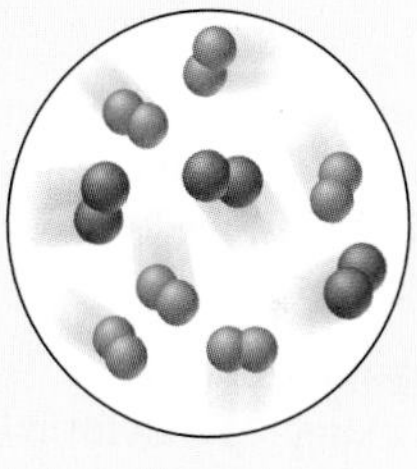

(a) What is the limiting reactant?
(b) How many molecules of product can form?

3.63 Percent yields are generally calculated from masses. Would the result be the same if amounts (mol) were used instead? Why?

Skill-Building Exercises (grouped in similar pairs)

3.64 Reactants A and B form product C. Draw a road map and write a plan to find the mass (g) of C when 25 g of A reacts with excess B.

3.65 Reactants D and E form product F. Draw a road map and write a plan to find the mass (g) of F when 27 g of D reacts with 31 g of E.

3.66 Chlorine gas can be made in the laboratory by the reaction of hydrochloric acid and manganese(IV) oxide:

$$4HCl(aq) + MnO_2(s) \longrightarrow MnCl_2(aq) + 2H_2O(g) + Cl_2(g)$$

When 1.82 mol of HCl reacts with excess MnO_2, (a) what amount (mol) of Cl_2 forms; (b) what mass (g) of Cl_2 forms?

3.67 Bismuth oxide reacts with carbon to form bismuth metal:

$$Bi_2O_3(s) + 3C(s) \longrightarrow 2Bi(s) + 3CO(g)$$

When 283 g of Bi_2O_3 reacts with excess carbon, (a) what amount (mol) of Bi_2O_3 reacts; (b) what amount (mol) of Bi forms?

3.68 Potassium nitrate decomposes on heating, producing potassium oxide and gaseous nitrogen and oxygen:

$$4KNO_3(s) \longrightarrow 2K_2O(s) + 2N_2(g) + 5O_2(g)$$

To produce 56.6 kg of oxygen, (a) what amount (mol) and (b) what mass (g) of KNO_3 must be heated?

3.69 Chromium(III) oxide reacts with hydrogen sulfide (H_2S) gas to form chromium(III) sulfide and water:

$$Cr_2O_3(s) + 3H_2S(g) \longrightarrow Cr_2S_3(s) + 3H_2O(l)$$

To produce 421 g of Cr_2S_3, (a) what amount (mol) and (b) what mass (g) of Cr_2O_3 are required?

3.70 Calculate the mass (g) of each product formed when 43.82 g of diborane (B_2H_6) reacts with excess water:

$$B_2H_6(g) + H_2O(l) \longrightarrow H_3BO_3(s) + H_2(g) \text{ (unbalanced)}$$

3.71 Calculate the mass (g) of each product formed when 174 g of silver sulfide reacts with excess hydrochloric acid:

$$Ag_2S(s) + HCl(aq) \longrightarrow AgCl(s) + H_2S(g) \text{ (unbalanced)}$$

3.72 Elemental phosphorus occurs as tetratomic molecules, P_4. What mass (g) of chlorine gas is needed to react completely with 455 g of phosphorus to form phosphorus pentachloride?

3.73 Elemental sulfur occurs as octatomic molecules, S_8. What mass (g) of fluorine gas is needed to react completely with 17.8 g of sulfur to form sulfur hexafluoride?

3.74 Solid iodine trichloride is prepared in two steps:
(1) A reaction between solid iodine and gaseous chlorine to form solid iodine monochloride
(2) Treatment with more chlorine
(a) Write a balanced equation for each step.
(b) Write a balanced equation for the overall reaction.
(c) What mass (g) of iodine is needed to prepare 2.45 kg of final product?

3.75 Lead can be prepared from galena, lead(II) sulfide, by first roasting the galena in oxygen gas to form lead(II) oxide and sulfur dioxide. Heating the metal oxide with more galena forms the molten metal and more sulfur dioxide.
(a) Write a balanced equation for each step.
(b) Write an overall balanced equation for the process.
(c) What mass (in tonnes, t) of sulfur dioxide forms for every tonne of lead obtained?

3.76 Many metals react with oxygen gas to form a metal oxide. For example, calcium reacts as follows:

$$2Ca(s) + O_2(g) \longrightarrow 2CaO(s)$$

Suppose that you wish to calculate the mass (g) of calcium oxide that can be prepared from 4.20 g of Ca and 2.80 g of O_2.
(a) What amount (mol) of CaO can be produced from the given mass of Ca?
(b) What amount (mol) of CaO can be produced from the given mass of O_2?
(c) Which reactant is the limiting reactant?
(d) What mass (g) of CaO can be produced?

3.77 Metal hydrides react with water to form hydrogen gas and a metal hydroxide. For example,

$$SrH_2(s) + 2H_2O(l) \longrightarrow Sr(OH)_2(s) + 2H_2(g)$$

You wish to calculate the mass (g) of hydrogen gas that can be prepared from 5.70 g of SrH_2 and 4.75 g of H_2O.
(a) What amount (mol) of H_2 can be produced from the given mass of SrH_2?
(b) What amount (mol) of H_2 can be produced from the given mass of H_2O?
(c) Which reactant is the limiting reactant?
(d) What mass (g) of H_2 can be produced?

3.78 Calculate the maximum amount (mol) and mass (g) of iodic acid (HIO_3) that can form when 635 g of iodine trichloride reacts with 118.5 g of water:

$$ICl_3(s) + H_2O(l) \longrightarrow ICl(g) + HIO_3(aq) + HCl(g) \text{ (unbalanced)}$$

What mass of the excess reactant remains?

3.79 Calculate the maximum amount (mol) and mass (g) of H_2S that can form when 158 g of aluminum sulfide reacts with 131 g of water:

$$Al_2S_3(s) + H_2O(l) \longrightarrow Al(OH)_3(aq) + H_2S(g) \text{ (unbalanced)}$$

What mass of the excess reactant remains?

3.80 When 0.100 mol of carbon is burned in a closed vessel with 8.00 g of oxygen, what mass (g) of carbon dioxide can form? Which reactant is in excess, and what mass of this reactant remains after the reaction?

3.81 A mixture of 0.0375 g of hydrogen and 0.0185 mol of oxygen, in a closed container, is sparked to initiate a reaction. What mass of water can form? Which reactant is in excess, and what mass of this reactant remains after the reaction?

3.82 Aluminum nitrite and ammonium chloride react to form aluminum chloride, nitrogen, and water. What mass (g) of each substance is present after 72.5 g of aluminum nitrite and 58.6 g of ammonium chloride react completely?

3.83 Calcium nitrate and ammonium fluoride react to form calcium fluoride, dinitrogen monoxide, and water vapour. What mass of each substance is present after 16.8 g of calcium nitrate and 17.50 g of ammonium fluoride react completely?

3.84 Two successive reactions, A $\longrightarrow$ B and B $\longrightarrow$ C, have yields of 73% and 68%, respectively. What is the overall percent yield for conversion of A to C?

3.85 Two successive reactions, D $\longrightarrow$ E and E $\longrightarrow$ F, have yields of 48% and 73%, respectively. What is the overall percent yield for the conversion of D to F?

3.86 What is the percent yield of a reaction in which 45.5 g of tungsten(VI) oxide (WO_3) reacts with excess hydrogen gas to produce metallic tungsten and 9.60 mL of water ($d = 1.00$ g/mL)?

3.87 What is the percent yield of a reaction in which 200. g of phosphorus trichloride reacts with excess water to form 128 g of HCl and aqueous phosphorous acid (H_3PO_3)?

3.88 When 20.5 g of methane and 45.0 g of chlorine gas undergo a reaction that has a 75.0% yield, what mass (g) of chloromethane (CH_3Cl) forms? Hydrogen chloride also forms.

3.89 When 56.6 g of calcium and 30.5 g of nitrogen gas undergo a reaction that has a 93.0% yield, what mass (g) of calcium nitride forms?

Problems in Context

3.90 Cyanogen, $(CN)_2$, has been observed in the atmosphere of Titan, Saturn's largest moon, and in the gases of interstellar nebulas. On Earth, it is used as a welding gas and a fumigant. In its reaction with fluorine gas, carbon tetrafluoride and nitrogen trifluoride gases are produced. What mass (g) of carbon tetrafluoride forms when 60.0 g of each reactant is used?

3.91 Gaseous dichlorine monoxide decomposes readily to form chlorine and oxygen gases.
(a) Which of the following circles best depicts the product mixture after the decomposition?

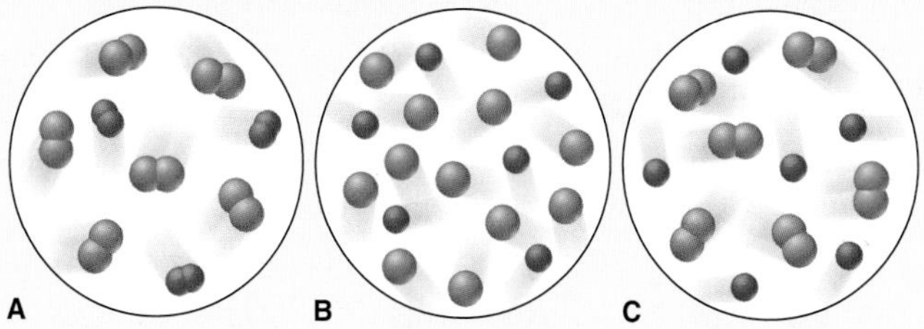

(b) Write the balanced equation for the decomposition.
(c) If each oxygen atom represents 0.050 mol, how many molecules of dichlorine monoxide were present before the decomposition?

3.92 An intermediate step in the production of nitric acid involves the reaction of ammonia with oxygen gas to form nitrogen monoxide and water. What mass (g) of nitrogen monoxide can form in the reaction of 485 g of ammonia with 792 g of oxygen?

3.93 Butane gas is compressed and used as a liquid fuel in disposable lighters and lightweight camping stoves. Suppose that a lighter contains 5.50 mL of butane ($d = 0.579$ g/mL).
(a) What mass (g) of oxygen is needed to burn the butane completely?
(b) What amount (mol) of H_2O forms when all the butane burns?
(c) How many total molecules of gas form when the butane burns completely?

3.94 Sodium borohydride ($NaBH_4$) is used industrially in many organic syntheses. One way to prepare it is by reacting sodium hydride with gaseous diborane (B_2H_6). Assuming an 88.5% yield, what mass of $NaBH_4$ can be prepared by reacting 7.98 g of sodium hydride with 8.16 g of diborane?

Fundamentals of Solution Stoichiometry
(Sample Problems 3.22 to 3.27)

Concept Review Questions

3.95 Box A represents a unit volume of a solution. Choose, from boxes B and C, the box representing the same unit volume of solution that has (a) more solute added; (b) more solvent added; (c) higher concentration; (d) lower concentration.

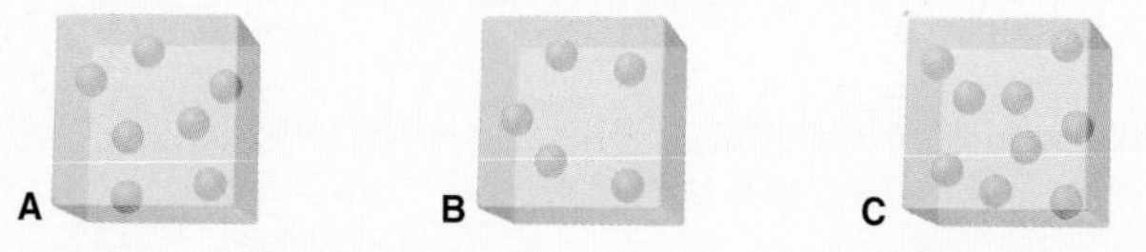

3.96 A mathematical equation useful for dilution calculations is $c_{dil} \times V_{dil} = c_{conc} \times V_{conc}$.
(a) What does each symbol mean, and why does the equation work?
(b) Given the volume and concentration of a $CaCl_2$ solution, how do you determine the amount (mol) and the mass (g) of solute?

3.97 Decide whether the following instructions for diluting a 10.0 mol/L solution to a 1.00 mol/L solution are correct: Take 100.0 mL of the 10.0 mol/L solution, and add 900.0 mL of water. Explain your decision.

3.98 Six different aqueous solutions (with solvent molecules omitted for clarity) are represented in the beakers below, and their total volumes are noted.

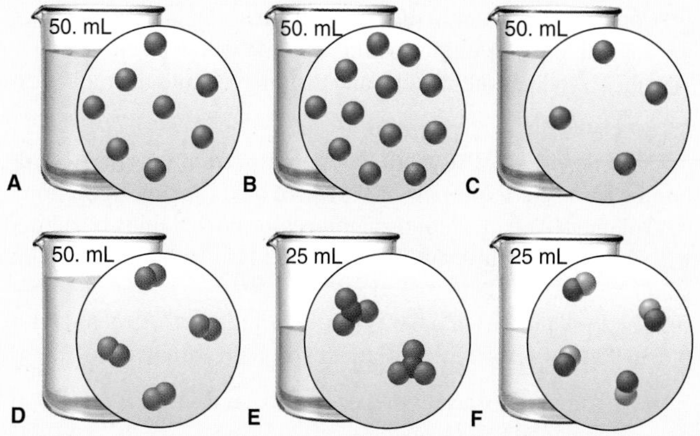

(a) Which solution has the highest concentration (mol/L)?
(b) Which solutions have the same concentration (mol/L)?
(c) If you mix solutions A and C, does the resulting solution have a concentration (mol/L) that is higher, lower, or the same as the concentration of solution B?
(d) After 50. mL of water is added to solution D, is its concentration (mol/L) higher, lower, or the same as the concentration of solution F after 75 mL of water is added to it?
(e) How much solvent must be evaporated from solution E for it to have the same concentration (mol/L) as solution A?

Skill-Building Exercises (grouped in similar pairs)

3.99 Calculate each quantity:
(a) Mass (g) of solute in 185.8 mL of 0.267 mol/L calcium acetate
(b) Concentration (mol/L) of 500. mL of solution containing 21.1 g of potassium iodide
(c) Amount (mol) of solute in 145.6 L of 0.850 mol/L sodium cyanide

3.100 Calculate each quantity:
(a) Volume (mL) of 2.26 mol/L potassium hydroxide that contains 8.42 g of solute
(b) Number of Cu^{2+} ions in 52 L of 2.3 mol/L copper(II) chloride
(c) Concentration (mol/L) of 275 mL of solution that contains 135 mmol of glucose

3.101 Calculate each quantity:
(a) Mass (g) of solute needed to make 475 mL of 5.62×10^{-2} mol/L potassium sulfate
(b) Concentration (mol/L) of a solution that contains 7.25 mg of calcium chloride in each millilitre
(c) Number of Mg^{2+} ions in each millilitre of 0.184 mol/L magnesium bromide

3.102 Calculate each quantity:
(a) Concentration (mol/L) of the solution that results from dissolving 46.0 g of silver nitrate in enough water to give a final volume of 335 mL
(b) Volume (L) of 0.385 mol/L manganese(II) sulfate that contains 63.0 g of solute
(c) Volume (mL) of 6.44×10^{-2} mol/L adenosine triphosphate (ATP) that contains 1.68 mmol of ATP

3.103 Calculate each quantity:
(a) Concentration (mol/L) of a solution prepared by diluting 37.00 mL of 0.250 mol/L potassium chloride to 150.00 mL
(b) Concentration (mol/L) of a solution prepared by diluting 25.71 mL of 0.0706 mol/L ammonium sulfate to 500.00 mL
(c) Concentration (mol/L) of sodium ion in a solution made by mixing 3.58 mL of 0.348 mol/L sodium chloride with 500. mL of 6.81×10^{-2} mol/L sodium sulfate (assume that the volumes are additive)

3.104 Calculate each quantity:
(a) Volume (L) of 2.050 mol/L copper(II) nitrate that must be diluted with water to prepare 750.0 mL of a 0.8543 mol/L solution
(b) Volume (L) of 1.63 mol/L calcium chloride that must be diluted with water to prepare 350. mL of a 2.86×10^{-2} mol/L chloride ion solution
(c) Final volume (L) of a 0.0700 mol/L solution prepared by diluting 18.0 mL of 0.155 mol/L lithium carbonate with water

3.105 A sample of concentrated nitric acid has a density of 1.41 g/mL and contains 70.0% HNO_3 by mass.
(a) What mass (g) of HNO_3 is present per litre of solution?
(b) What is the concentration (mol/L) of the solution?

3.106 Concentrated sulfuric acid (18.3 mol/L) has a density of 1.84 g/mL.
(a) What amount (mol) of H_2SO_4 is in each millilitre of solution?
(b) What is the mass percent of H_2SO_4 in the solution?

3.107 What volume (mL) of 0.383 mol/L HCl is needed to react with 16.2 g of $CaCO_3$?

$$2HCl(aq) + CaCO_3(s) \longrightarrow CaCl_2(aq) + CO_2(g) + H_2O(l)$$

3.108 What mass (g) of NaH_2PO_4 is needed to react with 43.74 mL of 0.285 mol/L NaOH?

$$NaH_2PO_4(s) + 2NaOH(aq) \longrightarrow Na_3PO_4(aq) + 2H_2O(l)$$

3.109 What mass (g) of solid barium sulfate forms when 35.0 mL of 0.160 mol/L barium chloride reacts with 58.0 mL of 0.065 mol/L sodium sulfate? Aqueous sodium chloride also forms.

3.110 What amount (mol) of excess reactant is present when 350. mL of 0.210 mol/L sulfuric acid reacts with 0.500 L of 0.196 mol/L sodium hydroxide to form water and aqueous sodium sulfate?

Problems in Context

3.111 Ordinary household bleach is an aqueous solution of sodium hypochlorite. What is the concentration (mol/L) of a bleach solution that contains 20.5 g of sodium hypochlorite in 375 mL?

3.112 Muriatic acid, an industrial grade of concentrated HCl, is used to clean masonry and cement. Its concentration is 11.7 mol/L.
(a) Write instructions for diluting the concentrated acid to make 3.0 L of 3.5 mol/L acid for routine use.
(b) What volume (mL) of the muriatic acid solution contains 9.66 g of HCl?

3.113 A sample of impure magnesium was analyzed by allowing it to react with excess HCl solution:

$$Mg(s) + 2HCl(aq) \longrightarrow MgCl_2(aq) + H_2(g)$$

After 1.32 g of the impure metal was treated with 0.100 L of 0.750 mol/L HCl, 0.0125 mol of HCl remained. Assuming that the impurities do not react, what is the mass percent of Mg in the sample?

The Role of Water as a Solvent

(Sample Problems 3.28 and 3.29)

Concept Review Questions

3.114 What two factors cause water to be polar?

3.115 What types of substances are most likely to be soluble in water?

3.116 What must be present in an aqueous solution for it to conduct an electric current? What general classes of compounds form solutions that conduct an electric current?

3.117 What occurs on the molecular level when an ionic compound dissolves in water?

3.118 Which scene best represents how the ions occur in an aqueous solution of (a) $CaCl_2$; (b) Li_2SO_4; (c) NH_4Br?

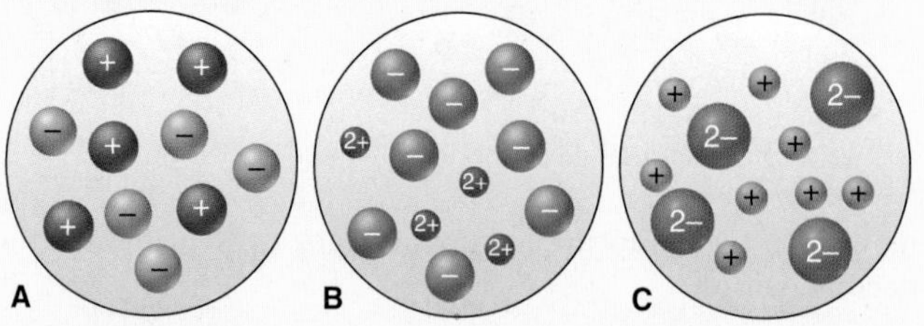

3.119 Which scene best represents a volume from a solution of magnesium nitrate?

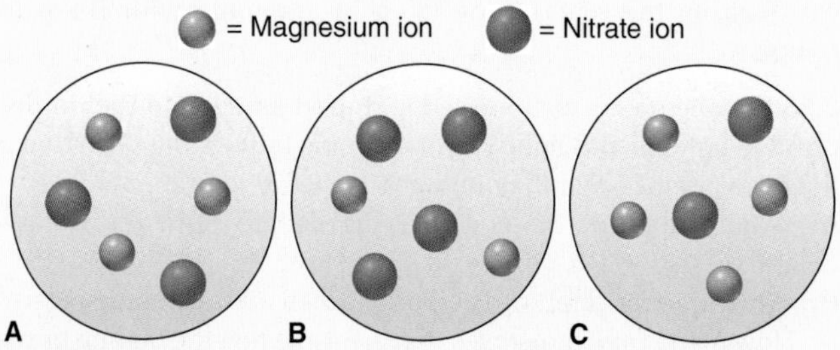

3.120 Why are some ionic compounds soluble in water, while others are not?

3.121 Why are some covalent compounds soluble in water, while others are not?

3.122 Some covalent compounds dissociate into ions in water. What atom do these compounds have in their structure? What type of solution do they form? Name three examples of this type of solution.

Skill-Building Exercises (grouped in similar pairs)

3.123 Is each compound very soluble in water? Explain.
(a) Benzene, C_6H_6
(c) Ethanol, CH_3CH_2OH
(b) Sodium hydroxide
(d) Potassium acetate

3.124 Is each compound very soluble in water? Explain.
(a) Lithium nitrate
(c) Pentane
(b) Glycine, H_2NCH_2COOH
(d) Ethane-1,2-diol, $HOCH_2CH_2OH$

3.125 Does an aqueous solution of each compound conduct an electric current? Explain.
(a) Cesium bromide (b) Hydrogen iodide

3.126 Does an aqueous solution of each compound conduct an electric current? Explain.
(a) Potassium sulfate (b) Sucrose, $C_{12}H_{22}O_{11}$

3.127 What total amount (mol) of ions is released when each compound dissolves in water?
(a) 0.32 mol of NH_4Cl
(b) 25.4 g of $Ba(OH)_2 \cdot 8H_2O$
(c) 3.55×10^{19} formula units of LiCl

3.128 What total amount (mol) of ions is released when each compound dissolves in water?
(a) 0.805 mol of Rb_2SO_4
(b) 3.85×10^{-3} g of $Ca(NO_3)_2$
(c) 4.03×10^{19} formula units of $Sr(HCO_3)_2$

3.129 What total amount (mol) of ions is released when each compound dissolves in water?
(a) 0.75 mol of K_3PO_4
(b) 6.88×10^{-3} g of $NiBr_2 \cdot 3H_2O$
(c) 2.23×10^{22} formula units of $FeCl_3$

3.130 What total amount (mol) of ions is released when each compound dissolves in water?
(a) 0.734 mol of Na_2HPO_4
(b) 3.86 g of $CuSO_4 \cdot 5H_2O$
(c) 8.66×10^{20} formula units of $NiCl_2$

3.131 What amount (mol) and how many ions of each type are present in solution?
(a) 130 mL of 0.45 mol/L aluminum chloride
(b) 9.80 mL of a solution containing 2.59 g of lithium sulfate per litre
(c) 245 mL of a solution containing 3.68×10^{22} formula units of potassium bromide per litre

3.132 What amount (mol) and how many ions of each type are present in solution?
(a) 88 mL of 1.75 mol/L magnesium chloride
(b) 321 mL of a solution containing 0.22 g of aluminum sulfate per litre
(c) 1.65 L of a solution containing 8.83×10^{21} formula units of cesium nitrate per litre

3.133 What amount (mol) of H^+ ions is present in each aqueous solution?
(a) 1.40 L of 0.25 mol/L perchloric acid
(b) 6.8 mL of 0.92 mol/L nitric acid
(c) 2.6 L of 0.085 mol/L hydrochloric acid

3.134 What amount (mol) of H^+ ions is present in each aqueous solution?
(a) 1.4 mL of 0.75 mol/L hydrobromic acid
(b) 2.47 mL of 1.98 mol/L hydriodic acid
(c) 395 mL of 0.270 mol/L nitric acid

Problems in Context

3.135 To study a marine organism, a biologist prepares a 1.00 kg sample to simulate the ion concentrations in seawater. She mixes 26.5 g of NaCl, 2.40 g of $MgCl_2$, 3.35 g of $MgSO_4$, 1.20 g of $CaCl_2$, 1.05 g of KCl, 0.315 g of $NaHCO_3$, and 0.098 g of NaBr in distilled water.
(a) If the density of the solution is 1.025 g/cm^3, what is the concentration (mol/L) of each ion?
(b) What is the total concentration (mol/L) of alkali metal ions?
(c) What is the total concentration (mol/L) of alkaline earth metal ions?
(d) What is the total concentration (mol/L) of anions?

3.136 Water "softeners" remove metal ions, such as Ca^{2+} and Fe^{3+}, by replacing them with enough Na^+ ions to maintain the same number of positive charges in the solution. If 1.0×10^3 L of "hard" water is 0.015 mol/L Ca^{2+} and 0.0010 mol/L Fe^{3+}, what amount (mol) of Na^+ is needed to replace these ions?

Writing Equations for Aqueous Ionic Reactions

Concept Review Questions

3.137 Which ions do not appear in a net ionic equation? Why?

3.138 Write two sets of equations (both molecular and total ionic) with different reactants that have the same net ionic equation as the following equation:

$$Ba(NO_3)_2(aq) + Na_2CO_3(aq) \longrightarrow BaCO_3(s) + 2NaNO_3(aq)$$

Comprehensive Problems

3.139 The mole is defined in terms of the carbon-12 atom. Use the definition to find (a) the mass (g), equal to one atomic mass unit (u); (b) the ratio of the gram to the atomic mass unit.

3.140 The first sulfur-nitrogen compound was prepared in 1835 and has been used to synthesize many others. In the early 1980s, researchers made another such compound that conducts electricity like a metal. Mass spectrometry of the compound shows that it has a molar mass of 184.27 g/mol, and analysis shows that it contains 2.288 g of S for every 1.000 g of N. What is its molecular formula?

3.141 Hydroxyapatite, $Ca_5(PO_4)_3(OH)$, is the main mineral component of dental enamel, dentin, and bone. Coating the compound on metallic implants (such as titanium alloys and stainless steels) helps the body accept the implant. When placed in bone voids, the powder encourages natural bone to grow into the void. Hydroxyapatite is prepared by adding aqueous phosphoric acid to a dilute slurry of calcium hydroxide.
(a) Write a balanced equation for this preparation.
(b) What mass (g) of hydroxyapatite could form from 100. g of 85% phosphoric acid and 100. g of calcium hydroxide?

3.142 Narceine is a narcotic in opium that crystallizes from solution as a hydrate. It contains 10.8 mass % water and has a molar mass of 499.52 g/mol. Determine x in narceine$\cdot xH_2O$.

3.143 Hydrogen-containing fuels have a "fuel value" based on their mass % H. Rank the following compounds from highest fuel value to lowest: ethane, propane, benzene, ethanol, cetyl palmitate (whale oil, $C_{32}H_{64}O_2$, not shown).

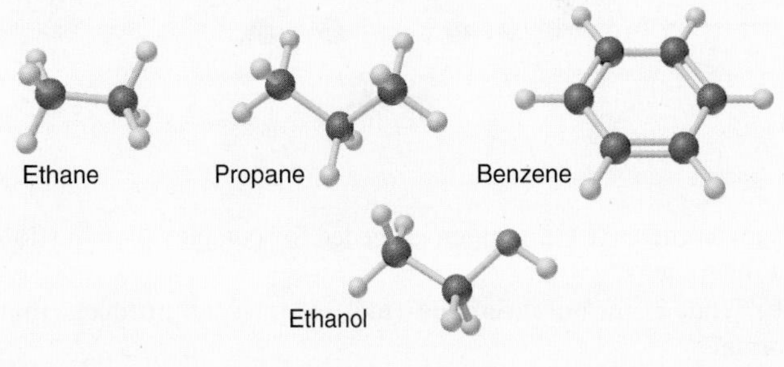

3.144 Serotonin ($\mathcal{M}$ = 176 g/mol) transmits nerve impulses between neurons. It contains 68.2% C, 6.86% H, 15.9% N, and 9.08% O by mass. What is its molecular formula?

3.145 In 1961, scientists agreed that the unified atomic mass unit (u) would be defined as $\frac{1}{12}$ the mass of an atom of ^{12}C. Before then, it was defined as $\frac{1}{16}$ the *average* mass of an atom of naturally occurring oxygen (a mixture of ^{16}O, ^{17}O, and ^{18}O). The current atomic mass of oxygen is 15.9994 u.
(a) Did Avogadro's number change after the definition of an atomic mass unit changed and, if so, in what direction?
(b) Did the definition of the mole change?
(c) Did the mass of 1 mol of a substance change?
(d) Before 1961, was Avogadro's number (6.02×10^{23} to three significant figures) the same as it is today?

3.146 Convert each description into a balanced equation:
(a) In a gaseous reaction, hydrogen sulfide burns in oxygen to form sulfur dioxide and water vapour.
(b) When crystalline potassium chlorate is heated to just above its melting point, it reacts to form two different crystalline compounds, potassium chloride and potassium perchlorate.
(c) When hydrogen gas is passed over powdered iron(III) oxide, iron metal and water vapour form.
(d) The combustion of gaseous ethane in air forms carbon dioxide and water vapour.
(e) Iron(II) chloride is converted to iron(III) fluoride by treatment with chlorine trifluoride gas. Chlorine gas is also formed.

3.147 Isobutylene is a hydrocarbon used in the manufacture of synthetic rubber. When 0.847 g of isobutylene was subjected to combustion analysis, the gain in mass of the CO_2 absorber was 2.657 g and that of the H_2O absorber was 1.089 g. What is the empirical formula for isobutylene?

3.148 The multistep smelting of ferric oxide to form elemental iron occurs at high temperatures in a blast furnace. In the first step, ferric oxide reacts with carbon monoxide to form Fe_3O_4. This substance reacts with more carbon monoxide to form iron(II) oxide, which reacts with still more carbon monoxide to form molten iron. Carbon dioxide is also produced in each step.
(a) Write an overall balanced equation for the iron-smelting process.
(b) What mass (g) of carbon monoxide is required to form 45.0 t (tonnes) of iron from ferric oxide?

3.149 One of the compounds that is used to increase the octane rating of gasoline is toluene (*below*). Suppose that 20.0 mL of toluene (d = 0.867 g/mL) is consumed when a sample of gasoline burns in air.

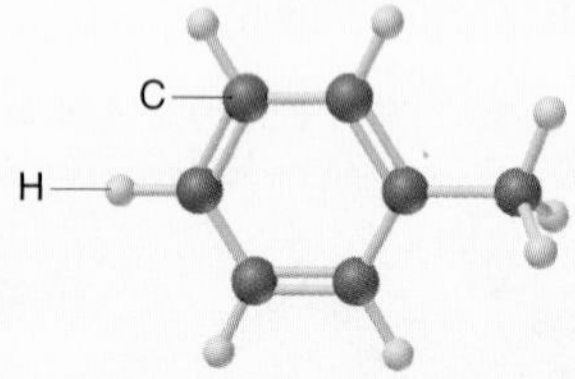

(a) What mass (g) of oxygen is needed for complete combustion of the toluene?
(b) What is the total amount (mol) of gaseous products that forms?
(c) How many molecules of water vapour form?

3.150 Recall the following reaction from Sample Problem 3.20:

$$2N_2H_4(l) + N_2O_4(l) \longrightarrow 3N_2(g) + 4H_2O(g)$$

During studies of this reaction, a chemical engineer measured a less-than-expected yield of N_2 and discovered that the following side reaction occurs:

$$N_2H_4(l) + 2N_2O_4(l) \longrightarrow 6NO(g) + 2H_2O(g)$$

In one experiment, 10.0 g of NO formed when 100.0 g of each reactant was used. What is the highest percent yield of N_2 that can be expected?

3.151 A 0.652 g sample of a pure strontium halide was reacted with excess sulfuric acid. The solid strontium sulfate that formed was separated, dried, and found to weigh 0.755 g. What is the formula for the original halide?

3.152 The following circles represent a chemical reaction between AB_2 and B_2:

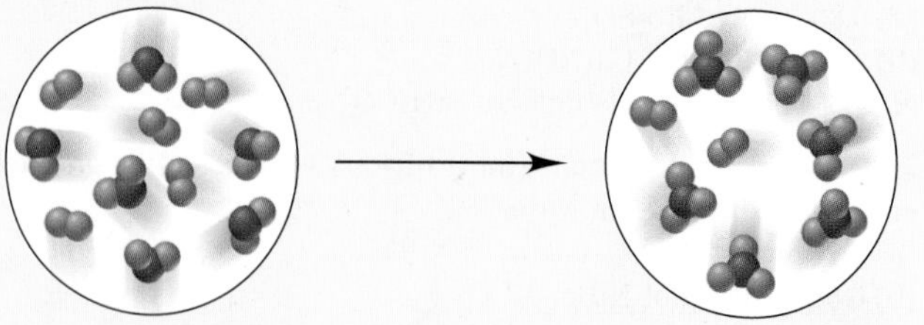

(a) Write a balanced equation for the reaction.
(b) What is the limiting reactant?
(c) What amount (mol) of product can be made from 3.0 mol of B_2 and 5.0 mol of AB_2?
(d) What amount (mol) of excess reactant remains after the reaction in part (c)?

3.153 Calculate each quantity:
(a) Volume (L) of 18.0 mol/L sulfuric acid that must be added to water to prepare 2.00 L of a 0.429 mol/L solution
(b) Concentration (mol/L) of the solution obtained by diluting 80.6 mL of 0.225 mol/L ammonium chloride to 0.250 L
(c) Volume (L) of water added to 0.130 L of 0.0372 mol/L sodium hydroxide to obtain a 0.0100 mol/L solution (assume that the volumes are additive at these low concentrations)
(d) Mass (g) of calcium nitrate in each millilitre of a solution prepared by diluting 64.0 mL of 0.745 mol/L calcium nitrate to a final volume of 0.100 L

3.154 Which models represent compounds that have the same empirical formula? What is the molecular mass of this common empirical formula?

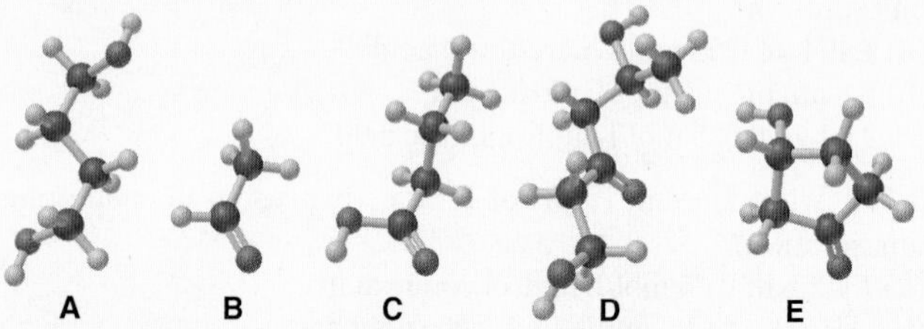

3.155 The zirconium oxalate $K_2Zr(C_2O_4)_3(H_2C_2O_4)\cdot H_2O$ was synthesized by mixing 1.68 g of $ZrOCl_2\cdot 8H_2O$ with 5.20 g of $H_2C_2O_4\cdot 2H_2O$ and an excess of aqueous KOH. After 2 months, 1.25 g of crystalline product was obtained, along with aqueous KCl and water. Calculate the percent yield.

3.156 Seawater is approximately 4.0% by mass dissolved ions, 85% of which are from NaCl.
(a) Find the mass percent of NaCl in seawater.
(b) Find the mass percent of Na^+ ions and of Cl^- ions in seawater.
(c) Find the concentration (mol/L) of NaCl in seawater at 15°C (d of seawater = 1.025 g/mL at 15°C).

3.157 Is each statement true or false? Correct any statements that are false.
(a) 1 mol of one substance has the same number of atoms as 1 mol of any other substance.
(b) The theoretical yield for a reaction is based on the balanced chemical equation.
(c) A limiting-reactant problem is being stated when the available quantity of one of the reactants is given in moles.
(d) To prepare 1.00 L of 3.00 mol/L NaCl, weigh 175.5 g of NaCl and dissolve it in 1.00 L of distilled water.
(e) The concentration of a solution is an intensive property, but the amount of solute in a solution is an extensive property.

3.158 Box A represents one unit volume of solution A. Which box—B, C, or D—represents one unit volume after adding enough solvent to solution A to (a) triple its volume; (b) double its volume; (c) quadruple its volume?

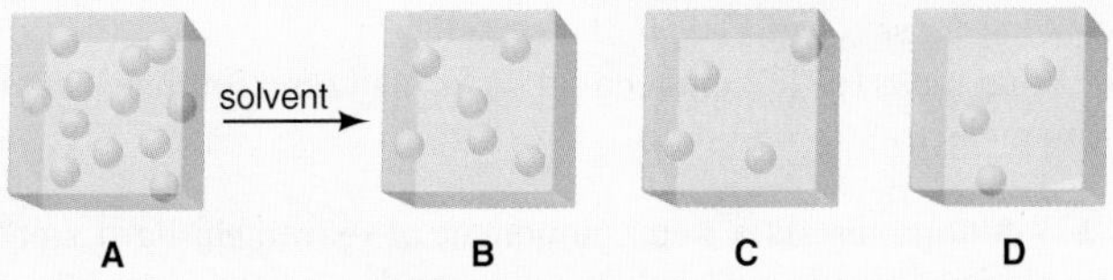

3.159 In each pair, choose the larger of the indicated quantities or state that the samples are equal:
(a) Entities: 0.4 mol of O_3 molecules or 0.4 mol of O atoms
(b) Mass (g): 0.4 mol of O_3 molecules or 0.4 mol of O atoms
(c) Amount (mol): 4.0 g of N_2O_4 or 3.3 g of SO_2
(d) Mass (g): 0.6 mol of C_2H_4 or 0.6 mol of F_2
(e) Total ions: 2.3 mol of sodium chlorate or 2.2 mol of magnesium chloride
(f) Molecules: 1.0 g of H_2O or 1.0 g of H_2O_2
(g) Na^+ ions: 0.500 L of 0.500 mol/L NaBr or 0.0146 kg of NaCl
(h) Mass (g): 6.02×10^{23} atoms of ^{235}U or 6.02×10^{23} atoms of ^{238}U

3.160 Write a balanced equation for the reaction between solid tetraphosphorus trisulfide and oxygen gas, which forms solid tetraphosphorus decoxide and sulfur dioxide gas. Show the equation in terms of (a) molecules; (b) amount (mol); (c) mass (g). (See Table 3.4.)

3.161 Hydrogen gas is considered a clean fuel because it produces only water vapour when it burns. If the reaction has a 98.8% yield, what mass (g) of hydrogen forms 105 kg of water?

3.162 Solar winds composed of free protons, electrons, and α particles bombard Earth constantly, knocking gas molecules out of the atmosphere. In this way, Earth loses about 3.0 kg of matter per second. It is estimated that the atmosphere will be gone in about 50 billion years. Use this estimate to calculate (a) the mass (kg) of Earth's atmosphere; (b) the amount (mol) of nitrogen, which makes up 75.5 mass % of the atmosphere.

3.163 Calculate each of the following quantities:
(a) Amount (mol) of 0.588 g of ammonium bromide
(b) Number of potassium ions in 88.5 g of potassium nitrate
(c) Mass (g) of 5.85 mol of glycerol ($C_3H_8O_3$)
(d) Volume (L) of 2.85 mol of chloroform ($CHCl_3$; $d = 1.48$ g/mL)
(e) Number of sodium ions in 2.11 mol of sodium carbonate
(f) Number of atoms in 25.0 mg of cadmium
(g) Number of atoms in 0.0015 mol of fluorine gas

3.164 Hydrocarbon mixtures are used as fuels.
(a) What mass (g) of $CO_2(g)$ is produced by the combustion of 200 g of a mixture that is 25.0% CH_4 and 75.0% C_3H_8 by mass?
(b) A 252 g gaseous mixture of CH_4 and C_3H_8 burns in excess O_2, and 748 g of CO_2 gas is collected. What is the mass percent of CH_4 in the mixture?

3.165 Suppose that you add 3.57 L of an HCl solution of unknown concentration to 1.35 L of 0.325 mol/L HCl. The resulting solution is 0.893 mol/L HCl. Assuming that the volumes are additive, calculate the concentration (mol/L) of the HCl solution you added.

3.166 Elements X (*green*) and Y (*purple*) react according to the following equation:

$$X_2 + 3Y_2 \longrightarrow 2XY_3$$

Which molecular scene represents the product of this reaction?

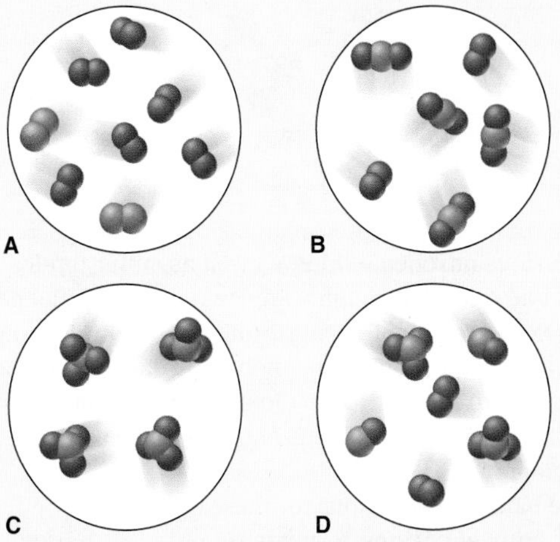

3.167 Nitrogen (N), phosphorus (P), and potassium (K) are the main nutrients in plant fertilizers. By industry convention, the numbers on a label refer to the mass percents of N, P_2O_5, and K_2O, in that order. Calculate the N/P/K ratio of a 30/10/10 fertilizer in terms of the amount (mol) of each element, and express this ratio as $x/y/1.0$.

3.168 What mass percents of ammonium sulfate, ammonium hydrogen phosphate, and potassium chloride would you use to prepare 10/10/10 plant fertilizer (see Problem 3.167)?

3.169 Ferrocene, synthesized in 1951, was the first organic iron compound with Fe—C bonds. An understanding of the structure of ferrocene gave rise to new ideas about chemical bonding and led to the preparation of many useful compounds. In the combustion analysis of ferrocene, which contains only Fe, C, and H, a 0.9437 g sample produced 2.233 g of CO_2 and 0.457 g of H_2O. What is the empirical formula for ferrocene?

3.170 When carbon-containing compounds are burned in a limited amount of air, $CO_2(g)$ and some $CO(g)$ are produced. A gaseous product mixture is 65.0 mass % CO_2 and 35.0 mass % CO. What is the mass percent of C in the mixture?

3.171 Write a balanced equation for the reaction depicted below:

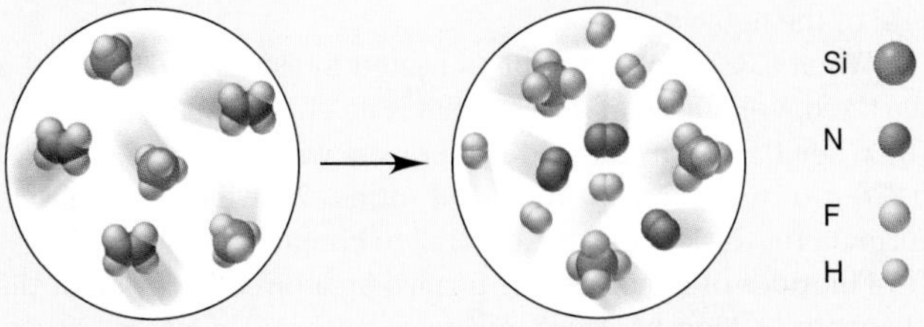

If each reactant molecule represents 1.25×10^{-2} mol and the reaction yield is 87%, what mass (g) of Si-containing product forms?

3.172 Citric acid (*below*) is concentrated in citrus fruits and plays a central metabolic role in nearly every animal and plant cell.
(a) What are the molar mass and formula for citric acid?
(b) What amount (mol) of citric acid is in 1.50 L of lemon juice (d = 1.09 g/mL) that is 6.82% citric acid by mass?

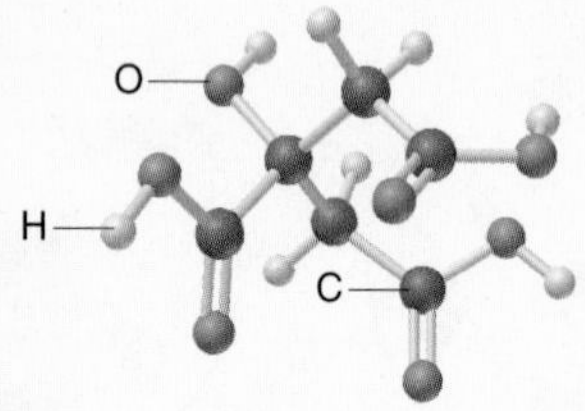

3.173 Various nitrogen oxides, as well as sulfur oxides, contribute to acidic rainfall through complex reaction sequences. Nitrogen and oxygen combine during the high-temperature combustion of fuels in air to form nitrogen monoxide gas, which reacts with more oxygen to form nitrogen dioxide gas. In contact with water vapour, nitrogen dioxide forms aqueous nitric acid and more nitrogen monoxide.
(a) Write balanced equations for these reactions.
(b) Use your equations to write one overall balanced equation that does *not* include nitrogen monoxide and nitrogen dioxide.
(c) What mass (tonnes, t) of nitric acid forms when 1350 t of atmospheric nitrogen is consumed (1 t = 1000 kg)?

3.174 Nitrogen monoxide reacts with elemental oxygen to form nitrogen dioxide. The scene shown below represents an initial mixture of reactants.

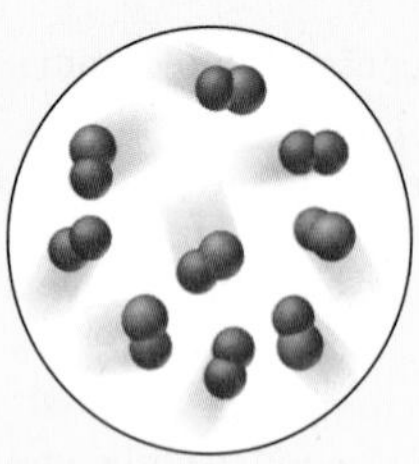

If the reaction has a 66% yield, which of the following scenes (A, B, or C) best represents the final product mixture?

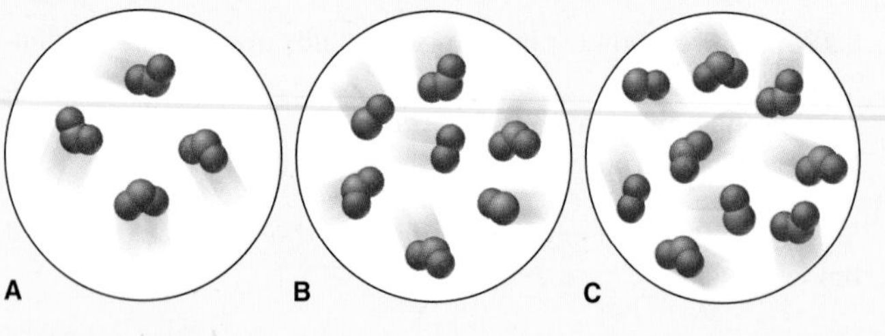

3.175 Fluorine is so reactive that it forms compounds with several of the noble gases.
(a) When 0.327 g of platinum is heated in fluorine, 0.519 g of a dark red, volatile solid forms. What is its empirical formula?
(b) When 0.265 g of this red solid reacts with excess xenon gas, 0.378 g of an orange-yellow solid forms. What is the empirical formula for this compound, the first to contain a noble gas?
(c) Fluorides of xenon can be formed by a direct reaction of the elements at high pressure and temperature. Under conditions that produce only tetrafluorides and hexafluorides, 1.85×10^{-4} mol of xenon reacts with 5.00×10^{-4} mol of fluorine, and 9.00×10^{-6} mol of xenon is found in excess. What is the mass percent of each xenon fluoride in the product mixture?

3.176 Hemoglobin is 6.0% heme ($C_{34}H_{32}FeN_4O_4$) by mass. To remove the heme, hemoglobin is treated with acetic acid and NaCl, which forms hemin ($C_{34}H_{32}N_4O_4FeCl$). A blood sample from a crime scene contains 0.65 g of hemoglobin.
(a) What mass (g) of heme is in the sample?
(b) What amount (mol) of heme is in the sample?
(c) What mass (g) of Fe is in the sample?
(d) What mass (g) of hemin could be formed for a forensic chemist to measure?

3.177 Manganese is a key component of extremely hard steel. The element occurs naturally in many oxides. A 542.3 g sample of a manganese oxide has an Mn/O ratio of 1.00/1.42 and consists of braunite (Mn_2O_3) and manganosite (MnO).
(a) What mass (g) of braunite and what mass (g) of manganosite are in the ore?
(b) What is the Mn^{3+}/Mn^{2+} ratio in the ore?

3.178 The human body excretes nitrogen in the form of urea, NH_2CONH_2. The key step in its biochemical formation is the reaction of water with arginine to produce urea and ornithine:

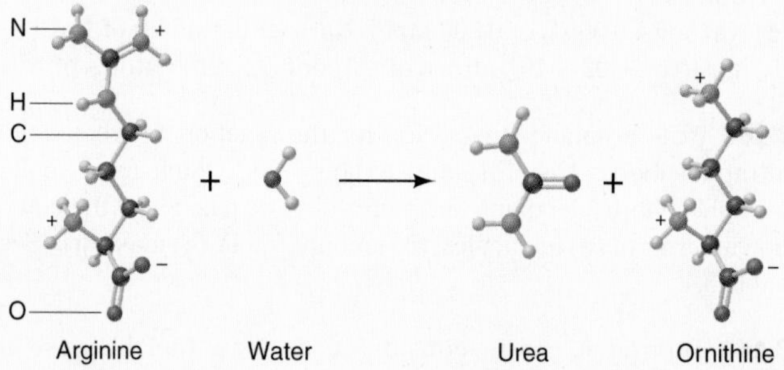

(a) What is the mass percent of nitrogen in urea, in arginine, and in ornithine?
(b) What mass (g) of nitrogen can be excreted as urea when 135.2 g of ornithine is produced?

3.179 Aspirin (acetylsalicylic acid, $C_9H_8O_4$) is made by reacting salicylic acid, $C_7H_6O_3$, with acetic anhydride, $(CH_3CO)_2O$:

$$C_7H_6O_3(s) + (CH_3CO)_2O(l) \longrightarrow C_9H_8O_4(s) + CH_3COOH(l)$$

In one preparation, 3.077 g of salicylic acid and 5.50 mL of acetic anhydride react to form 3.281 g of Aspirin.
(a) Which reactant is the limiting reactant? (d of acetic anhydride = 1.080 g/mL)
(b) What is the percent yield of this reaction?
(c) What is the percent atom economy of this reaction?

3.180 The rocket fuel hydrazine (N_2H_4) is made by the three-step Raschig process, which has the following overall equation:

$$NaOCl(aq) + 2NH_3(aq) \longrightarrow N_2H_4(aq) + NaCl(aq) + H_2O(l)$$

What is the percent atom economy of this process?

3.181 Lead(II) chromate ($PbCrO_4$) is used as the yellow pigment for marking traffic lanes, but it is banned from house paint because of the risk of lead poisoning. It is produced from chromite ($FeCr_2O_4$), an ore of chromium:

$$4FeCr_2O_4(s) + 8K_2CO_3(aq) + 7O_2(g) \longrightarrow 2Fe_2O_3(s) + 8K_2CrO_4(aq) + 8CO_2(g)$$

Lead(II) ion then replaces the K^+ ions. If a yellow paint needs to have 0.511% $PbCrO_4$ by mass, what mass (g) of chromite is needed per kilogram of paint?

3.182 Ethanol (CH_3CH_2OH), the intoxicant in alcoholic beverages, is also used to make other organic compounds. In concentrated sulfuric acid, ethanol forms diethyl ether and water:

$$2CH_3CH_2OH(l) \longrightarrow CH_3CH_2OCH_2CH_3(l) + H_2O(g)$$

In a side reaction, some ethanol forms ethylene and water:

$$CH_3CH_2OH(l) \longrightarrow CH_2CH_2(g) + H_2O(g)$$

(a) If 50.0 g of ethanol yields 35.9 g of diethyl ether, what is the percent yield of diethyl ether?
(b) If 45.0% of the ethanol that did not produce the ether reacts by the side reaction, what mass (g) of ethylene is produced?

3.183 When powdered zinc is heated with sulfur, a violent reaction occurs and zinc sulfide forms:

$$Zn(s) + S_8(s) \longrightarrow ZnS(s) \text{ (unbalanced)}$$

Some of the reactants also combine with oxygen in air to form zinc oxide and sulfur dioxide. When 83.2 g of Zn reacts with 52.4 g of S_8, 104.4 g of ZnS forms.
(a) What is the percent yield of ZnS?
(b) If all the remaining reactants combine with oxygen, what mass (g) of each of the two oxides forms?

3.184 Cocaine ($C_{17}H_{21}O_4N$) is a natural substance that is found in coca leaves, which have been used for centuries as a local anaesthetic and stimulant. Illegal cocaine arrives in Canada either as the pure compound or as the hydrochloride salt ($C_{17}H_{21}O_4NHCl$). At 25°C, the salt is very soluble in water (2.50 kg/L), but cocaine is much less so (1.70 g/L).
(a) What is the maximum mass (g) of the salt that can dissolve in 50.0 mL of water?
(b) If this solution is treated with NaOH, the salt is converted to cocaine. How much more water (L) is needed to dissolve it?

3.185 High-temperature superconducting oxides hold great promise in the utility, transportation, and computer industries.
(a) One superconductor is $La_{2-x}Sr_xCuO_4$. Calculate the molar masses of this oxide when $x = 0$, $x = 1$, and $x = 0.163$.
(b) Another common superconducting oxide is made by heating a mixture of barium carbonate, copper(II) oxide, and yttrium(III) oxide, followed by further heating in O_2:

$$4BaCO_3(s) + 6CuO(s) + Y_2O_3(s) \longrightarrow 2YBa_2Cu_3O_{6.5}(s) + 4CO_2(g)$$

$$2YBa_2Cu_3O_{6.5}(s) + \tfrac{1}{2}O_2(g) \longrightarrow 2YBa_2Cu_3O_7(s)$$

When equal masses of the three reactants are heated, which reactant is limiting?
(c) After the product in part (b) is removed, what is the mass percent of each reactant in the remaining solid mixture?

3.186 In a combustion analysis, 0.5431 g of sorbitol, an artificial sweetener containing C, H and O, was burned in an excess of oxygen to produce 0.7871 g of carbon dioxide and 0.3756 g of water. What is the empirical formula of sorbitol? If the molar mass of sorbitol is 182.17 g/mol, what is the molecular formula of sorbitol?

3.187 Industrially, when the reaction

$$PbS(s) + O_2(g) \longrightarrow PbO(s) + SO_2(g) \text{ (unbalanced)}$$

is carried out, 125 kg of PbS is reacted with 19.0 kg of oxygen gas.
(a) What is the balanced reaction?
(b) What is the limiting reactant?
(c) What mass of PbO should be produced?
(d) If 75.3 kg of PbO is produced by this reaction, what is the percent yield?

3.188 In the water gas shift reaction, $CH_4(g) + H_2O(g) \longrightarrow H_2(g) + CO_2(g)$ (unbalanced), equal masses of methane and water were placed in the reactor and the total mass of the reactants was 274 kg.
(a) What is the balanced reaction?
(b) What is the limiting reactant?
(c) What mass of hydrogen gas should be produced?
(d) If the reaction is found to have a 73.2% yield, what actual mass of hydrogen was obtained?

3.189 A sample of citronellal, the C—H—O compound that gives citronella its distinct lemony scent, with mass 1.8392 g was placed in a combustion chamber. Excess oxygen passing through the chamber at high temperatures resulted in complete combustion of the citronellal, forming 5.2469 g of carbon dioxide and 1.9318 g of water. What is the empirical formula of citronellal? Given that its molar mass is 154.25 g/mol, what is the molecular formula of citronellal?

Gases and the Kinetic-Molecular Theory

IN THIS CHAPTER . . . We explore the physical behaviour of gases and the theory that explains it. In the process, we see how scientists use mathematics to model nature. By the end of this chapter, you should be able to

- Differentiate between the behaviours of gases, liquids, and solids
- Describe laboratory methods for measuring gas pressure
- Describe the behaviour of a gas in terms of how its volume changes with a change in (1) pressure, (2) temperature, or (3) amount
- Derive the ideal gas law, which encompasses these three laws, and apply it to solve gas law problems
- Rearrange the ideal gas law to determine the density and molar mass of an unknown gas, the partial pressure of any gas in a mixture, and the amounts of gaseous reactants and products in a chemical change
- Apply the kinetic-molecular theory to explain the gas laws and account for other important behaviours of gas particles
- Apply key ideas about gas behaviour to Earth's atmosphere
- Describe the conditions under which gas behaviour may be non-ideal and explain some modifications and refinements to the ideal gas law and kinetic-molecular theory that better reflect on non-ideal gas behaviour

4

Every breath we take involves chemistry. From the application of the gas laws (Boyle's law and Charles's law) to the actual breathing process to the composition of the mixture of gases we breathe, chemistry is at work. In this chapter, we will examine the experimentally validated relationships between gas pressure, volume, temperature, and amount and show how these empirical laws led to the development of the ideal gas law. We will further connect the behaviour of gases to chemical stoichiometry, studied in Chapter 3, and see how we can predict the volume, temperature, pressure, or amount of a gas product given the other variables. We will study how gases behave in mixtures as well as other important properties of gases, such as effusion and diffusion. Finally, we will determine the factors that limit the ideality of gases and explore some of the ways to characterize non-ideal gases and their behaviour.

Concepts and Skills to Review before Studying This Chapter

- Physical states of matter (Section 1.1)
- SI unit conversions (Section 1.5)
- Amount-mass-number conversions (Section 3.1)

4.1 An Overview of the Physical States of Matter

As we have seen in Chapter 1, substances may exist as a Bose-Einstein condensate, a solid, a liquid, a gas, or a plasma depending on the conditions of pressure and temperature. In Chapter 1, we used the relative position and motion of the particles of a substance to distinguish how the three most common states fill a container (Figure 4.1):

- A gas adopts the shape of the container and fills the container because its particles are far apart and move randomly.
- A liquid adopts the shape of the container to the extent of the container's volume because its particles are close together but free to move around each other.
- A solid has a fixed shape regardless of the shape of the container because its particles are close together and held rigidly in place.

Figure 4.1 focuses on the three common states of bromine.

Several other aspects of their behaviour distinguish gases from liquids and solids:

1. *The volume of a gas changes significantly with pressure.* When a sample of gas is confined to a container of variable volume, such as a cylinder with a piston, *increasing* the force on the piston *decreases* the volume of the gas. Removing the external force allows the volume to increase again. Gases under pressure can do a lot of work: rapidly expanding compressed air in a jackhammer breaks rock and cement; the compression and expansion of high-temperature and high-pressure gases produced by the combustion of fuels in a car exert direct pressure on the pistons and convert chemical energy to mechanical energy. In contrast, the volume of a liquid or a solid does not change significantly under pressure.
2. *The volume of a gas changes significantly with temperature.* When a sample of gas is heated, it expands; when it is cooled, it shrinks. This volume change is 50 to 100 times that for liquids or solids. The expansion that occurs when gases are rapidly heated can have dramatic effects, such as lifting a rocket into space, and everyday effects, such as popping corn.
3. *Gases flow very freely.* Gases flow much more freely than liquids and solids. This behaviour allows gases to be transported more easily through pipes, but it also means that they leak more rapidly out of small holes and cracks.

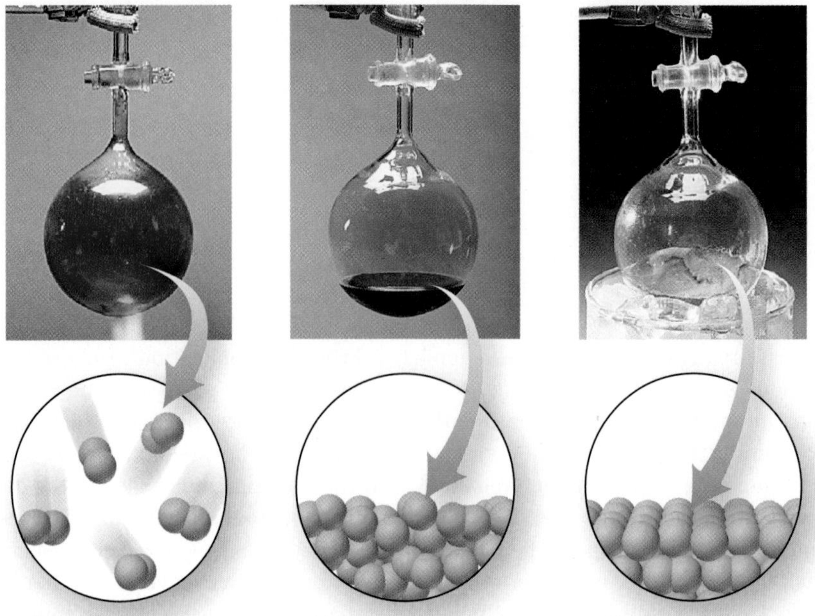
Gas: Particles are far apart, move freely, and fill the available space.
Liquid: Particles are close together but move around one another.
Solid: Particles are close together in a regular array and do not move around one another.

FIGURE 4.1 The three common states of matter. Many pure substances, such as bromine (Br_2), can exist under particular conditions of pressure and temperature as a gas, a liquid, or a solid.

4. *Gases have relatively low densities.* The density of a gas is usually measured in units of grams per litre (g/L), whereas the density of a liquid or a solid is usually measured in grams per millilitre (g/mL), which is about 1000 times as dense (see Table 1.4). For example, at standard temperature and pressure conditions (0°C and 1 bar), the density of dry air is 1.2754 kg/m^3, whereas the density of ethanol(l) is 0.806 g/mL and the density of NaCl(s) is 2.17 g/mL. When a gas cools, its density *increases* because its volume *decreases*: on cooling from 0°C to −20°C, the density of dry air(g) increases from 1.2754 kg/m^3 to 1.3943 kg/m^3.
5. *A gas forms a solution in any proportions.* Air is a solution of 18 gases. Two liquids, however, may or may not form a solution: water and ethanol do, but water and gasoline do not. Two solids generally do not form a solution unless they are melted and mixed while liquids and then allowed to solidify (as is done to make the alloy bronze from copper and tin).

Similar to a gas completely filling a container, these macroscopic properties—changing volume with pressure or temperature, greater ability to flow, low density, and ability to form solutions—arise because the particles in a gas are much farther apart than those in either a liquid or a solid.

SUMMARY OF SECTION 4.1

- The volume of a gas can be altered significantly by changing the applied force or the temperature. Corresponding changes for a liquid or a solid are *much* smaller.
- Gases flow more freely and have much lower densities than liquids and solids.
- Gases mix in any proportions to form solutions; liquids and solids generally do not.
- Differences in the physical states are due to the greater average distance between particles in a gas than in a liquid or a solid.

4.2 Gas Pressure and Its Measurement

You can blow up a balloon or pump up a tire because *a gas exerts pressure on the walls of its container.* **Pressure (p)** is defined as the force exerted per unit of surface area:

$$\text{Pressure} = \frac{\text{force}}{\text{area}}$$

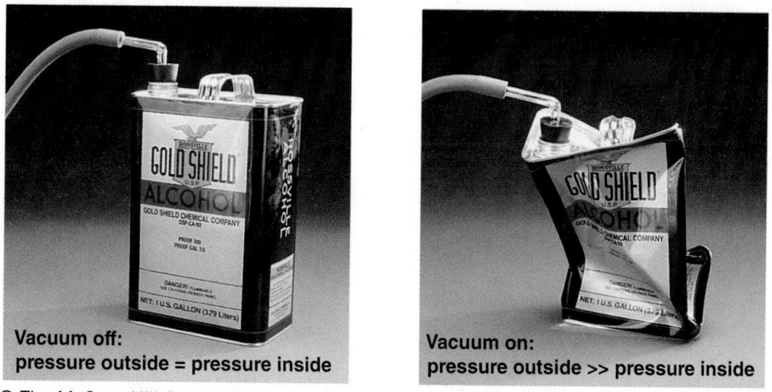

FIGURE 4.2 Effect of atmospheric pressure on a familiar object

Earth's gravity attracts the atmospheric gases, and they exert a force uniformly on *all* surfaces. The force, or *weight*, of these gases creates a pressure of about 10.1 N/cm^2 (newtons per square centimetre) on the surface. Thus, a pressure of 10.1 N/cm^2 exists on the outside of your room (or your body), and it equals the pressure on the inside. What would happen if the pressures were *not* equal? Consider the empty can in Figure 4.2. Under atmospheric pressure conditions, the can maintains its shape because the pressure on the outside is equal to the pressure on the inside of the can. When water is added to the can and it is heated and then inverted in an ice bath, the water condenses rapidly, creating a partial vacuum in the can. With the internal pressure of the can greatly decreased, the pressure of the atmosphere easily crushes the can. The vacuum-filtration flasks and tubing that you may have used in the laboratory have thick walls that withstand the relatively higher external pressure.

Laboratory Devices for Measuring Gas Pressure

The **barometer** *is used to measure atmospheric pressure.* This device is still essentially the same as it was when it was invented in 1643 by the Italian physicist Evangelista Torricelli: a tube about 1 m long, closed at one end, filled with mercury (atomic symbol, Hg), and inverted into a dish containing more mercury. When the tube is inverted, some of the mercury flows out into the dish, and a vacuum forms above the mercury remaining in the tube (Figure 4.3). At sea level, under ordinary atmospheric conditions, the mercury stops flowing out when the surface of the mercury in the tube is about 760 mm above the surface of the mercury in the dish. At that height, the column of mercury exerts the same pressure (weight/area) on the mercury surface in the dish as the atmosphere does: $p_{Hg} = p_{atm}$. Likewise, if you evacuate a closed tube and invert it into a dish of mercury, the atmosphere pushes the mercury up to a height of about 760 mm.

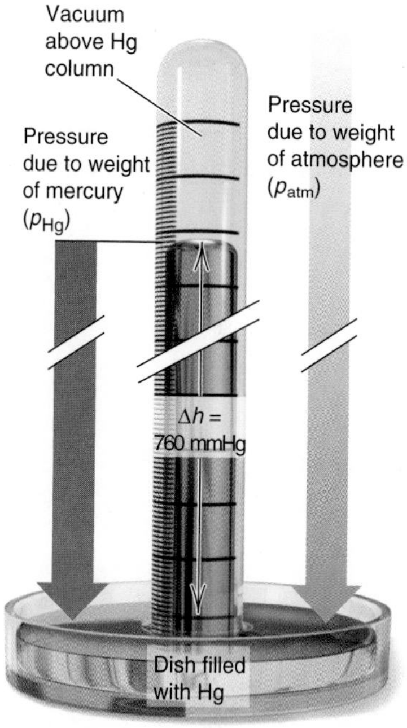

FIGURE 4.3 A mercury barometer. The pressure of the atmosphere, p_{atm}, balances the pressure of the mercury column, p_{Hg}.

Notice that we did not specify the diameter of the barometer tube. If the mercury in a 1 cm diameter tube rises to a height of 760 mm, the mercury in a 2 cm diameter tube will also rise to that height. The *weight* of mercury is greater in the wider tube, but so is the area; thus, the *pressure*, the *ratio* of weight to area, is the same.

Because the pressure of the mercury column is directly proportional to its height, a unit that has been commonly used for pressure is millimetres of mercury (mmHg). (We will discuss other units of pressure shortly.) At sea level and 0°C, atmospheric pressure is about 760 mmHg; at the top of Mount Robson in the Canadian Rockies (elevation 3954 m), the atmospheric pressure is only about 465 mmHg. Thus, *pressure decreases with altitude*: the column of air above the sea is taller, so it weighs more than the column of air above Mount Robson.

Laboratory barometers contain mercury because its high density allows a barometer to be a convenient size. If a barometer contained water instead, it would have to be more than 10.4 m high, because the pressure of the atmosphere equals the pressure of a column of water about 10 400 mm high. For a given pressure, the ratio of the heights (h) of the liquid columns is inversely related to the ratio of the densities (d) of the liquids:

$$\frac{h_{H_2O}}{h_{Hg}} = \frac{d_{Hg}}{d_{H_2O}}$$

Interestingly, several centuries ago, people thought a vacuum had mysterious "suction" powers, and they did not understand why a suction pump could remove water

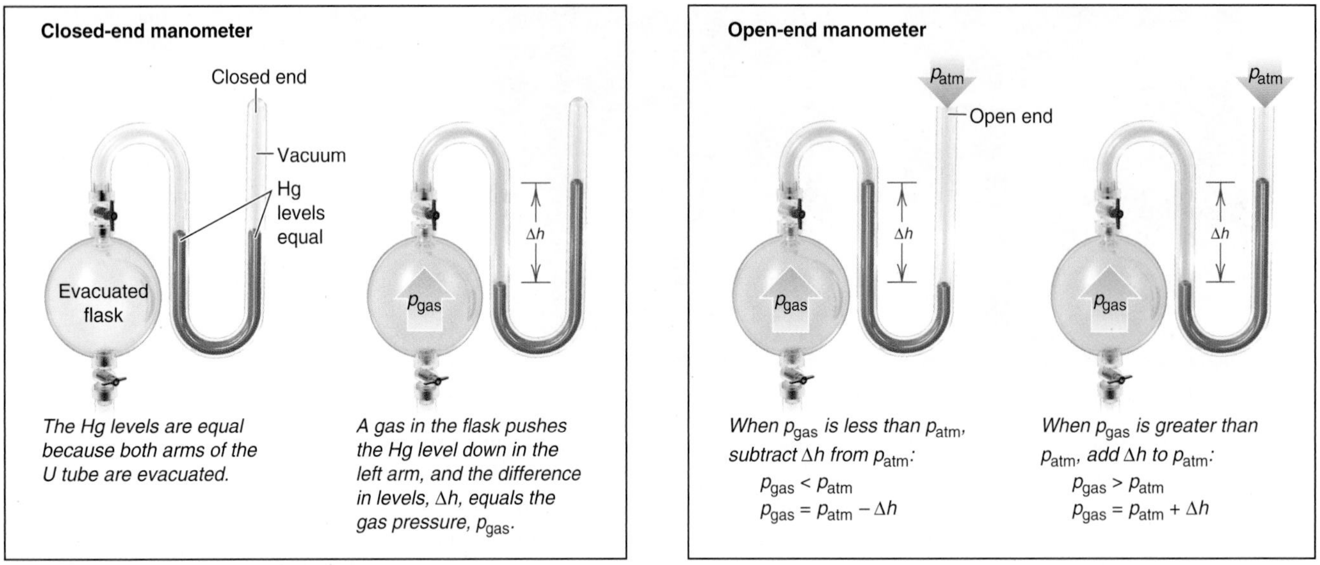
Closed-end manometer
Closed end
Vacuum
Hg levels equal
Evacuated flask
Δh
p_{gas}
The Hg levels are equal because both arms of the U tube are evacuated.
A gas in the flask pushes the Hg level down in the left arm, and the difference in levels, Δh, equals the gas pressure, p_{gas}.
Open-end manometer
p_{atm}
Open end
Δh
p_{gas}
When p_{gas} is less than p_{atm}, subtract Δh from p_{atm}:
$p_{gas} < p_{atm}$
$p_{gas} = p_{atm} - \Delta h$
When p_{gas} is greater than p_{atm}, add Δh to p_{atm}:
$p_{gas} > p_{atm}$
$p_{gas} = p_{atm} + \Delta h$

FIGURE 4.4 Two types of manometers

from a well only to a depth of 10.4 m. We know now, as the great 17th-century scientist Galileo explained, that a vacuum does not suck mercury up into a barometer tube, and a suction pump does not suck water up from a well. Similarly, the vacuum in Figure 4.2 does not suck in the walls of the crushed can, and the vacuum you create in a straw does not suck the drink into your mouth. Only matter—in this case, the atmospheric gases—can exert a force.

Manometers are devices that are used to measure the pressure of a gas in an experiment. Figure 4.4 shows two types of manometers. In the *closed-end manometer* (*left*), a mercury-filled, curved tube is *closed* at one end and attached to a flask at the other end. When the flask is evacuated, the mercury levels in the two arms of the tube are the same because no gas exerts pressure on either mercury surface. When a gas is in the flask, it pushes down the mercury level in the near arm, causing the level to rise in the far arm. The *difference* in the column heights (Δh) equals the gas pressure.

The *open-end manometer* (*right*) also consists of a curved tube filled with mercury, but one end of the tube is *open* to the atmosphere and the other end is connected to the gas sample. The atmosphere pushes on one mercury surface, and the gas pushes on the other. Again, Δh equals the difference between the two pressures. However, when using this type of manometer, we must measure the atmospheric pressure with a barometer and either add or subtract Δh from that value.

Units of Pressure

Pressure results from a force exerted on an area. The SI unit of force is the newton (N): 1 N = 1 kg • m/s^2 (about the weight of an apple). The SI unit of pressure is the **pascal (Pa)**, which equals a force of one newton exerted on an area of one square metre:

$$1 \text{ Pa} = 1 \text{ N/m}^2$$

A much larger unit is the **standard atmosphere (atm)**, the average atmospheric pressure measured at sea level and 0°C. It is an exact unit defined in terms of the pascal (Pa):

$$1 \text{ atm} = 101.325 \text{ kPa (kilopascal)} = 1.01325 \times 10^5 \text{ Pa}$$

Another unit is the **millimetre of mercury (mmHg)**, which was mentioned earlier; in honour of Torricelli, this unit has been renamed the **Torr**:

$$1 \text{ Torr} = 1 \text{ mmHg} = \frac{1}{760} \text{ atm} = \frac{101.325}{760} \text{ kPa} = 133.322 \text{ Pa}$$

The **bar** is currently used for defining **standard pressure** and when describing thermodynamic standards:

$$1 \text{ bar} = 1 \times 10^2 \text{ kPa} = 1 \times 10^5 \text{ Pa}$$

TABLE 4.1 Common Units of Pressure

Unit	Atmospheric Pressure at Sea Level and 15°C
pascal (Pa); kilopascal (kPa)	1.01325×10^5 Pa; 101.325 kPa*
atmosphere (atm)	1 atm*
millimetre of mercury (mmHg)	760 mmHg*
torr (Torr)	760 Torr*
bar	1.01325 bar

*This is an exact quantity.

Despite a gradual change to SI units, many chemists still express pressure in Torr and atmosphere; in this book, we will mainly use pascal and bar. Table 4.1 lists some important pressure units, with the corresponding values for atmospheric pressure.

Sample Problem 4.1 Converting Units of Pressure

Problem A geochemist heats a limestone ($CaCO_3$) sample and collects the CO_2 that is released in an evacuated flask attached to a closed-end manometer. After the system comes to room temperature, $\Delta h = 291.4$ mmHg. Calculate the CO_2 pressure in units of bar and kPa.

Plan The CO_2 pressure is given in units of mmHg, so we construct conversion factors from Table 4.1 to find the pressure in the other units.

Solution Convert from mmHg to bar:

$$p_{CO_2}\text{ (bar)} = 291.4\ \cancel{\text{mmHg}} \times \frac{1.013\,25\text{ bar}}{760\ \cancel{\text{mmHg}}} = 0.3885\text{ bar}$$

Convert from bar to kPa:

$$p_{CO_2}\text{ (kPa)} = 0.3885\ \cancel{\text{bar}} \times \frac{100\text{ kPa}}{1\ \cancel{\text{bar}}} = 38.85\text{ kPa}$$

Check There is 760 Torr in roughly 1 bar, so 300 Torr should be < 0.5 bar. There is 100 kPa in 1 bar, so < 0.5 bar should be < 50 kPa.

Comment **1.** In the conversion from mmHg to bar, we retained four significant figures because this unit conversion factor involves *exact* numbers; that is, 760 mmHg has as many significant figures as the calculation requires (see the footnote under Table 4.1).

2. From here on, except in particularly complex situations, *unit cancelling will no longer be shown.*

Follow-Up Problem 4.1 The CO_2 released from another mineral sample was collected in an evacuated flask connected to an open-end manometer. If the barometer reading is 753.6 mmHg, and p_{gas} is less than p_{atm} to give a Δh of 174.0 mmHg, calculate p_{CO_2} in units of Pa, kPa, and bar. (See Brief Solutions to Follow-Up Problems at the end of the chapter.)

SUMMARY OF SECTION 4.2

- Gases exert pressure (force/area) on all surfaces they contact.
- A barometer measures atmospheric pressure based on the height of a mercury column that the atmosphere can support (760 mmHg at sea level and 15°C).
- Closed-end and open-end manometers are used to measure the pressure of a gas sample.
- Pressure units include the SI unit pascal (Pa), the standard unit bar (bar), the atmosphere (atm), and the Torr (identical to mmHg).

4.3 The Gas Laws and Their Experimental Foundations

The physical behaviour of a sample of gas can be described completely by four variables: pressure (p), volume (V), temperature (T), and amount (mol, n). The variables are interdependent, which means that *any one of them can be determined by measuring the other three.* Three key relationships exist among the four gas variables: Boyle's law, Charles's law, and Avogadro's law. Each of these *gas laws expresses the effect of one variable on another, with the remaining two variables held constant.* Because the volume of a gas is so easy to measure, the laws are expressed as the effect on the volume of a gas, caused by a change in the pressure, temperature, or amount (mol) of the gas.

The individual gas laws are special cases of a unifying relationship called the *ideal gas law*, which quantitatively describes the behaviour of an **ideal gas**, *a gas that exhibits linear relationships among volume, pressure, temperature, and amount.* Although *no ideal gas actually exists*, most simple gases—such as N_2, O_2, H_2, and the noble gases—behave nearly ideally at ordinary temperatures and pressures. We will discuss the ideal gas law after discussing the three individual laws that led to its development.

The Relationship between Volume and Pressure: Boyle's Law

Following Torricelli's invention of the barometer, the great 17th-century English chemist Robert Boyle studied the effect of pressure on the volume of a sample of gas.

1. *The experiment.* Figure 4.5 illustrates the setup that Boyle might have used in his experiments (parts A and B), the data he might have collected (part C), and graphs of the data (parts D and E). Boyle sealed the shorter leg of a J-shaped glass tube and poured mercury into the longer open leg, thereby trapping some air (the gas in the experiment) in the shorter leg. He calculated the gas volume (V_{gas}) from the height of the trapped air and the diameter of the tube. The total pressure, p_{total}, applied to the trapped gas is the pressure of the atmosphere, p_{atm} (760 mm, measured with a barometer), plus the difference in the heights of the mercury columns (Δh) in the two legs of the J tube, 20 mm (Figure 4.5A); thus, p_{total} is 780 Torr. By adding mercury, Boyle increased p_{total}, and the gas volume decreased. In Figure 4.5B, more mercury has been added to the original Δh of 20 mm, so p_{total} doubles to 1560 Torr; note that V_{gas} is halved from 20 mL to 10 mL. In this way, by keeping the temperature and amount of the gas constant, Boyle was able to measure the effect of the applied pressure on the gas volume.

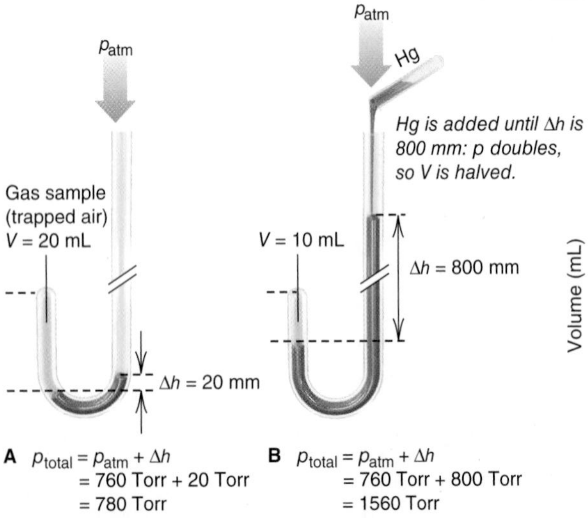

C

V (mL)	p (Torr) Δh	+ p_{atm}	= p_{total}	$\frac{1}{p_{total}}$	pV (Torr•mL)
20.0	20.0	760	780	0.001 28	1.56×10^4
15.0	278	760	1038	0.000 963	1.56×10^4
10.0	800	760	1560	0.000 641	1.56×10^4
5.0	2352	760	3112	0.000 321	1.56×10^4

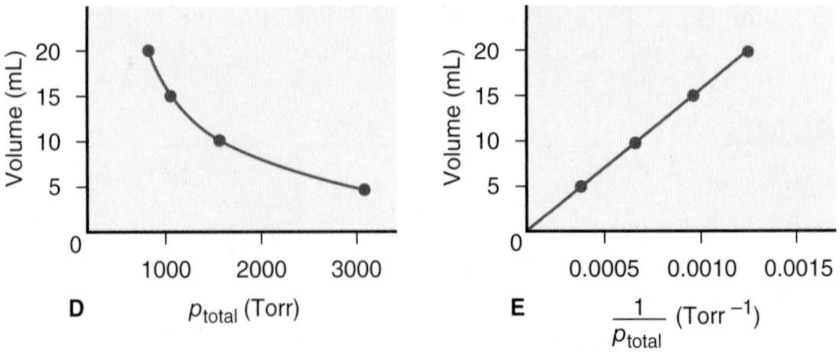

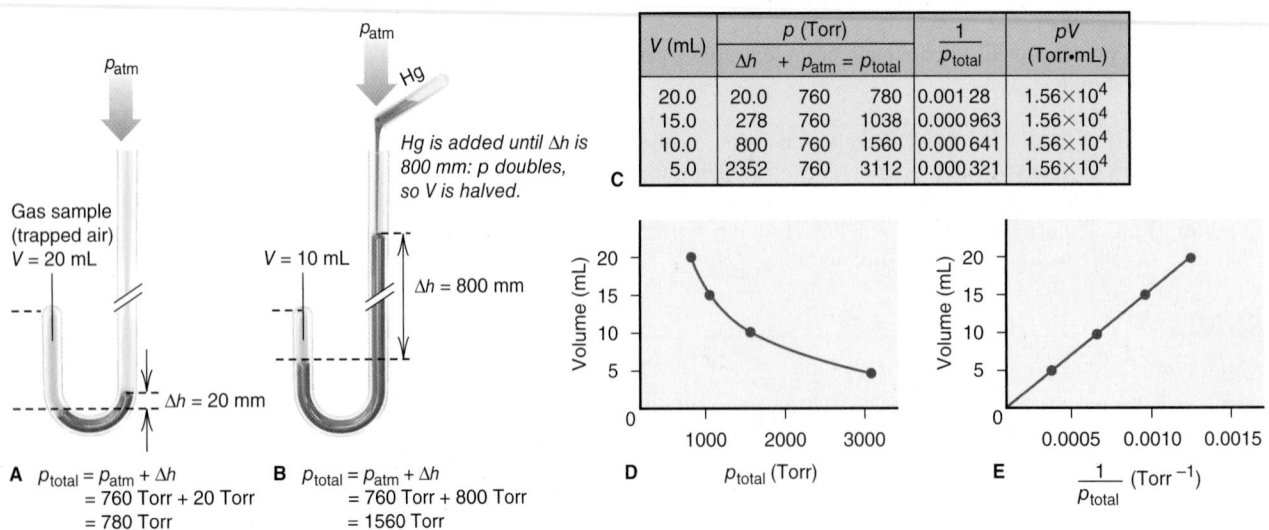

FIGURE 4.5 Boyle's law, the relationship between the volume and pressure of a gas

Note the following results in Figure 4.5:

- The product of corresponding p and V values is a constant (part C, rightmost column).
- V is *inversely* proportional to p (part D).
- V is *directly* proportional to $\frac{1}{p}$ (part E), and a plot of V versus $\frac{1}{p}$ is linear. This *linear relationship between two gas variables* is a hallmark of ideal gas behaviour.

2. *Conclusion and statement of the law.* The generalization of Boyle's observations is known as **Boyle's law**: *at constant temperature, the volume occupied by a fixed amount of gas is **inversely** proportional to the applied (external) pressure*, or

$$V \propto \frac{1}{p} \quad (T \text{ and } n \text{ fixed}) \tag{4.1}$$

The inverse proportionality indicates that the product of p and V is a constant. We can convert the proportionality to an equality as shown:

$$V = \frac{\text{constant}}{p} \quad \text{or} \quad pV = \text{constant} \quad (T \text{ and } n \text{ fixed})$$

This means that, at fixed T and n,

$$p\uparrow, V\downarrow \quad \text{and} \quad p\downarrow, V\uparrow$$

The constant is the same for most simple gases under ordinary conditions. Thus, tripling the external pressure reduces the volume of a gas to a third of its initial value, halving the pressure doubles the volume, and so on.

The wording of Boyle's law focuses on *external* pressure. Notice, however, that the mercury level rises as mercury is added, until the pressure of the trapped gas *on* the mercury increases enough to stop its rise. At this point, the pressure exerted *on* the gas equals the pressure exerted *by* the gas (p_{gas}). Thus, in general, if V_{gas} increases, p_{gas} decreases, and vice versa.

The Relationship between Volume and Temperature: Charles's Law

Boyle's work showed that the pressure-volume relationship holds only at constant temperature, but why should that be so? It would take more than a century, until the work of French scientists J. A. C. Charles and J. L. Gay-Lussac, for the relationship between gas volume and temperature to be understood.

1. *The experiment.* Let us examine this relationship by measuring the volume at different temperatures of a fixed amount of a gas under constant pressure. A straight tube, closed at one end, traps a fixed amount of gas (air) under a small mercury plug. The tube is immersed in a water bath that is warmed with a heater or cooled with ice. After each change in temperature, we measure the length of the gas column, which is proportional to its volume. The total pressure exerted on the gas is constant because the mercury plug and the atmospheric pressure do not change (Figures 4.6A and 4.6B).

Figure 4.6C shows some typical data. Consider the red line, which shows how the volume of 0.04 mol of gas at atmospheric pressure changes with temperature. Extrapolating this line to lower temperatures (dashed portion) shows that, in theory, the gas occupies zero volume at −273.15°C (the intercept on the temperature axis). Plots for a different amount of gas (*green*) or a different gas pressure (*blue*) have different slopes, but they all converge at −273.15°C. William Thomson (Lord Kelvin) later used this linear relation between gas volume and temperature to devise the absolute temperature scale (Section 1.5).

2. *Conclusion and statement of the law.* Above all, note that *the volume-temperature relationship is linear*, but, unlike volume and pressure, volume and temperature are *directly* proportional. This behaviour is incorporated into the modern statement of the volume-temperature relationship, which is known as **Charles's law**: *at constant pressure, the volume occupied by a fixed amount of gas is **directly** proportional to its absolute (Kelvin) temperature*, or

$$V \propto T \quad (p \text{ and } n \text{ fixed}) \tag{4.2}$$

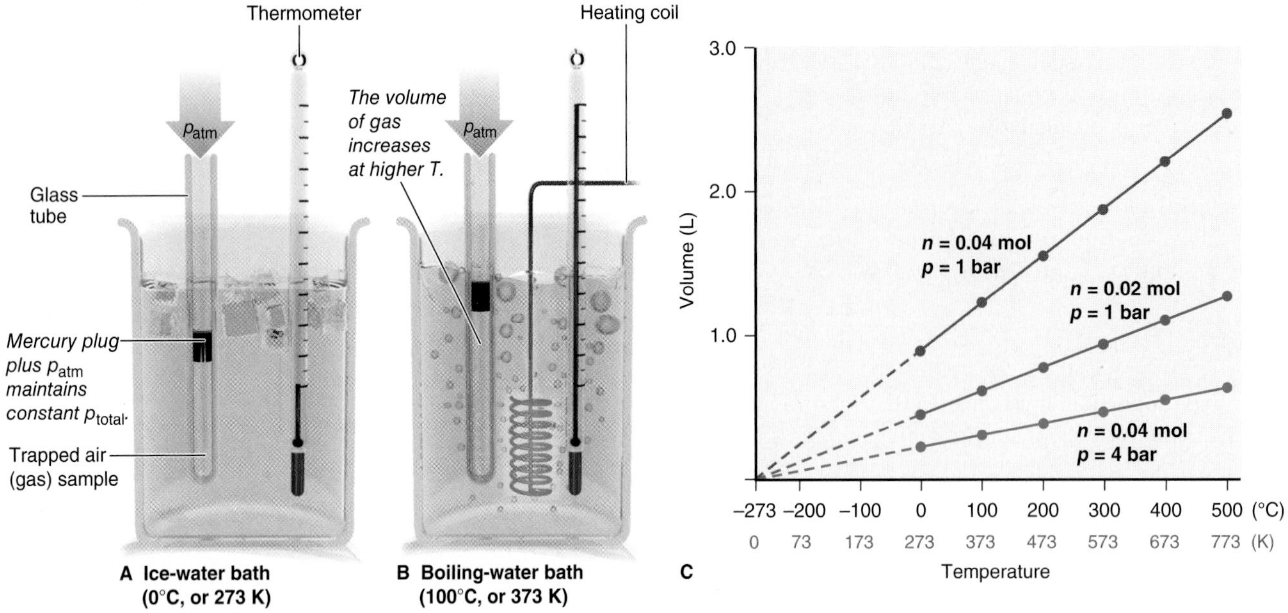

FIGURE 4.6 Charles's law, the relationship between the volume and temperature of a gas

The direct proportionality indicates that the quotient of V over T has a constant value. This relationship can be expressed in equation form as

$$V = T \times \text{constant} \quad \text{or} \quad \frac{V}{T} = \text{constant } (p \text{ and } n \text{ fixed})$$

This means that, at fixed p and n,

$$T\uparrow, V\uparrow \qquad \text{and} \qquad T\downarrow, V\downarrow$$

If T increases, V increases, and vice versa. Once again, for any given p and n, the constant is the same for most simple gases under ordinary conditions.

Absolute zero (0 K or −273.15°C) is the temperature at which an ideal gas would have zero volume. (Absolute zero has never been reached, but physicists have attained 10^{-9} K.) In reality, no sample of matter can have zero volume, and every non-ideal gas condenses to a liquid at some temperature higher than 0 K. Nevertheless, the linear dependence of volume on absolute temperature holds for most common gases over a wide temperature range. This dependence of gas volume on the *absolute* temperature adds a practical requirement for chemists (and chemistry students): *the Kelvin scale must be used in gas law calculations.* For instance, if the temperature changes from 200 K to 400 K, the volume of gas doubles. But, if the temperature changes from 200°C to 400°C, the volume increases by a factor of 1.42; that is,

$$\frac{400°\text{C} + 273.15}{200°\text{C} + 273.15} = \frac{673}{473} = 1.42$$

Other Relationships Based on Boyle's and Charles's Laws Two other important relationships arise from Boyle's and Charles's laws:

1. *The pressure-temperature relationship.* Charles's law is expressed as the effect of temperature on gas *volume* at constant pressure. However, volume and pressure are interdependent, so a similar relationship can be expressed for the effect of temperature on pressure (sometimes referred to as *Amontons's law*). Measure the pressure in car or bike tires before and after a long ride, and you will find that it increases. Heating due to friction between the tires and the road increases the air temperature inside the tires, but since the volume of a tire cannot increase very much, the air pressure does. Thus, *at constant volume, the pressure exerted by a fixed amount of gas is directly proportional to the absolute temperature*:

$$p \propto T \quad (V \text{ and } n \text{ fixed}) \tag{4.3}$$

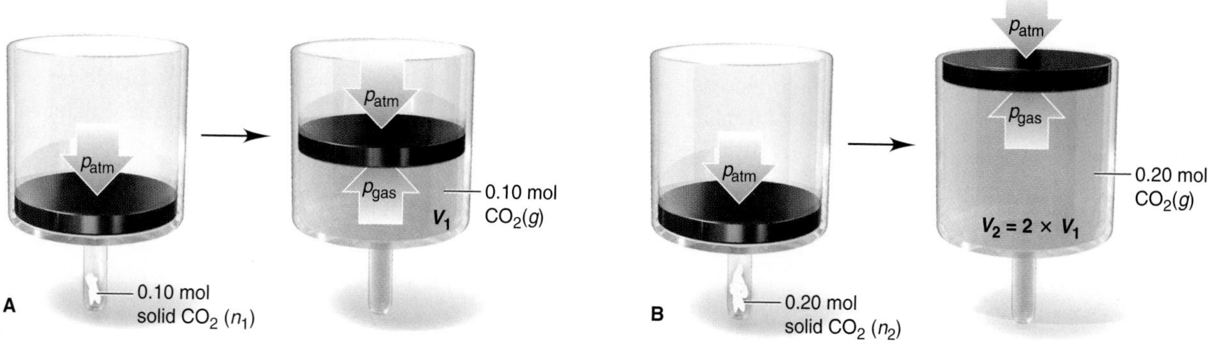

FIGURE 4.7 The relationship between the volume and amount of a gas

Or (similar to Charles's law)

$$p = \text{constant} \times T \quad \text{or} \quad \frac{p}{T} = \text{constant } (n, V \text{ constant})$$

This means that, at fixed V and n,

$$T\uparrow, p\uparrow \quad \text{and} \quad T\downarrow, p\downarrow$$

2. *The combined gas law.* Combining Boyle's and Charles's laws gives the *combined gas law*, which applies to situations in which changes in *two* of the three variables (V, p, T) affect the third:

$$V \propto \frac{T}{p} \quad \text{or} \quad V = \frac{T}{p} \times \text{constant} \quad \text{or} \quad \frac{pV}{T} = \text{constant}$$

The Relationship between Volume and Amount: Avogadro's Law

Let us see why both Boyle's and Charles's laws specify a fixed amount of gas.

1. *The experiment.* Figure 4.7 shows an experiment that involves two small test tubes, each fitted to a much larger piston-cylinder assembly. We add 0.10 mol (4.4 g) of dry ice (solid CO_2) to the first tube (A) and 0.20 mol (8.8 g) to the second tube (B). As the solid CO_2 warms to room temperature, it changes to gaseous CO_2, and the volume increases until $p_{gas} = p_{atm}$. At constant temperature, when all of the solid has changed to gas, cylinder B has twice the volume of cylinder A.

2. *Conclusion and statement of the law.* Thus, *at fixed temperature and pressure, the volume occupied by a gas is directly proportional to the amount (mol) of the gas:*

$$V \propto n \quad (p \text{ and } T \text{ fixed}) \tag{4.4}$$

In other words, as n increases, V increases, and vice versa. This relationship is also expressed as

$$V = n \times \text{constant} \quad \text{or} \quad \frac{V}{n} = \text{constant}$$

This means that, at fixed p and T,

$$n\uparrow, V\uparrow \quad \text{and} \quad n\downarrow, V\downarrow$$

The constant is the same for all simple gases at ordinary temperature and pressure. This relationship is another way to express **Avogadro's law**, which states that *at fixed temperature and pressure, equal volumes of* ***any*** *ideal gas contain equal numbers of particles (or moles).*

Familiar Applications of the Gas Laws The gas laws apply to countless familiar phenomena. In a car engine, a smaller amount (mol) of gasoline and O_2 react to form a larger amount (mol) of CO_2 and H_2O vapour, and heat is released. The increases in n (Avogadro's law) and T (Charles's law) increase V, and the piston is pushed back. Dynamite is a solid that forms more moles of hot gases very rapidly when it reacts (Avogadro's and Charles's laws). Dough rises because yeast digests sugar, which creates bubbles of CO_2 (Avogadro's law); the dough expands more as the bread bakes in a hot oven (Charles's law).

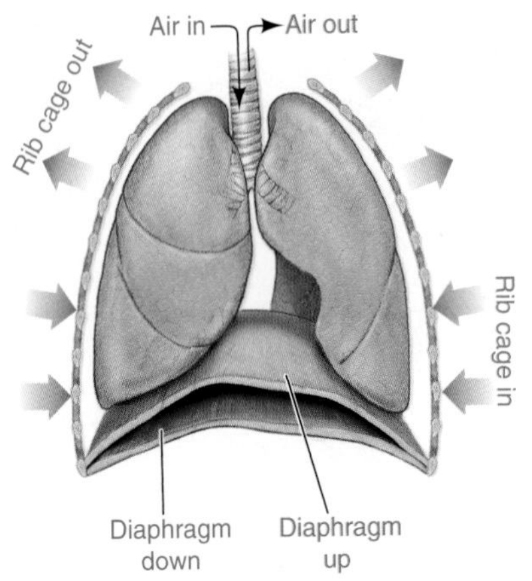

FIGURE 4.8 The process of breathing applies the gas laws

No application of the gas laws can be more vital or familiar than breathing (Figure 4.8). When you inhale, muscles move your diaphragm down and your rib cage out (*blue*). This coordinated movement increases the volume of your lungs, which decreases the air pressure inside them (Boyle's law). The inside pressure is 1.3 to 4.0 mbar *less* than atmospheric pressure, so air rushes in. The greater amount of air stretches the elastic tissue of the lungs and further expands the volume (Avogadro's law). The air also expands as it warms from the external temperature to your body temperature (Charles's law). When you exhale, the diaphragm moves up and the rib cage moves in, so your lung volume decreases (*red*). The inside pressure becomes 1.3 to 4.0 mbar *more* than the outside pressure (Boyle's law), so air rushes out.

Gas Behaviour at Standard Conditions

To better understand the factors that influence gas behaviour, chemists have assigned a baseline set of *standard conditions* called **standard temperature and pressure (STP)**:

$$\text{STP}: 0°\text{C } (273.15 \text{ K}) \text{ and } 1 \text{ bar } (10^5 \text{ Pa}) \tag{4.5}$$

Under these conditions, *the volume of 1 mol of an ideal gas* is called the **standard molar volume**:

$$\text{Standard molar volume} = 22.710\,953(21) \text{ L or } 22.7 \text{ L (3 sf)} \tag{4.6}$$

It is extremely important to note that, unlike in thermodynamics (Chapter 5), the standard temperature is *not* 25°C. At STP, helium (He), nitrogen (N_2), oxygen (O_2), and other simple gases behave nearly ideally (Figure 4.9). Note that the mass, and thus the density (d), depend on the specific gas, but 1 mol of any gas occupies 22.7 L at STP.

The Ideal Gas Law

Each of the three gas laws shows how one of the three other gas variables affects gas volume:

- Boyle's law focuses on pressure ($V \propto \frac{1}{p}$).
- Charles's law focuses on temperature ($V \propto T$).
- Avogadro's law focuses on the amount (mol) of gas ($V \propto n$).

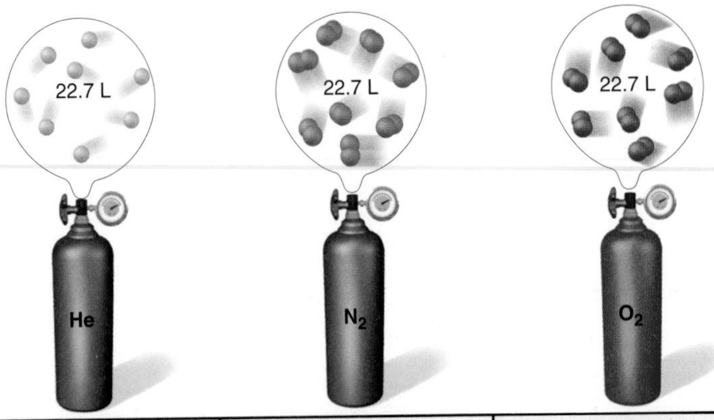

n = 1 mol	n = 1 mol	n = 1 mol
p = 10^5 Pa (1 bar)	p = 10^5 Pa (1 bar)	p = 10^5 Pa (1 bar)
T = 0°C (273 K)	T = 0°C (273 K)	T = 0°C (273 K)
V = 22.7 L	V = 22.7 L	V = 22.7 L
Number of gas particles = 6.022×10^{23}	Number of gas particles = 6.022×10^{23}	Number of gas particles = 6.022×10^{23}
Mass = 4.003 g	Mass = 28.02 g	Mass = 32.00 g
d = 0.179 g/L	d = 1.25 g/L	d = 1.43 g/L

FIGURE 4.9 Standard molar volume. One mole of an ideal gas occupies 22.7 L at STP (0°C and 10^5 Pa).

By combining these individual effects, we obtain the **ideal gas law** (or *ideal gas equation*):

$$V \propto \frac{nT}{p} \quad \text{or} \quad pV \propto nT \quad \text{or} \quad \frac{pV}{nT} = \text{constant}$$

where the proportionality constant is known as the **universal gas constant (*R*)**. Rearranging gives the most common form of the ideal gas law:

$$pV = nRT \tag{4.7}$$

We obtain a value of R by measuring the volume, temperature, and pressure of a given amount of gas and substituting the values into the ideal gas law. For example, using standard conditions for the gas variables and 1 mol of gas, we have

$$R = \frac{pV}{nT} = \frac{(1 \times 10^5\ \text{Pa})[22.710\,953(21) \times 10^{-3}\ \text{m}^3]}{(1\ \text{mol})(273.15\ \text{K})}$$
$$= 8.314\,462\ \frac{\text{Pa}\cdot\text{m}^3}{\text{mol}\cdot\text{K}} = 8.314\ \frac{\text{J}}{\text{mol}\cdot\text{K}}\ [4\ \text{sf}] \tag{4.8}$$

This numerical value of R corresponds to p, V, and T expressed *in these units*; *R has a different numerical value when different units are used (see margin)*. Figure 4.10 makes a central point: the ideal gas law *becomes* one of the individual gas laws when two of the four variables are kept constant. When initial conditions (subscript 1) change to final conditions (subscript 2), we have

$$p_1V_1 = n_1RT_1 \quad \text{and} \quad p_2V_2 = n_2RT_2$$

Thus,

$$\frac{p_1V_1}{n_1T_1} = R \quad \text{and} \quad \frac{p_2V_2}{n_2T_2} = R, \quad \text{so} \quad \frac{p_1V_1}{n_1T_1} = \frac{p_2V_2}{n_2T_2}$$

Notice that if, for example, the two variables p and T remain constant, then $p_1 = p_2$ and $T_1 = T_2$, and we obtain an expression for Avogadro's law:

$$\frac{\cancel{p_1}V_1}{n_1\cancel{T_1}} = \frac{\cancel{p_2}V_2}{n_2\cancel{T_2}} \quad \text{or} \quad \frac{V_1}{n_1} = \frac{V_2}{n_2}$$

As you will see next, you can use a similar approach to solve gas law problems. Thus, by keeping track of the initial and final values of the gas variables, you avoid the need to memorize the three individual gas laws.

Different Numerical Values and Their Units for *R*

8.314 462	$\frac{\text{Pa}\cdot\text{m}^3}{\text{mol}\cdot\text{K}}$
8.314 462	$\frac{\text{kPa}\cdot\text{L}}{\text{mol}\cdot\text{K}}$
8.314 462	$\frac{\text{J}}{\text{mol}\cdot\text{K}}$
0.083 146	$\frac{\text{bar}\cdot\text{L}}{\text{mol}\cdot\text{K}}$
0.082 06	$\frac{\text{L}\cdot\text{atm}}{\text{mol}\cdot\text{K}}$

Solving Gas Law Problems

Gas law problems are phrased in many ways, but they can usually be grouped into two types:

1. *A change in one of the four variables causes a change in another, while the two other variables remain constant.* In this type of problem, the ideal gas law

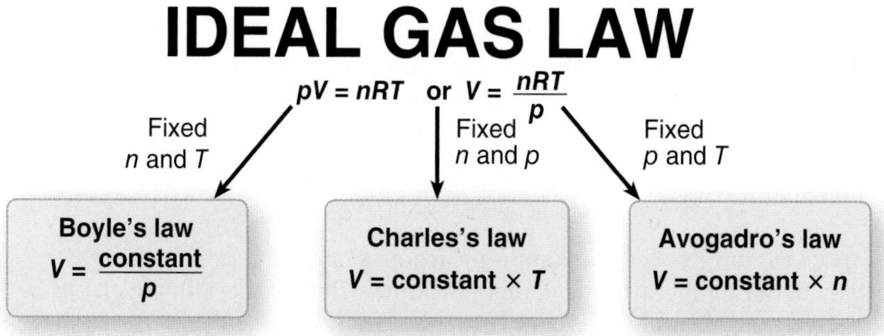

FIGURE 4.10 The individual gas laws as special cases of the ideal gas law

reduces to one of the individual gas laws, and you solve for the new value of the affected variable. Units must be consistent and T must always be in K, but R is not involved. Sample Problems 4.2, 4.3, 4.4, and 4.6 are this type of problem. (A variation in this type of problem involves the combined gas law, when simultaneous changes in two of the variables cause a change in a third.)

2. *One variable is unknown, but the other three are known and no change occurs.* In this type of problem, exemplified by Sample Problem 4.5, you apply the ideal gas law directly to find the unknown, and the units must conform to those in R.

Solving these problems requires a systematic approach:

- Summarize the changing gas variables—known and unknown—and the variables held constant.
- Convert units, if necessary.
- Rearrange the ideal gas law to obtain the needed relationship of variables, and solve for the unknown.

Sample Problem 4.2 Applying the Volume-Pressure Relationship

Problem Boyle's apprentice finds that the air trapped in a J tube occupies 24.8 cm³ at 1.13 bar. By adding mercury to the tube, he increases the pressure on the trapped air to 2.67 bar. Assuming constant temperature, what is the new volume of air (in litres)?

Plan We must find the final volume (V_2) in litres, given the initial volume (V_1), initial pressure (p_1), and final pressure (p_2). The temperature and amount of gas are fixed. We convert the units of V_1 from cm³ to mL and then to L, rearrange the ideal gas law to the appropriate form, and solve for V_2. (Note that the road map has two parts.)

Road Map

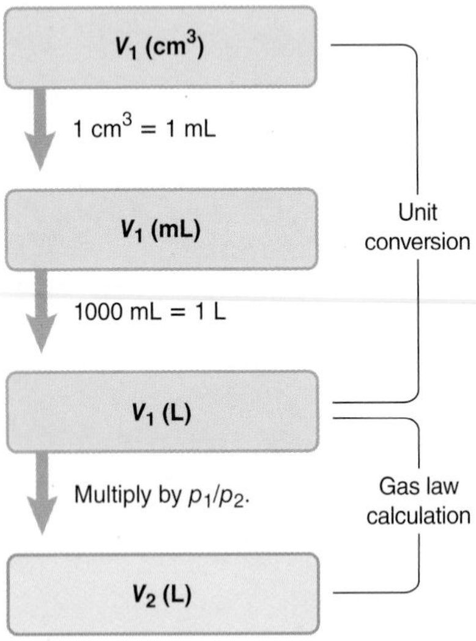

Solution Summarize the gas variables:

$p_1 = 1.13$ bar $\qquad p_2 = 2.67$ bar

$V_1 = 24.8$ cm³ (convert to L) $\qquad V_2 =$ unknown $\qquad T$ and n remain constant.

Convert V_1 from cm³ to L:

$$V_1 = 24.8 \text{ cm}^3 \times \frac{1 \text{ mL}}{1 \text{ cm}^3} \times \frac{1 \text{ L}}{1000 \text{ mL}} = 0.0248 \text{ L}$$

Now rearrange the ideal gas law and solve for V_2. At fixed n and T,

$$\frac{p_1V_1}{\cancel{n_1}\cancel{T_1}} = \frac{p_2V_2}{\cancel{n_2}\cancel{T_2}} \quad \text{or} \quad p_1V_1 = p_2V_2$$

$$V_2 = V_1 \times \frac{p_1}{p_2} = 0.0248 \text{ L} \times \frac{1.13 \text{ bar}}{2.67 \text{ bar}} = 0.0105 \text{ L}$$

Check The relative values of p and V can help us check the math: p more than doubled, so V_2 should be less than $\frac{1}{2}V_1$ $\left(\frac{0.0105}{0.0248} < \frac{1}{2}\right)$.

Comment Predicting the direction of the change provides another check on the problem setup: since p increases, V will decrease; thus, V_2 should be less than V_1. To make $V_2 < V_1$, we must multiply V_1 by a number *less than* 1. This means that the ratio of pressures must be *less than* 1, so the larger pressure (p_2) must be in the denominator, or $\frac{p_1}{p_2}$.

Follow-Up Problem 4.2 A sample of argon gas occupies 105 mL at 0.883 bar. If the temperature remains constant, what is the volume (L) at 26.3 kPa?

Sample Problem 4.3 Applying the Pressure-Temperature Relationship

Problem A steel tank used for fuel delivery is fitted with a safety valve that opens if the internal pressure exceeds 1.33 bar. The tank is filled with methane at 23°C and 100 kPa and placed in boiling water at 100.°C. Will the safety valve open?

Plan The question "Will the safety valve open?" translates to "Is p_2 greater than 1.33 bar at T_2?" Thus, p_2 is the unknown, and T_1, T_2, and p_1 are given, with V (steel tank) and n fixed. We convert both T values to K and compare p_1 with p_2 to determine whether the safety-limit pressure has been exceeded. We rearrange the ideal gas law and solve for p_2.

Road Map

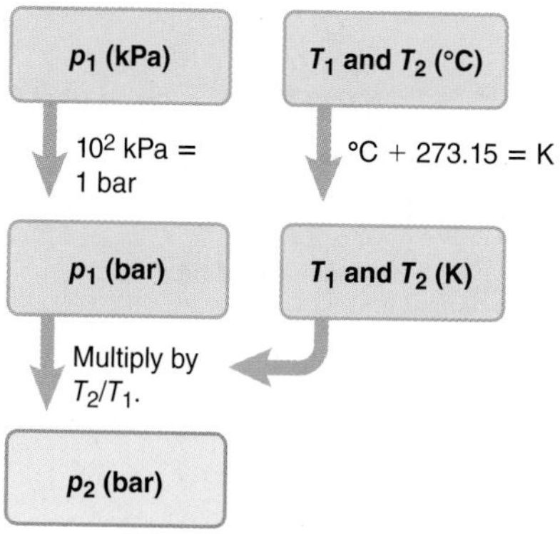

Solution Summarize the gas variables:

$p_1 = 100$ kPa (convert to bar) $\quad\quad p_2 =$ unknown

$T_1 = 23$°C (convert to K) $\quad\quad T_2 = 100.$°C (convert to K)

V and n remain constant.

Convert T from °C to K:

$$T_1\ (\text{K}) = 23 + 273.15 = 296\ \text{K} \qquad T_2\ (\text{K}) = 100. + 273.15 = 373\ \text{K}$$

Convert p from kPa to bar:

$$p_1\ (\text{bar}) = 100\ \text{kPa} \times \frac{1\ \text{bar}}{100\ \text{kPa}} = 1\ \text{bar (exact amount)}$$

Now rearrange the ideal gas law and solve for p_2. At fixed n and V,

$$\frac{p_1 \cancel{V_1}}{\cancel{n_1} T_1} = \frac{p_2 \cancel{V_2}}{\cancel{n_2} T_2} \quad \text{or} \quad \frac{p_1}{T_1} = \frac{p_2}{T_2}$$

$$p_2 = p_1 \times \frac{T_2}{T_1} = 1\ \text{bar} \times \frac{373\ \text{K}}{296\ \text{K}} = 1.26\ \text{bar}$$

p_2 is less than 1.33 bar, so the valve will *not* open.

Check Let us predict the change to check the math. Because $T_2 > T_1$, we expect $p_2 > p_1$. Thus, the temperature ratio should be > 1 (T_2 in the numerator). The T ratio is about 1.25 ($\frac{373}{296}$), so the p ratio should also be about 1.25 ($\frac{1.26}{1} \approx 1.25$).

Follow-Up Problem 4.3 An engineer pumps air at 0°C into a newly designed piston-cylinder assembly. The volume measures 6.83 cm^3. At what temperature (K) will the volume be 9.75 cm^3?

Sample Problem 4.4 Applying the Volume-Amount Relationship

Problem A scale model of a blimp rises when it is filled with helium to a volume of 55.0 dm^3. When 1.10 mol of He is added to the blimp, the volume is 26.2 dm^3. What additional mass of He must be added to make the blimp rise? Assume constant temperature and pressure.

Plan We are given the initial amount of helium (n_1), the initial volume of the blimp (V_1), and the volume needed for the blimp to rise (V_2), and we need the additional mass of helium to make it rise. So, we first need to find n_2. We rearrange the ideal gas law to the appropriate form, solve for n_2, subtract n_1 to find the additional amount ($n_{\text{add'l}}$), and then convert amount (mol) to mass.

Solution Summarize the gas variables:

$n_1 = 1.10$ mol $\quad n_2 =$ unknown (Find and then subtract n_1.)

$V_1 = 26.2$ dm^3 $\quad V_2 = 55.0$ dm^3

p and T remain constant.

Road Map

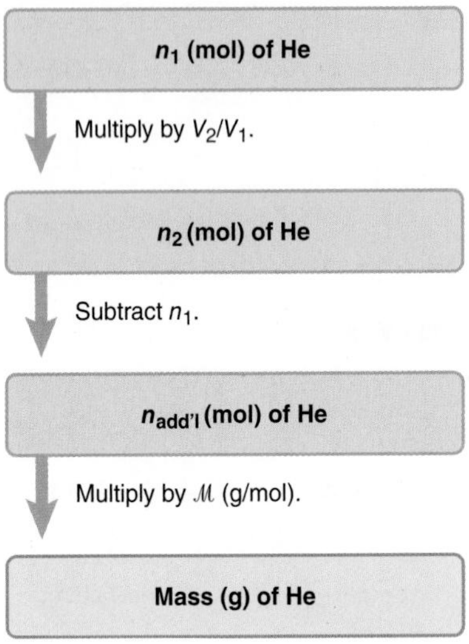

Next, rearrange the ideal gas law and solve for n_2. At fixed p and T,

$$\frac{p_1V_1}{n_1T_1} = \frac{p_2V_2}{n_2T_2} \quad \text{or} \quad \frac{V_1}{n_1} = \frac{V_2}{n_2}$$

$$n_2 = n_1 \times \frac{V_2}{V_1} = 1.10 \text{ mol He} \times \frac{55.0 \text{ dm}^3}{26.2 \text{ dm}^3}$$
$$= 2.31 \text{ mol He}$$

Find the additional amount of He:

$$n_{\text{add'l}} = n_2 - n_1 = 2.31 \text{ mol He} - 1.10 \text{ mol He}$$
$$= 1.21 \text{ mol He}$$

Convert amount (mol) of He to mass (g):

$$\text{Mass (g) of He} = 1.21 \text{ mol He} \times \frac{4.003 \text{ g He}}{1 \text{ mol He}}$$
$$= 4.84 \text{ g He}$$

Check We predict that $n_2 > n_1$ because $V_2 > V_1$: since V_2 is about twice V_1 ($\frac{55}{26} \approx 2$), n_2 should be about twice n_1 ($\frac{2.3}{1.1} \approx 2$). Since $n_2 > n_1$, we were right to multiply n_1 by a number > 1 (that is, $\frac{V_2}{V_1}$). About 1.2 mol × 4 g/mol ≈ 4.8 g.

Comment **1.** A different sequence of steps will give you the same answer: first find the additional volume ($V_{\text{add'l}} = V_2 - V_1$), and then solve directly for $n_{\text{add'l}}$. Try this yourself.
2. You saw that Charles's law ($V \propto T$ at fixed p and n) becomes a similar relationship between p and T at fixed V and n. The follow-up problem demonstrates that Avogadro's law ($V \propto n$ at fixed p and T) becomes a similar relationship at fixed V and T.

Follow-Up Problem 4.4 A rigid plastic container holds 35.0 g of ethylene gas (C_2H_4) at a pressure of 1.06 bar. What is the pressure if 5.0 g of ethylene is removed at constant temperature?

Sample Problem 4.5 Solving for an Unknown Gas Variable at Fixed Conditions

Problem A steel tank has a volume of 438 L and is filled with 0.885 kg of O_2. Calculate the pressure of O_2 at 21°C.

Plan We are given V, T, and the mass of O_2, and we must find p. Since conditions are not changing, we apply the ideal gas law without rearranging it. We convert V to m^3, T to K, and mass (kg) of O_2 to amount (mol) and solve for p.

Solution Summarize the gas variables:

$V = 438$ L (convert to m^3) $\qquad T = 21$°C (convert to K)
$n = 0.885$ kg O_2 (convert to mol) $\qquad p =$ unknown

Convert V from L to m^3:

$$V = 438 \text{ L} \times \frac{10^{-3} \text{ m}^3}{1 \text{ L}} = 0.438 \text{ m}^3$$

Convert T from °C to K:

$$T \text{ (K)} = 21 + 273.15 = 294 \text{ K}$$

Convert from mass (g) of O_2 to amount (mol):

$$n = \text{amount (mol) of O}_2 = 0.885 \text{ kg O}_2 \times \frac{1000 \text{ g}}{1 \text{ kg}} \times \frac{1 \text{ mol O}_2}{32.00 \text{ g O}_2} = 27.7 \text{ mol O}_2$$

Solve for p (note the units cancelling here):

$$p = \frac{nRT}{V}$$

$$= \frac{27.7\ \text{mol} \times 8.314\,472\ \frac{\text{Pa}\cdot\text{m}^3}{\text{mol}\cdot\text{K}} \times 294\ \text{K}}{0.438\ \text{m}^3}$$

$$= 1.55 \times 10^5\ \text{Pa}$$

Check The amount of O_2 seems correct: $\frac{\sim 900\ \text{g}}{30\ \text{g/mol}} = 30$ mol. To check the approximate size of the final calculation, round off the values, including the value for R:

$$p = \frac{30\ \text{mol}\ O_2 \times 8\ \frac{\text{Pa}\cdot\text{m}^3}{\text{mol}\cdot\text{K}} \times 300\ \text{K}}{450 \times 10^{-3}\ \text{m}^{-3}} = 2 \times 10^5\ \text{Pa}$$

The result is reasonably close to 1.55×10^5 Pa.

Follow-Up Problem 4.5 The tank in this sample problem develops a slow leak that is discovered and sealed. The new pressure is 1.39 bar. What mass of O_2 remains?

Finally, in a picture problem, we apply the gas laws to determine the balanced equation for a gaseous reaction.

Sample Problem 4.6 Using Gas Laws to Determine a Balanced Equation

Problem The piston-cylinder below is depicted before and after a gaseous reaction that is carried out inside it, at constant pressure. The temperature is 150 K before and 300 K after the reaction. (Assume that the cylinder is insulated.)

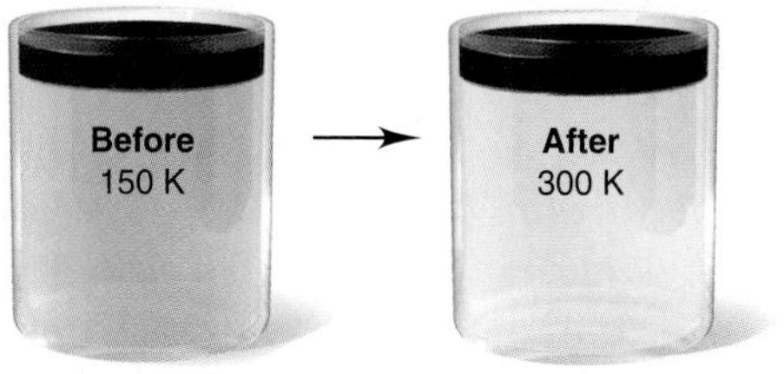

Which of the following balanced equations describes the reaction?

(1) $A_2(g) + B_2(g) \longrightarrow 2AB(g)$ (2) $2AB(g) + B_2(g) \longrightarrow 2AB_2(g)$

(3) $A(g) + B_2(g) \longrightarrow AB_2(g)$ (4) $2AB_2(g) \longrightarrow A_2(g) + 2B_2(g)$

Plan We are shown a depiction of the volume and temperature of a gas mixture before and after a reaction and must deduce the balanced equation. The problem says that p is constant, and the picture shows that, when T doubles, V stays the same. If n were also constant, Charles's law tells us that V should double when T doubles. However, since V does not change, n cannot be constant. From Avogadro's law, the only way to maintain V constant, with p constant and T doubling, is for n to be halved. So, we examine the four balanced equations and count the number of moles on each side to determine the equation in which n is halved.

Solution In equation (1), n does not change, so doubling T would double V.

In equation (2), n decreases from 3 mol to 2 mol, so doubling T would increase V by one-third.

In equation (3), n decreases from 2 mol to 1 mol. Doubling T would exactly balance the decrease from halving n, so V would stay the same.

In equation (4), n increases, so doubling T would more than double V.

Therefore, equation (3) is correct:

$$A(g) + B_2(g) \longrightarrow AB_2(g)$$

Follow-Up Problem 4.6 The piston-cylinder below shows the volumes of a gaseous reaction mixture before and after a reaction that takes place at constant pressure. The initial temperature is −73°C.

If the *unbalanced* equation is $CD(g) \longrightarrow C_2(g) + D_2(g)$, what is the final temperature in °C?

SUMMARY OF SECTION 4.3

- Four interdependent variables define the physical behaviour of an ideal gas: volume (V), pressure (p), temperature (T), and amount (mol, n).
- Most simple gases display nearly ideal behaviour at ordinary temperatures and pressures.
- Boyle's, Charles's, and Avogadro's laws refer to the linear relationships between the volume of a gas and the pressure, temperature, and amount of gas, respectively.
- At STP (0°C and 10^5 Pa, or 1 bar), 1 mol of an ideal gas occupies 22.7 L.
- The ideal gas law incorporates the individual gas laws into one equation, $pV = nRT$, where R is the universal gas constant.

4.4 Rearrangements of the Ideal Gas Law

In this section, we mathematically rearrange the ideal gas law to find gas density, molar mass, the partial pressure of each gas in a mixture, and the amount of gaseous reactant or product in a reaction.

The Density of a Gas

One mole of any gas behaving ideally occupies the same volume at a given temperature and pressure, so differences in gas density ($d = \frac{m}{V}$) depend on differences in molar mass (see Figure 4.9). For example, at STP, 1 mol of O_2 occupies the same volume as 1 mol of N_2; however, O_2 is denser because each O_2 molecule has a greater mass (32.00 u) than each N_2 molecule (28.02 u). Thus, d of O_2 is

$$\frac{32.00}{28.02} \times d \text{ of } N_2$$

We can rearrange the ideal gas law to calculate the density of a gas from its molar mass. Recall that the amount (mol) (n) is the mass (m) divided by the molar mass ($\mathcal{M}$), or

$$n = \frac{m}{\mathcal{M}}$$

Substituting for n in the ideal gas law gives

$$pV = \frac{m}{\mathcal{M}}RT$$

Rearranging to isolate $\frac{m}{V}$ gives

$$\frac{m}{V} = d = \frac{\mathcal{M} \times p}{RT} \quad (4.9)$$

Two important ideas are expressed by Equation 4.9:

- *The density of a gas is directly proportional to its molar mass.* The volume of a given amount of a heavier gas equals the volume of the same amount of a lighter gas (Avogadro's law), so the density of the heavier gas is higher (as you just saw for O_2 and N_2).

- *The density of a gas is inversely proportional to the temperature.* As the volume of a gas increases with temperature (Charles's law), the same mass occupies more space, so the density of the gas is lower.

We use Equation 4.9 to find the density of a gas at any temperature and pressure near standard conditions.

Sample Problem 4.7 Calculating Gas Density

Problem To apply a green chemistry approach, a chemical engineer uses waste CO_2 from a manufacturing process, instead of chlorofluorocarbons, as a "blowing agent" in the production of polystyrene. Find the density (g/L) of CO_2 and the number of molecules per litre **(a)** at STP (0°C and 100 kPa); **(b)** at room conditions (25.°C and 100 kPa).

Plan We must find the density (d) and the number of molecules of CO_2, given two sets of P and T data. We find m, convert T to K, and calculate d with Equation 4.9. Then we convert the mass per litre to molecules per litre with Avogadro's number.

Solution **(a)** We need to find the density and the number of molecules of CO_2 per litre at STP.

Summarize the gas properties:

$$T = 0 + 273.15 = 273\text{ K} \quad p = 1 \times 10^5\text{ Pa} \quad \mathcal{M}\text{ of }CO_2 = 44.01\text{ g/mol}$$

Calculate the density (note the unit cancelling here):

$$d = \frac{\mathcal{M} \times p}{RT} = \frac{(44.01\text{ g/}\cancel{\text{mol}})(1 \times 10^5\ \cancel{\text{Pa}})}{8.314\ \frac{\cancel{\text{Pa}}\cdot\text{m}^3}{\cancel{\text{mol}\cdot\text{K}}} \times 273\ \cancel{\text{K}}} = 1.94 \times \cancel{10^3}\ \frac{\text{g}}{\cancel{\text{m}^3}} \times \frac{1\ \cancel{\text{m}^3}}{\cancel{10^3}\text{ L}} = 1.94\text{ g/mol}$$

Convert from mass/L to molecules/L:

$$\text{Molecules }CO_2\text{/L} = \frac{1.94\text{ g }CO_2}{1\text{ L}} \times \frac{1\text{ mol }CO_2}{44.01\text{ g }CO_2} \times \frac{6.022 \times 10^{23}\text{ molecules }CO_2}{1\text{ mol }CO_2}$$

$$= 2.65 \times 10^{22}\text{ molecules }CO_2\text{/L}$$

Comment: Note that while calculating the density we need to use the molar mass of the gas. If a different gas is used, the density will change. However, when calculating the number of molecules of the gas, we divided by the molar mass of carbon dioxide. This is because the number of molecules does NOT depend on *what* gas is used, only on the amount (mol) of the gas.

(b) Now we need to find the density and the number of molecules of CO_2 per litre at room conditions.

Summarize the gas properties:

$$T = 25.°\text{C} + 273.15 = 298\text{ K} \quad p = 1 \times 10^5\text{ Pa} \quad \mathcal{M}\text{ of }CO_2 = 44.01\text{ g/mol}$$

Calculate the density:

$$d = \frac{\mathcal{M} \times p}{RT} = \frac{(44.01\text{ g/mol})(1 \times 10^5\text{ Pa})}{8.314\ \frac{\text{Pa}\cdot\text{m}^3}{\text{mol}\cdot\text{K}} \times 298\text{ K}} = \left(1.78 \times \cancel{10^3}\ \frac{\text{g}}{\cancel{\text{m}^3}}\right)\left(\frac{1\ \cancel{\text{m}^3}}{10^3\text{ L}}\right) = 1.78\text{ g/L}$$

Convert from mass/L to molecules/L:

$$\text{Molecules }CO_2\text{/L} = \frac{1.78\text{ g }CO_2}{1\text{ L}} \times \frac{1\text{ mol }CO_2}{44.01\text{ g }CO_2} \times \frac{6.022 \times 10^{23}\text{ molecules }CO_2}{1\text{ mol }CO_2}$$

$$= 2.44 \times 10^{22}\text{ molecules }CO_2\text{/L}$$

Check Round off to check the density values; for example, in part (a), at STP:

$$\frac{(50\text{ g/mol})(1 \times 10^5\text{ Pa})}{8\ \frac{\text{Pa}\cdot\text{m}^3}{\text{mol}\cdot\text{K}} \times 250\text{ K}} = \left(2 \times \cancel{10^3}\ \frac{\text{g}}{\cancel{\text{m}^3}}\right)\left(\frac{1\ \cancel{\text{m}^3}}{\cancel{10^3}\text{ L}}\right) = 2\text{ g/L} \approx 1.94\text{ g/L}$$

At the higher temperature in part (b), the density should decrease, which can happen only if there are fewer molecules per litre, so the answer is reasonable.

Comment An *alternative approach* for finding the density of most simple gases, but *at STP only*, is to divide the molar mass by the standard molar volume, 22.7 L:

$$d = \frac{\mathcal{M}}{V} = \frac{44.01\ \text{g/mol}}{22.7\ \text{L/mol}} = 1.94\ \text{g/L}$$

Once you know the density at one temperature (0°C), you can find the density at any other temperature with the following relationship:

$$\frac{d_1}{d_2} = \frac{T_2}{T_1}$$

Follow-Up Problem 4.7 Compare the density of CO_2 at 0°C and 0.507 bar with its density at STP.

Gas Density and the Human Condition A few applications demonstrate the wide-ranging relevance of gas density:

- *Engineering.* Architectural designers and heating engineers place heating ducts near the floor so that the warmer, and thus less dense, air coming from the ducts will rise and mix with the cooler room air.
- *Safety and air pollution.* In the absence of mixing, a less dense gas will lie above more dense air. Fire extinguishers that release CO_2 are effective because CO_2 is heavier than air: it sinks onto the fire and keeps more O_2 from reaching the fuel. The dense gases in the smog that blankets urban centres, such as Mexico City, Los Angeles, and Beijing, contribute to respiratory illnesses.
- *Toxic releases.* During World War I, poisonous phosgene gas ($COCl_2$) was used against ground troops because it was dense enough to sink into their trenches. In 1984, the accidental release of poisonous methylisocyanate gas from a Union Carbide plant in India killed thousands of people as it blanketed nearby neighbourhoods. In 1986, CO_2 released naturally from Lake Nyos in Cameroon suffocated thousands of people as it flowed down valleys into villages. Some paleontologists suggest that the release of CO_2 from volcanic lakes may have contributed to the widespread dying off of dinosaurs.
- *Ballooning.* When the gas in a hot-air balloon is heated, its volume increases and the balloon inflates. Further heating causes some of the gas to escape. Thus, the gas density decreases and the balloon rises. In 1783, Jacques Charles (of Charles's law) made one of the first balloon flights, and, 20 years later, Joseph Gay-Lussac (who studied the pressure-temperature relationship) set a solo altitude record that held for 50 years.

The Molar Mass of a Gas

Through another rearrangement of the ideal gas law, we can determine the molar mass of an unknown gas or a volatile liquid (a liquid that is easily vaporized):

$$n = \frac{m}{\mathcal{M}} = \frac{pV}{RT} \quad \text{so} \quad \mathcal{M} = \frac{mRT}{pV} \tag{4.10}$$

Notice that this equation is just a rearrangement of Equation 4.9.

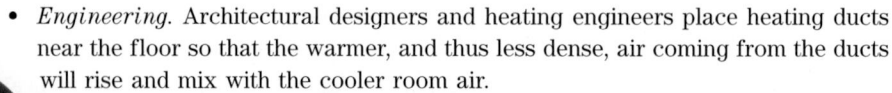

Sample Problem 4.8 Finding the Molar Mass of a Volatile Liquid

Problem An organic chemist isolates a colourless liquid from a petroleum sample. She places the liquid in a flask of known mass and puts the flask in boiling water, which vaporizes the liquid and fills the flask with gas. She closes the flask and again measures its mass. She obtains the following data:

Volume (V) of flask = 213 mL	T = 100.0°C p = 1.00 bar
Mass of flask + gas = 78.416 g	Mass of flask = 77.834 g

Calculate the molar mass of the liquid.

Plan We are given V, T, p, and the mass data, and we must find the molar mass ($\mathcal{M}$) of the liquid. We convert V to m^3, T to K, and p to Pa. Then we find the mass of the gas by subtracting the mass of the flask from the mass of the flask plus the gas and use Equation 4.10 to calculate $\mathcal{M}$.

Solution Summarize and convert the gas variables:

$$V\text{ (L)} = 213\text{ mL} \times \frac{1\text{ L}}{1000\text{ mL}} \times \frac{1\text{ m}^3}{1000\text{ L}} = 2.13 \times 10^{-4}\text{ m}^3$$

$$T\text{ (K)} = 100.0 + 273.15 = 373.2\text{ K}$$

$$p\text{ (Pa)} = 1.00\text{ bar} \times \frac{10^5\text{ Pa}}{1\text{ bar}} = 1.00 \times 10^5\text{ Pa} \quad m = 78.416\text{ g} - 77.834\text{ g} = 0.582\text{ g}$$

Calculate $\mathcal{M}$:

$$\mathcal{M} = \frac{mRT}{pV} = \frac{0.582\text{ g} \times 8.314\,\frac{\text{Pa}\cdot\text{m}^3}{\text{mol}\cdot\text{K}} \times 373.2\text{ K}}{(1.00 \times 10^5\text{ Pa})(2.13 \times 10^{-4}\text{ m}^3)} = 84.8\text{ g/mol}$$

Check Rounding to check the arithmetic, we have

$$\frac{0.6\text{ g} \times 8\,\frac{\text{Pa}\cdot\text{m}^3}{\text{mol}\cdot\text{K}} \times 375\text{ K}}{(1 \times 10^5\text{ Pa})(2 \times 10^{-4}\text{ m}^3)} = 90\text{ g/mol} \quad \text{(which is close to 84.8 g/mol)}$$

Follow-Up Problem 4.8 An empty 149 mL flask has a mass of 68.322 g before a sample of volatile liquid is added. The flask is then placed in a hot (95.0°C) water bath; the barometric pressure is 0.987 bar. The liquid vaporizes, and the gas fills the flask. After cooling, the flask and condensed liquid together have a mass of 68.697 g. What is the molar mass of the liquid?

The Partial Pressure of Each Gas in a Mixture of Gases

The gas behaviour we have discussed so far was observed in experiments with air, which is a mixture of gases; thus, the ideal gas law holds for virtually any gas at ordinary conditions, whether pure or a mixture, for the following reasons:

- Gases mix homogeneously (form a solution) in any proportions.
- Each gas in a mixture behaves as if it were the only gas present (assuming no chemical interactions).

Dalton's Law of Partial Pressures The second reason above was discovered by John Dalton during his lifelong study of humidity. He observed that, when water vapour is added to dry air, the total air pressure increases by the pressure of the water vapour:

$$p_{\text{humid air}} = p_{\text{dry air}} p_{\text{added water vapour}}$$

He concluded that each gas in the mixture exerts a **partial pressure** equal to the pressure it would exert *by itself*. This can be stated as **Dalton's law of partial pressures**: *in a mixture of gases that do not react, the total pressure is the sum of the partial pressures of the individual gases*:

$$p_{\text{total}} = p_1 + p_2 + p_3 + \cdots \tag{4.11}$$

As an example, suppose that we have a tank of fixed volume, containing nitrogen gas at a certain pressure, and we introduce a sample of hydrogen gas into the tank. The gases behave independently, so we can write an ideal gas law expression for each gas:

$$p_{N_2} = \frac{n_{N_2}RT}{V} \quad \text{and} \quad p_{H_2} = \frac{n_{H_2}RT}{V}$$

Because each gas occupies the same total volume and is at the same temperature, the pressure of each gas depends only on its amount, n. Thus, the total pressure is

$$p_{\text{total}} = p_{N_2} + p_{H_2} = \frac{n_{N_2}RT}{V} + \frac{n_{H_2}RT}{V} = \frac{(n_{N_2} + n_{H_2})RT}{V} = \frac{n_{\text{total}}RT}{V}$$

where $n_{\text{total}} = n_{N_2} + n_{H_2}$.

Each component in a mixture contributes a fraction of the total number of moles in the mixture; this portion is the **mole fraction (*X*)** of that component. Multiplying X by 100 gives the mole percent. The sum of the mole fractions of all the components must be 1, and the sum of the mole percents must be 100%. For N_2 in our mixture, the mole fraction is

$$X_{N_2} = \frac{n_{N_2}}{n_{\text{total}}} = \frac{n_{N_2}}{n_{N_2} + n_{H_2}}$$

If the total pressure is due to the total number of moles, the partial pressure of gas A is the total pressure multiplied by the mole fraction of A, or X_A:

$$p_A = X_A \times p_{\text{total}} \qquad (4.12)$$

Equation 4.12 is a very useful result. To see that it is valid for the mixture of N_2 and H_2, we recall that $X_{N_2} + X_{H_2} = 1$; then we obtain

$$p_{\text{total}} = p_{N_2} + p_{H_2} = (X_{N_2} \times p_{\text{total}}) + (X_{H_2} \times p_{\text{total}}) = (X_{N_2} + X_{H_2})p_{\text{total}} = 1 \times p_{\text{total}}$$

Sample Problem 4.9 Applying Dalton's Law of Partial Pressures

Problem In a study of O_2 uptake by muscle at high altitude, a physiologist prepares an atmosphere consisting of 79 mole % N_2, 17 mole % $^{16}O_2$, and 4.0 mole % $^{18}O_2$. (The isotope ^{18}O will be measured to determine the O_2 uptake.) The total pressure is 0.76 bar to simulate high altitude. Calculate the mole fraction and partial pressure of $^{18}O_2$ in the mixture.

Road Map

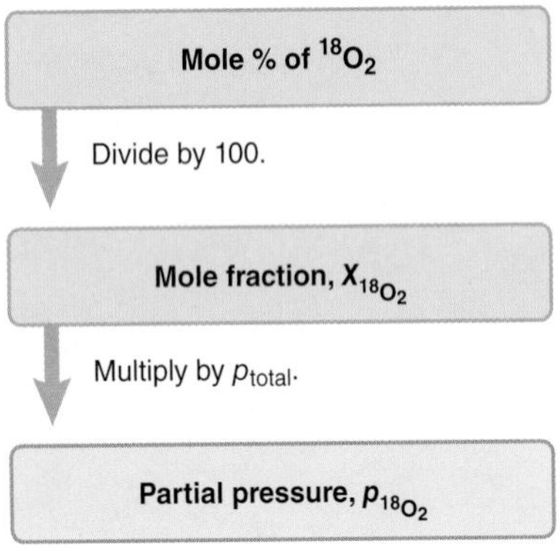

Plan We must find $X_{^{18}O_2}$ and $p_{^{18}O_2}$ from p_{total} (0.76 bar) and the mole % of $^{18}O_2$ (4.0). Dividing the mole % by 100 gives the mole fraction, $X_{^{18}O_2}$. Then, using Equation 4.12, we multiply $X_{^{18}O_2}$ by p_{total} to find $p_{^{18}O_2}$.

Solution Calculate the mole fraction of $^{18}O_2$:

$$X_{^{18}O_2} = \frac{4.0 \text{ mol}}{100} = 0.040$$

Solve for the partial pressure of $^{18}O_2$:

$$p_{^{18}O_2} = X_{^{18}O_2} \times p_{\text{total}} = 0.040 \times 0.76 \text{ bar} = 0.030 \text{ bar}$$

Check $X_{^{18}O_2}$ is small because the mole % is small, so $p_{^{18}O_2}$ should also be small.

Comment At high altitudes, specialized brain cells that are sensitive to O_2 and CO_2 levels in the blood trigger an increase in the rate and depth of breathing for several days, until a person becomes acclimatized.

Follow-Up Problem 4.9 To prevent the presence of air, noble gases are placed over highly reactive chemicals to act as inert "blanketing" gases. A chemical engineer places a mixture of noble gases consisting of 5.50 g of He, 15.0 g of Ne, and 35.0 g of Kr in a piston-cylinder assembly at STP. Calculate the partial pressure of each gas.

Collecting a Gas over Water Whenever a gas is in contact with water, some of the water vaporizes into the gas. The water vapour that mixes with the gas contributes the *vapour pressure*, a portion of the total pressure that depends only on the water temperature (Table 4.2). A common use of the law of partial pressures is to determine the yield of a water-insoluble gas formed in a reaction: the gaseous product bubbles through water, some water vaporizes into the bubbles, and the mixture of product gas and water vapour is collected into an inverted container (Figure 4.11).

To determine the yield, we look up the vapour pressure (p_{H_2O}) at the temperature of the experiment in Table 4.2 and subtract it from the total gas pressure (p_{total}, corrected for barometric pressure) to get the partial pressure of the gaseous product (p_{gas}). It should be noted here that the vapour pressure curve is actually exponential (see the graph accompanying Table 4.2). However, over very short ranges of temperature, such as 2°C to 5°C, it is not unreasonable to assume a linear relationship

TABLE 4.2 Vapour Pressure of Water (p_{H_2O}) at Different T

T (°C)	p_{H_2O} (kPa)	T (°C)	p_{H_2O} (kPa)
0	0.611 15	29	4.0092
5	0.872 58	30	4.247
10	1.2282	35	5.629
15	1.7058	40	7.3849
16	1.8188	45	9.595
17	1.9384	50	12.352
18	2.0647	55	15.762
19	2.1983	60	19.946
20	2.3393	65	25.042
21	2.4882	70	31.201
22	2.6453	75	38.595
23	2.8111	80	47.414
24	2.9858	85	57.867
25	3.1699	90	70.182
26	3.3639	95	84.608
27	3.5681	100	101.33
28	3.7831	105	120.8

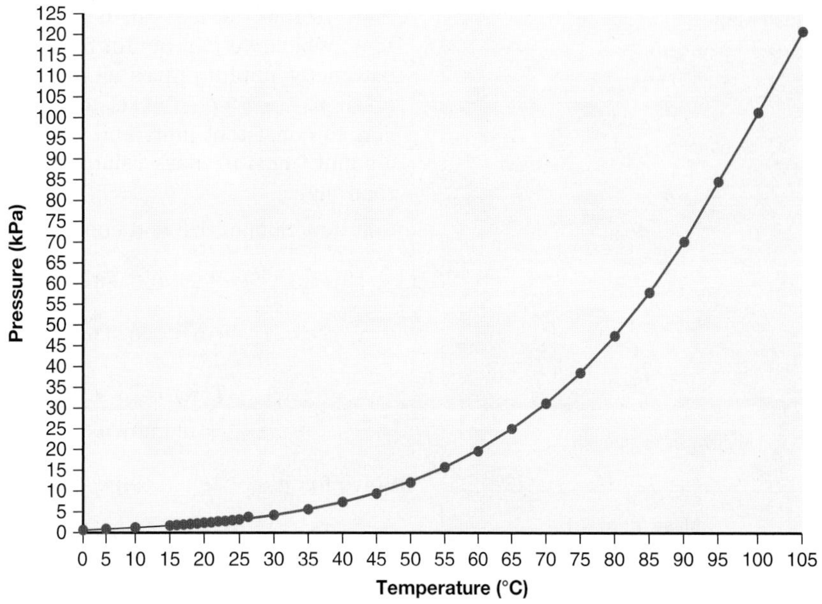

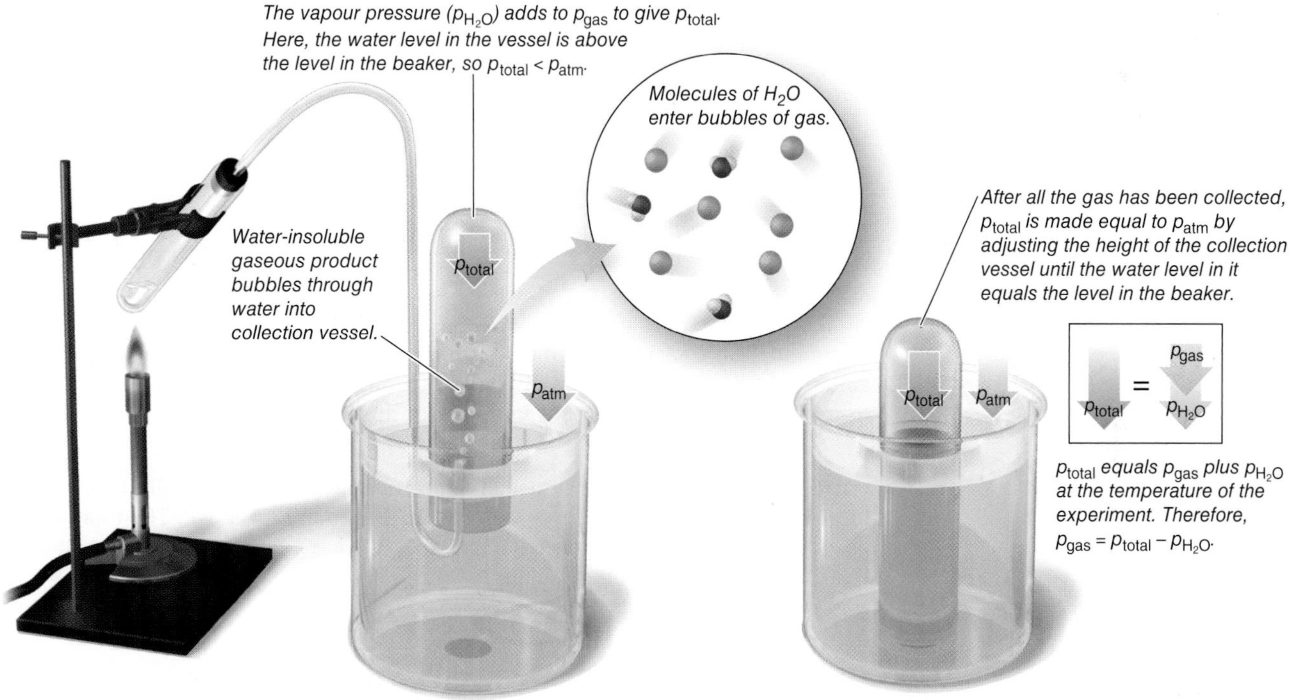

The vapour pressure (p_{H_2O}) adds to p_{gas} to give p_{total}. Here, the water level in the vessel is above the level in the beaker, so $p_{total} < p_{atm}$.
Molecules of H_2O enter bubbles of gas.
Water-insoluble gaseous product bubbles through water into collection vessel.
p_{total}
p_{atm}
After all the gas has been collected, p_{total} is made equal to p_{atm} by adjusting the height of the collection vessel until the water level in it equals the level in the beaker.
p_{total}
p_{atm}
p_{gas}
p_{total} = p_{H_2O}
p_{total} equals p_{gas} plus p_{H_2O} at the temperature of the experiment. Therefore, $p_{gas} = p_{total} - p_{H_2O}$.

FIGURE 4.11 Collecting a water-insoluble gaseous product and determining its pressure

between the points; this allows the extrapolation of vapour pressure data at temperatures not found on the curve or in the table. With V and T known, we can calculate the amount of product.

Sample Problem 4.10 Calculating the Amount of Gas Collected over Water

Problem Ethyne (or acetylene) (C_2H_2), an important fuel in welding, is produced in the laboratory when calcium carbide (CaC_2) reacts with water:

$$CaC_2(s) + 2H_2O(l) \longrightarrow C_2H_2(g) + Ca(OH)_2(aq)$$

For a sample of ethyne collected over water, the total gas pressure (adjusted to barometric pressure) is 0.984 bar and the volume is 523 mL. At the temperature of the gas (23°C), the vapour pressure of water is 2.8×10^3 Pa. What mass of ethyne is collected?

Road Map

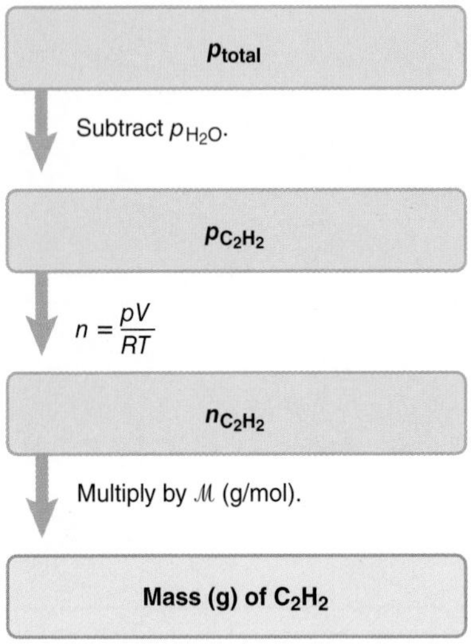

Plan To find the mass of C_2H_2, we first need to find the amount (mol) of C_2H_2, $n_{C_2H_2}$, which we can obtain from the ideal gas law by calculating $p_{C_2H_2}$. The barometer reading gives us p_{total}, which is the sum of $p_{C_2H_2}$ and p_{H_2O}, and we are given p_{H_2O}, so we subtract to find $p_{C_2H_2}$. We are also given V and T, so we convert to consistent units and find $n_{C_2H_2}$ from the ideal gas law. Then we convert amount (mol) to mass using the molar mass from the formula, as shown in the road map.

Solution Summarize and convert the gas variables:

$$p_{C_2H_2}\ (\text{Pa}) = p_{total} - p_{H_2O} = 9.84 \times 10^4\ \text{Pa} - 2.8 \times 10^3\ \text{Pa} = 9.56 \times 10^4\ \text{Pa}$$

$$V(\text{m}^3) = 523\ \text{ml} \times \frac{1\ \text{L}}{1000\ \text{mL}} \times \frac{1\ \text{m}^3}{1000\ \text{L}} = 5.23 \times 10^{-4}\ \text{m}^3$$

$$T(\text{K}) = 23 + 273.15 = 296\ \text{K}$$

$$n_{C_2H_2} = \text{unknown}$$

Solve for $n_{C_2H_2}$:

$$n_{C_2H_2} = \frac{pV}{RT} = \frac{(9.56 \times 10^4\ \text{Pa})(5.23 \times 10^{-4}\ \text{m}^3)}{8.314\ \frac{\text{Pa}\cdot\text{m}^3}{\text{mol}\cdot\text{K}} \times 296\ \text{K}} = 0.0203\ \text{mol}$$

Convert to mass (g):

$$\text{Mass (g) of } C_2H_2 = 0.0203\ \text{mol } C_2H_2 \times \frac{26.04\ \text{g } C_2H_2}{1\ \text{mol } C_2H_2}$$

$$= 0.529\ \text{g } C_2H_2$$

Check Rounding to one significant figure, a quick arithmetic check for n gives

$$n \approx \frac{(1 \times 10^5\ \text{Pa})(0.5 \times 10^{-3}\ \text{m}^3)}{8\ \frac{\text{Pa}\cdot\text{m}^3}{\text{mol}\cdot\text{K}} \times 300\ \text{K}} = 0.02\ \text{mol} \approx 0.0203\ \text{mol}$$

Comment The C_2^{2-} ion (called the *carbide*, or *acetylide, ion*) is an interesting anion. It is simply $^-C{\equiv}C^-$, which acts as a base in water, removing an H^+ ion from two H_2O molecules to form ethyne (or acetylene), $H{-}C{\equiv}C{-}H$.

Follow-Up Problem 4.10 A small piece of zinc reacts with dilute HCl to form H_2, which is collected over water at 16°C into a large flask. The total pressure is adjusted to barometric pressure (1.00 bar), and the volume is 1495 mL. Use Table 4.2 to help you calculate the partial pressure and mass of H_2.

The Ideal Gas Law and Reaction Stoichiometry

As you saw in Chapter 3, and in the preceding discussion about collecting a gas over water, many reactions involve gases as reactants or products. From the balanced equation for such a reaction, you can calculate the amounts (mol) of reactants and products and convert these amounts into masses or numbers of molecules. Figure 4.12 shows how you can use the ideal gas law to convert between gas variables (p, T, and V) and amounts (mol) of gaseous reactants and products. In effect, you combine a gas law problem with a stoichiometry problem, as you will see in Sample Problems 4.11 and 4.12.

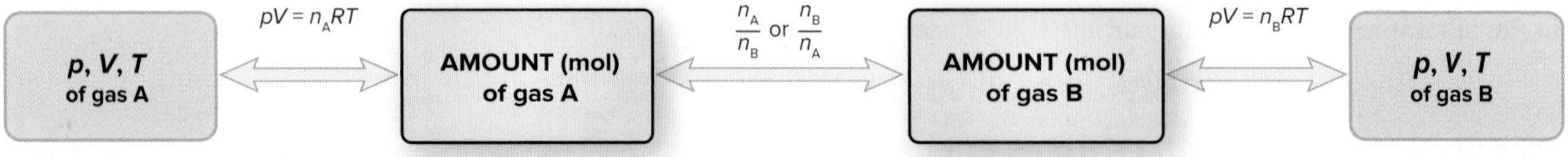

FIGURE 4.12 The relationships among the amount (mol, n) of gaseous reactant (or product) and the gas pressure (p), volume (V), and temperature (T)

Sample Problem 4.11 Using Gas Variables to Find Amounts of Reactants or Products I

Problem Engineers use copper in absorbent beds to react with and remove oxygen impurities in the ethylene used to make polyethylene. The beds are regenerated when hot H_2 reduces the copper(II) oxide, forming the pure metal and H_2O. On a laboratory scale, what volume of H_2 at 1.02 bar and 225°C is needed to reduce 35.5 g of copper(II) oxide?

Plan This is a stoichiometry *and* gas law problem. To find V_{H_2}, we first need n_{H_2}. We write and balance the equation. Next, we convert the given mass (35.5 g) of copper(II) oxide, CuO, to amount (mol) and use the molar ratio to find the amount (mol) of H_2 needed (stoichiometry portion). Then we use the ideal gas law to convert the amount of H_2 to volume (gas law portion). A road map is shown, but you are familiar with all the steps.

Road Map

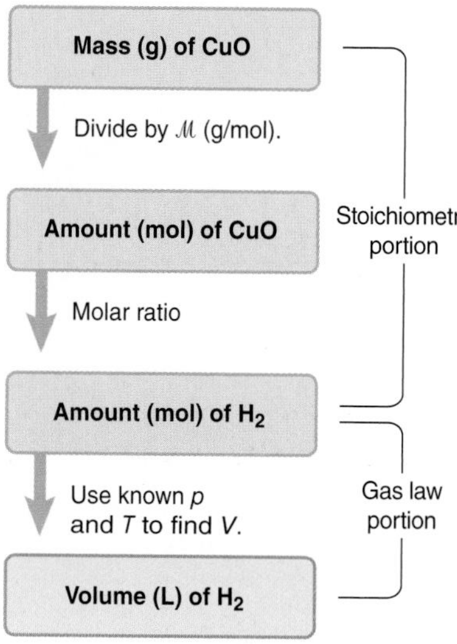

Solution Write the balanced equation:

$$\mathrm{CuO}(s) + \mathrm{H_2}(g) \longrightarrow \mathrm{Cu}(s) + \mathrm{H_2O}(g)$$

Calculate n_{H_2}:

$$n_{H_2} = 35.5\ \text{g CuO} \times \frac{1\ \text{mol CuO}}{79.55\ \text{g CuO}} \times \frac{1\ \text{mol H}_2}{1\ \text{mol CuO}} = 0.446\ \text{mol H}_2$$

Summarize and convert other gas variables:

$$V = \text{unknown} \qquad p\ (\text{Pa}) = 1.02\ \text{bar} \times \frac{10^5\ \text{Pa}}{1\ \text{bar}} = 1.02 \times 10^5\ \text{Pa}$$

$$T\ (\text{K}) = 225 + 273.15 = 498\ \text{K}$$

Solve for V_{H_2}:

$$V = \frac{nRT}{p} = \frac{0.446\ \text{mol} \times 8.314\ \frac{\text{Pa}\cdot\text{m}^3}{\text{mol}\cdot\text{K}} \times 498\ \text{K}}{1.02 \times 10^5\ \text{Pa}} = 18.1 \times 10^{-3}\ \text{m}^3 = 18.1\ \text{L}$$

Check One way to check the answer is to compare it with the molar volume of an ideal gas at STP (22.7 L at 273.15 K and 1 bar). Since 1 mol of H_2 at STP occupies about 23 L, less than 0.5 mol occupies less than 11 L. *T* is less than twice 273 K, so *V* should be less than twice 11 L.

Comment The main point here is that stoichiometry provides one gas variable (*n*), two more variables are given, and the ideal gas law is used to find the fourth.

Follow-Up Problem 4.11 Sulfuric acid reacts with sodium chloride to form aqueous sodium sulfate and hydrogen chloride gas. What volume (mL) of gas forms at STP when 0.117 kg of sodium chloride reacts with excess sulfuric acid?

Sample Problem 4.12 Using Gas Variables to Find Amounts of Reactants or Products II

Problem The alkali metals (group 1) react with the halogens (group 17) to form ionic metal halides. What mass of potassium chloride forms when 5.25 L of chlorine gas at 0.963 bar and 293 K reacts with 17.0 g of potassium (*see photo*)?

Chlorine gas reacting with potassium

Plan The amounts of two reactants are given, so this is a limiting-reactant problem. The only difference between this problem and previous limiting-reactant problems (such as Sample Problem 3.20) is that here we use the ideal gas law to find the amount (*n*) of gaseous reactant from the known *V*, *p*, and *T*. First, we write the balanced equation, and then we use it to find the limiting reactant and the amount and mass of product.

Solution Write the balanced equation:

$$2K(s) + Cl_2(g) \longrightarrow 2KCl(s)$$

Summarize the gas variables:

$$p = 0.963 \text{ bar} \qquad V = 5.25 \text{ L}$$
$$T = 293 \text{ K} \qquad n = \text{unknown}$$

Solve for n_{Cl_2}:

$$n_{Cl_2} = \frac{pV}{RT} = \frac{(0.963 \text{ bar})(5.25 \text{ L})}{0.08314 \dfrac{\text{bar·L}}{\text{mol·K}} \times 293 \text{ K}} = 0.2075 \text{ mol}$$

Convert from mass (g) of potassium (K) to amount (mol):

$$\text{Amount (mol) of K} = 17.0 \text{ g K} \times \frac{1 \text{ mol K}}{39.10 \text{ g K}} = 0.435 \text{ mol K}$$

Now determine the limiting reactant. If Cl_2 is limiting,

$$\text{Amount (mol) of KCl} = 0.2075 \text{ mol } Cl_2 \times \frac{2 \text{ mol KCl}}{1 \text{ mol } Cl_2} = 0.415 \text{ mol KCl}$$

If K is limiting,

$$\text{Amount (mol) of KCl} = 0.435 \text{ mol K} \times \frac{2 \text{ mol KCl}}{2 \text{ mol K}} = 0.435 \text{ mol KCl}$$

Cl_2 is the limiting reactant because it forms less KCl. Convert from amount (mol) of KCl to mass (g):

$$\text{Mass (g) of KCl} = 0.415 \text{ mol KCl} \times \frac{74.55 \text{ g KCl}}{1 \text{ mol KCl}} = 30.9 \text{ g KCl}$$

Check The gas law calculation seems correct. At STP, 22 L of Cl_2 gas contains about 1 mol, so a 5 L volume will contain a bit less than 0.25 mol of Cl_2. Moreover, since p (in the numerator) is slightly less than STP, and T (in the denominator) is slightly greater than STP, these values should lower the calculated n farther below the ideal value. The mass of KCl seems correct: less than 0.5 mol of KCl gives < 0.5 mol $\times$ $\mathcal{M}$ (~75 g/mol), and 31.0 g $<$ 0.5 mol $\times$ 75 g/mol.

Follow-Up Problem 4.12 Ammonia and hydrogen chloride gases react to form solid ammonium chloride. A 10.0 L reaction flask contains ammonia at 0.458 bar and 22°C, and 155 mL of hydrogen chloride gas at 7.60 bar and 271 K is introduced. After the reaction occurs and the temperature returns to 22°C, what is the pressure inside the flask? (Neglect the volume of the solid product.)

SUMMARY OF SECTION 4.4

- Gas density is inversely related to temperature: higher T causes lower d, and vice versa. At the same p and T, gases with larger m have higher d.
- In a mixture of gases, each component contributes its partial pressure to the total pressure (Dalton's law of partial pressures). The mole fraction of each component is the ratio of its partial pressure to the total pressure.
- When a gaseous reaction product is collected by bubbling it through water, the total pressure is the sum of the gas pressure and the vapour pressure of water at the given temperature.
- By converting the variables p, V, and T for a gaseous reactant (or product) to amount (n, mol), we can solve stoichiometry problems for gaseous reactions.

4.5 The Kinetic-Molecular Theory: A Model for Gas Behaviour

The **kinetic-molecular theory** is the model that accounts for macroscopic gas behaviour at the level of individual particles (atoms or molecules). Developed by some of the great scientists of the 19th century, most notably James Clerk Maxwell and Ludwig Boltzmann, the theory was able to explain the gas laws that some of the great scientists of the 18th century had arrived at empirically. The theory draws quantitative conclusions based on a few postulates (assumptions), but our discussion will be largely qualitative.

How the Kinetic-Molecular Theory Explains the Gas Laws

Let us address some questions that the theory must answer. Then we will state the postulates and draw conclusions that explain the gas laws and related phenomena.

Questions Concerning Gas Behaviour Observing gas behaviour at the macroscopic level, we must derive a molecular model that explains it:

1. *Origin of pressure.* Pressure is a measure of the force that a gas exerts on a surface. How do individual gas particles create this force?
2. *Boyle's law* ($V \propto \frac{1}{P}$). A change in gas pressure in one direction causes a change in gas volume in the other direction. What happens to the particles when external pressure compresses the gas volume? And why are liquids and solids much less compressible?
3. *Dalton's law* ($P_{total} = P_1 + P_2 + P_3 + \cdots$). The pressure of a gas mixture is the sum of the pressures of the individual gases. Why does each gas contribute to the total pressure in proportion to its number of particles?
4. *Charles's law* ($V \propto T$). A change in temperature causes a corresponding change in volume. What effect does higher temperature have on gas particles, causing an increase in gas volume? This question raises a more fundamental one: what does temperature measure on the molecular scale?
5. *Avogadro's law* ($V \propto n$). Gas volume depends on the amount (mol) present, not on the chemical nature of the gas. But should 1 mol of heavier particles not exert more pressure, and thus take up more space, than 1 mol of lighter particles?

Postulates of the Kinetic-Molecular Theory The theory is based on three postulates:

Postulate 1: *Particle volume.* A gas consists of a large collection of individual particles with empty space between them. The volume of each particle is so small compared with the volume of the whole sample that it is assumed to be zero; each particle is essentially a point of mass. In addition, since the particles are so small and there is so much space between them, there are essentially no interactions between the particles. The concept of ideality relies largely on the necessity that the particles move about independently without exerting any forces upon one another.

Postulate 2: *Particle motion.* The particles are in constant, random, straight-line motion, except when they collide with the container walls or with each other.

Postulate 3: *Particle collisions.* The collisions are *elastic*, which means that, like minute billiard balls, the colliding molecules exchange energy but do not lose any energy through friction. Thus, *their total kinetic energy* (E_k) *is constant.* Between collisions, the molecules do not influence each other by attractive or repulsive forces.

Imagine what a sample of gas in a container looks like. Countless minute particles move in every direction, smashing into the container walls and each other. Any given particle changes its speed often—at one moment standing still from a head-on collision and the next moment zooming away from a smash on the side.

In the sample as a whole, *each particle has a molecular speed* (u); *most are moving near the most probable speed, but some are much faster and others are*

FIGURE 4.13 Distribution of molecular speeds for N_2 at three temperatures

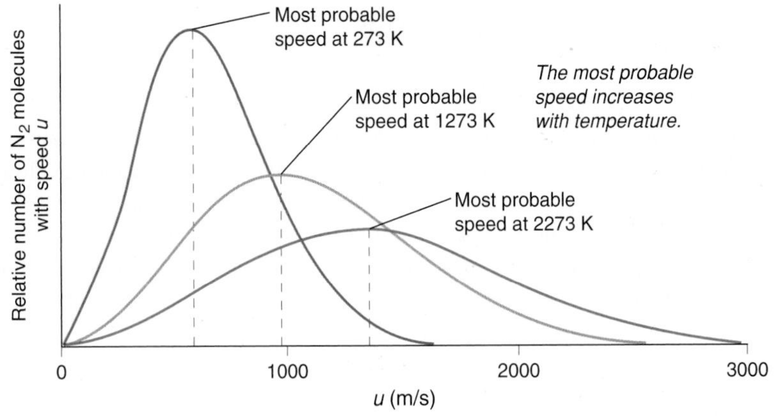
Most probable
speed at 273 K
Most probable
speed at 1273 K
The most probable
speed increases
with temperature.
Most probable
speed at 2273 K
Relative number of N2 molecules
with speed u
0
1000
2000
3000
u (m/s)

much slower. Figure 4.13 depicts this distribution of molecular speeds for N_2 gas at three temperatures.

Note that the curves flatten and spread at higher temperatures and that the *most probable speed (the peak of each curve) increases as the temperature increases*. This increase occurs because the average kinetic energy of the molecules, which is related to the most probable speed, is proportional to the absolute temperature: $\overline{E_k} \propto T$ or $\overline{E_k} = c \times T$, where $\overline{E_k}$ is the average kinetic energy of the molecules (an overbar indicates the average value of a quantity) and c is a constant that is the same for any gas. (We will return to this equation shortly.) Thus, a major conclusion based on the distribution of speeds, which arises directly from postulate 3, is that, *at a given temperature, all gases have the same average kinetic energy*.

A Molecular View of the Gas Laws Let us keep visualizing gas particles in a container to see how the theory explains the macroscopic behaviour of gases and answers the questions we posed above:

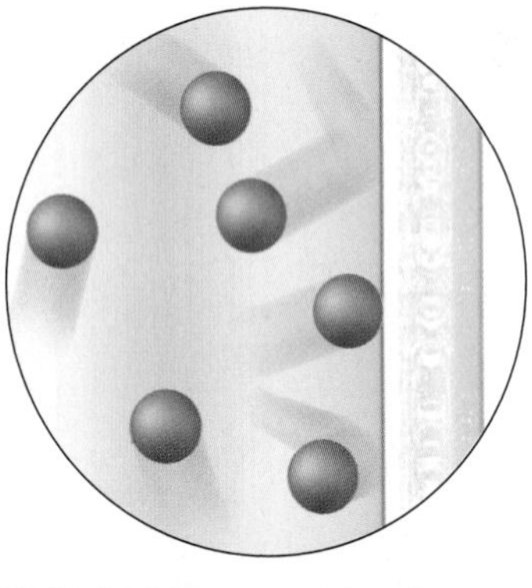

FIGURE 4.14 Pressure arises from countless collisions between gas particles and walls

1. *Origin of pressure* (Figure 4.14). From postulates 1 and 2, each gas particle (point of mass) colliding with the container walls (and the bottom of the piston) exerts a force. Countless collisions over the inner surface of the container result in a pressure. The greater the number of particles, the more frequently they collide with the container, and so the greater the pressure.
2. *Boyle's law* ($V \propto \frac{1}{p}$, shown in Figure 4.15). The particles in a gas are points of mass with empty space between them (postulate 1). Before any change in pressure, the pressure exerted *by* the gas (p_{gas}) equals the pressure exerted *on* the gas (p_{ext}), and there is some average distance (d_1) between the particles and the container walls. As p_{ext} increases at a constant temperature, the average distance (d_2) between the particles and the walls decreases (that is, $d_2 < d_1$), and so the sample volume decreases. Collisions of the particles with the walls become more frequent over the shorter average distance, which causes p_{gas} to increase until it again equals p_{ext}. The fact that liquids and solids cannot be compressed implies that there is little, if any, free space between their particles.

FIGURE 4.15 A molecular view of Boyle's law

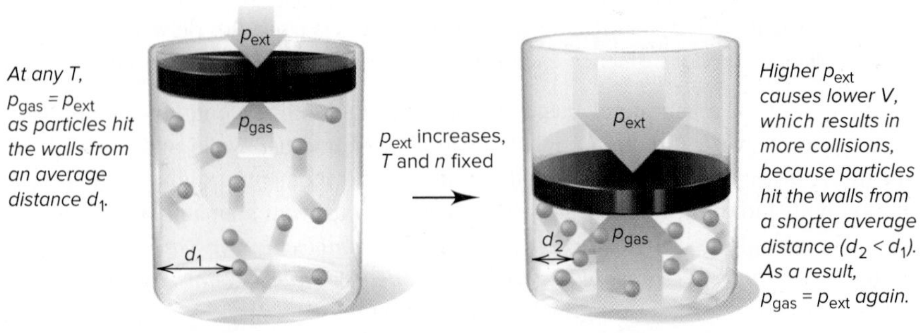
At any T,
pgas = pext
as particles hit
the walls from
an average
distance d1.
pext
pgas
d1
pext increases,
T and n fixed
pext
pgas
d2
Higher pext
causes lower V,
which results in
more collisions,
because particles
hit the walls from
a shorter average
distance (d2 < d1).
As a result,
pgas = pext again.

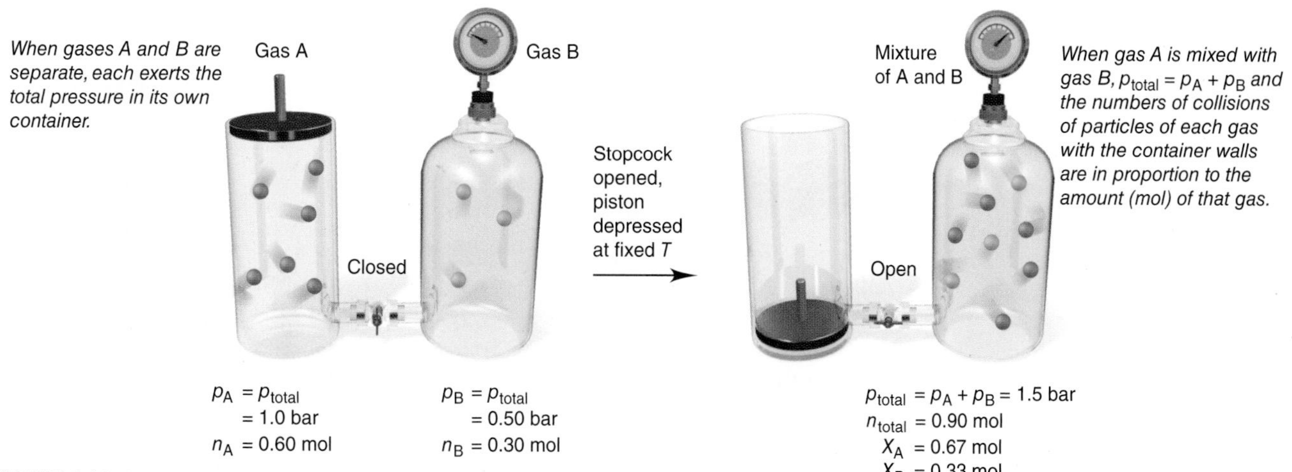

When gases A and B are separate, each exerts the total pressure in its own container.
Gas A
Gas B
Closed
Stopcock opened, piston depressed at fixed T
Mixture of A and B
Open
When gas A is mixed with gas B, $p_{total} = p_A + p_B$ and the numbers of collisions of particles of each gas with the container walls are in proportion to the amount (mol) of that gas.
$p_A = p_{total} = 1.0$ bar
$n_A = 0.60$ mol
$p_B = p_{total} = 0.50$ bar
$n_B = 0.30$ mol
$p_{total} = p_A + p_B = 1.5$ bar
$n_{total} = 0.90$ mol
$X_A = 0.67$ mol
$X_B = 0.33$ mol

FIGURE 4.16 A molecular view of Dalton's law

3. *Dalton's law of partial pressures* ($p_{total} = p_A + p_B$, shown in Figure 4.16). Adding a given amount (mol) of gas A to a given amount (mol) of gas B causes an increase in the total number of particles, in proportion to the particles of A added. This increase causes a corresponding increase in the total number of collisions with the walls per second (postulate 2), which causes a corresponding increase in the total pressure of the gas mixture (p_{total}). Each gas exerts a fraction of p_{total} in proportion to its fraction of the total number of particles (or, equivalently, its fraction of the total number of moles, that is, the mole fraction).
4. *Charles's law* ($V \propto T$, shown in Figure 4.17). At some starting temperature, T_1, the external (atmospheric) pressure (p_{atm}) equals the pressure of the gas (p_{gas}). When the gas is heated and the temperature increases to T_2, the most probable molecular speed and the average kinetic energy increase (postulate 3). Thus, the particles hit the walls more frequently *and* more energetically. This change temporarily increases p_{gas}. As a result, the piston moves up, which increases the volume and lowers the collision frequency until p_{atm} and p_{gas} are again equal.

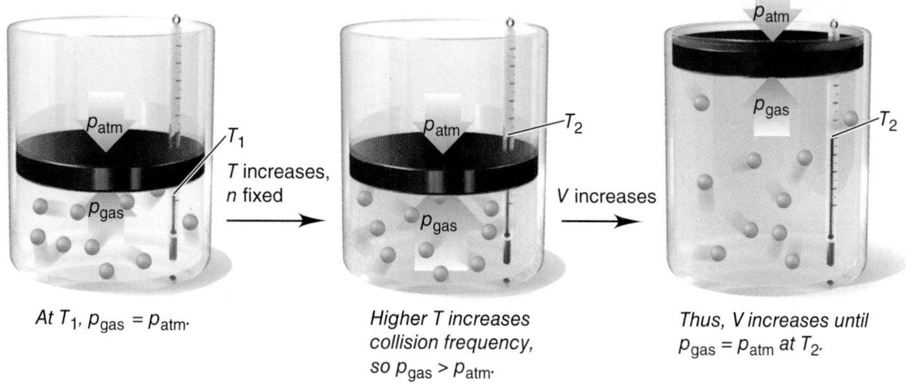

p_{atm}
p_{gas}
T_1
T increases, n fixed
T_2
V increases
At T_1, $p_{gas} = p_{atm}$.
Higher T increases collision frequency, so $p_{gas} > p_{atm}$.
Thus, V increases until $p_{gas} = p_{atm}$ at T_2.

FIGURE 4.17 A molecular view of Charles's law

5. *Avogadro's law* ($V \propto n$, shown in Figure 4.18). At some starting amount, n_1, of gas, p_{atm} equals p_{gas}. When more gas is added from the attached tank, the amount increases to n_2. Thus, more particles hit the walls more frequently, which temporarily increases p_{gas}. As a result, the piston moves up, which increases the volume and lowers the collision frequency until p_{atm} and p_{gas} are again equal.

The Central Importance of Kinetic Energy Recall, from Chapter 1, that the kinetic energy of an object is the energy that is associated with its motion. The kinetic energy of an object is key to explaining some of the implications of Avogadro's law and, most important, the meaning of temperature.

1. *Implications of Avogadro's law.* As we just saw, Avogadro's law says that, at any given T and p, the volume of a gas depends only on the number of moles—that is, the

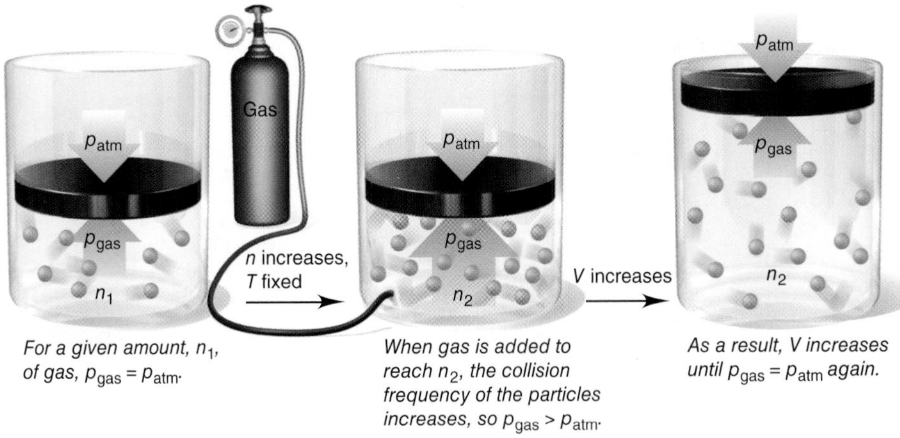

FIGURE 4.18 A molecular view of Avogadro's law

number of particles—in the sample. The law does not mention the chemical nature of the gas, so equal numbers of particles of any two gases, say O_2 and H_2, should occupy the same volume. However, why do the heavier O_2 molecules not exert more pressure on the container walls, and thus take up more volume, than the lighter H_2 molecules? To answer this question, we will show one way to express kinetic energy mathematically:

$$E_k = \frac{1}{2}\text{mass} \times \text{speed}^2 = \frac{1}{2}mu^2$$

The equation says that, for a given E_k, an object's mass and speed are inversely related, which means that *a heavier object moving slower can have the same kinetic energy as a lighter object moving faster.* Figure 4.19 shows that, for several gases, *the most probable speed (top of each curve) increases as the molar mass (the number in parentheses) decreases.*

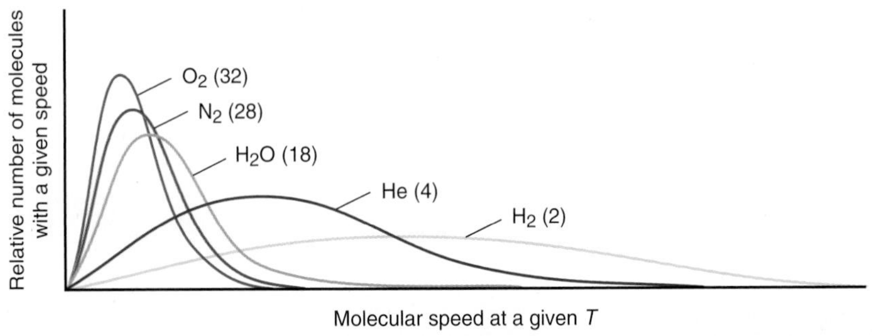

FIGURE 4.19 The relationship between molar mass and molecular speed. At a given temperature, gases with lower molar masses (the numbers in parentheses) have higher most probable speeds (the peaks of the curves).

As we saw earlier, postulate 3 of the kinetic-molecular theory directly implies that, at a given T, all gases have the same average kinetic energy. From Figure 4.19, we see that O_2 molecules move more slowly, on average, than H_2 molecules. With their higher most probable speed, H_2 molecules collide with the walls of a container more often than O_2 molecules do, but their lower mass means that each collision has less force. Therefore, at a given T, equimolar samples of H_2 and O_2 (or any other gas) exert the same pressure and, thus, occupy the same volume because, on average, *their molecules hit the walls with the same kinetic energy.*

2. *The meaning of temperature.* Closely related to these ideas is the central relation between kinetic energy and temperature. Earlier we said that the average kinetic energy of the particles ($\overline{E_k}$) equals the absolute temperature times a constant; that is, $\overline{E_k} = c \times T$. Using the definitions of velocity, momentum, force, and pressure, we can also express this relationship by the following equation: ■

$$\overline{E_k} = \frac{3}{2}\left(\frac{R}{N_A}\right)T$$

where R is the gas constant and N_A is the symbol for Avogadro's number. This equation makes the essential point that *temperature is a* ***measure*** *of the average*

■ We can derive the equation relating kinetic energy and temperature by using a series of equations that link together. The pressure of a gas can be expressed in terms of the frequency with which collisions occur and the momentum transferred when a molecule bounces back from a container wall. The collision frequency is a function of how fast a molecule is moving and how many molecules per unit volume are present. The momentum transfer is a function of the speed of the molecule and its mass. Putting this together, and taking into account the fact that molecules can move in three dimensions, we can express the pressure, in terms of number of molecules (N) per unit volume (V), as

$$p = \frac{1}{3}\frac{N}{V}m\overline{u^2}$$

Combining this equation with the ideal gas law (in which we take $n = 1$, thus making the number of molecules N_A) and expressing $\frac{1}{3}$ as the product of $\frac{2}{3}$ and $\frac{1}{2}$, we obtain $pV = \frac{2}{3}N_A(\frac{1}{2}m\overline{u^2}) = RT$. Substituting $\overline{E_k}$ for $\frac{1}{2}m\overline{u^2}$ and rearranging the expression, we obtain the expression that gives the average kinetic energy as a function of temperature.

kinetic energy of the particles: as T increases, $\overline{E_k}$ increases, and vice versa. Temperature is an intensive property (Section 1.5), so it is not related to the *total* energy of motion of the particles, which depends on the size of the sample, but to the *average* energy.

Thus, for example, in the macroscopic world, we heat a beaker of water over a flame and see the mercury rise inside a thermometer we put in the beaker. We see this because, in the molecular world, kinetic energy transfers, in turn, from the higher-energy gas particles in the flame to the lower-energy particles in the beaker glass, the water molecules, the particles in the thermometer glass, and the atoms of mercury.

Root-Mean-Square Speed Finally, let us derive an expression for the speed of a gas particle that has the average kinetic energy of the particles in a sample. From the general expression for the kinetic energy of an object,

$$E_k = \frac{1}{2}\text{mass} \times \text{speed}^2 = \frac{1}{2}mu^2$$

the average kinetic energy of each particle in a large population is

$$\overline{E_k} = \frac{1}{2}m\overline{u^2}$$

where m is the mass (atomic or molecular) of the particle and $\overline{u^2}$ is the average of the squares of the atomic or molecular speeds. Setting this expression for average kinetic energy equal to the earlier equation gives

$$\frac{1}{2}m\overline{u^2} = \frac{3}{2}\left(\frac{R}{N_A}\right)T$$

Multiplying through by Avogadro's number, N_A, gives the average kinetic energy for a mole of gas particles:

$$\frac{1}{2}N_A m\overline{u^2} = \frac{3}{2}RT$$

Avogadro's number times the molecular mass, $N_A \times m$, is the molar mass, $\mathcal{M}$, and solving for $\overline{u^2}$, we have

$$\overline{u^2} = \frac{3RT}{\mathcal{M}}$$

The square root of $\overline{u^2}$ is the root-mean-square speed, or **rms speed (u_{rms})**: *a particle moving at this speed has the average kinetic energy.** That is, taking the square root of both sides of the previous equation gives

$$u_{rms} = \sqrt{\frac{3RT}{\mathcal{M}}} \qquad (4.13)$$

where R is the gas constant, T is the absolute temperature, and $\mathcal{M}$ is the molar mass. (Because we want u in metres per second and R includes the joule, which has units of kg • m^2/s^2, we use the value 8.314 $\frac{J}{mol \cdot K}$ for R and express $\mathcal{M}$ in kg/mol.)

Thus, as an example, the root-mean-square speed of an O_2 molecule ($\mathcal{M}$ = 3.200 × 10^{-2} kg/mol) at 20°C, or 293 K, in the air you are breathing right now is

$$u_{rms} = \sqrt{\frac{3RT}{\mathcal{M}}} = \sqrt{\frac{3(8.314 \frac{J}{mol \cdot K})(293\ K)}{3.200 \times 10^{-2}\ kg/mol}}$$

$$= \sqrt{\frac{3(8.314 \frac{kg \cdot m^2/s^2}{\cancel{mol} \cdot \cancel{K}})(293\ \cancel{K})}{3.200 \times 10^{-2}\ \cancel{kg/mol}}}$$

$$= 478\ m/s$$

*The rms speed, u_{rms}, is equal to the square root of the average of the squares of all atomic or molecular speeds in a sample. It is proportional to, but slightly higher than, the most probable speed; for an ideal gas, u_{rms} = 1.09 × $\overline{u}$ (average speed of the molecules in a sample).

Effusion and Diffusion

The movement of a gas into a vacuum and the movement of gases through one another are phenomena with some vital applications.

The Process of Effusion One of the early triumphs of the kinetic-molecular theory was an explanation of **effusion**, the process by which a gas escapes through a tiny hole in its container into an evacuated space. In 1846, Thomas Graham studied the effusion rate of a gas, the number of molecules escaping per unit time, and found that it was inversely proportional to the square root of the density of the gas. However, density is directly related to molar mass, so **Graham's law of effusion** is stated as follows: *the rate of effusion of a gas is inversely proportional to the square root of its molar mass*, or

$$\text{Rate of effusion} \propto \frac{1}{\sqrt{\mathcal{M}}}$$

Argon (Ar) is lighter than krypton (Kr), so it effuses faster, assuming equal pressures of the two gases (Figure 4.20). Thus, the ratio of the rates is

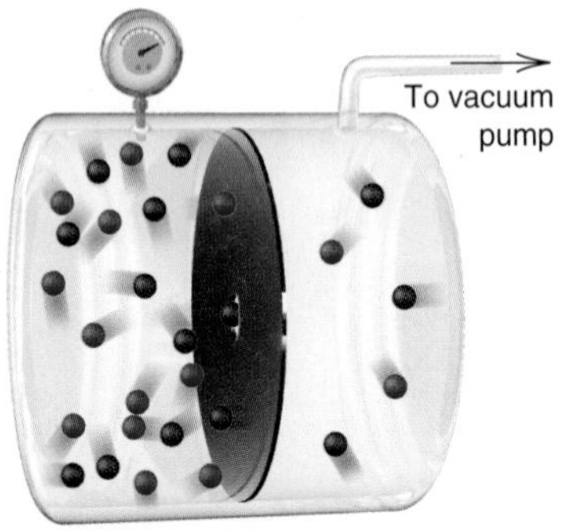

FIGURE 4.20 An example of effusion. Lighter (*black*) particles effuse faster than heavier (*red*) particles.

$$\frac{\text{Rate}_{\text{Ar}}}{\text{Rate}_{\text{Kr}}} = \frac{\sqrt{\mathcal{M}_{\text{Kr}}}}{\sqrt{\mathcal{M}_{\text{Ar}}}} \text{ or, in general, } \frac{\text{Rate}_{\text{A}}}{\text{Rate}_{\text{B}}} = \frac{\sqrt{\mathcal{M}_{\text{B}}}}{\sqrt{\mathcal{M}_{\text{A}}}} = \sqrt{\frac{\mathcal{M}_{\text{B}}}{\mathcal{M}_{\text{A}}}} \qquad (4.14)$$

The kinetic-molecular theory explains that, at a given temperature and pressure, *the gas with the lower molar mass effuses faster because the rms speed of its molecules is higher; therefore, more molecules reach the hole and escape per unit time.* This relationship allows us to derive a similar ratio for the relative times of effusion of two gases. A gas that effuses faster takes less time to effuse than a gas that effuses slowly. The ratio of the times of effusion of gases A and B would be the inverse of the ratio of the rates of effusion. Thus, the ratio of the relative times of effusion is

$$\frac{\text{Time of effusion}_{\text{Ar}}}{\text{Time of effusion}_{\text{Kr}}} = \frac{\text{Rate}_{\text{Kr}}}{\text{Rate}_{\text{Ar}}} = \frac{\sqrt{\mathcal{M}_{\text{Ar}}}}{\sqrt{\mathcal{M}_{\text{Kr}}}} \text{ or, in general, } \frac{t(\text{effusion})_{\text{A}}}{t(\text{effusion})_{\text{B}}} = \frac{\sqrt{\mathcal{M}_{\text{A}}}}{\sqrt{\mathcal{M}_{\text{B}}}} = \sqrt{\frac{\mathcal{M}_{\text{A}}}{\mathcal{M}_{\text{B}}}}$$

Sample Problem 4.13 Applying Graham's Law of Effusion

Problem A mixture of helium (He) and methane (CH_4) is placed in an effusion apparatus. Calculate the ratio of their effusion rates.

Plan Effusion rate is inversely proportional to $\sqrt{\mathcal{M}}$, so we find the molar mass of each substance from the formula and take its square root. The inverse of the ratio of the square roots is the ratio of the effusion rates.

Solution

$$\mathcal{M} \text{ of } CH_4 = 16.04 \text{ g/mol} \qquad \mathcal{M} \text{ of He} = 4.003 \text{ g/mol}$$

Calculate the ratio of the effusion rates:

$$\frac{\text{Rate}_{\text{He}}}{\text{Rate}_{CH_4}} = \sqrt{\frac{\mathcal{M}_{CH_4}}{\mathcal{M}_{\text{He}}}} = \sqrt{\frac{16.04 \text{ g/mol}}{4.003 \text{ g/mol}}} = \sqrt{4.007} = 2.002$$

Check A ratio > 1 makes sense because the lighter He should effuse faster than the heavier CH_4. Because the molar mass of CH_4 is about four times the molar mass of He, He should effuse about twice as fast as CH_4 ($\sqrt{4}$).

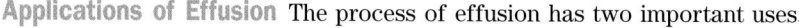

Follow-Up Problem 4.13 If it takes 1.25 min for 0.010 mol of He to effuse, how long will it take for the same amount of ethane (C_2H_6) to effuse?

Applications of Effusion The process of effusion has two important uses.

1. *Determination of molar mass.* We can use Graham's law to *determine the molar mass of an unknown gas.* By comparing the effusion rate of gas X with the effusion rate of a known gas, such as He, we can solve for the molar mass of X:

$$\frac{\text{Rate}_{\text{X}}}{\text{Rate}_{\text{He}}} = \sqrt{\frac{\mathcal{M}_{\text{He}}}{\mathcal{M}_{\text{X}}}}$$

Squaring both sides and solving for the molar mass of X gives

$$\mathcal{M}_{\text{X}} = \mathcal{M}_{\text{He}} \times \left(\frac{\text{rate}_{\text{He}}}{\text{rate}_{\text{X}}}\right)^2$$

2. *Preparation of nuclear fuel.* By far the most important application of Graham's law is in the preparation of fuel for nuclear energy reactors. The process of *isotope enrichment* increases the proportion of fissionable, but rarer, ^{235}U (only 0.7% by mass of naturally occurring uranium) to the non-fissionable, more abundant ^{238}U (99.3% by mass). Because the two isotopes have identical chemical properties, they are extremely difficult to separate chemically. However, one way to separate them takes advantage of a difference in a physical property—the effusion rate of gaseous compounds. Uranium ore is treated with fluorine to yield a gaseous mixture of $^{238}UF_6$ and $^{235}UF_6$ that is pumped through a series of chambers separated by porous barriers. Molecules of $^{235}UF_6$ are slightly lighter ($\mathcal{M}$ = 349.03) than molecules of $^{238}UF_6$ ($\mathcal{M}$ = 352.04), so they move slightly faster and effuse through each barrier 1.0043 times as fast. Many passes must be made, each one increasing the fraction of $^{235}UF_6$, until the mixture obtained is 3% to 5% by mass $^{235}UF_6$. This process was developed during the latter years of World War II and produced enough ^{235}U for two of the world's first atomic bombs. Today, a less expensive centrifuge process is used more often. The ability to enrich uranium has become a key international concern, as more countries aspire to develop nuclear energy and nuclear arms.

The Process of Diffusion Closely related to effusion is the process of gaseous **diffusion**, the movement of one gas through another. Graham's law can also be used to describe diffusion rates:

$$\text{Rate of diffusion} \propto \frac{1}{\sqrt{\mathcal{M}}}$$

For two gases at equal pressures, such as NH_3 and HCl, moving through another gas or a mixture of gases, such as air, we find

$$\frac{\text{Rate}_{\text{NH}_3}}{\text{Rate}_{\text{HCl}}} = \sqrt{\frac{\mathcal{M}_{\text{HCl}}}{\mathcal{M}_{\text{NH}_3}}}$$

The reason for this dependence on molar mass is the same as it is for effusion rates: *lighter molecules have higher average speeds than heavier molecules, so they move farther in a given time.*

If gas molecules move at hundreds of metres per second (see Figure 4.13), why does it take a second or two after you open a bottle of perfume to smell it? Although convection plays an important role in this process, another reason for the time lag is that a gas particle does not travel very far before it collides with another particle (Figure 4.21). Thus, a perfume molecule travels slowly because it collides with countless molecules in the air. The presence of so many other particles means that

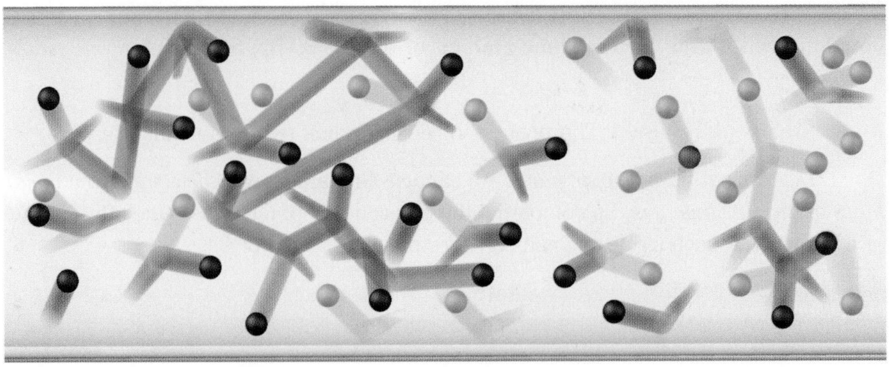

FIGURE 4.21 Diffusion of gases. When different gases (*black*, from the left, and *green*, from the right) move through each other, they mix. For simplicity, the complex path of only one black particle is shown (in *red*). In reality, all the particles have similar paths.

diffusion rates are much lower than effusion rates. Imagine how much quicker you can walk through an empty room than you can through a room crowded with other moving people.

Diffusion also occurs when a gas enters a liquid (and even, to a small extent, a solid). However, the average distances between molecules in a liquid are so much shorter that collisions are much more frequent; thus, diffusion of a gas through a liquid is *much* slower than it is through a gas. Nevertheless, this type of diffusion is a vital process in biological systems, for example, in the movement of O_2 from lungs to blood.

The Chaotic World of Gases: Mean Free Path and Collision Frequency

Refinements of the basic kinetic-molecular theory provide a view into the chaotic molecular world of gases. Try to visualize an "average" N_2 molecule in the room that you are now in. The N_2 molecule is continuously changing speed as it collides with other molecules—going 4000 km/h at one instant, and standing still at another. But these extreme speeds are *much* less likely than the most probable speed and those near it (see Figure 4.13). At 20°C and 1 bar pressure, the N_2 molecule is hurtling at an average speed of 470 m/s (rms speed = 510 m/s), or nearly 1700 km/h!

Mean Free Path From a particle's diameter, we can obtain the **mean free path**, *the average distance it travels between collisions at a given temperature and pressure.* An N_2 molecule (3.7×10^{-10} m in diameter) has a mean free path of 6.6×10^{-8} m, which means that it travels an average of 180 molecular diameters before smashing into a fellow traveller. (An N_2 molecule that is the size of a billiard ball would travel an average of about 9 m before hitting another.) Therefore, even though gas molecules are *not* points of mass, it is still valid to assume that a gas sample *is* nearly all empty space. Mean free path is a key factor in the rate of diffusion and the rate of heat flow through a gas.

Collision Frequency Divide the most probable speed (metres per second) by the mean free path (metres per collision) and you obtain the **collision frequency**, the average number of collisions per second that each particle undergoes. As you can see, the N_2 molecule experiences, on average, an enormous number of collisions every second: ■

$$\text{Collision frequency} = \frac{4.7 \times 10^2 \text{ m/s}}{6.6 \times 10^{-8} \text{ m/collision}} = 7.1 \times 10^9 \text{ collisions/s}$$

■ **Danger in a Molecular Amusement Park** To really appreciate the astounding events in the molecular world, let us use a two-dimensional analogy. Compare the moving N_2 molecule with a bumper car you are driving in an enormous amusement park ride. To match the collision frequency of the N_2 molecule, you would need to be travelling 4.5 billion km/s (much faster than the speed of light!), and you would smash into another bumper car every 640 m!

Distribution of speed (and kinetic energy) and collision frequency are essential ideas for understanding the speed of a reaction, as you will see in Chapter 14. As the Chemical Connections section shows, many of the concepts we have discussed so far apply directly to our planet's atmosphere.

An **atmosphere** is an envelope of gases that extends continuously from a planet's surface outward, thinning gradually until it is identical to outer space. A sample of clean, dry air at sea level on Earth contains 18 gases (Table B4.1). Under standard conditions, the gases behave nearly ideally, so volume percent equals mole percent (Avogadro's law), and the mole fraction of a component relates directly to its partial pressure (Dalton's law). Let us see how the gas laws and kinetic-molecular theory apply to our atmosphere, first with regard to variations in pressure and temperature, and then as explanations of some very familiar phenomena.

TABLE B4.1 Composition of Clean, Dry Air at Sea Level

Component	Mole Fraction
Nitrogen (N_2)	0.780 84
Oxygen (O_2)	0.209 46
Argon (Ar)	0.009 34
Carbon dioxide (CO_2)	0.000 318
Neon (Ne)	1.818×10^{-5}
Helium (He)	5.24×10^{-6}
Methane (CH_4)	2×10^{-6}
Krypton (Kr)	1.14×10^{-6}
Hydrogen (H_2)	5×10^{-7}
Dinitrogen monoxide (N_2O)	5×10^{-7}
Carbon monoxide (CO)	1×10^{-7}
Xenon (Xe)	8×10^{-8}
Ozone (O_3)	2×10^{-8}
Ammonia (NH_3)	6×10^{-9}
Nitrogen dioxide (NO_2)	6×10^{-9}
Nitrogen monoxide (NO)	6×10^{-10}
Sulfur dioxide (SO_2)	2×10^{-10}
Hydrogen sulfide (H_2S)	2×10^{-10}

The Smooth Variation in Pressure with Altitude Because gases are compressible (Boyle's law), the pressure of the atmosphere *increases* smoothly as we approach Earth's surface, with a more rapid increase at lower altitudes (Figure B4.1, *left*). No boundary delineates the beginning of the atmosphere from the end of outer space, but the densities and compositions are identical at an altitude of about 10 000 km. Yet, about 99% of the atmosphere's mass lies within 30 km of the surface, and 75% lies within the lowest 11 km.

The Zig-Zag Variation in Temperature with Altitude Unlike pressure, temperature does *not* change smoothly with altitude above Earth's surface. The atmosphere is classified into regions based on the direction of temperature change, and we will start at the surface (Figure B4.1, *right*).

1. *The troposphere.* In the troposphere (which extends from the surface to between 7 km at the poles and 17 km at the equator), the temperature *drops* 7°C per kilometre to −55°C (218 K). This region contains about 80% of the total mass of the atmosphere, with 50% of the mass in the lower 5.6 km. All weather occurs in this region, and nearly all aircraft, except supersonic aircraft, fly here.

2. *The stratosphere and the ozone layer.* In the stratosphere, the temperature *rises* from −55°C to about 7°C (280 K) at 50 km. This rise is due to a variety of complex reactions, mostly involving ozone, and is caused by the absorption of solar radiation. Most high-energy radiation is absorbed by the upper levels of the atmosphere, but some reaches the stratosphere and breaks O_2 into O atoms. The energetic O atoms collide with more O_2 to form ozone (O_3), another molecular form of oxygen:

$$O_2(g) \xrightarrow{\text{high-energy radiation}} 2O(g)$$

$$M + O(g) + O_2(g) \longrightarrow O_3(g) + M + \text{heat}$$

where M is any particle that can carry away excess energy. This reaction releases heat, which is why stratospheric temperatures increase with altitude. Over 90% of atmospheric ozone remains in a thin layer within the stratosphere, its thickness varying both geographically (greatest at the poles) and seasonally (greatest in the spring in the northern hemisphere and in the fall in the southern hemisphere). Stratospheric ozone is vital to life because it absorbs over 95% of the harmful ultraviolet (UV) solar radiation that would otherwise reach the surface. In Chapter 23, we will discuss the destructive effect of certain industrial chemicals on the ozone layer.

3. *The mesosphere.* In the mesosphere, the temperature *drops* again to −93°C (180 K) at around 80 km.

4. *The outer atmosphere.* Within the *thermosphere*, which extends to around 500 km, the temperature *rises* again, but varies between 700 and 2000 K, depending on the intensity of the solar radiation and sunspot activity. The *exosphere*, the outermost region, maintains these temperatures and merges with outer space.

What does it actually mean to have a temperature of 2000 K at 500 km above Earth's surface? Would a piece of iron (melting point = 1808 K) glow red-hot within a couple of minutes and melt in the thermosphere, as it does if heated to 2000 K in the troposphere? The answer involves the relation between temperature and the time taken to transfer kinetic energy. Our use of the words *hot* and *cold* refers to measurements near the surface of Earth. Here, the collision frequency of gas particles with a thermometer is enormous, and so the transfer of their kinetic energy is very fast. At an altitude of 500 km, however, where the density of gas particles is one-millionth that near the surface, collision frequency is extremely low, and a thermometer, or any object, experiences a *very slow transfer of kinetic energy*. Thus, the object would not become "hot" in the usual sense in any reasonable time. But the high-energy solar radiation that *is* transferred to the few particles present in these regions makes their average kinetic energy extremely high, as indicated by the high absolute temperature.

(Continued)

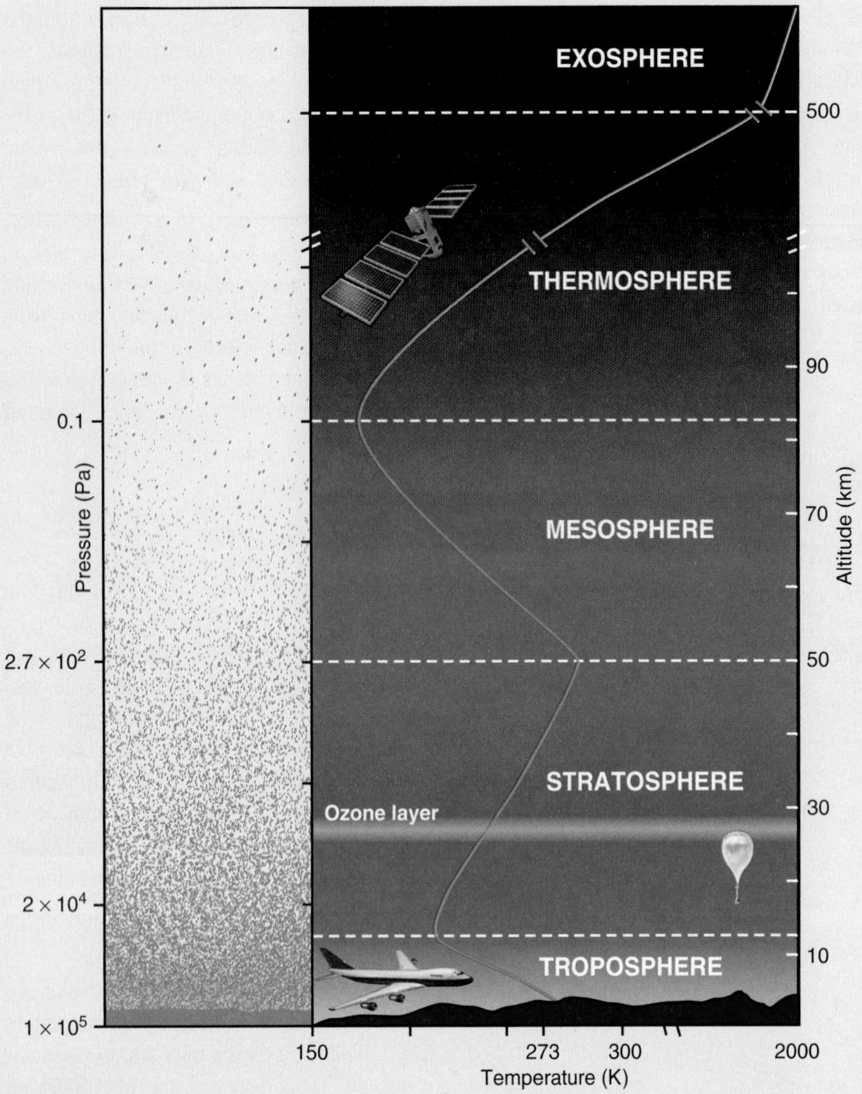

FIGURE B4.1 Variations in pressure and temperature with altitude in Earth's atmosphere

Convection in the Lower Atmosphere

Why must you take more breaths per minute on a high mountaintop than at sea level? At the higher elevation, there is a smaller *amount* of O_2 in each breath. However, the *proportion* (mole percent) of O_2 throughout the lower atmosphere remains about 21%. This uniform composition arises from *vertical (convective) mixing*, and the gas laws explain how it occurs.

Let us follow an air mass from ground level as solar heating of Earth's surface warms it. The warmer air mass expands (Charles's law), which makes it less dense, so it rises. As it does so, the pressure on it decreases, making it expand further (Boyle's law). Pushing against the surrounding air requires energy, so the temperature of the air mass decreases and thus the air mass shrinks slightly (Charles's law). But, as its temperature drops, its water vapour condenses (or solidifies), and these changes of state release heat; therefore, the air mass becomes warmer and rises higher. Meanwhile, the cooler, and thus denser, air that was above it sinks, becomes warmer through contact with the surface, and goes through the same process as the first air mass. As a result of this vertical mixing, the composition of the lower atmosphere remains uniform.

Warm air rising from the ground, called a *thermal*, is used by soaring birds and glider pilots to stay aloft. Convection helps to clean the air in urban areas, because the rising air carries up pollutants, which are dispersed by winds. Under certain conditions, however, a warm air mass remains stationary over a cool air mass. The resulting *temperature inversion* blocks normal convection, and harmful pollutants build up, causing severe health problems.

Problems

B4.1 Suggest a reason why supersonic aircraft are kept well below their maximum speeds until they reach their highest altitudes.

B4.2 Gases behave nearly ideally under Earth's conditions. Elsewhere in the Solar System, however, conditions are very different. On which planet would you expect atmospheric gases to deviate most from ideal behaviour: Saturn (4×10^6 bar and 130 K) or Venus (91 bar and 730 K)? Explain.

B4.3 What is the volume percent and partial pressure (Pa) of argon in a sample of dry air at sea level?

B4.4 Earth's atmosphere is estimated to have a mass of 5.14×10^{15} t (1 t = 1000 kg).

(a) If the average molar mass of air is 28.8 g/mol, what amount (mol) of gas is in the atmosphere?

(b) What volume (L) would the atmosphere occupy at 25°C and 1 bar?

SUMMARY OF SECTION 4.5

- The kinetic-molecular theory postulates that gas particles have no volume, move in straight-line paths between elastic (energy-conserving) collisions, and have average kinetic energies that are proportional to the absolute temperature of the gas.
- This theory explains the gas laws in terms of changes in distances between particles and the container walls, changes in molecular speed, and the energy of collisions.
- Temperature is a measure of the average kinetic energy of the particles.
- Effusion and diffusion rates are *inversely* proportional to the square root of the molar mass (Graham's law) because they are *directly* proportional to molecular speed.
- Molecular motion is characterized by a temperature-dependent, most probable speed (within a range of speeds), which affects mean free path and collision frequency.
- The atmosphere is a complex mixture of gases that exhibits variations in pressure and temperature with altitude. High temperatures in the upper atmosphere result from the absorption of high-energy solar radiation. The lower atmosphere has a uniform composition as a result of convective mixing.

4.6 Non-ideal Gases: Deviations from Ideal Behaviour

A fundamental principle of science is that simpler models are more useful than complex models, as long as they explain the data. With only a few postulates, the kinetic-molecular theory explains the behaviour of most gases under ordinary conditions. However, two of the postulates are useful approximations that do not reflect the following reality:

1. *Gas particles are **not** points of mass* but have volumes determined by the sizes of their atoms and the lengths and directions of their bonds.
2. *Attractive and repulsive forces **do** exist among gas particles* because atoms contain charged subatomic particles and many bonds are polar. (As you will see in Chapter 11, such forces lead to changes of physical state.)

These real situations cause deviations from ideal behaviour under *extreme conditions of low temperature and high pressure.* The deviations mean that we must alter the simple model and the ideal gas law to predict the behaviour of gases under extreme conditions.

Effects of Extreme Conditions on Gas Behaviour

At ordinary conditions—relatively high temperatures and low pressures—most gases exhibit nearly ideal behaviour. Yet, even at STP (0°C and 1 bar), gases deviate *slightly* from ideal behaviour. Table 4.3 shows that the standard molar volumes of several gases, when measured to five significant figures, do not equal the ideal values. Note that the deviations increase as the boiling point rises.

The phenomena that cause slight deviations under standard conditions exert more influence as temperature decreases and pressure increases. Figure 4.22 shows a plot of $\frac{pV}{RT}$ versus external pressure (p_{ext}) for 1 mol of several gases and an ideal gas. The $\frac{pV}{RT}$ values range from normal (at $p_{\text{ext}} = 1$ bar, $\frac{pV}{RT} = 1$) to very high (at $p_{\text{ext}} \approx 1013$ bar, $\frac{pV}{RT} \approx 1.6$ to 2.3). For the *ideal* gas, $\frac{pV}{RT}$ is 1 at any p_{ext}.

The $\frac{pV}{RT}$ curve for methane (CH_4) is typical of most gases: it decreases *below* the ideal value at moderately high p_{ext} and then rises *above* the ideal value as p_{ext} increases to very high values. This shape arises from two overlapping effects:

- At moderately high p_{ext}, $\frac{pV}{RT}$ values are lower than ideal values (less than 1) because of *interparticle attractions.*
- At very high p_{ext}, $\frac{pV}{RT}$ values are greater than ideal values (more than 1) because of *particle volume.*

TABLE 4.3 Molar Volume of Some Common Gases at 0°C and 1 bar

Gas	Molar Volume (L/mol)	Boiling Point (°C)
He	22.732	−268.9
H_2	22.729	−252.8
Ne	22.719	−246.1
Ideal gas	**22.711**	—
Ar	22.694	−185.9
N_2	22.693	−195.8
O_2	22.687	−183.0
CO	22.685	−191.5
Cl_2	22.478	−34.0
NH_3	22.372	−33.4

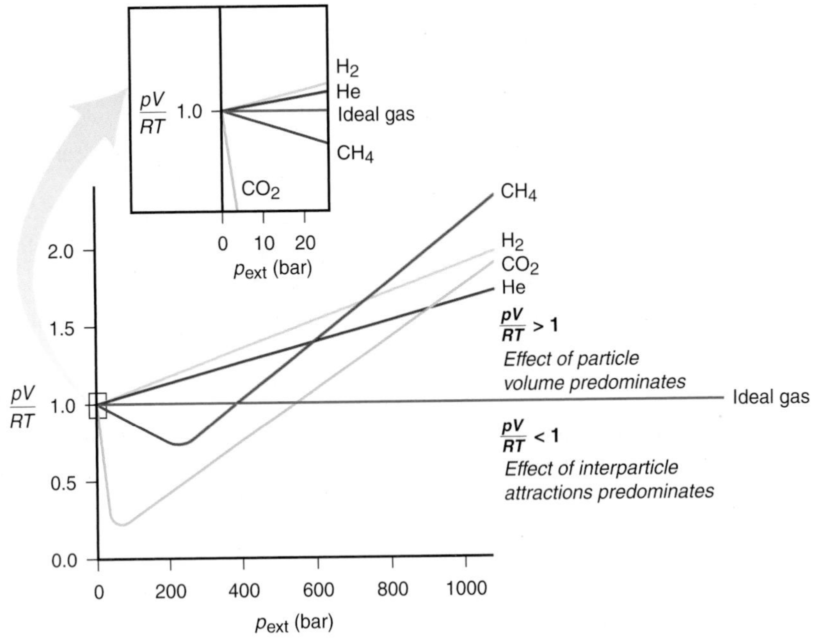

FIGURE 4.22 Deviations from ideal behaviour with increasing external pressure. The horizontal line shows that, for 1 mol of ideal gas, $\frac{pV}{RT} = 1$ at all p_{ext}. At very high p_{ext}, most gases deviate significantly from ideal behaviour, but small deviations appear even at ordinary pressures (expanded portion).

Let us examine these effects on the molecular level:

1. *Effect of interparticle attractions.* Interparticle attractions occur between separate atoms or molecules and are caused by imbalances in electron distributions. They are important only over *very* short distances and are *much* weaker than the covalent bonding forces that hold a molecule together. At normal p_{ext}, the spaces between the particles are so large that attractions are negligible and the gas behaves nearly ideally. As p_{ext} rises, the volume of the sample decreases and the particles get closer together, so interparticle attractions have a greater effect. As a particle approaches the container wall under these higher pressures, nearby particles attract it, lessening the force of its impact (Figure 4.23). *Repeated throughout the sample, this effect results in decreased gas pressure and, thus, a smaller numerator in* $\frac{pV}{RT}$. Similarly, lowering the temperature slows the particles, so they attract each other for a longer time.

2. *Effect of particle volume.* At normal p_{ext}, the space between particles (free volume) is enormous compared with the volume of the particles *themselves* (particle volume); thus, the free volume is essentially equal to V, the container volume in $\frac{pV}{RT}$. At *moderately* high p_{ext} and as free volume decreases, the particle volume makes up an increasing proportion of the container volume (Figure 4.24). At *extremely* high pressures, the space taken up by the particles themselves makes the free volume significantly *less* than the container volume. Nevertheless, we continue to use the container volume for V in $\frac{pV}{RT}$, which causes the numerator, and thus the ratio, to become artificially high. This particle volume effect increases as p_{ext} increases, eventually outweighing the effect of interparticle attractions and causing $\frac{pV}{RT}$ to rise above the ideal value.

FIGURE 4.23 The effect of interparticle attractions on measured gas pressure

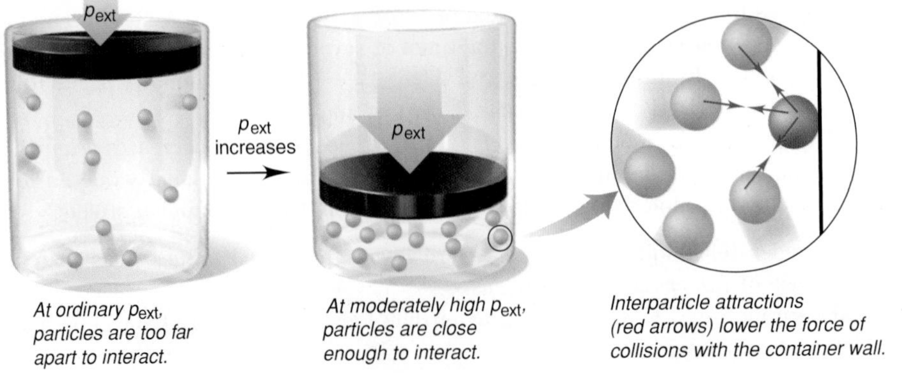

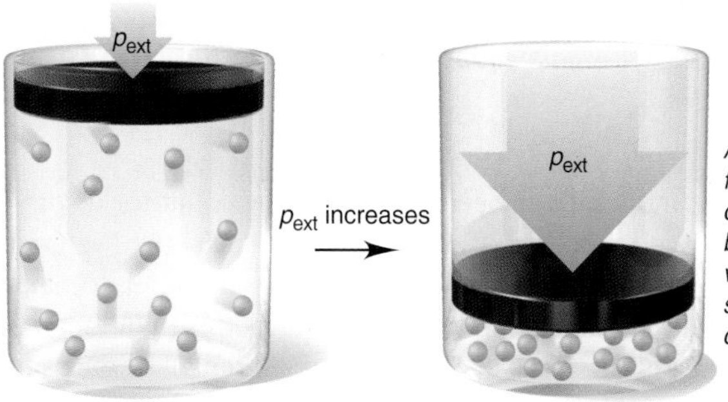

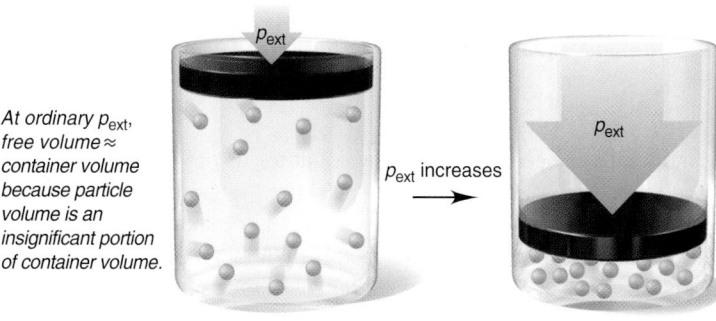

FIGURE 4.24 The effect of particle volume on measured gas volume

In Figure 4.22, note that the H_2 and He curves do not show the typical dip at moderate pressures. These gases consist of particles with such weak interparticle attractions that the particle volume effect predominates at all pressures.

Adjusting the Ideal Gas Law

The van der Waals Equation To describe non-ideal gas behaviour more accurately, we need to adjust the ideal gas equation in two ways:

1. Adjust *p up* by adding a factor that accounts for interparticle attractions.
2. Adjust *V down* by subtracting a factor that accounts for particle volume.

In 1873, Johannes van der Waals revised the ideal gas equation to account for the behaviour of non-ideal gases. The **van der Waals equation** for n mol of a non-ideal gas is

$$\underbrace{\left(p + \frac{n^2a}{V^2}\right)}_{\text{adjusts } p \text{ up}}\underbrace{(V - nb)}_{\text{adjusts } V \text{ down}} = nRT \qquad (4.15)$$

where p is the measured pressure, V is the known container volume, n and T have their usual meanings, and a and b are **van der Waals constants**, experimentally determined and specific for a given gas (Table 4.4). The constant a depends on the number and distribution of electrons, which relates to the complexity of a particle and the strength of its interparticle attractions. The constant b relates to the volume of the particle. For instance, CO_2 is both more complex and larger than H_2, and the values of their constants reflect this.

Here is a typical application of the van der Waals equation. A 1.98 L vessel contains 215 g (4.89 mol) of dry ice. After standing at 26°C (299 K), the $CO_2(s)$ changes to $CO_2(g)$. The pressure is measured (p_{meas}) and then calculated by the ideal gas law (p_{IGL}) and, using the appropriate values of a and b, by the van der Waals equation (p_{VDW}). The results are revealing:

$$p_{meas} = 45.4 \text{ bar} \qquad p_{IGL} = 61.4 \text{ bar} \qquad p_{VDW} = 46.5 \text{ bar}$$

Comparing the measured value with each calculated value shows that p_{IGL} is 35.3% greater than p_{meas}, but p_{VDW} is only 2.5% greater than p_{meas}. At these conditions, CO_2 deviates so much from ideal behaviour that the ideal gas law is not very useful.

It is important to realize that, according to kinetic-molecular theory, the constants a and b are zero for an ideal gas because the gas particles do not attract each other and have no volume. Yet, even in normal gases at ordinary pressures, the particles are very far apart. This large average interparticle distance has two consequences:

- Attractive forces are miniscule, so $p + \frac{n^2a}{V^2} \approx p$.
- The particle volume is a minute fraction of the container volume, so $V - nb \approx V$.

Table 4.4 Van der Waals Constants for Some Common Gases

Gas	$a\left(\frac{\text{bar}\cdot\text{L}^2}{\text{mol}^2}\right)$	$b\left(\frac{\text{L}}{\text{mol}}\right)$
He	0.0346	0.0238
Ne	0.208	0.016 72
Ar	1.355	0.032 01
Kr	2.325	0.0396
Xe	4.192	0.051 56
H_2	0.2453	0.026 51
N_2	1.370	0.0387
O_2	1.382	0.031 86
Cl_2	6.343	0.054 22
CH_4	2.300	0.043 01
CO	1.472	0.039 48
CO_2	3.658	0.042 86
NH_3	4.225	0.037 13
H_2O	5.537	0.030 49

Therefore, *at ordinary conditions, because the value of the correction factors becomes negligible, the van der Waals equation* **reduces to** *the ideal gas equation.*

Other Models for Non-ideal Gases

While many other models have been proposed to account for deviations from ideality, including the Berthelot and modified Berthelot models, the Dieterici model, the Clausius model, the Virial model, and the Peng-Robinson model, to name only a few, one model that actually predicts non-ideal behaviour better than the van der Waals model in almost all cases is the Redlich Kwong model. Similar to the van der Waals model in that it accounts for deviations from ideality using two parameters, it has the form

$$p = \frac{RT}{V_m - b} - \frac{a}{V_m(V_m + b)} \qquad (4.16)$$

where the values of a and b can be estimated using the values for critical temperature and critical pressure from the following relationships:

$$a = \frac{0.42748R^2T_c^{2.5}}{p_cT^{1/2}}, \quad b = \frac{0.0867RT_c}{p_c} \qquad (4.17)$$

V_m represents the molar volume of the gas and, as with the van der Waals model, a corrects for particle interactions while b corrects for particle volume. The Redlich Kwong model can be adjusted to determine the compressibility factor of a gas as well as to account for gas pressures in a mixture. What is most interesting about the Redlich Kwong model is not that it is a better model than van der Waals with the same number of parameters, but that it is better than many three-parameter models.

SUMMARY OF SECTION 4.6

- At very high p or low T, all gases deviate significantly from ideal behaviour.
- As external pressure increases, most gases exhibit first a lower and then a higher $\frac{pV}{RT}$; for 1 mol of an ideal gas, this ratio remains constant at 1.
- The deviations from ideal behaviour are due to (1) attractions between particles, which lower the pressure (and decrease $\frac{pV}{RT}$), and (2) the volume of the particles themselves, which takes up an increasingly larger fraction of the container volume (and increases $\frac{pV}{RT}$).
- The van der Waals equation includes constants that are specific for a given gas to correct for deviations from ideal behaviour. At ordinary p and T, the van der Waals equation reduces to the ideal gas equation.
- There are many other models for non-ideal gases, of which the Redlich Kwong is one of the better ones.

CHAPTER REVIEW GUIDE

Learning Objectives

Relevant section (§) and/or sample problem (SP) numbers appear in parentheses.

Concepts

1. Explain how the macroscopic properties of gases differ from those of liquids and solids. (§4.1)
2. Explain the meaning of pressure and the operation of a barometer and a manometer. (§4.2)
3. Describe the relations among gas variables expressed by Boyle's, Charles's, and Avogadro's laws. (§4.3)
4. Describe how the individual gas laws are incorporated into the ideal gas law. (§4.3)
5. Describe how the ideal gas law can be used to study gas density, molar mass, and amounts of gases in reactions. (§4.4)
6. Explore the relationship between the density and temperature of a gas. (§4.4)

7. Explain the meaning of Dalton's law, and describe the relationship between partial pressure and the mole fraction of a gas; describe how Dalton's law applies to collecting a gas over water. (§4.4)
8. Discuss how the postulates of the kinetic-molecular theory are applied to explain the origin of pressure and the gas laws. (§4.5)
9. Discuss the relationships among molecular speed, average kinetic energy, and temperature. (§4.5)
10. Differentiate between *effusion* and *diffusion* and how their rates and times are related to molar mass. (§4.5)
11. Discuss the relationships between mean free path, molecular speed, and collision frequency. (§4.5)
12. Explain why intermolecular attractions and molecular volume cause gases to deviate from ideal behaviour at low temperatures and high pressures. (§4.6)
13. Discuss how the van der Waals equation and other models that account for non-ideality correct the ideal gas law for extreme conditions. (§4.6)

Skills

1. Convert between the units of pressure (bar, Pa, kPa, atm, and mmHg or Torr). (SP 4.1)
2. Reduce the ideal gas law to the individual gas laws. (SPs 4.2–4.5)
3. Apply the gas laws to choose the correct chemical equation. (SP 4.6)
4. Rearrange the ideal gas law to calculate the density of a gas (SP 4.7) and the molar mass of a volatile liquid. (SP 4.8)
5. Calculate the mole fraction and the partial pressure of a gas. (SP 4.9)
6. Apply the vapour pressure of water to correct for the amount of a gas collected over water. (SP 4.10)
7. Apply stoichiometry and the gas laws to calculate amounts of reactants and products. (SPs 4.11, 4.12)
8. Use Graham's law to solve problems involving gaseous effusion. (SP 4.13)

Key Terms

Section 4.2
pressure (p)
barometer
manometers
pascal (Pa)
standard atmosphere (atm)
millimetre of mercury (mmHg)
Torr
bar
standard pressure

Section 4.3
ideal gas
Boyle's law
Charles's law
Avogadro's law
standard temperature and pressure (STP)
standard molar volume
ideal gas law
universal gas constant (R)

Section 4.4
partial pressure
Dalton's law of partial pressures
mole fraction (X)

Section 4.5
kinetic-molecular theory
rms speed (u_{rms})
effusion
Graham's law of effusion
diffusion
mean free path
collision frequency
atmosphere

Section 4.6
van der Waals equation
van der Waals constants

Key Equations and Relationships

4.1 Expressing the volume-pressure relationship (Boyle's law):

$$V \propto \frac{1}{p} \quad \text{or} \quad pV = \text{constant} \quad (T \text{ and } n \text{ fixed})$$

4.2 Expressing the volume-temperature relationship (Charles's law):

$$V \propto T \quad \text{or} \quad \frac{V}{T} = \text{constant} \quad (p \text{ and } n \text{ fixed})$$

4.3 Expressing the pressure-temperature relationship (Amontons's law):

$$p \propto T \quad \text{or} \quad \frac{p}{T} = \text{constant} \quad (V \text{ and } n \text{ fixed})$$

4.4 Expressing the volume-amount relationship (Avogadro's law):

$$V \propto n \quad \text{or} \quad \frac{V}{n} = \text{constant} \quad (p \text{ and } T \text{ fixed})$$

4.5 Defining standard temperature and pressure:

$$\text{STP}: \quad 0°\text{C } (273.15 \text{ K}) \text{ and } 1 \text{ bar } (10^5 \text{ Pa})$$

4.6 Defining the volume of 1 mol of an ideal gas at STP:

Standard molar volume = 22.710 953(21) L or 22.7 L (3 sf)

4.7 Relating volume to pressure, temperature, and amount (ideal gas law):

$$pV = nRT \quad \text{and} \quad \frac{p_1V_1}{n_1T_1} = \frac{p_2V_2}{n_2T_2}$$

4.8 Calculating the value of R:

$$R = \frac{pV}{nT} = \frac{(1 \times 10^5 \text{ Pa})[(22.710\,953(21) \times 10^{-3} \text{ m}^3)]}{(1 \text{ mol})(273.15 \text{ K})}$$

$$= 8.314\,462 \frac{\text{Pa}\cdot\text{m}^3}{\text{mol}\cdot\text{K}} = 8.314 \frac{\text{J}}{\text{mol}\cdot\text{K}} \text{ (4 sf)}$$

4.9 Rearranging the ideal gas law to find gas density:

$$pV = \frac{m}{\mathcal{M}}RT$$

so

$$\frac{m}{V} = d = \frac{\mathcal{M} \times p}{RT}$$

4.10 Rearranging the ideal gas law to find molar mass:

$$n = \frac{m}{\mathcal{M}} = \frac{pV}{RT} \quad \text{so} \quad \mathcal{M} = \frac{mRT}{pV}$$

4.11 Relating the total pressure of a gas mixture to the partial pressures of the components (Dalton's law of partial pressures):

$$p_{\text{total}} = p_1 + p_2 + p_3 + \cdots$$

4.12 Relating partial pressure to mole fraction:

$$p_A = X_A \times p_{\text{total}}$$

4.13 Defining rms speed as a function of molar mass and temperature:

$$u_{\text{rms}} = \sqrt{\frac{3RT}{\mathcal{M}}}$$

4.14 Applying Graham's law of effusion:

$$\frac{\text{Rate}_A}{\text{Rate}_B} = \frac{\sqrt{\mathcal{M}_B}}{\sqrt{\mathcal{M}_A}} = \sqrt{\frac{\mathcal{M}_B}{\mathcal{M}_A}}$$

$$\frac{t(\text{effusion})_A}{t(\text{effusion})_B} = \frac{\sqrt{\mathcal{M}_A}}{\sqrt{\mathcal{M}_B}} = \sqrt{\frac{\mathcal{M}_A}{\mathcal{M}_B}}$$

4.15 Applying the van der Waals equation to find the pressure or volume of a gas under extreme conditions:

$$\left(p + \frac{n^2a}{V^2}\right)(V - nb) = nRT$$

4.16, 4.17 Applying the Redlich Kwong equation to find the pressure of a gas under extreme conditions:

$$p = \frac{RT}{V_m - b} - \frac{a}{V_m(V_m + b)}$$

$$a = \frac{0.42748R^2T_c^{2.5}}{p_cT^{1/2}},\ b = \frac{0.0867RT_c}{p_c}$$

Brief Solutions to Follow-Up Problems

4.1 $p_{CO_2}\ (\text{Torr}) = (753.6\ \text{mmHg} - 174.0\ \text{mmHg}) \times \frac{1\ \text{Torr}}{1\ \text{mmHg}}$

$= 579.6\ \text{Torr}$

$p_{CO_2}\ (\text{Pa}) = 579.6\ \text{Torr} \times \frac{1.01325 \times 10^5\ \text{Pa}}{760\ \text{Torr}}$

$= 7.727 \times 10^4\ \text{Pa}$

$= 77.27\ \text{kPa}$

$p_{CO_2}\ (\text{bar}) = 579.6\ \text{Torr} \times \frac{1.01325\ \text{bar}}{760\ \text{Torr}} = 0.7727\ \text{bar}$

4.2 $p_2\ (\text{bar}) = 26.3\ \text{kPa} \times \frac{1\ \text{bar}}{100\ \text{kPa}} = 0.263\ \text{bar}$

$V_2\ (\text{L}) = 105\ \text{mL} \times \frac{1\ \text{L}}{1000\ \text{mL}} \times \frac{0.883\ \text{bar}}{0.263\ \text{bar}} = 0.353\ \text{L}$

4.3 $T_2\ (K) = 273\ \text{K} \times \frac{9.75\ \text{cm}^3}{6.83\ \text{cm}^3} = 390.\ \text{K}$

4.4 $p_2\ (\text{bar}) = 1.06\ \text{bar} \times \frac{35.0\ \text{g} - 5.0\ \text{g}}{35.0\ \text{g}} = 0.909\ \text{bar}$

(There is no need to convert mass to moles because the ratio of masses equals the ratio of moles.)

4.5 $p_{O_2} = 1.39\ \text{bar} \times \frac{100\ \text{kPa}}{1\ \text{bar}} = 139\ \text{kPa}$

$n = \frac{pV}{RT} = \frac{139\ \text{kPa} \times 438\ \text{L}}{8.314\ \frac{\text{kPa.L}}{\text{mol}\cdot\text{K}} \times 294\ \text{K}} = 24.9\ \text{mol}\ O_2$

$\text{Mass (g) of } O_2 = 24.9\ \text{mol}\ O_2 \times \frac{32.00\ \text{g}\ O_2}{1\ \text{mol}\ O_2} = 7.97 \times 10^2\ \text{g}\ O_2$

4.6 The balanced equation is $2CD(g) \longrightarrow C_2(g) + D_2(g)$, so n does not change. Therefore, given constant p, the temperature, T, must double: $T_1 = -73 + 273.15 = 200$ K, so $T_2 = 400$ K; $400 - 273.15 = 127°C$.

4.7 $d\ (\text{at } 0°\text{C and } 0.507\ \text{bar}) = \frac{44.01\ \text{g/mol} \times 0.507\ \text{bar}}{0.08314\ \frac{\text{bar}\cdot\text{L}}{\text{mol}\cdot\text{K}} \times 273\ \text{K}}$

$= 0.983\ \text{g/L}$

The density is lower at the smaller p because V is larger. In this case, d is lowered by one-half because p is about one-half as much.

4.8

$$\mathcal{M} = \frac{(68.697\ \text{g} - 68.322\ \text{g}) \times 0.08314\ \frac{\text{bar}\cdot\text{L}}{\text{mol}\cdot\text{K}} \times (273.15 + 95.0)\ \text{K}}{0.987\ \text{bar} \times \frac{149\ \text{mL}}{1000\ \text{mL/L}}}$$

$= 78.0\ \text{g/mol}$

4.9 $n_{\text{total}} = \left(5.50\ \text{g He} \times \frac{1\ \text{mol He}}{4.003\ \text{g He}}\right) + \left(15.0\ \text{g Ne} \times \frac{1\ \text{mol Ne}}{20.18\ \text{g Ne}}\right) + \left(35.0\ \text{g Kr} \times \frac{1\ \text{mol Kr}}{83.80\ \text{g Kr}}\right)$

$= 2.53\ \text{mol}$

$$p_{He} = \left(\frac{5.50\ \text{g He} \times \frac{1\ \text{mol He}}{4.003\ \text{g He}}}{2.53\ \text{mol}}\right) \times 1\ \text{bar} = 0.543\ \text{bar}$$

$p_{Ne} = 0.294\ \text{bar} \quad p_{Kr} = 0.165\ \text{bar}$

4.10 $p_{H_2} = 1.00\ \text{bar} - 0.018\ \text{bar} = 0.98\ \text{bar}$

$$\text{Mass (g) of } H_2 = \frac{(0.98\ \cancel{\text{bar}})(1.495\ \cancel{\text{L}})}{(0.08314\ \frac{\cancel{\text{bar}\cdot\text{L}}}{\cancel{\text{mol}\cdot\text{K}}})(289\ \cancel{\text{K}})} \times \frac{2.016\ \text{g}\ H_2}{1\ \cancel{\text{mol}}\ H_2}$$

$= 0.123\ \text{g}\ H_2$

4.11 $H_2SO_4(aq) + 2NaCl(s) \longrightarrow Na_2SO_4(aq) + 2HCl(g)$

$n_{HCl} = 0.117\ \text{kg NaCl} \times \frac{10^3\ \text{g}}{1\ \text{kg}} \times \frac{1\ \text{mol NaCl}}{58.44\ \text{g NaCl}} \times \frac{2\ \text{mol HCl}}{2\ \text{mol NaCl}}$

$= 2.00\ \text{mol HCl}$

At STP, $V\ (\text{mL}) = 2.00\ \text{mol} \times \frac{22.7\ \text{L}}{1\ \text{mol}} \times \frac{10^3\ \text{mL}}{1\ \text{L}} = 4.54 \times 10^4\ \text{mL}$

4.12 $NH_3(g) + HCl(g) \longrightarrow NH_4Cl(s)$

$n_{NH_3} = 0.187$ mol and $n_{HCl} = 0.0523$ mol; thus, HCl is the limiting reactant.

n_{NH_3} after reaction

$= 0.187\ \text{mol}\ NH_3 - \left(0.0523\ \text{mol HCl} \times \frac{1\ \text{mol}\ NH_3}{1\ \text{mol HCl}}\right)$

$= 0.135\ \text{mol}\ NH_3$

$$p = \frac{0.135\ \text{mol} \times 8.314\ \frac{\text{kPa}\cdot\text{L}}{\text{mol}\cdot\text{K}} \times 295\ \text{K}}{10.0\ \text{L}} = 33.1\ \text{kPa}$$

4.13 $\frac{\text{Rate of He}}{\text{Rate of } C_2H_6} = \sqrt{\frac{30.07\ \text{g/mol}}{4.003\ \text{g/mol}}} = 2.741$

Time for C_2H_6 to effuse $= 1.25\ \text{min} \times 2.741 = 3.43\ \text{min}$

PROBLEMS

Problems with **red** numbers are answered in Appendix G and worked in detail in the Student Solutions Manual. Problem sections match those in this book and provide the numbers of relevant sample problems. Most offer Concept Review Questions, Skill-Building Exercises (grouped in pairs covering the same concept), and Problems in Context. The Comprehensive Problems are based on material from any section or previous chapter.

An Overview of the Physical States of Matter

Concept Review Questions

4.1 How does a sample of gas differ in its behaviour from a sample of liquid in each situation?
(a) The sample is transferred from one container to a larger container.
(b) The sample is heated in an expandable container, but no change of state occurs.
(c) The sample is placed in a cylinder with a piston, and an external force is applied.

4.2 Are the particles of a gas farther apart or closer together than the particles of a liquid? Use your answer to explain each general observation:
(a) Gases are more compressible than liquids.
(b) Gases have lower viscosities than liquids.
(c) After thorough stirring, all gas mixtures are solutions.
(d) The density of a substance in the gas state is lower than its density in the liquid state.

Gas Pressure and Its Measurement

(Sample Problem 4.1)

Concept Review Questions

4.3 How does a barometer work? Is the column of mercury in a barometer shorter when the barometer is on a mountaintop or at sea level? Explain.

4.4 How can a unit of length, such as millimetre of mercury (mmHg), be used as a unit of pressure, which has the dimensions of force per unit area?

4.5 In a closed-end manometer, the mercury level in the arm attached to the flask can never be higher than the mercury level in the other arm. In an open-end manometer, however, it *can* be higher. Explain.

Skill-Building Exercises (grouped in similar pairs)

4.6 On a cool, rainy day, the barometric pressure is 730 mmHg. What is the barometric pressure in centimetres of water (cmH_2O) (*d* of Hg = 13.5 g/mL; *d* of H_2O = 1.00 g/mL)?

4.7 A long glass tube, sealed at one end, has an inner diameter of 10.0 mm. The tube is filled with water and inverted into a pail of water. If the atmospheric pressure is 755 mmHg, how high (in mmH_2O) is the column of water in the tube (*d* of Hg = 13.5 g/mL; *d* of H_2O = 1.00 g/mL)?

4.8 Convert each pressure:
(a) 0.745 atm to kPa (b) 992 Torr to bar
(c) 36.5 kPa to Pa (d) 804 mmHg to kPa

4.9 Convert each pressure:
(a) 768 mmHg to Pa (b) 27.5 atm to kPa
(c) 6.50 atm to bar (d) 0.937 kPa to bar

4.10 In Figure P4.10, what is the pressure of the gas in the flask (in bars) if the barometer reads 738.5 Torr?

4.11 In Figure P4.11, what is the pressure of the gas in the flask (in kPa) if the barometer reads 765.2 mmHg?

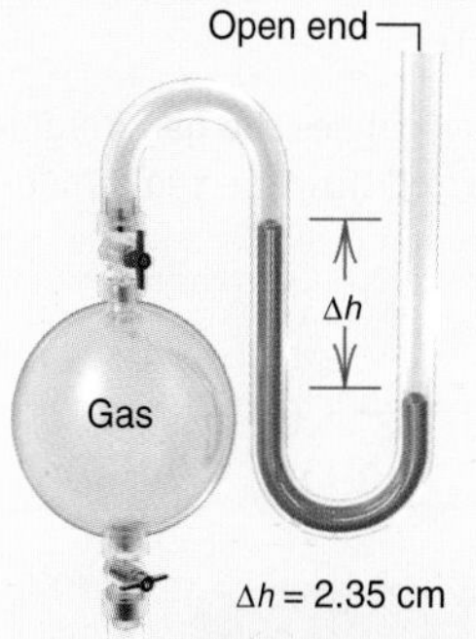

Figure P4.10

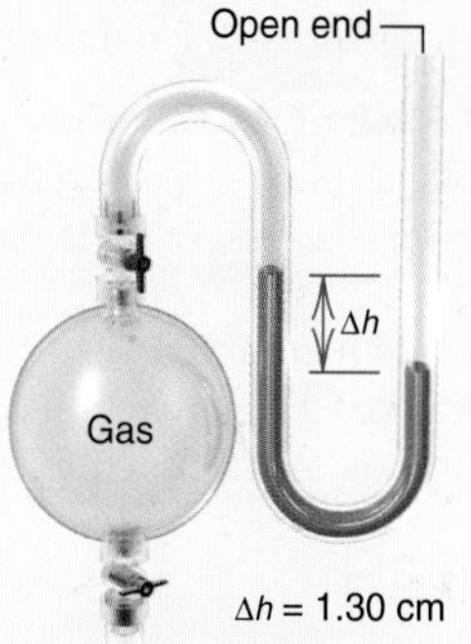

Figure P4.11

4.12 If the sample flask in Figure P4.12 is closed to the air, what is the atmospheric pressure (in bar)?

4.13 What is the pressure (in Pa) of the gas in the flask in Figure P4.13?

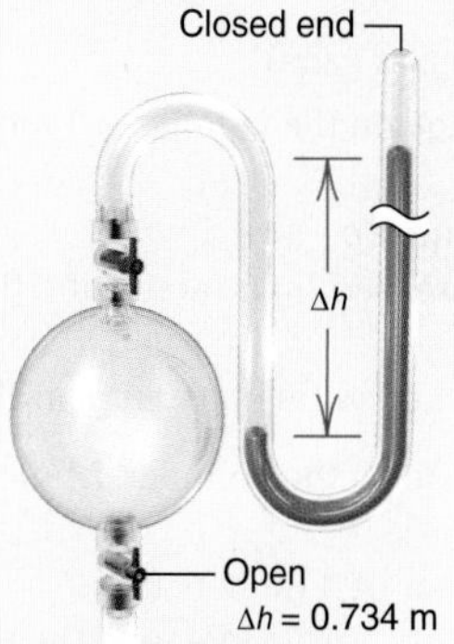

Figure P4.12

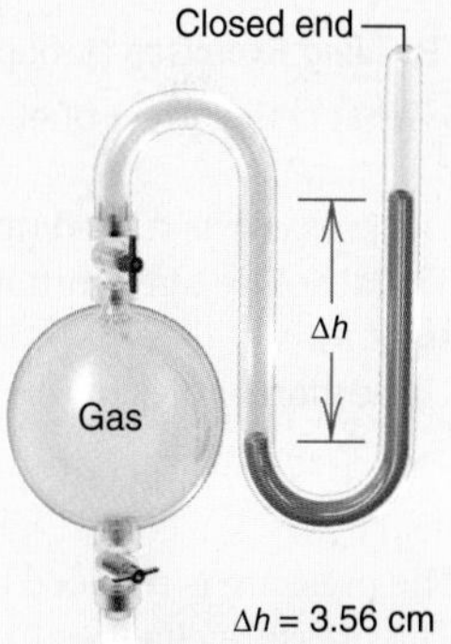

Figure P4.13

Problems in Context

4.14 Convert each of the pressures described below to Pa:
(a) At the peak of Mount Everest, atmospheric pressure is only 2.75×10^2 mmHg.
(b) The meteorologist at a local news station announces that the barometric pressure is 1.14 bar.
(c) The surface of Venus has an atmospheric pressure of 9.15×10^3 kPa.
(d) At 30.5 m below sea level, a scuba diver experiences a pressure of 2.54×10^4 Torr.

4.15 The gravitational force exerted by an object is given by $F = mg$, where F is the force in newton, m is the mass in kilogram, and g is the acceleration due to gravity (9.81 m/s^2).
(a) Use the definition of the pascal to calculate the mass (kg) of the atmosphere above 1 m^2 of ocean.
(b) Osmium ($Z = 76$) is a transition metal in group 8 and has the highest density of any element (22.6 g/mL). If an osmium column is 1 m^2 in area, what must its height be for its pressure to equal atmospheric pressure? [Use the answer from part (a) in your calculation.]

The Gas Laws and Their Experimental Foundations

(Sample Problems 4.2 to 4.6)

Concept Review Questions

4.16 A student states Boyle's law as follows: "The volume of a gas is inversely proportional to its pressure." How is this statement incomplete? Give a correct statement of Boyle's law.

4.17 In the following relationships, which quantities are variables and which are fixed: (a) Charles's law; (b) Avogadro's law; (c) Amontons's law?

4.18 Boyle's law relates gas volume to pressure, and Avogadro's law relates gas volume to amount (mol). State a relationship between gas pressure and amount (mol).

4.19 Each of the following processes causes the gas volume to double, as shown. For each process, tell how the remaining gas variable changes, or state that it remains fixed:
(a) *T* doubles at fixed *p*.
(b) *T* and *n* are fixed.
(c) At fixed *T*, the reaction is $CD_2(g) \longrightarrow C(g) + D_2(g)$.
(d) At fixed *p*, the reaction is $A_2(g) + B_2(g) \longrightarrow 2AB(g)$.

Skill-Building Exercises (grouped in similar pairs)

4.20 What is the effect of each change on the volume of 1 mol of an ideal gas?
(a) The pressure is tripled (at constant *T*).
(b) The absolute temperature is increased by a factor of 3.0 (at constant *p*).
(c) Three more moles of the gas are added (at constant *p* and *T*).

4.21 What is the effect of each change on the volume of 1 mol of an ideal gas?
(a) The pressure is reduced by a factor of 4 (at constant *T*).
(b) The pressure changes from 1.01 bar to 202 kPa, and the temperature changes from 37°C to 155 K.
(c) The temperature changes from 305 K to 32°C, and the pressure changes from 2 bar to 101 kPa.

4.22 What is the effect of each change on the volume of 1 mol of an ideal gas?
(a) Temperature decreases from 800 K to 400 K (at constant *p*).
(b) Temperature increases from 250°C to 500°C (at constant *p*).
(c) Pressure increases from 2 bar to 6 bar (at constant *T*).

4.23 What is the effect of each change on the volume of 1 mol of an ideal gas?
(a) Half the gas escapes (at constant *p* and *T*).
(b) The initial pressure is 72.2 kPa, and the final pressure is 0.950 bar; the initial temperature is 90.°C, and the final temperature is 273 K.
(c) Both the pressure and temperature decrease to one-fourth their initial values.

4.24 A sample of sulfur hexafluoride gas occupies 9.10 L at 198°C. Assuming that the pressure remains constant, what temperature (°C) is needed to reduce the volume to 2.50 L?

4.25 A 93 L sample of dry air cools from 145°C to −22°C while the pressure is maintained at 2.85 bar. What is the final volume?

4.26 A sample of Freon-12 (CF_2Cl_2) occupies 25.5 L at 298 K and 153.3 kPa. Find its volume at STP.

4.27 A sample of carbon monoxide occupies 3.65 L at 298 K and 74.5 kPa. Find its volume at −14°C and 36.7 kPa.

4.28 A sample of chlorine gas is confined in a 5.0 L container at 32.8 kPa and 37°C. What amount (mol) of gas is in the sample?

4.29 If 1.47×10^{-3} mol of argon occupies a 75.0 mL container at 26°C, what is the pressure (bar)?

4.30 You have 357 mL of chlorine trifluoride gas at 69.9 kPa and 45°C. What is the mass (g) of the sample?

4.31 A 75.0 g sample of dinitrogen monoxide is confined in a 3.1 L vessel. What is the pressure (bar) at 115°C?

Problems in Context

4.32 In preparation for a demonstration, a professor brings a 1.5 L bottle of sulfur dioxide into the lecture hall before class to allow the gas to reach room temperature. If the pressure gauge reads 5.9 bar and the temperature in the room is 23°C, what amount (mol) of sulfur dioxide is in the bottle? (*Hint*: The gauge reads zero when 1.013 bar of gas remains.)

4.33 A gas-filled weather balloon with a volume of 65.0 L is released at sea-level conditions of 0.993 bar and 25°C. The balloon can expand to a maximum volume of 835 L. When the balloon rises to an altitude at which the temperature is −5°C and the pressure is 6.9 kPa, will it reach its maximum volume?

Rearrangements of the Ideal Gas Law

(Sample Problems 5.7 to 5.12)

Concept Review Questions

4.34 Why is moist air less dense than dry air?

4.35 To collect a beaker of H_2 gas by displacing the air already in the beaker, would you hold the beaker upright or inverted? Why? How would you hold the beaker to collect CO_2?

4.36 Why can we use a gas mixture, such as air, to study the general behaviour of an ideal gas under ordinary conditions?

4.37 How does the partial pressure of gas A in a mixture compare with its mole fraction in the mixture? Explain.

4.38 The scene at the right represents a portion of a mixture of four gases: A (*purple*), B (*black*), C (*green*), and D_2 (*orange*).

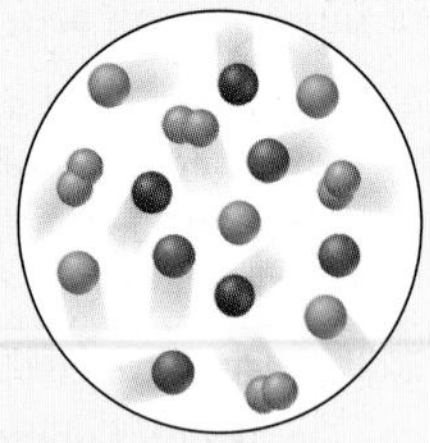

(a) Which gas has the highest partial pressure?
(b) Which gas has the lowest partial pressure?
(c) If the total pressure is 0.75 bar, what is the partial pressure of D_2?

Skill-Building Exercises (grouped in similar pairs)

4.39 What is the density of Xe gas at STP?

4.40 Find the density of Freon-11 ($CFCl_3$) at 120°C and 1.5 bar.

4.41 What amount (mol) of gaseous arsine (AsH_3) occupies 0.0400 L at STP? What is the density of gaseous arsine?

4.42 The density of a noble gas is 2.71 g/L at 3.00 bar and 0°C. Identify the gas.

4.43 Calculate the molar mass of a gas at 0.517 bar and 45°C if 206 ng occupies 0.206 mL.

4.44 When an evacuated 63.8 mL glass bulb is filled with a gas at 22°C and 9.96×10^4 Pa, the bulb gains 0.103 g in mass. Is the gas N_2, Ne, or Ar?

4.45 After 0.600 L of Ar at 1.22 bar and 227°C is mixed with 0.200 L of O_2 at 66.8 kPa and 127°C in a 400 mL flask at 27°C, what is the pressure in the flask?

4.46 A 355 mL container holds 0.146 g of Ne and an unknown amount of Ar at 35°C and a total pressure of 0.835 bar. Calculate the amount (mol) of Ar present.

4.47 What mass (g) of phosphorus reacts with 35.5 L of O_2 at STP to form tetraphosphorus decoxide?

$$P_4(s) + 5O_2(g) \longrightarrow P_4O_{10}(s)$$

4.48 What mass (g) of potassium chlorate decomposes to potassium chloride and 638 mL of O_2 at 128°C and 1.00 bar?

$$2KClO_3(s) \longrightarrow 2KCl(s) + 3O_2(g)$$

4.49 What mass (g) of phosphine (PH_3) can form when 37.5 g of phosphorus and 83.0 L of hydrogen gas react at STP?

$$P_4(s) + H_2(g) \longrightarrow PH_3(g)\text{(unbalanced)}$$

4.50 When 35.6 L of ammonia and 40.5 L of oxygen gas at STP burn, nitrogen monoxide and water form. After the products return to STP, what mass (g) of nitrogen monoxide is present?

$$NH_3(g) + O_2(g) \longrightarrow NO(g) + H_2O(l)\ \text{(unbalanced)}$$

4.51 Aluminum reacts with excess hydrochloric acid to form aqueous aluminum chloride and 35.8 mL of hydrogen gas over water at 27°C and 1.00 bar. What mass (g) of aluminum reacted?

4.52 What volume (L) of hydrogen gas is collected over water at 18°C and 1.00 bar when 0.84 g of lithium reacts with water? Aqueous lithium hydroxide also forms.

Problems in Context

4.53 The air in a hot-air balloon at 9.92×10^4 Pa is heated from 17°C to 60.0°C. Assuming that the amount (mol) of air and the pressure remain constant, what is the density of the air at each temperature? (The average molar mass of air is 28.8 g/mol.)

4.54 On a certain winter day in Alberta, the average atmospheric pressure is 86.7 kPa. What is the molar density (mol/L) of the air if the temperature is −25°C?

4.55 A sample of a liquid hydrocarbon known to consist of molecules with five carbon atoms is vaporized in a 0.204 L flask by immersion in a water bath at 101°C. The barometric pressure is 1.02 bar, and the remaining gas weighs 0.482 g. What is the molecular formula of the hydrocarbon?

4.56 A sample of air contains 78.08% nitrogen, 20.94% oxygen, 0.05% carbon dioxide, and 0.93% argon, by volume. How many molecules of each gas are present in 1.00 L of the sample at 25°C and 1.01 bar?

4.57 An environmental chemist is sampling industrial exhaust gases from a coal-burning plant. The chemist collects a CO_2-SO_2-H_2O mixture in a 21 L steel tank until the pressure reaches 1.13 bar at 45°C.

(a) What amount (mol) of gas is collected?

(b) If the SO_2 concentration in the mixture is 7.95×10^3 parts per million by volume (ppmv), what is its partial pressure? [*Hint*: ppmv = (volume of component/volume of mixture) $\times 10^6$.]

4.58 "Strike anywhere" matches contain the compound tetraphosphorus trisulfide, which burns to form tetraphosphorus decoxide and sulfur dioxide gas. What volume (mL) of sulfur dioxide, measured at 96.7 kPa and 32°C, can be produced from burning 0.800 g of tetraphosphorus trisulfide?

4.59 Freon-12 (CF_2Cl_2), widely used as a refrigerant and aerosol propellant, is a dangerous air pollutant. In the troposphere, it traps heat 25 times as effectively as CO_2. In the stratosphere, it participates in the breakdown of ozone. Freon-12 is prepared industrially by the reaction of gaseous carbon tetrachloride with hydrogen fluoride. Hydrogen chloride gas also forms. What mass (g) of carbon tetrachloride is required for the production of 16.0 dm^3 of Freon-12 at 27°C and 1.22 bar?

4.60 Xenon hexafluoride was one of the first noble gas compounds to be synthesized. The solid reacts rapidly with the silicon dioxide in glass or quartz containers to form liquid $XeOF_4$ and gaseous silicon tetrafluoride. What is the pressure in a 1.00 L container at 25°C after 2.00 g of xenon hexafluoride reacts? (Assume that silicon tetrafluoride is the only gas present and that it occupies the entire volume.)

4.61 In the four cylinder-piston assemblies below, the reactant in the left assembly is about to undergo a reaction at constant T and P:

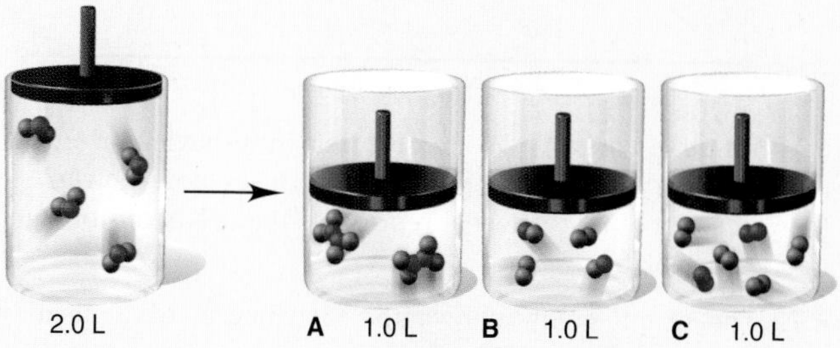

Which of the other three assemblies best represents the products of the reaction?

4.62 Roasting galena [lead(II) sulfide] is a step in the industrial isolation of lead. What volume of sulfur dioxide, measured at STP, is produced by the reaction of 3.75 kg of galena with 228 L of oxygen gas at 220°C and 202.7 kPa? Lead(II) oxide also forms.

4.63 In one of his most critical studies into the nature of combustion, Lavoisier heated mercury(II) oxide and isolated elemental mercury and oxygen gas. If 40.0 g of mercury(II) oxide is heated in a 502 mL vessel and 20.0% (by mass) decomposes, what is the pressure (in Pa) of the oxygen that forms at 25.0°C? (Assume that the gas occupies the entire volume.)

The Kinetic-Molecular Theory: A Model for Gas Behaviour

(Sample Problem 4.13)

Concept Review Questions

4.64 Use the kinetic-molecular theory to explain the change in gas pressure that results from warming a sample of gas.

4.65 How does the kinetic-molecular theory explain why 1 mol of krypton and 1 mol of helium have the same volume at STP?

4.66 Is the rate of effusion of a gas higher than, lower than, or equal to its rate of diffusion? Explain. For two gases with molecules of approximately the same size, is the ratio of their effusion rates higher than, lower than, or equal to the ratio of their diffusion rates? Explain.

4.67 Consider two 1 L samples of gas: one is H_2 and the other is O_2. Both are at 1 bar and 25°C. Compare the samples in terms of (a) mass; (b) density; (c) mean free path; (d) average molecular kinetic energy; (e) average molecular speed; (f) time for a given fraction of molecules to effuse.

4.68 Three 5 L flasks, fixed with pressure gauges and small valves, each contain 4 g of gas at 273 K. Flask A contains H_2, flask B contains He, and flask C contains CH_4. Rank the contents of the flasks in terms of (a) pressure; (b) average molecular kinetic energy; (c) diffusion rate after the valve is opened; (d) total kinetic energy of the molecules; (e) density; (f) collision frequency.

Skill-Building Exercises (grouped in similar pairs)

4.69 What is the ratio of effusion rates for the lightest gas, H_2, and the heaviest known gas, UF_6?

4.70 What is the ratio of effusion rates for O_2 and Kr?

4.71 The graph below shows the distribution of molecular speeds for argon and helium at the same temperature.

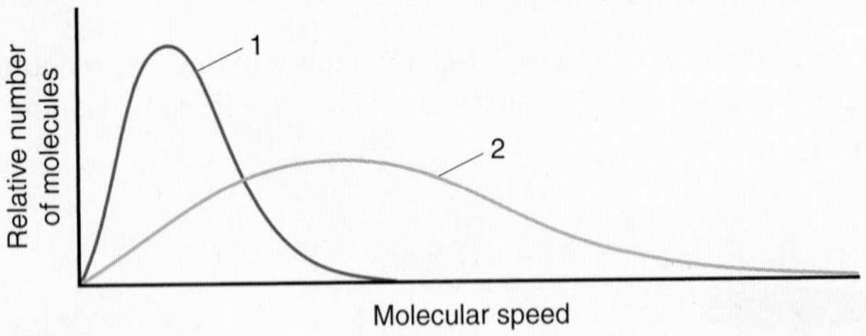

(a) Does curve 1 or curve 2 better represent the behaviour of argon?
(b) Which curve represents the gas that effuses more slowly?
(c) Which curve more closely represents the behaviour of fluorine gas? Explain.

4.72 The graph below shows the distribution of molecular speeds for a gas at two different temperatures.

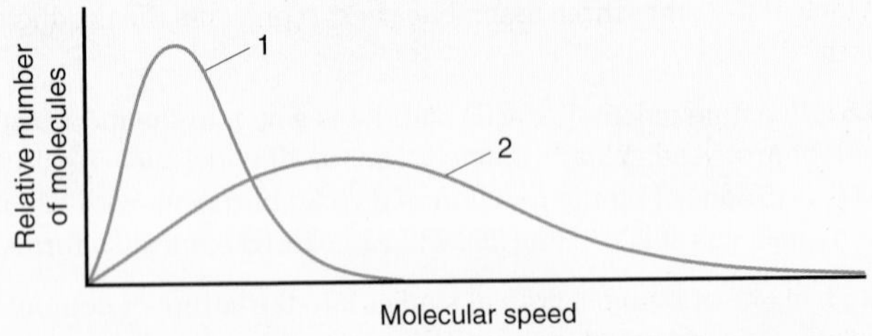

(a) Does curve 1 or curve 2 better represent the behaviour of the gas at the lower temperature?
(b) Which curve represents the gas when it has a higher $\overline{E_k}$?
(c) Which curve is consistent with a higher diffusion rate?

4.73 At a given pressure and temperature, it takes 4.85 min for a 1.5 L sample of He to effuse through a membrane. How long does it take for 1.5 L of F_2 to effuse under the same conditions?

4.74 A sample of an unknown gas effuses in 11.1 min. An equal volume of H_2 in the same apparatus, under the same conditions, effuses in 2.42 min. What is the molar mass of the unknown gas?

Problems in Context

4.75 White phosphorus melts and then vaporizes at a high temperature. The gas effuses at a rate that is 0.404 times the rate of neon in the same apparatus under the same conditions. How many atoms are in a molecule of gaseous white phosphorus?

4.76 Helium (He) is the lightest noble gas component of air, and xenon (Xe) is the heaviest. [For this problem, use $R = 8.314$ J/(mol • K) and $\mathcal{M}$ in kg/mol.]
(a) Find the rms speed of He in winter (0.°C) and in summer (30.°C).
(b) Compare the rms speed of He with that of Xe at 30.°C.
(c) Find the average kinetic energy per mole of He and of Xe at 30.°C.
(d) Find the average kinetic energy per molecule of He at 30.°C.

4.77 A mixture of gaseous disulfur difluoride, dinitrogen tetrafluoride, and sulfur tetrafluoride is placed in an effusion apparatus.
(a) Rank the gases in order of increasing effusion rate.
(b) Find the ratio of effusion rates of disulfur difluoride and dinitrogen tetrafluoride.
(c) If gas X is added, and it effuses at 0.935 times the rate of sulfur tetrafluoride, find the molar mass of X.

Non-ideal Gases: Deviations from Ideal Behaviour

Skill-Building Exercises (grouped in similar pairs)

4.78 Do interparticle attractions cause negative or positive deviations from the $\frac{pV}{RT}$ ratio of an ideal gas? Use Table 4.3 to rank Kr, CO_2, and N_2 in order of increasing magnitude of these deviations.

4.79 Does particle volume cause negative or positive deviations from the $\frac{pV}{RT}$ ratio of an ideal gas? Use Table 4.3 to rank Cl_2, H_2, and O_2 in order of increasing magnitude of these deviations.

4.80 Does N_2 behave more ideally at 1 bar or at 500 bar? Explain.

4.81 Does SF_6 (boiling point = 16°C at 1 bar) behave more ideally at 150°C or at 20°C? Explain.

Comprehensive Problems

4.82 An "empty" gasoline can, with dimensions 15.0 cm by 40.0 cm by 12.5 cm, is attached to a vacuum pump and evacuated. If the atmospheric pressure is 1.013 bar, what is the total force (in bar) on the outside of the can?

4.83 Hemoglobin is the protein that transports O_2 through the blood from the lungs to the rest of the body. In doing so, each molecule of hemoglobin combines with four molecules of O_2. If 1.00 g of hemoglobin combines with 1.53 mL of O_2 at 37°C and 97.8 kPa, what is the molar mass of hemoglobin?

4.84 A baker uses sodium hydrogen carbonate (baking soda) as the leavening agent in a banana-nut quick-bread. The baking soda decomposes in either of two possible reactions:
(1) $2NaHCO_3(s) \longrightarrow Na_2CO_3(s) + H_2O(l) + CO_2(g)$
(2) $NaHCO_3(s) + H^+(aq) \longrightarrow H_2O(l) + CO_2(g) + Na^+(aq)$
Calculate the volume (in mL) of CO_2 that forms at 200.°C and 0.988 bar per gram of $NaHCO_3$ in each reaction.

4.85 A weather balloon containing 600. L of He is released near the equator, where the atmospheric conditions are 1.02 bar and 305 K. It rises to a point where the conditions are 0.495 bar and 218 K, and it eventually lands in the northern hemisphere under the conditions 1.02 bar and 250 K. If one-fourth of the helium leaked out during this journey, what is the volume (L) of the balloon when it lands?

4.86 Chlorine is produced from sodium chloride by the electrochemical chlor-alkali process. During the process, the chlorine is collected in a container that is isolated from the other products to prevent unwanted (and explosive) reactions. If a 15.50 L container holds 0.5950 kg of Cl_2 gas at 225°C, calculate (a) p_{IGL}; (b) p_{VDW}.

4.87 In a certain experiment, magnesium boride (Mg_3B_2) reacted with acid to form a mixture of four boron hydrides (B_xH_y), three as liquids (labelled I, II, and III) and one as a gas (labelled IV).
(a) When a 0.1000 g sample of each liquid was transferred to an evacuated 750.0 mL container and volatilized at 70.00°C, sample I had a pressure of 6.030×10^3 Pa; sample II, 7.138×10^3 Pa; and sample III, 5.843×10^3 Pa. What is the molar mass of each liquid?
(b) Boron is 85.63% by mass in sample I, 81.10% in sample II, and 82.98% in sample III. What is the molecular formula for each sample?

(c) Sample IV was found to be 78.14% boron. Its rate of effusion was compared with that of sulfur dioxide; under identical conditions, 350.0 mL of sample IV effused in 12.00 min, and 250.0 mL of sulfur dioxide effused in 13.04 min. What is the molecular formula for sample IV?

4.88 Three equal volumes of gas mixtures, all at the same temperature, are depicted below (with gas A *red*, gas B *green*, and gas C *blue*):

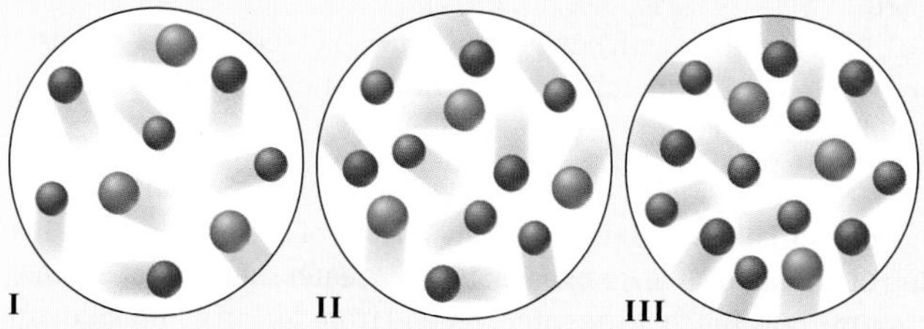

(a) Which sample, if any, has the highest partial pressure of A?
(b) Which sample, if any, has the lowest partial pressure of B?
(c) In which sample, if any, do the gas particles have the highest average kinetic energy?

4.89 Will the volume of a gas increase, decrease, or remain unchanged for each set of changes?
(a) The pressure is decreased from 2 bar to 1 bar, while the temperature is decreased from 200°C to 100°C.
(b) The pressure is increased from 1 bar to 3 bar, while the temperature is increased from 100°C to 300°C.
(c) The pressure is increased from 3 bar to 6 bar, while the temperature is increased from −73°C to 127°C.
(d) The pressure is increased from 0.2 bar to 0.4 bar, while the temperature is decreased from 300°C to 150°C.

4.90 When air is inhaled, it enters the alveoli of the lungs, and varying amounts of the component gases exchange with dissolved gases in the blood. The resulting alveolar gas mixture is quite different from the atmospheric mixture. The following table presents selected data on the composition and partial pressure of four gases in the atmosphere and in the alveoli:

	Atmosphere (Sea Level)		Alveoli	
Gas	**Mole %**	**Partial Pressure (kPa)**	**Mole %**	**Partial Pressure (kPa)**
N_2	78.6	—	—	74.9
O_2	20.9	—	—	13.7
CO_2	0.04	—	—	5.3
H_2O	0.46	—	—	6.2

If the total pressure of each gas mixture is 1.01 bar, calculate the following:
(a) The partial pressure (in kPa) of each gas in the atmosphere
(b) The mole percent of each gas in the alveoli
(c) The number of O_2 molecules in 0.50 L of alveolar air (volume of an average breath of a person at rest) at 37°C

4.91 Radon (Rn) is the heaviest, and only radioactive, member of group 18 (noble gases). It is a product of the disintegration of heavier radioactive nuclei found in minute concentrations in many common rocks used for building and construction. In recent years, health concerns about the cancers caused from inhaled residential radon have grown. If 1.0×10^{15} atoms of radium (Ra) produce an average of 1.373×10^4 atoms of Rn per second, what volume (L) of Rn, measured at STP, is produced per day by 1.0 g of Ra?

4.92 At 1.933 bar and 286 K, a skin diver exhales a 208 mL bubble of air that is 77% N_2, 17% O_2, and 6.0% CO_2 by volume.
(a) What would the volume (mL) of the bubble have been if it had been exhaled at the surface, at 1 bar and 298 K?
(b) What amount (mol) of N_2 is in the bubble?

4.93 Nitrogen dioxide is used industrially to produce nitric acid, but it contributes to acid rain and photochemical smog. What volume (L) of nitrogen dioxide is formed at 0.980 bar and 28.2°C by reacting 4.95 cm^3 of copper (d = 8.95 g/cm^3) with 230.0 mL of nitric acid (d = 1.42 g/cm^3, 68.0% HNO_3 by mass)?

$$Cu(s) + 4HNO_3(aq) \longrightarrow Cu(NO_3)_2(aq) + 2NO_2(g) + 2H_2O(l)$$

4.94 In the average adult male, the residual volume (RV) of the lungs, which is the volume of air remaining after a forced exhalation, is 1200 mL.
(a) What amount (mol) of air is present in the RV at 1.0 bar and 37°C?
(b) How many molecules of gas are present under these conditions?

4.95 In a bromine-producing plant, what volume (L) of gaseous elemental bromine at 300°C and 86.6 kPa is formed by the reaction of 275 g of sodium bromide and 175.6 g of sodium bromate in aqueous acid solution? (Assume that no Br_2 dissolves.)

$$5NaBr(aq) + NaBrO_3(aq) + 3H_2SO_4(aq) \longrightarrow 3Br_2(g) + 3Na_2SO_4(aq) + 3H_2O(g)$$

4.96 In a collision of sufficient force, automobile air bags respond by electrically triggering the explosive decomposition of sodium azide (NaN_3) to its elements. A 50.0 g sample of sodium azide was decomposed, and the nitrogen gas generated was collected over water at 26°C. The total pressure was 9.94×10^4 Pa. What volume (L) of dry N_2 was generated?

4.97 An anaesthetic gas contains 64.81% carbon, 13.60% hydrogen, and 21.59% oxygen by mass. If 2.00 L of the gas at 25°C and 0.426 bar had a mass of 2.57 g, what is the molecular formula for the anaesthetic?

4.98 Aluminum chloride is easily vaporized above 180°C. The gas escapes through a pinhole 0.122 times as fast as helium at the same conditions of temperature and pressure in the same apparatus. What is the molecular formula for aluminum chloride gas?

4.99 (a) What is the total volume (L) of gaseous *products*, measured at 350°C and 0.980 bar, when an automobile engine burns 100. g of C_8H_{18} (a typical component of gasoline)?
(b) For part (a), the source of O_2 is air, which is 78% N_2, 21% O_2, and 1.0% Ar by volume. Assuming that all the O_2 reacts, but no N_2 or Ar does, what is the total volume (L) of gaseous *exhaust*?

4.100 An atmospheric chemist studying the pollutant SO_2 places a mixture of SO_2 and O_2 in a 2.00 L container at 800. K and 192.5 kPa. When the reaction occurs, gaseous SO_3 forms, and the pressure falls to 167.2 kPa. What amount (mol) of SO_3 forms?

4.101 The thermal decomposition of ethene (ethylene) occurs during the compound's transit in pipelines and during the formation of polyethene (polyethylene). The decomposition reaction is

$$CH_2 = CH_2(g) \longrightarrow CH_4(g) + C(s, \text{graphite})$$

Suppose that the decomposition begins at 10°C and 50.7 bar, with a gas density of 0.215 g/mL, and the temperature increases by 950 K.
(a) What is the final pressure of the confined gas? (Ignore the volume of graphite, and use the van der Waals equation.)
(b) How does the $\frac{pV}{RT}$ value of CH_4 compare with that in Figure 4.22? Explain.

4.102 Ammonium nitrate, a common fertilizer, was used by terrorists to cause the tragic explosion in Oklahoma City in 1995 and is a strictly regulated product in Canada. What volume (L) of gas at 307°C and 1.01 bar is formed by the explosive decomposition of 15.0 kg of ammonium nitrate to nitrogen, oxygen, and water vapour?

4.103 An environmental engineer analyzes a sample of air contaminated with sulfur dioxide. To a 500. mL sample at 0.933 bar and 38°C, she adds 20.00 mL of 0.010 17 mol/L aqueous iodine, which reacts as follows:

$$SO_2(g) + I_2(aq) + H_2O(l) \longrightarrow HSO_4^-(aq) + I^-(aq) + H^+(aq) \text{ (unbalanced)}$$

Excess I_2 reacts with 11.37 mL of 0.0105 mol/L sodium thiosulfate:

$$I_2(aq) + S_2O_3^{2-}(aq) \rightarrow I^-(aq) + S_4O_6^{2-}(aq) \text{ (unbalanced)}$$

What is the volume percent of SO_2 in the air sample?

4.104 Canadian chemists have developed a modern variation of the 1899 Mond process for preparing extremely pure metallic nickel. A sample of impure nickel reacts with carbon monoxide at 50°C to form gaseous nickel carbonyl, $Ni(CO)_4$.
(a) What mass (g) of nickel can be converted to the carbonyl with 3.55 m^3 of CO at 100.7 kPa?
(b) The carbonyl is then decomposed at 21 bar and 155°C to pure (>99.95%) nickel. What mass (g) of nickel is obtained per cubic metre of the carbonyl?
(c) The released carbon monoxide is cooled and collected for reuse by passing it through water at 35°C. If the barometric pressure is 1.03 bar, what volume (m^3) of CO is formed per cubic metre of carbonyl?

4.105 Analysis of a newly discovered gaseous silicon-fluorine compound shows that it contains 33.01 mass % silicon. At 27°C, 2.60 g of the compound exerts a pressure of 1.52 bar in a 0.250 L vessel. What is the molecular formula for the compound?

4.106 A gaseous organic compound containing only carbon, hydrogen, and nitrogen is burned in oxygen gas, and the volume of each reactant and product is measured under the same conditions of temperature and pressure. Reaction of 4 volumes of the compound produces 4 volumes of CO_2, 2 volumes of N_2, and 10 volumes of water vapour.
(a) How many volumes of O_2 were required?
(b) What is the empirical formula for the compound?

4.107 Containers A, B, and C are attached by closed stopcocks of negligible volume. Each particle shown in the picture represents 10^6 particles.

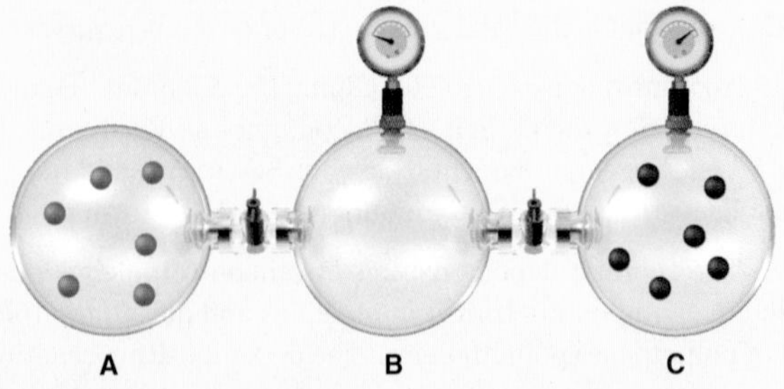

(a) How many blue particles and black particles are in B after the stopcocks are opened and the system reaches equilibrium?
(b) How many blue particles and black particles are in A after the stopcocks are opened and the system reaches equilibrium?
(c) If the pressure in C, p_C, is 1.00 bar before the stopcocks are opened, what is p_C afterward?
(d) What is p_B afterward?

4.108 Combustible vapour-air mixtures are flammable over a limited range of concentrations. The minimum volume percent of vapour that gives a combustible mixture is called the *lower flammable limit* (LFL). Generally, the LFL is about half the stoichiometric mixture, the concentration required for complete combustion of the vapour in air.
(a) If oxygen is 20.9 vol % of air, estimate the LFL for hexane, C_6H_{14}.
(b) What volume (mL) of hexane (d = 0.660 g/cm^3) is required to produce a flammable mixture of hexane in 1.000 m^3 of air at STP?

4.109 By what factor would a scuba diver's lungs expand if he ascended rapidly to the surface from a depth of 38.1 m without inhaling or exhaling? If an expansion factor greater than 1.5 causes lung rupture, how far could he safely ascend from 38.1 m without breathing? Assume constant temperature. (d of seawater = 1.04 g/mL; d of Hg = 13.5 g/mL.)

4.110 When 15.0 g of fluorite (CaF_2) reacts with excess sulfuric acid, hydrogen fluoride gas is collected at 97.9 kPa and 25.5°C. Solid calcium sulfate is the other product. What gas temperature is required to store the gas in an 8.63 L container at 1.17 bar?

4.111 Dilute aqueous hydrogen peroxide is used as a bleaching agent and for disinfecting surfaces and small cuts. Its concentration is sometimes given as a certain number of "volumes hydrogen peroxide," which refers to the number of volumes of O_2 gas, measured at STP, that a given volume of hydrogen peroxide solution will release when it decomposes to O_2 and liquid H_2O. What mass (g) of hydrogen peroxide is in 0.100 L of "20 volumes hydrogen peroxide" solution?

4.112 At a height of 300 km above Earth's surface, an astronaut finds that the atmospheric pressure is about 10^{-6} Pa and the temperature is 500 K. How many molecules of gas are present, per millilitre, at this altitude?

4.113 (a) What is the rms speed of O_2 at STP?
(b) If the mean free path of O_2 molecules at STP is 6.33×10^{-8} m, what is their collision frequency? [Use R = 8.314 J/(mol • K) and $\mathcal{M}$ in kg/mol.]

4.114 Prop-2-enoic acid ($CH_2{=}CHCOOH$) is used to prepare polymers, adhesives, and paints. The first step in making acrylic acid involves the vapour-phase oxidation of propene ($CH_2{=}CHCH_3$) to prop-2-enal ($CH_2{=}CHCHO$). This step is carried out at 330°C and 2.5 bar in a large bundle of tubes around which circulates a heat-transfer agent. The reactants spend an average of 1.8 s in the tubes, which have a void space of 2.8 m^3. What mass (kg) of propene must be added per hour in a mixture whose mole fractions are 0.07 propene, 0.35 steam, and 0.58 air?

4.115 Standard conditions are based on relevant environmental conditions. If normal average surface temperature and pressure on Venus are 730. K and 91 bar, respectively, what is the standard molar volume of an ideal gas on Venus?

4.116 A barometer tube is 1.00×10^2 cm long and has a cross-sectional area of 1.20 cm^2. The height of the mercury column is 74.0 cm, and the temperature is 24°C. A small amount of N_2 is introduced into the evacuated space above the mercury, which causes the mercury level to drop to a height of 64.0 cm. What mass (g) of N_2 was introduced?

4.117 What is the concentration (mol/L) of the cleaning solution formed when 10.0 L of ammonia gas at 33°C and 96.7 kPa dissolves in enough water to give a final volume of 0.750 L?

4.118 The Hawaiian volcano Kilauea emits an average of 1.5×10^3 m^3 of gas each day, when corrected to 298 K and 1.01 bar. The mixture contains gases that contribute to global warming and acid rain, and some are toxic. An atmospheric chemist analyzes a sample and finds the following mole fractions: 0.4896 CO_2, 0.0146 CO, 0.3710 H_2O, 0.1185 SO_2, 0.0003 S_2, 0.0047 H_2, 0.0008 HCl, and 0.0003 H_2S. What volume of each gas is emitted per year?

4.119 To study a key fuel-cell reaction, a chemical engineer has 20.0 L tanks of H_2 and O_2 and wants to use up both tanks to form 28.0 mol of water at 23.8°C.
(a) Use the ideal gas law to find the pressure needed in each tank.
(b) Use the van der Waals equation to find the pressure needed in each tank.
(c) Compare the results from the two equations.

4.120 For each pair, which gas shows the greater deviation from ideal behaviour at the same set of conditions? Explain.
(a) Argon or xenon (b) Water vapour or neon
(c) Mercury vapour or radon (d) Water vapour or methane

4.121 What volume (L) of gaseous hydrogen bromide at 29°C and 0.978 bar will a chemist need if she wishes to prepare 3.50 L of 1.20 mol/L hydrobromic acid?

4.122 A mixture consisting of 7.0 g of CO and 10.0 g of SO_2, two atmospheric pollutants, has a pressure of 0.33 bar when placed in a sealed container. What is the partial pressure of CO?

4.123 Sulfur dioxide is used to make sulfuric acid. One method of producing sulfuric acid is by roasting mineral sulfides:

$$FeS_2(s) + O_2(g) \longrightarrow SO_2(g) + Fe_2O_3(s) \text{ (unbalanced)}$$

A production error leads to the sulfide being placed in a 950 L vessel with insufficient oxygen. Initially, the partial pressure of O_2 is 0.65 bar, and the total pressure is 1.06 bar, with the balance due to N_2. The reaction is run until 85% of the O_2 is consumed, and the vessel is then cooled to its initial temperature. What are the total pressure and the partial pressure of each gas in the vessel?

4.124 A mixture of CO_2 and Kr weighs 35.0 g and exerts a pressure of 0.717 bar in its container. Since Kr is expensive, you wish to recover it from the mixture. After the CO_2 is completely removed by absorption with NaOH(s), the pressure in the container is 0.256 bar.
(a) What mass (g) of CO_2 was originally present?
(b) What mass of Kr can be recovered?

4.125 When a car accelerates quickly, the passengers feel a force that presses them back into their seats, but a balloon filled with helium floats forward. Why?

4.126 Gases such as CO are gradually oxidized in the atmosphere, not by O_2 but by the hydroxyl radical, OH, a species with one less electron than a hydroxide ion. At night, the OH concentration is nearly zero, but it increases to 2.5×10^{12} molecules/m^3 in polluted air during the day. At daytime conditions of 1.01 bar and 22°C, what is the partial pressure and mole percent of OH in air?

4.127 Aqueous sulfurous acid (H_2SO_3) was made by dissolving 0.200 L of sulfur dioxide gas at 19°C and 99.3 kPa in water to yield 500.0 mL of solution. The acid solution required 10.0 mL of sodium hydroxide solution to reach the titration end point. What was the concentration (mol/L) of the sodium hydroxide solution?

4.128 In the 19th century, J. B. A. Dumas devised a method for finding the molar mass of a volatile liquid from the volume, temperature, pressure, and mass of its vapour. He placed a sample of such a liquid in a flask that was closed with a stopper fitted with a narrow tube, immersed the flask in a hot water bath to vaporize the liquid, and then cooled the flask. Find the molar mass of a volatile liquid, given the following data:
Mass of empty flask = 65.347 g
Mass of flask filled with water at 25°C = 327.4 g
Density of water at 25°C = 0.997 g/mL
Mass of flask plus condensed unknown liquid = 65.739 g
Barometric pressure = 101.2 kPa
Temperature of water bath = 99.8°C

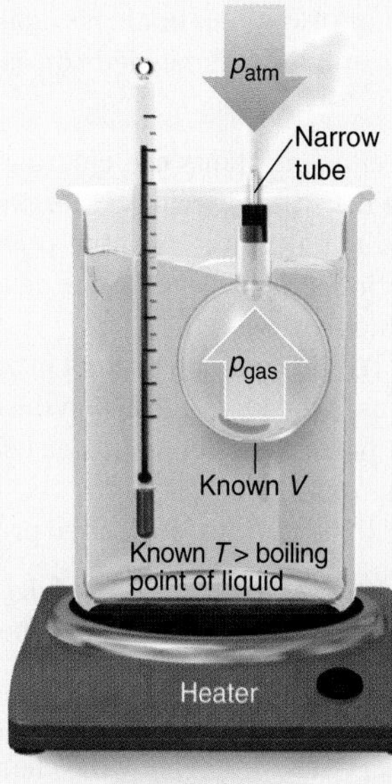

4.129 During World War II, a portable source of hydrogen gas was needed for weather balloons, and solid metal hydrides were the most convenient form. Many metal hydrides react with water to generate the metal hydroxide and hydrogen. Two candidates were lithium hydride and magnesium hydride. What volume (L) of gas is formed from 454 g of each hydride reacting with excess water at 1.00×10^5 Pa and 27°C?

4.130 The lunar surface reaches 370 K at midday. The atmosphere consists of neon, argon, and helium at a total pressure of only 2×10^{-9} Pa. Calculate the rms speed of each component in the lunar atmosphere. [Use $R = 8.314$ J/(mol • K) and $\mathcal{M}$ in kg/mol.]

4.131 A person inhales air richer in O_2 and exhales air richer in CO_2 and water vapour. During each hour of sleep, a person exhales a total of about 300 L of this CO_2-enriched and H_2O-enriched air.
(a) If the partial pressures of CO_2 and H_2O in exhaled air are both 4.00×10^3 Pa at 37.0°C, calculate the mass (g) of CO_2 and the mass of H_2O exhaled in 1 h of sleep.
(b) What amount (g) of body mass does the person lose in an 8 h sleep if all the CO_2 and H_2O exhaled comes from the metabolism of glucose?

$$C_6H_{12}O_6(s) + 6O_2(g) \longrightarrow 6CO_2(g) + 6H_2O(g)$$

4.132 Popcorn pops because the horny endosperm, a tough, elastic material, resists gas pressure within the heated kernel until it reaches explosive force. A 0.25 mL kernel has a water content of 1.6% by mass, and the water vapour reaches 170°C and 9.1 bar before the kernel ruptures. Assume that water vapour can occupy 75% of the kernel's volume.
(a) What is the mass (g) of the kernel?
(b) What volume (mL) would this amount of water vapour occupy at 25°C and 1.01 bar?

4.133 Sulfur dioxide emissions from coal-burning power plants are removed by *flue-gas desulfurization*. The flue gas passes through a scrubber, and a slurry of wet calcium carbonate reacts with it to form carbon dioxide and calcium sulfite. The calcium sulfite then reacts with oxygen to form calcium sulfate, which is sold as gypsum.
(a) If the sulfur dioxide concentration is 1000 times its mole fraction in clean dry air (2×10^{-10}), how much calcium sulfate (kg) can be made from scrubbing 4 GL of flue gas at 25°C (1 GL = 1×10^9 L)? A state-of-the-art scrubber removes at least 95% of the sulfur dioxide.
(b) If the mole fraction of oxygen in air is 0.209, what volume (L) of air at 1.01 bar and 25°C is needed to react with all the calcium sulfite?

4.134 Many water treatment plants use chlorine gas to kill micro-organisms before the water is released for residential use. A plant engineer has to maintain the chlorine pressure in a tank below the 86.1-bar rating and, to be safe, decides to fill the tank to 80.0% of this maximum pressure.
(a) What amount (mol) of Cl_2 gas can be kept in an 850. L tank at 298 K if the engineer uses the ideal gas law in the calculation?
(b) What is the tank pressure if the engineer uses the van der Waals equation for this amount of gas?
(c) Did the engineer fill the tank to the desired pressure?

4.135 At 10.0°C and 102.5 kPa, the density of dry air is 1.26 g/L. What is the average "molar mass" of dry air at these conditions?

4.136 Cylinder A contains 0.1 mol of a gas that behaves ideally. Choose the cylinder (B, C, or D) that correctly represents the volume of the gas after each of the following changes. If none of the cylinders is correct, specify "none."
(a) p is doubled at fixed n and T.
(b) T is reduced from 400 K to 200 K at fixed n and p.
(c) T is increased from 100°C to 200°C at fixed n and p.
(d) 0.1 mol of gas is added at fixed p and T.
(e) 0.1 mol of gas is added, and p is doubled at fixed T.

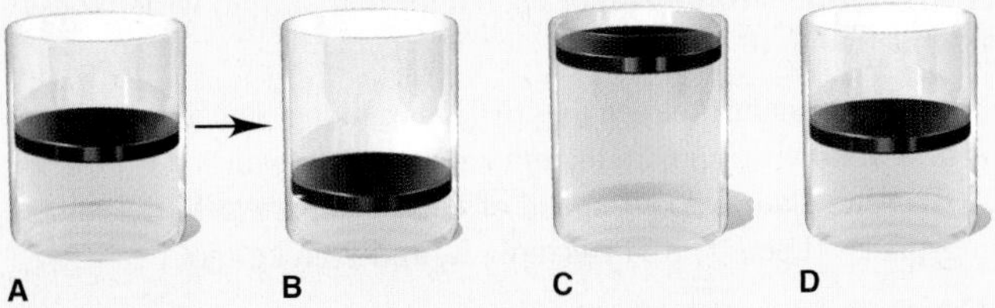

4.137 Ammonia is essential to so many industries that, on a molar basis, it is the most heavily produced substance in the world. Calculate p_{IGL} and p_{VDW} (bar) of 51.1 g of ammonia in a 3.000 L container at 0°C and 400.°C, the industrial temperature. (See Table 4.4 for the values of the van der Waals constants.)

4.138 A 6.0 L flask contains a mixture of methane (CH_4), argon, and helium at 45°C and 1.77 bar. If the mole fractions of helium and argon are 0.25 and 0.35, respectively, how many molecules of methane are present?

4.139 A large portion of metabolic energy arises from the biological combustion of glucose:

$$C_6H_{12}O_6(s) + 6O_2(g) \longrightarrow 6CO_2(g) + 6H_2O(g)$$

(a) If this reaction is carried out in an expandable container at 37°C and 104. kPa, what volume of CO_2 is produced from 20.0 g of glucose and excess O_2?
(b) If the reaction is carried out at the same conditions with the stoichiometric amount of O_2, what is the partial pressure of each gas when the reaction is 50% complete (10.0 g of glucose remains)?

4.140 What is the average kinetic energy and rms speed of N_2 molecules at STP? Compare these values with the values for H_2 molecules at STP. [Use $R = 8.314$ J/(mol • K) and $\mathcal{M}$ in kg/mol.]

4.141 According to government standards, the *8 h threshold limit value* is 5000 ppmv for CO_2 and 0.1 ppmv for Br_2 (1 ppmv is 1 part by volume in 10^6 parts by volume). Exposure to either gas for 8 h above these limits is unsafe. At STP, which of the following would be unsafe for 8 h of exposure?
(a) Air with a partial pressure of 30 Pa of Br_2
(b) Air with a partial pressure of 30 Pa of CO_2
(c) 1000 L of air containing 0.0004 g of Br_2 gas
(d) 1000 L of air containing 2.8×10^{22} molecules of CO_2

4.142 One way to prevent emission of the pollutant NO from industrial plants is by a catalyzed reaction with NH_3:

$$4NH_3(g) + 4NO(g) + O_2(g) \xrightarrow{\text{catalyst}} 4N_2(g) + 6H_2O(g)$$

(a) If NO has a partial pressure of 4.6 Pa in the flue gas, what volume (L) of NH_3 is needed per litre of flue gas at 1.01 bar?
(b) If the reaction takes place at 1.01 bar and 365°C, what mass (g) of NH_3 is needed per kilolitre of flue gas?

4.143 An equimolar mixture of Ne and Xe is accidentally placed in a container that has a tiny leak. After a short while, a very small proportion of the mixture has escaped. What is the mole fraction of Ne in the effusing gas?

4.144 From the relative rates of effusion of $^{235}UF_6$ and $^{238}UF_6$, find the number of steps needed to produce a sample of the enriched fuel used in many nuclear reactors, which is 3.0 mole % ^{235}U. The natural abundance of ^{235}U is 0.72%.

4.145 A slight deviation from ideal behaviour exists even at normal conditions. If it behaved ideally, 1 mol of CO would occupy 22.711 L and exert 1 bar pressure at 273.15 K. Calculate p_{VDW} for 1.000 mol of CO at 273.15 K.

4.146 In preparation for a combustion demonstration, a professor fills a balloon with equal molar amounts of H_2 and O_2, but the demonstration has to be postponed until the next day. During the night, both gases leak through pores in the balloon. If 35% of the H_2 leaks, what is the O_2/H_2 ratio in the balloon the next day?

4.147 Phosphorus trichloride is important in the manufacture of insecticides, fuel additives, and flame retardants. Phosphorus has only one naturally occurring isotope, ^{31}P, whereas chlorine has two, ^{35}Cl (75%) and ^{37}Cl (25%).
(a) What different molecular masses (u) can be found for PCl_3?
(b) Which is the most abundant?
(c) What is the ratio of the effusion rates of the heaviest and the lightest PCl_3 molecules?

4.148 A truck tire has a volume of 218 L and is filled with air to 241 kPa at 295 K. After a drive, the air heats up to 318 K.
(a) If the tire volume is constant, what is the pressure (kPa)?
(b) If the tire volume increases 2.0%, what is the pressure (kPa)?
(c) If the tire leaks 1.5 g of air per minute and the temperature is constant, how many minutes will it take for the tire to reach the original pressure of 241 kPa ($\mathcal{M}$ of air = 28.8 g/mol)?

4.149 Allotropes are different molecular forms of an element, such as dioxygen (O_2) and ozone (O_3).
(a) What is the density of each oxygen allotrope at 0°C and 1.01 bar?
(b) Calculate the ratio of densities, $\frac{d_{O_3}}{d_{O_2}}$, and explain the significance of this number.

4.150 When gaseous F_2 and solid I_2 are heated to high temperatures, the I_2 sublimes and gaseous iodine heptafluoride forms. If 46.7 kPa of F_2 and 2.50 g of solid I_2 are put into a 2.50 L container at 250. K and the container is heated to 550. K, what is the final pressure (bar)? What is the partial pressure of I_2 gas?

4.151 2-methoxy-2-methylpropane or methyl tert butyl ether, MTBE, is an additive that raises the octane number of gasoline but is controversial due to its persistence in groundwater.
(a) A 3.2914 g sample of MTBE, a C—H—O compound, was combusted and yielded 8.2145 g of carbon dioxide and 4.0326 g of water. What is the empirical formula of MTBE?
(b) A 0.317 g sample of MTBE was allowed to evaporate to fill a 250.0 mL flask at 24.7°C. When measured, the MTBE was found to have a pressure of 0.3561 bar. What is the molecular formula of MTBE?

4.152 The reaction of solid calcium carbide with liquid water to form solid calcium hydroxide and ethyne (C_2H_2) gas was used industrially to produce ethyne. Write the balanced reaction. What mass of calcium carbide is required to collect 325 mL of ethyne gas at 26.0°C and 0.971 bar if it is collected over water at the same temperature? The vapour pressure of water at 26.0°C is 3.3639 kPa.

4.153 Tetrachloroethene is a solvent commonly used for dry-cleaning.
(a) Complete combustion of 12.2324 g of this carbon-chlorine compound resulted in the production of 6.4913 g of carbon dioxide. What is the empirical formula of this compound?
(b) The mass of an empty flask was found to be 43.8879 g. When filled with water, the flask had a mass of 291.75 g. If 2.2323 g of tetrachloroethene at 30.7°C and 1.372 bar was allowed to fill the flask, what is the molecular formula of tetrachloroethene?

4.154 (a) A flask containing ethoxyethane (commonly known as diethyl ether), a C—H—O compound once used as a general anaesthetic, was filled to a pressure of 0.585 bar at 20.0°C. The density of the gas was measured and found to be 1.78 g/L. What is the molar mass of ethoxyethane?
(b) A 2.56 g sample of ethoxyethane was burned in the presence of excess oxygen, producing 6.076 g of carbon dioxide and 3.107 g of water. Determine the empirical formula and molecular formula of ethoxyethane.

4.155 Firefighters and mine rescue personnel are often in situations where they have a lack of light and oxygen. Under such circumstances, they often resort to using chlorate candles, which produce breathable oxygen. The chlorate candle consists of a core of sodium chlorate mixed with iron and a small amount of combustibles. When lit, the chlorate candle heats the iron sufficiently to cause decomposition of the sodium chlorate, forming sodium chloride and oxygen according to the reaction

$$NaClO_3(s) \longrightarrow NaCl(s) + O_2(g) \text{(unbalanced)}$$

(a) Balance the reaction.
(b) What mass of $NaClO_3$ is needed to generate 1355 L of oxygen gas at 31.5°C and 1.041 bar?
(c) If each kilogram of the sodium chlorate/iron candle provides enough oxygen for an adult to breathe normally for 6.5 hours, what is the minimum number of 3.5 kg candles that would be needed to allow 10 trapped miners to survive for 3 days?

Thermochemistry: Energy Flow and Chemical Change

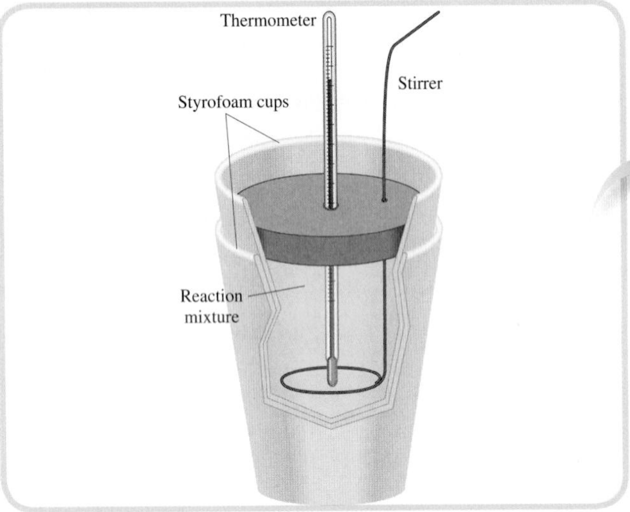

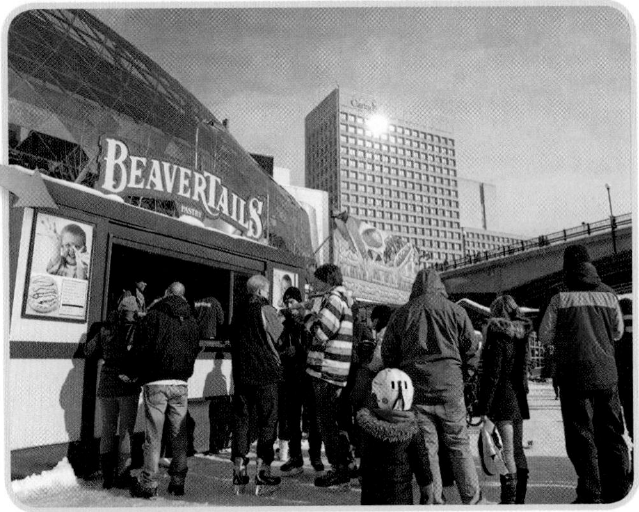

IN THIS CHAPTER . . . We see how heat, or thermal energy, flows when matter changes, how to measure the quantity of heat for a given change, and how to determine the heat flow accompanying any reaction. By the end of this chapter, you should be able to

- Show how energy always flows between a system and its surroundings in the form of heat or work, so that the total energy is conserved
- Differentiate between units of energy and show that the amount by which the energy changes does not depend on how the change occurs
- Identify the heat of a reaction in an open container as a change in enthalpy, and determine whether it is negative (exothermic reaction) or positive (endothermic reaction)
- Describe how a calorimeter measures heat and how the quantity of heat in a reaction is proportional to the amounts of substances
- Define the standard conditions used to compare enthalpies of reactions and obtain the change in enthalpy for any reaction
- Compare and contrast some current and future energy sources, bearing in mind the critical relation between energy demand and climate change

5

The chemistry of fire has fascinated humans since the dawn of time and was instrumental in changing the course of human development. Today, we have many ways of creating heat, some of which still rely on fire. What is this heat, and why do some substances and processes create heat while others require it? We are able to transfer heat easily to some materials, while others are difficult to heat. What makes heat transfer depend on the substance being heated? In this chapter, we will define energy; its conservation; and its conversion to other forms, such as heat and work. We will examine the properties of matter that help determine whether a substance heats rapidly or not. We will derive a quantity called enthalpy that will allow us to determine the amount of heat inherent in certain substances or the amount of heat required or released when a reaction or process occurs, and we will study the methods by which we can determine numerical values for enthalpy. Finally, we will look at energy now and, potentially, in the future.

Concepts and Skills to Review before Studying This Chapter

- Energy and its interconversion (Section 1.1)
- Distinction between heat and temperature (Section 1.5)
- Nature of chemical bonding (Section 2.7)
- Calculations of reaction stoichiometry (Section 3.4)
- Properties of the gaseous state (Section 4.1)
- Relation between kinetic energy and temperature (Section 4.5)

5.1 Forms of Energy and Their Interconversion

In Chapter 1, we saw that all energy is either potential or kinetic, and that these forms are interconvertible. An object has potential energy by virtue of its position and kinetic energy by virtue of its motion. Let us re-examine these ideas by considering a weight raised above the ground. As a motor or your muscles raise the weight, its potential energy increases; this energy is converted to kinetic energy as the weight falls (see Figure 1.3). When it hits the ground, some of that kinetic energy appears as *work* done when the weight moves the soil and pebbles slightly, and some appears as *heat* when it warms them slightly. Thus, in this situation, *potential energy is converted to kinetic energy, which appears as work and heat.*

Several other forms of energy—solar, electrical, nuclear, and chemical—are examples of potential and kinetic energy on the atomic scale. No matter what the form of energy or the situation, *when energy is transferred from one object to another, it appears as work and/or heat.* In this section, we examine this idea in terms of the release or absorption of energy during a chemical or physical change.

Defining the System and Its Surroundings

In any thermodynamic study, including measuring a change in energy, the first step is to define the **system**, *the part of the universe on which we are focused.* The moment that we define the system, *everything else* is defined as the **surroundings**.

Figure 5.1 shows a typical chemical system: the contents of a flask. The flask itself, other equipment, and perhaps the rest of the laboratory are the surroundings. In principle, the rest of the universe is the surroundings. In practice, however, we consider only the parts that are relevant to the system: it is not likely that a thunderstorm in central Asia or a methane blizzard on Neptune will affect the contents of the flask, but the temperature and pressure of the laboratory might. Thus, the experimenter defines the system and the relevant surroundings. An astronomer defines a galaxy as the system and nearby galaxies as the surroundings, a

FIGURE 5.1 A chemical system and its surroundings

microbiologist defines a given cell as the system and neighbouring cells and the extracellular fluid as the surroundings, and so on.

We can define three types of systems in a thermochemical sense: open systems, closed systems, and isolated systems. We can use a bowl of hot soup to demonstrate all three systems. An open system is a bowl of hot soup sitting on a counter. As the soup is in a bowl (our system), water vapour (steam) leaves the bowl, as does heat. Eventually, the soup gets thick and cold. An **open system** thus *allows for transfer of both matter and energy from the system to the surroundings.* If we have the same hot soup in a Styrofoam container with a lid (closed system), matter (steam) can no longer be transferred from the system, but the soup still gets cold, although it does take longer. Thus, a **closed system** still *allows transfer of heat from the system to the surroundings.* If we put the soup in a good insulated bottle, we have an **isolated system**, in which *neither matter nor heat is transferred* and the soup stays hot.

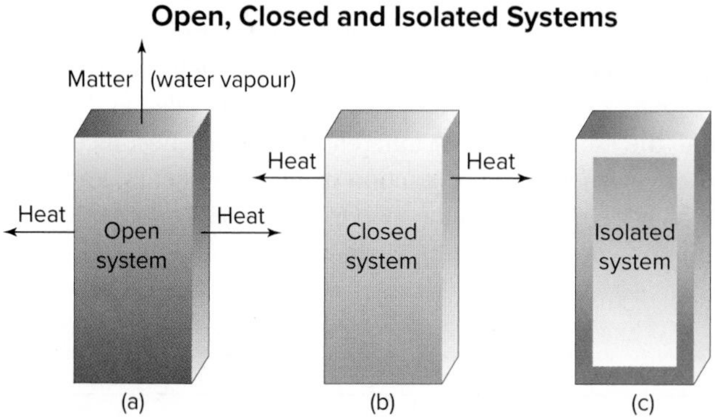

Energy Transfer to and from a System

Each particle in a system has potential energy and kinetic energy, and the sum of all these energies is the **internal energy (U)** of the system. When the reactants in a chemical system change to products, the system's internal energy has changed. This change, ΔU, is the difference between the internal energy *after* the change (U_{final}) and the internal energy *before* the change (U_{initial}):

$$\Delta U = U_{\text{final}} - U_{\text{initial}} = U_{\text{products}} - U_{\text{reactants}} \quad (5.1)$$

where Δ (Greek *delta*) means "a change (or difference) in an extensive thermodynamic property" and refers to the *final state* **minus** *the initial state.* Thus, ΔU is the final quantity of energy of the system *minus* the initial quantity.
In an *energy diagram*, the final and initial states are horizontal lines along a vertical energy axis, with ΔU the difference in the heights of the lines. A system can change its internal energy in one of two ways:

- By releasing some energy in a transfer *to* the surroundings (Figure 5.2A):

$$U_{\text{final}} < U_{\text{initial}} \quad \text{so} \quad \Delta U < 0$$

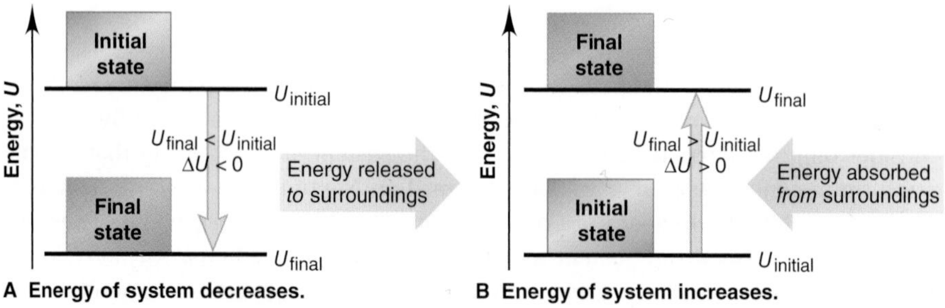

FIGURE 5.2 Energy diagrams for the transfer of internal energy (U) between a system and its surroundings. A. When the system releases energy, ΔU ($U_{\text{final}} - U_{\text{initial}}$) is negative. **B.** When the system absorbs energy, ΔU ($U_{\text{final}} - U_{\text{initial}}$) is positive. (The vertical *yellow* arrow always has its tail at the initial state.)

- By absorbing some energy in a transfer *from* the surroundings (Figure 5.2B):

$$U_{\text{final}} > U_{\text{initial}} \text{ so } \Delta U > 0$$

Thus, ΔU is a *transfer* of energy from system to surroundings, or vice versa.

Heat and Work: Two Forms of Energy Transfer

Energy that is transferred from system to surroundings, or vice versa, appears in two forms:

1. *Heat. Heat* (q) or *thermal energy* is *the energy transferred as a result of a difference in temperature between the system and the surroundings.* For example, energy in the form of heat is transferred from hot soup (system) to the bowl, air, and table (surroundings) because they are at a lower temperature.

2. *Work.* All other forms of energy transfer involve some type of **work (w)**, *the energy that is transferred when an object is moved by a force.* When you (system) kick a football, energy is transferred as work because the force of the kick moves the ball and the air (surroundings). When you pump up a ball, energy is transferred as work because the added air (system) exerts a force on the inner wall of the ball (surroundings) and moves it outward.

The total change in a system's internal energy is the sum of the energy transferred as heat and/or work:

$$\Delta U = q + w \tag{5.2}$$

The values of q and w (and, therefore, the value of ΔU) can have either a positive or a negative sign. *We define the sign of the energy change from the* ***system's*** *perspective:*

- Energy transferred *into* the system is *positive*, because the *system ends up with more* energy.
- Energy transferred *out from* the system is *negative*, because the *system ends up with less* energy.

Innumerable combinations of heat and/or work can change a system's internal energy. In the rest of this subsection, we will examine the four simplest cases: two that involve only heat and two that involve only work.

Energy Transferred as Heat Only For a system that transfers energy only as heat (q) and does no work ($w = 0$), we have, from Equation 5.2, $\Delta U = q + 0 = q$. There are two cases in which this transfer can happen:

1. *Heat flowing* ***out*** *from a system.* Suppose that hot water is the system, and the beaker holding it and the rest of the laboratory are the surroundings. The water transfers energy as heat outward until the temperature of the water and the surroundings are equal. Since heat flows *out* from the system, the final energy of the system is less than the initial energy. Heat was released, so *q is negative*, and therefore *ΔU is negative* (Figure 5.3A).

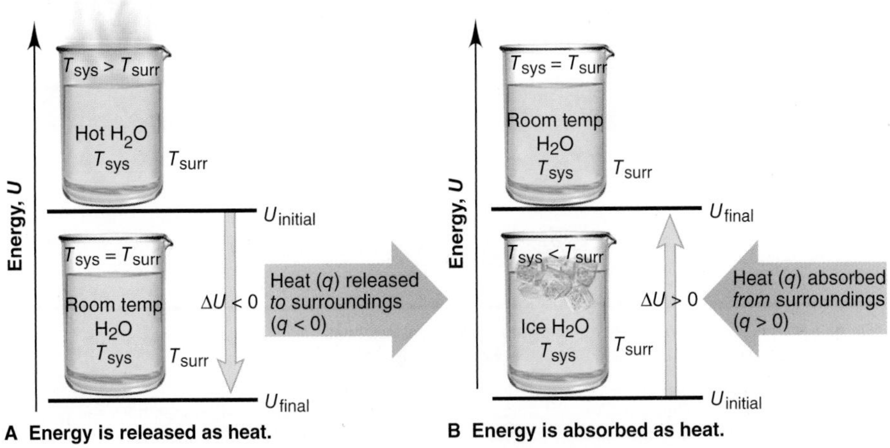

FIGURE 5.3 The two cases in which energy is transferred as heat only. **A.** The system releases heat. **B.** The system absorbs heat.

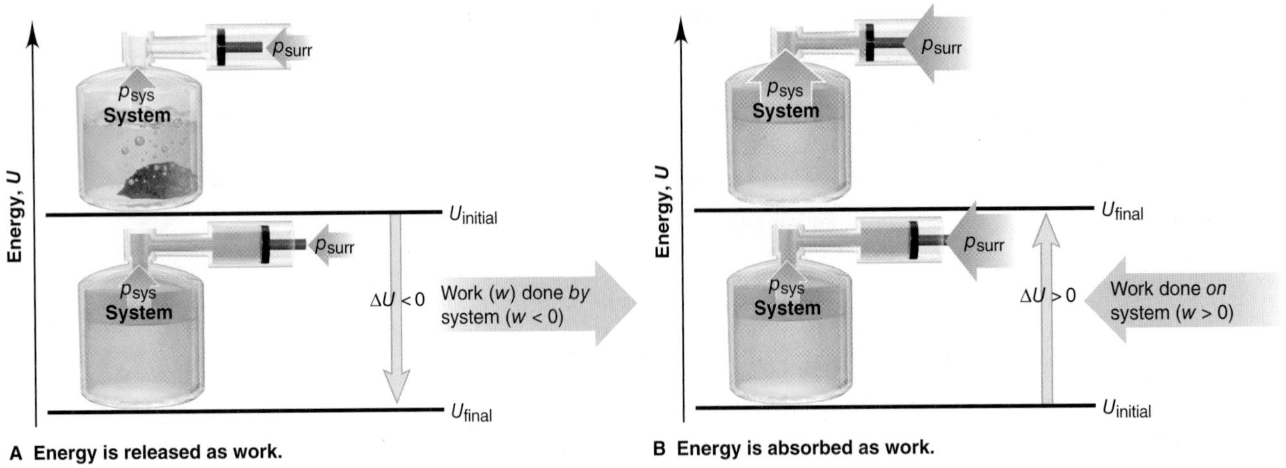

FIGURE 5.4 The two cases in which energy is transferred as work only. A. The system does work *on* the surroundings. **B.** The system has work done on it *by* the surroundings.

2. *Heat flowing* ***into*** *a system.* If the system consists of ice water, the surroundings transfer energy as heat *into* the system, once again until the ice melts and the temperature of the water and the surroundings becomes equal. In this case, heat flows *in*, so the final energy of the system is higher than its initial energy. Heat was absorbed, so q is *positive*, and therefore ΔU *is positive* (Figure 5.3B). ■

■ **Thermodynamics in the Kitchen** A new view of two familiar kitchen appliances can clarify the sign of q. The air in a refrigerator (surroundings) has a lower temperature than a newly added piece of food (system), so the food releases energy as heat to the refrigerator air, $q < 0$. The air in a hot oven (surroundings) has a higher temperature than a newly added piece of food (system), so the food absorbs energy as heat from the oven air, $q > 0$.

Energy Transferred as Work Only For a system that transfers energy only as work, $q = 0$; therefore, $\Delta U = 0 + w = w$. There are two cases in which this transfer can happen:

1. *Work done* ***by*** *a system* (Figure 5.4A). Consider an aqueous reaction of a grey metal (for example, Zn) and a strong acid (for example, HCl) that has a gas (for example, H_2) as one of the products. The reaction takes place in a nearly evacuated (narrow p_{sys} arrow, *top*), insulated container attached to a piston-cylinder assembly. (The container is insulated so that heat does not flow.) We define the system as the reaction mixture, and the container, piston-cylinder, outside air, and so on, as the surroundings. In the initial state, the internal energy is the energy of the reactants, and, in the final state, it is the energy of the products. As the gas forms, it pushes back the piston. Thus, energy is transferred as work done *by* the system *on* the surroundings. Since the system releases energy as work, w *is negative*; the final energy of the system is less than the initial energy, so ΔU *is negative.**

2. *Work done* ***on*** *a system* (Figure 5.4B). Suppose that, after the reaction is over, we increase the pressure of the surroundings (wider p_{surr} arrow) so that the piston moves in. Energy is transferred as work done *by* the surroundings *on* the system, so w *is positive.* The final energy of the system is greater than the initial energy, so ΔU *is positive.*

■ One easy way to remember the sign convention is to think of *yourself* as the system. When *you* have to *do* work, that is a *negative* feeling. When work is done *on you* (such as a nice massage!), that is a *positive* feeling. When you are cold (heat is *leaving you*), that is a *negative* feeling. When you are nice and warm (heat is *coming* in), that is a *positive* feeling.

Table 5.1 summarizes the sign conventions for q and w and their effect on the sign of ΔU. ■

Table 5.1 The Sign Conventions* for q, w, and ΔU

q	+	w	=	ΔU
+		+		+
+		−		Depends on the sizes of q and w
−		+		Depends on the sizes of q and w
−		−		−

*For q: + means system *absorbs* heat; − means system *releases* heat.
For w: + means work done *on* system; − means work done *by* system.

*The system shown in Figure 5.4A is not doing very useful work because it just pushes back the piston and outside air. However, if the reaction mixture is gasoline and oxygen, and the surroundings are an automobile engine, much of the internal energy is transferred as the work done to move a car.

The Law of Energy Conservation

As you have seen, when a system absorbs energy, the surroundings release it; when a system releases energy, the surroundings absorb it. Energy transferred between system and surroundings can be in the form of heat and/or various types of work—mechanical, electrical, and so on.

Indeed, energy is often converted from one form to another during transfers. For example, when gasoline burns in a car engine, the reaction releases energy that is transferred as heat and work. The heat warms the car parts, passenger compartment, and surrounding air. The work is done when mechanical energy turns the car's wheels and belts. That energy is converted into the electrical energy of the sound system, the radiant energy of the headlights, the chemical energy of the battery, and so on. The sum of all these forms equals the change in energy between reactants and products as the gasoline is burned.

Complex biological processes exhibit the same general pattern. During photosynthesis, green plants convert solar energy into the chemical energy of the bonds in starch and O_2; when you digest starch, the bonds in the compound are converted into the muscular (mechanical) energy needed to run a marathon.

The most important point to realize in these and all other situations is that the energy changes form but does not simply appear or disappear—energy cannot be created or destroyed. Put another way, *energy is conserved: the total energy of the system plus the surroundings remains constant.* The **law of conservation of energy (first law of thermodynamics)** restates this basic observation: *the total energy of the universe is constant.* It is expressed mathematically as

$$\Delta U_{\text{universe}} = \Delta U_{\text{system}} + \Delta U_{\text{surroundings}} = 0 \tag{5.3}$$

This law applies, as far as we know, to all systems—from a burning match to continental drift, from the pumping of your heart to the formation of the Solar System.

Units of Energy

The SI unit of energy is the **joule (J)**, a derived unit composed of three base units:

$$1\ \text{J} = 1\ \text{kg} \cdot \text{m}^2/\text{s}^2$$

Both heat and work are expressed in joules. Let us see why the joule is the unit for work. The work (w) done on a mass is the force (F) times the distance (d) that the mass moves: $w = F \times d$. A *force* changes the velocity of a mass over time; that is, a force *accelerates* a mass. Velocity has units of metres per second (m/s), so acceleration (a) has units of metres per second squared (m/s^2). Force, therefore, has units of mass (m, in kilograms) times acceleration:

$$F = m \times a \qquad \text{has units of} \qquad \text{kg} \cdot \text{m/s}^2$$

Therefore,

$$w = F \times d \qquad \text{has units of} \qquad (\text{kg} \cdot \text{m/s}^2) \times \text{m} = \text{kg} \cdot \text{m}^2/\text{s}^2 = \text{J}$$

The **calorie (cal)** is an older unit, defined originally as the quantity of energy needed to raise the temperature of 1 g of water by 1°C (from 14.5°C to 15.5°C). The calorie is now defined in terms of the joule:

$$1\ \text{cal} \equiv 4.184\ \text{J} \quad \text{or} \quad 1\ \text{J} = \frac{1}{4.184}\ \text{cal} = 0.2390\ \text{cal}$$

Since the quantities of energy involved in chemical reactions are usually quite large, chemists use the kilojoule (kJ), or, in earlier sources, the kilocalorie (kcal):

$$1\ \text{kJ} = 1000\ \text{J} = 0.2390\ \text{kcal} = 239.0\ \text{cal}$$

The nutritional Calorie (note the capital C), the unit that shows the energy available from food, is actually a kilocalorie. In general, the SI unit (J or kJ) is used in this book. Some interesting quantities of energy appear in Figure 5.5.

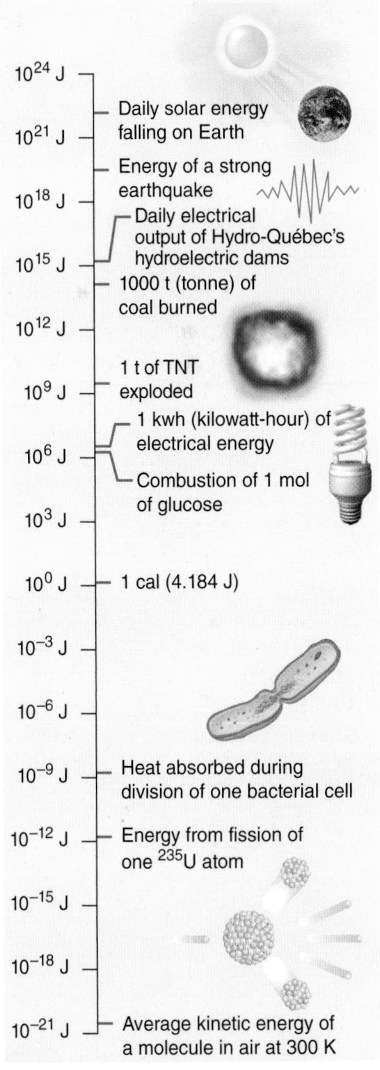

FIGURE 5.5 Some quantities of energy. The vertical scale is exponential.

Sample Problem 5.1 Determining the Change in Internal Energy of a System

Problem When gasoline burns in a car engine, the heat released causes the products CO_2 and H_2O to expand, which pushes the pistons outward. Excess heat is removed by the car's radiator. If the expanding gases do 451 J/mol of work on the pistons and the system releases 325 J/mol to the surroundings as heat, calculate the change in energy (ΔU) in J/mol, kJ/mol, and kcal/mol.

Plan We must define the system and surroundings to choose signs for q and w, and then we can calculate ΔU with Equation 5.2. The system is the reactants and products, and the surroundings are the pistons, the radiator, and the rest of the car. Heat is released by the system, so q is negative. Work is done by the system to push the pistons outward, so w is also negative. We obtain the answer in J/mol and then convert it to kJ/mol and kcal/mol.

Solution Calculate ΔU (from Equation 5.2) in J/mol:

$$q = -325 \text{ J/mol}$$
$$w = -451 \text{ J/mol}$$
$$\Delta U = q + w = -325 \text{ J/mol} + (-451 \text{ J/mol}) = -776 \text{ J/mol}$$

Convert from J to kJ:

$$\Delta U = -776 \text{ J/mol} \times \frac{1 \text{ kJ}}{1000 \text{ J}} = -0.776 \text{ kJ/mol}$$

Convert from kJ to kcal:

$$\Delta U = -0.776 \text{ kJ/mol} \times \frac{1 \text{ kcal}}{4.184 \text{ kJ}} = -0.185 \text{ kcal/mol}$$

Check The answer is reasonable: the combustion of gasoline releases energy from the system, so $U_{final} < U_{initial}$ and ΔU should be negative. Rounding shows that, since 4 kJ/mol ≈ 1 kcal/mol, nearly 0.8 kJ/mol should be nearly 0.2 kcal/mol.

Follow-Up Problem 5.1 In a reaction, gaseous reactants form a liquid product. The heat absorbed by the surroundings is 26.0 kcal/mol, and the work done on the system is 3.78 kcal/mol. Calculate ΔU in kJ/mol. (See Brief Solutions to Follow-Up Problems at the end of the chapter.)

State Functions and the Path Independence of the Energy Change

The internal energy (U) of a system is called a **state function**, a property that depends only on the *current* state of the system (its composition, volume, pressure, and temperature), *not* on the path that the system takes to reach this state. The energy change of a system can occur by countless combinations of heat (q) and/or work (w). However, because U is a state function, the overall ΔU is the same no matter what the specific combination is. That is, *ΔU does* ***not*** *depend on how the change takes place, but only on the* ***difference*** *between the final and initial states.* As an example, let us define a system in its initial state as 1 mol of octane (a component of gasoline) together with enough O_2 to burn it. In its final state, the system is the CO_2 and H_2O that form:

$$\underset{\text{initial state } (U_{initial})}{C_8H_{18}(l) + \frac{25}{2}O_2(g)} \longrightarrow \underset{\text{final state } (U_{final})}{8CO_2(g) + 9H_2O(g)}$$

Energy is transferred *out* from the system as heat and/or work, so ΔU is negative. Figure 5.6 shows just two of the many ways that the change can occur. If we burn the octane in an open container (*left*), ΔU is transferred almost completely as heat (with a small amount of work done to push back the atmosphere). If we burn the

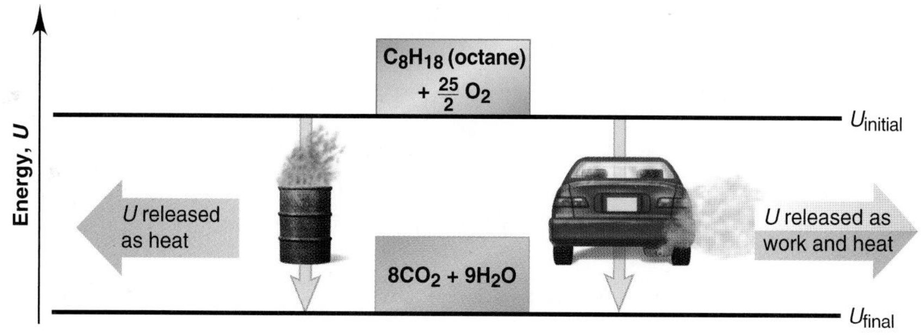

FIGURE 5.6 Two different paths for the energy change of a system. Even though q and w for the two paths are different, ΔU is the same.

octane in a car engine (*right*), a much larger portion (~30%) of ΔU is transferred as work that moves the car, with the rest released as heat that warms the car, exhaust gases, and surrounding air. If we burn the octane in a lawn mower or an airplane, ΔU appears as other combinations of work and heat. Thus, q and w are *not* state functions because their values *do* depend on the path that the system takes, but ΔU (the *sum* of q and w) *does not.* ■

Pressure (p), volume (V), and temperature (T) are some other state functions. Path independence means that *changes in state functions—ΔU, Δp, ΔV, and ΔT—depend only on the initial and final states.*

■ **Your Personal Financial State Function** The *balance* in your chequebook is a state function of your personal financial system. You can open a new chequing account by depositing a birthday gift of $50, or you can open one by depositing a $100 paycheque and then writing two $25 cheques. The two paths to the balance are different, but the balance (current state) is the same.

SUMMARY OF SECTION 5.1

- Internal energy (U) is transferred as heat (q) when the system and surroundings are at different temperatures or as work (w) when an object is moved by a force.
- Heat or work absorbed by a system ($q > 0$; $w > 0$) increases its U; heat or work released by a system ($q < 0$; $w < 0$) decreases its U. The change in the internal energy is the sum of the heat and work: $\Delta U = q + w$. Heat and work are measured in joules (J).
- Energy is always conserved: it can change from one form to another and move into or out from a system, but the total quantity of energy in the universe (system *plus* surroundings) is constant.
- The internal energy of a system is a state function and, thus, is independent of how the system attained this energy; therefore, the same overall ΔU can occur through any combination of q and w.

5.2 Enthalpy: Chemical Change at Constant Pressure

Most physical and chemical changes occur at nearly constant atmospheric pressure: a reaction in an open flask, the freezing of a lake, a biochemical process in an organism. In this section, we discuss *enthalpy*, a thermodynamic variable that relates directly to energy changes at constant pressure.

The Meaning of Enthalpy

To determine ΔU, we must measure both heat and work. The two most important types of chemical work are electrical work, done by moving charged particles (Chapter 19), and **pressure-volume work (pV work)**, *the mechanical work done when the volume of the system changes in the presence of an external pressure (p).* The quantity of pV work equals p times the change in volume (ΔV, or $V_{final} - V_{initial}$). In an open flask or in a cylinder with a weightless, frictionless piston (Figure 5.7), a system of an expanding gas does pV work *on* the surroundings, so it has a negative sign:

$$w = -p\Delta V \quad (5.4)$$

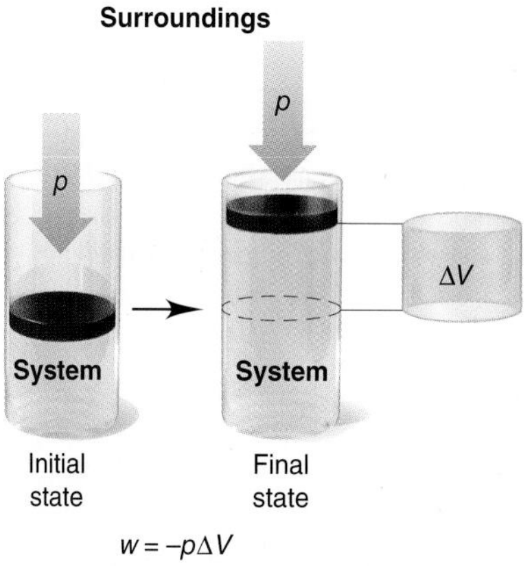

FIGURE 5.7 Pressure-volume work. An expanding gas pushing back the atmosphere does pV work ($w = -p\Delta V$).

For reactions *at constant pressure*, a thermodynamic variable called **enthalpy (H)** eliminates the need to measure pV work. The enthalpy of a system is defined as *the internal energy* ***plus*** *the product of the pressure and volume*:

$$H = U + pV$$

The **change in enthalpy (ΔH)** is *the change in internal energy* ***plus*** *the product of the pressure, which is constant, and the change in volume (ΔV)*:

$$\Delta H = \Delta U + p\Delta V \tag{5.5}$$

Combining Equations 5.2 ($\Delta U = q + w$) and 5.4 ($w = -p\Delta V$) gives

$$\Delta U = q + w = q + (-p\Delta V) = q - p\Delta V$$

At constant pressure, we denote q as q_p, giving $\Delta U = q_p - p\Delta V$, which we can use to solve for q_p:

$$q_p = \Delta U + p\Delta V$$

Notice that the right side of this equation and the right side of Equation 5.5 are the same:

$$q_p = \Delta U + p\Delta V = \Delta H \tag{5.6}$$

Thus, *the change in enthalpy equals the heat absorbed or released at* ***constant*** *pressure*. For most changes occurring at constant pressure, ΔH is more relevant than ΔU *and* easier to obtain: *to find ΔH, measure q_p*. We discuss the laboratory method in Section 5.3.

Comparing ΔU and ΔH

Knowing the *enthalpy* change of a system tells us a lot about its *energy* change as well. In fact, because many reactions involve little (if any) pV work, most (or all) of the energy change occurs as a transfer of heat. Here are three cases:

1. *Reactions that do not involve gases.* Gases do not appear in many reactions, such as precipitation reactions, many acid-base reactions, and many redox reactions. For example,

   $$2KOH(aq) + H_2SO_4(aq) \longrightarrow K_2SO_4(aq) + 2H_2O(l)$$

   Because liquids and solids undergo *very* small volume changes, $\Delta V \approx 0$; thus, $p\Delta V \approx 0$ and $\Delta H \approx \Delta U$.
2. *Reactions in which the amount (mol) of gas does* ***not*** *change.* When the total amount of gaseous reactants equals the total amount of gaseous products, the volume is constant, $\Delta V = 0$, so $p\Delta V = 0$ and $\Delta H = \Delta U$. For example,

   $$N_2(g) + O_2(g) \longrightarrow 2NO(g)$$

3. *Reactions in which the amount (mol) of gas* ***does*** *change.* In these reactions, $p\Delta V \neq 0$. However, q_p is usually *much* larger than $p\Delta V$, so ΔH is still very close to ΔU. For instance, in the combustion of H_2, 3 mol of gas yields 2 mol of gas:

   $$2H_2(g) + O_2(g) \longrightarrow 2H_2O(g)$$

   In this reaction, $\Delta H = -483.6$ kJ/mol and $p\Delta V = -2.5$ kJ/mol, ■ so (from Equation 5.5), $\Delta H = \Delta U + p\Delta V$ and

   $$\Delta U = \Delta H - p\Delta V = -483.6 \text{ kJ/mol} - (-2.5 \text{ kJ/mol}) = -481.1 \text{ kJ/mol}$$

Since most ΔU occurs as energy transferred as heat, $\Delta H \approx \Delta U$. While this is a valid approximation for many reactions, it is not always the case.

■ How do we know that $p\Delta V = -2.5$ kJ/mol? We make use of the ideal gas law, which states that $pV = nRT$. Thus, $p\Delta V = RT\Delta n$ (since R is a constant and does not change, and we are using constant temperature). In the equation, there are 3 mol of gas on the reactant side and 2 mol on the product side:

$$\Delta n = n_f - n_i = 2 - 3 = -1$$

Thus,

$$\begin{aligned} p\Delta V &= RT\Delta n \\ &= \left(8.314 \frac{\text{J}}{\text{mol}\cdot\text{K}}\right)(298.15\text{ K})(-1) \\ &= -2.5 \times 10^3 \text{ J/mol} = -2.5 \text{ kJ/mol} \end{aligned}$$

Exothermic and Endothermic Processes

Because H is a combination of the three state functions U, p, and V, it is also a state function. Therefore, ΔH equals H_{final} ***minus*** $H_{initial}$. For a reaction, H_{final} is $H_{products}$ and $H_{initial}$ is $H_{reactants}$, so the enthalpy change of a reaction is

$$\Delta H = H_{final} - H_{initial} = H_{products} - H_{reactants}$$

Since $H_{products}$ can be either more or less than $H_{reactants}$, the *sign* of ΔH indicates whether heat is absorbed or released during the reaction. We determine the sign of *ΔH by imagining the heat as "reactant" or "product."* There are two possibilities:

1. *Exothermic ("heat out") process.* An **exothermic process** *releases* heat and results in a *decrease* in the enthalpy of the system:

$$\text{Exothermic:} \quad H_{products} < H_{reactants} \quad \text{so} \quad \Delta H < 0$$

For example, when methane burns in air, heat flows *out from* the system into the surroundings, so we show it as a product:

$$CH_4(g) + 2O_2(g) \longrightarrow CO_2(g) + 2H_2O(g) + \textit{heat}$$

The reactants (1 mol of CH_4 and 2 mol of O_2) release heat during the reaction, so they originally had more enthalpy than the products (1 mol of CO_2 and 2 mol of H_2O):

$$H_{products} < H_{reactants} \quad \text{so} \quad \Delta H\ (= H_{products} - H_{reactants}) < 0$$

This exothermic change is shown in the **enthalpy diagram** in Figure 5.8A.

2. *Endothermic ("heat in") process.* An **endothermic process** *absorbs* heat and results in an *increase* in the enthalpy of the system:

$$\text{Endothermic:} \quad H_{products} > H_{reactants} \quad \text{so} \quad \Delta H > 0$$

When ice melts, for instance, heat flows *into* the ice from the surroundings, so we show the heat as a reactant:

$$\textit{heat} + H_2O(s) \longrightarrow H_2O(l)$$

Because heat is absorbed, the enthalpy of the product (water) must be higher than the enthalpy of the reactant (ice):

$$H_{water} > H_{ice} \quad \text{so} \quad \Delta H\ (= H_{water} - H_{ice}) > 0$$

This endothermic change is shown in Figure 5.8B. (In general, the value of an enthalpy change is determined with the reactants and the products at the same temperature.)

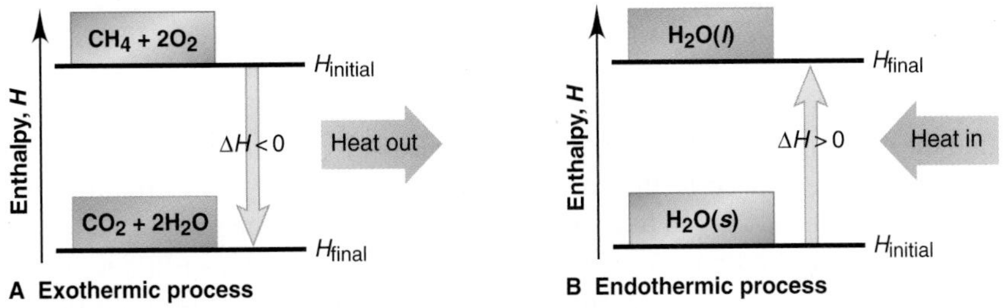

FIGURE 5.8 Enthalpy diagrams for exothermic and endothermic processes. **A.** The combustion of methane is exothermic: $\Delta H < 0$. **B.** The melting of ice is endothermic: $\Delta H > 0$.

Sample Problem 5.2 Drawing Enthalpy Diagrams and Determining the Sign of ΔH

Problem For each reaction, determine the sign of ΔH, state whether the reaction is exothermic or endothermic, and draw an enthalpy diagram:

(a) $H_2(g) + \frac{1}{2}O_2(g) \longrightarrow H_2O(l) + 285.8\ \text{kJ/mol}$

(b) $40.7\ \text{kJ/mol} + H_2O(l) \longrightarrow H_2O(g)$

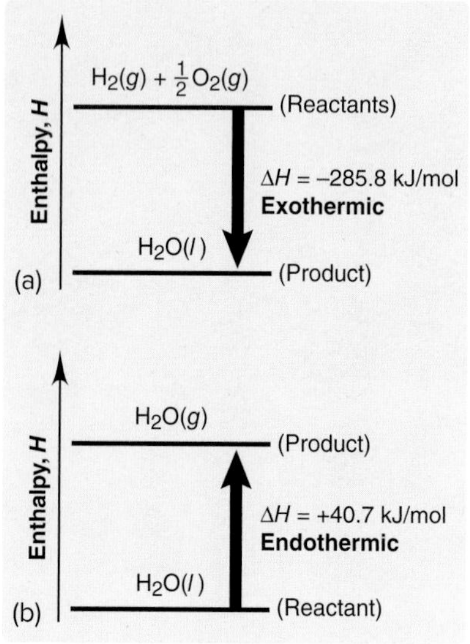

Plan From each equation, we note whether heat is a "product" (exothermic; $\Delta H < 0$) or a "reactant" (endothermic; $\Delta H > 0$). For exothermic reactions, reactants are above products on the enthalpy diagram; for endothermic reactions, reactants are below products. The ΔH arrow *always* points from reactants to products.

Solution (a) Heat is a product (on the right side of the equation), so $\Delta H < 0$ and the reaction is exothermic. The enthalpy diagram appears in the margin (*top*).

(b) Heat is a reactant (on the left side of the equation), so $\Delta H > 0$ and the reaction is endothermic. The enthalpy diagram appears in the margin (*bottom*).

Check Substances on the same side of the equation as the heat have less enthalpy than substances on the other side, so make sure that the substances on the same side of the equation as the heat (enthalpy) are placed on the lower line of the diagram.

Follow-Up Problem 5.2 When nitroglycerin decomposes, the reaction creates a violent explosion and releases 5.72×10^3 kJ of heat per mole:

$$C_3H_5(NO_3)_3(l) \longrightarrow 3CO_2(g) + \frac{5}{2}H_2O(g) + \frac{1}{4}O_2(g) + \frac{3}{2}N_2(g)$$

Is the decomposition of nitroglycerin exothermic or endothermic? Draw an enthalpy diagram for the process.

SUMMARY OF SECTION 5.2

- Enthalpy (H) is a state function, so any change in enthalpy (ΔH) is independent of how the change occurred. At constant p, the value of ΔH equals ΔU + pV work, which occurs when the volume of the system changes in the presence of an external pressure.
- ΔH equals q_p, the heat released or absorbed during a chemical or physical change that takes place at constant pressure.
- In most cases, ΔH is equal, or very close, to ΔU.
- A change that releases heat is exothermic ($\Delta H < 0$); a change that absorbs heat is endothermic ($\Delta H > 0$).

5.3 Calorimetry: Measuring the Heat of a Chemical or Physical Change

Data about energy content and usage are everywhere—the caloric value of a slice of bread, the energy efficiency of a dishwasher, or the fuel efficiency of a new car. In this section, we will see how these values are determined.

Specific Heat Capacity

To find the energy change during a process, we measure the quantity of heat released or absorbed by relating it to the change in temperature. You know, from everyday experience, that the more you heat an object, the higher its temperature will become. Similarly, the more you cool an object, the lower its temperature will get. In other words, the quantity of heat (q) absorbed or released by an object is proportional to its temperature change:

$$q \propto \Delta T \quad \text{or} \quad q = \text{constant} \times \Delta T \quad \text{or} \quad \frac{q}{\Delta T} = \text{constant}$$

What you might not know is that the temperature change depends on the object. Every object has its own **heat capacity**, the quantity of heat required to change its temperature by 1 K. Heat capacity is the proportionality constant in the preceding equation:

$$\text{Heat capacity} = \frac{q}{\Delta T} \quad \text{(in units of J/K)}$$

A related property is **specific heat capacity (*c*)**, the quantity of heat required to change the temperature of 1 g (gram) of a substance by 1 K:

$$\text{Specific heat capacity } (c) = \frac{q}{\text{mass} \times \Delta T} \quad [\text{in units of J/(g} \cdot \text{K)}]$$

If we know *c* of the object being heated (or cooled), we can measure the mass and temperature change and calculate the heat absorbed (or released):

$$q = c \times \text{mass} \times \Delta T \tag{5.7}$$

According to Equation 5.7, when an object gets hotter [that is, when ΔT (= $T_{final} - T_{initial}$) is positive], $q > 0$ (the sample absorbs heat). When an object gets cooler (that is, when ΔT is negative), $q < 0$ (the sample releases heat). Table 5.2 lists the specific heat capacities of some representative substances and materials. Notice that metals have relatively low values of *c* and water has a very high value: for instance, it takes over 30 times as much energy to increase the temperature of 1 g of water by 1 K as it does to increase the temperature of 1 g of gold by 1 K!

Closely related to the specific heat capacity (but reserved for substances) is the **molar heat capacity (C_m**; note the capital letter), *the quantity of heat required to change the temperature of **1 mol** of a substance by 1 K*:

$$\text{Molar heat capacity } (C_m) = \frac{q}{\text{amount (mol)} \times \Delta T} \quad [\text{in units of J/(mol} \cdot \text{K)}]$$

To find C_m of liquid H_2O, we multiply *c* of liquid H_2O [4.184 J/(g • K)] by the molar mass of H_2O (18.02 g/mol):

$$C_m \text{ of } H_2O(l) = 4.184 \frac{\text{J}}{\text{g} \cdot \text{K}} \times \frac{18.02 \text{ g}}{1 \text{ mol}} = 75.40 \frac{\text{J}}{\text{mol} \cdot \text{K}}$$

Liquid water has a *very* high specific heat capacity [~4.2 J/(g • K)], about six times the specific heat capacity of rock [~0.7 J/(g • K)].

Table 5.2 Specific Heat Capacities (c) of Some Elements, Compounds, and Materials

Substance	c [J/(g·K)]*
Elements	
Aluminum, Al	0.900
Graphite, C	0.711
Iron, Fe	0.450
Copper, Cu	0.387
Gold, Au	0.129
Compounds	
Water, $H_2O(l)$	4.184
Ethyl alcohol, $C_2H_5OH(l)$	2.46
Ethylene glycol, $(CH_2OH)_2(l)$	2.42
Carbon tetrachloride, $CCl_4(l)$	0.862
Solid Materials	
Wood	1.76
Cement	0.88
Glass	0.84
Granite	0.79
Steel	0.45

*At 298 K (25°C).

Sample Problem 5.3 Finding the Quantity of Heat from a Temperature Change

Problem A layer of copper welded to the bottom of a skillet has a mass of 125 g. How much heat is needed to raise the temperature of the copper layer from 25°C to 300.°C? The specific heat capacity (*c*) of Cu is given in Table 5.2.

Plan We know the mass (125 g) and *c* [0.387 J/(g • K)] of Cu and can find ΔT in °C, which equals ΔT in K. We then use Equation 5.7 to calculate the heat.

Solution Calculate ΔT and then *q*:

$$\Delta T = T_{final} - T_{initial} = 300.°\text{C} - 25°\text{C} = 275°\text{C} = 275 \text{ K}$$

$$q = c \times \text{mass (g)} \times \Delta T = 0.387 \frac{\text{J}}{\text{g} \cdot \text{K}} \times 125 \text{ g} \times 275 \text{ K} = 1.33 \times 10^4 \text{ J}$$

Check Heat is absorbed by the copper bottom (system), so *q* is positive. Rounding shows that the arithmetic is reasonable: 0.4 J/(g • K) × 100 g × 300 K = 1.2 × 10^4 J.

Note It is *very* important to remember *not to subtract* the initial temperature from the final temperature (with both in degrees Celsius) and then *add* 273.15. Because we are looking at a *difference* in temperature, this difference will be the same whether expressed in °C or K (as the intervals in both temperature scales are the same).

Follow-Up Problem 5.3 Find the heat released (kJ) when 5.50 L of ethylene glycol (*d* = 1.11 g/mL; see Table 5.2) in a car radiator cools from 37.0°C to 25.0°C.

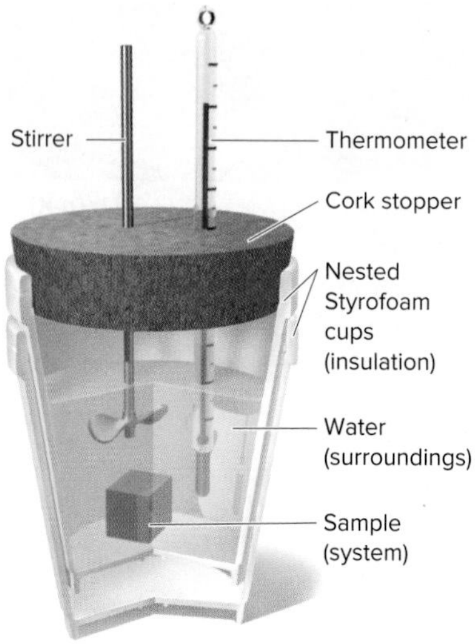

FIGURE 5.9 Coffee-cup calorimeter. This device measures the heat transferred at constant pressure (q_p).

The Two Major Types of Calorimetry

How do we know the amount of heat generated in an acid-base reaction or the caloric content in a teaspoon of sugar? We construct "surroundings" that retain the heat as reactants change to products and then note the temperature change. In the laboratory, these "surroundings" take the form of a **calorimeter**, a device used to measure the heat released (or absorbed) by a physical or chemical process. Let us look at two types of calorimeter—one designed to measure the heat at constant pressure and the other to measure it at constant volume.

Constant-Pressure Calorimetry For processes that take place at constant pressure, the heat transferred (q_p) is often measured in a *coffee-cup calorimeter* (Figure 5.9). This simple apparatus is often used to find the heat of an aqueous reaction or the heat absorbed or released in a process, such as dilution or dissolution of a salt.

A coffee-cup calorimeter can also be used to find the specific heat capacity of a solid, as long as the solid does not react with or dissolve in water. The solid (the system) is weighed, heated to some known temperature, and added to a known mass and temperature of water (surroundings) in the calorimeter. After stirring, the final water temperature is measured, which is also the final temperature of the solid. According to the laws of energy conservation, no heat can be lost or gained. Thus, we can write

$$q_{solid} + q_{water} + q_{calorimeter} = 0$$

In general, when we use a coffee-cup calorimeter, the heat capacity of the calorimeter is negligible and is thus ignored, although this may not always be the case. If it is necessary to take the heat capacity of the calorimeter into account, it will either be given to you or you will be given enough information to calculate it.

Assuming that the heat capacity of the coffee cup calorimeter is negligible and that no heat escapes the calorimeter, the heat released by the system ($-q_{sys}$, or $-q_{solid}$) is equal in magnitude but opposite in sign to the heat absorbed by the surroundings ($+q_{surr}$, or $+q_{H_2O}$):

$$-q_{solid} = q_{H_2O}$$

Substituting from Equation 5.7 on each side of this equation gives

$$-(c_{solid} \times \text{mass}_{solid} \times \Delta T_{solid}) = c_{water} \times \text{mass}_{water} \times \Delta T_{water}$$

All the quantities are known or measured except c_{solid}:

$$c_{solid} = -\frac{c_{H_2O} \times \text{mass}_{H_2O} \times \Delta T_{H_2O}}{\text{mass}_{solid} \times \Delta T_{solid}}$$

Sample Problem 5.4 Determining the Specific Heat Capacity of a Solid

Problem You heat 22.05 g of a solid in a test tube to 100.00°C and then add the solid to 50.00 g of water in a coffee-cup calorimeter. The water temperature changes from 25.10°C to 28.49°C. Find the specific heat capacity of the solid.

Plan We are given the masses of the solid (22.05 g) and H_2O (50.00 g), and we can find the temperature changes of the water and the solid by subtracting the given values, always using $T_{final} - T_{initial}$. Using Equation 5.7, we set the heat released by the solid ($-q_{solid}$) equal to the heat absorbed by the water (q_{water}). The specific heat of water is known, and we solve for c_{solid}.

Solution Find ΔT_{solid} and ΔT_{water}:

$$\Delta T_{water} = T_{final} - T_{initial} = 28.49°\text{C} - 25.10°\text{C} = 3.39°\text{C} = 3.39\text{ K}$$
$$\Delta T_{solid} = T_{final} - T_{initial} = 28.49°\text{C} - 100.00°\text{C} = -71.51°\text{C} = -71.51\text{ K}$$

Solve for c_{solid}:

$$c_{solid} = -\frac{c_{H_2O} \times mass_{H_2O} \times \Delta T_{H_2O}}{mass_{solid} \times \Delta T_{solid}} = -\frac{4.184 \frac{J}{g \cdot K} \times 50.00\ g \times 3.39\ K}{22.05\ g \times (-71.51\ K)} = 0.450 \frac{J}{g \cdot K}$$

Check Rounding gives $-\frac{4 \frac{J}{g \cdot K} \times 50\ g \times 3\ K}{20\ g \times (-70\ K)} = 0.4 \frac{J}{g \cdot K}$, so the answer seems correct.

Follow-Up Problem 5.4 A 12.18 g sample of a shiny, orange-brown metal is heated to 65.00°C in a controlled water bath. The metal is then added to 25.00 g of water in a coffee-cup calorimeter, and the water temperature changes from 25.55°C to 27.25°C. What is the unknown metal (see Table 5.2)?

In the next sample problem, the calorimeter is used to study the change in heat during an aqueous acid-base reaction. Recall that, if a reaction takes place at constant pressure, the heat of the reaction (q_{rxn}) can be used to find its enthalpy change ΔH, in this case the enthalpy of neutralization, $\Delta_{neut}H$. Similar methods can be used to find enthalpies of dilution and enthalpies of dissolution of salts.

Sample Problem 5.5 Determining the Enthalpy Change of an Aqueous Reaction

Problem You place 50.0 mL of 0.500 mol/L NaOH in a coffee-cup calorimeter at 25.00°C and add 25.0 mL of 0.500 mol/L HCl, also at 25.00°C. After stirring, the final temperature is 27.21°C. [Assume that the total volume is the sum of the individual volumes. Also assume that the final solution has the same density (1.00 g/mL) and specific heat capacity (Table 5.2) as water.] **(a)** Calculate q_{soln} (J). **(b)** Calculate the change in enthalpy, ΔH, of the reaction (kJ/mol of H_2O formed).

(a) Calculate q_{soln}.

Plan The solution is the surroundings, and, as the reaction takes place, heat flows into the solution. To find q_{soln}, we use Equation 5.7, so we need the mass of the solution, the change in temperature, and the specific heat capacity. We know the volumes of the solutions (25.0 mL and 50.0 mL), so we find their masses with the given density (1.00 g/mL). Then, to find q_{soln}, we multiply the total mass by the given c [4.184 J/(g • K)] and the change in T, which we find from $T_{final} - T_{initial}$.

Solution Find $m_{tot,\ soln}$ and ΔT_{soln}:

$$m_{tot,\ soln} = (25.0\ mL + 50.0\ mL) \times 1.00\ g/mL = 75.0\ g$$
$$\Delta T_{soln} = 27.21°C - 25.00°C = 2.21°C = 2.21\ K$$

Find q_{soln}:

$$q_{soln} = c_{soln} \times m_{tot,\ soln} \times \Delta T_{soln} = \left(4.184 \frac{J}{g \cdot K}\right)(75.0\ g)(2.21\ K) = 693\ J$$

Check Rounding to check q_{soln} gives 4 J/(g • K) × 75 g × 2 K = 600 J, which is close to the answer.

(b) Calculate the change in enthalpy for the neutralization ($\Delta_{neut}H$).

Plan To find $\Delta_{neut}H$, we write the balanced equation for the acid-base reaction and use the volumes and the concentrations (0.500 mol/L) to find the amount (mol) of each reactant (H^+ and OH^-). Since the amounts of two reactants are given, we determine which reactant is limiting, that is, which reactant gives less product (H_2O). The heat of the surroundings is q_{soln}, and it is the negative of the heat of the reaction (q_{rxn}), which equals $\Delta_{neut}H$. Dividing q_{rxn} by the amount (mol) of water formed gives $\Delta_{neut}H$ in kJ per mol of water formed.

Solution Write the balanced equation:

$$HCl(aq) + NaOH(aq) \longrightarrow H_2O(l) + NaCl(aq)$$

Find the amount (mol) of reactants:

$$\text{Amount (mol) of HCl} = n_{HCl} = 0.500 \text{ mol HCl/L} \times 0.0250 \text{ L} = 0.0125 \text{ mol HCl}$$

$$\text{Amount (mol) of NaOH} = n_{NaOH} = 0.500 \text{ mol NaOH/L} \times 0.0500 \text{ L} = 0.0250 \text{ mol NaOH}$$

Now we need to find the amount (mol) of product. All the coefficients in the equation are 1, which means that the amount (mol) of reactant yields that amount of product. Therefore, HCl is limiting because it yields less product: 0.0125 mol of H_2O.

Finally, we need to find $\Delta_{neut}H$. Heat absorbed by the solution was released by the reaction:

$$q_{soln} = -q_{rxn} = 693 \text{ J} \quad \text{so} \quad q_{rxn} = -693 \text{ J}$$

$$\Delta_{neut}H \text{ (kJ/mol)} = \frac{q_{rxn}}{\text{amount (mol) } H_2O} \times \frac{1 \text{ kJ}}{1000 \text{ J}}$$

$$= \frac{-693 \text{ J}}{0.0125 \text{ mol } H_2O} \times \frac{1 \text{ kJ}}{1000 \text{ J}} = -55.4 \text{ kJ/mol } H_2O$$

Check We check for the limiting reactant. The volume of H^+ is half the volume of OH^-, but they have the same concentration and a 1:1 stoichiometric ratio. Therefore, the amount (mol) of H^+ determines the amount of product. Rounding and taking the negative of q_{soln} to find ΔH gives $\frac{-600 \text{ J}}{0.012 \text{ mol}} = -5 \times 10^4$ J/mol, or −50 kJ/mol, so the answer seems correct.

Follow-Up Problem 5.5 After 50.0 mL of 0.500 mol/L $Ba(OH)_2$ and the same volume and concentration of HCl react in a coffee-cup calorimeter, you find q_{soln} to be 1.386 kJ. Calculate $\Delta_{neut}H$ of the reaction, in kJ/mol of H_2O formed.

Two other types of constant-pressure calorimetry that are encountered increasingly are ice calorimetry and microcalorimetry. In ice calorimetry, a known mass of ice in a constant-pressure calorimeter surrounds the sample or the reaction whose enthalpy is to be measured. The amount of heat generated by the sample or the reaction causes the ice to melt. The mass of ice melted is multiplied by the heat of fusion of ice to determine the amount of heat generated in the reaction or by the sample. Microcalorimetry is an increasingly useful technique that allows us to measure the amount of heat generated by processes that occur over long periods of time, from days to years. Microcalorimetry measurements are subject to stringent conditions, and samples must be carefully prepared and even more carefully introduced into the microcalorimeter in order to ensure the integrity of the measurements.

Constant-Volume Calorimetry Constant-volume calorimetry is often carried out in a *bomb calorimeter*, a device commonly used to measure the heat of combustion reactions, such as those for fuels and foods. In the coffee-cup calorimeter, we assume that all the heat is absorbed by the water. In reality, however, some must be absorbed by the stirrer, thermometer, and so on. With the much more precise bomb calorimeter, the *heat capacity of the entire calorimeter* is known (or can be determined).

Figure 5.10 depicts the pre-weighed combustible sample in a metal-walled chamber (the bomb), which is filled with oxygen gas and immersed in an insulated water bath fitted with a motorized stirrer and thermometer. A heating coil connected to an electrical source ignites the sample, and the heat released raises the temperature of the bomb, water, and other calorimeter parts. Because we know the mass of the sample and the heat capacity of the entire calorimeter, we can use the measured ΔT to calculate the heat released.

Note that the steel bomb is tightly sealed, not open to the atmosphere (like the coffee cup is), so the pressure is *not* constant. And the volume of the bomb is fixed, so $\Delta V = 0$ and, thus, $p\Delta V = 0$. Therefore, the energy change measured is the *heat released at constant volume* (q_V), which equals ΔU, not ΔH:

$$\Delta U = q + w = q_V + 0 = q_V$$

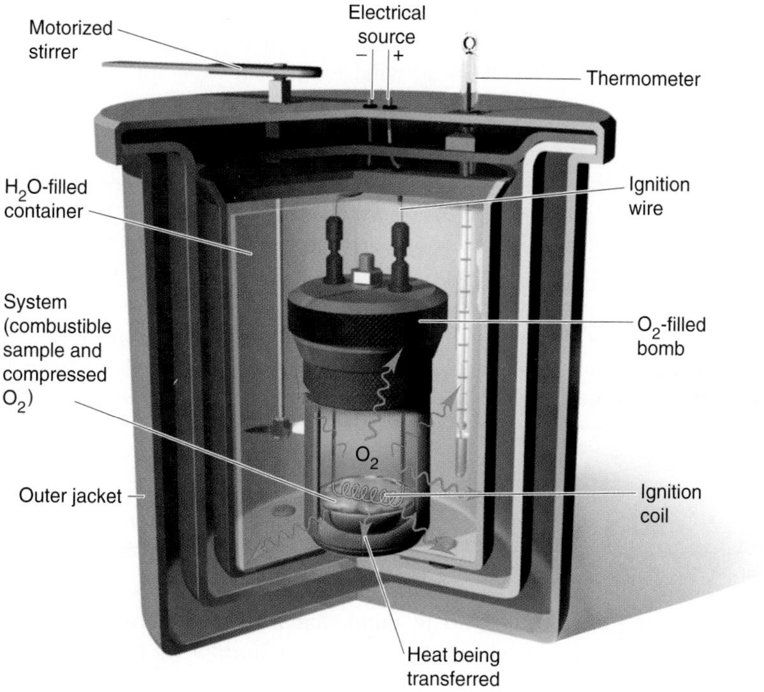

FIGURE 5.10 A bomb calorimeter. This device measures the heat released at constant volume (q_V). It is often used to study combustion reactions.

Recall, from Section 5.2, however, that even though the amount (mol) of gas may change, $p\Delta V$ is usually *much* less than ΔU, so ΔH is very close to ΔU. For example, ΔH is only 0.5% larger than ΔU for the combustion of H_2 and only 0.2% smaller than ΔU for the combustion of octane.

Sample Problem 5.6 Calculating the Heat of a Combustion Reaction

Problem A manufacturer claims that its new dietetic dessert has "less than 10 Cal per serving." To test this claim, a chemist at the Office of Consumer Affairs of Industry Canada places one serving in a bomb calorimeter and burns it in O_2. The initial temperature is 21.862°C, and the temperature rises to 26.799°C. If the heat capacity of the calorimeter is 8.151 kJ/K, is the manufacturer's claim correct?

Plan When the dessert (system) burns, the heat released is absorbed by the calorimeter:

$$-q_{\text{system}} = q_{\text{calorimeter}}$$

The claim is correct if the heat of the system is less than 10 Cal. To find the heat value, we multiply the given heat capacity, C, of the calorimeter (8.151 kJ/K) by ΔT and then convert the result to Cal. Note that the heat capacity, C, will have units of J/K rather than $\frac{\text{J}}{\text{mol}\cdot\text{K}}$ or $\frac{\text{J}}{\text{g}\cdot\text{K}}$ as it represents the amount of heat absorbed by the entire calorimeter ensemble.

Solution Find ΔT:

$$\Delta T = T_{\text{final}} - T_{\text{initial}} = 26.799°\text{C} - 21.862°\text{C} = 4.937°\text{C} = 4.937\ \text{K}$$

Calculate the heat absorbed by the calorimeter:

$$q_{\text{calorimeter}} = C \times \Delta T = 8.151\ \text{kJ/K} \times 4.937\ \text{K} = 40.24\ \text{kJ}$$

Recall that 1 Cal = 1 kcal = 4.184 kJ. Therefore, 10 Cal = 41.84 kJ, so the claim is correct.

Check A quick math check shows that the answer is reasonable: 8 kJ/K × 5 K = 40 kJ.

Follow-Up Problem 5.6 A chemist burns 0.8650 g of graphite (a form of carbon) in a new bomb calorimeter, and CO_2 forms. If 393.5 kJ of heat is released per mole of graphite, and ΔT is 2.613 K, what is the heat capacity of the bomb calorimeter?

SUMMARY OF SECTION 5.3

- We calculate ΔH of a process by measuring the energy transferred as heat at constant pressure (q_p). To do this, we determine ΔT and then multiply ΔT by the mass of the substance and by its specific heat capacity (c), which is the quantity of energy needed to raise the temperature of 1 g of the substance by 1 K.
- Calorimeters measure the heat released (or absorbed) during a process, either at constant pressure (coffee-cup calorimeter; $q_p = \Delta H$) or at constant volume (bomb calorimeter; $q_V = \Delta U$).

5.4 Stoichiometry of Thermochemical Equations

A **thermochemical equation** is *a balanced equation that includes the enthalpy change of the reaction (ΔH).* Keep in mind that a given ΔH refers only to the *amounts (mol) of substances* **and** *their states of matter in the equation.* The enthalpy change of any process has two aspects:

- *Sign.* The sign of ΔH depends on whether the reaction is exothermic (−) or endothermic (+). A forward reaction has the *opposite* sign of the reverse reaction.

  Decomposition of 2 mol of water to its elements (endothermic):

  $$2H_2O(l) \longrightarrow 2H_2(g) + O_2(g) \qquad \Delta H = 572 \text{ kJ/mol}$$

  Formation of 2 mol of water from its elements (exothermic):

  $$2H_2(g) + O_2(g) \longrightarrow 2H_2O(l) \qquad \Delta H = -572 \text{ kJ/mol}$$

  In these two reactions, notice that the units for the enthalpy are kilojoules per mole (kJ/mol). Here, the mole represents 1 mol *of the reaction as written.* In other words, for the reaction describing the decomposition of water to its elements, we can state that 572 kJ of energy is required to decompose 2 mol of liquid water, *or* 572 kJ of energy is required to form 2 mol of hydrogen gas, *or* 572 kJ of energy is required to form 1 mol of oxygen gas. All three statements are equivalent. Similarly, for the formation of water from its elements, 572 kJ of energy is released upon the reaction of 2 mol of hydrogen gas, the reaction of 1 mol of oxygen gas, and/or the formation of 2 mol of liquid water.
- *Magnitude.* The magnitude of ΔH is *proportional to the amount of substance.*

  Formation of 1 mol of water from its elements (half the preceding amount):

  $$H_2(g) + \frac{1}{2}O_2(g) \longrightarrow H_2O(l) \qquad \Delta H = -286 \text{ kJ/mol}$$

  Notice that, for this reaction, the formation of 1 mol of liquid H_2O releases 286 kJ of energy. This statement is in agreement with the statement derived for the doubled reactions.

Keep in mind the following key points about thermochemical equations:

1. *Balancing coefficients.* When necessary, we use fractional coefficients to balance an equation, because we are specifying the magnitude of ΔH for a *particular reaction as written:*

$$\frac{1}{8}S_8(s) + O_2(g) \longrightarrow SO_2(g) \qquad \Delta H = -296.8 \text{ kJ/mol}$$

for the reaction of $\frac{1}{8}$ mol of $S_8(s)$, the reaction of 1 mol of $O_2(g)$, or the formation of 1 mol of $SO_2(g)$.

2. *Thermochemical equivalence.* We can consider the above statement in a different way. For a *particular reaction*, a certain amount of substance is thermochemically equivalent to a certain quantity of energy. For example, in the preceding reaction,

296.8 kJ is thermochemically equivalent to $\frac{1}{8}$ mol of $S_8(s)$

296.8 kJ is thermochemically equivalent to 1 mol of $O_2(g)$

296.8 kJ is thermochemically equivalent to 1 mol of $SO_2(g)$

Just as we use stoichiometrically equivalent molar ratios to find amounts of substances, we use thermochemically equivalent quantities to find the ΔH of a reaction for a given amount of substance. Also, just as we use molar mass (g/mol) to convert an amount (mol) of a substance to mass (g), we use the ΔH (kJ/mol) to convert an amount of a substance to an equivalent quantity of heat (kJ). Figure 5.11 shows this new relationship, and Sample Problem 5.7 applies it.

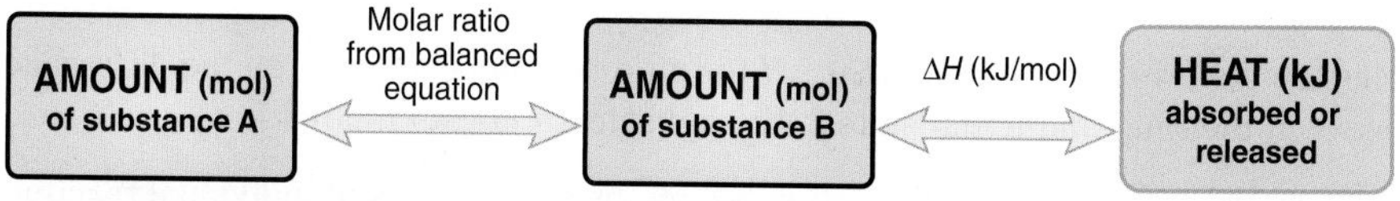

FIGURE 5.11 The relationship between amount (mol) of substance and the energy (kJ) transferred as heat during a reaction

Sample Problem 5.7 Using the Enthalpy Change of a Reaction (ΔH) to Find Amounts of Substance

Problem The major source of aluminum in the world is bauxite (mostly aluminum oxide). Its thermal decomposition can be written as follows:

$$Al_2O_2(s) \xrightarrow{\Delta} 2\,Al(s) + \frac{3}{2}O_2(g) \qquad \Delta H = 1676\ \text{kJ/mol}$$

If aluminum is produced this way (see Comment below), what mass (g) of aluminum can form when 1.000×10^3 kJ of heat is transferred?

Road Map

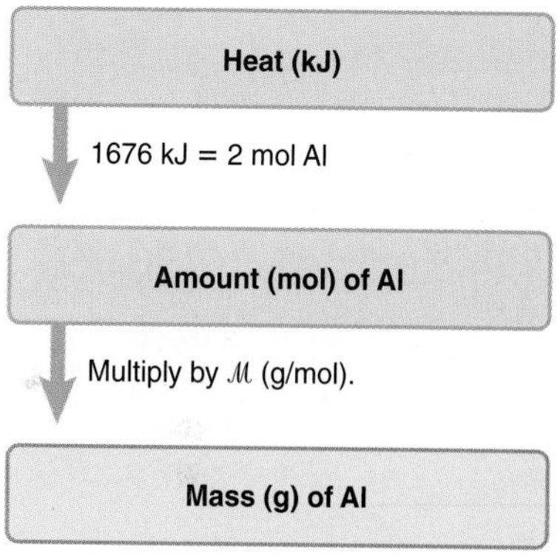

Plan From the balanced equation and the enthalpy change, we see that 2 mol of Al(*s*) forms when 1676 kJ of heat is absorbed. With this equivalent quantity, we convert the given heat energy (kJ) to amount (mol) formed and then convert amount to mass (g).

Solution Combine steps to convert from heat transferred to mass of Al:

$$\text{Mass (g) of Al} = (1.000 \times 10^3\ \text{kJ}) \times \frac{2\ \text{mol Al}}{1676\ \text{kJ}} \times \frac{26.98\ \text{g Al}}{1\ \text{mol Al}} = 32.20\ \text{g Al}$$

Check The mass of aluminum seems correct: ~1700 kJ forms about 2 mol of Al (54 g), so 1000 kJ should form a bit more than half this amount (27 g).

Comment In practice, aluminum is not obtained by heating bauxite but by supplying electrical energy (as you will see in Chapter 23). Because *H* is a state function, however, the total energy required for the change, ΔH, is the same no matter how it occurs.

Follow-Up Problem 5.7 Hydrogenation reactions, in which H_2 and an "unsaturated" organic compound combine, are used in the food, fuel, and polymer industries. In the simplest case, ethene (C_2H_4) and H_2 form ethane (C_2H_6). If 137 kJ is released per mole of C_2H_4 reacting, how much heat is released when 15.0 kg of C_2H_6 forms?

SUMMARY OF SECTION 5.4

- A thermochemical equation shows a balanced reaction *and* its ΔH value in kJ/mol. The sign of ΔH for a forward reaction is opposite the sign for the same reverse reaction. The magnitude of ΔH is specific for the given equation.
- The amount of a substance and the quantity of heat specified by the balanced equation are thermochemically equivalent and act as conversion factors to find the quantity of heat transferred when any amount of the substance reacts.

5.5 Hess's Law: Finding ΔH of Any Reaction

Some reactions are difficult, even impossible, to carry out individually: a reaction may be part of a complex biochemical process, or take place under extreme conditions, or require a change in conditions to occur. Even if we cannot run a reaction in a laboratory, we can still find its enthalpy change. In fact, the state-function property of enthalpy (H) allows us to find ΔH of *any* reaction for which we can write an equation.

This application is based on **Hess's law**: *the enthalpy change of an overall process is the sum of the enthalpy changes of its individual steps:*

$$\Delta_{\text{overall}}H = \Delta_1 H + \Delta_2 H + \cdots + \Delta_n H \qquad (5.8)$$

$$\Delta_r H = \sum_n \Delta_n H$$

This law follows from the fact that ΔH for a process depends only on the difference between the final and initial states. We apply Hess's law as follows:

- We imagine that an overall reaction occurs through a series of individual reaction steps, whether or not it actually does. Adding the steps must give the overall reaction.
- We choose individual reaction steps for which each step has a known ΔH.
- We add the known ΔH values for the steps to get the unknown ΔH for the overall reaction. We can also find an unknown ΔH value for any step by subtraction, if we know the ΔH values for the overall reaction and all the other steps.

Let us apply Hess's law to the oxidation of sulfur to sulfur trioxide, the key change in the industrial production of sulfuric acid and the formation of acid rain. When we burn S_8 in an excess of O_2, sulfur dioxide (SO_2) forms, *not* sulfur trioxide (SO_3) (Equation 1). After a change in conditions, we add more O_2 and oxidize SO_2 to SO_3 (Equation 2). Thus, we cannot put S_8 and O_2 in a calorimeter and find ΔH for the overall reaction of S_8 to SO_3 (Equation 3). However, we *can* find it with Hess's law. Here are the three equations:

$$\text{Equation 1: } \tfrac{1}{8}S_8(s) + O_2(g) \longrightarrow SO_2(g) \qquad \Delta_1 H = -296.8 \text{ kJ/mol}$$

$$\text{Equation 2: } 2SO_2(g) + O_2(g) \longrightarrow 2SO_3(g) \qquad \Delta_2 H = -198.4 \text{ kJ/mol}$$

$$\text{Equation 3: } \tfrac{1}{8}S_8(s) + \tfrac{3}{2}O_2(g) \longrightarrow SO_3(g) \qquad \Delta_3 H = ?$$

If we can manipulate Equation 1 and/or Equation 2, *along with their ΔH value(s)*, so that the equations add up to Equation 3, their ΔH values will add up to the unknown $\Delta_3 H$.

First, we identify our "target" equation, the one whose ΔH we want to find, and note the amount (mol) of each reactant and product; our "target" equation is Equation 3. Then we manipulate Equation 1 and/or Equation 2 to make them add up to Equation 3:

- Equations 3 and 1 have the same amount of S_8, so we do not change Equation 1.
- Equation 3 has half as much SO_3 as Equation 2, so we multiply Equation 2, *and* $\Delta_2 H$, by $\frac{1}{2}$; that is, we always treat the equation and its ΔH value in the same way.
- With the targeted amounts of reactants and products present, we add Equation 1 to the halved Equation 2 and cancel terms that appear on both sides:

$$\text{Equation 1: } \tfrac{1}{8}S_8(s) + O_2(g) \longrightarrow SO_2(g) \qquad \Delta_1 H = -296.8 \text{ kJ/mol}$$

$$\tfrac{1}{2}(\text{Equation 2}): \; SO_2(g) + \tfrac{1}{2}O_2(g) \longrightarrow SO_3(g) \qquad \tfrac{1}{2}(\Delta_2 H) = -99.2 \text{ kJ/mol}$$

$$\text{Equation 3: } \tfrac{1}{8}S_8(s) + O_2(g) + \cancel{SO_2(g)} + \tfrac{1}{2}O_2(g) \longrightarrow \cancel{SO_2(g)} + SO_3(g)$$

$$\tfrac{1}{8}S_8(s) + \tfrac{3}{2}O_2(g) \longrightarrow SO_3(g) \qquad \Delta_3 H = -396.0 \text{ kJ/mol}$$

Because ΔH depends only on the difference between H_{final} and H_{initial}, Hess's law tells us that the difference between the enthalpies of the reactants ($\frac{1}{8}$ mol of S_8 and $\frac{3}{2}$ mol of O_2) and the product (1 mol of SO_3) is the same, whether S_8 is oxidized directly to SO_3 (impossible) or through the intermediate formation of SO_2 (actual).

To summarize, calculating an unknown ΔH involves three steps:

1. Identify the target equation, the step whose ΔH is unknown, and note the amount (mol) of each reactant and product.
2. Manipulate each equation with known ΔH values so that the target amount (mol) of each substance is on the correct side of the equation. Remember:
   - Change the sign of ΔH when you reverse an equation.
   - Multiply amount (mol) and ΔH by the same factor.
3. Add the manipulated equations and their resulting ΔH values to get the target equation and its ΔH. All substances except those in the target equation must cancel.

Sample Problem 5.8 Using Hess's Law to Calculate an Unknown ΔH

Problem Two pollutants that form in auto exhaust are CO and NO. An environmental chemist must convert these pollutants to less harmful gases through the following reaction:

$$CO(g) + NO(g) \longrightarrow CO_2(g) + \frac{1}{2}N_2(g) \qquad \Delta_C H = ?$$

Given the following information, calculate the unknown $\Delta_C H$:

Equation A: $CO(g) + \frac{1}{2}O_2(g) \longrightarrow CO_2(g) \quad \Delta_A H = -283.0\ \text{kJ/mol}$

Equation B: $N_2(g) + O_2(g) \longrightarrow 2NO(g) \quad \Delta_B H = 180.6\ \text{kJ/mol}$

Plan We note the amount (mol) of each substance and the side on which each substance appears in the target equation. We manipulate Equation A and/or Equation B *and* their ΔH values as needed and add them together to obtain the target equation and the unknown ΔH.

Solution We start by looking at the substances in the target equation. For reactants, there is 1 mol of both CO and of NO; for products, there is 1 mol of CO_2 and $\frac{1}{2}$ mol of N_2.

Manipulate the given equations:

- Equation A has the same amounts of CO and CO_2 on the same sides of the arrow as the target equation, so we leave it as written.
- Equation B has twice as much N_2 and NO as the target equation, and they are on the opposite sides in the target. Thus, we reverse Equation B, change the sign of its ΔH, and multiply both by $\frac{1}{2}$:

$-\frac{1}{2}$(Equation B): $\frac{1}{2}[2NO(g) \longrightarrow N_2(g) + O_2(g)] \quad \Delta_{-\frac{1}{2}B}H = -\frac{1}{2}(\Delta_B H) = -\frac{1}{2}(180.6\ \text{kJ/mol})$

or $NO(g) \longrightarrow \frac{1}{2}N_2(g) + \frac{1}{2}O_2(g) \quad \Delta_{-\frac{1}{2}B}H = -90.3\ \text{kJ/mol}$

Add the manipulated equations to obtain the target equation:

Equation A: $CO(g) + \cancel{\frac{1}{2}O_2(g)} \longrightarrow CO_2(g) \quad \Delta_A H = -283.0\ \text{kJ/mol}$

$-\frac{1}{2}$(Equation B): $NO(g) \longrightarrow \frac{1}{2}N_2(g) + \cancel{\frac{1}{2}O_2(g)} \quad \Delta_{-\frac{1}{2}B}H = -90.3\ \text{kJ/mol}$

Target equation: $CO(g) + NO(g) \longrightarrow CO_2(g) + \frac{1}{2}N_2(g) \quad \Delta_C H = -373.3\ \text{kJ/mol}$

Check Obtaining the desired target equation is a sufficient check. Be sure to remember to change the *sign* of ΔH for any equation you reverse.

Follow-Up Problem 5.8 Nitrogen oxides undergo many reactions in the environment and in industry. Given the following information, calculate ΔH for the overall equation $2NO_2(g) + \frac{1}{2}O_2(g) \longrightarrow N_2O_5(s)$:

$$N_2O_5(s) \longrightarrow 2NO(g) + \frac{3}{2}O_2(g) \qquad \Delta_1 H = 223.7\ \text{kJ/mol}$$

$$NO(g) + \frac{1}{2}O_2(g) \longrightarrow NO_2(g) \qquad \Delta_2 H = -57.1\ \text{kJ/mol}$$

SUMMARY OF SECTION 5.5

- Because H is a state function, we can use Hess's law to determine ΔH of any reaction by assuming that it is the sum of other reactions.
- After manipulating the equations of those reactions and their ΔH values to match the substances in the target equation, we add the manipulated ΔH values to find the unknown ΔH.

5.6 Standard Enthalpies of Reaction ($\Delta_r H°$)

In this section, we see how Hess's law can be used to determine the ΔH values of an enormous number of reactions. Thermodynamic variables, such as ΔH, vary somewhat with conditions. Therefore, in order to study and compare reactions, chemists have established a set of specific conditions called **standard states**:

- For a *gas*, the standard state is 1 bar and ideal behaviour.
- For a substance in *aqueous solution*, the standard state is 1 mol/L concentration.
- For a *pure substance* (element or compound), the standard state is usually the most stable form of the substance at 1 bar and the temperature of interest. In this book (and in most thermodynamic tables), the temperature of interest is usually 25°C (298 K).
- For an element in its standard state, the standard enthalpy of formation is zero, $\Delta_f H° = 0$. Here, it is very important to remember that it is the standard state for which the enthalpy of formation is zero. For example, the standard state of bromine, Br_2, is a liquid. The enthalpy of formation of liquid bromine is zero. However, the enthalpy of formation of gaseous bromine is *not* zero. Another case where the standard state is important is for elements that exist in different allotropic forms in the same phase. One example is carbon, which exists as both diamond and graphite. Of these two solid forms, carbon (*s*, graphite) is the standard state and not carbon (*s*, diamond).

The standard-state symbol (shown as a degree sign) indicates that the variable has been measured with *all the substances in their standard states*. For example, *when the enthalpy change of a reaction is measured at the standard state*, it is the **standard enthalpy of reaction ($\Delta_r H°$**, where "r" stands for reaction; also called the *standard heat of reaction*).

Table 5.3 Selected Standard Enthalpies of Formation at 25°C (298.15 K)

Formula	$\Delta_f H°$ (kJ/mol)
Calcium	
Ca(*s*)	0
CaO(*s*)	−635.1
$CaCO_3$(*s*)	−1206.9
Carbon	
C(*s*, graphite)	0
C(*s*, diamond)	1.9
CO(*g*)	−110.5
CO_2(*g*)	−393.5
CH_4(*g*)	−74.9
CH_3OH(*l*)	−238.6
HCN(*g*)	135
CS_2(*l*)	87.9
Chlorine	
Cl(*g*)	121.0
Cl_2(*g*)	0
HCl(*g*)	−92.3
Hydrogen	
H(*g*)	218.0
H_2(*g*)	0
Nitrogen	
N_2(*g*)	0
NH_3(*g*)	−45.9
NO(*g*)	90.3
Oxygen	
O_2(*g*)	0
O_3(*g*)	143
H_2O(*g*)	−241.8
H_2O(*l*)	−285.8
Silver	
Ag(*s*)	0
AgCl(*s*)	−127.0
Sodium	
Na(*s*)	0
Na(*g*)	107.8
NaCl(*s*)	−411.1
Sulfur	
S_8(*s*, rhombic)	0
S_8(*s*, monoclinic)	0.3
SO_2(*g*)	−296.8
SO_3(*g*)	−396.0

Formation Equations and Their Standard Enthalpy Changes

The **formation equation** is the chemical equation that represents the formation of 1 mol of a pure substance from its elements. The **standard enthalpy of formation ($\Delta_f H°$)** (also called the *standard heat of formation*) is the enthalpy change for the formation equation when all the substances are in their standard states. For instance, the formation equation for methane (CH_4) is

$$C(s, \text{graphite}) + 2H_2(g) \longrightarrow CH_4(g) \qquad \Delta_f H° = -74.9 \text{ kJ/mol}$$

Fractional coefficients are often used with reactants to obtain 1 mol of the product:

$$Na(s) + \frac{1}{2}Cl_2(g) \longrightarrow NaCl(s) \qquad \Delta_f H° = -411.1 \text{ kJ/mol}$$

$$2C(s, \text{graphite}) + 3H_2(g) + \frac{1}{2}O_2(g) \longrightarrow C_2H_5OH(l) \qquad \Delta_f H° = -277.6 \text{ kJ/mol}$$

Standard enthalpies of formation have been tabulated for many substances. Table 5.3 shows several, and a much more extensive table appears in Appendix B. The values in Table 5.3 reiterate one of the points mentioned above and show us another interesting point:

1. *For an element in its standard state,* $\Delta_f H° = 0$.
   - The standard state for metals, such as sodium, is the solid ($\Delta_f H° = 0$); it takes 107.8 kJ of heat to form 1 mol of gaseous Na ($\Delta_f H° = 107.8$ kJ/mol).
   - The standard state for molecular elements, such as the halogens, is the molecular form, not separate atoms; for Cl_2, $\Delta_f H° = 0$, but for Cl, $\Delta_f H° = 121.0$ kJ/mol.
   - Some elements exist in different forms (called *allotropes*; Chapter 13), but only one is the standard state. The standard state of carbon is graphite ($\Delta_f H° = 0$), not diamond ($\Delta_f H° = 1.9$ kJ/mol); the standard state of oxygen is O_2 ($\Delta_f H° = 0$), not ozone (O_3; $\Delta_f H° = 143$ kJ/mol); and the standard state of sulfur is S_8 in its rhombic crystal form ($\Delta_f H° = 0$), not in its monoclinic form ($\Delta_f H° = 0.3$ kJ/mol).
2. *Most compounds have a negative* $\Delta_f H°$. That is, most compounds have exothermic formation reactions: *under standard conditions, heat is released when most compounds form from their elements.*

Sample Problem 5.9 Writing Formation Equations

Problem Write a balanced formation equation for each compound, including the value of $\Delta_f H°$:

(a) $AgCl(s)$ **(b)** $CaCO_3(s)$ **(c)** $HCN(g)$

Plan We write the elements as the reactants, and we write 1 mol of the compound as the product, being sure that all the substances are in their standard states. Then we balance the equations and find the $\Delta_f H°$ values in Table 5.3 or Appendix B.

Solution **(a)** $Ag(s) + \frac{1}{2}Cl_2(g) \longrightarrow AgCl(s) \quad \Delta_f H° = -127.0$ kJ/mol

(b) $Ca(s) + C(s, \text{graphite}) + \frac{3}{2}O_2(g) \longrightarrow CaCO_3(s) \quad \Delta_f H° = -1206.9$ kJ/mol

(c) $\frac{1}{2}H_2(g) + C(s, \text{graphite}) + \frac{1}{2}N_2(g) \longrightarrow HCN(g) \quad \Delta_f H° = 135$ kJ/mol

Follow-Up Problem 5.9 Write a balanced formation equation for each compound, including the value of $\Delta_f H°$:

(a) $CH_3OH(l)$ **(b)** $CaO(s)$ **(c)** $CS_2(l)$

Determining $\Delta_r H°$ from $\Delta_f H°$ Values for Reactants and Products

We can use $\Delta_f H°$ values to determine $\Delta_r H°$ for any reaction. By applying Hess's law, we can imagine the reaction occurring in two steps (Figure 5.12):

Step 1. Each reactant decomposes to its elements. This is the *reverse* of the formation reaction for the *reactant*, so the standard enthalpy change is $-\Delta_f H°$.

Step 2. Each product forms from its elements. This step is the formation reaction for the *product*, so the standard enthalpy change is $\Delta_f H°$.

According to Hess's law, we add the enthalpy changes for these steps to obtain the overall enthalpy change for the reaction ($\Delta_r H°$). Suppose that we want $\Delta_r H°$ for

$$TiCl_4(l) + 2H_2O(g) \longrightarrow TiO_2(s) + 4HCl(g)$$

We write this equation as though it were the sum of four individual equations, one for each compound. The first two equations show step 1, the decomposition of the reactants to their elements (*reverse* of their formation); the second two equations show step 2, the formation of the products from their elements:

$$\begin{array}{lr} TiCl_4(l) \longrightarrow Ti(s) + 2Cl_2(g) & -\Delta_f H°[TiCl_4(l)] \\ 2H_2O(g) \longrightarrow 2H_2(g) + O_2(g) & -2\Delta_f H°[H_2O(g)] \\ Ti(s) + O_2(g) \longrightarrow TiO_2(s) & \Delta_f H°[TiO_2(s)] \\ 2H_2(g) + 2Cl_2(g) \longrightarrow 4HCl(g) & 4\Delta_f H°[HCl(g)] \end{array}$$

$$TiCl_4(l) + 2H_2O(g) + \cancel{Ti(s)} + \cancel{O_2(g)} + \cancel{2H_2(g)} + \cancel{2Cl_2(g)} \longrightarrow \cancel{Ti(s)} + \cancel{2Cl_2(g)} + \cancel{2H_2(g)} + \cancel{O_2(g)} + TiO_2(s) + 4HCl(g)$$

or

$$TiCl_4(l) + 2H_2O(g) \longrightarrow TiO_2(s) + 4HCl(g)$$

It is important to realize that when titanium(IV) chloride and water react, the reactants do not *actually* decompose to their elements, which then recombine to form

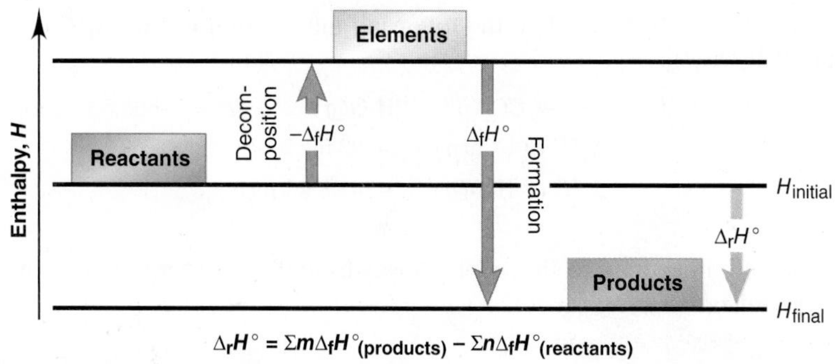

FIGURE 5.12 The two-step process for determining $\Delta_r H°$ from $\Delta_f H°$ values

the products. The great usefulness of Hess's law and the state-function concept is that $\Delta_r H°$ is the difference between two state functions, $H°_{products}$ minus $H°_{reactants}$, so how the reaction *actually* occurs does not matter. We add the individual enthalpy changes to find $\Delta_r H°$:

$$\Delta_r H° = \Delta_f H°[TiO_2(s)] + 4\Delta_f H°[HCl(g)] + \{-\Delta_f H°[TiCl_4(l)]\} + \{-2\Delta_f H°[H_2O(g)]\}$$
$$= \underbrace{\Delta_f H°[TiO_2(s)] + 4\Delta_f H°[HCl(g)]}_{\text{Products}} - \underbrace{\Delta_f H°[TiCl_4(l)] + 2\Delta_f H°[H_2O(g)]}_{\text{Reactants}}$$

The arithmetic in this case gives $\Delta_r H°$ + −25.39 kJ/mol. More important, however, when we generalize the pattern, we see that *the standard enthalpy of reaction is the sum of the standard enthalpies of formation of the* ***products*** *minus the sum of the standard enthalpies of formation of the* ***reactants*** (see Figure 5.12):

$$\Delta_r H° = \Sigma m\Delta_f H°_{(products)} - \Sigma n\Delta_f H°_{(reactants)} \quad (5.9)$$

where Σ means "sum of" and m and n are the amounts (mol) of the products and reactants given by the coefficients in the balanced equation.

Sample Problem 5.10 Calculating $\Delta_r H°$ from $\Delta_f H°$ Values

Problem Nitric acid is used to make many products, including fertilizers, dyes, and explosives. The first step in its production is the oxidation of ammonia:

$$4NH_3(g) + 5O_2(g) \longrightarrow 4NO(g) + 6H_2O(g)$$

Calculate $\Delta_r H°$ from $\Delta_f H°$ values.

Plan We use values from Table 5.3 (or Appendix B) and apply Equation 5.9.

Solution Calculate $\Delta_r H°$:

$$\begin{aligned}\Delta_r H° &= \Sigma m\Delta_f H°_{(products)} - \Sigma n\Delta_f H°_{(reactants)}\\ &= \{4\Delta_f H°[NO(g)] + 6\Delta_f H°[H_2O(g)]\} - \{4\Delta_f H°[NH_3(g)] + 5\Delta_f H°[O_2(g)]\}\\ &= [4(90.3\text{ kJ/mol}) + 6(-241.8\text{ kJ/mol})]\\ &\quad -[4(-45.9\text{ kJ/mol}) + 5(0\text{ kJ/mol})]\\ &= 361.2\text{ kJ/mol} - 1450.8\text{ kJ/mol} + 183.6\text{ kJ/mol} - 0\text{ kJ/mol} = -906\text{ kJ/mol}\end{aligned}$$

Check We write formation equations, with $\Delta_f H°$ values for the amounts of the compounds, in the correct direction (forward for products and reverse for reactants) and find the sum:

$$\begin{array}{llr}4NH_3(g) \longrightarrow \cancel{2N_2(g)} + \cancel{6H_2(g)} & -4\Delta_f H° = -4(-45.9\text{ kJ}) = & 183.6\text{ kJ/mol}\\ \cancel{2N_2(g)} + 2O_2(g) \longrightarrow 4NO(g) & 4\Delta_f H° = 4(90.3\text{ kJ}) = & 361.2\text{ kJ/mol}\\ \cancel{6H_2(g)} + 3O_2(g) \longrightarrow 6H_2O(g) & 6\Delta_f H° = 6(-241.8\text{ kJ}) = & -1450.8\text{ kJ/mol}\\ \hline 4NH_3(g) + 5O_2(g) \longrightarrow 4NO(g) + 6H_2O(g) & \Delta_r H° = & -906\text{ kJ/mol}\end{array}$$

The mol^{-1} in the unit attached to the enthalpy value indicates that the value of the enthalpy is for 1 mol of the reaction *as written*.

Comment In this problem, we know the individual $\Delta_f H°$ values and find the sum, $\Delta_r H°$. In the follow-up problem, we know the sum and want to find one of the $\Delta_f H°$ values.

Follow-Up Problem 5.10 Use the following information to find $\Delta_f H°$ of methanol $[CH_3OH(l)]$:

$$CH_3OH(l) + \tfrac{3}{2}O_2(g) \longrightarrow CO_2(g) + 2H_2O(g) \quad \Delta_r H° = -638.5\text{ kJ/mol}$$
$$\Delta_f H° \text{ of } CO_2(g) = -393.5\text{ kJ/mol}$$
$$\Delta_f H° \text{ of } H_2O(g) = -241.8\text{ kJ/mol}$$

The Chemical Connections section applies ideas from this chapter to new approaches for energy utilization.

Out of necessity, we must rethink our global use of energy. The dwindling supplies of our most common fuels and the environmental impact of their combustion products threaten the well-being of humans and the survival of many other species. Energy production presents scientists, engineers, and political leaders with some of the greatest challenges of our time.

A changeover in energy sources from wood to coal and then to petroleum took place over the past century. The **fossil fuels**—coal, petroleum, and natural gas—remain our major sources, but they are *non-renewable* because the natural processes that form them are many orders of magnitude slower than our rate of consuming them. Around the world, ongoing research seeks new alternatives to using coal and new approaches to using nuclear energy. The use of *renewable* sources—biomass, hydrogen, sunlight, wind, geothermal heat, and tides—is the subject of intensive efforts as well.

In this section, we focus on converting coal and biomass to cleaner fuels, understanding the effects of carbon-based fuels on climate, developing hydrogen as a fuel, using solar energy, and conserving energy.

Converting Coal to Cleaner Fuels Although Canada is the world's second-largest generator of hydroelectric power (one of the cleanest forms of energy), more than a quarter of our energy is still generated by thermal means. A thermal power plant uses the heat generated by burning oil, natural gas, or coal to heat water; the steam produced by the heated water is used to move turbines. Of the fossil fuels, coal is the most used, accounting for 19% of Canada's net electricity generation. Canada has an enormous supply of hydrocarbon-based energy in the form of oil, oil sands, coal, and natural gas, but the combustion of these fuels produces CO_2, SO_2, NO_x (various oxides of nitrogen), particulates, and volatile organic compounds (VOCs), as well as releasing toxins such as mercury. Exposure to SO_2 and particulates causes respiratory diseases, and SO_2 and NO_x can be oxidized to H_2SO_4 and HNO_3, key components of acid rain (Chapter 17). The trace amounts of Hg, a neurotoxin, spread as Hg vapour and bioaccumulate in fish. Although the amount of mercury found in coal is small (parts per million or less), the huge quantity of coal that is burned to produce heat and power (67 Mt in Canada alone in 2012, based on data found at coal.ca/production/) leads to the accumulation of a significant amount of mercury.

Two processes reduce the amounts of SO_2:

1. *Desulfurization.* The removal of sulfur dioxide from flue gases is done with devices called *scrubbers* that heat powdered limestone ($CaCO_3$) or spray lime-water slurries [$Ca(OH)_2$]:

$$CaCO_3(s) + SO_2(g) \xrightarrow{\Delta} CaSO_3(s) + CO_2(g) \quad \textbf{(B5.1)}$$

$$2CaSO_3(s) + O_2(g) + 4H_2O(l) \longrightarrow 2CaSO_4{\cdot}2H_2O(s;\ \text{gypsum}) \quad \textbf{(B5.2)}$$

$$2Ca(OH)_2(aq) + 2SO_2(g) + O_2(g) + 2H_2O(l) \longrightarrow 2CaSO_4 \cdot 2H_2O(s) \quad \textbf{(B5.3)}$$

A drywall plant built next to the power plant sells the gypsum produced by the desulfurization process (almost 1 t per customer each year); about 20% of the drywall produced each year is made with synthetic gypsum.

2. *Gasification.* In **coal gasification**, solid coal is converted to sulfur-free gaseous fuels. In this process, the sulfur in the coal is reduced to H_2S, which is removed through an acid-base reaction with a base such as ethanolamine ($HOCH_2CH_2NH_2$). The resulting salt is heated to release H_2S, which is converted to elemental sulfur by the Claus process (Chapter 23) and sold. Several reactions yield mixtures with increasing fuel value, that is, with a lower C/H ratio:

- Pulverized coal reacts with limited O_2 and water at 800°C to 1500°C to form an approximately 2/1 mixture of CO/H_2.
- Alternatively, in the *water-gas reaction* (or *steam-carbon reaction*), an exothermic oxidation of C to CO is followed by the endothermic reaction of C with steam to form a nearly 1/1 mixture of CO/H_2, called *water gas*:

$$C(s) + \frac{1}{2}O_2(g) \longrightarrow CO(g) \qquad \Delta_rH^\circ = -110\ \text{kJ/mol} \quad \textbf{(B5.4)}$$

$$C(s) + H_2O(g) \longrightarrow CO(g) + H_2(g) \quad \Delta_rH^\circ = 131\ \text{kJ/mol} \quad \textbf{(B5.5)}$$

However, water gas has a much lower fuel value than methane (CH_4). For example, a mixture of 0.5 mol of CO and 0.5 mol of H_2 releases about one-third as much energy as 1.0 mol of methane ($\Delta H = -802$ kJ/mol):

$$\frac{1}{2}H_2(g) + \frac{1}{4}O_2(g) \longrightarrow \frac{1}{2}H_2O(g) \quad \Delta_rH^\circ = -121\ \text{kJ/mol} \quad \textbf{(B5.6)}$$

$$\frac{1}{2}CO(g) + \frac{1}{4}O_2(g) \longrightarrow \frac{1}{2}CO_2(g) \quad \Delta_rH^\circ = -142\ \text{kJ/mol} \quad \textbf{(B5.7)}$$

$$\frac{1}{2}H_2(g) + \frac{1}{2}CO(g) + \frac{1}{2}O_2(g) \longrightarrow \frac{1}{2}H_2O(g) + \frac{1}{2}CO_2(g) \quad \Delta_rH^\circ = -263\ \text{kJ/mol} \quad \textbf{(B5.8)}$$

- In the *CO-shift* (or *water-gas shift*) reaction, the H_2 content of water gas is increased to produce synthesis gas (*syngas*), a fuel-gas mixture that consists mainly of carbon monoxide, hydrogen, and some carbon dioxide:

$$CO(g) + H_2O(g) \longrightarrow CO_2(g) + H_2(g) \qquad \Delta_rH^\circ = -41\ \text{kJ/mol} \quad \textbf{(B5.9)}$$

- To produce CH_4, a syngas that is a 3/1 mixture of H_2/CO, from which the CO_2 has been removed, is used:

$$CO(g) + 3H_2(g) \longrightarrow CH_4(g) + H_2O(g) \qquad \Delta_rH^\circ = -206\ \text{kJ/mol} \quad \textbf{(B5.10)}$$

Drying the product gives **synthetic natural gas (SNG)**.

Syngas is used in some newer methods as well. In the *integrated gasification combined cycle (IGCC)*, electricity is produced in two ways: first when syngas is burned in a combustion turbine, and then when the hot product gases generate steam to power a steam turbine. In the *Fischer-Tropsch process*, syngas is used to form liquid hydrocarbon fuels that have higher molar masses than methane:

$$nCO(g) + (2n+1)H_2(g) \longrightarrow C_nH_{2n+2}(l) + nH_2O(g) \quad \textbf{(B5.11)}$$

(Continued)

Ontario had set a goal to phase out the use of coal by 2014, and in April 2014 the Thunder Bay Generating Station was the final coal plant to stop burning coal. Across Canada, other coal generation facilities will be retired by 2025. Ontario intends to replace the energy generated by coal with energy generated by wind, solar, and other cleaner energy sources.

Converting Biomass to Fuels Nearly half the world's people rely on wood for energy. In principle, wood is renewable, but widespread deforestation has resulted from its use for fuel, lumber, and paper. Therefore, great emphasis has been placed on **biomass conversion** of vegetable and animal waste. In one process, chemical and/or microbial methods convert vegetable (sugarcane, corn, and switchgrass) and tree waste into fuel, mainly ethanol (C_2H_5OH).

Methanogenesis forms methane by anaerobic (oxygen-free) biodegradation of organic sources, such as manure and vegetable waste or wastewater. China's and India's biogas-generating facilities apply this technique, and similar facilities in the United States use garbage and sewage. In addition to producing methane, the process yields residues that improve soil; it also prevents the wastes from polluting natural waters.

Another biomass conversion process changes vegetable oils (soybean, cottonseed, sunflower, canola, and even used cooking oil) into a mixture called *biodiesel*, whose combustion produces less CO, SO_2, and particulate matter than fossil fuel–based diesel. The CO_2 produced in the combustion of biodiesel has minimal impact on the net atmospheric amount because the carbon is taken up by the plants grown as sources of the vegetable oils.

The Greenhouse Effect and Global Warming Combustion of all carbon-based fuels releases CO_2, and, over the past few decades, it has become clear that the atmospheric buildup of this gas from our increased use of these fuels is changing the climate. Throughout Earth's history, CO_2 has played a key temperature-regulating role in the atmosphere. Much of the sunlight that shines on Earth is absorbed by the land and oceans and converted to heat infrared (IR) radiation; see Figure B5.1. Like the glass of a greenhouse, CO_2 does not absorb visible light from the Sun, but it absorbs and re-emits some of the heat radiating from Earth's surface and, thus, helps to warm the atmosphere. This process is called the *natural greenhouse effect* (Figure B5.1, *left*).

Over several billion years, due to the spread of plant life, which uses CO_2 in photosynthesis, the amount of CO_2 originally present (from volcanic activity) decreased to 0.028% by volume. With this amount, Earth's average temperature was about 14°C (57°F); without CO_2, it would be −17°C (2°F)! However, as a result of intensive use of fossil fuels for the past 150 years, this amount has increased to 0.039%. Thus, although the same quantity of solar energy passes through the atmosphere, more is being trapped as heat, which has created an *enhanced greenhouse effect* that is changing the climate through *global warming* (Figure B5.1, *right*). Based on current trends, CO_2 amounts will increase to between 0.049% and 0.126% by 2100 (Figure B5.2).

If this projected increase in CO_2 occurs, two closely related questions arise: (1) How much will the temperature rise? (2) How will this rise affect life on Earth? Despite constantly improving models, answers are difficult to obtain. Natural fluctuations in temperature

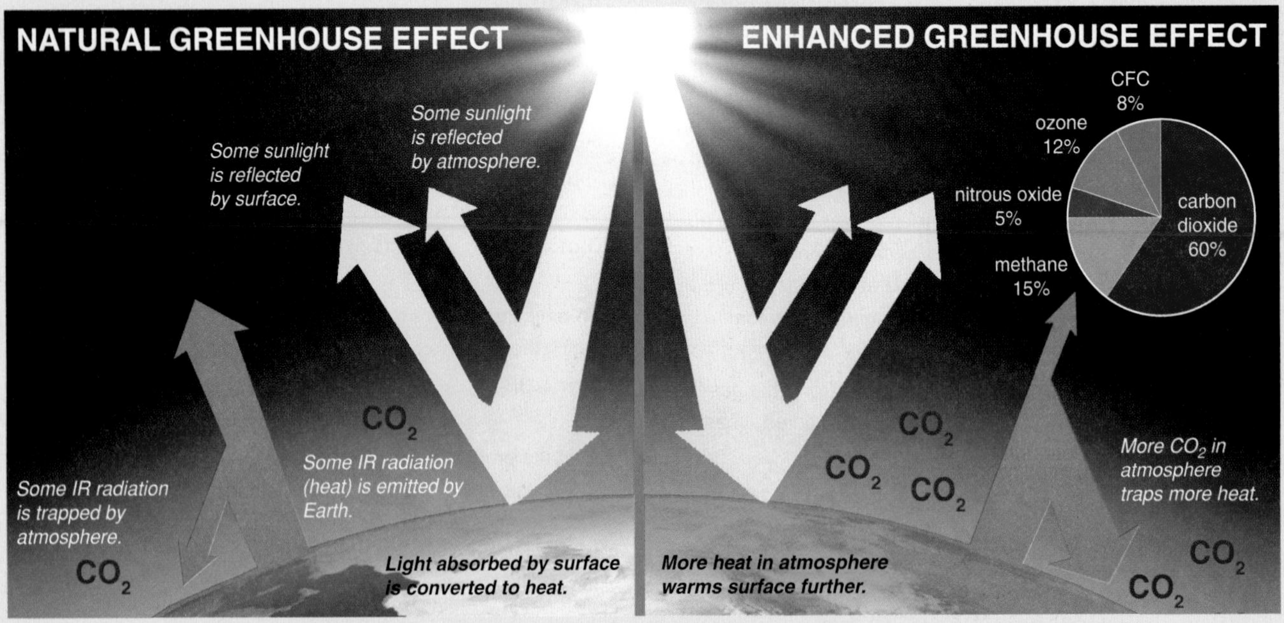

FIGURE B5.1 The trapping of heat by the atmosphere. Some of the sunlight that reaches Earth is reflected, and some is absorbed and converted to IR radiation (heat). Some of the heat that is emitted by the surface is trapped by atmospheric CO_2, creating a *natural* greenhouse effect (*left*) that has been essential to life. However, largely as a result of human activities in the past 150 years, and especially the past several decades, the buildup of CO_2 and several other greenhouse gases (*pie chart*) has created an *enhanced* greenhouse effect (*right*).

(Continued)

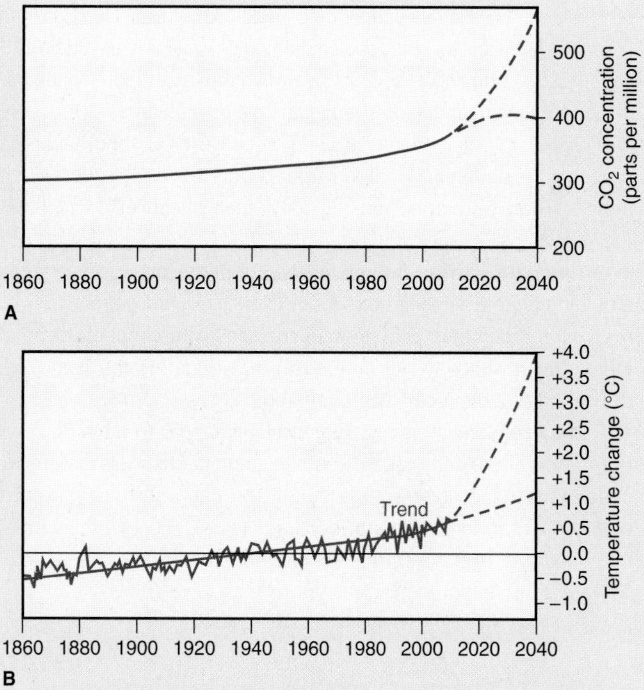

FIGURE B5.2 Evidence for the enhanced greenhouse effect. **A.** Since the mid-19th century, atmospheric CO_2 has increased. **B.** Since the mid-19th century, average global temperature has risen 0.6°C. (Zero is the average from 1957 to 1970.) The projections in the graphs (*dashed lines*) assume that current fossil fuel consumption and deforestation continue (*upper line*) or slow (*lower line*).

must be taken into account, as well as cyclic changes in solar activity. Moreover, as the amount of CO_2 increases from fossil-fuel burning, so do the amounts of particulate matter and SO_2, which may block sunlight and have a cooling effect. Water vapour also traps heat, and, as temperatures rise, more water evaporates. The increased amounts of water vapour may thicken the cloud cover and lead to cooling as well.

Despite such opposing factors, all the best models predict net warming, and scientists are observing the effects. The average temperature has increased by 0.6±0.2°C since the late 19th century and 0.2°C to 0.3°C over the past 25 years. Globally, the decade from 2001 to 2010 was the warmest on record. Snow cover and glacial size in the northern hemisphere and floating ice in the Arctic Ocean have decreased dramatically. Antarctica is also experiencing widespread breakup of icebound regions. Globally, sea level has risen an average of 17 cm over the past century, and flooding and other extreme weather events have increased.

About 15 years ago, the best models predicted a temperature rise of 1.0°C to 3.5°C; today, their predictions are more than 50% higher. Such increases would alter rainfall patterns and crop yields throughout the world and increase sea level as much as 1 m, thus flooding regions such as the Netherlands and affecting many island nations. To make matters worse, as we burn fossil fuels that *release* CO_2, we cut down forests that *absorb* it.

In addition to studying ways to reduce fossil-fuel consumption, researchers at CanmetENERGY, a Canadian leader in clean energy research and technology development, are studying methods of *carbon capture and storage (CCS)*, including *CO_2 sequestration*, both naturally by maintaining forests and jungles and industrially by liquefying CO_2 formed and either burying it underground or using it to produce mineral carbonates.

The 1997 United Nations Conference on Climate Change in Kyoto, Japan, created a treaty with legally binding targets to limit greenhouse gases. It was ratified by 189 countries, but not by the United States, the largest emitter. In 2009, Canada signed the Copenhagen Accord, which, unlike the Kyoto Accord, is a non-binding agreement. In December 2011, after an intense session in Durban, South Africa, in which representatives from 200 countries participated, Canada withdrew from the Kyoto protocol. The reasoning behind the withdrawal was that, in the absence of participation by two of the major contributors to greenhouse gas emissions (the United States and China), the Kyoto Accord could not function effectively.

Hydrogen The use of H_2 as an energy source holds great promise but presents several problems. The simplest element and most plentiful gas in the universe always occurs on Earth in compounds, most importantly with oxygen in water and carbon in hydrocarbons. Once freed, however, H_2 is an excellent energy source because of several properties:

- It has the highest energy content per *mass* unit of any fuel—nearly three times as much as gasoline.
- Its energy content per *volume* unit is low at STP but increases greatly when the H_2 is liquefied under extremely high p and low T.
- Its combustion does not produce CO_2, particulates, or sulfur oxides. It is such a clean energy source that H_2 fuel cells operated electrical systems aboard the space shuttles, and the crews drank the pure water product.

A hydrogen-based economy is the dream of many because, even though combustion of H_2 ($\Delta_r H^\circ = -242$ kJ/mol) produces less than one-third as much energy per mole as combustion of CH_4 ($\Delta_r H^\circ = -802$ kJ/mol), it yields nonpolluting water vapour. The keys to realizing the dream, however, involve major improvements in both the production of H_2 and its transportation and storage.

Production of H_2. Hydrogen is produced by a number of methods, some more widely used than others:

- *Fossil-fuel by-product.* Hydrogen can be produced by the steam-reforming process, in which methane or natural gas is treated with steam in high-temperature reactions (Chapter 23).
- *Electrolysis of liquid water.* Formation of H_2 from the decomposition of liquid water is endothermic ($\Delta_r H^\circ = 286$ kJ/mol), and most direct methods that use electricity are costly. However, energy from flowing water, wind, and geothermal sources can provide the needed electricity. An exciting new *photoelectrochemical approach* uses a photovoltaic panel coupled to an electrolysis unit: the energy of sunlight powers the decomposition of water.

(Continued)

- *Biological methods.* Green algae and cyanobacteria store the energy of sunlight in the form of starch or glycogen molecules for their own metabolism. Under certain conditions, however, these microbes can be made to produce H_2 gas instead of the large organic molecules. For example, eliminating sulfur from the diet of green algae activates a long-dormant gene, and the cells produce H_2. But major improvements in yields are needed to make this method cost effective.

Canada, the largest per-capita hydrogen producer in the Organisation for Economic Co-operation and Development (OECD), uses many of the above methods and many other novel and environmentally friendly methods to produce hydrogen. One such pilot project is being researched at the University of Regina in their Faculty of Engineering and Applied Science (uregina.ca/engineering/faculty-staff/index.html).

Transportation and storage of H_2. The absence of an infrastructure for transporting and storing large quantities of hydrogen is still a roadblock to realizing a hydrogen-based economy. Currently, the cost of energy to liquefy the gas by cooling and compressing it is very high. Hydrogen is carried in pipelines, but its ability to escape through metals and make them brittle requires more expensive piping. Storage in the form of solid metal hydrides has been explored, but heating the hydrides to release the stored hydrogen requires energy. The use of H_2 in fuel cells is a major area of electrochemical research (Chapter 19), and a significant leader in this technology is the Canadian company Ballard. Ballard currently offers fuel cell products for buses, cars (through a partnership with Daimler), backup power, and distributed generation of power.

Solar Energy The Sun's energy drives global winds and ocean currents; the cycle of evaporation and condensation; and many biological processes, especially photosynthesis. More energy falls on Earth's surface as sunlight in 1 h [1.2×10^5 TW (terawatts); 1 TW = 1×10^{12} W] than is used in all human activities in 1 year. Solar energy is, however, a dilute source (only 1 kW/m^2 at noon); in 1 h, about 60 to 1000 TW strikes land at sites that are suitable for collecting. Nevertheless, covering 0.16% of this land with today's photovoltaic systems would provide 20 TW/h, equivalent to the output of 20 000 nuclear power plants producing 1 GW (gigawatt) each. Clearly, solar energy is the largest carbon-free option among renewable sources.

Three important ways to use solar energy are with electronic materials, biomass, and thermal systems:

Electronic materials. A **photovoltaic cell** converts sunlight directly into electricity by employing certain combinations of elements (usually metalloids; we discuss the behaviour of elements with *semiconductor* properties in Chapter 11). Early cells relied primarily on crystalline silicon, but their *yield*, the proportion of electrical energy produced relative to the radiant energy supplied, was only about 10%. More recent devices have achieved much higher yields by using thin films of polycrystalline materials that incorporate combinations of other elements, such as cadmium telluride (CdTe cells, 16%), gallium arsenide (GaAs cells, 18%), and copper indium gallium selenide (CIGS cells, 20%). Ongoing research on nanometre-sized semiconductor devices, called *quantum dots*, suggests yields as high as 70%.

Biomass. Solar energy is used to convert CO_2 and water via photosynthesis to plant material (carbohydrates) and oxygen. Plant biomass can be burned directly as a fuel or converted to other fuels, such as ethanol or hydrogen.

Thermal systems. The operation of a photovoltaic cell creates a buildup of heat, which lowers its efficiency. A *photovoltaic thermal (pVT) hybrid cell* thus combines such a cell with conductive metal piping (filled with water or antifreeze) that transfers the heat from the cell to a domestic hot-water system.

Because sunlight is typically available only 6 to 8 h a day and is greatly reduced in cloudy weather, another thermal system focuses on storing solar energy during the day and releasing it at night. When an ionic hydrate, such as $Na_2SO_4 \cdot 10H_2O$, is warmed by sunlight to over 32°C, the 3 mol of ions dissolve in the 10 mol of water in an endothermic process:

$$Na_2SO_4 \cdot 10H_2O(s) \xrightarrow{>32°C} Na_2SO_4(aq) \qquad \Delta_r H° = 354 \text{ kJ/mol} \qquad \textbf{(B5.12)}$$

When cooled below 32°C after sunset, the solution recrystallizes, releasing the absorbed energy for heating:

$$Na_2SO_4(aq) \xrightarrow{<32°C} Na_2SO_4 \cdot 10H_2O(s) \qquad \Delta_r H° = -354 \text{ kJ/mol} \qquad \textbf{(B5.13)}$$

Unfortunately, as of the middle of the second decade of the 21st century, solar energy is not yet competitive with highly subsidized fossil fuels. Nevertheless, changes in economic policy, as well as improvements in photovoltaic technology and biomass processing, should greatly increase solar energy's contribution to a carbon-free energy future.

Nuclear Energy Despite ongoing problems with the disposal of radioactive waste, energy from nuclear *fission*—the splitting of large, unstable atomic nuclei—is used extensively, especially in Canada, France, and other parts of northern Europe. In 2014, five nuclear power plants in three provinces provided about 15% of Canada's electricity. Nuclear *fusion*—the combining of small atomic nuclei—avoids the problems of fission but has been achieved so far only with a net consumption of power. (We will discuss these processes in detail in Chapter 25.)

Energy Conservation: More from Less All systems, whether organisms or factories, waste energy. This is, in effect, a waste of fuel. For example, the production of one aluminum beverage can requires energy equivalent to burning 0.25 L of gasoline. Energy conservation lowers costs, extends our fuel supply, and reduces the effects of climate change. Following are three examples of conservation.

Passive solar design. Heating and cooling costs can be reduced by as much as 50% when buildings are constructed with large windows and materials that absorb the Sun's heat in the day and release

(Continued)

it slowly at night. This reduction is enhanced by the fact that new materials are capable of providing better insulation to the buildings, thus ensuring that heat entering the building remains there.

Residential heating systems. A high-efficiency gas-burning furnace channels hot waste gases through a system of baffles to transfer more heat to the room and then to an attached domestic hot-water system. While moving through the system, the gases cool to 100°C, so the water vapour condenses, thus releasing about 10% more heat:

$$CH_4(g) + 2O_2(g) \longrightarrow CO_2(g) + \cancel{2H_2O(g)} \quad \Delta_rH^\circ = -802 \text{ kJ/mol} \quad \textbf{(B5.14)}$$

$$\cancel{2H_2O(g)} \longrightarrow 2H_2O(l) \quad \Delta_rH^\circ = -88 \text{ kJ/mol} \quad \textbf{(B5.15)}$$

$$CH_4(g) + 2O_2(g) \longrightarrow CO_2(g) + 2H_2O(l) \quad \Delta_rH^\circ = -890 \text{ kJ/mol} \quad \textbf{(B5.16)}$$

Modern lighting. Incandescent bulbs give off light when the material in their filament reaches a high temperature. They are very inefficient because less than 7% of the electrical energy is converted to light, the rest being wasted as heat. *Compact fluorescent lamps (CFLs)* have an efficiency of about 18% and are having a major economic impact: in Canada, inefficient incandescent light bulbs were phased out in 2014. *Light-emitting diodes (LEDs)* are even more efficient (about 30%) and are already being used in car indicator lamps and tail lights, street lights, and household lighting, to name a few. Light-emitting diodes based on organic semiconductors (OLEDs) are used in flat-panel displays, and recent use of Al-In-Ga-P materials and Ga and In nitrides in LEDs has increased their brightness and improved their colour richness. Alternatives to incandescents—such as LEDs, compact fluorescents (CFL), and halogen infrared and enhanced incandescent lights—are available in many shapes, sizes, light outputs, and colour temperatures.

Engineers and chemists will continue to explore alternatives for energy production and use, but a more hopeful energy future ultimately depends on our dedication to conserving planetary resources.

Problems

B5.1 To make use of an ionic hydrate for storing solar energy, you place 500.0 kg of sodium sulfate decahydrate on the roof of your house. Assuming complete reaction and 100% efficiency of heat transfer, how much heat (kJ) is released to your house at night?

B5.2 In one step of coal gasification, coal reacts with superheated steam:

$$C(s, \text{coal}) + H_2O(g) \longrightarrow CO(g) + H_2(g) \quad \Delta_rH^\circ = 129.7 \text{ kJ/mol}$$

(a) Use this reaction and the following two reactions to write an overall reaction for the production of methane:

$$CO(g) + H_2O(g) \longrightarrow CO_2(g) + H_2(g) \quad \Delta_rH^\circ = -41 \text{ kJ/mol}$$

$$CO(g) + 3H_2(g) \longrightarrow CH_4(g) + H_2O(g) \quad \Delta_rH^\circ = -206 \text{ kJ/mol}$$

(b) Calculate Δ_rH° for the overall reaction.

(c) Using the value from part (b) and a calculated value of Δ_rH° for the combustion of methane, find the total ΔH for gasifying 1.00 kg of coal and burning the methane formed (assume that water forms as a gas and $\mathcal{M}$ of coal = 12.00 g/mol).

SUMMARY OF SECTION 5.6

- Standard states are a set of specific conditions used to determine thermodynamic variables for all substances.
- When 1 mol of a compound forms from its elements, with all the substances in their standard states, the enthalpy change is the standard enthalpy of formation, Δ_fH°.
- Hess's law allows us to picture a reaction as the decomposition of the reactants to their elements, followed by the formation of the products from their elements.
- We can use tabulated Δ_fH° values to find Δ_rH°, or we can use known Δ_rH° and Δ_fH° values to find an unknown Δ_fH°.
- Because of major concerns about climate change, chemists are developing energy alternatives, including coal and biomass conversion, hydrogen fuel, and non-combustible energy sources.

CHAPTER REVIEW GUIDE

Learning Objectives

Relevant section (§) and/or sample problem (SP) numbers appear in parentheses.

Concepts

1. Differentiate between a system and its surroundings. (§5.1)
2. Explain and discuss the transfer of energy to or from a system as heat and/or work. (§5.1)
3. Describe the relation between internal energy change, heat, and work. (§5.1)
4. Explain the meaning of energy conservation. (§5.1)
5. Explain the meaning of a state function, and describe why ΔU is constant even though q and w vary. (§5.1)
6. Explain the meaning of enthalpy, and describe the relation between ΔU and ΔH. (§5.2)
7. Explain the meaning of ΔH, and describe the distinction between exothermic and endothermic reactions. (§5.2)
8. Discuss the relation between specific heat capacity and heat. (§5.3)
9. Describe how constant-pressure (coffee-cup) and constant-volume (bomb) calorimeters work. (§5.3)
10. Explain the relation between ΔH and the amount of substance. (§5.4)
11. Calculate ΔH values with Hess's law. (§5.5)
12. Differentiate between the meanings of a formation equation and the standard enthalpy of formation. (§5.6)
13. Describe how a reaction can be viewed as the decomposition of reactants followed by the formation of products. (§5.6)

Skills

1. Calculate the change in a system's internal energy in different units. (SP 5.1)
2. Draw enthalpy diagrams for chemical and physical changes. (SP 5.2)
3. Solve problems involving specific heat capacity and heat transferred in a reaction. (SPs 5.3–5.6)
4. Relate the heat transferred in a reaction to the amounts of substances changing. (SP 5.7)
5. Use Hess's law to find an unknown ΔH. (SP 5.8)
6. Write formation equations and use $\Delta_f H°$ values to find $\Delta_r H°$. (SPs 5.9, 5.10)

Key Terms

Section 5.1
system
surroundings
open system
closed system
isolated system
internal energy (U)
work (w)
law of conservation of energy (first law of thermodynamics)
joule (J)
calorie (cal)
state function

Section 5.2
pressure-volume work (pV work)
enthalpy (H)
change in enthalpy (ΔH)
exothermic process
enthalpy diagram
endothermic process

Section 5.3
heat capacity
specific heat capacity (c)
molar heat capacity (C_m)
calorimeter

Section 5.4
thermochemical equation

Section 5.5
Hess's law

Section 5.6
standard states
standard enthalpy of reaction ($\Delta_r H°$)
formation equation
standard enthalpy of formation ($\Delta_f H°$)
fossil fuels
coal gasification
synthetic natural gas (SNG)
biomass conversion
methanogenesis
photovoltaic cell

Key Equations and Relationships

5.1 Defining the change in internal energy:

$$\Delta U = U_{\text{final}} - U_{\text{initial}} = U_{\text{products}} - U_{\text{reactants}}$$

5.2 Expressing the change in internal energy in terms of heat and work:

$$\Delta U = q + w$$

5.3 Stating the first law of thermodynamics (law of conservation of energy):

$$\Delta U_{\text{universe}} = \Delta U_{\text{system}} + \Delta U_{\text{surroundings}} = 0$$

5.4 Determining the work due to a change in volume at constant pressure (pV work):

$$w = -p\Delta V$$

5.5 Relating the enthalpy change to the internal energy change at constant pressure:

$$\Delta H = \Delta U + p\Delta V$$

5.6 Identifying the enthalpy change with the heat absorbed or released at constant pressure:

$$q_p = \Delta U + p\Delta V = \Delta H$$

5.7 Calculating the heat absorbed or released when a substance undergoes a temperature change or a reaction occurs:

$$q = c \times \text{mass} \times \Delta T$$

5.8 Calculating the overall enthalpy change of a reaction (Hess's law):

$$\Delta_{\text{overall}} H = \Delta_1 H + \Delta_2 H + \cdots + \Delta_n H$$

$$\Delta_r H = \sum_n \Delta_n H$$

5.9 Calculating the standard enthalpy of a reaction:

$$\Delta_r H° = \Sigma m \Delta_f H°_{(\text{products})} - \Sigma n \Delta_f H°_{(\text{reactants})}$$

Brief Solutions to Follow-Up Problems

5.1 $\Delta U = q + w = (-26.0 \text{ kcal/mol}) + (3.78 \text{ kcal/mol})$
$= -22.22 \text{ kcal/mol} \times 4.184 \text{ kJ/kcal}$
$= -93.0 \text{ kJ/mol}$

5.2 The reaction is exothermic.

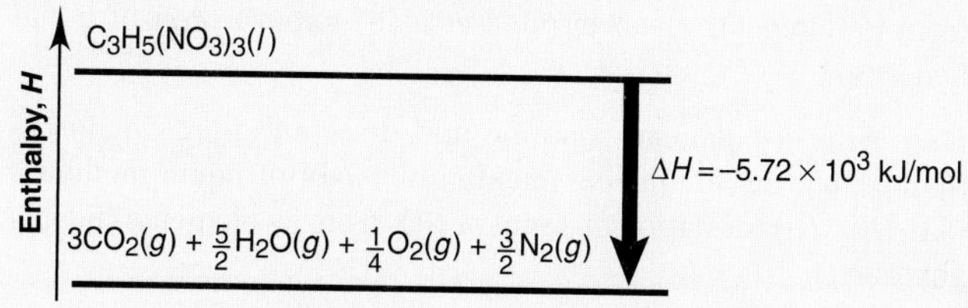

5.3 $\Delta T = 25.0°C - 37.0°C = -12.0°C = -12.0 \text{ K}$

$$\text{Mass (g)} = 1.11 \text{ g/mL} \times \frac{1000 \text{ mL}}{1 \text{ L}} \times 5.50 \text{ L} = 6.10 \times 10^3 \text{ g}$$

$q = c \times \text{mass} \times \Delta T$

$$= \left(2.42 \frac{\text{J}}{\text{g·K}}\right)\left(\frac{1 \text{ kJ}}{1000 \text{ J}}\right)(6.10 \times 10^3 \text{ g})(-12.0 \text{ K}) = -177 \text{ kJ}$$

5.4 $$c_{\text{solid}} = -\frac{4.184 \frac{\text{J}}{\text{g·K}} \times 25.00 \text{ g} \times 1.70 \text{ K}}{12.18 \text{ g} \times (-37.75 \text{ K})} = 0.387 \text{ J/(g·K)}$$

From Table 5.2, the metal is copper.

5.5 $2HCl(aq) + Ba(OH)_2(aq) \longrightarrow 2H_2O(l) + BaCl_2(aq)$
Amount (mol) of HCl = 0.500 mol HCl/L × 0.0500 L = 0.0250 mol HCl
Similarly, we have 0.0250 mol $Ba(OH)_2$.
Find the limiting reactant from the balanced equation:

$$\text{Amount (mol) of } H_2O = 0.0250 \text{ mol HCl} \times \frac{2 \text{ mol } H_2O}{2 \text{ mol HCl}}$$
$$= 0.0250 \text{ mol } H_2O$$

$$\text{Amount mol of } H_2O = 0.0250 \text{ mol Ba(OH)}_2 \times \frac{2 \text{ mol } H_2O}{1 \text{ mol Ba(OH)}_2}$$
$$= 0.0500 \text{ mol } H_2O$$

Thus, HCl is limiting.

$$\Delta_{\text{neut}}H \text{ (kJ/mol } H_2O) = \frac{q_{\text{rxn}}}{\text{amount (mol) } H_2O} = \frac{-1.386 \text{ kJ}}{0.0250 \text{ mol } H_2O}$$
$$= -55.4 \text{ kJ/mol } H_2O$$

5.6 Let x be the heat capacity.

$$-q_{\text{sample}} = q_{\text{calorimeter}}$$

$$-(0.8650 \text{ g·C})\left(\frac{1 \text{ mol C}}{12.01 \text{ g·C}}\right)[-393.5 \text{ kJ/(mol·C)}] = (2.613 \text{ K})x$$

$$x = 10.85 \text{ kJ/K}$$

5.7 $C_2H_4(g) + H_2(g) \longrightarrow C_2H_6(g) + 137 \text{ kJ}$

$$\text{Heat (kJ)} = 15.0 \text{ kg} \times \frac{1000 \text{ g}}{1 \text{ kg}} \times \frac{1 \text{ mol } C_2H_6}{30.07 \text{ g } C_2H_6} \times \frac{137 \text{ kJ}}{1 \text{ mol}}$$
$$= 6.83 \times 10^4 \text{ kJ}$$

5.8
$$2NO(g) + \tfrac{3}{2}O_2(g) \longrightarrow N_2O_5(s) \quad \Delta H = -223.7 \text{ kJ/mol}$$
$$2NO_2(g) \longrightarrow 2NO(g) + O_2(g) \quad \Delta H = 114.2 \text{ kJ/mol}$$
$$\cancel{2NO(g)} + \tfrac{3}{2}O_2(g) + 2NO_2(g) \longrightarrow N_2O_5(s) + \cancel{2NO(g)} + \cancel{O_2(g)}$$
$$2NO_2(g) + \tfrac{1}{2}O_2(g) \longrightarrow N_2O_5(s) \quad \Delta H = -109.5 \text{ kJ/mol}$$

5.9 (a) $C(s, \text{graphite}) + 2H_2(g) + \frac{1}{2}O_2(g) \longrightarrow CH_3OH(l)$ $\quad \Delta_fH° = -238.6 \text{ kJ/mol}$
(b) $Ca(s) + \frac{1}{2}O_2(g) \longrightarrow CaO(s)$ $\quad \Delta_fH° = -635.1 \text{ kJ/mol}$
(c) $C(s, \text{graphite}) + \frac{1}{4}S_8(s, \text{rhombic}) \longrightarrow CS_2(l)$ $\quad \Delta_fH° = 87.9 \text{ kJ/mol}$

5.10 $\Delta_fH°$ of $CH_3OH(l)$
$= -\Delta_rH° + 2\Delta_fH°[H_2O(g)] + \Delta_fH°[CO_2(g)]$
$= 638.5 \text{ kJ} + (2)(-241.8 \text{ kJ/mol}) + (1)(-393.5 \text{ kJ/mol})$
$= -238.6 \text{ kJ/mol}$

PROBLEMS

Problems with **red** numbers are answered in Appendix G and worked in detail in the Student Solutions Manual. Problem sections match those in this book and provide the numbers of relevant sample problems. Most offer Concept Review Questions, Skill-Building Exercises (grouped in pairs covering the same concept), and Problems in Context. The Comprehensive Problems are based on material from any section or previous chapter.

Forms of Energy and Their Interconversion

(Sample Problem 5.1)

Concept Review Questions

5.1 Why do heat (q) and work (w) have positive values when entering a system and negative values when leaving?

5.2 If you feel warm after exercising, have you increased the internal energy of your body? Explain.

5.3 An *adiabatic* process is a process that involves no heat transfer. What is the relationship between work and the change in internal energy in an adiabatic process?

5.4 State two ways that you increase the internal energy of your body and two ways that you decrease it.

5.5 Name a common device used to accomplish each change:
(a) Electrical energy to thermal energy
(b) Electrical energy to sound energy
(c) Electrical energy to light energy
(d) Mechanical energy to electrical energy
(e) Chemical energy to electrical energy

5.6 In winter, an electric heater uses a certain amount of electrical energy to heat a room to 20°C. In summer, an air conditioner uses the same amount of electrical energy to cool the room to 20°C. Is the change in internal energy of the heater larger, smaller, or the same as that of the air conditioner? Explain.

5.7 Suppose that you lift a heavy book and drop it onto a desk. Describe the energy transformations (from one form to another) that occur, moving backward in time from a moment after impact.

Skill-Building Exercises (grouped in similar pairs)

5.8 A system receives 425 J/mol of heat from and delivers 425 J/mol of work to its surroundings. What is the change in internal energy of the system (J/mol)?

5.9 A system conducts 255 cal/mol of heat to the surroundings and delivers 428 cal/mol of work. What is the change in internal energy of the system (cal/mol)?

5.10 What is the change in internal energy (J/mol) of a system that releases 675 J/mol of thermal energy to its surroundings and has 530 cal/mol of work done on it?

5.11 What is the change in internal energy (J/mol) of a system that absorbs 0.615 kJ/mol of heat from its surroundings and has 0.247 kcal/mol of work done on it?

5.12 Complete combustion of 2.0 t of coal to gaseous carbon dioxide releases 6.6×10^{10} J of heat. Convert this energy to (a) kilojoules; (b) kilocalories.

5.13 Thermal decomposition of 5.0 t of limestone to lime and carbon dioxide absorbs 9.0×10^{6} kJ of heat. Convert this energy to (a) joules; (b) calories.

Problems in Context

5.14 The nutritional calorie (Calorie) is equivalent to 1 kcal. One pound of body fat is equivalent to about 4.1×10^{3} Cal. Express this quantity of energy in joules and kilojoules.

5.15 If an athlete expends 1950 kJ/h, how long does it take her to work off 454 g of body fat? (See Problem 5.14.)

Enthalpy: Chemical Change at Constant Pressure

(Sample Problem 5.2)

Concept Review Questions

5.16 Why is the work that is done when a system expands against a constant external pressure assigned a negative sign?

5.17 Why is it often more convenient to measure ΔH than ΔU?

5.18 "Hot packs" used by winter athletes apply the crystallization of sodium acetate from a concentrated solution. What is the sign of ΔH for this crystallization? Is the reaction exothermic or endothermic?

5.19 Classify the following processes as exothermic or endothermic:
(a) Water freezing
(b) Water boiling
(c) Food being digested
(d) A person running
(e) A person growing
(f) Wood being chopped
(g) Heating with a furnace

5.20 What are the two main components of the internal energy of a substance? On what are they based?

5.21 For each process, state whether ΔH is less than (more negative), equal to, or greater than ΔU of the system. Explain.
(a) An ideal gas is cooled at constant pressure.
(b) A gas mixture reacts exothermically at a fixed volume.
(c) A solid reacts exothermically to yield a mixture of gases in a container of variable volume.

Skill-Building Exercises (grouped in similar pairs)

5.22 Draw an enthalpy diagram for a general exothermic reaction; label the axis, reactants, products, and ΔH (including its sign).

5.23 Draw an enthalpy diagram for a general endothermic reaction; label the axis, reactants, products, and ΔH (including its sign).

5.24 Write a balanced equation and draw an approximate enthalpy diagram for (a) combustion of 1 mol of ethane; (b) freezing of liquid water.

5.25 Write a balanced equation and draw an approximate enthalpy diagram for (a) formation of 1 mol of sodium chloride from its elements (heat is released); (b) vaporization of liquid benzene.

5.26 Write a balanced equation and draw an approximate enthalpy diagram for (a) combustion of 1 mol of liquid methanol (CH_3OH); (b) formation of 1 mol of NO_2 from its elements (heat is absorbed).

5.27 Write a balanced equation and draw an approximate enthalpy diagram for (a) sublimation of dry ice [conversion of $CO_2(s)$ directly to $CO_2(g)$]; (b) reaction of 1 mol of SO_2 with O_2.

5.28 The following circles represent a phase change at constant temperature:

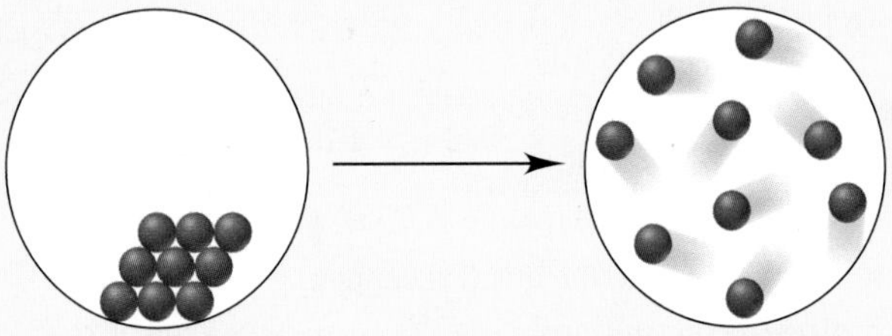

Is the value of each of the following positive (+), negative (−), or zero?
(a) q_{sys} (b) $\Delta_{sys}U$ (c) $\Delta_{univ}U$

5.29 The scenes below represent a physical change taking place in a piston-cylinder assembly:

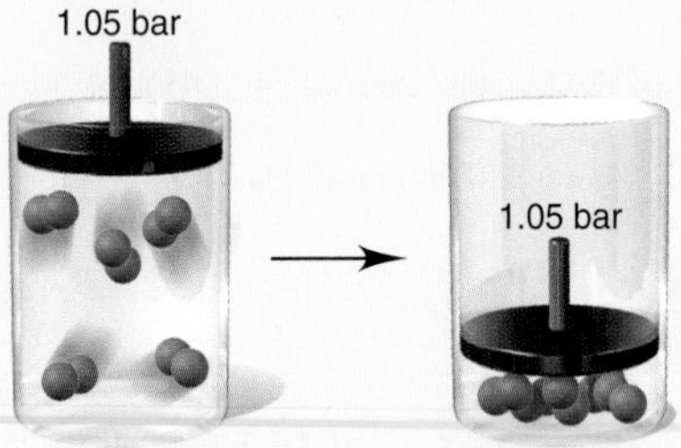

(a) Is w_{sys} positive (+), negative (−), or zero?
(b) Is $\Delta_{sys}H$ positive, negative, or zero?
(c) Can you determine whether $\Delta_{surr}U$ is positive, negative, or zero? Explain.

Calorimetry: Measuring the Heat of a Chemical or Physical Change

(Sample Problems 5.3 to 5.6)

Concept Review Questions

5.30 Which is larger, the specific heat capacity or the molar heat capacity of a substance? Explain.

5.31 What data do you need to determine the specific heat capacity of a substance?

5.32 Is the specific heat capacity of a substance an intensive or extensive property? Explain.

5.33 Distinguish between specific heat capacity, molar heat capacity, and heat capacity.

5.34 Both a coffee-cup calorimeter and a bomb calorimeter can be used to measure the heat transferred in a reaction. Which measures ΔU, and which measures ΔH? Explain.

Skill-Building Exercises (grouped in similar pairs)

5.35 Find q when 22.0 g of water is heated from 25.0°C to 100.°C.

5.36 Calculate q when 0.10 g of ice is cooled from 10.°C to −75°C [$c_{ice} = 2.087$ J/(g • K)].

5.37 A 295 g aluminum engine part at an initial temperature of 13.00°C absorbs 75.0 kJ of heat. What is the final temperature of the part [c of Al = 0.900 J/(g • K)]?

5.38 A 27.7 g sample of the radiator coolant ethylene glycol releases 688 J of heat. What was the initial temperature of the sample if the final temperature is 32.5°C [c of ethylene glycol = 2.42 J/(g • K)]?

5.39 Two iron bolts of equal mass—one at 100.°C, and the other at 55°C—are placed in an insulated container. Assuming that the heat capacity of the container is negligible, what is the final temperature inside the container [c of iron = 0.450 J/(g • K)]?

5.40 One piece of copper jewellery at 105°C has twice the mass of another piece at 45°C. Both are placed in a calorimeter of negligible heat capacity. What is the final temperature inside the calorimeter [c of copper = 0.387 J/(g • K)]?

5.41 When 155 mL of water at 26°C is mixed with 75 mL of water at 85°C, what is the final temperature? (Assume that no heat is released to the surroundings; d of water is 1.00 g/mL.)

5.42 An unknown volume of water at 18.2°C is added to 24.4 mL of water at 35.0°C. If the final temperature is 23.5°C, what was the unknown volume? (Assume that no heat is released to the surroundings; d of water is 1.00 g/mL.)

5.43 A 455 g piece of copper tubing is heated to 89.5°C and placed in an insulated vessel containing 159 g of water at 22.8°C. Assuming no loss of water and a heat capacity of 10.0 J/K for the vessel, what is the final temperature [c of copper = 0.387 J/(g • K)]?

5.44 A 30.5 g sample of an alloy at 93.0°C is placed into 50.0 g of water at 22.0°C in an insulated coffee cup with a heat capacity of 9.2 J/K. If the final temperature of the system is 31.1°C, what is the specific heat capacity of the alloy?

Problems in Context

5.45 High-purity benzoic acid (C_6H_5COOH; ΔH for combustion = −3227 kJ/mol) is used to calibrate bomb calorimeters. A 1.221 g sample burns in a calorimeter (heat capacity = 1365 J/°C) that contains 1.200 kg of water. What is the temperature change?

5.46 Two aircraft rivets, one iron and the other copper, are placed in a calorimeter that has an initial temperature of 20.°C. The data for the rivets are given in the following table:

	Iron	Copper
Mass (g)	30.0	20.0
Initial T (°C)	0.0	100.0
c [J/(g • K)]	0.450	0.387

(a) Will heat flow from Fe to Cu or from Cu to Fe?
(b) What other information is needed to correct any measurements in an actual experiment?
(c) What is the maximum final temperature of the system (assuming that the heat capacity of the calorimeter is negligible)?

5.47 A chemical engineer placed 1.520 g of a hydrocarbon in the bomb of a calorimeter (see Figure 5.10). The bomb was immersed in 2.550 L of water, and the sample was burned. The water temperature rose from 20.00°C to 23.55°C. If the calorimeter (excluding the water) had a heat capacity of 403 J/K, what was the heat released (q_V) per gram of hydrocarbon?

5.48 When 25.0 mL of 0.500 mol/L H_2SO_4 is added to 25.0 mL of 1.00 mol/L KOH in a coffee-cup calorimeter at 23.50°C, the temperature rises to 30.17°C. Calculate ΔH of this reaction. (Assume that the total volume is the sum of the volumes, and the density and specific heat capacity of the solution are the same as they are for water.)

Stoichiometry of Thermochemical Equations

(Sample Problem 5.7)

Concept Review Questions

5.49 Does a negative ΔH mean that the heat should be treated as a reactant or as a product?

5.50 Would you expect $O_2(g) \longrightarrow 2O(g)$ to have a positive or negative ΔH? Explain.

5.51 Is ΔH positive or negative when 1 mol of water vapour condenses to liquid water? Why? How does this value compare with ΔH for the vaporization of 2 mol of liquid water to water vapour?

Skill-Building Exercises (grouped in similar pairs)

5.52 Consider the following balanced thermochemical equation for a reaction sometimes used for H_2S production:

$$\frac{1}{8}S_8(s) + H_2(g) \longrightarrow H_2S(g) \quad \Delta H = -20.2 \text{ kJ/mol}$$

(a) Is this an exothermic or endothermic reaction?
(b) What is ΔH for the reverse reaction?
(c) What is q when 2.6 mol of S_8 reacts?
(d) What is q when 25.0 g of S_8 reacts?

5.53 Consider the following balanced thermochemical equation for the decomposition of the mineral magnesite:

$$MgCO_3(s) \longrightarrow MgO(s) + CO_2(g) \quad \Delta H = 117.3 \text{ kJ/mol}$$

(a) Is heat absorbed or released in the reaction?
(b) What is ΔH for the reverse reaction?
(c) What is ΔH when 5.35 mol of CO_2 reacts with excess MgO?
(d) What is ΔH when 35.5 g of CO_2 reacts with excess MgO?

5.54 When 1 mol of NO(g) forms from its elements, 90.29 kJ of heat is absorbed.
(a) Write a balanced thermochemical equation.
(b) What is ΔH when 3.50 g of NO decomposes to its elements?

5.55 When 1 mol of KBr(s) decomposes to its elements, 394 kJ of heat is absorbed.
(a) Write a balanced thermochemical equation.
(b) What is ΔH when 10.0 kg of KBr forms from its elements?

Problems in Context

5.56 Liquid hydrogen peroxide, an oxidizing agent in many rocket fuel mixtures, releases oxygen gas on decomposition:

$$2H_2O_2(l) \longrightarrow 2H_2O(l) + O_2(g) \quad \Delta H = -196.1 \text{ kJ/mol}$$

How much heat is released when 652 kg of H_2O_2 decomposes?

5.57 Compounds of boron and hydrogen are remarkable for their unusual bonding (described in Section 13.5) and also for their reactivity. With the more reactive halogens, for example, diborane (B_2H_6) forms trihalides even at low temperatures:

$$B_2H_6(g) + 6Cl_2(g) \longrightarrow 2BCl_3(g) + 6HCl(g) \quad \Delta H = -755.4 \text{ kJ/mol}$$

What is ΔH per kilogram of diborane that reacts?

5.58 Deterioration of buildings, bridges, and other structures through the rusting of iron costs millions of dollars a day. The actual process requires water, but a simplified equation is given below:

$$4Fe(s) + 3O_2(g) \longrightarrow 2Fe_2O_3(s) \quad \Delta H = -1.65 \times 10^3 \text{ kJ/mol}$$

(a) How much heat is released when 0.250 kg of iron rusts?
(b) How much rust forms when 4.85×10^3 kJ of heat is released?

5.59 A mercury mirror forms inside a test tube as a result of the thermal decomposition of mercury(II) oxide:

$$2HgO(s) \longrightarrow 2Hg(l) + O_2(g) \quad \Delta H = 181.6 \text{ kJ/mol}$$

(a) How much heat is absorbed to decompose 555 g of the oxide?
(b) If 275 kJ of heat is absorbed, what mass of Hg forms?

5.60 Most ethene (C_2H_4), the starting material for producing polyethene, comes from petroleum processing. It also occurs naturally as a fruit-ripening hormone and as a component of natural gas.
(a) The heat transferred during the combustion of C_2H_4 is -1411 kJ/mol. Write a balanced thermochemical equation.
(b) What mass of C_2H_4 must burn to give 70.0 kJ of heat?

5.61 Sucrose ($C_{12}H_{22}O_{11}$, table sugar) is oxidized in the body by O_2 via a complex set of reactions that produces $CO_2(g)$ and $H_2O(g)$ and releases 5.64×10^3 kJ/mol of energy.
(a) Write a balanced thermochemical equation for the overall process.
(b) How much heat is released per gram of sucrose oxidized?

Hess's Law: Finding ΔH of Any Reaction

(Sample Problem 5.8)

Concept Review Questions

5.62 Express Hess's law in your own words.

5.63 What is the main application of Hess's law?

5.64 When carbon burns in a deficiency of O_2, a mixture of CO and CO_2 forms. Carbon burns in excess O_2 to form only CO_2, and CO burns in excess O_2 to form only CO_2. Use ΔH values of the latter two reactions (from Appendix B) to calculate ΔH for the following reaction:

$$C(s, \text{graphite}) + \frac{1}{2}O_2(g) \longrightarrow CO(g)$$

Skill-Building Exercises (grouped in similar pairs)

5.65 Calculate ΔH for

$$Ca(s) + \frac{1}{2}O_2(g) + CO_2(g) \longrightarrow CaCO_3(s)$$

given the following reactions:

$$Ca(s) + \tfrac{1}{2}O_2(g) \longrightarrow CaO(s) \quad \Delta H = -635.1 \text{ kJ/mol}$$
$$CaCO_3(s) \longrightarrow CaO(s) + CO_2(g) \quad \Delta H = 178.3 \text{ kJ/mol}$$

5.66 Calculate ΔH for

$$2NOCl(g) \longrightarrow N_2(g) + O_2(g) + Cl_2(g)$$

given the following reactions:

$$\tfrac{1}{2}N_2(g) + \tfrac{1}{2}O_2(g) \longrightarrow NO(g) \quad \Delta H = 90.3 \text{ kJ/mol}$$
$$NO(g) + \tfrac{1}{2}Cl_2(g) \longrightarrow NOCl(g) \quad \Delta H = -38.6 \text{ kJ/mol}$$

5.67 Write the balanced overall equation (Equation 3) for the following process, calculate $\Delta_{\text{overall}}H$, and match the number of each equation with the letter of the appropriate arrow in Figure P5.67:

(1) $N_2(g) + O_2(g) \longrightarrow 2NO(g) \quad \Delta H = 180.6$ kJ/mol
(2) $2NO(g) + O_2(g) \longrightarrow 2NO_2(g) \quad \Delta H = -114.2$ kJ/mol
(3) $\Delta_{\text{overall}}H = ?$

5.68 Write the balanced overall equation (Equation 3) for the following process, calculate $\Delta_{\text{overall}}H$, and match the number of each equation with the letter of the appropriate arrow in Figure P5.68:

(1) $P_4(s) + 6Cl_2(g) \longrightarrow 4PCl_3(g) \quad \Delta H = -1148$ kJ/mol
(2) $4PCl_3(g) + 4Cl_2(g) \longrightarrow 4PCl_5(g) \quad \Delta H = -460$ kJ/mol
(3) $\Delta_{\text{overall}}H = ?$

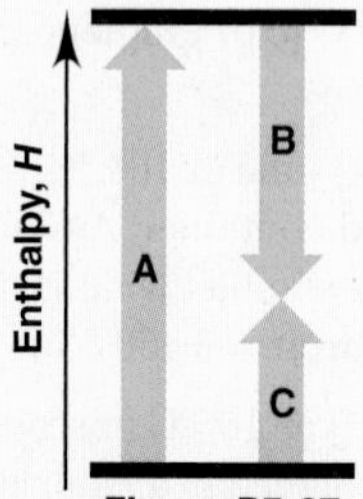

Figure P5.67

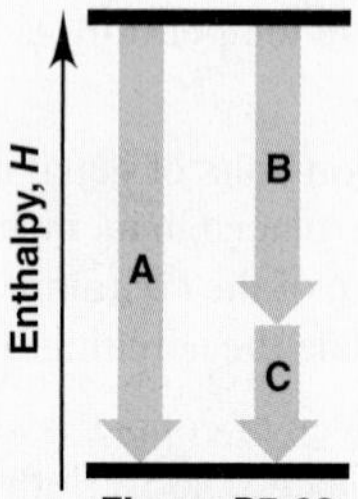

Figure P5.68

5.69 At a given set of conditions, 241.8 kJ of heat is released when 1 mol of $H_2O(g)$ forms from its elements. Under the same conditions, 285.8 kJ is released when 1 mol of $H_2O(l)$ forms from its elements. Find ΔH for the vaporization of water at these conditions.

5.70 When 1 mol of $CS_2(l)$ forms from its elements at 1 bar and 25°C, 89.7 kJ of heat is absorbed, and it takes 27.7 kJ to vaporize 1 mol of the liquid. How much heat is absorbed when 1 mol of $CS_2(g)$ forms from its elements at these conditions?

Problems in Context

5.71 Diamond and graphite are two crystalline forms of carbon. At 1 bar and 25°C, diamond changes to graphite so slowly that the enthalpy change of the process must be obtained indirectly. Using equations from the numbered list below, determine ΔH for this reaction:

$$C(s, \text{diamond}) \longrightarrow C(s, \text{graphite})$$

(1) $C(s, \text{diamond}) + O_2(g) \longrightarrow CO_2(g) \quad \Delta H = -395.4$ kJ/mol
(2) $2CO_2(g) \longrightarrow 2CO(g) + O_2(g) \quad \Delta H = 566.0$ kJ/mol
(3) $C(s, \text{graphite}) + O_2(g) \longrightarrow CO_2(g) \quad \Delta H = -393.5$ kJ/mol
(4) $2CO(g) \longrightarrow C(s, \text{graphite}) + CO_2(g) \quad \Delta H = -172.5$ kJ/mol

Standard Enthalpies of Reaction ($\Delta_r H°$)

(Sample Problems 5.9 and 5.10)

Concept Review Questions

5.72 What is the difference between the standard enthalpy of formation and the standard enthalpy of reaction?

5.73 How are $\Delta_f H°$ values used to calculate $\Delta_r H°$?

5.74 Make any changes needed in each equation to make the enthalpy change equal to $\Delta_f H°$ for the compound:
(a) $Cl(g) + Na(s) \longrightarrow NaCl(s)$
(b) $H_2O(g) \longrightarrow 2H(g) + \tfrac{1}{2}O_2(g)$
(c) $\tfrac{1}{2}N_2(g) + \tfrac{3}{2}H_2(g) \longrightarrow NH_3(g)$

Skill-Building Exercises (grouped in similar pairs)

5.75 Use Table 5.3 or Appendix B to write a balanced formation equation, at standard conditions, for each compound:
(a) $CaCl_2$ (b) $NaHCO_3$ (c) CCl_4 (d) HNO_3

5.76 Use Table 5.3 or Appendix B to write a balanced formation equation, at standard conditions, for each compound:
(a) HI (b) SiF_4 (c) O_3 (d) $Ca_3(PO_4)_2$

5.77 Calculate $\Delta_r H°$ for each reaction:
(a) $2H_2S(g) + 3O_2(g) \longrightarrow 2SO_2(g) + 2H_2O(g)$
(b) $CH_4(g) + Cl_2(g) \longrightarrow CCl_4(l) + HCl(g)$ (unbalanced)

5.78 Calculate $\Delta_r H°$ for each reaction:
(a) $SiO_2(s) + 4HF(g) \longrightarrow SiF_4(g) + 2H_2O(l)$
(b) $C_2H_6(g) + O_2(g) \longrightarrow CO_2(g) + H_2O(g)$ (unbalanced)

5.79 Copper(I) oxide can be oxidized to copper(II) oxide:

$$Cu_2O(s) + \tfrac{1}{2}O_2(g) \longrightarrow 2CuO(s) \quad \Delta_r H° = -146.0 \text{ kJ/mol}$$

Given $\Delta_f H°$ of $Cu_2O(s) = -168.6$ kJ/mol, find $\Delta_f H°$ of $CuO(s)$.

5.80 Acetylene burns in air according to the following equation:

$$C_2H_2(g) + \tfrac{5}{2}O_2(g) \longrightarrow 2CO_2(g) + H_2O(g) \quad \Delta_r H° = -1255.8 \text{ kJ/mol}$$

Given $\Delta_f H°$ of $CO_2(g) = -393.5$ kJ/mol and $\Delta_f H°$ of $H_2O(g) = -241.8$ kJ/mol, find $\Delta_f H°$ of $C_2H_2(g)$.

Problems in Context

5.81 The common lead-acid car battery produces a large burst of current, even at low temperatures, and is rechargeable. The following reaction occurs while a "dead" battery is being recharged:

$$2PbSO_4(s) + 2H_2O(l) \longrightarrow Pb(s) + PbO_2(s) + 2H_2SO_4(l)$$

(a) Use $\Delta_f H°$ values from Appendix B to calculate $\Delta_r H°$.
(b) Use the following equations to check your answer to part (a):
(1) $Pb(s) + PbO_2(s) + 2SO_3(g) \longrightarrow 2PbSO_4(s)$ $\quad \Delta_r H° = -768$ kJ/mol
(2) $SO_3(g) + H_2O(l) \longrightarrow H_2SO_4(l)$ $\quad \Delta_r H° = -132$ kJ/mol

Comprehensive Problems

5.82 Stearic acid ($C_{18}H_{36}O_2$) is a fatty acid, a molecule with a long hydrocarbon chain and an organic acid group (COOH) at the end. It is used to make cosmetics, ointments, soaps, and candles and is found in animal tissue as part of many saturated fats. In fact, when you eat meat, you are ingesting some fats that contain stearic acid.
(a) Write a balanced equation for the combustion of stearic acid to gaseous products.
(b) Calculate $\Delta_r H°$ for this combustion ($\Delta_f H°$ of $C_{18}H_{36}O_2 = -948$ kJ/mol).
(c) Calculate the heat (q) released (kJ and kcal) when 1.00 g of stearic acid is burned completely.
(d) A chocolate bar contains 11.0 g of fat and 100. Cal from fat. Is this consistent with your answer for part (c)?

5.83 Diluting sulfuric acid with water is highly exothermic:

$$H_2SO_4(l) \xrightarrow{H_2O} H_2SO_4(aq) + \text{heat}$$

(a) Use Appendix B to find $\Delta_r H°$ for diluting 1.00 mol of $H_2SO_4(l)$ ($d = 1.83$ g/mL) to 1 L of 1.00 mol/L $H_2SO_4(aq)$ ($d = 1.060$ g/mL).
(b) Suppose that you carry out the dilution in a calorimeter. The initial temperature is 25.0°C, and the specific heat capacity of the final solution is 3.50 J/(g • K). What is the final temperature?
(c) Use the ideas of density and heat capacity to explain why you should add acid to water rather than water to acid.

5.84 A balloonist begins a trip in a helium-filled balloon in the early morning, when the temperature is 15°C. By mid-afternoon, the temperature is 30.°C. Assuming that the pressure remains at 1.01 bar, for each mole of helium, calculate the following, and then explain the relationship between your answers to parts (d) and (e):
(a) The initial and final volumes
(b) The change in internal energy, ΔU (*Hint*: Helium behaves like an ideal gas, so $U = \frac{3}{2}nRT$. Be sure that the units of R are consistent with those of U.)
(c) The work (w) done by the helium (J/mol)
(d) The heat (q) transferred (J/mol)
(e) ΔH for the process (J/mol)

5.85 In winemaking, the sugars in grapes undergo *fermentation* by yeast to yield CH_3CH_2OH and CO_2. During cellular *respiration* (combustion), sugar and ethanol yield water vapour and CO_2.
(a) Using $C_6H_{12}O_6$ for sugar, calculate $\Delta_r H°$ of fermentation and respiration.
(b) Write a combustion reaction for ethanol. Which has a higher $\Delta_r H°$ for combustion per mole of C, sugar or ethanol?

5.86 Three of the reactions that occur when the paraffin of a candle (typical formula $C_{21}H_{44}$) burns are given below:
(1) Complete combustion forms CO_2 and water vapour.
(2) Incomplete combustion forms CO and water vapour.
(3) Some wax is oxidized to elemental C (soot) and water vapour.
(a) Find $\Delta_r H°$ of each reaction. ($\Delta_f H°$ of $C_{21}H_{44} = -476$ kJ/mol; use graphite for elemental carbon.)
(b) Find q (kJ) when a 254 g candle burns completely.
(c) Find q (kJ) when 8.00% by mass of the candle burns incompletely and 5.00% by mass of it undergoes soot formation.

5.87 Epoxyethane (EE) is prepared by the vapour-phase oxidation of ethene. Its main uses are in the preparation of the antifreeze ethane-1,2-diol, also known as ethylene glycol, and in the production of poly(ethene terephthalate), which is used to make beverage bottles and fibres. Pure EE vapour can decompose explosively:

Liquid EE has $\Delta_f H° = -77.4$ kJ/mol, and $\Delta H°$ for its vaporization = 569.4 J/g.
(a) Calculate $\Delta_r H°$ for the gas-phase reaction.
(b) External heating causes the vapour to decompose at 10 bar and 93°C in a distillation column. What is the final temperature if the average specific heat capacity of the products is 2.5 J/(g • °C)?

5.88 The following scenes represent a gaseous reaction between compounds of nitrogen (*blue*) and oxygen (*red*) at 298 K:

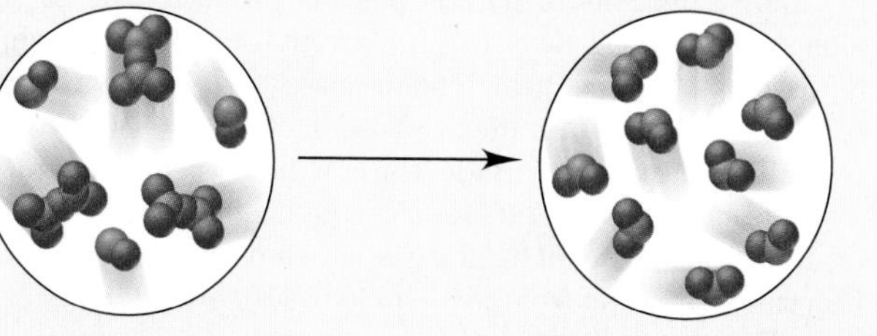

(a) Write a balanced equation, and use Appendix B to calculate $\Delta_r H°$.
(b) If each molecule of product represents 1.50×10^{-2} mol, what quantity of heat (J) is released or absorbed?

5.89 2,2,4-trimethylpentane, also known as isooctane (C_8H_{18}; d = 0.692 g/mL) is used as the fuel in a test of a new automobile drive train.
(a) How much energy (kJ) is released by combustion of 77.2 L of 2,2,4-trimethylpentane to gases ($\Delta_rH° = -5.44 \times 10^3$ kJ/mol)?
(b) The energy delivered to the wheels at 104 km/h is 5.5×10^4 kJ/h. Assuming that *all* the energy is transferred as work to the wheels, how far (km) can the car travel on the 77.2 L of fuel?
(c) If the actual range is 728 km, explain your answer to part (b).

5.90 Four 50. g samples of different colourless liquids are placed in beakers at $T_{initial}$ = 25.00°C. Each liquid is heated until 450. J of heat has been absorbed; T_{final} is shown on each beaker below. Rank the liquids in order of increasing specific heat capacity.

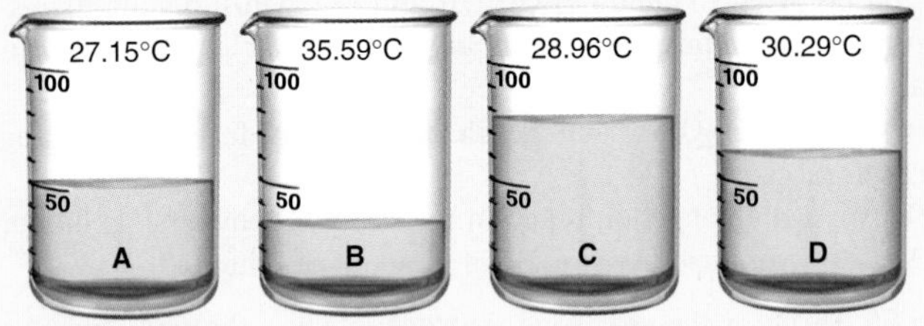

5.91 Reaction of gaseous ClF with F_2 yields liquid ClF_3, an important fluorinating agent. Use the following thermochemical equations to calculate $\Delta_rH°$ for this reaction:

(1) $3ClF(g) + O_2(g) \longrightarrow Cl_2O(g) + OF_2(g)$ $\quad \Delta_rH° = 167.5$ kJ/mol

(2) $2F_2(g) + O_2(g) \longrightarrow 2OF_2(g)$ $\quad \Delta_rH° = -43.5$ kJ/mol

(3) $2ClF_3(l) + 2O_2(g) \longrightarrow Cl_2O(g) + 3OF_2(g)$ $\quad \Delta_rH° = 394.1$ kJ/mol

5.92 Silver bromide is used to coat ordinary black-and-white photographic film, while high-speed film uses silver iodide.
(a) When 50.0 mL of 5.0 g/L $AgNO_3$ is added to a coffee-cup calorimeter containing 50.0 mL of 5.0 g/L NaI, with both solutions at 25°C, what mass of AgI forms?
(b) Use Appendix B to find $\Delta_rH°$.
(c) What is $\Delta_{soln}T$? (Assume that the volumes are additive and the solution has the density and specific heat capacity of water.)

5.93 (a) What amount (mol) of methane must be burned to give 1×10^8 J of energy? (Assume that water forms as a gas.)
(b) If natural gas costs $0.66 per 10^8 J, what is the cost per mole of methane? (Assume that natural gas is pure methane.)
(c) How much would it cost to warm 1202 L of water in a hot tub from 15.0°C to 42.0°C by burning methane?

5.94 When organic matter decomposes under oxygen-free (anaerobic) conditions, methane is one of the products. Thus, enormous deposits of natural gas, which is almost entirely methane, serve as a major source of fuel for home and industry.
(a) Known deposits of natural gas can produce 5600 EJ (exajoules) of energy (1 EJ = 10^{18} J). Current total global energy usage is 4.0×10^2 EJ per year. Find the mass (kg) of known deposits of natural gas ($\Delta_rH°$ for the combustion of CH_4 = −802 kJ/mol).
(b) At current rates of usage, for how many years could these deposits supply the world's total energy needs?
(c) What volume (m^3) of natural gas, measured at STP, is required to heat 0.946 L of water from 25.0°C to 100.0°C (d of H_2O = 1.00 g/mL; d of CH_4 at STP = 0.72 g/L)?
(d) The fission of 1 mol of uranium (about 0.01 L) in a nuclear reactor produces 2×10^{13} J. What volume (m^3) of natural gas would produce the same amount of energy?

5.95 A reaction takes place in a steel vessel within a chamber filled with argon gas. Shown below are molecular views of the argon adjacent to the surface of the reaction vessel before and after the reaction. Was the reaction exothermic or endothermic? Explain.

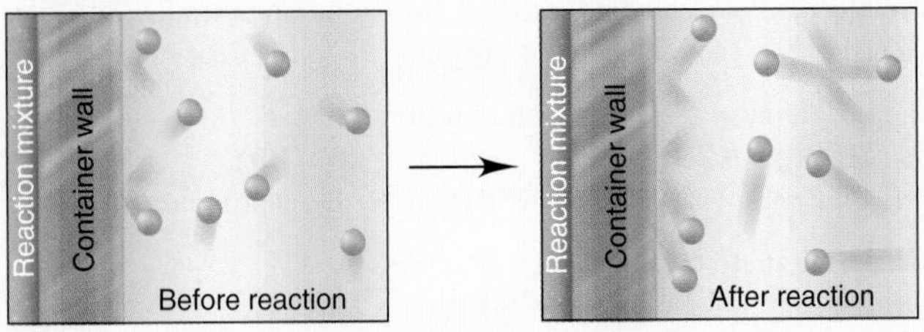

5.96 An aqueous waste stream with a maximum concentration of 0.50 mol/L H_2SO_4 (d = 1.030 g/mL at 25°C) is neutralized by controlled addition of 40% NaOH (d = 1.430 g/L) before it goes to the process sewer and then to the chemical plant's waste-treatment facility. A safety review finds that the waste stream could meet a small stream of an immiscible organic compound, which could form a flammable vapour in air at 40.°C. The maximum temperature reached by the NaOH solution and the waste stream is 31°C. Could the temperature increase due to the heat transferred by the neutralization cause the organic vapour to explode? [Assume that the specific heat capacity of each solution is 4.184 J/(g • K).]

5.97 Kerosene, a common space-heater fuel, is a mixture of hydrocarbons whose "average" formula is $C_{12}H_{26}$.
(a) Write a balanced equation, using the simplest whole-number coefficients, for the complete combustion of kerosene to gases.
(b) If $\Delta_rH° = -1.50 \times 10^4$ kJ/mol for the combustion equation as written in part (a), determine $\Delta_fH°$ of kerosene.
(c) Calculate the heat that is released by the combustion of 0.50 L of kerosene (d of kerosene = 0.749 g/mL).
(d) What volume (L) of kerosene must be burned for a kerosene furnace to produce 1320 kJ?

5.98 Silicon tetrachloride is produced annually on the multikiloton scale and used to make transistor-grade silicon. It can be produced directly from the elements (reaction 1) or, more cheaply, by heating sand and graphite with chlorine gas (reaction 2). If water is present in reaction 2, some tetrachloride may be lost in an unwanted side reaction (reaction 3):

(1) $Si(s) + 2Cl_2(g) \longrightarrow SiCl_4(g)$

(2) $SiO_2(s) + 2C(s, \text{graphite}) + 2Cl_2(g) \longrightarrow SiCl_4(g) + 2CO(g)$

(3) $SiCl_4(g) + 2H_2O(g) \longrightarrow SiO_2(s) + 4HCl(g)$ $\quad \Delta_rH° = -139.5$ kJ/mol

(a) Use reaction 3 to calculate the standard enthalpies of reaction of reactions 1 and 2.
(b) What is the standard enthalpy of reaction for a fourth reaction that is the sum of reactions 2 and 3?

5.99 One mole of nitrogen gas, confined within a cylinder by a piston, is heated from 0°C to 819°C at 1.01 bar.
(a) Calculate the work done by the expanding gas in joules. (1 J = 1.00×10^{-2} bar • L; assume that all the energy is used to do work.)
(b) What would be the temperature change if the gas were heated using the same amount of energy in a container of fixed volume? [Assume that the specific heat capacity of N_2 is 1.00 J/(g • K).]

5.100 The chemistry of nitrogen oxides is very versatile. Given the following numbered reactions and their standard enthalpy changes, calculate the standard enthalpy of reaction for

$$N_2O_3(g) + N_2O_5(s) \longrightarrow 2N_2O_4(g)$$

(1) $NO(g) + NO_2(g) \longrightarrow N_2O_3(g)$ $\Delta_r H° = -39.8$ kJ/mol

(2) $NO(g) + NO_2(g) + O_2(g) \longrightarrow N_2O_5(g)$ $\Delta_r H° = -112.5$ kJ/mol

(3) $2NO_2(g) \longrightarrow N_2O_4(g)$ $\Delta_r H° = -57.2$ kJ/mol

(4) $2NO(g) + O_2(g) \longrightarrow 2NO_2(g)$ $\Delta_r H° = -114.2$ kJ/mol

(5) $N_2O_5(s) \longrightarrow N_2O_5(g)$ $\Delta_r H° = 54.1$ kJ/mol

5.101 Electric generating plants transport large amounts of hot water through metal pipes, and oxygen dissolved in the water can cause a major corrosion problem. Hydrazine (N_2H_4) added to the water avoids the problem by reacting with the oxygen:

$$N_2H_4(aq) + O_2(g) \longrightarrow N_2(g) + 2H_2O(l)$$

About 4×10^7 kg of hydrazine is produced every year by reacting ammonia with sodium hypochlorite in the *Raschig process*:

$$2NH_3(aq) + NaOCl(aq) \longrightarrow N_2H_4(aq) + NaCl(aq) + H_2O(l) \quad \Delta_r H° = -151 \text{ kJ/mol}$$

(a) If $\Delta_f H°$ of $NaOCl(aq) = -346$ kJ/mol, find $\Delta_f H°$ of $N_2H_4(aq)$.
(b) What is the heat released when aqueous N_2H_4 is added to 5.00×10^3 L of water that is 2.50×10^{-4} mol/L O_2?

5.102 Liquid methanol (CH_3OH) can be used as an alternative fuel in pickup truck and SUV engines. An industrial method for preparing it involves the catalytic hydrogenation of carbon monoxide:

$$CO(g) + 2H_2(g) \xrightarrow{\text{catalyst}} CH_3OH(l)$$

How much heat (kJ) is released when 15.0 L of CO at 85°C and 112 kPa reacts with 18.5 L of H_2 at 75°C and 0.992 bar?

5.103 (a) How much heat is released when 25.0 g of methane burns in excess O_2 to form gaseous CO_2 and H_2O?
(b) Calculate the temperature of the product mixture if the methane and air are both at an initial temperature of 0.0°C. Assume a stoichiometric ratio of methane to oxygen from the air, with air being 21% O_2 by volume [c of CO_2 = 57.2 J/(mol • K); c of $H_2O(g)$ = 36.0 J/(mol • K); c of N_2 = 30.5 J/(mol • K)].

5.104 One industrially important reaction that is a step in the production of nitric acid is

$$4NH_3(g) + 5O_2(g) \longrightarrow 4NO(g) + 6H_2O(g)$$

(a) Calculate the enthalpy of the reaction.
(b) If 25.12 L of ammonia gas at 57.4°C and 2.45 bar reacts with 37.54 L of oxygen gas at 86.7°C and 139 kPa, what amount of NO (g) is formed and what amount of heat is released?
(c) If 57.9 L of NO (g) must be produced at 113.5°C and 1.789 bar, what amount of heat will be released?

5.105 An average person consumes 3.0 L of water per day. If a person goes on a two-day camping trip, what approximate volume of camping fuel (a mixture of hydrocarbons) needs to be taken to boil enough water for two days? Assume that the average temperature of the water is 22.5°C, the compound in the camping fuel is heptane (C_7H_{16}), the average density of the fuel is 0.7 g/mL, and only 18% of the heat generated from the fuel goes to heat the water (the rest is lost to the environment). The enthalpy of formation of heptane is −187.9 kJ/mol.

5.106 A woman expends 2092 kJ/hr in an aerobics class. (a) If the class is 45 min long, how many (rounded to the nearest whole number) chocolate-covered cherries would she have to eat to have enough energy for the class? One serving consists of 2 pieces (33 g) with an energy content of 586 kJ. (b) Assuming all the energy comes from sugar in the form of glucose, $C_6H_{12}O_6$, which has a standard enthalpy of formation of −1273.3 kJ/mol, what mass of sugar would have to be consumed by the woman to have enough energy for the class? The reaction for the metabolism of sugar is the same as the complete combustion reaction. (c) If 2 chocolate-covered cherries contain 20 g of sugar, how does this compare to the value obtained in part (b)?

5.107 A rectangular rhodium rod, 8.57 cm long, 9.32 mm wide, and 2.26 cm high, is heated to 141.3°C and dropped into 239.8 mL of water at 27.35°C. The temperature eventually stabilizes at 32.93°C. If the heat capacity of the calorimeter is 52.94 J/°C, what is the specific heat capacity of rhodium? The density of rhodium is 12.41 g/cm^3.

Quantum Theory and Atomic Structure

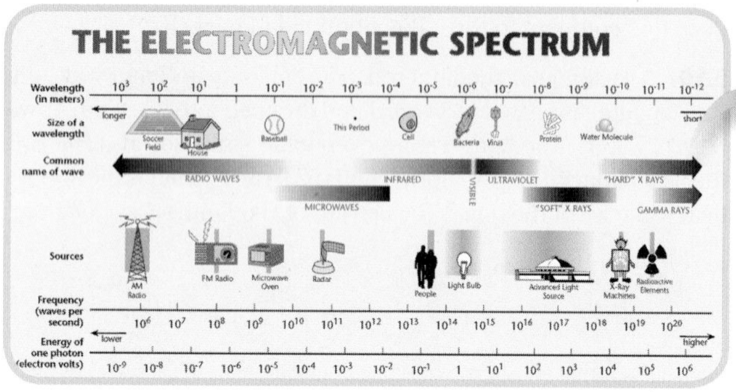

IN THIS CHAPTER . . . We discuss quantum mechanics, the theory that explains the fundamental nature of energy and matter and accounts for atomic structure. By the end of this chapter, you should be able to

- Distinguish between the classical wave properties of energy and the particle properties of matter
- Discuss the two observations—blackbody radiation and the photoelectric effect—whose explanations led to a *quantized*, or particulate, model of light
- Explain how light emitted by excited atoms—an *atomic spectrum*—suggests an atom with distinct energy levels, and apply spectra to chemical analysis
- Explain how wave-particle duality, which shows that matter and energy have similar properties, and the uncertainty principle, which proposes that electron behaviour can never be known exactly, led to the current model of the hydrogen atom
- Describe the quantum numbers, which specify the size, shape, and orientation of atomic orbitals, which are the regions that an electron occupies in the hydrogen atom (in Chapter 7, we will examine quantum numbers for atoms with more than one electron)

6

Your desktop computer works extraordinarily fast on one computation at a time. Now imagine a computer that could work just as fast, but on a million computations at once. This is the power of a quantum computer, imagined by scientists and engineers but not yet created, and the goal toward which powerful companies, such as Google, Microsoft, and IBM, are all working. While we are still a long way from being able to solve quantum-mechanical equations on our quantum computers while sipping coffee at home, our daily lives are nevertheless greatly affected by the amazing advances made in the understanding of the quantum nature of the atom. The incredible diagnostic equipment used in hospitals these days, the wonders of plasma television, and our incredibly complex cellular telephones/microcomputers, without which we are unable to function, are just a few examples of the effects of quantum science on our lives. In this chapter, we will start with light and determine its wave properties as compared to the particle properties of matter. We will look at the developments in science that led to the necessity of creating a quantum model of the atom. We will then determine how the idea of quantization allowed many other scientific puzzles to be solved. Finally, we will present the quantum model of the atom and some of the variables, such as the quantum numbers, that we use to describe this model.

Over a few remarkable decades—from around 1890 to 1930—a number of revolutions in science and culture took place worldwide. ■ One of these revolutions concerned our view of matter and energy, and you demonstrate this view with every flick of a light switch. This view is essential for understanding computers, television screens, fireworks, and neon signs, not to mention the structure of the atom! However, revolutions in science are not the violent upheavals of political overthrow. Flaws appear in a model due to conflicting evidence, a startling discovery widens the flaws into cracks, and the theoretical structure crumbles. Then new insight, verified by experiments, builds a model that is more consistent with reality. So it was when Lavoisier's theory of combustion overthrew the phlogiston model, when Dalton's atomic theory established the idea of individual units of matter, and when Rutherford's model substituted nuclear atoms for "billiard balls" or "plum puddings." You will see the process of scientific revolution unfold again as we discuss the development of modern atomic theory.

■ **A Time of Revolution** Here are some of the revolutionary discoveries and events that were witnessed by the science world between 1890 and 1930:

1895	Röntgen: discovery of X-rays	Nobel Prize in Physics: 1901
1896	Becquerel: discovery of radioactivity	Nobel Prize in Physics: 1903
1896	M. and P. Curie: discovery of radiation	Nobel Prize in Physics: 1903
1897	Thomson: discovery of the electron	Nobel Prize in Physics: 1906
1900	Planck: quantum theory	Nobel Prize in Physics: 1918
1901	Marconi: invention of the radio	Nobel Prize in Physics: 1909
1905	Rutherford: radioactivity explained	Nobel Prize in Chemistry: 1908
1905	Einstein: relativity and photon theories	Nobel Prize in Physics: 1921
1908	M. Curie: discovery of polonium and radium	Nobel Prize in Chemistry: 1911
1911	Rutherford: nuclear model of the atom	Nobel Prize in Chemistry: 1908
1913	Bohr: atomic spectra and quantum model of the atom	Nobel Prize in Physics: 1922
1923	Compton: photon momentum	Nobel Prize in Physics: 1927
1924	De Broglie: wave theory of matter	Nobel Prize in Physics: 1929
1926	Schrödinger: wave equation	Nobel Prize in Physics: 1933
1927	Heisenberg: uncertainty principle	Nobel Prize in Physics: 1932

Concepts and Skills to Review before Studying This Chapter

- Discovery of the electron and atomic nucleus (Section 2.4)
- Major features of atomic structure (Section 2.5)
- Changes in the energy state of a system (Section 5.1)

6.1 The Nature of Light

Visible light, X-rays, and microwaves are some of the types of **electromagnetic radiation** (also called *electromagnetic energy* or *radiant energy*). All electromagnetic radiation consists of energy propagated by electric and magnetic fields that increase and decrease in intensity as they move through space. This *classical wave model* explains why rainbows form, how magnifying glasses work, and many other familiar observations. However, it cannot explain observations of the very *un*familiar atomic scale. In this section, we describe some of the properties of electromagnetic radiation and note how they are distinguished from the properties of matter. Then we see that other properties blur this distinction and require a new model to explain them.

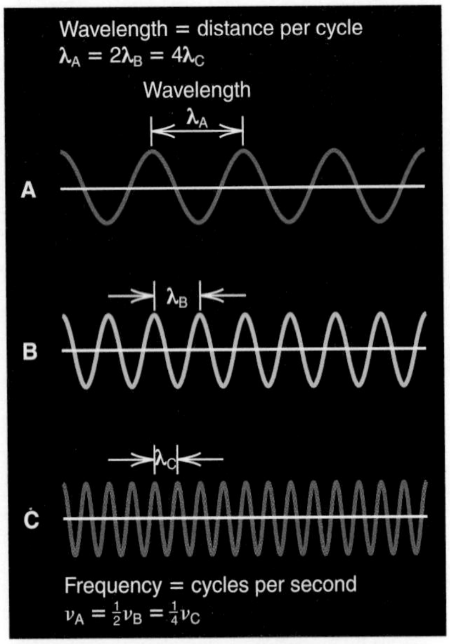

FIGURE 6.1 The reciprocal relationship of frequency and wavelength

The Wave Nature of Light

The wave properties of electromagnetic radiation are described by three variables and one constant (Figure 6.1):

1. *Frequency.* The **frequency (ν)** (ν is the Greek letter *nu*) of a wave is *the number of cycles that it undergoes per second,* expressed by the unit 1/s [s^{-1}; also called a *hertz* (Hz)].
2. *Wavelength.* The **wavelength (λ)** (λ is the Greek letter *lambda*) of a wave is *the distance between any point on the wave and the corresponding point on the next crest (or trough) of the wave*; in other words, the wavelength is the distance that the wave travels during one cycle. The units for wavelength are metres or, for very short wavelengths, nanometres (nm, 10^{-9} m) or picometres (pm, 10^{-12} m).
3. *Amplitude.* The **amplitude** of a wave is *the height of the crest (or depth of the trough).* For an electromagnetic wave, the amplitude is related to the *intensity* of the radiation or, in the case of visible light, its brightness. Light of a particular colour has a specific frequency (and thus a specific wavelength), but, as Figure 6.2 shows, it can be dimmer (lower amplitude, less intense) or brighter (higher amplitude, more intense).
4. *Speed.* The **speed** of a wave is *the distance that it moves per unit time* (metres per second), which is the product of its frequency (cycles per second) and its wavelength (metres per cycle):

$$\text{Units for speed of wave: } \frac{\cancel{\text{cycles}}}{\text{s}} \times \frac{\text{m}}{\cancel{\text{cycles}}} = \frac{\text{m}}{\text{s}}$$

In a vacuum, electromagnetic radiation moves at 2.997 924 58 × 10^8 m/s (3.00 × 10^8 m/s to three significant figures), a *physical constant* called the **speed of light (c)**:

$$c = \nu \times \lambda \tag{6.1}$$

Since the product of ν and λ is a constant, they have a reciprocal relationship: *radiation with a high frequency has a short wavelength, and vice versa:*

$$\nu\uparrow\lambda\downarrow \quad \text{and} \quad \nu\downarrow\lambda\uparrow$$

The Electromagnetic Spectrum Visible light represents a small region of the **electromagnetic spectrum** (Figure 6.3). *All waves in the spectrum travel at the same speed through a vacuum but differ in frequency and, therefore, wavelength.*

The spectrum is a *continuum of radiant energy*, so each region meets the next. For instance, the **infrared (IR)** region meets the *microwave* region on one end and the *visible* region on the other end. We perceive different wavelengths (or frequencies) of visible light as colours, from red ($\lambda \approx$ 750 nm) to violet ($\lambda \approx$ 400 nm). Light of a single wavelength is called *monochromatic* (Greek, meaning "one colour"), whereas light of many wavelengths is *polychromatic*. White light is polychromatic. The region adjacent to visible light on the short-wavelength end consists of **ultraviolet (UV)** radiation (also called *ultraviolet light*). Still shorter wavelengths (higher frequencies) make up the X-ray and gamma (γ) ray regions.

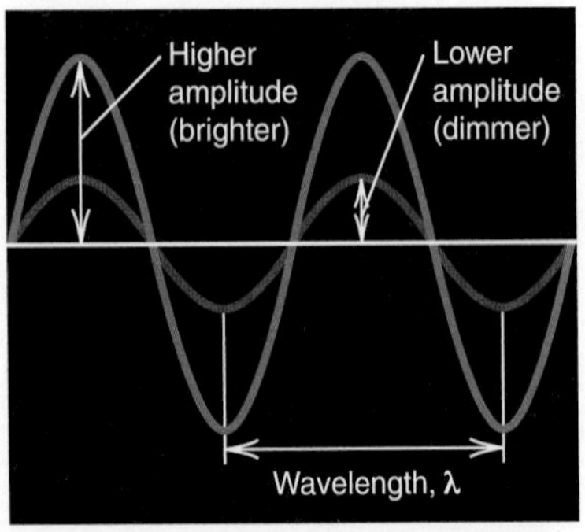

FIGURE 6.2 Differing amplitudes (brightness, or intensity) of a wave

Some types of electromagnetic radiation are used by familiar devices; for example, long-wavelength, low-frequency radiation is used by microwave ovens, radios,

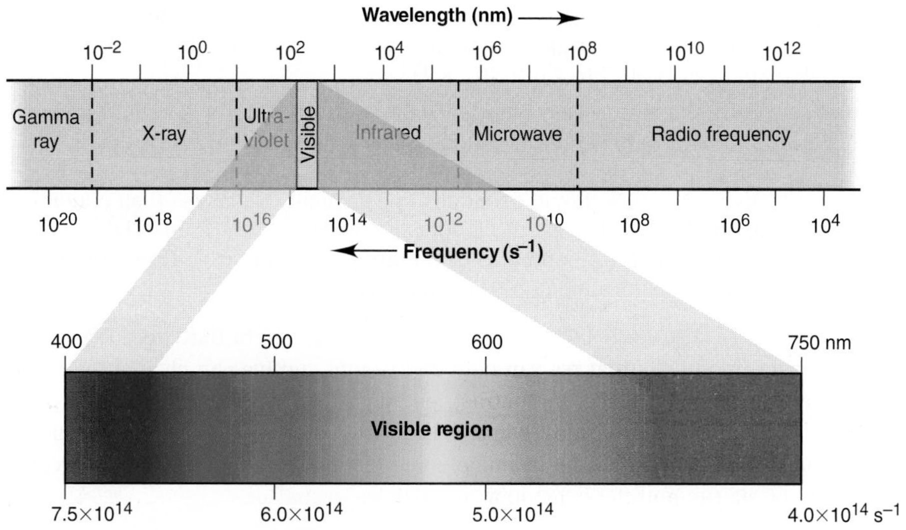

FIGURE 6.3 Regions of the electromagnetic spectrum. The visible region is expanded (and the scale made linear) to show the component colours.

and cellphones. Whether from human inventions such as lightbulbs, X-ray equipment, and car motors, or from natural sources such as the Sun, lightning, radioactivity, and even the glow of fireflies, electromagnetic emissions are everywhere! Our knowledge of the universe comes from the radiation that enters our eyes and the modern tools that we use (such as light, X-rays, and radio telescopes).

Sample Problem 6.1 Interconverting Wavelength and Frequency

Problem A dental hygienist uses X-rays ($\lambda = 100.$ pm) to take a series of dental radiographs while the patient listens to a radio station ($\lambda = 325$ cm) and looks out a window at the blue sky ($\lambda = 473$ nm). What is the frequency (in s^{-1}) of the electromagnetic radiation from each source? (Assume that the radiation travels at the speed of light, 3.00×10^8 m/s.)

Plan We are given the wavelengths, so we use Equation 6.1 to find the frequencies. However, we must first convert the wavelengths to metres because c has units of m/s.

Solution Convert the X-rays from picometres to metres:

$$\lambda = 100.\ \text{pm} \times \frac{10^{-12}\ \text{m}}{1\ \text{pm}} = 1.00 \times 10^{-10}\ \text{m}$$

Calculate the frequency:

$$\nu = \frac{c}{\lambda} = \frac{3.00 \times 10^{8}\ \text{ms}}{1.00 \times 10^{-10}\ \text{m}} = 3.00 \times 10^{18}\ \text{s}^{-1}$$

Combine steps to calculate the frequency of the radio signal:

$$\nu = \frac{c}{\lambda} = \frac{3.00 \times 10^{8}\ \text{m/s}}{325\ \text{cm} \times \dfrac{10^{-2}\ \text{m}}{1\ \text{cm}}} = 9.23 \times 10^{7}\ \text{s}^{-1}$$

Combine steps to calculate the frequency of the blue sky:

$$\nu = \frac{c}{\lambda} = \frac{3.00 \times 10^{8}\ \text{m/s}}{473\ \text{nm} \times \dfrac{10^{-9}\ \text{m}}{1\ \text{nm}}} = 6.34 \times 10^{14}\ \text{s}^{-1}$$

Road Map

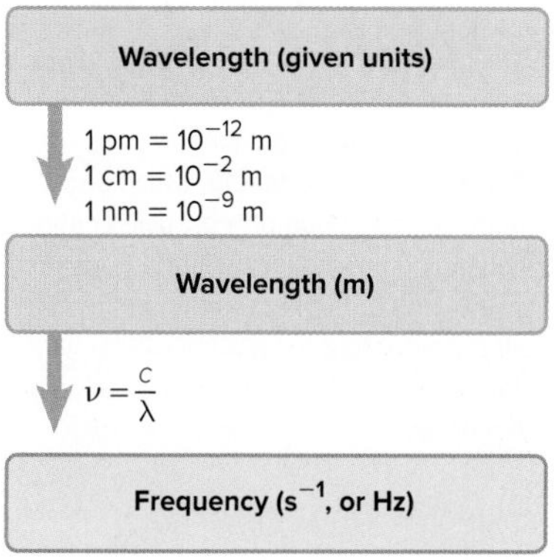

Check The orders of magnitude are correct for the regions of the electromagnetic spectrum (see Figure 6.3): X-rays (10^{19} to 10^{16} s^{-1}), radio waves (10^9 to 10^4 s^{-1}), and visible light (7.5×10^{14} to 4.0×10^{14} s^{-1}).

Comment The radio station here is broadcasting at 92.3×10^6 s^{-1}, or 92.3 million Hz (92.3 MHz), about midway in the FM (frequency modulation) range.

Follow-Up Problem 6.1 Some diamonds appear yellow because they contain nitrogen compounds that absorb purple light of frequency 7.23×10^{14} Hz. Calculate the wavelength (nm and pm) of the absorbed light. (See Brief Solutions to Follow-Up Problems at the end of the chapter.)

The Classical Distinction between Energy and Matter In our everyday world, matter comes in chunks that we can hold and weigh, and we can change the quantity of matter piece by piece. In contrast, energy is massless, and its quantity can change continuously. Matter moves in specific paths, whereas radiant energy (light) travels in diffuse waves. Let us examine some distinctions between the behaviour of waves of energy and the behaviour of particles of matter.

1. *Refraction and dispersion.* Light of a given wavelength travels at different speeds through various transparent media—vacuum, air, water, quartz, and so on. Therefore, when a light wave passes from one medium into another, the speed of the wave changes. Figure 6.4A shows the phenomenon known as **refraction**. If a wave strikes the boundary between media, say, between air and water, at an angle other than 90°, the change in speed causes a change in direction, and the wave continues at a different angle. The angle of refraction depends on the two media and the wavelength of the light. In the related process of *dispersion*, white light separates (disperses) into its component colours when it passes through a prism (or another refracting object) because each incoming wave is refracted at a slightly different angle.

In contrast to a wave of light, a particle of matter, like a pebble, does not undergo refraction. If you throw a pebble through the air into a pond, it continues to slow down gradually along a curved path after entering the water (Figure 6.4B).

2. *Diffraction and interference.* When a wave strikes the edge of an object, it bends around the object in a phenomenon called **diffraction**. If the wave passes through a slit about as wide as its wavelength, it bends around both edges of the slit and forms a semicircular wave on the other side of the opening (Figure 6.4C).

In contrast, when you throw a collection of particles, like a handful of sand, at a small opening, some particles hit the edge of the opening while others go through the opening and continue in a narrower group (Figure 6.4D).

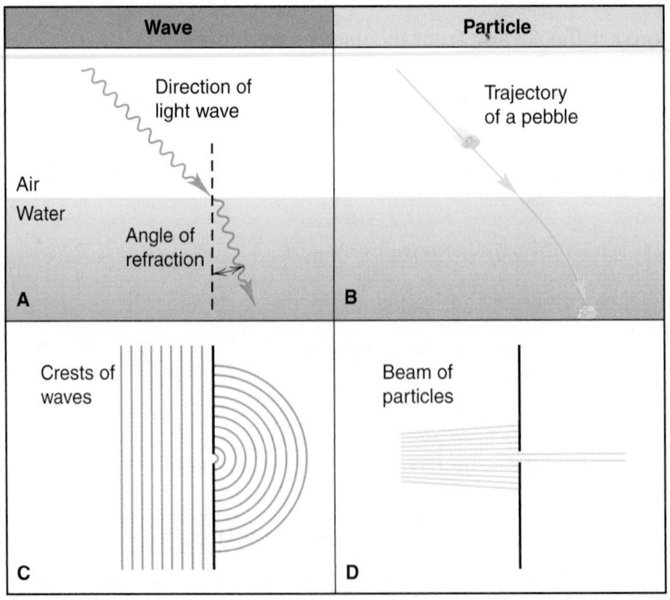

FIGURE 6.4 **Different behaviours of waves and particles. A.** Refraction: The speed of a light wave passing between media changes immediately, which bends its path. **B.** The speed of a particle continues changing gradually. **C.** Diffraction: A wave bends around both edges of a small opening, forming a semicircular wave. **D.** Particles either enter a small opening or do not enter.

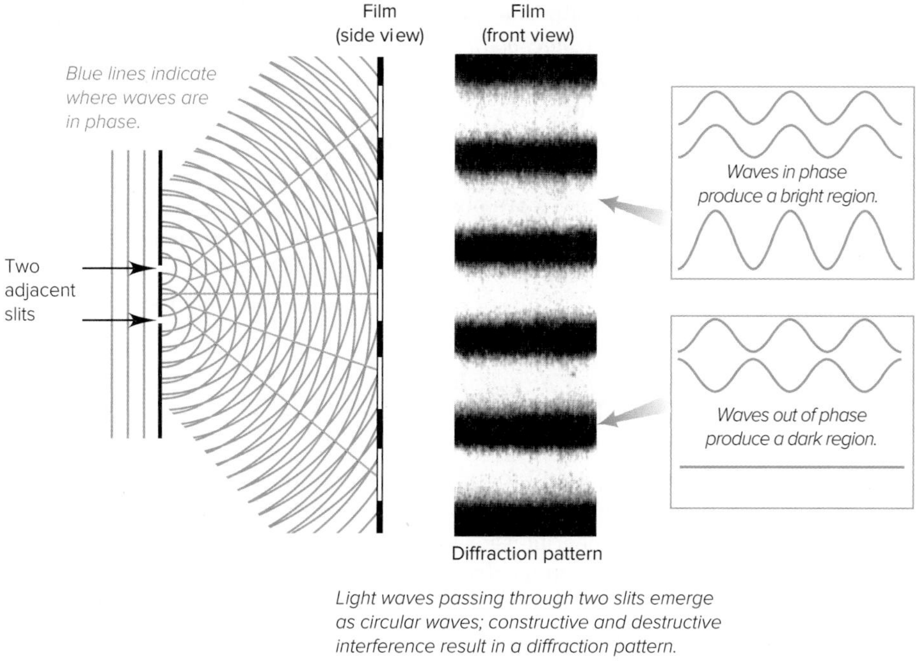
Film
(side view)
Film
(front view)
Blue lines indicate where waves are in phase.
Two adjacent slits
Waves in phase produce a bright region.
Waves out of phase produce a dark region.
Diffraction pattern
Light waves passing through two slits emerge as circular waves; constructive and destructive interference result in a diffraction pattern.

FIGURE 6.5 Formation of a diffraction pattern. Light waves passing through two slits emerge as circular waves and create a diffraction pattern due to constructive and destructive interference of the waves.

When waves of light pass through two adjacent slits, the nearby emerging circular waves interact through the process of *interference*. If the crests of the waves coincide (*in phase*), they interfere *constructively*—the amplitudes add together to form a brighter region. If the crests coincide with troughs (*out of phase*), they interfere *destructively*—the amplitudes cancel to form a darker region. The result is a *diffraction pattern* (Figure 6.5).

In contrast, particles passing through adjacent openings continue in straight paths, some colliding and moving at different angles.

The Particle Nature of Light

Three observations involving matter and light confounded physicists at the turn of the 20th century: blackbody radiation, the photoelectric effect, and atomic spectra. Explaining these phenomena required a radically new picture of energy. We discuss the first two of them here and the third in Section 6.2.

Blackbody Radiation and the Quantum Theory of Energy The first of the puzzling observations involved the light given off by an object being heated.

- *Observation: blackbody radiation.* When a solid object is heated to about 1000 K, it begins to emit visible light, as seen in the red glow of smouldering coal (Figure 6.6A). At about 1500 K, the light is brighter and more orange, like that from an electric heating coil (Figure 6.6B). At temperatures greater than 2000 K, the light is still brighter and whiter, like that emitted by the filament of a light bulb (Figure 6.6C).

A Smouldering coal **B** Electrical heating element **C** Light-bulb filament

FIGURE 6.6 Familiar examples of light emission related to blackbody radiation

Max Planck (1858–1947), was born in Kiel, Germany. His early research was in the field of thermodynamics, but his interests later turned to problems in radiation processes, which he showed to be electromagnetic in nature. He applied himself to one of the biggest scientific puzzles of the time, which was the difference between experimental observations in the field of blackbody radiation and the predicted results based on classical physics. He published a paper in 1900 that showed that the energy for a resonator of frequency ν is $h\nu$, where h, a universal constant whose value he determined, is now called Planck's constant. Planck was an accomplished pianist who at one time even considered a career in music.

These changes in the intensity and wavelength of emitted light as an object is heated are characteristic of *blackbody radiation*, light given off by a hot *blackbody*.* All attempts to account for these changes by applying classical electromagnetic theory failed. According to classical physics, the intensity should have been infinite at short wavelengths (or at high frequencies), as shown by the *pink* line at 5000 K in Figure 6.7. Experimentally, however, the *blue* line was observed at this temperature. (Similar curves are observed at other temperatures, as shown by the other lines in Figure 6.7.) Many (unsuccessful) attempts were made to reconcile classical theory in physics to what was experimentally observed. The disagreement between theoretical classical physics and experiment was called the *ultraviolet catastrophe*.

- *Explanation: the quantum theory.* In 1900, the German physicist Max Planck (1858–1947), while working on high-brightness low-energy light bulbs, developed a formula that fit the data perfectly. Planck's initial solution to the ultraviolet catastrophe did not, in fact, involve quantization at all when it was first presented. But Planck himself felt that his first solution was not ideal, since he could not explain the deeper physics involved. In general, Planck did not favour Boltzmann's use of statistical mechanics in the interpretation of the second law of thermodynamics (Chapter 18). When he redeveloped his formula, Planck not only made use of statistical mechanics but also made a significant deviation from accepted physics. He assumed that the frequencies could not simply assume any value (as had been previously supposed), just as a guitar string (which is fixed at both ends) makes only a fixed number of standing waves for a certain length of string,

Planck stated that the frequency could only assume certain values. This concept of limiting the frequency and thereby the energy to only certain values came to be known as *quantization*:

$$E = nh\nu$$

where E is the energy of the radiation, ν is its frequency, n is a positive integer (1, 2, 3, and so on) called a **quantum number**, and $\boldsymbol{h}$ is **Planck's constant**. With energy in joules (J) and frequency in s^{-1}, h has units of J•s:

$$h = 6.626\,068\,76 \times 10^{-34}\ \text{J}\cdot\text{s}$$

Since the hot object emits only certain quantities of energy, and the energy must be emitted by the object's atoms, this means that each atom *emits* only certain quantities of energy. It follows, then, that each atom *has* only certain quantities of energy. Thus, the energy of an atom is *quantized*: it occurs in fixed quantities, rather

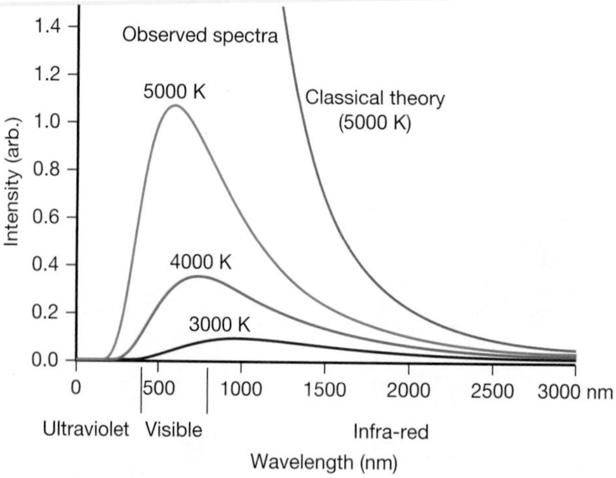

FIGURE 6.7 Variation of intensity with wavelength of light emitted by a blackbody, according to theoretical classical physics (pink) and based on experiment (other colours)

*A blackbody is an idealized object that absorbs all the radiation incident on it. A hollow cube with a small hole in one wall approximates a blackbody.

than being continuous. Each change in an atom's energy occurs when the atom absorbs or emits one or more "packets," or definite amounts, of energy. Each energy packet is called a **quantum** (meaning "fixed quantity"; plural: *quanta*). A quantum of energy is equal to $h\nu$. Thus, *an atom changes its energy state by emitting (or absorbing) one or more quanta*, and the energy of the emitted (or absorbed) radiation is equal to the *difference in the atom's energy states*:

$$\Delta_{\text{atom}}E = E_{\text{emitted(or absorbed) radiation}} = \Delta nh\nu$$

Because the atom can change its energy only by integer multiples of $h\nu$, the smallest change occurs when an atom in a given energy state changes to an adjacent state, that is, when $\Delta n = 1$:

$$\Delta E = h\nu = E_{\text{photon}} \quad (6.2)$$

Planck was so suspicious of the implications, both physical and philosophical, of the use of statistical mechanics that his availing himself of this theory was, in his own words, "an act of despair . . . I was ready to sacrifice any of my previous convictions about physics."*

The Photoelectric Effect and the Photon Theory of Light Despite the idea of quantization, physicists still pictured energy as travelling in waves. However, the wave model could not explain the second confusing observation, the flow of current when light strikes a metal.

- *Observation: the* **photoelectric effect**. When monochromatic light of sufficient frequency shines on a metal plate, a current flows (Figure 6.8). Scientists originally thought that the current arises because light transfers energy that frees outer electrons from the metal surface. However, the effect had two confusing features, the presence of a threshold frequency and the absence of a time lag:
  1. *Presence of a threshold frequency.* For current to flow, the light shining on the metal must have a minimum, or threshold, *frequency*, and different metals have different minimum frequencies. However, the wave theory associates the energy of light with its *intensity* (brightness), not its frequency (colour). Thus, the theory predicts that an outer electron will break free when it absorbed enough energy from light of *any* colour.
  2. *Absence of a time lag.* Current flows the moment that light of the minimum frequency shines on the metal, regardless of the light's intensity. However, the wave theory predicts that, with dim light, there will be a time lag before the current flowed, because the outer electrons will have to absorb enough energy to break free.
- *Explanation: the photon theory.* Building on Planck's ideas, Einstein proposed that light itself is particulate, quantized into tiny "bundles" of energy, later called **photons**. Each atom changes its energy, ΔE_{atom}, when it absorbs or emits one photon, or one "particle" of light, whose energy is related to its *frequency*, not its intensity:

$$E_{\text{photon}} = h\nu = \Delta_{\text{atom}}E$$

Let us see how the photon theory explains the two features of the photoelectric effect:

1. *Why is there a threshold frequency?* A beam of light consists of an enormous number of photons. The intensity is related to the *number* of photons, but *not* to the energy of each. Therefore, a photon of a certain *minimum* energy must be absorbed to free an outer electron from the surface (see Figure 6.8). Since energy depends on frequency ($h\nu$), the theory predicts a threshold frequency.
2. *Why is there no time lag?* An outer electron breaks free when it absorbs a photon of *enough* energy; it cannot break free by "saving up" energy from several

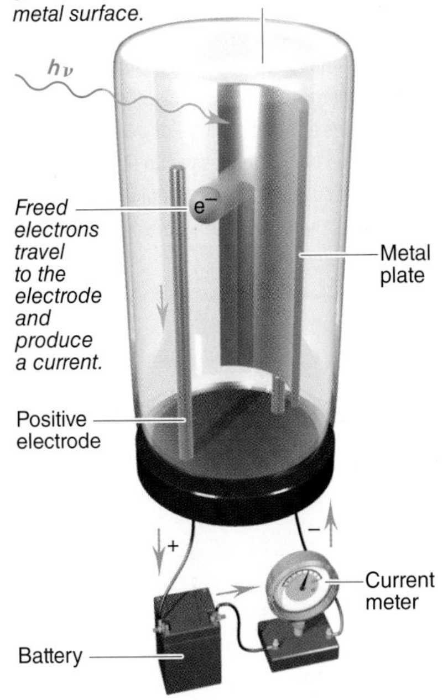

FIGURE 6.8 The photoelectric effect

Albert Einstein (1879–1955) was born in Ulm, Germany. While working as a patent clerk in Bern, Switzerland, in 1905, he not only obtained his doctorate in physics but also published a paper on special relativity, one of his four most influential research papers. He was famous for having determined the relationship between matter and energy, but his work on the photoelectric effect was no less influential on our modern lives. In fact, Einstein received the Nobel Prize in 1921, not for the theory of relativity but for his contributions to theoretical physics and most notably his discovery of the law of the photoelectric effect. He emigrated to Princeton, in the United States, in 1933, where he worked until he died. Einstein was a true believer in Gandhian principles and remained a steadfast pacifist. He was a gifted musician, playing both the piano and the violin with exceptional skill.

*Helge Kragh, "Max Planck: the reluctant revolutionary," *Physics World*. December 2000.

Ping-Pong Photons This analogy helps to explain why light of *insufficient* energy *cannot* free an outer electron from the metal surface. If one Ping-Pong ball does not have enough energy to knock a book off a shelf, neither does a series of Ping-Pong balls, because the book cannot save up the energy from the individual impacts. But one baseball travelling at the same speed does have enough energy.

photons, each having less than the minimum energy. The current is weak in dim light because fewer photons of enough energy can free fewer outer electrons per unit time, but some current flows *as soon as* light of sufficient energy (frequency) strikes the metal surface.

Essentially, then, the photoelectric effect shows that if light of insufficient energy (frequency too low) is applied to a metal surface, no current will be observed. (No outer electrons can be removed since the low-frequency/low-energy photons are not strong enough to overcome the energy binding the outer electron to the surface of the metal.) However, if the metal surface is exposed to light with a frequency that gives it exactly the same energy as the energy required to bind the outer electron to the surface of the metal, an outer electron will break free but it will have no remaining energy. This energy is also known as the *binding energy* or *work function*, ϕ. Here, ϕ equals $h\nu$, where ν is the frequency of the applied light. If the metal surface is exposed to light with a frequency that gives it energy greater than the energy binding the outer electron to the surface of the metal, the outer electron not only breaks free but also has sufficient kinetic energy to move away at high speed. We can express this as follows:

$$E_k = \tfrac{1}{2}mu_e^2 = h\nu - \phi \quad (6.3)$$

where E_k is the kinetic energy of the electron, m is the mass of the electron, u_e is the velocity of the removed electron, h is Planck's constant, ν is the frequency of the incident light, and ϕ is the binding energy of the electron (also called the work function of the metal).

Sample Problem 6.2 Calculating the Energy of Radiation from Its Wavelength

Problem A student uses a microwave oven to heat a meal. The wavelength of the radiation is 1.20 cm. What is the energy of one photon of this microwave radiation?

Plan We know λ in centimetres (1.20 cm), so we convert to metres, find the frequency with Equation 6.1, and then find the energy of one photon with Equation 6.2.

Solution Combine steps to find the energy of a photon:

$$E = h\nu = \frac{hc}{\lambda} = \frac{(6.626 \times 10^{-34}\ \text{J·s})(3.00 \times 10^{8}\ \text{m/s})}{(1.20\ \text{cm})\left(\dfrac{10^{-2}\ \text{m}}{1\ \text{cm}}\right)} = 1.66 \times 10^{-23}\ \text{J}$$

Check Checking the order of magnitude gives

$$\frac{10^{-33}\ \text{J·s} \times 10^{8}\ \text{m/s}}{10^{-2}\ \text{m}} = 10^{-23}\ \text{J}$$

Note: Although the value of h is 6.626 068 761 034 × 10^{-34} J·s, a value of h rounded to four significant figures (i.e., 6.626 × 10^{-34}) is sufficiently accurate for the majority of calculations we will perform.

Follow-Up Problem 6.2 Calculate the energy of one photon of **(a)** ultraviolet light ($\lambda = 1 \times 10^{-8}$ m); **(b)** visible light ($\lambda = 5 \times 10^{-7}$ m); **(c)** infrared light ($\lambda = 1 \times 10^{-4}$ m). What do the answers indicate about the relationship between the wavelength and energy of light?

Sample Problem 6.3 Finding the Velocity of an Electron Removed from a Metal Surface

Problem A particular metal has a work function of 166.0 kJ/mol. What is the maximum wavelength of light that will allow an outer electron to be removed? What will the velocity of the removed electron be if light of wavelength 538 nm is applied to the metal surface?

Plan We know ϕ in kJ/mol (166.0 kJ/mol), so we convert to J/photon using Avogadro's number and then find the wavelength using Equation 6.1; we then find the kinetic energy of one removed outer electron using the wavelength 538 nm and Equation 6.2. Finally, we use the expression for kinetic energy and the mass of an electron to find the velocity of the ejected electron.

Solution Find the work function in J for one photon:

$$E = \left(166.0\ \frac{\text{kJ}}{\text{mol}}\right)\left(\frac{10^3\ \text{J}}{1\ \text{kJ}}\right)\left(\frac{1\ \text{mol}}{6.022 \times 10^{23}}\right) = 2.757 \times 10^{-19}\ \text{J}$$

Find the wavelength of the photon:

$$E = h\nu = \frac{hc}{\lambda}$$

$$\lambda = \frac{hc}{E} = \frac{(6.626 \times 10^{-34}\ \text{J}{\cdot}\text{s})(3.00 \times 10^{8}\ \text{m}{\cdot}\text{s}^{-1})}{2.757 \times 10^{-19}\ \text{J}}$$

$$= 7.21 \times 10^{-7}\ \text{m} = 721\ \text{nm}$$

Check Checking the order of magnitude gives

$$\frac{(10^{-33}\ \text{J}{\cdot}\text{s})(10^{8}\ \text{m}{\cdot}\text{s}^{-1})}{10^{-19}\ \text{J}} = 10^{-6}\ \text{m}$$

Find the kinetic energy of the removed outer electron:

$$E_k = h\nu - \phi = h\left(\frac{c}{\lambda}\right) - \phi$$

$$= (6.626 \times 10^{-34}\ \text{J}{\cdot}\text{s})\left(\frac{3.00 \times 10^{8}\ \text{m}{\cdot}\text{s}^{-1}}{538 \times 10^{-9}\ \text{m}}\right) - 2.757 \times 10^{-19}\ \text{J} = 9.38 \times 10^{-20}\ \text{J}$$

Comment: Notice that the final value for kinetic energy contains only three significant figures. Remember to apply the rules for significant figures when subtracting numbers. In general, photon energies are in the range of 10^{-19} J to 10^{-20} J.

Find the velocity of the removed outer electron (1 J = 1 kg•m^2•s^{-2}):

$$E_k = \frac{1}{2}mu^2$$

$$u = \sqrt{\frac{2E_k}{m}} = \sqrt{\frac{(2)(9.38 \times 10^{-20}\ \text{kg}{\cdot}\text{m}^2{\cdot}\text{s}^{-2})}{9.109 \times 10^{-31}\ \text{kg}}} = 4.54 \times 10^{5}\ \frac{\text{m}}{\text{s}}$$

Comment: A joule is a unit of work, which is force times distance. Force is mass times acceleration. Hence, the base units for work are kg•m•s^{-2}•m, which simplifies to kg•m^2•s^{-2}. When divided by mass, we end up with m^2/s^2. So, when we take the square root, we get the correct unit of m/s for velocity. Here, 10^{-20} divided by 10^{-31} gives 10^{11}, for which the square root is 10^5, giving us the correct order of magnitude for the answer.

Note: Always be aware that Equations 6.1 and 6.2 are per *photon*, not per mole. If you are given quantities per mole, it is essential to convert them to quantities per photon using Avogadro's number to get the correct answer. Similarly, if asked to give an answer in kJ/mol, remember to multiply the quantity per photon by Avogadro's number.

Follow-Up Problem 6.3 Determine the wavelength of light that would produce a removed outer electron with a velocity of 829 km/s in a metal for which incident light, with a wavelength of 799 nm, was just able to remove an electron.

SUMMARY OF SECTION 6.1

- Electromagnetic radiation travels in waves, which are characterized by a given wavelength (λ) and frequency (ν).
- Electromagnetic waves travel through a vacuum at the speed of light, c (3.00×10^8 m/s), which equals $\nu \times \lambda$. Therefore, wavelength and frequency have a reciprocal relationship.
- The intensity (brightness) of light is related to the amplitude of its waves.
- The electromagnetic spectrum ranges from very long radio waves to very short gamma rays and includes the visible region between wavelengths 750 nm (red) and 400 nm (violet).
- Refraction (change in a wave's speed when entering a different medium) and diffraction (bending of a wave around an edge of an object) indicate that energy is wavelike, with properties distinct from the properties of particles of mass.
- Blackbody radiation and the photoelectric effect, however, are consistent with energy occurring in discrete packets, like particles.
- Light exists as photons (quanta), whose energy is proportional to the frequency.
- According to quantum theory, an atom has only certain quantities of energy ($E = nh\nu$), and it can change its energy only by absorbing or emitting a photon, whose energy equals the change in the atom's energy.

6.2 Atomic Spectra

The third observation involving matter and energy that confounded physicists at the beginning of the 20th century related to the light emitted when an element is vaporized and then excited electrically. In this section, we discuss the nature of this light and see why it created a problem for the existing atomic model and how a new model solved the problem.

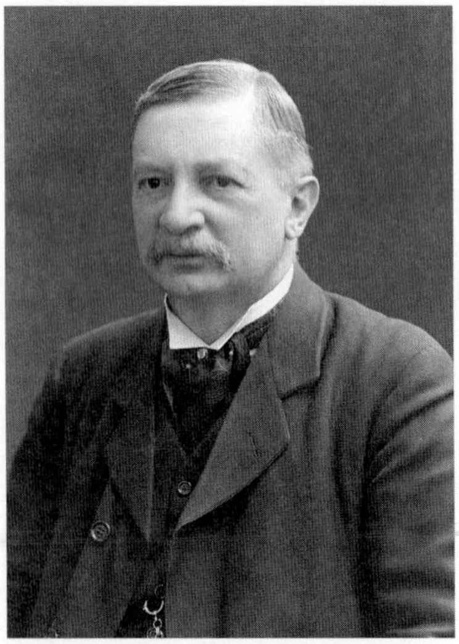

Johannes Rydberg (1854–1919) was born in Halmstad, Sweden. A mathematician who subsequently worked as a physics teacher at the University of Lund's physics institute, he had a deep desire to understand the workings of the periodic table. He turned to the mass of spectral data that had been collected at the time and devised the novel idea of using the wavenumber (the reciprocal of the wavelength) as a measure of the frequency. A pattern immediately began to emerge, and Rydberg formulated his now-famed equation. His work was validated when Johann Balmer's published results proved to be a special case of his own equation.

Line Spectra and the Rydberg Equation

When light from electrically excited gaseous atoms passes through a slit and is refracted by a prism, it does not create a *continuous spectrum*, or rainbow, as sunlight does. Instead, it creates a **line spectrum**, a series of fine lines at specific frequencies separated by black spaces. Figure 6.9A shows the apparatus and the line spectrum of atomic hydrogen. Figure 6.9B shows that each spectrum is *characteristic* of the element producing it.

Features of the Rydberg Equation Spectroscopists studying atomic hydrogen identified several series of spectral lines in different regions of the electromagnetic spectrum. Figure 6.10 shows three of these series. Working on the problem of relating spectral lines in alkali metals, the Swedish physicist Johannes Rydberg observed that plots of the wavenumbers ($\frac{1}{\lambda}$) as a function of the consecutive integers representing the lines in a particular series produced similar curves. Rydberg attempted, using empirical data, to develop a single equation that would, upon substitution of the correct constants, give the correct curve. He was unsuccessful until he became aware of the work being done by Johann Balmer on a series of lines in the hydrogen atom. Working with transitions from $n \geq 3$ down to $n = 2$, Balmer empirically developed a relationship that yielded the wavelength of the absorption/emission lines for hydrogen:

$$\lambda = B\left(\frac{m^2}{m^2 - n^2}\right)$$

where λ represented the wavelength of the spectral line, B was a constant that Balmer found to have a value of 364.56 nm, n equalled 2, and m was any integer greater than 2.

Rydberg modified the Balmer formula by using the wavenumber ($\frac{1}{\lambda}$) instead of the wavelength (λ), a modification that Niels Bohr later hailed as a key development.

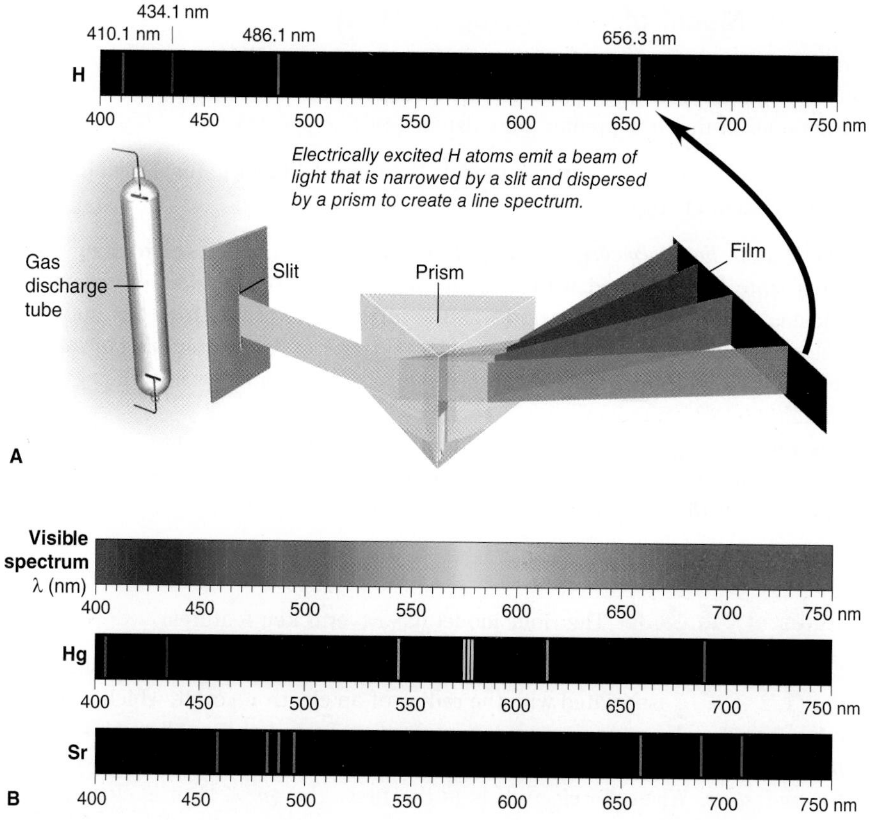

FIGURE 6.9 The line spectra of several elements. A. The line spectrum of atomic hydrogen. **B.** Unlike the continuous spectrum of white light, emission spectra of elements, such as mercury and strontium, appear as characteristic series of coloured lines.

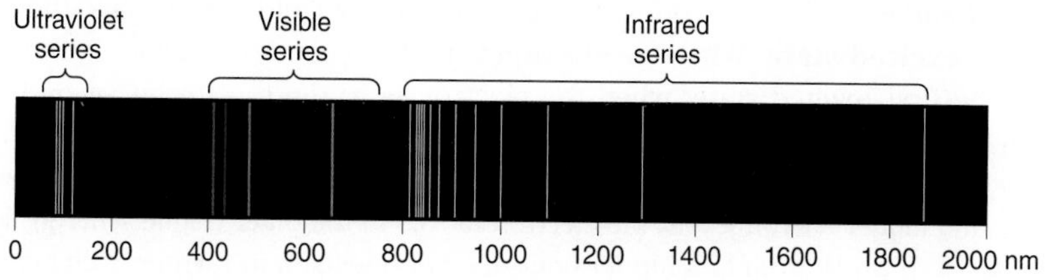

FIGURE 6.10 Three series of spectral lines of atomic hydrogen

The resulting *Rydberg equation* can be used to predict the position and wavelength of any line in a given series:

$$\frac{1}{\lambda} = R\left(\frac{1}{n_1^2} - \frac{1}{n_2^2}\right) \quad (6.4)$$

where λ is the wavelength of the line, n_1 and n_2 are positive integers with $n_2 > n_1$, and R is the Rydberg constant ($1.096\ 776 \times 10^7\ \text{m}^{-1}$). For the visible series, $n_1 = 2$,

$$\frac{1}{\lambda} = R\left(\frac{1}{2^2} - \frac{1}{n_2^2}\right), \text{ with } n_2 = 3, 4, 5, \ldots$$

Problems 6.23 and 6.24, at the end of this chapter, are two of several that apply the Rydberg equation.

Problems with Rutherford's Nuclear Model Almost as soon as Rutherford proposed his nuclear model (described in Chapter 2), a major problem arose. A positive nucleus and a negative electron attract each other, and, for them to stay apart, the kinetic energy of the electron's motion must counterbalance the potential energy of attraction. However, the laws of classical physics say that a negative particle moving in a curved path around a positive particle *must* emit radiation and, thus, lose energy. If the orbiting electrons behaved in this way, they would spiral into the nucleus, and all the atoms would collapse! The laws of classical physics also say that the frequency of the emitted radiation should change smoothly as a negative particle spirals inward and, thus, create a continuous spectrum, not a line spectrum. The behaviour of subatomic particles seemed to violate real-world experiences and accepted principles.

Niels Bohr (1885–1962) was born in Copenhagen, Denmark, into a family that highly valued education. His doctoral work, which was largely theoretical, was his first occasion to consider the work of Planck. His subsequent work in the labs of Sir. J. J. Thomson and Ernest Rutherford led to the eventual development of his model of the atom, incorporating Max Planck's idea of quantization. He continued atomic research at the Institute for Theoretical Physics at Copenhagen University and made many significant contributions not only to the field of physics but also to an understanding of how these advances in physics affected everyday life. He was a strong advocate of the peaceful uses of atomic energy and did much to further atomic research by encouraging young talent in the field of theoretical physics.

The Bohr Model of the Hydrogen Atom

Two years after the nuclear model was proposed, Niels Bohr (1885–1962), a young Danish physicist working in Rutherford's laboratory, suggested a model for the hydrogen atom that *did* predict the existence of line spectra.

Postulates of the Model In his model, Bohr used Planck's and Einstein's ideas about quantized energy and proposed three postulates:

1. *The H atom has only certain energy levels*, which Bohr called **stationary states**. Each state is associated with a fixed circular orbit of the electron around the nucleus. The higher the energy level, the farther the orbit is from the nucleus.
2. *The atom does* **not** *radiate energy while in one of its stationary states.* Even though it violates principles of classical physics, the atom does not change energy while the electron moves *within* an orbit.
3. *The atom changes to another stationary state* (the electron moves to another orbit) *only by absorbing or emitting a photon. The energy of the photon ($h\nu$) equals the difference in the energies of the two states:*

$$E_{\text{photon}} = \Delta_{\text{atom}}E = E_{\text{final}} - E_{\text{initial}} = h\nu$$

Features of the Model The Bohr model has several key features:

- *Quantum numbers and electron orbit.* The quantum number n is a positive integer (1, 2, 3, . . .) associated with the radius of an electron's orbit, which is directly related to the electron's energy: *the lower the n value, the smaller the radius of the orbit, and the lower the energy level.*
- *Ground state.* When the electron is in the first orbit ($n = 1$), it is closest to the nucleus, and the H atom is in its lowest (first) energy level, called the **ground state**.
- *Excited states.* If the electron is in any orbit farther from the nucleus, the atom is in an **excited state**. When the electron is in the second orbit ($n = 2$), the atom is in the first excited state; when the electron is in the third orbit ($n = 3$), the atom is in the second excited state; and so on.
- *Absorption.* If an H atom *absorbs* a photon whose energy equals the *difference* between lower and higher energy levels, the electron moves to the outer (higher energy) level.
- *Emission.* If an H atom in a higher energy level (electron in farther orbit) returns to a lower energy level (electron in closer orbit), the atom *emits* a photon whose energy equals the difference between the two levels. Figure 6.11 shows an analogy that illustrates absorption and emission.

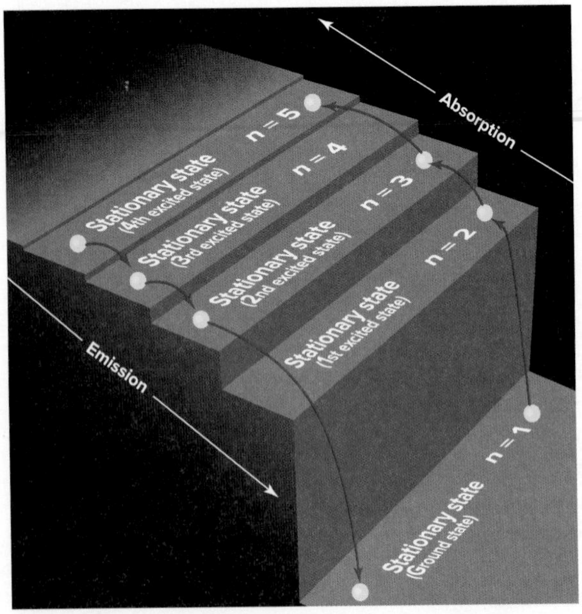

FIGURE 6.11 A quantum "staircase" as an analogy for atomic energy levels. Note that the electron can move up or down one or more steps at a time but cannot lie *between* steps.

How the Model Explains Line Spectra A spectral line results because a photon of specific energy (and thus frequency) is *emitted*. The emission occurs when the electron moves to an orbit closer to the nucleus as the atom's energy changes from a higher state to a lower state. Therefore, *an atomic spectrum is not continuous, because the atom's energy is not continuous, but rather has only certain states.*

Figure 6.12 shows how the Bohr model accounts for three series of spectral lines of hydrogen. When a sample of gaseous H atoms is excited, the atoms absorb different quantities of energy. Each atom has one electron, but there are so many atoms in the whole sample that all the energy levels (orbits) have electrons. When electrons drop from outer orbits to the $n = 3$ orbit (second excited state), the emitted photons create the *infrared* series of lines. The *visible* series is created when electrons drop to the $n = 2$ orbit (first excited state), and the *ultraviolet* series is created when electrons drop to the $n = 1$ orbit (ground state).

Limitations of the Model Despite its great success in accounting for the spectral lines of the H atom, the Bohr model failed to predict the spectrum of any other atom. The reason is that it is a *one-electron model*: it works beautifully for the H atom and for other one-electron species, such as He^+ ($Z = 2$), Li^{2+} ($Z = 3$), and Be^{3+} ($Z = 4$), but it fails completely for atoms with more than one electron because the electron-electron repulsions and additional nucleus-electron attractions that are present create much more complex interactions. Even more fundamentally, as we will see in Section 6.4, *electrons do not move in fixed, defined orbits.* However, even though the Bohr model is incorrect as a picture of the atom, we still use the terms *ground state* and *excited state*, and we still retain the central idea that *the energy of an atom occurs in discrete levels, and an atom changes energy by absorbing or emitting a photon of specific energy.*

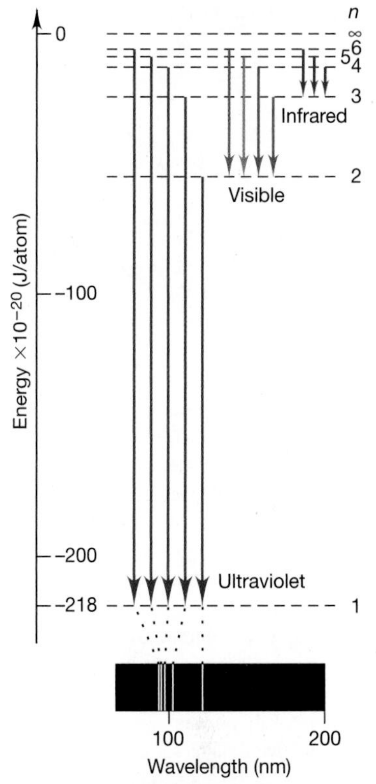

FIGURE 6.12 Bohr's explanation of three series of spectral lines emitted by the hydrogen atom. An energy diagram shows how the ultraviolet series arises.

The Energy Levels of the Hydrogen Atom

Bohr's work led to an equation for calculating the energy levels of an atom:

$$E = -2.18 \times 10^{-18}\,\text{J}\left(\frac{Z^2}{n^2}\right)$$

where Z is the charge of the nucleus. For the H atom, $Z = 1$, so we have

$$E = -2.18 \times 10^{-18}\,\text{J}\left(\frac{1^2}{n^2}\right) = -2.18 \times 10^{-18}\,\text{J}\left(\frac{1}{n^2}\right) \qquad (6.5)$$

Therefore, the energy of the ground state ($n = 1$) of the H atom is

$$E = -2.18 \times 10^{-18}\,\text{J}\left(\frac{1}{1^2}\right) = -2.18 \times 10^{-18}\,\text{J}$$

There is a negative sign for the energy (also used on the axis in Figure 6.12) because we *define* the zero point of the atom's energy when *the electron is completely removed from the nucleus.* Thus, $E = 0$ when $n = \infty$, so $E < 0$ for any smaller n. ■

■ A Tabletop Analogy for Defining the Energy of a System If we define the zero point of a book's potential energy as being when the book is on a table, then the energy is negative when the book is on the floor.

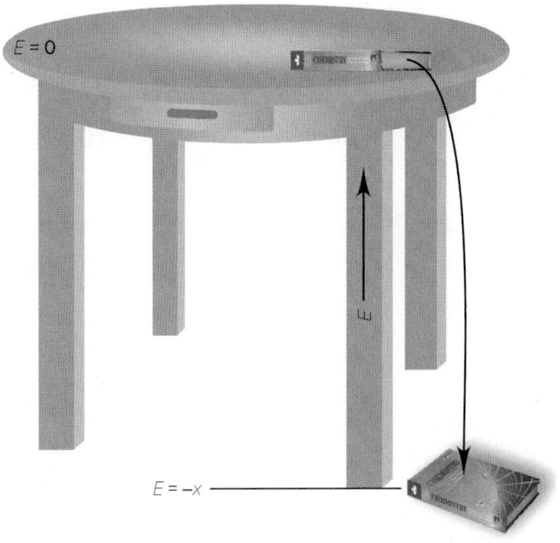

Applying Bohr's Equation for the Energy Levels of an Atom We can use the equation for the energy levels of an atom in several ways:

1. *Finding the difference in energy between two levels.* By subtracting the initial energy level of the H atom from its final energy level, we find the change in energy when the electron moves between the two levels in a hydrogen atom. To find the energy of the initial level, we use

$$E_{\text{initial}} = -2.18 \times 10^{-18}\,\text{J}\left(\frac{1}{n^2_{\text{initial}}}\right)$$

Similarly, we can find the energy of the final level. The change in the energy when the electron moves is simply $\Delta E = E_{\text{final}} - E_{\text{initial}}$. Note that, since n is in the denominator, the following is observed:

- *When the atom emits energy,* the electron moves closer to the nucleus ($n_{\text{final}} < n_{\text{initial}}$), so the atom's final energy is a *larger* negative number and ΔE is negative.
- *When the atom absorbs energy,* the electron moves away from the nucleus ($n_{\text{final}} > n_{\text{initial}}$), so the atom's final energy is a *smaller* negative number and ΔE is positive. (Analogously, in Chapter 5, you saw that ΔH is negative when the system releases heat and positive when the system absorbs heat.)

2. *Finding the energy needed to ionize the H atom.* We can also find the energy needed to remove the electron completely; that is, we can find ΔE for the following change:

$$H(g) \longrightarrow H^+(g) + e^-$$

We substitute $n_{\text{final}} = \infty$ and $n_{\text{initial}} = 1$ into Equation 6.5 and obtain

$$E_{\text{final}} = -2.18 \times 10^{-18}\,\text{J}\left(\frac{1}{\infty^2}\right) = 0$$

$$E_{\text{initial}} = -2.18 \times 10^{-18}\,\text{J}\left(\frac{1}{1^2}\right) = -2.18 \times 10^{-18}\,\text{J}$$

$$\Delta E = E_{\text{final}} - E_{\text{initial}} = 0 - (-2.18 \times 10^{-18}\,\text{J}) = 2.18 \times 10^{-18}\,\text{J}$$

Gerhard Herzberg was born in Hamburg, Germany, but lived in Canada for a large part of his life. He received the Nobel Prize in Chemistry in 1971 for "his contributions to the knowledge of electronic structure and geometry of molecules, particularly free radicals" (Nobel Media). He worked as a professor in the Department of Physics at the University of Saskatchewan for nine years and was the Chancellor at Carleton University (Ottawa) from 1973 to 1980. In 1968, he was named a Companion of the Order of Canada. Seven years before his death in 1999, he was sworn in to the Queen's Privy Council for Canada. In 2000, the Natural Sciences and Engineering Research Council (NSERC) named the Gerhard Herzberg Gold Medal for Science and Engineering in his honour, as they felt his research embodied two of the main attributes that the work of recipients should possess, namely the ability to influence other research and high quality. Herzberg wrote the authoritative texts *Atomic Spectra and Atomic Structure* and the four-volume *Molecular Spectra and Molecular Structure* that are still considered "bibles" by spectroscopists.

Energy must be *absorbed* to remove the electron from the nucleus, so ΔE is positive.

The *ionization energy* of hydrogen is the energy required to form 1 mol of gaseous H^+ ions from 1 mol of gaseous H atoms. Thus, for 1 mol of H atoms,

$$\Delta E = \left(2.18 \times 10^{-18}\ \frac{\text{J}}{\text{atom}}\right)\left(6.022 \times 10^{23}\ \frac{\text{atoms}}{\text{mol}}\right)\left(\frac{1\ \text{kJ}}{10^3\ \text{J}}\right) = 1.31 \times 10^3\ \text{kJ/mol}$$

Ionization energy is a key atomic property, and we will return to it in Chapter 7.

3. *Finding the wavelength of a spectral line.* Once we find ΔE using Equation 6.5, we can determine the wavelengths of the spectral lines of the H atom by combining the relation between frequency and wavelength (Equation 6.1) with Planck's expression for the change in energy of an atom (Equation 6.2) and solving for λ:

$$\Delta E = h\nu = \frac{hc}{\lambda} \quad \text{or} \quad \lambda = \frac{hc}{\Delta E}$$

In Sample Problem 6.4, we find the energy change when an H atom absorbs a photon.

Sample Problem 6.4 Determining ΔE and λ of an Electron Transition

Problem A hydrogen atom absorbs a photon of UV light (see Figure 6.12), and its electron enters the $n = 4$ energy level. Calculate **(a)** the change in energy of the atom; **(b)** the wavelength (nm) of the photon.

Plan (a) The H atom absorbs energy, so $E_{\text{final}} > E_{\text{initial}}$. We are given $n_{\text{final}} = 4$, and Figure 6.12 shows that $n_{\text{initial}} = 1$ because a UV photon is absorbed. We apply Equation 6.5 to find ΔE. **(b)** Once we know ΔE, we find the frequency with Equation 6.2 and the wavelength (m) with Equation 6.1. Then we convert from metres to nanometres.

Solution (a) Substitute the known values into Equation 6.5:

$$E_{\text{final}} = -2.18 \times 10^{-18}\ \text{J}\left(\frac{1}{n^2_{\text{final}}}\right) = -2.18 \times 10^{-18}\ \text{J}\left(\frac{1}{4^2}\right) = -1.36 \times 10^{-19}\ \text{J}$$

$$E_{\text{initial}} = -2.18 \times 10^{-18}\ \text{J}\left(\frac{1}{n^2_{\text{initial}}}\right) = -2.18 \times 10^{-18}\ \text{J}\left(\frac{1}{1^2}\right) = -2.18 \times 10^{-18}\ \text{J}$$

$$\Delta E = E_{\text{final}} - E_{\text{initial}} = -1.36 \times 10^{-19}\ \text{J} - (-2.18 \times 10^{-18}\ \text{J}) = 2.04 \times 10^{-18}\ \text{J}$$

(b) Use Equations 6.2 and 6.1 to solve for λ:

$$\Delta E = h\nu = \frac{hc}{\lambda}$$

Therefore,

$$\lambda = \frac{hc}{\Delta E} = \frac{(6.626 \times 10^{-34}\ \text{J·s})(3.00 \times 10^8\ \text{m/s})}{2.04 \times 10^{-18}\ \text{J}} = 9.74 \times 10^{-8}\ \text{m}$$

Convert m to nm:

$$\lambda = 9.74 \times 10^{-8}\ \text{m}\left(\frac{1\ \text{nm}}{10^{-9}\ \text{m}}\right) = 97.4\ \text{nm}$$

Check (a) The energy change is positive, which is consistent with absorption. **(b)** The wavelength is within the UV region (about 10 nm to 380 nm).

Comment In the follow-up problem, note that if ΔE is negative (the atom loses energy), we use its absolute value, $|\Delta E|$, because λ must have a positive value.

Follow-Up Problem 6.4 A hydrogen atom with its electron in the $n = 6$ energy level emits a photon of IR light. Calculate **(a)** the change in energy of the atom; **(b)** the wavelength (pm) of the photon.

Spectrometric analysis of the hydrogen atom led to the Bohr model, the first step toward our current model of the atom. From its use by 19th-century chemists to identify elements and compounds, spectrometry has developed into a major tool of modern chemistry (see Tools of the Laboratory).

Spectrometry in Chemical Analysis

The use of spectral data to identify and quantify substances is essential to modern chemical analysis. The terms *spectroscopy* and **spectrometry** refer to a large group of instrumental techniques that obtain spectra to gather data on a substance's atomic and molecular energy levels.

Types of Spectra

Emission and absorption spectra are two important types of spectra:

1. An **emission spectrum**, such as the H atom line spectrum, occurs when atoms in an excited state *emit* photons as they return to a lower energy state. Some elements produce an intense spectral line, or several closely spaced spectral lines, which can be used to indicate their presence. A **flame test**, performed by placing a granule of an ionic compound or a drop of its solution in a flame, relies on these intense emissions (Figure B6.1A), and some colours of fireworks are due to the same emissions (Figure B6.1B). Similarly, the colours of sodium-vapour and mercury-vapour streetlamps are due to the intense emission lines in the spectra of these elements.

2. An **absorption spectrum** is produced when atoms *absorb* photons of certain wavelengths and become excited. When white light passes through sodium vapour, for example, the absorption spectrum shows dark lines at the same wavelengths as the yellow-orange lines in sodium's emission spectrum (Figure B6.2).

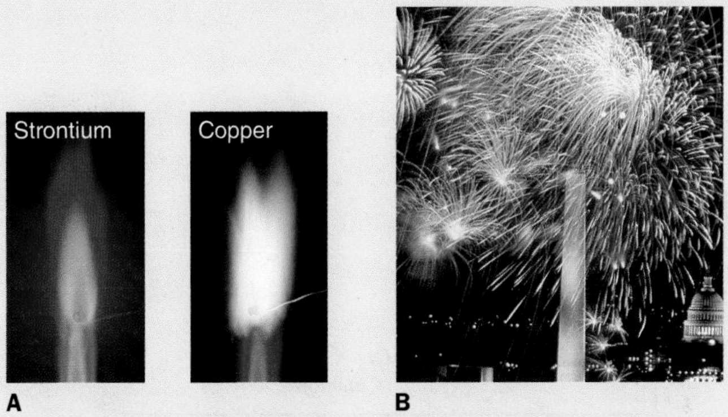
Strontium
Copper
A
B

FIGURE B6.1 Flame tests and fireworks. A. The flame's colour is due to intense emission by the element of light of a particular wavelength. **B.** Fireworks display emissions similar to those seen in flame tests.

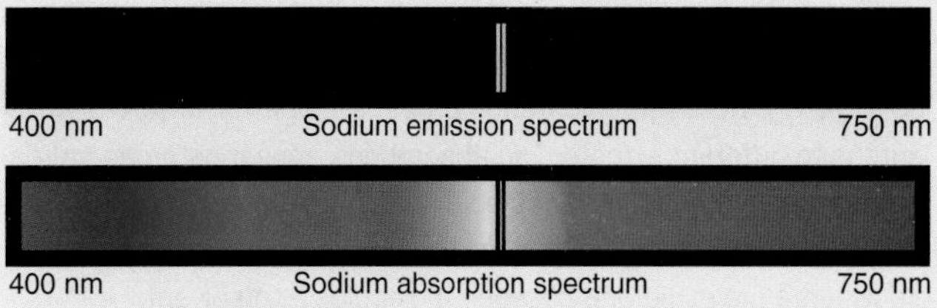
400 nm
Sodium emission spectrum
750 nm
400 nm
Sodium absorption spectrum
750 nm

FIGURE B6.2 Emission and absorption spectra of sodium atoms. The wavelengths of the bright emission lines correspond to those of the dark absorption lines because both are created by the same energy change: $\Delta E_{emission} = -\Delta E_{absorption}$. (Only the two most intense lines in the Na spectra are shown.)

Basic Instrumentation

Instruments based on absorption spectra are much more common than those based on emission spectra because many substances absorb relatively few wavelengths, so their spectra are more characteristic. As well, absorption is less destructive of fragile molecules.

Design differences depend on the region of the electromagnetic spectrum used to irradiate the sample, but all modern spectrometers have components that perform the same basic functions (Figure B6.3). (We will discuss infrared spectroscopy and nuclear magnetic resonance spectroscopy in Chapter 22.)

Identifying and Quantifying a Substance

In chemical analysis, spectra are used to identify a substance and/or quantify the amount in a sample. Visible light is often used for coloured substances, because they absorb only some of the wavelengths. A leaf looks green, for example, because its chlorophyll absorbs red and blue light strongly but green light weakly, so most of the green light is reflected.

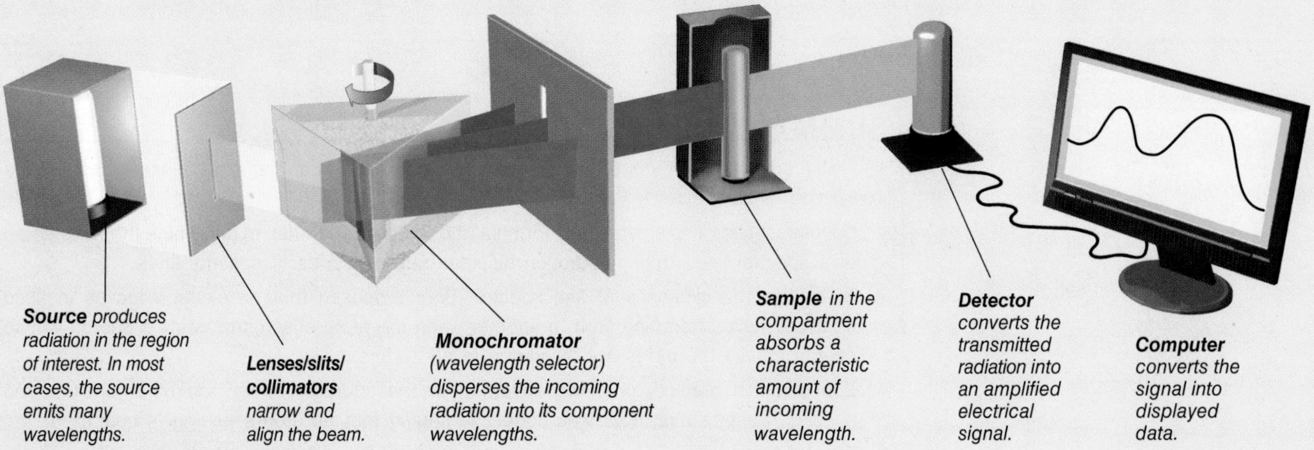
Source produces radiation in the region of interest. In most cases, the source emits many wavelengths.
Lenses/slits/collimators narrow and align the beam.
Monochromator (wavelength selector) disperses the incoming radiation into its component wavelengths.
Sample in the compartment absorbs a characteristic amount of incoming wavelength.
Detector converts the transmitted radiation into an amplified electrical signal.
Computer converts the signal into displayed data.

FIGURE B6.3 Components of a typical spectrometer

(Continued)

Figure B6.4 shows the visible absorption spectrum of chlorophyll *a* in ether solution. The shape of the curve and the wavelengths of the major peaks are characteristic of chlorophyll *a*. The curve varies in height because chlorophyll *a* absorbs light of different wavelengths to different extents. The absorptions appear as broad bands, rather than as the distinct lines seen in the absorption spectra of elements, because there are greater numbers and types of energy levels within a molecule and between molecules and solvent.

In addition to identifying a substance, a spectrometer can be used to measure its concentration because *absorbance*, the quantity of light of a given wavelength absorbed by a substance, is *proportional to the number of molecules*. Suppose that you want to determine the concentration of chlorophyll *a* in a leaf extract. You select a strongly absorbed wavelength from the compound's spectrum (such as 663 nm in Figure B6.4A), measure the absorbance of the *unknown* solution, and compare it with the absorbances of solutions of *known* concentration (Figure B6.4B).

Problems

B6.1 The sodium salt of 2-quinizarinsulfonic acid forms a complex with Al^{3+} that absorbs strongly at 560 nm. (a) Use the data below to draw a plot of absorbance versus concentration of the complex in solution, and find the slope and *y*-intercept:

Concentration (mol/L)	Absorbance (560 nm)
1.0×10^{-5}	0.131
1.5×10^{-5}	0.201
2.0×10^{-5}	0.265
2.5×10^{-5}	0.329
3.0×10^{-5}	0.396

(b) When 20.0 mL of a solution of this complex is diluted with water to 150. mL, its absorbance is 0.236. Find the concentrations of the diluted solution and the original solution.

B6.2 In fireworks displays, flame tests, and other emission events, light of a given wavelength often indicates the presence of a particular element or ion. What are the frequency and colour of the light associated with each of the following?

(a) Li, $\lambda = 671$ nm

(b) Cs^+, $\lambda = 453$ nm

(c) Na, $\lambda = 589$ nm

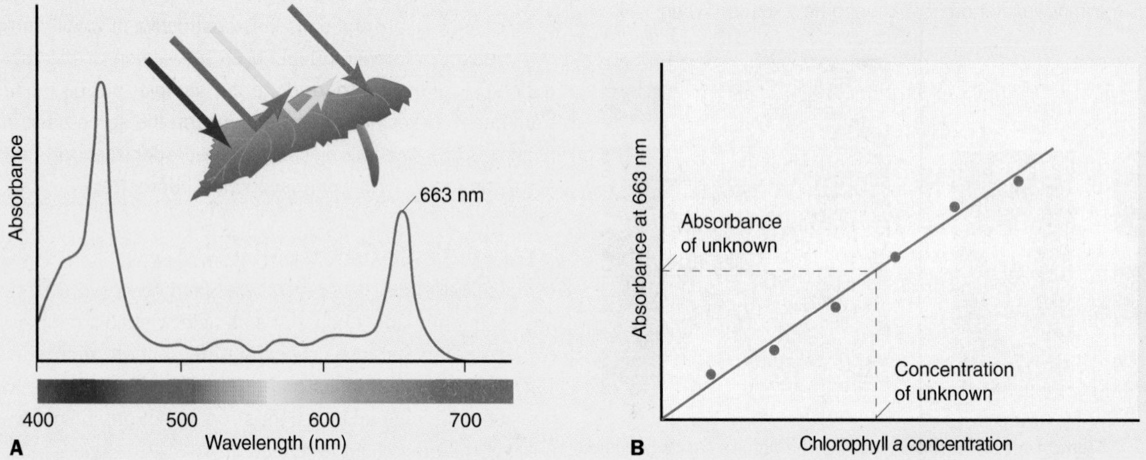

FIGURE B6.4 Measuring the chlorophyll *a* concentration in a leaf extract. A. The spectrometer is set to measure the strong absorption at 663 nm in the chlorophyll *a* spectrum. **B.** The absorbance from the leaf extract is compared with the absorbances of known standards.

SUMMARY OF SECTION 6.2

- Unlike sunlight, light emitted by electrically excited atoms of elements appears as separate spectral lines.
- Spectroscopists use an empirical formula (the Rydberg equation) to determine the wavelength of a spectral line. Atomic hydrogen displays several series of spectral lines.
- To explain the existence of line spectra, Bohr proposed that an electron moves in fixed orbits. It moves from one orbit to another when the atom absorbs or emits a photon whose energy equals the difference in energy levels (orbits).
- Bohr's model predicts only the spectrum of the hydrogen atom and other one-electron species. Despite this, Bohr was correct in saying that an atom's energy is quantized.
- Spectrometry is an instrumental technique that uses emission and absorption spectra to identify substances and measure their concentrations.

6.3 The Wave-Particle Duality of Matter and Energy

The year 1905 was a busy one for Albert Einstein. He had just presented the photon theory and explained the photoelectric effect. A friend remembered him in his small apartment, rocking his baby in a carriage with one hand, while scribbling ideas for a new branch of physics with the other hand. The new branch of physics was the theory of relativity. One of its revelations was that *matter and energy are alternative forms of the same entity.* This revelation is embodied in Einstein's famous equation $E = mc^2$, which relates a specific mass to an equivalent amount of energy. Certain scientific results that showed energy to be particle-like were being developed at the same time as others that showed matter to be wavelike. How these remarkable ideas became entwined is the key to understanding our modern atomic model.

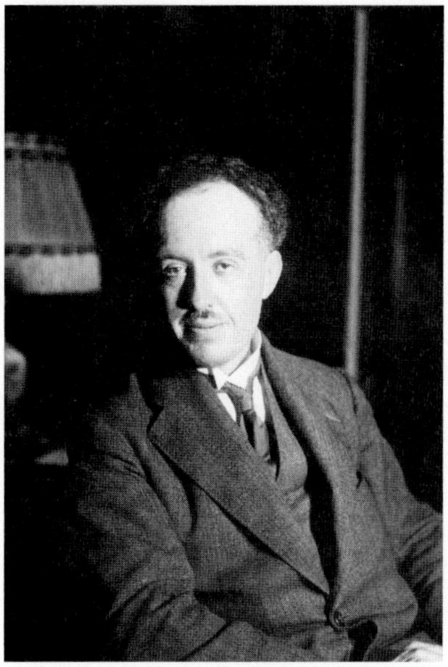

Louis de Broglie (1892–1987) was born in Dieppe, France. He obtained his first degree in history, and then, as his interest in the sciences grew, he took a second degree in science. After World War I, de Broglie resumed his interest and studies in physics and in 1924 published his doctoral thesis, *Recherches sur la Théorie des Quanta* (Research on the Theory of Quanta). In his thesis, he first proposed the ideas that later led to the development of wave mechanics. His ideas were experimentally confirmed by the results of Davisson and Germer in 1927. He continued his research in wave mechanics and was a much published researcher who had a number of graduate students who went on to become famous. De Broglie was a strong advocate for international scientific cooperation.

The Wave Nature of Electrons and the Particle Nature of Photons

Bohr's model was a perfect case of fitting theory to data: he *assumed* that an atom has only certain energy levels to *explain* line spectra. However, Bohr had no theoretical basis for his assumption. Several breakthroughs in the early 1920s provided this basis and blurred the distinction between matter (chunky and massive) and energy (diffuse and massless).

The Wave Nature of Electrons Attempting to explain why an atom has fixed energy levels, a French physics student, Louis de Broglie, considered other systems that display only certain allowed motions, such as the vibrations of a plucked guitar string. Figure 6.13 shows that, because the ends of the string are fixed, only certain vibrational frequencies (and wavelengths) are allowable to create a note. De Broglie proposed that *if energy is particle-like, perhaps matter is wavelike.* He reasoned that *if electrons have wavelike motion* in orbits of fixed radii, they would have only certain allowable frequencies and energies.

Combining the equations for mass-energy equivalence ($E = mc^2$) and energy of a photon ($E = h\nu = \frac{hc}{\lambda}$), de Broglie derived an equation for the wavelength of any particle of mass m—whether planet, baseball, or electron—moving at speed u:

$$\lambda = \frac{h}{mu} \qquad (6.6)$$

The term mu (mass times velocity) in the denominator is also called the *momentum (p)* of the particle. According to this equation for the **de Broglie wavelength**, *matter behaves as though it moves in a wave.* An object's wavelength is *inversely*

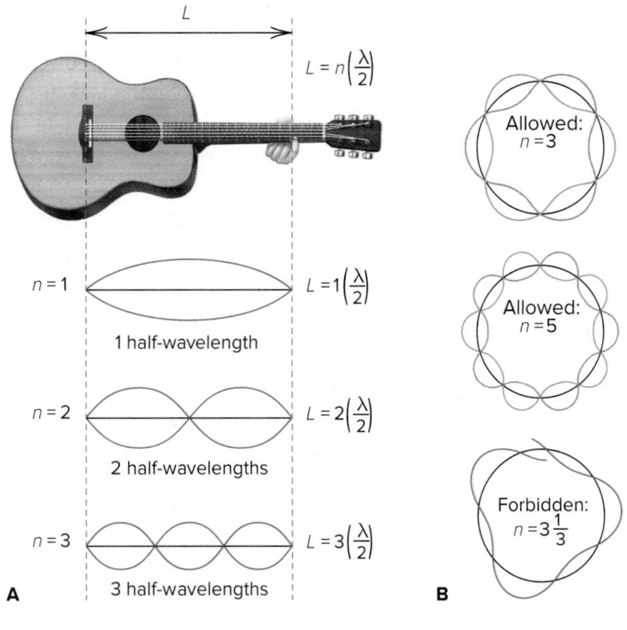

FIGURE 6.13 Wave motion in restricted systems. A. One half-wavelength ($\frac{\lambda}{2}$) is the "quantum" of the guitar string's vibration. With string length L fixed by a finger on the fret, allowed vibrations occur when L is a whole-number multiple (n) of $\frac{\lambda}{2}$. **B.** In a circular electron orbit, only whole numbers of wavelengths are allowed ($n = 3$ and $n = 5$ are shown). A wave with a fractional number of wavelengths (such as $n = 3\frac{1}{3}$) is "forbidden" because it dies out through overlap of crests and troughs.

TABLE 6.1 The de Broglie Wavelengths of Several Objects

Substance	Mass (kg)	Speed (m/s)	λ (m)
Slow electron	9×10^{-31}	1.0	7×10^{-4}
Fast electron	9×10^{-31}	5.9×10^{6}	1×10^{-10}
Alpha particle	6.6×10^{-27}	1.5×10^{7}	7×10^{-15}
1 g mass	1.0×10^{-3}	0.01	7×10^{-29}
Baseball	0.142	40.0	1×10^{-34}
Earth	6.0×10^{24}	3.0×10^{4}	4×10^{-63}

proportional to its mass, so a heavy object, such as a planet or a baseball, has a wavelength *many* orders of magnitude smaller than the object itself (Table 6.1).

Sample Problem 6.5 Calculating the de Broglie Wavelength of an Electron

Problem Find the de Broglie wavelength of an electron with a speed of 1.00×10^6 m/s. (Electron mass $= 9.109 \times 10^{-31}$ kg; $h = 6.626 \times 10^{-34}$ kg·m²/s.)

Plan We know the speed (1.00×10^6 m/s) and mass (9.109×10^{-31} kg) of the electron, so we substitute these into Equation 6.6 to find λ.

Solution

$$\lambda = \frac{h}{mu} = \frac{6.626 \times 10^{-34}\ \text{kg·m}^2/\text{s}}{(9.109 \times 10^{-31}\ \text{kg})(1.00 \times 10^6\ \text{m/s})} = 7.27 \times 10^{-10}\ \text{m}$$

Check The order of magnitude and units seem correct:

$$\lambda \approx \frac{10^{-33}\ \text{kg·m}^2/\text{s}}{(10^{-30}\ \text{kg})(10^6\ \text{m/s})} = 10^{-9}\ \text{m}$$

Comment As you will see in the upcoming discussion, such fast-moving electrons, with wavelengths in the range of atomic sizes, exhibit remarkable properties.

Follow-Up Problem 6.5 What is the speed of an electron that has a de Broglie wavelength of 100. nm?

If electrons travel in waves, they should exhibit the properties of waves. In other words, electrons should show diffraction and interference patterns, just as light does. This characteristic is a corollary of de Broglie's work. A fast-moving electron has a wavelength of about 10^{-10} m, so a beam of such electrons should be diffracted by the spaces between atoms in a crystal—about 10^{-10} m. In 1927, C. Davisson and L. Germer guided a beam of X-rays and then a beam of electrons at a nickel crystal and obtained two diffraction patterns; Figure 6.14 shows these patterns for aluminum. Thus, electrons—particles with mass and charge—create diffraction patterns, just as electromagnetic waves do.

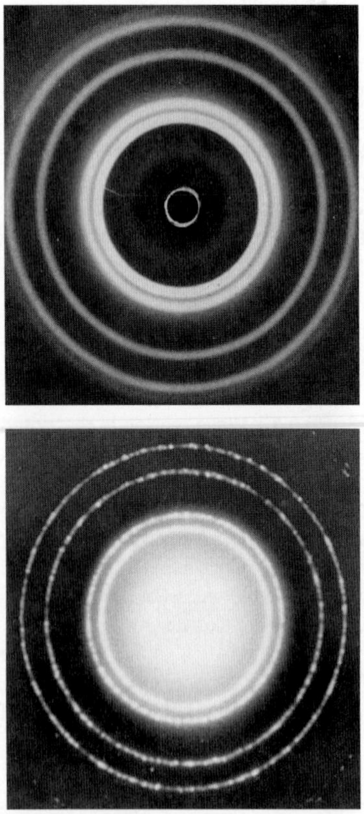

FIGURE 6.14 Diffraction patterns of aluminum with X-rays (*top*) and electrons (*bottom*). The diffraction pattern observed with X-rays (waves) is very similar to that obtained using electrons (particles), supporting the idea of wave-particle duality.

A major application of electrons travelling in waves is the *electron microscope.* Its great advantage over light microscopes is that high-speed electrons have much smaller wavelengths than visible light, which allow much higher resolution. A transmission electron microscope focuses a beam of electrons through a lens, and the beam then passes through a thin section of the specimen to a second lens and then a third lens. These "lenses" are electromagnetic fields, which can result in up to a 200 000-fold magnification. In a scanning electron microscope, the beam scans the specimen, knocking electrons from the specimen to create a current, which generates an image that looks like the surface of the specimen (Figure 6.15).

The Particle Nature of Photons If electrons have properties of energy, do photons have properties of matter? The de Broglie equation suggests that we can calculate the momentum (p), the product of mass and speed, for a photon. Substituting the speed of light (c) for speed u in Equation 6.6 and then solving for p gives

$$\lambda = \frac{h}{mc} = \frac{h}{p} \quad \text{and} \quad p = \frac{h}{\lambda}$$

The inverse relationship between p and λ in this equation means that shorter wavelength (higher energy) photons have greater momentum. Thus, a decrease in a photon's momentum should appear as an increase in its wavelength. In 1923, Arthur Compton directed a beam of X-ray photons at graphite and observed an increase in the wavelength of the reflected photons. Thus, just as billiard balls transfer momentum when they collide, the photons transferred momentum to the electrons in the carbon atoms of the graphite. In this experiment, the photons behaved as particles.

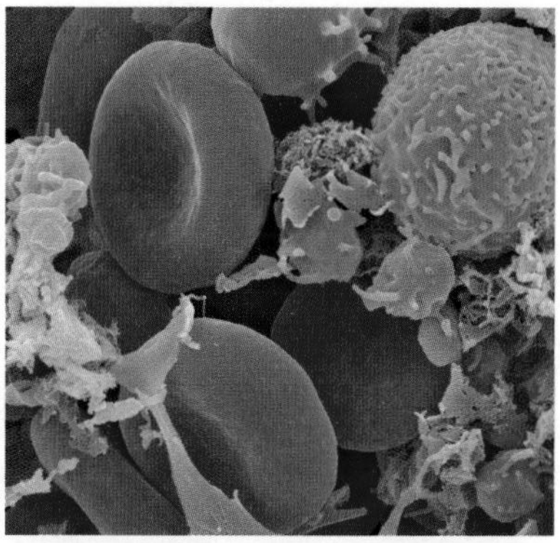

FIGURE 6.15 False-colour scanning electron micrograph of blood cells (×1200)

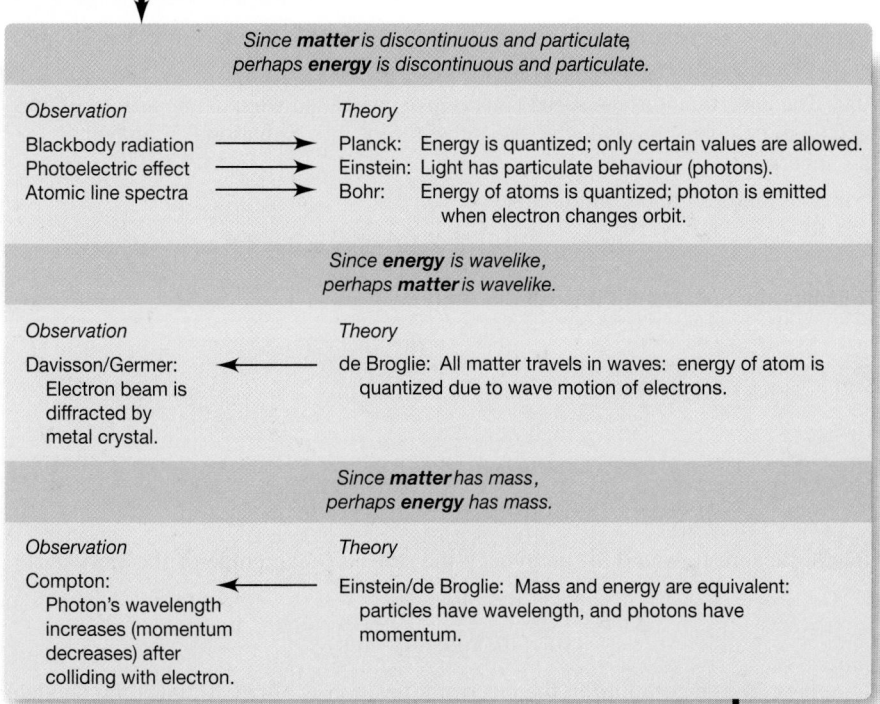

Wave-Particle Duality Classical experiments had shown matter to be particle-like and energy to be wavelike. However, results on the atomic scale show electrons moving in waves and photons having momentum. Thus, every property of matter was also a property of energy. The truth is that *both* matter and energy show *both* behaviours: each possesses both "faces." In some experiments, we observe one face; in other experiments, we observe the other face. Our everyday distinction between matter and energy is meaningful in the macroscopic world but *not* in the atomic world. The distinction is in our minds and the limited definitions we have created, not inherent in nature. This dual character of matter and energy is known as the **wave-particle duality**. It is important to understand that when we use the term "dual character," we are not saying that the particle is sometimes a wave, and the wave is sometimes a particle. Rather, we are saying that all substances have *both* wave *and* particle characteristics. For macroscopic objects, the particle nature is predominant due to the large mass. For energy, the wave nature is predominant due to the insignificant mass. This duality becomes interesting when we consider particles, such as electrons, that have a measurable but very small mass and, hence, measurable particle and wave properties. Figure 6.16 summarizes the theories and observations that led to this new understanding.

CLASSICAL THEORY

Matter	**Energy**
particulate, massive	continuous, wavelike

↓

*Since **matter** is discontinuous and particulate, perhaps **energy** is discontinuous and particulate.*

Observation		*Theory*
Blackbody radiation	→	Planck: Energy is quantized; only certain values are allowed.
Photoelectric effect	→	Einstein: Light has particulate behaviour (photons).
Atomic line spectra	→	Bohr: Energy of atoms is quantized; photon is emitted when electron changes orbit.

*Since **energy** is wavelike, perhaps **matter** is wavelike.*

Observation		*Theory*
Davisson/Germer: Electron beam is diffracted by metal crystal.	←	de Broglie: All matter travels in waves: energy of atom is quantized due to wave motion of electrons.

*Since **matter** has mass, perhaps **energy** has mass.*

Observation		*Theory*
Compton: Photon's wavelength increases (momentum decreases) after colliding with electron.	←	Einstein/de Broglie: Mass and energy are equivalent: particles have wavelength, and photons have momentum.

↓

QUANTUM THEORY

Energy *and* Matter
particulate, massive, wavelike

FIGURE 6.16 Major observations and theories leading from classical theory to quantum theory

Werner Heisenberg (1901–1976) was born in Würtzburg, Germany. He worked with Niels Bohr for two years and, at the young age of 26, was appointed Professor of Theoretical Physics at the University of Leipzig. He became famous for his doctoral dissertation, published when he was only 23 years old and due to which the field of quantum mechanics was born. Heisenberg's work applied matrices to define quantities such as position and velocity and stated that the uncertainties in the determination of these quantities necessarily led to a minimum error that could not be less than the value of the Planck constant, h. Famously, Einstein took issue with this idea and spent many years trying to disprove what came to be known as the Heisenberg uncertainty principle. Heisenberg was a gifted classical pianist.

Heisenberg's Uncertainty Principle

In classical physics, a moving particle has a definite location at any instant, whereas a wave is spread out in space. If an electron has the properties of *both* a particle *and* a wave, can we determine its position in the atom? In 1927, the German physicist Werner Heisenberg postulated the **uncertainty principle**, which states that it is impossible to know simultaneously the position *and* momentum (mass times speed) of a particle. For a particle with constant mass m, the uncertainty principle is expressed mathematically as

$$\Delta x \cdot \Delta p \geq \frac{h}{4\pi} \quad \text{or} \quad \Delta x \cdot \Delta(mu) \geq \frac{h}{4\pi} \tag{6.7}$$

where Δx is the uncertainty in position, Δp is the uncertainty in momentum, and h is Planck's constant. The more accurately we know the position of the particle (smaller Δx), the less accurately we know its momentum [larger Δp or $\Delta(mu)$], and vice versa. The best-case scenario is that we know the product of these uncertainties, which is equal to $\frac{h}{4\pi}$.

For a macroscopic object like a baseball, Δx and Δu are generally insignificant because the mass is enormous compared with $\frac{h}{4\pi}$. It is important to realize that when you have a large object, the mass can be determined so precisely that Δm is essentially zero and, thus, we can reduce $\Delta(mu)$ to $m\Delta u$. Therefore, by knowing the position and speed of a pitched baseball, we can use the laws of motion to predict its trajectory and whether it will be a ball or a strike. However, using the position and speed of an electron to find its trajectory is a very different proposition, as Sample Problem 6.6 demonstrates.

Sample Problem 6.6 Applying the Uncertainty Principle

Problem An electron moving near an atomic nucleus has a speed of 6×10^6 m/s $\pm$ 1%. What is the uncertainty in its position (Δx)?

Plan The uncertainty in the speed (Δu) is given as 1%, so we multiply u (6×10^6 m/s) by 0.01 to calculate the value of Δu, substitute it into Equation 6.7, and solve for the uncertainty in position (Δx).

Solution Find the uncertainty in speed, Δu:

$$\Delta u = 1\% \text{ of } u = 0.01(6 \times 10^6 \text{ m/s}) = 6 \times 10^4 \text{ m/s}$$

Calculate the uncertainty in position, Δx:

$$\Delta x \cdot m\Delta u \geq \frac{h}{4\pi}$$

Thus,

$$\Delta x \geq \frac{h}{4\pi\, m\Delta u} = \frac{6.626 \times 10^{-34} \text{ kg·m}^2\text{/s}}{4\pi(9.109 \times 10^{-31} \text{ kg})(6 \times 10^4 \text{ m/s})} = 1 \times 10^{-9} \text{ m}$$

Check Be sure to round off and check the order of magnitude of the answer:

$$\Delta x \geq \frac{10^{-33} \text{ kg·m}^2\text{/s}}{(10^1)(10^{-30} \text{ kg})(10^5 \text{ m/s})} = 10^{-9} \text{ m}$$

Comment The uncertainty in the electron's position is about 10 times the diameter of the entire atom (10^{-10} m)! Therefore, we have no precise idea where in the atom the electron is located. In the follow-up problem, you will see if an umpire has any better idea about the position of a baseball.

Follow-Up Problem 6.6 How accurately can an umpire know the position of a baseball (mass = 0.142 kg) moving at 44.7 m/s $\pm$ 1.00%?

SUMMARY OF SECTION 6.3

- As a result of Planck's quantum theory and Einstein's theory of relativity, we no longer view matter and energy as distinct entities.
- The de Broglie wavelength is based on the idea that an electron (or any object) has wavelike motion. Allowed atomic energy levels are related to allowed wavelengths of the electron's motion.
- Electrons exhibit diffraction, just as light waves do, and photons exhibit transfer of momentum, just as objects do. This wave-particle duality of matter and energy is observable only on the atomic scale.
- According to the uncertainty principle, we can never know the position and momentum of an electron simultaneously.

6.4 The Quantum-Mechanical Model of the Atom

Acceptance of the dual nature of matter and energy, and the uncertainty principle, culminated in the field of **quantum mechanics**, which examines the wave nature of objects on the atomic scale. In 1926, Erwin Schrödinger derived an equation that is the basis for the *quantum-mechanical model* of the hydrogen atom. This model describes an atom with specific quantities of energy that result from allowed frequencies of its electron's wavelike motion. The electron's position can only be known within a certain probability. Key features of the quantum-mechanical model are described in the following subsections.

The Atomic Orbital and the Probable Location of the Electron

Two central aspects of the quantum-mechanical model concern the atomic orbital and the electron's probable location.

The Schrödinger Equation and the Atomic Orbital The electron's matter-wave occupies the space near the nucleus and is continuously influenced by it. The **Schrödinger equation** is quite complex but can be represented in simpler form as

$$\mathcal{H}\psi = E\psi$$

where E is the energy of the atom. The symbol ψ (Greek *psi*, pronounced "sigh") is called a **wave function (atomic orbital)**, a mathematical description of the electron's matter-wave in three dimensions. The symbol $\mathcal{H}$, called the Hamiltonian operator, represents a set of mathematical operations that, when carried out with a particular ψ, yields one of the allowed energy states of the atom.* Thus, *each solution to the equation gives an energy state associated with a given atomic orbital.*

An important point to keep in mind, throughout this discussion, is that an "orbital" in the quantum-mechanical model *bears no resemblance* to an "orbit" as described in the Bohr model. An *orbital* is a mathematical function that describes the electron's matter-wave (the three-dimensional space in which the highest probability exists of finding an electron). ■

■ **Visualizing the Idea of an Orbital and What It Represents** For the *s* orbital, which is spherical, imagine a spherical room with no doors. Now imagine that a fly is trapped in this room. How would you describe where you might find the fly? Flies are creatures that flit from place to place, so, at any time, the fly could be anywhere! Now, imagine that there is an invisible string with a drop of honey at the tip, placed so the drop of honey is at the centre of the spherical room. How would you describe where you might find the fly now? It will *still* not stay anywhere for too long, but you might find that it tends to hover more around where the honey is. It might flit on and then around the honey, and then flit back. In other words, the probability of finding the fly is now greater around the centre of the room and likely decreases as you head toward the borders of the room. In a sense, this is what the orbital represents—the probability of where you are most likely to find the electron at a given time.

The Probable Location of the Electron While we cannot know *exactly* where the electron is at any moment, we can know where it *probably* is, that is, where it spends most of its time. We get this information by squaring the wave function. Thus, even though ψ has no physical meaning, ψ^2 does and is called the *probability density*, a measure of the probability of finding the electron in some tiny volume of the atom.

*The complete form of the Schrödinger equation in terms of the three linear axes is

$$\left[-\frac{h^2}{8\pi^2 m_e}\left(\frac{\partial^2}{\partial x^2}+\frac{\partial^2}{\partial y^2}+\frac{\partial^2}{\partial z^2}\right)+V(x, y, z)\right]\psi(x, y, z) = E\psi(x, y, z)$$

where h is Planck's constant, ψ is the wave function, m_e is the mass of the electron, E is the total quantized energy of the atomic system, and V is the potential energy at point (x, y, z).

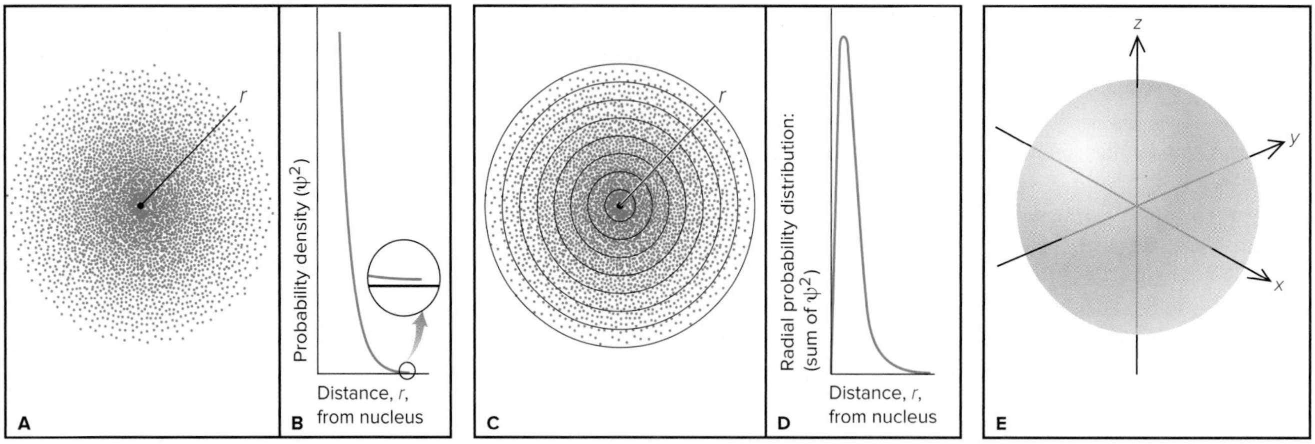

FIGURE 6.17 Electron probability density in the ground-state hydrogen atom. A. In the electron density diagram, the density of dots represents the probability of the electron within a tiny volume and decreases with distance, *r*, from the nucleus. **B.** The probability density (ψ^2) decreases with *r* but does not reach zero (*blow-up circle*). **C.** Counting dots within each layer gives the total probability of the electron being in that layer. **D.** A radial probability distribution plot shows that total electron density peaks *near*, but not *at*, the nucleus. **E.** A 90% probability contour for the ground state of the hydrogen atom.

We depict the electron's probable location in several ways, which we will look at first for the hydrogen atom's *ground state*:

1. *Probability of the electron being in some tiny volume of the atom.* For each energy level, we can create an *electron probability density diagram*, or, more simply, an **electron density diagram**. The value of ψ^2 for a given volume is shown with dots: the greater the density of the dots, the higher the probability of finding the electron within this volume. Note that, for the ground state of the H atom, *the electron probability density decreases with distance from the nucleus* along a line, *r* (Figure 6.17A).

An electron density diagram is also called an **electron cloud depiction** because, if we *could* take a time-exposure photograph of the electron in wavelike motion around the nucleus, it would appear as a "cloud" of positions. Keep in mind that the electron cloud depiction is an *imaginary* picture of the electron changing its position rapidly over time; it does *not* mean that the electron is a diffuse cloud of charge.

Figure 6.17B shows a plot of ψ^2 versus *r*. Due to the thickness of the printed line, the curve appears to touch the axis; in the blow-up circle, however, we see that *the probability of the electron being far from the nucleus is very small, but not zero.*

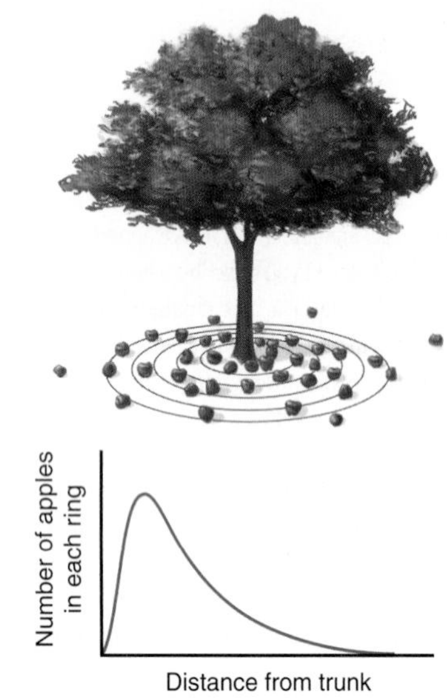

A Radial Probability Distribution of Apples An analogy might clarify why the curve in the radial probability distribution plot peaks and then falls off. Picture fallen apples around the base of an apple tree: the density of the apples is greatest near the trunk and decreases with distance. Divide the ground under the tree into 30-cm-wide concentric rings, and collect the apples within each ring. Apple density is greatest in the first ring, but the area of the second ring is larger, and so it contains a greater *total* number of apples. Farther out, near the edge of the tree, rings have more area but lower apple "density," so the total number of apples decreases. A plot of "number of apples in each ring" versus "distance from the trunk" shows a peak at some distance close to the trunk, as shown in Figure 6.17D.

2. *Total probability density at some distance from the nucleus.* To find the *radial probability distribution* (that is, the *total* probability of finding the electron at some distance, *r*, from the nucleus), we first mentally divide the volume around the nucleus into thin, concentric, spherical layers, like the layers of an onion (shown in cross section in Figure 6.17C). Then we find the *sum of the* ψ^2 *values* in each layer to see which layer is most likely to contain the electron.

The fall-off in probability density with distance has an important effect. Near the nucleus, *the volume of each layer increases faster than its density of dots decreases.* The result of these opposing effects is that the *total* probability peaks in a layer *near*, but not *at*, the nucleus. For example, the total probability in the second layer is higher than the total probability in the first layer, but this result disappears with greater distance. Figure 6.16D shows this result as a **radial probability distribution plot**. ■

3. *Probability contour and the size of the atom.* How far away from the nucleus can we find the electron? This is the same as asking, "How big is the H atom?" Recall, from Figure 6.17B, that the probability of finding the electron far from the nucleus is not zero. Therefore, we *cannot* assign a definite volume to an atom. However, we can visualize an atom with a 90% **probability contour**: the electron is somewhere within that volume 90% of the time (Figure 6.17E).

As you will see later in this section, each atomic orbital has a distinctive radial probability distribution and 90% probability contour.

Table 6.2 The Hierarchy of Quantum Numbers for Atomic Orbitals

Name, Symbol (Property)	Allowed Values	Quantum Numbers					
Principal, n (size, energy)	Positive integer (1, 2, 3, . . .)	1	2		3		
Angular momentum, l (shape)	0 to $n - 1$	0	0	1	0	1	2
Magnetic, m_l (orientation)	$-l, \ldots, 0, \ldots, +l$	0	0	−1 0 +1	0	−1 0 +1	−2 −1 0 +1 +2

Quantum Numbers of an Atomic Orbital

An atomic orbital is specified by three quantum numbers (Table 6.2), which are part of the solution to the Schrödinger equation and indicate the size, shape, and orientation in space of the orbital:*

1. The **principal quantum number (n)** is a *positive integer* (1, 2, 3, and so on). It indicates the relative *size* of the orbital and therefore the relative *distance from the nucleus* of the peak in the radial probability distribution plot. The principal quantum number specifies the *energy level, called the shell,* of the H atom: *the higher the n value, the higher the energy level.* When the electron occupies an orbital with $n = 1$, the H atom is in its ground state and has its lowest energy. When the electron occupies an orbital with $n = 2$ (first excited state), the atom has more energy.

2. The **angular momentum quantum number (l)** is an *integer from* 0 *to* $n - 1$. It is related to the *shape* of the orbital. Note that the principal quantum number sets a limit on the angular momentum quantum number: n limits l. For an orbital with $n = 1$, l can have only one value: 0. For orbitals with $n = 2$, l can have two values: 0 or 1. For orbitals with $n = 3$, l can have three values: 0, 1, or 2; and so on. Thus, the number of possible l values equals the value of n.

3. The **magnetic quantum number (m_l)** is an *integer from* $-l$ *through* 0 *to* $+l$. It prescribes the three-dimensional *orientation* of the orbital in the space around the nucleus. The angular momentum quantum number sets a limit on the magnetic quantum number: l limits m_l. An orbital with $l = 0$ can have only $m_l = 0$. However, an orbital with $l = 1$ can have one of three m_l values: -1, 0, or $+1$; that is, there are three possible orbitals with $l = 1$, each with its own orientation. The number of values that m_l can take is called the multiplicity or degeneracy. All the orbitals have the same energy, but they differ in their three-dimensional orientation. Note that the number of m_l values *equals* $2l + 1$, which is the number of orbitals for a given l. The total number of m_l values (that is, the total number of orbitals) for a given n value is n^2.

Sample Problem 6.7 Determining Quantum Numbers for an Energy Level

Problem What values of the angular momentum (l) and magnetic (m_l) quantum numbers are allowed for a principal quantum number (n) of 3? How many orbitals are allowed?

Plan We determine the allowable quantum numbers with the rules from this book: l values are integers from 0 to $n - 1$, and m_l values are integers from $-l$ to 0 to

*For ease in discussion, chemists often refer to the size, shape, and orientation of an "atomic orbital," although we really mean the size, shape, and orientation of an "atomic orbital's radial probability distribution."

Erwin Schrödinger (1887–1961) was born in Vienna, Austria. From a young age, he displayed tremendous interest in science and mathematics. He worked with Boltzmann's successor, Fritz Hasenöhrl, and mastered eigenvalue problems, a skill that was to serve him well (eigenvalues are a special set of scalars, quantities that are fully described by a magnitude, associated with linear sygtems of equations such as matrices). He worked with a number of famous scientists, during which time he published numerous papers, including several on atomic spectra. It was this work that led to the development of the wave equation for which he became famous. He was dissatisfied with Bohr's use of the quantum condition as applied to orbit theory and felt that it should be approached as an eigenvalue problem. He shared the Nobel prize for this work, in 1933, with Dirac. Schrödinger remained interested in atomic physics and attempted to create a wave theory exclusively because he disliked the idea of a wave and particle theory. He also was very interested in a unified field theory, an idea that Einstein pursued until the end.

$+l$. One m_l value is assigned to each orbital, so the number of m_l values gives the number of orbitals.

Solution Determine the l values:

$$\text{For } n = 3,\ l = 0,\ 1,\ 2.$$

Determine m_l for each l value:

$$\text{For } l = 0,\ m_l = 0.$$
$$\text{For } l = 1,\ m_l = -1,\ 0,\ +1.$$
$$\text{For } l = 2,\ m_l = -2,\ -1,\ 0,\ +1,\ +2.$$

There are nine m_l values, so there are nine orbitals with $n = 3$.

Check Table 6.2 shows that we are correct. As we saw, the total number of orbitals for a given n value is n^2; for $n = 3$, $n^2 = 9$.

Follow-Up Problem 6.7 What are the possible l and m_l values for $n = 4$?

Quantum Numbers and Shells

The energy states and orbitals of the atom are described with specific terms and are associated with one or more quantum numbers:

1. *Shell.* The atom's energy levels, or **shells**, are given by the n value: the smaller the n value, the lower the shell and the greater the probability that the electron is closer to the nucleus.

2. *Subshell.* The atom's levels are divided into **subshells**, which are given by the l value. We use letters to describe the orbital shapes that correspond to a particular l value:

$l = 0$ is called the s subshell.
$l = 1$ is called the p subshell.
$l = 2$ is called the d subshell.
$l = 3$ is called the f subshell.

(The letters derive from names of spectroscopic lines: *s*harp, *p*rincipal, *d*iffuse, and *f*undamental.) Subshells with l values greater than 3 are designated by consecutive letters after f: g subshell, h subshell, and so on. A subshell is named with its n value and letter designation; for example, the subshell with $n = 2$ and $l = 0$ is called the 2s subshell. We discuss orbital shapes below.

3. *Orbital.* Each combination of n, l, and m_l specifies the size (energy), shape, and spatial orientation of one of the atom's orbitals. We know the quantum numbers of the orbitals in a subshell from the name of the subshell and the quantum-number hierarchy. For example, any orbital in the 2s subshell has $n = 2$ and $l = 0$; given that l value, the 2s subshell can only have $m_l = 0$. Thus, the 2s subshell contains only one orbital. Any orbital in the 3p subshell has $n = 3$ and $l = 1$; given this l value, the 3p subshell can have three orbitals with m_l values corresponding to -1, 0, and $+1$. Thus, the 3p subshell has three orbitals.

Sample Problem 6.8 Determining Subshell Names and Orbital Quantum Numbers

Problem Give the name, magnetic quantum numbers, and number of orbitals for each subshell with the given n and l quantum numbers:

(a) $n = 3$, $l = 2$

(b) $n = 2$, $l = 0$

(c) $n = 5$, $l = 1$

(d) $n = 4$, $l = 3$

Plan We name the subshell with the n value and the letter designation of the l value. From the l value, we find the number of possible m_l values, which equals the number of orbitals in the subshell.

Solution

	n	l	Name of Subshell	Possible m_l Values	Number of Orbitals
(a)	3	2	3d	−2, −1, 0, +1, +2	5
(b)	2	0	2s	0	1
(c)	5	1	5p	−1, 0, +1	3
(d)	4	3	4f	−3, −2, −1, 0, +1, +2, +3	7

Check Check the number of orbitals in each subshell:

$$\text{No. of orbitals} = \text{no. of } m_l \text{ values} = 2l + 1$$

Follow-Up Problem 6.8 What are the n, l, and possible m_l values for the $2p$ and $5f$ subshells?

Dr. Tom Woo, at the University of Ottawa, is a physical chemist who studies technologically important issues, such as energy storage and conversion and the catalysts used in the synthesis of new compounds using computational chemistry. The computational simulations are used to obtain important information about processes that are difficult or impossible to study experimentally. These simulations allow the study of the quantum-mechanical effects in molecules at the atomic and electronic levels. For more information about the work being done by Dr. Woo and his group, please go to science.uottawa.ca/chemistry/people/woo-tom.

Sample Problem 6.9 Identifying Incorrect Quantum Numbers

Problem What is wrong with each quantum number designation and/or subshell name?

	n	l	m_l	Name
(a)	1	1	0	1p
(b)	4	3	+1	4d
(c)	3	1	−2	3p

Solution (a) A subshell with $n = 1$ can have only $l = 0$, not $l = 1$. The only possible subshell name is 1s.

(b) A subshell with $l = 3$ is an f subshell, not a d subshell. The name should be $4f$.

(c) A subshell with $l = 1$ can have only −1, 0, or +1 for m_l, not −2.

Check Check that l is always less than n, and m_l is always $\geq -l$ and $\leq +l$.

Follow-Up Problem 6.9 Supply the missing quantum numbers and subshell names.

	n	l	m_l	Name
(a)	?	?	0	4p
(b)	2	1	0	?
(c)	3	2	−2	?
(d)	?	?	?	2s

Shapes of Atomic Orbitals

Each subshell of the hydrogen atom consists of a set of orbitals with characteristic shapes. As you will see in Chapter 7, orbitals for the other atoms have similar shapes.

The s Orbital An orbital with $l = 0$ has a *spherical* shape with the nucleus at its centre and is called an **s orbital**. Because a sphere has only one orientation, an s orbital has only one m_l value: for any s orbital, $m_l = 0$.

1. *The 1s orbital* holds the electron in the H atom's ground state. *The electron probability density is highest at the nucleus.* Figure 6.18A shows this graphically (*top*) and has an electron density *relief map* that depicts the graph's curve in three dimensions (*inset*). Note that the quarter-section of a three-dimensional electron cloud depiction (*middle*) has the darkest shading at the nucleus. For reasons discussed earlier (see Figure 6.17D and its related margin note), the radial probability distribution plot (*bottom*) is highest slightly out from the nucleus. Both plots fall off smoothly with distance.

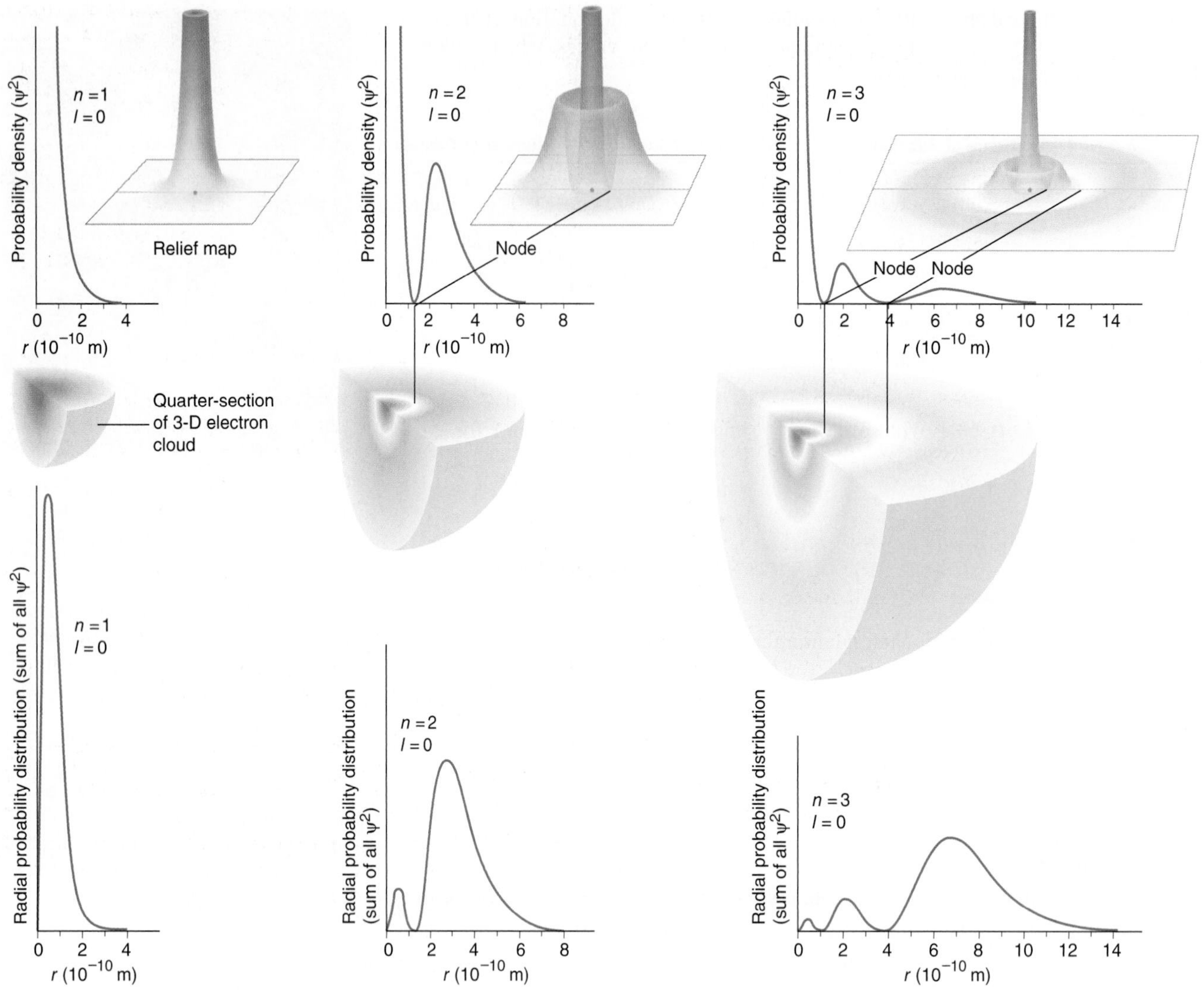

A **1*s* orbital**

B **2*s* orbital**

C **3*s* orbital**

FIGURE 6.18 Representations of the 1*s*, 2*s*, and 3*s* orbitals. For each of the s orbitals, there is a plot of probability density versus distance (*top*), with a relief map (*inset*) showing the plot in three dimensions. Below the plot (*middle*), there is a quarter-section of an electron cloud depiction of the 90% probability contour. Below this (*bottom*), there is a radial probability distribution plot.

2. *The 2s orbital* (Figure 6.18B) has two regions of higher electron density. The radial probability distribution (Figure 6.18B, *bottom*) of the more distant region is *higher* than that of the closer region because the sum of ψ^2 for it is taken over a much larger volume. Between the two regions is a spherical **node**, where the probability of finding the electron drops to zero ($\psi^2 = 0$ at the node, analogous to zero amplitude of a wave exactly between the peak and the trough). Because the 2*s* orbital is larger than the 1*s* orbital, an electron in the 2*s* orbital spends more time farther from the nucleus (in the larger of the two regions) than it does when it occupies the 1*s* orbital.

3. *The 3s orbital* (Figure 6.18C) has three regions of high electron density and two nodes. Here again, the highest radial probability is at the greatest distance from the nucleus. This pattern of more nodes and higher probability with distance from the nucleus continues with the 4*s*, 5*s*, and so on.

The *p* Orbital An orbital with $l = 1$ is called a ***p* orbital** and has two regions (lobes) of high probability, one on *either side* of the nucleus (Figure 6.19). The *nucleus lies at the nodal plane* of this dumbbell-shaped orbital. Since the maximum value of l is $n - 1$, only levels with $n = 2$ or higher have a *p* orbital: the lowest energy *p* orbital (the one closest to the nucleus) is the 2*p* orbital. *One p* orbital consists of *two* lobes,

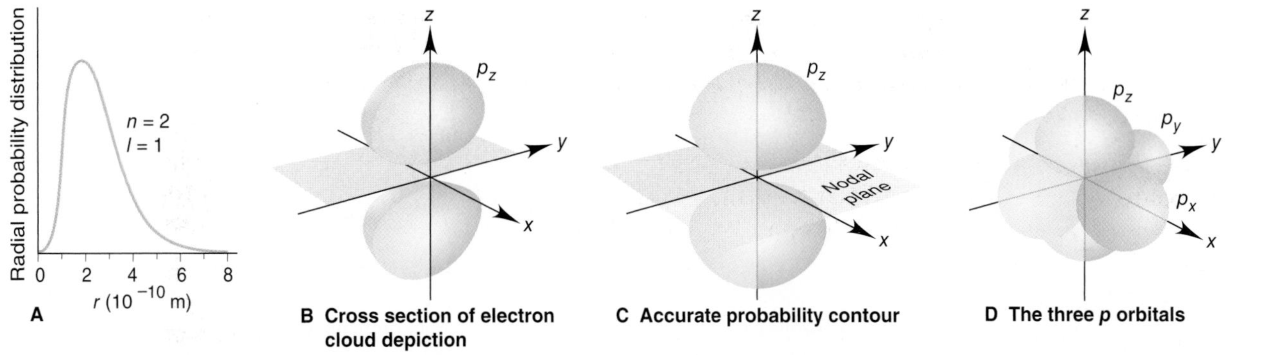

FIGURE 6.19 The 2*p* orbitals. A. A radial probability distribution plot of the 2*p* orbital shows a peak much farther from the nucleus than the peak for the 1*s* orbital. **B.** Cross section of an electron cloud depiction of the 90% probability contour of the $2p_z$ orbital shows a nodal plane. **C.** An accurate representation of the $2p_z$ probability contour. **D.** The three 2*p* orbitals occupy mutually perpendicular regions of space, contributing to the atom's overall spherical shape.

and the electron spends *equal* time in both. Similar to the pattern for *s* orbitals, a 3*p* orbital is larger than a 2*p* orbital, a 4*p* orbital is larger than a 3*p* orbital, and so on.

Unlike *s* orbitals, *p* orbitals *have* different spatial orientations. The three possible m_l values of -1, 0, and $+1$ refer to three orbitals that differ in the direction and amount of angular momentum associated with them; that is, while identical in size, shape, and energy, the three *p* orbitals differ in orientation. We associate *p* orbitals with the *x*, *y*, and *z* axes: the p_x orbital lies along the *x* axis, the p_y along the *y* axis, and the p_z along the *z* axis. (There is no relationship between a particular axis and a given m_l value.)

It is important for us to remember, at all times, that atomic orbitals are really mathematical solutions to the Schrödinger wave equation. Many mathematical equations have solutions that may have positive values at times and may have negative values at times, depending on the values of the variables. This sign of the solution to an equation is known as the *phase* of the orbital. The *s* orbital has a positive phase at all times in all three dimensions because it is a spherically symmetric orbital. The *p* orbital, however, has a positive phase and a negative phase (like all waves do). For example, the p_x orbital has a positive phase whenever *x* has a positive value and a negative phase whenever *x* has a negative value (the different phases are shown as different colours in the depictions of the orbitals). The *d* and *f* orbitals also have positive and negative phases. This particular property of atomic orbitals will have an important impact when we consider bonding.

The *d* Orbital An orbital with $l = 2$ is called a ***d* orbital**. There are five possible m_l values for $l = 2$: $-2, -1, 0, +1$, and $+2$. Thus, a *d* orbital has any one of five orientations (Figure 6.20). Four of the five *d* orbitals have four lobes (a cloverleaf shape), with two mutually perpendicular nodal planes between them and the nucleus at the junction of the lobes (Figure 6.20C). (The orientation of the nodal planes is always between the orbital lobes.) Three of these orbitals lie in the *xy*, *xz*, and *yz* planes, with their lobes *between* the axes, and are called the d_{xy}, d_{xz}, and d_{yz} orbitals. A fourth, the $d_{x^2-y^2}$ orbital, also lies in the *xy* plane, but its lobes are *along* the axes. The fifth *d* orbital, the d_{z^2}, has two major lobes *along* the *z* axis and a doughnut-shaped region that girds the centre. An electron in a *d* orbital spends equal time in all of its lobes.

In keeping with the quantum-number hierarchy, a *d* orbital ($l = 2$) must have a principal quantum number of $n = 3$ or higher, so 3*d* is the lowest energy *d* subshell. Orbitals in the 4*d* subshell are larger (extend farther from the nucleus) than those in the 3*d* subshell, and those in the 5*d* subshell are larger still.

Orbitals with Higher *l* Values Orbitals with $l = 3$ are *f* orbitals and have a principal quantum number of at least $n = 4$. Figure 6.21 shows one of the seven *f* orbitals ($2l + 1 = 7$); each *f* orbital has a complex, multi-lobed shape with several nodal planes. Orbitals with $l = 4$ are *g* orbitals, but they play no known role in chemical bonding.

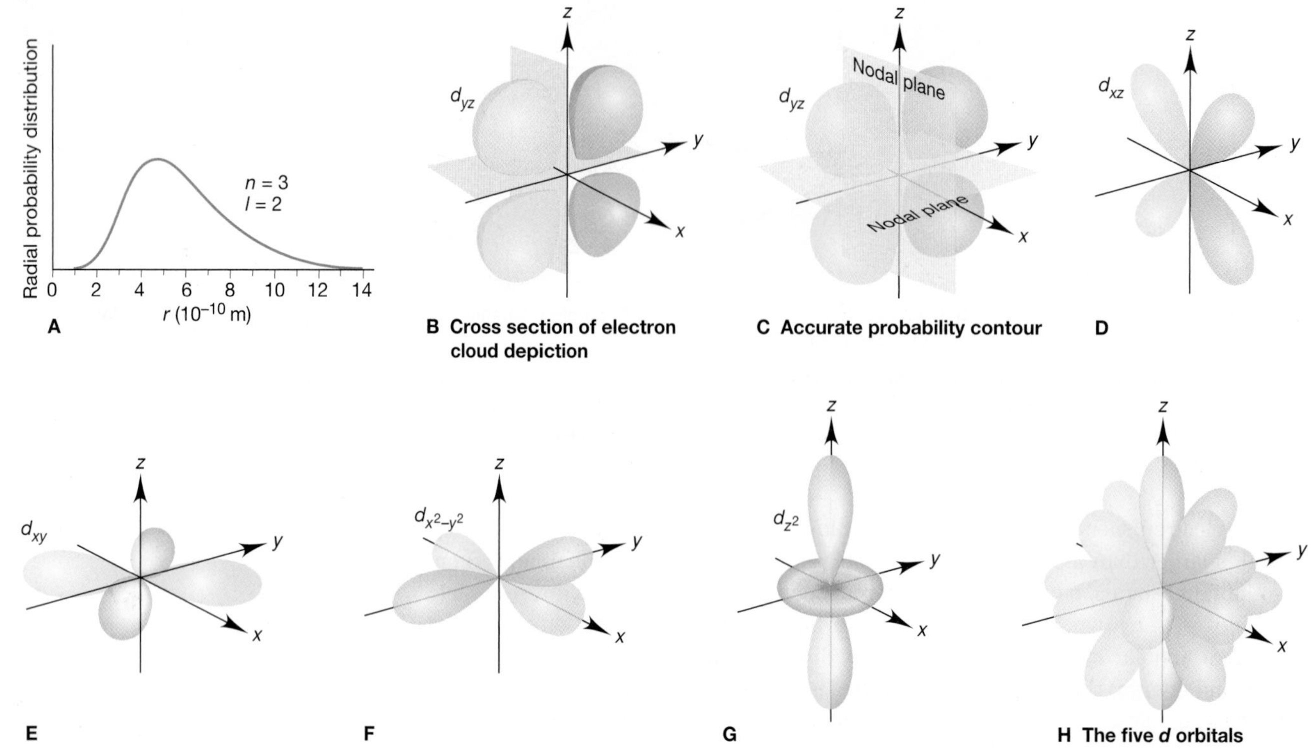

FIGURE 6.20 The 3*d* orbitals. **A.** A radial probability distribution plot. **B.** Cross section of an electron cloud depiction of the $3d_{yz}$ orbital probability contour shows two mutually perpendicular nodal planes and lobes lying *between* the axes. **C.** An accurate representation of the $3d_{yz}$ orbital probability contour. **D.** The (stylized) $3d_{xz}$ orbital. **E.** The $3d_{xy}$ orbital. **F.** The lobes of the $d_{x^2-y^2}$ orbital lie *along* the *x* and *y* axes. **G.** The d_{z^2} orbital has two lobes and a central, doughnut-shaped region. **H.** A composite of the five 3*d* (stylized) orbitals shows how they contribute to the atom's spherical shape.

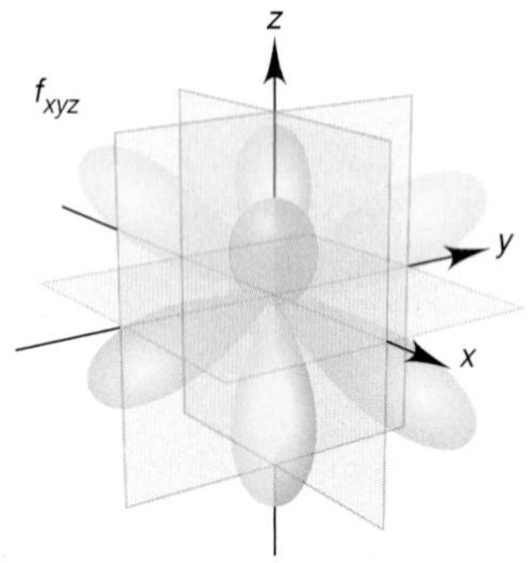

FIGURE 6.21 The $4f_{xyz}$ orbital (stylized), one of the seven 4*f* orbitals

The Special Case of Shells in the Hydrogen Atom

With regard to shells and subshells, the hydrogen atom is a special case. When a hydrogen atom gains energy, its electron occupies an orbital of higher n value, which is (on average) farther from the nucleus. However, because it has just one electron, *hydrogen is the only atom whose energy state depends completely on the principal quantum number, n.* As you will see in Chapter 7, because of additional nucleus-electron attractions and electron-electron repulsions, the energy states of all the other atoms depend on the n *and* l values of the occupied orbitals. Thus, *for the hydrogen atom only,* all four $n = 2$ orbitals (one 2*s* and three 2*p*) have the same energy, all nine $n = 3$ orbitals (one 3*s*, three 3*p*, and five 3*d*) have the same energy (Figure 6.22), and so on.

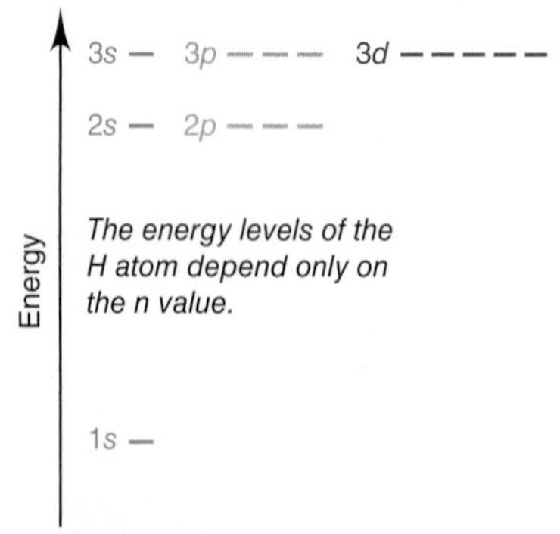

FIGURE 6.22 Energy levels of the hydrogen atom

SUMMARY OF SECTION 6.4

- The atomic orbital (ψ, wave function) is a mathematical description of the electron's wavelike behaviour in an atom. The Schrödinger equation converts each allowed wave function to one of the atom's energy states.
- The probability density of finding the electron at a particular location is represented by ψ^2. For a given energy level, an electron density diagram and a radial probability distribution plot show how the electron occupies the space near the nucleus.
- An atomic orbital is described by three quantum numbers: size (n), shape (l), and orientation (m_l): n limits l to $n - 1$ values, and l limits m_l to $2l + 1$ values.
- A shell has subshells with the same n value; a subshell has orbitals with the same n and l values but different m_l values.
- A subshell with $l = 0$ is described by a spherical (*s*) orbital, a subshell with $l = 1$ is described by three two-lobed (*p*) orbitals, and a subshell with $l = 2$ is described by five multi-lobed (*d*) orbitals.
- In the special case of the hydrogen atom, the shells depend only on the n value.

CHAPTER REVIEW GUIDE

Learning Objectives

Relevant section (§) and/or sample problem (SP) numbers appear in parentheses.

Concepts

1. Describe the wave characteristics of light (the interrelations of frequency, wavelength, and speed and the meaning of amplitude) and the general regions of the electromagnetic spectrum. (§6.1)
2. Differentiate between particles and waves in terms of the phenomena of refraction, diffraction, and interference. (§6.1)
3. Explain the quantization of energy and how an atom changes its energy by emitting or absorbing quanta of radiation. (§6.1)
4. Describe how the photon theory explains the photoelectric effect. (§6.1)
5. Describe how Bohr's theory explained the line spectra of the hydrogen atom; explain why the theory is wrong and which ideas we retain. (§6.2)
6. Describe the wave-particle duality of matter and energy, and the relevant theories and experiments that led to it (de Broglie wavelength, electron diffraction, and photon momentum). (§6.3)
7. Explain the meaning of the uncertainty principle and how uncertainty limits our knowledge of electron properties. (§6.3)
8. Distinguish between ψ (wave function, or atomic orbital) and ψ^2 (probability density). (§6.4)
9. Explain how electron density diagrams and radial probability distribution plots depict the electron's location within the atom. (§6.4)
10. Categorize the hierarchy of the quantum numbers that describe the size (n, energy), shape (l), and orientation (m_l) of an orbital. (§6.4)
11. Explain the distinction between a shell, a subshell, and an orbital. (§6.4)
12. Describe the shapes of *s*, *p*, and *d* orbitals. (§6.4)

Skills

1. Interconvert wavelength and frequency. (SP 6.1)
2. Calculate the energy of a photon from its wavelength. (SP 6.2)
3. Find the maximum wavelength of light that will allow an outer electron to be removed from a metal surface, as well as the velocity of the removed electron. (SP 6.3)
4. Find the energy change and wavelength of the photon absorbed or emitted when an H atom changes its energy level. (SP 6.4)
5. Apply de Broglie's equation to find the wavelength of an electron. (SP 6.5)
6. Apply the uncertainty principle to see that the location and momentum of a particle cannot be determined simultaneously. (SP 6.6)
7. Determine quantum numbers and subshell designations. (SPs 6.7–6.9)

Key Terms

Section 6.1
electromagnetic radiation
frequency (ν)
wavelength (λ)
amplitude
speed
speed of light (c)
electromagnetic spectrum
infrared (IR)
ultraviolet (UV)
refraction
diffraction
quantum number
Planck's constant (h)
quantum
photoelectric effect
photons

Section 6.2
line spectrum
stationary states
ground state
excited state
spectrometry
emission spectrum
flame test
absorption spectrum

Section 6.3
de Broglie wavelength
wave-particle duality
uncertainty principle

Section 6.4
quantum mechanics
Schrödinger equation
wave function (atomic orbital)
electron density diagram
electron cloud depiction
radial probability distribution plot
probability contour
principal quantum number (n)
angular momentum quantum number (l)
magnetic quantum number (m_l)
shells
subshells
s orbital
node
p orbital
d orbital

Key Equations and Relationships

6.1 Relating the speed of light to its frequency and wavelength:

$$c = \nu \times \lambda$$

6.2 Determining the smallest change in an atom's energy:

$$\Delta E = h\nu = E_{\text{photon}}$$

6.3 Relating the kinetic energy of an outer electron removed from a metal surface to the frequency of the incident light:

$$E_k = \tfrac{1}{2}mu_e^2 = h\nu - \phi$$

where E_k is the kinetic energy, m is the mass of the removed outer electron, v_e is the velocity of the removed outer electron, h is Planck's constant, ν is the frequency of the incident light, and ϕ is the binding energy or work function

6.4 Calculating the wavelength of any line in the spectrum of the hydrogen atom (Rydberg equation):

$$\frac{1}{\lambda} = R\left(\frac{1}{n_1^2} - \frac{1}{n_2^2}\right)$$

where n_1 and n_2 are positive integers and $n_2 > n_1$

6.5 Finding the energy of levels in the hydrogen atom:

$$E_n = -2.18 \times 10^{-18}\,\text{J}\left(\frac{1}{n^2}\right)$$

Finding the difference between two energy levels in any one-electron species:

$$\Delta E = E_{\text{final}} - E_{\text{initial}} = -(2.18 \times 10^{-18}\,\text{J})(Z^2)\left(\frac{1}{n^2_{\text{final}}} - \frac{1}{n^2_{\text{initial}}}\right)$$

6.6 Calculating the wavelength of any moving particle (de Broglie wavelength):

$$\lambda = \frac{h}{mu} = \frac{h}{p}$$

6.7 Finding the uncertainty in position or speed of a particle (Heisenberg's uncertainty principle):

$$\Delta x \cdot \Delta p \geq \frac{h}{4\pi} \quad \text{or} \quad \Delta x \cdot \Delta(mu) \geq \frac{h}{4\pi}$$

Brief Solutions to Follow-Up Problems

6.1 $\lambda\,(\text{nm}) = \dfrac{3.00 \times 10^8\,\text{m/s}}{7.23 \times 10^{14}\,\text{s}^{-1}} \times \dfrac{10^9\,\text{nm}}{1\,\text{m}} = 415\,\text{nm}$

$\lambda\,(\text{pm}) = 415\,\text{nm} \times \dfrac{1000\,\text{pm}}{1\,\text{nm}} = 415\,000\,\text{pm}$

6.2 (a) UV: $E = \frac{hc}{\lambda}$

$$= \frac{(6.626 \times 10^{-34}\,\text{J·s})(3.00 \times 10^8\,\text{m/s})}{1 \times 10^{-8}\,\text{m}}$$

$$= 2 \times 10^{-17}\,\text{J}$$

(b) Visible: $E = 4 \times 10^{-19}$ J
(c) IR: $E = 2 \times 10^{-21}$ J
As λ increases, E decreases.

6.3 $\phi = h\nu = \dfrac{hc}{\lambda} = \dfrac{(6.626 \times 10^{-34}\,\text{J·s})\left(3.00 \times 10^8\,\dfrac{\text{m}}{\text{s}}\right)}{799 \times 10^{-9}\,\text{m}}$

$$= 2.49 \times 10^{-19}\,\text{J}$$

$$E_k = \frac{1}{2}mu^2 = \frac{1}{2}(9.109 \times 10^{-31}\,\text{kg})\left(829 \times 10^3\,\frac{\text{m}}{\text{s}}\right)^2$$

$$= 3.13 \times 10^{-19}\,\text{J}$$

$E_k = h\nu - \phi$

$\Rightarrow h\nu = E_k + \phi$

$= (3.13 \times 10^{-19}\,\text{J}) + (2.49 \times 10^{-19}\,\text{J}) = 5.62 \times 10^{-19}\,\text{J}$

$$\nu = \frac{5.62 \times 10^{-19}\,\text{J}}{6.626 \times 10^{-34}\,\text{J·s}} = 8.48 \times 10^{14}\,\text{s}^{-1} = \frac{c}{\lambda}$$

$$\lambda = \frac{3.00 \times 10^8\,\text{m/s}}{8.48 \times 10^{14}\,\text{s}^{-1}} = 3.54 \times 10^{-7}\,\text{m} = 354\,\text{nm}$$

6.4 (a) With $n_{\text{final}} = 3$ for an IR photon,

$$\Delta E = -2.18 \times 10^{-18}\,\text{J}\left(\frac{1}{n^2_{\text{final}}} - \frac{1}{n^2_{\text{initial}}}\right)$$

$$= -2.18 \times 10^{-18}\,\text{J}\left(\frac{1}{3^2} - \frac{1}{6^2}\right)$$

$$= -2.18 \times 10^{-18}\,\text{J}\left(\frac{1}{9} - \frac{1}{36}\right) = -1.82 \times 10^{-19}\,\text{J}$$

(b) $\lambda = \dfrac{hc}{|\Delta E|} = \dfrac{(6.626 \times 10^{-34}\,\text{J·s})(3.00 \times 10^8\,\text{m/s})}{1.82 \times 10^{-19}\,\text{J}} \times \dfrac{1\,\text{pm}}{10^{-12}\,\text{m}}$

$$= 1.09 \times 10^6\,\text{pm}$$

6.5 $u = \dfrac{h}{m\lambda} = \dfrac{6.626 \times 10^{-34}\,\text{kg·m}^2\text{/s}}{(9.109 \times 10^{-31}\,\text{kg})\left(100.\,\text{nm} \times \dfrac{1\,\text{m}}{10^9\,\text{nm}}\right)}$

$$= 7.27 \times 10^3\,\text{m/s}$$

6.6 $\Delta x \geq \dfrac{6.626 \times 10^{-34}\,\text{kg·m}^2\text{/s}}{4\pi(0.142\,\text{kg})(0.447\,\text{m/s})} = 8.31 \times 10^{-34}\,\text{m}$

6.7 $n = 4$, so $l = 0, 1, 2, 3$; in addition to the nine m_l values in Sample Problem 6.7, there are those for $l = 3$:

$$m_l = -3, -2, -1, 0, +1, +2, +3$$

6.8 For 2*p*: $n = 2, l = 1, m_l = -1, 0, +1$
For 5*f*: $n = 5, l = 3, m_l = -3, -2, -1, 0, +1, +2, +3$

6.9 (a) $n = 4$, $l = 1$; (b) name is 2*p*; (c) name is 3*d*; (d) $n = 2$, $l = 0$, $m_l = 0$

PROBLEMS

Problems with red numbers are answered in Appendix G and worked in detail in the Student Solutions Manual. Problem sections match those in this book and provide the numbers of relevant sample problems. Most offer Concept Review Questions, Skill-Building Exercises (grouped in pairs covering the same concept), and Problems in Context. Comprehensive Problems are based on material from any section or previous chapter.

The Nature of Light

(Sample Problems 6.1 to 6.3)

Concept Review Questions

6.1 In what ways are microwave and ultraviolet radiation the same? In what ways are they different?

6.2 Consider the following types of electromagnetic radiation:
(1) Microwave (2) Ultraviolet (3) Radio waves
(4) Infrared (5) X-ray (6) Visible
Arrange them in order of (a) increasing wavelength; (b) increasing frequency; (c) increasing energy.

6.3 Define each wave phenomenon, and give an example of where each occurs:
(a) Refraction (b) Diffraction
(c) Dispersion (d) Interference

6.4 In the 17th century, Newton proposed that light was a stream of particles. The wave-particle debate continued for over 250 years, until Planck and Einstein presented their ideas. Give two pieces of evidence for the wave model and two pieces of evidence for the particle model.

6.5 Portions of electromagnetic waves A, B, and C are represented below:

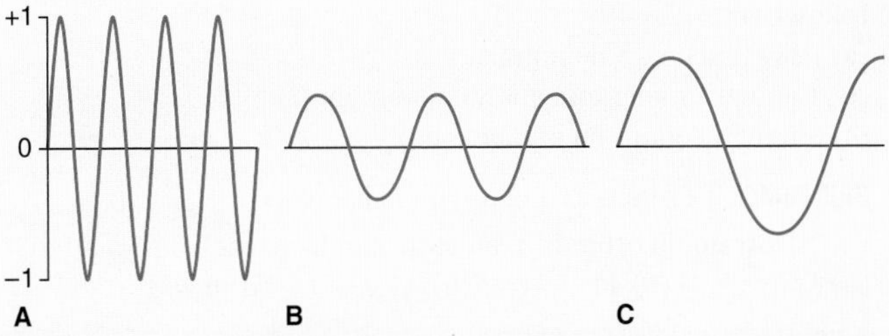

(a) Rank the waves in order of (i) increasing frequency; (ii) increasing energy; (iii) increasing amplitude.
(b) If wave B just barely fails to cause a current when shining on a metal, is wave A or wave C more likely to do so?
(c) If wave B represents visible radiation, is wave A or wave C more likely to be IR radiation?

6.6 (a) What new idea about light did Einstein use to explain the photoelectric effect?
(b) Why does the photoelectric effect exhibit a threshold frequency but not a time lag?

Skill-Building Exercises (grouped in similar pairs)

6.7 An AM station broadcasts rock music at "950 on your radio dial." Units for AM frequencies are given in kilohertz (kHz). Find the wavelength of the station's radio waves in metres (m) and nanometres (nm).

6.8 An FM station broadcasts music at 93.5 MHz (megahertz, or 10^6 Hz). Find the wavelength (in m and nm) of these waves.

6.9 A radio wave has a frequency of 3.8×10^{10} Hz. What is the energy (J) of one photon of this radiation?

6.10 An X-ray has a wavelength of 13 nm. Calculate the energy (J) of one photon of this radiation.

6.11 Rank these photons in terms of increasing energy:
(a) Blue ($\lambda = 453$ nm) (b) Red ($\lambda = 660$ nm)
(c) Yellow ($\lambda = 595$ nm)

6.12 Rank these photons in terms of decreasing energy:
(a) IR ($\nu = 6.5 \times 10^{13}\ s^{-1}$) (b) Microwave ($\nu = 9.8 \times 10^{11}\ s^{-1}$)
(c) UV ($\nu = 8.0 \times 10^{15}\ s^{-1}$)

Problems in Context

6.13 Police often monitor traffic with "K-band" radar guns, which operate in the microwave region at 22.235 GHz (1 GHz = 10^9 Hz). Find the wavelength (nm) of this radiation.

6.14 Covalent bonds in a molecule absorb radiation in the IR region and vibrate at characteristic frequencies.
(a) The C—O bond absorbs radiation of wavelength 9.6 μm. What frequency (s^{-1}) corresponds to this wavelength?
(b) The H—Cl bond has a vibration frequency of 8.652×10^{13} Hz. What wavelength (μm) corresponds to this frequency?

6.15 Cobalt-60 is a radioactive isotope used to treat cancers. A gamma ray emitted by this isotope has an energy of 1.33 MeV (million electron volts; 1 eV = 1.602×10^{-19} J). What are the frequency (Hz) and the wavelength (m) of this gamma ray?

6.16 (a) Ozone formation in the upper atmosphere starts when oxygen molecules absorb UV radiation of wavelengths ≤ 242 nm. Find the frequency and energy of the least energetic of these photons.
(b) Ozone absorbs radiation of wavelengths 220 nm to 290 nm, thus protecting organisms from this radiation. Find the frequency and energy of the most energetic of these photons.

Atomic Spectra

(Sample Problem 6.4)

Concept Review Questions

6.17 How is n_1 in the Rydberg equation (Equation 6.4) related to the quantum number, n, in the Bohr model?

6.18 What key assumption of the Bohr model would a "Solar System" model of the atom violate? What was the theoretical basis for this assumption?

6.19 Distinguish between an absorption spectrum and an emission spectrum. With which did Bohr work?

6.20 Which of these electron transitions correspond to absorption of energy, and which correspond to emission of energy?
(a) $n = 2$ to $n = 4$ (b) $n = 3$ to $n = 1$
(c) $n = 5$ to $n = 2$ (d) $n = 3$ to $n = 4$

6.21 Why could the Bohr model not predict spectra for atoms other than hydrogen?

6.22 The H atom and the Be^{3+} ion each have one electron. Would you expect the Bohr model to predict their spectra accurately? Would you expect their spectra to be identical? Explain.

Skill-Building Exercises (grouped in similar pairs)

6.23 Use the Rydberg equation to find the wavelength (nm) of the photon emitted when an H atom undergoes a transition from $n = 5$ to $n = 2$.

6.24 Use the Rydberg equation to find the wavelength (nm) of the photon absorbed when an H atom undergoes a transition from $n = 1$ to $n = 3$.

6.25 What is the wavelength (nm) of the least energetic spectral line in the infrared series of the H atom?

6.26 What is the wavelength (nm) of the least energetic spectral line in the visible series of the H atom?

6.27 Calculate the energy difference (ΔE) for the transition in Problem 6.23 for 1 mol of H atoms.

6.28 Calculate the energy difference (ΔE) for the transition in Problem 6.24 for 1 mol of H atoms.

6.29 Arrange these H atom electron transitions in order of *increasing* frequency of the photon absorbed or emitted:
(a) $n = 2$ to $n = 4$ (b) $n = 2$ to $n = 1$
(c) $n = 2$ to $n = 5$ (d) $n = 4$ to $n = 3$

6.30 Arrange these H atom electron transitions in order of *decreasing* wavelength of the photon absorbed or emitted:
(a) $n = 2$ to $n = \infty$ (b) $n = 4$ to $n = 20$
(c) $n = 3$ to $n = 10$ (d) $n = 2$ to $n = 1$

6.31 The electron in a ground-state H atom absorbs a photon of wavelength 97.20 nm. To what energy level does it move?

6.32 An electron in the $n = 5$ level of an H atom emits a photon of wavelength 1281 nm. To what energy level does it move?

Problems in Context

6.33 In addition to continuous radiation, fluorescent lamps emit some visible lines from mercury. A prominent line has a wavelength of 436 nm. What is the energy (J) of one photon of this line?

6.34 A Bohr-model representation of the H atom is shown below, with several electron transitions depicted by arrows:

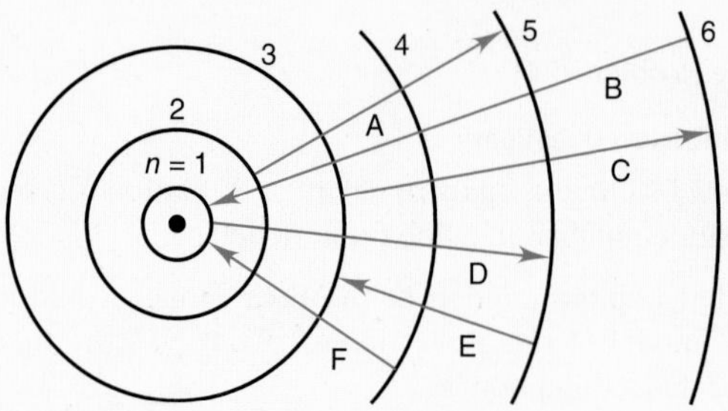

(a) Which transitions are absorptions, and which are emissions?
(b) Rank the emissions in terms of increasing energy.
(c) Rank the absorptions in terms of increasing wavelength of light emitted.

The Wave-Particle Duality of Matter and Energy

(Sample Problems 6.5 and 6.6)

Concept Review Questions

6.35 In what sense is the wave motion of a guitar string analogous to the motion of an electron in an atom?

6.36 What experimental support did de Broglie's concept receive?

6.37 If particles have wavelike motion, why do we not observe this motion in the macroscopic world?

6.38 Why can we not overcome the uncertainty predicted by Heisenberg's principle by building more precise instruments to reduce the error in measurements below the $\frac{h}{4\pi}$ limit?

Skill-Building Exercises (grouped in similar pairs)

6.39 A 105 kg football fullback runs 37 m at 19.8 ± 0.1 km/h.
(a) What is his de Broglie wavelength in metres?
(b) What is the uncertainty in his position?

6.40 An alpha particle (mass = 6.6×10^{-24} g) emitted by a radium isotope travels at $3.4 \times 10^7 \pm 0.1 \times 10^7$ m/s.
(a) What is its de Broglie wavelength in metres?
(b) What is the uncertainty in its position?

6.41 How fast must a 56.5 g tennis ball travel to have a de Broglie wavelength equal to that of a photon of green light (540 nm)?

6.42 How fast must a 142 g baseball travel to have a de Broglie wavelength equal to that of an X-ray photon with λ = 100. pm?

6.43 A sodium flame has a characteristic yellow colour due to emissions of wavelength 589 nm. What is the mass equivalence of one photon of this wavelength (1 J = 1 kg • m^2/s^2)?

6.44 A lithium flame has a characteristic red colour due to emissions of wavelength 671 nm. What is the mass equivalence of 1 mol of photons of this wavelength (1 J = 1 kg • m^2/s^2)?

The Quantum-Mechanical Model of the Atom

(Sample Problems 6.7 to 6.9)

Concept Review Questions

6.45 What physical meaning is attributed to ψ^2?

6.46 What does "electron density in a tiny volume of space" mean?

6.47 Explain what it means for the peak in the radial probability distribution plot for the $n = 1$ level of an H atom to be at 52.9 pm. Is the probability of finding an electron at 52.9 pm from the nucleus greater for the 1*s* orbital or the 2*s* orbital?

6.48 What feature of an orbital is related to each of the following?
(a) Principal quantum number (n)
(b) Angular momentum quantum number (l)
(c) Magnetic quantum number (m_l)

Skill-Building Exercises (grouped in similar pairs)

6.49 How many orbitals in an atom can have each designation?
(a) 1*s* (b) 4*d* (c) 3*p* (d) $n = 3$

6.50 How many orbitals in an atom can have each designation?
(a) 5*f* (b) 4*p* (c) 5*d* (d) $n = 2$

6.51 Give all possible m_l values for orbitals that have each designation:
(a) $l = 2$ (b) $n = 1$ (c) $n = 4, l = 3$

6.52 Give all possible m_l values for orbitals that have each designation:
(a) $l = 3$ (b) $n = 2$ (c) $n = 6, l = 1$

6.53 Draw 90% probability contours (with axes) for each orbital:
(a) s (b) p_x

6.54 Draw 90% probability contours (with axes) for each orbital:
(a) p_z (b) d_{xy}

6.55 For each of the following, give the subshell designation, the allowable m_l values, and the number of orbitals:
(a) $n = 4, l = 2$ (b) $n = 5, l = 1$ (c) $n = 6, l = 3$

6.56 For each of the following, give the subshell designation, the allowable m_l values, and the number of orbitals:
(a) $n = 2, l = 0$ (b) $n = 3, l = 2$
(c) $n = 5, l = 1$

6.57 For each subshell, give the n and l values and the number of orbitals:
(a) 5*s* (b) 3*p* (c) 4*f*

6.58 For each subshell, give the n and l values and the number of orbitals:
(a) 6*g* (b) 4*s* (c) 3*d*

6.59 Is each combination allowed? If not, show two ways to correct it:
(a) $n = 2, l = 0, m_l = -1$ (b) $n = 4, l = 3, m_l = -1$
(c) $n = 3, l = 1, m_l = 0$ (d) $n = 5, l = 2, m_l = +3$

6.60 Is each combination allowed? If not, show two ways to correct it:
(a) $n = 1, l = 0, m_l = 0$ (b) $n = 2, l = 2, m_l = +1$
(c) $n = 7, l = 1, m_l = +2$ (d) $n = 3, l = 1, m_l = -2$

Comprehensive Problems

6.61 The orange colour of carrots and orange peels is due mostly to β-carotene, an organic compound that is insoluble in water, but soluble in benzene and chloroform. Describe an experiment to determine the concentration of β-carotene in the oil from an orange peel.

6.62 The quantum-mechanical treatment of the H atom gives the energy, E, of the electron as a function of n:

$$E = -\frac{h^2}{8\pi^2 m_e a_0^2 n^2} \quad (n = 1, 2, 3, \ldots)$$

where h is Planck's constant, m_e is the electron's mass, and a_0 is 52.92×10^{-12} m.
(a) Write the expression in the form $E = -(\text{constant})(\frac{1}{n^2})$, evaluate the constant (J), and compare the expression with the corresponding expression from Bohr's theory.
(b) Use the expression to find ΔE between $n = 2$ and $n = 3$.
(c) Calculate the wavelength of the photon that corresponds to this energy change.

6.63 The photoelectric effect is illustrated in the following plot of the kinetic energies of electrons ejected from the surface of potassium metal and silver metal at different frequencies of incident light.

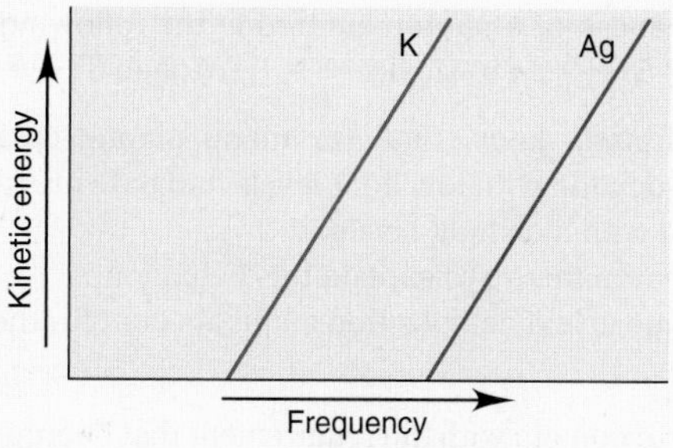

(a) Why do the lines not begin at the origin?
(b) Why do the lines not begin at the same point?
(c) From which metal will light of shorter wavelength eject an electron?
(d) Why are the slopes equal?

6.64 The optic nerve needs to have a minimum of 2.0×10^{-17} J of energy to trigger a series of impulses that will eventually reach the brain.
(a) How many photons of red light (700. nm) are needed?
(b) How many photons of blue light (475 nm) are needed?

6.65 One reason carbon monoxide (CO) is toxic is that it binds to the blood protein hemoglobin more strongly than oxygen does. The bond between hemoglobin and CO absorbs radiation of 1953 cm^{-1}. (The units are the reciprocal of the wavelength in centimetres.) Calculate the wavelength (nm) and the frequency (Hz) of the absorbed radiation.

6.66 A metal ion, Mn^+, has a single electron. The highest energy line in its emission spectrum has a frequency of 2.961×10^{16} Hz. Identify the ion.

6.67 Compare the wavelengths of an electron (mass = 9.109×10^{-31} kg) and a proton (mass = 1.67×10^{-27} kg), each having (a) a speed of 3.4×10^6 m/s; (b) a kinetic energy of 2.7×10^{-15} J.

6.68 Five lines in the H atom spectrum have these wavelengths: (a) 121 nm; (b) 434 nm; (c) 486 nm; (d) 656 nm; (e) 1094 nm. Three lines result from transitions to $n_{\text{final}} = 2$ (visible series). The other two result from transitions in different series, one with $n_{\text{final}} = 1$ and the other with $n_{\text{final}} = 3$. Identify n_{initial} for each line.

6.69 In Einstein's explanation of the threshold frequency in the photoelectric effect, he reasoned that the absorbed photon must have a minimum energy to dislodge an electron from the metal surface. This energy is called the work function (ϕ) of the metal. What is the longest wavelength of radiation (nm) that could cause the photoelectric effect in each of these metals?
(a) Calcium, $\phi = 4.60 \times 10^{-19}$ J
(b) Titanium, $\phi = 6.94 \times 10^{-19}$ J
(c) Sodium, $\phi = 4.41 \times 10^{-19}$ J

6.70 Refractometry is based on the difference in the speed of light through a substance (u) and through a vacuum (c). In the procedure, light of known wavelength passes through a fixed thickness of the substance at a known temperature. The index of refraction equals $\frac{c}{u}$. Using yellow light (λ = 589 nm) at 20°C, for example, the index of refraction of water is 1.33 and the index of refraction of diamond is 2.42. Calculate the speed of light in (a) water; (b) diamond.

6.71 A laser (*l*ight *a*mplification by *s*timulated *e*mission of *r*adiation) provides nearly monochromatic high-intensity light. Lasers are used in eye surgery, CD and DVD players, basic research, and many other areas. Some dye lasers can be "tuned" to emit a desired wavelength. Fill in the blanks in the following table of the properties of some common lasers:

Type	λ (nm)	ν (s^{-1})	E (J)	Colour
He-Ne	632.8	?	?	?
Ar	?	6.148×10^{14}	?	?
Ar-Kr	?	?	3.499×10^{-19}	?
Dye	663.7	?	?	?

6.72 The following combinations are not allowed. If n and m_l are correct, change the l value to create an allowable combination:
(a) $n = 3, l = 0, m_l = -1$ (b) $n = 3, l = 3, m_l = +1$
(c) $n = 7, l = 2, m_l = +3$ (d) $n = 4, l = 1, m_l = -2$

6.73 A ground-state H atom absorbs a photon of wavelength 94.91 nm, and its electron attains a higher energy level. The atom then emits two photons: one photon of wavelength 1281 nm to reach an intermediate energy level, and a second photon to return to the ground state.
(a) What higher level did the electron reach?
(b) What intermediate level did the electron reach?
(c) What was the wavelength of the second photon that was emitted?

6.74 Ground-state ionization energies of three one-electron species are given below:
H = 1.31×10^3 kJ/mol
He^+ = 5.24×10^3 kJ/mol
Li^{2+} = 1.18×10^4 kJ/mol
(a) Write a general expression for the ionization energy of any one-electron species.
(b) Use your expression to calculate the ionization energy of B^{4+}.
(c) What is the minimum wavelength required to remove the electron from the $n = 3$ shell of He^+?
(d) What is the minimum wavelength required to remove the electron from the $n = 2$ shell of Be^{3+}?

6.75 Use the relative size of the 3s orbital below to answer the following questions about orbitals A to D.

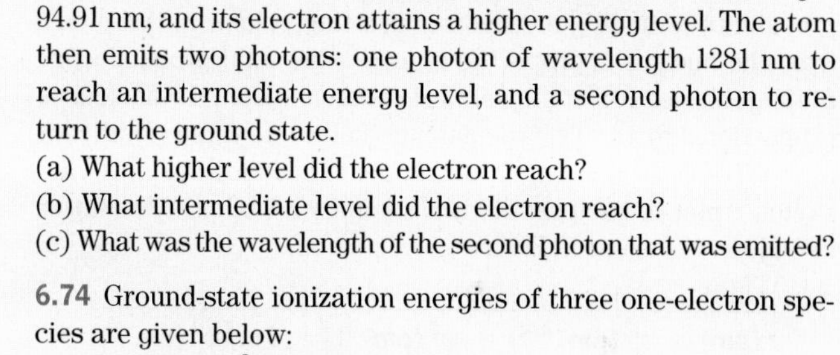

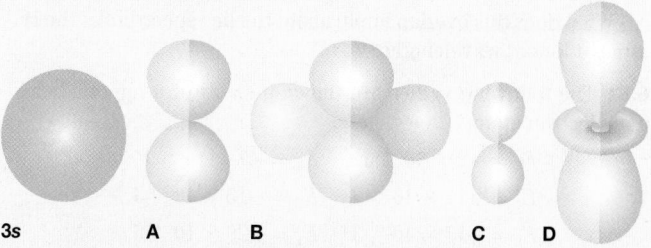

(a) Which orbital has the highest value of n?
(b) Which orbital(s) has (have) a value of $l = 1$? a value of $l = 2$?
(c) How many other orbitals with the same value of n have the same shape as orbital B? orbital C?
(d) Which orbital has the highest energy? lowest energy?

6.76 In the course of developing his model, Bohr arrived at the following formula for the radius of the electron's orbit: $r_n = \frac{n^2h^2\epsilon_0}{\pi m_e e^2}$, where m_e is the electron's mass, e is its charge, and ϵ_0 is a constant related to charge attraction in a vacuum. Given that $m_e = 9.109 \times 10^{-31}$ kg, $e = 1.602 \times 10^{-19}$ C, and $\epsilon_0 = 8.854 \times 10^{-12}$ $C^2/(J \cdot m)$, calculate the following:
(a) The radius of the first ($n = 1$) orbit in the H atom
(b) The radius of the tenth ($n = 10$) orbit in the H atom

6.77 (a) Find the Bohr radius of an electron in the $n = 3$ orbit of an H atom (see Problem 6.76).
(b) What is the energy (J) of the atom in part (a)?
(c) What is the energy of an Li^{2+} ion with its electron in the $n = 3$ orbit?
(d) Why are the answers to parts (b) and (c) different?

6.78 Enormous numbers of microwave photons are needed to warm macroscopic samples of matter. A portion of soup containing 252 g of water is heated in a microwave oven from 20.°C to 98°C, with radiation of wavelength 1.55×10^{-2} m. How many photons are absorbed by the water in the soup?

6.79 The quantum-mechanical treatment of the hydrogen atom gives this expression for the wave function, ψ, of the 1s orbital:

$$\psi = \frac{1}{\sqrt{\pi}}\left(\frac{1}{a_0}\right)^{3/2} e^{-\frac{r}{a_0}}$$

where r is the distance from the nucleus and a_0 is 52.92 pm. The probability of finding the electron in a tiny volume at distance r from the nucleus is proportional to ψ^2. The total probability of finding the electron at all points at distance r from the nucleus is proportional to $4\pi r^2\psi^2$. Calculate the values (to three significant figures) of ψ, ψ^2, and $4\pi r^2\psi^2$ to complete the following table, and sketch a plot of each set of values versus r. Compare the latter two plots with those in Figure 6.18A.

r (pm)	ψ ($pm^{-3/2}$)	ψ^2 (pm^{-3})	$4\pi r^2\psi^2$ (pm^{-1})
0			
50			
100			
200			

6.80 Lines in one spectral series can overlap lines in another.
(a) Does the range of wavelengths in the $n_1 = 1$ series for the H atom overlap the range in the $n_1 = 2$ series?
(b) Does the range in the $n_1 = 3$ series overlap the range in the $n_1 = 4$ series?
(c) How many lines in the $n_1 = 4$ series lie in the range of the $n_1 = 5$ series?
(d) What does this overlap imply about the line spectrum of the H atom at longer wavelengths?

6.81 The following values are the only shells of a hypothetical one-electron atom:

$$E_6 = -2 \times 10^{-19}\text{ J} \qquad E_5 = -7 \times 10^{-19}\text{ J}$$
$$E_4 = -11 \times 10^{-19}\text{ J} \qquad E_3 = -15 \times 10^{-19}\text{ J}$$
$$E_2 = -17 \times 10^{-19}\text{ J} \qquad E_1 = -20 \times 10^{-19}\text{ J}$$

(a) If the electron were in the $n = 3$ shell, what would be the highest frequency (and minimum wavelength) of radiation that could be emitted?
(b) What is the ionization energy (kJ/mol) of the atom in its ground state?
(c) If the electron were in the $n = 4$ shell, what would be the shortest wavelength (nm) of radiation that could be absorbed without causing ionization?

6.82 Photoelectron spectroscopy applies the principle of the photoelectric effect to study orbital energies of atoms and molecules. High-energy radiation (usually UV or X-ray) is absorbed by a sample, and an electron is ejected. The orbital energy can be calculated from the known energy of the radiation and the measured energy of the electron lost. The following energy differences were determined for several electron transitions:

$$\Delta E_{2\longrightarrow1} = 4.098 \times 10^{-17}\text{ J} \qquad \Delta E_{3\longrightarrow1} = 4.854 \times 10^{-17}\text{ J}$$
$$\Delta E_{5\longrightarrow1} = 5.242 \times 10^{-17}\text{ J} \qquad \Delta E_{4\longrightarrow2} = 1.024 \times 10^{-17}\text{ J}$$

Calculate ΔE and λ of a photon emitted in the following transitions: (a) shell 3 $\longrightarrow$ 2; (b) shell 4 $\longrightarrow$ 1; (c) shell 5 $\longrightarrow$ 4.

6.83 Horticulturists know that, for many plants, dark-green leaves are associated with low light levels and pale-green leaves are associated with high light levels.
(a) Use the photon theory to explain this behaviour.
(b) What change in leaf composition might account for the difference in colour?

6.84 In order to comply with the requirement that energy be conserved, Einstein showed, in the photoelectric effect, that the energy of a photon ($h\nu$) absorbed by a metal is the sum of the work function (ϕ), the minimum energy needed to dislodge an electron from the metal's surface, and the kinetic energy (E_k) of the electron: $h\nu = \phi + E_k$. When light of wavelength 358.1 nm falls on the surface of potassium metal, the speed (u) of the dislodged electron is 6.40×10^5 m/s.
(a) What is E_k ($\frac{1}{2}mu^2$) of the dislodged electron?
(b) What is ϕ (J) of potassium?

6.85 For any microscope, the smallest object observable is one-half the wavelength of the radiation used. For example, the smallest object observable with light of 400 nm is 2×10^{-7} m. What is the smallest object observable with an electron microscope, using electrons moving at (a) 5.5×10^4 m/s; (b) 3.0×10^7 m/s?

6.86 In fireworks, the heat of the reaction of an oxidizing agent, such as $KClO_4$, with an organic compound excites certain salts, which emit specific colours. Strontium salts have an intense emission at 641 nm, and barium salts have an intense emission at 493 nm.
(a) What colours do these emissions produce?
(b) What is the energy (kJ) of these emissions for 5.00 g each of the chloride salts of Sr and Ba? (Assume that all the heat produced is converted to emitted light.)

6.87 Atomic hydrogen produces several series of spectral lines. Each series fits the Rydberg equation with its own particular n_1 value. Calculate the value of n_1 (by trial and error, if necessary) that would produce a series of lines in which (a) the *highest* energy line has a wavelength of 3282 nm; (b) the *lowest* energy line has a wavelength of 7460 nm.

6.88 Fish-liver oil is a good source of vitamin A, whose concentration is measured spectrometrically at a wavelength of 329 nm.
(a) Suggest a reason for using this wavelength.
(b) In what region of the spectrum does this wavelength lie?
(c) When 0.1232 g of fish-liver oil is dissolved in 500. mL of solvent, the absorbance is 0.724 units. When 1.67×10^{-3} g of vitamin A is dissolved in 250. mL of solvent, the absorbance is 1.018 units. Calculate the vitamin A concentration in the fish-liver oil.

6.89 Many calculators use photocells as their energy source. Find the maximum wavelength needed to remove an electron from silver ($\phi = 7.59 \times 10^{-19}$ J). Is silver a good choice for a photocell that uses visible light?

6.90 In a game of Clue, Ms. White is murdered in the conservatory. A spectrometer in each room records who is present to help find the murderer. For example, if someone wearing yellow is in a room, light at 580 nm is reflected. The suspects are Col. Mustard, Prof. Plum, Mr. Green, Ms. Peacock (blue), and Ms. Scarlet. At the time of the murder, the spectrometer in the dining room shows a reflection at 520 nm, the spectrometers in the lounge and study record lower frequencies, and the spectrometer in the library records the shortest possible wavelength. Who killed Ms. White? Explain.

6.91 Technetium (Tc; $Z = 43$) is a synthetic element that is used as a radioactive tracer in medical studies. A Tc atom emits a beta particle (electron) with a kinetic energy ($E_k = \frac{1}{2}mu^2$) of 4.71×10^{-15} J. What is the de Broglie wavelength of this electron?

6.92 Electric power is measured in watts (1 W = 1 J/s). About 95% of the power output of an incandescent bulb is converted to heat, and 5% is converted to light. If 10% of this light shines on your chemistry textbook, how many photons per second shine on the book from a 75 W bulb? (Assume that the photons have a wavelength of 550 nm.)

6.93 The flame tests for sodium and potassium are based on the emissions at 589 nm and 404 nm, respectively. When both elements are present, the Na^+ emission is so strong that the K^+ emission can only be seen by looking through a cobalt-glass filter.
(a) What are the colours of these Na^+ and K^+ emissions?
(b) What does the cobalt-glass filter do?
(c) Why is $KClO_4$, rather than $NaClO_4$, used as an oxidizing agent in fireworks?

6.94 The net change during photosynthesis involves CO_2 and H_2O forming glucose ($C_6H_{12}O_6$) and O_2. Chlorophyll absorbs light in the 600 nm to 700 nm region.
(a) Write a balanced thermochemical equation for the formation of 1.00 mol of glucose.
(b) What minimum number of photons with $\lambda = 680.$ nm is needed to form 1.00 mol of glucose?

6.95 Only certain transitions are allowed from one energy level to another. In one-electron species, the change in l of an allowed transition is 61. For example, a $3p$ electron can move to a $2s$ orbital, but not to a $2p$ orbital. Thus, in the UV series, where $n_{final} =$ 1, allowed transitions can start in a p orbital ($l = 1$) of $n = 2$ or higher but not in an s ($l = 0$) or d ($l = 2$) orbital of $n = 2$ or higher. From what orbital does each of the allowed transitions start for the first four emission lines in the visible series ($n_{final} = 2$)?

6.96 The discharge of phosphate in detergents to the environment has led to imbalances in the life cycle of freshwater lakes. A chemist uses a spectrometric method to measure the total phosphate in lakes and obtains the following data for known standards:

Absorbance (880 nm)	Concentration (mol/L)
0	0.0×10^{-5}
0.10	2.5×10^{-5}
0.16	3.2×10^{-5}
0.20	4.4×10^{-5}
0.25	5.6×10^{-5}
0.38	8.4×10^{-5}
0.48	10.5×10^{-5}
0.62	13.8×10^{-5}
0.76	17.0×10^{-5}
0.88	19.4×10^{-5}

(a) Draw a curve of absorbance versus phosphate concentration.
(b) If a sample of lake water has an absorbance of 0.55, what is its phosphate concentration?

6.97 How many photons are required to just melt 675 g of ice in a microwave oven emitting radiation of wavelength 13.25 cm? The enthalpy of fusion of water is 6.02 kJ/mol. If the power output of the microwave oven is 1.5 kW (1 W = 1 J/s), how long will it take for the ice to melt?

6.98 In a photoelectric experiment,* the work function (binding energy, ϕ) for samarium was found to be 260.5 kJ/mol.
(a) What is the maximum wavelength of light that will cause an electron to be ejected from the surface of the samarium metal?
(b) What is the kinetic energy of the electron that is ejected if light of wavelength 325 nm is directed at the surface of the samarium metal?
(c) What is the velocity of the electron ejected in part (b)?

6.99 An electron ejected from the surface of a sheet of lanthanum metal has a velocity of 8.76×10^5 m/s. If the binding energy (work function) of the lanthanum metal is 337.7 kJ/mol, what was the wavelength of the incident light?

6.100 (a) A microwave oven emitting radiation of wavelength 14.47 cm is used to heat 137 mL of water (about half a cup) originally at 23.2°C by *only* 1°C. How many photons are required to heat the water by 1°C?
(b) A bedroom lamp has a CFL bulb that is 9.8% energy efficient. If its power output is 15 W and the lamp emits light with an average wavelength of 550 nm, how long will the lamp have to be on continuously to emit the same number of photons as in part (a)?

*Values obtained from B. E. Nieuwenhuys, O. G. Van Aardenne, and W. M. H. Sachtler, "Adsorption of xenon on group VIII and ib metals studied by photoelectric work function measurements," *Chemical Physics*. September 1974.

Electron Configuration and Chemical Periodicity

IN THIS CHAPTER . . . We explore recurring patterns of electron distributions in atoms to see how they account for the recurring behaviour of the elements. By the end of this chapter, you should be able to

- Describe a new quantum number and explain the restriction it places on the number of electrons in an orbital, thus defining a unique set of quantum numbers for each electron in an atom of any element
- Discuss the electrostatic effects that lead to the splitting of atomic shells into subshells and give rise to the order in which orbitals fill with electrons
- Explain how this filling order correlates with the order of the elements in the periodic table
- Discuss with examples the reasons for periodic trends in atomic properties
- Explain how these trends account for chemical reactivity related to metallic and redox behaviour, ion formation, and magnetic behaviour

7

Concepts and Skills to Review before Studying This Chapter

- Format of the periodic table (Section 2.6)
- Characteristics of metals and nonmetals (Section 2.6)
- Attractions, repulsions, and Coulomb's law (Section 2.7)
- Rules for assigning quantum numbers (Section 6.4)

Who would have believed it possible to predict not only the existence but also the properties, both physical and chemical, of something that no one even knew existed? Dmitri Ivanovich Mendeleev, Дми́трий Ива́нович Менделе́ев, did exactly that. He took the elements known in his time (which we now know were in no particular order) and put them into a table ordered by mass so that elements in columns (called *groups*) and elements in rows (called *periods*) had recurring chemical and physical properties. Because of the periodic occurrence of these properties, his table came to be called the *periodic table of the elements*. What made his table so powerful was his ability to foretell the numerical values of many of the properties before the element was even discovered. Today, we know that the periodic table is ordered based on the number of electrons in each atom and that the periodic properties of elements are due to the similarity in the *electron configurations* of these elements. In this chapter, we will learn how to determine the electron configurations of multi-electron atoms using the quantum numbers we learned in Chapter 6. Then we will see how the electron configurations lead us to see trends in the periodic properties of the elements that form our universe.

We have become expert at measuring recurring patterns in nature—day and night, the seasons, solar and lunar eclipses, and even the swirl in a shell and the rhythm of a heartbeat. Elements exhibit recurring patterns in properties because their atoms do, and we can measure these patterns.

The outpouring of scientific creativity by early-20th-century physicists, which led to the new quantum-mechanical model of the atom, was preceded by countless hours of laboratory work by 19th-century chemists who were exploring the nature of electrolytes, the kinetic-molecular theory, and chemical thermodynamics. The fields of organic chemistry and biochemistry were born in the 19th century, as were the fertilizer, explosives, glassmaking, soap-making, bleaching, and dyestuff industries. As well, for the first time, chemistry became a subject in universities in Europe and North America.

Condensed from all these efforts, an enormous body of facts emerged about the elements. These facts became organized into the periodic table:

- *The original periodic table.* In 1870, the Russian chemist Dmitri Mendeleev arranged the 65 elements known at the time into a table and summarized their behaviour in the **periodic law**: when arranged by *atomic mass*, the elements exhibit a periodic recurrence of similar properties. Mendeleev left blank spaces in his table and was even able to *predict* the properties of several elements, such as germanium, that were not discovered until later.
- *The modern periodic table.* Today's table (*inside front cover*) includes 53 elements not known in 1870 and, most important, arranges the elements by *atomic number* (number of protons), *not* atomic mass. This change is based on the work of the British physicist Henry G. J. Moseley, who obtained X-ray spectra of various metals by bombarding the metals with electrons. Moseley found that the largest X-ray peak for each metal was related to the nuclear charge, which increased by one for each successive element.

The great test for the new atomic model was to answer one of the central questions in chemistry: *why* do the elements behave as they do? Or, rephrasing to fit the main topic of this chapter, how does the **electron configuration** of an element—*the distribution of electrons within the shells and subshells of its atoms*—relate to its chemical and physical properties?

7.1 Characteristics of Many-Electron Atoms

Unlike the Bohr model, the Schrödinger equation (introduced in Chapter 6) gives excellent *approximate* solutions for the shells of *many-electron atoms*, atoms with more than one electron. However, it does not give *exact* solutions. Three additional features become important in many-electron atoms: (1) a fourth quantum number, (2) a limit on the number of electrons in an orbital, and (3) a splitting of shells into subshells.

The Electron-Spin Quantum Number

The three quantum numbers—n, l, and m_l—describe the size (energy), shape, and orientation, respectively, of an atomic orbital. An additional quantum number describes a property called *spin*, which is a property of the electron, not the orbital.

In 1922, Otto Stern and Walther Gerlach performed what is now known as the Stern-Gerlach experiment using silver atoms. When a beam of atoms that have one or more lone electrons passes through a nonuniform magnetic field (created by magnet faces with different shapes), it splits into two beams; Figure 7.1 shows this for a beam of Ag atoms. Each electron behaves like a spinning charge and generates a tiny magnetic field, which can have one of two values of *spin*. The two electron fields have opposing directions, so half of the electrons are *attracted* by the large external magnetic field, while the other half are *repelled* by it.

Corresponding to the two directions of the electron's field, the **spin quantum number (m_s)** can take one of two possible values, $+\frac{1}{2}$ or $-\frac{1}{2}$. Thus, *each electron in an atom is described completely by a set of* ***four*** *quantum numbers: the first three describe its orbital, and the fourth describes its spin.* The quantum numbers are summarized in Table 7.1.

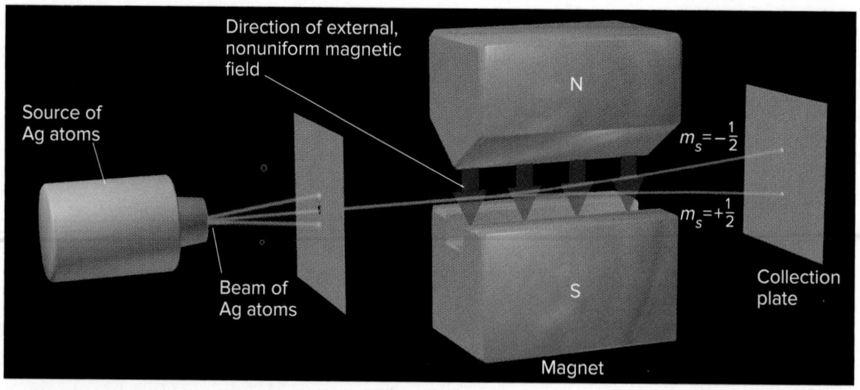

FIGURE 7.1 The effect of electron spin. A beam of Ag atoms splits because each atom's electron has one of the two possible values of spin.

TABLE 7.1 Summary of Quantum Numbers of Electrons in Atoms

Name	Symbol	Permitted Values	Property
Principal	n	Positive integers (1, 2, 3, . . .)	Orbital energy (size)
Angular momentum	l	Integers from 0 to $n - 1$	Orbital shape (the l values 0, 1, 2, and 3 are described as s, p, d, and f orbitals, respectively)
Magnetic	m_l	Integers from $-l$ to 0 to $+l$	Orbital orientation
Spin	m_s	$+\frac{1}{2}$ or $-\frac{1}{2}$	Direction of e^- spin

Now we can write a set of four quantum numbers for any electron in the ground state of any atom. For example, the set of quantum numbers for the lone electron in hydrogen (H; $Z = 1$) is $n = 1$, $l = 0$, $m_l = 0$, and $m_s = +\frac{1}{2}$. (The spin quantum number could just as well have been $-\frac{1}{2}$, but, by convention, we assign $+\frac{1}{2}$ to the first electron in an orbital.)

The Exclusion Principle

The element after hydrogen is helium (He; $Z = 2$), the first with atoms that have more than one electron. The first electron in the He ground state has the same set of quantum numbers as the electron in the H atom, but the second He electron does not. Based on observations of excited states, the Austrian physicist Wolfgang Pauli formulated the **exclusion principle**: *no two electrons in the same atom can have the same four quantum numbers.* Therefore, the second He electron occupies the same orbital as the first, but has an opposite spin: $n = 1$, $l = 0$, $m_l = 0$, and $m_s = -\frac{1}{2}$. ■

The major consequence of the exclusion principle is that *an atomic orbital can hold a maximum of two electrons and they must have opposing spins.* We say that the 1*s* orbital in He is *filled*, and that the electrons have *paired spins.* Thus, a beam of He atoms is not split in an experiment like the one shown in Figure 7.1.

■ **Hockey Quantum Numbers** The unique set of quantum numbers that describes an electron is analogous to the unique location of a box seat at a hockey game. The arena (atom) is arranged by section (n, shell), box (l, subshell or orbital), row (m_l, multiplicity), and seat (m_s, spin). Only one person (electron) can have this particular set of arena "quantum numbers."

Electrostatic Effects and Energy-Level Splitting

Electrostatic effects—the attraction of opposite charges and the repulsion of like charges—play a major role in determining the energy states of many-electron atoms. Unlike the H atom, in which there is only the attraction between nucleus and electron, and the energy state is determined *only* by the n value, the energy states of many-electron atoms are also affected by electron-electron repulsions. You will see shortly how *these additional interactions give rise to the splitting of shells into subshells of differing energies*, and how *the energy of an orbital in a many-electron atom depends mostly on its n value (size) and to a lesser extent on its l value (shape).*

Our first encounter with shell splitting occurs with lithium (Li; $Z = 3$). The first two electrons of Li fill its 1*s* orbital, so the third Li electron must go into the $n = 2$ level. However, this level has 2*s and* 2*p* subshells. Which subshell does the third electron enter? For reasons we will discuss below, the 2*s* subshell is lower in energy than the 2*p* subshell, so the ground state of Li has its third electron in the 2*s* subshell.

This energy difference arises from three factors: *nuclear attraction, electron repulsions*, and *orbital shape* (that is, radial probability distribution). The interplay among these factors leads to two phenomena—*shielding* and *penetration*—that occur in all atoms *except* hydrogen. [In the following discussion, keep in mind that more energy is needed to remove an electron from a more stable (lower energy) subshell than from a less stable (higher energy) subshell.]

The Effect of Nuclear Charge (Z) on Subshell Energy Higher charges interact more strongly than lower charges (Coulomb's law, Section 2.7). Therefore, *a higher nuclear charge increases nucleus-electron attractions and, thus, lowers subshell energy (stabilizes the atom).* We see this effect by comparing the 1*s* subshell energies of three species with one electron: the H atom ($Z = 1$), the He^+ ion ($Z = 2$), and the Li^{2+} ion ($Z = 3$). Figure 7.2 shows that the 1*s* subshell in H is the least stable (highest energy), so the least energy is needed to remove its electron. In comparison, the 1*s* subshell in Li^{2+} is the most stable, so the most energy is needed to remove its electron.

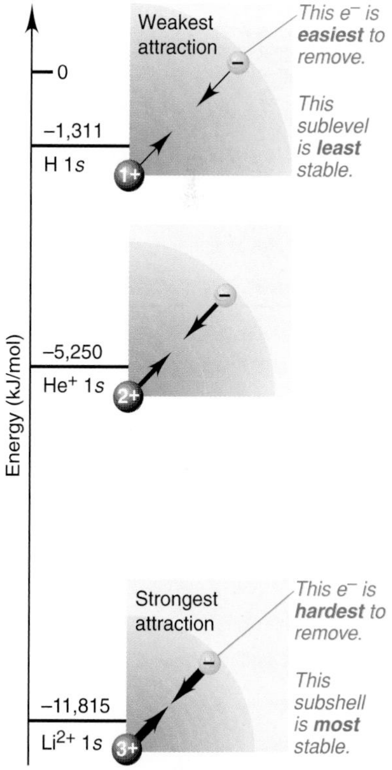

FIGURE 7.2 The effect of nuclear charge on subshell energy. Greater nuclear charge lowers subshell energy (to a more negative number), which makes the electron harder to remove. (The strength of attraction is indicated by the thickness of the black arrows.)

Shielding: The Effect of Electron Repulsions on Subshell Energy In many-electron atoms, each electron "feels" not only the attraction to the nucleus but also repulsions from other electrons. Repulsions counteract the nuclear attraction somewhat, making each electron easier to remove by, in effect, helping to push it away. We speak of each electron "shielding" the other electrons, to some extent, from the nuclear charge. **Shielding** (also called *screening*) reduces the full nuclear charge to an **effective nuclear charge (Z_{eff})**, the nuclear charge that an electron *actually experiences*, and *this lower nuclear charge makes the electron easier to remove.*

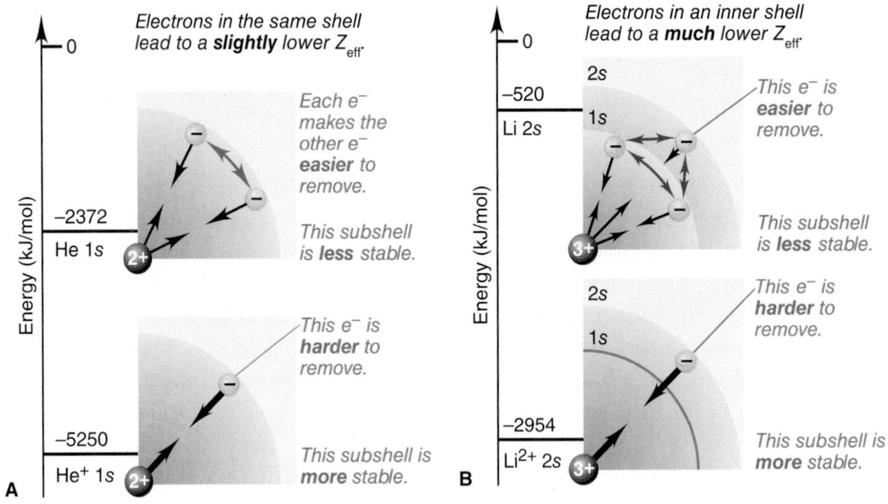

FIGURE 7.3 Shielding and shells.
A. Within a shell, each electron shields (*red arrows*) other electrons from the full nuclear charge (*black arrows*), so they experience a lower Z_{eff}. **B.** Core electrons shield outer electrons *much* more effectively than electrons in the same shell.

There are two types of shielding:

1. *Shielding by other electrons in a given shell.* Electrons in the *same* shell shield each other somewhat. Compare the He atom with the He^+ ion: both have a 2+ nuclear charge, but He has two electrons in the 1*s* subshell and He^+ has only one (Figure 7.3A). It takes less than half as much energy to remove an electron from He (2372 kJ/mol) as it takes to remove an electron from He^+ (5250 kJ/mol), because the second electron in He repels the first, in effect causing a lower Z_{eff}.

2. *Shielding by electrons in inner shells.* Because core electrons spend nearly all their time *between* the outer electrons and the nucleus, they cause a *much* lower Z_{eff} than electrons in the same shell do. We can see this by comparing two atomic systems with the same nucleus, one *with* core electrons and the other *without*. The ground-state Li atom has two core (1*s*) electrons and one valence (2*s*) electron, whereas the Li^{2+} ion has only one electron, which occupies the 2*s* orbital in the first excited state (Figure 7.3B). It takes about one-sixth as much energy to remove the 2*s* electron from the Li atom (520 kJ/mol) as it takes to remove this electron from the Li^{2+} ion (2954 kJ/mol), because the *core electrons shield very effectively*.

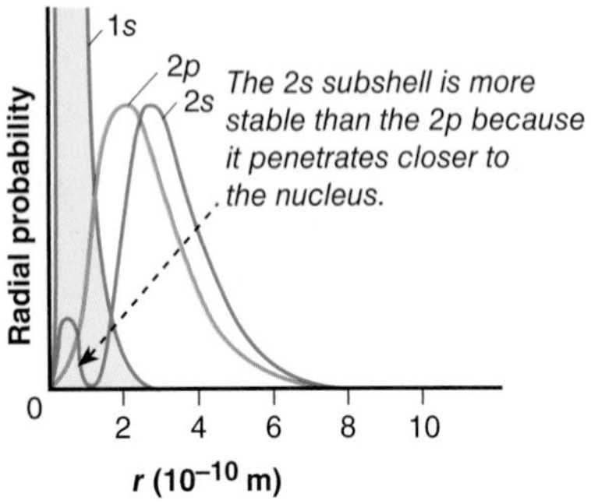

FIGURE 7.4 Penetration and subshell energy

Penetration: The Effect of Orbital Shape on Subshell Energy To see why the third Li electron occupies the 2*s* subshell rather than the 2*p* subshell, we have to consider orbital shapes, that is, radial probability distributions (Figure 7.4). A 2*p* orbital (*orange curve*) is slightly closer to the nucleus, on average, than the major portion of the 2*s* orbital (*blue curve*). However, a small portion of the 2*s* radial probability distribution peaks within the 1*s* region. Thus, an electron in the 2*s* orbital spends part of its time "penetrating" very close to the nucleus. **Penetration** has two effects:

- It *increases the nuclear attraction* for a 2*s* electron over that for a 2*p* electron.
- It *decreases the shielding* of a 2*s* electron by the 1*s* electrons.

Therefore, since it takes more energy to remove a 2*s* electron (520 kJ/mol) than it takes to remove a 2*p* electron (341 kJ/mol), the 2*s* subshell is lower in energy than the 2*p* subshell.

Splitting of Shells into Subshells In general, *penetration and the resulting effects on shielding cause a shell to split into subshells of differing energies.* The lower the *l* value of a subshell, the more its electrons penetrate, and so the greater their attraction to the nucleus. Therefore, *for a given n value, a lower l value indicates a more stable (lower energy) subshell*:

$$\text{Order of subshell energies: } s < p < d < f \qquad (7.1)$$

Thus, the 2*s* subshell ($l = 0$) is lower in energy than the 2*p* subshell ($l = 1$), the 3*p* subshell ($l = 1$) is lower than the 3*d* subshell ($l = 2$), and so on.

Figure 7.5 shows the general energy order of the shells (n value) and how they are split into subshells (l values) of differing energies. (Compare this with the H atom shells in Figure 6.22.) In the next section, we will use the energy order to construct a periodic table of ground-state atoms.

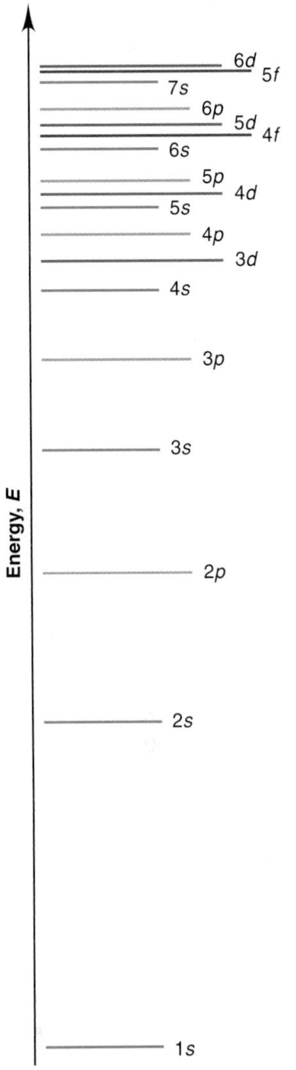

FIGURE 7.5 The order for filling shells with electrons. In general, the energies of the subshells increase with the principal quantum number n ($1 < 2 < 3$, and so on) and the angular momentum quantum number l ($s < p < d < f$). As n increases, some subshells overlap; for example, the 4s subshell is lower in energy than the 3d subshell. (Line colour indicates subshell type.)

SUMMARY OF SECTION 7.1

- Identifying electrons in many-electron atoms requires four quantum numbers: three (n, l, m_l) describe the orbital, and a fourth (m_s) describes the electron spin.
- The exclusion principle requires each electron to have a unique set of four quantum numbers; therefore, an orbital can hold no more than two electrons, and their spins must be paired (opposite).
- Electrostatic interactions determine subshell energies as follows:
  1. Greater nuclear charge lowers subshell energy, making electrons harder to remove.
  2. Electron-electron repulsions raise subshell energy, making electrons easier to remove. Repulsions shield electrons from the full nuclear charge, reducing it to an effective nuclear charge, Z_{eff}. Core electrons shield outer electrons very effectively.
  3. Penetration makes an electron harder to remove because nuclear attraction increases and shielding decreases. As a result, a shell is split into subshells with the energy order $s < p < d < f$.

7.2 The Quantum-Mechanical Model and the Periodic Table

Quantum mechanics provides the theoretical foundation for the experimentally based periodic table. In this section, we fill the table by determining the ground-state electron configuration of each element, that is, *the lowest energy distribution of electrons in the subshells of its atoms.* Note especially the *recurring pattern in electron configurations, which is the basis for recurring patterns in chemical behaviour.*

A useful way to determine electron configurations is based on the **Aufbau principle** (German *aufbauen*, meaning "to build up"). We start at the beginning of the periodic table and add one proton to the nucleus and one electron to the *lowest energy subshell available.* (Of course, one or more neutrons are also added to the nucleus.)

There are two common ways to indicate the distribution of electrons:

- *The electron configuration.* This shorthand notation consists of the principal shell (n value), the letter designation of the subshell (l value), and the number of electrons (#) in the subshell, written as a superscript: $nl^{\#}$.
- *The orbital diagram.* An **orbital diagram** consists of a box (or circle, or just a line) for each orbital in a given shell, grouped by subshell (with nl designation shown beneath), with an arrow representing an electron *and* its spin: ↑ is $+\frac{1}{2}$ and ↓ is $-\frac{1}{2}$. (Throughout this book, orbital occupancy is also indicated by colour intensity: no colour is empty, pale colour is half-filled, and full colour is filled.)

Building Up Period 1

Let us begin by applying the Aufbau principle to period 1, whose ground-state elements have only the $n = 1$ shell and, thus, only the 1s subshell, which consists of only the 1s orbital. We will also assign a set of four quantum numbers to each element's *last added* electron.

1. *Hydrogen.* For the electron in H, as you have seen, the set of quantum numbers is H ($Z = 1$): $n = 1$, $l = 0$, $m_l = 0$, $m_s = +\frac{1}{2}$. The electron configuration (spoken "one-ess-one") and orbital diagram are

H ($Z = 1$) $1s^1$ [↑]
1s

2. *Helium.* Recall that the first electron in He has the same quantum numbers as the electron in H, but the second He electron has opposing spin (exclusion principle): He ($Z = 2$): $n = 1, l = 0, m_l = 0, m_s = -\frac{1}{2}$. The electron configuration (spoken "one-ess-two," *not* "one-ess-squared") and orbital diagram are

He ($Z = 2$) $1s^2$

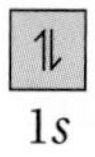

It should be noted here that the electron in H (which is shown as pointing up or having positive spin) could *just* as easily have been shown pointing down (negative spin). Then the second electron, added to form He, would have been shown pointing *up*. The net effect would have been the same; in other words, there would have been a maximum of two electrons with opposite spins in the 1*s* orbital. This will be true *whenever* we add electrons to any orbital in the rest of the chapter, as shown in the next subsection, when we find the electron configurations of Li and Be.

Building Up Period 2

The exclusion principle says an orbital can hold no more than two electrons. Therefore, with He, the 1*s* orbital (which can also be called the 1*s* subshell, the $n = 1$ shell, or period 1) is filled. Filling the $n = 2$ shell builds up period 2 and begins with the 2*s* subshell, which is the next lowest in energy (see Figure 7.5) and consists of only the 2*s* orbital. When the 2*s* subshell is filled, we proceed to fill the 2*p* orbital.

1. *Lithium.* The first two electrons in Li fill the 1*s* subshell, and the last added Li electron has quantum numbers $n = 2$, $l = 0$, and $m_l = 0$. As it is the first electron in the 2*s* orbital, it can have a spin of $+\frac{1}{2}$ or $-\frac{1}{2}$. For the example below, we will take $m_s = -\frac{1}{2}$. The electron configuration and orbital diagram then resemble

Li ($Z = 3$) $1s^22s^1$

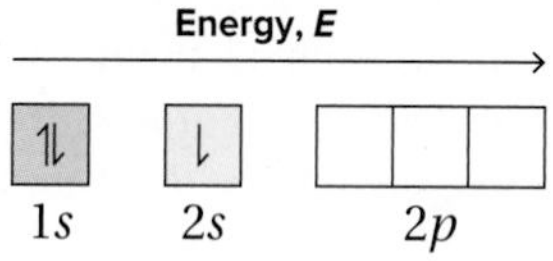

(Note that a complete orbital diagram shows all the orbitals for the given *n* value, whether or not they are occupied.) To save space on a page, orbital diagrams are written horizontally, with *the subshell energy increasing left to right.* However, Figure 7.6 highlights the energy increase with a vertical orbital diagram for lithium.

2. *Beryllium.* The 2*s* orbital is only half-filled in Li, and the fourth electron of beryllium fills it with the electron's spin paired: $n = 2, l = 0, m_l = 0, m_s = +\frac{1}{2}$.

Be ($Z = 4$) $1s^22s^2$

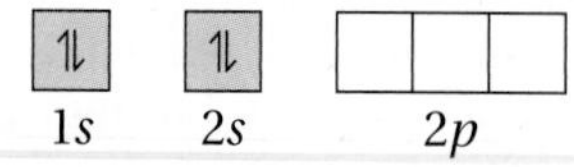

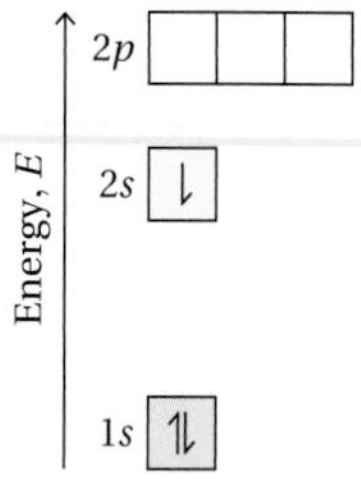

FIGURE 7.6 A vertical orbital diagram for the Li ground state

3. *Boron.* The next lowest energy subshell is the 2*p* subshell. A *p* subshell has $l = 1$, so we can assume m_l (orientation) values of −1, 0, or +1. The three orbitals in the 2*p* subshell have *equal energy* (same *n* and *l* values), which means that the fifth electron of boron can go into *any one of the 2p orbitals.* For convenience, let us label the boxes from left to right: −1, 0, +1. By convention, we start on the left and place the fifth electron in the $m_l = -1$ orbital: $n = 2, l = 1, m_l = -1, m_s = +\frac{1}{2}$. However, it is important to understand that the electron in the 2*p* orbital could have been placed in any of the three boxes and that the values of m_l shown above the boxes in the 2*p* orbital have been randomly assigned.

B ($Z = 5$) $1s^22s^22p^1$

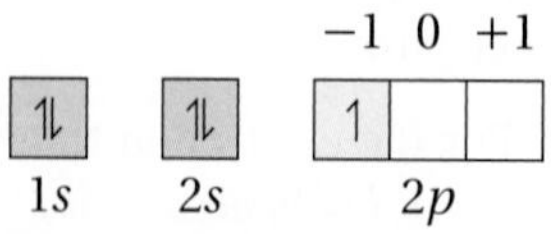

4. *Carbon.* To minimize electron-electron repulsions, the sixth electron of carbon enters one of the *unoccupied* 2*p* orbitals; by convention, we place it in the $m_l = 0$ orbital. Experiment shows that the spin of this electron is *parallel* to (the same as) the spin of the other 2*p* electron. This result exemplifies **Hund's rule**: *when orbitals*

of equal energy are available, the electron configuration of lowest energy has the maximum number of unpaired electrons with parallel spins. Thus, the sixth electron has $n = 2$, $l = 1$, $m_l = 0$, and $m_s = +\frac{1}{2}$. Again, it is important to understand two things:

(1) *Both* of the spins could have been $-\frac{1}{2}$ as well.
(2) The second electron could have been placed in either of the two empty boxes in the diagram since all p orbitals are equivalent in energy.

C ($Z = 6$) $1s^22s^22p^2$

1s	2s	2p		
↑↓	↑↓	↑	↑	

5. *Nitrogen.* Based on Hund's rule, nitrogen's seventh electron enters the last empty $2p$ orbital, with its spin parallel to the other two: $n = 2$, $l = 1$, $m_l = +1$, $m_s = +\frac{1}{2}$. Note that, in the case of nitrogen, while all three spins could be $-\frac{1}{2}$, there is no other choice about where to place the electron.

N ($Z = 7$) $1s^22s^22p^3$

1s	2s	2p		
↑↓	↑↓	↑	↑	↑

6. *Oxygen.* The eighth electron in oxygen must enter one of the three half-filled $2p$ orbitals and "pair up" with (oppose the spin of) the electron present. We place the electron in a half-filled $2p$ orbital: $n = 2$, $l = 1$, $m_l = -1$, $m_s = -\frac{1}{2}$. As in the case of boron, we could place the fourth p electron in any of the three boxes. However, it *must* have a spin opposite to the spin of the other electron in the orbital.

O ($Z = 8$) $1s^22s^22p^4$

1s	2s	2p		
↑↓	↑↓	↑↓	↑	↑

7. *Fluorine.* Fluorine's ninth electron enters the next of the two remaining half-filled $2p$ orbitals: $n = 2$, $l = 1$, $m_l = 0$, $m_s = -\frac{1}{2}$.

F ($Z = 9$) $1s^22s^22p^5$

1s	2s	2p		
↑↓	↑↓	↑↓	↑↓	↑

8. *Neon.* Only one unfilled $2p$ orbital remains, so the 10th electron of neon occupies it: $n = 2$, $l = 1$, $m_l = +1$, $m_s = -\frac{1}{2}$. With neon, the $n = 2$ shell is filled.

Ne ($Z = 10$) $1s^22s^22p^6$

1s	2s	2p		
↑↓	↑↓	↑↓	↑↓	↑↓

Sample Problem 7.1 Determining Quantum Numbers from Orbital Diagrams

Problem Use the orbital diagram shown above for fluorine to write the set of quantum numbers for (a) the third electron and (b) the eighth electron of the F atom.

Plan Referring to the orbital diagram, we identify the electron of interest and note its shell (n), subshell (l), number of equivalent orbitals (m_l), and spin (m_s).

Solution **(a)** The third electron is in the $2s$ orbital. The upward arrow indicates a spin of $+\frac{1}{2}$:

$$n = 2, l = 0, m_l = 0, m_s = +\tfrac{1}{2}$$

(b) The eighth electron is in the first $2p$ orbital, which is designated $m_l = -1$, and has a downward arrow:

$$n = 2, l = 1, m_l = -1, m_s = -\tfrac{1}{2}$$

Follow-Up Problem 7.1 Use the periodic table to identify the element with the electron configuration $1s^22s^22p^4$. Write its orbital diagram and the set of quantum numbers for its sixth electron. (See Brief Solutions to Follow-Up Problems at the end of the chapter.)

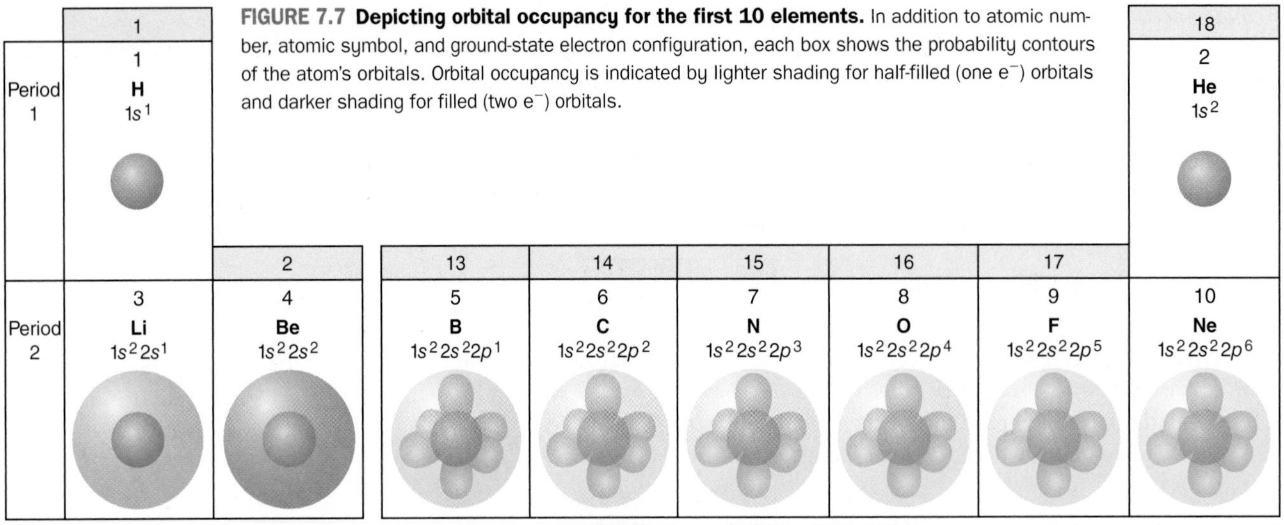

FIGURE 7.7 Depicting orbital occupancy for the first 10 elements. In addition to atomic number, atomic symbol, and ground-state electron configuration, each box shows the probability contours of the atom's orbitals. Orbital occupancy is indicated by lighter shading for half-filled (one e^-) orbitals and darker shading for filled (two e^-) orbitals.

With our attention on these notations, it is important to keep in mind that atoms are real objects and electrons occupy volumes with specific shapes and orientations. Figure 7.7 shows orbital contours for the first 10 elements, arranged in periodic table format.

Even now, we can make an important correlation: *elements in the same group have similar valence electron configurations and similar patterns of reactivity.* As an example, helium (He) and neon (Ne) in group 18 both have filled outer subshells—$1s^2$ for helium and $2s^22p^6$ for neon—and neither element forms compounds. Filled outer subshells make elements much more stable and unreactive.

Building Up Period 3

The period 3 elements, Na through Ar, lie directly under the period 2 elements, Li through Ne. That is, even though the $n = 3$ shell splits into $3s$, $3p$, and $3d$ subshells, period 3 fills only $3s$ and $3p$; as you will see shortly, the $3d$ subshell is filled in period 4. Table 7.2 introduces three ways to present electron distributions more concisely:

- *Partial orbital diagrams* show only the subshells being filled, here the $3s$ and $3p$ subshells.

TABLE 7.2 Partial Orbital Diagrams and Electron Configurations* for the Elements in Period 3

Atomic Number	Element	Partial Orbital Diagram (3s and 3p Subshells Only)		Full Electron Configuration†	Condensed Electron Configuration
		3s	3p		
11	Na	[↑]	[ ][ ][ ]	$[1s^22s^22p^6]\ 3s^1$	$[Ne]\ 3s^1$
12	Mg	[↑↓]	[ ][ ][ ]	$[1s^22s^22p^6]\ 3s^2$	$[Ne]\ 3s^2$
13	Al	[↑↓]	[ ][ ][↑]	$[1s^22s^22p^6]\ 3s^23p^1$	$[Ne]\ 3s^23p^1$
14	Si	[↑↓]	[↑][ ][↑]	$[1s^22s^22p^6]\ 3s^23p^2$	$[Ne]\ 3s^23p^2$
15	P	[↑↓]	[↑][↑][↑]	$[1s^22s^22p^6]\ 3s^23p^3$	$[Ne]\ 3s^23p^3$
16	S	[↑↓]	[↑][↑↓][↑]	$[1s^22s^22p^6]\ 3s^23p^4$	$[Ne]\ 3s^23p^4$
17	Cl	[↑↓]	[↑][↑↓][↑↓]	$[1s^22s^22p^6]\ 3s^23p^5$	$[Ne]\ 3s^23p^5$
18	Ar	[↑↓]	[↑↓][↑↓][↑↓]	$[1s^22s^22p^6]\ 3s^23p^6$	$[Ne]\ 3s^23p^6$

*Coloured type indicates the subshell to which the last electron is added.
†The full configuration is not usually written with square brackets, but they are included here to show how the [Ne] designation arises.

Period	1	2	13	14	15	16	17	18
1	1 **H** $1s^1$							2 **He** $1s^2$
2	3 **Li** [He] $2s^1$	4 **Be** [He] $2s^2$	5 **B** [He] $2s^22p^1$	6 **C** [He] $2s^22p^2$	7 **N** [He] $2s^22p^3$	8 **O** [He] $2s^22p^4$	9 **F** [He] $2s^22p^5$	10 **Ne** [He] $2s^22p^6$
3	11 **Na** [Ne] $3s^1$	12 **Mg** [Ne] $3s^2$	13 **Al** [Ne] $3s^23p^1$	14 **Si** [Ne] $3s^23p^2$	15 **P** [Ne] $3s^23p^3$	16 **S** [Ne] $3s^23p^4$	17 **Cl** [Ne] $3s^23p^5$	18 **Ar** [Ne] $3s^23p^6$

FIGURE 7.8 Condensed electron configurations in the first three periods. Elements in a group have similar valence electron configurations (*colour*).

- *Condensed electron configurations (rightmost column)* have the element symbol of the previous noble gas in brackets, to stand for its configuration, followed by the electron configuration of filled inner subshells and the shell being filled. For example, the condensed electron configuration of sulfur is [Ne] $3s^23p^4$, where [Ne] stands for $1s^22s^22p^6$.
- *Valence shell electron configurations* show the shells after the noble gas core in order of increasing energy. As a result, the final shell shown is normally the valence shell.

In Na (the second alkali metal) and Mg (the second alkaline earth metal), electrons are added to the 3*s* subshell, which contains only the 3*s* orbital; this is directly comparable to the filling of the 2*s* subshell in Li and Be in period 2. Next, in the same way as the 2*p* orbitals of B, C, and N in period 2 are half-filled, the last electrons added to Al, Si, and P in period 3 half-fill successive 3*p* orbitals with spins parallel (Hund's rule). Then the last electrons added to S, Cl, and Ar successively pair up to fill those 3*p* orbitals, thus filling the 3*p* subshell.

Similar Electron Configurations within Groups

One of the central concepts in chemistry is that *similar valence electron configurations correlate with similar chemical behaviour.* Figure 7.8 shows the condensed electron configurations of the first 18 elements. Note the similarities within each group. Here are examples from three of the groups:

- In group 1, Li and Na have the valence electron configuration ns^1 (where *n* is the quantum number of the highest energy shell), as do the other alkali metals (K, Rb, Cs, and Fr). All are highly reactive metals whose atoms lose the valence electron when they form ionic compounds with nonmetals, and all react vigorously with water to displace H_2 (Figure 7.9A).
- In group 17, F and Cl have the valence electron configuration ns^2np^5, as do the other halogens (Br, I, and At). All are reactive nonmetals that occur as diatomic molecules (X_2), and all form ionic compounds with metals (KX, MgX_2) in which

FIGURE 7.9 Similar reactivities in a group. A. Potassium reacting with water. **B.** Chlorine reacting with potassium.

the ion charge is 1− (Figure 7.9B), covalent compounds with hydrogen (HX) that yield acidic solutions in water, and covalent compounds with carbon (CX_4).

- In group 18, He has the electron configuration ns^2, and all the other elements in the group have the valence configuration ns^2np^6. Consistent with their *filled* shells, all the members are very unreactive monatomic gases.

Summarizing the connection between quantum mechanics and chemical periodicity, *subshells are filled in order of increasing energy, which leads to valence electron configurations that recur periodically, which leads to chemical properties that recur periodically.*

Building Up Period 4: The First Transition Series

Period 4 contains the first series of **transition elements**, those in which *d* orbitals are being filled. Let us examine three factors that affect the filling pattern in a period with a transition series. For the sake of simplicity, the orbital diagrams in Table 7.3 are filled from left to right using first up and then down arrows; however, the filling could be completely random (as shown in Table 7.2), as long as Hund's rule and the Pauli exclusion principle are observed.

Table 7.3 Partial Orbital Diagrams and Electron Configurations* for the Elements in Period 4

Atomic Number	Element	Partial Orbital Diagram (4s, 3d, and 4p Subshells Only): 4s	3d	4p	Full Electron Configuration	Condensed Electron Configuration	Valence Shell Electron Configuration
19	K	[↑]	[ ][ ][ ][ ][ ]	[ ][ ][ ]	$1s^22s^22p^63s^23p^64s^1$	[Ar] $4s^1$	[Ar]$4s^1$
20	Ca	[↑↓]	[ ][ ][ ][ ][ ]	[ ][ ][ ]	$1s^22s^22p^63s^23p^64s^2$	[Ar] $4s^2$	[Ar]$4s^2$
21	Sc	[↑↓]	[↑][ ][ ][ ][ ]	[ ][ ][ ]	$1s^22s^22p^63s^23p^64s^23d^1$	[Ar] $4s^23d^1$	[Ar]$3d^14s^2$
22	Ti	[↑↓]	[↑][↑][ ][ ][ ]	[ ][ ][ ]	$1s^22s^22p^63s^23p^64s^23d^2$	[Ar] $4s^23d^2$	[Ar]$3d^24s^2$
23	V	[↑↓]	[↑][↑][↑][ ][ ]	[ ][ ][ ]	$1s^22s^22p^63s^23p^64s^23d^3$	[Ar] $4s^23d^3$	[Ar]$3d^34s^2$
24	Cr	[↑]	[↑][↑][↑][↑][↑]	[ ][ ][ ]	$1s^22s^22p^63s^23p^64s^13d^5$	[Ar] $4s^13d^5$	[Ar]$3d^54s^1$
25	Mn	[↑↓]	[↑][↑][↑][↑][↑]	[ ][ ][ ]	$1s^22s^22p^63s^23p^64s^23d^5$	[Ar] $4s^23d^5$	[Ar]$3d^54s^2$
26	Fe	[↑↓]	[↑↓][↑][↑][↑][↑]	[ ][ ][ ]	$1s^22s^22p^63s^23p^64s^23d^6$	[Ar] $4s^23d^6$	[Ar]$3d^64s^2$
27	Co	[↑↓]	[↑↓][↑↓][↑][↑][↑]	[ ][ ][ ]	$1s^22s^22p^63s^23p^64s^23d^7$	[Ar] $4s^23d^7$	[Ar]$3d^74s^2$
28	Ni	[↑↓]	[↑↓][↑↓][↑↓][↑][↑]	[ ][ ][ ]	$1s^22s^22p^63s^23p^64s^23d^8$	[Ar] $4s^23d^8$	[Ar]$3d^84s^2$
29	Cu	[↑]	[↑↓][↑↓][↑↓][↑↓][↑↓]	[ ][ ][ ]	$1s^22s^22p^63s^23p^64s^13d^{10}$	[Ar] $4s^13d^{10}$	[Ar]$3d^{10}4s^1$
30	Zn	[↑↓]	[↑↓][↑↓][↑↓][↑↓][↑↓]	[ ][ ][ ]	$1s^22s^22p^63s^23p^64s^23d^{10}$	[Ar] $4s^23d^{10}$	[Ar]$3d^{10}4s^2$
31	Ga	[↑↓]	[↑↓][↑↓][↑↓][↑↓][↑↓]	[↑][ ][ ]	$1s^22s^22p^63s^23p^64s^23d^{10}4p^1$	[Ar] $4s^23d^{10}4p^1$	[Ar]$3d^{10}4s^24p^1$
32	Ge	[↑↓]	[↑↓][↑↓][↑↓][↑↓][↑↓]	[↑][↑][ ]	$1s^22s^22p^63s^23p^64s^23d^{10}4p^2$	[Ar] $4s^23d^{10}4p^2$	[Ar]$3d^{10}4s^24p^2$
33	As	[↑↓]	[↑↓][↑↓][↑↓][↑↓][↑↓]	[↑][↑][↑]	$1s^22s^22p^63s^23p^64s^23d^{10}4p^3$	[Ar] $4s^23d^{10}4p^3$	[Ar]$3d^{10}4s^24p^3$
34	Se	[↑↓]	[↑↓][↑↓][↑↓][↑↓][↑↓]	[↑↓][↑][↑]	$1s^22s^22p^63s^23p^64s^23d^{10}4p^4$	[Ar] $4s^23d^{10}4p^4$	[Ar]$3d^{10}4s^24p^4$
35	Br	[↑↓]	[↑↓][↑↓][↑↓][↑↓][↑↓]	[↑↓][↑↓][↑]	$1s^22s^22p^63s^23p^64s^23d^{10}4p^5$	[Ar] $4s^23d^{10}4p^5$	[Ar]$3d^{10}4s^24p^5$
36	Kr	[↑↓]	[↑↓][↑↓][↑↓][↑↓][↑↓]	[↑↓][↑↓][↑↓]	$1s^22s^22p^63s^23p^64s^23d^{10}4p^6$	[Ar] $4s^23d^{10}4p^6$	[Ar]$3d^{10}4s^24p^6$

*Coloured type indicates a subshell whose occupancy changes when the last electron is added.

1. *Effects of shielding and penetration on subshell energy.* The $3d$ subshell is filled in period 4, but *the 4s subshell is filled first.* This switch in filling order is due to shielding and penetration effects. Based on the $3d$ radial probability distribution (see Figure 6.20), a $3d$ electron spends most of its time outside the filled inner $n = 1$ and $n = 2$ shells, so it is shielded very effectively from the nuclear charge. However, the outermost $4s$ electron penetrates close to the nucleus part of its time, so it is subject to a greater attraction. As a result, the $4s$ orbital is slightly *lower* in energy than the $3d$ orbital and fills first. In any period, *the ns subshell fills before the $(n - 1)d$ subshell.* Other variations in the filling pattern occur at higher values of n because subshell energies become very close together (see Figure 7.5).

2. *Filling the 4s and 3d subshells.* Table 7.3 shows the partial orbital diagrams and the full, condensed, and valence shell electron configurations for the 18 elements in period 4. The first two elements, K and Ca, are the next alkali and alkaline earth metals, respectively; the last electron of K half-fills the $4s$ subshell, and the last electron of Ca fills the $4s$ subshell. The last electron of scandium (Sc; $Z = 21$), the first transition element, occupies any one of the five $3d$ orbitals because they are equal in energy; Sc has the valence shell electron configuration [Ar] $3d^14s^2$. Filling of the $3d$ orbitals proceeds one electron at a time, as does filling of the p orbitals, except in two cases, chromium (Cr; $Z = 24$) and copper (Cu; $Z = 29$), which are discussed next.

3. *Stability of half-filled and filled subshells.* Vanadium (V; $Z = 23$) has three half-filled d orbitals ([Ar] $3d^34s^2$). However, the last electron of the next element, Cr, does not enter a fourth empty d orbital to give [Ar] $3d^44s^2$; instead, Cr has one electron in the $4s$ subshell and five electrons in the $3d$ subshell, making both subshells half-filled: [Ar] $3d^54s^1$ *(see margin)*. In the next element, manganese (Mn; $Z = 25$), the $4s$ subshell is filled again ([Ar] $3d^54s^2$).

Cr ($Z = 24$) [Ar] $3d^54s^1$

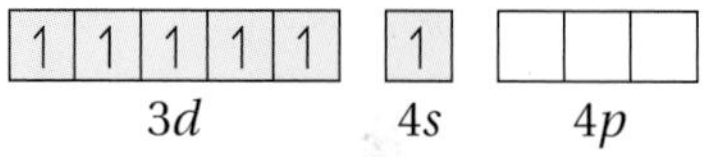

Cu ($Z = 29$) [Ar] $3d^{10}4s^1$

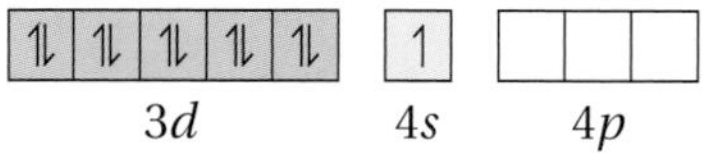

Because copper follows after nickel (Ni; [Ar] $3d^84s^2$), it would be expected to have the configuration [Ar] $3d^94s^2$. Instead, the $4s$ subshell of Cu is *half-filled* with one electron, and the $3d$ subshell is *filled* with 10 *(see margin)*. From these two exceptions, Cr and Cu, we conclude that *half-filled and filled subshells are unexpectedly stable (low in energy)*; we see this pattern in many other elements.

With zinc (Zn; $Z = 30$), the $4s$ subshell is filled ([Ar] $3d^{10}4s^2$) and the first transition series ends. The $4p$ subshell is filled by the next six elements, and period 4 ends with the noble gas krypton (Kr; $Z = 36$).

General Principles of Electron Configurations

Figure 7.10 shows the partial (highest energy subshells filled) ground-state electron configurations of the 118 known elements. Let us highlight some key relationships among them:

- *Similar valence electron configurations within a group.* Among the main-group elements (groups 1, 2, and 13 to 18)—the s-block and p-block elements—valence electron configurations within a group are identical. Some variations in the transition elements (groups 3 to 12, d block) and inner transition elements (f block) occur, as we will see.
- *Orbital filling order.* When the elements are "built up" by filling shells and subshells in order of increasing energy, we obtain the sequence in the periodic table. Reading the table from left to right, like words on a page, gives the energy order of the shells and subshells (Figure 7.11); the margin note is a memory aid for subshell filling order when a periodic table is not available. ■

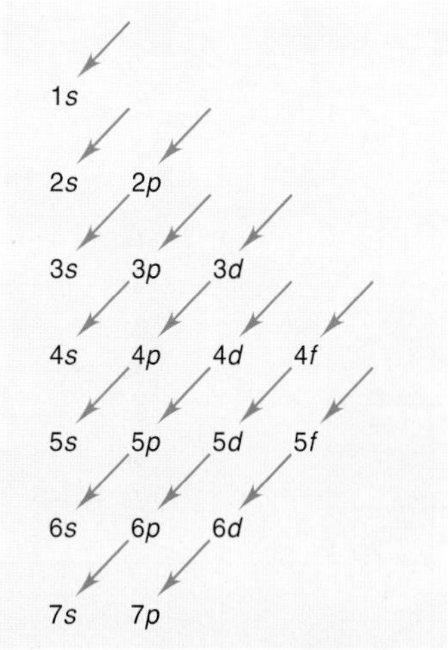

■ **Aid to Memorizing Subshell Filling Order** List the subshells as shown, and read from 1s, following the direction of the arrows. Note:

- The n value is constant horizontally.
- The l value is constant vertically.
- The value of $n + l$ is constant diagonally.

Categories of Electrons

Atoms have three categories of electrons:

1. **Core electrons** are those that an atom has in common with the previous noble gas and any *completed* transition series. They fill all the *lower energy shells* of an atom.
2. **Outer electrons** are those in the *highest energy shell* (highest n value). They spend most of their time farthest from the nucleus.

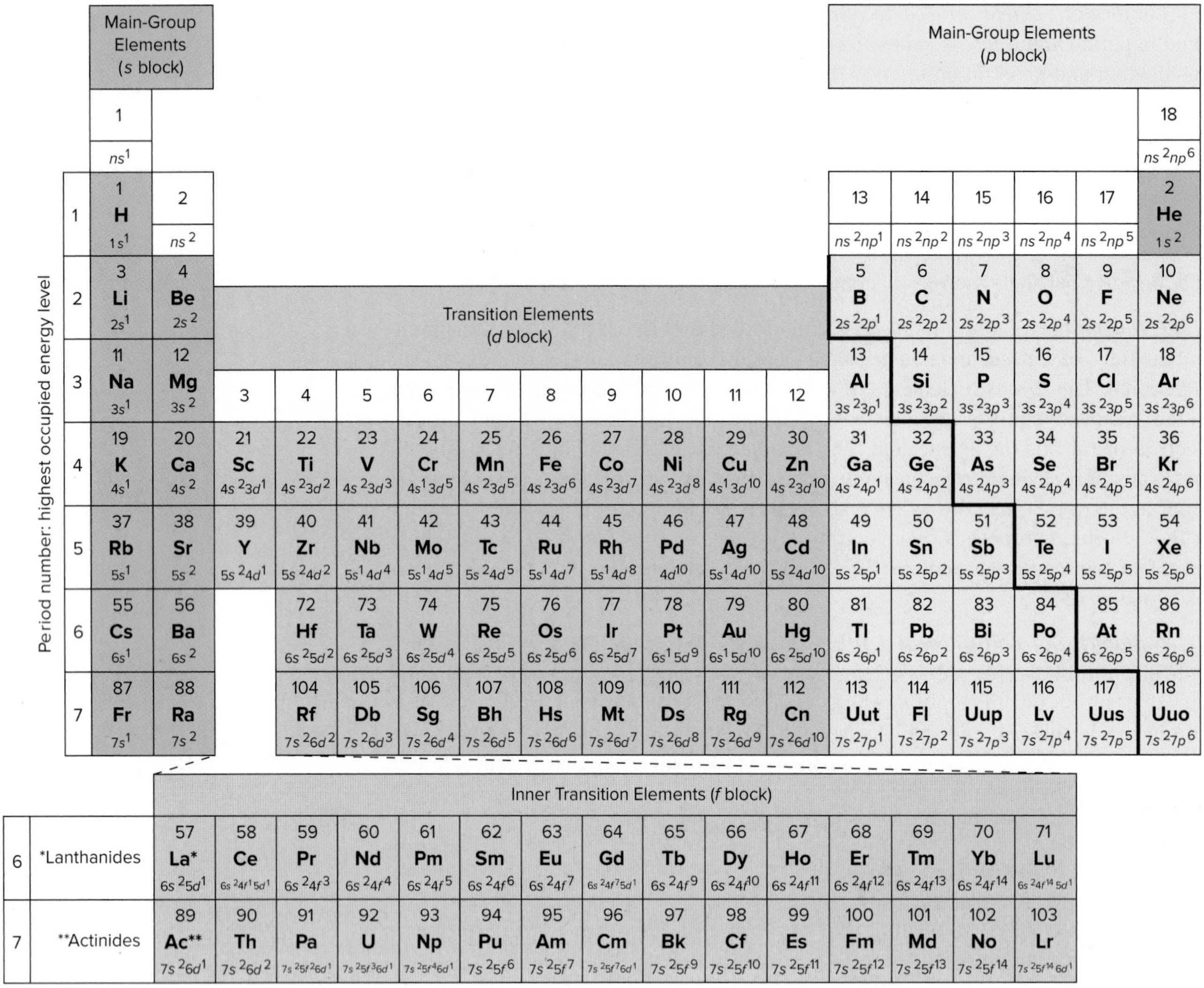
Main-Group Elements (s block)
Main-Group Elements (p block)
Transition Elements (d block)
Inner Transition Elements (f block)
Period number: highest occupied energy level
1 ns^1
2 ns^2
13 ns^2np^1 14 ns^2np^2 15 ns^2np^3 16 ns^2np^4 17 ns^2np^5 18 ns^2np^6
3 4 5 6 7 8 9 10 11 12
1: 1 H $1s^1$; 2 He $1s^2$
2: 3 Li $2s^1$; 4 Be $2s^2$; 5 B $2s^22p^1$; 6 C $2s^22p^2$; 7 N $2s^22p^3$; 8 O $2s^22p^4$; 9 F $2s^22p^5$; 10 Ne $2s^22p^6$
3: 11 Na $3s^1$; 12 Mg $3s^2$; 13 Al $3s^23p^1$; 14 Si $3s^23p^2$; 15 P $3s^23p^3$; 16 S $3s^23p^4$; 17 Cl $3s^23p^5$; 18 Ar $3s^23p^6$
4: 19 K $4s^1$; 20 Ca $4s^2$; 21 Sc $4s^23d^1$; 22 Ti $4s^23d^2$; 23 V $4s^23d^3$; 24 Cr $4s^13d^5$; 25 Mn $4s^23d^5$; 26 Fe $4s^23d^6$; 27 Co $4s^23d^7$; 28 Ni $4s^23d^8$; 29 Cu $4s^13d^{10}$; 30 Zn $4s^23d^{10}$; 31 Ga $4s^24p^1$; 32 Ge $4s^24p^2$; 33 As $4s^24p^3$; 34 Se $4s^24p^4$; 35 Br $4s^24p^5$; 36 Kr $4s^24p^6$
5: 37 Rb $5s^1$; 38 Sr $5s^2$; 39 Y $5s^24d^1$; 40 Zr $5s^24d^2$; 41 Nb $5s^14d^4$; 42 Mo $5s^14d^5$; 43 Tc $5s^24d^5$; 44 Ru $5s^14d^7$; 45 Rh $5s^14d^8$; 46 Pd $4d^{10}$; 47 Ag $5s^14d^{10}$; 48 Cd $5s^24d^{10}$; 49 In $5s^25p^1$; 50 Sn $5s^25p^2$; 51 Sb $5s^25p^3$; 52 Te $5s^25p^4$; 53 I $5s^25p^5$; 54 Xe $5s^25p^6$
6: 55 Cs $6s^1$; 56 Ba $6s^2$; 72 Hf $6s^25d^2$; 73 Ta $6s^25d^3$; 74 W $6s^25d^4$; 75 Re $6s^25d^5$; 76 Os $6s^25d^6$; 77 Ir $6s^25d^7$; 78 Pt $6s^15d^9$; 79 Au $6s^15d^{10}$; 80 Hg $6s^25d^{10}$; 81 Tl $6s^26p^1$; 82 Pb $6s^26p^2$; 83 Bi $6s^26p^3$; 84 Po $6s^26p^4$; 85 At $6s^26p^5$; 86 Rn $6s^26p^6$
7: 87 Fr $7s^1$; 88 Ra $7s^2$; 104 Rf $7s^26d^2$; 105 Db $7s^26d^3$; 106 Sg $7s^26d^4$; 107 Bh $7s^26d^5$; 108 Hs $7s^26d^6$; 109 Mt $7s^26d^7$; 110 Ds $7s^26d^8$; 111 Rg $7s^26d^9$; 112 Cn $7s^26d^{10}$; 113 Uut $7s^27p^1$; 114 Fl $7s^27p^2$; 115 Uup $7s^27p^3$; 116 Lv $7s^27p^4$; 117 Uus $7s^27p^5$; 118 Uuo $7s^27p^6$
6 *Lanthanides: 57 La* $6s^25d^1$; 58 Ce $6s^24f^15d^1$; 59 Pr $6s^24f^3$; 60 Nd $6s^24f^4$; 61 Pm $6s^24f^5$; 62 Sm $6s^24f^6$; 63 Eu $6s^24f^7$; 64 Gd $6s^24f^75d^1$; 65 Tb $6s^24f^9$; 66 Dy $6s^24f^{10}$; 67 Ho $6s^24f^{11}$; 68 Er $6s^24f^{12}$; 69 Tm $6s^24f^{13}$; 70 Yb $6s^24f^{14}$; 71 Lu $6s^24f^{14}5d^1$
7 **Actinides: 89 Ac** $7s^26d^1$; 90 Th $7s^26d^2$; 91 Pa $7s^25f^26d^1$; 92 U $7s^25f^36d^1$; 93 Np $7s^25f^46d^1$; 94 Pu $7s^25f^6$; 95 Am $7s^25f^7$; 96 Cm $7s^25f^76d^1$; 97 Bk $7s^25f^9$; 98 Cf $7s^25f^{10}$; 99 Es $7s^25f^{11}$; 100 Fm $7s^25f^{12}$; 101 Md $7s^25f^{13}$; 102 No $7s^25f^{14}$; 103 Lr $7s^25f^{14}6d^1$

FIGURE 7.10 A periodic table of partial ground-state electron configurations

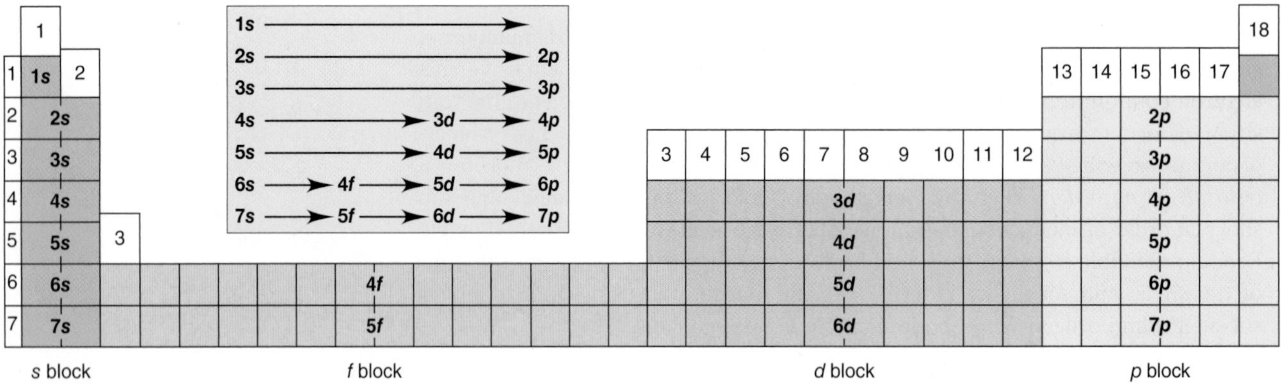

FIGURE 7.11 Orbital filling and the periodic table. This form of the periodic table shows the subshell blocks. The inset box (*pale green*) summarizes the subshell filling order.

3. **Valence electrons** are those that are involved in forming compounds:
   - For main-group elements, the *valence electrons* ***are*** *the outer electrons.*
   - For transition elements, in addition to the outer *ns* electrons, the $(n - 1)d$ electrons are valence electrons. However, the metals Fe ($Z = 26$) through Zn ($Z = 30$) may use only a few, if any, of their *d* electrons in bonding.

Group and Period Numbers Key information is embedded in the periodic table:

- Among the main-group elements (groups 1, 2, and 13 to 18), *the number of valence electrons* increases from one to eight.
- *The period number is the n value of the highest energy shell.*
- For an energy shell, the *n* value squared (n^2) is the number of *orbitals*, and $2n^2$ is the maximum number of *electrons* (or elements). For example, consider the $n = 3$ shell. The number of orbitals is $n^2 = 9$: one 3*s*, three 3*p*, and five 3*d*. The number of electrons is $2n^2 = 18$: two 3*s* and six 3*p* electrons for the eight elements of period 3, and ten 3*d* electrons for the ten transition elements of period 4.

Intervening Series: Transition and Inner Transition Elements

As Figure 7.11 shows, the *d* block and *f* block occur between the main-group *s* and *p* blocks.

1. *Transition series.* Periods 4, 5, 6, and 7 incorporate the 3*d*, 4*d*, 5*d*, and 6*d* subshells, respectively. As you have seen, the general pattern is the $(n - 1)d$ subshell being filled between the *ns* and *np* subshells. Thus, in period 5, the filling order is 5*s*, then 4*d*, and then 5*p*.

2. *Inner transition series.* In period 6, there is the first of two series of **inner transition elements**, those in which *f* orbitals are being filled (Figure 7.11). The *f* orbitals have $l = 3$, so the possible m_l values are −3, −2, −1, 0, +1, +2, and +3; that is, there are seven *f* orbitals, for a total of 14 elements in *each* of the two inner transition series:

- The period 6 inner transition series, called the **lanthanides (rare earths)**, occurs after barium (Ba; $Z = 56$) and involves the 4*f* orbitals being filled.
- The period 7 inner transition series, called the **actinides**, occurs after radium (Ra; $Z = 88$) and involves the 5*f* orbitals being filled.
- When we are filling levels above $n = 5$, the difference in energy between *d* and *f* orbitals becomes blurred. Although, traditionally, periodic tables showed element 57 as being part of the 5*d* level and element 58 as beginning the 4*f* level, based on the properties and behaviour of the elements, it is more appropriate to begin the 4*f* level at element 57. Although this would technically make element 71 the first element in the 5*d* level, it is still shown as part of the 4*f* block; the most recent versions of the periodic table thus show 15 elements in the *f* block rather than 14. The same principle is applied for the 6*d* and 5*f* elements.

Thus, in periods 6 and 7, the filling sequence is *ns*, all $(n - 2)f$, $(n - 1)d$, and *np*.

Period 6 ends with the 6*p* subshell. Period 7 was completed in 2010 with the creation and detection of element 117. This element will be given a name once IUPAC (the International Union for Pure and Applied Chemistry) recognizes its discovery officially. The discovery of element 117 is remarkable in that it completes the seventh period, meaning that any new elements created will belong to an entirely new shell with $n = 8$.

3. *Irregular filling patterns.* Irregularities in a filling pattern, such as those for Cr and Cu in period 4, occur in the *d* and *f* blocks because the subshell energies in these larger atoms differ very little. Even though occasional deviations occur in the *d* block, the sum of *ns* electrons and $(n - 1)d$ electrons always equals the group number. For instance, despite variations in group *6*—Cr, Mo, W, and Sg—the sum of *ns* and $(n - 1)d$ electrons is *6*; for group *10*—Ni, Pd, Pt, and Ds—the sum is *10*.

Sample Problem 7.2 Determining Electron Configurations

Problem Using the periodic table (not Figure 7.10 or Table 7.3) and assuming a regular filling pattern, give the full and valence shell electron configurations, partial orbital diagrams showing the valence electrons only, and the number of core electrons for each element:

(a) Potassium (K; $Z = 19$) **(b)** Technetium (Tc; $Z = 43$) **(c)** Lead (Pb; $Z = 82$)

Plan The atomic number tells us the number of electrons, and the periodic table shows the order for filling the subshells. In the partial orbital diagrams, we include all the electrons added after the previous noble gas *except* those in *filled* inner subshells. The number of core electrons is the sum of those in the previous noble gas and in the filled *d* and *f* subshells.

Solution **(a)** For K ($Z = 19$), the full electron configuration is $1s^2 2s^2 2p^6 3s^2 3p^6 4s^1$.
The valence shell electron configuration is [Ar] $4s^1$.
The partial orbital diagram, showing the valence electrons, is

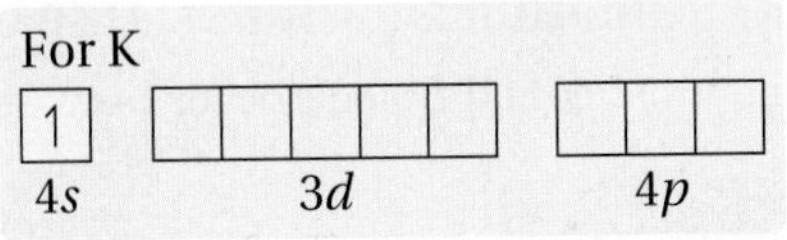

K is a main-group element in group 1 and period 4, so there are 18 core electrons.

(b) For Tc ($Z = 43$), assuming the expected pattern, the full electron configuration is $1s^2 2s^2 2p^6 3s^2 3p^6 4s^2 3d^{10} 4p^6 5s^2 4d^5$.
The valence shell electron configuration is [Kr] $4d^5 5s^2$.
The partial orbital diagram, showing the valence electrons, is

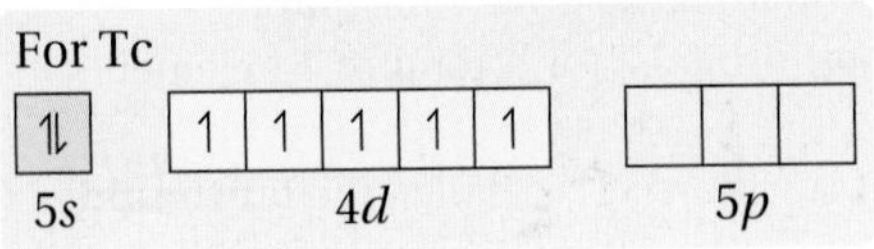

Tc is a transition element in group 7 and period 5, so there are 36 core electrons.

(c) For Pb ($Z = 82$), the full electron configuration is
$1s^2 2s^2 2p^6 3s^2 3p^6 4s^2 3d^{10} 4p^6 5s^2 4d^{10} 5p^6 6s^2 4f^{14} 5d^{10} 6p^2$.
The valence shell electron configuration is [Xe] $4f^{14} 5d^{10} 6s^2 6p^2$.
The partial orbital diagram, showing the valence electrons (no filled inner subshells), is

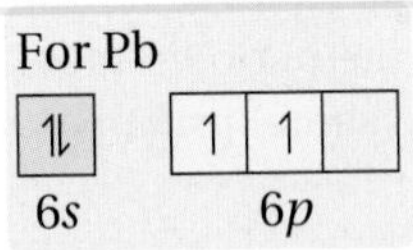

Pb is a main-group element in group 14 and period 6, so there are 54 (in Xe) + 14 (in the 4*f* subshell) + 10 (in the 5*d* subshell) = 78 core electrons.

Check Be sure that the sum of the superscripts (the numbers of electrons) in the full electron configuration equals the atomic number. Also be sure that the number of *valence* electrons in the condensed configuration equals the number of electrons in the partial orbital diagram.

Follow-Up Problem 7.2 Without referring to Figure 7.10 or Table 7.3, give the full and valence shell electron configurations, partial orbital diagrams showing the valence electrons only, and the number of core electrons for each element:

(a) Ni ($Z = 28$) **(b)** Sr ($Z = 38$) **(c)** Po ($Z = 84$)

SUMMARY OF SECTION 7.2

- By the Aufbau principle, one electron is added to an atom of each successive element in accordance with the exclusion principle (no two electrons can have the same set of quantum numbers) and Hund's rule (orbitals of equal energy become half-filled, with electron spins parallel, before any pairing of spins occurs).
- The elements of a group have similar valence electron configurations and similar chemical behaviour.
- For the main-group elements, valence electrons (those involved in reactions) are in the outer (highest energy) shell only. For transition elements, $(n - 1)d$ electrons are also considered valence electrons.
- Because of shielding of *d* electrons by electrons in inner subshells, as well as penetration by the *ns* electron, the $(n - 1)d$ subshell fills after the *ns* subshell and before the *np* subshells.
- In periods 6 and 7, $(n - 2)f$ orbitals fill between the first and second $(n - 1)d$ orbitals.

7.3 Periodic Trends

In this section, we focus on three atomic properties that are directly influenced by electron configuration and effective nuclear charge: atomic size, ionization energy, and electron affinity. We also touch briefly on the property of electronegativity. Most important, these properties are *periodic*, which means that they generally exhibit consistent changes, or *trends*, within a group or period.

Trends in Atomic Size

Recall, from Chapter 6, that we often represent atoms with spherical contours in which the electrons spend 90% of their time. We *define* atomic size (the extent of the contour) in terms of how close one atom is to another. However, in practice, as we will discuss in Chapter 11, we measure the distance between atomic nuclei in a sample of an element and divide this distance in half. Because atoms do not have hard surfaces, the size of an atom in a given compound somewhat depends on the atoms near it. In other words, *atomic size varies slightly from substance to substance.*

Figure 7.12 shows two common definitions of atomic size:

1. Used mostly for *metals*, the **metallic radius** is *one-half the shortest distance between the nuclei of adjacent, individual atoms in a crystal of an element* (Figure 7.12A).
2. Used for elements occurring as molecules, mostly *nonmetals*, the **covalent radius** is *one-half the shortest distance between the nuclei of bonded atoms* (Figure 7.12B).

Radii measured for some elements are used to determine the radii of other elements from distances between atoms in compounds. For instance, in a carbon-chlorine compound, the distance between nuclei in a C—Cl bond is 177 pm. Using the known covalent radius of Cl (100 pm), we can find the covalent radius of C (Figure 7.12C): 177 pm − 100 pm = 77 pm.

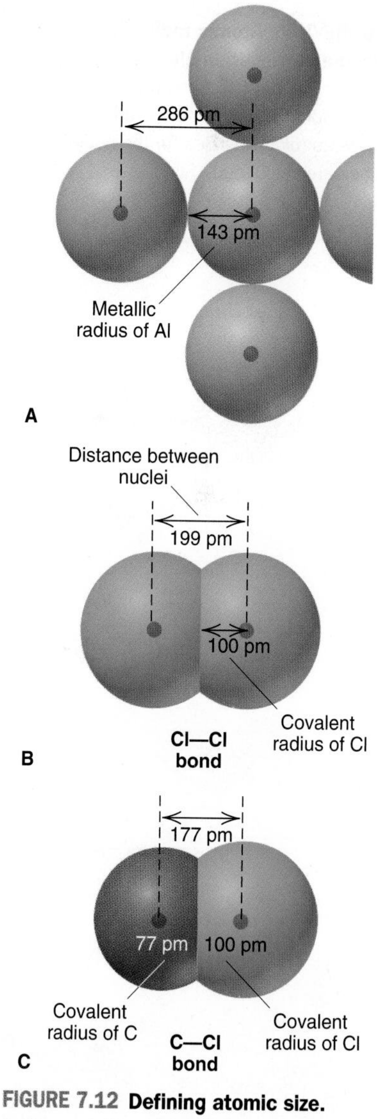

FIGURE 7.12 Defining atomic size. **A.** The metallic radius of aluminum. **B.** The covalent radius of chlorine. **C.** Known covalent radii and the distances between nuclei can be used to find unknown radii.

Main-Group Elements Figure 7.13 shows the atomic radii of the main-group elements and most of the transition elements. The atomic size of the main-group elements varies within both groups and periods as a result of two opposing influences:

1. *Changes in n.* As the principal quantum number (n) increases, the probability that outer electrons spend most of their time farther from the nucleus increases as well; thus, the atomic size increases.

2. *Changes in Z_{eff}.* As the effective nuclear charge (Z_{eff}) increases, outer electrons are pulled closer to the nucleus; thus, the atomic size decreases.

The net effect of these influences depends on how effectively the core electrons shield the increasing nuclear charge:

1. *Down a group, n dominates.* As we move down a main group, each member has *one more shell of core electrons that shields the outer electrons very effectively,* and the atoms get larger as a result of the increasing *n* value:

   Atomic radius generally ***increases*** *down a group.*

2. *Across a period, Z_{eff} dominates.* Across a period from left to right, electrons are added to the *same* outer shell, so the shielding by core electrons does not change. Despite greater electron repulsions, outer electrons shield each other only slightly, so Z_{eff} rises significantly, and the outer electrons are pulled closer to the nucleus:

   Atomic radius generally ***decreases*** *across a period.*

Transition Elements As Figure 7.13 shows, size trends are *not* as consistent for the transition elements:

1. *Down a transition group,* *n* increases, but shielding by an additional shell of core electrons results in only a small size increase from period 4 to period 5 and no size increase from period 5 to period 6.

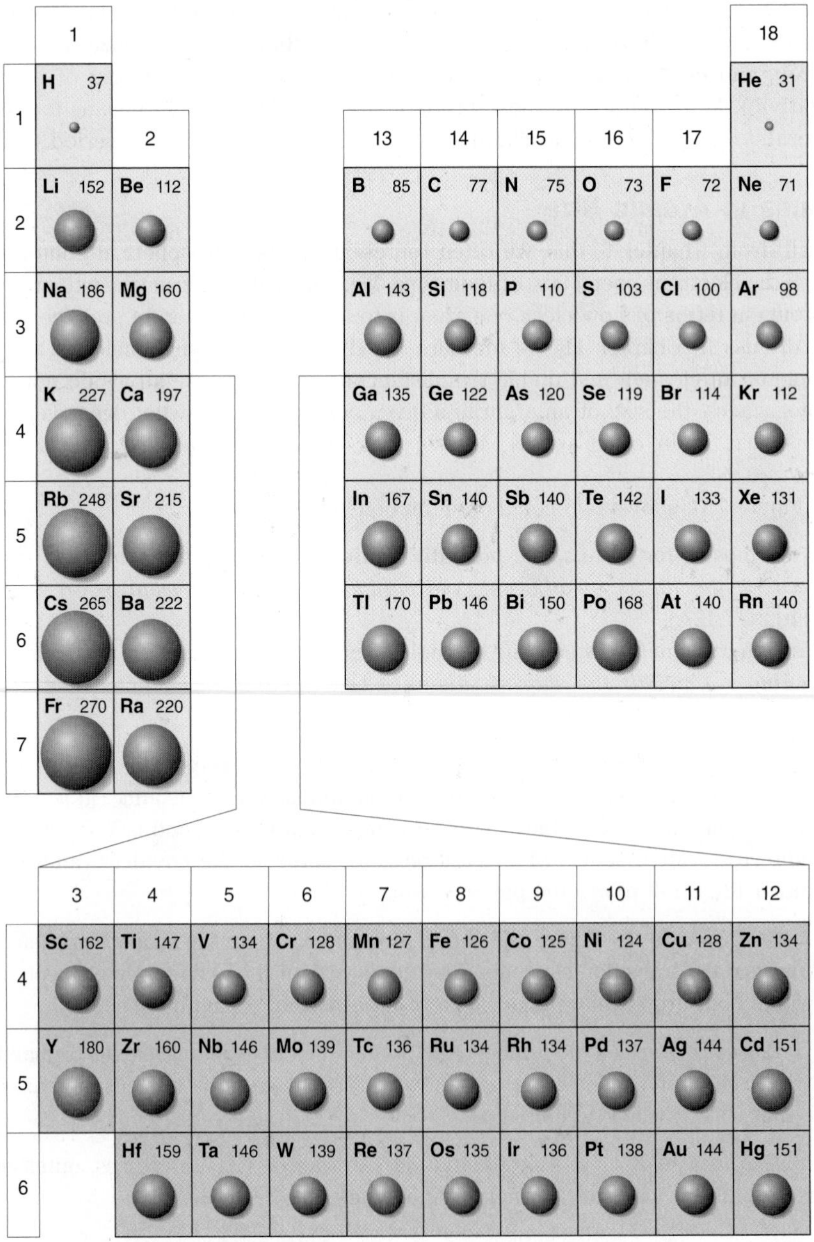

FIGURE 7.13 Atomic radii of the main-group and transition elements. Atomic radii (in picometres) are shown for the main-group elements (*tan*) and the transition elements (*blue*). (Values for the noble gases are calculated.)

2. *Across a transition series*, atomic size shrinks through the first two or three elements because of the increasing nuclear charge. From then on, however, *size remains relatively constant* because shielding by the core *d* electrons counteracts the increase in Z_{eff}. Thus, for example, in period 4, the third transition element, vanadium (V; $Z = 23$), has the same radius as the last, zinc (Zn; $Z = 30$). This pattern also appears in periods 5 and 6 in the transition series and both inner transition series.

3. *A transition series affects atomic size in neighbouring main groups*. Shielding by *d* electrons causes a *major size decrease from group 2 to group 13* in periods 4 through 6. Because the *np* subshell has more penetration than the $(n - 1)d$ subshell, the first *np* electron (added in group 13) "feels" a much greater Z_{eff} due to all the protons added in the intervening transition elements. The greatest decrease occurs in period 4: calcium (Ca; $Z = 20$) in group 2 is nearly 50% larger than gallium (Ga; $Z = 31$) in group 13. In fact, *d*-orbital shielding causes gallium to be slightly *smaller* than aluminum (Al; $Z = 13$), the element above it!

Sample Problem 7.3 Ranking Elements by Atomic Size

Problem Using only the periodic table (not Figure 7.13), rank each set of main-group elements in order of *decreasing* atomic size:

(a) Ca, Mg, Sr **(b)** K, Ga, Ca **(c)** Br, Rb, Kr **(d)** Sr, Ca, Rb

Plan To rank the elements by atomic size, we find them in the periodic table. They are all main-group elements, so size increases down a group and decreases across a period.

Solution (a) These three elements are in group 2, and size decreases up the group: Sr > Ca > Mg

(b) These three elements are in period 4, and size decreases across a period: K > Ca > Ga

(c) Rb is largest because it has one more energy shell (period 5) and is farthest to the left; Kr is smaller than Br because Kr is farther to the right in period 4: Rb > Br > Kr

(d) Ca is smallest because it has one less energy shell; Sr is smaller than Rb because it is farther to the right: Rb > Sr > Ca

Check From Figure 7.13, we see that the rankings are correct.

Follow-Up Problem 7.3 Using only the periodic table, rank the elements in each set in order of *increasing* size:

(a) Se, Br, Cl **(b)** I, Xe, Ba

Periodicity of Atomic Size Figure 7.14 shows the variation in atomic size with atomic number. Note the up-and-down pattern as size drops across a period to the noble gas (*purple*) and then leaps up to the alkali metal (*brown*) that begins the next period. Also note the deviations from the smooth size decrease in each transition (*blue*) and inner transition (*green*) series. Although it is difficult to detect on this scale, there is a similar but smaller deviation between groups 15 and 16, when the electron-electron repulsion due to the first pairing of electrons causes the radius to increase marginally. This deviation becomes more evident as the value of *n* increases.

Trends in Ionization Energy

The **ionization energy (IE)** is the energy required for the *complete removal* of 1 mol of electrons from 1 mol of gaseous atoms or ions. Pulling an electron away from a nucleus *requires* energy to overcome their electrostatic attraction. Because energy flows *into* a system, the ionization energy is always positive (like ΔH of an endothermic reaction). (Chapter 6 viewed the ionization energy of the H atom as the

FIGURE 7.14 Periodicity of atomic radius

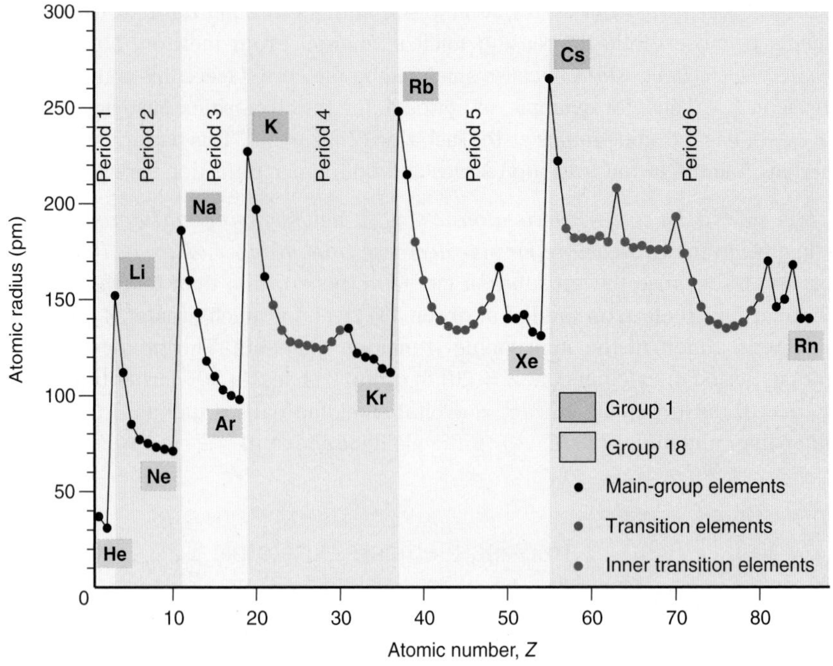

energy difference between $n = 1$ and $n = \infty$, where the electron is completely removed.) The ionization energy is a key factor in determining an element's reactivity.

Many-electron atoms can lose more than one electron. The *first* ionization energy (IE_1) removes an outermost electron (highest energy subshell) from a gaseous atom:

$$\text{atom}(g) \longrightarrow \text{ion}^+(g) + e^- \qquad \Delta E = IE_1 > 0 \qquad (7.2)$$

The *second* ionization energy (IE_2) removes a second electron. Since this electron is pulled away from a positive ion, IE_2 is always larger than IE_1:

$$\text{ion}^+(g) \longrightarrow \text{ion}^{2+}(g) + e^- \qquad \Delta E = IE_2 \text{ (always} > IE_1)$$

Periodicity of First Ionization Energy Figure 7.15 shows the variation in first ionization energy with atomic number. This up-and-down pattern—IE_1 rising across a period

FIGURE 7.15 Periodicity of first ionization energy (IE_1). This trend is the *inverse* of the trend in atomic size (see Figure 7.14).

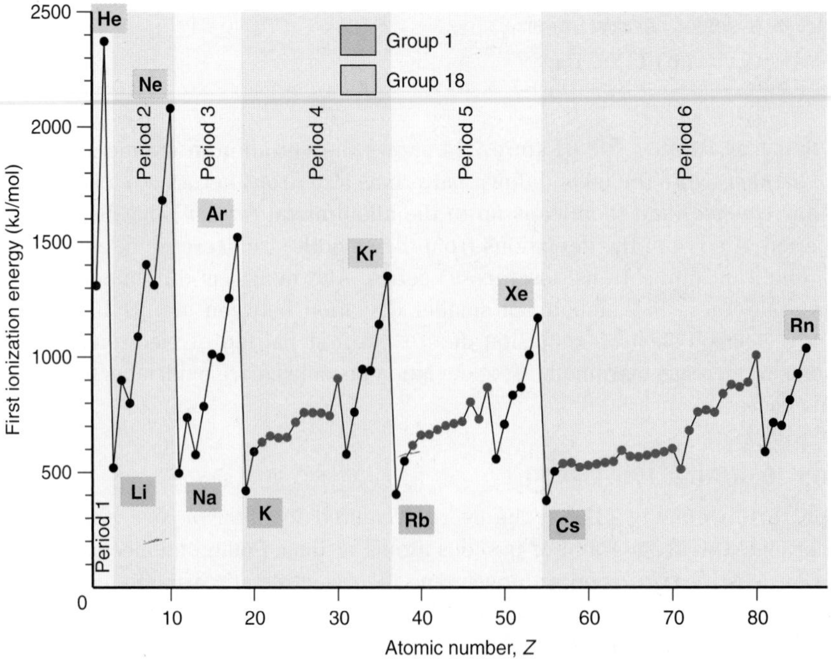

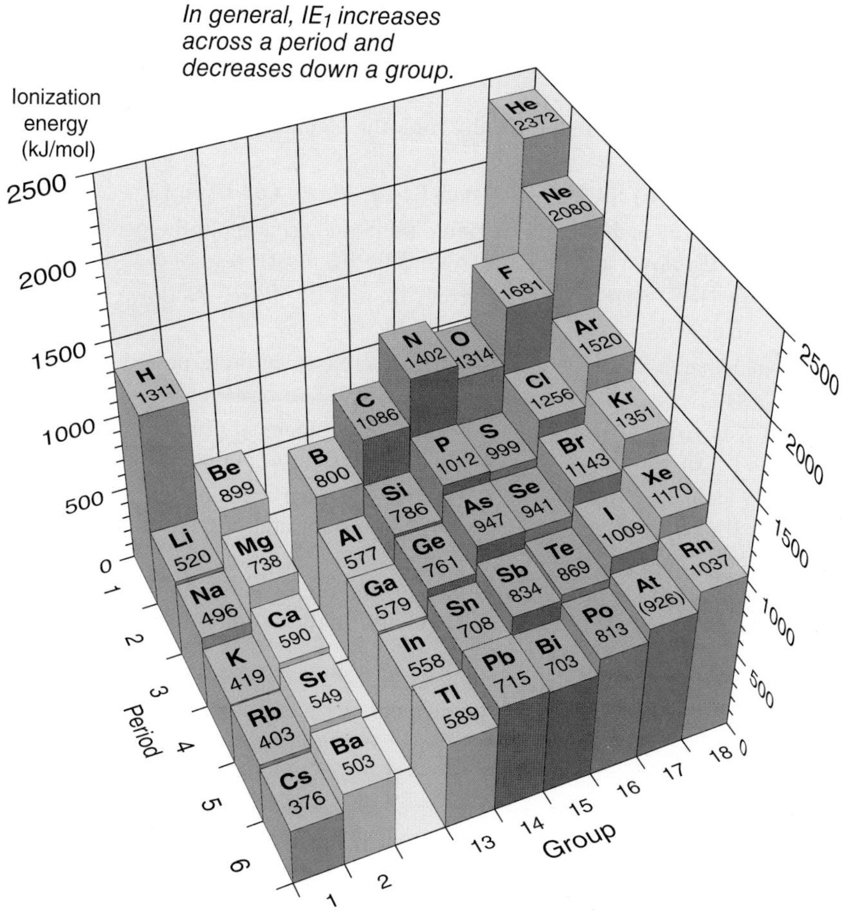

FIGURE 7.16 First ionization energies of the main-group elements

to the noble gas (*purple*) and then dropping down to the next alkali metal (*brown*)—is the inverse of the variation in atomic size (Figure 7.14): *as size decreases, more energy is needed to remove an electron because the nucleus is closer, so IE_1 increases.*

Let us examine the group and period trends, and their exceptions:

1. *Down a group.* As we move *down* a main group, the n value increases, so atomic size increases as well. As the distance from nucleus to valence electron increases, their attraction lessens, so the electron is easier to remove (Figure 7.16):

Ionization energy generally ***decreases*** *down a group.*

The only significant exception occurs in group 13: IE_1 decreases from boron (B) to aluminum (Al), but not in the rest of the group. Filling the transition series in periods 4, 5, and 6 causes a much higher Z_{eff} and an unusually small change in size, so valence electrons in the larger group 13 elements are held tighter.

2. *Across a period.* As we move left to right across a period, Z_{eff} increases and atomic size decreases. The attraction between nucleus and valence electron increases, so the electron is harder to remove:

Ionization energy generally ***increases*** *across a period.*

There are two exceptions to the otherwise smooth increase in IE_1 across periods:

- In periods 2 and 3, there are dips at two group 13 elements, B and Al. These elements have the first *np* electrons, which are removed more easily because the resulting ions have a filled (stable) *ns* subshell.
- In periods 2 and 3, once again, there are dips at two group 16 elements, O and S. These elements have a fourth *np* electron, the first to pair up with another *np* electron, and electron-electron repulsions raise the orbital energy. The fourth *np* electron is easier to remove because doing so relieves the repulsions and leaves a half-filled (stable) *np* subshell.

Sample Problem 7.4 Ranking Elements by First Ionization Energy

Problem Using the periodic table only, rank the elements in each set in order of *decreasing* IE_1:

(a) Kr, He, Ar **(b)** Sb, Te, Sn **(c)** K, Ca, Rb **(d)** I, Xe, Cs

Plan We find the elements in the periodic table and then apply the general trends of decreasing IE_1 down a group and increasing IE_1 across a period.

Solution (a) These elements are in group 18, and IE_1 decreases down a group: He > Ar > Kr

(b) These elements are in period 5, and IE_1 increases across a period: Te > Sb > Sn

(c) IE_1 of K is larger than IE_1 of Rb because K is higher in group 1; IE_1 of Ca is larger than IE_1 of K because Ca is farther to the right in period 4: Ca > K > Rb

(d) IE_1 of I is smaller than IE_1 of Xe because I is farther to the left; IE_1 of I is larger than IE_1 of Cs because I is farther to the right and in the previous period: Xe > I > Cs

Check Because trends in IE_1 are generally the opposite of the trends in size, you can rank the elements by size and check that you obtain the reverse order.

Follow-Up Problem 7.4 Rank the elements in each set in order of *increasing* IE_1:

(a) Sb, Sn, I **(b)** Sr, Ca, Ba

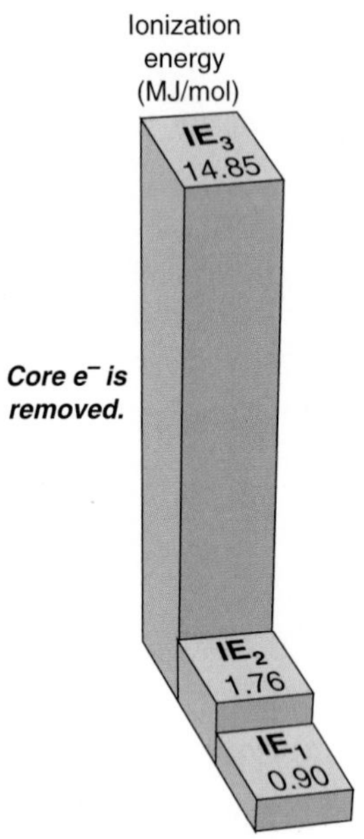

FIGURE 7.17 The first three ionization energies of beryllium. Beryllium has two valence electrons, so IE_3 is much larger than IE_2.

Successive Ionization Energies For a given element, IE_1, IE_2, and so on, increase because each electron is pulled away from a species with a higher positive charge. This increase includes an enormous jump *after* the valence electrons have been removed because *much* more energy is needed to remove an inner electron (Figure 7.17).

Table 7.4 shows the successive ionization energies for period 2 and the first element in period 3. If we move across the values for any element, we reach a point that separates relatively low IE values from relatively high IE values (*shaded area*). For example, follow the values for boron (B): IE_1 (0.80 MJ) is lower than IE_2 (2.43 MJ), which is lower than IE_3 (3.66 MJ), which is *much* lower than IE_4 (25.02 MJ). From this jump, we know that boron has three electrons in the highest energy shell ($1s^22s^22p^1$). Because they are so difficult to remove, *core electrons are not involved in reactions.*

TABLE 7.4 Successive Ionization Energies of the Elements Lithium through Sodium

Z	Element	Number of Valence Electrons	IE_1	IE_2	IE_3	IE_4	IE_5	IE_6	IE_7	IE_8	IE_9	IE_{10}
			Ionization Energy (MJ/mol)*									
3	Li	1	0.52	7.30	11.81							
4	Be	2	0.90	1.76	14.85	21.01						
5	B	3	0.80	2.43	3.66	25.02	32.82					
6	C	4	1.09	2.35	4.62	6.22	37.83	47.28				
7	N	5	1.40	2.86	4.58	7.48	9.44	53.27	64.36			
8	O	6	1.31	3.39	5.30	7.47	10.98	13.33	71.33	84.08		
9	F	7	1.68	3.37	6.05	8.41	11.02	15.16	17.87	92.04	106.43	
10	Ne	8	2.08	3.95	6.12	9.37	12.18	15.24	20.00	23.07	115.38	131.43
11	Na	1	0.50	4.56	6.91	9.54	13.35	16.61	20.11	25.49	28.93	141.37

CORE ELECTRONS

*MJ/mol, or megajoules per mole = 10^3 kJ/mol.

Sample Problem 7.5 Identifying an Element from Its Ionization Energies

Problem Name the period 3 element with the following ionization energies (kJ/mol), and write its full electron configuration:

IE_1	IE_2	IE_3	IE_4	IE_5	IE_6
1012	1903	2910	4956	6278	22 230

Plan We look for a large jump in the IE values, which occurs after all the valence electrons have been removed. Then we refer to the periodic table to find the period 3 element with this number of valence electrons and write its electron configuration.

Solution The large jump occurs after IE_5, indicating that the element has five valence electrons and, thus, is in group 15. This period 3 element is phosphorus (P; $Z = 15$). Its electron configuration is $1s^22s^22p^63s^23p^3$.

Follow-Up Problem 7.5 Element Q is in period 3 and has the following ionization energies (in kJ/mol):

IE_1	IE_2	IE_3	IE_4	IE_5	IE_6
577	1816	2744	11 576	14 829	18 375

Name element Q, and write its full electron configuration.

Trends in Electron Affinity

The **electron affinity (EA)** is the energy change (kJ/mol) that accompanies the *addition* of 1 mol of electrons to 1 mol of gaseous atoms or ions. The *first electron affinity* (EA_1) refers to the formation of 1 mol of monovalent (1−) gaseous anions:

$$\text{atom}(g) + e^- \longrightarrow \text{ion}^-(g) \qquad \Delta E = EA_1$$

As with ionization energy, there is a first electron affinity, a second, and so on. The first electron is *attracted* by the atom's nucleus, so, in most cases, *EA_1 is negative* (energy is released), analogous to the negative ΔH for an exothermic reaction.* But the second electron affinity (EA_2) is always positive because energy must be *absorbed* to overcome electrostatic repulsions and add another electron to a negative ion.

Factors other than Z_{eff} and atomic size affect electron affinities, so trends are not regular, as are the trends for size and IE_1. The many exceptions arise from changes in subshell energy and electron-electron repulsion:

- *Down a group.* We might expect a smooth decrease (smaller negative numbers) down a group because size increases, so the nucleus is farther away from an electron being added. However, only group 1 elements exhibit this behaviour (Figure 7.18).
- *Across a period.* We might expect a regular increase (larger negative numbers) across a period because size decreases, so higher Z_{eff} should more strongly attract the electron being added. However, although there is an overall left-to-right increase, it is not at all regular.

1	2	13	14	15	16	17	18
H −72.8							He (0.0)
Li −59.6	Be ≤0	B −26.7	C −122	N +7	O −141	F −328	Ne (+29)
Na −52.9	Mg ≤0	Al −42.5	Si −134	P −72.0	S −200	Cl −349	Ar (+35)
K −48.4	Ca −2.37	Ga −28.9	Ge −119	As −78.2	Se −195	Br −325	Kr (+39)
Rb −46.9	Sr −5.03	In −28.9	Sn −107	Sb −103	Te −190	I −295	Xe (+41)
Cs −45.5	Ba −13.95	Tl −19.3	Pb −35.1	Bi −91.3	Po −183	At −270	Rn (+41)

FIGURE 7.18 Electron affinities of the main-group elements (kJ/mol). Values for group 18 are estimates, indicated in parentheses.

Despite these irregularities, relative values of IE and EA show three general patterns in behaviour:

1. *Anions (negative ions).* Elements, such as those in group 16 and especially group 17 (the halogens), that have high IEs and highly negative (exothermic) EAs tend to lose electrons with difficulty but attract them strongly. Therefore, *in their ionic compounds, they have a tendency to form negative ions.*

*Some tables of EA_1 list them as positive values because energy would be *absorbed* to remove an electron from the anion.

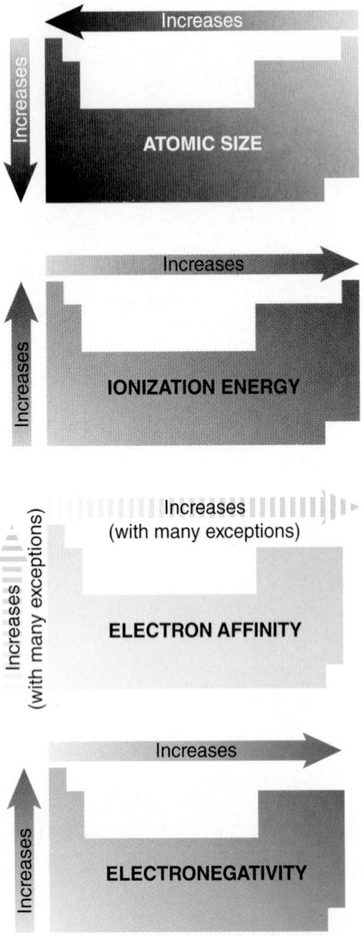

FIGURE 7.19 Trends in atomic properties and the similarity to electronegativity. The dashed arrows for electron affinity indicate that there are numerous exceptions to the expected trends.

2. *Cations (positive ions).* Elements, such as those in groups 1 and 2, that have low IEs and slightly negative (exothermic) EAs tend to lose electrons easily but attract them weakly, if at all. Therefore, *in their ionic compounds, they have a tendency to form positive ions.*

3. *Non-ions.* Elements, such as those in group 18, that have very high IEs and slightly positive (endothermic) EAs *tend **not** to lose or gain electrons.* In fact, only the larger members of the group (Kr, Xe, and Rn) form compounds at all.

Trends in Electronegativity

Although it is not a property of individual atoms, another quantity that merits discussion in this section about periodic trends is electronegativity. **Electronegativity** (χ) (χ is the Greek letter *chi*) is the ability of an atom *in a bond* to attract electron density toward itself. From the description of electronegativity, we can see that it must be the property of an atom that is bonded to at least one other atom; that is, an atom in a molecule or an ion. If we consider water, H_2O, for example, we know that the hydrogen atoms are attached in a V shape below the oxygen atom (Figure 3.17). We also know that the hydrogen atoms are slightly positive and the oxygen atom is slightly negative due to an unequal sharing of the electrons. The oxygen atom pulls the electron density more strongly toward itself, since it is more *electronegative* than the hydrogen atom. Although electronegativity and electron affinity are different properties, they essentially follow the same trend. In general, atoms that have a high electron affinity tend to be very electronegative. We will explore electronegativity in more detail in Chapter 8. For the purposes of this chapter, we can simply say that, in the periodic table, elements that tend to form anions are more electronegative than elements that tend to form cations, and smaller atoms are generally more electronegative than larger atoms.

SUMMARY OF SECTION 7.3

- Trends in three atomic properties are summarized in Figure 7.19.
- Atomic size (half the distance between the nuclei of adjacent atoms) increases down a main group and decreases across a period. In a transition series, atomic size remains relatively constant.
- First ionization energy (the energy required to remove the outermost electron from a mole of gaseous atoms) is inversely related to atomic size: IE_1 decreases down a main group and increases across a period.
- Successive ionization energies of an element show a very large increase after all the valence electrons have been removed, because the first inner electron is in an orbital of much lower energy and so is held very tightly.
- Electron affinity (the energy involved in adding an electron to a mole of gaseous atoms) shows many variations from expected trends.
- Electronegativity (the ability of an atom in a bond to attract electron density toward itself) tends to follow the same trends as electron affinity.
- Based on the relative sizes of IEs and EAs, group 1 and 2 elements tend to form cations and group 16 and 17 elements tend to form anions in ionic compounds. Group 18 elements are very unreactive.

7.4 Atomic Properties and Chemical Reactivity

All the physical and chemical behaviours of the elements and their compounds are based on electron configuration and effective nuclear charge. In this section, we will see how atomic properties determine metallic behaviour.

Trends in Metallic Behaviour

The three general classes of elements have distinguishing properties:

- *Metals*, found in the left and lower three-quarters of the periodic table, are typically shiny solids, have moderate to high melting points, are good conductors of heat and electricity, can be machined into wires and sheets, and lose electrons to nonmetals.

- *Nonmetals*, found in the upper right quarter of the table, are typically not shiny, have relatively low melting points, are poor conductors, are mostly crumbly solids or gases, and tend to gain electrons from metals.
- *Metalloids*, found between the other two classes, have intermediate properties.

Thus, *metallic behaviour decreases from left to right across a period and increases down a group in the periodic table* (Figure 7.20).

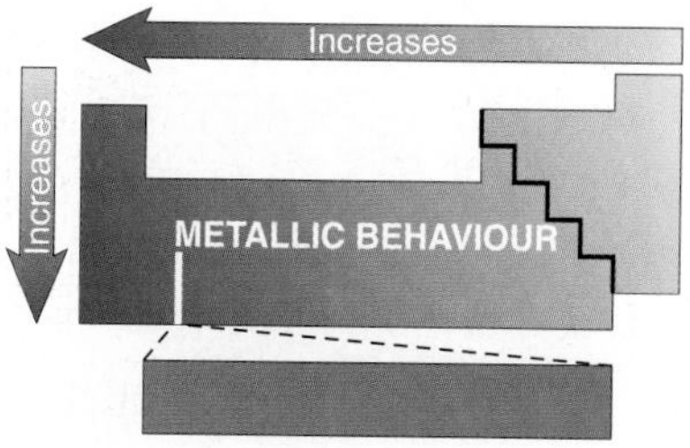

FIGURE 7.20 Trends in metallic behaviour. Hydrogen appears next to helium.

Remember, however, that some elements do not fit into these categories: for example, nonmetallic carbon, as graphite, is a good electrical conductor; the nonmetal iodine is shiny; metallic gallium melts in your hand; mercury is a liquid; and iron is brittle. Despite such exceptions, in this discussion, we will make several generalizations about metallic behaviour.

Relative Tendency to Lose or Gain Electrons Metals tend to lose electrons to nonmetals during reactions:

1. *Down a main group.* The increase in metallic behaviour down a group is consistent with an increase in size and a decrease in IE and is most obvious in groups with more than one class of element, such as group 15 (Figure 7.21, *vertical*): *elements at the top can form anions, and elements at the bottom can form cations.* Nitrogen (N) is a gaseous nonmetal, and phosphorus (P) is a soft nonmetal; both occur occasionally as 3− anions in their compounds. Arsenic (As) and antimony (Sb) are metalloids, with Sb the more metallic, and neither forms ions readily. Bismuth (Bi) is a typical metal, forming a 3+ cation in its mostly ionic compounds. Groups 13, 14, and 16 show a similar trend. However, even in group 2, which contains only metals, the tendency to form cations increases down the group: beryllium (Be) forms covalent compounds with nonmetals, whereas all the compounds of barium (Ba) are ionic.

2. *Across a period.* The decrease in metallic behaviour across a period is consistent with a decrease in size, an increase in IE, and a more favourable (more negative) EA. Consider period 3 (Figure 7.21, *horizontal*): *elements at the left tend to form cations, and those at the right tend to form anions.* Sodium and magnesium are metals that occur as Na^+ and Mg^{2+} in seawater, minerals, and organisms.

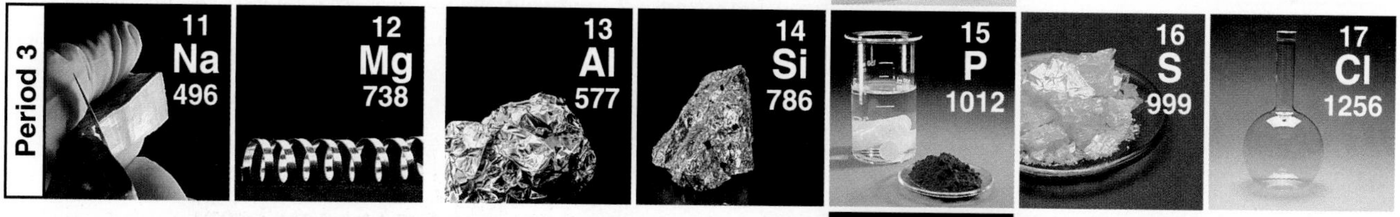

FIGURE 7.21 Metallic behaviour in group 15 and period 3. Moving down from N to Bi, there is an *increase* in metallic behaviour. Moving across from Na to Cl, there is a *decrease* in metallic behaviour.

Aluminum is metallic physically and occurs as Al^{3+} in some compounds, but it bonds covalently in most. Silicon (Si) is a shiny metalloid that does not occur as a monatomic ion. Phosphorus is a white, waxy nonmetal that occurs rarely as P^{3-}, whereas crumbly, yellow sulfur forms S^{2-} in many compounds, and gaseous, yellow-green chlorine almost always occurs in nature as Cl^-.

Properties of Monatomic Ions

So far we have focused on the reactants—the atoms—in the process of electron loss and gain. Now we focus on the products—the ions—as we consider their electron configurations, magnetic properties, and sizes.

Electron Configurations of Main-Group Ions Why does an ion have a particular charge: Na^+ not Na^{2+}, or F^- not F^{2-}? Why do some metals form two ions, such as Sn^{2+} and Sn^{4+}? The answer relates to the

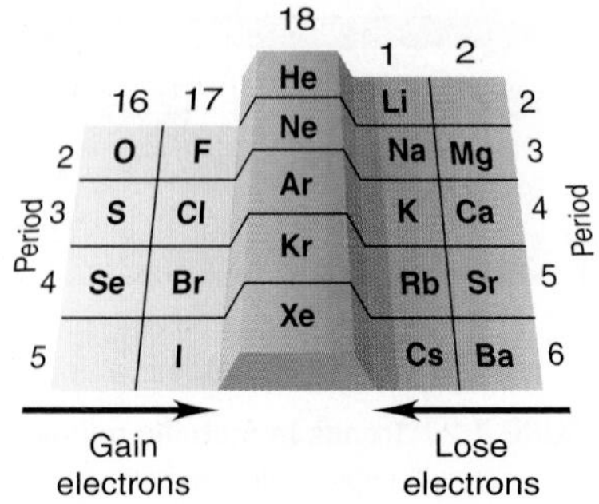

FIGURE 7.22 Main-group elements whose ions have noble gas electron configurations

location of the element in the periodic table and the energy associated with losing or gaining electrons.

1. *Ions with a noble gas configuration.* Atoms of the noble gases have very low reactivity because their highest energy shell is filled (ns^2np^6). Thus, *when elements at either end of a period form ions, they attain a filled valence shell—a noble gas configuration.* These elements lie on either side of group 18, and their ions are **isoelectronic** (Greek *iso*, meaning "same") with the nearest noble gas (Figure 7.22; see also Figure 2.13).

- Elements in groups 1 and 2 *lose* electrons and become isoelectronic with the *previous* noble gas. The Na^+ ion, for example, is isoelectronic with neon (Ne):

$$\text{Na } (1s^22s^22p^63s^1) \longrightarrow e^- + \text{Na}^+([\text{He}]\ 2s^22p^6)\ [\text{isoelectronic with Ne } ([\text{He}]\ 2s^22p^6)]$$

- Elements in groups 16 and 17 *gain* electrons and become isoelectronic with the *next* noble gas. The Br^- ion, for example, is isoelectronic with krypton (Kr):

$$\text{Br } ([\text{Ar}]\ 3d^{10}4s^24p^5) + e^- \longrightarrow \text{Br}^- \ ([\text{Ar}]\ 3d^{10}4s^24p^6)$$
$$[\text{isoelectronic with Kr } ([\text{Ar}]\ 3d^{10}4s^24p^6)]$$

The energy needed to remove electrons from metals or add them to nonmetals determines the charges of the resulting ions:

- *Cations.* Removing another electron from Na^+ or Mg^{2+} means removing a core electron, which requires too much energy; thus, $NaCl_2$ and MgF_3 do *not* exist.
- *Anions.* Similarly, adding another electron to F^- or O^{2-} means putting the atom into the next higher energy shell ($n = 3$). With 10 electrons ($1s^22s^22p^6$) acting as core electrons, the nuclear charge would be shielded very effectively, and adding a valence electron would require too much energy; thus, we never see Na_2F or Mg_3O_2.

2. *Ions without a noble gas configuration.* Except for aluminum, the metals of groups 13 to 15 do not form ions with noble gas configurations. Instead, they form cations with two different stable configurations:

- *Pseudo–noble gas configuration.* If the metal atom empties its highest energy shell, it attains the stability of empty *ns* and *np* subshells and a filled inner $(n - 1)d$ subshell. This $(n - 1)d^{10}$ configuration is called a **pseudo–noble gas configuration**. For example, tin (Sn; $Z = 50$) loses four electrons to form the tin(IV) ion (Sn^{4+}), which has empty 5*s* and 5*p* subshells and a filled inner 4*d* subshell:

$$\text{Sn } ([\text{Kr}]\ 4d^{10}5s^25p^2) \longrightarrow \text{Sn}^{4+}\ ([\text{Kr}]\ 4d^{10}) + 4e^-$$

- *Inert pair configuration.* Alternatively, the metal atom may lose just its *np* electrons and attain a stable configuration with filled *ns* and $(n - 1)d$ subshells. The retained ns^2 electrons are sometimes called an *inert pair*. For example, in the more common tin(II) ion (Sn^{2+}), the atom loses the two 5*p* electrons and has filled 5*s* and 4*d* subshells:

$$\text{Sn } ([\text{Kr}]\ 4d^{10}5s^25p^2) \longrightarrow \text{Sn}^{2+}\ ([\text{Kr}]\ 4d^{10}5s^2) + 2e^-$$

Thallium, lead, and bismuth, the largest and thus most metallic atoms in groups 13 to 15, form ions that retain the ns^2 pair: Tl^+, Pb^{2+}, and Bi^{3+}.

Once again, energy considerations explain these configurations. It would be energetically impossible for metals in groups 13 to 15 to achieve noble gas configurations: tin, for example, would have to lose fourteen electrons—ten 4*d* in addition to the two 5*p* and two 5*s*—to be isoelectronic with krypton (Kr; $Z = 36$), the previous noble gas.

Sample Problem 7.6 Writing Electron Configurations of Main-Group Ions

Problem Using valence shell electron configurations, write equations that represent the formation of the ion(s) of the following elements:

(a) Iodine ($Z = 53$) **(b)** Potassium ($Z = 19$) **(c)** Indium ($Z = 49$)

Plan We identify the element's position in the periodic table and recall the following:

- Ions of elements in groups 1, 2, 16, and 17 are isoelectronic with the nearest noble gas.

• Metals in groups 13 to 15 lose their *ns* and *np* electrons or just their *np* electrons.

Solution **(a)** Iodine is in group 17, so it gains one electron, and I^- is isoelectronic with xenon:

$$I\ ([Kr]\ 4d^{10}5s^25p^5) + e^- \longrightarrow I^-\ ([Kr]\ 4d^{10}5s^25p^6) \qquad \text{(same as Xe)}$$

(b) Potassium is in group 1, so it loses one electron; K^+ is isoelectronic with argon:

$$K\ ([Ar]\ 4s^1) \longrightarrow K^+([Ar]) + e^-$$

(c) Indium is in group 13, so it loses either three electrons to form In^{3+} (with a pseudo–noble gas configuration) or one electron to form In^+ (with an inert pair):

$$In\ ([Kr]\ 4d^{10}5s^25p^1) \longrightarrow In^{3+}([Kr]\ 4d^{10}) + 3e^-$$

$$In\ ([Kr]\ 4d^{10}5s^25p^1) \longrightarrow In^{+}([Kr]\ 4d^{10}5s^2) + e^-$$

Check Be sure that the number of electrons in the ion's electron configuration, plus the number gained or lost to form the ion, equals *Z*.

Follow-Up Problem 7.6 Using valence shell electron configurations, write equations that represent the formation of the ion(s) of the following elements: **(a)** Ba ($Z = 56$) **(b)** O ($Z = 8$) **(c)** Pb ($Z = 82$)

Electron Configurations of Transition Metal Ions In contrast to many main-group ions, *transition metal ions rarely attain a noble gas configuration.* Aside from the period 4 elements scandium, which forms Sc^{3+}, and titanium, which occasionally forms Ti^{4+}, *a transition element typically forms more than one cation by losing all of its ns and some of its (n − 1)d electrons.*

The reason, once again, is that the energy costs are too high. Let us consider the filling of period 4. At the beginning of period 4 (and other periods), penetration makes the 4*s* subshell *more stable* than the 3*d* subshell. Therefore, the first and second electrons that are added enter the 4*s* subshell, which is the outer subshell. However, the 3*d* subshell is an *inner* subshell, so, as it begins to fill, its electrons are not well shielded from the increasing nuclear charge.

A *crossover of subshell energies* results: the 3*d* subshell becomes *more stable* than the 4*s* subshell in the transition series (Figure 7.23). This crossover has a major effect on the formation of period 4 transition metal ions: because the 3*d* electrons are held tightly and shield those in the outer subshell, *the 4s electrons of a transition metal are lost* ***before*** *the 3d electrons.* Thus, 4*s* electrons are *added* before 3*d* electrons to form the *atom* and are *lost* before 3*d* electrons to form the *ion*, the so-called "first-in, first-out" rule.

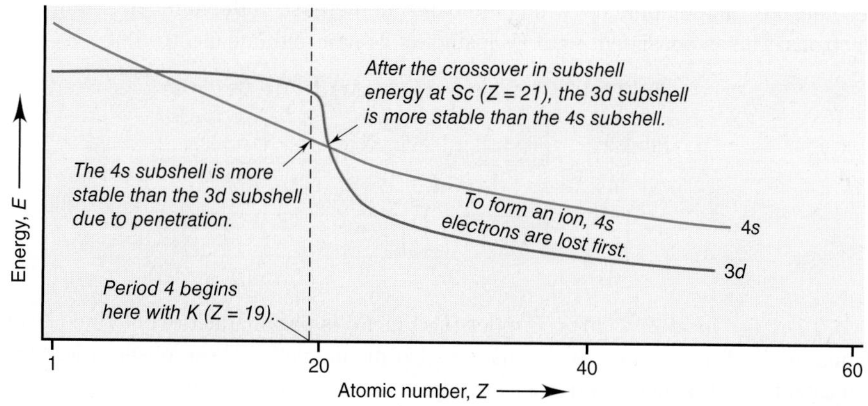
After the crossover in subshell energy at Sc (Z = 21), the 3d subshell is more stable than the 4s subshell.
The 4s subshell is more stable than the 3d subshell due to penetration.
To form an ion, 4s electrons are lost first.
4s
3d
Period 4 begins here with K (Z = 19).
Energy, E
1
20
40
60
Atomic number, Z

FIGURE 7.23 The crossover of subshell energies in period 4

Ion Formation: A Summary of Electron Loss or Gain The various ways that cations form have one characteristic in common—*the valence electrons are removed first.* Here is a summary of the rules for the formation of any main-group or transition metal ion:

- Main-group *s*-block metals lose all electrons with the highest *n* value.
- Main-group *p*-block metals lose *np* electrons before *ns* electrons.

- Transition (d-block) metals lose ns electrons before $(n - 1)d$ electrons.
- Nonmetals gain electrons in the p orbitals of highest n value.

Magnetic Properties of Transition Metal Ions We can learn a great deal about an element's electron configuration from atomic spectra, and magnetic studies provide additional evidence.

Recall that electron spin generates a tiny magnetic field, which causes a beam of H atoms to split in an external magnetic field (see Figure 7.1). Only a beam of a species (atoms, ions, or molecules) with *unpaired* electrons will split. A beam of silver atoms (Ag; $Z = 47$) was used in the original 1922 experiment by Stern and Gerlach:

Ag ($Z = 47$) [Kr] $4d^{10}5s^1$

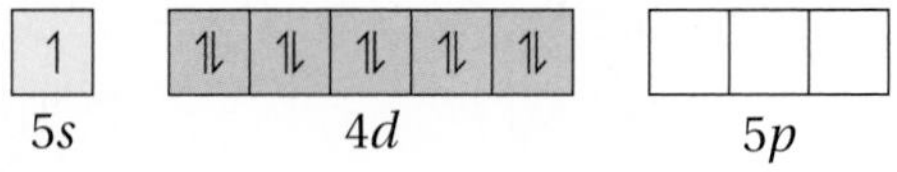

Note the unpaired 5s electron. A beam of cadmium atoms (Cd; $Z = 48$) is not split because their 5s electrons are *paired* ([Kr] $4d^{10}5s^2$).

A species that contains one or more unpaired electrons exhibits **paramagnetism**: it is attracted by an external field. A species with all of its electrons paired exhibits **diamagnetism**: it is slightly repelled by the field (Figure 7.24). Many transition metals and their compounds are paramagnetic because their atoms and ions have unpaired electrons.

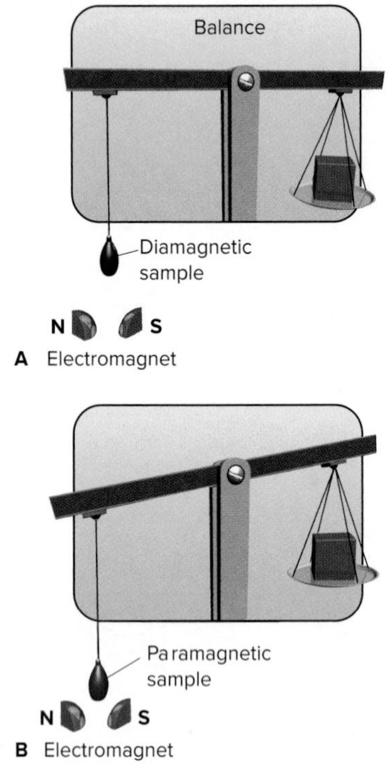

FIGURE 7.24 Measuring the magnetic behaviour of a sample. The substance is weighed with the external magnetic field "off." **A.** If the substance is diamagnetic (has all *paired* electrons), its apparent mass is unaffected (or slightly reduced) with the field "on." **B.** If the substance is paramagnetic (has *unpaired* electrons), its apparent mass increases.

Let us see three examples of how magnetic studies might provide evidence for a proposed electron configuration:

1. *The Ti^{2+} ion.* Spectral analysis of titanium metal yields the electron configuration [Ar] $3d^24s^2$, and experiments show that the metal is paramagnetic, which indicates the presence of unpaired electrons. Spectral analysis shows that the Ti^{2+} ion is [Ar] $3d^2$, indicating loss of the 4s electrons. In support of the spectra, magnetic studies show that Ti^{2+} compounds are paramagnetic. If Ti had lost its 3d electrons to form Ti^{2+}, its compounds would be diamagnetic:

$$\text{Ti ([Ar] }3d^24s^2) \longrightarrow Ti^{2+}\text{ ([Ar] }3d^2) + 2e^-$$

The partial orbital diagrams are

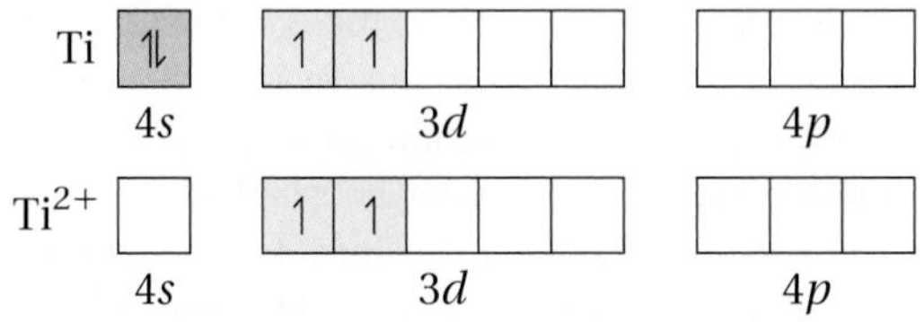

2. *The Fe^{3+} ion.* An increase in paramagnetism occurs when iron metal (Fe) becomes Fe^{3+} in compounds, which indicates an increase in the number of unpaired electrons. This is consistent with Fe losing its 4s pair and one electron of a 3d pair:

$$\text{Fe ([Ar] }3d^64s^2) \longrightarrow Fe^{3+}\text{([Ar] }3d^5) + 3e^-$$

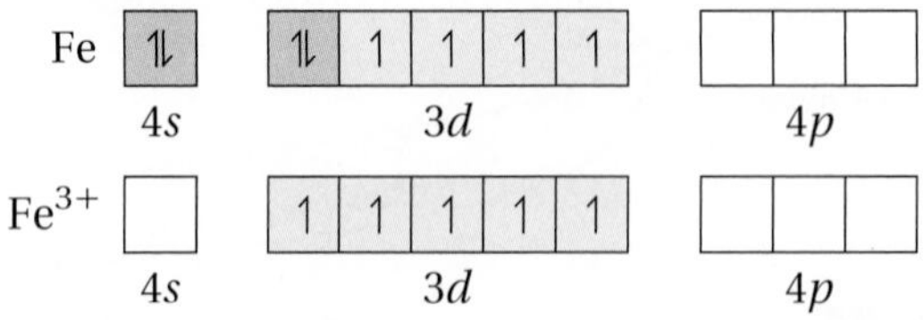

3. *The Cu^+ and Zn^{2+} ions.* Copper (Cu) metal is paramagnetic, but zinc (Zn) is diamagnetic. The Cu^+ and Zn^{2+} ions are also diamagnetic. These observations are consistent with the ions being isoelectronic, which means that 4s electrons were lost:

$$\text{Cu ([Ar] }3d^{10}4s^1) \longrightarrow Cu^+\text{([Ar] }3d^{10}) + e^-$$

$$\text{Zn ([Ar] }3d^{10}4s^2) \longrightarrow Zn^{2+}\text{([Ar] }3d^{10}) + 2e^-$$

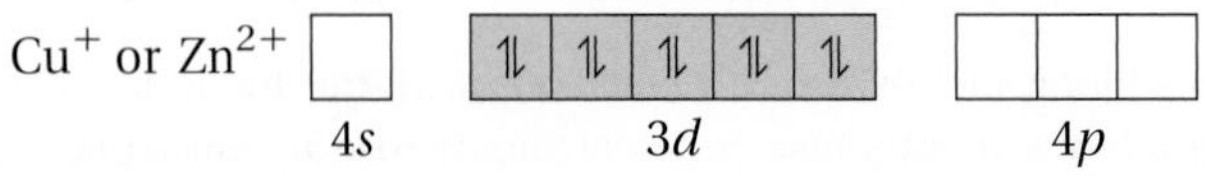

Sample Problem 7.7 Writing Electron Configurations and Predicting the Magnetic Behaviour of Transition Metal Ions

Problem Use valence shell electron configurations to write an equation for the formation of each transition metal ion, and predict whether it is paramagnetic:

(a) Mn^{2+} ($Z = 25$) **(b)** Cr^{3+} ($Z = 24$) **(c)** Hg^{2+} ($Z = 80$)

Plan We first write the valence shell electron configuration of the atom, recalling the irregularity for Cr. Then we remove electrons, beginning with *ns* electrons, to attain the ion charge. If unpaired electrons are present, the ion is paramagnetic.

Solution (a) Mn ([Ar] $3d^5 4s^2$) $\longrightarrow$ Mn^{2+} ([Ar] $3d^5$) + $2e^-$

There are five unpaired e^-, so Mn^{2+} is paramagnetic.

(b) Cr ([Ar] $3d^5 4s^1$) $\longrightarrow$ Cr^{3+} ([Ar] $3d^3$) + $3e^-$

There are three unpaired e^-, so Cr^{3+} is paramagnetic.

(c) Hg ([Xe] $4f^{14} 5d^{10} 6s^2$) $\longrightarrow$ Hg^{2+} ([Xe] $4f^{14} 5d^{10}$) + $2e^-$

The 4*f* and 5*d* subshells are filled, so there are no unpaired e^-. Therefore, Hg^{2+} is *not* paramagnetic.

Check We removed the *ns* electrons first, and the sum of the lost electrons and the electrons in the electron configuration of the ion equals *Z*.

Follow-Up Problem 7.7 Write the valence shell electron configuration of each transition metal ion, and predict whether it is paramagnetic:

(a) V^{3+} ($Z = 23$) **(b)** Ni^{2+} ($Z = 28$) **(c)** La^{3+} ($Z = 57$)

Ionic Size versus Atomic Size The **ionic radius** is *a measure of the size of an ion* and is obtained from the distance between the nuclei of adjacent ions in a crystalline ionic compound (Figure 7.25). From the relation between effective nuclear charge (Z_{eff}) and atomic size, we can predict the size of an ion relative to its parent atom:

- *Cations are smaller than parent atoms.* When a cation forms, electrons are *removed from* the outer shell. The resulting decrease in shielding and value of *nl* allows the nucleus to pull the remaining electrons closer.
- *Anions are larger than parent atoms.* When an anion forms, electrons are *added to* the outer shell. The increases in shielding and electron repulsions mean that the electrons occupy more space.

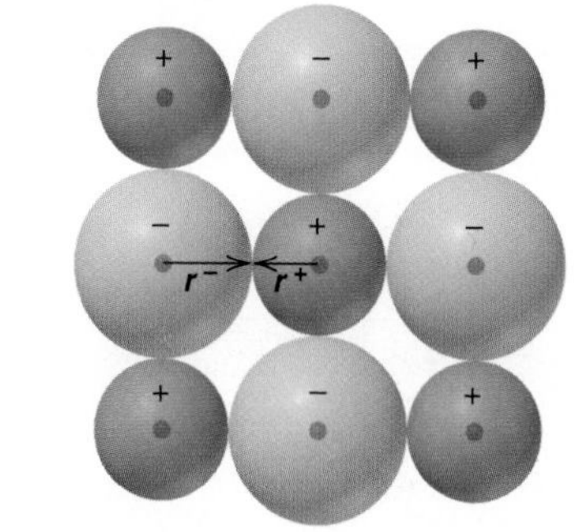

FIGURE 7.25 Ionic radius. Together, the cation radius (r^+) and the anion radius (r^-) make up the distance between nuclei.

Figure 7.26 shows the radii of some main-group ions and their parent atoms:

1. *Down a group, ionic size increases* because *n* increases.
2. *Across a period*, the pattern is complex. For example, consider period 3:
   - *Among cations*, the increase in Z_{eff} from left to right makes Na^+ larger than Mg^{2+}, which is larger than Al^{3+}.
   - *From last cation to first anion*, a great jump in size occurs. We are *adding* electrons rather than removing them, so repulsions increase sharply: P^{3-} has eight more electrons than Al^{3+}.
   - *Among anions*, the increase in Z_{eff} from left to right makes P^{3-} larger than S^{2-}, which is larger than Cl^-.
   - *Within an isoelectronic series*, these factors have striking results. The ions within the dashed outline in Figure 7.26 are isoelectronic with neon. Period 2 anions are much larger than period 3 cations because the same number of electrons are attracted by an increasing nuclear charge. The pattern is

$$3- > 2- > 1- > 1+ > 2+ > 3+$$

3. *Cation size decreases with charge.* When a metal forms more than one cation, *the greater the ionic charge, the smaller the ionic radius.* With the two ions of iron, for example, Fe^{3+} has one less electron, so shielding is reduced somewhat, and the same nucleus is attracting fewer electrons. As a result, Z_{eff} increases, so Fe^{3+} is smaller than Fe^{2+}.

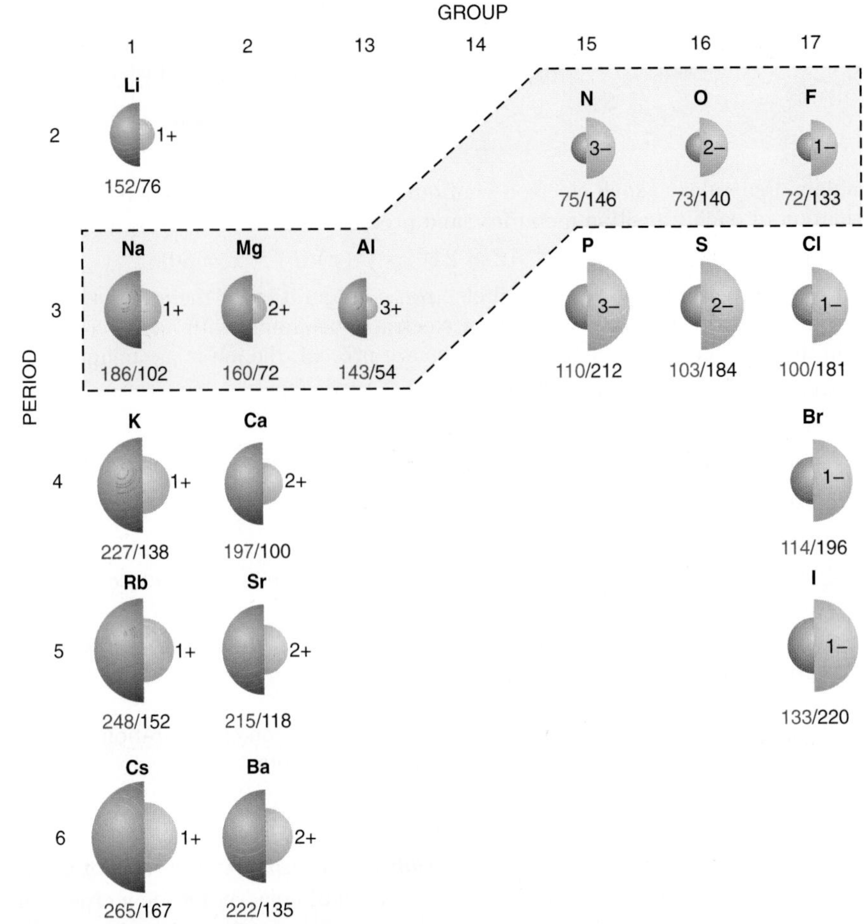

FIGURE 7.26 Ionic versus atomic radii. Atomic radii (*colour*) and ionic radii (*grey*) are given in picometres. Metal atoms (*blue*) form *smaller* positive ions, and nonmetal atoms (*red*) form *larger* negative ions. Ions in the dashed outline are *isoelectronic* with neon.

Sample Problem 7.8 Ranking Ions by Size

Problem Rank each set of ions in order of *decreasing* size, and explain your ranking:
(a) Ca^{2+}, Sr^{2+}, Mg^{2+} **(b)** K^+, S^{2-}, Cl^- **(c)** Au^+, Au^{3+}

Plan We find the position of each element in the periodic table and apply the ideas just discussed.

Solution **(a)** Mg^{2+}, Ca^{2+}, and Sr^{2+} are all from group 2, so their sizes decrease up the group: $Sr^{2+} > Ca^{2+} > Mg^{2+}$

(b) The ions K^+, S^{2-}, and Cl^- are isoelectronic. S^{2-} has a lower Z_{eff} than Cl^-, so it is larger. K^+ is a cation and has the highest Z_{eff}, so it is the smallest: $S^{2-} > Cl^- > K^+$

(c) Au^+ has a lower charge than Au^{3+}, so it is larger: $Au^+ > Au^{3+}$

Follow-Up Problem 7.8 Rank each set of ions in order of *increasing* size:
(a) Cl^-, Br^-, F^- **(b)** Na^+, Mg^{2+}, F^- **(c)** Cr^{2+}, Cr^{3+}

SUMMARY OF SECTION 7.4

- Metallic behaviour correlates with large atomic size and low ionization energy. Thus, metallic behaviour increases down a group and decreases across a period.
- Many main-group elements form ions that are isoelectronic with the nearest noble gas. Removing (or adding) more electrons than needed to attain the noble gas configuration requires a prohibitive amount of energy.
- Metals in groups 13 to 15 lose either their *np* electrons or both their *ns* and *np* electrons.
- Transition metals lose *ns* electrons before $(n - 1)d$ electrons and commonly form more than one ion.
- Many transition metals and their compounds are paramagnetic because their atoms (or ions) have unpaired electrons.
- Cations are smaller than their parent atoms, and anions are larger than their parent atoms. Ionic radii increase down a group. Across a period, ionic radii generally decrease, but a large increase occurs from the last cation to the first anion.

CHAPTER REVIEW GUIDE

Learning Objectives

Relevant section (§) and/or sample problem (SP) numbers appear in parentheses.

Concepts

1. Explain the meaning of the periodic law and the arrangement of elements by atomic number. (Introduction)
2. Describe the reason for the spin quantum number and its two possible values. (§7.1)
3. Explain how the exclusion principle applies to orbital filling. (§7.1)
4. Describe the effects of nuclear charge, shielding, and penetration on the splitting of energy shells; explain the meaning of effective nuclear charge. (§7.1)
5. Describe how the arrangement of the periodic table is based on the order of subshell energies. (§7.2)
6. Describe how subshells are filled in main-group and transition elements; explain the importance of Hund's rule. (§7.2)
7. Show how valence electron configuration within a group is related to chemical behaviour. (§7.2)
8. Distinguish between core and valence electrons. (§7.2)
9. Differentiate between the meanings of *atomic radius*, *ionization energy*, and *electron affinity*. (§7.3)
10. Explain how the n value and the effective nuclear charge give rise to the periodic trends of atomic size and ionization energy. (§7.3)
11. Discuss the importance of core electrons to the pattern of successive ionization energies. (§7.3)
12. Discuss how atomic properties relate to the tendency to form ions. (§7.3)
13. Describe the general properties of metals and nonmetals. (§7.4)
14. Show how vertical and horizontal trends in metallic behaviour are related to ion formation. (§7.4)
15. Explain why main-group ions are isoelectronic with the nearest noble gas or have a pseudo–noble gas electron configuration. (§7.4)
16. Describe why transition elements lose *ns* electrons first. (§7.4)
17. Explain the origin of paramagnetic and diamagnetic behaviour. (§7.4)
18. Discuss the relationship between ionic and atomic size, and the trends in ionic size. (§7.4)

Skills

1. Use orbital diagrams to determine the set of quantum numbers for any electron in an atom. (SP 7.1)
2. Write full and condensed electron configurations for an element. (SP 7.2)
3. Use periodic trends to rank elements by atomic size and first ionization energy. (SPs 7.3, 7.4)
4. Identify an element from its successive ionization energies. (SP 7.5)
5. Write the electron configurations of main-group and transition metal ions. (SPs 7.6, 7.7)
6. Use periodic trends to rank ions by size. (SP 7.8)

Key Terms

periodic law
electron configuration

Section 7.1
spin quantum number (m_s)
exclusion principle
shielding
effective nuclear charge (Z_{eff})
penetration

Section 7.2
Aufbau principle
orbital diagram
Hund's rule
transition elements
core electrons
outer electrons
valence electrons
inner transition elements
lanthanides (rare earths)
actinides

Section 7.3
metallic radius
covalent radius
ionization energy (IE)
electron affinity (EA)
electronegativity (χ)

Section 7.4
isoelectronic
pseudo–noble gas configuration
paramagnetism
diamagnetism
ionic radius

Key Equations and Relationships

7.1 Defining the energy order of subshells in terms of the angular momentum quantum number (l value):

Order of subshell energies: $s < p < d < f$

7.2 Meaning of the first ionization energy:

$$\text{atom}(g) \longrightarrow \text{ion}^+(g) + e^- \qquad \Delta E = \text{IE}_1 > 0$$

Brief Solutions to Follow-Up Problems

7.1 The element has eight electrons, so $Z = 8$: oxygen.

1s	2s	2p		
⇅	⇅	⇅	↑	↑

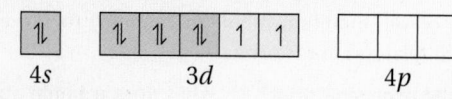

Quantum numbers for sixth electron: $n = 2$, $l = 1$, $m_l = 0$, $m_s = +\frac{1}{2}$

7.2 (a) Ni: $1s^22s^22p^63s^23p^64s^23d^8$; [Ar] $3d^84s^2$

4s	3d					4p		
⇅	⇅	⇅	⇅	↑	↑			

Ni has 18 core electrons.

(b) Sr: $1s^22s^22p^63s^23p^64s^23d^{10}4p^65s^2$; [Kr] $5s^2$

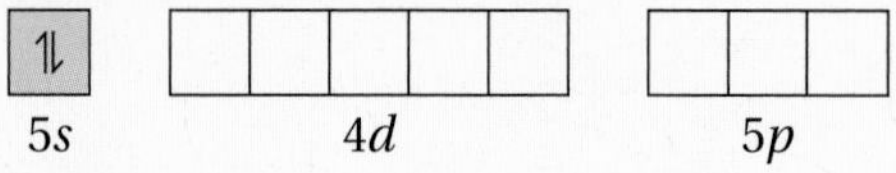

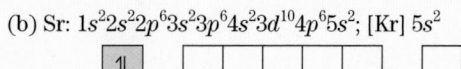

Sr has 36 core electrons.
(c) Po: $1s^22s^22p^63s^23p^64s^23d^{10}4p^65s^24d^{10}5p^66s^24f^{14}5d^{10}6p^4$; [Xe] $4f^{14}5d^{10}6s^26p^4$

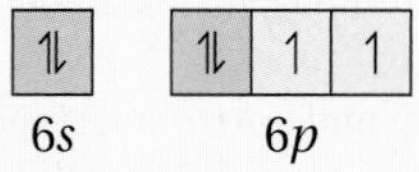

Po has 78 core electrons.

7.3 (a) Cl < Br < Se; (b) Xe < I < Ba

7.4 (a) Sn < Sb < I; (b) Ba < Sr < Ca

7.5 Q is aluminum: $1s^22s^22p^63s^23p^1$

7.6 (a) Ba ([Xe] $6s^2$) $\longrightarrow$ Ba^{2+} ([Xe]) + $2e^-$
(b) O ([He] $2s^22p^4$) + $2e^-$ $\longrightarrow$ O^{2-} + ([He] $2s^22p^6$) (same as Ne)
(c) Pb ([Xe] $4f^{14}5d^{10}6s^26p^2$) $\longrightarrow$ Pb^{2+} ([Xe] $4f^{14}5d^{10}6s^2$) + $2e^-$
Pb ([Xe] $4f^{14}5d^{10}6s^26p^2$) $\longrightarrow$ Pb^{4+} ([Xe] $4f^{14}5d^{10}$) + $4e^-$

7.7 (a) V^{3+}: [Ar] $3d^2$; paramagnetic
(b) Ni^{2+}: [Ar] $3d^8$; paramagnetic
(c) La^{3+}: [Xe]; not paramagnetic (diamagnetic)

7.8 (a) $F^- < Cl^- < Br^-$ (b) $Mg^{2+} < Na^+ < F^-$
(c) $Cr^{3+} < Cr^{2+}$

PROBLEMS

Problems with **red** numbers are answered in Appendix G and worked in detail in the Student Solutions Manual. Problem sections match those in this book and provide the numbers of relevant sample problems. Most offer Concept Review Questions, Skill-Building Exercises (grouped in pairs covering the same concept), and Problems in Context. The Comprehensive Problems are based on material from any section or previous chapter.

Introduction

Concept Review Questions

7.1 What would be your reaction to a claim that a new element had been discovered and it fit between tin (Sn) and antimony (Sb) in the periodic table?

7.2 Based on the results of his study of atomic X-ray spectra, Moseley discovered a relationship that replaced atomic mass as the criterion for ordering the elements. By what criterion are the elements now ordered in the periodic table? Give an example of a sequence of elements that was confirmed by Moseley's findings.

Skill-Building Exercises (grouped in similar pairs)

7.3 Before Mendeleev published his periodic table, Döbereiner grouped elements with similar properties into triads, in which the unknown properties of one element could be predicted by averaging known values of the properties of the others. To test Döbereiner's idea, predict the value of each quantity:
(a) The atomic mass of K from the atomic masses of Na and Rb
(b) The melting point of Br_2 from the melting points of Cl_2 (−101.0°C) and I_2 (113.6°C) (actual value = −7.2°C)

7.4 To test Döbereiner's idea (Problem 7.3), predict each boiling point:
(a) The boiling point of HBr from the boiling points of HCl (−84.9°C) and HI (−35.4°C) (actual value = −67.0°C)
(b) The boiling point of AsH_3 from the boiling points of PH_3 (−87.4°C) and SbH_3 (−17.1°C) (actual value = −55°C)

Characteristics of Many-Electron Atoms

Concept Review Questions

7.5 Summarize the rules for the allowable values of the four quantum numbers of an electron in an atom.

7.6 Which of the quantum numbers relate(s) to the electron only? Which relate(s) to the orbital?

7.7 State the exclusion principle. What does it imply about the number and spin of electrons in an atomic orbital?

7.8 What is the key distinction between subshell energies in one-electron species, such as the H atom, and those in many-electron species, such as the C atom? What factors led to this distinction? Would you expect the pattern of subshell energies in Be^{3+} to be more like the pattern in H or the pattern in C? Explain.

7.9 Define *shielding* and *effective nuclear charge*. What is the connection between the two?

7.10 What is penetration? How is it related to shielding? Use the penetration effect to explain the difference in relative orbital energies of a 3*p* electron and a 3*d* electron in the same atom.

Skill-Building Exercises (grouped in similar pairs)

7.11 How many electrons in an atom can have each quantum-number or subshell designation?
(a) $n = 2, l = 1$ (b) 3*d* (c) 4*s*

7.12 How many electrons in an atom can have each quantum-number or subshell designation?
(a) $n = 2, l = 1, m_l = 0$ (b) 5*p* (c) $n = 4, l = 3$

7.13 How many electrons in an atom can have each quantum-number or subshell designation?
(a) 4*p* (b) $n = 3, l = 1, m_l = +1$ (c) $n = 5, l = 3$

7.14 How many electrons in an atom can have each quantum-number or subshell designation?
(a) 2*s* (b) $n = 3, l = 2$ (c) 6*d*

The Quantum-Mechanical Model and the Periodic Table

(Sample Problems 7.1 and 7.2)

Concept Review Questions

7.15 State the periodic law, and explain its relation to electron configuration. (Use Na and K in your explanation.)

7.16 State Hund's rule in your own words, and show its application in the orbital diagram of the nitrogen atom.

7.17 How did the Aufbau principle, in connection with the periodic law, lead to the format of the periodic table?

7.18 For main-group elements, are valence electron configurations similar or different within a group? within a period? Explain.

7.19 For which blocks of elements are the outer electrons the same as the valence electrons? For which blocks are the *d* electrons often included among the valence electrons?

7.20 What is the electron capacity of the *n*th energy shell? What is the electron capacity of the fourth energy shell?

Skill-Building Exercises (grouped in similar pairs)

7.21 Write a full set of quantum numbers for each electron:
(a) The outermost electron in an Rb atom
(b) The electron gained when an S^- ion becomes an S^{2-} ion
(c) The electron lost when an Ag atom ionizes
(d) The electron gained when an F^- ion forms from an F atom

7.22 Write a full set of quantum numbers for each electron:
(a) The outermost electron in an Li atom
(b) The electron gained when a Br atom becomes a Br^- ion
(c) The electron lost when a Cs atom ionizes
(d) The highest energy electron in the ground-state B atom

7.23 Write the full ground-state electron configuration for each element:
(a) Rb (b) Ge (c) Ar

7.24 Write the full ground-state electron configuration for each element:
(a) Br (b) Mg (c) Se

7.25 Write the full ground-state electron configuration for each element:
(a) Cl (b) Si (c) Sr

7.26 Write the full ground-state electron configuration for each element:
(a) S^+ (b) Kr (c) Cs

7.27 Draw a partial (valence-shell) orbital diagram, and write the condensed ground-state electron configuration for each element:
(a) Ti (b) Cl (c) V

7.28 Draw a partial (valence-shell) orbital diagram, and write the condensed ground-state electron configuration for each element:
(a) Ba (b) Co (c) Ag

7.29 Draw a partial (valence-shell) orbital diagram, and write the condensed ground-state electron configuration for each element:
(a) Mn (b) P (c) Fe

7.30 Draw a partial (valence-shell) orbital diagram, and write the condensed ground-state electron configuration for each element:
(a) Ga (b) Zn (c) Sc

7.31 Draw a partial (valence-shell) orbital diagram, and write the symbol, group number, and period number of each element:
(a) [He] $2s^22p^4$ (b) [Ne] $3s^23p^3$

7.32 Draw a partial (valence-shell) orbital diagram, and write the symbol, group number, and period number of each element:
(a) [Kr] $4d^{10}5s^2$ (b) [Ar] $3d^84s^2$

7.33 Draw a partial (valence-shell) orbital diagram, and write the symbol, group number, and period number of each element:
(a) [Ne] $3s^23p^5$ (b) [Ar] $3d^{10}4s^24p^3$

7.34 Draw a partial (valence-shell) orbital diagram, and write the symbol, group number, and period number of each element:
(a) [Ar] $3d^54s^2$ (b) [Kr] $4d^25s^2$

7.35 From each partial (valence-shell) orbital diagram, write the condensed electron configuration and group number:

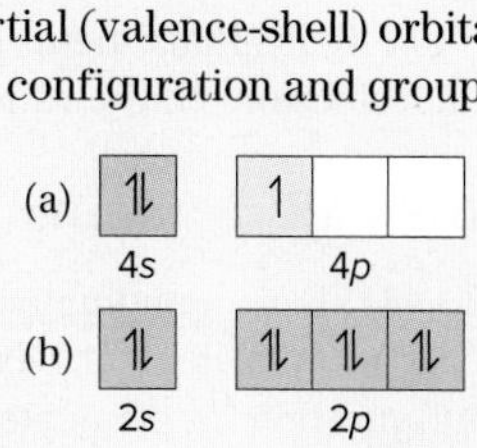

7.36 From each partial (valence-shell) orbital diagram, write the condensed electron configuration and group number:

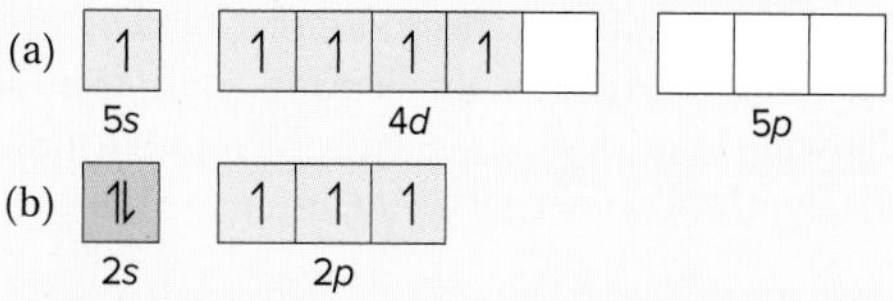

7.37 How many core and valence electrons are present in an atom of each element?
(a) O (b) Sn (c) Ca (d) Fe (e) Se

7.38 How many core and valence electrons are present in an atom of each element?
(a) Br (b) Cs (c) Cr (d) Sr (e) F

7.39 Identify each element below, and give the symbols of the other elements in its group:
(a) [He] $2s^22p^1$ (b) [Ne] $3s^23p^4$ (c) [Xe] $5d^16s^2$

7.40 Identify each element below, and give the symbols of the other elements in its group:
(a) [Ar] $3d^{10}4s^24p^4$ (b) [Xe] $4f^{14}5d^26s^2$ (c) [Ar] $3d^54s^2$

7.41 Identify each element below, and give the symbols of the other elements in its group:
(a) [He] $2s^22p^2$ (b) [Ar] $3d^34s^2$ (c) [Ne] $3s^23p^3$

7.42 Identify each element below, and give the symbols of the other elements in its group:
(a) [Ar] $3d^{10}4s^24p^2$ (b) [Ar] $3d^74s^2$ (c) [Kr] $4d^55s^2$

Problems in Context

7.43 After an atom in its ground state absorbs energy, it exists in an excited state. Spectral lines are produced when the atom returns to its ground state. The yellow-orange line in the sodium spectrum, for example, is produced by the emission of energy when excited sodium atoms return to their ground state. Write the electron configuration and the orbital diagram of the first excited state of sodium. (*Hint:* The outermost electron is excited.)

7.44 One reason spectroscopists study excited states is to gain information about the energies of orbitals that are unoccupied in an atom's ground state. Each electron configuration represents an atom in an excited state. Identify the element, and write its condensed ground-state configuration:
(a) $1s^22s^22p^63s^13p^1$ (b) $1s^22s^22p^63s^23p^44s^1$
(c) $1s^22s^22p^63s^23p^64s^23d^44p^1$ (d) $1s^22s^22p^53s^1$

Periodic Trends

(Sample Problems 7.3 to 7.5)

Concept Review Questions

7.45 If the exact outer limit of an isolated atom cannot be measured, what criterion can we use to determine its atomic radius? What is the difference between a covalent radius and a metallic radius?

7.46 Given the following partial (valence-shell) electron configurations, (a) identify each element; (b) rank the four elements in order of increasing atomic size; (c) rank the four elements in order of increasing ionization energy:

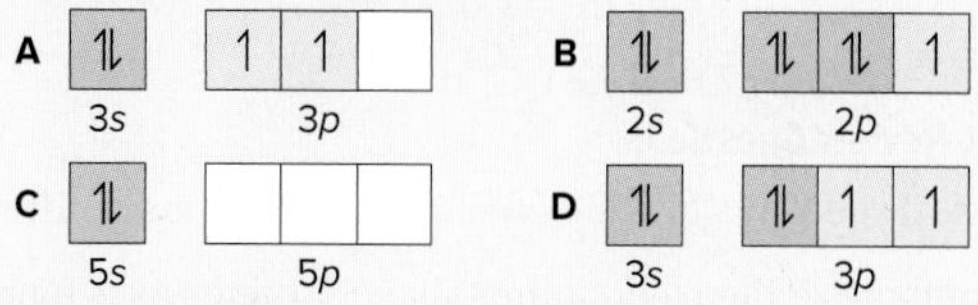

7.47 In what region of the periodic table will you find elements with relatively high IEs? with relatively low IEs?

7.48 (a) Why do successive IEs of a given element always increase? (b) When the difference between successive IEs of a given element is exceptionally large (for example, between IE_1 and IE_2 of K), what do we learn about its electron configuration?
(c) The following bar graph represents the relative magnitudes of the first five ionization energies of an atom:

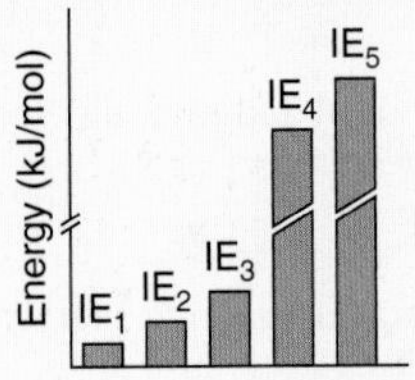

Identify the element, and write its complete electron configuration, assuming that it comes from (i) period 2; (ii) period 3; (iii) period 4.

7.49 In a plot of IE_1 for the period 3 elements (see Figure 7.15), why do the values for elements in groups 13 and 16 drop slightly below the generally increasing trend?

7.50 Which group in the periodic table has elements with high (endothermic) IE_1 and very negative (exothermic) first electron affinities (EA_1)? Give the charge on the ions that these atoms form.

7.51 The EA_2 of an oxygen atom is positive, even though its EA_1 is negative. Why does this change of sign occur? Which other elements exhibit a positive EA_2? Explain.

7.52 How does *d*-electron shielding influence atomic size among the period 4 transition elements?

Skill-Building Exercises (grouped in similar pairs)

7.53 Arrange each set of atoms in order of *increasing* atomic size:
(a) Rb, K, Cs (b) C, O, Be (c) Cl, K, S (d) Mg, K, Ca

7.54 Arrange each set of atoms in order of *decreasing* atomic size:
(a) Ge, Pb, Sn (b) Sn, Te, Sr (c) F, Ne, Na (d) Be, Mg, Na

7.55 Arrange each set of atoms in order of *increasing* IE_1:
(a) Sr, Ca, Ba (b) N, B, Ne (c) Br, Rb, Se (d) As, Sb, Sn

7.56 Arrange each set of atoms in order of *decreasing* IE_1:
(a) Na, Li, K (b) Be, F, C (c) Cl, Ar, Na (d) Cl, Br, Se

7.57 Write the full electron configuration of the period 2 element with the following successive IEs:
$IE_1 = 801$ kJ/mol $IE_2 = 2427$ kJ/mol $IE_3 = 3659$ kJ/mol
$IE_4 = 25\,022$ kJ/mol $IE_5 = 32\,822$ kJ/mol

7.58 Write the full electron configuration of the period 3 element with the following successive IEs:
$IE_1 = 738$ kJ/mol $IE_2 = 1450$ kJ/mol $IE_3 = 7732$ kJ/mol
$IE_4 = 10\,539$ kJ/mol $IE_5 = 13\,628$ kJ/mol

7.59 Which element in each set would you expect to have the *highest* IE_2?
(a) Na, Mg, Al (b) Na, K, Fe (c) Sc, Be, Mg

7.60 Which element in each set would you expect to have the *lowest* IE_3?
(a) Na, Mg, Al (b) K, Ca, Sc (c) Li, Al, B

Atomic Properties and Chemical Reactivity

(Sample Problems 7.6 to 7.8)

Concept Review Questions

7.61 List three ways in which metals and nonmetals differ.

7.62 Summarize the trend in metallic character as a function of position in the periodic table. Is it the same as the trend in atomic size? the trend in ionization energy?

7.63 What ions are possible for the two largest stable elements in group 14? How does each ion arise?

7.64 What is a pseudo–noble gas configuration? Give an example of one ion from group 13 that has this configuration.

7.65 How are measurements of paramagnetism used to support electron configurations that have been derived spectroscopically? Use Cu(I) and Cu(II) chlorides as examples.

7.66 The charges of a set of isoelectronic ions vary from 3+ to 3–. Place the ions in order of increasing size.

Skill-Building Exercises (grouped in similar pairs)

7.67 Which element would you expect to be *more* metallic?
(a) Ca or Rb (b) Mg or Ra (c) Br or I

7.68 Which element would you expect to be *more* metallic?
(a) S or Cl (b) In or Al (c) As or Br

7.69 Which element would you expect to be *less* metallic?
(a) Sb or As (b) Si or P (c) Be or Na

7.70 Which element would you expect to be *less* metallic?
(a) Cs or Rn (b) Sn or Te c) Se or Ge

7.71 Write the charge and full ground-state electron configuration of the monatomic ion most likely to be formed by each atom:
(a) Cl (b) Na (c) Ca

7.72 Write the charge and full ground-state electron configuration of the monatomic ion most likely to be formed by each atom:
(a) Rb (b) N (c) Br

7.73 Write the charge and full ground-state electron configuration of the monatomic ion most likely to be formed by each atom:
(a) Al (b) S (c) Sr

7.74 Write the charge and full ground-state electron configuration of the monatomic ion most likely to be formed by each atom:
(a) P (b) Mg (c) Se

7.75 How many unpaired electrons are present in a ground-state atom from each group?
(a) 2 (b) 15 (c) 18 (d) 13

7.76 How many unpaired electrons are present in a ground-state atom from each group?
(a) 14 (b) 17 (c) 1 (d) 16

7.77 Which of these atoms are paramagnetic in their ground state?
(a) Ga (b) Si (c) Be (d) Te

7.78 Are compounds of these ground-state ions paramagnetic?
(a) Ti^{2+} (b) Zn^{2+} (c) Ca^{2+} (d) Sn^{2+}

7.79 Write the condensed ground-state electron configurations of these transition metal ions, and state which are paramagnetic:
(a) V^{3+} (b) Cd^{2+} (c) Co^{3+} (d) Ag^{+}

7.80 Write the condensed ground-state electron configurations of these transition metal ions, and state which are paramagnetic:
(a) Mo^{3+} (b) Au^{+} (c) Mn^{2+} (d) Hf^{2+}

7.81 Palladium (Pd; $Z = 46$) is diamagnetic. Draw partial orbital diagrams to show which electron configuration is consistent with this fact:
(a) [Kr] $4d^8 5s^2$ (b) [Kr] $4d^{10}$ (c) [Kr] $4d^9 5s^1$

7.82 Niobium (Nb; $Z = 41$) has an anomalous ground-state electron configuration for a group 5 element: [Kr] $4d^4 5s^1$. What is the expected electron configuration for elements in this group? Draw partial orbital diagrams to show how paramagnetic measurements could support niobium's actual configuration.

7.83 Rank the ions in each set in order of *increasing* size, and explain your ranking:
(a) Li^+, K^+, Na^+ (b) Se^{2-}, Rb^+, Br^- (c) O^{2-}, F^-, N^{3-}

7.84 Rank the ions in each set in order of *decreasing* size, and explain your ranking:
(a) Se^{2-}, S^{2-}, O^{2-} (b) Te^{2-}, Cs^+, I^- (c) Sr^{2+}, Ba^{2+}, Cs^+

Comprehensive Problems

7.85 Name the element that matches each description:
(a) Smallest atomic radius in group 16
(b) Largest atomic radius in period 6
(c) Smallest metal in period 3
(d) Highest IE_1 in group 14
(e) Lowest IE_1 in period 5
(f) Most metallic in group 15
(g) Period 4 element with the highest energy shell filled
(h) Condensed ground-state electron configuration of [Ne] $3s^23p^2$
(i) Condensed ground-state electron configuration of [Kr] $4d^65s^2$
(j) Element that forms a 2+ ion with the electron configuration [Ar] $3d^3$
(k) Period 5 element that forms a 3+ ion with a pseudo–noble gas configuration
(l) Period 4 transition element that forms a 3+ diamagnetic ion
(m) Period 4 transition element that forms a 2+ ion with a half-filled *d* subshell
(n) Heaviest lanthanide
(o) Period 3 element whose 2− ion is isoelectronic with Ar
(p) Alkaline earth metal whose cation is isoelectronic with Kr

7.86 Use electron configurations to account for the stability of the lanthanide ions Ce^{4+} and Eu^{2+}.

7.87 A fundamental relationship of electrostatics states that the energy required to separate opposite charges of magnitudes Q_1 and Q_2 that are a distance d apart is proportional to $\frac{Q_1 \times Q_2}{d}$. Use this relationship, and any other factors, to explain why (a) the IE_2 of He ($Z = 2$) is *more* than twice the IE_1 of H ($Z = 1$); (b) why the IE_1 of He is *less* than twice the IE_1 of H.

7.88 The energy difference between the 5*d* and 6*s* subshells in gold accounts for its colour. Assuming that this energy difference is about 2.7 eV (electron volts), explain why gold has a warm yellow colour (1 eV = 1.602×10^{-19} J).

7.89 Write the formula and name of the compound formed from each ionic interaction:
(a) The 2+ ion and the 1− ion are both isoelectronic with the atoms of a chemically unreactive period 4 element.
(b) The 2+ ion and the 2− ion are both isoelectronic with the period 3 noble gas.
(c) The 2+ ion is the smallest, with a filled *d* subshell; the anion forms from the smallest halogen.
(d) The ions form from the largest and smallest ionizable atoms in period 2.

7.90 The energy changes for many unusual reactions can be determined using Hess's law (Section 5.5).
(a) Calculate ΔE for the conversion of $F^-(g)$ into $F^+(g)$.
(b) Calculate ΔE for the conversion of $Na^+(g)$ into $Na^-(g)$.

7.91 The hot glowing gases around the Sun, the *corona*, can reach millions of degrees Celsius, high enough to remove many electrons from gaseous atoms. Iron ions with charges as high as 14+ have been observed in the corona. Which ions from Fe^+ to Fe^{14+} are paramagnetic? Which would be most strongly attracted to a magnetic field?

7.92 There are some exceptions to the trends of first and successive ionization energies. For each pair, explain which ionization energy would be higher:
(a) IE_1 of Ga or IE_1 of Ge (b) IE_2 of Ga or IE_2 of Ge
(c) IE_3 of Ga or IE_3 of Ge (d) IE_4 of Ga or IE_4 of Ge

7.93 Use Figure 7.16 to find (a) the longest wavelength of electromagnetic (EM) radiation that can ionize an alkali metal atom; (b) the longest wavelength of EM radiation that can ionize an alkaline earth metal atom; (c) the elements, other than the alkali and alkaline earth metals, that could also be ionized by the radiation described in part (b); (d) the region of the EM spectrum in which these photons are found.

7.94 Rubidium and bromine atoms are depicted below.
(a) What monatomic ions do they form?
(b) What electronic feature characterizes this pair of ions, and which noble gas are they related to?
(c) Which pair best represents their relative ionic sizes?

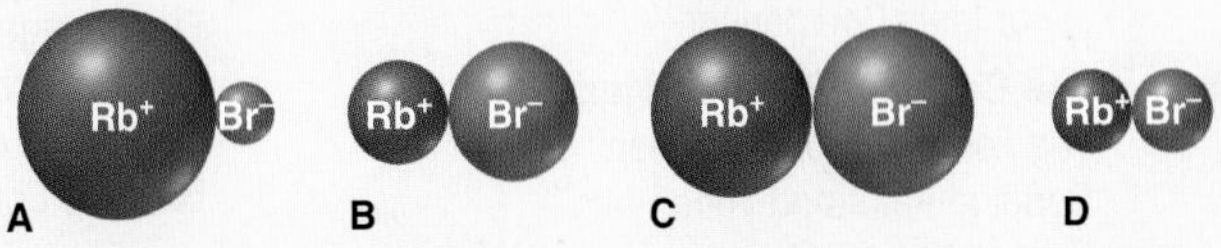

7.95 Partial (valence-shell) electron configurations for four different ions are shown below:

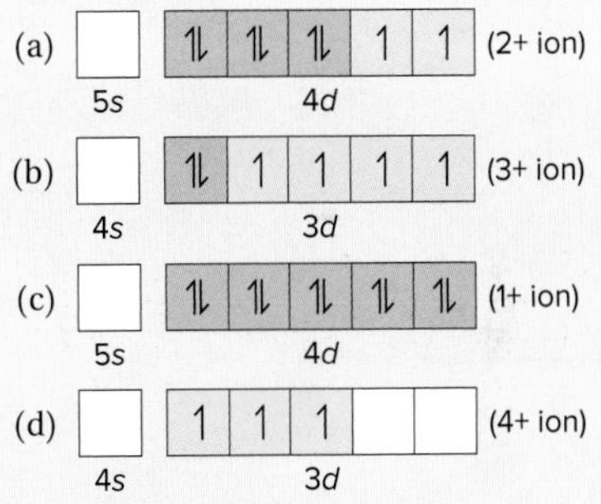

Identify the elements from which these ions are derived, and write the formula for the oxide that each ion forms.

7.96 Data for some main-group elements from the planet Zog are given in the table below. (Zoggian units are linearly related to Earth units, but are not shown.) Radio signals from Zog reveal that balloonium is a monatomic gas with two positive nuclear charges. Use the data to deduce the names that Earthlings give to these elements:

Name	Atomic Radius	IE_1	EA_1
Balloonium	10	339	0
Inertium	24	297	+4.1
Allotropium	34	143	−28.6
Brinium	63	70.9	−7.6
Canium	47	101	−15.3
Fertilium	25	200	0
Liquidium	38	163	−46.4
Utilium	48	82.4	−6.1
Crimsonium	72	78.4	−2.9

Models of Chemical Bonding

IN THIS CHAPTER . . . We examine how atomic properties give rise to three models of chemical bonding—ionic, covalent, and metallic—and how each model explains the behaviour of substances. By the end of this chapter, you should be able to

- Differentiate between the three types of bonding that metals and nonmetals use to combine and depict atoms and ions with Lewis symbols
- Describe the steps in the formation of an ionic solid and explain the importance of lattice energy
- Describe how a bond forms and explain the relationships among bond order, energy, and length
- Explore the relationship between bond energy and the enthalpy change of a reaction, with a focus on fuels and foods
- Compare periodic trends in electronegativity and explain its role in the range of bonding, from pure covalent to ionic, as well as its role in bond polarity
- Apply the octet rule to convert a molecular formula into a flat structural formula that shows atom attachments and electron-pair locations

8

Concepts and Skills to Review before Studying This Chapter

- Characteristics of ionic and covalent compounds; Coulomb's law (Section 2.7)
- Polar covalent bonds and the polarity of water (Section 3.6)
- Hess's law, ΔH, $\Delta_r H°$, and $\Delta_f H°$ (Sections 5.5 and 5.6)
- Atomic and ionic electron configurations (Sections 7.2 and 7.4)
- Trends in atomic properties and metallic behaviour (Sections 7.3 and 7.4)

- Apply the concept of resonance in cases where electron delocalization limits our ability to depict a molecule with a single structural formula
- Propose a simple bonding model that explains the properties of metals.

From a distance, a mound of snow and a mound of salt look very similar. Both are white and look as if they are made of individual granules. If we look more closely, however, they are very different. We could, in theory, isolate a single molecule of the water that composes snow. We can never, however, isolate a single unit of sodium chloride that forms salt. What makes these two substances that look similar in appearance so different in their properties? Sodium chloride and water form entirely different types of bonds, the former being ionic and the latter covalent. How does the type of bond affect the properties of the material? Why does heat affect substances with different types of bonds in different ways?

This chapter provides insight into the nature of chemical bonds and the different types of bonding that can occur between atoms. How these bonds can be characterized, quantified, and represented will also be shown.

8.1 Atomic Properties and Chemical Bonds

Before we examine the types of chemical bonding, we should start with the most fundamental question: why do atoms bond? In general, *bonding lowers the potential energy between positive and negative particles* (see Figure 1.3), whether the particles are oppositely charged ions or nuclei and electron pairs. Just as the electron configuration and the strength of nucleus-electron attractions determine the properties of an atom, the type and strength of chemical bonds determine the properties of a substance. Richard F. W. Bader, one of Canada's most prominent theoretical chemists (Figure 8.1), confirmed that there is a balance of the repulsive and attractive forces by the accumulation of electron density in the binding region. Bonding is the accumulation of electron density between the nuclei, which lowers the potential energy. It has the effect of increasing the kinetic energy.

Types of Bonding: Three Ways That Metals and Nonmetals Combine

In general, there is a gradation from atoms of more metallic elements to atoms of more nonmetallic elements across a period *and* up a group (Figure 8.2). Three types of bonding result from the three ways that these two types of atoms can combine:

1. *Metal with nonmetal: electron transfer and ionic bonding* (Figure 8.3A). We observe **ionic bonding** between atoms with large differences in their tendencies to lose or gain electrons. Such differences occur between reactive metals (groups 1 and 2) and nonmetals (group 17 and the top of group 16). A metal atom (low ionization energy, IE) loses its one, two, or three valence electrons, and a nonmetal atom (highly negative electron affinity, EA) gains the electron(s). *Electron transfer* from metal to nonmetal occurs, and each atom forms an ion with a noble gas electron configuration. The electrostatic attractions between these positive and negative ions

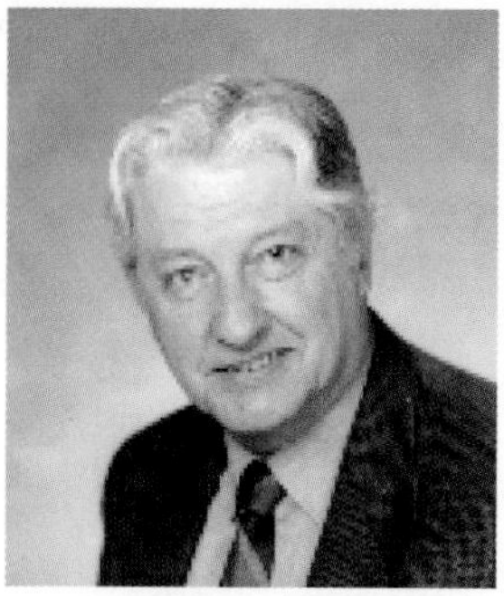

FIGURE 8.1 Richard F. W. Bader (1931–2012) was a Canadian quantum chemist (McMaster University, Hamilton, Ontario) noted for his work on the "atoms in molecules" approach. His research focused on the concept that all bonding is a result of the accumulation of electron density in the binding region. His work, the *Quantum Theory of Atoms in Molecules* (QTAIM), is applied today by thousands of chemists, physicists, material scientists, molecular biologists, crystallographers, and others all over the world and has found applications in the physics of crystals, in chemical biology, and in the formulation of new strategies to design drugs. He was among the first to ask seriously and answer a number of fundamental questions about the physical basis for such powerful chemical concepts as atoms in molecules, functional groups, bonding, electron pairs, and electron localization and delocalization—concepts that chemists and other scientists use every day.

draw them into a three-dimensional array to form an ionic solid. Note that the chemical formula for an ionic compound is the *empirical formula* because it gives the cation-to-anion ratio.

2. *Nonmetal with nonmetal: electron sharing and covalent bonding* (Figure 8.3B). When two atoms differ little, or not at all, in their tendencies to lose or gain electrons, we observe *electron sharing* and **covalent bonding**, which occurs most commonly between nonmetal atoms. Each atom holds onto its own electrons tightly (high IE) and attracts other electrons (highly negative EA). The nucleus of each atom attracts the valence electrons of the other, which draws the atoms together. The shared electron pair is typically *localized* between the two atoms, linking them in a covalent bond of a particular length and strength. In most cases, separate molecules result when atoms bond covalently. Note that the chemical formula for a covalent compound is the *molecular formula* because it gives the actual numbers of atoms in each molecule.

3. *Metal with metal: electron pooling and metallic bonding* (Figure 8.3C). Metal atoms are relatively large, and their few outer electrons are well shielded by filled inner levels (core electrons). Thus, they lose outer electrons easily (low IE) and do not gain outer electrons readily (slightly negative or positive EA). These properties lead metal atoms to share their valence electrons, but not by covalent bonding. In the simplest model of **metallic bonding**, the enormous number of atoms in a sample of a metal *pool* their valence electrons into a "sea" of electrons that "flows" between and around each metal-ion core (nucleus plus inner electrons), thereby attracting them together. Unlike the localized electrons in covalent bonding, electrons in metallic bonding are *delocalized*, moving freely throughout the entire piece of metal.

In the world of real substances, there are exceptions to these idealized models, so you cannot always predict the type of bond from the positions of the elements in the periodic table. As just one example, when the metal beryllium (group 2) combines with the nonmetal chlorine (group 17), there is more electron sharing than transferring—that is, the bonding is more covalent than ionic. Thus, just as we see

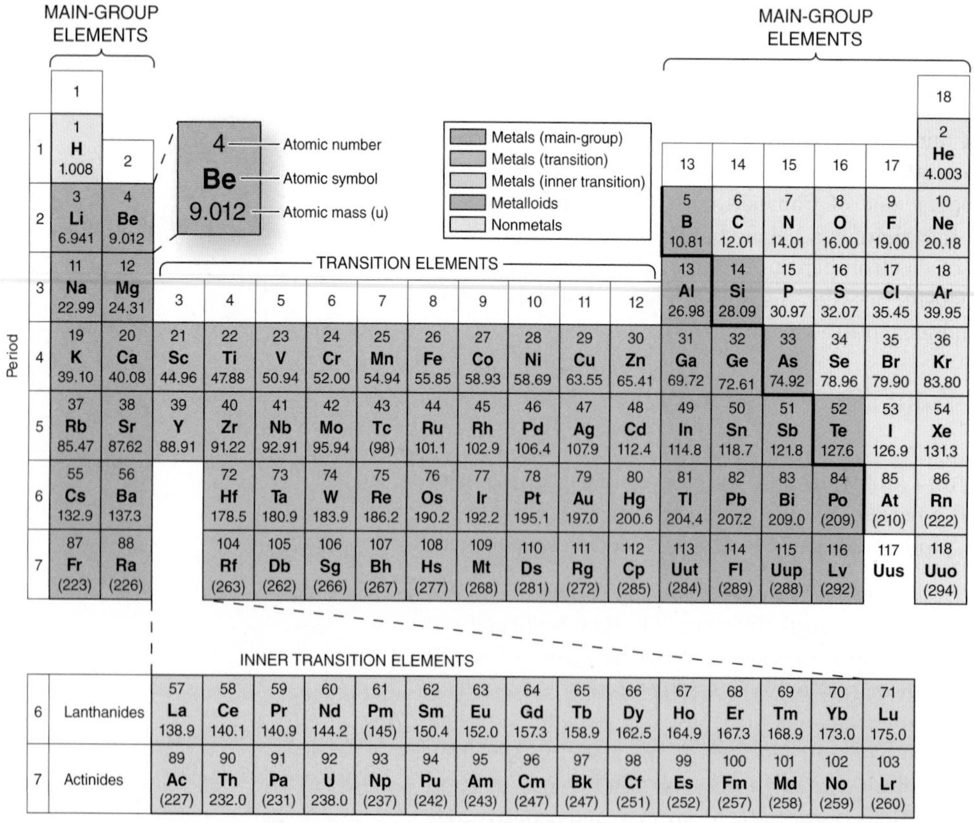

PROPERTY	METAL ATOM	NONMETAL ATOM
Atomic size	Larger	Smaller
Z_{eff}	Lower	Higher
IE	Lower	Higher
EA	Less negative	More negative

B

FIGURE 8.2 A comparison of metals and nonmetals. A. Location within the periodic table. **B.** Relative magnitudes of some atomic properties across a period.

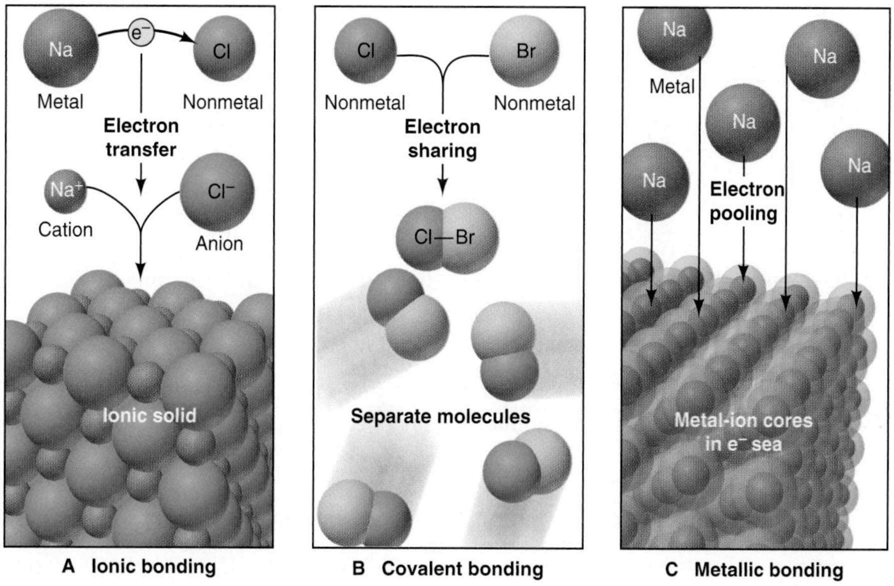

FIGURE 8.3 Three models of chemical bonding

gradations in atomic behaviour within a group or period, we see gradations in the type of bonding between atoms from different groups and periods (Figure 8.4). However, a 100 percent ionic bond or a 100 percent covalent bond does not exist, as we will see in Section 8.5.

Lewis Symbols and the Octet Rule

Before examining each of the three models, let us learn a method for depicting the valence electrons of interacting atoms, which predicts how they bond. In a **Lewis electron-dot symbol** (named for the American chemist G. N. Lewis; *see photo* ■), the element symbol represents the nucleus *and* inner electrons, and dots around the symbol represent the valence electrons (Figure 8.5). You have already learned how to work out the number of valence electrons from electron configurations. Note that the pattern of dots is the same for elements within a group.

We use these steps to write the Lewis symbol for any main-group element:

1. Find the number of valence electrons by writing the electron configuration of the element. Elements in groups 1 and 2 have one and two valence electrons, respectively. The number of valence electrons for elements in groups 13 to 18 can be calculated by subtracting 10 from their group number.

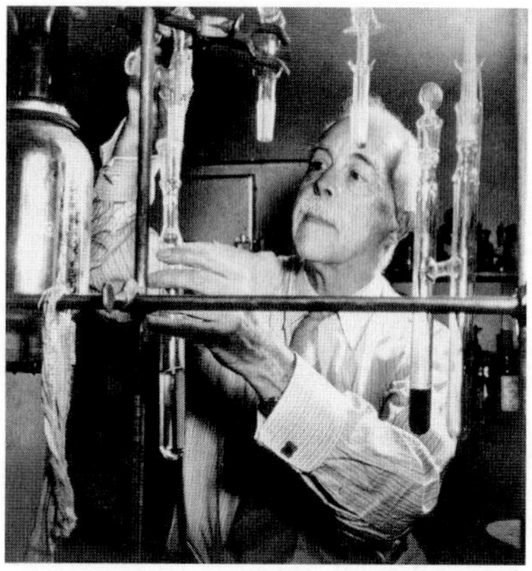

■ **Gilbert Newton Lewis (1875–1946)** was an American physical chemist who developed the concept of the covalent bond and the symbols that are still used to describe ways in which the atoms bond. He was the first scientist to produce *heavy water* and the one who coined the term *photon*. He was nominated for the Nobel Prize in chemistry more than 30 times, but surprisingly was not a recipient.

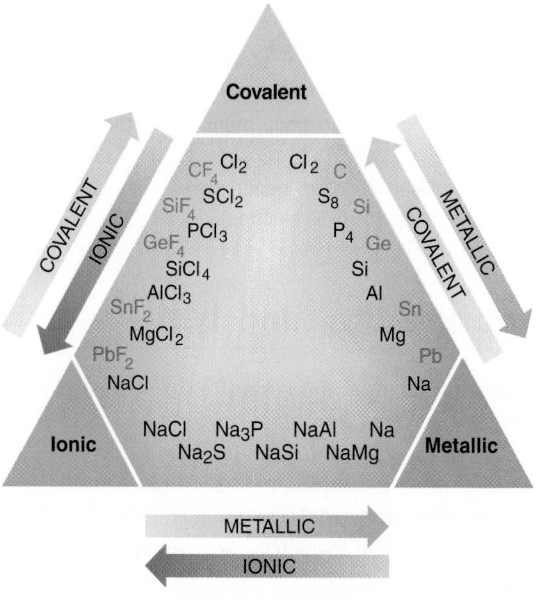

FIGURE 8.4 Gradations in bond type among elements in period 3 (*black type*) and group 4 (*red type*)

FIGURE 8.5 Lewis electron-dot symbols for elements in periods 2 and 3

Group	1	2	13	14	15	16	17	18
Valence electrons	ns^1	ns^2	ns^2np^1	ns^2np^2	ns^2np^3	ns^2np^4	ns^2np^5	ns^2np^6
Period 2	• Li	• Be •	• B •	• C •	• N •	: O •	: F :	: Ne :
Period 3	• Na	• Mg •	• Al •	• Si •	• P •	: S •	: Cl :	: Ar :

2. Place one dot at a time on each of the four sides (left, right, top, and bottom) of the element symbol.
3. Keep adding dots, and pairing them, until all are used up.

The specific placement of the dots is not important; that is, in addition to the format shown in Figure 8.5, the Lewis symbol for nitrogen can *also* be written as follows:

·N: or ·N· or :N·

The Lewis symbol provides information about an element's bonding behaviour:

- For a *metal*, the *total* number of dots is the number of electrons that an atom loses to form a cation.
- For a *nonmetal*, the number of *unpaired* dots equals either the number of electrons that an atom *gains* to form an anion or the number that it *shares* to form covalent bonds.

The Lewis symbol for carbon illustrates the last point. Rather than one pair of dots and two unpaired dots, as its electron configuration seems to call for ([He]$2s^22p^2$), carbon has four unpaired dots because it forms four bonds. Larger nonmetals can form as many bonds as the number of dots in their Lewis symbol, as we will see later in this chapter. In Lewis's pioneering studies, he generalized much of bonding behaviour into the **octet rule**: *when atoms bond, they lose, gain, or share electrons to attain a filled outer level of eight electrons* (or *two* electrons, for atoms whose outer level resembles He). The octet rule holds for nearly all the compounds of period 2 elements and a large number of other compounds as well. However, its use is restricted to elements in groups 1 and 2 and groups 13 to 18.

SUMMARY OF SECTION 8.1

- Nearly all naturally occurring substances consist of atoms or ions bonded to other atoms or ions. Chemical bonding allows atoms to lower their energy.
- Ionic bonding occurs when metal atoms transfer electrons to nonmetal atoms, and the resulting ions attract each other and form an ionic solid.
- Covalent bonding is most common between nonmetal atoms and usually results in individual molecules. Bonded atoms share one or more pairs of electrons that are localized between them.
- Metallic bonding occurs when many metal atoms pool their valence electrons into a delocalized electron "sea," which holds together all the atoms in the sample.
- The Lewis electron-dot symbol of a main-group atom shows the valence electrons as dots surrounding the element symbol.
- According to the octet rule, when bonding, many atoms lose, gain, or share electrons to attain a filled outer level of eight (or two) electrons.

8.2 The Ionic Bonding Model

The central idea of the ionic bonding model is the *transfer of electrons from metal atoms to nonmetal atoms to form ions that attract each other into a solid compound*. In most cases, for the main groups, the ion that forms has a filled outer level of either two or eight electrons, the number in the nearest noble gas (octet rule).

The transfer of an electron from a lithium atom to a fluorine atom is depicted in three ways in Figure 8.6. In each, Li loses its single outer electron and is left with a

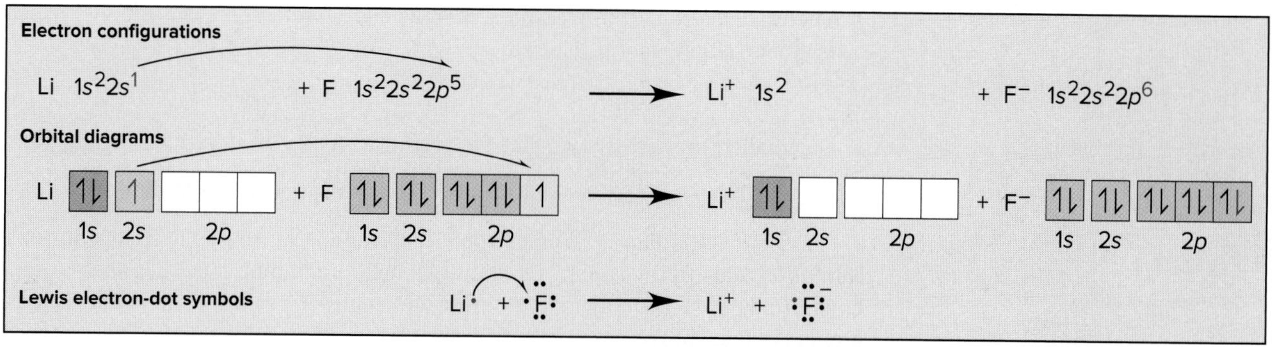

FIGURE 8.6 Three ways to depict electron transfer in the formation of Li^+ and F^-. The electron being transferred is shown in *red*.

filled $n = 1$ level (two e^-), while F gains a single electron to fill its $n = 2$ level (eight e^-). Each of these atoms is one electron away from the configuration of its nearest noble gas, so the number of electrons lost by each Li equals the number gained by each F. Therefore, equal numbers of Li^+ and F^- ions form, as the formula LiF indicates. In other words, in ionic bonding, *the total number of electrons lost by the metal atom(s) equals the total number of electrons gained by the nonmetal atom(s).*

Sample Problem 8.1 Depicting Ion Formation

Problem Use partial orbital diagrams and Lewis symbols to depict the formation of Na^+ and S^{2-} ions from the atoms, and give the formula for the compound formed.

Plan First we draw the orbital diagrams and Lewis symbols for Na and S atoms. To attain filled outer levels, Na loses one electron and S gains two. To make the number of electrons lost equal the number of electrons gained, two Na atoms are needed for each S atom.

Solution

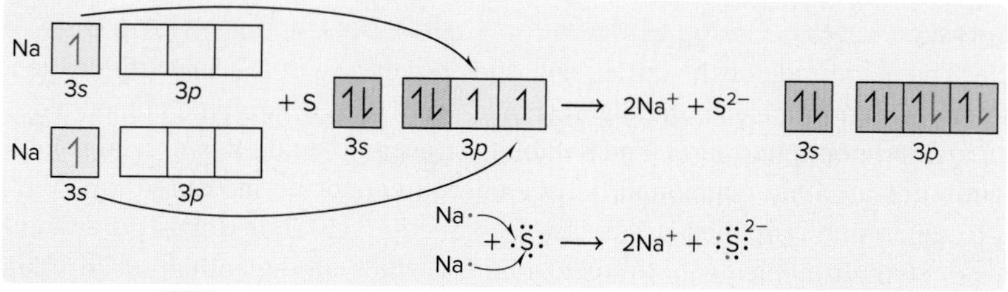

The formula is Na_2S.

Follow-Up Problem 8.1 Use condensed electron configurations and Lewis symbols to depict the formation of Mg^{2+} and Cl^- ions from the atoms, and give the formula for the compound formed. (See Brief Solutions to Follow-Up Problems at the end of the chapter.)

Why Ionic Compounds Form: The Importance of Lattice Energy

You may be surprised to learn that energy is *absorbed* during electron transfer. So, why does electron transfer occur? As well, in view of this absorption of energy, why do ionic substances exist at all? As you will see, the answer involves the enormous quantity of energy that is *released* after electron transfer, as the ions form a solid.

1. *The electron-transfer process.* Consider the electron-transfer process for the formation of lithium fluoride, which involves a gaseous Li atom losing an electron and a gaseous F atom gaining this electron:

- The first ionization energy (IE_1) of Li is the energy absorbed when 1 mol of gaseous Li atoms loses 1 mol of valence electrons:

$$Li(g) \longrightarrow Li^+(g) + e^- \qquad IE_1 = 520 \text{ kJ/mol}$$

- The first electron affinity (EA_1) of F is the energy released when 1 mol of gaseous F atoms gains 1 mol of electrons:

$$F(g) + e^- \longrightarrow F^-(g) \qquad EA_1 = -328 \text{ kJ/mol}$$

- Calculating the sum shows that electron transfer *by itself* requires energy:

$$Li(g) + F(g) \longrightarrow Li^+(g) + F^-(g) \qquad IE_1 + EA_1 = 192 \text{ kJ/mol}$$

2. *Other steps that absorb energy.* The total energy that is needed prior to ion formation adds to the sum of IE_1 and EA_1: metallic lithium must be made into gaseous atoms (161 kJ/mol), and fluorine molecules must be broken into separate atoms (79.5 kJ/mol).

3. *Steps that release energy.* Despite these endothermic steps, the standard enthalpy of formation ($\Delta_f H°$) of solid LiF is −617 kJ/mol; that is, 617 kJ is *released* when 1 mol of LiF(*s*) forms from its elements. Formation of LiF is typical of reactions between active metals and nonmetals: ionic solids form readily (Figure 8.7).

If the overall reaction releases energy, there must be some step that is exothermic enough to outweigh the endothermic steps. This step involves the *strong attraction between pairs of oppositely charged ions.* When 1 mol of $Li^+(g)$ and 1 mol of $F^-(g)$ form 1 mol of gaseous LiF molecules, a large quantity of heat is released:

$$Li^+(g) + F^-(g) \longrightarrow LiF(g) \qquad \Delta H° = -755 \text{ kJ/mol}$$

As we know, however, LiF does not exist as gaseous molecules under ordinary conditions: *even more energy is released when the separate gaseous ions coalesce into a crystalline solid* because each ion attracts *several* oppositely charged ions:

$$Li^+(g) + F^-(g) \longrightarrow LiF(s) \qquad \Delta H° = -1050 \text{ kJ/mol}$$

The negative of this enthalpy change is 1050 kJ/mol, the lattice energy of LiF. The *enthalpy of lattice destruction* is known as lattice energy. Therefore, the **lattice energy ($\Delta_{lattice}H°$)** is the enthalpy change that accompanies the reverse of this equation—1 mol of ionic solid separating into gaseous ions.

Determining Lattice Energy (Enthalpy of Lattice Destruction) with a Born-Haber Cycle The magnitude of the lattice energy is a measure of the strength of the ionic interactions and influences the macroscopic properties of an ionic compound, such as its melting point, hardness, and solubility. Despite playing this crucial role in the formation of an ionic compound, lattice energy cannot be measured directly. One way to determine it applies Hess's law (see Section 5.5) in a **Born-Haber cycle**, a series of steps from elements to ionic solid for which all the enthalpies* are known except the lattice energy.

Let us go through a Born-Haber cycle for the formation of lithium fluoride to calculate $\Delta_{lattice}H°$. Figure 8.8 shows two possible paths—a direct combination reaction ($\Delta_f H°$; *black arrow*) and a multistep path (*orange arrows*) in which one step is

A

B

FIGURE 8.7 The exothermic formation of sodium bromide. A. Sodium (in the beaker under mineral oil) and bromine. **B.** The reaction is rapid and vigorous.

*Strictly speaking, ionization energy (IE) and electron affinity (EA) are internal energy changes (ΔU), not enthalpy changes (ΔH). In these steps, however, $\Delta H = \Delta U$ because $\Delta V = 0$ (see Section 5.2).

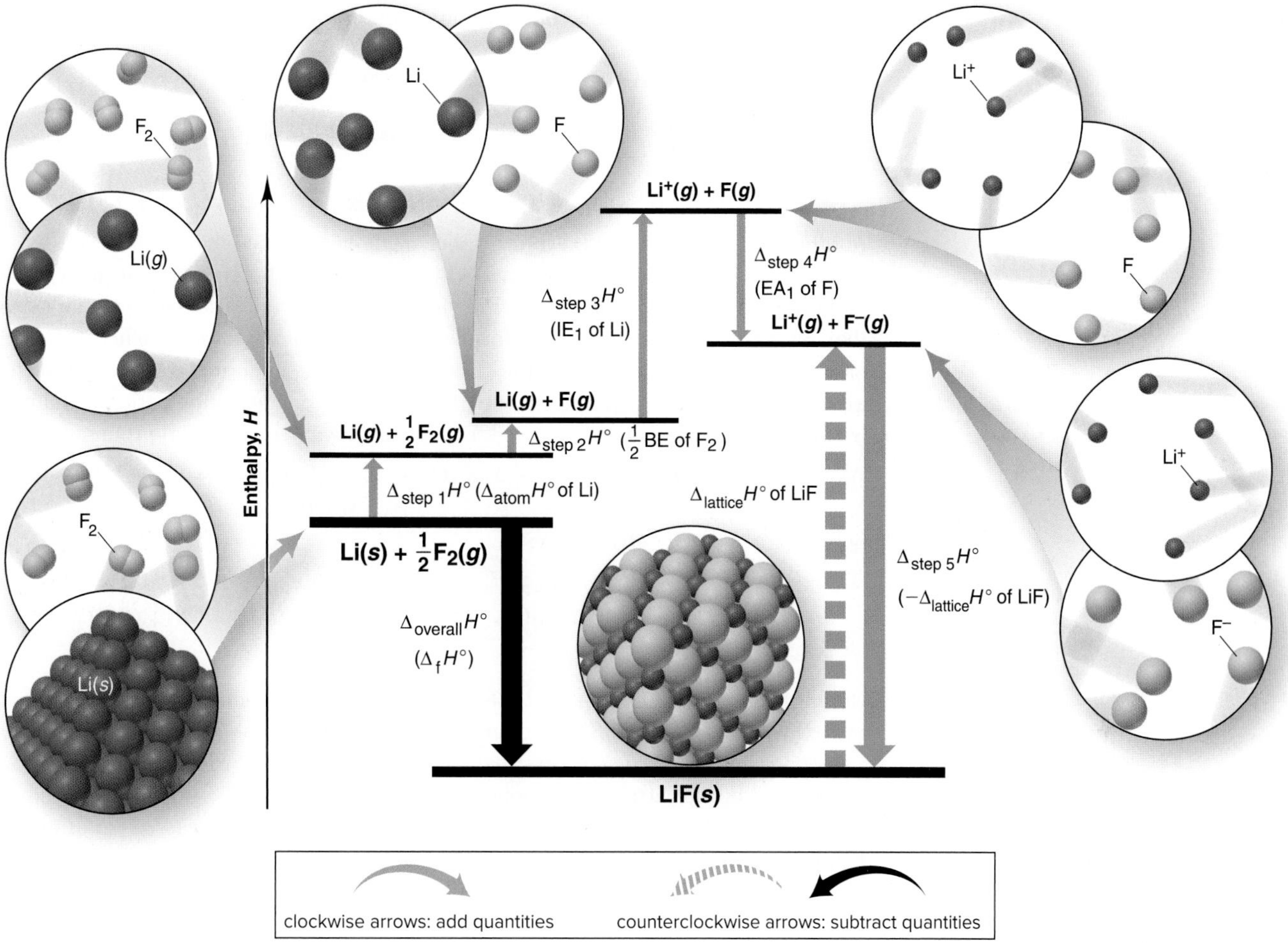

FIGURE 8.8 The Born-Haber cycle for lithium fluoride. The formation of LiF(*s*) from its elements can happen in one combination reaction (*black arrow*) or in five steps (*orange arrows*). The unknown enthalpy change is $\Delta_{\text{step 5}}H°$ ($-\Delta_{\text{lattice}}H°$ of LiF). The dashed orange arrow in step 5 represents $\Delta_{\text{lattice}}H°$.

the unknown $\Delta_{\text{lattice}}H°$ (*orange dashed arrow*). Hess's law tells us that both paths involve the same overall enthalpy change:

$$\Delta_f H° \text{ of LiF}(s) = \text{sum of } \Delta H° \text{ values for multistep path}$$

Hess's law lets us choose *hypothetical* steps whose enthalpy changes we can measure, even though *they are **not** the actual steps that occur when lithium reacts with fluorine.* We identify each $\Delta H°$ by its step number in Figure 8.8:

Step 1. From solid Li to Li atoms. This step, called *atomization,* has the enthalpy change $\Delta_{\text{atom}}H°$. It involves breaking metallic bonds, so it absorbs energy:

$$\text{Li}(s) \longrightarrow \text{Li}(g) \qquad \Delta_{\text{step 1}}H° = \Delta_{\text{atom}}H° = 161 \text{ kJ/mol}$$

Step 2. From F_2 molecules to F atoms. This step involves breaking a covalent bond, so it absorbs energy; as we will discuss later, this is the *bond energy* (BE) of F_2. Since we need 1 mol of F atoms to make 1 mol of LiF, we start with $\frac{1}{2}$ mol of F_2:

$$\frac{1}{2}\text{F}_2(g) \longrightarrow \text{F}(g) \qquad \Delta_{\text{step 2}}H° = \frac{1}{2}(\text{BE of F}_2) = \frac{1}{2}(159 \text{ kJ/mol}) = 79.5 \text{ kJ/mol}$$

Step 3. From Li to Li^+. Removing the 2*s* electron from Li absorbs energy:

$$\text{Li}(g) \longrightarrow \text{Li}^+(g) + e^- \qquad \Delta_{\text{step 3}}H° = \text{IE}_1 = 520. \text{ kJ/mol}$$

Step 4. From F to F^-. Adding an electron to F releases energy:

$$\text{F}(g) + e^- \longrightarrow \text{F}^-(g) \qquad \Delta_{\text{step 4}}H° = \text{EA}_1 = -328 \text{ kJ/mol}$$

Step 5. From gaseous ions to ionic solid. Forming solid LiF from gaseous Li^+ and F^- releases a lot of energy. The enthalpy change for this step is unknown but is, by definition, the negative of the lattice energy:

$$\text{Li}^+(g) + \text{F}^-(g) \longrightarrow \text{LiF}(s) \qquad \Delta_{\text{step 5}}H° = -\Delta_{\text{lattice}}H° \text{ of LiF} = ?$$

The enthalpy change of the combination reaction (*black arrow*) is

$$\text{Li}(s) + \frac{1}{2}\text{F}_2(g) \longrightarrow \text{LiF}(s) \qquad \Delta_{\text{overall}}H° = \Delta_f H° = -617\ \text{kJ/mol}$$

We set $\Delta_f H°$ equal to the sum of the $\Delta H°$ values for the steps and solve for $\Delta_{\text{lattice}}H°$ (*orange dashed arrow*):

$$\Delta_f H° = \Delta_{\text{step 1}}H° + \Delta_{\text{step 2}}H° + \Delta_{\text{step 3}}H° + \Delta_{\text{step 4}}H° + (-\Delta_{\text{lattice}}H° \text{ of LiF})$$

Another way to approach the Born-Haber cycle is to notice that, in general (as we can see in Figure 8.8), there are two kinds of arrows: the arrows that go with the definitions of the steps (*all straight orange arrows going clockwise*) and the arrow that goes with the path taken in forming the cycle (*black arrow*). When the cycle path is in the same direction as the definition path, the quantity (*all straight orange arrows*) is added in. When the cycle path is opposite to a definition path, the quantity is subtracted. These all sum to zero. Doing this always gives the right answer:

$$\Delta_{\text{step 1}}H° + \Delta_{\text{step 2}}H° + \Delta_{\text{step 3}}H° + \Delta_{\text{step 4}}H° + \Delta_{\text{step 5}}H° - \Delta_f H° = 0$$
$$\text{where } \Delta_{\text{step 5}}H° = -\Delta_{\text{lattice}}H° \text{ of LiF}$$

Solving for $\Delta_{\text{lattice}}H°$ of LiF gives

$$\begin{aligned}\Delta_{\text{lattice}}H° \text{of LiF} &= -\Delta_f H° + (\Delta_{\text{step 1}}H° + \Delta_{\text{step 2}}H° + \Delta_{\text{step 3}}H° + \Delta_{\text{step 4}}H°)\\ &= -(-617\ \text{kJ/mol}) + 161\ \text{kJ/mol} + 79.5\ \text{kJ/mol} + 520.\ \text{kJ/mol} - 328\ \text{kJ/mol}\\ &= 1050\ \text{kJ/mol}\end{aligned}$$

Note that *the lattice energy is, by far, the largest component of the multistep process.* (Problems 8.30 and 8.31 are among several that focus on Born-Haber cycles.)

The Born-Haber cycle shows that the energy *required* for elements to form ions is *supplied* by the attraction among the ions in the solid. The take-home lesson is that *ionic solids exist* ***only*** *because the lattice energy far exceeds the total energy needed to form the ions.*

A simplified Born-Haber cycle can be represented as follows:

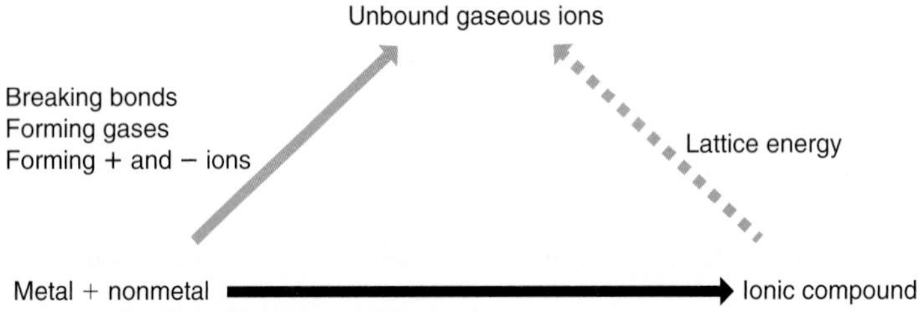

Sample Problem 8.2 Constructing a Born-Haber Cycle

Problem Construct a Born-Haber cycle, and use it to calculate the first electron affinity of chlorine, given the following information: $\Delta_f H°$ $MgCl_2$ = −641 kJ/mol, $\Delta_{\text{atom}}H°$ Mg = +148 kJ/mol, BE of Cl_2 = +244 kJ/mol, IE_1 of Mg = +738 kJ/mol, IE_2 of Mg = +1451 kJ/mol, $\Delta_{\text{lattice}}H°$ of $MgCl_2$ = +2526 kJ/mol.

Plan First we write and balance an equation with the reactants and the product. The metal (Mg) and the nonmetal (Cl_2) will react to form the ionic compound $MgCl_2$. Then we draw an arrow on top of the metal showing its atomization (see Figure 8.8) and, on top of it, another arrow showing the first and second ionizations. On top of the nonmetal, we draw an arrow showing the breaking of the bond (bond energy) and, on top of that arrow, another arrow corresponding to the formation of the anion (which is the unknown electron affinity).

Solution Keeping in mind that the sum of all the quantities (taking into account the direction of the arrows) should be zero, we add the values going clockwise and subtract the values going counterclockwise.

$$\Delta_{\text{atom}}H° \text{ of Mg} + \text{IE}_1 \text{ of Mg} + \text{IE}_2 \text{ of Mg} + \text{BE of Cl}_2 + 2\text{EA of Cl}_2$$
$$- \Delta_{\text{lattice}}H° \text{ of MgCl}_2 - \Delta_f H° \text{ of MgCl}_2 = 0$$

The answer is −348 kJ/mol. Did you get this answer? If not, try again! Remember that the algebraic addition of all the enthalpies results in a value of EA for 2 mol of Cl, which means that you need to divide by 2 to get the EA of 1 mol of chlorine.

Follow-Up Problem 8.2 Use the data given below to construct a Born-Haber cycle to determine the electron affinity of Br. (The energy associated with each step is given in parentheses.)

$K(s) \longrightarrow K(g)$	(89 kJ/mol)
$K(g) \longrightarrow K^+(g) + e^-$	(419 kJ/mol)
$\frac{1}{2}Br_2(l) \longrightarrow Br(g)$	(112 kJ/mol)
$K(s) + \frac{1}{2}Br_2(l) \longrightarrow KBr(s)$	(−394 kJ/mol)
$KBr(s) \longrightarrow K^+(g) + Br^-(g)$	(674 kJ/mol)

Periodic Trends in Lattice Energy

The lattice energy results from electrostatic interactions among ions, so its magnitude depends on ionic size, ionic charge, and ionic arrangement in the solid. Therefore, we expect to see periodic trends in lattice energy.

Explaining the Trends with Coulomb's Law Recall, from Chapter 2, that **Coulomb's law** states that the electrostatic energy between particles A and B is directly proportional to the product of their charges and inversely proportional to the distance between them:

$$\text{Electrostatic energy} \propto \frac{\text{charge A} \times \text{charge B}}{\text{distance}}$$

Lattice energy is directly proportional to electrostatic energy. In an ionic solid, cations and anions lie as close to each other as possible, so the distance between them is the sum of the ionic radii (see Figure 7.25):

$$\text{Electrostatic energy} \propto \frac{\text{cation charge} \times \text{anion charge}}{\text{cation radius} + \text{anion radius}} \propto \Delta_{\text{lattice}}H^\circ \qquad (8.1)$$

This relationship helps to explain the effects of ionic size and charge on trends in lattice energy:

1. *Effect of ionic size.* As we move down a group, the ionic radius increases, so the electrostatic energy between the cations and anions decreases; thus, the lattice energy should decrease as well. Figure 8.9 shows that, for the alkali-metal halides,

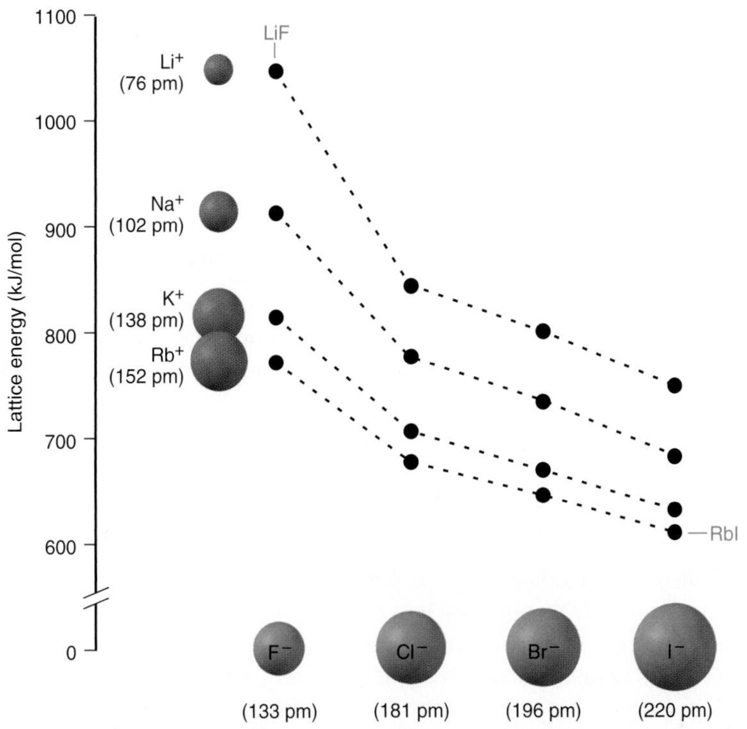

FIGURE 8.9 Trends in lattice energy. The lattice energies are shown for compounds formed from a given group 1 cation (*left side*) and one of the group 17 anions (*bottom*). LiF (smallest ions) has the highest lattice energy, and RbI (largest ions) has the lowest lattice energy.

lattice energy decreases down the group, whether we hold the cation constant (LiF to LiI) or the anion constant (LiF to RbF).

2. *Effect of ionic charge.* Across a period, the ionic charge changes. For example, lithium fluoride and magnesium oxide have cations and anions of about equal radii (Li^+ = 76 pm and Mg^{2+} = 72 pm; F^- = 133 pm and O^{2-} = 140 pm). The major difference is between the singly charged Li^+ and F^- ions and doubly charged Mg^{2+} and O^{2-} ions. The difference in the lattice energies of the two compounds is striking:

$$\Delta_{\text{lattice}}H° \text{ of LiF} = 1050 \text{ kJ/mol} \quad \text{and} \quad \Delta_{\text{lattice}}H° \text{ of MgO} = 3923 \text{ kJ/mol}$$

This nearly fourfold increase in $\Delta_{\text{lattice}}H°$ reflects the fourfold increase in the product of the charges (1 × 1 versus 2 × 2) in the numerator of Equation 8.1.

Sample Problem 8.3 Predicting Relative Lattice Energy from Ionic Properties

Problem Use ionic properties to explain which compound in each pair has the larger lattice energy: **(a)** RbI or NaBr; **(b)** KCl or CaS.

Plan To choose the compound with the larger lattice energy, we apply Coulomb's law (Equation 8.1) and periodic trends in ionic radius and charge (see Figure 2.12). We examine the ions in each compound: for ions of similar size, higher charge leads to a larger lattice energy; for ions with the same charge, smaller size leads to larger lattice energy because the ions can get closer together.

Solution (a) NaBr. All the ions have single charges, so charge is not a factor. Size increases down a group, so Rb^+ is larger than Na^+, and I^- is larger than Br^-. Therefore, NaBr has the larger lattice energy because it consists of smaller ions.

(b) CaS. Size decreases from left to right, so K^+ is slightly larger than Ca^{2+}, and S^{2-} is slightly larger than Cl^-. However, these small differences are not nearly as important as the charges: Ca^{2+} and S^{2-} have twice the charge of K^+ and Cl^-, so CaS has the larger lattice energy.

Check The actual lattice energies are **(a)** RbI = 598 kJ/mol and NaBr = 719 kJ/mol; **(b)** KCl = 676 kJ/mol and CaS = 3039 kJ/mol.

Follow-Up Problem 8.3 **(a)** Use ionic properties to explain which compound has the *larger* lattice energy: BaF_2 or SrF_2. **(b)** Use ionic properties to explain which compound has the *smaller* lattice energy: SrS or RbCl.

Why Does MgO Exist? We might ask how ionic solids, such as MgO, with its doubly charged ions, could even form. After all, forming 1 mol of Mg^{2+} involves the sum of the first *and* second ionization energies:

$$Mg(g) \longrightarrow Mg^{2+}(g) + 2e^- \quad \Delta H° = IE_1 + IE_2 = 738 \text{ kJ/mol} + 1450 \text{ kJ/mol} = 2188 \text{ kJ/mol}$$

As well, while forming 1 mol of O^- ions is exothermic (first electron affinity, EA_1), adding a second mole of electrons (second electron affinity, EA_2) is endothermic because the electron is added to a negative ion. The overall formation of O^{2-} ions is endothermic:

$$\begin{aligned} O(g) + e^- &\longrightarrow O^-(g) & \Delta H° &= EA_1 & &= -141 \text{ kJ/mol} \\ O^-(g) + e^- &\longrightarrow O^{2-}(g) & \Delta H° &= EA_2 & &= 878 \text{ kJ/mol} \\ \hline O(g) + 2e^- &\longrightarrow O^{2-}(g) & \Delta H° &= EA_1 + EA_2 & &= 737 \text{ kJ/mol} \end{aligned}$$

There are also the endothermic steps for converting Mg(*s*) to Mg(*g*) (148 kJ/mol) and breaking $\frac{1}{2}$ mol of O_2 molecules into O atoms (498 kJ/mol). Nevertheless, the 2+ and 2− ionic charges make the lattice energy so large ($\Delta_{\text{lattice}}H°$ = 3923 kJ/mol) that solid MgO forms readily whenever Mg burns in air ($\Delta_f H°$ = −601 kJ/mol).

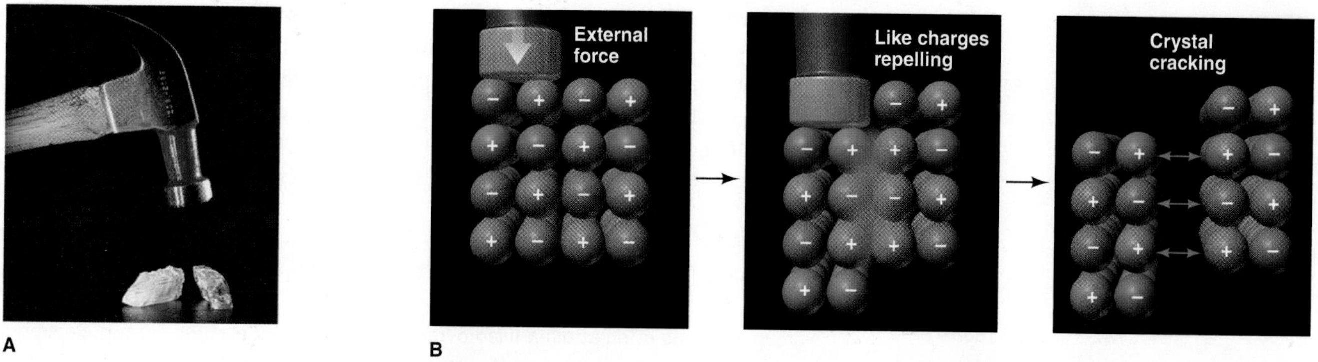

FIGURE 8.10 Why ionic compounds crack. **A.** Ionic compounds crack when struck with enough force. **B.** When a force moves like charges near each other, the repulsions cause a crack.

How the Model Explains the Properties of Ionic Compounds

The central role of any model is to explain the facts. With atomic-level views, we can see how the ionic bonding model accounts for the properties of ionic solids:

1. *Physical behaviour.* As a typical ionic compound, a piece of rock salt (NaCl) is *hard* (does not dent), *rigid* (does not bend), and *brittle* (cracks without deforming). These properties arise from the strong attractive forces that hold the ions *in specific positions.* Moving the ions out of position requires overcoming these forces, so rock salt does not dent or bend. If enough force is applied, ions of like charge are brought next to each other, and repulsions between them crack the sample suddenly (Figure 8.10).

2. *Electrical conductivity.* Ionic compounds typically *do not* conduct electricity in the solid state, but *do* conduct electricity when melted or dissolved. According to the model, the solid consists of fixed ions, but when it melts or dissolves, the ions can move and carry a current (Figure 8.11).

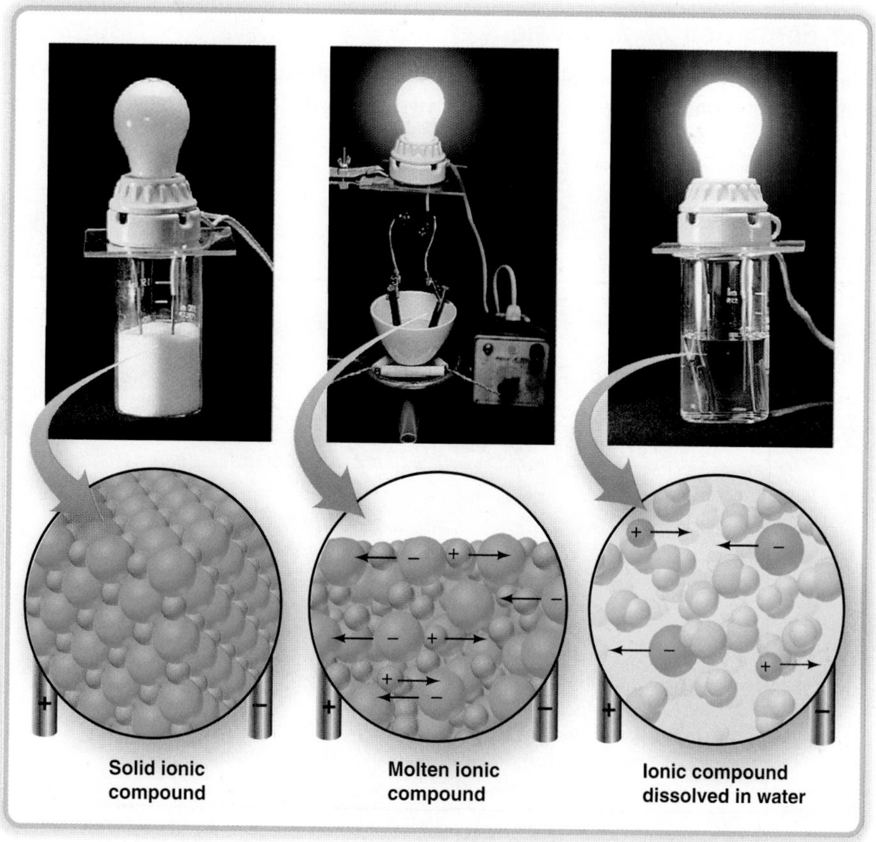

FIGURE 8.11 Electrical conductance and ion mobility

TABLE 8.1 Melting and Boiling Points of Some Ionic Compounds

Compound	Melting Point (°C)	Boiling Point (°C)
CsBr	636	1300
NaI	661	1304
$MgCl_2$	714	1412
KBr	734	1435
$CaCl_2$	782	>1600
NaCl	801	1413
LiF	845	1676
KF	858	1505
MgO	2852	3600

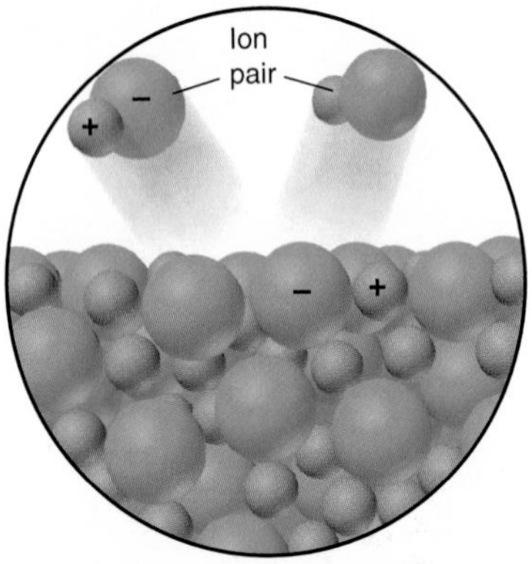

FIGURE 8.12 Ion pairs that are formed when an ionic compound vaporizes

3. *Thermal conductivity.* Large amounts of energy are needed to free the ions from their positions and separate them. Thus, we expect ionic compounds to have high melting points and much higher boiling points (Table 8.1). In fact, the interionic attraction is so strong that the vapour consists of **ion pairs**, or gaseous ionic molecules, rather than individual ions (Figure 8.12). In their normal state, as you know, ionic compounds are solid arrays of ions, and *no separate molecules exist.*

SUMMARY OF SECTION 8.2

- In ionic bonding, a metal transfers electrons to a nonmetal, and the resulting ions attract each other to form a solid.
- Main-group elements often attain a filled outer level (either eight electrons or two electrons) by forming ions with the electron configuration of the nearest noble gas.
- Ion formation by itself *absorbs* energy, but more than enough energy is *released* when the ions form a solid. The lattice energy, which is the energy required to separate the solid into gaseous ions, is the reason ionic solids exist.
- The lattice energy is determined by applying Hess's law in a Born-Haber cycle.
- The lattice energy increases with higher ionic charge and decreases with larger ionic radius.
- According to the ionic bonding model, the strong electrostatic attractions that keep ions in position explain why ionic solids are hard, conduct a current only when melted or dissolved, and have high melting and boiling points.
- Ion pairs form when an ionic compound vaporizes.

8.3 The Covalent Bonding Model

Look through the *Handbook of Chemistry and Physics*, and you will find that the number of covalent compounds dwarfs the number of ionic compounds. Molecules that are held together by covalent bonds range from tiny, diatomic hydrogen to biological and synthetic macromolecules with thousands of atoms, and covalent bonds occur in all polyatomic ions, too. Without doubt, *sharing electrons is the main way that atoms interact.*

The Formation of a Covalent Bond

Why does hydrogen gas consist of H_2 molecules and not separate H atoms? Figure 8.13 plots the potential energy of a system of two isolated H atoms versus the distance

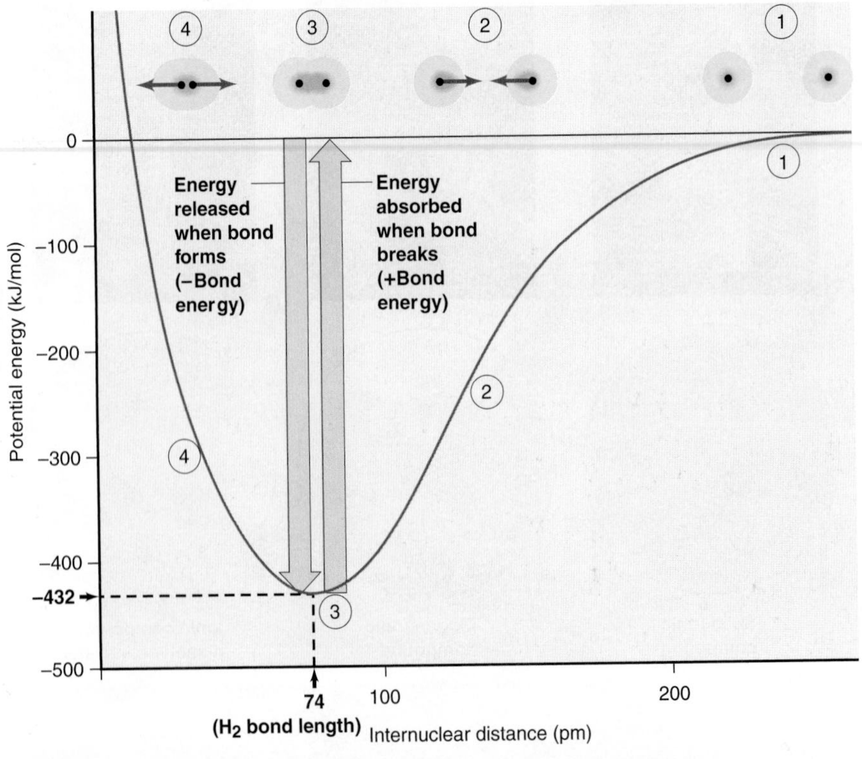

FIGURE 8.13 Covalent bond formation in H_2. The energy difference between points 1 and 3 is the H_2 bond energy (432 kJ/mol): it is released when the bond forms (the kinetic energy transforms into potential energy) and absorbed to break the bond (energy from the outside is required to drive the nuclei apart). The absorbing energy comes from being violently hit by another atom to break the bond. The formed bond will fall apart immediately, unless this bond loses energy to the surroundings. The internuclear distance at point 3 is the H_2 bond length (74 pm). A similar trend is observed for the formation of an ionic bond.

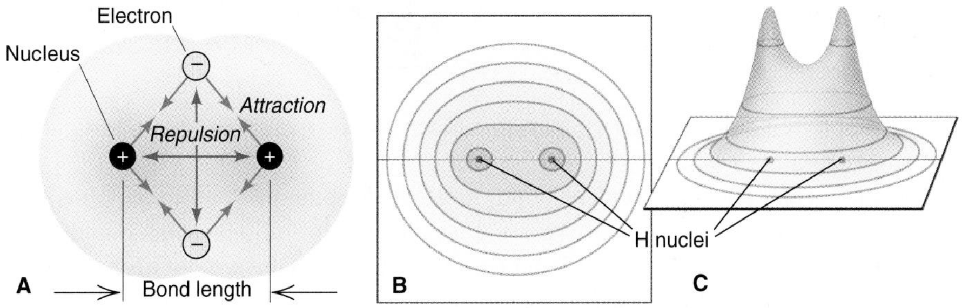

FIGURE 8.14 Distribution of electron density in H_2. **A.** At some distance (bond length), attractions balance repulsions. Electron density (*blue shading*) is high around and between the nuclei. **B.** Electron density doubles with each concentric curve. **C.** The highest regions of electron density are shown as peaks.

between their nuclei (see also Figure 2.14). Let us start at the right end of the curve and move along it as the atoms get closer:

- *At point 1*, the atoms are far apart, and each atom acts as though the other atom is not present.
- *At point 2*, the distance between the atoms has decreased enough for each nucleus to start attracting the other atom's electron, which lowers the potential energy. As the atoms get closer, these attractions increase, but so do the repulsions between the nuclei and between the electrons.
- *At point 3* (bottom of the energy "well"), the maximum attraction is achieved in the face of the increasing repulsion, and the system has its minimum energy.
- *At point 4*, if it were reached, the atoms would be too close, and the rise in potential energy from increasing repulsions would push the atoms apart, toward point 3 again.

Thus, a **covalent bond** arises from the balance between the nuclei attracting the electrons and the electrons repelling each other and nuclei repelling each other. (We will return to Figure 8.13 shortly.)

Formation of a covalent bond always results in greater electron density ***between*** *the nuclei.* Figure 8.14 depicts this fact with a cross section of a space-filling model (A), an *electron density contour map* (B), and an *electron density relief map* (C).

Paul G. Mezey, Canada Research Chair in Scientific Modelling and Simulation at Memorial University of Newfoundland and Labrador (Figure 8.15A), and Russell J. Boyd, professor of Chemistry at Dalhousie University in Halifax (Figure 8.15B), are two Canadian scientists who are studying electron density. Dr. Mezey is building a state-of-the-art Scientific Modelling and Simulation Laboratory (SMSL) to develop fundamental methods and computer software for a broad range of scientific modelling and simulation applications. Dr. Boyd has earned an international reputation for his research in theoretical chemistry, with an emphasis on the analysis of electron correlation and electron density distributions. As well, he is recognized for his applications of contemporary computational methods to the study of biological systems.

A.

B.

FIGURE 8.15 A. Paul G. Mezey (Memorial University of Newfoundland and Labrador, St. John's) is the author of more than 340 papers. Dr Mezey proposed the first ab initio quality linear-scaling macromolecular quantum chemistry method and holographic electron density theorem. His research deals with high-quality biomolecular modelling and simulation, which provide powerful tools for molecule design, molecular engineering, and molecular-level biotechnology. **B.** Russell J. Boyd (Dalhousie University, Halifax, Nova Scotia) has published more than 220 peer-reviewed papers in computational and theoretical chemistry. Dr. Boyd has earned an international reputation for his research in theoretical chemistry, with an emphasis on the analysis of electron correlation and electron density distributions. More recently, he and his students have gained much recognition for their applications of contemporary computational methods to the study of biological systems. This is a rapidly developing field of research because there is often little information about the identity of short-lived biomolecules, and even less information about the mechanisms by which they affect the physiology of living systems.

Bonding Pairs and Lone Pairs

To achieve a full outer (valence) level of electrons, *each atom in a covalent bond "counts" the shared electrons as belonging entirely to itself.* Thus, the two shared electrons in H_2 simultaneously fill the outer level of *both* H atoms, as clarified by the blue circles added below (which are *not* part of the Lewis structures). The **bonding pair (shared pair)**, is represented by a pair of dots or a line:

An outer-level electron pair that is *not* involved in bonding is called a **lone pair (unshared pair)**. The bonding pair in HF fills the outer level of the H atom *and*, together with three lone pairs, fills the outer level of the F atom as well:

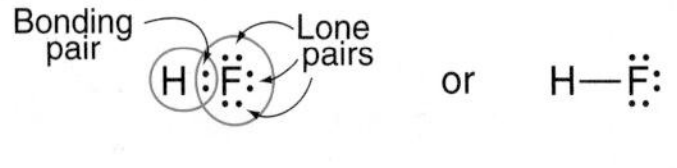

In F_2, the bonding pair and three lone pairs fill the outer level of *each* F atom:

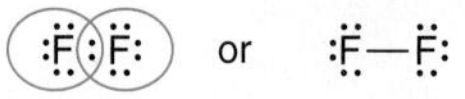

(In this book, bonding pairs are generally shown as lines, and lone pairs are generally shown as dots.)

In the formation of a covalent bond, most of the time each atom supplies one electron to the bond; however, there are cases in which both electrons come from the same atom, forming what is called a **coordinate bond (dative covalent bond)**, such as the N—B bond in BF_3NH_3:

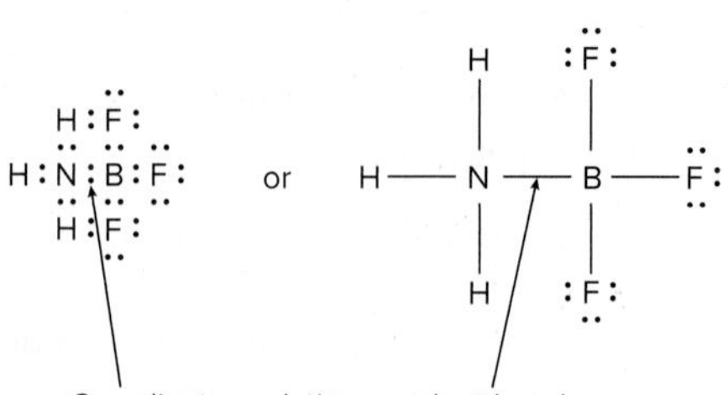

Properties of a Covalent Bond: Bond Order, Energy, and Length

A covalent bond has three important properties that are closely related to one another and to the compound's reactivity: bond order, bond energy, and bond length.

1. *Bond order.* The **bond order** is the number of electron pairs being shared by a given pair of atoms:

- A **single bond**, as shown above, in H_2, HF, and F_2, is the most common bond and consists of one bonding pair of electrons. Thus, a *single bond has a bond order of 1.*
- Many molecules (and ions) contain *multiple bonds*, in which more than one pair is shared between two atoms. Multiple bonds usually involve C, O, and/or N atoms. A **double bond** consists of two bonding electron pairs, four electrons shared between two atoms, so *the bond order is* 2. Ethene (C_2H_4) contains a carbon-carbon double bond and four carbon-hydrogen single bonds:

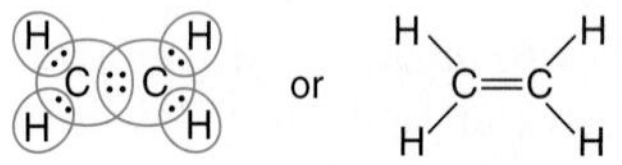

  To attain an octet, *each* carbon "counts" the four electrons in the double bond and the four electrons in its two single bonds to hydrogen atoms.
- A **triple bond** consists of three shared pairs: two atoms share six electrons, so *the bond order is* 3. The N_2 molecule has a triple bond, and each N atom also has a lone pair. Six shared and two unshared electrons give *each* N atom an octet:

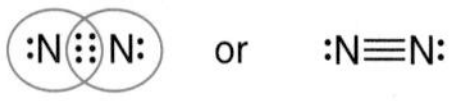

2. *Bond energy.* The strength of a covalent bond depends on the magnitude of the attraction between the nuclei and shared electrons. The **bond energy (BE)** (also called *bond enthalpy* or *bond strength*) is the energy needed to overcome this attraction and is defined as the standard enthalpy change for breaking the bond in 1 mol of *gaseous* molecules. Bond breakage is an *endothermic* process, so *bond energy is always positive*:

$$\text{A—B}(g) \longrightarrow \text{A}(g) + \text{B}(g) \qquad \Delta_{\text{bond breaking}}H^\circ = \text{BE}_{\text{A—B}} \text{ (always} > 0)$$

The bond energy is the difference in energy between separated and bonded atoms (the potential energy difference between points 1 and 3, the energy "well" in Figure 8.13). *The same quantity of energy that is absorbed to break the bond is released when the bond forms.* Bond formation is an *exothermic* process, so *the sign of its enthalpy change is always negative*:

$$\text{A}(g) + \text{B}(g) \longrightarrow \text{A—B}(g) \qquad \Delta_{\text{bond forming}}H^\circ = -\text{BE}_{\text{A—B}} \text{ (always} < 0)$$

Table 8.2 lists the energies of some common bonds. By definition,

- *Stronger bonds are lower in energy (have a deeper energy well)*
- *Weaker bonds are higher in energy (have a shallower energy well)*

The energy of a given bond varies slightly from molecule to molecule, and even within the same molecule, so each value is an *average* bond energy.

3. *Bond length.* A covalent bond has a **bond length**, which is the distance between the nuclei of two bonded atoms. Thus, the bond length is related to the sum of the radii of the bonded atoms. In Figure 8.13, the bond length is the distance between the nuclei at the point of minimum energy (the bottom of the "well"), and Table 8.2 shows the lengths of some covalent bonds. Like bond energies, these lengths are *average* bond lengths in different substances. In fact, most atomic radii are calculated from measured bond lengths (see Figure 7.12C). Bond lengths for a series of similar bonds, such as the bonds in the halogens, increase with atomic size (Figure 8.16).

The order, energy, and length of a covalent bond are interrelated. Two nuclei are more strongly attracted to two shared pairs than to one shared pair, so double-bonded atoms are drawn closer together *and* are more difficult to pull apart than single-bonded atoms: *for a given pair of atoms, a higher bond order results in a shorter bond length and a higher bond energy.* Thus, as Table 8.3 shows, for a given pair of atoms, *a shorter bond is a stronger bond.*

In some cases, we can see a relation among atomic size, bond length, and bond energy by varying one of the atoms in a single bond while holding the other atom constant:

- *Variation within a group.* The trend in carbon-halogen single bond lengths, C—I > C—Br > C—Cl, parallels the trend in atomic size, I > Br > Cl, and is opposite to the trend in bond energy, C—Cl > C—Br > C—I.
- *Variation within a period.* Looking again at single bonds involving carbon, the trend in bond lengths, C—N > C—O > C—F, is opposite to the trend in bond energy, C—F > C—O > C—N.

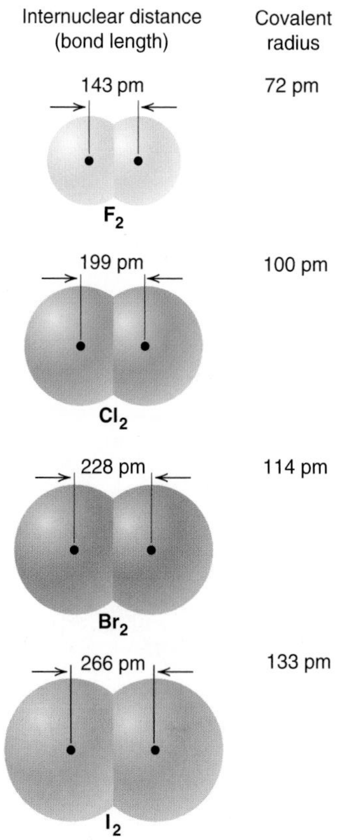

FIGURE 8.16 Bond length and covalent radius

TABLE 8.2 Average Bond Energies (kJ/mol) and Bond Lengths (pm)

Bond	Energy	Length	Bond	Energy	Length	Bond	Energy	Length	Bond	Energy	Length
Single Bonds											
H—H	432	74	N—H	391	101	Si—H	323	148	S—H	347	134
H—F	565	92	N—N	160	146	Si—Si	226	234	S—S	266	204
H—Cl	427	127	N—P	209	177	Si—O	368	161	S—F	327	158
H—Br	363	141	N—O	201	144	Si—S	226	210	S—Cl	271	201
H—I	295	161	N—F	272	139	Si—F	565	156	S—Br	218	225
			N—Cl	200	191	Si—Cl	381	204	S—I	~170	234
C—H	413	109	N—Br	243	214	Si—Br	310	216			
C—C	347	154	N—I	159	222	Si—I	234	240	F—F	159	143
C—Si	301	186							F—Cl	193	166
C—N	305	147	O—H	467	96	P—H	320	142	F—Br	212	178
C—O	358	143	O—P	351	160	P—Si	213	227	F—I	263	187
C—P	264	187	O—O	204	148	P—P	200	221	Cl—Cl	243	199
C—S	259	181	O—S	265	151	P—F	490	156	Cl—Br	215	214
C—F	453	133	O—F	190	142	P—Cl	331	204	Cl—I	208	243
C—Cl	339	177	O—Cl	203	164	P—Br	272	222	Br—Br	193	228
C—Br	276	194	O—Br	234	172	P—I	184	246	Br—I	175	248
C—I	216	213	O—I	234	194				I—I	151	266
Multiple Bonds											
C=C	614	134	N=N	418	122	C≡C	839	121	N≡N	945	110
C=N	615	127	N=O	607	120	C≡N	891	115	N≡O	631	106
C=O	745 (799 in CO_2)	123	O=O	498	121	C≡O	1070	113			

TABLE 8.3 The Relation of Bond Order, Bond Length, and Bond Energy

Bond	Bond Order	Average Bond Length (pm)	Average Bond Energy (kJ/mol)
C—O	1	143	358
C=O	2	123	745
C≡O	3	113	1070
C—C	1	154	347
C=C	2	134	614
C≡C	3	121	839
N—N	1	146	160
N=N	2	122	418
N≡N	3	110	945

Sample Problem 8.4 Comparing Bond Length and Bond Strength

Problem Without referring to Table 8.2, rank the bonds in each set in order of *decreasing* bond length and *decreasing* bond strength:

(a) S—F, S—Br, S—Cl **(b)** C=O, C—O, C≡O

Plan **(a)** S is singly bonded to three different halogen atoms, so the bond order is the same. Bond length increases and bond strength decreases as the halogen's atomic radius increases. **(b)** The same two atoms are bonded, but the bond orders differ. In this case, bond strength increases and bond length decreases as bond order increases.

Solution **(a)** Atomic size increases down a group, so F < Cl < Br:

Bond length: S—Br > S—Cl > S—F

Bond strength: S—F > S—Cl > S—Br

(b) By ranking the bond orders, C≡O > C=O > C—O, we obtain the following bond length and strength:

Bond length: C—O > C=O > C≡O

Bond strength: C≡O > C=O > C—O

Check From Table 8.2, we see that the rankings are correct.

Comment For bonds involving pairs of different atoms, as in part (a), remember that *the relationship between length and strength holds* ***only*** *for single bonds* and not in every case, so apply this relationship carefully.

Follow-Up Problem 8.4 Rank the bonds in each set in order of *increasing* bond length and *increasing* bond strength:

(a) Si—F, Si—C, Si—O **(b)** N=N, N—N, N≡N

How the Model Explains the Properties of Covalent Substances

The covalent bonding model proposes that electron sharing between pairs of atoms leads to *strong, localized bonds.* Most, but not all, covalent substances consist of individual molecules. However, there are chemical compounds in which the atoms are joined by covalent bonds in a continuous network, forming giant molecular lattice structures. These macromolecules are called **network covalent solids**. The bonding between the atoms in a network covalent solid extends infinitely in three dimensions. In contrast, *molecular covalent substances* have very different physical properties, because different types of forces give rise to them:

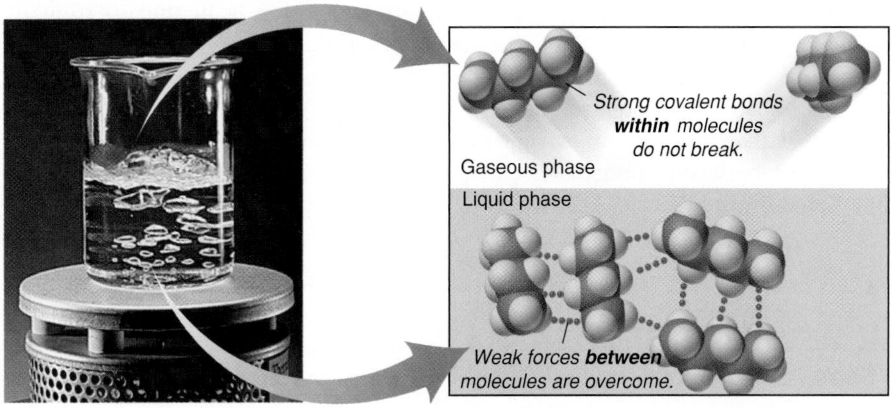
Strong covalent bonds
within molecules
do not break.
Gaseous phase
Liquid phase
Weak forces **between**
molecules are overcome.

FIGURE 8.17 Strong forces within molecules and weak forces between them

1. *Physical properties of **molecular** covalent substances.* At first glance, the model seems inconsistent with the physical properties of covalent substances. Most are gases (such as methane and ammonia), liquids (such as benzene and water), or low-melting solids (such as sulfur and paraffin wax). If covalent bonds are so strong (~200 to 500 kJ/mol), why do covalent substances melt and boil at such low temperatures?

To answer this, we consider two different forces: (1) *strong bonding forces*, which hold the atoms together within the molecule, and (2) *weak intermolecular forces*, which act between separate molecules in the sample. It is the weak forces *between* the molecules that account for the physical properties of *molecular* covalent substances. For example, look what happens when pentane (C_5H_{12}) boils (Figure 8.17): the weak forces *between* the pentane molecules are overcome, but not the strong C—C and C—H bonds *within* each pentane molecule.

2. *Physical properties of **network** covalent solids.* As mentioned before, some covalent substances do not consist of separate molecules. Rather, these *network* covalent solids are held together by covalent bonds *between atoms throughout the sample*, and their properties *do* reflect the strength of the covalent bonds. Some examples are quartz and diamond (Figure 8.18), as well as graphite. Quartz (SiO_2; *top*) has silicon-oxygen covalent bonds in three dimensions; no separate SiO_2 molecules exist. Quartz is very hard and melts at 1550°C. Diamond (*bottom*) has covalent bonds connecting each carbon atom to four others. It is the hardest natural substance known and melts at around 3550°C. In general, *network* covalent solids have the following properties:

- They have a high melting point, due to the large amount of energy required to rearrange the covalent bonds.
- They are hard, due to the strong covalent bonds throughout the lattice. (However, the layers of carbon atoms in graphite can be easily displaced, allowing the substance to be malleable.)
- They are generally insoluble in any solvent, due to the difficulty of solvating a very large molecule.

As mentioned before, covalent bonds *are* strong, but most covalent substances consist of separate molecules with weak forces between them. (We will discuss intermolecular forces in detail in Chapter 11.)

3. *Electrical conductivity.* An electric current is carried by either mobile electrons or mobile ions. Most covalent substances are poor electrical conductors, whether melted or dissolved, because their electrons are localized as either shared or unshared pairs, and no ions are present. Infrared spectroscopy, described in Chapter 22, is a tool that is used widely to study the types of bonds in covalent substances.

International Symbols Labels—An **explosive hazard** (e.g., TNT, nitroglycerin, RDX) involves reactions containing reactants with weak chemical bonds and products with very strong bonds. These reactions will be the most exothermic.

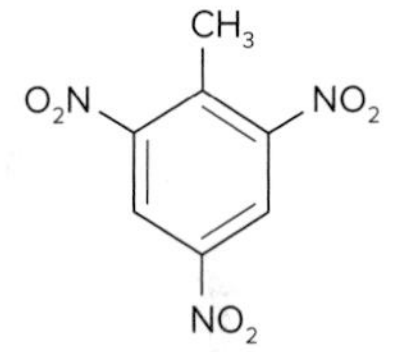

TNT (trinitrotoluene or 2-methyl-1,3,5-trinitrobenzene)

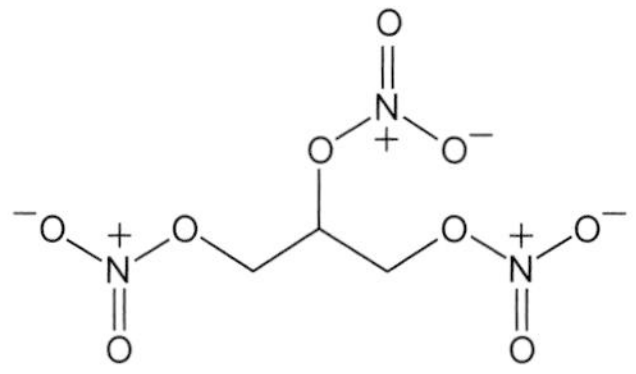

Nitroglycerin (1,2,3-trinitroxypropane)

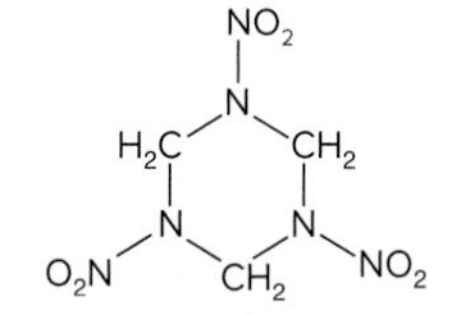

RDX (Research Department explosive or 1,3,5-trinitroperhydro-1,3,5-triazine)

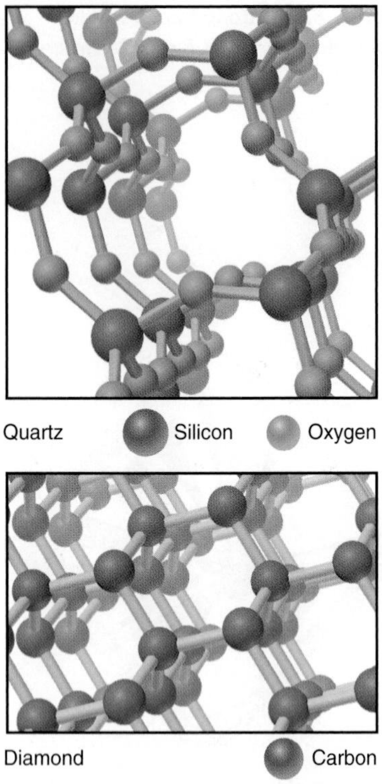

FIGURE 8.18 Covalent bonds of two network covalent solids: quartz and diamond

One of the many applications of bonding theories is in the creation of designer drug molecules. When the drugs enter the human body, they might cause it to react in a certain way by bonding at receptors, ion channels, enzymes, and cell transporter proteins at specific binding sites. These interactions are very basic, just like those of other types of chemical bonding. They occur through attractions between opposite charges, or between hydrogen atoms and polar functional groups.

SUMMARY OF SECTION 8.3

- A shared, localized pair of valence electrons holds together the nuclei of two atoms in a covalent bond, filling each atom's outer level.
- Bond order is the number of shared pairs between two atoms. Bond energy is the energy that is absorbed to separate the atoms; the same quantity of energy is released when the bond forms. Bond length is the distance between the nuclei of the atoms.
- For a given pair of atoms, bond order is directly related to bond energy and inversely related to bond length.
- Molecular covalent substances are soft and have low melting points because of the weak forces *between* the molecules, not the strong bonding forces *within* them. Network covalent solids are hard and have high melting points because covalent bonds join all the atoms in the sample.
- Most covalent substances have low electrical conductivity because their electrons are localized and ions are absent.

8.4 Bond Energy and Chemical Change

The relative strengths of the bonds in the reactants and products determine whether heat is released or absorbed in a chemical reaction. The change in *bond energy* is one of two factors that determine whether the reaction occurs at all (see Chapters 5 and 18). In this section, we discuss the origin of the enthalpy of reaction ($\Delta_r H^\circ$), use bond energies to calculate it, and look at the energy available from fuels and foods.

Changes in Bond Energy: Where Does $\Delta_r H^\circ$ Come From?

In Chapter 5, we discussed the heat involved in a chemical change, but we never asked a central question: where does the enthalpy of reaction ($\Delta_r H^\circ$) come from? For example, when 1 mol of H_2 and 1 mol of F_2 react to form 2 mol of HF at 101.3 kPa and 298 K, where does the 546 kJ/mol come from?

$$H_2(g) + F_2(g) \longrightarrow 2HF(g) + 546\ \text{kJ/mol}$$

We find the answer by looking closely at the energies of the molecules involved. A system's total internal energy is composed of its kinetic energy and its potential energy. Let us see how these change during the formation of HF:

- *Kinetic energy.* The most important contributions to the kinetic energy are the molecules' movements in space, as well as their rotations and vibrations. However, since kinetic energy is proportional to temperature, which is constant at 298.15 K, it does not change during the reaction.
- *Potential energy.* The most important contributions to the potential energy are phase changes and changes in the attraction between vibrating atoms, between the nucleus and electrons (and between the electrons) in each atom, between the protons and neutrons in each nucleus, and between the nuclei and the shared

electron pair in each bond. However, there are no phase changes, vibrational forces vary only slightly as the bonded atoms change, and forces within the atoms and nuclei do not change at all. The only significant change in potential energy comes from changes in the attraction between the nuclei and the shared electron pair—the bond energy.

Thus, our answer to "Where does $\Delta_r H°$ come from?" is that it does not really "come from" anywhere: *the heat released or absorbed during a chemical change is due to differences between reactant bond energies and product bond energies.*

Using Bond Energies to Calculate $\Delta_r H°$

Hess's law allows us to think of any reaction as a two-step process, whether or not it actually occurs this way:

1. A quantity of heat is *absorbed* ($\Delta H° > 0$) to break the reactant bonds and form separate atoms.
2. A different quantity of heat is then *released* ($\Delta H° < 0$) when the atoms form product bonds.

The sum (symbolized by Σ) of these enthalpy changes is the enthalpy of reaction, $\Delta_r H°$:

$$\Delta_r H° = \Sigma\Delta_{\text{reactant bonds broken}}H° + \Sigma\Delta_{\text{product bonds formed}}H° \tag{8.2}$$

- In an exothermic reaction, the magnitude of $\Delta_{\text{product bonds formed}}H°$ is *greater* than the magnitude of $\Delta_{\text{reactant bonds broken}}H°$, so the sum, $\Delta_r H°$, is *negative* (heat is released).
- In an endothermic reaction, the opposite situation is true. The magnitude of $\Delta_{\text{product bonds formed}}H°$ is *smaller* than the magnitude of $\Delta_{\text{reactant bonds broken}}H°$, so $\Delta_r H°$ is *positive* (heat is absorbed).

An equivalent form of Equation 8.2 uses bond energies:

$$\Delta_r H° = \Sigma BE_{\text{reactant bonds broken}} - \Sigma BE_{\text{product bonds formed}}$$

(We need the minus sign because all bond energies are positive.)

Typically, only certain bonds break and form during a reaction. However, with Hess's law, the following method is simpler for calculating $\Delta_r H°$:

1. Break *all* the reactant bonds to obtain the individual atoms.
2. Use the atoms to form *all* the product bonds.
3. Add the bond energies, with the appropriate signs, to obtain the enthalpy of reaction.

(This method assumes that the reactants and products do not change their physical state, and it is *only valid for gas-phase reactions.* Additional heat is involved when phase changes occur, as we will discuss in Chapter 11.)

Let us use this method to calculate $\Delta_r H°$ for two reactions:

1. *Formation of HF.* When 1 mol of H—H bonds and 1 mol of F—F bonds absorb energy and break, the 2 mol of H atoms and 2 mol of F atoms form 2 mol of H—F bonds, which releases energy (Figure 8.19). We find the bond energy values in Table 8.2 and use a positive sign for bonds broken and a negative sign for bonds formed.

Bonds broken:

$$1 \times \text{H—H} = (1)(432\ \text{kJ/mol}) = 432\ \text{kJ/mol}$$
$$1 \times \text{F—F} = (1)(159\ \text{kJ/mol}) = 159\ \text{kJ/mol}$$
$$\Sigma\Delta_{\text{reactant bonds broken}}H° = 591\ \text{kJ/mol}$$

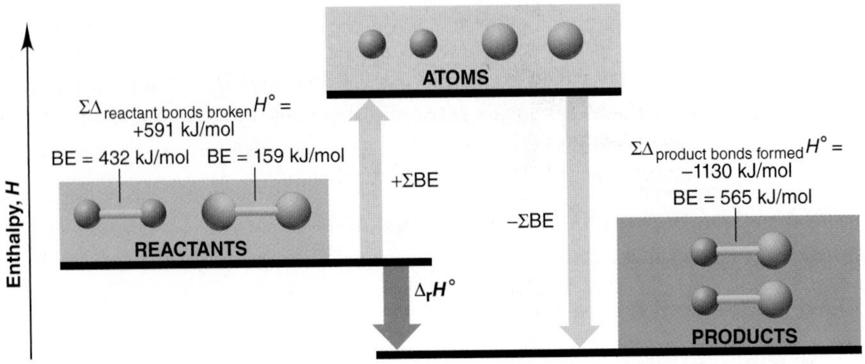

FIGURE 8.19 Using bond energies to calculate $\Delta_r H°$ for HF formation

Bonds formed:

$$2 \times \text{H—F} = (2)(-565\text{ kJ/mol}) = \Sigma\Delta_{\text{product bonds formed}}H° = -1130\text{ kJ/mol}$$

Applying Equation 8.2 gives

$$\Delta_r H° = \Sigma\Delta_{\text{reactant bonds broken}}H° + \Sigma\Delta_{\text{product bonds formed}}H°$$
$$= 591\text{ kJ/mol} + (-1130\text{ kJ/mol}) = -539\text{ kJ/mol}$$

The small discrepancy between this bond energy value (−539 kJ/mol) and the value from tabulated $\Delta H°$ values (−546 kJ/mol) is due to variations in the experimental method.

2. *Combustion of CH_4.* In this more complicated reaction, all the bonds in CH_4 and O_2 break, and the atoms form all the bonds in CO_2 and H_2O (Figure 8.20). Once again, we use Table 8.2 and appropriate signs for bonds broken and bonds formed.

Bonds broken:

$$4 \times \text{C—H} = (4)(413\text{ kJ/mol}) = 1652\text{ kJ/mol}$$
$$2 \times O_2 = (2)(498\text{ kJ/mol}) = 996\text{ kJ/mol}$$
$$\Sigma\Delta_{\text{reactant bonds broken}}H° = 2648\text{ kJ/mol}$$

FIGURE 8.20 Using bond energies to calculate $\Delta_r H°$ for the combustion of methane

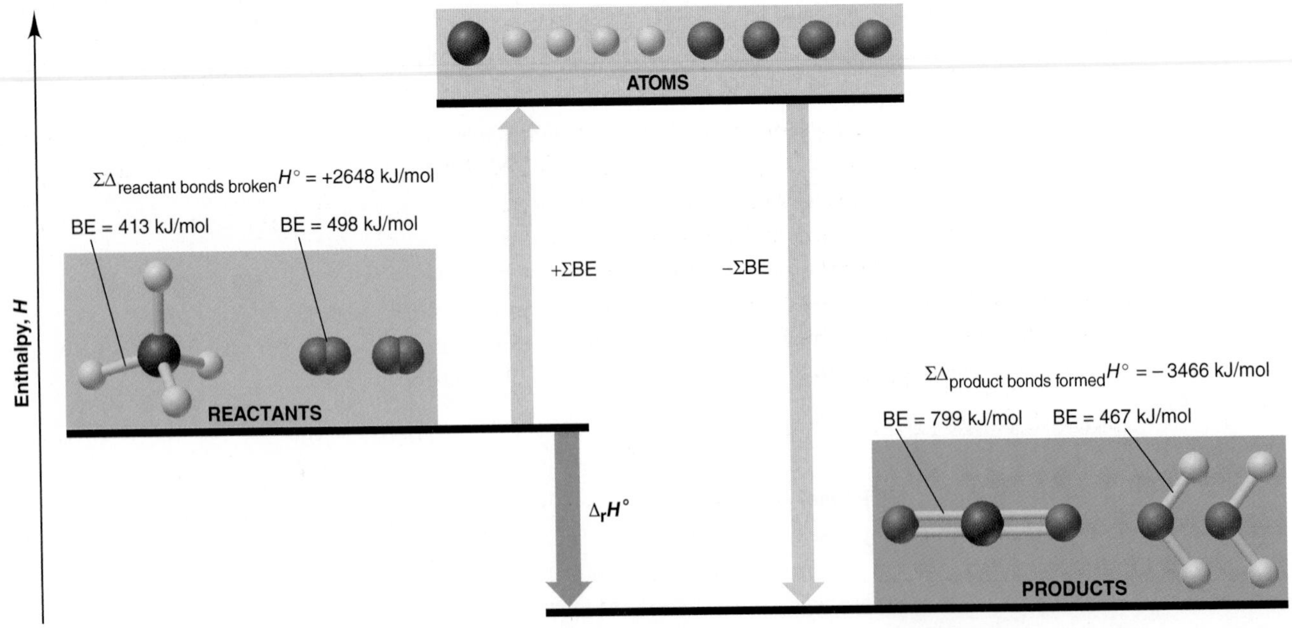

Bonds formed:

$$2 \times \text{C}{=}\text{O} = (2)(-799\ \text{kJ/mol}) = -1598\ \text{kJ/mol}$$
$$4 \times \text{O—H} = (4)(-467\ \text{kJ/mol}) = -1868\ \text{kJ/mol}$$
$$\Sigma\Delta_{\text{product bonds formed}}H° = -3466\ \text{kJ/mol}$$

Applying Equation 8.2 gives

$$\Delta_r H° = \Sigma\Delta_{\text{reactant bonds broken}}H° + \Sigma\Delta_{\text{product bonds formed}}H°$$
$$= 2648\ \text{kJ/mol} + (-3466\ \text{kJ/mol}) = -818\ \text{kJ/mol}$$

In addition to variations in the experimental method, there is a more basic reason for the discrepancy between the $\Delta_r H°$ obtained from bond energies (−818 kJ/mol) and the value obtained by calorimetry (−802 kJ/mol; Section 5.3). A bond energy is an *average* value for a given bond in many compounds. The value *in a particular substance* is usually close, but not equal, to the average value. For example, 413 kJ/mol is the average value for the C—H bond energy in many molecules. In fact, 415 kJ is actually required to break 1 mol of C—H bonds in methane, and 1660 kJ is required to break 4 mol of these bonds, which gives a $\Delta_r H°$ closer to the calorimetric value. Thus, it is not surprising to find small discrepancies between the $\Delta_r H°$ values obtained in different ways.

Sample Problem 8.5 Using Bond Energies to Calculate $\Delta_r H°$

Problem (a) Calculate $\Delta_r H°$ for the chlorination of methane to form trichloromethane (chloroform):

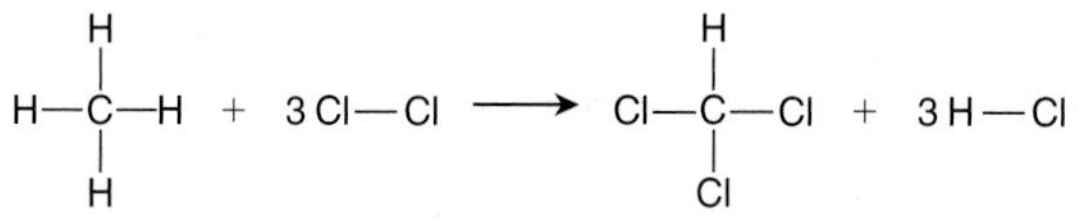

(b) Calculate $\Delta_r H°$ for the reaction of ethene with hydrobromic acid to form bromoethane:

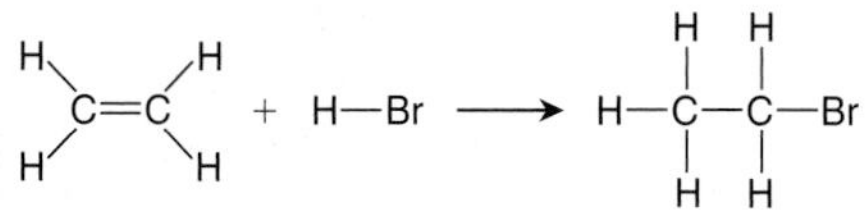

Plan *All* the reactant bonds break, and *all* the product bonds form. We find the bond energies in Table 8.2 and substitute the two sums, with correct signs, into Equation 8.2.

Solution First we need to find the standard enthalpy changes for bonds broken and bonds formed.

(a) For bonds broken, the bond energy values are

$$4 \times \text{C—H} = (4)(413\ \text{kJ/mol}) = 1652\ \text{kJ/mol}$$
$$3 \times \text{Cl—Cl} = (3)(243\ \text{kJ/mol}) = 729\ \text{kJ/mol}$$
$$\Sigma\Delta_{\text{bonds broken}}H° = 2381\ \text{kJ/mol}$$

For bonds formed, the bond energy values are

$$3 \times \text{C—Cl} = (3)(-339\ \text{kJ/mol}) = -1017\ \text{kJ/mol}$$
$$1 \times \text{C—H} = (1)(-413\ \text{kJ/mol}) = -413\ \text{kJ/mol}$$
$$3 \times \text{H—Cl} = (3)(-427\ \text{kJ/mol}) = -1281\ \text{kJ/mol}$$
$$\Sigma\Delta_{\text{bonds formed}}H° = -2711\ \text{kJ/mol}$$

Calculate $\Delta_r H°$:

$$\begin{aligned}\Delta_r H° &= \Sigma\Delta_{\text{bonds broken}}H° + \Sigma\Delta_{\text{bonds formed}}H°\\ &= 2381 \text{ kJ/mol} + (-2711 \text{ kJ/mol})\\ &= -330 \text{ kJ/mol}\end{aligned}$$

Check The signs of the enthalpy changes are correct: $\Sigma\Delta_{\text{bonds broken}}H° > 0$ and $\Sigma\Delta_{\text{bonds formed}}H° < 0$. More energy is released than absorbed, so $\Delta_r H°$ is negative:

$$\sim 2400 \text{ kJ/mol} + [\sim(-2700 \text{ kJ/mol})] = -300 \text{ kJ/mol}$$

(b) For bonds broken, the bond energy values are

$$\begin{aligned}4 \times \text{C—H} &= (4)(413 \text{ kJ/mol}) = 1652 \text{ kJ/mol}\\ 1 \times \text{C═C} &= (1)(614 \text{ kJ/mol}) = 614 \text{ kJ/mol}\\ 1 \times \text{H—Br} &= (1)(363 \text{ kJ/mol}) = 363 \text{ kJ/mol}\\ \Sigma\Delta_{\text{bonds broken}}H° &= 2629 \text{ kJ/mol}\end{aligned}$$

For bonds formed, the bond energy values are

$$\begin{aligned}5 \times \text{C—H} &= (5)(-413 \text{ kJ/mol}) = -2065 \text{ kJ/mol}\\ 1 \times \text{C—Br} &= (1)(-276 \text{ kJ/mol}) = -276 \text{ kJ/mol}\\ 1 \times \text{C—C} &= (1)(-347 \text{ kJ/mol}) = -347 \text{ kJ/mol}\\ \Sigma\Delta_{\text{bonds formed}}H° &= -2688 \text{ kJ/mol}\end{aligned}$$

Calculate $\Delta_r H°$:

$$\begin{aligned}\Delta_r H° &= \Sigma\Delta_{\text{bonds broken}}H° + \Sigma\Delta_{\text{bonds formed}}H°\\ &= 2629 \text{ kJ/mol} + (-2688 \text{ kJ/mol})\\ &= -59 \text{ kJ/mol}\end{aligned}$$

Check The signs of the enthalpy changes are correct: $\Sigma\Delta_{\text{bonds broken}}H° > 0$ and $\Sigma\Delta_{\text{bonds formed}}H° < 0$. More energy is released than absorbed, so $\Delta_r H°$ is negative.

Follow-Up Problem 8.5 One of the most important industrial reactions is the formation of ammonia from its elements:

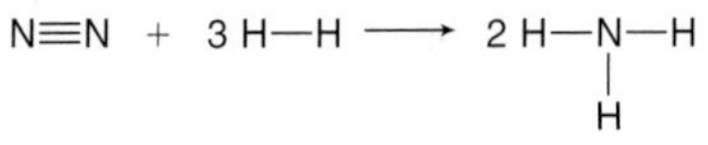

Use bond energies to calculate $\Delta_r H°$.

Bond Strengths and the Heat Released from Fuels and Foods

A *fuel* is a material that reacts with atmospheric oxygen to release energy *and* is available at a reasonable cost. The most common fuels for machines are hydrocarbons and coal, and the most common fuels for organisms are fats and carbohydrates. All of these fuels are composed of large organic molecules with many C—C and C—H bonds and fewer C—O and O—H bonds. According to our two-step approach, when the fuel reacts with O_2, all the bonds break, and the C, H, and O atoms form C═O and O—H bonds in the products, CO_2 and H_2O. Because their combustion is exothermic, the total of the bond energies in the products is *greater* than the total in the reactants. *Weaker bonds (less stable, more reactive) are easier to break than stronger bonds (more stable, less reactive) because they are already higher in energy.* Therefore, the bonds in CO_2 and H_2O are stronger (lower energy, more stable) than the bonds in gasoline (or cooking oil) and O_2 (weaker, higher energy, less stable).

Fuels with more weak bonds yield more energy than fuels with fewer weak bonds. When a hydrocarbon burns, C—C and C—H bonds break; when an alcohol burns, C—O and O—H bonds also break. Table 8.2 shows that the sum for C—C and C—H bonds (760 kJ/mol) is less than the sum for C—O and O—H bonds (825 kJ/mol).

Therefore, it takes more energy to break the bonds of a fuel with a lot of C—O and O—H bonds. In general, *a fuel with fewer bonds to O releases more energy* (Figure 8.21).

Both fats and carbohydrates serve as high-energy foods and consist of chains or rings of C atoms attached to H atoms, with some C—O, C═O, and O—H bonds (*shown in red below*):

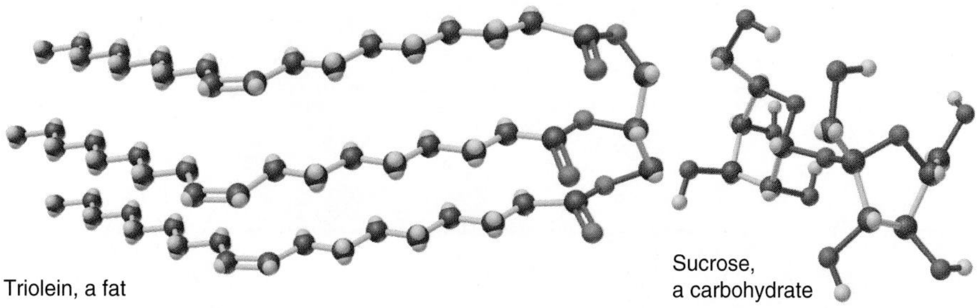

Carbohydrates have fewer chains of C atoms and bonds to H, and more bonds to O. Fats contain more Calories per gram than carbohydrates because fats have fewer bonds to O (Table 8.4).

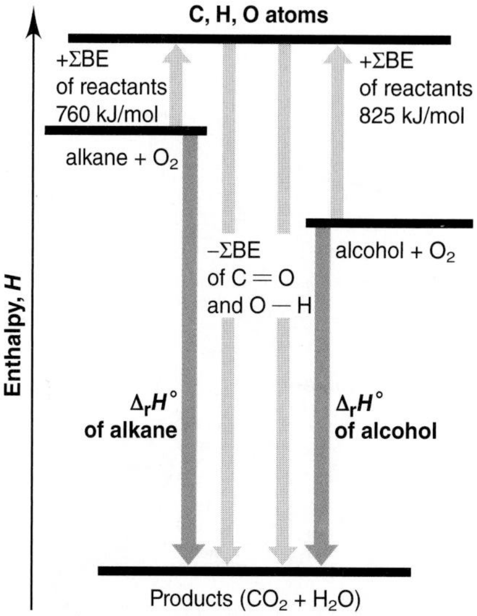

FIGURE 8.21 Relative bond strength and energy from fuels

TABLE 8.4 Enthalpies of Reaction for the Combustion of Some Foods

Substance	$\Delta_rH°$ (kJ/g)
Fats	
Vegetable oil	−37.0
Margarine	−30.1
Butter	−30.0
Carbohydrates	
Table sugar (sucrose)	−16.2
Brown rice	−14.9
Maple syrup	−10.4

SUMMARY OF SECTION 8.4

- The only component of internal energy that changes significantly during a reaction is the bond energies of the reactants and products, and this change appears as the enthalpy of reaction, $\Delta_rH°$.
- A reaction involves breaking reactant bonds and forming product bonds. Applying Hess's law, we use tabulated bond energies to calculate $\Delta_rH°$.
- Bonds in fuels are weaker (less stable, higher energy) than bonds in the combustion products. Fuels with more weak bonds release more energy than fuels with fewer weak bonds.

8.5 Between the Extremes: Electronegativity and Bond Polarity

Scientific models are idealized descriptions of reality. The ionic and covalent bonding models portray compounds as formed by *either* complete electron transfer *or* complete electron sharing. However, in real substances, most atoms are joined by *polar covalent bonds*—partly ionic and partly covalent (Figure 8.22). As mentioned before, we are unlikely to find a bond that is either 100% covalent or 100% ionic. In this section, we explore the "in-between" nature of these bonds and its importance in the properties of substances. To simplify calculations, we will consider that the oxidation number assumes the bond is 100 percent ionic, and that the formal charges assume 100 percent covalent bonding.

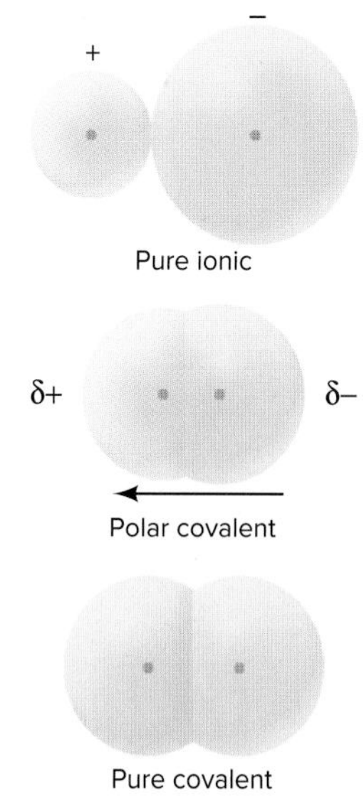

FIGURE 8.22 Bonding between the models. Pure ionic bonding (*top*) and pure covalent bonding (*bottom*) are far less common than polar covalent bonding (*middle*).

Electronegativity

As discussed in Section 7.3, electronegativity (χ) is the relative ability of a bonded atom to attract shared electrons.* We might expect the H—F bond energy to be the average of an H—H bond (432 kJ/mol) and an F—F bond (159 kJ/mol), or 296 kJ/mol. However, the actual HF bond energy is 565 kJ/mol, or 269 kJ/mol *higher*! To explain this difference, the American chemist Linus Pauling reasoned as follows: If F attracted the shared electron pair more strongly than H (that is, if F were more *electronegative* than H), the electrons would spend more time closer to F. This

*Electronegativity refers to a *bonded* atom attracting a shared pair of electrons; electron affinity refers to a gaseous atom gaining an electron to form an anion. Elements with a high χ also have a highly negative EA.

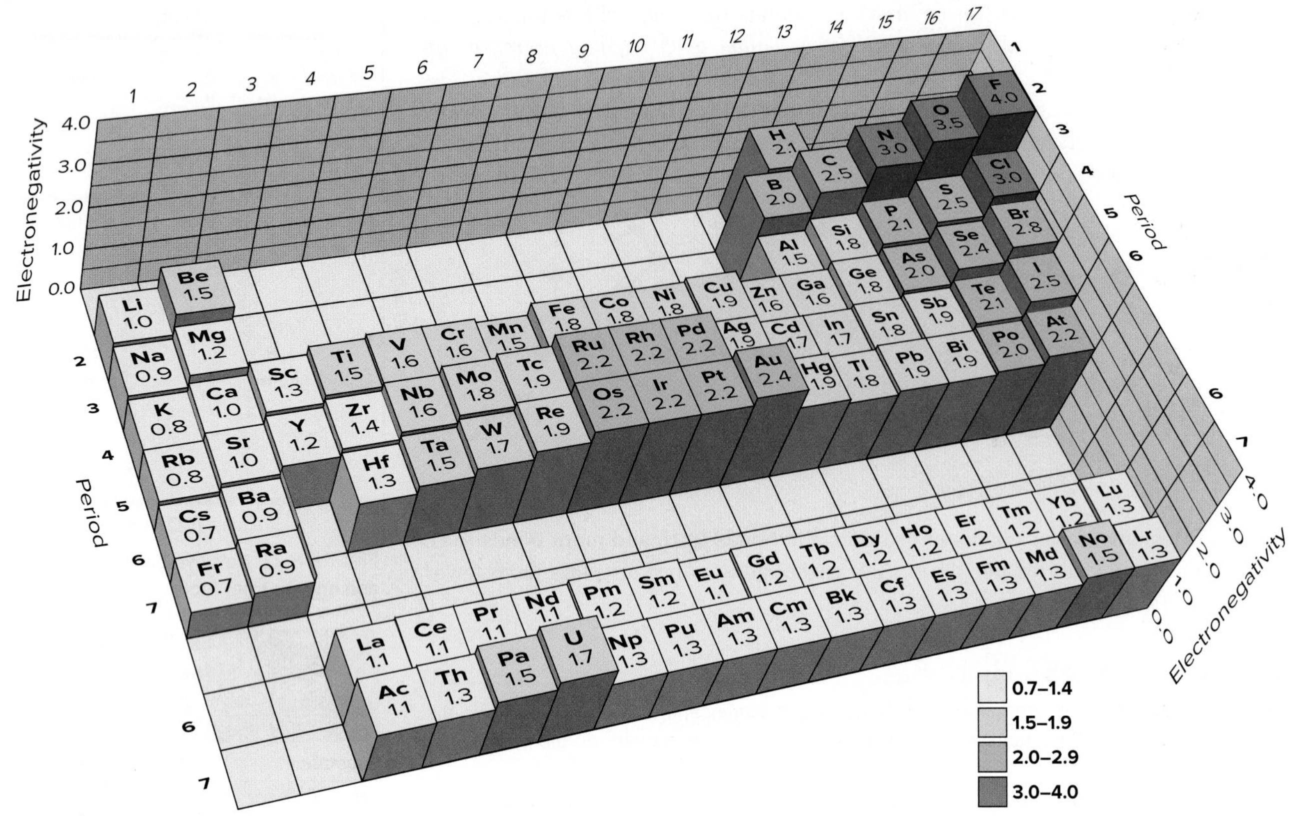

FIGURE 8.23 The Pauling electronegativity (χ) scale. The height of each post is proportional to the χ value, which is shown on top. The key has several χ cutoffs. In the main groups, χ *increases* across and *decreases* down the table. The transition and inner transition elements show little change in χ. Here, hydrogen is placed near elements with similar χ values.

unequal sharing would make the F end of the bond partially negative and the H end partially positive. The electrostatic attraction between these partially charged bond "poles" would *increase* the energy required to break the bond.

From studies with many other compounds, Pauling derived a scale of *relative χ values* based on fluorine having the highest χ value, 4.0 (Figure 8.23). Other chemists, such as Allred and Rochow, interpreted the definition of electronegativity as corresponding to the electrostatic force of attraction exerted by an atom on its valence electrons, suggesting a scale of electronegativity based on this definition. The Allred-Rochow electronegativity is often denoted as χ_{AR}. The force of attraction is given by force $= \frac{e^2 Z_{eff}}{r^2}$, where r is the distance between the nucleus and the electron (covalent radius), e is the charge of an electron, and Z_{eff} is the effective charge at the electron due to the nucleus and its surrounding electrons. The quantity $\frac{Z_{eff}}{r^2}$ correlates well with the Pauling electronegativity scale, and the two scales can be made to coincide by expressing the Allred-Rochow electronegativity as

$$\chi_{AR} = 0.744 + \frac{0.359 Z_{eff}}{r^2}$$

Trends in Electronegativity In general, *electronegativity is inversely related to atomic size* because the nucleus of a smaller atom is closer to the shared pair than the nucleus of a larger atom is, so it attracts the electrons more strongly (Figure 8.24):

- *Down a main group*, electronegativity (the height of a post) decreases as size (the hemisphere on top of the post) increases.
- *Across a period of main-group elements*, electronegativity increases.
- Nonmetals are *more* electronegative than metals.

The most electronegative element is fluorine, with oxygen a close second. Thus, except when it bonds with fluorine, oxygen always pulls bonding electrons toward itself. The least electronegative element is francium, in the lower left corner of the

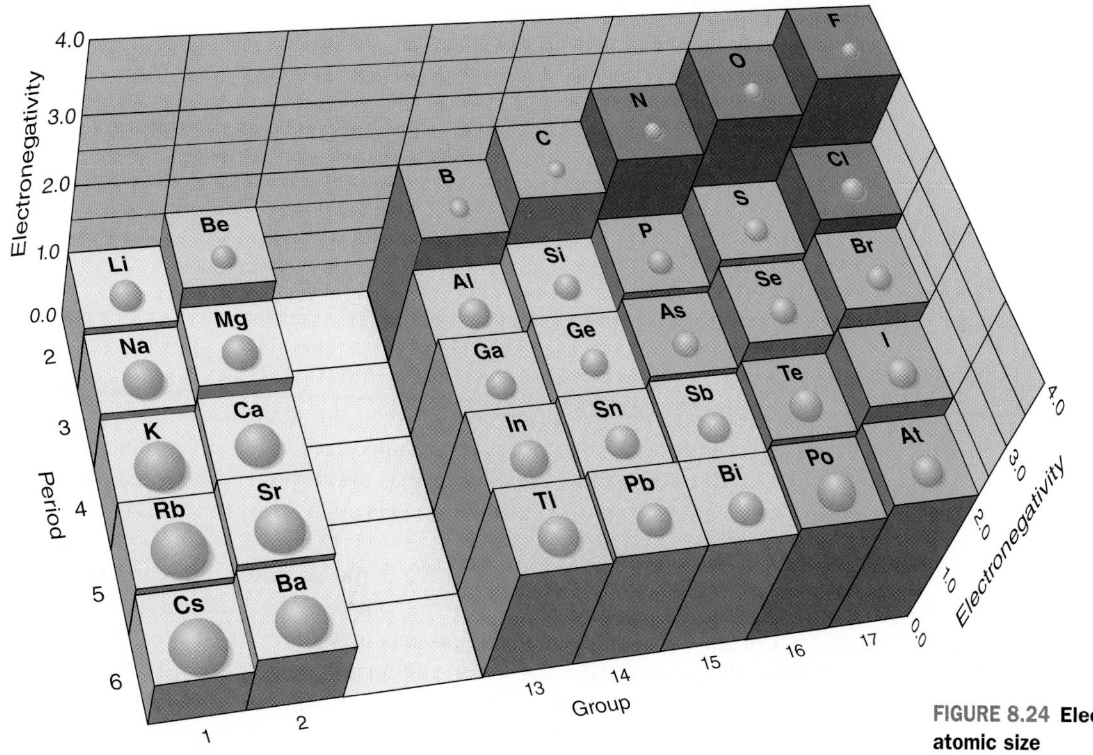

FIGURE 8.24 Electronegativity and atomic size

periodic table, but it is radioactive and extremely rare, so, for all practical purposes, cesium is the least electronegative.*

Electronegativity and Oxidation Number The oxidation number is the charge that an atom would have if all the ligands (the atoms attached to it) were removed, along with the electron pairs that it is sharing with other atom(s). An important use of electronegativity is in determining an atom's oxidation number (O.N.):

1. The more electronegative atom in a bond is assigned *all* the *shared* electrons; the less electronegative atom is assigned *none.*
2. Each atom in a bond is assigned *all* of its *unshared* electrons.
3. The oxidation number is given by

$$\text{O.N.} = \text{no. of valence e}^- - (\text{no. of shared e}^- + \text{no. of unshared e}^-)$$

In HCl, for example, Cl is more electronegative than H. Cl has 7 valence electrons and is assigned 8 (2 shared + 6 unshared), so its O.N. is 7 − 8 = −1. The H atom has 1 valence electron and is assigned none, so its O.N. is 1 − 0 = +1.

Bond Polarity and Partial Ionic Character

Whenever atoms of different electronegativities form a bond, such as H (2.1) and F (4.0) in HF, the bonding pair is shared *unequally.* This unequal distribution of electron density results in a **polar covalent bond**. A polar covalent bond is depicted by a polar arrow (⟶) pointing toward the partially positive pole (to be consistent with the IUPAC statement that the direction of the dipole moment is from the negative charge to the positive charge) or by δ+ and δ− symbols (see Figure 3.17):

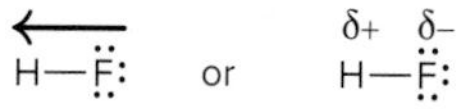

In the H—H and F—F bonds, where the atoms are identical, the bonding pair is shared *equally*, and a **nonpolar covalent bond** results. In Figure 8.25, relief maps show the distribution of the electron density in H_2, F_2, and HF.

*In 1934, the American physicist Robert S. Mulliken developed electronegativity values based on atomic properties: $\chi = \frac{(\text{IE} - \text{EA})}{2}$. By this approach as well, fluorine, with a high ionization energy (IE) and a large negative electron affinity (EA), has a high χ, and cesium, with a low IE and a small EA, has a low χ.

FIGURE 8.25 Electron density distributions in H_2, F_2, and HF. In HF, the electron density shifts from H to F. (The electron density peak for F has been cut off to limit the height of the figure.)

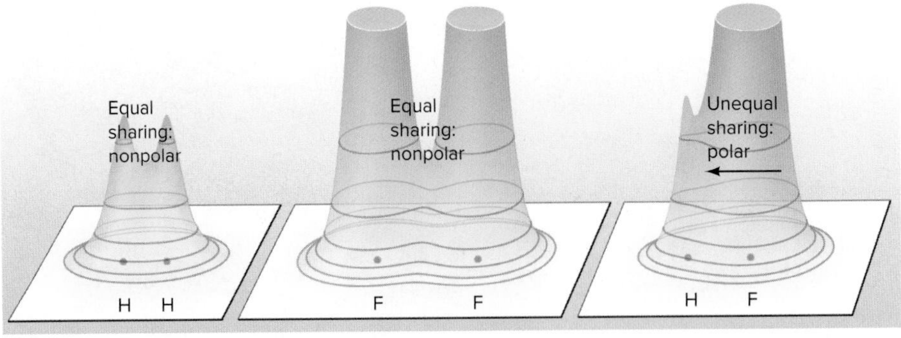

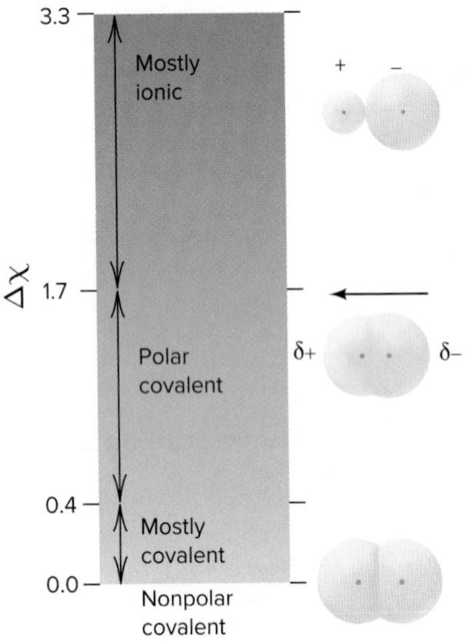

FIGURE 8.26 Δχ ranges for classifying the partial ionic character of bonds

The Importance of the Electronegativity Difference (Δχ) The **electronegativity difference (Δχ)**, the difference between the χ values of bonded atoms, is directly related to a bond's polarity. It ranges from 0.0 in a diatomic element, such as H_2, O_2, or Cl_2, all the way up to 3.3, the difference between the more electronegative atom, F (4.0), and the less electronegative, Cs (0.7), in the ionic compound CsF.

Another parameter closely related to Δχ is the **partial ionic character** of a bond: *a greater Δχ results in larger partial charges and a higher partial ionic character.* Consider three Cl-containing gaseous molecules: for LiCl(*g*), Δχ is 3.0 − 1.0 = 2.0; for HCl(*g*), it is 3.0 − 2.1 = 0.9; and for $Cl_2(g)$, it is 3.0 − 3.0 = 0.0. Thus, the bond in LiCl has more ionic character than the bond in HCl, which has more ionic character than the bond in Cl_2.

Here are two approaches that quantify ionic character. Both use arbitrary cutoffs, which is not really consistent with the actual gradation in bonding:

1. *Δχ range.* This approach divides bonds into mostly ionic, polar covalent, mostly covalent, and nonpolar covalent based on a range of Δχ values (Figure 8.26).

2. *Percent ionic character.* This approach is based on the behaviour of a diatomic molecule in an electric field. A plot of *percent ionic character* versus Δχ for several gaseous molecules shows that, as expected, *percent ionic character generally increases with Δχ* (Figure 8.27A). A value of 50% divides ionic bonds from covalent bonds. Note that a substance like $Cl_2(g)$ has 0% ionic character, but no substances have 100% ionic character: *electron sharing occurs to some extent in every bond,* even in an alkali halide (Figure 8.27B).

FIGURE 8.27 Percent ionic character as a function of Δχ. A. Δχ correlates with ionic character. **B.** Even in highly ionic LiF (Δχ = 3.0), the relief map shows some electron sharing between the ions.

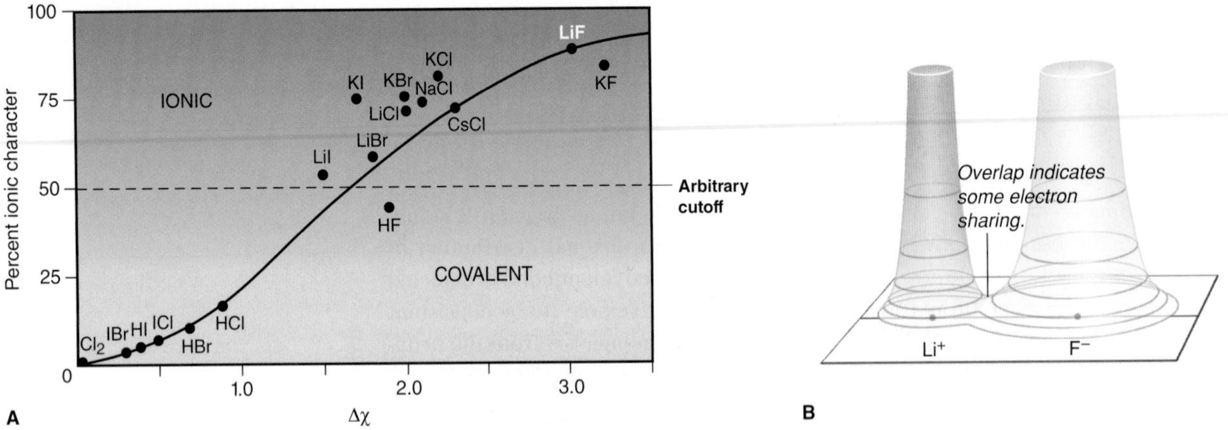

Sample Problem 8.6 Determining Bond Polarity from χ Values

Problem **(a)** Use a polar arrow to indicate the polarity of each bond: N—H, F—N, I—Cl.

(b) Rank the following bonds in order of *increasing* polarity and *decreasing* percent ionic character: H—N, H—O, H—C.

Plan **(a)** We use Figure 8.23 to find the χ values for the atoms and point the polar arrow toward the less electronegative atom. **(b)** To rank the bond polarity, we determine $\Delta\chi$: the higher the value, the greater the polarity. Percent ionic character is also directly related to $\Delta\chi$ (and bond polarity); it decreases in the opposite order that polarity increases.

Solution **(a)** χ of N = 3.0 and χ of H = 2.1, so $\overrightarrow{\text{N—H}}$

χ of F = 4.0 and χ of N = 3.0, so $\overrightarrow{\text{F—N}}$

χ of I = 2.5 and χ of Cl = 3.0, so $\overleftarrow{\text{I—Cl}}$

(b) The $\Delta\chi$ values are 0.9 for H—N, 1.4 for H—O, and 0.4 for H—C.

The order of *increasing* bond polarity is H—C < H—N < H—O.

The order of *decreasing* percent ionic character is H—O > H—N > H—C.

Check In (b), we can check the order of bond polarity using periodic trends. Each bond involves H and a period 2 atom. Since size decreases and χ increases across a period, the polarity is greatest for the bond to O (farthest to the right in period 2).

Comment In the following section, you will see that bond polarity contributes to the overall polarity of a molecule, which is a major factor for determining behaviour.

Follow-Up Problem 8.6 Arrange each set of bonds in order of increasing polarity, and indicate bond polarity with $\delta+$ and $\delta-$ symbols:

(a) Cl—F, Br—Cl, Cl—Cl **(b)** Si—Cl, P—Cl, S—Cl, Si—Si

Bonding Changes across a Period

A metal and a nonmetal—elements from the left and right sides of the periodic table—have a relatively large $\Delta\chi$ and typically form an ionic compound. Two nonmetals—both from the right side of the periodic table—have a small $\Delta\chi$ and form a covalent compound. When we combine chlorine with each of the other period 3 elements, starting with sodium, we observe a steady decrease in $\Delta\chi$ and a gradation in bond type from ionic through polar covalent to nonpolar covalent.

Figure 8.28 shows an electron density relief map of a bond in each of the common period 3 chlorides. Note the steady increase in the height of electron density *between* the peaks—the bonding region—which indicates an *increase in electron sharing*. Figure 8.29 shows samples of common period 3 chlorides—NaCl, $MgCl_2$, $AlCl_3$, $SiCl_4$, PCl_3, and SCl_2, as well as Cl_2—along with the change in $\Delta\chi$ and two physical properties:

- *NaCl.* Sodium chloride is a white (colourless) crystalline solid with a $\Delta\chi$ of 2.1, a high melting point, and high electrical conductivity when molten—ionic by any criteria. However, just as for LiF (Figure 8.27B), a small but significant region of electron sharing appears in the relief map.
- *$MgCl_2$.* With a $\Delta\chi$ of 1.8, magnesium chloride is still ionic, but it has a lower melting point and lower conductivity, as well as slightly more electron sharing.
- *$AlCl_3$.* Rather than being a three-dimensional lattice of Al^{3+} and Cl^- ions, aluminum chloride, with a $\Delta\chi$ value of 1.5, consists of layers of highly polar Al—Cl bonds. Weak forces between layers result in a much lower melting point, and the low conductivity implies few free ions. As well, electron density between the nuclei is higher.
- *$SiCl_4$, PCl_3, SCl_2, and Cl_2.* The trend toward more covalent bonding continues through the remaining substances. Each occurs as separate molecules, which have no conductivity and such weak forces *between* them that the melting point is below 0°C. In Cl_2, the bond is nonpolar ($\Delta\chi$ = 0.0). The relief maps show the increasing height of the electron density in the bonding region.

Thus, *as $\Delta\chi$ decreases, the bond becomes more covalent,* and the character of the substance changes from ionic solid to covalent gas.

FIGURE 8.28 Electron density distributions in bonds of the period 3 chlorides. Note the steady increase in electron sharing from left to right.

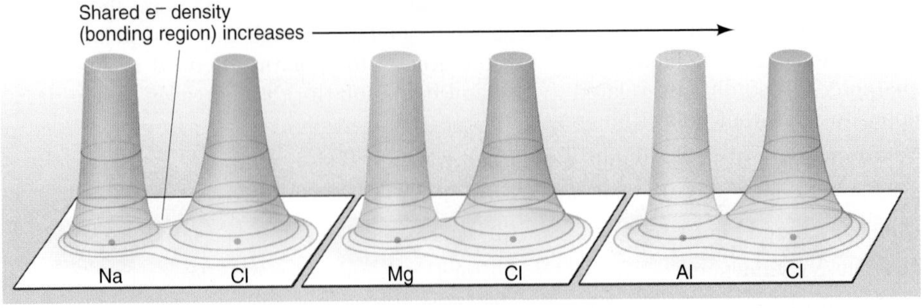

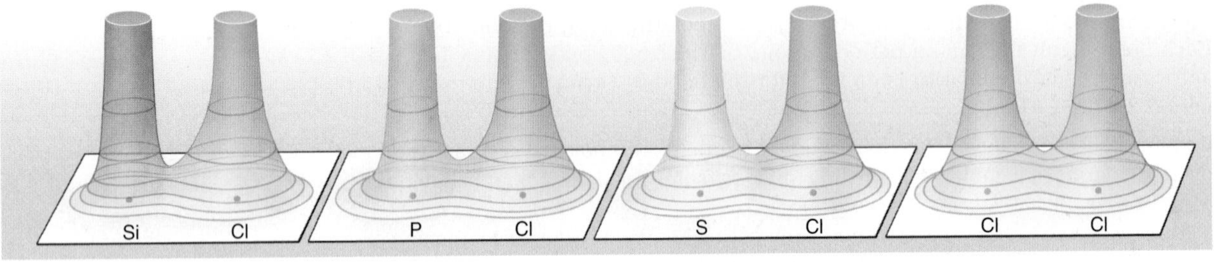

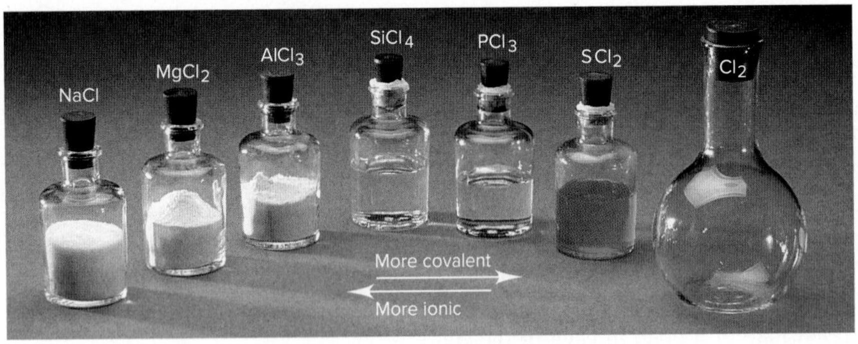

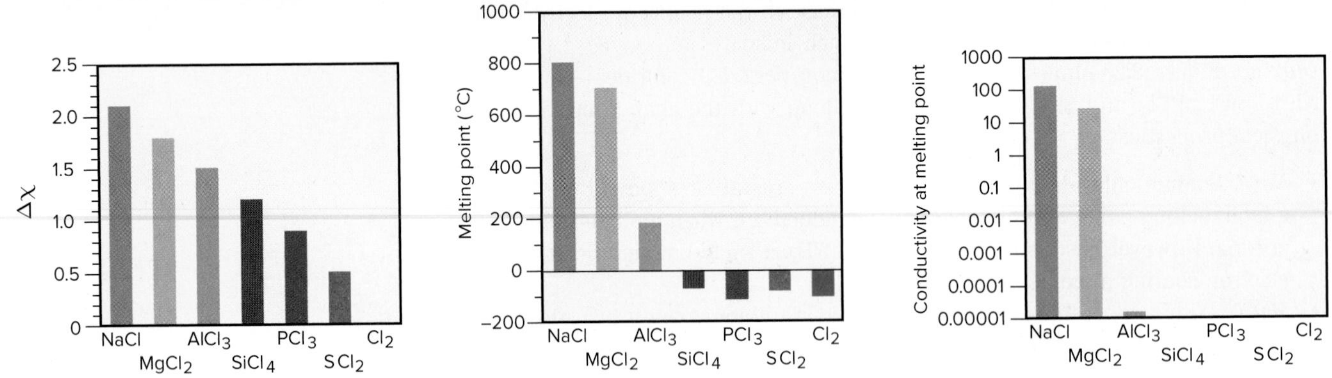

FIGURE 8.29 Properties of the period 3 chlorides. As $\Delta\chi$ decreases, melting point and electrical conductivity decrease because the bond type changes from ionic to polar covalent to nonpolar covalent.

SUMMARY OF SECTION 8.5

- Electronegativity is the ability of a bonded atom to attract shared electrons, which generates opposite partial charges at the ends of the bond and contributes to the bond energy.
- Electronegativity increases across a period and decreases down a group, the reverse of the trends in atomic size.
- The larger the $\Delta\chi$ for two bonded atoms, the more polar the bond and the greater its ionic character.
- For period 3 chlorides, there is a gradation in bond type from ionic to polar covalent to nonpolar covalent.

8.6 Depicting Molecules and Ions with Lewis Structures

The first step toward visualizing a molecule is to convert its molecular formula to its **Lewis structure (Lewis formula*)**, which shows electron-dot symbols for the atoms, the bonding pairs as lines, and the lone pairs that fill each atom's outer level (valence shell) as pairs of dots.

Applying the Octet Rule to Write Lewis Structures

To write a Lewis structure, we decide on the relative placement of the atoms in the molecule or polyatomic ion and then distribute the total number of valence electrons as bonding and lone pairs. In many, but not all, cases, the octet rule (Section 8.1) guides us in distributing the electrons. We begin with species that "obey" the octet rule, in which each atom fills the outer level with eight electrons (or two for hydrogen).

FIGURE 8.30 The steps for converting a molecular formula into a Lewis structure

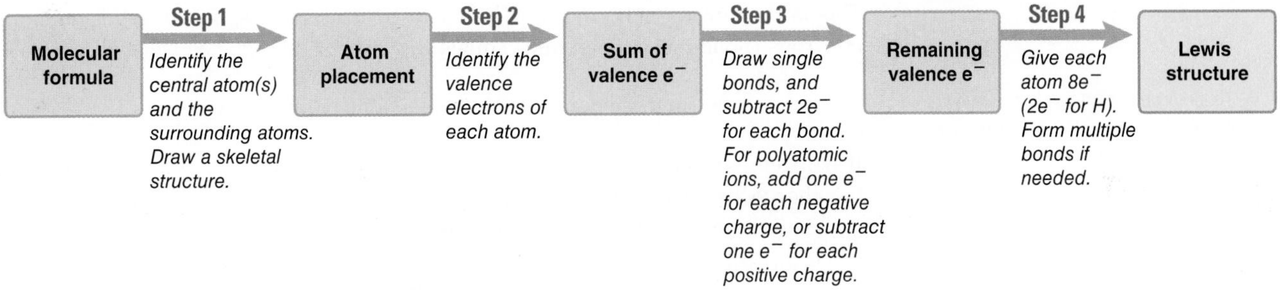

Molecules with Single Bonds Figure 8.30 outlines the steps for writing a Lewis structure for species with only single bonds. Let us use nitrogen trifluoride, NF_3, to illustrate the steps:

Step 1. Place the atoms relative to each other. For compounds with the general molecular formula AB_n, place the odd atom out in the middle. This atom is usually the one that is able to form more bonds. In NF_3, for example, the atom with the *lower group number* is placed at the centre because it needs more electrons to attain an octet; in NF_3, this is also the atom with the *lower electronegativity*. The N (group 15; $\chi = 3.0$) has five electrons and so needs three, whereas each F (group 17; $\chi = 4.0$) has seven and needs only one; thus, N goes in the centre with the three F atoms around it:

If the atoms have the same group number, as in SO_3, place the atom with the *higher period number* (also lower χ) at the centre. H can form only one bond, so it is *never* a central atom.

Step 2. Determine the total number of valence electrons.

- For molecules, add up the valence electrons of the atoms. (Recall that you need to determine the electron configuration to find out the number of valence electrons.) In NF_3, N has five valence electrons, and each F has seven:

$$[1 \times \text{N}\ (5e^-)] + [3 \times \text{F}\ (7e^-)] = 5e^- + 21e^- = 26 \text{ valence } e^-$$

- For polyatomic ions, *add* one e^- for each negative charge, or *subtract* one e^- for each positive charge. By convention, we write the structure in square brackets, placing the charge outside the upper left corner.

*A Lewis *structure* does *not* indicate the three-dimensional shape, so it may be more correct to call it a Lewis *formula*, but we follow convention and use the term *structure*.

Step 3. Draw a single bond from each surrounding atom to the central atom, and subtract two e^- from the total for each bond to find the number of e^- remaining:

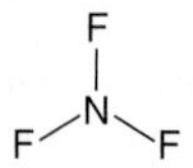

$$3 \text{ N—F bonds} \times 2e^- = 6e^- \quad \text{so} \quad 26e^- - 6e^- = 20e^- \text{ remaining}$$

Step 4. Distribute the remaining electrons in pairs so that each atom ends up with eight e^- (or two e^- for H). First, place lone pairs on the *surrounding (more electronegative) atoms* to give each atom an octet. If any electrons remain, place them around the central atom—each F gets three pairs ($3 \times 6e^- = 18e^-$) and the N gets one pair ($2e^-$), for a total of $20e^-$. Then check that each atom has eight e^-:

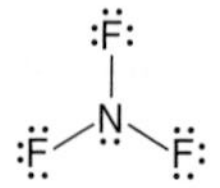

This is the Lewis structure for NF_3. It is a neutral species, so the total number of electrons (bonds plus lone pairs) equals the sum of the valence electrons:

$$6e^- \text{ in three bonds} + 20e^- \text{ in ten lone pairs} = 26 \text{ valence } e^-$$

If we were writing a Lewis structure for a polyatomic ion, we would also consider the charge, as described in step 2. ■

■ The total number of bonds in a Lewis structure can also be calculated by dividing the total number of shared electrons (S) by 2, where

$$S = N - A$$

N is the total number of valence electrons needed by all atoms to achieve noble gas configurations. If the structure is a cation, subtract an electron for each positive charge; if it is an anion, add electrons for each negative charge. A is the total number of valence electrons (calculated in step 2).

Since Lewis structures do not indicate shape, an equally correct depiction of NF_3 is

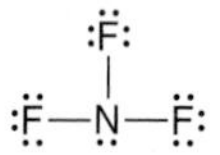

or any other depiction that retains the *same connections among the atoms*—a central N atom connected by single bonds to each of three surrounding F atoms.

Using these four steps, you can write a Lewis structure for any singly bonded species with a central C, N, or O atom, as well as for some species with central atoms from higher periods. Nearly all of their *neutral* compounds have the following characteristics:

- Hydrogen atoms form one bond. This means that H is *always* a terminal or surrounding atom.
- Carbon atoms usually form four bonds. C is usually a central atom.
- Nitrogen atoms usually form three bonds.
- Oxygen atoms usually form two bonds. If there are more than two oxygen atoms in the structure, generally they will not be written next to each other. Some exceptions are oxygen gas (O_2), ozone (O_3), peroxides (such as H_2O_2), and ethers (see Follow-Up Problem 8.8).
- Halogens are *often* terminal atoms. This means that surrounding halogens form one bond; fluorine is *always* a surrounding atom.

Sample Problem 8.7 Writing a Lewis Structure for a Molecule with One Central Atom

Problem Write a Lewis structure for CCl_2F_2, one of the compounds responsible for the depletion of stratospheric ozone.

Solution *Step 1.* Place the atoms relative to each other. In CCl_2F_2, carbon has the lowest group number and χ, so it is the central atom (*see margin*). The halogen atoms surround it, but their specific positions are not important.

Step 1.

Cl
F C F
Cl

Step 2. Determine the total number of valence electrons. C is in group 14; F and Cl are in group 17. Therefore, we have

$$[1 \times C\ (4e^-)] + [2 \times F\ (7e^-)] + [2 \times Cl\ (7e^-)] = 32 \text{ valence } e^-$$

Step 3. Draw single bonds to the central atom (*see margin*) and subtract $2e^-$ for each bond:

$$4 \text{ bonds} \times 2e^- = 8e^- \quad \text{so} \quad 32e^- - 8e^- = 24e^- \text{ remaining}$$

Step 3.

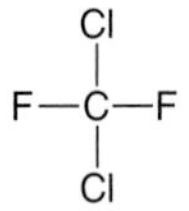

Step 4. Distribute the remaining electrons in pairs, beginning with the surrounding atoms, so that each atom has an octet. Each surrounding halogen gets three pairs (*see margin*).

Step 4.

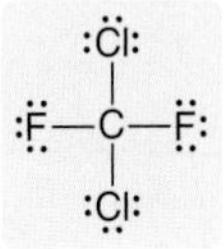

Check Always check that each atom has an octet. Bonding electrons belong to each atom in the bond. The total number in the bonds ($8e^-$) and the lone pairs ($24e^-$) equals 32 valence e^-. As expected, C has four bonds and each of the surrounding halogens forms one bond.

Follow-Up Problem 8.7 Write a Lewis structure for **(a)** H_2S; **(b)** OF_2; **(c)** $SOCl_2$.

In molecules with two or more central atoms bonded to each other, it is usually clear which atoms are central and which are surrounding.

Sample Problem 8.8 Writing a Lewis Structures for a Molecule with More Than One Central Atom

Problem Write the Lewis structure for methanol (molecular formula: CH_4O), an important industrial alcohol that can be used as a gasoline alternative in race cars.

Solution *Step 1.* Place the atoms relative to each other. The H atoms can form only one bond, so all the H atoms will be surrounding or terminal atoms. On the other hand, C and O must be central and adjacent to each other. Since C can form four bonds and O can form two, we arrange the H atoms accordingly (*see margin*).

Step 1.

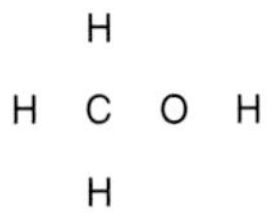

Step 2. Find the sum of the valence electrons (C is in group 14, and O is in group 16):

$$[1 \times C\ (4e^-)] + [1 \times O\ (6e^-)] + [4 \times H\ (1e^-)] = 14e^-$$

Step 3. Add single bonds (*see margin*) and subtract $2e^-$ for each bond:

$$5 \text{ bonds} \times 2e^- = 10e^- \quad \text{so} \quad 14e^- - 10e^- = 4e^- \text{ remaining}$$

Step 3.

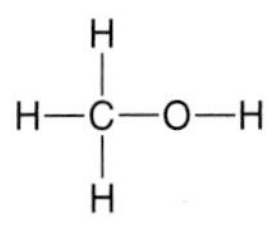

Step 4. Add the remaining electrons, in pairs, to fill each valence level after counting how many electrons each atom has so far. C already has an octet, and each H shares $2e^-$ with the C, so the remaining $4e^-$ form two lone pairs on O to give the Lewis structure for methanol (*see margin*). In this particular molecule, O is the only atom to which you can add the remaining electrons. However, if you have a molecule with several atoms that are lacking electrons to complete the octet, you should start adding electrons to the atoms in the order of decreasing electronegativity.

Step 4.

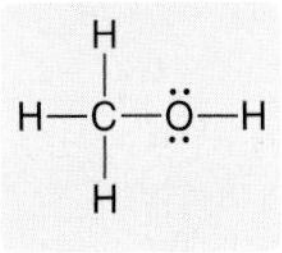

Check Each H atom has $2e^-$, and C and O each have $8e^-$. The total number of valence electrons is $14e^-$, which equals $10e^-$ in bonds plus $4e^-$ in two lone pairs. Each H has one bond, C has four bonds, and O has two bonds.

Follow-Up Problem 8.8 Write a Lewis structure for **(a)** hydroxylamine (NH_3O); **(b)** methoxymethane (dimethyl ether, C_2H_6O; no O—H bonds).

Molecules with Multiple Bonds In most cases, if there are not enough electrons for the central atom(s) to attain an octet, a multiple bond is present. We need to add the following step to the procedure for writing a Lewis structure:

Step 5: Cases involving multiple bonds. If a central atom does not end up with an octet, change a lone pair on a surrounding atom into another bonding pair to the central atom, thus forming a multiple bond. For example, let us write the Lewis structure for sulfur dioxide (SO_2), a colourless gas used as a bleaching agent and during the production of wood pulp for the manufacture of paper.

First, we need to decide which atom will be the central atom. Both O and S have six valence electrons. However, since S has a lower χ, we will place it in the centre when drawing the skeletal structure:

O—S—O

Then we determine the total number of valence electrons:

$$[1 \times \text{O}\ (6e^-)] + [2 \times \text{S}\ (6e^-)] = 18e^-$$

To determine the number of electrons to distribute, we subtract $2e^-$ for each bond that is drawn on the skeletal structure:

$$18e^- - (2 \times 2e^-) = 14e^- \text{ to distribute}$$

To distribute the valence electrons around the atoms until each atom has a complete valence shell (eight valence electrons), we start with the most electronegative atoms:

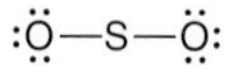

Then we add the remaining electrons to the central atom (S):

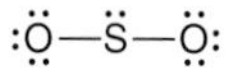

Now we need to check to make sure that the central atom has a complete octet. In SO_2, the S atom has only six of the eight required electrons. Therefore, we must remove one of the nonbonding electrons around one of the O atoms and create a double bond between that O atom and the S atom. This, however, raises one question: from which O atom do we remove the nonbonding electrons to form the double bond? The following two structures are basically the same, except for the choice of the oxygen atom with which the S atom forms the double bond. They are resonance structures. (We will study resonance structures in the following subsection.)

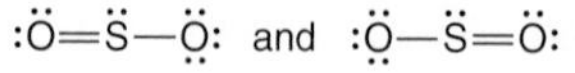

Let us write the Lewis structure for hydrogen cyanide (HCN), an extremely poisonous colourless liquid that is a very important precursor to many chemical compounds, ranging from polymers to pharmaceuticals. First, we identify the central atom. We know that H can form only one bond, so it cannot be the central atom. C and N are in the same period, but since C has a lower χ, it will be the central atom. So, we write the skeletal structure as follows:

H—C—N

Then we identify the valence electrons of each atom by determining the electron configuration, and we calculate the total number of valence electrons:

$$[1 \times \text{H}\ (1e^-)] + [1 \times \text{C}\ (4e^-)] + [1 \times \text{N}\ (5e^-)] = 10e^-$$

The number of electrons to distribute is

$$10e^- - (2 \times 2e^-) = 6e^-$$

Now we proceed as in the previous example, distributing the valence electrons around the atoms. We start with the most electronegative atom first (recall that H will not need any additional electrons):

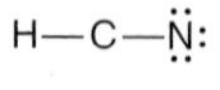

Since the C atom does not have a complete octet, we must remove two pairs of nonbonding electrons around the N atom and form a triple bond between the C atom and the N atom:

H—C≡N:

Sample Problem 8.9 Writing Lewis Structures for Molecules with Multiple Bonds

Problem Write Lewis structures for the following molecules:

(a) Ethene (ethylene, C_2H_4), the most important reactant in the manufacture of polymers

(b) Nitrogen (N_2), the most abundant atmospheric gas (about 78%)

Plan We show the structure that results from steps 1 to 4: placing the atoms, counting the total valence electrons, making single bonds, and distributing the remaining valence electrons in pairs to attain octets. Then we continue with step 5, if needed.

Solution **(a)** For C_2H_4, we have the following structure after steps 1 to 4:

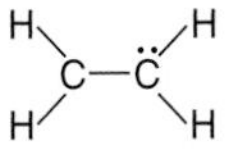

Step 5. Change a lone pair to a bonding pair. The right C has an octet, but the left C has only $6e^-$, so we change the lone pair to another bonding pair between the two C atoms:

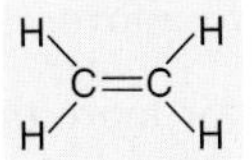

(b) For N_2, we have this structure after steps 1 to 4:

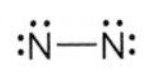

Step 5. Neither N has an octet, so we change a lone pair to a bonding pair. In this case, moving one lone pair to make a double bond still does not give the N atoms an octet, so we move a lone pair from each N to make a triple bond:

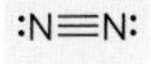

Check **(a)** Each C has four bonds and counts the $4e^-$ in the double bond as part of its own octet. The valence electron total is $12e^-$, all in six bonds. **(b)** Each N counts the $6e^-$ in the triple bond as part of its own octet. The valence electron total is $10e^-$, which equals the electrons in three bonds and two lone pairs.

Follow-Up Problem 8.9 Write Lewis structures for **(a)** CO (the only common molecule in which C has three bonds); **(b)** C_2H_2; **(c)** CO_2.

Resonance: Delocalized Electron-Pair Bonding

Today, scientists realize that bonding electron pairs in many molecules are not as well localized as Lewis believed. We can often write more than one Lewis structure for a molecule or polyatomic ion with *double bonds next to single bonds*. Which structure, if any, is correct?

The Need for Resonance Structures To understand this issue, consider ozone (O_3), an air pollutant at ground level but an absorber of harmful ultraviolet (UV) radiation in the stratosphere. Two Lewis structures (with lettered O atoms for clarity) are

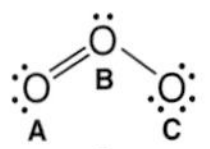

I

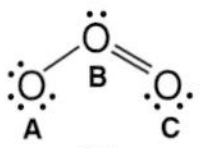

II

In structure I, oxygen B has a double bond to oxygen A and a single bond to oxygen C. In structure II, the single and double bonds are reversed. You can rotate I to get II, so these are *not* different types of ozone molecules, but different Lewis structures for the *same* molecule.

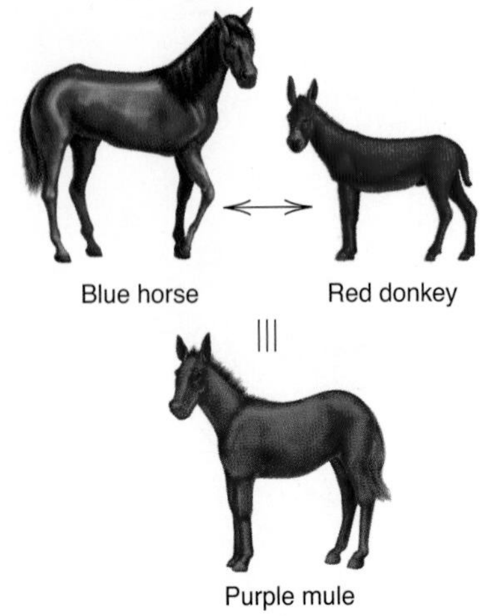

FIGURE 8.31 A purple mule, not a blue horse and a red donkey! A mule is a genetic mix, or a hybrid, of a horse and a donkey; it is not a horse one instant and a donkey the next. Similarly, the colour purple is a mix of red and blue, not red one instant and blue the next. In the same sense, a resonance hybrid is one molecular species, not one resonance form this instant and another resonance form the next. The problem is that we cannot accurately depict the actual species, the hybrid, with a single Lewis structure.

In fact, *neither* Lewis structure depicts O_3 accurately, because the two oxygen-oxygen bonds in O_3 are actually identical in length and energy. The bonds in O_3 have properties between an O—O bond and an O=O bond, something like a "one-and-a-half" bond. The molecule is shown more correctly with two Lewis structures, called **resonance structures (resonance forms),** and a two-headed resonance arrow (⟷) between them. Resonance structures *have the same relative placement of atoms, but different locations of bonding and lone electron pairs.* You can convert one resonance form to another by moving lone pairs to bonding positions, and vice versa (by convention, a double hook arrow denotes the movement of two electrons, while a single arrow hook is used for one-electron movement):

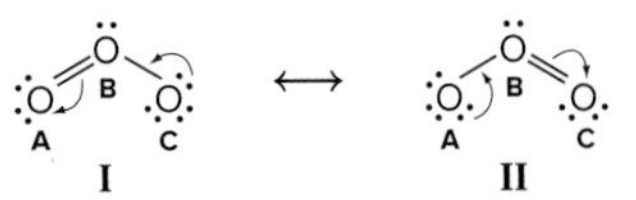

Keep in mind, however, that *resonance structures are not real bonding depictions.* O_3 does *not* change back and forth quickly from structure I to structure II. The actual molecule is a **resonance hybrid**, an average of the resonance forms (Figure 8.31).

Note: If you are not sure where to place a double bond when you are writing a Lewis structure, then you need to write as many resonance structures as you have possible options for the double bond. Connect the different resonance structures with double-headed arrows, as shown above.

Electron Delocalization Our need for more than one Lewis structure to depict O_3 is due to **electron-pair delocalization**. In a single, double, or triple bond, each electron pair is *localized* between the bonded atoms. In a resonance hybrid, two of the electron pairs (one bonding pair and one lone pair) are *delocalized*: their density is "spread" over a few adjacent atoms. (This delocalization involves just a few e^- pairs, so it is *much* less extensive than the electron delocalization in metals, which we will discuss in Section 8.7.)

In O_3, the result is two identical bonds, each consisting of a single bond (the localized pair) and a *partial bond* (the contribution from one of the delocalized pairs). We draw the resonance hybrid with a curved dashed line to show the delocalized pairs:

Resonance hybrid 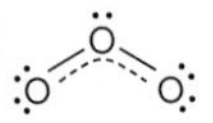

Resonance is very common. For example, benzene (C_6H_6, *shown below*) has two important resonance forms in which alternating single and double bonds have different positions. The actual molecule is an average of the two forms with six C—C bonds and three electron pairs delocalized over all six C atoms. The delocalized pairs are often shown as a dashed circle (or simply a circle):

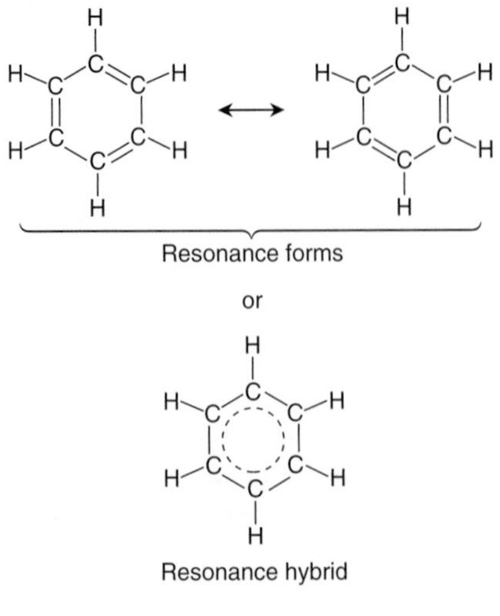

Fractional Bond Orders Partial bonding, as in resonance hybrids, often leads to fractional bond orders. For O_3, we have the following bond order:

$$\text{Bond order} = \frac{3 \text{ electron pairs}}{2 \text{ bonded-atom pairs}} = 1\frac{1}{2}$$

The carbon-to-carbon bond order in benzene is $\frac{9 \text{ electron pairs}}{6 \text{ bonded-atom pairs}}$, which is also $1\frac{1}{2}$. For the carbonate ion, CO_3^{2-}, three resonance structures can be drawn. Each has four electron pairs shared among three bonded-atom pairs, so the bond order is $\frac{4}{3}$, or $1\frac{1}{3}$. One of the three resonance structures for CO_3^{2-} is

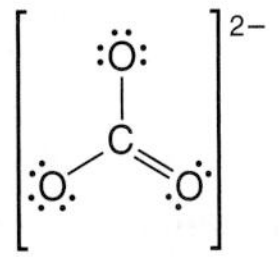

Note here, and in Sample Problem 8.10, that the Lewis structure of a polyatomic ion is written in *square brackets with the ionic charge outside the brackets.* These two polyatomic ions, CO_3^{2-} and NO^{3-}, are examples of fully deprotonated oxoacids, which have lost their acidic hydrogen atoms (each H atom was originally bonded to an O atom), resulting in their ionic charges.

Sample Problem 8.10 Writing Resonance Structures

Problem Write resonance structures for the nitrate ion, NO_3^-, and find the bond order.

Plan We write a Lewis structure, remembering to add one e^- to the total number of valence electrons because of the 1− ionic charge. Then we move the lone pairs and the bonding pairs to write other resonance forms and connect them with the resonance arrow. The bond order is the number of shared electron pairs divided by the number of atom pairs.

Solution After steps 1 to 4, we have the following structure:

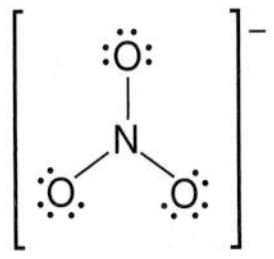

Step 5. Since N has only six e^-, we change a lone pair on one of the O atoms to a bonding pair to form a double bond, which gives each atom an octet. All the O atoms are equivalent, however, so we can move a lone pair from any one of the three and obtain three resonance structures:

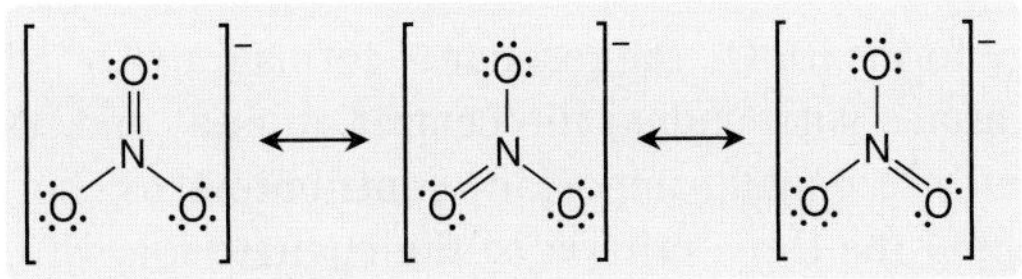

The bond order is

$$\frac{4 \text{ shared electron pairs}}{3 \text{ bonded-atom pairs}} = 1\frac{1}{3}$$

Check Each structure has the same relative placement of atoms, an octet around each atom, and $24e^-$ (the sum of the valence electron total and $1e^-$ from the ionic charge, distributed in four bonds and eight lone pairs).

Comment These three resonance forms contribute equally to the resonance hybrid because all the surrounding atoms are identical. This is not always the case, as you will see next.

Follow-Up Problem 8.10 One of the three resonance structures for CO_3^{2-} was shown just before Sample Problem 8.10. Draw the other two.

Formal Charge: Selecting the More Important Resonance Structure

If one resonance form "looks" more like the resonance hybrid than the other forms do, it "weights" the average in its favour. One way to select the most important resonance form is to determine each atom's **formal charge**, the charge that the atom would have *if the bonding electrons were shared equally*. We will examine this concept and then see how the formal charge compares with the oxidation number.

Determining Formal Charge An atom's formal charge is its total number of valence electrons minus *all* of its unshared valence electrons and *half* of its shared valence electrons. Thus,

$$\text{Formal charge of atom} = \text{no. of valence e}^- - \left(\text{no. of unshared valence e}^- + \frac{1}{2}\text{no. of shared valence e}^-\right) \quad (8.3)$$

For example, in O_3, the formal charge of oxygen A in resonance form I is

$$6 \text{ valence e}^- - \left(4 \text{ unshared e}^- + \frac{1}{2} \text{ of } 4 \text{ shared e}^-\right) = 6 - 4 - 2 = 0$$

The formal charges of all the atoms in the two O_3 resonance forms are

$O_A[6 - 4 - \frac{1}{2}(4)] = 0$

$O_B[6 - 2 - \frac{1}{2}(6)] = +1$

$O_C[6 - 6 - \frac{1}{2}(2)] = -1$

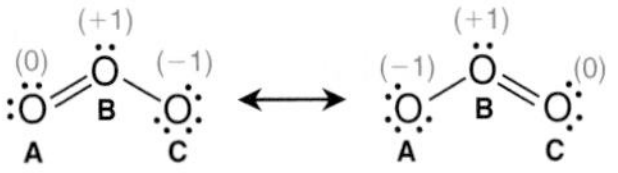

I ⟷ II

$O_A[6 - 6 - \frac{1}{2}(2)] = -1$

$O_B[6 - 2 - \frac{1}{2}(6)] = +1$

$O_C[6 - 4 - \frac{1}{2}(4)] = 0$

Forms I and II have the same formal charges, but on different O atoms, so they contribute equally to the resonance hybrid. *Formal charges must sum to the actual charge on the species*: zero for a molecule or the ionic charge for an ion.

In form I, note that, instead of oxygen's usual two bonds, O_B has three bonds and O_C has one. Only when an atom has a zero formal charge does it have its usual number of bonds; the same holds for C in CO_3^{2-} and N in NO_3^-.

Choosing the More Important Resonance Form Three criteria help us choose the more important resonance structure:

- Smaller formal charges (positive *or* negative) are preferable to larger formal charges.
- The *same* nonzero formal charges on adjacent atoms are not preferred.
- A more negative formal charge should reside on a more electronegative atom.

As in the resonance forms for O_3, the resonance forms for CO_3^{2-}, NO_3^-, and benzene all have identical atoms surrounding the central atom(s) and, thus, have identical formal charges and are equally important contributors to the resonance hybrid. However, let us apply the three criteria to the cyanate ion, NCO^-, which has two *different* atoms around the central atom. Three resonance forms, with formal charges, are given below:

Formal charges:

Resonance forms:

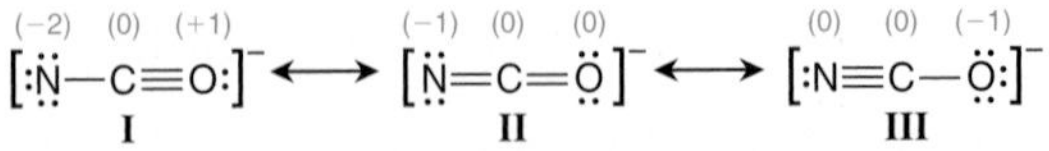

I II III

Form I is not an important contributor to the hybrid because it has a larger formal charge on N and a positive formal charge on the more electronegative O. Forms II and III have the same magnitude of charges, but III has a 1− charge on O, the more electronegative atom. Therefore, II and III are more important than I, and III is more important than II.

Formal Charge versus Oxidation Number The formal charge (used to examine resonance structures) is *not* the same as the oxidation number (used to monitor redox reactions, as we will study in Section 19.1):

- For a *formal charge*, the bonding electrons are *shared equally* by the atoms (as if the bonding were *nonpolar covalent*), so that each atom has half of them:

$$\text{Formal charge} = \text{valence e}^- - \left(\text{lone pair e}^- + \frac{1}{2}\text{bonding e}^-\right)$$

- For an *oxidation number*, the bonding electrons are *transferred completely* to the more electronegative atom (as if the bonding were *pure ionic*):

$$\text{Oxidation number} = \text{valence e}^- - (\text{lone pair e}^- + \text{bonding e}^-)$$

Let us consider the three resonance structures for the cyanate ion:

Formal charges:

Oxidation numbers:

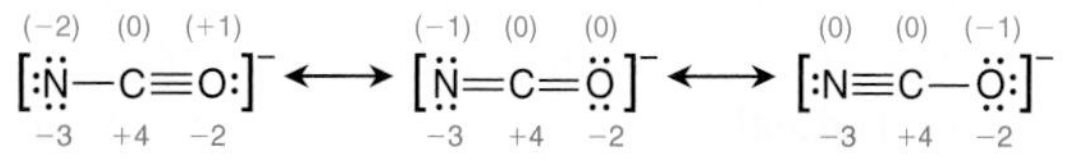

Notice that the oxidation numbers *do not* change from one resonance form to another (because the electronegativities *do not* change), but the formal charges *do* change (because the numbers of bonding and lone pairs *do* change).

Lewis Structures: Exceptions to the Octet Rule

The octet rule applies to most molecules (and ions) with period 2 central atoms, but not to all of them, and not to many molecules with central atoms from period 3 and higher. We now know about many more exceptions to the octet rule than Lewis did. Three important exceptions occur in molecules with (1) electron-deficient atoms, (2) odd-electron atoms, and (3) atoms with expanded valence shells. In this discussion, you will also see that the formal charge has limitations for selecting the best resonance form. Today, scientists realize that bonding electron pairs in many molecules are not as well localized as Lewis believed. Nevertheless, resonance structures (that is, plausible alternative Lewis structures) are still often used to describe such molecules. Scientists also realize that electrons are not always found in pairs. The electron density distribution in a molecule can now be analyzed using functions such as the ELF (electron localization function). There are also other functions, which use electron density analysis, that can show where electron pairs are most likely to be found in a molecule.

Molecules with Electron-Deficient Atoms Gaseous molecules containing either beryllium or boron as the central atom are often **electron deficient**: they have *fewer* than eight electrons around the central atom. The Lewis structures, with formal charges, of gaseous beryllium chloride* and boron trifluoride are

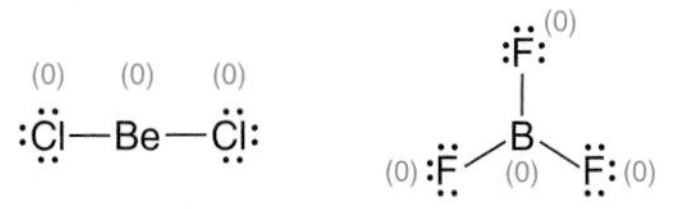

There are only four electrons around Be and six electrons around B. Surrounding halogen atoms do not form multiple bonds to the central atoms to give them an octet, because the halogens are much more electronegative. Formal charges make the following structures unlikely:

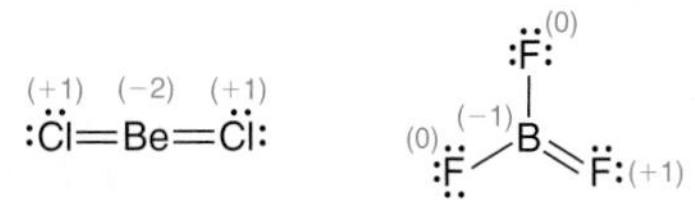

(Some data for BF_3 show a shorter than expected B—F bond. Shorter bonds indicate double-bond character, so the structure with the B═F bond may be a minor contributor to a resonance hybrid.) Electron-deficient atoms often attain an octet by forming additional bonds in reactions. When BF_3 reacts with ammonia, for instance, a compound in which boron attains an octet is formed:[†]

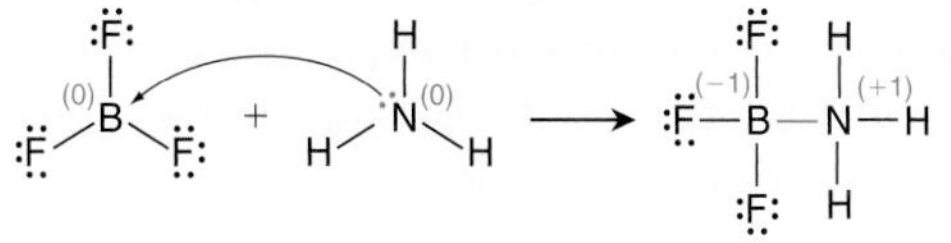

*Even though beryllium is in group 2, most Be compounds have considerable covalent bonding. For example, molten $BeCl_2$ does not conduct electricity, indicating a lack of ions.

[†]Reactions in which one species "donates" an electron pair to another to form a covalent bond are Lewis acid-base reactions, which we discuss fully in Chapter 16.

Aluminum also forms many covalent compounds. Another example of a molecule with electron-deficient atoms is $AlCl_3$:

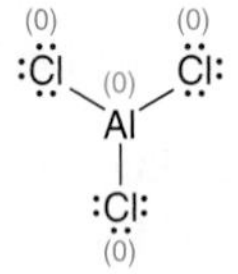

Molecules with Odd-Electron Atoms A few molecules contain a central atom with an odd number of valence electrons, so they cannot have all of their electrons in pairs. Any species (molecules, atoms, or ions) that contain unpaired electrons are called **free radicals**. Most free radicals have a central atom from an odd-numbered group, such as N (group 15) or Cl (group 17). Because they contain a lone (unpaired) electron, they are paramagnetic (Section 7.4) and extremely reactive. Free radicals are dangerous because they can bond to an H atom in a biomolecule and extract it, forming a new free radical. This step repeats and can disrupt genes and membranes. Recent studies suggest that free radicals may be involved in cancer and even aging. Antioxidants, such as vitamin E, interrupt free-radical proliferation.

Consider the free radical nitrogen dioxide, NO_2, a major contributor to urban smog. NO_2 is formed when the NO in auto exhaust is oxidized, and it has several resonance forms. Two differ in terms of which O atom is doubly bonded, as in the case of ozone. Two others have the lone electron residing on the N or on an O, so the resonance hybrid has the lone electron delocalized over these two atoms:

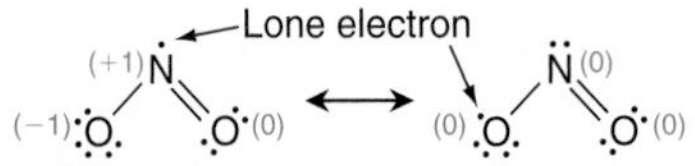

Let us see if formal charge considerations help us decide where the electron resides most of the time. The form of NO_2 with the electron on the singly bonded O has zero formal charges (*right*), while the form with the electron on N (*left*) has some nonzero charges. Thus, based on formal charge, the form on the right is more important. However, chemical facts suggest otherwise. Free radicals often react with each other to pair their lone electrons. When two NO_2 molecules react, the lone electrons pair up to form the N—N bond in dinitrogen tetroxide (N_2O_4) and each N attains an octet:

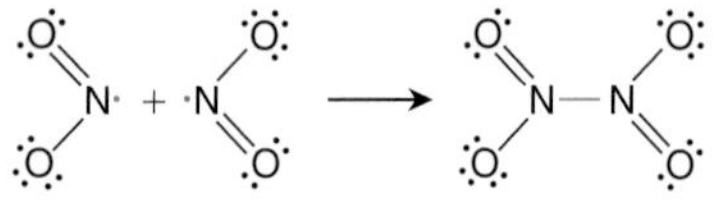

Thus, given the way that NO_2 reacts, the lone electron may spend most of its time on N, making *that* form more important. Apparently, in this case, formal charge is not very useful for picking the more important resonance form; we will see other cases below.

Atoms with Expanded Valence Shells Many molecules (and ions) have more than eight valence electrons around the central atom. *An atom expands its valence shell to form more bonds, which releases energy.* The central atom must be large and have empty orbitals that can hold the additional pairs. Therefore, **expanded valence shells** occur only with *nonmetals from period 3 or higher because they have d orbitals available.* Such a central atom may be bonded to more than four atoms, or to four or fewer:

1. *Central atom bonded to more than four atoms.* Phosphorus pentachloride, PCl_5, is a fuming yellow-white solid that is used to manufacture lacquers and films. It forms when phosphorus trichloride, PCl_3, reacts with chlorine gas. The P in PCl_3 has an octet, but two more bonds to chlorine form and P expands its valence shell to 10 electrons in PCl_5. Note that, when PCl_5 forms, *one* Cl—Cl bond breaks (*left side of the equation*), and *two* P—Cl bonds form (*right side*), for a net increase of one bond:

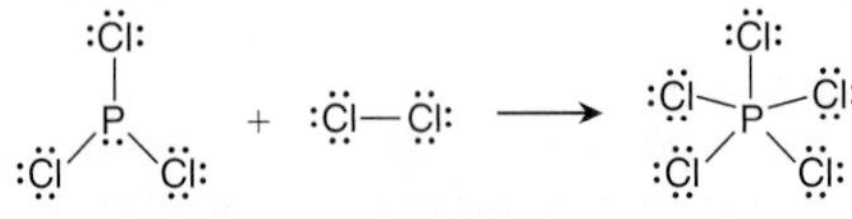

Sulfur hexafluoride, SF_6, a dense and inert gas, is used as an electrical insulator. Like PCl_5, it forms when sulfur tetrafluoride, SF_4, reacts with F_2. The S in SF_4 already has an expanded valence shell of 10 electrons, but 2 more bonds to F expand the valence shell further, to 12 electrons:

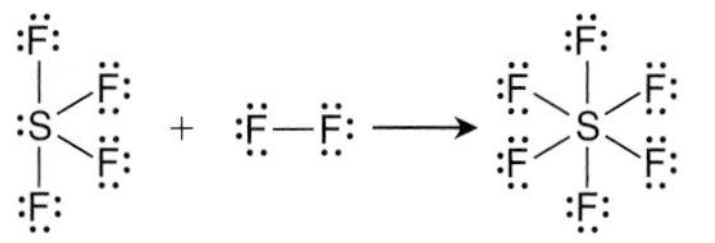

2. *Central atoms bonded to four or fewer atoms.* By applying the concept of formal charge, we can draw Lewis structures with expanded valence shells of central atoms bonded to *four or fewer* atoms. The S in SF_4 is one example. Some others are as follows:

- *Sulfuric acid.* Two resonance forms of H_2SO_4, with formal charges, are given below:

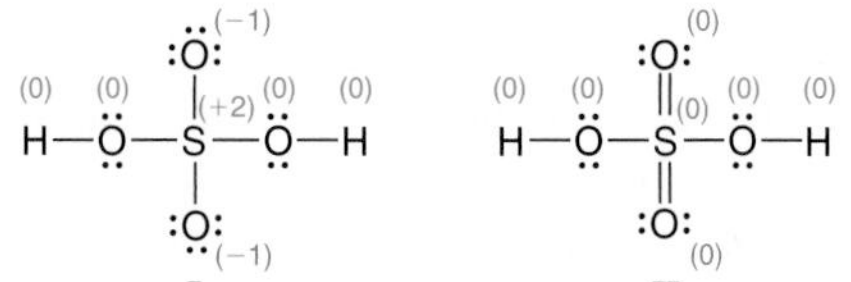

I II

Form I obeys the octet rule, but it has several nonzero formal charges. In form II, sulfur has 12 electrons (6 bonds) around it, but all zero formal charges. Thus, based on the formal charge rules alone, form II contributes more than form I to the resonance hybrid. More important than whether the rules are followed, however, form II is consistent with observations. In gaseous H_2SO_4, the two sulfur-oxygen bonds *with* an H atom attached to the O atom are 157 pm long, whereas the two sulfur-oxygen bonds *without* an H atom are 142 pm long. This shorter bond indicates double-bond character, and other measurements indicate greater electron density in the bonds without the attached H.

- *Sulfate ion.* When sulfuric acid loses two H^+ ions, it forms the sulfate ion, SO_4^{2-}. Measurements indicate that all the bonds in SO_4^{2-} are 149 pm long, between the length of an S═O bond (~142 pm) and the length of an S—O bond (~157 pm). Six of the seven resonance forms consistent with these data have an expanded valence shell and zero formal charges. The Lewis structures below show two of these six forms (*left*) and the form that obeys the octet rule (*right*):

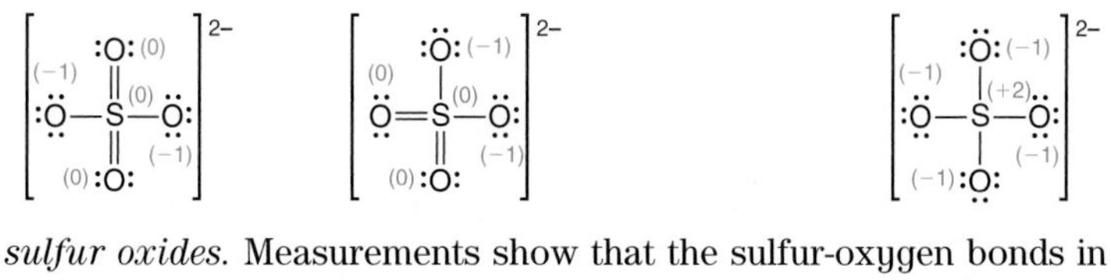

- *Two sulfur oxides.* Measurements show that the sulfur-oxygen bonds in SO_2 and SO_3 are all approximately 142 pm long, indicating S═O bonds. Lewis structures consistent with these data have zero formal charges (*two at the left*), but others (*two at the right*) obey the octet rule:

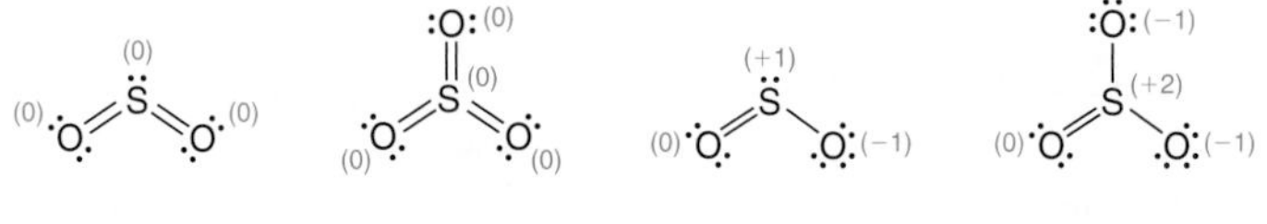

3. *Limitations of Lewis structures and formal charge.* Chemistry has been a central science for well over two centuries, yet controversies often arise over interpretations of data, even in established areas such as bonding and structure. We have seen that a single Lewis structure often cannot accurately depict a molecule, so we need several resonance forms to do this.

Formal charge rules have limitations, too. They were not useful for choosing the correct location of the lone electron in NO_2. As well, based on quantum-mechanical calculations, resonance forms that have expanded valence shells with zero formal charges look *less* like the actual species than forms that follow the octet rule but have higher formal charges. These findings suggest that shorter bonds arise not from double-bond character, but because the higher formal charges draw the bonded atoms closer. Such considerations favour the octet-rule forms for H_2SO_4, SO_4^{2-}, SO_2, and SO_3. Thus, formal charge rules may be useful for selecting the more important resonance form, but they are far from perfect. Nevertheless, while keeping these contrary findings in mind, we will continue to draw structures based on formal charge rules because they provide a simple approach, consistent with experimental data.

Sample Problem 8.11 Writing Lewis Structures for Octet-Rule Exceptions

Problem Write a Lewis structure and identify the octet-rule exception for **(a)** $SClF_5$; **(b)** H_3PO_4 (draw two resonance forms and select the more important form); **(c)** $BFCl_2$.

Plan We write each Lewis structure and examine it for exceptions to the octet rule. **(a)** and **(b)** The central atoms are in period 3, so they can have more than an octet. **(c)** The central atom is B, which can have less than an octet of electrons.

Solution **(a)** $SClF_5$ has an *expanded valence shell.* The Lewis structure is

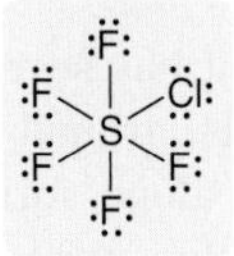

(b) H_3PO_4 has two resonance forms. The structures, with formal charges, are

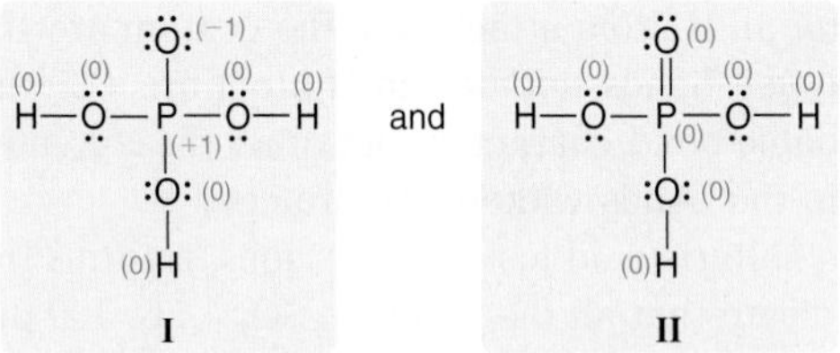

Structure I obeys the octet rule but has nonzero formal charges. Structure II has an expanded valence shell with zero formal charges. According to the formal charge rules, structure II is the more important form.

(c) $BFCl_2$ is an *electron-deficient molecule*; B has only six electrons surrounding it:

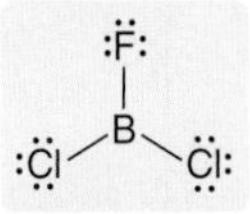

Comment In (b), structure II is consistent with bond-length measurements, which show one shorter (152 pm) and three longer (157 pm) phosphorus-oxygen bonds. Nevertheless, as for H_2SO_4, calculations show that structure I may be more important.

Follow-Up Problem 8.11 Write a Lewis structure with minimal formal charges for **(a)** $POCl_3$; **(b)** ClO_2; **(c)** XeF_4.

SUMMARY OF SECTION 8.6

- A step-by-step process can be used to convert a molecular formula into a Lewis structure, which is a two-dimensional representation of a molecule (or ion) that shows the placement of the atoms and the distribution of the valence electrons among bonding and lone pairs.
- When two or more Lewis structures can be drawn for the same relative placement of the atoms, the actual structure is a hybrid of those resonance forms.
- Formal charges can be useful for choosing the more important contributor to the hybrid, but experimental data always determine the choice.
- Molecules with an electron-deficient atom (central Be or B) or an odd-electron atom (free radicals) have less than an octet around the central atom but often attain an octet in reactions.
- In a molecule (or ion) with a central atom from period 3 or higher, this atom can have more than eight valence electrons because it is larger and has empty *d* orbitals for expanding its valence shell.

8.7 An Introduction to Metallic Bonding

In this section, you will see how a simple, qualitative model for metallic bonding accounts for the properties of metals; a more detailed model is presented in Chapter 11.

The Electron-Sea Model

Metals can transfer electrons to nonmetals and form ionic solids, such as NaCl. As well, experiments with metals in the gas phase show that two metal atoms can even share their valence electrons to form gaseous, diatomic molecules, such as Na_2. But what holds the atoms together in a sample of Na metal? The **electron-sea model** of metallic bonding proposes that all the metal atoms in the sample contribute their valence electrons to form a delocalized electron "sea" throughout the sample, with the metal ions (nuclei and core electrons) lying in an orderly array (see Figure 8.3C). *All the atoms in the sample share the electrons*, and the sample is held together by the mutual attraction of the metal cations for the mobile valence electrons. Thus, bonding in metals is fundamentally different from the other two types of bonding:

- *In contrast to ionic bonding*, the metal ions are not held in place as rigidly.
- *In contrast to covalent bonding*, no particular pair of metal atoms is bonded through a localized electron pair.
- *Instead of forming compounds*, two or more metals typically form **alloys**, solid mixtures of variable composition. Alloys appear in car and airplane bodies, bridges, coins, jewellery, dental fillings, and many other familiar objects.

How the Model Explains the Properties of Metals

The physical properties of metals vary over a wide range. Two features of the electron-sea model that account for these properties are (1) the *regularity*, but not rigidity, of the metal-ion array and (2) the *number* and *mobility* of the valence electrons.

1. *Melting and boiling points*. Nearly all metals are solids with moderate to high melting points and much higher boiling points (Table 8.5). These properties are related to the energy of the metallic bonding. Melting points are only moderately high because the cations can move without breaking the attraction to the surrounding electrons. Boiling points are very high because each cation and its valence electron(s) must break away from the others. Gallium is a striking example: it can melt in your hand (mp 29.8°C) but does not boil until over 2400°C.

Periodic trends are consistent with the strength of the bonding:

- *Down a group*, melting points decrease because the larger metal ions have a weaker attraction to the electron sea.
- *Across a period*, melting points increase. Alkaline earth metals (group 2) have higher melting points than alkali metals (group 1) because their 2+ cations have stronger attractions to twice as many valence electrons (Figure 8.32).

2. *Mechanical properties*. When a piece of metal is deformed by a hammer, the metal ions do not repel each other, but rather slide past each other through the

TABLE 8.5 Melting and Boiling Points of Some Metals

Element	Melting Point (°C)	Boiling Point (°C)
Lithium (Li)	180	1347
Tin (Sn)	232	2623
Aluminum (Al)	660	2467
Barium (Ba)	727	1850
Silver (Ag)	961	2155
Copper (Cu)	1083	2570
Uranium (U)	1130	3930

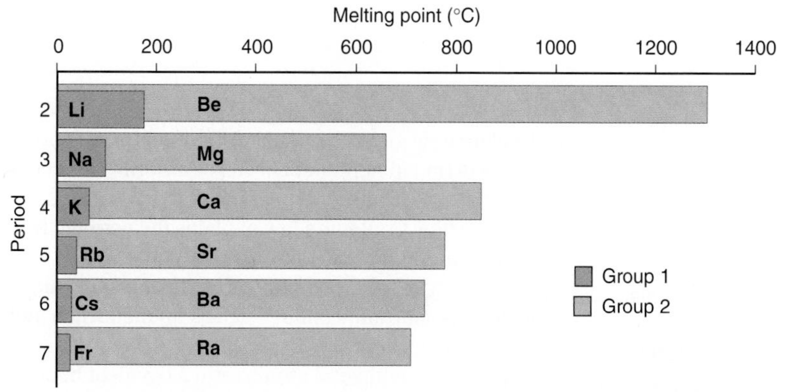

FIGURE 8.32 Melting points of group 1 and group 2 metals

FIGURE 8.33 Why metals dent and bend rather than crack

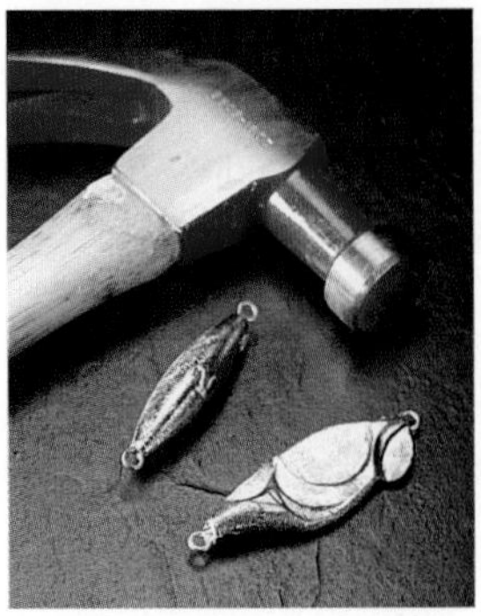

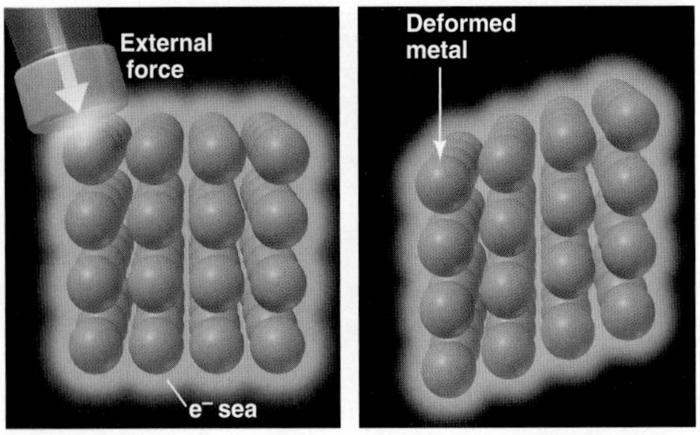

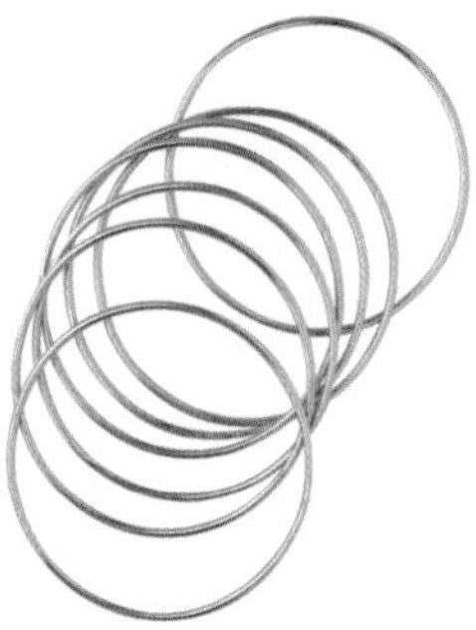

Ductility (ability of a material to deform under tensile stress) and malleability (ability of a material to be deformed by compression without cracking or rupturing) are properties of metals that are widely used in different industries.

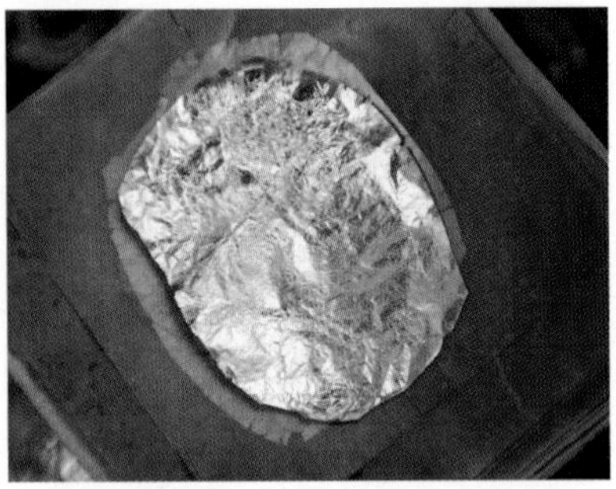

How thin is gold leaf?

electron sea and end up in new positions. Thus, metals dent and bend, as shown in Figure 8.33. Compare this behaviour with the interionic repulsions that occur when an ionic solid is struck (see Figure 8.10).

All the group 11 metals—copper, silver, and gold—are soft enough to be machined into sheets (malleable) and wires (ductile), but gold is in a class by itself. One gram of gold, about the size of a small ball bearing, can be drawn into a wire that is 20 μm thick and 165 m long or hammered into a sheet that is 70 nm (about 230 atoms) thick and has an area of 1.0 m^2.

3. *Electrical conductivity.* Unlike ionic and covalent substances, metals are good conductors of electricity in both their solid and liquid states because of their mobile electrons. When a piece of metal wire is attached to a battery, electrons flow from one terminal into the wire, replacing the electrons that flow from the wire into the other terminal. Foreign atoms disrupt the array of metal atoms and reduce conductivity. Copper used in electrical wiring is over 99.99% pure because traces of other atoms drastically restrict electron flow.

4. *Thermal conductivity.* Mobile electrons also make metals good conductors of heat. Place your hand on a piece of metal and a piece of wood that are both at room temperature. The metal feels colder because it conducts heat away from your hand much faster than the wood does. The mobile delocalized electrons in the metal disperse the heat from your hand more quickly than the localized electron pairs in the covalent bonds of the wood.

SUMMARY OF SECTION 8.7

- According to the electron-sea model, the valence electrons of the metal atoms in a sample are highly delocalized and attract all the metal cations, holding them together.
- Metals have only moderately high melting points because the metal ions remain attracted to the electron sea even if their relative positions change.
- Boiling involves complete separation of individual cations with their valence electrons from all the others, so metals have very high boiling points.
- Metals can be deformed because the electron sea prevents repulsions among the cations.
- Metals conduct electricity and heat because their electrons are mobile.

CHAPTER REVIEW GUIDE

Learning Objectives

Relevant section (§) and/or sample problem (SP) numbers appear in parentheses.

Concepts

1. Explain how differences in atomic properties lead to differences in bond type, and describe the basic distinctions among the three types of bonding. (§8.1)
2. Explain the essential features of ionic bonding: electron transfer to form ions and the electrostatic attraction of ions to form a solid. (§8.2)
3. Explain how lattice energy is ultimately responsible for the formation of ionic compounds. (§8.2)
4. Explain how ionic compound formation is conceptualized as occurring in hypothetical steps (a Born-Haber cycle) to calculate the lattice energy. (§8.2)
5. Explain how Coulomb's law explains the periodic trends in lattice energy. (§8.2)
6. Describe why ionic compounds are brittle, have high melting points, and conduct electricity only when molten or dissolved in water. (§8.2)
7. Explain how nonmetal atoms form a covalent bond. (§8.3)

8. Describe how bonding and lone electron pairs fill the outer level (valence shell) of each atom in a molecule. (§8.3)
9. Explain the interrelationships among bond order, bond length, and bond energy. (§8.3)
10. Describe how the distinction between bonding and nonbonding forces explains the properties of covalent molecules and network covalent solids. (§8.3)
11. Explain how changes in bond strength account for the enthalpy of reaction. (§8.4)
12. Explain how a reaction can be divided conceptually into bond-breaking and bond-forming steps. (§8.4)
13. Describe the periodic trends in electronegativity, and explain the inverse relationship of χ values to atomic sizes. (§8.5)
14. Explain how bond polarity arises from differences in the electronegativities of bonded atoms, and describe the direction of bond polarity. (§8.5)
15. Explain the change in partial ionic character with $\Delta\chi$ and the change from ionic to polar covalent to nonpolar covalent bonding across a period. (§8.5)
16. Describe how Lewis structures depict the atoms, bonding pairs, and lone electron pairs in a molecule or polyatomic ion. (§8.6)
17. Discuss how both resonance and electron-pair delocalization explain bond properties in many compounds that have double bonds adjacent to single bonds. (§8.6)
18. Discuss the meaning of *formal charge* and how it is used to select the more important resonance structure. Explain the difference between formal charge and oxidation number. (§8.6)
19. Apply the octet rule and its three major exceptions: a molecule with a central atom that has an electron deficiency, a molecule with an odd number of electrons, and a molecule with an expanded valence shell. (§8.6)
20. Analyze the role of delocalized electrons in metallic bonding. (§8.7)
21. Discuss how the electron-sea model explains why metals bend, have very high boiling points, and conduct electricity in solid or molten form. (§8.7)

Skills

1. Use Lewis electron-dot symbols to depict main-group atoms. (§8.1)
2. Depict the formation of ions with electron configurations, orbital diagrams, and Lewis symbols, and write the formula for the ionic compound. (SP 8.1)
3. Calculate lattice energy from the enthalpies of the steps to ionic compound formation. (§8.2 and SPs 8.2 and 8.3)
4. Rank similar covalent bonds according to their length and strength. (SP 8.4)
5. Use bond energies to calculate $\Delta_r H°$. (SP 8.5)
6. Determine bond polarity from χ values. (SP 8.6)
7. Use a step-by-step method for writing a Lewis structure from a molecular formula. (SPs 8.7–8.9, 8.11)
8. Write resonance structures for molecules and ions. (SP 8.10)
9. Calculate the formal charge of any atom in a molecule or an ion. (SP 8.11)

Key Terms

Section 8.1
ionic bonding
covalent bonding
metallic bonding
Lewis electron-dot symbol
octet rule

Section 8.2
lattice energy ($\Delta_{lattice}H°$)
Born-Haber cycle
Coulomb's law
ion pairs

Section 8.3
covalent bond
bonding pair (shared pair)
lone pair (unshared pair)
coordinate bond (dative covalent bond)
bond order
single bond
double bond
triple bond
bond energy (BE)
bond length
network covalent solids

Section 8.5
polar covalent bond
nonpolar covalent bond
electronegativity difference ($\Delta\chi$)
partial ionic character

Section 8.6
Lewis structure (Lewis formula)
resonance structures (resonance forms)
resonance hybrid
electron-pair delocalization
formal charge
electron deficient
free radicals
expanded valence shells

Section 8.7
electron-sea model
alloys

Key Equations and Relationships

8.1 Relating the energy of attraction to the lattice energy:

$$\text{Electrostatic energy} \propto \frac{\text{cation charge} \times \text{anion charge}}{\text{cation radius} + \text{anion radius}} \propto \Delta_{lattice}H°$$

8.2 Calculating enthalpy of reaction from bond enthalpies or bond energies:

$$\Delta_r H° = \Sigma\Delta_{\text{reactant bonds broken}}H° + \Sigma\Delta_{\text{product bonds formed}}H°$$

or

$$\Delta_r H° = \Sigma BE_{\text{reactant bonds broken}} - \Sigma BE_{\text{product bonds formed}}$$

8.3 Determining the formal charge of an atom:

$$\text{Formal charge of atom} = \text{no. of valence e}^- - \left(\text{no. of unshared valence e}^- + \frac{1}{2}\text{no. of shared valence e}^-\right)$$

Brief Solutions to Follow-Up Problems

8.1 $Mg([Ne]3s^2) + 2Cl([Ne]3s^23p^5) \longrightarrow Mg^{2+}([Ne]) + 2Cl^-([Ne]3s^23p^6)$

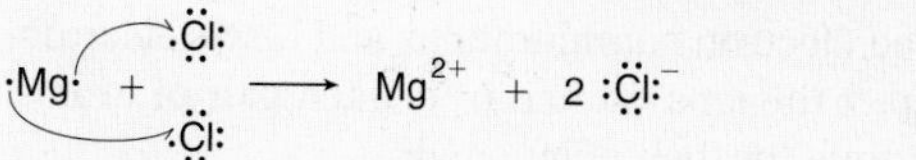

Formula: $MgCl_2$

8.2 $Br(g) + e^- \longrightarrow Br^-(g)$ EA = −324 kJ/mol

8.3 (a) SrF_2. The only difference between these compounds is the size of the cation: the Sr^{2+} ion is smaller than the Ba^{2+} ion. (b) RbCl. The sizes of the cations and of the anions are nearly the same, but the charges of Rb^+ and $Cl^{2-}$are half as much as the charges of Sr^{2+} and S^{2+}. In nearly every case, charge is more important than size.

8.4 (a) Bond length: Si—F < Si—O < Si—C
Bond strength: Si—C < Si—O < Si—F
(b) Bond length: N≡N < N=N < N—N
Bond strength: N—N < N=N < N≡N

8.5 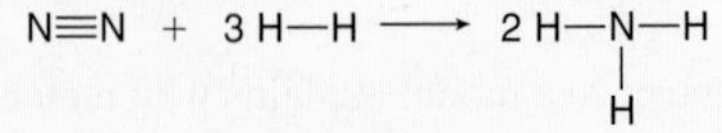

$\Sigma\Delta_{\text{bonds broken}}H° = 1\ \text{N}{\equiv}\text{N} + 3\ \text{H—H}$
$= 945\ \text{kJ/mol} + 1296\ \text{kJ/mol} = 2241\ \text{kJ/mol}$
$\Sigma\Delta_{\text{bonds formed}}H° = 6\ \text{N—H} = -2346\ \text{kJ/mol}$
$\Delta_r H° = -105\ \text{kJ/mol}$

8.6 (a) Cl—Cl < $\overset{\delta+}{\text{Br}}$—$\overset{\delta-}{\text{Cl}}$ < $\overset{\delta+}{\text{Cl}}$—$\overset{\delta-}{\text{F}}$

(b) Si—Si < $\overset{\delta+}{\text{S}}$—$\overset{\delta-}{\text{Cl}}$ < $\overset{\delta+}{\text{P}}$—$\overset{\delta-}{\text{Cl}}$ < $\overset{\delta+}{\text{Si}}$—$\overset{\delta-}{\text{Cl}}$

8.7 (a) 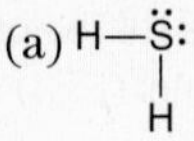 (b) 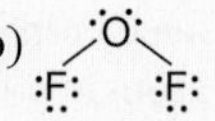 (c) 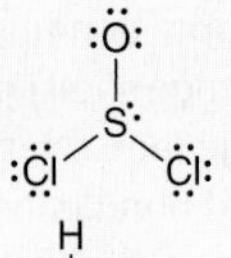

8.8 (a) 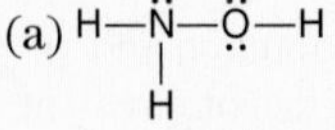 (b) 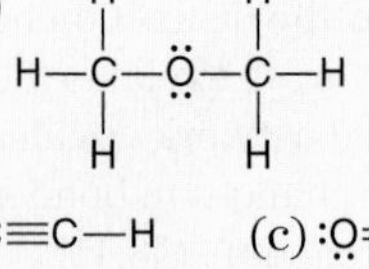

8.9 (a) :C≡O: (b) H—C≡C—H (c) :O=C=O:

8.10 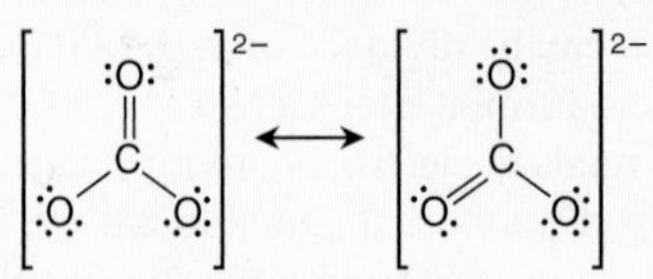

8.11 (a) 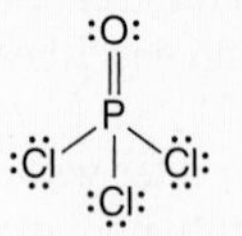 (b) 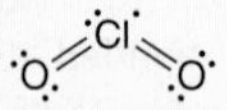 (c) 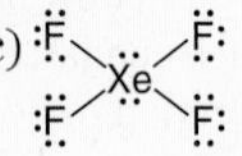

PROBLEMS

Problems with **red** numbers are answered in Appendix G and worked in detail in the Student Solutions Manual. Problem sections match those in this book and provide the numbers of relevant sample problems. Most offer Concept Review Questions, Skill-Building Exercises (grouped in pairs covering the same concept), and Problems in Context. The Comprehensive Problems are based on material from any section or previous chapter.

Atomic Properties and Chemical Bonds

Concept Review Questions

8.1 In general terms, how does each atomic property influence the metallic character of the main-group elements in a period?
(a) Ionization energy (b) Atomic radius
(c) Number of outer electrons (d) Effective nuclear charge

8.2 Three solids are represented below. What is the predominant type of intramolecular bonding in each?

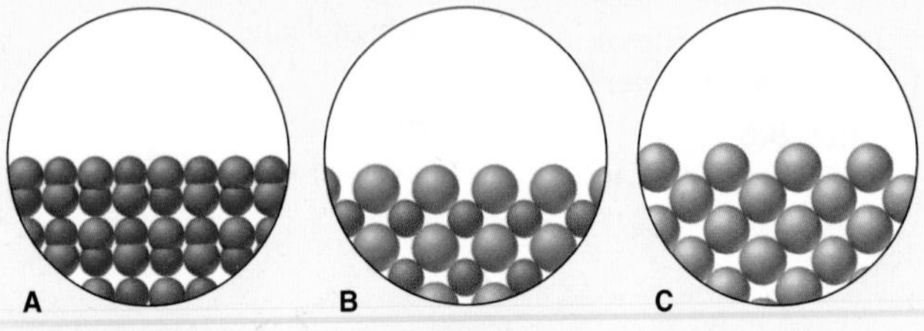

8.3 What is the relationship between the tendency of a main-group element to form a monatomic ion and its position in the periodic table? In what part of the table are the main-group elements that typically form cations? Anions?

Skill-Building Exercises (grouped in similar pairs)

8.4 Which member of each pair is *more* metallic?
(a) Na or Cs (b) Mg or Rb (c) As or N

8.5 Which member of each pair is *less* metallic?
(a) I or O (b) Be or Ba (c) Se or Ge

8.6 State the type of bonding—ionic, covalent, or metallic—you would expect in (a) CsF(*s*); (b) N_2(*g*); (c) Na(*s*).

8.7 State the type of bonding—ionic, covalent, or metallic—you would expect in (a) ICl_3(*g*); (b) N_2O(*g*); (c) LiCl(*s*).

8.8 State the type of bonding—ionic, covalent, or metallic—you would expect in (a) O_3(*g*); (b) $MgCl_2$(*s*); (c) BrO_2(*g*).

8.9 State the type of bonding—ionic, covalent, or metallic—you would expect in (a) Cr(*s*); (b) H_2S(*g*); (c) CaO(*s*).

8.10 Draw a Lewis electron-dot symbol for (a) Rb; (b) Si; (c) I.

8.11 Draw a Lewis electron-dot symbol for (a) Ba; (b) Kr; (c) Br.

8.12 Draw a Lewis electron-dot symbol for (a) Sr; (b) P; (c) S.

8.13 Draw a Lewis electron-dot symbol for (a) As; (b) Se; (c) Ga.

8.14 Give the group number and general electron configuration of an element with each electron-dot symbol:
(a) ·Ẍ: (b) Ẋ·

8.15 Give the group number and general electron configuration of an element with each electron-dot symbol:
(a) ·Ẋ: (b) ·Ẋ·

The Ionic Bonding Model
(Sample Problems 8.1 to 8.3)

Concept Review Questions

8.16 If energy is required to form monatomic ions from metals and nonmetals, why do ionic compounds exist?

8.17 (a) In general, how does the lattice energy of an ionic compound depend on the charges and sizes of the ions? (b) Ion arrangements of three general salts are represented below. Rank the arrangements in order of increasing lattice energy.

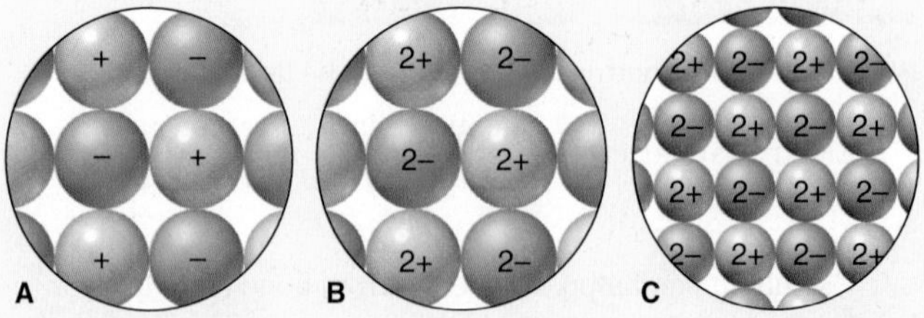

8.18 When gaseous Na^+ and Cl^- ions form gaseous NaCl ion pairs, 548 kJ/mol of energy is released. Why, then, does NaCl occur as a solid under ordinary conditions?

8.19 To form S^{2-} ions from gaseous sulfur atoms requires 214 kJ/mol, but these ions exist in solids such as K_2S. Explain.

Skill-Building Exercises (grouped in similar pairs)

8.20 Use condensed electron configurations and Lewis electron-dot symbols to depict the ions formed from each pair of atoms, and predict the formula for their compound:
(a) Ba and Cl (b) Sr and O (c) Al and F (d) Rb and O

8.21 Use condensed electron configurations and Lewis electron-dot symbols to depict the ions formed from each pair of atoms, and predict the formula for their compound:
(a) Cs and S (b) O and Ga (c) N and Mg (d) Br and Li

8.22 For each ionic compound formula, identify the main group to which X belongs:
(a) XF_2 (b) MgX (c) X_2SO_4

8.23 For each ionic compound formula, identify the main group to which X belongs:
(a) X_3PO_4 (b) $X_2(SO_4)_3$ (c) $X(NO_3)_2$

8.24 For each ionic compound formula, identify the main group to which X belongs:
(a) X_2O_3 (b) XCO_3 (c) Na_2X

8.25 For each ionic compound formula, identify the main group to which X belongs:
(a) CaX_2 (b) Al_2X_3 (c) XPO_4

8.26 For each pair, choose the compound with the higher lattice energy, and explain your choice:
(a) BaS or CsCl (b) LiCl or CsCl

8.27 For each pair, choose the compound with the higher lattice energy, and explain your choice:
(a) CaO or CaS (b) BaO or SrO

8.28 For each pair, choose the compound with the lower lattice energy, and explain your choice:
(a) CaS or BaS (b) NaF or MgO

8.29 For each pair, choose the compound with the lower lattice energy, and explain your choice:
(a) NaF or NaCl (b) K_2O or K_2S

8.30 Use the following information to calculate $\Delta_{lattice}H°$ of NaCl:

$Na(s) \longrightarrow Na(g)$ $\Delta H° = 109$ kJ/mol
$Cl_2(g) \longrightarrow 2Cl(g)$ $\Delta H° = 243$ kJ/mol
$Na(g) \longrightarrow Na^+(g) + e^-$ $\Delta H° = 496$ kJ/mol
$Cl(g) + e^- \longrightarrow Cl^-(g)$ $\Delta H° = -349$ kJ/mol
$Na(s) + \frac{1}{2}Cl_2(g) \longrightarrow NaCl(s)$ $\Delta H° = -411$ kJ/mol

Compared with the lattice energy of LiF (1050 kJ/mol), is the magnitude of the value of NaCl what you expected? Explain.

8.31 Use the following information to calculate $\Delta_{lattice}H°$ of MgF_2:

$Mg(s) \longrightarrow Mg(g)$ $\Delta H° = 148$ kJ/mol
$F_2(g) \longrightarrow 2F(g)$ $\Delta H° = 159$ kJ/mol
$Mg(g) \longrightarrow Mg^+(g) + e^-$ $\Delta H° = 738$ kJ/mol
$Mg^+(g) \longrightarrow Mg^{2+}(g) + e^-$ $\Delta H° = 1450$ kJ/mol
$F(g) + e^- \longrightarrow F^-(g)$ $\Delta H° = -328$ kJ/mol
$Mg(s) + F_2(g) \longrightarrow MgF_2(s)$ $\Delta H° = -1123$ kJ/mol

Compared with the lattice energy of LiF (1050 kJ/mol) or the lattice energy you calculated for NaCl in Problem 8.30, does the relative magnitude of the value of MgF_2 surprise you? Explain.

Problems in Context

8.32 Aluminum oxide (Al_2O_3) is a widely used industrial abrasive (emery, corundum), for which the specific application depends on the hardness of the crystal. What does this hardness imply about the magnitude of the lattice energy? Would you have predicted, from the chemical formula, that Al_2O_3 is hard? Explain.

8.33 Born-Haber cycles were used to obtain the first reliable values for electron affinity by considering the EA value as the unknown and using a theoretically calculated value for the lattice energy. Use a Born-Haber cycle for KF and the following values to calculate a value for the electron affinity of fluorine:

$K(s) \longrightarrow K(g)$ $\Delta H° = 90$ kJ/mol
$K(g) \longrightarrow K^+(g) + e^-$ $\Delta H° = 419$ kJ/mol
$F_2(g) \longrightarrow 2F(g)$ $\Delta H° = 159$ kJ/mol
$K(s) + \frac{1}{2}F_2(g) \longrightarrow KF(s)$ $\Delta H° = -569$ kJ/mol
$K^+(g) + F^-(g) \longrightarrow KF(s)$ $\Delta H° = -821$ kJ/mol

The Covalent Bonding Model

(Sample Problem 8.4)

Concept Review Questions

8.34 Describe the interactions that occur between individual chlorine atoms as they approach each other and form Cl_2. What combination of forces gives rise to the energy that holds the atoms together and to the final internuclear distance?

8.35 Define *bond energy* using the H—Cl bond as an example. When this bond breaks, is energy absorbed or released? Is the accompanying ΔH value positive or negative? How do the magnitude and sign of this ΔH value relate to the value that accompanies H—Cl bond formation?

8.36 For single bonds between similar types of atoms, how does the strength of the bond relate to the sizes of the atoms? Explain.

8.37 How does the energy of the bond between a given pair of atoms relate to the bond order? Why?

8.38 When liquid benzene (C_6H_6) boils, does the gas consist of molecules, ions, or separate atoms? Explain.

Skill-Building Exercises (grouped in similar pairs)

8.39 Using only the periodic table, arrange the bonds in each set in order of increasing bond *strength*:
(a) Br—Br, Cl—Cl, I—I (b) S—H, S—Br, S—Cl
(c) C═N, C—N, C≡N

8.40 Using only the periodic table, arrange the bonds in each set in order of increasing bond *length*:
(a) H—F, H—I, H—Cl (b) C—S, C═O, C—O
(c) N—H, N—S, N—O

Problem in Context

8.41 Methanoic acid (formic acid, HCOOH; structural formula shown *below*) is secreted by certain species of ants when they bite:

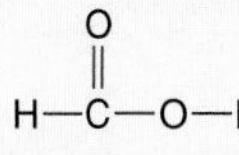

Rank the relative strengths of (a) the C—O and C═O bonds; (b) the H—C and H—O bonds. Explain your rankings.

Bond Energy and Chemical Change

(Sample Problem 8.5)

Concept Review Questions

8.42 Write a solution plan for calculating the total enthalpy change of the following reaction:

$$H_2(g) + O_2(g) \longrightarrow H_2O_2(g) \quad (\text{H—O—O—H})$$

Do not include actual numbers in your solution plan, but include the bond energies you would use and explain how you would combine them algebraically.

8.43 As discussed, for similar types of substances, the substance with weaker bonds is usually more reactive than the substance with stronger bonds. Why is this generally true?

8.44 Why is there a discrepancy between a heat of reaction obtained from calorimetry and a heat of reaction obtained from bond energies?

Skill-Building Exercises (grouped in similar pairs)

8.45 Which of the following gases would you expect to have the greater heat of reaction per mole for combustion? Why?

Methane or Methanal (formaldehyde)

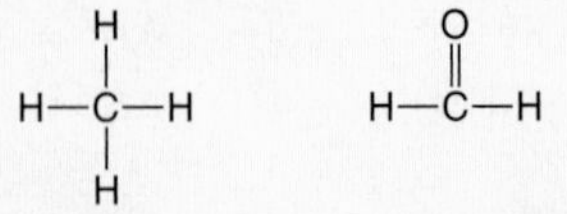

8.46 Which of the following gases would you expect to have the greater heat of reaction per mole for combustion? Why?

Ethanol or Methanol

8.47 Use bond energies to calculate the heat of reaction:

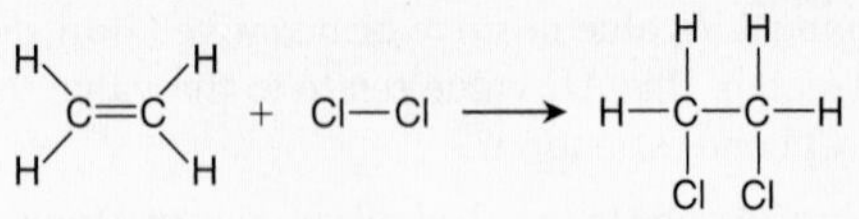

8.48 Use bond energies to calculate the heat of reaction:

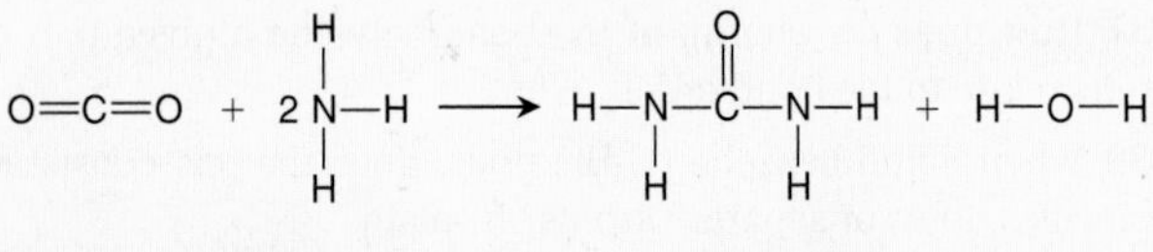

Problems in Context

8.49 An important industrial route to extremely pure ethanoic acid (acetic acid) is the reaction of methanol with carbon monoxide:

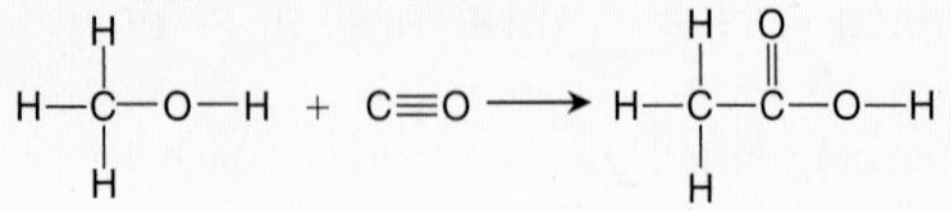

Use bond energies to calculate the enthalpy of reaction.

8.50 Sports trainers treat sprains and soreness with bromoethane (ethyl bromide). It is manufactured by reacting ethene (ethylene) with hydrogen bromide:

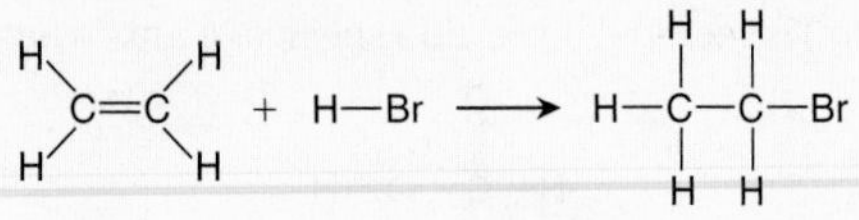

Use bond energies to find the enthalpy of reaction.

Between the Extremes: Electronegativity and Bond Polarity

(Sample Problem 8.6)

Concept Review Questions

8.51 Describe the vertical and horizontal trends in electronegativity (χ) among the main-group elements. According to the Pauling electronegativity scale, what are the two most electronegative elements? The two least electronegative elements?

8.52 What is the general relationship between IE_1 and χ for the elements? Why?

8.53 Is the H—O bond in water nonpolar covalent, polar covalent, or ionic? Define each term, and explain your choice.

8.54 How does electronegativity differ from electron affinity?

8.55 How is the partial ionic character of a bond in a diatomic molecule related to $\Delta\chi$ for the bonded atoms? Why?

Skill-Building Exercises (grouped in similar pairs)

8.56 Using only the periodic table, arrange the elements in each set in order of *increasing* χ:
(a) S, O, Si (b) Mg, P, As

8.57 Using only the periodic table, arrange the elements in each set in order of *increasing* χ:
(a) I, Br, N (b) Ca, H, F

8.58 Using only the periodic table, arrange the elements in each set in order of *decreasing* χ:
(a) N, P, Si (b) Ca, Ga, As

8.59 Using only the periodic table, arrange the elements in each set in order of *decreasing* χ:
(a) Br, Cl, P (b) I, F, O

8.60 Use Figure 8.23 to indicate the polarity of each bond with a *polar arrow*:
(a) N—B (b) N—O (c) C—S
(d) S—O (e) N—H (f) Cl—O

8.61 Use Figure 8.23 to indicate the polarity of each bond with *partial charges*:
(a) Br—Cl (b) F—Cl (c) H—O
(d) Se—H (e) As—H (f) S—N

8.62 Which is the more polar bond in each pair from Problem 8.60?
(a) N—B or N—O (b) C—S or S—O (c) N—H or Cl—O

8.63 Which is the more polar bond in each pair from Problem 8.61?
(a) Br—Cl or F—Cl (b) H—O or Se—H (c) As—H or S—N

8.64 Are the bonds in each substance ionic, nonpolar covalent, or polar covalent? Arrange the substances with polar covalent bonds in order of increasing bond polarity:
(a) S_8 (b) RbCl (c) PF_3 (d) SCl_2 (e) F_2 (f) SF_2

8.65 Are the bonds in each substance ionic, nonpolar covalent, or polar covalent? Arrange the substances with polar covalent bonds in order of increasing bond polarity:
(a) KCl (b) P_4 (c) BF_3 (d) SO_2 (e) Br_2 (f) NO_2

8.66 Rank the compounds in each set in order of *increasing* ionic character of their bonds. Use a *polar arrow* to indicate the bond polarity of each compound:
(a) HBr, HCl, HI (b) H_2O, CH_4, HF (c) SCl_2, PCl_3, $SiCl_4$

8.67 Rank the compounds in each set in order of *decreasing* ionic character of their bonds. Use *partial charges* to indicate the bond polarity of each compound:
(a) PCl_3, PBr_3, PF_3 (b) BF_3, NF_3, CF_4 (c) SeF_4, TeF_4, BrF_3

Problem in Context

8.68 The energy of the C—C bond is 347 kJ/mol, and the energy of the Cl—Cl bond is 243 kJ/mol. Which of the following values might you expect for the C—Cl bond energy? Explain.
(a) 590 kJ/mol (sum of the values given)
(b) 104 kJ/mol (difference of the values given)
(c) 295 kJ/mol (average of the values given)
(d) 339 kJ/mol (greater than the average of the values given)

Depicting Molecules and Ions with Lewis Structures

(Sample Problems 8.7 to 8.11)

Concept Review Questions

8.69 Which of these atoms *cannot* serve as a central atom in a Lewis structure? Explain.
(a) O (b) He (c) F (d) H (e) P

8.70 When is a resonance hybrid needed to depict the bonding in a molecule adequately? Using NO_2 as an example, explain how a resonance hybrid is consistent with the actual bond length, bond strength, and bond order.

8.71 In which of these bonding patterns does X obey the octet rule?

(a) (b) (c) (d) (e) (f) (g) (h)

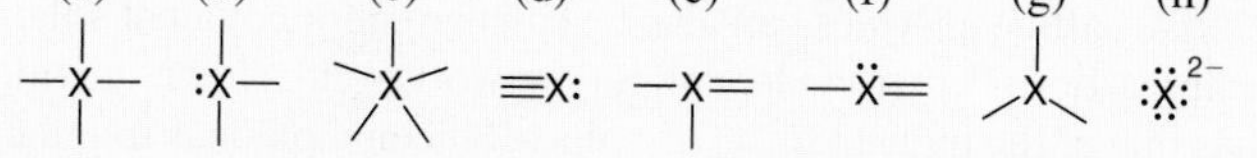

8.72 (a) What is required for an atom to expand its valence shell? (b) Which atom can expand its valence shell?
(i) F (ii) S (iii) H (iv) Al (v) Se (vi) Cl

Skill-Building Exercises (grouped in similar pairs)

8.73 Draw a Lewis structure for (a) SiF_4; (b) $SeCl_2$; (c) COF_2 (C is central).

8.74 Draw a Lewis structure for (a) PH_4^+; (b) C_2F_4; (c) SbH_3.

8.75 Draw a Lewis structure for (a) PF_3; (b) H_2CO_3 (both H atoms are attached to O atoms); (c) CS_2.

8.76 Draw a Lewis structure for (a) CH_4S; (b) S_2Cl_2; (c) $CHCl_3$.

8.77 Draw Lewis structures for all the important resonance forms of (a) NO_2^+; (b) NO_2F (N is central).

8.78 Draw Lewis structures for all the important resonance forms of (a) HNO_3 ($HONO_2$); (b) $HAsO_4^{2-}$ ($HOAsO_3^{2-}$).

8.79 Draw Lewis structures for all the important resonance forms of (a) N_3^-; (b) NO_2^-.

8.80 Draw Lewis structures for all the important resonance forms of (a) HCO_2^- (H is attached to C); (b) $HBrO_4$ ($HOBrO_3$).

8.81 Draw the Lewis structure with the lowest formal charges and determine the charge of each atom in (a) IF_5; (b) AlH_4^-.

8.82 Draw the Lewis structure with the lowest formal charges and determine the charge of each atom in (a) OCS; (b) NO.

8.83 Draw the Lewis structure with the lowest formal charges and determine the charge of each atom in (a) CN^-; (b) ClO^-.

8.84 Draw the Lewis structure with the lowest formal charges and determine the charge of each atom in (a) BF_4^-; (b) ClNO.

8.85 Draw a Lewis structure for the resonance form of each ion with the lowest possible formal charges, show the charges, and give oxidation numbers of the atoms:
(a) BrO_3^- (b) SO_3^{2-}

8.86 Draw a Lewis structure for the resonance form of each ion with the lowest possible formal charges, show the charges, and give oxidation numbers of the atoms:
(a) AsO_4^{3-} (b) ClO_2^-

8.87 These species do not obey the octet rule. Draw a Lewis structure for each, and state the type of octet-rule exception:
(a) BH_3 (b) AsF_4^- (c) $SeCl_4$

8.88 These species do not obey the octet rule. Draw a Lewis structure for each, and state the type of octet-rule exception:
(a) PF_6^- (b) ClO_3 (c) H_3PO_3 (one P—H bond)

8.89 These species do not obey the octet rule. Draw a Lewis structure for each, and state the type of octet-rule exception:
(a) BrF_3 (b) ICl_2^- (c) BeF_2

8.90 These species do not obey the octet rule. Draw a Lewis structure for each, and state the type of octet-rule exception:
(a) O_3^- (b) XeF_2 (c) SbF_4^-

Problems in Context

8.91 Molten beryllium chloride reacts with chloride ion from molten NaCl to form the $BeCl_4^{2-}$ ion, in which the Be atom attains an octet. Show the net ionic reaction with Lewis structures.

8.92 Despite many attempts, the perbromate ion (BrO_4^-) was not prepared in a laboratory until about 1970. (In fact, articles were published to explain theoretically why it could never be prepared!) Draw a Lewis structure for BrO_4^- in which all the atoms have the lowest formal charges.

8.93 Cryolite (Na_3AlF_6) is an indispensable component in the electrochemical production of aluminum. Draw a Lewis structure for the AlF_6^{3-} ion.

8.94 Phosgene is a colourless, highly toxic gas that was employed against troops in World War I and is used today as a key reactant in organic syntheses. From the following resonance structures, select the one with the lowest formal charges:

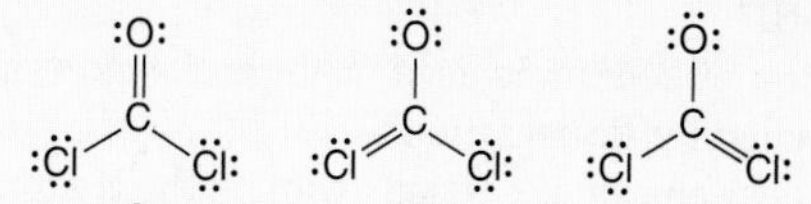

A B C

An Introduction to Metallic Bonding

Concept Review Questions

8.95 (a) List four physical characteristics of a solid metal.
(b) List two chemical characteristics of a metallic element.

8.96 Briefly account for the following relative values:
(a) The melting points of Na and K are 89°C and 63°C, respectively.
(b) The melting points of Li and Be are 180°C and 1287°C, respectively.
(c) The boiling point of Li is more than 1100°C higher than its melting point.

8.97 Magnesium metal is easily deformed by an applied force, whereas magnesium fluoride is shattered. Why do these two solids behave so differently?

Comprehensive Problems

8.98 Geologists have the following rule: when molten rock cools and solidifies, crystals of compounds with the smallest lattice energies appear at the bottom of the mass. Suggest a reason for this rule.

8.99 Ethyne (acetylene gas; HC≡CH) burns in an oxyethyne torch to produce carbon dioxide and water vapour. The heat of reaction for the combustion of ethyne is 1259 kJ/mol.
(a) Calculate the C≡C bond energy, and compare your value with the value in Table 8.2.
(b) When 500.0 g of ethyne (acetylene) burns, what amount of heat (kJ) is given off?
(c) What mass (g) of CO_2 forms?
(d) What volume (L) of O_2 at 298 K and 1823 kPa is consumed?

8.100 Use Lewis electron-dot symbols to represent the formation of (a) BrF_3 from bromine and fluorine atoms; (b) AlF_3 from aluminum and fluorine atoms.

8.101 Even though so much energy is required to form a metal cation with a 2+ charge, the alkaline earth metals form halides with general formula MX_2, rather than MX.
(a) Use the following data to calculate $\Delta_f H°$ of MgCl:

$Mg(s) \longrightarrow Mg(g)$ $\Delta H° = 148$ kJ/mol
$Cl_2(g) \longrightarrow 2Cl(g)$ $\Delta H° = 243$ kJ/mol
$Mg(g) \longrightarrow Mg^+(g) + e^-$ $\Delta H° = 738$ kJ/mol
$Cl(g) + e^- \longrightarrow Cl^-(g)$ $\Delta H° = -349$ kJ/mol
$\Delta_{lattice}H°$ of MgCl = 783.5 kJ/mol

(b) Is MgCl favoured energetically relative to Mg and Cl_2? Explain.
(c) Use Hess's law to calculate $\Delta H°$ for the conversion of MgCl to $MgCl_2$ and Mg ($\Delta_f H°$ of $MgCl_2 = -641.6$ kJ/mol).
(d) Is MgCl favoured energetically relative to $MgCl_2$? Explain.

8.102 Gases react explosively if the heat released when the reaction begins is sufficient to cause more reaction, which leads to a rapid expansion of the gases. Use bond energies to calculate $\Delta H°$ of the following reactions, and predict which reaction occurs explosively:
(a) $H_2(g) + Cl_2(g) \longrightarrow 2HCl(g)$
(b) $H_2(g) + I_2(g) \longrightarrow 2HI(g)$
(c) $2H_2(g) + O_2(g) \longrightarrow 2H_2O(g)$

8.103 By using photons of specific wavelengths, chemists can dissociate gaseous HI to produce H atoms with certain speeds. When HI dissociates, the H atoms move away rapidly, whereas the heavier I atoms move away more slowly.
(a) What is the longest wavelength (nm) that can dissociate a molecule of HI?
(b) If a photon of 254 nm is used, what is the excess energy (J) over that needed for dissociation?
(c) If this excess energy is carried away by the H atom as kinetic energy, what is its speed (m/s)?

8.104 In developing the concept of electronegativity, Pauling used the term *excess bond energy* for the difference between the actual bond energy of X—Y and the average bond energies of X—X and Y—Y (see discussion about HF in Section 8.4). Based on the values in Figure 8.23, which substance contains bonds with no excess bond energy?
(a) PH_3 (b) CS_2 (c) BrCl (d) BH_3 (e) Se_8

8.105 Use condensed electron configurations to predict the relative hardnesses and melting points of rubidium (Z = 37), vanadium (Z = 23), and cadmium (Z = 48).

8.106 Without stratospheric ozone (O_3), harmful solar radiation would cause gene alterations. Ozone forms when the bond in O_2 breaks and each O atom reacts with another O_2 molecule. Ozone is destroyed by its reaction with Cl atoms that are formed when the C—Cl bond in synthetic chemicals breaks. Find the wavelengths of light that can break the C—Cl bond and the bond in O_2.

8.107 "Inert" xenon actually forms many compounds, especially with highly electronegative fluorine. The $\Delta_f H°$ values for xenon difluoride, tetrafluoride, and hexafluoride are −105 kJ/mol, −284 kJ/mol, and −402 kJ/mol, respectively. Find the average bond energy of the Xe—F bonds in each fluoride.

8.108 The HF bond length is 92 pm, 16% shorter than the sum of the covalent radii of H (37 pm) and F (72 pm). Suggest a reason for this difference. Similar data show that the difference becomes smaller down the group, from HF to HI. Explain.

8.109 There are two main types of covalent bond breakage. In homolytic breakage (as in Table 8.2), each atom in the bond gets one of the shared electrons. In some cases, the electronegativity of adjacent atoms affects the bond energy. In heterolytic breakage, one atom gets both electrons and the other atom gets none; thus, a cation and an anion form.
(a) Why is the C—C bond in $H_3C—CF_3$ (423 kJ/mol) stronger than the C—C bond in $H_3C—CH_3$ (376 kJ/mol)?
(b) Use bond energy and any other data to calculate the enthalpy of reaction for the heterolytic cleavage of O_2.

8.110 Find the longest wavelengths of light that can cleave the bonds in elemental nitrogen, oxygen, and fluorine.

8.111 The work function (ϕ) of a metal is the minimum energy needed to remove an electron from its surface.
(a) Is it easier to remove an electron from a gaseous silver atom or from the surface of solid silver ($\phi = 7.59 \times 10^{-19}$ J; IE = 731 kJ/mol)?
(b) Explain your answer to part (a) in terms of the electron-sea model of metallic bonding.

8.112 Lattice energies can also be calculated for covalent solids using a Born-Haber cycle, and the network solid silicon dioxide has one of the highest $\Delta_{lattice}H°$ values. Silicon dioxide is found in pure crystalline form as transparent rock quartz. Much harder than glass, silicon dioxide was once prized for making lenses for optical devices and expensive eyeglasses. Use Appendix B and the following data to calculate $\Delta_{lattice}H°$ of SiO_2:

$Si(s) \longrightarrow Si(g)$ $\Delta H° = 454$ kJ/mol
$Si(g) \longrightarrow Si^{4+}(g) + 4e^-$ $\Delta H° = 9949$ kJ/mol
$O_2(g) \longrightarrow 2O(g)$ $\Delta H° = 498$ kJ/mol
$O(g) + 2e^- \longrightarrow O^{2}(g)$ $\Delta H° = 737$ kJ/mol

8.113 The average C—H bond energy in CH_4 is 415 kJ/mol. Use Table 8.2 and the following information to calculate the average C—H bond energy in (a) ethane (C_2H_6; C—C bond); (b) ethene (C_2H_4; C=C bond); (c) ethyne (C_2H_2; C≡C bond):

$C_2H_6(g) + H_2(g) \longrightarrow 2CH_4(g)$ $\Delta_r H° = -65.07$ kJ/mol
$C_2H_4(g) + 2H_2(g) \longrightarrow 2CH_4(g)$ $\Delta_r H° = -202.21$ kJ/mol
$C_2H_2(g) + 3H_2(g) \longrightarrow 2CH_4(g)$ $\Delta_r H° = -376.74$ kJ/mol

8.114 Carbon-carbon bonds form the "backbone" of nearly every organic and biological molecule. The average bond energy of the C—C bond is 347 kJ/mol. Calculate the frequency and wavelength of the least energetic photon that can break an average C—C bond. In what region of the electromagnetic spectrum is this radiation?

8.115 In a future hydrogen-fuel economy, the cheapest source of H_2 will certainly be water. It takes 467 kJ to produce 1 mol of H atoms from water. What are the frequency, wavelength, and minimum energy of a photon that can free an H atom from water?

8.116 Methoxymethane (dimethyl ether, CH_3OCH_3) and ethanol (CH_3CH_2OH) are constitutional isomers (see Table 3.3).
(a) Use Table 8.2 to calculate $\Delta_r H°$ for the formation of each compound as a gas from methane and oxygen; water vapour also forms.
(b) State which reaction is more exothermic.
(c) Calculate $\Delta_r H°$ for the conversion of ethanol to methoxymethane.

8.117 Enthalpies of reaction calculated from bond energies and from enthalpies of formation are often, but not always, close to each other.
(a) Industrial ethanol (CH_3CH_2OH) is produced by a catalytic reaction of ethene (ethylene, $CH_2{=}CH_2$) with water at high pressures and temperatures. Calculate $\Delta_r H°$ for this gas-phase hydration of ethene to ethanol, using bond energies and then using enthalpies of formation.
(b) Ethane-1,2-diol (ethylene glycol) is produced by the catalytic oxidation of ethene (ethylene) to epoxyethane (ethylene oxide), which then reacts with water to form ethane-1,2-diol:

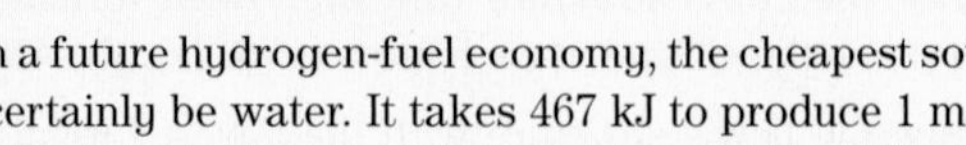

The $\Delta_r H°$ for this hydrolysis step, based on enthalpies of formation, is −97 kJ/mol. Calculate $\Delta_r H°$ for the hydrolysis using bond energies.
(c) Why are the two values relatively close for the hydration in part (a), but not close for the hydrolysis in part (b)?

8.118 In addition to ammonia, nitrogen forms three other hydrides: hydrazine (N_2H_4), diazene (N_2H_2), and tetrazene (N_4H_4).

(a) Use Lewis structures to compare the strength, length, and order of the nitrogen-nitrogen bonds in hydrazine, diazene, and N_2.
(b) Tetrazene (atom sequence H_2NNNNH_2) decomposes above 0°C to hydrazine and nitrogen gas. Draw a Lewis structure for tetrazene, and calculate $\Delta_r H°$ for its decomposition.

8.119 Draw a Lewis structure for each species:

(a) PF_5	(b) CCl_4	(c) H_3O^+	(d) ICl_3
(e) BeH_2	(f) PH_2^-	(g) $GeBr_4$	(h) CH_3^-
(i) BCl_3	(j) BrF_4^+	(k) XeO_3	(l) TeF_4

8.120 Like several other bonds, carbon-oxygen bonds have lengths and strengths that depend on the bond order. Draw Lewis structures for the following species, and arrange them in order of increasing carbon-oxygen bond length and then by increasing carbon-oxygen bond strength:
(a) CO (b) CO_3^{2-} (c) H_2CO (d) CH_4O
(e) HCO_3^- (H attached to O)

8.121 In the 1980s, there was an international agreement to destroy all stockpiles of mustard gas, $ClCH_2CH_2SCH_2CH_2Cl$. When this gas contacts the moisture in eyes, nasal passages, and skin, the —OH groups of water replace the Cl atoms and create high local concentrations of hydrochloric acid, which causes severe blistering and tissue destruction. Write a balanced equation for this reaction, and calculate $\Delta_r H°$.

8.122 Ethanol (CH_3CH_2OH) is being used as a gasoline additive or alternative in many parts of the world.
(a) Use bond energies to find $\Delta_r H°$ for the combustion of gaseous ethanol. (Assume that H_2O forms as a gas.)
(b) In its standard state at 25°C, ethanol is a liquid. Its vaporization requires 40.5 kJ/mol. Correct the value from part (a) to find the enthalpy of reaction for the combustion of liquid ethanol.
(c) Calculate the enthalpy of reaction for the combustion of liquid ethanol using standard enthalpies of formation (Appendix B). How does this value compare with the value you calculated in part (b)?
(d) "Greener" methods produce ethanol from corn and other plant material, but the main industrial method involves hydrating ethene (ethylene) from petroleum. Use Lewis structures and bond energies to calculate $\Delta_r H°$ for the formation of gaseous ethanol from ethene (ethylene) gas with water vapour.

8.123 In the following compounds, the C atoms form a single ring. Draw a Lewis structure for each compound, identify cases for which resonance exists, and determine the carbon-carbon bond order(s):
(a) C_3H_4 (b) C_3H_6 (c) C_4H_6 (d) C_4H_4 (e) C_6H_6

8.124 An experiment requires 50.0 mL of 0.040 mol/L NaOH for the titration of 1.00 mmol of acid. Mass analysis of the acid shows that its composition is 2.24% hydrogen, 26.7% carbon, and 71.1% oxygen. Draw the Lewis structure for the acid.

8.125 A gaseous compound has a composition by mass of 24.8% carbon, 2.08% hydrogen, and 73.1% chlorine. At STP, the gas has a density of 4.3 g/L. Draw a Lewis structure that matches this information. Would another structure be equally satisfactory? Explain.

8.126 Perchlorates are powerful oxidizing agents that are used in fireworks, in flares, and formerly in the booster rockets of space shuttles. Lewis structures for the perchlorate ion (ClO_4^-) can be drawn with all single bonds or with one, two, or three double bonds. Draw each of these possible resonance forms, use formal charges to determine the most important form, and calculate the average bond order of this form.

8.127 Methane burns in oxygen to form carbon dioxide and water vapour. Hydrogen sulfide burns in oxygen to form sulfur dioxide and water vapour. Use bond energies (Table 8.2) to determine the enthalpy of each reaction per mole of O_2. (Assume Lewis structures with zero formal charges; BE of S═O is 552 kJ/mol.)

8.128 Use Lewis structures to determine which *two* of the following are unstable:
(a) SF_2 (b) SF_3 (c) SF_4 (d) SF_5 (e) SF_6

8.129 A major short-lived neutral species in flames is OH.
(a) What is unusual about the electronic structure of OH?
(b) Use the standard enthalpy of formation of OH(*g*) (39.0 kJ/mol) and bond energies to calculate the O—H bond energy in OH(*g*).
(c) From the average value for the O—H bond energy (Table 8.2) and your value for the O—H bond energy in OH(*g*), find the energy needed to break the first O—H bond in water.

8.130 Pure HN_3 (atom sequence HNNN) is explosive. In aqueous solution, it is a weak acid that yields the azide ion, N_3^-. Draw resonance structures to explain why the nitrogen-nitrogen bond lengths are equal in N_3^- but unequal in HN_3.

8.131 Dinitrogen monoxide (N_2O) supports combustion in a manner similar to oxygen, with the nitrogen atoms forming N_2. Draw three resonance structures for N_2O (one N is central), and use formal charges to decide the relative importance of each. What correlation can you suggest between the most important structure and the observation that N_2O supports combustion?

8.132 Ethanedioic acid (oxalic acid, $H_2C_2O_4$) is found in toxic concentrations in rhubarb leaves. This acid forms two ions, $HC_2O_4^-$ and $C_2O_4^{2-}$, by the sequential loss of H^+ ions. Draw Lewis structures for the three species, and comment on the relative lengths and strengths of their carbon-oxygen bonds. The connections among the atoms are shown below with single bonds only.

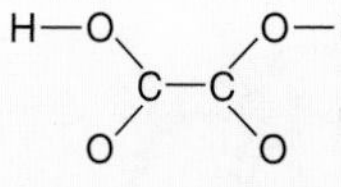

8.133 The Murchison meteorite that landed in Australia in 1969 contained 92 different amino acids, including 21 found in Earth organisms. A skeleton structure (single bonds only) of one of the extraterrestrial amino acids is shown below:

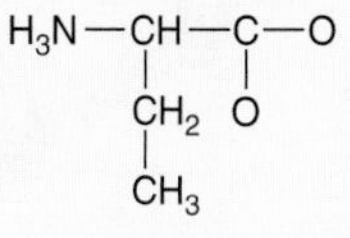

Draw a Lewis structure, and identify any atoms that have a nonzero formal charge.

8.134 Hydrazine (N_2H_4) is used as a rocket fuel because it reacts very exothermically with oxygen to form nitrogen gas and water vapour. The heat released and the increase in the number of moles of gas provide the thrust. Calculate the enthalpy of reaction.

8.135 When gaseous sulfur trioxide is dissolved in concentrated sulfuric acid, disulfuric acid forms:

$$SO_3(g) + H_2SO_4(l) \longrightarrow H_2S_2O_7(l)$$

Use bond energies (Table 8.2) to determine $\Delta_r H°$. (The S atoms in $H_2S_2O_7$ are bonded through an O atom. Assume Lewis structures with zero formal charges; BE of S═O is 552 kJ/mol.)

8.136 Hydrogen cyanide can be catalytically reduced with hydrogen to form methylamine. Use Lewis structures and bond energies to determine $\Delta_r H°$ for the following reaction:

$$HCN(g) + 2H_2(g) \longrightarrow CH_3NH_2(g)$$

The Shapes of Molecules

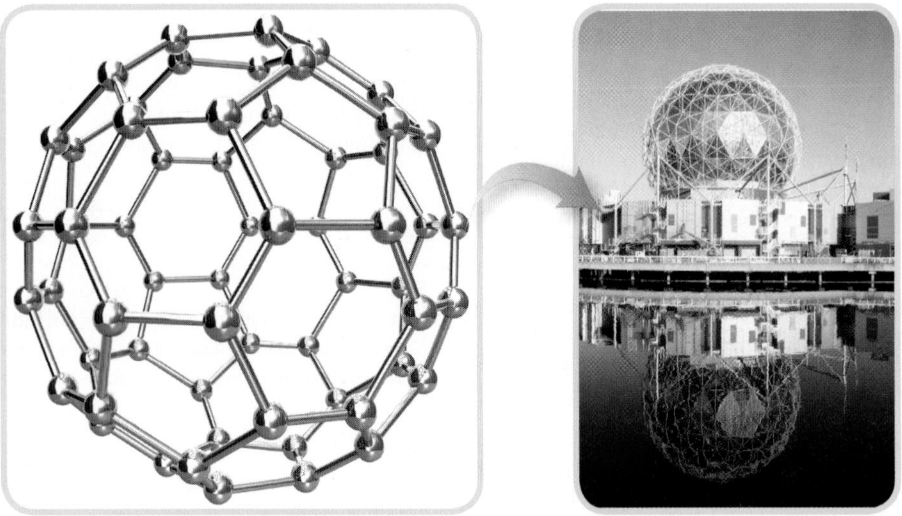

IN THIS CHAPTER . . . We learn how to picture molecules by applying two approaches that explain their shapes, and we examine some effects of molecular shape on physical and biochemical behaviour. By the end of this chapter, you will be able to

- Determine bond angles and apply a theory that converts a two-dimensional formula into a three-dimensional shape
- Describe five basic classes of shapes that many molecules adopt, explain how multiple bonds and lone pairs affect these shapes, and determine how smaller molecular portions can be combined into the shapes of more complex molecules
- Explain the relationships between bond polarity, shape, and molecular polarity and describe the effect of polarity on behaviour
- Discuss, with examples, how shape influences biological function

9

Carbon dioxide, nitrogen dioxide, and sulfur dioxide: CO_2, NO_2, and SO_2. At first glance, it would appear that, aside from the first element, these molecules should be identical in every way. What is most interesting is, in fact, that all three have very different structures that result in their having very different shapes, properties, and reactivities. Carbon dioxide is linear and nontoxic in small doses; nitrogen dioxide is bent with a lone electron, irritating at low doses, and potentially toxic at higher doses; and sulfur dioxide is bent and highly toxic. What makes these three substances so different?

This chapter describes the valence-shell electron-pair repulsion (VSEPR) theory. VSEPR can be used to predict how electrons are distributed in a substance; to determine how this distribution affects the structure and shape of the resulting molecules; and to obtain information about bond angles, lengths, and polarities.

Concepts and Skills to Review before Studying This Chapter

- Electron configurations of main-group elements (Section 7.2)
- Electron-dot symbols (Section 8.1)
- Octet rule (Section 8.1)
- Bond order, bond length, and bond energy (Sections 8.3 and 8.4)
- Polar covalent bonds and bond polarity (Section 8.5)
- Depictions of molecules and ions with Lewis structures (Section 8.6)

9.1 Valence-Shell Electron-Pair Repulsion Theory

Virtually every biochemical process hinges, to a great extent, on the shapes of interacting molecules. Every medicine you take, odour you smell, and flavour you taste depends on part or all of one molecule fitting together with another. Biologists have found that complex behaviours in many organisms, such as mating, defence, navigation, and feeding, often depend on one molecule's shape matching the shape of another. In this section, we discuss **valence-shell electron-pair repulsion (VSEPR) theory**, a model for predicting the shape of a molecule. It is also named *Gillespie-Nyholm theory* after its two main developers, the Canadian chemist Ronald James Gillespie (Figure 9.1A) and the Australian chemist Sir Ronald Sydney Nyholm (Figure 9.1B).

To obtain the molecular shape, chemists start with the Lewis structure and apply VSEPR theory. Its basic principle is that, *to minimize repulsions, each group of valence electrons around a central atom is located as far as possible from the others.* A "group" of electrons is any number that occupies a localized region around an atom: single bond, double bond, triple bond,* lone pair, or even lone electron. The **molecular shape** is the three-dimensional arrangement of nuclei joined by the bonding groups. The names of the molecular shapes, as you will see in the following subsections, are directly related to the positions of the atoms.

Electron-Group Arrangements and Molecular Shapes

When two, three, four, five, or six objects attached to a central point maximize the space between them, five geometric patterns result, which Figure 9.2A shows with balloons. If the objects are valence-electron groups, repulsions maximize the space that each group occupies around the central atom, and we obtain the five *electron-group arrangements* seen in the great majority of molecules and polyatomic ions.

Classifying Molecular Shapes *The electron-group arrangement* is defined by the bonding *and* nonbonding electron groups, but the *molecular shape* is defined by the relative positions of the nuclei, which are connected by the bonding groups only.

*The two electron pairs in a double bond (or the three pairs in a triple bond) occupy separate orbitals, so they remain near each other and act as one electron group (see Chapter 10).

A.

B.

FIGURE 9.1 **A.** Ronald J. Gillespie (McMaster University, Hamilton, Ontario) was the first person to formulate and develop the VSEPR model for predicting the shape of a molecule based on the number of electron pairs in its outer shells. Dr. Gillespie's distinguished career has also been highlighted by the discovery and characterization of super acid media, the earliest identification of many polyatomic cations of nonmetals, and early studies of noble gas fluorocations. In 2007, Dr. Gillespie was awarded the Order of Canada. **B.** Sir Ronald Sydney Nyholm (1917–1971) was an Australian chemist. Together with Dr. Gillespie, he developed the VSEPR theory.

FIGURE 9.2 **Electron-group repulsions and molecular shapes. A.** Five geometric orientations arise when each balloon occupies as much space as possible. **B.** Mutually repelling bonding groups (*grey sticks*) attach a surrounding atom (*dark grey*) to the central atom (*red*). The name is the electron-group arrangement.

Figure 9.2B shows the molecular shapes that occur when *all* the surrounding electron groups are *bonding* groups. When some of the electron groups are *nonbonding* groups, no nucleus is attached, so different molecular shapes occur. Thus, *the same electron-group arrangement can give rise to different molecular shapes*: some with all bonding groups (as in Figure 9.2B) and others with bonding and nonbonding groups. To classify molecular shapes, we assign each shape a specific AX_mE_n designation, where m and n are integers, A is the central atom, X is a surrounding atom, and E is a nonbonding valence-electron group (usually a lone pair, but can also be a single electron).

The Importance of Bond Angle The **bond angle** is the angle that is formed by the nuclei of two surrounding atoms with the nucleus of the central atom at the vertex. The angles shown for the shapes in Figure 9.2B are *ideal* bond angles, determined by basic geometry alone. We observe these angles when all the bonding groups are the same and are connected to the same type of atom. When this is not the case, the real bond angles deviate from the ideal angles:

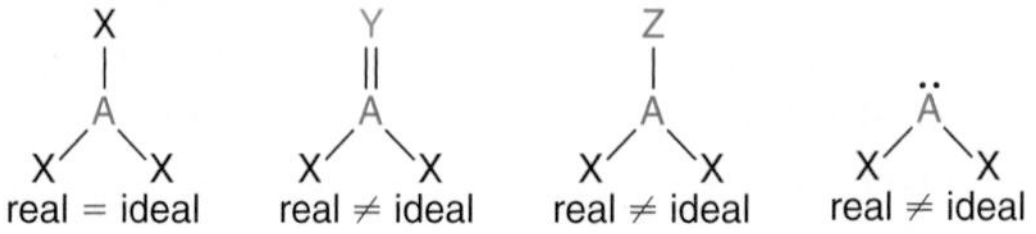

The Molecular Shape with Two Electron Groups (Linear Arrangement)

Two electron groups attached to a central atom point in opposite directions. This **linear arrangement** of electron groups results in a molecule with a **linear shape** and a bond angle of 180°. Figure 9.3 shows the general form (*top*) and shape (*middle*), as well as the VSEPR shape class (AX_2), and gives the formulas for three linear molecules.

Another linear molecule is gaseous beryllium chloride ($BeCl_2$). Recall that gaseous Be compounds are electron deficient, with two electron pairs around the central Be:

In carbon dioxide, the central C atom forms two double bonds with the O atoms:

Each double bond acts as one electron group and is 180° away from the other. The lone pairs on the O atoms of CO_2 or the Cl atoms of $BeCl_2$ are *not* involved in the molecular shape: only electron groups around the *central* atom affect the shape.

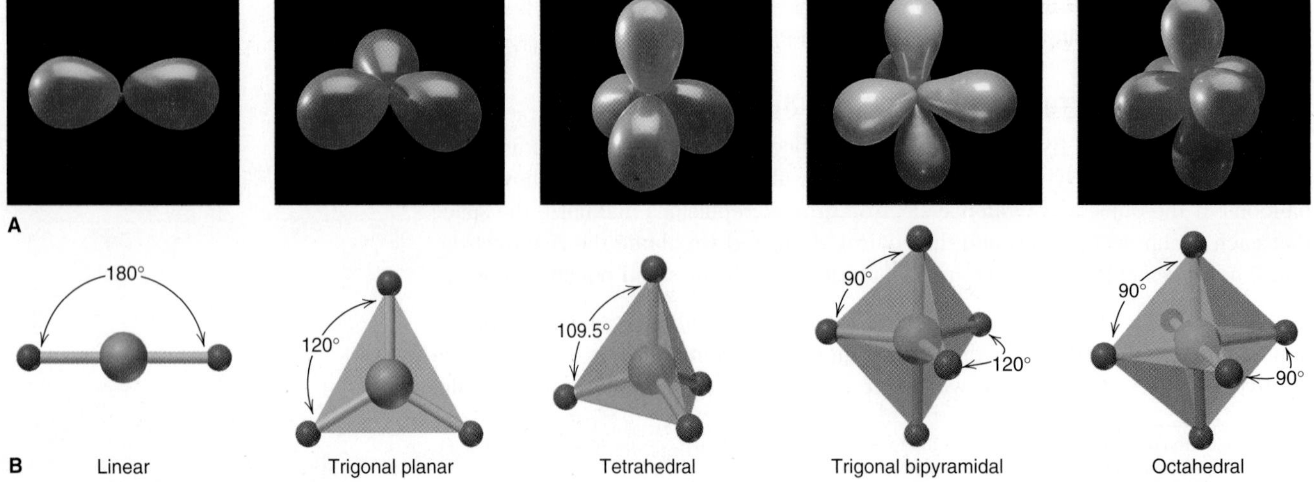

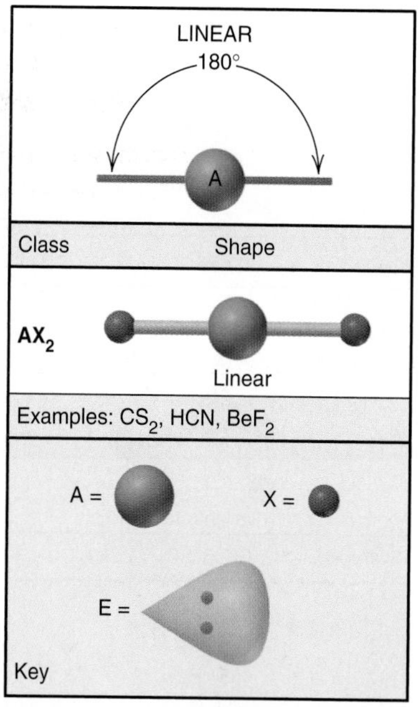

FIGURE 9.3 The single molecular shape of the linear electron-group arrangement. The key (*bottom*) for A, X, and E also applies to Figures 9.4, 9.5, 9.7, 9.8, and 9.10. Recall that A is the central atom, X is a surrounding atom, and E is a nonbonding valence-electron group (usually a lone pair, but can also be a single electron).

Molecular Shapes with Three Electron Groups (Trigonal Planar Arrangement)

Three electron groups around a central atom point to the corners of an equilateral triangle, which gives the **trigonal planar arrangement** and an ideal bond angle of 120° (Figure 9.4). This arrangement has two molecular shapes: one with all bonding groups and the other with one lone pair. It allows us to see *the effects of lone pairs and double bonds on bond angles.*

1. *All bonding groups: trigonal planar shape (AX_3).* Boron trifluoride (BF_3), another electron-deficient molecule, is one example of a molecule with the trigonal planar shape. It has six electrons around the central B atom, in three single bonds to F atoms. The four nuclei lie in a plane, and each F—B—F angle is 120°:

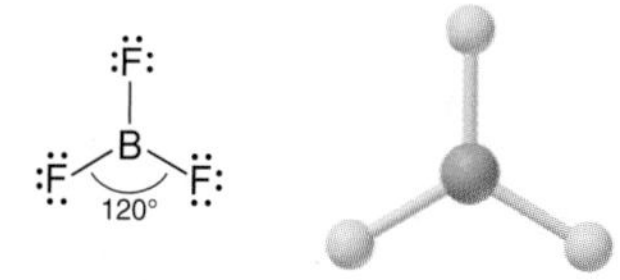

The nitrate ion (NO_3^-) is one of several polyatomic ions with the trigonal planar shape. One of three resonance forms of the nitrate ion (Sample Problem 8.10) is given below:

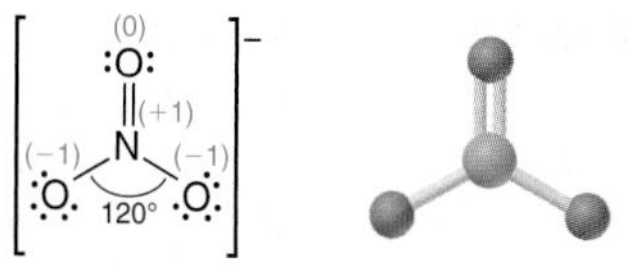

The resonance hybrid has three identical bonds of bond order $1\frac{1}{3}$, so the ideal bond angle is observed.

2. *One lone pair: bent or V shape (AX_2E_1).* Gaseous tin(II) chloride is a molecule with a **bent shape (V shape)**, with three electron groups in a trigonal plane and a lone pair at one of the triangle's corners. A lone pair often has a major effect on bond angle. Because it is held by only one nucleus, a lone pair is less confined than a bonding pair, so it exerts stronger repulsions. In general, *a lone pair repels bonding pairs more than bonding pairs repel each other, so it decreases the angle between bonding pairs.* Note the 95° bond angle in $SnCl_2$, which is considerably less than the ideal 120°:

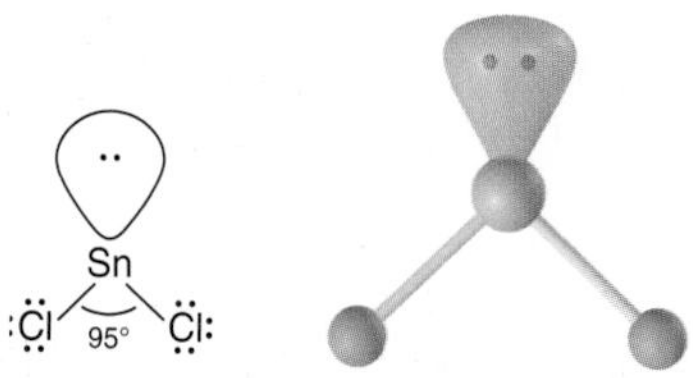

The Effect of Double Bonds on Bond Angle When the surrounding atoms and electron groups are not identical, the bond angles may also be affected. Consider methanal (formaldehyde, CH_2O), uses of which include the manufacture of countertops and the production of methanol. Its trigonal planar shape is due to two types of surrounding atoms (O and H) and two types of electron groups (single and double bonds):

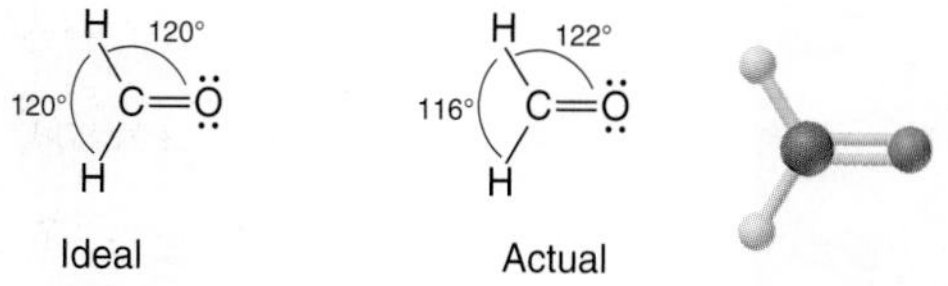

The actual H—C—H bond angle is less than the ideal 120° because *the greater electron density of a double bond repels electrons in single bonds more than the single bonds repel each other.*

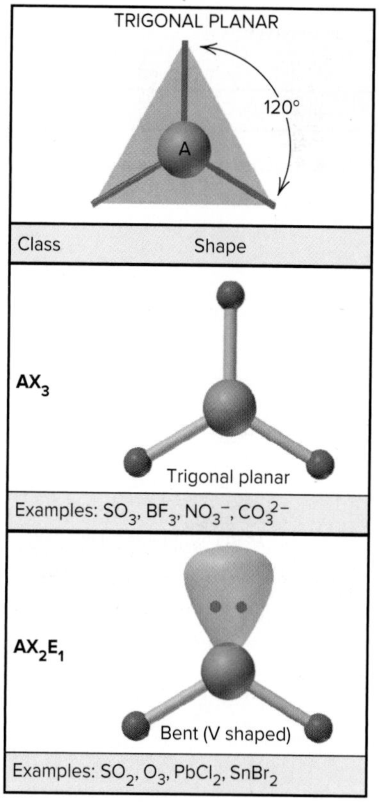

FIGURE 9.4 The two molecular shapes of the trigonal planar electron-group arrangement

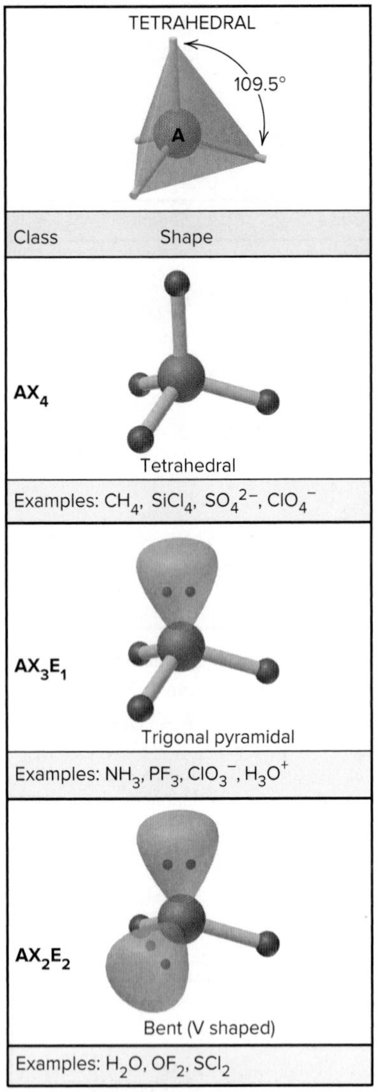

FIGURE 9.5 The three molecular shapes of the tetrahedral electron-group arrangement

Molecular Shapes with Four Electron Groups (Tetrahedral Arrangement)

Shapes based on two or three electron groups lie in a plane, but four electron groups require three dimensions to maximize separation. Consider methane, whose Lewis structure (*below, left*) shows four bonds pointing to the corners of a square, suggesting 90° bond angles. However, *Lewis structures do **not** depict shape.* In three dimensions, the four electron groups lie at the corners of a *tetrahedron,* a polyhedron with four faces made of equilateral triangles, giving bond angles of 109.5° (Figure 9.5):

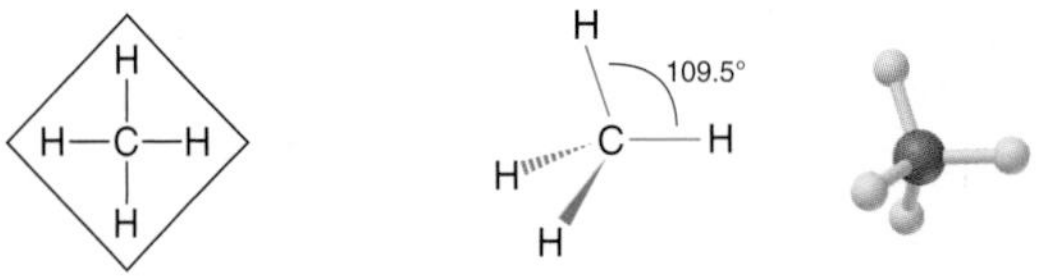

A *perspective drawing* for methane (*above, middle*) indicates depth by using solid and dashed wedges for bonds out of the plane of the page. The normal bond lines (*blue*) are in the plane of the page; the solid wedge (*green*) is the bond from the C atom in the plane of the page to the H above the plane; and the dashed wedge (*red*) is the bond from the C to the H below the plane of the page. The ball-and-stick model (*above, right*) shows the tetrahedral shape more clearly.

All molecules or ions with four electron groups around a central atom adopt the **tetrahedral arrangement**. There are three shapes with this arrangement:

1. *All bonding groups: tetrahedral shape* (AX_4). Methane has a tetrahedral shape, a very common geometry in organic molecules. In Sample Problem 8.7, we drew the Lewis structure for CCl_2F_2, without considering the relative placement of the four halogen atoms around the carbon atom. Because Lewis structures are flat, it may seem like we can write two different structures for CCl_2F_2. However, they represent the same molecule, as a twist of the wrist reveals (Figure 9.6).

2. *One lone pair: trigonal pyramidal shape* (AX_3E_1). Ammonia (NH_3) is an example of a molecule with a **trigonal pyramidal shape**, a tetrahedron with one vertex "missing." Stronger repulsions by the lone pair make the H—N—H bond angle slightly less than the ideal 109.5°. The lone pair forces the N—H pairs closer to each other, and the bond angle is 107.3°. NF_3 also has a trigonal pyramidal shape.

Picturing shapes is a great way to visualize a reaction. For instance, when ammonia reacts with an acid, the lone pair on N forms a bond to the H^+ and yields the ammonium ion (NH_4^+), one of many tetrahedral polyatomic ions. As the lone pair becomes a bonding pair, the H—N—H angle expands from 107.3° to 109.5°:

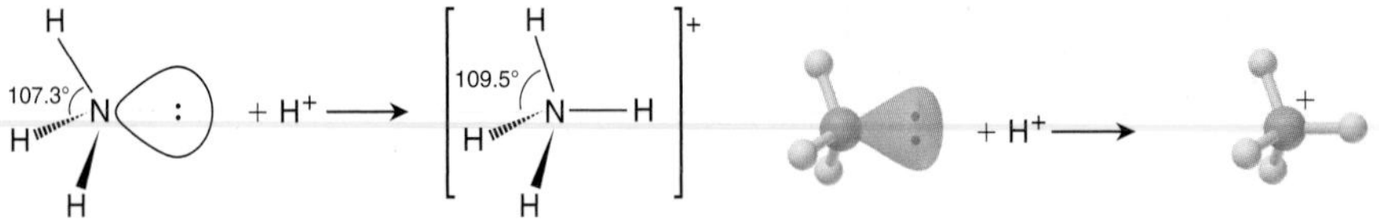

3. *Two lone pairs: bent or V shape* (AX_2E_2). Water is the most important V-shaped molecule with the tetrahedral arrangement. [Note that, in the trigonal

FIGURE 9.6 Lewis structures do not indicate molecular shape. In this model, Cl is green and F is yellow.

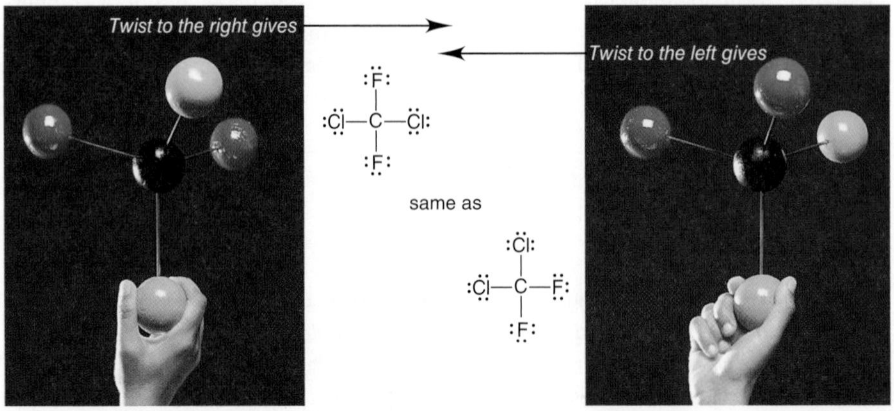

planar arrangement, the V shape has two bonding groups and *one* lone pair (AX_2E_1), and its ideal bond angle is 120°, not 109.5°.] Repulsions from two lone pairs are greater than repulsions from one lone pair, and the H—O—H bond angle is 104.5°, less than the H—N—H angle in NH_3:

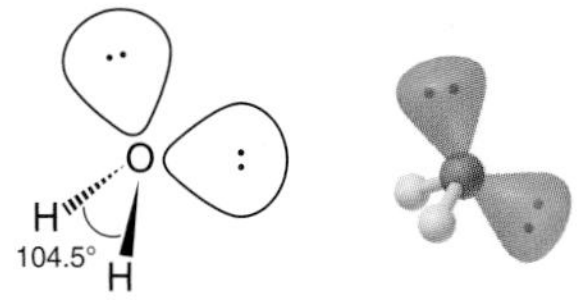

Thus, for similar molecules within a given electron-group arrangement, electron-electron repulsions cause deviations from ideal bond angles in the following order:

lone pair–lone pair > lone pair–bonding pair > bonding pair–bonding pair (9.1)

Molecular Shapes with Five Electron Groups (Trigonal Bipyramidal Arrangement)

All molecules with five or six electron groups have a central atom from period 3 or higher because only those atoms have *d* orbitals available to expand the valence shell.

Relative Positions of Electron Groups Five mutually repelling electron groups form the **trigonal bipyramidal arrangement**, in which two trigonal pyramids have a common base (Figure 9.7). This is the only case in which *there are two different positions for electron groups and two ideal bond angles.* Three **equatorial groups** lie in a trigonal plane that includes the central atom, and two **axial groups** lie above and below this plane. Therefore, a 120° bond angle separates the equatorial groups, and a 90° angle separates the axial groups from the equatorial groups. Two factors need to be considered:

- The greater the bond angle, the weaker the repulsions, so *equatorial-equatorial (120°) repulsions are weaker than axial-equatorial (90°) repulsions.*
- The stronger repulsions from lone pairs mean that, when possible, *lone pairs occupy equatorial positions.*

Shapes for the Trigonal Bipyramidal Arrangement The tendency for lone pairs to occupy equatorial positions, and thus minimize stronger axial-equatorial repulsions, governs three of the four shapes for this arrangement.

1. *All bonding groups: trigonal bipyramidal shape (AX_5).* Phosphorus pentachloride (PCl_5) has a trigonal bipyramidal shape. With five identical surrounding atoms, the bond angles are ideal:

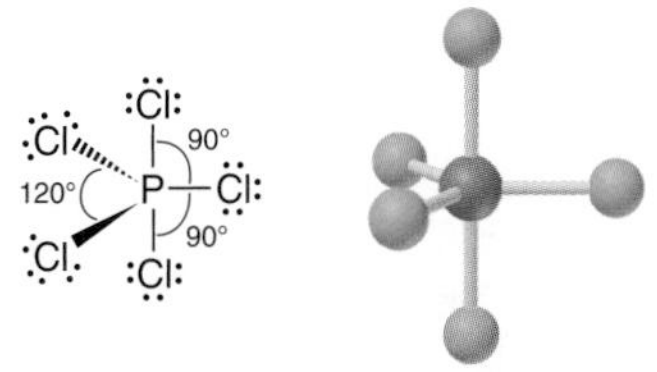

2. *One lone pair: seesaw shape (AX_4E_1).* Sulfur tetrafluoride (SF_4), a strong fluorinating agent, has the **seesaw shape**; in Figure 9.7, the "seesaw" is tipped on an end. This is the first example of *lone pairs occupying equatorial positions* to minimize repulsions. The lone pair repels all four bonding pairs, reducing the bond angles to 101.5° and 86.8°:

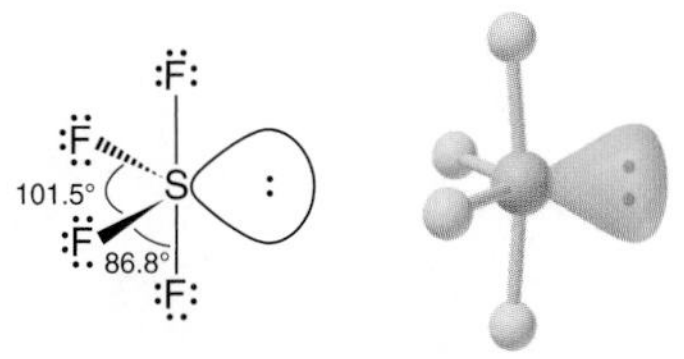

3. *Two lone pairs: T shape (AX_3E_2).* Bromine trifluoride (BrF_3), one of many compounds with fluorine bonded to a larger halogen, has a **T shape**. Since both

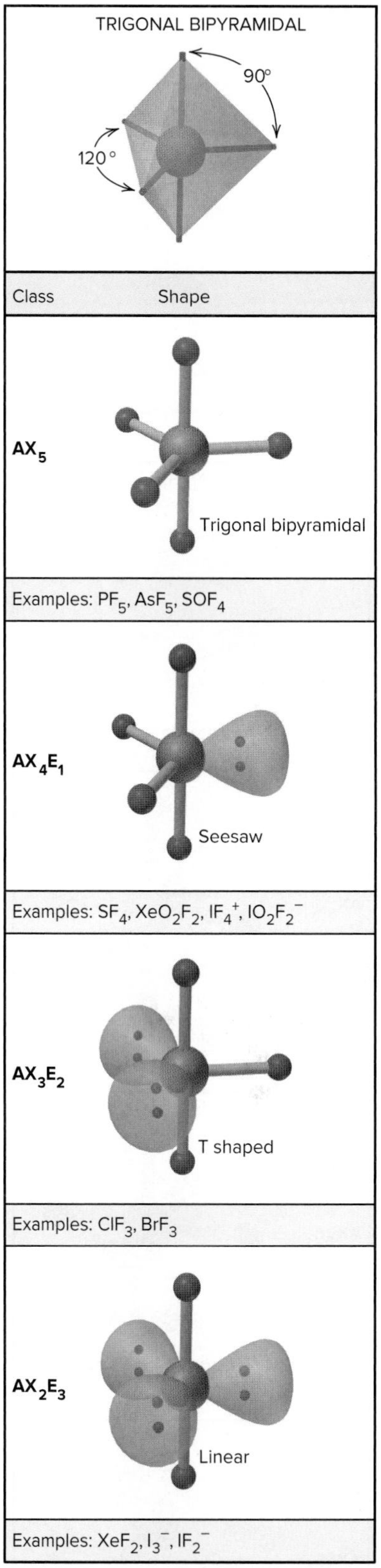

FIGURE 9.7 The four molecular shapes of the trigonal bipyramidal electron-group arrangement

lone pairs occupy equatorial positions, we see a greater decrease in the axial-equatorial bond angle, down to 86.2°:

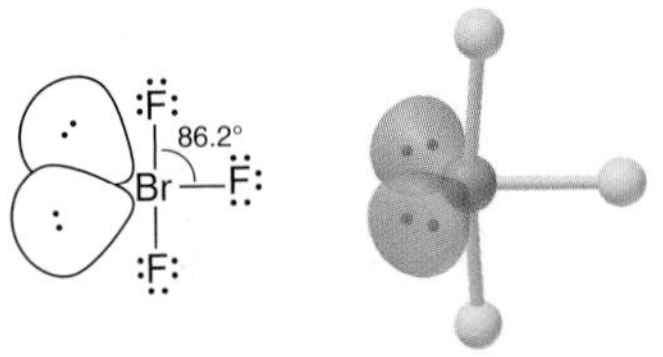

4. *Three lone pairs: linear shape (AX_2E_3).* The triiodide ion (I_3^{-0}), which forms when I_2 dissolves in aqueous I^- solution, is linear. With three equatorial lone pairs and two axial bonding pairs, the three nuclei form a straight line and a 180° X—A—X bond angle:

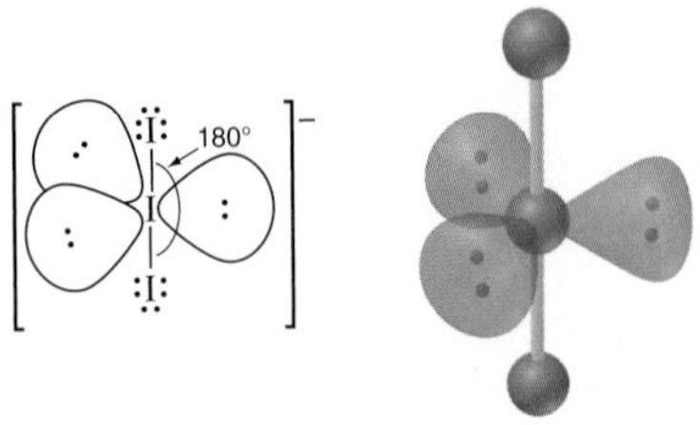

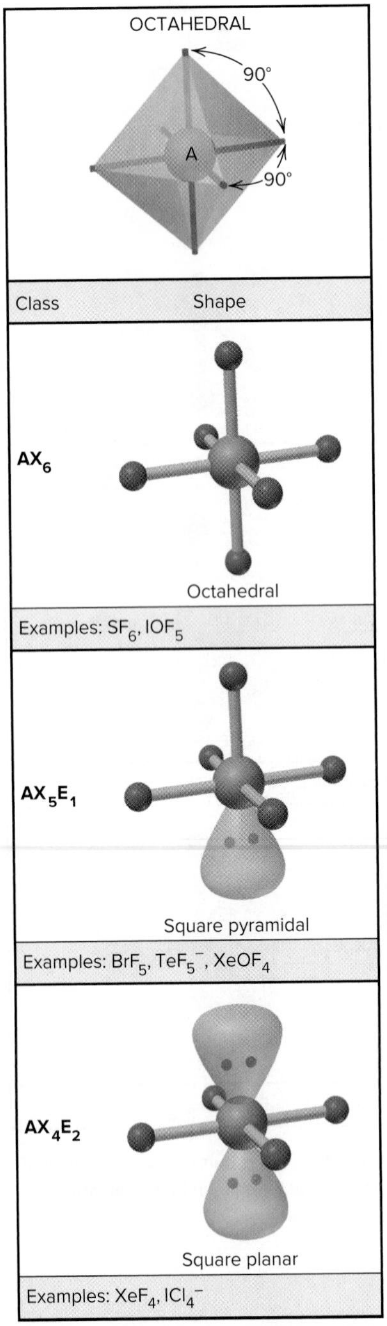

FIGURE 9.8 The three molecular shapes of the octahedral electron-group arrangement

Molecular Shapes with Six Electron Groups (Octahedral Arrangement)

Six electron groups form the **octahedral arrangement**. An *octahedron* is a polyhedron with eight equilateral triangles for faces and six identical vertices (Figure 9.8). Each of the six groups points to a corner, which gives a 90° ideal bond angle.

1. *All bonding groups: octahedral shape (AX_6).* When seesaw-shaped SF_4 reacts with more F_2, the central S atom expands its valence shell further to form octahedral sulfur hexafluoride (SF_6):

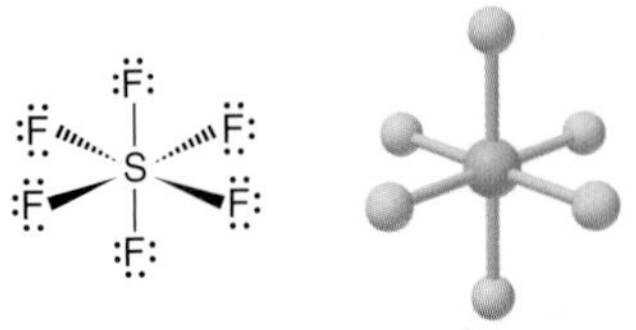

2. *One lone pair: square pyramidal shape (AX_5E_1).* Iodine pentafluoride (IF_5) has a **square pyramidal shape**. Note that it makes no difference where the one lone pair resides because all the bond angles are 90°. The lone pair reduces the bond angles to 81.9°:

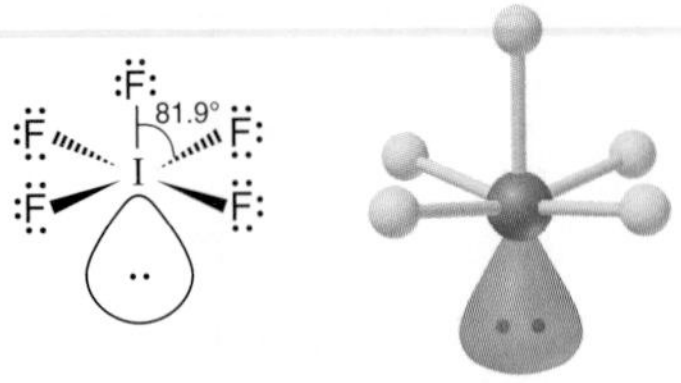

3. *Two lone pairs: square planar shape (AX_4E_2).* Xenon tetrafluoride (XeF_4) has a **square planar shape**. To avoid stronger 90° lone pair–lone pair repulsions, two lone pairs lie *opposite* each other:

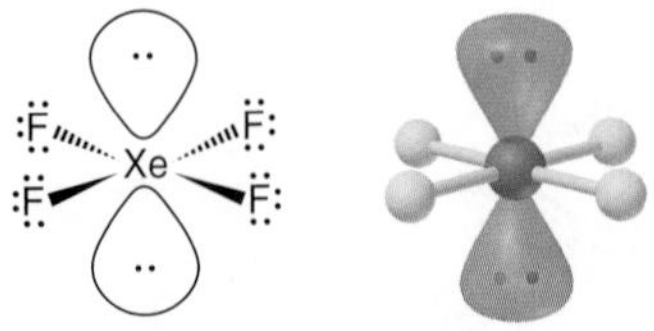

Figure 9.9 displays the shapes that are possible for elements in different periods, and Figure 9.10 summarizes the molecular shapes we have discussed.

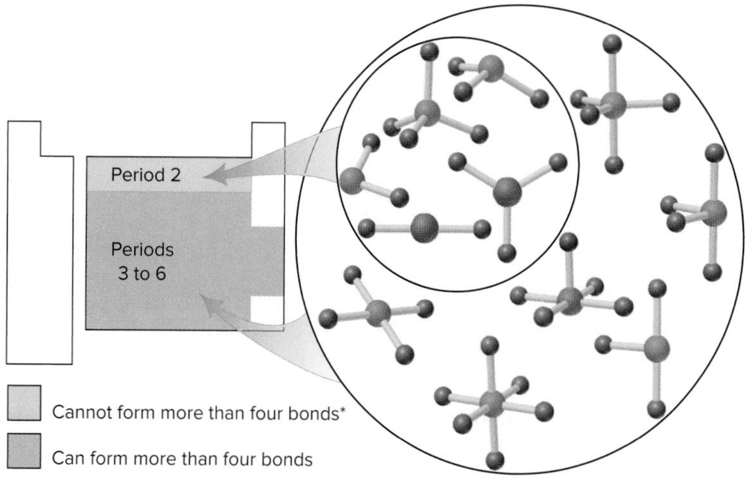

FIGURE 9.9 Molecular shapes for central atoms in period 2 and higher periods

*Although C usually forms four bonds, it can make stable compounds from two to six bonds, as in carbenes (two bonds).

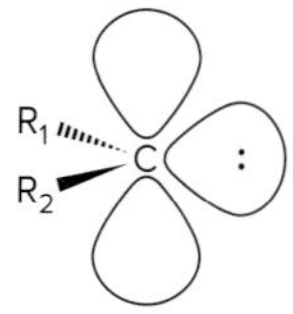

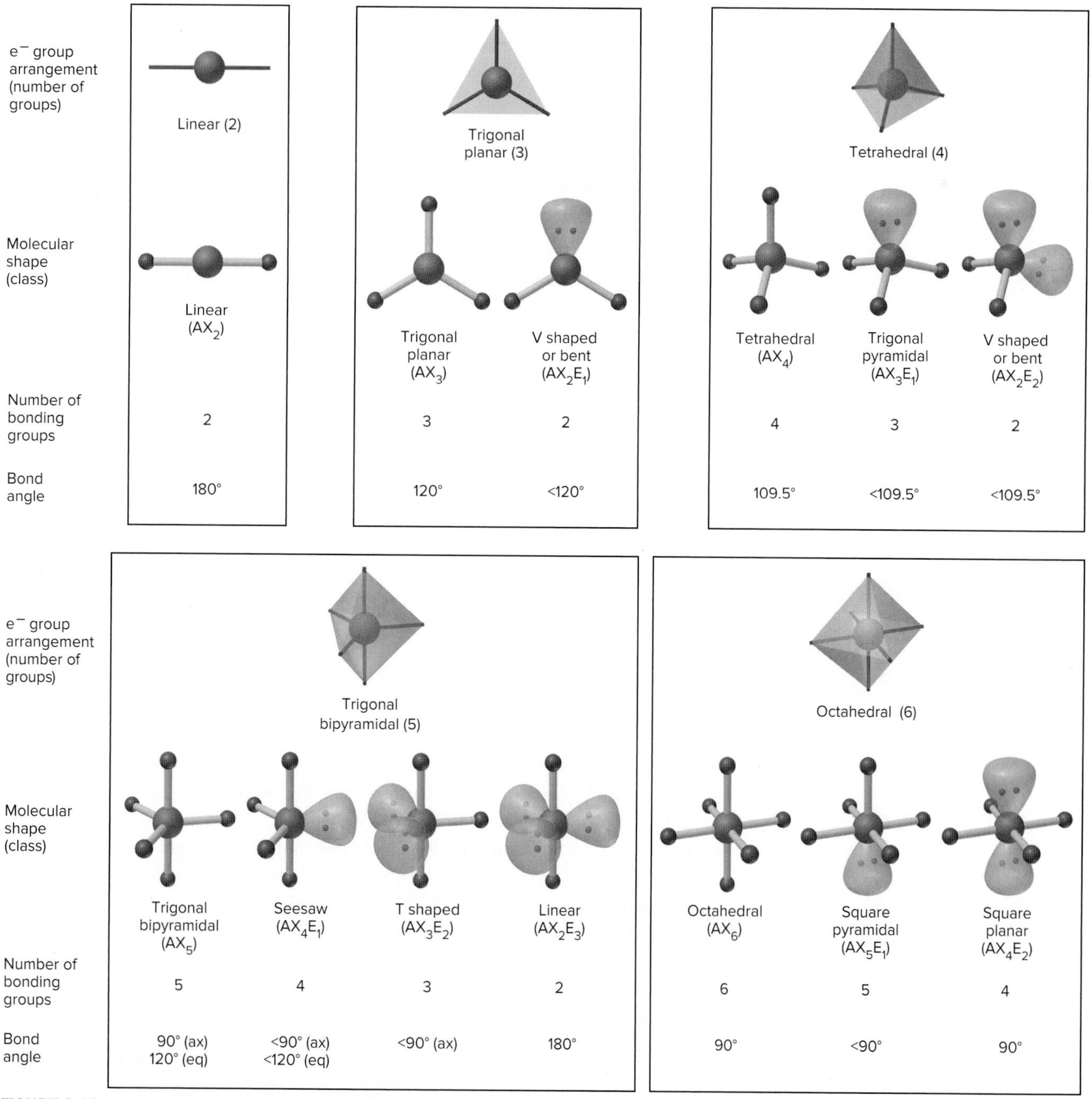

FIGURE 9.10 A summary of common molecular shapes with two to six electron groups

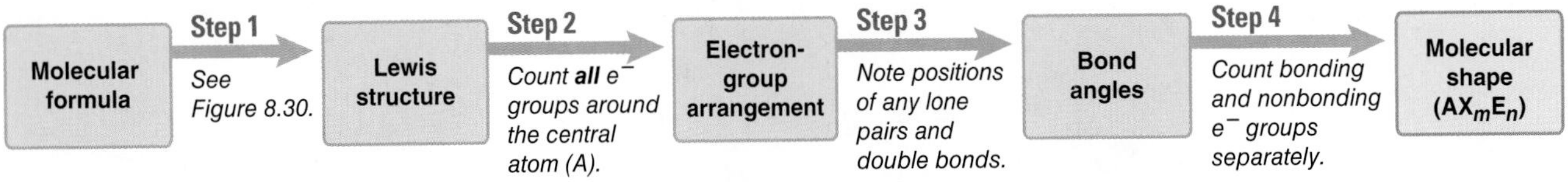

FIGURE 9.11 The four steps for converting a molecular formula to a molecular shape

Using Valence-Shell Electron-Pair Repulsion Theory to Determine Molecular Shape

We can use the following steps, based on VSEPR theory, to determine a molecular shape from a molecular formula (Figure 9.11):

Step 1. Write the Lewis structure from the molecular formula (Figure 8.30) to see the relative placement of the atoms and the number of electron groups.

Step 2. Assign an electron-group arrangement by counting *all* the electron groups (bonding plus nonbonding) around the central atom.

Step 3. Predict the ideal bond angle from the electron-group arrangement and *the effect of any deviation* caused by lone pairs or double bonds.

Step 4. Draw and name the molecular shape by counting bonding groups and nonbonding groups separately.

The next two sample problems show how these steps can be used.

Sample Problem 9.1 Examining Shapes with Two, Three, or Four Electron Groups

Problem Draw the molecular shapes and predict the bond angles (relative to the ideal angles) of **(a)** PF_3; **(b)** $COCl_2$.

Solution (a) PF_3:

Step 1. Write the Lewis structure from the formula. See below, left.

Step 2. Assign the electron-group arrangement. Three bonding groups and one lone pair give four electron groups around P and the *tetrahedral arrangement.*

Step 3. Predict the bond angle. The ideal bond angle is 109.5°. There is one lone pair, so the actual bond angle will be less than 109.5°.

Step 4. Draw and name the molecular shape. With one lone pair, PF_3 has a trigonal pyramidal shape (AX_3E_1):

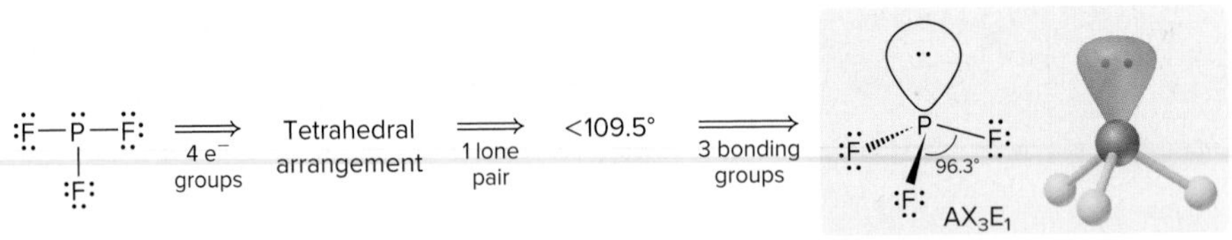

(b) $COCl_2$:

Step 1. Write the Lewis structure from the formula. See below, left.

Step 2. Assign the electron-group arrangement. Two single bonds and one double bond give three electron groups around C and the *trigonal planar arrangement.*

Step 3. Predict the bond angles. The ideal bond angle is 120°, but the double bond between C and O will compress the Cl—C—Cl angle to less than 120°.

Step 4. Draw and name the molecular shape. With three electron groups and no lone pairs, $COCl_2$ has a trigonal planar shape (AX_3):

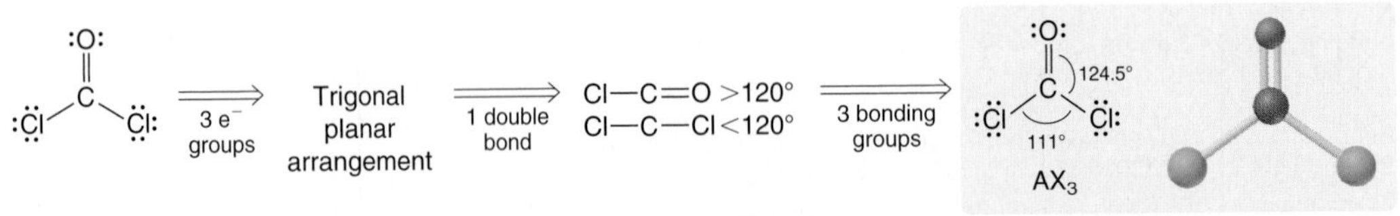

Check We compare the answers with the general information in Figure 9.10.

Comment We need to be sure that the Lewis structure is correct because it determines the other steps.

Follow-Up Problem 9.1 Draw the molecular shapes and predict the bond angles (relative to the ideal angles) of **(a)** CS_2; **(b)** $PbCl_2$; **(c)** CBr_4; **(d)** SF_2. (See Brief Solutions to Follow-Up Problems at the end of the chapter.)

Sample Problem 9.2 Examining Shapes with Five or Six Electron Groups

Problem Draw the molecular shapes and predict the bond angles (relative to the ideal angles) of **(a)** SbF_5; **(b)** BrF_5.

Plan We proceed as in Sample Problem 9.1, being sure to minimize the number of axial-equatorial repulsions.

Solution **(a)** SbF_5:

Step 1. Lewis structure. See below, left.

Step 2. Electron-group arrangement. With five electron groups, this is the *trigonal bipyramidal* arrangement.

Step 3. Bond angles. All the groups and surrounding atoms are identical, so the bond angles are ideal: 120° between equatorial groups and 90° between axial and equatorial groups.

Step 4. Molecular shape. Five electron groups and no lone pairs give the trigonal bipyramidal shape (AX_5):

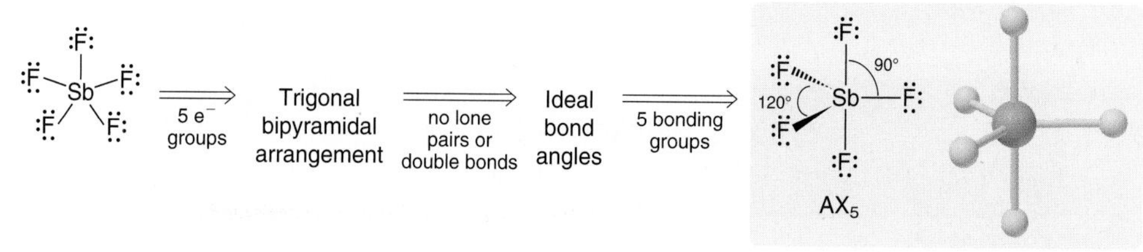

(b) BrF_5:

Step 1. Lewis structure. See below, left.

Step 2. Electron-group arrangement. Six electron groups give the *octahedral* arrangement.

Step 3. Bond angles. The lone pair will make all bond angles less than the ideal 90°.

Step 4. Molecular shape. With one lone pair, BrF_5 has the square pyramidal shape (AX_5E_1):

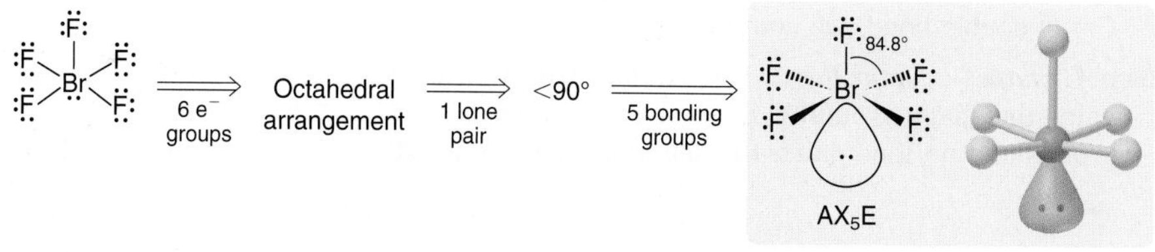

Check We can compare our answers with the information in Figure 9.10.

Comment We will also see the linear, tetrahedral, square planar, and octahedral shapes in an important group of substances called *coordination compounds* in Chapter 24.

Follow-Up Problem 9.2 Draw the molecular shapes and predict the bond angles (relative to the ideal angles) of **(a)** ICl_2^-; **(b)** ClF_3; **(c)** SOF_4.

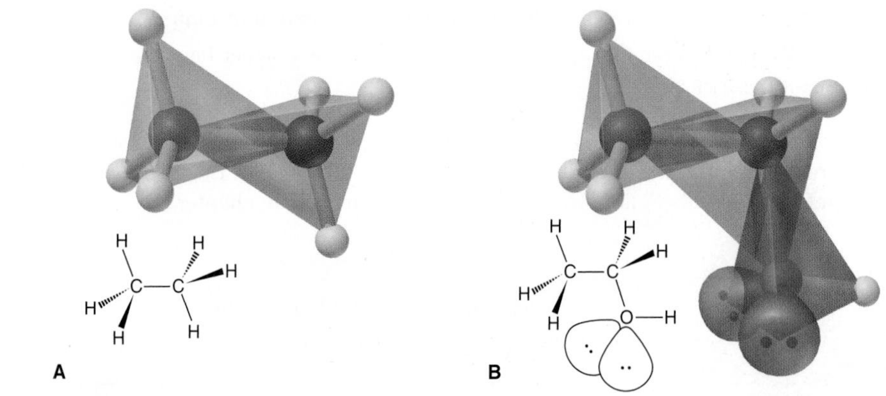

FIGURE 9.12 **The tetrahedral shapes around the central atoms and the overall shapes of ethane (A) and ethanol (B)**

Molecular Shapes with More Than One Central Atom

The shapes of molecules with more than one central atom are composites of the shapes around each of the atoms. Here are two examples:

1. Ethane (CH_3CH_3; molecular formula C_2H_6) is a component of natural gas. With four bonding groups and no lone pairs around the two central carbons, ethane is shaped like two overlapping tetrahedrons (Figure 9.12A).
2. Ethanol (CH_3CH_2OH; molecular formula C_2H_6O), the central nervous system depressant in beer, wine, and whisky, has three central atoms (Figure 9.12B). The CH_3— group is tetrahedrally shaped. The —CH_2— group has four bonding groups around its central C atom, so it is also tetrahedrally shaped. The O atom has two bonding groups and two lone pairs around it, so the —OH group has a V shape (AX_2E_2).

Sample Problem 9.3 Predicting Molecular Shapes with More Than One Central Atom

Problem Determine the shape around each central atom in propanone (acetone, $[(CH_3)_2CO]$).

Plan There are three central C atoms, two of which are in the CH_3— groups. We determine the shape around one central atom at a time.

Solution *Step 1. Lewis structure.* See below, left.

Step 2. Electron-group arrangement. Each CH_3— group has four electron groups around its central C, so its electron-group arrangement is *tetrahedral.* The third C atom has three electron groups around it, so it has the *trigonal planar arrangement.*

Step 3. Bond angles. The H—C—H angle in CH_3— should be near the ideal 109.5°. The C=O double bond will compress the C—C—C angle to less than the ideal 120°.

Step 4. Shapes around the central atoms. With four electron groups and no lone pairs, the shape around C in each CH_3— is tetrahedral (AX_4). With three electron groups and no lone pairs, the shape around the middle C is trigonal planar (AX_3):

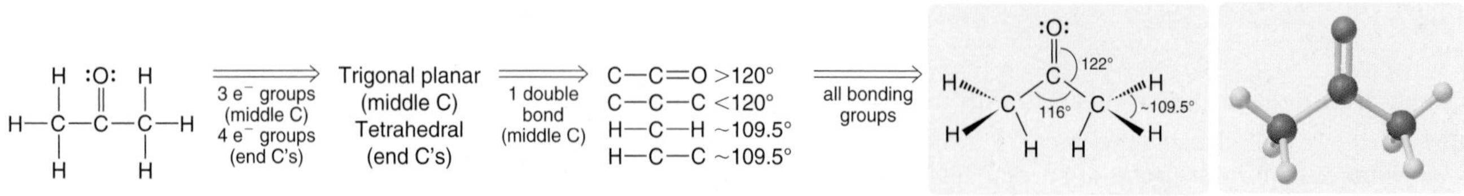

Follow-Up Problem 9.3 Determine the shape around each central atom and predict any deviations from the ideal bond angles in **(a)** H_2SO_4; **(b)** propyne (C_3H_4; there is one C≡C bond); **(c)** S_2F_2.

SUMMARY OF SECTION 9.1

- VSEPR theory proposes that each electron group (single bond, multiple bond, lone pair, or lone electron) around a central atom remains as far from the others as possible.
- Five electron-group arrangements are possible when two, three, four, five, or six electron groups surround a central atom. Each arrangement is associated with one or more molecular shapes, depending on the numbers of bonding and lone pairs.
- The ideal bond angles are based on the regular geometric arrangements. Deviations from the ideal bond angles occur when surrounding atoms and/or electron groups are not identical.
- Lone pairs and double bonds exert stronger repulsions than single bonds.
- Shapes of larger molecules are composites of the shapes around each central atom.

9.2 Molecular Shape and Molecular Polarity

Knowing the shape of the molecules in a substance is key to understanding the physical and chemical behaviour of the substance. One of the most far-reaching effects of molecular shape is molecular polarity, which can influence melting and boiling points, solubility, reactivity, and even biological function.

Recall, from Chapter 8, that a covalent bond is *polar* when the atoms have different electronegativities and, thus, share the electrons unequally. In diatomic molecules, such as HF, the only bond is polar, so the molecule is polar. In larger molecules, *both shape and bond polarity determine* **molecular polarity**, an imbalance of charge over the whole molecule or a large portion of it. Polar molecules become oriented in an electric field, with their partially charged ends pointing toward the oppositely charged plates (Figure 9.13). **Dipole moment (μ)** is a measure of molecular polarity, given in the unit *debye* (D)* derived from SI units of charge (coulomb, C) and length (metre, m): 1 D = 3.34×10^{-30} C·m. **Electrostatic potential maps**, or electron-density models, which illustrate the charge distributions of molecules in 3-D, will allow us to visualize variably charged regions of the molecules.

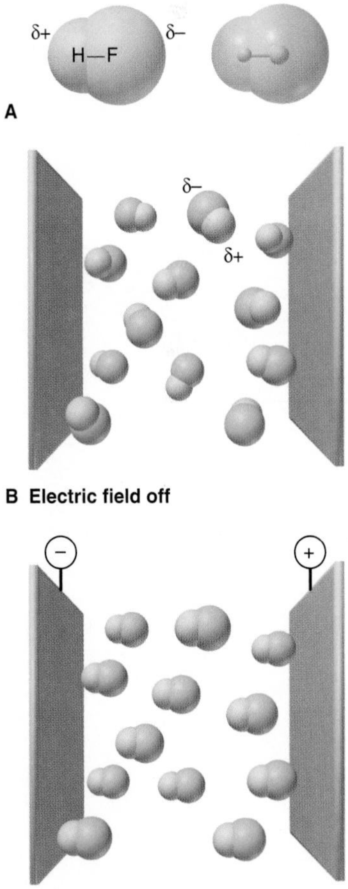

FIGURE 9.13 The orientation of polar molecules in an electric field. A. Space-filling (*left*) and electron-density (*right*) models of the polar HF molecule. **B.** With the external electric field off, HF molecules are oriented randomly. **C.** With the electric field on, the molecules, on average, become oriented.

Bond Polarity, Bond Angle, and Dipole Moment

The presence of polar bonds does not *always* result in a polar molecule; we must also consider shape and the atoms surrounding the central atom. Here are three cases:

1. *CO_2: polar bonds, nonpolar molecule.* In carbon dioxide, the electronegativity difference between C (χ = 2.5) and O (χ = 3.5) makes each C═O bond polar. But CO_2 is linear, so the bonds point 180° from each other. The two bond polarities are counterbalanced, and the molecule has *no net dipole moment* (μ = 0 D). The electron density model shows regions of high negative charge (*red*) distributed equally on either side of the central region of high positive charge (*blue*). According to IUPAC, the direction of the dipole moment is from the negative charge to the positive charge. We can represent the orientation of the dipole with vectors, as follows:

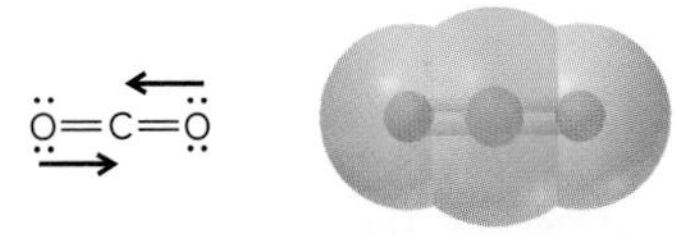

2. *H_2O: polar bonds, polar molecule.* Water also has two polar bonds, but it *is* polar (μ = 1.85 D). In each O—H bond, electron density is pulled from H (χ = 2.1) toward O (χ = 3.5). Bond polarities are *not* counterbalanced because the molecule is V shaped (see also Figure 3.17). The bond polarities are partially reinforced,

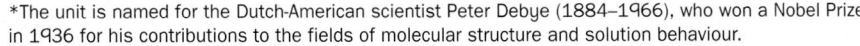

*The unit is named for the Dutch-American scientist Peter Debye (1884–1966), who won a Nobel Prize in 1936 for his contributions to the fields of molecular structure and solution behaviour.

making the O end partially negative and the other end (the region between the H atoms) partially positive:

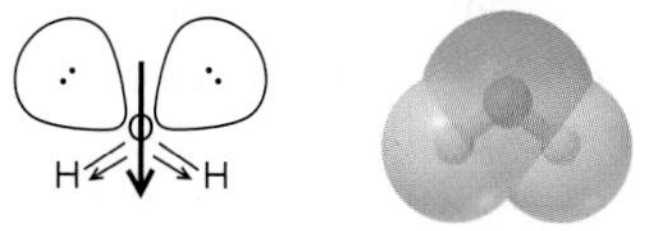

(The molecular polarity of water has some amazing effects, from determining the composition of the oceans to supporting life itself, as you will see in Chapter 11.)

3. *Same shapes, different polarities.* When different molecules have the same shape, the identities of the surrounding atoms affect polarity. Tetrachloromethane (carbon tetrachloride, CCl_4) and trichloromethane (chloroform, $CHCl_3$) are tetrahedral molecules with very different polarities. In CCl_4, all the surrounding atoms are Cl atoms. Each C—Cl bond is polar ($\Delta\chi = 0.5$), but the molecule is nonpolar ($\mu = 0$ D) because the bond polarities counterbalance each other. In $CHCl_3$, an H replaces one Cl, disrupting the balance and giving trichloromethane a significant dipole moment ($\mu = 1.01$ D):

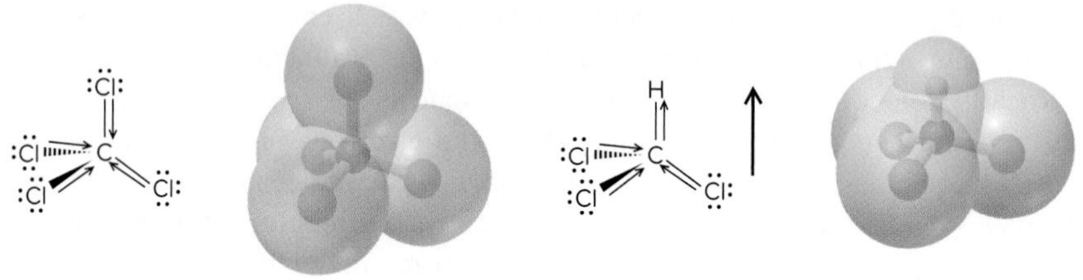

Sample Problem 9.4 Predicting the Polarity of Molecules

Problem For each molecule, use the molecular shape and χ values and trends (Figure 8.23) to predict the direction of bond polarities and molecular polarity, if present:

(a) Ammonia, NH_3

(b) Boron trifluoride, BF_3

(c) Carbonyl sulfide, COS (atom sequence SCO)

Plan We draw and name the molecular shape and point a polar arrow toward the atom with higher χ in each bond. If the bond polarities balance one another, the molecule is nonpolar; if they reinforce each other, we show the direction of the molecular polarity.

Solution (a) The molecular shape of NH_3 is trigonal pyramidal. N ($\chi = 3.0$) is more electronegative than H ($\chi = 2.1$), so the bond polarities are directed toward N and partially reinforce each other; thus, the molecular polarity is shifted toward N:

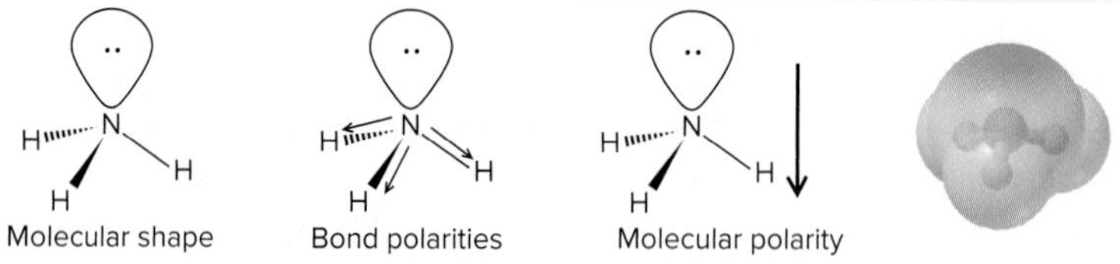

Therefore, ammonia is polar.

(b) The molecular shape of BF_3 is trigonal planar. F ($\chi = 4.0$) is farther to the right in period 2 than B ($\chi = 2.0$), so it is more electronegative; thus, each bond polarity is shifted toward F. However, the bond angle is 120°, so the three bond polarities balance each other, and BF_3 has no molecular polarity:

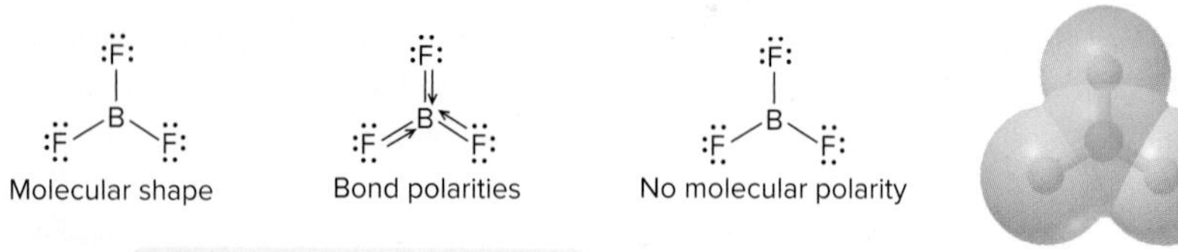

Therefore, boron trifluoride is nonpolar.

(c) The molecular shape of COS is linear. Since C and S have the same χ, the C=S bond is nonpolar. However, the C=O bond is quite polar ($\Delta\chi = 1.0$), so there is a net molecular polarity toward the O:

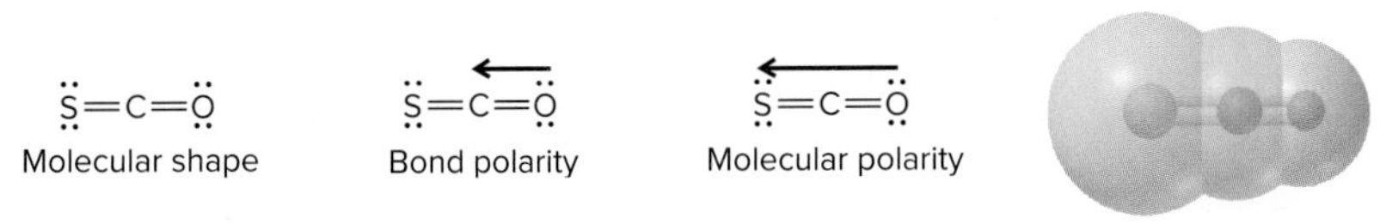

Therefore, carbonyl sulfide is polar.

Check The electron density models confirm our conclusions. Note that, in part (b), the negative (*red*) regions surround the central B (*blue*) symmetrically.

Follow-Up Problem 9.4 Show the bond polarities and molecular polarity, if any, for each molecule:

(a) Dichloromethane (CH_2Cl_2)

(b) Iodine oxide pentafluoride (IOF_5)

(c) Nitrogen tribromide (NBr_3)

The Effect of Molecular Polarity on Behaviour

Earlier we mentioned that molecular polarity influences physical behaviour. Let us see how a molecular property, such as dipole moment, affects a macroscopic property, such as boiling point. Consider the two dichloroethene molecules shown below. They have the *same* molecular formula ($C_2H_2Cl_2$), but *different* physical and chemical properties; that is, these stereoisomers are geometric isomers (Sections 3.2 and 20.4). Both molecules are planar, with a trigonal planar shape around each C atom. The *trans* isomer is nonpolar ($\mu = 0$ D) because the polar C—Cl bonds balance each other. The *cis* isomer is polar ($\mu = 1.90$ D) because the bond polarities partially reinforce each other, with the molecular polarity pointing between the Cl atoms.

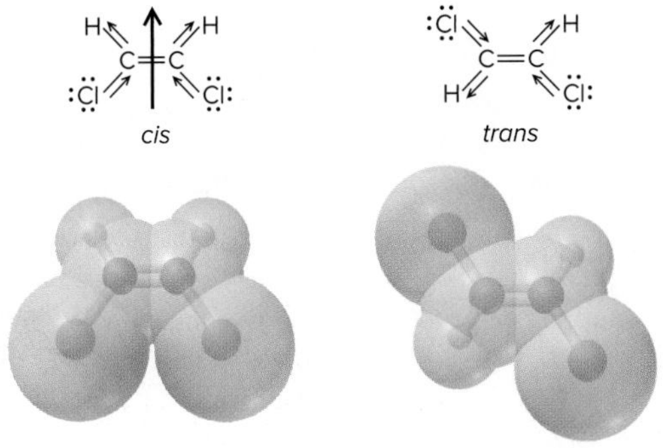

A liquid boils when it forms bubbles against the atmospheric pressure. To enter the bubble, the molecules in the liquid must overcome the weak attractive forces *between* them. Because of their polarity, the *cis* molecules attract each other more strongly than the *trans* molecules do. Since more energy is needed to overcome the stronger attractions, we expect the *cis* isomer to have a higher boiling point. In fact, *cis*-1,2-dichloroethene boils at 13°C higher than *trans*-1,2-dichloroethene.

Figure 9.14 shows the chain of influences of atomic properties on the behaviour of substances. We will extend these relationships in Chapter 11, and the upcoming Chemical Connections section discusses some biological effects of molecular shape and polarity.

FIGURE 9.14 The influences of atomic properties on macroscopic behaviour

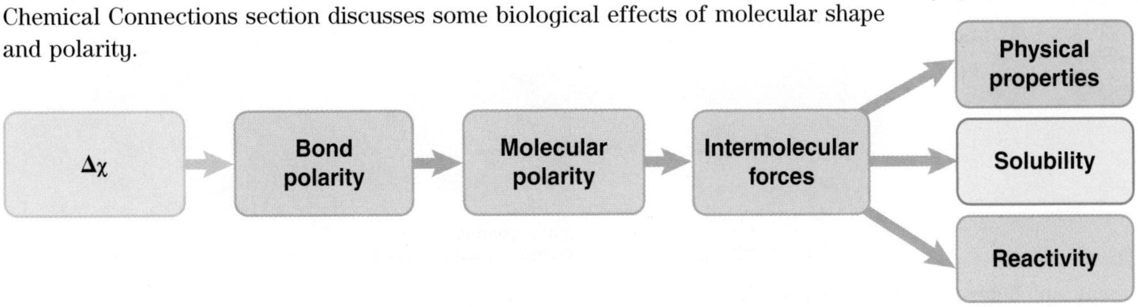

CHEMICAL CONNECTIONS TO SENSORY PHYSIOLOGY

Molecular Shape, Biological Receptors, and the Sense of Smell

A biological cell can be thought of as a membrane-bound sack that is filled with an aqueous fluid, which contains many molecules of various shapes. Many complex processes begin when a molecular "key" fits into a molecular "lock" with a complementary shape. Typically, the key is a small molecule circulating in cellular or other bodily fluid, and the lock, the *biological receptor*, is a large molecule often embedded in a membrane. The *receptor site* is a small region of the receptor with a shape that matches the molecular key. Thousands of molecules per second collide with the receptor site. When a molecule with the correct shape collides with the receptor site, the receptor "grabs" it through intermolecular attractions, and the biological response begins.

Molecular Shape and the Sense of Smell Molecular shapes fitting together is crucial to the sense of smell (olfaction). To have an odour, a substance must be a gas or a volatile liquid or solid and must be at least slightly soluble in the aqueous film of the nasal passages. As well, the odorous molecule, or a portion of it, must fit into one of the receptor sites on nerve endings in the area. When this happens, nerve impulses travel to the brain, which interprets them as a specific odour.

In the mid-twentieth century, it was proposed that molecular shape (and sometimes polarity), but *not* composition, determines odour. The theory stated that any molecule producing one of seven primary odours—camphor-like, musky, floral, minty, ethereal, pungent, or putrid—matches a receptor site of a particular shape. Figure B9.1 shows three of the seven sites holding a molecule with that odour. Several predictions of the theory proved to be correct:

- If different substances fit a given receptor, they have the same odour. The four molecules in Figure B9.2 fit the camphor-like receptor and smell like moth repellent.
- If different parts of a molecule fit different receptors, it has a mixed odour. Portions of benzaldehyde fit the camphor-like, floral, and minty receptors, and it smells like almonds; other molecules with this odour fit the same three receptors.

However, other predictions were not confirmed. For example, an odour predicted from the shape was often not the odour smelled. The reason is that molecules may not have the same shape in solution at the receptor as they have in the gas phase.

Evidence from the 1990s suggests that the original model is far too simple. The 2004 Nobel Prize in Medicine or Physiology was given to Richard Axel and Linda B. Buck for showing that olfaction involves about 1000 different receptors, and molecules fitting various combinations of them produce the over 10 000 odours that humans smell.

Thus, although the process is much more complex than originally thought, the central idea that odour depends on molecular shape is valid and is being actively researched in the food, cosmetics, and insecticide industries. The last is of vital importance in Canada and the United States, due to the insect epidemic that is affecting and killing the trees of our national forests (Figure B9.3). One interesting application of this idea is being studied by Erika Plettner (Figure B9.4), whose research at Simon Fraser University focuses on studying proteins in the olfactory system of insects that interact with pheromones, using various biological and biophysical techniques. One of her goals is to develop a natural pesticide that will block or confuse the pest insects' olfactory receptors, which allow the insects to find each other to reproduce; this will help to control and reduce their population.

The Biological Significance of Molecular Shape *Countless other examples show that no chemical property is more crucial to living systems than molecular shape.* Here are just a few of the many biochemical processes that are controlled by one molecule fitting into a receptor site on another:

- Enzymes are proteins that bind cellular reactants and speed their reaction. An early step in energy metabolism involves glucose binding to the "active" site of the enzyme hexokinase (Figure B9.5).

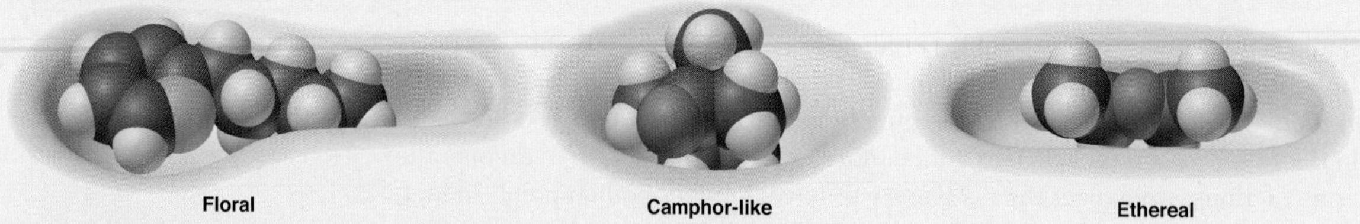

FIGURE B9.1 Shapes of some olfactory receptor sites

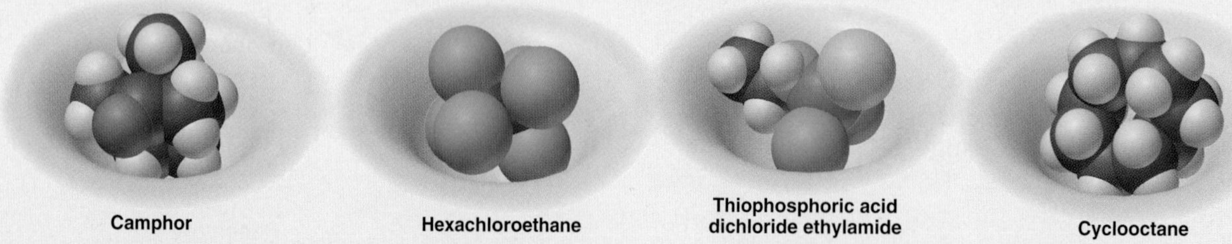

FIGURE B9.2 Different molecules with the same odour

(Continued)

- Nerve impulses are transmitted when small molecules released from one nerve fit into receptors on the next. Mind-altering drugs act by chemically disrupting the molecular fit at such nerve receptors in the brain.
- One type of immune response is triggered when a molecule on a bacterial surface binds to the receptors on "killer" cells in the bloodstream.
- Hormones regulate processes by fitting into and activating specific receptors on target tissues and organs.
- Genes function when certain nucleic acid molecules fit into specific regions on other molecules.

A.

B.

FIGURE B9.3 Destructive effects of plant insect pests. A. Trees killed by the mountain pine beetle. **B.** This insect has decimated forests in central British Columbia and continues to spread.

FIGURE B9.4 Erika Plettner, professor at Simon Fraser University in Vancouver. As part of Dr. Plettner's research, she is interested in designing pheromone mimics that will specifically interfere with pheromone olfaction. Such compounds could be used in pest control. Dr. Plettner obtained her Ph.D. with Dr. Keith N. Slessor, working on the biosynthesis of functionalized fatty acids in honeybee queens and workers. She then pursued postdoctoral studies in protease enzymology and insect olfaction. Since joining the faculty at Simon Fraser University in 1999, she has continued her studies in insect olfaction, with particular emphasis on the mechanisms of molecular recognition of chiral insect pheromones. Other interests of her research group include the biosynthesis of honeybee ester pheromones, the design of insect deterrents, and reactions of cytochromes. Practical applications that could come from this research include the development of new insect control methods (for example, using novel deterrents in integrated pest management schemes) and new ways to degrade hydrophobic pollutants that accumulate in the environment, using mutant cytochromes.

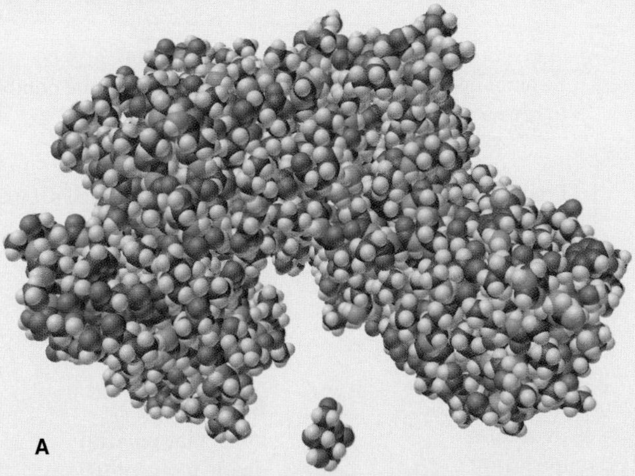

A

B

FIGURE B9.5 Molecular shape and enzyme action. A. A small sugar molecule (*bottom*) is shown near a specific region of an enzyme molecule. **B.** When the sugar lands in this region, the reaction begins.

(Continued)

Problems

B9.1 As you will learn in Chapter 10, groups joined by a single bond rotate freely around the bond, but groups joined by double bonds do not. Peptide bonds make up a major portion of protein chains. Determine the molecular shape around the central C and N atoms in the following two resonance forms of the peptide bond:

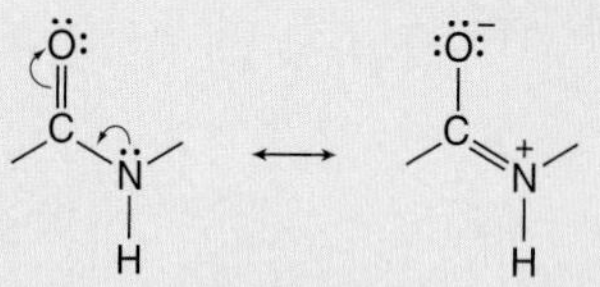

B9.2 Lewis structures of mescaline, a hallucinogenic compound in peyote cactus, and dopamine, a neurotransmitter in the mammalian brain, are shown below. Suggest a reason for mescaline's ability to disrupt nerve impulses.

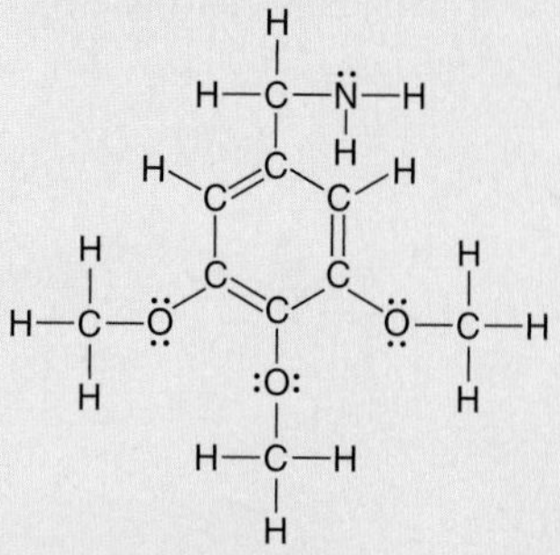

Mescaline

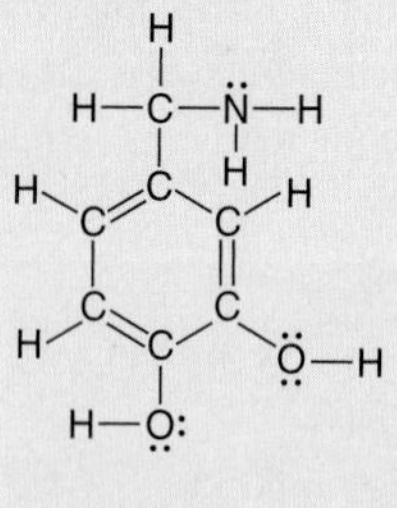

Dopamine

SUMMARY OF SECTION 9.2

- Bond polarity and molecular shape determine molecular polarity, which is measured as a dipole moment.
- A molecule with polar bonds is not necessarily a polar molecule. When bond polarities counterbalance each other, the molecule is nonpolar; when bond polarities reinforce each other, the molecule is polar.
- Molecular shape and polarity can affect physical properties, such as boiling point, and they play a central role in biological function.

CHAPTER REVIEW GUIDE

Learning Objectives

Relevant section (§) and/or sample problem (SP) numbers appear in parentheses.

Concepts

1. Explain how electron-group repulsions lead to molecular shapes. (§9.1)
2. Describe the five electron-group arrangements and their associated molecular shapes. (§9.1)
3. Explain why double bonds and lone pairs cause deviations from ideal bond angles. (§9.1)
4. Explain how bond polarities and molecular shape combine to give a molecule polarity. (§9.2)

Skills

1. Predict molecular shapes from Lewis structures. (SPs 9.1–9.3)
2. Use molecular shape and electronegativity values to predict the polarity of a molecule. (SP 9.4)

Key Terms

Section 9.1
valence-shell electron-pair repulsion (VSEPR) theory
molecular shape
bond angle
linear arrangement
linear shape
trigonal planar arrangement
bent shape (V shape)
tetrahedral arrangement
trigonal pyramidal shape
trigonal bipyramidal arrangement
equatorial groups
axial groups
seesaw shape
T shape
octahedral arrangement
square pyramidal shape
square planar shape

Section 9.2
molecular polarity
dipole moment (μ)
electrostatic potential maps

Key Equations and Relationships

9.1 Ranking the effect of electron-pair repulsions on bond angle:

lone pair–lone pair > lone pair–bonding pair > bonding pair–bonding pair

Brief Solutions to Follow-Up Problems

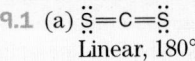

9.1 (a) S=C=S Linear, 180°

(b) 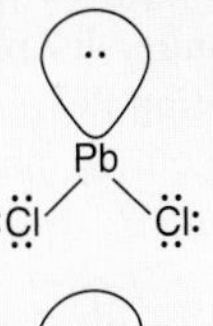 V shaped, <120°

(c) 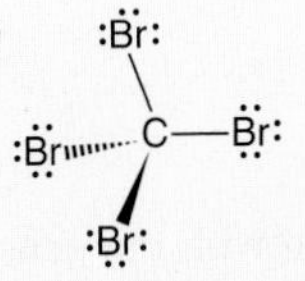 Tetrahedral, 109.5°

(d) 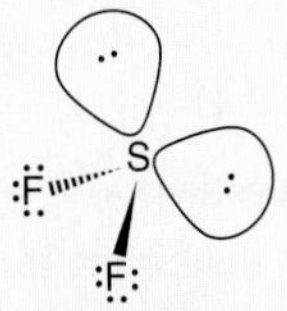 V shaped, <109.5°

9.2 (a) 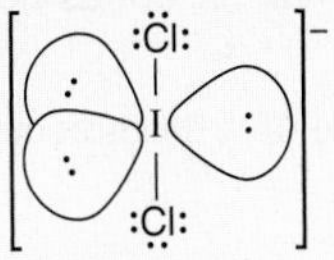 Linear; 180°

(b) 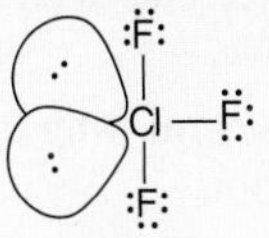 T shaped, <90°

(c) 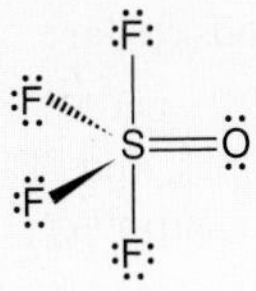 Trigonal bipyramidal, F_{eq}—S—F_{eq} angle <120° and F_{ax}—S—F_{eq} angle <90°

9.3 (a) 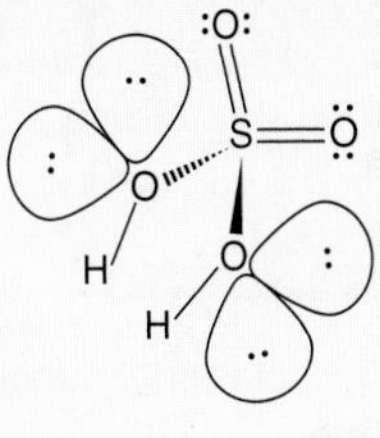 S is tetrahedral; double bonds compress O—S—O angle to <109.5°. Shape around each O in an —OH is V shape; lone pairs compress H—O—S angle to <109.5°.

(b) 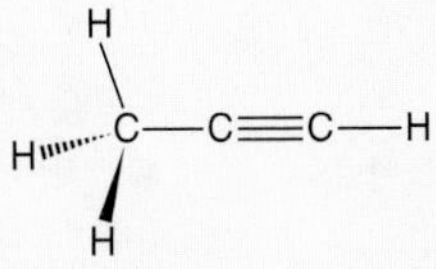 Shape around C in CH_3— is tetrahedral, with angles ~109.5°; other C atoms are linear, 180°.

(c) 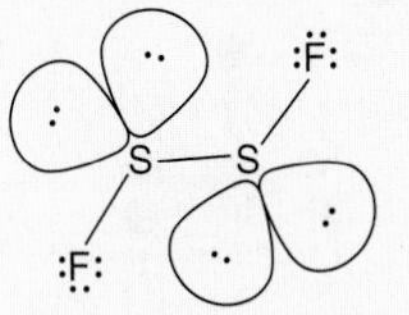 V shape around each S; F—S—S angle <109.5°.

9.4

(a) 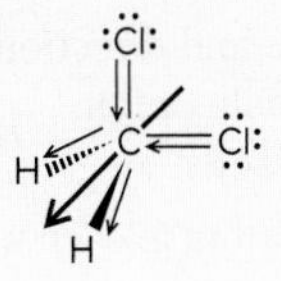

(b) 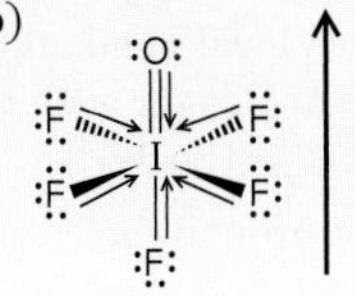

(c) 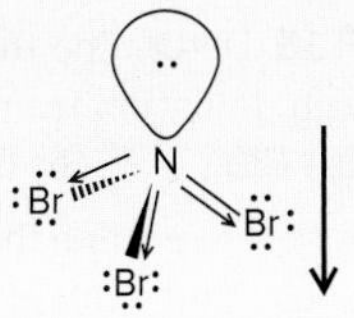

PROBLEMS

Problems with **red** numbers are answered in Appendix G and worked in detail in the Student Solutions Manual. Problem sections match those in this book and provide the numbers of relevant sample problems. Most offer Concept Review Questions, Skill-Building Exercises (grouped in pairs covering the same concept), and Problems in Context. The Comprehensive Problems are based on material from any section or previous chapter.

Valence-Shell Electron-Pair Repulsion Theory

(Sample Problems 9.1 to 9.3)

Concept Review Questions

9.1 If you know the formula for a molecule or an ion, what is the first step in predicting its shape?

9.2 In what situation is the name of the molecular shape the same as the name of the electron-group arrangement?

9.3 Which of the following numbers of electron groups can give rise to a bent (V-shaped) molecule: two, three, four, five, or six? Draw an example for each number of electron groups, showing the shape classification (AX_mE_n) and the ideal bond angle.

9.4 Name all the molecular shapes that have a tetrahedral electron-group arrangement.

9.5 Consider the following molecular shapes:

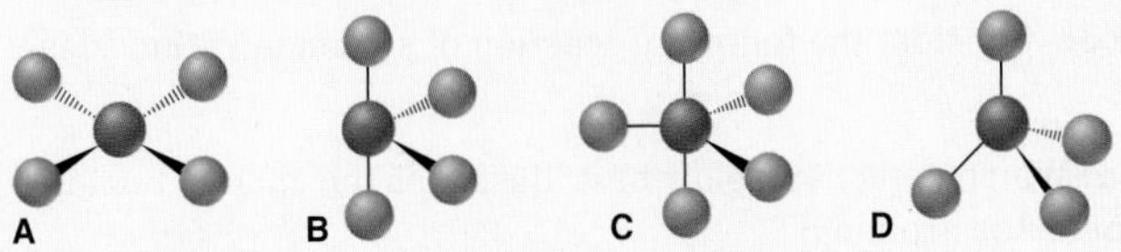

A B C D

(a) Which has the most electron pairs (both shared and unshared) around the central atom?
(b) Which has the most unshared pairs around the central atom?
(c) Do any have only shared pairs around the central atom?

9.6 Use wedge-bond perspective drawings (if necessary) to sketch the positions of the atoms in a general molecule of formula (not shape class) AX_n that has each shape:
(a) V shape (b) trigonal planar (c) trigonal bipyramidal
(d) T shape (e) trigonal pyramidal (f) square pyramidal

9.7 What would you expect to be the electron-group arrangement around atom A in each molecule or ion? For each arrangement, give the ideal bond angle and the direction of any expected deviation:

(a) 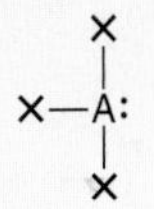 (b) X—A≡X (c) 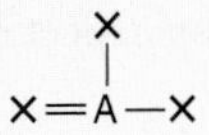 (d) X—Ä—X

(e) X=A=X (f) 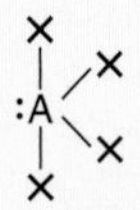

Skill-Building Exercises (grouped in similar pairs)

9.8 Determine the electron-group arrangement, molecular shape, and ideal bond angle(s) for each molecule or ion:
(a) O_3 (b) H_3O^+ (c) NF_3

9.9 Determine the electron-group arrangement, molecular shape, and ideal bond angle(s) for each molecule or ion:
(a) SO_4^{2-} (b) NO_2^- (c) PH_3

9.10 Determine the electron-group arrangement, molecular shape, and ideal bond angle(s) for each molecule or ion:
(a) CO_3^{2-} (b) SO_2 (c) CF_4

9.11 Determine the electron-group arrangement, molecular shape, and ideal bond angle(s) for each molecule:
(a) SO_3 (b) N_2O (N is central) (c) CH_2Cl_2

9.12 Name the shape and give the AX_mE_n classification and ideal bond angle(s) for each general molecule:

(a) (b) (c)

9.13 Name the shape and give the AX_mE_n classification and ideal bond angle(s) for each general molecule:

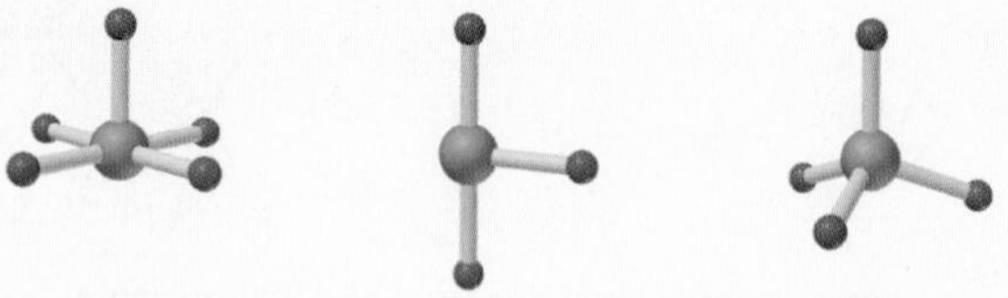

(a) (b) (c)

9.14 Determine the shape, ideal bond angle(s), and direction of any deviation from those angles for each molecule or ion:
(a) ClO_2^- (b) PF_5 (c) SeF_4 (d) KrF_2

9.15 Determine the shape, ideal bond angle(s), and direction of any deviation from those angles for each molecule or ion:
(a) ClO_3^- (b) IF_4^- (c) $SeOF_2$ (d) TeF_5^-

9.16 Determine the shape around each central atom in each molecule, and explain any deviation from the ideal bond angles:
(a) CH_3OH (b) N_2O_4 (O_2NNO_2)

9.17 Determine the shape around each central atom in each molecule, and explain any deviation from the ideal bond angles:
(a) H_3PO_4 (no H—P bond) (b) $CH_3—O—CH_2CH_3$

9.18 Determine the shape around each central atom in each molecule, and explain any deviation from the ideal bond angles:
(a) CH_3COOH (b) H_2O_2

9.19 Determine the shape around each central atom in each molecule, and explain any deviation from the ideal bond angles:
(a) H_2SO_3 (no H—S bond) (b) N_2O_3 ($ONNO_2$)

9.20 Arrange the following AF_n species in order of *increasing* F—A—F bond angles: BF_3, BeF_2, CF_4, NF_3, OF_2.

9.21 Arrange the following ACl_n species in order of *decreasing* Cl—A—Cl bond angles: SCl_2, OCl_2, PCl_3, $SiCl_4$, $SiCl_6^{2-}$.

9.22 State an ideal value for each bond angle in each molecule, and note where you expect deviations:

(a) (b) (c)

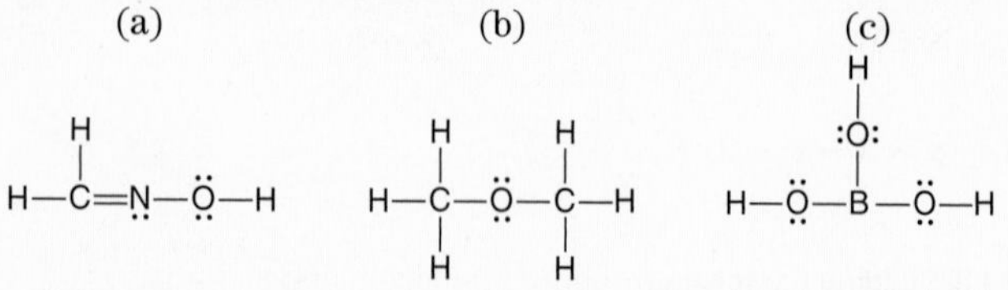

9.23 State an ideal value for each bond angle in each molecule, and note where you expect deviations:

(a) (b) (c)

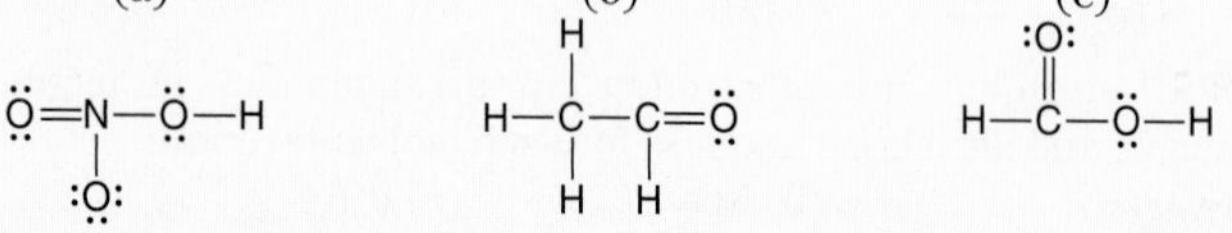

Problems in Context

9.24 Because both tin and carbon are members of group 14, they form structurally similar compounds. However, tin exhibits a greater variety of structures because it forms several ionic species. Predict the shapes and ideal bond angles, including any deviations, for each molecule or ion:
(a) $Sn(CH_3)_2$ (b) $SnCl_3^-$ (c) $Sn(CH_3)_4$
(d) SnF_5^- (e) SnF_6^{2-}

9.25 In the gas phase, phosphorus pentachloride exists as separate molecules. In the solid phase, however, the compound is composed of alternating PCl_4^+ and PCl_6^- ions. What change(s) in molecular shape occur(s) as PCl_5 solidifies? How does the Cl—P—Cl angle change?

Molecular Shape and Molecular Polarity

(Sample Problem 9.4)

Concept Review Questions

9.26 How do you determine if a molecule with the general formula AX_n (where $n > 2$) is polar?

9.27 How can a molecule with polar covalent bonds not be polar? Give an example.

9.28 Explain, in general, why the shape of a biomolecule is important to its function.

Skill-Building Exercises (grouped in similar pairs)

9.29 Consider the molecules SCl_2, F_2, CS_2, CF_4, and BrCl.
(a) Which molecule has bonds that are the most polar?
(b) Which molecules have a dipole moment?

9.30 Consider the molecules BF_3, PF_3, BrF_3, SF_4, and SF_6.
(a) Which molecule has bonds that are the most polar?
(b) Which molecules have a dipole moment?

9.31 Which molecule in each pair has the greater dipole moment? Give the reason for your choice.
(a) SO_2 or SO_3 (b) ICl or IF
(c) SiF_4 or SF_4 (d) H_2O or H_2S

9.32 Which molecule in each pair has the greater dipole moment? Give the reason for your choice.
(a) ClO_2 or SO_2 (b) HBr or HCl
(c) $BeCl_2$ or SCl_2 (d) AsF_3 or AsF_5

Problems in Context

9.33 There are three different dichloroethene isomers (molecular formula $C_2H_2Cl_2$), which we can designate X, Y, and Z. Compound X has no dipole moment, but compound Z does. Compounds X and Z each combine with hydrogen to give the same product:

$$C_2H_2Cl_2 \text{ (X or Z)} + H_2 \longrightarrow ClCH_2 - CH_2Cl$$

What are the structures of X, Y, and Z? Would you expect compound Y to have a dipole moment?

9.34 Dinitrogen difluoride, N_2F_2, is the only stable, simple inorganic molecule with an N═N bond. It occurs in *cis* and *trans* forms.
(a) Draw the molecular shapes of the two forms of N_2F_2.
(b) Predict the direction of the polarity, if any, of each form.

Comprehensive Problems

9.35 Give the molecular shape of each species in Problem 8.119 (Chapter 8).

9.36 Consider the following reaction of silicon tetrafluoride:

$$SiF_4 + F^- \longrightarrow SiF_5^-$$

(a) Which depiction below best illustrates the change in molecular shape around Si?

(b) Give the name and AX_mE_n designation of each shape in the depiction you chose in part (a).

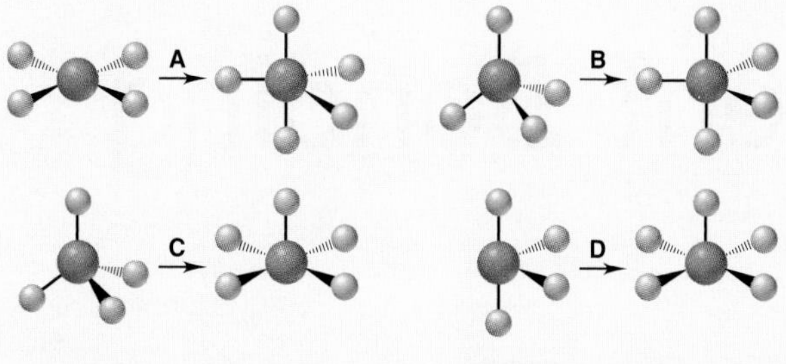

9.37 Both aluminum and iodine form chlorides, Al_2Cl_6 and I_2Cl_6, with "bridging" Cl atoms. Their Lewis structures are given below:

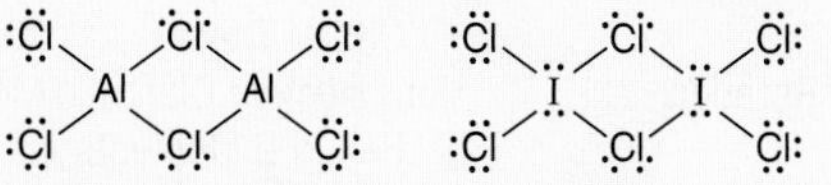

(a) What is the formal charge on each atom?
(b) Which of these molecules has a planar shape? Explain.

9.38 VSEPR theory was developed before any xenon compounds had been prepared. Thus, these compounds provided an excellent test of the theory's predictive power. What would you have predicted for the shapes of XeF_2, XeF_4, and XeF_6?

9.39 When SO_3 gains two electrons, SO_3^{2-} forms.
(a) Which depiction below best illustrates the change in molecular shape around S?
(b) Does molecular polarity change during this reaction?

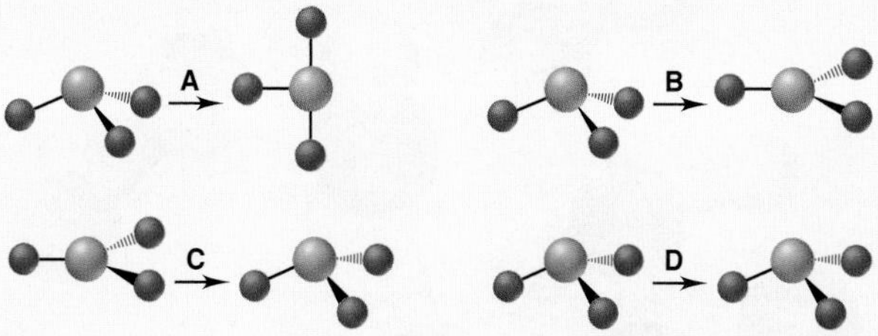

9.40 The actual bond angle in NO_2 is 134.3°, and the actual bond angle in NO_2^- is 115.4°, although the ideal bond angle is 120° in both. Explain.

9.41 "Inert" xenon actually forms several compounds, especially with the highly electronegative elements oxygen and fluorine. The simple fluorides XeF_2, XeF_4, and XeF_6 are all formed by direct reaction of the elements. As you might expect from the size of the xenon atom, the Xe—F bond is not a strong one. Calculate the Xe—F bond energy in XeF_6, given that the enthalpy of formation is −402 kJ/mol.

9.42 Epoxypropane (propylene oxide) is used to make many products, including plastics such as polyurethane. One method for synthesizing it involves oxidizing propene with hydrogen peroxide:

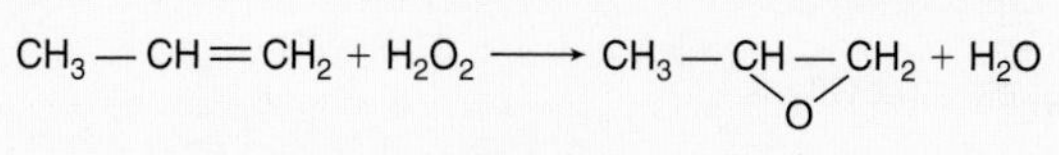

(a) What are the molecular shape and ideal bond angle around each carbon atom in epoxypropane?
(b) Predict any deviation from the ideal bond angles for the actual C—C—C bond angles. (Assume that the three atoms in the ring form an equilateral triangle.)

9.43 Chloral, $Cl_3C—CH{=}O$, reacts with water to form the sedative and hypnotic agent chloral hydrate, $Cl_3C—CH(OH)_2$. Draw Lewis structures for these substances, and describe the change in molecular shape, if any, that occurs around each carbon atom during the reaction.

9.44 The four bonds of tetrachloromethane (carbon tetrachloride, CCl_4) are polar, but the molecule is nonpolar because the bond polarity is cancelled by the symmetrical tetrahedral shape. When other atoms are substituted for some of the Cl atoms, the symmetry is broken and the molecule becomes polar. Use Figure 8.23 to rank the following molecules from the least polar to the most polar: CH_2Br_2, CF_2Cl_2, CH_2F_2, CH_2Cl_2, CBr_4, CF_2Br_2.

9.45 Except for nitrogen, all of the elements in group 15 form pentafluorides, and most form pentachlorides. The chlorine atoms of PCl_5 can be replaced with fluorine atoms one at a time to give, successively, PCl_4F, PCl_3F_2, . . . , PF_5.
(a) Given the sizes of F and Cl, would you expect the first two F substitutions to be at axial or equatorial positions? Explain.
(b) Which of the five fluorine-containing molecules have no dipole moment?

9.46 A student isolates a product with the molecular shape shown below. The orange spheres represent F.
(a) If the species is a neutral compound, can the black sphere represent selenium (Se)?
(b) If the species is an anion, can the black sphere represent N?
(c) If the black sphere represents Br, what is the charge of the species?

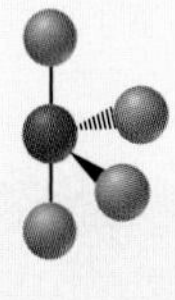

9.47 A molecule of formula AY_3 is found experimentally to be polar. Which molecular shapes are possible and which are impossible for AY_3?

9.48 Consider the following molecular shapes:

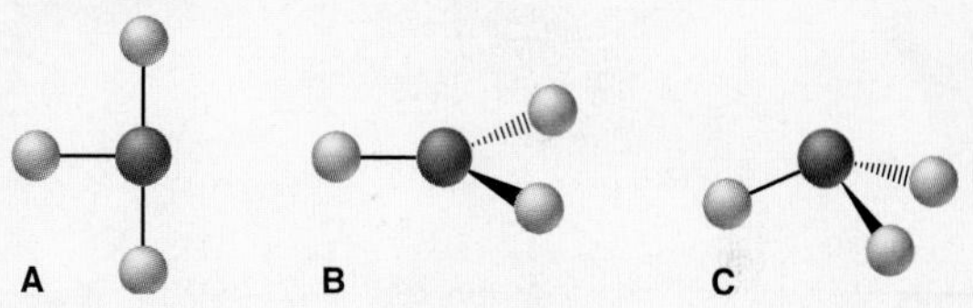

(a) Match each shape with one of the following species: XeF_3^+, $SbBr_3$, $GaCl_3$.
(b) Which species, if any, is polar?
(c) Which species has the most valence electrons around the central atom?

9.49 Ethene (ethylene, C_2H_4) and tetrafluoroethene (tetrafluoroethylene, C_2F_4) are used to make the polymers polyethylene and polytetrafluoroethylene (Teflon), respectively.
(a) Draw the Lewis structures for C_2H_4 and C_2F_4, and give the ideal H—C—H and F—C—F bond angles.
(b) The actual H—C—H and F—C—F bond angles are 117.4° and 112.4°, respectively. Explain these deviations.

9.50 Using bond lengths in Table 8.2, and assuming ideal geometry, calculate each distance:
(a) Between the H atoms in C_2H_2
(b) Between the F atoms in SF_6 (two answers)
(c) Between the equatorial F atoms in PF_5

9.51 Phosphorus pentachloride, a key industrial compound with annual world production of about 2×10^7 kg, is used to make other compounds. It reacts with sulfur dioxide to produce phosphorus oxychloride ($POCl_3$) and thionyl chloride ($SOCl_2$). Draw a Lewis structure and name the molecular shape of each product.

Theories of Covalent Bonding

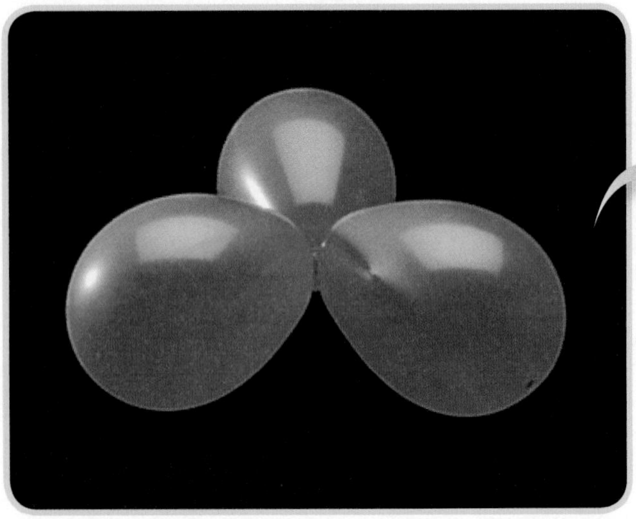

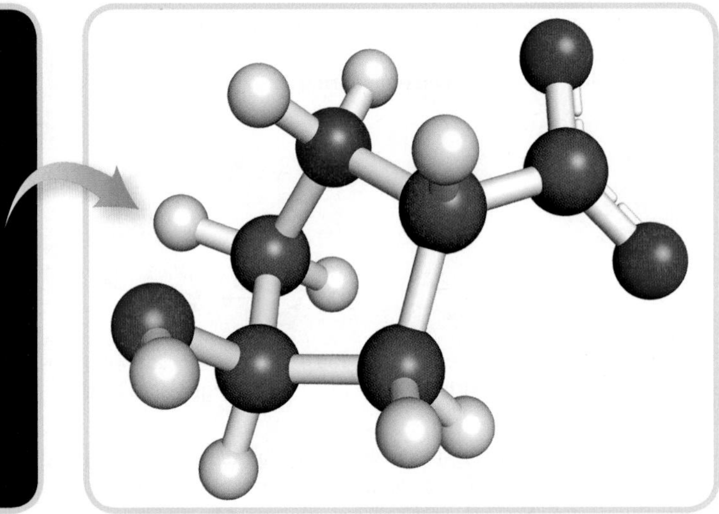

IN THIS CHAPTER . . . We introduce two theories of bonding in molecules, based on the interactions of the orbitals of their atoms. By the end of this chapter, you should be able to

- Explain the valence bond theory, which rationalizes molecular shapes through interactions of atomic orbitals during bonding to form hybrid orbitals
- Apply valence bonding theory to sigma (σ) and pi (π) bonds, the two types of covalent bonds
- Discuss why parts of molecules do not rotate freely around multiple bonds, and explain why this has major effects on reactivity, physical properties, and biological behaviour
- Discuss molecular orbital theory, which explains molecular energy levels and properties through the formation of orbitals that spread over the whole molecule
- Apply molecular orbital theory to understand key properties of some diatomic molecules and simple polyatomic molecules

10

Why do atoms bond the way they do? Why do some atoms bond by sharing electrons while others do not? We looked at valence-shell electron-pair repulsion (VSEPR) theory in the previous chapter as a simple way of explaining the shapes of molecules for which we had drawn Lewis structures. While this is a very good model for simple molecules, it is not necessarily the best model for more complex molecules. In this chapter, other bonding models will be introduced that provide better explanations for certain types of molecules. Each model has its limitations, which must be kept in mind when deciding which model will be used for a particular structure. We will also discuss whether physical properties, chemical properties, and biological behaviour can be predicted using these models.

Concepts and Skills to Review before Studying This Chapter

- Atomic orbital shapes (Section 6.4)
- Exclusion principle (Section 7.1)
- Hund's rule (Section 7.2)
- Lewis structures (Section 8.6)
- Resonance in covalent bonding (Section 8.6)
- Molecular shapes (Section 9.1)
- Molecular polarity (Section 9.2)

10.1 Valence Bond Theory and Orbital Hybridization

What *is* a covalent bond, and what characteristic gives it strength? How can we explain *molecular* shapes based on the interactions of *atomic* orbitals? The most useful approach for answering these questions is based on quantum mechanics (Chapter 6) and is called **valence bond (VB) theory**.

The Central Themes of Valence Bond Theory

The basic principle of VB theory is that *a covalent bond forms when orbitals of two atoms overlap and a pair of electrons occupy the overlap region.* In the terminology of quantum mechanics, overlap of the two orbitals (formation of bonding orbitals) means that their wave functions are *in phase* (constructive interference), so the amplitude between the nuclei increases (see Figure 6.5). On the other hand, antibonding orbitals form when *out of phase* orbitals combine (destructive interference). The four central themes of VB theory derive from this principle:

1. *Opposing spins of the electron pair.* As the exclusion principle (Section 7.1) prescribes, the space formed by the overlapping orbitals *has a maximum capacity for two electrons that have opposite (paired) spins.* In the simplest case, a molecule of H_2 forms when the 1*s* orbitals of two H atoms overlap, and the electrons, with their spins paired, spend more time in the overlap region. (See the up and down arrows in Figure 10.1.)

2. *Maximum overlap of bonding orbitals.* Bond strength depends on the attraction between nuclei and shared electrons, so *the greater the orbital overlap, the stronger the bond.* The extent of the overlap depends on the orbital shape and direction. An *s* orbital is spherical, so its orientation is the same in any direction, but the *p* and *d* orbitals have specified directions. Thus, a *p* or *d* orbital involved in a bond is oriented to maximize overlap. In HF, for example, the 1*s* orbital of H overlaps a half-filled 2*p* orbital of F *along its long axis* (Figure 10.2A). In F_2, the two half-filled 2*p* orbitals interact end to end, that is, *along the long axes* of the orbitals (Figure 10.2B).

3. *Hybridization of atomic orbitals.* To account for the bonding in diatomic molecules, we picture direct overlap of the *s* and/or *p* orbitals of isolated atoms. But how can we account for the shape of a molecule like methane from the shapes and

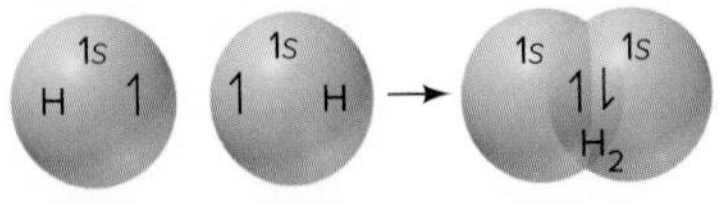

FIGURE 10.1 Orbital overlap and spin pairing in H_2

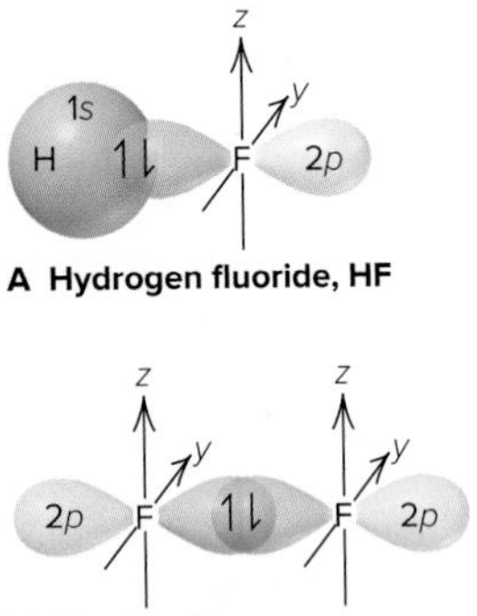

A Hydrogen fluoride, HF

B Fluorine, F_2

FIGURE 10.2 Orbital orientation and maximum overlap. The $2p_x$ orbital is shown bonding; the other two 2p orbitals of F are omitted for clarity.

orientations of C and H atomic orbitals (AOs)? A C atom ([He]$2s^22p^2$) has two valence electrons in the spherical 2s orbital and one each in two of the three mutually perpendicular 2p orbitals. If the half-filled p orbitals overlapped the 1s orbitals of two H atoms, *two* C—H bonds would form with a 90° H—C—H bond angle. However, methane has the formula CH_4, not CH_2, and its bond angle is 109.5°.

To explain such facts, Linus Pauling proposed that, during bonding, *the valence AOs in the isolated atoms become* ***different*** *when they are in the molecule.* Quantum-mechanical calculations show that if we mathematically "mix" certain combinations of orbitals, we form new orbitals whose spatial orientations *do* match the observed molecular shapes. The process of orbital mixing is called **hybridization**, and the new AOs are called **hybrid orbitals**.

4. *Features of hybrid orbitals.* Here are some central points about hybrid orbitals that arise during bonding:

- The *number* of hybrid orbitals formed *equals* the number of AOs mixed. We only need to know the total number of AOs involved to know how many hybrid orbitals are formed.
- The *type* of hybrid orbitals formed *varies* with the types of AOs mixed. The name given to the hybrid orbitals indicates which AOs were mixed and how many were mixed. As a convention, we indicate all the AOs involved in the hybridization with a superscript that states how many of each are involved in the mix (but only if it is a number greater than one, as we will discuss in the next section).
- The *shape* and *orientation* of a hybrid orbital *maximize* overlap with the orbital of the other atom in the bond. The hybrid orbitals will always be as far away from each other as possible.

You can think of hybridization as a process in which AOs mix, hybrid orbitals form and overlap other orbitals, and electrons enter the overlap region with opposing spins, thus forming stable bonds. In truth, hybridization is a concept used to explain the results we calculate, which predict the molecular shapes we observe.

Types of Hybrid Orbitals

It is interesting to note that, according to VB theory, the type of hybrid orbitals in a molecule is postulated after observing its shape. Note that the orientations of the five types of hybrid orbitals we discuss in this section correspond to the five electron-group arrangements in VSEPR theory, which we explored in Chapter 9.

sp Hybridization Looking at the name given to the *sp* hybridization, we can tell that only one *s* AO and one *p* AO were mixed. We can also tell that, since the number of hybrid orbitals formed equals the number of AOs mixed, the number of *sp* hybrid orbitals formed is two. Therefore, the maximum separation between them will occur when they are opposite each other. In general, when two electron groups surround the central atom, we observe a linear shape, which means that the bonding orbitals must have a linear orientation.

1. *AOs mixed to form hybrid orbitals.* VB theory proposes that two *nonequivalent* orbitals of a central atom, one *s* and one *p*, mix and form two *equivalent* ***sp*** **hybrid orbitals** that are oriented 180° apart (Figure 10.3A). The shape of these hybrid orbitals, with one large and one small lobe, differs markedly from the shapes of the AOs. The orbital orientations increase electron density *in the bonding direction.*

2. *Overlap of orbitals from central and surrounding atoms: $BeCl_2$.* In beryllium chloride, the Be atom is *sp* hybridized. How can we use VB theory to explain this hybridization? The electron configuration for the Be atom (Z = 4) shows its two valence electrons in 2*s*. However, to form two covalent bonds with two Cl atoms, the Be atom needs to have two unpaired electrons. For that to happen, one electron from the orbital 2*s* has to be promoted to 2*p*. This is possible because the energy absorbed in the process is less than the energy released when the bonds are formed. Now these two AOs can mix and form two *sp* hybrid orbitals. Figure 10.3B depicts the hybridization of Be in an orbital box diagram, and Figure 10.3C shows an orbital box diagram with shaded contours instead of arrows. Bond formation with Cl is shown in Figure 10.3D.

For Be ($Z = 4$), the full electron configuration is $1s^2 2s^2$.

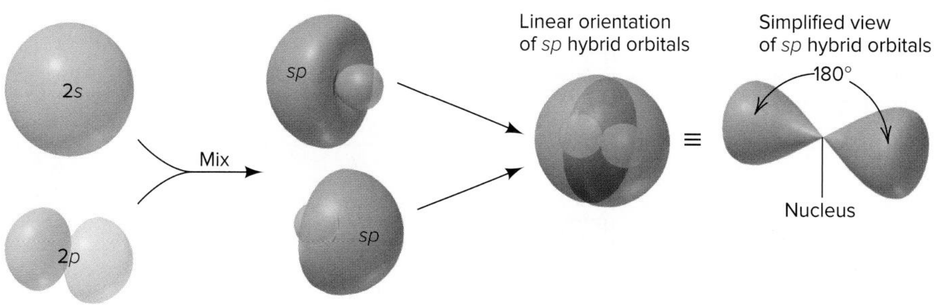

A Formation of *sp* hybrid orbitals

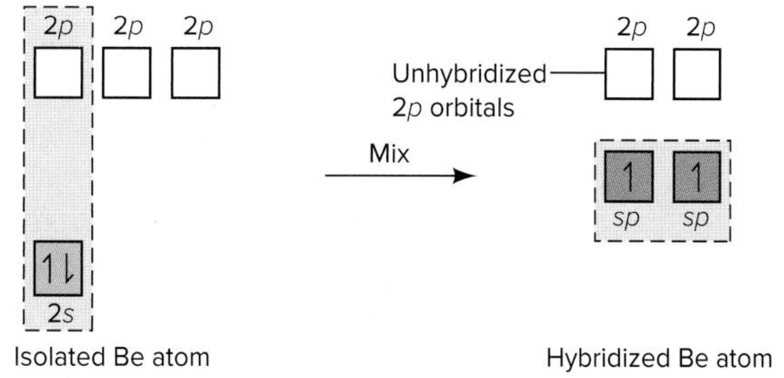

B Orbital box diagram

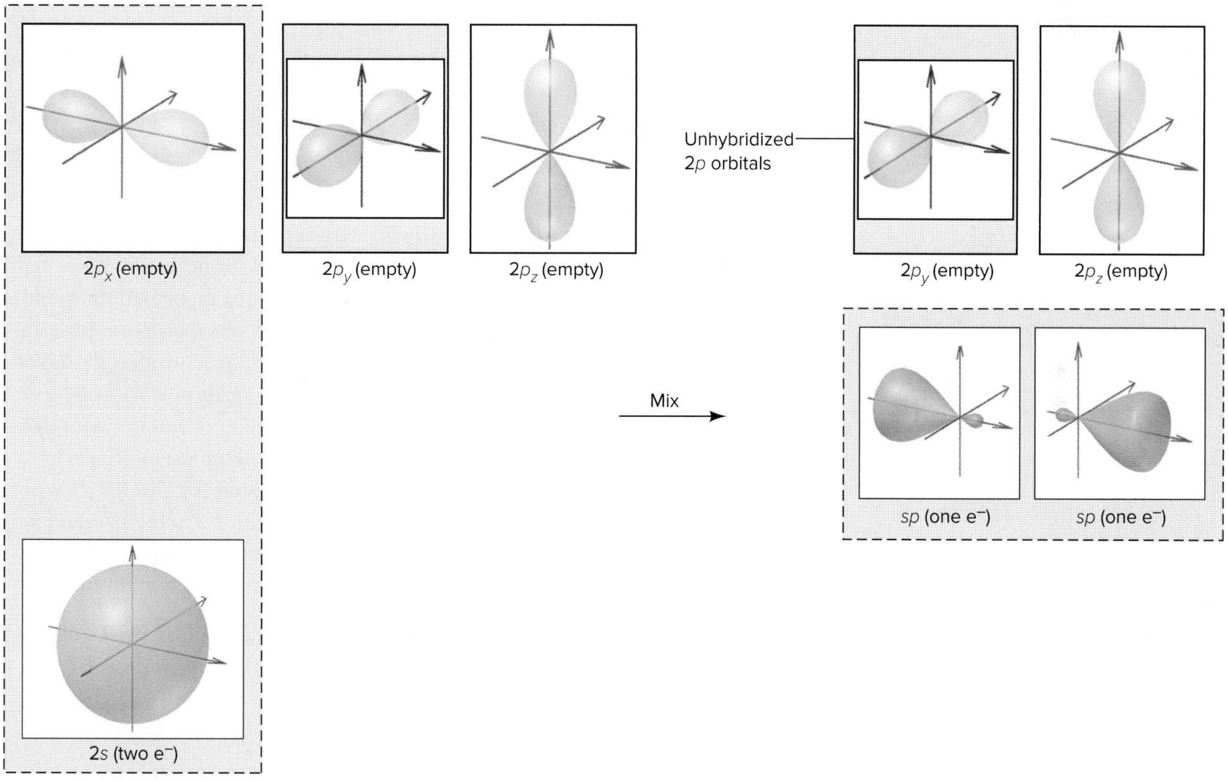

C Box diagram with orbital contours

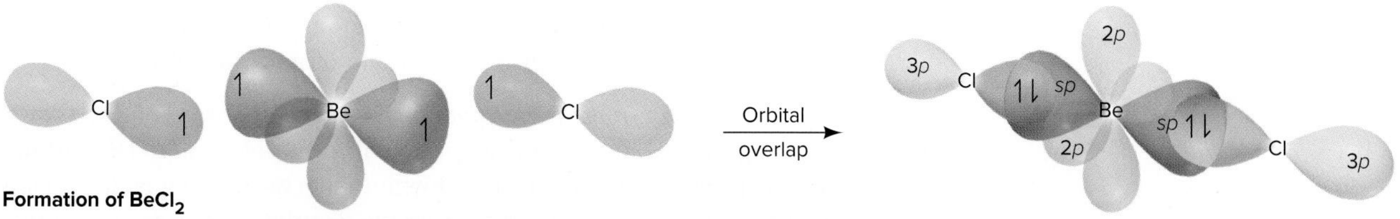

D Formation of $BeCl_2$

FIGURE 10.3 Formation and orientation of the *sp* hybrid orbitals and the bonding in $BeCl_2$. A. One 2s AO and one 2*p* AO mix to form two *sp* hybrid orbitals. (The simplified hybrid orbitals at the far right are used elsewhere, often even without the small lobe.) **B.** The orbital box diagram for the hybridization of Be, drawn vertically. **C.** The orbital box diagram with orbital contours. **D.** Overlap of Be and Cl orbitals to form $BeCl_2$. (Only the Cl 3*p* orbital involved in bonding is shown.)

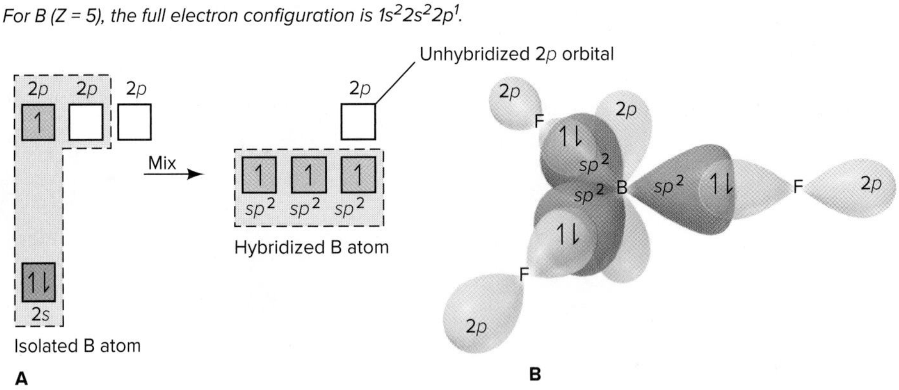

FIGURE 10.4 The sp^2 hybrid orbitals in BF_3. A. The orbital box diagram shows the formation of three sp^2 hybrid orbitals. One $2p$ orbital is unhybridized and empty. **B.** Contour depiction of BF_3. (Only the F $2p$ orbital involved in bonding is shown.)

The filled $2s$ orbital and one of the three empty $2p$ orbitals of Be mix and form two half-filled *sp* orbitals. Two empty unhybridized $2p$ orbitals of Be lie perpendicular to each other and to the *sp* hybrids. The hybrid orbitals overlap the half-filled $3p$ orbital in each of two Cl atoms. The four electrons—two from Be and one from each Cl—appear as pairs with opposite spins in the two overlap regions. (The $3p$ and *sp* hybrid orbitals that are partially coloured on the left become fully coloured on the right, after the bonds form and each orbital is filled with two electrons.)

sp^2 Hybridization We use the sp^2 hybridization to rationalize the two shapes that are possible for the trigonal planar electron-group arrangement.

1. *Orbitals mixed and orbitals formed.* Mixing one *s* and two *p* orbitals gives three **sp^2 hybrid orbitals** that point to the corners of an equilateral triangle, with their axes 120° apart. (In hybrid orbitals, unlike in electron configurations, superscripts refer to the number of *AOs* of a given type, *not* to the number of *electrons* in the orbital; thus, one *s* and two *p* orbitals give s^1p^2, or sp^2.) The third $2p$ orbital remains unhybridized.

2. *Overlap of orbitals from central and surrounding atoms: BF_3.* Looking at the shape of this molecule, we can infer that the central B atom in BF_3 is sp^2 hybridized, with the three sp^2 orbitals in a trigonal plane and the third $2p$ orbital unhybridized and perpendicular to this plane (Figure 10.4). If we do the electron configuration for the B atom ($Z = 5$), we find out that its three valence electrons are supposed to be distributed such that there are two in $2s$ and one in $2p$. We know by now that there is only a small energy gap between the $2s$ and $2p$ orbitals, and that the energy absorbed to promote one electron from $2s$ to an empty $2p$ is less than the energy released when forming the bonds with the F atoms, which compensates for the initial input. These three AOs can then mix and form three sp^2 hybrid orbitals Each half-filled sp^2 orbital overlaps the half-filled $2p$ orbital of an F atom, and the six valence electrons—three from B and one from each of the three F atoms—form three bonding pairs.

3. *Placement of lone pairs.* To account for other molecular shapes within a given electron-group arrangement, one or more hybrid orbitals contain a lone pair. In ozone (O_3), for example, the central O is sp^2 hybridized and a lone pair fills one of the three sp^2 orbitals.

sp^3 Hybridization The sp^3 hybridization, which accounts for the shape of methane, applies to any species with a tetrahedral electron-group arrangement.

1. *Orbitals mixed and orbitals formed.* Mixing one *s* and three *p* orbitals gives four **sp^3 hybrid orbitals** that point to the corners of a tetrahedron.

2. *Overlap of orbitals from central and surrounding atoms: CH_4.* The C atom in methane is sp^3 hybridized. But why is it sp^3 hybridized if the C atom ($Z = 6$) has its four valence electrons supposedly arranged with two in $2s$ and two in $2p$? Again, we can assume a promotion of one electron from the $2s$ orbital to an empty $2p$ orbital. However, if the four bonds in methane are equivalent, this means that the C atom uses four similar hybrid orbitals (each of them with one electron) to bond to the H atoms. Therefore, its four valence electrons half-fill the four sp^3 hybrids, which overlap the half-filled $1s$ orbitals of four H atoms to form four C—H bonds (Figure 10.5).

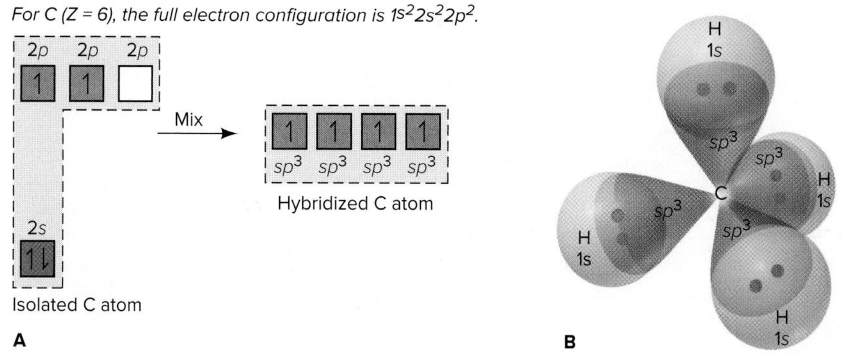

FIGURE 10.5 The sp^3 hybrid orbitals in CH_4. A. The orbital box diagram shows the formation of four sp^3 hybrids. **B.** Contour depiction of CH_4, with the electron pairs shown as dots.

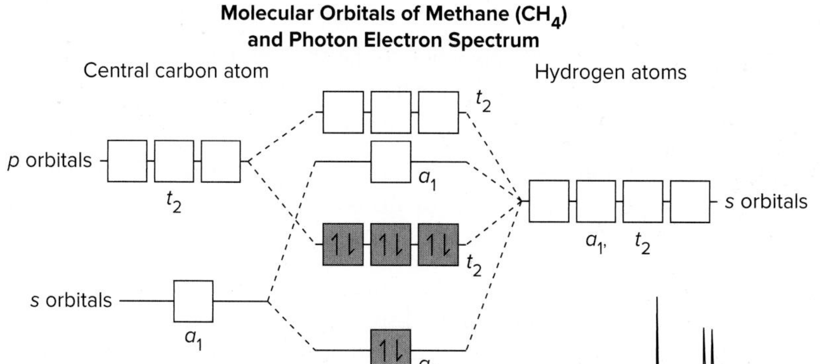

Comment The hybrid orbital theory predicts four equivalent bonds in methane (CH_4). Therefore, the PES (photon electron spectrum) of methane should show a single peak; however, as we can see below, this is not the case (two peaks are clearly present), and the integrated intensities of those peaks are very close to 3:1. This shows that the orbitals are not all equivalent and are not sp^3 orbitals. The orbitals can only be explained via molecular orbital (MO) theory, which we will study in more detail in Section 10.3.

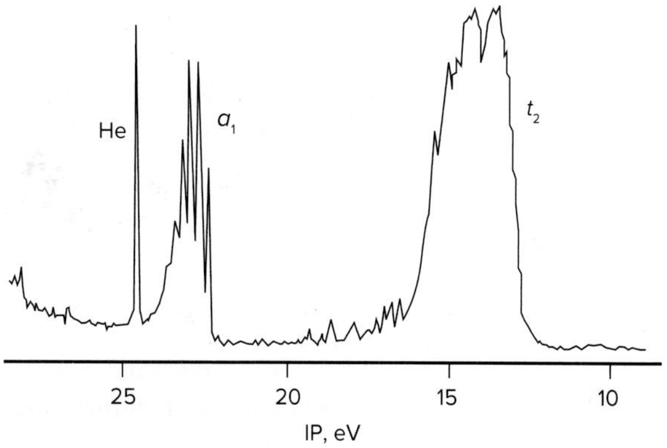

3. *Placement of lone pairs.* The trigonal pyramidal shape of NH_3 arises when a lone pair fills any one of the four sp^3 orbitals of N. The bent shape of H_2O arises when lone pairs fill any two of the sp^3 orbitals of O (Figure 10.6).

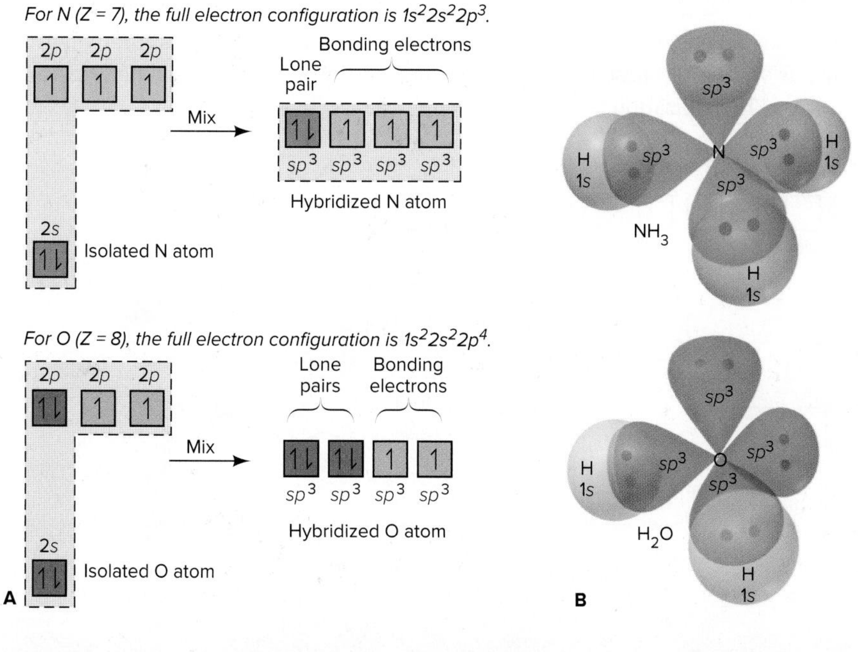

FIGURE 10.6 The sp^3 hybrid orbitals in NH_3 and H_2O. A. The orbital box diagrams show sp^3 hybridization, with lone pairs filling one (NH_3) or two (H_2O) hybrid orbitals. **B.** Contour depictions of NH_3 and H_2O.

FIGURE 10.7 The sp^3d hybrid orbitals in PCl_5. **A.** The orbital box diagram shows the formation of five half-filled sp^3d orbitals. Four 3*d* orbitals are unhybridized and empty. **B.** Contour depiction of PCl_5. (For clarity, unhybridized 3*d* orbitals, the other two Cl 3*p* orbitals, and the five bonding P—Cl pairs are not shown.)

For P (Z = 15), the full electron configuration is $1s^22s^22p^63s^23p^3$.

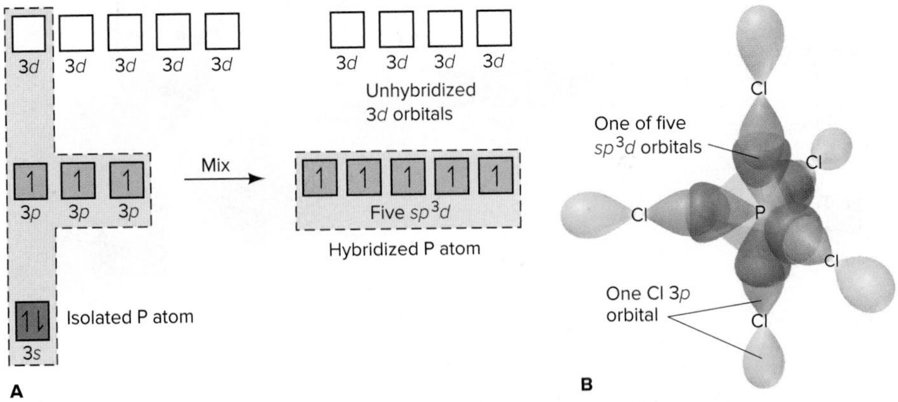

sp^3d Hybridization Molecules with shapes due to the trigonal bipyramidal electron-group arrangement have central atoms from period 3 or higher. VB theory proposes that *d* orbitals, as well as *s* and *p* orbitals, are mixed to form hybrid orbitals.

1. *Orbitals mixed and orbitals formed.* Mixing one 3*s*, the three 3*p*, and one of the five 3*d* orbitals gives five ***sp^3d* hybrid orbitals**, which point to the corners of a trigonal bipyramid.

2. *Overlap of orbitals from central and surrounding atoms: PCl_5.* The P atom in PCl_5 is sp^3d hybridized. Each hybrid orbital overlaps a 3*p* orbital of a Cl atom, and ten valence electrons—five from P and one from each of the five Cl atoms—form five P—Cl bonds (Figure 10.7).

3. *Placement of lone pairs.* Seesaw, T-shaped, and linear molecules have lone pairs in, respectively, one, two, or three of the central atom's sp^3d orbitals.

sp^3d^2 Hybridization According to VB theory, molecules with shapes that have the octahedral electron-group arrangement also use *d* orbitals to form hybrids.

1. *Orbitals mixed* and *orbitals formed.* Mixing one 3*s*, the three 3*p*, and two 3*d* orbitals gives six ***sp^3d^2* hybrid orbitals**, which point to the corners of an octahedron.

2. *Overlap of orbitals from central and surrounding atoms: SF_6.* The S atom in SF_6 is sp^3d^2 hybridized. Each half-filled hybrid orbital overlaps a half-filled 2*p* orbital of an F atom, and twelve valence electrons—six from S and one from each of the six F atoms—form six S—F bonds (Figure 10.8).

3. *Placement of lone pairs.* Square pyramidal and square planar molecules have lone pairs in one and two of the central atom's sp^3d^2 orbitals, respectively.

There is some controversy regarding the use of *d* orbitals in hybrids for main-group atoms, as will be discussed further at the end of this section, in "Limitations to the Concept of Hybridization."

FIGURE 10.8 The sp^3d^2 hybrid orbitals in SF_6. **A.** The orbital box diagram shows the formation of six half-filled sp^3d^2 orbitals; three 3*d* orbitals remain unhybridized and empty. **B.** Contour depiction of SF_6. (For clarity, unhybridized 3*d* orbitals, the other two F 2*p* orbitals, and the six bonding S—F pairs are not shown.)

For S (Z = 16), the full electron configuration is $1s^22s^22p^63s^23p^4$.

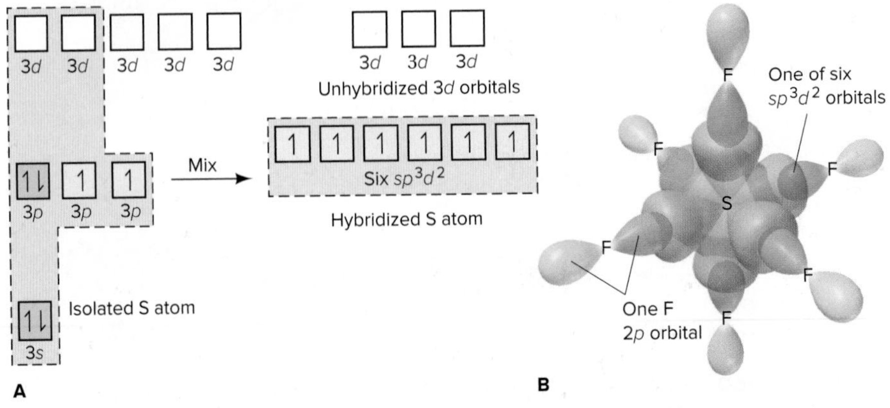

TABLE 10.1 Composition and Orientation of Hybrid Orbitals

	Linear	Trigonal Planar	Tetrahedral	Trigonal Bipyramidal	Octahedral
Atomic orbitals mixed	one *s* one *p*	one *s* two *p*	one *s* three *p*	one *s* three *p* one *d*	one *s* three *p* two *d*
Hybrid orbitals formed	two *sp*	three sp^2	four sp^3	five sp^3d	six sp^3d^2
Unhybridized orbitals remaining	two *p*	one *p*	none	four *d*	three *d*
Orientation		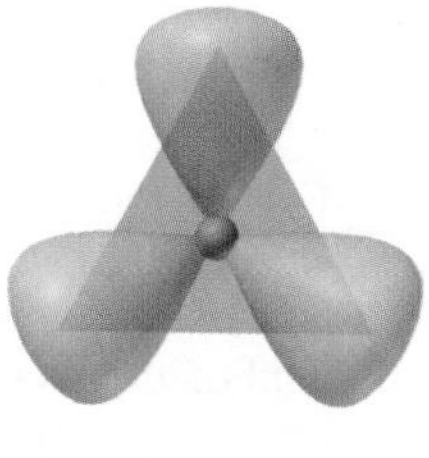	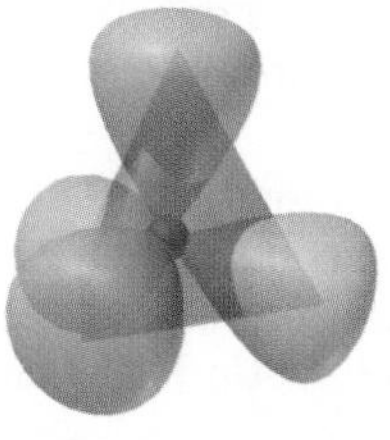	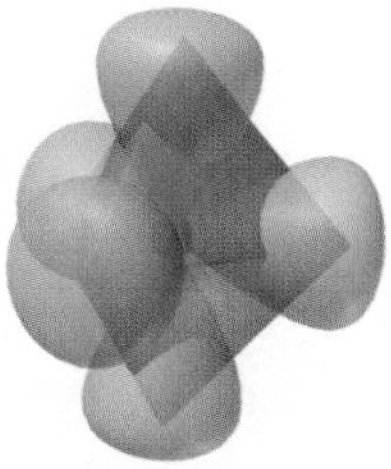	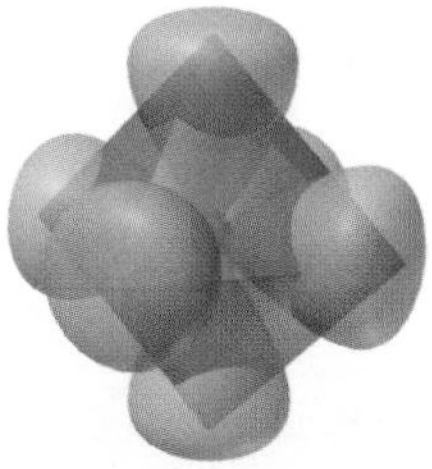

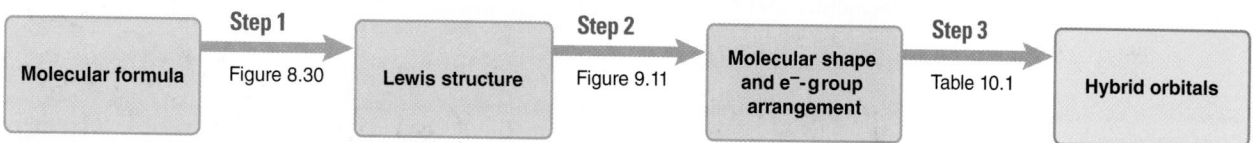

Table 10.1 summarizes the numbers and types of AOs that mix to form the five types of hybrid orbitals. Figure 10.9 shows three conceptual steps, from a molecular formula to the hybrid orbitals in the molecule, and Sample Problem 10.1 focuses on the third step.

Molecular formula → **Step 1** (Figure 8.30) → **Lewis structure** → **Step 2** (Figure 9.11) → **Molecular shape and e⁻-group arrangement** → **Step 3** (Table 10.1) → **Hybrid orbitals**

FIGURE 10.9 From molecular formula to hybrid orbitals. See Figures 8.30 and 9.11, as well as Table 10.1.

Sample Problem 10.1 Postulating Hybrid Orbitals in a Molecule

Problem Use partial orbital diagrams to describe how mixing the AOs of the central atom(s) leads to the hybrid orbitals in each molecule:

(a) Methanol, CH_3OH **(b)** Sulfur tetrafluoride, SF_4

Plan Prior to the steps described below, we have used the molecular formula to draw the Lewis structure and determined the electron-group arrangement of each central atom. Then we can use Table 10.1 to postulate the type of hybrid orbitals. We write the partial orbital diagram for each central atom before and after the orbitals are hybridized.

Solution (a) In CH_3OH, the electron-group arrangement is tetrahedral around both the C atom and the O atom, so each mixed one 2*s* orbital and three 2*p* orbitals to become sp^3 hybridized. The C atom has four half-filled sp^3 orbitals:

↿	↿		2*p*
⇅			2*s*

Isolated C atom —Mix→ | ↿ | ↿ | ↿ | ↿ | sp^3 Hybridized C atom

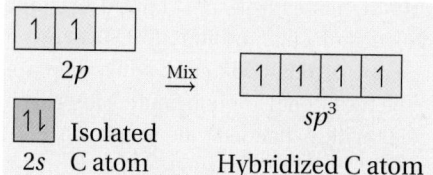

The O atom has two half-filled sp^3 orbitals and two orbitals filled with lone pairs:

⇅	↿	↿	2*p*
⇅			2*s*

Isolated O atom —Mix→ | ⇅ | ⇅ | ↿ | ↿ | sp^3 Hybridized O atom

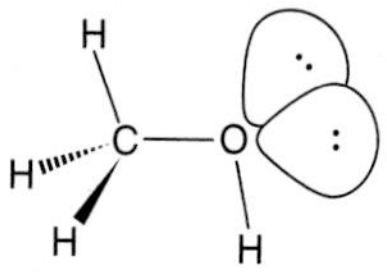

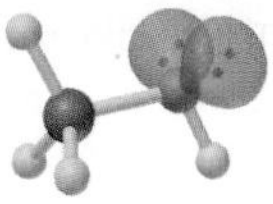

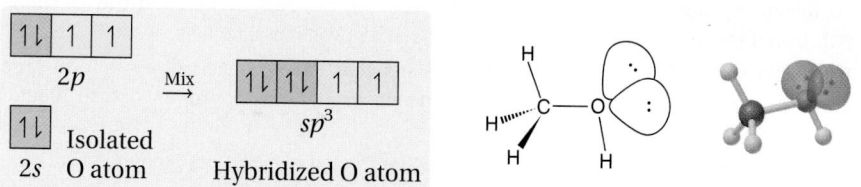

(b) In SF_4, the electron-group arrangement is trigonal bipyramidal, so the central S atom is sp^3d hybridized, which means that one 3*s* orbital, three 3*p* orbitals, and one $3d$ orbital are mixed. One hybrid orbital is filled with a lone pair, and four orbitals are half-filled. Four unhybridized $3d$ orbitals remain empty:

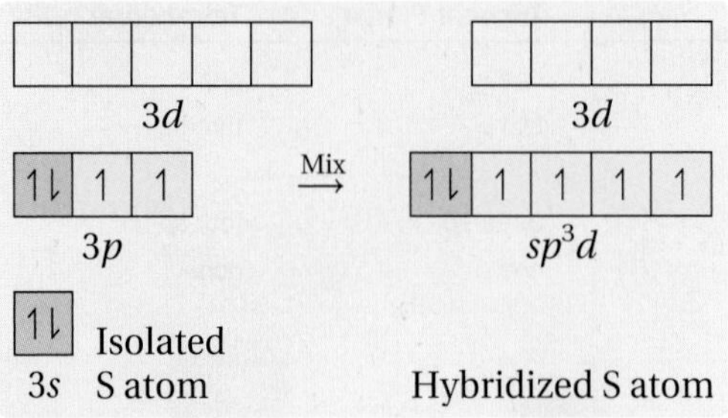

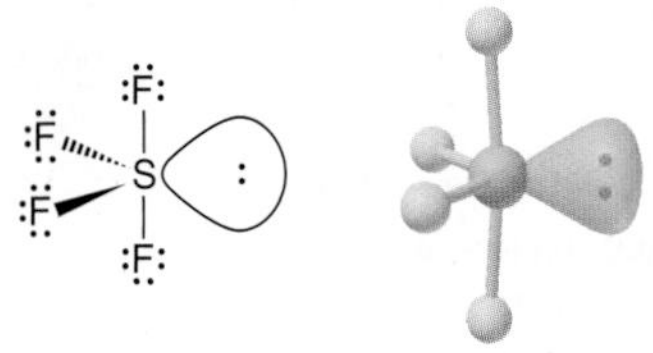

Follow-Up Problem 10.1 Use partial orbital diagrams to show how the AOs of the central atom mix to form hybrid orbitals in **(a)** beryllium fluoride, BeF_2; **(b)** silicon tetrachloride, $SiCl_4$; **(c)** xenon tetrafluoride, XeF_4. (See Brief Solutions to Follow-Up Problems at the end of the chapter.)

Limitations to the Concept of Hybridization VSEPR and VB theories can be used to rationalize an observed molecular shape. In some cases, however, these theories may not be consistent with other findings.

1. *Hybridization does not apply: large nonmetal hydrides.* Consider the Lewis structure and bond angle of H_2S:

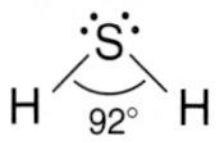

Based on VSEPR theory, we would predict that, as in H_2O, the four electron groups around H_2S point to the corners of a tetrahedron, and the two lone pairs compress the H—S—H bond angle below the ideal 109.5°. Based on VB theory, we would predict that the $3s$ and $3p$ orbitals of the S atom mix and form four sp^3 hybrids. Two of these hybrids are filled with lone pairs, while the other two overlap $1s$ orbitals of two H atoms and are filled with bonding pairs.

However, observations do *not* support these predictions. The bond angle of 92° is close to the 90° angle between *unhybridized p* orbitals. Similar angles occur in the hydrides of other large nonmetals of groups 15 and 16. Why apply a theory if the facts do not support it? Real factors—bond length, atomic size, and electrostatic repulsions—influence shape. Larger atoms form longer bonds to H, which decreases repulsions; thus, overlap of *unhybridized* orbitals explains these shapes.

2. *d-orbital hybridization is less important: rationalizing shapes with expanded valence shells.* Quantum-mechanical calculations show that d orbitals have such high energies that they do not hybridize effectively with the *much* more stable s and p orbitals of a given n value. Thus, for example, some scientists have proposed that SF_6 is most stable when the bonding orbitals of the central S atom use a combination of sp hybrid orbitals and unhybridized $3p$ orbitals instead of sp^3d^2 hybrid orbitals. Other scientists prefer explanations that involve molecular orbitals (MOs) or even ionic structures. These topics are beyond the scope of this book, and, while they are being actively debated, we will continue to use the traditional, though limited, approach of including d-orbital hybridization for molecules with expanded valence shells.

SUMMARY OF SECTION 10.1

- VB theory explains that a covalent bond forms when two AOs overlap and two electrons with paired (opposite) spins spend more time in the overlapped regions.
- To explain molecular shape, VB theory proposes that, during bonding, AOs mix to form hybrid orbitals with a different shape and direction. This process gives rise to greater overlap and, thus, stronger bonds.
- Based on the observed molecular shape (and the related electron-group arrangement), we can predict the type of hybrid orbital that accounts for the shape. In many cases, especially molecules with larger central atoms, the concept of hybridization has major limitations.

10.2 Modes of Orbital Overlap and the Types of Covalent Bonds

Orbitals can overlap by two *modes*—end to end or side to side—which give rise to two types of covalent bonds—*sigma* (σ) and *pi* (π). In this section, we will use VB theory to describe these two types of covalent bonds. As you will see, they can also be described by MO theory.

Orbital Overlap in Single and Multiple Bonds

Ethane (C_2H_6), ethene (C_2H_4), and ethyne (C_2H_2) have different shapes. Ethane is tetrahedral at both carbons, with bond angles near the ideal 109.5°. Ethene is trigonal planar at both carbons, with bond angles near the ideal 120°. Ethyne is linear, with bond angles of 180°:

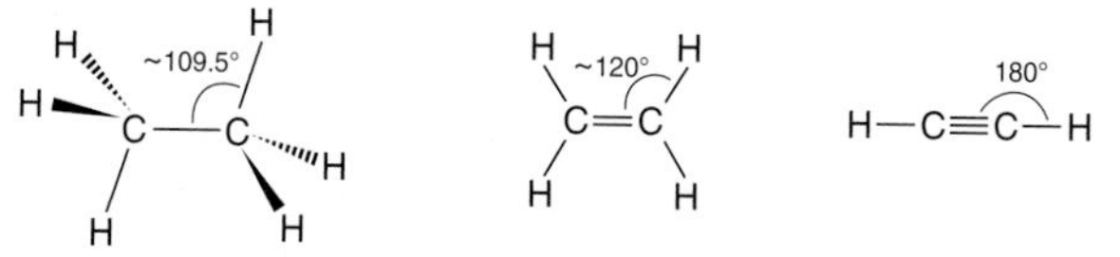

Ethane Ethene Ethyne

In these molecules, two modes of orbital overlap result in two types of bonds.

End-to-End Overlap and Sigma (σ) Bonding Since both C atoms of ethane are tetrahedral, they are sp^3 hybridized (Figure 10.10). The C—C bond arises from the overlap of the end of one sp^3 orbital with the end of the other. *End-to-end* overlap forms a **sigma (σ) bond**, which has its *highest electron density along the bond axis* and is shaped like an ellipse rotated about its long axis (like a football). *All single bonds are σ bonds*, including the six C—H bonds in ethane.

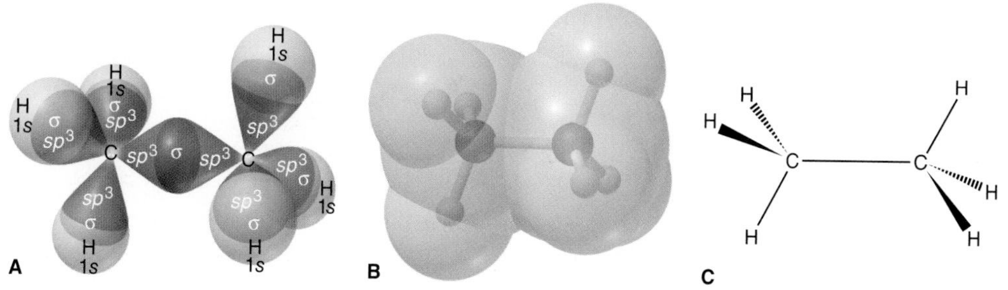

FIGURE 10.10 The σ bonds in ethane (C_2H_6). A. Depiction using atomic contours. **B.** An electron density model shows very slightly positive (*blue*) and negative (*red*) regions. **C.** Wedge-bond perspective drawing.

Side-to-Side Overlap and Pi (π) Bonding The pi (π) MOs arise any time there are contiguous arrays of unhybridized *p* orbitals. If the arrays are *in phase*, they will form a **pi (π) bond**. For this type of bond, let us examine ethene and ethyne:

1. *In ethene*, each C atom is sp^2 hybridized. The four valence electrons of C half-fill the three sp^2 orbitals *and* the unhybridized 2*p* orbital, which lies perpendicular to the sp^2 plane (Figure 10.11). Two sp^2 orbitals of each C form C—H σ bonds. The third sp^2 orbital forms a σ bond with the other C. With the σ-bonded C atoms near each other, their half-filled 2*p* orbitals overlap *side to side*, forming a π bond. The π bond has *two regions (lobes) of electron density*, one above and one below the σ-bond axis. *The two electrons in one π bond occupy both lobes. A double bond*

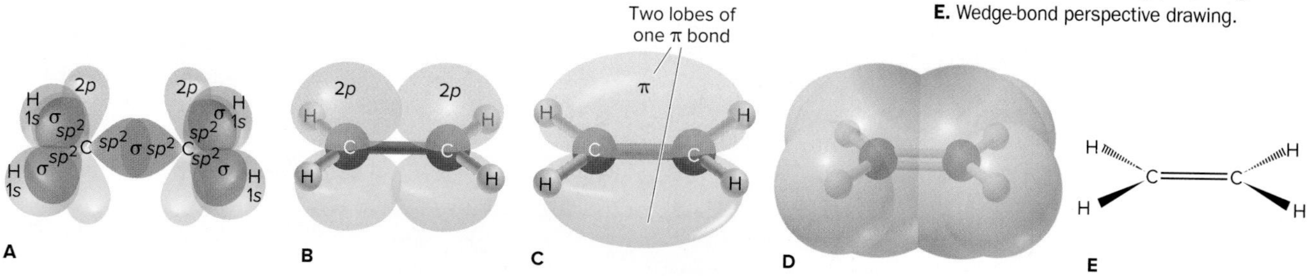

FIGURE 10.11 The σ and π bonds in ethene (C_2H_4). A. The C—C σ bond and the four C—H σ bonds are shown, as well as the unhybridized 2*p* orbitals. **B.** An accurate depiction of the 2*p* orbitals shows the side-to-side overlap; σ bonds are shown in ball-and-stick form. **C.** Two overlapping regions make up *one* π bond, which is occupied by two electrons. **D.** With four electrons (one σ bond and one π bond) between the C atoms, electron density (*red*) is higher. **E.** Wedge-bond perspective drawing.

FIGURE 10.12 The σ bonds and π bonds in ethyne (C_2H_2). **A.** A contour depiction shows the C—C σ bond, the two C—H σ bonds, and two unhybridized 2*p* orbitals. **B.** The 2*p* orbitals (accurate form) overlap side to side; σ bonds are shown by a ball-and-stick model. **C.** Overlapping regions of two π bonds perpendicular to each other. **D.** The molecule has cylindrical symmetry. Six electrons (one σ bond and two π bonds) create even higher electron density (*red*) between the C atoms. **E.** Bond-line drawing.

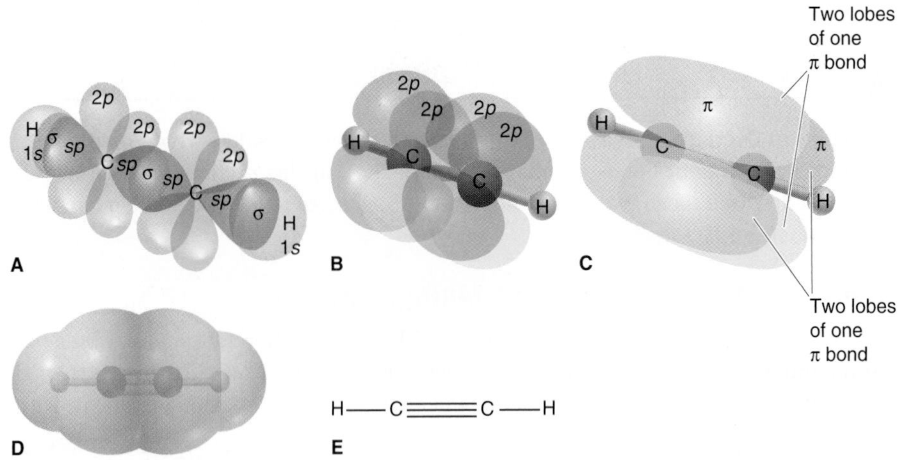

consists of one σ bond and one π bond, which increases electron density between the nuclei (Figure 10.11D). Notice that we can have an overlap of the two positive lobes (phases) of the *p* orbitals *or* the two negative lobes (phases) of the *p* orbitals, but *we cannot have* a negative lobe of one with a positive lobe of the other. (For example, in Figure 10.11, we *cannot* have a *pink* with a *blue*.)

The two electron pairs act as one electron group because each pair occupies a different orbital, which reduces repulsions. The following orbital box diagram shows the sp^2 hybridization:

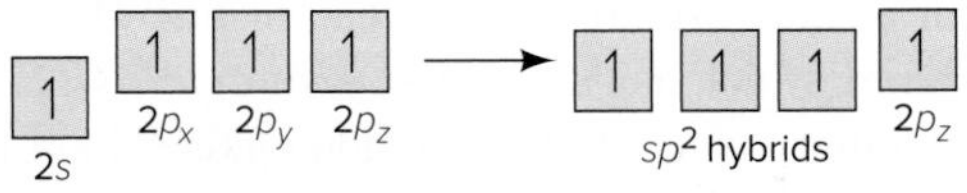

The sp^2 hybridization is also shown in the energy diagram below:

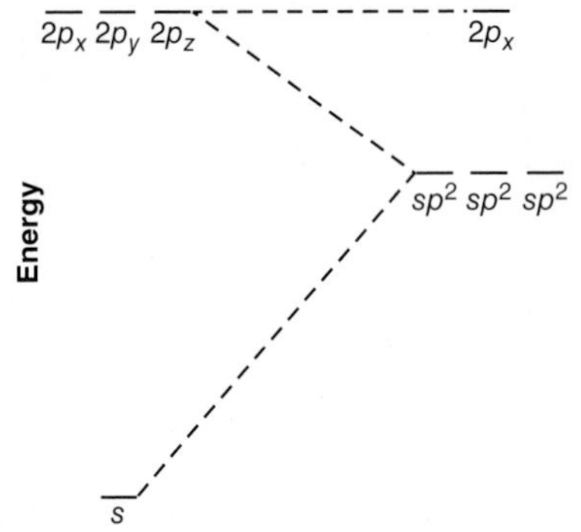

2. *In ethyne*, each C atom is *sp* hybridized, and its four valence electrons half-fill the two *sp* hybrids *and* the two unhybridized 2*p* orbitals (Figure 10.12). Each C forms a C—H σ bond with one *sp* orbital and a C—C σ bond with the other *sp* orbital. The side-to-side overlap of one pair of 2*p* orbitals gives one π bond, with electron density above and below the σ bond. The side-to-side overlap of the other pair of 2*p* orbitals gives another π bond, 90° away from the first, with electron density in the front and back of the σ bond. The result is a *cylindrically symmetrical* H—C≡C—H molecule. Note the greater electron density between the C atoms, which is created by the six bonding electrons. Any triple bond *consists of one σ and two π bonds*. The *sp* hybridization is shown in the following diagram:

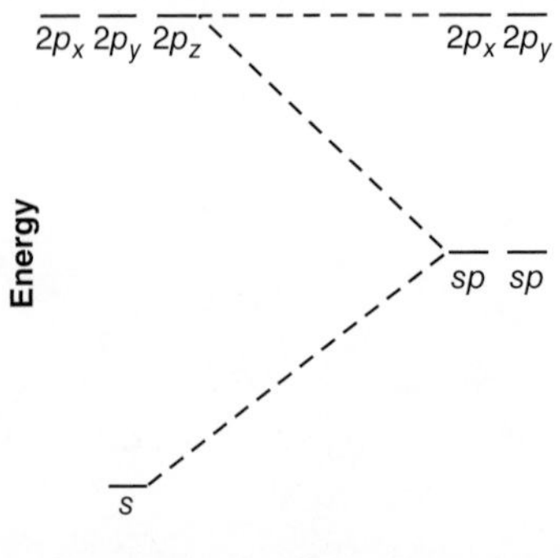

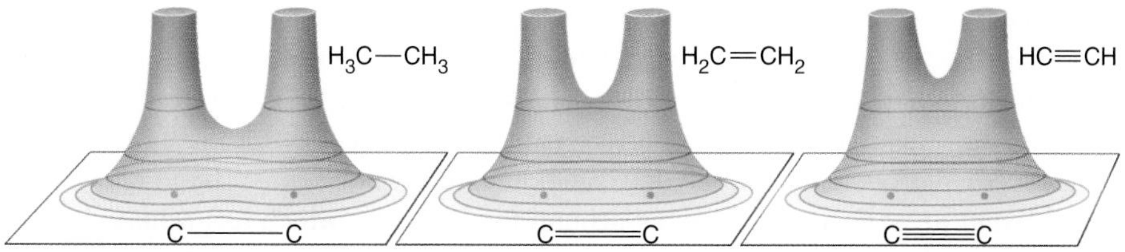

FIGURE 10.13 Electron density and bond order in ethane, ethene, and ethyne

Mode of Overlap, Bond Strength, and Bond Order Because orbitals overlap less side to side than end to end, a π bond is weaker than a σ bond; thus, for carbon-carbon bonds, a double bond is less than twice as strong as a single bond (Table 8.2). Figure 10.13 shows electron density relief maps of the three types of carbon-carbon bonds; note the increasing electron density between the nuclei from single to double to triple bond.

Lone-pair repulsions, bond polarities, and other factors affect the overlap between other pairs of atoms. Nevertheless, as a rough approximation, in terms of bond order (BO), a double bond (BO = 2) is slightly less than twice as strong as a single bond (BO = 1), and a triple bond (BO = 3) is less than three times as strong as a single bond.

Sample Problem 10.2 Describing the Types of Bonds in Molecules

Problem Describe the types of bonds and orbitals in propanone (acetone, $[(CH_3)_2CO]$).

Plan We use the shape around each central atom to predict the hybrid orbitals, and we use unhybridized orbitals to form the C=O bond.

Solution The shapes are tetrahedral around each C of the two CH_3 (methyl) groups and trigonal planar around the middle C (see Sample Problem 9.3). Thus, the middle C has three sp^2 orbitals and one unhybridized p orbital. Each of the two methyl C atoms has four sp^3 orbitals. Three of these orbitals form σ bonds with the 1s orbitals of H atoms; the fourth forms a σ bond with an sp^2 orbital of the middle C. Thus, two of the three sp^2 orbitals of the middle C form σ bonds with the other two C atoms.

The O atom is also sp^2 hybridized and has an unhybridized p orbital that can form a π bond. Two of the O atom's sp^2 orbitals hold lone pairs, and the third forms a σ bond with the third sp^2 orbital of the middle C atom. The unhybridized, half-filled 2p orbitals of C and O form a π bond. The σ and π bonds constitute the C=O bond:

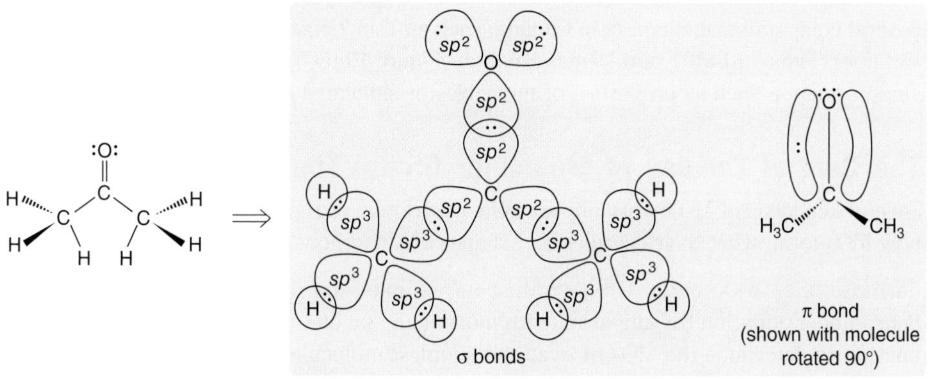

Comment Why would the O atom in propanone be hybridized? After all, it could use two perpendicular p orbitals for the σ and π bonds with C and leave the other p orbital and the s orbital to hold the two lone pairs. However, having each lone pair in an sp^2 orbital oriented away from the C=O bond lowers electron-electron repulsions. Note that, since the shape is used in most cases to determine the form of hybridization, it might not be needed to determine the terminal atom's hybridization.

Follow-Up Problem 10.2 Describe the types of bonds and orbitals in **(a)** hydrogen cyanide, HCN; **(b)** carbon dioxide, CO_2.

Orbital Overlap and Molecular Rotation

The type of overlap—end-to-end or side-to-side—affects rotation around the bond:

- *Sigma bond. A σ bond allows free rotation* because the extent of overlap is not affected. If you could hold one CH_3 group of ethane, the other CH_3 group could spin without affecting the overlap of the C—C σ bond (see Figure 10.10).

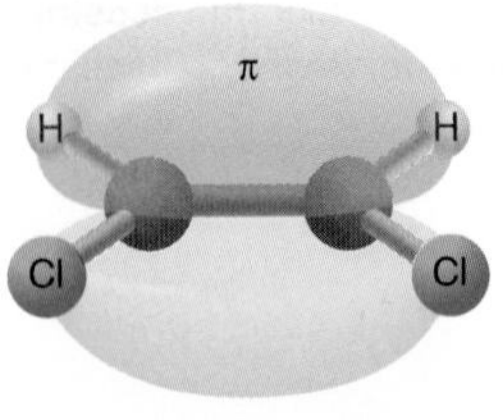

Cis-1,2-dichloroethene

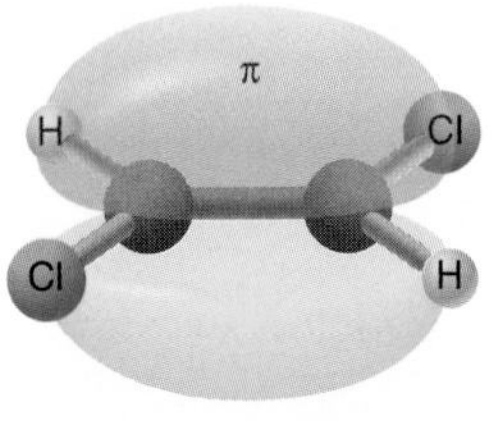

Trans-1,2-dichloroethene

FIGURE 10.14 Restricted rotation around a π bond. *Cis-* and *trans*-1,2-dichloroethene are different molecules because the π bond restricts rotation.

- *Pi bond.* A π *bond restricts rotation* because *p* orbitals must be parallel to each other to overlap most effectively. Holding one CH_2 group in ethene and trying to spin the other CH_2 group decreases the side-to-side overlap and breaks the π bond. For this reason, distinct *cis* and *trans* structures exist for compounds such as 1,2-dichloroethene (Section 9.2). As Figure 10.14 shows, the π bond allows two *different* arrangements of atoms around the C atoms, which has a major effect on molecular polarity. Rotation around a triple bond is not meaningful: each triple-bonded C atom is bonded to one other group in a linear arrangement, so there can be no difference in the relative positions of the attached groups.

SUMMARY OF SECTION 10.2

- End-to-end overlap of AOs forms a σ bond, which allows free rotation of the bonded parts of the molecule.
- Side-to-side overlap forms a π bond, which restricts rotation.
- A multiple bond consists of a σ bond and either one π bond (double bond) or two π bonds (triple bond). Multiple bonds have greater electron density between the nuclei than single bonds do and, thus, have higher bond energies.

10.3 Molecular Orbital Theory and Electron Delocalization

Scientists choose the theory that best answers a question: VSEPR theory for a question about molecular shape and VB theory for a question about orbital overlap. However, neither adequately explains magnetic and spectral properties, and both understate the importance of electron delocalization. To deal with phenomena like these, which involve molecular energy levels, chemists use **molecular orbital (MO) theory**. MO theory is a quantum-mechanical model for molecules, similar to the model for atoms (Chapter 7): just as an atom has AOs of given energies and shapes that are occupied by the atom's electrons, a molecule has **molecular orbitals** of given energies and shapes that are occupied by the molecule's electrons.

There is a key distinction between the VB and MO theories:

- VB theory pictures a molecule as a group of atoms bonded through *localized* overlapping of valence-shell atomic and/or hybrid orbitals occupied by electrons.
- MO theory pictures a molecule as a collection of nuclei, with the orbitals *delocalized* over the whole molecule and occupied by electrons.

Several computational chemists in Canada, such as Tom Ziegler (Figure 10.15A), Stacey Wetmore (Figure 10.15B), and Dennis Salahub (Figure 10.15C), use computers to generate information such as properties of molecules or simulated experimental results.

The Central Themes of Molecular Orbital Theory

Several key ideas of MO theory appear in its description of H_2 and other simple species: how MOs form, what their energies and shapes are, and how they fill with electrons.

Formation of Molecular Orbitals Just as we need approximations to solve the Schrödinger equation for any atom with more than one electron, we need an approximation to determine the MOs of even the simplest molecule, H_2. The approximation is needed to *combine* mathematically (add or subtract) AOs (atomic wave functions) of nearby atoms to form MOs (molecular wave functions). Thus, when two H nuclei lie near each other, their AOs overlap and combine in two ways, depending on their phase relationship, where the phase of an orbital is a direct consequence of the wavelike properties of electrons:

- *Adding the wave functions together.* This combination, a constructive overlap, forms a **bonding molecular orbital**, which has *a region of high electron density between the nuclei* and energy that is lower than the energy of the original AOs.

A B C

FIGURE 10.15 A. Tom Ziegler (Canada Research Chair, University of Calgary) makes extensive use of computational methods (MO theory) as a tool to study catalytic processes and reactive intermediates. Using quantum-mechanical computer calculations that provide information about energetically favourable reaction pathways and insights into electronic structure, Dr. Ziegler's group performs theoretical studies of fuel oxidation on the anode and electrolyte surfaces, anode degradation as a result of coking and sulfur poisoning, and mechanisms to enhance fuel cell performance. **B.** Stacey Wetmore (Canada Research Chair, University of Lethbridge, Alberta) uses computer modelling to investigate systems related to DNA damage and repair. Her research is primarily focused on the structure of damaged DNA components to understand their influence on the structure and function of DNA and the mechanism of action of DNA repair enzymes to understand how nature repairs DNA damage. **C.** Dennis Salahub (Professor, University of Calgary) has improved density functional theory (DFT) methods and software, which has helped to extend the range of applications in materials and biomolecular modelling. Dr. Salahub and his research group proposed, tested, and implemented new improved functionals in the code suite deMon, developed in Montréal and now used in dozens of labs around the world. A fusion of DFT-deMon with other techniques (such as reaction fields and molecular dynamics) is under way. His overall research goal for the next decade is to develop the theoretical, computational, and conceptual expertise necessary to attain a detailed microscopic understanding of chemical reactions that take place in real, complex environments.

Additive overlap is analogous to light waves reinforcing each other, which makes the amplitude higher and the light brighter. For electron waves, the overlap *increases* the probability that the electrons are between the nuclei (Figure 10.16A). A bond involving MOs that are symmetrical with respect to rotation around the bond axis (no change) is a sigma (σ) bond. In the event of a phase change, the bond becomes a pi (π) bond.

- *Subtracting the wave functions from each other.* AOs can also interact with each other out of phase, which leads to destructive cancellation. This combination forms an **antibonding molecular orbital (MO)**, which has *a node, a region of zero electron density, between the nuclei* (Figure 10.16B). Subtractive overlap is analogous to light waves cancelling each other, causing the light to disappear. With electron waves, the probability that the electrons lie between the nuclei *decreases* to zero. In the antibonding MO, which has much higher energy than the original AOs, any electrons present are located in lobes pointing away from the central internuclear axis.

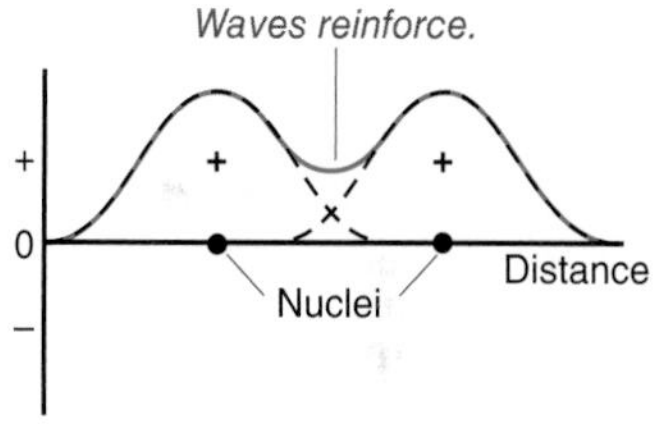

A Amplitudes of wave functions added

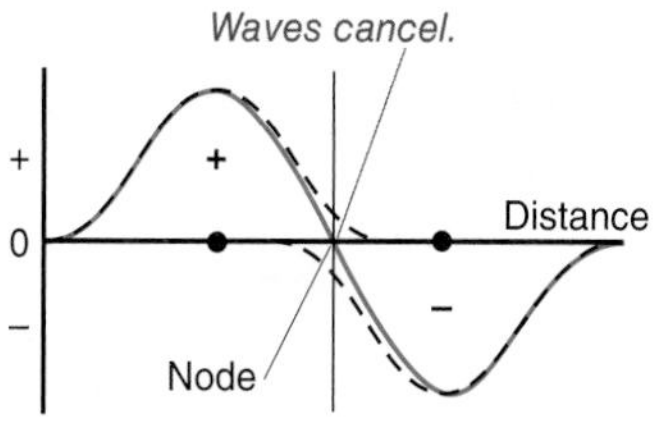

B Amplitudes of wave functions subtracted

FIGURE 10.16 An analogy between light waves and atomic wave functions

In graphical representations of orbitals, the orbital phase is depicted either by shading one lobe, as shown in Figure 10.26, or by a plus or minus sign. The latter representation, which we will not use here, might be confusing because there is no relationship to electrical charge, and the sign of the phase itself does not have physical meaning except when mixing orbitals to form MOs.

The two possible combinations for hydrogen atoms H_A and H_B are

$$\text{AO of } H_A + \text{AO of } H_B = \text{bonding MO of } H_2 \text{ (more } e^- \text{ density between nuclei)}$$

$$\text{AO of } H_A - \text{AO of } H_B = \text{antibonding MO of } H_2 \text{ (less } e^- \text{ density between nuclei)}$$

Notice that *the number of AOs combined always equals the number of MOs formed*: two H AOs combine to form two H_2 MOs.

Shape and Energy of H_2 Molecular Orbitals Bonding and antibonding MOs have different shapes and energies. Figure 10.17 shows these orbitals for H_2.

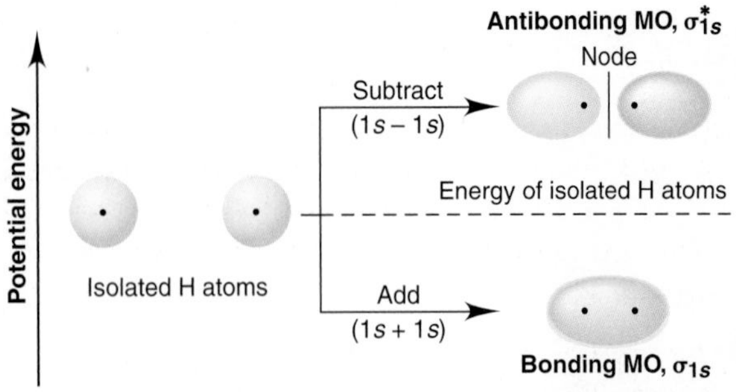

FIGURE 10.17 Contours and energies of H_2 bonding and antibonding MOs

Note: Even though it might not be appreciable in Figure 10.17 and following figures, the energy gap from AO to bonding MO is slightly smaller than the energy gap from AO to antibonding MOs.

- *Bonding MO.* A bonding MO is *lower in energy* than the AOs that form it. Because it is spread mostly *between* the nuclei, nuclear repulsions decrease while nucleus-electron attractions increase. Moreover, two electrons in this MO can delocalize their charges over a larger volume than in nearby, separate AOs, which lowers electron repulsions. Because of these electrostatic effects, when electrons occupy this orbital, the H_2 molecule is *more stable* than the separate H atoms.
- *Antibonding MO.* An antibonding MO is *higher in energy* than the AOs that form it. With most of its electron density *outside* the internuclear region, it has a node between the nuclei, and nuclear repulsions increase. Therefore, when electrons occupy this orbital, the H_2 molecule is *less stable* than the separate H atoms.

Both the bonding and antibonding MOs of H_2 are **sigma (σ) molecular orbitals** because they are cylindrically symmetrical about an imaginary line between the nuclei. The bonding MO is denoted by σ_{1s}, that is, a σ MO derived from 1*s* AOs. Antibonding orbitals are denoted with a superscript star: the antibonding MO derived from 1*s* AOs is σ^*_{1s} (read as "sigma, one ess, star").

For AOs to interact enough to form MOs, they must be similar in *energy* and *orientation*. The 1*s* orbitals of two H atoms have identical energy and orientation, so they interact strongly. We will revisit this requirement for molecules composed of many-electron atoms later.

Electrons in Molecular Orbitals Several aspects of MO theory—filling of MOs with electrons, energy-level diagrams, electron configurations, and bond order—relate to earlier ideas:

1. *Filling of MOs with electrons.* Electrons enter MOs just as they enter AOs:
- MOs are filled in order of increasing energy (Aufbau principle).
- An MO can hold a maximum of two electrons with opposite spins (exclusion principle).
- Orbitals of equal energy are half-filled, with spins parallel, before any of them are filled (Hund's rule).

2. *MO energy-level diagrams.* A **molecular orbital diagram** shows the relative energy and number of electrons in each MO, as well as the AOs from which they formed. In the MO diagram for H_2 (Figure 10.18), two electrons, one from the AO of each H, fill the H_2 bonding MO. The antibonding MO remains empty.

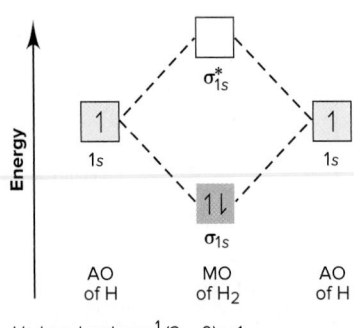

FIGURE 10.18 MO diagram for H_2. The vertical placement of the boxes indicates the relative energies. Orbital occupancy is shown with arrows and shading (*dark* = full, *pale* = half-filled, *none* = empty).

3. *Electron configuration.* Just as we write an electron configuration for an atom, we can write an electron configuration for a molecule. The symbol of each occupied MO is written in parentheses, with the number of electrons in the MO as a superscript outside: for example, the electron configuration for H_2 is $(\sigma_{1s})^2$.

4. *Bond order.* In a Lewis structure, bond order is the number of electron pairs per atom-to-atom linkage. The **molecular orbital bond order** is the number of electrons in bonding MOs minus the number in antibonding MOs, divided by 2:

$$\text{Bond order} = \frac{1}{2}[(\text{no. of } e^- \text{ in bonding MO}) - (\text{no. of } e^- \text{ in antibonding MO})]$$

$$\text{Bond order} = e_b^- - e_{ab}^- \qquad (10.1)$$

Keep in mind these four key points about MO bond order:

- *Bond order > 0.* The molecule is more stable than the separate atoms, so it *will* form. For H_2, the bond order is $\frac{1}{2}(2 - 0) = 1$.
- *Bond order = 0.* The molecule is slightly less stable than the separate atoms, so it will *not* form. (This occurs when equal numbers of electrons occupy bonding and antibonding MOs.)

- The *bond order* only deals with the covalent portion of the bond. For example, the bond order of HF by this equation does not reflect the actual one because the bond has a large ionic character.
- *Bond strength.* In general, the *higher* the bond order, the *stronger* the bond is.

Do He_2^+ and He_2 Exist? One of the early triumphs of MO theory was its ability to *predict* the existence of He_2^+, the helium molecule ion, which consists of two He nuclei and three electrons. Let us use MO diagrams to see why He_2^+ exists, but He_2 does not:

- In He_2^+, the 1*s* AOs form MOs (Figure 10.19A). The three electrons are distributed as a pair in the σ_{1s} MO and a lone electron in the σ^*_{1s} MO. The bond order is $\frac{1}{2}(2-1)=\frac{1}{2}$. Thus, He_2^+ has a relatively weak bond, but it should exist. Indeed, this species has been observed frequently when He atoms collide with He^+ ions. Its electron configuration is $(\sigma_{1s})^2(\sigma^*_{1s})^1$.
- In He_2, with four electrons in the σ_{1s} and (σ^*_{1s}) MOs, both the bonding and antibonding orbitals are filled (Figure 10.19B). Stabilization from the electron pair in the bonding MO is cancelled by destabilization from the electron pair in the antibonding MO. With a zero bond order [$\frac{1}{2}(2-2)=0$], we predict, and experiment has so far confirmed, that a covalent He_2 molecule does not exist.

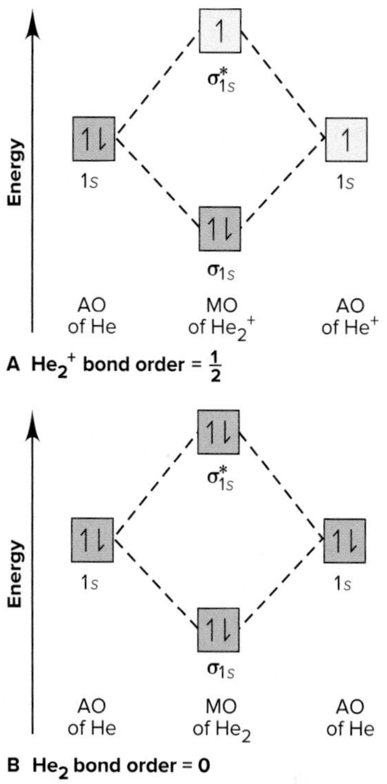

FIGURE 10.19 **MO diagrams for He_2^+ and He_2**

Sample Problem 10.3 Predicting Stability of Species Using MO Diagrams

Problem Use MO diagrams to find the bond orders and predict whether **(a)** H_2^+ and **(b)** H_2^- exist. If either exists, write its electron configuration.

Plan Since the 1*s* AOs form the MOs, the MO diagrams are similar to the MO diagram for H_2. We find the number of electrons in each species and distribute them, one at a time, to the MOs in order of increasing energy. We obtain the bond order with Equation 10.1 and write the electron configuration as described in this book.

Solution (a) H_2 has two e^-, so H_2^+ has only one, which enters the bonding MO (*see diagram, below left*). The bond order is $\frac{1}{2}(1-0)=\frac{1}{2}$, so we can predict that H_2^+ exists. The electron configuration is $(\sigma_{1s})^1$. **(b)** H_2 has two e^-, so H_2^- has three. We place two paired e^- in the bonding MO and one unpaired e^- in the antibonding MO (*see diagram, below right*). The bond order is $\frac{1}{2}(2-1)=\frac{1}{2}$, so we can predict that H_2^- exists. The electron configuration is $(\sigma_{1s})^2(\sigma^*_{1s})^1$.

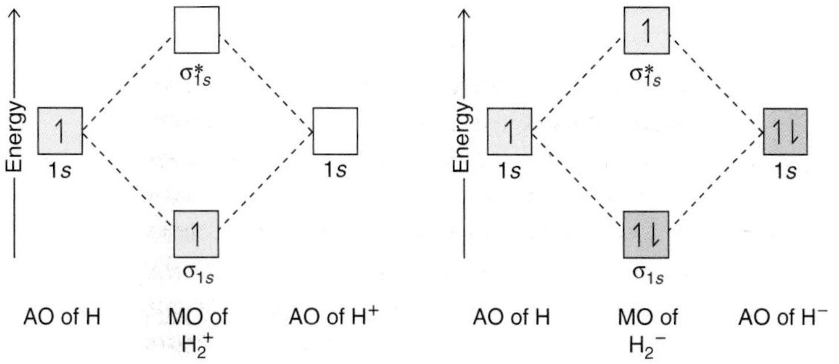

Check The number of electrons in the MOs equals the number of electrons in the AOs.

Comment Both species have been detected spectroscopically—H_2^+ in the material around stars and H_2^- in the laboratory.

Follow-Up Problem 10.3 Use an MO diagram to find the bond order to predict whether two H^- ions could form H_2^{2-}. If so, write the electron configuration for H_2^{2-}.

Homonuclear Diatomic Molecules of Period 2 Elements

Homonuclear diatomic molecules are composed of two identical atoms. In addition to H_2 from period 1, you are familiar with N_2, O_2, and F_2 from period 2 as the elemental forms under standard conditions. Others from period 2—Li_2, Be_2, B_2, C_2, and Ne_2—may be observed at high temperatures. We will divide homonuclear diatomic molecules into molecules from the *s* block (groups 1 and 2) and the *p* block (groups 13 through 18).

Bonding in s-Block Homonuclear Diatomic Molecules Both Li and Be occur as metals under normal conditions, but MO theory can examine their stability as diatomic gases, dilithium (Li_2) and diberyllium (Be_2).

These atoms have electrons in inner (1*s*) and outer (2*s*) AOs, but we ignore the inner AOs because, in general, *only outer (valence) AOs interact enough to form MOs*. Like those formed from 1*s* AOs, these 2*s* AOs form σ MOs, which are cylindrically symmetrical around the internuclear axis.

- In Li_2, the two valence electrons fill the bonding (σ_{2s}) MO, with opposing spins, leaving the antibonding (σ^*_{2s}) MO empty (Figure 10.20A). The bond order is $\frac{1}{2}(2 - 0) = 1$. In fact, Li_2 *has* been observed; the electron configuration is $(\sigma_{2s})^2$.
- In Be_2, the four valence electrons fill the σ_{2s} and σ^*_{2s} MOs (Figure 10.20B), giving an orbital occupancy similar to that in He_2. The bond order is $\frac{1}{2}(2 - 2) = 0$, and the ground state of Be_2 has never been observed.

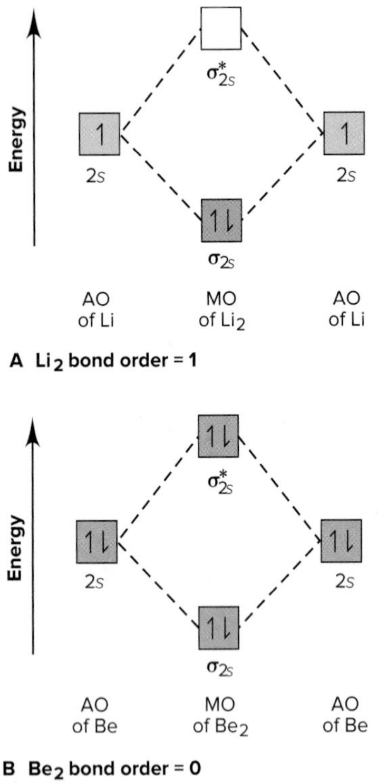

FIGURE 10.20 Bonding in *s*-block homonuclear diatomic molecules. Only outer (valence) AOs interact enough to form MOs.

Shape and Energy of Molecular Orbitals from Atomic *p*-Orbital Combinations As we move to boron, the atomic 2*p* orbitals are occupied. Recall that *p* orbitals can overlap by two modes, which correspond to two ways that their wave functions combine (Figure 10.21):

- End-to-end combination gives a pair of σ MOs, the σ_{2p} and σ^*_{2p}.
- Side-to-side combination gives a pair of **pi (π) molecular orbitals,** π_{2p} and π^*_{2p}.

Despite the different shapes, MOs derived from *p* orbitals are like those derived from *s* orbitals. Bonding MOs have most of the electron density *between* the nuclei, and antibonding MOs have most *outside* the internuclear region, with a node between the nuclei.

The order of energy levels for MOs, whether bonding or antibonding, is based on the order of AO energy levels *and* on the mode of the *p*-orbital overlap:

- MOs formed from 2*s* orbitals are *lower in energy* than MOs formed from 2*p* orbitals because 2*s* AOs are lower in energy than 2*p* AOs.

FIGURE 10.21 Shapes and energies of σ and π MOs from combinations of 2*p* AOs. A. The *p* orbitals that lie along the internuclear axis (designated p_x) overlap end to end and form σ_{2p} and σ^*_{2p} MOs. **B.** The *p* orbitals that are perpendicular to the internuclear axis overlap side to side and form two π MOs. (The p_y interactions, shown here, are the same as the p_z interactions, for a total of four π MOs.)

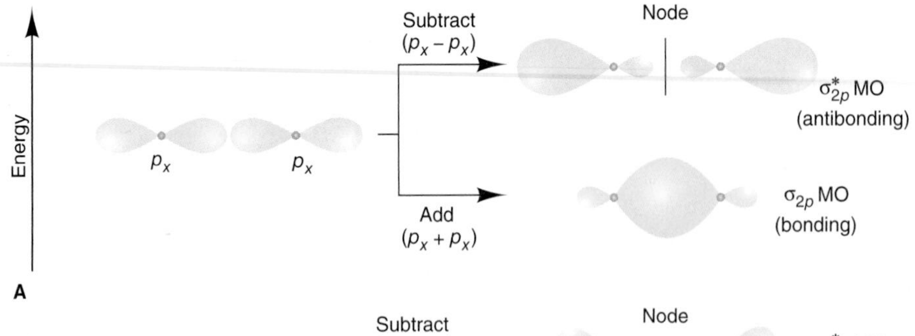

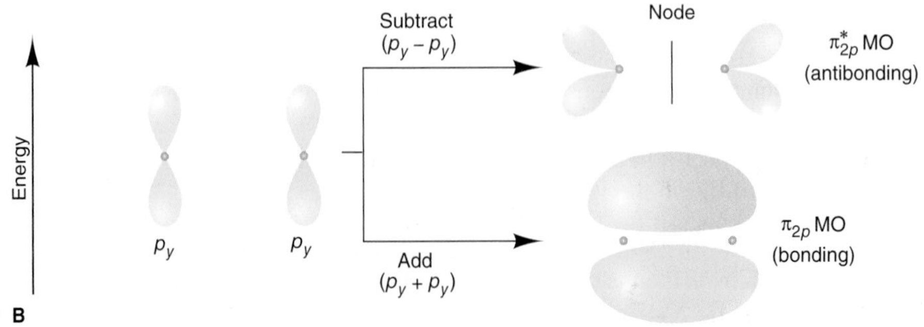

- Bonding MOs are *lower in energy* than antibonding MOs: σ_{2p} is lower in energy than σ^*_{2p}, and π_{2p} is lower in energy than π^*_{2p}.
- Atomic p orbitals overlap more extensively end to end than side to side. Thus, the σ_{2p} MO is usually lower in energy than the π_{2p} MO. We also find that the destabilizing effect of the σ^*_{2p} MO is greater than that of the π^*_{2p} MO.

Thus, the energy order for MOs derived from $2p$ orbitals is typically

$$\sigma_{2p} < \pi_{2p} < \pi^*_{2p} < \sigma^*_{2p}$$

Each atom has three mutually perpendicular $2p$ orbitals. When the six p orbitals in two atoms combine, the two orbitals that interact end to end form one σ MO and one σ^* MO, and the two pairs of orbitals that interact side to side form two π MOs and two π^* MOs. Placing these orientations within the energy order gives the *expected* MO diagram for the p-block period 2 homonuclear diatomic molecules (Figure 10.22A).

Recall that only AOs of similar energy interact enough to form MOs. This leads to two energy orders:

1. ***Without** s and p orbital mixing: O_2, F_2, and Ne_2.* The order in Figure 10.22A assumes that s and p AOs are so different in energy that they do not interact; we say that the orbitals do not *mix*. Lying at the right of period 2, O, F, and Ne are relatively small. Thus, as electrons start to pair up in the half-filled $2p$ orbitals, strong repulsions raise the energy of the $2p$ orbitals high enough above the $2s$ orbitals to prevent orbital mixing.

2. ***With** s and p orbital mixing: B_2, C_2, and N_2.* B, C, and N atoms are relatively large, with $2p$ AOs only half-filled, so repulsions are weaker. As a result, orbital energies are close enough for some mixing to occur between the $2s$ of one atom and the end-on $2p$ of the other. The effect is to *lower* the energy of the σ_{2s} and (σ^*_{2s}) MOs and *raise* the energy of the σ_{2p} and σ^*_{2p} MOs; the π MOs are not affected. The MO diagram for B_2, C_2, and N_2 reflects this mixing (Figure 10.22B). The only difference from the MO diagram for O_2, F_2, and Ne_2 is the *reverse in energy order* of the σ_{2p} and π_{2p} MOs.

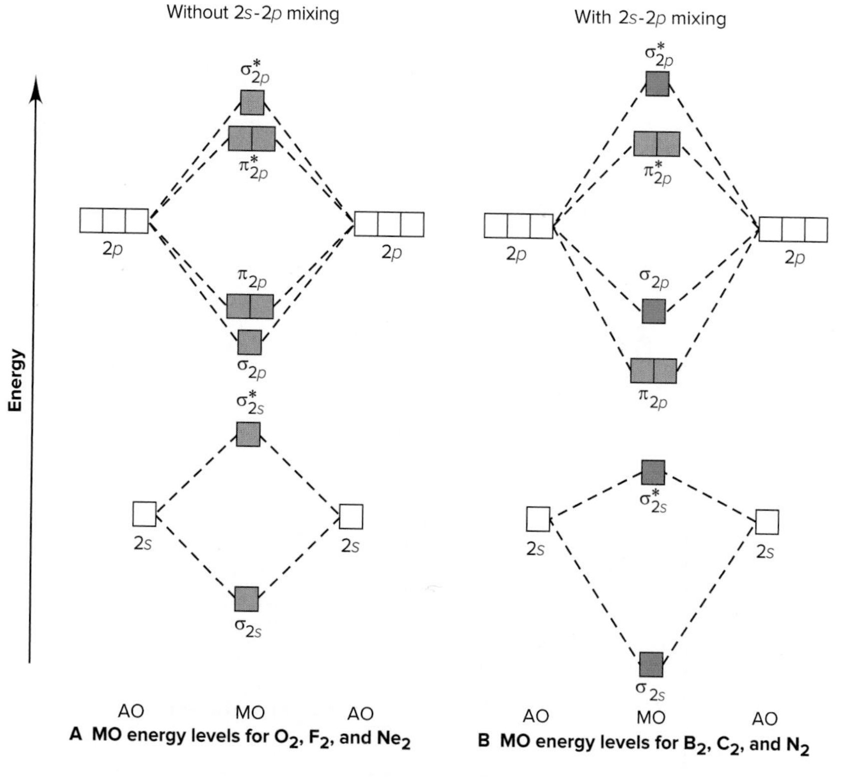

FIGURE 10.22 Relative MO energy levels for period 2 homonuclear diatomic molecules. For clarity, the MOs that are affected by mixing are shown in *purple*.

Bonding in the *p*-Block Homonuclear Diatomic Molecules Figure 10.23 shows the MOs, electron occupancy, and some other properties of homonuclear diatomic molecules from B_2 through Ne_2. Note the following:

1. *Higher* bond order correlates with *greater* bond energy and *shorter* bond length.
2. Orbital occupancy correlates with magnetic properties. Recall, from Chapter 7, that a substance is *paramagnetic* if it has unpaired electrons, and it is *diamagnetic* if all the electrons are paired. The same applies to molecules. Let us examine the MO occupancy and some properties of paramagnetic and diamagnetic molecules:

- *B_2*. The B_2 molecule has six outer electrons: four fill the σ_{2s} and σ^*_{2s} MOs. The remaining two electrons occupy the two π_{2p} MOs, one in each orbital, in keeping with Hund's rule. With four electrons in bonding MOs and two electrons in antibonding MOs, the bond order of B_2 is $\frac{1}{2}(4 - 2) = 1$. As expected from the two lone electrons, B_2 is paramagnetic.
- *C_2*. Two additional electrons in C_2 fill the two π_{2p} MOs. With two more bonding electrons than B_2, the bond order of C_2 is 2 and the bond is stronger and shorter. But with all the electrons paired, C_2 is diamagnetic.
- *N_2*. Two more electrons in N_2 fill the σ_{2p} MO, so this molecule is also diamagnetic. The bond order of 3 is consistent with the triple bond in the Lewis structure and with a stronger, shorter bond.
- *O_2*. In this molecule, we see the power of MO theory over VB theory and others, based on electrons in localized orbitals. It is impossible to write one Lewis structure consistent with the fact that O_2 is double bonded and paramagnetic. We can write one Lewis structure with a double bond and paired electrons, and another with a single bond and two unpaired electrons:

Ö=Ö or :Ö—Ö:

MO theory resolves this paradox beautifully: with eight electrons in bonding MOs and four electrons in antibonding MOs, the bond order is $\frac{1}{2}(8 - 4) = 2$. O_2 is paramagnetic because *one* electron occupies each of *two* π^*_{2p} MOs, with unpaired

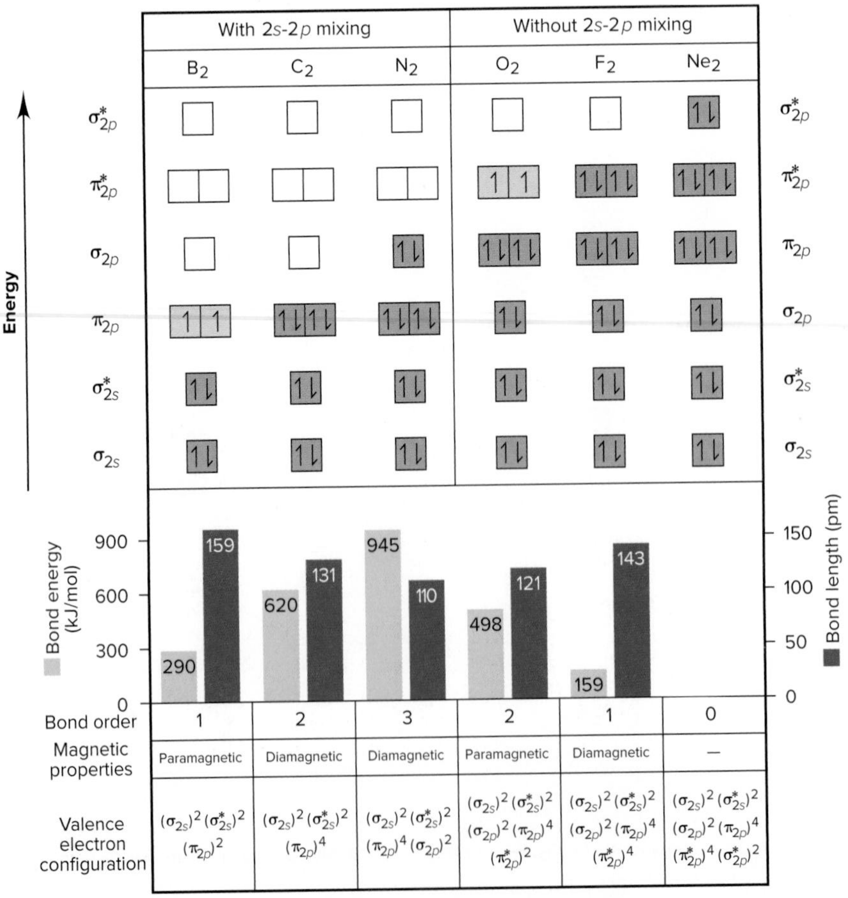

FIGURE 10.23 MO occupancy and some properties of homonuclear diatomic molecules from B_2 through Ne_2. Energy order and occupancy of MOs are shown above the bar graphs for bond energy and length. Bond order, magnetic properties, and valence (outer) electron configuration are given below the bar graphs.

(parallel) spins. A thin stream of liquid O_2 will remain suspended between the poles of a powerful magnet (Figure 10.24). As expected, the bond is weaker and longer than the bond in N_2.

- F_2. Two more electrons in F_2 fill the π^*_{2p} orbitals, so F_2 is diamagnetic. The bond order is 1, and the bond is weaker and longer than the bond in O_2. Note that this bond is shorter, yet about half as strong as the single bond in B_2. We might have expected it to be stronger because F is smaller than B. However, 18 electrons in the smaller volume of F_2 cause greater repulsions than the 10 electrons in B_2, so the F_2 bond is weaker.
- Ne_2. The final member of the period 2 series does not exist for the same reason that He_2 does not exist: all the MOs are filled, which gives a bond order of zero.

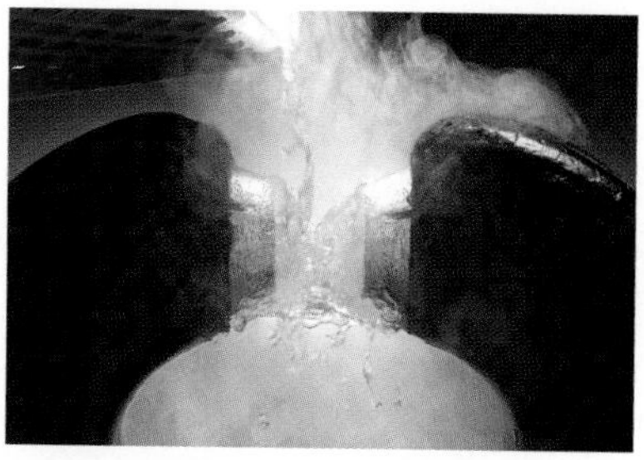

FIGURE 10.24 **The paramagnetic properties of O_2**

Sample Problem 10.4 Using MO Theory to Explain Bond Properties

Problem Explain the following data with diagrams that show the occupancy of MOs:

	N_2	N_2^+	O_2	O_2^+
Bond energy (kJ/mol)	945	841	498	623
Bond length (pm)	110	112	121	112

Plan The data show that removing an electron from each parent molecule has opposite effects: N_2^+ has a weaker, longer bond than N_2, but O_2^+ has a stronger, shorter bond than O_2. We determine the valence electrons in each species, draw the sequence of MO energy levels (showing orbital mixing in N_2 but not in O_2), and fill them with electrons. To explain the data, we calculate bond orders, which relate directly to bond energy and inversely to bond length.

Solution Determine the valence electrons:

N has 5 valence e^-, so N_2 has 10 and N_2^+ has 9.

O has 6 valence e^-, so O_2 has 12 and O_2^+ has 11.

Draw and fill MO diagrams:

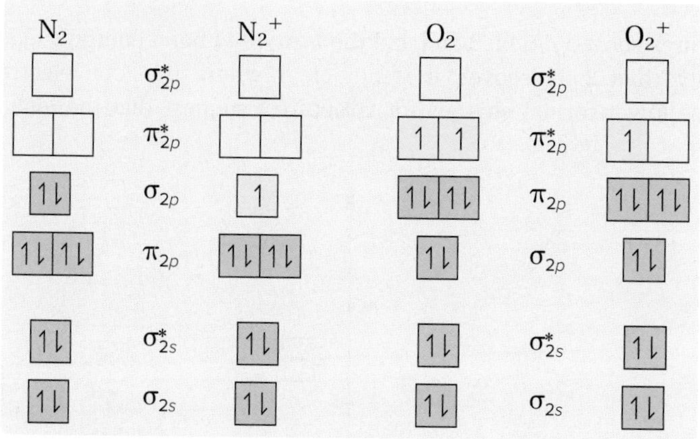

Calculate bond orders:

$$\frac{1}{2}(8-2)=3 \qquad \frac{1}{2}(7-2)=2.5 \qquad \frac{1}{2}(8-4)=2 \qquad \frac{1}{2}(8-3)=2.5$$

Explain the data:

1. When N_2 becomes N_2^+, a *bonding* electron is removed, so the bond order decreases. Thus, N_2^+ has a weaker, longer bond than N_2.
2. When O_2 becomes O_2^+, an *antibonding* electron is removed, so the bond order increases. Thus, O_2^+ has a stronger, shorter bond than O_2.

Check The answers make sense in terms of bond order, bond energy, and bond length. Check that the total number of bonding and antibonding electrons equals the number of valence electrons calculated.

Follow-Up Problem 10.4 Determine the bond orders for the following species: F_2^{2-}, F_2^-, F_2, F_2^+, F_2^{2+}. List the species in order of increasing bond energy and in order of increasing bond length.

Two Heteronuclear Diatomic Molecules: HF and NO

Heteronuclear diatomic molecules have asymmetric MO diagrams because the AOs of the *different* atoms have unequal energies. Atoms with greater effective nuclear charge (Z_{eff}) pull their electrons closer, so they have more stable (lower energy) AOs (Section 7.1) and higher χ values (Section 8.5). Let us examine the bonding in HF and NO.

Bonding in HF To form the MOs in HF, we need to decide which AOs will combine. The high Z_{eff} of F holds its electrons so tightly that the 1*s*, 2*s*, and 2*p* orbitals of F have lower energy than the 1*s* of H. The half-filled F 2*p* orbital interacts with the H 1*s* orbital through end-on overlap, which forms σ and σ^* MOs. The two filled 2*p* orbitals of F are called **nonbonding molecular orbitals**. Because they are not involved in bonding, they have the same energy as the isolated AOs (Figure 10.25A). The AOs with closer energies will combine more readily than the AOs with disparate energies. This is why the 1*s* orbital of H combines with the 2*p* orbital of F, not the 2*s* orbital of F.

The bonding MO is closer in energy to the AOs of F, so the F 2*p* orbital contributes more to the HF bond than the H 1*s* orbital does. In polar covalent molecules, *bonding MOs are closer in energy to the AOs of the more electronegative atom.* In effect, F's greater electronegativity lowers the energy of the bonding MO and draws the bonding electrons closer. The H—F bonding MO looks more like the F AO because it is closer in energy to the F AO. Since the effective nuclear charge controls the orbital energies, it also controls where the electrons mostly reside (electronegativity). Therefore, the electrons in the bond spend more time near the F than the H.

Bonding in NO Nitrogen monoxide (nitric oxide) is highly reactive because of its lone electron. Two possible Lewis structures for NO, with formal charges (Section 8.6), are given below:

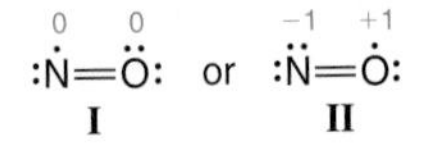

Both structures show a double bond, but the *measured* bond energy suggests a bond order *higher* than 2. Moreover, it is not clear where the lone electron resides, although the lower formal charges for structure I suggest that the lone electron is on the N.

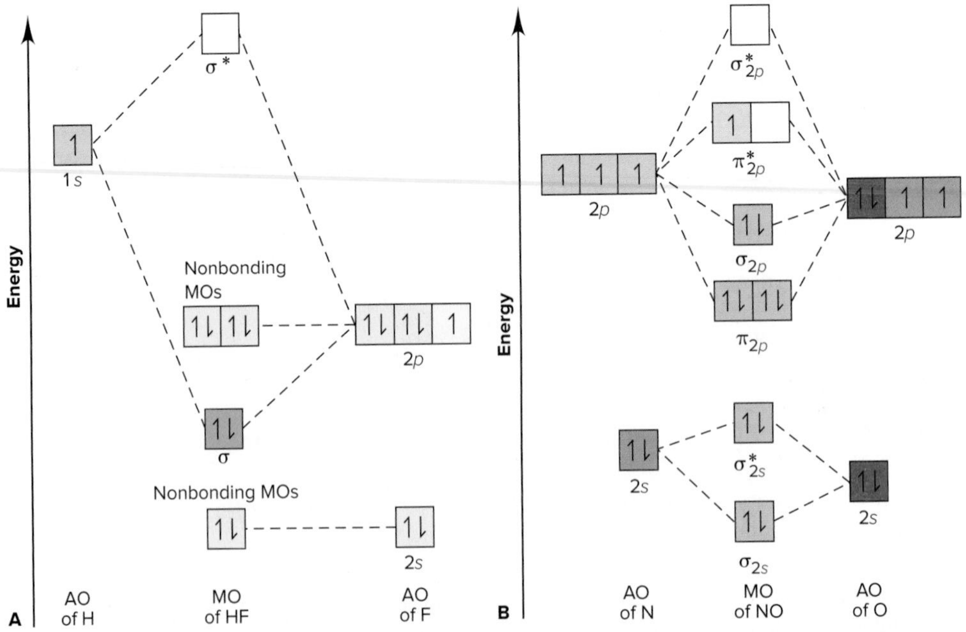

FIGURE 10.25 **MO diagrams for HF and NO. A.** In HF, the bonding MO is closer in energy to the 2*p* orbital of F. (The 2*s* AO of F is not shown.) **B.** In NO, the lone electron occupies an MO closer in energy to the 2*p* of N.

The MO diagram is asymmetrical, with the AOs of the more electronegative O lower in energy (Figure 10.25B). When the 11 valence electrons fill the MOs of NO, the lone electron occupies one of the π^*_{2p} orbitals. Eight bonding and three antibonding electrons give a bond order of $\frac{1}{2}(8 - 3) = 2.5$, more in keeping with the data than either Lewis structure. Bonding electrons lie in MOs closer in energy to the AOs of the O atom. The $2p$ orbitals of N contribute more to the orbital that holds the lone electron, so it spends more time closer to N. (This diagram shows the energy order *with* $2s$-$2p$ mixing, as in Figure 10.22, but the order is the same as the order without mixing.)

Two Polyatomic Molecules: Benzene and Ozone

The orbital shapes and MO diagrams for polyatomic molecules are too complex for a detailed discussion here. However, we will briefly discuss how the model eliminates the need for resonance forms and helps to explain the effects of the absorption of energy.

Recall that we cannot draw one Lewis structure for either benzene or ozone, so we draw separate forms of a resonance hybrid. VB theory also uses resonance because it relies on *localized* bonds. In contrast, MO theory pictures a structure of delocalized σ and π MOs. Figure 10.26 shows the lowest energy π-bonding MOs in benzene and ozone. The extended areas of electron density allow delocalization of one π-bonding electron pair over the entire molecule, thus eliminating the need for separate resonance forms:

- *In benzene,* the upper and lower hexagonal lobes of this MO lie above and below the σ plane of the six C nuclei.
- *In ozone,* the two lobes of this MO extend over and under the three O nuclei.

The full MO diagrams for these molecules help us rationalize how bonding electrons in O_3 become excited and occupy empty antibonding orbitals when the molecule absorbs UV radiation in the stratosphere, and why the UV spectrum of benzene has its characteristic absorption bands.

Figure 10.27 represents an MO diagram for benzene, where all of benzene's electrons are in MOs that are lower in energy than the atomic p orbitals of carbon ("zero" of energy). This makes the molecule very stable. Notice that all of the six p AOs of the C atoms are in phase. Figure 10.28 shows that there is a pure $2p$ AO at each oxygen, perpendicular to the plane of the molecule. These $2p$ orbitals lead, via linear combinations, to three delocalized π MOs.

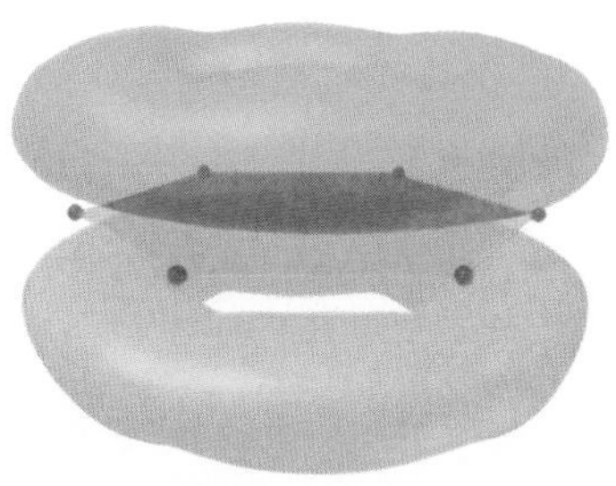

Benzene, C_6H_6

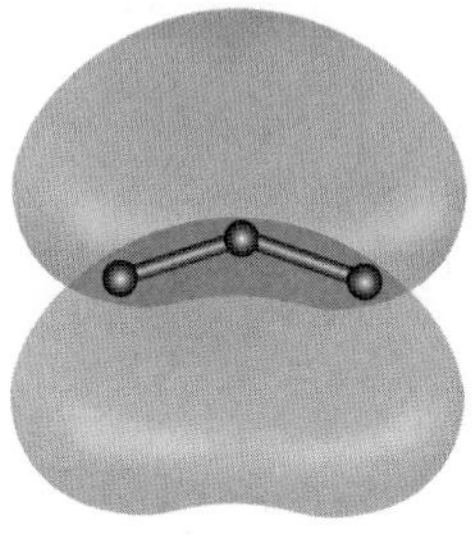

Ozone, O_3

FIGURE 10.26 The lowest energy π-bonding MOs in benzene and ozone

Metal-Metal Multiple Bonds

Although double and triple bonds are common in organic chemistry (see Chapters 13 and 20), the existence of multiple metal-metal bonding is an interesting topic of study in inorganic chemistry. We can find, for example, bond orders of 3.0, 3.5, and 4.0 between Re-Re in phosphine compounds such as $[Re_2Cl_4(PMe_2Ph)_4]^{n+}$ ($n = 0, 1, 2$) or a quadruple bond between the super-short Cr(II)-Cr(II) bond.

In general, metals with a low d electron count have a high oxidation state. The low d electron count allows the formation of many bonds. A d^0 metal centre can accommodate up to nine bonds without violating the 18-electron rule, whereas a d^6 species can only accommodate six bonds. The 18-electron rule was first proposed by N. V. Sigwick, who extended the Lewis octet theory to explain coordination compounds. It is used mainly to predict the formulas for stable metal complexes. It relies on the fact that the valence shell of a transition metal consists of nine AOs that give rise to nine MOs, which can accommodate up to 18 electrons as either bonding or nonbonding electron pairs. In some cases, the bond order–bond distance relationship does not follow the trend we have already studied. This abnormality may be caused by the contraction of the metal d orbitals due to an increase in the oxidation number.

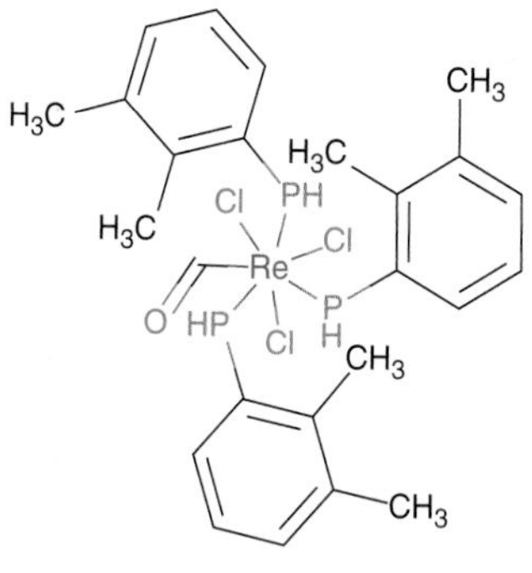

$Re_2Cl_4(PMe_2Ph)_4$

Most of the compounds containing metals consist of a coordination complex, or metal complex, where an atom (usually metallic) is bonded to a surrounding array of molecules or anions known as ligands or complexing agents. These compounds are of both scholarly and practical interest, and they have many different and important applications in industry.

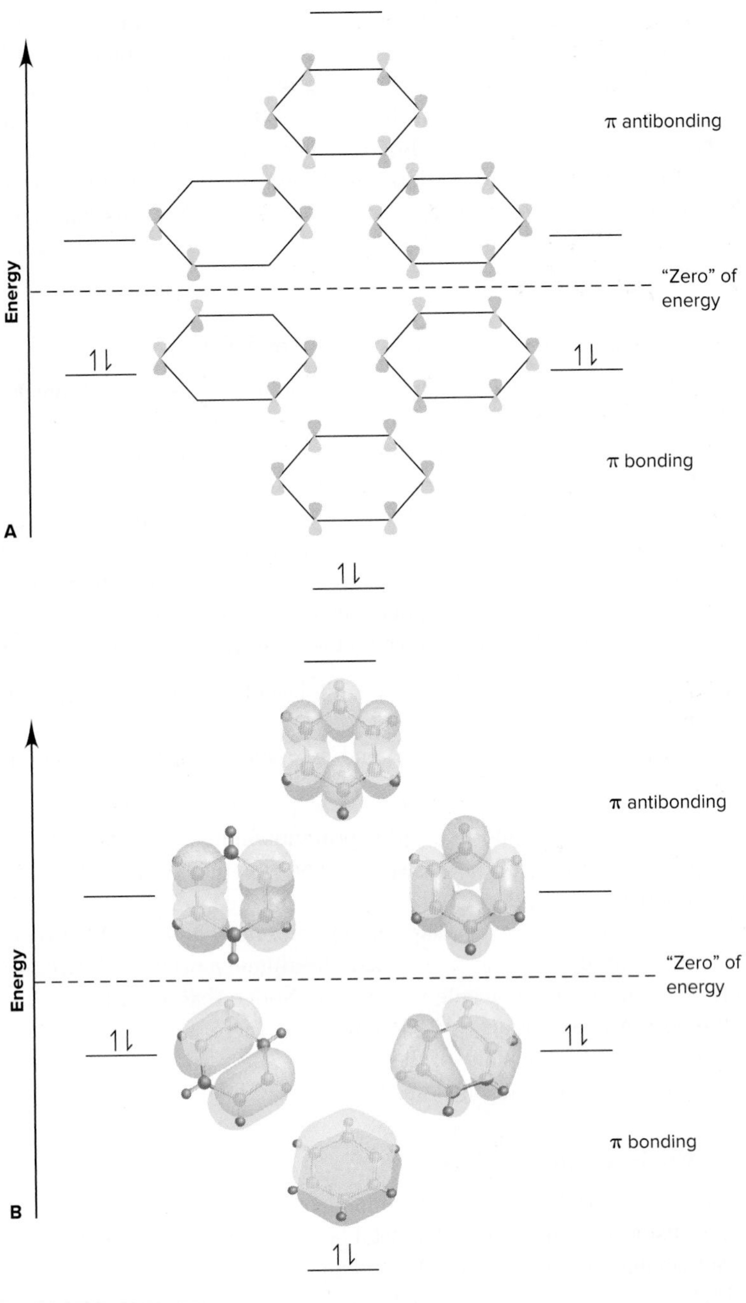

FIGURE 10.27 A. MO diagram for benzene. **B.** An alternative MO diagram for benzene with a different perspective.

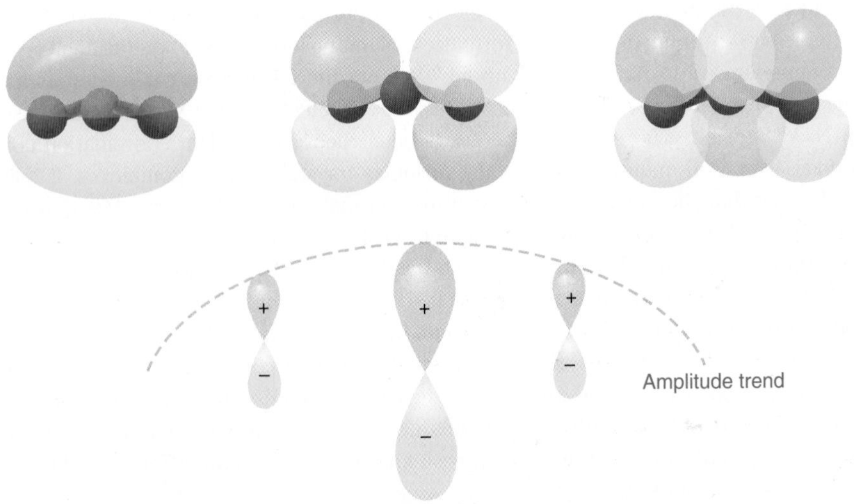

FIGURE 10.28 Three combinations of the three 2*p* AOs at each O in ozone.

SUMMARY OF SECTION 10.3

- MO theory treats a molecule as a collection of nuclei, with MOs delocalized over the entire structure.
- AOs of comparable energy can be added or subtracted to obtain bonding or antibonding MOs, respectively.
- Bonding MOs, whether σ or π, have most of the electron density between the nuclei and are lower in energy than the AOs. Most of the electron density in antibonding MOs does not lie between the nuclei, so these MOs are higher in energy.
- MOs are filled in the order of their energy, with paired electrons having opposite spins.
- MO diagrams show energy levels and orbital occupancy. Diagrams for the period 2 homonuclear diatomic molecules explain bond energy, bond length, and magnetic behaviour.
- In heteronuclear diatomic molecules, the more electronegative atom contributes more to the bonding MOs.
- MO theory eliminates the need for resonance forms to depict larger molecules.

CHAPTER REVIEW GUIDE

Learning Objectives

Relevant section (§) and/or sample problem (SP) numbers appear in parentheses.

Concepts

1. Describe the main ideas of valence bond (VB) theory—orbital overlap, opposing electron spins, and hybridization—as a means of rationalizing molecular shapes. (§10.1)
2. Explain how orbitals mix to form hybrid orbitals with different spatial orientations. (§10.1)
3. Explain the distinction between end-to-end and side-to-side overlap and the origin of sigma (σ) and pi (π) bonds in simple molecules. (§10.2)
4. Explain how these two modes of orbital overlap lead to single, double, and triple bonds. (§10.2)
5. Explain why π bonding restricts rotation around double bonds. (§10.2)
6. Distinguish between the localized bonding of VB theory and the delocalized bonding of molecular orbital (MO) theory. (§10.3)
7. Explain how the addition or subtraction of atomic orbitals (AOs) forms bonding or antibonding MOs. (§10.3)
8. Describe the shapes of MOs formed from combinations of two *s* orbitals and combinations of two *p* orbitals. (§10.3)
9. Explain how MO bond order predicts the stability of molecular species. (§10.3)
10. Describe how MO theory explains the bonding and magnetic properties of homonuclear and heteronuclear diatomic molecules of period 2 elements. (§10.3)

Skills

1. Use molecular shape to postulate the hybrid orbitals used by a central atom. (SP 10.1)
2. Describe the types of bonds and orbitals in a molecule. (SP 10.2)
3. Draw MO diagrams, calculate bond orders, and write electron configurations of molecular species. (SP 10.3)
4. Explain bond and magnetic properties with MO theory. (SP 10.4)

Key Terms

Section 10.1

valence bond (VB) theory
hybridization
hybrid orbitals
sp hybrid orbitals
sp^2 hybrid orbitals
sp^3 hybrid orbitals
sp^3d hybrid orbitals
sp^3d^2 hybrid orbitals

Section 10.2

sigma (σ) bond
pi (π) bond

Section 10.3

molecular orbital (MO) theory
molecular orbitals
bonding molecular orbital
antibonding molecular orbital
sigma (σ) molecular orbitals
molecular orbital diagram
molecular orbital bond order
homonuclear diatomic molecules
pi (π) molecular orbitals
nonbonding molecular orbitals

Key Equations and Relationships

10.1 Calculating the MO bond order:

$$\text{Bond order} = \frac{1}{2}[(\text{no. of e}^- \text{ in bonding MO}) - (\text{no. of e}^- \text{ in antibonding MO})]$$

$$= e^-_b - e^-_{ab}$$

Brief Solutions to Follow-Up Problems

10.1 (a) The shape is linear, so Be is *sp* hybridized.

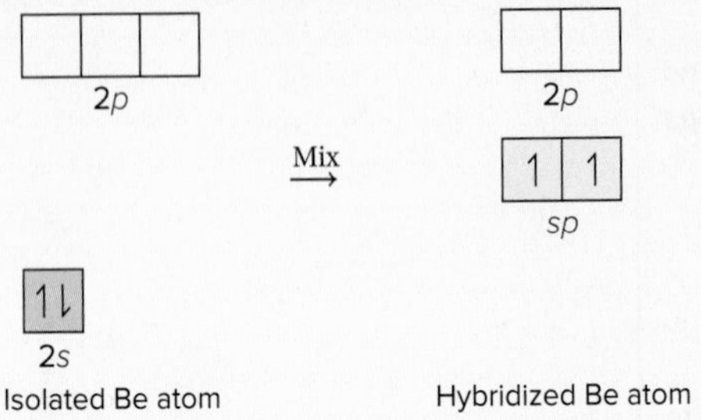

(b) The shape is tetrahedral, so Si is sp^3 hybridized.

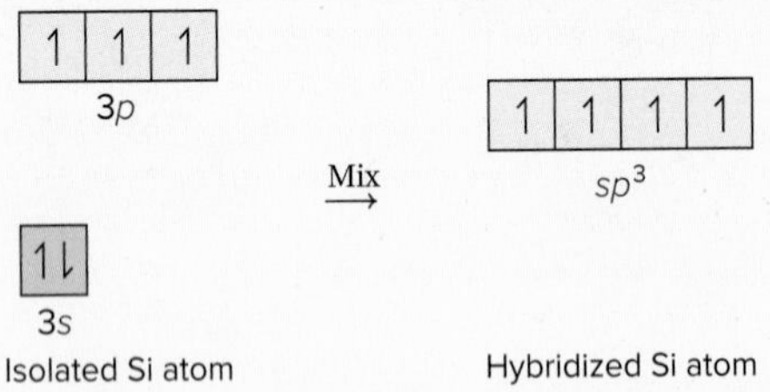

(c) The shape is square planar, so Xe is sp^3d^2 hybridized.

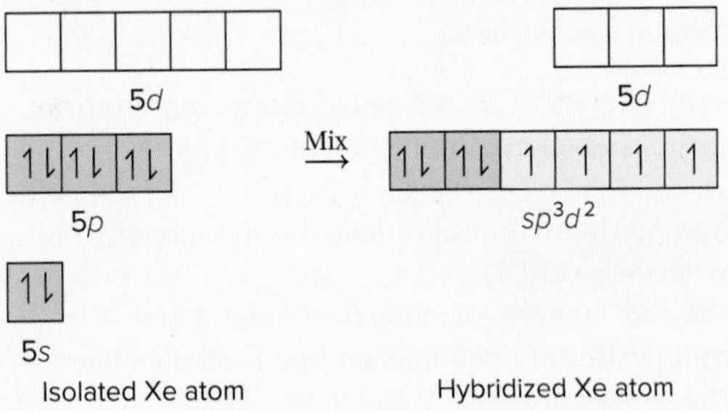

10.2 (a) H—C≡N:

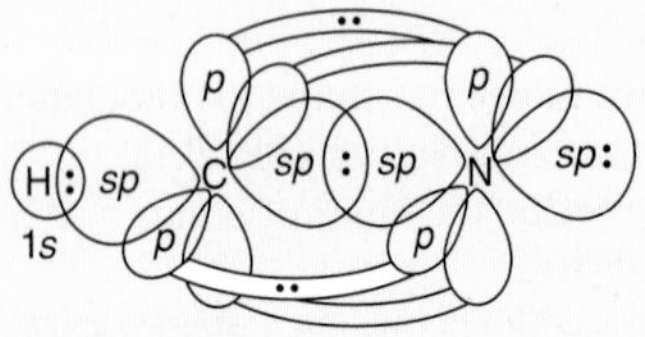

HCN is linear, so C is *sp* hybridized. N is also *sp* hybridized. One *sp* of C overlaps the 1*s* of H to form a σ bond. The other *sp* of C overlaps one *sp* of N to form a σ bond. The other *sp* of N holds a lone pair. Two unhybridized *p* orbitals of N and two unhybridized *p* orbitals of C overlap to form two π bonds.

(b) $\ddot{\text{O}}$=C=$\ddot{\text{O}}$

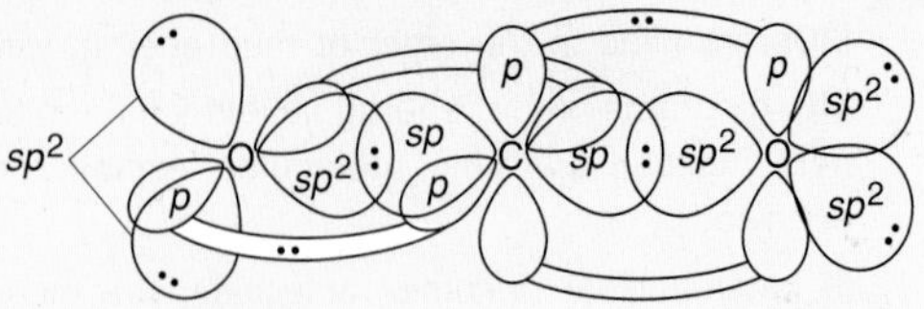

CO_2 is linear, so C is *sp* hybridized. Both O atoms are sp^2 hybridized. Each *sp* of C overlaps one sp^2 of an O to form two σ bonds. Each of the two unhybridized *p* orbitals of C forms a π bond with the unhybridized *p* orbital of one of the two O atoms. Two sp^2 orbitals of each O hold lone pairs.

10.3

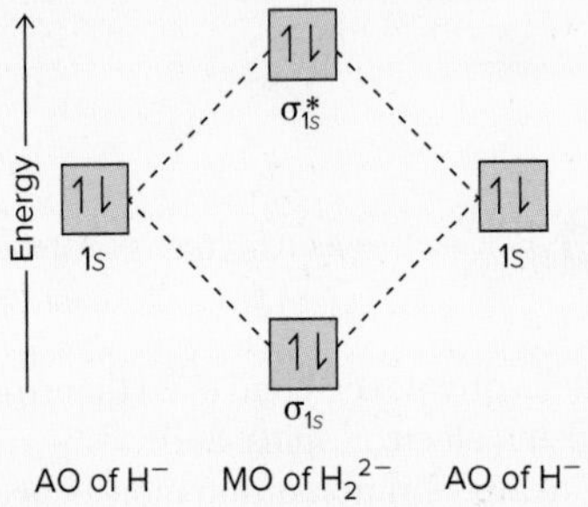

Does not exist: bond order $= \frac{1}{2}(2 - 2) = 0$.

10.4 Bond orders: $F_2^{2-} = 0$; $F_2^- = \frac{1}{2}$; $F_2 = 1$; $F_2^+ = 1\frac{1}{2}$
$F_2^{2+} = 2$

Bond energy: $F_2^{2-} < F_2^- < F_2 < F_2^+ < F_2^{2+}$
Bond length: $F_2^{2+} < F_2^+ < F_2 < F_2^-$; F_2^{2-} does not exist.

PROBLEMS

Problems with **red** numbers are answered in Appendix G and worked in detail in the Student Solutions Manual. Problem sections match those in this book and provide the numbers of relevant sample problems. Most offer Concept Review Questions, Skill-Building Exercises (grouped in pairs covering the same concept), and Problems in Context. The Comprehensive Problems are based on material from any section or previous chapter.

Valence Bond Theory and Orbital Hybridization
(Sample Problem 10.1)

Concept Review Questions

10.1 What type of central-atom orbital hybridization corresponds to each electron-group arrangement?
(a) Trigonal planar (b) Octahedral (c) Linear
(d) Tetrahedral (e) Trigonal bipyramidal

10.2 What is the orbital hybridization of a central atom that has one lone pair and bonds to (a) two other atoms; (b) three other atoms; (c) four other atoms; (d) five other atoms?

10.3 How do carbon and silicon differ with regard to the *types* of orbitals that are available for hybridization? Explain.

10.4 How many hybrid orbitals form when four AOs of a central atom mix? Explain.

Skill-Building Exercises (grouped in similar pairs)

10.5 Give the number and type of hybrid orbital that forms when each set of AOs mixes:
(a) Two *d*, one *s*, and three *p* (b) Three *p* and one *s*

10.6 Give the number and type of hybrid orbital that forms when each set of AOs mixes:
(a) One *p* and one *s* (b) Three *p*, one *d*, and one *s*

10.7 What is the hybridization of nitrogen in (a) NO; (b) NO_2; (c) NO_2^-?

10.8 What is the hybridization of carbon in (a) CO_3^{2-}; (b) $C_2O_4^{2-}$; (c) NCO^-?

10.9 What is the hybridization of chlorine in (a) ClO_2; (b) ClO_3^-; (c) ClO_4^-?

10.10 What is the hybridization of bromine in (a) BrF_3; (b) BrO_2^-; (c) BrF_5?

10.11 Which types of AOs of the central atom mix to form hybrid orbitals in (a) $SiClH_3$; (b) CS_2; (c) SCl_3F; (d) NF_3?

10.12 Which types of AOs of the central atom mix to form hybrid orbitals in (a) Cl_2O; (b) $BrCl_3$; (c) PF_5; (d) SO_3^{2-}?

10.13 Phosphine (PH_3) reacts with borane (BH_3) as follows:

$$PH_3 + BH_3 \longrightarrow H_3P\text{—}BH_3$$

(a) Which illustration depicts the change, if any, in the orbital hybridization of P during this reaction?
(b) Which depicts the change, if any, in the orbital hybridization of B?

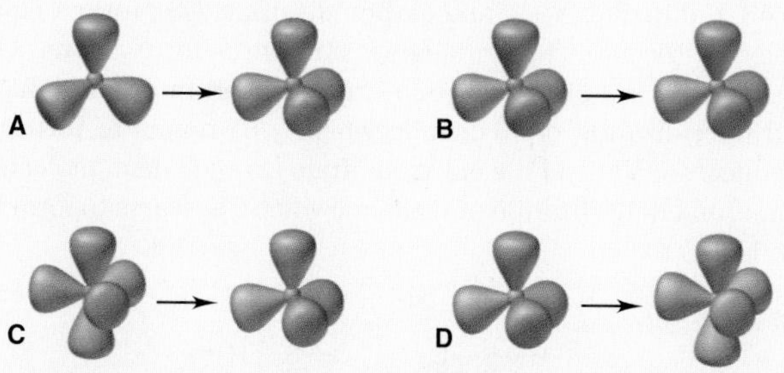

10.14 The following illustrations depict differences in the orbital hybridization of some tellurium (Te) fluorides. (a) Which illustration depicts the difference, if any, between TeF_6 (*left*) and TeF_5^- (*right*)? (b) Which illustration depicts the difference, if any, between TeF_4 (*left*) and TeF_6 (*right*)?

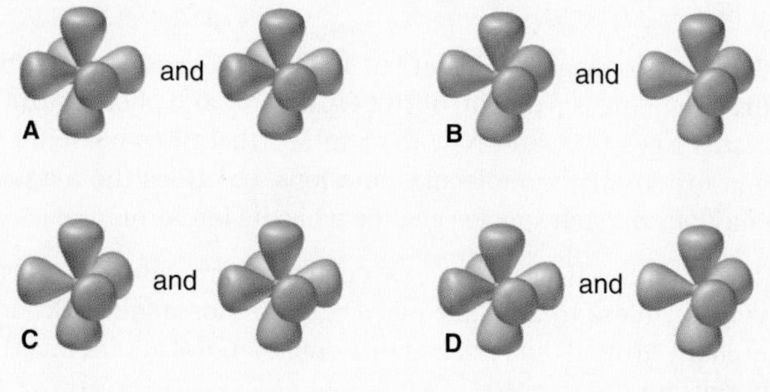

10.15 Use partial orbital diagrams to show how the AOs of the central atom lead to hybrid orbitals in (a) $GeCl_4$; (b) BCl_3; (c) CH_3^+.

10.16 Use partial orbital diagrams to show how the AOs of the central atom lead to hybrid orbitals in (a) BF_4^-; (b) PO_4^{3-}; (c) SO_3.

10.17 Use partial orbital diagrams to show how the AOs of the central atom lead to hybrid orbitals in (a) $SeCl_2$; (b) H_3O^+; (c) IF_4^-.

10.18 Use partial orbital diagrams to show how the AOs of the central atom lead to hybrid orbitals in (a) $AsCl_3$; (b) $SnCl_2$; (c) PF_6^-.

Problem in Context

10.19 Methyl isocyanate, $CH_3\text{—}\ddot{N}\text{=}C\text{=}\ddot{O}$:, is an intermediate in the manufacture of many pesticides. In 1984, a leak from a manufacturing plant resulted in the immediate deaths of more than 2000 people in Bhopal, India. What are the hybridizations of the N atom and the two C atoms in methyl isocyanate? Sketch the molecular shape.

Modes of Orbital Overlap and the Types of Covalent Bonds

(Sample Problem 10.2)

Concept Review Question

10.20 Are these statements true or false? Correct any statements that are false.
(a) Two σ bonds form a double bond.
(b) A triple bond consists of one π bond and two σ bonds.
(c) Bonds formed from atomic *s* orbitals are always σ bonds.
(d) A π bond restricts rotation about the σ-bond axis.
(e) A π bond consists of two pairs of electrons.
(f) End-to-end overlap results in a bond with electron density above and below the bond axis.

Skill-Building Exercises (grouped in similar pairs)

10.21 Describe the hybrid orbitals used by the central atom and the type(s) of bonds formed in (a) NO_3^-; (b) CS_2; (c) CH_2O.

10.22 Describe the hybrid orbitals used by the central atom and the type(s) of bonds formed in (a) O_3; (b) I_3^-; (c) $COCl_2$ (C is central).

10.23 Describe the hybrid orbitals used by the central atom(s) and the type(s) of bonds formed in (a) FNO; (b) C_2F_4; (c) $(CN)_2$.

10.24 Describe the hybrid orbitals used by the central atom(s) and the type(s) of bonds formed in (a) BrF_3; (b) $CH_3C\equiv CH$; (c) SO_2.

Problem in Context

10.25 The molecule but-2-ene ($CH_3CH\text{=}CHCH_3$) is a starting material in the manufacture of lubricating oils and many other compounds. Draw two different structures for but-2-ene, indicating the σ and π bonds in each.

Molecular Orbital Theory and Electron Delocalization

(Sample Problems 10.3 and 10.4)

Concept Review Questions

10.26 Two *p* orbitals from one atom and two *p* orbitals from another atom are combined to form MOs for the joined atoms. How many MOs result from this combination? Explain.

10.27 Certain AOs on two atoms were combined to form the following MOs. Name the AOs used and the MOs formed, and explain which MO has higher energy:

10.28 How do the bonding and antibonding MOs formed from a given pair of AOs compare with each other with regard to (a) energy; (b) presence of nodes; (c) internuclear electron density?

10.29 Antibonding MOs always have at least one node. Can a bonding MO have a node? If so, draw an example.

Skill-Building Exercises (grouped in similar pairs)

10.30 How many electrons does it take to fill (a) a σ bonding MO; (b) a π antibonding MO; (c) the MOs formed from a combination of the 1*s* orbitals of two atoms?

10.31 How many electrons does it take to fill (a) the MOs formed from a combination of the 2*p* orbitals of two atoms; (b) a σ^*_{2p} MO; (c) the MOs formed from a combination of the 2*s* orbitals of two atoms?

10.32 The MOs shown below are derived from the 2*p* AOs in F_2^+. (a) Give the orbital designations. (b) Which orbital is occupied by at least one electron in F_2^+? (c) Which orbital is occupied by only one electron in F_2^+?

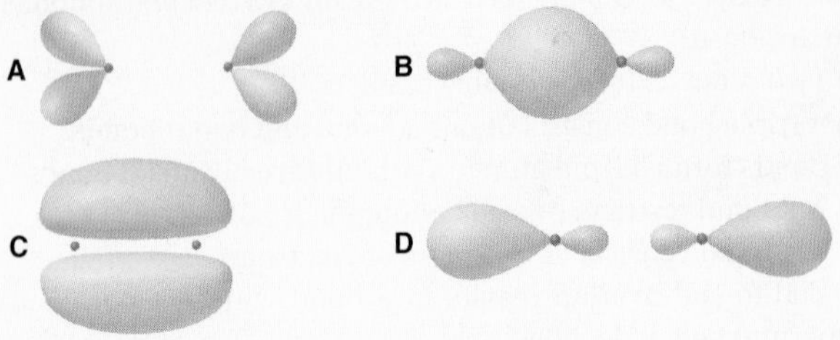

10.33 The MOs shown below are derived from $n = 2$ AOs. (a) Give the orbital designations. (b) Which orbital is highest in energy? (c) Which orbital is lowest in energy? (d) Rank the MOs in order of increasing energy for B_2.

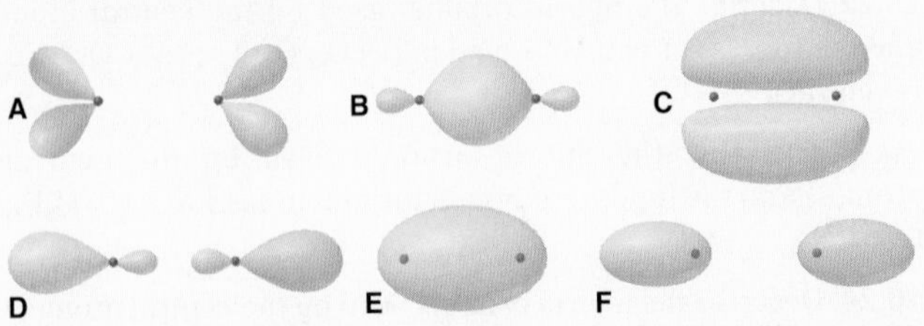

10.34 Use MO diagrams, and the bond orders you obtain from them, to answer the following questions: (a) Is Be_2^+ stable? (b) Is Be_2^+ diamagnetic? (c) What is the valence (outer) electron configuration of Be_2^+?

10.35 Use MO diagrams, and the bond orders you obtain from them, to answer the following questions: (a) Is O_2^- stable? (b) Is O_2^- paramagnetic? (c) What is the valence (outer) electron configuration of O_2^-?

10.36 Use MO diagrams to place C_2^-, C_2, and C_2^+ in order of (a) increasing bond energy; (b) increasing bond length.

10.37 Use MO diagrams to place B_2^+, B_2, and B_2^- in order of (a) decreasing bond energy; (b) decreasing bond length.

Comprehensive Problems

10.38 Predict the shape, state the hybridization of the central atom, and give the ideal bond angle(s) and any expected deviations for each ion:
(a) BrO_3^- (b) $AsCl_4^-$ (c) SeO_4^{2-} (d) BiF_5^{2-} (e) SbF_4^+ (f) AlF_6^{3-} (g) IF_4^+

10.39 Buta-1.3-diene (*below*) is a colourless gas that is used to make synthetic rubber and many other compounds. (a) How many σ bonds and π bonds does the molecule have? (b) Are *cis-trans* arrangements around the double bonds possible? Explain.

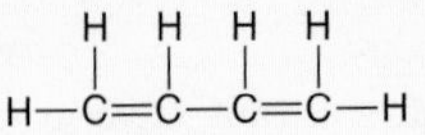

10.40 Epinephrine (or adrenaline; *below*) is a naturally occurring hormone that is also manufactured commercially for use as a heart stimulant, a nasal decongestant, and a glaucoma treatment:

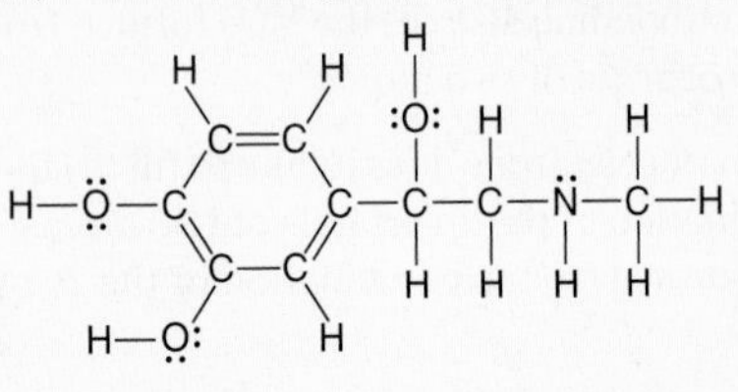

(a) What is the hybridization of each C, O, and N atom? (b) How many σ bonds does the molecule have? (c) How many π electrons are delocalized in the ring?

10.41 Use partial orbital diagrams to show how the AOs of the central atom lead to the hybrid orbitals in (a) IF_2^-; (b) ICl_3; (c) $XeOF_4$; (d) BHF_2.

10.42 Isoniazid (*below*) is an antibacterial agent that is very useful against many common strains of tuberculosis. (a) How many σ bonds are in the molecule? (b) What is the hybridization of each C and N atom?

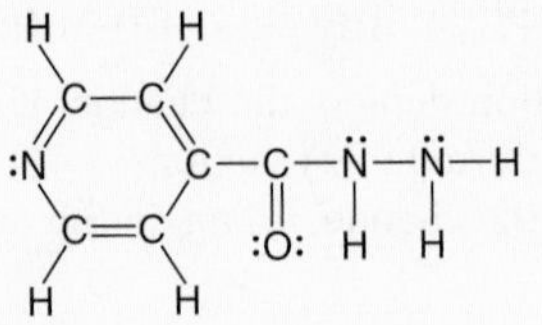

10.43 Hydrazine, N_2H_4, and carbon disulfide, CS_2, form a cyclic molecule (*below*). (a) Draw Lewis structures for N_2H_4 and CS_2. (b) How do the electron-group arrangement, molecular shape, and hybridization of N change when N_2H_4 reacts to form the product? (c) How do the electron-group arrangement, molecular shape, and hybridization of C change when CS_2 reacts to form the product?

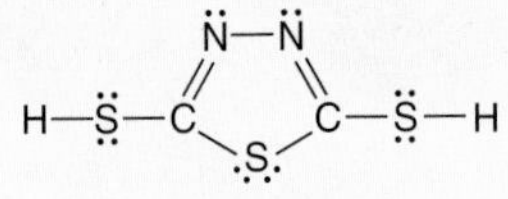

10.44 What hybridization change, if any, occurs for the underlined atom in each equation?
(a) $\underline{B}F_3 + NaF \longrightarrow Na^+ + BF_4^-$
(b) $\underline{P}Cl_3 + Cl_2 \longrightarrow PCl_5$
(c) $H\underline{C}{\equiv}CH + H_2 \longrightarrow H_2C{=}CH_2$
(d) $\underline{Si}F_4 + 2F^- \longrightarrow SiF_6^{2-}$
(e) $\underline{S}O_2 + \frac{1}{2}O_2 \longrightarrow SO_3$

10.45 The ionosphere lies about 100 km above Earth's surface. This layer consists mostly of NO, O_2, and N_2, and photoionization creates NO^+, O_2^+, and N_2^+. (a) Use MO theory to compare the bond orders of the molecules and ions. (b) Does the magnetic behaviour of each species change when its ion forms?

10.46 Glyphosate (*below*) is a common herbicide that is relatively harmless to animals, but deadly to most plants. Describe the shape around the P, N, and three numbered C atoms, and their hybridization.

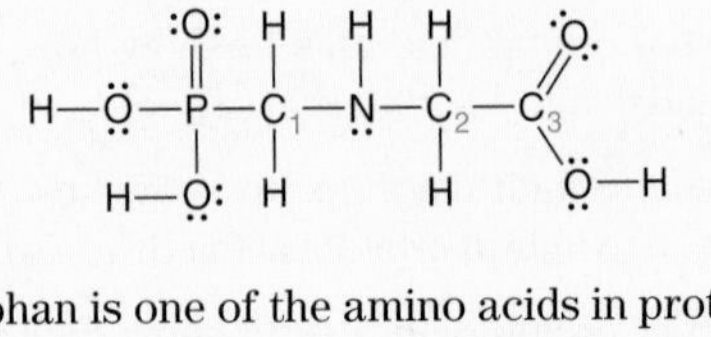

10.47 Tryptophan is one of the amino acids in proteins:

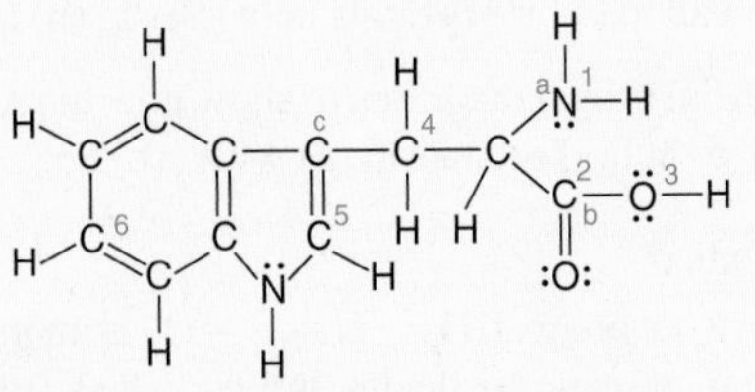

(a) What is the hybridization of each of the numbered C, N, and O atoms? (b) How many σ bonds are present in tryptophan? (c) Predict the bond angles at points a, b, and c.

10.48 Some species with only two oxygen atoms are the oxygen molecule, O_2, the peroxide ion, O_2^{2-}, the superoxide ion, O_2^-, and the dioxygenyl ion, O_2^+. Draw MO diagrams, rank the species in order of increasing bond length, and find the number of unpaired electrons in each.

10.49 Molecular nitrogen, carbon monoxide, and cyanide ion are isoelectronic. (a) Draw an MO diagram for each species. (b) CO and CN^- are toxic. What property may explain why N_2 is not?

10.50 There is concern in health-related government agencies that the Canadian diet contains too much meat. Numerous recommendations have been made, urging people to consume more fruit and vegetables. One of the richest sources of vegetable protein is soy, available in many forms. One of these forms is soybean curd, or tofu, which is a staple of many Asian diets. Chemists have isolated an anticancer agent called *genistein* from tofu, which may explain the much lower incidence of cancer among people in Asia. A valid Lewis structure for genistein is given below:

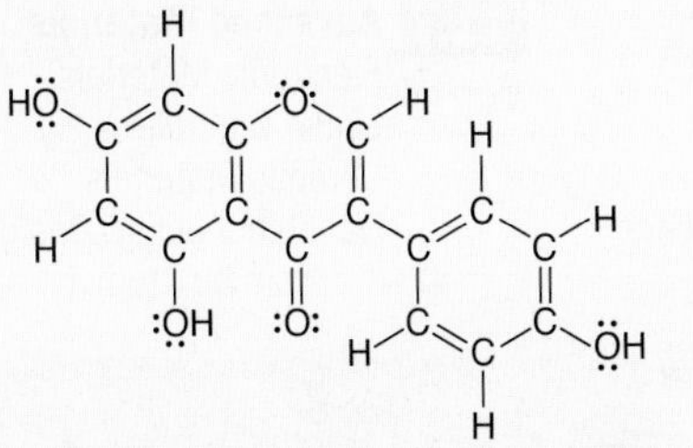

(a) Is the hybridization of each C in the right-hand ring the same? Explain. (b) Is the hybridization of the O atom in the centre ring the same as the hybridization of the O atoms in the OH groups? Explain. (c) How many carbon-oxygen σ bonds are present in genistein? How many carbon-oxygen π bonds are present? (d) Do all the lone pairs on oxygens occupy the same type of hybrid orbital? Explain.

10.51 An organic chemist synthesizes the molecule below:

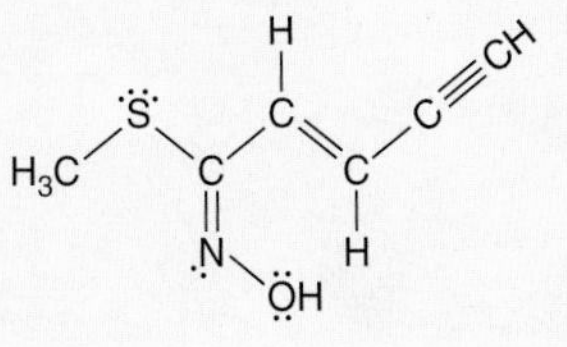

(a) Which of the following orientations of hybrid orbitals are present in the molecule? (b) Are there any present that are not shown below? If so, what are they? (c) How many of each type of hybrid orbital are present?

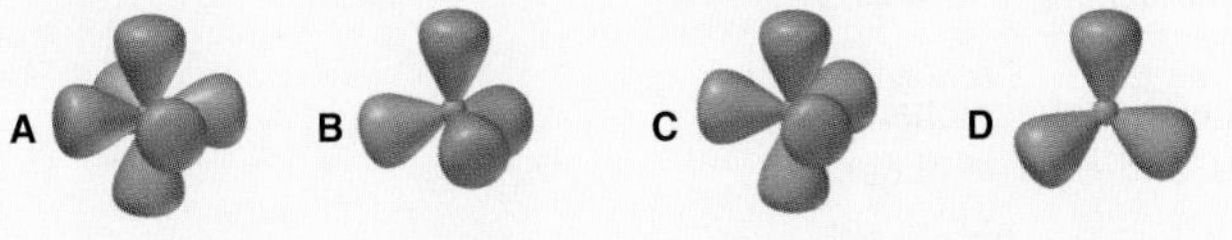

10.52 Simple proteins consist of amino acids linked together in a long chain. A small portion of such a chain is shown below:

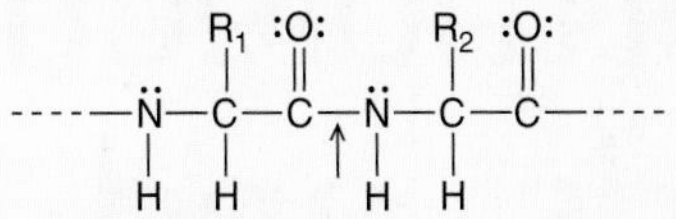

Experiments show that rotation about the C—N bond (indicated by the arrow) is somewhat restricted. Explain why, using resonance structures, and show the types of bonding involved.

10.53 Sulfur forms oxides, oxoanions, and halides. What is the hybridization of the central S in SO_2, SO_3, SO_3^{2-}, SCl_4, SCl_6, and S_2Cl_2 (atom sequence Cl—S—S—Cl)?

10.54 The compound 2,6-dimethylpyrazine (*below*) gives chocolate its odour and is used in flavourings. (a) Which AOs mix to form the hybrid orbitals of N? (b) In what type of hybrid orbital do the lone pairs of N reside? (c) Is C in CH_3 hybridized the same as any C in the ring? Explain.

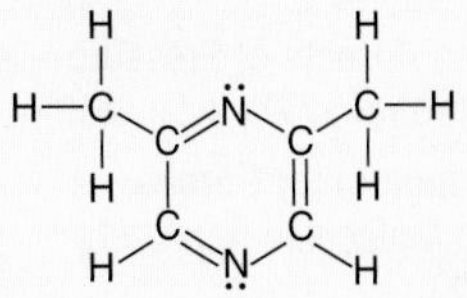

10.55 Acetylsalicylic acid (Aspirin), the most widely used medicine in the world, has the Lewis structure shown below. (a) What is the hybridization of each C and each O atom? (b) How many localized π bonds are present? (c) How many C atoms have a trigonal planar shape around them? A tetrahedral shape?

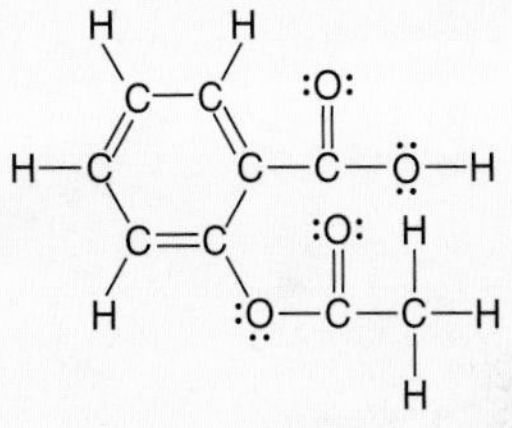

10.56 Linoleic acid is an essential fatty acid that is found in many vegetable oils, such as soy, peanut, and cottonseed. A key structural feature of the molecule is the *cis* orientation around its two double bonds, where R_1 and R_2 represent two different groups that form the rest of the molecule:

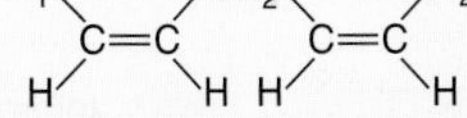

(a) How many different compounds are possible, changing only the *cis-trans* arrangements around these two double bonds?
(b) How many different compounds are possible for a similar molecule with three double bonds?

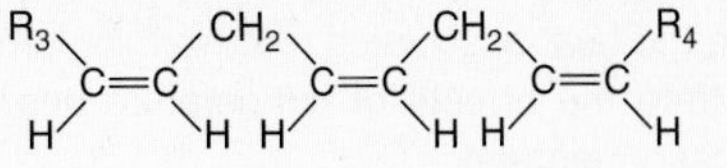

Intermolecular Forces: Liquids, Solids, and Phase Changes

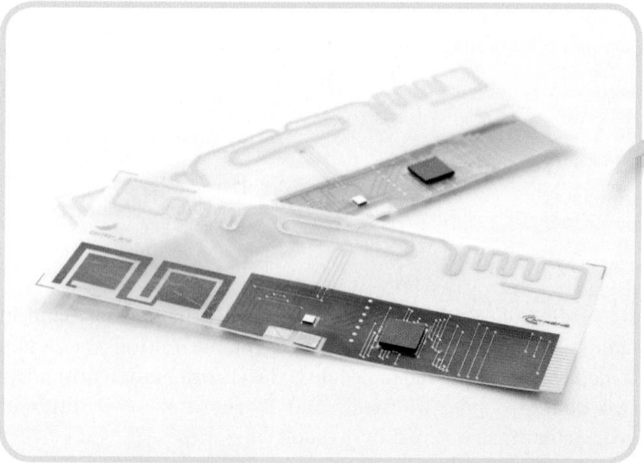

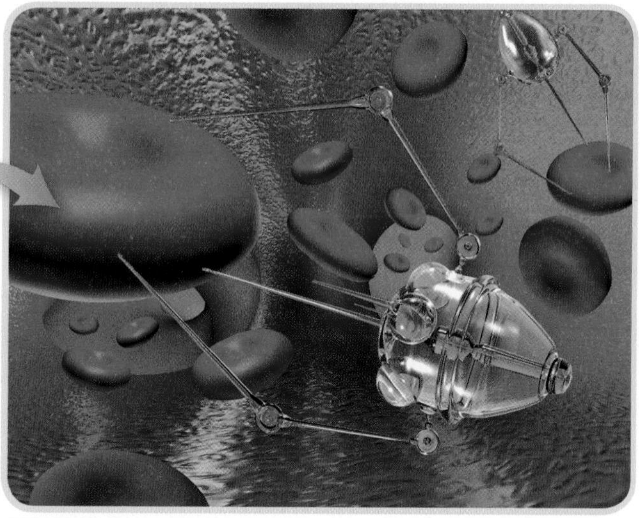

IN THIS CHAPTER . . . We explore the interplay of forces that give rise to the three states of matter and their changes, with special attention to liquids and solids. By the end of this chapter, you should be able to

- Apply the kinetic-molecular theory to determine the relative magnitudes of potential energy and kinetic energy and relate these values to the behaviour of gases, liquids, and solids
- Demonstrate how temperature is related to phase changes
- Calculate the heat associated with warming or cooling a phase and with a phase change
- Demonstrate the effects of temperature and pressure on phases and their changes, using phase diagrams
- Describe the types and relative strengths of the intermolecular forces that occur within pure substances
- Discuss how intermolecular forces give rise to the properties of liquids
- Correlate the unique and vital physical properties of water to the electron configurations of its atoms
- Discuss the properties of solids, emphasizing the relationship between type of bonding and predominant intermolecular force, and compare and contrast the methods used to study crystal structures
- Explain the key features of several kinds of advanced materials: semiconductors, liquid crystals, ceramics, and nanostructures

11

Concepts and Skills to Review before Studying This Chapter

- Properties of gases, liquids, and solids (Section 4.1)
- Kinetic-molecular theory of gases (Section 4.5)
- Kinetic energy and potential energy (Section 5.1)
- Enthalpy change, heat capacity, and Hess's law (Sections 5.2, 5.3, and 5.5)
- Diffraction of light (Section 6.1)
- Coulomb's law (Section 8.2)
- Chemical bonding models (Chapter 8)
- Molecular polarity (Section 9.2)
- Molecular orbital treatment of diatomic molecules (Section 10.3)

What could sitting in a hot springs bath, watching the steam rise up, and looking at the snowy peaks surrounding you at Le Spa Nordik in Québec have to do with chemistry? In this chapter, we will see how temperature and pressure affect the states of matter and how the kinetic molecular theory, first seen in Chapter 4, can to some extent be used to explain the behaviour of solids and liquids. We will learn about phase diagrams and how they relate to states of matter for a substance at different temperatures and pressures. We will determine the relationship between physical properties of many substances and the intermolecular forces at play in these molecules. The unique properties of water, whose importance in our world and in chemistry were first seen in Chapter 3, will be examined in detail and should explain how the Spa Nordik experience is very much related to chemistry.

11.1 An Overview of Physical States and Phase Changes

Each physical state, or **phase**, is a physically distinct, homogeneous part of a system. The water in a closed container is one phase; the water vapour above the liquid is a second phase; then add some ice, and there are three phases.

In this section, you will see that interactions between the potential energy and the kinetic energy of the particles give rise to the properties of each phase:

- *The potential energy*, manifested in the form of **intermolecular (interparticle) forces**, has the tendency to draw molecules together. Note, however, that the potential energy is not the only source of intermolecular forces. According to Coulomb's law, the electrostatic potential energy depends on the charges of the particles and the distances between them (Section 8.2).
- The *kinetic energy* associated with the random motion of the molecules has the tendency to disperse the molecules. It is related to their average speed and is proportional to the absolute temperature (Section 4.5).

These interactions also explain **phase changes**, changes in physical state from one phase to another—liquid to solid, solid to gas, and so on.

A Kinetic-Molecular View of the Three States Imagine yourself among the particles in any of the three states of water. Look closely, and you will discover two types of electrostatic forces at work:

1. *Intramolecular forces* exist *within* each molecule. The *chemical* behaviour of the three states is identical because each state consists of the same bent polar H—O—H molecules, held together by identical covalent bonding forces.

2. *Intermolecular* forces exist *between* the molecules. The *physical* behaviour of the states is different because the strengths of these forces differ from state to state.

Whether a substance occurs as a gas, a liquid, or a solid depends on the interplay of the potential energy and kinetic energy:

- *In a gas*, the potential energy (energy of attraction) is small relative to the kinetic energy (energy of motion); thus, on average, the particles are far apart. This large

TABLE 11.1 A Macroscopic Comparison of Gases, Liquids, and Solids

State	Shape and Volume	Compressibility	Ability to Flow
Gas	Conforms to the shape and volume of the container	High	High
Liquid	Conforms to the shape of the container; volume limited by the surface	Very low	Moderate
Solid	Maintains its own shape and volume	Almost none	Almost none

distance has several macroscopic consequences: a gas fills its container, is highly compressible, and flows easily through another gas (Table 11.1).

- *In a liquid*, the attractions are stronger because the particles are touching, but the particles have enough kinetic energy to move randomly around one another. Thus, a liquid conforms to the shape of its container but has a surface; it resists an applied force and, thus, compresses very slightly; and it flows, but *much* more slowly than does a gas.
- *In a solid*, the attractions dominate the motion so much that the particles are fixed in position relative to one another, just jiggling in place. Thus, a solid has its own shape, compresses even less than a liquid does, and does not flow significantly.

How does the ability to flow influence composition in nature?

The environment provides a perfect demonstration of these differences in ability to flow (*see margin photo*). Atmospheric gases mix so well that the lowest 80 km of air has a uniform composition. Much less mixing in the oceans allows the composition at various depths to support different species. Rocks intermingle so little that adjacent strata remain separated for millions of years.

Types of Phase Changes and Their Enthalpies When we consider phase changes, an understanding of the effects of temperature is critical:

- As *temperature increases*, the average kinetic energy does too, so the faster-moving particles overcome attractions more easily.
- As *temperature decreases*, particles slow, so attractions can pull them together.

Each phase change has a name and an associated enthalpy change:

1. *Gas to liquid, and liquid to gas.* As the temperature drops, the molecules in the gas phase come together and form a liquid in the process of **condensation**; the opposite process, changing from a liquid to a gas, is called **vaporization** and occurs when heat is added to a system. The boiling of water resulting in the formation of water vapour (steam) is one such example (Figure 11.1A).

2. *Liquid to solid, and solid to liquid.* As the temperature drops further, the particles move slower and become fixed in position in the process of **freezing**; the opposite change is called **melting (fusion)**. In common speech, *freezing* implies low temperature because we think of water. However, molten metals, for example, freeze (solidify) at much higher temperatures and have medical, industrial, and artistic applications, such as gold dental crowns, steel auto bodies, and bronze statues.

3. *Solid to gas, and gas to solid.* All three common states of water are familiar because they are stable under ordinary conditions. Carbon dioxide (CO_2), on the other hand, is familiar as a gas and a solid (dry ice), but liquid CO_2 occurs only at pressures of 516.7 kPa or greater. Under ordinary conditions, solid CO_2 changes directly to a gas in a process called **sublimation**. Freeze-dried foods are prepared by sublimation. The opposite process, changing from a gas directly into a solid, is called **deposition**: ice crystals form on a cold window from the deposition of water vapour (Figure 11.1B).

The accompanying enthalpy changes are either exothermic or endothermic:

- *Exothermic changes.* As the molecules of a gas attract each other into a liquid and then become fixed in a solid, the system of particles *loses* energy, which is released as heat. Thus, *condensing, freezing, and depositing are exothermic changes.*
- *Endothermic changes.* Heat must be absorbed by the system to overcome the attractive forces that keep the particles fixed in place in a solid or near each other in a liquid. Thus, *melting, vaporizing, and subliming are endothermic changes.*

A Vaporization

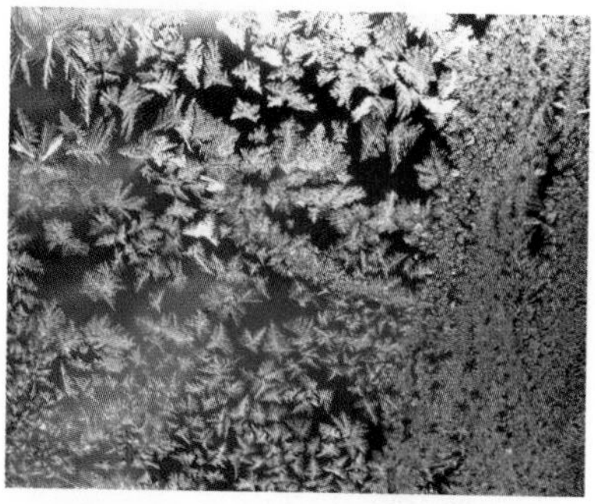

B Deposition

FIGURE 11.1 Two familiar phase changes

We see these phase changes in our everyday lives as well. Sweating has a cooling effect because heat from our bodies vaporize the water. To achieve this same cooling effect, cats lick themselves and dogs pant.

For a pure substance, each phase change is accompanied by a standard enthalpy change, given in units of *kilojoules per mole* (measured at 101.3 kPa and the temperature of the change). For vaporization, it is the **heat (enthalpy) of vaporization ($\Delta_{vap}H°$)**. For fusion (melting), it is the **heat (enthalpy) of fusion ($\Delta_{fus}H°$)**. In the case of water, we have

$$H_2O(l) \longrightarrow H_2O(g) \qquad \Delta H = \Delta_{vap}H° = 40.7 \text{ kJ/mol (at 100°C)}$$
$$H_2O(s) \longrightarrow H_2O(l) \qquad \Delta H = \Delta_{fus}H° = 6.02 \text{ kJ/mol (at 0°C)}$$

The reverse processes, condensing and freezing, have enthalpy changes of the *same magnitude but opposite sign*:

$$H_2O(g) \longrightarrow H_2O(l) \qquad \Delta H = -\Delta_{vap}H° = -40.7 \text{ kJ/mol}$$
$$H_2O(l) \longrightarrow H_2O(s) \qquad \Delta H = -\Delta_{fus}H° = -6.02 \text{ kJ/mol}$$

Water behaves typically in that much less energy is needed to melt the solid than to vaporize the liquid: $\Delta_{fus}H° < \Delta_{vap}H°$; that is, less energy is needed to reduce the intermolecular forces enough for the molecules to move out of their fixed positions (melt a solid) than to separate them completely (vaporize a liquid) (Figure 11.2).

The **heat (enthalpy) of sublimation ($\Delta_{subl}H°$)** is the enthalpy change when 1 mol of a substance sublimes, and the negative of this value is the change when 1 mol of the substance deposits. Since sublimation can be considered a combination of melting and vaporizing, Hess's law (Section 5.5) says that the heat of sublimation equals the sum of the heats of fusion and vaporization:

$$\begin{array}{lc} \text{Solid} \longrightarrow \text{liquid} & \Delta_{fus}H° \\ \text{Liquid} \longrightarrow \text{gas} & \Delta_{vap}H° \\ \hline \text{Solid} \longrightarrow \text{gas} & \Delta_{subl}H° \end{array}$$

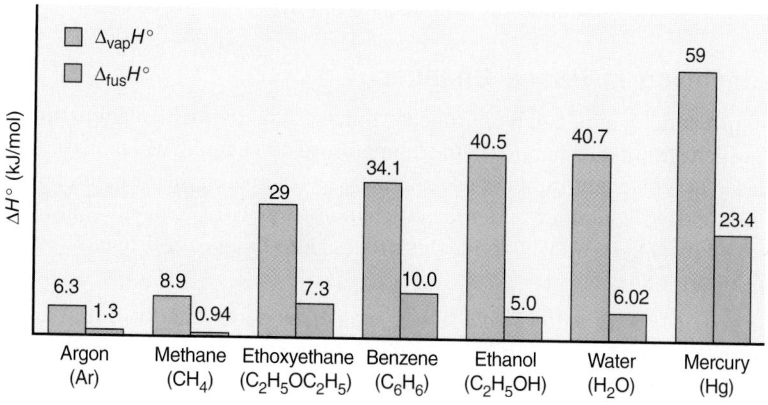

FIGURE 11.2 Heats of vaporization and fusion for several common substances

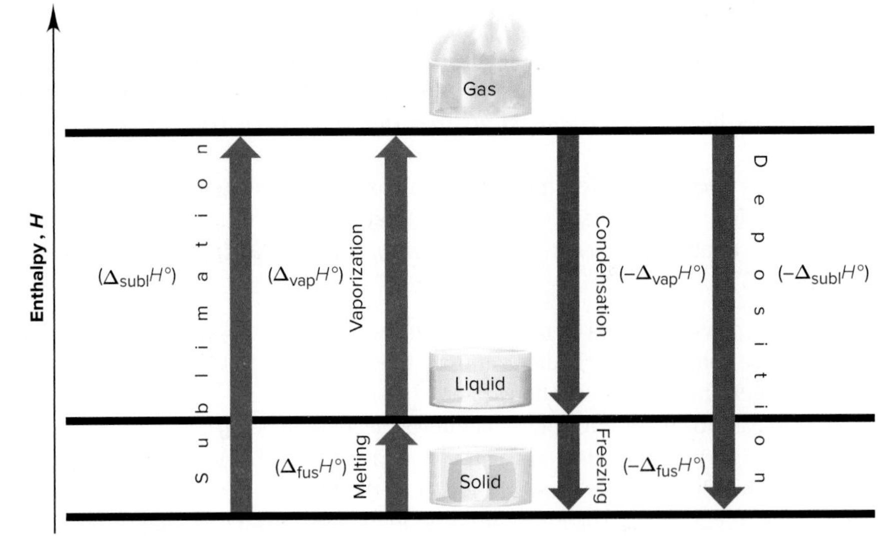

FIGURE 11.3 Phase changes and their enthalpy changes. Fusion (or melting), vaporization, and sublimation are endothermic changes (positive $\Delta H°$), whereas freezing, condensation, and deposition are exothermic changes (negative $\Delta H°$).

Figure 11.3 summarizes the phase changes and their enthalpy changes.

SUMMARY OF SECTION 11.1

- Because of the relative magnitudes of intermolecular forces (potential energy) and average speed (kinetic energy), the particles in a gas are far apart and moving randomly, the particles in a liquid are in contact and moving relative to one another, and the particles in a solid are in contact and in fixed positions. These molecular-level differences account for the macroscopic differences in shape, compressibility, and ability to flow.
- When a solid becomes a liquid (melting, or fusion), a liquid becomes a gas (vaporization), or a solid becomes a gas (sublimation), energy is absorbed to overcome the intermolecular forces and increase the average distance between particles. As particles come closer together in the reverse changes (freezing, condensation, and deposition), energy is released. Each phase change is associated with a given enthalpy change under specified conditions.

11.2 Quantitative and Qualitative Aspects of Phase Changes

Many phase changes occur around you every day, accompanied by the release or absorption of heat. When it rains, water vapour has condensed to a liquid, which changes back to a gas as puddles dry up. Solid water melts in the spring and then freezes again in the winter. The same phase changes take place whenever you make a pot of tea or a tray of ice cubes. In this section, we quantify the heat involved in a phase change and examine the equilibrium nature of the process.

Heat Involved in Phase Changes

We apply a kinetic-molecular approach to phase changes with a **heating-cooling curve**, which shows the changes in the temperature of a sample when heat is absorbed or released at a constant rate. Let us examine what happens when 2.50 mol of gaseous water in a closed container undergoes a change from 130°C to −40°C at a constant pressure of 101.3 kPa. We will divide this process into five heat-releasing (exothermic) stages (Figure 11.4):

Stage 1: Gaseous water cools. Water molecules zoom chaotically at a range of speeds, smashing into one another and the container walls. At the starting temperature, the most probable speed of the molecules, and thus their average kinetic energy (E_k), is high enough to overcome the potential energy (E_p) of attractions. As the

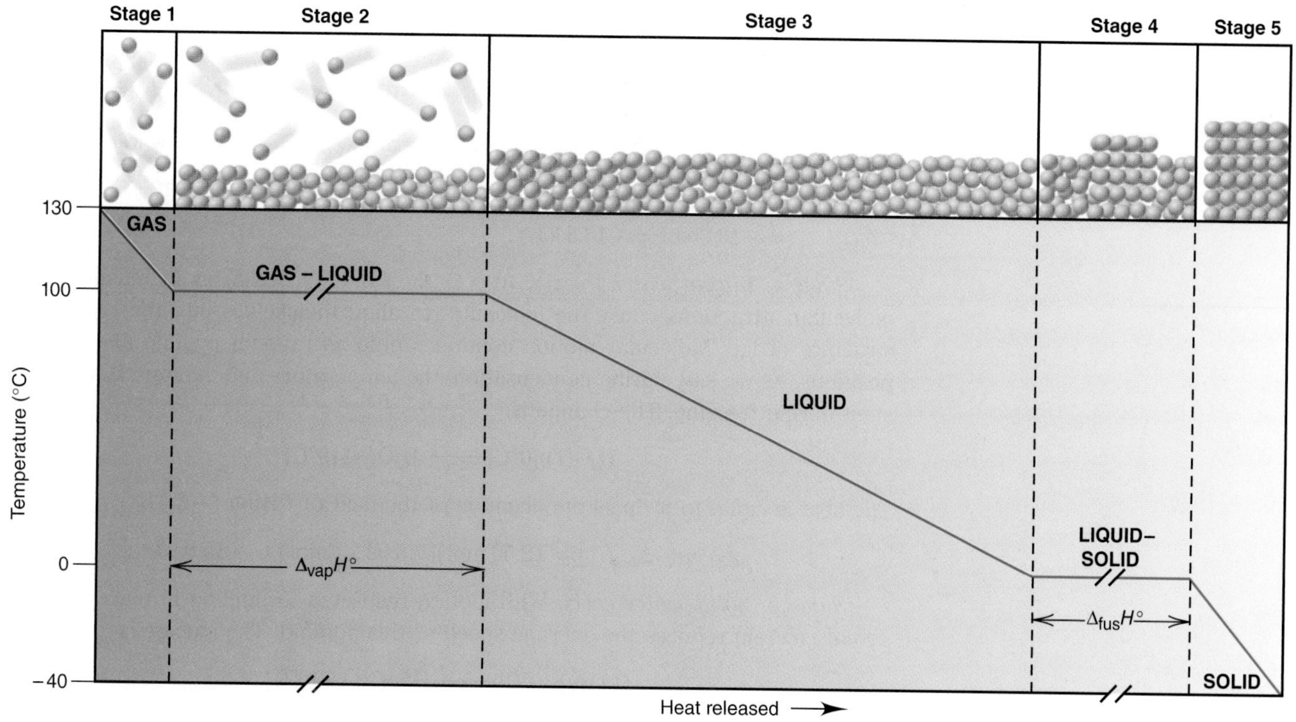

FIGURE 11.4 A cooling curve for the conversion of gaseous water to ice. A plot of temperature versus heat released as gaseous water changes to ice is shown, with a molecular-level depiction for each stage. The slopes of the lines in stages 1, 3, and 5 reflect the molar heat capacities of the phases. Although not drawn to scale, the line in stage 2 is longer than the line in stage 4 because $\Delta_{vap}H°$ of water is greater than $\Delta_{fus}H°$. A plot of temperature versus heat absorbed, starting at −40°C, would have the same steps but in reverse order. *Note:* The result of the calculation for each stage is shown with three significant figures. Recall that you do not round off until the very end, which, in this example, would be the total heat released in the process.

temperature falls, the average E_k decreases and attractions become more important. The change is

$$H_2O(g)\ [130°C] \longrightarrow H_2O(g)\ [100°C]$$

The heat (q) is the product of the amount (number of moles, n) of water; the molar heat capacity of *gaseous* water, $C_{water(g)}$; and the temperature change, ΔT ($T_{final} - T_{initial}$):

$$q = n \times C_{water(g)} \times \Delta T = (2.50\ mol)[33.1\ J/(mol{\cdot}°C)](100°C - 130°C)$$
$$= -2482\ J = -2.48\ kJ$$

The minus sign indicates that heat is released.*

Stage 2: Gaseous water condenses. At the condensation point, intermolecular attractions cause the slowest of the molecules to aggregate into microdroplets and then a bulk liquid. Note the following observation during the phase change:

- The temperature of the sample, and thus its average E_k, is constant. At the same temperature, molecules move farther between collisions in a gas than they do in a liquid, but their *average* speed is the same.
- Releasing heat from the sample decreases the average E_p as the molecules approach each other.

Thus, at 100°C, gaseous and liquid water have the same average E_k, but the liquid has lower average E_p. The change is

$$H_2O(g)\ [100°C] \longrightarrow H_2O(l)\ [100°C]$$

The heat is the amount (n) times the negative of the heat of vaporization ($-\Delta_{vap}H°$):

$$q = n(-\Delta_{vap}H°) = (2.50\ mol)(-40.7\ kJ/mol) = -102\ kJ$$

This stage contributes the *greatest portion of the total heat released* because of the large decrease in E_p as the molecules become so much closer in the liquid than they were in the gas.

Stage 3: Liquid water cools. The molecules in the liquid state continue to lose heat, which appears as a decrease in temperature, that is, as a decrease in the most

*For the purposes of cancelling, the units for molar heat capacity, C, include °C, rather than K, but this does not affect the magnitude of C because we are using ΔT and not T. Since the size of the temperature increments is the same in the Kelvin and Celsius scales, ΔT has the same magnitude in both scales.

probable molecular speed and, thus, the average E_k. The temperature decreases as long as the sample remains liquid. The change is

$$H_2O(l)\ [100°C] \longrightarrow H_2O(l)\ [0°C]$$

The heat depends on amount (n), the molar heat capacity of *liquid* water, and ΔT:

$$q = n \times C_{water(l)} \times \Delta T = (2.50 \text{ mol})[75.4 \text{ J/(mol·°C)}](0°C - 100°C)$$
$$= -18\,850 \text{ J} = -18.8 \text{ kJ}$$

Stage 4: Liquid water freezes. At 0°C, the sample loses E_p as increasing intermolecular attractions cause the molecules to align themselves into the crystalline structure of ice. Molecular motion continues only as random jiggling about fixed positions. As we saw during condensation, the temperature and average E_k are constant during freezing. The change is

$$H_2O(l)\ [0°C] \longrightarrow H_2O(s)\ [0°C]$$

The heat is equal to n times the negative of the heat of fusion ($-\Delta_{fus}H°$):

$$q = n(-\Delta_{fus}H°) = (2.50 \text{ mol})(-6.02 \text{ kJ/mol}) = -15.0 \text{ kJ}$$

Stage 5: Solid water cools. With motion restricted to jiggling in place, further cooling merely reduces the average speed of this jiggling. The change is

$$H_2O(s)\ [0°C] \longrightarrow H_2O(s)\ [-40°C]$$

The heat depends on n, the molar heat capacity of *solid* water, and ΔT:

$$q = n \times C_{water(s)} \times \Delta T = (2.50 \text{ mol})[37.6 \text{ J/(mol·°C)}](-40°C - 0°C)$$
$$= -3760 \text{ J} = -3.76 \text{ kJ}$$

According to Hess's law, the total heat released is the sum of the heats released for the individual stages.* The sum of q for stages 1 to 5 is −142 kJ. Two key points stand out for this or any similar process, whether exothermic or endothermic:

- *Within a phase,* heat flow is accompanied by *a change in temperature,* which is associated with a change in average E_k, as *the most probable speed of the molecules changes.* The heat released or absorbed depends on the amount of substance, the molar heat capacity *for that phase,* and the change in temperature.
- *During a phase change,* heat flow occurs at a *constant temperature,* which is associated with a change in average E_p, as *the average distance between molecules changes.* Both phases are present and (as you will see below) are in equilibrium during the change. The heat released or absorbed depends on the amount of substance and the enthalpy change for that phase change.

Let us work a molecular-scene sample problem to illustrate these ideas.

Sample Problem 11.1 Finding the Heat of a Phase Change Depicted by Molecular Scenes

Problem The scenes below represent a phase change of water. Select data from the previous discussion to find the heat (kJ) released or absorbed when 24.3 g of H_2O undergoes this phase change.

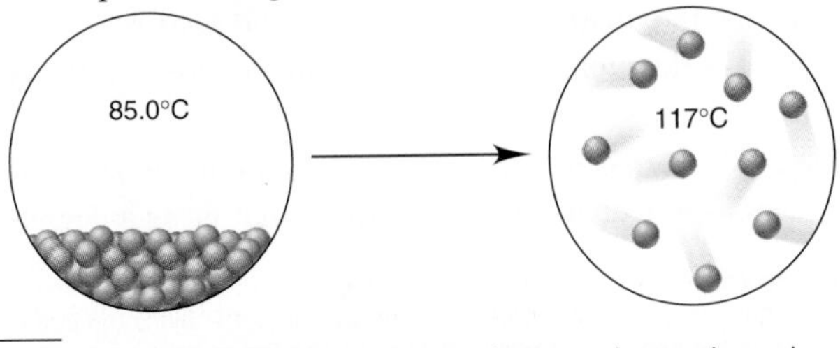

*Recall that Hess's law is a law for enthalpy. In this case, however, since work equals zero, the heat is mathematically equal to the enthalpy.

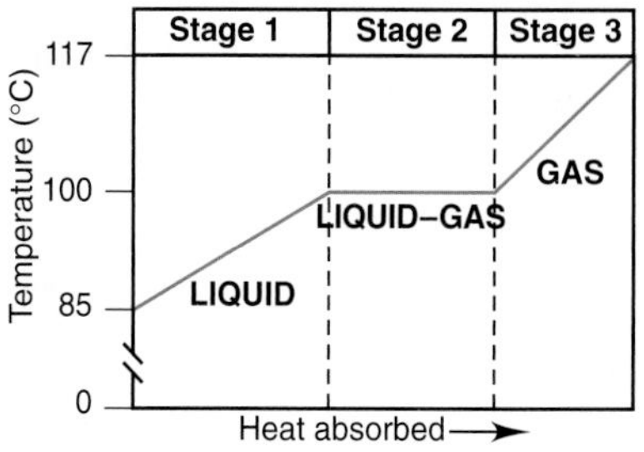

Plan From the molecular scenes, data from the discussion, and the given mass (24.3 g) of water, we have to find the heat that accompanies this change. The scenes show a disorderly condensed phase at 85.0°C changing to separate molecules at 117°C. Thus, the phase change they depict is vaporization, an endothermic process. The data in the discussion are given per mole, so we first convert the mass (g) of water to amount (mol). There are three stages: (1) heating the liquid from 85.0°C to 100.°C, (2) converting liquid water at 100.°C to gaseous water at 100.°C, and (3) heating the gas from 100.°C to 117°C (*see margin*). We add the values of q for these stages to obtain the total heat.

Solution Convert from mass (g) of H_2O to amount (mol):

$$\text{Amount (mol) of } H_2O = 24.3 \text{ g } H_2O \times \frac{1 \text{ mol}}{18.02 \text{ g } H_2O} = 1.349 \text{ mol}$$

Find the flow accompanying stage 1, $H_2O(l)$ [85.0°C] $\longrightarrow$ $H_2O(l)$ [100.°C]:

$$q = n \times C_{\text{water}(l)} \times \Delta T = (1.349 \text{ mol})[75.4 \text{ J/(mol·°C)}](100.°\text{C} - 85.0°\text{C})$$
$$= 1526 \text{ J} = 1.526 \text{ kJ}$$

Find the heat accompanying stage 2, $H_2O(l)$ [100.°C] $\longrightarrow$ $H_2O(g)$ [100.°C]:

$$q = n(\Delta_{\text{vap}}H°) = (1.349 \text{ mol})(40.7 \text{ kJ/mol}) = 54.90 \text{ kJ}$$

Find the heat accompanying stage 3, $H_2O(g)$ [100.0°C] $\longrightarrow$ $H_2O(g)$ [117.°C]:

$$q = n \times C_{\text{water}(g)} \times \Delta T = (1.349 \text{ mol})[35.1 \text{ J/(mol·°C)}](117°\text{C} - 100.°\text{C})$$
$$= 804.9 \text{ J} = 0.805 \text{ kJ}$$

Add the three heats together to find the total heat of the process:

$$\text{Total heat (kJ)} = 1.526 \text{ kJ} + 54.90 \text{ kJ} + 0.805 \text{ kJ} = 57.2 \text{ kJ}$$

Check The heat should have a positive value because it is absorbed. Be sure to round to check each value of q; for example, in stage 1, 1.33 mol × 75 J/(mol·°C) × 15°C = 1500 J. Note that the phase change itself (stage 2) requires the most energy and, thus, dominates the final answer. The $\Delta_{\text{vap}}H°$ units include kJ, whereas the molar heat capacity units include J, which is a thousandth as large.

Follow-Up Problem 11.1 The scenes below represent a phase change of water. Select data from the discussion to find the heat (kJ) released or absorbed when 2.25 mol of H_2O undergoes this change. (See Brief Solutions to Follow-Up Problems at the end of the chapter.)

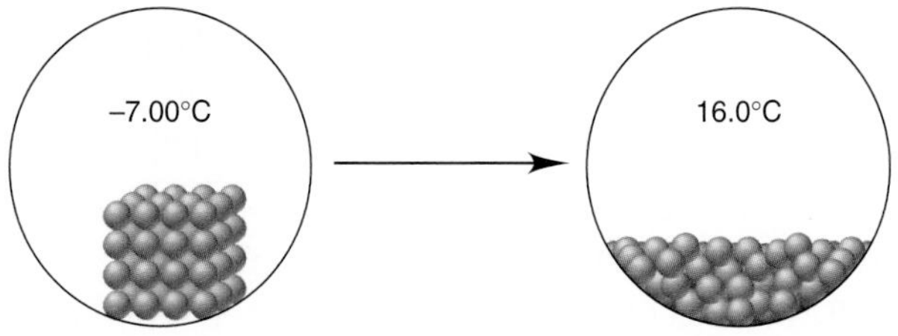

The Equilibrium Nature of Phase Changes

In everyday experiences, phase changes take place in *open* containers—the outdoors, a pot on a stove, the freezer compartment of a refrigerator—so they are not reversible. In a *closed* container, however, *phase changes **are** reversible and reach equilibrium*, just as chemical changes do. In this discussion, we examine the three phase equilibria.

Liquid-Gas Equilibria Vaporization and condensation are familiar events. Let us see how these processes differ in open and closed systems of a liquid in a flask:

1. *Open system: nonequilibrium process.* Picture an *open* flask containing a pure liquid at constant temperature. Within their range of speeds, some molecules at the surface have a high enough E_k to overcome attractions and vaporize. Nearby molecules fill the gap, and, with heat supplied by the constant-temperature surroundings, the process continues until the entire liquid phase is gone.

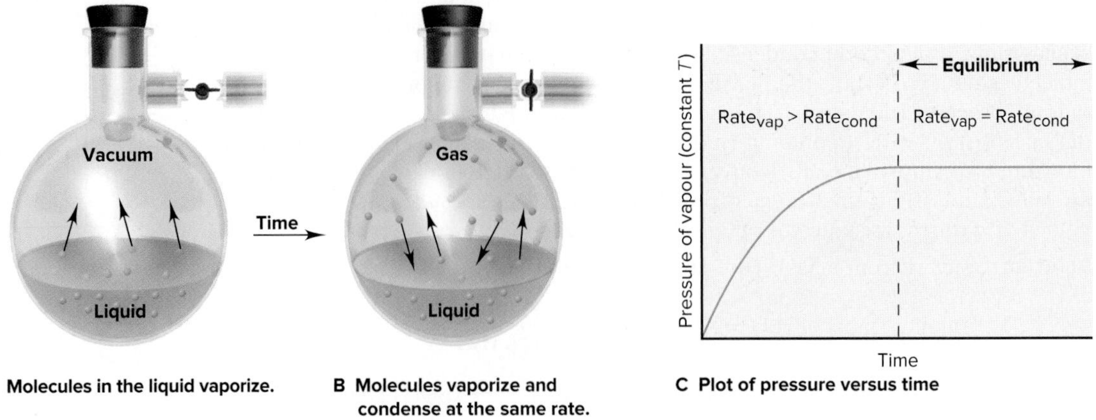

FIGURE 11.5 Liquid-gas equilibrium. A. Molecules leave the surface at a constant rate, and the pressure rises. **B.** At equilibrium, the same number of molecules leave and enter the liquid in a given time. **C.** The pressure increases until, at equilibrium, it is constant.

2. *Closed system: equilibrium process.* Now picture a *closed* flask at constant temperature and assume that a vacuum exists above a liquid (Figure 11.5A). Two processes take place: (1) Some molecules at the surface have a high enough E_k to *vaporize.* (2) After a short time, molecules in the vapour collide with the surface, and the slower molecules are attracted strongly enough to *condense.*

At first, these two processes occur at different rates. The number of molecules in a given surface area is constant, so the number of molecules leaving the surface area per unit of time—the rate of vaporization—is also constant, and the pressure increases. Over time, the number of molecules colliding with and entering the surface area—the rate of condensation—increases as the vapour becomes more populated, so the increase in pressure slows. Eventually, the rate of condensation equals the rate of vaporization; from this time onward, *the pressure is constant* (Figure 11.5B).

Macroscopically, the situation at this point seems static, but, at the molecular level, molecules are entering and leaving the liquid at equal rates. The system has reached a state of *dynamic equilibrium:*

$$\text{liquid} \rightleftharpoons \text{gas}$$

The pressure exerted by the vapour at equilibrium is called the *equilibrium vapour pressure*, or just the **vapour pressure**, of the liquid at that temperature. Figure 11.5C depicts the entire process graphically. If we started with a larger flask, the number of molecules in the vapour would be greater at equilibrium. However, as long as the temperature is constant and some liquid is present, the vapour pressure will be the same.

3. *Disturbing a system at equilibrium.* Let us see what happens if we alter certain conditions. This is called "disturbing" the system:

- *Decrease in pressure.* Suppose that we pump some vapour out of the flask, immediately lowering the pressure. (In a cylinder fitted with a piston, we lower the pressure by moving the piston outward, thus increasing the volume.) The rate of condensation temporarily falls below the rate of vaporization (the forward process is faster) because fewer molecules enter the liquid than leave it. The pressure rises until, after a short time, the condensation rate increases enough for equilibrium to be reached again.
- *Increase in pressure.* Suppose that we pump more vapour in (or move the piston inward, thus decreasing the volume), thereby immediately raising the pressure. The rate of condensation temporarily exceeds the rate of vaporization because more molecules enter the liquid than leave it (the reverse process is faster). Soon, however, the condensation rate decreases until the pressure again reaches the equilibrium value.

This general behaviour of a liquid and its vapour is seen in any system at equilibrium: *when a system at equilibrium is disturbed, it counteracts the disturbance until it reestablishes equilibrium.* We will return to this key idea often in later chapters.

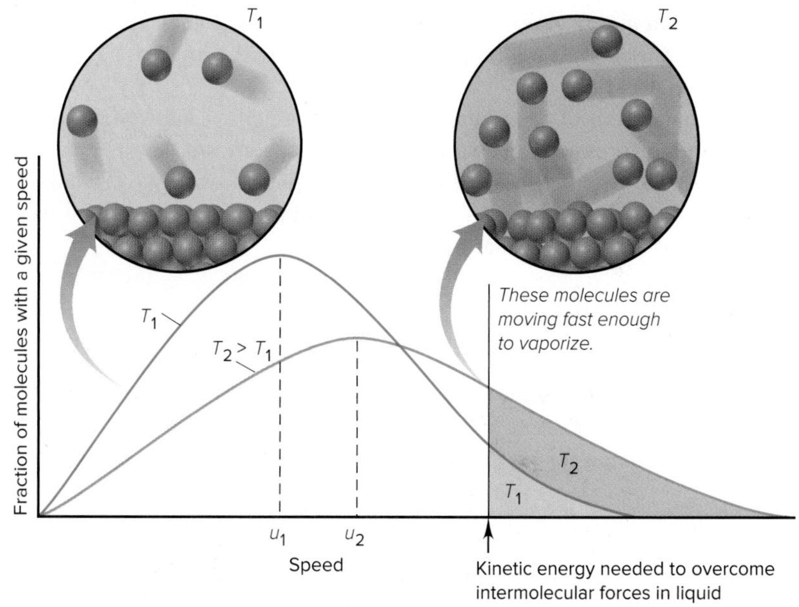

FIGURE 11.6 The effect of temperature on the distribution of molecular speeds

The Effects of Temperature and Intermolecular Forces on Vapour Pressure While the vapour pressure is not affected by the removal or addition of gas, it is affected by two other factors: a change in temperature and a change in the gas, that is, in the types of intermolecular forces.

1. *Effect of temperature.* Temperature has a major effect on vapour pressure because it changes the fraction of molecules moving fast enough to escape the liquid and, by the same token, the fraction moving slowly enough to be recaptured. In Figure 11.6, we see the familiar skewed bell-shaped curve of the distribution of molecular speeds (also see Figure 4.14). At the higher temperature, T_2, more molecules have enough energy to leave the surface. Thus, in general, ***the higher the temperature, the higher the vapour pressure***:

$$\text{higher } T \Longrightarrow \text{higher } p$$

2. *Effect of intermolecular forces.* At a given T, all substances have the same average E_k. Therefore, molecules with weaker intermolecular forces are held less tightly at the surface and vaporize more easily. In general, ***the weaker the intermolecular forces, the higher the vapour pressure***:

$$\text{weaker forces} \Longrightarrow \text{higher } p$$

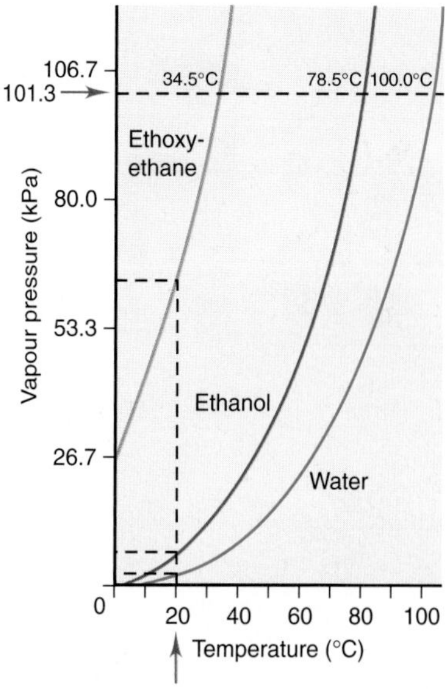

FIGURE 11.7 Vapour pressure as a function of temperature and intermolecular forces

Figure 11.7 shows the vapour pressure of three liquids as a function of temperature:

- The effect of temperature is seen in the steeper rise as the temperature increases.
- The effect of intermolecular forces is seen in the values of the vapour pressure, the short horizontal dashed lines intersecting the vertical (pressure) axis at a given temperature *(vertical dashed line at 20°C)*: the intermolecular forces in ethoxyethane (highest vapour pressure) are weaker than those in ethanol, which are weaker than those in water (lowest vapour pressure).

Quantifying the Effect of Temperature The nonlinear relationship between p and T is converted into a linear relationship with the **Clausius-Clapeyron equation**:

$$\ln p = \frac{-\Delta_{vap}H}{R}\left(\frac{1}{T}\right) + C$$
$$y = \quad m \quad x \quad + b$$

where $\ln p$ is the natural logarithm of the vapour pressure, $\Delta_{vap}H$ is the heat of vaporization, R is the universal gas constant [8.314 J/(mol•K)], T is the absolute temperature, and C is a constant (not related to heat capacity). The equation is often used to find the heat of vaporization. The equation for a straight line is shown under it

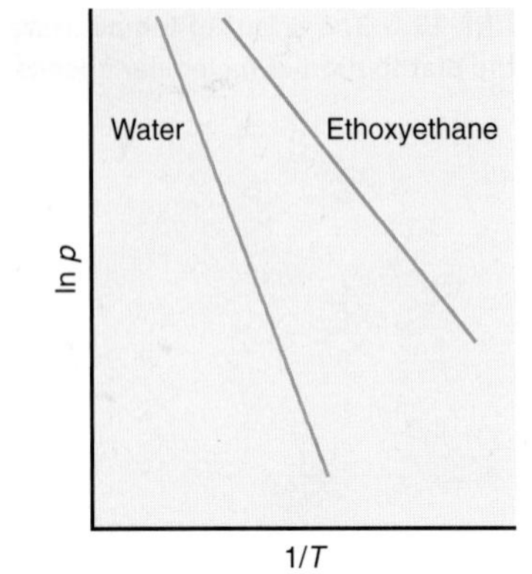

FIGURE 11.8 Linear plots of the relationship between vapour pressure and temperature. The slope is steeper for water because its $\Delta_{vap}H$ is greater.

in blue, with $y = \ln p$, $x = \frac{1}{T}$, m (the slope) $= \frac{-\Delta_{vap}H}{R}$, and b (the y-axis intercept) $= C$. A plot of ln p versus $\frac{1}{T}$ gives a straight line, as shown in the graph for ethoxyethane and water in Figure 11.8.

A two-point version of the Clausius-Clapeyron equation allows us to calculate $\Delta_{vap}H$ if the vapour pressures at two temperatures are known. Let us consider two points on one of the straight lines in Figure 11.8: at point 1, we have pressure p_1 and temperature T_1; at point 2, we have pressure p_2 and temperature T_2. We can then write the Clausius-Clapeyron equation for each point:

$$\text{Point 1:}\quad \ln p_1 = \frac{-\Delta_{vap}H}{R}\left(\frac{1}{T_1}\right) + C$$

$$\text{Point 2:}\quad \ln p_2 = \frac{-\Delta_{vap}H}{R}\left(\frac{1}{T_2}\right) + C$$

Subtracting the equation for point 1 from the equation for point 2, we get

$$\ln p_2 - \ln p_1 = \left[\frac{-\Delta_{vap}H}{R}\left(\frac{1}{T_1}\right) + C\right] + \left[\frac{\Delta_{vap}H}{R}\left(\frac{1}{T_2}\right) - C\right]$$

Simplifying and reorganizing this equation, we get the two-point Clausius-Clapeyron equation:

$$\ln\frac{p_2}{p_1} = \frac{-\Delta_{vap}H}{R}\left(\frac{1}{T_2} - \frac{1}{T_1}\right) \tag{11.1}$$

If $\Delta_{vap}H$ and p_1 at T_1 are known, we can also calculate the vapour pressure (p_2) at any other temperature (T_2) or the temperature at any other pressure.

Sample Problem 11.2 Applying the Clausius-Clapeyron Equation

Problem The vapour pressure of ethanol is 15.33 kPa at 34.9°C. If $\Delta_{vap}H$ of ethanol is 40.5 kJ/mol, calculate the temperature (°C) when the vapour pressure is 101.3 kPa.

Plan We are given $\Delta_{vap}H$, p_1, p_2, and T_1. We substitute them into Equation 11.1 to solve for T_2. Here, the value of R is 8.314 J/(mol•K), so we must convert T_1 to K to obtain T_2, and then convert T_2 back to °C.

Solution Substitute the values into Equation 11.1 and solve for T_2:

$$T_1 = 34.9°\text{C} + 273.15 = 308.0 \text{ K}$$

$$\ln\frac{p_2}{p_1} = \frac{-\Delta_{vap}H}{R}\left(\frac{1}{T_2} - \frac{1}{T_1}\right)$$

$$\ln\frac{101.3 \text{ kPa}}{15.33 \text{ kPa}} = \left(-\frac{40.5\times10^3 \text{ J/mol}}{8.314 \text{ J/[mol•K]}}\right)\left(\frac{1}{T_2} - \frac{1}{308.0 \text{ K}}\right)$$

$$1.888 = (-4.871\times10^3 \text{ K})\left[\frac{1}{T_2} - (3.247\times10^{-3}\text{ K}^{-1})\right]$$

$$T_2 = 350. \text{ K}$$

Convert T_2 from K to °C:

$$T_2 = 350. \text{ K} - 273.15 = 77°\text{C}$$

Check Round off to check the math. The change is in the right direction: higher p should occur at higher T. As we discuss next, a substance has a vapour pressure of 101.3 kPa at its *normal boiling point*. Checking the *CRC Handbook of Chemistry and Physics* shows that the boiling point of ethanol is 78.5°C, very close to our answer.

Follow-Up Problem 11.2 At 34.1°C, the vapour pressure of water is 5.346 kPa. What is the vapour pressure at 85.5°C? The $\Delta_{vap}H$ of water is 40.7 kJ/mol.

Vapour Pressure and Boiling Point Let us discuss what is happening when a liquid boils and then see the effect of pressure on boiling point.

1. *How a liquid boils.* In an *open* container, the weight of the atmosphere bears down on a liquid surface. As the temperature rises, molecules move more quickly throughout the liquid. At some temperature, the average E_k of the molecules in the liquid is great enough for them to form bubbles of vapour *in the interior*, and the liquid boils. At any lower temperature, the bubbles collapse as soon as they start to form because the external pressure is greater than the vapour pressure inside the bubbles. Thus, the **boiling point** *is the temperature at which the vapour pressure equals the external pressure,* which is usually the pressure of the atmosphere. As in condensation and freezing, once boiling begins, the temperature of the liquid remains constant until all of the liquid is gone.

2. *Effect of pressure on boiling point.* The boiling point of a liquid varies with elevation. At high elevations, a lower atmospheric pressure is exerted on the liquid surface, so molecules in the interior need less kinetic energy to form bubbles. Thus, *the boiling point depends on the applied pressure.*

In mountainous regions, food takes *more* time to cook because the boiling point is lower and the boiling liquid is not as hot; for instance, in Lake Louise, Alberta (elevation 1534 m), water boils at 94.7°C. On the other hand, in a pressure cooker, food takes *less* time to cook because the boiling point is higher at the higher pressure. The *normal boiling point* is observed at standard atmospheric pressure (101.3 kPa; *long, horizontal dashed line* in Figure 11.7).

Solid-Liquid Equilibria The particles in a solid are continuously jiggling about their fixed positions. As the temperature rises, the particles jiggle more rapidly, until some have enough kinetic energy to break free of their positions. At this point, melting begins. As more molecules enter the liquid (molten) phase, some collide with the solid and become fixed in position again. Because the phases remain in contact, a dynamic equilibrium is established when the melting rate equals the freezing rate. The temperature at which this occurs is called the **melting point**. The temperature remains fixed at the melting point until all the solid melts.

Because liquids and solids are nearly incompressible, pressure has little effect on the rates of melting and freezing: a plot of pressure versus temperature for a solid-liquid phase change is typically a *nearly* vertical straight line.

Solid-Gas Equilibria Sublimation is not very familiar because solids have *much* lower vapour pressures than liquids. A substance sublimes rather than melts because the intermolecular attractions are not great enough to keep the molecules near one another when they leave the solid state. Some solids *do* have high enough vapour pressures to sublime at ordinary conditions. Three of these solids are dry ice (carbon dioxide), iodine (Figure 11.9), and moth repellant, all nonpolar molecules with weak intermolecular forces.

The plot of pressure versus temperature for a solid-gas phase change reflects the large effect of temperature on vapour pressure; thus, it resembles the liquid-gas curve because it rises steeply with higher temperatures.

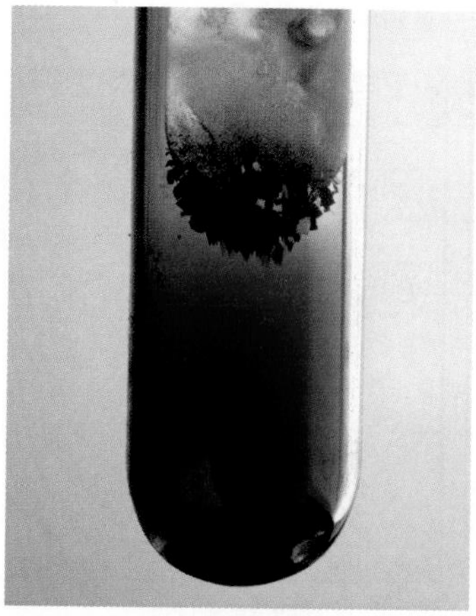

FIGURE 11.9 Iodine subliming. After the solid sublimes, I_2 vapour deposits on a cold surface (water-filled inner test tube).

Phase Diagrams: Effects of Pressure and Temperature on Physical State

The **phase diagram** of a substance combines the liquid-gas, solid-liquid, and solid-gas curves and gives the conditions of temperature and pressure at which each phase is stable and where phase changes occur.

The Phase Diagram for CO_2 and Most Substances The diagram for CO_2, which is typical of most substances, has four general features (Figure 11.10):

1. *Regions of the diagram.* Each region presents the conditions of pressure and temperature for which the phase is stable. If another phase is placed under those conditions, it will change to the stable phase. In general, the solid is stable at low

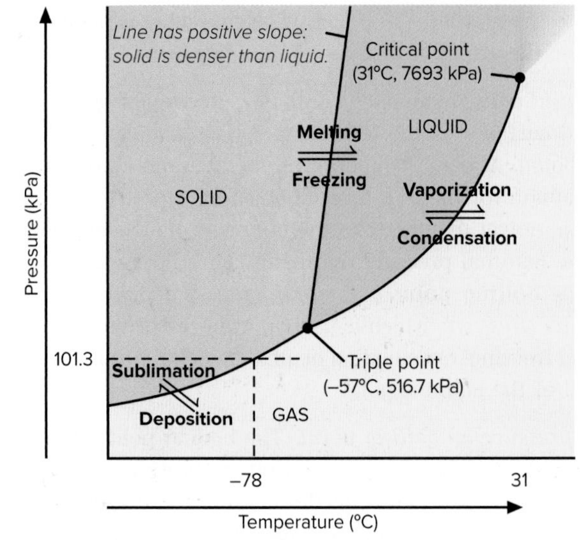

FIGURE 11.10 Phase diagram for CO_2. (The slope of the solid-liquid line is exaggerated, and the axes are not linear in both this figure and Figure 11.13.)

temperature and high pressure, the gas is stable at high temperature and low pressure, and the liquid is stable at intermediate conditions.

2. *Lines between regions.* The lines are the phase-transition curves discussed earlier. Any point along a line shows the pressure and temperature at which the phases are in equilibrium. The solid-liquid line has a slightly *positive* slope (slants to the *right* with increasing pressure) because, for most substances, the solid is more dense than the liquid: an increase in pressure converts the liquid to the solid. (Water is *the* major exception.)

3. *The triple point.* The three phase-transition curves meet at the **triple point**, at which all three phases are in equilibrium. As strange as it sounds, at the triple point in Figure 11.10, CO_2 is subliming and depositing, melting and freezing, and vaporizing and condensing simultaneously! Substances with several solid and/or liquid forms can have more than one triple point.

The CO_2 phase diagram shows why dry ice (solid CO_2) does not melt under ordinary conditions. The triple-point pressure is 516.8 kPa, so liquid CO_2 does not occur at 101.3 kPa because it is not stable. The horizontal dashed line at 101.3 kPa crosses the solid-gas line. Therefore, when solid CO_2 is heated, it sublimes at −78°C rather than melts. If our normal atmospheric pressure were 526.9 kPa, liquid CO_2 *would* occur.

4. *The critical point.* Heat a liquid in a closed container, and its density decreases. At the same time, more of the liquid vaporizes, so the density of the vapour increases. At the **critical point**, the two densities become equal and the phase boundary disappears. The temperature at the critical point is the *critical temperature* (T_c), and the pressure is the *critical pressure* (p_c). The average E_k is so high at this point that the vapour cannot be condensed at any pressure. The two most common gases in air have critical temperatures far below room temperature: O_2 cannot be condensed above −119°C, and N_2 cannot be condensed above −147°C.

Beyond the critical temperature, a *supercritical fluid* (SCF) exists, rather than separate liquid and gaseous phases. An SCF expands and contracts like a gas and has unusual solvent properties. Since an SCF has the low viscosity of a gas and the high density of a liquid, it cannot be liquefied using any amount of pressure. However, an SCF can be liquefied by lowering the temperature, as shown in Figure 11.11.

Supercritical CO_2 is used to extract caffeine from coffee beans, nicotine from tobacco, and fats from potato chips, and it is used as a dry-cleaning agent. Lower the pressure, and the SCF disperses as a harmless gas. Supercritical H_2O dissolves nonpolar substances, even though liquid water cannot. Studies are under way to use supercritical H_2O to remove nonpolar organic toxins, such as PCBs, from

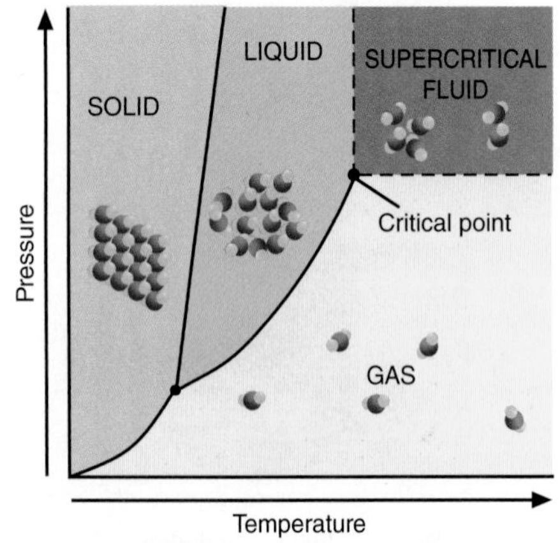

FIGURE 11.11 Phase diagram showing an SCF

FIGURE 11.12 Robert C. Burk (Carleton University, Ottawa) currently performs research related to SCF extraction from solid and liquid matrices, cloud point extraction of hydrophilic analyses from water and biological matrices, the uses of carbon nanotubes for water remediation and analysis, and stir bar sorption extraction for water analysis.

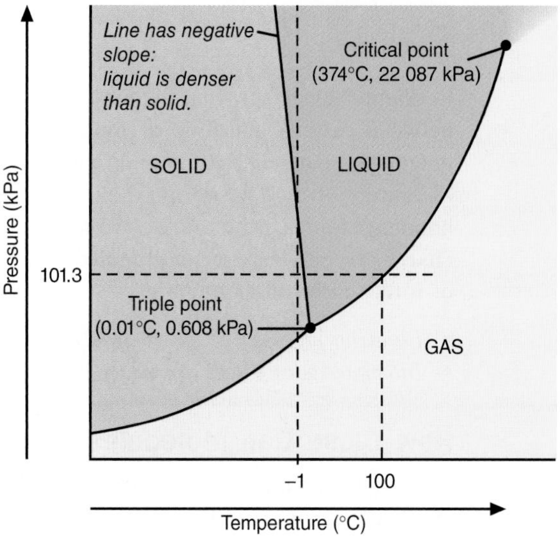

FIGURE 11.13 Phase diagram for H_2O

industrial waste. Several Canadian researchers are studying SCFs. One of them is Robert C. Burk (Figure 11.12), who is interested in the physical chemistry underlying the transfer of organic and inorganic solutes from various matrices to supercritical carbon dioxide.

The Solid-Liquid Line for Water The phase diagram for water differs from other phase diagrams in one major way, which reveals a key property. Unlike almost any other substance, the solid form is *less dense* than the liquid form; that is, *water expands upon freezing*. Thus, the solid-liquid line has a *negative* slope (slants to the *left* with increasing pressure): an increase in pressure converts the solid to the liquid; the higher the pressure, the lower the temperature at which water freezes (Figure 11.13). The vertical dashed line at −1°C crosses the solid-liquid line, which means that ice melts with only an increase in pressure.

The triple point of water occurs at a low pressure, 0.608 kPa. Therefore, when solid water is heated at 101.3 kPa (*horizontal dashed line*), the solid-liquid line is crossed at 0°C, the normal melting point. Thus, ice melts rather than sublimes. The horizontal dashed line then crosses the liquid-gas curve at 100°C, the normal boiling point. The unique properties of water will be discussed in more detail in Section 11.5.

SUMMARY OF SECTION 11.2

- A heating-cooling curve depicts the change in temperature when a substance absorbs or releases heat at a constant rate. Within a phase, temperature (and average E_k) change. During a phase change, temperature (and average E_k) are constant, but E_p changes. The total enthalpy change for the system is found using Hess's law.
- In a closed container, the liquid and gas phases of a substance reach equilibrium. The vapour pressure, the pressure of the gas at equilibrium, is related directly to temperature and inversely to the strength of the intermolecular forces.
- The Clausius-Clapeyron equation relates vapour pressure to temperature and is often used to find $\Delta_{vap}H$.
- A liquid in an open container boils when its vapour pressure equals the external pressure.
- Solid-liquid equilibrium occurs at the melting point. Some solids sublime because they have very weak intermolecular forces.
- The phase diagram of a substance shows the phase that is stable at any p and T, the conditions at which phase changes occur, and the conditions at the critical point and the triple point. Water differs from most substances in that its solid phase is less dense than its liquid phase, so its solid-liquid line has a negative slope.

11.3 Types of Intermolecular Forces

In Chapter 8, we saw that bonding (*intra*molecular) forces are due to the attraction between cations and anions (ionic bonding), nuclei and electron pairs (covalent bonding), or metal cations and delocalized electrons (metallic bonding). However, the nature of the phases and their changes is due primarily to *inter*molecular (nonbonding) forces, which arise from the attraction between molecules with partial charges or between ions and molecules. Coulomb's law explains why the two types of forces differ in magnitude:

- *Bonding forces are relatively strong* because larger charges are closer together.
- *Intermolecular forces are relatively weak* because smaller charges are farther apart.

How Close Can Molecules Approach Each Other?

To see the minimum distance *between* molecules, consider solid Cl_2. When we measure the distances between two Cl nuclei, we obtain two different values (Figure 11.14A):

- *Bond length and covalent radius.* The shorter distance, called the *bond length*, is between *two bonded Cl atoms in the* ***same*** *molecule.* One-half this distance is the *covalent radius.*
- *Van der Waals distance and radius.* The longer distance is between *two nonbonded Cl atoms in* ***adjacent*** *molecules.* It is called the *van der Waals (VDW) distance.* At this distance, intermolecular attractions balance electron-cloud repulsions; thus, the VDW distance is as close as one Cl_2 molecule can approach another. The **van der Waals radius** is one-half the closest distance between the nuclei of identical *nonbonded* atoms. The VDW radius of an atom is *always larger than its covalent radius.* Like covalent radii, VDW radii decrease across a period and increase down a group (Figure 11.14B).

As we discuss intermolecular forces (also called *van der Waals forces*), consult Table 11.2, which compares intermolecular forces with bonding forces.

Ion-Dipole Forces

When an ion and a nearby polar molecule (dipole) attract each other, an **ion-dipole force** results. The most important example takes place when an ionic compound dissolves in water. The ions become separated because the attractions between the ions and the oppositely charged poles of the H_2O molecules are stronger than the attractions between the ions themselves. Ion-dipole forces in solutions, and their associated energy, will be discussed fully in Chapter 12.

FIGURE 11.14 Covalent and VDW radii and their periodic trends. A. The VDW radius is one-half the distance between adjacent *nonbonded* atoms ($\frac{1}{2}$ × VDW distance). **B.** Like covalent radii (*blue quarter-circles and numbers*), VDW radii (*red quarter-circles and numbers*) increase down a group and decrease across a period.

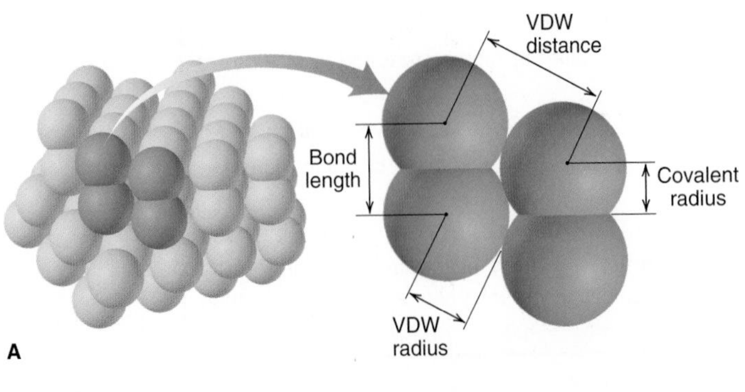

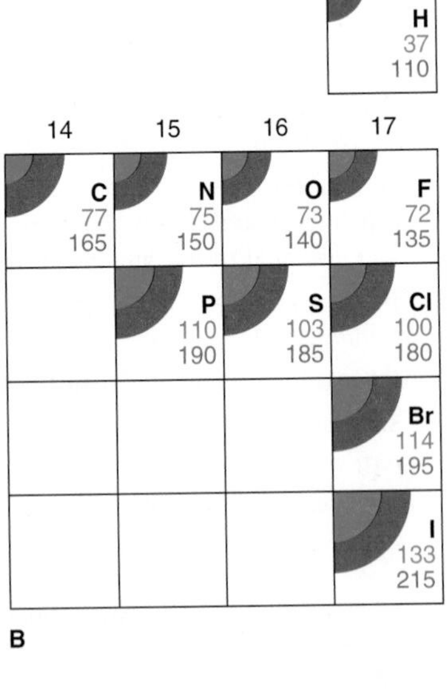

TABLE 11.2 Comparison of Bonding and Nonbonding (Intermolecular) Forces

Force	Model	Basis of Attraction	Energy (kJ/mol)	Example
Bonding				
Ionic	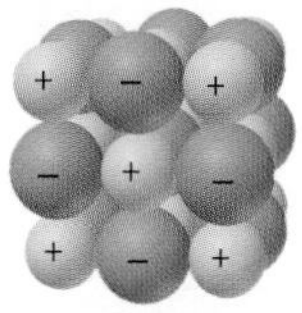	Cation-anion	400–4000	NaCl
Covalent	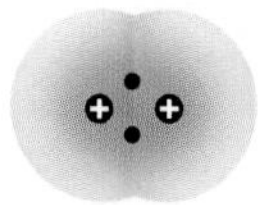	Nuclei–shared e^- pair	150–1100	H—H
Metallic	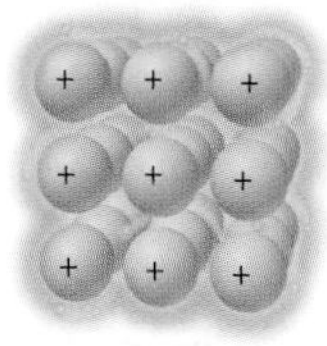	Cations–delocalized electrons	75–1000	Fe
Nonbonding (Intermolecular)				
Ion-dipole		Ion charge–dipole charge	40–600	Na^+···OH_2
Hydrogen bond	δ− δ+ δ− —A—H·······:B—	Polar bond to H–dipole charge (high χ of N, O, F)	10–40	:Ö—H···:Ö—H (each O bonded to H)
Dipole-dipole	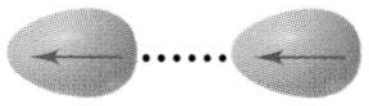	Dipole charges	5–25	I—Cl···I—Cl
Ion–induced dipole		Ion charge–polarizable e^- cloud	3–15	Fe^{2+}···O_2
Dipole–induced dipole		Dipole charge–polarizable e^- cloud	2–10	H—Cl···Cl—Cl
Induced dipole–induced dipole (dispersion or London)	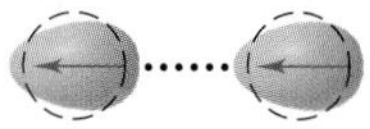	Polarizable e^- clouds	0.05–40	F—F···F—F

Dipole-Dipole Forces

In Figure 9.13, an external electric field orients gaseous polar molecules. The polar molecules in liquids and solids lie near each other, and their partial charges act as tiny electric fields and give rise to **dipole-dipole forces**: the positive pole of one molecule attracts the negative pole of another (Figure 11.15). The orientation is more orderly in a solid than in a liquid because the average kinetic energy is lower.

These forces depend on the magnitude of the molecular dipole moment. For compounds of similar molar masses, the greater the molecular dipole moment, the greater the dipole-dipole forces, so the more energy it takes to separate the molecules; thus, the boiling point is higher. Chloromethane, for instance, has a smaller dipole moment than ethanal (acetaldehyde) and boils at a lower temperature (Figure 11.16).

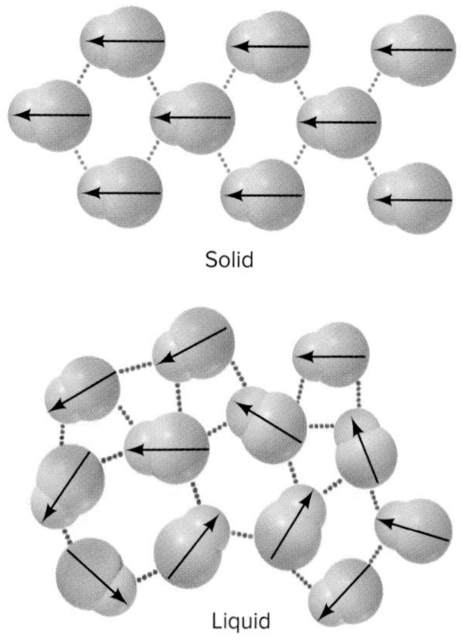

FIGURE 11.15 Polar molecules and dipole-dipole forces. Spaces between the molecules are exaggerated.

The Hydrogen Bond

A special type of dipole-dipole force arises between molecules that have *an H atom bonded to a small, highly electronegative atom with lone electron pairs, specifically N, O, or F.* The H—N, H—O, and H—F bonds are very polar. When the partially

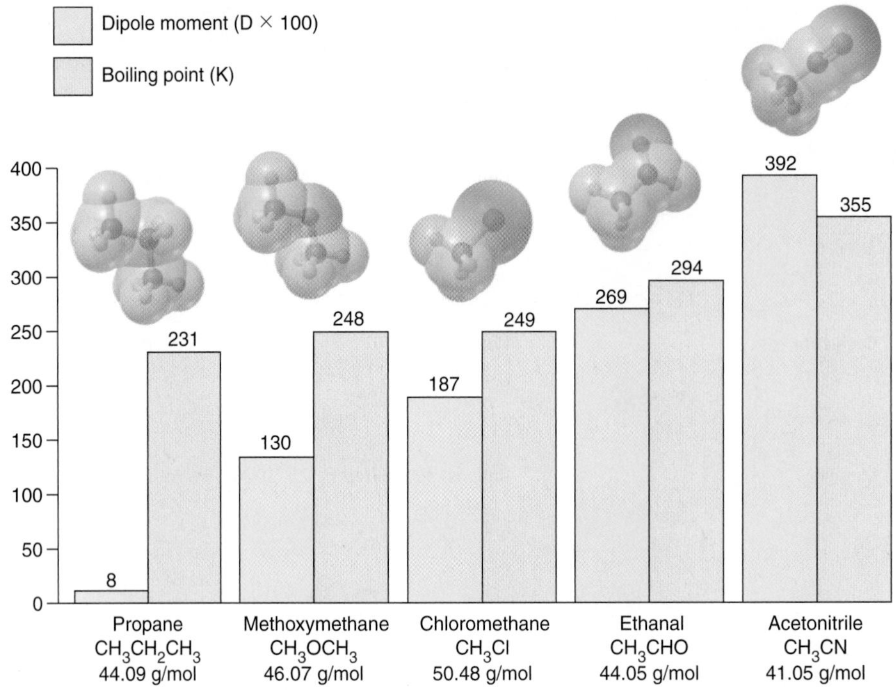

FIGURE 11.16 Dipole moment and boiling point. Relative strengths of the dipole moments are indicated by colour intensities in the electron-density models.

positive H of one molecule is attracted to the partially negative lone pair on the N, O, or F of another molecule, a **hydrogen bond** forms. Thus, the atom sequence of a hydrogen bond (dotted line) is —B:····H—A—, where A *and* B are N, O, or F. Three examples are shown below:

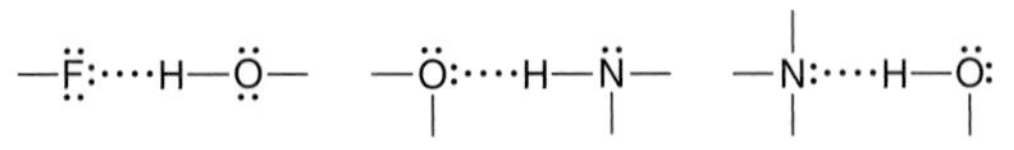

The small sizes of N, O, and F are essential to hydrogen bonding for two reasons:

1. The atoms are so electronegative that their covalently bonded H is highly positive.
2. The lone pair on the N, O, or F of the other molecule can come close to the H.

The only electron in the hydrogen atom is strongly bonded to the nucleus. The net force on the electron in the hydrogen atom is just as large as the electrical attraction from the nucleus, because there are no other electrons that might *shield* or decrease the interaction between the nucleus and the outermost electron. Thus, the electron will not be removed from the hydrogen atom in the previously mentioned molecules. This lack of *shielding effect* is a key factor in the formation of the hydrogen bond. Hydrogen bonds are basically electrostatic interactions and are weaker than covalent bonds. They are, however, the strongest kind of dipole-dipole interaction. The electronegative atom to which the hydrogen atom is bonded pulls electron density away from the hydrogen atom, developing a partially positive charge. Therefore, the hydrogen atom can interact with a partially negatively charged atom (F, O, or N) through an electrostatic interaction.

The Significance of Hydrogen Bonding Hydrogen bonding has a profound impact in many systems. We will examine one effect on physical properties and preview its importance in biological systems, which we will address in Chapter 22.

Figure 11.17 shows the effect of hydrogen bonding on the boiling points of the binary hydrides of groups 14 through 17. For reasons we will discuss shortly, boiling point rises with molar mass, as the group 14 hydrides show. However, the first member in each of the other groups—NH_3, H_2O, and HF—deviates enormously from this expected trend. Within samples of these substances, the molecules form strong hydrogen bonds, so it takes more energy for the molecules to separate and enter the gas phase. For example, on the basis of molar mass alone,

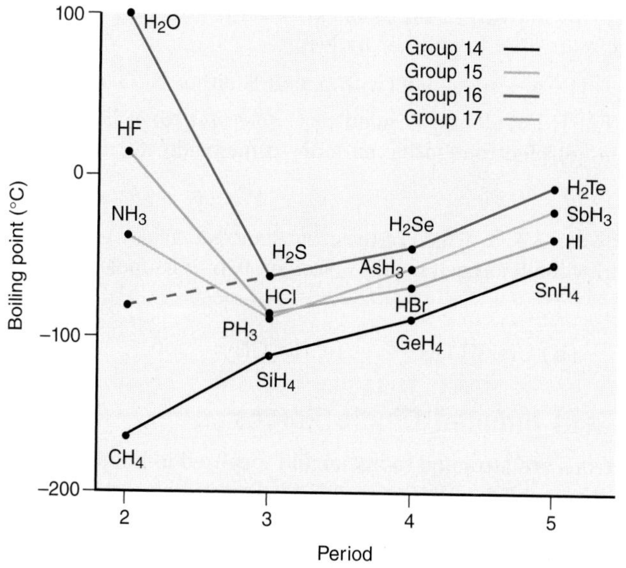

FIGURE 11.17 Hydrogen bonding and boiling point. NH_3, H_2O, and HF have exceptionally high boiling points because they form hydrogen bonds.

we would expect water to boil about 200°C lower than it actually does (*red dashed line*). (In Section 11.5, we will discuss the effects that the hydrogen bonds in water have in nature.)

The significance of hydrogen bonding in biological systems cannot be emphasized too strongly. In Chapter 22 you will see that it is a key feature in the structure and function of the biological macromolecules—proteins and nucleic acids. It is also responsible for the action of many *enzymes*, the proteins that speed metabolic reactions.

Sample Problem 11.3 Drawing Hydrogen Bonds between Molecules of a Substance

Problem Which substance(s) exhibit(s) hydrogen bonding? For any that do, draw the hydrogen bonds between two of its molecules.

(a) C_2H_6 **(b)** CH_3OH **(c)** $CH_3\overset{\overset{O}{\|}}{C}—NH_2$

Plan If the molecule does *not* contain N, O, or F, it cannot form hydrogen bonds. If it contains any of these atoms covalently bonded to hydrogen, we draw two molecules in the pattern.

Solution (a) C_2H_6: There is no N, O, or F, so no hydrogen bonds can form.

(b) CH_3OH: The H covalently bonded to the O in one molecule forms a hydrogen bond to the lone pair on the O of an adjacent molecule.

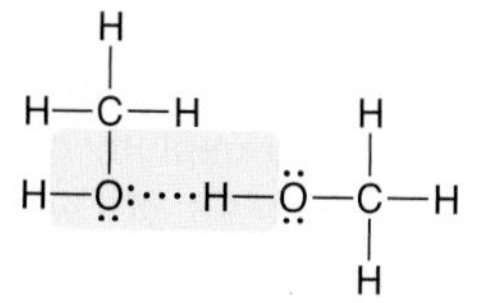

(c) $CH_3\overset{\overset{O}{\|}}{C}—NH_2$: Two of these molecules can form one hydrogen bond between an H bonded to N and the O, or they can form two such hydrogen bonds.

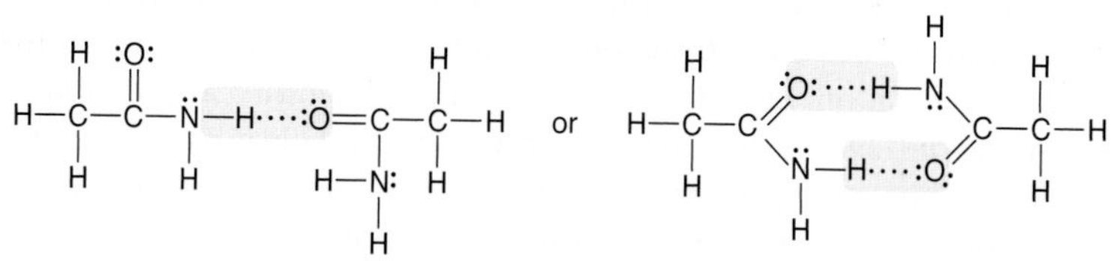

A third possibility (not shown) could be between an H attached to N in one molecule and the lone pair of N in another molecule.

Check The —B:····H—A— sequence (with A and B either N, O, or F) is present.

Comment Note that H covalently bonded to C *does not form hydrogen bonds* because carbon is not electronegative enough to make the C—H bond sufficiently polar.

Follow-Up Problem 11.3 Which of these substances exhibit(s) hydrogen bonding? For any that do, draw the hydrogen bond(s) between two of its molecules.

(a) $CH_3C(=O)—OH$ **(b)** CH_3CH_2OH **(c)** $CH_3C(=O)CH_3$

Polarizability and Induced Dipole Forces

Even though electrons are attracted to nuclei and localized in bonding and lone pairs, we often picture them as "clouds" of negative charge because they are in constant motion. A nearby electric field can *induce* a distortion in the cloud, pulling electron density toward a positive pole of a field or pushing it away from a negative pole:

- *For a nonpolar molecule,* the distortion induces a temporary dipole moment.
- *For a polar molecule,* it enhances the dipole moment already present.

In addition to charged plates connected to a battery, the source of the electric field can be the charge of an ion or the partial charges of a polar molecule.

How easily an electron cloud can be distorted is called its **polarizability**. Smaller atoms (or ions) are less polarizable than larger atoms (or ions) because their electrons are closer to the nucleus and therefore held more tightly. Thus, we observe the following trends:

- Polarizability *increases down a group* because atomic size increases and larger electron clouds are easier to distort.
- Polarizability *decreases across a period* because increasing Z_{eff} makes the atoms smaller and holds the electrons more tightly.
- Cations are *less* polarizable than their parent atoms because they are smaller; anions are *more* polarizable because they are larger.

Ion–induced dipole and dipole–induced dipole forces are the two types of charge-induced dipole forces; they are most important in solution, so we will focus on them in Chapter 12. Nevertheless, *polarizability affects all intermolecular forces.*

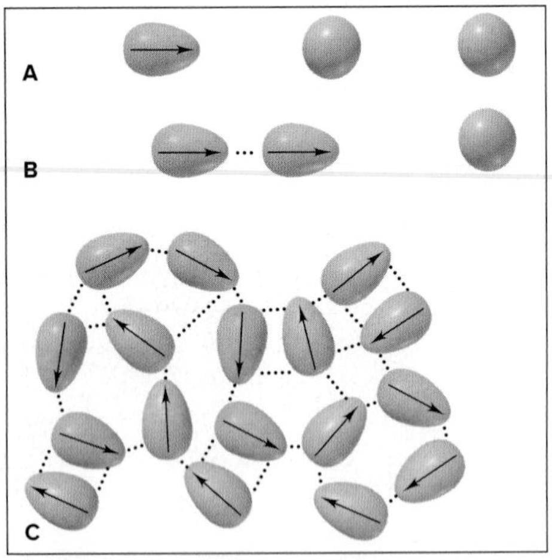

FIGURE 11.18 Dispersion forces among nonpolar particles. A. When atoms are far apart, an instantaneous dipole in one atom (*left*) does not influence the others. **B.** When atoms get closer together, the instantaneous dipole in one atom (*left*) induces a dipole in another (*centre*). **C.** The process occurs throughout the sample.

Dispersion (London) Forces

So far, we have discussed forces that depend on the existing charge of an ion or a polar molecule. But what forces cause nonpolar substances such as octane, chlorine, and argon to condense and solidify? As you will see, polarizability plays a central role in the most universal intermolecular force.

The intermolecular force responsible for the condensed states of nonpolar substances is the **dispersion force (London force)**. London forces are named for Fritz London, the physicist who explained their quantum-mechanical basis. Dispersion forces *are present between all particles (atoms, ions, and molecules)* because they are caused by the *motion of electrons in atoms.* Let us examine key aspects of this force:

1. *Source.* Picture one atom in a sample of, for example, argon gas. Over time, its 18 electrons are distributed uniformly, so the atom is nonpolar. However, at any instant, there may be more electrons on one side of the nucleus than on the other side, which gives the atom an *instantaneous dipole.* When two argon atoms are far apart, they do not influence each other, but when they are close together, *the instantaneous dipole in one atom induces a dipole in its neighbour,* and they attract each other. This process spreads to other atoms and throughout the sample. At low

temperatures, these attractions keep the atoms together (Figure 11.18). Thus, dispersion forces are *instantaneous induced dipole–induced dipole forces.*

2. *Prevalence.* Although they are the *only* force existing between nonpolar particles, *dispersion forces contribute to the energy of attraction in all substances* because they exist between *all* particles. In fact, except for the forces between small, highly polar molecules or between molecules forming hydrogen bonds, the *dispersion force is the dominant intermolecular force.* Calculations show, for example, that 85% of the attraction between HCl molecules is due to dispersion forces and only 15% to dipole-dipole forces. Even for water, 75% of the attraction comes from hydrogen bonds and 25% from dispersion forces.

3. *Relative strength.* The relative strength of dispersion forces depends on the polarizability of the particles, so they are weak for small particles, such as H_2 and He, but stronger for larger particles, such as I_2 and Xe. *Polarizability depends on the number of electrons, which correlates closely with molar mass* because heavier particles are either larger atoms or molecules with more atoms and, thus, more electrons. For this reason, as molar mass increases down the group 14 hydrides (see Figure 11.17), or down the halogens or the noble gases, dispersion forces increase and so do boiling points (Figure 11.19).

4. *Effect of molecular shape.* For a pair of nonpolar substances with the same molar mass, a molecular shape that has more area over which electron clouds can be distorted allows stronger attractions. For example, the two five-carbon alkanes, pentane and 2,2-dimethylpropane (neopentane), are structural isomers—the same molecular formula (C_5H_{12}) but different properties. In comparison, pentane is more cylindrical and 2,2-dimethylpropane is more spherical (Figure 11.20). Thus, two pentane molecules make more contact than two 2,2-dimethylpropane molecules do, so dispersion forces act at more points, and pentane has a higher boiling point.

Figure 11.21 shows how to determine the intermolecular forces in a sample.

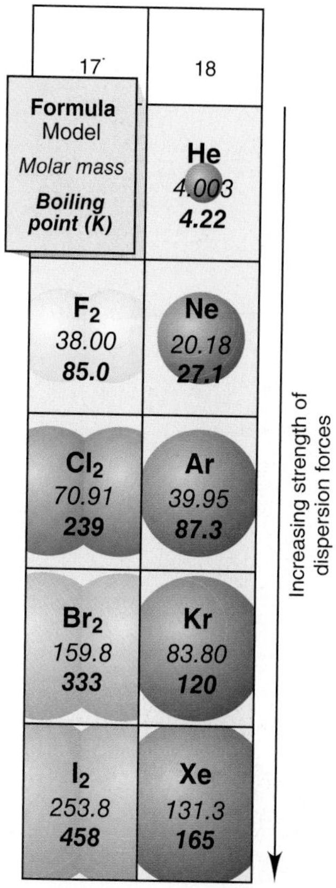

FIGURE 11.19 Molar mass and trends in boiling point

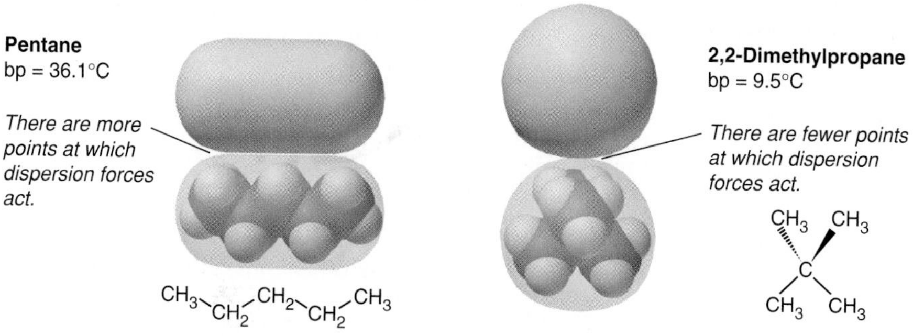

FIGURE 11.20 Molecular shape, intermolecular contact, and boiling point

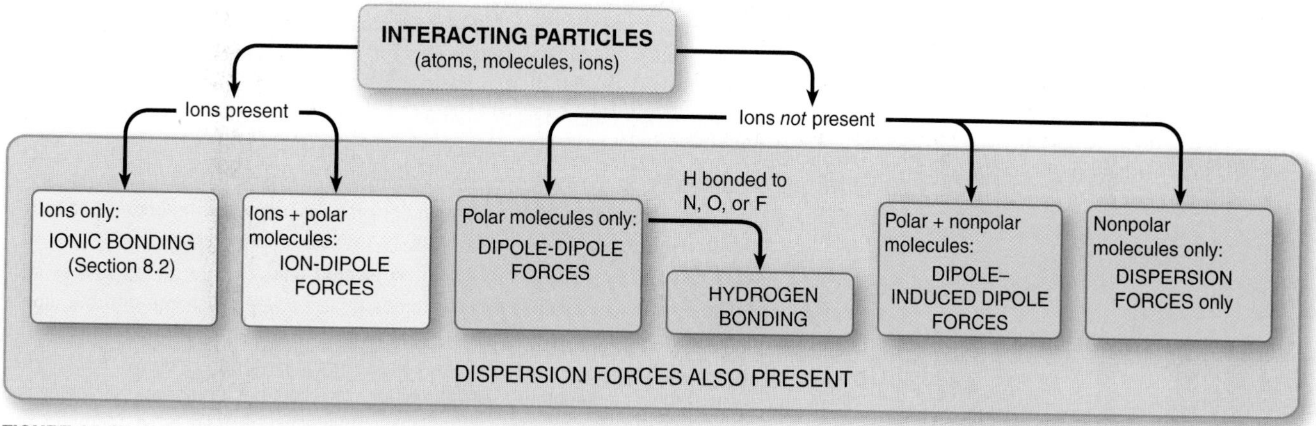

FIGURE 11.21 Determining the intermolecular forces in a sample

Sample Problem 11.4 Predicting the Types of Intermolecular Forces

Problem Identify the key bonding and/or intermolecular force(s) in each substance. Then predict which substance in each pair has the higher boiling point:

(a) $MgCl_2$ or PCl_3
(b) CH_3NH_2 or CH_3F
(c) CH_3OH or CH_3CH_2OH
(d) Hexane ($CH_3CH_2CH_2CH_2CH_2CH_3$) or 2,2-dimethylbutane 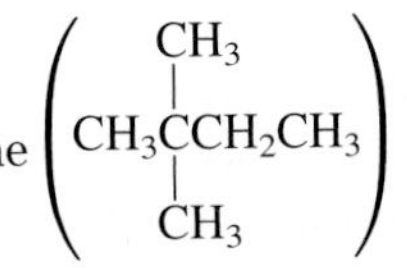

Plan We examine the formulas and structures to look for key differences between the substances in each pair: Are ions present? Are the molecules polar or nonpolar? Is N, O, or F bonded to H? Do the molecules have different masses or shapes? To rank the boiling points, we consult Figure 11.19 and Table 11.2. Remember the following:

- Bonding forces are stronger than intermolecular forces.
- Hydrogen bonding is a strong type of dipole-dipole force.
- Dispersion forces are decisive when the difference is molar mass or molecular shape.

Solution (a) $MgCl_2$ consists of Mg^{2+} and Cl^- ions held together by ionic bonding forces; PCl_3 consists of polar molecules, so intermolecular dipole-dipole forces are present. The forces in $MgCl_2$ are stronger, so it should have a higher boiling point.
(b) CH_3NH_2 and CH_3F both consist of polar molecules with about the same molar mass. CH_3NH_2 has N—H bonds, so it can form hydrogen bonds (*see margin*). CH_3F contains a C—F bond but no H—F bond, so dipole-dipole forces occur but not hydrogen bonds. Therefore, CH_3NH_2 should have the higher boiling point.
(c) CH_3OH and CH_3CH_2OH molecules both contain an O—H bond, so they can form hydrogen bonds (*see margin*). CH_3CH_2OH has an additional $—CH_2—$ group and thus a larger molar mass, which correlates with stronger dispersion forces; therefore, it should have a higher boiling point.
(d) Hexane and 2,2-dimethylbutane are nonpolar molecules with the same molar mass but different molecular shapes (*see margin*). Cylindrical hexane molecules make more intermolecular contact than more compact 2,2-dimethylbutane molecules do, so hexane should have stronger dispersion forces and a higher boiling point.

(b)
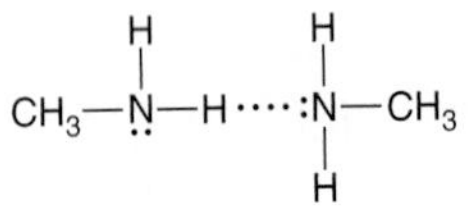

(c)
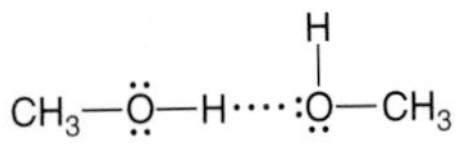

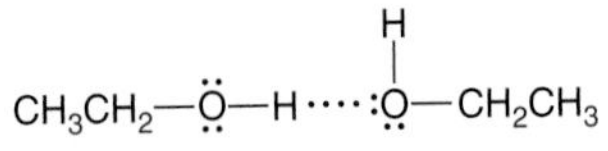

(d)
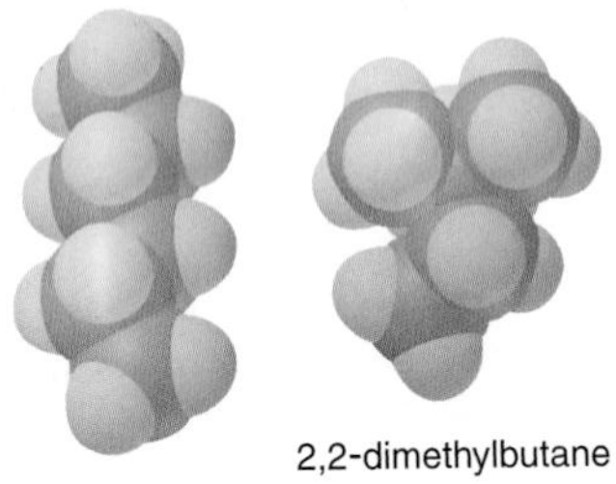

2,2-dimethylbutane

Hexane

Check The actual boiling points show that our predictions are correct:
(a) $MgCl_2$ (1412°C) and PCl_3 (76°C)
(b) CH_3NH_2 (−6.3°C) and CH_3F (−78.4°C)
(c) CH_3OH (64.7°C) and CH_3CH_2OH (78.5°C)
(d) Hexane (69°C) and 2,2-dimethylbutane (49.7°C)

Comment Dispersion forces are *always* present, but, in the substances in parts (a) and (b), they are much less significant than the other forces that occur.

Follow-Up Problem 11.4 Identify the intermolecular forces in each substance. Then predict which substance in each pair has the higher boiling point:
(a) CH_3Br or CH_3F **(b)** $CH_3CH_2CH_2OH$ or $CH_3CH_2OCH_3$ **(c)** C_2H_6 or C_3H_8

SUMMARY OF SECTION 11.3

- The van der Waals radius determines the shortest distance over which intermolecular forces operate; it is always larger than the covalent radius.
- Intermolecular forces are much weaker than bonding (intramolecular) forces.
- Ion-dipole forces occur between ions and polar molecules.
- Dipole-dipole forces occur between oppositely charged poles on polar molecules.
- Hydrogen bonding, a special type of dipole-dipole force, occurs when hydrogen bonded to N, O, or F is attracted to the lone pair of N, O, or F in another molecule.
- Electron clouds can be distorted (polarized) in an electric field.
- Ion–induced dipole and dipole–induced dipole forces arise between a charge and the dipole it induces in another molecule.
- Dispersion (London) forces are instantaneous induced dipole–induced dipole forces that occur among all particles and increase with the number of electrons (molar mass). Molecular shape determines the extent of contact between molecules and can be a factor in the strength of dispersion forces.

11.4 Properties of the Liquid State

Of the three states, only liquids combine the ability to flow with the effects of strong intermolecular forces. We understand this state the least at the molecular level. Because of the *random* arrangement of the particles in a gas, any region of a sample is virtually identical to any other region. Also, different regions of a crystalline solid are identical because of the *orderly* arrangement of the particles (Section 11.6). Liquids, however, have regions that are orderly one moment and random the next. Despite this complexity, many macroscopic properties, such as surface tension, capillarity, and viscosity, are well understood.

Surface Tension

Intermolecular forces have different effects on a molecule at the surface compared with a molecule in the interior (Figure 11.22):

- An interior molecule is attracted by other molecules on all sides.
- A surface molecule is only attracted by other molecules below and to the sides, so it experiences a *net attraction downward.*

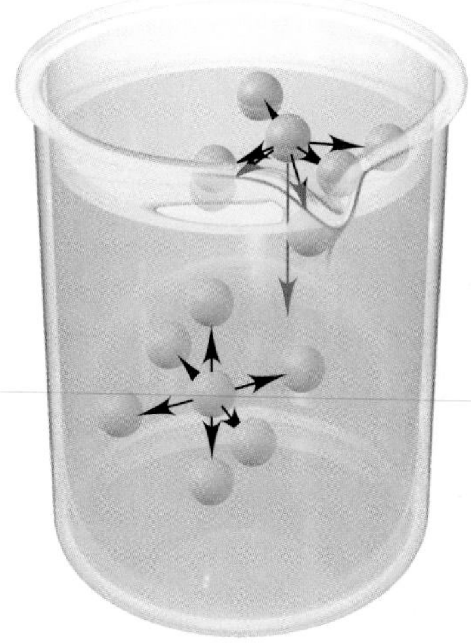

FIGURE 11.22 The molecular basis of surface tension

Therefore, to increase attractions and become more stable, a surface molecule tends to move into the interior. For this reason, *a liquid surface has the fewest molecules and, thus, the smallest area possible.* In effect, the surface behaves like a taut skin covering the interior.

The only way to increase the surface area is for molecules to move up by breaking attractions in the interior, which requires energy. The **surface tension** is the energy required to increase the surface area and has units of J/m^2 (Table 11.3). In general, *the stronger the forces between particles, the more energy is needed to increase the surface area, so the greater the surface tension.* Water has a high surface tension because its molecules form multiple hydrogen bonds. *Surfactants* (*surf*ace-*act*ive *a*ge*nts*), such as soaps, petroleum recovery agents, and fat emulsifiers, decrease the surface tension of water by congregating at the surface and disrupting the hydrogen bonds.

Surface tension is the cause of many familiar phenomena. For example, bubbles are round because this shape minimizes the surface area around the gas. Surface tension explains why some insects can walk on water (*see margin photograph*).

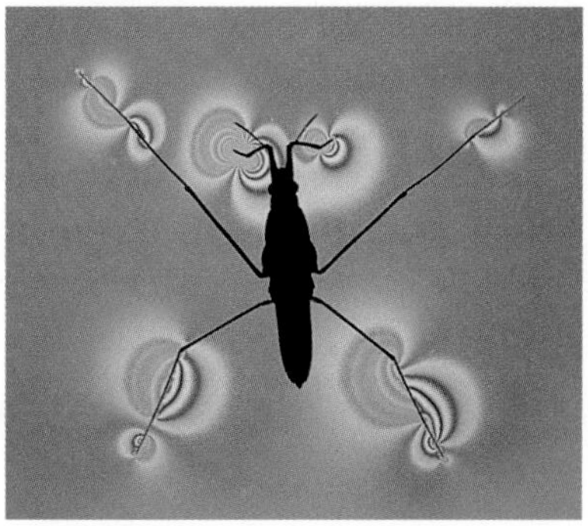

A water strider's widespread legs do not exert enough pressure to exceed the surface tension.

Capillarity

The rising of a liquid against the pull of gravity through a narrow space, such as a thin tube, is called *capillary action*, or **capillarity**. Capillarity results from a competition between the intermolecular forces within the liquid (cohesive forces) and those between the liquid and the tube walls (adhesive forces). Let us look at the difference between the capillarities of water and mercury in glass:

1. *Water in glass.* When you place a narrow glass tube in water, why does the liquid rise up the tube and form a concave meniscus? Glass is mostly silicon dioxide (SiO_2), so water molecules form adhesive hydrogen bonding forces with the O atoms of the glass. As a result, a thin film of water creeps up the wall. At the same

TABLE 11.3 Surface Tension and Forces between Particles

Substance	Formula	Surface Tension (J/m^2) at 20°C	Major Force(s)
Ethoxyethane	$CH_3CH_2OCH_2CH_3$	1.7×10^{-2}	Dipole-dipole; dispersion
Ethanol	CH_3CH_2OH	2.3×10^{-2}	Hydrogen bonding
Butanol	$CH_3CH_2CH_2CH_2OH$	2.5×10^{-2}	Hydrogen bonding; dispersion
Water	H_2O	7.3×10^{-2}	Hydrogen bonding
Mercury	Hg	48×10^{-2}	Metallic bonding

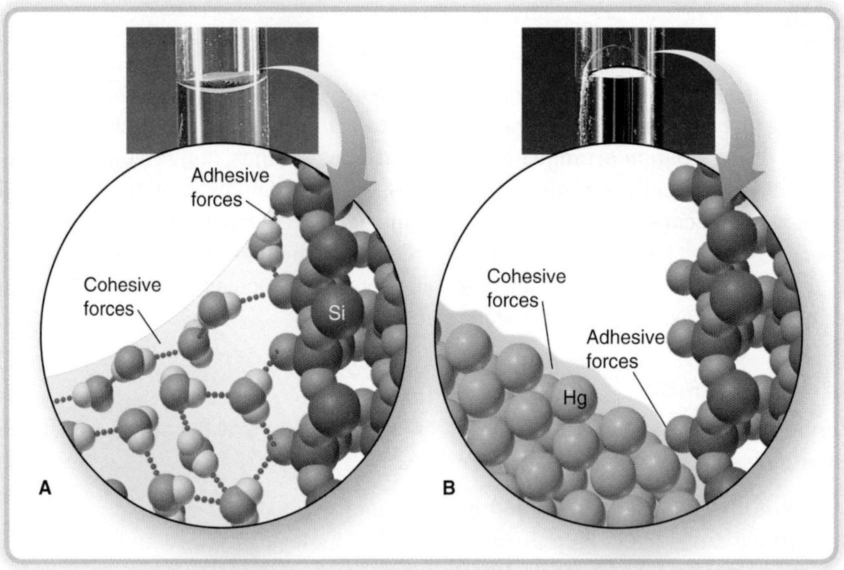

FIGURE 11.23 Capillary action and the shape of the water or mercury meniscus in glass. A. Water displays a concave meniscus. **B.** Mercury displays a convex meniscus.

time, cohesive hydrogen bonding forces between water molecules, which give rise to surface tension, make the surface taut. These adhesive and cohesive forces combine to raise the water level and produce the concave meniscus (Figure 11.23A). The liquid rises until the gravity pulling down is balanced by the adhesive forces pulling up.

2. *Mercury in glass.* When you place a glass tube in a dish of mercury, why does the liquid drop below the level in the dish and form a convex meniscus? The cohesive forces among the mercury atoms are metallic bonds, so they are *much* stronger than the mostly dispersion adhesive forces between the mercury and the glass. As a result, the liquid pulls away from the walls. At the same time, the surface atoms are being pulled toward the interior by mercury's high surface tension, so the level drops. These combined forces produce the convex meniscus seen in a laboratory barometer (Figure 11.23B).

Capillarity plays a key role in many everyday events. The adhesive (dipole–induced dipole) forces between water and a nonpolar surface are much weaker than the cohesive (hydrogen bond) forces within water. As a result, water pulls away from a nonpolar surface and forms beaded droplets, as on a freshly waxed car or on the waxy coating of a leaf after a rainfall. Even more familiar, after a shower, capillary action draws water away from your body through the closely spaced fibres in a cotton towel, which is made of cellulose molecules, so the water molecules also form hydrogen bonds with the —OH groups of cellulose.

Viscosity

Viscosity is the resistance of a fluid to flow, and it results from intermolecular attractions that impede the movement of molecules around and past one another. Both gases and liquids flow, but liquid viscosities are *much* higher because there are many more points for intermolecular forces to act due to the much shorter distances. Let us examine two factors—temperature and molecular shape—that influence viscosity:

TABLE 11.4 Viscosity of Water at Several Temperatures

Temperature (°C)	Viscosity (N·s/m^2)*
20	1.00×10^{-3}
40	0.65×10^{-3}
60	0.47×10^{-3}
80	0.35×10^{-3}

*The units of viscosity are newton-second per square metre.

- *Effect of temperature.* Viscosity decreases with heating (Table 11.4). Faster-moving molecules overcome intermolecular forces more easily, so the resistance to flow decreases. The next time you heat cooking oil, watch the oil flow more easily and spread out in the pan as it warms. When automobiles were first invented, petroleum oil was used to lubricate the engine components. However, in cold weather, petroleum motor oil thickened to the point where it became

semisolid. For this reason, synthetic motor oils, which perform much better in cold weather, were developed. Their viscosity is not just based on additive systems, but also on their molecular structure.
- *Effect of molecular shape.* Small, spherical molecules make little contact and pour easily, like peas from a bowl. Long molecules make more contact and become entangled and pour slowly, like cooked spaghetti from a bowl. Thus, given the same types of intermolecular forces, liquids that consist of longer molecules have higher viscosities.

To protect engine parts during long drives, motor oils contain *polymeric viscosity improvers*. As the oil warms, these additive molecules change from compact spheres to long strands that become tangled with the long hydrocarbon molecules of the oil. Greater dispersion forces increase the viscosity to compensate for the oil thinning due to heating.

SUMMARY OF SECTION 11.4

- Surface tension is a measure of the energy required to increase a liquid's surface area. Greater intermolecular forces within a liquid create higher surface tension.
- Capillarity, the rising of a liquid through a narrow space, occurs when the forces between a liquid and a surface (adhesive) are greater than the forces within the liquid (cohesive).
- Viscosity, the resistance to flow, depends on molecular shape and decreases with temperature. Stronger intermolecular forces create higher viscosity.

11.5 The Uniqueness of Water

Water is absolutely amazing stuff, with some of the most unusual properties of any substance. However, it is so familiar that we take it for granted. Like any substance, its properties arise inevitably from those of its atoms. Each O atom and H atom attains a filled outer level by sharing electrons in single bonds. With two bonding pairs and two lone pairs around O and a large electronegativity difference in each O—H bond, the H_2O molecule is bent and highly polar. This arrangement is crucial because it allows each molecule to engage in four hydrogen bonds with its neighbours (Figure 11.24). From these basic atomic and molecular facts emerges unique and remarkable macroscopic behaviour.

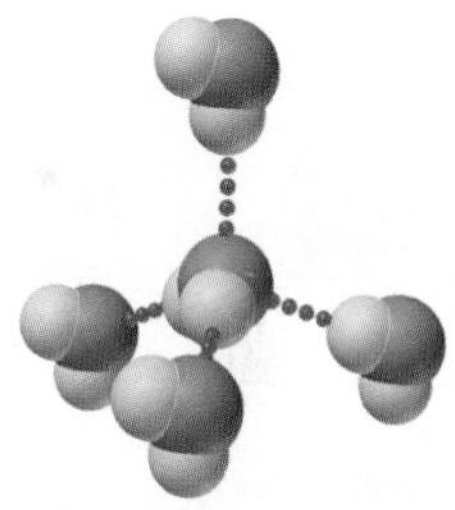

FIGURE 11.24 Hydrogen-bonding ability of water. One H_2O molecule can form four hydrogen bonds to other molecules, resulting in a tetrahedral arrangement.

Solvent Properties of Water

The *great solvent power* of water results from its polarity and its hydrogen-bonding ability:

- Water dissolves ionic compounds through ion-dipole forces that separate the ions from the solid and keep them in solution (see Figure 3.18).
- It dissolves polar nonionic substances, such as ethanol (CH_3CH_2OH) and glucose ($C_6H_{12}O_6$), by hydrogen bonding.
- It dissolves nonpolar atmospheric gases to a limited extent through dipole–induced dipole and dispersion forces.

Water is the environmental and biological solvent, forming the complex solutions we know as oceans, lakes, and cytoplasm. Aquatic animals could not survive without dissolved O_2, and aquatic plants could not survive without dissolved CO_2. Tiny marine animals form coral reefs made of carbonates from dissolved CO_2 and HCO_3^-. Life began in a "primordial soup," an aqueous mixture of simple molecules from which emerged larger molecules capable of self-sustaining reactions. From a chemical point of view, all organisms, from bacteria to humans, are highly organized systems of membranes enclosing and compartmentalizing complex aqueous solutions.

Thermal Properties of Water

When a substance is heated, some of the added energy increases average molecular speed, some increases molecular vibration and rotation, and some is used to overcome intermolecular forces.

1. *Specific heat capacity.* Because water has so many strong hydrogen bonds, its *specific heat capacity* is higher than the specific heat capacity of any common liquid. With oceans covering 70% of Earth's surface, daytime energy from the Sun causes relatively small changes in temperature, allowing life to survive. On the waterless, airless Moon, temperatures range from 100°C to −150°C during a complete lunar day. Even in Earth's deserts, day-night temperature differences of 40°C are common.

2. *Heat of vaporization.* Numerous strong hydrogen bonds give water a very *high heat of vaporization.* Two examples show why this is crucial. The average adult has 40 kg of body water and generates about 10 000 kJ of heat each day from metabolism. If this heat were used to increase the average E_k of water molecules in the body, the rise in body temperature of tens of degrees would mean immediate death. Instead, the heat is converted to E_p as it breaks hydrogen bonds and evaporates sweat, resulting in a stable body temperature and minimal loss of body fluid. On a planetary scale, the Sun's energy vaporizes ocean water in warm latitudes, and the potential energy is released as heat to warm cooler regions when the vapour condenses to rain. This global-scale cycling of water powers many weather patterns.

Surface Properties of Water

Hydrogen bonding is also responsible for water's *high surface tension* and *high capillarity.* Except for some molten metals and salts, water has the highest surface tension of any liquid. It keeps plant debris resting on a pond surface, providing shelter and nutrients for fish and insects. High capillarity means that water rises through the tiny spaces between soil particles, so plant roots can absorb deep groundwater during dry periods.

The Unusual Density of Solid Water

In the solid state, the tetrahedral pattern of hydrogen bonding (see Figure 11.24) leads to the hexagonal *open structure* of ice (Figure 11.25A), and the symmetrical beauty of snowflakes (Figure 11.25B) reflects this hexagonal organization. The large spaces within ice make *the solid less dense than the liquid* and explain the negative slope of the solid-liquid line in the phase diagram for water (see Figure 11.13). As pressure is applied, some hydrogen bonds break, so the ordered crystal structure is disrupted and the ice liquefies. When ice melts at 0°C, the loosened molecules pack much more closely, filling spaces in the collapsing solid structure. As a result, liquid water is most dense (1.000 g/mL) at around 4°C (3.98°C). With more heating, the density decreases through normal thermal expansion. This behaviour has important effects in nature:

- *Surface ice of lakes.* When the surface of a lake freezes in the winter, the ice floats. If the solid were denser than the liquid, as is true for nearly every other substance, the surface water would freeze and sink until the entire lake became solid. Aquatic life would not survive from year to year.
- *Nutrient turnover.* As lake water becomes colder in the early winter, it becomes more dense *before* it freezes. Similarly, in the spring, less-dense ice thaws to form more-dense water *before* the water expands. During both of these seasonal density changes, the top layer of water reaches the maximum density first and sinks. The next layer of water rises because it is slightly less dense, reaches 4°C, and likewise sinks. This alternation of sinking and rising distributes nutrients and dissolved oxygen.
- *Soil formation.* When rain fills the crevices in rocks and freezes, an outward force is applied that is relieved when the ice melts. In time, the repeated freeze-thaw stress cracks the rock. Over eons, this effect helps to produce sand and soil.

The far-reaching consequences of the properties of water illustrate chemistry's central theme: the macroscopic world we know is the cumulative outcome of the atomic world we seek to know (Figure 11.26).

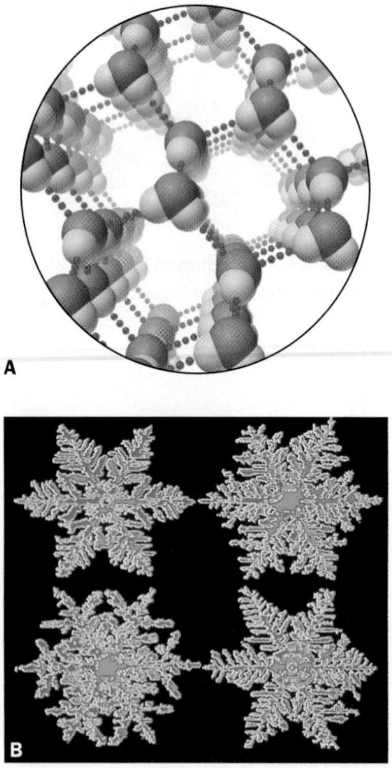

FIGURE 11.25 The hexagonal structure of ice. **A.** The open, hexagonal molecular structure of ice. **B.** The beauty of six-pointed snowflakes reflects this hexagonal structure.

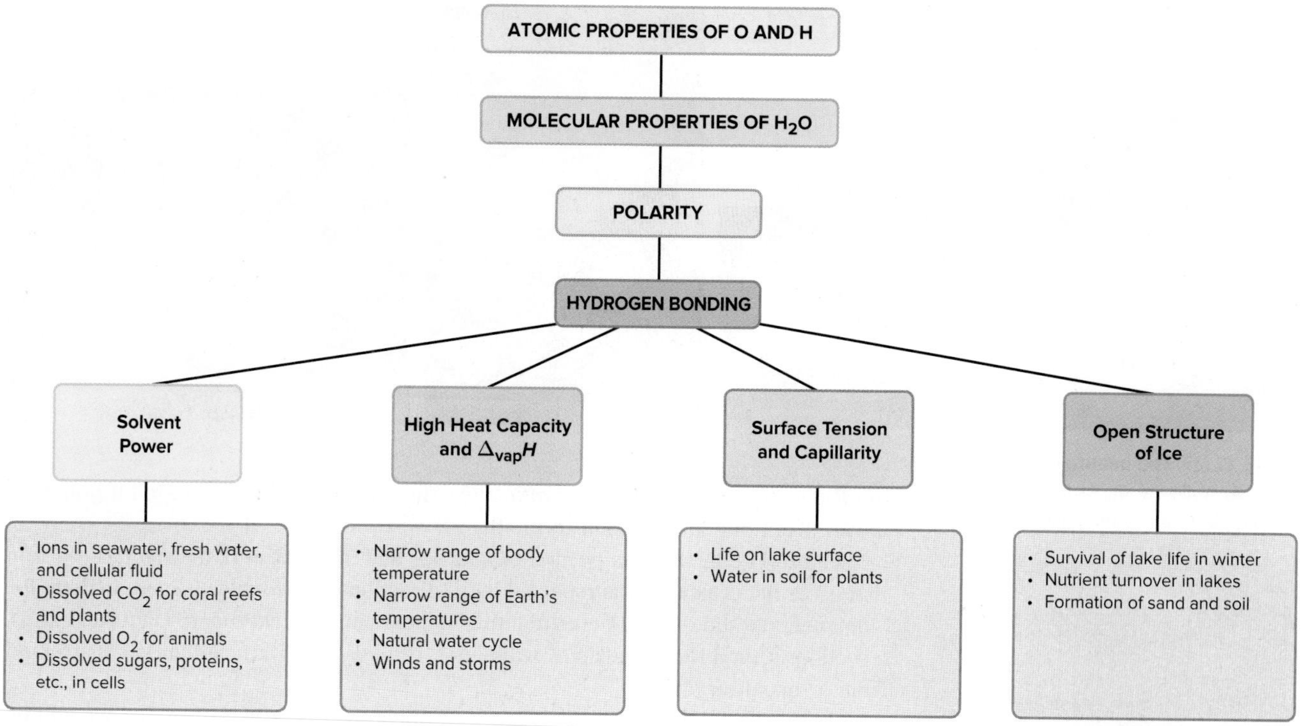

FIGURE 11.26 The unique macroscopic behaviour of water that emerges from its atomic and molecular properties

SUMMARY OF SECTION 11.5

- The atomic properties of H and O result in water's bent molecular shape, polarity, and hydrogen-bonding ability.
- These properties give water the ability to dissolve many ionic and polar compounds.
- Water's high specific heat capacity and heat of vaporization give Earth and its organisms a narrow temperature range.
- The high surface tension and capillarity of water are essential to plants and animals.
- Because water expands when it freezes, lake life survives in winter, nutrients mix due to seasonal density changes, and soil forms through freeze-thaw stress on rocks.

11.6 The Solid State: Structure, Properties, and Bonding

Stroll through a museum's mineral collection, and you will be struck by the variety and beauty of crystalline solids. In this section, we discuss the structural features of these and other solids and the intermolecular forces that create them. We also consider the main bonding model that explains many properties of solids.

Structural Features of Solids

We can divide solids into two broad categories:

- **Crystalline solids** have well-defined shapes because their particles—atoms, molecules, or ions—occur in an orderly arrangement (Figure 11.27).
- **Amorphous solids** have poorly defined shapes because their particles lack an orderly arrangement.

The Crystal Lattice and the Unit Cell The particles in a crystal are packed tightly in an orderly, three-dimensional array. As the simplest case, consider the particles as *identical* spherical atoms, and imagine a point at the centre of each. The collection of points forms a regular pattern called a crystal **lattice**. The lattice consists

FIGURE 11.27 The beauty of crystalline solids. **A.** Wulfanite. **B.** Barite. **C.** Beryl (emerald). **D.** Quartz (amethyst).

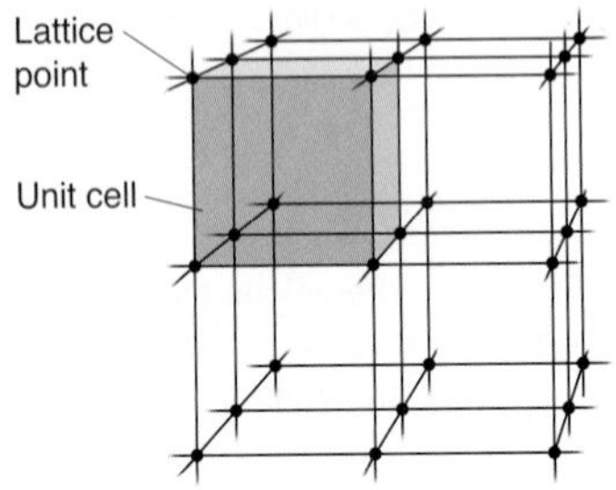

A Portion of 3-D lattice

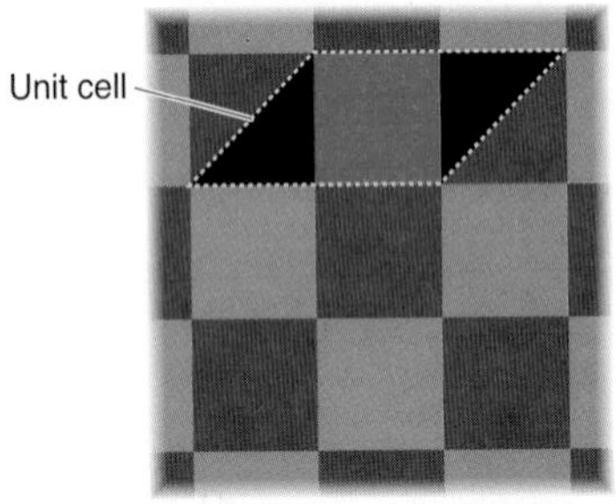

B 2-D analogy for unit cell and lattice

FIGURE 11.28 The crystal lattice and the unit cell. **A.** A small portion of a lattice is shown as points connected by lines, with a unit cell (*coloured*). **B.** A checkerboard analogy for a lattice.

The efficient packing of fruits and vegetables

of *all points with identical surroundings*; that is, there is no way to tell if you have moved from one lattice point to another.

Figure 11.28A shows a portion of a lattice and the **unit cell**, the *smallest* portion that gives the crystal if it is repeated in all directions. A two-dimensional analogy for a unit cell and the resulting crystal lattice appears in a checkerboard (Figure 11.28B), a section of tiled floor, a strip of wallpaper, or any other pattern that is constructed from a repeating unit.

Seven crystal systems and 14 types of unit cells occur in nature, but we will be concerned primarily with the *cubic system*. The solid states of a majority of metallic elements, some covalent compounds, and many ionic compounds occur as cubic lattices. A key parameter of any lattice is the **coordination number**, the number of *nearest* neighbours of a particle. There are three types of cubic unit cells within the cubic system:

1. In the **simple cubic unit cell** (Figure 11.29A), the centres of eight identical particles define the corners of a cube (shown in the expanded view, *top row*). The particles touch along the cube edges (see the space-filling view, *second row*), but they do not touch diagonally along the faces of the cube or through its centre. An expanded portion of the crystal (*third row*) shows that the coordination number of each particle is 6: four neighbours in its own layer, one neighbour in the layer above, and one neighbour in the layer below.
2. In the **body-centred cubic unit cell** (Figure 11.29B), identical particles lie at each corner *and* in the centre of the cube. Those at the corners do not touch each other, but they all touch the particle in the centre. Each particle is surrounded by eight nearest neighbours, four above and four below, so the coordination number is 8.
3. In the **face-centred cubic unit cell** (Figure 11.29C), identical particles lie at each corner *and* in the centre of each face but not in the centre of the cube. Particles at the corners touch those in the faces but not each other. The coordination number is 12.

How many particles make up a unit cell? For particles of the same size, *the higher the coordination number, the greater the number of particles in a given volume.* Since one unit cell touches another, with no gaps, a particle at a corner or face is *shared* by adjacent cells. In the cubic unit cells, the particle at each corner is part of eight adjacent cells (Figure 11.29, *third row*), so one-eighth of each particle belongs to each cell (*bottom row*). There are eight corners in a cube, so

- A simple cubic unit cell contains $8 \times \frac{1}{8}$ particle = 1 particle
- A body-centred cubic unit cell contains $8 \times \frac{1}{8}$ particle = 1 particle plus 1 particle in the centre, for a total of 2 particles
- A face-centred cubic unit cell contains $8 \times \frac{1}{8}$ particle = 1 particle plus $\frac{1}{2}$ particle in each of the six faces, or $6 \times \frac{1}{2}$ particle = 3 particles, for a total of 4 particles

Packing Efficiency and the Creation of Unit Cells Unit cells result from the ways that atoms pack together, which are similar to the ways that macroscopic spheres—such as marbles, golf balls, and fruits and vegetables—are packed (*see photo*). Let

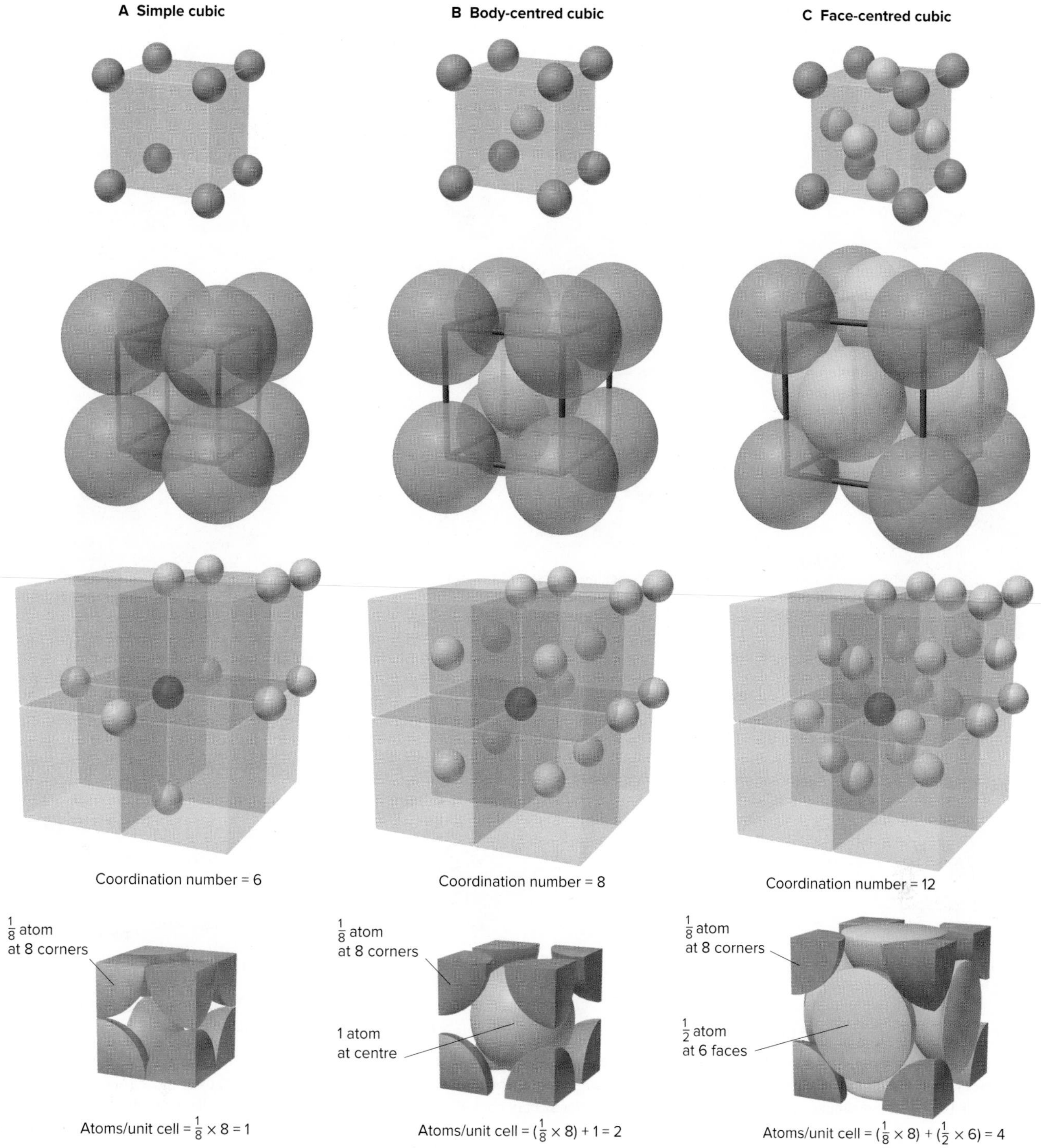

FIGURE 11.29 The three cubic unit cells. A. Simple cubic unit cell. **B.** Body-centred cubic unit cell. **C.** Face-centred cubic unit cell. *Top row:* Cubic arrangements of atoms in expanded view. *Second row:* Space-filling view of these cubic arrangements. All atoms are identical but, for clarity, the corner atoms are blue, the body-centred atoms are pink, and the face-centred atoms are yellow. *Third row:* A unit cell (*shaded blue*) in an expanded portion of the crystal. The number of nearest neighbours around one particle (*dark blue in centre*) is the coordination number. *Bottom row:* The total number of atoms in the actual unit cell. The simple cube has one atom, the body-centred cube has two, and the face-centred cube has four.

us pack *identical* spheres to create the three cubic unit cells and the hexagonal unit cell and determine the **packing efficiency**, the percentage of the total volume that is occupied by the spheres themselves:

1. *Simple cubic unit cell.* When we arrange the first layer of spheres in vertical and horizontal rows, large diamond-shaped spaces are formed (Figure 11.30A, *cutaway portion*). If we place the next layer of spheres *directly above* the first, we

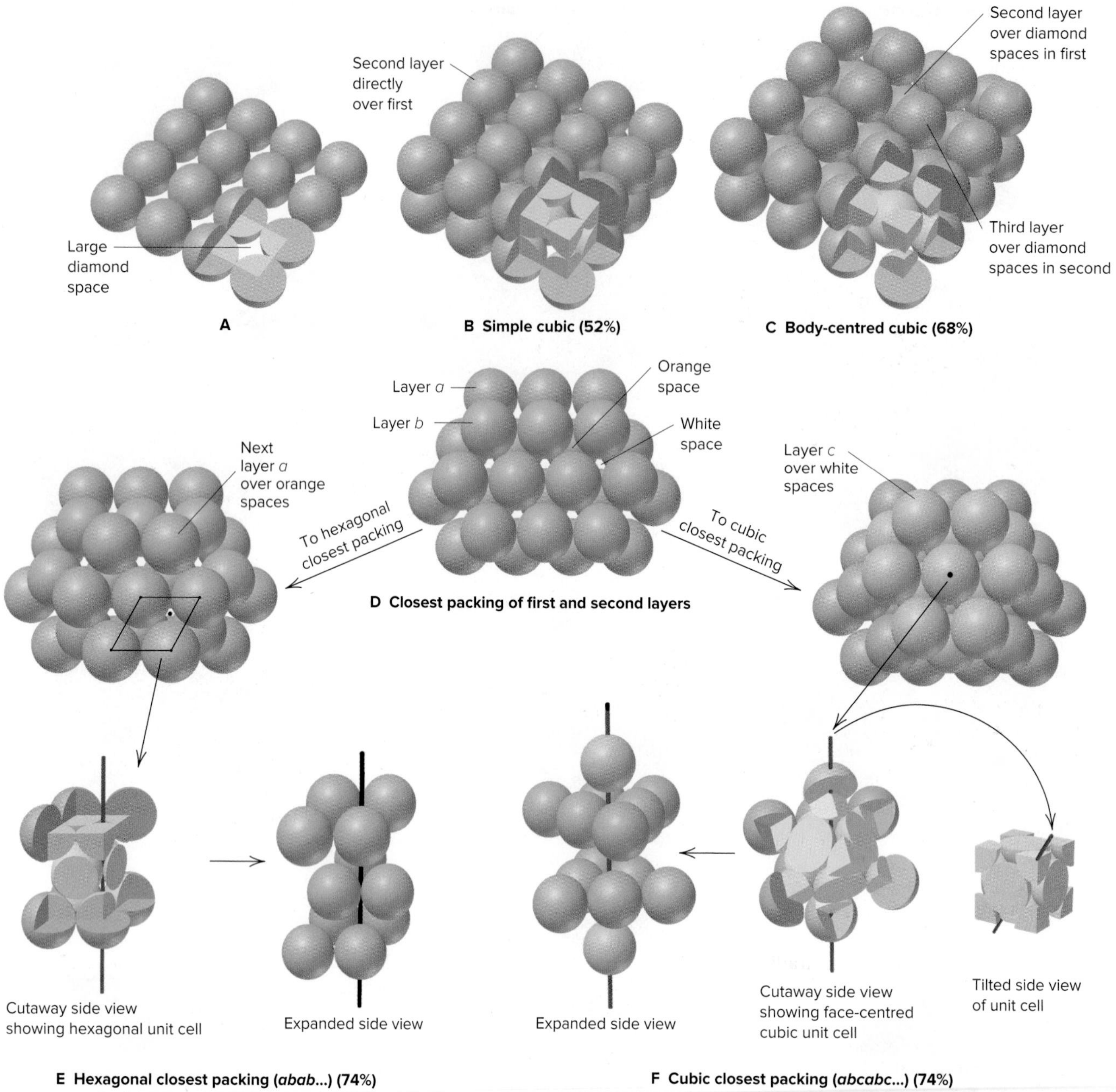

FIGURE 11.30 Packing spheres to obtain three cubic and hexagonal unit cells. A. In the first layer, each sphere lies next to another horizontally and vertically. Notice the large diamond-shaped spaces (*see cutaway*). **B.** If the spheres in the next layer lie directly over those in the first layer, the packing is based on the simple cubic unit cell (*pale orange cube, lower right corner*). **C.** If the spheres in the next layer lie in the diamond-shaped spaces of the first layer, the packing is based on the *body-centred cubic* unit cell (*lower right corner*). **D.** The closest possible packing of the first layer (*layer a, orange*) is obtained by shifting every other row in part A to obtain smaller triangular spaces. The spheres in the second layer (*layer b, green*) are placed above these spaces. Notice the orange and white spaces that result. **E.** When the third layer (*next layer a, orange*) is placed over the orange spaces, we obtain an *abab*... pattern and the hexagonal unit cell. **F.** When the third layer (*layer c, blue*) covers the white spaces, we get an *abcabc*... pattern and the face-centred cubic unit cell.

obtain an arrangement based on the *simple* cubic unit cell (Figure 11.30B). The spheres occupy only 52% of the unit-cell volume, so 48% is empty space between them. This is a very inefficient way to pack spheres, so neither fruit nor atoms are usually packed this way.

2. *Body-centred cubic unit cell.* Rather than placing the second layer (coloured *green* for clarity) directly above the first layer, we use space more efficiently by placing it on the diamond-shaped spaces in the first layer (Figure 11.30C). Then we pack the third layer onto the diamond-shaped spaces in the second, which makes

the first and third layers line up vertically. This arrangement is based on the *body-centred* cubic unit cell, and its packing efficiency is much higher, at 68%. Several metallic elements, including chromium, iron, and all the group 1 elements, have a crystal structure based on this unit cell.

3. *Hexagonal and face-centred cubic unit cells.* Spheres are packed the most efficiently in these cells. First, in the bottom layer (labelled *a*, *orange*), we shift every other row laterally so that the large diamond-shaped spaces become smaller triangular spaces. Then we place the second layer (labelled *b*, *green*) over these spaces (Figure 11.30D).

In layer *b*, notice that some spaces are orange because they lie above *spheres* in layer *a*, whereas other spaces are white because they lie above *spaces* in layer *a*. We can place the third layer in either of two ways, which gives rise to two different unit cells:

- *Hexagonal unit cell.* If we place the third layer of spheres (*orange*) over the orange spaces (look down and left to Figure 11.30E), they lie directly over the spheres in layer *a*. Every other layer is placed identically (an *abab*... layering pattern), and we obtain **hexagonal closest packing**, which is based on the *hexagonal* unit cell.
- *Face-centred unit cell.* If we place the third layer of spheres (*blue*) over the white spaces in layer *b* (look down and right to Figure 11.30F), the placement is different from the placement in layers *a* and *b* (an *abcabc*... pattern) and we obtain **cubic closest packing**, which is based on the *face-centred cubic* unit cell.

The packing efficiency of both hexagonal and cubic closest packing is 74%, and the coordination number of both is 12. Most metallic elements crystallize in one of these arrangements. Magnesium, titanium, and zinc are three elements that adopt the hexagonal structure. Nickel, copper, and lead adopt the cubic structure, as do many ionic compounds and other substances, such as CO_2, CH_4, and most noble gases.

In Sample Problem 11.5, we use density and our knowledge of unit cells to find an atomic radius.

Sample Problem 11.5 Determining Atomic Radius

Problem Barium is the largest nonradioactive alkaline earth metal. It has a body-centred cubic unit cell and a density of 3.62 g/cm^3. What is the atomic radius of barium (Volume of a sphere: $V = \frac{4}{3}\pi r^3$)?

Plan An atom is spherical, so we can find its radius from its volume. If we multiply the reciprocal of density (volume/mass) by the molar mass (mass/mole), we find the volume/mole of Ba metal. The metal crystallizes in the body-centred cubic structure, so 68% of this volume is occupied by 1 mol of the Ba atoms themselves (see Figure 11.29C). Dividing by Avogadro's number gives the volume of one Ba atom, from which we find the radius.

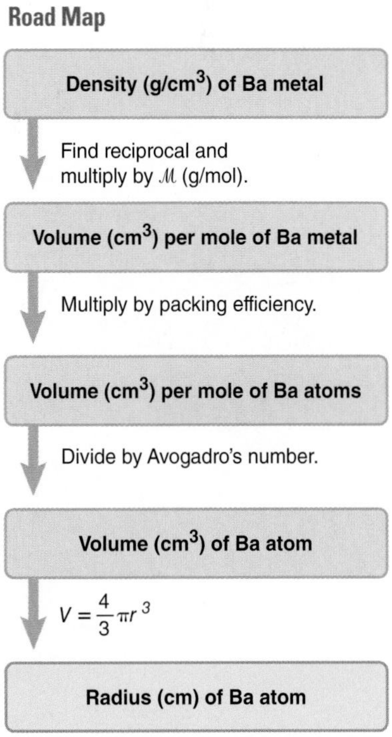

Solution Combine steps to find the volume of 1 mol of Ba metal:

$$\text{Volume/mole of Ba metal} = \frac{1}{\text{density}} \times \mathcal{M} = \frac{1\ \text{cm}^3}{3.62\ \text{g Ba}} \times \frac{137.3\ \text{g Ba}}{1\ \text{mol Ba}} = 37.93\ \text{cm}^3/\text{mol Ba}$$

Find the volume (cm^3) of 1 mol of Ba *atoms*:

$$\text{Volume/mole of Ba atoms} = \frac{\text{cm}^3}{\text{mol Ba}} \times \text{packing efficiency}$$

$$= \frac{37.93\ \text{cm}^3}{\text{mol Ba}} \times 0.68 = 25.79\ \frac{\text{cm}^3}{\text{mol Ba atoms}}$$

Find the volume of one Ba atom:

$$\text{Volume of Ba atom} = \frac{25.79\ \text{cm}^3}{1\ \text{mol Ba atoms}} \times \frac{1\ \text{mol Ba atoms}}{6.022 \times 10^{23}\ \text{Ba atoms}} = 4.28 \times 10^{-23}\text{cm}^3/\text{Ba atom}$$

Find the atomic radius of Ba from the volume of a sphere:

$$V \text{ of Ba atom} = \frac{4}{3}\pi r^3 \qquad \text{so} \qquad r^3 = \frac{3V}{4\pi}$$

Thus, $$r = \sqrt[3]{\frac{3V}{4\pi}} = \sqrt[3]{\frac{3(4.28 \times 10^{-23}\ \text{cm}^3)}{4 \times 3.142}} = 2.2 \times 10^{-8}\ \text{cm}$$

Check The order of magnitude is correct for an atom (~10^{-8} cm = 10^{-10} m). The actual value for barium is, in fact, 2.22×10^{-8} cm (see Figure 7.13).

Follow-Up Problem 11.5 Iron crystallizes in a body-centred cubic structure. The volume of one Fe atom is 8.38×10^{-24} cm^3, and the density of Fe is 7.874 g/cm^3. Calculate an approximate value for Avogadro's number.

Our understanding of crystal structures comes from our ability to "see" them. Two techniques for doing this are described in the Tools of the Laboratory feature below.

The edge length of the unit cell is obtained from X-ray crystallography. We can then apply the Pythagorean theorem and some arithmetic to find atomic (or ionic) radii in the three cubic unit cells (Figures 11.29 and 11.31). Sample Problem 11.6 shows this approach.

FIGURE 11.31 Edge length and atomic (ionic) radius in the three cubic unit cells

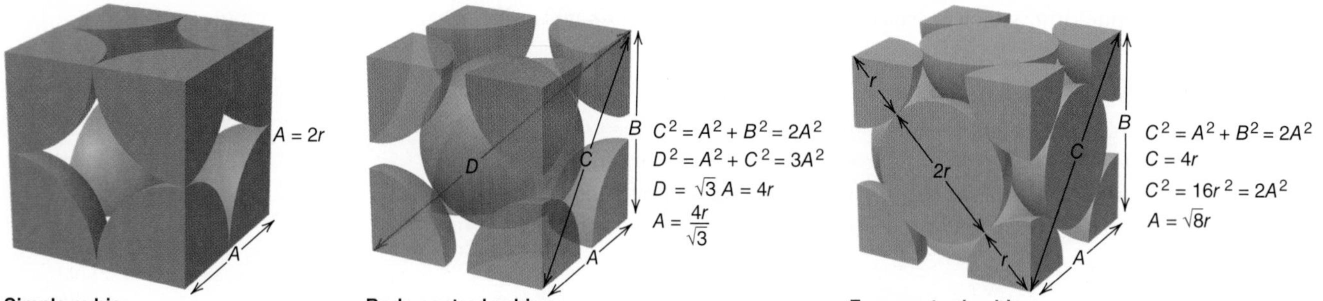

Sample Problem 11.6 Determining Atomic Radius from the Unit Cell

Problem Copper adopts cubic closest packing, and the edge length of the unit cell is 361.5 pm. What is the atomic radius of copper?

Plan Cubic closest packing has a face-centred cubic unit cell, and we know the edge length. With Figure 11.31 and A = 361.5 pm, we can solve for r (*see margin*).

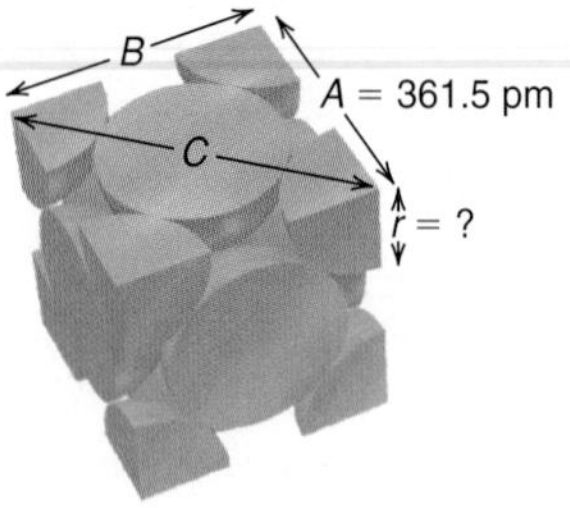

Solution Use the Pythagorean theorem to find C, the diagonal of the cell's face:

$$C = \sqrt{A^2 + B^2}$$

The unit cell is a cube, so $A = B$. Therefore,

$$C = \sqrt{2A^2} = \sqrt{2(361.5\ \text{pm})^2} = 511.2\ \text{pm}$$

Find r:

$$C = 4r$$

$$r = \frac{511.2\ \text{pm}}{4} = 127.8\ \text{pm}$$

Check Rounding and quickly checking the math gives

$$C = \sqrt{2(4 \times 10^2\ \text{pm})^2} = \sqrt{2(16 \times 10^4\ \text{pm}^2)} \text{ or } {\sim}500-600\ \text{pm}$$

Thus, $r \approx$ 125–150 pm. The actual value for copper is 128 pm (see Figure 7.13).

Follow-Up Problem 11.6 Iron crystallizes in a body-centred cubic structure. If the atomic radius of Fe is 126 pm, find the edge length of the unit cell.

X-Ray Diffraction Analysis and Scanning Tunnelling Microscopy

Various tools exist for measuring atomic-scale dimensions, and two of the most powerful are **X-ray diffraction analysis** and **scanning tunnelling microscopy**.

X-Ray Diffraction Analysis

In Chapter 6, we saw how diffraction patterns of bright and dark regions appear when light passes through slits spaced as far apart as the wavelength (see Figure 6.5). X-ray wavelengths are about the same size as the spaces between the layers of spheres in a crystal, so we use the spaces as "slits" to diffract X-rays.

As one example, let us use the X-ray diffraction analysis approach to measure the distance (d) between two adjacent layers of atoms in a simplified lattice (Figure B11.1). Two waves impinge on the crystal at an angle θ and are diffracted at that angle by the layers. When the first wave strikes the top layer and the second wave strikes the next layer, they are *in phase* (peaks aligned with peaks, and troughs aligned with troughs). If they are still in phase after being diffracted, they form a spot on a photographic plate. This will occur only if the additional distance that is travelled by the second wave ($DE + EF$ in the figure) is a whole number of wavelengths, $n\lambda$, where n is a positive integer referred to as the *order* of the diffraction. From trigonometry, we find that

$$n\lambda = 2d \sin \theta$$

where θ is the angle of incoming light and d, the unknown, is the distance between the layers. This is the *Bragg equation*, named for W. H. Bragg and his son W. L. Bragg, who shared the Nobel Prize in Physics in 1915 for their work on crystal structure analysis.

Rotating the crystal changes θ and produces different sets of spots. Eventually, the complete diffraction pattern reveals distances and angles in the lattice (Figure B11.2). X-ray diffraction analysis is used in many studies, but its greatest recent impact is in uncovering the structures of DNA and proteins.

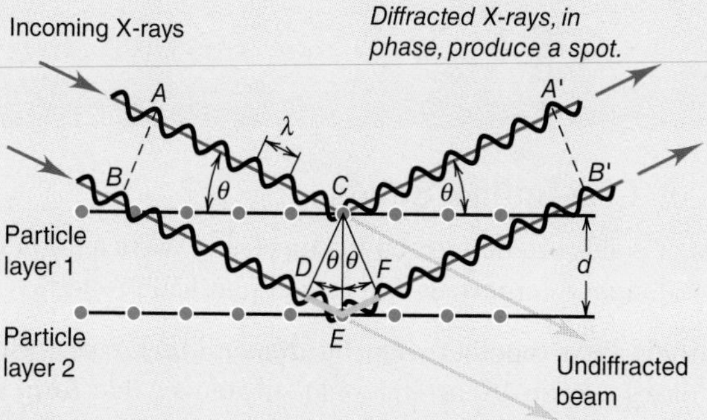

FIGURE B11.1 Diffraction of X-rays by crystal planes

Scanning Tunnelling Microscopy

The scanning tunnelling microscopy technique, invented by Gerd Binnig and Heinrich Rohrer, who shared the Nobel Prize in Physics in 1986 with Ernst Ruska, is used to observe surfaces. It is based on the idea that an electron has a small, but finite, probability of being so far from its nucleus that it can move ("tunnel") closer to another atom.

In practice, an extremely sharp tungsten-tipped probe, the source of the tunnelling electrons, is placed very close (about 0.5 nm) to the surface under study. A small applied electric potential increases the probability that the electrons will tunnel across this minute gap. The probe moves tiny distances up and

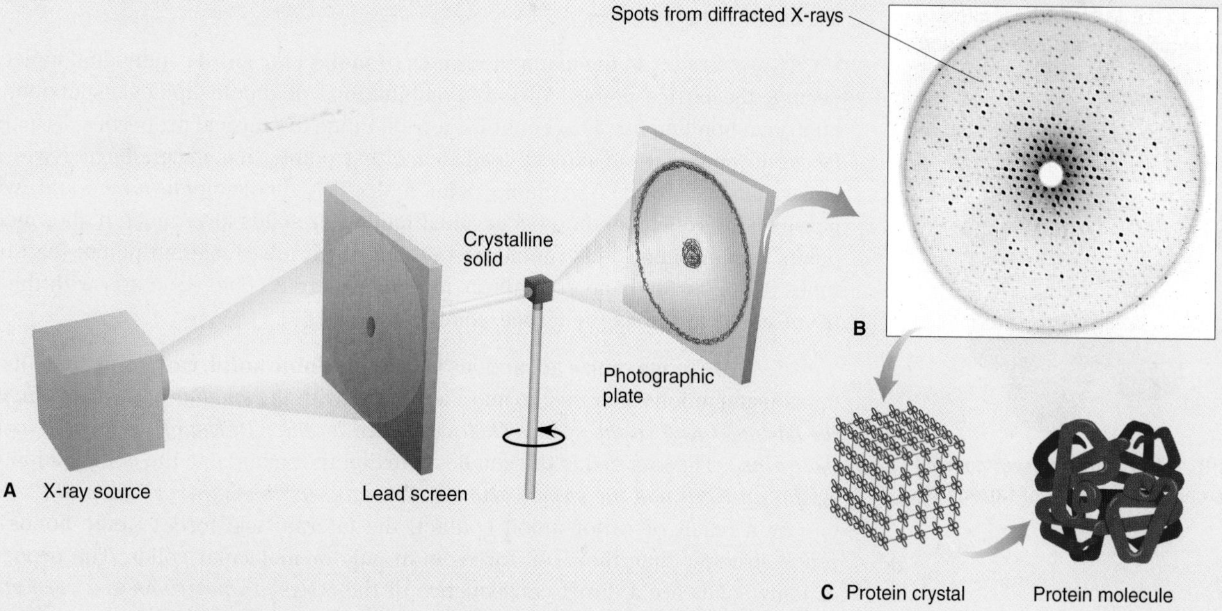

FIGURE B11.2 Formation of an X-ray diffraction pattern of the protein hemoglobin. **A.** A sample of crystalline protein is rotated to obtain different angles of incoming and diffracted X-rays. **B.** The diffraction pattern is a complex series of spots. **C.** Computerized analysis provides a picture of the molecule.

(Continued)

down to maintain the constant current across the gap, thus following the surface at the atomic scale. This movement is monitored electronically, and, after many scans, a three-dimensional map of the surface is obtained. Scanning tunnelling microscopy has revealed images of atoms and molecules coated on surfaces and is being used to study the nature of defects and many other surface features (Figure B11.3).

FIGURE B11.3 A scanning tunnelling micrograph of cesium atoms (*red*) on gallium arsenide

Problems

B11.1 X-rays of wavelength $\lambda = 0.709 \times 10^{-10}$ m undergo a first-order ($n = 1$) diffraction of 11.6° when aimed at crystalline nickel. What is the spacing between layers of Ni atoms?

B11.2 A first-order ($n = 1$) diffraction of X-rays aimed at a crystal of NaCl occurs at 15.9°. (See Figure 11.34, below.)
(a) What is the wavelength (pm) of the X-rays?
(b) At what angle would a second-order ($n = 2$) diffraction appear?

FIGURE 11.32 Cubic closest packing of frozen argon (face-centred cubic unit cell)

FIGURE 11.33 Cubic closest packing (face-centred unit cell) of frozen CH_4

Types and Properties of Crystalline Solids

The five most important types of solids are defined by the type(s) of particle(s) in their crystals (Table 11.5). We will highlight interparticle forces and physical properties.

Atomic Solids Individual atoms held together only by *dispersion forces* form an **atomic solid**, and the noble gases (group 18) are the only substances that form such solids. The very weak forces among the atoms mean that melting and boiling points and heats of vaporization and fusion are all very low, rising smoothly with increasing molar mass. Argon crystallizes in a cubic closest packing structure (Figure 11.32), as do the other noble gases.

Molecular Solids In the many thousands of **molecular solids**, individual molecules occupy the lattice points. Various combinations of dipole-dipole, dispersion, and hydrogen-bonding forces account for a wide range of physical properties. Dispersion forces in nonpolar substances lead to melting points that generally increase with molar mass (Table 11.5). Among polar molecules, dipole-dipole forces and, where possible, hydrogen bonding occur. Most molecular solids have much higher melting points than atomic solids (noble gases), but much lower melting points than other types of solids. Methane crystallizes in a face-centred cubic structure, with the centre of each carbon as the lattice point (Figure 11.33).

Ionic Solids To maximize attractions in a binary **ionic solid**, cations are surrounded by as many anions as possible, and vice versa, with *the smaller of the ions (usually the cation) lying in the spaces (holes) formed by the packing of the larger (usually the anion)*. The unit cell is the smallest portion that maintains the composition; that is, *the unit cell has the same cation/anion ratio as the empirical formula.*

As a result of cation-anion contact, the interparticle forces (ionic bonds) are *much* stronger than the VDW forces in atomic or molecular solids. The properties of ionic solids are a direct consequence of the *fixed ion positions* and *very strong attractive forces*, which create a high lattice energy. Thus, ionic solids typically have high melting points and low electrical conductivities. When a large quantity of heat is supplied and the ions gain enough kinetic energy to break free of their positions, the solid melts and the mobile ions conduct a current. Ionic compounds are hard

TABLE 11.5 Characteristics of the Major Types of Crystalline Solids

Type	Particle(s)	Interparticle Forces	Physical Properties	Examples [melting point in °C]
Atomic	Atoms	Dispersion	Soft, very low melting point, poor thermal and electrical conductors	Group 18: Ne [−249] to Rn [−71]
Molecular	Molecules	Dispersion, dipole-dipole, hydrogen bonds	Fairly soft, low to moderate melting point, poor thermal and electrical conductors	Nonpolar:* O_2 [−219] C_4H_{10} [−138] Cl_2 [2101] C_6H_{14} [−95] P_4 [44.1] Polar: SO_2 [−73] $CHCl_3$ [−64] HNO_3 [−42] H_2O [0.0] CH_3COOH [17]
Ionic	Positive and negative ions	Ion-ion attraction	Hard and brittle, high melting point, good thermal and electrical conductors when molten	CsCl [645] NaCl [801] CaF_2 [1423] MgO [2852]
Metallic	Atoms	Metallic bond	Soft to hard, low to very high melting point, excellent thermal and electrical conductors, malleable and ductile	Na [97.8] Zn [420] Fe [1535]
Network covalent	Atoms	Covalent bond	Very hard, very high melting point, usually poor thermal and electrical conductors	SiO_2 (quartz) [1610] C (diamond) [~4000]

*Nonpolar molecular solids are arranged in order of increasing molar mass. Notice the correlation with increasing melting point.

because only a strong external force can change the relative positions of the huge number of ions that are attracting one another. If enough force is applied to move them, ions of like charge are brought near each other, and their repulsions crack the crystal (see Figure 8.10).

Ionic solids often adopt cubic closest packing crystal structures. We will describe two examples with a 1/1 ratio of ions and then two examples with a 2/1 (or 1/2) ratio:

1. The *sodium chloride structure* is found in many ionic compounds, including most alkali (group 1) halides and hydrides, alkaline earth (group 2) oxides and sulfides, and several transition metal oxides and sulfides. To visualize this structure, imagine Cl^- anions and Na^+ cations in separate face-centred cubic arrays. Now imagine the two arrays penetrating each other, with the smaller Na^+ ions ending up in the holes between the larger Cl^- ions (Figure 11.34A). Thus, each Na^+ is

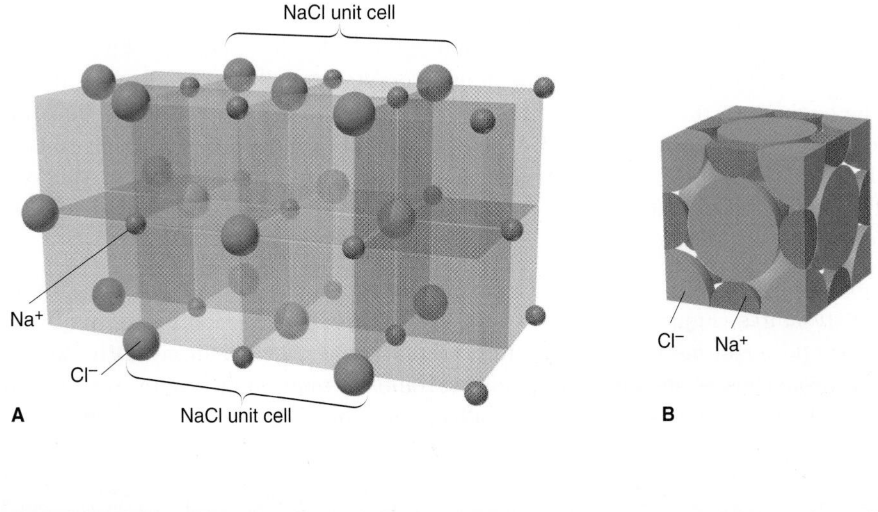

FIGURE 11.34 The sodium chloride structure. A. Expanded view. **B.** Space-filling model of the NaCl unit cell.

surrounded by six Cl^-, and vice versa (coordination number = 6). Figure 11.34B is a space-filling depiction of the unit cell: four Cl^- $[(8 \times \frac{1}{8}) + (6 \times \frac{1}{2}) = 4\ Cl^-]$ and four Na^+ $[(12 \times \frac{1}{4}) + 1$ in the centre $= 4\ Na^+]$ give a 1/1 ion ratio.

2. The *zinc blende (ZnS) structure* can be pictured as two face-centred cubic arrays, one of Zn^{2+} ions (*grey*) and the other of S^{2-} ions (*yellow*), penetrating each other such that each ion is tetrahedrally surrounded by four of the other ions (coordination number = 4) (Figure 11.35). There are four $[(8 \times \frac{1}{8}) + (6 \times \frac{1}{2}) = 4]$ S^{2-} ions and four Zn^{2+} ions for a 1/1 ion ratio. Many other compounds, including AgI, CdS, and the copper(I) halides, adopt the zinc blende structure.

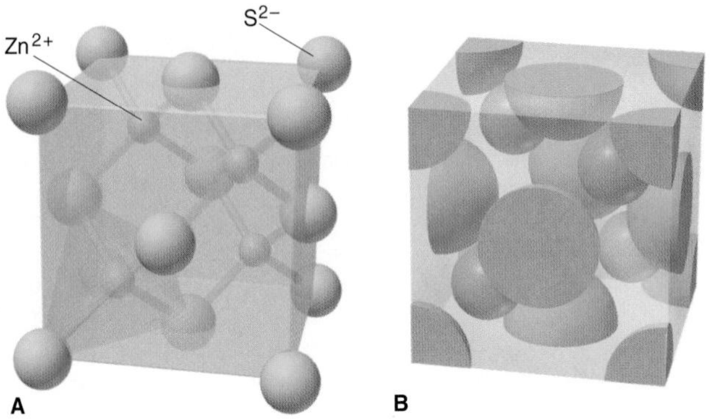

FIGURE 11.35 The zinc blende structure. A. Expanded view (with bonds shown for clarity). **B.** The unit cell is expanded a bit to show the interior ions.

3. The *fluorite (CaF_2) structure* is common among salts with a 1/2 cation/anion ratio and relatively large cations and small anions, such as SrF_2 and $BaCl_2$. In CaF_2, the unit cell is a face-centred cubic array of Ca^{2+} ions, with F^- ions occupying *all* eight available holes (Figure 11.36). This results in a Ca^{2+}/F^- ratio of 4/8, or 1/2.

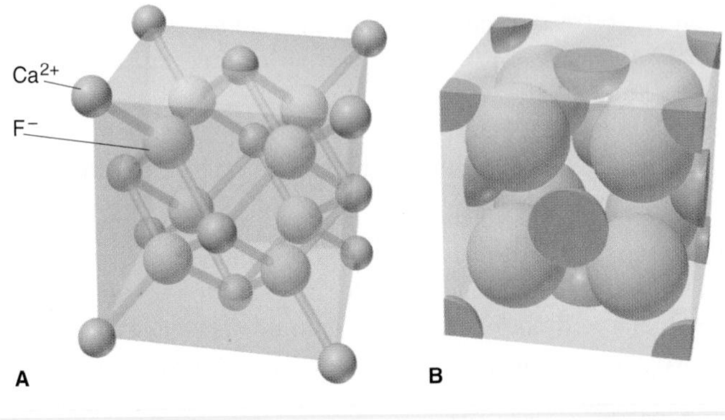

FIGURE 11.36 The fluorite structure. A. Expanded view (with bonds shown for clarity). **B.** The unit cell is expanded a bit to show the interior ions.

4. The *antifluorite structure* is common in compounds with a 2/1 cation/anion ratio and a relatively large anion, such as K_2S. The ion arrangement is the opposite of that in the fluorite structure: the cations occupy all eight holes formed by the cubic closest packing of the anions.

There are also tetrahedral and octahedral holes in closest packing, which can be occupied by other atoms or ions in crystal structures of salts and alloys. The packing of the spheres and the formation of the tetrahedral and octahedral holes are shown in Figure 11.37. Whenever you put four spheres together so that they are touching one another, you get a tetrahedral arrangement of spheres. The space in the centre is called a **tetrahedral hole**. The **octahedral hole** is formed by six spheres.

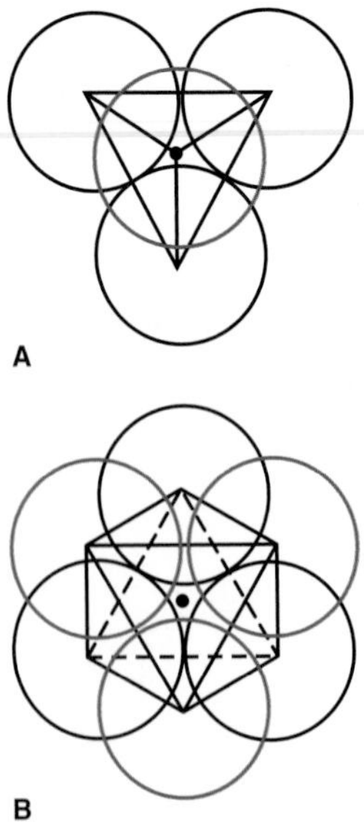

FIGURE 11.37 Tetrahedral (A) and octahedral (B) holes in closest packing

Metallic Solids Most metallic elements crystallize in one of the two closest packing structures (Figure 11.38). In contrast to the weak dispersion forces in atomic solids, powerful metallic bonding forces hold atoms together in **metallic solids**. The properties of metals—high electrical and thermal conductivity, lustre, and malleability—result from their delocalized electrons (Section 8.7). Melting point and hardness are related to the packing efficiency and number of valence electrons. For example, group 2 metals are harder and melt at higher temperatures than group 1 metals (see Figure 8.32), because the group 2 metals have twice as many delocalized valence electrons.

In addition to metallic solids, it is important to mention intermetallic compounds, such as the ones studied by Arthur Mar (Figure 11.39). They are hard, shiny, electrical conductors and are often magnetic, the properties we typically associate with metals. Intermetallic compounds have many applications, including structural materials in aircraft, superconductors in magnetic resonance imaging (MRI) instruments, permanent magnets in computer disk drives, and thermoelectric materials in solid state refrigerators.

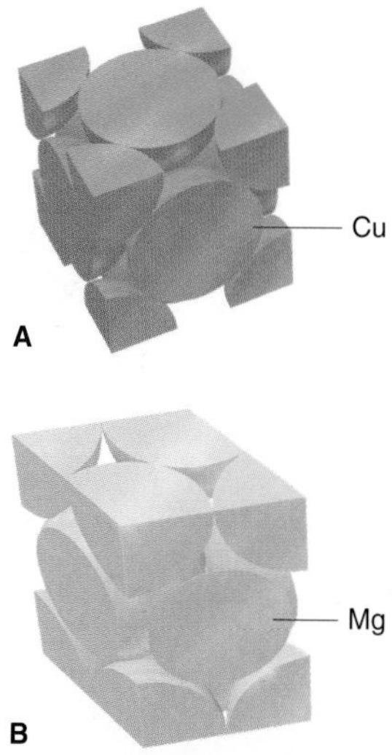

FIGURE 11.38 Crystal structures of metals. **A.** Copper adopts cubic closest packing. **B.** Magnesium adopts hexagonal closest packing.

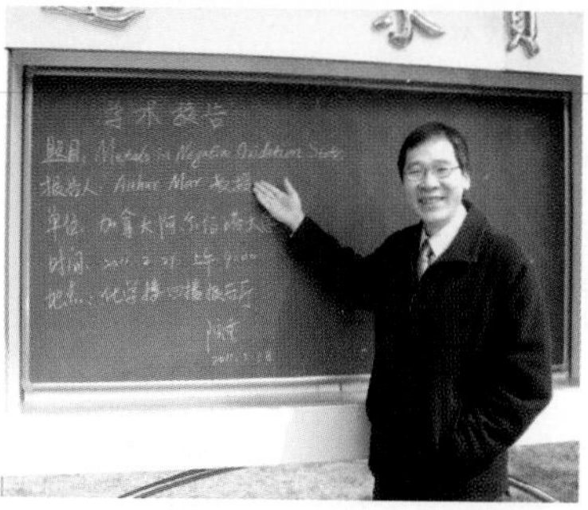

FIGURE 11.39 Arthur Mar is a professor at the University of Alberta, in Edmonton. His research focuses on intermetallic compounds involving combinations of two or more metals or metalloids from the left side of the periodic table (such as alkaline earth, rare-earth, and early transition metals) with elements from the right side of the periodic table (such as Ge, Sb, Bi, Au, and Hg). His long-term goal is to relate the crystal structures of these compounds to their physical properties so that he can understand how to design better materials for applications.

Network Covalent Solids Strong covalent bonds link the atoms together in **network covalent solids**; thus, separate particles are not present. These substances adopt a variety of crystal structures, depending on the details of their bonding.

As a consequence of strong bonding, all network covalent solids have extremely high melting and boiling points, but their conductivity and hardness vary. Two examples with the same composition but strikingly different properties are the two common crystalline forms of elemental carbon, graphite and diamond:

- *Graphite* occurs as stacked flat sheets of hexagonal carbon rings with a strong σ-bond framework and delocalized π bonds, reminiscent of benzene; the arrangement looks like chicken wire or a honeycomb. Whereas the π-bonding electrons of benzene are delocalized over one ring, those of graphite are delocalized over the entire sheet. Thus, graphite conducts electricity well—it is a common electrode material—but only in the plane of the sheets. The sheets interact via dispersion forces. Impurities, such as O_2, between the sheets allow them to slide past each other, which explains why graphite is soft and used as a lubricant.
- *Diamond* has a face-centred cubic unit cell, with each C tetrahedrally bonded to four others in an endless array. Throughout the crystal, strong single bonds make diamond the hardest natural substance known. Like most network covalent solids, diamond does not conduct electricity because the bonding electrons are localized.

These properties are compared in Table 11.6.

The most important network covalent solids are the *silicates*, which consist of extended arrays of covalently bonded silicon and oxygen atoms. Quartz (SiO_2) is a common example. We will discuss silicates, which form the structure of clays, rocks, and many minerals, when we consider the chemistry of silicon in Chapter 13.

Amorphous Solids

Amorphous solids are noncrystalline. Many have small, somewhat ordered regions interspersed among large disordered regions. Charcoal, rubber, and glass are examples.

The process that forms quartz glass is typical of the processes that form many amorphous solids. Crystalline quartz (SiO_2), which adopts cubic closest packing, is

TABLE 11.6 Comparison of the Properties of Diamond and Graphite

Property	Graphite	Diamond
Density (g/cm^3)	2.27	3.51
Hardness	<1 (very soft)	10 (hardest)
Melting point (K)	4100	4100
Colour	Shiny black	Colourless transparent
Electrical conductivity	High (along sheet)	None
$\Delta_r H°$ for combustion (kJ/mol)	−393.5	−395.4
$\Delta_f H°$ (kJ/mol)	0 (standard state)	1.90

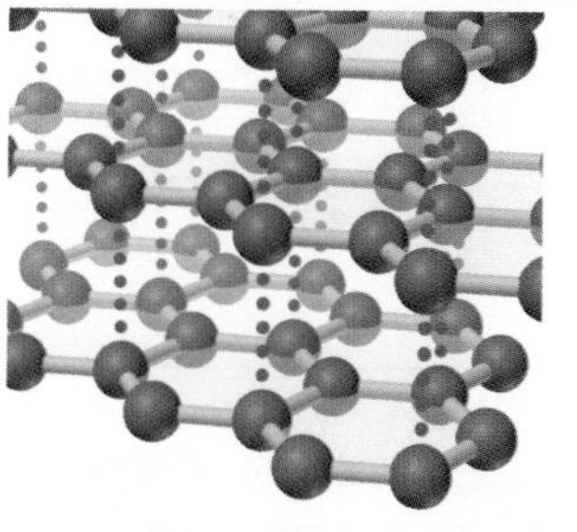

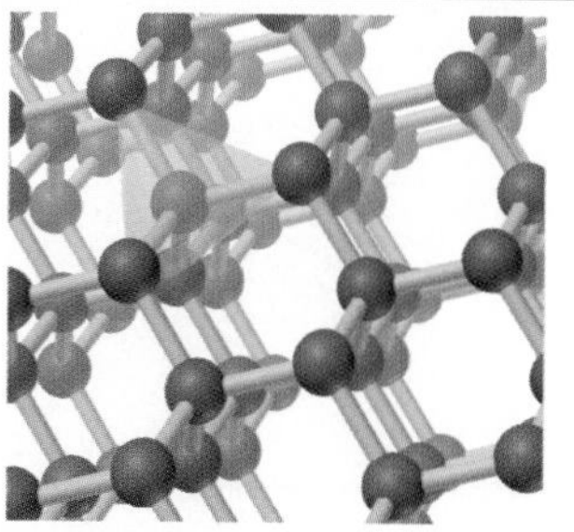

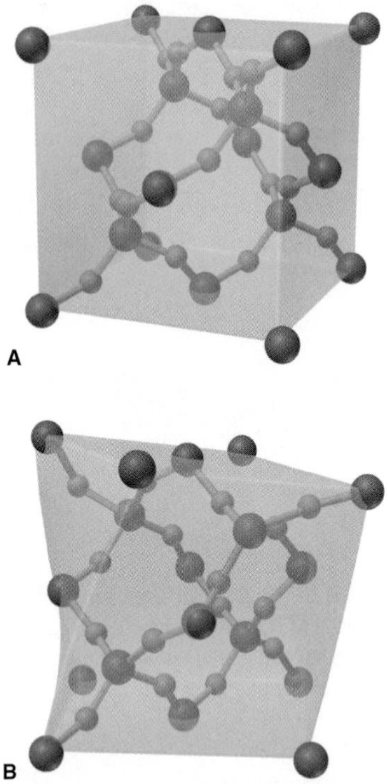

FIGURE 11.40 Crystalline and amorphous silicon dioxide. A. Cristobalite, a crystalline form of silica (SiO_2), has cubic closest packing. **B.** Quartz glass is amorphous with a generally disordered structure.

melted, and the viscous liquid is cooled rapidly to prevent it from recrystallizing. The chains of Si and O atoms cannot orient themselves quickly enough, so they solidify in a distorted jumble containing gaps and misaligned rows (Figure 11.40). The absence of regularity confers some properties of a liquid; in fact, glasses are sometimes referred to as *supercooled liquids*.

Bonding in Solids: Molecular Orbital Band Theory

Chapter 8 introduced a qualitative model of metallic bonding, with metal ions submerged in a "sea" of mobile, delocalized valence electrons. Molecular orbital (MO) theory offers a more quantitative, and therefore more useful, model called **band theory**. We will focus on the bonding in metals and the conductivity of metals, metalloids, and nonmetals.

Formation of Valence and Conduction Bands Recall, from Section 10.3, that when two atoms form a diatomic molecule, their atomic orbitals (AOs) combine to form an equal number of MOs. Figure 11.41 shows the formation of MOs in lithium. To form Li_2, each Li atom has four valence orbitals (one 2*s* and three 2*p*) that combine to form eight MOs, four bonding and four antibonding. The order of the MOs shows 2*s*-2*p* mixing (Figure 10.22B). Two more Li atoms form Li_4, a slightly larger aggregate, with 16 delocalized MOs. As more Li atoms join the cluster, more MOs are created, and their energy levels lie closer and closer together. Extending this process to 7 g (1 mol) of lithium results in 6×10^{23} Li atoms (Li_{N_A}) combining to form an extremely large number (4 × Avogadro's number) of delocalized MOs. *The energies of the MOs are so close that they form a continuum, or band, of MOs.*

The lower energy MOs are occupied by the $2s^1$ valence electrons and make up the **valence band**. The empty MOs that are higher in energy make up the **conduction band**. In Li metal, the valence band is derived from the 2*s* AOs, and the conduction band is derived from 2*s* and mostly 2*p* AOs. In Li_2, two valence electrons fill the lowest energy bonding MO and leave the antibonding MO empty. In Li metal, 1 mol of valence electrons fills the valence band and leaves the conduction band empty.

How Band Theory Explains Metallic Properties The key to understanding the properties of metals is knowing that *the valence and conduction bands are contiguous*; that is, the highest level of one touches the lowest of the other. This means

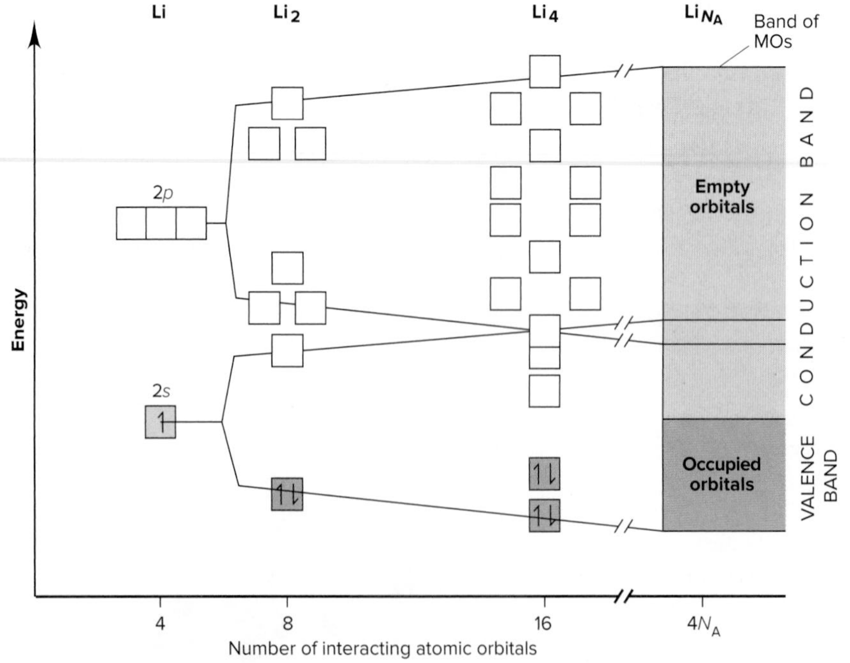

FIGURE 11.41 The band of MOs in lithium metal

that, given an infinitesimal quantity of energy, electrons jump from the filled valence band to the unfilled conduction band: the electrons are completely delocalized.

- *Electrical conductivity.* Metals conduct so well because an applied field easily excites the highest energy valence electrons into empty conduction orbitals, allowing them to move through the sample.
- *Lustre.* With so many closely spaced levels available, electrons absorb and release photons of many frequencies as they move between the valence and conduction bands.
- *Malleability.* Under an applied force, layers of positive metal ions move past each other, always protected from mutual repulsions by the delocalized electrons.
- *Thermal conductivity.* When a metal wire is heated, the highest energy electrons are excited and their extra energy is transferred as kinetic energy along the length of the wire.

Conductivity of Solids and the Size of the Energy Gap Like metal atoms, large numbers of nonmetal and metalloid atoms can form bands of MOs. Band theory explains differences in electrical conductivity and the effect of temperature among these three classes of substances in terms of the presence of an energy gap between their valence and conduction bands (Figure 11.42):

1. *Conductors (metals).* The valence and conduction bands of a **conductor** have *no energy gap* between them, so electrons flow when a tiny electric potential difference is applied. When the temperature is raised, greater random motion of the atoms hinders electron movement. Therefore, conductivity *decreases* when a metal is heated.

2. *Semiconductors (metalloids).* In a **semiconductor**, a *small energy gap* exists between the valence and conduction bands. Thermally excited electrons can cross the gap, allowing a small current to flow. In contrast to a conductor, conductivity *increases* when a semiconductor is heated.

3. *Insulators (nonmetals).* In an **insulator**, a *large energy gap* exists between the bands. No current is observed, even when the substance is heated.

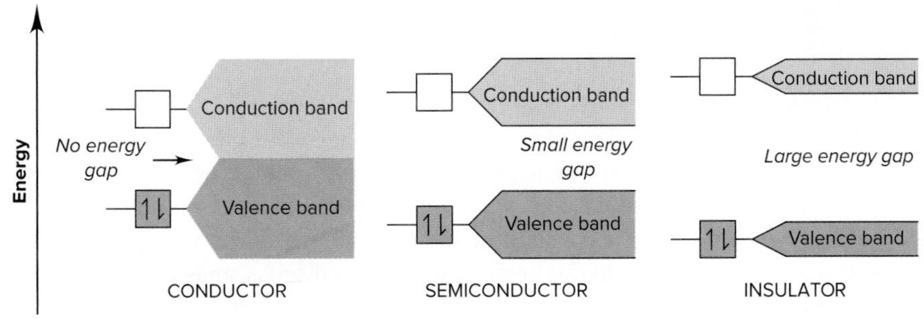

FIGURE 11.42 Electrical conductivity in a conductor, a semiconductor, and an insulator

Another type of electrical conductivity, called **superconductivity**, has been the focus of intensive research for the past few decades. When metals conduct at ordinary temperatures, moving electrons collide with vibrating atoms; the reduction in their flow appears as resistive heating and represents a loss of energy.

For many years, to conduct with no energy loss—to superconduct—required minimizing atom movement by cooling with liquid helium (boiling point = 4 K; price = \$11/L). Then, in 1986, certain ionic oxides that superconduct near the boiling point of liquid nitrogen (boiling point = 77 K; price = \$0.20/L) were prepared. Like metal conductors, oxide superconductors, such as $YBa_2Cu_3O_7$, have no gap between bands. In 1989, oxides with Bi and Tl instead of Y and Ba were synthesized and found to superconduct at 125 K. In 1993, an oxide with Hg, Ba, and Ca, in addition to Cu and O, was found to superconduct at 133 K. Such materials could transmit electricity with no loss of energy, allowing power plants to be located far from cities. They could be part of ultrasmall microchips for ultrafast computers, electromagnets to levitate superfast trains, and inexpensive medical diagnostic equipment with superb image clarity. However, the oxides are brittle and not easy to machine, and the superconductivity may disappear when the oxides are warmed and may not return on cooling. Addressing these and related problems will involve chemists, physicists, and engineers for many years.

SUMMARY OF SECTION 11.6

- Particles in crystalline solids lie at points that form a structure of repeating unit cells.
- The three types of unit cells of the cubic system are simple, body-centred, and face-centred. The highest packing efficiency occurs with cubic (face-centred) and hexagonal closest packing.
- Bond angles and distances in a crystal are determined with X-ray diffraction analysis and scanning tunnelling microscopy. These data are used to determine atomic radii.
- Atomic (group 18) solids adopt cubic closest packing, with atoms held together by weak dispersion forces.
- Molecular solids have molecules at the lattice points and often adopt cubic closest packing. Combinations of intermolecular forces (dispersion, dipole-dipole, and hydrogen bonding) result in physical properties that vary greatly.
- Ionic solids crystallize with one ion filling holes in the cubic closest packing array of the other. High melting points, hardness, and low conductivity arise from strong ionic attractions.
- Most metals have a closest packing structure. Their physical properties result from the high packing efficiency and the presence of delocalized electrons.
- Atoms of network covalent solids are covalently bonded throughout the sample, so these substances have very high melting and boiling points.
- Amorphous solids have little regularity in their structure.
- Band theory proposes that atomic orbitals (AOs) of many atoms combine to form a continuum, or band, of molecular orbitals (MOs). Metals are electrical conductors because electrons move freely from the filled (valence) band to the empty (conduction) band. Insulators have a large energy gap between the two portions, and semiconductors have a small gap, which can be bridged by heating.

11.7 Advanced Materials

In the last few decades, the exciting field of materials science has grown from solid-state chemistry, physics, and engineering. Objects that were once considered futuristic fantasies are becoming realities: powerful, ultrafast computers smaller than this book; cars powered by sunlight and made of nonmetallic parts stronger than steel and lighter than aluminum; ultrasmall machines constructed by manipulating individual atoms and molecules. In this section, we briefly discuss some of these remarkable materials. Polymeric materials will be discussed in Chapter 22.

Electronic Materials

The ideal of a perfectly ordered crystal is attainable only if the crystal is grown very slowly under carefully controlled conditions. When crystals form more rapidly, **crystal defects** inevitably form. Planes of particles are misaligned, particles are out of place or missing entirely, and foreign particles replace those that belong in the lattice. Although defects usually weaken a substance, they can be introduced intentionally to improve materials, giving them increased strength, hardness, or conductivity. Centuries of metal-working exemplify the traditional importance of defects, which are the basis of modern electronic materials.

Introducing Defects in Welding and Alloying In the process of *welding* two metals together, *vacancies* form near the surface when atoms vaporize, and then these vacancies move deeper as atoms from lower rows rise to fill the gaps. Welding causes the two types of metal atoms to intermingle and fill each other's vacancies. Metal *alloying* introduces several kinds of crystal defects, as when some atoms of a second metal occupy lattice sites of the first. Often, the alloy is harder than the pure metal; an example is brass, an alloy of copper with zinc. One reason the welded

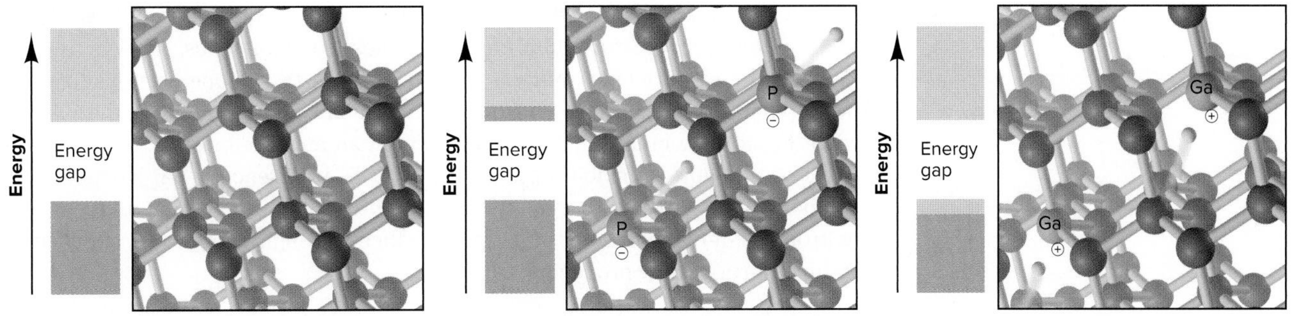

A Pure silicon crystal B n-type doping with phosphorus C p-type doping with gallium

FIGURE 11.43 **Crystal structures and band representations of doped semiconductors. A.** Pure silicon has an energy gap between its valence and conduction bands, which keeps conductivity low at room temperature. **B.** Doping silicon with phosphorus (*purple*) adds additional valence electrons. **C.** Doping silicon with gallium (*orange*) removes electrons from the valence band and introduces positive holes.

metals are stronger and the alloy is harder is that the second metal contributes additional valence electrons for metallic bonding.

Doped Semiconductors By controlling the number of valence electrons through the creation of specific types of crystal defects, chemists and engineers can greatly increase the conductivity of a semiconductor. Pure silicon (Si; group 14) conducts poorly at room temperature because an energy gap separates its filled valence band from its conduction band (Figure 11.43A). Its conductivity can be greatly enhanced by **doping**, adding small amounts of other elements to increase or decrease the number of valence electrons in the bands:

- *n-type semiconductor: increasing the number of valence electrons.* When Si is doped with phosphorus (or another group 15 element), P atoms occupy some of the lattice sites. Since P has one more valence electron than Si, this additional electron must enter an empty orbital in the conduction band, thus bridging the energy gap and increasing conductivity. Such doping creates an *n-type semiconductor,* so called because extra *n*egative charges (electrons) are present (Figure 11.43B).
- *p-type semiconductor: decreasing the number of valence electrons.* When Si is doped with gallium (or another group 13 element), Ga atoms occupy some sites (Figure 11.43C). Since Ga has one less valence electron than Si, some of the orbitals in the valence band are empty, which creates a positive site. Si electrons can migrate to these empty orbitals, thereby increasing conductivity. Such doping creates a *p-type semiconductor,* so called because the empty orbitals act as *p*ositive holes.

In contact with each other, an n-type semiconductor and a p-type semiconductor form a *p-n junction.* When the negative terminal of a battery is connected to the n-type portion and the positive terminal is connected to the p-type portion, electrons (*yellow, negative spheres*) flow freely in the n-to-p direction, which has the simultaneous effect of moving holes (*white, positive spheres*) in the p-to-n direction (Figure 11.44A). No current flows if the terminals are reversed (Figure 11.44B). Such unidirectional current flow makes a p-n junction act as a *rectifier,* a device that converts alternating current into direct current. A p-n junction in a modern integrated circuit can be made smaller than a square with 10 μm sides.

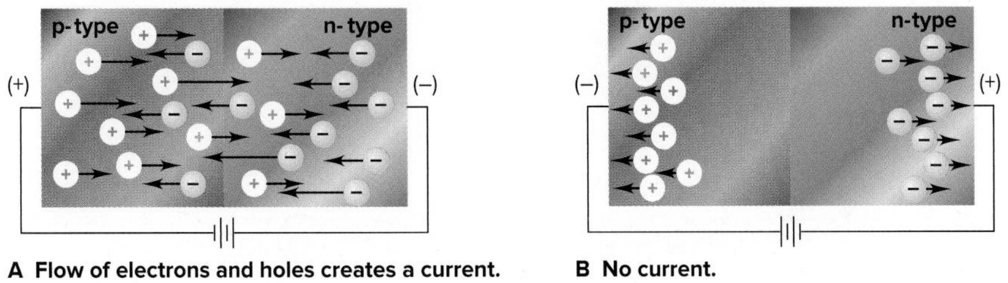

A Flow of electrons and holes creates a current. B No current.

FIGURE 11.44 **The p-n junction.** Placing a p-type semiconductor adjacent to an n-type semiconductor creates a p-n junction.

A modern computer chip the size of a nickel may incorporate millions of p-n junctions in the form of *transistors*. One of the most common types, an n-p-n transistor, is made by sandwiching a p-type portion between two n-type portions to form adjacent p-n junctions. The current flowing through one junction controls the current flowing through the other junction and results in an amplified signal.

One of the more common types of *solar cell* is, in essence, a p-n junction with an n-type surface exposed to a light source. Light provides the energy to free electrons from the n-type portion and accelerate them through an external circuit into the p-type portion, thus producing a current.

Liquid Crystals

Incorporated in the membrane of every cell in your body and in the display of every electronic device are **liquid crystals**. These materials flow like liquids but, like crystalline solids, pack at the molecular level with a high degree of order.

Ordering the Particles Let us examine how particles are ordered in the three common physical states and how this affects their properties. Both gases and liquids are *isotropic*: their physical properties are the same in every direction within the phase. For example, the viscosity of a gas or a liquid is the same regardless of direction. Glasses and other amorphous solids are also isotropic because they have no regular lattice structure.

In contrast, crystalline solids have a high degree of order among their particles. The properties of a crystal *do* depend on direction, so a crystal is *anisotropic*. The facets in a cut diamond, for instance, arise because the crystal cracks in one direction more easily than in another direction. Liquid crystals are anisotropic in several physical properties, including electrical and optical properties, which differ with direction through the phase.

Molecular Characteristics Liquid crystal phases consist of individual molecules with two characteristics: (1) a long, cylindrical shape and (2) a structure that fosters intermolecular attractions but inhibits perfect crystalline packing. Molecules that form liquid crystal phases have rodlike shapes and certain groups (for example, flat benzene-like ring systems in the molecules shown in Figure 11.45) that keep the molecules extended. Many of these molecules also have polarity associated with the long molecular axis. A strong electric field can orient these polar molecules like compass needles in a magnetic field.

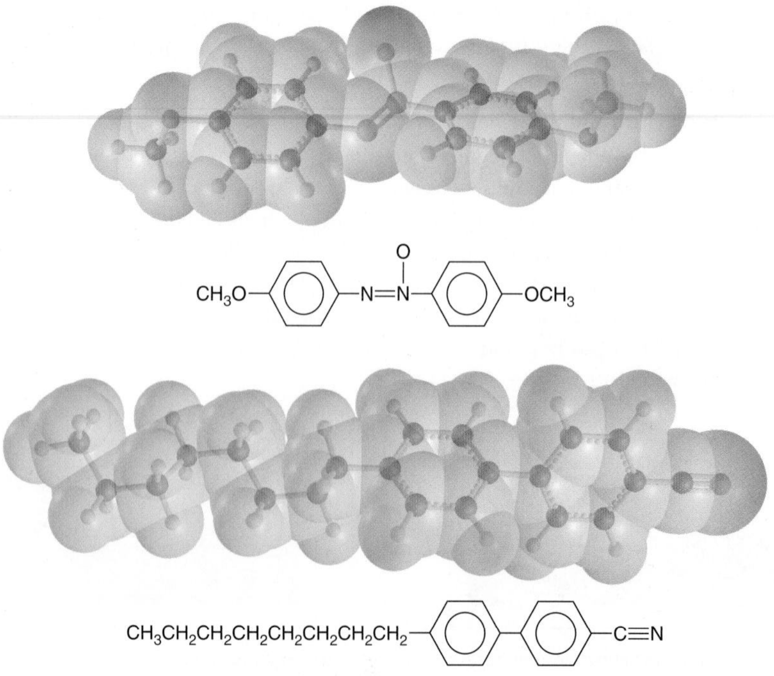

FIGURE 11.45 Structures of two typical molecules that form liquid crystal phases. Notice the long extended shapes and the regions of high (*red*) and low (*blue*) electron density.

The viscosity of a liquid crystal phase is lowest in the direction parallel to the long axis. Like moistened microscope slides, it is easier for the molecules to slide along each other (because the total attractive force remains the same) than it is for them to pull straight apart from each other. As a result, the molecules tend to align while the phase flows.

Controlling Conditions to Form Phases Liquid crystal phases can arise in two general ways, and, sometimes, either way can occur in the same substance:

1. *Thermotropic phases* develop as a result of a change in temperature. As a crystalline solid is heated, the molecules leave their lattice sites, but the intermolecular interactions are still strong enough to keep the molecules aligned with each other along their long axes. Like any other phase, the liquid crystal phase has sharp transition temperatures, but they occur over a relatively small temperature range. Further heating provides enough kinetic energy for the molecules to become disordered, as in a normal liquid. The typical range for liquid crystal phases of pure substances is from $<1°C$ to as much as 10°C, but mixing phases of two or more substances can greatly extend the range. For this reason, the liquid crystal phases used within display devices, as well as those within cell membranes, consist of mixtures of molecules.

2. *Lyotropic phases* occur in solution as a result of changes in concentration, but the conditions for forming these phases vary for different substances. For example, when purified, some biomolecules that exist in cell membranes form lyotropic phases in water at the temperature that occurs within the organism. At the other extreme, Kevlar, a fibre used in bulletproof vests and high-performance sports equipment, forms a lyotropic phase in a concentrated H_2SO_4 solution.

In some cases, a substance that forms a given liquid crystal phase under one set of conditions forms other phases under different conditions. Thus, a given thermotropic liquid crystal substance can change from a disordered liquid, through a series of distinct liquid crystal phases, to an ordered crystal as a result of a decrease in temperature. A lyotropic substance can undergo similar changes as a result of an increase in concentration.

Types of Order in Liquid Crystal Phases Molecules that form liquid crystal phases can exhibit various types of order. Three common types are nematic, cholesteric, and smectic:

1. In a *nematic phase*, the molecules lie in the same direction but their ends are not aligned, much like a school of fish swimming in synchrony (Figure 11.46A). The nematic phase is the least ordered type of liquid crystal phase.

2. In a *cholesteric phase*, which is somewhat more ordered, the molecules lie in layers. Each layer exhibits nematic-type ordering. Rather than lying in a parallel fashion, however, each layer is rotated by a fixed angle with respect to the next layer, giving a helical (corkscrew) arrangement. A cholesteric phase is often called a *twisted nematic phase* (Figure 11.46B).

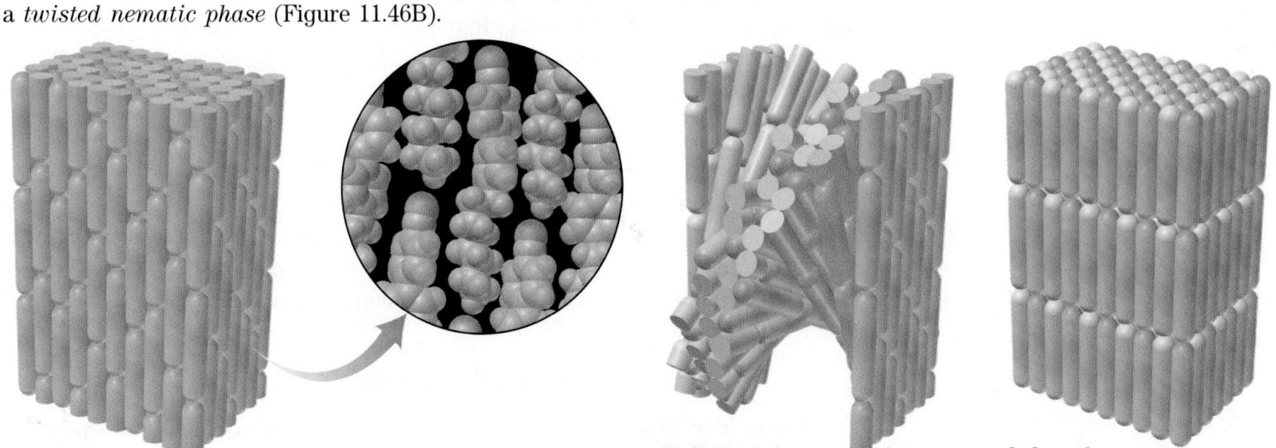

FIGURE 11.46 The three common types of ordering in liquid crystal phases

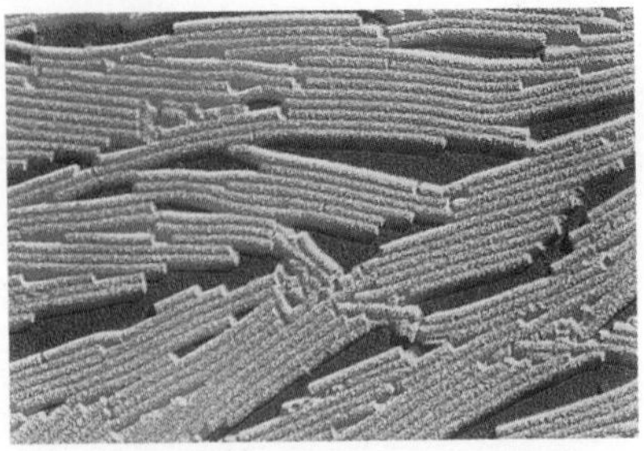
A

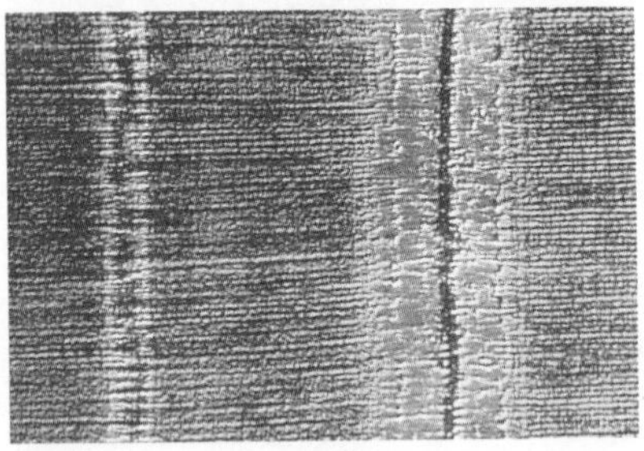
B

FIGURE 11.47 Liquid crystal–type phases in biological systems. **A.** Nematic arrays of tobacco mosaic virus particles within the fluid of a tobacco leaf. **B.** The smectic-like arrangement of actin and myosin protein filaments in voluntary muscle cells.

3. In a *smectic phase*, which is the most ordered, the molecules lie parallel to each other, *with* their ends aligned, in layers that are stacked directly over each other (Figure 11.46C). The long molecular axis has a well-defined angle (shown in the figure as 90°) with respect to the plane of the layer. The molecules in Figure 11.45 form nematic or smectic phases. Liquid crystal–type phases appear in many biological systems (Figure 11.47).

Applications of Liquid Crystals The ability to control the orientation of the molecules in a liquid crystal allows us to produce materials with high strength or unique optical properties:

1. *High-strength applications* involve the use of extremely long molecules called *polymers*. While in a thermotropic liquid crystal phase and during their flow through the processing equipment, these molecules become highly aligned, like the fibres in wood. Cooling solidifies them into fibres, rods, and sheets that can be shaped into materials with superior mechanical properties. Sporting equipment, supersonic aircraft parts, and the sails used in the America's Cup races are fabricated from these polymeric materials. (We will discuss the structure and physical behaviour of polymers later, in Chapter 22.)

2. *Optical applications* include the liquid crystal displays (LCDs) used in countless devices, such as watches, calculators, cellphones, and computers. All of these devices depend on *changes in molecular orientation in an electric field.* Figure 11.48 shows a small portion of a wristwatch LCD. Layers of nematic phases are sandwiched between thin glass plates that incorporate transparent electrodes. The long molecular axis lies parallel to the plane of the plates. The distance between plates (6 μm to 8 μm) allows the molecular axis within each succeeding layer to twist just enough for the molecular orientation at the bottom plate to be 90° from the molecular orientation at the top plate. Above and below this "sandwich" are thin polarizing filters that allow light waves oriented in only one direction to pass through. The filters are placed in a "crossed" arrangement, so that light passing through the top filter must twist 90° to pass through the bottom filter. This whole grouping of filters, plates, and liquid crystal phase lies on a mirror.

A current generated by the watch battery controls the orientation of the molecules. With the current on in one region of the display, the molecules become oriented *toward* the field and, thus, block the light from passing through to the bottom filter, so that this region appears dark. With the current off in another region, light passes through the molecules and bottom filter to the mirror and back again, so that this region appears bright.

Cholesteric liquid crystals are used in applications that involve colour changes with temperature. The twisted molecular orientation "unwinds" with heating, and the extent of the unwinding determines the colour. Liquid crystal thermometers include a mixture of substances to widen their range of temperatures. Newer uses include "mapping" the area of a tumour, detecting faulty connections in electronic circuit boards, and nondestructively testing materials under stress. Another application is in "mood rings" or "mood strips," in which a liquid crystal changes colour to indicate the "mood" of a person.

Ceramic Materials

First developed by Stone Age people, **ceramics** are nonmetallic, nonpolymeric solids that are hardened by high temperature. Clay ceramics consist of silicate microcrystals suspended in a glassy cementing medium. For example, to "fire" a ceramic pot made of an aluminosilicate clay, such as kaolinite, a kiln heats the pot to 1500°C and the clay loses water:

$$Si_2Al_2O_5(OH)_4(s) \longrightarrow Si_2Al_2O_7(s) + 2H_2O(g)$$

During the heating process, the structure rearranges to an extended network of Si-centred and Al-centred tetrahedrons of O atoms (Section 13.6).

TABLE 11.7 Some Uses of Modern Ceramics and Ceramic Mixtures

Ceramic	Applications
SiC, Si_3N_4, TiB_2, Al_2O_3	Whiskers (fibres) to strengthen Al and other ceramics
Si_3N_4	Car engine parts; turbine rotors for "turbo" cars; electronic sensor units
Si_3N_4, BN, Al_2O_3	Support or layering materials (as insulators) in electronic microchips
SiC, Si_3N_4, TiB_2, ZrO_2, Al_2O_3, BN	Cutting tools; edge sharpeners (as coatings and whole devices); scissors; surgical tools; industrial "diamond"
BN, SiC	Armour-plating reinforcement fibres (as in Kevlar composites)
ZrO_2, Al_2O_3	Surgical implants (hip and knee joints)

Bricks, porcelain, glazes, and other clay ceramics are hard and resistant to heat and chemicals. Today's high-tech ceramics have these characteristics in addition to superior electrical and magnetic properties (Table 11.7). As just one example, consider the unusual electrical behaviour of certain zinc oxide (ZnO) composites. Ordinarily a semiconductor, ZnO can be doped so that it becomes a conductor. Embedding particles of the doped oxide into an insulating ceramic produces a variable resistor: the material conducts poorly at low voltage but conducts well at high voltage. The changeover voltage can be preset by controlling the size of the ZnO particles and the thickness of the insulating medium.

Preparing Modern Ceramics Among the important modern ceramics are silicon carbide (SiC) and silicon nitride (Si_3N_4), boron nitride (BN), and the superconducting oxides. They are prepared by standard chemical methods that involve driving off a volatile component during the reaction:

1. *SiC ceramics* are made from compounds used in silicone polymer manufacture (we will discuss these polymers in Section 22.1):

$$n(CH_3)_2SiCl_2(l) + 2nNa(s) \longrightarrow 2nNaCl(s) + [(CH_3)_2Si]_n(s)$$

This product is heated to 800°C to form the ceramic:

$$[(CH_3)_2Si]_n \longrightarrow nCH_4(g) + nH_2(g) + nSiC(s)$$

Silicon carbide can also be prepared by the direct reaction of silicon and graphite in a vacuum:

$$Si(s) + C(s, \text{graphite}) \xrightarrow{\sim 1500°C} SiC(s)$$

The nitride is also prepared by reaction of the elements:

$$3Si(s) + 2N_2(g) \xrightarrow{>1300°C} Si_3N_4(s)$$

2. *BN ceramics* are produced through the reaction of boron trichloride or boric acid with ammonia:

$$B(OH)_3(s) + 3NH_3(g) \longrightarrow B(NH_2)_3(s) + 3H_2O(g)$$

Heat drives off some of the bound nitrogen as NH_3 to yield a ceramic:

$$B(NH_2)_3(s) \xrightarrow{\Delta} 2NH_3(g) + BN(s)$$

3. One type of *superconducting oxide* is made by heating a mixture of barium carbonate with copper and yttrium oxides, followed by further heating in the presence of O_2:

$$4BaCO_3(s) + 6CuO(s) + Y_2O_3(s) \xrightarrow{\Delta} 2YBa_2Cu_3O_{6.5}(s) + 4CO_2(g)$$

$$YBa_2Cu_3O_{6.5}(s) + \frac{1}{4}O_2(g) \xrightarrow{\Delta} YBa_2Cu_3O_7(s)$$

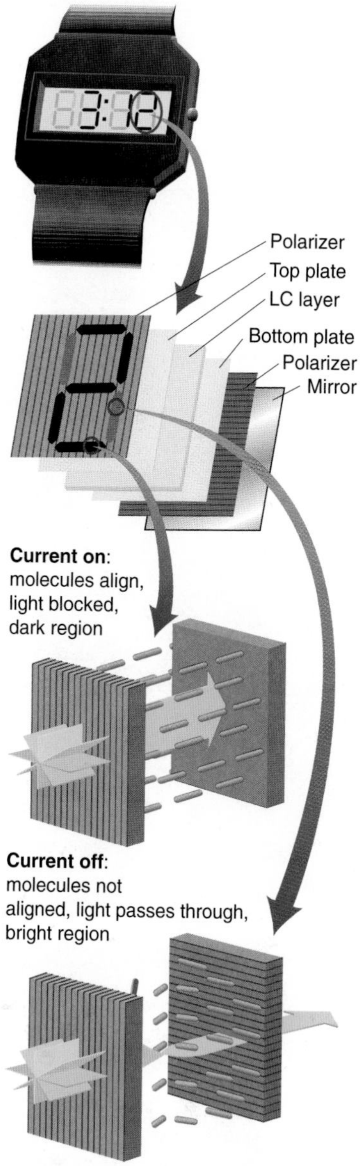

FIGURE 11.48 An LCD. A close-up of the "2" on a wristwatch LCD reveals two polarizers sandwiching two glass plates, which sandwich a liquid crystal (LC) layer, all lying on a mirror. When light waves enter the first polarizer, waves oriented in one direction emerge to enter the LC layer. Enlarging a dark region (*top blow-up*) shows the LC molecules aligned, keeping light waves from passing through; thus, the viewer sees no light. Enlarging a bright region (*bottom blow-up*) shows the LC molecules rotating the plane of the light waves, which can then pass through and reach the viewer.

Ceramic Structures and Uses The structures of several ceramic materials are shown in Figure 11.49.

1. *Superconducting oxides* often contain copper in an unusual oxidation state. In $YBa_2Cu_3O_7$, for instance (Figure 11.49A), assuming oxidation states of +3 for Y, +2 for Ba, and −2 for O, the three Cu atoms have a total oxidation state of +7. This is allocated as $Cu(II)_2Cu(III)$, with one Cu in the unusual +3 state. X-ray diffraction analysis indicates that a distortion in the structure makes four of the oxide ions unusually close to the Y^{3+} ion, which aligns the Cu ions into chains within the crystal. It is suspected that a specific half-filled 3*d* orbital in Cu oriented toward a neighbouring O^{2-} ion may be associated with superconductivity.

2. *Silicon carbide* (Figure 11.49B) has a diamond-like structure. Network covalent bonding gives this material great strength. It is made into thin fibres, called *whiskers*, to reinforce other ceramics and prevent cracking in a composite structure, much like steel rods reinforcing concrete.

3. *Silicon nitride* is virtually inert chemically, retains its strength and wear resistance for extended periods above 1000°C, is dense and hard, and acts as an electrical insulator. Many automakers are testing it in high-efficiency car and truck engines because of its low weight, tolerance to high operating temperature, and little need for lubrication.

4. *BN ceramics* (Figure 11.49C) exist in two structures, analogous to the common crystalline forms of carbon. In the graphite-like form, BN has extraordinary properties as an electrical insulator. At high temperature and very high pressure (1800°C and 8.6×10^4 kPa), it converts to a diamond-like structure, which is extremely hard and durable. Both forms are virtually invisible to radar.

Because of brittleness, these ceramics are difficult to fashion into wires, but methods for making films and ribbons have been developed. The brittleness arises from the strength of the ionic-covalent bonding and the resulting inability to deform. Under stress, a microfine defect widens until the ceramic cracks. One new method forms defect-free superconducting oxide ceramics using controlled packing and heat treatment of extremely small, uniform oxide particles coated with organic polymers. Another method is aimed at arresting a widening crack. The ceramics are embedded with zirconia (ZrO_2), whose crystal structure expands up to 5% under mechanical stress. When an advancing crack reaches the zirconia particles, they pinch it shut. A third method prepares the material with precisely grown crystals of Al_2O_3 and $GdAlO_3$, which become entangled during the solidification process. The resulting material bends without cracking above 1800 K.

Nanotechnology: Designing Materials Atom by Atom

Nanotechnology is the science and engineering of nanoscale systems, whose sizes range from 1 nm to 100 nm. (Recall that 1 nm is 10^{-9} m—about one-billionth the width of a desk or one-millionth the width of the tip of a pen. A nanometre is to a

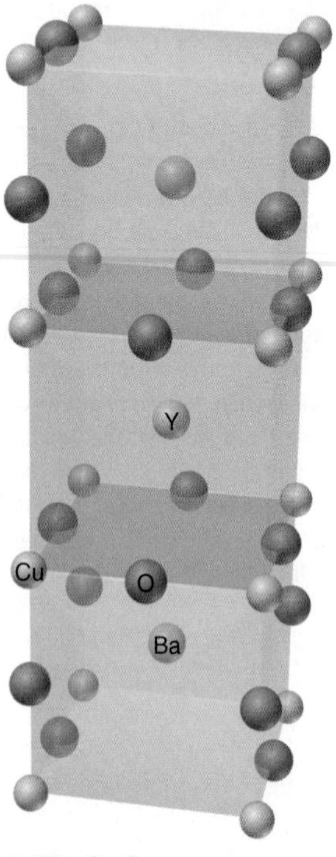

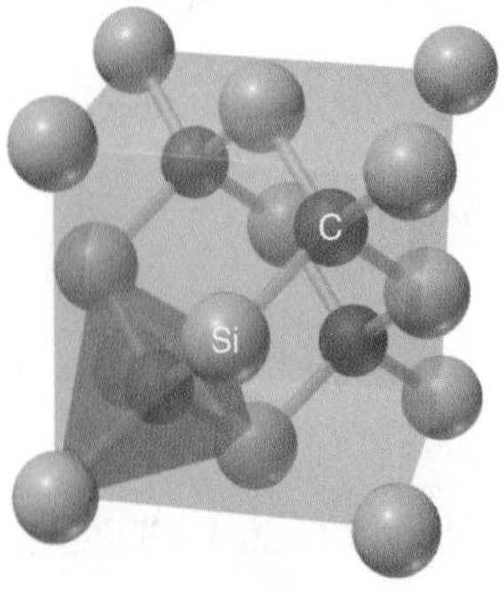

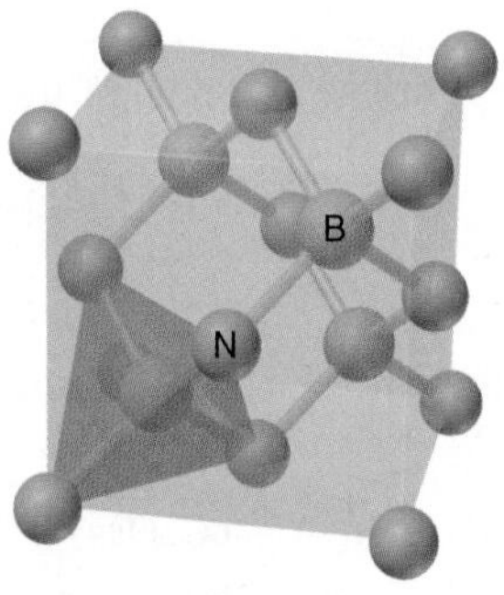

A $YBa_2Cu_3O_7$ **B** Silicon carbide **C** Cubic BN (borazon)

FIGURE 11.49 Expanded view of the atom arrangements in some modern ceramic materials

metre as a marble is to Earth!) Nanotechnology joins scientists from physics, materials science, chemistry, biology, environmental science, medicine, and many branches of engineering.

Nanoscale materials behave like neither atoms, which are smaller (1×10^1 nm), nor crystals, which are at least 1×10^5 nm. The different behaviour is due to a surface area that is similar to their interior volume: for example, a 5-nm particle has about half of its atoms on its surface. Cut a piece of ordinary aluminum foil into pieces just large enough to see with a light microscope and they still behave like the original piece. However, if you could cut the original piece into 20-nm to 30-nm pieces, they would explode. Indeed, aluminum nanoparticles are used in fireworks and propellants.

As so often happens in science, a technological advance allows a new field to blossom. Nanotechnologists can use the scanning tunnelling microscope (see the Tools of the Laboratory feature above) and similar scanning probes, such as atomic force microscopes, to examine the chemical and physical properties of nanostructures and build them up, in many cases, atom by atom.

Two features of nanoscale engineering occur routinely in nature:

1. *Self-assembly* is the ability of smaller, simpler parts to organize themselves into a larger, more complex whole. On the molecular scale, self-assembly refers to atoms or small molecules aggregating through intermolecular forces. Oppositely charged regions on two such particles make contact to form a larger particle, which, in turn, aggregates with another to form a still larger particle.

2. *Controlled orientation* is the positioning of two molecules near each other long enough for intermolecular forces to result in a change, often a bond breaking or forming. All enzymes function that way.

Because this new field is changing many times faster than a textbook does, we can only provide a glimpse of a few exciting research directions in nanotechnology.

Nanoscale Optical Materials *Quantum dots* are nanoparticles of semiconducting materials, such as gallium arsenide (GaAs) and gallium selenide (GaSe), that are smaller than 10 nm. Rather than having a band of energy levels, a quantum dot has discrete energy levels, much like a single atom. As in atoms, the energy levels can be studied with spectroscopy; unlike atoms, quantum dots can be modified chemically. When high-energy (short-wavelength) UV light irradiates a suspension of quantum dots, they become excited and emit radiation of lower energy (longer wavelength). Most important, the emitted wavelengths depend on the size of the dots: the larger the particles, the longer the wavelength emitted. Thus, the colour of the emitted light can be "tuned" by varying the dimensions of the nanoparticles (Figure 11.50).

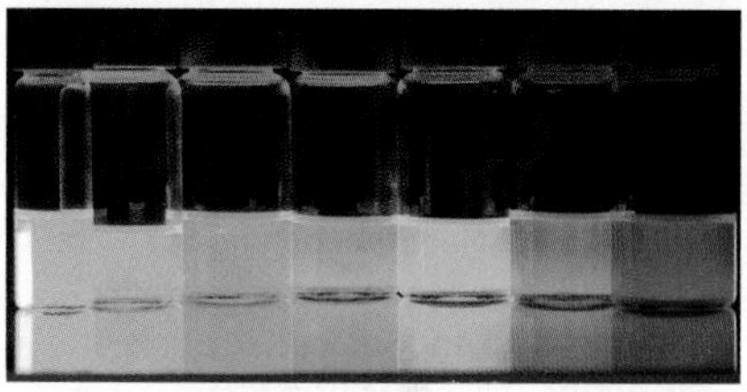

FIGURE 11.50 The colours of quantum dots. The smaller the dots, the shorter the emitted wavelengths.

One use for quantum dots is the imaging of specific cells. A quantum dot can be encapsulated in a coating and then bonded to a particular protein used by a target organ. The cells in the target organ incorporate the protein, along with its attached quantum dot, and the location of the protein is viewed spectroscopically.

Nanostructured Materials *Nanostructuring*, the construction of bulk materials from nanoscale building blocks, increases strength, ductility, plasticity, and many other properties. The resulting *nanocomposites* behave as liquid magnets, ductile cements, conducting elastomers, and many other unique materials.

One example mimics a familiar natural material. Bones are natural nanocomposites of hydroxyapatite (a type of calcium phosphate) and other minerals. Synthetic nanocrystalline hydroxyapatite has been made with weight-bearing properties identical to bone and, thus, can provide a framework for natural bone tissue to grow into and heal a fracture. Another material, called a *ferrofluid*, consists of magnetic nanoparticles (usually magnetite, Fe_3O_4) dispersed in a viscous fluid. When placed in a magnetic field, the particles align to conform to the shape of the field (Figure 11.51). Ferrofluids find uses in automobile shock absorbers, magnetic plastics, audio speakers, computer hard drives, high-vacuum valves, and many other devices.

FIGURE 11.51 The magnetic behaviour of a ferrofluid. Nanoparticles of magnetite (Fe_3O_4) dispersed in a viscous fluid are suspended between the poles of a magnet.

High-Surface Area Materials Carbon nanotubes have remarkable mechanical and electrical properties. A team of researchers at the National Research Council's Steacie

FIGURE 11.52 Hockey sticks with carbon nanotubes

Institute for Molecular Sciences in Ottawa is currently using the cutting-edge science of nanotechnology to make better hockey sticks with carbon nanotubes (Figure 11.52). Despite their size, carbon nanotubes have 100 times the strength of steel and only one-sixth the weight. Adding carbon nanotubes to the composites used in today's hockey sticks can dramatically improve their durability, resulting in lighter, tougher, and more flexible sticks that will not break at a crucial moment in a game.

A possible future application for carbon nanotubes involves storing H_2 gas for hydrogen-powered vehicles. Currently, H_2 is stored in metal cylinders at high pressures (>7092 kPa). Porous solids, such as carbon nanotubes, can store the same amount of gas at much lower pressures because they have surface areas of 4500 m^2/g—about four football fields per gram of material! Thus, a medium-size container of nanotubes could store a large amount of H_2 fuel under conditions safe enough for a family car.

Other high–surface area materials being developed are porous membranes for water purification or batteries and multilayer films that incorporate photosynthetic molecules for high-efficiency solar cells. A very exciting development is a molecule-specific biosensor, which can detect as little as 10^{-14} mol of DNA using colour changes in gold nanoparticles. Such biosensors could circulate freely in the bloodstream to measure levels of specific disease-related molecules, deliver drugs to individual cells, and even alter individual genes.

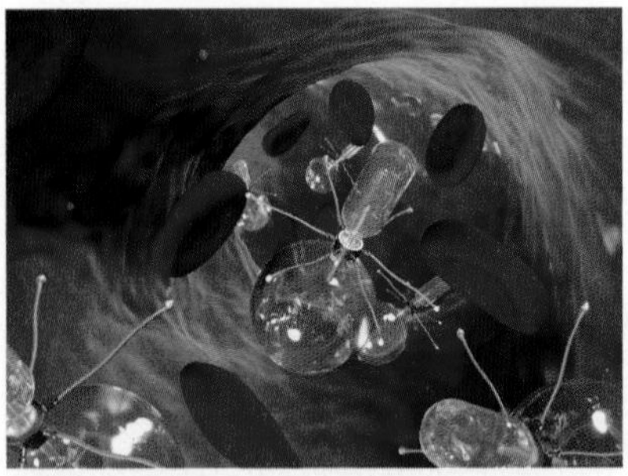

FIGURE 11.53 Nanotechnology is an area of science concerned with producing mechanical entities on the scale of nanometres (billionths of a metre). It is hoped that robots of this size will have medical applications, such as being used at a cellular level to remove cancerous cells, to remove the plaque that builds up in arteries causing atherosclerosis, or to fight illness.

Nanomachines In nature, a virus is a marvel of nanoscale engineering, designed to deliver genetic material to infect a host cell. For example, the T4 virus latches onto a bacterial cell with its tail fibres, bores through the bacterial membrane with its end plate, and then injects the payload of DNA, packaged in its head, into the bacterium. These complex functions are performed by a biological "machine" that measures 60 nm by 200 nm. Researchers are exploring ways for viruses to deliver medicinal agents to specific cells.

In other feats of nanoscale engineering, researchers have created machines such as nanovalves and nanopropellers (Figure 11.53). These **nanomachines** are small computers that are measured in nanometres and can perform tasks on a molecular level. Experts have predicted that, within one or two decades, nanomachines will make it possible to do many new things, such as decipher the human genome, change the flavour of any food on demand, and detect toxic chemicals and measure their concentrations in the environment. The possible applications of nanomachines are limitless, making nanotechnology one of the most profitable and advanced industries in the near future, especially in the fields of health, biological enhancements, food, crime, sports, and household appliances.

SUMMARY OF SECTION 11.7

- Doping increases the conductivity of semiconductors and is essential to modern electronic materials. Doping silicon with group 15 atoms introduces negative sites (n-type semiconductors) by adding valence electrons to the conduction band, whereas doping with group 13 atoms adds positive holes (p-type semiconductors) by emptying some orbitals in the valence band. Placing these two types of semiconductors in contact with one another forms a p-n junction. Sandwiching a p-type portion between two n-type portions forms a transistor.
- Liquid crystal phases flow like liquids but have molecules ordered like crystalline solids. Typically, the molecules have rodlike shapes, and their intermolecular forces keep them aligned. Thermotropic phases are prepared by heating the solid; lyotropic phases form when the solvent concentration is varied. The nematic, cholesteric, and smectic phases of liquid crystals differ in their molecular order. Liquid crystal applications depend on controlling the orientation of the molecules.
- Ceramics are very resistant to heat and chemicals. Most are network covalent solids formed at high temperature from simple reactants. They add lightweight strength to other materials.
- Nanoscale materials can be made through construction processes involving self-assembly and controlled orientation of molecules.

Research and development (R&D) activities in nanotechnology in Canada are spearheaded by the federal and provincial governments, as well as universities and national institutes. At the federal level, nine institutes of the National Research Council (NRC) are conducting R&D in nanotechnology, with the major concentrations of both research and industry in Alberta, British Columbia, Ontario, and Québec.

Many notable researchers work in this field. Geoffrey Ozin (Figure B11.4A) is a professor at the University of Toronto and a pioneer in the field of nanochemistry. His research involves synthetic strategies, self-assembly protocols, and experimental and theoretical methods for making and understanding the structure, property, and function relations of new classes of nanomaterials, mesoporous materials, photonic crystals, and nanomachines. He has developed nanomaterial-based platforms with properties and functions designed, for example, to control molecules and materials, and electrons and photons, in new ways that are proving to be useful for a wide range of nanotechnology applications. As well, he has invented two new and exciting photonic crystal technologies: photonic ink (Figure B11.4B) and elastic ink. Dr. Ozin is co-founder of OpaluxInc.

Robert Burrell, Canada Research Chair in Nanostructured Biomaterials at the University of Alberta, in Edmonton, has helped to design more than 15 medical processes and products. He is named as the inventor on over 290 patents and patent applications. He made an important breakthrough in dressing design using nanocrystalline silver technology, making one of the most significant advances in wound-care history by inventing Acticoat (Figure B11.5), a silver-based wound dressing that has antimicrobial properties and speeds healing. The dressing is often used in burn units and is now sold around the world. He has been awarded the World Union of Wound Healing Society Lifetime Achievement Award in recognition of his contributions to wound healing around the world, the 2009 ASM Engineering Materials Achievement Award, and the 2009 Manning Innovation Award.

Neil Branda (Figure B11.6A) is a Professor of Chemistry and a Canada Research Chair at Simon Fraser University (SFU) in Vancouver, British Columbia. He is also the Executive Director of 4D LABS, a research centre at SFU for advanced materials and nanoscale devices. His research program lies at the interface of

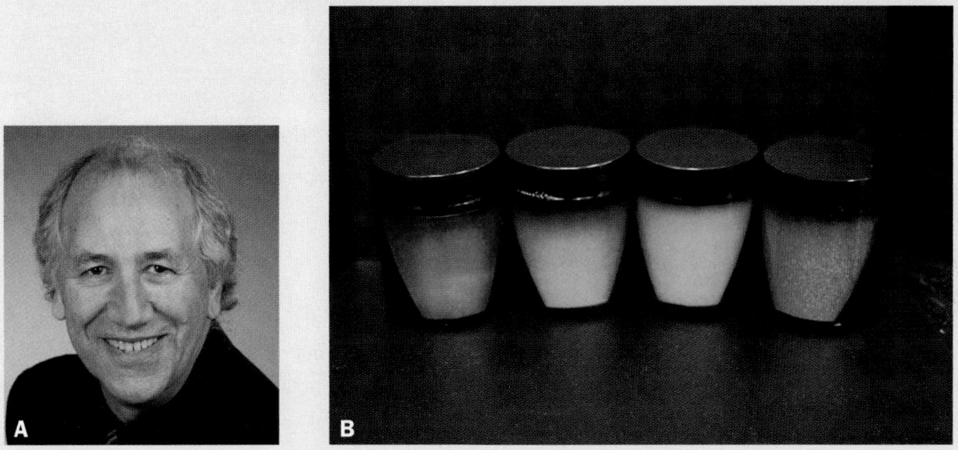

FIGURE B11.4 **A.** Geoffrey Ozin. **B.** Photonic ink invented by Dr. Ozin

FIGURE B11.5 Acticoat, a nanocrystalline antimicrobial dressing invented by Dr. Burrell.

(Continued)

organic chemistry and materials science, with a focus on designing and synthesizing molecular switches (molecules that change their structure and function when triggered with light, electricity, or chemical stimuli), integrating photo- (and electro-) responsive molecules into digital data storage systems (Figure B11.6B), synthetic reagents and catalysts, sensors and dosimeters, and drug delivery systems (using light as a trigger to selectively unmask known drug architectures and using optics to control important metabolic intermediates and enzyme cofactors). Dr. Branda was among Canada's Top 40 under 40 in 2006.

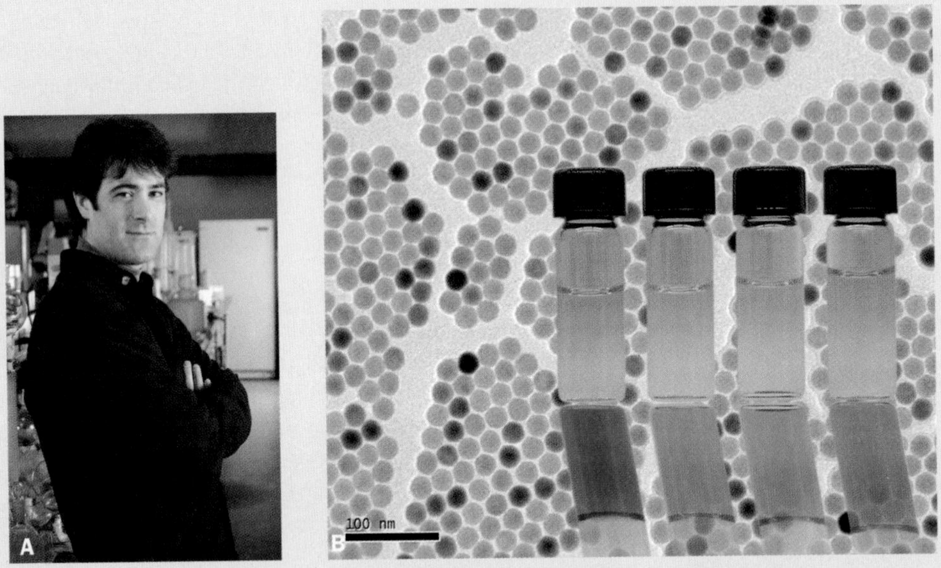

FIGURE B11.6 **A.** Neil Branda. **B.** The Branda group decorates nanoparticles with photoresponsive compounds. The compounds change colour when exposed to light.

Vladimir Kitaev (Figure B11.7A) is an Associate Professor at Wilfrid Laurier University, in Waterloo, Ontario. After working with Prof. Ozin on the synthesis of well-defined monodisperse nanoparticles, their self-assembly for functional applications, and colloidal photonic crystals, Dr. Kitaev is focusing his research in control of size, shape, and chirality on the nanoscale, including plasmonic nanoparticles of silver and gold for optical and sensor applications (Figure B11.7A).

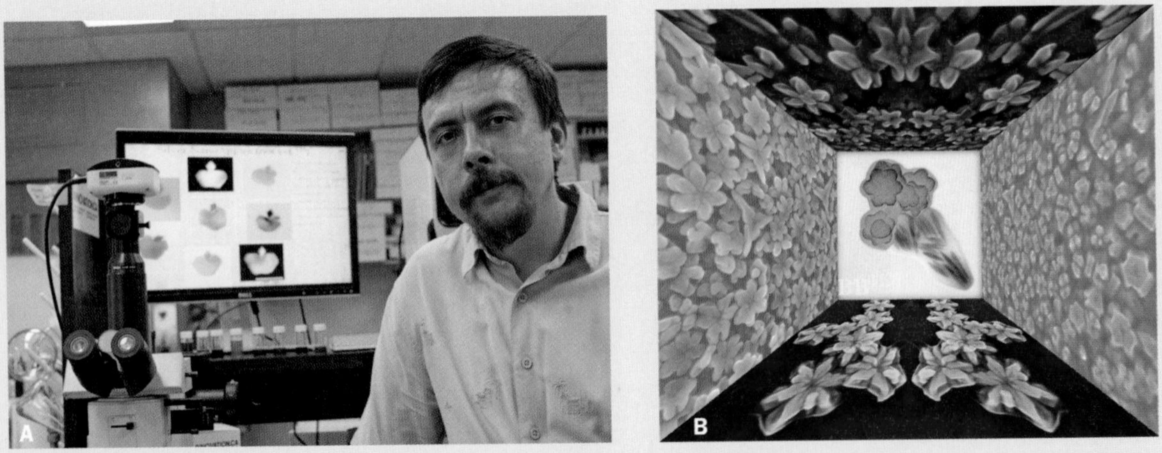

FIGURE B11.7 **A.** Vladimir Kitaev. **B.** Dr. Kitaev transformed the images he got by electron microscopy into modern art.

CHAPTER REVIEW GUIDE

Learning Objectives

Relevant section (§) and/or sample problem (SP) numbers appear in parentheses.

Concepts

1. Explain how the interplay between kinetic energy and potential energy underlies the properties of the three states of matter and their phase changes. (§11.1)
2. Explain the processes involved, both within a phase and during a phase change, when heat is added or removed from a pure substance. (§11.2)
3. Explain the meaning of *vapour pressure*, and discuss how phase changes are dynamic equilibrium processes. (§11.2)
4. Explain how temperature and intermolecular forces influence vapour pressure. (§11.2)
5. Explain the relationship between vapour pressure and boiling point. (§11.2)
6. Explain how a phase diagram shows the phases of a substance at differing conditions of pressure and temperature. (§11.2)
7. Distinguish between bonding and intermolecular forces on the basis of Coulomb's law and the meaning of the van der Waals (VDW) radius of an atom. (§11.3)
8. Differentiate between and discuss the types and relative strengths of intermolecular forces acting in a substance (dipole-dipole, hydrogen-bonding, dispersion), the impact of hydrogen bonding on physical properties, and the meaning of *polarizability*. (§11.3)
9. Explain the meanings of *surface tension*, *capillarity*, and *viscosity*, and describe how intermolecular forces influence their magnitudes. (§11.4)
10. Explain how the important macroscopic properties of water arise from atomic and molecular properties. (§11.5)
11. Explain the meaning of *crystal lattice* and the characteristics of the three types of cubic unit cells. (§11.6)
12. Describe how the packing of spheres gives rise to hexagonal and cubic unit cells. (§11.6)
13. Describe the types of crystalline solids, and explain how their intermolecular forces give rise to their properties. (§11.6)
14. Explain how band theory accounts for the properties of metals and the relative conductivities of metals, nonmetals, and metalloids. (§11.6)
15. Explain the structures, properties, and functions of modern materials (doped semiconductors, liquid crystals, ceramics, and nanostructures). (§11.7)

Skills

1. Calculate the overall enthalpy change when heat is gained or lost by a pure substance. (§11.2 and SP 11.1)
2. Use the Clausius-Clapeyron equation to examine the relationship between vapour pressure and temperature. (SP 11.2)
3. Use a phase diagram to predict the physical state and/or phase change of a substance. (§11.2)
4. Determine whether a substance can form hydrogen bonds, and draw the hydrogen-bonded structures. (SP 11.3)
5. Predict the types and relative strength of the bonding and intermolecular forces acting within a substance, based on its structure. (SP 11.4)
6. Find the number of particles in a unit cell. (§11.6)
7. Calculate the atomic radius of an element from its crystal structure. (SP 11.5, 11.6)

Key Terms

Section 11.1
phase
intermolecular (interparticle) forces
phase changes
condensation
vaporization
freezing
melting (fusion)
sublimation
deposition
heat (enthalpy) of vaporization ($\Delta_{vap}H°$)
heat (enthalpy) of fusion ($\Delta_{fus}H°$)
heat (enthalpy) of sublimation ($\Delta_{subl}H°$)

Section 11.2
heating-cooling curve
vapour pressure
Clausius-Clapeyron equation
boiling point
melting point
phase diagram
triple point
critical point

Section 11.3
van der Waals radius
ion-dipole force
dipole-dipole forces
hydrogen bond
polarizability
dispersion force (London force)

Section 11.4
surface tension
capillarity
viscosity

Section 11.6
crystalline solids
amorphous solids
lattice
unit cell
coordination number
simple cubic unit cell
body-centred cubic unit cell
face-centred cubic unit cell
packing efficiency
hexagonal closest packing
cubic closest packing
X-ray diffraction analysis
scanning tunnelling microscopy
atomic solid
molecular solids
ionic solid
tetrahedral hole
octahedral hole
metallic solids
network covalent solid
band theory
valence band
conduction band
conductor
semiconductor
insulator
superconductivity

Section 11.7
crystal defects
doping
liquid crystals
ceramics
nanotechnology
nanomachines

Key Equations and Relationships

11.1 Using the vapour pressure at one temperature to find the vapour pressure at another temperature (two-point form of the Clausius-Clapeyron equation):

$$\ln\frac{p_2}{p_1} = \frac{-\Delta_{vap}H}{R}\left(\frac{1}{T_2} - \frac{1}{T_1}\right)$$

Brief Solutions to Follow-Up Problems

11.1 The scenes represent solid water at −7.00°C melting to liquid water at 16.0°C, so there are three stages:

Stage 1, $H_2O(s)[-7.00°C] \longrightarrow H_2O(s)\ [0.00°C]$:

$$q = n \times C_{water(s)} \times \Delta T = (2.25\ \text{mol})[37.6\ \text{J/(mol·°C)}](7.00°C) = 592\ \text{J} = 0.592\ \text{kJ}$$

Stage 2, $H_2O(s)\ [0.00°C] \longrightarrow H_2O(l)\ [0.00°C]$:

$$q = n(\Delta_{fus}H°) = (2.25\ \text{mol})(6.02\ \text{kJ/mol}) = 13.5\ \text{kJ}$$

Stage 3, $H_2O(l)\ [0.00°C] \longrightarrow H_2O(l)\ [16.0°C]$:

$$q = n \times C_{water(l)} \times \Delta T = (2.25\ \text{mol})[75.4\ \text{J/(mol·°C)}](16.0°C) = 2714\ \text{J} = 2.71\ \text{kJ}$$

Total heat (kJ) = 0.592 kJ + 13.5 kJ + 2.71 kJ = 16.8 kJ

11.2
$$\ln\frac{p_2}{p_1} = \left(\frac{-40.7 \times 10^3\ \text{J/mol}}{8.314\ \text{J/[mol·K]}}\right) \times \left(\frac{1}{273.15 + 85.5\ \text{K}} - \frac{1}{273.15 + 34.1\ \text{K}}\right)$$
$$= (-4.90 \times 10^3\ \text{K})(-4.6 \times 10^{-4}\ \text{K}^{-1}) = 2.28$$
$$\frac{p_2}{p_1} = 9.8$$

Thus, $p_2 = 5.346\ \text{kPa} \times 9.8 = 52\ \text{kPa}$.

11.3 (a)

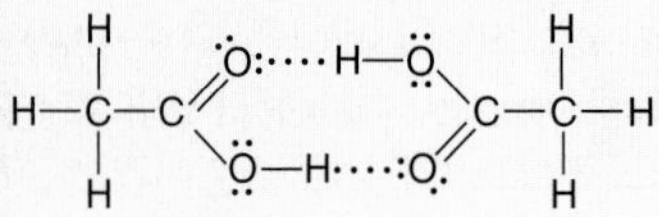

(b)

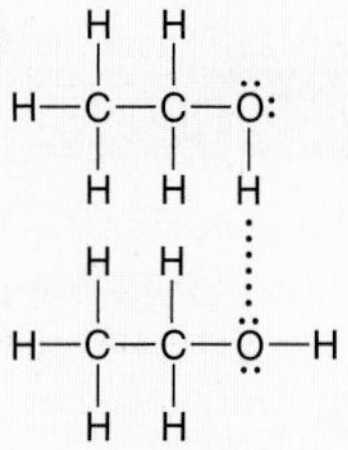

(c) There is no hydrogen bonding.

11.4 (a) Dipole-dipole, dispersion; CH_3Br
(b) Hydrogen bonds, dipole-dipole, dispersion; $CH_3CH_2CH_2OH$
(c) Dispersion; C_3H_8

11.5
$$\text{Avogadro's no.} = \frac{1\ \text{cm}^3}{7.874\ \text{g Fe}} \times \frac{55.85\ \text{g Fe}}{1\ \text{mol Fe}} \times 0.68 \times \frac{1\ \text{Fe atom}}{8.38 \times 10^{-24}\ \text{cm}^3} = 5.8 \times 10^{23}\ \text{Fe atoms/mol Fe}$$

11.6 From Figure 11.31B,

$$A = \frac{4r}{\sqrt{3}} = \frac{4(126\ \text{pm})}{\sqrt{3}} = 291\ \text{pm}$$

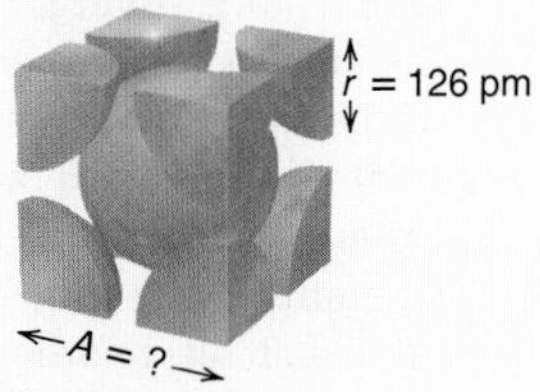

PROBLEMS

Problems with **red** numbers are answered in Appendix G and worked in detail in the Student Solutions Manual. Problem sections match those in this book and provide the numbers of relevant sample problems. Most offer Concept Review Questions, Skill-Building Exercises (grouped in pairs covering the same concept), and Problems in Context. The Comprehensive Problems are based on material from any section or previous chapter.

An Overview of Physical States and Phase Changes

Concept Review Questions

11.1 How does the energy of attraction between particles compare with their energy of motion in a gas and a solid? As part of your answer, identify two macroscopic properties that are different in a gas and a solid.

11.2 (a) Why are gases more easily compressed than liquids?
(b) Why do liquids have a greater ability to flow than solids?

11.3 State which type of force, intramolecular or intermolecular, does each of the following:
(a) prevents ice cubes from adopting the shape of their container
(b) is overcome when ice melts
(c) is overcome when liquid water is vaporized
(d) is overcome when gaseous water is converted to hydrogen gas and oxygen gas

11.4 (a) Why is the heat of fusion ($\Delta_{fus}H$) of a substance smaller than its heat of vaporization ($\Delta_{vap}H$)?
(b) Why is the heat of sublimation ($\Delta_{subl}H$) of a substance greater than its heat of vaporization?
(c) At a given temperature and pressure, how does the magnitude of the heat of vaporization of a substance compare with the magnitude of its heat of condensation?

Skill-Building Exercises (grouped in similar pairs)

11.5 Which forces are intramolecular, and which are intermolecular?
(a) Those preventing oil from evaporating at room temperature
(b) Those preventing butter from melting in a refrigerator
(c) Those allowing silver to tarnish
(d) Those preventing O_2 in air from forming O atoms

11.6 Which forces are intramolecular, and which are intermolecular?
(a) Those allowing fog to form on a cool, humid evening
(b) Those allowing water to form when H_2 is sparked
(c) Those allowing liquid benzene to crystallize when cooled
(d) Those responsible for the low boiling point of hexane

11.7 Name the phase change in each situation:
(a) Dew appears on a lawn in the morning.
(b) Icicles change into liquid water.
(c) Wet clothes dry on a summer day.

11.8 Name the phase change in each situation:
(a) A diamond film forms on a surface from gaseous carbon atoms in a vacuum.
(b) Mothballs in a bureau drawer disappear over time.
(c) Molten iron from a blast furnace is cast into ingots ("pigs").

Problems in Context

11.9 Liquid propane, a widely used fuel, is produced by compressing gaseous propane. During the process, approximately 15 kJ of energy is released for each mole of gas that is liquefied. Where does this energy come from?

11.10 Many heat-sensitive and oxygen-sensitive solids, such as camphor, are purified by warming them in a vacuum. The solid vaporizes directly, and the vapour crystallizes on a cool surface. What phase changes are involved in this process?

Quantitative and Qualitative Aspects of Phase Changes

(Sample Problems 11.1 and 11.2)

Concept Review Questions

11.11 Describe the changes (if any) in potential energy and kinetic energy among the molecules when gaseous PCl_3 condenses to a liquid at a fixed temperature.

11.12 When benzene is at its melting point, two processes occur simultaneously and balance each other. Describe these processes on the macroscopic and molecular levels.

11.13 Liquid hexane (boiling point = 69°C) is placed in a closed container at room temperature. At first, the pressure of the vapour phase increases. After a short time, however, it stops changing. Why?

11.14 Explain the effect of strong intermolecular forces on (a) critical temperature; (b) boiling point; (c) vapour pressure; (d) heat of vaporization.

11.15 Match each numbered point in the following phase diagram for compound Q with the correct molecular depiction at the right:

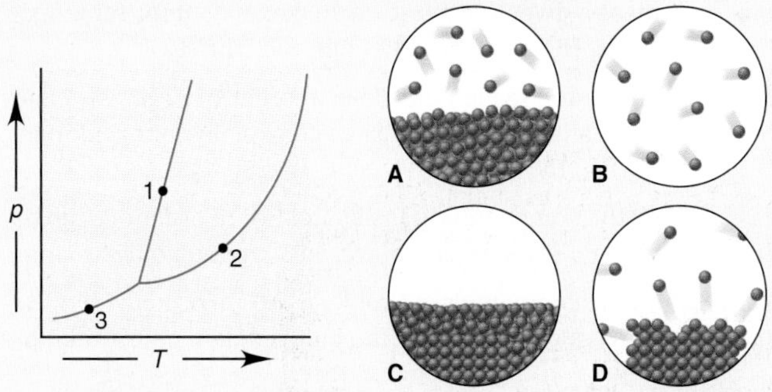

11.16 A liquid is in equilibrium with its vapour in a closed vessel at a fixed temperature. The vessel is connected by a stopcock to an evacuated vessel. When the stopcock is opened, will the final pressure of the vapour be different from the initial pressure if (a) some liquid remains; (b) all the liquid is first removed? Explain.

11.17 The phase diagram for substance A has a solid-liquid line with a positive slope, and the phase diagram for substance B has a solid-liquid line with a negative slope. What macroscopic property can distinguish substance A from substance B?

11.18 Why does water vapour at 100°C cause a more severe burn than liquid water at 100°C?

Skill-Building Exercises (grouped in similar pairs)

11.19 From the data below, calculate the total heat (J) needed to convert 22.00 g of ice at −6.00°C to liquid water at 0.500°C:
melting point at 101.3 kPa = 0.0°C $\Delta_{fus}H° = 6.02$ kJ/mol
$c_{liquid} = 4.21$ J/(g•°C) $c_{solid} = 2.09$ J/(g•°C)

11.20 From the data below, calculate the total heat (J) needed to convert 0.333 mol of gaseous ethanol at 300°C and 101.3 kPa to liquid ethanol at 25.0°C and 101.3 kPa:
boiling point at 101.3 kPa = 78.5°C $\Delta_{vap}H° = 40.5$ kJ/mol
$c_{gas} = 1.43$ J/(g•°C) $c_{liquid} = 2.45$ J/(g•°C)

11.21 A liquid has a $\Delta_{vap}H°$ of 35.5 kJ/mol and a boiling point of 122°C at 101.3 kPa. What is its vapour pressure at 113°C?

11.22 Ethoxyethane has a $\Delta_{vap}H°$ of 29.1 kJ/mol and a vapour pressure of 71 226 Pa at 25.0°C. What is its vapour pressure at 95.0°C?

11.23 What is the $\Delta_{vap}H°$ of a liquid that has a vapour pressure of 82.8 kPa at 85.2°C and a boiling point of 95.6°C at 101.3 kPa?

11.24 Methane (CH_4) has a boiling point of −164°C at 101.3 kPa and a vapour pressure of 4336 kPa at −100°C. What is the heat of vaporization of CH_4?

11.25 Use the following data to draw a qualitative phase diagram for ethene (C_2H_4). Is $C_2H_4(s)$ more or less dense than $C_2H_4(l)$?
Boiling point at 1 bar = −103.7°C
Melting point at 1 bar = −169.16°C
Critical point = 9.9°C and 51.2 bar
Triple point = −169.17°C and 1.22×10^{-3} bar

11.26 Use the following data to draw a qualitative phase diagram for H_2. Does H_2 sublime at 0.05 bar? Explain.
Melting point at 1 bar = 13.96 K
Boiling point at 1 bar = 20.39 K
Triple point = 13.95 K and 0.07 bar
Critical point = 33.2 K and 13 bar
Vapour pressure of solid at 10 K = 0.001 bar

Problems in Context

11.27 Sulfur dioxide is produced in enormous amounts for sulfuric acid production. It melts at −73°C and boils at −10.°C. Its $\Delta_{fus}H°$ is 8.619 kJ/mol, and its $\Delta_{vap}H°$ is 25.73 kJ/mol. The specific heat capacities of the liquid and the gas are 0.995 J/(g•K) and 0.622 J/(g•K), respectively. How much heat is required to convert 2.500 kg of solid SO_2 at its melting point to a gas at 60.°C?

11.28 Butane is a common fuel used in cigarette lighters and camping stoves. Normally supplied in metal containers under pressure, the fuel exists as a mixture of liquid and gas, so high temperature may cause the container to explode. At 25.0°C, the vapour pressure of butane is 233 kPa. What is the pressure in the container at 135°C ($\Delta_{vap}H° = 24.3$ kJ/mol)?

11.29 Use Figure 11.10 to answer the following questions:
(a) Carbon dioxide is sold in steel cylinders under a pressure of approximately 2026 kPa. Is there liquid CO_2 in the cylinder at (i) room temperature (~20°C); (ii) 40°C; (iii) −40°C; (iv) −120°C?
(b) Carbon dioxide is also sold as solid chunks, called *dry ice*, in insulated containers. If the chunks are warmed by leaving them in an open container at room temperature, will they melt?
(c) If a container is nearly filled with dry ice and then sealed and warmed to room temperature, will the dry ice melt?
(d) If dry ice is compressed at a temperature below its triple point, will it melt?

Types of Intermolecular Forces

(Sample Problems 11.3 and 11.4)

Concept Review Questions

11.30 Why are covalent bonds typically much stronger than intermolecular forces?

11.31 (a) Name the type of force depicted in each scene below.
(b) Rank the forces in order of increasing strength.

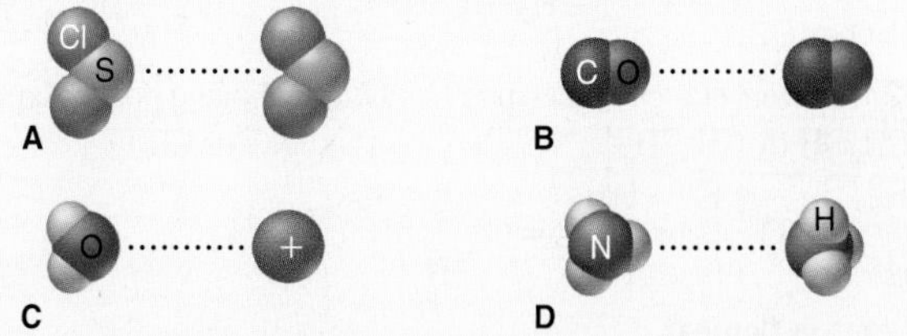

11.32 Oxygen and selenium are members of group 16. Water forms hydrogen bonds, but H_2Se does not. Explain.

11.33 In solid I_2, is the distance between the two I nuclei of one I_2 molecule longer or shorter than the distance between two I nuclei of adjacent I_2 molecules? Explain.

11.34 Polar molecules exhibit dipole-dipole forces. Do they also exhibit dispersion forces? Explain.

11.35 Distinguish between *polarizability* and *polarity*. How does each influence intermolecular forces?

11.36 How can one nonpolar molecule induce a dipole in a nearby nonpolar molecule?

Skill-Building Exercises (grouped in similar pairs)

11.37 What is the strongest interparticle force in each substance?
(a) CH_3OH (b) CCl_4 (c) Cl_2

11.38 What is the strongest interparticle force in each substance?
(a) H_3PO_4 (b) SO_2 (c) $MgCl_2$

11.39 What is the strongest interparticle force in each substance?
(a) CH_3Cl (b) CH_3CH_3 (c) NH_3

11.40 What is the strongest interparticle force in each substance?
(a) Kr (b) BrF (c) H_2SO_4

11.41 Which compound in each pair forms intermolecular hydrogen bonds? Draw the hydrogen-bonded structures.
(a) $CH_3\underset{\substack{|\\ OH}}{C}HCH_3$ or CH_3SCH_3 (b) HF or HBr

11.42 Which compound in each pair forms intermolecular hydrogen bonds? Draw the hydrogen-bonded structures.
(a) $(CH_3)_2NH$ or $(CH_3)_3N$ (b) $HOCH_2CH_2OH$ or FCH_2CH_2F

11.43 Which forces oppose vaporization of each substance?
(a) Hexane (b) Water (c) $SiCl_4$

11.44 Which forces oppose vaporization of each substance?
(a) Br_2 (b) SbH_3 (c) CH_3NH_2

11.45 Which ion or molecule has greater polarizability? Explain.
(a) Br^- or I^- (b) $CH_2{=}CH_2$ or $CH_3{-}CH_3$ (c) H_2O or H_2Se

11.46 Which ion or molecule has greater polarizability? Explain.
(a) Ca^{2+} or Ca (b) CH_3CH_3 or $CH_3CH_2CH_3$ (c) CCl_4 or CF_4

11.47 Which liquid in each pair has a *higher* vapour pressure at a given temperature? Explain.
(a) C_2H_6 or C_4H_{10} (b) CH_3CH_2OH or CH_3CH_2F (c) NH_3 or PH_3

11.48 Which liquid in each pair has a *lower* vapour pressure at a given temperature? Explain.
(a) $HOCH_2CH_2OH$ or $CH_3CH_2CH_2OH$
(b) CH_3COOH or $(CH_3)_2\,C{=}O$ (c) HF or HCl

11.49 Which substance in each pair has a *lower* boiling point? Explain.
(a) LiCl or HCl (b) NH_3 or PH_3 (c) Xe or I_2

11.50 Which substance in each pair has a *higher* boiling point? Explain.
(a) CH_3CH_2OH or $CH_3CH_2CH_3$ (b) NO or N_2
(c) H_2S or H_2Te

11.51 Which substance in each pair has a *lower* boiling point? Explain.
(a) $CH_3CH_2CH_2CH_3$ or $\begin{matrix} CH_2{-}CH_2 \\ | \quad\;\; | \\ CH_2{-}CH_2 \end{matrix}$ (b) NaBr or PBr_3
(c) H_2O or HBr

11.52 Which substance in each pair has a *higher* boiling point? Explain.
(a) CH_3OH or CH_3CH_3 (b) FNO or ClNO
(c) 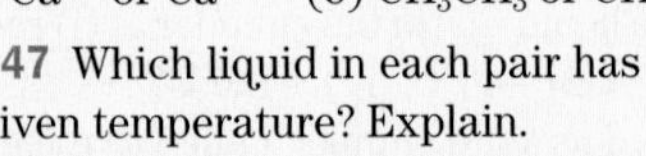

Problems in Context

11.53 For pairs of molecules in the gas phase, average hydrogen bond dissociation energies are 17 kJ/mol for NH_3, 22 kJ/mol for H_2O, and 29 kJ/mol for HF. Explain this increase in hydrogen bond strength.

11.54 Dispersion forces are the only intermolecular forces present in motor oil, yet motor oil has a high boiling point. Explain.

11.55 Why does ethane-1,2-diol (ethylene glycol, $HOCH_2CH_2OH$; $\mathcal{M} = 62.07$ g/mol), an ingredient in antifreeze, have a boiling point of 197.6°C, whereas propanol ($CH_3CH_2CH_2OH$; $\mathcal{M} = 60.09$ g/mol), a compound with a similar molar mass, has a boiling point of only 97.4°C?

Properties of the Liquid State

Concept Review Questions

11.56 Before the phenomenon of surface tension was understood, physicists described the surface of water as being covered with a "skin." What causes this skinlike phenomenon?

11.57 Small equal-sized drops of oil, water, and mercury lie on a waxed floor. How does each liquid behave? Explain.

11.58 Why does an aqueous solution of ethanol (CH_3CH_2OH) have a lower surface tension than water?

11.59 Why are units of energy per area (J/m^2) used for surface tension values?

11.60 Does the *strength* of the intermolecular forces in a liquid change as the liquid is heated? Explain. Why does liquid viscosity decrease with rising temperature?

Skill-Building Exercises (grouped in similar pairs)

11.61 Rank the following molecules in order of *increasing* surface tension at a given temperature, and explain your ranking: $CH_3CH_2CH_2OH$, $HOCH_2CH(OH)CH_2OH$, $HOCH_2CH_2OH$

11.62 Rank the following molecules in order of *decreasing* surface tension at a given temperature, and explain your ranking: CH_3OH, CH_3CH_3, $H_2C{=}O$

11.63 Rank the compounds in Problem 11.61 in order of *decreasing* viscosity at a given temperature, and explain your ranking.

11.64 Rank the compounds in Problem 11.62 in order of *increasing* viscosity at a given temperature, and explain your ranking.

Problems in Context

11.65 Soil vapour extraction (SVE) is used to remove volatile organic pollutants, such as chlorinated solvents, from the soil at hazardous waste sites. Vent wells are drilled, and a vacuum pump is applied to the subsurface.
(a) How does this process remove pollutants?
(b) Why does heating combined with SVE speed up this process?

11.66 Use Figure 11.2 to answer the following questions:
(a) Does it take more heat to melt 12.0 g of CH_4 or 12.0 g of Hg?
(b) Does it take more heat to vaporize 12.0 g of CH_4 or 12.0 g of Hg?
(c) What is the principal intermolecular force in each sample?

11.67 Pentanol ($C_5H_{11}OH$; $\mathcal{M} = 88.15$ g/mol) has nearly the same molar mass as hexane (C_6H_{14}; $\mathcal{M} = 86.17$ g/mol), but it is more than 12 times as viscous at 20°C. Explain.

The Uniqueness of Water

Concept Review Questions

11.68 For what types of substances is water a good solvent? For what types of substances is it a poor solvent? Explain.

11.69 A water molecule can engage in as many as four hydrogen bonds. Explain.

11.70 Warm-blooded animals have a narrow range for body temperature because their bodies have a high water content. Explain.

11.71 What property of water keeps plant debris on the surface of lakes and ponds? What is the ecological significance of this?

11.72 A drooping plant may become upright if the ground around it is watered. Explain.

11.73 Describe the molecular basis of the property of water that is responsible for the presence of ice on the surface of a frozen lake.

11.74 Describe, in molecular terms, what occurs when ice melts.

The Solid State: Structure, Properties, and Bonding

(Sample Problems 11.5 and 11.6)

Concept Review Questions

11.75 What is the difference between an amorphous solid and a crystalline solid on the macroscopic and molecular levels? Give an example of each.

11.76 How are the unit cell and crystal structure of a solid related?

11.77 For structures consisting of identical atoms, how many atoms are contained in the simple, body-centred, and face-centred cubic unit cells? Explain how you obtained these values.

11.78 An element has a crystal structure in which the width of the cubic unit cell equals the diameter of an atom. What type of unit cell does the element have?

11.79 What specific difference in the positioning of spheres gives a crystal structure based on the face-centred cubic unit cell less empty space than a crystal structure based on the body-centred cubic unit cell?

11.80 Both solid Kr and solid Cu consist of individual atoms. Why do their physical properties differ so much?

11.81 What is the energy gap in band theory? Compare its size in superconductors, conductors, semiconductors, and insulators.

11.82 Predict the effect (if any) of an increase in temperature on the electrical conductivity of (a) a conductor; (b) a semiconductor; (c) an insulator.

11.83 Besides the type of unit cell, what information is needed to find the density of a solid consisting of identical atoms?

Skill-Building Exercises (grouped in similar pairs)

11.84 What type of crystal lattice does each metal form? (The number of atoms per unit cell is given in parentheses.)
(a) Ni (4) (b) Cr (2) (c) Ca (4)

11.85 For each metal, what is the number of atoms per unit cell?
(a) Polonium, Po (b) Manganese, Mn (c) Silver, Ag

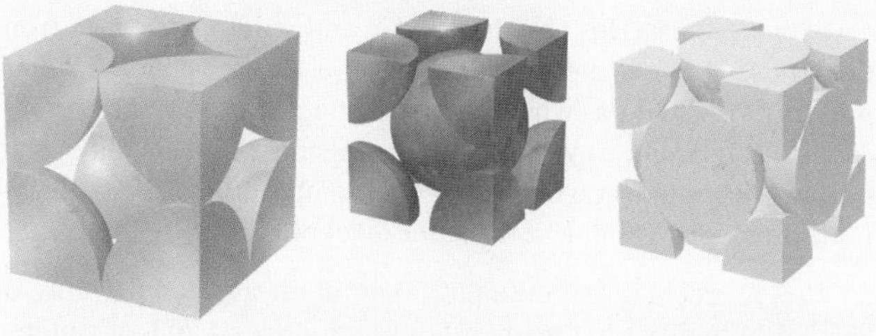

11.86 When cadmium oxide reacts to form cadmium selenide, a change in the unit cell occurs, as depicted in A and B below:
(a) What is the change in the unit cell?
(b) Does the coordination number of cadmium change? Explain.

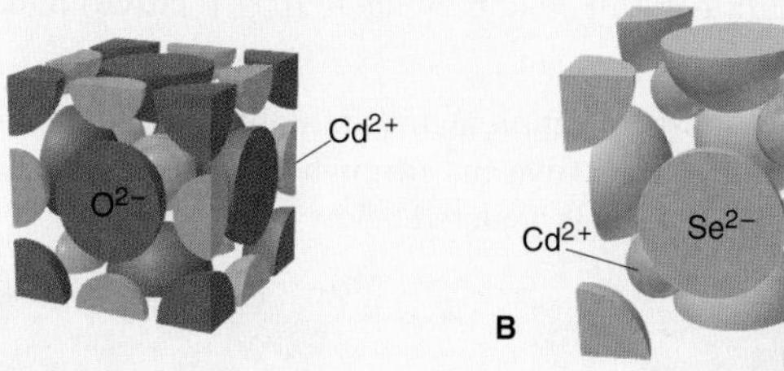

11.87 As molten iron cools to 1674 K, it adopts one type of cubic unit cell, shown in diagram A. Then, as the temperature drops below 1181 K, it changes to another type of unit cell, shown in diagram B.
(a) What is the change in the unit cell?
(b) Which crystal structure has greater packing efficiency?

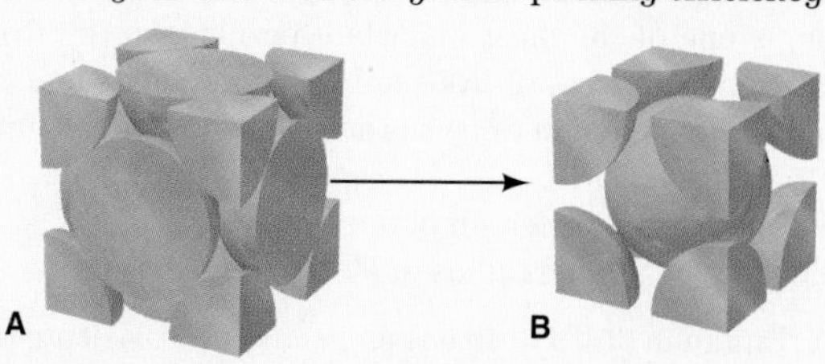

11.88 Which of the five major types of crystalline solid does each substance form, and why?
(a) Ni (b) F_2 (c) CH_3OH (d) Sn (e) Si (f) Xe

11.89 Which of the five major types of crystalline solid does each substance form, and why?
(a) SiC (b) Na_2SO_4 (c) SF_6
(d) cholesterol ($C_{27}H_{45}OH$) (e) KCl (f) BN

11.90 Zinc oxide adopts the zinc blende crystal structure (Figure P11.90 below). How many Zn^{2+} ions are in the ZnO unit cell?

11.91 Calcium sulfide adopts the sodium chloride crystal structure (Figure P11.91). How many S^{2-} ions are in the CaS unit cell?

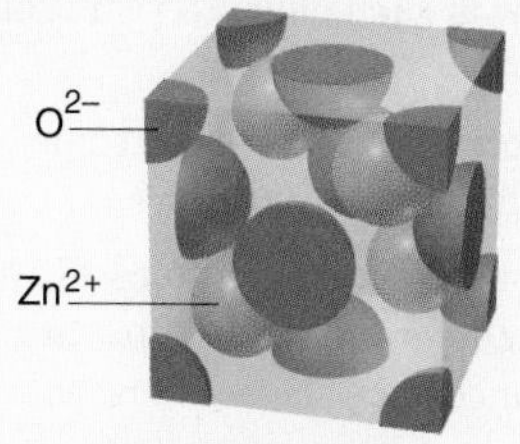

Figure P11.90

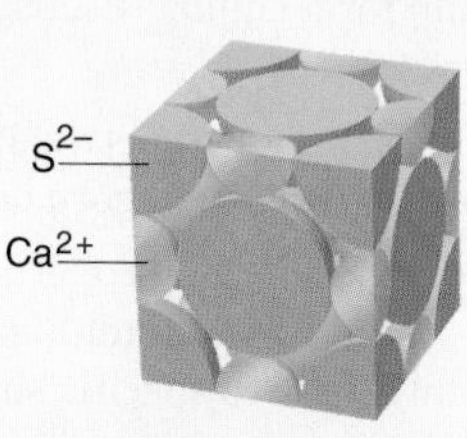

Figure P11.91

11.92 Zinc selenide (ZnSe) crystallizes in the zinc blende structure (see Figure P11.90) and has a density of 5.42 g/cm^3.
(a) How many Zn and Se ions are in each unit cell?
(b) What is the mass of a unit cell?
(c) What is the volume of a unit cell?
(d) What is the edge length of a unit cell?

11.93 An element crystallizes in a face-centred cubic lattice and has a density of 1.45 g/cm^3. The edge length of its unit cell is 4.52×10^{-8} cm.
(a) How many atoms are in each unit cell?
(b) What is the volume of a unit cell?
(c) What is the mass of a unit cell?
(d) Calculate an approximate atomic mass for the element.

11.94 Classify each substance as a conductor, an insulator, or a semiconductor:
(a) Phosphorus (b) Mercury (c) Germanium

11.95 Classify each substance as a conductor, an insulator, or a semiconductor:
(a) Carbon (graphite) (b) Sulfur (c) Platinum

11.96 Predict the effect (if any) of an increase in temperature on the electrical conductivity of (a) antimony; (b) tellurium; (c) bismuth.

11.97 Predict the effect (if any) of a decrease in temperature on the electrical conductivity of (a) silicon; (b) lead; (c) germanium.

Problems in Context

11.98 Polonium, the period 6 element in group 16, is a rare radioactive metal that is the only element with a crystal structure based on the simple cubic unit cell. If its density is 9.142 g/cm^3, calculate an approximate atomic radius for polonium.

11.99 The coinage metals—copper, silver, and gold—crystallize in a cubic closest packing structure. Use the density of copper (8.95 g/cm^3) and its molar mass (63.55 g/mol) to calculate an approximate atomic radius for copper.

11.100 Nitrogenase, the plant protein that catalyzes nitrogen fixation, is one of the most important enzymes in the world. It contains active clusters of iron, sulfur, and molybdenum atoms. Crystalline molybdenum (Mo) has a body-centred cubic unit cell (d of Mo = 10.28 g/cm^3).
(a) Determine the edge length of the unit cell.
(b) Calculate the atomic radius of Mo.

11.101 Tantalum (Ta; d = 16.634 g/cm^3; $\mathcal{M}$ = 180.9479 g/mol) has a body-centred cubic structure. The edge length of its unit cell is 3.3058 × 10^{-10} m. Use these data to calculate Avogadro's number.

Advanced Materials

Concept Review Questions

11.102 When tin is added to copper, the resulting alloy (bronze) is much harder than copper. Explain.

11.103 In the process of doping a semiconductor, certain impurities are added to increase its electrical conductivity. Explain this process for an n-type semiconductor and a p-type semiconductor.

11.104 State two molecular characteristics of substances that typically form liquid crystals. How is each characteristic related to function?

11.105 Distinguish between isotropic and anisotropic substances. To which category do liquid crystals belong?

11.106 How are the properties of high-tech ceramics the same as the properties of traditional clay ceramics, and how are they different? Refer to specific substances in your answer.

Skill-Building Exercises (grouped in similar pairs)

11.107 Silicon and germanium are both semiconducting elements from group 14 that can be doped to improve their conductivity. Would each of the following form an n-type semiconductor or a p-type semiconductor?
(a) Ge doped with P (b) Si doped with In

11.108 Silicon and germanium are both semiconducting elements from group 14 that can be doped to improve their conductivity. Would each of the following form an n-type semiconductor or a p-type semiconductor?
(a) Ge doped with As (b) Si doped with B

Comprehensive Problems

11.109 A 0.75 L bottle is cleaned, dried, and closed in a room where the air temperature is 22°C and the relative humidity is 44% (that is, the water vapour in the air is 0.44 of the equilibrium vapour pressure at 22°C). The bottle is then brought outside and stored at 0.0°C.
(a) What mass of liquid water condenses inside the bottle?
(b) Would liquid water condense at 10°C? (See Table 4.2.)

11.110 In an experiment, 5.00 L of N_2 is saturated with water vapour at 22°C and then compressed to half its volume at constant T.
(a) What is the partial pressure of H_2O in the compressed gas mixture?
(b) What mass of water vapour condenses to a liquid?

11.111 Two important characteristics that are used to evaluate the risk of fire or explosion are a compound's *lower flammable limit* (LFL) and *flash point*. The LFL is the minimum percentage, by volume, in air that is ignitable. Below the LFL, the mixture is too "lean" to burn. The flash point is the temperature at which the air over a confined liquid becomes ignitable. The boiling point of hexane is 68.7°C at 101.3 kPa, and its vapour pressure is 16.1 kPa at 20.0°C. The LFL of n-hexane is 1.1%. Calculate the flash point of n-hexane.

11.112 Bismuth is used to calibrate instruments that are used in high-pressure studies because it has several well-characterized crystalline phases. Its phase diagram (*below*) shows the liquid phase and five solid phases that are stable above 101.3 × 10^3 kPa and up to 300°C.
(a) Which solid phases are stable at 25°C?
(b) Which phase is stable at 506.6 × 10^4 kPa and 175°C?
(c) As the pressure is reduced from 101.3 × 10^5 to 101.3 × 10^3 kPa at 200°C, what phase transitions does bismuth undergo?
(d) What phases are present at each of the triple points?

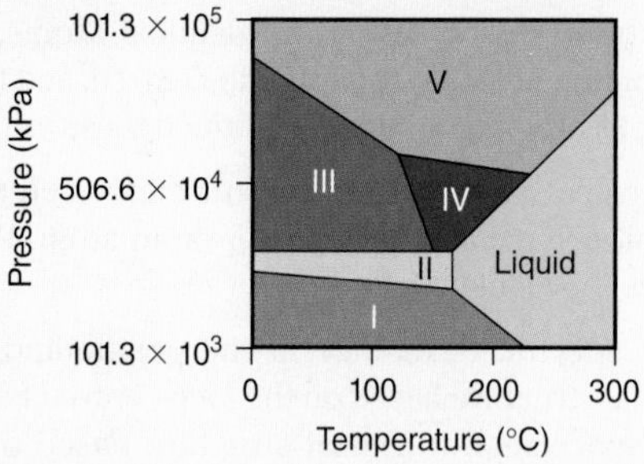

11.113 When making computer chips, a 4.00 kg cylindrical ingot of ultrapure n-type doped silicon, 13.208 cm in diameter, is sliced into wafers that are 1.12 × 10^{-4} m thick.
(a) Assuming no waste, how many wafers can be made?
(b) What is the mass of a wafer (d of Si = 2.34 g/cm^3; V of a cylinder = $\pi r^2 h$)?
(c) A key step in making p-n junctions for a chip is the chemical removal of the oxide layer on the wafer through treatment with gaseous HF. Write a balanced equation for this reaction.
(d) If 0.750% of the Si atoms are removed during the treatment in part (c), what amount (mol) of HF is required per wafer, assuming 100% reaction yield?

11.114 Methyl 2-hydroxybenzoate (methyl salicylate, $C_8H_8O_3$), the odorous constituent of oil of wintergreen, has a vapour pressure of 133 Pa at 54.3°C and a vapour pressure of 1333 Pa at 95.3°C.
(a) What is its vapour pressure at 25°C?
(b) What is the minimum volume of air that must pass over a sample of the compound at 25°C to vaporize 1.0 mg of the sample?

11.115 Mercury (Hg) vapour is toxic and readily absorbed from the lungs. At 20.°C, mercury ($\Delta_{vap}H$ = 59.1 kJ/mol) has a vapour pressure of 0.160 Pa, which is high enough to be hazardous. To reduce the danger to workers in processing plants, Hg is cooled to lower its vapour pressure. At what temperature would the vapour pressure of Hg be at the safer level of 6.67 × 10^{-3} Pa?

11.116 A greenhouse contains 256 m^3 of air at a temperature of 26°C. A humidifier in the greenhouse vaporizes 4.20 L of water.
(a) What is the pressure of water vapour in the greenhouse, assuming that none escapes and that the air was originally completely dry (d of H_2O = 1.00 g/mL)?
(b) What total volume of liquid water would have to be vaporized to saturate the air (that is, achieve 100% relative humidity)? (See Table 4.2.)

11.117 Like most transition metals, tantalum (Ta) exhibits several oxidation states. Give the formula of each tantalum compound whose unit cell is depicted below:

(a) (b)

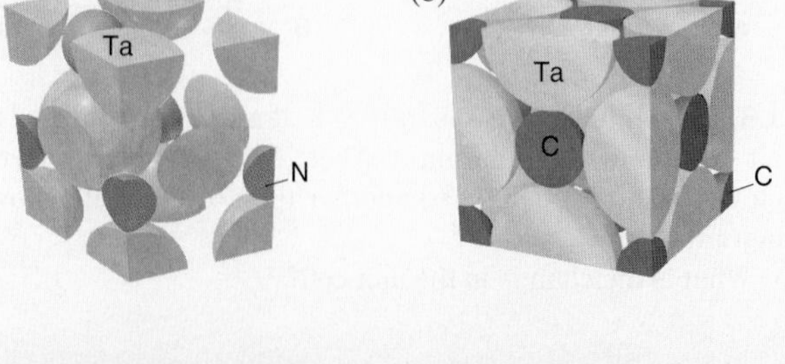

11.118 KF has the same type of crystal structure as NaCl. The unit cell of KF has an edge length of 5.39×10^{-8} cm. Find the density of KF.

11.119 Furfural, which is prepared from corncobs, is an important solvent in synthetic rubber manufacturing. It is reduced to furfuryl alcohol, which is used to make polymer resins. Furfural can also be oxidized to 2-furoic acid.

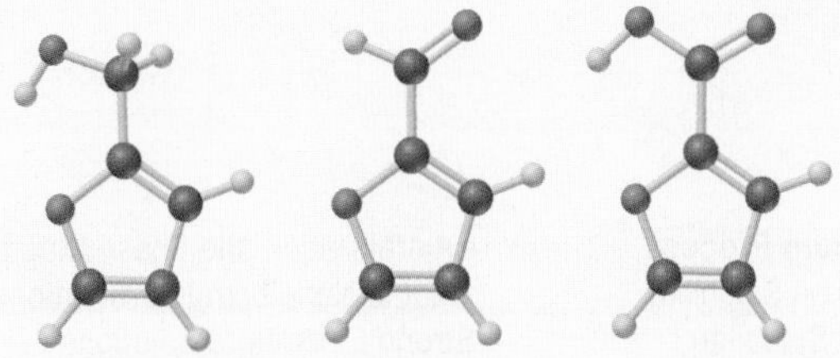

Furfuryl alcohol Furfuryl 2-furoic acid

(a) Which of these compounds can form hydrogen bonds? Draw the hydrogen-bonded structures.
(b) The molecules of some substances can form an "internal" hydrogen bond, that is, a hydrogen bond *within* a molecule. An internal hydrogen bond takes the form of a polygon, with atoms as corners and bonds as sides and with a hydrogen bond as one of the sides. Which of these molecules is (are) likely to form a stable internal hydrogen bond? Draw the structure. (*Hint:* Structures with five or six atoms as corners are the most stable.)

11.120 On a humid day in Toronto, the temperature is 22.0°C, and the partial pressure of water vapour in the air is 4.133 kPa. A 9000 t (tonne) air-conditioning system in a large building maintains an inside air temperature of 22.0°C, but the partial pressure of water vapour is 1.333 kPa. The volume of air in the building is 2.4×10^6 m^3, and the total pressure is 101.3 kPa, both inside and outside the building.
(a) What mass of water (t) must be removed every time the inside air is completely replaced with outside air? (*Hint:* What amount (mol) of gas is in the building? Of water vapour? Of dry air? What amount (mol) of outside air must be added to the air in the building to simulate the composition of the outside air?)
(b) Find the heat that is released when this mass of water condenses.

11.121 The boiling point of amphetamine, $C_9H_{13}N$, is 201°C at 101.3 kPa and 83°C at 1.733 kPa. What is the concentration (g/m^3) of amphetamine when it is in contact with 20.°C air?

11.122 Diamonds have a face-centred cubic unit cell, with four more C atoms in tetrahedral holes within the cell. The densities of diamonds vary from 3.01 g/cm^3 to 3.52 g/cm^3 because C atoms are missing from some holes.
(a) Calculate the edge length of the unit cell of the densest diamond.
(b) Assuming that the cell dimensions are fixed, how many C atoms are in the unit cell of the diamond with the lowest density?

11.123 Is it possible for a salt with the formula AB_3 to have a face-centred cubic unit cell of anions, with cations in all of the eight available holes? Explain.

11.124 The density of solid gallium at its melting point is 5.9 g/cm^3, whereas the density of liquid gallium is 6.1 g/cm^3. Is the temperature at the triple point higher or lower than the normal melting point? Is the slope of the solid-liquid line for gallium positive or negative?

11.125 A 4.7 L sealed bottle containing 0.33 g of liquid ethanol, C_2H_6O, is placed in a refrigerator and reaches equilibrium with its vapour at −11°C.
(a) What mass of ethanol is present in the vapour?
(b) When the container is removed and warmed to room temperature, 20.°C, will all the ethanol vaporize?
(c) How much liquid ethanol would be present at 0.0°C? The vapour pressure of ethanol is 1.33 kPa at −2.3°C and 5.33 kPa at 19°C.

11.126 Substance A has the following properties:

Melting point at 101.3 kPa = −20.°C	Boiling point at 101.3 kPa = 85°C
$\Delta_{fus}H$ = 180. J/g	$\Delta_{vap}H$ = 500. J/g
c_{solid} = 1.0 J/(g•°C)	c_{liquid} = 2.5 J/(g•°C)
c_{gas} = 0.5 J/(g•°C)	

At 101.3 kPa, a 25 g sample of A is heated from −40.°C to 100.°C at a constant rate of 450. J/min.
(a) How many minutes does the sample take to reach its melting point?
(b) How many minutes does the sample take to melt?
(c) Perform any other necessary calculations, and draw a curve of temperature versus time for the entire heating process.

11.127 An aerospace manufacturer is building a prototype experimental aircraft that cannot be detected by radar. Boron nitride is chosen for incorporation into the body parts, and the boric acid/ammonia method is used to prepare the ceramic material. Given 85.5% and 86.8% yields for the two reaction steps, how much boron nitride can be prepared from 1.00 t (tonne) of boric acid and 12.5 m^3 of ammonia at 275 K and 3.07×10^3 kPa? Assume that ammonia does not behave ideally under these conditions and is recycled completely in the reaction process.

11.128 The following ball-and-stick models represent three compounds with the same molecular formula, $C_4H_8O_2$:

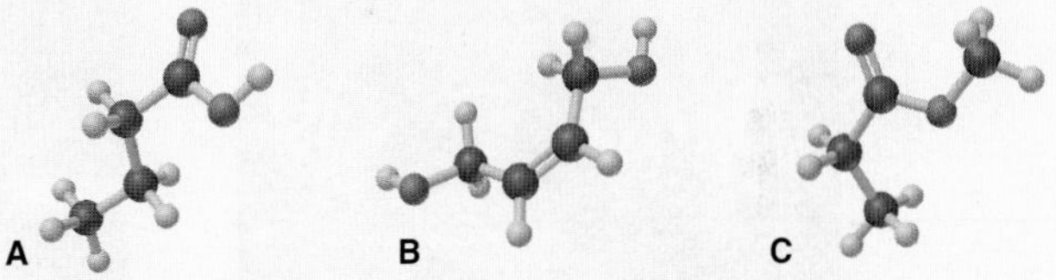

A B C

(a) Which compound(s) can form intermolecular hydrogen bonds?
(b) Which has the highest viscosity?

11.129 The $\Delta_f H°$ of gaseous ethoxyethane (CH_3OCH_3) is −185.4 kJ/mol. Its vapour pressure is 101.3 kPa at −23.7°C and 53.3 kPa at −37.8°C.
(a) Calculate $\Delta_{vap}H°$ of ethoxyethane.
(b) Calculate $\Delta_f H°$ of liquid ethoxyethane.

11.130 The crystal structure of sodium is based on the body-centred cubic unit cell. What is the mass of one unit cell of Na?

11.131 One way to purify gaseous H_2 is to pass it, under high pressure, through the holes of a metal's crystal structure. Palladium, which adopts a cubic closest packing structure, absorbs more H_2 than any other element and is one of the metals used for this purpose. How the metal and H_2 interact is unclear, but scientists estimate that the density of absorbed H_2 approaches the density of liquid hydrogen (70.8 g/L). What volume (L) of gaseous H_2 (at STP) can be packed into the spaces of 1 dm^3 of palladium metal?

The Properties of Mixtures: Solutions and Colloids

IN THIS CHAPTER . . . We focus on how intermolecular forces and other energy considerations affect a solute dissolving in a solvent, how we can calculate concentrations, and how solutions differ from pure substances. We also briefly consider the behaviour and applications of colloids. By the end of this chapter, you should be able to

- Describe the intermolecular forces between solute and solvent, and explain why substances with similar types of forces form a solution
- Draw a step-by-step cycle to explain why a substance dissolves, and examine the roles of the heat of solution and the dispersal of matter (for which we will introduce the concept of entropy)
- Discuss the equilibrium nature of solubility and describe how temperature and pressure affect it
- Differentiate between various solution concentration units and convert from any one unit to any other
- Differentiate between the physical properties of solutions and pure substances and apply those differences
- Describe colloids and explain how solution and colloid chemistry can be applied to the purification of water.

12

Anyone who lives in Halifax can easily imagine this scenario. Imagine hunching down in the middle of winter, walking from campus to the nearest place you can buy lunch. You have to walk carefully to avoid the icy patches and are extremely thankful for the salt on the sidewalk. You stamp your feet when you get in and shake off the snow. A hot plate of pasta with sweet or spicy pickles on the side looks really inviting! As you wait, you watch the cook liberally salt the water in which he is going to cook the pasta. After eating lunch, you step back out, and, luckily, one of your friends, driving back from the hospital where her mother is going through dialysis, pulls up, offering you a ride. What does any of this have to do with chemistry? In fact, many things in this scenario are related to a single theme, namely the colligative properties. In this chapter, we will learn how four important properties of solutions are related only to the number of particles in solution and not what the particles themselves are. In order to better understand these properties, we will first review the nature of solutions and the intermolecular forces that lead to their formation. We will examine the different units we can use to describe solution concentration. Eventually, we will be able to see how salting roads, pickles, dialysis, salting boiling water, and antifreeze all relate to colligative properties.

Recall, from Chapter 2, that a mixture has two defining characteristics: *its composition is variable,* and *it retains some properties of its components.* We focus here on two common types of mixtures—solutions and colloids—whose main differences relate to particle size and number of phases:

- A *solution* is a *homogeneous* mixture; thus, it exists as one phase. In a solution, the particles are individual atoms, ions, or small molecules.
- A *colloid* is a type of *heterogeneous* mixture. A heterogeneous mixture has two or more phases. They may be visibly distinct, such as pebbles in concrete, or not visibly distinct, such as the much smaller particles in the colloids smoke and milk. In a colloid, the particles are typically macromolecules or aggregations of small molecules that are dispersed so finely they do not settle out.

Concepts and Skills to Review before Studying This Chapter

- Separation of mixtures (Section 2.9)
- Calculations involving mass percent (Section 3.1) and amount of substance concentration (Section 3.5)
- Electrolytes and water as a solvent (Sections 3.6 and 11.5)
- Mole fraction and Dalton's law (Section 4.4)
- Intermolecular forces and polarizability (Section 11.3)
- Vapour pressure of liquids (Section 11.2)

12.1 Types of Solutions: Intermolecular Forces and Solubility

A **solute** dissolves in a **solvent** to form a solution. In general, *the solvent is the most abundant component.* In some cases, however, the substances are **miscible**—soluble in each other in any proportion—so the terms *solute* and *solvent* lose their meanings. *The physical state of the solvent usually determines the physical state of the solution.* Solutions can be gaseous, liquid, or solid, but we focus mostly on liquid solutions because they are by far the most important.

The **solubility (s)** of a solute is the maximum amount that dissolves in a fixed quantity of a given solvent at a given temperature, where an excess of the solute is present. Different solutes have different solubilities:

- Sodium chloride (NaCl), s = 39.12 g/100. mL water at 100.°C
- Silver chloride (AgCl), s = 0.0021 g/100. mL water at 100.°C

Solubility is a *quantitative* term, but *dilute* and *concentrated* are qualitative, referring to the *relative* amounts of dissolved solute. The NaCl solution above is concentrated, and the AgCl solution is dilute.

A given solute may dissolve in one solvent, but not in another. The explanation lies in the relative strengths of the intermolecular forces within both solute and solvent, and between them. A useful rule of thumb called the **like-dissolves-like rule** says that *substances with similar types of intermolecular forces dissolve in each other.* Thus, by knowing the forces, we can often predict whether a solute will dissolve in a solvent.

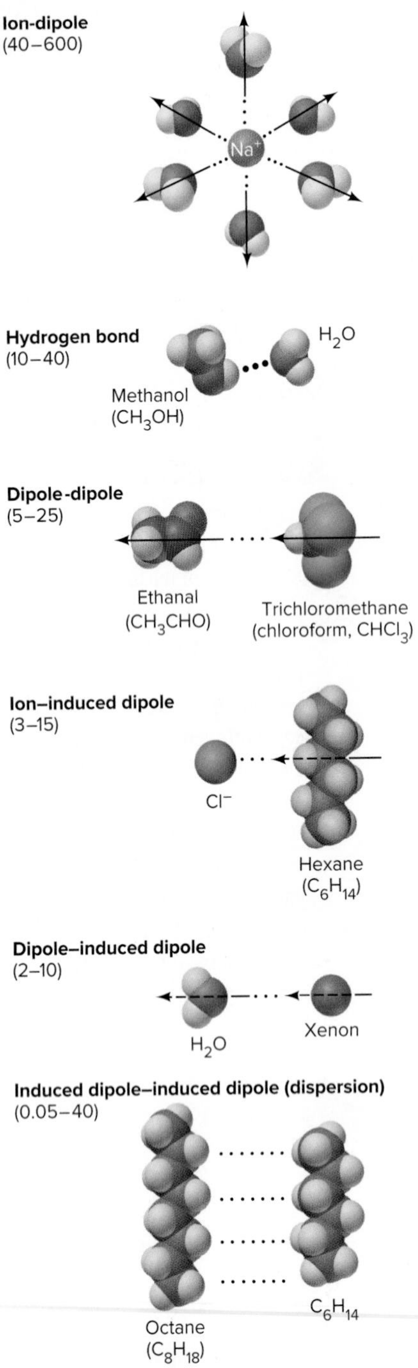

FIGURE 12.1 Types of intermolecular forces in solutions. The forces are listed in order of decreasing strength (kJ/mol), with an example of each.

Intermolecular Forces in Solution

All the intermolecular forces we discussed for pure substances also occur in solutions (Figure 12.1; see also Section 11.2):

1. *Ion-dipole forces* are the principal forces involved when an ionic compound dissolves in water. Two events occur simultaneously:
   - *Forces compete.* When a salt is added to water, an ion and the oppositely charged pole of a water molecule attract each other. These attractions compete with and overcome attractions between the ions, and the crystal structure breaks down.
   - *Hydration shells form.* As an ion separates, water molecules cluster around it in **hydration shells**. The number of water molecules in the closest shell depends on the size of the ion: four water molecules fit tetrahedrally around small ions, such as Li^+, while the larger Na^+ and F^- ions have six water molecules surrounding them octahedrally (Figure 12.2). In the innermost shell, normal hydrogen bonding is disrupted to form the ion-dipole forces. However, these water molecules are hydrogen-bonded to others in the next shell, and those water molecules are hydrogen-bonded to others still farther away.
2. *Hydrogen bonding* is the principal force in solutions of polar O-containing and N-containing organic and biological compounds, such as alcohols, amines, and amino acids.
3. *Dipole-dipole forces,* in the absence of hydrogen bonding, allow polar molecules such as propanal (CH_3CH_2CHO) to dissolve in polar solvents such as dichloromethane (CH_2Cl_2).
4. **Ion–induced dipole forces**, one type of *charge-induced dipole force*, rely on polarizability. They arise when an ion's charge distorts the electron cloud of a nearby nonpolar molecule. This type of force initiates the binding of the Fe^{2+} ion in hemoglobin to an O_2 molecule entering a red blood cell.

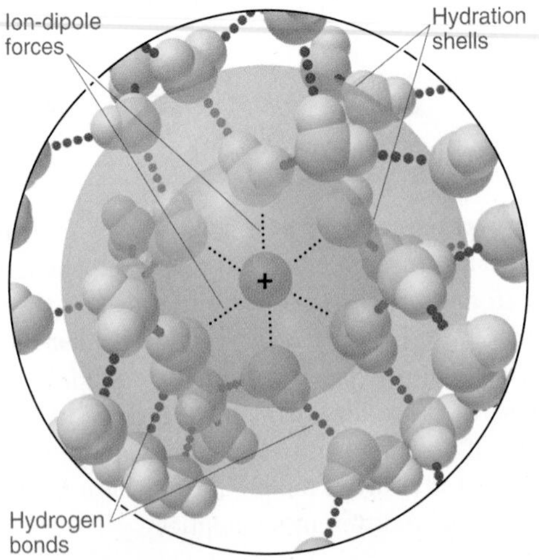

FIGURE 12.2 Hydration shells around an Na+ ion. Ion-dipole forces orient water molecules around an ion. Here, in the innermost shell, six water molecules surround the cation octahedrally.

5. **Dipole–induced dipole forces**, also based on polarizability, arise when a polar molecule distorts the electron cloud of a nonpolar molecule. They are weaker than ion–induced dipole forces because the charge of each pole is less than an ion's charge (Coulomb's law). The solubility of atmospheric O_2, N_2, and noble gases in water, while limited, is due in part to these forces. Paint thinners and grease solvents also use these forces.
6. *Dispersion forces*, or *induced dipole–induced dipole forces*, contribute to the solubility of all solutes in all solvents, but they are the *principal* intermolecular forces in solutions of nonpolar substances, such as the substances in petroleum or gasoline.

The same forces keep biological macromolecules in their active shapes (Section 22.4).

Liquid Solutions and the Role of Molecular Polarity

From cytoplasm to tree sap, gasoline to cleaning fluid, and iced tea to urine, liquid solutions are very familiar. Water is the most prominent solvent, but there are many other liquid solvents, with polarities ranging from very polar to nonpolar.

Applying the Like-Dissolves-Like Rule The like-dissolves-like rule says that when the forces *within* a solute are *similar* to those *within* a solvent, the forces *replace* each other and a solution forms. Thus, the following properties are observed:

- *Salts are soluble in water* because the strong ion-dipole attractions between the ions and water are so similar to the strong attractions between the ions and the strong hydrogen bonds between water molecules that they *can* replace each other.
- *Salts are insoluble in hexane* (C_6H_{14}) because the weak ion–induced dipole forces between the ions and nonpolar hexane *cannot* replace attractions between the ions.
- *Oil is insoluble in water* because the weak dipole–induced dipole forces between the oil and water molecules cannot replace the strong hydrogen bonds within the water molecules and the extensive dispersion forces within the oil.
- *Oil is soluble in hexane* because the dispersion forces in one readily replace the dispersion forces in the other.

Dual Polarity and Effects on Solubility To examine these properties further, let us compare the solubilities of a series of alcohols in water and hexane, solvents with very different intermolecular forces. Alcohols are organic compounds that have a dual polarity, a polar hydroxyl (—OH) group bonded to a nonpolar hydrocarbon group:

- The —OH portion interacts through strong hydrogen bonds with water and through weak dipole–induced dipole forces with hexane.
- The hydrocarbon portion interacts through many dispersion forces with hexane and through weak dipole–induced dipole forces with water.

The general formula for an alcohol is $CH_3(CH_2)_nOH$. We will look at straight-chain examples with one to six carbons ($n = 0$ to 5):

1. *Solubility in water is high for smaller alcohols.* From the models in Table 12.1, we see that the —OH group is a relatively large portion of alcohols with one to three carbons ($n = 0$ to $n = 2$). These molecules interact with each other through hydrogen bonding, just as water molecules do. When they mix with water, hydrogen bonding in solute and in solvent is replaced by hydrogen bonding *between* solute and solvent (Figure 12.3). As a result, these smaller alcohols are miscible with water.

2. *Solubility in water is low for larger alcohols.* Solubility decreases dramatically for alcohols larger than three carbons ($n > 2$); in fact, alcohols with chains longer than six carbons are insoluble in water. For larger alcohols to dissolve, the nonpolar chains have to move among the water molecules, replacing strong hydrogen bonds between water molecules with weak attractions to water. While the —OH

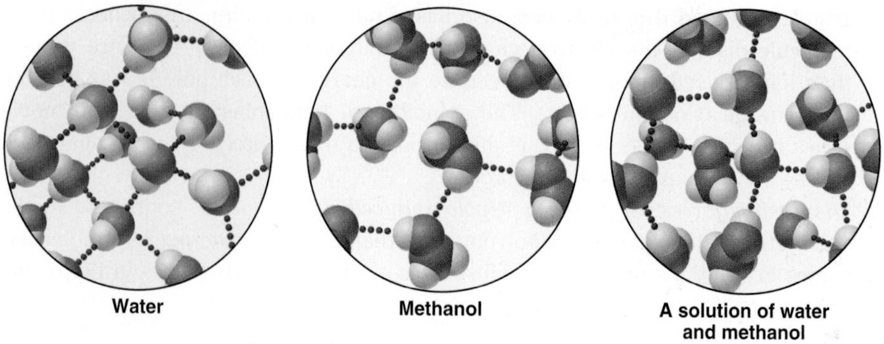

FIGURE 12.3 Like dissolves like: the solubility of methanol in water. The hydrogen bonds in water and in methanol replace one another when the two substances form a solution.

portion of such an alcohol forms hydrogen bonds to water, these bonds cannot make up for all the other hydrogen bonds between water molecules that have to break to make room for the hydrocarbon portion.

Table 12.1 shows that the opposite trend occurs with hexane:

1. *Solubility in hexane is low for smaller alcohols.* For alcohols in hexane, in addition to dispersion forces, weak dipole–induced dipole forces exist between the —OH of methanol (CH_3OH) and hexane. These forces cannot replace the strong hydrogen bonding between CH_3OH molecules, so solubility is low.

2. *Solubility in hexane is high for larger alcohols.* In any larger alcohol ($n > 0$), dispersion forces between the hydrocarbon portion and hexane *can* replace the dispersion forces between hexane molecules. With only weak forces within the solvent to be replaced, even ethanol, with a two-carbon chain, has enough dispersion forces between it and hexane to be miscible.

Many organic molecules have polar and nonpolar portions, which determine their solubility. For example, carboxylic acids and amines behave like alcohols: methanoic acid (formic acid, HCOOH) and methanamine (CH_3NH_2) are miscible with water and slightly soluble in hexane, whereas hexanoic acid [$CH_3(CH_2)_4COOH$] and hexan-1-amine [$CH_3(CH_2)_5NH_2$] are slightly soluble in water and very soluble in hexane.

TABLE 12.1 Solubility* of a Series of Alcohols in Water and in Hexane

Alcohol	Model	Solubility in Water	Solubility in Hexane
CH_3OH (methanol)		∞	1.2
CH_3CH_2OH (ethanol)		∞	∞
$CH_3(CH_2)_2OH$ (propan-1-ol)		∞	∞
$CH_3(CH_2)_3OH$ (butan-1-ol)		1.1	∞
$CH_3(CH_2)_4OH$ (pentan-1-ol)		0.30	∞
$CH_3(CH_2)_5OH$ (hexan-1-ol)		0.058	∞

*Expressed in mol alcohol/1000 g solvent at 20°C.

Sample Problem 12.1 Predicting Relative Solubilities

Problem Predict which solvent will dissolve more of the given solute:

(a) Sodium chloride in methanol (CH_3OH) or in propan-1-ol ($CH_3CH_2CH_2OH$)

(b) Ethylene glycol (ethane-1,2-diol, $HOCH_2CH_2OH$) in hexane ($CH_3CH_2CH_2CH_2CH_2CH_3$) or in water

(c) Ethoxyethane (diethyl ether, $CH_3CH_2OCH_2CH_3$) in water or in ethanol (CH_3CH_2OH)

Plan We examine the formulas of the solute and solvent to determine the forces in and between the solute and solvent. A solute is more soluble in a solvent whose intermolecular forces are similar to, and therefore can replace, its own.

Solution (a) Methanol can dissolve more sodium chloride. NaCl is ionic, so it dissolves through ion-dipole forces. Both methanol and propan-1-ol have a polar —OH group, but the hydrocarbon portion of each alcohol interacts only weakly with the ions, and propan-1-ol has a longer hydrocarbon portion than methanol.

(b) Water can dissolve more ethylene glycol. Ethylene glycol molecules have two —OH groups, so they interact with each other through hydrogen bonding. Hydrogen bonds formed with H_2O can replace these hydrogen bonds between solute molecules better than dipole–induced dipole forces with hexane can.

(c) Ethanol can dissolve more ethoxyethane. Ethoxyethane (diethyl ether) molecules interact through dipole-dipole and dispersion forces. They can form hydrogen bonds to water or to ethanol. However, ethanol can also interact with the ether effectively through dispersion forces because it has a hydrocarbon chain.

Follow-Up Problem 12.1 Which solute is more soluble in the given solvent? **(a)** Butan-1-ol ($CH_3CH_2CH_2CH_2OH$) or butane-1,4-diol ($HOCH_2CH_2CH_2CH_2OH$) in water; **(b)** trichloromethane (chloroform, $CHCl_3$) or tetrachloromethane (carbon tetrachloride, CCl_4) in water. (See Brief Solutions to Follow-Up Problems at the end of the chapter.)

Gas-Liquid Solutions A substance with very weak intermolecular attractions has a low boiling point and is a gas under ordinary conditions. Likewise, it is not very soluble in water because solute-solvent forces are weak. Thus, for nonpolar or slightly polar gases, boiling point generally correlates with solubility in water (Table 12.2).

The small amount of a nonpolar gas that *does* dissolve may be vital. At 25°C and 101.3 kPa, the solubility of O_2 is only 3.2 mL/100. mL of water, but aquatic animal life requires it. Sometimes, the solubility of a nonpolar gas may *seem* high because it is also *reacting* with solvent. Oxygen seems more soluble in blood than in water because it bonds to hemoglobin in red blood cells. Carbon dioxide, which is essential for aquatic plants and coral-reef growth, seems very soluble in water (~81 mL of CO_2/100. mL of H_2O at 25°C and 101.3 kPa) because it is dissolving *and* reacting:

$$CO_2(g) + H_2O(l) \longrightarrow H^+(aq) + HCO_3^-(aq)$$

TABLE 12.2 Correlation between Boiling Point and Solubility in Water

Gas	Solubility (mol/L)*	Boiling Point (K)
He	4.2×10^{-4}	4.2
Ne	6.6×10^{-4}	27.1
N_2	10.4×10^{-4}	77.4
CO	15.6×10^{-4}	81.6
O_2	21.8×10^{-4}	90.2
NO	32.7×10^{-4}	121.4

*At 273 K and 101.3 kPa.

Gas Solutions and Solid Solutions

Gas solutions and solid solutions also have vital importance and numerous applications.

Gas-Gas Solutions *All gases are miscible with each other.* Air is the classic example of a gaseous solution, consisting of about 18 gases in widely differing proportions. The proportions of anaesthetic gases are finely adjusted to the needs of each patient and the length of the surgical procedure. The proportions of many industrial gas mixtures, such as CO/H_2 in syngas production and N_2/H_2 in ammonia production, are controlled to optimize product yield under varying conditions of temperature and pressure.

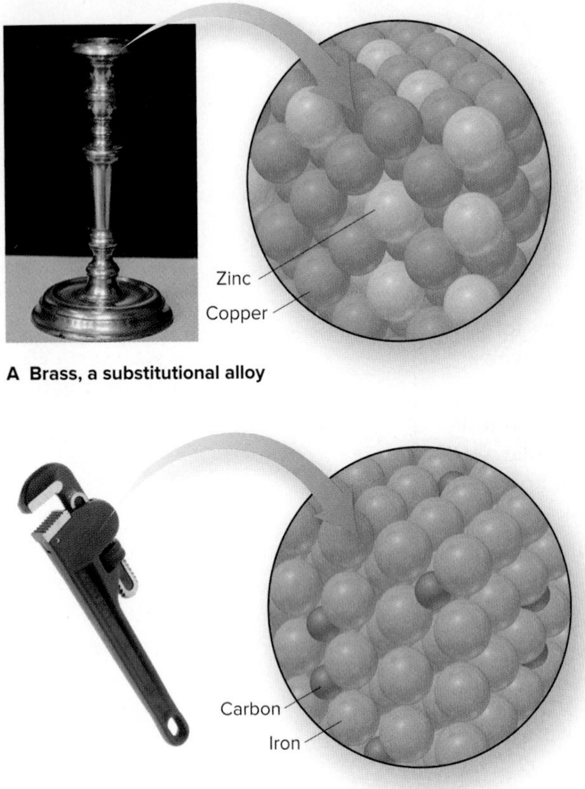

A Brass, a substitutional alloy

B Carbon steel, an interstitial alloy

FIGURE 12.4 The arrangement of atoms in two types of alloys

Gas-Solid Solutions *When a gas dissolves in a solid, it occupies the spaces between the closely packed particles.* Hydrogen gas can be purified by passing an impure sample through palladium. Only H_2 molecules are small enough to fit between the Pd atoms, where they form Pd—H bonds. The H atoms move from one Pd atom to another and emerge from the metal as H_2 molecules (see Figure 13.2).

The ability of gases to dissolve in solids has disadvantages, however. For example, the electrical conductivity of copper is greatly reduced by the presence of O_2, which dissolves into the crystal structure and reacts to form copper(I) oxide. High-conductivity copper is prepared by melting and recasting the metal in an O_2-free atmosphere.

Solid-Solid Solutions Solids diffuse so little that their mixtures are usually heterogeneous. Some solid-solid solutions can be formed by melting the solids and then mixing them and allowing them to freeze. Many **alloys**, mixtures of elements that have a metallic character, are solid-solid solutions (although several have microscopic heterogeneous regions). In a *substitutional alloy*, such as brass (Figure 12.4A), atoms of zinc replace atoms of the main element, copper, at some sites in the cubic closest packing array. In an *interstitial alloy*, such as carbon steel (Figure 12.4B), atoms of carbon (a nonmetal is typical in this type of alloy) fill some of the spaces (*interstices*) between atoms of the main element, iron, in the body-centred array.

Waxes are also solid-solid solutions. Most are amorphous solids with some small regions of crystalline regularity. A natural *wax* is defined as a solid of biological origin that is insoluble in water and soluble in nonpolar solvents. Beeswax, which bees secrete to build their honeycombs, is a homogeneous mixture of fatty acids, long-chain carboxylic acids, and hydrocarbons. Some of the molecules contain chains that are more than 40 carbon atoms long! Carnauba wax, from a South American palm, is a mixture of compounds, each consisting of a fatty acid bound to a long-chain alcohol. It is hard but forms a thick gel in nonpolar solvents, so it is perfect for waxing cars.

SUMMARY OF SECTION 12.1

- A solution is a homogeneous mixture of a solute dissolved in a solvent through the action of intermolecular forces.
- Ion-dipole, ion–induced dipole, and dipole–induced dipole forces occur in solutions, in addition to all the intermolecular forces that occur in pure substances.
- If similar intermolecular forces occur in a solute and a solvent, they replace each other when the substances mix, and a solution is likely to form (the like-dissolves-like rule).
- When ionic compounds dissolve in water, the ions become surrounded by hydration shells of hydrogen-bonded water molecules.
- The solubility of organic molecules in various solvents depends on the relative sizes of their polar and nonpolar portions.
- The solubility of nonpolar gases in water is low because of weak intermolecular forces. Gases are miscible with one another and dissolve in solids by fitting into spaces in the crystal structure.
- Solid-solid solutions include alloys and waxes, and some are formed by mixing molten components.

12.2 Why Substances Dissolve: Understanding the Solution Process

The qualitative *macroscopic* rule "like dissolves like" is based on *molecular* interactions between solute and solvent. To see *why* like dissolves like, we will break down the solution process conceptually into steps and examine them quantitatively.

Heat of Solution: Solution Cycles

Before a solution forms, the solute particles are attracting each other, and the solvent particles are also attracting each other. For the solute to dissolve in the solvent, the following three steps must take place. Each step is accompanied by an enthalpy change:

Step 1. Solute particles separate from each other. This step involves overcoming intermolecular attractions, so it is *endothermic*:

$$\text{Solute (aggregated)} + heat \longrightarrow \text{solute (separated)} \qquad \Delta_{\text{solute}}H > 0$$

Step 2. Solvent particles separate from each other. This step also involves overcoming attractions, so it is also *endothermic*:

$$\text{Solvent (aggregated)} + heat \longrightarrow \text{solvent (separated)} \qquad \Delta_{\text{solvent}}H > 0$$

Step 3. Solute and solvent particles mix and form a solution. The different particles attract each other and come together, so this step is *exothermic*:

$$\text{Solute (separated)} + \text{solvent (separated)} \longrightarrow \text{solution} + heat \qquad \Delta_{\text{mix}}H < 0$$

The overall process is called a *thermochemical solution cycle*. In yet another application of Hess's law, we combine the three individual enthalpy changes to find the **heat of solution ($\Delta_{\text{soln}}H$)**, the total enthalpy change that occurs when a solute and a solvent form a solution:

$$\Delta_{\text{soln}}H = \Delta_{\text{solute}}H + \Delta_{\text{solvent}}H + \Delta_{\text{mix}}H \qquad (12.1)$$

Overall solution formation is either exothermic or endothermic, and $\Delta_{\text{soln}}H$ is either positive or negative, depending on the relative sizes of the individual ΔH values:

- *Exothermic process: $\Delta_{soln}H < 0$.* If the sum of the endothermic terms ($\Delta_{\text{solute}}H$ + $\Delta_{\text{solvent}}H$) is *smaller* than the exothermic term ($\Delta_{\text{mix}}H$), the process is exothermic and $\Delta_{\text{soln}}H$ is negative (Figure 12.5A).
- *Endothermic process: $\Delta_{soln}H > 0$.* If the sum of the endothermic terms is *larger* than the exothermic term, the process is endothermic and $\Delta_{\text{soln}}H$ is positive (Figure 12.5B). If $\Delta_{\text{soln}}H$ is highly positive, the solute may not dissolve significantly in that solvent.

Heat of Hydration: Ionic Solids in Water

The $\Delta_{\text{solvent}}H$ and $\Delta_{\text{mix}}H$ components of the solution cycle are difficult to measure individually. Combined, they equal the enthalpy change for **solvation**, the process of surrounding a solute particle with solvent particles:

$$\Delta_{\text{solvation}}H = \Delta_{\text{solvent}}H + \Delta_{\text{mix}}H$$

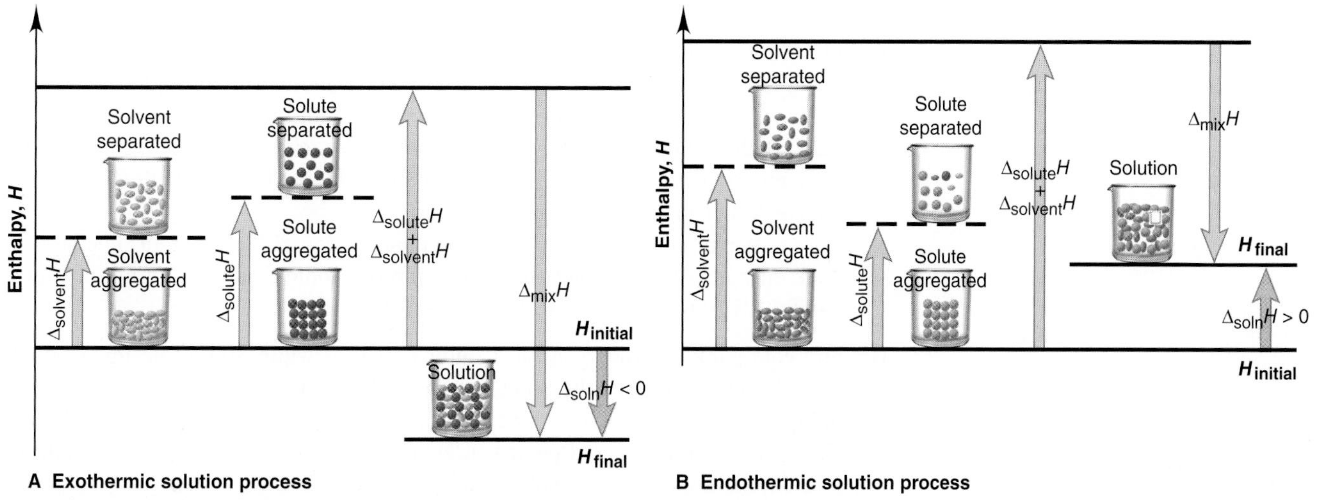

FIGURE 12.5 Enthalpy components of the heat of solution. A. $\Delta_{\text{mix}}H$ is larger than the sum of $\Delta_{\text{solute}}H$ and $\Delta_{\text{solvent}}H$, so $\Delta_{\text{soln}}H$ is negative. **B.** $\Delta_{\text{mix}}H$ is smaller than the sum of $\Delta_{\text{solute}}H$ and $\Delta_{\text{solvent}}H$, so $\Delta_{\text{soln}}H$ is positive.

Solvation in water is called **hydration**. Thus, enthalpy changes for separating the water molecules ($\Delta_{\text{solvent}}H$) and mixing the separated solute with them ($\Delta_{\text{mix}}H$) are combined into the **heat of hydration ($\Delta_{\text{hydr}}H$)**. In water, Equation 12.1 becomes

$$\Delta_{\text{soln}}H = \Delta_{\text{solute}}H + \Delta_{\text{hydr}}H$$

The heat of hydration is a key factor in dissolving an ionic solid. Breaking hydrogen bonds in water is more than compensated for by forming the stronger ion-dipole forces, so hydration of an ion is *always* exothermic. The $\Delta_{\text{hydr}}H$ of an ion is defined as the enthalpy change for the hydration of 1 mol of separated (gaseous) ions:

$$M^+(g)\ [\text{or } X^-(g)] \xrightarrow{H_2O} M^+(aq)\ [\text{or } X^-(aq)] \qquad \Delta_{\text{hydr of the ion}}H\ (\text{always} < 0)$$

Importance of Charge Density Heats of hydration exhibit trends based on an ion's **charge density**, the ratio of its charge to its volume. In general, the higher the charge density, the more negative $\Delta_{\text{hydr}}H$ is. Coulomb's law explains why: the higher the charge of an ion and the smaller its radius, the closer it gets to the oppositely charged pole of an H_2O molecule (see Figure 2.12), and the stronger the attraction. Thus,

- A 2+ ion attracts H_2O molecules more strongly than a 1+ ion of similar size
- A small 1+ ion attracts H_2O molecules more strongly than a large 1+ ion

Periodic trends in $\Delta_{\text{hydr}}H$ values are based on trends in charge density (Table 12.3):

- *Down a group*, the charge stays the same and the size increases; thus, the charge densities decrease, as do the $\Delta_{\text{hydr}}H$ values.
- *Across a period*, for example, from group 1 to group 2, the group 2 ion has a smaller radius *and* a higher charge, so its charge density and $\Delta_{\text{hydr}}H$ are greater.

TABLE 12.3 Trends in Heats of Hydration

Ion	Ionic Radius (pm)	$\Delta_{\text{hydr}}H$ (kJ/mol)
Group 1		
Na^+	102	−410
K^+	138	−336
Rb^+	152	−315
Cs^+	167	−282
Group 2		
Mg^{2+}	72	−1903
Ca^{2+}	100	−1591
Sr^{2+}	118	−1424
Ba^{2+}	135	−1317
Group 17		
F^-	133	−431
Cl^-	181	−313
Br^-	196	−284
I^-	220	−247

Components of Aqueous Heats of Solution To separate an ionic solute, MX, into gaseous ions ($\Delta_{\text{solute}}H$) requires a lot of energy. Recall, from Chapter 8, that this is the lattice energy, and it is highly positive:

$$MX(s) \longrightarrow M^+(g) + X^-(g) \qquad \Delta_{\text{solute}}H\ (\text{always} > 0) = \Delta_{\text{lattice}}H$$

Thus, for ionic compounds in water, the heat of solution is the lattice energy (always positive) plus the combined heats of hydration of the ions (always negative):

$$\Delta_{\text{soln}}H = \Delta_{\text{lattice}}H + \Delta_{\text{hydr of the ions}}H \tag{12.2}$$

Once again, the sizes of the individual terms determine the sign of $\Delta_{\text{soln}}H$.

Figure 12.6 shows enthalpy diagrams for three ionic solutes dissolving in water:

- *NaCl.* Sodium chloride has a small positive $\Delta_{\text{soln}}H$ (3.9 kJ/mol) because its lattice energy is only slightly greater than the combined ionic heats of hydration. If you dissolve NaCl in water in a flask, you do not feel any temperature change.
- *NaOH.* Sodium hydroxide has a large negative $\Delta_{\text{soln}}H$ (−44.5 kJ/mol) because its lattice energy is much *smaller* than the combined ionic heats of hydration. If you dissolve NaOH in water, the flask feels hot.
- *NH_4NO_3.* Ammonium nitrate has a large positive $\Delta_{\text{soln}}H$ (25.7 kJ/mol) because its lattice energy is much *larger* than the combined ionic heats of hydration. If you dissolve NH_4NO_3 in water, the flask feels cold.

Hot and cold packs consist of a thick outer pouch of water and a thin inner pouch of a salt. A squeeze breaks the inner pouch, and the salt dissolves. Most hot packs use anhydrous $CaCl_2$ ($\Delta_{\text{soln}}H = -82.8$ kJ/mol), and most cold packs use NH_4NO_3 ($\Delta_{\text{soln}}H = 25.7$ kJ/mol). A cold pack can keep the solution at 0°C for about half an hour. In Japan, some soup is sold in a double-walled can, with a packet of salt immersed in water between the walls. Open the can and the packet breaks, the salt dissolves, and the soup quickly warms to about 90°C.

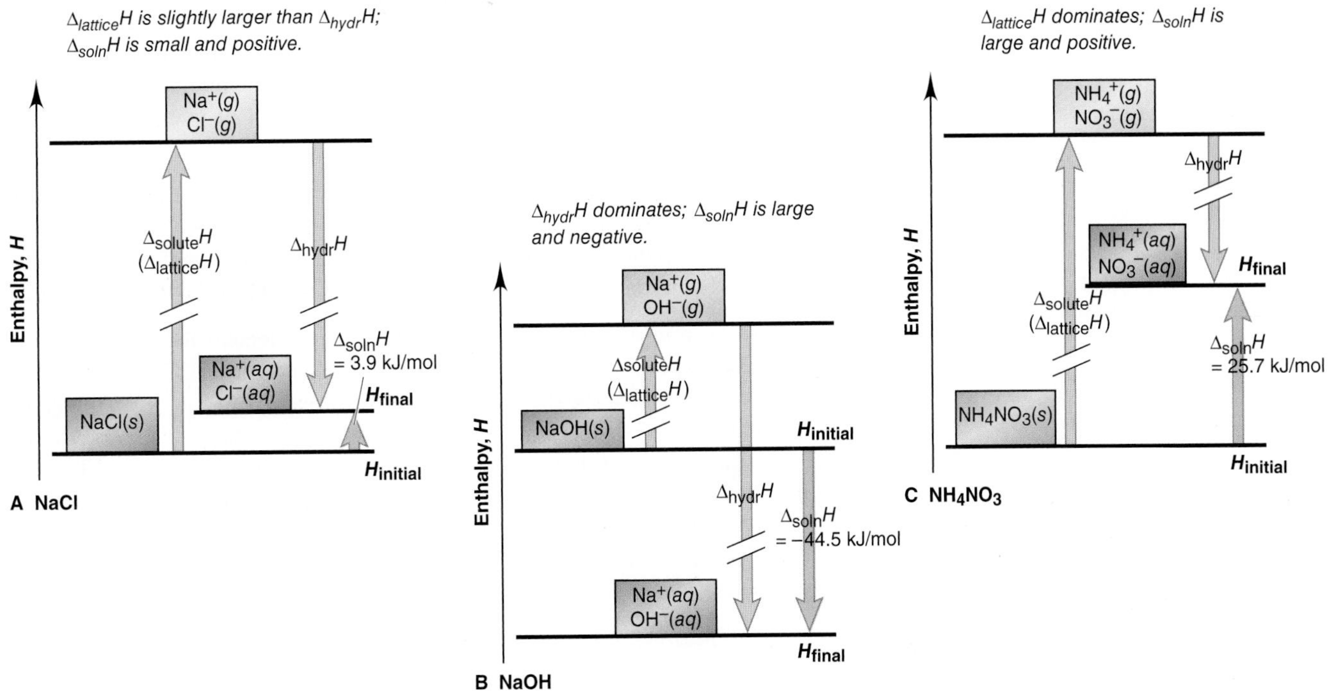

FIGURE 12.6 Enthalpy diagrams for three ionic compounds dissolving in water

The Solution Process and the Change in Entropy

The heat of solution ($\Delta_{soln}H$) is one of two factors that determine whether a solute dissolves. The other factor involves the natural tendency of a system of particles to spread out, and the system's kinetic energy to become more dispersed or distributed. A thermodynamic variable called **entropy (S)** is directly related to the number of ways that a system can disperse its energy, which is closely related to the freedom of motion of the particles. (Entropy will be discussed in more detail in Chapter 18).

Let us see what it means for a system to distribute its energy. We will first compare the three physical states and then compare a solute and a solvent with a solution.

Entropy and the Three Physical States The states of matter differ significantly in their entropy:

- In a solid, the particles are fixed in their positions with little freedom of motion. In a liquid, the particles can move around one another and so have greater freedom of motion. In a gas, the particles have little restriction and much more freedom of motion.
- The more freedom of motion the particles have, the more ways that they can distribute their kinetic energy. Thus, a liquid has higher entropy than a solid, and a gas has higher entropy than a liquid:

$$S_{gas} > S_{liquid} > S_{solid}$$

- We can also look at the *change* in entropy (ΔS) of a phase change. Thus, for example, the change in entropy when a liquid vaporizes ($\Delta_{vap}S = S_{gas} - S_{liquid}$) is positive ($\Delta_{vap}S > 0$), the change in entropy when a liquid freezes (fusion; $\Delta_{fus}S = S_{solid} - S_{liquid}$) is negative ($\Delta_{fus}S < 0$), and so on.

Entropy and the Formation of Solutions The formation of solutions also involves a change in entropy. *A solution usually has higher entropy than the pure solute and pure solvent* because the number of ways to distribute the energy is related to the number of interactions between different molecules. There are far more interactions possible when a solute and a solvent are mixed than when they are pure. Thus,

$$S_{soln} > (S_{solute} + S_{solvent}) \quad \text{or} \quad \Delta_{soln}S > 0$$

You know from everyday experience that solutions form naturally, but pure solutes and solvents do not. You have probably seen sugar dissolve in water, but you have never seen a sugar solution separate into pure sugar and water. In Chapter 18, you will see that energy is needed to reverse the natural tendency of systems to distribute their energy—to get "mixed up." Water treatment plants, oil refineries, steel mills, and many other industrial facilities expend a lot of energy to separate mixtures into pure components.

Enthalpy versus Entropy Changes in Solution Formation Solution formation involves the interplay of two factors: systems change toward a state of *lower enthalpy* and *higher entropy*, so the relative sizes of $\Delta_{soln}H$ and $\Delta_{soln}S$ determine whether a solution forms. Let us consider three solute-solvent pairs to see which factor dominates:

1. *NaCl in hexane.* Given their very different intermolecular forces, we predict that sodium chloride does *not* dissolve in hexane (C_6H_{14}) (Figure 12.7A). Separating the solvent is easy because the dispersion forces are weak ($\Delta_{solvent} H \geq 0$), but separating the solute requires supplying the very large $\Delta_{lattice}H$ ($\Delta_{solute}H >> 0$). Mixing releases little heat because ion–induced dipole forces between Na^+ (or Cl^-) and hexane are weak ($\Delta_{mix}H \leq 0$). Because the sum of the endothermic terms is *much* larger than the exothermic term, $\Delta_{soln}H >> 0$. *A solution does not form because the entropy increase from mixing solute and solvent would be much smaller than the enthalpy increase required to separate the solute:* $\Delta_{mix}S << \Delta_{solute}H$.

2. *Octane in hexane.* We predict that octane (C_8H_{18}) is soluble in hexane because both are held together by dispersion forces of similar strength; in fact, they are miscible. That is, both $\Delta_{solute}H$ and $\Delta_{solvent}H$ are around zero. The similar forces mean that $\Delta_{mix}H$ is also around zero. As well, not a lot of heat is released; in fact, $\Delta_{soln}H$ is around zero (Figure 12.7B). So, why does a solution form so readily? *With no enthalpy change driving the process, octane dissolves in hexane because the entropy increases greatly when the pure substances mix:* $\Delta_{mix}S >> \Delta_{soln}H$.

3. *NH_4NO_3 in water.* A large enough increase in entropy can sometimes cause a solution to form even when the enthalpy increase is large ($\Delta_{soln}H >> 0$). As shown in Figure 12.6C, when ammonium nitrate (NH_4NO_3) dissolves in water, the process is highly endothermic; that is, $\Delta_{lattice}H >> \Delta_{hydr\ of\ the\ ions}H$. Nevertheless, *the increase in entropy that occurs when the crystal breaks down and the ions mix with water molecules is greater than the increase in enthalpy:* $\Delta_{soln}S > \Delta_{soln}H$.

In Chapter 18, we will examine the relation between enthalpy and entropy in depth to understand physical and chemical systems.

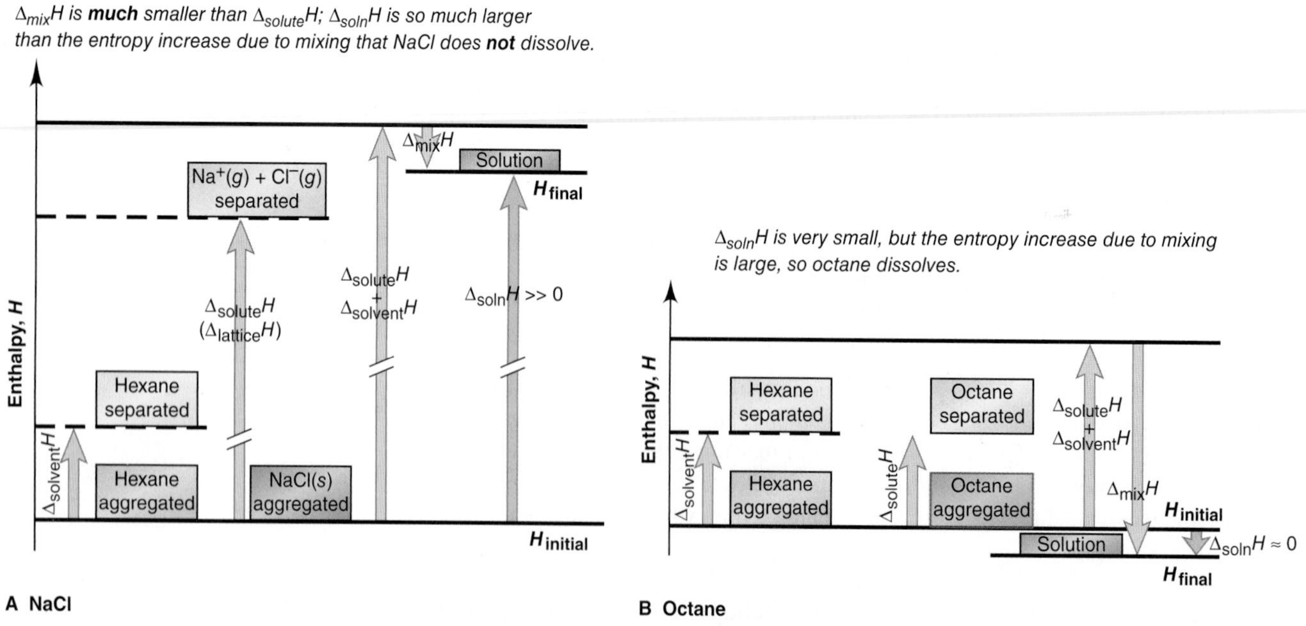

FIGURE 12.7 Enthalpy diagrams for dissolving NaCl (A) and octane (B) in hexane

SUMMARY OF SECTION 12.2

- In a thermochemical solution cycle, the heat of solution is the sum of the endothermic separations of solute and solvent and the exothermic mixing of their particles.
- In water, the combination of solvent separation and mixing with solute particles is hydration. For ions, heats of hydration depend on the ion's charge density but are always negative because ion-dipole forces are strong.
- Systems naturally increase their entropy (distribute their energy in more ways). A gas has higher entropy than a liquid, which has higher entropy than a solid, and a solution has higher entropy than the pure solute and solvent.
- Relative sizes of enthalpy and entropy changes determine solution formation. A substance with a positive $\Delta_{soln}H$ dissolves *only* if $\Delta_{soln}S$ is larger than $\Delta_{soln}H$.

FIGURE 12.8 Equilibrium in a saturated solution. At some temperature, the number of solute particles dissolving (*white arrows*) per unit time equals the number of solute particles recrystallizing (*black arrows*).

12.3 Solubility as an Equilibrium Process

When an excess amount of solid is added to a solvent, particles leave the crystal, are surrounded by solvent, and move away. Some dissolved solute particles collide with undissolved solute and recrystallize, but, as long as the rate of dissolving is greater than the rate of recrystallizing, the concentration rises. At a given temperature, when solid is dissolving at the same rate as dissolved particles are recrystallizing, the concentration remains constant and *undissolved solute is in equilibrium with dissolved solute:*

$$\text{Solute (undissolved)} \rightleftharpoons \text{solute (dissolved)}$$

Figure 12.8 shows an ionic solid in equilibrium, with dissolved cations and anions.

The following three terms express the extent of the solution process:

- A **saturated solution** is at equilibrium and contains the maximum amount of dissolved solute at a given temperature in the presence of undissolved solute. Therefore, if we filter off the solution and add more solute, the additional solute does not dissolve.
- An **unsaturated solution** contains *less* than the equilibrium concentration of dissolved solute. If we add more solute, more dissolves until the solution is saturated.
- A **supersaturated solution** contains *more* than the equilibrium concentration and is unstable relative to the saturated solution. If we add a "seed" crystal of solute or tap the container, the excess solute crystallizes immediately, leaving a saturated solution (Figure 12.9). We can often prepare a supersaturated solution if the solute is more soluble at a higher temperature. While heating, we dissolve more than the amount required for saturation at a lower temperature, and then slowly cool the solution. If the excess solute remains dissolved, the solution is supersaturated.

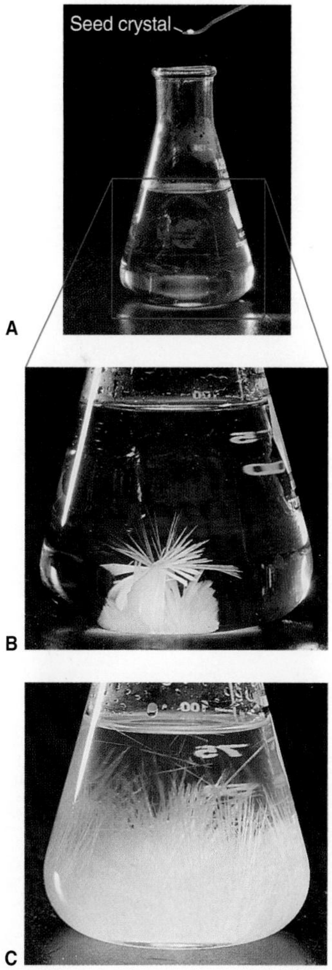

FIGURE 12.9 Sodium acetate crystallizing from a supersaturated solution. When a seed crystal of sodium acetate is added to a supersaturated solution of sodium acetate **(A)**, solute begins to crystallize **(B)** and continues to crystallize until the remaining solution is saturated **(C)**.

The Effect of Temperature on Solubility

We know that more sugar dissolves in hot tea than in iced tea; in fact, temperature affects the solubility of most substances. Let us examine the effects of temperature on the solubility of solids and gases.

Temperature and the Solubility of Solids in Water Like sugar, *most solids are more soluble at higher temperatures* (Figure 12.10). Note that cerium sulfate is one of several exceptions, most of which are other sulfates. Some salts have higher solubility up to a certain temperature and then lower solubility at still higher temperatures.

Unfortunately, the effect of temperature on solubility is a complex phenomenon, and the sign of $\Delta_{soln}H$ does not reflect this complexity. Tabulated $\Delta_{soln}H$ values give the enthalpy change for making a solution at the standard state of 1 mol/L. To understand the effect of temperature, however, we need to know the sign of the enthalpy change very close to the point of saturation, which may differ from the sign of the tabulated value. For example, tables give a negative

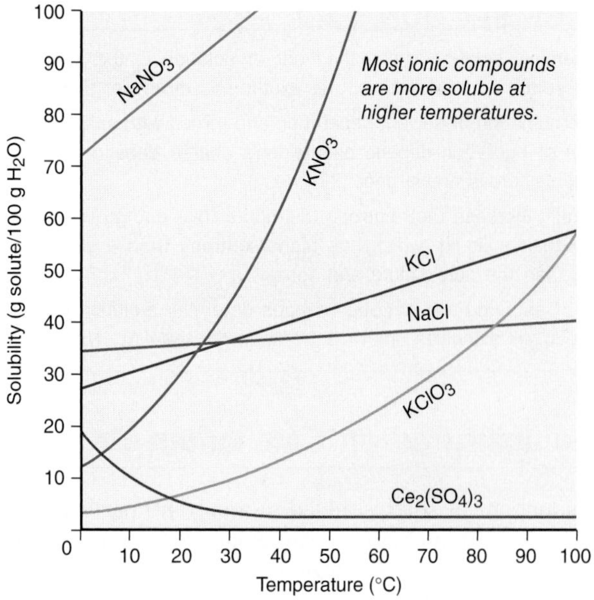

FIGURE 12.10 **Relationship between solubility and temperature for several ionic compounds**

$\Delta_{soln}H$ for NaOH and a positive $\Delta_{soln}H$ for NH_4NO_3, yet both compounds are more soluble at higher temperatures. The point is that, although the effect of temperature reflects the equilibrium nature of solubility, no single measure can predict this effect for a given solute.

Temperature and the Solubility of Gases in Water The effect of temperature on *gas* solubility is much more predictable. When a solid dissolves in a liquid, the solute particles must separate, so $\Delta_{solute}H > 0$. In contrast, gas particles are already separated, so $\Delta_{solute}H \approx 0$. Because hydration is exothermic ($\Delta_{hydr}H < 0$), the sum of these two terms is negative. Thus, $\Delta_{soln}H < 0$ for all gases in water:

$$\text{solute}(g) + \text{water}(l) \rightleftharpoons \text{saturated solution}(aq) + \text{heat}$$

Thus, *gas solubility in water* ***decreases*** *with rising temperature (the addition of heat).* Gases have weak intermolecular forces with water. When the temperature rises, the average kinetic energy increases, allowing the gas particles to easily overcome these forces and reenter the gas phase.

This behaviour leads to *thermal pollution.* Many electric power plants use large amounts of water from a nearby river or lake for cooling, and the warmed water is returned to the source. The metabolic rates of fish and other aquatic animals increase in this warmer water, increasing their need for O_2. However, the concentration of dissolved O_2 is lower in warm water, so the aquatic animals become oxygen deprived. Also, the less dense warm water floats and prevents O_2 from reaching the cooler water below. Thus, even creatures at deeper levels become oxygen deprived. Farther from the plant, the water temperature and O_2 solubility return to normal. To mitigate the problem, cooling towers lower the temperature of the water before it exits the plant (Figure 12.11); nuclear power plants use a similar approach (Section 25.7).

FIGURE 12.11 **Preventing thermal pollution with cooling towers**

The Effect of Pressure on Solubility

Pressure has little effect on the solubility of liquids and solids because they are almost incompressible. Pressure has a *major* effect, however, on the solubility of gases. Consider a piston-cylinder assembly with a gas above a saturated aqueous solution of the gas (Figure 12.12A). At equilibrium, at a given pressure, the same amount of gas molecules enter and leave the solution per unit time:

$$\text{gas} + \text{solvent} \rightleftharpoons \text{saturated solution}$$

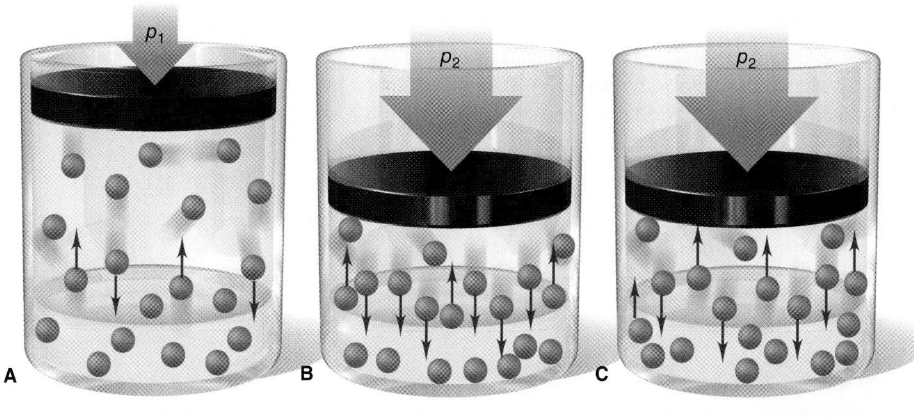

FIGURE 12.12 The effect of pressure on gas solubility

Push down on the piston, and we disturb the equilibrium: gas volume decreases, so gas pressure (and concentration) increase, and gas particles collide with the liquid surface more often. Thus, more particles enter than leave the solution per unit time (Figure 12.12B). More gas dissolves to reduce this disturbance (a shift to the right in the preceding equation) until the system reestablishes equilibrium (Figure 12.12C).

The relation between gas pressure and solubility has many familiar applications. In a closed can of cola, dissolved CO_2 is in equilibrium with 405.3 kPa of CO_2 in the small volume above the solution. Open the can, and the dissolved CO_2 bubbles out of solution until the drink goes "flat" because it reaches equilibrium with CO_2 in the air ($p_{CO_2} = 3 \times 10^{-2}$ kPa). In a different situation (*see photo*), scuba divers who breathe compressed air have a high concentration of N_2 dissolved in their blood. If they ascend too quickly, they may suffer decompression sickness (the "bends"): lower external pressure allows the dissolved N_2 to form bubbles in the blood, which block capillaries.

What happens to gases that are dissolved in the blood of a scuba diver?

Henry's law expresses the quantitative relationship between gas pressure and solubility: *the solubility of a gas (s_{gas}) is directly proportional to the partial pressure of the gas (p_{gas}) above the solution:*

$$s_{gas} = k_H \times p_{gas} \quad (12.3)$$

where k_H is the *Henry's law constant* (an equilibrium constant) and is specific for a given gas-solvent combination at a given temperature. With s_{gas} in mol/L and p_{gas} in kPa, the units of k_H are mol/(L•kPa).

William Henry (1774–1836) studied the quantity of a gas dissolved in water under different pressures.

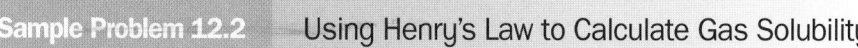

Sample Problem 12.2 Using Henry's Law to Calculate Gas Solubility

Problem The partial pressure of carbon dioxide gas inside a bottle of cola is 4×10^2 kPa at 25°C. What is the solubility of CO_2? The Henry's law constant for CO_2 in water is 3.3×10^{-4} mol/(L•kPa) at 25°C.

Plan We know p_{CO_2} (4×10^2 kPa) and the value of k_H [3.3×10^{-4} mol/(L•kPa)], so we can substitute them into Equation 12.3 to find s_{CO_2}.

Solution $s_{CO_2} = k_H \times p_{CO_2} = [3.3 \times 10^{-4}\,\text{mol/(L•kPa)}](4 \times 10^2\,\text{kPa}) = 0.1\,\text{mol/L}$

Check The units are correct. We rounded to one significant figure because there is one significant figure in the pressure value. A 0.5 L bottle of cola has about 0.2 g of dissolved CO_2.

Follow-Up Problem 12.2 If air contains 78% N_2 by volume, what is the solubility of N_2 in water at 25°C and 101.3 kPa [k_H for N_2 in H_2O at 25°C = 7×10^{-6} mol/(L•kPa)]?

SUMMARY OF SECTION 12.3

- A solution that contains the maximum amount of dissolved solute in the presence of excess undissolved solute is saturated. A saturated solution is in equilibrium with excess solute, because solute particles are entering and leaving the solution at the same rate.
- Most solids are more soluble at higher temperatures.
- All gases have a negative $\Delta_{soln}H$ in water, so heating lowers gas solubility in water.
- Henry's law says that the solubility of a gas is directly proportional to its partial pressure above the solution.

12.4 Concentration Terms

Concentration is the *proportion* of a substance in a mixture, so it is an *intensive* property (like density and temperature), a property that does not depend on the quantity of mixture; for example, 1.0 L and 1.0 mL of 0.1 mol/L NaCl have the same concentration. Concentration is a ratio of quantities (Table 12.4), most often solute to *solution*, but sometimes solute to *solvent*. Both parts of the ratio can be given in units of mass, volume, or amount (mol), and chemists express concentration using several terms, including amount of substance concentration in mol/L, molality, and various expressions of "parts of solute per part by solution."

The Amount of Substance Concentration and Molality

Two very common concentration terms are concentration in mol/L (c) and molality (m):

1. Amount of substance concentration, amount concentration, or concentration (mol/L), also called molarity (M), is the *number of moles of solute dissolved in 1 L of solution*:

$$c = \text{concentration (mol/L)} = \frac{\text{amount (mol) of solute}}{\text{volume (L) of solution}} = \frac{n}{V} \qquad (12.4)$$

In Chapter 3, we used concentration (mol/L) to convert litres of solution into the amount of dissolved solute. The amount of substance concentration, or concentration (mol/L), has two drawbacks that affect its use in precise work:

- *Effect of temperature.* A liquid expands when heated, so a unit volume of hot solution contains less solute than a unit volume of cold solution; thus, the concentration (mol/L) is different.
- *Effect of mixing.* Because of solute-solvent interactions that are difficult to predict, *volumes may not be additive*; for example, adding 500. mL of one solution to 500. mL of another solution may not give 1000. mL of final solution.

TABLE 12.4 Concentration Definitions

Concentration Term	Ratio
Concentration (mol/L) (c)	$\frac{\text{amount (mol) of solute}}{\text{volume (L) of solution}}$
Molality (m)	$\frac{\text{amount (mol) of solute}}{\text{mass (kg) of solvent}}$
Parts by mass	$\frac{\text{mass of solute}}{\text{mass of solution}}$
Parts by volume	$\frac{\text{volume of solute}}{\text{volume of solution}}$
Mole fraction (X)	$\frac{\text{amount (mol) of solute}}{\text{amount (mol) of solute + amount (mol) of solvent}}$

2. **Molality (m)** does not contain volume in its ratio; it is *the amount (mol) of solute dissolved in 1000 g (1 kg) of solvent*:

$$\text{Molality}\ (m) = \frac{\text{amount (mol) of solute}}{\text{mass (kg) of solvent}} = \frac{n_{\text{solute}}}{kg_{\text{solvent}}} \qquad (12.5)$$

Note that molality includes the quantity of *solvent*, not solution. For precise work, molality has two advantages over the concentration expressed in mol/L:

- *Effect of temperature.* Molal solutions are based on *masses* of components, not *volume*. Thus, since mass does not change with temperature, neither does molality.
- *Effect of mixing.* Unlike volumes, masses *are* additive: adding 500. g of one solution to 500. g of another *does* give 1000. g of final solution.

For these reasons, molality is the preferred term when temperature, and hence density, may change, as in a study of physical properties. Note that, in the case of water, 1 L has a mass of 1 kg, so *molality and concentration (mol/L) are nearly the same for dilute aqueous solutions.*

Sample Problem 12.3 Calculating Molality

Problem What is the molality of a solution prepared by dissolving 32.0 g of $CaCl_2$ in 271 g of water?

Plan To use Equation 12.5, we convert mass of $CaCl_2$ (32.0 g) to amount (mol) with the molar mass (g/mol) and then divide by the mass of water (271 g), being sure to convert from grams to kilograms (*see road map*).

Road Map

Mass (g) of $CaCl_2$

↓ Divide by $\mathcal{M}$ (g/mol).

Amount (mol) of $CaCl_2$

↓ Divide by kg of water.

Molality (m) of $CaCl_2$ solution

Solution Convert from grams of solute to moles:

$$\text{Amount (mol) of } CaCl_2 = 32.0\ \text{g}\ CaCl_2 \times \frac{1\ \text{mol}\ CaCl_2}{110.98\ \text{g}\ CaCl_2} = 0.2883\ \text{mol}\ CaCl_2$$

Find molality:

$$\text{Molality} = \frac{\text{mol solute}}{\text{kg solvent}} = \frac{0.2883\ \text{mol}\ CaCl_2}{271\ \text{g} \times \dfrac{1\ \text{kg}}{10^3\ \text{g}}} = 1.06\ \text{mol/kg}\ CaCl_2$$

Check The answer seems reasonable: the given amount (mol) of $CaCl_2$ and mass (kg) of H_2O are about the same, so their ratio is about 1.

Follow-Up Problem 12.3 What mass (in grams) of glucose ($C_6H_{12}O_6$) must be dissolved in 563 g of ethanol (C_2H_5OH) to prepare a 2.40×10^{-2} mol/kg solution?

Parts of Solute by Parts of Solution

Several concentration terms relate the amount of solute (or solvent) parts to the amount of *solution* parts. The amount of parts can be expressed in units of mass, volume, or amount (mol).

Parts by Mass The most common of these concentration terms is **mass percent [% (w/w)]**, which you encountered in Chapter 3. The word *percent* means "per hundred," so *mass percent* means "mass of solute dissolved in 100. parts by mass of solution":

$$\text{Mass percent} = \frac{\text{mass of solute}}{\text{mass of solute} + \text{mass of solvent}} \times 100 = \frac{\text{mass of solute}}{\text{mass of solution}} \times 100 \qquad (12.6)$$

Mass percent values appear on jars of solid chemicals to indicate impurities.

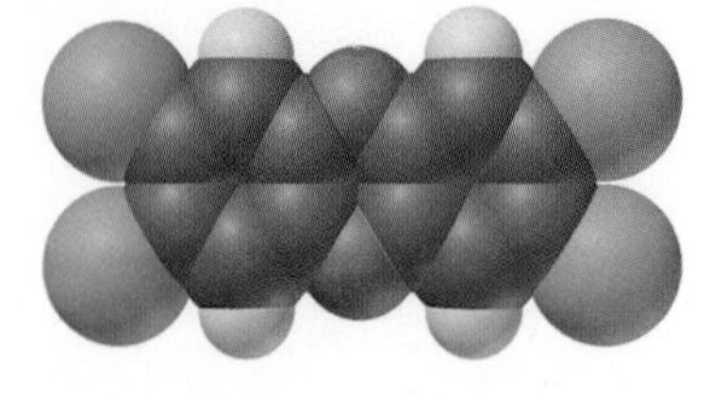

Is this deadly toxin in your body?

Two very similar terms are parts per million (ppm) by mass and parts per billion (ppb) by mass, or grams of solute per million or billion grams of solution. (In Equation 12.6, you multiply by 10^6 or by 10^9, respectively, instead of by 100.) Environmental toxicologists use these units to measure pollutants. For example, TCDD (*tetrachlorodibenzodioxin*) (*see margin*), a by-product of paper bleaching, is unsafe at soil levels above 1 ppb. From normal contact with air, water, and soil, North Americans have an average of 0.01 ppb TCDD in their tissues.

Parts by Volume The most common parts-by-volume term is **volume percent [% (v/v)]**, the volume of solute in 100. volumes of solution:

$$\text{Volume percent} = \frac{\text{volume of solute}}{\text{volume of solution}} \times 100 \qquad (12.7)$$

For example, rubbing alcohol is an aqueous solution of propan-2-ol (also known as isopropanol, a three-carbon alcohol) that contains 70 volumes of alcohol in 100 volumes of solution, or 70% (v/v).

Parts by volume is often used to express tiny concentrations of liquids or gases. Minor atmospheric components occur in parts per million by volume (ppmv). For example, about 0.05 ppmv of the toxic gas carbon monoxide (CO) is in clean air, 1000 times as much (50 ppmv of CO) in air over urban traffic, and 10 000 times as much (500 ppmv of CO) in cigarette smoke. *Pheromones* are compounds that are secreted to signal food, danger, sexual readiness, and so on. Many organisms, including dogs and monkeys, release pheromones, and researchers suspect that humans do as well. Some insect pheromones, such as the sexual attractant of the gypsy moth, are active at a few hundred molecules per millilitre of air, 100 parts per quadrillion by volume.

A concentration term often used in health-related facilities for aqueous solutions is % (w/v), a ratio of solute *weight* (actually mass) to solution *volume*. Thus, a 1.5% (w/v) NaCl solution contains 1.5 g of NaCl per 100. mL of *solution*.

Mole Fraction The **mole fraction (X)** of a solute is the ratio of the amount (mol) of solute to the total amount (mol) (solute plus solvent), that is, parts by mole:

$$\text{Mole fraction } (X) = \frac{\text{amount (mol) of solute}}{\text{amount (mol) of solute} + \text{amount (mol) of solvent}}$$

$$= \frac{n_{\text{solute}}}{n_{\text{solute}} + n_{\text{solvent}}} \qquad (12.8)$$

Put another way, the mole fraction gives the proportion of solute (or solvent) particles in solution. The *mole percent* is the mole fraction expressed as a percentage:

$$\text{Mole percent (mol \%)} = \text{mole fraction} \times 100$$

Sample Problem 12.4 Expressing Concentrations in Parts by Mass, Parts by Volume, and Mole Fraction

Problem (a) Find the concentration of calcium ion (ppm) in a 3.50 g pill that contains 40.5 mg of Ca^{2+}.

(b) The label on a 0.750 L bottle of Chianti indicates "11.5% alcohol by volume." How many litres of alcohol does the wine contain?

(c) A sample of rubbing alcohol contains 142 g of propan-2-ol (isopropyl alcohol, C_3H_7OH) and 58.0 g of water. What are the mole fractions of alcohol and water?

Plan (a) We know the mass of Ca^{2+} (40.5 mg) and the mass of the pill (3.50 g). We convert the mass of Ca^{2+} from mg to g, find the mass ratio of Ca^{2+} to pill, and multiply by 10^6 to obtain ppm. **(b)** We know the volume % (11.5%, or 11.5 parts by volume of alcohol to 100. parts of chianti) and the total volume (0.750 L), so we use Equation 12.7 to find litres of alcohol. **(c)** We know the mass and formula for each component, so we convert masses to amounts (mol) and apply Equation 12.8 to find the mole fractions.

Solution **(a)** Combine the steps to find parts per million by mass of Ca^{2+}:

$$\text{ppm } Ca^{2+} = \frac{\text{mass of } Ca^{2+}}{\text{mass of pill}} \times 10^6 = \frac{40.5 \text{ mg } Ca^{2+} \times \frac{1 \text{ g}}{10^3 \text{mg}}}{3.50 \text{ g}} \times 10^6$$

$$= 1.16 \times 10^4 \text{ ppm } Ca^{2+}$$

(b) Find the volume (L) of alcohol:

$$\text{Volume (L) of alcohol} = 0.750 \text{ L chianti} \times \frac{11.5 \text{ L alcohol}}{100. \text{ L chianti}}$$

$$= 0.0862 \text{ L}$$

(c) To find the mole fractions, first we convert from mass (g) to amount (mol):

$$\text{Amount (mol) of } C_3H_7OH = 142 \text{ g } C_3H_7OH \times \frac{1 \text{ mol } C_3H_7OH}{60.09 \text{ g } C_3H_7OH} = 2.363 \text{ mol } C_3H_7OH$$

$$\text{Amount (mol) of } H_2O = 58.0 \text{ g } H_2O \times \frac{1 \text{ mol } H_2O}{18.02 \text{ g } H_2O} = 3.219 \text{ mol } H_2O$$

Calculate the mole fractions:

$$X_{C_3H_7OH} = \frac{\text{amount (mol) of } C_3H_7OH}{\text{total amount (mol)}} = \frac{2.363 \text{ mol}}{2.363 \text{ mol} + 3.219 \text{ mol}} = 0.423$$

$$X_{H_2O} = \frac{\text{amount (mol) of } H_2O}{\text{total amount (mol)}} = \frac{3.219 \text{ mol}}{2.363 \text{ mol} + 3.219 \text{ mol}} = 0.577$$

Check **(a)** The mass ratio is about 0.04 g/4 g $= 10^{-2}$, and $10^{-2} \times 10^6 = 10^4$ ppm, so it seems correct. **(b)** The volume % is a bit more than 10%, so the volume of alcohol should be a bit more than 75 mL (0.075 L). **(c)** Always check that the *mole fractions add up to 1.* Here, $0.423 + 0.577 = 1.000$.

Follow-Up Problem 12.4 An alcohol solution contains 35.0 g of propan-1-ol (C_3H_7OH) and 150. g of ethanol (C_2H_5OH). Calculate the mass percent and the mole fraction of each alcohol.

Interconverting Concentration Terms

All the terms we have just discussed represent different ways to express concentration, so they are interconvertible. Keep the following ideas in mind:

- To convert a term based on amount to a term based on mass, you need the molar mass. These conversions are similar to the mass-mole conversions you have done earlier.
- To convert a term based on mass to a term based on volume, you need the solution *density.* Given the mass of a solution, the density (mass/volume) gives the volume, or vice versa.
- Molality includes the quantity of *solvent*; the other terms include the quantity of *solution.*

Sample Problem 12.5 Interconverting Concentration Terms

Problem Hydrogen peroxide is a powerful oxidizing agent; it is used in a concentrated solution in rocket fuel, but in a dilute solution in hair bleach. An aqueous solution of H_2O_2 is 30.0% by mass and has a density of 1.11 g/mL. Calculate **(a)** the molality; **(b)** the mole fraction of H_2O_2; **(c)** the concentration in mol/L.

Plan We know the mass % (30.0) and the density (1.11 g/mL). **(a)** For molality, we need the amount (mol) of solute and the mass (kg) of *solvent.* If we assume 100.0 g of solution, the mass % equals the mass (g) of H_2O_2, which we subtract to obtain the mass (g) of solvent. To find molality, we convert the mass (g) of H_2O_2 to amount (mol) and divide by the mass of solvent (converting g to kg). **(b)** To find the mole fraction, we use the amount (mol) of H_2O_2 [from part (a)] and convert the mass (g) of H_2O to amount (mol). Then we divide the amount (mol) of H_2O_2 by the total amount (mol). **(c)** To find concentration in mol/L, we assume 100.0 g of solution and use the solution density to find the volume. Then we divide the amount (mol) of H_2O_2 [from part (a)] by the *solution* volume (in L).

Solution **(a)** We need to convert from mass % to molality.

Find the mass of solvent (assuming 100.0 g of solution):

$$\text{Mass (g) of } H_2O = 100.0 \text{ g solution} - 30.0 \text{ g } H_2O_2 = 70.0 \text{ g } H_2O$$

Convert from mass (g) of H_2O_2 to amount (mol):

$$\text{Amount (mol) of } H_2O_2 = 30.0 \text{ g } H_2O_2 \times \frac{1 \text{ mol } H_2O_2}{34.02 \text{ g } H_2O_2} = 0.8818 \text{ mol } H_2O_2$$

Calculate the molality:

$$\text{Molality of } H_2O_2 = \frac{0.8818 \text{ mol } H_2O_2}{70.0 \text{ g} \times \dfrac{1 \text{ kg}}{10^3 \text{ g}}} = 12.6 \text{ mol/kg } H_2O$$

(b) We need to convert from mass % to mole fraction:

$$\text{Amount (mol) of } H_2O_2 = 0.8818 \text{ mol } H_2O_2 \text{ [from part (a)]}$$

$$\text{Amount (mol) of } H_2O = 70.0 \text{ g } H_2O \times \frac{1 \text{ mol } H_2O}{18.02 \text{ g } H_2O} = 3.885 \text{ mol } H_2O$$

$$X_{H_2O_2} = \frac{0.8818 \text{ mol}}{0.8818 \text{ mol} + 3.885 \text{ mol}} = 0.185$$

(c) We need to convert from mass % and density to concentration (mol/L).

Convert from solution mass to volume:

$$\text{Volume (mL) of solution} = 100.0 \text{ g} \times \frac{1 \text{ mL}}{1.11 \text{ g}} = 90.09 \text{ mL}$$

Calculate c, or concentration (mol/L):

$$c = \frac{\text{mol } H_2O_2}{\text{L soln}} = \frac{0.8818 \text{ mol } H_2O_2}{90.09 \text{ mL} \times \dfrac{1 \text{ L soln}}{10^3 \text{ mL}}} = 9.79 \text{ mol/L } H_2O_2$$

Check Rounding shows that the answers are reasonable: **(a)** The ratio of ~0.9 mol/0.07 kg is greater than 10. **(b)** $\frac{\sim 0.9 \text{ mol } H_2O_2}{1 \text{ mol} + 4 \text{ mol}} \approx 0.2$. **(c)** The ratio of amount (mol) to volume (L) (0.9/0.09) is around 10.

Follow-Up Problem 12.5 Concentrated hydrochloric acid is 11.8 mol/L HCl and has a density of 1.190 g/mL. Calculate the mass percent, molality, and mole fraction of HCl.

SUMMARY OF SECTION 12.4

- The concentration of a solution is independent of the quantity of solution and can be expressed as the amount of substance concentration (mol solute/L solution), molality (mol solute/kg solvent), parts by mass (mass solute/mass solution), parts by volume (volume solute/volume solution), or mole fraction [mol solute/(mol solute + mol solvent)] (Table 12.4).
- Molality is based on mass, so it is independent of temperature. The mole fraction gives the proportion of dissolved particles.
- If, in addition to the quantities of solute and solution, the solution density is known, the various ways of expressing the concentration are interconvertible.

12.5 Colligative Properties of Solutions

The presence of a solute gives a solution different physical properties than the pure solvent. For four important properties, however, it is the *number* of solute particles, *not* their chemical identity, that makes the difference. These **colligative properties** (*colligative* means "collective") are vapour pressure lowering, boiling point elevation, freezing point depression, and osmotic pressure. Most of the effects are small, but they have many applications, including some that are vital to organisms.

We predict the magnitude of a colligative property from the solute formula, which shows the number of particles in solution and is closely related to our classification of solutes by their ability to conduct an electric current (Chapter 3; Figure 12.13):

1. *Electrolytes.* An aqueous solution of an **electrolyte** conducts electricity because the solute separates into ions as it dissolves.
   - *Strong electrolytes*—soluble salts, strong acids, and strong bases—dissociate completely, so their solutions conduct well.
   - *Weak electrolytes*—weak acids and weak bases—dissociate very little, so their solutions conduct poorly.

2. *Nonelectrolytes.* Compounds such as sugar and alcohol do not dissociate into ions at all. They are **nonelectrolytes** because their solutions do not conduct a current.

Thus, we can make the following predictions:

- *For nonelectrolytes,* 1 mol of compound yields 1 mol of particles in solution. For example, 0.35 mol/L glucose contains 0.35 mol of solute particles per litre.
- *For strong electrolytes,* 1 mol of compound yields the amount (mol) of ions in the formula unit: 0.4 mol/L Na_2SO_4 has 0.8 mol of Na^+ ions and 0.4 mol of SO_4^{2-} ions, or 1.2 mol of ion particles, per litre (see Sample Problem 3.28).
- *For weak electrolytes,* the calculation is complicated because the solution reaches equilibrium. We will examine the properties of systems containing weak electrolytes in Chapters 16 and 17.

In this section, we discuss colligative properties of three types of solutes: nonvolatile nonelectrolytes, volatile nonelectrolytes, and strong electrolytes.

Nonvolatile Nonelectrolyte Solutions

We start with solutions of *nonvolatile nonelectrolytes* because they provide the clearest examples of the colligative properties. These solutes do not dissociate and have negligible vapour pressure at the boiling point of the solvent; sucrose (table sugar) dissolved in water is an example.

Vapour Pressure Lowering *The vapour pressure of a nonvolatile nonelectrolyte solution is always lower than the vapour pressure of the pure solvent.* The difference in vapour pressures is the **vapour pressure lowering (Δp)**.

1. *Why the vapour pressure of a solution is lower.* We can explain this lowering in terms of maintaining equilibrium or increasing entropy (Figure 12.14):

- *Maintaining equilibrium.* In pure solvent, the opposing rates of vaporization (molecules leaving the liquid) and condensation (molecules entering the liquid) are equal at equilibrium. In a solution, some solute molecules are on the surface, so the number of solvent molecules on the surface is lower; therefore, fewer vaporize per unit time. To maintain equilibrium, fewer gas molecules must enter the liquid, which happens if the vapour pressure is lower.
- *Increasing entropy.* A solvent vaporizes because the gaseous state has higher entropy than the liquid state. However, with dissolved particles present, the solution already has higher entropy, so the solvent has less tendency to vaporize.

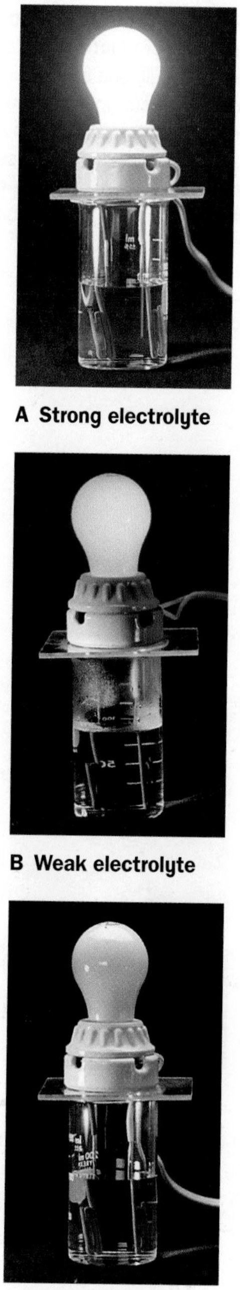

A Strong electrolyte

B Weak electrolyte

C Nonelectrolyte

FIGURE 12.13 Conductivity of three types of electrolyte solutions. A. For strong electrolytes, the number of particles equals the number of ions in a formula unit. **B.** Weak electrolytes form few ions. **C.** For nonelectrolytes, the number of particles equals the number of molecules.

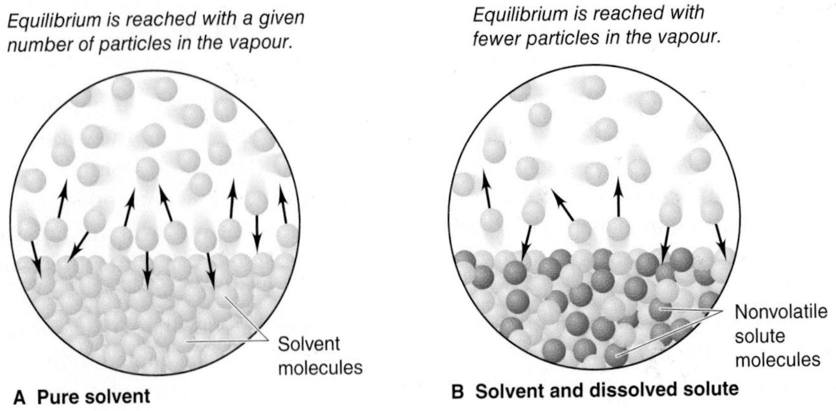
Equilibrium is reached with a given number of particles in the vapour.
Equilibrium is reached with fewer particles in the vapour.
Solvent molecules
Nonvolatile solute molecules
A Pure solvent
B Solvent and dissolved solute

FIGURE 12.14 Effect of solute on the vapour pressure of solution. A. Equilibrium between a liquid and its vapour occurs when equal numbers of molecules vaporize and condense in a given time. **B.** In many solutions, solute molecules decrease the number of solvent molecules at the surface *and* increase the entropy, so equilibrium occurs at a lower vapour pressure.

2. *Quantifying vapour pressure lowering.* **Raoult's law** says that the vapour pressure of a solvent above a solution ($p_{solvent}$) equals the mole fraction of the solvent ($X_{solvent}$) times the vapour pressure of the pure solvent ($p^{\circ}_{solvent}$):

$$p_{solvent} = X_{solvent} \times p^{\circ}_{solvent} \tag{12.9}$$

Since $X_{solvent}$ is less than 1 in a solution, $p_{solvent}$ is less than $p^{\circ}_{solvent}$. An **ideal solution** is a solution that follows Raoult's law at any concentration. However, just as real gases deviate from ideality, so do real solutions. In practice, Raoult's law works reasonably well for *dilute* solutions and becomes exact at infinite dilution.

Let us see how the *amount* of dissolved solute affects Δp. A solution consists of a solvent and a solute, so the sum of their mole fractions equals 1:

$$X_{solvent} + X_{solute} = 1; \qquad \text{thus,} \qquad X_{solvent} = 1 - X_{solute}$$

From Raoult's law, we have

$$p_{solvent} = X_{solvent} \times p^{\circ}_{solvent} = (1 - X_{solute}) \times p^{\circ}_{solvent}$$

Multiplying through on the right side gives

$$p_{solvent} = p^{\circ}_{solvent} - (X_{solute} \times p^{\circ}_{solvent})$$

Rearranging and introducing Δp gives

$$p^{\circ}_{solvent} - p_{solvent} = \Delta p = X_{solute} \times p^{\circ}_{solvent} \tag{12.10}$$

Thus, Δp equals the mole fraction of the solute times the vapour pressure of the pure solvent—a relationship applied in the next sample problem.

François-Marie Raoult (1830–1901) studied depression of a solvent's vapour pressure and its freezing point, due to a solute.

Sample Problem 12.6 Using Raoult's Law to Find Δp

Problem Find the vapour pressure lowering, Δp, when 10.0 mL of glycerol (propane-1,2,3-triol, $C_3H_8O_3$) is added to 500. mL of water at 50.°C. At this temperature, the vapour pressure of pure water is 12.3 kPa and its density is 0.988 g/mL. The density of glycerol is 1.26 g/mL.

Plan To calculate Δp, we use Equation 12.10. We are given the vapour pressure of pure water, $p^{\circ}_{H_2O} = 12.3$ kPa, so we just need the mole fraction of glycerol, $X_{glycerol}$. We convert the given volume of glycerol (10.0 mL) to mass using the given density (1.26 g/L), find the molar mass from the formula, and convert the mass (g) to the amount (mol). The same procedure gives the amount of H_2O. From these amounts, we find $X_{glycerol}$ and Δp.

Solution Calculate the amount (mol) of glycerol and of water:

$$\text{Amount (mol) of glycerol} = 10.0 \text{ mL glycerol} \times \frac{1.26 \text{ g glycerol}}{1 \text{ mL glycerol}} \times \frac{1 \text{ mol glycerol}}{92.09 \text{ g glycerol}}$$
$$= 0.1368 \text{ mol glycerol}$$

$$\text{Amount (mol) of } H_2O = 500. \text{ mL } H_2O \times \frac{0.988 \text{ g } H_2O}{1 \text{ mL } H_2O} \times \frac{1 \text{ mol } H_2O}{18.02 \text{ g } H_2O} = 27.41 \text{ mol } H_2O$$

Calculate the mole fraction of glycerol:

$$X_{\text{glycerol}} = \frac{0.1368 \text{ mol}}{0.1368 \text{ mol} + 27.41 \text{ mol}} = 0.004966$$

Find the vapour pressure lowering:

$$\Delta p = X_{\text{glycerol}} \times p^{\circ}_{\text{H}_2\text{O}} = 0.004966 \times 12.3 \text{ kPa} = 6.11 \times 10^{-2} \text{ kPa}$$

Check The amount of each component seems correct: for glycerol, ~10 mL × 1.25 g/mL ÷ 100 g/mol = 0.125 mol; for H_2O, ~500 mL × 1 g/mL ÷ 20 g/mol = 25 mol. The small Δp is reasonable because the mole fraction of the solute is small.

Comment **1.** The calculation assumes that glycerol is nonvolatile. At 101.3 kPa, glycerol boils at 290.0°C, so the vapour pressure of glycerol at 50°C is negligible. **2.** To introduce the calculation, we assume that the solution is close to ideal and Raoult's law holds.

Road Map

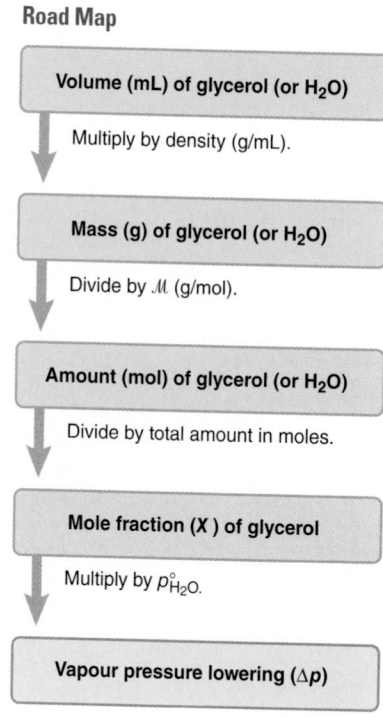

Volume (mL) of glycerol (or H_2O)

↓ Multiply by density (g/mL).

Mass (g) of glycerol (or H_2O)

↓ Divide by $\mathcal{M}$ (g/mol).

Amount (mol) of glycerol (or H_2O)

↓ Divide by total amount in moles.

Mole fraction (X) of glycerol

↓ Multiply by $p^{\circ}_{\text{H}_2\text{O}}$.

Vapour pressure lowering (Δp)

Follow-Up Problem 12.6 Calculate the vapour pressure lowering of a solution of 2.00 g of Aspirin ($\mathcal{M}$ = 180.15 g/mol) in 50.0 g of methanol (CH_3OH) at 21.2°C. Pure methanol has a vapour pressure of 12.5 kPa at this temperature.

Boiling Point Elevation *A solution boils at a higher temperature than the pure solvent.* This colligative property results from the previous vapour pressure lowering:

1. *Why a solution boils at a higher T.* Recall that the boiling point, T_b, of a liquid is the temperature at which its vapour pressure equals the external pressure, p_{ext}. However, the vapour pressure of a solution is always lower than the vapour pressure of the pure solvent. Therefore, it is lower than p_{ext} at T_b of the solvent, so the solution does not boil. Thus, the **boiling point elevation ($\Delta_b T$)** results because a higher temperature is needed to raise the solution's vapour pressure to equal p_{ext}.

We superimpose a phase diagram for the solution on a phase diagram for the solvent to see $\Delta_b T$ (Figure 12.15): the gas-liquid line for the solution lies *below* the line for the solvent at any T and, to the right of it, at any p, and the line crosses 101.3 kPa (p_{ext}, or p_{atm}) at a higher T.

2. *Quantifying boiling point elevation.* Like vapour pressure lowering, boiling point elevation is proportional to the concentration of solute:

$$\Delta_b T \propto m \text{ or } \Delta_b T = K_b m \qquad (12.11)$$

where m is the solution molality and K_b is the *molal boiling point elevation constant.* Since $\Delta_b T > 0$, we subtract the lower solvent T_b from the higher solution T_b:

$$\Delta_b T = T_{\text{b (solution)}} - T_{\text{b(solvent)}}$$

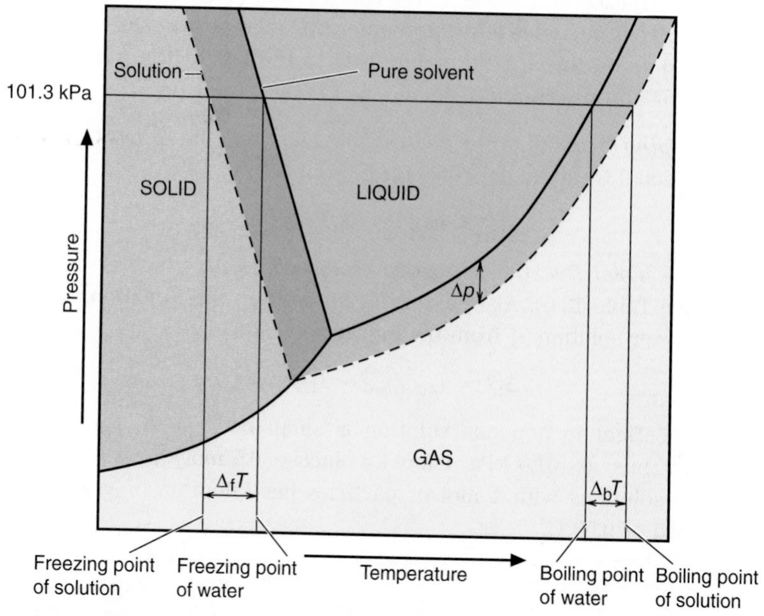

FIGURE 12.15 Boiling and freezing points of a solvent and a solution. Phase diagrams of an aqueous solution (*dashed lines*) and pure water (*solid lines*) show that, by lowering the vapour pressure (Δp), a solute elevates the boiling point ($\Delta_b T$) and depresses the freezing point ($\Delta_f T$). (The slope of the solid-liquid line is exaggerated.)

TABLE 12.5 Molal Boiling Point Elevation and Freezing Point Depression Constants of Several Solvents

Solvent	Boiling Point (°C)*	K_b (°C·kg/mol)	Melting Point (°C)	K_f (°C·kg/mol)
Ethanoic acid (acetic acid)	117.9	3.07	16.6	3.90
Benzene	80.1	2.53	5.5	4.90
Carbon disulfide	46.2	2.34	−111.5	3.83
Tetrachloromethane (carbon tetrachloride)	76.5	5.03	−23	30.
Trichloromethane (chloroform)	61.7	3.63	−63.5	4.70
Ethoxyethane (diethyl ether)	34.5	2.02	−116.2	1.79
Ethanol	78.5	1.22	−117.3	1.99
Water	100.0	0.512	0.0	1.86

*At 101.3 kPa.

Molality is used because it relates to mole fraction and, thus, to particles of solute, and it is not affected by temperature.

The constant K_b has units of °C per molal unit (°C/(mol/kg) or °C•kg/mol) and is specific for a given solvent (Table 12.5). The K_b for water is 0.512°C•kg/mol, so $\Delta_b T$ for aqueous solutions is quite small. For example, if you dissolve 1.00 mol of glucose (180. g; 1.00 mol of particles) or 0.500 mol of NaCl (29.2 g; also 1.00 mol of particles) in 1.00 kg of water at 101.3 kPa, the solutions will boil at 100.512°C instead of 100.000°C.

Freezing Point Depression *A solution freezes at a lower temperature than the pure solvent*, and this colligative property also results from vapour pressure lowering.

1. *Why a solution freezes at a lower T.* Only the solvent vaporizes from a solution, so the solute molecules are left behind. Similarly, *only the solvent freezes*, again leaving the solute molecules behind. The freezing point of a solution is the temperature at which its vapour pressure equals the vapour pressure of the pure solvent, that is, when the solid solvent and the liquid solution are in equilibrium. The **freezing point depression ($\Delta_f T$)** occurs because the vapour pressure of the solution is always lower than the vapour pressure of the solvent, so the solution freezes at a lower temperature; that is, only at a lower temperature will the solvent particles leave and enter the solid at the same rate. In Figure 12.15, the solid-liquid line for the solution is to the left of the pure solvent line at 101.3 kPa, or any pressure.

2. *Quantifying freezing point depression.* Like $\Delta_b T$, the freezing point depression is proportional to the molal concentration of solute:

$$\Delta_f T \propto m \quad \text{or} \quad \Delta_f T = K_f m \tag{12.12}$$

where K_f is the *molal freezing point depression constant*, which also has units of °C•mol/kg. (see Table 12.5). Also, like $\Delta_b T$, $\Delta_f T > 0$. In this equation, however, we subtract the lower solution T_f from the higher solvent T_f:

$$\Delta_f T = T_{f\,(solvent)} - T_{f\,(solution)}$$

Here, too, the effect in aqueous solution is small because K_f for water is just 1.86°C•kg/mol. Thus, at 101.3 kPa, 1 mol/kg glucose, 0.5 mol/kg NaCl, and 0.33 mol/kg K_2SO_4—all solutions with 1 mol of particles per kilogram of water—freeze at −1.86°C instead of 0.00°C.

Sample Problem 12.7 Determining Boiling and Freezing Points of a Solution

Problem Suppose that you add 1.00 kg of ethylene glycol (ethane-1,2-diol, $C_2H_6O_2$) antifreeze to 4450 g of water in the radiator of a car. What are the boiling and freezing points of the solution?

Plan To find the boiling and freezing points, we need $\Delta_b T$ and $\Delta_f T$. First, we find the molality by converting the mass of the solute (1.00 kg) to amount (mol) and then dividing by the mass of the solvent (4450 g, converted to kg). Next, we calculate $\Delta_b T$ and $\Delta_f T$ from Equations 12.11 and 12.12 (using constants from Table 12.5). We add $\Delta_b T$ to the solvent boiling point and subtract $\Delta_f T$ from its freezing point. The road map shows these steps.

Road Map

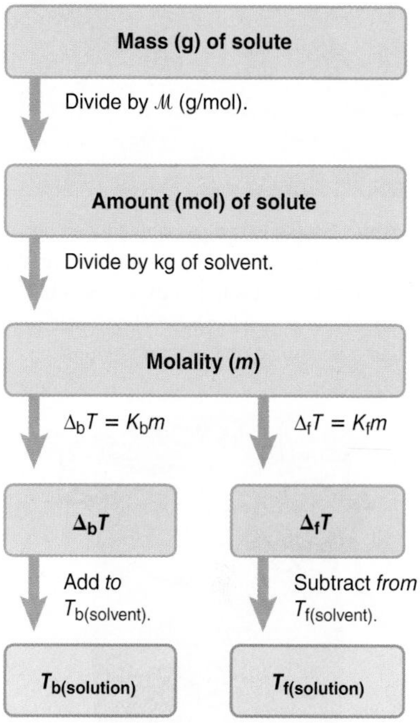

Solution Calculate the molality:

$$\text{Amount (mol) of } C_2H_6O_2 = 1.00 \text{ kg } C_2H_6O_2 \times \frac{10^3 \text{ g}}{1 \text{ kg}} \times \frac{1 \text{ mol } C_2H_6O_2}{62.07 \text{ g } C_2H_6O_2}$$

$$= 16.11 \text{ mol } C_2H_6O_2$$

$$\text{Molality} = \frac{\text{amount (mol) solute}}{\text{mass (kg) solvent}} = \frac{16.11 \text{ mol } C_2H_6O_2}{4450 \text{ g } H_2O \times \dfrac{1 \text{ kg}}{10^3 \text{ g}}} = 3.620 \text{ mol/kg } C_2H_6O_2$$

Find the boiling point elevation and $T_{b(\text{solution})}$, with $K_b = 0.512°\text{C}{\cdot}\text{kg/mol}$:

$$\Delta_b T = 0.512°\text{C}{\cdot}\text{kg/mol} \times 3.620 \text{ mol/kg} = 1.85°\text{C}$$

$$T_{b\,(\text{solution})} = T_{b\,(\text{solvent})} + \Delta_b T = 100.00°\text{C} + 1.85°\text{C} = 101.85°\text{C}$$

Find the freezing point depression and $T_{f(\text{solution})}$, with $K_f = 1.86°\text{C}{\cdot}\text{kg/mol}$:

$$\Delta_f T = 1.86°\text{C}{\cdot}\text{kg/mol} \times 3.62 \text{ mol/kg} = 6.73°\text{C}$$

$$T_{f(\text{solution})} = T_{f(\text{solvent})} - \Delta_f T = 0.00°\text{C} - 6.73°\text{C} = -6.73°\text{C}$$

Check The changes in the boiling and freezing points should be in the same proportion as the constants used; that is, $\frac{\Delta_b T}{\Delta_f T}$ should equal $\frac{K_b}{K_f}$: $\frac{1.85}{6.73} = 0.275 = \frac{0.512}{1.86}$

Comment These answers are approximate because the concentration far exceeds the concentration of a *dilute* solution, for which Raoult's law is most accurate.

Follow-Up Problem 12.7 What is the lowest molality of ethylene glycol solution that will protect your car's coolant from freezing at –17.8°C? (Assume that the solution is ideal.)

Osmotic Pressure The colligative property of osmotic pressure is observed when solutions of higher and lower concentrations are separated by a **semipermeable membrane**, a membrane that allows solvent, but *not* solute, to pass through. The phenomenon is called **osmosis**: a net flow of solvent into the more concentrated solution causes a pressure difference known as *osmotic pressure*. Many organisms regulate internal concentrations by osmosis.

1. *Why osmotic pressure arises.* Consider a simple apparatus in which a semipermeable membrane lies at the curve of a U tube and separates an aqueous sugar solution from pure water. Water molecules pass in *either* direction, but the larger sugar molecules do not. Because sugar molecules are on the solution side of the membrane, fewer water molecules touch that side, so fewer leave the solution than enter it in a given time (Figure 12.16A). The *net flow of water into the solution* increases its volume and, thus, decreases its concentration.

As the height of the solution rises and the height of the water falls, a pressure difference forms. This pressure difference resists more water entering and pushes some water back through the membrane. When water is being pushed out of the solution at

Biological examples of osmosis:

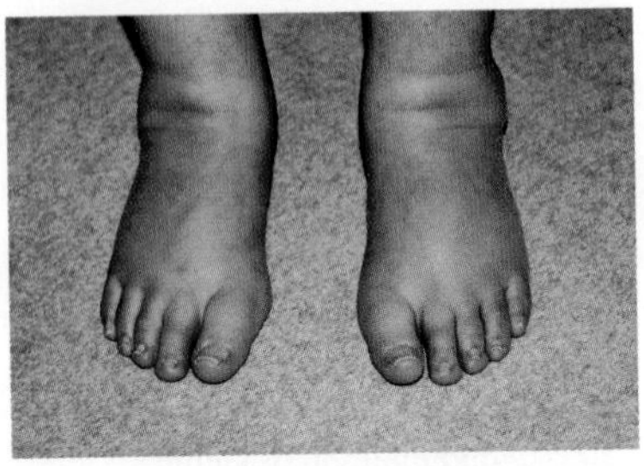

Edema: Edema is the swelling caused by fluid retention. People who eat a lot of salty food might retain water in tissue cells and intercellular space because of osmosis. Edema is very common during pregnancy.

From cucumber to pickle: A cucumber placed in concentrated brine (aqueous NaCl) loses water via osmosis and shrivels into a pickle

Preserving food: Bacteria on candied fruit or salted meat lose water through osmosis, shrivel, and die.

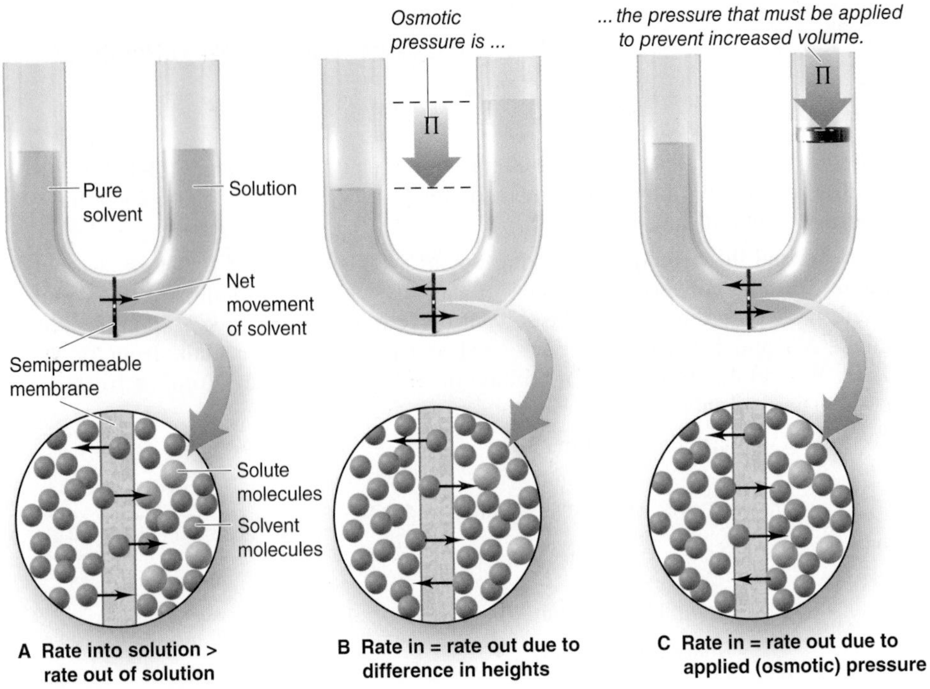

FIGURE 12.16 The development of osmotic pressure. A. In a given time, more solvent enters the solution through the membrane than leaves the solution. **B.** At equilibrium, solvent flow is equalized. **C.** The *osmotic pressure* (Π) *prevents* the volume change.

the same rate that it is entering, the system is at equilibrium (Figure 12.16B). The pressure difference at this point is the **osmotic pressure (Π)**, which is the same as the pressure that must be applied to *prevent* net movement of water from solvent to solution (or from lower to higher concentration, Figure 12.16C).

2. *Quantifying osmotic pressure.* The osmotic pressure is proportional to the amount of solute particles in a given solution *volume*, that is, to the concentration (mol/L):

$$\Pi \propto \frac{n_{\text{solute}}}{V_{\text{soln}}} \quad \text{or} \quad \Pi \propto c$$

where c is concentration (mol/L). The proportionality constant is R times the absolute temperature T. Thus,

$$\Pi = \frac{n_{\text{solute}}}{V_{\text{soln}}} RT = cRT \qquad \textbf{(12.13)}$$

The similarity of Equation 12.13 to the ideal gas law ($p = \frac{nRT}{V}$) is not surprising, because both relate the pressure of a system to its concentration and temperature.

The Underlying Theme of Colligative Properties A common theme runs through our explanations of the four colligative properties of nonvolatile solutes. Each property arises because solute particles cannot move between two phases:

- They cannot enter the gas phase, which leads to *vapour pressure lowering* and *boiling point elevation.*
- They cannot enter the solid phase, which leads to *freezing point depression.*
- They cannot cross a semipermeable membrane, which leads to *osmotic pressure.*

In each situation, the presence of solute decreases the mole fraction of solvent, which lowers the number of solvent particles leaving the solution per unit time. This lowering requires a new balance in numbers of particles moving between two phases per unit time, and the new balance results in the measured colligative property.

Using Colligative Properties to Find Solute Molar Mass

Each colligative property is proportional to solute concentration. Thus, by measuring the property—lower freezing point, higher boiling point, and so on—we determine the amount (mol) of solute particles and, given the mass of solute, the molar mass.

In principle, any of the colligative properties can be used, but, of the four, osmotic pressure creates the largest changes and, thus, the most precise measurements. Polymer chemists and biochemists estimate molar masses as great as 10^5 g/mol by measuring osmotic pressure. Because only a tiny fraction of a mole of a macromolecular solute dissolves, the change in the other colligative properties would be too small.

Sample Problem 12.8 Determining Molar Mass from Osmotic Pressure

Problem Biochemists have discovered more than 400 mutant varieties of hemoglobin, the blood protein that carries O_2. A physician dissolves 21.5 mg of one variety in water to make 1.50 mL of solution at 5.0°C. She measures an osmotic pressure of 481 Pa. What is the molar mass of the protein?

Plan We know the osmotic pressure (Π = 481 Pa), R, and T (5.0°C). We convert Π from Pa to kPa, and T from °C to K, and we use Equation 12.13 to solve for concentration (mol/L). Then we calculate the amount (mol) of hemoglobin from the known volume (1.50 mL) and use the known mass (21.5 mg) to find $\mathcal{M}$.

Road Map

Π (kPa)

$c = \Pi/RT$

c (mol/L)

Multiply by volume (L) of solution.

Amount (mol) of solute

Divide *into* mass (g) of solute.

$\mathcal{M}$ (g/mol)

Solution Combine unit conversion steps and solve for concentration:

$$c = \frac{\Pi}{RT} = \frac{\dfrac{481\ \text{Pa}}{1 \times 10^3\ \text{Pa/1 kPa}}}{\left(8.314\ \dfrac{\text{kPa}\cdot\text{L}}{\text{mol}\cdot\text{K}}\right)(273.15\ \text{K} + 5.0)} = 2.080 \times 10^{-4}\ \text{mol/L}$$

Find amount (mol) of solute (after changing mL to L):

$$\text{Amount (mol) of solute} = c \times V = \frac{2.080 \times 10^{-4}\ \text{mol}}{1\ \text{L soln}} \times 0.00150\ \text{L soln} = 3.120 \times 10^{-7}\ \text{mol}$$

Calculate molar mass of hemoglobin (after changing mg to g):

$$\mathcal{M} = \frac{0.0215\ \text{g}}{3.120 \times 10^{-7}\ \text{mol}} = 6.89 \times 10^4\ \text{g/mol}$$

Check The small osmotic pressure implies a very low concentration. Hemoglobin is a biopolymer, so we expect a small amount (mol) [(~2×10^{-4} mol/L)(1.5×10^{-3} L) = 3×10^{-7} mol] and a high $\mathcal{M}$ ($\frac{\sim 21 \times 10^{-3}\ \text{g}}{3 \times 10^{-7}\ \text{mol}} = 7 \times 10^4$ g/mol). Mammalian hemoglobin has a molar mass of 64 500 g/mol.

Follow-Up Problem 12.8 At 37°C, 0.30 mol/L sucrose has about the same osmotic pressure as blood. What is the osmotic pressure of blood?

Crystallization and Phase Diagrams

Phase diagrams serve as a useful framework to guide crystallization trials, especially for proteins. The **crystallization** process can be seen as three different processes:

1. *Nucleation.* This is the initial process that occurs in the formation of a crystal from a solution, a liquid, or a vapour. A small number of ions become arranged in a pattern that is characteristic of a crystalline solid, forming a site upon which additional particles are deposited as the crystal grows.

2. *Growth.* The constituent atoms, molecules, or ions are arranged in an orderly repeating pattern extending in all three spatial dimensions.

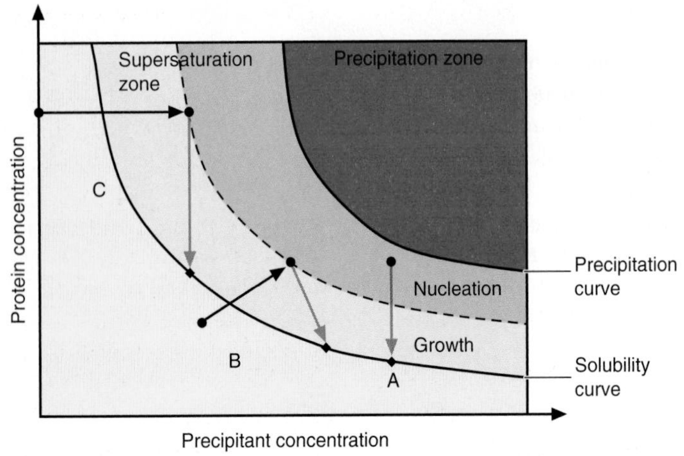

FIGURE 12.17 Protein crystallization phase diagram schema. The crystals grow under conditions of supersaturation only when nuclei have been formed. In this diagram, yellow and red indicate no crystal growth, dark blue indicates the target area to land for crystallization screening, and light blue is the area to be for crystal growth. [Source: http://researchblogging.org/blogger/home/id/2079, by Peter Nollert]

3. *Cessation of growth.* A crystal will only grow when the surrounding solution is saturated with solute. When the solution is saturated, no more material will be deposited on the crystal. This is an equilibrium condition that depends on the temperature. A solution may go from saturated to unsaturated if the temperature of the saturated solution increases, even by only a few degrees, depending on the nature of the solute.

It is important to keep in mind that crystallization occurs only from supersaturated solutions, where the substance concentration exceeds its solubility in a given solution. Figure 12.17 is a protein crystallization phase diagram. Notice that crystal nucleation occurs only in an area of the supersaturation zone. We can distinguish three different pathways:

Pathway A: Batch-type crystallization. By mixing with the precipitant solution, the protein becomes supersaturated. The crystal nuclei form and the crystals grow until the protein concentration in the solution is saturated.

Pathway B: Vapour diffusion–type crystallization. The concentration occurs during vapour diffusion, following mixing of the protein with the precipitant solution, causing the protein to become supersaturated. During vapour diffusion, the precipitant concentration increases and extends the crystal growth process.

Pathway C: Dialysis-type crystallization. As the precipitant diffuses into the protein-holding chamber, the protein supersaturates. Once the nuclei have formed, protein crystals grow as long as the protein concentration remains supersaturated.

Volatile Nonelectrolyte Solutions

What is the effect on vapour pressure when the solute *is* volatile, that is, when the vapour consists of solute *and* solvent molecules? From Raoult's law (Equation 12.9),

$$p_{\text{solvent}} = X_{\text{solvent}} \times p^{\circ}_{\text{solvent}} \qquad \text{and} \qquad p_{\text{solute}} = X_{\text{solute}} \times p^{\circ}_{\text{solute}}$$

where X_{solvent} and X_{solute} are the mole fractions in the *liquid* phase. According to Dalton's law, the total vapour pressure is the sum of the partial vapour pressures:

$$p_{\text{total}} = p_{\text{solvent}} + p_{\text{solute}} = (X_{\text{solvent}} \times p^{\circ}_{\text{solvent}}) + (X_{\text{solute}} \times p^{\circ}_{\text{solute}})$$

Thus, just as a nonvolatile solute lowers the vapour pressure of the solvent by making the solvent's mole fraction less than 1, *the presence of each volatile component lowers the vapour pressure of the other* by making each mole fraction less than 1.

Let us examine this idea using a solution of benzene (C_6H_6) and toluene (C_7H_8), which are miscible. The mole fractions in the liquid are equal: $X_{\text{ben}} = X_{\text{tol}} = 0.500$. At 25°C, the vapour pressures of the pure substances are 12.7 kPa for benzene (p°_{ben}) and 3.79 kPa for toluene (p°_{tol}). Note that benzene is *more volatile* than toluene. We find the partial pressures from Raoult's law:

$$p_{\text{ben}} = X_{\text{ben}} \times p^{\circ}_{\text{ben}} = 0.500 \times 12.7 \text{ kPa} = 6.35 \text{ kPa}$$
$$p_{\text{tol}} = X_{\text{tol}} \times p^{\circ}_{\text{tol}} = 0.500 \times 3.79 \text{ kPa} = 1.90 \text{ kPa}$$

Thus, the presence of benzene in the liquid lowers the vapour pressure of toluene, and vice versa.

Now let us calculate the mole fraction of each substance *in the vapour* by applying Dalton's law. Recall, from Section 4.4, that $X_A = \frac{p_A}{p_{\text{total}}}$. Therefore, for benzene and toluene in the vapour,

$$X_{\text{ben}} = \frac{p_{\text{ben}}}{p_{\text{total}}} = \frac{6.35 \text{ kPa}}{6.35 \text{ kPa} + 1.90 \text{ kPa}} = 0.770$$

$$X_{\text{tol}} = \frac{p_{\text{tol}}}{p_{\text{total}}} = \frac{1.90 \text{ kPa}}{6.35 \text{ kPa} + 1.90 \text{ kPa}} = 0.230$$

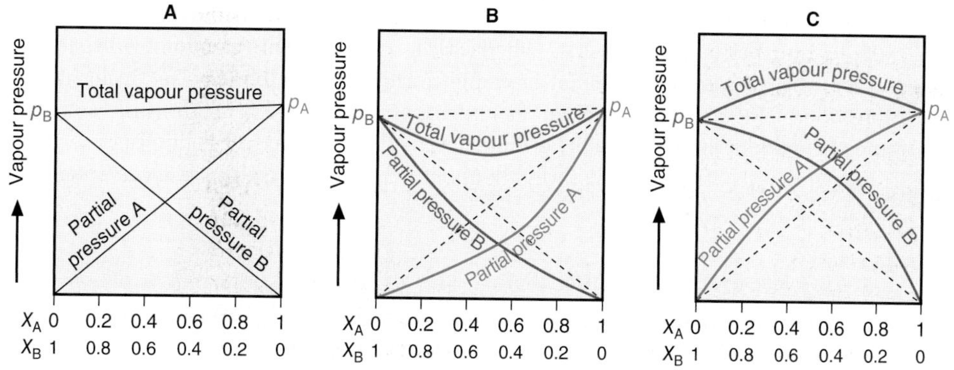

FIGURE 12.18 Deviations from Raoult's law for a solution of two volatile liquids. A. When Raoult's law is obeyed. **B.** A negative deviation from Raoult's law. **C.** A positive deviation from Raoult's law. (p_A and p_B are pure vapour pressures; X_A and X_B are molar fractions.)

The key point is that *the vapour has a higher mole fraction of the* ***more*** *volatile component.* Through a single vaporization-condensation step, a 50/50 *liquid* ratio of benzene-to-toluene created a 77/23 *vapour* ratio. Condense this vapour into a separate container, and the new *liquid* will have this 77/23 composition, and the new *vapour* above it will be enriched still further in the more volatile benzene.

In the laboratory method of **fractional distillation**, a solution of two or more volatile components is attached to a *fractionating column* packed with glass beads (or ceramic chunks) and connected to a condenser and collection flask. As the solution is heated and the vapour mixture meets the beads, numerous vaporization-condensation steps enrich the mixture until the vapour leaving the column, and thus the liquid finally collected, consists of only the most volatile component (see Figure 12.21). Fractional distillation can be used to separate mixtures of volatile liquids with different boiling points, as long as they do not form an **azeotrope**. An azeotrope is a solution with a constant boiling point, which can be higher (negative azeotrope) or lower (positive azeotrope) than the boiling points of its volatile components. It has the characteristic of distilling without changing its composition. When an azeotrope boils, the resulting vapour has the same ratio of components as the original liquid mixture. Azeotropes consisting of two components—such as ethanol-water, trichloromethane (chloroform)-water, and methanoic acid (formic acid)–benzene (positive azeotropes), or mixtures of water with hydrochloric acid, nitric acid, or sulfuric acid (negative azeotropes)—are called *binary* azeotropes. Those consisting of three components—mixtures of ethanol-water with ethyl ethanoate (ethyl acetate), benzene, or trichloromethane, or mixtures of methanol-propanone (acetone) with trichloromethane, cyclohexane, or methyl ethanoate (methyl acetate)—are called *ternary* azeotropes. There are also azeotropes of more than three components.

Non-ideal Solutions: Deviations from Raoult's Law Very few liquid mixtures obey Raoult's law exactly. Instead of getting a graph similar to Figure 12.18A, *when Raoult's law is obeyed,* we could get graphs similar to Figures 12.18B and 12.18C. How, why, and when does this happen? Let us analyze a solution formed by mixing two liquids, A and B. If the attractive forces between the molecules of A and B are the same as when the molecules of A and the molecules of B are by themselves (pure solvents), then we say that the solution follows Raoult's law, and we get a straight line (Figure 12.18A) when we graph vapour pressure versus molar fraction. However, if the attractive forces between the molecules of A and B are greater than the attractive forces between A and A, or B and B, then the vapour pressure of the solution is less than the expected vapour pressure from Raoult's law. This is called a *negative deviation from Raoult's law.* The greater attraction between different molecules in a solution (that is, through the formation of hydrogen bonds) will reduce the tendency of A and B to evaporate. An example of a negative deviation from Raoult's law (Figure 12.18B) can be found when mixing propanone (acetone, CH_3COCH_3) and trichloromethane (chloroform, $CHCl_3$):

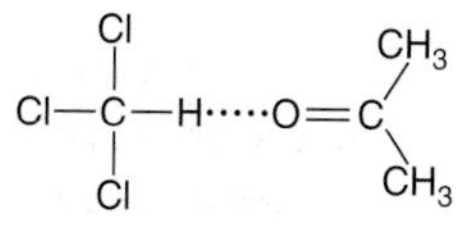

The opposite behaviour is shown in a solution of trichloromethane (chloroform, $CHCl_3$) and methanol (CH_3OH). When $CHCl_3$ molecules are randomly distributed among CH_3OH molecules, the latter cannot form hydrogen bonds effectively. Therefore, the molecules of both components can escape more easily from the solution, and the vapour pressure will be higher than the expected vapour pressure from Raoult's law. In cases like this, we talk about a *positive deviation from Raoult's law* (Figure 12.18C).

Strong Electrolyte Solutions

For colligative properties of strong electrolyte solutions, the solute formula tells us the amount of particles. For instance, the boiling point elevation ($\Delta_b T$) of 0.050 mol/kg NaCl should be 2 × $\Delta_b T$ of 0.050 mol/kg glucose ($C_6H_{12}O_6$), because NaCl dissociates into two particles per formula unit. Thus, we use a multiplying factor called the *van't Hoff factor* (i), named after the Dutch chemist Jacobus van't Hoff (1852–1911):

$$i = \frac{\text{measured value for electrolyte solution}}{\text{expected value for nonelectrolyte solution}}$$

To calculate colligative properties for strong electrolyte solutions, we include i:

For vapour pressure lowering:	$\Delta p = i(X_{\text{solute}} \times p^{\circ}_{\text{solvent}})$
For boiling point elevation:	$\Delta_b T = i(K_b m)$
For freezing point depression:	$\Delta_f T = i(K_f m)$
For osmotic pressure:	$\Pi = i(cRT)$

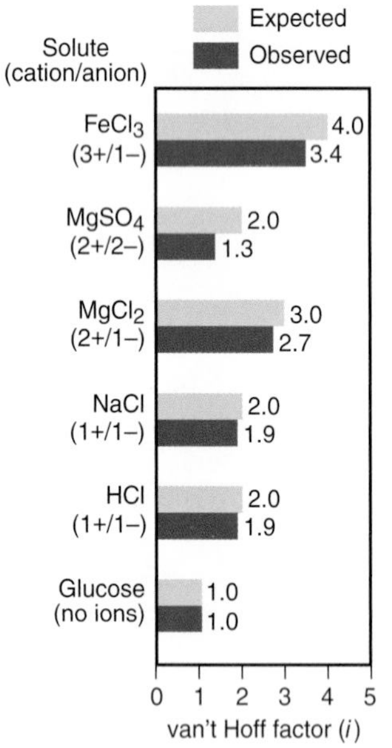

FIGURE 12.19 Non-ideal behaviour of strong electrolyte solutions. van't Hoff factors (i) for 0.050 *mol/kg* solutions show the largest deviation for salts with multiply-charged ions. Glucose (a nonelectrolyte) behaves as expected.

Non-ideal Solutions and Ionic Atmospheres *If* strong electrolyte solutions behave ideally, the factor i is the amount (mol) of particles in solution divided by the amount (mol) of dissolved solute; that is, i is 2 for KBr, 3 for $Mg(NO_3)_2$, and so forth. However, *most strong electrolyte solutions are* ***not*** *ideal*, and the measured value of i is typically *lower* than the value expected from the formula. For example, for the boiling point elevation of 0.050 mol/kg NaCl, the expected value is 2.0, but, from experiment, we have

$$i = \frac{\Delta_b T \text{ of } 0.050 \text{ mol/kg NaCl}}{\Delta_b T \text{ of } 0.050 \text{ mol/kg glucose}} = \frac{0.049°\text{C}}{0.026°\text{C}} = 1.9$$

It seems as though the ions are not behaving independently, even though other evidence indicates that soluble salts dissociate completely. One clue to understanding the results is that multiply charged ions cause a larger deviation (Figure 12.19).

To explain this non-ideal behaviour, we picture positive ions clustered, on average, near negative ions, and vice versa, to form an **ionic atmosphere** of net opposite charge (Figure 12.20). In effect, each type of ion acts "tied up," so its actual concentration seems *lower*. The *effective* concentration is the *stoichiometric* concentration, which is based on the formula multiplied by i. The greater the charge, the stronger the attractions, which explains the larger deviation for salts with multiply charged ions.

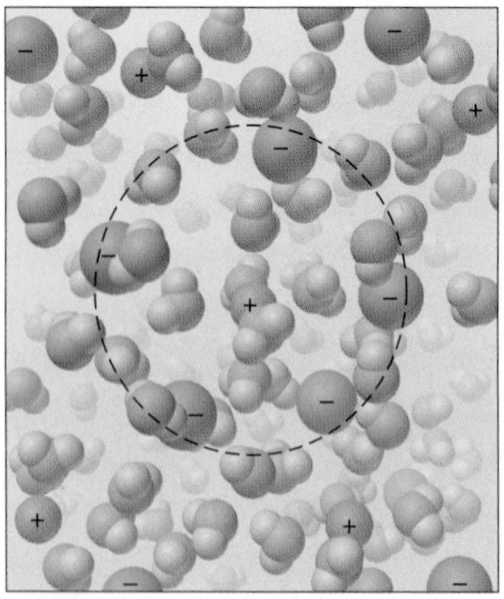

FIGURE 12.20 An ionic atmosphere model for non-ideal behaviour of electrolyte solutions

Comparing Real Solutions and Real Gases At ordinary conditions and concentrations, particles are *much* closer together in liquids than they are in gases. As a result, non-ideal behaviour is much more common, and the deviations are much larger. Nevertheless, the two systems have similarities:

- Gases display nearly ideal behaviour at low pressures because the distances between particles are large. Similarly, the van't Hoff factor (i) approaches the ideal value as the solution becomes more dilute, that is, as the distance between ions increases.
- Attractions between particles cause deviations from the expected pressure in gases and from the expected size of a colligative property in ionic solutions.
- For both real gases and real solutions, we use empirically determined numbers (van der Waals constants or van't Hoff factors) to transform theories (the ideal gas law or Raoult's law) into more useful relations.

Sample Problem 12.9 Depicting Strong Electrolyte Solutions

Problem A 0.952 g sample of magnesium chloride dissolves in 100. g of water in a flask.

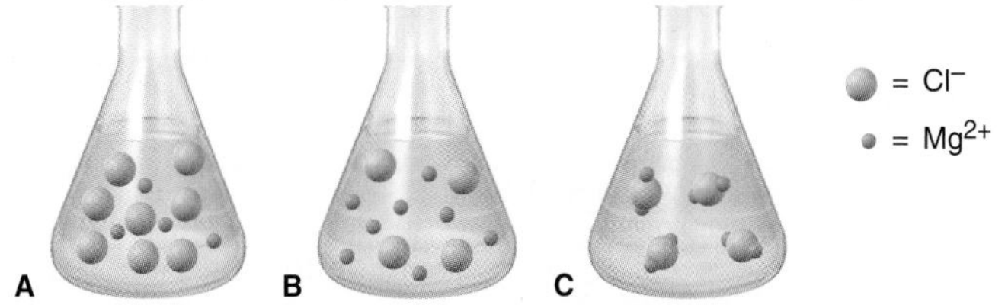

(a) Which scene depicts the solution best?

(b) What is the amount (mol) represented by each green sphere?

(c) Assuming that the solution is ideal, what is its freezing point (at 101.3 kPa)?

Plan (a) We find the amounts of cations and anions per formula unit from the name and compare these numbers with the three scenes. **(b)** We convert the given mass to amount (mol), use the answer from part (a) to find amount (mol) of chloride ions (*green spheres*), and divide by the number of green spheres to get mol/sphere. **(c)** We find the molality (m) from the amount (mol) of solute divided by the given mass of water (changed to kg). We multiply K_f for water (Table 12.5) by m to get $\Delta_f T$ and then subtract $\Delta_f T$ from 0.000°C to get the solution freezing point.

Solution (a) The formula is $MgCl_2$; only scene A has 1 Mg^{2+} for every 2 Cl^-.

(b) $$\text{Amount (mol) of } MgCl_2 = \frac{0.952 \text{ g } MgCl_2}{95.21 \text{ g/mol } MgCl_2} = 0.01000 \text{ mol } MgCl_2$$

Therefore,

$$\text{Amount (mol) of } Cl^- = 0.01000 \text{ mol } MgCl_2 \times \frac{2\ Cl^-}{1\ MgCl_2} = 0.02000 \text{ mol } Cl^-$$

$$\text{Mole/sphere} = \frac{0.02000 \text{ mol } Cl^-}{8 \text{ spheres}} = 2.50 \times 10^{-3} \text{ mol/sphere}$$

(c) $$\text{Molality } (m) = \frac{\text{mol of solute}}{\text{kg of solvent}} = \frac{0.01000 \text{ mol } MgCl_2}{100. \text{ g} \times \dfrac{1 \text{ kg}}{1000 \text{ g}}} = 0.1000 \text{ mol/kg } MgCl_2$$

Assuming an ideal solution, $i = 3$ for $MgCl_2$ (3 ions per formula unit), so we have

$$\Delta_f T = i(K_f m) = 3(1.86°\text{C·kg/mol} \times 0.1000 \text{ mol/kg}) = 0.558°\text{C}$$

and

$$T_f = 0.000°\text{C} - 0.558°\text{C} = -0.558°\text{C}$$

Check Let us quickly check the $\Delta_f T$ in part (c): we have 0.01 mol dissolved in 0.1 kg, or 0.1 mol/kg; then, rounding K_f, we have about 3(2°C·kg/mol × 0.1 mol/kg) = 0.6°C.

Follow-Up Problem 12.9 The $MgCl_2$ solution in the sample problem has a density of 1.006 g/mL at 20.0°C.

(a) What is the osmotic pressure of the solution?

(b) A U tube, with a semipermeable membrane, is filled with the $MgCl_2$ solution in the left arm and a glucose solution of equal concentration (mol/L) in the right arm. After time, which scene depicts the U tube best?

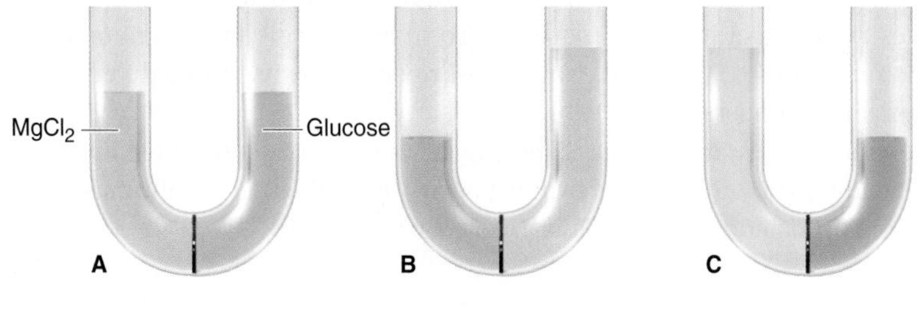

Applications of Colligative Properties

Two colligative properties in particular, freezing point depression and osmotic pressure, have key applications that involve all three types of solutes we have discussed, nonvolatile and volatile nonelectrolytes and strong electrolytes.

Uses of Freezing Point Depression Applications of this property appear in everyday life, nature, and industry:

An Air Canada Jazz airplane has to have its wings sprayed with de-icer on the tarmac of the John G. Diefenbaker International Airport in Saskatoon, Saskatchewan.

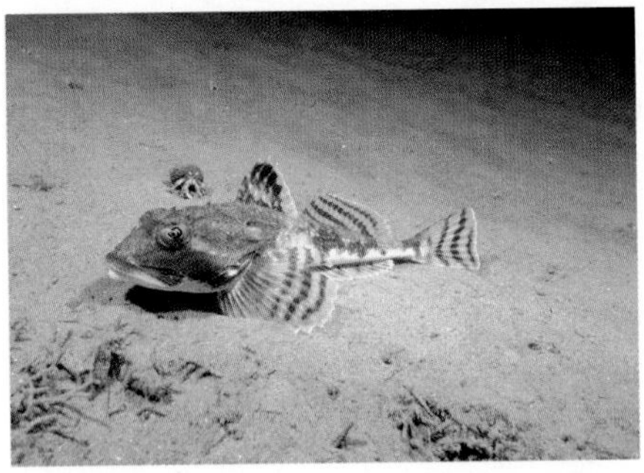

Scientists at the University of Guelph, in Ontario, study antifreeze proteins (AFPs) in the longhorn sculpin. The AFPs enable these fish to survive in subzero environments.

- *Plane de-icer and car antifreeze.* The main ingredient in plane de-icer and car antifreeze, ethylene glycol (ethane-1,2-diol, $C_2H_6O_2$), lowers the freezing point of water in the winter and raises its boiling point in the summer. Due to extensive hydrogen bonding, ethylene glycol is miscible with water and nonvolatile at 100°C.
- *Biological antifreeze.* Structurally similar to ethylene glycol and also miscible with water, glycerol (propane-1,2,3-triol, $C_3H_8O_3$) is produced by some fish and insects, including the common housefly, to lower the freezing point of their blood. This allows them to survive the winter.
- *De-icing sidewalks and roads.* NaCl, or a mixture of NaCl and $CaCl_2$, is used to melt ice on roads. A small amount dissolves in the ice by lowering its freezing point and melting it. More salt dissolves, more ice melts, and so on. An advantage of $CaCl_2$ is that it has a highly negative $\Delta_{soln}H$, so heat is released when it dissolves, which melts more ice.
- *Refining petroleum and silicon.* Countless vaporization-condensation steps within a 30 m fractionating tower separate the hundreds of volatile compounds in crude oil into a few "fractions" based on boiling point range (Figure 12.21). ■ Impure silicon is refined by continuous melting and refreezing into a sample pure enough for use in computer chips (Chapter 23).

■ Kerosene is a Canadian discovery. Abraham Gesner, a Nova Scotia doctor, was searching for a cheaper, brighter lamp oil when he tried distilling New Brunswick coal. The process resulted in burning oils that he called "kerosene." By 1850, kerosene was lighting up homes in Dartmouth and Halifax, Nova Scotia. When petroleum was discovered a few years later (in Pennsylvania in 1859 and in Ontario in 1860), it soon became the major source for the production of kerosene, the C12–C15 distillate fraction. Gesner's work led to the use of petroleum for lighting and gave birth to the modern petroleum refining industry. Kerosene, originally developed for heating and lighting, is even more important today as a fuel for jet aircraft. [Source: biographi.ca/en/bio/gesner_abraham_9E.html.]

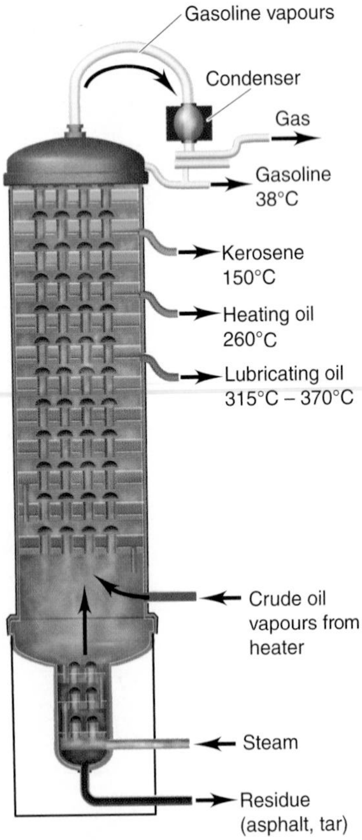

FIGURE 12.21 Fractional distillation in petroleum refining

Uses of Osmotic Pressure Of the four colligative properties, the osmotic pressure property has the most applications that are vital to organisms:

- *Tonicity and cell shape.* The term *tonicity* refers to the tone, or firmness, of a cell. Placing the cell in an *isotonic* solution, a solution that has the same concentration of particles as the cell fluid, maintains the cell's normal shape because water enters and leaves the cell at the same rate (Figure 12.22, *top*). A *hypotonic* solution has a lower concentration of particles, so water enters the cell faster than it leaves, causing the cell to burst (*middle*). In contrast, a *hypertonic* solution has a higher concentration, so the cell shrinks because water leaves faster than it enters (*bottom*). To maintain cell shape, contact-lens rinse is isotonic saline (0.15 mol/L NaCl), as are solutions used to deliver nutrients or drugs intravenously.
- *Absorption of water by trees.* Tree sap is a more concentrated solution than the surrounding groundwater, so water passes through root membranes into a tree, creating an osmotic pressure that can exceed 2×10^3 kPa in the tallest trees.

- *Regulation of water volume in the body.* Of the four major biological cations—Na^+, K^+, Mg^{2+}, and Ca^{2+}—the sodium ion has the primary role of regulating water volume. It accounts for over 90 mol % of cations *outside* a cell: high Na^+ draws water out of the cell and low Na^+ leaves more water inside. The primary role of the kidneys is to regulate Na^+ concentration.
- *Food preservation.* Before refrigeration, salt was valued as a preservative. Packed onto food, salt causes microbes on the surface to shrivel as they lose water. In fact, salt was so highly prized for this purpose that Roman soldiers were paid in salt, from which comes the word *salary*.

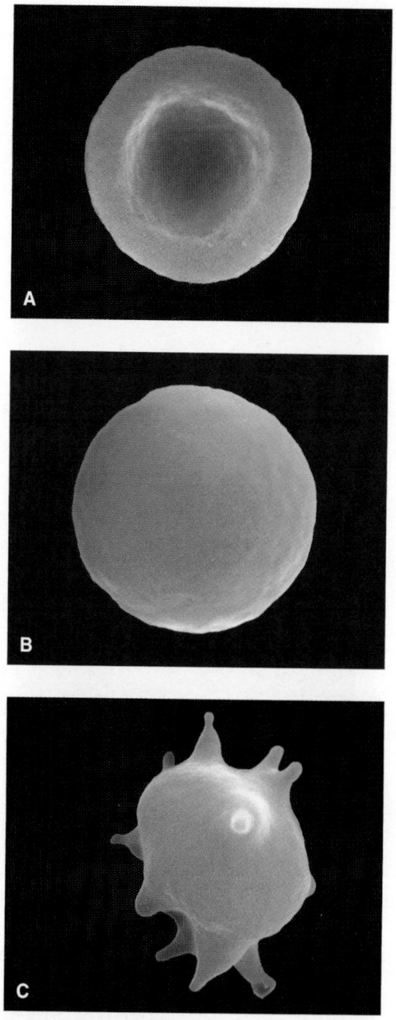

FIGURE 12.22 Osmotic pressure and cell shape. Isotonic (*top*), hypotonic (*middle*), and hypertonic (*bottom*) solutions influence the shape of a red blood cell.

SUMMARY OF SECTION 12.5

- Colligative properties arise from the number, not type, of dissolved solute particles.
- Compared with pure solvent, a solution has lower vapour pressure (Raoult's law), elevated boiling point, and depressed freezing point, and it gives rise to osmotic pressure.
- Colligative properties are used to determine solute molar mass; osmotic pressure gives the most precise measurements.
- When both solute *and* solvent are volatile, the vapour pressure of the more volatile component is greater. When the vapour is condensed, the new solution is richer in this component than the original solution was.
- Calculating colligative properties of electrolyte solutions requires a factor (*i*) that adjusts for the amount of ions per formula unit. These solutions exhibit non-ideal behaviour because charge attractions effectively reduce the concentration of ions.

12.6 The Structure and Properties of Colloids

Particle size plays a defining role in three types of mixtures:

- *Suspensions.* Stir a handful of fine sand into a glass of water, and the particles are suspended at first but gradually settle to the bottom. Sand in water is a **suspension**, a *heterogeneous* mixture containing particles large enough (>1000 nm) to be visibly distinct from the surrounding fluid.
- *Solutions.* In contrast, stirring sugar into water forms a solution, a *homogeneous* mixture in which the particles are invisible, individual sugar molecules (around 1 nm) distributed evenly throughout the surrounding fluid.
- *Colloids.* Between these extremes is a large group of mixtures called *colloidal dispersions*, or **colloids**, in which a dispersed (solute-like) substance is distributed throughout a dispersing (solvent-like) substance. The particles are larger than simple molecules but too small to settle out.

In this section, we examine the classification and key features of colloids:

1. *Particle size and surface area.* Colloidal particles range in diameter from 1 nm to 1000 nm (10^{-9} m to 10^{-6} m). A colloid may be a single macromolecule (natural or synthetic) or an aggregate of many atoms, ions, or molecules. As a result of the small particle size, a colloid has a *very large total surface area*. Think about a cube with 1 cm sides, which has a total surface area of 6 cm^2. If the cube is divided equally into 10^{12} cubes, the cubes are the size of large colloidal particles and have a total surface area of 60 000 cm^2, or 6 m^2. The enormous surface area of a colloid attracts other particles through various intermolecular forces.

2. *Classification of colloids.* Colloids are commonly classified by the physical state of the dispersed and dispersing substances (Table 12.6). Many familiar commercial products and natural objects are colloids. Whipped cream and shaving cream are *foams*, a gas dispersed in a liquid. Styrofoam is a *solid foam*, a gas dispersed in a solid. Most biological fluids are aqueous *sols*, solids dispersed in water: proteins and nucleic acids are colloidal-size particles dispersed in an

TABLE 12.6 Types of Colloids

Colloid Type	Dispersed Substance	Dispersing Medium	Example(s)
Aerosol	Liquid	Gas	Fog
Aerosol	Solid	Gas	Smoke
Foam	Gas	Liquid	Whipped cream
Solid foam	Gas	Solid	Marshmallow
Emulsion	Liquid	Liquid	Milk
Solid emulsion	Liquid	Solid	Butter
Sol	Solid	Liquid	Paint, cell fluid
Solid sol	Solid	Solid	Opal

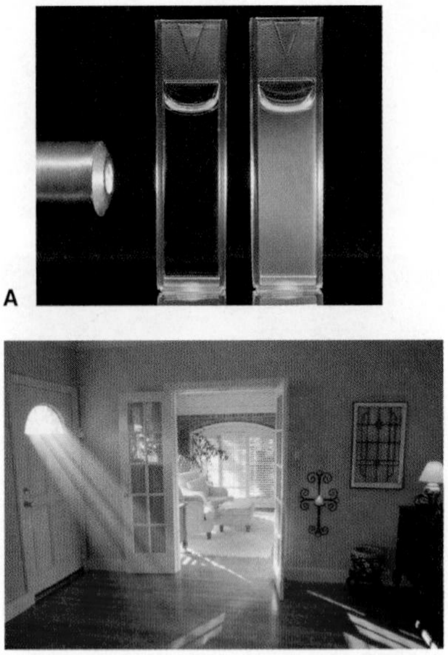

A

B

FIGURE 12.23 Light scattering and the Tyndall effect. A. The narrow, barely visible light beam that passes through a solution (*left*) is scattered and broadened by passing through a colloid (*right*). **B.** Sunlight is scattered by dust in air.

aqueous fluid of ions and small molecules within a cell. Soaps and detergents work by forming an *emulsion*, a liquid dispersed in another liquid. Other emulsions are mayonnaise and hand cream. Bile salts convert fats to an emulsion in the watery fluid of the small intestine.

3. *Tyndall effect and Brownian motion.* Light passing through a colloid is scattered randomly because the dispersed particles have sizes that are similar to the wavelengths of visible light (400 nm to 750 nm). The scattered light beam appears broader than a light beam passing through a solution, an example of the **Tyndall effect** (Figure 12.23). Dust scatters sunlight shining through it, as does mist pierced by headlights.

Under low magnification, colloidal particles exhibit *Brownian motion*, an erratic change of speed and direction. Brownian motion results from collisions with molecules of the dispersing medium. Einstein's explanation of Brownian motion in 1905 led many scientists to accept the molecular nature of matter.

4. *Stabilizing and destabilizing colloids.* Why do colloidal particles not aggregate and settle out? Colloidal particles dispersed in water have charged surfaces that stabilize the colloid through ion-dipole forces. Molecules with dual polarities, such as lipids and soaps, form spherical *micelles*, with charged heads on the exterior and hydrocarbon tails in the interior. Aqueous proteins mimic this arrangement, with charged amino acid groups facing the water and nonpolar groups buried within the molecule. Oily particles can be dispersed in water by adding ions, which are adsorbed onto their surface. Repulsions between the ions on the oil and the partial charges on the water molecules prevent the particles from aggregating.

Despite these forces, various methods that coagulate the particles can destabilize a colloid. Heating makes the particles collide more often and with enough force to coalesce and settle out. Addition of an electrolyte solution containing oppositely charged ions neutralizes the surface charges, so that the particles coagulate and settle. In smokestack gases from a coal-burning power plant, ions become adsorbed onto uncharged colloidal particles, which are then attracted to the charged plates of a device called a *Cotrell precipitator* installed in the stack. At the mouths of rivers, where salt concentrations increase near an ocean or a sea, colloidal clay particles coalesce into muddy deltas, like those of the Mississippi, Nile, and Fraser rivers (Figure 12.24). Thus, the city of New Orleans and the ancient Egyptian empire were made possible by large-scale colloid chemistry.

The following Chemical Connections section describes the application of solution and colloid chemistry to the purification of water for residential and industrial uses.

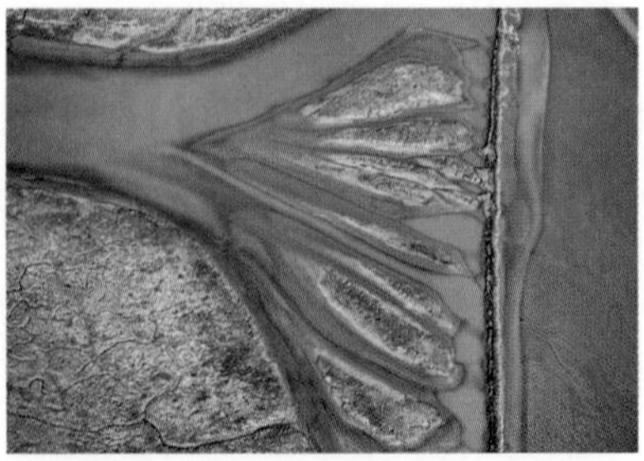

FIGURE 12.24 The Fraser river delta as it enters the Pacific Ocean in British Columbia

Solutions and Colloids in Water Purification

Most water destined for human use comes from lakes, rivers, reservoirs, or groundwater. Present in this essential resource may be soluble toxic organic compounds and high concentrations of NO_3^- and Fe^{3+}, colloidal clay and microbes, and suspended debris. According to a 2013 study, Canada's water ranks 4th out of 17 countries, surpassed only by Sweden, Austria, and Norway. Our water quality is most affected by industrial effluent, agricultural runoff, and municipal sewage pollution. Let us see how water is treated to remove these dissolved, dispersed, and suspended particles.

Water Treatment Plants Treating water involves several steps (Figure B12.1):

Step 1: Screening and settling. As water enters the facility, screens remove debris, and settling removes particles such as sand.

Step 2: Coagulating. This step, and the next two, remove colloids. Colloid particles have negative surfaces that repel each other. Added aluminum sulfate [cake alum; $Al_2(SO_4)_3$] or iron(III) chloride ($FeCl_3$), which supplies Al^{3+} or Fe^{3+} ions to neutralize the charges, coagulates the particles through intermolecular forces.

Step 3: Flocculating and sedimenting. Mixing water and flocculating agents in large basins causes a fluffy *floc* to form. Added cationic polymers form long-chain bridges between floc particles, which grow bigger and heavier and flow into other basins, where they form a sediment and are removed. Some plants use *dissolved air flotation* instead: bubbles forced through the water attach to the floc, and the floating mass is skimmed.

Step 4: Filtering. Various filters remove the remaining particles. In *slow sand filters*, the water passes through sand and/or gravel of increasing particle size. In *rapid sand filters*, the sand is backwashed with water, and the colloidal mass is removed. Membrane filters (*not shown*), with pore sizes of 0.1 μm to 10 μm, are thin tubes bundled together inside a vessel. The water is forced into these tubes, and the colloid-free filtrate is collected from a large central tube. Filtration is very effective at removing microorganisms that are resistant to disinfecting.

Step 5: Disinfecting. Water sources often contain harmful microorganisms that are killed by one of three agents:

- Chlorine, as an aqueous bleach (ClO^-) or Cl_2, is the most common, but carcinogenic chlorinated organic compounds can form.
- UV light, emitted by high-intensity fluorescent tubes, disinfects by disrupting the micro-organisms' DNA.
- Ozone (O_3) gas is a powerful oxidizing agent.

Sodium fluoride (NaF), to prevent tooth decay, and phosphate salts, to prevent the leaching of lead from pipes, may then be added.

Step 6 (not shown): Adsorbing onto granular activated carbon (GAC). Petroleum and other organic contaminants are removed by adsorption. GAC is a highly porous agent formed by "activating" wood, coal, or coconut shells with steam: 1 kg of GAC has a surface area of 1.11×10^6 m^2!

Water Softening via Ion Exchange Water with large amounts of 2+ ions, such as Ca^{2+} and Mg^{2+}, is called *hard water*. Combined with fatty-acid anions in soap, these cations form solid deposits on clothes, washing machines, and sinks:

$$Ca^{2+}(aq) + 2\underset{\text{soap}}{C_{17}H_{35}COONa}(aq) \longrightarrow \underset{\text{insoluble deposit}}{(C_{17}H_{35}COO)_2Ca(s)} + 2\,Na^+(aq)$$

When a large amount of HCO_3^- is present, the cations form *scale*, a carbonate deposit in boilers and hot-water pipes that interferes with the transfer of heat:

$$Ca^{2+}(aq) + 2\,HCO_3^-(aq) \longrightarrow CaCO_3(s) + CO_2(g) + H_2O(l)$$

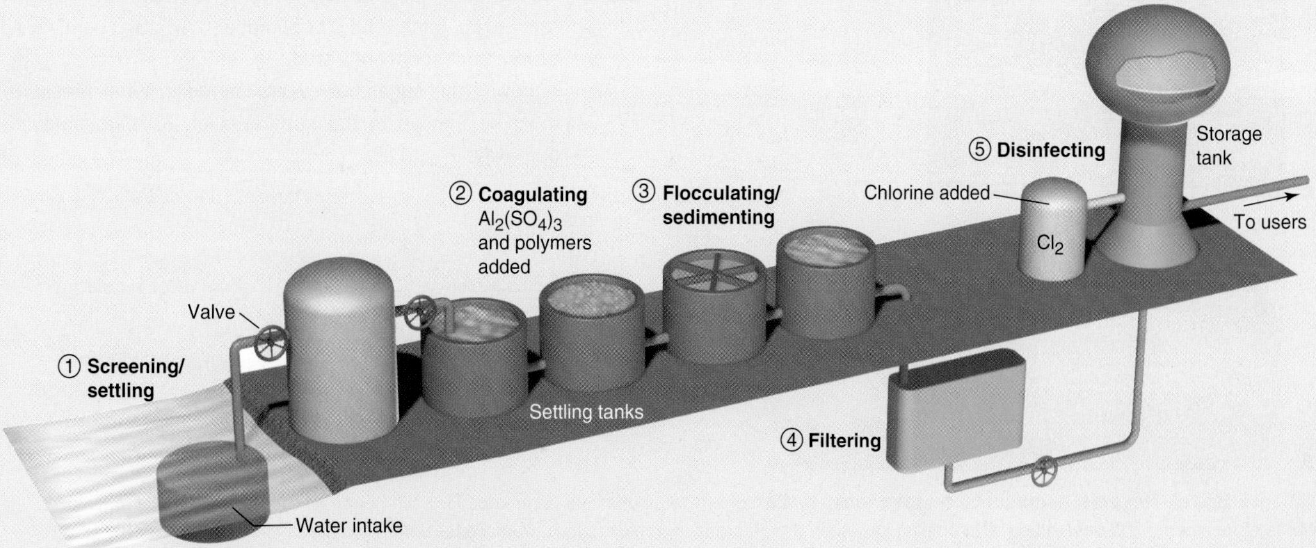

FIGURE B12.1 The typical steps in municipal water treatment

(Continued)

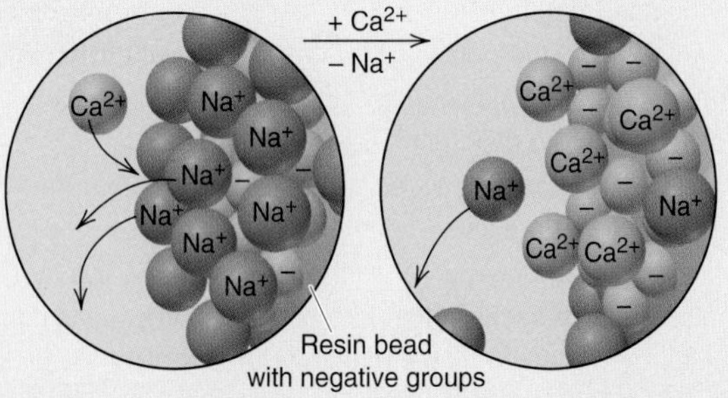

FIGURE B12.2 Ion exchange to remove hard-water cations

Removing hard-water cations, called **water softening**, is done by exchanging Na^+ ions for Ca^{2+} and Mg^{2+} ions. A domestic **ion exchange** system contains an insoluble polymer *resin* with bonded anionic groups, such as $—SO_3^-$ or $—COO^-$, and Na^+ ions for charge balance (Figure B12.2). The hard-water cations displace the Na^+ ions and bind to the anionic groups. When the resin sites are occupied, the resin is regenerated with a concentrated Na^+ solution, which exchanges Na^+ ions for bound Ca^{2+} and Mg^{2+} ions.

Membrane Processes and Reverse Osmosis Membranes with 0.0001 μm to 0.01 μm pores can remove unwanted ions from water. Recall that solutions of different concentrations, separated by a semipermeable membrane, create osmotic pressure. In **reverse osmosis**, a pressure *greater* than the osmotic pressure is *applied* to the more concentrated solution to force water back through the membrane and filter out ions. In homes, toxic *heavy-metal ions*, such as Pb^{2+}, Cd^{2+}, and Hg^{2+}, are removed this way. On a large scale, reverse osmosis is used for **desalination**, which can convert seawater (40 000 ppm of ions) to drinking water (400 ppm) (Figure B12.3).

Wastewater Treatment **Wastewater**, which is used domestic or industrial water, is treated in several ways before being returned to a natural source:

- In *primary* treatment, the wastewater enters a settling basin to remove particles.
- In *biological* treatment, bacteria metabolize organic compounds, which are then removed by settling.
- In *advanced* treatment, a process is tailored to remove a specific pollutant. For example, ammonia, which causes the excessive growth of plants and algae, is removed in two steps:

1. *Nitrification.* Certain bacteria oxidize ammonia (electron donor) with O_2 (electron acceptor) to form nitrate (NO_3^-) ions:

$$NH_4^+ + 2O_2 \longrightarrow NO_3^- + 2H^+ + H_2O$$

2. *Denitrification.* Other bacteria oxidize an added compound, such as methanol (CH_3OH), using the NO_3^- ions:

$$5CH_3OH + 6NO_3^- \longrightarrow 3N_2 + 5CO_2 + 7H_2O + 6OH^-$$

Thus, the process converts NH_3 in wastewater to N_2, which is released to the atmosphere.

Problems

B12.1 Briefly answer each question:

(a) Why is cake alum, $Al_2(SO_4)_3$, added during water purification?

(b) Why is water that contains large amounts of Ca^{2+} and Mg^{2+} difficult to use for cleaning?

(c) What is the meaning of *reverse* in *reverse osmosis*?

(d) Why might a water treatment plant use ozone as a disinfectant, instead of chlorine?

(e) How does passing a saturated NaCl solution through a "spent" ion exchange resin regenerate the resin?

B12.2 Wastewater that is discharged into a stream by a sugar refinery contains 3.55 g of sucrose ($C_{12}H_{22}O_{11}$) per litre. A government-sponsored study is testing the feasibility of removing the sugar by reverse osmosis. What pressure must be applied to the apparatus at 20.°C to produce pure water?

A

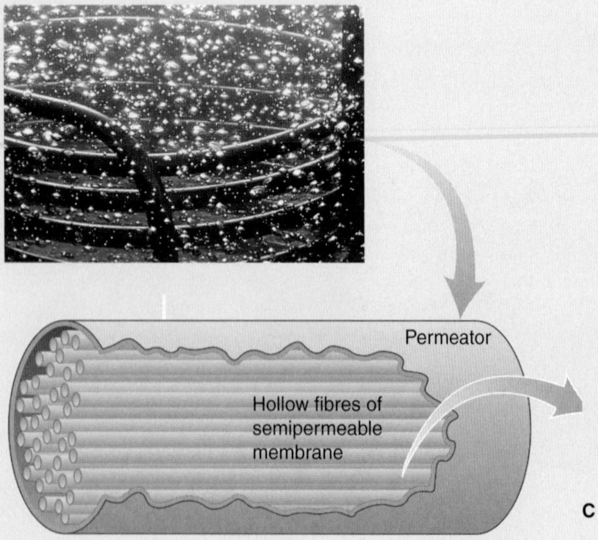

B

C

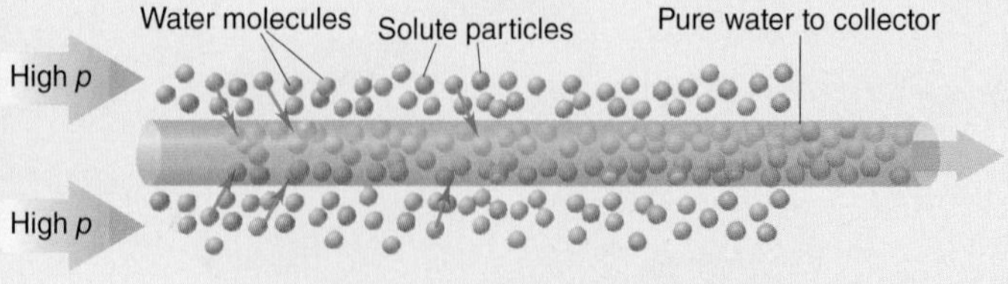

FIGURE B12.3 Reverse osmosis to remove ions. A. Part of a reverse-osmosis permeator. **B.** Each permeator contains a bundle of hollow fibres of semipermeable membrane. **C.** Pumping seawater at high pressure removes ions. Purer water enters the fibres and is collected.

SUMMARY OF SECTION 12.6

- Particles in a colloid are smaller than particles in a suspension and larger than particles in a solution.
- Colloids are classified by the physical states of the dispersed and dispersing substances and involve many combinations of gas, liquid, and/or solid.
- Colloids have extremely large surface areas, scatter incoming light (Tyndall effect), and exhibit random (Brownian) motion.
- Colloidal particles in water are stabilized by charged surfaces that keep them dispersed, but they can be coagulated by heating or by the addition of ions.
- Solution behaviour and colloid chemistry are applied to water treatment and purification.

CHAPTER REVIEW GUIDE

Learning Objectives

Relevant section (§) and/or sample problem (SP) numbers appear in parentheses.

Concepts

1. Explain the quantitative meaning of *solubility*. (§12.1)
2. Describe the major types of intermolecular forces in solution and their relative strengths. (§12.1)
3. Explain how the like-dissolves-like rule depends on intermolecular forces. (§12.1)
4. Explain why gases have relatively low solubilities in water. (§12.1)
5. Describe the general characteristics of solutions formed by various combinations of gases, liquids, and solids. (§12.1)
6. Explain the enthalpy components of a solution cycle and their effect on $\Delta_{soln}H$. (§12.2)
7. Discuss the dependence of $\Delta_{hydr}H$ on ionic charge density and the factors that determine whether ionic solution processes are exothermic or endothermic. (§12.2)
8. Explain the meaning of *entropy* and how the balance between the change in enthalpy and the change in entropy governs the solution process. (§12.2)
9. Describe the differences among saturated, unsaturated, and supersaturated solutions, and explain the equilibrium nature of a saturated solution (§12.3)
10. Explain the relation between temperature and the solubility of solids. (§12.3)
11. Explain why the solubility of gases in water decreases with a rise in temperature. (§12.3)
12. Explain the effect of gas pressure on solubility and its quantitative expression as Henry's law. (§12.3)
13. Differentiate between amount of substance concentration in mol/L, molality, mole fraction, and parts by mass or by volume of a solution, and show how to convert among them. (§12.4)
14. Differentiate between electrolytes and nonelectrolytes in solution. (§12.5)
15. Describe the four colligative properties, and explain their dependence on the number of dissolved particles. (§12.5)
16. Describe ideal solutions, and explain the importance of Raoult's law. (§12.5)
17. Explain how the phase diagram of a solution differs from the phase diagram of the pure solvent. (§12.5)
18. Explain how phase diagrams can be used to guide crystallization trials. (§12.5)
19. Explain why the vapour over a solution of a volatile nonelectrolyte is richer in the more volatile component. (§12.5)
20. Describe non-ideal solutions, and explain deviations from Raoult's law. (§12.5)
21. Explain why electrolyte solutions are not ideal, and discuss the meanings of the *van't Hoff factor* and *ionic atmosphere*. (§12.5)
22. Explain how particle size distinguishes suspensions, colloids, and solutions. (§12.6)
23. Explain how colloidal behaviour is demonstrated by the Tyndall effect and Brownian motion. (§12.6)

Skills

1. Predict relative solubilities from intermolecular forces. (SP 12.1)
2. Use Henry's law to calculate the solubility of a gas. (SP 12.2)
3. Express concentration in terms of molality, parts by mass, parts by volume, and mole fraction. (SPs 12.3, 12.4)
4. Interconvert among the various terms for expressing concentration. (SP 12.5)
5. Use Raoult's law to calculate the vapour pressure lowering of a solution. (SP 12.6)
6. Determine the boiling point elevation and freezing point depression of a solution. (SP 12.7)
7. Use a colligative property to calculate the molar mass of a solute. (SP 12.8)
8. Calculate the composition of vapour over a solution of volatile nonelectrolyte. (§12.5)
9. Calculate the van't Hoff factor (i) from the magnitude of a colligative property. (§12.5)
10. Use a solution depiction to determine colligative properties. (SP 12.9)

Key Terms

Section 12.1
solute
solvent
miscible
solubility (s)
like-dissolves-like rule
hydration shells
ion–induced dipole forces
dipole–induced dipole forces
alloys

Section 12.2
heat of solution ($\Delta_{soln}H$)
solvation
hydration
heat of hydration ($\Delta_{hydr}H$)
charge density
entropy (S)

Section 12.3
saturated solution
unsaturated solution
supersaturated solution
Henry's law

Section 12.4
molality (m)
mass percent [% (w/w)]
volume percent [% (v/v)]
mole fraction (X)

Section 12.5
colligative properties
electrolyte
nonelectrolytes
vapour pressure lowering (Δp)
Raoult's law
ideal solution
boiling point elevation ($\Delta_b T$)
freezing point depression ($\Delta_f T$)
semipermeable membrane
osmosis
osmotic pressure (Π)
crystallization
fractional distillation
azeotrope
ionic atmosphere

Section 12.6
suspension
colloids
Tyndall effect
water softening
ion exchange
reverse osmosis
desalination
wastewater

Key Equations and Relationships

12.1 Dividing the general heat of solution into component enthalpies:

$$\Delta_{soln}H = \Delta_{solute}H + \Delta_{solvent}H + \Delta_{mix}H$$

12.2 Dividing the heat of solution of an ionic compound in water into component enthalpies:

$$\Delta_{soln}H = \Delta_{lattice}H + \Delta_{hydr\ of\ the\ ions}H$$

12.3 Relating gas solubility to its partial pressure (Henry's law):

$$s_{gas} = k_H \times p_{gas}$$

12.4 Defining amount of substance concentration in terms of amount concentration (also called molarity):

$$\text{Concentration } (c, \text{mol/L}) = \frac{\text{amount (mol) of solute}}{\text{volume (L) of solution}} = \frac{n}{V}$$

12.5 Defining concentration in terms of molality:

$$\text{Molality}(m) = \frac{\text{amount (mol) of solute}}{\text{mass (kg) of solvent}} = \frac{n_{solute}}{kg_{solvent}}$$

12.6 Defining concentration in terms of mass percent:

$$\text{Mass percent } [\%(\text{w/w})] = \frac{\text{mass of solute}}{\text{mass of solution}} \times 100$$

12.7 Defining concentration in terms of volume percent:

$$\text{Volume percent } [\%(\text{v/v})] = \frac{\text{volume of solute}}{\text{volume of solution}} \times 100$$

12.8 Defining concentration in terms of mole fraction:

Mole fraction (X)

$$= \frac{\text{amount (mol) of solute}}{\text{amount (mol) of solute} + \text{amount (mol) of solvent}}$$

$$= \frac{n_{solute}}{n_{solute} + n_{solvent}}$$

12.9 Expressing the relationship between the vapour pressure of a solvent above a solution and its mole fraction in the solution (Raoult's law):

$$p_{solvent} = X_{solvent} \times p^{\circ}_{solvent}$$

12.10 Calculating the vapour pressure lowering due to a solute:

$$\Delta p = X_{solute} \times p^{\circ}_{solvent}$$

12.11 Calculating the boiling point elevation of a solution:

$$\Delta_b T = K_b m$$

12.12 Calculating the freezing point depression of a solution:

$$\Delta_f T = K_f m$$

12.13 Calculating the osmotic pressure of a solution:

$$\Pi = \frac{n_{solute}}{V_{soln}} RT = cRT$$

Brief Solutions to Follow-Up Problems

12.1 (a) Butane-1,4-diol is more soluble in water because it can form more hydrogen bonds.
(b) Trichloromethane (chloroform) is more soluble in water because of dipole-dipole forces.

12.2 $s_{N_2} = [7 \times 10^{-6} \text{mol/(L·kPa)}](79.014 \text{ kPa})$
$= 6 \times 10^{-4} \text{mol/L}$

12.3 Mass (g) of glucose

$$= 563 \text{ g ethanol} \times \frac{1 \text{ kg}}{10^3 \text{g}} \times \frac{2.40 \times 10^{-2} \text{ mol glucose}}{1 \text{ kg ethanol}} \times \frac{180.16 \text{ g glucose}}{1 \text{ mol glucose}}$$

$= 2.43$ g glucose

12.4 Mass % $C_3H_7OH = \dfrac{35.0\ g}{35.0\ g + 150.\ g} \times 100$

$= 18.9$ mass

Mass % $C_2H_5OH = 100.0 - 18.9$

$= 81.1$ mass %

$$X_{C_3H_7OH} = \frac{35.0\ g\ C_3H_7OH \times \dfrac{1\ mol\ C_3H_7OH}{60.09\ g\ C_3H_7OH}}{\left(35.0\ g\ C_3H_7OH \times \dfrac{1\ mol\ C_3H_7OH}{60.09\ g\ C_3H_7OH}\right) + \left(150.\ g\ C_2H_5OH \times \dfrac{1\ mol\ C_2H_5OH}{46.07\ g\ C_2H_5OH}\right)} = 0.152$$

$X_{C_2H_5OH} = 1.000 - 0.152 = 0.848$

12.5 Mass % HCl $= \dfrac{\text{mass of HCl}}{\text{mass of soln}} \times 100$

$$= \frac{\dfrac{11.8\ mol\ HCl}{1\ L\ soln} \times \dfrac{36.46\ g\ HCl}{1\ mol\ HCl}}{\dfrac{1.190\ g}{1\ mL\ soln} \times \dfrac{10^3\ mL}{1\ L}} \times 100$$

$= 36.2$ mass % HCl

Mass (kg) of soln $= 1\ L\ soln \times \dfrac{1.190 \times 10^{-3}\ kg\ soln}{1 \times 10^{-3}\ L\ soln}$

$= 1.190$ kg soln

Mass (kg) of HCl $= 11.8\ mol\ HCl \times \dfrac{36.46\ g\ HCl}{1\ mol\ HCl} \times \dfrac{1\ kg}{10^3\ g}$

$= 0.430$ kg HCl

Molality of HCl $= \dfrac{mol\ HCl}{kg\ water} = \dfrac{mol\ HCl}{kg\ soln - kg\ HCl}$

$= \dfrac{11.8\ mol\ HCl}{0.760\ kg\ H_2O} = 15.5$ mol/kg HCl

$$X_{HCl} = \frac{mol\ HCl}{mol\ HCl + mol\ H_2O}$$

$$= \frac{11.8\ mol}{11.8\ mol + \left(760\ g\ H_2O \times \dfrac{1\ mol}{18.02\ g\ H_2O}\right)} = 0.219$$

12.6 $\Delta p = X_{Aspirin} \times p^\circ_{methanol}$

$$= \frac{\dfrac{2.00\ g}{180.15\ g/mol}}{\dfrac{2.00\ g}{180.15\ g/mol} + \dfrac{50.0\ g}{32.04\ g/mol}} \times 12.5\ kPa$$

$= 0.0883$kPa

12.7 Molality of $C_2H_6O_2 = \dfrac{0°C - (-17.8°C)}{1.86°C\cdot kg/mol} = 9.57$ mol/kg

12.8 $\Pi = cRT = (0.30\ mol/L)\left(8.314\ \dfrac{kPa\cdot L}{mol\cdot K}\right)(37°C + 273.15)$

$= 7.7 \times 10^2$ kPa

12.9 (a) Mass of 0.100 mol/kg solution

= mass of water + mass of $MgCl_2$

$= 1\ kg \times \dfrac{1000\ g}{1\ kg} + 0.100\ mol\ MgCl_2 \times 95.205\ \dfrac{g}{mol\ MgCl_2}$

$= 1000\ g + 9.52\ g = 1009.52\ g$

Volume of solution $= 1009.52\ g \times \dfrac{1\ mL}{1.006\ g} = 1003$ mL

Concentration (mol/L) $= \dfrac{9.52\ g\ MgCl_2}{1003\ mL\ soln} \times \dfrac{1\ mol}{95.21\ g\ MgCl_2} \times \dfrac{10^3\ mL}{1\ L}$

$= 9.97 \times 10^{-2}$ mol/L

Osmotic pressure (Π)

$= i(cRT)$

$= 3(9.97 \times 10^{-2}\ mol/L)\left(8.314\ \dfrac{kPa\cdot L}{mol\cdot K}\right)(293\ K)$

$= 7.29 \times 10^2$ kPa

(b) Scene C, because i for sucrose equals 1, which makes the osmotic pressure of the sucrose solution less than that of $MgCl_2$.

PROBLEMS

Problems with **red** numbers are answered in Appendix G and worked in detail in the Student Solutions Manual. Problem sections match those in this book and provide the numbers of relevant sample problems. Most offer Concept Review Questions, Skill-Building Exercises (grouped in pairs covering the same concept), and Problems in Context. The Comprehensive Problems are based on material from any section or previous chapter.

Types of Solutions: Intermolecular Forces and Solubility

(Sample Problem 12.1)

Concept Review Questions

12.1 Describe how properties of seawater illustrate the two characteristics that define mixtures.

12.2 What types of intermolecular forces give rise to hydration shells in an aqueous solution of sodium chloride?

12.3 Ethanoic acid (acetic acid) is miscible with water. Would you expect carboxylic acids with the general formula $CH_3(CH_2)_nCOOH$ to become more or less water soluble as n increases? Explain.

12.4 Which would you expect to be more effective as a soap: sodium ethanoate (sodium acetate) or sodium octadecanoate (sodium stearate)? Explain.

12.5 Hexane and methanol are miscible as gases but only slightly soluble in each other as liquids. Explain.

12.6 Hydrogen chloride (HCl) gas is much more soluble than propane gas (C_3H_8) in water, even though HCl has a lower boiling point. Explain.

Skill-Building Exercises (grouped in similar pairs)

12.7 Which gives the more concentrated solution: (a) KNO_3 in H_2O or (b) KNO_3 in carbon tetrachloride (CCl_4)? Explain.

12.8 Which gives the more concentrated solution: (a) stearic acid [octadecanoic acid, $CH_3(CH_2)_{16}COOH$] in H_2O or (b) stearic acid in CCl_4? Explain.

12.9 What is the strongest type of intermolecular force between the solute and the solvent in each solution?

(a) $CsCl(s)$ in $H_2O(l)$ (b) $CH_3C(=O)CH_3(l)$ in $H_2O(l)$
(c) $CH_3OH(l)$ in $CCl_4(l)$

12.10 What is the strongest type of intermolecular force between the solute and the solvent in each solution?
(a) $Cu(s)$ in $Ag(s)$ (b) $CH_3Cl(g)$ in $CH_3OCH_3(g)$
(c) $CH_3CH_3(g)$ in $CH_3CH_2CH_2NH_2(l)$

12.11 What is the strongest type of intermolecular force between the solute and the solvent in each solution?
(a) $CH_3OCH_3(g)$ in $H_2O(l)$ (b) $Ne(g)$ in $H_2O(l)$
(c) $N_2(g)$ in $C_4H_{10}(g)$

12.12 What is the strongest type of intermolecular force between the solute and the solvent in each solution?
(a) $C_6H_{14}(l)$ in $C_8H_{18}(l)$ (b) $H_2C{=}O(g)$ in $CH_3OH(l)$
(c) $Br_2(l)$ in $CCl_4(l)$

12.13 Which substance in each pair is more soluble in ethoxyethane (diethyl ether)? Why?

(a) $NaCl(s)$ or $HCl(g)$ (b) $H_2O(l)$ or $CH_3CH(=O)(l)$
(c) $MgBr_2(s)$ or $CH_3CH_2MgBr(s)$

12.14 Which substance in each pair is more soluble in water? Why?
(a) $CH_3CH_2OCH_2CH_3(l)$ or $CH_3CH_2OCH_3(g)$
(b) $CH_2Cl_2(l)$ or $CCl_4(l)$
(c)

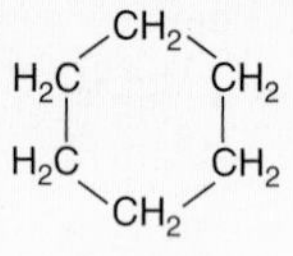

Cyclohexane

or

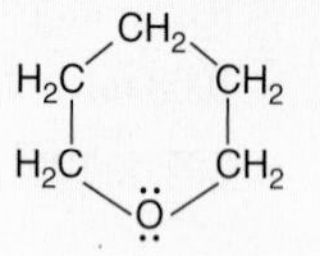

Oxacyclohexane (tetrahydropyran)

Problems in Context

12.15 The dictionary defines *homogeneous* as "uniform in composition throughout." River water is a mixture of dissolved compounds, such as calcium bicarbonate, and suspended soil particles. Is river water homogeneous? Explain.

12.16 Gluconic acid (2,3,4,5,6-pentahydroxyhexanoic acid) is a derivative of glucose that is used in cleaners and in the dairy and brewing industries. Caproic acid (hexanoic acid) is a carboxylic acid that is used in the flavouring industry. Although both are six-carbon acids (*see the structures below*), gluconic acid is soluble in water and nearly insoluble in hexane, whereas caproic acid has the opposite solubility behaviour. Explain.

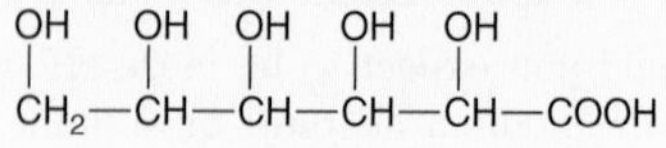

Gluconic acid

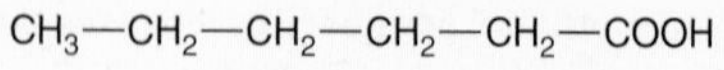

Caproic acid

Why Substances Dissolve: Understanding the Solution Process

Concept Review Questions

12.17 What is the relationship between solvation and hydration?

12.18 For a general solvent, which enthalpy terms in the thermochemical solution cycle are combined to obtain $\Delta_{solvation}H$?

12.19 (a) What is the charge density of an ion, and what two properties of an ion affect it?
(b) Arrange the following ions in order of increasing charge density:

(c) How do the two properties you named in part (a) affect the ionic heat of hydration, $\Delta_{hydr}H$?

12.20 For $\Delta_{soln}H$ to be very small, what quantities must be nearly equal in magnitude? Will their signs be the same or opposite?

12.21 Water is added to a flask containing solid NH_4Cl. As the salt dissolves, the solution becomes colder.
(a) Is the dissolving of NH_4Cl exothermic or endothermic?
(b) Is the magnitude of $\Delta_{lattice}H$ of NH_4Cl larger or smaller than the combined $\Delta_{hydr}H$ of the ions? Explain.
(c) Based on your answer to part (a), why does NH_4Cl dissolve in water?

12.22 An ionic compound has a highly negative $\Delta_{soln}H$ in water. Would you expect the compound to be very soluble or nearly insoluble in water? Explain your answer in terms of enthalpy and entropy changes.

Skill-Building Exercises (grouped in similar pairs)

12.23 Sketch an enthalpy diagram for the process of dissolving $KCl(s)$ in H_2O (endothermic).

12.24 Sketch an enthalpy diagram for the process of dissolving $NaI(s)$ in H_2O (exothermic).

12.25 Which ion in each pair has the greater charge density? Explain.
(a) Na^+ or Cs^+ (b) Sr^{2+} or Rb^+ (c) Na^+ or Cl^-
(d) O^{2-} or F^- (e) OH^- or SH^- (f) Mg^{2+} or Ba^{2+}
(g) Mg^{2+} or Na^+ (h) NO_3^- or CO_3^{2-}

12.26 Which ion in each pair has the lower ratio of charge to volume? Explain.
(a) Br^- or I^- (b) Sc^{3+} or Ca^{2+} (c) Br^- or K^+
(d) S^{2-} or Cl^- (e) Sc^{3+} or Al^{3+} (f) SO_4^{2-} or ClO_4^-
(g) Fe^{3+} or Fe^{2+} (h) Ca^{2+} or K^+

12.27 Which ion in each pair in Problem 12.25 has the *larger* ΔH?

12.28 Which ion in each pair in Problem 12.26 has the *smaller* $\Delta_{hydr}H$?

12.29 (a) Use the following data to calculate the combined heat of hydration for the ions in potassium bromate ($KBrO_3$):

$$\Delta_{lattice}H = 745 \text{ kJ/mol} \quad \Delta_{soln}H = 41.1 \text{ kJ/mol}$$

(b) Which ion contributes more to your answer to part (a)? Why?

12.30 (a) Use the following data to calculate the combined heat of hydration for the ions in sodium acetate ($NaC_2H_3O_2$):

$$\Delta_{lattice}H = 763 \text{ kJ/mol} \quad \Delta_{soln}H = 17.3 \text{ kJ/mol}$$

(b) Which ion contributes more to your answer to part (a)? Why?

12.31 State whether the entropy of the system increases or decreases in each process:
(a) Gasoline burns in a car engine.
(b) Gold is extracted and purified from its ore.
(c) Ethanol (CH_3CH_2OH) dissolves in propan-1-ol ($CH_3CH_2CH_2OH$).

12.32 State whether the entropy of the system increases or decreases in each process:
(a) Pure gases are mixed to prepare an anaesthetic.
(b) Electronic-grade silicon is prepared from sand.
(c) Dry ice (solid CO_2) sublimes.

Problem in Context

12.33 Silver nitrate ($AgNO_3$) is used in photography to make black-and-white film. It is used in a similar way in forensic science. The NaCl left behind in the sweat of a fingerprint is treated with $AgNO_3$ solution to form AgCl. This precipitate is developed to show the black-and-white fingerprint pattern. Given $\Delta_{lattice}H$ of $AgNO_3$ = 822 kJ/mol and $\Delta_{hydr}H$ = −799 kJ/mol, calculate the $\Delta_{soln}H$ of $AgNO_3$.

Solubility as an Equilibrium Process

(Sample Problem 12.2)

Concept Review Questions

12.34 You are given a bottle of solid X and three aqueous solutions of X: one saturated, one unsaturated, and one supersaturated. How would you determine which solution is which?

12.35 Potassium permanganate ($KMnO_4$) has a solubility of 6.4 g/100 g of H_2O at 20°C and a curve of solubility versus temperature that slopes upward to the right. How would you prepare a supersaturated solution of $KMnO_4$?

12.36 Why does the solubility of any gas in water decrease with rising temperature?

Skill-Building Exercises (grouped in similar pairs)

12.37 For a saturated aqueous solution of each gas at 20°C and 101.3 kPa, will the solubility increase, decrease, or stay the same when the indicated change occurs?
(a) $O_2(g)$, increase p (b) $N_2(g)$, increase V

12.38 For a saturated aqueous solution of each gas at 20°C and 101.3 kPa, will the solubility increase, decrease, or stay the same when the indicated change occurs?
(a) He(g), decrease T (b) RbI(s), increase p

Problems in Context

12.39 The Henry's law constant (k_H) for O_2 in water at 20°C is 1.26×10^{-5} mol/(L•kPa).
(a) What mass of O_2 will dissolve in 2.50 L of H_2O that is in contact with pure O_2 at 101.3 kPa?
(b) What mass of O_2 will dissolve in 2.50 L of H_2O that is in contact with air if the partial pressure of O_2 is 21.2 kPa?

12.40 Argon makes up 0.93% by volume of air. Calculate its solubility (mol/L) in water at 20°C and 101.3 kPa. The Henry's law constant for Ar under these conditions is 1.5×10^{-5} mol/(L•kPa).

12.41 Caffeine is about 10 times as soluble in hot water as it is in cold water. A chemist puts a hot-water extract of caffeine into an ice bath, and some caffeine crystallizes. Is the remaining solution saturated, unsaturated, or supersaturated?

12.42 The partial pressure of CO_2 gas above the liquid in a bottle of champagne at 20°C is 557.2 kPa. What is the solubility of CO_2 in champagne? Assume that the Henry's law constant is the same for champagne as it is for water: at 20°C, $k_H = 3.7 \times 10^{-4}$ mol/(L•kPa).

12.43 Respiratory problems are treated with devices that deliver air with a higher partial pressure of O_2 than normal air. Why?

Concentration Terms

(Sample Problems 12.3 to 12.5)

Concept Review Questions

12.44 Explain the difference between concentration (mol/L) and molality. Under what circumstances would molality be a more accurate measure of the concentration of a prepared solution than amount of solute (mol)/volume of solution (L)? Why?

12.45 Which way of expressing concentration includes (a) volume of solution; (b) mass of solution; (c) mass of solvent?

12.46 A solute has a solubility of 21 g/kg solvent in water. Is this value the same as 21 g/kg solution? Explain.

12.47 You want to convert among concentration (mol/L), molality, and mole fraction of a solution. You know the masses of the solute and solvent and the volume of the solution. Is this enough information to carry out all the conversions? Explain.

12.48 When a solution is heated, which ways of expressing concentration change in value? Which remain unchanged? Explain.

Skill-Building Exercises (grouped in similar pairs)

12.49 Calculate the concentration (mol/L) of each aqueous solution:
(a) 32.3 g of table sugar ($C_{12}H_{22}O_{11}$) in 100. mL of solution
(b) 5.80 g of $LiNO_3$ in 505 mL of solution

12.50 Calculate the concentration (mol/L) of each aqueous solution:
(a) 0.82 g of ethanol (C_2H_5OH) in 10.5 mL of solution
(b) 1.27 g of gaseous NH_3 in 33.5 mL of solution

12.51 Calculate the concentration (mol/L) of each aqueous solution:
(a) 78.0 mL of 0.240 mol/L NaOH diluted to 0.250 L with water
(b) 38.5 mL of 1.2 mol/L HNO_3 diluted to 0.130 L with water

12.52 Calculate the concentration (mol/L) of each aqueous solution:
(a) 25.5 mL of 6.25 mol/L HCl diluted to 0.500 L with water
(b) 8.25 mL of 2.00×10^{-2} mol/L KI diluted to 12.0 mL with water

12.53 How would you prepare each aqueous solution?
(a) 365 mL of 8.55×10^{-2} mol/L KH_2PO_4 from solid KH_2PO_4
(b) 465 mL of 0.335 mol/L NaOH from 1.25 mol/L NaOH

12.54 How would you prepare each aqueous solution?
(a) 2.5 L of 0.65 mol/L NaCl from solid NaCl
(b) 15.5 L of 0.3 mol/L urea, $(NH_2)_2C{=}O$, from 2.1 mol/L urea

12.55 How would you prepare each aqueous solution?
(a) 1.40 L of 0.288 mol/L KBr from solid KBr
(b) 255 mL of 0.0856 mol/L $LiNO_3$ from 0.264 mol/L $LiNO_3$

12.56 How would you prepare each aqueous solution?
(a) 57.5 mL of 1.53×10^{-3} mol/L $Cr(NO_3)_3$ from solid $Cr(NO_3)_3$
(b) 5.8×10^3 m^3 of 1.45 mol/L NH_4NO_3 from 2.50 mol/L NH_4NO_3

12.57 Calculate the molality of each solution:
(a) A solution containing 85.4 g of glycine (2-aminoethanoic acid, NH_2CH_2COOH) in 1.270 kg of H_2O
(b) A solution containing 8.59 g of glycerol (propane-1,2,3-triol, $C_3H_8O_3$) in 77.0 g of ethanol (C_2H_5OH)

12.58 Calculate the molality of each solution:
(a) A solution containing 174 g of HCl in 757 g of H_2O
(b) A solution containing 16.5 g of naphthalene ($C_{10}H_8$) in 53.3 g of benzene (C_6H_6)

12.59 What is the molality of a solution consisting of 44.0 mL of benzene (C_6H_6; d = 0.877 g/mL) in 167 mL of hexane (C_6H_{14}; d = 0.660 g/mL)?

12.60 What is the molality of a solution consisting of 2.66 mL of tetrachloromethane (carbon tetrachloride) (CCl_4; $d = 1.59$ g/mL) in 76.5 mL of dichloromethane (CH_2Cl_2; $d = 1.33$ g/mL)?

12.61 How would you prepare the following aqueous solutions?
(a) 3.10×10^2 g of 0.125 mol/kg ethylene glycol (ethane-1,2-diol, $C_2H_6O_2$) from ethylene glycol and water
(b) 1.20 kg of 2.20 mass % HNO_3 from 52.0 mass % HNO_3

12.62 How would you prepare the following aqueous solutions?
(a) 1.50 kg of 0.0355 mol/kg ethanol (C_2H_5OH) from ethanol and water
(b) 445 g of 13.0 mass % HCl from 34.1 mass % HCl

12.63 A solution contains 0.35 mol of propan-2-ol (isopropanol, C_3H_7OH) dissolved in 0.85 mol of water. What is (a) the mole fraction of isopropanol; (b) the mass percent; (c) the molality?

12.64 A solution contains 0.100 mol of NaCl dissolved in 8.60 mol of water. What is (a) the mole fraction of NaCl; (b) the mass percent; (c) the molality?

12.65 What mass of cesium bromide must be added to 0.500 L of water ($d = 1.00$ g/mL) to produce a 0.400 mol/kg solution? What are the mole fraction and the mass percent of CsBr?

12.66 Calculate the mole fraction and the mass percent of a solution made by dissolving 0.30 g of KI in 0.400 L of water ($d = 1.00$ g/mL).

12.67 Calculate the molality, concentration (mol/L), and mole fraction of NH_3 in an 8.00 mass % aqueous solution ($d = 0.9651$ g/mL).

12.68 Calculate the molality, concentration (mol/L), and mole fraction of $FeCl_3$ in a 28.8 mass % aqueous solution ($d = 1.280$ g/mL).

Problems in Context

12.69 Wastewater from a cement factory contains 0.25 g of Ca^{2+} ion and 0.056 g of Mg^{2+} ion per 100.0 L of solution. The solution density is 1.001 g/mL. Calculate the Ca^{2+} and Mg^{2+} concentrations in ppm (by mass).

12.70 An automobile antifreeze mixture is made by mixing equal volumes of ethylene glycol (ethane-1,2-diol, $d = 1.114$ g/mL; $\mathcal{M} = 62.07$ g/mol) and water ($d = 1.00$ g/mL) at 20°C. The density of the mixture is 1.070 g/mL. Express the concentration of ethylene glycol as (a) a volume percent; (b) a mass percent; (c) a concentration (mol/L); (d) a molality; (e) a mole fraction.

Colligative Properties of Solutions

(Sample Problems 12.6 to 12.9)

Concept Review Questions

12.71 The chemical formula for a solute does *not* affect the extent of the solution's colligative properties. What characteristic of a solute *does* affect these properties? Name a physical property of a solution that is affected by the chemical formula for the solute.

12.72 What is a nonvolatile nonelectrolyte? Why is this type of solute the simplest case for examining colligative properties?

12.73 In what sense is a strong electrolyte "strong"? What property of the substance makes it a strong electrolyte?

12.74 Express Raoult's law in words. Is Raoult's law valid for a solution of a volatile solute? Explain.

12.75 What are the most important differences between the phase diagram of a pure solvent and the phase diagram of a solution of this solvent?

12.76 Is the composition of the vapour at the top of a fractionating column different from the composition at the bottom? Explain.

12.77 Is the boiling point of 0.01 mol/kg KF(*aq*) higher or lower than the boiling point of 0.01 mol/kg glucose(*aq*)? Explain.

12.78 Which aqueous solution has a boiling point closer to its predicted value: 0.050 mol/kg NaF or 0.50 mol/kg KCl? Explain.

12.79 Which aqueous solution has a freezing point closer to its predicted value: 0.01 mol/kg NaBr or 0.01 mol/kg $MgCl_2$? Explain.

12.80 The freezing point depression constants of the solvents cyclohexane and naphthalene are 20.1°C•mol/kg and 6.94°C•mol/kg, respectively. Which solvent would give a more accurate result if you were using freezing point depression to determine the molar mass of a substance that is soluble in either solvent? Why?

Skill-Building Exercises (grouped in similar pairs)

12.81 Classify each substance as a strong electrolyte, a weak electrolyte, or a nonelectrolyte:
(a) Hydrogen chloride (HCl) (b) Potassium nitrate (KNO_3)
(c) Glucose ($C_6H_{12}O_6$) (d) Ammonia (NH_3)

12.82 Classify each substance as a strong electrolyte, a weak electrolyte, or a nonelectrolyte:
(a) Sodium permanganate ($NaMnO_4$)
(b) Ethanoic acid (acetic acid, CH_3COOH)
(c) Methanol (CH_3OH)
(d) Calcium ethanoate [calcium acetate, $Ca(C_2H_3O_2)_2$]

12.83 What amount of solute particles is present in 1 L of each aqueous solution?
(a) 0.3 mol/L KBr
(b) 0.065 mol/L HNO_3
(c) 10^{-4} mol/L $KHSO_4$
(d) 0.06 mol/L ethanol (C_2H_5OH)

12.84 What amount of solute particles is present in 1 mL of each aqueous solution?
(a) 0.02 mol/L $CuSO_4$ (b) 0.004 mol/L $Ba(OH)_2$
(c) 0.08 mol/L pyridine (C_5H_5N) (d) 0.05 mol/L $(NH_4)_2CO_3$

12.85 Which solution in each pair has the lower freezing point?
(a) 11.0 g of CH_3OH in 100. g of H_2O *or* 22.0 g of CH_3CH_2OH in 200. g of H_2O
(b) 20.0 g of H_2O in 1.00 kg of CH_3OH *or* 20.0 g of CH_3CH_2OH in 1.00 kg of CH_3OH

12.86 Which solution in each pair has the higher boiling point?
(a) 38.0 g of $C_3H_8O_3$ in 250. g of ethanol *or* 38.0 g of $C_2H_6O_2$ in 250. g of ethanol
(b) 15 g of $C_2H_6O_2$ in 0.50 kg of H_2O *or* 15 g of NaCl in 0.50 kg of H_2O

12.87 Rank the following aqueous solutions in order of increasing (a) osmotic pressure; (b) boiling point; (c) freezing point; (d) vapour pressure at 50°C:
(I) 0.100 mol/kg $NaNO_3$
(II) 0.100 mol/kg glucose
(III) 0.100 mol/kg $CaCl_2$

12.88 Rank the following aqueous solutions in order of decreasing (a) osmotic pressure; (b) boiling point; (c) freezing point; (d) vapour pressure at 298 K:
(I) 0.04 mol/kg urea, $(NH_2)_2C{=}O$
(II) 0.01 mol/kg $AgNO_3$
(III) 0.03 mol/kg $CuSO_4$

12.89 Calculate the vapour pressure of a solution of 34.0 g of glycerol (propane-1,2,3-triol, $C_3H_8O_3$) in 500.0 g of water at 25°C. The vapour pressure of water at 25°C is 3168 Pa. (Assume ideal behaviour.)

12.90 Calculate the vapour pressure of a solution of 0.39 mol of cholesterol in 5.4 mol of toluene (methylbenzene) at 32°C. Pure toluene has a vapour pressure of 5467 Pa at 32°C. (Assume ideal behaviour.)

12.91 What is the freezing point of 0.251 mol/kg urea in water?

12.92 What is the boiling point of 0.200 mol/kg lactose in water?

12.93 The boiling point of ethanol (C_2H_5OH) is 78.5°C. What is the boiling point of a solution of 6.4 g of vanillin ($\mathcal{M}$ = 152.14 g/mol) in 50.0 g of ethanol (K_b of ethanol = 1.22°C·mol/kg)?

12.94 The freezing point of benzene is 5.5°C. What is the freezing point of a solution of 5.00 g of naphthalene ($C_{10}H_8$) in 444 g of benzene (K_f of benzene = 4.90°C·mol/kg)?

12.95 What is the minimum mass of ethylene glycol (ethane-1,2-diol, $C_2H_6O_2$) that must be dissolved in 14.5 kg of water to prevent the solution from freezing at −24.4°C? (Assume ideal behaviour.)

12.96 What is the minimum mass of glycerol (propane-1,2,3-triol, $C_3H_8O_3$) that must be dissolved in 11.0 mg of water to prevent the solution from freezing at −15°C? (Assume ideal behaviour.)

12.97 Calculate the molality and van't Hoff factor (i) for each aqueous solution:
(a) 1.00 mass % NaCl, freezing point = −0.593°C
(b) 0.500 mass % CH_3COOH, freezing point = −0.159°C

12.98 Calculate the molality and van't Hoff factor (i) for each aqueous solution:
(a) 0.500 mass % KCl, freezing point = −0.234°C
(b) 1.00 mass % H_2SO_4, freezing point = −0.423°C

Problems in Context

12.99 In a study designed to prepare new gasoline-resistant coatings, a polymer chemist dissolves 6.053 g of poly(vinyl alcohol) in enough water to make 100.0 mL of solution. At 25°C, the osmotic pressure of this solution is 27.6 kPa. What is the molar mass of the polymer sample?

12.100 Health Canada lists dichloromethane (CH_2Cl_2) and tetrachloromethane (CCl_4) among the many cancer-causing chlorinated organic compounds. What are the partial pressures of these substances in the vapour above a solution of 1.60 mol of CH_2Cl_2 and 1.10 mol of CCl_4 at 23.5°C? The vapour pressures of pure CH_2Cl_2 and CCl_4 at 23.5°C are 46.9 kPa and 15.7 kPa, respectively. (Assume ideal behaviour.)

The Structure and Properties of Colloids

Concept Review Questions

12.101 Is the fluid inside a bacterial cell considered a solution, a colloid, or both? Explain.

12.102 What type of colloid is each of the following?
(a) Milk (b) Fog (c) Shaving cream

12.103 What is Brownian motion, and what causes it?

12.104 In a movie theatre, you can see the beam of projected light. What phenomenon does this exemplify? Why does it occur?

12.105 Why do soap micelles not coagulate and form large globules? Is soap more effective in fresh water or in seawater? Why?

Comprehensive Problems

12.106 The three aqueous ionic solutions shown have total volumes of 25. mL for A, 50. mL for B, and 100. mL for C. If each sphere represents 0.010 mol of ions, calculate (a) the total concentration (mol/L) of ions for each solution; (b) the highest concentration (mol/L) of solute; (c) the lowest molality of solute (assuming the solution densities are equal); (d) the highest osmotic pressure (assuming ideal behaviour).

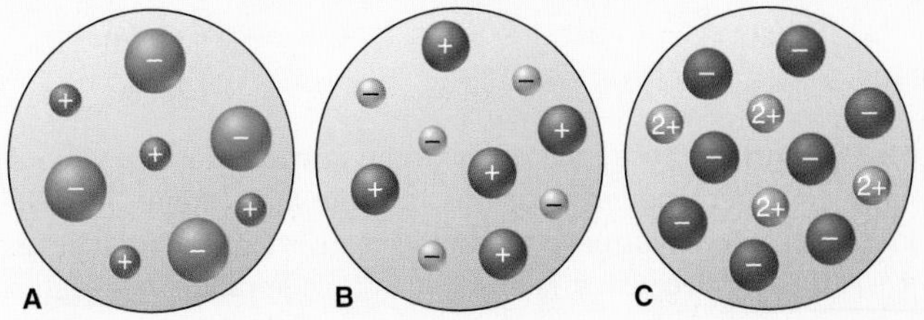

12.107 An aqueous solution is 10.% glucose by mass (d = 1.039 g/mL at 20°C). Calculate its freezing point, boiling point at 101.3 kPa, and osmotic pressure.

12.108 Because zinc has nearly the same atomic radius as copper (d = 8.95 g/cm^3), zinc atoms substitute for some copper atoms in the many types of brass. Calculate the density of the brass with (a) 10.0 atom % Zn; (b) 38.0 atom % Zn.

12.109 Gold occurs in seawater at an average concentration of 1.1×10^{-2} ppb. What volume (L) of seawater must be processed to recover 1 oz t (troy ounce) of gold, assuming 81.5% efficiency? (d of seawater = 1.025 g/mL; 1 oz t = 31.1 g)

12.110 Use atomic properties to explain why xenon is 11 times as soluble as helium in water at 0°C on an amount (mol) basis.

12.111 Which scene best represents a molecular-scale view of an ionic compound in aqueous solution? Explain.

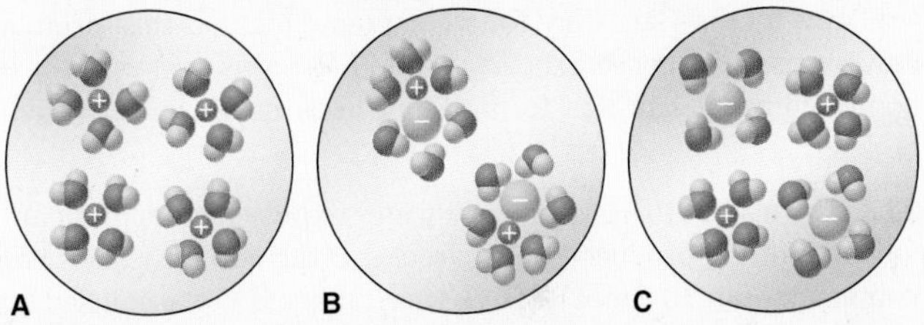

12.112 Four 0.50 mol/kg aqueous solutions are depicted below. Assume that the solutions behave ideally.
(a) Which has the highest boiling point?
(b) Which has the lowest freezing point?
(c) Can you determine which has the highest osmotic pressure? Explain.

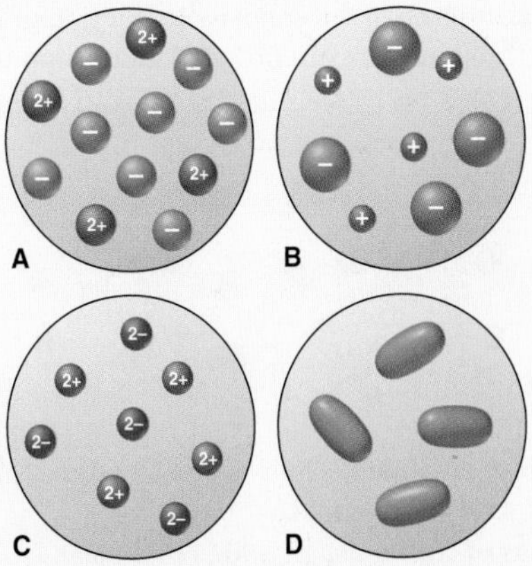

12.113 Thermal pollution from industrial wastewater causes the temperature of river or lake water to increase, which can affect the survival of fish as the concentration of dissolved O_2 decreases. Use the following data to find the concentration (mol/L) of O_2 at each temperature. (Assume that the solution density is the same as the density of water.)

Temperature (°C)	Solubility of O_2 (mg/kg H_2O)	Density of H_2O (g/mL)
0.0	14.5	0.99987
20.0	9.07	0.99823
40.0	6.44	0.99224

12.114 Pyridine (*below*) is an essential portion of many biologically active compounds, such as nicotine and vitamin B_6. Like ammonia, it has a nitrogen with a lone pair, which makes it act as a weak base. Because it is miscible in a wide range of solvents, from water to benzene, pyridine is one of the most important bases and solvents in organic syntheses. Account for its solubility behaviour in terms of intermolecular forces.

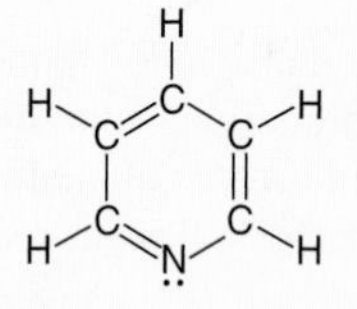

12.115 A chemist is studying small organic compounds for their potential use as an antifreeze. When 0.243 g of a compound is dissolved in 25.0 mL of water, the freezing point of the solution is −0.201°C.
(a) Calculate the molar mass of the compound (*d* of water = 1.00 g/mL).
(b) Analysis shows that the compound is 53.31 mass % C and 11.18 mass % H, the remainder being O. Calculate the empirical and molecular formulas for the compound.
(c) Draw a Lewis structure for a compound that has this formula and forms hydrogen bonds. Draw another Lewis structure for a compound that has this formula but does not form hydrogen bonds.

12.116 Until 2010, bartenders were exposed to significant amounts of CO produced by cigarettes. If the air in a smoky bar contained 4.0×10^{-6} mol/L of CO, what mass of CO was inhaled by a bartender who respired at a rate of 11 L/min during an 8.0 h shift?

12.117 Is 50% by mass of methanol dissolved in ethanol different from 50% by mass of ethanol dissolved in methanol? Explain.

12.118 Three gaseous mixtures of N_2 (*blue*), Cl_2 (*green*), and Ne (*purple*) are depicted below.
(a) Which mixture has the smallest mole fraction of N_2?
(b) Which mixtures have the same mole fraction of Ne?
(c) Rank all three mixtures in order of increasing mole fraction of Cl_2.

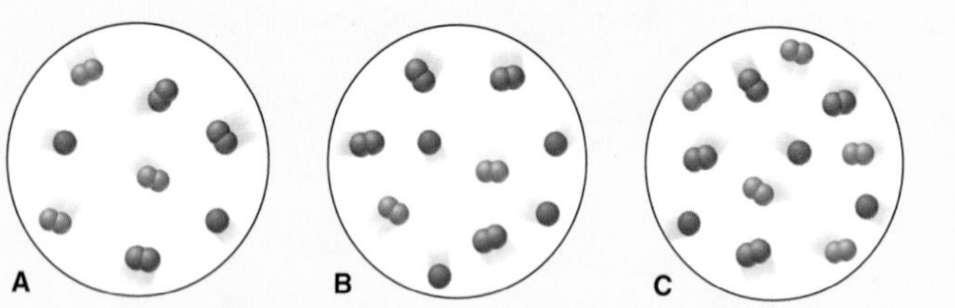

12.119 A water treatment plant needs to attain a fluoride concentration of 4.50×10^{-5} mol/L.
(a) What mass of NaF must be added to a 5000. L blending tank of water?
(b) What mass (per day) of fluoride is ingested by a person who drinks 2.0 L of this water per day?

12.120 Four U tubes each have distilled water in the right arm, a solution in the left arm, and a semipermeable membrane between the arms.
(a) If the solute is KCl, which solution is most concentrated?
(b) If each solute is different, but all the solutions have the same amount of substance concentration, which contains the smallest amount of dissolved ions?

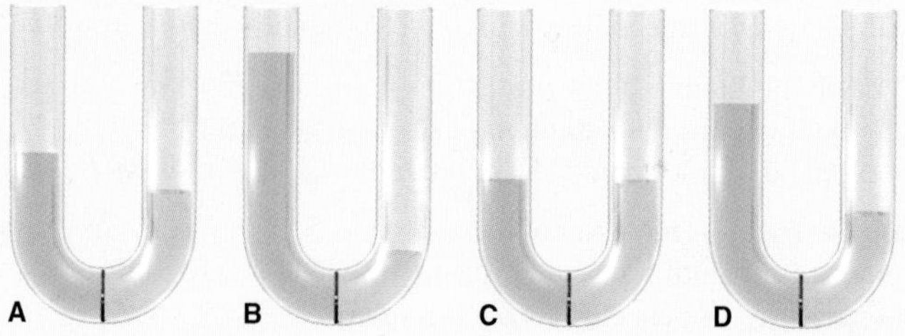

12.121 β-pinene ($C_{10}H_{16}$) and α-terpineol ($C_{10}H_{18}O$) are used in cosmetics to provide a "fresh pine" scent. At 367 K, the pure substances have vapour pressures of 13.4 kPa and 1.31 kPa, respectively. What is the composition of the vapour (in terms of mole fractions) above a solution containing equal masses of these compounds at 367 K? (Assume ideal behaviour.)

12.122 A solution of 1.50 g of solute dissolved in 25.0 mL of H_2O at 25°C has a boiling point of 100.45°C.
(a) What is the molar mass of the solute if it is a nonvolatile nonelectrolyte and the solution behaves ideally (*d* of H_2O at 25°C = 0.997 g/mL)?
(b) Conductivity measurements show the solute to be ionic of general formula AB_2 or A_2B. What is the molar mass if the solution behaves ideally?
(c) Analysis indicates an empirical formula of CaN_2O_6. Explain the difference between the actual formula mass and the formula mass calculated from the boiling point elevation.
(d) Find the van't Hoff factor (*i*) for this solution.

12.123 A pharmaceutical preparation made with ethanol (C_2H_5OH) is contaminated with methanol (CH_3OH). A sample of vapour above the liquid mixture contains a 97/1 mass ratio of C_2H_5OH to CH_3OH. What is the mass ratio of these alcohols in the liquid? At the temperature of the liquid, the vapour pressures of C_2H_5OH and CH_3OH are 8.07 kPa and 16.8 kPa, respectively.

12.124 Water treatment plants commonly use chlorination to destroy bacteria. A by-product is trichloromethane (chloroform, $CHCl_3$), a suspected carcinogen, produced when HOCl, formed by reaction of Cl_2 and water, reacts with dissolved organic matter. Canada, the United States, and the World Health Organization have set a limit of 100. ppb of $CHCl_3$ in drinking water. Convert this concentration into (a) amount of substance concentration (mol/L); (b) molality; (c) mole fraction; (d) mass percent.

12.125 A saturated Na_2CO_3 solution is prepared, and a small excess of solid is present (white pile in the beaker). A seed crystal of $Na_2{}^{14}CO_3$ (^{14}C is a radioactive isotope of ^{12}C) is added (small red piece), and the radioactivity is measured over time.

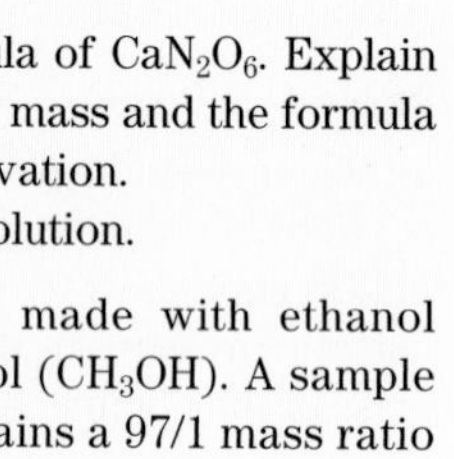

(a) Would you expect radioactivity in the solution? Explain.
(b) Would you expect radioactivity in all the solid or just in the seed crystal? Explain.

12.126 A biochemical engineer isolates a bacterial gene fragment and dissolves a 10.0 mg sample in enough water to make

30.0 mL of solution. The osmotic pressure of the solution is 45.3 Pa at 25°C.
(a) What is the molar mass of the gene fragment?
(b) If the solution density is 0.997 g/mL, what is the freezing point depression for this solution (K_f of water = 1.86°C·kg/mol)?

12.127 A river is contaminated with 0.65 mg/L of 1,1-dichloroethene ($C_2H_2Cl_2$). What is the concentration (ng/L) of 1,1-dichloroethene at 21°C in the air breathed by a person swimming in the river [k_H for $C_2H_2Cl_2$ in water is 0.00033 mol/(L·kPa)]?

12.128 At an air-water interface, fatty acids such as oleic acid lie in a one-molecule-thick layer (*monolayer*), with the heads in the water and the tails perpendicular in the air. When 2.50 mg of oleic acid is placed on a water surface, it forms a circular monolayer, 38.6 cm in diameter. Find the surface area (cm^2) occupied by one molecule ($\mathcal{M}$ of oleic acid = 283 g/mol).

12.129 A simple device used for estimating the concentration of total dissolved solids in an aqueous solution works by measuring the electrical conductivity of the solution. The method assumes that equal concentrations of different solids give approximately the same conductivity and that the conductivity is proportional to the concentration. The table below gives some actual electrical conductivities (in arbitrary units) for solutions of selected solids at the indicated concentrations (ppm by mass):

	Conductivity		
Sample	**0 ppm**	**5.00 × 10^3 ppm**	**10.00 × 10^3 ppm**
$CaCl_2$	0.0	8.0	16.0
K_2CO_3	0.0	7.0	14.0
Na_2SO_4	0.0	6.0	11.0
Seawater (dil)	0.0	8.0	15.0
Sucrose, $C_{12}H_{22}O_{11}$	0.0	0.0	0.0
Urea, $(NH_2)_2C{=}O$	0.0	0.0	0.0

(a) How reliable are these measurements for estimating concentrations of dissolved solids?
(b) For what types of substances might this method have a large error? Why?
(c) Based on this method, an aqueous $CaCl_2$ solution has a conductivity of 14.0 units. Calculate its mole fraction and molality.

12.130 Two beakers are placed in a closed container (*left*). One beaker contains water, and the other contains a concentrated aqueous sugar solution. With time, the solution volume increases and the water volume decreases (*right*). Explain this on the molecular level.

12.131 The release of volatile organic compounds into the atmosphere is regulated to limit ozone formation. In a laboratory simulation, 5% of the ethanol in a liquid detergent is released. Thus, a "down-the-drain" factor of 0.05 is used to estimate ethanol emissions from the detergent. The k_H values for ethanol and 2-butoxyethanol ($C_4H_9OCH_2CH_2OH$) are 5 × 10^{-4} kPa·m^3/mol and 1.6 × 10^{-4} kPa·m^3/mol, respectively.
(a) Estimate a "down-the-drain" factor for 2-butoxyethanol in the detergent.
(b) What is the k_H for ethanol in units of L·kPa/mol?
(c) Is this value consistent with a value given as 0.64 Pa·m^3/mol?

12.132 Although other solvents are available, dichloromethane (CH_2Cl_2) is still often used to "decaffeinate" drinks because the solubility of caffeine in CH_2Cl_2 is 8.35 times its solubility in water.
(a) A 100.0 mL sample of cola containing 10.0 mg of caffeine is extracted with 60.0 mL of CH_2Cl_2. What mass of caffeine remains in the aqueous phase?
(b) A second identical cola sample is extracted with two successive 30.0 mL portions of CH_2Cl_2. What mass of caffeine remains in the aqueous phase after each extraction?
(c) Which approach extracts more caffeine?

12.133 How do you prepare 250. g of 0.150 mol/kg aqueous $NaHCO_3$?

12.134 Tartaric acid (2,3-dihydroxybutanedioic acid) occurs in crystalline residues found in wine vats. It is used in baking powders and as an additive in foods. It contains 32.3% by mass carbon and 3.97% by mass hydrogen; the balance is oxygen. When 0.981 g of tartaric acid is dissolved in 11.23 g of water, the solution freezes at −1.26°C. Find the empirical and molecular formulas for tartaric acid.

12.135 Methanol (CH_3OH) and ethanol (C_2H_5OH) are miscible because the major intermolecular force for each is hydrogen bonding. In some methanol-ethanol solutions, the mole fraction of methanol is higher, but the mass percent of ethanol is higher. What is the range of mole fraction of methanol for these solutions?

12.136 A solution of 5.0 g of benzoic acid (C_6H_5COOH) in 100.0 g of tetrachloromethane has a boiling point of 77.5°C.
(a) Calculate the molar mass of benzoic acid in the solution.
(b) Suggest a reason for the difference between the molar mass based on the formula and the molar mass you calculated in part (a).

12.137 Derive a general equation that expresses the relationship between the amount of substance concentration (mol/L) and the molality of a solution. Why are the numerical values of these two terms approximately equal for very dilute aqueous solutions?

12.138 A florist prepares a solution of nitrogen-phosphorus fertilizer by dissolving 5.66 g of NH_4NO_3 and 4.42 g of $(NH_4)_3PO_4$ in enough water to make 20.0 L of solution. What are the concentrations (mol/L) of NH_4^+ and of PO_4^{3-} in the solution?

12.139 Suppose that coal-fired power plants used water in scrubbers to remove SO_2 from smokestack gases (see Chemical Connections in Chapter 5).
(a) If the partial pressure of SO_2 in the stack gases is 203 Pa, what is the solubility of SO_2 in the scrubber liquid [k_H for SO_2 in water is 1.21 × 10^{-2} mol/(L·kPa) at 200.°C]?
(b) From your answer to part (a), why are basic solutions, such as limewater slurries, $Ca(OH)_2$, used in scrubbers?

12.140 Urea is a white crystalline solid used in the pharmaceutical industry, in the manufacture of certain polymer resins, and as a fertilizer. Analysis of urea reveals that, by mass, it is 20.1% carbon, 6.7% hydrogen, and 46.5% nitrogen, and the rest is oxygen.
(a) Find the empirical formula for urea.
(b) A 5.0 g/L solution of urea in water has an osmotic pressure of 206.7 kPa, measured at 25°C. What are the molar mass and molecular formula for urea?

12.141 The total concentration of dissolved particles in blood is 0.30 mol/L. An intravenous (IV) solution must be isotonic with blood, which means that it must have the same concentration.
(a) To relieve dehydration, a patient is given 100. mL/h of IV glucose ($C_6H_{12}O_6$) for 2.5 h. What mass (g) of glucose did she receive?
(b) If isotonic saline (NaCl) is used, what is the concentration (mol/L) of the solution?
(c) If the patient is given 150. mL/h of IV saline for 1.5 h, what mass of NaCl did she receive?

12.142 Deviations from Raoult's law lead to the formation of *azeotropes*, constant boiling mixtures that cannot be separated by distillation, making industrial separations difficult. For components A and B, there is a positive deviation if the A-B attraction is less than A-A and B-B attractions (A and B reject each other) and a negative deviation if the A-B attraction is greater than A-A and B-B attractions. If the A-B attraction is nearly equal to the A-A and B-B attractions, the solution obeys Raoult's law. Explain whether the behaviour of each pair will be nearly ideal, have a positive deviation, or have a negative deviation:
(a) Benzene (C_6H_6) and methanol
(b) Water and ethyl ethanoate (ethyl acetate)
(c) Hexane and heptane
(d) Methanol and water
(e) Water and hydrochloric acid

12.143 Acrylic acid (prop-2-enoic acid, $H_2C{=}CHCOOH$) is a monomer used to make superabsorbent polymers and various compounds for paint and adhesive production. At 101.3 kPa, it boils at 141.5°C but is prone to polymerization. Its vapour pressure at 25°C is 4.1 mbar. What pressure (kPa) is needed to distill the pure acid at 65°C?

12.144 To effectively stop polymerization, certain inhibitors require the presence of a small amount of O_2. At equilibrium with 101.3 kPa of air, the concentration of O_2 dissolved in the monomer acrylic acid (prop-2-enoic acid, $H_2C{=}CHCOOH$) is 1.64×10^{-3} mol/L. (Pure acrylic acid is 14.6 mol/L; p_{O_2} in air is 21.2 kPa.)
(a) What is the k_H [mol/(L•kPa)] for O_2 in acrylic acid?
(b) If 0.5 kPa of O_2 is sufficient to stop polymerization, what is the concentration in mol/L of O_2?
(c) What is the mole fraction?
(d) What is the concentration in ppm?

12.145 Volatile organic solvents have been implicated in adverse health effects in industrial workers. Greener methods are phasing these solvents out. Rank the solvents in Table 12.5 in terms of increasing volatility.

12.146 At ordinary temperatures, water is a poor solvent for organic substances. At high pressure and above 200°C, however, water develops many properties of organic solvents. Find the minimum pressure needed to maintain water as a liquid at 200.°C. ($\Delta_{vap}H$ = 40.7 kJ/mol at 100°C and 101.3 kPa; assume that it remains constant with temperature.)

12.147 In ice cream making, the ingredients are kept below 0.0°C in an ice-salt bath.
(a) Assuming that NaCl dissolves completely and forms an ideal solution, what mass of NaCl is needed to lower the melting point of 5.5 kg of ice to −5.0°C?
(b) Given the same assumptions as in part (a), what mass of $CaCl_2$ is needed?

12.148 Perfluorocarbons (PFCs), hydrocarbons with all the H atoms replaced by F atoms, have very weak cohesive forces. One unusual application has a live mouse able to breathe normally while submerged in O_2-saturated PFCs.
(a) At 298 K, perfluorohexane (C_6F_{14}; $\mathcal{M}$ = 338 g/mol, d = 1.674 g/mL) in equilibrium with 101 325 Pa of O_2 has a mole fraction of O_2 of 4.28×10^{-3}. What is the k_H in mol/(L•kPa)?
(b) According to one source, the k_H for O_2 in water at 25°C is 7.67×10^4 L•kPa/mol. What is the solubility of O_2 in water at 25°C in ppm?
(c) Rank the k_H values in descending order for O_2 in water, ethanol, C_6F_{14}, and C_6H_{14}. Explain your ranking.

12.149 The solubility of N_2 in blood is a serious problem for divers breathing compressed air (78% N_2 by volume) at depths greater than 15 m. (k_H for N_2 in water is 6.9×10^{-6} mol/(L•kPa) at 25°C and 6.1×10^{-6} mol/(L•kPa) at 37°C; assume that d of water is 1.00 g/mL.)
(a) What is the concentration (mol/L) of N_2 in blood at 101.3 kPa?
(b) What is the concentration (mol/L) of N_2 in blood at a depth of 15 m?
(c) Find the volume (mL) of N_2, measured at 25°C and 101.3 kPa, released per litre of blood when a diver at a depth of 15 m rises to the surface.

12.150 Figure 11.13 shows the phase changes of pure water. Consider how the phase diagram would change if air were present at 101.3 kPa and dissolved in the water.
(a) Would the three phases of water still attain equilibrium at some temperature? Explain.
(b) In principle, would this temperature be higher, lower, or the same as the triple point for pure water? Explain.
(c) Would ice sublime at a few degrees below the freezing point under this pressure? Explain.
(d) Would the liquid have the same vapour pressure as that shown in Figure 11.7 (i) at 80°C; (ii) at 100°C?

12.151 KNO_3, $KClO_3$, KCl, and NaCl are recrystallized as follows:
Step 1. A saturated aqueous solution of the compound is prepared at 50°C.
Step 2. The mixture is filtered to remove any undissolved compound.
Step 3. The filtrate is cooled to 0°C.
Step 4. The crystals that form are filtered, dried, and weighed.
(a) Which compound has the highest percent recovery, and which the lowest (see Figure 12.10)?
(b) Starting with 100. g of each compound, what mass of each can be recovered?

12.152 Eighty proof whiskey is 40% ethanol (C_2H_5OH) by volume. A man has 7.0 L of blood and drinks 28 mL of the whiskey, of which 22% of the ethanol goes into his blood.
(a) What concentration (g/mL) of ethanol is in his blood (d of ethanol = 0.789 g/mL)?
(b) What volume (mL) of whiskey would raise his blood alcohol level to 8.0×10^{-4} g/mL, the level at which a person is considered intoxicated, if he drank it all at once?

12.153 Soft drinks are canned under 405.3 kPa of CO_2 and release CO_2 when opened [k_H for CO_2 at 25°C is 3.3×10^{-4} mol/(L•kPa); p_{CO_2} in air is 3×10^{-2} kPa].
(a) What amount (mol) of CO_2 is dissolved in a 355 mL can of pop (i) before it is opened; (ii) after it has gone flat?
(b) What volume (in L) would the released CO_2 occupy at 101.3 kPa and 25°C?

12.154 Gaseous O_2 in equilibrium with O_2 dissolved in water at 283 K is depicted at the right.

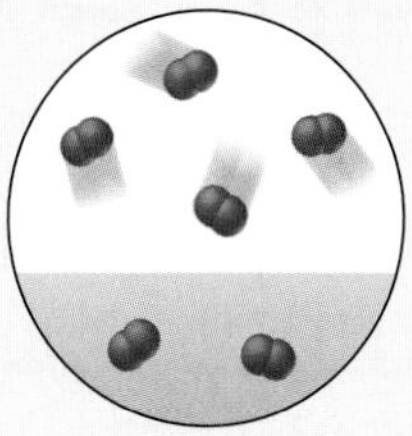

(a) Which scene below (A, B, or C) represents the system at 298 K?
(b) Which scene represents the system when the pressure of O_2 is increased by half?

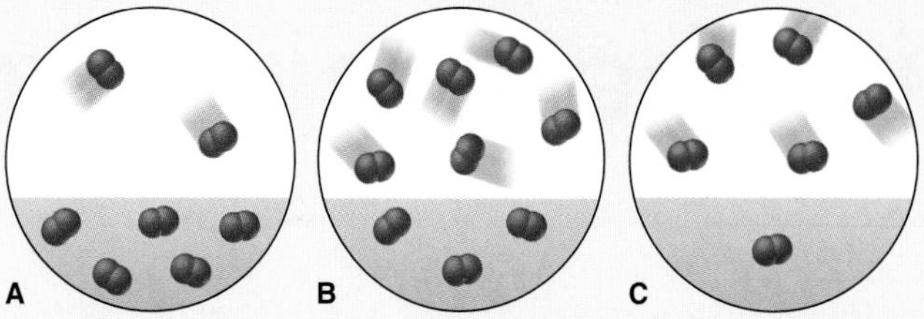

Periodic Patterns in the Main-Group Elements

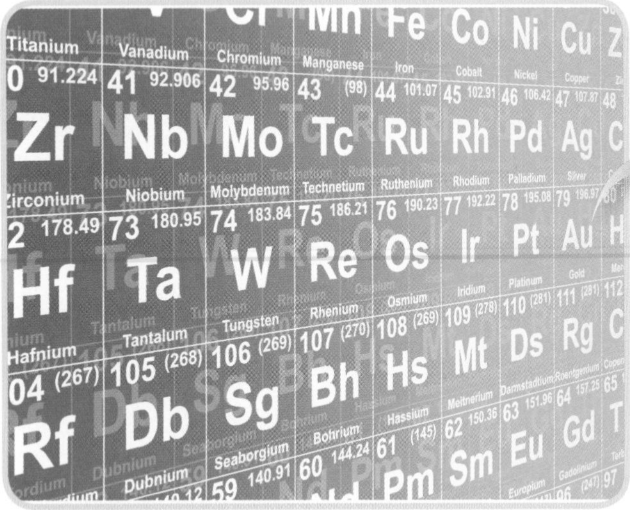

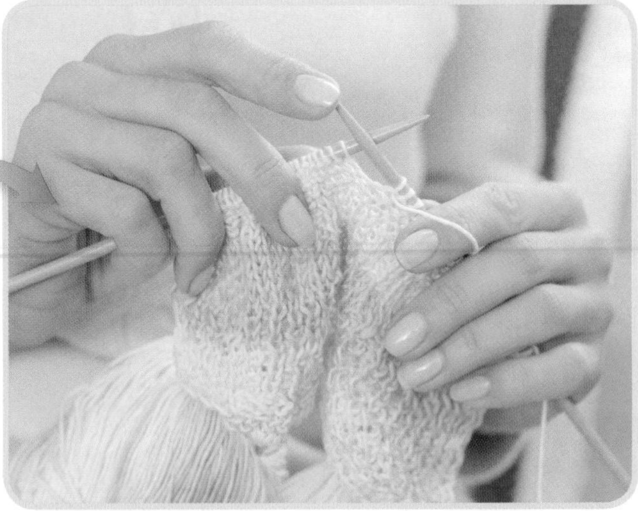

IN THIS CHAPTER . . . We apply general ideas of bonding, structure, and reactivity (from Chapters 6 to 11) to the main-group elements (also referred to as representative elements) and see how their behaviour correlates with their position in the periodic table. By the end of this chapter, you should be able to

- Use the periodic table to differentiate between the three types of hydrides
- Describe the general changes in chemical and physical properties from left to right across the periodic table
- Compare each of the eight families of main-group elements, exploring vertical trends in physical and chemical properties in each group and focusing on some especially important elements: boron, carbon, silicon, nitrogen, phosphorus, oxygen, sulfur, and the halogens

13

From an early age, we learn to look for ideas that repeat as a way to structure our learning. Whether it is in the repetition of scales to learn an instrument, or footwork to learn dance, we see patterns emerge in all our daily activities. Statisticians look for patterns, as do artists. Chemists are no different. Our most helpful tool, the periodic table, is named for the fact that patterns recur *periodically*. In this chapter, we will learn to use the periodic table to elicit the patterns it so elegantly lays out for us and determine the trends in physical and chemical properties that allowed Mendeleev to lay the groundwork for the design of this chemical map.

Concepts and Skills to Review before Studying This Chapter

- Redox behaviour and oxidation states (Section 8.5)
- Electron configurations (Section 7.2)
- Trends in atomic size, ionization energy, metallic behaviour, and electronegativity (Sections 7.3, 7.4, and 8.5)
- Trends in element properties and types of bonding (Sections 7.4 and 8.5)
- Models of ionic, covalent, and metallic bonding (Sections 8.2, 8.3, 8.7, and 11.6)
- Resonance and formal charge (Section 8.6)
- Molecular shape and polarity (Sections 9.1 and 9.2)
- Orbital hybridization and modes of orbital overlap (Sections 10.1 and 10.2)
- Phase changes and phase diagrams, intermolecular forces, and crystalline solids (Sections 11.2, 11.3, and 11.6)

13.1 Hydrogen, the Simplest Atom

A hydrogen atom consists of a nucleus with a single positive charge, surrounded by a single electron. Despite this simple structure, or perhaps because of it, hydrogen may be the most important element of all. In the Sun, hydrogen (H) nuclei combine to form helium (He) nuclei in a process that provides nearly all of Earth's energy. About 90% of the atoms in the universe are H atoms, making hydrogen the most abundant element by far. On Earth, only tiny amounts of the free, diatomic element occur naturally, but hydrogen is abundant in combination with oxygen in water. Because of its simple structure and low molar mass, nonpolar gaseous H_2 is colourless and odourless, and its extremely weak dispersion forces result in its very low melting point (−259°C) and boiling point (−253°C). Whereas the chemical properties of hydrogen isotopes are nearly the same, their physical properties are more dramatically different than those of most other elements because hydrogen is so small.

Where Does Hydrogen Fit in the Periodic Table?

Hydrogen has no perfectly suitable position in the periodic table (Figure 13.1). Depending on the property, hydrogen may fit better in group 1, 14, or 17:

- *Like the group 1 elements*, hydrogen has an outer electron configuration of ns^1, a single valence electron, and a common +1 oxidation state. However, unlike the alkali metals, hydrogen *shares* its electron with nonmetals rather than transferring its electron to them. Moreover, hydrogen has a much higher ionization energy (IE = 1311 kJ/mol) and electronegativity (χ = 2.1) than the alkali metals. By comparison, lithium has an ionization energy of only 520 kJ/mol and an electronegativity of 1.0, the highest of the alkali metals.
- *Like the group 14 elements*, hydrogen has a half-filled valence level (but with only one electron). Hydrogen also has similar ionization energy, electron affinity, electronegativity, and bond energies.
- *Like the group 17 elements*, hydrogen occurs as diatomic molecules and fills its outer (valence) level either by electron sharing or by gaining one electron from a metal to form an anion (hydride, H^-) with a 1− charge. However, whereas the monatomic halide ions (X^-) are common and stable, the H^- ion is rare and reactive. Moreover, hydrogen has a lower electronegativity (χ = 2.1) than any of the halogens (whose values of χ range from 2.2 to 4.0) and lacks their three valence electron pairs.

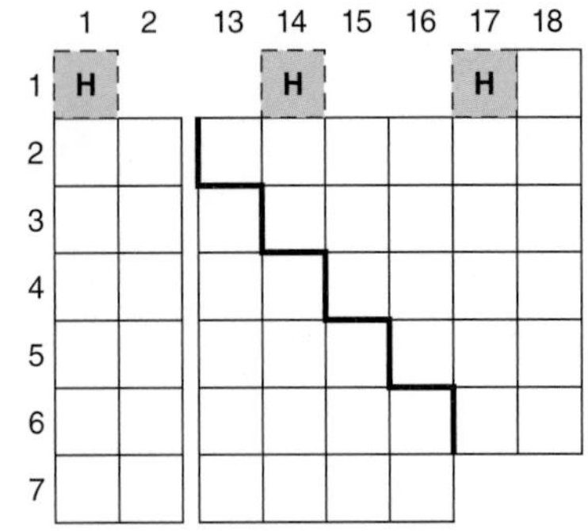

FIGURE 13.1 Where does hydrogen belong?

Hydrogen's unique behaviour is attributable to its tiny size. Hydrogen has a high IE because its electron is very close to the nucleus, with no inner electrons to shield it from the positive charge. It has a low χ (for a nonmetal), because it has only one proton to attract bonding electrons. In this chapter, hydrogen will appear in either group 1 or group 17, depending on the property being considered.

Highlights of Hydrogen Chemistry

In Chapters 11 and 12, we discussed hydrogen bonding and its impact on physical properties (melting and boiling points, heats of fusion and vaporization, and specific heat capacity) and solubility. You learned that hydrogen bonding plays a critical role in stabilizing Earth's climate, as well as your own body temperature, and you glimpsed its role in the functional structures of biomolecules. In Chapter 5 (Chemical Connections), we considered the possible future use of hydrogen as a fuel. In Chapter 23, we will see how hydrogen is produced industrially and how it is used to recover some metals from their ores. Hydrogen is very reactive, combining with nearly every element, so we focus here on the three types of hydrides.

Ionic (Saltlike) Hydrides With very reactive metals, such as those in group 1 and the larger members of group 2 (Ca, Sr, and Ba), hydrogen forms *saltlike hydrides*: white, crystalline solids composed of the metal cation and the hydride ion:

$$2Li(s) + H_2(g) \longrightarrow 2LiH(s)$$
$$Ca(s) + H_2(g) \longrightarrow CaH_2(s)$$

In water, H^- is a strong base that pulls H^+ from surrounding H_2O molecules to form H_2 and OH^-:

$$NaH(s) + H_2O(l) \longrightarrow Na^+(aq) + OH^-(aq) + H_2(g)$$

The hydride ion is also a powerful reducing agent. For example, it reduces Ti(IV) to the free metal:

$$TiCl_4(l) + 4LiH(s) \longrightarrow Ti(s) + 4LiCl(s) + 2H_2(g)$$

Covalent (Molecular) Hydrides Hydrogen reacts with nonmetals to form many *covalent hydrides*, such as CH_4, PH_3, H_2S, and HCl. Most are gases consisting of small molecules, but many hydrides of boron and carbon are liquids or solids that consist of much larger molecules. In most covalent hydrides, hydrogen has an oxidation number of +1 because the other nonmetal has a higher electronegativity.

Conditions for preparing the covalent hydrides depend on the reactivity of the other nonmetal. For example, hydrogen reacts at high temperatures (~400°C) and pressures (~25 × 10^3 kPa) with stable, triple-bonded N_2. The reaction needs a catalyst to proceed at any practical speed:

$$N_2(g) + 3H_2(g) \xrightarrow{\text{catalyst}} 2NH_3(g) \qquad \Delta_r H° = -91.8 \text{ kJ/mol}$$

Industrial facilities throughout the world use this reaction to produce millions of tonnes of ammonia each year for fertilizers, explosives, and synthetic fibres.

In comparison, hydrogen combines rapidly with reactive, single-bonded F_2, even at extremely low temperatures (−196°C):

$$F_2(g) + H_2(g) \longrightarrow 2HF(g) \qquad \Delta_r H° = -546 \text{ kJ/mol}$$

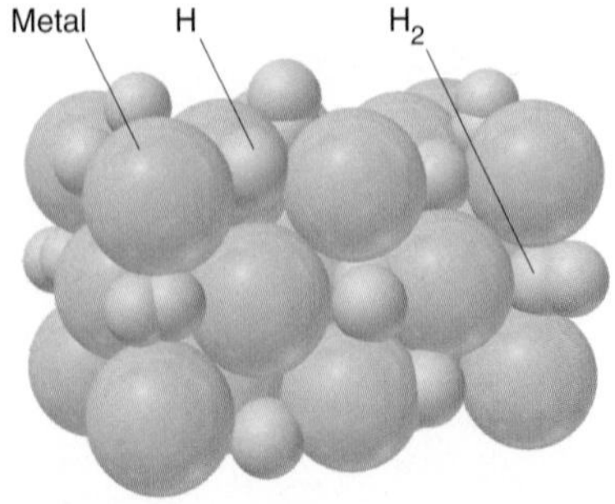

FIGURE 13.2 A metallic (interstitial) hydride

Metallic (Interstitial) Hydrides Many transition elements form *metallic (interstitial) hydrides*, in which H_2 molecules (and H atoms) occupy the holes in the metal's crystal structure (Figure 13.2). Thus, such hydrides are *not* compounds but gas-solid solutions. Also, unlike ionic and covalent hydrides, interstitial hydrides, such as $TiH_{1.7}$, typically lack a single, stoichiometric formula because the metal can incorporate variable amounts of hydrogen, depending on the pressure and temperature of the gas.

Scientists have considered using metallic hydrides as "storage containers" for hydrogen fuel in cars. Unfortunately, the metals that store the most hydrogen are expensive and heavy and release hydrogen only at very high temperatures. As pointed out by the

Canadian Hydrogen and Fuel Cell Association, a successful hydrogen storage system would have low-cost recyclable/rechargeable vessels, near-ambient temperature and pressure operation, fast recharge/discharge kinetics, impact safety, and tolerance to trace poisoning. The Waterloo Institute for Sustainable Energy (WISE), at the University of Waterloo, is developing various hydrides that have gravimetric hydrogen capacities higher than ~5 wt.% (a borderline for the FreedomCar program), with much improved hydrogen desorption properties. So far, a number of metal/nonmetal-hydrogen systems, based mainly on light elements (such as Mg, Li, Na, Al, N, and B) and sometimes with added transition metals (such as Fe, Co, and Mn), have been extensively studied.

13.2 Trends across the Periodic Table: The Period 2 Elements

Table 13.1 presents the trends in the atomic properties of the period 2 elements, lithium through neon, and the physical and chemical properties that emerge from them. In general, these trends apply to the other periods as well. Note the following properties:

- Electrons fill the one *ns* and the three *np* orbitals according to Pauli's exclusion principle and Hund's rule.
- As a result of increasing nuclear charge and the addition of electrons to orbitals of the same energy level (same *n* value), atomic size generally decreases, whereas first ionization energy and electronegativity generally increase.
- Metallic character decreases with increasing nuclear charge as period members change from metals to metalloids to nonmetals.
- Reactivity is highest at the left and right ends of the period, except for the inert noble gas, because Li and F are only one electron away from attaining a filled outer level.
- Bonding between atoms of an element changes from metallic to covalent in networks, to covalent in individual molecules, to none (noble gases exist as separate atoms). As expected, physical properties, such as melting point, change abruptly at the network/molecule boundary, which occurs in period 2 between carbon (solid) and nitrogen (gas).
- Bonding between each element and an active nonmetal changes from ionic to polar covalent to covalent. Bonding between each element and an active metal changes from metallic to polar covalent to ionic.
- The acid-base behaviour of the common element oxide in water changes from basic to amphoteric to acidic as the bond between the element and oxygen changes from ionic to covalent.
- Reducing strength decreases through the metals, and oxidizing strength increases through the nonmetals. In period 2, common oxidation numbers equal the group number for Li and Be and the group number minus 18 for O and F. Boron has several oxidation numbers, Ne has none, and C and N show all possible oxidation numbers for their groups.

The Anomalous Behaviour of Period 2 Members One property that is not shown in Table 13.1 is the *anomalous (unrepresentative) behaviour of the period 2 elements within their groups*. This behaviour arises from the relatively *small size* and *small number of orbitals* in the outer energy level of these elements.

1. *Anomalous properties of lithium.* In group 1, Li is the only member that forms a simple oxide and nitride, Li_2O and Li_3N, on reaction with O_2 and N_2 in air. Li is also the only member that forms molecular compounds with organic halides:

$$2Li(s) + CH_3CH_2Cl(g) \longrightarrow CH_3CH_2Li(s) + LiCl(s)$$

Because of its small size, Li^+ has a relatively high charge density. Therefore, it can deform nearby electron clouds to a much greater extent than the other group 1 ions can, which increases orbital overlap and gives many lithium salts significant covalent character. Thus, LiCl, LiBr, and LiI are much more soluble in polar organic solvents, such as ethanol and acetone, than the halides of Na and K are, because

TABLE 13.1 Trends in Atomic, Physical, and Chemical Properties of the Period 2 Elements

Group: **Element/Atom. No.:**	**1** **Lithium (Li) $Z = 3$**	**2** **Beryllium (Be) $Z = 4$**	**13** **Boron (B) $Z = 5$**	**14** **Carbon (C) $Z = 6$**
Atomic Properties				
Condensed electron configuration; partial orbital diagram	[He] $2s^1$ 2s: [↑] 2p: [ ][ ][ ]	[He] $2s^2$ 2s: [↑↓] 2p: [ ][ ][ ]	[He] $2s^2 2p^1$ 2s: [↑↓] 2p: [↑][ ][ ]	[He] $2s^2 2p^2$ 2s: [↑↓] 2p: [↑][↑][ ]
Physical Properties				
Appearance				
Metallic character	Metal	Metal	Metalloid	Nonmetal
Hardness	Soft	Hard	Very hard	Graphite: soft Diamond: extremely hard
Melting point/ boiling point	Low melting point for a metal	High melting point	Extremely high melting point	Extremely high melting point
Chemical Properties				
General reactivity	Reactive	Low reactivity at room temperature	Low reactivity at room temperature	Low reactivity at room temperature; graphite more reactive
Bonding among atoms of element	Metallic	Metallic	Network covalent	Network covalent
Bonding with nonmetals	Ionic	Polar covalent	Polar covalent	Covalent (π bonds common)
Bonding with metals	Metallic	Metallic	Polar covalent	Polar covalent
Acid-base behaviour of common oxide	Strongly basic	Amphoteric	Very weakly acidic	Very weakly acidic
Redox behaviour (oxidation number)	Strong reducing agent (+1)	Moderately strong reducing agent (+2)	Complex hydrides good reducing agents (+3, −3)	Every oxidation state from +4 to −4
Relevance/Uses of Element and Compounds	Li soaps for auto grease; thermonuclear bombs; high-voltage, low-weight batteries; treatment of bipolar disorders (Li_2CO_3)	Rocket nose cones; alloys for springs and gears; nuclear reactor parts; X-ray tubes	Cleaning agent (borax); eyewash, antiseptic (boric acid); armour (B_4C); borosilicate glass; plant nutrient	Graphite: lubricant, structural fibre Diamond: jewellery, cutting tools, protective films Limestone ($CaCO_3$): building material, component of cement Organic compounds: drugs, fuels, textiles, biomolecules

Group:	15	16	17	18
Element/Atom. No.:	**Nitrogen (N) $Z = 7$**	**Oxygen (O) $Z = 8$**	**Fluorine (F) $Z = 9$**	**Neon (Ne) $Z = 10$**
Atomic Properties				
Condensed electron configuration; partial orbital diagram	[He] $2s^2 2p^3$ 2s: ↑↓ 2p: ↑ ↑ ↑	[He] $2s^2 2p^4$ 2s: ↑↓ 2p: ↑↓ ↑ ↑	[He] $2s^2 2p^5$ 2s: ↑↓ 2p: ↑↓ ↑↓ ↑	[He] $2s^2 2p^6$ 2s: ↑↓ 2p: ↑↓ ↑↓ ↑↓
Physical Properties				
Appearance				
Metallic character	Nonmetal	Nonmetal	Nonmetal	Nonmetal
Hardness	—	—	—	—
Melting point/ boiling point	Very low melting point and boiling point	Very low melting point and boiling point	Very low melting point and boiling point	Extremely low melting point and boiling point
Chemical Properties				
General reactivity	Inactive at room temperature	Very reactive	Extremely reactive	Chemically inert
Bonding among atoms of element	Covalent N_2 molecules	Covalent O_2 (or O_3) molecules	Covalent F_2 molecules	None; separate atoms
Bonding with nonmetals	Covalent (π bonds common)	Covalent (π bonds common)	Covalent	None
Bonding with metals	Ionic/polar covalent; anions with active metals	Ionic	Ionic	None
Acid-base behaviour of common oxide	Strongly acidic (NO_2)	—	Acidic	None
Redox behaviour (oxidation number)	Every oxidation state from +5 to −3	O_2 (and O_3) very strong oxidizing agents (−2)	Strongest oxidizing agent (−1)	None
Relevance/Uses of Element and Compounds	Component of proteins, nucleic acids; ammonia for fertilizers, explosives; oxides involved in manufacturing and air pollution (smog, acid rain)	Component of biological macromolecules; final oxidizer in residential, industrial, and biological energy production	Manufacture of coatings (Teflon); glass etching (HF); refrigerants involved in ozone depletion (CFCs); dental protection (NaF, SnF_2)	Electrified gas in advertising signs

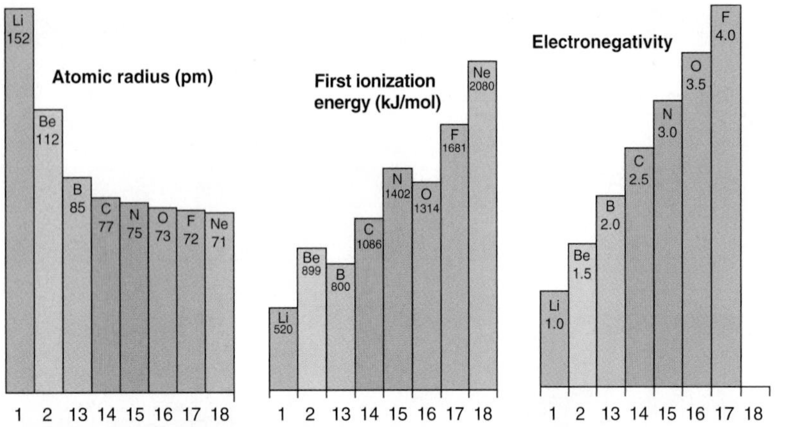

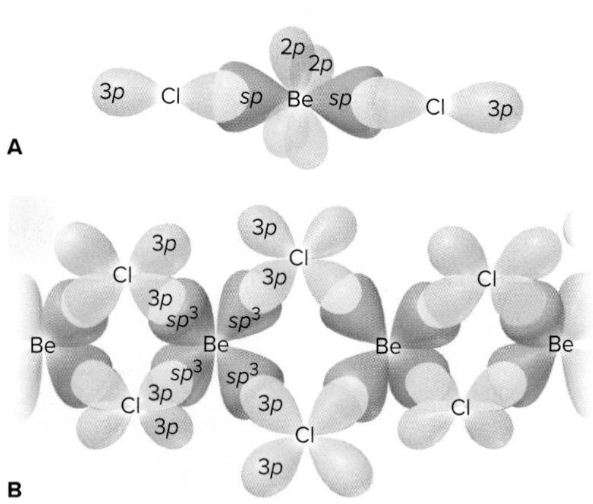

FIGURE 13.3 Overcoming electron deficiency in beryllium chloride

the lithium halide bond polarity interacts with these solvents through dipole-dipole forces. The small, highly positive Li^+ makes dissociation of Li salts into ions more difficult in water; thus, the fluoride, carbonate, hydroxide, and phosphate of Li are much less soluble in water than those of Na and K.

2. *Anomalous properties of beryllium.* In group 2, beryllium displays especially anomalous behaviour. Because of the extremely high charge density of Be^{2+}, the discrete ion does not exist, and *all Be compounds exhibit covalent bonding.* With only two valence electrons, Be does not attain an octet in its simple gaseous compounds (Section 8.6). When it bonds to an electron-rich atom, however, this electron deficiency is overcome as the gas condenses. Consider beryllium chloride ($BeCl_2$). At temperatures greater than 900°C, it consists of linear molecules in which two *sp* hybrid orbitals hold four electrons around the central Be (Figure 13.3A). As it cools, the molecules bond together, solidifying in long chains, with each Cl bridging two Be atoms. Each Be is sp^3 hybridized to finally attain an octet (Figure 13.3B).

3. *Anomalous properties of other period 2 elements.* In group 13, boron is the only member to form a complex family of compounds with metals, as well as covalent compounds with hydrogen (boranes). Carbon, in group 14, shows extremely unusual behaviour: it bonds to itself (and a small number of other elements) so extensively and diversely that it gives rise to countless organic compounds. In group 15, triple-bonded, unreactive, gaseous nitrogen is dramatically different from its reactive, solid family members. Oxygen, the only gas in group 16, is much more reactive than sulfur and the other members. In group 17, fluorine is so electronegative that it reacts violently with water, and it is the only member that forms a weak hydrohalic acid, HF.

The chemistry of a first-row (second-period) element often has similarities to the chemistry of the second-row (third-period) element that is one column to the right of it in the periodic table. This is called a *diagonal relationship.* Diagonal adjacent pairs, such as Li and Mg, Be and Al, and B and Si, exhibit similar properties, and charge density has been found to be a factor for diagonal relationships to exist. We will discuss these relationships further in subsequent sections.

13.3 Group 1: The Alkali Metals

The first group of elements in the periodic table is named for the alkaline (basic) nature of their oxides and for the basic solutions that the elements form in water. Group 1 provides the best example of regular trends with no significant exceptions. All the elements in the group—lithium (Li); sodium (Na); potassium (K); rubidium (Rb); cesium (Cs); and rare, radioactive francium (Fr)*—are very reactive metals.

*Francium is so rare (estimates indicate only 15 g of francium in the top kilometre of Earth's crust) that its properties are largely unknown. Therefore, we will mention it only occasionally in the discussion.

The Family Portrait of Group 1 is the first in a series that provides an overview of each of the main groups, summarizing the key atomic, physical, and chemical properties.

Lithium floating in oil floating on water

Why Are the Alkali Metals Soft, Low Melting, and Lightweight?

Alkali metals have some properties that are unique for metals:

- They are unusually soft and can be easily cut with a knife. Na has the consistency of cold butter, and K can be squeezed like clay.
- Alkali metals have lower melting and boiling points than any other group of metals. Li is the only alkali metal that has a melting point above 100°C, and Cs melts only a few degrees above room temperature.
- Alkali metals also have lower densities than most metals. Li floats on lightweight household oil (*see margin photo*).

The unusual physical behaviour of the alkali metals can be traced to their atomic size—the largest in their respective periods—and to the ns^1 valence electron configuration. Because the single valence electron is relatively far from the nucleus, there is only a weak attraction between the delocalized electrons and the metal-ion cores in the crystal structure. This weak metallic attraction means that the alkali metal crystal structure can be easily deformed or broken down, which results in a soft consistency and low melting point. The low densities of the alkali metals result from their having the lowest molar masses and largest atomic radii (and thus volumes) in their periods.

Why Are the Alkali Metals So Reactive?

The alkali metals are extremely reactive elements. They are *powerful reducing agents* (as we will study in Section 19.1), always occurring in nature as 1+ cations rather than free metals. (As we will discuss in Section 23.4, highly endothermic reduction processes are required to prepare the free metals industrially from their molten salts.) Some examples of their reactivity follow:

- The alkali metals (E) reduce halogens to form ionic solids in highly exothermic reactions:

$$2E(s) + X_2 \longrightarrow 2EX(s) \qquad (X = F, Cl, Br, I)$$

- They reduce hydrogen in water, reacting vigorously (Rb and Cs explosively) to form H_2 and a metal hydroxide solution (*see margin photo*):

$$2E(s) + 2H_2O(l) \longrightarrow 2E^+(aq) + 2OH^-(aq) + H_2(g)$$

Potassium reacting with water

- They reduce molecular hydrogen to form ionic hydrides:

$$2E(s) + H_2(g) \longrightarrow 2EH(s)$$

- They reduce O_2 in air and, thus, tarnish rapidly. Because of this reactivity, Na and K are usually kept under mineral oil (an unreactive liquid) in the laboratory, and Rb and Cs are handled with gloves under an inert argon atmosphere.

The ns^1 configuration, which is the basis for their physical properties, is also why alkali metals form salts so readily. Properties based on the ns^1 configuration are involved in each of the following steps that take place in the reaction between an alkali metal and a nonmetal:

1. *The solid metal separates into gaseous atoms.* Consistent with the alkali metals' low melting and boiling points, their weak metallic bonding leads to *low values for* $\Delta_{atom}H$ (the heat needed to convert the solid into individual gaseous atoms), which decrease down the group:

$$E(s) \longrightarrow E(g) \qquad \Delta_{atom}H \text{ (Li > Na > K > Rb > Cs)}$$

2. *The metal atom becomes a cation after transferring its outer electron to the nonmetal.* Alkali metals have *low ionization energies* (the lowest in their periods)

Family Portrait Group 1: The Alkali Metals

Key Atomic Properties, Physical Properties, and Reactions

KEY	Atomic no. / Symbol / Atomic mass / Valence e^- configuration / (Common oxidation states)
3	Li, 6.941, $2s^1$, (+1)
11	Na, 22.99, $3s^1$, (+1)
19	K, 39.10, $4s^1$, (+1)
37	Rb, 85.47, $5s^1$, (+1)
55	Cs, 132.9, $6s^1$, (+1)
87	Fr, (223), $7s^1$, (+1)

No sample available

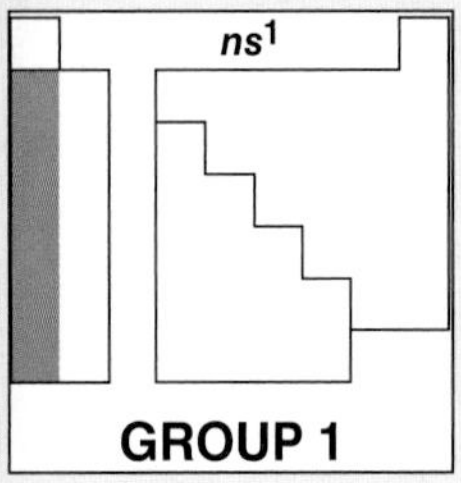

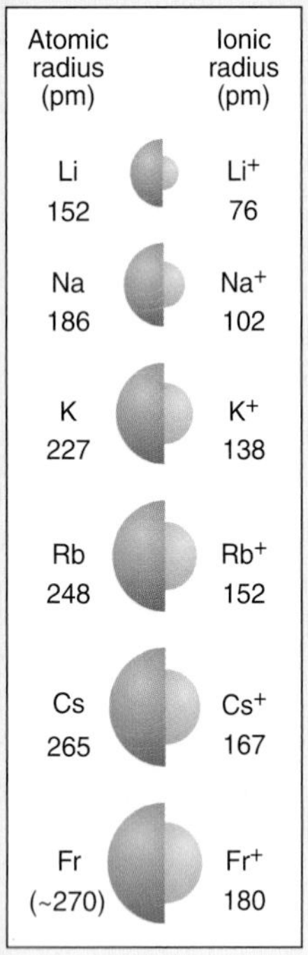

Atomic Properties

The group electron configuration is ns^1. All members have the +1 oxidation state and form an E^+ ion. Atoms have the largest size and lowest IE and χ in their periods. Down the group, atomic and ionic sizes increase, whereas IE and χ decrease.

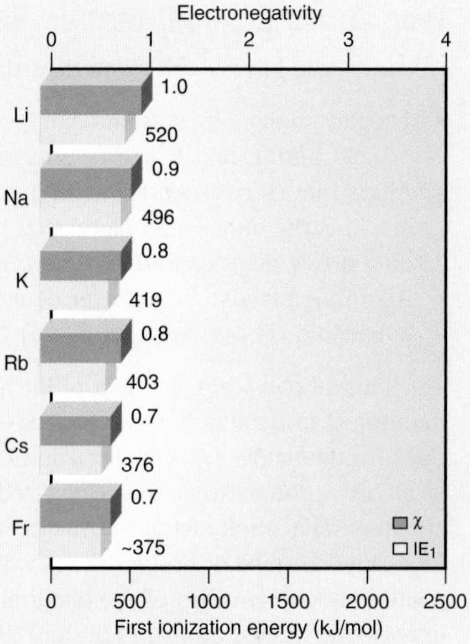

Physical Properties

Metallic bonding is relatively weak because there is only one valence electron. Therefore, these metals are soft with relatively low melting and boiling points. These values decrease down the group because larger atom cores attract delocalized electrons less strongly. Large atomic size and low atomic mass result in low density; thus, density generally increases down the group because mass increases more than size.

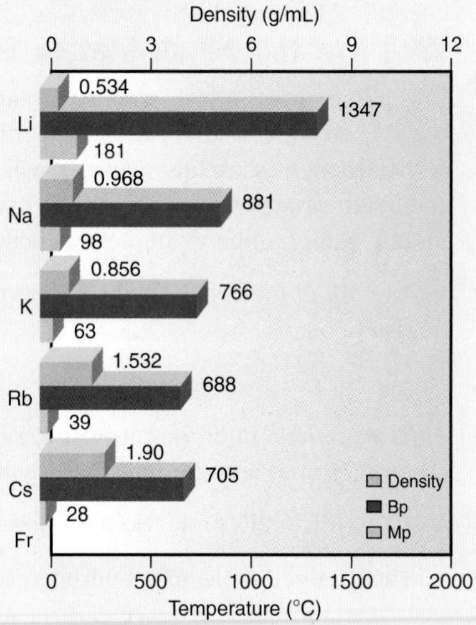

Reactions

1. The alkali metals (E)* reduce H in H_2O from the +1 to the 0 oxidation state:

$$2E(s) + 2H_2O(l) \longrightarrow 2E^+(aq) + 2OH^-(aq) + H_2(g)$$

The reaction becomes more vigorous down the group.

2. The alkali metals reduce oxygen, but the product depends on the metal. Li forms the oxide, Li_2O; Na forms the peroxide, Na_2O_2 (oxidation number of O = −1); K, Rb, and Cs form the superoxide, EO_2 (oxidation number of O = $-\frac{1}{2}$):

$$4Li(s) + O_2(g) \longrightarrow 2Li_2O(s)$$

$$2Na(s) + O_2(g) \longrightarrow Na_2O_2(s)$$

$$K(s) + O_2(g) \longrightarrow KO_2(s)$$

In emergency breathing units, KO_2 reacts with H_2O and CO_2 in exhaled air to release O_2 (Section 23.4).

3. The alkali metals reduce hydrogen to form ionic (saltlike) hydrides:

$$2E(s) + H_2(g) \longrightarrow 2EH(s)$$

NaH is an industrial base and reducing agent that is used to prepare other reducing agents, such as $NaBH_4$.

4. The alkali metals reduce halogens to form ionic halides:

$$2E(s) + X_2 \longrightarrow 2EX(s) \qquad (X = F, Cl, Br, I)$$

*Throughout the chapter, we use E to represent any element in a group.

and form *cations with small radii*, since a great decrease in size occurs when the outer electron is lost: the volume of the Li^+ ion is less than 13% of the volume of the Li atom! Thus, group 1 ions are small spheres with considerable charge density.

3. *The resulting cations and anions attract one another to form an ionic solid.* Group 1 salts have *high lattice energies* because the small cations lie close to the anions. Thus, the endothermic atomization and ionization steps are easily outweighed by the highly exothermic formation of the solid. For a given anion, the trend in lattice energy is the inverse of the trend in cation size: *as the cation becomes larger, the lattice energy becomes smaller.* This steady decrease in lattice energy within the group 1 and 2 chlorides is shown in Figure 13.4.

Despite the strong ionic attractions in the solid, *nearly all group 1 salts are water soluble.* The attraction between the ions and the water molecules creates a highly exothermic heat of hydration ($\Delta_{hydr}H$), and a large increase in entropy occurs when ions in the organized crystal become dispersed and hydrated in solution; together, these factors outweigh the high lattice energy.

The magnitude of the hydration energy *decreases* as the ionic size increases:

$$E^+(g) \longrightarrow E^+(aq) \qquad \Delta H = -\Delta_{hydr}H \;(Li^+ < Na^+ < K^+ < Rb^+ < Cs^+)$$

Interestingly, the *smaller* ions attract water molecules strongly enough to form larger *hydrated ions.* This size trend has a major effect on the function of nerves, kidneys, and cell membranes, because the *sizes* of $Na^+(aq)$ and $K^+(aq)$, the most common cations in cell fluids, influence their movement in and out of cells.

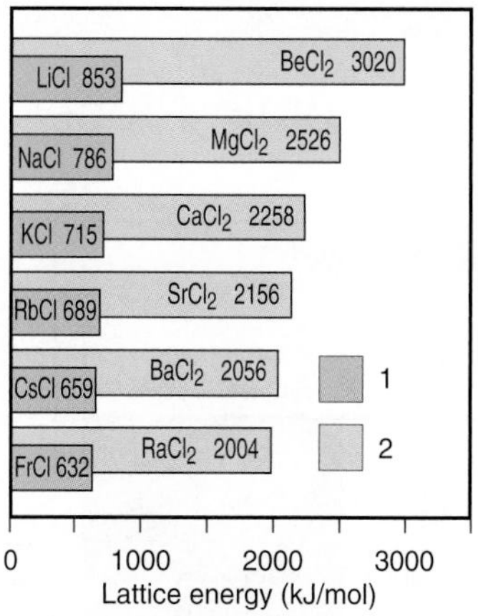

FIGURE 13.4 Lattice energies of the group 1 and 2 chlorides

13.4 Group 2: The Alkaline Earth Metals

The group 2 elements are called *alkaline earth metals* because their oxides give basic (alkaline) solutions and melt at such high temperatures that they remained as solids ("earths") in the alchemists' fires. This group is a fascinating collection of elements: rare beryllium (Be), common magnesium (Mg) and calcium (Ca), less familiar strontium (Sr) and barium (Ba), and radioactive radium (Ra). The Group 2 Family Portrait presents an overview of these elements.

How Do the Physical Properties of the Alkaline Earth and Alkali Metals Compare?

In general terms, the elements in groups 1 and 2 behave as close cousins. Whatever differences occur between the groups are those of degree, not kind, and are due to the change in outer electron configuration: ns^2 versus ns^1. Two electrons are available from each group 2 atom for metallic bonding, and the nucleus contains one additional positive charge. These factors strengthen the attraction between delocalized electrons and atom cores. Consequently, melting and boiling points are much higher for group 2 metals than for the corresponding group 1 metals; in fact, the group 2 elements melt at around the same temperatures as the group 1 elements boil! Compared with transition metals such as iron and chromium, the alkaline earth metals are soft and lightweight, but their stronger metallic bonding and smaller atomic sizes make them harder and denser than the alkali metals.

How Do the Chemical Properties of the Alkaline Earth and Alkali Metals Compare?

Group 2 elements have smaller atomic radii and higher ionization energies than group 1 elements. This is because the second valence electron in an alkaline earth metal lies in the same sublevel as the first. Therefore, it is not shielded from the additional nuclear charge very well, so Z_{eff} is greater. This greater attraction of the

Family Portrait Group 2: The Alkaline Earth Metals

Key Atomic Properties, Physical Properties, and Reactions

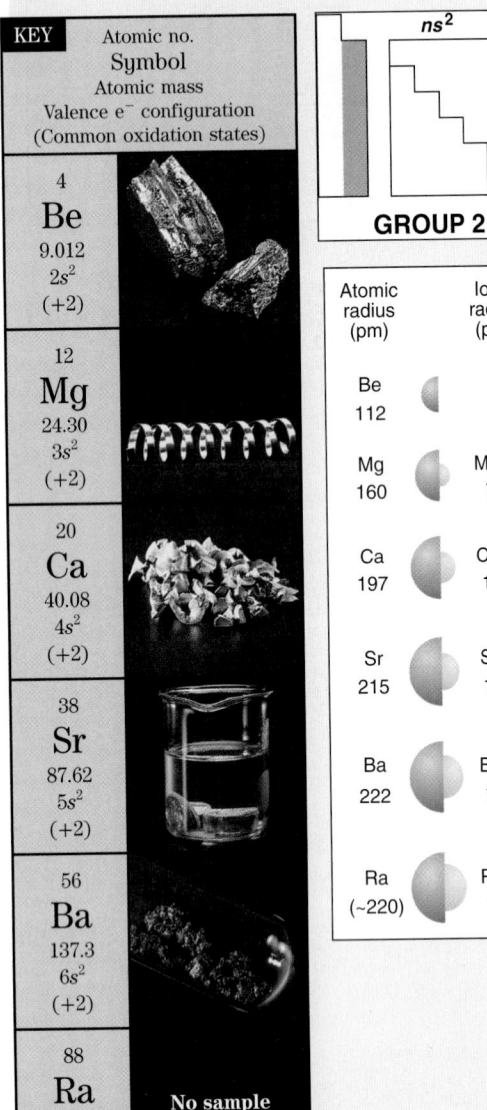

KEY: Atomic no. / Symbol / Atomic mass / Valence e$^-$ configuration / (Common oxidation states)
4 **Be** 9.012 $2s^2$ (+2)
12 **Mg** 24.30 $3s^2$ (+2)
20 **Ca** 40.08 $4s^2$ (+2)
38 **Sr** 87.62 $5s^2$ (+2)
56 **Ba** 137.3 $6s^2$ (+2)
88 **Ra** (226) $7s^2$ (+2)

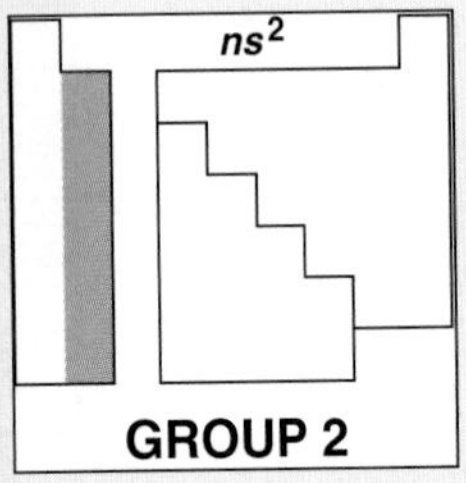

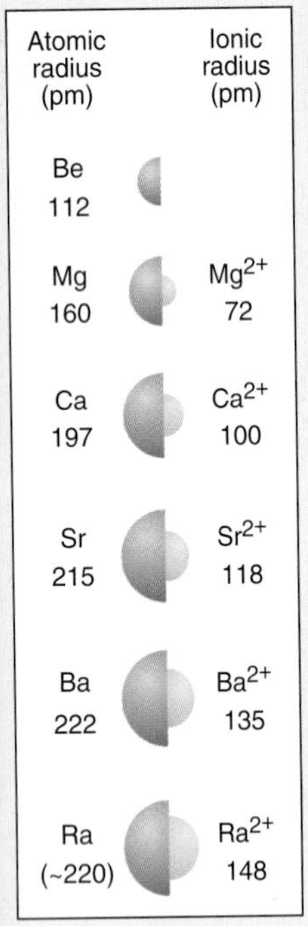

Atomic Properties

The group electron configuration is ns^2 (filled *ns* sublevel). All members have the +2 oxidation state and, except for Be, form compounds with an E^{2+} ion. Atomic and ionic sizes increase down the group but are smaller than for the corresponding group 1 elements. IE and χ decrease down the group but are higher than for the corresponding group 1 elements.

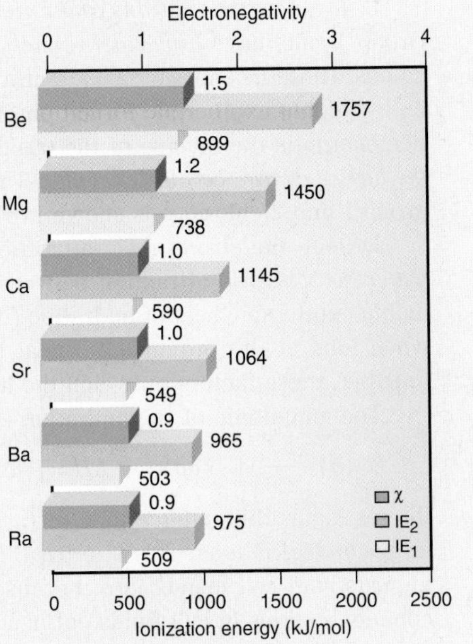

Physical Properties

Metallic bonding involves two valence electrons. These metals are still relatively soft but are much harder than the group 1 metals. Melting and boiling points generally decrease, and densities generally increase, down the group. These values are much higher than for group 1 elements, and the trend is not as regular.

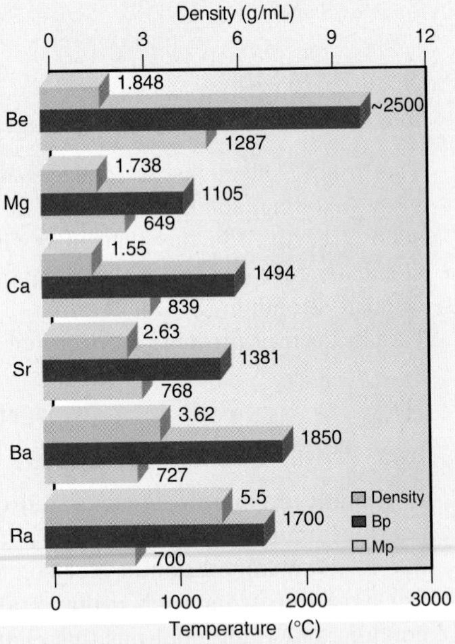

Reactions

1. The metals reduce O_2 to form the oxides:

$$2E(s) + O_2(g) \longrightarrow 2EO(s)$$

Ba also forms the peroxide BaO_2.

2. The larger metals reduce water to form hydrogen gas:

$$E(s) + 2H_2O(l) \longrightarrow E^{2+}(aq) + 2OH^-(aq) + H_2(g) \quad (E = Ca, Sr, Ba)$$

Be and Mg form an oxide coating that allows only a slight reaction.

3. The metals reduce halogens to form ionic halides:

$$E(s) + X_2 \longrightarrow EX_2(s) \quad [X = F \text{ (except with Be)}, Cl, Br, I]$$

4. Most of the elements reduce hydrogen to form ionic hydrides:

$$E(s) + H_2(g) \longrightarrow EH_2(s) \quad (E = \text{all except Be})$$

5. The elements reduce nitrogen to form ionic nitrides:

$$3E(s) + N_2(g) \longrightarrow E_3N_2(s)$$

6. Except for amphoteric BeO, the element oxides are basic:

$$EO(s) + H_2O(l) \longrightarrow E^{2+}(aq) + 2OH^-(aq)$$

$Ca(OH)_2$ is a component of cement and mortar.

7. All carbonates undergo thermal decomposition to the oxide:

$$ECO_3(s) \xrightarrow{\Delta} EO(s) + CO_2(g)$$

This reaction is used to produce CaO (lime) in huge amounts from naturally occurring limestone.

nucleus for the valence electrons makes those electrons more difficult to remove. Despite the higher IEs, *all the alkaline earth metals (except Be) form ionic compounds* as 2+ cations. Beryllium behaves differently because so much energy is needed to remove two electrons from this tiny atom that it never forms a discrete Be^{2+} ion, and its bonds are polar covalent.

Like the alkali metals, the alkaline earth metals are *strong reducing agents*:

- Each alkaline earth metal reduces O_2 in air to form the oxide. (Ba also forms the peroxide BaO_2.)
- Except for Be and Mg, which form adherent oxide coatings, the alkaline earth metals reduce H_2O at room temperature to form H_2.
- Except for Be, the alkaline earth metals reduce the halogens, N_2, and H_2 to form ionic compounds.

The group 2 oxides are strongly basic (except for amphoteric BeO) and react with acidic oxides to form salts, such as sulfites and carbonates, for example,

$$SrO(s) + CO_2(g) \longrightarrow SrCO_3(s)$$

Natural carbonates, such as limestone and marble, are major structural materials and the commercial sources for most group 2 compounds. Calcium carbonate is heated to obtain calcium oxide (lime). This important industrial compound has essential roles in steelmaking, water treatment, and smokestack scrubbing, and it is used to make glass, whiten paper, and neutralize acidic soil.

The alkaline earth metals are reactive because the high lattice energies of their compounds more than compensate for the large total IE required to form 2+ cations (Section 8.2). Group 2 salts have much higher lattice energies than group 1 salts (see Figure 13.4) because the group 2 cations are smaller and doubly charged, resulting in much higher charge densities, which lead to the lower solubility of group 2 salts in water. Although higher charge density increases the heat of hydration, it increases lattice energy even more. In fact, most group 2 fluorides, carbonates, phosphates, and sulfates have very low solubility, unlike the corresponding group 1 compounds. Nevertheless, the ion-dipole attractions between 2+ ions and water molecules are so strong that many slightly soluble group 2 salts crystallize as hydrates; two examples are Epsom salt ($MgSO_4 \cdot 7H_2O$), which is used as a soak for inflammations, and gypsum ($CaSO_4 \cdot 2H_2O$), which is used as the bonding material between the paper sheets in drywall and as the cement in surgical casts.

Diagonal Relationships: Lithium and Magnesium

One of the clearest ways in which atomic properties influence chemical behaviour appears in the **diagonal relationships**, the similarities between a period 2 element and an element diagonally down and to the right in period 3 (Figure 13.5).

The first of three such relationships occurs between Li and Mg, which have similar atomic and ionic sizes. Note that *one period down increases atomic (or ionic) size and one group to the right decreases it.* Thus, the radius of Li is 152 pm, and the radius of Mg is 160 pm; the radius of Li^+ is 76 pm, and the radius of Mg^{2+} is 72 pm. From similar atomic properties emerge similar chemical properties. Both elements form nitrides with N_2, hydroxides and carbonates that decompose easily with heat, organic compounds with a polar covalent metal-carbon bond, and salts with similar solubilities. We will discuss the diagonal relationships between Be and Al and between B and Si in upcoming sections.

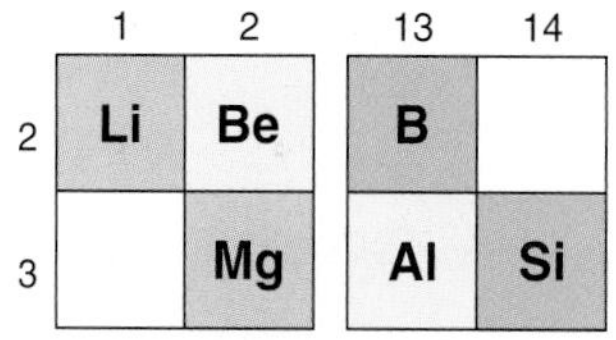

FIGURE 13.5 Three diagonal relationships in the periodic table

13.5 Group 13: The Boron Family

The third family of main-group elements contains both familiar and unusual members, which engage in some exotic bonding and have strange physical properties. Boron (B) heads the family, but, as we will see, its properties are not representative. Metallic aluminum (Al) has properties that are more typical of the group, but its great abundance and importance contrast with the rareness of gallium (Ga); indium (In); thallium (Tl); and, of course, the recently synthesized Element 113. The atomic, physical, and chemical properties of these elements are summarized in the Group 13 Family Portrait.

How Do Transition Elements Influence Group 13 Properties?

Group 13 is the first of the *p* block. If you look at the main groups only, the elements in this group seem to be just one group away from those in group 2. In period 4 and higher, however, a large gap separates the two groups (see Figure 7.11). The gap holds 10 transition elements (*d* block) in each of periods 4 to 7 and an additional 14 inner transition elements (*f* block) in periods 6 and 7. Recall, from Section 7.1, that *d* and *f* electrons penetrate very little and so spend very little time near the nucleus. Thus, the heavier group 13 members—Ga, In, and Tl (and presumably Element 113)—have nuclei with many more protons, but their outer (*s* and *p*) electrons are only partially shielded from the much higher positive charge; as a result, these elements have greater Z_{eff} values than the two lighter members of the group. This stronger nuclear attraction affects many properties of the heavier group 13 elements. For example, Ga, In, and Tl have smaller atomic radii and larger ionization energies and electronegativities than expected. The deviations for Ga and In reflect the *d*-block *contraction in size* and can be explained by the *d* electrons' limited shielding of the 10 additional protons in the first transition series. Similarly, the deviations for Tl reflect the *f*-block (lanthanide) contraction and can be explained by the *f* electrons' limited shielding of the 14 additional protons in the first inner transition series.

Physical properties are influenced by the type of bonding in the element. Boron is a network covalent metalloid—black, hard, and with very high melting point. The other group members are metals—shiny and relatively soft and with low melting point. Aluminum's low density and three valence electrons make it an exceptional conductor: for a given mass, aluminum conducts a current twice as effectively as copper. Gallium has the largest liquid temperature range of any element: it melts at skin temperature (*see photo in Family Portrait*) but does not boil until 2403°C. Its metallic bonding is too weak to keep the Ga atoms fixed when the solid is warmed, but strong enough to keep them from escaping the molten metal until it is very hot. An important use of gallium is in the production of gallium arsenide (GaAs) semiconductors. Since GaAs chips create a current when they absorb light and emit light when a current is supplied, they are used in light-powered calculators, wrist watches, and solar panels.

What New Features Appear in the Chemical Properties of Group 13?

Looking down group 13, we see a wide range of chemical behaviour. Boron, the anomalous member from period 2, is the first metalloid we have encountered so far and the only metalloid in the group. It is much less reactive at room temperature than the other members and forms covalent bonds exclusively. Although aluminum acts like a metal physically, its halides exist in the gas phase as covalent *dimers*—molecules formed by joining two identical smaller molecules (Figure 13.6)—and its oxide is amphoteric rather than basic. Most of the other group 13 compounds are ionic, but they have more covalent character than similar group 2 compounds. Because the group 13 cations are smaller and more highly charged than the group 2 cations, they polarize an anion's electron cloud more effectively.

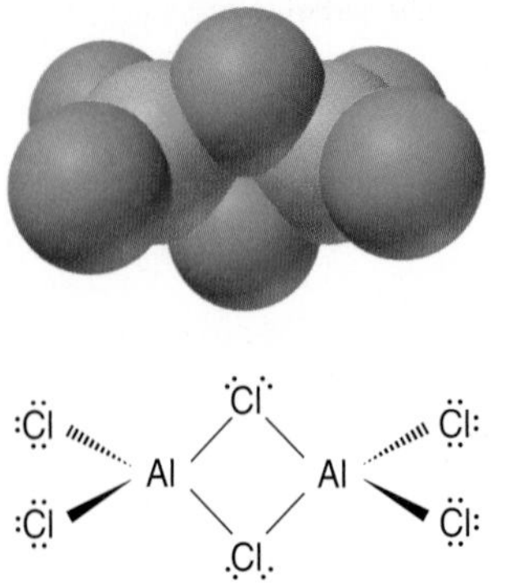

FIGURE 13.6 The dimeric structure of gaseous aluminum chloride

The redox behaviour of the elements in this group provides a chance to note three general principles that appear in groups 13 to 16:

1. *Presence of multiple oxidation states.* Many of the larger elements in these groups also have an important oxidation state *12 lower than the group number.* The lower state occurs when the atoms lose their *np* electrons only, not their two *ns* electrons. This fact is often called the *inert pair effect* (Section 7.4).

2. *Increasing prominence of the lower oxidation state.* When a group exhibits more than one oxidation state, *the lower state becomes more prominent going down the group.* In group 13, for instance, all the members exhibit the +3 state, but the

Group 13: The Boron Family

Key Atomic Properties, Physical Properties, and Reactions

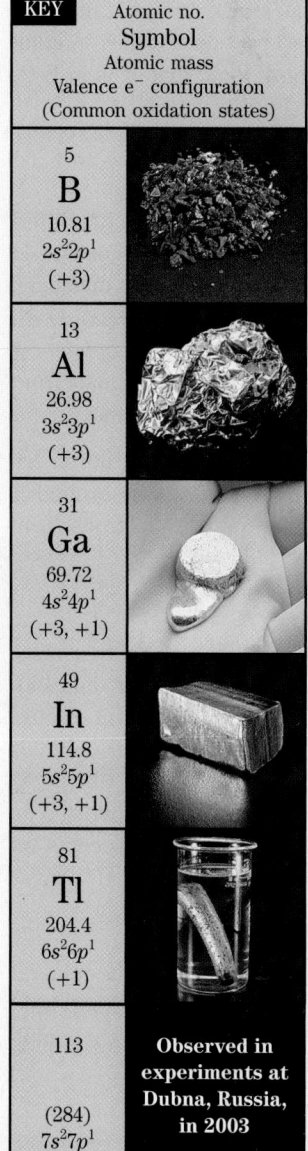

KEY: Atomic no. / Symbol / Atomic mass / Valence e^- configuration / (Common oxidation states)

Atomic no.	Symbol	Atomic mass	Valence e^- configuration	(Common oxidation states)	
5	B	10.81	$2s^22p^1$	(+3)	
13	Al	26.98	$3s^23p^1$	(+3)	
31	Ga	69.72	$4s^24p^1$	(+3, +1)	
49	In	114.8	$5s^25p^1$	(+3, +1)	
81	Tl	204.4	$6s^26p^1$	(+1)	
113		(284)	$7s^27p^1$		Observed in experiments at Dubna, Russia, in 2003

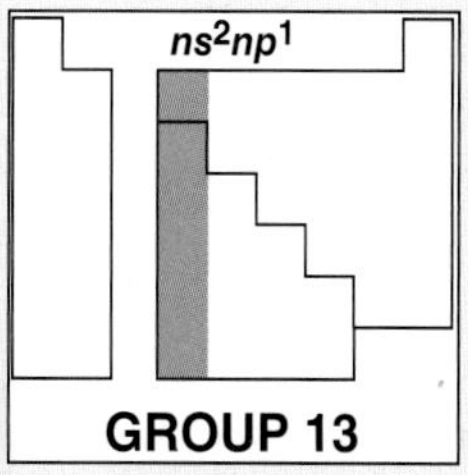

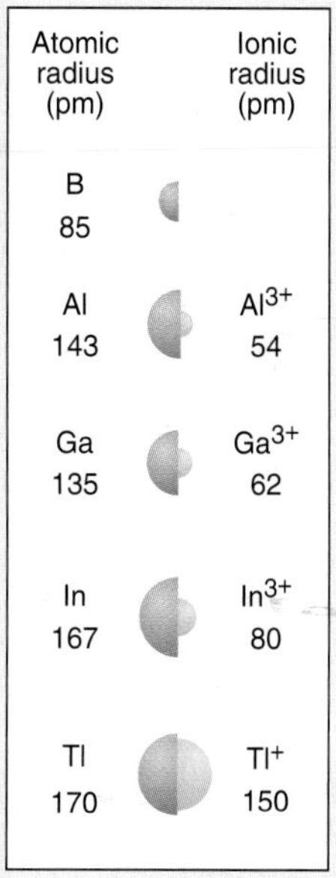

Atomic Properties

The group electron configuration is ns^2np^1. All except Tl commonly display the +3 oxidation state. The +1 state becomes more common down the group. Atomic size is smaller and χ is higher than for group 2 elements; IE is lower, however, because it is easier to remove an electron from the higher-energy *p* sublevel. Atomic size, IE, and χ do not change as expected down the group because there are intervening transition and inner transition elements.

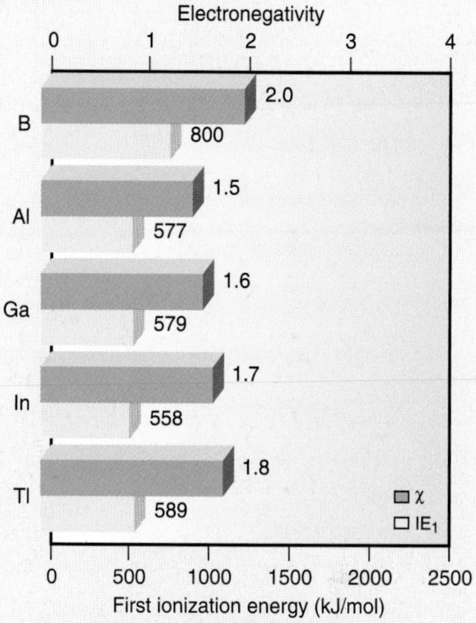

Physical Properties

Bonding changes from network covalent in B to metallic in the rest of the group. Thus, B has a much higher melting point than the others, but there is no overall trend. Boiling points decrease down the group. Densities increase down the group.

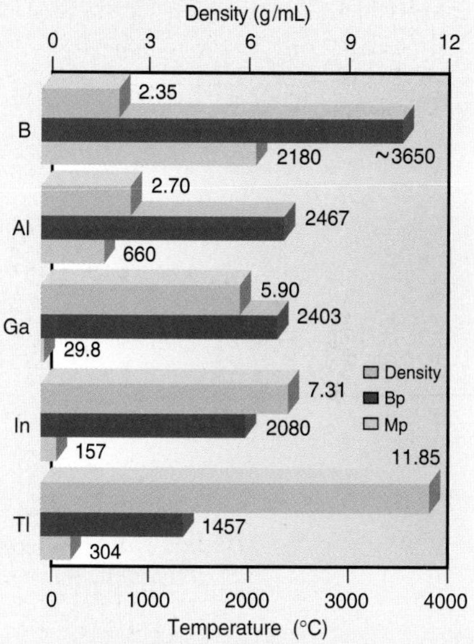

Reactions

1. The elements react sluggishly, if at all, with water:

$$2Ga(s) + 6H_2O(hot) \longrightarrow 2Ga^{3+}(aq) + 6OH^-(aq) + 3H_2(g)$$

$$2Tl(s) + 2H_2O(steam) \longrightarrow 2T^+(aq) + 2OH^-(aq) + H_2(g)$$

Al becomes covered with a layer of Al_2O_3 that prevents further reaction.

2. When strongly heated in pure O_2, all members form oxides:

$$4E(s) + 3O_2(g) \xrightarrow{\Delta} 2E_2O_3(s) \qquad (E = B, Al, Ga, In)$$

$$4Tl(s) + O_2(g) \xrightarrow{\Delta} 2Tl_2O(s)$$

Oxide acidity decreases down the group: B_2O_3 (weakly acidic) > Al_2O_3 > Ga_2O_3 > In_2O_3 > Tl_2O (strongly basic). The +1 oxide is more basic than the +3 oxide.

3. All the members reduce halogens (X_2):

$$2E(s) + 3X_2 \longrightarrow 2EX_3 \qquad (E = B, Al, Ga, In)$$

$$2Tl(s) + X_2 \longrightarrow 2TlX(s)$$

The BX_3 compounds are volatile and covalent. Trihalides of Al, Ga, and In are (mostly) ionic solids.

+1 state first appears with some of the compounds of gallium and becomes the only important state of thallium.

3. *Relative basicity of oxides. In general, oxides of an element in the lower oxidation state are more basic than oxides in the higher oxidation state.* For example, in group 13, In_2O is more basic than In_2O_3. In general, when an element has more than one oxidation state, *it acts more like a metal in its lower state,* and this too is related to ionic charge density. For example, the lower charge of In^+ does not polarize the O^{2-} ion as much as the higher charge of In^{3+} does. Thus, in compounds of general formula E_2O, the E-to-O bonding is more ionic than it is in E_2O_3 compounds, so the O^{2-} ion is more available to act as a base.

Highlights of Boron Chemistry

Like the other period 2 elements, the chemical behaviour of boron is strikingly different from the chemical behaviour of the other members of its group. As was discussed earlier, *all boron compounds are covalent,* and, unlike the other group 13 members, boron forms network covalent compounds or large molecules with metals, H, O, N, and C. The unifying feature of many boron compounds is boron's *electron deficiency,* but boron adopts two strategies to fill its outer level: accepting a bonding pair from an electron-rich atom and forming bridge bonds with an electron-poor atom.

Accepting a Bonding Pair from an Electron-Rich Atom In gaseous boron trihalides (BX_3), the B atom is electron deficient, with only six electrons around it (Section 8.6). To attain an octet, the B atom accepts a lone pair of electrons from an electron-rich atom (*blue*) and forms a covalent bond:

$$BF_3(g) + :NH_3(g) \longrightarrow F_3B—NH_3(g)$$

(Such reactions, in which one reactant accepts an electron pair from another reactant to form a covalent bond, are very widespread in inorganic, organic, and biochemical processes. They are known as *Lewis acid-base reactions.* We will discuss them in Chapters 16 and 24 and see examples of them throughout the second half of this book.)

Similarly, B has only six electrons in boric acid, $B(OH)_3$ (sometimes written as H_3BO_3). In water, the acid itself does not release a proton. Rather, it accepts an electron pair from the O in H_2O, forming a fourth bond and releasing an H^+ ion:

$$B(OH)_3(s) + H_2O(l) \rightleftharpoons B(OH)_4^-(aq) + H^+(aq)$$

Boron's outer shell is filled in the wide variety of borate salts, such as the mineral borax (sodium borate), $Na_2[B_4O_5(OH)_4] \cdot 8H_2O$, used for decades as a household cleaning agent. Strong heating of boric acid (or borate salts) drives off water molecules and gives molten boron oxide:

$$2B(OH)_3(s) \xrightarrow{\Delta} B_2O_3(l) + 3H_2O(g)$$

Labware made of borosilicate glass

When mixed with silica (SiO_2), the molten oxide forms borosilicate glass. Its high transparency and small change in size when heated or cooled make borosilicate glass useful in cookware and in the glassware used in laboratories (*see margin photo*).

Forming Bridge Bonds with an Electron-Poor Atom In elemental boron and its many hydrides (boranes), there is no electron-rich atom to supply boron with electrons. In these substances, boron attains an octet through some unusual bonding. In diborane (B_2H_6) and many larger boranes, for example, two types of B—H bonds exist. The first type is a normal electron-pair bond. The valence bond picture in Figure 13.7 shows an sp^3 orbital of B overlapping a 1*s* orbital of H in each of the four terminal B—H bonds, using two of the three electrons in the valence level of each B atom.

The other type of bond is a hydride **bridge bond** (or *three-centre, two-electron bond*), in which *each B—H—B grouping is held together by only two electrons.* Two sp^3 orbitals, one from *each* B, overlap an H 1*s* orbital between them. Two electrons move through this extended bonding orbital—one from one of the B atoms and the other from the H atom—and join the two B atoms via the H atom bridge. Notice

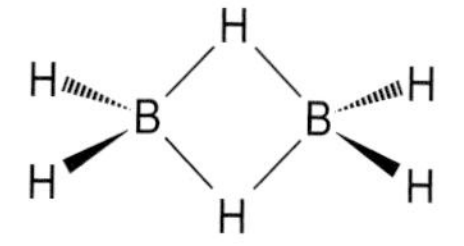

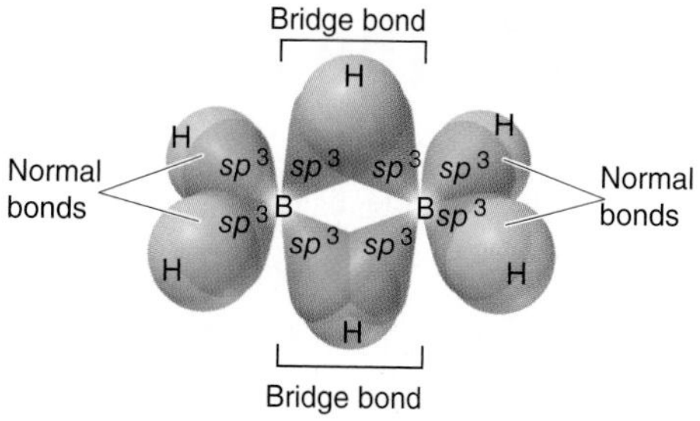

FIGURE 13.7 The two types of covalent bonding in diborane

that *each B atom is surrounded by eight electrons*: four from the two normal B—H bonds and four from the two B—H—B bridge bonds with a tetrahedral arrangement around each B atom. In many boranes and in elemental boron (Figure 13.8), one B atom bridges two others in a three-centre, two-electron B—B—B bond.

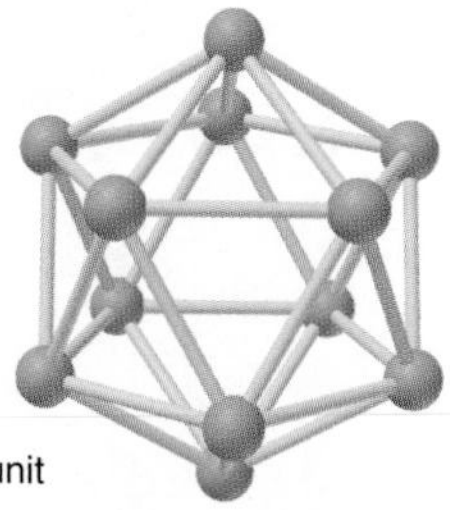

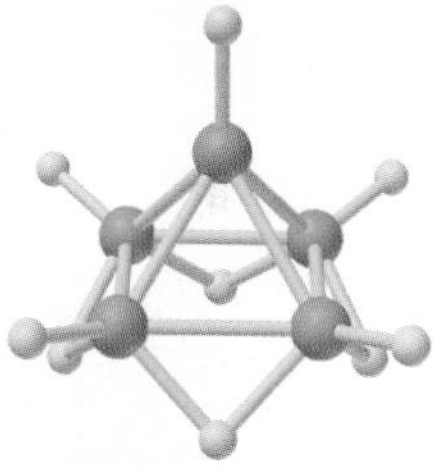

FIGURE 13.8 The boron icosahedron and one of the boranes

Diagonal Relationships: Beryllium and Aluminum

Beryllium in group 2 and aluminum in group 13 are another pair of diagonally related elements. Both form oxoanions in a strong base: beryllate, $Be(OH)_4^{2-}$, and aluminate, $Al(OH)_4^-$. Both have bridge bonds in their hydrides and chlorides. Both form oxide coatings impervious to reaction with water (which explains aluminum's great use as a structural metal), and both oxides are amphoteric, extremely hard, and high melting. Although the atomic and ionic sizes of these elements differ, the small, highly charged Be^{2+} and Al^{3+} ions polarize nearby electron clouds strongly. Therefore, some Al compounds and all Be compounds have significant covalent character.

13.6 Group 14: The Carbon Family

The whole range of elemental behaviour occurs within group 14: nonmetallic carbon (C) leads off, followed by the metalloids silicon (Si) and germanium (Ge), with metallic tin (Sn) and lead (Pb) next, and recently synthesized Element 114 at the bottom of the group. Information about the compounds of C and of Si fills libraries: organic chemistry, most polymer chemistry, and biochemistry are based on carbon, whereas geochemistry and some extremely important polymer and electronic technologies are based on silicon. The Group 14 Family Portrait summarizes its atomic, physical, and chemical properties.

How Does the Bonding in an Element Affect Physical Properties?

The elements of group 14 and their neighbours in groups 13 and 15 illustrate how physical properties, such as melting point and heat of fusion ($\Delta_{fus}H$), depend on the type of bonding in an element (Table 13.2). Within group 14, the large decrease in

TABLE 13.2 Bond Type and the Melting Process in Groups 13 to 15

	Group 13				Group 14				Group 15			
Period	Element	Bond Type	Melting Point (°C)	$\Delta_{fus}H$ (kJ/mol)	Element	Bond Type	Melting Point (°C)	$\Delta_{fus}H$ (kJ/mol)	Element	Bond Type	Melting Point (°C)	$\Delta_{fus}H$ (kJ/mol)
2	B	Covalent network	2180	23.6	C	Covalent network	4100	Very high	N	Covalent molecule	−210	0.7
3	Al	Metallic	660	10.5	Si	Covalent network	1420	50.6	P	Covalent molecule	44.1	2.5
4	Ga	Metallic	30	5.6	Ge	Covalent network	945	36.8	As	Covalent network	816	27.7
5	In	Metallic	157	3.3	Sn	Metallic	232	7.1	Sb	Covalent network	631	20.0
6	Tl	Metallic	304	4.3	Pb	Metallic	327	4.8	Bi	Metallic	271	10.5

Key: Metallic; Covalent network; Covalent molecule; Metal; Metalloid; Nonmetal

Family Portrait Group 14: The Carbon Family

Key Atomic Properties, Physical Properties, and Reactions

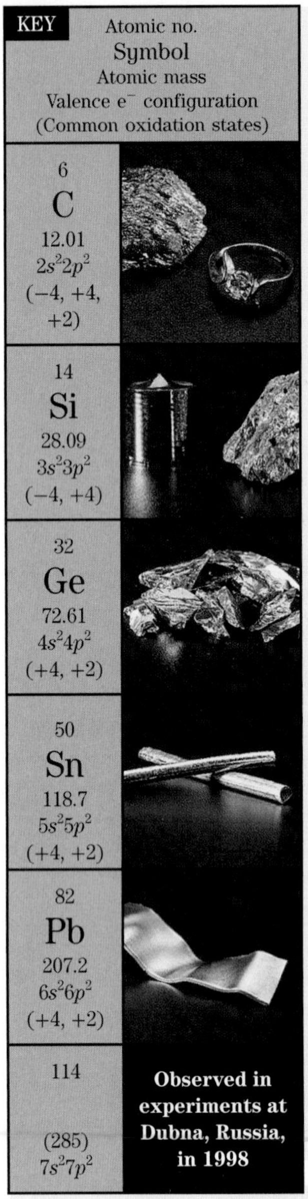

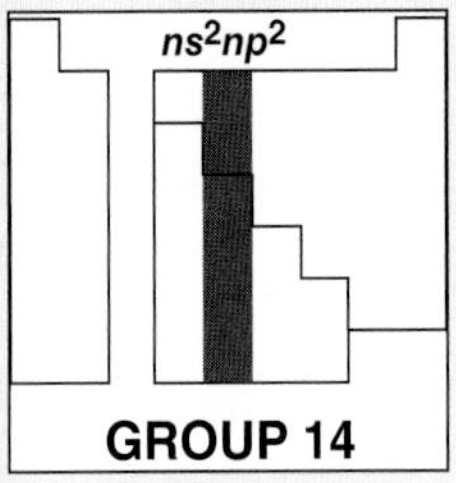

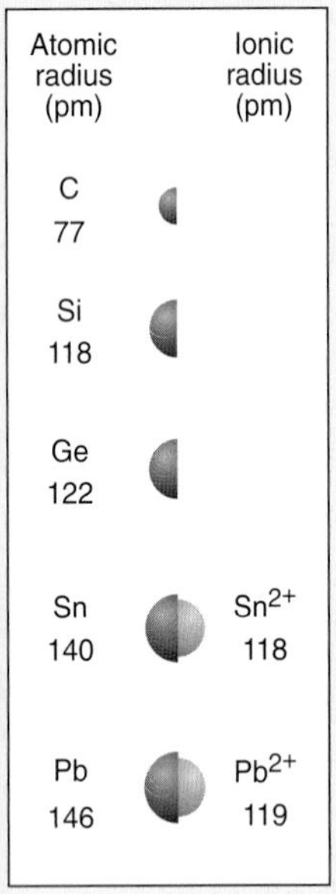

Atomic Properties

The group electron configuration is ns^2np^2. Down the group, the number of oxidation states decreases, and the lower (+2) state becomes more common. Down the group, size increases. Because transition and inner transition elements intervene, IE and χ do not decrease smoothly.

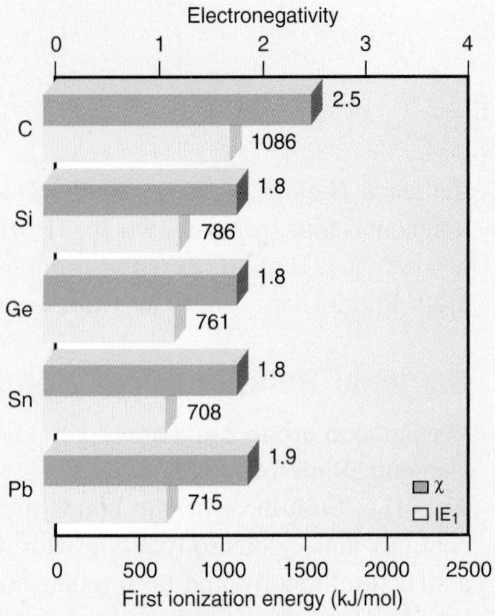

Physical Properties

Trends in properties, such as decreasing hardness and melting point, are due to changes in types of bonding within the solid: covalent network in C, Si, and Ge; metallic in Sn and Pb (*see main text*). Down the group, density increases because of several factors, including differences in crystal packing.

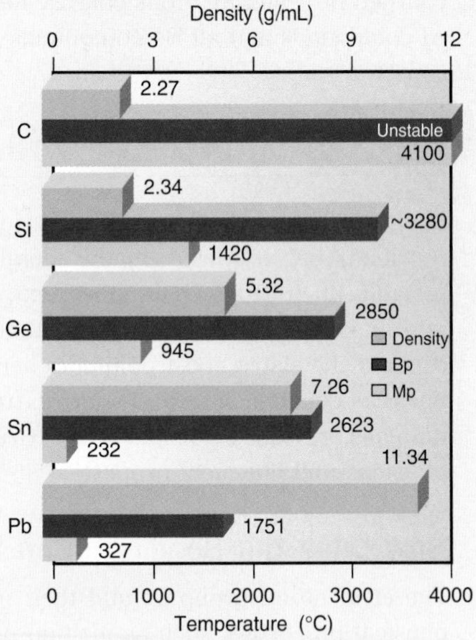

Reactions

1. The elements are oxidized by halogens:

$$E(s) + 2X_2 \longrightarrow EX_4 \qquad (E = C, Si, Ge)$$

The +2 halides are more stable for tin and lead, SnX_2 and PbX_2.

2. The elements are oxidized by O_2:

$$E(s) + O_2(g) \longrightarrow EO_2 \qquad (E = C, Si, Ge, Sn)$$

Pb forms the +2 oxide, PbO. Oxides become more basic down the group. The reaction of CO_2 and H_2O provides the weak acidity of natural unpolluted waters:

$$CO_2(g) + H_2O(l) \rightleftharpoons H_2CO_3(aq) \rightleftharpoons H^+(aq) + HCO_3^-(aq)$$

3. Hydrocarbons react with O_2 to form CO_2 and H_2O. The reaction for methane is adapted to yield heat or electricity:

$$CH_4(g) + 2O_2(g) \longrightarrow CO_2(g) + 2H_2O(g)$$

4. Silica is reduced to form elemental silicon:

$$SiO_2(s) + 2C(s) \longrightarrow Si(s) + 2CO(g)$$

This crude silicon is made ultrapure through zone refining for use in the manufacture of computer chips.

melting point between the network covalent solids C and Si is due to longer, weaker bonds in the Si structure; the large decrease between Ge and Sn is due to the change from covalent network to metallic bonding. Similarly, looking at horizontal trends, the large increases in melting point and $\Delta_{fus}H$ across a period between Al and Si and between Ga and Ge reflect the change from metallic to covalent network bonding. Notice the abrupt rises in the values for these properties from metallic Al, Ga, and Sn to the network covalent metalloids Si, Ge, and Sb. Also notice the abrupt drops from the covalent networks of C and Si to the individual molecules of N and P in group 15.

Allotropism: Different Forms of an Element Striking variations in physical properties often appear among **allotropes**, different crystalline or molecular forms of a substance. One allotrope is usually more stable than another at a particular pressure and temperature. Group 14 provides the first important examples of allotropism, in the forms of carbon and tin.

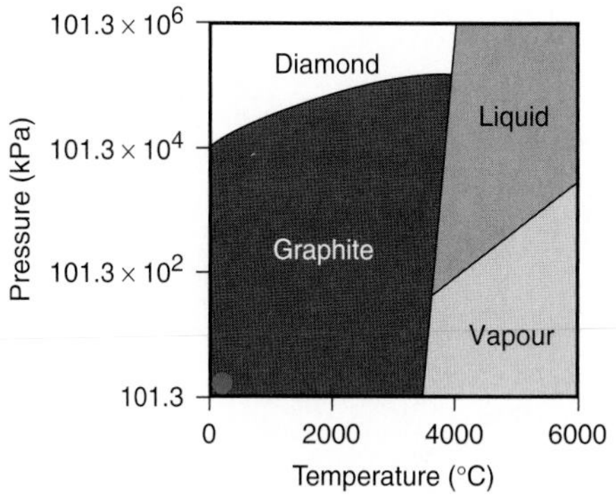

FIGURE 13.9 Phase diagram of carbon

1. *Allotropes of carbon.* It is difficult to imagine two substances, made entirely of the same atom, that are more different than graphite and diamond. Graphite is a black electrical conductor that is soft and "greasy," whereas diamond is a colourless electrical insulator that is extremely hard. Graphite is the standard state of carbon, the more stable form at ordinary temperature and pressure, conditions indicated by the red dot in the phase diagram in Figure 13.9. Fortunately for jewellery owners, diamond changes to graphite at a negligible rate under normal conditions.

In the mid-1980s, a newly discovered allotrope of carbon began generating great interest. Mass spectrometric analysis of soot had shown evidence for a molecule of formula C_{60}, shaped like a soccer ball (Figure 13.10A). More recently, the molecule has also been found in geological samples formed by meteorite impacts, even the impact that occurred around the time when the dinosaurs became extinct. The molecule has been dubbed *buckminsterfullerene* (informally called a "buckyball") after the architect-engineer R. Buckminster Fuller, who designed structures with similar shapes. Excitement rose in 1990, when scientists learned how to prepare multigram quantities of C_{60} and related fullerenes, enough to study macroscopic behaviour and possible applications. Since then, metal atoms have been incorporated into the structure, and many different atoms and groups (such as fluorine, hydroxyl groups, and sugars) have been attached, resulting in compounds with a range of useful properties.

In 1991, scientists passed an electric discharge through graphite rods surrounded by helium and sealed in a container. They obtained extremely thin (~1 nm in diameter) graphite-like tubes with fullerene ends (Figure 13.10B). These *nanotubes* are rigid and, on a mass basis, much stronger than steel along their long axis. They also conduct electricity along this axis because of the delocalized electrons. With potential applications in nanoscale electronics, energy storage, catalysis, polymers, and medicine, fullerene and nanotube chemistry is an active area in materials research.

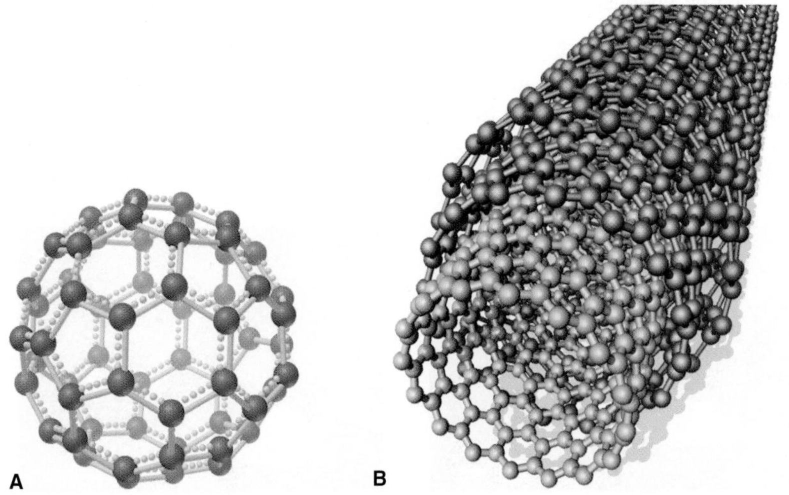

FIGURE 13.10 A. A buckyball. **B.** Model of concentric nanotubes, in different colours for clarity, without the usual fullerene ends.

(The 2010 Nobel Prize in Physics was awarded for research into a new form of carbon called *graphene*. It has remarkable conductivity and strength, yet exists as extended sheets only one atom thick.)

2. *Allotropes of tin.* Tin has two allotropes. White β-tin is stable at room temperature and above, whereas grey α-tin is more stable below 13°C. When white tin is kept for long periods at a low temperature, some converts to microcrystals of grey tin. The random formation and growth of these regions of grey tin, which has a different crystal structure, weaken the metal and make it crumble. In the unheated cathedrals of medieval northern Europe, tin pipes of magnificent organs were sometimes destroyed by the "tin disease" caused by this allotropic transition.

How Does the Type of Bonding Change in Group 14 Compounds?

The group 14 elements display a wide range of chemical behaviour, from the covalent compounds of carbon to the ionic compounds of lead. Carbon's intermediate χ of 2.5 ensures that it virtually always forms covalent bonds, but the larger members of the group form bonds with increasing ionic character. With nonmetals, Si and Ge form strong polar covalent bonds, such as the Si—O bond, which is one of the strongest of any period 3 element (BE = 368 kJ/mol) and is responsible for the physical and chemical stability of Earth's solid surface, as we discuss later in this section. Although individual Sn or Pb ions rarely exist, the bonding of either element with a nonmetal has considerable ionic character.

The pattern of elements having more than one oxidation state, observed in group 13, also appears here. Thus, compounds with Si in the +4 state are much more stable than compounds with Si in the +2 state, whereas compounds with Pb in the +2 state are more stable than those with Pb in the +4 state. These elements behave more like metals in the lower oxidation state. For example, $SnCl_2$ and $PbCl_2$ are white, relatively high-melting, water-soluble crystals—typical properties of a salt (Figure 13.11)—whereas $SnCl_4$ is a volatile, benzene-soluble liquid, and $PbCl_4$ is a thermally unstable oil. Similarly, SnO and PbO are more basic than SnO_2 and PbO_2: because the +2 metals are less able to polarize the O^{2-} ion, the E-to-O bonding is more ionic.

Highlights of Carbon Chemistry

Like the other period 2 elements, carbon is an anomaly in its group; indeed, it may be an anomaly in the entire periodic table. Carbon forms bonds with the smaller group 1 and 2 metals, many transition metals, the halogens, and many other metalloids and nonmetals. As well, it exhibits every oxidation state possible for its group, from +4 in CO_2 and halides such as CCl_4 through −4 in CH_4.

Organic Compounds Two major properties of carbon give rise to the enormous field of organic chemistry.

- Carbon has the ability to bond to itself—a process known as *catenation*. As a result of its small size and its capacity for four bonds,* carbon can form chains, branches, and rings that lead to myriad structures. Add a lot of H, some O and N, a bit of S, P, halogens, and a few metals, and you have the whole organic world! Figure 13.12 shows three of the several million organic compounds known.

FIGURE 13.11 Saltlike +2 chlorides and oily +4 chlorides show the greater metallic character of tin and lead in the lower oxidation state

*Although carbon usually forms four bonds, in compounds such as carbenes, carbon forms only two bonds.

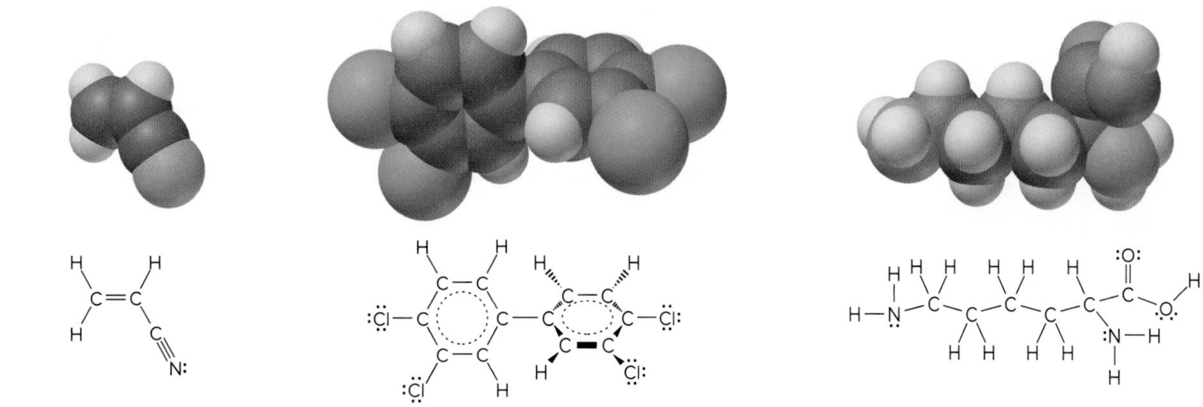

Prop-2-enenitrile (acrylonitrile) PCB Lysine

FIGURE 13.12 Three of the several million known organic compounds of carbon: prop-2-enenitrile (acrylonitrile), a precursor of acrylic fibres; a typical PCB (one of the polychlorinated biphenyls); and lysine, one of about 20 amino acids that occur in proteins.

- Carbon can form multiple bonds. Multiple bonds are common in carbon structures because the C—C bond is short enough for side-to-side overlap of two half-filled $2p$ orbitals to form π bonds. (In Chapters 20 and 21, we will discuss, in detail, how the atomic properties of carbon give rise to the diverse structures and reactivities of organic compounds.)

Because the other group 14 members are larger, E—E bonds become longer and weaker, so catenation and multiple bonding are much less important down the group.

Inorganic Compounds In contrast to its organic compounds, carbon's inorganic compounds are simple.

1. *Carbonates.* Metal carbonates are the main mineral form. Marble, limestone, chalk, coral, and several other types are found in enormous deposits throughout the world. Many of these compounds are remnants of fossilized marine organisms. Carbonates are used in several common antacids because they react with the HCl in stomach acid:

$$CaCO_3(s) + 2HCl(aq) \longrightarrow CaCl_2(aq) + CO_2(g) + H_2O(l)$$

Identical net ionic reactions with sulfuric and nitric acids protect lakes bounded by limestone deposits from the harmful effects of acid rain.

2. *Oxides.* Unlike the other group 14 members, which form only solid network covalent or ionic oxides, carbon forms two common gaseous oxides, CO_2 and CO.

- Carbon dioxide is essential to all life. It is the primary source of the carbon in plants through photosynthesis—and animals eat the plants. Its aqueous solution is the cause of mild acidity in natural waters. However, its atmospheric buildup from deforestation and the excessive use of fossil fuels is severely affecting the global climate.
- Carbon monoxide forms when carbon or its compounds burn in an inadequate supply of O_2:

$$2C(s) + O_2(g) \longrightarrow 2CO(g)$$

Carbon monoxide is a key component of syngas fuels (see Chemical Connections in Chapter 5) and is widely used in the production of methanol, methanal (formaldehyde), and other major industrial compounds. CO binds strongly to many transition metals. When inhaled in cigarette smoke or polluted air, it enters the blood and binds strongly to the Fe(II) in hemoglobin, preventing the normal binding of O_2, and to other iron-containing proteins. The cyanide ion (CN^-) is *isoelectronic* with CO:

$$[:C{\equiv}N:]^- \quad \text{same electronic structure as} \quad :C{\equiv}O:$$

Cyanide binds to many of the same iron-containing proteins and is also toxic.

3. *Halides.* Carbon halides have major uses as solvents and in structural plastics. The chlorofluorocarbons (CFCs, or Freons) have short, strong carbon-halogen bonds that make them both useful and harmful (Figure 13.13). Chemically and thermally stable, nontoxic, and nonflammable, they are excellent as cleaners for electronic

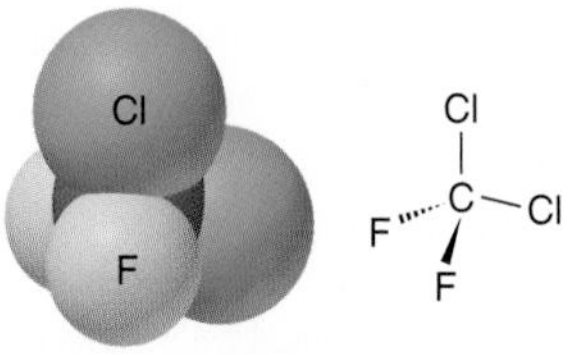

FIGURE 13.13 Freon-12 (CCl_2F_2), a chlorofluorocarbon

"Frustrated Lewis Pairs" In 1923, G. N. Lewis described electron donors as "bases" and electron acceptors as "acids." Such Lewis acids and bases can combine to share electrons, forming a Lewis acid-base adduct. In 2006, Doug Stephan and coworkers, then at the University of Windsor, in Ontario, uncovered a perturbation to this long-standing principle. Combining electron-rich and electron-poor species where sterically demanding molecular structures preclude direct interaction of the electron acceptor and donor, they showed that such "frustrated" Lewis pairs (FLPs) provide a unique approach to new reactivity. In a truly remarkable and unprecedented finding, exposure of a solution of neutral sterically demanding combinations of phosphines and borane were capable of splitting H_2 at 25°C. For example, the simple combinations of $B(C_6F_5)_3$ with a variety of sterically frustrated phosphines R_3P ($R = t$Bu, $C_6H_2Me_3$) do not react with the Lewis acid $B(C_6F_5)_3$, confirming that the mixtures constitute FLPs. However, upon exposure to 1 bar H_2, there was an immediate reaction to form the corresponding salts $[R_3PH][HB(C_6F_5)_3]$.

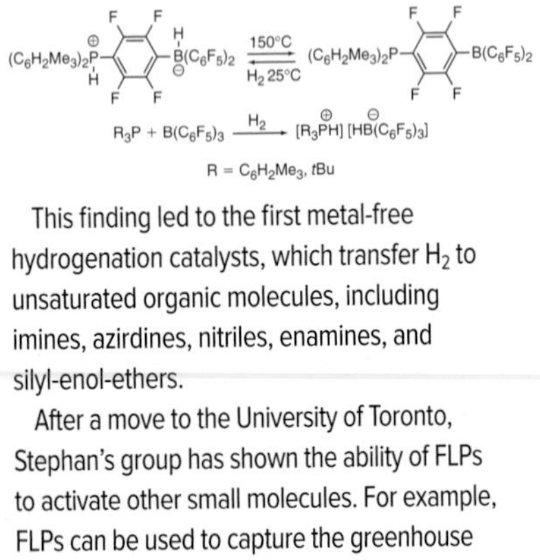

This finding led to the first metal-free hydrogenation catalysts, which transfer H_2 to unsaturated organic molecules, including imines, azirdines, nitriles, enamines, and silyl-enol-ethers.

After a move to the University of Toronto, Stephan's group has shown the ability of FLPs to activate other small molecules. For example, FLPs can be used to capture the greenhouse gases CO_2, N_2O, NO, and SO_2. In the case of CO_2, metal-free mediated reduction suggests new strategies to combatting global warming.

In summary, the concept of combining Lewis acids and Lewis bases where steric demands preclude classical Lewis acid-base adduct formation provides a new strategy for the catalysis and activation of a variety of small molecules. Intense research efforts are underway to exploit this fundamental finding of the reactivity of main-group compounds for applications in chemical synthesis and transformation.

[Source: Text by Dr. Doug Stephan.]

parts, coolants in refrigerators and air conditioners, and propellants in aerosol cans. However, their bond strengths mean that they decompose *very* slowly near Earth's surface. In the stratosphere, they are bombarded by ultraviolet (UV) radiation, which breaks the otherwise stable C—Cl bonds, releasing free Cl atoms that initiate ozone-destroying reactions (Chapter 14). Legal production of CFCs has ended in Canada and the United States, but international production and smuggling are widespread.

Highlights of Silicon Chemistry

To a great extent, the chemistry of silicon is the chemistry of the *silicon-oxygen bond.* Just as carbon forms unending C—C chains, the —Si—O— grouping repeats itself endlessly in a wide variety of **silicates**, the *most important minerals on the planet*, and in **silicones**, *synthetic polymers that have many applications*:

1. *Silicate minerals.* From common sand and clay to semiprecious amethyst and carnelian, silicate minerals are the dominant form of matter in the nonliving world. Oxygen, the most abundant element on Earth, and silicon, the next most abundant, compose these minerals and account for four of every five atoms on the surface of the planet!

The silicate building unit is the *orthosilicate grouping,* —SiO_4—, a tetrahedral arrangement of four oxygen atoms around a central silicon. Several well-known minerals contain SiO_4^{4-} ions or small groups of these ions, linked together. The gemstone zircon ($ZrSiO_4$) contains one unit; hemimorphite [$Zn_4(OH)_2Si_2O_7{\cdot}H_2O$] contains two units linked through an oxygen corner; and beryl ($Be_3Al_2Si_6O_{18}$), the major source of beryllium, contains six units joined into a cyclic ion (Figure 13.14). In extended structures, one of the O atoms links the next Si—O group to form chains, a second O atom forms crosslinks to neighbouring chains to form sheets, and the third O atom forms more crosslinks to create three-dimensional frameworks. Chains of silicate groups compose the asbestos minerals, sheets give rise to talc and mica, and frameworks occur in feldspar and quartz (Figure 13.15).

2. *Silicone polymers.* Unlike the naturally occurring silicates, silicone polymers are manufactured substances, consisting of alternating Si and O atoms with two organic groups also bonded to each Si atom in a very long Si—O chain, as in *poly(dimethyl siloxane):*

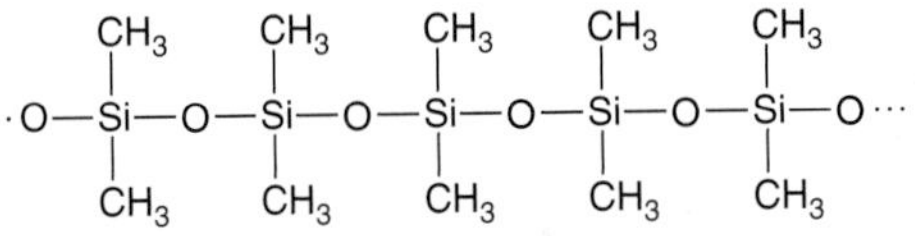

Silicones have properties of both plastics and minerals. The organic groups give them the flexibility and weak intermolecular forces between chains that are characteristic of a plastic, whereas the O—Si—O backbone confers the thermal stability and nonflammability of a mineral. Structures similar to those of the silicates can be

FIGURE 13.14 Structures of the silicate anions in some minerals

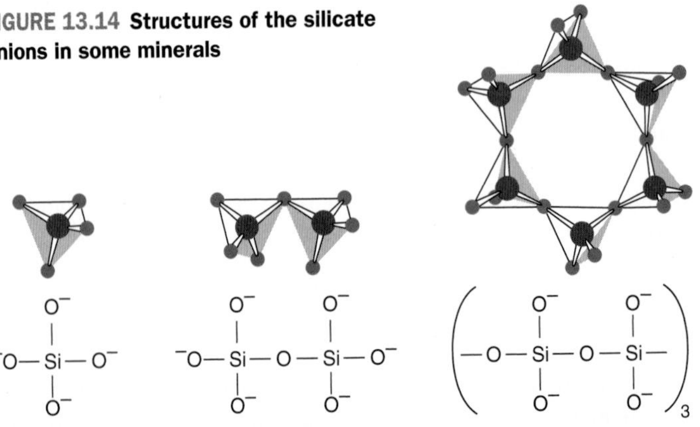

SiO_4^{4-} — **Silicate ion in zircon**

$Si_2O_7^{6-}$ — **Silicate ion in hemimorphite**

$Si_6O_{18}^{12-}$ — **Silicate ion in beryl**

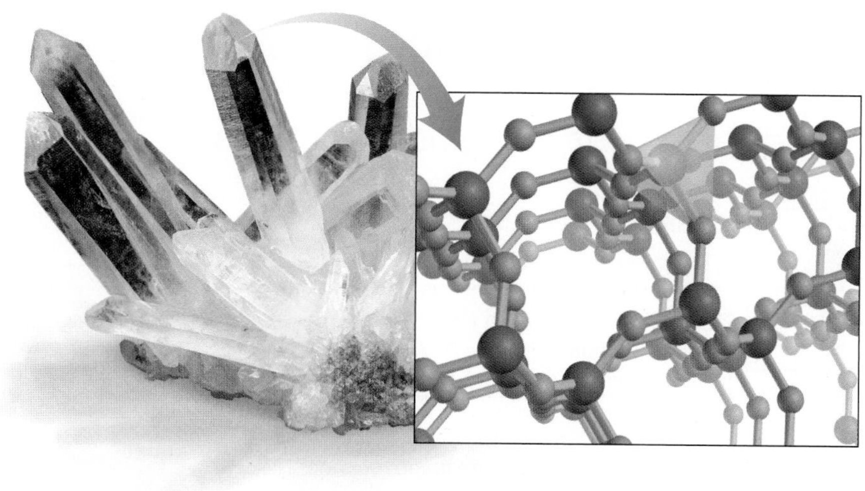

FIGURE 13.15 Quartz is a three-dimensional framework silicate

created by adding various reactants to form silicone chains, sheets, and frameworks. The chains are oily liquids used as lubricants and as components of car polish and makeup. The sheets are components of gaskets, space suits, and contact lenses. The frameworks are used as laminates on circuit boards, in nonstick cookware, and in artificial skin and bone.

Diagonal Relationships: Boron and Silicon

The final diagonal relationship that we consider occurs between the metalloids boron and silicon, which are both semiconductors (Section 11.6). Both B and Si and their mineral oxoanions—borates and silicates—occur in extended covalent networks. Both boric acid [$B(OH)_3$] and silicic acid [$Si(OH)_4$] are weakly acidic solids that occur as layers held together by widespread hydrogen bonding. Their hydrides—the compact boranes and the extended silanes—are flammable, low-melting compounds that act as reducing agents.

13.7 Group 15: The Nitrogen Family

The first two elements in group 15, gaseous nonmetallic nitrogen (N) and solid nonmetallic phosphorus (P), play major roles in both nature and industry. Below these nonmetals are two metalloids, arsenic (As) and antimony (Sb), followed by the metal bismuth (Bi) and the recently synthesized Element 115. The Group 15 Family Portrait provides an overview. The cycling of nitrogen and phosphorus through the environment will be discussed in Section 23.2.

What Accounts for the Wide Range of Physical Behaviour in Group 15?

Group 15 displays the widest range of physical behaviour we have seen so far because of large changes in bonding and intermolecular forces:

- *Nitrogen* occurs as a gas consisting of N_2 molecules with such weak intermolecular forces that the element *boils* more than 200°C below room temperature.
- *Phosphorus* exists most commonly as tetrahedral P_4 molecules in the solid phase. Because P is heavier and more polarizable than N, it has stronger dispersion forces and melts about 25°C above room temperature.
- *Arsenic* consists of extended, puckered sheets in which each As atom is covalently bonded to three others and has nonbonding interactions with its three nearest neighbours in adjacent sheets, giving As the highest melting point in the group.
- *Antimony* has a similar covalent network, also resulting in a high melting point.
- *Bismuth* has metallic bonding and, thus, a lower melting point than As and Sb.

Family Portrait Group 15: The Nitrogen Family

Key Atomic Properties, Physical Properties, and Reactions

KEY
Atomic no.
Symbol
Atomic mass
Valence e^- configuration
(Common oxidation states)

7
N
14.01
$2s^22p^3$
(−3, +5, +4, +3, +2, +1)

15
P
30.97
$3s^23p^3$
(−3, +5, +3)

33
As
74.92
$4s^24p^3$
(−3, +5, +3)

51
Sb
121.8
$5s^25p^3$
(−3, +5, +3)

83
Bi
209.0
$6s^26p^3$
(+3)

115
(288)
$7s^27p^3$
Observed in experiments at Dubna, Russia, in 2003

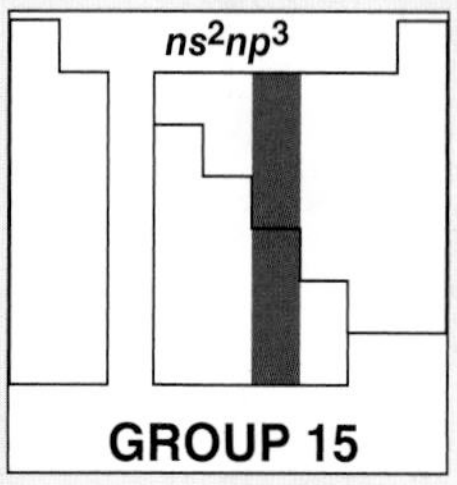

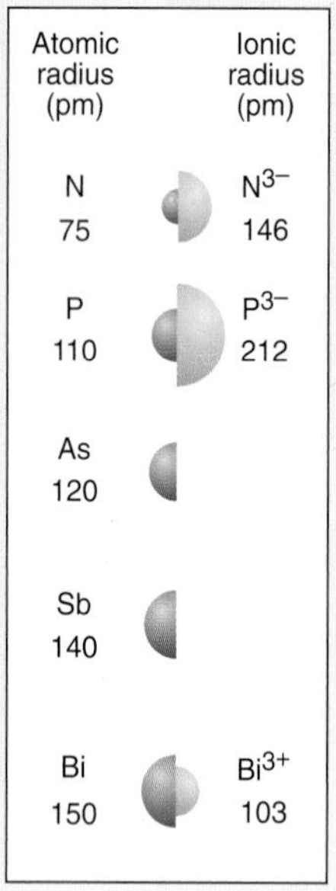

Atomic Properties

The group electron configuration is ns^2np^3. The np sublevel is half-filled, with each p orbital containing one electron (parallel spin). The number of oxidation states decreases down the group, and the lower (+3) state becomes more common. Atomic properties follow generally expected trends. The large (~50%) increase in size from N to P correlates with the much lower IE and χ of P.

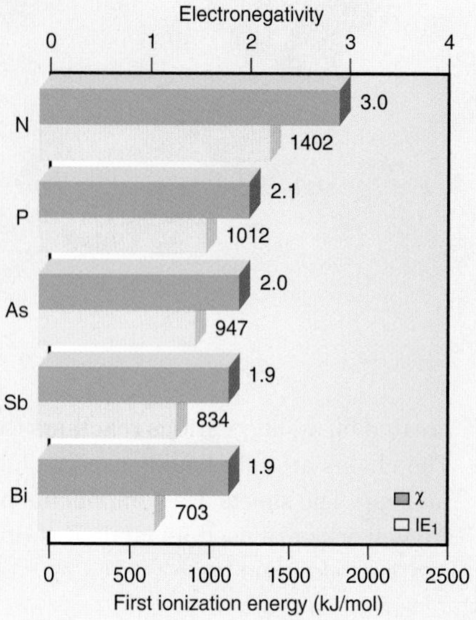

Physical Properties

Physical properties reflect the change from individual molecules (N, P) to network covalent solid (As, Sb) to metal (Bi). Thus, melting points increase and then decrease. Large atomic size and low atomic mass result in low density. Because mass increases more than size down the group, the density of the elements as solids increases. The dramatic increase in density from P to As is due to the intervening transition elements.

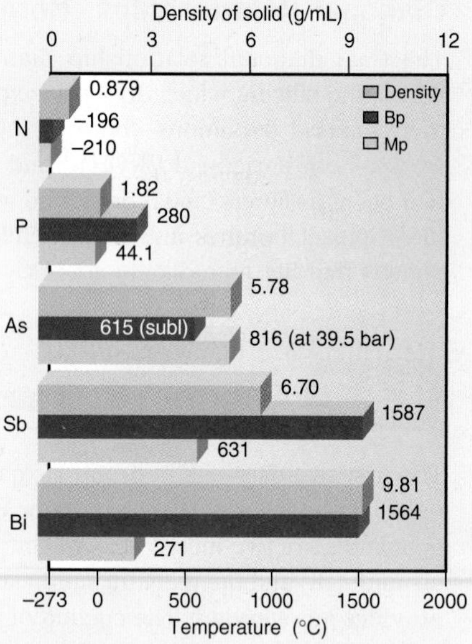

Reactions

1. Nitrogen is "fixed" industrially in the Haber process (Section 15.6):

$$N_2(g) + 3H_2(g) \rightleftharpoons 2NH_3(g)$$

Further reactions convert NH_3 to NO, NO_2, and HNO_3 (*see discussion*). Hydrides of some other group members are formed from reaction in water (or with H_3O^+) of a metal phosphide or arsenide, or a similar compound:

$$Ca_3P_2(s) + 6H_2O(l) \longrightarrow 2PH_3(g) + 3Ca(OH)_2(aq)$$

2. Halides are formed by direct combination of the elements:

$$2E(s) + 3X_2 \longrightarrow 2EX_3 \quad (E = \text{all except N})$$

$$EX_3 + X_2 \longrightarrow EX_5 \quad (\text{with } X = F \text{ and Cl, but no } BiCl_5;\ E = P \text{ for } X = Br)$$

3. Oxoacids are formed from the halides in a reaction with water that is common to many nonmetal halides:

$$EX_3 + 3H_2O(l) \longrightarrow H_3EO_3(aq) + 3HX(aq) \quad (E = \text{all except N})$$

$$EX_5 + 4H_2O(l) \longrightarrow H_3EO_4(aq) + 5HX(aq)$$

(E = all except N and Bi)

Notice that the oxidation number of E does *not* change.

The emerging importance of the pnictogen elements (arsenic, antimony, and bismuth) in the development of new thermoelectric materials, electronic materials, pharmaceuticals, and polymeric materials requires a more extensive understanding of their fundamental chemistry. Neil Burford (Figure 13.16A), for example, investigates the potential antimicrobial as well as other bioactivities of new bismuth compounds, related to the active ingredient in Pepto-Bismol, as well as other pnictogen elements. On the other hand, one of Michael D. Fryzuk's (Figure 13.16B) research areas is the activation and functionalization of molecular nitrogen.

Two Allotropes of Phosphorus Phosphorus has several allotropes, which have very different properties. Two major allotropes are white and red phosphorus:

- *White phosphorus* consists of individual tetrahedral molecules (Figure 13.17A), making it a low-melting (mp = 44.5°C), whitish, waxy solid that is soluble in nonpolar solvents. Each P atom uses its half-filled $3p$ orbitals to bond to the other three; with a small 60° bond angle and, thus, weak P—P bonds, it is highly reactive (Figure 13.17B). White phosphorus is poisonous and cannot be purchased legally. It burns fiercely and can set ammunition, cloth, fuel, and other combustibles on fire, causing serious burns or death. It was used as early as World War I and is even used today as an incendiary weapon.
- *Red phosphorus* is formed by heating the white form in the absence of air. One of the P—P bonds in each tetrahedron breaks, and these $3p$ orbitals overlap with others to form chains of P_4 units (Figure 13.17C). The chains make the red allotrope much less reactive, high-melting, and insoluble.

A

B

FIGURE 13.16 **A.** Neil Burford (University of Victoria) investigates efficient and effective preparative or synthetic routes to new, fundamentally important molecules containing P, As, Sb, or Bi. In such molecules, the pnictogen centre exhibits an unusual local structure, is engaged in a new connectivity, or has relevance in established bioactivity. As well, it may provide materials with new spectroscopic, physical, or reactivity properties. **B.** Michael D. Fryzuk (University of British Columbia, Vancouver) and his group study the coordination chemistry of dinitrogen (N_2). The Fryzuk lab centres the strategy for ligand design around the combination of soft and hard donor types arranged in a chelating array to facilitate changes in oxidation state at the metal centre. The ultimate goal of this research is to discover a homogeneous catalyst that can take N_2, which is readily available in the atmosphere, and convert it into useful, higher-value organonitrogen-containing materials.

What Patterns Appear in the Chemical Behaviour of Group 15?

The same general pattern of chemical behaviour that we discussed for group 14 appears again in this group, reflected in the change from nonmetallic N to metallic Bi. The overwhelming majority of group 15 compounds have *covalent bonds.* Whereas N can form no more than four bonds, the next three members can expand their valence shells by using empty d orbitals.

1. *Formation of ions.* For a group 15 element to form an ion with a noble gas electron configuration, it must *gain* three electrons, the last two in endothermic steps. Nevertheless, the enormous lattice energy released when such highly charged anions attract cations drives their formation. However, the 3− anion of N occurs only in compounds with active metals, such as Li_3N and Mg_3N_2. (The 3− anion of P may occur in Na_3P.) Metallic Bi forms mostly covalent compounds but exists as a cation in a few compounds, such as BiF_3 and $Bi(NO_3)_3 \cdot 5H_2O$, through *loss* of its three valence p electrons.

2. *Oxidation states and oxides.* As in group 13 and group 14, fewer oxidation states occur down the group, with the lower state becoming more prominent: N exhibits every state possible for a group 15 element, from +5 to −3; only the +5 and +3 states are common for P, As, and Sb; and +3 is the only common state of Bi. The oxides change from acidic to amphoteric to basic, reflecting the increase in the metallic character of the elements. In addition, the lower oxide of an element is more basic than the higher oxide, reflecting the greater ionic character of the E-to-O bonding in the lower oxide.

3. *Formation of hydrides.* All the group 15 elements form gaseous hydrides with the formula EH_3. Except for NH_3, these are extremely reactive and poisonous and are synthesized by reaction of a metal phosphide or arsenide, or a similar compound, which acts as a strong base in water or aqueous acid, for example,

$$Ca_3As_2(s) + 6H_2O(l) \longrightarrow 2AsH_3(g) + 3Ca(OH)_2(aq)$$

Ammonia is made industrially by direct combination of the elements at high pressure and moderately high temperature:

$$N_2(g) + 3H_2(g) \rightleftharpoons 2NH_3(g)$$

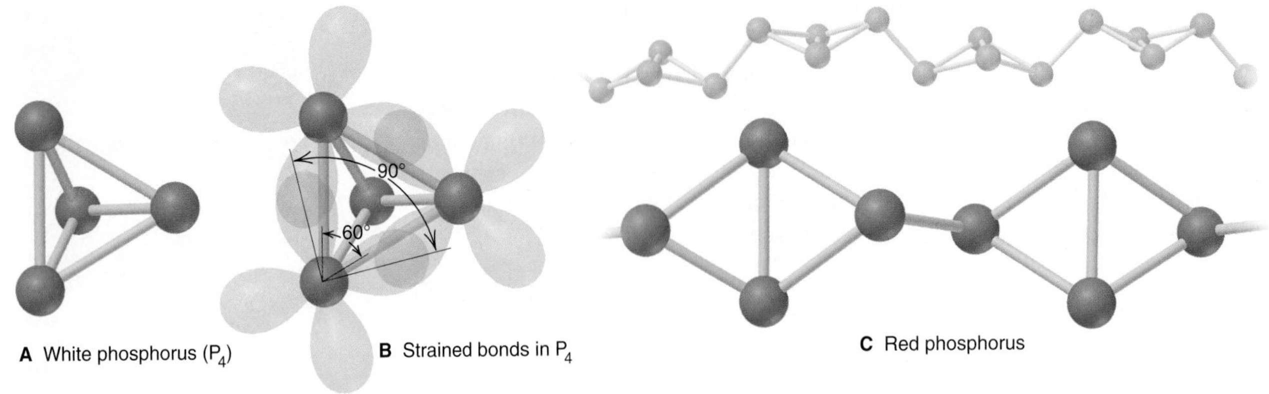

FIGURE 13.17 Two allotropes of phosphorus

Nitrogen forms a second hydride, hydrazine (N_2H_4). Like NH_3, hydrazine is a weak base; it is used to make antituberculin drugs, plant growth regulators, and fungicides.

Molecular properties of the group 15 hydrides reveal some interesting bonding and structural patterns:

- Despite its much lower molar mass, NH_3 melts and boils at higher temperatures than the other group 15 hydrides as a result of hydrogen *bonding.*
- Bond angles decrease from 107.3° for NH_3 to around 90° for the other hydrides, which suggests that the larger atoms use unhybridized *p* orbitals.
- E—H bond lengths increase down the group, so bond strength and thermal stability decrease: AsH_3 decomposes at 250°C, SbH_3 at 20°C, and BiH_3 at −45°C.

We will see these properties—hydrogen bonding for the smallest member, change in bond angles, and change in bond energies—in the hydrides of group 16 as well.

4. *Types and properties of halides.* All the group 15 elements form trihalides (EX_3). All except nitrogen form pentafluorides (EF_5), but only a few other pentahalides (PCl_5, PBr_5, $AsCl_5$, and $SbCl_5$) are known. Nitrogen cannot form pentahalides because it cannot expand its valence shell. Most trihalides are prepared by direct combination:

$$P_4(s) + 6Cl_2(g) \longrightarrow 4PCl_3(l)$$

The pentahalides form with excess halogen:

$$PCl_3(l) + Cl_2(g) \longrightarrow PCl_5(s)$$

As with the hydrides, the thermal stability of the halides decreases as the E—X bond becomes longer. Among the nitrogen halides, for example, NF_3 is a stable, rather unreactive gas. NCl_3 is explosive and reacts rapidly with water. (The chemist who first prepared it lost three fingers and an eye!) NBr_3 can only be made below −87°C. NI_3 has never been prepared, but an ammoniated product ($NI_3 \cdot NH_3$) explodes at the slightest touch.

5. *Reaction of halides in water.* In a reaction pattern *typical of many nonmetal halides,* each group 15 halide reacts with water to yield the hydrogen halide and the oxoacid, in which E has the *same* oxidation number as it does in the original halide. For example, PX_5 (oxidation number of P = +5) produces phosphoric acid (oxidation number of P = +5) and HX:

$$PCl_5(s) + 4H_2O(l) \longrightarrow H_3PO_4(l) + 5HCl(g)$$

Highlights of Nitrogen Chemistry

Surely the most striking highlight of nitrogen chemistry is the inertness of N_2 itself. Nearly four-fifths of the atmosphere consists of N_2, and the other fifth is nearly all O_2, a very strong oxidizing agent. Nevertheless, the searing temperature of lightning is required for significant amounts of atmospheric nitrogen oxides to form. Although

TABLE 13.3 Structures and Properties of the Nitrogen Oxides

Formula	Name	Space-Filling Model	Lewis Structure	Oxidation State of N	$\Delta_f H$ (kJ/mol) at 298 K	Comment
N_2O	Dinitrogen monoxide (dinitrogen oxide, nitrous oxide)		$:N\equiv N-\ddot{\underset{..}{O}}:$	+1 (0, +2)	82.0	Colourless gas; used as dental anaesthetic ("laughing gas") and aerosol propellant
NO	Nitrogen monoxide (nitrogen oxide, nitric oxide)		$:\dot{N}=\ddot{O}:$	+2	90.3	Colourless paramagnetic gas; biochemical messenger; air pollutant
N_2O_3	Dinitrogen trioxide	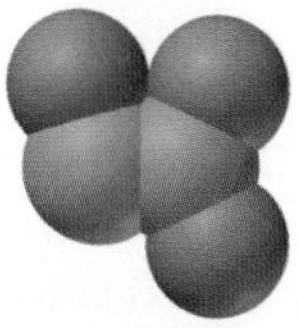	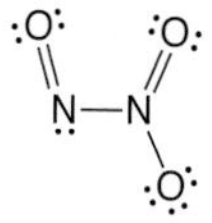	+3 (+2, +4)	83.7	Reddish-brown gas (reversibly dissociates to NO and NO_2)
NO_2	Nitrogen dioxide		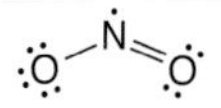	+4	33.2	Orange-brown paramagnetic gas formed during HNO_3 manufacture; poisonous air pollutant
N_2O_4	Dinitrogen tetroxide	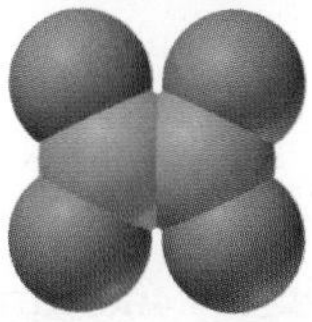	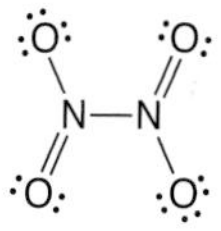	+4	9.16	Colourless to yellow liquid (reversibly dissociates to NO_2)
N_2O_5	Dinitrogen pentoxide	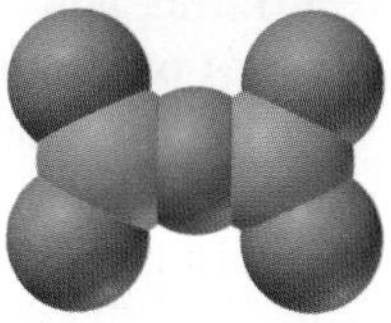	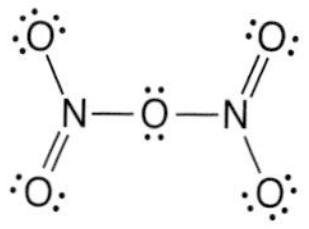	+5	11.3	Colourless volatile solid consisting of NO_2^+ and NO_3^-; gas consists of N_2O_5 molecules

N_2 is inert at moderate temperatures, it reacts at high temperatures with H_2, Li, group 2 members, B, Al, C, Si, Ge, O_2, and many transition elements. In fact, nearly every element in the periodic table forms bonds to N. Here we focus on the oxides and oxoacids and their salts.

Nitrogen Oxides Nitrogen is remarkable for having six stable oxides, each with a *positive* enthalpy of formation because of the great strength of the N≡N bond (BE = 945 kJ/mol). The structures of the nitrogen oxides and some of their properties are shown in Table 13.3. Unlike the hydrides and halides of nitrogen, the nitrogen oxides are planar. Nitrogen displays all of its positive oxidation states in these compounds. In N_2O and N_2O_3, the two N atoms have different states. Let us highlight the three most important nitrogen oxides: dinitrogen monoxide, nitrogen monoxide, and nitrogen dioxide.

1. *Dinitrogen monoxide* (N_2O; also called *dinitrogen oxide* or *nitrous oxide*) is the dental anaesthetic known as "laughing gas" and the propellant in canned whipped cream. It is a linear molecule with an electronic structure that is best described by three resonance forms (notice the formal charges):

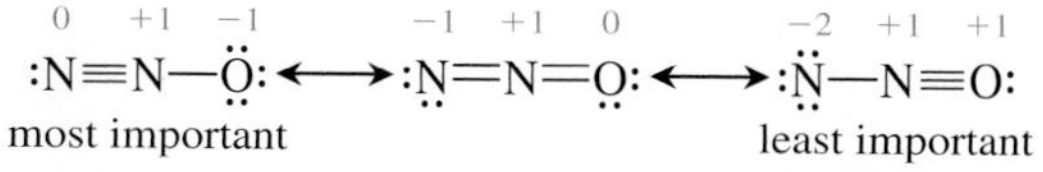

2. *Nitrogen monoxide* (NO; also called *nitrogen oxide* or *nitric oxide*) is an odd-electron molecule with biochemical functions ranging from neurotransmission to the control of blood flow. In Section 10.3, we used molecular orbital theory to

explain its bonding. The commercial preparation of NO through the oxidation of ammonia occurs as a first step in the production of nitric acid:

$$4NH_3(g) + 5O_2(g) \longrightarrow 4NO(g) + 6H_2O(g)$$

Nitrogen monoxide is also produced whenever air is heated to high temperatures, as in a car engine or a lightning storm:

$$N_2(g) + O_2(g) \xrightarrow{\text{high } T} 2NO(g)$$

Heating converts NO to two other oxides:

$$3NO(g) \xrightarrow{\Delta} N_2O(g) + NO_2(g)$$

This type of redox reaction is called a **disproportionation reaction**. It occurs when a substance *acts as both the oxidizing agent and the reducing agent in a reaction.* In the process, an atom with an intermediate oxidation state in the reactant occurs in both the lower and the higher states in the products: the oxidation state of N in NO (+2) is intermediate between its oxidation state in N_2O (+1) and NO_2 (+4).

3. *Nitrogen dioxide* (NO_2), a brown poisonous gas, forms to a small extent when NO reacts with additional oxygen:

$$2NO(g) + O_2(g) \rightleftharpoons 2NO_2(g)$$

Like NO, NO_2 is an odd-electron molecule with the electron (*red*) more localized on the N atom. Thus, NO_2 dimerizes reversibly to *dinitrogen tetroxide*:

$$O_2N\cdot(g) + \cdot NO_2(g) \rightleftharpoons O_2N\text{—}NO_2(g) \qquad (\text{or } N_2O_4)$$

Photochemical smog over St. Joseph's Oratory (Montréal, Québec)

Thunderstorms form NO and NO_2 and carry them down to the soil, where they act as natural fertilizers. In urban traffic, however, their formation leads to *photochemical smog* (*see photo*) in a series of reactions also involving sunlight, ozone (O_3), unburned gasoline, and various other substances. Although air pollution in Canada is not as severe as it is in some other areas in the world, it can still be a problem. For example, the southeastern parts of Canada along the Windsor-Québec City corridor of Ontario and Québec, as well as southern Nova Scotia, are renowned for smog and acid rain in the summer months.

Nitrogen Oxoacids and Oxoanions There are two common oxoacids of nitrogen: nitric acid and nitrous acid (Figure 13.18).

1. *Nitric acid* (HNO_3) is produced in the *Ostwald process.* We have already seen the first two steps: the oxidation of NH_3 to NO and the oxidation of NO to NO_2. The final step is a *disproportionation reaction*, as the oxidation numbers show:

$$3\overset{+4}{N}O_2(g) + H_2O(l) \longrightarrow 2H\overset{+5}{N}O_3(aq) + \overset{+2}{N}O(g)$$

The NO is recycled to make more NO_2.

In nitric acid, as in all oxoacids, *the acidic H is attached to one of the O atoms. When the proton is lost, the trigonal planar nitrate ion is formed* (Figure 13.18A).

FIGURE 13.18 The structures of nitric and nitrous acids and their oxoanions

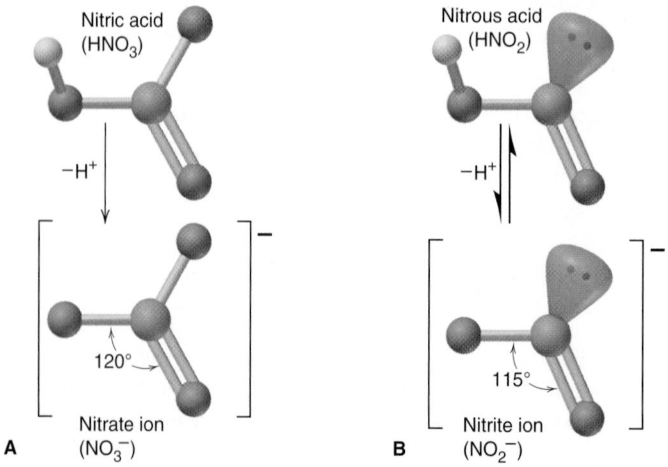

In the laboratory, nitric acid is used as a strong oxidizing acid. The products of its reactions with metals vary with the metal's reactivity and the acid's concentration. Notice in the following examples, from the net ionic equations, that *the* NO_3^- *ion is the oxidizing agent.* The nitrate ion, which is not reduced, is a spectator ion and does not appear in the net ionic equations.

- With an active metal, such as Al, and a dilute acid, N is reduced from the +5 state all the way to the −3 state in the ammonium ion, NH_4^+:

$$8Al(s) + 30HNO_3(aq; 1\ mol/L) \longrightarrow 8Al(NO_3)_3(aq) + 3NH_4NO_3(aq) + 9H_2O(l)$$
$$8Al(s) + 30H^+(aq) + 3NO_3^-(aq) \longrightarrow 8Al^{3+}(aq) + 3NH_4^+(aq) + 9H_2O(l)$$

- With a less reactive metal, such as Cu, and a more concentrated acid, N is reduced to the +2 state in NO:

$$3Cu(s) + 8HNO_3(aq; 3\ to\ 6\ mol/L) \longrightarrow 3Cu(NO_3)_2(aq) + 4H_2O(l) + 2NO(g)$$
$$3Cu(s) + 8H^+(aq) + 2NO_3^-(aq) \longrightarrow 3Cu^{2+}(aq) + 4H_2O(l) + 2NO(g)$$

- With a still more concentrated acid, N is reduced only to the +4 state in NO_2:

$$Cu(s) + 4HNO_3(aq; 12\ mol/L) \longrightarrow Cu(NO_3)_2(aq) + 2H_2O(l) + 2NO_2(g)$$
$$Cu(s) + 4H^+(aq) + 2NO_3^-(aq) \longrightarrow Cu^{2+}(aq) + 2H_2O(l) + 2NO_2(g)$$

Nitrates form when HNO_3 reacts with metals or with their hydroxides, oxides, or carbonates. *All nitrates are soluble in water.*

2. *Nitrous acid* (HNO_2), a much weaker acid than HNO_3, forms when metal nitrites are treated with a strong acid:

$$NaNO_2(aq) + HCl_2 \longrightarrow HNO_2(aq) + NaCl(aq)$$

This acid forms the planar nitrite ion (Figure 13.18B), in which nitrogen's lone pair reduces the ideal 120° bond angle to 115°.

These two acids reveal a *general pattern in relative acid strength among oxoacids: the more O atoms are bonded to the central nonmetal, the stronger the acid.* The O atoms pull electron density from the N atom, which in turn pulls electron density from the O of the O—H bond, facilitating the release of the H^+ ion to H_2O. The O atoms also act to stabilize the resulting oxoanion by delocalizing its negative charge. The same pattern occurs in the oxoacids of sulfur and the halogens; we will discuss this pattern quantitatively in Chapter 16.

Highlights of Phosphorus Chemistry

Like nitrogen, phosphorus forms important oxides (although not as many) and oxoacids. Here we focus on these as well as some other important phosphorus compounds.

Phosphorus Oxides Phosphorus forms two important oxides: tetraphosphorus hexoxide and tetraphosphorus decoxide.

1. *Tetraphosphorus hexoxide* (P_4O_6) has P in its +3 oxidation state. It forms when white P_4 reacts with limited oxygen:

$$P_4(s) + 3O_2(g) \longrightarrow P_4O_6(s)$$

P_4O_6 has the same tetrahedral orientation of the P atoms in P_4, with an O atom between each pair of P atoms (Figure 13.19A).

2. *Tetraphosphorus decoxide* (P_4O_{10}) has P in the +5 oxidation state. Commonly known as *phosphorus pentoxide*, from the empirical formula (P_2O_5), it forms when P_4 burns in excess O_2:

$$P_4(s) + 5O_2(g) \longrightarrow P_4O_{10}(s)$$

Its structure can be viewed as the structure of P_4O_6 with another O atom bonded to each of the four corner P atoms (Figure 13.19B). P_4O_{10} is a powerful drying agent.

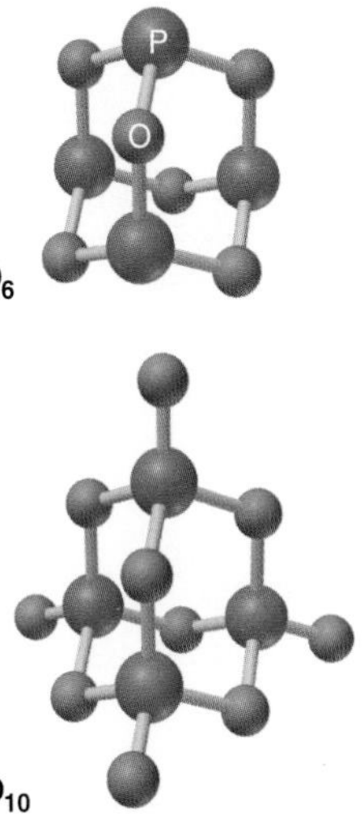

A P_4O_6

B P_4O_{10}

FIGURE 13.19 Important oxides of phosphorus

Phosphorus Oxoacids and Oxoanions The two common phosphorus oxoacids are phosphorous acid (note the change in spelling) and phosphoric acid.

1. *Phosphorous acid* (H_3PO_3) is formed when P_4O_6 reacts with water:

$$P_4O_6(s) + 6H_2O(l) \longrightarrow 4H_3PO_3(l)$$

The formula H_3PO_3 is misleading because the acid has only two acidic H atoms; the third is bonded to the central P and does not dissociate. Phosphorous acid is a weak acid in water but reacts completely in two steps with excess strong base:

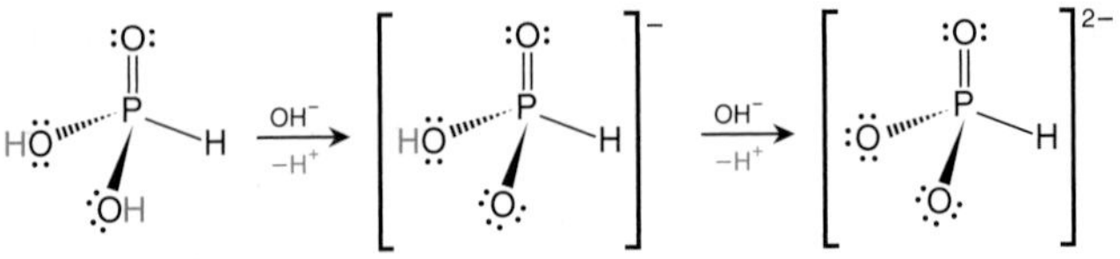

Salts of phosphorous acid contain the phosphite ion, HPO_3^{2-}.

2. *Phosphoric acid* (H_3PO_4), one of the 10 most important compounds in chemical manufacturing, is formed in a vigorous exothermic reaction of P_4O_{10} with water:

$$P_4O_{10}(s) + 6H_2O(l) \longrightarrow 4H_3PO_4(l)$$

The presence of many hydrogen bonds makes pure H_3PO_4 syrupy, more than 75 times as viscous as water. The laboratory-grade concentrated acid is an 85% by mass aqueous solution. H_3PO_4 is a weak triprotic acid; in water, it loses one proton in the following equilibrium reaction:

$$H_3PO_4(l) + H_2O(l) \rightleftharpoons H_2PO_4^-(aq) + H_3O^+(aq)$$

In excess strong base, however, the three protons dissociate completely in three steps to give the three phosphate oxoanions:

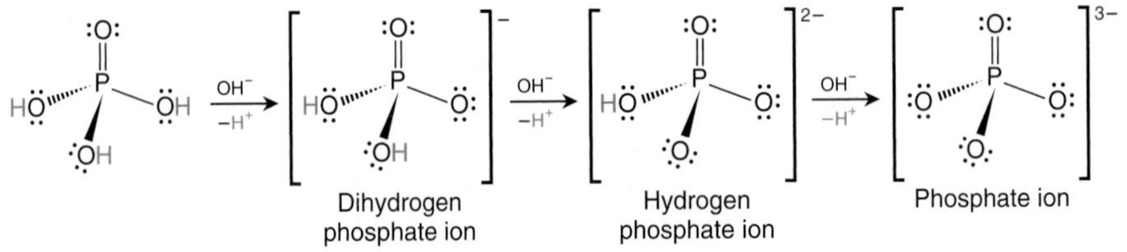

Phosphoric acid has a central role in fertilizer production and is also used as an additive in soft drinks, to give a touch of tartness. The various phosphate salts have numerous essential applications: Na_3PO_4 is a paint stripper and grease remover, K_3PO_4 is used to stabilize latex for synthetic rubber, and K_2HPO_4 is a radiator corrosion inhibitor. Ammonium phosphates are used as fertilizers and as flame retardants on curtains. Calcium phosphates are used in baking powders and toothpastes, as mineral supplements in livestock feed, and as fertilizers. (Phosphates are a nonrenewable resource, as we will see in Section 23.2).

Polyphosphates Hydrogen phosphates lose water when heated, as they form P—O—P linkages in the formation of compounds called *polyphosphates*. This type of reaction, in which *an H_2O molecule is lost for every pair of groups that join*, is called **dehydration-condensation**; it occurs frequently in the formation of polyoxoanion chains and other polymeric structures, both synthetic and natural. For example, sodium diphosphate, $Na_4P_2O_7$, is prepared by heating sodium hydrogen phosphate:

$$2Na_2HPO_4(s) \xrightarrow{\Delta} Na_4P_2O_7(s) + H_2O(g)$$

The diphosphate ion, $P_2O_7^{4-}$, the smallest of the polyphosphates, consists of two PO_4 units linked through a common oxygen corner (Figure 13.20A). Its reaction with water, the reverse of the previous reaction, generates heat:

$$P_2O_7^{4-}(aq) + H_2O(l) \longrightarrow 2HPO_4^-(aq) + heat$$

A similar process is put to vital use by organisms, when a third PO_4 unit linked to diphosphate creates the triphosphate grouping, part of the all-important high-energy

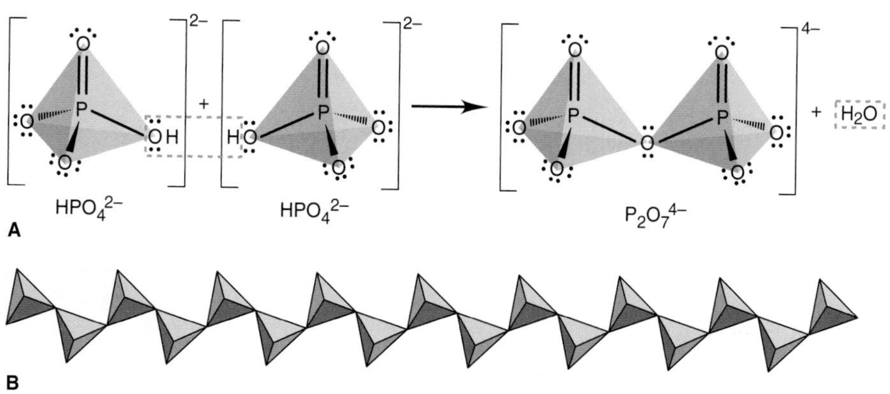

FIGURE 13.20 The diphosphate ion and polyphosphates

biomolecule adenosine triphosphate (ATP). In Chapters 18 and 19, we will discuss the central role of ATP in biological energy production. Extended polyphosphate chains consist of many tetrahedral PO_4 units (Figure 13.20B) and are structurally similar to silicate chains.

Phosphorus Compounds with Sulfur and Nitrogen Phosphorus forms many sulfides and nitrides. P_4S_3 is used in "strike anywhere" match heads, and P_4S_{10} is used in the manufacture of organophosphorus pesticides, such as malathion. Compounds of phosphorus and nitrogen, called *polyphosphazenes*, have properties similar to those of silicones. The $—(R_2)P{=}N—$ unit is isoelectronic with the silicone unit, $—(R_2)Si—N—$. Sheets, films, fibres, and foams of polyphosphazene are water repellent, flame resistant, solvent resistant, and flexible at low temperatures—perfect for the gaskets and O-rings in spacecraft and polar vehicles.

13.8 Group 16: The Oxygen Family

The first two members of this family—gaseous nonmetallic oxygen (O) and solid nonmetallic sulfur (S)—are among the most important elements in industry, the environment, and living things. Two metalloids, selenium (Se) and tellurium (Te), appear below them, followed by the lone metal—radioactive polonium (Po)—and recently synthesized Element 116. The Group 16 Family Portrait displays the features of these elements.

How Do the Oxygen and Nitrogen Families Compare Physically?

Group 16 resembles group 15 in many respects, so let us look at some common themes. The pattern of physical properties we saw in group 15 appears again in this group.

- *Oxygen*, like nitrogen, occurs as a low-boiling diatomic gas.
- *Sulfur*, like phosphorus, occurs as a polyatomic molecular solid.
- *Selenium*, like arsenic, commonly occurs as a grey metalloid.
- *Tellurium*, like antimony, is slightly more metallic than the preceding group member but still displays network covalent bonding.
- *Polonium*, like bismuth, has a metallic crystal structure.

As in group 15, electrical conductivity increases steadily down group 16 as bonding changes from nonmetal molecules (insulators) to metalloid networks (semiconductors), to a metallic solid (conductor).

Allotropism in the Oxygen Family Allotropism is more common in group 16 than it is in group 15.

1. *Oxygen.* Oxygen has two allotropes: life-giving dioxygen (O_2) and poisonous triatomic ozone (O_3). Oxygen gas is colourless, odourless, paramagnetic, and thermally stable. In contrast, ozone gas is bluish, has a pungent odour, is diamagnetic, and decomposes in heat and especially in ultraviolet (UV) light:

$$2O_3(g) \xrightarrow{\text{UV}} 3O_2(g)$$

Family Portrait Group 16: The Oxygen Family

Key Atomic Properties, Physical Properties, and Reactions

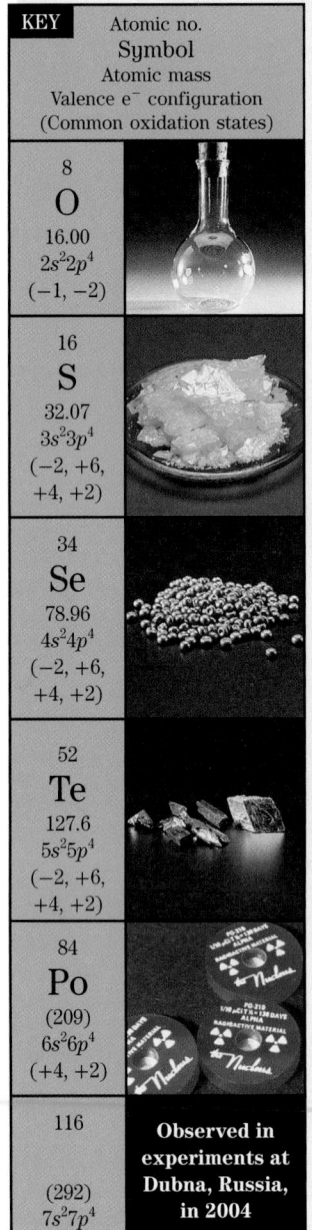

KEY
Atomic no.
Symbol
Atomic mass
Valence e^- configuration
(Common oxidation states)

8
O
16.00
$2s^22p^4$
(−1, −2)

16
S
32.07
$3s^23p^4$
(−2, +6, +4, +2)

34
Se
78.96
$4s^24p^4$
(−2, +6, +4, +2)

52
Te
127.6
$5s^25p^4$
(−2, +6, +4, +2)

84
Po
(209)
$6s^26p^4$
(+4, +2)

116

(292)
$7s^27p^4$

Observed in experiments at Dubna, Russia, in 2004

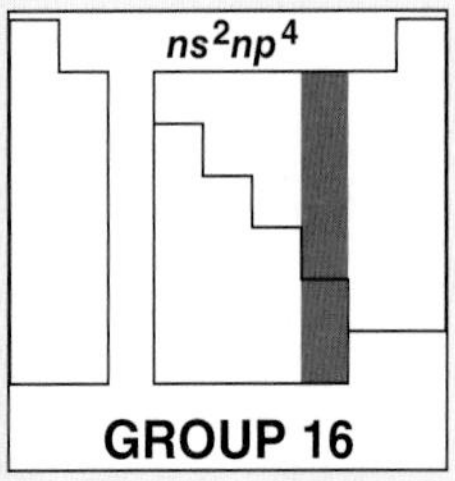

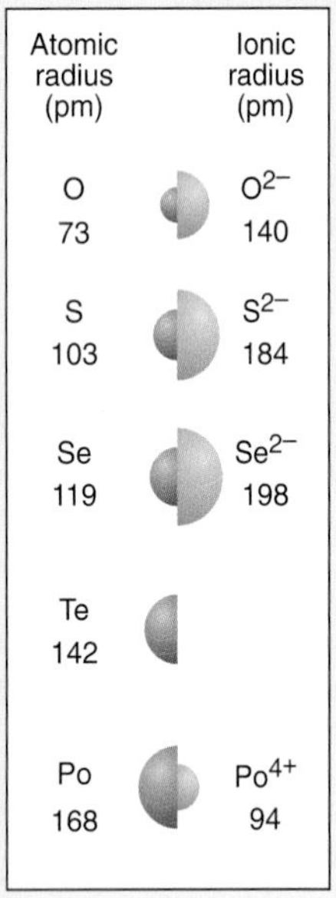

Atomic Properties

The group electron configuration is ns^2np^4. As in groups 13 and 15, a lower (+4) oxidation state becomes more common down the group. Down the group, atomic and ionic sizes increase, and IE and χ decrease.

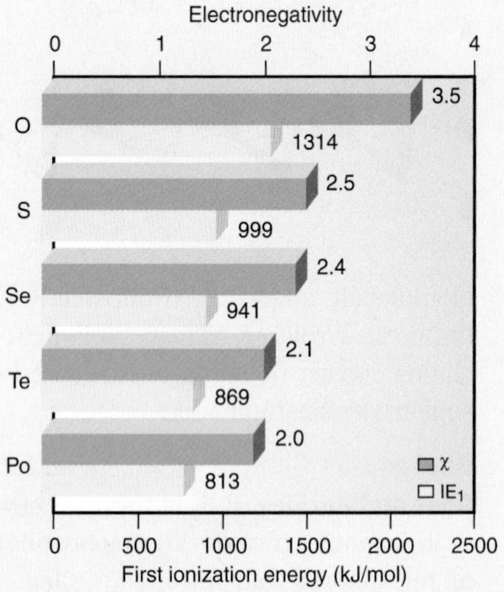

Physical Properties

Melting points increase through Te, which has covalent bonding, and then decrease for Po, which has metallic bonding. Densities of the elements as solids increase steadily.

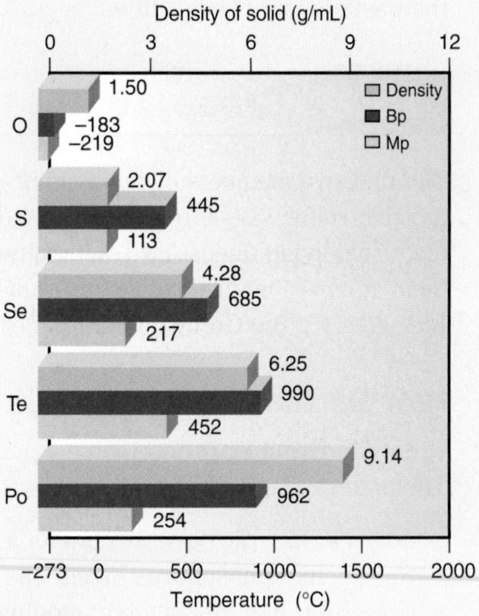

Reactions

1. Halides are formed by direct combination:

$$E(s) + X_2(g) \longrightarrow \text{various halides} \quad (E = S, Se, Te; X = F, Cl)$$

2. The other elements in the group are oxidized by O_2:

$$E(s) + O_2(g) \longrightarrow EO_2 \quad (E = S, Se, Te, Po)$$

SO_2 is oxidized further, and the product is used in the final step of H_2SO_4 manufacture (*see discussion*):

$$2SO_2(g) + O_2(g) \longrightarrow 2SO_3(g)$$

3. The thiosulfate ion is formed when an alkali metal sulfite reacts with sulfur, as in the preparation of "hypo," a developing solution used by photographers:

$$S_8(s) + 8Na_2SO_3(s) \longrightarrow 8Na_2S_2O_3(aq)$$

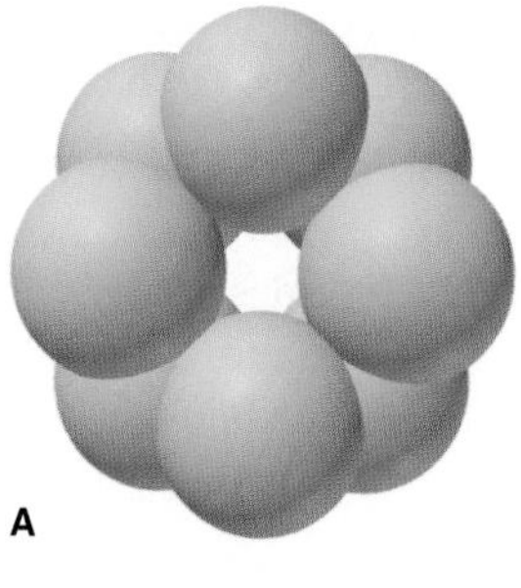

A

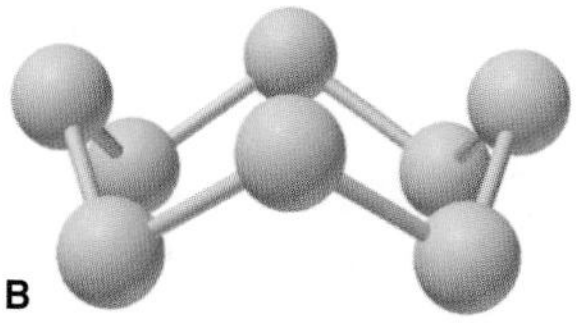

B

FIGURE 13.21 The cyclo-S_8 molecule.
A. Top view of a space-filling model.
B. Side view of a ball-and-stick model; notice the crownlike shape.

Ozone's ability to absorb high-energy photons makes stratospheric ozone vital to life. A thinning of the ozone layer, observed above the North Pole and especially above the South Pole, means that more UV light is reaching Earth's surface, with potentially hazardous effects. (We will discuss the chemical causes of ozone depletion in Chapter 14.)

2. *Sulfur.* Sulfur is the allotrope "champion" of the periodic table, with more than 10 forms. The S atom's ability to bond to other S atoms (catenate) creates rings and chains, many with S—S bond lengths that range from 180 pm to 260 pm and bond angles that range from 90° to 180°. At room temperature, the sulfur molecule is a crown-shaped ring of eight atoms, called *cyclo-S_8* (Figure 13.21). The most stable allotrope is orthorhombic α-S_8, which consists of cyclo-S_8; all other S allotropes eventually revert to this one.

The Athabasca Oil Sands in Alberta have a high sulfur content. Stockpiles of elemental sulfur, which have been recovered from these hydrocarbons, now exist throughout Alberta and British Columbia (*see margin photo*). Sulfur has many different uses. It is utilized in the steel, petroleum, rubber, and sugar refining industry, as well as in the production of sulfuric acid, inorganic chemicals, glass, matches, explosives, fireworks, cement, adhesives, and fertilizers. It is also used as a fungicide, fumigant, and insecticide.

Industrial sulfur piles, North Vancouver, British Columbia

3. *Selenium.* Selenium also has several allotropes, some consisting of crown-shaped Se_8 molecules. Grey Se is composed of layers of helical chains. When molten glass, cadmium sulfide, and grey Se are mixed and heated in the absence of air, a ruby-red glass forms. You may still see this glass when you stop at a traffic light. The ability of grey Se to conduct an electric current when illuminated is used by the photocopying industry. A film of amorphous Se is deposited on an aluminum drum and electrostatically charged. Exposure to a document produces an "image" of low and high positive charges, which correspond to the document's bright and dark areas. Negatively charged black dry-ink (toner) particles are attracted to regions of high charge more than they are attracted to regions of low charge. This pattern of black particles is transferred electrostatically to paper, and the particles are fused to the paper's surface by heat or solvent. Excess toner is removed from the Se film, the charges are "erased" by exposure to light, and the film is ready for the next page.

How Do the Oxygen and Nitrogen Families Compare Chemically?

Changes in chemical behaviour in group 16 are also similar to those in group 15. Even though O and S occur as anions much more often than N and P do, they bond covalently with almost every other nonmetal, like N and P do. Covalent bonds appear in the compounds of Se and Te (as in the compounds of As and Sb), whereas Po behaves like a metal (as does Bi) in some of its saltlike compounds. In contrast to nitrogen, oxygen has few common oxidation states, but the earlier pattern returns with the other group 16 members: the +6, +4, and −2 states occur most often, with the lower positive (+4) state becoming more common in Te and Po [as the lower positive (+3) state does in Sb and Bi].

The range in atomic properties is wider in this group than in group 15 because of oxygen's high χ (3.5) and great oxidizing strength, second only to that of fluorine. However, the other members of group 16 behave very little like oxygen: they are much less electronegative, they form anions much less often (S^{2-} occurs with active metals), and their hydrides exhibit no hydrogen bonding.

FIGURE 13.22 Tris Chivers (University of Calgary) is interested in the general area of inorganic chemistry of the main-group elements, particularly boron, nitrogen, phosphorus, sulfur, selenium, and tellurium. His work involves a combination of synthesis and structural studies, aided by techniques such as multinuclear NMR, IR/Raman, ESR, and UV-visible spectroscopies in conjunction with X-ray crystallography. Dr. Chivers and his group evaluate the applications of new inorganic materials in the context of their unusual reactivities or unique electronic, optical, or magnetic properties. He and his group have co-authored more than 300 publications, including reviews, books, and chapters of books.

Professor Tristram "Tris" Chivers (Figure 13.22) is one of several researchers in Canada who are studying the oxygen and the nitrogen families. He has more than 40 years of experience in the chemistry of inorganic ring systems, with an emphasis on groups 15 and 16.

Types and Properties of Hydrides Oxygen forms two hydrides, water and hydrogen peroxide (H_2O_2). Both have relatively high boiling points and viscosities due to hydrogen bonding. In peroxides, O is in the -1 oxidation state, midway between its oxidation state in O_2 (0) and oxides (-2); thus, H_2O_2 readily disproportionates:

$$H_2O_2(l) \longrightarrow H_2O(l) + \frac{1}{2}O_2(g)$$

Aside from its familiar use as a hair bleach and topical disinfectant, much more H_2O_2 is used to bleach paper pulp, textiles, straw, and leather, and to oxidize bacteria in sewage treatment.

The other group 16 elements form foul-smelling, poisonous, gaseous hydrides (H_2E) when treated with acids of the metal sulfide, selenide, and so on, for example,

$$FeSe(s) + 2HCl(aq) \longrightarrow H_2Se(g) + FeCl_2(aq)$$

Hydrogen sulfide also forms naturally in swamps from the breakdown of organic matter. It is as toxic as HCN; even worse, it anaesthetizes your olfactory nerves, so that you smell it less as its concentration increases! The other hydrides are about 100 times *more* toxic.

In their bonding and thermal stability, these group 16 hydrides have several features in common with those of group 15:

- Only water and H_2O_2 can form hydrogen bonds, so they melt and boil at much higher temperatures than the other H_2E compounds (see Figure 11.17).
- Bond angles drop from the nearly tetrahedral value for H_2O (104.5°) to around 90° for the larger group 16 hydrides, suggesting that the central atom uses unhybridized *p* orbitals.
- E—H bond length increases (bond energy decreases) down the group. Thus, H_2Te decomposes above 0°C (273.15 K), and H_2Po can be made only in extreme cold because thermal energy from the radioactive Po decomposes it. Another result of longer (weaker) bonds is that the group 16 hydrides are acids in water, and their acidity increases from H_2S to H_2Po.

Types and Properties of Halides Except for O, the group 16 elements form a wide range of halides. The structure and reactivity patterns of these halides depend on the *sizes of the central atom and the surrounding halogens:*

- Sulfur forms many fluorides, a few chlorides, one bromide, but no stable iodides.
- As the central atom becomes larger, the halides become more stable. Thus, tetrachlorides and tetrabromides of Se, Te, and Po are known, as are tetraiodides of Te and Po. Hexafluorides are known only for S, Se, and Te.

The inverse relationship between bond length and bond strength that we have previously seen does not account for this pattern. Rather, it is based on the effect of electron repulsions due to the crowding of lone pairs and halogen (X) atoms around the central group 16 atom. With S, the larger X atoms would be too crowded, which explains why sulfur iodides do not occur. However, with increasing size of E, and therefore increasing length of E—X bonds, lone pairs and X atoms do not crowd each other as much, so a greater number of stable halides form.

Highlights of Oxygen Chemistry: Range of Oxide Properties

Oxygen is the most abundant element on Earth's surface, occurring both as a free element and in innumerable oxides, silicates, carbonates, and phosphates, as well as in water. Virtually all free O_2 has a biological origin, having been formed for billions of years by photosynthetic algae and multicellular plants in an overall equation that looks deceptively simple:

$$nH_2O(l) + nCO_2(g) \xrightarrow{\text{light}} nO_2(g) + (CH_2O)_n \text{ (carbohydrates)}$$

The reverse process occurs during combustion and respiration. Through these O_2-forming and O_2-utilizing processes, the 1.5×10^9 km^3 of water on Earth is, on average, used and remade every 2 million years!

Every element except He, Ne, and Ar forms at least one oxide, many by direct combination. A broad spectrum of properties characterizes these compounds. Some oxides are gases that condense at very low temperatures, such as CO (bp = −192°C); others are solids that melt at extremely high temperatures, such as BeO (mp = 2530°C). Oxides cover the full range of conductivity: insulators (MgO), semiconductors (NiO), conductors (ReO_3), and superconductors ($YBa_2Cu_3O_7$). They may be thermally stable (CaO) or unstable (HgO), as well as chemically reactive (Li_2O) or inert (Fe_2O_3).

Given this vast range of behaviour, another useful way to classify element oxides is by their acid-base properties. The oxides of group 16 exhibit the expected trends in acidity, with SO_3 being the most acidic and PoO_2 the most basic.

Highlights of Sulfur Chemistry

Like phosphorus, sulfur forms two common oxides and two oxoacids, one of which is essential to a wide variety of industries. There are also several important metal sulfides.

Sulfur Oxides Sulfur forms two important oxides: sulfur dioxide and sulfur trioxide.

1. *Sulfur dioxide* (SO_2) has S in its +4 oxidation state. It is a colourless, choking gas that forms when S, H_2S, or a metal sulfide burns in air:

$$2H_2S(g) + 3O_2(g) \longrightarrow 2H_2O(g) + 2SO_2(g)$$
$$4FeS_2(s) + 11O_2(g) \longrightarrow 2Fe_2O_3(s) + 8SO_2(g)$$

2. *Sulfur trioxide* (SO_3), which has S in the +6 oxidation state, is produced when sulfur dioxide reacts in oxygen. A catalyst (Chapter 14) must be used to speed up this very slow reaction. En route to the production of sulfuric acid, a vanadium(V) oxide catalyst is used:

$$SO_2(g) + \frac{1}{2}O_2(g) \xrightarrow{V_2O_5/K_2O \text{ catalyst}} SO_3(g)$$

Sulfur Oxoacids Sulfur forms two important oxoacids: sulfurous acid and sulfuric acid.

1. *Sulfurous acid* (H_2SO_3), formed when sulfur dioxide dissolves in water, exists in equilibrium with hydrated SO_2 rather than as stable H_2SO_3 molecules:

$$SO_2(aq) + H_2O(l) \rightleftharpoons H_2SO_3(aq) \rightleftharpoons H^+(aq) + HSO_3^-(aq)$$

Sulfurous acid is weak and has two acidic protons, forming the hydrogen sulfite (bisulfite, HSO_3^-) and sulfite (SO_3^{2-}) ions with a strong base. Because the S in SO_3^{2-} is in the +4 state and is easily oxidized to the +6 state, sulfites are good reducing agents and are used to preserve food and wine by eliminating undesirable products of air oxidation.

2. *Sulfuric acid* (H_2SO_4) is produced when SO_2 is oxidized catalytically to SO_3, which is then absorbed into concentrated H_2SO_4 and treated with additional H_2O:

$$SO_3(\text{in concentrated } H_2SO_4) + H_2O(l) \longrightarrow H_2SO_4(l)$$

With more than 4.3 million t produced each year in Canada alone, H_2SO_4 ranks first among all industrial chemicals. It is vital to fertilizer production; to metal, pigment, and textile processing; and to soap and detergent manufacturing. (The production of H_2SO_4 will be discussed in detail in Chapter 23.)

Concentrated laboratory-grade sulfuric acid is a viscous, colourless liquid that is 98% H_2SO_4 by mass. Like other strong acids, H_2SO_4 dissociates completely in water, forming the hydrogen sulfate (or bisulfate) ion, a much weaker acid:

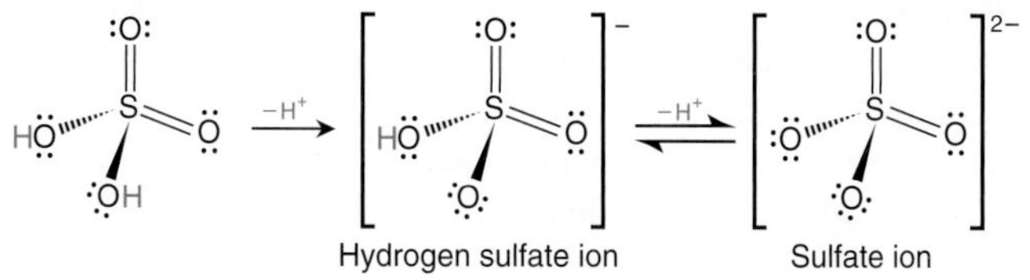

Hydrogen sulfate ion

Sulfate ion

FIGURE 13.23 The dehydration of sugar by sulfuric acid

Industrial sources produce the sulfur oxides that lead to acid rain.

Most common hydrogen sulfates and sulfates are water soluble, but those of group 2 members (except $MgSO_4$), Ag^+, and Pb^{2+} are not.

Concentrated sulfuric acid is an excellent dehydrating agent. Its loosely held proton transfers to water in a highly exothermic formation of hydronium (H_3O^+) ions. This process can occur even when the reacting substance contains no free water. For example, H_2SO_4 dehydrates wood, natural fibres, and many other organic substances, such as table sugar $(CH_2O)_n$, by removing the components of water from the molecular structure, leaving behind a carbonaceous mass (Figure 13.23).

Sulfuric acid is one of the components of acid rain. Enormous amounts of SO_2 are emitted by coal-burning power plants, petroleum refineries, and metal-ore smelters (*see photo*). In contact with rain, the SO_2 and its oxidation product, SO_3, form H_2SO_3 and H_2SO_4 in the atmosphere. The H_2SO_3 and H_2SO_4 then fall in rain, snow, and dust on animals, plants, buildings, and lakes (see Chemical Connections in Chapter 17).

Metal Sulfides Many metals combine directly with S to form *metal sulfides.* Sulfide ores are mined for the extraction of many metals, including copper, zinc, lead, and silver. Aside from the sulfides of groups 1 and 2, most metal sulfides do not have discrete S^{2-} ions. Several transition metals, such as chromium, iron, and nickel, form covalent, alloy-like, nonstoichiometric compounds with S, such as $Cr_{0.88}S$ or $Fe_{0.86}S$. Some important minerals contain S_2^{2-} ions; an example is iron pyrite, or "fool's gold" (FeS_2). We will discuss the metallurgy of ores in Chapter 23.

13.9 Group 17: The Halogens

The last elements of great reactivity are found in group 17. The halogens begin with fluorine (F), the strongest electron "grabber" of all. Chlorine (Cl, by far the most important industrially), bromine (Br), and iodine (I) also form compounds with most elements. Even rare radioactive astatine (At) is thought to be reactive. The key features of group 17 are presented in the Family Portrait.

What Accounts for the Regular Changes in the Halogens' Physical Properties?

Like the alkali metals at the other end of the periodic table, the halogens display regular trends in their physical properties. However, whereas the melting and boiling points and the heats of fusion and vaporization *decrease* down group 1, these properties *increase* down group 17. The reason for the opposite trends is the different type of bonding in the elements. The alkali metals consist of atoms held together by metallic bonding, which *decreases* in strength as the atoms become larger. The halogens, on the other hand, exist as diatomic molecules that interact through dispersion forces, which *increase* in strength as the atoms become larger and more easily polarized. Thus, F_2 is a very pale yellow gas, Cl_2 is a yellow-green gas, Br_2 is a brown-orange liquid, and I_2 is a purple-black solid.

Group 17: The Halogens

Family Portrait

Key Atomic Properties, Physical Properties, and Reactions

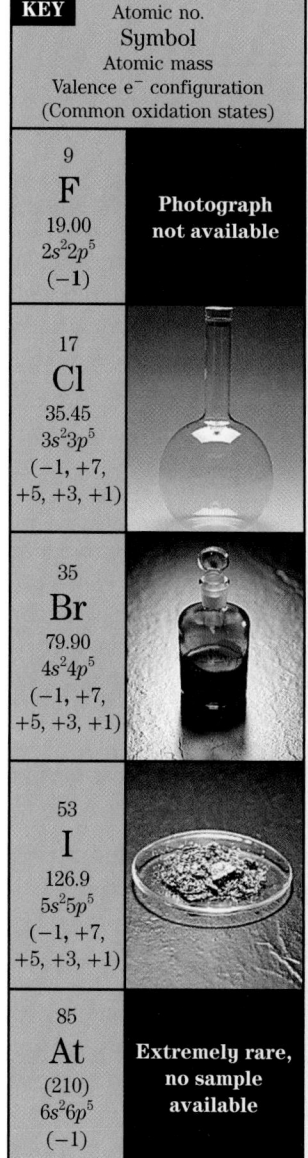

KEY	
Atomic no. Symbol Atomic mass Valence e⁻ configuration (Common oxidation states)	
9 **F** 19.00 $2s^22p^5$ (−1)	**Photograph not available**
17 **Cl** 35.45 $3s^23p^5$ (−1, +7, +5, +3, +1)	
35 **Br** 79.90 $4s^24p^5$ (−1, +7, +5, +3, +1)	
53 **I** 126.9 $5s^25p^5$ (−1, +7, +5, +3, +1)	
85 **At** (210) $6s^26p^5$ (−1)	**Extremely rare, no sample available**

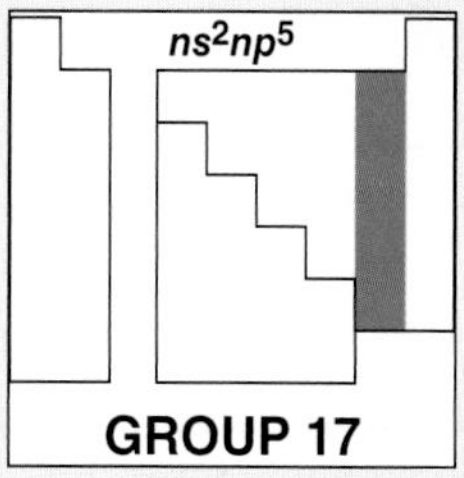

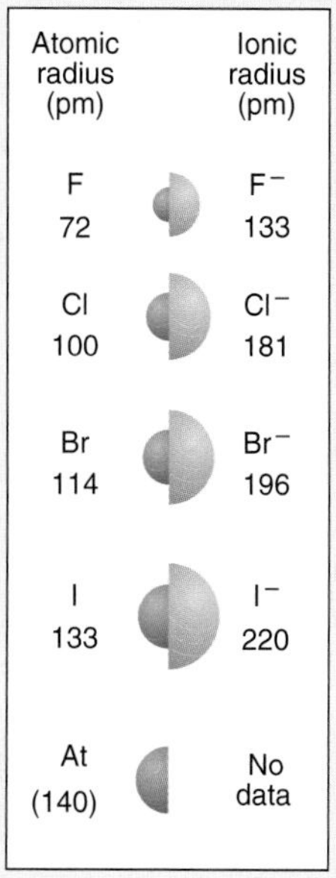

Atomic Properties

The group electron configuration is ns^2np^5; elements lack one electron to complete their outer level. The −1 oxidation state is the most common for all members. Except for F, the halogens exhibit all odd-numbered states (+7 through −1). Down the group, atomic and ionic sizes increase steadily, as IE and χ decrease.

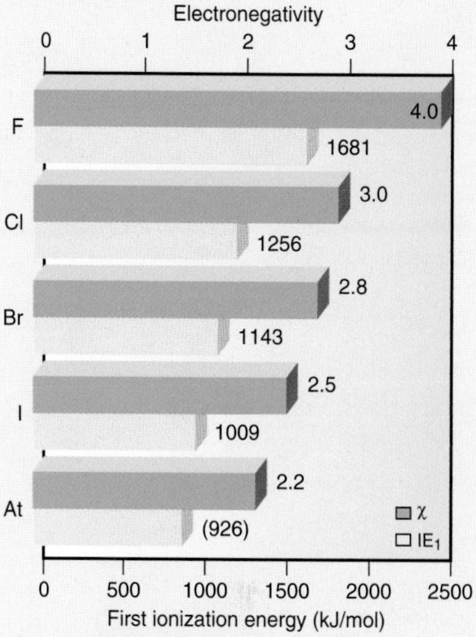

Physical Properties

Down the group, melting and boiling points increase smoothly as a result of stronger dispersion forces between larger molecules. The densities of the elements as liquids (at the given T) increase steadily with molar mass.

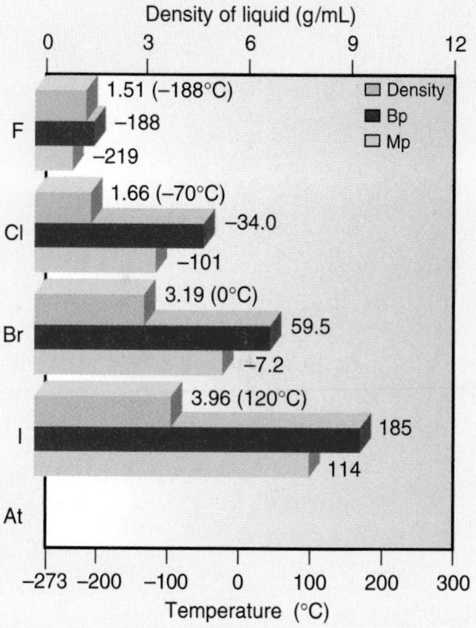

Reactions

1. The halogens (X_2) oxidize many metals and nonmetals. The reaction with hydrogen, although not used commercially for HX production (except for high-purity HCl), is characteristic of these strong oxidizing agents:

$$X_2 + H_2(g) \longrightarrow 2HX(g)$$

2. The halogens undergo disproportionation in water:

$$X_2 + H_2O(l) \rightleftharpoons HX(aq) + HXO(aq)$$

(X = Cl, Br, I)

In an aqueous base, the reaction goes to completion to form hypohalites (*see discussion*) and, at higher temperatures, halates, for example,

$$3Cl_2(g) + 6OH^-(aq) \xrightarrow{\Delta} ClO_3^-(aq) + 5Cl^-(aq) + 3H_2O(l)$$

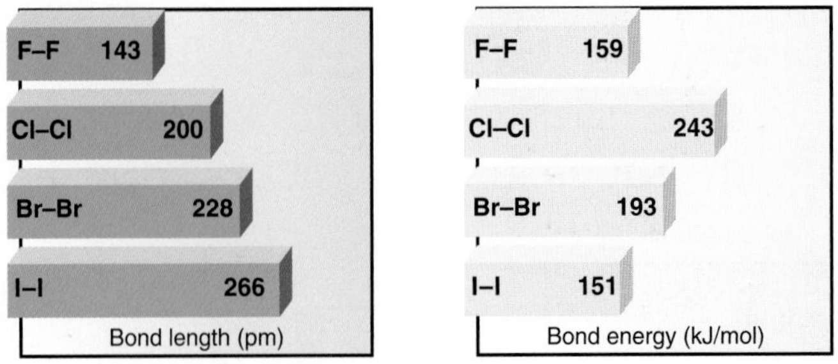

FIGURE 13.24 Bond energies and bond lengths of the halogens

Why Are the Halogens So Reactive?

The group 17 elements react with most metals and nonmetals to form many ionic and covalent compounds: metal and nonmetal halides, halogen oxides, and oxoacids. The main reason for halogen reactivity is the same as for alkali metal reactivity: an electron configuration that is one electron away from the electron configuration of a noble gas. Whereas a group 1 metal atom must lose one electron to attain a filled outer level, *a group 17 nonmetal atom must gain one electron to fill its outer level.* It accomplishes this filling in one of two ways:

1. Gaining an electron from a metal atom, thus forming a negative ion as the metal forms a positive ion
2. Sharing an electron pair with a nonmetal atom, thus forming a covalent bond

Electronegativity and Bond Properties The halogens display the largest range in electronegativity of any group, but all the halogens are electronegative enough to behave as nonmetals. Down the group, reactivity reflects the decrease in electronegativity: F_2 is the most reactive, and I_2 is the least reactive. The exceptional reactivity of elemental F_2 is also related to the weakness of the F—F bond. Although bond energy generally decreases as atomic size increases down the group (Figure 13.24), F_2 deviates from this trend. The short F—F bond is weaker than expected because lone pairs on each small F atom repel those on the other. As a result of these factors, F_2 reacts with every element (except He, Ne, and Ar), in many cases explosively.

Redox Behaviour The halogens act as *oxidizing agents* in the majority of their reactions, and halogens higher in the group can oxidize halide ions lower down:

$$F_2(g) + 2X^-(aq) \longrightarrow 2F^-(aq) + X_2(aq) \qquad (X = Cl, Br, I)$$

Thus, the oxidizing ability of X_2 increases *up* the group: the higher the χ, the more strongly each X atom pulls electrons away. Similarly, the reducing ability of X^- increases *down* the group: the larger the ion, the more easily it gives up its electron (Figure 13.25A). Aqueous Cl_2 added to a solution of I^- (Figure 13.25B, *top layer*) oxidizes the I^- to I_2, which dissolves in the CCl_4 solvent (*bottom layer*) to give a purple solution.

The halogens undergo some important aqueous redox chemistry. Fluorine is such a powerful oxidizing agent that it reacts vigorously with water, oxidizing the

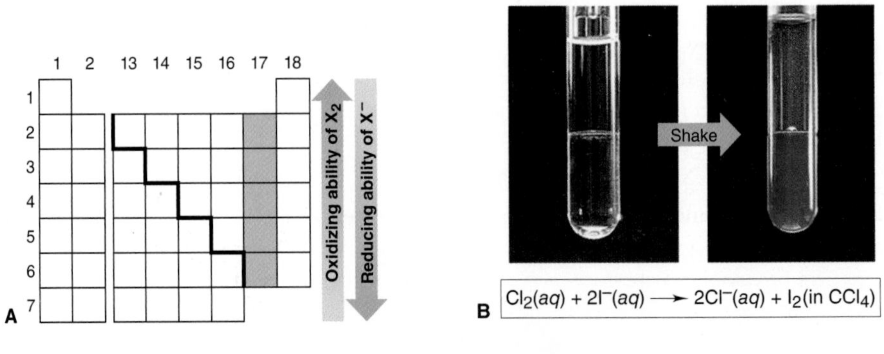

FIGURE 13.25 The relative oxidizing ability of the halogens

O to produce O_2, some O_3, and HFO (hypofluorous acid). The other halogens undergo disproportionations (note the oxidation numbers):

$$\overset{0}{X_2} + H_2O(l) \rightleftharpoons H\overset{-1}{X}(aq) + H\overset{+1}{X}O(aq) \qquad (X = Cl, Br, I)$$

At equilibrium, very little product is present unless excess OH^- ions are added. The excess OH^- ions react with the HX and HXO and drive the reaction to completion:

$$X_2 + 2OH^-(aq) \longrightarrow X^-(aq) + XO^-(aq) + H_2O(l)$$

When X is Cl, the product mixture acts as a bleach: household bleach is a dilute solution of sodium hypochlorite (NaClO). Heating causes XO^- to disproportionate further, creating oxoanions with X in a higher oxidation state:

$$3\overset{+1}{X}O^-(aq) \xrightarrow{\Delta} 2\overset{-1}{X}^-(aq) + \overset{+5}{X}O_3^-(aq)$$

Highlights of Halogen Chemistry

In this section, we examine the compounds that the halogens form with hydrogen and with each other, as well as their oxides, oxoanions, and oxoacids.

The Hydrogen Halides The halogens form gaseous hydrogen halides (HX) through direct combination with H_2 or through the action of a concentrated acid on the metal halide (a nonoxidizing acid is used for HBr and HI):

$$CaF_2(s) + H_2SO_4(l) \longrightarrow CaSO_4(s) + 2HF(g)$$

$$3NaBr(s) + H_3PO_4(l) \longrightarrow Na_3PO_4(s) + 3HBr(g)$$

Commercially, most HCl is formed as a by-product in the chlorination of hydrocarbons for plastics production:

$$CH_2{=}CH_2(g) + Cl_2(g) \longrightarrow ClCH_2CH_2Cl(l) \xrightarrow{500°C} \underset{\text{Chloroethene (vinyl chloride)}}{CH_2{=}CHCl(g)} + HCl(g)$$

The chloroethene (vinyl chloride) reacts in a separate process to form poly(vinyl chloride), or PVC, a polymer used extensively in plumbing and other piping needs.

In water, gaseous HX molecules form a *hydrohalic acid.* Only HF, with its relatively short strong bond, forms a weak acid:

$$HF(g) + H_2O(l) \rightleftharpoons H_3O^+(aq) + F^-(aq)$$

HF has many uses, including the synthesis of cryolite (Na_3AlF_6) for aluminum production (Chapter 23), fluorocarbons for refrigeration, and NaF for water fluoridation. HF is also used in nuclear fuel processing and for glass etching.

The other hydrohalic acids dissociate completely to form the stoichiometric amount of H_3O^+ ions:

$$HBr(g) + H_2O(l) \longrightarrow H_3O^+(aq) + Br^-(aq)$$

These reactions involve the *transfer* of a proton from an acid to H_2O and are classified as *Brønsted-Lowry acid-base reactions.* In Chapter 16, we will discuss these reactions thoroughly and examine the relation between bond length and acidity of the larger HX molecules.

HCl, a common laboratory reagent, is used in the "pickling" of steel to remove adhering oxides and in the production of syrups, rayon, and plastic. HCl(*aq*) occurs naturally in the stomach fluids of mammals.

Interhalogen Compounds: The "Halogen Halides" *Halogens react exothermically with one another to form many* **interhalogen compounds**. The simplest are diatomic molecules, such as ClF and BrCl. Every binary combination of the four common halogens is known. The more electronegative halogen is in the −1 oxidation state, and the less electronegative halogen is in the +1 state. Interhalogens of general formula XY_n (n = 3, 5, 7) form through a variety of reactions, including direct reaction of the elements. In every case, the central atom has the lower *electronegativity* and a positive oxidation state.

FIGURE 13.26 Molecular shapes of the main types of interhalogen compounds (as seen in Section 9.1)

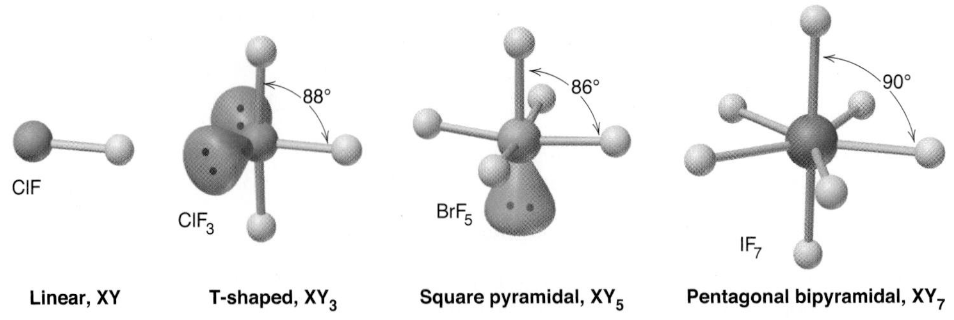

Some interhalogens are used commercially as powerful *fluorinating agents*, which react with metals, nonmetals, and oxides—even wood and asbestos:

$$Sn(s) + ClF_3(l) \longrightarrow SnF_2(s) + ClF(g)$$

$$P_4(s) + 5ClF_3(l) \longrightarrow 4PF_3(g) + 3ClF(g) + Cl_2(g)$$

$$2B_2O_3(s) + 4BrF_3(l) \longrightarrow 4BF_3(g) + 2Br_2(l) + 3O_2(g)$$

Their reactions with water are nearly explosive and yield HF and the *oxoacid with the central halogen in the same oxidation state,* for example,

$$3H_2O(l) + \overset{+5}{Br}F_5(l) \longrightarrow 5HF(g) + H\overset{+5}{Br}O_3(aq)$$

The Oddness and Evenness of Oxidation States *Almost all stable molecules have paired electrons,* either as bonding or lone pairs. Therefore, *when bonds form or break, two electrons are involved, so the oxidation state changes by 2.* For this reason, odd-numbered groups exhibit odd-numbered oxidation states, and even-numbered groups exhibit even-numbered oxidation states.

1. *Odd-numbered oxidation states.* Consider the interhalogens. Four general formulas are XY, XY_3, XY_5, and XY_7; examples are shown in Figure 13.26. With Y in the -1 state, X must be in the $+1$, $+3$, $+5$, and $+7$ state, respectively. The -1 state arises when Y fills its valence level; the $+7$ state arises when the central halogen (X) is completely oxidized, that is, when all seven valence electrons have shifted away from it to the more electronegative Y atoms around it.

Let us examine the iodine fluorides to see why the oxidation states jump by two units. When I_2 reacts with F_2, IF forms (note the oxidation number of I):

$$I_2 + F_2 \longrightarrow 2\overset{+1}{I}F$$

In IF_3, I uses *two* more valence electrons to form *two* more bonds:

$$\overset{+1}{I}F + F_2 \longrightarrow \overset{+3}{I}F_3$$

Otherwise, an unstable lone-electron species containing two fluorines would form. With more fluorine, another jump of two units occurs and the pentafluoride forms:

$$\overset{+3}{I}F_3 + F_2 \longrightarrow \overset{+5}{I}F_5$$

With still more fluorine, the heptafluoride forms:

$$\overset{+5}{I}F_5 + F_2 \longrightarrow \overset{+7}{I}F_7$$

2. *Even-numbered oxidation states.* An element in an even-numbered group, such as sulfur in group 16, shows the same tendency to have paired electrons in its compounds. Elemental sulfur (oxidation number = 0) gains or shares two electrons to complete its shell (oxidation number = -2). It uses two electrons when it reacts with fluorine to form SF_2 (oxidation number = $+2$), two more electrons to form SF_4 (oxidation number = $+4$), and another two electrons to form SF_6 (oxidation number = $+6$).

Thus, an element with one even state typically has all even states, and an element with one odd state typically has all odd states. To reiterate the main point, *successive oxidation states differ by two units because stable molecules have electrons in pairs around their atoms.*

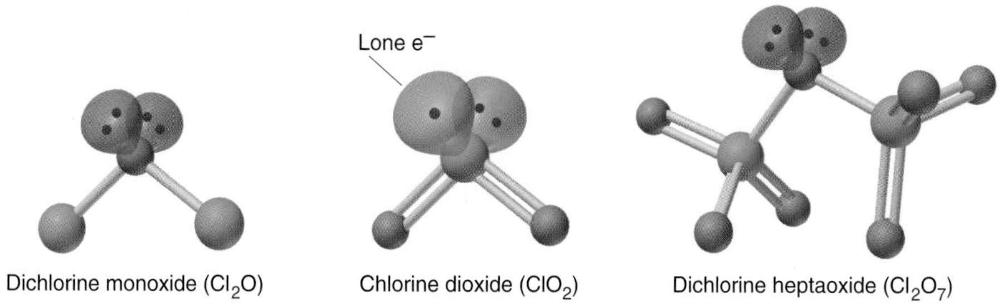

FIGURE 13.27 Chlorine oxides. These structures show each central Cl atom with its lowest formal charge.

Halogen Oxides, Oxoacids, and Oxoanions The group 17 elements form many oxides that are *powerful oxidizing agents and acids in water.* Dichlorine monoxide (Cl_2O) and especially chlorine dioxide (ClO_2) are used to bleach paper (Figure 13.27). ClO_2 is unstable to heat and shock, so it is prepared on site, and more than 100 000 t is used annually:

$$2NaClO_3(s) + SO_2(g) + H_2SO_4(aq) \longrightarrow 2ClO_2(g) + 2NaHSO_4(aq)$$

The dioxide has an unpaired electron and Cl in the unusual +4 oxidation state.

Chlorine is in its highest (+7) oxidation state in dichlorine heptoxide, Cl_2O_7, which is a symmetrical molecule formed when two $HClO_4$ ($HO—ClO_3$) molecules undergo a dehydration-condensation reaction:

$$O_3Cl—O\boxed{H + HO}—ClO_3 \longrightarrow O_3Cl—O—ClO_3(l) + H_2O(l)$$

The halogen oxoacids and oxoanions are produced by reaction of the halogens and their oxides with water. Most of the oxoacids are stable only in solution. Table 13.4 shows ball-and-stick models of the acids in which each atom has its lowest formal charge; note the formulas, which emphasize that H is bonded to O. The hypohalites (XO^-), halites (XO_2^-), and halates (XO_3^-) are oxidizing agents formed by aqueous disproportionation reactions. (See the Group 17 Family Portrait, reaction 2.) You may have heated solid alkali chlorates in the laboratory to form small amounts of O_2:

$$2MClO_3(s) \xrightarrow{\Delta} 2MCl(s) + 3O_2(g)$$

Potassium chlorate is the oxidizer in "safety" matches.

Several perhalates (XO_4^-) are also strong oxidizing agents. Thousands of tonnes of perchlorates are made each year for use in explosives and fireworks. Ammonium perchlorate, prepared from sodium perchlorate, is the oxidizing agent for the aluminum powder that was used in the solid-fuel booster rockets of space shuttles; each launch used more than 700 t of NH_4ClO_4:

$$10Al(s) + 6NH_4ClO_4(s) \longrightarrow 4Al_2O_3(s) + 12H_2O(g) + 3N_2(g) + 2AlCl_3(g)$$

TABLE 13.4 The Known Halogen Oxoacids*

Central Atom	Hypohalous Acid (HOX)	Halous Acid (HOXO)	Halic Acid ($HOXO_2$)	Perhalic Acid ($HOXO_3$)
	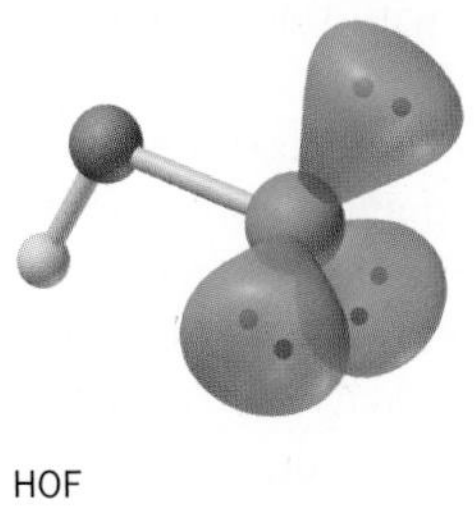	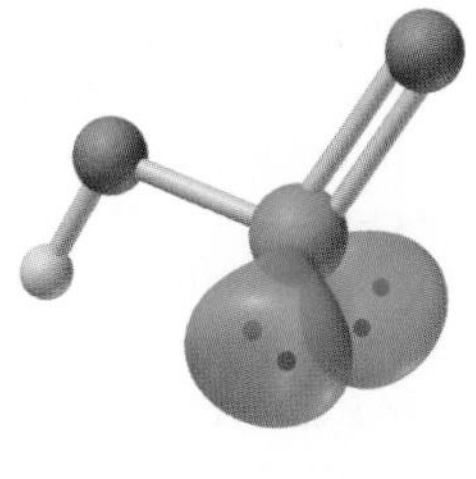	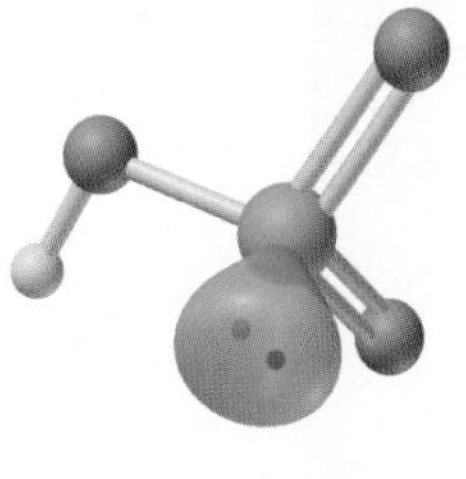	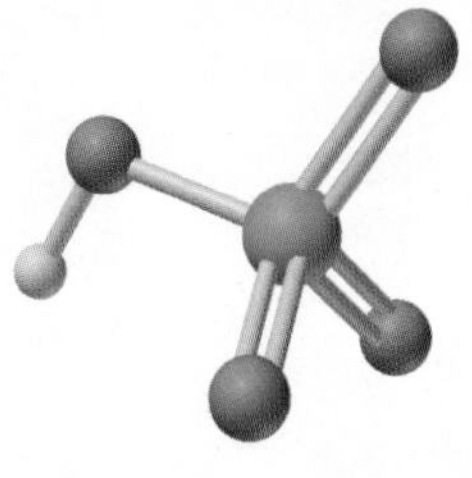
Fluorine	HOF	—	—	—
Chlorine	HOCl	HOClO	$HOClO_2$	$HOClO_3$
Bromine	HOBr	(HOBrO?)	$HOBrO_2$	$HOBrO_3$
Iodine	HOI	—	$HOIO_2$	$HOIO_3$, $(HO)_5IO$
Oxoanion	Hypohalite	Halite	Halate	Perhalate

*Lone pairs are shown only on the halogen atom, and each atom has its lowest formal charge.

The relative strengths of the halogen oxoacids depend on two factors:

1. *Electronegativity of the halogen.* Among oxoacids with the halogen in the same oxidation state, such as the halic acids, HXO_3 (or $HOXO_2$), acid strength decreases as the halogen's χ decreases:

$$HOClO_2 > HOBrO_2 > HOIO_2$$

The more electronegative the halogen, the more electron density it removes from the O—H bond, and the more easily the proton is lost.

2. *Oxidation state of the halogen.* Among oxoacids of a given halogen, such as chlorine, acid strength decreases as the oxidation state of the halogen decreases:

$$HOClO_3 > HOClO_2 > HOClO > HOCl$$

The higher the oxidation state (number of attached O atoms) of the halogen, the more electron density it pulls from the O—H bond. We will consider these trends quantitatively in Chapter 16.

FIGURE 13.28 Neil Bartlett (1932–2008) was most famous for the work he did while at the University of British Columbia, in Vancouver, in 1962. On May 23, 2006, the Canadian Society for Chemistry (CSC) and the American Chemical Society (ACS) designated the work of Neil Bartlett and the reactive noble gases an International Historic Chemical Landmark at the University of British Columbia:

There was a young man of Vancouver
Who devised a clever manoeuvre.
He showed that a gas he was keen on
Could even react with xenon.
Thus he greatly enhanced his whole oeuvre.

13.10 Group 18: The Noble Gases

The last main group consists of individual atoms too "noble" to interact much with others. The group 18 elements display regular trends in physical properties and very low, if any, reactivity. The group consists of the following elements: helium (He), the second most abundant element in the universe; neon (Ne); argon (Ar); krypton (Kr) and xenon (Xe), the only members for which compounds have been well studied; and, finally, radioactive radon (Rn). The noble gases make up about 1% by volume of the atmosphere, primarily due to the abundance of Ar. The Group 18 Family Portrait presents an overview of these elements.

Physical Properties of the Noble Gases

Lying at the far right side of the periodic table, the group 18 elements consist of individual atoms with filled outer levels and the smallest radii in their periods. Even Li, the smallest alkali metal (152 pm), is bigger than Rn, the largest noble gas (140 pm). These elements come as close to behaving as ideal gases as any other substances. Only at very low temperatures do they condense and solidify. In fact, He is the only substance that does *not* solidify by a reduction in temperature alone; it requires an increase in pressure as well. Helium has the lowest melting point known (−272.2°C at 2533 kPa), only one degree above 0 K (−273.15°C), and it boils only about three degrees higher. Weak dispersion forces hold these elements in condensed states, with melting and boiling points that increase, as expected, with molar mass.

How Can Noble Gases Form Compounds?

Ever since their discovery in the late 19th century, these elements had been considered, and even formerly named, the "inert" gases. Atomic theory and, more important, all experiments had supported this idea. Then, in 1962, this changed when the first noble gas compound was prepared. How, with filled outer levels and extremely high ionization energies, *can* noble gases react?

The discovery of noble gas reactivity is a classic example of clear thinking in the face of an unexpected event. At the time, a young inorganic chemist named Neil Bartlett (Figure 13.28) was studying platinum fluorides, known to be strong oxidizing agents. When he accidentally exposed PtF_6 to air, its deep-red colour lightened slightly, and analysis showed that the PtF_6 had oxidized O_2 to form the ionic compound $[O_2]^+[PtF_6]^-$. Knowing that the ionization energy of the oxygen molecule ($O_2 \longrightarrow O_2^+ + e^-$; IE = 1175 kJ/mol) is very close to IE_1 of xenon (1170 kJ/mol), Bartlett reasoned that PtF_6 might be able to oxidize xenon. Shortly thereafter, he prepared $XePtF_6$, an orange-yellow solid. Within a few months, the white crystalline XeF_2 and XeF_4 (Figure 13.29) were also prepared. In addition to its +2 and +4 oxidation states, Xe has the +6 state in several compounds, such as XeF_6, and the +8 state in the unstable oxide XeO_4. A few compounds of Kr and Rn have also been made.

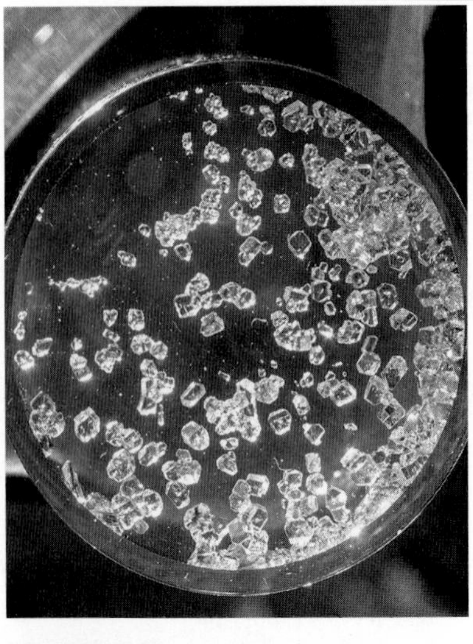

FIGURE 13.29 Crystals of xenon tetrafluoride (XeF_4)

Group 18: The Noble Gases

Family Portrait

Key Atomic and Physical Properties

KEY
Atomic no. Symbol Atomic mass Valence e^- configuration (Common oxidation states)
2 **He** 4.003 $1s^2$ **(none)**
10 **Ne** 20.18 $2s^22p^6$ **(none)**
18 **Ar** 39.95 $3s^23p^6$ **(none)**
36 **Kr** 83.80 $4s^24p^6$ **(+2)**
54 **Xe** 131.3 $5s^25p^6$ (+8, **+6**, **+4**, +2)
86 **Rn** (222) $6s^26p^6$ (+2)

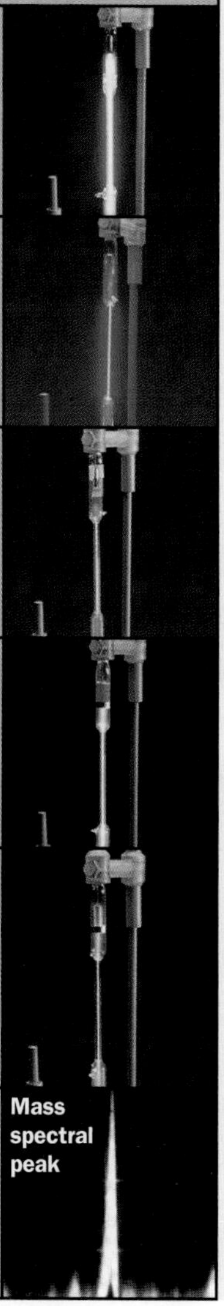

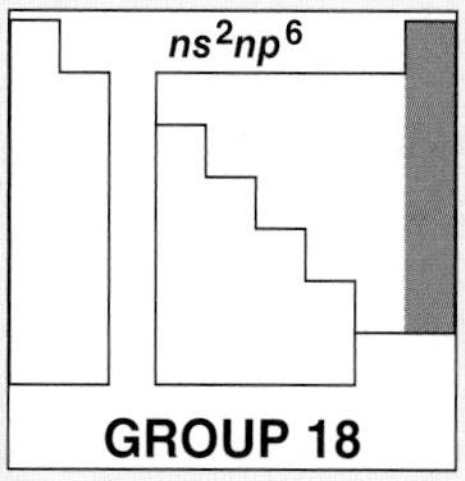

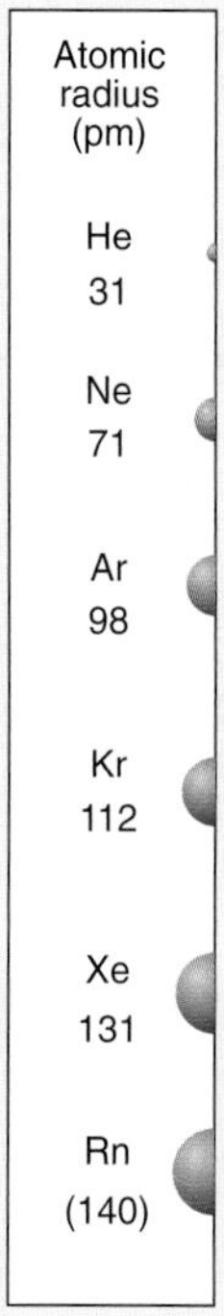

Atomic Properties

The group electron configuration is $1s^2$ for He and ns^2np^6 for the others. The valence shell is filled. Only Kr, Xe, and Rn are known to form compounds. The more reactive Xe exhibits all even oxidation states (+2 to +8). This group contains the smallest atoms with the highest IEs in their periods. Down the group, atomic size increases and IE decreases steadily. (χ values are given only for Kr and Xe.)

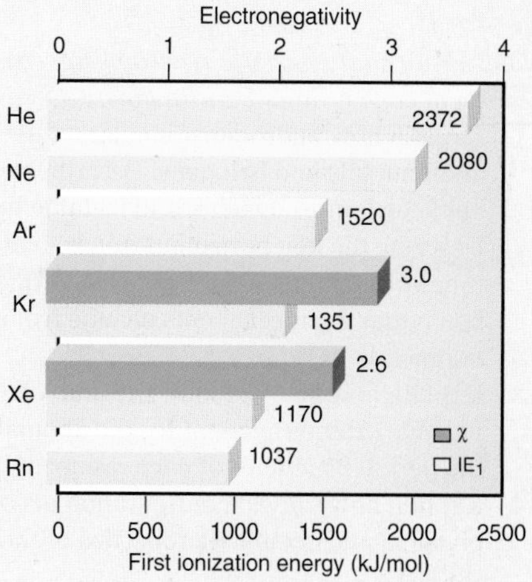

Physical Properties

Melting and boiling points of these gaseous elements are extremely low but increase down the group because of stronger dispersion forces. Note the extremely small liquid ranges. Densities (at STP) increase steadily, as expected.

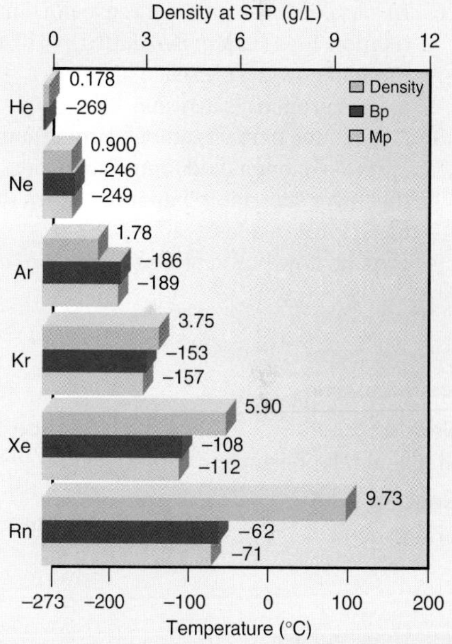

Nowadays, more research is being conducted to study the reactivity of noble gases. For example, Gary Schrobilgen (Figure 13.30) is widely regarded as one of the world leaders in noble gas–halogen chemistry.

FIGURE 13.30 Gary J. Schrobilgen (McMaster University, Hamilton, Ontario). Dr. Schrobilgen's achievements include the syntheses and structural characterizations of a significant number of the known noble-gas compounds, as well as fluoro- and oxofluoro-derivatives of the main-group and transition elements in their highest oxidation states and at the limits of coordination.

CHAPTER REVIEW GUIDE

Learning Objectives

Relevant section (§) numbers appear in parentheses.

Concepts

Note: Many characteristic reactions appear in the "Reactions" section of each group's Family Portrait.

1. Explain how hydrogen is similar to, yet different from, alkali metals and halogens. Describe the differences between ionic, covalent, and metallic hydrides. (§13.1)
2. Describe the key horizontal trends in atomic properties, types of bonding, oxide acid-base properties, and redox behaviour as the elements change from metals to nonmetals. (§13.2)
3. Explain how small atomic size and a limited number of valence orbitals account for the anomalous behaviour of the period 2 member of each group. (§13.2)
4. Explain how the ns^1 configuration accounts for the physical and chemical properties of the alkali metals. (§13.3)
5. Explain how the ns^2 configuration accounts for the key differences between groups 1 and 2. (§13.4)
6. Discuss the basis of the three important diagonal relationships (Li/Mg, Be/Al, B/Si). (§13.4–13.6)
7. Explain how the presence of inner $(n - 1)d$ electrons affects properties in group 13. (§13.5)
8. Explain the patterns among larger members of groups 13 to 16: two common oxidation states (inert pair effect), with the lower state more important down the group, and more basic lower oxides. (§13.5–13.8)
9. Explain how boron attains an octet of electrons. (§13.5)
10. Explain the effect of bonding on the physical behaviour of groups 14 to 16. (§13.6–13.8)
11. Describe allotropism in carbon, phosphorus, and sulfur. (§13.6–13.8)
12. Explain how atomic properties lead to catenation and multiple bonds in organic compounds. (§13.6)
13. Describe the structures and explain the properties of the silicates and silicones. (§13.6)
14. Analyze the patterns of behaviour among the hydrides and halides of groups 15 and 16. (§13.7, 13.8)
15. Explain the meaning of *disproportionation*. (§13.7)
16. Analyze the structure and chemistry of the nitrogen oxides and oxoacids. (§13.7)
17. Analyze the structure and chemistry of the phosphorus oxides and oxoacids. (§13.7)
18. Explain dehydration-condensation reactions and polyphosphate structures. (§13.7)
19. Analyze the structure and chemistry of the sulfur oxides and oxoacids. (§13.8)
20. Explain how the ns^2np^5 configuration accounts for the reactivity of halogens with metals. (§13.9)
21. Explain why the oxidation states of an element change by two units. (§13.9)
22. Analyze the structure and chemistry of the halogen oxides and oxoacids. (§13.9)
23. Explain how the ns^2np^6 configuration accounts for the relative inertness of the noble gases. (§13.10)

Key Terms

Section 13.4
diagonal relationships

Section 13.5
bridge bond

Section 13.6
allotropes
silicates
silicones

Section 13.7
disproportionation reaction
dehydration-condensation

Section 13.9
interhalogen compounds

PROBLEMS

Problems with **red** numbers are answered in Appendix G and worked in detail in the Student Solutions Manual. Problem sections match those in this book. Most offer Concept Review Questions, Skill-Building Exercises (grouped in pairs covering the same concept), and Problems in Context. The Comprehensive Problems are based on material from any section or previous chapter.

Hydrogen, the Simplest Atom

Concept Review Questions

13.1 Hydrogen has only one proton, but its IE_1 is much greater than the IE_1 of lithium, which has three protons. Explain.

13.2 Sketch a periodic table, and label the areas containing elements that give rise to the three types of hydrides.

Skill-Building Exercises (grouped in similar pairs)

13.3 Draw Lewis structures for each pair of compounds, and predict which member of each pair will form hydrogen bonds:
(a) NF_3 or NH_3 (b) CH_3OCH_3 or CH_3CH_2OH

13.4 Draw Lewis structures for each pair of compounds, and predict which member of each pair will form hydrogen bonds:
(a) NH_3 or AsH_3 (b) CH_4 or H_2O

13.5 Complete and balance each equation:
(a) An active metal reacting with acid:

$$Al(s) + HCl(aq) \longrightarrow$$

(b) A saltlike (alkali metal) hydride reacting with water:

$$LiH(s) + H_2O(l) \longrightarrow$$

13.6 Complete and balance each equation:
(a) A saltlike (alkaline earth metal) hydride reacting with water:

$$CaH_2(s) + H_2O(l) \longrightarrow$$

(b) Reduction of a metal halide by hydrogen to form a metal:

$$PdCl_2(aq) + H_2(g) \longrightarrow$$

Problems in Context

13.7 Compounds such as $NaBH_4$, $Al(BH_4)_3$, and $LiAlH_4$ are complex hydrides used as reducing agents in many syntheses.
(a) Give the oxidation state of each element in these compounds.
(b) Write a Lewis structure for the polyatomic anion in $NaBH_4$, and predict its shape.

13.8 Unlike the F^- ion, which has an ionic radius close to 133 pm in all alkali metal fluorides, the ionic radius of H^- varies from 137 pm in LiH to 152 pm in CsH. Suggest an explanation for the large variability in the size of H^- but not the size of F^-.

Trends across the Periodic Table: The Period 2 Elements

Concept Review Questions

13.9 How does the maximum oxidation number vary across a period in the main groups? Is the pattern in period 2 different?

13.10 Each of the chemically active period 2 elements forms stable compounds in which it has bonds to fluorine.
(a) What are the names and formulas for these compounds?
(b) Does $\Delta\chi$ increase or decrease left to right across the period?
(c) Does percent ionic character increase or decrease left to right?
(d) Draw Lewis structures for these compounds.

13.11 Period 6 is unusual in several ways:
(a) It is the longest period in the table. How many elements belong to period 6? How many metals?
(b) It contains no metalloids. Where is the metal/nonmetal boundary in period 6?

13.12 An element forms an oxide, E_2O_3, and a fluoride, EF_3.
(a) Of which two groups might E be a member?
(b) How does the group to which E belongs affect the properties of the oxide and the fluoride?

13.13 Fluorine lies between oxygen and neon in period 2. Whereas atomic sizes and ionization energies of these three elements change smoothly, their electronegativities display a dramatic change. What is this change, and how do their electron configurations explain it?

Group 1: The Alkali Metals

Concept Review Questions

13.14 Lithium salts are often much less soluble in water than the corresponding salts of other alkali metals. For example, at 18°C, the concentration of a saturated LiF solution is 1.0×10^{-2} mol/L, whereas the concentration of a saturated KF solution is 1.6 mol/L. How would you explain this behaviour?

13.15 The alkali metals play virtually the same general chemical role in all their reactions.
(a) What is this role?
(b) How is this role based on atomic properties?
(c) Using sodium, write two balanced equations that illustrate this role.

13.16 How do atomic properties account for the low densities of the group 1 elements?

Skill-Building Exercises (grouped in similar pairs)

13.17 Each property shows a regular trend in group 1. Predict whether each property increases or decreases *down* the group:
(a) Density (b) Ionic size
(c) E—E bond energy (d) IE_1
(e) Magnitude of $\Delta_{hydr}H$ of E^+ ion

13.18 Each property shows a regular trend in group 1. Predict whether each property increases or decreases *up* the group:
(a) Melting point (b) E—E bond length
(c) Hardness (d) Molar volume
(e) Lattice energy of EBr

13.19 Write a balanced equation for the formation of sodium peroxide, an industrial bleach, from its elements.

13.20 Write a balanced equation for the formation of rubidium bromide through a reaction of a strong acid and a strong base.

Problems in Context

13.21 Although the alkali metal halides can be prepared directly from the elements, the far less expensive industrial route is treatment of the carbonate or hydroxide with aqueous hydrohalic acid (HX) followed by recrystallization. Balance the reaction between potassium carbonate and aqueous hydroiodic acid.

13.22 Lithium forms several useful organolithium compounds. Calculate the mass percent of Li in each compound:
(a) Lithium stearate ($C_{17}H_{35}COOLi$), a water-resistant grease that is used in cars because it does not harden at cold temperatures
(b) Butyllithium (LiC_4H_9), a reagent in organic syntheses

Group 2: The Alkaline Earth Metals

Concept Review Questions

13.23 How do groups 1 and 2 compare with respect to the reaction of the metals with water?

13.24 Alkaline earth metals are involved in two key diagonal relationships in the periodic table.
(a) Give the two pairs of elements in these diagonal relationships.
(b) For each pair, cite two similarities that demonstrate the relationship.
(c) Why are the members of each pair so similar in behaviour?

13.25 The melting points of alkaline earth metals are many times higher than the melting points of the alkali metals. Explain this difference on the basis of atomic properties. Name three other physical properties for which group 2 metals have higher values than the corresponding group 1 metals.

Skill-Building Exercises (grouped in similar pairs)

13.26 Write a balanced equation for each reaction:
(a) "Slaking" of lime (treatment with water)
(b) Combustion of calcium in air

13.27 Write a balanced equation for each reaction:
(a) Thermal decomposition of witherite (barium carbonate)
(b) Neutralization of stomach acid (HCl) by milk of magnesia (magnesium hydroxide)

Problems in Context

13.28 Lime (CaO) is one of the most abundantly produced chemicals in the world. Write balanced equations for these reactions:
(a) The preparation of lime from natural sources
(b) The use of slaked lime to remove SO_2 from flue gases
(c) The reaction of lime with arsenic acid (H_3AsO_4) to manufacture the insecticide calcium arsenate
(d) The regeneration of NaOH in the paper industry by the reaction of lime with aqueous sodium carbonate

13.29 In some reactions, Be behaves like a typical alkaline earth metal; in other reactions, it does not. Complete and balance each reaction. In which reaction does Be behave like the other group 2 members?
(a) $BeO(s) + H_2O(l) \longrightarrow$
(b) $BeCl_2(l) + Cl^-(l$; from molten NaCl$) \longrightarrow$

Group 13: The Boron Family

Concept Review Questions

13.30 How do the transition metals in period 4 affect the pattern of ionization energies in group 13? How does this pattern compare with the pattern in group 3?

13.31 How do the acidities of aqueous solutions of Tl_2O and Tl_2O_3 compare with each other? Explain.

13.32 Despite the expected decrease in atomic size, there is an unexpected drop in the first ionization energy between groups 2 and 13 in periods 2 through 4. Explain this pattern in terms of electron configurations and orbital energies.

13.33 Many compounds of group 13 elements have chemical behaviour that reflects an electron deficiency.
(a) What is the meaning of *electron deficiency*?
(b) Give two reactions that illustrate this behaviour.

13.34 Boron's chemistry is not typical of its group.
(a) Cite three ways in which boron and its compounds differ significantly from the other group 13 elements and their compounds.
(b) What is the reason for these differences?

Skill-Building Exercises (grouped in similar pairs)

13.35 Rank the following oxides in order of increasing aqueous *acidity*: Ga_2O_3, Al_2O_3, In_2O_3.

13.36 Rank the following hydroxides in order of increasing aqueous *basicity*: $Al(OH)_3$, $B(OH)_3$, $In(OH)_3$.

13.37 Thallium forms the compound TlI_3.
(a) What is the apparent oxidation state of Tl in this compound?
(b) Given that the anion is I_3^-, what is the actual oxidation state of Tl?
(c) Draw the shape of the anion, giving its VSEPR class and bond angles. Propose a reason that the compound does not exist as $(Tl^{3+})(I^-)_3$.

13.38 Very stable dihalides of the group 13 metals are known.
(a) What is the apparent oxidation state of Ga in $GaCl_2$?
(b) Given that $GaCl_2$ consists of a Ga^+ cation and a $GaCl_4^-$ anion, what are the actual oxidation states of Ga?
(c) Draw the shape of the anion, giving its VSEPR class and bond angles.

Problems in Context

13.39 Give the name and symbol or formula for a group 13 element or compound that fits each description or use:
(a) Component of heat-resistant glass (such as Pyrex)
(b) Largest temperature range for the liquid state of an element
(c) Elemental substance with three-centre, two-electron bonds
(d) Metal protected from oxidation by an adherent oxide coat
(e) Toxic metal that lies between two other toxic metals

13.40 Indium (In) reacts with HCl to form a diamagnetic solid with the formula $InCl_2$.
(a) Write condensed electron configurations for In, In^+, In^{2+}, and In^{3+}.
(b) Which of these species is (are) diamagnetic, and which is (are) paramagnetic?
(c) What is the apparent oxidation state of In in $InCl_2$?
(d) Given your answers to parts (b) and (c), explain how $InCl_2$ can be diamagnetic.

13.41 Use VSEPR theory to draw structures, with ideal bond angles, for boric acid and the anion it forms when it reacts with water.

13.42 Boron nitride (BN) has a structure similar to graphite, but it is a white insulator rather than a black conductor. It is synthesized by heating diboron trioxide with ammonia at about 1000°C.
(a) Write a balanced equation for the formation of BN; water also forms.
(b) Calculate $\Delta_r H°$ for the production of BN ($\Delta_f H°$ of BN = −254 kJ/mol).
(c) Boron is obtained from the mineral borax, $Na_2B_4O_7{\cdot}10H_2O$. How much borax is needed to produce 1.0 kg of BN, assuming 72% yield?

Group 14: The Carbon Family

Concept Review Questions

13.43 How does the basicity of SnO_2 in water compare with that of CO_2? Explain.

13.44 Nearly every compound of silicon has the element in the +4 oxidation state. In contrast, most compounds of lead have the element in the +2 state.
(a) What general observation do these facts illustrate?
(b) Explain this observation in terms of atomic and molecular properties.
(c) Give an analogous example from group 13.

13.45 The sum of IE_1 through IE_4 for group 14 elements shows a decrease from C to Si, a slight increase from Si to Ge, a decrease from Ge to Sn, and an increase from Sn to Pb.
(a) What is the expected trend for IEs down a group?
(b) Suggest a reason for the deviations in group 14.
(c) Which group might show even greater deviations?

13.46 Explain the large drops in melting point from C to Si and from Ge to Sn.

13.47 What is an allotrope? Name two group 14 elements that exhibit allotropism, and identify two of their allotropes.

13.48 Even though χ values vary relatively little down group 14, the elements change from nonmetal to metal. Explain.

13.49 How do atomic properties account for the enormous number of carbon compounds? Why do other group 14 elements not behave similarly?

Skill-Building Exercises (grouped in similar pairs)

13.50 Draw a Lewis structure for each species:
(a) The cyclic silicate ion $Si_4O_{12}^{8-}$
(b) A cyclic hydrocarbon with formula C_4H_8

13.51 Draw a Lewis structure for each species:
(a) The cyclic silicate ion $Si_6O_{18}^{12-}$
(b) A cyclic hydrocarbon with formula C_6H_{12}

Problems in Context

13.52 Zeolite A, $Na_{12}[(AlO_2)_{12}(SiO_2)_{12}] \cdot 27H_2O$, is used to soften water by replacing Ca^{2+} and Mg^{2+} with Na^+. Hard water from a certain source is 4.5×10^{-3} mol/L Ca^{2+} and 9.2×10^{-4} mol/L Mg^{2+}, and a pipe delivers 25 000 L of this hard water per day. What mass (kg) of zeolite A is needed to soften a week's supply of the water? (Assume that zeolite A loses its capacity to exchange ions when 85 mol % of its Na^+ has been lost.)

13.53 Give the name and symbol or formula for a group 14 element or compound that fits each description or use:
(a) Hardest known natural substance
(b) Medicinal antacid
(c) Atmospheric gas implicated in the greenhouse effect
(d) Gas that binds to Fe(II) in blood
(e) Element used in the manufacture of computer chips

13.54 One similarity between B and Si is the explosive combustion of their hydrides in air. Write balanced equations for the combustion of B_2H_6 and the combustion of Si_4H_{10}.

Group 15: The Nitrogen Family

Concept Review Questions

13.55 Which group 15 elements form trihalides? Which form pentahalides? Explain.

13.56 As you move down group 15, the melting points of the elements increase and then decrease. Explain.

13.57 (a) What is the range of oxidation states shown by the elements of group 15 as you move down the group? (b) How does this range illustrate the general rule for the range of oxidation states in groups on the right side of the periodic table?

13.58 Bismuth(V) compounds are such powerful oxidizing agents that they have not been prepared in pure form. How is this consistent with the location of Bi in the periodic table?

13.59 Rank the following oxides in order of increasing acidity in water: Sb_2O_3, Bi_2O_3, P_4O_{10}, Sb_2O_5.

Skill-Building Exercises (grouped in similar pairs)

13.60 Assuming that acid strength relates directly to the electronegativity of the central atom, rank the following compounds in order of *increasing* acid strength: H_3PO_4, HNO_3, H_3AsO_4.

13.61 Assuming that acid strength relates directly to the number of O atoms bonded to the central atom, rank the following compounds in order of *decreasing* acid strength: $H_2N_2O_2$ [or $(HON)_2$]; HNO_3 (or $HONO_2$); HNO_2 (or HONO).

13.62 Complete and balance each equation:
(a) $As(s) + \text{excess } O_2(g) \longrightarrow$
(b) $Bi(s) + \text{excess } F_2(g) \longrightarrow$
(c) $Ca_3As_2(s) + H_2O(l) \longrightarrow$

13.63 Complete and balance each equation:
(a) Excess $Sb(s) + Br_2(l) \longrightarrow$
(b) $HNO_3(aq) + MgCO_3(s) \longrightarrow$
(c) $K_2HPO_4(s) \xrightarrow{\Delta}$

13.64 Complete and balance each equation:
(a) $N_2(g) + Al(s) \xrightarrow{\Delta}$
(b) $PF_5(g) + H_2O(l) \longrightarrow$

13.65 Complete and balance each equation:
(a) $AsCl_3(l) + H_2O(l) \longrightarrow$
(b) $Sb_2O_3(s) + NaOH(aq) \longrightarrow$

13.66 Based on the relative sizes of F and Cl, predict the structure of PF_2Cl_3.

13.67 Use the VSEPR model to predict the structure of the cyclic ion $P_3O_9^{3-}$.

Problems in Context

13.68 The pentafluorides of the larger members of group 15 have been prepared, but N can have only eight electrons. A claim has been made that, at low temperatures, a compound with the empirical formula NF_5 forms. Draw a possible Lewis structure for this compound. (*Hint*: NF_5 is ionic.)

13.69 Give the name and symbol or formula for a group 15 element or compound that fits each description or use:
(a) Hydride that exhibits hydrogen bonding
(b) Compound used in "strike anywhere" match heads
(c) Oxide used as a laboratory drying agent
(d) Odd-electron molecule (two examples)
(e) Compound used as an additive in soft drinks

13.70 In addition to the nitrogen oxides in Table 13.3, other less-stable nitrogen oxides exist. Draw a Lewis structure for each of the following:
(a) N_2O_2, a dimer of nitrogen monoxide with an N—N bond
(b) N_2O_2, a dimer of nitrogen monoxide with no N—N bond
(c) N_2O_3, a nitrogen oxide with no N—N bond
(d) NO^+ and NO_3^-, products of the ionization of liquid N_2O_4

13.71 Nitrous oxide (N_2O), the "laughing gas" that is used as an anaesthetic by dentists, is made by the thermal decomposition of solid NH_4NO_3. Write a balanced equation for this reaction. What are the oxidation states of N in NH_4NO_3 and N_2O?

13.72 Write balanced equations for the thermal decomposition of potassium nitrate (a) at a low temperature to the nitrite; (b) at a high temperature to the metal oxide and nitrogen. (O_2 is also formed in both reactions.)

Group 16: The Oxygen Family

Concept Review Questions

13.73 Rank the following elements in order of increasing electrical conductivity, and explain your ranking: Po, S, Se.

13.74 The oxygen and nitrogen families have some obvious similarities and differences.
(a) State two general physical similarities between group 15 and 16 elements.
(b) State two general chemical similarities between group 15 and 16 elements.
(c) State two chemical similarities between P and S.
(d) State two physical similarities between N and O.
(e) State two chemical differences between N and O.

13.75 A molecular property of the group 16 hydrides changes abruptly down the group. This change has been explained in terms of a change in orbital hybridization.
(a) Between what periods does this change occur? (b) What is the change in the molecular property? (c) What is the change in hybridization? (d) What other group displays a similar change?

Skill-Building Exercises (grouped in similar pairs)

13.76 Complete and balance each equation:
(a) $NaHSO_4(aq) + NaOH(aq) \longrightarrow$
(b) $S_8(s) + \text{excess } F_2(g) \longrightarrow$
(c) $FeS(s) + HCl(aq) \longrightarrow$
(d) $Te(s) + I_2(s) \longrightarrow$

13.77 Complete and balance each equation:
(a) $H_2S(g) + O_2(g) \longrightarrow$
(b) $SO_3(g) + H_2O(l) \longrightarrow$
(c) $SF_4(g) + H_2O(l) \longrightarrow$
(d) $Al_2Se_3(s) + H_2O(l) \longrightarrow$

13.78 Is each oxide basic, acidic, or amphoteric in water?
(a) SeO_2 (b) N_2O_3 (c) K_2O
(d) BeO (e) BaO

13.79 Is each oxide basic, acidic, or amphoteric in water?
(a) MgO (b) N_2O_5 (c) CaO
(d) CO_2 (e) TeO_2

13.80 Rank the following hydrides in order of *increasing* acid strength: H_2S, H_2O, H_2Te.

13.81 Rank the following species in order of *decreasing* acid strength: H_2SO_4, H_2SO_3, HSO_3^-.

Problems in Context

13.82 (a) Describe the physical changes that are observed when solid sulfur is heated from room temperature to 440°C and then poured quickly into cold water.
(b) Explain the molecular changes that are responsible for these macroscopic changes.

13.83 Give the name and symbol or formula for a group 16 element or compound that fits each description or use:
(a) Unstable allotrope of oxygen
(b) Oxide with sulfur in which sulfur has the same oxidation number as it does in sulfuric acid
(c) Air pollutant produced by burning sulfur-containing coal
(d) Powerful dehydrating agent
(e) Compound used in solution in the photographic process

13.84 Give the oxidation state of sulfur in (a) S_8; (b) SF_4; (c) SF_6; (d) H_2S; (e) FeS_2; (f) H_2SO_4; (g) $Na_2S_2O_3 \cdot 5H_2O$.

13.85 Disulfur decafluoride is intermediate in reactivity between SF_4 and SF_6. It disproportionates at 150°C to these monosulfur fluorides. Write a balanced equation for this reaction, and give the oxidation state of S in each compound.

Group 17: The Halogens

Concept Review Questions

13.86 (a) Give the physical state and colour of each halogen at STP.
(b) Explain the change in physical state down group 17 in terms of molecular properties.

13.87 (a) What are the common oxidation states of the halogens?
(b) Explain the range and values of the oxidation states of chlorine, based on electron configuration.
(c) Why is fluorine an exception to the pattern of oxidation states of the other group members?

13.88 (a) How many electrons does a halogen atom need to complete its octet? (b) Give examples of the ways that a Cl atom can complete its orbit.

13.89 (a) Select the stronger bond in each pair:
(i) Cl—Cl or Br—Br; (ii) Br—Br or I—I; (iii) F—F or Cl—Cl.
(b) Why does the F—F bond strength *not* follow the group trend?

13.90 In addition to interhalogen compounds, many interhalogen ions exist. Would you expect interhalogen ions with a 1+ charge or a 1− charge to have an even or odd number of atoms? Explain.

13.91 (a) A halogen (X_2) disproportionates in a base, in several steps, to X^- and XO_3^-. Write the overall equation for the disproportionation of Br_2 to Br^- and BrO_3^-.
(b) Write a balanced equation for the reaction of ClF_5 with an aqueous base. (*Hint*: See the reaction of BrF_5 shown in Section 13.9.)

Skill-Building Exercises (grouped in similar pairs)

13.92 Complete and balance each equation. If no reaction occurs, write NR:
(a) $Rb(s) + Br_2(l) \longrightarrow$
(b) $I_2(s) + H_2O(l) \longrightarrow$
(c) $Br_2(l) + I^-(aq) \longrightarrow$
(d) $CaF_2(s) + H_2SO_4(l) \longrightarrow$

13.93 Complete and balance each equation. If no reaction occurs, write NR:
(a) $H_3PO_4(l) + NaI(s) \longrightarrow$
(b) $Cl_2(g) + I^-(aq) \longrightarrow$
(c) $Br_2(l) + Cl^-(aq) \longrightarrow$
(d) $ClF(g) + F_2(g) \longrightarrow$

13.94 Rank the following acids in order of *increasing* acid strength: HClO, $HClO_2$, HBrO, HIO.

13.95 Rank the following acids in order of *decreasing* acid strength: $HBrO_3$, $HBrO_4$, HIO_3, $HClO_4$.

Problems in Context

13.96 Give the name and symbol or formula for a group 17 element or compound that fits each description or use:
(a) Substance used for etching glass
(b) Compound used in household bleach
(c) Weakest hydrohalic acid
(d) Element that is a liquid at room temperature
(e) Organic chloride used to make PVC

13.97 An industrial chemist treats solid NaCl with concentrated H_2SO_4 and obtains gaseous HCl and $NaHSO_4$. When the chemist substitutes solid NaI for NaCl, gaseous H_2S, solid I_2, and S_8 are obtained, but no HI.
(a) What type of reaction did the H_2SO_4 undergo with NaI?
(b) Why does NaI, but not NaCl, cause this type of reaction?
(c) To produce HI(*g*) by the reaction of NaI with an acid, how does the acid have to differ from sulfuric acid?

13.98 Rank the halogens Cl_2, Br_2, and I_2 in order of increasing oxidizing strength, based on their products with metallic Re ($ReCl_6$, $ReBr_5$, and ReI_4). Explain your ranking.

Group 18: The Noble Gases

Concept Review Questions

13.99 (a) Which noble gas is the most abundant in the universe?
(b) Which noble gas is the most abundant in Earth's atmosphere?

13.100 What oxidation states does Xe show in its compounds?

13.101 Why do the noble gases have such low boiling points?

13.102 Explain why Xe, and to a limited extent Kr, form compounds, whereas He, Ne, and Ar do not.

13.103 (a) Why do stable xenon fluorides have an even number of F atoms?
(b) Why do the ionic species XeF_3^+ and XeF_7^- have odd numbers of F atoms?
(c) Predict the shape of XeF_3^+.

Comprehensive Problems

13.104 Xenon tetrafluoride reacts with antimony pentafluoride to form the ionic complex $[XeF_3]^+[SbF_6]^-$.
(a) Which depiction shows the molecular shapes of the reactants and product?
(b) How, if at all, does the hybridization of xenon change in the reaction?

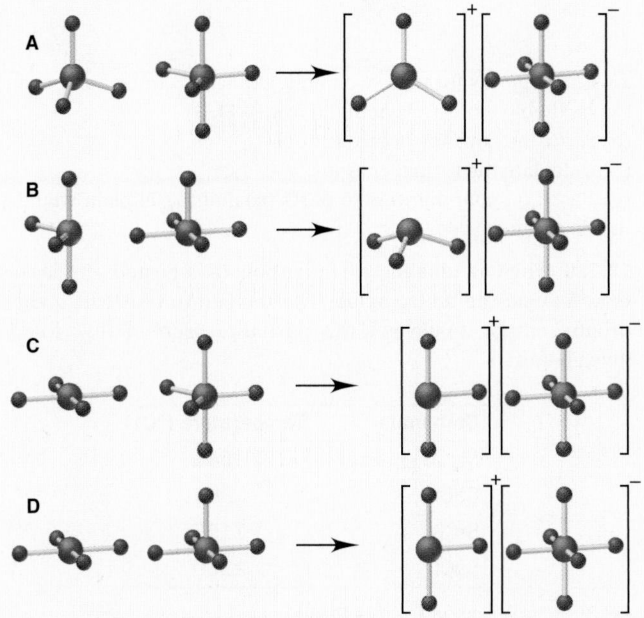

13.105 Consider the following information:

$$H^+(g) + H_2O(g) \longrightarrow H_3O^+(g) \quad \Delta H = -720 \text{ kJ/mol}$$
$$H^+(g) + H_2O(l) \longrightarrow H_3O^+(aq) \quad \Delta H = -1090 \text{ kJ/mol}$$
$$H_2O(l) \longrightarrow H_2O(g) \quad \Delta H = 40.7 \text{ kJ/mol}$$

Use this information to calculate the heat of solution of the hydronium ion:

$$H_3O^+(g) \xrightarrow{H_2O} H_3O^+(aq)$$

13.106 The electronic transition in Na from $3p^1$ to $3s^1$ gives rise to a bright yellow-orange emission at 589.2 nm. What is the energy of this transition?

13.107 Unlike other group 2 metals, beryllium reacts like aluminum and zinc with a concentrated aqueous base to release hydrogen gas and form oxoanions with the general formula $M(OH)_4{}^{n-}$. Write equations for the reactions of these three metals with NaOH.

13.108 The interhalogen IF undergoes the reaction depicted below (I is *purple*, and F is *green*):

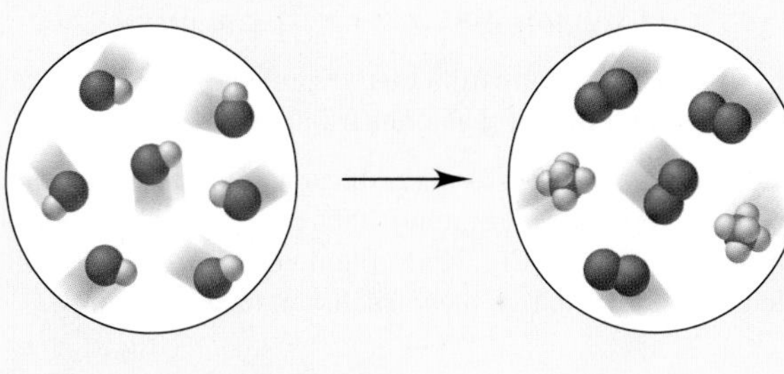

(a) Write the balanced equation.
(b) Name the interhalogen product.
(c) What type of reaction is shown?
(d) If each molecule of IF represents 2.50×10^{-3} mol, what mass of each product forms?

13.109 The main reason alkali metal dihalides (MX_2) do *not* form is the high IE_2 of the metal.
(a) Why is IE_2 so high for alkali metals?
(b) The IE_2 for Cs is 2255 kJ/mol, low enough for CsF_2 to form exothermically ($\Delta_f H° = -125$ kJ/mol). This compound cannot be synthesized, however, because CsF forms with a much greater release of heat ($\Delta_f H° = -530$ kJ/mol). Thus, the breakdown of CsF_2 to CsF happens readily. Write the equation for this breakdown, and calculate the enthalpy of reaction per mole of CsF.

13.110 Semiconductors made from elements in groups 13 and 15 are typically prepared by the direct reaction of the elements at high temperatures. An engineer treats 32.5 g of molten gallium with 20.4 L of white phosphorus vapour at 515 K and 195 kPa. If purification losses are 7.2% by mass, what mass of gallium phosphide is prepared?

13.111 Two substances with the empirical formula HNO are hyponitrous acid ($\mathcal{M} = 62.04$ g/mol) and nitroxyl ($\mathcal{M} = 31.02$ g/mol).
(a) What is the molecular formula for each species?
(b) For each species, draw the Lewis structure that has the lowest formal charges. (*Hint*: Hyponitrous acid has an N=N bond.)
(c) Predict the shape around the N atoms of each species.
(d) When hyponitrous acid loses two protons, it forms the hyponitrite ion. Draw the *cis* and *trans* forms of this ion.

13.112 The species CO, CN^-, and $C_2{}^{2-}$ are isoelectronic.
(a) Draw their Lewis structures.
(b) Draw their molecular orbital (MO) diagrams (assume 2*s*-2*p* mixing, as in N_2), and give the bond order and electron configuration for each.

13.113 The Ostwald process is a series of three reactions used for the industrial production of nitric acid from ammonia.
(a) Write a series of balanced equations for the Ostwald process.
(b) If NO is *not* recycled, what amount (mol) of NH_3 is consumed per mole of HNO_3 produced?
(c) In a typical industrial unit, the process is very efficient, with a 96% yield for the first step. Assuming 100% yields for the subsequent steps, what volume of concentrated aqueous nitric acid (60.% by mass; $d = 1.37$ g/mL) can be prepared for each cubic metre of a gas mixture that is 90.% air and 10.% NH_3 by volume at the industrial conditions of 507 kPa and 850.°C?

13.114 All common plant fertilizers contain nitrogen compounds. Determine the mass % of N in (a) ammonia; (b) ammonium nitrate; (c) ammonium hydrogen phosphate.

13.115 Producer gas is a fuel that is formed by passing air over red-hot coke (amorphous carbon). What mass of a producer gas that consists of 25% CO, 5.0% CO_2, and 70.% N_2 by mass can be formed from 1.75 t of coke, assuming an 87% yield?

13.116 Gaseous F_2 reacts with water to form HF and O_2. In NaOH solution, F_2 forms F^-, water, and oxygen difluoride (OF_2), a highly toxic gas and powerful oxidizing agent. The OF_2 reacts with excess OH^-, forming O_2, water, and F^-.
(a) For each reaction, write a balanced equation, give the oxidation state of O in all the compounds, and identify the oxidizing and reducing agents.
(b) Draw a Lewis structure for OF_2, and predict its shape.

13.117 What is a disproportionation reaction, and which of these reactions fit(s) this description?
(a) $I_2(s) + KI(aq) \longrightarrow KI_3(aq)$
(b) $2ClO_2(g) + H_2O(l) \longrightarrow HClO_3(aq) + HClO_2(aq)$
(c) $Cl_2(g) + 2NaOH(aq) \longrightarrow NaCl(aq) + NaClO(aq) + H_2O(l)$
(d) $NH_4NO_2(s) \longrightarrow N_2(g) + 2H_2O(g)$
(e) $3MnO_4^{2-}(aq) + 2H_2O(l) \longrightarrow 2MnO_4^-(aq) + MnO_2(s) + 4OH^-(aq)$
(f) $3AuCl(s) \longrightarrow AuCl_3(s) + 2Au(s)$

13.118 Explain the following observations:
(a) In reactions with Cl_2, phosphorus forms PCl_5 in addition to the expected PCl_3, but nitrogen forms only NCl_3.
(b) Tetrachloromethane (carbon tetrachloride) is unreactive toward water, but tetrachlorosilane (silicon tetrachloride) reacts rapidly and completely. (What is produced?)
(c) The sulfur-oxygen bond in SO_4^{2-} is shorter than expected for an S—O single bond.
(d) Chlorine forms ClF_3 and ClF_5, but ClF_4 is unknown.

13.119 Which group(s) in the periodic table is (are) described by each general statement?
(a) The elements form compounds of VSEPR class AX_3E_1.
(b) The free elements are strong oxidizing agents and form monatomic ions and oxoanions.
(c) The atoms form compounds by combining with two other atoms, which donate one electron each.
(d) The free elements are strong reducing agents, show only one nonzero oxidation state, and form mainly ionic compounds.
(e) The elements can form stable compounds with only three bonds. As a central atom, however, they can accept a pair of electrons from a fourth atom without expanding their valence shell.
(f) Only larger members of the group are chemically active.

13.120 Diiodine pentoxide (I_2O_5) was discovered by Joseph Gay-Lussac in 1813, but its structure was unknown until 1970! Like Cl_2O_7, it can be prepared by the dehydration-condensation of the corresponding oxoacid.
(a) Name the precursor oxoacid, write a reaction for the formation of the oxide, and draw a likely Lewis structure.
(b) Data show that the bonds to the terminal O are shorter than the bonds to the bridging O. Why?
(c) I_2O_5 is one of the few chemicals that can oxidize CO rapidly and completely; elemental iodine forms in the process. Write a balanced equation for this reaction.

13.121 Bromine monofluoride (BrF) disproportionates to bromine gas and bromine tri- and pentafluorides. Use the following to find $\Delta_r H^\circ$ for the decomposition of BrF to its elements:

$$3BrF(g) \longrightarrow Br_2(g) + BrF_3(l) \quad \Delta_r H = -125.3 \text{ kJ/mol}$$
$$5BrF(g) \longrightarrow 2Br_2(g) + BrF_5(l) \quad \Delta_r H = -166.1 \text{ kJ/mol}$$
$$BrF_3(l) + F_2(g) \longrightarrow BrF_5(l) \quad \Delta_r H = -158.0 \text{ kJ/mol}$$

13.122 White phosphorus is prepared by heating phosphate rock [principally $Ca_3(PO_4)_2$] with sand and coke:

$$Ca_3(PO_4)_2(s) + SiO_2(s) + C(s) \longrightarrow CaSiO_3(s) + CO(g) + P_4(g) \text{ (unbalanced)}$$

What mass (in kg) of phosphate rock is needed to produce 315 mol of P_4, assuming that the conversion is 90.% efficient?

13.123 Element E forms an oxide of general structure A and a chloride of general structure B, as shown below. For the anion EF_5^-, what is (a) the molecular shape; (b) the hybridization of E; (c) the oxidation number of E?

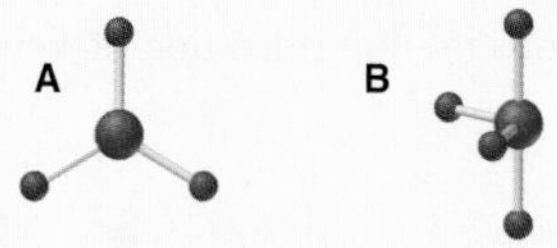

13.124 From its formula, we might expect CO to be quite polar, but its dipole moment is actually low (0.11 D).
(a) Draw the Lewis structure for CO.
(b) Calculate the formal charges.
(c) Based on your answers to parts (a) and (b), explain why the dipole moment is so low.

13.125 When an alkaline earth carbonate is heated, it releases CO_2, leaving the metal oxide. The temperature at which each group 2 carbonate yields a CO_2 partial pressure of 101.3 kPa is given below:

Carbonate	Temperature (°C)
$MgCO_3$	542
$CaCO_3$	882
$SrCO_3$	1155
$BaCO_3$	1360

(a) Suggest a reason for this trend.
(b) Mixtures of $CaCO_3$ and MgO are used to absorb dissolved silicates from boiler water. How would you prepare a mixture of $CaCO_3$ and MgO from dolomite, which contains $CaCO_3$ and $MgCO_3$?

13.126 The bond angles in the nitrite ion, nitrogen dioxide, and the nitronium ion (NO_2^+) are 115°, 134°, and 180°, respectively. Explain these values using Lewis structures and VSEPR theory.

13.127 A common method used to produce a gaseous hydride is to treat a salt containing the anion of the volatile hydride with a strong acid.
(a) Write an equation for (i) the production of HF from CaF_2; (ii) the production of HCl from NaCl; (iii) the production of H_2S from FeS.
(b) In some cases, even a weak acid, such as water, will suffice if the anion of the salt has a sufficiently strong attraction for protons. An example is the production of PH_3 from Ca_3P_2 and water. Write the equation for this reaction.
(c) By analogy, predict the products and write the equation for the reaction of Al_4C_3 with water.

13.128 Chlorine trifluoride was formerly used in the production of uranium hexafluoride for the U.S. nuclear industry:

$$U(s) + 3ClF_3(l) \longrightarrow UF_6(l) + 3ClF(g)$$

What mass of UF_6 can form from 1.00 t of uranium ore that is 1.55% by mass uranium and 12.75 L of chlorine trifluoride (d = 1.88 g/mL)?

13.129 Chlorine is used to make bleach solutions that contain 5.25% NaClO by mass. Assuming 100% yield in the reaction that produces NaClO from Cl_2, what volume of $Cl_2(g)$ at STP will be needed to make 1000. L of bleach solution (d = 1.07 g/mL)?

13.130 The triatomic molecular ion H_3^+ was first detected and characterized by J. J. Thomson using mass spectrometry. Use the bond energy of H_2 (432 kJ/mol) and the proton affinity of H_2 ($H_2 + H^+ \longrightarrow H_3^+$; $\Delta H = -337$ kJ/mol) to calculate the enthalpy of reaction for $H + H + H^+ \longrightarrow H_3^+$.

13.131 An atomic hydrogen torch is used to cut and weld thick sheets of metal. When H_2 passes through an electric arc, the molecules decompose into atoms, which react with O_2. Temperatures over 5000°C are reached, which can melt all metals. Write equations for the breakdown of H_2 to H atoms and for the subsequent overall reaction of the H atoms with oxygen. Use Appendix B to find the standard heat of each reaction, per mole of product.

13.132 Which of the following oxygen ions are paramagnetic: O^+, O^-, O^{2-}, O^{2+}?

13.133 Copper(II) hydrogen arsenite ($CuHAsO_3$) is a green pigment that was once used in wallpaper. In damp conditions, mould metabolizes this pigment to trimethylarsine, $(CH_3)_3As$, a highly toxic gas.

(a) Calculate the mass percent of As in each compound.

(b) If arsenic is toxic at 0.50 mg/m^3, how much $CuHAsO_3$ must react to reach a toxic level in a room that measures 12.35 m × 7.52 m × 2.98 m?

13.134 Hydrogen peroxide can act as either an oxidizing agent or a reducing agent.

(a) When H_2O_2 is treated with aqueous KI, I_2 forms. In which role is H_2O_2 acting? What oxygen-containing product is formed?

(b) When H_2O_2 is treated with aqueous $KMnO_4$, the purple colour of MnO_4^- disappears and a gas forms. In which role is H_2O_2 acting? What oxygen-containing product is formed?

Kinetics: Rates and Mechanisms of Chemical Reactions

IN THIS CHAPTER . . . We examine the rate of a reaction, the factors that affect it, the theories that explain these effects, and the step-by-step changes that reactants undergo as they transform into products. By the end of this chapter, you should be able to

- Differentiate between different reaction rates and discuss how three key factors, concentration, physical state, and temperature, affect them
- Express rate through a rate law and determine its components
- Calculate how concentrations change as a reaction proceeds and determine the half-life of a reaction
- Determine the effects of concentration and temperature on reaction rates and discuss two related theories of chemical kinetics
- Explain reaction mechanisms, noting the steps that a reaction goes through and picturing the chemical species that exist as reactant bonds are breaking and product bonds are forming
- Describe how catalysts increase reaction rates, highlighting two vital examples: the reactions in a living cell and the depletion of stratospheric ozone

14

Apply a hot glass rod to a piece of treated cotton and the entire piece of cotton disappears in a flash. Expose metal to water and oxygen and eventually, over time, it will rust. Put hydrogen and oxygen in a balloon and they might stay there forever. Bring a flame near that balloon and it will explode with a huge noise and an even bigger flame! Why do some reactions happen so quickly and others so slowly? The answer to this question is not simple, but some explanations are offered in this chapter as we study chemical kinetics, the rate at which reactions occur. In this chapter, we will look at what the rate of a reaction is and at some of the factors that affect the rate of a reaction, including concentration, time, and temperature. Kinetics is currently a field that is entirely empirical, or determined by experimental evidence. Nevertheless, we will try to use some theoretical models to explain why certain factors affect reaction rates as they do and then see how these models give us the ability to determine the possible mechanisms by which reactions occur.

Concepts and Skills to Review before Studying This Chapter

- Influence of temperature on molecular speed and collision frequency (Section 4.5)

14.1 Focusing on Reaction Rate

By definition, in a chemical reaction, reactants change into products. **Chemical kinetics**, the study of how fast this change occurs, focuses on the **reaction rate**, the change in the concentrations of reactants (or products) as a function of time. Different reactions have different rates: *in a faster reaction (higher rate), the reactant concentration decreases quickly; in a slower reaction (lower rate), the reactant concentration decreases slowly* (Figure 14.1). Under any given set of conditions, the rate is determined by the nature of the reactants. At room temperature, for example, hydrogen reacts explosively with fluorine but extremely slowly with nitrogen:

$$H_2(g) + F_2(g) \longrightarrow 2HF(g) \quad \text{(very fast)}$$
$$3H_2(g) + N_2(g) \longrightarrow 2NH_3(g) \quad \text{(very slow)}$$

Furthermore, any given reaction has a different rate under different conditions.

Chemical processes occur over a wide range of rates (Figure 14.2). Some processes—such as neutralization, precipitation, or explosion—may take a second or less. Processes that include many reactions—such as the ripening of fruit—take days to months. Human aging continues for decades, and the formation of coal from dead plants takes hundreds of millions of years. Knowing the reaction rate can be essential: how quickly a medicine acts can make the difference between life and death, and how long a polymer takes to form can make the difference between profit and loss.

We can control four factors that affect rate: the concentrations of the reactants, their physical state, the temperature of the reaction, and the use of a catalyst. We consider the first three here and the fourth in Section 14.7.

1. *Concentration: Molecules must collide to react.* A major factor that influences reaction rate is reactant concentration. Consider the single-step reaction between ozone and nitrogen monoxide (nitric oxide):

$$NO(g) + O_3(g) \longrightarrow NO_2(g) + O_2(g)$$

This reaction occurs in the stratosphere, where the nitric oxide is released in the exhaust gases of supersonic aircraft, but it can be simulated in a laboratory. In a

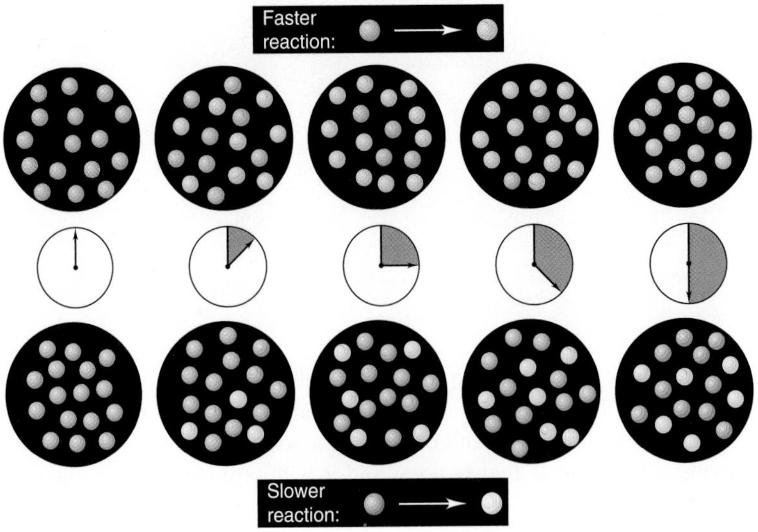

FIGURE 14.1 A faster reaction (*top*) and a slower reaction (*bottom*). As time elapses, the reactant *decreases* and the product *increases*.

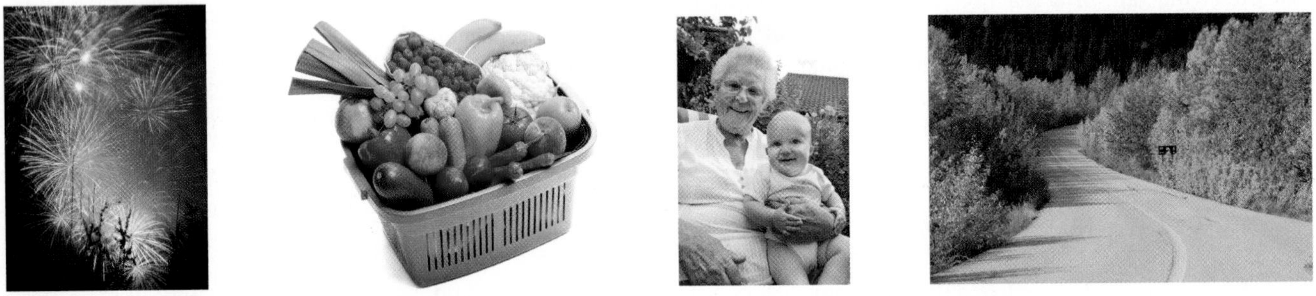

FIGURE 14.2 The wide range of reaction rates

reaction vessel, the molecules zoom every which way, crashing into one another and the vessel walls, but a reaction can occur only when NO and O_3 molecules collide. The more molecules that are present, the more frequently they collide, and the more often they react. Thus, *reaction rate is proportional to the number of collisions, which depends on the concentration of the reactants*:

$$\text{rate} \propto \text{collision frequency} \propto \text{concentration}$$

2. *Physical state: Molecules must mix to collide.* Collision frequency also depends on physical state, which determines how easily the reactants mix. When the reactants are in the same phase, as in an aqueous solution, random thermal motion brings them into contact, but gentle stirring mixes them further. When the reactants are in different phases, contact occurs only at the interface between the phases, so vigorous stirring or even grinding may be needed. Thus, *the more finely divided a solid or liquid reactant is, the greater its surface area, the more contact it makes with the other reactant, and the faster the reaction occurs.* In Figure 14.3, a hot steel nail (*left*) placed in oxygen gas glows feebly, but the same mass of steel wool (*right*) bursts into flame. For the same reason, you start a campfire with twigs, not logs.

3. *Temperature: Molecules must collide with enough energy.* Temperature usually has a major effect on the rate of a reaction. Two kitchen appliances employ this effect: a refrigerator slows down chemical processes that spoil food; an oven speeds up other chemical processes that cook it. Temperature affects reaction rate by increasing the *frequency* and, more important, the *average energy* of collisions:

- *Frequency of collisions.* Recall that molecules in a sample of gas have a range of speeds, with the most probable speed being a function of the temperature

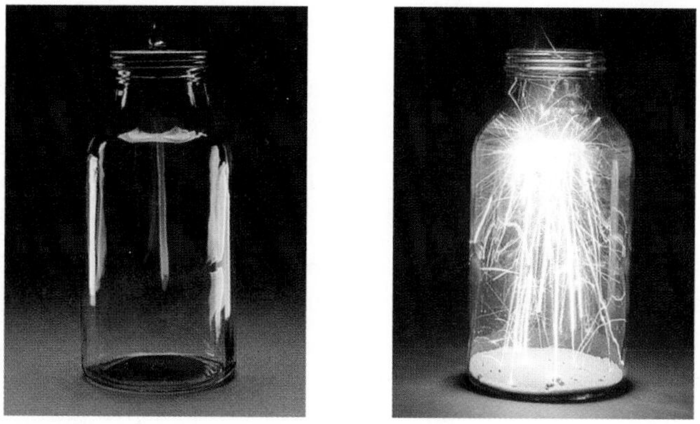

FIGURE 14.3 The effect of surface area on reaction rate

(see Figure 4.14). Thus, *at a higher temperature, collisions occur more frequently, and so more molecules react:*

$$\text{rate} \propto \text{collision } \textit{frequency} \propto \text{temperature}$$

- *Energy of collisions.* Even more important, *temperature affects the kinetic energy* of the molecules. In the jumble of NO and O_3 molecules in the reaction vessel, most collisions have only enough energy for the molecules to bounce off each other. However, some collisions occur with sufficient energy for the molecules to react (Figure 14.4). *At a higher temperature, more sufficiently energetic collisions occur, and so more molecules react:*

$$\text{rate} \propto \text{collision } \textit{energy} \propto \text{temperature}$$

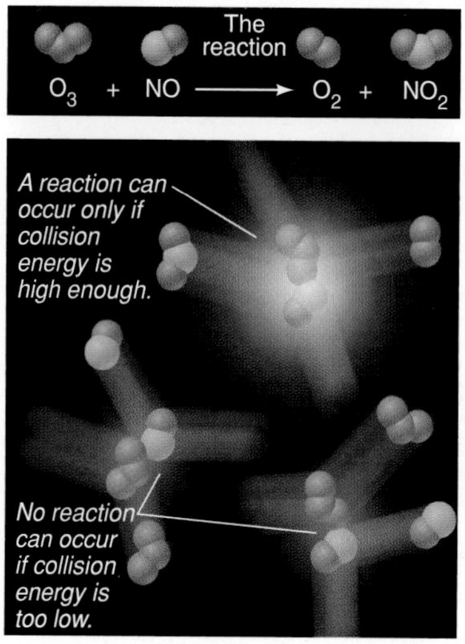
The reaction
O_3 + NO ⟶ O_2 + NO_2
A reaction can occur only if collision energy is high enough.
No reaction can occur if collision energy is too low.

FIGURE 14.4 Sufficient collision energy is required for a reaction to occur

SUMMARY OF SECTION 14.1

- Chemical kinetics focuses on the reaction rate and the factors that affect it.
- Under a given set of conditions, each reaction has its own rate.
- Concentration affects the reaction rate by influencing the frequency of collisions between the reactant molecules.
- Physical state affects the reaction rate by determining how well the reactants can mix.
- Temperature affects the reaction rate by influencing the frequency and, more important, the energy of the collisions between the reactant molecules.

14.2 Expressing the Reaction Rate

In general terms, a *rate* is a change in some variable per unit of time. The most familiar examples relate to speed (*see margin photo*), the change in position of an object divided by the change in time. For instance, if we measure a runner's initial position, x_1, at time t_1, and final position, x_2, at time t_2, the average speed is as follows:

Over the course of a race, the runner is never in the same spot. The runner moves physically over a period of time.

$$\text{Rate of motion (speed)} = \frac{\text{change in position}}{\text{change in time}} = \frac{x_2 - x_1}{t_2 - t_1} = \frac{\Delta x}{\Delta t}$$

For the rate of a *reaction*, we measure the changes in concentrations of reactants or products per unit time: *reactant concentrations decrease while product concentrations increase.* For the general reaction A ⟶ B, we measure the initial reactant concentration (conc A_1) at t_1, allow the reaction to proceed, and then quickly measure the final reactant concentration (conc A_2) at t_2. The change in concentration divided by the change in time gives the rate:

$$\text{Rate} = -\left(\frac{\text{change in concentration of A}}{\text{change in time}}\right) = -\left(\frac{\text{conc A}_2 - \text{conc A}_1}{t_2 - t_1}\right) = -\left(\frac{\Delta(\text{conc A})}{\Delta t}\right)$$

The negative sign is important because, by convention, the reaction rate is a *positive* number. However, since conc A_2 must be *lower* than conc A_1, the *change in concentration (final − initial) of reactant A is negative.* Therefore, we use the negative sign to convert the negative change in reactant concentration to a positive value for the rate.

Suppose that the concentration of A changes from 1.2 mol/L (conc A_1) to 0.75 mol/L (conc A_2) over a 125 s period. The rate is as follows:

$$\text{Rate} = -\left(\frac{0.75\ \text{mol/L} - 1.2\ \text{mol/L}}{125\ \text{s} - 0\ \text{s}}\right) = 3.6 \times 10^{-3}\ \text{mol/(L·s)}$$

The *square brackets,* [ ], *indicate a concentration in moles per litre.* For example, [A] is the concentration of A in mol/L, and the rate expressed in terms of A is

$$\text{Rate} = -\frac{\Delta[\text{A}]}{\Delta t} \qquad (14.1)$$

The units for the rate are moles per litre per second [mol L^{-1} s^{-1}, or mol/(L•s)], or any time unit convenient for the reaction (minutes, years, and so on).

If, instead, we measure the *product* concentrations to determine the rate, we find that conc B_2 is always *higher* than conc B_1. Thus, the *change* in product concentration, Δ[B], is *positive*, and the reaction rate for A ⟶ B expressed in terms of B is

$$\text{Rate} = +\frac{\Delta[\text{B}]}{\Delta t}$$

The plus sign is usually understood and not shown.

Average, Instantaneous, and Initial Reaction Rates

In most cases, *the rate varies as the reaction proceeds.* Consider the reversible gas-phase reaction between ethene and ozone, one of the many reactions that may be involved in the formation of smog:

$$C_2H_4(g) + O_3(g) \rightleftharpoons C_2H_4O(g) + O_2(g)$$

The equation shows that, for every molecule of C_2H_4 that reacts, a molecule of O_3 reacts. Thus, $[O_3]$ and $[C_2H_4]$ decrease at the same rate:

$$\text{Rate} = -\frac{\Delta[C_2H_4]}{\Delta t} = -\frac{\Delta[O_3]}{\Delta t}$$

When we start with a known $[O_3]$ in a closed vessel at 30°C (303 K) and measure $[O_3]$ at 10.0 s intervals during the first minute after adding C_2H_4, we obtain the data in Table 14.1 and the plot of $[O_3]$ versus t in Figure 14.5 (*red curve*). Consider the following two key points:

- The data points in Figure 14.5 result in a curved line, which means that the rate is changing. (A straight line would mean that the rate is constant.)
- The rate *decreases* during the course of the reaction because we are plotting *reactant* concentration versus time. As O_3 molecules react, fewer are present to collide with C_2H_4 molecules. Therefore, the rate, the change in $[O_3]$ over time, decreases.

TABLE 14.1 Concentration of O_3 at Various Times during Its Reaction with C_2H_4 at 303 K

Time (s)	Concentration of O_3 (mol/L)
0.0	3.20×10^{-5}
10.0	2.42×10^{-5}
20.0	1.95×10^{-5}
30.0	1.63×10^{-5}
40.0	1.40×10^{-5}
50.0	1.23×10^{-5}
60.0	1.10×10^{-5}

Three types of reaction rates are shown in Figure 14.5:

1. *Average rate.* Over a given period of time, the **average rate** is the slope of the line joining two points along the curve. The rate over the entire 60.0 s is the total change in concentration divided by the total change in time (*line* ***a***):

$$\text{Rate} = -\frac{\Delta[O_3]}{\Delta t} = -\frac{(1.10 \times 10^{-5}\ \text{mol/L}) - (3.20 \times 10^{-5}\ \text{mol/L})}{60.0\ \text{s} - 0.0\ \text{s}} = 3.50 \times 10^{-7}\ \text{mol/(L·s)}$$

This quantity, which is the slope of line **a** (that is, $\frac{\Delta[O_3]}{\Delta t}$), is the average rate over the entire period: during the first 60.0 s of the reaction, $[O_3]$ decreases an *average* of 3.50×10^{-7} mol/L each second.

When you drive a car or ride a bike for a few kilometres, your speed over shorter distances may be lower or higher than your average speed. Similarly, the decrease in $[O_3]$ over the whole time period does not show the rate over any shorter

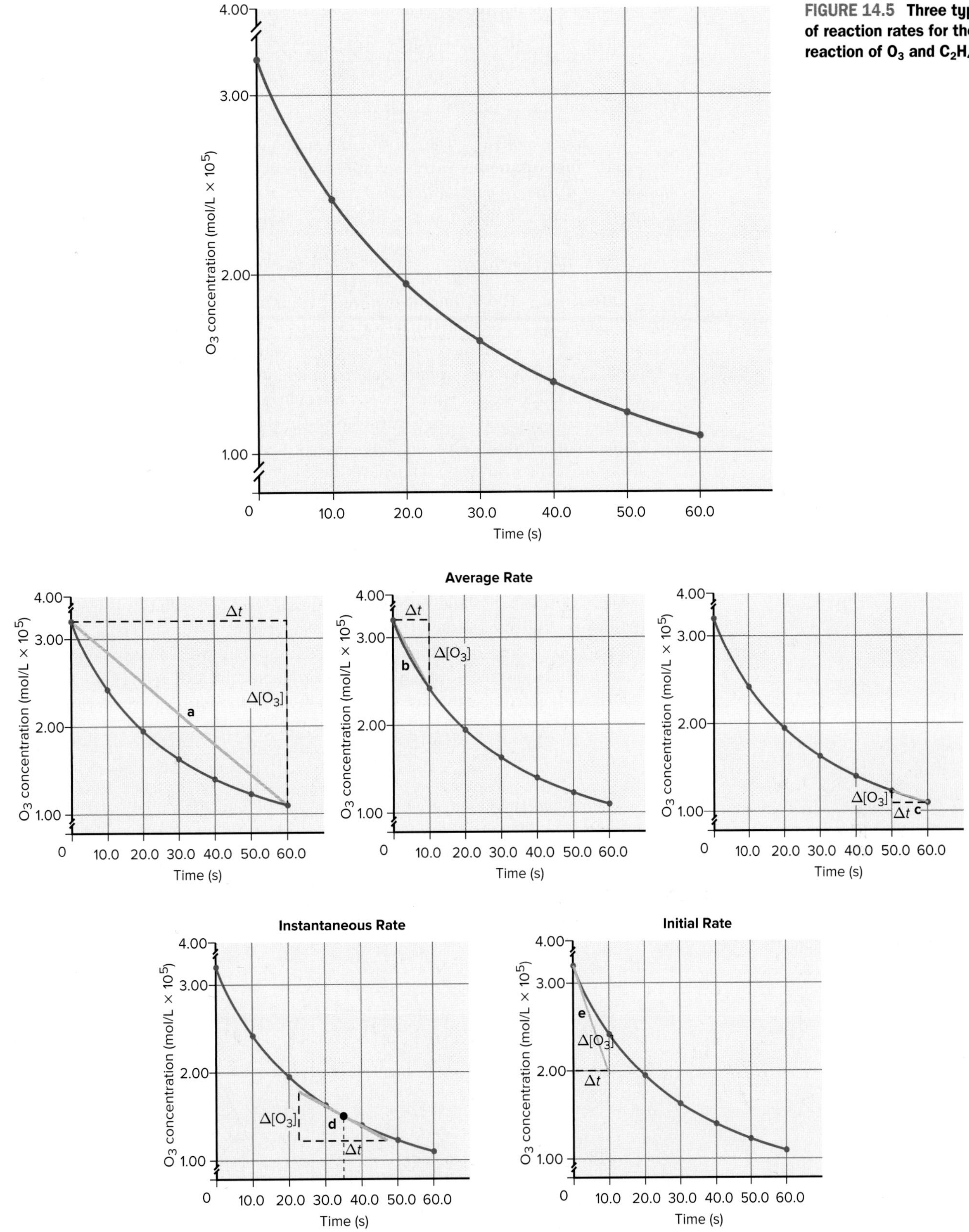

FIGURE 14.5 Three types of reaction rates for the reaction of O_3 and C_2H_4

time period. This *change* in reaction rate is evident when we calculate the average rate over two shorter time periods. For the first 10.0 s, between 0.0 s and 10.0 s, the average rate (*line* ***b***) is

$$\text{Rate} = -\frac{\Delta[\text{O}_3]}{\Delta t} = -\frac{(2.42 \times 10^{-5}\ \text{mol/L}) - (3.20 \times 10^{-5}\ \text{mol/L})}{10.0\ \text{s} - 0.0\ \text{s}} = 7.80 \times 10^{-7}\ \text{mol/(L·s)}$$

For the last 10.0 s, between 50.0 s and 60.0 s, the average rate (*line* **c**) is

$$\text{Rate} = -\frac{\Delta[O_3]}{\Delta t} = -\frac{(1.10 \times 10^{-5}\ \text{mol/L}) - (1.23 \times 10^{-5}\ \text{mol/L})}{60.0\ \text{s} - 50.0\ \text{s}} = 1.30 \times 10^{-7}\ \text{mol/(L·s)}$$

The earlier rate is six times the later rate.

2. *Instantaneous rate.* The shorter the time period we choose, the closer we come to the **instantaneous rate**, the rate at a particular instant during the reaction. *The slope of a line tangent to the curve at any point gives the instantaneous rate at that time.* For example, the rate at 35.0 s is 2.50×10^{-7} mol/(L·s), the slope of the line tangent to the curve through the point at $t = 35.0$ s (*line* **d**). In general, we use the term *reaction rate* to mean *instantaneous* reaction rate.

3. *Initial rate.* The instantaneous rate at the moment when the reactants are mixed (that is, at $t = 0$) is the **initial rate**. We use this rate to avoid a complication: as a reaction proceeds in the *forward* direction (reactants $\longrightarrow$ products), the product increases, causing the *reverse* reaction (reactants $\longleftarrow$ products) to occur more quickly. To find the overall (net) rate, we would have to calculate the difference between the forward and reverse rates. However, the initial rate occurs at $t = 0$, so the product concentrations are negligible, as is the reverse rate. We find the initial rate from the slope of the line tangent to the curve at $t = 0$ s (*line* **e**). We typically use initial rates to find other kinetic parameters.

Expressing Rate in Terms of Reactant and Product Concentrations

So far, for the reaction between C_2H_4 and O_3, we have expressed the rate in terms of $[O_3]$, which is *decreasing*. The rate would be expressed the same way in terms of $[C_2H_4]$. However, the rate is exactly the opposite in terms of the product concentrations because they are *increasing*. From the balanced equation, we see that one molecule of C_2H_4O and one molecule of O_2 appear for every molecule of C_2H_4 and O_3 that disappear. Thus, we can express the rate in terms of any of the four substances:

$$\text{Rate} = -\frac{\Delta[C_2H_4]}{\Delta t} = -\frac{\Delta[O_3]}{\Delta t} = \frac{\Delta[C_2H_4O]}{\Delta t} = \frac{\Delta[O_2]}{\Delta t}$$

Figure 14.6A plots the changes in concentrations of one reactant (C_2H_4) and one product (O_2) simultaneously. The curves have the same shape but are inverted relative to each other because, *for this reaction*, the product appears at the same rate that the reactant disappears.

FIGURE 14.6 Plots of [reactant] and [product] versus time. C_2H_4 and O_2 **(A)** and H_2 and HI **(B)**.

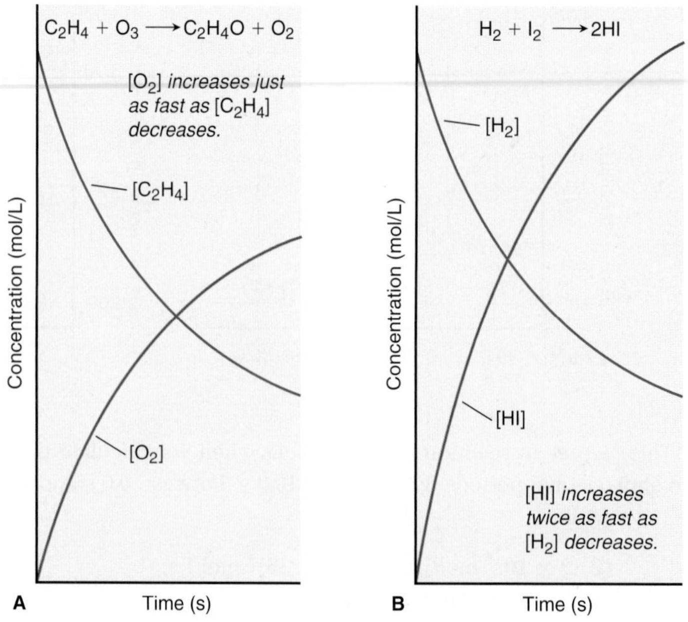

For many other reactions, however, reactants disappear and products appear at different rates. Consider the reaction between hydrogen and iodine to form hydrogen iodide:

$$H_2(g) + I_2(g) \longrightarrow 2HI(g)$$

From the balancing coefficients, we see that, for every molecule of H_2 that disappears, *one* molecule of I_2 disappears and *two* molecules of HI appear. In other words, the rate of $[H_2]$ decrease is the same as the rate of $[I_2]$ decrease, but both are only half the rate of [HI] increase. Thus, in Figure 14.6B, the [HI] curve rises twice as fast as the $[H_2]$ curve drops. If we reference the changes in $[I_2]$ and [HI] to the change in $[H_2]$, we obtain

$$\text{Rate} = -\frac{\Delta[H_2]}{\Delta t} = -\frac{\Delta[I_2]}{\Delta t} = \frac{1}{2}\frac{\Delta[HI]}{\Delta t}$$

If, instead, we reference the changes in $[H_2]$ and $[I_2]$ to the change in [HI], we obtain

$$\text{Rate} = \frac{\Delta[HI]}{\Delta t} = -2\frac{\Delta[H_2]}{\Delta t} = -2\frac{\Delta[I_2]}{\Delta t}$$

Note that this expression gives a rate that is double the previous rate. Thus, *the expression for the rate of a reaction and its numerical value depend on which substance serves as the reference.*

We can summarize these results for any reaction,

$$aA + bB \longrightarrow cC + dD$$

where a, b, c, and d are coefficients of the balanced equation, as follows:

$$\text{Rate} = -\frac{1}{a}\frac{\Delta[A]}{\Delta t} = -\frac{1}{b}\frac{\Delta[B]}{\Delta t} = \frac{1}{c}\frac{\Delta[C]}{\Delta t} = \frac{1}{d}\frac{\Delta[D]}{\Delta t} \quad (14.2)$$

Sample Problem 14.1 Expressing Rate in Terms of Changes in Concentration with Time

Problem Hydrogen gas has a nonpolluting combustion product, water vapour. H_2 was used as a fuel in the space shuttle program and is used for earthbound cars with prototype engines:

$$2H_2(g) + O_2(g) \longrightarrow 2H_2O(g)$$

(a) Express the rate in terms of changes in $[H_2]$, $[O_2]$, and $[H_2O]$ with time.

(b) When $[O_2]$ is decreasing at 0.23 mol/(L•s), at what rate is $[H_2O]$ increasing?

Plan **(a)** Of the three substances in the equation, let us choose O_2 as the reference because its coefficient is 1. For every molecule of O_2 that disappears, two molecules of H_2 disappear. Thus, the rate of $[O_2]$ decrease is one-half the rate of $[H_2]$ decrease. By similar reasoning, the rate of $[O_2]$ decrease is one-half the rate of $[H_2O]$ increase. **(b)** Because $[O_2]$ is decreasing, the change in its concentration must be negative. We substitute the given rate as a negative value [−0.23 mol/(L•s)] into the expression and solve for $\frac{\Delta[H_2O]}{\Delta t}$.

Solution **(a)** Express the rate in terms of each component:

$$\text{Rate} = -\frac{\Delta[O_2]}{\Delta t} = -\frac{1}{2}\frac{\Delta[H_2]}{\Delta t} = \frac{1}{2}\frac{\Delta[H_2O]}{\Delta t}$$

(b) Calculate the rate of change of $[H_2O]$:

$$\frac{1}{2}\frac{\Delta[H_2O]}{\Delta t} = -\frac{\Delta[O_2]}{\Delta t} = -[-0.23\ \text{mol/(L•s)}]$$

$$\frac{\Delta[H_2O]}{\Delta t} = 2[0.23\ \text{mol/(L•s)}] = 0.46\ \text{mol/(L•s)}$$

Check **(a)** A good check involves using the rate expression to obtain the balanced equation: $[H_2]$ changes twice as fast as $[O_2]$, so two H_2 molecules react for each O_2. Since $[H_2O]$ changes twice as fast as $[O_2]$, two H_2O molecules are formed from each O_2. Thus, we get $2H_2 + O_2 \longrightarrow 2H_2O$. The values of $[H_2]$ and $[O_2]$ decrease, so they have minus signs; the value of $[H_2O]$ increases, so it has a plus sign. Another check involves using Equation 14.2, with $A = H_2$, $a = 2$; $B = O_2$, $b = 1$; and $C = H_2O$, $c = 2$:

$$\text{Rate} = -\frac{1}{a}\frac{\Delta[\text{A}]}{\Delta t} = -\frac{1}{b}\frac{\Delta[\text{B}]}{\Delta t} = \frac{1}{c}\frac{\Delta[\text{C}]}{\Delta t}$$

or

$$\text{Rate} = -\frac{1}{2}\frac{\Delta[\text{H}_2]}{\Delta t} = -\frac{\Delta[\text{O}_2]}{\Delta t} = \frac{1}{2}\frac{\Delta[\text{H}_2\text{O}]}{\Delta t}$$

(b) Given the rate expression, it makes sense that the numerical value of the rate of $[H_2O]$ increase is twice the numerical value of the rate of $[O_2]$ decrease.

Comment Thinking through this type of problem at the molecular level is the best approach, but use Equation 14.2 to confirm your answer.

Follow-Up Problem 14.1 **(a)** Balance the following equation, and express the rate in terms of the change in concentration with time for each substance:

$$\text{NO}(g) + \text{O}_2(g) \longrightarrow \text{N}_2\text{O}_3(g)$$

(b) How fast is $[O_2]$ decreasing when [NO] is decreasing at a rate of 1.60×10^{-4} mol/(L•s)? (See Brief Solutions to Follow-Up Problems at the end of the chapter.)

SUMMARY OF SECTION 14.2

- The average reaction rate is the change in reactant (or product) concentration over a change in time, Δt. The rate slows as the reaction proceeds because the reactants are being used up.
- The instantaneous rate at time *t* is the slope of the tangent to a curve that plots concentration versus time.
- The initial rate, the instantaneous rate at $t = 0$, occurs when the reactants have just been mixed and before any product accumulates.
- The expression for a reaction rate and its numerical value depend on which reaction component is being referenced.

14.3 The Rate Law and Its Components

The centrepiece of any kinetic study of a reaction is the **rate law (rate equation)**, which expresses the rate as a function of concentrations and temperature. The rate law is based on experiment, so any hypothesis about how the reaction occurs on the molecular level must conform to it.

In this discussion, we generally consider reactions for which the products do not appear in the rate law, so the rate depends only on *reactant* concentrations and temperature. For a general reaction occurring at a fixed temperature,

$$a\text{A} + b\text{B} + \cdots \longrightarrow c\text{C} + d\text{D} + \cdots$$

the rate law is

$$\text{Rate} = k[\text{A}]^m[\text{B}]^n \cdots \tag{14.3}$$

The term k is a proportionality constant, called the **rate constant**, which is specific for a given reaction at a given temperature and does *not* change as the reaction proceeds. (As we will see in Section 14.5, *k does* change with temperature.) The exponents m and n, called the **reaction orders**, define how the rate is affected by reactant concentration; we will see how to determine them shortly. Two key points to remember are the following:

- *The balancing coefficients a and b in the reaction equation are* ***not*** *necessarily related in any way to the reaction orders m and n.*
- *The components of the rate law—rate, reaction orders, and rate constant—**must** be found by experiment.*

In the remainder of this section, we will find the components of the rate law by measuring concentrations to determine the *initial rate*, using initial rates to determine the *reaction orders*, and using these values to calculate the *rate constant*. With the rate law for a reaction, we can predict the rate for any initial concentrations.

Some Laboratory Methods for Determining the Initial Rate

We determine an initial rate from a plot of concentration versus time, so we need a quick, accurate method for measuring concentration. Let us briefly discuss three common types of methods:

1. *Spectrometric methods* measure the concentration of a component that absorbs (or emits) characteristic wavelengths of light. For example, in the reaction of NO and O_3, only NO_2 has colour:

$$NO(g, \text{colourless}) + O_3(g, \text{colourless}) \longrightarrow O_2(g, \text{colourless}) + NO_2(g, \text{brown})$$

Known amounts of reactants are injected into a tube of known volume within a spectrometer (see Tools of the Laboratory at the end of Section 6.2), which is set to measure the wavelength and intensity of the brown colour. Since the intensity of the brown colour is directly proportional to the concentration of NO_2 gas, the rate of NO_2 formation is proportional to the increase in this intensity over time.

2. *Conductometric methods* rely on the change in the electrical conductivity of the reaction solution when nonionic reactants form ionic products, and vice versa. Consider the reaction between an organic halide, such as 2-bromo-2-methylpropane, and water:

$$(CH_3)_3C\text{—}Br(l) + H_2O(l) \longrightarrow (CH_3)_3C\text{—}OH(l) + H^+(aq) + Br^-(aq)$$

The HBr that forms is a strong acid, so it dissociates completely in the water. As time passes, more ions form, so the conductivity of the reaction mixture increases.

3. *Manometric methods* employ a manometer attached to a reaction vessel of fixed volume and temperature. The manometer measures the pressure, over time, of a reaction that involves a change in the amount (mol) of gas. Consider the reaction between zinc and acetic acid:

$$Zn(s) + 2CH_3COOH(aq) \longrightarrow Zn^{2+}(aq) + 2CH_3COO^-(aq) + H_2(g)$$

The rate is directly proportional to the increase in H_2 gas pressure.

Determining Reaction Orders

With the initial rate, we can determine reaction orders. Let us start by discussing what reaction orders are, and then we will see how to determine them by controlling reactant concentrations.

Meaning and Terminology A reaction has an *individual* order "with respect to" or "in" each reactant, as well as an *overall* order, the sum of the individual orders. First, let us consider the simplest case, a reaction with only one reactant, A:

$$A \longrightarrow \text{products}$$

- *First order.* If the rate doubles when [A] doubles, the rate depends on [A] raised to the first power, $[A]^1$ (the ^{1} is generally omitted). Thus, the reaction is *first order in* (or *with respect to*) A and *first order* overall:

$$\text{Rate} = k[A]^1 = k[A]$$

- *Second order.* There are two ways a reaction can be second order. The first case is when we have a single reactant, A. If the rate quadruples when [A] doubles, the rate depends on [A] squared, $[A]^2$. In this case, the reaction is *second order* in A and *second order* overall:

$$\text{Rate} = k[A]^2$$

A reaction can also be second order if it is first order in two different reactants, A and B. Then the rate is determined as follows:

$$\text{Rate} = k[A]^1[B]^1 = k[A][B]$$

Pseudo first-order reactions There are situations in which a second-order reaction might appear (or might be *made* to appear) to be first order in an experiment. This type of situation occurs if the amount of one of the reactants in the reaction mixture greatly exceeds the amount of the other reactant. In general, this type of situation is more complicated kinetically, but more realistic. In the laboratory, we often create such a situation by adding an excess of one of the reactants. For example, in the example above, if we added an excess of reactant B (where an excess could be 100 to 1000 times reactant A, or even more than that, depending on the reaction and the circumstances) over the course of the reaction, the concentrations of *both* A and B would change. However, because there is so much more of B, the change in its concentration over time (which defines the rate as well) would be negligible. In other words, over the course of the reaction, while the concentration of A changes, the concentration of B remains essentially constant. Under these circumstances, we could write a modified version of the equation above:

$$\text{When } [B] \gg [A], \text{ Rate} = k[B]\cdot[A] = k'[A]$$

The reaction now *appears* to be first order, because the concentration of [B] (essentially a constant) has been incorporated into the value of k, creating a new rate constant, k'. We call such an experimentally created situation a *pseudo first-order rate expression*. Under certain conditions, a pseudo first-order rate expression can be used to find the true second-order rate constant.

Why might we consider measuring a pseudo first-order rate constant? One reason might involve the cost of the reactants. If one of the reactants is particularly expensive and the other reactant is much less expensive, we could adjust the kinetics by placing an excess of the inexpensive reactant in the system and adding only a small quantity of the expensive reactant. ■ Toxicity and cost of disposal are two other reasons we might consider using pseudo first-order reactions.

■ A somewhat simplified analogy might be if you (with a limited budget) were to go out for an evening with a very rich friend. If you played laser tag (3 games for \$20), had dinner at a nice restaurant (\$50 per person) and went for a movie with popcorn and drinks after (\$20), the evening might put a significant dent in your budget. For your rich friend, however, it would not make a dent in her budget. Similarly, in the case of a pseudo first-order reaction, even if the concentration of both reactants changes by the same amount, in the case of the reactant present in smaller quantities, this results in a large change in concentration; for the reactant present in much larger quantity, the change is negligible.

- *Zero order.* If the rate does not change when [A] doubles, the rate does *not* depend on [A]. We express this mathematically by saying that the rate depends on [A] raised to the zero power, $[A]^0$. The reaction is *zero order* in A and *zero order* overall:

$$\text{Rate} = k[A]^0 = k(1) = k$$

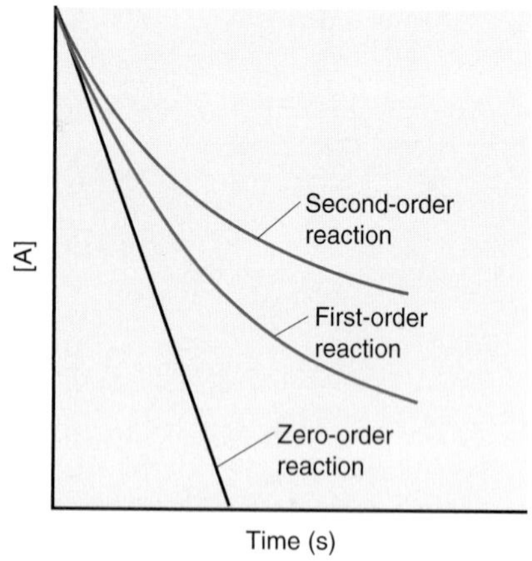

FIGURE 14.7 Plots of reactant concentration, [A], versus time for first-order, second-order, and zero-order reactions

Figure 14.7 shows plots of [A] versus time for first-order, second-order, and zero-order reactions. (In all the plots, the value of k was assumed to be the same.) Note the observations below:

- The decrease in [A] does not change as time goes on for a zero-order reaction.
- The decrease slows down as time goes on for a first-order reaction.
- The decrease slows even more for a second-order reaction.

These results are reflected in Figure 14.8, which shows plots of rate versus [A] for the same reaction orders. Note the following observations:

- The plot is a horizontal line for the zero-order reaction because the rate does not change, no matter what the value of [A].
- The plot is an upward-sloping *line* for the first-order reaction because the rate is directly proportional to [A].
- The plot is an upward-sloping *curve* for the second-order reaction because the rate is proportional to $[A]^2$.

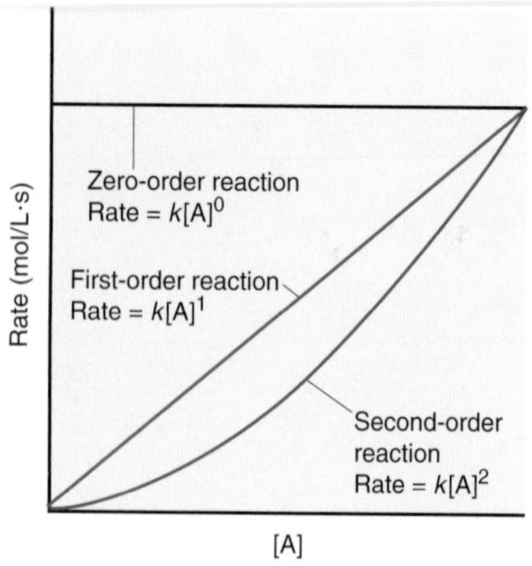

FIGURE 14.8 Plots of rate versus reactant concentration, [A], for first-order, second-order, and zero-order reactions

Let us look at some examples of observed rate laws and note the reaction orders. For the reaction between NO and hydrogen gas,

$$2NO(g) + 2H_2(g) \longrightarrow N_2(g) + 2H_2O(g)$$

the rate law is as follows:

$$\text{Rate} = k[NO]^2[H_2]$$

This reaction is second order in NO. Even though H_2 has a coefficient of 2 in the balanced equation, the reaction is first order in H_2. It is third order overall $(2 + 1 = 3)$.

The reaction between nitrogen monoxide and ozone,

$$NO(g) + O_3(g) \longrightarrow NO_2(g) + O_2(g)$$

has the following rate law:

$$\text{Rate} = k[NO][O_3]$$

This reaction is first order with respect to NO and first order with respect to O_3, so it is second order overall $(1 + 1 = 2)$.

Finally, for the hydrolysis of 2-bromo-2-methylpropane,

$$(CH_3)_3C\text{—}Br(l) + H_2O(l) \longrightarrow (CH_3)_3C\text{—}OH(l) + H^+(aq) + Br^-(aq)$$

the rate law is

$$\text{Rate} = k[(CH_3)_3CBr]$$

This reaction is first order in 2-bromo-2-methylpropane and zero order with respect to H_2O, despite its coefficient of 1 in the balanced equation. If we want to note that water is a reactant, we can write

$$\text{Rate} = k[(CH_3)_3CBr][H_2O]^0$$

Overall, this is a first-order reaction $(1 + 0 = 1)$.

These examples reiterate an important point mentioned earlier: *reaction orders* ***cannot*** *be deduced from the balanced equation but* ***must*** *be determined from experimental data.*

Although reaction orders are usually positive integers or zero, they can also be fractional or negative. For the reaction

$$CHCl_3(g) + Cl_2(g) \longrightarrow CCl_4(g) + HCl(g)$$

a fractional order appears in the rate law:

$$\text{Rate} = k[CHCl_3][Cl_2]^{1/2}$$

This reaction order means that if, for example, $[Cl_2]$ increases by a factor of 4, the rate increases by a factor of 2, the square root of the change in $[Cl_2]$. The overall order of this reaction is $\frac{3}{2}$.

A negative reaction order means that the rate *decreases* when the concentration of the reference component increases. Negative orders are often seen when the rate law includes products. For example, for the atmospheric reaction

$$2O_3(g) \rightleftharpoons 3O_2(g)$$

the rate law is

$$\text{Rate} = k[O_3]^2[O_2]^{-1} = k\frac{[O_3]^2}{[O_2]}$$

If $[O_2]$ doubles, the reaction proceeds half as fast. This reaction is second order in O_3 and negative first order in O_2, so it is first order overall $[2 + (-1) = 1]$.

Sample Problem 14.2 Determining Reaction Orders from Rate Laws

Problem For each reaction, use the given rate law to determine the reaction order with respect to each reactant, as well as the overall order:

(a) $2NO(g) + O_2(g) \longrightarrow 2NO_2(g)$; Rate $= k[NO]^2[O_2]$

(b) $CH_3CHO(g) \longrightarrow CH_4(g) + CO(g)$; Rate $= k[CH_3CHO]^{3/2}$

(c) $H_2O_2(aq) + 3I^-(aq) + 2H^+(aq) \longrightarrow I_3^-(aq) + 2H_2O(l)$; Rate $= k[H_2O_2][I^-]$

Plan We inspect the exponents in the rate law, *not* the coefficients in the balanced equation, to find the individual orders. Then we take their sum to find the overall reaction order.

Solution **(a)** The exponent of [NO] is 2, so the reaction is second order with respect to NO, first order with respect to O_2, and third order overall.

(b) The reaction is $\frac{3}{2}$ order in CH_3CHO and $\frac{3}{2}$ order overall.

(c) The reaction is first order in H_2O_2, first order in I^-, and second order overall. The reactant H^+ does not appear in the rate law, so the reaction is zero order in H^+.

Check Be sure that each reactant has an order and that the sum of the individual orders gives the overall order.

Follow-Up Problem 14.2 Experiment shows that the reaction

$$5Br^-(aq) + BrO_3^-(aq) + 6H^+(aq) \longrightarrow 3Br_2(l) + 3H_2O(l)$$

obeys the following rate law:

$$\text{Rate} = k[Br^-][BrO_3^-][H^+]^2$$

What are the reaction order in each reactant and the overall reaction order?

Determining Reaction Orders by Changing Reactant Concentrations Now let us see how reaction orders are found *before* the rate law is known. Before looking at a real reaction, let us go through the process for substances A and B in this reaction:

$$A + 2B \rightarrow C + D$$

The rate law, expressed in general terms, is

$$\text{Rate} = k[A]^m[B]^n$$

To find the values of m and n, *we run a series of experiments in which one reactant concentration changes while the other is kept constant, and we measure the effect on the initial rate in each experiment.* Table 14.2 shows the results.

1. *Finding m, the order with respect to A.* By comparing experiments 1 and 2, in which [A] doubles and [B] is constant, we can obtain m. First, we take the ratio of the general rate laws for these two experiments:

$$\frac{\text{Rate 2}}{\text{Rate 1}} = \frac{k[A]_2^m[B]_2^n}{k[A]_1^m[B]_1^n}$$

where $[A]_2$ is the concentration of A in experiment 2, $[B]_1$ is the concentration of B in experiment 1, and so forth. Because k is a constant and [B] does not change between the two experiments, these quantities cancel:

$$\frac{\text{Rate 2}}{\text{Rate 1}} = \frac{\cancel{k}[A]_2^m\cancel{[B]_2^n}}{\cancel{k}[A]_1^m\cancel{[B]_1^n}} = \frac{[A]_2^m}{[A]_1^m} = \left(\frac{[A]_2}{[A]_1}\right)^m$$

Substituting the values from Table 14.2, we have

$$\frac{3.50 \times 10^{-3}\ \text{mol/(L·s)}}{1.75 \times 10^{-3}\ \text{mol/(L·s)}} = \left(\frac{5.00 \times 10^{-2}\ \text{mol/L}}{2.50 \times 10^{-2}\ \text{mol/L}}\right)^m$$

TABLE 14.2 Initial Rates for the Reaction between A and B

Experiment	Initial Rate [mol/(L·s)]	Initial [A] (mol/L)	Initial [B] (mol/L)
1	1.75×10^{-3}	2.50×10^{-2}	3.00×10^{-2}
2	3.50×10^{-3}	5.00×10^{-2}	3.00×10^{-2}
3	3.50×10^{-3}	2.50×10^{-2}	6.00×10^{-2}
4	7.00×10^{-3}	5.00×10^{-2}	6.00×10^{-2}

Dividing, we obtain

$$2.00 = (2.00)^m, \text{ so } m = 1$$

Thus, the reaction is first order in A: when [A] doubles, the rate doubles.

2. *Finding n, the order with respect to B.* To find n, we compare experiments 3 and 1, in which [A] is held constant and [B] doubles:

$$\frac{\text{Rate 3}}{\text{Rate 1}} = \frac{k[\text{A}]_3^m\,[\text{B}]_3^n}{k[\text{A}]_1^m[\text{B}]_1^n}$$

As before, k is a constant. In this pair of experiments, [A] does not change. Thus, k and [A] cancel, and we have

$$\frac{\text{Rate 3}}{\text{Rate 1}} = \frac{\cancel{k[\text{A}]_3^m}[\text{B}]_3^n}{\cancel{k[\text{A}]_1^m}[\text{B}]_1^n} = \frac{[\text{B}]_3^n}{[\text{B}]_1^n} = \left(\frac{[\text{B}]_3}{[\text{B}]_1}\right)^n$$

The actual values give

$$\frac{3.50 \times 10^{-3}\ \text{mol/(L·s)}}{1.75 \times 10^{-3}\ \text{mol/(L·s)}} = \left(\frac{6.00 \times 10^{-2}\ \text{mol/L}}{3.00 \times 10^{-2}\ \text{mol/L}}\right)^n$$

Dividing, we obtain

$$2.00 = (2.00)^n, \text{ so } n = 1$$

Thus, the reaction is also first order in B: when [B] doubles, the rate doubles. We can check this conclusion from experiment 4: when *both* [A] and [B] double, the rate should quadruple, and it does. Thus, the rate law, with m and n equal to 1, is

$$\text{Rate} = k[\text{A}][\text{B}]$$

Note, especially, that while the order with respect to B is 1, the coefficient of B in the balanced equation is 2. Thus, as we mentioned earlier, *reaction orders must be determined from experiment.*

Next, let us go through this process for a real reaction, the reaction between oxygen and nitrogen monoxide, a key step in the formation of acid rain and in the industrial production of nitric acid:

$$O_2(g) + 2NO(g) \longrightarrow 2NO_2(g)$$

The general rate law is

$$\text{Rate} = k[O_2]^m[\text{NO}]^n$$

Table 14.3 shows experiments that change one reactant concentration while keeping the other constant. If we compare experiments 1 and 2, we see the effect of doubling $[O_2]$ on the rate. First, we take the ratio of their rate laws:

$$\frac{\text{Rate 2}}{\text{Rate 1}} = \frac{k[O_2]_2^m[\text{NO}]_2^n}{k[O_2]_1^m[\text{NO}]_1^n}$$

As before, the constant quantities—k and [NO]—cancel:

$$\frac{\text{Rate 2}}{\text{Rate 1}} = \frac{k[O_2]_2^m\cancel{[\text{NO}]_2^n}}{k[O_2]_1^m\cancel{[\text{NO}]_1^n}} = \frac{[O_2]_2^m}{[O_2]_1^m} = \left(\frac{[O_2]_2}{[O_2]_1}\right)^m$$

TABLE 14.3 Initial Rates for the Reaction between O_2 and NO

Experiment	Initial Rate [mol/(L·s)]	Initial Reactant Concentrations (mol/L) $[O_2]$	[NO]
1	3.21×10^{-3}	1.10×10^{-2}	1.30×10^{-2}
2	6.40×10^{-3}	2.20×10^{-2}	1.30×10^{-2}
3	12.8×10^{-3}	1.10×10^{-2}	2.60×10^{-2}
4	9.60×10^{-3}	3.30×10^{-2}	1.30×10^{-2}
5	28.8×10^{-3}	1.10×10^{-2}	3.90×10^{-2}

Substituting the values from Table 14.3, we obtain

$$\frac{6.40 \times 10^{-3}\ \text{mol/(L·s)}}{3.21 \times 10^{-3}\ \text{mol/(L·s)}} = \left(\frac{2.20 \times 10^{-2}\ \text{mol/L}}{1.10 \times 10^{-2}\ \text{mol/L}}\right)^m$$

Dividing, we obtain

$$1.99 = (2.00)^m$$

Rounding to one significant figure gives

$$2 = 2^m, \text{ so } m = 1$$

Sometimes, the exponent is not as easy to find by inspection as it is here. In these cases, we solve for m with an equation of the form $a = b^m$. If we take the base 10 logarithm of both sides and use the log rules (Appendix A), we get $\log a = m \log b$, which can be rearranged to give

$$m = \frac{\log a}{\log b} = \frac{\log 1.99}{\log 2.00} = 0.993, \text{ which rounds to } 1$$

Thus, the reaction is first order in O_2: when $[O_2]$ doubles, the rate doubles.

To find the order with respect to NO, we compare experiments 3 and 1, in which $[O_2]$ is held constant and [NO] is doubled:

$$\frac{\text{Rate 3}}{\text{Rate 1}} = \frac{k[O_2]_3^m[\text{NO}]_3^n}{k[O_2]_1^m[\text{NO}]_1^n}$$

Cancelling the constant k and unchanging $[O_2]$, we have

$$\frac{\text{Rate 3}}{\text{Rate 1}} = \left(\frac{[\text{NO}]_3}{[\text{NO}]_1}\right)^n$$

The actual values give

$$\frac{12.8 \times 10^{-3}\ \text{mol/(L·s)}}{3.21 \times 10^{-3}\ \text{mol/(L·s)}} = \left(\frac{2.60 \times 10^{-2}\ \text{mol/L}}{1.30 \times 10^{-2}\ \text{mol/L}}\right)^n$$

Dividing, we obtain

$$3.99 = (2.00)^n$$

Solving for n:

$$n = \frac{\log 3.99}{\log 2.00} = 2.00 \text{ (or 2)}$$

The reaction is second order in NO: when [NO] doubles, the rate quadruples. Thus, the actual rate law is

$$\text{Rate} = k[O_2][\text{NO}]^2$$

In this case, the reaction orders happen to be the same as the equation coefficients; nevertheless, the reaction orders must *always* be determined by experiment.

The next two sample problems offer practice with this approach. The first problem is based on data, and the second problem is based on molecular scenes.

Sample Problem 14.3 Determining Reaction Orders from Rate Data

Problem Many gaseous reactions occur in car engines and exhaust systems. One of these reactions is

$$NO_2(g) + CO(g) \longrightarrow NO(g) + CO_2(g) \qquad \text{Rate} = k[NO_2]^m[CO]^n$$

Use the following data to determine the individual and overall reaction orders:

Experiment	Initial Rate [mol/(L·s)]	Initial $[NO_2]$ (mol/L)	Initial [CO] (mol/L)
1	0.0050	0.10	0.10
2	0.080	0.40	0.10
3	0.0050	0.10	0.20

Plan We need to solve the general rate law for m and n and then add these orders to get the overall order. To solve for each exponent, we proceed as discussed, taking the ratio of the rate laws for two experiments in which only the reference reactant changes.

Solution First, we need to calculate m in $[NO_2]^m$.

Take the ratio of the rate laws for experiments 1 and 2, in which $[NO_2]$ varies but [CO] is constant:

$$\frac{\text{Rate 2}}{\text{Rate 1}} = \frac{k[NO_2]_2^m[CO]_2^n}{k[NO_2]_1^m[CO]_1^n} = \left(\frac{[NO_2]_2}{[NO_2]_1}\right)^m \quad \text{or} \quad \frac{0.080 \text{ mol/(L·s)}}{0.0050 \text{ mol/(L·s)}} = \left(\frac{0.40 \text{ mol/L}}{0.10 \text{ mol/L}}\right)^m$$

This gives $16 = (4.0)^m$, which can be written as $4^2 = (4.0)^m$, so by inspection, $m = 2$. We can also use the log method to give the same answer: $m = \frac{\log 16}{\log 4.0} = 2.0$. The reaction is second order in NO_2.

Next, we need to calculate n in $[CO]^n$.

Taking the ratio of the rate laws for experiments 1 and 3, in which [CO] varies but $[NO_2]$ is constant:

$$\frac{\text{Rate 3}}{\text{Rate 1}} = \frac{k[NO_2]_3^m[CO]_3^n}{k[NO_2]_1^m[CO]_1^n} = \left(\frac{[CO]_3}{[CO]_1}\right)^n \quad \text{or} \quad \frac{0.0050 \text{ mol/(L·s)}}{0.0050 \text{ mol/(L·s)}} = \left(\frac{0.20 \text{ mol/L}}{0.10 \text{ mol/L}}\right)^n$$

We have $1.0 = (2.0)^n$, so $n = 0$. The rate does not change when [CO] varies, so the reaction is zero order in CO.

Therefore, the rate law is

$$\text{Rate} = k[NO_2]^2[CO]^0 = k[NO_2]^2(1) = k[NO_2]^2$$

The reaction is second order overall.

Check We can check by reasoning through the orders. If $m = 1$, quadrupling $[NO_2]$ would quadruple the rate; but the rate *more* than quadruples, so $m > 1$. If $m = 2$, quadrupling $[NO_2]$ would increase the rate by a factor of 16 (4^2). The ratio of rates is $\frac{0.080}{0.005} = 16$, so $m = 2$. In contrast, increasing [CO] has no effect on the rate, which can happen only if $[CO]^n = 1$, so $n = 0$.

Follow-Up Problem 14.3 Find the rate law and the individual and overall reaction orders for the reaction

$$H_2 + I_2 \longrightarrow 2HI$$

using the following data at 450°C:

Experiment	Initial Rate [mol/(L·s)]	Initial $[H_2]$ (mol/L)	Initial $[I_2]$ (mol/L)
1	1.9×10^{-23}	0.0113	0.0011
2	1.1×10^{-22}	0.0220	0.0033
3	9.3×10^{-23}	0.0550	0.0011
4	1.9×10^{-22}	0.0220	0.0056

Sample Problem 14.4 Determining Reaction Orders from Molecular Scenes

Problem At a particular temperature and volume, two gases, A (*red*) and B (*blue*), react. The following molecular scenes represent starting mixtures for four experiments:

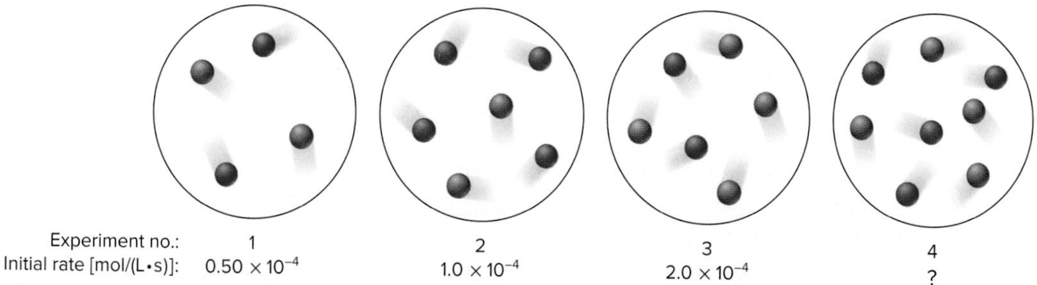

(a) What is the reaction order with respect to A and with respect to B? What is the overall order?

(b) Write the rate law for the reaction.

(c) Predict the initial rate of experiment 4.

Plan **(a)** As before, we find the individual reaction orders by seeing how a change in each reactant changes the rate. In this problem, however, instead of using concentration data, we count numbers of particles. The sum of the individual orders is the overall order. **(b)** To write the rate law, we use the orders from part (a) as exponents in the general rate law. **(c)** Using the results from experiments 1 through 3 and the rate law from part (b), we find the unknown initial rate of experiment 4.

Solution **(a)** We need to find the individual and overall orders.

Reactant A (*red*): Based on experiments 1 and 2, when the number of particles of A doubles (from 2 to 4) and the number of particles of B remains constant (at 2), the rate doubles [from 0.5×10^{-4} mol/(L·s) to 1.0×10^{-4} mol/(L·s)]. Thus, the order with respect to A is 1.

Reactant B (*blue*): Based on experiments 1 and 3, when the number of particles of B doubles (from 2 to 4) and the number of particles of A remains constant (at 2), the rate quadruples [from 0.5×10^{-4} mol/(L·s) to 2.0×10^{-4} mol/(L·s)]. Thus, the order with respect to B is 2. The overall order is $1 + 2 = 3$.

(b) Next, we need to write the rate law. The general rate law is

$$\text{Rate} = k[\text{A}]^m[\text{B}]^n$$

so we have

$$\text{Rate} = k[\text{A}][\text{B}]^2$$

(c) Finally, we need to find the initial rate of experiment 4. There are several possibilities, but let us compare experiments 3 and 4, in which the number of particles of A doubles (from 2 to 4) and the number of particles of B does not change. Since the rate law shows that the reaction is first order in A, the initial rate in experiment 4 should be double the initial rate in experiment 3, or 4.0×10^{-4} mol/(L·s).

Check We can check by comparing other pairs of experiments. **(a)** Comparing experiments 2 and 3 shows that the number of B doubles, which causes the rate to quadruple, and the number of A decreases by half, which causes the rate to halve. Therefore, the overall rate change should double [from 1.0×10^{-4} mol/(L·s) to 2.0×10^{-4} mol/(L·s)], which it does. **(c)** Comparing experiments 2 and 4, in which the number of A is constant and the number of B doubles, the rate should quadruple. This means that the initial rate of experiment 4 should be 4.0×10^{-4} mol/(L·s), as we found.

Follow-Up Problem 14.4 The scenes below show three experiments at a given temperature and volume involving reactants X (*black*) and Y (*green*):

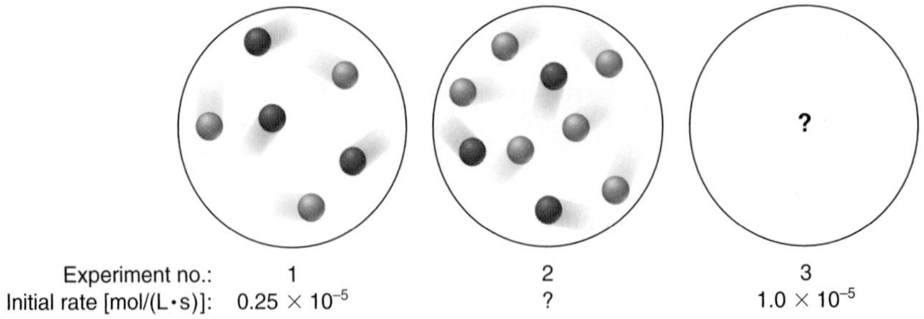

The rate law for the reaction is

$$\text{Rate} = k[\text{X}]^2$$

(a) What is the initial rate of experiment 2?

(b) Draw a scene for experiment 3 that involves a single change in the scene for experiment 1.

Determining the Rate Constant

Let us find the rate constant for the reaction of O_2 and NO. Since we know the rate, reactant concentrations, and reaction orders, the sole remaining unknown in the rate law is the rate constant, k. We can use data from any of the experiments in Table 14.3 to solve for k. From experiment 1, for instance, we have

$$\text{Rate} = k[O_2]_1[NO]_1^2$$

$$k = \frac{\text{rate 1}}{[O_2]_1[NO]_1^2} = \frac{3.21 \times 10^{-3}\ \text{mol/(L·s)}}{(1.10 \times 10^{-2}\ \text{mol/L})(1.30 \times 10^{-2}\ \text{mol/L})^2}$$

$$= \frac{3.21 \times 10^{-3}\ \text{mol/(L·s)}}{1.86 \times 10^{-6}\ \text{mol}^3/\text{L}^3} = 1.73 \times 10^3\ \text{L}^2/(\text{mol}^2\text{·s})$$

Always check that the values of k for a series of experiments are constant within experimental error. To three significant figures, the average value of k for the five experiments in Table 14.3 is 1.72×10^3 $L^2/(mol^2·s)$.

With concentrations in mol/L and the reaction rate in units of mol/(L·time), the units for k depend on the order of the reaction and, of course, the unit for time. For this reaction, the units for k, $L^2/(mol^2·s)$, are required to give a rate with units of mol/(L·s):

$$\frac{\text{mol}}{\text{L·s}} = \frac{\text{L}^2}{\text{mol}^2\text{·s}} \times \frac{\text{mol}}{\text{L}} \times \left(\frac{\text{mol}}{\text{L}}\right)^2$$

The rate constant will *always* have these units for an overall third-order reaction with the time in seconds. Table 14.4 shows the units of k for common integer overall orders, but you can always determine the units mathematically.

Figure 14.9 summarizes the steps for studying the kinetics of a reaction.

TABLE 14.4 Units of the Rate Constant k for Several Overall Reaction Orders

Overall Reaction Order	Units of k (t in seconds)
0	mol/(L·s) (or mol L^{-1} s^{-1})
1	1/s (or s^{-1})
2	L/(mol·s) (or L mol^{-1} s^{-1})
3	$L^2/(mol^2·s)$ (or L^2 mol^{-2} s^{-1})

General formula:

$$\text{Units of } k = \frac{\left(\frac{\text{L}}{\text{mol}}\right)^{\text{order}-1}}{\text{unit of } t}$$

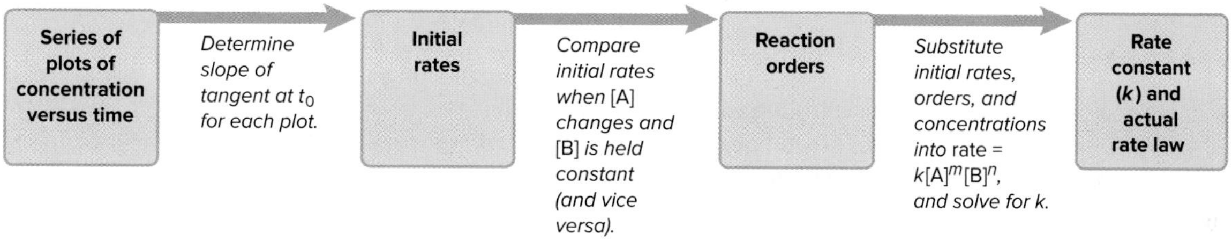
Series of plots of concentration versus time
Determine slope of tangent at t_0 for each plot.
Initial rates
Compare initial rates when [A] changes and [B] is held constant (and vice versa).
Reaction orders
Substitute initial rates, orders, and concentrations into rate = $k[A]^m[B]^n$, and solve for k.
Rate constant (k) and actual rate law

FIGURE 14.9 Information sequence to determine the kinetic parameters of a reaction.

SUMMARY OF SECTION 14.3

- An experimentally determined rate law shows how the rate of a reaction depends on concentration. Considering only initial rates (that is, no products), the expression for the general rate law is

$$\text{Rate} = k[A]^m[B]^n\cdots$$

  This reaction is mth order with respect to A and nth order with respect to B. The overall reaction order is $m + n$.
- With an accurate method for obtaining initial rates, reaction orders are determined experimentally by varying the concentration of one reactant at a time to see its effect on the rate.
- By substituting the known rate, concentrations, and reaction orders into the rate law, we can solve for the rate constant, k.

14.4 Integrated Rate Laws: Concentration Changes over Time

The rate laws that we have developed so far tell us the rate or concentration at a given instant, allowing us to answer questions such as "How fast is the reaction proceeding at the moment y moles per litre of A is mixed with z moles per litre of

B?" and "What is [B], when [A] is x moles per litre?" By using different forms of the rate laws, called **integrated rate laws**, we can include time as a variable and answer questions such as "How long will it take to use up x moles per litre of A?" and "What is [A] after y minutes of reaction?"

Integrated Rate Laws for First-Order, Second-Order, and Zero-Order Reactions

As we have seen, for a *general first-order reaction*, the rate is the negative of the change in [A] divided by the change in time:

$$\text{Rate} = -\frac{\Delta[\text{A}]}{\Delta t}$$

The rate can also be expressed in terms of the rate law:

$$\text{Rate} = k[\text{A}]$$

Setting these two expressions equal to each other gives

$$-\frac{\Delta[\text{A}]}{\Delta t} = k[\text{A}] \quad \text{or} \quad -\frac{\Delta[\text{A}]}{[\text{A}]} = k\Delta t$$

Using calculus methods, we can integrate over time to obtain the integrated rate law for a first-order reaction:

$$\ln\frac{[\text{A}]_0}{[\text{A}]_t} = kt \qquad (\text{first-order reaction; Rate} = k[\text{A}]) \tag{14.4}$$

From the calculus derivation, we can see that this equation can also be written as

$$\ln\,[\text{A}]_0 - \ln\,[\text{A}]_t = kt$$

where ln is the natural logarithm, $[\text{A}]_0$ is the concentration of A at $t = 0$, and $[\text{A}]_t$ is the concentration of A at any time t.

The integrated rate law is exactly what it says. We start with the equation just before Equation 14.4 and integrate both sides:

$$\text{Rate} = \frac{-d[\text{A}]}{dt} = k[\text{A}]$$

If we rearrange the equation and bring the terms containing A to the same side of the equation, we obtain

$$\frac{-d[\text{A}]}{[\text{A}]} = kdt$$

Now, we integrate both sides. On the left, we integrate from our initial concentration, $[\text{A}]_0$, to our final concentration, $[\text{A}]_t$. On the right, we integrate from our initial time, $t = 0$, to our final time, t:

$$\int_{[\text{A}]_0}^{[\text{A}]_t} -\frac{d[\text{A}]}{[\text{A}]} = \int_0^t kdt = k\int_0^t dt$$

Since k is a constant, it can be factored out of the right side of the equation. On the right, the integral form of dt is just t, from time 0 to time t, or $t - 0 = t$. Thus, the right-hand side is simply kt. The integral form of $\frac{-1}{[\text{A}]}d[\text{A}]$ is $-\ln[\text{A}]$ from $[\text{A}]_0$ to $[\text{A}]_t$; or $-\ln[\text{A}]_t - (-\ln[\text{A}]_0)$ or $\ln[\text{A}]_0 - \ln[\text{A}]_t$. According to the rules of logarithms, $\log a - \log b = \log\frac{a}{b}$. Thus, we arrive at our integrated first-order rate expression:

$$\ln\frac{[\text{A}]_0}{[\text{A}]_t} = kt$$

For a *general second-order reaction* involving two reactants, A and B, the expression including time is complex. In the simpler case, the rate law contains only reactant A. Setting the two rate expressions equal to each other gives

$$\text{Rate} = -\frac{\Delta[\text{A}]}{\Delta t} = k[\text{A}]^2 \quad \text{or} \quad -\frac{\Delta[\text{A}]}{[\text{A}]^2} = k\Delta t$$

Integrating over time gives the integrated rate law for a second-order reaction involving one reactant:

$$\frac{1}{[\text{A}]_t} - \frac{1}{[\text{A}]_0} = kt \quad \text{(second-order reaction; Rate} = k[\text{A}]^2\text{)} \qquad (14.5)$$

For a *general zero-order reaction*, setting the two rate expressions equal to each other gives

$$\text{Rate} = -\frac{\Delta[\text{A}]}{\Delta t} = k[\text{A}]^0 = k \quad \text{or} \quad -\Delta[\text{A}] = k\Delta t$$

Integrating over time gives the integrated rate law for a zero-order reaction:

$$[\text{A}]_t - [\text{A}]_0 = -kt \quad \text{(zero-order reaction; Rate} = k[\text{A}]^0 = k\text{)} \qquad (14.6)$$

Sample Problem 14.5 shows one way that integrated rate laws are applied.

Sample Problem 14.5 Determining the Reactant Concentration at a Given Time

Problem At 1000°C, cyclobutane (C_4H_8) decomposes in a first-order reaction, with the very high rate constant of 87 s^{-1}, to two molecules of ethene (C_2H_4).

(a) The initial C_4H_8 concentration is 2.00 mol/L. What is the concentration after 0.010 s?

(b) What fraction of C_4H_8 has decomposed in this time?

Plan (a) We must find the concentration of cyclobutane at time t, $[C_4H_8]_t$. The problem tells us that this is a first-order reaction, so we use the integrated first-order rate law:

$$\ln\frac{[C_4H_8]_0}{[C_4H_8]_t} = kt$$

We know k (87 s^{-1}), t (0.010 s), and $[C_4H_8]_0$ (2.00 mol/L), so we can solve for $[C_4H_8]_t$.

(b) The fraction that has decomposed is the concentration that has decomposed divided by the initial concentration:

$$\text{Fraction decomposed} = \frac{[C_4H_8]_0 - [C_4H_8]_t}{[C_4H_8]_0}$$

Solution (a) Substitute the data into the integrated rate law:

$$\ln\frac{2.00\text{ mol/L}}{[C_4H_8]_t} = (87\text{ s}^{-1})(0.010\text{ s}) = 0.87$$

Take the antilog of both sides:

$$\frac{2.00\text{ mol/L}}{[C_4H_8]_t} = e^{0.87} = 2.39$$

Solve for $[C_4H_8]_t$:

$$[C_4H_8]_t = \frac{2.00\text{ mol/L}}{2.39} = 0.84\text{ mol/L}$$

(b) Find the fraction that has decomposed after 0.010 s:

$$\frac{[C_4H_8]_0 - [C_4H_8]_t}{[C_4H_8]_0} = \frac{2.00\text{ mol/L} - 0.84\text{ mol/L}}{2.00\text{ mol/L}} = 0.58$$

Check The concentration remaining after 0.010 s (0.84 mol/L) is less than the starting concentration (2.00 mol/L), which makes sense. Raising e to an exponent slightly less than 1 should give a number (2.39) slightly less than the value of e (2.718). Moreover, the final result makes sense: a high rate constant indicates a fast reaction, so it is not surprising that so much decomposes in such a short time.

Follow-Up Problem 14.5 At 25°C, hydrogen iodide breaks down very slowly to hydrogen and iodine: Rate $= k[\text{HI}]^2$. The rate constant at 25°C is 2.4×10^{-21} L/(mol·s). If 0.0100 mol of HI(g) is placed in a 1.0 L container, how long will it take for the concentration of HI to reach 0.00900 mol/L (10.0% reacted)?

Determining Reaction Orders from an Integrated Rate Law

In Sample Problem 14.3, we found the reaction orders using rate data. If rate data are not available, we can rearrange the integrated rate law into an equation for a straight line, $y = mx + b$, where m is the slope and b is the y intercept. We can then use a graphical method to find the order:

- For a *first-order reaction*, we have

$$\ln \frac{[\text{A}]_0}{[\text{A}]_t} = kt$$

From Appendix A, we know that $\ln \frac{a}{b} = \ln a - \ln b$, so we have

$$\ln [\text{A}]_0 - \ln [\text{A}]_t = kt$$

Rearranging gives

$$\ln [\text{A}]_t = -kt + \ln [\text{A}]_0$$
$$y = mx + b$$

Therefore, a plot of $\ln [\text{A}]_t$ versus t gives a straight line with slope $= -k$ and y intercept $= \ln [\text{A}]_0$ (Figure 14.10A).

- For a *second-order reaction* with one reactant, we have

$$\frac{1}{[\text{A}]_t} - \frac{1}{[\text{A}]_0} = kt$$

Rearranging gives

$$\frac{1}{[\text{A}]_t} = kt + \frac{1}{[\text{A}]_0}$$
$$y = mx + b$$

Here, a plot of $\frac{1}{[\text{A}]_t}$ versus t gives a straight line with slope $= k$ and y intercept $= \frac{1}{[\text{A}]_0}$ (Figure 14.10B).

- For a *zero-order reaction*, we have

$$[\text{A}]_t - [\text{A}]_0 = -kt$$

Rearranging gives

$$[\text{A}]_t = -kt + [\text{A}]_0$$
$$y = mx + b$$

A plot of $[\text{A}]_t$ versus t gives a straight line with slope $= -k$ and y intercept $= [\text{A}]_0$ (Figure 14.10C).

Some trial-and-error graphical plotting is required to find the reaction order from the concentration and time data:

- If we obtain a straight line when we plot ln [reactant] versus t, the reaction is *first order* with respect to the reactant.
- If we obtain a straight line when we plot $\frac{1}{[\text{reactant}]}$ versus t, the reaction is *second order* with respect to the reactant.
- If we obtain a straight line when we plot [reactant] versus t, the reaction is *zero order* with respect to the reactant.

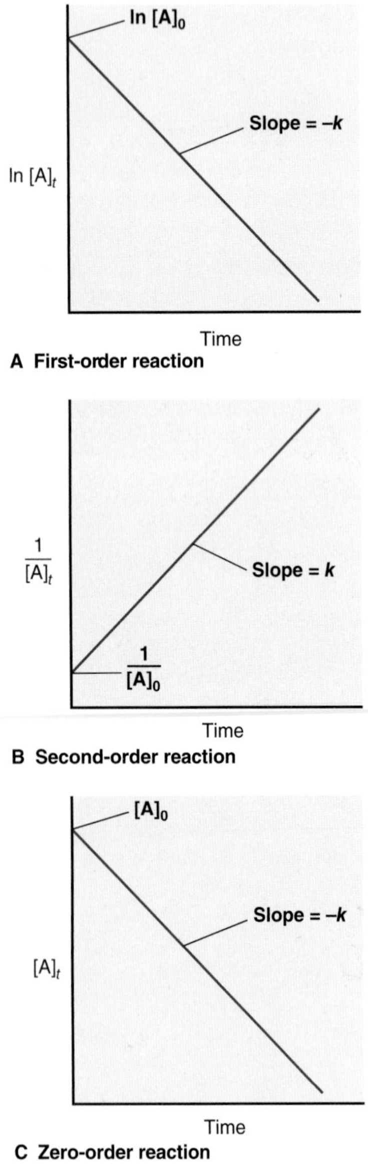

FIGURE 14.10 Graphical method for finding the reaction order from the integrated rate law

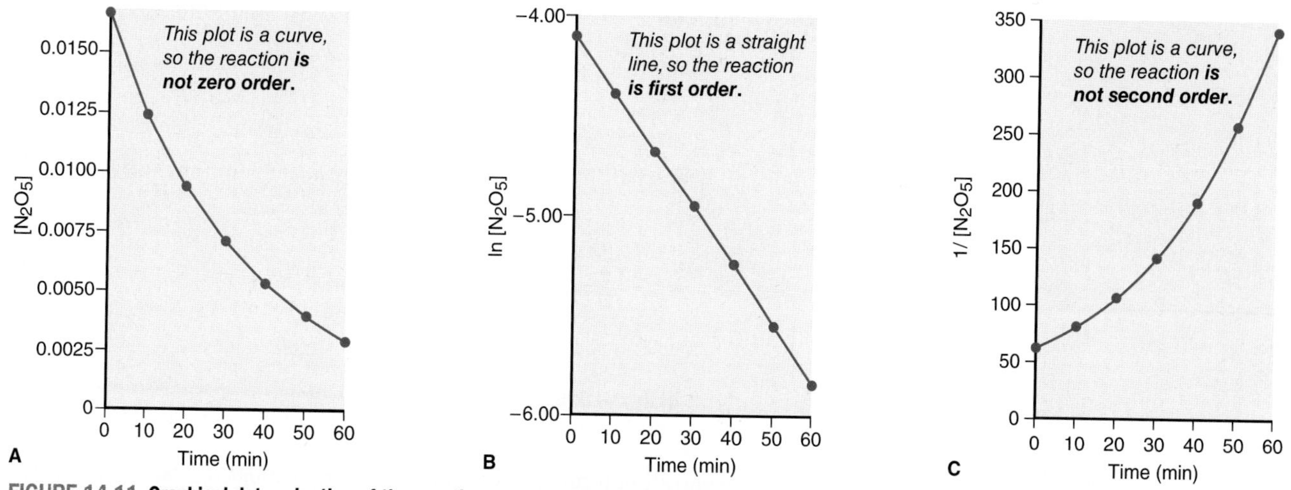

FIGURE 14.11 Graphical determination of the reaction order for the decomposition of N_2O_5. The time and concentration data in the table are used to obtain the three plots, A, B, and C.

Time (min)	$[N_2O_5]$	ln $[N_2O_5]$	$1/[N_2O_5]$
0	0.0165	–4.104	60.6
10	0.0124	–4.390	80.6
20	0.0093	–4.68	1.1×10^2
30	0.0071	–4.95	1.4×10^2
40	0.0053	–5.24	1.9×10^2
50	0.0039	–5.55	2.6×10^2
60	0.0029	–5.84	3.4×10^2

Figure 14.11 shows how to use this approach to determine the order for the decomposition of N_2O_5. When we plot the data from each column in the table versus time, we find that the plot of ln $[N_2O_5]$ versus time is *linear* (plot B), whereas the plot of $[N_2O_5]$ versus time (plot A) and the plot of $\frac{1}{[N_2O_5]}$ versus t (plot C) are *not linear.* Therefore, the decomposition is first order in N_2O_5.

Reaction Half-Life

The **half-life ($t_{1/2}$)** of a reaction is the time that the reactant concentration takes to reach *half its initial value.* A half-life has time units appropriate for the reaction and is characteristic of the reaction at a given temperature. For example, at 45°C, the half-life for the decomposition of N_2O_5, which we know is first order, is 24.0 min. Therefore, if we start with 0.0600 mol/L of N_2O_5 at 45°C, after 24 min (one half-life), 0.0300 mol/L will have reacted and 0.0300 mol/L will remain; after 48 min (two half-lives), 0.0150 mol/L will remain; after 72 min (three half-lives), 0.0075 mol/L will remain, and so on (Figure 14.12). The mathematical expression for the half-life depends on the overall order of the reaction.

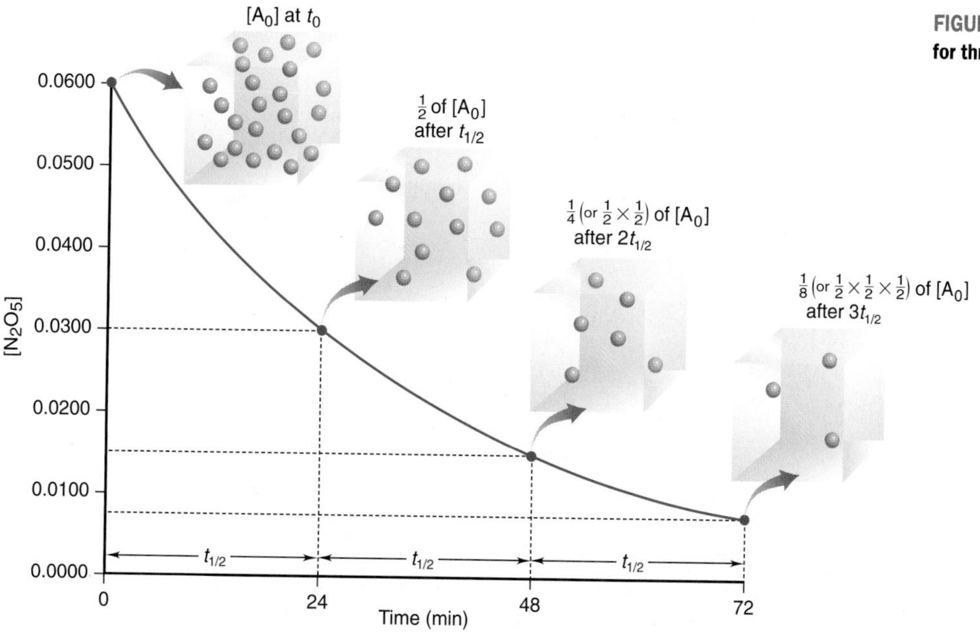

FIGURE 14.12 A plot of $[N_2O_5]$ versus time for three reaction half-lives

First-Order Reactions We can derive an expression for the half-life of a first-order reaction from the integrated rate law, which is

$$\ln \frac{[A]_0}{[A]_t} = kt$$

By definition, after one half-life, $t = t_{1/2}$ and $[A]_t = \frac{1}{2}[A]_0$. Substituting and cancelling $[A]_0$ gives

$$\ln \frac{\cancel{[A]_0}}{\frac{1}{2}\cancel{[A]_0}} = kt_{1/2} \qquad \text{or} \qquad \ln 2 = kt_{1/2}$$

Then, solving for $t_{1/2}$, we have

$$t_{1/2} = \frac{\ln 2}{k} = \frac{0.693}{k} \qquad \text{(first-order process; Rate} = k[\text{A}]) \qquad \textbf{(14.7)}$$

Because no concentration term appears, *for a first-order reaction, the time it takes to reach one-half the starting concentration is a constant and, thus, independent of the reactant concentration.*

An interesting example of the practical application of first-order processes is in the area of pharmaceuticals. The amount of medicine that must be administered (dosing) to ensure that the correct quantity enters the bloodstream is determined using first-order kinetics. Similarly, the time it takes for the medicine to be absorbed or to react in the body also follows first-order kinetics. Finally, the amount of time we must wait to take the next dose of medicine (to ensure that we have enough but not too much medication in our bodies) is also determined by first-order kinetics. The field of pharmacokinetics is vital and requires a thorough understanding of how a particular medication reacts with, remains in, and is removed from the body.

Decay of an unstable, radioactive nucleus is an example of a first-order process that does not involve a *chemical* change. For example, the half-life for the decay of uranium-235 is 7.1×10^8 years. Thus, a sample of ore containing uranium-235 will have one-half the original mass of uranium-235 after 7.1×10^8 years: a sample containing 1 kg will contain 0.5 kg of uranium-235, a sample containing 1 mg will contain 0.5 mg, and so on. (We will discuss radioactive decay thoroughly in Chapter 25, with an emphasis on the kinetics of radioactive decay in Section 25.2).

The next two sample problems show some ways to use half-life in calculations.

Sample Problem 14.6 Using Molecular Scenes to Find Quantities at Various Times

Problem Substance A (*green*) decomposes into two other substances, B (*blue*) and C (*yellow*), in a first-order gaseous reaction. The molecular scenes below show a portion of the reaction mixture at two different times:

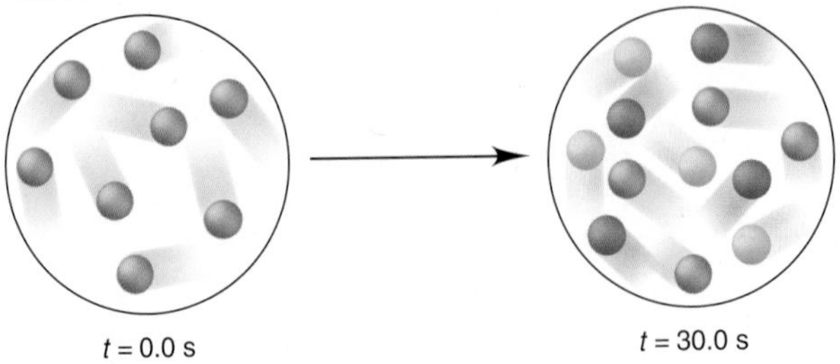

(a) Draw a similar molecular scene of the reaction mixture at t = 60.0 s.

(b) Find the rate constant of the reaction.

(c) If the total pressure (P_{total}) of the mixture is 5.00 bar at 90.0 s, what is the partial pressure of substance B (P_B)?

Plan We are shown molecular scenes of a reaction at two times, with various numbers of reactant and product particles, and we have to predict quantities at two later times. **(a)** We count the number of particles of A and see that A has decreased by half after 30.0 s; thus, the half-life is 30.0 s. The time t = 60.0 s represents two half-lives, so the number of particles of A will decrease by half again, and each A forms one B and one C. **(b)** We substitute the value of the half-life in Equation 14.7 to find k. **(c)** First, we find the numbers of particles at t = 90.0 s, which represents three half-lives. To find p_B, we multiply the mole fraction of B, X_B, by p_{total} (5.00 bar) (Chapter 4). To find X_B, we know that the number of particles is equivalent to the amount (mol), so we divide the number of particles of B by the total number of particles.

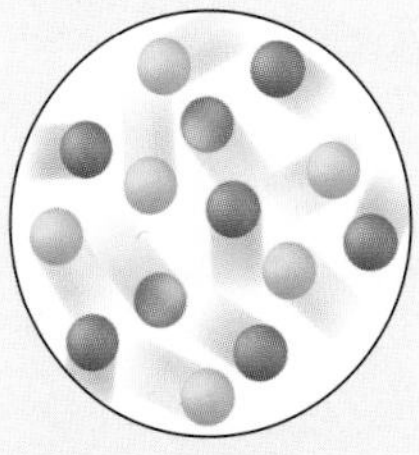

t = 60.0 s

Solution **(a)** The number of particles of A decreased from 8 to 4 in one half-life (30.0 s). Therefore, after two half-lives (60.0 s), there would be 2 (or $\frac{1}{2}$ of 4) particles of A. Each A decomposes to 1 B and 1 C, so 6 (= 8 − 2) particles of A form 6 particles of B and 6 particles of C (*see margin*).

(b) Find the rate constant, k:

$$t_{1/2} = \frac{0.693}{k} \quad \text{so} \quad k = \frac{0.693}{t_{1/2}} = \frac{0.693}{30.0 \text{ s}} = 2.31 \times 10^{-2} \text{s}^{-1}$$

(c) After 90.0s, or three half-lives, there would be 1 A, 7 B, and 7 C particles.

Find the mole fraction of B, X_B:

$$X_B = \frac{7}{1 + 7 + 7} = \frac{7}{15} = 0.467$$

Find the partial pressure of B, p_B:

$$p_B = X_B \times p_{total} = 0.467 \times 5.00 \text{ bar} = 2.33 \text{ bar}$$

Check For (b), rounding gives $\frac{0.7}{30}$, which is a bit over 0.02, so the answer seems correct. For (c), X_B is almost 0.5, so P_B is a bit less than half of 5 bar, or < 2.5 bar.

Follow-Up Problem 14.6 Substance X (*black*) changes to substance Y (*red*) in a first-order gaseous reaction. The scenes below represent the reaction mixture in cubic containers at two different times:

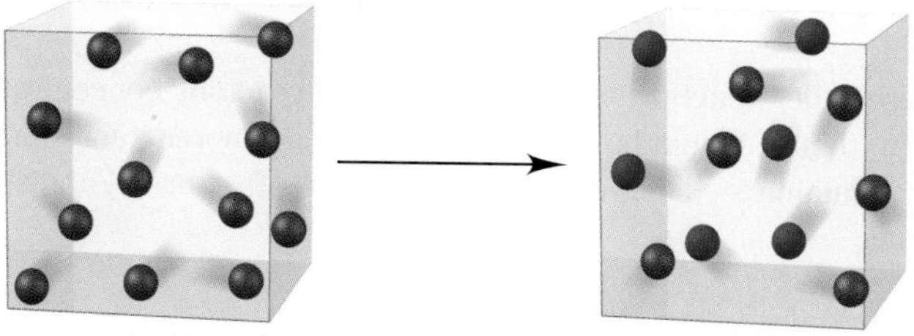

t = 0.0 min t = 2.5 min

(a) Draw a scene that represents the mixture at 5.0 min.

(b) If each sphere represents 0.20 mol of particles and the volume of the cubic container is 0.50 L, what is the concentration (mol/L) of X at 10.0 min?

Sample Problem 14.7 Determining the Half-Life of a First-Order Reaction

Problem Cyclopropane is the smallest cyclic hydrocarbon. Because its 60° bond angles reduce orbital overlap, its bonds are weak. As a result, it is thermally unstable and rearranges to propene at 1000°C via the following first-order reaction:

$$\text{H}_2\text{C}\!\overset{\text{CH}_2}{\overset{/\backslash}{}}\!\!-\text{CH}_2(g) \xrightarrow{\Delta} \text{CH}_3\text{—CH=CH}_2(g)$$

The rate constant is 9.2 s^{-1}.

(a) What is the half-life of the reaction?

(b) How long does it take for the concentration of cyclopropane to reach one-quarter of the initial value?

Plan (a) The cyclopropane rearrangement is first order, so we use Equation 14.7 to find $t_{1/2}$ and substitute for k ($9.2\ s^{-1}$). **(b)** Each half-life decreases the concentration to one-half of its initial value, so two half-lives decrease it to one-quarter.

Solution (a) Solve for $t_{1/2}$:

$$t_{1/2} = \frac{\ln 2}{k} = \frac{0.693}{9.2\ s^{-1}} = 0.075\ s$$

It takes 0.075 s for half of the cyclopropane to form propene at this temperature.

(b) Find the time to reach one-quarter of the initial concentration:

$$\text{Time} = 2(t_{1/2}) = 2(0.075\ s) = 0.15\ s$$

Check For part (a), rounding gives $\frac{0.7}{9\ s^{-1}} = 0.08$ s, so the answer seems correct.

Follow-Up Problem 14.7 Iodine-123 is used to study thyroid gland function. This radioactive isotope breaks down in a first-order process with a half-life of 13.1 h. What is the rate constant for the process?

Second-Order Reactions In contrast to the half-life of a first-order reaction, the half-life of a second-order reaction *does* depend on reactant concentration:

$$t_{1/2} = \frac{1}{k[A]_0} \quad \text{(second-order process; Rate} = k[A]^2\text{)}$$

Note that, here, *the half-life is* ***inversely*** *proportional to the initial reactant concentration.* This relationship means that a second-order reaction with a high initial reactant concentration has a *shorter* half-life, and a second-order reaction with a low initial reactant concentration has a *longer* half-life.

Zero-Order Reactions In contrast to the half-life of a second-order reaction, *the half-life of a zero-order reaction is* ***directly*** *proportional to the initial reactant concentration*:

$$t_{1/2} = \frac{[A]_0}{2k} \quad \text{(zero-order process; Rate} = k\text{)}$$

Thus, if a zero-order reaction begins with a high reactant concentration, it has a longer half-life than it would have if it began with a low reactant concentration.

Table 14.5 summarizes the features of zero-order, first-order, and simple second-order reactions.

TABLE 14.5 An Overview of Zero-Order, First-Order, and Simple Second-Order Reactions

	Zero Order	First Order	Second Order
Rate law	Rate = k	Rate = $k[A]$	Rate = $k[A]^2$
Units for k	mol/(L•s)	1/s	L/(mol•s)
Half-life	$\frac{[A]_0}{2k}$	$\frac{\ln 2}{k}$	$\frac{1}{k[A]_0}$
Integrated rate law in straight-line form	$[A]_t = -kt + [A]_0$	$\ln [A]_t = -kt + \ln [A]_0$	$1/[A]_t = kt + 1/[A]_0$
Plot for straight line	$[A]_t$ versus t	$\ln [A]_t$ versus t	$1/[A]_t$ versus t
Slope, y intercept	$-k$, $[A]_0$	$-k$, $\ln [A]_0$	k, $1/[A]_0$
Graph	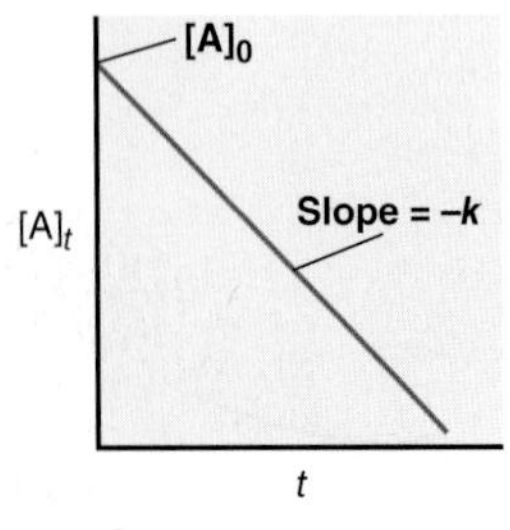	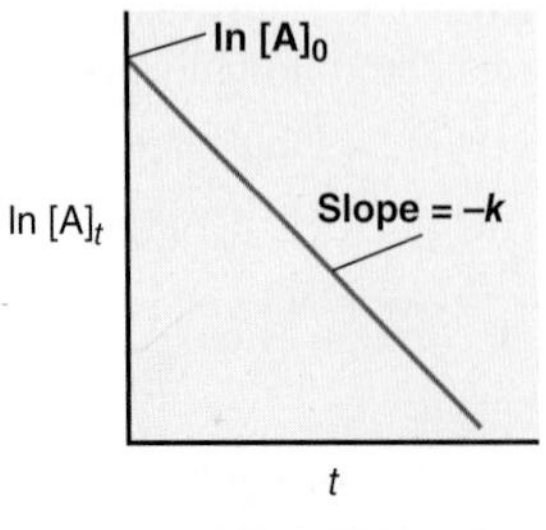	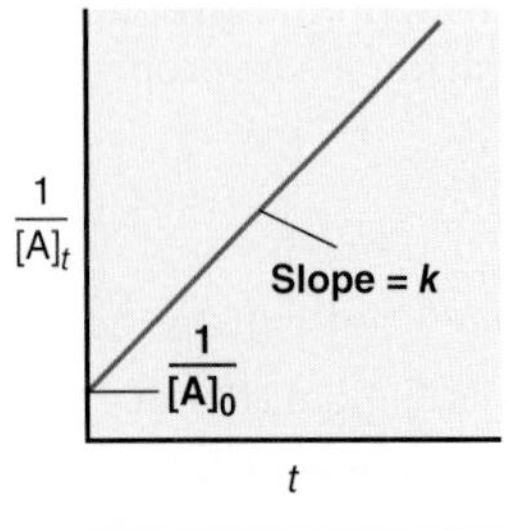

SUMMARY OF SECTION 14.4

- Integrated rate laws are used to find either the time needed to reach a certain concentration of reactant or the concentration present after a given time.
- Rearrangements of the integrated rate laws that give equations in the form of a straight line allow us to determine reaction orders and rate constants graphically.
- The half-life is the time needed for the reactant concentration to reach half its initial value. For first-order reactions, the half-life is constant; that is, it is independent of concentration.

14.5 Theories of Chemical Kinetics

As discussed at the beginning of this chapter, concentration and temperature have major effects on reaction rate. Chemists employ two models—*collision theory* and *transition state theory*—to explain these effects.

Collision Theory: Basis of the Rate Law

The basic tenet of **collision theory** is that particles—atoms, molecules, or ions—must collide to react. However, the number of collisions cannot be the only factor that determines rate; if it were, every reaction would be over in an instant. For example, at 101.3 kPa and 20°C, the N_2 and O_2 molecules in 1 mL of air experience about 10^{27} collisions per second. If all that was needed for a reaction to occur was an N_2 molecule colliding with an O_2 molecule, our atmosphere would consist of almost all NO; in fact, only traces of NO molecules are present. Thus, collision theory also relies on the concepts of collision energy and molecular structure to explain the effects of concentration and temperature on rate.

Why Concentrations Are Multiplied in the Rate Law Particles must collide to react, so the collision frequency, the number of collisions per unit time, provides an *upper limit* on how fast a reaction can take place. In its basic form, collision theory deals with one-step reactions, those in which two particles collide and form products:

$$A + B \longrightarrow \text{products}$$

Suppose that we have only two particles of A and two particles of B confined in a vessel. Figure 14.13 shows that four A-B collisions are possible. The laws of probability tell us that the number of collisions depends on the *product* of the numbers of reactant particles, not their sum. Thus, when we add another particle of A, six A-B collisions (3 × 2) are possible, not just five (3 + 2). Similarly, when we add another particle of B, nine A-B collisions (3 × 3) are possible, not just six (3 + 3). Thus, collision theory explains why we *multiply* the concentrations in the rate law to obtain the observed rate.

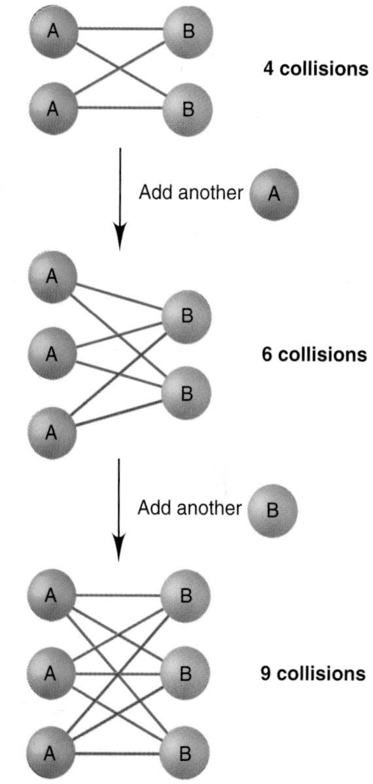

FIGURE 14.13 The number of possible collisions is the product, not the sum, of the reactant concentrations.

The Effect of Temperature on the Rate Constant and the Rate Temperature typically has a dramatic effect on reaction rate: for many reactions near room temperature, an increase of 10 K (10°C) doubles or triples the reaction rate. Figure 14.14A shows kinetic data for an organic reaction—the hydrolysis, or reaction with water, of the ester ethyl acetate. To understand the effect of temperature, we measure concentrations and times for the reaction run at different temperatures. Solving each rate expression for k and plotting the results (Figure 14.14B), we find that *k increases exponentially as T increases.*

FIGURE 14.14 Increase of the rate constant with temperature for the hydrolysis of an ester

A

Expt	[Ester]	$[H_2O]$	T (K)	Rate [mol/(L•s)]	k [L/(mol•s)]
1	0.100	0.200	288	1.040×10^{-3}	0.0521
2	0.100	0.200	298	2.02×10^{-3}	0.101
3	0.100	0.200	308	3.68×10^{-3}	0.184
4	0.100	0.200	318	6.64×10^{-3}	0.332

B

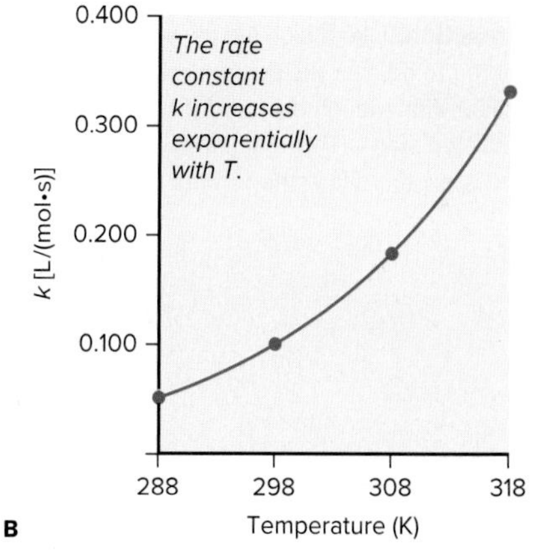

These results are consistent with the findings obtained in 1889 by the Swedish chemist Svante Arrhenius. In its modern form, the **Arrhenius equation** is

$$k = Ae^{-\frac{E_a}{RT}} \tag{14.8}$$

where k is the rate constant, e is the base of natural logarithms, T is the absolute temperature, and R is the universal gas constant. (We will focus on the term E_a in the next subsection, and on the term A a bit later.) Notice the relationship between k and T: especially notice that T is in the denominator of a negative exponent. Thus, *as T increases, the value of the negative exponent becomes smaller, which means that k becomes larger, so the rate increases*:

$$\text{higher } T \Longrightarrow \text{larger } k \Longrightarrow \text{increased rate}$$

The Central Importance of Activation Energy The effect of temperature on k is closely related to the **activation energy (E_a)** of a reaction, an energy *threshold* that the colliding molecules must exceed to react. As an analogy, to be successful, a high jumper must exert at least enough energy to get over the bar. Similarly, if reactant molecules collide with a certain minimum energy, they form an activated complex from which they can change to products (Figure 14.15).

As you can see from Figure 14.15, a reversible reaction has two activation energies. The activation energy for the forward reaction, $E_{a(fwd)}$, is the energy difference between the activated state and the reactants. The activation energy for the reverse reaction, $E_{a(rev)}$, is the energy difference between the activated state and the products. The reaction represented in the diagram is exothermic ($\Delta_r H < 0$) in the forward direction. Thus, the products are at a lower energy than the reactants, and $E_{a(fwd)}$ is less than $E_{a(rev)}$. This difference equals the heat of reaction, $\Delta_r H$:

$$\Delta_r H = E_{a(fwd)} - E_{a(rev)} \tag{14.9}$$

The Effect of Temperature on Collision Energy A rise in temperature has two effects on moving particles: it causes a higher collision *frequency* and a higher collision *energy*. Let us see how each affects rate.

- *Collision frequency*. If particles move faster, they collide more often. Calculations show that a 10 K rise in temperature, from 288 K to 298 K, for example, increases the average molecular speed by 2%. This would lead to, at most, a 4% increase in rate. Thus, higher collision frequency cannot possibly account for the doubling or tripling of rates observed with a 10 K rise. Indeed, the effect of temperature on collision *frequency* is only minor.
- *Collision energy*. On the other hand, the effect of temperature on collision *energy* is major. At a given temperature, the fraction, f, of collisions with energy equal to or greater than E_a is

$$f = e^{-\frac{E_a}{RT}}$$

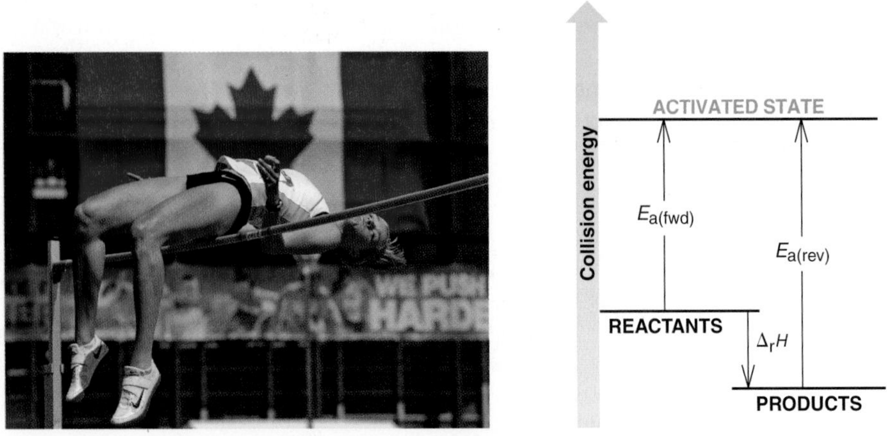

FIGURE 14.15 Energy-level diagram for a reaction. Like a high jumper with enough energy to go over the bar, molecules must collide with enough energy, E_a, to reach an activated state. (This reaction is reversible and is exothermic in the forward direction.)

where e is the base of natural logarithms, T is the absolute temperature, and R is the universal gas constant. The right side of this equation appears in the Arrhenius equation (Equation 14.8), which shows that a rise in T causes a larger k. We now see why—because *a rise in temperature enlarges the fraction of collisions with enough energy to exceed* E_a (Figure 14.16).

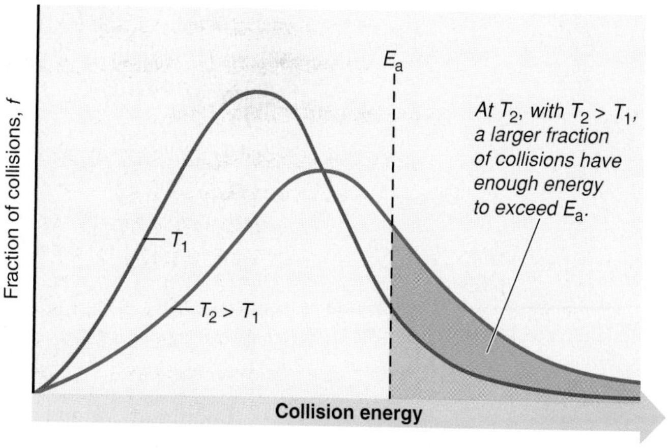

FIGURE 14.16 The effect of temperature on the distribution of collision energies

Table 14.6 shows that the magnitudes of *both* E_a *and* T *affect the size of this fraction* for the hydrolysis of the ester we discussed above. In the top portion, temperature is held constant, and the fraction of sufficiently energetic collisions shrinks several orders of magnitude with each 25 kJ/mol increase in E_a. (To extend the high jump analogy, as the height of the bar is raised, a smaller fraction of high jumpers have enough energy to jump over it.) In the bottom portion, E_a is held constant, and the fraction nearly doubles for each 10 K (10°C) rise in temperature.

Therefore, *the smaller the activation energy (or the higher the temperature), the larger the fraction of sufficiently energetic collisions, the larger the value of k, and the higher the reaction rate*:

$$\text{smaller } E_a \text{ (or higher } T) \Longrightarrow \text{larger } f \Longrightarrow \text{larger } k \Longrightarrow \text{higher rate}$$

TABLE 14.6 The Effects of E_a and T on the Fraction (f) of Collisions with Sufficient Energy to Allow a Reaction

E_a (kJ/mol)	f (at T = 298 K)
50	1.70×10^{-9}
75	7.03×10^{-14}
100	2.90×10^{-18}
T	**f (at E_a = 50 kJ/mol)**
25°C (298 K)	1.70×10^{-9}
35°C (308 K)	3.29×10^{-9}
45°C (318 K)	6.12×10^{-9}

Calculating the Activation Energy We can calculate E_a from the Arrhenius equation by taking the natural logarithm of both sides of the equation, which recasts the equation into the form of an equation for a straight line:

$$k = Ae^{-\frac{E_a}{RT}}$$

$$\ln k = \ln A - \frac{E_a}{R}\left(\frac{1}{T}\right)$$

$$y = b + mx$$

A plot of ln k versus $1/T$ gives a straight line whose slope is $-E_a/R$ and whose y intercept is ln A (Figure 14.17). We know the constant R, so we can determine E_a graphically from a series of k values at different temperatures.

We can find E_a in another way if we know the rate constants at two temperatures, T_2 and T_1:

$$\ln k_2 = \ln A - \frac{E_a}{R}\left(\frac{1}{T_2}\right) \qquad \ln k_1 = \ln A - \frac{E_a}{R}\left(\frac{1}{T_1}\right)$$

When we subtract ln k_1 from ln k_2, the term ln A drops out and the other terms can be rearranged to give

$$\ln\frac{k_2}{k_1} = -\frac{E_a}{R}\left(\frac{1}{T_2} - \frac{1}{T_1}\right) \qquad (14.10)$$

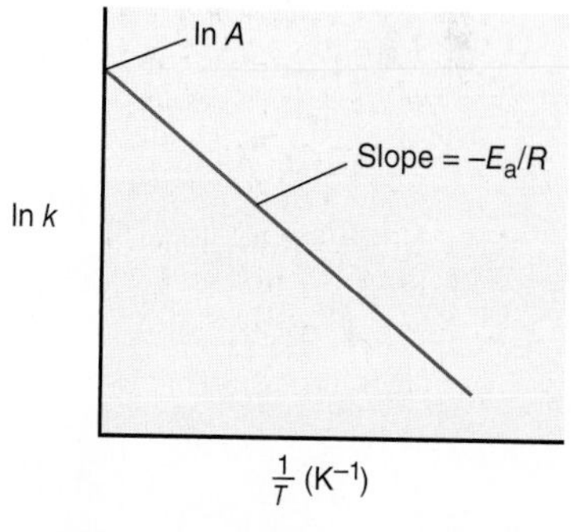

FIGURE 14.17 Graphical determination of the activation energy

Then we can solve for E_a, as in the next sample problem. This derivation is valid as long as the pre-exponential factor (A) is temperature independent. In the examples in this chapter, we will assume that the value of A changes negligibly with temperature. However, in fact, A does have a slight dependence on temperature.

Sample Problem 14.8 Determining the Activation Energy

Problem The decomposition of hydrogen iodide is shown below:

$$2HI(g) \longrightarrow H_2(g) + I_2(g)$$

This reaction has rate constants of 9.51×10^{-9} L/(mol·s) at 500. K and 1.10×10^{-5} L/(mol·s) at 600. K. Find E_a.

Plan We are given the rate constants, k_1 and k_2, at two temperatures, T_1 and T_2, so we can substitute into Equation 14.10 and solve for E_a.

Solution Rearrange Equation 14.10 to solve for E_a:

$$\ln\frac{k_2}{k_1} = -\frac{E_a}{R}\left(\frac{1}{T_2} - \frac{1}{T_1}\right)$$

$$E_a = -R\left(\ln\frac{k_2}{k_1}\right)\left(\frac{1}{T_2} - \frac{1}{T_1}\right)^{-1}$$

$$= -[8.314\ \text{J/(mol·K)}]\left[\ln\frac{1.10\times10^{-5}\ \text{L/(mol·s)}}{9.51\times10^{-9}\ \text{L/(mol·s)}}\right]\left(\frac{1}{600.\ \text{K}} - \frac{1}{500.\ \text{K}}\right)^{-1}$$

$$= 1.76\times10^{5}\ \text{J/mol} = 1.76\times10^{2}\ \text{kJ/mol}$$

Comment We need to retain the same number of significant figures in $\frac{1}{T}$ as there are in T, to avoid introducing a significant error. We round to the correct number of significant figures only at the final answer.

Follow-Up Problem 14.8 The following reaction has an E_a of 1.00×10^2 kJ/mol and a rate constant of 0.286 L/(mol•s) at 500. K:

$$2NOCl(g) \longrightarrow 2NO(g) + Cl_2(g)$$

What is the rate constant at 490. K?

The Effect of Molecular Structure on Rate At ordinary temperatures, the enormous number of collisions per second between reactant particles is reduced six or more orders of magnitude by counting only those collisions with enough energy to react. Even this tiny fraction of collisions is typically much larger than the number of **effective collisions**, those that actually lead to product because *the atoms that become bonded in the product make contact.* Thus, to be effective, a collision must have enough energy *and* the appropriate *molecular orientation.*

In the Arrhenius equation, molecular orientation is contained in the term A:

$$k = Ae^{-\frac{E_a}{RT}}$$

This term is the **pre-exponential factor**, *the product of the collision frequency (Z) and an orientation probability factor (p) that is specific to each reaction:*

$$A = pZ$$

The factor p is related to the structural complexity of the reactants. You can think of p as the ratio of effectively oriented collisions to all possible collisions. Figure 14.18 shows five of the collision orientations for the following reaction:

$$NO(g) + NO_3(g) \longrightarrow 2NO_2(g)$$

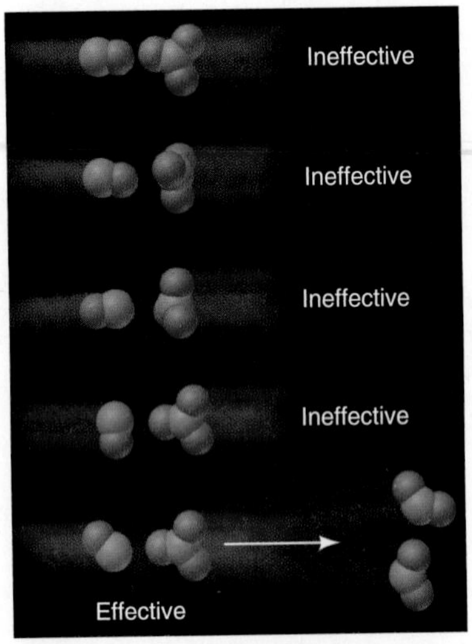

FIGURE 14.18 The importance of molecular orientation to an effective collision. In the one effective orientation (*bottom*), contact occurs between the atoms that become bonded in the product.

Of the five orientations shown, only one has the effective orientation, with the N of NO making contact with an O of NO_3. Actually, the p value for this reaction is 0.006: only 6 collisions in every 1000 (1 in 167) have the correct orientation.

The more complex the molecular structure, the smaller the p value. Individual atoms are spherical, so reactions between them have p values near 1: as long as reacting atoms collide with enough energy, the product forms. At the other extreme are biochemical reactions, in which two small molecules (or portions of larger molecules) react only when they collide with enough energy on a specific tiny region of a giant protein. The p value for these reactions is often much less than 10^{-6}, less than one in a million. The fact that countless such biochemical reactions are occurring in you right now attests to the astounding number of collisions per second.

Transition State Theory: What the Activation Energy Is Used For

Collision theory explains the importance of effective collisions, and **transition state theory** focuses on the high-energy species that exists at the moment of an effective collision when reactants are becoming product.

Visualizing the Transition State As two molecules approach each other, repulsions between their electron clouds continuously increase, so they slow down as some of their kinetic energy is converted to potential energy. If they collide but the energy of the collision is *less* than the activation energy, the molecules bounce off each other.

However, in a tiny fraction of collisions in which the molecules are moving fast enough, *their kinetic energies push them together with enough force to overcome the repulsions and surpass the activation energy.* In an even tinier fraction of these sufficiently energetic collisions, the molecules are oriented effectively. In these collisions, nuclei in one molecule attract electrons in the other, atomic orbitals overlap, electron densities shift, and some bonds lengthen and weaken while other bonds shorten and strengthen. At some point during this smooth transformation, there is *a species with partial bonds* that is neither reactant nor product. This very unstable species, called the **transition state (activated complex)** exists only at the instant of highest potential energy. Thus, *the activation energy of a reaction is used to reach the transition state.*

The scientist who was responsible for the development of transition state theory, Henry Eyring (1901–1981), wrote his major work on rates of chemical reactions in 1935. He was a physical chemist who was greatly influenced by the work of G. N. Lewis, who was quoted as saying "Physical chemistry is everything that is interesting." It was a great puzzle to many scientists that Eyring did not receive the Nobel prize for his groundbreaking calculations of electronic potential energy surfaces for chemical reactions and the use of statistical mechanical tools to develop his absolute rate theory, which became the basis for transition state theory. It is believed that the Nobel committee did not understand his brilliant work until it was much too late; the Royal Swedish Academy of Sciences awarded him the next-highest prize, the Berzelius Medal in gold, in 1977, in recognition of his outstanding contributions. One of Eyring's graduate students was Dr. Keith J. Laidler, who was one of the first professors in the Department of Chemistry at the University of Ottawa, where he remained even after retirement as Professor Emeritus, until his death in 2003. (References: Wikipedia, Nobelprize.org, and "Book reviews," *C&EN News*, Volume 86, Issue 23, pp. 46–49, June 9, 2008.)

Transition states cannot be isolated, but the work of Ahmed H. Zewail, who received the 1999 Nobel Prize in Chemistry, greatly expanded our knowledge of them. Using lasers pulsing at the time scale of bond vibrations (10^{-15} s), he observed transition states forming and decomposing. A few have been well studied, such as the transition state that forms when methyl bromide reacts with hydroxide ion:

$$BrCH_3 + OH^- \longrightarrow Br^- + CH_3OH$$

The electronegative bromine makes the carbon in $BrCH_3$ partially positive, and the carbon attracts the negatively charged oxygen in OH^-. As a C—O bond begins to form, the Br—C bond begins to weaken. In the transition state (Figure 14.19), the C atom is surrounded by five atoms (trigonal bipyramidal; Section 9.1), which never occurs in its stable compounds. This high-energy species has three normal C—H bonds and two partial bonds, one from C to O and the other from Br to C.

Reaching the transition state does not guarantee that a reaction will proceed to products, because *a transition state can change in either direction.* In this case, if the C—O bond continues to strengthen, products form, but, if the Br—C bond becomes stronger again, the transition state reverts to reactants.

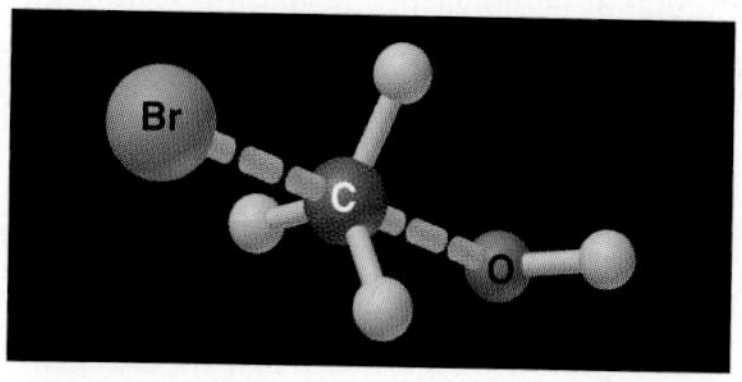

FIGURE 14.19 The transition state of the reaction between $BrCH_3$ and OH^-. Notice the partial (*dashed*) C—O and Br—C bonds and the trigonal bipyramidal shape.

Depicting the Change with a Reaction Energy Diagram A useful way to depict these events is to use a **reaction energy diagram**, which plots how potential energy changes as the reaction proceeds from reactants to products (the *reaction progress*). A reaction energy diagram shows the relative energy levels of reactants, products, and transition state, as well as the forward and reverse activation energies and the enthalpy of reaction.

A diagram for the reaction of $BrCH_3$ and OH^- is shown at the bottom of Figure 14.20. This reaction is exothermic, so the reactants are higher in energy than the products, which means that $E_{a(fwd)}$ is less than $E_{a(rev)}$. Above the diagram

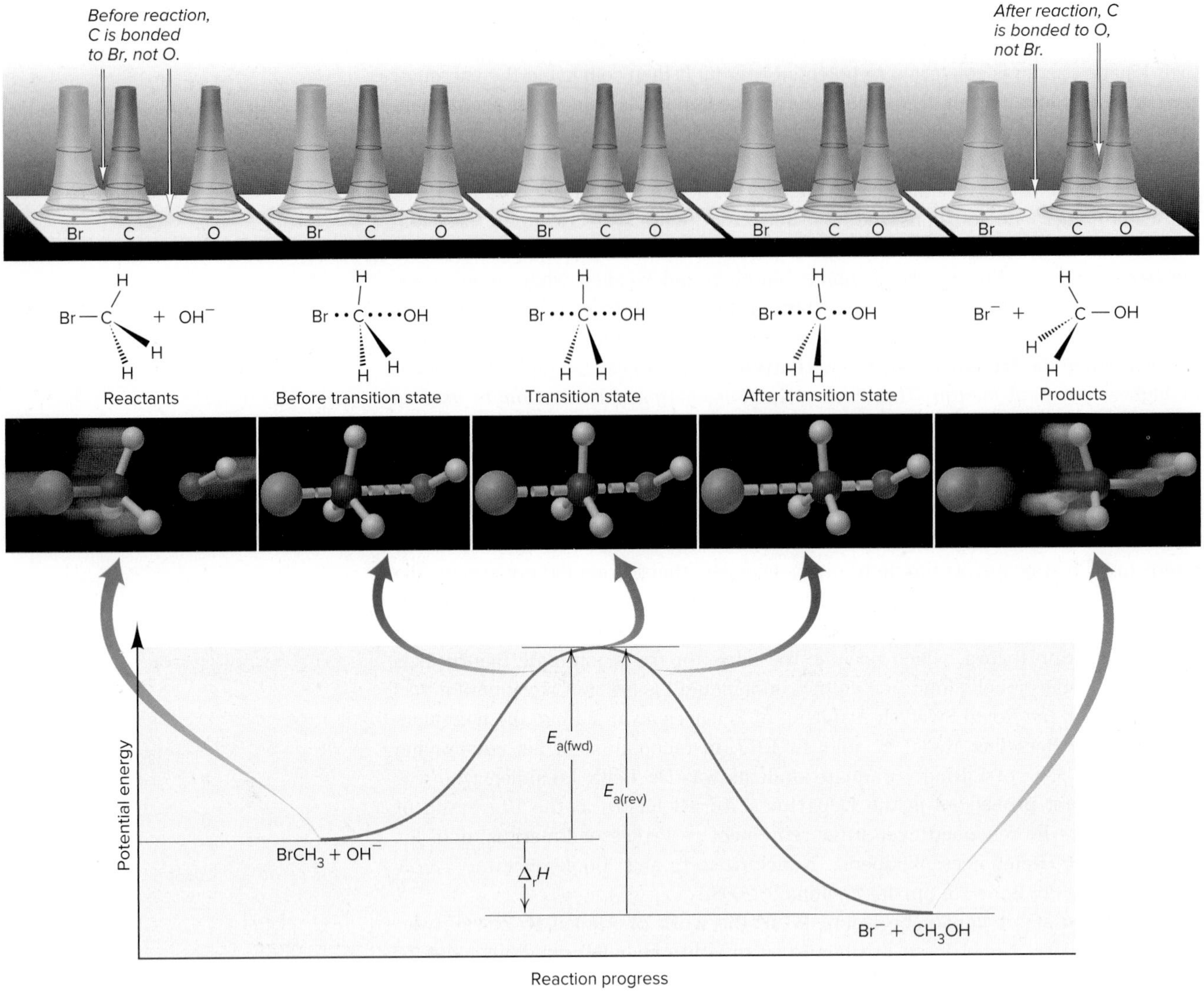

FIGURE 14.20 Depicting the reaction between $BrCH_3$ and OH^-. From *bottom* to *top*, the reaction energy diagram with blow-up arrows at five points during the reaction, molecular-scale views, structural formulas, and electron density relief maps. Notice the gradual, simultaneous C—O bond forming and Br—C bond breaking.

are molecular-scale views at various points during the reaction, as well as the corresponding structural formulas. At the top of Figure 14.20 are electron density relief maps of the Br, C, and O atoms; notice the gradual change in electron density from C overlapping Br (*left*) to C overlapping O (*right*).

Transition state theory proposes that *every reaction (or every step in an overall reaction) goes through its own transition state.* Figure 14.21 presents reaction energy diagrams for two gas-phase reactions. For each reaction, the structure of the transition state is predicted from the orientations of the reactant atoms that must become bonded in the product. Henry Eyring developed the **Eyring equation** (sometimes called the Eyring-Polanyi equation, as it was almost simultaneously developed by Eyring; Michael Polanyi, father of John Charles Polanyi (see margin below); and Meredith Gwynne Evans), an equation that allowed for the determination of the rate constant of a reaction using transition state theory:

$$k = \frac{k_B T}{h} e^{-\frac{\Delta G^{\ddagger}}{RT}}$$

where k is the rate constant, k_B represents the Boltzmann constant (Chapter 18), T is the temperature (K), h is Planck's constant, R is the ideal gas law constant, and $\Delta G^{\ddagger}$ represents the free energy of activation (Chapter 18). Interestingly, the Eyring equation resembles the Arrhenius equation quite closely; it is very important

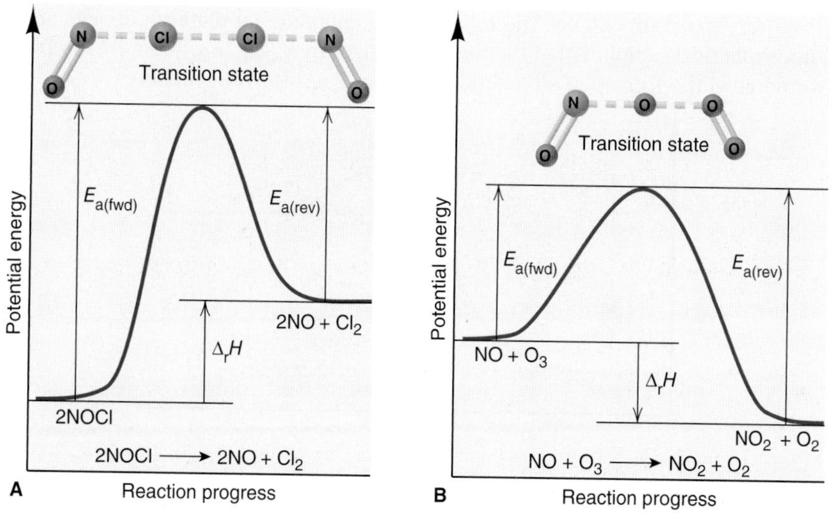

FIGURE 14.21 Reaction energy diagrams and possible transition states for two reactions. A. Endothermic reaction. **B.** Exothermic reaction.

to note, however, that the Eyring equation was *derived* from the transition state theory, whereas the Arrhenius equation was determined *empirically* (from experiment).

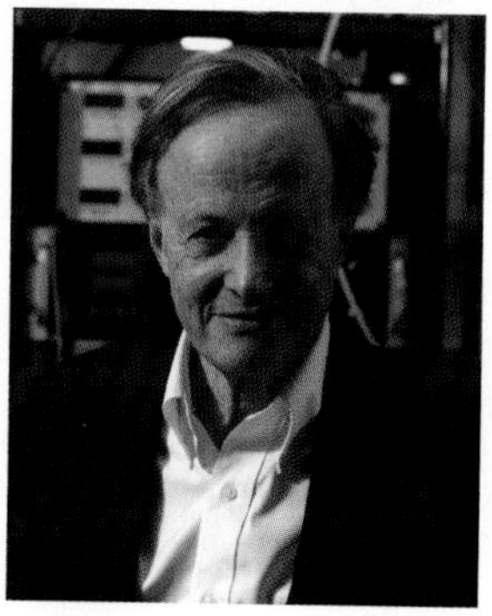

John Charles Polanyi (a professor at the University of Toronto) received the Nobel Prize in Chemistry in 1986 for his work on reaction dynamics, an area of chemical kinetics. He studied the dynamics of elementary chemical processes using infrared chemiluminescence, a technique that he developed. In particular, he studied weak infrared emissions from newly formed molecules to determine energy disposal during the chemical reaction. While in Sweden receiving the Nobel Prize, he learned about the technique of scanning tunnelling microscopy. He now uses this technique to study surfaces and surface molecules. [SOURCE: Information from Wikipedia and the Press Release for the 1986 Nobel Prize in Chemistry.]

Sample Problem 14.9 Drawing Reaction Energy Diagrams and Transition States

Problem The following is a key reaction in the upper atmosphere:

$$O_3(g) + O(g) \longrightarrow 2O_2(g)$$

The $E_{a(fwd)}$ is 19 kJ/mol, and the $\Delta_r H$ for the reaction as written is −392 kJ/mol. Draw a reaction energy diagram, predict a structure for the transition state, and calculate $E_{a(rev)}$.

Plan The reaction is highly exothermic ($\Delta_r H = -392$ kJ/mol), so the products are much lower in energy than the reactants. The small $E_{a(fwd)}$ (19 kJ/mol) means that the energy of the reactants lies slightly below the energy of the transition state. We use Equation 14.9 to calculate $E_{a(rev)}$. To predict the transition state, we sketch the species. Note that one of the bonds in O_3 weakens, and this partially bonded O begins forming a bond to the separate O atom.

Solution Solve for $E_{a(rev)}$:

$$\Delta_r H = E_{a(fwd)} - E_{a(rev)}$$

So,

$$E_{a(rev)} = E_{a(fwd)} - \Delta_r H = 19 \text{ kJ/mol} - (-392 \text{ kJ/mol}) = 411 \text{ kJ/mol}$$

The reaction energy diagram (not drawn to scale), with the transition state, is shown below:

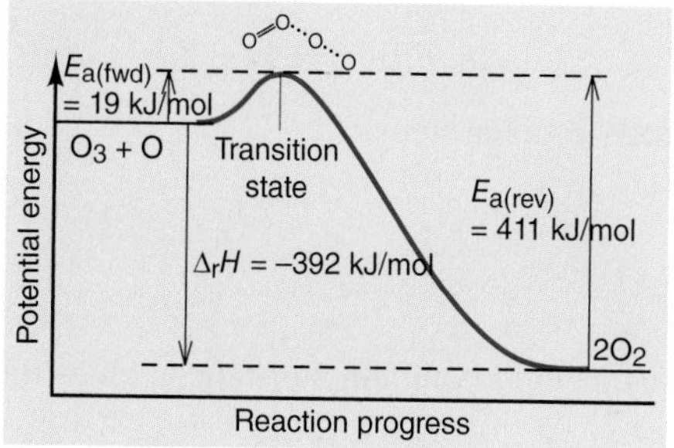

Check Rounding to find $E_{a(rev)}$ gives ~ 20 + 390 = 410.

Follow-Up Problem 14.9 The following reaction energy diagram depicts another key atmospheric reaction. Label the axes, and identify $E_{a(fwd)}$, $E_{a(rev)}$, and $\Delta_r H$. Then draw and label the transition state, and calculate $E_{a(rev)}$ for the reaction.

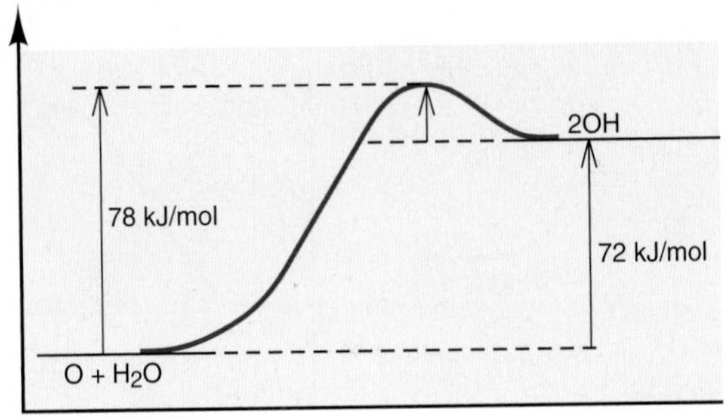

SUMMARY OF SECTION 14.5

- According to collision theory, reactant particles must collide to react, and the number of possible collisions is found by multiplying the number of reactant particles.
- As the Arrhenius equation shows, a rise in temperature increases the reaction rate because it increases the rate constant.
- The activation energy, E_a, is the minimum energy needed for colliding particles to react.
- The relative E_a values for forward and reverse reactions depend on whether the overall reaction is exothermic or endothermic.
- At higher temperatures, more collisions have enough energy to exceed E_a.
- E_a can be determined graphically from k values obtained at different T values.
- Molecules must collide with an effective orientation for them to react, so structural complexity decreases the reaction rate.
- Transition state theory focuses on the change of kinetic energy to potential energy as reactant particles collide and form an unstable transition state.
- Given a sufficiently energetic collision and an effective molecular orientation, the reactant species form the transition state, which either continues toward product(s) or reverts to reactant(s).
- A reaction energy diagram depicts the changing potential energy throughout a reaction's progress from reactants through transition states to products.

14.6 Reaction Mechanisms: The Steps from Reactant to Product

We cannot understand how a car works by examining the body, wheels, and dashboard, or even the engine as a whole. We need to look inside the engine to see how its parts fit together and function. Similarly, we cannot understand how a reaction works by examining the overall balanced equation. We must look "inside the reaction" to see how the reactants change into products.

Most reactions occur through a **reaction mechanism**, a sequence of single reaction steps that sum to the overall equation. For example, a possible mechanism for the overall reaction

$$2A + B \longrightarrow E + F$$

might involve these three simpler steps:

(1) $A + B \longrightarrow C$
(2) $C + A \longrightarrow D$
(3) $D \longrightarrow E + F$

Adding the steps and cancelling common substances gives the overall equation:

$$A + B + \cancel{C} + A + \cancel{D} \longrightarrow \cancel{C} + \cancel{D} + E + F \quad \text{or} \quad 2A + B \longrightarrow E + F$$

A mechanism is a hypothesis about how a reaction occurs. Chemists *propose* a mechanism and then *test* it to see if it fits with the observed rate law.

TABLE 14.7 Rate Laws for General Elementary Steps

Elementary Step	Molecularity	Rate Law
$A \longrightarrow$ product	Unimolecular	Rate $= k[A]$
$2A \longrightarrow$ product	Bimolecular	Rate $= k[A]^2$
$A + B \longrightarrow$ product	Bimolecular	Rate $= k[A][B]$
$2A + B \longrightarrow$ product	Termolecular	Rate $= k[A]^2[B]$

Elementary Reactions and Molecularity

The individual steps that make up a reaction mechanism are called **elementary reactions (elementary steps)**. Each describes a *single molecular event*, such as one particle decomposing or two particles combining.

An elementary step is characterized by its **molecularity**, the number of *reactant* particles in the step. Consider the mechanism for the breakdown of ozone in the stratosphere. The overall equation is

$$2O_3(g) \longrightarrow 3O_2(g)$$

A two-step mechanism has been proposed:

(1) $O_3(g) \longrightarrow O_2(g) + O(g)$
(2) $O_3(g) + O(g) \longrightarrow 2O_2(g)$

The first step is a **unimolecular reaction**, a reaction that involves the decomposition or rearrangement of a single particle (O_3). The second step is a **bimolecular reaction**, a reaction in which two particles (O_3 and O) react. Some *termolecular* elementary steps occur, but they are extremely rare because the probability of three particles colliding simultaneously with enough energy and an effective orientation is very small. Higher molecularities are not known. Therefore, in general, *we propose unimolecular and/or bimolecular reactions as reasonable steps in a mechanism.*

Because an elementary reaction occurs in one step, its rate law, unlike the rate law for an overall reaction, *can* be deduced from the reaction stoichiometry: *reaction order equals molecularity*. Therefore, *only* for an elementary step, *we use the equation coefficients as the reaction orders in the rate law* (Table 14.7).

Sample Problem 14.10 Determining Molecularity and Rate Laws for Elementary Steps

Problem The following elementary steps are proposed for a reaction mechanism:
(1) $NO_2Cl(g) \longrightarrow NO_2(g) + Cl(g)$
(2) $NO_2Cl(g) + Cl(g) \longrightarrow NO_2(g) + Cl_2(g)$

(a) Write the overall balanced equation.

(b) Determine the molecularity of each step.

(c) Write the rate law for each step.

Plan We find the overall equation from the sum of the elementary steps. The molecularity of each step equals the total number of *reactant* particles. We write the rate law for each step using the molecularities as reaction orders.

Solution (a) Write the overall balanced equation:

$$NO_2Cl(g) \longrightarrow NO_2(g) + Cl(g)$$
$$NO_2Cl(g) + Cl(g) \longrightarrow NO_2(g) + Cl_2(g)$$
$$NO_2Cl(g) + NO_2Cl(g) + \cancel{Cl(g)} \longrightarrow NO_2(g) + \cancel{Cl(g)} + NO_2(g) + Cl_2(g)$$
$$2NO_2Cl(g) \longrightarrow 2NO_2(g) + Cl_2(g)$$

(b) We need to determine the molecularity of each step. The first step has one reactant, NO_2Cl, so it is unimolecular. The second step has two reactants, NO_2Cl and Cl, so it is bimolecular.

(c) Write the rate laws for the elementary steps:
(1) $Rate_1 = k_1[NO_2Cl]$
(2) $Rate_2 = k_2[NO_2Cl][Cl]$

Check In part (a), be sure that the equation is balanced. In part (c), be sure that the substances in brackets are the *reactants* of the elementary steps.

Follow-Up Problem 14.10 These elementary steps are proposed for a mechanism:

(1) $2NO(g) \longrightarrow N_2O_2(g)$
(2) $2H_2(g) \longrightarrow 4H(g)$
(3) $N_2O_2(g) + H(g) \longrightarrow N_2O(g) + HO(g)$
(4) $HO(g) + H(g) \longrightarrow H_2O(g)$
(5) $H(g) + N_2O(g) \longrightarrow HO(g) + N_2(g)$

(a) Write the balanced equation for the overall reaction.

(b) Determine the molecularity of each step.

(c) Write the rate law for each step.

The Rate-Determining Step of a Reaction Mechanism

All the elementary steps in a mechanism have their own rates. However, one step is usually *much* slower than the others. This step, called the **rate-determining step (rate-limiting step)**, limits how fast the overall reaction can proceed. Therefore, *the rate law of the rate-determining step becomes the rate law for the overall reaction.* ■

■ **A Rate-Determining Step for Traffic Flow** Imagine driving back to Canada after a visit to the United States. Traffic flows smoothly until you come to the border crossing. At the border crossing, there is a huge delay, causing traffic to come to a stop for an extended period of time. Once you cross the border, traffic returns to normal and the rest of your trip home is without delay. Getting through the bottleneck of the border crossing takes longer than the rest of the trip combined and, therefore, determines the time for the whole trip.

Consider the reaction between nitrogen dioxide and carbon monoxide:

$$NO_2(g) + CO(g) \longrightarrow NO(g) + CO_2(g)$$

If this reaction were an elementary step—that is, if the mechanism consisted of only one step—we could immediately write the overall rate law as

$$\text{Rate} = k[NO_2][CO]$$

However, as you saw in Sample Problem 14.3, experimental data show that the rate law is

$$\text{Rate} = k[NO_2]^2$$

Thus, the overall reaction cannot be elementary.

A proposed two-step mechanism is

(1) $NO_2(g) + NO_2(g) \longrightarrow NO_3(g) + NO(g)$ (slow; rate-determining)
(2) $NO_3(g) + CO(g) \longrightarrow NO_2(g) + CO_2(g)$ (fast)

In this mechanism, NO_3 functions as a **reaction intermediate**, a substance that is formed and used up during the reaction. Even though it does not appear in the overall balanced equation, a reaction intermediate is essential for the reaction to occur. Intermediates are less stable than the reactants and products, but, unlike the *much* less stable transition states, they have normal bonds and can sometimes be isolated.

The rate laws for the two elementary steps above are

(1) $\text{Rate}_1 = k_1[NO_2][NO_2] = k_1[NO_2]^2$
(2) $\text{Rate}_2 = k_2[NO_3][CO]$

Notice the following three key points about this mechanism:

- If $k_1 = k$, the rate law for the rate-determining step (step 1) becomes identical to the observed rate law.
- Because the first step is slow, $[NO_3]$ is low. As soon as any NO_3 forms, it is consumed by the fast second step, so the reaction takes as long as the first step does.
- CO does not appear in the rate law (reaction order = 0) because it takes part in the mechanism *after* the rate-determining step.

Correlating the Mechanism with the Rate Law

Coming up with a reasonable reaction mechanism is a classic demonstration of the scientific method. Using observations and data from rate experiments, we hypothesize the individual steps and then test our hypothesis with further evidence. If the evidence supports our mechanism, we can accept it; if not, we must propose a different mechanism. We can never *prove* that a mechanism represents the *actual* chemical change, only that it is consistent with the data.

A valid mechanism must meet three criteria:

1. *The elementary steps must add up to the overall balanced equation.*
2. *The elementary steps must be reasonable.* They should generally involve one reactant particle (unimolecular) or two (bimolecular).
3. *The mechanism must correlate with the rate law*, not the other way around.

Mechanisms with a Slow Initial Step The reaction between NO_2 and CO that we considered earlier has a mechanism with a slow initial step; this means that the first step is rate-determining. The reaction between nitrogen dioxide and fluorine is another example:

$$2NO_2(g) + F_2(g) \longrightarrow 2NO_2F(g)$$

The experimental rate law is first order in NO_2 and F_2:

$$\text{Rate} = k[NO_2][F_2]$$

The accepted mechanism is given below. Notice that the free fluorine atom is a reaction intermediate.

(1) $NO_2(g) + F_2(g) \longrightarrow NO_2F(g) + F(g)$ (slow; rate-determining)
(2) $NO_2(g) + F(g) \longrightarrow NO_2F(g)$ (fast)

Let us see how this mechanism meets the three criteria.

1. The elementary reactions sum to the balanced equation:

$$NO_2(g) + NO_2(g) + F_2(g) + \cancel{F(g)} \longrightarrow NO_2F(g) + NO_2F(g) + \cancel{F(g)}$$

or

$$2NO_2(g) + F_2(g) \longrightarrow 2NO_2F(g)$$

2. Both steps are bimolecular and, thus, reasonable.
3. To determine whether the mechanism is consistent with the observed rate law, we first write the rate laws for the elementary steps:
   (1) $\text{Rate}_1 = k_1[NO_2][F_2]$
   (2) $\text{Rate}_2 = k_2[NO_2][F]$
   Step 1 is the rate-determining step. With $k_1 = k$, it is the same as the overall rate law, so the third criterion is met. Note that the second molecule of NO_2 is involved *after* the rate-determining step, so it does not appear in the overall rate law. Thus, as in the mechanism for the reaction of NO_2 and CO, *the overall rate law includes all the reactants involved in the rate-determining step.*

Mechanisms with a Fast Initial Step If the rate-limiting step in a mechanism is *not* the initial step, the product of the fast initial step builds up and starts reverting to the reactant. With time, that *fast, reversible step reaches equilibrium* (we will learn more about equilibrium in the next chapter, Chapter 15), as the product changes to the reactant as fast as it forms. As you will see, this situation allows us to fit the mechanism to the overall rate law.

Consider, once again, the oxidation of nitrogen monoxide:

$$2NO(g) + O_2(g) \longrightarrow 2NO_2(g)$$

The observed rate law is

$$\text{Rate} = k[NO]^2[O_2]$$

and a proposed mechanism is

(1) $NO(g) + O_2(g) \rightleftharpoons NO_3(g)$ (fast; reversible)
(2) $NO_3(g) + NO(g) \longrightarrow 2NO_2(g)$ (slow; rate-determining)

Let us go through the three criteria to see if this mechanism is valid. With cancellation of the reaction intermediate, NO_3, the sum of the steps gives the overall equation, so the first criterion is met:

$$NO(g) + O_2(g) + \cancel{NO_3(g)} + NO(g) \longrightarrow \cancel{NO_3(g)} + 2NO_2(g)$$

or

$$2NO(g) + O_2(g) \longrightarrow 2NO_2(g)$$

Both steps are bimolecular, so the second criterion is met.

Now we need to see whether the third criterion—that the mechanism is consistent with the observed rate law—is met. First, we write rate laws for the elementary steps:

(1) $\text{Rate}_{1(\text{fwd})} = k_1[\text{NO}][\text{O}_2]$
$\text{Rate}_{1(\text{rev})} = k_{-1}[\text{NO}_3]$

where k_{-1} is the rate constant and NO_3 is the reactant for the reverse reaction.

(2) $\text{Rate}_2 = k_2[\text{NO}_3][\text{NO}]$

Next, we show that the rate law for the rate-determining step (step 2) gives the overall rate law. As written, it does not, because it contains the intermediate NO_3, and *an overall rate law includes only reactants (and products)*. We eliminate $[NO_3]$ from the rate law for step 2 by expressing it in terms of reactants, as follows. Step 1 reaches equilibrium when the forward and reverse rates are equal:

$$\text{Rate}_{1(\text{fwd})} = \text{Rate}_{1(\text{rev})} \quad \text{or} \quad k_1[\text{NO}][\text{O}_2] = k_{-1}[\text{NO}_3]$$

To express $[NO_3]$ in terms of reactants, we isolate it algebraically:

$$[\text{NO}_3] = \frac{k_1}{k_{-1}}[\text{NO}][\text{O}_2]$$

Then, substituting for $[NO_3]$ in the rate law for the slow step, step 2, we obtain

$$\text{Rate}_2 = k_2[\text{NO}_3][\text{NO}] = k_2\Bigg(\underbrace{\frac{k_1}{k_{-1}}[\text{NO}][\text{O}_2]}_{[\text{NO}_3]}\Bigg)[\text{NO}] = \frac{k_2k_1}{k_{-1}}[\text{NO}]^2[\text{O}_2]$$

With $k = \dfrac{k_2k_1}{k_{-1}}$, this rate law is identical to the overall rate law.

To summarize, we assess the validity of a mechanism with a fast initial step as follows:

1. We write rate laws for the fast step (both directions) and for the slow step.
2. We express [intermediate] in terms of [reactant] by setting the forward rate law of the reversible step equal to the reverse rate law. Then we solve for [intermediate].
3. We substitute the expression for [intermediate] into the rate law for the slow step to obtain the overall rate law.

Several end-of-chapter problems, including Problems 14.72 and 14.73, provide additional examples of this approach.

It is important to note that, *for any mechanism, only the reactants involved up to and including the slow (rate-determining) step appear in the overall rate law.*

Steady State Approximation: Mechanisms in Which the Nature of the Elementary Steps Is Unknown In the previous two examples, we knew not only the elementary steps of the proposed mechanism, but also which step was fast or slow. Many reactions, however, have more than two elementary steps in the proposed mechanism, and we may not know whether these steps are slow rate-determining steps or fast equilibrium steps. How can we determine whether the mechanism is valid under such conditions? One method that we can use is called the **steady state approximation**. In this method, we use an intermediate in the mechanism to determine the rate law, and then we use our known, experimentally determined rate law to establish whether the step was a fast equilibrium step or a slow rate-determining step.

Let us use the previous reaction and the proposed mechanism to illustrate the steady state approximation. Since we have already looked at the reaction, we will not check to ensure that the mechanism yields the correct reaction. The mechanism, once again, is given below:

(1) $\text{NO}(g) + \text{O}_2(g) \longrightarrow \text{NO}_3(g)$
(2) $\text{NO}_3(g) + \text{NO}(g) \longrightarrow 2\text{NO}_2(g)$

Notice that we have not indicated whether the elementary steps are fast or slow or whether there are any equilibria. In the steady state approximation, we use the intermediate to establish the rate law from the elementary steps. In the mechanism above, the intermediate species is $NO_3(g)$. This species is formed in the forward reaction (1), used up in the reverse of reaction (1), and again used up in the forward reaction (2). The steady state approximation is based on the concept that *the concentration of the intermediate species remains roughly constant throughout the reaction.* In other words, if the concentration of the intermediate species does not change, then the rate at which it is produced must equal the rate at which it is consumed. This is also true mathematically, since the *rate* of change of a species must be zero if its concentration is constant:

$$\frac{d[NO_3]}{dt} = \text{rate of appearance of } NO_3 - \text{rate of disappearance of } NO_3 = 0$$

$$\text{rate of appearance of } NO_3 = \text{rate of disappearance of } NO_3$$

From (1), the rate of appearance of NO_3 is $k_1[NO][O_2]$. From the reverse of (1), the rate of disappearance of NO_3 is $k_{-1}[NO_3]$. From (2), the rate of disappearance is $k_2[NO_3][NO]$. Substituting these terms into the above equivalence, we obtain

$$k_1[NO][O_2] = k_{-1}[NO_3] + k_2[NO_3][NO]$$

Solving algebraically for the concentration of the intermediate, we find that

$$[NO_3] = \frac{k_1[NO][O_2]}{k_{-1} + k_2[NO]}$$

Now that we have defined the steady state, we can approach the rate-determining step, which will determine the rate law for the reaction. Of the two steps, (1) and (2), we do not know which is the fast step and which is the slow step (theoretically, of course). From the experimental rate law, we know that the rate is a function of $[NO]^2$ and $[O_2]$. In (1), only one molecule of NO reacts with one molecule of O_2. For this reason, (1) *cannot* be the rate-determining step since it would not yield a rate law expression with a square dependence on [NO]. The rate-determining step must be the second step (because the second molecule of NO is introduced in this step) and so, using (2), we can establish the rate law as follows:

$$\text{Rate}_2 = k_2[NO_3][NO]$$

Now, we can substitute the expression from the steady state approximation into the rate law to replace the concentration of the intermediate, NO_3, to obtain

$$\text{Rate}_2 = k_2 \cdot \underbrace{\frac{k_1[NO][O_2]}{k_{-1} + k_2[NO]}}_{[NO_3]} \cdot [NO]$$

Although this equation looks *similar* to the equation derived previously, notice that it is not the same. It is, in fact, more complicated. However, we can now determine the relative rates of reactions (1) and (2) by making certain assumptions:

A. If the rate at which NO_3 is consumed in (2) is greater than the rate at which NO_3 is consumed in the reverse of (1), we can say that

$$k_2[NO_3][NO] > k_{-1}[NO_3]$$

or, by simplifying,

$$k_2[NO] > k_{-1}$$

Using this simplification, the denominator in the expression for $[NO_3]$ becomes $k_2[NO]$, the expression for $[NO_3]$ simplifies to $\frac{k_1[O_2]}{k_2}$, and the expression for the rate law becomes

$$\text{Rate}_2 = k_1[NO][O_2]$$

However, this rate law does *not* match the experimentally observed rate law. Therefore, we can conclude that the assumption that the rate at which NO_3 is consumed in (2) is greater than the rate at which it is consumed in the reverse of (1) is *not* valid.

B. If we invert the statement above and assume that the rate at which NO_3 is consumed by the reverse of (1) is greater than the rate of consumption in (2), this would mean that

$$k_{-1}[NO_3] > k_2[NO_3][NO]$$

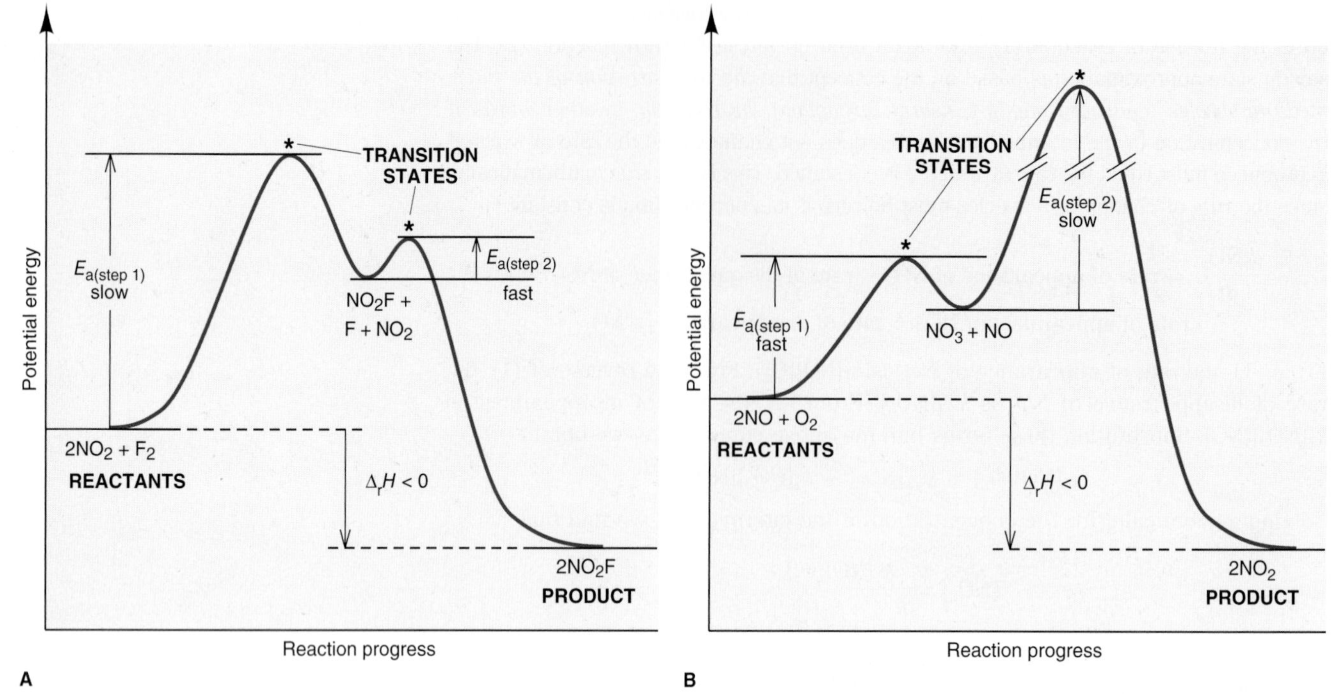

FIGURE 14.22 Reaction energy diagrams for the two-step reaction of NO_2 and F_2 (A) and NO and O_2 (B)

or, by simplifying,

$$k_{-1} > k_2[\text{NO}]$$

Substituting into the expression for the steady state concentration of NO_3 gives us

$$[\text{NO}_3] = \frac{k_1[\text{NO}][\text{O}_2]}{k_{-1}}$$

Substituting again for $[NO_3]$ in the rate law, we obtain

$$\text{Rate}_2 = k_2 \cdot \underbrace{\frac{k_1[\text{NO}][\text{O}_2]}{k_{-1}}}_{[\text{NO}_3]} \cdot [\text{NO}] = \frac{k_2 k_1}{k_{-1}}[\text{NO}]^2[\text{O}_2] = k'[\text{NO}]^2[\text{O}_2]$$

where $k' = \dfrac{k_2 k_1}{k_{-1}}$.

Notice that the derived rate law and the experimentally obtained rate law now agree. This tells us that (1) is a fast equilibrium and (2) is the slow rate-determining step, since NO_3 is consumed more quickly in the reverse of (1) than it is in (2). Our conclusion is consistent with having used (2) to determine the rate law, although we did so for mechanistic reasons. In general, when dealing with complex reaction mechanisms, the steady state approximation is useful for determining which reactions are more likely to contribute to the rate law.

Depicting a Multistep Mechanism with a Reaction Energy Diagram Figure 14.22 shows reaction energy diagrams for two reactions, each with a two-step mechanism. The reaction of NO_2 and F_2 (part A) starts with a slow step, and the reaction of NO and O_2 (part B) starts with a fast step. Both overall reactions are exothermic, so the product is lower in energy than the reactants. Note these key points:

- *Each step in the mechanism has its own peak, with the transition state at the top.*
- The intermediates (F in part A and NO_3 in part B) are reactive, unstable species, so they are higher in energy than the reactants or product.
- The slow rate-determining step (step 1 in A and step 2 in B) has a *larger* E_a than the other step.

SUMMARY OF SECTION 14.6

- The mechanisms of most common reactions consist of two or more elementary steps, each of which shows a single molecular event.
- The molecularity of an elementary step equals the number of reactant particles and is the same as the total reaction order of the step. Only unimolecular and bimolecular steps are reasonable.
- The rate-determining (slowest) step in a mechanism determines how fast the overall reaction occurs, and its rate law is equivalent to the overall rate law.
- Reaction intermediates are species that form in one step and react in a later step.
- For a mechanism to be valid, (1) the elementary steps must add up to the overall balanced equation, (2) the steps must be reasonable, and (3) the mechanism must correlate with the rate law.
- If a mechanism begins with a slow step, only the reactants that are involved in the slow step appear in the overall rate law.
- If a mechanism begins with a fast step, the product of the fast step accumulates as an intermediate, and the step reaches equilibrium. To show that the mechanism is valid, we express [intermediate] in terms of [reactant].
- Only reactants that are involved in steps up to and including the slow rate-determining step appear in the overall rate law.
- Each step in a mechanism has its own transition state, which appears at the top of a peak in the reaction energy diagram. A slower step has a higher peak (larger E_a).

14.7 Catalysis: Speeding Up a Reaction

Increasing the rate of a reaction has countless applications, in both engineering and biology. Higher temperatures can speed up a reaction, but energy for industrial processes is costly, and many organic and biological substances are heat sensitive. More commonly, by far, a reaction is accelerated by a **catalyst**, a substance that increases the rate *without* being consumed. Thus, only a small, nonstoichiometric amount of a catalyst is required to speed up a reaction. Despite this, catalysts are used in so many processes that several million tonnes are produced annually in North America alone. Nature is the master designer and user of catalysts: every organism relies on protein catalysts, known as *enzymes*, to speed up life-sustaining reactions, and even the simplest bacterium employs thousands of them.

The Basis of Catalytic Action

Each catalyst has its own specific way of functioning, but, in general, *a catalyst causes a lower activation energy, which, in turn, makes the rate constant larger and, thus, the reaction rate higher*:

$$\text{catalyst} \Longrightarrow \text{lower } E_a \Longrightarrow \text{larger } k \Longrightarrow \text{higher rate}$$

Consider a general *uncatalyzed* reaction that proceeds by a one-step mechanism involving a bimolecular collision between the reactants A and B (Figure 14.23). The activation energy is relatively large, so the rate is relatively low:

$$\text{A} + \text{B} \longrightarrow \text{product} \qquad (\text{larger } E_a \Longrightarrow \text{lower rate})$$

In the *catalyzed* reaction, reactant A interacts with the catalyst in one step to form the intermediate C, and then C reacts with B in a second step to form the product and regenerate the catalyst:

$$(1)\ \text{A} + \text{catalyst} \longrightarrow \text{C} \qquad (\text{smaller } E_a \Longrightarrow \text{higher rate})$$
$$(2)\ \text{C} + \text{B} \longrightarrow \text{product} + \text{catalyst} \qquad (\text{smaller } E_a \Longrightarrow \text{higher rate})$$

Notice the following three points in Figure 14.23:

- A catalyst speeds up the forward *and* reverse reactions. Thus, a reaction has the *same yield* with or without a catalyst, but the product forms *faster*.

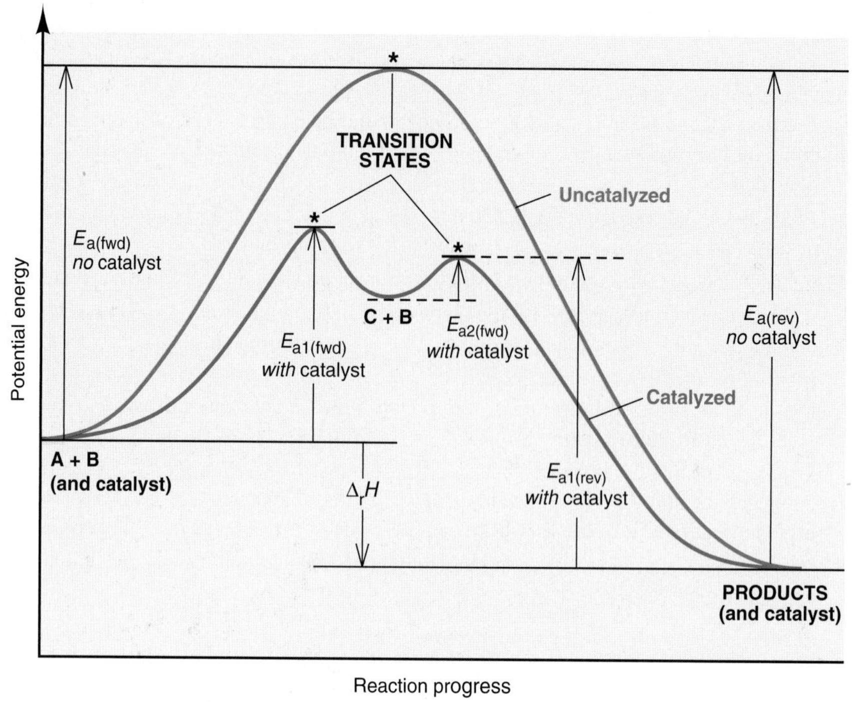

FIGURE 14.23 Reaction energy diagram for a catalyzed (*green*) and an uncatalyzed (*red*) process. A catalyst speeds a reaction by providing a different, lower-energy pathway (*green*). (Only the first step in the catalyzed reverse reaction is labelled.)

- A catalyst causes a *lower total activation energy* by providing a *different mechanism* for the reaction. The total of the activation energies for both steps of the catalyzed pathway $[E_{a1(fwd)} + E_{a2(fwd)}]$ is less than the forward activation energy of the uncatalyzed pathway.
- The catalyst is *not* consumed but rather used and then regenerated.

Homogeneous Catalysis

Chemists classify catalysts based on whether or not they act in the same phase as the reactants and products. A **homogeneous catalyst** exists in solution with the reaction mixture, so it must be a gas, a liquid, or a soluble solid.

A thoroughly studied example of homogeneous catalysis in the gas phase was formerly used in sulfuric acid manufacture. The key step, the oxidation of sulfur dioxide to sulfur trioxide, occurs so slowly that it is not economical:

$$SO_2(g) + \frac{1}{2}O_2(g) \longrightarrow SO_3(g)$$

In the presence of nitrogen monoxide, however, the reaction speeds up dramatically:

$$NO(g) + \frac{1}{2}O_2(g) \longrightarrow NO_2(g)$$
$$NO_2(g) + SO_2(g) \longrightarrow NO(g) + SO_3(g)$$

Notice that NO and NO_2 cancel to give the overall reaction. Also notice that NO_2 acts as an intermediate (formed and then consumed) and NO acts as a catalyst (used and then regenerated).

Another well-studied example of homogeneous catalysis involves the decomposition of hydrogen peroxide in aqueous solution:

$$2H_2O_2(aq) \longrightarrow 2H_2O(l) + O_2(g)$$

Commercial H_2O_2 decomposes in light and in the presence of the small amounts of ions dissolved from glass, but it is quite stable in dark plastic containers. Many other substances speed its decomposition, including the bromide ion, Br^- (Figure 14.24). The catalyzed process is thought to occur in two steps:

$$2Br^-(aq) + H_2O_2(aq) + 2H^+(aq) \longrightarrow Br_2(aq) + 2H_2O(l)$$
$$Br_2(aq) + H_2O_2(aq) \longrightarrow 2Br^-(aq) + 2H^+(aq) + O_2(g)$$

In this process, Br_2, Br^-, and H^+ cancel to give the overall balanced equation. Br^- (in the presence of H^+) is the catalyst, and Br_2 is the intermediate.

A

B

FIGURE 14.24 The catalyzed decomposition of H_2O_2. A. A small amount of NaBr is added to a solution of H_2O_2. **B.** Oxygen gas forms quickly as $Br^-(aq)$ catalyzes the H_2O_2 decomposition. The intermediate, Br_2, turns the solution orange.

Heterogeneous Catalysis

A **heterogeneous catalyst** speeds up a reaction in a different phase. Very early in a reaction, the rate depends on the reactant concentration, but, almost immediately, the reaction becomes zero order: the rate-determining step occurs on the catalyst's surface, so once the reactant covers it, adding more reactant cannot increase the rate further. Heterogeneous catalysts are most often solids interacting with gaseous or liquid reactants, and they have enormous surface areas, sometimes as much as 500 m^2/g.

A very important organic example of heterogeneous catalysis is **hydrogenation**, the addition of H_2 to C=C bonds to form C—C bonds. The petroleum, plastics, and food industries employ this process on an enormous scale. The simplest hydrogenation converts ethene to ethane:

$$H_2C{=}CH_2(g) + H_2(g) \longrightarrow H_3C{-}CH_3(g)$$

In the absence of a catalyst, the reaction is very slow. However, at high H_2 pressure (high $[H_2]$) and in the presence of finely divided Ni, Pd, or Pt metal, the reaction is rapid even at ordinary temperatures. The group 10 metals catalyze by *chemically adsorbing the reactants onto their surface* (Figure 14.25). In the rate-determining step, the adsorbed H_2 splits into two H atoms that become weakly bonded to the catalyst's surface (catM):

$$H{-}H(g) + 2\text{catM}(s) \longrightarrow 2\text{catM}{-}H \text{ (H atoms bound to metal surface)}$$

C_2H_4 then adsorbs and reacts with the H atoms, one at a time, to form C_2H_6. Thus, the catalyst acts by lowering the activation energy of the slow step as part of a different mechanism.

A solid mixture of transition metals and their oxides forms a heterogeneous catalyst in a car's exhaust system. The catalytic converter speeds both the oxidation of toxic CO and unburned gasoline to CO_2 and H_2O *and* the reduction of the pollutant NO to N_2. As in the hydrogenation mechanism, the catalyst adsorbs the molecules, weakening and splitting their bonds, thus allowing the atoms to form new bonds more quickly. The process is extremely efficient: for example, an NO molecule can be split into catalyst-bound N and O atoms in less than 2×10^{-12} s.

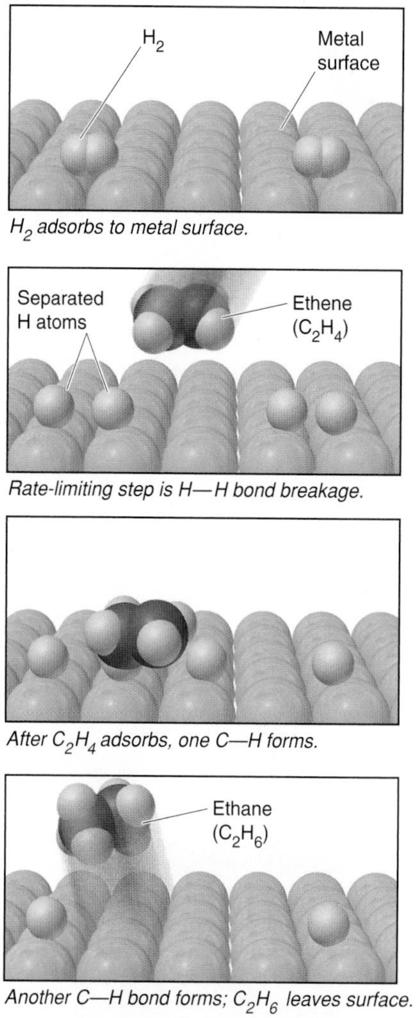
H2
Metal surface
H2 adsorbs to metal surface.
Separated H atoms
Ethene (C2H4)
Rate-limiting step is H—H bond breakage.
After C2H4 adsorbs, one C—H forms.
Ethane (C2H6)
Another C—H bond forms; C2H6 leaves surface.

FIGURE 14.25 The metal-catalyzed hydrogenation of ethene

Kinetics and Function of Biological Catalysts

Whereas most industrial chemical reactions occur under extreme conditions with high concentrations, thousands of complex reactions occur in every living cell in dilute solutions at ordinary temperatures and pressures. Moreover, the rate of each reaction responds to the rates of other reactions, chemical signals from other cells, and environmental stress. In this marvellous chemical harmony, each rate is controlled by an **enzyme**, a protein catalyst whose function has been perfected through evolution.

Enzymes are globular proteins with complex shapes (Section 20.4) and molar masses ranging from about 15 000 to 1 000 000 g/mol. A small part of an enzyme's surface—like a hollow carved into a mountainside—is the **active site**, a region whose shape results from the amino acid side chains involved in catalyzing the reaction. Although active site groups are often far apart in the sequence of the polypeptide chain, they lie near each other because of the chain's three-dimensional folding. When the reactant molecules, called **substrates**, collide effectively with the active site, they become attached through *intermolecular forces*, and the chemical change begins.

Characteristics of Enzyme Action The catalytic behaviour of enzymes has several common features:

1. *Catalytic activity.* An enzyme behaves like both types of catalysts. Enzymes are typically *much* larger than the substrate, and many enzymes are embedded within cell membranes. Thus, like a heterogeneous catalyst, an enzyme provides a surface on which one reactant is temporarily immobilized to wait until the other reactant lands nearby. However, like many homogeneous catalysts, active site groups interact with substrate(s) in multistep sequences in the presence of solvents and other species.

2. *Catalytic efficiency.* Enzymes are incredibly *efficient* in terms of the number of reactions catalyzed per unit of time. For example, consider the hydrolysis of urea:

$$(NH_2)_2C{=}O(aq) + 2H_2O(l) + H^+(aq) \longrightarrow 2NH_4^+(aq) + HCO_3^-(aq)$$

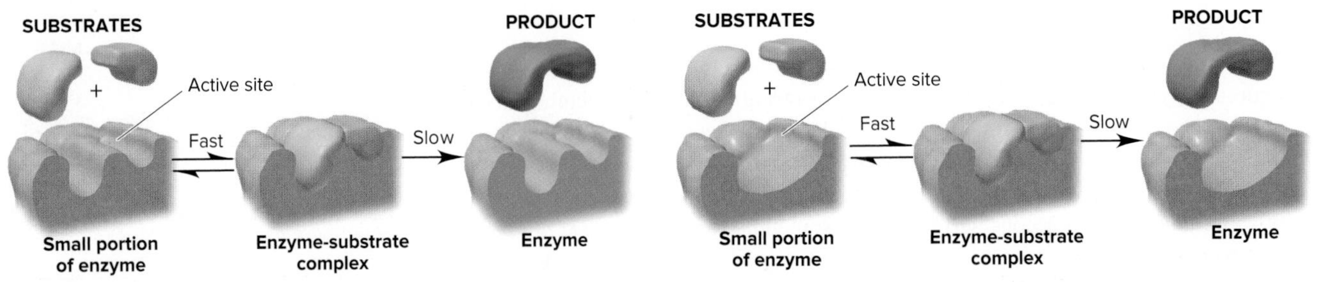

A Lock-and-key model: fixed shape of active site matches shape of substrate(s)

B Induced-fit model: active site changes shape to bind substrate(s) more effectively

FIGURE 14.26 Two models of enzyme action

The rate constant is 3×10^{-10} s^{-1} for the uncatalyzed reaction in water at room temperature. Under the same conditions in the presence of the enzyme *urease* (pronounced "*yur*-ee-ase"), the rate constant is 3×10^4 s^{-1}, a 10^{14}-fold increase. Enzymes typically increase rates by 10^8 to 10^{20} times, values that industrial chemists who use catalysts can only dream about.

3. *Catalytic specificity.* As a result of the particular groups at an active site, enzymes are also highly *specific*: each enzyme generally catalyzes only one reaction. Urease catalyzes *only* the hydrolysis of urea, and no other enzyme does this.

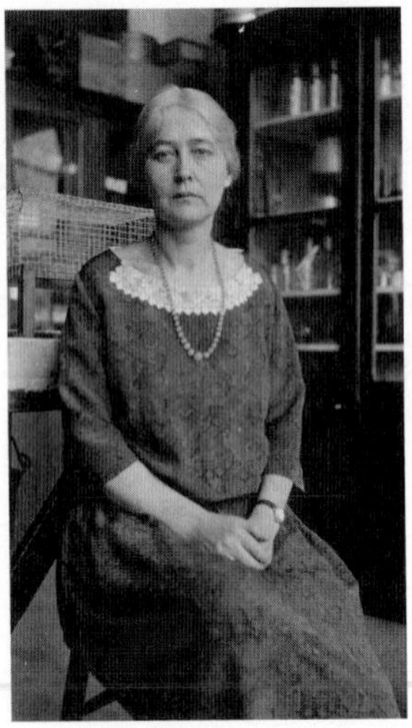

Maud Leonora Menten (1879–1960) was born in Port Lambton, Ontario. One of the first women to receive the degree of MD in 1911, she moved abroad because women were not permitted to do research in Canada at the time. She worked with Leonor Michaelis to develop the Michaelis-Menten equation, which demonstrated the relationship between the rate of an enzyme-catalyzed reaction and the amount of enzyme-substrate complex. Subsequently, she completed her Ph.D. at the University of Chicago and went on to author or co-author over 100 papers. She eventually returned to Canada. Dr. Menten was an accomplished artist and clarinet player and was best known for her enthusiasm and love of research. [Source: Wikipedia.]

Models of Enzyme Action Two models of enzyme action have been proposed. According to the earlier **lock-and-key model** (Figure 14.26A), when the "key" (substrate) fits the "lock" (active site), the chemical change begins. However, X-ray crystallographic and spectroscopic analyses show that, in most cases, *an enzyme changes shape when the substrate lands at the active site.* Thus, according to the more current **induced-fit model** (Figure 14.26B), the substrate induces the active site to adopt a perfect fit. Rather than a rigid key in a lock, we picture a flexible hand in a glove; the "glove" (active site) does not attain its functional shape until the "hand" (substrate) moves into it.

Kinetics of Enzyme Action Kinetic studies of enzyme catalysis show that the substrate (S) and the enzyme (E) form an intermediate **enzyme-substrate complex (ES)**, whose concentration determines the rate of formation of the product (P). In other words, the rate of an enzyme-catalyzed reaction is proportional to the concentration of ES, that is, of the reactant bound to the catalyst. The following steps are common to virtually all enzyme-catalyzed reactions:

(1) $E + S \rightleftharpoons ES$ (fast; reversible)
(2) $ES \longrightarrow E + P$ (slow; rate-determining)

Thus,

$$\text{Rate} = k[\text{ES}]$$

As in the heterogeneous catalysis of hydrogenation (Figure 14.25), [S] is much greater than [E], and, once all the enzyme molecules are bound to the substrate, [ES] is as high as possible. At this point, increasing [S] has no effect on the rate, and the process is zero order. Many biochemical processes are assumed to follow *Michaelis-Menten kinetics*, given by the **Michaelis-Menten equation**,

$$\text{Rate} = \frac{d[P]}{dt} = \frac{V_{\text{max}}[S]}{K_{\text{M}} + [S]}$$

where V_{max} represents the maximum rate, $[S]$ represents the concentration of the substrate, and K_{M} is the **Michaelis constant**, which represents the substrate concentration at which the reaction rate is half of V_{max}.

CHEMICAL CONNECTIONS
TO CATALYSIS AND THE OIL SANDS

From the Lab of Dr. Kevin J. Smith, University of British Columbia

About 1 million barrels per day (bpd) of oil equivalent is currently produced from Canada's oil sands, and this is expected to reach 3.8 million bpd by 2022 (energy.alberta.ca/oilsands/oilsands.asp). The oil sands' recoverable reserves are estimated at 300 billion barrels, and continuing development of these reserves is expected to provide much of Canada's future energy needs and contribute to North America's future energy and economic security. Oil sands consist mainly of bitumen, sand, and water. The bitumen, a viscous mix of polyaromatic hydrocarbons with high sulfur levels and an American Petroleum Institute (API) gravity less than 10, is separated from the sand before being upgraded to a synthetic crude oil that can be refined into commercial products, such as gasoline and diesel. An API gravity greater than 10 indicates that the liquid is lighter than water and floats, whereas an API gravity less than 10 indicates that the liquid is heavier and sinks.

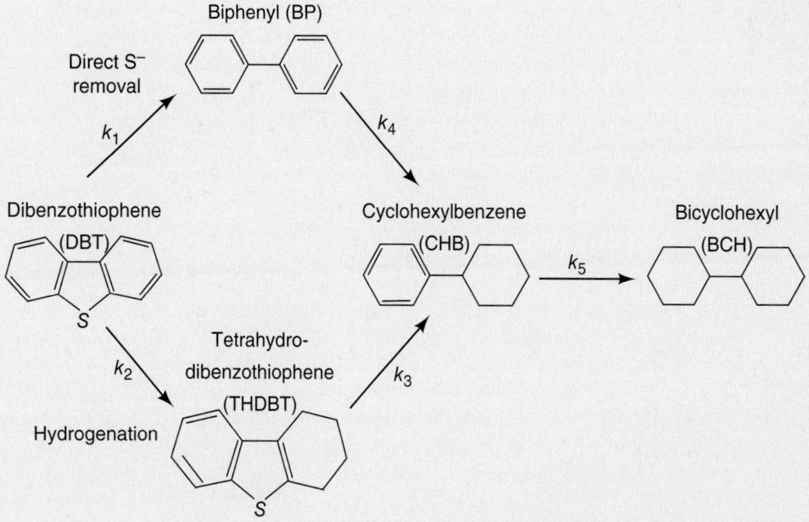

FIGURE B14.2 Reaction network for the hydrodesulfurization of dibenzothiophene

Dr. Smith and his group are interested in two areas of bitumen upgrading. The first is the primary upgrading of bitumen by hydroconversion in the presence of a catalyst, producing liquid product and small amounts of coke. One approach that is used to limit the effect of catalyst deactivation by the coke is to perform the reaction in slurry hydrocracking reactors, with unsupported metal sulfide catalysts dispersed in the bitumen. Although the catalysts may be used on a once-through basis, the cost of the catalysts dictates the need for catalyst recycling and/or regeneration, especially if molybdenum catalysts with concentrations above 200 ppm are used in the feed oil. Although dispersed catalyst hydroconversion technologies have been developed, none have been commercialized, in part because of the cost of the catalysts. The catalyst that is used in bitumen hydroconversion influences the coke yield and the quality of the liquid product.

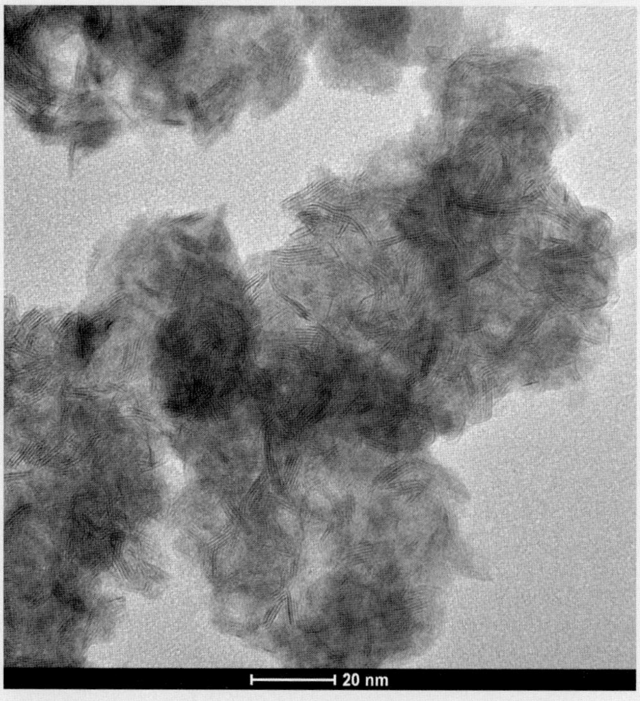

FIGURE B14.1 Transmission electron microscopy (TEM) of the catalyst MoS_2, which is derived from ammonium heptamolybdate and used for bitumen hydroconversion

A second area of interest is in hydrotreating the upgraded bitumen to remove sulfur and nitrogen from the product fuel. Dr. Smith's group has focused on the influence of the properties of catalysts and on new types of catalysts (metal phosphides) for hydrodesulfurization (HDS) and hydrodenitrogenation (HDN) reactions. As an example, they have examined the kinetics of the HDS of benzothiophene on MoS_2 catalysts (Figure B14.1). Figure B14.2 shows the complexity of the reaction for a relatively simple molecule. The accompanying graph (Figure B14.3) shows the product yields as a function of the conversion of dibenzothiophene.

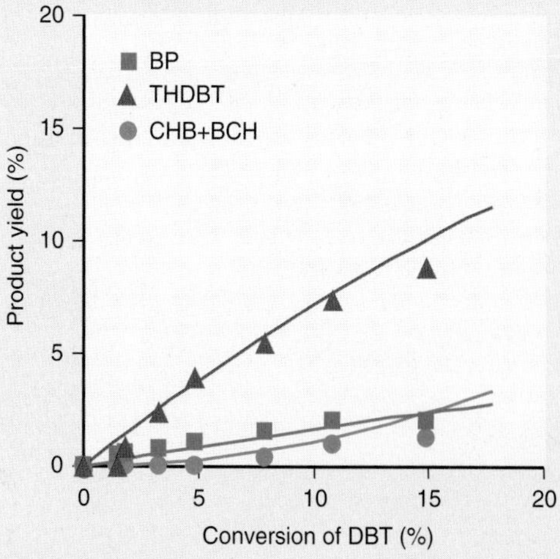

FIGURE B14.3 Product yields as a function of the conversion of dibenzothiophene (DBT)

(Continued)

References to relevant papers:

1. R. Wang and K. J. Smith, "The effect of preparation conditions on the properties of high-surface area Ni_2P catalysts," *Applied Catalysis A: General* 380 (2010), pp. 149–164.
2. H. Rezaei, X. Liu, S. Jooyaa, K. J. Smith, and M. Bricker, "A study of Cold Lake Vacuum Residue hydroconversion in batch and semi-batch reactors using dispersed catalysts," *Catalysis Today* 150 (2010), pp. 244–254.

Problems

B14.1 The decomposition of ethane is a complex first-order process whose mechanism contains many steps. However, we can consider the first step to be the following reaction:

$$C_2H_6(g) \longrightarrow CH_3\cdot + CH_3\cdot$$

This reaction has been found to have $A = 2.5 \times 10^{16}\ s^{-1}$ and $\frac{E_a}{R} = 44\ 210\ K$ in the temperature range of 200 K to 2500 K.

(a) Determine the value of k at 300 K.
(b) Determine the value of k at 2500 K.
(c) Calculate E_a (J/mol) from these two values.

B14.2 The entire mechanism, proposed by Rice-Herzfeld, contains two initiation steps, two propagation steps, and a termination step:

(1) $C_2H_6 \xrightarrow{k_1} 2CH_3$
(2) $CH_3 + C_2H_6 \longrightarrow C_2H_5 + CH_4$
(3) $C_2H_5 \longrightarrow C_2H_4 + H$
(4) $H + C_2H_6 \longrightarrow C_2H_5 + H_2$
(5) $H + C_2H_5 \longrightarrow C_2H_6$

(a) What is the molecularity of each reaction?
(b) What are the intermediate species in the reaction mechanism?

(http://www.cosbkup.gatech.edu/group/chem780/CHAPT2.pdf: source of numeric data and mechanism)

Mechanisms of Enzyme Action Enzymes employ a variety of reaction mechanisms. In some cases, the active site groups bring specific atoms of the substrate close together. In other cases, the groups bound to the substrate move apart, stretching the bond to be broken. Many *hydrolases* have acidic groups that provide H^+ ions to speed bond cleavage. Two examples are lysozyme and chymotrypsin. Lysozyme, an enzyme found in tears, hydrolyzes a polysaccharide in bacterial cell walls to protect the eyes from infection. Chymotrypsin, an enzyme found in the small intestine, hydrolyzes proteins during digestion.

No matter what their mode of action, *all enzymes catalyze by stabilizing the reaction's transition state.* For instance, in the lysozyme-catalyzed reaction, the transition state is a portion of the polysaccharide whose bonds have been twisted and stretched by the active site groups until it fits the active site well. Stabilizing the transition state lowers the activation energy and, thus, increases the rate.

SUMMARY OF SECTION 14.7

- A catalyst increases the rate of a reaction without being consumed. It accomplishes this by providing another mechanism with lower activation energy.
- Homogeneous catalysts function in the same phase as the reactants. Heterogeneous catalysts function in a different phase from the reactants.
- The hydrogenation of carbon-carbon double bonds takes place on a solid metal catalyst, which speeds up the rate-determining step, the breakage of the H_2 bond.
- Enzymes are protein catalysts of high efficiency and specificity that act by stabilizing the transition state and, thus, lowering the activation energy.

CHAPTER REVIEW GUIDE

Learning Objectives

Relevant section (§) and/or sample problem (SP) numbers appear in parentheses.

Concepts

1. Describe how the rate of a reaction depends on concentration, physical state, and temperature. (§14.1)
2. Explain the meaning of *reaction rate* in terms of changing concentrations over time. (§14.2)
3. Describe how the reaction rate can be expressed in terms of reactant or product concentration. (§14.2)
4. Differentiate between the average rate and the instantaneous rate, and show why the instantaneous rate changes during a reaction. (§14.2)
5. Describe the interpretation of the reaction rate in terms of reactant and product concentrations. (§14.2)
6. Explain the experimental basis of the rate law, and describe the information needed to determine it: initial rate data, reaction orders, and rate constant. (§14.3)
7. Discuss the importance of reaction orders when determining the reaction rate. (§14.3)
8. Explain how reaction orders are determined from initial rates at different concentrations. (§14.3)
9. Explain how integrated rate laws show the dependence of concentration on time. (§14.4)
10. Describe what the term *half-life* means and explain why the half-life is constant for a first-order reaction. (§14.4)
11. Show why concentrations are multiplied in the rate law. (§14.5)
12. Describe activation energy and the effect of temperature on the rate constant (Arrhenius equation). (§14.5)
13. Show how temperature affects rate by influencing collision energy and, thus, the fraction of collisions with energy that exceeds the activation energy. (§14.5)
14. Explain why molecular orientation and complexity influence the number of effective collisions and the rate. (§14.5)
15. Describe the transition state as the momentary species between reactants and products, whose formation requires the activation energy. (§14.5)
16. Show why an elementary reaction represents a single molecular event and how its molecularity equals the number of colliding particles. (§14.6)
17. Show how a reaction mechanism consists of several elementary steps, with the slowest step determining the overall rate. (§14.6)
18. Describe the criteria for a valid reaction mechanism. (§14.6)
19. Explain how a catalyst speeds up a reaction by lowering the activation energy. (§14.7)
20. Differentiate between homogeneous and heterogeneous catalysis. (§14.7)

Skills

1. Calculate instantaneous rate from the slope of a tangent to a concentration versus time plot. (§14.2)
2. Express the reaction rate in terms of changes in concentration over time. (SP14.1)
3. Determine reaction orders from a known rate law. (SP14.2)
4. Determine reaction orders from changes in initial rate with concentration. (SPs 14.3, 14.4)
5. Calculate the rate constant and its units. (§14.3)
6. Use an integrated rate law to find the concentration at a given time or the time to reach a given concentration. (SP 14.5)
7. Determine reaction orders graphically with a rearranged integrated rate law. (§14.4)
8. Determine the half-life of a reaction. (SPs 14.6, 14.7)
9. Use a form of the Arrhenius equation to calculate the activation energy. (SP 14.8)
10. Use reaction energy diagrams to depict the energy changes during a reaction. (SP 14.9)
11. Predict a transition state for a simple reaction. (SP 14.9)
12. Determine the molecularity and rate law for elementary steps. (SP 14.10)
13. Construct a mechanism with either a slow initial step or a fast initial step. (§14.6)

Key Terms

Section 14.1
chemical kinetics
reaction rate

Section 14.2
average rate
instantaneous rate
initial rate

Section 14.3
rate law (rate equation)
rate constant
reaction orders

Section 14.4
integrated rate laws
half-life ($t_{1/2}$)

Section 14.5
collision theory
Arrhenius equation
activation energy (E_a)
effective collisions
pre-exponential factor
transition state theory
transition state (activated complex)
reaction energy diagram
Eyring equation

Section 14.6
reaction mechanism
elementary reactions (elementary steps)
molecularity
unimolecular reaction
bimolecular reaction
rate-determining step (rate-limiting step)
reaction intermediate
steady state approximation

Section 14.7
catalyst
homogeneous catalyst
heterogeneous catalyst
hydrogenation
enzyme
active site
substrates
lock-and-key model
induced-fit model
enzyme-substrate complex (ES)
Michaelis-Menten equation
Michaelis constant

Key Equations and Relationships

14.1 Expressing reaction rate in terms of reactant A:

$$\text{Rate} = -\frac{\Delta[A]}{\Delta t}$$

14.2 Expressing the rate of a general reaction of the form $aA + bB \longrightarrow cC + dD$:

$$\text{Rate} = -\frac{1}{a}\frac{\Delta[A]}{\Delta t} = -\frac{1}{b}\frac{\Delta[B]}{\Delta t} = \frac{1}{c}\frac{\Delta[C]}{\Delta t} = \frac{1}{d}\frac{\Delta[D]}{\Delta t}$$

14.3 Writing a general rate law (in which products do not appear):

$$\text{Rate} = k[A]^m[B]^n \cdots$$

14.4 Calculating the time to reach a given [A] in a first-order reaction (Rate = k[A]):

$$\ln\frac{[A]_0}{[A]_t} = kt \quad \text{or} \quad \ln[A]_0 - \ln[A]_t = kt$$

14.5 Calculating the time to reach a given [A] in a simple second-order reaction (Rate = $k[A]^2$):

$$\frac{1}{[A]_t} - \frac{1}{[A]_0} = kt$$

14.6 Calculating the time to reach a given [A] in a zero-order reaction (Rate = k):

$$[A]_t - [A]_0 = -kt$$

14.7 Finding the half-life of a first-order process (Rate = k[A]):

$$t_{1/2} = \frac{\ln 2}{k} = \frac{0.693}{k}$$

14.8 Relating the rate constant to the temperature (Arrhenius equation):

$$k = Ae^{-\frac{E_a}{RT}}$$

14.9 Relating the heat of reaction to the forward and reverse activation energies:

$$\Delta_r H = E_{a(\text{fwd})} - E_{a(\text{rev})}$$

14.10 Calculating the activation energy (rearranged form of Arrhenius equation):

$$\ln\frac{k_2}{k_1} = -\frac{E_a}{R}\left(\frac{1}{T_2} - \frac{1}{T_1}\right)$$

Brief Solutions to Follow-Up Problems

14.1 (a) $4NO(g) + O_2(g) \longrightarrow 2N_2O_3(g)$

$$\text{Rate} = -\frac{\Delta[O_2]}{\Delta t} = -\frac{1}{4}\frac{\Delta[NO]}{\Delta t} = \frac{1}{2}\frac{\Delta[N_2O_3]}{\Delta t}$$

(b) $-\frac{\Delta[O_2]}{\Delta t} = -\frac{1}{4}\frac{\Delta[NO]}{\Delta t} = -\frac{1}{4}[-1.60 \times 10^{-4}\ \text{mol/(L·s)}]$

$$= 4.00 \times 10^{-5}\ \text{mol/(L·s)}$$

14.2 First order in Br^-, first order in BrO_3^-, second order in H^+, and fourth order overall

14.3 Rate = $k[H_2]^m[I_2]^n$
From experiments 1 and 3, $m = 1$. From experiments 2 and 4, $n = 1$. Rate = $k[H_2][I_2]$; second order overall.

14.4 (a) The rate law shows the reaction is zero order in Y, so the rate is not affected by doubling Y. Rate of experiment 2 = 0.25×10^{-5} mol/(L·s).
(b) The rate of experiment 3 is four times the rate of experiment 1, so [X] doubles.

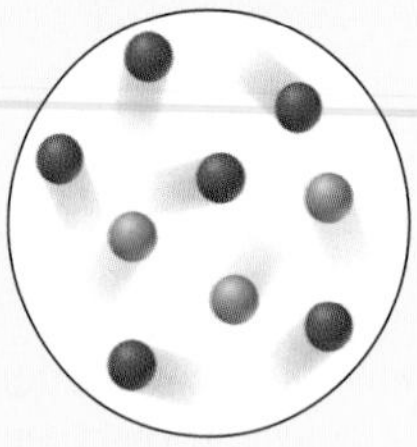

14.5 $\frac{1}{[HI]_t} - \frac{1}{[HI]_0} = kt$

111 L/mol − 100. L/mol = [2.4×10^{-21} L/(mol·s)](t)
$t = 4.6 \times 10^{21}$ s (or 1.5×10^{14} years)

14.6 (a)

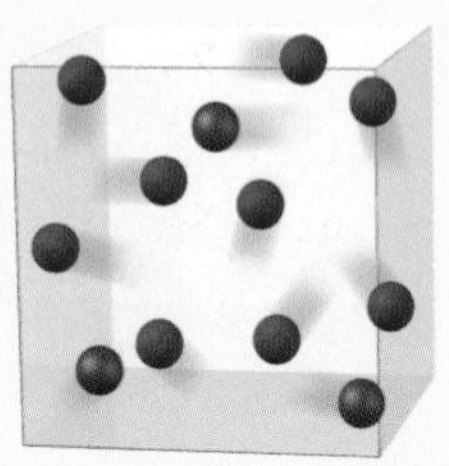

(b) At 10.0 min (four half-lives), there are 0.75 particles of X.

$$\text{Amount(mol)} = 0.75\ \text{particles} \times \frac{0.20\ \text{mol X}}{1\ \text{particle}} = 0.15\ \text{mol X}$$

$$c = \frac{0.15\ \text{mol X}}{0.50\ \text{L}} = 0.30\ \text{mol/L}$$

14.7 $t_{1/2} = \frac{\ln 2}{k}$; $k = \frac{0.693}{13.1\ \text{h}} = 5.29 \times 10^{-2}\ \text{h}^{-1}$

14.8 $\ln\frac{0.286\ \text{L/(mol·s)}}{k_1} = -\frac{1.00 \times 10^5\ \text{J/mol}}{8.314\ \text{J/(mol·K)}} \times \left(\frac{1}{500.\ \text{K}} - \frac{1}{490.\ \text{K}}\right) = 0.491$

$$k_1 = 0.175\ \text{L/(mol·s)}$$

14.9

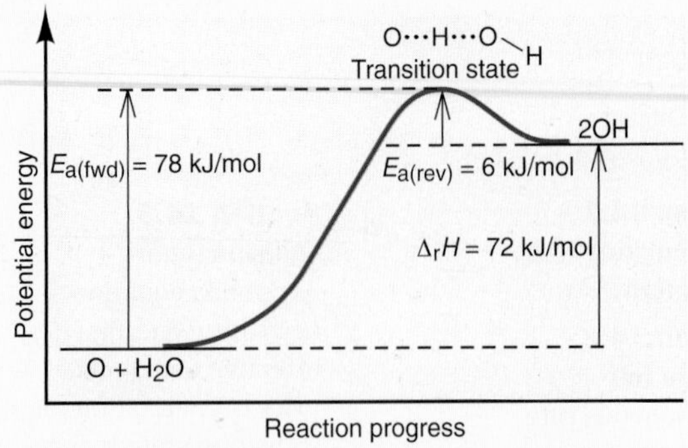

14.10 (a) Balanced equation (after doubling step 4):

$$2NO(g) + 2H_2(g) \longrightarrow N_2(g) + 2H_2O(g)$$

(b) Step 2 is unimolecular since the equation can be reduced to $H_2(g) \longrightarrow 2H(g)$. All the other steps are bimolecular.
(c) Rate$_1$ = $k_1[NO]^2$; Rate$_2$ = $k_2[H_2]$; Rate$_3$ = $k_3[N_2O_2][H]$; Rate$_4$ = $k_4[HO][H]$; Rate$_5$ = $k_5[H][N_2O]$

PROBLEMS

Problems with **red** numbers are answered in Appendix G and worked in detail in the Student Solutions Manual. Problem sections match those in this book and provide the numbers of relevant sample problems. Most offer Concept Review Questions, Skill-Building Exercises (grouped in pairs covering the same concept), and Problems in Context. The Comprehensive Problems are based on material from any section or previous chapter.

Focusing on Reaction Rate

Concept Review Questions

14.1 What variable of a chemical reaction is measured over time to obtain the reaction rate?

14.2 How does an increase in pressure affect the rate of a gas-phase reaction? Explain.

14.3 A reaction is carried out with water as the solvent. How does the addition of more water to the reaction vessel affect the rate of the reaction? Explain.

14.4 A gas reacts with a solid that is present in large chunks. Then the reaction is run again with the solid pulverized. How does the increase in the surface area of the solid affect the rate of its reaction with the gas? Explain.

14.5 How does an increase in temperature affect the rate of a reaction? Explain the two factors involved.

14.6 In a kinetics experiment, a chemist places crystals of iodine in a closed reaction vessel, introduces a given quantity of H_2 gas, and obtains data to calculate the rate of HI formation. In a second experiment, the chemist uses the same amounts of iodine and hydrogen, but first warms the flask to 130°C, a temperature above the sublimation point of iodine. In which of these experiments does the reaction proceed at a higher rate? Explain.

Expressing the Reaction Rate

(Sample Problem 14.1)

Concept Review Questions

14.7 Define *reaction rate*. Assuming constant temperature and a closed reaction vessel, why does the reaction rate change with time?

14.8 (a) What is the difference between an average rate and an instantaneous rate?
(b) What is the difference between an initial rate and an instantaneous rate?

14.9 Give two reasons to measure initial rates in a kinetics study.

14.10 For the reaction $A(g) \longrightarrow B(g)$, sketch two curves on the same set of axes that show (a) the formation of the product as a function of time; (b) the consumption of the reactant as a function of time.

14.11 For the reaction $C(g) \longrightarrow D(g)$, [C] versus time is plotted:

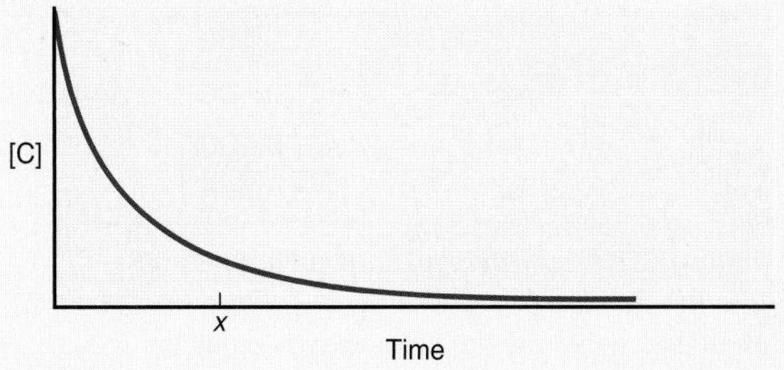

(a) How do you determine (i) the average rate over the entire experiment; (ii) the reaction rate at time *x*; (iii) the initial reaction rate?
(b) Would the values in part (a) be different if you plotted [D] versus time? Explain.

Skill-Building Exercises (grouped in similar pairs)

14.12 The compound AX_2 decomposes according to the following equation:

$$2AX_2(g) \longrightarrow 2AX(g) + X_2(g)$$

In one experiment, $[AX_2]$ was measured at various times and these data were obtained:

Time (s)	$[AX_2]$ (mol/L)
0.0	0.0500
2.0	0.0448
6.0	0.0300
8.0	0.0249
10.0	0.0209
20.0	0.0088

(a) Find the average rate over the entire experiment.
(b) Is the initial rate higher or lower than the rate in part (a)? Use graphical methods to estimate the initial rate.

14.13 (a) Use the data from Problem 14.12 to calculate the average rate from 8.0 s to 20.0 s.
(b) Is the rate at exactly 5.0 s higher or lower than the rate in part (a)? Use graphical methods to estimate the rate at 5.0 s.

14.14 (a) Express the rate of reaction in terms of the change in concentration of each reactant and product:

$$2A(g) \longrightarrow B(g) + C(g)$$

(b) When [C] is increasing at 2 mol/(L•s), how fast is [A] decreasing?

14.15 (a) Express the rate of reaction in terms of the change in concentration of each reactant and product:

$$D(g) \longrightarrow \frac{3}{2}E(g) + \frac{5}{2}F(g)$$

(b) When [E] is increasing at 0.25 mol/(L•s), how fast is [F] increasing?

14.16 (a) Express the rate of reaction in terms of the change in concentration of each reactant and product:

$$A(g) + 2B(g) \longrightarrow C(g)$$

(b) When [B] is decreasing at 0.5 mol/(L•s), how fast is [A] decreasing?

14.17 (a) Express the rate of reaction in terms of the change in concentration of each reactant and product:

$$2D(g) + 3E(g) + F(g) \longrightarrow 2G(g) + H(g)$$

(b) When [D] is decreasing at 0.1 mol/(L•s), how fast is [H] increasing?

14.18 Reaction rate is expressed in terms of changes in concentration of reactants and products. Write a balanced equation for the following reaction:

$$\text{Rate} = -\frac{1}{2}\frac{\Delta[N_2O_5]}{\Delta t} = \frac{1}{4}\frac{\Delta[NO_2]}{\Delta t} = \frac{\Delta[O_2]}{\Delta t}$$

14.19 Reaction rate is expressed in terms of changes in concentration of reactants and products. Write a balanced equation for the following reaction:

$$\text{Rate} = -\frac{\Delta[CH_4]}{\Delta t} = -\frac{1}{2}\frac{\Delta[O_2]}{\Delta t} = \frac{1}{2}\frac{\Delta[H_2O]}{\Delta t} = \frac{\Delta[CO_2]}{\Delta t}$$

Problems in Context

14.20 The decomposition of NOBr is studied manometrically because the amount (mol) of gas changes; it cannot be studied spectrometrically because both NOBr and Br_2 are reddish brown:

$$2NOBr(g) \longrightarrow 2NO(g) + Br_2(g)$$

Use the data in the table below to answer the following questions:

Time (s)	[NOBr] (mol/L)
0.00	0.0100
2.00	0.0071
4.00	0.0055
6.00	0.0045
8.00	0.0038
10.00	0.0033

(a) Determine the average rate over the entire experiment.
(b) Determine the average rate between 2.00 s and 4.00 s.
(c) Use graphical methods to estimate the initial reaction rate.
(d) Use graphical methods to estimate the rate at 7.00 s.
(e) At what time does the instantaneous rate equal the average rate over the entire experiment?

14.21 The formation of ammonia is one of the most important processes in the chemical industry:

$$N_2(g) + 3H_2(g) \longrightarrow 2NH_3(g)$$

Express the rate in terms of changes in $[N_2]$, $[H_2]$, and $[NH_3]$.

14.22 Although the depletion of stratospheric ozone threatens life on Earth today, its accumulation was one of the crucial processes that allowed life to develop in prehistoric times:

$$3O_2(g) \longrightarrow 2O_3(g)$$

(a) Express the reaction rate in terms of $[O_2]$ and $[O_3]$.
(b) At a given instant, the reaction rate in terms of $[O_2]$ is 2.17×10^{-5} mol/(L·s). What is the reaction rate in terms of $[O_3]$?

The Rate Law and Its Components

(Sample Problems 14.2 to 14.4)

Concept Review Questions

14.23 The rate law for the general reaction

$$aA + bB + \cdots \longrightarrow cC + dD + \cdots$$

is Rate $= k[A]^m[B]^n\cdots$.
(a) Explain the meaning of k.
(b) Explain the meanings of m and n. Does $m = a$ and $n = b$? Explain.
(c) If the reaction is first order in A and second order in B, and time is measured in minutes, what are the units for k?

14.24 You are studying the following reaction to determine its rate law:

$$A_2(g) + B_2(g) \longrightarrow 2AB(g)$$

Assuming that you have a valid experimental procedure for obtaining $[A_2]$ and $[B_2]$ at various times, explain how you determine (a) the initial rate; (b) the reaction orders; (c) the rate constant.

14.25 By what factor does the rate change in each reaction? (Assume constant temperature.)
(a) A reaction is first order in reactant A, and [A] is doubled.
(b) A reaction is second order in reactant B, and [B] is halved.
(c) A reaction is second order in reactant C, and [C] is tripled.

Skill-Building Exercises (grouped in similar pairs)

14.26 Give the individual reaction orders for all the substances in the following rate law, as well as the overall reaction order:

$$\text{Rate} = k[BrO_3^-][Br^-][H^+]^2$$

14.27 Give the individual reaction orders for all the substances in the following rate law, as well as the overall reaction order:

$$\text{Rate} = k\frac{[O_3]^2}{[O_2]}$$

14.28 By what factor does the rate in Problem 14.26 change if each change occurs?
(a) $[BrO_3^-]$ is doubled.
(b) $[Br^-]$ is halved.
(c) $[H^+]$ is quadrupled.

14.29 By what factor does the rate in Problem 14.27 change if each change occurs?
(a) $[O_3]$ is doubled.
(b) $[O_2]$ is doubled.
(c) $[O_2]$ is halved.

14.30 Give the individual reaction orders for all the substances in this rate law, as well as the overall reaction order:

$$\text{Rate} = k[NO_2]^2[Cl_2]$$

14.31 Give the individual reaction orders for all the substances in this rate law, as well as the overall reaction order:

$$\text{Rate} = k\frac{[HNO_2]^4}{[NO]^2}$$

14.32 By what factor does the rate in Problem 14.30 change if each change occurs?
(a) $[NO_2]$ is tripled.
(b) $[NO_2]$ and $[Cl_2]$ are doubled.
(c) $[Cl_2]$ is halved.

14.33 By what factor does the rate in Problem 14.31 change if each change occurs?
(a) $[HNO_2]$ is doubled.
(b) [NO] is doubled.
(c) $[HNO_2]$ is halved.

14.34 For the reaction

$$4A(g) + 3B(g) \longrightarrow 2C(g)$$

the following data were obtained at constant temperature:

Experiment	Initial Rate [mol/(L·min)]	Initial [A] (mol/L)	Initial [B] (mol/L)
1	5.00	0.100	0.100
2	45.0	0.300	0.100
3	10.0	0.100	0.200
4	90.0	0.300	0.200

(a) What is the order with respect to each reactant?
(b) Write the rate law.
(c) Calculate k using the data from experiment 1.

14.35 For the reaction

$$A(g) + B(g) + C(g) \longrightarrow D(g)$$

the following data were obtained at constant temperature:

Experiment	Initial Rate [mol/(L·s)]	Initial [A] (mol/L)	Initial [B] (mol/L)	Initial [C] (mol/L)
1	6.25×10^{-3}	0.0500	0.0500	0.0100
2	1.25×10^{-2}	0.1000	0.0500	0.0100
3	5.00×10^{-2}	0.1000	0.1000	0.0100
4	6.25×10^{-3}	0.0500	0.0500	0.0200

(a) What is the order with respect to each reactant?
(b) Write the rate law.
(c) Calculate k using the data from experiment 1.

14.36 Without consulting Table 14.4, give the units of the rate constant for a reaction with each overall order:
(a) First order (b) Second order
(c) Third order (d) $\frac{5}{2}$ order

14.37 Give the overall reaction order that corresponds to a rate constant with each unit: (a) mol/(L•s); (b) year^{-1}; (c) $(\text{mol/L})^{1/2}\cdot\text{s}^{-1}$; (d) $(\text{mol/L})^{-5/2}\cdot\text{min}^{-1}$.

Problem in Context

14.38 Phosgene is a toxic gas prepared by the reaction of carbon monoxide with chlorine:

$$CO(g) + Cl_2(g) \longrightarrow COCl_2(g)$$

These data were obtained in a kinetics study of its formation:

Experiment	Initial Rate [mol/(L·s)]	Initial [CO] (mol/L)	Initial [Cl_2] (mol/L)
1	1.29×10^{-29}	1.00	0.100
2	1.33×10^{-30}	0.100	0.100
3	1.30×10^{-29}	0.100	1.00
4	1.32×10^{-31}	0.100	0.0100

(a) Write the rate law for the formation of phosgene.
(b) Calculate the average value of the rate constant.

Integrated Rate Laws: Concentration Changes over Time

(Sample Problems 14.5 to 14.7)

Concept Review Questions

14.39 (a) How are integrated rate laws used to determine reaction order?
(b) For each plot described below, what is the order in the reactant?
(i) The natural logarithm of [reactant] versus time is linear.
(ii) The inverse of [reactant] versus time is linear.
(iii) [Reactant] versus time is linear.

14.40 Define the *half-life* of a reaction. Explain, on the molecular level, why the half-life of a first-order reaction is constant.

Skill-Building Exercises (grouped in similar pairs)

14.41 For the simple decomposition reaction below, Rate = $k[AB]^2$ and $k = 0.2$ L/(mol•s):

$$AB(g) \longrightarrow A(g) + B(g)$$

How long will it take for [AB] to reach one-third of its initial concentration of 1.50 mol/L?

14.42 For the reaction in Problem 14.41, what is [AB] after 10.0 s?

14.43 In a first-order decomposition reaction, 50.0% of a compound decomposes in 10.5 min.
(a) What is the rate constant of the reaction?
(b) How long does it take for 75.0% of the compound to decompose?

14.44 Consider a decomposition reaction that has a rate constant of 0.0012 year^{-1}.
(a) What is the half-life of the reaction?
(b) How long does it take for [reactant] to reach 12.5% of its original value?

Problem in Context

14.45 In a study of ammonia production, an industrial chemist discovers that the compound decomposes to its elements N_2 and H_2 in a first-order process. She collects the following data:

Time (s)	0	1.000	2.000
[NH_3] (mol/L)	4.000	3.986	3.974

(a) Use graphical methods to determine the rate constant.
(b) What is the half-life for ammonia decomposition?

Theories of Chemical Kinetics

(Sample Problems 14.8 and 14.9)

Concept Review Questions

14.46 What is the central idea of collision theory? How does this model explain the effect of concentration on reaction rate?

14.47 Is collision frequency the only factor that affects rate? Explain.

14.48 Arrhenius proposed that each reaction has an energy threshold that must be reached for the particles to react. The kinetic theory of gases proposes that the average kinetic energy of the particles is proportional to the absolute temperature. How do these concepts relate to the effect of temperature on rate?

14.49 Use the exponential term in the Arrhenius equation to explain how temperature affects reaction rate.

14.50 How is the activation energy determined from the Arrhenius equation?

14.51 (a) Graph the relationship between k (vertical axis) and T (horizontal axis).
(b) Graph the relationship between ln k (vertical axis) and $\frac{1}{T}$ (horizontal axis). How is the activation energy determined from this graph?

14.52 (a) For a reaction with a given E_a, how does an increase in T affect the rate?
(b) For a reaction at a given T, how does a decrease in E_a affect the rate?

14.53 In the following reaction, 4×10^{-5} mol of AB molecules collide with 4×10^{-5} mol of CD molecules:

$$AB + CD \rightleftharpoons EF$$

Will 4×10^{-5} mol of EF form? Explain.

14.54 Assuming that the activation energies are equal, which reaction will proceed at a higher rate at 50°C? Explain.

$$NH_3(g) + HCl(g) \longrightarrow NH_4Cl(s)$$
$$N(CH_3)_3(g) + HCl(g) \longrightarrow (CH_3)_3NHCl(s)$$

Skill-Building Exercises (grouped in similar pairs)

14.55 For the following reaction, how many unique collisions between A and B are possible if there are four particles of A and three particles of B in the vessel?

$$A(g) + B(g) \longrightarrow AB(g)$$

14.56 For the following reaction, how many unique collisions between A and B are possible if the vessel contains 1.01 mol of A(*g*) and 2.12 mol of B(*g*)?

$$A(g) + B(g) \longrightarrow AB(g)$$

14.57 At 25°C, what is the fraction of collisions with energy equal to or greater than an activation energy of 100. kJ/mol?

14.58 If the temperature in Problem 14.57 is increased to 50.°C, by what factor does the fraction of collisions with energy equal to or greater than the activation energy change?

14.59 The rate constant of a reaction is $4.7 \times 10^{-3}\ s^{-1}$ at 25°C, and the activation energy is 33.6 kJ/mol. What is k at 75°C?

14.60 The rate constant of a reaction is 4.50×10^{-5} L/(mol•s) at 195°C and 3.20×10^{-3} L/(mol•s) at 258°C. What is the activation energy of the reaction?

14.61 For the following reaction, $\Delta_r H° = -55$ kJ/mol and $E_{a(fwd)} = 215$ kJ/mol:

$$ABC + D \rightleftharpoons AB + CD$$

Assuming a one-step reaction, (a) draw a reaction energy diagram; (b) calculate $E_{a(rev)}$; (c) sketch a possible transition state if ABC is V-shaped.

14.62 For the following reaction, $E_{a(fwd)} = 125$ kJ/mol and $E_{a(rev)} = 85$ kJ/mol:

$$A_2 + B_2 \longrightarrow 2AB$$

Assuming that the reaction occurs in one step, (a) draw a reaction energy diagram; (b) calculate $\Delta_r H°$; (c) sketch a possible transition state.

Problems in Context

14.63 Understanding the high-temperature formation and breakdown of the nitrogen oxides is essential for controlling the pollutants generated from power plants and cars. The first-order breakdown of dinitrogen monoxide to its elements has rate constants of $0.76\ s^{-1}$ at 727°C and $0.87\ s^{-1}$ at 757°C. What is the activation energy of this reaction?

14.64 Aqua regia, a mixture of HCl and HNO_3, has been used since alchemical times to dissolve many metals, including gold. Its orange colour is due to the presence of nitrosyl chloride. Consider this one-step reaction for the formation of aqua regia:

$$NO(g) + Cl_2(g) \longrightarrow NOCl(g) + Cl(g) \quad \Delta H° = 83 \text{ kJ/mol}$$

(a) Draw a reaction energy diagram, given $E_{a(fwd)} = 86$ kJ/mol.
(b) Calculate $E_{a(rev)}$.
(c) Sketch a possible transition state for the reaction. (*Note*: The atom sequence of nitrosyl chloride is Cl—N—O.)

Reaction Mechanisms: The Steps from Reactant to Product

(Sample Problem 14.10)

Concept Review Questions

14.65 Is the rate of an overall reaction lower than, higher than, or equal to the average rate of the individual steps? Explain.

14.66 Explain why the coefficients of an elementary step equal the reaction orders of its rate law but the coefficients of an overall reaction do not.

14.67 Is it possible for more than one mechanism to be consistent with the rate law of a given reaction? Explain.

14.68 What is the difference between a reaction intermediate and a transition state?

14.69 Why is a bimolecular step more reasonable physically than a termolecular step?

14.70 If a slow step precedes a fast step in a two-step mechanism, do the substances in the fast step appear in the rate law? Explain.

14.71 If a fast step precedes a slow step in a two-step mechanism, how is the fast step affected? How is this effect used to determine the validity of the mechanism?

Skill-Building Exercises (grouped in similar pairs)

14.72 The proposed mechanism for a reaction is given below:

(1) $A(g) + B(g) \rightleftharpoons X(g)$ (fast)
(2) $X(g) + C(g) \longrightarrow Y(g)$ (slow)
(3) $Y(g) \longrightarrow D(g)$ (fast)

(a) What is the overall equation?
(b) Identify the intermediate(s), if any.
(c) What are the molecularity and the rate law for each step?
(d) Is the mechanism consistent with the actual rate law?

$$\text{Rate} = k[A][B][C]$$

(e) Is the following one-step mechanism equally valid?

$$A(g) + B(g) + C(g) \longrightarrow D(g)$$

14.73 Consider the following mechanism:

(1) $ClO^-(aq) + H_2O(l) \rightleftharpoons HClO(aq) + OH^-(aq)$ (fast)
(2) $I^-(aq) + HClO(aq) \longrightarrow HIO(aq) + Cl^-(aq)$ (slow)
(3) $OH^-(aq) + HIO(aq) \longrightarrow H_2O(l) + IO^-(aq)$ (fast)

(a) What is the overall equation?
(b) Identify the intermediate(s), if any.
(c) What are the molecularity and the rate law for each step?
(d) Is the mechanism consistent with the actual rate law?

$$\text{Rate} = k[ClO^-][I^-]$$

Problems in Context

14.74 In a study of nitrosyl halides, a chemist proposes the following mechanism for the synthesis of nitrosyl bromide:

$NO(g) + Br_2(g) \rightleftharpoons NOBr_2(g)$ (fast)
$NOBr_2(g) + NO(g) \longrightarrow 2NOBr(g)$ (slow)

If the rate law is Rate $= k[NO]^2[Br_2]$, is the proposed mechanism valid? If so, show that it satisfies the three criteria for validity.

14.75 The rate law for the reaction below is Rate $= k[NO]^2[O_2]$:

$$2NO(g) + O_2(g) \longrightarrow 2NO_2(g)$$

In addition to the mechanism presented in Section 14.6, the following mechanisms have been proposed:

(1) $2NO(g) + O_2(g) \longrightarrow 2NO_2(g)$
(2) $2NO(g) \rightleftharpoons N_2O_2(g)$ (fast)
$N_2O_2(g) + O_2(g) \longrightarrow 2NO_2(g)$ (slow)
(3) $2NO(g) \rightleftharpoons N_2(g) + O_2(g)$ (fast)
$N_2(g) + 2O_2(g) \longrightarrow 2NO_2(g)$ (slow)

(a) Which of these mechanisms is consistent with the rate law?
(b) Which of these mechanisms is most reasonable? Why?

Catalysis: Speeding Up a Reaction

Concept Review Questions

14.76 Consider the following reaction:

$$N_2O(g) \xrightarrow{Au} N_2(g) + \frac{1}{2}O_2(g)$$

(a) Does the gold catalyst (Au, above the arrow) act as a homogeneous catalyst or a heterogeneous catalyst?
(b) On the same set of axes, sketch the reaction energy diagrams for the catalyzed reaction and the uncatalyzed reaction.

14.77 Does a catalyst increase the reaction rate by the same means as a rise in temperature does? Explain.

14.78 In a classroom demonstration, hydrogen gas and oxygen gas are mixed in a balloon. The mixture is stable under normal conditions. However, if a spark is applied to the mixture or some powdered metal is added, the mixture explodes.
(a) Is the spark acting as a catalyst? Explain.
(b) Is the metal acting as a catalyst? Explain.

14.79 A principle of green chemistry is that the energy needs of industrial processes should have minimal environmental impact. How can the use of catalysts lead to greener technologies?

14.80 Enzymes are remarkably efficient catalysts that can increase reaction rates by as many as 20 orders of magnitude.
(a) How does an enzyme affect the transition state of a reaction, and how does this effect increase the reaction rate?
(b) What characteristics of enzymes give them this effectiveness as catalysts?

Comprehensive Problems

14.81 Experiments show that each redox reaction is second order overall:

Reaction 1: $NO_2(g) + CO(g) \longrightarrow NO(g) + CO_2(g)$

Reaction 2: $NO(g) + O_3(g) \longrightarrow NO_2(g) + O_2(g)$

(a) When $[NO_2]$ in reaction 1 is doubled, the rate quadruples. Write the rate law for this reaction.
(b) When [NO] in reaction 2 is doubled, the rate doubles. Write the rate law for this reaction.
(c) In each reaction, the initial concentrations of the reactants are equal. For each reaction, what is the ratio of the initial rate to the rate when the reaction is 50% complete?
(d) In reaction 1, the initial $[NO_2]$ is twice the initial [CO]. What is the ratio of the initial rate to the rate at 50% completion?
(e) In reaction 2, the initial [NO] is twice the initial $[O_3]$. What is the ratio of the initial rate to the rate at 50% completion?

14.82 Consider the following reaction energy diagram:

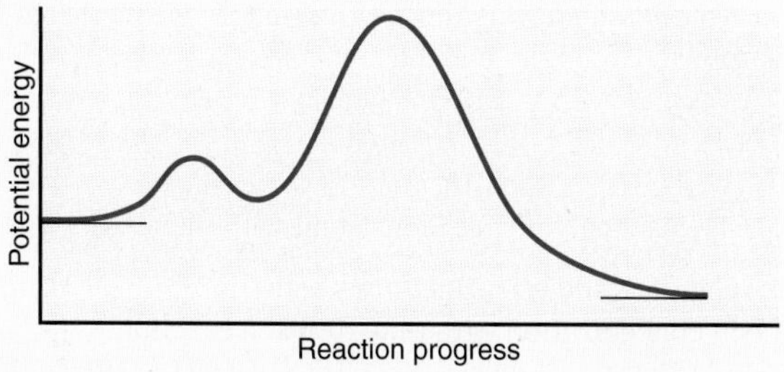

(a) How many elementary steps are in the reaction mechanism?
(b) Which step is rate-limiting?
(c) Is the overall reaction exothermic or endothermic?

14.83 Reactions between certain organic (alkyl) halides and water produce alcohols. Consider the overall reaction for *t*-butyl bromide (2-bromo-2-methylpropane):

$$(CH_3)_3CBr(aq) + H_2O(l) \longrightarrow (CH_3)_3COH(aq) + H^+(aq) + Br^-(aq)$$

The experimental rate law is Rate $= k[(CH_3)_3CBr]$. The accepted mechanism for the reaction is given below:

(1) $(CH_3)_3C{-}Br(aq) \longrightarrow (CH_3)_3C^+(aq) + Br^-(aq)$ (slow)
(2) $(CH_3)_3C^+(aq) + H_2O(l) \longrightarrow (CH_3)_3C{-}OH_2^+(aq)$ (fast)
(3) $(CH_3)_3C{-}OH_2^+(aq) \longrightarrow H^+(aq) + (CH_3)_3C{-}OH(aq)$ (fast)

(a) Why does H_2O not appear in the rate law?
(b) Write rate laws for the elementary steps.
(c) What reaction intermediates appear in the mechanism?
(d) Show that the mechanism is consistent with the experimental rate law.

14.84 Archaeologists can determine the age of an artifact made of wood or bone by measuring the amount of the radioactive isotope ^{14}C that is present in the artifact. The amount of isotope decreases in a first-order process. If 15.5% of the original amount of ^{14}C is present in a wooden tool at the time of analysis, what is the age of the tool? The half-life of ^{14}C is 5730 years.

14.85 A slightly bruised apple will rot extensively in about 4 days at room temperature (20°C). If the apple is kept in the refrigerator at 0°C, the same extent of rotting takes about 16 days. What is the activation energy for the rotting reaction?

14.86 Benzoyl peroxide, the substance most widely used against acne, has a half-life of 9.8×10^3 days when refrigerated. How long will it take to lose 5% of its potency (95% remaining)?

14.87 The rate law for the following reaction is Rate $= k[NO_2]^2$:

$$NO_2(g) + CO(g) \longrightarrow NO(g) + CO_2(g)$$

(a) One possible mechanism is presented in Section 14.6. Draw a reaction energy diagram for this mechanism, given that $\Delta_{overall}H° = -226$ kJ/mol.
(b) Consider the alternative mechanism below:
(1) $2NO_2(g) \longrightarrow N_2(g) + 2O_2(g)$ (slow)
(2) $2CO(g) + O_2(g) \longrightarrow 2CO_2(g)$ (fast)
(3) $N_2(g) + O_2(g) \longrightarrow 2NO(g)$ (fast)
Is the alternative mechanism consistent with the rate law? Is one mechanism more reasonable physically? Explain.

14.88 Consider the following general reaction and data:

$$2A + 2B + C \longrightarrow D + 3E$$

Experiment	Initial Rate [mol/(L·s)]	Initial [A] (mol/L)	Initial [B] (mol/L)	Initial [C] (mol/L)
1	6.0×10^{-6}	0.024	0.085	0.032
2	9.6×10^{-5}	0.096	0.085	0.032
3	1.5×10^{-5}	0.024	0.034	0.080
4	1.5×10^{-6}	0.012	0.170	0.032

(a) What is the reaction order with respect to each reactant?
(b) Calculate the rate constant.
(c) Write the rate law for this reaction.
(d) Express the rate in terms of changes in concentration with time for each of the components.

14.89 In an acidic solution, the breakdown of sucrose into glucose and fructose has this rate law: Rate $= k[H^+]$[sucrose]. The initial rate of sucrose breakdown is measured in a solution that is 0.01 mol/L H^+, 1.0 mol/L sucrose, 0.1 mol/L fructose, and 0.1 mol/L glucose. How does the rate change in each situation?
(a) [Sucrose] is changed to 2.5 mol/L.
(b) [Sucrose], [fructose], and [glucose] are all changed to 0.5 mol/L.
(c) $[H^+]$ is changed to 0.0001 mol/L.
(d) [Sucrose] and $[H^+]$ are both changed to 0.1 mol/L.

14.90 The citric acid cycle is the central reaction sequence in human metabolism. One of the key steps is catalyzed by the enzyme isocitrate dehydrogenase and the oxidizing agent NAD^+. In yeast, the reaction is 11th order:

$$\text{Rate} = k[\text{enzyme}][\text{isocitrate}]^4[\text{AMP}]^2[\text{NAD}^+]^m[\text{Mg}^{2+}]^2$$

What is the order with respect to NAD^+?

14.91 The following molecular scenes represent starting mixtures I and II for the reaction of A (*black*) with B (*orange*):

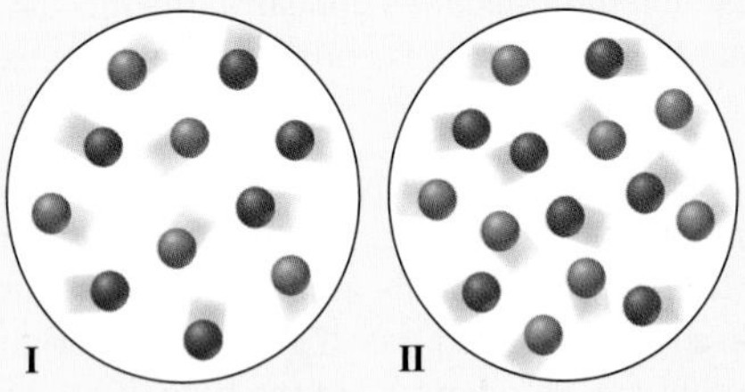

Each sphere represents 0.010 mol, and the volume is 0.50 L. If the reaction is first order in A and first order in B, and the initial rate for I is 8.3×10^{-4} mol/(L•min), what is the initial rate for II?

14.92 Experiments show that the rate of formation of carbon tetrachloride from chloroform, shown below, is first order in $CHCl_3$, $\frac{1}{2}$ order in Cl_2, and $\frac{3}{2}$ order overall:

$$CHCl_3(g) + Cl_2(g) \longrightarrow CCl_4(g) + HCl(g)$$

Show that the following mechanism is consistent with the rate law:

(1) $Cl_2(g) \rightleftharpoons 2Cl(g)$ (fast)
(2) $Cl(g) + CHCl_3(g) \longrightarrow HCl(g) + CCl_3(g)$ (slow)
(3) $CCl_3(g) + Cl(g) \longrightarrow CCl_4(g)$ (fast)

14.93 A biochemist who is studying the breakdown of the insecticide DDT finds that it decomposes by a first-order reaction with a half-life of 12 years. How long does it take DDT in a soil sample to decompose from 275 ppbm (parts per billion by mass) to 10. ppbm?

14.94 Insulin is a polypeptide hormone that is released into the blood from the pancreas and stimulates fat and muscle to take up glucose; the insulin is used up in a first-order process. In a certain patient, the insulin has a half-life of 3.5 min. To maintain an adequate blood concentration of insulin, the insulin must be replenished in a time interval equal to $\frac{1}{k}$. How long is the time interval for this patient?

14.95 For the following reaction, the rate is 0.20 mol/(L•s) when $[A]_0 = [B]_0 = 1.0$ mol/L:

$$A(g) + B(g) \longrightarrow AB(g)$$

If the reaction is first order in B and second order in A, what is the rate when $[A]_0 = 2.0$ mol/L and $[B]_0 = 3.0$ mol/L?

14.96 The acid-catalyzed hydrolysis of sucrose occurs by the following overall reaction, whose kinetic data are given below:

$$\underset{\text{Sucrose}}{C_{12}H_{22}O_{11}(s)} + H_2O(l) \longrightarrow \underset{\text{Glucose}}{C_6H_{12}O_6(aq)} + \underset{\text{Fructose}}{C_6H_{12}O_6(aq)}$$

[Sucrose] (mol/L)	Time (h)
0.501	0.00
0.451	0.50
0.404	1.00
0.363	1.50
0.267	3.00

(a) Determine the rate constant and the half-life of the reaction.
(b) How long does it take to hydrolyze 75% of the sucrose?
(c) Other studies have shown that this reaction is actually second order overall but appears to follow first-order kinetics. (Such a reaction is called a *pseudo first-order reaction*; see Section 14.3.) Suggest a reason for this apparent first-order behaviour.

14.97 At body temperature (37°C), the rate constant of an enzyme-catalyzed decomposition is 2.3×10^{14} times the rate constant of the uncatalyzed reaction. If the pre-exponential factor, *A*, is the same for both processes, by how much does the enzyme lower the E_a?

14.98 Is each of these statements true? If not, explain why.
(a) At a given *T*, all molecules have the same kinetic energy.
(b) Halving the *P* of a gaseous reaction doubles the rate.
(c) A higher activation energy gives a lower reaction rate.
(d) A temperature rise of 10°C doubles the rate of any reaction.
(e) If reactant molecules collide with greater energy than the activation energy, they change into product molecules.
(f) The activation energy of a reaction depends on temperature.
(g) The rate of a reaction increases as the reaction proceeds.
(h) Activation energy depends on collision frequency.
(i) A catalyst increases the rate by increasing the collision frequency.
(j) Exothermic reactions are faster than endothermic reactions.
(k) Temperature has no effect on the pre-exponential factor (*A*).
(l) The activation energy of a reaction is lowered by a catalyst.
(m) For most reactions, $\Delta_r H$ is lowered by a catalyst.
(n) The orientation probability factor (*p*) is near 1 for reactions between single atoms.
(o) The initial rate of a reaction is its maximum rate.
(p) A bimolecular reaction is generally twice as fast as a unimolecular reaction.
(q) The molecularity of an elementary reaction is proportional to the molecular complexity of the reactant(s).

14.99 For the decomposition of gaseous dinitrogen pentoxide, shown below, the rate constant is $k = 2.8 \times 10^{-3}$ s^{-1} at 60°C:

$$2N_2O_5(g) \longrightarrow 4NO_2(g) + O_2(g)$$

The initial concentration of N_2O_5 is 1.58 mol/L.
(a) What is $[N_2O_5]$ after 5.00 min?
(b) What fraction of the N_2O_5 has decomposed after 5.00 min?

14.100 Even when a mechanism is consistent with the rate law, later work may show it to be incorrect. For example, the reaction between hydrogen and iodine has this rate law: Rate = $k[H_2][I_2]$. The long-accepted mechanism had a single bimolecular step; that is, the overall reaction was thought to be elementary:

$$H_2(g) + I_2(g) \longrightarrow 2HI(g)$$

In the 1960s, however, spectroscopic evidence showed the presence of free iodine atoms during the reaction. Kineticists have since proposed a three-step mechanism:

(1) $I_2(g) \rightleftharpoons 2I(g)$ (fast)
(2) $H_2(g) + I(g) \rightleftharpoons H_2I(g)$ (fast)
(3) $H_2I(g) + I(g) \longrightarrow 2HI(g)$ (slow)

Show that this mechanism is consistent with the rate law.

14.101 Suggest an experimental method for measuring the change in concentration with time for each reaction:
(a) $CH_3CH_2Br(l) + H_2O(l) \longrightarrow CH_3CH_2OH(l) + HBr(aq)$
(b) $2NO(g) + Cl_2(g) \longrightarrow 2NOCl(g)$

14.102 An atmospheric chemist fills a container with gaseous N_2O_5 to a pressure of 125 kPa, and the gas decomposes to NO_2 and O_2. What is the partial pressure of NO_2, P_{NO_2} (kPa), when the total pressure is 178 kPa?

14.103 Many drugs decompose in blood by a first-order process.
(a) Two tablets of Aspirin supply 0.60 g of the active compound. After 30 min, this compound reaches a maximum concentration of 2 mg/100 mL of blood. If the half-life for its breakdown is 90 min, what is its concentration (mg/100 mL) 2.5 h after it reaches its maximum concentration?
(b) For the decomposition of an antibiotic in a person with a normal temperature (37.0°C), $k = 3.1 \times 10^{-5}$ s^{-1}; for the decomposition in a person with a fever (38.8°C), $k = 3.9 \times 10^{-5}$ s^{-1}. If the person with a fever must take another pill when $\frac{2}{3}$ of the first pill

has decomposed, how many hours should the person wait to take (i) a second pill; (ii) a third pill? (Assume that the pill is effective immediately.)
(c) Calculate E_a for decomposition of the antibiotic in part (b).

14.104 While developing a catalytic process to make ethane-1,2-diol (ethylene glycol) from synthesis gas ($CO + H_2$), a chemical engineer finds that the rate is fourth order with respect to gas pressure. The uncertainty in the pressure reading is 5%. When the catalyst is modified, the rate increases by 10%. If you were the company's patent attorney, would you file for a patent on this catalyst modification? Explain.

14.105 Iodide ion reacts with chloromethane to displace chloride ion in a common organic substitution reaction:

$$I^- + CH_3Cl \longrightarrow CH_3I + Cl^-$$

(a) Draw a wedge-bond structure of chloroform, and indicate the most effective direction of I^- attack.
(b) The analogous reaction with 2-chlorobutane [Figure P14.105(b)] results in a major change in specific rotation as measured by polarimetry. Explain why, showing a wedge-bond structure of the product.
(c) Under different conditions, 2-chlorobutane loses Cl^- in a rate-determining step to form a planar intermediate [Figure P14.105(c)]. This cationic species reacts with HI and then loses H^+ to form a product that exhibits no optical activity. Explain why, showing a wedge-bond structure.

Figure P14.105(b)

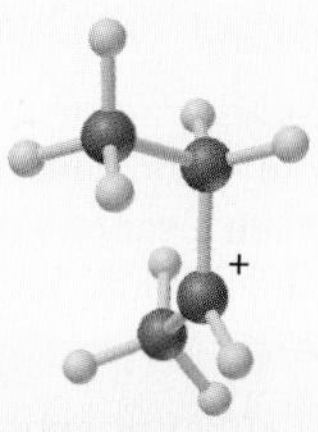

Figure P14.105(c)

14.106 Assume that water boils at 100.0°C in St. John's, Newfoundland and Labrador (near sea level), and at 90.0°C in Chatter Creek, British Columbia (near 2896 m). If it takes 4.8 min to cook an egg in Chatter Creek and 4.5 min in St. John's, what is E_a for this process?

14.107 Sulfonation of benzene has the following mechanism:

(1) $2H_2SO_4 \longrightarrow H_3O^+ + HSO_4^- + SO_3$ (fast)
(2) $SO_3 + C_6H_6 \longrightarrow H(C_6H_5^+)SO_3^-$ (slow)
(3) $H(C_6H_5^+)SO_3^- + HSO_4^- \longrightarrow C_6H_5SO_3^- + H_2SO_4$ (fast)
(4) $C_6H_5SO_3^- + H_3O^+ \longrightarrow C_6H_5SO_3H + H_2O$ (fast)

(a) Write an overall equation for the reaction.
(b) Write the overall rate law in terms of the initial rate of the reaction.

14.108 In the lower troposphere, ozone is one of the components of photochemical smog. It is generated in air when nitrogen dioxide, formed by the oxidation of nitrogen monoxide from car exhaust, reacts by the following mechanism:

(1) $NO_2(g) \xrightarrow[h\nu]{k_1} NO(g) + O(g)$

(2) $O(g) + O_2(g) \xrightarrow{k_2} O_3(g)$

Assuming that the rate of formation of atomic oxygen in step 1 equals the rate of its consumption in step 2, use the data below to calculate (a) the concentration of atomic oxygen, [O]; (b) the rate of ozone formation:

$k_1 = 6.0 \times 10^{-3}\,s^{-1}$ $[NO_2] = 4.0 \times 10^{-9}$ mol/L
$k_2 = 1.0 \times 10^6$ L/(mol•s) $[O_2] = 1.0 \times 10^{-2}$ mol/L

14.109 Chlorine is commonly used to disinfect drinking water, and inactivation of pathogens by chlorine follows first-order kinetics. The following data show *E. coli* inactivation:

Contact Time (min)	Percent (%) Inactivation
0.00	0.0
0.50	68.3
1.00	90.0
1.50	96.8
2.00	99.0
2.50	99.7
3.00	99.9

(a) Determine the first-order inactivation constant, k. [*Hint:* % inactivation = $100 \times (\frac{1 - [A]_t}{[A]_0})$.]
(b) How much contact time is required for 95% inactivation?

14.110 The reaction and rate law for the gas-phase decomposition of dinitrogen pentoxide are given below:

$$2N_2O_5(g) \longrightarrow 4NO_2(g) + O_2(g) \quad \text{Rate} = k[N_2O_5]$$

Which of the following can be considered valid mechanisms for the reaction?
(a) One-step collision
(b) $2N_2O_5(g) \longrightarrow 2NO_3(g) + 2NO_2(g)$ (slow)
$2NO_3(g) \longrightarrow 2NO_2(g) + 2O(g)$ (fast)
$2O(g) \longrightarrow O_2(g)$ (fast)
(c) $N_2O_5(g) \rightleftharpoons NO_3(g) + NO_2(g)$ (fast)
$NO_2(g) + N_2O_5(g) \longrightarrow 3NO_2(g) + O(g)$ (slow)
$NO_3(g) + O(g) \longrightarrow NO_2(g) + O_2(g)$ (fast)
(d) $2N_2O_5(g) \rightleftharpoons 2NO_2(g) + N_2O_3(g) + 3O(g)$ (fast)
$N_2O_3(g) + O(g) \longrightarrow 2NO_2(g)$ (slow)
$2O(g) \longrightarrow O_2(g)$ (fast)
(e) $2N_2O_5(g) \longrightarrow N_4O_{10}(g)$ (slow)
$N_4O_{10}(g) \longrightarrow 4NO_2(g) + O_2(g)$ (fast)

14.111 Nitrification is a biological process for removing NH_3 from wastewater as NH_4^+:

$$NH_4^+ + 2O_2 \longrightarrow NO_3^- + 2H^+ + H_2O$$

The first-order rate constant is given as

$$k_1 = 0.47e^{0.095(T - 15°C)}$$

where k_1 is in day^{-1}.
(a) If the initial concentration of NH_3 is 3.0 mol/m^3, how long will it take to reduce the concentration to 0.35 mol/m^3 (i) in the early summer ($T = 20°C$); (ii) in the early spring ($T = 10°C$)?
(b) Using your answer for the early summer in part (a), what is the rate of O_2 consumption?

14.112 Carbon disulfide, a poisonous flammable liquid, is an excellent solvent for phosphorus, sulfur, and some other nonmetals. A kinetic study of its gaseous decomposition reveals these data:

Experiment	Initial Rate [mol/(L•s)]	Initial $[CS_2]$ (mol/L)
1	2.7×10^{-7}	0.100
2	2.2×10^{-7}	0.080
3	1.5×10^{-7}	0.055
4	1.2×10^{-7}	0.044

(a) Write the rate law for the decomposition of CS_2.
(b) Calculate the average value of the rate constant.

14.113 Like any catalyst, palladium, platinum, and nickel catalyze both directions of a reaction: addition of hydrogen to carbon double bonds (hydrogenation) and its elimination from carbon double bonds (dehydrogenation).
(a) Which variable determines whether an alkene will be hydrogenated or dehydrogenated?
(b) Which reaction requires a higher temperature?
(c) How can all-*trans* fats arise during the hydrogenation of fats that contain some *cis* double bonds?

14.114 In a *clock reaction*, a dramatic colour change occurs at a time determined by concentration and temperature. Consider the iodine clock reaction, whose overall equation follows:

$$2I^-(aq) + S_2O_8^{2-}(aq) \longrightarrow I_2(aq) + 2SO_4^{2-}(aq)$$

As I_2 forms, it is immediately consumed by its reaction with a fixed amount of added $S_2O_3^{2-}$:

$$I_2(aq) + 2S_2O_3^{2-}(aq) \longrightarrow 2I^-(aq) + S_4O_6^{2-}(aq)$$

Once the $S_2O_3^{2-}$ is consumed, the excess I_2 forms a blue-black product with the starch that is present in solution:

$$I_2 + \text{starch} \longrightarrow \text{starch}\cdot I_2 \text{ (blue-black)}$$

The rate of the reaction is also influenced by the total concentration of ions, so KCl and $(NH_4)_2SO_4$ are added to maintain a constant value. Use the data below, obtained at 23°C, to determine (a) the average rate for each trial; (b) the order with respect to each reactant; (c) the rate constant; (d) the rate law for the overall reaction.

	Experiment		
	1	2	3
0.200 mol/L KI (mL)	10.0	20.0	20.0
0.100 mol/L $Na_2S_2O_8$ (mL)	20.0	20.0	10.0
0.0050 mol/L $Na_2S_2O_3$ (mL)	10.0	10.0	10.0
0.200 mol/L KCl (mL)	10.0	0.0	0.0
0.100 mol/L $(NH_4)_2SO_4$ (mL)	0.0	0.0	10.0
Time to colour (s)	29.0	14.5	14.5

14.115 Heat transfer to and from a reaction flask is often a critical factor in controlling reaction rate. The heat transferred (q) depends on the heat transfer coefficient (h) for the flask material; the temperature difference (ΔT) across the flask wall; and the commonly "wetted" area (A) of the flask and bath, $q = hA\Delta T$. When an exothermic reaction is run at a given T, there is a bath temperature at which the reaction can no longer be controlled, and the reaction "runs away" suddenly. A similar problem is often seen when a reaction is "scaled up" from, for example, a half-filled small flask to a half-filled large flask. Explain these behaviours.

14.116 The molecular scenes below represent the first-order reaction in which cyclopropane (*red*) is converted to propene (*green*):

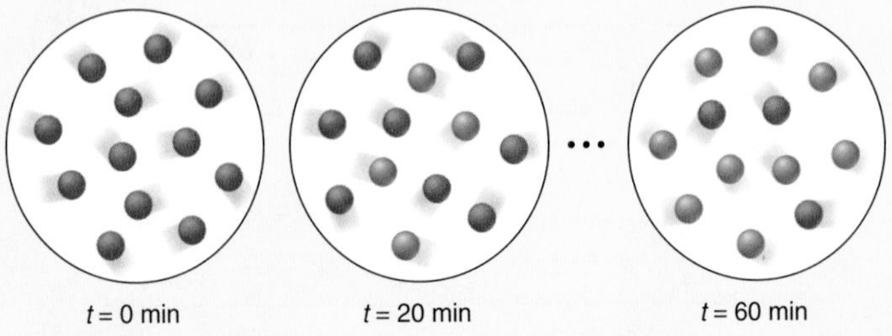

Determine (a) the half-life; (b) the rate constant.

14.117 The growth of *Pseudomonas* bacteria is modelled as a first-order process with $k = 0.035\ \text{min}^{-1}$ at 37°C. The initial *Pseudomonas* population density is 1.0×10^3 cells/L.
(a) What is the population density after 2 h?
(b) What is the time required for the population to go from 1.0×10^3 cells/L to 2.0×10^3 cells/L?

14.118 Consider the following organic reaction, in which one halogen replaces another in an alkyl halide:

$$CH_3CH_2Br + KI \longrightarrow CH_3CH_2I + KBr$$

In acetone, this particular reaction goes to completion because KI is soluble in acetone but KBr is not. In the mechanism, I^- approaches the carbon *opposite* to the Br (see Figure 14.20, with I^- instead of OH^-). After Br^- has been replaced by I^- and precipitates as KBr, other I^- ions react with the ethyl iodide by the same mechanism.
(a) If we designate the carbon bonded to the halogen as C-1, what is the shape around C-1 and the hybridization of C-1 in ethyl iodide?
(b) In the transition state, one of the two lobes of the unhybridized $2p$ orbital of C-1 overlaps a p orbital of I, while the other lobe overlaps a p orbital of Br. What are the shape around C-1 and the hybridization of C-1 in the transition state?
(c) The deuterated reactant, CH_3CHDBr (where D is deuterium, ^2H), has two optical isomers because C-1 is chiral. If the reaction is run with one of the isomers, the ethyl iodide is *not* optically active. Explain.

14.119 Another radioisotope of iodine, ^{131}I, is also used to study thyroid gland function (see Follow-Up Problem 14.7). A patient is given a sample that is 1.7×10^{-4} mol/L ^{131}I. If the half-life is 8.04 days, what fraction of the radioactivity remains after 30. days?

14.120 The effect of substrate concentration on the first-order growth rate of a microbial population follows the Monod equation (similar in form to the Michaelis-Menten equation, but different in that it was *empirically* derived):

$$\mu = \frac{\mu_{max} S}{K_s + S}$$

where μ is the first-order growth rate (s^{-1}), μ_{max} is the maximum growth rate (s^{-1}), S is the substrate concentration (kg/m^3), and K_s is the value of S that gives one-half the maximum growth rate (kg/m^3). For $\mu_{max} = 1.5 \times 10^{-4}\ s^{-1}$, $K_s = 0.03\ kg/m^3$.
(a) Plot μ versus S for S between 0.0 and 1.0 kg/m^3.
(b) The initial population density is 5.0×10^3 cells/m^3. What is the density after 1.0 h if the initial S is 0.30 kg/m^3?
(c) What is the density after 1.0 h if the initial S is 0.70 kg/m^3?

14.121 The following scenes depict four initial reaction mixtures for the reaction of A (*blue*) and B (*yellow*), with and without a solid present (*grey cubes*). The initial rate, $\frac{-\Delta[A]}{\Delta t}$ [mol/(L•s)], is shown, with each sphere representing 0.010 mol and the container volume at 0.50 L.

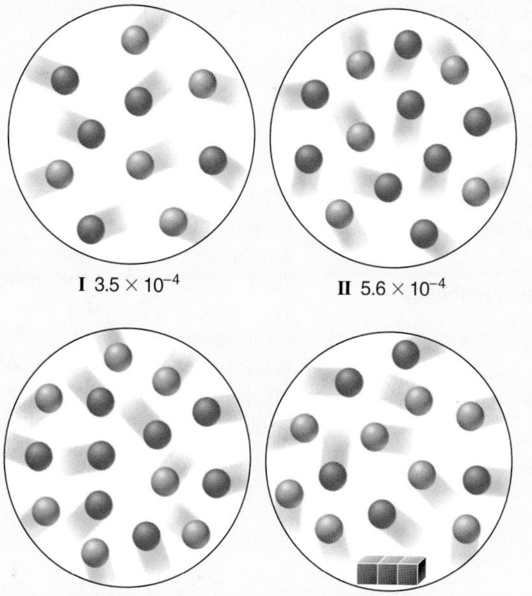

I 3.5×10^{-4} II 5.6×10^{-4}

III 5.6×10^{-4} IV 4.9×10^{-4}

(a) What is the rate law in the absence of a catalyst?
(b) What is the overall reaction order?
(c) Find the rate constant.
(d) Do the grey cubes have a catalytic effect? Explain.

14.122 The mathematics of the first-order rate law can be applied to any situation in which a quantity decreases by a constant fraction per unit of time (or unit of any other variable).
(a) As light moves through a solution, its intensity decreases per unit distance travelled in the solution. Show that

$$\ln\left(\frac{\text{intensity of light leaving the solution}}{\text{intensity of light entering the solution}}\right) = -\text{fraction of light removed per unit of length} \times \text{distance travelled in solution}$$

(b) The value of your savings declines under conditions of constant inflation. Show that

$$\ln\left(\frac{\text{value remaining}}{\text{initial value}}\right) = -\text{fraction lost per unit of time} \times \text{savings time interval}$$

14.123 Figure 14.25 shows key steps in the metal-catalyzed (M) hydrogenation of ethene:

$$C_2H_4(g) + H_2(g) \xrightarrow{M} C_2H_6(g)$$

Use the following symbols to write a mechanism that gives the overall equation:

Symbol	Meaning
$H_2(ads)$	Adsorbed hydrogen molecules
M—H	Hydrogen atoms bonded to metal atoms
$C_2H_4(ads)$	Adsorbed ethene molecules
$C_2H_5(ads)$	Adsorbed ethyl radicals

14.124 Human liver enzymes catalyze the degradation of ingested toxins. By what factor is the rate of a detoxification changed if an enzyme lowers the E_a by 5 kJ/mol at 37°C?

14.125 Acetone is one of the most important solvents in organic chemistry; it is used to dissolve everything from fats and waxes to airplane glue and nail polish. At high temperatures, it decomposes in a first-order process to methane and ethenone ($CH_2{=}C{=}O$). At 600°C, the rate constant is $8.7 \times 10^{-3}\ s^{-1}$.
(a) What is the half-life of the reaction?
(b) How long does it take for 40.% of a sample of acetone to decompose?
(c) How long does it take for 90.% of a sample of acetone to decompose?

14.126 A (*green*), B (*blue*), and C (*red*) are structural isomers. The molecular filmstrip depicts them undergoing a chemical change as time proceeds:

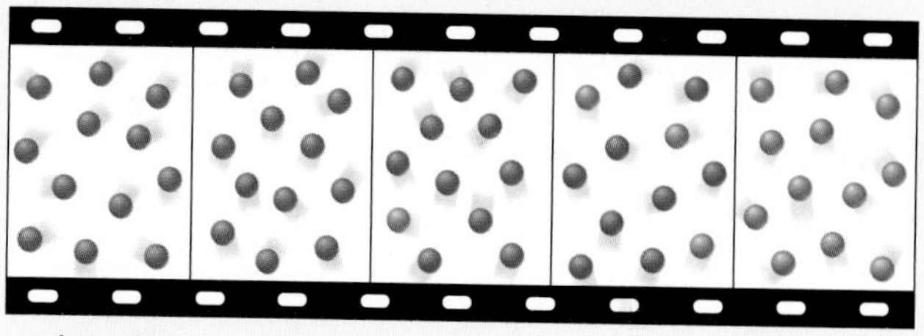

1 $t = 0$ 2 3 4 5 $t = \infty$

(a) Write a mechanism for this reaction.
(b) What role does C play?

14.127 The reaction $2N_2O_5(g) \longrightarrow 4NO_2(g) + O_2(g)$ is first order in N_2O_5. In a particular experiment, 5.750 g of N_2O_5 was placed in a 755 mL reaction vessel at 647°C and left until 3.250 g of NO_2 was formed. If the half-life of the reaction is 11.2 min, how long was the reaction allowed to proceed?

14.128 For a certain reaction, the first-order rate constant was found to be $3.71\ min^{-1}$ at 35°C and $8.32 \times 10^{-4}\ s^{-1}$ at 14°C. What is the value of the rate constant at 55°C?

14.129 A new medication that was prescribed to a patient contained 200 mg of the active ingredient in the tablet. The pharmacokinetics of the medication was first order and the half-life of the medication was found to be 1.08×10^4 s. If the minimum required amount of the medication in the body should be 50 mg for the medication to remain effective, when should the next dose be taken?

14.130 For the reaction cyclopropane $\longrightarrow$ propene, the frequency factor, A, was found to be $2.9 \times 10^{14}\ s^{-1}$. If the activation energy is 261 kJ/mol, what is the rate constant at 496.9°C? If the initial concentration of cyclopropane is 13.74 mol/L, what will the concentration be after 26 min? What are the initial and final rates of reaction?

Equilibrium: The Extent of Chemical Reactions

$$H_2O(l) \rightleftharpoons H_2O(g)$$

IN THIS CHAPTER . . . We consider the principles of equilibrium in systems of gases and pure liquids and solids. By the end of this chapter, you should be able to

- Describe the equilibrium state at the macroscopic and molecular levels and explain that equilibrium occurs when the forward and reverse reaction rates are equal
- Compare and contrast the reaction quotient (*Q*), which changes as the reaction proceeds, with the equilibrium constant (*K*), which applies to the reaction at equilibrium
- Express the equilibrium condition in terms of concentrations or partial pressures and see how the two expressions are related
- Develop a systematic approach to solving a variety of equilibrium problems
- Compare values of *Q* and *K* to determine the direction in which a system must proceed to reach equilibrium
- Explain Le Châtelier's principle, which describes how a change in conditions—concentration, pressure, temperature, or catalyst—affects the equilibrium state, and apply this principle to a major industrial process
- Discuss how equilibrium concepts operate in metabolic pathways

15

Your boss at the chemical plant comes to you and asks you, as the chemist/chemical engineer in charge of the process, to boost the yield of the product you are making. How can you do this? Depending on the type of reaction taking place, you may need to have an understanding of equilibrium, how it affects reaction yields, and what can be done to alter the equilibrium state that will allow you to obtain more of your product. In this chapter, we will look at what equilibrium is and how it relates to rates from Chapter 14. We will also see how equilibrium affects almost all reactions that take place and also many processes that occur in our daily lives. We will examine the factors that affect reactions already at equilibrium and then consider equilibrium in a biological system.

Just as reactions vary greatly in their speed, they also vary greatly in their extent. Indeed, kinetics and equilibrium apply to different aspects of a reaction:

- Kinetics applies to the *speed* (or rate) of a reaction, the concentration of reactant that disappears (or the concentration of product that appears) per unit time.
- Equilibrium applies to the *extent* of a reaction, the concentrations of reactant and product that are present after an unlimited time, when no further change occurs.

At equilibrium, no further *net* change occurs because the forward and reverse reactions reach a *balance*. Like balls being juggled, traffic going back and forth across a bridge, or people going up and down an escalator, the change in one direction is balanced by the change in the other. A fast reaction may proceed almost completely, just partially, or only slightly toward products before this balance is reached. Consider acid dissociation in water. In 1 mol/L HCl, virtually all the hydrogen chloride molecules are dissociated into ions; in 1 mol/L CH_3COOH, less than 1% of the ethanoic (acetic) acid molecules are dissociated. Yet both reactions take less than a second. Similarly, some slow reactions yield a large amount of product, whereas others yield very little. A steel water-storage tank will rust after a few years, and it will rust completely given enough time. However, no matter how long we wait, the water inside the tank will not decompose to hydrogen and oxygen.

Knowing the extent of a given reaction is crucial. How much product—medicine, polymer, fuel—can you obtain from a particular reaction mixture? How can you adjust the conditions to obtain more? If a reaction is slow but has a good yield, will a catalyst speed up the reaction enough to make it useful?

Concepts and Skills to Review before Studying This Chapter

- Equilibrium vapour pressure (Section 11.2)
- Equilibrium nature of a saturated solution (Section 12.3)
- Dependence of rate on concentration (Sections 14.2 and 14.5)
- Rate laws for elementary reactions (Section 14.6)
- Function of a catalyst (Section 14.7)

15.1 The Equilibrium State and the Equilibrium Constant

Experimental results from countless reactions have shown that, *given sufficient time, the concentrations of reactants and products no longer change.* This apparent cessation of chemical change occurs because *all reactions are reversible and reach a state of equilibrium.* Let us examine a chemical system at the macroscopic and molecular levels to see how equilibrium arises and then consider some quantitative aspects of the process.

1. *A macroscopic view of equilibrium.* The system we will consider is the reversible gaseous reaction between colourless dinitrogen tetroxide and brown nitrogen dioxide:

$$N_2O_4(g;\ \text{colourless}) \rightleftharpoons 2NO_2(g;\ \text{brown})$$

As soon as we introduce some liquid N_2O_4 [boiling point (bp) = 21°C] into a sealed container kept at 200°C, it vaporizes, and the gas begins to turn pale brown. As time passes, the brown darkens, until, after more time, the colour stops changing. The first three photos in Figure 15.1 show the colour change, and the last photo shows no further change.

2. *A molecular view of equilibrium.* On the molecular level, as shown in the blow-up circles in Figure 15.1, a dynamic scene unfolds. The N_2O_4 molecules fly wildly throughout the container, a few splitting into two NO_2 molecules. As time passes, more N_2O_4 molecules decompose and the concentration of NO_2 rises. As the number of N_2O_4 molecules decreases, N_2O_4 decomposition slows down. At the same time, as the number of NO_2 molecules increases, more collide, and the formation of N_2O_4 speeds up. Eventually, the system reaches equilibrium: N_2O_4 molecules are decomposing into NO_2 molecules just as fast as NO_2 molecules are forming N_2O_4.

Thus, at equilibrium, *reactant and product concentrations or partial pressures are constant because a change in one direction is balanced by a change in the other direction as the forward and reverse rates become equal:*

$$\text{At equilibrium: rate}_{\text{fwd}} = \text{rate}_{\text{rev}} \tag{15.1}$$

3. *A quantitative view of equilibrium: a constant ratio of constants.* Let us see how reactant and product amounts affect the equilibrium process. At a particular temperature, when the system reaches equilibrium, we have

$$\text{rate}_{\text{fwd}} = \text{rate}_{\text{rev}}$$

In this reaction system, both forward and reverse reactions are elementary steps (Section 14.6), so we can write their rate laws directly from the balanced equation:

$$k_{\text{fwd}} \times p_{N_2O_4(\text{eq})} = k_{\text{rev}} \times p^2_{NO_2(\text{eq})}$$

where k_{fwd} and k_{rev} are the forward and reverse rate constants, respectively, and the subscript "eq" refers to the concentrations or partial pressures at equilibrium. By rearranging, we set the ratio of the rate constants equal to the ratio of the concentration or partial pressure terms:

$$\frac{k_{\text{fwd}}}{k_{\text{rev}}} = \frac{p^2_{NO_2(\text{eq})}}{p_{N_2O_4(\text{eq})}}$$

FIGURE 15.1 Reaching equilibrium on the macroscopic and molecular levels. **A.** The reaction mixture consists mostly of colourless N_2O_4. **B.** As N_2O_4 decomposes to NO_2, the mixture becomes pale brown. **C.** At equilibrium, the colour and concentrations of NO_2 and N_2O_4 no longer change. **D.** The reaction continues in both directions at equal rates, so the concentrations (and colour) remain constant.

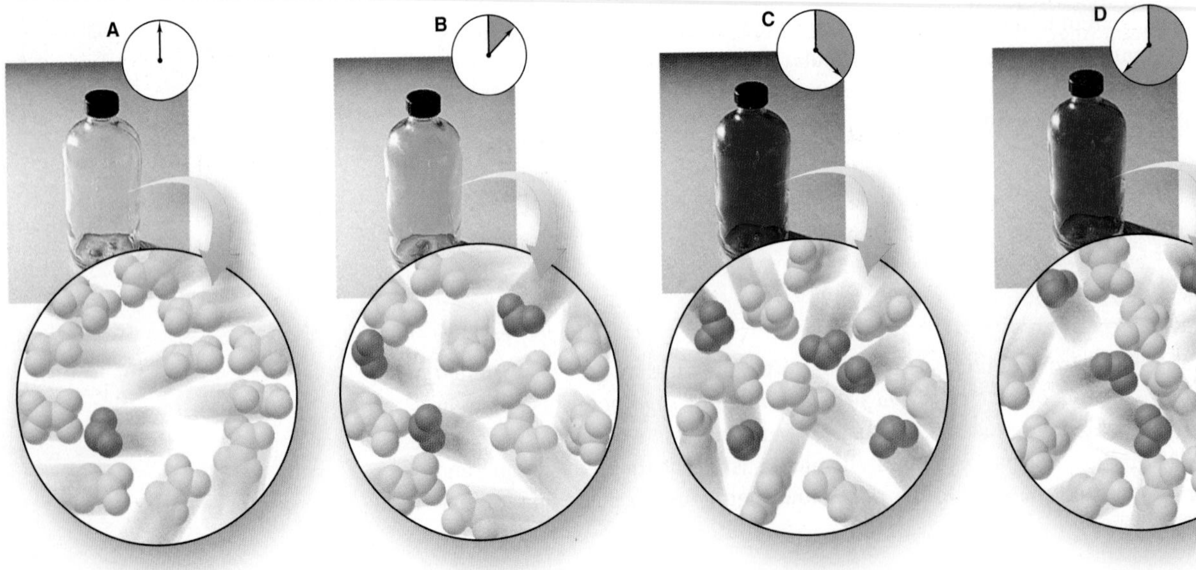

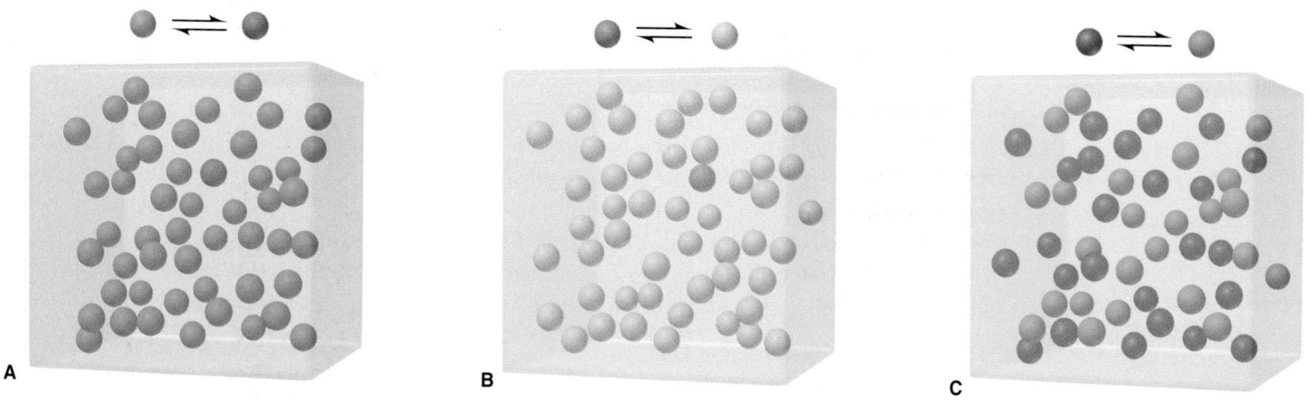

FIGURE 15.2 **The range of equilibrium constants. A.** For this reaction, $K = \frac{1}{49} = 0.020$. **B.** For this reaction, $K = \frac{49}{1} = 49$. **C.** For this reaction, $K = \frac{25}{25} = 1.0$.

The ratio of constants creates a new constant called the **equilibrium constant (*K*)**:

$$K = \frac{k_{\text{fwd}}}{k_{\text{rev}}} = \frac{p^2_{\text{NO}_2(\text{eq})}}{p_{\text{N}_2\text{O}_4(\text{eq})}} \qquad (15.2)$$

Thus, *K is a number equal to a particular ratio of equilibrium amounts of product(s) and reactant(s) at a particular temperature.* In this example, we used the forward and reverse rates of the reaction (which we defined based on concentrations or partial pressures) to establish the equilibrium constant. In the next few sections, we will see how we can use thermodynamics to create an even broader definition of the equilibrium constant, *K*.

4. *K as a measure of reaction extent.* The *magnitude* of *K* is an indication of *how far a reaction proceeds toward product at a given temperature.* Different reactions, even at the same temperature, have a wide range of concentrations or partial pressures at equilibrium—from almost all reactant to almost all product—and, therefore, have a wide range of equilibrium constants. Here are three examples of different magnitudes of *K*:

- *Small K* (Figure 15.2A). If a reaction yields little product before reaching equilibrium, it has a small *K*; if *K* is very small ($K < 10^{-5}$), we may say that there is "very little reaction." For example, there is very little reaction between nitrogen and oxygen at 1000 K:*

$$N_2(g) + O_2(g) \rightleftharpoons 2NO(g) \qquad K = 1 \times 10^{-30}$$

- *Large K* (Figure 15.2B). Conversely, if a reaction reaches equilibrium with little reactant remaining, it has a large *K*; if *K* is very large ($K > 10^5$), we say that the reaction "essentially goes to completion." The oxidation of carbon monoxide essentially goes to completion at 1000 K:

$$2CO(g) + O_2(g) \rightleftharpoons 2CO_2(g) \qquad K = 2.2 \times 10^{22}$$

- *Intermediate K* (Figure 15.2C). When significant amounts of both reactant and product are present at equilibrium, *K* has an intermediate value ($10^{-5} < K < 10^5$), as when bromine monochloride breaks down to its elements at 1000 K:

$$2BrCl(g) \rightleftharpoons Br_2(g) + Cl_2(g) \qquad K = 5$$

SUMMARY OF SECTION 15.1

- Kinetics and equilibrium are distinct aspects of a reaction system, and the rate and extent of a reaction are not necessarily related.
- When the forward and reverse reactions occur at the same rate, concentrations or partial pressures no longer change and the system has reached equilibrium.
- The equilibrium constant (*K*) is a number based on a particular ratio of product and reactant concentrations or partial pressures: *K* is small if there is a high concentration or partial pressure of reactant(s) at equilibrium, and it is large if there is a high concentration or partial pressure of product(s) at equilibrium.

*To distinguish the equilibrium constant from the temperature unit, the kelvin, the equilibrium constant is represented by a capital italic *K*, whereas the kelvin is a capital roman K. Also, since the kelvin is a unit, it always follows a number.

15.2 The Reaction Quotient and the Equilibrium Constant

We have introduced the equilibrium constant in terms of a ratio of rate constants, but the original research on chemical equilibrium was developed many years before the principles of kinetics. In 1864, two Norwegian chemists, Cato Guldberg and Peter Waage, observed that, *at a given temperature, a chemical system reaches a state in which a particular ratio of reactant and product concentrations or partial pressures has a constant value.* This is a statement of the **law of chemical equilibrium (law of mass action).**

Changing Value of the Reaction Quotient

The particular ratio of concentration terms that we write for a given reaction is called the **reaction quotient (*Q*)**, also known as the *mass-action expression.* For the reversible breakdown of N_2O_4 to NO_2, the reaction quotient is

$$N_2O_4(g) \rightleftharpoons 2NO_2(g) \qquad Q = \frac{p^2_{NO_2}}{p_{N_2O_4}}$$

As the reaction proceeds toward equilibrium, the concentrations or partial pressures of reactants and products change continuously, and so does their ratio, the value of Q: at a given temperature, at the beginning of the reaction, the pressures have initial values, and Q has an initial value; a moment later, the pressures have slightly different values, and so does Q; after another moment, the pressures and the value of Q change further. These changes continue *until the system reaches equilibrium.* At this point, reactant and product concentrations or partial pressures have their equilibrium values and no longer change. Thus, the value of Q no longer changes and equals K at that temperature:

$$\text{At equilibrium: } Q = K \qquad (15.3)$$

In formulating the law of mass action, Guldberg and Waage found that, *for a particular system and temperature, the same equilibrium state is attained regardless of the starting concentrations or partial pressures.* For example, consider the data in Table 15.1, collected during four experiments for the N_2O_4-NO_2 system at 200°C. Three essential points stand out:

- The ratio of *initial* partial pressures and concentrations (*fourth column*) varies widely but always gives the same ratio of *equilibrium* partial pressures and concentrations (*rightmost column*).
- The *individual* equilibrium partial pressures and concentrations are different for each experiment, but the *ratio* of these equilibrium partial pressures and concentrations is constant.
- The ratio of the *partial pressures* and the ratio of the *concentrations* are *not* the same. In other words, K calculated using pressures does not equal K calculated using concentrations. The reason for this difference will be discussed in Section 15.3.

Thus, monitoring Q tells whether a system has reached equilibrium or, if it has not, how far away it is from equilibrium and in which direction it is changing.

The curves in Figure 15.3 show the experiments in Table 15.1. Note that $[N_2O_4]$ and $[NO_2]$ change smoothly during the course of the reaction (as indicated by the changing brown colour at the top), and so does the value of Q. Once the system reaches equilibrium (constant brown colour), the concentrations no longer change and Q equals K. In other words, for any given chemical system, *K is a special value of Q that occurs when the reactant and product concentrations have reached their equilibrium values.*

TABLE 15.1 Initial and Equilibrium Partial Pressure and Concentration Ratios for the N_2O_4/NO_2 System at 200°C (473 K)

	Initial Partial Pressures (bar) and Q			Equilibrium Partial Pressure (bar) and K		
Experiment	$p_{N_2O_4}$	p_{NO_2}	Q, $(p_{NO_2})^2/p_{N_2O_4}$	$p_{N_2O_4,\,eq}$	$p_{NO_2,eq}$	K, $(p_{NO_{2,eq}})^2/p_{N_2O_{4,eq}}$
1	3.933	0.0000	0.000 0	0.1404	7.590	410.3
2	0.0000	3.93300	∞	0.03634	3.862	410.4
3	1.966	1.966	0.050 0	0.08022	5.741	410.9
4	2.949	0.9831	0.3277	0.1081	6.685	413.3
	Initial Concentrations (mol/L) and Q			**Initial Concentrations (mol/L) and K**		
Experiment	$[N_2O_4]$	$[NO_2]$	Q, $[NO_2]^2/[N_2O_4]$	$[N_2O_4]_{eq}$	$[NO_2]_{eq}$	K, $[NO_2]^2_{eq}/[N_2O_4]_{eq}$
1	0.1000	0.0000	0.000 0	0.003 57	0.193	10.4
2	0.0000	0.1000	∞	0.000 924	0.0982	10.4
3	0.0500	0.0500	0.050 0	0.002 04	0.146	10.4
4	0.0750	0.0250	0.008 33	0.002 75	0.170	10.5

Writing the Reaction Quotient in Its Various Forms

The reaction quotient *must* be written directly from the balanced net equation. In contrast, as discussed in Chapter 14, the rate law for an overall reaction *cannot* be written from the balanced equation but must be determined from rate data.

Constructing the Reaction Quotient The most common form of the reaction quotient shows the reactant and product terms as molar concentrations, which are designated by square brackets, [ ]. In general, however, the terms used in the reaction quotient and in K are *effective concentrations.* If we use effective concentrations, the expressions we formulate for K and Q will always be applicable. How, then, do we determine the effective concentration of a particular species? It can be determined using the *activity* of the species. The **activity** of a species in solution is defined as

$$a_X = \gamma_X \frac{[X]}{c^\circ} \qquad (15.4)$$

where γ_X is known as the *activity coefficient*; [X] represents the concentration of species X in units of mol/L; and c° represents a standard reference concentration, which is usually 1 mol/L. Note that the activity becomes a *unitless quantity* due to the ratio of the concentration of the solution to the reference concentration. We take the activity coefficient γ_X to be 1, although this is not strictly true. So, to a reasonable degree in a solution that is not very concentrated and behaves ideally, the activity a_X equals the concentration of the solution in mol/L.

When dealing with reactions involving gases, the activity is expressed as

$$a_Y = \gamma_Y \frac{p_Y}{p^\circ} \qquad (15.5)$$

where γ_Y is once again the activity coefficient; p_Y represents the partial pressure of species Y; and p° represents a standard reference pressure, which is usually 1 bar. Note that, again, the activity is unitless and that, although not strictly correct, we take the activity coefficient to be 1. So, to a reasonable degree in a gas that behaves ideally, the activity of the gas equals the partial pressure of the gas in bar.

Now that we have established the identity of the terms that we will put in the expression for Q (or K), we can construct the expression from the balanced chemical equation. For the general chemical equation

$$aA + bB \rightleftharpoons cC + dD$$

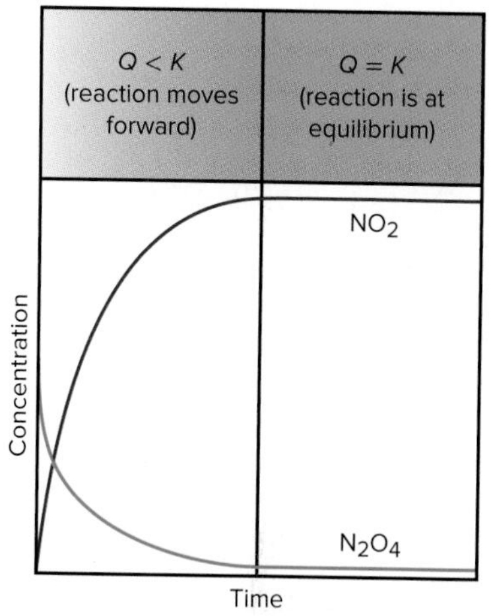

FIGURE 15.3 The change in Q during the N_2O_4-NO_2 reaction, based on data from Table 15.1

where a, b, c, and d are the stoichiometric coefficients, the reaction quotient is

$$Q = \frac{a_C^c a_D^d}{a_A^a a_B^b} \tag{15.6}$$

Thus, *Q is a ratio of product activity terms multiplied together and divided by reactant activity terms multiplied together, with each term raised to the power of its balancing or stoichiometric coefficient.*

Two steps are needed to construct a reaction quotient for any chemical system:

1. *Start with the balanced equation and carefully note the phases of both the reactants and the products.* For example, in the formation of ammonia from its elements, the balanced equation (with coloured coefficients for reference) is

$$N_2(g) + 3H_2(g) \rightleftharpoons 2NH_3(g)$$

2. *Arrange the terms and exponents.* Place the product terms in the numerator and the reactant terms in the denominator, multiplied by each other, and raise each term to the power of its balancing coefficient (coloured as in the balanced equation). *Remember*: If all the reactants and products are in the gas phase, the activities of the reactants and products can be approximated by their partial pressures; if all the reactants and products are in aqueous solution, the activities of the reactants and products can be approximated by their concentrations in mol/L. When the phases of all the reactants and products are the same, the resulting Q or K expression is known as a homogeneous reaction quotient or equilibrium constant. If, however, the chemical equation contains both gaseous and aqueous terms, the resulting Q or K expression is known as a heterogeneous reaction quotient or equilibrium constant, since both pressure and concentration terms appear in the expression.

$$Q = \frac{p_{NH_3}^2}{p_{N_2} p_{H_2}^3}$$

Sample Problem 15.1 Writing the Reaction Quotient from the Balanced Equation

Problem Write the reaction quotient, Q, for each reaction:

(a) The decomposition of dinitrogen pentoxide, $N_2O_5(g) \rightleftharpoons NO_2(g) + O_2(g)$

(b) The formation of cobalt hexamine in solution,
$Co^{3+}(aq) + 6\,NH_3(aq) \rightleftharpoons Co(NH_3)_6^{3+}(aq)$

(c) The reaction between fluorine gas and tin ion in solution,
$F_2(g) + Sn^{2+}(aq) \rightleftharpoons 2F^-(aq) + Sn^{4+}(aq)$

Plan We balance the equations, note the phase of each reactant and product, and then construct the reaction quotient (Equation 15.6).

Solution **(a)** $2N_2O_5(g) \rightleftharpoons 4NO_2(g) + O_2(g)$

Notice that all the reactants and products are in the gas phase. Thus, in the expression for Q, we will use the partial pressure of each reactant and product to represent the activity.

$$Q = \frac{p_{NO_2}^4 p_{O_2}}{p_{N_2O_5}^2}$$

(b) $Co^{3+}(aq) + 6NH_3(aq) \rightleftharpoons Co(NH_3)_6^{3+}(aq)$

All the products and reactants are in the aqueous phase, so we will use concentrations (mol/L) instead of activities.

$$Q = \frac{[Co(NH_3)_6^{3+}]}{[Co^{3+}][NH_3]^6}$$

(c) $F_2(g) + Sn^{2+}(aq) \rightleftharpoons 2F^-(aq) + Sn^{4+}(aq)$
Fluorine is a gas, while all the other species are aqueous. This is an example of a heterogeneous reaction quotient. We will use concentrations (mol/L) for all the aqueous terms and pressure for the fluorine gas, instead of their activities.

$$Q = \frac{[F^-]^2[Sn^{4+}]}{p_{F_2}[Sn^{2+}]}$$

Check Be sure that the exponents in Q are the same as the balancing coefficients. A good check is to reverse the process and see if you obtain the balanced equation: change the numerator to products, the denominator to reactants, and the exponents to coefficients. Check the phase of each reactant and product to determine whether you will use partial pressure or concentration to approximate the activity.

Follow-Up Problem 15.1 Write the reaction quotient, Q, for each unbalanced reaction:

(a) The first step in nitric acid production, $NH_3(g) + O_2(g) \rightleftharpoons NO(g) + H_2O(g)$

(b) The disproportionation of nitric oxide, $NO(g) \rightleftharpoons N_2O(g) + NO_2(g)$

(See Brief Solutions to Follow-Up Problems at the end of the chapter.)

Why Q and K Are Unitless In this book (and most others), *the values of Q and K are unitless numbers*. The reason, as previously explained, is that each term in Q represents the activity, which is the *ratio* of the quantity of the substance (molar concentration or partial pressure) to its thermodynamic standard-state quantity. Recall, from Section 5.6, that the standard states are 1 mol/L for a substance in solution, 1 bar for a gas, and the pure substance for a liquid or a solid. Thus, a concentration of 1.20 mol/L becomes

$$\frac{1.20 \text{ mol/L (measured quantity)}}{1 \text{ mol/L (standard-state quantity)}} = 1.20$$

Similarly, a pressure of 0.53 bar becomes $\frac{0.53 \text{ bar}}{1 \text{ bar}} = 0.53$. Since the quantity terms are unitless, their ratio, which gives the value of Q (or K), is also unitless.

Form of Q for an Overall Reaction We follow the same procedure for writing a reaction quotient, regardless of whether the equation represents an individual reaction step or an overall multistep reaction. *If an overall reaction is the* ***sum*** *of two or more reactions, the overall reaction quotient (or equilibrium constant) is the* ***product*** *of the reaction quotients (or equilibrium constants) for the steps:*

$$Q_{overall} = Q_1 \times Q_2 \times Q_3 \times \cdots$$

and

$$K_{overall} = K_1 \times K_2 \times K_3 \times \cdots \qquad (15.7)$$

Sample Problem 15.2 demonstrates this procedure.

Sample Problem 15.2 Writing the Reaction Quotient and Finding K for an Overall Reaction

Problem Nitrogen dioxide is a toxic pollutant that contributes to photochemical smog at 25°C. The following sequence is one way that it forms:

(1) $N_2(g) + O_2(g) \rightleftharpoons 2NO(g)$ $\qquad K_1 = 4.3 \times 10^{-25}$

(2) $2NO(g) + O_2(g) \rightleftharpoons 2NO_2(g)$ $\qquad K_2 = 2.6 \times 10^{8}$

(a) Show that the overall Q for this reaction sequence is the same as the product of the Q's for the individual reactions.

(b) Given that both reactions occur at the same temperature, find K for the overall reaction.

Plan **(a)** We first write the overall reaction by adding the individual reactions, and then we write the overall Q. Next, we write the Q for each step. We *add* the steps and *multiply* their Q's, cancelling common terms, to obtain the overall Q. **(b)** We know the individual K's (4.3×10^{-25} and 2.6×10^{8}), so we multiply them to find $K_{overall}$.

Solution **(a)** Write the overall reaction and its reaction quotient:

$$\begin{array}{ll} (1) & N_2(g) + O_2(g) \rightleftharpoons \cancel{2NO(g)} \\ (2) & \cancel{2NO(g)} + O_2(g) \rightleftharpoons 2NO_2(g) \\ \hline \text{Overall:} & N_2(g) + 2O_2(g) \rightleftharpoons 2NO_2(g) \end{array}$$

$$Q = \frac{p^2_{NO_2}}{p_{N_2}p^2_{O_2}}$$

Write the reaction quotients for the individual steps:

For step 1, $$Q_1 = \frac{p^2_{NO}}{p_{N_2}p_{O_2}}$$

For step 2, $$Q_2 = \frac{p^2_{NO_2}}{p^2_{NO}p_{O_2}}$$

Multiply the individual reaction quotients and cancel:

$$(Q_1)(Q_2) = Q = \left(\frac{\cancel{p^2_{NO}}}{p_{N_2}p_{O_2}}\right)\left(\frac{p^2_{NO_2}}{\cancel{p^2_{NO}}p_{O_2}}\right) = \frac{p^2_{NO_2}}{p_{N_2}p^2_{O_2}} = Q_{overall}$$

(b) Calculate the overall K:

$$K_{overall} = K_1 \times K_2 = (4.3 \times 10^{-25})(2.6 \times 10^{8}) = 1.2 \times 10^{-16}$$

Check Round off and check the calculation in part (b):

$$K \approx (4 \times 10^{-25})(3 \times 10^{8}) = 1 \times 10^{-16}$$

Follow-Up Problem 15.2 The following sequence of steps has been proposed for the overall reaction between H_2 and Br_2 to form HBr:

(1) $Br_2(g) \rightleftharpoons 2Br(g)$
(2) $Br(g) + H_2(g) \rightleftharpoons HBr(g) + H(g)$
(3) $H(g) + Br(g) \rightleftharpoons HBr(g)$

Write the overall balanced equation, and show that the overall Q is the product of the Q's for the individual steps.

Form of Q for a Forward Reaction and a Reverse Reaction The form of the reaction quotient depends on the *direction* in which the balanced equation is written. Consider, for example, the oxidation of sulfur dioxide to sulfur trioxide. This reaction is a key step in acid rain formation and sulfuric acid production. The balanced equation is

$$2SO_2(g) + O_2(g) \rightleftharpoons 2SO_3(g)$$

The reaction quotient for this equation, *as written*, is

$$Q_{fwd} = \frac{p^2_{SO_3}}{p^2_{SO_2}p_{O_2}}$$

If we had written the reverse reaction, the decomposition of sulfur trioxide,

$$2SO_3(g) \rightleftharpoons 2SO_2(g) + O_2(g)$$

the reaction quotient would have been the *reciprocal* of Q_{fwd}:

$$Q_{rev} = \frac{p^2_{SO_2}p_{O_2}}{p^2_{SO_3}} = \frac{1}{Q_{fwd}}$$

Thus, *a reaction quotient (or equilibrium constant) for a forward reaction is the* ***reciprocal*** *of the reaction quotient (or equilibrium constant) for the reverse reaction*:

$$Q_{\text{fwd}} = \frac{1}{Q_{\text{rev}}} \quad \text{and} \quad K_{\text{fwd}} = \frac{1}{K_{\text{rev}}} \tag{15.8}$$

The K values for the forward and reverse reactions shown above, at 1000 K, are

$$K_{\text{fwd}} = 3.18 \quad \text{and} \quad K_{\text{rev}} = \frac{1}{K_{\text{fwd}}} = \frac{1}{3.18} = 0.314$$

These values make sense: if the forward reaction goes to the right (shifts toward the products; high K), the reverse reaction does not (lower K).

Form of Q for a Reaction with Coefficients Multiplied by a Common Factor

Multiplying all the coefficients of the equation by a factor also changes the form of Q. For example, multiplying all the coefficients in the equation for the formation of SO_3 by $\frac{1}{2}$ gives

$$SO_2(g) + \frac{1}{2}O_2(g) \rightleftharpoons SO_3(g)$$

For this equation, the reaction quotient is

$$Q'_{\text{fwd}} = \frac{p_{SO_3}}{p_{SO_2}p_{O_2}^{1/2}}$$

Notice that Q for the halved equation equals Q for the original equation raised to the power of $\frac{1}{2}$:

$$Q'_{\text{fwd}} = Q_{\text{fwd}}^{1/2} = \left(\frac{p_{SO_3}^2}{p_{SO_2}^2 p_{O_2}}\right)^{1/2} = \frac{p_{SO_3}}{p_{SO_2}p_{O_2}^{1/2}}$$

Once again, the same property holds for the equilibrium constants. Relating the halved reaction to the original reaction, we have

$$K'_{\text{fwd}} = K_{\text{fwd}}^{1/2} = (3.18)^{1/2} = 1.78$$

It may seem that we have changed the extent of the reaction, as indicated by a different K, just by changing the coefficients of the equation. However, this cannot be true. *A particular K has meaning only in relation to a particular equation.* If K_{fwd} and K'_{fwd} relate to different equations, we cannot compare them directly.

In general, *if all the coefficients of the balanced equation are multiplied by a factor, that factor becomes the exponent for relating the reaction quotients and the equilibrium constants.* For a multiplying factor n, which we can write as

$$n(a\text{A} + b\text{B} \rightleftharpoons c\text{C} + d\text{D})$$

the reaction quotient and the equilibrium constant are

$$Q' = Q^n = \left(\frac{a_C^c a_D^d}{a_A^a a_B^b}\right)^n \quad \text{and} \quad K' = K^n \tag{15.9}$$

Sample Problem 15.3 Finding the Equilibrium Constant for an Equation Multiplied by a Common Factor

Problem For the ammonia-formation reaction, the reference (ref) equation is

$$N_2(g) + 3H_2(g) \rightleftharpoons 2NH_3(g)$$

K is 3.6×10^{-7} at 1000 K. What are the values of K for the following balanced equations?

(a) $\frac{1}{3}N_2(g) + H_2(g) \rightleftharpoons \frac{2}{3}NH_3(g)$ **(b)** $NH_3(g) \rightleftharpoons \frac{1}{2}N_2(g) + \frac{3}{2}H_2(g)$

Plan We compare each equation with the reference equation to see how the direction and coefficients have changed.

Solution The reaction quotient for the reference equation is

$$Q_{ref} = \frac{p_{NH_3}^2}{p_{N_2}p_{H_2}^3}$$

(a) This equation is the reference equation multiplied by $\frac{1}{3}$, so K equals $K_{ref} = 3.6 \times 10^{-7}$ raised to the power of $\frac{1}{3}$.

$$Q = Q_{ref}^{1/3} = \left(\frac{p_{NH_3}^2}{p_{N_2}p_{H_2}^3}\right)^{1/3} = \frac{p_{NH_3}^{2/3}}{p_{N_2}^{1/3}p_{H_2}}$$

Thus,

$$K = K_{ref}^{1/3} = (3.6 \times 10^{-7})^{1/3} = 7.1 \times 10^{-3}$$

(b) This equation is one-half the *reverse* of the reference equation, so K is the reciprocal of K_{ref} raised to the power of $\frac{1}{2}$.

$$Q = \left(\frac{1}{Q_{ref}}\right)^{1/2} = \left(\frac{1}{\left[\frac{p_{NH_3}^2}{p_{N_2}p_{H_2}^3}\right]}\right)^{1/2} = \frac{p_{N_2}^{1/2}p_{H_2}^{3/2}}{p_{NH_3}}$$

Thus,

$$Q = \left(\frac{1}{K_{ref}}\right)^{1/2} = \left(\frac{1}{3.6 \times 10^{-7}}\right)^{1/2} = 1.7 \times 10^3$$

Check A good check is to work the math backwards. **(a)** $(7.1 \times 10^{-3})^3 = 3.6 \times 10^{-7}$, our original value of K. The reaction goes in the same direction, so there should be mostly reactants, as $K < 1$ indicates. **(b)** $\frac{1}{(1.7 \times 10^3)^2} = 3.5 \times 10^{-7}$, within rounding. At equilibrium, the reverse reaction should yield mostly products, as $K < 1$ indicates.

Follow-Up Problem 15.3 At 1200 K, the reaction of hydrogen and chlorine to form hydrogen chloride is

$$H_2(g) + Cl_2(g) \rightleftharpoons 2HCl(g) \qquad K = 7.6 \times 10^8$$

Calculate K for the following reactions:

(a) $\frac{1}{2}H_2(g) + \frac{1}{2}Cl_2(g) \rightleftharpoons HCl(g)$ **(b)** $\frac{4}{3}HCl(g) \rightleftharpoons \frac{2}{3}H_2(g) + \frac{2}{3}Cl_2(g)$

Form of Q for a Reaction Involving Pure Liquids or Solids Until now, we have looked at reactions involving only gases. These reactions are *homogeneous* equilibria, systems in which all the components are in the same phase. When the components are in different phases, the system reaches *heterogeneous* equilibrium. Consider the decomposition of limestone to lime and carbon dioxide:

$$CaCO_3(s) \rightleftharpoons CaO(s) + CO_2(g)$$

Based on the rules for writing the reaction quotient, we have

$$Q = \frac{a_{CaO}a_{CO_2}}{a_{CaCO_3}}$$

However, the activity of a pure solid is 1. We can explain this by thinking about density. As with density, a pure solid always has the same "concentration" at a given temperature, that is, the same amount (mol) per litre of solid. Therefore, the "concentration" of a pure solid is constant, as is the "concentration" of a pure liquid. Since the pure solid or pure liquid in the standard state is being compared to the pure substance in the standard state, the quotient must be 1.

Essentially, we are only concerned with quantities that *change* as they approach equilibrium; the fact that the activities of pure liquids and solids are 1 effectively eliminates *the terms for pure liquids and solids from the reaction quotient.* Our reaction quotient is

$$Q = p_{CO_2}$$

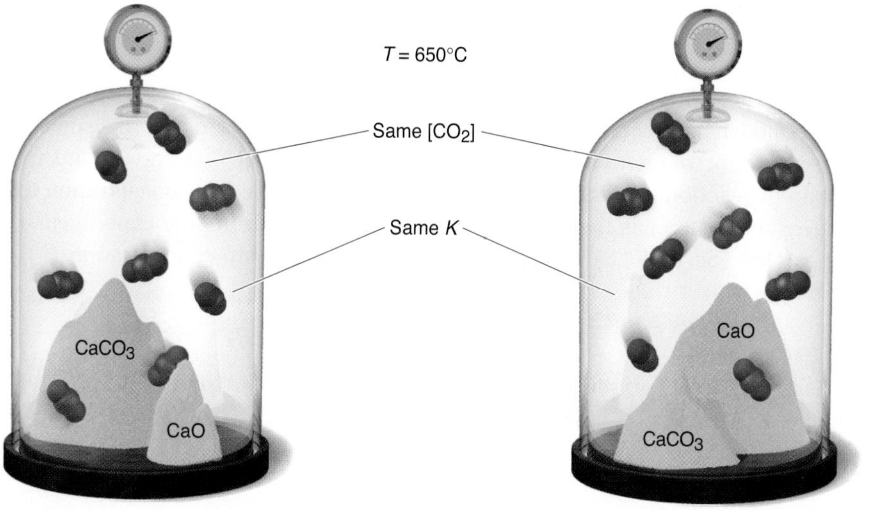

FIGURE 15.4 The reaction quotient for a heterogeneous system depends only on concentrations that change.

Thus, only a change in the CO_2 pressure has an effect on Q. It is *vital*, however, that at least some CaO and/or $CaCO_3$ be present for the reaction to proceed, but the reaction quotient itself depends only on the CO_2 pressure (Figure 15.4).

Table 15.2 summarizes the ways of writing Q and calculating K.

TABLE 15.2 Ways of Expressing Q and Calculating K

Form of Chemical Equations	Form of Q	Value of K
Reaction I: A $\rightleftharpoons$ B	$Q_{ref} = \frac{a_B}{a_A}$	$K_{ref} = \frac{a_{B,eq}}{a_{A,eq}}$
Reverse of Reaction I: B $\rightleftharpoons$ A	$Q = \frac{1}{Q_{ref}} = \frac{a_A}{a_B}$	$K = \frac{1}{K_{ref}}$
Reaction as a sum of two steps:		
(1) A $\rightleftharpoons$ C	$Q_1 = \frac{a_C}{a_A}; Q_2 = \frac{a_B}{a_C}$	
(2) C $\rightleftharpoons$ B	$Q_{overall} = Q_1Q_2 = Q_{ref} = \frac{a_C}{a_A} \times \frac{a_B}{a_C} = \frac{a_B}{a_A}$	$K_{overall} = K_1 \times K_2 = K_{ref}$
Coefficients multiplied by n	$Q = Q_{ref}^n$	$K = K_{ref}^n$
Reaction with pure solid or liquid component, such as A(s)	$Q = a_B$	$K = a_B$

SUMMARY OF SECTION 15.2

- The reaction quotient, Q, is a particular ratio of product to reactant activities. The value of Q changes as the reaction proceeds. When the system reaches equilibrium at a particular temperature, $Q = K$.
- For a reaction involving gaseous terms, the activities may be replaced by the pressures of the gaseous terms in bar. For a reaction involving aqueous terms, the activities may be replaced by the concentrations (mol/L) of the aqueous terms.
- The activities of pure liquids and solids do not appear in Q because they are equal to 1.
- If a reaction is the sum of two or more steps, the overall Q (or K) is the product of the individual Q's (or K's).
- The *form* of Q is based on the balanced equation exactly as written, so it changes if the equation is reversed or multiplied by a factor, and K changes accordingly.
- Three criteria define a system at equilibrium:
  1. Reactant and product concentrations are constant over time.
  2. The opposing reaction rates are equal: $rate_{fwd} = rate_{rev.}$
  3. The reaction quotient equals the equilibrium constant: $Q = K$.

15.3 Expressing Equilibria with Pressure Terms: Relation between K and K_c

The thermodynamic equilibrium constant K requires that, for gases, we substitute the pressure for the activity (if not using the activity itself). However, under the majority of reaction conditions, it is more convenient to use gas concentration than gas pressure. As long as the gases behave nearly ideally during the experiment, the ideal gas law (Section 4.3) allows us to relate pressure (p) to concentration ($\frac{n}{V}$):

$$pV = nRT, \quad \text{so} \quad p = \frac{n}{V}RT \quad \text{or} \quad \frac{p}{RT} = \frac{n}{V}$$

Thus, at constant T, *pressure is directly proportional to molar concentration.* For example, in the reaction between gaseous NO and O_2,

$$2NO(g) + O_2(g) \rightleftharpoons 2NO_2(g)$$

the *reaction quotient*, Q, is

$$Q = \frac{p^2_{NO_2}}{p^2_{NO}p_{O_2}}$$

If the partial pressures represent equilibrium pressures, then this expression becomes the expression for the equilibrium constant, K. If, however, we use concentrations instead of the partial pressures, we can form an expression for the equilibrium constant with all the terms expressed as concentrations, in mol/L. However, our new expression does *not* represent the thermodynamic equilibrium constant and, for the sake of clarity, may be referred to as K_c. In many cases, K has a value different from K_c. But if we know one value, the *change in amount (mol) of gas*, Δn_{gas}, from the balanced equation allows us to calculate the other. Let us see how this conversion works for the oxidation of NO:

$$2NO(g) + O_2(g) \rightleftharpoons 2NO_2(g)$$

As the balanced equation shows, there is a total of 3 mol of gas on the reactant side (2 mol of NO and 1 mol of O_2) and 2 mol of gas on the product side (2 mol of NO_2):

$$3 \text{ mol gaseous reactants} \rightleftharpoons 2 \text{ mol gaseous products}$$

With Δ meaning final *minus* initial (products *minus* reactants), we have

$$\Delta n_{gas} = \text{moles of gaseous product} - \text{moles of gaseous reactant} = 2 - 3 = -1$$

Keep this value of Δn_{gas} in mind, because it appears in the conversion that follows.

The reaction quotient based on concentrations is

$$Q_c = \frac{[NO_2]^2}{[NO]^2[O_2]}$$

Rearranging the ideal gas law to $\frac{n}{V} = \frac{p}{RT}$, we write the terms in square brackets as $\frac{n}{V}$ and convert them to partial pressures, p; then we collect the RT terms and cancel:

$$Q_c = \frac{\dfrac{n^2_{NO_2}}{V^2}}{\dfrac{n^2_{NO}}{V^2} \times \dfrac{n_{O_2}}{V}} = \frac{\dfrac{p^2_{NO_2}}{(RT)^2}}{\dfrac{p^2_{NO}}{(RT)^2} \times \dfrac{p_{O_2}}{RT}} = \frac{p^2_{NO_2}}{p^2_{NO} \times p_{O_2}} \times \frac{\dfrac{1}{\cancel{(RT)^2}}}{\dfrac{1}{\cancel{(RT)^2}} \times \dfrac{1}{RT}} = \frac{p^2_{NO_2}}{p^2_{NO} \times p_{O_2}} \times RT$$

Notice that the far right side of the previous expression is Q multiplied by RT. Thus,

$$Q_c = Q(RT)$$

Also, at equilibrium, $K_c = K(RT)$. Thus,

$$K = \frac{K_c}{RT} \quad \text{or} \quad K = K_c(RT)^{-1}$$

Notice, especially, that *the exponent of the RT term equals the change in the amount (mol) of gas* (Δn_{gas}) *from the balanced equation*, −1. Thus, in general, we have

$$K = K_c(RT)^{\Delta n_{gas}} \qquad (15.10)$$

Based on Equation 15.10, *if the amount (mol) of gas does not change in the reaction*, $\Delta n_{gas} = 0$, so the RT term drops out (that is, equals 1) and $K = K_c$. (In calculations, be sure that the units for R are consistent with the units for pressure. This means that the value of R used must always be 0.083 14 $bar{\cdot}L{\cdot}mol^{-1}{\cdot}K^{-1}$.)

Sample Problem 15.4 Converting between K and K_c

Problem A chemical engineer injects limestone ($CaCO_3$) into the hot flue gas of a coal-burning power plant to form lime (CaO), which scrubs SO_2 from the gas and forms gypsum ($CaSO_4{\cdot}2H_2O$). Find K_c for the following reaction:

$$CaCO_3(s) \rightleftharpoons CaO(s) + CO_2(g) \qquad K = 2.1 \times 10^{-4} \text{ (at 1000. K)}$$

Plan We know K (2.1×10^{-4}), so, to convert between K and K_c, we must first determine Δn_{gas} from the balanced equation. Then we rearrange Equation 15.10. (If any pressure unit other than bar is used, we need to convert the pressure to bar so that we can use $R = 0.083\ 14\ bar{\cdot}L{\cdot}mol^{-1}{\cdot}K^{-1}$.)

Solution We need to determine Δn_{gas}. There is 1 mol of gaseous product and no gaseous reactant, so $\Delta n_{gas} = 1 - 0 = 1$.

Rearrange Equation 15.10 and calculate K_c:

$$K = K_c(RT)^1$$

So,

$$K_c = K(RT)^{-1} = (2.1 \times 10^{-4})\left(0.083\ 14\ \frac{bar{\cdot}L}{mol{\cdot}K} \times 1000.\ K\right)^{-1}$$
$$= 2.5 \times 10^{-6}$$

Check Rounding gives

$$(2 \times 10^{-4})(0.1 \times 10^3)^{-1} = (2 \times 10^{-4})(10^{-2})$$
$$= 2 \times 10^{-6}$$

Comment Because we are making a connection between the thermodynamic equilibrium constant K and another equilibrium constant, K_c, we see that the units do not cancel entirely when calculating K_c. For the sake of convenience, we take the K_c value to be unitless as well. For this reason, however, it is always preferable to use the thermodynamic equilibrium constant.

Follow-Up Problem 15.4 Calculate K for the following reaction:

$$PCl_3(g) + Cl_2(g) \rightleftharpoons PCl_5(g) \qquad K_c = 1.67 \text{ (at 500. K)}$$

SUMMARY OF SECTION 15.3

- For a gaseous reaction, the reaction quotient and the equilibrium constant are normally expressed in terms of partial pressures (Q and K), but they can occasionally be expressed in terms of concentration, with units of mol/L (Q_c and K_c).
- If you know K, you can find K_c, and vice versa: $K = K_c(RT)^{\Delta n_{gas}}$, where $R = 0.083\ 14\ bar{\cdot}L{\cdot}mol^{-1}{\cdot}K^{-1}$.

15.4 Comparing Q and K to Determine Reaction Direction

Suppose that you have a mixture of reactants and products and you know K at the temperature of the reaction. By comparing the values of Q and K, you can tell whether the reaction has reached equilibrium and, if not, in which direction it is progressing. With product terms in the numerator of Q and reactant terms in the denominator, *more product makes Q larger, and more reactant makes Q smaller.* There are three possibilities for the relative sizes of Q and K (Figure 15.5):

- $Q < K$. If Q is smaller than K, the denominator (reactants) is large relative to the numerator (products). For Q to equal K, the reactants must decrease and the products must increase. The reaction will progress to the right, toward the products:

$$\text{If } Q < K, \text{reactants} \longrightarrow \text{products}$$

- $Q < K$. If Q is larger than K, the numerator (products) will decrease and the denominator (reactants) will increase. The reaction will progress to the left, toward the reactants:

$$\text{If } Q > K, \text{reactants} \longleftarrow \text{products}$$

- $Q = K$. This situation occurs when the reactant and product terms equal their equilibrium values. No further net change takes place:

$$\text{If } Q = K, \text{reactants} \rightleftharpoons \text{products}$$

Sample Problem 15.5 relies on molecular scenes to determine reaction direction, and Sample Problem 15.6 relies on concentration data.

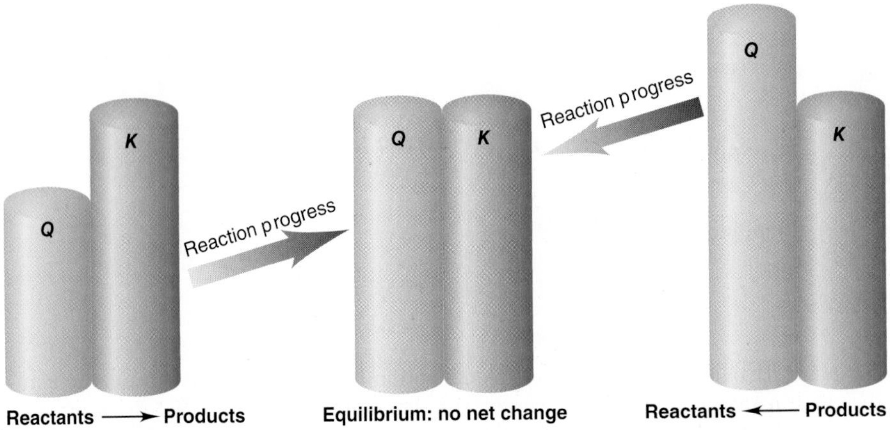

FIGURE 15.5 Reaction direction and the relative sizes of Q and K. When Q is smaller or larger than K, the reaction continues until $Q = K$. Note that K remains the same throughout.

Sample Problem 15.5 Using Molecular Scenes to Determine Reaction Direction

Problem For the reaction $A(g) \rightleftharpoons B(g)$, the equilibrium mixture at 175°C is $[A] = 2.8 \times 10^{-4}$ mol/L and $[B] = 1.2 \times 10^{-4}$ mol/L. The molecular scenes below represent mixtures at various times during runs 1 to 4 of the reaction (A is *red* and B is *blue*). Does the reaction progress to the right or left, or not at all, for each mixture to reach equilibrium?

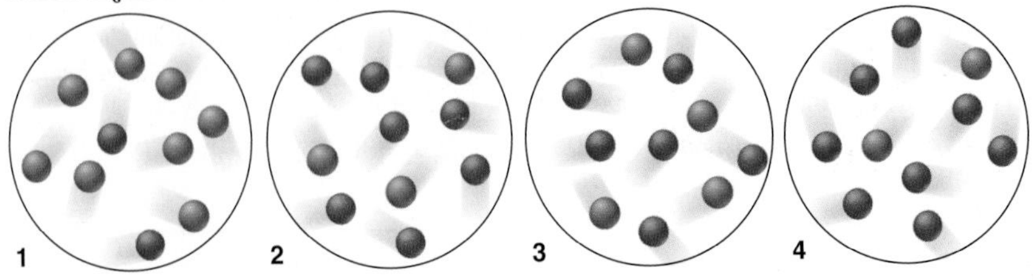

Plan We must compare Q_c with K_c to determine the reaction direction, so we first use the given equilibrium concentrations to find K_c. Then we count spheres and calculate Q_c for each mixture. If $Q_c < K_c$, the reaction progresses to the right (reactants to products); if $Q_c < K_c$, the reaction progresses to the left (products to reactants); and if $Q_c = K_c$, there is no further net change.

Solution Write the reaction quotient and use the data to find K_c:

$$Q_c = \frac{[B]}{[A]} = \frac{1.2 \times 10^{-4}}{2.8 \times 10^{-4}} = 0.43 = K_c$$

Count *red* (A) and *blue* (B) spheres to calculate Q_c for each mixture:

1. $Q_c = \frac{8}{2} = 4.0$ 2. $Q_c = \frac{3}{7} = 0.43$ 3. $Q_c = \frac{4}{6} = 0.67$ 4. $Q_c = \frac{2}{8} = 0.25$

Compare Q_c with K_c to determine the direction of the reaction:

1. $Q_c < K_c$: left 2. $Q_c = K_c$: no net change 3. $Q_c < K_c$: left 4. $Q_c < K_c$: right

Check Making an error in the calculation for K_c would lead to incorrect conclusions throughout, so check this step: the exponents are the same, and $\frac{1.2}{2.8}$ is a bit less than 0.5, as is the calculated K_c. You can check the final answers by inspection; for example, for the number of B (8) in mixture 1 to equal the number at equilibrium (3), more B must change to A, so the reaction must proceed to the left.

Follow-Up Problem 15.5 At 338 K, the reaction $X(g) \rightleftharpoons Y(g)$ has a K_c of 1.4. The scenes below represent different mixtures at 338 K (X is *orange* and Y is *green*). In which direction does the reaction proceed (if at all) for each mixture to reach equilibrium?

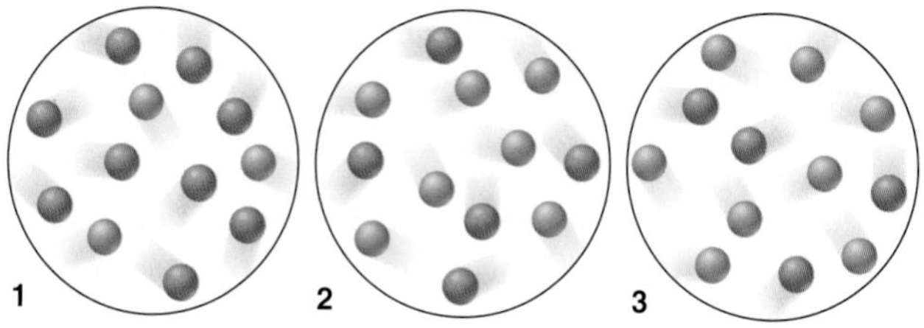

Sample Problem 15.6 Using Concentrations to Determine Reaction Direction

Problem For the following reaction, $K_c = 0.21$ at 100°C:

$$N_2O_4(g) \rightleftharpoons 2NO_2(g)$$

At a point during the reaction, $[N_2O_4] = 0.12$ mol/L and $[NO_2] = 0.55$ mol/L. Is the reaction at equilibrium? If not, in which direction is it progressing?

Plan We write Q_c, find its value by substituting the given concentrations, and compare its value with the given K_c.

Solution Write the reaction quotient and solve for Q_c:

$$Q_c = \frac{[NO_2]^2}{[N_2O_4]} = \frac{0.55^2}{0.12} = 2.5$$

With $Q_c < K_c$, the reaction is not at equilibrium and will proceed to the left until $Q_c = K_c$.

Check With $[NO_2] < [N_2O_4]$, we expect to obtain a value of Q_c that is greater than 0.21. If $Q_c < K_c$, the numerator will decrease and the denominator will increase until $Q_c = K_c$; that is, the reaction will proceed toward reactants.

Follow-Up Problem 15.6 Chloromethane forms by the following reaction:

$$CH_4(g) + Cl_2(g) \rightleftharpoons CH_3Cl(g) + HCl(g)$$

At 1500 K, $K = 1.6 \times 10^4$. In the reaction mixture, $p_{CH_4} = 0.13$ bar, $p_{Cl_2} = 0.035$ bar, $p_{CH_3Cl} = 0.24$ bar, and $p_{HCl} = 0.47$ bar. Is CH_3Cl or CH_4 forming?

SUMMARY OF SECTION 15.4

- We compare the values of Q and K to determine the direction in which a reaction will proceed toward equilibrium:
  - If $Q < K$, more product forms.
  - If $Q > K$, more reactant forms.
  - If $Q = K$, there is no net change.

15.5 How to Solve Equilibrium Problems

Many kinds of equilibrium problems arise in the real world—and on chemistry exams—but we can group most of them into two types:

1. We know the equilibrium quantities, and we solve for K.
2. We know K and the initial quantities, and we solve for the equilibrium quantities.

Using Quantities to Find the Equilibrium Constant

There are two common variations on this type of problem: one involves substituting quantities to solve for K, and the other requires finding some of the quantities first.

Substituting Equilibrium Quantities into Q to Find K In this type of problem, we use given equilibrium quantities to calculate K.

Suppose, for example, that equal amounts of gaseous hydrogen and iodine are injected into a 1.50 L flask at 375 K. In time, the following equilibrium is attained:

$$H_2(g) + I_2(g) \rightleftharpoons 2HI(g)$$

At equilibrium, the flask contains 1.80 mol of H_2, 1.80 mol of I_2, and 0.520 mol of HI. Since all the species in the chemical equation are gases, we calculate K by finding the partial pressures from the amounts, temperature, and flask volume and then substituting these pressures into the reaction quotient from the balanced equation:

$$Q = \frac{p_{HI}^2}{p_{H_2}p_{I_2}}$$

We multiply each amount (mol) by R (equal to 0.083 14 bar•L•mol^{-1}•K^{-1}) and the temperature (K) and then divide by the volume (L) to find each pressure (bar):

$$p_{H_2} = \frac{(1.80\ \text{mol})(0.083\ 14\ \text{bar·L·mol}^{-1}\text{·K}^{-1})(375\ \text{K})}{1.50\ \text{L}} = 37.41\ \text{bar}$$

Similarly, $p_{I_2} = 37.41$ bar and $p_{HI} = 10.81$ bar (carrying an extra decimal place). Substituting these values into the expression for Q gives K:

$$K = \frac{(10.81)^2}{(37.41)(37.41)} = 0.0835$$

Using a Reaction or ICE Table to Find Equilibrium Quantities and K When some of the quantities are not given, we determine them from the reaction stoichiometry, using a *reaction table or* ***I****nitial* ***C****hange* ***E****quilibrium (ICE) table*, and then find K.

For example, in a study of carbon oxidation, an evacuated vessel containing a small amount of powdered graphite is heated to 1080 K. Gaseous CO_2 is added to a pressure of 0.458 bar, and CO forms. At equilibrium, the total pressure is 0.757 bar. Suppose that we need to calculate K.

As always, we start by writing the balanced equation and the reaction quotient:

$$CO_2(g) + C(\text{graphite}) \rightleftharpoons 2CO(g)$$

The data are given in bar and we must find K, so we write the expression for Q. Notice that the expression does *not* include a term for the solid, C(graphite):

$$Q = \frac{p_{CO}^2}{p_{CO_2}}$$

We are given the initial p_{CO_2} and the p_{total} at equilibrium. To find K, we must find the equilibrium pressures of CO_2 and CO and then substitute them into Q.

Let us think through what happened in the vessel. An unknown portion of the CO_2 reacted with graphite to form an unknown amount of CO. We already know the *relative* amounts of CO_2 and CO from the balanced equation: for each mole of CO_2 that reacts, 2 mol of CO forms, which means that when x bar of CO_2 reacts, $2x$ bar of CO forms:

$$x \text{ bar } CO_2 \longrightarrow 2x \text{ bar CO}$$

The pressure of CO_2 at equilibrium, $p_{CO_2(eq)}$, is the initial pressure, $p_{CO_2(init)}$, *minus* x, the CO_2 that reacts (the change in p_{CO_2} due to the reaction):

$$p_{CO_2(init)} - x = p_{CO_2(eq)} = 0.458 - x$$

Similarly, the pressure of CO at equilibrium, $p_{CO(eq)}$, is the initial pressure, $p_{CO(init)}$, *plus* $2x$, the CO that forms (the change in p_{CO} due to the reaction). $p_{CO(init)}$ is zero, so

$$p_{CO(init)} + 2x = 0 + 2x = 2x = p_{CO(eq)}$$

We summarize this information in an ICE table, similar to those in Chapter 3, but the "Final" quantities are the quantities at equilibrium. The ICE table shows the balanced equation and the following information:

- The *initial* quantities (concentrations or pressures) of reactants and products
- The *changes* in these quantities during the reaction
- The *equilibrium* quantities

Pressure (bar)	**$CO_2(g)$**	+	**C(graphite)**	$\rightleftharpoons$	**$2CO(g)$**
Initial	0.458		—		0
Change	$-x$		—		$+2x$
Equilibrium	$0.458 - x$		—		$2x$

We treat each column like a list of numbers to add together: *the initial quantity plus the change in this quantity gives the equilibrium quantity.* Notice that we *only* include substances whose concentrations change; thus, the column for C(graphite) is blank.

To solve for K, we need to substitute the equilibrium values into Q, so we first have to find x. To do this, we use the other piece of data given, $p_{total(eq)}$. According to Dalton's law of partial pressures and using the equilibrium quantities from the ICE table,

$$p_{total(eq)} = 0.757 \text{ bar} = p_{CO_2(eq)} + p_{CO(eq)} = (0.458 \text{ bar} - x) + 2x$$

Thus,

$$0.757 \text{ bar} = 0.458 \text{ bar} + x \text{ and } x = 0.299 \text{ bar}$$

With x known, we determine the equilibrium partial pressures:

$$p_{CO_2(eq)} = 0.458 \text{ bar} - x = 0.458 \text{ bar} - 0.299 \text{ bar} = 0.159 \text{ bar}$$

$$p_{CO(eq)} = 2x = 2(0.299 \text{ bar}) = 0.598 \text{ bar}$$

Then we substitute these pressures into Q to find K:

$$Q = \frac{p_{CO(eq)}^2}{p_{CO_2(eq)}} = \frac{0.598^2}{0.159} = 2.25 = K$$

(From here on, the subscripts "init" and "eq" appear only when it is not clear whether a quantity is an initial value or an equilibrium value.)

Sample Problem 15.7 Calculating K_c from Concentration Data

Problem To study hydrogen halide decomposition, a researcher fills an evacuated 2.00 L flask with 0.200 mol of HI gas and allows the reaction to proceed at 453°C:

$$2HI(g) \rightleftharpoons H_2(g) + I_2(g)$$

At equilibrium, [HI] = 0.078 mol/L. Calculate K_c.

Plan To calculate K_c, we need the equilibrium concentrations. We can find the initial [HI] from the amount (0.200 mol) and the flask volume (2.00 L), and we are given [HI] at equilibrium (0.078 mol/L). From the balanced equation, when $2x$ mol of HI reacts, x mol of H_2 and x mol of I_2 form. We set up an ICE table, use the known [HI] at equilibrium to solve for x (the change in $[H_2]$ or $[I_2]$), and substitute the concentrations into K_c.

Solution Calculate initial [HI]:

$$[HI] = \frac{0.200 \text{ mol}}{2.00 \text{ L}} = 0.100 \text{ mol/L}$$

Set up the ICE table, with $x = [H_2]$ or $[I_2]$ that forms and $2x = [HI]$ that reacts:

Concentration (mol/L)	2HI(g)	$\rightleftharpoons$	$H_2(g)$	+	$I_2(g)$
Initial	0.100		0		0
Change	$-2x$		$+x$		$+x$
Equilibrium	$0.100 - 2x$		x		x

Solve for x, using the known [HI] at equilibrium:

$$[HI] = 0.100 \text{ mol/L} - 2x = 0.078 \text{ mol/L} \quad \text{so} \quad x = 0.011 \text{ mol/L}$$

Therefore, the equilibrium concentrations are

$$[H_2] = [I_2] = 0.011 \text{ mol/L and we are given } [HI] = 0.078 \text{ mol/L}$$

Substitute into the expression for K_c:

$$K_c = \frac{(0.011)(0.011)}{0.078^2} = 0.020$$

Check Rounding gives $\sim\frac{0.01^2}{0.08^2} = 0.02$. Because the initial [HI] of 0.100 mol/L fell slightly at equilibrium to 0.078 mol/L, relatively little product formed, so we expect $K_c < 1$.

Follow-Up Problem 15.7 The atmospheric oxidation of nitrogen monoxide was studied at 184°C, with initial pressures of both NO and O_2 equal to 1.000 bar:

$$2NO(g) + O_2(g) \rightleftharpoons 2NO_2(g)$$

At equilibrium, $p_{O_2} = 0.506$ bar. Calculate K.

Using the Equilibrium Constant to Find Quantities

The type of problem that involves finding equilibrium quantities also has several variations. Sample Problem 15.8 is one variation, in which we know K and some of the equilibrium concentrations and must find another equilibrium concentration.

Sample Problem 15.8 Determining Equilibrium Concentrations from K_c

Problem In a study of the conversion of methane to other fuels, a chemical engineer mixes gaseous CH_4 and H_2O in a 0.32 L flask at 1200 K. At equilibrium, the flask contains 0.26 mol of CO, 0.091 mol of H_2, and 0.041 mol of CH_4. What is $[H_2O]$ at equilibrium? $K_c = 0.26$ for this process at 1200 K.

Plan First, we write the balanced equation and the reaction quotient. We calculate the equilibrium concentrations from the given numbers of moles and the volume of the flask (0.32 L). Substituting these values into the equation for Q_c and setting the equation equal to the given K_c (0.26), we solve for the unknown equilibrium concentration, $[H_2O]$.

Solution Write the balanced equation and the reaction quotient:

$$CH_4(g) + H_2O(g) \rightleftharpoons CO(g) + 3H_2(g) \qquad Q_c = \frac{[CO][H_2]^3}{[CH_4][H_2O]}$$

Determine the equilibrium concentrations:

$$[CH_4] = \frac{0.041\ \text{mol}}{0.32\ \text{L}} = 0.13\ \text{mol/L}$$

Similarly, $[CO] = 0.81$ mol/L and $[H_2] = 0.28$ mol/L.

We need to calculate $[H_2O]$ at equilibrium. Since $Q_c = K_c$, rearranging gives

$$[H_2O] = \frac{[CO][H_2]^3}{[CH_4]K_c} = \frac{(0.81)(0.28)^3}{(0.13)(0.26)} = 0.53\ \text{mol/L}$$

Check Always check by substituting the concentrations into Q_c to confirm that the result is equal to K_c:

$$Q_c = \frac{[CO][H_2]^3}{[CH_4][H_2O]} = \frac{(0.81)(0.28)^3}{(0.13)(0.53)} = 0.26 = K_c$$

Follow-Up Problem 15.8 At 298 K, nitrogen monoxide decomposes according to the following equation:

$$2NO(g) \rightleftharpoons N_2(g) + O_2(g) \qquad K_c = 2.3 \times 10^{30}$$

In the atmosphere, $p_{O_2} = 0.209$ bar and $p_{N_2} = 0.781$ bar. What is the equilibrium partial pressure of NO in the air we breathe? (*Hint*: You need K to find the partial pressure.)

In a somewhat more involved variation, we know K and the *initial* quantities, and we must find the *equilibrium* quantities using an ICE table. Sample Problem 15.9 focuses on this variation. But first a math review.

Using the Quadratic Formula to Find the Unknown Suppose, for example, that we start with 2.00 mol/L CO and 1.00 mol/L H_2O. An ICE table can be completed as follows:

Concentration (mol/L)	$CO(g)$	+	$H_2O(g)$	$\rightleftharpoons$	$CO_2(g)$	+	$H_2(g)$
Initial	2.00		1.00		0		0
Change	$-x$		$-x$		$+x$		$+x$
Equilibrium	$2.00 - x$		$1.00 - x$		x		x

Substituting these values into Q_c, we obtain

$$Q_c = \frac{[CO_2][H_2]}{[CO][H_2O]} = \frac{(x)(x)}{(2.00 - x)(1.00 - x)} = \frac{x^2}{x^2 - 3.00x + 2.00}$$

At equilibrium, we have

$$1.56 = \frac{x^2}{x^2 - 3.00x + 2.00}$$

Cross multiplying, we obtain

$$1.56x^2 - 4.68x + 3.12 = x^2$$

Bringing all the variable terms to the left, the equation becomes

$$1.56x^2 - 4.68x + 3.12 - x^2 = 0$$

By combining like terms and simplifying, we get

$$0.56x^2 - 4.68x + 3.12 = 0$$

which has the form of a *quadratic equation*:

$$\begin{aligned} a\,x^2 &+ \quad b\,x \quad + \quad c = 0 \\ 0.56x^2 &+ (-4.68)x + 3.12 = 0 \end{aligned}$$

where $a = 0.56$, $b = -4.68$, and $c = 3.12$. Then we find x with the quadratic formula (Appendix A):

$$x = \frac{-b \pm \sqrt{b^2 - 4ac}}{2a}$$

The ± sign means that we have two possible values for x:

$$x = \frac{4.68 \pm \sqrt{(-4.68)^2 - 4(0.56)(3.12)}}{2(0.56)}$$

$$x = 7.6 \text{ mol/L} \qquad \text{and} \qquad x = 0.73 \text{ mol/L}$$

Only one of these values makes sense chemically. The positive root gives the larger value and a negative concentration at equilibrium (for example, 2.00 mol/L − 7.6 mol/L = −5.6 mol/L), which has no meaning. Therefore, x = 0.73 mol/L, and we have

$$[CO] = 2.00 \text{ mol/L} - x = 2.00 \text{ mol/L} - 0.73 \text{ mol/L} = 1.27 \text{ mol/L}$$

$$[H_2O] = 1.00 \text{ mol/L} - x = 0.27 \text{ mol/L}$$

$$[CO_2] = [H_2] = x = 0.73 \text{ mol/L}$$

Checking to see if these values give the known K_c, we get

$$K_c = \frac{(0.73)(0.73)}{(1.27)(0.27)} = 1.6 \text{ (within rounding of 1.56)}$$

Sample Problem 15.9 Determining Equilibrium Concentrations from Initial Concentrations and K_c

Problem Fuel engineers use the extent of the change from CO and H_2O to CO_2 and H_2 to regulate the proportions of synthetic fuel mixtures. If 0.250 mol of CO gas and 0.150 mol of H_2O gas are placed in a 125 mL flask at 900 K, what is the composition of the equilibrium mixture? At this temperature, K_c is 1.56.

Plan We have to find the "composition" of the equilibrium mixture, in other words, the equilibrium quantities. As always, we write the balanced equation and use it to write the reaction quotient. Note that, in this particular question, $\Delta n = 0$ and so $K = K_c$. We find the initial [CO] and [H_2O] from the amounts (0.250 mol and 0.150 mol) and volume (0.125 L); use the balanced equation to define x and set up an ICE table; substitute the values into Q_c; and solve for x, from which we calculate the concentrations.

Solution Write the balanced equation and the reaction quotient:

$$CO(g) + H_2O(g) \rightleftharpoons CO_2(g) + H_2(g) \qquad Q_c = \frac{[CO_2][H_2]}{[CO][H_2O]}$$

Calculate the initial reactant concentrations:

$$[CO] = \frac{0.250 \text{ mol}}{0.125 \text{ L}} = 2.00 \text{ mol/L}$$

$$[H_2O] = \frac{0.150 \text{ mol}}{0.125 \text{ L}} = 1.20 \text{ mol/L}$$

Set up an ICE table, with x = [CO] and [H_2O] that react:

Concentration (mol/L)	CO(*g*)	+	$H_2O(g)$	$\rightleftharpoons$	$CO_2(g)$	+	$H_2(g)$
Initial	2.00		1.20		0		0
Change	$-x$		$-x$		$+x$		$+x$
Equilibrium	$2.00 - x$		$1.20 - x$		x		x

Substitute into the reaction quotient and solve for x:

$$Q_c = \frac{[CO_2][H_2]}{[CO][H_2O]} = \frac{(x)(x)}{(2.00 - x)(1.20 - x)} = \frac{x^2}{(2.40 - 3.20x + x^2)}$$

At equilibrium, we have

$$Q_c = K_c = 1.56 = \frac{x^2}{(2.40 - 3.20x + x^2)}$$

Cross multiplying, we get

$$x^2 = 1.56(2.40 - 3.20x + x^2) = 3.744 - 4.992x + 1.56x^2$$

Collecting terms, we get

$$0.56x^2 - 4.992x + 3.744 = 0$$

We solve the quadratic to find the root:

$$x = \frac{-b \pm \sqrt{b^2 - 4ac}}{2a}$$

$$= \frac{4.992 \pm \sqrt{(-4.992)^2 - 4(0.56)(3.744)}}{2(0.56)}$$

$$= 0.827 \text{ mol/L or } 8.09 \text{ mol/L}$$

We can discard the second root, 8.09 mol/L, as it would mean we would have *negative* amounts of reactant, which is *physically impossible.*

Calculate the equilibrium concentrations:

$$[CO] = 2.00 \text{ mol/L} - x = 2.00 \text{ mol/L} - 0.827 \text{ mol/L} = 1.17 \text{ mol/L}$$
$$[H_2O] = 1.20 \text{ mol/L} - x = 1.20 \text{ mol/L} - 0.827 \text{ mol/L} = 0.37 \text{ mol/L}$$
$$[CO_2] = [H_2] = x = 0.827 \text{ mol/L}$$

Check Given the intermediate size of K_c (1.56), it makes sense that the changes in concentration are moderate. It is a good idea to check that the sign of x in the ICE table is correct: only reactants were initially present, so x has a negative sign for reactants and a positive sign for products. Also check that the equilibrium concentrations give the known K_c:

$$\frac{(0.827)(0.827)}{(1.17)(0.37)} = 1.58 \approx K_c$$

Follow-Up Problem 15.9 The decomposition of HI at a low temperature was studied by injecting 2.50 mol of HI into a 10.32 L vessel at 25°C:

$$2HI(g) \rightleftharpoons H_2(g) + I_2(g) \qquad K_c = 1.26 \times 10^{-3}$$

What is [H_2] at equilibrium for the reaction?

A Simplifying Assumption for Finding the Unknown In many problems, we can use chemical common sense to make an assumption that simplifies the math by avoiding the need to use the quadratic formula to find x. In general, *if a reaction has a relatively small K and a relatively large initial reactant concentration, the concentration change (x) is very small relative to the original concentration. Thus, the equilibrium concentration is often equal (to the number of significant figures given) to the original concentration.* This assumption does not mean that $x = 0$, because then there would be no reaction. It means that, if a reaction starts with a high $[\text{reactant}]_{\text{init}}$ and proceeds very little to reach equilibrium (small K), the reactant concentration at equilibrium, $[\text{reactant}]_{\text{eq}}$, will be nearly the same as $[\text{reactant}]_{\text{init}}$.

Here is an everyday analogy for this assumption. If you stand on a bathroom scale, you might see that your mass is 72 kg. If you take off your wristwatch, your mass is *still* 72 kg. Within the accuracy of the scale, the mass of the watch is so small compared with your body mass that it can be neglected:

$$\text{initial body mass} - \text{mass of watch} = \text{final body mass} \approx \text{initial body mass}$$

Thus, let us say that the initial concentration of A is 0.500 mol/L and, because of a small K_c, the concentration of A that reacts is 0.002 mol/L. We can then assume that

$$0.500 \text{ mol/L} - 0.002 \text{ mol/L} = 0.498 \text{ mol/L} \approx 0.500 \text{ mol/L}$$

That is,

$$[\text{A}]_{\text{init}} - [\text{A}]_{\text{reacting}} = [\text{A}]_{\text{eq}} \approx [\text{A}]_{\text{init}} \qquad (15.11)$$

To justify the assumption that x is negligible, we make sure that the error introduced is not significant. One common criterion for "significant" is the 5% rule: *if the assumption results in a change that is less than 5% of the initial concentration, the error is not significant, and the assumption is justified.* Let us see how making this assumption simplifies the math and whether it is justified if there are two different $[\text{reactant}]_{\text{init}}$ values.

Sample Problem 15.10 Making a Simplifying Assumption to Calculate Equilibrium Concentrations

Problem Phosgene is a potent chemical warfare agent that is now outlawed by international agreement. It decomposes by the following reaction:

$$COCl_2(g) \rightleftharpoons CO(g) + Cl_2(g) \qquad K_c = 8.3 \times 10^{-4} \text{ (at 360°C)}$$

Calculate [CO], $[Cl_2]$, and $[COCl_2]$ when each amount of phosgene decomposes and reaches equilibrium in a 10.0 L flask:

(a) 5.00 mol of $COCl_2$ **(b)** 0.100 mol of $COCl_2$

Plan We know, from the balanced equation, that when x mol of $COCl_2$ decomposes, x mol of CO and x mol of Cl_2 form. We use the volume (10.0 L) to convert the amount (5.00 mol or 0.100 mol) to the molar concentration, define x and set up an ICE table, and substitute the values into Q_c. Before using the quadratic formula, we assume that x is negligibly small. After solving for x, we check our assumption and find the equilibrium concentrations. If our assumption is not justified, we use the quadratic formula to find x.

Solution (a) Write the reaction quotient:

$$Q_c = \frac{[CO][Cl_2]}{[COCl_2]}$$

Calculate the initial reactant concentration, $[COCl_2]_{\text{init}}$, for 5.00 mol of $COCl_2$:

$$[COCl_2]_{\text{init}} = \frac{5.00 \text{ mol}}{10.0 \text{ L}} = 0.500 \text{ mol/L}$$

Set up an ICE table, with x equal to $[COCl_2]_{\text{reacting}}$:

Concentration (mol/L)	$COCl_2(g)$	$\rightleftharpoons$	$CO(g)$	+	$Cl_2(g)$
Initial	0.500		0		0
Change	$-x$		$+x$		$+x$
Equilibrium	$0.500 - x$		x		x

If we use the equilibrium values in Q_c with the given K_c, we obtain

$$Q_c = \frac{[CO][Cl_2]}{[COCl_2]} = \frac{x^2}{0.500 - x} = K_c = 8.3 \times 10^{-4}$$

Because K_c is small, the reaction does not proceed very far to the right, so let us assume that x ($[COCl_2]_{reacting}$) can be neglected. In other words, we assume that the equilibrium concentration is nearly the same as the initial concentration, 0.500 mol/L:

$$[COCl_2]_{init} - [COCl_2]_{reacting} \approx [COCl_2]_{eq}$$
$$0.500 \text{ mol/L} - x \approx 0.500 \text{ mol/L}$$

Using this assumption, we substitute and solve for x:

$$K_c = 8.3 \times 10^{-4} \approx \frac{x^2}{0.500}$$

$$x^2 \approx (8.3 \times 10^{-4})(0.500) \quad \text{so} \quad x \approx 2.0 \times 10^{-2}$$

Check the assumption by seeing if the error is <5%:

$$\frac{[\text{change}]}{[\text{initial}]} \times 100 = \frac{2.0 \times 10^{-2}}{0.500} \times 100 = 4\% \; (<5\%, \text{ so the assumption is justified})$$

Solve for the equilibrium concentrations:

$$[CO] = [Cl_2] = x = 2.0 \times 10^{-2} \text{ mol/L}$$
$$[COCl_2] = 0.500 \text{ mol/L} - x = 0.480 \text{ mol/L}$$

(b) The calculation for 0.100 mol of $COCl_2$ is the same as the calculation in part (a), except that $[COCl_2]_{init} = 0.100 \text{ mol}/10.0 \text{ L} = 0.0100 \text{ mol/L}$. Thus, at equilibrium,

$$Q_c = \frac{[CO][Cl_2]}{[COCl_2]} = \frac{x^2}{0.0100 - x} = K_c = 8.3 \times 10^{-4}$$

Assume that $0.0100 \text{ mol/L} - x \approx 0.0100 \text{ mol/L}$ and solve for x:

$$K_c = 8.3 \times 10^{-4} \approx \frac{x^2}{0.0100} \quad \text{so} \quad x \approx 2.9 \times 10^{-3}$$

Check the assumption:

$$\frac{2.9 \times 10^{-3}}{0.0100} \times 100 = 29\% \; (>5\%, \text{ so the assumption is } \textit{not} \text{ justified})$$

We must solve the quadratic equation:

$$x^2 + (8.3 \times 10^{-4})x - (8.3 \times 10^{-6}) = 0$$

The only meaningful value of x is 2.5×10^{-3}. Solve for the equilibrium concentrations:

$$[CO] = [Cl_2] = 2.5 \times 10^{-3} \text{ mol/L}$$
$$[COCl_2] = (1.00 \times 10^{-2} \text{ mol/L}) - x = 7.5 \times 10^{-3} \text{ mol/L}$$

Check Once again, we use the calculated values to make sure that we obtain the given K_c.

Comment Notice that, in this problem, the simplifying assumption was justified at the high $[COCl_2]_{init}$ but *not* at the low $[COCl_2]_{init}$.

Follow-Up Problem 15.10 In a study of the effect of temperature on halogen decomposition, 0.50 mol of I_2 was heated in a 2.5 L vessel, and the following reaction occurred:

$$I_2(g) \rightleftharpoons 2I(g)$$

(a) Calculate $[I_2]$ and [I] at equilibrium at 600 K if $K_c = 2.94 \times 10^{-10}$.

(b) Calculate $[I_2]$ and [I] at equilibrium at 2000 K if $K_c = 0.209$.

Predicting When the Assumption Will Be Justified To summarize, we assume that x ($[A]_{reacting}$) can be neglected if K_c is relatively small and/or $[A]_{init}$ is relatively large. The same holds for K and $p_{A(init)}$. But *how* small or large must these variables be? Here is a benchmark for deciding when to make the assumption:

- If $\dfrac{[A]_{init}}{K_c} > 400$, the assumption is justified. Neglecting x introduces an error that is *less than* 5%.

- If $\dfrac{[A]_{init}}{K_c} < 400$, the assumption is *not* justified. Neglecting x introduces an error that is *greater than* 5%, so we solve a quadratic equation to find x.

For example, let us consider the values from Sample Problem 15.10:

(a) For $[A]_{init} = 0.500$ mol/L, $\dfrac{0.500}{8.3 \times 10^{-4}} = 6.0 \times 10^2$, which is greater than 400.

(b) For $[A]_{init} = 0.0100$ mol/L, $\dfrac{0.0100}{8.3 \times 10^{-4}} = 12$, which is less than 400.

We will make a similar assumption in many problems in Chapters 16 and 17. While assumptions help us to simplify the mathematics of these chemistry problems, the most accurate answer can always be found by solving the quadratic equation.

Problems Involving Mixtures of Reactants and Products

In the problems so far, the reaction *had* to go toward products because it started with only reactants. And, therefore, in the ICE tables, we knew that the unknown change in reactant concentration had a negative sign ($-x$) and the change in product concentration had a positive sign ($+x$). If, however, we start with a *mixture* of reactants and products, the reaction direction is not obvious. In those cases, we apply the idea from Section 15.4 and first *compare the values of Q and K to find the direction the reaction is proceeding to reach equilibrium.* To focus on this idea, Sample Problem 15.11 uses concentrations that avoid the need for the quadratic formula.

Sample Problem 15.11 Predicting Reaction Direction and Calculating Equilibrium Concentrations

Problem The research and development unit of a chemical company is studying the reaction of CH_4 and H_2S, two components of natural gas:

$$CH_4(g) + 2H_2S(g) \rightleftharpoons CS_2(g) + 4H_2(g)$$

In one experiment, 1.00 mol of CH_4, 1.00 mol of CS_2, 2.00 mol of H_2S, and 2.00 mol of H_2 are mixed in a 250. mL vessel at 960°C. At this temperature, $K_c = 0.036$.

(a) In which direction will the reaction proceed to reach equilibrium?

(b) If $[CH_4] = 5.56$ mol/L at equilibrium, what are the equilibrium concentrations of the other substances?

Plan (a) To find the direction, we convert the given initial amounts and volume (0.250 L) to concentrations, calculate Q_c, and compare it with K_c. **(b)** Based on the information from part (a), we determine the sign of each concentration change for an ICE table and then use the known $[CH_4]$ at equilibrium (5.56 mol/L) to determine x and the other equilibrium concentrations.

Solution (a) Calculate the initial concentrations:

$$[CH_4] = \frac{1.00 \text{ mol}}{0.250 \text{ L}} = 4.00 \text{ mol/L}$$

Similarly, $[H_2S] = 8.00$ mol/L, $[CS_2] = 4.00$ mol/L, and $[H_2] = 8.00$ mol/L.

Calculate the value of Q_c:

$$Q_c = \frac{[CS_2][H_2]^4}{[CH_4][H_2S]^2} = \frac{(4.00)(8.00)^4}{(4.00)(8.00)^2} = 64.0$$

$Q_c > K_c$ (64.0 > 0.036), so the reaction proceeds to the left. Therefore, concentrations of reactants increase and concentrations of products decrease.

(b) Set up an ICE table, with x = [CS_2] that reacts, which equals [CH_4] that forms:

Concentration (mol/L)	$CH_4(g)$	+	$2H_2S(g)$	$\rightleftharpoons$	$CS_2(g)$	+	$4H_2(g)$
Initial	4.00		8.00		4.00		8.00
Change	$+x$		$+2x$		$-x$		$-4x$
Equilibrium	$4.00 + x$		$8.00 + 2x$		$4.00 - x$		$8.00 - 4x$

Solve for x. At equilibrium,

$$[CH_4] = 5.56\ \text{mol/L} = 4.00\ \text{mol/L} + x \quad \text{so} \quad x = 1.56\ \text{mol/L}$$

Thus,

$$[H_2S] = 8.00\ \text{mol/L} + 2x = 8.00\ \text{mol/L} + 2(1.56\ \text{mol/L}) = 11.12\ \text{mol/L}$$
$$[CS_2] = 4.00\ \text{mol/L} - x = 2.44\ \text{mol/L}$$
$$[H_2] = 8.00\ \text{mol/L} - 4x = 1.76\ \text{mol/L}$$

Check The comparison of Q_c and K_c showed the reaction proceeding to the left. The given data from part (b) confirm this because [CH_4] increases from 4.00 mol/L to 5.56 mol/L during the reaction. Check that the concentrations give the known K_c:

$$\frac{(2.44)(1.76)^4}{(5.56)(11.12)^2} = 0.0341, \text{ which is close to } 0.036$$

Comment Note that even though the reaction is proceeding to the left (that is, backwards) to produce reactants, we did not reverse the expression for Q_c, nor did we reverse the equation. Instead, we indicated that the products were decreasing by putting $-x$ on the product side and that the reactants were increasing by putting $+x$ on the reactant side.

Follow-Up Problem 15.11 An inorganic chemist who is studying the reactions of phosphorus halides mixes 0.1050 mol of PCl_5 with 0.0450 mol of Cl_2 and 0.0450 mol of PCl_3 in a 0.5000 L flask at 250°C:

$$PCl_5(g) \rightleftharpoons PCl_3(g) + Cl_2(g)$$

At this temperature, $K_c = 4.2 \times 10^{-2}$.

(a) In which direction will the reaction proceed?

(b) If [PCl_5] = 0.2065 mol/L at equilibrium, what are the equilibrium concentrations of the other components?

Figure 15.6 groups the steps for solving equilibrium problems in which you know K and some initial quantities and must find the equilibrium quantities.

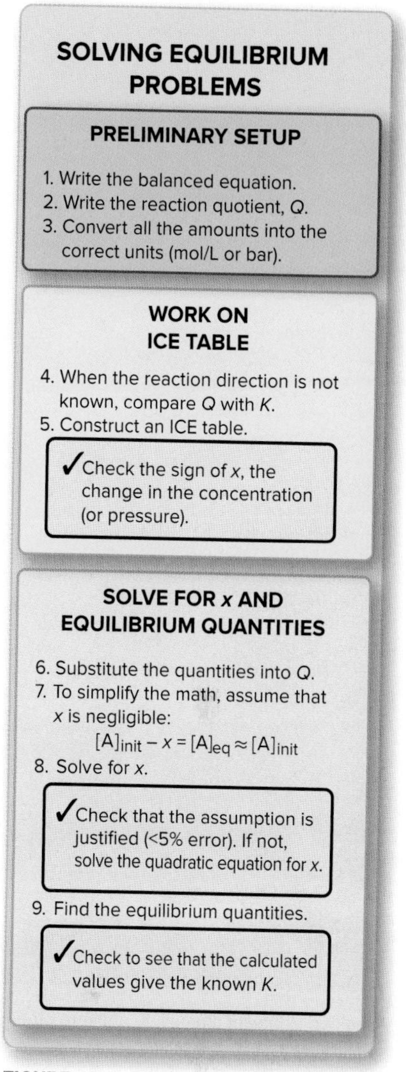
SOLVING EQUILIBRIUM PROBLEMS
PRELIMINARY SETUP
1. Write the balanced equation.
2. Write the reaction quotient, Q.
3. Convert all the amounts into the correct units (mol/L or bar).
WORK ON ICE TABLE
4. When the reaction direction is not known, compare Q with K.
5. Construct an ICE table.
✓Check the sign of x, the change in the concentration (or pressure).
SOLVE FOR x AND EQUILIBRIUM QUANTITIES
6. Substitute the quantities into Q.
7. To simplify the math, assume that x is negligible:
[A]init − x = [A]eq ≈ [A]init
8. Solve for x.
✓Check that the assumption is justified (<5% error). If not, solve the quadratic equation for x.
9. Find the equilibrium quantities.
✓Check to see that the calculated values give the known K.

FIGURE 15.6 Steps in solving equilibrium problems

SUMMARY OF SECTION 15.5

- In equilibrium problems, we typically use quantities (concentrations or partial pressures) of reactants and products to find K, or we use K to find quantities.
- ICE tables summarize the initial quantities, their changes during the reaction, and the equilibrium quantities. To simplify calculations, we assume that if K is small and the initial quantity of reactant is large, the unknown change in reactant (x) can be neglected. If this assumption is not justified (that is, if the error is greater than 5%), we use the quadratic formula to find x.
- For reactions that start with a mixture of reactants and products, we first determine the reaction direction by comparing Q and K. This allows us to determine the sign of x.

15.6 Reaction Conditions and Equilibrium: Le Châtelier's Principle

If conditions are changed so that a system is no longer at equilibrium, the system has a remarkable ability to adjust itself and regain equilibrium. This phenomenon is described by **Le Châtelier's principle**: when a chemical system at equilibrium is disturbed, it regains equilibrium by undergoing a net reaction that reduces the effect of the disturbance.

Two phrases in this statement need further explanation:

1. How is a system "disturbed"? At equilibrium, Q equals K. The system is disturbed when a change in conditions forces it temporarily out of equilibrium ($Q \neq K$). Three common disturbances are a change in concentration, a change in pressure (caused by a change in volume), and a change in temperature. We will discuss each of these changes in this section.

2. What is a "net reaction"? This term refers to a shift in the *equilibrium position* to the right or left. The *equilibrium position* is defined by the specific equilibrium concentrations (or pressures). A shift to the right is a net reaction from reactant to product until equilibrium is regained; a shift to the left is a net reaction from product to reactant. Thus, *concentrations (or pressures) change in a way that reduces the effect of the change in conditions, and the system attains a new equilibrium position.*

For the remainder of this section, we will examine a system at equilibrium to see how it responds to changes in concentration, pressure (volume), and temperature; then we will see what happens when we add a catalyst. As our example, we will use the gaseous reaction between phosphorus trichloride and chlorine to produce phosphorus pentachloride:

$$PCl_3(g) + Cl_2(g) \rightleftharpoons PCl_5(g)$$

Does Le Châtelier's principle apply to other systems?

Le Châtelier's principle holds for many systems at equilibrium in both the natural and social sciences: the effects of disease or drought on populations of predators and prey living on the African plain (*see margin*), a change in the cost of petroleum based on a balance of supply and demand, and even the formation of certain elements in the core of a star.

The Effect of a Change in Concentration

When a system at equilibrium is disturbed by a change in concentration of one of the components, it reacts in the direction that reduces the change:

- If the concentration of A increases, the system reacts to consume some of it.
- If the concentration of B decreases, the system reacts to produce some of it.

Only components that appear in Q can have an effect, so changes in the amounts of pure liquids and solids cannot disturb the equilibrium.

A Qualitative View of a Concentration Change At 523 K, the PCl_3-Cl_2-PCl_5 system reaches equilibrium when

$$PCl_3(g) + Cl_2(g) \rightleftharpoons PCl_5(g) \qquad Q_c = \frac{[PCl_5]}{[PCl_3][Cl_2]} = 24.0 = K_c$$

Starting with Q_c equal to K_c, let us think through some possible changes in concentration:

1. *Adding a reactant.* What happens if we disturb the system by adding some Cl_2 gas? To reduce the disturbance, the system will consume some of the added Cl_2 by shifting toward the product. What happens to the reaction quotient? When the $[Cl_2]$ term increases, the value of Q_c decreases; thus, $Q_c \neq K_c$ and, in fact, $Q < K$. As some of the added Cl_2 reacts with some of the PCl_3 to form more PCl_5, the denominator becomes smaller again and the numerator larger, until eventually $Q_c = K_c$ again. Notice the changes in the new equilibrium concentrations: $[Cl_2]$ and $[PCl_5]$ are higher than they are in the original equilibrium position, and $[PCl_3]$ is lower. Nevertheless, the ratio of values gives the same K_c. Thus, *the equilibrium position shifts to the right when a component on the left is added:*

$$PCl_3 + Cl_2\,(\text{added}) \longrightarrow PCl_5$$

2. *Removing a reactant.* What happens if we disturb the system by removing some PCl_3? To reduce this disturbance, the system will replace the PCl_3 by consuming some PCl_5

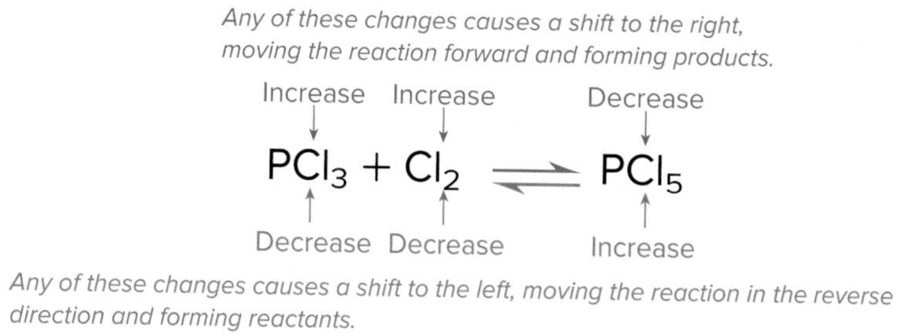

FIGURE 15.7 The effect of a change in concentration on a system at equilibrium

and proceeding toward reactants. What happens to the reaction quotient? When the $[PCl_3]$ term decreases, Q_c increases, so $Q_c \neq K_c$ and, in fact, $Q > K$. As some PCl_5 decomposes to PCl_3 and Cl_2, the numerator decreases and the denominator increases until $Q_c = K_c$ again. Once again, the new and old equilibrium concentrations are different, but the value of K_c is not. Thus, *the equilibrium position shifts to the left when a component on the left is removed:*

$$PCl_3\,(\text{removed}) + Cl_2 \longleftarrow PCl_5$$

3. *Adding or removing a product.* The effects of adding or removing a product are similar to the effects of adding or removing a reactant. If we add PCl_5, the equilibrium position shifts to the left; if we remove some PCl_5, the equilibrium position shifts to the right.

In other words, no matter how the disturbance in concentration comes about, *the system reacts to consume some of the added substance or produce some of the removed substance* to make $Q_c = K_c$ again (Figure 15.7):

- The equilibrium position shifts to the *right* if a reactant is added or a product is removed: [reactant] increases or [product] decreases.
- The equilibrium position shifts to the *left* if a reactant is removed or a product is added: [reactant] decreases or [product] increases.

A Quantitative View of a Concentration Change As discussed, a system reacts to reduce the effect of a disturbance. However, the effect is not completely eliminated, as shown by a quantitative comparison of original and new equilibrium positions.

Consider what happens when we add Cl_2 to a system whose original equilibrium position was established with $[PCl_3] = 0.200$ mol/L, $[Cl_2] = 0.125$ mol/L, and $[PCl_5] = 0.600$ mol/L, that is,

$$Q_c = \frac{[PCl_5]}{[PCl_3][Cl_2]} = \frac{0.600}{(0.200)(0.125)} = 24.0 = K_c$$

Suppose that we add enough Cl_2 to increase its concentration by 0.075 mol/L to a new $[Cl_2]_{init}$ of 0.200 mol/L. The reaction proceeds, and the system comes to a new equilibrium position. From Le Châtelier's principle, we predict that adding more reactant will shift the equilibrium position to the right. Experimental measurement shows that the new $[PCl_5]_{eq}$ is 0.637 mol/L.

Table 15.3 is an ICE table of the entire process: the original equilibrium position, the disturbance, the (new) initial concentrations, the direction of x (the change needed to re-establish equilibrium), and the new equilibrium position. Figure 15.8 depicts the process.

Let us determine the new equilibrium concentrations. From Table 15.3,

$$[PCl_5] = 0.600\text{ mol/L} + x = 0.637\text{ mol/L} \quad \text{so} \quad x = 0.037\text{ mol/L}$$

Thus,

$$[PCl_3] = [Cl_2] = 0.200\text{ mol/L} - x = 0.200\text{ mol/L} - 0.037\text{ mol/L} = 0.163\text{ mol/L}$$

Therefore, at equilibrium,

$$K_{c(\text{original})} = \frac{0.600}{(0.200)(0.125)} = 24.0$$

$$K_{c(\text{new})} = \frac{0.637}{(0.163)(0.163)} = 24.0$$

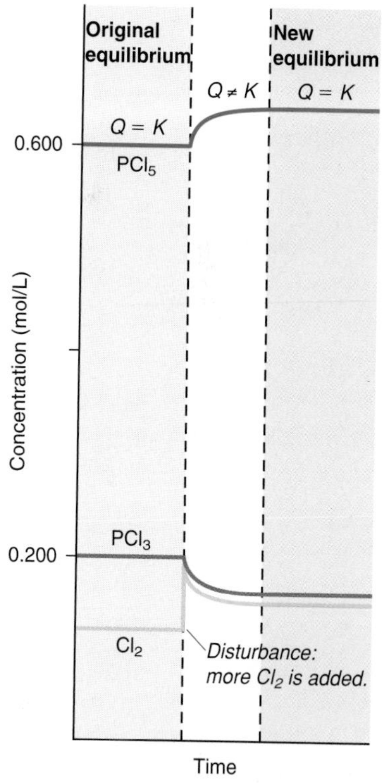

FIGURE 15.8 The effect of added Cl_2 on the PCl_3-Cl_2-PCl_5 system. The original equilibrium concentrations are shown at left (*grey region*). When Cl_2 (*yellow curve*) is added, its concentration increases instantly (*vertical part of yellow curve*) and then falls gradually as it reacts with PCl_3 to form more PCl_5. Soon, equilibrium is re-established at new concentrations (*blue region*), but with the same K.

Several key observations are related to the new equilibrium concentrations:

- As we predicted, $[PCl_5]$ (0.637 mol/L) is higher than the original concentration (0.600 mol/L).
- $[Cl_2]$ (0.163 mol/L) is higher than the *original* equilibrium concentration (0.125 mol/L), but lower than the new initial concentration (0.200 mol/L); thus, the disturbance is *reduced but not eliminated.*
- $[PCl_3]$ (0.163 mol/L), the concentration of the other reactant, is lower than the original equilibrium concentration (0.200 mol/L) because some of the PCl_3 reacted with the added Cl_2.
- Most important, although the position of equilibrium shifted to the right, *at a given temperature, K_c does* ***not*** *change with a change in concentration.*

TABLE 15.3 The Effect of Added Cl_2 on the PCl_3-Cl_2-PCl_5 System

Concentration (mol/L)	$PCl_3(g)$	+	$Cl_2(g)$	$\rightleftharpoons$	$PCl_5(g)$
Original equilibrium	0.200		0.125		0.600
Disturbance			+0.075		
New initial	0.200		0.200		0.600
Change	$-x$		$-x$		$+x$
New equilibrium	$0.200 - x$		$0.200 - x$		$0.600 + x$ (0.637)*

*Experimentally determined value.

Sample Problem 15.12 Predicting the Effect of a Change in Concentration on the Equilibrium Position

Problem To improve air quality and obtain a useful product, chemists often remove sulfur from coal and natural gas by treating the contaminant, hydrogen sulfide, with O_2:

$$2H_2S(g) + O_2(g) \rightleftharpoons 2S(s) + 2H_2O(g)$$

What happens to each concentration if the given change is made?

(a) $[H_2O]$ if O_2 is added
(b) $[H_2S]$ if O_2 is added
(c) $[O_2]$ if H_2S is removed
(d) $[H_2S]$ if sulfur is added

Plan We write the reaction quotient to see how Q_c is affected by each disturbance, relative to K_c. This effect tells us the direction in which the reaction proceeds for the system to re-establish equilibrium and the way that each concentration changes.

Solution Write the reaction quotient:

$$Q_c = \frac{[H_2O]^2}{[H_2S]^2[O_2]}$$

(a) When O_2 is added, the denominator of Q_c increases, so $Q_c < K_c$. The reaction proceeds to the right until $Q_c = K_c$ again, so $[H_2O]$ increases.

(b) As in part (a), when O_2 is added, $Q_c < K_c$. Some H_2S reacts with the added O_2 as the reaction proceeds to the right, so $[H_2S]$ decreases.

(c) When H_2S is removed, the denominator of Q_c decreases, so $Q_c > K_c$. As the reaction proceeds to the left to reform H_2S, more O_2 forms as well, so $[O_2]$ increases.

(d) The concentration of solid S is unchanged as long as some is present, so it does not appear in the reaction quotient. Adding more S has no effect, so $[H_2S]$ is unchanged (but see Comment 2 below).

Check Apply Le Châtelier's principle to see that the reaction proceeds in the direction that lowers the increased concentration or raises the decreased concentration.

Comment 1. Sulfur exists most commonly as S_8. How would this change in the formula affect the answers? The balanced equation and Q_c would be

$$8H_2S(g) + 4O_2(g) \rightleftharpoons S_8(s) + 8H_2O(g) \qquad Q_c = \frac{[H_2O]^8}{[H_2S]^8[O_2]^4}$$

The value of K_c is different for this equation, but the changes described in the problem have the same effects. Thus, shifts in equilibrium position predicted by Le Châtelier's principle are not affected by a change in the balancing coefficients. **2.** In (d), you saw that adding a solid has no effect on the concentrations of other components. Because *the activity of a solid is 1*, it does not appear in Q. However, *the* ***amount*** *of a solid can change.* Adding H_2S shifts the reaction to the right, so more S forms.

Follow-Up Problem 15.12 In a study of glass etching, a chemist examines the reaction between sand (SiO_2) and hydrogen fluoride at 150°C:

$$SiO_2(s) + 4HF(g) \rightleftharpoons SiF_4(g) + 2H_2O(g)$$

Predict the effect on $[SiF_4]$ when **(a)** $H_2O(g)$ is removed; **(b)** some liquid water is added; **(c)** HF is removed; **(d)** some sand is removed.

The Effect of a Change in Pressure (Volume)

Changes in pressure can have a large effect on equilibrium systems that contain gaseous components. (A change in pressure has a negligible effect on liquids and solids because they are nearly incompressible.) Pressure changes can occur in three ways:

1. *Changing the concentration of a gaseous component.* We just considered the effect of changing the concentration of a component, and the same reasoning applies here.

2. *Adding an inert gas (a gas that does not take part in the reaction).* As long as the volume of the system is constant, adding an inert gas has no effect on the equilibrium position because *all concentrations, and thus partial pressures, remain the same.* Moreover, the inert gas does not appear in Q, so it cannot have an effect.

3. *Changing the volume of the reaction vessel.* This change can cause a large shift in equilibrium position, but only for reactions in which the amount (mol) of gas, n_{gas}, changes.

Let us consider the two possible situations for the third way: changing the volume of the reaction vessel.

(a) *Reactions in which n_{gas} changes.* Suppose that the PCl_3-Cl_2-PCl_5 system is in a cylinder-piston assembly. We press down on the piston to halve the volume, so the gas pressure doubles. To reduce this disturbance, the system responds by *reducing the number of gas molecules.* The only way to do this is through a net reaction toward the side with *a smaller amount (mol) of gas*, in this case, toward the product:

$$PCl_3(g) + Cl_2(g) \longrightarrow PCl_5(g)$$
$$2 \text{ mol gas} \longrightarrow 1 \text{ mol gas}$$

Recall that $Q_c = \frac{[PCl_5]}{[PCl_3][Cl_2]}$. When the volume is halved, the concentrations double, but the denominator of Q_c is the product of two concentrations, so it quadruples while the numerator only doubles. Therefore, Q_c becomes less than K_c. As a result, the system forms more PCl_5 and a new equilibrium position is reached.

Thus, for a system that consists of gases at equilibrium, in which the amount (mol) of gas, n_{gas}, changes during the reaction (Figure 15.9), we can conclude the following:

- If the volume becomes smaller (pressure is higher), the reaction shifts so that the total number of gas molecules decreases.
- If the volume becomes larger (pressure is lower), the reaction shifts so that the total number of gas molecules increases.

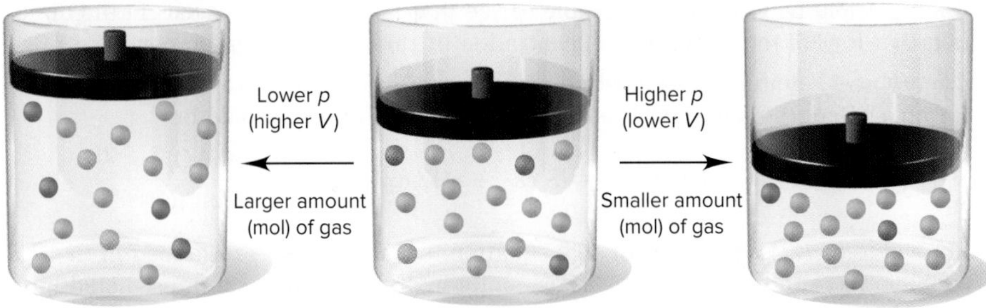

FIGURE 15.9 The effect of a change in pressure (volume) on a system at equilibrium. The system of gases (*centre*) is at equilibrium. For the reaction

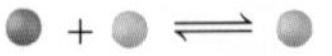

an increase in pressure (*right*) decreases the volume, so the equilibrium shifts to form *fewer* molecules. A decrease in pressure (*left*) increases the volume, so the equilibrium shifts to form *more* molecules.

(b) *Reactions in which n_{gas} does **not** change.* For the formation of hydrogen iodide from its elements, we have the same amount (mol) of gas on both sides:

$$H_2(g) + I_2(g) \rightleftharpoons 2HI(g)$$
$$2 \text{ mol gas} \longrightarrow 2 \text{ mol gas}$$

Therefore, Q_c has the same number of terms in the numerator and the denominator:

$$Q_c = \frac{[HI]^2}{[H_2][I_2]} = \frac{[HI][HI]}{[H_2][I_2]}$$

Because a change in volume has the same effect on the numerator and the denominator, *there is **no** effect on the equilibrium position.*

In terms of the equilibrium constant, *a change in volume is, in effect, a change in concentration*: a decrease in volume raises the concentration, and vice versa. Thus, like other changes in concentration, *a change in pressure due to a change in volume does **not** alter K_c.*

Sample Problem 15.13 Predicting the Effect of a Change in Volume (Pressure) on the Equilibrium Position

Problem How would you change the volume of each reaction to *increase* the product yield?

(a) $CaCO_3(s) \rightleftharpoons CaO(s) + CO_2(g)$

(b) $S(s) + 3F_2(g) \rightleftharpoons SF_6(g)$

(c) $Cl_2(g) + I_2(g) \rightleftharpoons 2ICl(g)$

Plan Whenever gases are present, a change in volume causes a change in concentration. For reactions in which the amount (mol) of gas changes, if the volume decreases (pressure increases), the equilibrium position shifts to lower the pressure by reducing the amount (mol) of gas. If the volume increases (pressure decreases), the opposite effect occurs.

Solution (a) The only gas is the product CO_2. To make the system produce more molecules of gas (that is, more CO_2), we increase the volume (decrease the pressure).

(b) With 3 mol of gas on the left and only 1 mol on the right, we decrease the volume (increase the pressure) to form fewer molecules of gas and, thus, more SF_6.

(c) The amount (mol) of gas is the same on both sides of the equation, so a change in volume (pressure) will have no effect on the yield of ICl.

Check Let us predict the relative values of Q_c and K_c.

(a) $Q_c = [CO_2]$, so increasing the volume will make $Q_c < K_c$ and the system will yield more CO_2.

(b) $Q_c = \frac{[SF_6]}{[F_2]^3}$, so lowering the volume will increase $[F_2]$ and $[SF_6]$ proportionately. However, Q_c will decrease because of the exponent 3 in the denominator. To make $Q_c = K_c$ again, $[SF_6]$ must increase.

(c) $Q_c = \frac{[ICl]^2}{[Cl_2][I_2]}$, so a change in the volume (pressure) will affect the numerator (2 mol) and the denominator (2 mol) equally; thus, it will have no effect.

Follow-Up Problem 15.13 Would you increase or decrease the pressure (via a volume change) of each reaction mixture to *decrease* the product yield?

(a) $2SO_2(g) + O_2(g) \rightleftharpoons 2SO_3(g)$

(b) $4NH_3(g) + 5O_2(g) \rightleftharpoons 4NO(g) + 6H_2O(g)$

(c) $CaC_2O_4(s) \rightleftharpoons CaCO_3(s) + CO(g)$

The Effect of a Change in Temperature

Of the three types of disturbances that may occur—changes in concentration, pressure, and temperature—*only temperature changes alter K*. To see why, let us focus on the sign of $\Delta_r H°$:

$$PCl_3(g) + Cl_2(g) \rightleftharpoons PCl_5(g) \qquad \Delta_r H° = -111\text{ kJ/mol}$$

The forward reaction is exothermic (releases heat; $\Delta_r H° < 0$), so the reverse reaction is endothermic (absorbs heat; $\Delta_r H° > 0$):

$$PCl_3(g) + Cl_2(g) \longrightarrow PCl_5(g) + \textit{heat}\text{ (exothermic)}$$
$$PCl_5(g) + \textit{heat}\text{ (endothermic)} \longrightarrow PCl_3(g) + Cl_2(g)$$

If we consider *heat as a component of the equilibrium system*, a rise in temperature occurs when heat is "added" to the system, and a drop in temperature occurs when heat is "removed" from the system. As with a change in any other component, the system shifts to reduce the effect of the change. Therefore, *a temperature increase (adding heat) favours the endothermic (heat-absorbing) direction, and a temperature decrease (removing heat) favours the exothermic (heat-releasing) direction.*

If we start with the system at equilibrium, Q equals K. If we increase the temperature, the system absorbs the added heat by decomposing some PCl_5 to PCl_3 and Cl_2. The denominator of Q becomes larger and the numerator becomes smaller, so the system reaches a new equilibrium position at a smaller ratio of concentration terms, that is, a lower K. Similarly, if we decrease the temperature, the system releases more heat by forming more PCl_5 from some PCl_3 and Cl_2. The numerator of Q becomes larger and the denominator becomes smaller, so the new equilibrium position has a higher K. Thus, we can make the following conclusions:

- *A temperature rise will increase K for a system with a positive $\Delta_r H°$.*
- *A temperature rise will decrease K for a system with a negative $\Delta_r H°$.*

Let us review these conclusions with a sample problem.

Sample Problem 15.14 Predicting the Effect of a Change in Temperature on the Equilibrium Position

Problem How does an *increase* in temperature affect the equilibrium concentration of the underlined substance and the value of K in each reaction?

(a) $CaO(s) + H_2O(l) \longrightarrow \underline{Ca(OH)_2(aq)}$ $\qquad \Delta_r H° = -82\text{ kJ/mol}$

(b) $CaCO_3(s) \rightleftharpoons CaO(s) + \underline{CO_2(g)}$ $\qquad \Delta_r H° = 178\text{ kJ/mol}$

(c) $\underline{SO_2(g)} \rightleftharpoons S(s) + O_2(g)$ $\qquad \Delta_r H° = 297\text{ kJ/mol}$

Plan We write each equation to show heat as a reactant or a product. The temperature increases when we add heat, so the system shifts to absorb the heat; that is, the endothermic reaction occurs. Thus, K will increase if the forward reaction is endothermic and decrease if the forward reaction is exothermic.

Solution (a) $CaO(s) + H_2O(l) \rightleftharpoons Ca(OH)_2(aq) + \textbf{heat}$

Adding heat shifts the equilibrium to the left (toward reactants): $[Ca(OH)_2]$ and K will decrease.

(b) $CaCO_3(s) + \textit{heat} \rightleftharpoons CaO(s) + CO_2(g)$

Adding heat shifts the equilibrium to the right (toward products): $[CO_2]$ and K will increase.

(c) $SO_2(g) + heat \rightleftharpoons S(s) + O_2(g)$

Adding heat shifts the equilibrium to the right (toward products): $[SO_2]$ will decrease and K will increase.

Check Check your answers by reasoning through a *decrease* in temperature: heat is removed and the exothermic direction is favoured. All the answers should be opposite.

Comment Note that, in part (a), our conclusions about K_c hold for solutions as well.

Follow-Up Problem 15.14 How does a *decrease* in temperature affect the partial pressure of the underlined substance and the value of K in each reaction?

(a) $C(\text{graphite}) + \underline{2H_2(g)} \rightleftharpoons CH_4(g)$ $\quad \Delta_rH° = -75$ kJ/mol

(b) $\underline{N_2(g)} + O_2(g) \rightleftharpoons 2NO(g)$ $\quad \Delta_rH° = 181$ kJ/mol

(c) $P_4(s) + 10Cl_2(g) \rightleftharpoons \underline{4PCl_5(g)}$ $\quad \Delta_rH° = -1528$ kJ/mol

The van't Hoff Equation: The Effect of T on K The *van't Hoff equation* shows quantitatively how the equilibrium constant is affected by changes in temperature:

$$\ln\frac{K_2}{K_1} = -\frac{\Delta_rH°}{R}\left(\frac{1}{T_2} - \frac{1}{T_1}\right) \tag{15.12}$$

where K_1 is the equilibrium constant at T_1, K_2 is the equilibrium constant at T_2, and R is the universal gas constant [8.314 J/(mol•K)]. If we know $\Delta_rH°$ and K at one temperature, the van't Hoff equation allows us to find K at any other temperature (or to find $\Delta_rH°$, given the two K's at two T's).

Equation 15.12 confirms the qualitative prediction from Le Châtelier's principle: for a temperature rise, we have

$$T_2 > T_1 \quad \text{and} \quad \frac{1}{T_2} < \frac{1}{T_1}$$

So,

$$\frac{1}{T_2} - \frac{1}{T_1} < 0$$

Therefore, we can make the following conclusions:

- For an endothermic reaction ($\Delta_rH° > 0$), the $-(\frac{\Delta_rH°}{R})$ term is less than zero. With $\frac{1}{T_2} - \frac{1}{T_1} < 0$, the right side of the equation is greater than zero. Thus, $\ln(\frac{K_2}{K_1}) > 0$, so $K_2 > K_1$.
- For an exothermic reaction ($\Delta_rH° < 0$), the $-(\frac{\Delta_rH°}{R})$ term is greater than zero. With $\frac{1}{T_2} - \frac{1}{T_1} < 0$, the right side of the equation is less than zero. Thus, $\ln(\frac{K_2}{K_1}) < 0$, so $K_2 < K_1$.

The van't Hoff equation is useful in many real-life situations. For example, several coal gasification processes begin with the formation of syngas from carbon and steam:

$$C(s) + H_2O(g) \rightleftharpoons CO(g) + H_2(g) \qquad \Delta_rH° = 131 \text{ kJ/mol}$$

An engineer knows that K is only 9.36×10^{-17} at 25°C and, therefore, wants to find a temperature that allows a much higher yield. The engineer decides to calculate K at 700.°C:

$$\ln\frac{K_2}{K_1} = -\frac{\Delta_rH°}{R}\left(\frac{1}{T_2} - \frac{1}{T_1}\right)$$

The temperatures must be in kelvin, and the units of $\Delta_rH°$ and R must be made consistent:

$$\ln\left(\frac{K_2}{9.36 \times 10^{-17}}\right) = -\frac{131 \times 10^3 \text{ J/mol}}{8.314 \text{ J/(mol•K)}}\left(\frac{1}{973 \text{ K}} - \frac{1}{298 \text{ K}}\right)$$

$$\frac{K_2}{9.36 \times 10^{-17}} = 8.51 \times 10^{15}$$

$$K_2 = 0.797 \quad \text{(a much higher yield)}$$

(For further practice with the van't Hoff equation, see Problems 15.73 and 15.74.)

The Lack of Effect of a Catalyst

Let us briefly consider what effect, if any, adding a catalyst would have on an equilibrium system. Recall, from Chapter 14, that a catalyst speeds up a reaction by lowering the activation energy, thereby increasing the forward *and* reverse rates to the same extent. Thus, *a catalyst shortens the time it takes to reach equilibrium, but has* ***no*** *effect on the equilibrium position.* That is, if we add a catalyst to a mixture of PCl_3 and Cl_2 at 523 K, the system attains the *same* equilibrium concentrations of PCl_3, Cl_2, and PCl_5 *more quickly* than it does without the catalyst. Figure 15.10 shows an engine in an imaginary world where catalysts speed up a reaction in only one direction. As we will see in a moment, however, catalysts play key roles in optimizing reaction systems.

Table 15.4 summarizes the effects of changing conditions. Many changes alter the equilibrium *position*, but only temperature changes alter the equilibrium *constant*. Sample Problem 15.15 shows how to visualize equilibrium at the molecular level.

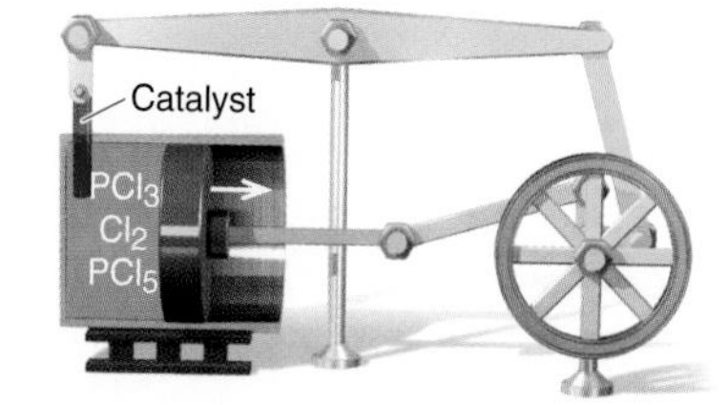

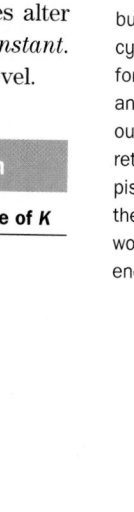

FIGURE 15.10 Catalyzed perpetual motion? This imaginary engine has a piston attached to a flywheel, whose rocker arm holds a catalyst that speeds PCl_5 breakdown but not formation. With the catalyst in the cylinder, PCl_5 breaks down faster than it forms, so the total gas pressure increases and the piston moves out. With the catalyst out of the cylinder, the old equilibrium returns, the gas pressure lowers, and the piston moves in. If a catalyst *could* change the rate in only one direction, this machine would supply power with no input of external energy!

TABLE 15.4 Effects of Various Disturbances on a System at Equilibrium

Disturbance	Effect on Equilibrium Position	Effect on Value of K
Concentration		
Increase [reactant]	Toward formation of product	None
Decrease [reactant]	Toward formation of reactant	None
Increase [product]	Toward formation of reactant	None
Decrease [product]	Toward formation of product	None
Pressure		
Increase p (decrease V)	Toward formation of fewer moles of gas	None
Decrease p (increase V)	Toward formation of more moles of gas	None
Increase p (add inert gas, no change in V)	None: concentrations unchanged	None
Temperature		
Increase T	Toward absorption of heat	Increases if $\Delta_r H° > 0$ Decreases if $\Delta_r H° < 0$
Decrease T	Toward release of heat	Increases if $\Delta_r H° < 0$ Decreases if $\Delta_r H° > 0$
Catalyst added	None: forward and reverse rates increase equally; equilibrium attained sooner	None

Sample Problem 15.15 Determining Equilibrium Parameters from Molecular Scenes

Problem Consider the following reaction:

$$X(g) + Y_2(g) \rightleftharpoons XY(g) + Y(g) \quad \Delta H > 0$$

These molecular scenes depict three different reaction mixtures (X is *green*, and Y is *purple*):

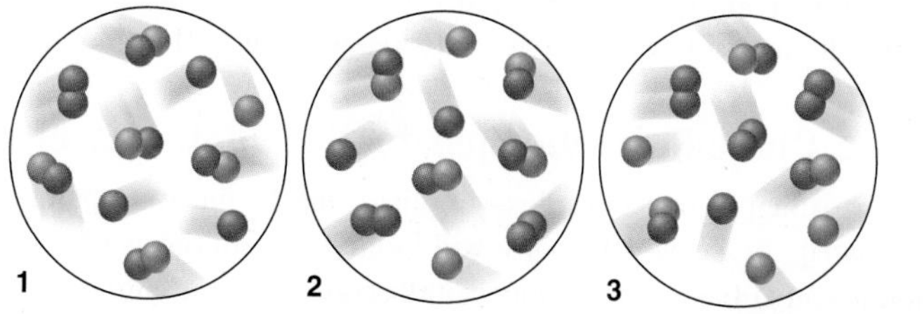

(a) If $K = 2$ at the temperature of the reaction, which scene represents the mixture at equilibrium?

(b) Will the reaction mixtures in the other two scenes proceed toward reactants or toward products to reach equilibrium?

(c) For the mixture at equilibrium, how will a rise in temperature affect $[Y_2]$?

Plan (a) We are given the balanced equation and K, and we must choose the scene that represents the mixture at equilibrium. We write Q and, for each scene, count particles and find the value of Q. Whichever scene gives a Q equal to K (2) will be the scene that represents the mixture at equilibrium. **(b)** For each of the other two reaction mixtures, we compare the value of Q with 2. If $Q > K$, the numerator (product side) is too high, so the reaction proceeds toward the reactants; if $Q < K$, the reaction proceeds toward the products. **(c)** We know that $\Delta H > 0$, so we must see whether a rise in T increases or decreases $[Y_2]$, one of the reactants.

Solution (a) For the reaction, $Q = \frac{[XY][Y]}{[X][Y_2]}$. Thus, we can calculate Q for each scene:

$$\text{Scene 1: } Q = \frac{5 \times 3}{1 \times 1} = 15 \qquad \text{Scene 2: } Q = \frac{4 \times 2}{2 \times 2} = 2 \qquad \text{Scene 3: } Q = \frac{3 \times 1}{3 \times 3} = \frac{1}{3}$$

For scene 2, $Q = K$, so scene 2 represents the mixture at equilibrium.

(b) For scene 1, Q (15) $> K$ (2), so the reaction will proceed toward the reactants.

For scene 3, Q ($\frac{1}{3}$) $< K$ (2), so the reaction will proceed toward the products.

(c) The reaction is endothermic, so heat acts as a reactant:

$$X(g) + Y_2(g) + heat \rightleftharpoons XY(g) + Y(g)$$

Therefore, adding heat to the left shifts the reaction to the right, so $[Y_2]$ decreases.

Check (a) Remember that quantities in the numerator (or denominator) of Q are multiplied, not added. For example, the denominator for scene 1 is $1 \times 1 = 1$, not $1 + 1 = 2$.

(c) A good check is to imagine that $\Delta H < 0$ and see if you get the opposite result:

$$X(g) + Y_2(g) \rightleftharpoons XY(g) + Y(g) + heat$$

If $\Delta H < 0$, adding heat would shift the reaction to the left and increase $[Y_2]$.

Follow-Up Problem 15.15 Consider the following reaction:

$$C_2(g) + D_2(g) \rightleftharpoons 2CD(g) \quad \Delta H < 0$$

These molecular scenes depict three different reaction mixtures (C is *red*, and D is *blue*):

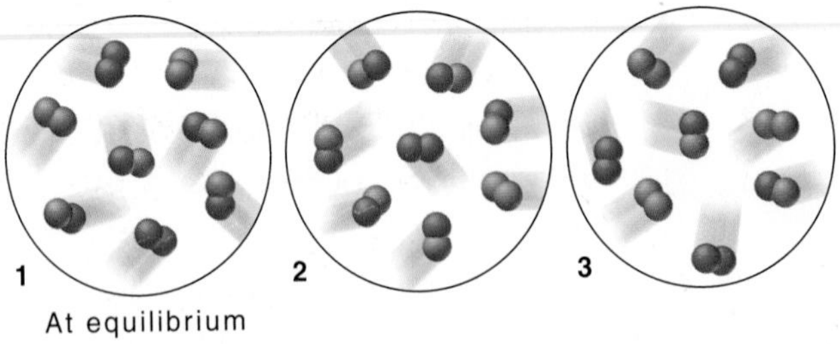

(a) Calculate the value of K.

(b) In which direction will the reaction proceed for the mixtures *not* at equilibrium?

(c) For the mixture at equilibrium, what effect will a rise in temperature have on the total amount (mol) of gas (increase, decrease, or no effect)? Explain.

Applying Le Châtelier's Principle to the Synthesis of Ammonia

Le Châtelier's principle has countless applications in natural systems and in the chemical industry. As a case study, we will look at the synthesis of ammonia, which, on an amount (mol) basis, is produced industrially in a greater amount than any other compound. Then, in the Chemical Connections that follows, we will consider how Le Châtelier's principle is applied to metabolism.

Even though four out of every five molecules in the atmosphere are N_2, the supply of *usable* nitrogen is limited because the strong triple bond in N_2 lowers its reactivity. Thus, the N atom is very difficult to "fix," that is, to combine with other atoms into useful compounds. Natural nitrogen fixation occurs through the fine-tuned activity of enzymes found in bacteria that live on plant roots and through the brute force of lightning. However, nearly 13% of all nitrogen fixation is done industrially via the **Haber process**:

$$N_2(g) + 3H_2(g) \rightleftharpoons 2NH_3(g) \qquad \Delta_r H^\circ = -91.8 \text{ kJ/mol}$$

Developed by the German chemist Fritz Haber in 1913 and first used in a plant making 12 000 t of ammonia a year, the process now yields over 110 million t a year. Over 80% of this ammonia is used as fertilizer, with most of the remainder used to make explosives and nylons and other polymers, and smaller amounts going into the production of refrigerants, rubber stabilizers, household cleaners, and pharmaceuticals.

Optimizing Reaction Conditions: Yield versus Rate The Haber process applies equilibrium *and* kinetics principles to achieve a compromise that makes the process economical. From the balanced equation, we see three ways to maximize NH_3 yield:

1. *Decrease [NH_3].* Removing NH_3, as it forms, makes the system shift toward producing more to regain equilibrium.

2. *Decrease volume (increase pressure).* Because 4 mol of gas reacts to form 2 mol of gas, decreasing the volume shifts the system toward making a smaller amount (mol) of gas.

3. *Decrease temperature.* Because the formation of NH_3 is exothermic, decreasing the temperature (removing heat) shifts the equilibrium position toward product, thereby increasing K_c (Table 15.5).

Therefore, the conditions for maximizing the yield of product are continuous removal of NH_3, high pressure, and low temperature. Figure 15.11 shows the percent yield of NH_3 at various combinations of pressure and temperature. Notice the almost complete conversion (98.3%) to product at 1000 bar and 473 K (200.°C).

Although the *yield* is favoured at this relatively low temperature, the *rate* of formation is so slow that the process is uneconomical. In practice, a compromise optimizes yield *and* rate. High pressure and continuous removal are used to increase the yield, but the temperature is raised to a moderate level and *a catalyst is used to increase the rate.* Achieving the same rate without a catalyst would require much higher temperatures and, thus, result in a much lower yield.

TABLE 15.5 Effect of Temperature on K_c for Ammonia Synthesis

T (K)	K_c
200.	7.17×10^{15}
300.	2.69×10^{8}
400.	3.94×10^{4}
500.	1.72×10^{2}
600.	4.53×10^{0}
700.	2.96×10^{-1}
800.	3.96×10^{-2}

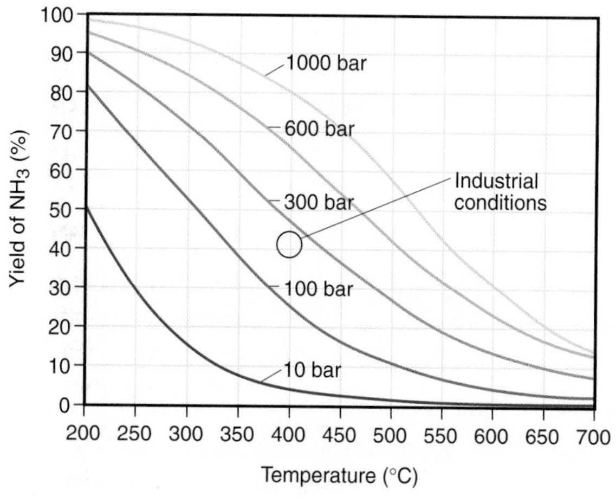

FIGURE 15.11 Percent yield of ammonia versus temperature at five different pressures. At very high p and low T (*top left*), the yield is high, but the rate is low. Industrial conditions (*circle*) are between 200 and 300 bar at about 400°C.

CHEMICAL CONNECTIONS TO CELLULAR METABOLISM

Design and Control of a Metabolic Pathway

From the simplest bacterium to the most specialized neuron, every cell performs thousands of reactions that allow it to grow and reproduce, feed and excrete, and move and communicate. Taken together, these myriad feats of breakdown, synthesis, and energy flow constitute the cell's *metabolism* and are organized into reaction sequences called **metabolic pathways**.

Continuous Shift toward Product In principle, each step in a metabolic pathway is a reversible reaction that is catalyzed by a specific enzyme (Section 14.7). However, equilibrium is never reached in a pathway, because the product of each reaction becomes the reactant of the next reaction. Consider the five-step pathway by which the amino acid threonine is converted into the amino acid isoleucine (Figure B15.1). Threonine, supplied from a different region of the cell, forms ketobutyrate through the catalytic action of enzyme 1. The equilibrium position of reaction 1 shifts to the right because the product, ketobutyrate, is the reactant in reaction 2. Similarly, reaction 2 shifts to the right as its product, acetohydroxybutyrate, is used up in reaction 3. In this way, each subsequent reaction shifts the previous reaction to the right. The final product, isoleucine, is removed to make proteins elsewhere in the cell. Thus, the entire pathway operates in one direction.

Creation of a Steady State This continuous shift in equilibrium position has two consequences for metabolic pathways:

1. *Each step proceeds with nearly 100% yield.* Virtually every molecule of threonine that enters this region of the cell eventually changes to ketobutyrate, every molecule of ketobutyrate eventually changes to the next product, and so on.
2. *Reactant and product concentrations remain nearly constant,* because they reach a *steady state*.
   - In an equilibrium system, equal rates in *opposing* directions create constant concentrations of reactants and products.
   - In a steady state system, the rates of reactions in *one* direction—into, through, and out of the system—create constant concentrations of intermediates.

Ketobutyrate, for example, is formed in reaction 1 as fast as it is consumed in reaction 2, so its concentration is constant. (A steady state amount of water results if you fill a sink and then adjust the flow of the faucet and drain so that the water enters as fast as it leaves.)

Regulation by Feedback Inhibition Recall, from Chapter 14, that substrate concentrations are *much* higher than enzyme concentrations. If the active sites on all the enzyme molecules were always occupied, all cellular reactions would occur at their maximum rates.

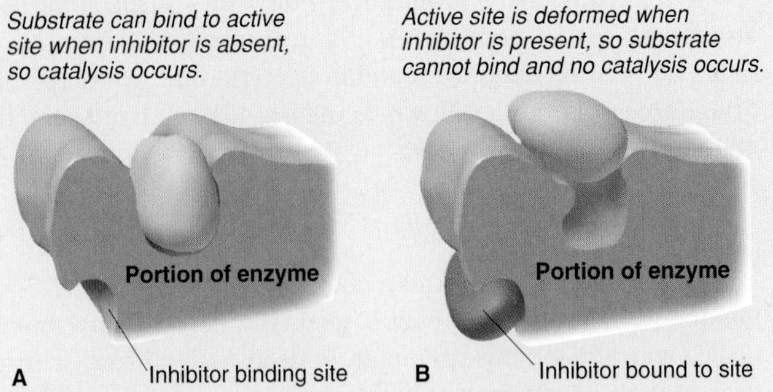

FIGURE B15.2 Effect of inhibitor binding on the shape of the active site. A. If the inhibitor site is not occupied, the enzyme catalyzes the reaction. **B.** If the inhibitor site is occupied, the enzyme does not function.

This might be ideal for an industrial process, but an organism needs to control amounts carefully. To regulate product formation, certain key steps are catalyzed by *regulatory enzymes*, which contain an *inhibitor site* in addition to an active site: when the inhibitor site is occupied, the shape of the active site is deformed and the reaction is not catalyzed (Figure B15.2).

In the simplest case of metabolic regulation (Figure B15.1), *the final product of a pathway is the inhibitor, and the regulatory enzyme catalyzes the first step*. Suppose, for instance, that a cell is temporarily making less protein, so isoleucine is not being removed as quickly. As its concentration rises, isoleucine lands on the inhibitor site of threonine dehydratase, the first enzyme in the pathway, thereby inhibiting its own production. This process is called *end-product feedback inhibition*. More complex pathways have more elaborate regulatory schemes.

Problem

B15.1 Many metabolites are products in branched pathways. In the pathway below, the letters represent compounds, and the numbers represent enzymes:

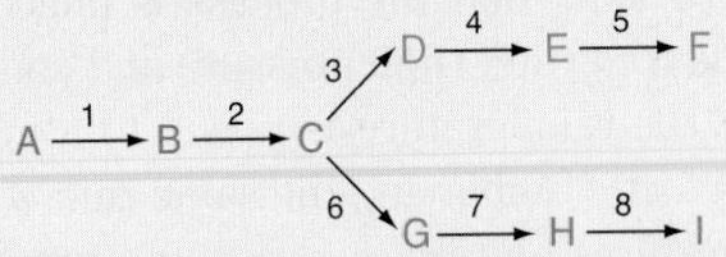

One method of control of these pathways occurs through inhibition of the first enzyme specific for a branch.
(a) Which enzyme is inhibited by F?
(b) Which enzyme is inhibited by I?
(c) What would be the disadvantage if F inhibited enzyme 1?
(d) What would be the disadvantage if enzyme 6 was inhibited by F?

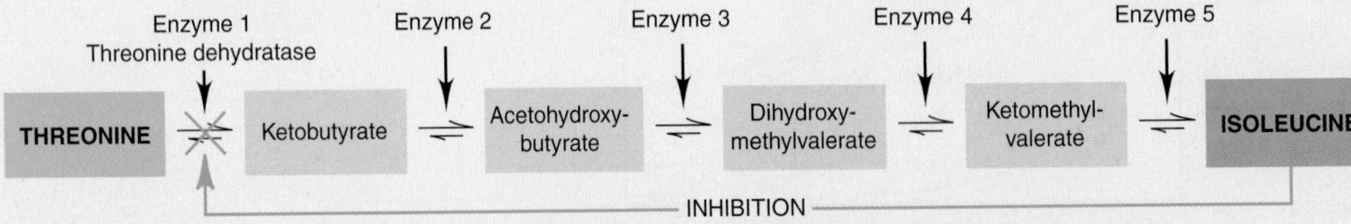

FIGURE B15.1 The biosynthesis of isoleucine from threonine. Isoleucine is synthesized from threonine in a sequence of five enzyme-catalyzed reactions. Once enough isoleucine is present, its concentration builds up and inhibits threonine dehydratase, the first enzyme in the pathway.

The Industrial Process The key stages in the industrial production of NH_3 are shown in Figure 15.12. To extend equipment life and minimize cost, modern plants operate at about 200 to 300 bar and around 673 K (400.°C). The stoichiometric ratio of reactant gases (N_2/H_2 = 1/3 by volume) is injected into the heated, pressurized reaction chamber. Some of the needed heat is supplied by $\Delta_r H°$. The gases flow over catalyst beds that consist of 5 mm to 10 mm chunks of iron crystals embedded in a fused mixture of MgO, Al_2O_3, and SiO_2. The emerging equilibrium mixture contains about 35% NH_3 and is cooled by refrigeration until the NH_3 (bp = −33.4°C) condenses; it is then removed and stored. Because N_2 and H_2 have much lower boiling points, they are recycled as gases by pumping them back into the reaction chamber.

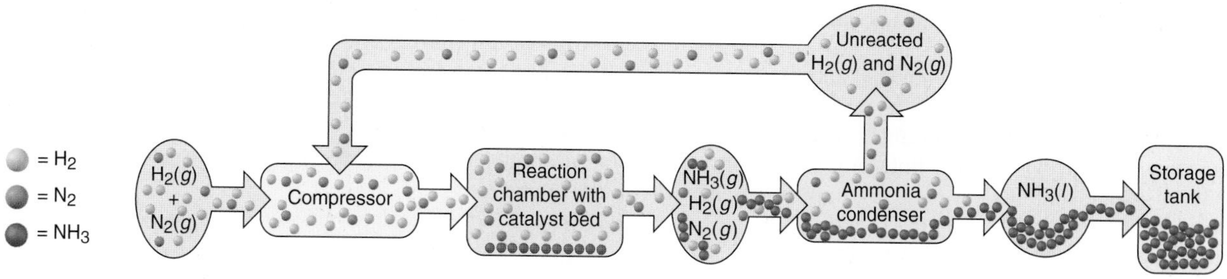

FIGURE 15.12 Key stages in the Haber process for synthesizing ammonia

SUMMARY OF SECTION 15.6

- If a system at equilibrium is disturbed, it undergoes a net reaction that reduces the disturbance and returns the system to equilibrium.
- Changes in concentration cause a net reaction to consume the added component or produce the removed component.
- For a reaction that involves a change in amount (mol) of gas, an increase in pressure (decrease in volume) causes a net reaction toward fewer moles of gas, and a decrease in pressure causes the opposite change.
- Although the equilibrium position changes as a result of a concentration or volume change, K does not change.
- A temperature change affects K: higher T increases K for an endothermic reaction (positive $\Delta_r H°$) and decreases K for an exothermic reaction (negative $\Delta_r H°$).
- A catalyst causes a system to reach equilibrium more quickly by speeding up the forward and reverse reactions equally, but it does not affect the equilibrium position.
- Ammonia is produced in a process that is favoured by high pressure, low temperature, and continuous removal of product. To make the process economical, intermediate temperature and pressure and a catalyst are used.
- A metabolic pathway is a cellular reaction sequence that has each step shifted completely toward product. Its overall yield is controlled by a feedback system that inhibits the activity of certain key enzymes.

CHAPTER REVIEW GUIDE

Learning Objectives

Relevant section (§) and/or sample problem (SP) numbers appear in parentheses.

Concepts

1. Distinguish between the speed (rate) and the extent of a reaction. (Introduction)
2. Describe why a system attains dynamic equilibrium when the forward and reverse reaction rates are equal. (§15.1)
3. Express the equilibrium constant as a number that is equal to a ratio of rate constants and a ratio of concentration terms. (§15.1)
4. Explain how the magnitude of K is related to the extent of the reaction. (§15.1)
5. Describe why the same equilibrium state is reached no matter what the starting concentrations or pressures of the reacting system. (§15.2)
6. Explain how the reaction quotient (Q) changes continuously until the system reaches equilibrium, at which point $Q = K$. (§15.2)
7. Explain why the form of Q is based exactly on the balanced equation *as written*. (§15.2)
8. Describe how the *sum* of reaction steps gives the overall reaction, and the *product* of Q's (or K's) gives the overall Q (or K). (§15.2)
9. Explain why pure solids and liquids do not appear in Q. (§15.2)
10. Describe how the interconversion of K_c and K is based on the ideal gas law and Δn_{gas}. (§15.3)
11. Explain why the reaction direction depends on the relative values of Q and K. (§15.4)
12. Describe how an ICE table is used to find an unknown quantity (concentration or pressure). (§15.5)
13. Explain how assuming that the change in [reactant] is relatively small can simplify finding equilibrium quantities. (§15.5)
14. Describe how Le Châtelier's principle explains the effects of a change in concentration, pressure (volume), or temperature on a system at equilibrium and on K. (§15.6)
15. Explain why a change in temperature *does* affect K. (§15.6)
16. Explain why the addition of a catalyst does *not* affect K. (§15.6)
17. Describe how adjusting reaction conditions and using a catalyst optimizes the synthesis of ammonia. (§15.6)

Skills

1. Write a reaction quotient (Q) from a balanced equation. (SP 15.1)
2. Write Q and calculate K for a reaction consisting of more than one step. (SP 15.2)
3. Write Q and find K for an equation multiplied by a common factor. (SP 15.3)
4. Write Q for heterogeneous equilibria. (§15.2)
5. Convert between K_c and K. (SP 15.4)
6. Compare Q and K to determine reaction direction. (SPs 15.5, 15.6)
7. Substitute quantities (concentrations or pressures) into Q to find K. (§15.5)
8. Use an ICE table to determine quantities and find K. (SP 15.7)
9. Calculate one equilibrium quantity from other equilibrium quantities and K. (SP 15.8)
10. Calculate an equilibrium quantity from initial quantities and K. (SP 15.9)
11. Solve a quadratic equation for an unknown equilibrium quantity. (§15.5)
12. Assume that the change in [reactant] is relatively small to find equilibrium quantities, and check this assumption. (SP 15.10)
13. Compare the values of Q and K to find the direction of the reaction and x, the unknown change in a quantity. (SP 15.11)
14. Using the relative values of Q and K, predict the effect of a change in concentration on the equilibrium position and K. (SP 15.12)
15. Using Le Châtelier's principle and Δn_{gas}, predict the effect of a change in volume (pressure) on the equilibrium position. (SP 15.13)
16. Using Le Châtelier's principle and $\Delta_r H°$, predict the effect of a change in temperature on the equilibrium position and K. (SP 15.14)
17. Using the van't Hoff equation, calculate K at one temperature given K at another temperature. (§15.6)
18. Using molecular scenes, determine equilibrium parameters. (SP 15.15)

Key Terms

Section 15.1
equilibrium constant (K)

Section 15.2
law of chemical equilibrium (law of mass action)
reaction quotient (Q)
activity

Section 15.6
Le Châtelier's principle
Haber process
metabolic pathways

Key Equations and Relationships

15.1 Defining equilibrium in terms of reaction rates:

$$\text{At equilibrium: rate}_{fwd} = \text{rate}_{rev}$$

15.2 Defining the equilibrium constant for the reaction $A \rightleftharpoons 2B$:

$$K = \frac{k_{fwd}}{k_{rev}} = \frac{[B]^2_{eq}}{[A]_{eq}}$$

15.3 Defining the equilibrium constant in terms of the reaction quotient:

$$\text{At equilibrium: } Q = K$$

15.4 Defining the activity of a species in solution:

$$a_X = \gamma_X \frac{[X]}{c^\circ}$$

15.5 Expressing the activity for reactions that involve gases:

$$a_Y = \gamma_Y \frac{p_Y}{p^\circ}$$

15.6 Expressing Q for the reaction $aA + bB \rightleftharpoons cC + dD$:

$$Q = \frac{a_C^c a_D^d}{a_A^a a_B^b}$$

15.7 Finding the overall K for a reaction sequence:

$$K_{overall} = K_1 \times K_2 \times K_3 \times \cdots$$

15.8 Finding Q of a reaction from Q of the reverse reaction and finding K of a reaction from K of the reverse reaction:

$$Q_{fwd} = \frac{1}{Q_{rev}} \qquad K_{fwd} = \frac{1}{K_{rev}}$$

15.9 Finding K of a reaction multiplied by a factor n:

$$Q' = Q^n = \left(\frac{a_C^c a_D^d}{a_A^a a_B^b}\right)^n \qquad K' = K^n$$

15.10 Relating K based on pressures to K based on concentrations:

$$K = K_c(RT)^{\Delta n_{gas}}, \quad R = 0.083\ 14\ \text{bar·L·mol}^{-1}\text{·K}^{-1}$$

15.11 Assuming that ignoring the concentration that reacts introduces no significant error:

$$[A]_{init} - [A]_{reacting} = [A]_{eq} \approx [A]_{init}$$

15.12 Finding K at one temperature given K at another temperature (van't Hoff equation):

$$\ln\frac{K_2}{K_1} = -\frac{\Delta_r H^\circ}{R}\left(\frac{1}{T_2} - \frac{1}{T_1}\right)$$

Brief Solutions to Follow-Up Problems

15.1 (a) $Q = \dfrac{p_{NO}^4 p_{H_2O}^6}{p_{NH_3}^4 p_{O_2}^5}$

(b) $Q = \dfrac{p_{N_2O}\,p_{NO_2}}{p_{NO}^3}$

15.2 $H_2(g) + Br_2(g) \rightleftharpoons 2HBr(g)$

$$Q_{overall} = \frac{p_{HBr}^2}{p_{H_2}p_{Br_2}}$$

$$Q_{overall} = Q_1 \times Q_2 \times Q_3$$

$$= \frac{p_{Br}^2}{p_{Br_2}} \times \frac{p_{HBr}p_{H}}{p_{Br}\,p_{H_2}} \times \frac{p_{HBr}}{p_{H}p_{Br}} = \frac{p_{HBr}^2}{p_{H_2}p_{Br_2}}$$

15.3 (a) $K = K_{ref}^{1/2} = 2.8 \times 10^4$

(b) $K = \left(\dfrac{1}{K_{ref}}\right)^{2/3} = 1.2 \times 10^{-6}$

15.4 $K = K_c(RT)^{-1} = 1.67\left(0.083\ 14\ \dfrac{\text{bar·L}}{\text{mol·K}} \times 500.\ \text{K}\right)^{-1}$

$= 4.02 \times 10^{-2}$

15.5 $K_c = \dfrac{[Y]}{[X]} = 1.4$

1. $Q = 0.33$: right; 2. $Q = 1.4$: no net change; 3. $Q = 2.0$: left

15.6 $Q = \dfrac{(p_{CH_3Cl})(p_{HCl})}{(p_{CH_4})(p_{Cl_2})} = \dfrac{(0.24)(0.47)}{(0.13)(0.035)} = 25;$

$Q < K$, so CH_3Cl is forming.

15.7 From the ICE table for $2NO + O_2 \rightleftharpoons 2NO_2$,

$$p_{O_2} = 1.000\ \text{bar} - x = 0.506\ \text{bar} \qquad x = 0.494\ \text{bar}$$

Also, $p_{NO} = 0.012$ bar and $p_{NO_2} = 0.988$ bar, so

$$K = \frac{(0.988)^2}{(0.012)^2(0.506)} = 1.3 \times 10^4$$

15.8 Since $\Delta n_{gas} = 0$, $K = K_c = 2.3 \times 10^{30} = \dfrac{(0.781)(0.209)}{p_{NO}^2}$;

$p_{NO} = 2.7 \times 10^{-16}$ bar

15.9 From the ICE table, $[H_2] = [I_2] = x$; $[HI] = 0.242 - 2x$;

thus, $K_c = 1.26 \times 10^{-3} = \dfrac{x^2}{(0.242 - 2x)^2}$

Taking the square root of both sides, ignoring the negative root, and solving gives $x = [H_2] = 8.02 \times 10^{-3}$ mol/L.

15.10 (a) Based on the ICE table, and assuming that 0.20 mol/L $- x \approx$ 0.20 mol/L,

$$K_c = 2.94 \times 10^{-10} \approx \frac{4x^2}{0.20} \qquad x \approx 3.8 \times 10^{-6}\ \text{mol/L}$$

The error is 1.9×10^{-3}%, so the assumption is justified. Therefore, at equilibrium, $[I_2] = 0.20$ mol/L and $[I] = 7.6 \times 10^{-6}$ mol/L. (b) Based on the same ICE table and assumption, $x < 0.10$. The error is 50%, so the assumption is *not* justified. Solve the equation:

$$4x^2 + 0.209x - 0.042 = 0 \qquad x = 0.080\ \text{mol/L}$$

Therefore, at equilibrium, $[I_2] = 0.12$ mol/L and $[I] = 0.16$ mol/L.

15.11 (a) $Q_c = \dfrac{(0.0900)(0.0900)}{0.2100} = 3.86 \times 10^{-2}$

$Q_c < K_c$, so the reaction proceeds to the right.
(b) From the ICE table,
$[PCl_5] = 0.2100$ mol/L $- x = 0.2065$ mol/L so $x = 0.0035$ mol/L
Thus, $[Cl_2] = [PCl_3] = 0.0900$ mol/L $+ x = 0.0935$ mol/L.

15.12 (a) $[SiF_4]$ increases; (b) $[SiF_4]$ decreases; (c) $[SiF_4]$ decreases (pure liquids and solids do not affect equilibrium pressures or concentrations); (d) no effect

15.13 (a) Decrease p; (b) increase p; (c) increase p

15.14 (a) p_{H_2} will decrease, and K will increase; (b) p_{N_2} will increase, and K will decrease; (c) both p_{PCl_5} and K will increase.

15.15 (a) Since $p = \frac{n}{V}RT$ and, in this case, V, R, and T cancel,

$$K = \frac{n_{CD}^2}{n_{C_2} \times n_{D_2}} = \frac{16}{(2)(2)} = 4$$

(b) For scene 2, the reaction will proceed to the left. For scene 3, the reaction will proceed to the right.
(c) There are 2 mol of gas on each side of the balanced equation, so increasing the temperature has no effect on the total amount of gas.

PROBLEMS

Problems with **red** numbers are answered in Appendix G and worked in detail in the Student Solutions Manual. Problem sections match those in this book and provide the numbers of relevant sample problems. Most offer Concept Review Questions, Skill-Building Exercises (grouped in pairs covering the same concept), and Problems in Context. The Comprehensive Problems are based on material from any section or previous chapter.

The Equilibrium State and the Equilibrium Constant

Concept Review Questions

15.1 A change in reaction conditions increases the rate of a certain forward reaction more than the rate of the reverse reaction. What are the effects on the equilibrium constant and the concentrations of reactants and products at equilibrium?

15.2 When a chemical company uses a new reaction to manufacture a product, the chemists consider its rate (kinetics) and yield (equilibrium). How does each of these affect the usefulness of the manufacturing process?

15.3 If there is no change in concentrations, why is the equilibrium state considered dynamic?

15.4 Is K very large or very small for a reaction that goes essentially to completion? Explain.

15.5 White phosphorus, P_4, is produced by the reduction of phosphate rock, $Ca_3(PO_4)_2$. If exposed to oxygen, the waxy white solid smokes, bursts into flames, and releases a large quantity of heat:

$$P_4(g) + 5O_2(g) \rightleftharpoons P_4O_{10}(s) + heat$$

Does this reaction have a large or a small equilibrium constant? Explain.

The Reaction Quotient and the Equilibrium Constant
(Sample Problems 15.1 to 15.3)

Concept Review Questions

15.6 For a given reaction at a given temperature, the value of K is constant. Is the value of Q also constant? Explain.

15.7 A chemist is studying the thermal decomposition of lithium peroxide:

$$2Li_2O_2(s) \rightleftharpoons 2Li_2O(s) + O_2(g)$$

The chemist finds that, as long as some Li_2O_2 is present at the end of the experiment, the amount of O_2 obtained in a given container at a given temperature is the same. Explain.

15.8 A chemist is studying the formation of HI from its elements:

$$H_2(g) + I_2(g) \rightleftharpoons 2HI(g)$$

The chemist placed equal amounts of H_2 and I_2 in a container, which was then sealed and heated.
(a) On one set of axes, sketch concentration versus time curves for H_2 and HI, and explain how Q changes as a function of time.
(b) Is the value of Q different if $[I_2]$ is plotted instead of $[H_2]$?

15.9 Explain the difference between a heterogeneous equilibrium and a homogeneous equilibrium. Give an example of each.

15.10 Does Q for the formation of 1 mol of NO from its elements differ from Q for the decomposition of 1 mol of NO to its elements? Explain, and give the relationship between the two Q's.

15.11 Does Q for the formation of 1 mol of NH_3 from H_2 and N_2 differ from Q for the formation of NH_3 from H_2 and 1 mol of N_2? Explain, and give the relationship between the two Q's.

Skill-Building Exercises (grouped in similar pairs)

15.12 Balance each reaction, and write its reaction quotient, Q:
(a) $NO(g) + O_2(g) \rightleftharpoons N_2O_3(g)$
(b) $SF_6(g) + SO_3(g) \rightleftharpoons SO_2F_2(g)$
(c) $SClF_5(g) + H_2(g) \rightleftharpoons S_2F_{10}(g) + HCl(g)$

15.13 Balance each reaction, and write its reaction quotient, Q:
(a) $C_2H_6(g) + O_2(g) \rightleftharpoons CO_2(g) + H_2O(g)$
(b) $CH_4(g) + F_2(g) \rightleftharpoons CF_4(g) + HF(g)$
(c) $SO_3(g) \rightleftharpoons SO_2(g) + O_2(g)$

15.14 Balance each reaction, and write its reaction quotient, Q:
(a) $NO_2Cl(g) \rightleftharpoons NO_2(g) + Cl_2(g)$
(b) $POCl_3(g) \rightleftharpoons PCl_3(g) + O_2(g)$
(c) $NH_3(g) + O_2(g) \rightleftharpoons N_2(g) + H_2O(g)$

15.15 Balance each reaction, and write its reaction quotient, Q:
(a) $O_2(g) \rightleftharpoons O_3(g)$
(b) $NO(g) + O_3(g) \rightleftharpoons NO_2(g) + O_2(g)$
(c) $N_2O(g) + H_2(g) \rightleftharpoons NH_3(g) + H_2O(g)$

15.16 At a particular temperature, $K = 1.6 \times 10^{-2}$ for the following reaction:

$$2H_2S(g) \rightleftharpoons 2H_2(g) + S_2(g)$$

Calculate K for each of these reactions:
(a) $\frac{1}{2}S_2(g) + H_2(g) \rightleftharpoons H_2S(g)$
(b) $5H_2S(g) \rightleftharpoons 5H_2(g) + \frac{5}{2}S_2(g)$

15.17 At a particular temperature, $K = 6.5 \times 10^2$ for the following reaction:

$$2NO(g) + 2H_2(g) \rightleftharpoons N_2(g) + 2H_2O(g)$$

Calculate K for each of these reactions:
(a) $NO(g) + H_2(g) \rightleftharpoons \frac{1}{2}N_2(g) + H_2O(g)$
(b) $2N_2(g) + 4H_2O(g) \rightleftharpoons 4NO(g) + 4H_2(g)$

15.18 Balance each example of heterogeneous equilibrium, and write each reaction quotient, Q:
(a) $Na_2O_2(s) + CO_2(g) \rightleftharpoons Na_2CO_3(s) + O_2(g)$
(b) $H_2O(l) \rightleftharpoons H_2O(g)$
(c) $NH_4Cl(s) \rightleftharpoons NH_3(g) + HCl(g)$

15.19 Balance each example of heterogeneous equilibrium, and write each reaction quotient, Q:
(a) $H_2O(l) + SO_3(g) \rightleftharpoons H_2SO_4(aq)$
(b) $KNO_3(s) \rightleftharpoons KNO_2(s) + O_2(g)$
(c) $S_8(s) + F_2(g) \rightleftharpoons SF_6(g)$

15.20 Balance each example of heterogeneous equilibrium, and write each reaction quotient, Q:
(a) $NaHCO_3(s) \rightleftharpoons Na_2CO_3(s) + CO_2(g) + H_2O(g)$
(b) $SnO_2(s) + H_2(g) \rightleftharpoons Sn(s) + H_2O(g)$
(c) $H_2SO_4(l) + SO_3(g) \rightleftharpoons H_2S_2O_7(l)$

15.21 Balance each example of heterogeneous equilibrium, and write each reaction quotient, Q:
(a) $Al(s) + NaOH(aq) + H_2O(l) \rightleftharpoons Na[Al(OH)_4](aq) + H_2(g)$
(b) $CO_2(s) \rightleftharpoons CO_2(g)$
(c) $N_2O_5(s) \rightleftharpoons NO_2(g) + O_2(g)$

Problems in Context

15.22 Write Q for each reaction:
(a) Hydrogen chloride gas reacts with oxygen gas to produce chlorine gas and water vapour.
(b) Solid diarsenic trioxide reacts with fluorine gas to produce liquid arsenic pentafluoride and oxygen gas.
(c) Gaseous sulfur tetrafluoride reacts with liquid water to produce gaseous sulfur dioxide and hydrogen fluoride gas.
(d) Solid molybdenum(VI) oxide reacts with gaseous xenon difluoride to form liquid molybdenum(VI) fluoride, xenon gas, and oxygen gas.

15.23 The interhalogen ClF_3 is prepared in a two-step fluorination of chlorine gas:

$$Cl_2(g) + F_2(g) \rightleftharpoons ClF(g)$$
$$ClF(g) + F_2(g) \rightleftharpoons ClF_3(g)$$

(a) Balance each step, and write the overall equation.
(b) Show that the overall Q equals the product of the Q's for the individual steps.

Expressing Equilibria with Pressure Terms: Relation between K and K_c

(Sample Problem 15.4)

Concept Review Questions

15.24 Guldberg and Waage proposed the definition of the equilibrium constant as a certain ratio of *concentrations*. What relationship allows us to use a particular ratio of *partial pressures* (for a gaseous reaction) to express an equilibrium constant? Explain.

15.25 When are K_c and K equal, and when are they not?

15.26 A certain reaction at equilibrium has a larger amount (mol) of gaseous products than of gaseous reactants.
(a) Is K_c larger or smaller than K?
(b) Write a statement about the relative sizes of K_c and K for any gaseous equilibrium.

Skill-Building Exercises (grouped in similar pairs)

15.27 Determine Δn_{gas} for each reaction:
(a) $2KClO_3(s) \rightleftharpoons 2KCl(s) + 3O_2(g)$
(b) $2PbO(s) + O_2(g) \rightleftharpoons 2PbO_2(s)$
(c) $I_2(s) + 3XeF_2(s) \rightleftharpoons 2IF_3(s) + 3Xe(g)$

15.28 Determine Δn_{gas} for each reaction:
(a) $MgCO_3(s) \rightleftharpoons MgO(s) + CO_2(g)$
(b) $2H_2(g) + O_2(g) \rightleftharpoons 2H_2O(l)$
(c) $HNO_3(l) + ClF(g) \rightleftharpoons ClONO_2(g) + HF(g)$

15.29 Calculate K_c for each equilibrium:
(a) $CO(g) + Cl_2(g) \rightleftharpoons COCl_2(g)$; $K = 3.9 \times 10^{-2}$ at 1000. K
(b) $S_2(g) + C(s) \rightleftharpoons CS_2(g)$; $K = 28.5$ at 500. K

15.30 Calculate K_c for each equilibrium:
(a) $H_2(g) + I_2(g) \rightleftharpoons 2HI(g)$; $K = 49$ at 730. K
(b) $2SO_2(g) + O_2(g) \rightleftharpoons 2SO_3(g)$; $K = 2.5 \times 10^{10}$ at 500. K

15.31 Calculate K for each equilibrium:
(a) $N_2O_4(g) \rightleftharpoons 2NO_2(g)$; $K_c = 6.1 \times 10^{-3}$ at 298 K
(b) $N_2(g) + 3H_2(g) \rightleftharpoons 2NH_3(g)$; $K_c = 2.4 \times 10^{-3}$ at 1000. K

15.32 Calculate K for each equilibrium:
(a) $H_2(g) + CO_2(g) \rightleftharpoons H_2O(g) + CO(g)$; $K_c = 0.77$ at 1020. K
(b) $3O_2(g) \rightleftharpoons 2O_3(g)$; $K_c = 1.8 \times 10^{-56}$ at 570. K

Comparing Q and K to Determine Reaction Direction

(Sample Problems 15.5 and 15.6)

Concept Review Questions

15.33 When the numerical value of Q is less than the numerical value of K, in which direction does the reaction proceed to reach equilibrium? Explain.

15.34 The following molecular scenes depict the aqueous reaction $2D \rightleftharpoons E$ (D is *red*, and E is *blue*). Each sphere represents 0.0100 mol, but the volume is 1.00 L in scene A and 0.500 L in scenes B and C.

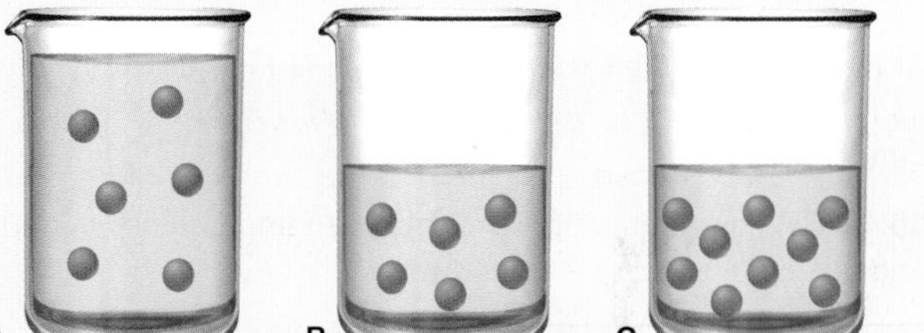
A B C

(a) If the reaction in scene A is at equilibrium, calculate K_c.
(b) Are the reactions in scenes B and C at equilibrium? Which, if either, is not, and in which direction will it proceed?

Skill-Building Exercises (grouped in similar pairs)

15.35 At 425°C, $K = 4.18 \times 10^{-9}$ for the following reaction:

$$2HBr(g) \rightleftharpoons H_2(g) + Br_2(g)$$

In one experiment, 0.20 bar of HBr(g), 0.010 bar of $H_2(g)$, and 0.010 bar of $Br_2(g)$ is introduced into a container. Is the system at equilibrium? If not, in which direction will the reaction proceed?

15.36 At 100°C, $K = 60.6$ for the following reaction:

$$2NOBr(g) \rightleftharpoons 2NO(g) + Br_2(g)$$

In a given experiment, 0.10 bar of each component is placed in a container. Is the system at equilibrium? If not, in which direction will the reaction proceed?

Problem in Context

15.37 The water-gas shift reaction plays a central role in the chemical methods for obtaining cleaner fuels from coal:

$$CO(g) + H_2O(g) \rightleftharpoons CO_2(g) + H_2(g)$$

At a given temperature, $K = 2.7$. If 0.13 mol of CO, 0.56 mol of H_2O, 0.62 mol of CO_2, and 0.43 mol of H_2 are put in a 2.0 L flask, in which direction will the reaction proceed?

How to Solve Equilibrium Problems
(Sample Problems 15.7 to 15.11)

Concept Review Questions

15.38 In the 1980s, CFC-11 was one of the most heavily produced chlorofluorocarbons. The last step in its formation is given below:

$$CCl_4(g) + HF(g) \rightleftharpoons CFCl_3(g) + HCl(g)$$

If you start the reaction with equal concentrations of CCl_4 and HF, you obtain equal concentrations of $CFCl_3$ and HCl at equilibrium. Are the final concentrations of $CFCl_3$ and HCl equal if you start with unequal concentrations of CCl_4 and HF? Explain.

15.39 For a problem involving the catalyzed reaction of methane and steam, the following ICE table was prepared:

Pressure (bar)	$CH_4(g)$	+ $2H_2O(g)$	$\rightleftharpoons$ $CO_2(g)$	+ $4H_2(g)$
Initial	0.30	0.40	0	0
Change	$-x$	$-2x$	$+x$	$+4x$
Equilibrium	$0.30 - x$	$0.40 - 2x$	x	$4x$

Explain the entries in the "Change" and "Equilibrium" rows.

15.40 (a) What is the basis of the approximation that avoids using the quadratic formula to find an equilibrium concentration? (b) When should this approximation *not* be made?

Skill-Building Exercises (grouped in similar pairs)

15.41 In an experiment to study the formation of HI(*g*), $H_2(g)$ and $I_2(g)$ were placed in a sealed container at a certain temperature:

$$H_2(g) + I_2(g) \rightleftharpoons 2HI(g)$$

At equilibrium, $[H_2] = 6.50 \times 10^{-5}$ mol/L, $[I_2] = 1.06 \times 10^{-3}$ mol/L, and $[HI] = 1.87 \times 10^{-3}$ mol/L. Calculate K_c for the reaction at this temperature.

15.42 Gaseous ammonia was introduced into a sealed container and heated to a certain temperature:

$$2NH_3(g) \rightleftharpoons N_2(g) + 3H_2(g)$$

At equilibrium, $[NH_3] = 0.0225$ mol/L, $[N_2] = 0.114$ mol/L, and $[H_2] = 0.342$ mol/L. Calculate K_c for the reaction at this temperature.

15.43 Gaseous PCl_5 decomposes according to the following reaction:

$$PCl_5(g) \rightleftharpoons PCl_3(g) + Cl_2(g)$$

In one experiment, 0.15 mol of $PCl_5(g)$ was introduced into a 2.0 L container. Construct the ICE table for this experiment.

15.44 Hydrogen fluoride, HF, can be made from the following reaction:

$$H_2(g) + F_2(g) \rightleftharpoons 2HF(g)$$

In one experiment, 0.10 mol of $H_2(g)$ and 0.050 mol of $F_2(g)$ are added to a 0.50 L flask. Write an ICE table for this experiment.

15.45 For the following reaction, $K = 6.5 \times 10^4$ at 308 K:

$$2NO(g) + Cl_2(g) \rightleftharpoons 2NOCl(g)$$

At equilibrium, $p_{NO} = 0.35$ bar and $p_{Cl_2} = 0.10$ bar. What is the equilibrium partial pressure of NOCl(*g*)?

15.46 For the following reaction, $K = 0.262$ at 1000°C:

$$C(s) + 2H_2(g) \rightleftharpoons CH_4(g)$$

At equilibrium, p_{H_2} is 1.22 bar. What is the equilibrium partial pressure of $CH_4(g)$?

15.47 Ammonium hydrogen sulfide decomposes according to the following reaction, for which $K = 0.11$ at 250°C:

$$NH_4HS(s) \rightleftharpoons H_2S(g) + NH_3(g)$$

If 55.0 g of $NH_4HS(s)$ is placed in a sealed 5.0 L container, what is the partial pressure of $NH_3(g)$ at equilibrium?

15.48 Hydrogen sulfide decomposes according to the following reaction, for which $K_c = 9.30 \times 10^{-8}$ at 700°C:

$$2H_2S(g) \rightleftharpoons 2H_2(g) + S_2(g)$$

If 0.45 mol of H_2S is placed in a 3.0 L container, what is the equilibrium concentration of $H_2(g)$ at 700°C?

15.49 Even at high *T*, the formation of NO is not favoured:

$$N_2(g) + O_2(g) \rightleftharpoons 2NO(g) \qquad K_c = 4.10 \times 10^{-4} \text{ at } 2000°C$$

What is [NO] when a mixture of 0.20 mol of $N_2(g)$ and 0.15 mol of $O_2(g)$ reaches equilibrium in a 1.0 L container at 2000°C?

15.50 Nitrogen dioxide decomposes according to the following reaction, where $K = 4.48 \times 10^{-13}$ at a certain temperature:

$$2NO_2(g) \rightleftharpoons 2NO(g) + O_2(g)$$

If 0.75 bar of NO_2 is added to a container and allowed to come to equilibrium, what are the equilibrium partial pressures of NO(*g*) and $O_2(g)$?

15.51 Hydrogen iodide decomposes according to the following reaction:

$$2HI(g) \rightleftharpoons H_2(g) + I_2(g)$$

A sealed 1.50 L container initially holds 0.006 23 mol of H_2, 0.004 14 mol of I_2, and 0.0244 mol of HI at 703 K. When equilibrium is reached, the concentration of $H_2(g)$ is 0.004 67 mol/L. What are the concentrations of HI(*g*) and $I_2(g)$?

15.52 Compound A decomposes according to the following equation:

$$A(g) \rightleftharpoons 2B(g) + C(g)$$

A sealed 1.00 L container initially contains 1.75×10^{-3} mol of A(*g*), 1.25×10^{-3} mol of B(*g*), and 6.50×10^{-4} mol of C(*g*) at 100°C. At equilibrium, [A] is 2.15×10^{-3} mol/L. Find [B] and [C].

Problems in Context

15.53 In an analysis of interhalogen reactivity, 0.500 mol of ICl was placed in a 5.00 L flask, where it decomposed at a high *T*:

$$2ICl(g) \rightleftharpoons I_2(g) + Cl_2(g)$$

Calculate the equilibrium concentrations of I_2, Cl_2, and ICl. (K_c = 0.110 at this temperature.)

15.54 A toxicologist studying mustard gas, $S(CH_2CH_2Cl)_2$, a blistering agent, prepares a mixture of 0.675 mol/L SCl_2 and 0.973 mol/L C_2H_4 and allows it to react at room temperature (20.0°C):

$$SCl_2(g) + 2C_2H_4(g) \rightleftharpoons S(CH_2CH_2Cl)_2(g)$$

At equilibrium, $[S(CH_2CH_2Cl)_2] = 0.350$ mol/L. Calculate *K*.

15.55 The first step in HNO_3 production is the catalyzed oxidation of NH_3. Without a catalyst, a different reaction predominates:

$$4NH_3(g) + 3O_2(g) \rightleftharpoons 2N_2(g) + 6H_2O(g)$$

When 0.0150 mol of $NH_3(g)$ and 0.0150 mol of $O_2(g)$ are placed in a 1.00 L container at a certain temperature, the N_2 concentration at equilibrium is 1.96×10^{-3} mol/L. Calculate K_c.

15.56 A key step in the extraction of iron from its ore is given below:

$$FeO(s) + CO(g) \rightleftharpoons Fe(s) + CO_2(g) \qquad K = 0.403 \text{ at } 1000°C$$

This step occurs in the 700°C to 1200°C temperature range within a blast furnace. What are the equilibrium partial pressures of

$CO(g)$ and $CO_2(g)$ when 1.00 bar of $CO(g)$ and excess $FeO(s)$ react in a sealed container at 1000°C?

Reaction Conditions and Equilibrium: Le Châtelier's Principle

(Sample Problems 15.12 to 15.15)

Concept Review Questions

15.57 What does the word *disturbance* mean in Le Châtelier's principle?

15.58 What is the difference between the equilibrium position and the equilibrium constant of a reaction? Which changes as a result of a change in reactant concentration?

15.59 Scenes A, B, and C (*below*) depict the following reaction at three temperatures:

$$NH_4Cl(s) \rightleftharpoons NH_3(g) + HCl(g) \qquad \Delta_rH^\circ = 176\ \text{kJ/mol}$$

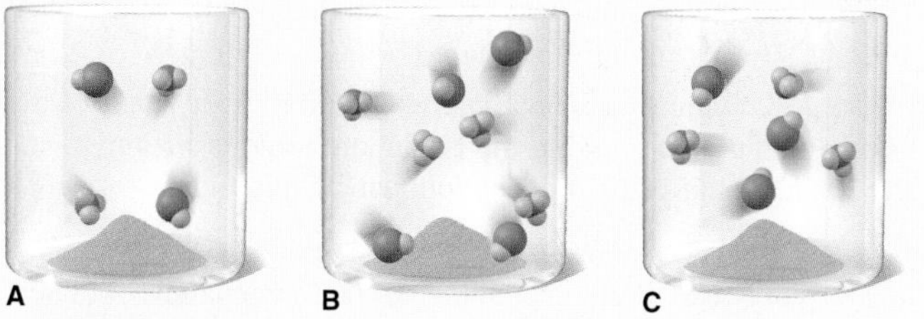

(a) Which scene best represents the reaction mixture at the highest temperature? Explain.
(b) Which scene best represents the reaction mixture at the lowest temperature? Explain.

15.60 What is implied by the word *constant* in the term *equilibrium constant*? Give two reaction parameters that can be changed without changing the value of an equilibrium constant.

15.61 Le Châtelier's principle is related ultimately to the rates of the forward and reverse steps in a reaction. Explain (a) why an increase in reactant concentration shifts the equilibrium position to the right, but does not change K; (b) why a decrease in V shifts the equilibrium position toward fewer moles of gas, but does not change K; (c) why a rise in T shifts the equilibrium position of an exothermic reaction toward reactants and also changes K; and (d) why a rise in temperature of an endothermic reaction from T_1 to T_2 results in K_2 being larger than K_1.

15.62 An equilibrium mixture of two solids and a gas, in the following reaction, is depicted below (X is *green*, and Y is *black*):

$$XY(s) \rightleftharpoons X(g) + Y(s)$$

Does scene A, B, or C best represent the system at equilibrium after two formula units of Y(*s*) is added? Explain.

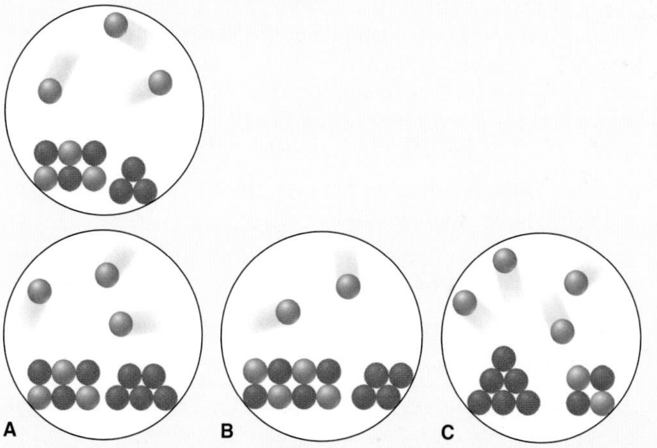

Skill-Building Exercises (grouped in similar pairs)

15.63 Consider this equilibrium system:

$$CO(g) + Fe_3O_4(s) \rightleftharpoons CO_2(g) + 3FeO(s)$$

How does the equilibrium position shift as a result of each disturbance?
(a) CO is added.
(b) CO_2 is removed by adding solid NaOH.
(c) Additional $Fe_3O_4(s)$ is added to the system.
(d) Dry ice is added at constant temperature.

15.64 Sodium bicarbonate undergoes thermal decomposition according to the following reaction:

$$2NaHCO_3(s) \rightleftharpoons Na_2CO_3(s) + CO_2(g) + H_2O(g)$$

How does the equilibrium position shift as a result of each disturbance?
(a) 0.20 bar of argon gas is added.
(b) $NaHCO_3(s)$ is added.
(c) $Mg(ClO_4)_2(s)$ is added as a drying agent to remove H_2O.
(d) Dry ice is added at constant temperature.

15.65 Predict the effect of *increasing* the container volume on the amount of each reactant and product in each reaction:
(a) $F_2(g) \rightleftharpoons 2F(g)$
(b) $2CH_4(g) \rightleftharpoons C_2H_2(g) + 3H_2(g)$

15.66 Predict the effect of *increasing* the container volume on the amount of each reactant and product in each reaction:
(a) $CH_3OH(l) \rightleftharpoons CH_3OH(g)$
(b) $CH_4(g) + NH_3(g) \rightleftharpoons HCN(g) + 3H_2(g)$

15.67 Predict the effect of *decreasing* the container volume on the amount of each reactant and product in each reaction:
(a) $H_2(g) + Cl_2(g) \rightleftharpoons 2HCl(g)$
(b) $2H_2(g) + O_2(g) \rightleftharpoons 2H_2O(l)$

15.68 Predict the effect of *decreasing* the container volume on the amount of each reactant and product in each reaction:
(a) $C_3H_8(g) + 5O_2(g) \rightleftharpoons 3CO_2(g) + 4H_2O(l)$
(b) $4NH_3(g) + 3O_2(g) \rightleftharpoons 2N_2(g) + 6H_2O(g)$

15.69 How would you adjust the *volume* of the container to maximize the product yield in each reaction?
(a) $Fe_3O_4(s) + 4H_2(g) \rightleftharpoons 3Fe(s) + 4H_2O(g)$
(b) $2C(s) + O_2(g) \rightleftharpoons 2CO(g)$

15.70 How would you adjust the *volume* of the container to maximize the product yield in each reaction?
(a) $Na_2O_2(s) \rightleftharpoons 2Na(l) + O_2(g)$
(b) $C_2H_2(g) + 2H_2(g) \rightleftharpoons C_2H_6(g)$

15.71 Predict the effect of *increasing* the temperature on the amount of each product in each reaction:
(a) $CO(g) + 2H_2(g) \rightleftharpoons CH_3OH(g) \qquad \Delta_rH^\circ = -90.7$ kJ/mol
(b) $C(s) + H_2O(g) \rightleftharpoons CO(g) + H_2(g) \qquad \Delta_rH^\circ = 131$ kJ/mol
(c) $2NO_2(g) \rightleftharpoons 2NO(g) + O_2(g)$ (endothermic)
(d) $2C(s) + O_2(g) \rightleftharpoons 2CO(g)$ (exothermic)

15.72 Predict the effect of *decreasing* the temperature on the amount of each reactant in each reaction:
(a) $C_2H_2(g) + H_2O(g) \rightleftharpoons CH_3CHO(g) \qquad \Delta_rH^\circ = 2151$ kJ/mol
(b) $CH_3CH_2OH(l) + O_2(g) \rightleftharpoons CH_3CO_2H(l) + H_2O(g)$
$\Delta_rH^\circ = -451$ kJ/mol
(c) $2C_2H_4(g) + O_2(g) \rightleftharpoons 2CH_3CHO(g)$ (exothermic)
(d) $N_2O_4(g) \rightleftharpoons 2NO_2(g)$ (endothermic)

15.73 The molecule D_2 (where D, deuterium, is ^2H) undergoes a reaction with ordinary H_2 that leads to isotopic equilibrium:

$$D_2(g) + H_2(g) \rightleftharpoons 2DH(g) \qquad K = 1.80 \text{ at } 298\ \text{K}$$

If Δ_rH° is 0.32 kJ/mol DH, calculate K at 500. K.

15.74 The formation of methanol is important to the processing of new fuels. At 298 K, $K = 2.25 \times 10^4$ for the following reaction:

$$CO(g) + 2H_2(g) \rightleftharpoons CH_3OH(l)$$

If $\Delta_r H° = -128$ kJ/mol CH_3OH, calculate K at 0°C.

Problems in Context

15.75 The minerals hematite (Fe_2O_3) and magnetite (Fe_3O_4) exist in equilibrium with atmospheric oxygen:

$$4Fe_3O_4(s) + O_2(g) \rightleftharpoons 6Fe_2O_3(s) \qquad K = 2.5 \times 10^{87} \text{ at } 298 \text{ K}$$

(a) Determine p_{O_2} at equilibrium.
(b) Given that p_{O_2} in air is 0.21 bar, in which direction will the reaction proceed to reach equilibrium?
(c) Calculate K_c at 298 K.

15.76 The oxidation of SO_2 is the key step in H_2SO_4 production:

$$SO_2(g) + \frac{1}{2}O_2(g) \rightleftharpoons SO_3(g) \qquad \Delta_r H° = -99.2 \text{ kJ/mol}$$

(a) What qualitative combination of temperature and pressure maximizes SO_3 yield?
(b) How does the addition of O_2 affect Q? How does it affect K?
(c) Why is catalysis used for this reaction?

15.77 A mixture of 3.00 volumes of H_2 and 1.00 volume of N_2 reacts at 344°C to form ammonia. The equilibrium mixture at 110. bar contains 41.49% NH_3 by volume. Calculate K for the reaction, assuming that the gases behave ideally.

15.78 You are a member of a research team of chemists discussing plans for a plant to produce ammonia:

$$N_2(g) + 3H_2(g) \rightleftharpoons 2NH_3(g)$$

(a) The plant will operate at close to 700 K, at which K is 1.00×10^{-4}, and use the stoichiometric 1/3 ratio of N_2/H_2. At equilibrium, the partial pressure of NH_3 is 50. bar. Calculate the partial pressure of each reactant and p_{total}.
(b) One member of the team has the following suggestion: since the partial pressure of H_2 is cubed in the reaction quotient, the plant could produce the same amount of NH_3 if the reactants were in a 1/6 ratio of N_2/H_2 and could do so at a lower pressure, which would cut operating costs. Calculate the partial pressure of each reactant and p_{total} under these conditions, assuming an unchanged partial pressure of 50. bar for NH_3. Is the suggestion valid?

Comprehensive Problems

15.79 Which changes will form more $CaCO_3$ in this equilibrium system?

$$CO_2(g) + CaOH_2(s) \rightleftharpoons CaCO_3(s) + H_2O(l) \qquad \Delta_r H° = -113 \text{ kJ/mol}$$

(a) Decrease temperature at constant pressure (no phase change).
(b) Increase volume at constant temperature.
(c) Increase partial pressure of CO_2.
(d) Remove one-half of the initial $CaCO_3$.

15.80 A gaseous mixture reaches equilibrium over time according to the following reaction:

$$X_2(g) + Y_2(g) \rightleftharpoons 2XY(g)$$

The "filmstrip" represents five molecular scenes of this reaction (X is *purple*, and Y is *orange*):

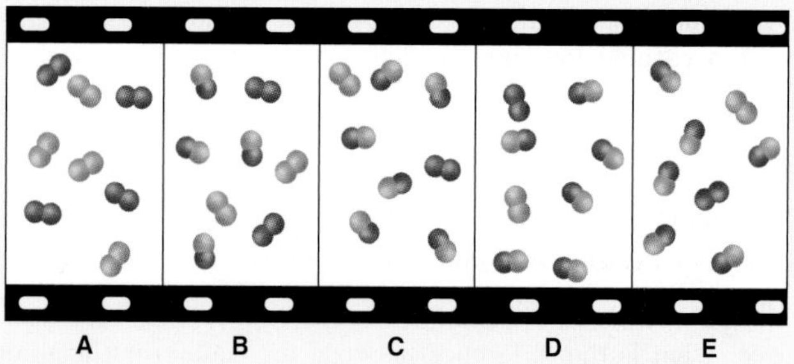

(a) Write the reaction quotient, Q, for this reaction.
(b) If each particle represents 0.1 mol, find Q for each scene.
(c) If $K < 1$, is time progressing to the right or to the left? Explain.
(d) Calculate K at this temperature.
(e) If $\Delta_r H° < 0$, which scene, if any, best represents the mixture at a higher temperature? Explain.
(f) Which scene, if any, best represents the mixture at a higher pressure (lower volume)? Explain.

15.81 Ammonium carbamate (NH_2COONH_4) is a salt of carbamic acid that is found in the blood and urine of mammals. At 250.°C, $K_c = 1.58 \times 10^{-8}$ for the following equilibrium:

$$NH_2COONH_4(s) \rightleftharpoons 2NH_3(g) + CO_2(g)$$

If 7.80 g of NH_2COONH_4 is put into a 0.500 L evacuated container, what is the total pressure at equilibrium?

15.82 Isolation of group 10 elements, used as industrial catalysts, involves a series of steps. For nickel, the sulfide ore is roasted in air:

$$Ni_3S_2(s) + O_2(g) \rightleftharpoons NiO(s) + SO_2(g)$$

The metal oxide is reduced by the H_2 in water gas ($CO + H_2$) to impure Ni:

$$NiO(s) + H_2(g) \rightleftharpoons Ni(s) + H_2O(g)$$

The CO in water gas then reacts with the metal in the Mond process to form gaseous nickel carbonyl, which is subsequently decomposed to the metal:

$$Ni(s) + CO(g) \rightleftharpoons NiCO_4(g)$$

(a) Balance each of the three steps, and obtain an overall balanced equation for the conversion of Ni_3S_2 to $Ni(CO)_4$.
(b) Show that the overall Q_c is the product of the Q_c's for the individual reactions.

15.83 Consider the formation of ammonia in two experiments.
(a) To a 1.00 L container at 727°C, 1.30 mol of N_2 and 1.65 mol of H_2 is added. At equilibrium, 0.100 mol of NH_3 is present. Calculate the equilibrium concentrations of N_2 and H_2, and find K_c for the following reaction:

$$2NH_3(g) \rightleftharpoons N_2(g) + 3H_2(g)$$

(b) In a different 1.00 L container at the same temperature, equilibrium is established with 8.34×10^{-2} mol of NH_3, 1.50 mol of N_2, and 1.25 mol of H_2. Calculate K_c for this reaction:

$$NH_3(g) \rightleftharpoons \frac{1}{2}N_2(g) + \frac{3}{2}H_2(g)$$

(c) What is the relationship between the K_c values in parts (a) and (b)? Why are these values not the same?

15.84 An important industrial source of ethanol is the reaction, catalyzed by H_3PO_4, of steam with ethylene derived from oil:

$$C_2H_4(g) + H_2O(g) \rightleftharpoons C_2H_5OH(g)$$

$$\Delta_r H° = -47.8 \text{ kJ/mol} \qquad K_c = 9 \times 10^3 \text{ at } 600. \text{ K}$$

(a) At equilibrium, $p_{C_2H_5OH} = 200.$ bar and $p_{H_2O} = 400.$ bar. Calculate $p_{C_2H_4}$.

(b) Is the highest yield of ethanol obtained at high or low pressure? Is it obtained at high or low temperature?
(c) Calculate K_c at 450. K.
(d) In NH_3 manufacture, the yield is increased by condensing the NH_3 to a liquid and removing it. Would condensing the C_2H_5OH have the same effect in ethanol production? Explain.

15.85 An industrial chemist introduces 2.0 bar of H_2 and 2.0 bar of CO_2 into a 1.00 L container at 25.0°C and then raises the temperature to 700.°C, at which $K_c = 0.534$:

$$H_2(g) + CO_2(g) \rightleftharpoons H_2O(g) + CO(g)$$

What mass of H_2 is present at equilibrium?

15.86 As an Environment Canada scientist studying catalytic converters and urban smog, you want to find K_c for the following reaction:

$$2NO_2(g) \rightleftharpoons N_2(g) + 2O_2(g) \qquad K_c = ?$$

Use the following data to find the unknown K_c:

$$\frac{1}{2}N_2(g) + \frac{1}{2}O_2(g) \rightleftharpoons NO(g) \qquad K_c = 4.8 \times 10^{-10}$$

$$2NO_2(g) \rightleftharpoons 2NO(g) + O_2(g) \quad K_c = 1.1 \times 10^{-5}$$

15.87 An engineer who is examining the oxidation of SO_2 in the manufacture of sulfuric acid determines that $K_c = 1.7 \times 10^8$ at 600. K:

$$2SO_2(g) + O_2(g) \rightleftharpoons 2SO_3(g)$$

(a) At equilibrium, $p_{SO_3} = 300.$ bar and $p_{O_2} = 100.$ bar. Calculate p_{SO_2}.
(b) The engineer places a mixture of 0.0040 mol of $SO_2(g)$ and 0.0028 mol of $O_2(g)$ in a 1.0 L container and raises the temperature to 1000 K. At equilibrium, 0.0020 mol of $SO_3(g)$ is present. Calculate K_c and p_{SO_2} for this reaction at 1000. K.

15.88 Phosgene ($COCl_2$) is a toxic substance that forms readily from carbon monoxide and chlorine at elevated temperatures:

$$CO(g) + Cl_2(g) \rightleftharpoons COCl_2(g)$$

If 0.350 mol of each reactant is placed in a 0.500 L flask at 600 K, what are the concentrations of all three substances at equilibrium? ($K_c = 4.95$ at this temperature.)

15.89 When 0.100 mol of $CaCO_3(s)$ and 0.100 mol of $CaO(s)$ are placed in an evacuated sealed 10.0 L container and heated to 385 K, $p_{CO_2} = 0.220$ bar after equilibrium is established:

$$CaCO_3(s) \rightleftharpoons CaO(s) + CO_2(g)$$

An additional 0.300 bar of $CO_2(g)$ is pumped in. What is the total mass (g) of $CaCO_3$ after equilibrium is re-established?

15.90 Use each reaction quotient to write a balanced equation:

(a) $Q = \dfrac{[CO_2]^2[H_2O]^2}{[C_2H_4][O_2]^3}$

(b) $Q = \dfrac{[NH_3]^4[O_2]^7}{[NO_2]^4[H_2O]^6}$

15.91 Hydrogenation of carbon-carbon π bonds is important in the petroleum and food industries. The conversion of acetylene to ethylene is a simple example of the process:

$$C_2H_2(g) + H_2(g) \rightleftharpoons C_2H_4(g)$$

The calculated K_c at 2000. K is 2.9×10^8. However, the process is run at lower temperatures with the aid of a catalyst to prevent decomposition. Use $\Delta H°$ values to calculate K_c at 300. K.

15.92 Consider the following reaction:

$$M_2 + N_2 \rightleftharpoons 2MN$$

Scene A (*below*) represents the mixture at equilibrium (M is *black*, and N is *orange*). If each molecule represents 0.10 mol and the volume is 1.0 L, what amount (mol) of each substance will be present in scene B, when the mixture reaches equilibrium?

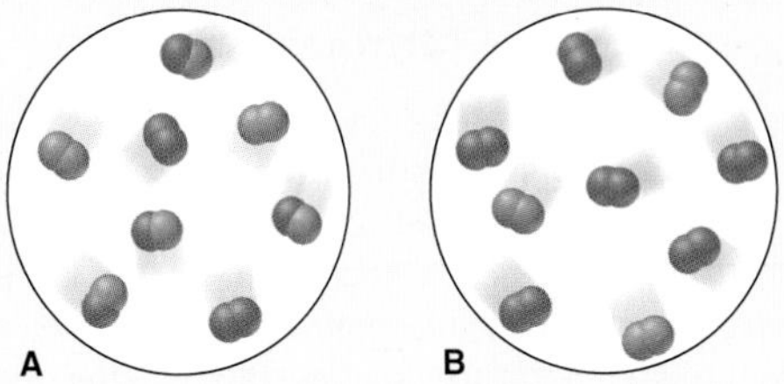

15.93 Highly toxic disulfur decafluoride decomposes by a free-radical process:

$$S_2F_{10}(g) \rightleftharpoons SF_4(g) + SF_6(g)$$

In a study of this decomposition, S_2F_{10} was placed in a 2.0 L flask and heated to 100°C; $[S_2F_{10}]$ was 0.50 mol/L at equilibrium. More S_2F_{10} was added and, when equilibrium was re-established, $[S_2F_{10}]$ was 2.5 mol/L. How did $[SF_4]$ and $[SF_6]$ change from the original to the new equilibrium position after the addition of more S_2F_{10}?

15.94 In a study of the water-gas shift reaction (see Problem 15.37), equilibrium was reached with $[CO] = [H_2O] = [H_2] = 0.10$ mol/L and $[CO_2] = 0.40$ mol/L. After 0.60 mol of H_2 was added to the 2.0 L container and equilibrium was re-established, what were the new concentrations of all the components?

15.95 A gaseous mixture of 10.0 volumes of CO_2, 1.00 volume of unreacted O_2, and 50.0 volumes of unreacted N_2 leaves an engine at 4.0 bar and 800. K. Assuming that the mixture reaches equilibrium, what are (a) the partial pressure and (b) the concentration (in picograms per litre, pg/L) of CO in this exhaust gas?

$$2CO_2(g) \rightleftharpoons 2CO(g) + O_2(g) \qquad K = 1.4 \times 10^{-28} \text{ at 800. K}$$

(The actual concentration of CO in exhaust gas is much higher because the gases do *not* reach equilibrium in the short transit time through the engine and exhaust system.)

15.96 When ammonia is made industrially, the mixture of N_2, H_2, and NH_3 that emerges from the reaction chamber is far from equilibrium. Why does the plant supervisor use reaction conditions that produce less than the maximum yield of ammonia?

15.97 The following reaction can be used to make H_2 for the synthesis of ammonia from the greenhouse gases carbon dioxide and methane:

$$CH_4(g) + CO_2(g) \rightleftharpoons 2CO(g) + 2H_2(g)$$

(a) An equimolar mixture of CH_4 and CO_2, with a total pressure of 20.0 bar, reaches equilibrium at 1200. K. If $K = 3.548 \times 10^6$, what is the percent yield of H_2?
(b) What is the percent yield of H_2 for this system at 1300. K, if $K = 2.626 \times 10^7$?
(c) Use the van't Hoff equation to find $\Delta_r H°$.

15.98 The methane that is used to obtain H_2 for NH_3 manufacture is impure and usually contains other hydrocarbons, such as propane, C_3H_8. Imagine the reaction of propane occurring in two steps, as follows:

(1) $C_3H_8(g) + 3H_2O(g) \rightleftharpoons 3CO(g) + 7H_2(g)$
$K = 8.175 \times 10^{15}$ at 1200. K

(2) $CO(g) + H_2O(g) \rightleftharpoons CO_2(g) + H_2(g)$
$K = 0.6944$ at 1200. K

(a) Write the overall equation for the reaction of propane and steam to produce carbon dioxide and hydrogen.

(b) Calculate K for the overall reaction at 1200. K.
(c) When 1.00 volume of C_3H_8 and 4.00 volumes of H_2O, each at 1200. K and 5.0 bar, are mixed in a container, what is the final pressure? Assume that the total volume remains constant, the reaction is essentially complete, and the gases behave ideally.
(d) What percentage of the C_3H_8 remains unreacted?

15.99 Using CH_4 and steam as a source of H_2 for NH_3 synthesis requires high temperatures. Rather than burning CH_4 separately to heat the mixture, it is more efficient to inject some O_2 into the reaction mixture. All the H_2 is, thus, released for the synthesis, and the heat of reaction for the combustion of CH_4 helps to maintain the required temperature. Imagine the reaction occurring in two steps, as follows:

(1) $2CH_4(g) + O_2(g) \rightleftharpoons 2CO(g) + 4H_2(g)$
$K = 9.34 \times 10^{28}$ at 1000. K

(2) $CO(g) + H_2O(g) \rightleftharpoons CO_2(g) + H_2(g)$
$K = 1.374$ at 1000. K

(a) Write the overall equation for the reaction of methane, steam, and oxygen to form carbon dioxide and hydrogen.
(b) What is K for the overall reaction?
(c) What is K_c for the overall reaction?
(d) A mixture of 2.0 mol of CH_4, 1.0 mol of O_2, and 2.0 mol of steam, with a total pressure of 30. bar, reacts at 1000. K at constant volume. Assuming that the reaction is complete and the ideal gas law is a valid approximation, what is the final pressure?

15.100 One mechanism for the synthesis of ammonia proposes that N_2 and H_2 molecules catalytically dissociate into atoms:

$$N_2(g) \rightleftharpoons 2N(g) \quad \log K = -43.10$$
$$H_2(g) \rightleftharpoons 2H(g) \quad \log K = -17.30$$

(a) Find the partial pressure of N in N_2 at 1000. K and 200. bar.
(b) Find the partial pressure of H in H_2 at 1000. K and 600. bar.
(c) How many N atoms and H atoms are present per litre?
(d) Based on these answers, which of the following is a more reasonable step to continue the mechanism after the catalytic dissociation? Explain.
(1) $N(g) + H(g) \longrightarrow NH(g)$
(2) $N_2(g) + H(g) \longrightarrow NH(g) + N(g)$

15.101 The following molecular scenes depict the reaction $Y \rightleftharpoons 2Z$ at four different times, out of sequence, as the reaction reaches equilibrium. Each sphere (Y is *red*, and Z is *green*) represents 0.025 mol, and the volume is 0.40 L.

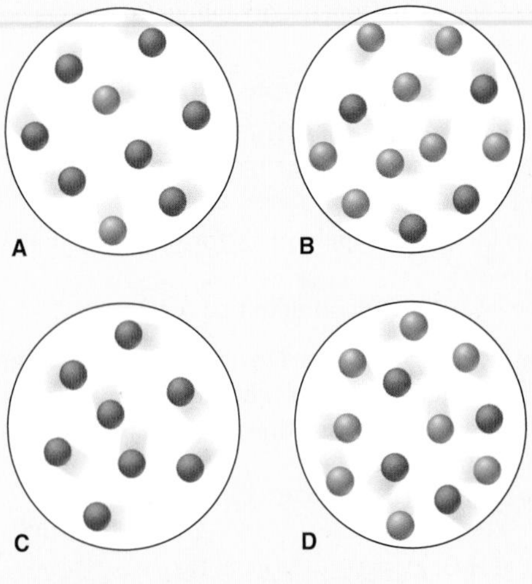

(a) Which scene(s) represent(s) equilibrium?
(b) List the scenes in the correct sequence.
(c) Calculate K_c.

15.102 For the following equilibrium, the initial concentrations of the three gases are 0.300 mol/L H_2S, 0.300 mol/L H_2, and 0.150 mol/L S_2:

$$2H_2S(g) \rightleftharpoons 2H_2(g) + S_2(g) \qquad K_c = 9.0 \times 10^{-8} \text{ at } 700°C$$

Determine the equilibrium concentrations of the gases.

15.103 The two most abundant atmospheric gases react, to a tiny extent, at 298 K in the presence of a catalyst:

$$N_2(g) + O_2(g) \rightleftharpoons 2NO(g) \qquad K = 4.35 \times 10^{-31}$$

(a) What are the equilibrium pressures of the three gases when the atmospheric partial pressures of O_2 (0.210 bar) and of N_2 (0.780 bar) are put into an evacuated 1.00 L flask at 298 K with the catalyst?
(b) What is p_{total} in the container?
(c) Find K_c at 298 K.

15.104 The oxidation of nitrogen monoxide is favoured at 457 K:

$$2NO(g) + O_2(g) \rightleftharpoons 2NO_2(g) \qquad K = 1.3 \times 10^4$$

(a) Calculate K_c at 457 K.
(b) Find $\Delta_r H°$ from the standard heats of formation.
(c) At what temperature does K_c equal 6.4×10^9?

15.105 The kinetics and equilibrium of the decomposition of hydrogen iodide have been studied extensively:

$$2HI(g) \rightleftharpoons H_2(g) + I_2(g)$$

(a) At 298 K, $K_c = 1.26 \times 10^{-3}$ for this reaction. Calculate K.
(b) Calculate K_c for the *formation* of HI at 298 K.
(c) Calculate $\Delta_r H°$ for the *decomposition* of HI from the $\Delta_f H°$ values.
(d) At 729 K, $K_c = 2.0 \times 10^{-2}$ for the decomposition of HI. Calculate $\Delta_r H°$ for this reaction from the van't Hoff equation.

15.106 Isopentyl alcohol reacts with pure acetic acid to form isopentyl acetate, the essence of banana oil:

$$C_5H_{11}OH + CH_3COOH \rightleftharpoons CH_3COOC_5H_{11} + H_2O$$

A student adds a drying agent to remove H_2O and, thus, increase the yield of banana oil. Is this approach reasonable? Explain.

15.107 Isomers Q (*blue*) and R (*yellow*) interconvert. They are depicted in an equilibrium mixture in scene A. Scene B represents the mixture after the addition of more Q. How many molecules of each isomer are present when the mixture in scene B re-establishes equilibrium?

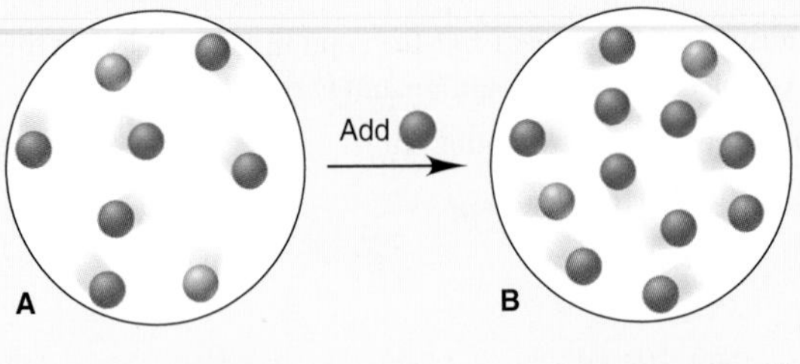

15.108 Glauber's salt, $Na_2SO_4 \cdot 10H_2O$, was used by J. R. Glauber in the 17th century as a medicinal agent. At 25°C, $K = 4.08 \times 10^{-25}$ for the loss of waters of hydration from Glauber's salt:

$$Na_2SO_4 \cdot 10H_2O(s) \rightleftharpoons Na_2SO_4(s) + 10H_2O(g)$$

(a) What is the vapour pressure of water at 25°C in a closed container holding a sample of $Na_2SO_4 \cdot 10H_2O(s)$?
(b) How do the following changes affect the ratio (higher, lower, or the same) of hydrated form to anhydrous form for the system above?
(i) Add more $Na_2SO_4(s)$. (ii) Reduce the container volume.
(iii) Add more water vapour. (iv) Add N_2 gas.

15.109 In a study of synthetic fuels, 0.100 mol of CO and 0.100 mol of water vapour is added to a 20.00 L container at 900.°C, and they react to form CO_2 and H_2. At equilibrium, [CO] is 2.24×10^{-3} mol/L.
(a) Calculate K_c at this temperature.
(b) Calculate p_{total} in the flask at equilibrium.
(c) What amount (mol) of CO must be added to double p_{total}?
(d) After p_{total} is doubled and the system regains equilibrium, what is $[CO]_{eq}$?

15.110 Synthetic diamonds are made under conditions of high temperature (2000 K) and high pressure (10^{10} Pa; 10^5 bar) in the presence of catalysts. Carbon's phase diagram is useful for finding the best conditions for the formation of natural and synthetic diamonds. Along the diamond-graphite line, the two allotropes are in equilibrium.
(a) At point A, what is the sign of ΔH for the formation of diamond from graphite? Explain.
(b) Which allotrope is denser? Explain.

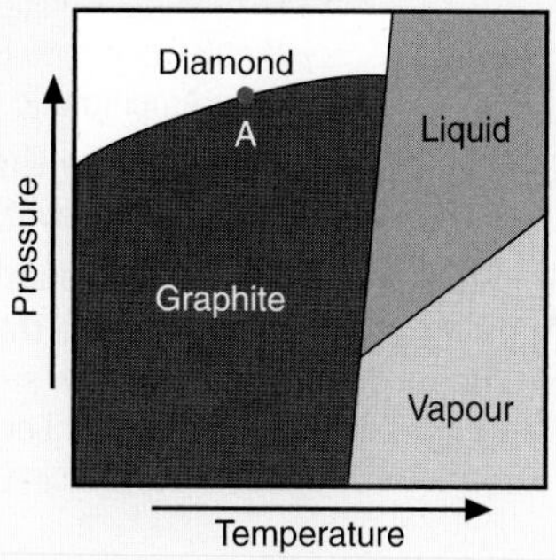

15.111 At 500.°C, the reaction $3H_2(g) + N_2(g) \rightleftharpoons 2NH_3(g)$ has $K = 3.5 \times 10^{-7}$. If the initial pressure of hydrogen is 1.5 bar and the initial pressure of nitrogen is 1.00 bar in a 2.00 L vessel, what are the pressures of all reactants and products at equilibrium? What assumptions, if any, did you need to make? Are they justifiable? Explain.

15.112 A certain endothermic reaction, $A(g) + B(g) \rightleftharpoons C(g) + D(g)$, has an equilibrium constant, K, of 1.83 at 627°C.
(a) If the reaction mixture originally contains 0.250 mol A, 0.500 mol B, 0.750 mol C, and 0.450 mol D in a 5.00 L vessel, in which direction will the reaction proceed?
(b) Determine the equilibrium pressures of all of species A, B, C, and D.
(c) Once the reaction has reached equilibrium, which way will the reaction shift if (i) the volume of the container is doubled; (ii) more C is added to the system; (iii) D is removed from the system; (iv) the temperature is raised?
Justify your answers.

Acid-Base Equilibria

IN THIS CHAPTER . . . We develop three definitions of acids and bases that explain an expanded range of reactions, and we apply equilibrium principles to understand acid-base behaviour. By the end of this chapter, you should be able to

- Describe the *Arrhenius* acid-base definition, which relies on formulas and behaviour in water
- Differentiate between the Arrhenius acid-base and Brønsted-Lowry acid-base definitions, using the idea of proton transfer, which expands the meaning of *base*, along with the scope of acid-base reactions
- Discuss acid dissociation and explain how variation in acid strength is expressed by the equilibrium constant
- Describe the pH scale used to measure the acidity of aqueous solutions
- Discuss weak bases and their interdependence with weak acids
- Apply a systematic approach to solving acid-base equilibrium problems
- Examine the molecular structures of acids to rationalize their relative strengths
- Determine the relative acidity of salts from the reactions of their conjugate acids and conjugate bases with water
- Describe and discuss how the designations "acid" and "base" depend on the relative strengths of the substances *and* on the solvents
- Differentiate between the Lewis acid-base definition, which greatly expands the meaning of the terms *acid* and *acid-base reaction*, and previous definitions

16

Acids are an inescapable part of our lives. Most of the food that we consume is acidic, our stomach uses acid to digest our food, and we have developed a preference for acidic soft drinks instead of water. We use acids to run our cars and acids to create some of the more important chemicals we use on a daily basis. Even our building blocks, DNA, have an acidic component, namely amino acids. Although we use bases almost as much as acids, we know much less about them. For instance, most people know that acids taste sour, but many people do not know that bases taste bitter. With the exception of a few common bases, such as sodium hydrogen carbonate (baking soda), ammonia (window cleaners), and sodium hydroxide (drain and oven cleaners), most bases go unnoticed in our daily lives. What is the difference between an acid and a base, and why is it important? In this chapter, we will look at how we define acidic substances and how we differentiate them from bases. We will also look at what happens when they react and how to characterize them.

Acids and bases played a central role in the laboratories of alchemists, and they remain indispensable today, not only in modern academic and industrial research, but also in many consumer products (Table 16.1).

The slice of lemon (citric acid) on the side of a glass of water and the vinegar (ethanoic acid, also known as acetic acid) in a vinaigrette salad dressing both have a sour taste. In fact, sourness has been a defining property since the 17th century: an acid was known as any substance that tasted sour; reacted with active metals, such as aluminum and zinc, to produce hydrogen gas; and turned certain organic compounds specific colours. (We discuss *indicators* in this chapter and in Chapter 17.) Similarly, a base (such as the amines in fish) was known as any substance that had a bitter taste and slippery feel and turned the same organic compounds different colours. Moreover, it was known that *when acids and bases react, each cancels the properties of the other in a process called neutralization.* Although these early definitions described distinctive properties, they gave way to other definitions based on molecular behaviour. As science progresses, limited definitions are replaced by broader definitions that explain more phenomena.

TABLE 16.1 Some Common Acids and Bases and Their Household Uses

Substance	Use
Acids	
Ethanoic (acetic) acid, CH_3COOH	Flavouring, preservative
Citric acid, $H_3C_6H_5O_7$	Flavouring
Ascorbic acid, $H_2C_6H_6O_6$	Vitamin C, nutritional supplement
Aluminum salts, $NaAl(SO_4)_2 \cdot 12H_2O$	Component of baking powder (with sodium hydrogen carbonate)
Bases	
Sodium hydroxide (lye), NaOH	Oven and drain cleaners
Ammonia, NH_3	Household cleaner
Sodium carbonate, Na_2CO_3	Water softener, grease remover
Sodium hydrogen carbonate, $NaHCO_3$	Fire extinguisher, rising agent in cake mixes (baking soda), mild antacid
Sodium phosphate, Na_3PO_4	Cleaner for surfaces before painting or wallpapering

Concepts and Skills to Review before Studying This Chapter

- Role of water as a solvent (Section 3.6)
- Rules for writing ionic equations (Section 3.7)
- Properties of an equilibrium constant (Section 15.2)
- Solutions to equilibrium problems (Section 15.5)

16.1 Acids and Bases

Most laboratory work with acids and bases involves water, as do most environmental, biological, and industrial applications. *Water is always one of the products in reactions between acids and bases*, as shown by the molecular and net ionic equations for any such reaction:

$$HX(aq) + MOH(aq) \longrightarrow MX(aq) + H_2O(l) \qquad \text{(molecular)}$$

and

$$H^+(aq) + OH^-(aq) \longrightarrow H_2O(l) \qquad \text{(net ionic)}$$

Furthermore, whenever an acid dissociates in water, solvent molecules participate in the reaction:

$$HA(g \text{ or } l) + H_2O(l) \longrightarrow A^-(aq) + H_3O^+(aq)$$

When the reaction above proceeds largely in the direction written, we say the acid is a *strong* acid. The same is true for bases that dissociate to a large extent. Acids that dissociate very little (or in which the majority of molecules remain in the undissociated state) are known as *weak* acids. Similarly, bases that remain largely undissociated are known as weak bases. When acids are added to water, water molecules surround the proton to form hydrogen-bonded species with the general formula $H(H_2O)_n^+$. The charge density of the proton is so high that it attracts water especially strongly. The proton bonds covalently to one of the lone electron pairs of the O atom of a water molecule to form a **hydronium ion** (**H_3O^+**), which forms hydrogen bonds to several other water molecules (see Figure 16.1). To emphasize the active role of water and the proton-water interaction, the hydrated proton is usually shown as $H_3O^+(aq)$, although, for simplicity, it is sometimes shown as $H^+(aq)$. In this book, H^+ and H_3O^+ will be used interchangeably, depending on the situation.

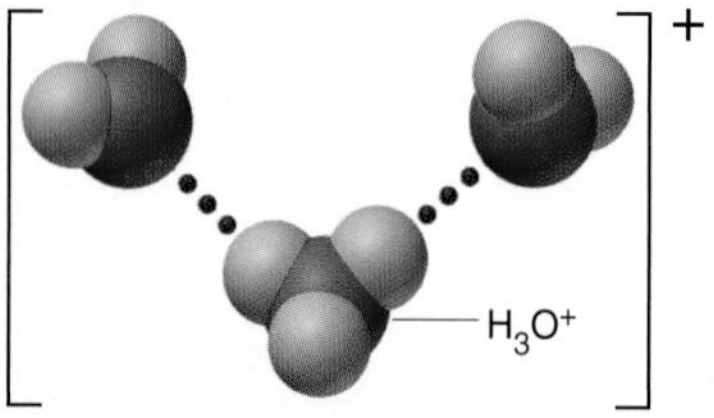

FIGURE 16.1 The H^+ ion as a solvated hydronium ion

Release of H^+ or OH^- and the Arrhenius Acid-Base Definition

The earliest definition that highlighted the molecular nature of acids and bases was the **Arrhenius acid-base definition**, which classified these substances in terms of their formulas and behaviour *in water*:

- An *acid* is a substance that has H in its formula and dissociates in water to yield H^+ (which may form H_3O^+).
- A *base* is a substance that has OH in its formula and dissociates in water to yield OH^-.

Some typical Arrhenius acids are HCl, HNO_3, and HCN, and some typical Arrhenius bases are NaOH, KOH, and $Ba(OH)_2$. Whereas Arrhenius bases contain ionically bonded OH^- ions in their structures, Arrhenius acids *never* contain discrete H^+ ions. Instead, they contain *covalently bonded H atoms that ionize when a molecule dissolves in water.*

When an acid and a base react, they undergo **neutralization**. The meaning of this term has changed, but, in the Arrhenius sense, neutralization occurs when *the H^+ from an acid and the OH^- from a base form H_2O.* A key point about neutralization that Arrhenius was able to explain is that no matter which strong acid and strong base react, and no matter which salt results, $\Delta_r H°$ is −55.9 kJ per mole of water formed. Arrhenius suggested that the enthalpy change is always the same because the reaction is always the same—a hydrogen ion and a hydroxide ion form water:

$$H^+(aq) + OH^-(aq) \longrightarrow H_2O(l) \qquad \Delta_r H° = -55.9 \text{ kJ/mol}$$

The dissolved salt that is present, such as NaCl in the reaction of sodium hydroxide with hydrochloric acid, exists as hydrated spectator ions and does not affect $\Delta_r H°$:

$$Na^+(aq) + OH^-(aq) + H^+(aq) + Cl^-(aq) \longrightarrow Na^+(aq) + Cl^-(aq) + H_2O(l)$$

Despite its importance at the time, limitations in the Arrhenius definition soon became apparent. Arrhenius and many others realized that, even though some substances do *not* have discrete OH^- ions, they still behave as bases. For example, NH_3 and K_2CO_3 also yield OH^- in water. As we will see shortly, broader acid-base definitions are required to include these species.

Proton Transfer and the Brønsted-Lowry Acid-Base Definition

Earlier, we noted one limitation of the Arrhenius definition: many substances that yield OH^- ions in water do not contain OH in their formulas. Examples are ammonia, the amines, and many salts of weak acids, such as NaF. Another limitation is that water has to be the solvent for acid-base reactions. In the early 20th century, J. N. Brønsted and T. M. Lowry suggested definitions that remove these limitations.

According to the **Brønsted-Lowry acid-base definition**, an acid and a base can be defined as follows:

- *An acid is a* ***proton donor****, any species that donates an* H^+ *ion.* An acid must contain H in its formula; HNO_3 and $H_2PO_4^-$ are two of many examples. All Arrhenius acids are also Brønsted-Lowry acids.
- *A base is a* ***proton acceptor****, any species that accepts an* H^+ *ion.* A base must contain a lone pair of electrons to bind H^+; a few examples are NH_3, CO_3^{2-}, and F^-, as well as OH^- itself. Brønsted-Lowry bases are not Arrhenius bases, but all Arrhenius bases contain the Brønsted-Lowry base OH^-.

Thus, an acid-base reaction occurs when *one species donates a proton and another species simultaneously accepts it: an acid-base reaction is, therefore, a proton-transfer process.* An acid-base reaction can occur between gases, in a nonaqueous solution, and in a heterogeneous mixture, as well as in an aqueous solution.

According to this definition, an acid-base reaction occurs even when an acid (or a base) just dissolves in water, because water acts as the proton acceptor (or donor):

1. *The acid donates a proton to water* (Figure 16.2A). When HCl dissolves in water, an H^+ ion (a proton) is transferred from HCl to H_2O, where it becomes attached to a lone pair of electrons on the O atom, forming H_3O^+. Thus, HCl (the acid) has *donated* the H^+, and H_2O (the base) has *accepted* it:

$$HCl(g) + H_2\ddot{O}(l) \longrightarrow Cl^-(aq) + H_3\ddot{O}^+(aq)$$

2. *The base accepts a proton from water* (Figure 16.2B). When ammonia dissolves in water, an H^+ from H_2O is transferred to the lone pair of N, forming NH_4^+, and the H_2O becomes an OH^- ion:

$$\ddot{N}H_3(aq) + H_2O(l) \rightleftharpoons NH_4^+(aq) + OH^-(aq)$$

In this reaction, H_2O (the acid) has *donated* the H^+, and NH_3 (the base) has *accepted* it.

Note that H_2O is *amphiprotic*: it acts as a base (accepts an H^+) in one reaction and as an acid (donates an H^+) in the other. Many other species are amphiprotic as well.

Stomach acid is roughly 1 mol/L HCl, a strong acid that is needed to digest the food we eat. Our stomach, where the digestion process takes place, is lined with a special mucous membrane that keeps the acid from damaging the soft tissue. The stomach produces stomach acid constantly throughout the day. When we eat food, the stomach is stimulated to produce a greater amount of stomach acid. After we eat, the stomach acid rises to cover the food. Then it digests, or breaks down, the food to provide us with the nutrients we need. The stomach acid also activates various digestive enzymes and kills potentially harmful bacteria. If, however, we lie down immediately after eating, or we have a lot of excess weight (which pushes the stomach higher up in the body), the acid can "spill" or leak out of the stomach. There is a valve between the stomach and the esophagus that prevents the backwash of stomach content into the esophagus. However, with repeated exposure to stomach acid, the valve may lose its ability to completely prevent stomach acid from leaking out. The soft tissue in the esophagus and just above the stomach is not designed to be in contact with the acid. The sensation we feel when the acid comes in contact with this soft tissue is called heartburn. Much chemistry goes into the development of antacids to treat heartburn.

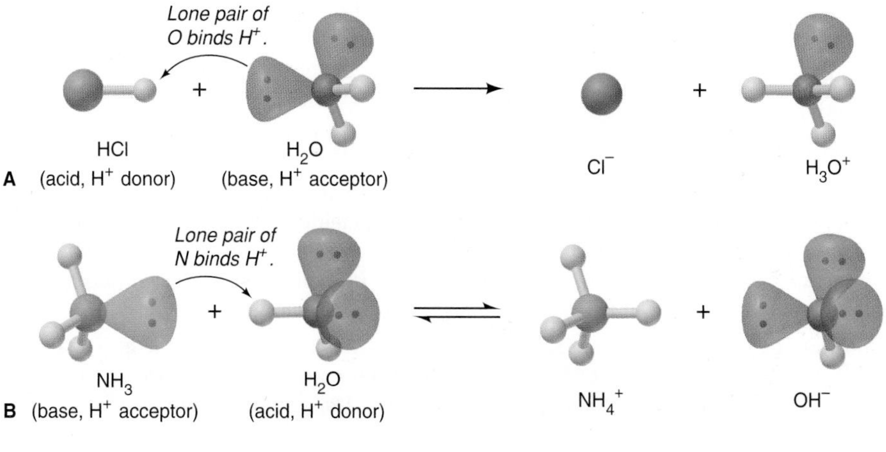

FIGURE 16.2 Dissolving of an acid or a base in water as a Brønsted-Lowry acid-base reaction. A. The acid HCl dissolving in the base water. **B.** The base NH_3 dissolving in the acid water.

Conjugate Acid-Base Pairs

The Brønsted-Lowry definition provides a new way to look at these reactions because it focuses on the reactants *and* the products as acids and bases. For example, let us examine the reaction between hydrogen sulfide and ammonia:

$$H_2S\,(aq) + NH_3(aq) \rightleftharpoons HS^-(aq) + NH_4^+(aq)$$

In the forward reaction, H_2S acts as an acid by donating an H^+ to NH_3, which acts as a base by accepting it. In the reverse reaction, the ammonium ion, NH_4^+, acts as an acid by donating an H^+ to the hydrogen sulfide ion, HS^-, which acts as a base. Notice that the acid, H_2S, becomes a base, HS^-, and the base, NH_3, becomes an acid, NH_4^+.

In Brønsted-Lowry terminology, H_2S and HS^- are a **conjugate acid-base pair**: HS^- is the conjugate base of the acid H_2S. Similarly, NH_3 and NH_4^+ are a conjugate acid-base pair: NH_4^+ is the conjugate acid of the base NH_3. *Every acid has a conjugate base, and every base has a conjugate acid.* For any conjugate acid-base pair,

- the conjugate base has one *less* H and one *more* minus charge than the acid
- the conjugate acid has one *more* H and one *less* minus charge than the base

A Brønsted-Lowry acid-base reaction occurs when *an acid and a base react to form their conjugate base and conjugate acid, respectively*:

$$\text{acid}_1 + \text{base}_2 \rightleftharpoons \text{base}_1 + \text{acid}_2$$

Table 16.2 shows some Brønsted-Lowry acid-base reactions. Note these points:

- Each reaction has an acid and a base as reactants *and* as products, comprising two conjugate acid-base pairs.
- Acids and bases can be neutral, cationic, or anionic.
- The same species can be an acid or a base (amphiprotic), depending on the other species reacting. Water behaves this way in reactions 1 and 4 in Table 16.2, and HPO_4^{2-} does so in reactions 4 and 6.

TABLE 16.2 The Conjugate Pairs in Some Acid-Base Reactions

(Conjugate Pair: Acid (left) and Base (right); Conjugate Pair: Base (left) and Acid (right))

	Acid	+	Base	⇌	Base	+	Acid
Reaction 1	$HF(aq)$	+	$H_2O(l)$	⇌	$F^-(aq)$	+	$H_3O^+(aq)$
Reaction 2	$HCOOH(aq)$	+	$CN^-(aq)$	⇌	$HCOO^-(aq)$	+	$HCN(aq)$
Reaction 3	$NH_4^+(aq)$	+	$CO_3^{2-}(aq)$	⇌	$NH_3(aq)$	+	$HCO_3^-(aq)$
Reaction 4	$H_2PO_4^-(aq)$	+	$OH^-(aq)$	⇌	$HPO_4^{2-}(aq)$	+	$H_2O(l)$
Reaction 5	$H_2SO_4(aq)$	+	$N_2H_5^+(aq)$	⇌	$HSO_4^-(aq)$	+	$N_2H_6^{2+}(aq)$
Reaction 6	$HPO_4^{2-}(aq)$	+	$SO_3^{2-}(aq)$	⇌	$PO_4^{3-}(aq)$	+	$HSO_3^-(aq)$

Sample Problem 16.1 Identifying Conjugate Acid-Base Pairs

Problem The following reactions are important environmental processes. Identify the conjugate acid-base pairs.

(a) $H_2PO_4^-(aq) + CO_3^{2-}(aq) \rightleftharpoons HCO_3^-(aq) + HPO_4^{2-}(aq)$

(b) $H_2O(l) + SO_3^{2-}(aq) \rightleftharpoons OH^-(aq) + HSO_3^-(aq)$

Plan To find the conjugate pairs, we find the species that donated an H^+ (acid) and the species that accepted it (base). The acid (or base) on the left becomes the conjugate base (or conjugate acid) on the right. Remember that the conjugate acid has one more H and one less minus charge than its conjugate base.

Solution (a) $H_2PO_4^-$ has one more H^+ than HPO_4^{2-}; CO_3^{2-} has one less H^+ than HCO_3^-. Therefore, the acids are $H_2PO_4^-$ and HCO_3^-, and the bases are HPO_4^{2-} and CO_3^{2-}. The conjugate acid-base pairs are $H_2PO_4^-/HPO_4^{2-}$ and HCO_3^-/CO_3^{2-}.

(b) H_2O has one more H^+ than OH^-; SO_3^{2-} has one less H^+ than HSO_3^-. The acids are H_2O and HSO_3^-, and the bases are OH^- and SO_3^{2-}. The conjugate acid-base pairs are H_2O/OH^- and HSO_3^-/SO_3^{2-}.

Follow-Up Problem 16.1 Identify the conjugate acid-base pairs in each reaction:

(a) $CH_3COOH(aq) + H_2O(l) \rightleftharpoons CH_3COO^-(aq) + H_3O^+(aq)$

(b) $H_2O(l) + F^-(aq) \rightleftharpoons OH^-(aq) + HF(aq)$

(See Brief Solutions to Follow-Up Problems at the end of the chapter.)

Relative Acid-Base Strength and the Net Direction of Reaction

The *net* direction of an acid-base reaction depends on the relative acid and base strengths: *a reaction proceeds to the greater extent in the direction in which a stronger acid and a stronger base form a weaker acid and a weaker base.*

Competition for the Proton The net direction of the reaction of H_2S and NH_3 is to the right ($K > 1$) because H_2S is a stronger acid than NH_4^+, the other acid, and NH_3 is a stronger base than HS^-, the other base:

$$\underset{\text{stronger acid}}{H_2S(aq)} + \underset{\text{stronger base}}{NH_3(aq)} \rightleftharpoons \underset{\text{weaker base}}{HS^-(aq)} + \underset{\text{weaker acid}}{NH_4^+(aq)} \quad (\longrightarrow)$$

We can think of the process as *a competition for the proton between the two bases*, NH_3 and HS^-, in which NH_3 wins.

The extent of acid (HA) dissociation in water depends on a competition for the proton between the two bases, A^- and H_2O. Strong and weak acids give different results:

1. *Strong acids.* When the strong acid HNO_3 dissolves, it completely transfers an H^+ to the base, H_2O, forming the conjugate base NO_3^- and the conjugate acid H_3O^+:

$$\underset{\text{stronger acid}}{HNO_3(aq)} + \underset{\text{stronger base}}{H_2O(l)} \rightleftharpoons \underset{\text{weaker base}}{NO_3^-(aq)} + \underset{\text{weaker acid}}{H_3O^+(aq)} \quad (\longrightarrow)$$

The net direction is so far to the right that $K >> 1$ and the reaction is essentially complete. HNO_3 is a stronger acid than H_3O^+, and H_2O is a stronger base than NO_3^-. Thus, with a strong HA, H_2O wins the competition for the proton because A^- is a *much* weaker base. In fact, *the only acidic species that remains in a strong-acid solution is* H_3O^+.

2. *Weak acids.* On the other hand, in a weak acid such as HF, the A^- (F^-) wins the competition because it is a stronger base than H_2O ($K < 1$):

$$\underset{\text{weaker acid}}{HF(aq)} + \underset{\text{weaker base}}{H_2O(l)} \rightleftharpoons \underset{\text{stronger base}}{F^-(aq)} + \underset{\text{stronger acid}}{H_3O^+(aq)} \quad (\longleftarrow)$$

Whether the acid or base is strong or weak, in *any* acid-base reaction, the *strongest acid present* will first react with the *strongest base present* to form a weaker acid species and a weaker base species. This is true in any solvent. If there is also a weaker acid or base species present in the solution, then it may react once the strong species is completely consumed, but not while the strong species is present.

Ranking Conjugate Pairs Based on evidence from many such reactions, we can rank conjugate pairs in terms of the ability of the acid to transfer its proton (Figure 16.3). Note that *a weaker acid has a stronger conjugate base*: the acid cannot give up its proton very readily because its conjugate base is holding it too strongly.

We can use Figure 16.3 to predict the net direction of a reaction between any two pairs—whether the equilibrium position lies to the right ($K > 1$) or to the left ($K < 1$). *A reaction proceeds to the right if an acid reacts with a base lower on the list.* Notice that both the pH and the pK_a decrease as the acid strength increases, and both the pOH and the pK_b decrease as the base strength increases. Here are two sample problems that demonstrate this key idea.

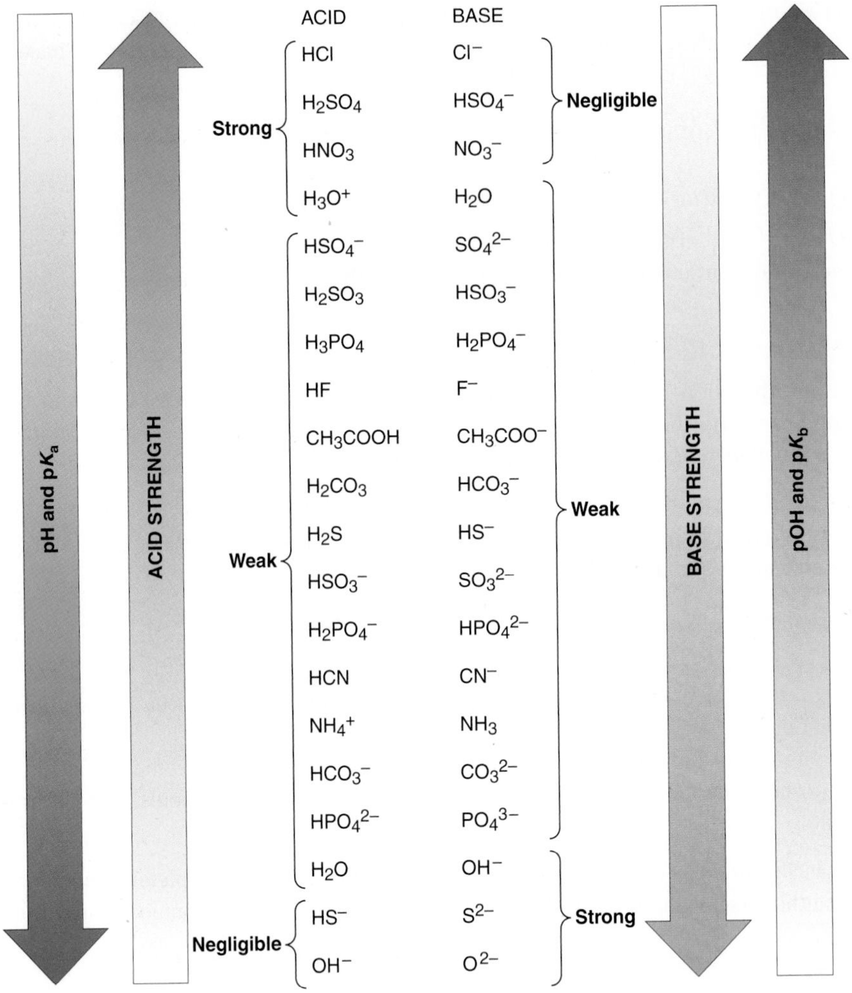

FIGURE 16.3 Strengths of conjugate acid-base pairs. The stronger the acid, the weaker its conjugate base. The strongest acid is at the top left, and the strongest base is at the bottom right. When an acid reacts with a base farther down the list, the reaction proceeds to the right ($K > 1$).

Sample Problem 16.2 Predicting the Net Direction of an Acid-Base Reaction

Problem Predict the net direction of each reaction and whether K is greater or less than 1. (Assume equal initial concentrations of all species.)

(a) $H_2PO_4^-(aq) + NH_3(aq) \rightleftharpoons NH_4^+(aq) + HPO_4^{2-}(aq)$

(b) $H_2O(l) + HS^-(aq) \rightleftharpoons OH^-(aq) + H_2S(aq)$

Plan We identify the conjugate acid-base pairs and consult Figure 16.3 to see which acid and base are stronger. The reaction proceeds in the direction in which the stronger acid and base form the weaker acid and base. If the reaction *as written* proceeds to the right, then [products] is higher than [reactants], so $K > 1$.

Solution (a) The conjugate pairs are $H_2PO_4^-/HPO_4^{2-}$ and NH_4^+/NH_3. Since $H_2PO_4^-$ is higher on the list of acids, it is stronger than NH_4^+; since NH_3 is lower on the list of bases, it is stronger than HPO_4^{2-}. Therefore,

$$\underset{\text{stronger acid}}{H_2PO_4^-(aq)} + \underset{\text{stronger base}}{NH_3(aq)} \rightleftharpoons \underset{\text{weaker base}}{HPO_4^{2-}(aq)} + \underset{\text{weaker acid}}{NH_4^+(aq)}$$

(stronger acid + stronger base ⟶ weaker base + weaker acid)

The net direction is to the right, so $K > 1$.

(b) The conjugate pairs are H_2O/OH^- and H_2S/HS^-. Since H_2S is higher on the list of acids, and OH^- is lower on the list of bases, we have

$$\underset{\text{weaker acid}}{H_2O(l)} + \underset{\text{weaker base}}{HS^-(aq)} \rightleftharpoons \underset{\text{stronger base}}{OH^-(aq)} + \underset{\text{stronger acid}}{H_2S(aq)}$$

(weaker acid + weaker base ⟵ stronger base + stronger acid)

The net direction is to the left, so $K < 1$.

Follow-Up Problem 16.2 Use balanced equations that show the net direction of the reaction to explain each observation:

(a) You smell ammonia when NH_3 dissolves in water.

(b) The odour goes away when you add an excess of HCl to the solution in part (a).

(c) The odour returns when you add an excess of NaOH to the solution in part (b).

Sample Problem 16.3 Using Molecular Scenes to Predict the Net Direction of an Acid-Base Reaction

Problem Given that 0.10 mol/L HX (*blue and green*) has a pH of 2.88, and 0.10 mol/L HY (*blue and orange*) has a pH of 3.52, which scene best represents the final mixture after equimolar solutions of HX and Y^- are mixed?

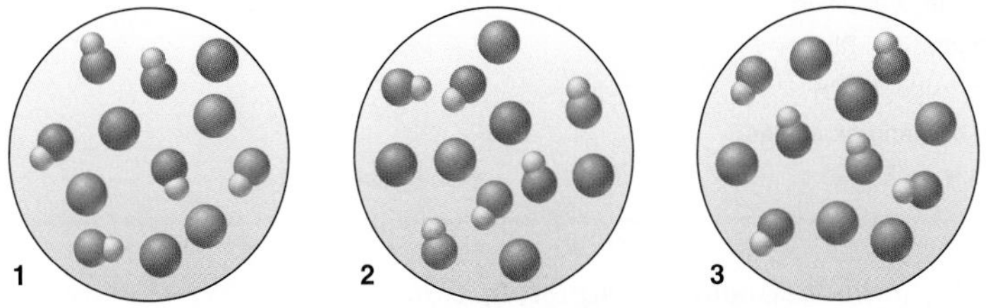

Plan A stronger acid and base yield a weaker acid and base, so we have to determine the relative strengths of the acids HX and HY to choose the correct molecular scene. The concentrations of the acid solutions are equal, so we can pick the stronger acid directly from the pH values of the two acid solutions. Because the stronger acid reacts to a greater extent, there should be fewer molecules of this acid than the weaker acid in the scene.

Solution Both solutions have the same concentration; however, the HX solution has a lower pH (2.88) than the HY solution (3.52), so we know right away that HX is the stronger acid and Y^- is the stronger base. Therefore, the reaction of HX and Y^- has a $K > 1$, which means that the equilibrium mixture will have more HY than HX. Scene 1 has equal numbers of HX and HY, which would occur if the acids were of equal strength. Scene 2 shows fewer HY than HX, which would occur if HY were stronger. Therefore, only scene 3 is consistent with the relative acid strengths.

Follow-Up Problem 16.3 The scene below represents the equilibrium mixture after 0.10 mol/L solutions of HA (*blue and red*) and B^- (*black*) react. Does their reaction have a K greater or less than 1? Which acid is stronger, HA or HB?

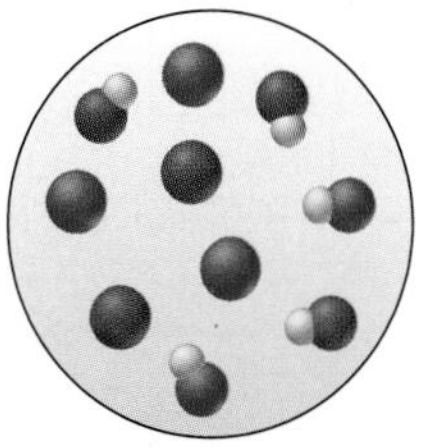

Variation in Acid Strength: The Acid-Dissociation Constant (K_a)

Acids (and bases) are classified by their *strength*, the amount of H_3O^+ (or OH^-) produced per mole of substance dissolved, or, in other words, by the extent of their dissociation into ions (see Table 16.3). Because acids and bases are electrolytes, their strength correlates with electrolyte strength: *strong electrolytes dissociate completely, and weak electrolytes dissociate slightly.*

TABLE 16.3 Strong and Weak Acids and Bases

Acids	
Strong	
Hydrochloric acid, HCl	Nitric acid, HNO_3
Hydrobromic acid, HBr	Sulfuric acid, H_2SO_4 (first proton only)
Hydriodic acid, HI	Perchloric acid, $HClO_4$
Weak (a few of many examples)	
Hydrofluoric acid, HF	Ethanoic acid, CH_3COOH (or $HC_2H_3O_2$)
Phosphoric acid, H_3PO_4	
Bases	
Strong	
Group 1 hydroxides:	*Heavy group 2 hydroxides:*
Lithium hydroxide, LiOH	Calcium hydroxide, $Ca(OH)_2$
Sodium hydroxide, NaOH	Strontium hydroxide, $Sr(OH)_2$
Potassium hydroxide, KOH	Barium hydroxide, $Ba(OH)_2$
Rubidium hydroxide, RbOH	
Cesium hydroxide, CsOH	
Weak (two of many examples)	
Ammonia, NH_3	Bicarbonate ion, HCO_3^-

1. *Strong acids dissociate* ***completely*** *into ions in water* (Figure 16.4A):

$$\mathrm{HA}(g \text{ or } l) \xrightarrow{\mathrm{H_2O}(l)} \mathrm{H^+}(aq) + \mathrm{A^-}(aq)$$

In a dilute solution of a strong acid, *HA molecules are no longer present*: $[\mathrm{H^+}] \approx [\mathrm{HA}]_{\mathrm{init}}$. In other words, $[\mathrm{HA}]_{\mathrm{eq}} \approx 0$, so the value of the equilibrium constant for acid dissociation, K_a, is extremely large:

$$K_a = \frac{a_{\mathrm{H^+}}a_{\mathrm{A^-}}}{a_{\mathrm{HA}}} \quad (\text{At equilibrium, } K_a \gg 1.)$$

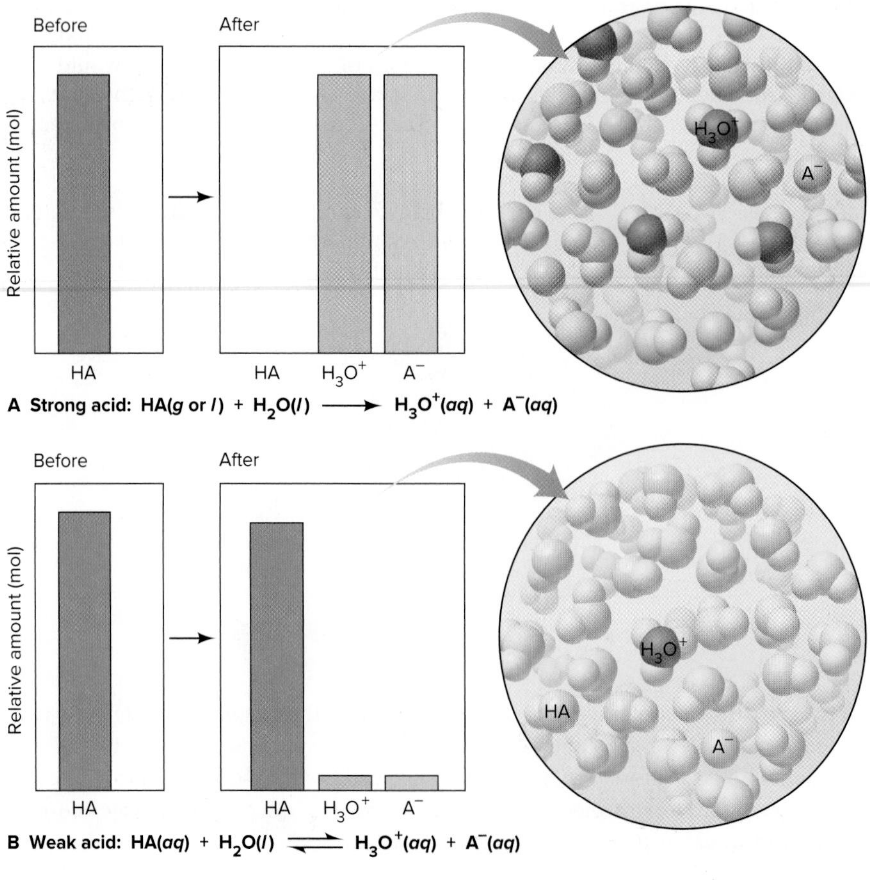

FIGURE 16.4 The extent of dissociation for strong acids and weak acids. The bar graphs show the relative numbers of moles of species before (*left*) and after (*right*) acid dissociation. **A.** A strong acid dissociates completely; virtually no HA molecules are present. **B.** A weak acid dissociates very little, remaining mostly as intact HA molecules.

Remember that the activities of aqueous species are approximated by concentrations in mol/L, and that the activity of a pure liquid is 1. Thus, for reasonably dilute solutions or for solutions that behave ideally, we can write

$$K_a = \frac{[H^+][A^-]}{[HA]}$$

Because the reaction is essentially complete, we usually do not express it as an equilibrium process. In dilute aqueous nitric acid, for example, there are virtually no undissociated nitric acid molecules:

$$HNO_3(aq) + H_2O(l) \longrightarrow H_3O^+(aq) + NO_3^-(aq)$$

2. *Weak acids dissociate* ***slightly*** *into ions in water* (Figure 16.4B):

$$HA(aq) + H_2O(l) \rightleftharpoons H_3O^+(aq) + A^-(aq)$$

In a dilute solution of a weak acid, *the great majority of HA molecules are undissociated.* Thus, $[H_3O^+] << [HA]_{init}$, and $[HA]_{eq} \approx [HA]_{init}$, so K_a is very small. Hydrocyanic acid is an example of a weak acid:

$$HCN(aq) + H_2O(l) \rightleftharpoons H_3O^+(aq) + CN^-(aq)$$

$$K_a = \frac{[H_3O]^+[CN^-]}{[HCN]} \quad \text{(At equilibrium, } K_a \ll 1.)$$

(As in Chapter 15, brackets with no subscript mean concentration in mol/L *at equilibrium*; that is, [X] means $[X]_{eq}$. In this chapter, we are dealing with systems *at equilibrium*, so we express K directly as a collection of equilibrium concentration terms.)

The difference in $[H^+]$ causes a much higher rate for the reaction of a strong acid with an active metal, such as zinc, compared with the reaction of a weak acid with an active metal (Figure 16.5):

$$Zn(s) + 2H^+(aq) \longrightarrow Zn^{2+}(aq) + H_2(g)$$

In a strong acid, with its much higher $[H^+]$, zinc reacts rapidly, vigorously forming bubbles of H_2. In a weak acid, $[H^+]$ is much lower, so zinc reacts slowly.

FIGURE 16.5 Reaction of zinc with a strong acid (left) and a weak acid (right)

The Meaning of K_a The equilibrium expression for the dissociation of a general *weak acid*, HA, in water is

$$K_a = \frac{[H^+][A^-]}{[HA]} \qquad (16.1)$$

The equilibrium constant for the dissociation of an acid is called the **acid-dissociation constant (acid-ionization constant), K_a**.

Like any equilibrium constant, K_a is a number whose magnitude is temperature dependent and tells how far to the right the reaction has proceeded to reach equilibrium. Thus, *the stronger the acid, the higher the $[H_3O^+]$ at equilibrium and the larger the K_a*:

$$\text{stronger acid} \Longrightarrow \text{higher } [H^+] \Longrightarrow \text{larger } K_a$$

The Range of K_a Values Acid-dissociation constants of weak acids range over many orders of magnitude. Some benchmark K_a values for typical weak acids give an idea of the fraction of HA molecules that dissociate into ions:

- For a weak acid with a relatively high K_a ($\sim 10^{-2}$), a 1 mol/L solution has ~10% of the HA molecules dissociated. The K_a of chlorous acid ($HClO_2$) is 1.1×10^{-2}, and 1 mol/L $HClO_2$ is 10.% dissociated.
- For a weak acid with a moderate K_a ($\sim 10^{-5}$), a 1 mol/L solution has ~0.3% of the HA molecules dissociated. The K_a of ethanoic acid (CH_3COOH) is 1.8×10^{-5}, and 1 mol/L CH_3COOH is 0.42% dissociated.
- For a weak acid with a relatively low K_a ($\sim 10^{-10}$), a 1 mol/L solution has ~0.001% of the HA molecules dissociated. The K_a of HCN is 6.2×10^{-10}, and 1 mol/L HCN is 0.0025% dissociated.

Anorexia is a psychological disorder that causes people to think that they are overweight, even if they are normal or below normal weight. People who suffer from anorexia may eat either extremely limited quantities of food or, after eating normal amounts of food, will feel that they have eaten too much. Bulimia, also known as binge eating, is another psychological disorder, where large amounts of food are consumed and then purged from the body. One similarity between people who are anorexic and people who are bulimic is that they often force themselves to vomit food that they have ingested. Since the food has not yet been digested, it has a high acid content and the vomiting process brings the stomach acid into direct contact with the teeth. The acid reacts with the calcium in the teeth, causing cavities, discoloration, and tooth loss, symptoms that doctors often check for when they suspect that a patient is anorexic or bulimic.

TABLE 16.4 K_a Values for Some Monoprotic Acids at 25°C

Name (Formula)	Lewis Structure*	K_a	pK_a
Chlorous acid ($HClO_2$)	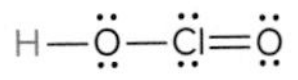	1.1×10^{-2}	1.96
Nitrous acid (HNO_2)	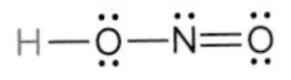	7.1×10^{-4}	3.15
Hydrofluoric acid (HF)	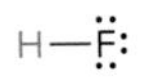	6.8×10^{-4}	3.17
Methanoic (formic) acid (HCOOH)	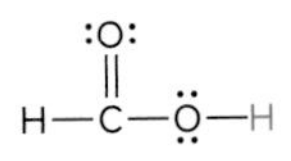	1.8×10^{-4}	3.74
Ethanoic (acetic) acid (CH_3COOH)	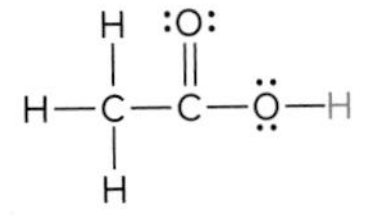	1.8×10^{-5}	4.74
Propanoic acid (CH_3CH_2COOH)	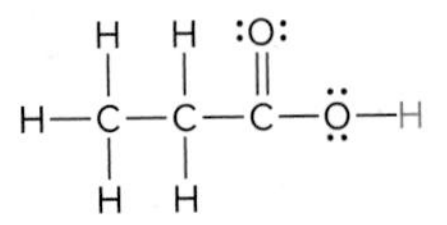	1.3×10^{-5}	4.89
Hypochlorous acid (HClO)	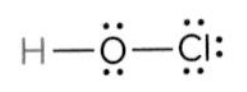	2.9×10^{-8}	7.54
Hydrocyanic acid (HCN)	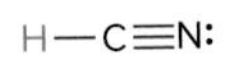	6.2×10^{-10}	9.21

*Red type indicates the ionizable proton; all atoms have zero formal charge.

Thus, for solutions with the same initial HA concentration, *the smaller the K_a, the lower the percent dissociation of HA*:

$$\text{weaker acid} \Longrightarrow \text{lower \% dissociation of HA} \Longrightarrow \text{smaller } K_a$$

Table 16.4 lists K_a values of some weak *monoprotic* acids, those with one ionizable proton. (A more extensive list is in Appendix C.) Notice that the ionizable proton in organic acids is bound to the O in —COOH; H atoms bonded to C do *not* ionize.

Classifying the Relative Strengths of Acids and Bases

Using a table of K_a values is the surest way to quantify the strengths of weak acids, but you can classify acids and bases qualitatively as strong or weak from their formulas.

- *Strong acids.* Two types of strong acids, with examples *you should memorize*, are given below:
  1. The hydrohalic acids HCl, HBr, and HI
  2. Oxoacids in which the number of O atoms exceeds the number of ionizable protons by two or more, such as HNO_3, H_2SO_4, and $HClO_4$; for example, in the case of H_2SO_4, 4 O's − 2 H's = 2

- *Weak acids.* There are many *more* weak acids than strong acids. Here are four types:
  1. The hydrohalic acid HF
  2. Acids in which H is not bonded to O or to a halogen, such as HCN and H_2S
  3. Oxoacids in which the number of O atoms equals or exceeds by one the number of ionizable protons, such as HClO, HNO_2, and H_3PO_4
  4. Carboxylic acids (general formula RCOOH, with the ionizable proton shown in red), such as CH_3COOH and C_6H_5COOH

- *Strong bases.* Water-soluble compounds that contain O^{2-} or OH^- ions are strong bases. The cations are usually those of the most active metals:
  1. M_2O or MOH, where M = group 1 metal (Li, Na, K, Rb, Cs)
  2. MO or $M(OH)_2$, where M = group 2 metal (Ca, Sr, Ba) [MgO and $Mg(OH)_2$ are only slightly soluble in water, but the soluble portion dissociates completely]

- *Weak bases.* Many compounds with an electron-rich nitrogen atom are weak bases. (None behave as Arrhenius bases; they are mostly Brønsted-Lowry bases.) The common structural feature is an N atom with a lone electron pair (shown in blue):
  1. Ammonia, $\ddot{N}H_3$
  2. Amines (general formula $R\ddot{N}H_2$, $R_2\ddot{N}H$, or $R_3\ddot{N}$), such as $CH_3CH_2\ddot{N}H_2$, $(CH_3)_2\ddot{N}H$, and $(C_3H_7)_3\ddot{N}$

Sample Problem 16.4 Classifying Acid and Base Strength from Chemical Formulas

Problem Classify each compound as a strong acid, a weak acid, a strong base, or a weak base:

(a) KOH **(b)** $(CH_3)_2CHCOOH$

(c) H_2SeO_4 **(d)** $(CH_3)_2CHNH_2$

Plan We examine each formula and classify the acid or base, using the given descriptions. For acids, notice the number of O atoms relative to the number of ionizable H atoms and the presence of the —COOH group. For bases, notice the nature of the cation or the presence of an N atom that has a lone pair.

Solution **(a)** Strong base: KOH is one of the group 1 hydroxides.

(b) Weak acid: $(CH_3)_2CHCOOH$ is a carboxylic acid, as indicated by the —COOH group.

(c) Strong acid: H_2SeO_4 is an oxoacid in which the number of O atoms exceeds the number of ionizable protons by two.

(d) Weak base: $(CH_3)_2CHNH_2$ has a lone pair on the N and is an amine.

Follow-Up Problem 16.4 Which compound in each pair is the stronger acid or base?

(a) HClO or $HClO_3$ **(b)** HCl or CH_3COOH **(c)** NaOH or CH_3NH_2

SUMMARY OF SECTION 16.1

- In aqueous solution, water binds the proton released from an acid to form a hydrated species represented by $H^+(aq)$ or $H_3O^+(aq)$.
- According to the Arrhenius definition, acids contain the element H and yield H^+ (sometimes written as H_3O^+) in water, bases contain OH and yield OH^- in water, and an acid-base reaction (neutralization) is the reaction of H^+ and OH^- to form H_2O.
- According to the Brønsted-Lowry definition, acids behave as proton donors and bases behave as proton acceptors. This leads to a different way of viewing acid-base reactions. The Brønsted-Lowry acid-base definition does not require a base to contain OH in its formula or an acid-base reaction to occur in aqueous solution.
- Each acid-base reaction produces a set of conjugate species. The stronger acid and base pair react to form the weaker acid and base pair.
- When an acid donates a proton, it becomes the conjugate base; when a base accepts a proton, it becomes the conjugate acid. In an acid-base reaction, acids and bases form their conjugates. A stronger acid has a weaker conjugate base, and vice versa.
- Conjugate species can be recognized as follows: the conjugate base of an acid has one hydrogen atom *less* and one negative charge *more* than the acid; the conjugate acid of a base has one hydrogen atom *more* and one negative charge *less* than the base.
- Acid strength depends on $[H^+]$ relative to [HA] in aqueous solution. Strong acids dissociate completely and weak acids dissociate slightly.
- The extent of dissociation is expressed by the acid-dissociation constant, K_a. Weak acids have K_a values lower than 1. Strong acids have K_a values larger than 1, and very strong acids sometimes have undefined K_a values.
- Many acids and bases can be classified as strong or weak based on their formulas.

16.2 Autoionization of Water and the pH Scale

Before we discuss the next major acid-base definition, let us examine a crucial property of water that enables us to quantify $[H_3O^+]$: *water dissociates very slightly into ions* in an equilibrium process known as **autoionization** (or self-ionization):

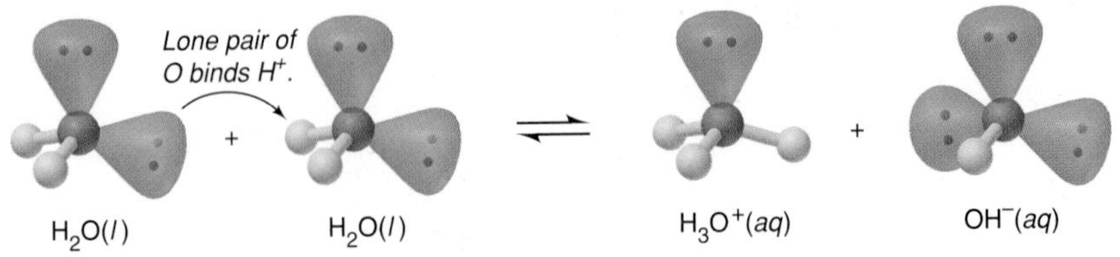

The Equilibrium Nature of Autoionization: The Ion-Product Constant for Water (K_w)

Like any equilibrium process, the autoionization of water is described quantitatively by an equilibrium constant, called the **ion-product constant for water (K_w)**:

$$K_w = [H^+][OH^-] = 1.0 \times 10^{-14} \text{ (at 25°C)} \quad (16.2)$$

Notice that *one H^+ ion and one OH^- ion form for each H_2O molecule that dissociates*. Therefore, in pure water, we find that

$$[H^+] = [OH^-] = \sqrt{1.0 \times 10^{-14}} = 1.0 \times 10^{-7} \text{ mol/L (at 25°C)}$$

It is important to remember that the terms H^+ and H_3O^+ are often used interchangeably in this chapter, and that, as with all K values, K_w is temperature sensitive.

Since pure water has a concentration of about 55.5 mol/L, these equilibrium concentrations are attained when only one in 555 million water molecules dissociates reversibly into ions.

The autoionization of water affects aqueous acid-base chemistry in two major ways:

1. *A change in $[H^+]$ causes an inverse change in $[OH^-]$*, and vice versa:

$$\text{higher } [H^+] \Longrightarrow \text{lower } [OH^-] \quad \text{and} \quad \text{higher } [OH^-] \Longrightarrow \text{lower } [H^+]$$

Recall, from Le Châtelier's principle (Section 15.6), that a change in concentration shifts the equilibrium position but does *not* change the equilibrium constant. Therefore, if some acid is added, $[H_3O^+]$ increases and $[OH^-]$ decreases as the ions react to form water; similarly, if some base is added, $[OH^-]$ increases and $[H_3O^+]$ decreases. In both cases, as long as the temperature is constant, the value of K_w is constant.

2. *Both ions are present in all aqueous systems.* Thus, all acidic solutions contain a low $[OH^-]$, and all basic solutions contain a low $[H_3O^+]$. The equilibrium nature of autoionization allows us to define *acidic* and *basic* solutions in terms of the relative magnitudes of $[H_3O^+]$ and $[OH^-]$:

In an *acidic* solution,	$[H_3O^+] > [OH^-]$
In a *neutral* solution,	$[H_3O^+] = [OH^-]$
In a *basic* solution,	$[H_3O^+] < [OH^-]$

Figure 16.6 summarizes these relationships. Moreover, if you know the value of K_w at a particular temperature and the concentration of one of the two ions, you can find the concentration of the other ion:

$$[H^+] = \frac{K_w}{[OH^-]} \quad \text{or} \quad [OH^-] = \frac{K_w}{[H^+]}$$

There is one very important point to note here: the concentration of the hydrogen ion is not an indicator of the strength of the acid. A very dilute solution of a strong acid could have the same concentration of the hydrogen ion as a concentrated solution of a weak acid.

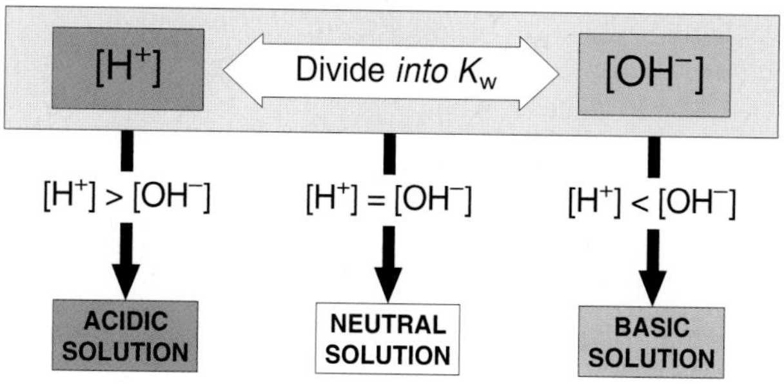

FIGURE 16.6 The relationship between $[H^+]$ and $[OH^-]$ and the relative acidity of solutions

Sample Problem 16.5 Calculating $[H^+]$ or $[OH^-]$ in Aqueous Solution

Problem A research chemist adds a measured amount of HCl gas to pure water at 25°C and obtains a solution with $[H^+] = 3.0 \times 10^{-4}$ mol/L. Calculate $[OH^-]$. Is the solution neutral, acidic, or basic?

Plan We use the known value of K_w at 25°C (1.0×10^{-14}) and the given $[H^+]$ (3.0×10^{-4} mol/L) to solve for $[OH^-]$. Then we compare $[H^+]$ with $[OH^-]$ to determine whether the solution is acidic, basic, or neutral (see Figure 16.6).

Solution Calculate $[OH^-]$:

$$[OH^-] = \frac{K_w}{[H^+]} = \frac{1.0 \times 10^{-14}}{3.0 \times 10^{-4}}$$
$$= 3.3 \times 10^{-11} \text{ mol/L}$$

Because $[H^+] > [OH^-]$, the solution is acidic.

Check It makes sense that adding an acid to water results in an acidic solution. Also, since $[H^+]$ is greater than 10^{-7} mol/L, $[OH^-]$ must be less than 10^{-7} mol/L to give a constant K_w.

Follow-Up Problem 16.5 Calculate $[H^+]$ in a solution that has $[OH^-] = 6.7 \times 10^{-2}$ mol/L at 25°C. Is the solution neutral, acidic, or basic?

Expressing the Hydronium Ion Concentration: The pH Scale

In aqueous solutions, $[H^+]$ or $[H_3O^+]$ can vary from about 10 mol/L to 10^{-15} mol/L. To handle numbers with negative exponents more conveniently in calculations, we convert them to positive numbers using a numerical system called a *p-scale*, the negative of the common (base-10) logarithm of the number. Applying this numerical system to $[H^+]$ gives **pH**, the negative logarithm of $[H^+]$ (or $[H_3O^+]$):

$$pH = -\log [H^+] \quad (16.3)$$

What is the pH of a 10^{-12} mol/L H^+ solution?

$$pH = -\log [H^+] = -\log 10^{-12} = (-1)(-12) = 12$$

Similarly, a 10^{-3} mol/L H^+ solution has a pH of 3, and a 5.4×10^{-4} mol/L H_3O^+ solution has a pH of 3.27:

$$pH = -\log [H^+] = (-1)\log (5.4 \times 10^{-4}) = 3.27$$

As with any measurement, the number of significant figures in a pH value reflects the precision with which the concentration is known. However, a pH value is a logarithm, so the number of significant figures in the concentration equals the number of digits *to the right of the decimal point in the pH value* (see Appendix A). In the preceding example, 5.4×10^{-4} mol/L has two significant figures, so its negative logarithm, 3.27, has two digits to the right of the decimal point.

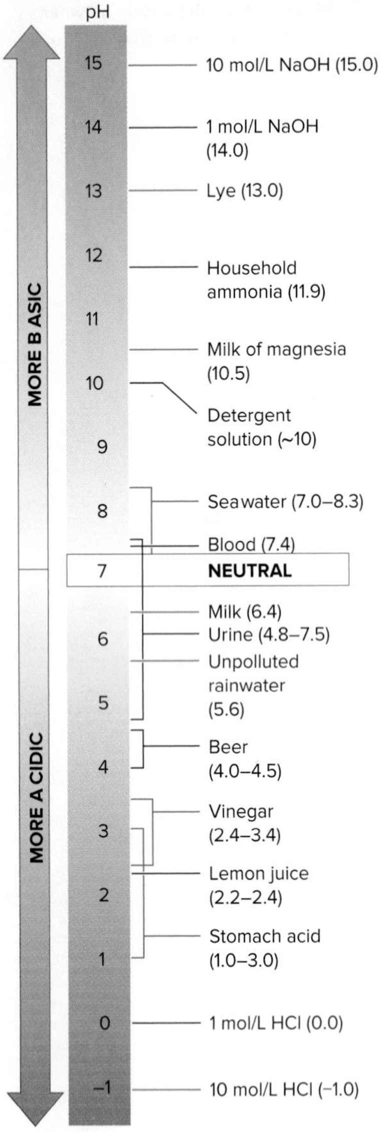

FIGURE 16.7 The pH values of some familiar aqueous solutions

Note, in particular, that *the higher the pH, the lower the [H^+]*. Therefore, *an acidic solution has a lower pH (higher [H^+]) than a basic solution.* At 25°C in pure water, [H_3O^+] is 1.0×10^{-7} mol/L. Thus,

$$\text{pH of an acidic solution} < 7.00$$
$$\text{pH of a neutral solution} = 7.00$$
$$\text{pH of a basic solution} > 7.00$$

Figure 16.7 shows that the pH values of some familiar aqueous solutions fall within a range of −1 to 15.

Because the pH scale is logarithmic, a solution with pH 1.0 has an [H^+] that is 10 times the [H^+] of a solution with pH 2.0, 100 times the [H^+] of a solution with pH 3.0, and so on. To find the [H^+] from the pH, you perform the opposite arithmetic process; that is, you find the negative antilog of pH:

$$[H^+] = 10^{-\text{pH}}$$

A p-scale (−log X) can be used to express other quantities as well:

- Hydroxide ion concentration can be expressed as pOH:

$$\text{pOH} = -\log\,[OH^-]$$

Acidic solutions have a higher pOH (lower [OH^-]) than basic solutions.

- Equilibrium constants can be expressed as pK:

$$\text{p}K = -\log K$$

A low pK corresponds to a high K. A reaction that reaches equilibrium with mostly products present (proceeds far to the right) has a low pK (high K), whereas a reaction that has mostly reactants present at equilibrium has a high pK (low K). Table 16.5 shows this relationship for the aqueous equilibria of some weak acids.

The Relationships among pH, pOH, and pK_w Taking the negative log of both sides of the K_w expression gives a very useful relationship among pH, pOH, and pK_w:

$$K_w = [H_3O^+][OH^-] = 1.0 \times 10^{-14} \text{ (at 25°C)}$$
$$-\log K_w = (-\log\,[H_3O^+]) + (-\log\,[OH^-]) = -\log\,(1.0 \times 10^{-14})$$

$$\text{p}K_w = \text{pH} + \text{pOH} = 14.00 \quad \text{(at 25°C)} \qquad (16.4)$$

Note these important points:

1. The sum of pH and pOH is pK_w for any aqueous solution at any temperature, and pK_w equals 14.00 at 25°C.
2. *Because K_w is constant, pH, pOH, [H_3O^+], and [OH^-] are interrelated:*
   - [H_3O^+] and [OH^-] change in opposite directions.
   - pH and pOH also change in opposite directions.
   - At 25°C, the product of [H_3O^+] and [OH^-] is 1.0×10^{-14}, and the sum of pH and pOH is 14.00 (Figure 16.8).

TABLE 16.5 The Relationship between K_a and pK_a

Acid Name (Formula)	K_a at 25°C	pK_a
Hydrogen sulfate ion (HSO_4^-)	1.0×10^{-2}	1.99
Nitrous acid (HNO_2)	7.1×10^{-4}	3.15
Ethanoic acid (CH_3COOH)	1.8×10^{-5}	4.75
Hypobromous acid (HBrO)	2.3×10^{-9}	8.64
Phenol (C_6H_5OH)	1.0×10^{-10}	10.00

	$[H_3O^+]$	pH	$[OH^-]$	pOH
	1.0×10^{-15}	15.00	1.0×10^{1}	−1.00
	1.0×10^{-14}	14.00	1.0×10^{0}	0.00
	1.0×10^{-13}	13.00	1.0×10^{-1}	1.00
BASIC	1.0×10^{-12}	12.00	1.0×10^{-2}	2.00
	1.0×10^{-11}	11.00	1.0×10^{-3}	3.00
	1.0×10^{-10}	10.00	1.0×10^{-4}	4.00
	1.0×10^{-9}	9.00	1.0×10^{-5}	5.00
	1.0×10^{-8}	8.00	1.0×10^{-6}	6.00
NEUTRAL	1.0×10^{-7}	7.00	1.0×10^{-7}	7.00
	1.0×10^{-6}	6.00	1.0×10^{-8}	8.00
	1.0×10^{-5}	5.00	1.0×10^{-9}	9.00
	1.0×10^{-4}	4.00	1.0×10^{-10}	10.00
ACIDIC	1.0×10^{-3}	3.00	1.0×10^{-11}	11.00
	1.0×10^{-2}	2.00	1.0×10^{-12}	12.00
	1.0×10^{-1}	1.00	1.0×10^{-13}	13.00
	1.0×10^{0}	0.00	1.0×10^{-14}	14.00
	1.0×10^{1}	−1.00	1.0×10^{-15}	15.00

FIGURE 16.8 The relationships among $[H_3O^+]$, pH, $[OH^-]$, and pOH at 25°C

Sample Problem 16.6 Calculating $[H^+]$, pH, $[OH^-]$, and pOH

Problem In an art restoration project, a conservator prepares copper-plate etching solutions by diluting concentrated HNO_3 to 2.0 mol/L, 0.30 mol/L, and 0.0063 mol/L HNO_3. Calculate $[H^+]$, pH, $[OH^-]$, and pOH of the three solutions at 25°C.

Plan We know, from its formula, that HNO_3 is a strong acid, so it dissociates completely; thus, $[H^+] = [HNO_3]_{init}$. We use the given concentrations and the value of K_w at 25°C (1.0×10^{-14}) to find $[H^+]$ and $[OH^-]$ and then use them to calculate pH and pOH.

Solution Calculate the values for 2.0 mol/L HNO_3:

$$[H^+] = 2.0 \text{ mol/L}$$
$$pH = -\log [H^+] = -\log 2.0 = -0.30$$
$$[OH^-] = \frac{K_w}{[H^+]} = \frac{1.0 \times 10^{-14}}{2.0} = 5.0 \times 10^{-15} \text{ mol/L}$$
$$pOH = -\log (5.0 \times 10^{-15}) = 14.30$$

Calculate the values for 0.30 mol/L HNO_3:

$$[H^+] = 0.30 \text{ mol/L}$$
$$pH = -\log [H^+] = -\log 0.30 = 0.52$$
$$[OH^-] = \frac{K_w}{[H^+]} = \frac{1.0 \times 10^{-14}}{0.30} = 3.3 \times 10^{-14} \text{ mol/L}$$
$$pOH = -\log (3.3 \times 10^{-14}) = 13.48$$

Calculate the values for 0.0063 mol/L HNO_3:

$$[H^+] = 6.3 \times 10^{-3} \text{ mol/L}$$
$$pH = -\log [H^+] = -\log(6.3 \times 10^{-3}) = 2.20$$
$$[OH^-] = \frac{K_w}{[H^+]} = \frac{1.0 \times 10^{-14}}{6.3 \times 10^{-3}} = 1.6 \times 10^{-12} \text{ mol/L}$$
$$pOH = -\log (1.6 \times 10^{-12}) = 11.80$$

Check As the solution becomes more dilute, $[H^+]$ decreases, so pH increases, as we expect. An $[H^+] > 1.0$ mol/L, as in 2.0 mol/L HNO_3, gives a positive log, so it results in a negative pH. The arithmetic seems correct because pH + pOH = 14.00 for each solution.

Follow-Up Problem 16.6 A solution of NaOH has a pH of 9.52. What are its pOH, $[H^+]$, and $[OH^-]$ at 25°C?

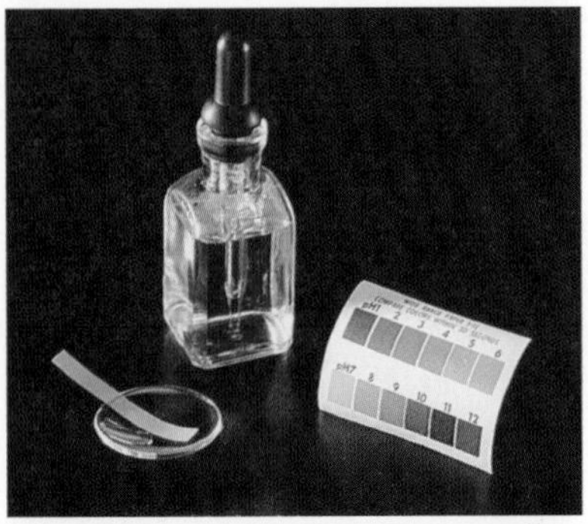

A

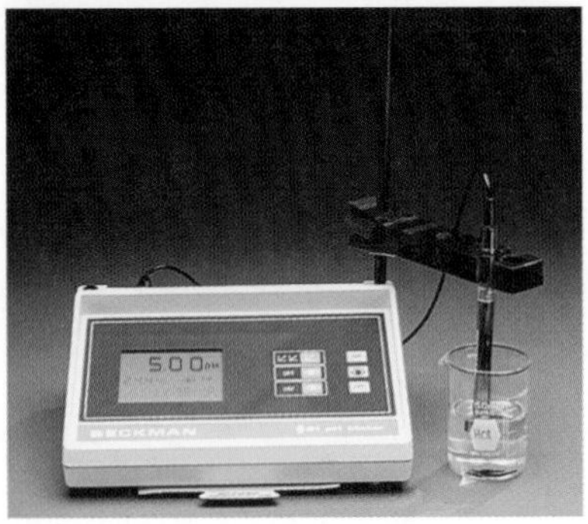

B

FIGURE 16.9 Methods for measuring the pH of an aqueous solution: pH paper **(A)** and pH meter **(B)**

Measuring pH In a laboratory, pH values are usually obtained in one of two ways:

1. **Acid-base indicators** are organic molecules whose colours depend on the acidity of the solution in which they are dissolved. A pH value can be estimated quickly with *pH paper*, a paper strip impregnated with an indicator or a mixture of indicators. A drop of solution is placed on the strip, and the colour is compared with the colours in a chart (Figure 16.9A).

2. A *pH meter* measures $[H_3O^+]$ using two electrodes immersed in the test solution. One electrode supplies a reference system; the other electrode consists of a very thin glass membrane that separates a known internal $[H_3O^+]$ from the unknown external $[H_3O^+]$. The difference in $[H_3O^+]$ creates a voltage difference across the membrane, which is displayed as a pH (Figure 16.9B).

SUMMARY OF SECTION 16.2

- Pure water has a low conductivity because it autoionizes to a small extent in a process whose equilibrium constant is the ion-product constant for water, K_w (1.0×10^{-14} at 25°C).
- $[H^+]$ and $[OH^-]$ are inversely related: $[H^+]$ is greater than $[OH^-]$ in an acidic solution, the reverse is true in a basic solution, and the two concentrations are equal in a neutral solution.
- To express small values of $[H^+]$, we use the pH scale (pH = $-\log [H^+]$). Similarly, pOH = $-\log [OH^-]$, and pK = $-\log K$.
- A high pH represents a low $[H^+]$. In acidic solutions, pH < 7.00; in basic solutions, pH > 7.00; and in neutral solutions, pH = 7.00. The sum of pH and pOH equals pK_w (14.00 at 25°C).
- The pH is typically measured with either an acid-base indicator or a pH meter.

16.3 Weak Bases and Their Relation to Weak Acids

The Brønsted-Lowry concept expands the definition of a base to encompass a host of species that the Arrhenius definition excludes: *to accept a proton, a base needs only a lone electron pair.*

Let us examine the equilibrium system of a weak base (B) as it dissolves in water. In the following reaction, B accepts a proton from H_2O, which acts as an acid, leaving behind an OH^- ion:

$$B(aq) + H_2O(l) \rightleftharpoons BH^+(aq) + OH^-(aq)$$

This general reaction for a base in water is described by the equilibrium expression for base dissociation in water,

$$K_b = \frac{a_{BH^+}a_{OH^-}}{a_B a_{H_2O}}$$

Since the activity of pure water is 1, if we are dealing with fairly dilute or reasonably ideal solutions, we can write the **base-dissociation constant (base-ionization constant), K_b**, as follows:

$$K_b = \frac{[BH^+][OH^-]}{[B]} \qquad (16.5)$$

Despite the name "base-dissociation constant," *no base dissociates in the process.* As in the relation between pK_a and K_a, we know that a lower pK_b indicates a higher K_b, that is, a stronger base. In aqueous solution, the two large classes of weak bases are (1) ammonia and the amines and (2) the conjugate bases of weak acids.

Molecules as Weak Bases: Ammonia and the Amines

Ammonia is the simplest N-containing compound that acts as a weak base in water:

$$NH_3(aq) + H_2O(l) \rightleftharpoons NH_4^+(aq) + OH^-(aq) \qquad K_b = 1.76 \times 10^{-5} \text{ (at 25°C)}$$

TABLE 16.6 K_b **Values for Some Molecular (Amine) Bases at 25°C**

Name (Formula)	Lewis Structure*	K_b	pK_b
N-ethylethanamine (diethylamine), $(CH_3CH_2)_2NH$	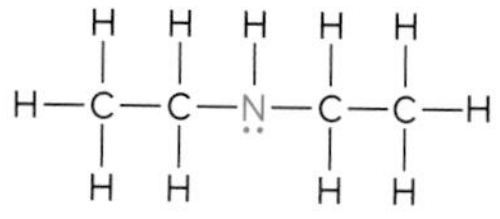	8.6×10^{-4}	3.07
N-methylmethanamine (dimethylamine), $(CH_3)_2NH$	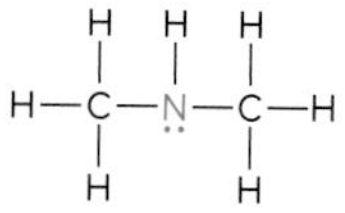	5.9×10^{-4}	3.23
Methanamine (methylamine), CH_3NH_2	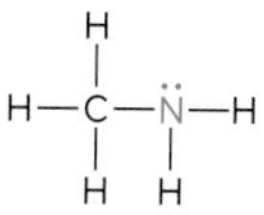	4.4×10^{-4}	3.36
Ammonia, NH_3	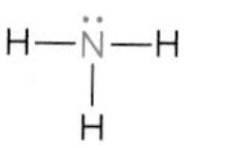	1.76×10^{-5}	4.75
Pyridine, C_5H_5N		1.7×10^{-9}	8.77
Phenylamine (aniline), $C_6H_5NH_2$	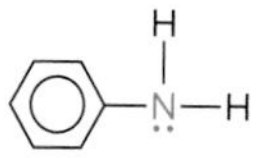	4.0×10^{-10}	9.40

*Blue type indicates the basic nitrogen and its lone pair.

Despite labels on reagent bottles that read "ammonium hydroxide," an aqueous solution of ammonia consists largely of water and *unprotonated* NH_3 molecules, as its small K_b indicates. In a 1.0 mol/L NH_3 solution, for example, $[OH^-] = [NH_4^+] = 4.2 \times 10^{-3}$ mol/L, so about 99.6% of the NH_3 is not protonated. Table 16.6 shows the K_b values for a few molecular bases. (A more extensive list appears in Appendix C.)

If one or more of the H atoms in ammonia is replaced by an organic group (designated as R), an *amine* results: RNH_2, R_2NH, or R_3N (see Chapter 20, Figure 20.25). The key structural feature of these organic compounds, as in all Brønsted-Lowry bases, is *a lone pair of electrons that can bind the proton donated by the acid*. Figure 16.10 shows this process for methanamine, the simplest amine.

Weak Bases That Are Conjugates of Weak Acids

The other large group of Brønsted-Lowry bases consists of conjugate bases of weak acids. The conjugate base of a weak acid, HA, is A^-, and the dissociation of the acid in water can be written as follows:

$$A^-(aq) + H_2O(l) \rightleftharpoons HA(aq) + OH^-(aq) \qquad K_b = \frac{[HA][OH^-]}{[A^-]}$$

This reaction is sometimes referred to as a *hydrolysis reaction* because water is dissociated (hydrolyzed). Actually, except for the charge on the base, this reaction

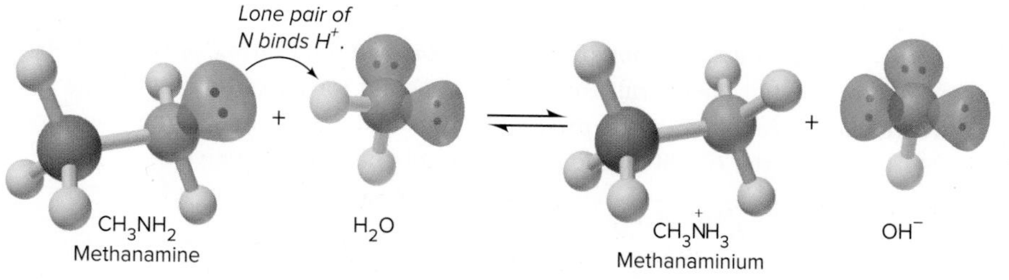

FIGURE 16.10 Abstraction of a proton from water by the base methanamine

is the same as the proton-abstraction process by ammonia and the amines. For example, F^-, the conjugate base of the weak acid HF, is a weak base:

$$F^-(aq) + H_2O(l) \rightleftharpoons HF(aq) + OH^-(aq) \qquad K_b = \frac{[HF][OH^-]}{[F^-]}$$

Why is a solution of HA acidic and a solution of A^- basic? We can find the answer from the relative concentrations of the species in 1 mol/L HF and in 1 mol/L NaF:

1. *The acidity of HA(aq).* HF is a weak acid, so the equilibrium position of the acid dissolving in water lies far to the left:

$$HF(aq) + H_2O(l) \rightleftharpoons H_3O^+(aq) + F^-(aq)$$

Water also contributes H_3O^+ and OH^-, but their concentrations are extremely small:

$$2H_2O(l) \rightleftharpoons H_3O^+(aq) + OH^-(aq)$$

Of all the species present—HF, H_2O, H_3O^+, F^-, and OH^-—the two that can influence the acidity of the solution are H_3O^+, predominantly from HF, and OH^-, from water. Thus, the HF solution is acidic because $[H_3O^+]_{\text{from HF}} \gg [OH^-]_{\text{from } H_2O}$.

2. *The basicity of $A^-(aq)$.* Now consider the species in 1 mol/L NaF. The salt dissociates completely to yield 1 mol/L Na^+ and 1 mol/L F^-. The Na^+ behaves as a spectator ion (it is too weak a conjugate acid to participate in any acid-base reaction), while some F^- reacts as a weak base to produce small amounts of HF and OH^-:

$$F^-(aq) + H_2O(l) \rightleftharpoons HF(aq) + OH^-(aq)$$

As before, water dissociation contributes minute amounts of H_3O^+ and OH^-. Thus, in addition to the Na^+ ion, the species that are present are the same as those in the HF solution: HF, H_2O, H_3O^+, F^-, and OH^-. The two species that affect the acidity are OH^-, predominantly from F^- reacting with water, and H_3O^+ from water. In this reaction, $[OH^-]_{\text{from } F^-} \gg [H_3O^+]_{\text{from } H_2O}$, so the solution is basic.

We can summarize as follows:

- In an HA solution, $[HA] \gg [A^-]$ and $[H_3O^+]_{\text{from HA}} \gg [OH^-]_{\text{from } H_2O}$, so the solution is acidic.
- In an A^- solution, $[A^-] \gg [HA]$ and $[OH^-]_{\text{from } A^-} \gg [H_3O^+]_{\text{from } H_2O}$, so the solution is basic.

While the above relationship is true for most solutions of acids and bases that are used in a lab setting, it is nevertheless important to verify that it is true each time any reaction is performed. If the concentration of a particular acid, whether strong or weak, is very low (such as 1.0×10^{-8} mol/L), the amount of hydrogen ion contributed by water (1.0×10^{-7} mol/L) becomes significant and must be taken into consideration. The same is true if the concentration of a particular base, whether strong or weak, is very low; then the amount of hydroxide ion contributed by water becomes significant.

The Relationship between K_a and K_b of a Conjugate Acid-Base Pair

A key relationship exists between the K_a of HA and the K_b of A^-, which we can see by writing the two reactions as a reaction sequence and adding them:

$$\begin{array}{l} \cancel{HA(aq)} + H_2O\ (l) \rightleftharpoons H_3O^+(aq) + \cancel{A^-(aq)} \\ \cancel{A^-(aq)} + H_2O\ (l) \rightleftharpoons \cancel{HA(aq)} + OH^-(aq) \\ \hline 2H_2O\ (l) \rightleftharpoons H_3O^+(aq) + OH^-(aq) \end{array}$$

The sum of the two dissociation reactions is the autoionization of water. Recall, from Chapter 15, that the overall equilibrium constant for a reaction that is the *sum* of two or more reactions is the *product* of the individual equilibrium constants. Therefore, writing the equilibrium expressions for the reactions gives

$$\frac{[H_3O^+]\cancel{[A^-]}}{\cancel{[HA]}} \times \frac{\cancel{[HA]}[OH^-]}{\cancel{[A^-]}} = [H_3O^+][OH^-]$$

or

$$K_a \times K_b = K_w \quad (16.6)$$

This relationship allows us to find the K_a of the acid in a conjugate pair, given the K_b of the base, and vice versa. Reference tables typically have K_a and K_b values for *molecular species only*. The K_b of F^- and the K_a of $CH_3NH_3^+$, for example, do not appear in standard tables, but you can calculate them by looking up the values for the molecular conjugate species and relating these values to K_w. To find the K_b of F^-, for instance, we look up the K_a of HF and apply Equation 16.6:

$$K_a \text{ of HF} = 6.8 \times 10^{-4} \text{ (from Appendix C)}$$

So, we have

$$K_a \text{ of HF} \times K_b \text{ of } F^- = K_w$$

or

$$K_b \text{ of } F^- = \frac{K_w}{K_a \text{ of HF}} = \frac{1.0 \times 10^{-14}}{6.8 \times 10^{-4}} = 1.5 \times 10^{-11}$$

SUMMARY OF SECTION 16.3

- The extent to which a weak base accepts a proton from water to form OH^- is expressed by a base-dissociation constant, K_b.
- Brønsted-Lowry bases include NH_3 and amines, as well as the conjugate bases of weak acids. All produce basic solutions by accepting H^+ from water, which yields OH^-, thus making $[H_3O^+] < [OH^-]$.
- A solution of HA is acidic because $[HA] >> [A^-]$, so $[H_3O^+] > [OH^-]$. A solution of A^- is basic because $[A^-] >> [HA]$, so $[OH^-] > [H_3O^+]$.
- By multiplying the expressions for K_a of HA and K_b of A^-, we obtain K_w. This relationship allows us to calculate either K_a of BH^+ or K_b of A^-.

16.4 Solving Problems Involving Weak-Acid or Weak-Base Equilibria

Just as you saw in Chapter 15 for equilibrium problems in general, there are two general types of equilibrium problems involving weak acids and their conjugate bases or weak bases and their conjugate acids:

1. Given equilibrium concentrations, find K_a or K_b.
2. Given K_a or K_b and some concentrations, find other equilibrium concentrations.

For both types of problems, we will apply the same problem-solving approach, notation system, and assumptions:

- *Problem-solving approach.* Start with what is given in the problem and move toward what you want to find. Make a habit of applying the following steps:
  1. Write the balanced equation and the K_a or K_b expression; these tell you what to find.
  2. Define x as the unknown change in concentration that occurs during the reaction. Frequently, x equals $[HA]_{dissoc}$, the concentration of HA that dissociates, which (through the use of certain assumptions) may also equal $[H^+]$ and $[A^-]$ at equilibrium in an acid-dissociation problem. In a base-dissociation problem, often x equals $[A^-]_{dissoc}$, the concentration of A^- that dissociates, which (again through the use of certain assumptions) may also equal [HA] and $[OH^-]$ at equilibrium.
  3. For most problems, construct an ICE (**I**nitial, **C**hange, **E**quilibrium) table that incorporates x.
  4. If warranted, make assumptions (usually that x is very small relative to the initial concentration; see below) that simplify the calculations.

5. Substitute the values into the K_a or K_b expression, and solve for x.
6. Check that the assumptions are justified with the 5% test first used in Sample Problem 15.10. If they are not justified, use the quadratic formula to find x. Although assumptions are sometimes helpful to simplify the arithmetic, it is important to understand the 3W rule: *why* they are being made, *what* they represent, and *when* you can use them.

- *Notation system.* As always, concentration in mol/L is indicated with brackets and is used to approximate the activity of aqueous species. A subscript refers to where the species comes from or when it occurs in the reaction process. For example, $[H^+]_{\text{from HA}}$ is the molar concentration of H^+ that comes from the dissociation of HA, $[HA]_{\text{init}}$ is the initial molar concentration of HA (that is, before dissociation), $[HA]_{\text{dissoc}}$ is the molar concentration of HA that dissociates, and so on. A bracketed formula with *no* subscript represents the molar concentration of the species *at equilibrium*.
- *Assumptions.* At times, we may make two assumptions to simplify the arithmetic:

1. $[H_3O^+]$ from the autoionization of water is so much smaller than $[H_3O^+]$ from the dissociation of HA that we can assume it does not contribute significantly in these problems:

$$[H_3O^+] = [H_3O^+]_{\text{from HA}} + [H_3O^+]_{\text{from } H_2O} \approx [H_3O^+]_{\text{from HA}}$$

Note that each molecule of HA that dissociates forms one H^+ and one A^-:

$$[HA]_{\text{dissoc}} = [H^+] = [A^-]$$

Before making any assumption, it is very important to ensure that the assumption is valid under the circumstances being studied. For example, the assumption above may not be valid if the K_a of the acid is in the range of 10^{-5} and the initial concentration of the acid is very small, such as 10^{-3} mol/L. If the initial concentration of the acid is lower than 1.0×10^{-7} mol/L, then the concentration of hydrogen ion in water must be taken into consideration.

2. A weak acid has a small K_a and, therefore, it dissociates to a very small extent. Such a small change in the initial concentration is observed that we can effectively say that the equilibrium concentration equals the initial concentration:

$$[HA] = [HA]_{\text{init}} - \underbrace{[HA]_{\text{dissoc}}}_{\text{mathematically insignificant}} \approx [HA]_{\text{init}}$$

Finding K_a or K_b Given Concentrations

This type of problem involves finding the K_a of a weak acid from the concentration of one of the species in solution, usually $[H_3O^+]$ from a given pH:

$$HA(aq) + H_2O(l) \rightleftharpoons H_3O^+(aq) + A^-(aq) \qquad K_a = \frac{[H_3O^+][A^-]}{[HA]}$$

or finding the K_b of a weak base from the concentration of one of the species in solution, usually $[OH^-]$ from a given pH or pOH:

$$BOH(aq) \rightleftharpoons B^+(aq) + OH^-(aq) \qquad K_b = \frac{[B^+][OH^-]}{[BOH]}$$

Some bases abstract a proton from water:

$$B(aq) + H_2O(l) \rightleftharpoons BH^+(aq) + OH^-(aq) \qquad K_b = \frac{[BH^+][OH^-]}{[B]}$$

You prepare an aqueous solution of HA or B (or BOH) and measure its pH. Thus, you know $[HA]_{\text{init}}$ or $[B]_{\text{init}}$ (or $[BOH]_{\text{init}}$); using the pH, you can find the $[H^+]$ or $[OH^-]$ at equilibrium, which equals $[A^-]$ or $[BH^+]$ (or $[B^+]$) at equilibrium as well. Using these concentrations, you can determine the $[HA]_{\text{eq}}$ or $[B]_{\text{eq}}$ (or $[BOH]_{\text{eq}}$). These values can be substituted into the appropriate expressions to determine either K_a or K_b. Let us go through this approach in Sample Problem 16.7.

Sample Problem 16.7 Finding K_a of a Weak Acid from the Solution pH

Problem A substance called 2-phenylethanoic (or phenylacetic) acid ($C_6H_5CH_2COOH$, simplified here to HPOAc; *see model*) builds up in the blood of people with phenylketonuria, an inherited disorder that, if untreated, causes decreased brain function and death. A study of the acid shows that the pH of 0.12 mol/L HPOAc is 2.62. What is the K_a of 2-phenylacetic acid?

Plan We are given $[\text{HPOAc}]_{\text{init}}$ (0.12 mol/L) and the pH (2.62), and we must find K_a. As always, we first write the equation for HPOAc dissociation and the expression for K_a to see which values we need. We assume that $[H^+]_{\text{from } H_2O}$ is negligible, so we can use the given pH to find $[H^+]$, which equals $[\text{POAc}^-]$ and $[\text{HPOAc}]_{\text{dissoc}}$. To find [HPOAc], we assume that very little dissociates (because HPOAc is a weak acid), so $[\text{HPOAc}]_{\text{init}} - [\text{HPOAc}]_{\text{dissoc}} = [\text{HPOAc}] \approx [\text{HPOAc}]_{\text{init}}$. We make these assumptions, substitute the equilibrium values, solve for K_a, and then check the assumptions. Note that, although we do not need an ICE table here (because the pH will give us $[\text{HPOAc}]_{\text{dissoc}}$), there is no harm in drawing one if it helps us visualize what is happening.

Solution Write the dissociation equation and K_a expression:

$$\text{HPOAc}(aq) + H_2O(l) \rightleftharpoons H_3O^+(aq) + \text{POAc}^-(aq) \qquad K_a = \frac{[H_3O^+][\text{POAc}^-]}{[\text{HPOAc}]}$$

Calculate $[H^+]$:

$$[H^+] = 10^{-\text{pH}} = 10^{-2.62} = 2.4 \times 10^{-3} \text{ mol/L}$$

Make the assumptions:

1. The calculated $[H_3O^+]$ (2.4×10^{-3} mol/L) $\gg$ $[H_3O^+]_{\text{from } H_2O}$ (1.0×10^{-7} mol/L), so we assume that $[H_3O^+] \approx [H_3O^+]_{\text{from HPOAc}} = [\text{POAc}^-] = x$ (the change in [HPOAc], or $[\text{HPOAc}]_{\text{dissoc}}$).
2. HPOAc is a weak acid, so we assume that $[\text{HPOAc}] = 0.12 \text{ mol/L} - x \approx 0.12$ mol/L.

Solve for the equilibrium concentrations:

$$x \approx [H_3O^+] = [\text{POAc}^-] = 2.4 \times 10^{-3} \text{ mol/L}$$

$$[\text{HPOAc}] = 0.12 \text{ mol/L} - x = 0.12\text{mol/L} - (2.4 \times 10^{-3} \text{ mol/L}) \approx 0.12 \text{ mol/L (to 2 sf)}$$

Substitute these values into K_a:

$$K_a = \frac{[H_3O^+][\text{POAc}^-]}{[\text{HPOAc}]} \approx \frac{(2.4 \times 10^{-3})(2.4 \times 10^{-3})}{0.12} = 4.8 \times 10^{-5}$$

Check the assumptions by finding the percent error in concentration:

1. For $[H_3O^+]_{\text{from } H_2O}$: $\dfrac{1 \times 10^{-7} \text{ mol/L}}{2.4 \times 10^{-3} \text{ mol/L}} \times 100 = 4 \times 10^{-3}\%$ (<5%; assumption is justified)
2. For $[\text{HPOAc}]_{\text{dissoc}}$: $\dfrac{2.4 \times 10^{-3} \text{ mol/L}}{0.12 \text{ mol/L}} \times 100 = 2.0\%$ (<5%; assumption is justified)

Check The $[H_3O^+]$ makes sense: pH 2.62 should give $[H_3O^+]$ between 10^{-2} and 10^{-3} mol/L. The value for the K_a calculation also seems in the correct range: $\frac{(10^{-3})^2}{10^{-1}} = 10^{-5}$, and the result seems reasonable for a weak acid.

Follow-Up Problem 16.7 The conjugate acid of ammonia is the weak acid NH_4^+. If a 0.2 mol/L NH_4Cl solution has a pH of 5.0, what is the K_a of NH_4^+?

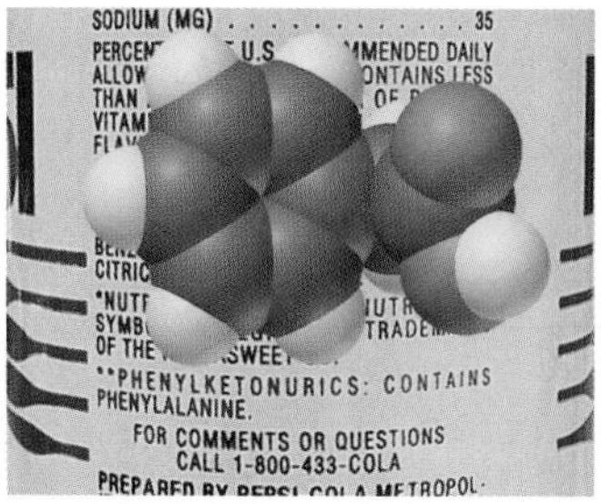

Phenylalanine, one of the amino acids that make up aspartame, is metabolized to 2-phenylethanoic (or phenylacetic) acid (model).

Finding Concentrations Given K_a or K_b

The second type of equilibrium problem gives some concentration data and either K_a or K_b; you are asked to solve for the equilibrium concentration of some component. Such problems are very similar to those we solved in Chapter 15, in which a substance with a given initial concentration reacted to an unknown extent (see Sample Problems 15.9 to 15.11). We *will* use an ICE table in these problems to solve for the missing values; where $[H_3O^+]_{\text{from } H_2O}$ is very small relative to $[H_3O^+]_{\text{from HA}}$, we will enter the initial $[H_3O^+]$ in the reaction table as zero.

Sample Problem 16.8 Determining Concentration from K_a and Initial [HA]

Problem Propanoic acid (CH_3CH_2COOH, which we simplify as EtCOOH) is a carboxylic acid whose salts are used to retard mould growth in foods. What is the $[H_3O^+]$ of 0.10 mol/L EtCOOH ($K_a = 1.3 \times 10^{-5}$)?

Plan We know the initial concentration (0.10 mol/L) and K_a (1.3×10^{-5}) of EtCOOH, and we need to find $[H_3O^+]$. First, we write the balanced equation and the expression for K_a. We know $[\text{EtCOOH}]_{\text{init}}$ but not [EtCOOH] (that is, the concentration at equilibrium). If we let $x = [\text{EtCOOH}]_{\text{dissoc}}$, x is also $[H_3O^+]_{\text{from EtCOOH}}$ and $[\text{EtCOO}^-]$ because each EtCOOH dissociates into one H_3O^+ and one EtCOO^-. With this information, we can set up an ICE table. We assume that, because EtCOOH has a small K_a, it dissociates very little. After solving for x, we check the assumption.

Solution Write the balanced equation and expression for K_a:

$$\text{EtCOOH}(aq) + H_2O(l) \rightleftharpoons H_3O^+(aq) + \text{EtCOO}^-(aq)$$

$$K_a = \frac{[H_3O^+][\text{EtCOO}^-]}{[\text{EtCOOH}]} = 1.3 \times 10^{-5}$$

Set up an ICE table, with $x = [\text{EtCOOH}]_{\text{dissoc}} = [\text{EtCOO}^-] = [H_3O^+]$:

Concentration (mol/L)	EtCOOH(*aq*)	+	$H_2O(l)$	⇌	$H_3O^+(aq)$	+	$\text{EtCOO}^-(aq)$
Initial	0.10		—		0		0
Change	−*x*		—		+*x*		+*x*
Equilibrium	0.10 − *x*		—		*x*		*x*

Since K_a is small, x is small compared with $[\text{EtCOOH}]_{\text{init}}$. Therefore, we can assume that $[\text{EtCOOH}]_{\text{init}} - x = [\text{EtCOOH}] \approx [\text{EtCOOH}]_{\text{init}}$, or $0.10 \text{ mol/L} - x \approx 0.10 \text{ mol/L}$.

Substitute into the K_a expression and solve for x:

$$K_a = \frac{[H_3O^+][\text{EtCOO}^-]}{[\text{EtCOOH}]} = 1.3 \times 10^{-5} \approx \frac{(x)(x)}{0.10}$$

$$x \approx \sqrt{(0.10)(1.3 \times 10^{-5})} = 1.1 \times 10^{-3} \text{ mol/L} = [H_3O^+]$$

We need to check our assumption.

$$\text{For } [\text{EtCOOH}]_{\text{dissoc}}: \frac{1.1 \times 10^{-3} \text{ mol/L}}{0.10 \text{ mol/L}} \times 100 = 1.1\% \ (<5\%; \text{ assumption is justified})$$

Check The $[H_3O^+]$ seems reasonable for a dilute solution of a weak acid with a small K_a. By reversing the calculation, we can check the math: $\frac{(1.1 \times 10^{-3})^2}{0.10} = 1.2 \times 10^{-5}$, which is within rounding of the given K_a.

Comment In Chapter 15 we introduced another benchmark, aside from the 5% rule, that we can use to see if an assumption is justified. (See the discussion following Sample Problem 15.10.)

- If $\frac{[\text{HA}]_{\text{init}}}{K_a} > 400$, the assumption is justified: neglecting x introduces an error < 5%.
- If $\frac{[\text{HA}]_{\text{init}}}{K_a} < 400$, the assumption is *not* justified; neglecting x introduces an error > 5%, so we solve a quadratic equation to find x.

In this sample problem, we have $\frac{0.10}{1.3 \times 10^{-5}} = 7.7 \times 10^3$, which is greater than 400. The alternative situation occurs in the follow-up problem.

Follow-Up Problem 16.8 Cyanic acid (HOCN) is an extremely acrid, unstable substance. What is the $[H_3O^+]$ of 0.10 mol/L HOCN ($K_a = 3.5 \times 10^{-4}$)?

To find the pH of a solution of a molecular weak base, we use an approach very similar to the approach used for a weak acid: write the equilibrium expression, set up an ICE table to find $[B]_{reacting}$, make the usual assumptions, and then solve for $[OH^-]$. The only additional step is converting $[OH^-]$ to $[H^+]$ to calculate pH.

Sample Problem 16.9 Determining pH from K_b and Initial [B]

Problem N-methylmethanamine (dimethylamine), $(CH_3)_2NH$ (*see margin*), a key intermediate in detergent manufacture, has a K_b of 5.9×10^{-4}. What is the pH of 1.5 mol/L $(CH_3)_2NH$?

Methylmethanamine (dimethylamine)

Plan We know the initial concentration (1.5 mol/L) and K_b (5.9×10^{-4}) of $(CH_3)_2NH$ and have to find the pH. The amine reacts with water to form OH^-, so we have to find $[OH^-]$ and then calculate $[H_3O^+]$ and pH. We first write the balanced equation and K_b expression. Because $K_b \gg K_w$, the $[OH^-]$ from the autoionization of water is small compared with the $[OH^-]$ from the amine reacting with water, so we can disregard it and assume that all the $[OH^-]$ comes from the amine reacting with water. Because K_b is small, the amount of amine reacting, $[(CH_3)_2NH]_{reacting}$, can be neglected. We set up an ICE table, make the assumption, and solve for x. Then we check the assumption, convert $[OH^-]$ to $[H_3O^+]$ using K_w, and calculate pH.

Solution Write the balanced equation and K_b expression:

$$(CH_3)_2NH(aq) + H_2O(l) \rightleftharpoons (CH_3)_2NH_2^+(aq) + OH^-(aq)$$

$$K_b = \frac{[(CH_3)_2NH_2^+][OH^-]}{[(CH_3)_2NH]}$$

Set up the ICE table, with $x = [(CH_3)_2NH]_{reacting} = [(CH_3)_2NH_2^+] = [OH^-]$:

Concentration (mol/L)	$(CH_3)_2NH(aq)$	+	$H_2O(l)$	$\rightleftharpoons$	$(CH_3)_2NH_2^+(aq)$	+	$OH^-(aq)$
Initial	1.5		—		0		0
Change	−x		—		+x		+x
Equilibrium	1.5 − x		—		x		x

Because K_b is small, we can assume that

$$[(CH_3)_2NH]_{init} - [(CH_3)_2NH]_{reacting} = [(CH_3)_2NH] \approx [(CH_3)_2NH]_{init}$$

Thus, 1.5 mol/L − $x \approx$ 1.5 mol/L.

Substitute into the K_b expression and solve for x:

$$K_b = \frac{[(CH_3)_2NH_2^+][OH^-]}{[(CH_3)_2NH]} = 5.9 \times 10^{-4} \approx \frac{x^2}{1.5}$$

$$x = [OH^-] \approx 3.0 \times 10^{-2}\ \text{mol/L}$$

Check the assumption:

$$\frac{3.0 \times 10^{-2}\ \text{mol/L}}{1.5\ \text{mol/L}} \times 100 = 2.0\%\ (< 5\%;\ \text{assumption is justified})$$

Note that the comment in Sample Problem 16.8 applies here as well:

$$\frac{[B]_{init}}{K_b} = \frac{1.5}{5.9 \times 10^{-4}} = 2.5 \times 10^3 > 400$$

Calculate pH:

$$[H^+] = \frac{K_w}{[OH^-]} = \frac{1.0 \times 10^{-14}}{3.0 \times 10^{-2}} = 3.3 \times 10^{-13} \text{ mol/L}$$

$$pH = -\log(3.3 \times 10^{-13}) = 12.48$$

Check The value of x seems reasonable: $\sqrt{(\sim 6 \times 10^{-4})(1.5)} = \sqrt{9 \times 10^{-4}} = 3 \times 10^{-2}$. Because $(CH_3)_2NH$ is a weak base, the pH should be several pH units above 7.

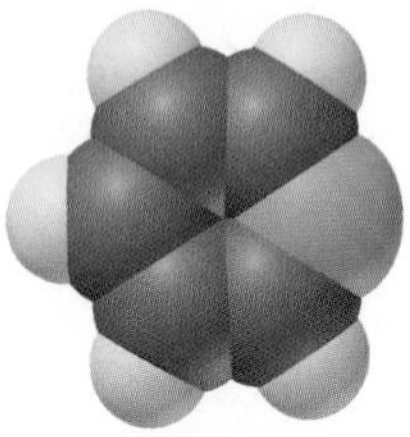

Pyridine

Follow-Up Problem 16.9 Pyridine (C_5H_5N, *see margin*) serves as a solvent *and* a base in many organic syntheses. It has a pK_b of 8.77. What is the pH of 0.10 mol/L pyridine?

The Effect of Concentration on the Extent of Acid or Base Dissociation

If we repeat the calculation in Sample Problem 16.8, but start with a lower [EtCOOH], we can make a very interesting observation about the **extent of dissociation** of a weak acid. Suppose that the initial concentration of EtCOOH is one-tenth as much, 0.010 mol/L rather than 0.10 mol/L. After filling in the ICE table and making the same assumptions, we find that

$$x = [H_3O^+] = [EtCOOH]_{dissoc} = 3.6 \times 10^{-4} \text{ mol/L}$$

Now let us compare the percentages of EtCOOH molecules dissociated at the two different initial acid concentrations, using the following relationship:

$$\text{Percent HA dissociated} = \frac{[HA]_{dissoc}}{[HA]_{init}} \times 100 \quad (16.7)$$

Case 1: $[EtCOOH]_{init} = 0.10$ mol/L:

$$\text{Percent EtCOOH dissociated} = \frac{1.1 \times 10^{-3} \text{ mol/L}}{1.0 \times 10^{-1} \text{ mol/L}} \times 100 = 1.1\%$$

Case 2: $[EtCOOH]_{init} = 0.010$ mol/L:

$$\text{Percent EtCOOH dissociated} = \frac{3.6 \times 10^{-4} \text{ mol/L}}{1.0 \times 10^{-2} \text{ mol/L}} \times 100 = 3.6\%$$

As the initial acid concentration decreases, the **percent dissociation** *of the acid increases.* Do not confuse the *concentration* of HA dissociated with the *percent* of HA dissociated. The concentration, $[HA]_{dissoc}$, is lower in the diluted HA solution because the actual *number* of dissociated HA molecules is smaller. It is the *fraction* (or the *percent*) of dissociated HA molecules that increases with dilution. The entire discussion in this section can be extended to bases as well, and the percent of a base that has dissociated can be determined by an equation analogous to Equation 16.7, that is, by simply replacing the concentrations of the acid with the concentrations of the base.

This phenomenon is analogous to a change in container volume (pressure) for a gas at equilibrium (Section 15.6). For a gas, an increase in volume shifts the equilibrium position to favour a larger amount (mol) of gas. For HA or B dissociation, a lower HA or B concentration, which is the same as an increase in volume, shifts the equilibrium position to favour a larger amount (mol) of ions. Sample Problem 16.10 uses molecular scenes to highlight this idea for an acid. (Note that, to depict the scenes practically, the acid has a much higher percent dissociation than any real weak acid would.)

Sample Problem 16.10 Using Molecular Scenes to Determine the Extent of HA Dissociation

Problem A 0.15 mol/L solution of HA (*blue and green*) is 33% dissociated. Which scene represents a sample of this solution after it is diluted with water?

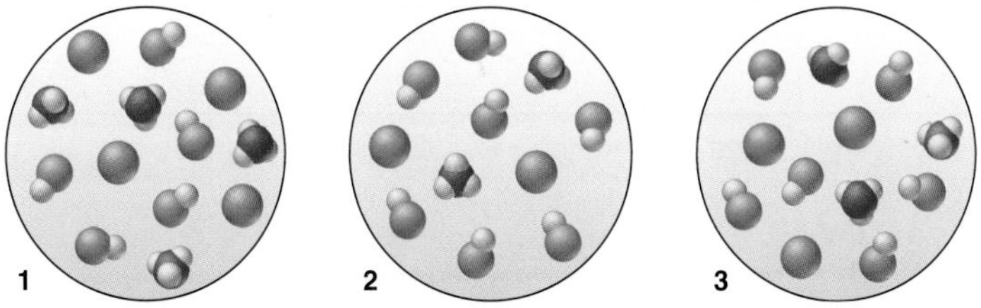

Plan We are given the percent dissociation of the original HA solution (33%), and we know that the percent dissociation increases as the acid is diluted. Thus, we calculate the percent dissociations of the diluted samples, represented by the given scenes, and see which is greater than 33%. To determine percent dissociation, we use Equation 16.7, with HA_{dissoc} equal to the number of H_3O^+ (or A^-) and HA_{init} equal to the number of HA *plus* the number of H_3O^+ (or A^-).

Solution Calculate the percent dissociation of each diluted sample using Equation 16.7:

$$\text{Percent dissociated in scene 1} = \frac{4}{5+4} \times 100 = 44\%$$

$$\text{Percent dissociated in scene 2} = \frac{2}{7+2} \times 100 = 22\%$$

$$\text{Percent dissociated in scene 3} = \frac{3}{6+3} \times 100 = 33\%$$

Therefore, scene 1 represents the diluted sample.

Check Let us confirm our choice by examining the other scenes: in scene 2, HA is *less* dissociated than originally, so this scene must represent a more concentrated HA sample; scene 3 represents another sample with the same percent dissociation as the original solution.

Follow-Up Problem 16.10 The scene in the margin represents a sample of a weak acid HB (*blue and purple*) dissolved in water. Draw a scene that represents the same volume after the solution has been diluted with water.

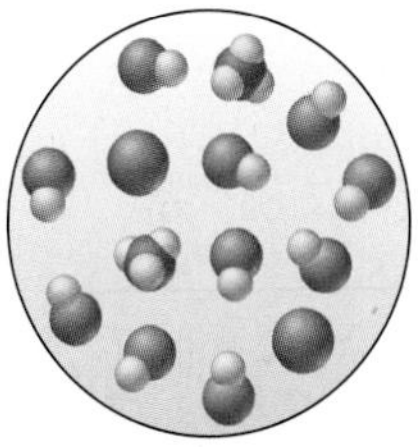

The next sample problem looks at how we deal with the dissociation of a salt of the conjugate base in water. Remember that once the salt is placed in water, it dissociates immediately. The cation behaves as a spectator ion and does not contribute to the pH. We only need to look at the chemistry that occurs between the conjugate base (the anion) and water.

Sample Problem 16.11 Determining the pH of a Solution of A^-

Problem Sodium ethanoate (or sodium acetate) (CH_3COONa, or NaOAc in this problem) is used in textile dyeing. What is the pH of 0.25 mol/L NaOAc at 25°C? K_a of ethanoic (or acetic) acid (HOAc) is 1.8×10^{-5}.

Plan We know the initial concentration of OAc^- (0.25 mol/L) and the K_a of HOAc (1.8×10^{-5}), and we have to find the pH of the OAc^- solution, which acts as a base in water. We write the base-dissociation equation and K_b expression. If we can find $[OH^-]$, we can use K_w to find $[H_3O^+]$ and convert it to pH. To solve for $[OH^-]$, we need the K_b of OAc^-, which we obtain from the K_a of HOAc by applying Equation 16.7. We set up an ICE table to find $[OH^-]$ and make the usual assumption that K_b is small, so $[OAc^-]_{init} \approx [OAc^-]$.

Solution Write the base-dissociation equation and K_b expression:

$$OAc^-(aq) + H_2O(l) \rightleftharpoons HOAc(aq) + OH^-(aq) \qquad K_b = \frac{[HOAc][OH^-]}{[OAc^-]}$$

Set up the ICE table, with $x = [OAc^-]_{reacting} = [HOAc] = [OH^-]$:

Concentration (mol/L)	$OAc^-(aq)$	+	$H_2O(l)$	$\rightleftharpoons$	$HOAc(aq)$	+	$OH^-(aq)$
Initial	0.25		—		0		0
Change	$-x$		—		$+x$		$+x$
Equilibrium	$0.25 - x$		—		x		x

Solve for K_b of OAc^-:

$$K_b = \frac{K_w}{K_a} = \frac{1.0 \times 10^{-14}}{1.8 \times 10^{-5}} = 5.6 \times 10^{-10}$$

Because K_b is small, we can assume that $0.25 \text{ mol/L} - x \approx 0.25\text{mol/L}$.

Substitute into the expression for K_b and solve for x:

$$K_b = \frac{[HOAc][OH^-]}{[OAc^-]} = 5.6 \times 10^{-10} \approx \frac{x^2}{0.25} \quad \text{so} \quad x = [OH^-] \approx 1.2 \times 10^{-5} \text{ mol/L}$$

Check the assumption:

$$\frac{1.2 \times 10^{-5} \text{ mol/L}}{0.25 \text{ mol/L}} \times 100 = 4.8 \times 10^{-3}\% \text{ } (<5\%;\text{ assumption is justified})$$

Also note that

$$\frac{0.25}{5.6 \times 10^{-10}} = 4.5 \times 10^8 > 400$$

Solve for pH:

$$[H^+] = \frac{K_w}{[OH^-]} = \frac{1.0 \times 10^{-14}}{1.2 \times 10^{-5}} = 8.3 \times 10^{-10} \text{ mol/L}$$

$$pH = -\log(8.3 \times 10^{-10}) = 9.08$$

Check The K_b calculation seems reasonable: $\frac{\sim 10 \times 10^{-15}}{2 \times 10^{-5}} = 5 \times 10^{-10}$. Because OAc^- is a weak base, $[OH^-] > [H_3O^+]$; thus, pH > 7, which makes sense.

Follow-Up Problem 16.11 Sodium hypochlorite (NaClO) is the active ingredient in household laundry bleach. What is the pH of 0.20 mol/L NaClO?

The Behaviour of Polyprotic Acids

An acid with more than one ionizable proton is a **polyprotic acid**. In solution, each dissociation step has a different K_a. For example, phosphoric acid is a triprotic acid (three ionizable protons), so it has three K_a values:

$$H_3PO_4(aq) + H_2O(l) \rightleftharpoons H_2PO_4^-(aq) + H_3O^+(aq)$$

$$K_{a1} = \frac{[H_2PO_4^-][H_3O^+]}{[H_3PO_4]} = 7.2 \times 10^{-3}$$

$$H_2PO_4^-(aq) + H_2O(l) \rightleftharpoons HPO_4^{2-}(aq) + H_3O^+(aq)$$

$$K_{a2} = \frac{[HPO_4^{2-}][H_3O^+]}{[H_2PO_4^-]} = 6.3 \times 10^{-8}$$

$$HPO_4^{2-}(aq) + H_2O(l) \rightleftharpoons PO_4^{3-}(aq) + H_3O^+(aq)$$

$$K_{a3} = \frac{[PO_4^{3-}][H_3O^+]}{[HPO_4^{2-}]} = 4.2 \times 10^{-13}$$

TABLE 16.7 Successive K_a Values for Some Polyprotic Acids at 25°C

Name (Formula)	Lewis Structure*	K_{a1}	K_{a2}	K_{a3}
Ethanedioic (oxalic) acid ($H_2C_2O_4$)	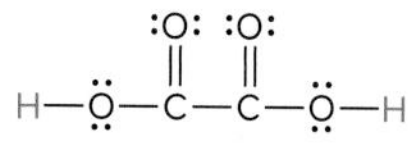	5.6×10^{-2}	5.4×10^{-5}	
Sulfurous acid (H_2SO_3)	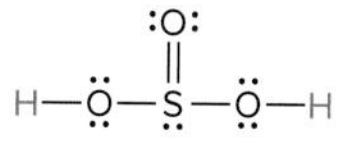	1.4×10^{-2}	6.5×10^{-8}	
Phosphoric acid (H_3PO_4)	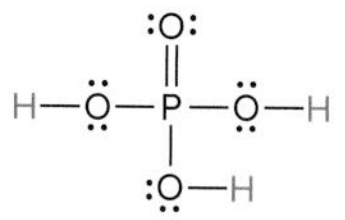	7.2×10^{-3}	6.3×10^{-8}	4.2×10^{-13}
Arsenic acid (H_3AsO_4)	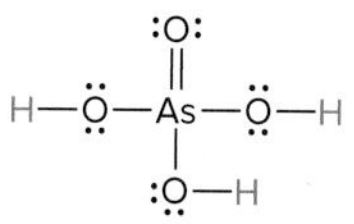	6×10^{-3}	1.1×10^{-9}	3×10^{-12}
Carbonic acid (H_2CO_3)	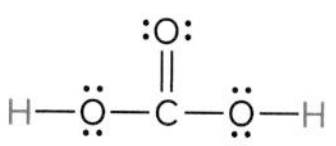	4.5×10^{-7}	4.7×10^{-11}	
Hydrosulfuric acid (H_2S)	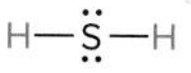	9×10^{-8}	1×10^{-17}	

*Red type indicates the ionizable protons.

The relative K_a values show that H_3PO_4 is a much stronger acid than $H_2PO_4^-$, which is a much stronger acid than HPO_4^{2-}.

Table 16.7 lists some common polyprotic acids and their K_a values. (More are listed in Appendix C.) Note that the general pattern for H_3PO_4 occurs for all polyprotic acids:

$$K_{a1} \gg K_{a2} \gg K_{a3}$$

This trend occurs because it is more difficult for an H^+ ion to leave a singly charged anion (such as $H_2PO_4^-$) than to leave a neutral molecule (such as H_3PO_4), and more difficult still for an H^+ ion to leave a doubly charged anion (such as HPO_4^{2-}). Successive K_a values typically differ by several orders of magnitude. This simplifies calculations because *we usually neglect the H_3O^+ coming from the subsequent dissociations.* While the simplification is valid in many cases, remember that you should *always* check whether an approximation is valid before using it in a calculation.

Many soft drinks, especially the "cola" (dark brown) variety, contain orthophosphoric acid. This acid reacts with the calcium in teeth and bones and then leaches the calcium from them. As a result, bone and tooth density decreases. Coupled with the tendency of young people to drink soft drinks instead of milk (which provides needed calcium), there is a growing concern among doctors that there will be signs of bone brittleness and osteoporosis at much younger ages in the future.

Sample Problem 16.12 Calculating Equilibrium Concentrations for a Polyprotic Acid

Problem Ascorbic acid ($H_2C_6H_6O_6$, or H_2Asc in this problem), known as vitamin C, is a diprotic acid ($K_{a1} = 1.0 \times 10^{-5}$ and $K_{a2} = 5 \times 10^{-12}$) found in citrus fruit. Calculate [$HAsc^-$], [Asc^{2-}], and the pH of 0.050 mol/L H_2Asc.

Plan We know the initial concentration (0.050 mol/L) and both K_a values for H_2Asc, and we have to calculate the equilibrium concentrations of all the species and convert [H_3O^+] to pH. We first write the equations and K_a expressions. Because $K_{a1} \gg K_{a2}$, we can assume that the first dissociation produces almost all the H_3O^+: $[H_3O^+]_{\text{from } H_2Asc} \gg [H_3O^+]_{\text{from } HAsc^-}$. Also, because K_{a1} is small, the amount of H_2Asc that dissociates can be neglected. We set up an ICE table for the first dissociation, with x equal to $[H_2Asc]_{\text{dissoc}}$, and then we solve for [H_3O^+] and [$HAsc^-$]. Because the second dissociation occurs to a much lesser extent, we can substitute values from the first dissociation directly to find [Asc^{2-}] from the second dissociation.

Solution Write the equations and K_a expressions:

$$H_2Asc(aq) + H_2O(l) \rightleftharpoons HAsc^-(aq) + H_3O^+(aq)$$

$$K_{a1} = \frac{[HAsc^-][H_3O^+]}{[H_2Asc]} = 1.0 \times 10^{-5}$$

Set up an ICE table with $x = [H_2Asc]_{dissoc} = [HAsc^-] \approx [H_3O^+]$:

Concentration (mol/L)	$H_2Asc(aq)$	+	$H_2O(l)$	$\rightleftharpoons$	$H_3O(aq)$	+	$HAsc^-(aq)$
Initial	0.050		—		0		0
Change	$-x$		—		$+x$		$+x$
Equilibrium	$0.050 - x$		—		x		x

Make the assumptions:

1. Because $K_{a2} \ll K_{a1}$, $[H_3O^+]_{\text{from }HAsc^-} \ll [H_3O^+]_{\text{from }H_2Asc}$, we can assume that

$$[H_3O^+]_{\text{from }H_2Asc} \approx [H_3O^+]$$

2. Because K_{a1} is small, $[H_2Asc]_{init} - x = [H_2Asc] \approx [H_2Asc]_{init}$, we can assume that

$$[H_2Asc] = 0.050\text{mol/L} - x \approx 0.050 \text{ mol/L}$$

Substitute into the expression for K_{a1} and solve for x:

$$K_{a1} = \frac{[H_3O^+][HAsc^-]}{[H_2Asc]} = 1.0 \times 10^{-5} = \frac{x^2}{0.050 - x} \approx \frac{x^2}{0.050}$$

$$x = [HAsc^-] \approx [H_3O^+] \approx 7.1 \times 10^{-4} \text{ mol/L}$$

$$\text{pH} = -\log [H_3O^+] = -\log (7.1 \times 10^{-4}) = 3.15$$

Check the assumptions:

1. $[H_3O^+]_{\text{from }HAsc^-} \ll [H_3O^+]_{\text{from }H_2Asc}$: For any second dissociation that does occur,

$$[H_3O^+]_{\text{from }HAsc^-} \approx \sqrt{[HAsc^-](K_{a2})} = \sqrt{(7.1 \times 10^{-4})(5 \times 10^{-12})} = 6 \times 10^{-8} \text{ mol/L}$$

This is even less than $[H_3O^+]_{\text{from }H_2O}$, so the assumption is justified.

2. $[H_2Asc]_{dissoc} \ll [H_2Asc]_{init}$:

$$\frac{7.1 \times 10^{-4} \text{ mol/L}}{0.050 \text{ mol/L}} \times 100 = 1.4\% \ (< 5\%; \text{ assumption is justified})$$

Also, note that

$$\frac{[H_3Asc]_{init}}{K_{a1}} = \frac{0.050}{1.0 \times 10^{-5}} = 5000 > 400$$

Use the equilibrium concentrations from the first dissociation to calculate $[Asc^{2-}]$:

$$K_{a2} = \frac{[H_3O^+][Asc^{2-}]}{[HAsc^-]} \quad \text{so} \quad [Asc^{2-}] = \frac{(K_{a2})[HAsc^-]}{[H_3O^+]}$$

$$[Asc^{2-}] = \frac{(5 \times 10^{-12})(7.1 \times 10^{-4})}{7.1 \times 10^{-4}} = 5 \times 10^{-12} \text{ mol/L}$$

Check $K_{a1} \gg K_{a2}$, so it makes sense that $[HAsc^-] \gg [Asc^{2-}]$ because Asc^{2-} is produced only in the second (much weaker) dissociation. Both K_a values are small, so all concentrations except $[H_2Asc]$ should be much lower than the original 0.050 mol/L.

Follow-Up Problem 16.12 Oxalic acid (HOOC—COOH, or $H_2C_2O_4$) is the simplest diprotic carboxylic acid. Its commercial uses include bleaching straw and leather and removing rust and ink stains. Calculate the equilibrium values of $[H_2C_2O_4]$, $[HC_2O_4^-]$, and $[C_2O_4^{2-}]$, and find the pH of a 0.150 mol/L $H_2C_2O_4$ solution. Use K_a values from Appendix C. (*Hint*: First check whether you need to use the quadratic equation to find x.)

SUMMARY OF SECTION 16.4

- Two common types of weak-acid or weak-base equilibrium problems involve finding K_a or K_b from a given concentration and finding a concentration from a given K_a or K_b.
- We simplify the arithmetic by assuming (1) that $[H_3O^+]_{\text{from } H_2O}$ and $[OH^-]_{\text{from } H_2O}$ are much smaller than $[OH^-]_{\text{from B}}$, so they can be neglected, and (2) that weak acids dissociate so little that $[H^+]_{\text{from HA}} \ll [HA]_{\text{init}}$. Thus, $[HA]_{\text{eq}} = [HA]_{\text{init}} - [H^+]_{\text{from HA}} \approx [HA]_{\text{init}}$. Similarly, $[B^-]_{\text{init}} \approx [B^-]_{\text{eq}}$ for weak bases.
- If the initial concentration of the acid or base is smaller than 1.0×10^{-7} mol/L, then the concentration of hydrogen ion or hydroxide ion from water must be taken into consideration.
- The *fraction* of weak acid or base molecules that dissociates is greater in a more dilute solution, even though the total $[H_3O^+]$ or $[OH^-]$ is lower.
- The salt of a conjugate acid or base, when dissolved in water, may cause the pH of the resulting solution to be acidic or basic because of hydrolysis reactions.
- Polyprotic acids have more than one ionizable proton, but we assume that the first dissociation provides virtually all the H_3O^+.

16.5 Molecular Properties and Acid Strength

The strength of an acid depends on its ability to donate a proton, which depends on the strength of the bond to the acidic proton. In this section, we apply trends in atomic and bond properties to determine the trends in acid strength of nonmetal hydrides and oxoacids and then discuss the acidity of hydrated metal ions.

Acid Strength of Nonmetal Hydrides

Two factors determine how easily a proton is released from a nonmetal hydride:

- The electronegativity of the central nonmetal (E)
- The strength of the E—H bond

Figure 16.11 displays the following two periodic trends among the nonmetal hydrides:

1. *Across a period, the acid strength increases.* The electronegativity of the nonmetal E determines the trend. From left to right, as E becomes more electronegative, it withdraws electron density from H, and the E—H bond becomes more polar. As a result, H^+ is pulled away more easily by an O atom of a water molecule. In aqueous solution, the hydrides of groups 13 to 15 do not behave as acids, but an increase in acid strength is seen from group 16 to group 17.

2. *Down a group, the acid strength increases.* As we go down a group, the size of the ion increases and so the conjugate base becomes more stable (weaker). This stability arises because the negative charge of the anion is spread over a larger surface area. As a result, charge density of I^- < charge density of Br^- < charge density of Cl^- $\ll$ charge density of F^-. In fact, F^- is so small that its large charge density causes it to be a strong enough base to make HF a weak acid instead of a strong acid (such as HCl):

$$\text{Acid strength: HF} \ll \text{HCl} < \text{HBr} < \text{HI}$$

(This trend is not observed in aqueous solution, where HCl, HBr, and HI are all equally strong. We will discuss how it *is* observed in Section 16.7.)

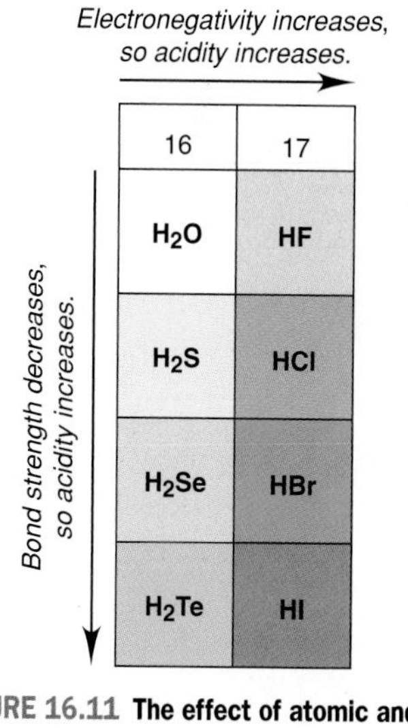

FIGURE 16.11 The effect of atomic and molecular properties on nonmetal hydride acidity

Acid Strength of Oxoacids

All oxoacids have the acidic H atom bonded to an O atom, so bond length is not involved. Two other factors determine the acid strength of oxoacids (Section 8.5):

- The electronegativity of the central nonmetal (E)
- The number of O atoms around E (oxidation number of E)

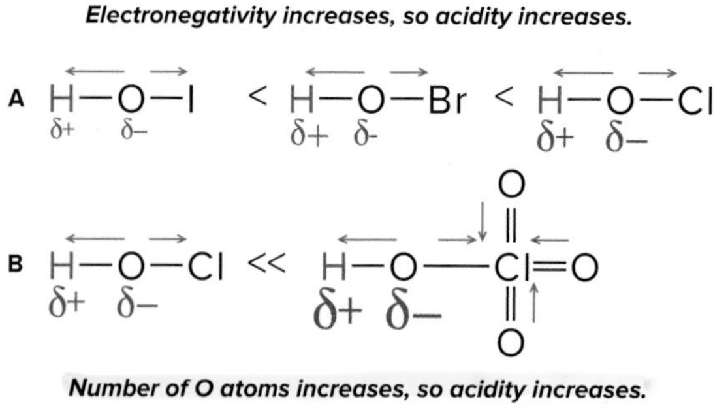

FIGURE 16.12 The relative strengths of oxoacids. A. Cl withdraws electron density from the O—H bond most effectively, making that bond most polar (relative size of δ symbols). **B.** Additional O atoms pull more electron density from the O—H bond.

Figure 16.12 summarizes the following trends:

1. *For oxoacids with the* ***same*** *number of O atoms, the acid strength increases with the electronegativity of E.* Consider the hypohalous acids (HOE, where E is a halogen atom). The more electronegative E is, the more polar the O—H bond becomes and the more easily H^+ is lost (Figure 16.12A). Electronegativity decreases down a group, as does acid strength:

$$K_a \text{ of HOCl} = 2.9 \times 10^{-8} \qquad K_a \text{ of HOBr} = 2.3 \times 10^{-9} \qquad K_a \text{ of HOI} = 2.3 \times 10^{-11}$$

Similarly, in group 16, H_2SO_4 is stronger than H_2SeO_4; in group 15, H_3PO_4 is stronger than H_3AsO_4; and so on.

2. *For oxoacids with the same central atom but* ***different*** *numbers of O atoms, the acid strength increases with the number of O atoms (or oxidation number of the central nonmetal).* The electronegative O atoms pull electron density away from E, which makes the O—H bond more polar. The more O atoms that are present, the greater the shift in electron density, and the more easily the H^+ ion comes off (Figure 16.12B). Therefore, the chlorine oxoacids ($HOClO_n$, with n from 0 to 3) increase in strength with the number of O atoms (and oxidation number of Cl):

$$K_a \text{ of HO}\overset{+1}{\text{Cl}} = 2.9 \times 10^{-8} \qquad K_a \text{ of HO}\overset{+3}{\text{Cl}}\text{O} = 1.12 \times 10^{-2} \qquad K_a \text{ of HO}\overset{+5}{\text{Cl}}\text{O}_2 \approx 1 \qquad K_a \text{ of HO}\overset{+7}{\text{Cl}}\text{O}_3 > 10^7$$

Similarly, HNO_3 is stronger than HNO_2, H_2SO_4 is stronger than H_2SO_3, and so on.

Acidity of Hydrated Metal Ions

The aqueous solutions of certain metal ions are acidic because the *hydrated* metal ion transfers an H^+ ion to water. Consider a general metal nitrate, $M(NO_3)_n$, as it dissolves in water. The ions separate and the metal ion becomes bonded to some number of H_2O molecules. The following equation shows the hydration of the cation (M^{n+}) with H_2O molecules and (aq); hydration of the anion (NO_3^-) is indicated by just (aq):

$$M(NO_3)_n(s) + xH_2O(l) \longrightarrow [M(H_2O)_x]^{n+}(aq) + nNO_3^-(aq)$$

If the metal ion, M^{n+}, is *small and highly charged*, its high charge density withdraws sufficient electron density from the O—H bonds of the bound water molecules for an H^+ to be released. Thus, the hydrated cation, $[M(H_2O)_x]^{n+}$, is a typical Brønsted-Lowry acid. The bound H_2O that releases the H^+ becomes a bound OH^- ion:

$$[M(H_2O)_x]^{n+}(aq) + H_2O(l) \rightleftharpoons [M(H_2O)_{x-1}OH]^{(n-1)+}(aq) + H_3O^+(aq)$$

The salts of most M^{2+} and M^{3+} ions yield acidic aqueous solutions. The K_a values for some acidic hydrated metal ions are given in Appendix C.

Consider the small, highly charged Al^{3+} ion. When an aluminum salt, such as $Al(NO_3)_3$, dissolves in water, the following steps occur:

(1) $Al(NO_3)_3(s) + 6H_2O(l) \longrightarrow [Al(H_2O)_6]^{3+}(aq) + 3NO_3^-(aq)$
(dissolution and hydration)

(2) $[Al(H_2O)_6]^{3+}(aq) + H_2O(l) \rightleftharpoons [Al(H_2O)_5OH]^{2+}(aq) + H_3O^+(aq)$
(dissociation of weak acid)

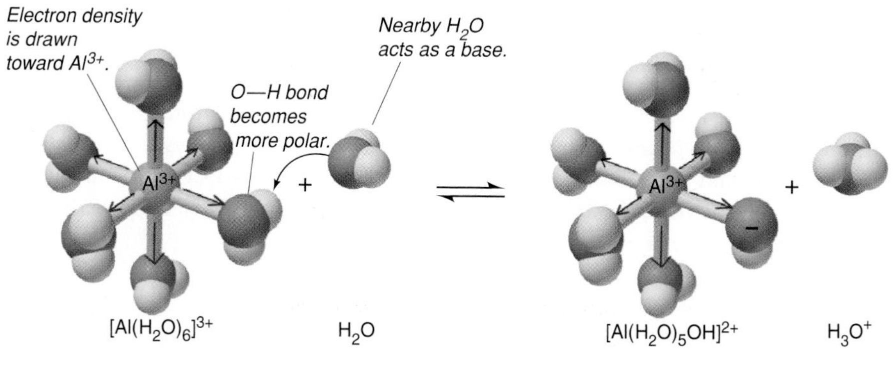

FIGURE 16.13 The acidic behaviour of the hydrated Al^{3+} ion. The hydrated Al^{3+} ion is small and multiply charged, so it pulls electron density from the O—H bonds, allowing an H^+ ion to be transferred to a nearby water molecule.

Notice the formulas of the hydrated metal ions in the second step. When H^+ is released, the number of bound H_2O molecules decreases by 1 (from 6 to 5) and the number of bound OH^- ions increases by 1 (from 0 to 1), which reduces the positive charge of the ion by 1 (from 3 to 2) (Figure 16.13). This pattern of changes in the formula of the hydrated metal ion before and after it loses a proton occurs for any highly charged metal ion in water.

SUMMARY OF SECTION 16.5

- For nonmetal hydrides, the acid strength increases across a period (with the electronegativity of the nonmetal, E) and down a group (with the length of the E—H bond).
- For oxoacids with the same number of O atoms, the acid strength increases with the electronegativity of E. For oxoacids with the same E, the acid strength increases with the number of O atoms (or oxidation number of E).
- Small, highly charged metal ions are acidic in water because they withdraw electron density from the O—H bonds of bound H_2O molecules, releasing an H^+ ion to the solution.

16.6 Acid-Base Properties of Salt Solutions

Often, when a salt dissolves, one or both of its ions may react with water and affect the pH of the solution. You have seen that conjugate acids of weak bases (such as NH_4^+) are acidic, conjugate bases of weak acids (such as CN^-) are basic, and small, highly charged metal cations (such as Al^{3+}) are acidic. In addition, certain ions (such as $H_2PO_4^-$ and HCO_3^-) can act as an acid or a base. In this section, we classify the acid-base behaviour of the various types of salt solutions.

Salts That Yield Neutral Solutions

A salt consisting of the conjugate base of a strong acid and the conjugate acid of a strong base yields a neutral solution because the ions do not react with water. When a strong acid such as HNO_3 dissolves, the reaction goes essentially to completion because *the conjugate base of a strong acid is a much weaker base than water.* The conjugate base is hydrated, but *it does not react with water*:

$$HNO_3(l) + H_2O(l) \longrightarrow NO_3^-(aq) + H_3O^+(aq) \quad \text{(dissolution and hydration)}$$

Similarly, a strong base, such as NaOH, dissolves completely. The cation, in this case Na^+, is hydrated, but *it is not small or charged enough to react with water*:

$$NaOH(s) \xrightarrow{H_2O} Na^+(aq) + OH^-(aq) \quad \text{(dissolution and hydration)}$$

The conjugate bases of strong acids are the halide ions (except F^-) and the ions of strong oxoacids, such as NO_3^- and ClO_4^-. The conjugate acids of strong bases are group 1 ions and Ca^{2+}, Sr^{2+}, and Ba^{2+} (group 2). *Salts containing only these ions yield neutral solutions.*

Salts That Yield Acidic Solutions

There are two types of acidic salts:

1. *A salt consisting of the conjugate acid of a weak base and the conjugate base of a strong acid yields an acidic solution because the conjugate acid acts as a weak acid,* and the anion does not react. For example, NH_4Cl yields an acidic solution because NH_4^+, the conjugate acid of the weak base NH_3, is a weak acid; Cl^-, the conjugate base of the strong HCl, does not react:

$$NH_4Cl(s) \xrightarrow{H_2O} NH_4^+(aq) + Cl^-(aq) \quad \text{(dissolution and hydration)}$$

$$NH_4^+(aq) + H_2O(l) \rightleftharpoons NH_3(aq) + H_3O^+(aq) \quad \text{(dissociation of weak acid)}$$

2. *A salt consisting of a small, highly charged metal cation and the conjugate base of a strong acid yields an acidic solution because the cation acts as a weak acid,* and the conjugate base does not react. For example, $Fe(NO_3)_3$ yields an acidic solution because the hydrated Fe^{3+} ion is a weak acid; NO_3^-, the conjugate base of the strong acid HNO_3, does not react:

$$Fe(NO_3)_3(s) + 6H_2O(l) \xrightarrow{H_2O} [Fe(H_2O)_6]^{3+}(aq) + 3NO_3^-(aq) \quad \text{(dissolution and hydration)}$$

$$[Fe(H_2O)_6]^{3+}(aq) + H_2O(l) \rightleftharpoons [Fe(H_2O)_5OH]^{2+}(aq) + H_3O^+(aq) \quad \text{(dissociation of weak acid)}$$

Salts That Yield Basic Solutions

A salt consisting of the conjugate base of a weak acid and the conjugate acid of a strong base yields a basic solution because the conjugate base acts as a weak base, and the conjugate acid does not react. Sodium ethanoate (sodium acetate), for example, yields a basic solution because the CH_3COO^- ion, the conjugate base of the weak acid CH_3COOH, acts as a weak base; Na^+, the conjugate acid of the strong base NaOH, does not react:

$$CH_3COONa(s) \xrightarrow{H_2O} Na^+(aq) + CH_3COO^-(aq) \quad \text{(dissolution and hydration)}$$

$$CH_3COO^-(aq) + H_2O(l) \rightleftharpoons CH_3COOH(aq) + OH^-(aq) \quad \text{(reaction of weak base)}$$

Sample Problem 16.13 Predicting Relative Acidity of Salt Solutions from Reactions of the Ions with Water

Problem Predict whether the aqueous solution of each salt is acidic, basic, or neutral, and write an equation for the reaction of any ion with water:

(a) Potassium perchlorate, $KClO_4$

(b) Sodium benzoate, C_6H_5COONa

(c) Chromium(III) nitrate, $Cr(NO_3)_3$

Plan The formula shows the conjugate acid and conjugate base. Depending on the ability of an ion to react with water, the solution will be neutral (strong-acid conjugate base and strong-base conjugate acid), acidic (weak-base conjugate acid or highly charged metal cation with strong-acid conjugate base), or basic (weak-acid conjugate base and strong-base conjugate acid).

Solution (a) Neutral: The ions are K^+ and ClO_4^-. The K^+ is from the strong base KOH, and the ClO_4^- is from the strong acid $HClO_4$. Neither ion reacts with water.

(b) Basic: The ions are Na^+ and $C_6H_5COO^-$. The Na^+ is the conjugate acid of the strong base NaOH, so it does not react with water. The benzoate ion, $C_6H_5COO^-$, is the conjugate base of the weak acid benzoic acid, so it reacts with water to produce OH^-:

$$C_6H_5COO^-(aq) + H_2O(l) \rightleftharpoons C_6H_5COOH(aq) + OH^-(aq)$$

(c) Acidic: The ions are Cr^{3+} and NO_3^-. The NO_3^- is the conjugate base of the strong acid HNO_3, so it does not react with water. Cr^{3+} is small and highly charged, so the hydrated ion, $Cr(H_2O)_6^{3+}$, reacts with water to produce H_3O^+:

$$[Cr(H_2O)_6]^{3+}(aq) + H_2O(l) \rightleftharpoons [Cr(H_2O)_5OH_3]^{2+}(aq) + H_3O^+(aq)$$

Follow-Up Problem 16.13 Predict, using an equation if necessary, whether the solution of each salt is acidic, basic, or neutral:

(a) $KClO_2$ **(b)** $CH_3NH_3NO_3$ **(c)** CsI

Salts of Weak Conjugate Acids and Weak Conjugate Bases

If a salt consists of a cation that is a weak acid *and* an anion that is a weak base, the overall acidity of the solution depends on the relative acid strength (K_a) or base strength (K_b) of the separated ions. Consider a solution of ammonium cyanide, NH_4CN, and the reactions that occur between the separated ions and water. The ammonium ion is the conjugate acid of a weak base, so it is a weak acid:

$$NH_4^+(aq) + H_2O(l) \rightleftharpoons NH_3(aq) + H_3O^+(aq)$$

The cyanide ion is the conjugate base of a weak acid, so it is a weak base:

$$CN^-(aq) + H_2O(l) \rightleftharpoons HCN(aq) + OH^-(aq)$$

The reaction that goes farther to the right determines the pH of the solution, so we compare the K_a of NH_4^+ with the K_b of CN^-. Only molecular compounds are listed in K_a and K_b tables, so we calculate these values for the ions using Equation 16.6:

$$K_a \text{ of } NH_4^+ = \frac{K_w}{K_b \text{ of } NH_3} = \frac{1.0\times10^{-14}}{1.76\times10^{-5}} = 5.7\times10^{-10}$$

$$K_b \text{ of } CN^- = \frac{K_w}{K_a \text{ of } HCN} = \frac{1.0\times10^{-14}}{6.2\times10^{-10}} = 1.6\times10^{-5}$$

Because K_b of CN^- > K_a of NH_4^+, we know that CN^- is a stronger base than NH_4^+ is an acid. Thus, the NH_4CN solution is basic.

Salts of Amphiprotic Conjugate Bases

The only salts left to consider are those in which the conjugate acid comes from a strong base and the conjugate base comes from a polyprotic acid with one or more ionizable protons still attached. These conjugate bases are amphiprotic: they can act as an acid and release a proton *to* water, or they can act as a base and abstract a proton *from* water. We can determine the overall acidity of their solutions the same way we determined the acidity of the solutions of the salts of weak conjugate acids and weak conjugate bases: we compare the magnitudes of K_a and K_b. Here, however, we compare both the K_a and the K_b of the same species, the conjugate base.

For example, Na_2HPO_4 consists of Na^+, the cation of a strong base (which does not react with water), and HPO_4^{2-}, the second conjugate base of the weak polyprotic acid H_3PO_4. In water, the salt undergoes three steps:

(1) $Na_2HPO_4(s) \xrightarrow{H_2O} 2Na^+(aq) + HPO_4^{2-}(aq)$ (dissolution and hydration)

(2) $HPO_4^{2-}(aq) + H_2O(l) \rightleftharpoons PO_4^{3-}(aq) + H_3O^+(aq)$ (acting as a weak acid)

(3) $HPO_4^{2-}(aq) + H_2O(l) \rightleftharpoons H_2PO_4^-(aq) + OH^-(aq)$ (acting as a weak base)

We must decide whether step 2 or step 3 goes farther to the right. Appendix C lists the K_a of HPO_4^{2-} as 4.2×10^{-13}, but the K_b of HPO_4^{2-} is not given, so we use Equation 16.6 to find it:

$$K_b \text{ of } HPO_4^{2-} = \frac{K_w}{K_a \text{ of } H_2PO_4^-} \quad \text{or} \quad \frac{1.0\times10^{-14}}{6.3\times10^{-8}} = 1.6\times10^{-7}$$

Because K_b (1.6×10^{-7}) > K_a (4.2×10^{-13}), HPO_4^{2-} is stronger as a base than as an acid, so a solution of Na_2HPO_4 is basic.

TABLE 16.8 The Acid-Base Behaviour of Salts in Water

Relative Acidity: Examples	pH*	Nature of Ions	Ion That Reacts with Water: Examples
Neutral: NaCl, KBr, $Ba(NO_3)_2$	7.0	Conjugate acid of strong base Conjugate base of strong acid	None
Acidic: NH_4Cl, NH_4NO_3, CH_3NH_3Br	<7.0	Conjugate acid of weak base Conjugate base of strong acid	Conjugate acid: $NH_4^+ + H_2O \rightleftharpoons NH_3 + H_3O^+$
Acidic: $Al(NO_3)_3$, $CrBr_3$, $FeCl_3$	<7.0	Small, highly charged cation Conjugate base of strong acid	Cation: $Al(H_2O)_6^{3+} + H_2O \rightleftharpoons Al(H_2O)_5OH^{2+} + H_3O^+$
Acidic/basic: NH_4ClO_2, NH_4CN, $Pb(CH_3COO)_2$	<7.0 if $K_{a(ca)} > K_{b(cb)}$ >7.0 if $K_{b(cb)} > K_{a(ca)}$	Conjugate acid of weak base (or small, highly charged cation) Conjugate base of weak acid	Cation *and* conjugate base: $CH_3NH_3^+ + H_2O \rightleftharpoons CH_3NH_2 + H_3O^+$ $F^- + H_2O \rightleftharpoons HF + OH^-$
Acidic/basic: NaH_2PO_4, $KHCO_3$, $NaHSO_3$	<7.0 if $K_{a(cb)} > K_{b(ca)}$ >7.0 if $K_{b(cb)} > K_{a(ca)}$	Conjugate acid of strong base Conjugate base of polyprotic acid	Conjugate base: $HSO_3^- + H_2O \rightleftharpoons SO_3^{2-} + H_3O^+$ $HSO_3^- + H_2O \rightleftharpoons H_2SO_3 + OH^-$

*$K_{a(cb)} = K_{a(\text{conjugate base})}$; $K_{a(ca)} = K_{a(\text{conjugate acid})}$.

Table 16.8 displays the acid-base behaviour of the various types of salts in water.

Sample Problem 16.14 Predicting the Relative Acidity of a Salt Solution from K_a and K_b of the Ions

Problem Determine whether an aqueous solution of zinc formate, $Zn(HCOO)_2$, at 25°C is acidic, basic, or neutral.

Plan The formula consists of the small, highly charged, and therefore weakly acidic Zn^{2+} cation and the weakly basic $HCOO^-$ conjugate base of the weak acid HCOOH. To determine the relative acidity of the solution, we write equations that show the reactions of the ions with water. Then we find K_a of Zn^{2+} (from Appendix C) and calculate K_b of $HCOO^-$ (from K_a of HCOOH in Appendix C) to see which ion reacts with water to a greater extent.

Solution Write the reactions with water:

$$[Zn(H_2O)_6]^{2+}(aq) + H_2O(l) \rightleftharpoons [Zn(H_2O)_5OH]^+(aq) + H_3O^+(aq)$$
$$HCOO^-(aq) + H_2O(l) \rightleftharpoons HCOOH(aq) + OH^-(aq)$$

From Appendix C, the K_a of $[Zn(H_2O)_6]^{2+}(aq)$ is 1×10^{-9} and the K_a of HCOOH is 1.8×10^{-4}. We use the K_a of HCOOH to solve for the K_b of $HCOO^-$:

$$K_b \text{ of } HCOO^- = \frac{K_w}{K_a \text{ of HCOOH}} = \frac{1.0 \times 10^{-14}}{1.8 \times 10^{-4}} = 5.6 \times 10^{-11}$$

K_a of $Zn(H_2O)_6^{2+}$ > K_b of $HCOO^-$, so the solution is acidic.

Follow-Up Problem 16.14 Determine whether an aqueous solution of each salt is acidic, basic, or neutral at 25°C:

(a) $Cu(CH_3COO)_2$ **(b)** NH_4F **(c)** $KHSO_3$

SUMMARY OF SECTION 16.6

- Salts that yield a neutral solution consist of ions that do not react with water.
- Salts that yield an acidic solution contain an unreactive conjugate base and a conjugate acid that releases a proton to water.
- Salts that yield a basic solution contain an unreactive conjugate acid and a conjugate base that accepts a proton from water.
- If both a conjugate acid and a conjugate base react with water, the ion that reacts to the greater extent (higher K) determines the acidity or basicity of the salt solution.
- If a conjugate base is amphiprotic (from a polyprotic acid), its strength as an acid (K_a) or as a base (K_b) determines the acidity of the salt.

16.7 Generalizing the Brønsted-Lowry Concept: The Levelling Effect

In general, *an acid yields the conjugate acid and a base yields the conjugate base of solvent autoionization.* For example, in H_2O, all Brønsted-Lowry acids yield H_3O^+ and all Brønsted-Lowry bases yield OH^-, which are the ions that form when water autoionizes. Similarly, when an acid and a base react in water, the acid yields the conjugate base of the acid, and the base yields the conjugate acid of the base.

This idea lets us examine a key question: Why are all strong acids and strong bases *equally* strong in water? The answer is that, *in water, the strongest acid possible is H_3O^+ and the strongest base possible is OH^-*:

- *Strong acids.* The moment we put gaseous HCl in water, it reacts with the base H_2O to form H_3O^+. The same holds for any strong acid, because it dissociates *completely* to form H_3O^+. Thus, we are actually observing the acid strength of H_3O^+.
- *Strong bases.* A strong base, such as $Ba(OH)_2$, dissociates completely in water to yield OH^-. Even strong bases that do not contain hydroxide ions in the solid, such as K_2O, do this. The oxide ion, which is a stronger base than OH^-, immediately takes a proton from water to form OH^-:

$$O^{2}(aq) + H_2O(l) \longrightarrow 2OH^-(aq)$$

Thus, water exerts a **levelling effect** on any strong acid or base by reacting with it to form the products of water's autoionization. Acting as a base, water levels the strength of all strong acids by making them appear equally strong. Acting as an acid, water levels the strength of all strong bases.

Therefore, to rank the relative strengths of strong acids, we must dissolve them in a solvent that is a *weaker* base than water, so it accepts their protons less readily. For example, the hydrohalic acids increase in strength as the halogen becomes larger, due to the longer, weaker H—X bond (see Figure 16.11). In water, HF is weak, and HCl, HBr, and HI appear equally strong because they dissociate completely. When we dissolve them in ethanoic (or acetic) acid, however, *the ethanoic acid acts as the base* and accepts a proton:

acid base base acid

$$HCl(g) + CH_3COOH(l) \rightleftharpoons CH(eth) + CH_3COOH_2^+(eth)$$

$$HBr(g) + CH_3COOH(l) \rightleftharpoons Br^-(eth) + CH_3COOH_2^+(eth)$$

$$HI(g) + CH_3COOH(l) \rightleftharpoons I^-(eth) + CH_3COOH_2^+(eth)$$

[The use of (*eth*) instead of (*aq*) indicates solvation by CH_3COOH.] Because ethanoic acid is a *weaker base* than water, the three acids donate their protons to *different* extents. Measurements show that HI protonates the solvent to a greater extent than HBr, and HBr does so more than HCl; that is, in pure ethanoic acid, $K_{HI} > K_{HBr} > K_{HCl}$. Similarly, the relative strengths of strong bases are determined in a solvent that is a weaker acid than H_2O, such as liquid NH_3.

SUMMARY OF SECTION 16.7

- Strong acids (or strong bases) dissociate completely to yield H_3O^+ (or OH^-) in water; in effect, water equalizes (levels) their strengths.
- Strong acids show differences in strength when dissolved in a weaker base than water, such as ethanoic acid.

16.8 Electron-Pair Donation and the Lewis Acid-Base Definition

The final acid-base concept we consider in this chapter was developed by Gilbert N. Lewis, whose contribution to understanding valence electron pairs in bonding was discussed in Chapter 8. Whereas the Brønsted-Lowry concept focuses on the proton in defining a species as an acid or a base, the Lewis concept highlights the role of the *electron pair.* The **Lewis acid-base definition** describes an acid and a base as follows:

- A *base* is any species that *donates* an electron pair to form a bond.
- An *acid* is any species that *accepts* an electron pair to form a bond.

The Lewis definition, like the Brønsted-Lowry definition, requires that a base have an electron pair to donate, so it does not expand the classes of bases. However, *it greatly expands the classes of acids.* Many species, such as CO_2 and Cu^{2+}, that do not contain H in their formula (and thus cannot be Brønsted-Lowry acids) are Lewis acids because they accept an electron pair in reactions. Thus, the proton itself is a Lewis acid because it accepts the electron pair donated by a base:

$$B: + H^+ \qquad B{-}H^+$$

Thus, *all Brønsted-Lowry acids donate H^+, a Lewis acid.*

The product of a Lewis acid-base reaction is an **adduct**, *a single species that contains a **new** covalent bond*:

$$A + :B \qquad A{-}B$$

As we will discuss in Chapter 21, a curved double-headed arrow represents the movement of an electron pair from its original position to a new position. The tail of the arrow shows where the electron pair has come from, and the head of the arrow shows where the electron pair is going.

The Lewis concept radically broadens the idea of an acid-base reaction:

- To Arrhenius, an acid-base reaction is the formation of H_2O from H^+ and OH^-.
- To Brønsted and Lowry, it is H^+ transfer from a stronger acid to a stronger base to form a weaker base and weaker acid.
- To Lewis, it is *the donation and acceptance of an electron pair to form a covalent bond in an adduct.*

By definition, then, the following apply:

- *A Lewis base must have a lone pair of electrons to donate.*
- *A Lewis acid must have a vacant orbital* (or the ability to rearrange its bonds to form a vacant orbital) to accept a lone pair and form a new bond.

In this section, we discuss molecules and positive ions that act as Lewis acids.

Molecules as Lewis Acids

Many molecules act as Lewis acids. In every case, the atom accepting the electron pair has low electron density due to either an electron deficiency or a polar multiple bond.

Lewis Acids with Electron-Deficient Atoms The most important of the *electron-deficient* Lewis acids are compounds of the group 13 elements boron and aluminum. Recall from Chapters 7 and 13 that these compounds have fewer than eight electrons around the central atom, so they react to complete the octet of that atom. For example, boron trifluoride accepts an electron pair from ammonia to form a covalent bond:

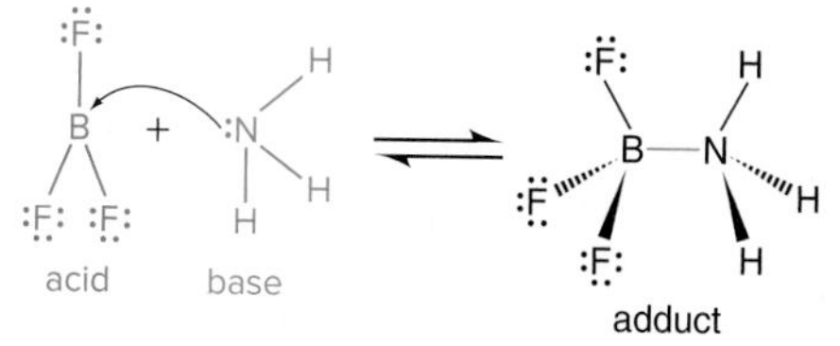

Aluminum chloride dissolves freely in relatively nonpolar diethyl ether when the O atom of the ether donates an electron pair to Al to form a covalent bond:

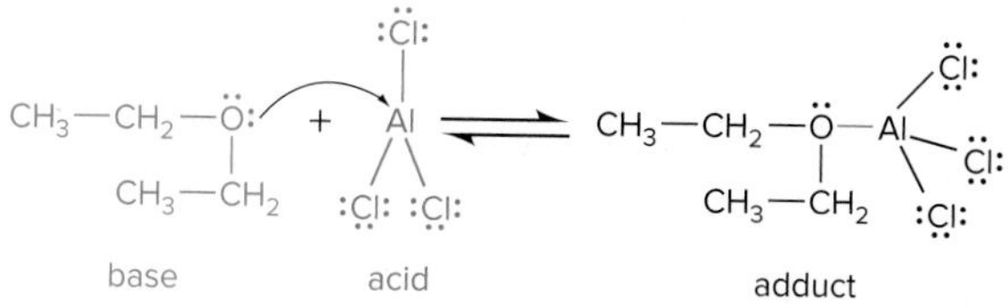

This acidic behaviour of boron and aluminum halides is often used in organic syntheses. For example, a methyl group is added to the benzene ring by the action of CH_3Cl in the presence of $AlCl_3$. The Lewis acid $AlCl_3$ abstracts the Lewis base Cl^- from CH_3Cl to form an adduct that has a reactive CH_3^+ group, which then attacks the electron-rich benzene ring:

$$\underset{\text{base}}{CH_3Cl} + \underset{\text{acid}}{AlCl_3} \rightleftharpoons \underset{\text{adduct}}{[CH_3]^+[Cl{-}AlCl_3]^-}$$

$$\underset{\text{benzene}}{C_6H_6} + [CH_3]^+[Cl{-}AlCl_3]^- \rightleftharpoons \underset{\text{toluene}}{C_6H_5CH_3} + AlCl_3 + HCl$$

Lewis Acids with Polar Multiple Bonds Molecules with a polar double bond also function as Lewis acids. An electron pair on the Lewis base approaches the partially positive end of the double bond to form the new bond in the adduct as the π bond breaks. For example, consider the reaction of SO_2 in water. The electronegative O atoms in SO_2 make the central S partially positive. The O atom of water donates a lone pair to the S, thus forming an S—O bond and breaking one of the π bonds. Then a proton is transferred from water to the O that was part of the π bond. The resulting adduct is sulfurous acid:

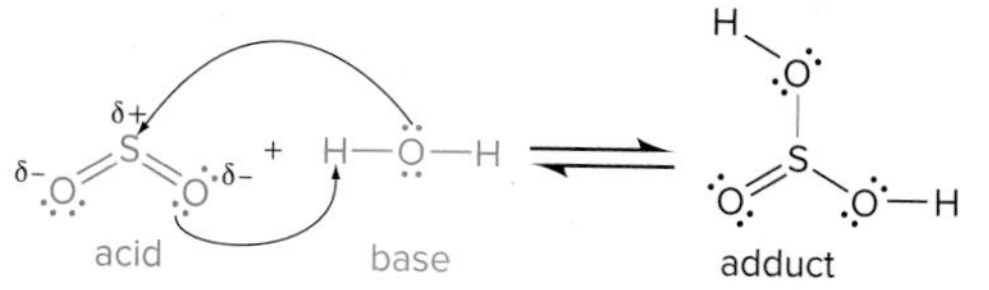

The analogous formation of a carbonate from a metal oxide and carbon dioxide occurs in a nonaqueous system. The O^{2-} ion (shown coming from CaO) donates an electron pair to the partially positive C in CO_2, a π bond breaks, and the CO_3^{2-} ion (shown as part of $CaCO_3$) is the adduct:

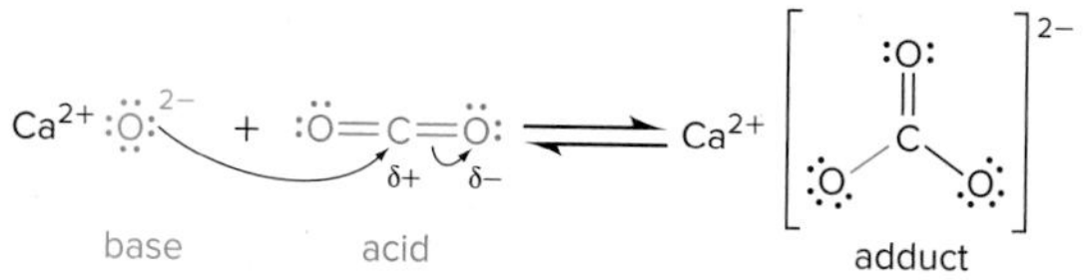

Metal Cations as Lewis Acids

In the Lewis sense, hydration of a metal ion is itself an acid-base reaction. As electron pairs on the O atoms of H_2O molecules form covalent bonds, the hydrated cation is the adduct; thus, *a metal ion acts as a Lewis acid when it dissolves in water*:

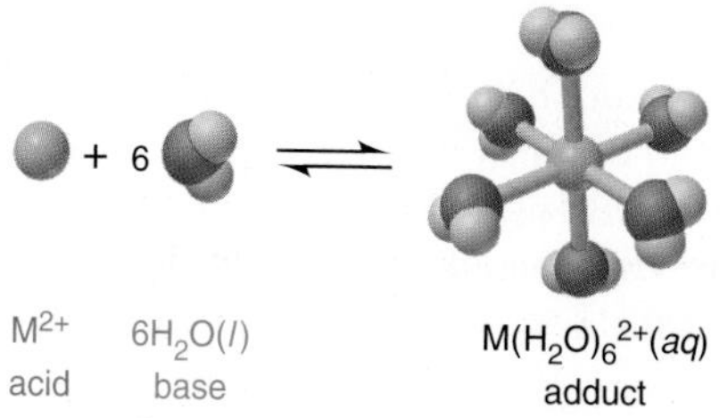

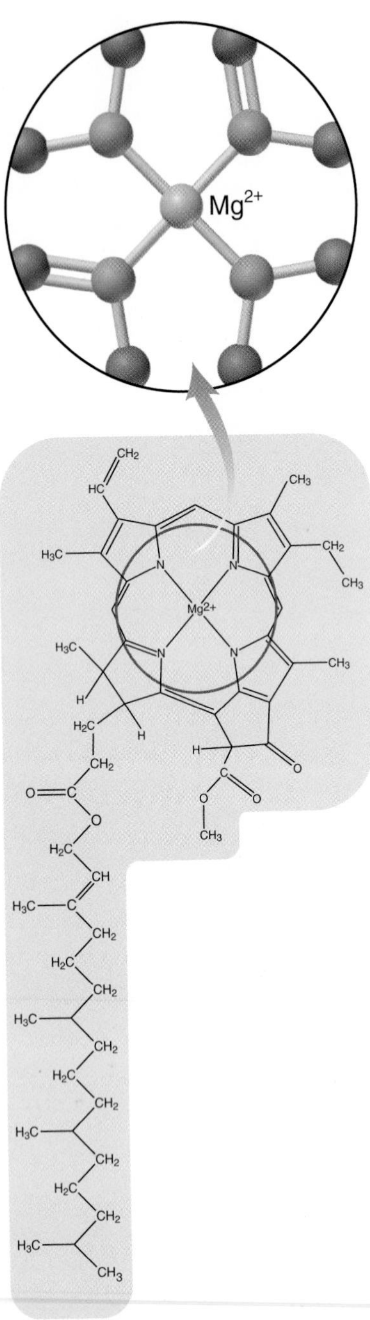

FIGURE 16.14 The Mg^{2+} ion as a Lewis acid in chlorophyll. Mg^{2+} accepts electron pairs from surrounding N atoms.

Ammonia is a stronger Lewis base than water because, when it is added to the aqueous solution of the hydrated cation, it displaces H_2O, with $K >> 1$:

$$\underset{\text{hydrated adduct}}{Ni(H_2O)_6^{2+}(aq)} + \underset{\text{base}}{6NH_3(aq)} \rightleftharpoons \underset{\text{ammoniated adduct}}{Ni(NH_3)_6^{2+}(aq)} + 6H_2O(l)$$

We will discuss the equilibrium nature of these acid-base reactions in greater detail in Chapter 17, and we will investigate the structures of these ions in Chapter 24.

Many biomolecules with central metal ions are Lewis adducts. Most often, O and N atoms of organic groups donate their lone pairs as the Lewis bases. Chlorophyll is a Lewis adduct of a central Mg^{2+} and the four N atoms of a tetrapyrrole (porphin) ring system (Figure 16.14). Vitamin B_{12} has a similar structure, with a central Co^{3+}, and so does heme, with a central Fe^{2+}. Several other metal ions, such as Zn^{2+}, Mo^{2+}, and Cu^{2+}, are bound at the active sites of enzymes and function as Lewis acids in the catalytic action.

Sample Problem 16.15 Identifying Lewis Acids and Bases

Problem Identify the Lewis acids and Lewis bases in each reaction:

(a) $H^+ + OH^- \rightleftharpoons H_2O$ **(b)** $Cl^- + BCl_3 \rightleftharpoons BCl_4^-$

(c) $K^+ + 6H_2O \rightleftharpoons K(H_2O)_6^+$

Plan We examine the formulas to see which species accepts the electron pair (Lewis acid) and which species donates it (Lewis base) when forming the adduct.

Solution **(a)** The H^+ ion accepts an electron pair from the OH^- ion when forming a bond. H^+ is the acid, and OH^- is the base.

(b) The Cl^- ion has four lone pairs and uses one to form a new bond to the central B. BCl_3 is the acid, and Cl^- is the base.

(c) The K^+ ion does not have any valence electrons to provide, so the bond is formed when electron pairs from the O atoms of water enter empty orbitals on K^+. K^+ is the acid, and H_2O is the base.

Check Each of the Lewis acids (H^+, BCl_3, and K^+) has an unfilled valence shell that can accept an electron pair from the Lewis bases (OH^-, Cl^-, and H_2O).

Follow-Up Problem 16.15 Identify the Lewis acids and Lewis bases in each reaction:

(a) $OH^- + Al(OH)_3 \rightleftharpoons Al(OH)_4^-$ **(b)** $SO_3 + H_2O \rightleftharpoons H_2SO_4$

(c) $Co^{3+} + 6NH_3 \rightleftharpoons Co(NH_3)_6^{3+}$

An Overview of Acid-Base Definitions

From a broader chemical perspective, the diversity of acid-base reactions takes on more unity. Chemists see a common theme in reactions as diverse as a standardized base being used to analyze an unknown fatty acid, baking soda being used to make bread rise, and even oxygen binding to hemoglobin in a blood cell. Let us see how the three acid-base definitions fit together.

1. The *Arrhenius definition*, which was the first attempt to describe acids and bases on the molecular level, is the most limited and narrow of the three definitions. It applies only to species with an H atom or an OH group that is released as an ion when the species dissolves in water. Because relatively few species have these structural prerequisites, Arrhenius acid-base reactions are relatively few in number, and all occur in H_2O and result in the formation of H_2O (for example, strong acids plus strong bases).

2. The *Brønsted-Lowry definition* sees acid-base reactions as proton-transfer processes that do not need to occur in water. A Brønsted-Lowry acid, like an Arrhenius acid, must have an H, but a Brønsted-Lowry base is any species with an electron pair available to accept a proton. This definition includes many more species as bases, including OH^-. It defines the acid-base reaction in terms of conjugate acid-base pairs,

with an acid and a base on both sides of the reaction. The system reaches equilibrium based on the relative strengths of the acid, the base, and their conjugates.

3. The *Lewis definition* has the widest scope and includes the other two definitions. The main event in a Lewis acid-base reaction is the donation and acceptance of an electron pair to form a new covalent bond in an adduct. Lewis bases must still have an electron pair to donate, but Lewis acids—the electron-pair acceptors—include many species not encompassed by the other definitions, including molecules with electron-deficient atoms or polar double bonds, metal ions, and even H^+ itself.

SUMMARY OF SECTION 16.8

- The Lewis acid-base definition focuses on the donation or acceptance of an electron pair to form a new covalent bond in an adduct, the product of an acid-base reaction. A Lewis base donates the electron pair, and a Lewis acid accepts it.
- Many species that do not contain H are Lewis acids. Molecules with polar double bonds act as Lewis acids, as do molecules with electron-deficient atoms.
- Metal ions act as Lewis acids when they dissolve in water, which acts as a Lewis base, to form the adduct, a hydrated cation.
- Many metal ions function as Lewis acids in biomolecules.

CHAPTER REVIEW GUIDE

Learning Objectives

Relevant section (§) and/or sample problem (SP) numbers appear in parentheses.

Concepts

1. Explain why the proton is bonded to a water molecule as H_3O^+ in all aqueous acid-base systems. (§16.1)
2. Differentiate between the Arrhenius definition of an acid and the Arrhenius definition of a base. (§16.1)
3. Explain why all reactions of a strong acid and a strong base have the same $\Delta_r H°$. (§16.1)
4. Reiterate the Brønsted-Lowry definitions of an acid and a base, and describe how an acid-base reaction can be viewed as a proton-transfer process. (§16.1)
5. Explain how water acts as a base (or as an acid) when an acid (or a base) dissolves in it. (§16.1)
6. Describe how a conjugate acid-base pair differs by one proton. (§16.1)
7. Explain how a Brønsted-Lowry acid-base reaction involves two conjugate acid-base pairs. (§16.1)
8. Explain why a stronger acid and base react ($K > 1$) to form a weaker base and acid. (§16.1)
9. Describe how the strength of an acid (or base) relates to the extent of its dissociation into ions in water. (§16.1)
10. Describe how relative acid strength is expressed by the acid-dissociation constant, K_a. (§16.1)
11. Explain why water is a very weak electrolyte and how its autoionization is expressed by K_w. (§16.2)
12. Explain why $[H_3O^+]$ is inversely related to $[OH^-]$ in any aqueous solution. (§16.2)
13. Describe how the relative magnitudes of $[H_3O^+]$ and $[OH^-]$ define whether a solution is acidic, basic, or neutral. (§16.2)
14. Explain how weak bases in water accept a proton rather than dissociate, and explain the meaning of K_b and pK_b. (§16.3)
15. Relate how ammonia, amines, and weak-acid conjugate bases act as weak bases in water. (§16.3)
16. Explain why relative concentrations of HA and A^- determine the acidity or basicity of their solution. (§16.3)
17. Describe the relationship of the K_a and K_b of a conjugate acid-base pair to K_w. (§16.3)
18. Explain how the percent dissociation of a weak acid increases as its concentration decreases. (§16.4)
19. Describe how a polyprotic acid dissociates in two or more steps and why only the first step supplies significant $[H_3O^+]$. (§16.4)
20. Explain the effects of electronegativity, bond polarity, and bond strength on acid strength. (§16.5)
21. Explain why aqueous solutions of small, highly charged metal ions are acidic. (§16.5)
22. Describe the various combinations of conjugate acids and conjugate bases that lead to acidic, basic, or neutral salt solutions. (§16.6)
23. Explain why the strengths of strong acids are levelled in water, but differentiated in a less basic solvent. (§16.7)
24. Describe the Lewis definitions of an acid and a base, and explain how a Lewis acid-base reaction involves the donation and acceptance of an electron pair to form a covalent bond. (§16.8)
25. Describe how molecules with electron-deficient atoms, molecules with polar multiple bonds, and metal cations act as Lewis acids. (§16.8)

Skills

1. Identify conjugate acid-base pairs. (SP 16.1)
2. Use relative acid strengths to predict the net direction of an acid-base reaction. (SPs 16.2, 16.3)
3. Classify strong and weak acids and bases from their formulas. (SP 16.4)

4. Use K_w to calculate $[H^+]$ or $[OH^-]$ in an aqueous solution. (SP 16.5)
5. Use p-scales to express $[H_3O^+]$, $[OH^-]$, and K. (§16.2)
6. Calculate $[H_3O^+]$, pH, $[OH^-]$, and pOH. (SP 16.6)
7. Calculate K_a of a weak acid from pH. (SP 16.7)
8. Calculate $[H_3O^+]$ (and thus pH) from K_a and $[HA]_{init}$. (SP 16.8)
9. Apply the quadratic equation to find a concentration. (Follow-Up Problem 16.8)
10. Calculate pH from K_b and $[B]_{init}$. (SP 16.9)
11. Calculate the percent dissociation of a weak acid. (§16.4 and SP 16.10)
12. Find K_b of A^- from K_a of HA and K_w. (§16.4 and SP 16.11)
13. Calculate pH from K_b of A^- and $[A^-]_{init}$. (SP 16.11)
14. Calculate $[H_3O^+]$ and other concentrations for a polyprotic acid. (SP 16.12)
15. Predict relative acid strengths of nonmetal hydrides and oxoacids. (§16.5)
16. Predict the relative acidity of a salt solution from the nature of the conjugate acid and conjugate base. (SPs 16.13, 16.14)
17. Identify Lewis acids and bases. (SP 16.15)

Key Terms

Section 16.1
hydronium ion (H_3O^+)
Arrhenius acid-base definition
neutralization
Brønsted-Lowry acid-base definition
proton donor
proton acceptor
conjugate acid-base pair
acid-dissociation constant (acid-ionization constant), K_a

Section 16.2
autoionization
ion-product constant for water (K_w)
pH
acid-base indicators

Section 16.3
base-dissociation constant (base-ionization constant), K_b

Section 16.4
extent of dissociation
percent dissociation
polyprotic acid

Section 16.7
levelling effect

Section 16.8
Lewis acid-base definition
adduct

Key Equations and Relationships

16.1 Defining the acid-dissociation constant:

$$K_a = \frac{[H^+][A^-]}{[HA]}$$

16.2 Defining the ion-product constant for water:

$$K_w = [H^+][OH^-] = 1.0 \times 10^{-14} \text{ (at 25°C)}$$

16.3 Defining pH:

$$pH = -\log [H^+]$$

16.4 Relating pK_w to pH and pOH:

$$pK_w = pH + pOH = 14.00 \text{ (at 25°C)}$$

16.5 Defining the base-dissociation constant:

$$K_b = \frac{[BH^+][OH^-]}{[B]}$$

16.6 Expressing the relationship among K_a, K_b, and K_w:

$$K_a \times K_b = K_w$$

16.7 Finding the percent dissociation of HA:

$$\text{Percent HA dissociated} = \frac{[HA]_{dissoc}}{[HA]_{init}} \times 100$$

Brief Solutions to Follow-Up Problems

16.1 (a) CH_3COOH/CH_3COO^- and H_3O^+/H_2O
(b) H_2O/OH^- and HF/F^-

16.2 (a) $NH_3(g) + H_2O(l) \rightleftharpoons NH_4^+(aq) + OH^-(aq)$
(b) $NH_3(g) + H_3O^+(aq\text{; from HCl}) \longrightarrow NH_4^+(aq) + H_2O(l)$
(c) $NH_4^+(aq) + OH^-(aq\text{; from NaOH}) \longrightarrow NH_3(g) + H_2O(l)$

16.3 There are more HB molecules than HA, so $K > 1$ and HA is the stronger acid.

16.4 (a) $HClO_3$ (b) HCl (c) NaOH

16.5 $[H^+] = \dfrac{1.0 \times 10^{-14}}{6.7 \times 10^{-2}} = 1.5 \times 10^{-13}$ mol/L; basic

16.6 $pOH = 14.00 - 9.52 = 4.48$
$[H^+] = 10^{-9.52} = 3.0 \times 10^{-10}$ mol/L
$[OH^-] = \dfrac{1.0 \times 10^{-14}}{3.0 \times 10^{-10}} = 3.3 \times 10^{-5}$ mol/L

16.7 $NH_4^+(aq) + H_2O(l) \rightleftharpoons NH_3(aq) + H_3O^+(aq)$
$[H_3O^+] = 10^{-pH} = 10^{-5.0} = 1 \times 10^{-5}$ mol/L $= [NH_3]$
$[NH_4^+] = 0.2$ mol/L $- (1 \times 10^{-5}$ mol/L$) = 0.2$ mol/L
$$K_a = \frac{[NH_3][H_3O^+]}{[NH_4^+]} \approx \frac{(1 \times 10^{-5})^2}{0.2} = 5 \times 10^{-10}$$

16.8 $K_a = \dfrac{[H_3O^+][OCN^-]}{[HOCN]} = \dfrac{(x)(x)}{0.10 - x} = 3.5 \times 10^{-4}$

Since $\frac{[HOCN]_{init}}{K_a} = \frac{0.10}{3.5 \times 10^{-4}} = 286 < 400$, you must solve a quadratic equation: $x^2 + (3.5 \times 10^{-4})x - (3.5 \times 10^{-5}) = 0$; $x = [H_3O^+] = 5.7 \times 10^{-3}$ mol/L.

16.9 $K_b = \frac{[C_5H_5NH^+][OH^-]}{[C_5H_5N]} = 10^{-8.77} = 1.7 \times 10^{-9}$

Assuming that 0.10 mol/L − $x \approx$ 0.10 mol/L,

$K_b = 1.7 \times 10^{-9} \approx \frac{(x)(x)}{0.10}$

$x = [OH^-] \approx 1.3 \times 10^{-5}$ mol/L

$[H_3O^+] = 7.7 \times 10^{-10}$ mol/L

pH = 9.11

16.10 There is no single correct scene. Any scene in which the total number of HB + H_3O^+ (or HB + B^-) is less than the total number in the original solution, yet the number of HB dissociated is greater, is correct. One example is given below:

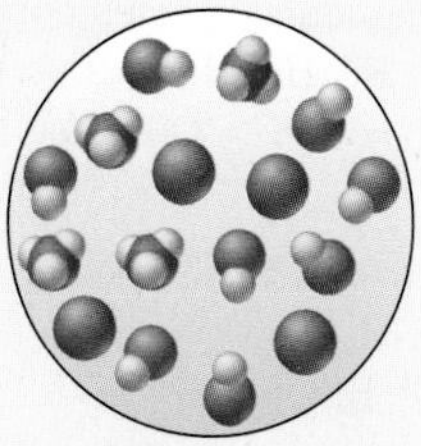

16.11 K_b of $ClO^- = \frac{K_w}{K_a \text{ of HClO}} = \frac{1.0 \times 10^{-14}}{2.9 \times 10^{-8}} = 3.4 \times 10^{-7}$

Assuming that 0.20 mol/L − $x \approx$ 0.20 mol/L,

$K_b = 3.4 \times 10^{-7} = \frac{[HClO][HO^-]}{[ClO^-]} \approx \frac{x^2}{0.20}$

$x = [OH^-] \approx 2.6 \times 10^{-4}$ mol/L

$[H_3O^+] = 3.8 \times 10^{-11}$ mol/L

pH = 10.42

16.12 $K_{a1} = \frac{[HC_2O_4^-][H_3O^+]}{[H_2C_2O_4]} = \frac{x^2}{0.150 - x} = 5.6 \times 10^{-2}$

Since $\frac{[H_2C_2O_4]_{init}}{K_{a1}} < 400$, you must solve a quadratic equation:

$x^2 + (5.6 \times 10^{-2})x - (8.4 \times 10^{-3}) = 0$

$x = [H_3O^+] = 0.068$ mol/L

pH = 1.17

$x = [HC_2O_4^-] = 0.068$ mol/L

$[H_2C_2O_4] = 0.150$ mol/L − x = 0.082 mol/L

$[C_2O_4^{2-}] = \frac{(K_{a2})[HC_2O_4^-]}{[H_3O^+]} = \frac{(5.4 \times 10^{-5})(0.068)}{0.068}$

$= 5.4 \times 10^{-5}$ mol/L

16.13 (a) Basic:

$ClO_2^-(aq) + H_2O(l) \rightleftharpoons HClO(aq) + OH^-(aq)$

K^+ is from the strong base KOH.

(b) Acidic:

$CH_3NH_3^+(aq) + H_2O(l) \rightleftharpoons CH_3NH_2(aq) + H_3O^+(aq)$

NO_3^- is from the strong acid HNO_3.

(c) Neutral: Cs^+ is from the strong base CsOH; I^- is from the strong acid HI.

16.14 (a) K_a of $Cu(H_2O)_6^{2+} = 3 \times 10^{-8}$

K_b of $CH_3COO^- = \frac{K_w}{K_a \text{ of } CH_3COOH} = 5.6 \times 10^{-10}$

Since $K_a > K_b$, $Cu(CH_3COO)_2(aq)$ is acidic.

(b) K_a of $NH_4^+ = \frac{K_w}{K_b \text{ of } NH_3} = 5.7 \times 10^{-10}$

K_b of $F^- = \frac{K_w}{K_a \text{ of HF}} = 1.5 \times 10^{-11}$

Because $K_a > K_b$, $NH_4F(aq)$ is acidic.

(c) From Appendix C, K_a of HSO_3^- is 6.5×10^{-8}.

K_b of $HSO_3^- = \frac{K_w}{K_a \text{ of } H_2SO_3} = 7.1 \times 10^{-13}$

Because $K_a > K_b$, $KHSO_3(aq)$ is acidic.

16.15 (a) OH^- is the Lewis base, and $Al(OH)_3$ is the Lewis acid.
(b) H_2O is the Lewis base, and SO_3 is the Lewis acid.
(c) NH_3 is the Lewis base, and Co^{3+} is the Lewis acid.

PROBLEMS

Problems with **red** numbers are answered in Appendix G and worked in detail in the Student Solutions Manual. Problem sections match those in this book and provide the numbers of relevant sample problems. Most offer Concept Review Questions, Skill-Building Exercises (grouped in pairs covering the same concept), and Problems in Context. The Comprehensive Problems are based on material from any section or previous chapter.

Note: Unless stated otherwise, all problems refer to aqueous solutions at 298 K (25°C).

Acids and Bases

(Sample Problems 16.1 to 16.4)

Concept Review Questions

16.1 What is the role of water in the Arrhenius acid-base definition?

16.2 (a) What do Arrhenius acids have in common? What do Arrhenius bases have in common?
(b) Explain neutralization in terms of the Arrhenius acid-base definition. What data led Arrhenius to propose this idea of neutralization?

16.3 Why is the Arrhenius acid-base definition too limited? Give an example of an acid to which the Arrhenius definition does not apply.

16.4 (a) How are the Arrhenius and Brønsted-Lowry acid-base definitions different? How are they similar?
(b) Name two Brønsted-Lowry bases that are not Arrhenius bases. Can you do the same for acids? Explain.

16.5 What is a conjugate acid-base pair? What is the relationship between the two members of the pair?

16.6 (a) A Brønsted-Lowry acid-base reaction proceeds in the net direction in which a stronger acid and a stronger base form a weaker acid and a weaker base. Explain.
(b) The molecular scene below depicts an aqueous solution of two conjugate acid-base pairs: HA/A^- and HB/B^-. The base in the first pair is represented by red spheres and the base in the second pair is represented by green spheres; solvent molecules are omitted for clarity. Which is the stronger acid? Which is the stronger base? Explain.

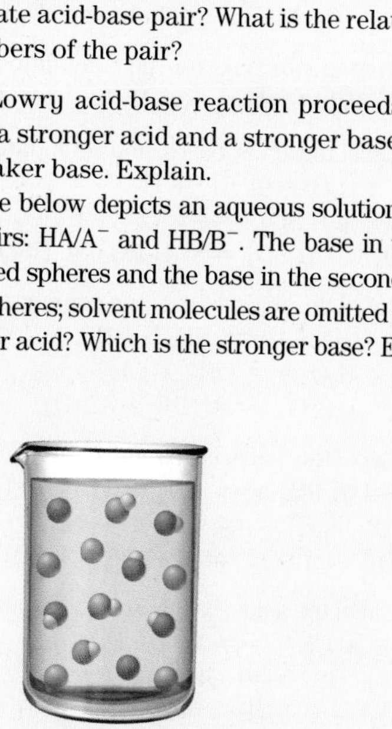

16.7 What is an amphiprotic species? Name one, and write balanced equations to show why it is amphiprotic.

16.8 (a) What do the words *strong* and *weak* mean for acids and bases?
(b) K_a values of weak acids vary over more than 10 orders of magnitude. What do the acids have in common that makes them "weak"?

Skill-Building Exercises (grouped in similar pairs)

16.9 Which are Arrhenius acids?
(a) H_2O (b) $Ca(OH)_2$ (c) H_3PO_3 (d) HI

16.10 Which are Arrhenius acids?
(a) $NaHSO_4$ (b) CH_4 (c) NaH (d) H_3N

16.11 Which are Arrhenius bases?
(a) H_3AsO_4 (b) $Ba(OH)_2$ (c) HClO (d) KOH

16.12 Which are Arrhenius bases?
(a) CH_3COOH (b) HOH (c) CH_3OH (d) H_2NNH_2

16.13 Write the K_a expression for each of the following in water:
(a) HCN (b) HCO_3^- (c) HCOOH

16.14 Write the K_a expression for each of the following in water:
(a) $CH_3NH_3^+$ (b) HClO (c) H_2S

16.15 Write the K_a expression for each of the following in water:
(a) HNO_2 (b) CH_3COOH (c) $HBrO_2$

16.16 Write the K_a expression for each of the following in water:
(a) $H_2PO_4^-$ (b) H_3PO_2 (c) HSO_4^-

16.17 Write balanced equations and K_a expressions for each Brønsted-Lowry acid in water:
(a) H_3PO_4 (b) C_6H_5COOH (c) HSO_4^-

16.18 Write balanced equations and K_a expressions for each Brønsted-Lowry acid in water:
(a) HCOOH (b) $HClO_3$ (c) $H_2AsO_4^-$

16.19 Give the formula for the conjugate base:
(a) HCl (b) H_2CO_3 (c) H_2O

16.20 Give the formula for the conjugate base:
(a) HPO_4^{2-} (b) NH_4^+ (c) HS^-

16.21 Give the formula for the conjugate acid:
(a) NH_3 (b) NH_2^- (c) nicotine, $C_{10}H_{14}N_2$

16.22 Give the formula for the conjugate acid:
(a) O^{2-} (b) SO_4^{2-} (c) H_2O

16.23 Label the acids, bases, and conjugate pairs in each equation:
(a) $HCl + H_2O \rightleftharpoons Cl^- + H_3O^+$
(b) $HClO_4 + H_2SO_4 \rightleftharpoons ClO_4^- + H_3SO_4^+$
(c) $HPO_4^{2-} + H_2SO_4 \rightleftharpoons H_2PO_4^- + HSO_4^-$

16.24 Label the acids, bases, and conjugate pairs in each equation:
(a) $NH_3 + HNO_3 \rightleftharpoons NH_4^+ + NO_3^-$
(b) $O^{2-} + H_2O \rightleftharpoons OH^- + OH^-$
(c) $NH_4^+ + BrO_3^- \rightleftharpoons NH_3 + HBrO_3$

16.25 Label the acids, bases, and conjugate pairs in each equation:
(a) $NH_3 + H_3PO_4 \rightleftharpoons NH_4^+ + H_2PO_4^-$
(b) $CH_3O^- + NH_3 \rightleftharpoons CH_3OH + NH_2^-$
(c) $HPO_4^{2-} + HSO_4^- \rightleftharpoons H_2PO_4^- + SO_4^{2-}$

16.26 Label the acids, bases, and conjugate pairs in each equation:
(a) $NH_4^+ + CN^- \rightleftharpoons NH_3 + HCN$
(b) $H_2O + HS^- \rightleftharpoons OH^- + H_2S$
(c) $HSO_3^- + CH_3NH_2 \rightleftharpoons SO_3^{2-} + CH_3NH_3^+$

16.27 Write a balanced net ionic equation for each reaction, and label the conjugate acid-base pairs:
(a) $NaOH(aq) + NaH_2PO_4(aq) \rightleftharpoons H_2O(l) + Na_2HPO_4(aq)$
(b) $KHSO_4(aq) + K_2CO_3(aq) \rightleftharpoons K_2SO_4(aq) + KHCO_3(aq)$

16.28 Write a balanced net ionic equation for each reaction, and label the conjugate acid-base pairs:
(a) $HNO_3(aq) + Li_2CO_3(aq) \rightleftharpoons LiNO_3(aq) + LiHCO_3(aq)$
(b) $2NH_4Cl(aq) + Ba(OH)_2(aq) \rightleftharpoons 2H_2O(l) + BaCl_2(aq) + 2NH_3(aq)$

16.29 The following aqueous species constitute two conjugate acid-base pairs: HS^-, Cl^-, HCl, H_2S. Use these species to write one acid-base reaction with $K > 1$ and another acid-base reaction with $K < 1$.

16.30 The following aqueous species constitute two conjugate acid-base pairs: NO_3^-, F^-, HF, HNO_3. Use these species to write one acid-base reaction with $K > 1$ and another acid-base reaction with $K < 1$.

16.31 Use Figure 16.3 to determine whether $K > 1$ for each reaction:
(a) $HCl + NH_3 \rightleftharpoons NH_4^+ + Cl^-$
(b) $H_2SO_3 + NH_3 \rightleftharpoons HSO_3^- + NH_4^+$

16.32 Use Figure 16.3 to determine whether $K > 1$ for each reaction:
(a) $OH^- + HS^- \rightleftharpoons H_2O + S^{2-}$
(b) $HCN + HCO_3^- \rightleftharpoons H_2CO_3 + CN^-$

16.33 Use Figure 16.3 to determine whether $K < 1$ for each reaction:
(a) $NH_4^+ + HPO_4^{2-} \rightleftharpoons NH_3 + H_2PO_4^-$
(b) $HSO_3^- + HS^- \rightleftharpoons H_2SO_3 + S^{2-}$

16.34 Use Figure 16.3 to determine whether $K < 1$ for each reaction:
(a) $H_2PO_4^- + F^- \rightleftharpoons HPO_4^{2-} + HF$
(b) $CH_3COO^- + HSO_4^- \rightleftharpoons CH_3COOH + SO_4^{2-}$

16.35 Use the chapter content and Appendix C to rank the following acids in order of *increasing* strength: HIO_3, HI, CH_3COOH, HF.

16.36 Use the chapter content and Appendix C to rank the following acids in order of *decreasing* strength: HClO, HCl, HCN, HNO_2.

16.37 Classify each species as a strong or weak acid or base:
(a) H_3AsO_4 (b) $Sr(OH)_2$ (c) HIO (d) $HClO_4$

16.38 Classify each species as a strong or weak acid or base:
(a) CH_3NH_2 (b) K_2O (c) HI (d) HCOOH

16.39 Classify each species as a strong or weak acid or base:
(a) RbOH (b) HBr (c) H_2Te (d) HClO

16.40 Classify each species as a strong or weak acid or base:
(a) $HOCH_2CH_2NH_2$ (b) H_2SeO_4 (c) HS^- (d) $B(OH)_3$

Autoionization of Water and the pH Scale

(Sample Problems 16.5 and 16.6)

Concept Review Questions

16.41 (a) What is an autoionization reaction?
(b) Write equations for the autoionization reactions of H_2O and H_2SO_4.

16.42 What is the difference between K and K_w for the autoionization of water?

16.43 (a) What is the change in pH when $[OH^-]$ increases by a factor of 10?
(b) What is the change in $[H^+]$ when the pH decreases by 3 units?

16.44 Which solution in each pair has the higher pH? Explain.
(a) A 0.1 mol/L solution of an acid with $K_a = 1 \times 10^{-4}$ or one with $K_a = 4 \times 10^{-5}$
(b) A 0.1 mol/L solution of an acid with $pK_a = 3.0$ or one with $pK_a = 3.5$
(c) A 0.1 mol/L solution of a weak acid or a 0.01 mol/L solution of a weak acid
(d) A 0.1 mol/L solution of a weak acid or a 0.1 mol/L solution of a strong acid
(e) A 0.1 mol/L solution of an acid or a 0.01 mol/L solution of a base
(f) A solution with a pOH of 6.0 or a solution with a pOH of 8.0

Skill-Building Exercises (grouped in similar pairs)

16.45 (a) What is the pH of 0.0111 mol/L NaOH? Is the solution neutral, acidic, or basic?
(b) What is the pOH of 1.35×10^{-3} mol/L HCl? Is the solution neutral, acidic, or basic?

16.46 (a) What is the pH of 0.0333 mol/L HNO_3? Is the solution neutral, acidic, or basic?
(b) What is the pOH of 0.0347 mol/L KOH? Is the solution neutral, acidic, or basic?

16.47 (a) What is the pH of 6.14×10^{-3} mol/L HI? Is the solution neutral, acidic, or basic?
(b) What is the pOH of 2.55 mol/L $Ba(OH)_2$? Is the solution neutral, acidic, or basic?

16.48 (a) What is the pH of 7.52×10^{-4} mol/L CsOH? Is the solution neutral, acidic, or basic?
(b) What is the pOH of 1.59×10^{-3} mol/L $HClO_4$? Is the solution neutral, acidic, or basic?

16.49 (a) What are $[H_3O^+]$, $[OH^-]$, and pOH in a solution with a pH of 9.85?
(b) What are $[H_3O^+]$, $[OH^-]$, and pH in a solution with a pOH of 9.43?

16.50 (a) What are $[H_3O^+]$, $[OH^-]$, and pOH in a solution with a pH of 3.47?
(b) What are $[H_3O^+]$, $[OH^-]$, and pH in a solution with a pOH of 4.33?

16.51 (a) What are $[H_3O^+]$, $[OH^-]$, and pOH in a solution with a pH of 4.77?
(b) What are $[H_3O^+]$, $[OH^-]$, and pH in a solution with a pOH of 5.65?

16.52 (a) What are $[H_3O^+]$, $[OH^-]$, and pOH in a solution with a pH of 8.97?
(b) What are $[H_3O^+]$, $[OH^-]$, and pH in a solution with a pOH of 11.27?

16.53 What amount (mol) of H_3O^+ or OH^-, and which one, must you add per litre of HA solution to adjust its pH from 3.15 to 3.65? Assume a negligible volume change.

16.54 What amount (mol) of H_3O^+ or OH^-, and which one, must you add per litre of HA solution to adjust its pH from 9.33 to 9.07? Assume a negligible volume change.

16.55 What amount (mol) of H_3O^+ or OH^-, and which one, must you add to 5.6 L of HA solution to adjust its pH from 4.52 to 5.25? Assume a negligible volume change.

16.56 What amount (mol) of H_3O^+ or OH^-, and which one, must you add to 87.5 mL of HA solution to adjust its pH from 8.92 to 6.33? Assume a negligible volume change.

Problems in Context

16.57 The following two molecular scenes depict the relative concentrations of H_3O^+ (*purple*) in solutions of the same volume (with counterions and solvent molecules omitted for clarity). If the pH in scene A is 4.8, what is the pH in scene B?

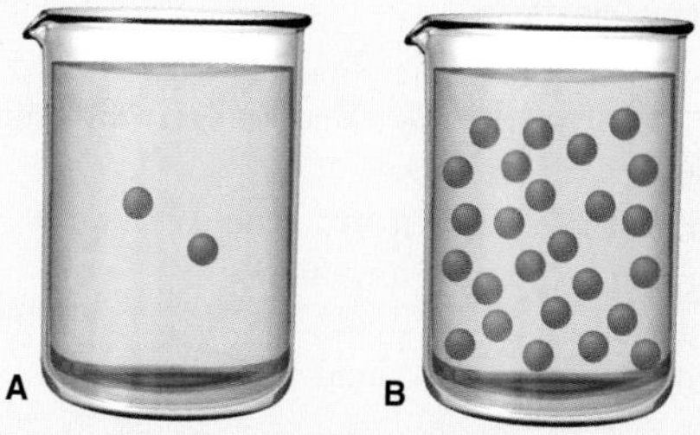

16.58 Like any equilibrium constant, K_w changes with temperature.
(a) Given that autoionization is endothermic, how does K_w change with rising temperature? Explain with a reaction that includes heat as a reactant or a product.
(b) In many medical applications, the value of K_w at 37°C (body temperature) may be more appropriate than the value at 25°C, 1.0×10^{-14}. The pH of pure water at 37°C is 6.80. Calculate K_w, pOH, and $[OH^-]$ at this temperature.

Weak Bases and Their Relation to Weak Acids

Concept Review Questions

16.59 What is the key structural feature of all Brønsted-Lowry bases? How does this feature function in an acid-base reaction?

16.60 Why are most conjugate bases basic in H_2O? Give formulas of four conjugate bases that are not basic.

16.61 Except for the Na^+ spectator ion, aqueous solutions of CH_3COOH and CH_3COONa contain the same species. (a) What are the species (other than H_2O)?
(b) Why is 0.1 mol/L CH_3COOH acidic and 0.1 mol/L CH_3COONa basic?

Skill-Building Exercises (grouped in similar pairs)

16.62 Write balanced equations and K_b expressions for each Brønsted-Lowry base in water:
(a) Pyridine, C_5H_5N (b) CO_3^{2-}

16.63 Write balanced equations and K_b expressions for each Brønsted-Lowry base in water:
(a) Benzoate ion, $C_6H_5COO^-$ (b) $(CH_3)_3N$

16.64 Write balanced equations and K_b expressions for each Brønsted-Lowry base in water:
(a) Hydroxylamine, HO—NH_2 (b) HPO_4^{2-}

16.65 Write balanced equations and K_b expressions for each Brønsted-Lowry base in water:
(a) Guanidine, $(H_2N)_2$ C=NH (double-bonded N is most basic)
(b) Acetylide ion, HC≡C^-

16.66 (a) What is the K_b of the ethanoate (acetate) ion?
(b) What is the K_a of the anilinium ion, $C_6H_5NH_3^+$?

16.67 (a) What is the K_b of the benzoate ion, $C_6H_5COO^-$?
(b) What is the K_a of the 2-hydroxyethanaminium ion, $HOCH_2CH_2NH_3^+$ (pK_b of $HOCH_2CH_2NH_2 = 4.49$)?

16.68 (a) What is the pK_b of ClO_2^-?
(b) What is the pK_a of the N-methylmethanaminium ion, $(CH_3)_2NH_2^+$?

16.69 (a) What is the pK_b of NO_2^-?
(b) What is the pK_a of the hydrazinium ion, $H_2N—NH_3^+$ (K_b of hydrazine = 8.5×10^{-7})?

Solving Problems Involving Weak-Acid or Weak-Base Equilibria

(Sample Problems 16.7 to 16.12)

Concept Review Questions

16.70 In each solution, is the concentration of acid before and after dissociation nearly the same or very different? Explain your reasoning.
(a) Concentrated solution of a strong acid
(b) Concentrated solution of a weak acid
(c) Dilute solution of a weak acid
(d) Dilute solution of a strong acid

16.71 A sample of 0.0001 mol/L HCl has [H_3O^+] close to that of a sample of 0.1 mol/L CH_3COOH. Are ethanoic acid and hydrochloric acid equally strong in these samples? Explain.

16.72 In which solutions will [H_3O^+] be approximately equal to [CH_3COO^-]? Explain.
(a) 0.1 mol/L CH_3COOH
(b) 1×10^{-7} mol/L CH_3COOH
(c) Solution containing both 0.1 mol/L CH_3COOH and 0.1 mol/L CH_3COONa

16.73 Why do successive K_a's decrease for all polyprotic acids?

Skill-Building Exercises (grouped in similar pairs)

16.74 A 0.15 mol/L solution of butanoic acid, $CH_3CH_2CH_2COOH$, contains 1.51×10^{-3} mol/L H_3O^+. What is the K_a of butanoic acid?

16.75 A 0.035 mol/L solution of a weak acid (HA) has a pH of 4.88. What is the K_a of the acid?

16.76 What is the pH of 0.070 mol/L dimethylamine?

16.77 What is the pH of 0.12 mol/L diethylamine?

16.78 What is the pH of 0.25 mol/L ethanolamine?

16.79 What is the pH of 0.26 mol/L aniline?

16.80 Nitrous acid, HNO_2, has a K_a of 7.1×10^{-4}. What are [H_3O^+], [NO_2^-], and [OH^-] in 0.60 mol/L HNO_2?

16.81 Hydrofluoric acid, HF, has a K_a of 6.8×10^{-4}. What are [H_3O^+], [F^-], and [OH^-] in 0.75 mol/L HF?

16.82 Chloroethanoic (chloroacetic) acid, $ClCH_2COOH$, has a pK_a of 2.87. What are [H_3O^+], pH, [$ClCH_2COO^-$], and [$ClCH_2COOH$] in 1.25 mol/L $ClCH_2COOH$?

16.83 Hypochlorous acid, HClO, has a pK_a of 7.54. What are [H_3O^+], pH, [ClO^-], and [HClO] in 0.115 mol/L HClO?

16.84 (a) What is the pH of 0.150 mol/L KCN?
(b) What is the pH of 0.40 mol/L triethylammonium chloride, $(CH_3CH_2)_3NHCl$?

16.85 (a) What is the pH of 0.100 mol/L sodium phenolate, C_6H_5ONa, the sodium salt of phenol?
(b) What is the pH of 0.15 mol/L methylammonium bromide, CH_3NH_3Br (K_b of CH_3NH_2 = 4.4×10^{-4})?

16.86 (a) What is the pH of 0.65 mol/L potassium formate, HCOOK?
(b) What is the pH of 0.85 mol/L NH_4Br?

16.87 (a) What is the pH of 0.75 mol/L NaF?
(b) What is the pH of 0.88 mol/L pyridinium chloride, C_5H_5NHCl?

16.88 In a 0.20 mol/L solution, a weak acid is 3.0% dissociated.
(a) Calculate the [H_3O^+], pH, [OH^-], and pOH of the solution.
(b) Calculate K_a of the acid.

16.89 In a 0.735 mol/L solution, a weak acid is 12.5% dissociated.
(a) Calculate the [H_3O^+], pH, [OH^-], and pOH of the solution.
(b) Calculate K_a of the acid.

16.90 A 0.250 mol sample of HX is dissolved in enough H_2O to form 655 mL of solution. If the pH of the solution is 3.54, what is the K_a of HX?

16.91 A 4.85×10^{-3} mol sample of HY is dissolved in enough H_2O to form 0.095 L of solution. If the pH of the solution is 2.68, what is the K_a of HY?

16.92 The weak acid HZ has a K_a of 2.55×10^{-4}.
(a) Calculate the pH of 0.075 mol/L HZ.
(b) Calculate the pOH of 0.045 mol/L HZ.

16.93 The weak acid HQ has a pK_a of 4.89.
(a) Calculate the [H_3O^+] of 3.5×10^{-2} mol/L HQ.
(b) Calculate the [OH^-] of 0.65 mol/L HQ.

16.94 (a) Calculate the pH of 0.175 mol/L HY if $K_a = 1.50 \times 10^{-4}$.
(b) Calculate the pOH of 0.175 mol/L HX if $K_a = 2.00 \times 10^{-2}$.

16.95 (a) Calculate the pH of 0.55 mol/L HCN if $K_a = 6.2 \times 10^{-10}$.
(b) Calculate the pOH of 0.044 mol/L HIO_3 if $K_a = 0.16$.

16.96 Use Appendix C to calculate the percent dissociation of 0.55 mol/L benzoic acid, C_6H_5COOH.

16.97 Use Appendix C to calculate the percent dissociation of 0.050 mol/L CH_3COOH.

16.98 Use Appendix C to calculate [H_2S], [HS^-], [S^{2-}], [H_3O^+], pH, [OH^-], and pOH in a 0.10 mol/L solution of the diprotic acid hydrosulfuric acid.

16.99 Use Appendix C to calculate [$H_2C_3H_2O_4$], [$HC_3H_2O_4^-$], [$C_3H_2O_4^{2-}$], [H_3O^+], pH, [OH^-], and pOH in a 0.200 mol/L solution of the diprotic acid malonic acid.

Problems in Context

16.100 Acetylsalicylic acid (Aspirin), $HC_9H_7O_4$, is the most widely used pain reliever and fever reducer. Find the pH of 0.018 mol/L aqueous Aspirin at body temperature (K_a at 37°C = 3.6×10^{-4}).

16.101 Methanoic (formic) acid, HCOOH, the simplest carboxylic acid, is used in the textile and rubber industries and is secreted as a defence by many species of ants (family *Formicidae*). Calculate the percent dissociation of 0.75 mol/L HCOOH.

16.102 Sodium hypochlorite solution, sold as chlorine bleach, is potentially dangerous because of the basicity of ClO^-, the active bleaching ingredient. (Assume that d of solution = 1.0 g/mL.)
(a) What is [OH^-] in an aqueous solution that is 6.5% NaClO by mass?
(b) What is the pH of the solution?

16.103 Codeine ($C_{18}H_{21}NO_3$) is a narcotic pain reliever that forms a salt with HCl. What is the pH of 0.050 mol/L codeine hydrochloride (pK_b of codeine = 5.80)?

Molecular Properties and Acid Strength

Concept Review Questions

16.104 Across a period, how does the electronegativity of a nonmetal affect the acidity of its binary hydride?

16.105 How does the atomic size of a nonmetal affect the acidity of its binary hydride?

16.106 A strong acid has a weak bond to its acidic proton, whereas a weak acid has a strong bond to its acidic proton. Explain.

16.107 Perchloric acid, $HClO_4$, is the strongest of the halogen oxoacids, and hypoiodous acid, HIO, is the weakest. What two factors govern this difference in acid strength?

Skill-Building Exercises (grouped in similar pairs)

16.108 Which is the *stronger* acid in each pair?
(a) H_2SeO_3 or H_2SeO_4 (b) H_3PO_4 or H_3AsO_4 (c) H_2S or H_2Te

16.109 Which is the *weaker* acid in each pair?
(a) HBr or H_2Se (b) $HClO_4$ or H_2SO_4 (c) H_2SO_3 or H_2SO_4

16.110 Which is the *stronger* acid in each pair?
(a) H_2Se or H_3As (b) $B(OH)_3$ or $Al(OH)_3$ (c) $HBrO_2$ or HBrO

16.111 Which is the *weaker* acid in each pair?
(a) HI or HBr (b) H_3AsO_4 or H_2SeO_4 (c) HNO_3 or HNO_2

16.112 Use Appendix C to choose the solution with the *lower* pH:
(a) 0.5 mol/L $CuBr_2$ or 0.5 mol/L $AlBr_3$
(b) 0.3 mol/L $ZnCl_2$ or 0.3 mol/L $SnCl_2$

16.113 Use Appendix C to choose the solution with the *lower* pH:
(a) 0.1 mol/L $FeCl_3$ or 0.1 mol/L $AlCl_3$
(b) 0.1 mol/L $BeCl_2$ or 0.1 mol/L $CaCl_2$

16.114 Use Appendix C to choose the solution with the *higher* pH:
(a) 0.2 mol/L $Ni(NO_3)_2$ or 0.2 mol/L $Co(NO_3)_2$
(b) 0.35 mol/L $Al(NO_3)_3$ or 0.35 mol/L $Cr(NO_3)_3$

16.115 Use Appendix C to choose the solution with the *higher* pH:
(a) 0.1 mol/L $NiCl_2$ or 0.1 mol/L NaCl
(b) 0.1 mol/L $Sn(NO_3)_2$ or 0.1 mol/L $Co(NO_3)_2$

Acid-Base Properties of Salt Solutions

(Sample Problems 16.13 and 16.14)

Concept Review Questions

16.116 What determines whether an aqueous solution of a salt will be acidic, basic, or neutral? Give an example of each type of salt.

16.117 Why is aqueous NaF basic but aqueous NaCl neutral?

16.118 The NH_4^+ ion forms acidic solutions, and the CH_3COO^- ion forms basic solutions. However, a solution of ammonium ethanoate (ammonium acetate) is almost neutral. Do all the ammonium salts of weak acids form neutral solutions? Explain your answer.

Skill-Building Exercises (grouped in similar pairs)

16.119 Explain, using equations and calculations when necessary, whether an aqueous solution of each salt is acidic, basic, or neutral:
(a) KBr (b) NH_4I (c) KCN

16.120 Explain, using equations and calculations when necessary, whether an aqueous solution of each salt is acidic, basic, or neutral:
(a) $Cr(NO_3)_3$ (b) NaHS (c) $Zn(CH_3COO)_2$

16.121 Explain, using equations and calculations when necessary, whether an aqueous solution of each salt is acidic, basic, or neutral:
(a) Na_2CO_3 (b) $CaCl_2$ (c) $Cu(NO_3)_2$

16.122 Explain, using equations and calculations when necessary, whether an aqueous solution of each salt is acidic, basic, or neutral:
(a) CH_3NH_3Cl (b) $KClO_4$ (c) CoF_2

16.123 Explain, using equations and calculations when necessary, whether an aqueous solution of each salt is acidic, basic, or neutral:
(a) $SrBr_2$ (b) $Ba(CH_3COO)_2$ (c) $(CH_3)_2NH_2Br$

16.124 Explain, using equations and calculations when necessary, whether an aqueous solution of each salt is acidic, basic, or neutral:
(a) $Fe(HCOO)_3$ (b) $KHCO_3$ (c) K_2S

16.125 Explain, using equations and calculations when necessary, whether an aqueous solution of each salt is acidic, basic, or neutral:
(a) $(NH_4)_3PO_4$ (b) Na_2SO_4 (c) LiClO

16.126 Explain, using equations and calculations when necessary, whether an aqueous solution of each salt is acidic, basic, or neutral:
(a) $Pb(CH_3COO)_2$ (b) $Cr(NO_2)_3$ (c) CsI

16.127 Rank the salts in each group in order of *increasing* pH of their 0.1 mol/L aqueous solutions:
(a) KNO_3, K_2SO_3, K_2S, $Fe(NO_3)_2$
(b) NH_4NO_3, $NaHSO_4$, $NaHCO_3$, Na_2CO_3

16.128 Rank the salts in each group in order of *decreasing* pH of their 0.1 mol/L aqueous solutions:
(a) $FeCl_2$, $FeCl_3$, $MgCl_2$, $KClO_2$
(b) NH_4Br, $NaBrO_2$, NaBr, $NaClO_2$

Generalizing the Brønsted-Lowry Concept: The Levelling Effect

Concept Review Questions

16.129 (a) Methoxide ion, CH_3O^-, and amide ion, NH_2^-, are very strong bases that are "levelled" by water. What does this mean?
(b) Write the reactions that occur in the levelling process described in part (a). What species do the two levelled solutions have in common?

16.130 Explain the differing extents of dissociation of H_2SO_4 in CH_3COOH, H_2O, and NH_3.

16.131 In H_2O, HF is weak and the other hydrohalic acids are equally strong. In NH_3, however, all the hydrohalic acids are equally strong. Explain.

Electron-Pair Donation and the Lewis Acid-Base Definition

(Sample Problem 16.15)

Concept Review Questions

16.132 What feature must a molecule or ion have for it to act as a Lewis base? What feature must it have to act as a Lewis acid? Explain the roles of these features.

16.133 (a) How do Lewis acids differ from Brønsted-Lowry acids? How are they similar?
(b) Do Lewis bases differ from Brønsted-Lowry bases? Explain.

16.134 (a) Is a weak Brønsted-Lowry base necessarily a weak Lewis base? Explain with an example.
(b) Identify the Lewis bases in the following reaction:

$$Cu(H_2O)_4^{2+}(aq) + 4CN^-(aq) \rightleftharpoons Cu(CN)_4^{2-}(aq) + 4H_2O(l)$$

(c) Given that $K > 1$ for the reaction in part (b), which Lewis base is stronger?

16.135 (a) In which of the three acid-base definitions can water be a product of an acid-base reaction?
(b) In which definition is water the only product?

16.136 (a) Give an example of a *substance* that is a base in two of the three acid-base definitions, but not in the third.
(b) Give an example of a *substance* that is an acid in one of the three acid-base definitions, but not in the other two.

Skill-Building Exercises (grouped in similar pairs)

16.137 Which species are Lewis acids, and which are Lewis bases?
(a) Cu^{2+} (b) Cl^- (c) $SnCl_2$ (d) OF_2

16.138 Which species are Lewis acids, and which are Lewis bases?
(a) Na^+ (b) NH_3 (c) CN^- (d) CO_2

16.139 Which species are Lewis acids, and which are Lewis bases?
(a) BF_3 (b) S^{2-} (c) SO_3^{2-} (d) SO_3

16.140 Which species are Lewis acids, and which are Lewis bases?
(a) Mg^{2+} (b) OH^- (c) SiF_4 (d) $BeCl_2$

16.141 Identify the Lewis acid and Lewis base in each equation:
(a) $Na^+ + 6H_2O \rightleftharpoons Na(H_2O)_6^+$
(b) $CO_2 + H_2O \rightleftharpoons H_2CO_3$
(c) $F^- + BF_3 \rightleftharpoons BF_4^-$

16.142 Identify the Lewis acid and Lewis base in each equation:
(a) $Fe^{3+} + 2H_2O \rightleftharpoons FeOH^{2+} + H_3O^+$
(b) $H_2O + H^- \rightleftharpoons OH^- + H_2$
(c) $4CO + Ni \rightleftharpoons Ni(CO)_4$

16.143 Classify each reaction as an Arrhenius, Brønsted-Lowry, or Lewis acid-base reaction. A reaction may fit all, two, one, or none of these classifications.
(a) $Ag^+ + 2NH_3 \rightleftharpoons Ag(NH_3)_2^+$
(b) $H_2SO_4 + NH_3 \rightleftharpoons HSO_4^- + NH_4^+$
(c) $2HCl \rightleftharpoons H_2 + Cl_2$
(d) $AlCl_3 + Cl^- \rightleftharpoons AlCl_4^-$

16.144 Classify each reaction as an Arrhenius, Brønsted-Lowry, or Lewis acid-base reaction. A reaction may fit all, two, one, or none of these classifications.
(a) $Cu^{2+} + 4Cl^- \rightleftharpoons CuCl_4^{2-}$
(b) $Al(OH)_3 + 3HNO_3 \rightleftharpoons Al^{3+} + 3H_2O + 3NO_3^-$
(c) $N_2 + 3H_2 \rightleftharpoons 2NH_3$
(d) $CN^- + H_2O \rightleftharpoons HCN + OH^-$

Comprehensive Problems

16.145 Chloral ($Cl_3C—CH═O$) forms a monohydrate, chloral hydrate, the sleep-inducing depressant called "knockout drops" in old movies.
(a) Write two possible structures for chloral hydrate, one that involves hydrogen bonding and one that is a Lewis adduct.
(b) What spectroscopic method could be used to identify the real structure? Explain.

16.146 In humans, blood pH is maintained within a narrow range: *acidosis* occurs if the blood pH is below 7.35, and *alkalosis* occurs if the pH is above 7.45. Given that the pK_w of blood is 13.63 at 37°C (body temperature), what is the normal range of $[H_3O^+]$ and the normal range of $[OH^-]$ in blood?

16.147 The disinfectant phenol, C_6H_5OH, has a pK_a of 10.0 in water, but 14.4 in methanol.
(a) Why are the values different?
(b) Is methanol a stronger or weaker base than water?
(c) Write the dissociation reaction of phenol in methanol.
(d) Write an expression for the autoionization constant of methanol.

16.148 When carbon dioxide dissolves in water, it undergoes a multistep equilibrium process, which can be simplified as follows:

(1) $CO_2(g) + H_2O(l) \rightleftharpoons H_2CO_3(aq)$

(2) $H_2CO_3(aq) + H_2O(l) \rightleftharpoons HCO_3^-(aq) + H_3O^+(aq)$

$K_{overall} = 4.5 \times 10^{-7}$

(a) Classify each step as a Lewis reaction or a Brønsted-Lowry reaction.
(b) What is the pH of nonpolluted rainwater in equilibrium with clean air [(P_{CO_2} in clean air $= 3.2 \times 10^{-4}$ bar; Henry's law constant for CO_2 at 25°C = 0.033 mol/(L•bar)]?
(c) What is $[CO_3^{2-}]$ in rainwater (K_a of $HCO_3^- = 4.7 \times 10^{-11}$)?
(d) If the partial pressure of CO_2 in clean air doubles in the next few decades, what will the pH of rainwater become?

16.149 Seashells are mostly calcium carbonate, which reacts with H_3O^+ according to the following equation:

$CaCO_3(s) + H_3O^+(aq) \rightleftharpoons Ca^{2+}(aq) + HCO_3^-(aq) + H_2O(l)$

If K_w increases at higher pressure, will seashells dissolve more rapidly near the surface of the ocean or at great depths? Explain.

16.150 Many molecules with a central atom from period 3 or higher take part in Lewis acid-base reactions in which the central atom expands to accommodate additional electrons in its valence level. $SnCl_4$ reacts with $(CH_3)_3N$ as follows:

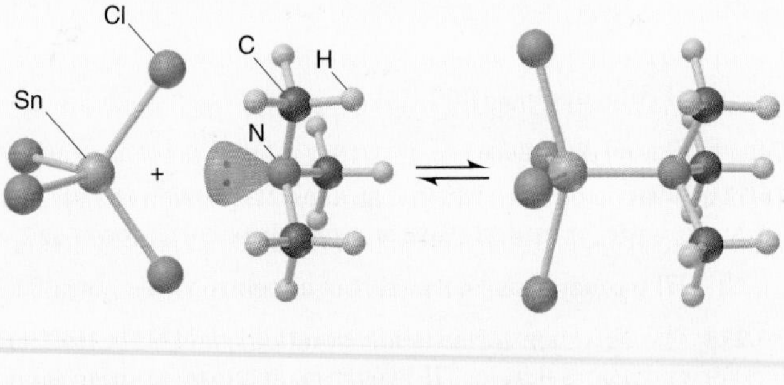

(a) Identify the Lewis acid and the Lewis base in the reaction.
(b) Give the *nl* designation of the sublevel of the central atom in the acid before it accepts the lone pair.

16.151 A chemist makes four successive 10-fold dilutions of 1.0×10^{-5} mol/L HCl. Calculate the pH of the original solution and the pH of each diluted solution (through 1.0×10^{-9} mol/L HCl).

16.152 Chlorobenzene, C_6H_5Cl, is a key intermediate in the manufacture of dyes and pesticides. It is made by the chlorination of benzene, catalyzed by $FeCl_3$, in this series of steps:

(1) $Cl_2 + FeCl_3 \rightleftharpoons FeCl_5$ (or $Cl^+FeCl_4^-$)

(2) $C_6H_6 + Cl^+FeCl_4^- \rightleftharpoons C_6H_6Cl^+ + FeCl_4^-$

(3) $C_6H_6Cl^+ \rightleftharpoons C_6H_5Cl + H^+$

(4) $H^+ + FeCl_4^- \rightleftharpoons HCl + FeCl_3$

(a) Which step(s) can be classified as a Lewis acid-base reaction?
(b) Identify the Lewis acids and bases in each step you named in part (a).

16.153 The beakers shown contain 0.300 L of aqueous solutions of a moderately weak acid, HY. Each particle represents 0.010 mol; the solvent molecules are omitted for clarity.
(a) The reaction in beaker A is at equilibrium. Calculate Q for the reactions in beakers B, C, and D to determine which, if any, is also at equilibrium.
(b) For any reaction that is not at equilibrium, in which direction does the reaction proceed?
(c) Does dilution affect the extent of dissociation of a weak acid? Explain.

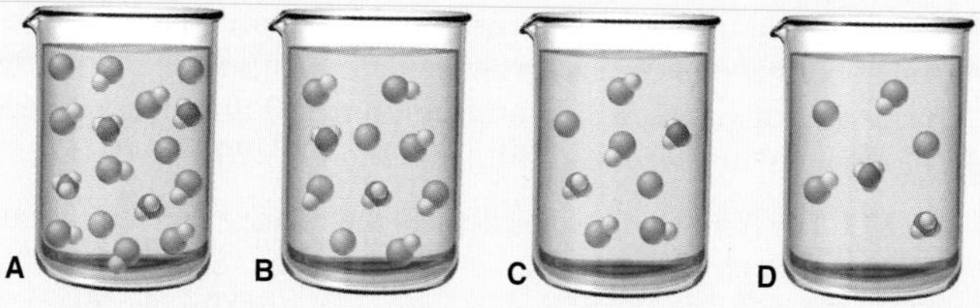

16.154 The strength of an acid or a base is related to its strength as an electrolyte.
(a) Is the electrical conductivity of 0.1 mol/L HCl higher than, lower than, or the same as the electrical conductivity of 0.1 mol/L CH_3COOH? Explain.
(b) Is the electrical conductivity of 1×10^{-7} mol/L HCl higher than, lower than, or the same as the electrical conductivity of 1×10^{-7} mol/L CH_3COOH? Explain.

16.155 Esters, RCOOR′, are formed by the reaction of a carboxylic acid, RCOOH, and an alcohol, R′OH, where R and R′ are hydrocarbon groups. Many esters are responsible for the odours of fruit and, thus, have important uses in the food and cosmetics industries. The first two steps in the mechanism of ester formation are given below:

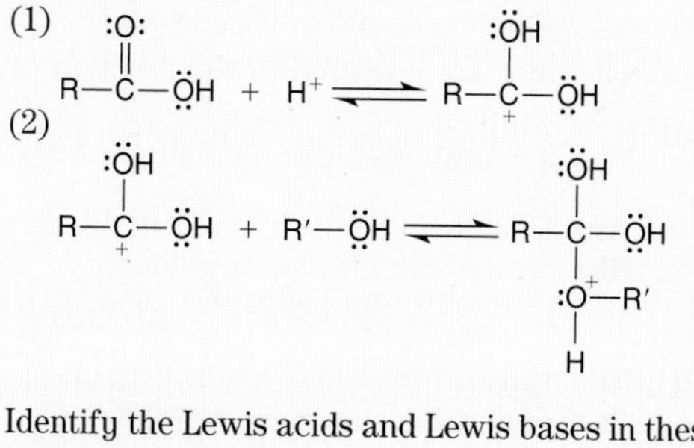

Identify the Lewis acids and Lewis bases in these steps.

16.156 Three beakers contain 100. mL of 0.10 mol/L HCl, $HClO_2$, and HClO, respectively.
(a) Find the pH of each solution.
(b) Describe quantitatively how to make the pH equal in the solutions through the addition of water only.

16.157 Human urine has a normal pH of 6.2. If a person eliminates an average of 1250. mL of urine per day, how many H^+ ions are eliminated per week?

16.158 Liquid ammonia autoionizes like water, as shown below:
$2NH_3(l) \longrightarrow NH_4^+(am) + NH_2^-(am)$, where (am) represents solvation by NH_3.
(a) Write the ion-product constant expression, K_{am}.
(b) What are the strongest acid and base that can exist in $NH_3(l)$?
(c) HNO_3 and HCOOH are levelled in $NH_3(l)$. Explain with equations.
(d) At the boiling point (−233°C), $K_{am} = 5.1 \times 10^{-27}$. Calculate $[NH_4^+]$ at this temperature.
(e) Pure sulfuric acid also autoionizes. Write the ion-product constant expression, K_{sulf}, and find the concentration of the conjugate base at 20°C ($K_{sulf} = 2.7 \times 10^{-4}$ at 20°C).

16.159 Thiamine hydrochloride ($C_{12}H_{18}ON_4SCl_2$) is a water-soluble form of thiamine (vitamin B_1; $K_a = 3.37 \times 10^{-7}$). What mass of the hydrochloride must be dissolved in 10.00 mL of water to give a pH of 3.50?

16.160 Tris(hydroxymethyl)aminomethane, known as TRIS or THAM, is a water-soluble base used to synthesize surfactants and pharmaceuticals, as an emulsifying agent in cosmetics, and in cleaning mixtures for textiles and leather. In biomedical research, solutions of TRIS are used to maintain nearly constant pH for the study of enzymes and other cellular components. Given that the pK_b is 5.91, calculate the pH of 0.075 mol/L TRIS.

16.161 When an Fe^{3+} salt is dissolved in water, the solution becomes acidic due to the formation of $Fe(H_2O)_5OH^{2+}$ and H_3O^+. The overall process involves both Lewis and Brønsted-Lowry acid-base reactions. Write the equations for the process.

16.162 What is the pH of a vinegar with 5.0% (w, v) ethanoic acid in water?

16.163 How would you differentiate between a strong and a weak monoprotic acid from the results of the following procedures?
(a) Electrical conductivity of an equimolar solution of each acid is measured.
(b) Equal concentrations of each acid are tested with pH paper.
(c) Zinc metal is added to solutions of equal concentration.

16.164 The catalytic efficiency of an enzyme is called its *activity* and refers to the rate at which it catalyzes the reaction. Most enzymes have optimum activity over a relatively narrow pH range, which is related to the pH of the local cellular fluid. The pH profiles of three digestive enzymes are shown:

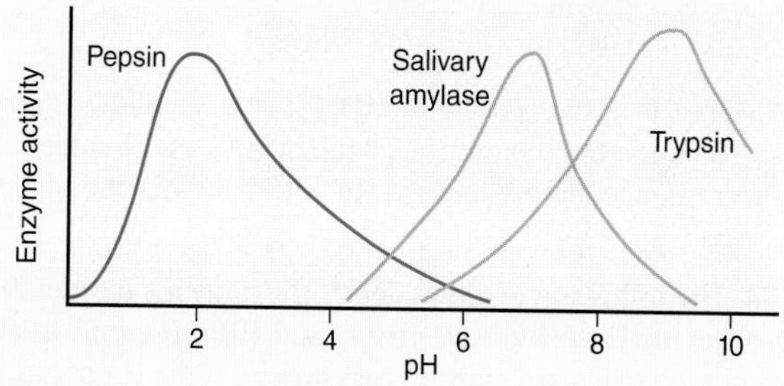

Salivary amylase begins the digestion of starches in the mouth and has optimum activity at a pH of 6.8. Pepsin begins protein digestion in the stomach and has optimum activity at a pH of 2.0. Trypsin, released in pancreatic juices, continues protein digestion in the small intestine and has optimum activity at a pH of 9.5. Calculate $[H_3O^+]$ in the local cellular fluid for each enzyme.

16.165 Ethanoic acid has a K_a of 1.8×10^{-5}, and ammonia has a K_b of 1.8×10^{-5}. Find $[H_3O^+]$, $[OH^-]$, pH, and pOH for (a) 0.240 mol/L ethanoic acid; (b) 0.240 mol/L ammonia.

16.166 The uses of sodium phosphate include clarifying crude sugar, manufacturing paper, removing boiler scale, and washing concrete. What is the pH of a solution that contains 33 g of Na_3PO_4 per litre? What is $[OH^-]$ of this solution?

16.167 The group 15 hydrides react with boron trihalides in a reversible Lewis acid-base reaction. When 0.15 mol of $PH_3BCl_3(s)$ is introduced into a 3.0 L container at a certain temperature, 8.4×10^{-3} mol of PH_3 is present at equilibrium:

$$PH_3BCl_3(s) \rightleftharpoons PH_3(g) + BCl_3(g)$$

(a) Find K_c for the reaction at this temperature.
(b) Draw a Lewis structure for the reactant.

16.168 A 1.000 mol/kg solution of chloroethanoic acid ($ClCH_2COOH$) freezes at −1.93°C. Find the K_a of chloroethanoic acid. (Assume that the molarities equal the molalities.)

16.169 Sodium stearate ($C_{17}H_{35}COONa$) is a major component of bar soap. The K_a of stearic acid is 1.3×10^{-5}. What is the pH of 10.0 mL of a solution containing 0.42 g of sodium stearate?

16.170 Calcium propionate [$Ca(CH_3CH_2COO)_2$; calcium propanoate] is a mould inhibitor that is used in food, tobacco, and pharmaceuticals.
(a) Use balanced equations to show whether aqueous calcium propionate is acidic, basic, or neutral.
(b) Use Appendix C to find the resulting pH when 8.75 g of $Ca(CH_3CH_2COO)_2$ dissolves in enough water to give 0.500 L of solution.

16.171 A site in Alberta receives a total annual deposition of 2.688 g/m^2 of sulfate from fertilizer and acid rain. The ratio by mass of ammonium sulfate/ammonium bisulfate/sulfuric acid is 3.0/5.5/1.0.
(a) How much acid, expressed as kilograms of sulfuric acid, is deposited over an area of 10. km^2?
(b) What mass of $CaCO_3$ is needed to neutralize this acid?
(c) If the area of an unpolluted lake, 3 m deep, is 10. km^2 and there is no loss of acid, what pH would be attained in the year? (Assume that the volume is constant.)

16.172 (a) If K_w is 1.139×10^{-15} at 0°C and 5.474×10^{-14} at 50°C, find the [H_3O^+] and pH of water at 0°C and 50°C.
(b) The autoionization constant for heavy water (deuterium oxide, D_2O) is 3.64×10^{-16} at 0°C and 7.89×10^{-15} at 50°C. Find [D_3O^+] and pD of heavy water at 0°C and 50°C.
(c) Suggest a reason for these differences.

16.173 HX ($\mathcal{M}$ = 150. g/mol) and HY ($\mathcal{M}$ = 50.0 g/mol) are weak acids. A solution that contains 12.0 g/L of HX has the same pH as a solution that contains 6.00 g/L of HY. Which is the stronger acid? Why?

16.174 The following beakers depict the aqueous dissociations of weak acids HA (*blue and green*) and HB (*blue and yellow*). (Solvent molecules are omitted for clarity.) If the HA solution is 0.50 L, and the HB solution is 0.25 L, and each particle represents 0.010 mol, find the K_a of each acid. Which acid, if either, is stronger?

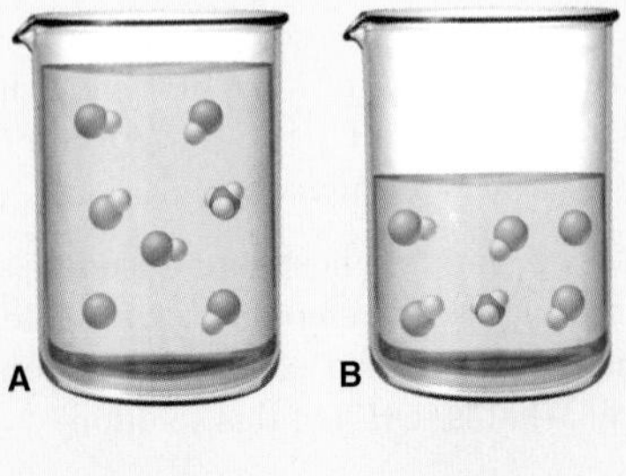

16.175 In his acid-base studies, Arrhenius discovered an important fact involving reactions like the following:

$$KOH(aq) + HNO_3(aq) \longrightarrow ?$$

$$NaOH(aq) + HCl(aq) \longrightarrow ?$$

(a) Complete the reactions, and use the data for the individual ions in Appendix B to calculate each $\Delta_r H°$.
(b) Explain your results, and use them to predict $\Delta_r H°$ for this reaction:

$$KOH(aq) + HCl(aq) \longrightarrow ?$$

16.176 Putrescine [$NH_2(CH_2)_4NH_2$], found in rotting animal tissue, is now known to be in all cells and to be essential for normal and abnormal (cancerous) growth. It also plays a key role in the formation of GABA, a neurotransmitter. A 0.10 mol/L aqueous solution of putrescine has [OH^-] = 2.1×10^{-3}. What is the K_b?

16.177 The following beaker depicts the relative concentrations of H_3O^+ (*purple*) and OH^- (*green*) in an aqueous solution at 25°C. (Counterions and solvent molecules are omitted for clarity.)
(a) Calculate the pH.
(b) How many H_3O^+ ions would you have to draw for every OH^- ion to depict a solution with a pH of 4?

16.178 Polymers are not very soluble in water, but their solubility increases if they have charged groups.
(a) Casein, a milk protein, contains many $—COO^-$ groups on its side chains. How does the solubility of casein vary with pH?
(b) Histones are proteins that are essential to the function of DNA. They are weakly basic due to the presence of side chains with $—NH_2$ and $=NH$ groups. How does the solubility of a histone vary with pH?

16.179 Hemoglobin (Hb) transports oxygen in the blood:

$$HbH^+(aq) + O_2(aq) + H_2O(l) \longrightarrow HbO_2(aq) + H_3O^+(aq)$$

[H_3O^+] in the blood is held nearly constant, at 4×10^{-8} mol/L.
(a) How does the equilibrium position change in the lungs?
(b) How does it change in O_2-deficient cells?
(c) Excessive vomiting may lead to metabolic *alkalosis*, in which [H_3O^+] in the blood *decreases*. How does this condition affect the ability of Hb to transport O_2?
(d) Diabetes mellitus may lead to metabolic *acidosis*, in which [H_3O^+] in the blood *increases*. How does this condition affect the ability of Hb to transport O_2?

16.180 Nitrogen is discharged from wastewater treatment facilities into rivers and streams, usually as NH_3 and NH_4^+:

$$NH_3(aq) + H_2O(l) \rightleftharpoons NH_4^+(aq) + OH^-(aq) \quad K_b = 1.76 \times 10^{-5}$$

One strategy for removing nitrogen is to raise the pH and "strip" the NH_3 from solution by bubbling air through the water.
(a) At a pH of 7.00, what fraction of the total nitrogen in solution is NH_3, defined as $\frac{[NH_3]}{[NH_3] + [NH_4^+]}$?
(b) What is the fraction at a pH of 10.00?
(c) Explain the basis of ammonia stripping.

16.181 A solution of propanoic acid (CH_3CH_2COOH), made by dissolving 7.500 g in sufficient water to make 100.0 mL, has a freezing point of $-1.890°C$.
(a) Calculate the molarity of the solution.
(b) Calculate the molarity of the propanoate ion. (Assume that the molarity of the solution equals the molality.)
(c) Calculate the percent dissociation of propanoic acid.

16.182 The antimalarial properties of quinine ($C_{20}H_{24}N_2O_2$) saved thousands of lives during the construction of the Panama Canal. Quinine is a classic example of the medicinal wealth of tropical forests. Both N atoms are basic, but the N (*blue*) of the 3° amine group is far more basic ($pK_b = 5.1$) than the N within the aromatic ring system ($pK_b = 9.7$).

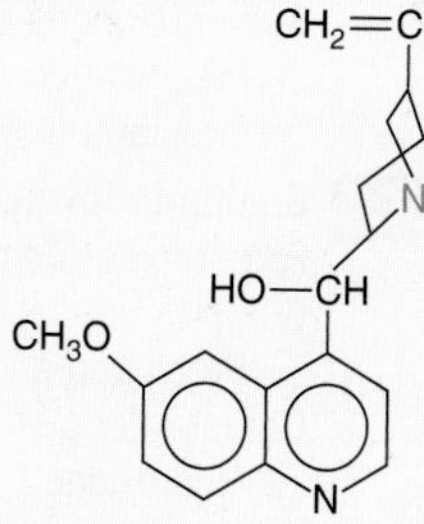

(a) A saturated solution of quinine in water is only 1.6×10^{-3} mol/L. What is the pH of the solution?
(b) Show that the aromatic N contributes negligibly to the pH of the solution.
(c) Because of its low solubility, quinine is given as the salt quinine hydrochloride ($C_{20}H_{24}N_2O_2 \cdot HCl$), which has 120 times the solubility of quinine. What is the pH of 0.33 mol/L quinine hydrochloride?
(d) An antimalarial concentration in water is 1.5% quinine hydrochloride by mass ($d = 1.0$ g/mL). What is the pH?

16.183 Drinking water is often disinfected with Cl_2, which hydrolyzes to form HClO, a weak acid but a powerful disinfectant:

$$Cl_2(aq) + 2H_2O(l) \longrightarrow HClO(aq) + H_3O^+(aq) + Cl^-(aq)$$

The fraction of HClO in solution is defined as follows:

$$\frac{[HClO]}{[HClO] + [ClO^-]}$$

(a) What is the fraction of HClO at a pH of 7.00 (K_a of HClO = 2.9×10^{-8})?
(b) What is the fraction at a pH of 10.00?

16.184 The following scenes represent three weak acids, HA (where A = X, Y, or Z), dissolved in water. (H_2O is not shown.)

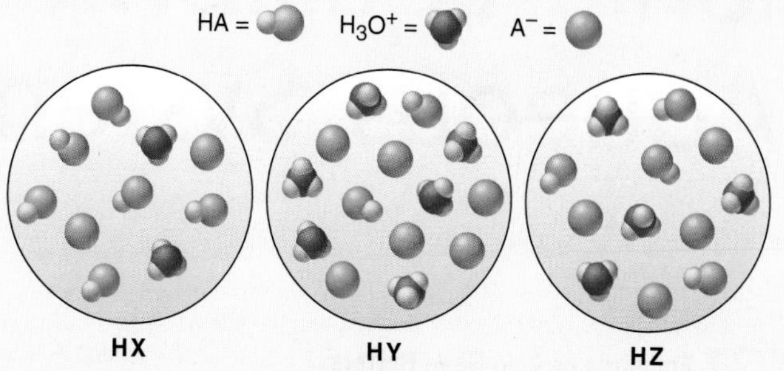

(a) Rank the acids in order of increasing K_a.
(b) Rank the acids in order of increasing pK_a.
(c) Rank the conjugate bases in order of increasing pK_b.
(d) What is the percent dissociation of HX?
(e) If equimolar amounts of the sodium salts of the acids (NaX, NaY, and NaZ) were dissolved in water, which solution would have the highest pOH? Which solution would have the lowest pH?

16.185 What concentration of ethanoic acid would have the same pH as a solution of 0.0035 mol/L HCl?

16.186 What concentration of ammonia would have the same pH as a solution of 0.0079 mol/L NaOH?

16.187 In what volume of water would 2.86 mg of CH_3COOK (potassium ethanoate) have to be dissolved to have a pH of 7.235?

16.188 If water is added to 0.150 mol/L ammonia, will the pH increase or decrease? Explain. What volume of water would have to be added to 500.0 mL of 0.150 mol/L $NH_3(aq)$ to change its pH by 1?

Ionic Equilibria in Aqueous Systems

IN THIS CHAPTER . . . We define and quantify three aqueous ionic equilibrium systems: acid-base buffers, slightly soluble salts, and complex ions. By the end of this chapter, you should be able to

- Explain how buffers work using the common-ion effect, describe how buffers are prepared, and discuss why buffers are important
- Describe several types of acid-base titrations and the role of buffers in these titrations
- Explain why slightly soluble salts have their own equilibrium constant and how a common ion and pH influence the solubility of these compounds
- Apply concepts of aqueous ionic equilibria to understand how a cave forms
- Describe how complex ions form and change from one type to another

17

Concepts and Skills to Review before Studying This Chapter

- Effect of concentration on equilibrium position (Section 15.6)
- Conjugate acid-base pairs (Section 16.1)
- Calculations for weak-acid and weak-base equilibria (Sections 16.3 and 16.4)
- Acid-base properties of salt solutions (Section 16.6)
- Lewis acids and bases (Section 16.8)

Whether it is part of the food we eat, the medications we use, or the environment in which we live, buffers play a significant role in our lives. Soft drinks, yogourt, and even beer are buffered to maintain their freshness and properties. Many foods prepared today are buffered to ensure that they have a particular taste or do not spoil. Many medications are buffered to ensure that the medication is in the proper form to be effective, to enhance its effectiveness, or to prevent damage to other tissue. Our environment and our bodies contain buffers to prevent the damage that can be inflicted by acids and bases. In this chapter, we will learn the definition of buffers, their properties, and the different ways they can be prepared. We will also look at quantitative reactions between acids and bases and see how a knowledge of buffers helps us treat these reactions. We will learn about the equilibria that exist between solids and solutions and how these equilibria are simply an extension of the other equilibria we have studied in the previous two chapters. Finally, we will learn about complexes and their properties in aqueous systems.

Expanding our view of the universal nature of equilibrium, we now examine aqueous ionic systems. The unique formations in limestone caves and the vast expanses of oceanic coral reefs arise from subtle shifts in carbonate solubility equilibria. Similar interactions prevent the acidification of lakes. Organisms survive by maintaining cellular pH within narrow limits through complex carbonate and phosphate equilibria. In soils, equilibria involving clays control the availability of ionic nutrients for plants. In industrial processes, the same principles govern the softening of water and the purification of products by the precipitation of unwanted ions. In cooking, these principles even explain how the weak acids in wine and vinegar influence the delicate taste of a fine sauce.

17.1 Equilibria of Acid-Base Buffers

Why do some lakes become acidic when showered by acid rain, while others remain unaffected? How does blood maintain a constant pH in contact with countless cellular acid-base reactions? How can a chemist sustain a nearly constant $[H_3O^+]$ in reactions that consume or produce H_3O^+ or OH^-? The answer in each case depends on the action of a buffer, and in this section we discuss how buffers work and how to prepare them.

What a Buffer Is and How It Works: The Common-Ion Effect

In everyday language, a buffer is something that lessens the impact of an external force. An **acid-base buffer** is a solution that *lessens the impact on pH from the addition of acid or base.* If a small amount of H_3O^+ or OH^- is added to an unbuffered solution, the pH changes by several units; thus, $[H^+]$ changes by *several orders of magnitude* (Figure 17.1). (As in Chapter 16, $H^+(aq)$ and $H_3O^+(aq)$ will be used interchangeably throughout this chapter.)

The same addition of a strong acid or a strong base to a buffered solution causes only a minor change in the pH (Figure 17.2). To withstand these additions, a buffer must have an acidic component that reacts with the added OH^- *and* a basic component that reacts with the added H_3O^+. However, these components cannot be just any acid and base because they would neutralize each other. Most often, *the*

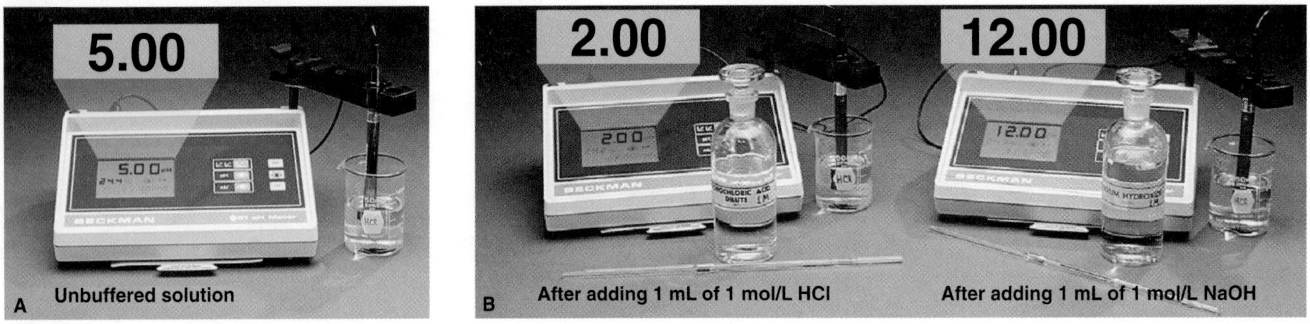

FIGURE 17.1 The effect of adding acid or base to an unbuffered solution. A. A 100 mL sample of dilute HCl is adjusted to pH 5.00. **B.** Adding 1 mL of strong acid (*left*) or strong base (*right*) changes the pH by several units.

components of a buffer are a conjugate acid-base pair (weak acid and conjugate base or weak base and conjugate acid). The buffer in Figure 17.2, for example, is a mixture of ethanoic (acetic) acid (CH_3COOH) and ethanoate (acetate) ion (CH_3COO^-).

Presence of a Common Ion Buffers work through the **common-ion effect**. When you dissolve ethanoic (acetic) acid in water, the acid dissociates slightly:

$$CH_3COOH(aq) + H_2O(l) \rightleftharpoons H_3O^+(aq) + CH_3COO^-(aq)$$

What happens if you now introduce the ethanoate (acetate) ion by adding the soluble salt sodium ethanoate (acetate)? From Le Châtelier's principle (Section 15.6), we know that adding the CH_3COO^- ion will shift the equilibrium position to the left; thus, $[H_3O^+]$ decreases, in effect lowering the extent of acid dissociation:

$$CH_3COOH(aq) + H_2O(l) \leftharpoondown\!\!\!\rightharpoonup H_3O^+(aq) + CH_3COO^-(aq;\text{ added})$$

We get the same result when we add ethanoic (acetic) acid to a sodium ethanoate (acetate) solution instead of water. The ethanoate (acetate) ion that is already present suppresses the acid from dissociating as much as it does in water, thus keeping the $[H_3O^+]$ lower. In either case, the effect is less acid dissociation. The ethanoate (acetate) ion is called *the common ion* because it is "common" to both the ethanoic (acetic) acid and sodium acetate solutions. *The common-ion effect occurs when a given ion is added to an equilibrium mixture that already contains this ion, and the position of equilibrium shifts away from forming it.*

Table 17.1 shows that the percent dissociation (and the $[H_3O^+]$) of an ethanoic (acetic) acid solution decreases as the concentration of ethanoate ion [supplied by dissolving sodium ethanoate (acetate)] increases. Thus, the *common ion, A^-, suppresses the dissociation of HA*, which makes the solution less acidic (higher pH).

Relative Concentrations of Buffer Components A buffer works because *large amounts of the acidic component (HA) and basic component (A^-) consume small amounts of the added OH^- and H_3O^+, respectively.* Consider what happens in a solution containing high $[CH_3COOH]$ and high $[CH_3COO^-]$ when we add small amounts of a strong acid or base. The equilibrium expression for HA dissociation is

$$K_a = \frac{[CH_3COO^-][H_3O^+]}{[CH_3COOH]}$$

FIGURE 17.2 The effect of adding acid or base to a buffered solution. A. A 100 mL sample of an ethanoate buffer, made by mixing 1 mol/L CH_3COOH with 1 mol/L CH_3COONa, is adjusted to pH 5.00. **B.** Adding 1 mL of strong acid (*left*) or strong base (*right*) changes the pH negligibly.

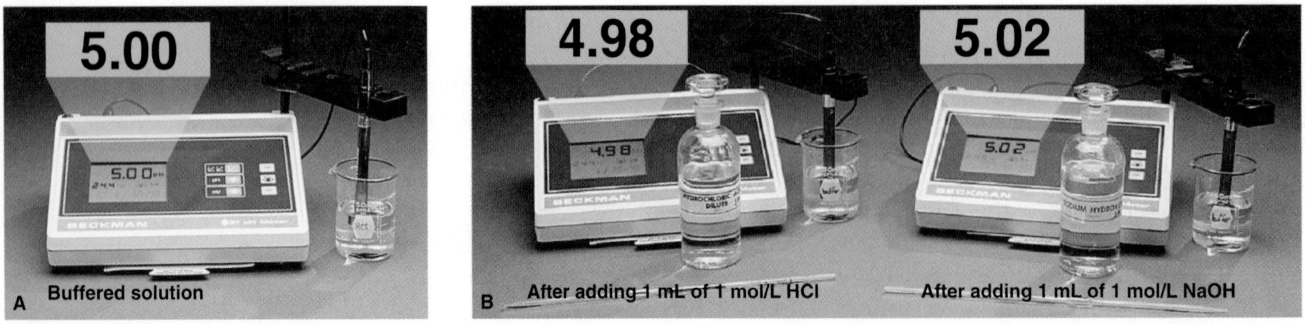

TABLE 17.1 The Effect of Added Ethanoate (Acetate) Ion on the Dissociation of Ethanoic (Acetic) acid

$[CH_3COOH]_{init}$	$[CH_3COO^-]_{added}$	% Dissociation*	H_3O^+	pH
0.10	0.00	1.3	1.3×10^{-3}	2.89
0.10	0.050	0.036	3.6×10^{-5}	4.44
0.10	0.10	0.018	1.8×10^{-5}	4.74
0.10	0.15	0.012	1.2×10^{-5}	4.92

*% Dissociation $= \frac{[CH_3COOH]_{dissoc}}{[CH_3COOH]_{init}} \times 100$.

Solving for $[H_3O^+]$ gives

$$[H_3O^+] = K_a \times \frac{[CH_3COOH]}{[CH_3COO^-]}$$

Since K_a is constant, *the $[H_3O^+]$ of the solution depends on the buffer-component concentration ratio,* $\frac{[CH_3COOH]}{[CH_3COO^-]}$:

- If the ratio $\frac{[HA]}{[A^-]}$ goes up, $[H_3O^+]$ goes up.
- If the ratio $\frac{[HA]}{[A^-]}$ goes down, $[H_3O^+]$ goes down.

Let us track this ratio as we add a strong acid or a strong base to a buffer in which $[HA] = [A^-]$ (Figure 17.3, *middle*):

1. *Strong acid.* When we add a small amount of strong acid, the H_3O^+ ions react with an *equal (stoichiometric) amount* of ethanoate (acetate) ion from the buffer to form more ethanoic (acetic) acid:

$$H_3O^+(aq;\ \text{added}) + CH_3COO^-(aq;\ \text{from buffer}) \longrightarrow CH_3COOH(aq) + H_2O(l)$$

As a result, $[CH_3COO^-]$ goes down by that small amount and $[CH_3COOH]$ goes up by that amount, which increases the buffer-component concentration ratio (Figure 17.3, *to the left*). The $[H_3O^+]$ increases *very* slightly.

2. *Strong base.* The addition of a small amount of strong base produces the opposite result. The OH^- ions react with an *equal (stoichiometric) amount* of CH_3COOH from the buffer to form that much more CH_3COO^- (Figure 17.3, *to the right*):

$$CH_3COOH(aq;\ \text{from buffer}) + OH^-(aq;\ \text{added}) \longrightarrow CH_3COO^-(aq) + H_2O(l)$$

This time, the buffer-component concentration ratio decreases, causing $[H_3O^+]$ to decrease *very* slightly.

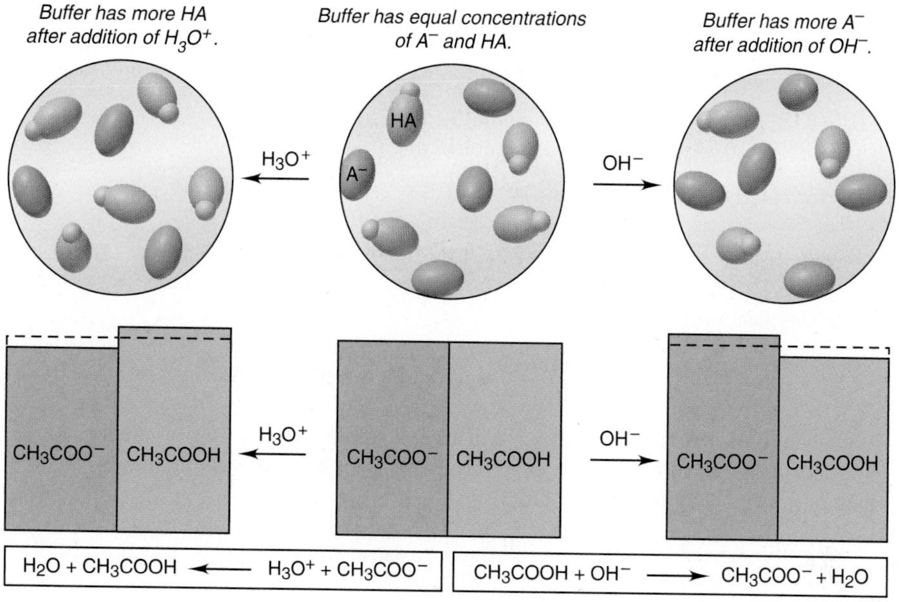

Buffer has more HA after addition of H3O+.
Buffer has equal concentrations of A− and HA.
Buffer has more A− after addition of OH−.
HA
A−
H3O+
OH−
CH3COO−
CH3COOH
H2O + CH3COOH ← H3O+ + CH3COO−
CH3COOH + OH− → CH3COO− + H2O

FIGURE 17.3 How a buffer works. The relative concentrations of the buffer components, ethanoic (acetic) acid (CH_3COOH, *pink*) and ethanoate (acetate) ion (CH_3COO^-, *purple*), are indicated by the heights of the bars.

Thus, the buffer components consume *nearly all* the added H_3O^+ or OH^-. To reiterate, as long as the amount of added H_3O^+ or OH^- is small compared to the amounts of the buffer components, *the conversion of one component into the other produces a small change in the buffer-component concentration ratio and, consequently, a small change in [H_3O^+] and pH.* Sample Problem 17.1 demonstrates how small these pH changes typically are. Note that parts (a) and (b) of the problem combine a stoichiometry portion, like the problems in Chapter 3, and a weak-acid dissociation portion, like those in Chapter 16.

Sample Problem 17.1 Calculating the Effect of Added H_3O^+ or OH^- on Buffer pH

Problem Calculate the following pH values (K_a of $CH_3COOH = 1.8 \times 10^5$; assume that the additions cause negligible volume changes):

(a) pH of a buffer solution consisting of 0.50 mol/L CH_3COOH and 0.50 mol/L CH_3COONa

(b) pH after adding 0.020 mol of solid NaOH to 1.0 L of the buffer solution in part (a)

(c) pH after adding 0.020 mol of HCl to 1.0 L of the buffer solution in part (a)

Plan For each part, we know or can find $[CH_3COOH]_{init}$ and $[CH_3COO^-]_{init}$. We know the K_a of CH_3COOH (1.8×10^{-5}) and need to find [H_3O^+] at equilibrium and convert it to pH. **(a)** We use the given concentrations of buffer components (each 0.50 mol/L) as the initial values. As in earlier problems, we assume that x, the [CH_3COOH] that dissociates, which equals [H_3O^+], is so small relative to $[CH_3COOH]_{init}$ that it can be neglected. We set up an ICE table, solve for x, and check the assumption. For **(b)** and **(c)**, we assume that the added OH^- or H_3O^+ reacts completely with the buffer components to yield new $[CH_3COOH]_{init}$ and $[CH_3COO^-]_{init}$, and then the acid dissociates to an unknown extent. We set up two reaction tables. The first table summarizes the stoichiometry of adding strong base (0.020 mol) or acid (0.020 mol). The second table summarizes the dissociation of the new $[CH_3COOH]_{init}$, and we proceed as in part (a) to find the new [H_3O^+].

Solution **(a)** The original pH is [H^+] in the original buffer.

We start by setting up an ICE table with $x = [CH_3COOH]_{dissoc} = [H_3O^+]$. (As in Chapter 16, we assume that [H_3O^+] from H_2O is very small relative to the [H_3O^+] in solution that comes from the presence of the acid, so it can be neglected.)

Concentration (mol/L)	$CH_3COOH(aq)$	+ $H_2O(l)$	$\rightleftharpoons$ $CH_3COO^-(aq)$	+ $H_3O^+(aq)$
Initial	0.50	—	0.50	0
Change	$-x$	—	$+x$	$+x$
Equilibrium	$0.50 - x$	—	$0.50 + x$	x

Then we need to find the equilibrium [CH_3COOH] and [CH_3COO^-]. With K_a being small, x is small relative to the initial concentration of 0.50 mol/L, so we can assume that

$$[CH_3COOH] = 0.50\text{ mol/L} - x \approx 0.50\text{ mol/L} \quad \text{and}$$
$$[CH_3COO^-] = 0.50\text{ mol/L} + x \approx 0.50\text{ mol/L}$$

Solve for x ([H_3O^+] at equilibrium):

$$x = [H_3O^+] = K_a \times \frac{[CH_3COOH]}{[CH_3COO^-]} \approx (1.8 \times 10^{-5}) \times \frac{0.50}{0.50} = 1.8 \times 10^{-5}\text{ mol/L}$$

Check the assumption:

$$\frac{1.8 \times 10^{-5}\text{ mol/L}}{0.50\text{ mol/L}} \times 100 = 3.6 \times 10^{-3}\% < 5\%$$

The assumption is justified, and we will use the same assumption in parts (b) and (c). Also, according to the other criterion (see Discussion following Sample Problem 15.10 or within Sample Problem 16.8),

$$\frac{[HA]_{init}}{K_a} = \frac{0.50}{1.8 \times 10^{-5}} = 2.8 \times 10^4 > 400$$

Calculate pH:

$$pH = -\log [H_3O^+] = -\log (1.8 \times 10^{-5}) = 4.74$$

(b) We need to find the pH after adding base (0.020 mol of NaOH to 1.0 L of buffer). Find $[OH^-]_{added}$:

$$[OH^-]_{added} = \frac{0.020 \text{ mol } OH^-}{1.0 \text{ L soln}} = 0.020 \text{ mol/L } OH^-$$

Set up an ICE table for the *stoichiometry* of adding OH^- to CH_3COOH:

Concentration (mol/L)	$CH_3COOH(aq)$ +	$OH^-(aq)$ ⟶	$CH_3COO^-(aq)$ +	$H_2O(l)$
Initial	0.50	0.020	0.50	—
Change	−0.020	−0.020	+0.020	—
Final	0.48	0	0.52	—

Using these new initial concentrations, we set up an ICE table for the *acid dissociation*. As in part (a), we assume that $x = [CH_3COOH]_{dissoc} = [H_3O^+]$.

Concentration (mol/L)	$CH_3COOH(aq)$ +	$H_2O(l)$ ⇌	$CH_3COO^-(aq)$ +	$H_3O^+(aq)$
Initial	0.48	—	0.52	0
Change	−x	—	+x	+x
Equilibrium	0.48 − x	—	0.52 + x	x

Assume that x is small and solve for x:

$[CH_3COOH] = 0.48 \text{ mol/L} - x \approx 0.48 \text{ mol/L}$ and
$[CH_3COO^-] = 0.52 \text{ mol/L} + x \approx 0.52 \text{ mol/L}$

$$x = [H_3O^+] = K_a \times \frac{[CH_3COOH]}{[CH_3COO^-]} \approx (1.8 \times 10^{-5}) \times \frac{0.48}{0.52} = 1.7 \times 10^{-5} \text{ mol/L}$$

Calculate the pH:

$$pH = -\log [H_3O^+] = -\log (1.7 \times 10^{-5}) = 4.77$$

The addition of a strong base increased the concentration of the basic buffer component at the expense of the acidic buffer component. Notice, especially, that the pH *increased only slightly*, from 4.74 to 4.77.

(c) We need to find the pH after adding acid (0.020 mol of HCl to 1.0 L of buffer). Find $[H_3O^+]_{added}$:

$$[H_3O^+]_{added} = \frac{0.020 \text{ mol } H_3O^+}{1.0 \text{ L soln}} = 0.020 \text{ mol/L } H_3O^+$$

Now we proceed as in part (b), by first setting up an ICE table for the *stoichiometry* of adding H_3O^+ to CH_3COO^-:

Concentration (mol/L)	$CH_3COO^-(aq)$ +	$H_3O^+(aq)$ ⟶	$CH_3COOH(aq)$ +	$H_2O(l)$
Initial	0.50	0.020	0.50	—
Change	−0.020	−0.020	+0.020	—
Final	0.48	0	0.52	—

Then we set up an ICE table for the *acid dissociation*, with $x = [CH_3COOH]_{dissoc} = [H_3O^+]$:

Concentration (mol/L)	$CH_3COOH(aq)$ +	$H_2O(l)$ ⇌	$CH_3COO^-(aq)$ +	$H_3O^+(aq)$
Initial	0.52	—	0.48	0
Change	−x	—	+x	+x
Equilibrium	0.52 − x	—	0.48 + x	x

Assume that x is small and solve for x:

$[CH_3COOH] = 0.52 \text{ mol/L} - x \approx 0.52 \text{ mol/L}$ and
$[CH_3COO^-] = 0.48\text{mol/L} + x \approx 0.48 \text{ mol/L}$

$$x = [H_3O^+] = K_a \times \frac{[CH_3COOH]}{[CH_3COO^-]} \approx (1.8 \times 10^{-5}) \times \frac{0.52}{0.48} = 2.0 \times 10^{-5} \text{ mol/L}$$

Calculate the pH:

$$pH = -\log [H_3O^+] = -\log (2.0 \times 10^{-5}) = 4.70$$

The addition of a strong acid increased the concentration of the acidic buffer component at the expense of the basic buffer component and *lowered* the pH only slightly, from 4.74 to 4.70.

Check The changes in $[CH_3COOH]$ and $[CH_3COO^-]$ occur in opposite directions in parts (b) and (c), which makes sense. The additions were equal amounts, so the pH increase in (b) should equal the pH decrease in (c), within rounding.

Comment In part (a), we justified our assumption that x can be neglected. Therefore, in parts (b) and (c), we could have used the "Final" values from the last line of the stoichiometry reaction tables directly for the ratio of buffer components; that would have allowed us to dispense with an ICE table for the acid dissociation. In subsequent problems in the chapter, we will follow this more straightforward approach.

Follow-Up Problem 17.1 Calculate the pH of a buffer consisting of 0.50 mol/L HF and 0.45 mol/L F^- **(a)** before and **(b)** after the addition of 0.40 g of NaOH to 1.0 L of the buffer (K_a of HF = 6.8×10^{-4}).
(See Brief Solutions to Follow-Up Problems at the end of the chapter.)

The Henderson-Hasselbalch Equation

For any weak acid, HA, the dissociation equation and K_a expression are

$$HA(aq) + H_2O(l) \rightleftharpoons H_3O^+(aq) + A^-(aq)$$

$$K_a = \frac{[H_3O^+][A^-]}{[HA]}$$

The key variable that determines $[H_3O^+]$ is the concentration *ratio* of acid species to base species, so, as before, rearranging to isolate $[H_3O^+]$ gives

$$[H_3O^+] = K_a \times \frac{[HA]}{[A^-]}$$

Taking the negative common (base 10) logarithm of both sides gives

$$-\log [H_3O^+] = -\log K_a - \log \left(\frac{[HA]}{[A^-]}\right)$$

Using our definitions, this reduces to

$$pH = pK_a - \log \left(\frac{[HA]}{[A^-]}\right)$$

Then, because of the nature of logarithms, when we invert the buffer-component concentration ratio, the sign of the logarithm changes to give

$$pH = pK_a + \log \left(\frac{[A^-]}{[HA]}\right)$$

Generalizing the previous equation for any conjugate acid-base pair gives the **Henderson-Hasselbalch equation**:

$$pH = pK_a + \log \left(\frac{[\text{base}]}{[\text{acid}]}\right) \qquad (17.1)$$

This relationship allows us to solve directly for pH instead of having to calculate $[H^+]$ first. For example, using the Henderson-Hasselbalch equation in part (b) of Sample Problem 17.1, the pH of the buffer after the addition of NaOH is

$$pH = pK_a + \log \left(\frac{[CH_3COO^-]}{[CH_3COOH]}\right) = 4.74 + \log \left(\frac{0.52}{0.48}\right) = 4.77$$

Buffer Capacity and Buffer Range

Let us consider two key aspects of a buffer: its capacity and the closely related range.

Buffer Capacity **Buffer capacity** is a *measure of the "strength" of the buffer, its ability to maintain the pH following the addition of a strong acid or base.* Capacity depends ultimately on the concentrations of the buffer components, both the absolute and relative concentrations:

1. In terms of *absolute* concentrations, *the more concentrated the buffer components, the greater the capacity.* Thus, for a given amount of added H^+ or OH^-, the pH of a higher-capacity buffer changes less than the pH of a lower-capacity buffer (Figure 17.4). Note that *buffer pH is independent of buffer capacity.* A buffer made of equal volumes of 1.0 mol/L CH_3COOH and 1.0 mol/L CH_3COO^- has the same pH (4.74) as a buffer made of equal volumes of 0.10 mol/L CH_3COOH and 0.10 mol/L CH_3COO^-, but the more concentrated buffer has a greater capacity.

2. In terms of *relative* concentrations, *the closer the concentrations of the components are to each other, the greater the capacity.* As a buffer functions, the concentration of one component increases relative to the other. Because the concentration ratio determines the pH, the less the ratio changes, the less the pH changes. Let us compare the percent change in the component concentration ratio for a buffer at two different initial ratios:

- Add 0.010 mol of OH^- to 1.00 L of buffer with initial concentrations [HA] = [A^-] = 1.000 mol/L. [A^-] becomes 1.010 mol/L, and [HA] becomes 0.990 mol/L:

$$\frac{[A^-]_{init}}{[HA]_{init}} = \frac{1.000 \text{ mol/L}}{1.000 \text{ mol/L}} = 1.000 \qquad \frac{[A^-]_{final}}{[HA]_{final}} = \frac{1.010 \text{ mol/L}}{0.990 \text{ mol/L}} = 1.02$$

$$\text{Percent change} = \frac{1.02 - 1.000}{1.000} \times 100 = 2\%$$

- Add 0.010 mol of OH^- to 1.00 L of buffer with initial concentrations [HA] = 0.250 mol/L and [A^-] = 1.750 mol/L. [A^-] becomes 1.760 mol/L, and [HA] becomes 0.240 mol/L:

$$\frac{[A^-]_{init}}{[HA]_{init}} = \frac{1.750 \text{ mol/L}}{0.250 \text{ mol/L}} = 7.00 \qquad \frac{[A^-]_{final}}{[HA]_{final}} = \frac{1.760 \text{ mol/L}}{0.240 \text{ mol/L}} = 7.33$$

$$\text{Percent change} = \frac{7.33 - 7.00}{7.00} \times 100 = 4.7\%$$

Note that the change in the concentration ratio is more than twice as large when the initial component concentrations are different than when they are the same. Thus, *a buffer has the highest capacity when the component concentrations are equal.* In the Henderson-Hasselbalch equation, when [A^-] = [HA], their ratio is 1, the log term is 0, and so pH = pK_a:

$$pH = pK_a + \log\left(\frac{[A^-]}{[HA]}\right) = pK_a + \log 1 = pK_a + 0 = pK_a$$

Therefore, *a buffer whose pH is equal to or near the pK_a of its acid component has the highest capacity* for a given concentration.

Buffer Range **Buffer range** is the *pH range over which the buffer is effective.* It is also related to the *relative* buffer-component concentrations. The farther the concentration ratio is from 1, the less effective the buffer is (the lower the buffer capacity). In practice, if the [A^-]/[HA] ratio is greater than 10 or less than 0.1—that is, if one component concentration is more than 10 times the other—the buffering action is poor. Since log 10 = +1 and log 0.1 = −1, *buffers have a usable range within ±1 of the pK_a of the acid component*:

$$pH = pK_a + \log\left(\frac{10}{1}\right) = pK_a + 1 \quad \text{and} \quad pH = pK_a + \log\left(\frac{1}{10}\right) = pK_a - 1$$

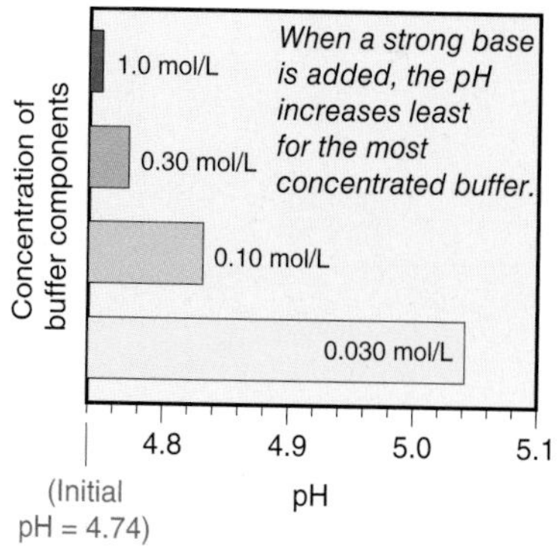

FIGURE 17.4 The relation between buffer capacity and pH change. The bars indicate the final pH values, after a strong base was added, for four CH_3COOH/CH_3COO^- buffers with the same initial pH (4.74) and different concentrations of components.

Sample Problem 17.2 Using Molecular Scenes to Examine Buffers

Problem The following molecular scenes represent samples of four HA/A^- buffers. (HA is *blue and green*, A^- is *green*, and the other ions and water are not shown.)

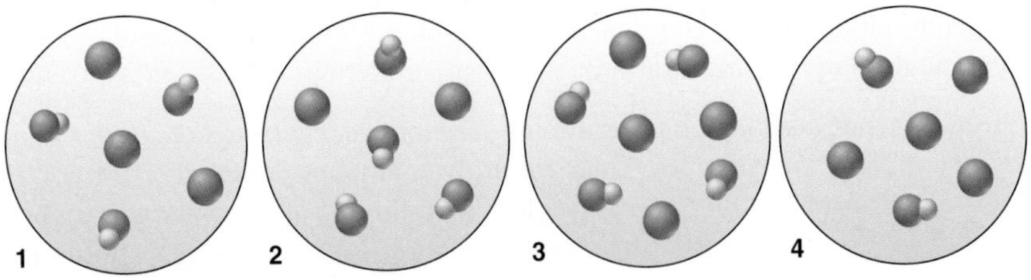

(a) Which buffer has the highest pH?

(b) Which buffer has the greatest capacity?

(c) To convert sample 1 to sample 2, should a small amount of concentrated strong acid or a small amount of concentrated strong base be added (assuming no volume change)?

Plan The molecular scenes show varying numbers of weak acid molecules (HA) and the conjugate base (A^-). Because the volumes are equal, the scenes represent molarities as well as numbers. **(a)** As the pH rises, more HA loses its H^+ and becomes A^-, so the [A^-]/[HA] ratio will increase. We examine the scenes to see which has the highest ratio. **(b)** Buffer capacity depends on buffer-component concentration *and* ratio. We examine the scenes to see which has a high concentration and a ratio close to 1. **(c)** Adding strong acid converts some A^- to HA, and adding strong base does the opposite. Comparing the [A^-]/[HA] ratios in samples 1 and 2 tells us which to add.

Solution **(a)** The [A^-]/[HA] ratio is 3/3 = 1 for sample 1, 0.5 for sample 2, 1 for sample 3, and 2 for sample 4. Sample 4 has the highest pH because it has the highest [A^-]/[HA] ratio.

(b) Samples 1 and 3 have a [A^-]/[HA] ratio of 1, but sample 3 has the greater capacity because it has a higher concentration.

(c) Sample 2 has a lower [A^-]/[HA] ratio than sample 1, so strong acid should be added to sample 1 to convert some A^- to HA.

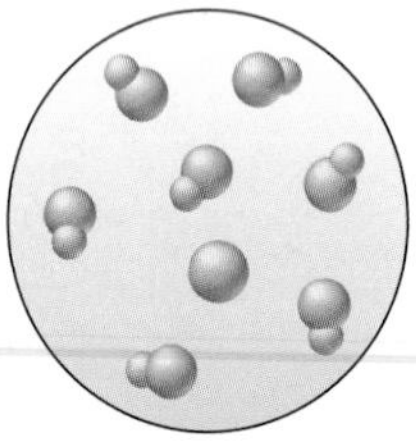

Follow-Up Problem 17.2 The molecular scene (*see margin*) shows a sample of an HB/B^- buffer. (HB is *blue and yellow*, B^- is *yellow*, and the other ions and water are not shown.)

(a) To increase the buffer capacity, should a small amount of concentrated strong acid or a small amount of concentrated strong base be added?

(b) Assuming no volume change, draw a scene that represents the buffer with the highest possible capacity after the addition in part (a).

Preparing a Buffer

Even though chemical supply companies offer buffers in a variety of pH values and concentrations, a specific buffer may have to be prepared in, for example, an environmental or biomedical application. Several steps are required to prepare a buffer:

1. *Choose the conjugate acid-base pair.* Deciding on the chemical composition is based, to a large extent, on the desired pH. Suppose that, for a biochemical experiment, we need a buffer whose pH is 3.90. Therefore, to maximize the capacity, the pK_a of the acid component should be close to 3.90, or $K_a = 10^{-3.90} = 1.3 \times 10^{-4}$. A table of K_a values (see Appendix C) shows that lactic acid ($pK_a = 3.86$), glycolic acid ($pK_a = 3.83$), and formic acid ($pK_a = 3.74$) are possibilities. To avoid substances that are common in biochemical systems, we choose formic acid, HCOOH, and formate ion, $HCOO^-$, supplied by a soluble salt such as sodium formate, HCOONa, as the basic component.

2. *Calculate the ratio of buffer-component concentrations.* To find the ratio $[A^-]/[HA]$ that gives the desired pH, we have

$$pH = pK_a + \log\left(\frac{[A^-]}{[HA]}\right) \quad \text{or} \quad 3.90 = 3.74 + \log\left(\frac{[HCOO^-]}{[HCOOH]}\right)$$

$$\log\left(\frac{[HCOO^-]}{[HCOOH]}\right) = 0.16 \quad \text{so} \quad \left(\frac{[HCOO^-]}{[HCOOH]}\right) = 10^{0.16} = 1.4$$

Thus, for every 1.0 mol of HCOOH in a given volume of solution, we need 1.4 mol of HCOONa.

3. *Determine the buffer concentration.* For most laboratory-scale applications, concentrations of about 0.5 mol/L are suitable, but the decision is often based on the availability of stock solutions. Suppose that we have a large stock of 0.40 mol/L HCOOH and we need approximately 1.0 L of final buffer. First, we find the amount (mol) of sodium formate that will give the needed 1.4/1.0 ratio, and then we convert the amount to mass (g):

$$\text{Amount (mol) of HCOOH} = 1.0 \text{ L soln} \times \frac{0.40 \text{ mol HCOOH}}{1.0 \text{ L soln}} = 0.40 \text{ mol HCOOH}$$

$$\text{Amount (mol) of HCOONa} = 0.40 \text{ mol HCOOH} \times \frac{1.4 \text{ mol HCOONa}}{1.0 \text{ mol HCOOH}} = 0.56 \text{ mol HCOONa}$$

$$\text{Mass (g) of HCOONa} = 0.56 \text{ mol HCOONa} \times \frac{68.01 \text{ g HCOONa}}{1 \text{ mol HCOONa}} = 38 \text{ g HCOONa}$$

4. *Mix the solution and correct the pH.* We dissolve 38 g of sodium formate in the stock 0.40 mol/L HCOOH to a total volume of 1.0 L. Because of non-ideal behaviour (Section 12.5), the buffer may vary slightly from the desired pH, so we adjust with a strong acid or strong base, drop by drop, while monitoring the solution with a pH meter.

Sample Problem 17.3 Preparing a Buffer

Problem An environmental chemist needs a carbonate buffer of pH 10.00 to study the effects of acid rain on limestone-rich soils. What mass of Na_2CO_3 must the chemist add to 1.5 L of 0.20 mol/L $NaHCO_3$ to make the buffer (K_a of $HCO_3^- = 4.7 \times 10^{-11}$)?

Plan The conjugate pair is HCO_3^- (acid) and CO_3^{2-} (base); we know the volume of the buffer (1.5 L) and the concentration (0.20 mol/L) of HCO_3^-, so we need the buffer-component concentration ratio that gives pH 10.00 to find the mass of Na_2CO_3 that we need to dissolve. We use the Henderson-Hasselbalch equation to find $[CO_3^{2-}]$. Multiplying by the volume of solution gives the amount (mol) of CO_3^{2-} needed, which we convert to mass (g) of Na_2CO_3.

Solution

$$pK_a = -\log K_a = -\log(4.7 \times 10^{-11}) = 10.33$$

$$pH = pK_a + \log\frac{[A^-]}{[HA]}$$

$$10.00 = 10.33 + \log\frac{[CO_3^{2-}]}{0.20 \text{ mol/L}}$$

$$\frac{[CO_3^{2-}]}{0.20 \text{ mol/L}} = 10^{-0.33}$$

$$[CO_3^{2-}] = 0.094 \text{ mol/L}$$

Calculate the amount (mol) of CO_3^{2-} needed for the given volume:

$$\text{Amount (mol) of } CO_3^{2-} = 1.5 \text{ L soln} \times \frac{0.094 \text{ mol } CO_3^{2-}}{1 \text{ L soln}} = 0.14 \text{ mol } CO_3^{2-}$$

Calculate the mass (g) of Na_2CO_3 needed:

$$\text{Mass (g) of } Na_2CO_3 = 0.14 \text{ mol } Na_2CO_3 \times \frac{105.99 \text{ g } Na_2CO_3}{1 \text{ mol } Na_2CO_3} = 15 \text{ g } Na_2CO_3$$

The chemist should dissolve 15 g of Na_2CO_3 into about 1.3 L of 0.20 mol/L $NaHCO_3$ and then add more 0.20 mol/L $NaHCO_3$ to make up 1.5 L. Using a pH meter, the chemist can adjust the pH to 10.00 by adding, drop by drop, concentrated strong acid or base.

Check For a useful buffer range, the concentration of the acidic component, $[HCO_3^-]$, must be within a factor of 10 of the concentration of the basic component, $[CO_3^{2-}]$. We have (1.5 L)(0.20 mol/L HCO_3^-), or 0.30 mol of HCO_3^-, and 0.14 mol of CO_3^{2-}; $\frac{0.30}{0.14} = 2.1$. Make sure that the relative amounts of the components are reasonable: we want a pH below the pK_a of HCO_3^- (10.33), so we want more of the acidic than the basic species.

Follow-Up Problem 17.3 How would a benzoic acid/benzoate buffer with pH = 4.25 be prepared, starting with 5.0 L of 0.050 mol/L sodium benzoate (C_6H_5COONa) solution and adding the acidic component [K_a of benzoic acid (C_6H_5COOH) = 6.3×10^{-5}]?

Another way to prepare a buffer is to form one of the components by *partial neutralization* of the other component. For example, an $HCOOH/HCOO^-$ buffer can be prepared by mixing aqueous solutions of HCOOH and NaOH. As OH^- reacts with HCOOH, neutralization of some of the HCOOH produces the $HCOO^-$ needed:

$$\text{HCOOH (HA total)} + \text{OH}^-\text{(amt added)} \longrightarrow \text{HCOOH (HA total} - \text{OH}^-\text{amt added)} + \text{HCOO}^- \text{(OH}^- \text{amt added)} + \text{H}_2\text{O}$$

This method is based on the same chemical process that occurs when a weak acid is titrated with a strong base, as we will discuss in Section 17.2. A sample problem that illustrates this method of making a buffer is given below.

Sample Problem 17.4 Calculating the pH of a Buffer

Problem A first-year university student, working in a lab, is asked to prepare a buffer solution by adding 5.00 mL of 5.0 mol/L NaOH to 100.0 mL of 0.50 mol/L HCOOH. What will the pH of the resulting buffer solution be (K_a of HCOOH = 1.8×10^{-4})?

Plan The conjugate pair is HCOOH (acid) and $HCOO^-$ (base), and we know the total volume (105.0 mL) and the concentrations of the acid and the added base. We need to write the equation for the reaction between the weak acid and the strong base, treating it as a limiting reactant problem. If all the strong base is consumed, then the resulting concentrations of weak acid and weak conjugate base can be used, along with the pK_a of the acid, to determine the pH of the resulting buffer.

Solution Write the balanced reaction between the weak acid and strong base:

$$\text{HCOOH}(aq) + \text{NaOH}(aq) \longrightarrow \text{Na}^+\text{HCOO}^-(aq) + \text{H}_2\text{O}(l)$$

Calculate the initial concentrations after dilution has occurred:

$$[\text{NaOH}]_{\text{init}} = 5.0 \text{ mol/L}\left(\frac{5.00 \text{ mL}}{105.0 \text{ mL}}\right) = 0.238 \text{ mol/L}$$

$$[\text{HCOOH}]_{\text{init}} = 0.50 \text{ mol/L}\left(\frac{100.0 \text{ mL}}{105.0 \text{ mL}}\right) = 0.476 \text{ mol/L}$$

Note that one extra significant figure has been carried in the calculations above.

We can now set up an ICE table. (We will not include water, since its activity is 1.)

Concentration (mol/L)	HCOOH(*aq*)	+	NaOH(*aq*)	⟶	Na^+HCOO^-(*aq*)	+	$H_2O(l)$
Initial	0.476		0.238		0		—
Change	−0.238		−0.238		+0.238		—
Equilibrium	+0.238		0		+0.238		—

From the reaction table, we see that the NaOH is completely consumed. It is the limiting reactant, as well as the driving force for the reaction. As long as there is NaOH, the reaction moves strongly in the forward direction. Once the NaOH is completely consumed, we have only a weak acid and its weak conjugate base in solution in equal concentrations. This is the ideal buffer situation. We can now use the Henderson-Hasselbalch equation to determine the pH of the final solution.

Calculate the pH of the final solution:

$$pH = pK_a + \log\left(\frac{[HCOO^-]}{[HCOOH]}\right) = -\log(1.8 \times 10^{-4}) + \log\left(\frac{0.238\ mol/L}{0.238\ mol/L}\right) = 3.74$$

The pH of the buffer produced by adding a small amount of NaOH to a solution of HCOOH is 3.74.

Check For a useful buffer range, the concentration of the acidic component, [HCOOH], must be within a factor of 10 of the concentration of the basic component, [$HCOO^-$]. Here, we have equal amounts of both. This is the ideal case for a buffer since it maximizes the buffer capacity. When the concentrations of the acid and conjugate base are equal, the pH of the buffer equals the pK_a.

Follow-Up Problem 17.4 A student is asked to prepare an NH_3-NH_4^+ buffer by adding 4.0 mL of 5.0 mol/L HCl to a beaker containing 150 mL of 0.25 mol/L $NH_3(aq)$. What is the pH of the resulting buffer solution [K_b of ammonia (NH_3) = 1.8×10^{-5}]?

The methods outlined in the preceding problems are only two of the six ways to prepare buffer solutions. The six methods are summarized in Figure 17.5. The ovals simply indicate the chemicals that need to be combined to form the buffer. In each case, the chemicals must be combined in the proper stoichiometric ratio

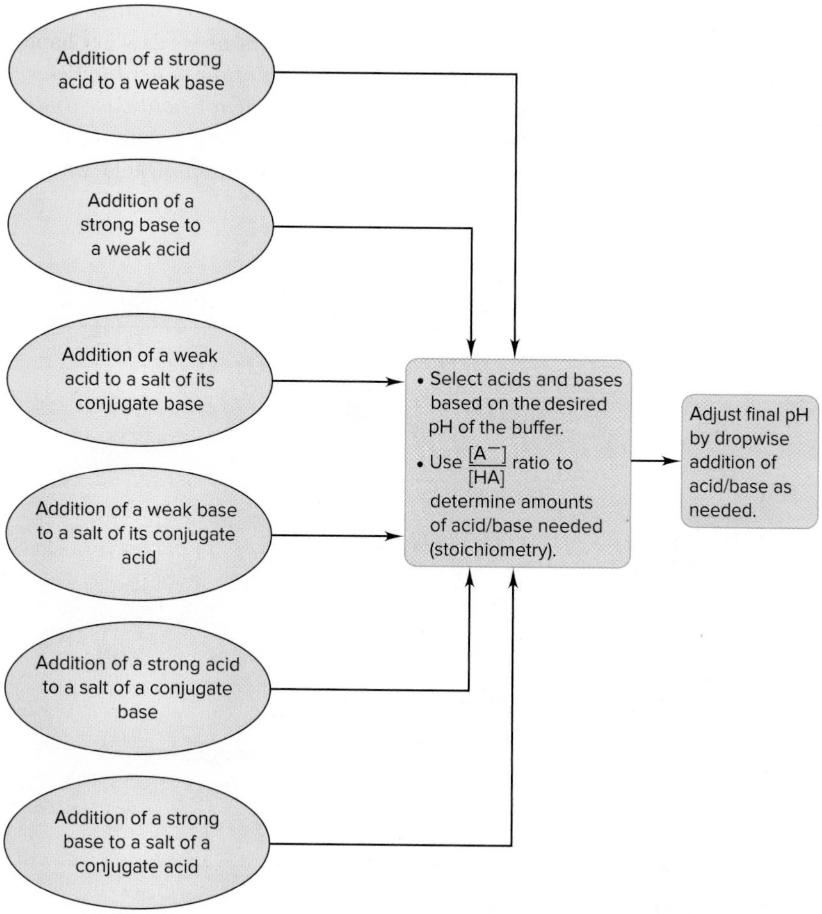

FIGURE 17.5 Six ways to make a buffer solution

(using the ratio of $[A^-]/[HA]$) to prepare the buffer at the desired pH, and the final pH must be carefully adjusted using a pH meter and by dropwise addition of either strong acid or base as needed.

SUMMARY OF SECTION 17.1

- The pH of a buffered solution changes much less than the pH of an unbuffered solution when H^+ or OH^- is added.
- A buffer consists of a weak acid and its conjugate base, or a weak base and its conjugate acid. To be effective, the amounts of the buffer components must be *much* greater than the amount of H^+ or OH^- added.
- The buffer-component concentration ratio determines the pH; the ratio and the pH are related by the Henderson-Hasselbalch equation.
- When H^+ or OH^- is added to a buffer, one component reacts to form the other; thus, $[H_3O^+]$ and pH change only slightly.
- A concentrated (higher-capacity) buffer undergoes smaller changes in pH than a dilute buffer. When the buffer pH equals the pK_a of the acid component, the buffer has its highest capacity.
- A buffer has an effective pH range of $pK_a \pm 1$.
- To prepare a buffer, choose the conjugate acid-base pair, using one of the methods in Figure 17.5; calculate the ratio of the components; determine the buffer concentration; and adjust the final solution to the desired pH.

17.2 Acid-Base Titrations and Curves

Aqueous acid-base reactions occur in processes as diverse as the metabolic action of proteins, the industrial production of fertilizer, and the revitalization of lakes damaged by acid rain. *These reactions involve water as a reactant or product*, in addition to its common role as a solvent. Of course, an **acid-base reaction** or **neutralization reaction** *occurs when an acid reacts with a base*, but the definitions of these terms and the scope of this reaction class have changed over the years, as we saw in Chapter 16.

In any **titration**, *the known concentration of one solution is used to determine the unknown concentration of another, with the aid of an acid-base indicator* (Figure 17.6). In a typical acid-base titration, a *standardized* solution of base (a solution whose concentration is *known*) is added to a solution of acid whose concentration is *unknown* (or vice versa).

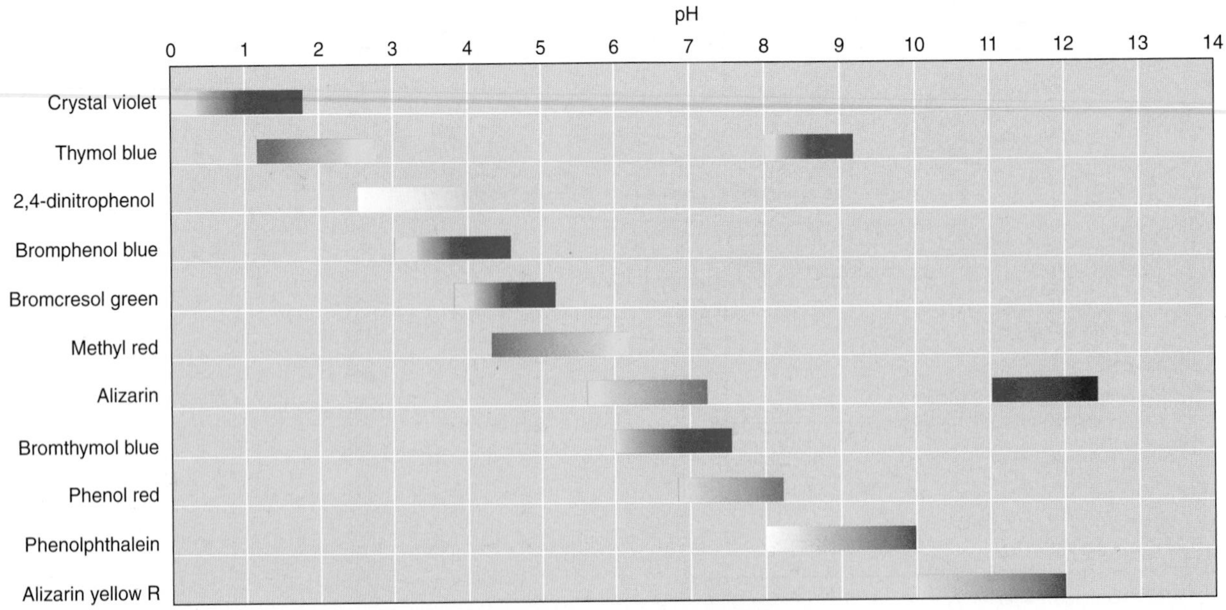

FIGURE 17.6 Colours and approximate pH ranges of some common acid-base indicators

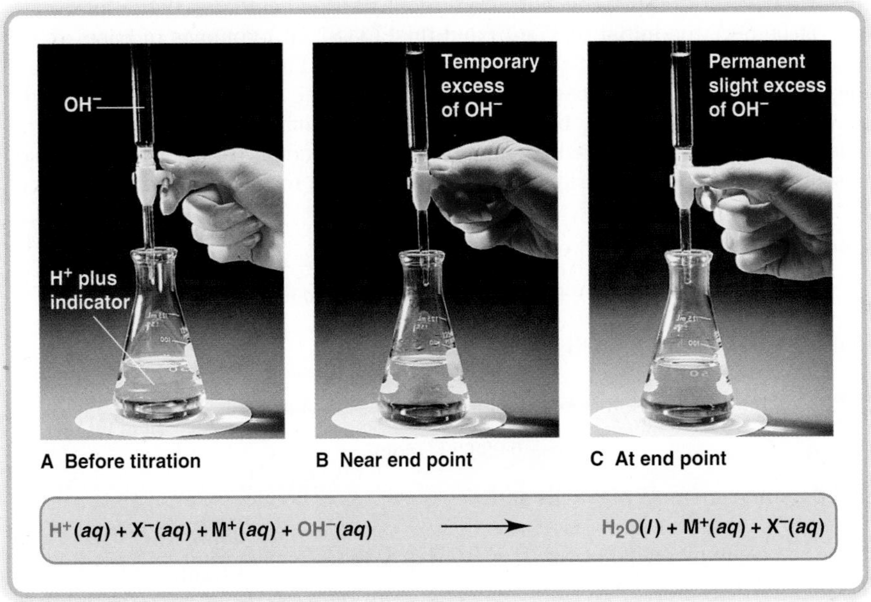

FIGURE 17.7 Stages in an acid-base titration

Figure 17.7A shows the laboratory setup for an acid-base titration with a known volume of acid and a few drops of indicator in a flask. An *acid-base indicator* is a substance whose colour is different in an acid than it is in a base; the indicator used in Figure 17.7 is phenolphthalein, which is pink in a base and colourless in an acid. (We will examine indicators in more detail in the next section.) The base is added from a burette, and the OH^- ions react with the H^+ ions. In general, the flask should always be swirled with the strong hand (right or left) to ensure that the solution is continuously homogeneous and that even the most minute addition of base will cause the colour to change as the end point of the titration approaches. The stopcock should be controlled by the weaker hand (left or right) and the hand should be placed around the back of the burette to control the movement of the stopcock with finesse. An experienced scientist can add a quarter of a drop in this manner. As the titration nears its end (Figure 17.7B), the drop of added base creates a temporary excess of OH^-, causing some indicator molecules to change to the basic colour; they return to the acidic colour when the flask is swirled. There are two key stages in the titration:

- The **equivalence point**, which occurs when *the amount (mol) of H^+ ions in the original volume of acid has reacted with the same amount (mol) of OH^- ions from the burette*:

$$\text{amount (mol) of } H^+ \text{(originally in flask)} = \text{amount (mol) of } OH^- \text{(added from burette)}$$

- The **end point**, which occurs when *a tiny excess of OH^- ions changes the indicator permanently to its basic colour* (Figure 17.7C)

In calculations such as those in Sample Problem 17.5, we assume that this tiny excess of OH^- ions is insignificant, and that *the amount of base needed to reach the end point is the same as the amount needed to reach the equivalence point.*

Sample Problem 17.5 Finding the Concentration of an Acid from a Titration

Problem To standardize an HCl solution, 50.00 mL of the solution is placed in a flask with a few drops of indicator. A 0.1524 mol/L NaOH solution is placed in a burette. The burette reads 0.55 mL at the start and 33.87 mL at the end point. Find the concentration (mol/L) of the HCl solution.

Road Map

Volume (L) of base (difference in burette readings)

↓ Multiply by concentration (mol/L) of base.

Amount (mol) of base

↓ Molar ratio

Amount (mol) of acid

↓ Divide by volume (L) of acid.

Concentration (mol/L) of acid

Plan We have to find the concentration (mol/L) of the acid from the volume of acid (50.00 mL), the initial (0.55 mL) and final (33.87 mL) volumes of base, and the concentration (mol/L) of the base (0.1524 mol/L). First, we balance the equation. The volume of added base is the difference in the burette readings, and we use the concentration of the base to calculate its amount (mol). Then, we use the molar ratio from the balanced equation to find the amount (mol) of acid originally present and divide by the original volume of the acid to find the concentration of the acid (*see road map*).

Solution Write the balanced equation:

$$\text{NaOH}(aq) + \text{HCl}(aq) \longrightarrow \text{NaCl}(aq) + \text{H}_2\text{O}(l)$$

Find the volume (L) of NaOH solution added:

$$\text{Volume (L) of solution} = (33.87 \text{ mL soln} - 0.55 \text{ mL soln}) \times \frac{1 \text{ L}}{1000 \text{ mL}}$$
$$= 0.033\ 32 \text{ L soln}$$

Find the amount (mol) of NaOH added:

$$\text{Amount (mol) of NaOH} = 0.033\ 32 \text{ L soln} \times \frac{0.1524 \text{ mol NaOH}}{1 \text{ L soln}}$$
$$= 5.078 \times 10^{-3} \text{ mol NaOH}$$

Since the molar ratio is 1/1, we can find the amount (mol) of HCl originally present as follows:

$$\text{Amount (mol) of HCl} = 5.078 \times 10^{-3} \text{ mol NaOH} \times \frac{1 \text{ mol HCl}}{1 \text{ mol NaOH}} = 5.078 \times 10^{-3} \text{ mol HCl}$$

Calculate the concentration of HCl:

$$\text{Concentration of HCl} = \frac{5.078 \times 10^{-3} \text{ mol HCl}}{50.00 \text{ mL}} \times \frac{1000 \text{ mL}}{1 \text{ L}} = 0.1016 \text{ mol/L HCl}$$

Check The answer makes sense: a large volume of less concentrated acid was neutralized by a small volume of more concentrated base. With rounding, the amounts (mol) of H^+ and OH^- are about equal: 50 mL × 0.1 mol/L H^+ = 0.005 mol = 33 mL × 0.15 mol/L OH^-.

Follow-Up Problem 17.5 What volume of 0.1292 mol/L $Ba(OH)_2$ would neutralize 50.00 mL of the HCl solution standardized in Sample Problem 17.5?

We will now look at the **acid-base titration curve**, a *plot of pH versus volume of titrant added*. We will discuss curves for four types of titrations: strong acid–strong base, weak acid–strong base, weak base–strong acid, and polyprotic acid–strong base. Running a titration is an activity for a lab, but understanding the roles of acid-base indicators, salt solutions (Section 16.6), and buffers applies key principles of acid-base equilibria.

Monitoring pH with Acid-Base Indicators

An *acid-base indicator* is a weak organic acid (denoted here as HIn) whose colour differs from the colour of its conjugate base (In^-) and whose colour change occurs over a specific, narrow pH range. Typically, one or both forms are intensely coloured, so only a tiny amount of indicator is needed, far too little to affect the pH during the titration.

Figure 17.6 shows the colour change(s) and pH range(s) of some acid-base indicators. To select an indicator, you must know the approximate pH of the titration end point, which means that you must know the ionic species that are present. Because the indicator is a weak acid, the [HIn]/[In^-] ratio is governed by the [H^+] of the solution:

$$\text{HIn}(aq) + \text{H}_2\text{O}(l) \rightleftharpoons \text{H}_3\text{O}^+(aq) + \text{In}^-(aq) \qquad K_a \text{ of HIn} = \frac{[\text{H}_3\text{O}^+][\text{In}^-]}{[\text{HIn}]}$$

Therefore,

$$\frac{[HIn]}{[In^-]} = \frac{[H_3O^+]}{K_a}$$

The colour of an indicator reflects the concentration ratio:

- We see the HIn colour if the $[HIn]/[In^-]$ ratio is 10/1 or greater.
- We see the In^- colour if the $[HIn]/[In^-]$ ratio is 1/10 or less.
- Between these extremes, the two colours merge into an intermediate hue.

Thus, an indicator has a *colour range* equal to a 10^2-fold range in the [HIn]/[In] ratio: an *indicator changes colour in the range of about ±1 of the pK_a value of the indicator.* For example, bromthymol blue has a pH range of about 6.0 to 7.6. As Figure 17.8 shows, it is yellow below this range (*left*), blue above it (*right*), and greenish in between (*centre*).

FIGURE 17.8 The colour change of the indicator bromthymol blue

Strong Acid–Strong Base Titration Curves

Figure 17.9 shows a typical curve for the titration of a strong acid with a strong base, the data used to construct it, and molecular scenes of the key species in solution at various points during the titration.

Features of the Curve There are three distinct regions of the titration curve, which correspond to three major changes in slope:

1. *The pH starts out low*, reflecting the high $[H^+]$ of the strong acid, and increases slowly as the acid is gradually neutralized by the added base.

2. *The pH rises 6 to 8 units very rapidly.* This steep increase begins when the amount (mol) of OH^- added nearly equals the amount (mol) of H^+ originally present in the acid. One or two more drops of base neutralize the remaining tiny excess of acid and introduce a tiny excess of base.

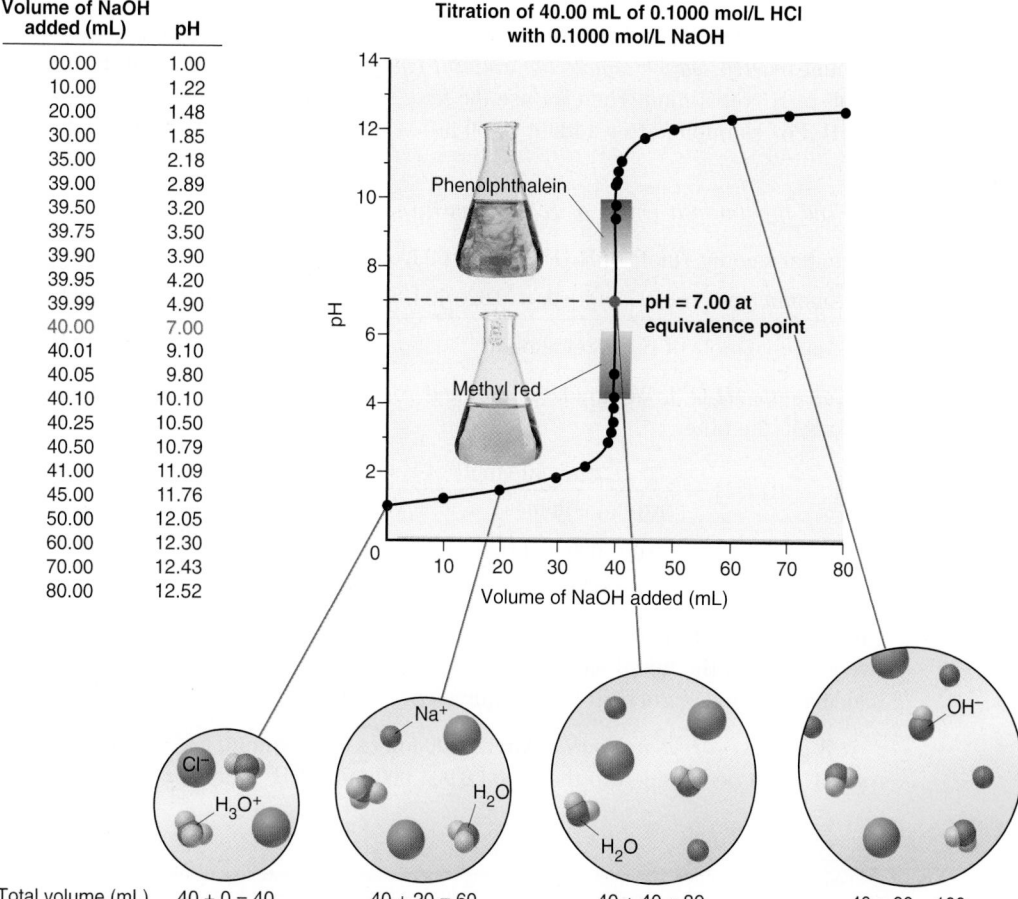

Volume of NaOH added (mL)	pH
00.00	1.00
10.00	1.22
20.00	1.48
30.00	1.85
35.00	2.18
39.00	2.89
39.50	3.20
39.75	3.50
39.90	3.90
39.95	4.20
39.99	4.90
40.00	7.00
40.01	9.10
40.05	9.80
40.10	10.10
40.25	10.50
40.50	10.79
41.00	11.09
45.00	11.76
50.00	12.05
60.00	12.30
70.00	12.43
80.00	12.52

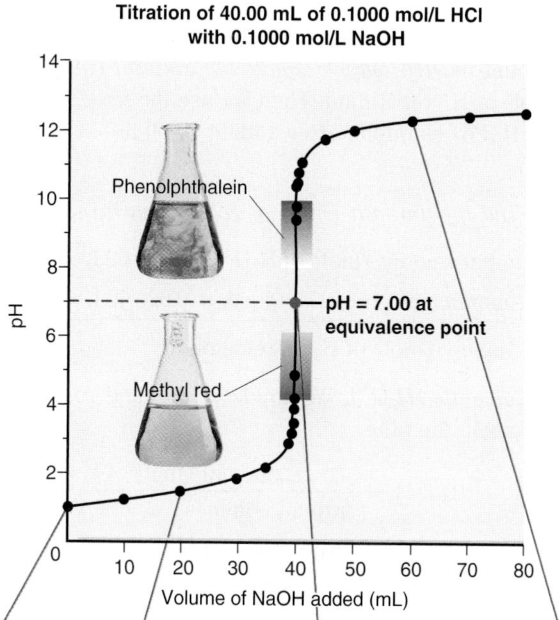

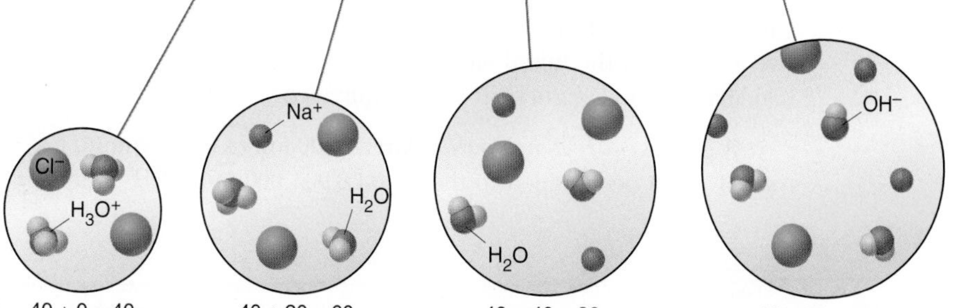

FIGURE 17.9 Curve for a strong acid–strong base titration. Data (*at left*) for the titration of 40.00 mL of 0.1000 mol/L HCl with 0.1000 mol/L NaOH. In the titration, the pH increases gradually, then rapidly near and at the equivalence point, and then gradually again. In a strong acid–strong base titration, pH = 7.00 at the equivalence point. Each molecular view shows the relative numbers of species, other than solvent, in a portion of the solution at one point during the titration. The circles increase in size in proportion to the increase in the total volume of the solution.

3. *The pH increases slowly* beyond the steep rise as more base is added.

Let us distinguish between the *equivalence point* and the *end point*, both of which occur on the steep portion of the curve:

- The equivalence point occurs when the *amount (mol) of added* OH^- *equals the amount (mol) of* H^+ originally present. *The solution consists of the conjugate base of the strong acid and the conjugate acid of the strong base.* Recall, from Chapter 16, that *these ions do not react with water, so the solution is neutral: pH = 7.00.*
- The end point occurs when the indicator, which was added before the titration, changes colour. *We choose an indicator with a colour change close to the pH of the equivalence point.* Figure 17.6 includes two indicators that are suitable for a strong acid–strong base titration. Methyl red changes from red at pH 4.2 to yellow at pH 6.3, and phenolphthalein changes from colourless at pH 8.3 to pink at pH 10.0. Neither colour change occurs *at* the equivalence point (pH 7.00), but both occur on the vertical portion of the curve. Therefore, when methyl red turns yellow or phenolphthalein turns pink, we are within a drop or two of the equivalence point. In practice, then, the *visible* end point signals the *invisible* equivalence point.

Calculating the pH during This Titration By knowing the chemical species that are present during a titration, we can calculate the pH at various points along the way:

1. *Initial solution of strong HA.* In Figure 17.9, 40.00 mL of 0.1000 mol/L HCl is titrated with 0.1000 mol/L NaOH. Because a strong acid is completely dissociated, $[HCl] = [H^+] = 0.1000$ mol/L. Therefore, the initial pH is*

$$pH = -\log [H_3O^+] = -\log (0.1000) = 1.00$$

2. *Before the equivalence point.* As we start adding base, some acid is neutralized and the volume of the solution increases. To find the pH at various points up to the equivalence point, we find the *initial* amount (mol) of H^+ and subtract the amount *reacted, which equals the amount (mol) of* OH^- *added,* to find the amount (mol) of H^+ remaining. Then we use the *total* volume to calculate $[H^+]$ and convert to pH. For example, after adding 20.00 mL of 0.1000 mol/L NaOH, we follow these steps:

- *Find the amount (mol) of* H_3O^+ *remaining:*

$$\begin{array}{lll} \text{Initial amount (mol) of } H_3O^+ & = 0.040\,00 \text{ L} \times 0.1000 \text{ mol/L} & = 0.004\,000 \text{ mol } H_3O^+ \\ -\text{Amount (mol) of } OH^- \text{ added} & = 0.020\,00 \text{ L} \times 0.1000 \text{ mol/L} & = 0.002\,000 \text{ mol } OH^- \\ \hline \text{Amount (mol) of } H_3O^+ \text{ remaining} & & = 0.002\,000 \text{ mol } H_3O^+ \end{array}$$

- *Calculate* $[H_3O^+]$. We divide by the *total volume* because one solution dilutes the ions in the other:

$$[H_3O^+] = \frac{\text{amount (mol) of } H_3O^+ \text{remaining}}{\text{original volume of acid + volume of added base}}$$

$$= \frac{0.002\,000 \text{ mol } H_3O^+}{0.040\,00 \text{ L} + 0.020\,00 \text{ L}} = 0.033\,33 \text{ mol/L} \quad \text{so} \quad pH = 1.48$$

Given the amount of OH^- added, we are halfway to the equivalence point. However, we are still on the initial slow rise of the curve, so the pH is still very low. Similar calculations give values up to the equivalence point.

3. *At the equivalence point.* After 40.00 mL of 0.1000 mol/L NaOH (0.004 000 mol of OH^-) has been added to the initial 0.004 000 mol of H^+, the equivalence point is

*In an acid-base titration, volumes and concentrations are usually known to four significant figures, and pH is usually reported to two digits to the right of the decimal point, unless a high-precision pH meter is used.

reached. The solution contains Na^+ and Cl^-, neither of which reacts with water. Because of the autoionization of water, however,

$$H_2O(l) \rightleftharpoons H^+(aq) + OH^-(aq)$$
$$K_w = [H^+][OH^-] = x^2 = 1.0 \times 10^{-14}$$
$$x = [H_3O^+] = [OH^-] = 1.0 \times 10^{-7}\ \text{mol/L} \quad \text{so} \quad pH = 7.00$$

4. *After the equivalence point.* From the equivalence point onward, the pH calculation is based on the amount (mol) of *excess* OH^- present. For example, after adding 50.00 mL of NaOH, we have the following:

Total amount (mol) of OH^- added	$= 0.050\,00\ \text{L} \times 0.1000\ \text{mol/L} = 0.005\,000\ \text{mol}\ OH^-$
−Amount (mol) of H_3O^+ consumed	$= 0.040\,00\ \text{L} \times 0.1000\ \text{mol/L} = 0.004\,000\ \text{mol}\ H_3O^+$
Amount (mol) of excess OH^-	$= 0.001\,000\ \text{mol}\ OH^-$

$$[OH^-] = \frac{0.001\,000\ \text{mol}\ OH^-}{0.040\,00\ \text{L} + 0.050\,00\ \text{L}} = 0.011\,11\ \text{mol/L} \quad \text{so} \quad pOH = 1.95$$
$$pH = pK_w - pOH = 14.00 - 1.95 = 12.05$$

Weak Acid–Strong Base Titration Curves

Figure 17.10 shows a curve for the titration of a weak acid with a strong base: a titration of 40.00 mL of 0.1000 mol/L propanoic acid ($K_a = 1.3 \times 10^{-5}$) with 0.1000 mol/L NaOH. (We have abbreviated the propanoic acid, CH_3CH_2COOH, as EtCOOH; we have abbreviated the conjugate base, $CH_3CH_2COO^-$, as $EtCOO^-$.)

Features of the Curve The dotted portion of the curve in Figure 17.10 corresponds to the bottom half of the strong acid–strong base curve (Figure 17.9). There are four

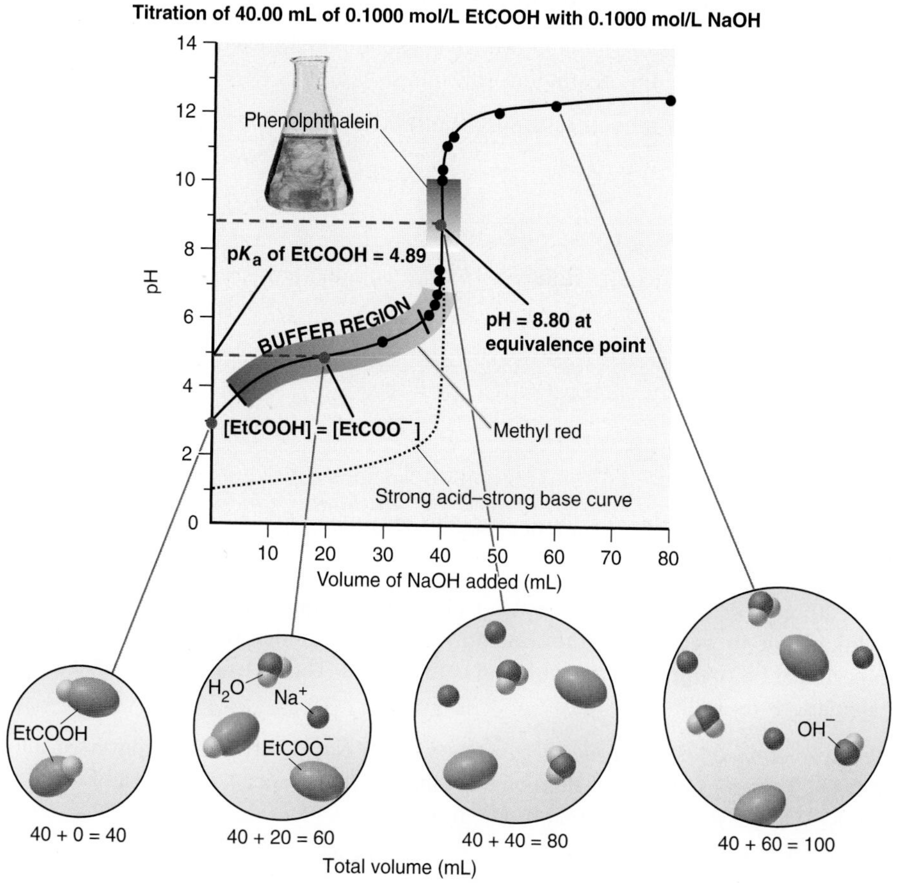

FIGURE 17.10 Curve for a weak acid–strong base titration

key features to note for the weak-acid curve; the first three differ from the features of the strong-acid curve:

1. *The initial pH is higher.* Because the weak acid (EtCOOH) dissociates slightly, there is much less H_3O^+ in the weak acid than there is in the strong acid.

2. *The curve rises gradually in the so-called buffer region before the steep rise to the equivalence point.* As EtCOOH reacts with strong base, more $EtCOO^-$ forms, which creates an $EtCOOH/EtCOO^-$ buffer. At the midpoint of the buffer region, half the initial EtCOOH has reacted (that is, half the OH^- needed to reach the equivalence point has been added), so $[EtCOOH] = [EtCOO^-]$, or $[EtCOO^-]/[EtCOOH] = 1$. Therefore, *the pH equals the* pK_a at the midpoint of the buffer region:

$$pH = pK_a + \log\left(\frac{[EtCOO^-]}{[EtCOOH]}\right) = pK_a + \log 1 = pK_a + 0 = pK_a$$

The pH observed at this point is used to estimate the pK_a of an unknown acid.

3. *The pH at the equivalence point is above 7.00.* The solution contains the strong-base conjugate acid Na^+, which does not react with water, and the weak-acid conjugate base $EtCOO^-$, which acts as a weak conjugate base to accept a proton from H_2O and yield OH^-.

4. *The pH increases slowly* beyond the equivalence point as excess OH^- is added.

The choice of indicator is more limited for a weak acid–strong base titration than it is for a strong acid–strong base titration because the steep rise occurs over a smaller pH range. Phenolphthalein works because it changes colour within this range. However, methyl red does not work because its colour change requires about 30 mL of titrant, rather than just a drop or two.

Calculating the pH during This Titration During the weak acid–strong base titration, we must take into account the partial dissociation of the weak acid, the presence of the buffer, and the reaction of the conjugate base with water. Thus, each of the four regions of the curve requires its own calculation to find $[H^+]$:

1. *Initial HA solution.* Initially, we have a solution of a weak acid in water. The pH can be determined by creating an ICE table:

Concentration (mol/L)	EtCOOH(*aq*)	+	$H_2O(l)$	⇌	$EtCOO^-(aq)$	+	$H_3O^+(aq)$
Initial	0.1000		—		0		0
Change	−x		—		+x		+x
Equilibrium	0.1000 − x		—		+x		+x

Since the acid is weak, and we can assume that $x \ll 0.1000$ so that $0.1000 - x \approx 0.1000$, we have

$$K_a = \frac{[H_3O^+][EtCOO^-]}{[EtCOOH]} = \frac{x^2}{0.1000 - x} \approx \frac{x^2}{0.1000}$$

Therefore,

$$x^2 = (0.1000)(1.3 \times 10^{-5}) = 1.3 \times 10^{-6} \quad \text{so} \quad x = 1.1 \times 10^{-3}\ \text{mol/L} = [H^+]$$

(We must now verify our assumption that $x \ll 0.1000$, and, indeed, it is only 1% of 0.1000 mol/L.)

Although it is sometimes possible to use the assumption $[H^+] = \sqrt{K_a \cdot [HA]_i}$, it is important to verify that the assumption is valid under the conditions of the experiment before proceeding with any calculations. This is true for the subsequent calculations as well.

2. *Solution of HA and added base.* As we add NaOH, EtCOOH is converted into $EtCOO^-$. Thus, for much of the titration up to the equivalence point, we have an $EtCOOH/EtCOO^-$ buffer. Therefore, we find $[H_3O^+]$ from

$$[H_3O^+] = K_a \times \frac{[EtCOOH]}{[EtCOO^-]}$$

(Alternatively, we can find pH directly with the Henderson-Hasselbalch equation.) Because *the volumes cancel in the concentration ratio,* that is, $[EtCOOH]/[EtCOO^-]$ = amount (mol) of EtCOOH/amount (mol) of $EtCOO^-$, we do not need the volumes or have to calculate the concentrations.

3. *Equivalent amounts of HA and added base.* At the equivalence point, all the EtCOOH has reacted, so the solution contains $EtCOO^-$, which reacts with water to form OH^-:

$$EtCOO^-(aq) + H_2O(l) \rightleftharpoons EtCOOH(aq) + OH^-(aq)$$

This explains why, in a weak acid–strong base titration, pH > 7.00 at the equivalence point. To calculate $[H_3O^+]$ (see Section 16.3), we first find K_b of $EtCOO^-$ from K_a of EtCOOH, set up an ICE table (assume $[EtCOO^-] >> [EtCOO^-]_{reacting}$), and solve for $[OH^-]$. Since we use a single concentration, $[EtCOO^-]$, to solve for $[OH^-]$, we *do* need the total volume. Then, we convert $[OH^-]$ to $[H^+]$. These two steps are shown below:

(a) $[OH^-] \approx \sqrt{K_b \times [EtCOO^-]}$

$$\text{where} \quad K_b = \frac{K_w}{K_a} \quad \text{and} \quad [EtCOO^-] = \frac{\text{amount (mol) of EtCOOH}_{init}}{\text{total volume}}$$

(b) $[H_3O^+] = \dfrac{K_w}{[OH^-]}$

Combining them into one step gives

$$[H_3O^+] \approx \frac{K_w}{\sqrt{K_b \times [EtCOO^-]}}$$

Once again, it is essential that the validity of the assumption be verified before proceeding with the calculations.

4. *Solution of excess base.* Beyond the equivalence point, as in the strong acid–strong base titration, excess OH^- is being added:

$$[H_3O^+] = \frac{K_w}{[OH^-]} \quad \text{where} \quad [OH^-] = \frac{\text{amount (mol) of excess OH}^-}{\text{total volume}}$$

Sample Problem 17.6 shows the overall approach.

Sample Problem 17.6 Finding the pH during a Weak Acid–Strong Base Titration

Problem Calculate the pH during the titration of 40.00 mL of 0.1000 mol/L propanoic acid (EtCOOH; $K_a = 1.3 \times 10^{-5}$) after each addition of 0.1000 mol/L NaOH:

(a) 0.00 mL **(b)** 30.00 mL **(c)** 40.00 mL **(d)** 50.00 mL

Plan **(a)** 0.00 mL: No base has been added yet, so this is a weak-acid dissociation. We calculate the pH as we did in Section 16.4. **(b)** 30.00 mL: We find the amount (mol) of $EtCOO^-$ and the amount (mol) of EtCOOH; then we substitute into the K_a expression to solve for $[H_3O^+]$ and convert to pH. **(c)** 40.00 mL: The amount (mol) of NaOH added equals the initial amount (mol) of EtCOOH, so a solution of Na^+ and $EtCOO^-$ exists. We calculate the pH as we did in Section 16.4, except that we need the *total* volume to find $[EtCOO^-]$. **(d)** 50.00 mL: We calculate the amount (mol) of excess OH^- in the total volume, convert to $[H_3O^+]$, and then convert to pH.

Solution **(a)** 0.00 mL of 0.1000 mol/L NaOH is added.

Following the approach used in Sample Problem 16.8 and just described, we obtain

$$[H_3O^+] \approx \sqrt{K_a \times [EtCOOH]_{init}} = \sqrt{(1.3 \times 10^{-5})(0.1000)} = 1.1 \times 10^{-3} \text{ mol/L}$$

$$pH = 2.96$$

(b) 30.00 mL of 0.1000 mol/L NaOH is added.

Calculate the ratio of amount (mol) of EtCOOH to amount (mol) of $EtCOO^-$:

Initial amount (mol) of EtCOOH = 0.040 00 L × 0.1000 mol/L = 0.004 000 mol EtCOOH

Amount (mol) of NaOH added = 0.030 00 L × 0.1000 mol/L = 0.003 000 mol OH^-

For every mole of NaOH, 1 mol of $EtCOO^-$ forms, so we can set up this stoichiometry reaction table:

Amount (mol)	EtCOOH(*aq*)	+	OH^-(*aq*)	⟶	$EtCOO^+$(*aq*)	+	H_2O(*l*)
Initial	0.004 000		0.003 000		0		—
Change	−0.003 000		−0.003 000		+0.003 000		—
Final	0.001 000		0		0.003 000		—

The last line of the table gives the new initial amounts of EtCOOH and $EtCOO^-$ that react to attain a new equilibrium. Since x is very small relative to [EtCOOH], we assume that the [EtCOOH]/[$EtCOO^-$] ratio at equilibrium is essentially equal to the ratio of these new initial amounts (see the Comment at the end of Sample Problem 17.1). Thus,

$$\frac{[\text{EtCOOH}]}{[\text{EtCOO}^-]} = \frac{0.001\ 000\ \text{mol}}{0.003\ 000\ \text{mol}} = 0.3333$$

Solve for $[H_3O^+]$:

$$[\text{H}_3\text{O}^+] = K_a \times \frac{[\text{EtCOOH}]}{[\text{EtCOO}^-]} = (1.3 \times 10^{-5})(0.3333) = 4.3 \times 10^{-6}\ \text{mol/L}$$

$$\text{pH} = 5.37$$

(c) 40.00 mL of 0.1000 mol/L NaOH is added. Calculate [$EtCOO^-$] after all EtCOOH has reacted:

$$[\text{EtCOO}^-] = \frac{0.004\ 000\ \text{mol}}{0.040\ 00\ \text{L} + 0.04\ 000\ \text{L}} = 0.050\ 00\ \text{mol/L}$$

Calculate K_b using the given K_a (see Sample Problem 17.5 for similar steps):

$$K_b = \frac{K_w}{K_a} = \frac{1.0 \times 10^{-14}}{1.3 \times 10^{-5}} = 7.7 \times 10^{-10}$$

Solve for $[H_3O^+]$ as described above for region 3 of the curve:

$$[\text{H}_3\text{O}^+] = \frac{K_w}{\sqrt{K_b \times [\text{EtCOO}^-]}} = \frac{1.0 \times 10^{-14}}{\sqrt{(7.7 \times 10^{-10})(0.05000)}} = 1.6 \times 10^{-9}\ \text{mol / L}$$

$$\text{pH} = 8.80$$

(d) 50.00 mL of 0.1000 mol/L NaOH is added.

Amount (mol) of excess OH^- = (0.1000 mol/L)(0.050 00 L − 0.040 00 L) = 0.001 000 mol

$$[\text{OH}^-] = \frac{\text{amount (mol) of excess OH}^-}{\text{total volume}} = \frac{0.001\ 000\ \text{mol}}{0.090\ 00\ \text{L}} = 0.011\ 11\ \text{mol/L}$$

$$[\text{H}_3\text{O}^+] = \frac{K_w}{[\text{OH}^-]} = \frac{1.0 \times 10^{-14}}{0.011\ 11} = 9.0 \times 10^{-13}\ \text{mol/L}$$

$$\text{pH} = 12.05$$

Check As expected, the pH increases through the four regions of the titration. Be sure to round only at the end of the calculations (to avoid rounding errors) and to check the arithmetic along the way.

Follow-Up Problem 17.6 A chemist titrates 20.00 mL of 0.2000 mol/L HBrO ($K_a = 2.3 \times 10^{-9}$) with 0.1000 mol/L NaOH.

(a) What is the pH **(i)** before any base is added; **(ii)** when [HBrO] = [BrO^-]; **(iii)** at the equivalence point; **(iv)** when the amount (mol) of OH^- added is twice the amount of HBrO present initially?

(b) Sketch the titration curve, and label the pK_a and equivalence point.

Weak Base–Strong Acid Titration Curves

The opposite of a weak acid–strong base titration is the titration of a weak base (NH_3) with a strong acid (HCl). This titration curve, shown in Figure 17.11, has the *same shape as the weak acid–strong base curve, but it is inverted.*

This titration curve also has the same features as the weak acid–strong base curve, but *the pH decreases* throughout the titration:

1. *The initial weak-base solution has a pH well above 7.00* (see Sample Problem 16.9).
2. *The pH decreases gradually in the buffer region,* where significant amounts of NH_3 and its conjugate acid, NH_4^+, are present. At the midpoint of this region, *the pH equals the* pK_a of NH_4^+.
3. *The curve drops steeply to the equivalence point.* All the NH_3 has reacted with added HCl, and the solution contains only NH_4^+ and Cl^-. Notice that *the pH at the equivalence point is below 7.00* because Cl^- does not react with water and NH_4^+ is acidic:

$$NH_4^+(aq) + H_2O(l) \rightleftharpoons NH_3(aq) + H_3O^+(aq)$$

4. *The pH decreases slowly beyond the equivalence point as excess H_3O^+ is added.*

For this titration, phenolphthalein changes colour too slowly. However, the change in methyl red occurs on the steep portion of the curve and straddles the equivalence point, so it is the perfect choice.

Titration Curves for Polyprotic Acids

Except for sulfuric acid, the common polyprotic acids are all weak. The successive K_a values for a polyprotic acid differ by several orders of magnitude, as these values for sulfurous acid show:

$$H_2SO_3(aq) + H_2O(l) \rightleftharpoons HSO_3^-(aq) + H_3O^+(aq)$$

$$K_{a1} = 1.4 \times 10^{-2} \quad \text{and} \quad pK_{a1} = 1.85$$

$$HSO_3^-(aq) + H_2O(l) \rightleftharpoons SO_3^{2-}(aq) + H_3O^+(aq)$$

$$K_{a2} = 6.5 \times 10^{-8} \quad \text{and} \quad pK_{a2} = 7.19$$

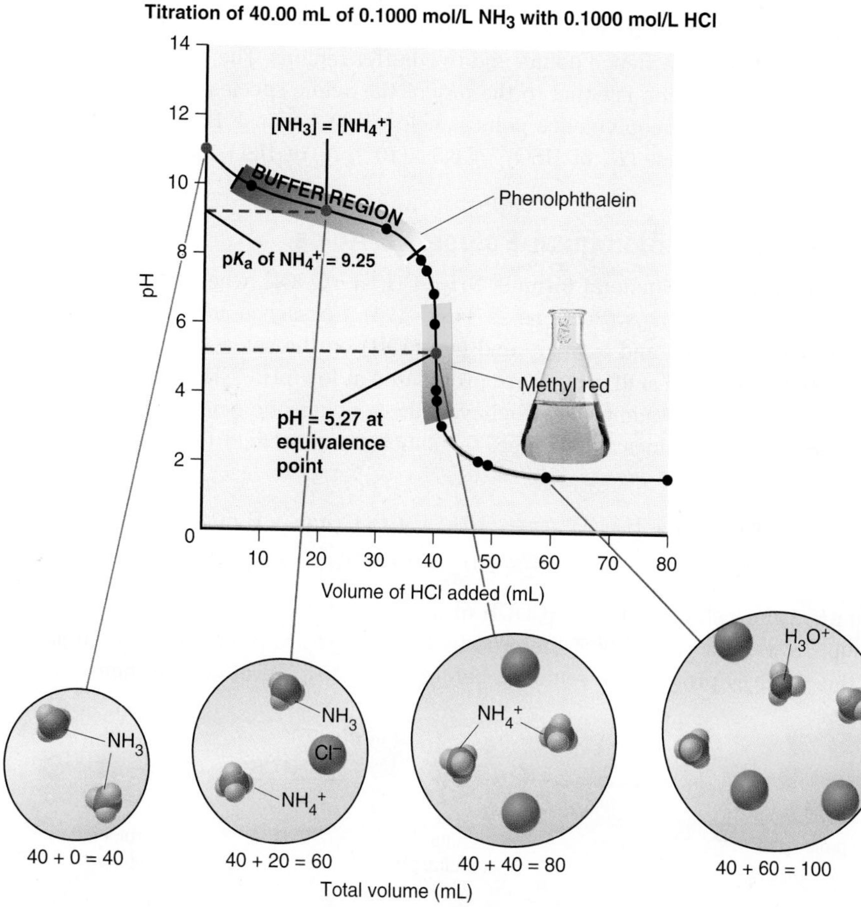

FIGURE 17.11 Curve for a weak base–strong acid titration

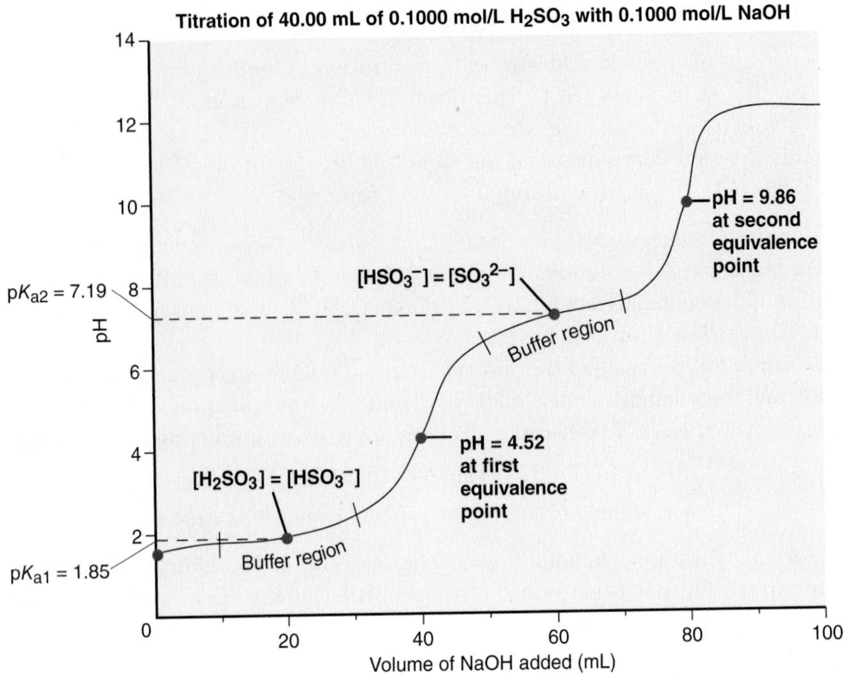

FIGURE 17.12 Curve for the titration of a weak polyprotic acid. Because the K_a values are separated by several orders of magnitude, the titration curve looks like two weak acid–strong base curves joined.

In a titration of a diprotic acid, such as H_2SO_3, two OH^- ions are required to react with the two H^+ ions of each acid molecule. Figure 17.12 shows the titration curve for sulfurous acid with a strong base. Because of the large difference in K_a values, each mole of H^+ is titrated separately, so the H_2SO_3 molecules lose one H^+ before any HSO_3^- ions do:

$$H_2SO_3 \xrightarrow{\text{1 mol OH}^-} HSO_3^- \xrightarrow{\text{1 mol OH}^-} SO_3^{2-}$$

Several features of the curve are described below:

- The same amount of base (0.004 000 mol OH^-) is required per mole of H^+.
- There are two equivalence points and two buffer regions. The pH at the midpoint of each buffer region is equal to the pK_a of the acidic species.
- The pH of the first equivalence point is below 7.00, because HSO_3^- is a stronger acid than it is a base (K_a of $HSO_3^- = 6.5 \times 10^{-8}$; K_b of $HSO_3^- = 7.1 \times 10^{-13}$).

Amino Acids as Biological Polyprotic Acids

Amino acids have the general formula NH_2—CH(R)—COOH, where R can be one of about 20 different groups (Sections 22.3 and 22.4). In essence, amino acids contain a weak base (—NH_2) and a weak acid (—COOH) on the same molecule. Both the amino and carboxylic acid groups are protonated at low pH: $^+NH_3$—CH(R)—COOH. Thus, in this form, an amino acid behaves like a polyprotic acid. The dissociation reactions and pK_a values for glycine, the simplest amino acid (R = H), are given below:

$$^+NH_3CH_2COOH(aq) + H_2O(l) \rightleftharpoons {}^+NH_3CH_2COO^-(aq) + H_3O^+(aq) \quad pK_{a1} = 2.35$$
$$^+NH_3CH_2COO^-(aq) + H_2O(l) \rightleftharpoons NH_2CH_2COO^-(aq) + H_3O^+(aq) \quad pK_{a2} = 9.78$$

The pK_a values show that the —COOH group is *much* more acidic than the —NH_3^+ group. As we saw with H_2SO_3, the protons (*black circles*) are titrated separately, so all the —COOH protons are removed before any —NH_3^+ protons are removed:

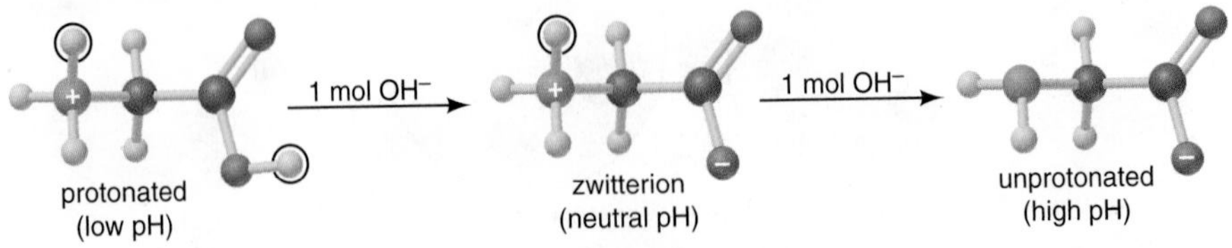

Thus, at physiological pH (~7), which is between the two pK_a values, *an amino acid exists as a zwitterion* (from the German word *zwitter*, meaning "double"), a species with opposite charges on the same molecule. For glycine, the zwitterion is $^+NH_3CH_2COO^-$.

Of the 20 different R groups of amino acids, several have *additional* $—COO^-$ or $—NH_3^+$ groups at pH 7 (see Figure 22.8). When amino acids link to form a protein, their charged R groups give the protein its overall charge, which is often related to its function. A widely studied example of this relationship occurs in the hereditary disease sickle cell anemia. Normal red blood cells contain hemoglobin molecules that have two glutamic acid R groups ($—CH_2CH_2COO^-$), each providing a negative charge at a critical region. Abnormal hemoglobin molecules in sickle cell anemia have the valine R group ($—CH(CH_3)_2$), which is uncharged, at the same region. This change in just two of hemoglobin's 574 amino acids lowers the charge repulsions between the hemoglobin molecules. As a result, they clump together in fibre-like structures, which leads to the sickle shape of the red blood cells (Figure 17.13). Since the misshapen cells block capillaries, sickle cell anemia is painful and usually causes an early death.

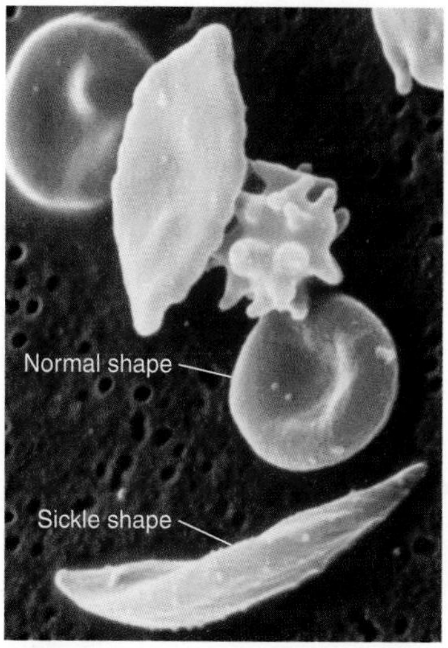

FIGURE 17.13 Abnormal shape of red blood cells in sickle cell anemia

SUMMARY OF SECTION 17.2

- An acid-base (pH) indicator is a weak acid that has a different-colour conjugate base form and changes colour over about ±1 of its pK_a value.
- An acid-base titration is the stoichiometric reaction of an acid and a base to yield a salt and water. Such a reaction is also called a neutralization reaction.
- The equivalence point of a titration is the point at which the amount (mol) of the acid equals the amount (mol) of the base (for strong acids and bases, we would use H^+ and OH^- as the acid and base, respectively). The end point of a titration is the point at which a small amount of added base causes the colour of the indicator to change from its colour in an acidic solution to its colour in a basic solution.
- In a strong acid–strong base titration, the pH starts low, rises slowly, and then shoots up near the equivalence point (pH = 7).
- In a weak acid–strong base titration, the pH starts higher, rises slowly in the buffer region (pH = pK_a at the midpoint), and then rises quickly near the equivalence point (pH > 7).
- A weak base–strong acid titration curve has a shape that is the inverse of the weak acid–strong base curve, with the pH decreasing to the equivalence point (pH < 7).
- Polyprotic acids have two or more acidic protons, each with its own K_a value. Because the K_a values differ by several orders of magnitude, each proton is titrated separately.
- Amino acids exist in charged forms that depend on the pH of the solution and determine the overall charge of a protein.

17.3 Equilibria of Slightly Soluble Ionic Compounds

In this section, we explore an equilibrium system that involves the solubility of ionic compounds. Recall, from Chapter 12, that most solutes, even those we call "soluble," have a limited solubility in a particular solvent. In a saturated solution at a particular temperature, equilibrium exists between dissolved and undissolved solute. Slightly soluble ionic compounds, which we have been calling "insoluble," reach equilibrium with very little solute dissolved. In this introductory section, we *assume* that, like a soluble ionic compound, the small amount of a slightly soluble ionic compound that does dissolve dissociates completely into ions.

The Ion-Product Expression (Q_{sp}) and the Solubility-Product Constant (K_{sp})

For a slightly soluble ionic compound, *equilibrium exists between solid solute and aqueous ions*. Thus, for example, we can write the equilibrium reaction for a saturated solution of lead(II) fluoride as follows:

$$PbF_2(s) \rightleftharpoons Pb^{2+}(aq) + 2F^-(aq)$$

As for any equilibrium system, we can write an equilibrium expression:

$$Q = \frac{a_{Pb^{2+}}a_{F^-}^2}{a_{PbF_2}}$$

Since the activity of a pure solid is 1, and we will deal mostly with solutions that are fairly dilute and behave reasonably ideally, we can substitute concentrations for the activities of the ions in solution, giving a relationship called the *ion-product expression*, Q_{sp}:

$$Q_{sp} = [Pb^{2+}][F^-]^2$$

When the solution is saturated, the numerical value of Q_{sp} attains a constant value called the **solubility-product constant (K_{sp})**. The K_{sp} for PbF_2 at 25°C, for example, is 3.6×10^{-8}. Like other equilibrium constants, K_{sp} depends *only* on the temperature, not on the individual ion concentrations.

Writing the Ion-Product Expression The form of Q_{sp} is identical to the form of other reaction quotients: each concentration is raised to an exponent equal to the coefficient in the balanced equation, which in this case also *equals the subscript of each ion in the formula of the compound.* At saturation, the concentration terms have their equilibrium values. Thus, in general, for a slightly soluble ionic compound, M_pX_q, composed of the ions M^{n+} and X^{z-}, the ion-product expression at equilibrium is

$$Q_{sp} = [M^{n+}]^p[X^{z-}]^q = K_{sp} \quad (17.2)$$

Since the expressions are identical at equilibrium, we can write that Q_{sp} is equal to K_{sp}. ■

■ We have seen different *K* values in Chapters 15, 16, and 17. It is important not to consider them all as separate from one another. *Any* reaction that reaches a state of equilibrium has a *K* value. For some reactions, such as those with gases where the *K* value is between 10^{-4} and 10^{+4}, we cannot make any approximations and we always have to solve the equation that results. In the case of acid-base reactions, water (a pure liquid) is always part of the reaction. This simplifies the chemistry and the mathematics because pure liquids are not part of the *K* expression. In solubility equilibria, the left-hand side of the chemical equation will always be a solid and the right-hand side will be the ions formed when the solid dissociates in water. The chemistry and mathematics are even more simplified here as there will not be a denominator in the *K* expression. Nevertheless, the approach to solving all these problems is exactly the same: use an ICE table, substitute the known values into the ICE table, solve for the unknown variable, and determine the answer to the question.

The Special Case of Metal Sulfides The aqueous equilibria of metal sulfides have a special form of the ion-product expression. The sulfide ion, S^{2-}, is so basic that it reacts completely with water to form the hydrogen sulfide ion (HS^-) and OH^-:

$$S^{2-}(aq) + H_2O(l) \longrightarrow HS^-(aq) + OH^-(aq)$$

Although S^{2-} is not stable in water, we can think of the dissolution process as the sum of two steps, with S^{2-} formed in the first step and consumed immediately in the second step. Thus, for example, the equilibrium reaction for manganese(II) sulfide can be written as follows:

$$MnS(s) \rightleftharpoons Mn^{2+}(aq) + \cancel{S^{2-}(aq)}$$
$$\cancel{S^{2-}(aq)} + H_2O(l) \longrightarrow HS^-(aq) + OH^-(aq)$$
$$MnS(s) + H_2O(l) \rightleftharpoons Mn^{2+}(aq) + HS^-(aq) + OH^-(aq)$$

Note that MnS is a solid and water is a liquid (both with activities of 1), so this gives the following ion-product expression at equilibrium:

$$Q_{sp} = K_{sp} = [Mn^{2+}][HS^-][OH^-]$$

Sample Problem 17.7 Writing Ion-Product Expressions

Problem Write the ion-product expression at equilibrium for **(a)** magnesium carbonate; **(b)** iron(II) hydroxide; **(c)** calcium phosphate; **(d)** silver sulfide.

Plan We write an equation for a saturated solution and then write the ion-product expression at equilibrium, K_{sp} (Equation 17.2), noting the sulfide in part (d).

Solution **(a)** Magnesium carbonate:

$$MgCO_3(s) \rightleftharpoons Mg^{2+}(aq) + CO_3^{2-}(aq) \qquad K_{sp} = [Mg^{2+}][CO_3^{2-}]$$

(b) Iron(II) hydroxide:

$$Fe(OH)_2(s) \rightleftharpoons Fe^{2+}(aq) + 2OH^-(aq) \qquad K_{sp} = [Fe^{2+}][OH^-]^2$$

(c) Calcium phosphate:

$$Ca_3(PO_4)_2(s) \rightleftharpoons 3Ca^{2+}(aq) + 2PO_4^{3-}(aq) \qquad K_{sp} = [Ca^{2+}]^3[PO_4^{3-}]^2$$

(d) Silver sulfide:

$$Ag_2S(s) \rightleftharpoons 2Ag^+(aq) + \cancel{S^{2-}(aq)}$$
$$\cancel{S^{2-}(aq)} + H_2O(l) \longrightarrow HS^-(aq) + OH^-(aq)$$
$$Ag_2S(s) + H_2O(l) \rightleftharpoons 2Ag^+(aq) + HS^-(aq) + OH^-(aq) \quad K_{sp} = [Ag^+]^2[HS^-][OH^-]$$

Check Except for part (d), we can check by reversing the process to see if we obtain the formula for the compound from the K_{sp}.

Follow-Up Problem 17.7 Write the ion-product expression at equilibrium for **(a)** calcium sulfate; **(b)** chromium(III) carbonate; **(c)** magnesium hydroxide; **(d)** arsenic(III) sulfide.

Calculations Involving the Solubility-Product Constant

In Chapters 15 and 16, we described two types of equilibrium problems. In one type, we used concentrations or pressure to find K; in the other type, we used K to find equilibrium pressures or concentrations. Here, we encounter the same two types of problems.

The Complication from Assuming Complete Dissociation Before we focus on calculations, let us address a complication that affects accuracy and results in approximate answers. Our assumption that the small dissolved amount of a slightly soluble ionic compound dissociates completely into separate ions is an oversimplification. Many slightly soluble salts have polar covalent metal-nonmetal bonds, and partially dissociated or even undissociated species occur in solution. Here are two of many examples:

- With slightly soluble lead(II) chloride in water, the solution contains not only the separate $Pb^{2+}(aq)$ and $Cl^-(aq)$ ions we expect from complete dissociation, but also undissociated $PbCl_2(aq)$ molecules and $PbCl^+(aq)$ ions.
- In an aqueous solution of $CaSO_4$, undissociated ion pairs, $Ca^{2+}SO_4^{2-}(aq)$, are present.

These partly dissociated and undissociated species increase the *apparent* solubility of a slightly soluble salt above the value obtained by assuming complete dissociation. More advanced courses discuss this complication, but we simply mention it in the Comments at the ends of several sample problems. Thus, we should view the problem answers as first approximations.

The Meaning of K_{sp} and How to Determine It from Solubility Values The K_{sp} value indicates *how far the dissolution proceeds at equilibrium (saturation).* Table 17.2 presents a few K_{sp} values, and Appendix C lists many more. Note that all the values are low, but they range over many orders of magnitude. In Sample Problem 17.8, we find K_{sp} from the solubility of a compound.

TABLE 17.2 Solubility-Product Constants (K_{sp}) of Selected Ionic Compounds at 25°C

Name, Formula	K_{sp}
Aluminum hydroxide, $Al(OH)_3$	3×10^{-34}
Cobalt(II) carbonate, $CoCO_3$	1.0×10^{-10}
Iron(II) hydroxide, $Fe(OH)_2$	4.1×10^{-15}
Lead(II) fluoride, PbF_2	3.6×10^{-8}
Lead(II) sulfate, $PbSO_4$	1.6×10^{-8}
Mercury(I) iodide, Hg_2I_2	4.7×10^{-29}
Silver sulfide, Ag_2S	8×10^{-48}
Zinc iodate, $Zn(IO_3)_2$	3.9×10^{-6}

Sample Problem 17.8 Determining K_{sp} from Solubility

Problem (a) Lead(II) sulfate ($PbSO_4$) is a key component in lead-acid car batteries. Its solubility in water at 25°C is 4.25×10^{-3} g/100 mL solution. What is the K_{sp} of $PbSO_4$?

(b) When lead(II) fluoride (PbF_2) is shaken with pure water at 25°C, its solubility is found to be 0.64 g/L. Calculate the K_{sp} of PbF_2.

Plan We are given the solubilities in various units and must find K_{sp}. As always, we write a dissolution equation and the ion-product expression for each compound. This tells us the amount (mol) of each ion. We use the molar mass to convert the solubility of the compound from the given mass units to *molar solubility* (mol/L). Then we use the molar mass to find the concentration of each ion and substitute the concentrations into the ion-product expression to calculate K_{sp}.

Solution (a) Write the equation and ion-product (K_{sp}) expression for $PbSO_4$:

$$PbSO_4(s) \rightleftharpoons Pb^{2+}(aq) + SO_4^{2-}(aq) \qquad K_{sp} = [Pb^{2+}][SO_4^{2-}]$$

Convert solubility to molar solubility:

$$\text{Molar solubility of } PbSO_4 = \frac{0.004\,25 \text{ g } PbSO_4}{100 \text{ mL soln}} \times \frac{1000 \text{ mL}}{1 \text{ L}} \times \frac{1 \text{ mol } PbSO_4}{303.3 \text{ g } PbSO_4}$$

$$= 1.40 \times 10^{-4} \text{ mol/L } PbSO_4$$

Finally, we need to determine the concentrations of the ions. Because 1 mol of Pb^{2+} and 1 mol of SO_4^{2-} form when 1 mol of $PbSO_4$ dissolves, we have

$$[Pb^{2+}] = [SO_4^{2-}] = 1.40 \times 10^{-4} \text{ mol/L}$$

Substitute these values into the ion-product expression to calculate K_{sp}:

$$K_{sp} = [Pb^{2+}][SO_4^{2-}] = (1.40 \times 10^{-4})^2 = 1.96 \times 10^{-8}$$

(b) Write the equation and K_{sp} expression for PbF_2:

$$PbF_2(s) \rightleftharpoons Pb^{2+}(aq) + 2F^-(aq) \qquad K_{sp} = [Pb^{2+}][F^-]^2$$

Convert solubility to molar solubility:

$$\text{Molar solubility of } PbF_2 = \frac{0.64 \text{ g } PbF_2}{1 \text{ L soln}} \times \frac{1 \text{ mol } PbF_2}{245.2 \text{ g } PbF_2} = 2.6 \times 10^{-3} \text{ mol/L } PbF_2$$

Then we determine the concentrations of the ions. Since 1 mol of Pb^{2+} and 2 mol of F^- form when 1 mol of PbF_2 dissolves, we have

$$[Pb^{2+}] = 2.6 \times 10^{-3} \text{ mol/L} \quad \text{and} \quad [F^-] = 2(2.6 \times 10^{-3} \text{ mol/L}) = 5.2 \times 10^{-3} \text{ mol/L}$$

Substitute these values into the ion-product expression to calculate K_{sp}:

$$K_{sp} = [Pb^{2+}][F^-]^2 = (2.6 \times 10^{-3})(5.2 \times 10^{-3})^2 = 7.0 \times 10^{-8}$$

Check The low solubilities are consistent with the K_{sp} values being small. **(a)** The molar solubility seems about right: $\sim\frac{4 \times 10^{-2} \text{g/L}}{3 \times 10^{2} \text{g/mol}} \approx 1.3 \times 10^{-4}$ mol/L. Squaring this number gives 1.7×10^{-8}, close to the calculated K_{sp}. **(b)** We can check the math in the final step as follows: $\sim(3 \times 10^{-3})(5 \times 10^{-3})^2 = 7.5 \times 10^{-8}$, close to the calculated K_{sp}.

Comment 1. In part (b), the formula PbF_2 means that $[F^-]$ is twice $[Pb^{2+}]$. Then we follow the ion-product expression exactly and square this value of $[F^-]$.

2. The tabulated K_{sp} values for these compounds (Table 17.2) are lower than our calculated values. For PbF_2, for instance, the tabulated value is 3.6×10^{-8}, but we calculated 7.0×10^{-8} from solubility data. The discrepancy arises because we assumed that PbF_2 in solution dissociates completely to Pb^{2+} and F^-. Here is an example of the complication pointed out earlier. Actually, about one-third of the PbF_2 dissolves as $PbF^+(aq)$ and a small amount dissolves as undissociated $PbF_2(aq)$. The solubility given in the problem statement (0.64 g/L) is determined experimentally and includes these other species, which we did *not* include in our calculation.

Fluorite, the mineral form of CaF_2

Follow-Up Problem 17.8 When fluorite (CaF_2; *see photo*) is pulverized and shaken in water at 18°C, 10.0 mL of solution contains 1.5×10^{-4} g of solute. Find the K_{sp} of CaF_2 at 18°C.

Determining Solubility from K_{sp} The reverse of Sample Problem 17.7 involves finding the solubility of a compound based on its formula and K_{sp} value. We will use an approach similar to the one we used for weak acids in Sample Problem 16.8: we define the unknown amount dissolved—the molar solubility—as *s*, include ion concentrations in terms of this unknown in an ICE table, and solve for *s*.

Sample Problem 17.9 Determining Solubility from K_{sp}

Problem Calcium hydroxide (slaked lime) is a major component of mortar, plaster, and cement, and solutions of $Ca(OH)_2$ are used in industries as strong, inexpensive bases. Calculate the molar solubility of $Ca(OH)_2$ in water if K_{sp} is 6.5×10^{-6}.

Plan We write the dissolution equation and the ion-product expression. We know K_{sp} (6.5×10^{-6}). To find molar solubility (s), we set up an ICE table that expresses $[Ca^{2+}]$ and $[OH^-]$ in terms of s, substitute into the ion-product expression, and solve for s.

Solution Write the equation and ion-product expression:

$$CaOH_2(s) \rightleftharpoons Ca^{2+}(aq) + 2OH^-(aq) \qquad K_{sp} = [Ca^{2+}][OH^-]^2 = 6.5 \times 10^{-6}$$

Set up an ICE table, with s = molar solubility:

Concentration (mol/L)	$Ca(OH)_2(s)$	$\rightleftharpoons$	$Ca^{2+}(aq)$	+	$2OH^-(aq)$
Initial	—		0		0
Change	—		+s		+2s
Equilibrium	—		s		2s

Substitute into the ion-product expression and solve for s:

$$K_{sp} = [Ca^{2+}][OH^-]^2 = (s)(2s)^2 = (s)(4s^2) = 4s^3 = 6.5 \times 10^{-6}$$

$$s = \sqrt[3]{\frac{6.5 \times 10^{-6}}{4}} = 1.2 \times 10^{-2} \text{ mol/L}$$

Check We expect a low solubility from a slightly soluble salt. If we reverse the calculation, we should obtain the given K_{sp}: $4(1.2 \times 10^{-2})^3 = 6.9 \times 10^{-6}$, which is close to 6.5×10^{-6}.

Comment 1. Note that we did not double and *then* square $[OH^-]$. $2s$ *is* the $[OH^-]$, so we just squared it, as the ion-product expression required.
2. Once again, we assumed that the solid dissociates completely. In fact, the solubility is increased to about 2.0×10^{-2} mol/L by the presence of $CaOH^+(aq)$ formed in the following reaction:

$$Ca(OH)_2(s) \rightleftharpoons CaOH^+(aq) + OH^-(aq)$$

Our calculated answer is only approximate because we did not take this other species into account.

Follow-Up Problem 17.9 Milk of magnesia, a suspension of $Mg(OH)_2$ in water, relieves indigestion by neutralizing stomach acid. What is the molar solubility of $Mg(OH)_2$ in water if K_{sp} is 6.3×10^{-10}?

Using K_{sp} Values to Compare Solubilities As long as we compare compounds with the *same total number of ions* in their formulas, the K_{sp} values indicate *relative* solubility: *the higher the K_{sp}, the greater the solubility* (Table 17.3). Note that the relationship holds for compounds that form three ions, whether the cation/anion ratio is 1/2 or 2/1, because the mathematical expression containing s is the same ($4s^3$) in the calculation (see Sample Problem 17.9).

TABLE 17.3 Relationship between K_{sp} and Solubility at 25°C

No. of Ions	Formula	Cation/Anion	K_{sp}	Solubility (mol/L)
2	$MgCO_3$	1/1	3.5×10^{-8}	1.9×10^{-4}
2	$PbSO_4$	1/1	1.6×10^{-8}	1.3×10^{-4}
2	$BaCrO_4$	1/1	2.1×10^{-10}	1.4×10^{-5}
3	$Ca(OH)_2$	1/2	6.5×10^{-6}	1.2×10^{-2}
3	BaF_2	1/2	1.5×10^{-6}	7.2×10^{-3}
3	CaF_2	1/2	3.2×10^{-11}	2.0×10^{-4}
3	Ag_2CrO_4	2/1	2.6×10^{-12}	8.7×10^{-5}

$PbCrO_4(s) \rightleftharpoons Pb^{2+}(aq) + CrO_4^{2-}(aq)$ | $PbCrO_4(s) \rightleftharpoons Pb^{2+}(aq) + CrO_4^{2-}(aq;\ \text{added})$

FIGURE 17.14 The effect of a common ion on solubility. A. Lead(II) chromate, a slightly soluble salt, forms a saturated solution. **B.** When Na_2CrO_4 solution is added, the amount of $PbCrO_4(s)$ increases.

Effect of a Common Ion on Solubility

From Le Châtelier's principle, we know that *adding a common ion decreases the solubility of a slightly soluble ionic compound.* Consider a saturated solution of lead(II) chromate:

$$PbCrO_4(s) \rightleftharpoons Pb^{2+}(aq) + CrO_4^{2-}(aq) \qquad K_{sp} = [Pb^{2+}][CrO_4^{2-}] = 2.3 \times 10^{-13}$$

At a given temperature, K_{sp} depends on the product of the ion concentrations. If the concentration of either ion goes up, the concentration of the other goes down to maintain K_{sp}. Suppose that we add Na_2CrO_4, a soluble salt, to the saturated $PbCrO_4$ solution. The concentration of the common ion, CrO_4^{2-}, increases, and some of it combines with the Pb^{2+} ion to form more solid $PbCrO_4$ (Figure 17.14). In other words, the equilibrium position shifts to the left:

$$PbCrO_4(s) \rightleftharpoons Pb^{2+}(aq) + CrO_4^{2-}\ (aq;\ \text{added})$$

As a result of the addition, $[Pb^{2+}]$ is lower. As well, since the dissociation of $PbCrO_4$ produces equimolar amounts of Pb^{2+} and CrO_4^{2-}, we can say that $[Pb^{2+}]$ defines the solubility of $PbCrO_4$; in effect, the solubility of $PbCrO_4$ has decreased. Note that we would get the same result if the method of addition were reversed, that is, $PbCrO_4$ were more soluble in water than in aqueous Na_2CrO_4.

Sample Problem 17.10 Calculating the Effect of a Common Ion on Solubility

Problem In Sample Problem 17.9, we calculated the solubility of $Ca(OH)_2$ in water. What is its solubility in 0.10 mol/L $Ca(NO_3)_2$ (K_{sp} of $Ca(OH)_2 = 6.5 \times 10^{-6}$)?

Plan The addition of Ca^{2+}, the common ion, should lower the solubility. We write the equation and ion-product expression and set up an ICE table, with $[Ca^{2+}]_{init}$ reflecting the 0.10 mol/L $Ca(NO_3)_2$ and s equal to $[Ca^{2+}]_{from\ Ca(OH)_2}$. To simplify the math, we assume that, because K_{sp} is low, s is so small relative to $[Ca^{2+}]_{init}$ that $[Ca^{2+}]_{init} + s \approx [Ca^{2+}]_{init}$. Then we solve for s and check the assumption.

Solution Write the equation and ion-product expression:

$$CaOH_2(s) \rightleftharpoons Ca^{2+}(aq) + 2OH^-(aq) \qquad K_{sp} = [Ca^{2+}][OH^-]^2 = 6.5 \times 10^{-6}$$

Set up the ICE table, with $s = [Ca^{2+}]_{from\ Ca(OH)_2}$:

Concentration (mol/L)	$Ca(OH)_2(s)$	$\rightleftharpoons$	$Ca^{2+}(aq)$	+	$2OH^-(aq)$
Initial	—		0.10		0
Change	—		+s		+2s
Equilibrium	—		0.10 + s		2s

Since K_{sp} is small, $s << 0.10$ mol/L; thus, we can assume that $0.10 \text{ mol/L} + s \approx 0.10$ mol/L. Substitute into the ion-product expression and solve for s:

$$K_{sp} = [Ca^{2+}][OH^-]^2 = 6.5 \times 10^{-6} \approx (0.10)(2s)^2$$

Therefore,

$$4s^2 \approx \frac{6.5 \times 10^{-6}}{0.10} \quad \text{so} \quad s \approx \sqrt{\frac{6.5 \times 10^{-5}}{4}} = 4.0 \times 10^{-3} \text{ mol/L}$$

Check the assumption:

$$\frac{4.0 \times 10^{-3} \text{ mol/L}}{0.10 \text{ mol/L}} \times 100 = 4.0\% < 5\%$$

Check In Sample Problem 17.9, the solubility of $Ca(OH)_2$ was 0.012 mol/L; here, it is 0.0040 mol/L, one-third as much. As expected, the solubility *decreased* in the presence of added Ca^{2+}, the common ion.

Follow-Up Problem 17.10 To improve the X-ray image used to diagnose intestinal disorders, patients drink an aqueous suspension of $BaSO_4$ before the procedure. Ba^{2+} is opaque to X-rays (*see margin photo*). However, Ba^{2+} is also toxic; thus, the $[Ba^{2+}]$ is lowered by adding dilute Na_2SO_4. What is the solubility of $BaSO_4$ in **(a)** water; **(b)** 0.10 mol/L Na_2SO_4 (K_{sp} of $BaSO_4 = 1.1 \times 10^{-10}$)?

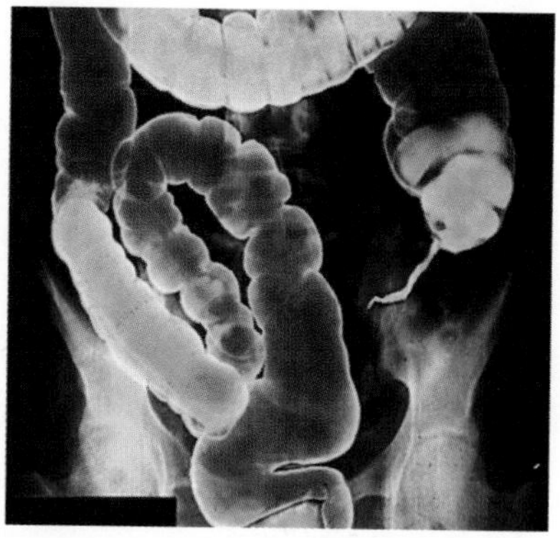

$BaSO_4$ imaging of a large intestine

Effect of pH on Solubility

If a slightly soluble ionic compound contains the conjugate base of a weak acid, *the addition of H^+ (from a strong acid) increases its solubility.* Once again, Le Châtelier's principle explains why. Consider a saturated solution of calcium carbonate:

$$CaCO_3(s) \rightleftharpoons Ca^{2+}(aq) + CO_3^{2-}(aq)$$

Adding strong acid introduces H^+, which reacts with the conjugate base of a weak acid, CO_3^{2-}, to form the conjugate base of another weak acid, HCO_3^-:

$$CO_3^{2-}(aq) + H_3O^+(aq) \longrightarrow HCO_3^-(aq) + H_2O(l)$$

If enough H_3O^+ is added, carbonic acid forms and then decomposes to H_2O and CO_2, and the CO_2 gas escapes the container:

$$HCO_3^-(aq) + H_3O^+(aq) \longrightarrow H_2CO_3(aq) + H_2O(l) \longrightarrow CO_2(g) + 2H_2O(l)$$

The net effect of adding H^+ is a shift in the equilibrium position to the right, and more $CaCO_3$ dissolves:

$$CaCO_3(s) \rightleftharpoons Ca^{2+}(aq) + CO_3^{2-}(aq) \xrightarrow{H^+(aq)} HCO_3^-(aq) \xrightarrow{H^+(aq)} H_2CO_3(aq) \longrightarrow CO_2(g) + H_2O(l) + Ca^{2+}(aq)$$

This example illustrates a qualitative field test for carbonate minerals, because the CO_2 bubbles vigorously (Figure 17.15).

FIGURE 17.15 Test for the presence of a carbonate. When a carbonate mineral is treated with HCl, bubbles of CO_2 form.

In contrast, adding H_3O^+ to a saturated solution of a slightly soluble ionic compound with a strong-acid conjugate base, such as AgCl, has no effect on its solubility:

$$AgCl(s) \rightleftharpoons Ag^+(aq) + Cl^-(aq)$$

The Cl^- ion can coexist with high $[H_3O^+]$, so the equilibrium position is not affected.

Sample Problem 17.11 Predicting Whether the Addition of a Strong Acid Affects Solubility

Problem Write a balanced equation to show whether the addition of H_3O^+ from a strong acid affects the solubility of **(a)** lead(II) bromide; **(b)** copper(II) hydroxide; **(c)** iron(II) sulfide.

Plan We write the balanced dissolution equation and note the anion:

- Weak-acid conjugate bases react with H_3O^+ and increase solubility when a strong acid is added.
- Strong-acid conjugate bases do not react with H_3O^+, so the addition of a strong acid has no effect.

Solution (a) $PbBr_2(s) \rightleftharpoons Pb^{2+}(aq) + 2Br^-(aq)$

There is no effect. Br^- is the conjugate base of HBr, a strong acid, so it does not react with H^+.

(b) $Cu(OH)_2(s) \rightleftharpoons Cu^{2+}(aq) + 2OH^-(aq)$

The addition of H_3O^+ increases solubility. OH^- is the conjugate base of H_2O, a very weak acid, so it reacts with the added H_3O^+:

$$OH^-(aq) + H_3O^+(aq) \longrightarrow 2H_2O(l)$$

(c) $FeS(s) + H_2O(l) \rightleftharpoons Fe^{2+}(aq) + HS^-(aq) + OH^-(aq)$

The addition of H_3O^+ increases solubility. The S^{2-} ion reacts completely with water to form HS^- and OH^-. The added H_3O^+ reacts with both of these weak-acid conjugate bases:

$$HS^-(aq) + H_3O^+(aq) \longrightarrow H_2S(aq) + H_2O(l)$$
$$OH^-(aq) + H_3O^+(aq) \longrightarrow 2H_2O(l)$$

Follow-Up Problem 17.11 Write a balanced equation to show whether the addition of $HNO_3(aq)$ affects the solubility of **(a)** calcium fluoride; **(b)** zinc sulfide; **(c)** silver iodide.

FIGURE 17.16 Limestone cave in Nerja, Málaga, Spain

Applying Ionic Equilibria to the Formation of a Limestone Cave

Limestone caves, and the remarkable structures within them, provide striking evidence of the results of aqueous ionic equilibria involving carbonate rocks and the carbon dioxide and water that have flowed through them for many hundreds of millennia (Figure 17.16).

The Cave-Forming Reactions Limestone is mostly calcium carbonate ($CaCO_3$; $K_{sp} = 3.3 \times 10^{-9}$). Two key reactions help us understand how limestone caves form:

1. Gaseous CO_2 in air is in equilibrium with aqueous CO_2 in natural water:

$$CO_2(g) \overset{H_2O(l)}{\rightleftharpoons} CO_2(aq) \qquad \textbf{(Equation 1)}$$

The concentration of aqueous CO_2 is proportional to the partial pressure of $CO_2(g)$ in contact with the water (Henry's law; Section 12.3):

$$[CO_2(aq)] \propto p_{CO_2}$$

Due to the continuous release of CO_2 from within Earth (outgassing), p_{CO_2} in soil-trapped air is *higher* than p_{CO_2} in the atmosphere.

2. The reaction of CO_2 with water produces H_3O^+:

$$CO_2(aq) + 2H_2O(l) \rightleftharpoons H_3O^+(aq) + HCO_3^-(aq)$$

Thus, since $CaCO_3$ contains the conjugate base of a weak acid, this formation of H_3O^+ increases the solubility of $CaCO_3$:

$$CaCO_3(s) + CO_2(aq) + H_2O(l) \rightleftharpoons Ca^{2+}(aq) + 2HCO_3^-(aq) \qquad \textbf{(Equation 2)}$$

The Cave-Forming Process Here is an overview of the process that forms most limestone caves:

1. As surface water trickles through cracks in the ground, it meets soil-trapped air with a high p_{CO_2}. As a result, $[CO_2(aq)]$ increases (Equation 1 shifts to the right), and the solution becomes more acidic.

2. When this CO_2-rich water contacts $CaCO_3$, more $CaCO_3$ dissolves (Equation 2 shifts to the right). As a result, more rock is carved out, more water flows in, and so on. Centuries pass, and a cave slowly begins to form.

3. Some of the aqueous solution, dilute $Ca(HCO_3)_2$, passes through the ceiling of the growing cave. As it drips, it meets air, which has a lower p_{CO_2} than the soil, and some $CO_2(aq)$ comes out of solution (Equation 1 shifts to the left).

4. Consequently, some $CaCO_3$ precipitates on the ceiling and on the floor below, where the drops land (Equation 2 shifts to the left). After many decades, the ceiling bears a *stalactite*, and a spike, called a *stalagmite*, grows up from the floor. Eventually, the stalactite and stalagmite meet to form a column of precipitated limestone.

The same chemical process can lead to different shapes. Standing pools of $Ca(HCO_3)_2$ solution form limestone "lily pads" or "corals." Cascades of solution form delicate limestone "draperies" on a cave wall, with fabulous colours arising from trace metal ions, such as iron (reddish brown) and copper (bluish green).

Predicting the Formation of a Precipitate: Q_{sp} versus K_{sp}

As in Chapter 15, we are now going to compare the values of Q_{sp} and K_{sp} to see if a reaction has reached equilibrium and, if not, in which net direction the reaction will move until it does. Using solutions of *soluble* salts that contain the ions of *slightly* soluble salts, we can calculate the ion concentrations and predict the result when we mix the solutions:

- If $Q_{sp} = K_{sp}$, the solution is saturated and no change will occur.
- If $Q_{sp} > K_{sp}$, a precipitate will form until the remaining solution is saturated.
- If $Q_{sp} < K_{sp}$, no precipitate will form because the solution is unsaturated.

To help us determine which ions are soluble and which are slightly soluble, general solubility rules for some common ions are given in Table 17.4.

TABLE 17.4 Solubility Rules for Ionic Compounds in Water

Soluble Ionic Compounds	Insoluble Ionic Compounds
1. All common compounds of group 1 ions (such as Li^+, Na^+, and K^+) and the ammonium ion (NH_4^+) are soluble.	1. All common metal hydroxides are insoluble, except those of group 1 ions and the larger members of group 2 (beginning with Ca^{2+}).
2. All common nitrates (NO_3^-), ethanoates (acetates) (CH_3COO^-), and most perchlorates (ClO_4^-) are soluble.	2. All common carbonates (CO_3^{2-}) and phosphates (PO_4^{3-}) are insoluble, except those of group 1 ions and NH_4^+.
3. All common chlorides (Cl^-), bromides (Br^-), and iodides (I^-) are soluble, except those of Ag^+, Pb^{2+}, Cu^+, and Hg_2^{2+}. All common fluorides (F^-) are soluble, except those of Pb^{2+} and group 2 ions.	3. All common sulfides are insoluble, except those of group 1 ions, group 2 ions, and NH_4^+.
4. All common sulfates (SO_4^{2-}) are soluble, except those of Ca^{2+}, Sr^{2+}, Ba^{2+}, Ag^+, and Pb^{2+}.	

Sample Problem 17.12 Predicting Whether a Precipitate Will Form

Precipitation of CaF_2

Problem A common laboratory method for preparing a precipitate is to mix solutions containing the component ions. Does a precipitate form when 0.100 L of 0.30 mol/L $Ca(NO_3)_2$ is mixed with 0.200 L of 0.060 mol/L NaF?

Plan First, we decide which slightly soluble salt could form, look up its K_{sp} value in Appendix C, and write a dissolution equation and ion-product expression. To see whether mixing these solutions forms a precipitate, we find the initial ion concentrations by calculating the amount (mol) of each ion from its concentration and volume and then dividing by the *total* volume because each solution dilutes the other. Finally, we substitute these concentrations to calculate Q_{sp}, and compare Q_{sp} with K_{sp}.

Solution The ions present are Ca^{2+}, Na^+, F^-, and NO_3^-. All sodium and all nitrate salts are soluble (Table 17.4), so the only possibility is CaF_2 ($K_{sp} = 3.2 \times 10^{-11}$).

Write the equation and ion-product expression:

$$CaF_2(s) \rightleftharpoons Ca^{2+}(aq) + 2F^-(aq) \qquad Q_{sp} = [Ca^{2+}][F^-]^2$$

Calculate the ion concentrations:

$$\text{Amount (mol) of } Ca^{2+} = 0.30 \text{ mol/L } Ca^{2+} \times 0.100 \text{ L} = 0.030 \text{ mol } Ca^{2+}$$

$$[Ca^{2+}]_{init} = \frac{0.030 \text{ mol } Ca^{2+}}{0.100 \text{ L} + 0.200 \text{ L}} = 0.10 \text{ mol/L } Ca^{2+}$$

$$\text{Amount (mol) of } F^- = 0.060 \text{ mol/L } F^- \times 0.200 \text{ L} = 0.012 \text{ mol } F^-$$

$$[F^-]_{init} = \frac{0.012 \text{ mol } F^-}{0.100 \text{ L} + 0.200 \text{ L}} = 0.040 \text{ mol/L } F^-$$

Substitute into the ion-product expression and compare Q_{sp} with K_{sp}:

$$Q_{sp} = [Ca^{2+}]_{init}[F^-]^2_{init} = (0.10)(0.040)^2 = 1.6 \times 10^{-4}$$

Because $Q_{sp} > K_{sp}$, CaF_2 will precipitate until $Q_{sp} = 3.2 \times 10^{-11}$.

Check Make sure that you round off and quickly check the math. For example, $Q_{sp} = (1 \times 10^{-1})(4 \times 10^{-2})^2 = 1.6 \times 10^{-4}$. With K_{sp} so low, CaF_2 must have a low solubility. Given the sizable concentrations being mixed, we would expect CaF_2 to precipitate.

Follow-Up Problem 17.12 As a result of mineral erosion and biological activity, the phosphate ion is common in natural water, where it often precipitates as insoluble salts, such as $Ca_3(PO_4)_2$. If $[Ca^{2+}]_{init} = [PO_4^{3-}]_{init} = 1.0 \times 10^{-9}$ mol/L in a given river, will $Ca_3(PO_4)_2$ precipitate [K_{sp} of $Ca_3(PO_4)_2 = 1.2 \times 10^{-29}$]?

Sample Problem 17.13 Using Molecular Scenes to Predict Whether a Precipitate Will Form

Problem These four scenes represent solutions of silver (*grey*) and carbonate (*black and red*) ions above solid silver carbonate. (The solid, other ions, and water are not shown.)

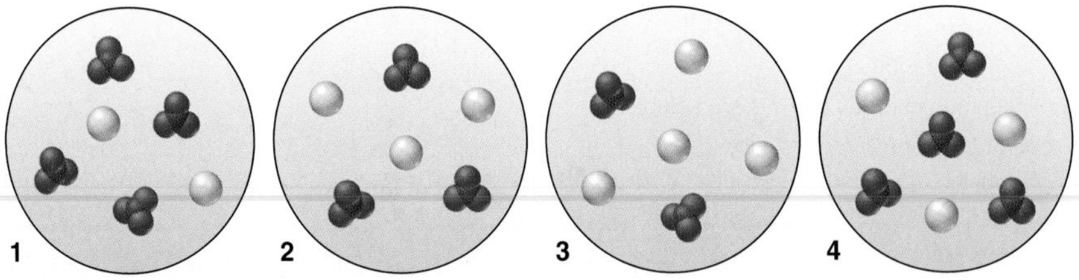

(a) Which scene best represents the solution in equilibrium with the solid?

(b) In which, if any, other scene(s) will additional solid silver carbonate form?

(c) Explain how, if at all, the addition of a small volume of a concentrated strong acid will affect the $[Ag^+]$ and the mass of solid present in scene 4.

Plan (a) The solution of silver and carbonate ions in equilibrium with the solid (Ag_2CO_3) should have the same relative numbers of cations and conjugate bases as there are in the formula. We examine the scenes to see which has a ratio of 2 Ag^+ to 1 CO_3^{2-}. **(b)** A solid forms if the value of Q_{sp} exceeds the value of K_{sp}. We write the dissolution equation and the Q_{sp} expression. Then we count ions to calculate Q_{sp} in each scene and see which, if any, exceeds the value for the solution from part (a). **(c)** The CO_3^{2-} ion reacts with added H_3O^+, so adding strong acid will shift the equilibrium to the right. We write the equations and determine how a shift to the right affects $[Ag^+]$ and the mass of solid Ag_2CO_3.

Solution (a) Scene 3 is the only scene with an Ag^+/CO_3^{2-} ratio of 2/1, as in the formula of the solid.

(b) Calculate the ion products:

$$Ag_2CO_3(s) \rightleftharpoons 2Ag^+(aq) + CO_3^{2-}(aq) \qquad Q_{sp} = [Ag^+]^2[CO_3^{2-}]$$

Scene 1: $Q_{sp} = (2)^2(4) = 16$ Scene 2: $Q_{sp} = (3)^2(3) = 27$

Scene 3: $Q_{sp} = (4)^2(2) = 32$ Scene 4: $Q_{sp} = (3)^2(4) = 36$

Therefore, from scene 3, $K_{sp} = 32$. The Q_{sp} value for scene 4 is the only other value that equals or exceeds 32, so a precipitate of Ag_2CO_3 will form in scene 4.

(c) Write the equations:

(1) $Ag_2CO_3(s) \rightleftharpoons 2Ag^+(aq) + CO_3^{2-}(aq)$

(2) $CO_3^{2-}(aq) + 2H_3O^+(aq) \longrightarrow H_2CO_3(aq) + 2H_2O(l) \longrightarrow 3H_2O(l) + CO_2(g)$

The CO_2 leaves as a gas, so adding H_3O^+ shifts the equilibrium position of reaction 2 to the right. This change lowers the $[CO_3^{2-}]$ in reaction 1, thereby causing more CO_3^{2-} to form. As a result, more solid dissolves, which means that $[Ag^+]$ increases and the mass of Ag_2CO_3 decreases.

Check (a) In scene 1, the formula has two CO_3^{2-} per formula unit, not two Ag^+. **(b)** Even though scene 4 has fewer Ag^+ ions than scene 3, its Q_{sp} value is higher and exceeds K_{sp}.

Follow-Up Problem 17.13 The following scenes represent solutions of nickel(II) (*black*) and hydroxide (*red and blue*) ions above solid nickel(II) hydroxide. (For clarity, the solid, other ions, and water are not shown.)

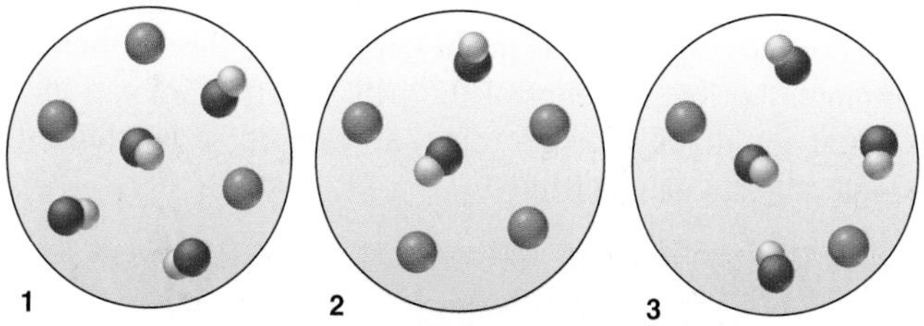

(a) Which scene best depicts the solution at equilibrium with the solid?

(b) In which, if any, other scene(s) will additional solid form?

(c) Will addition of a small amount of concentrated strong acid or strong base affect the mass of solid present in any scene? Explain.

Separating Ions by Selective Precipitation and Simultaneous Equilibria

Let us consider two ways to separate one ion from another by reacting them with a given precipitating ion to form compounds with different solubilities.

Selective Precipitation In the process of **selective precipitation**, a solution of one precipitating ion is added to a solution of two ionic compounds until the Q_{sp} of the *more soluble* compound is almost equal to its K_{sp}. This method ensures that the K_{sp} of the *less soluble* compound is exceeded as much as possible. As a result, the maximum amount of the less soluble compound precipitates, but none of the more soluble compound does.

Sample Problem 17.14 Separating Ions by Selective Precipitation

Problem A solution consists of 0.20 mol/L $MgCl_2$ and 0.10 mol/L $CuCl_2$. Calculate the $[OH^-]$ that will separate the metal ions (K_{sp} of $Mg(OH)_2 = 6.3 \times 10^{-10}$; K_{sp} of $Cu(OH)_2 = 2.2 \times 10^{-20}$).

Plan Because both compounds have 1/2 ratios of cation/anion, when we compare their K_{sp} values, we find that $Mg(OH)_2$ is about 10^{10} times more soluble than $Cu(OH)_2$; thus, $Cu(OH)_2$ precipitates first. (See "Using K_{sp} Values to Compare Solubilities" earlier in this section.) We write the dissolution equations and ion-product expressions. We are given both cation concentrations, so we solve for the $[OH^-]$ that gives a saturated solution of $Mg(OH)_2$, because this $[OH^-]$ will precipitate the most Cu^{2+}. Then we calculate the $[Cu^{2+}]$ remaining to see if the separation was accomplished.

Solution Write the equations and ion-product expressions:

$$Mg(OH)_2(s) \rightleftharpoons Mg^{2+}(aq) + 2OH^-(aq) \quad K_{sp} = [Mg^{2+}][OH^-]^2 = 6.3 \times 10^{-10}$$
$$Cu(OH)_2(s) \rightleftharpoons Cu^{2+}(aq) + 2OH^-(aq) \quad K_{sp} = [Cu^{2+}][OH^-]^2 = 2.2 \times 10^{-20}$$

Calculate the $[OH^-]$ that gives a saturated $Mg(OH)_2$ solution:

$$[OH^-] = \sqrt{\frac{K_{sp}}{[Mg^{2+}]}} = \sqrt{\frac{6.3 \times 10^{-10}}{0.20}} = 5.6 \times 10^{-5} \text{ mol/L}$$

This is the maximum $[OH^-]$ that will *not* precipitate Mg^{2+} ion.

Calculate the $[Cu^{2+}]$ remaining in the solution with this $[OH^-]$:

$$[Cu^{2+}] = \frac{K_{sp}}{[OH^-]^2} = \frac{2.2 \times 10^{-20}}{(5.6 \times 10^{-5})^2} = 7.0 \times 10^{-12} \text{ mol/L}$$

Since the initial $[Cu^{2+}]$ is 0.10 mol/L, virtually all the Cu^{2+} ion is precipitated.

Check Rounding, we find that $[OH^-]$ seems correct: $\approx \sqrt{(6 \times 10^{-10})/0.2} = 5 \times 10^{-5}$.
The $[Cu^{2+}]$ remaining also seems correct: $\frac{(200 \times 10^{-22})}{(5 \times 10^{-5})^2} = 8 \times 10^{-12}$.

Follow-Up Problem 17.14 A solution contains 0.050 mol/L $BaCl_2$ and 0.025 mol/L $CaCl_2$. What concentration of SO_4^{2-} will separate the cations in solution (K_{sp} of $BaSO_4$ = 1.1×10^{-10}; K_{sp} of $CaSO_4 = 2.4 \times 10^{-5}$)?

Simultaneous Equilibria Sometimes two or more equilibrium systems are controlled simultaneously to separate one metal ion from another as their sulfides, using HS^- as the precipitating ion. We control the $[HS^-]$ so that we can increase the K_{sp} value of one metal sulfide but not the other, and we do so by shifting the H_2S dissociation through adjustments of $[H_3O^+]$:

$$H_2S(aq) + H_2O(l) \rightleftharpoons H_3O^+(aq) + HS^-(aq)$$

These adjustments are controlled as follows:

- If we add strong acid, the high $[H^+]$ shifts H_2S dissociation to the left, which decreases $[HS^-]$, so the *less* soluble sulfide precipitates.
- If we add strong base, the low $[H^+]$ shifts H_2S dissociation to the right, which increases $[HS^-]$, so the *more* soluble sulfide precipitates.

Thus, we shift one equilibrium system (H_2S dissociation) by adjusting a second (H_2O ionization) to control a third (metal sulfide solubility).

As the following Chemical Connections section demonstrates, the principles of ionic equilibria often help us understand the chemical basis of complex environmental problems and may provide ways to solve these problems.

SUMMARY OF SECTION 17.3

- Only as a first approximation does the dissolved portion of a slightly soluble salt dissociate completely into ions.
- In a saturated solution, dissolved ions and the undissolved solid salt are in equilibrium. The product of the ion concentrations, each raised to the power of its subscript in the formula, has a constant value ($Q_{sp} = K_{sp}$).
- The value of K_{sp} can be obtained from the solubility, and vice versa.
- Adding a common ion lowers the solubility of a compound.
- Adding H^+ (lowering the pH) increases the solubility of a compound if the anion of the compound is also the conjugate base of a weak acid.
- Limestone caves result from shifts in the $CaCO_3/CO_2$ equilibrium system.
- An ionic solid forms if $Q_{sp} > K_{sp}$ when solutions that each contain one of the ions are mixed.
- Ions are precipitated selectively from a solution of two compounds by adding a precipitating ion until the K_{sp} of one compound is increased as much as possible without exceeding the K_{sp} of the other. An extension of this approach uses simultaneous control of three equilibrium systems to separate metal ions as their sulfides.
- Lakes bounded by limestone-rich soils form buffer systems that prevent acidification.

Acid rain, the deposition of acids in wet form as rain, snow, or fog, or in dry form as solid particles, is an environmental problem that has been observed in Canada, the United States, Mexico, the Amazon basin, Europe, Russia, many parts of Asia, and even at the North and South Poles. Let us see how it arises and how some of its harmful effects can be prevented.

Origins of Acid Rain The strong acids H_2SO_4 and HNO_3 cause the greatest concern, so let us consider how they form (Figure B17.1):

1. *Sulfuric acid.* Sulfur dioxide (SO_2), formed mostly by the burning of high-sulfur coal, forms sulfurous acid (H_2SO_3) in contact with water. The atmospheric pollutants hydrogen peroxide (H_2O_2) and ozone (O_3) dissolve in the water in clouds and oxidize the sulfurous acid to sulfuric acid:

$$H_2O_2(aq) + H_2SO_3(aq) \longrightarrow H_2SO_4(aq) + H_2O(l)$$

Alternatively, SO_2 is oxidized by atmospheric hydroxyl radicals ($HO\cdot$) to sulfur trioxide (SO_3), which forms H_2SO_4 with water.

2. *Nitric acid.* Nitrogen oxides (NO_x) form from N_2 and O_2 during high-temperature combustion in the engines of cars and trucks and in electrical power plants. Once formed, NO forms NO_2 and HNO_3 in a process that creates smog. (See "Highlights of Nitrogen Chemistry" in Section 13.7.) At night, NO_x is converted to N_2O_5, which reacts with water to form HNO_3:

$$N_2O_5(g) + H_2O(l) \longrightarrow 2HNO_3(aq)$$

Ammonium salts, deposited as NH_4HSO_4 or NH_4NO_3, produce HNO_3 in soil through biochemical oxidation. In Canada, the major sources of sulfur dioxide emissions are non-ferrous metal smelters, followed by coal-fired generators. Motor vehicles and, to a lesser extent, coal-fired generators, are the major sources of nitrogen oxides. [Source: http://www.ec.gc.ca/eau-water/default.asp?lang=En&n=FDF30C16-1 (2010).]

The pH of Acid Rain Normal, clean rainwater is weakly acidic from the reaction of atmospheric CO_2 with water (see Problem 17.151) and should have a pH value of about 5.6:

$$CO_2(g) + 2H_2O(l) \rightleftharpoons H_3O^+(aq) + HCO_3^-(aq)$$

Between 1980 and 1984, the pH of rain in parts of southern Ontario, Québec, and eastern Canada was as low as 4.2. Today, due to reductions in the emissions of SO_2, the rain is less acidic. In eastern Canada, however, the average pH of rain is still 4.5 in some places. Rain in Wheeling, West Virginia, has had a pH of 1.8, lower than the pH of lemon juice. Rain in industrial parts of Sweden once had a pH of 2.7, about the same as vinegar.

Effects of Acid Rain Some effects of the increased acidity of water are listed in Table B17.1. With tens of thousands of rivers and lakes around the world becoming or already acidified, the loss of freshwater fish became a major concern long ago. Although some ecosystems in eastern Canada are slowly recovering due to the large decrease in industrial emissions, many systems have not recovered and some may take a very long time to return to normal. [REFERENCE: http://www.ec.gc.ca/air/default.asp?lang=En&n=7E5E9F00-1]

In addition, acres of forest have been affected by the acid, which removes nutrients and releases toxic substances from the soil. The aluminosilicates that make up most soils are extremely insoluble in water. Through a complex series of simultaneous equilibria, contact with water at pH < 5 causes these materials to release some bound Al^{3+}, which is toxic to fish and plants. At the same time, acid rain dissolves Ca^{2+} and Mg^{2+} ions, which are nutrients for plants and animals, from the soil.

Marble and limestone (both primarily $CaCO_3$) in buildings and monuments react with sulfuric acid to form gypsum ($CaSO_4\cdot 2H_2O$), which flakes off. Interestingly, the same process that destroys these structures rescues lakes that are bounded by limestone and calcite-rich soil. The minerals dissolve sufficiently in lake water to form an HCO_3^-/CO_3^{2-} buffer, which is capable of absorbing the incoming H_3O^+ and maintaining a mildly basic pH:

$$CO_3^{2-}(aq) + H_3O^+(aq) \rightleftharpoons HCO_3^-(aq) + H_2O(l)$$

Many thousands of Canada's lakes, however, lie on the Precambrian Shield, which contains little limestone or calcite. As a result, these lakes exhibit greater effects from acid rain, including decreased pH level of the water, increased sulfate concentrations, and larger amounts of metals such as aluminum and manganese. Acidified lakes and rivers that are in contact with granite and other weathering-resistant bedrock can be remediated by *liming* (treating with

TABLE B17.1 The Effects of Decreasing pH on Aquatic Life

As Water pH Approaches ...	Effects
6.0	• Crustaceans, insects, and some plankton species begin to disappear.
5.0	• Major changes in the makeup of the plankton community occur. • Less-desirable species of mosses and plankton may begin to invade. • The progressive loss of some fish populations is likely, with the more highly valued species being generally the least tolerant of acidity.
Less than 5.0	• The water is largely devoid of fish. • The bottom is covered with undecayed material. • Areas near the shore may be dominated by mosses. • Terrestrial animals that depend on aquatic ecosystems are affected. Waterfowl, for example, depend on aquatic organisms for nourishment and nutrients. As these food sources are reduced or eliminated, the quality of the waterfowl's habitat declines and their reproductive success is affected.

(Continued)

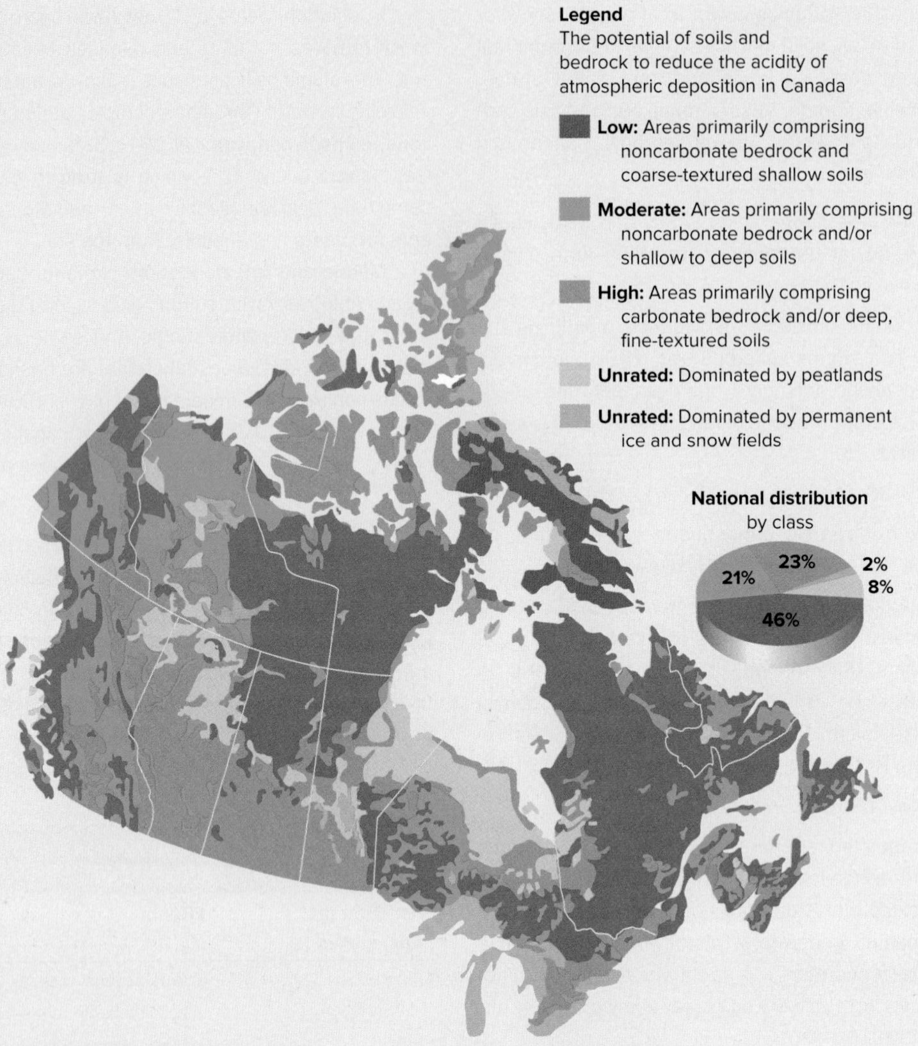

FIGURE B17.1 Neutralizing potential of Canada's soil and bedrock

limestone). This approach is expensive, however, and only a stopgap because the lakes are acidic again within several years.

Remarkably, in parts of western Canada, notably the Prairies, there are alkaline ponds and lakes. These bodies of water are considered wetlands; the water is not deep and not always present. Acid rain has less effect on these alkaline waters, since they are capable of neutralizing the pH. Western Canada has the added benefit of weather patterns that tend to move eastward, as well as groundrock that has a higher capacity for neutralizing acid rain. In recent years, however, there has been some concern that the development of the Alberta Oil Sands will lead to an increase in the observed effects of acid rain due to higher emissions of nitrogen oxides and sulfur dioxide.

Preventing Acid Rain The effective prevention of acid rain has to address the sources of the sulfur and nitrogen pollutants:

1. *Sulfur pollutants.* As was pointed out in the Chemical Connections in Chapter 5, the principal method for minimizing sulfur dioxide release is by "scrubbing" power-plant emissions with limestone in both dry and wet forms. Another method reduces SO_2 with methane, coal, or H_2S, and the mixture is converted catalytically to sulfur, which is sold as a by-product:

$$16H_2S(g) + 8SO_2(g) \xrightarrow{\text{catalyst}} 3S_8(s) + 16H_2O(l)$$

Although low-sulfur coal is rare and expensive to mine, coal that contains sulfur can be converted into gaseous and liquid low-sulfur fuels. The sulfur is removed (as H_2S) in an acid-gas scrubber after gasification.

2. *Nitrogen pollutants.* Through the use of a catalytic converter in an auto exhaust system, NO_x species are reduced to N_2. In power-plant emissions, NO_x is decreased by adjusting the conditions and by treating hot stack gases with ammonia in the presence of a heterogeneous catalyst:

$$4NO(g) + 4NH_3(g) + O_2(g) \xrightarrow{\text{catalyst}} 4N_2(g) + 6H_2O(g)$$

Reducing the power-plant emissions in North America and Europe has increased the pH of rainfall and some surface waters (Figure B17.2). Lime (CaO) is routinely used to react with acid rain falling on cropland. Further progress is expected under current legislation, but in much of eastern North America and northern Europe, additional measures are needed for full recovery.

(Continued)

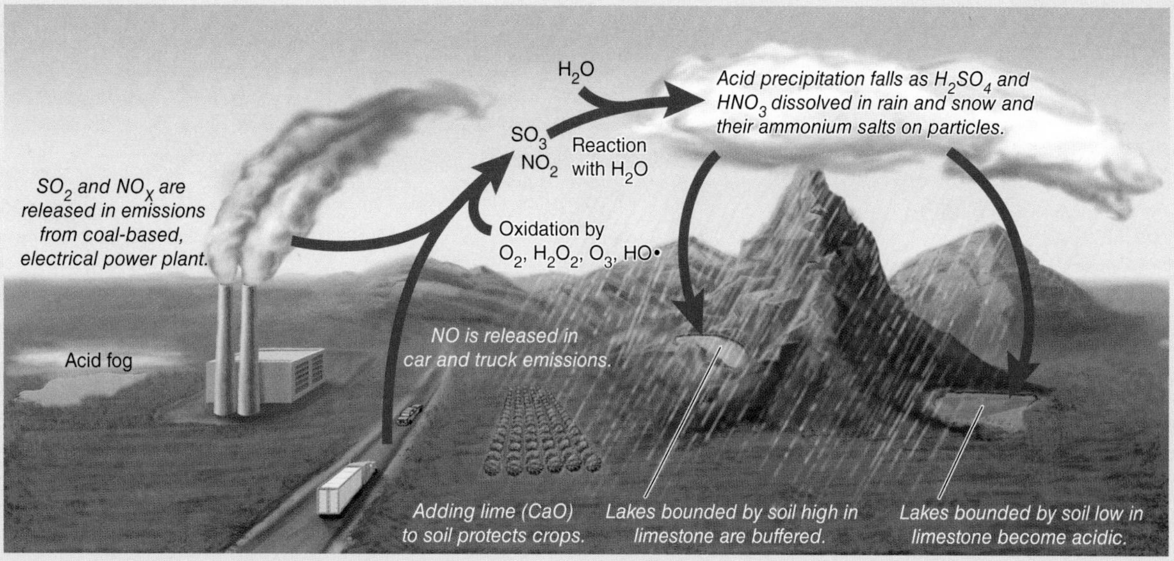

FIGURE B17.2 Formation of acidic precipitation. A complex interplay of human activities, atmospheric chemistry, and environmental distribution leads to acidic precipitation and its harmful effects.

Problems

B17.1 An environmental technician collects a sample of rainwater. Back in the lab, her pH meter is not working, so she uses indicator solutions to estimate the pH. A piece of litmus paper turns red, indicating acidity, so she divides the sample into thirds and obtains the following results: thymol blue turns yellow, bromphenol blue turns green, and methyl red turns red. Estimate the pH of the rainwater.

B17.2 A lake that has a surface area of 0.040 km^2 receives 25.4 mm of rain with a pH of 4.20. (Assume that the acidity of the rain is due to a strong, monoprotic acid.)

(a) What amount (mol) of H^+ is in the rain falling on the lake?

(b) If the lake is unbuffered and its average depth is 3.05 m before the rain, find the pH after the rain has been mixed with lake water. (Assume that the initial pH is 7.00, and ignore runoff from the surrounding land.)

(c) If the lake contains hydrogen carbonate ions (HCO_3^-), what mass of HCO_3^- would neutralize the acid in the rain?

17.4 Equilibria Involving Complex Ions

A third kind of aqueous ionic equilibrium involves a type of ion that we mentioned briefly in Section 16.8. A *simple ion,* such as Na^+ or CH_3COO^-, consists of one or a few bonded atoms, with an excess or deficit of electrons. A **complex ion** consists of a central metal ion covalently bonded to two or more anions or molecules, called **ligands**. Hydroxide, chloride, and cyanide ions are examples of ionic ligands; water, carbon monoxide, and ammonia are examples of molecular ligands. In a complex ion, the central metal is placed first, followed by the ligand, and the entire ion is placed inside square brackets. In the complex ion $[Cr(NH_3)_6]^{3+}$, for example, the central Cr^{3+} is surrounded by six NH_3 ligands (Figure 17.17). Hydrated metal ions are complex ions with water as ligands (Section 16.5). In Chapter 24, we will discuss the transition metals and the structures and properties of the numerous complex ions they form. Here, we focus on the equilibria of hydrated ions with ligands other than water.

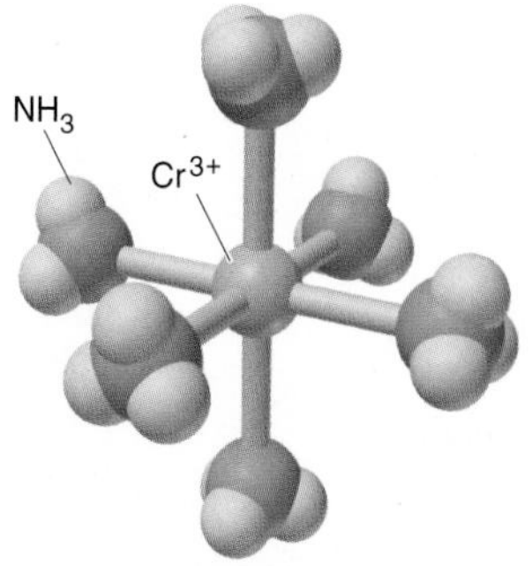

FIGURE 17.17 $[Cr(NH_3)_6]^{3+}$**, a typical complex ion**

Formation of Complex Ions

When a salt dissolves in water, a complex ion is formed, with water as ligands around the metal ion. In many cases, when we treat this hydrated cation with a

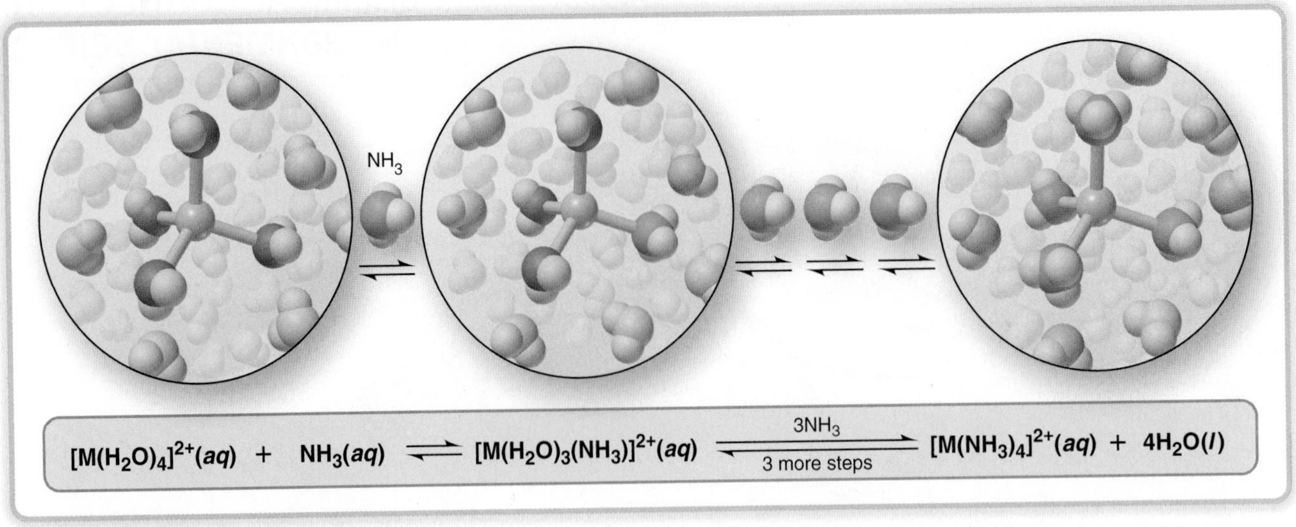

FIGURE 17.18 The step-by-step exchange of NH_3 for H_2O in $[M(H_2O)_4]^{2+}$. The molecular views show the first exchange and the fully ammoniated ion.

solution of another ligand, the bound water molecules are replaced by the other ligand. For example, a hydrated M^{2+} ion, $[M(H_2O)_4]^{2+}$, forms the ammoniated ion, $[M(NH_3)_4]^{2+}$, in aqueous NH_3:

$$[M(H_2O)_4]^{2+}(aq) + 4NH_3(aq) \rightleftharpoons [M(NH_3)_4]^{2+}(aq) + 4H_2O(l)$$

At equilibrium, this system is expressed by a ratio of concentration terms whose form follows that of any other equilibrium expression:

$$K = \frac{[[M(NH_3)_4]^{2+}]}{[[M(H_2O)_4]^{2+}][NH_3]^4} = K_f$$

Once again, we do not include the term for water because its activity is 1.

At the molecular level, depicted in Figure 17.18, the actual process occurs step by step, with ammonia molecules replacing water molecules one at a time. This process yields a series of intermediate species, each with its own formation constant:

$$[M(H_2O)_4]^{2+}(aq) + NH_3(aq) \rightleftharpoons [M(H_2O)_3(NH_3)]^{2+}(aq) + H_2O(l)$$

$$K_{f_1} = \frac{[[M(H_2O)_3(NH_3)]^{2+}]}{[[M(H_2O)_4]^{2+}][NH_3]}$$

$$[M(H_2O)_3(NH_3)]^{2+}(aq) + NH_3(aq) \rightleftharpoons [M(H_2O)_2(NH_3)_2]^{2+}(aq) + H_2O(l)$$

$$K_{f_2} = \frac{[[M(H_2O)_2(NH_3)_2]^{2+}]}{[[M(H_2O)_3(NH_3)]^{2+}][NH_3]}$$

$$[M(H_2O)_2(NH_3)_2]^{2+}(aq) + NH_3(aq) \rightleftharpoons [M(H_2O)(NH_3)_3]^{2+}(aq) + H_2O(l)$$

$$K_{f_3} = \frac{[[M(H_2O)(NH_3)_3]^{2+}]}{[[M(H_2O)_2(NH_3)_2]^{2+}][NH_3]}$$

$$[M(H_2O)(NH_3)_3]^{2+}(aq) + NH_3(aq) \rightleftharpoons [M(NH_3)_4]^{2+}(aq) + H_2O(l)$$

$$K_{f_4} = \frac{[[M(NH_3)_4]^{2+}]}{[[M(H_2O)(NH_3)_3]^{2+}][NH_3]}$$

The *sum* of the equations gives the overall equation, so the *product* of the individual formation constants gives the overall **formation constant (K_f)**:

$$K_f = K_{f_1} \times K_{f_2} \times K_{f_3} \times K_{f_4}$$

Recall that *all complex ions are Lewis adducts* (Section 16.8). The metal ion acts as a Lewis acid (accepts an electron pair), and the ligand acts as a Lewis base (donates an electron pair). In the formation of $[M(NH_3)_4]^{2+}$, the K_f for each step is much larger than 1 because ammonia is a stronger Lewis base than water. Therefore, if we add excess ammonia to the $[M(H_2O)_4]^{2+}$ solution, nearly all the M^{2+} ions exist as $[M(NH_3)_4]^{2+}(aq)$.

Table 17.5 and Appendix C provide the K_f values of some complex ions. Notice that these values are all 10^6 or greater, which means that the ions form readily from hydrated

TABLE 17.5 Formation Constants (K_f) of Some Complex Ions at 25°C

Complex Ion	K_f
$[Ag(CN)_2]^-$	3.0×10^{20}
$[Ag(NH_3)_2]^+$	1.7×10^7
$[Ag(S_2O_3)_2]^{3-}$	4.7×10^{13}
$[AlF_6]^{3-}$	4×10^{19}
$[Al(OH)_4]^-$	3×10^{33}
$[Be(OH)_4]^{2-}$	4×10^{18}
$[CdI_4]^{2-}$	1×10^6
$[Co(OH)_4]^{2-}$	5×10^9
$[Cr(OH)_4]^-$	8.0×10^{29}
$[Cu(NH_3)_4]^{2+}$	5.6×10^{11}
$[Fe(CN)_6]^{4-}$	3×10^{35}
$[Fe(CN)_6]^{3-}$	4.0×10^{43}
$[Hg(CN)_4]^{2-}$	9.3×10^{38}
$[Ni(NH_3)_6]^{2+}$	2.0×10^8
$[Pb(OH)_3]^-$	8×10^{13}
$[Sn(OH)_3]^-$	3×10^{25}
$[Zn(CN)_4]^{2-}$	4.2×10^{19}
$[Zn(NH_3)_4]^{2+}$	7.8×10^8
$[Zn(OH)_4]^{2-}$	3×10^{15}

ions. Because of this behaviour, some uses of complex-ion formation are to retrieve a metal from its ore; eliminate a toxic or unwanted metal ion from a solution; or convert a metal ion to a different form, as Sample Problem 17.15 shows for the zinc ion.

Sample Problem 17.15 Calculating the Concentration of a Complex Ion

Problem An industrial chemist converts $[Zn(H_2O)_4]^{2+}$ to the more stable $[Zn(NH_3)_4]^{2+}$ by mixing 50.0 L of 0.0020 mol/L $[Zn(H_2O)_4]^{2+}$ and 25.0 L of 0.15 mol/L NH_3. What is $[[Zn(H_2O)_4]^{2+}]$ at equilibrium (K_f of $[Zn(NH_3)_4]^{2+} = 7.8 \times 10^8$)?

Plan We write the equation and the K_f expression and use an ICE table to calculate the equilibrium concentrations. To set up the reaction table, we must first find $[[Zn(H_2O)_4]^{2+}]_{init}$ and $[NH_3]_{init}$. We are given the individual volumes and molar concentrations, so we find the number of moles and divide by the *total* volume because the solutions are mixed. With the large excess of NH_3 and the high K_f, we assume that almost all the $[Zn(H_2O)_4]^{2+}$ is converted to $[Zn(NH_3)_4]^{2+}$. Because $[[Zn(H_2O)_4]^{2+}]$ at equilibrium is very small, we use x to represent it.

Solution Write the equation and the K_f expression:

$$[Zn(H_2O)_4]^{2+}(aq) + 4NH_3(aq) \rightleftharpoons [Zn(NH_3)_4]^{2+}(aq) + 4H_2O(l)$$

$$K_f = \frac{[[Zn(NH_3)_4]^{2+}]}{[[Zn(H_2O)_4]^{2+}][NH_3]^4}$$

Find the initial reactant concentrations:

$$[[Zn(H_2O)_4]^{2+}]_{init} = \frac{50.0\ L \times 0.0020\ mol/L}{50.0\ L + 25.0\ L} = 1.3 \times 10^{-3}\ mol/L$$

$$[NH_3]_{init} = \frac{25.0\ L \times 0.15\ mol/L}{50.0\ L + 25.0\ L} = 5.0 \times 10^{-2}\ mol/L$$

We assume that nearly all the $[Zn(H_2O)_4]^{2+}$ is converted to $[Zn(NH_3)_4]^{2+}$, so we set up the ICE table with $x = [[Zn(H_2O)_4]^{2+}]$ at equilibrium. Because 4 mol of NH_3 is needed per mole of $Zn(H_2O)_4^{2+}$, the change in $[NH_3]$ is

$$[NH_3]_{reacted} \approx 4(1.3 \times 10^{-3}\ mol/L) = 5.2 \times 10^{-3}\ mol/L$$

and

$$[[Zn(NH_3)_4]^{2+}] \approx 1.3 \times 10^{-3}\ mol/L$$

Concentration (mol/L)	$[Zn(H_2O)_4]^{2+}(aq)$	$+ 4NH_3(aq)$	$\rightleftharpoons$	$[Zn(NH_3)_4]^{2+}$	+	$4H_2O(l)$
Initial	1.3×10^{-3}	5.0×10^{-2}		0		—
Change	$\sim(-1.3 \times 10^{-3})$	$\sim(-5.2 \times 10^{-3})$		$\sim(+1.3 \times 10^{-3})$		—
Equilibrium	x	4.5×10^{-2}		1.3×10^{-3}		—

Solve for x, the $[[Zn(H_2O)_4]^{2+}]$ remaining at equilibrium:

$$K_f = \frac{[[Zn(NH_3)_4]^{2+}]}{[[Zn(H_2O)_4]^{2+}][NH_3]^4} = 7.8 \times 10^8 \approx \frac{1.3 \times 10^{-3}}{x(4.5 \times 10^{-2})^4}$$

$$x = [[Zn(H_2O)_4]^{2+}] \approx 4.1 \times 10^{-7}\ mol/L$$

Check The K_f is large, so we expect the remaining $[[Zn(H_2O)_4]^{2+}]$ to be very low.

Follow-Up Problem 17.15 The cyanide ion is toxic because it forms stable complex ions with the Fe^{3+} ion in certain proteins that are involved in cellular energy production. To study this effect, a biochemist mixes 25.5 mL of 3.1×10^{-2} mol/L $[Fe(H_2O)_6]^{3+}$ with 35.0 mL of 1.5 mol/L NaCN. What is the final $[[Fe(H_2O)_6]^{3+}]$ (K_f of $[Fe(CN)_6]^{3-} = 4.0 \times 10^{43}$)?

Complex Ions and the Solubility of Precipitates

In Section 17.3, we saw that H_3O^+ increases the solubility of a slightly soluble ionic compound if its conjugate base is that of a weak acid. Similarly, *a ligand increases the solubility of a slightly soluble ionic compound if it forms a complex ion with the cation.* For example, iron(II) sulfide is very slightly soluble:

$$FeS(s) + H_2O(l) \rightleftharpoons Fe^{2+}(aq) + HS^-(aq) + OH^-(aq) \qquad K_{sp} = 8 \times 10^{-16}$$

When we add some 1.0 mol/L NaCN, the CN^- ions act as ligands and react with the small amount of $Fe^{2+}(aq)$ to form the complex ion $Fe(CN)_6^{\ 4-}$:

$$Fe^{2+}(aq) + 6CN^-(aq) \rightleftharpoons [Fe(CN)_6]^{4-}(aq) \qquad K_f = 3 \times 10^{35}$$

To see the effect of complex-ion formation on the solubility of FeS, we add the equations and, therefore, multiply their equilibrium constants:

$$FeS(s) + 6CN^-(aq) + H_2O(l) \rightleftharpoons [Fe(CN)_6]^{4-}(aq) + HS^-(aq) + OH^-(aq)$$
$$K_{overall} = K_{sp} \times K_f = (8 \times 10^{-16})(3 \times 10^{35}) = 2 \times 10^{20}$$

The overall dissociation of FeS into ions increased by more than a factor of 10^{35} in the presence of the ligand.

Sample Problem 17.16 Calculating the Effect of Complex-Ion Formation on Solubility

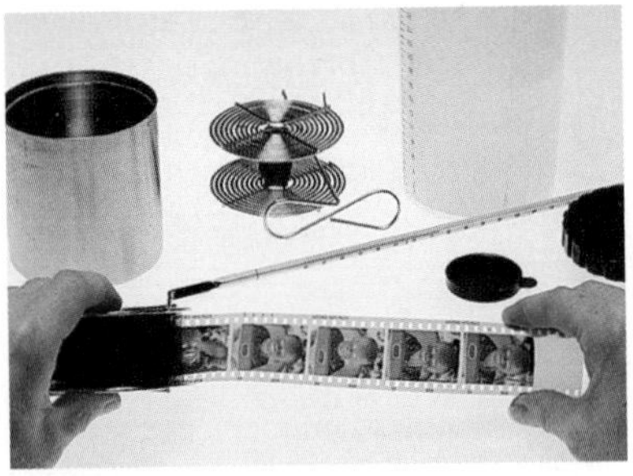

Developing an image in hypo

Problem In black-and-white film developing (*see photo*), excess AgBr is removed from a film negative with "hypo," an aqueous solution of sodium thiosulfate ($Na_2S_2O_3$), which forms the complex ion $[Ag(S_2O_3)_2]^{3-}$. Calculate the solubility of AgBr in **(a)** H_2O; **(b)** 1.0 mol/L hypo (K_f of $[Ag(S_2O_3)_2]^{3-} = 4.7 \times 10^{13}$; K_{sp} of AgBr $= 5.0 \times 10^{-13}$).

Plan **(a)** After writing the equation and the ion-product expression, we use the given K_{sp} to solve for *s*, the molar solubility of AgBr. **(b)** In hypo, Ag^+ forms a complex ion with $S_2O_3^{\ 2-}$, which shifts the equilibrium and dissolves more AgBr. We write the complex-ion equation and add it to the equation for dissolving AgBr to obtain the overall equation for dissolving AgBr in hypo. We multiply K_{sp} by K_f to find $K_{overall}$. To find the solubility of AgBr in hypo, we set up an ICE table, with $s = [[Ag(S_2O_3)_2]^{3-}]$; substitute the values into the expression for $K_{overall}$; and solve for *s*.

Solution **(a)** To find the solubility of AgBr in water, we start by writing the equation for the saturated solution and the ion-product expression:

$$AgBr(s) \rightleftharpoons Ag^+(aq) + Br^-(aq) \qquad K_{sp} = [Ag^+][Br^-]$$

Then we solve for solubility (*s*) directly from the equation. We know that

$$s = [AgBr]_{dissolved} = [Ag^+] = [Br^-]$$

Thus,

$$K_{sp} = [Ag^+][Br^-] = s^2 = 5.0 \times 10^{-13}$$

So,

$$s = 7.1 \times 10^{-7}\ mol/L$$

(b) To find the solubility of AgBr in 1.0 mol/L hypo, we write the overall equation:

$$AgBr(s) \rightleftharpoons \cancel{Ag^+(aq)} + Br^-(aq)$$
$$\cancel{Ag^+(aq)} + 2S_2O_3^{\ 2-}(aq) \rightleftharpoons [Ag(S_2O_3)_2]^{3-}(aq)$$
$$AgBr(s) + 2S_2O_3^{\ 2-}(aq) \rightleftharpoons [Ag(S_2O_3)_2]^{3-}(aq) + Br^-(aq)$$

Calculate $K_{overall}$:

$$K_{overall} = \frac{[[Ag(S_2O_3)_2]^{3-}][Br^-]}{[S_2O_3^{\ 2-}]^2} = K_{sp} \times K_f = (5.0 \times 10^{-13})(4.7 \times 10^{13}) = 24$$

Set up an ICE table, with $s = [AgBr]_{dissolved} = [[Ag(S_2O_3)_2]^{3-}]$:

Concentration (mol/L)	AgBr(s) +	$2S_2O_3^{2-}(aq)$	⇌	$[Ag(S_2O_3)_2]^{3-}(aq)$ +	$Br^-(aq)$
Initial	—	1.0		0	0
Change	—	−2s		+s	+s
Equilibrium	—	1.0 − 2s		s	s

Substitute the values into the expression for $K_{overall}$ and solve for s:

$$K_{overall} = \frac{[[Ag(S_2O_3)_2]^{3-}][Br^-]}{[S_2O_3^{2-}]^2} = \frac{s^2}{(1.0 \text{ mol/L} - 2s)^2} = 24$$

Take the square root of both sides:

$$\frac{s}{1.0 \text{ mol/L} - 2s} = \sqrt{24} = 4.9 \quad \text{so} \quad s = 4.9 \text{ mol/L} - 9.8s \quad \text{and} \quad 10.8s = 4.9 \text{ mol/L}$$

$$[[Ag(S_2O_3)_2]^{3-}] = s = 0.45 \text{ mol/L}$$

Check **(a)** From the number of ions in the formula of AgBr, we know that $s = \sqrt{K_{sp}}$, so the order of magnitude seems right: $\sim\sqrt{10^{-14}} \approx 10^{-7}$. **(b)** The $K_{overall}$ seems correct: the exponents cancel, and $5 \times 5 = 25$. Most important, the answer makes sense because the photographic process requires the remaining AgBr to be washed off the film, and the large $K_{overall}$ confirms this. We can check s by rounding and working backwards to find $K_{overall}$. From the ICE table, we find that

$$[S_2O_3^{2-}] = 1.0 \text{ mol/L} - 2s = 1.0 \text{ mol/L} - 2(0.45 \text{ mol/L})$$
$$= 1.0 \text{ mol/L} - 0.90 \text{ mol/L} = 0.1 \text{ mol/L}$$

So, $K_{overall} \approx \frac{(0.45)^2}{(0.1)^2} = 20$, within rounding of the calculated value.

Follow-Up Problem 17.16 How does the solubility of AgBr in 1.0 mol/L NH_3 compare with its solubility in 1.0 mol/L hypo (K_f of $[Ag(NH_3)_2]^+ = 1.7 \times 10^7$)?

Complex Ions of Amphoteric Hydroxides

Many of the same metals that form amphoteric oxides also form slightly soluble *amphoteric hydroxides*. These compounds dissolve very little in water, but they dissolve to a much greater extent in both acidic and basic solutions. Aluminum hydroxide is one of several examples:

$$Al(OH)_3(s) \rightleftharpoons Al^{3+}(aq) + 3OH^-(aq)$$

It is sparingly soluble in water ($K_{sp} = 3 \times 10^{-34}$), but it dissolves in an acid or a base:

- It dissolves in an acid because H_3O^+ reacts with the OH^- anion [Sample Problem 17.11, part (b)],

$$3H_3O^+(aq) + 3OH^-(aq) \longrightarrow 6H_2O(l)$$

giving the overall equation

$$Al(OH_3)(s) + 3H_3O^+(aq) \longrightarrow Al^{3+}(aq) + 6H_2O(l)$$

- It dissolves in a base through the formation of a complex ion:

$$Al(OH)_3(s) + OH^-(aq) \longrightarrow AlOH_4^-(aq)$$

Let us look more closely at this behaviour. When we dissolve a soluble aluminum salt, such as $Al(NO_3)_3$, in water and then slowly add a strong base, a white precipitate first forms and then dissolves as more base is added. What reactions are occurring? The formula for the hydrated Al^{3+} ion is $[Al(H_2O)_6]^{3+}(aq)$. It acts as a weak polyprotic acid and reacts with added OH^- ions in a step-by-step process. In each step, one of the bound H_2O molecules loses a proton and becomes a bound OH^- ion, so the number of bound H_2O molecules is reduced by 1:

$$[Al(H_2O)_6]^{3+}(aq) + OH^-(aq) \rightleftharpoons [Al(H_2O)_5OH]^{2+}(aq) + H_2O(l)$$
$$[Al(H_2O)_5OH]^{2+}(aq) + OH^-(aq) \rightleftharpoons [Al(H_2O)_4(OH)_2]^+(aq) + H_2O(l)$$
$$[Al(H_2O)_4(OH)_2]^+(aq) + OH^-(aq) \rightleftharpoons Al(H_2O)_3(OH)_3(s) + H_2O(l)$$

After three protons have been removed from each $[Al(H_2O)_6]^{3+}$, the white precipitate has formed. The white precipitate is the insoluble hydroxide $Al(H_2O)_3(OH)_3(s)$,

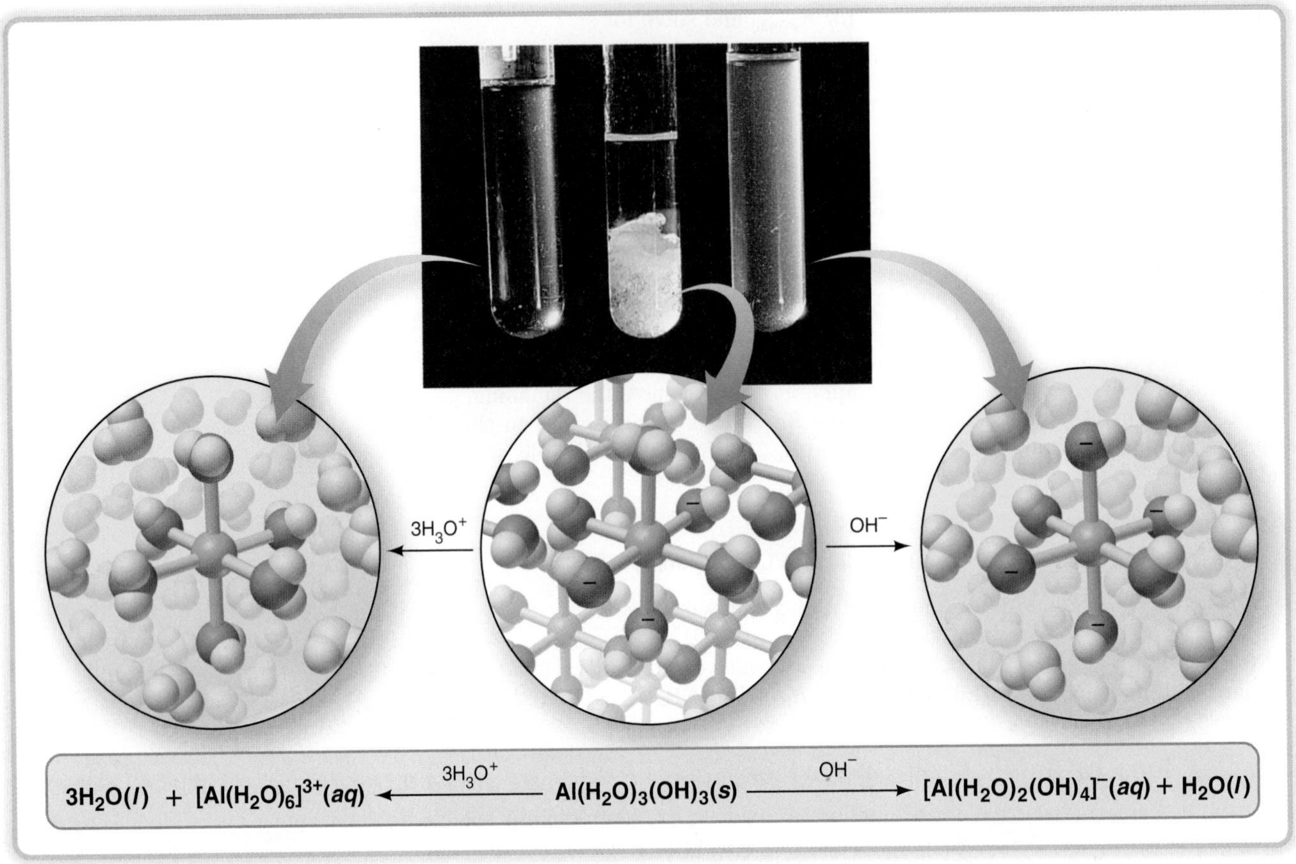

FIGURE 17.19 The amphoteric behaviour of aluminum hydroxide. When solid $Al(OH)_3$ is treated with H_3O^+ (*left*) or with OH^- (*right*), it dissolves as a result of the formation of soluble complex ions.

often written more simply as $Al(OH)_3(s)$. Now we can see that the precipitate actually consists of the hydrated Al^{3+} ion with an H^+ removed from each of three bound H_2O molecules (Figure 17.19, *centre*). The addition of H_3O^+ protonates the OH^- ions and re-forms the hydrated Al^{3+} ion (Figure 17.19, *left*).

Further addition of OH^- removes a fourth H^+, and the precipitate dissolves as the soluble ion $[Al(H_2O)_2(OH)_4]^-(aq)$ forms (Figure 17.19, *right*). [We usually write the formula for this ion as $Al(OH)_4^-(aq)$.]

$$Al(H_2O)_3(OH)_2(s) + OH^-(aq) \rightleftharpoons [Al(H_2O)_2(OH)_4]^-(aq) + H_2O(l)$$

In other words, this complex ion is not created by ligands substituting for bound water molecules, but through an acid-base reaction in which added OH^- ions titrate bound water molecules.

Several other slightly soluble hydroxides, including those of cadmium, chromium (III), cobalt(III), lead(II), tin(II), and zinc, are amphoteric and exhibit similar reactions:

$$Zn(H_2O)_2(OH)_2(s) + OH^-(aq) \rightleftharpoons [Zn(H_2O)(OH)_3]^-(aq) + H_2O(l)$$

In contrast, the slightly soluble hydroxides of iron(II), iron(III), and calcium dissolve in an acid but do *not* dissolve in a base, because the three remaining bound water molecules are not acidic enough to lose any of their protons:

$$Fe(H_2O)_3(OH)_3(s) + 3H_3O^+(aq) \longrightarrow [Fe(H_2O)_6]^{3+}(aq) + 3H_2O(l)$$
$$Fe(H_2O)_3(OH)_3(s) + OH^-(aq) \longrightarrow \text{no reaction}$$

This difference between the solubilities of $Al(OH)_3$ and $Fe(OH)_3$ in a base is the key to an important separation step in the production of aluminum metal, so we will consider it again in Section 23.4.

SUMMARY OF SECTION 17.4

- A complex ion consists of a central metal ion covalently bonded to two or more negatively charged or neutral ligands. Its formation is described by a formation constant, K_f.
- A hydrated metal ion is a complex ion with water molecules as ligands. Other ligands (stronger Lewis bases) can displace the water in a step-by-step process. In most cases, the K_f value of each step is large, so the fully substituted complex ion forms almost completely with excess ligand.
- Adding a solution containing a ligand increases the solubility of an ionic precipitate if the cation forms a complex ion with the ligand.
- Amphoteric metal hydroxides dissolve in an acid or a base as a result of reactions that involve complex ions.

CHAPTER REVIEW GUIDE

Learning Objectives

Concepts

1. Describe how the presence of a common ion suppresses the reaction that forms it. (§17.1)
2. Explain why the concentrations of buffer components must be high to minimize the change in pH from the addition of small amounts of H^+ or OH^-. (§17.1)
3. Describe how buffer capacity depends on the buffer concentration and the pK_a of the acid component, and explain why the buffer range is within ± 1 of the pK_a. (§17.1)
4. Discuss the nature of an acid-base indicator as a conjugate acid-base pair with different-colour acidic and basic forms. (§17.2)
5. Describe the process of titration. (§17.2)
6. Distinguish between the equivalence point and the end point in an acid-base titration. (§17.2)
7. Explain why the shapes of strong acid–strong base, weak acid–strong base, and weak base–strong acid titration curves differ. (§17.2)
8. Discuss how the pH at the equivalence point is determined by the species present, and explain why the pH at the midpoint of the buffer region equals the pK_a of the acid. (§17.2)
9. Show how the titration curve of a polyprotic acid has a buffer region and equivalence point for each ionizable proton. (§17.2)
10. Explain how a slightly soluble ionic compound reaches equilibrium in water, expressed by an equilibrium (solubility-product) constant, K_{sp}. (§17.3)
11. Describe why the incomplete dissociation of an ionic compound leads to approximate calculated values for K_{sp} and solubility. (§17.3)
12. Explain why a common ion in a solution decreases the solubility of its compounds. (§17.3)
13. Discuss how pH affects the solubility of a compound that contains a weak-acid conjugate base. (§17.3)
14. Describe how precipitate formation depends on the relative values of Q_{sp} and K_{sp}. (§17.3)
15. Distinguish between selective precipitation and simultaneous equilibria as they are used to separate ions. (§17.3)
16. Describe how complex-ion formation is a step-by-step process, expressed by an overall equilibrium (formation) constant, K_f. (§17.4)
17. Explain why the addition of a ligand increases the solubility of a compound whose metal ion forms a complex ion. (§17.4)
18. Describe how the aqueous chemistry of amphoteric hydroxides involves precipitation, complex-ion formation, and acid-base equilibria. (§17.4)

Skills

1. Use stoichiometry and equilibrium problem-solving techniques to calculate the effect of added H_3O^+ or OH^- on buffer pH. (SP 17.1)
2. Use the Henderson-Hasselbalch equation to calculate buffer pH. (§17.1)
3. Use molecular scenes to depict buffers. (SP 17.2)
4. Choose the components of a buffer with a given pH, and calculate their quantities. (SP 17.3)
5. Calculate the pH of a buffer. (SP 17.4)
6. Find the concentration of an acid from a titration. (SP 17.5)
7. Calculate the pH at any point in an acid-base titration. (§17.2 and SP 17.6)
8. Choose an appropriate indicator based on the pH at various points in a titration. (§17.2)
9. Write K_{sp} expressions for slightly soluble ionic compounds. (SP 17.7)
10. Use K_{sp} values to compare solubilities for compounds with the same total number of ions. (§17.3)
11. Calculate K_{sp} values from solubility data. (SP 17.8)
12. Calculate solubility from a K_{sp} value. (SP 17.9)
13. Calculate the decrease in solubility caused by the presence of a common ion. (SP 17.10)
14. Predict whether the addition of H_3O^+ affects solubility. (SP 17.11)
15. Use ion concentrations to calculate Q_{sp} and compare it with K_{sp} to predict whether a precipitate will form. (SPs 17.12, 17.13)
16. Compare K_{sp} values to calculate the $[OH^-]$ that will separate ions by selective precipitation. (SP 17.14)
17. Calculate the concentration of metal ion remaining after the addition of excess ligand forms a complex ion. (SP 17.15)
18. Use an overall equilibrium constant ($K_{sp} \times K_f$) to calculate the effect of complex-ion formation on solubility. (SP 17.16)

Key Terms

Section 17.1
acid-base buffer
common-ion effect
Henderson-Hasselbalch equation
buffer capacity
buffer range

Section 17.2
acid-base reaction
neutralization reaction
titration
equivalence point
end point
acid-base titration curve

Section 17.3
solubility-product constant (K_{sp})
selective precipitation

Section 17.4
complex ion
ligands
formation constant (K_f)

Key Equations and Relationships

17.1 Finding the pH from known concentrations of a conjugate acid-base pair (Henderson-Hasselbalch equation):

$$pH = pK_a + \log\left(\frac{[base]}{[acid]}\right)$$

17.2 Defining the equilibrium condition for a saturated solution of a slightly soluble compound, M_pX_q, composed of Mn^{n+} and X^{z-} ions:

$$Q_{sp} = [M^{n+}]^p[X^{z-}]^q = K_{sp}$$

Brief Solutions to Follow-Up Problems

17.1 (a) Before the addition of 0.40 g of NaOH (0.010 mol of NaOH) to 1.0 L of buffer (assuming that x is small enough to be neglected):

$[HF] = 0.50$ mol/L and $[F^-] = 0.45$ mol/L

$$[H_3O^+] = K_a \times \frac{[HF]}{[F^-]} \approx (6.8 \times 10^{-4})\left(\frac{0.50}{0.45}\right) = 7.6 \times 10^{-4} \text{ mol/L}$$

$pH = 3.12$

(b) After the addition of 0.40 g of NaOH (0.010 mol of NaOH) to 1.0 L of buffer:

$[HF] = 0.49$ mol/L and $[F^-] = 0.46$ mol/L

$$[H_3O^+] \approx (6.8 \times 10^{-4})\left(\frac{0.49}{0.46}\right) = 7.2 \times 10^{-4} \text{ mol/L}$$

$pH = 3.14$

17.2 (a) Adding a strong base would convert HB to B^-, thereby making the ratio closer to 1.

(b)

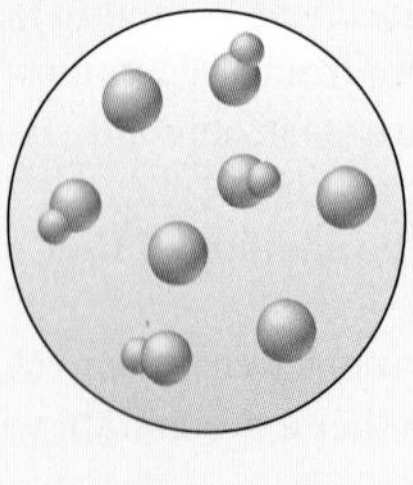

17.3 $[H_3O^+] = 10^{-pH} = 10^{-4.25} = 5.6 \times 10^{-5}$

$$[C_6H_5COOH] = \frac{[H_3O^+][C_6H_5COO^-]}{K_a} = \frac{(5.6 \times 10^{-5})(0.050)}{6.3 \times 10^{-5}} = 0.044 \text{ mol/L}$$

Mass (g) of C_6H_5COOH

$$= 5.0 \text{ L soln} \times \frac{0.044 \text{ mol } C_6H_5COOH}{1 \text{ L soln}} \times \frac{122.12 \text{ g } C_6H_5COOH}{1 \text{ mol } C_6H_5COOH}$$

$= 27$ g C_6H_5COOH

Dissolve 27 g of C_6H_5COOH in 4.9 L of 0.050 mol/L C_6H_5COONa, and add solution to make up 5.0 L. Adjust the pH to 4.25 with a strong acid or base.

17.4 $$[NH_3]_{init} = 0.25 \text{ mol/L}\left(\frac{150.0 \text{ mL}}{154.0 \text{ mL}}\right) = 0.244 \text{ mol/L}$$

$$[HCl]_{init} = 5.0 \text{ mol/L}\left(\frac{4.0 \text{ mL}}{154.0 \text{ mL}}\right) = 0.130 \text{ mol/L}$$

From an ICE table, $[NH_3] = 0.114$ mol/L and $[NH_4^+] = 0.130$ mol/L.

$$K_a = \frac{K_w}{K_b} = \frac{1.0 \times 10^{-14}}{1.8 \times 10^{-5}} = 5.6 \times 10^{-10}$$

$$pK_a = -\log(5.6 \times 10^{-10}) = 9.25$$

$$pH = pK_a + \log\left(\frac{[NH_3]}{[NH_4^+]}\right) = 9.25 + \log\left(\frac{0.114 \text{ mol/L}}{0.130 \text{ mol/L}}\right) = 9.19$$

17.5 $Ba(OH)_2(aq) + 2HCl(aq) \longrightarrow BaCl_2(aq) + 2H_2O(l)$

$$\text{Volume (L) of soln} = 50.00 \text{ mL HCl soln} \times \frac{1 \text{ L}}{10^3 \text{ mL}} \times \frac{0.1016 \text{ mol HCl}}{1 \text{ L soln}} \times \frac{1 \text{ mol } Ba(OH)_2}{2 \text{ mol HCl}} \times \frac{1 \text{ L soln}}{0.1292 \text{ mol } Ba(OH)_2}$$

$= 0.019\,66$ L

17.6 (a) (i) $[H_3O^+] \approx \sqrt{(2.3 \times 10^{-9})(0.2000)} = 2.1 \times 10^{-5}$ mol/L

pH = 4.68

(ii) $[H_3O^+] = K_a \times \frac{[HBrO]}{[BrO^-]} = (2.3 \times 10^{-9})(1) = 2.3 \times 10^{-9}$ mol/L

pH = 8.64

(iii)

$$[BrO^-] = \frac{\text{amount (mol) of BrO}^-}{\text{total volume (L)}} = \frac{0.004\,000 \text{ mol}}{0.060\,00 \text{ L}} = 0.066\,67 \text{ mol/L}$$

$$K_b \text{ of BrO}^- = \frac{K_w}{K_a \text{ of HBrO}} = 4.3 \times 10^{-6}$$

$$[H_3O^+] = \frac{K_w}{\sqrt{K_b \times [BrO^-]}}$$

$$\approx \frac{1.0 \times 10^{-14}}{\sqrt{(4.3 \times 10^{-6})(0.066\,67)}} = 1.9 \times 10^{-11} \text{ mol/L}$$

pH = 10.72

(iv) Amount (mol) of OH^- added = 0.008 000 mol

Volume (L) of OH^- soln = 0.080 00 L

$$[OH^-] = \frac{\text{amount (mol) of OH}^- \text{ unreacted}}{\text{total volume (L)}}$$

$$= \frac{0.008\,000 \text{ mol} - 0.004\,000 \text{ mol}}{(0.020\,00 + 0.080\,00) \text{ L}} = 0.040\,00 \text{ mol/L}$$

$$[H_3O^+] = \frac{K_w}{[OH^-]} = 2.5 \times 10^{-13}$$

pH = 12.60

(b)

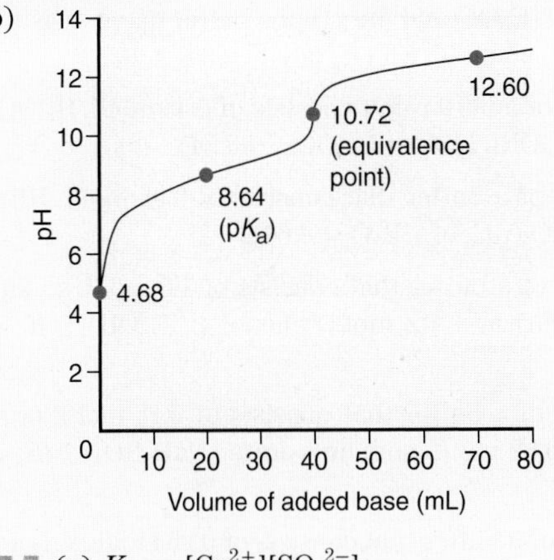

17.7 (a) $K_{sp} = [Ca^{2+}][SO_4^{2-}]$
(b) $K_{sp} = [Cr^{3+}]^2[CO_3^{2-}]^3$
(c) $K_{sp} = [Mg^{2+}][OH^-]^2$
(d) $K_{sp} = [As^{3+}]^2[HS^-]^3[OH^-]^3$

17.8 $$[CaF_2] = \frac{1.5 \times 10^{-4} \text{ g CaF}_2}{10.0 \text{ mL soln}} \times \frac{1000 \text{ mL}}{1 \text{ L}} \times \frac{1 \text{ mol CaF}_2}{78.08 \text{ g CaF}_2}$$

$= 1.9 \times 10^{-4}$ mol/L

$CaF_2(s) \rightleftharpoons Ca^2(aq) + 2F^-(aq)$

$[Ca^{2+}] = 1.9 \times 10^{-4}$ mol/L and $[F^-] = 3.8 \times 10^{-4}$ mol/L

$K_{sp} = [Ca^{2+}][F^-]^2 = (1.9 \times 10^{-4})(3.8 \times 10^{-4})^2 = 2.7 \times 10^{-11}$

17.9 From the reaction table, $[Mg^{2+}] = s$ and $[OH^-] = 2s$.

$K_{sp} = [Mg^{2+}][OH^-]^2 = 4s^3 = 6.3 \times 10^{-10}$; $s = 5.4 \times 10^{-4}$ mol/L

17.10 (a) In water: $K_{sp} = [Ba^{2+}][SO_4^{2-}] = s^2 = 1.1 \times 10^{-10}$
$s = 1.0 \times 10^{-5}$
(b) In 0.10 mol/L Na_2SO_4: $[SO_4^{2-}] = 0.10$ mol/L
$K_{sp} = 1.1 \times 10^{-10} \approx s \times 0.10$; $s = 1.1 \times 10^{-9}$ mol/L
(s decreases in the presence of the common ion SO_4^{2-}.)

17.11 (a) Increases solubility:

$$CaF_2(s) \rightleftharpoons Ca^{2+}(aq) + 2F^-(aq)$$
$$F^-(aq) + H_3O^+(aq) \longrightarrow HF(aq) + H_2O(l)$$

(b) Increases solubility:

$$ZnS(s) + H_2O(l) \rightleftharpoons Zn^2(aq) + HS^-(aq) + OH^-(aq)$$
$$HS^-(aq) + H_3O^+(aq) \longrightarrow H_2S(aq) + H_2O(l)$$
$$OH^-(aq) + H_3O^+(aq) \longrightarrow 2H_2O(l)$$

(c) No effect; $I^-(aq)$ is the conjugate base of the strong acid HI.

17.12 $Ca_3(PO_4)_2(s) \rightleftharpoons 3Ca^{2+}(aq) + 2PO_4^{3-}(aq)$
$Q_{sp} = [Ca^{2+}]^3[PO_4^{3-}]^2 = (1.0 \times 10^{-9})^5 = 1.0 \times 10^{-45}$
$Q_{sp} < K_{sp}$, so $Ca_3(PO_4)_2$ will not precipitate.

17.13 (a) Scene 3 has the same relative numbers of ions as in the formula.
(b) Based on

$$Ni(OH)_2(s) \rightleftharpoons Ni^{2+}(aq) + 2OH^-(aq)$$

and $Q_{sp} = [Ni^{2+}][OH^-]^2$, the ion products are $(3)(4)^2 = 48$ in scene 1, $(4)(2)^2 = 16$ in scene 2, and $(2)(4)^2 = 32 = K_{sp}$ in scene 3. Therefore, Q_{sp} of scene 1 exceeds K_{sp} of scene 3.
(c) The addition of acid will decrease the mass of $Ni(OH)_2(s)$ by reacting with OH^-, thereby causing more solid to dissolve. The addition of base will increase the mass of $Ni(OH)_2(s)$ due to the common-ion effect.

17.14 Both are 1/1 salts, and the K_{sp} values show that $CaSO_4$ is more soluble:

$$[SO_4^{2-}] = \frac{K_{sp}}{[Ca^{2+}]} = \frac{2.4 \times 10^{-5}}{0.025} = 9.6 \times 10^{-4} \text{ mol/L}$$

17.15 $$[[Fe(H_2O)_6]^{3+}]_{init} = \frac{(0.0255 \text{ L})(3.1 \times 10^{-2} \text{ mol/L})}{0.0255 \text{ L} + 0.0350 \text{ L}}$$

$= 1.3 \times 10^{-2}$ mol/L

Similarly, $[CN^-]_{init} = 0.87$ mol/L. From the ICE table,

$$K_f = \frac{[[Fe(CN)_6]^{3-}]}{[[Fe(H_2O)_6]^{3+}][CN^-]^6} = 4.0 \times 10^{43} \approx \frac{1.3 \times 10^{-2}}{x(0.79)^6}$$

$x = [[Fe(H_2O)_6]^{3+}] \approx 1.3 \times 10^{-45}$

17.16 $AgBr(s) + 2NH_3(aq) \rightleftharpoons [Ag(NH_3)_2]^+(aq) + Br^-(aq)$
$K_{overall} = K_{sp}$ of AgBr $\times K_f$ of $[Ag(NH_3)_2]^+$
$= 8.5 \times 10^{-6}$
From the ICE table,

$$\frac{s}{1.0 - 2s} = \sqrt{8.5 \times 10^{-6}} = 2.9 \times 10^{-3}$$

$s = [[Ag(NH_3)_2]^+] = 2.9 \times 10^{-3}$ mol/L

The solubility of AgBr is greater in 1.0 mol/L hypo than in 1.0 mol/L NH_3.

PROBLEMS

Problems with red numbers are answered in Appendix G and worked in detail in the Student Solutions Manual. Problem sections match those in this book and provide the numbers of relevant sample problems. Most offer Concept Review Questions, Skill-Building Exercises (grouped in pairs covering the same concept), and Problems in Context. The Comprehensive Problems are based on material from any section or previous chapter.

Note: Unless stated otherwise, all problems refer to aqueous solutions at 298 K (25°C).

Equilibria of Acid-Base Buffers

(Sample Problems 17.1 to 17.4)

Concept Review Questions

17.1 What is the purpose of an acid-base buffer?

17.2 How do the acid and base components of a buffer function? Why are they often a conjugate acid-base pair of a weak acid?

17.3 What is the common-ion effect? How is it related to Le Châtelier's principle? Explain with equations that include HF and NaF.

17.4 The scenes below depict solutions of the same HA/A^- buffer (with counterions and water molecules omitted for clarity).

(a) Which solution has the greatest buffer capacity?
(b) Explain how the pH ranges of the buffers compare.
(c) Which solution can react with the largest amount of added strong acid?

17.5 (a) What is the difference between a buffer with a high capacity and a buffer with a low capacity?
(b) Will adding 0.01 mol of HCl produce a greater pH change in a buffer with a high capacity or a buffer with a low capacity? Explain.

17.6 Which of these factors influence buffer capacity? How?
(a) Conjugate acid-base pair
(b) pH of the buffer
(c) Concentration of buffer components
(d) Buffer range
(e) pK_a of the acid component

17.7 What is the relationship between the buffer range and the buffer-component concentration ratio?

17.8 (a) A chemist needs a buffer with a pH of 3.5. Should the chemist use NaOH with formic acid (K_a = 1.8 × 10^{-4}) or NaOH with ethanoic (acetic) acid (K_a =1.8 × 10^{-5})? Why?
(b) What is the disadvantage of choosing the other acid?
(c) What is the role of the NaOH?

17.9 State and explain the relative change in the pH and in the buffer-component concentration ratio, [NaA]/[HA], when each addition is made:
(a) Add 0.1 mol/L NaOH to the buffer.
(b) Add 0.1 mol/L HCl to the buffer.
(c) Dissolve pure NaA in the buffer.
(d) Dissolve pure HA in the buffer.

17.10 Does the pH increase or decrease, and does it do so to a large or small extent, when each addition is made?
(a) Five drops of 0.1 mol/L NaOH to 100 mL of 0.5 mol/L ethanoate (acetate) buffer
(b) Five drops of 0.1 mol/L HCl to 100 mL of 0.5 mol/L ethanoate (acetate) buffer
(c) Five drops of 0.1 mol/L NaOH to 100 mL of 0.5 mol/L HCl
(d) Five drops of 0.1 mol/L NaOH to distilled water

Skill-Building Exercises (grouped in similar pairs)

17.11 What are the [H_3O^+] and the pH of a propanoic acid–propanoate buffer that consists of 0.35 mol/L CH_3CH_2COONa and 0.15 mol/L CH_3CH_2COOH (K_a of propanoic acid = 1.3 × 10^{-5})?

17.12 What are the [H_3O^+] and the pH of a benzoic acid–benzoate buffer that consists of 0.33 mol/L C_6H_5COOH and 0.28 mol/L C_6H_5COONa (K_a of benzoic acid = 6.3 × 10^{-5})?

17.13 What are the [H_3O^+] and the pH of a buffer that consists of 0.55 mol/L HNO_2 and 0.75 mol/L KNO_2 (K_a of HNO_2 = 7.1 × 10^{-4})?

17.14 What are the [H_3O^+] and the pH of a buffer that consists of 0.20 mol/L HF and 0.25 mol/L KF (K_a of HF = 6.8 × 10^{-4})?

17.15 Find the pH of a buffer that consists of 0.45 mol/L HCOOH and 0.63 mol/L HCOONa (pK_a of HCOOH = 3.74).

17.16 Find the pH of a buffer that consists of 0.95 mol/L HBrO and 0.68 mol/L KBrO (pK_a of HBrO = 8.64).

17.17 Find the pH of a buffer that consists of 1.3 mol/L sodium phenolate (C_6H_5ONa) and 1.2 mol/L phenol (C_6H_5OH) (pK_a of phenol = 10.00).

17.18 Find the pH of a buffer that consists of 0.12 mol/L boric acid (H_3BO_3) and 0.82 mol/L sodium borate (NaH_2BO_3) (pK_a of boric acid = 9.24).

17.19 Find the pH of a buffer that consists of 0.25 mol/L NH_3 and 0.15 mol/L NH_4Cl (pK_b of NH_3 = 4.75).

17.20 Find the pH of a buffer that consists of 0.50 mol/L methanamine (CH_3NH_2) and 0.60 mol/L CH_3NH_3Cl (pK_b of CH_3NH_2 = 3.35).

17.21 A buffer consists of 0.22 mol/L $KHCO_3$ and 0.37 mol/L K_2CO_3. Carbonic acid is a diprotic acid with K_{a1} = 4.5 × 10^{-7} and K_{a2} = 4.7 × 10^{-11}.
(a) Which K_a value is more important to this buffer?
(b) What is the buffer pH?

17.22 A buffer consists of 0.50 mol/L NaH_2PO_4 and 0.40 mol/L Na_2HPO_4. Phosphoric acid is a triprotic acid with K_{a1} = 7.2 × 10^{-3}, K_{a2} = 6.3 × 10^{-8}, and K_{a3} = 4.2 × 10^{-13}.
(a) Which K_a value is most important to this buffer?
(b) What is the buffer pH?

17.23 What is the component concentration ratio, [$EtCOO^-$]/[EtCOOH], of a buffer that has a pH of 5.44 (K_a of EtCOOH = 1.3 × 10^{-5})?

17.24 What is the component concentration ratio, $[NO_2^-]/[HNO_2]$, of a buffer that has a pH of 2.95 (K_a of $HNO_2 = 7.1 \times 10^{-4}$)?

17.25 What is the component concentration ratio, $[BrO^-]/[HBrO]$, of a buffer that has a pH of 7.95 (K_a of $HBrO = 2.3 \times 10^{-9}$)?

17.26 What is the component concentration ratio, $[CH_3COO^-]/[CH_3COOH]$, of a buffer that has a pH of 4.39 (K_a of $CH_3COOH = 1.8 \times 10^{-5}$)?

17.27 A buffer that contains 0.2000 mol/L of acid HA and 0.1500 mol/L of its conjugate base, A^-, has a pH of 3.35. What is the pH after 0.0015 mol of NaOH is added to 0.5000 L of this solution?

17.28 A buffer that contains 0.40 mol/L of base B and 0.25 mol/L of its conjugate acid, BH^+, has a pH of 8.88. What is the pH after 0.0020 mol of HCl is added to 0.25 L of this solution?

17.29 A buffer that contains 0.110 mol/L HY and 0.220 mol/L Y^- has a pH of 8.77. What is the pH after 0.0015 mol of $Ba(OH)_2$ is added to 0.350 L of this solution?

17.30 A buffer that contains 1.05 mol/L B and 0.750 mol/L BH^+ has a pH of 9.50. What is the pH after 0.0050 mol of HCl is added to 0.500 L of this solution?

17.31 A buffer is prepared by mixing 204 mL of 0.452 mol/L HCl and 0.500 L of 0.400 mol/L sodium ethanoate. (See Appendix C.)
(a) What is the pH?
(b) What mass of KOH must be added to 0.500 L of the buffer to change the pH by 0.15 units?

17.32 A buffer is prepared by mixing 50.0 mL of 0.050 mol/L sodium hydrogen carbonate and 10.7 mL of 0.10 mol/L NaOH. (See Appendix C.)
(a) What is the pH?
(b) What mass of HCl must be added to 25.0 mL of the buffer to change the pH by 0.07 units?

17.33 Choose specific acid-base conjugate pairs to make a buffer with (a) pH $\approx$ 4.5; (b) pH $\approx$ 7.0. (See Appendix C.)

17.34 Choose specific acid-base conjugate pairs to make a buffer with (a) $[H_3O^+] \approx 1 \times 10^{-9}$ mol/L; (b) $[OH^-] \approx 3 \times 10^{-5}$ mol/L. (See Appendix C.)

17.35 Choose specific acid-base conjugate pairs to make a buffer with (a) pH $\approx$ 3.5; (b) pH $\approx$ 5.5. (See Appendix C.)

17.36 Choose specific acid-base conjugate pairs to make a buffer with (a) $[OH^-] \approx 1 \times 10^{-6}$ mol/L; (b) $[H_3O^+] \approx 4 \times 10^{-4}$ mol/L. (See Appendix C.)

Problems in Context

17.37 An industrial chemist who is studying bleaching and sterilizing prepares several hypochlorite buffers. Find the pH of (a) 0.100 mol/L HClO and 0.100 mol/L NaClO; (b) 0.100 mol/L HClO and 0.150 mol/L NaClO; (c) 0.150 mol/L HClO and 0.100 mol/L NaClO; (d) 1.0 L of the solution in part (a) after 0.0050 mol of NaOH has been added.

17.38 Oxoanions of phosphorus are buffer components in blood. For a KH_2PO_4/Na_2HPO_4 solution with pH = 7.40 (pH of normal arterial blood), what is the buffer-component concentration ratio?

Acid-Base Titrations and Curves

(Sample Problems 17.5 and 17.6)

Concept Review Questions

17.39 State a general equation for a neutralization reaction.

17.40 (a) The net ionic equation for the aqueous neutralization reaction between ethanoic (acetic) acid and sodium hydroxide is different from the net ionic equation for the reaction between hydrochloric acid and sodium hydroxide. Explain why, by writing balanced net ionic equations.
(b) For a solution of ethanoic acid in water, list the major species in decreasing order of concentration.

17.41 (a) How can you estimate the pH range of an indicator's colour change?
(b) Why do some indicators have two separate pH ranges?

17.42 Why does the colour change of an indicator take place over a range of about ± 1 of the pK_a?

17.43 Why does the addition of an acid-base indicator not affect the pH of the test solution?

17.44 What is the difference between the end point of a titration and the equivalence point? Is the equivalence point always reached first? Explain.

17.45 The scenes below depict the relative concentrations of H_3PO_4, $H_2PO_4^-$, and HPO_4^{2-} during a titration with aqueous NaOH, but they are out of order. (Phosphate groups are *purple*, hydrogens are *blue*, and Na^+ ions and water molecules are not shown.)

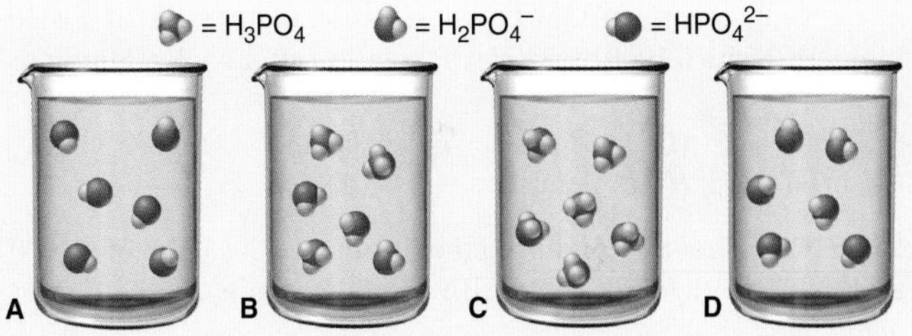

(a) List the scenes in the correct order.
(b) What is the pH in the correctly ordered second scene (see Appendix C)?
(c) If 10.00 mL of the NaOH solution is required to reach the correctly ordered second scene, how much more is needed to reach the last scene?

17.46 Explain how *strong acid*–strong base, *weak acid*–strong base, and *weak base*–strong acid titrations using the same concentrations differ in terms of (a) the initial pH; (b) the pH at the equivalence point. (The component in italics is in the flask.)

17.47 (a) What species are in the buffer region of a weak acid–strong base titration?
(b) How are they different from the species at the equivalence point?
(c) How are they different from the species in the buffer region of a weak base–strong acid titration?

17.48 Why is the centre of the buffer region of a weak acid–strong base titration significant?

17.49 How does the titration curve of a monoprotic acid differ from the titration curve of a diprotic acid?

Skill-Building Exercises (grouped in similar pairs)

17.50 If 25.98 mL of 0.1180 mol/L KOH solution reacts with 52.50 mL of CH_3COOH solution, what is the concentration (mol/L) of the acid solution?

17.51 If 26.25 mL of 0.1850 mol/L NaOH solution reacts with 25.00 mL of H_2SO_4, what is the concentration (mol/L) of the acid solution?

17.52 The indicator cresol red has $K_a = 3.5 \times 10^{-9}$. Over what approximate pH range does it change colour?

17.53 The indicator ethyl red has $K_a = 3.8 \times 10^{-6}$. Over what approximate pH range does it change colour?

17.54 Use Figure 17.6 to find an indicator for each titration:
(a) 0.10 mol/L HCl with 0.10 mol/L NaOH
(b) 0.10 mol/L HCOOH (Appendix C) with 0.10 mol/L NaOH

17.55 Use Figure 17.6 to find an indicator for each titration:
(a) 0.10 mol/L CH_3NH_2 (Appendix C) with 0.10 mol/L HCl
(b) 0.50 mol/L HI with 0.10 mol/L KOH

17.56 Use Figure 17.6 to find an indicator for each titration:
(a) 0.5 mol/L $(CH_3)_2NH$ (Appendix C) with 0.5 mol/L HBr
(b) 0.2 mol/L KOH with 0.2 mol/L HNO_3

17.57 Use Figure 17.6 to find an indicator for each titration:
(a) 0.25 mol/L C_6H_5COOH (Appendix C) with 0.25 mol/L KOH
(b) 0.50 mol/L NH_4Cl (Appendix C) with 0.50 mol/L NaOH

17.58 Calculate the pH during the titration of 40.00 mL of 0.1000 mol/L HCl with 0.1000 mol/L NaOH solution after each addition of base:
(a) 0 mL (b) 25.00 mL (c) 39.00 mL (d) 39.90 mL
(e) 40.00 mL (f) 40.10 mL (g) 50.00 mL

17.59 Calculate the pH during the titration of 30.00 mL of 0.1000 mol/L KOH with 0.1000 mol/L HBr solution after each addition of acid:
(a) 0 mL (b) 15.00 mL (c) 29.00 mL (d) 29.90 mL
(e) 30.00 mL (f) 30.10 mL (g) 40.00 mL

17.60 Find the pH during the titration of 20.00 mL of 0.1000 mol/L butanoic acid, $CH_3CH_2CH_2COOH$ ($K_a = 1.54 \times 10^{-5}$), with 0.1000 mol/L NaOH solution after each addition of titrant:
(a) 0 mL (b) 10.00 mL (c) 15.00 mL (d) 19.00 mL
(e) 19.95 mL (f) 20.00 mL (g) 20.05 mL (h) 25.00 mL

17.61 Find the pH during the titration of 20.00 mL of 0.1000 mol/L triethylamine, $(CH_3CH_2)_3N$ ($K_b = 5.2 \times 10^{-4}$), with 0.1000 mol/L HCl solution after each addition of titrant:
(a) 0 mL (b) 10.00 mL (c) 15.00 mL (d) 19.00 mL
(e) 19.95 mL (f) 20.00 mL (g) 20.05 mL (h) 25.00 mL

17.62 Find the pH at the equivalence point(s) and the volume (mL) of 0.0372 mol/L NaOH needed to reach the equivalence point(s) in titrations of (a) 42.2 mL of 0.0520 mol/L CH_3COOH; (b) 28.9 mL of 0.0850 mol/L H_2SO_3 (two equivalence points).

17.63 Find the pH at the equivalence point(s) and the volume (mL) of 0.0588 mol/L KOH needed to reach the equivalence point(s) in titrations of (a) 23.4 mL of 0.0390 mol/L HNO_2; (b) 17.3 mL of 0.130 mol/L H_2CO_3 (two equivalence points).

17.64 Find the pH at the equivalence point(s) and the volume (mL) of 0.125 mol/L HCl needed to reach the equivalence point(s) in titrations of (a) 65.5 mL of 0.234 mol/L NH_3; (b) 21.8 mL of 1.11 mol/L CH_3NH_2.

17.65 Find the pH at the equivalence point(s) and the volume (mL) of 0.447 mol/L HNO_3 needed to reach the equivalence point(s) in titrations of (a) 2.65 L of 0.0750 mol/L pyridine (C_5H_5N); (b) 0.188 L of 0.250 mol/L ethylenediamine ($H_2NCH_2CH_2NH_2$).

Problems in Context

17.66 A mechanic spills 88 mL of 2.6 mol/L H_2SO_4 solution from an automobile battery. What volume (mL) of 1.6 mol/L $NaHCO_3$ must be poured on the spill to react completely with the sulfuric acid?

17.67 Sodium hydroxide is used extensively in acid-base titrations because it is a strong, inexpensive base. A sodium hydroxide solution was standardized by titrating 25.00 mL of 0.1528 mol/L standard hydrochloric acid. The initial burette reading of the sodium hydroxide was 2.24 mL, and the final reading was 39.21 mL. What was the concentration (mol/L) of the base solution?

17.68 An unknown amount of acid can often be determined by adding an excess of base and then "back-titrating" the excess. A 0.3471 g sample of a mixture of oxalic acid (which has two ionizable protons) and benzoic acid (which has one ionizable proton) is treated with 100.0 mL of 0.1000 mol/L NaOH. The excess NaOH is titrated with 20.00 mL of 0.2000 mol/L HCl. Find the mass percent of benzoic acid.

17.69 One of the first steps in the enrichment of uranium for use in nuclear power plants involves a displacement reaction between UO_2 and aqueous HF:

$$UO_2(s) + HF(aq) \longrightarrow UF_4(s) + H_2O(l) \text{(unbalanced)}$$

What volume (L) of 2.40 mol/L HF will react with 2.15 kg of UO_2?

17.70 A mixture of bases can sometimes be the active ingredient in an antacid tablet. If 0.4826 g of a mixture of $Al(OH)_3$ and $Mg(OH)_2$ is neutralized with 17.30 mL of 1.000 mol/L HNO_3, what is the mass percent of $Al(OH)_3$ in the mixture?

Equilibria of Slightly Soluble Ionic Compounds
(Sample Problems 17.7 to 17.14)

Concept Review Questions

17.71 The molar solubility (s) of M_2X is 5×10^{-5} mol/L.
(a) Find s of each ion.
(b) How do you set up the calculation to find K_{sp}? What assumption must you make about the dissociation of M_2X into ions?
(c) Why is the calculated K_{sp} higher than the actual value?

17.72 Why does pH affect the solubility of BaF_2 but not $BaCl_2$?

17.73 A list of K_{sp} values, like that in Appendix C, can be used to compare the solubility of silver chloride directly with the solubility of silver bromide but not with the solubility of silver chromate. Explain.

17.74 In a gaseous equilibrium, the reverse reaction occurs when $Q > K$. What occurs in aqueous solution when $Q_{sp} > K_{sp}$?

Skill-Building Exercises (grouped in similar pairs)

17.75 Write the ion-product expressions for (a) silver carbonate; (b) barium fluoride; (c) copper(II) sulfide.

17.76 Write the ion-product expressions for (a) iron(III) hydroxide; (b) barium phosphate; (c) tin(II) sulfide.

17.77 Write the ion-product expressions for (a) calcium chromate; (b) silver cyanide; (c) nickel(II) sulfide.

17.78 Write the ion-product expressions for (a) lead(II) iodide; (b) strontium sulfate; (c) cadmium sulfide.

17.79 The solubility of silver carbonate is 0.032 mol/L at 20°C. Calculate its K_{sp}.

17.80 The solubility of zinc oxalate is 7.9×10^{-3} mol/L at 18°C. Calculate its K_{sp}.

17.81 The solubility of silver dichromate is 8.3×10^{-3} g/100 mL solution at 15°C. Calculate its K_{sp}.

17.82 The solubility of calcium sulfate at 30°C is 0.209 g/100 mL solution. Calculate its K_{sp}.

17.83 Find the molar solubility of $SrCO_3$ in (a) pure water; (b) 0.13 mol/L $Sr(NO_3)_2$ (K_{sp} of $SrCO_3 = 5.4 \times 10^{-10}$).

17.84 Find the molar solubility of $BaCrO_4$ in (a) pure water; (b) 1.5×10^{-3} mol/L Na_2CrO_4 (K_{sp} of $BaCrO_4 = 2.1 \times 10^{-10}$).

17.85 Calculate the molar solubility of $Ca(IO_3)_2$ in (a) 0.060 mol/L $Ca(NO_3)_2$; (b) 0.060 mol/L $NaIO_3$. (See Appendix C.)

17.86 Calculate the molar solubility of Ag_2SO_4 in (a) 0.22 mol/L $AgNO_3$; (b) 0.22 mol/L Na_2SO_4. (See Appendix C.)

17.87 Which compound in each pair is more soluble in water?
(a) Magnesium hydroxide or nickel(II) hydroxide
(b) Lead(II) sulfide or copper(II) sulfide
(c) Silver sulfate or magnesium fluoride

17.88 Which compound in each pair is more soluble in water?
(a) Strontium sulfate or barium chromate
(b) Calcium carbonate or copper(II) carbonate
(c) Barium iodate or silver chromate

17.89 Which compound in each pair is more soluble in water?
(a) Barium sulfate or calcium sulfate
(b) Calcium phosphate or magnesium phosphate
(c) Silver chloride or lead(II) sulfate

17.90 Which compound in each pair is more soluble in water?
(a) Manganese(II) hydroxide or calcium iodate
(b) Strontium carbonate or cadmium sulfide
(c) Silver cyanide or copper(I) iodide

17.91 Write equations to show whether the solubility of either (a) AgCl or (b) $SrCO_3$ is affected by pH.

17.92 Write equations to show whether the solubility of either (a) CuBr or (b) $Ca_3(PO_4)_2$ is affected by pH.

17.93 Write equations to show whether the solubility of either (a) $Fe(OH)_2$ or (b) CuS is affected by pH.

17.94 Write equations to show whether the solubility of either (a) PbI_2 or (b) $Hg_2(CN)_2$ is affected by pH.

17.95 Does any solid $Cu(OH)_2$ form when 0.075 g of KOH is dissolved in 1.0 L of 1.0×10^{-3} mol/L $Cu(NO_3)_2$?

17.96 Does any solid $PbCl_2$ form when 3.5 mg of NaCl is dissolved in 0.250 L of 0.12 mol/L $Pb(NO_3)_2$?

17.97 Does any solid $Ba(IO_3)_2$ form when 7.5 mg of $BaCl_2$ is dissolved in 500. mL of 0.023 mol/L $NaIO_3$?

17.98 Does any solid Ag_2CrO_4 form when 2.7×10^{-5} g of $AgNO_3$ is dissolved in 15.0 mL of 4.0×10^{-4} mol/L K_2CrO_4?

Problems in Context

17.99 When blood is donated, sodium oxalate solution is used to precipitate Ca^{2+}, which triggers clotting. A 104 mL sample of blood contains 9.7×10^{-5} g Ca^{2+}/mL. A technologist treats the sample with 100.0 mL of 0.1550 mol/L $Na_2C_2O_4$. Calculate $[Ca^{2+}]$ after the treatment. (See Appendix C for the K_{sp} of $CaC_2O_4 \cdot H_2O$.)

17.100 A 50.0 mL volume of 0.50 mol/L $Fe(NO_3)_3$ is mixed with 125 mL of 0.25 mol/L $Cd(NO_3)_2$.
(a) If aqueous NaOH is added, which ion precipitates first? (See Appendix C.)
(b) Describe how the metal ions can be separated using NaOH.
(c) Calculate the $[OH^-]$ that will accomplish the separation.

Equilibria Involving Complex Ions

(Sample Problems 17.15 and 17.16)

Concept Review Questions

17.101 How can a metal cation be at the centre of a complex anion?

17.102 Write equations to show the step-by-step reaction of $[Cd(H_2O)_4]^{2+}$ in an aqueous solution of KI to form $[CdI_4]^{2-}$. Show that $K_{f(overall)} = K_{f_1} \times K_{f_2} \times K_{f_3} \times K_{f_4}$.

17.103 Consider the dissolution of PbS in water:

$$PbS(s) + H_2O(l) \rightleftharpoons Pb^{2+}(aq) + HS^-(aq) + OH^-(aq)$$

Adding aqueous NaOH causes more PbS to dissolve. Does this violate Le Châtelier's principle? Explain.

Skill-Building Exercises (grouped in similar pairs)

17.104 Write a balanced equation for the reaction of $[Hg(H_2O)_4]^{2+}$ in aqueous KCN.

17.105 Write a balanced equation for the reaction of $[Zn(H_2O)_4]^{2+}$ in aqueous NaCN.

17.106 Write a balanced equation for the reaction of $[Ag(H_2O)_2]^+$ in aqueous $Na_2S_2O_3$.

17.107 Write a balanced equation for the reaction of $[Al(H_2O)_6]^{3+}$ in aqueous KF.

17.108 What is $[Ag^+]$ when 25.0 mL of 0.044 mol/L $AgNO_3$ and 25.0 mL of 0.57 mol/L $Na_2S_2O_3$ are mixed (K_f of $[Ag(S_2O_3)_2]^{3-} = 4.7 \times 10^{13}$)?

17.109 Potassium thiocyanate, KSCN, is often used to detect the presence of Fe^{3+} ions in solution through the formation of the red $[Fe(H_2O)_5SCN]^{2+}$ (or, more simply, $[FeSCN]^{2+}$). What is $[Fe^{3+}]$ when 0.50 L of 0.0015 mol/L $Fe(NO_3)_3$ and 0.50 L of 0.20 mol/L KSCN are mixed (K_f of $[FeSCN]^{2+} = 8.9 \times 10^2$)?

17.110 Find the solubility of $Cr(OH)_3$ in a buffer of pH 13.0 (K_{sp} of $Cr(OH)_3 = 6.3 \times 10^{-31}$; K_f of $[Cr(OH)_4]^- = 8.0 \times 10^{29}$).

17.111 Find the solubility of AgI in 2.5 mol/L NH_3 (K_{sp} of AgI $= 8.3 \times 10^{-17}$; K_f of $[Ag(NH_3)_2]^+ = 1.7 \times 10^7$).

17.112 When 0.84 g of $ZnCl_2$ is dissolved in 245 mL of 0.150 mol/L NaCN, what are $[Zn^{2+}]$, $[[Zn(CN)_4]^{2-}]$, and $[CN^-]$ (K_f of $[Zn(CN)_4]^{2-} = 4.2 \times 10^{19}$)?

17.113 When 2.4 g of $Co(NO_3)_2$ is dissolved in 0.350 L of 0.22 mol/L KOH, what are $[Co^{2+}]$, $[[Co(OH)_4]^{2-}]$, and $[OH^-]$ (K_f of $[Co(OH)_4]^{2-} = 5 \times 10^9$)?

Comprehensive Problems

17.114 What volumes of 0.200 mol/L HCOOH and 2.00 mol/L NaOH would make 500. mL of a buffer with the pH of a buffer made from 475 mL of 0.200 mol/L benzoic acid and 25 mL of 2.00 mol/L NaOH?

17.115 A microbiologist is preparing a medium on which to culture *E. coli* bacteria. He buffers the medium at a pH of 7.00 to minimize the effect of acid-producing fermentation. What volumes of equimolar aqueous solutions of K_2HPO_4 and KH_2PO_4 must he combine to make 100. mL of the pH 7.00 buffer?

17.116 As a Health Canada physiologist, you need 0.700 L of methanoic acid–methanoate buffer with a pH of 3.74.
(a) What is the required buffer-component concentration ratio?
(b) How do you prepare this solution from stock solutions of 1.0 mol/L HCOOH and 1.0 mol/L NaOH?
(c) What is the final concentration of HCOOH in this solution?

17.117 Tris(hydroxymethyl)aminomethane, $(HOCH_2)_3CNH_2$, known as TRIS, is a weak base used in biochemical experiments to make buffer solutions in the pH range of 7 to 9. A certain TRIS buffer has a pH of 8.10 at 25°C and a pH of 7.80 at 37°C. Why does the pH change with temperature?

17.118 Water flowing through pipes of carbon steel must be kept at a pH of 5 or greater to limit corrosion. If a 3.6×10^3 kg/h stream of water contains 10 ppm sulfuric acid and 0.015% ethanoic (acetic) acid, at what rate (kg/h) must sodium ethanoate trihydrate be added to maintain this pH?

17.119 Gout is caused by an error in metabolism that leads to a buildup of uric acid in body fluids, which is deposited as slightly soluble sodium urate ($C_5H_3N_4O_3Na$) in the joints. If the extracellular $[Na^+]$ is 0.15 mol/L and the solubility of sodium urate is 0.085 g/100. mL, what is the minimum urate ion concentration (abbreviated $[Ur^-]$) that will cause a deposit of sodium urate?

17.120 In the process of cave formation (Section 17.3), the dissolution of CO_2 (Equation 1) has a K_{eq} of 3.1×10^{-2}, and the formation of aqueous $Ca(HCO_3)_2$ (Equation 2) has a K_{eq} of 1×10^{-12}. The fraction by volume of atmospheric CO_2 is 3×10^{-4}.
(a) Find $[CO_2(aq)]$ in equilibrium with atmospheric CO_2.
(b) Determine $[Ca^{2+}]$ arising from Equation 2, given current levels of atmospheric CO_2.
(c) Calculate $[Ca^{2+}]$ if atmospheric CO_2 doubles.

17.121 Phosphate systems form essential buffers in organisms. Calculate the pH of a buffer made by dissolving 0.80 mol of NaOH in 0.50 L of 1.0 mol/L H_3PO_4.

17.122 The solubility of KCl is 3.7 mol/L at 20°C. Two beakers contain 100. mL of saturated KCl solution; 100. mL of 6.0 mol/L HCl is added to the first beaker, and 100. mL of 12 mol/L HCl is added to the second.
(a) Find the ion-product constant of KCl at 20°C.
(b) What mass, if any, of KCl will precipitate from each beaker?

17.123 It is possible to detect NH_3 gas over 10^{-2} mol/L NH_3. To what pH must 0.15 mol/L NH_4Cl be raised to form detectable NH_3?

17.124 Manganese(II) sulfide is one of the compounds that is found in nodules on the ocean floor. These nodules may eventually be a primary source of many transition metals. The solubility of MnS is 4.7×10^{-4} g/100 mL solution. Estimate the K_{sp} of MnS.

17.125 The normal pH of blood is 7.40 ± 0.05 and is controlled, in part, by the H_2CO_3/HCO_3^- buffer system.
(a) Assuming that the K_a value for carbonic acid at 25°C applies to blood, what is the $[H_2CO_3]/[HCO_3^-]$ ratio in normal blood?
(b) In a condition called *acidosis*, the blood is too acidic. What is the $[H_2CO_3]/[HCO_3^-]$ ratio in a patient whose blood pH is 7.20?

17.126 A bioengineer preparing cells for cloning bathes a small piece of rat epithelial tissue in a TRIS buffer (see Problem 17.117). The buffer is made by dissolving 43.0 g of TRIS ($pK_b = 5.91$) in enough 0.095 mol/L HCl to make 1.00 L of solution. What are the concentration (mol/L) of TRIS and the pH of the buffer?

17.127 Sketch a qualitative curve for the titration of 1,2-ethanediamine, $H_2NCH_2CH_2NH_2$, with 0.1 mol/L HCl.

17.128 A solution contains 0.10 mol/L $ZnCl_2$ and 0.020 mol/L $MnCl_2$. Given the following information, how would you adjust the pH to separate the ions as their sulfides? ($[H_2S]$ of a saturated aqueous solution at 25°C = 0.10 mol/L; $K_w = 1.0 \times 10^{-14}$ at 25°C.)

$$MnS + H_2O \rightleftharpoons Mn^{2+} + HS^- + OH^- \quad K_{sp} = 3 \times 10^{-11}$$
$$ZnS + H_2O \rightleftharpoons Zn^{2+} + HS^- + OH^- \quad K_{sp} = 2 \times 10^{-22}$$
$$H_2S + H_2O \rightleftharpoons H_3O^+ + HS^- \quad K_{a1} = 9 \times 10^{-8}$$

17.129 Amino acids [general formula $NH_2CH(R)COOH$] can be considered polyprotic acids. In many cases, the R group contains additional amine and carboxyl groups.
(a) Can an amino acid dissolved in pure water have a carboxyl (COOH) group and an amine (NH_2) group? Use glycine, NH_2CH_2COOH, to explain why (K_a of COOH group = 4.47×10^{-3}; K_b of NH_2 group = 6.03×10^{-5}).
(b) Calculate $[^+NH_3CH_2COO^-]/[^+NH_3CH_2COOH]$ at pH 5.5.
(c) The R group of lysine is $-CH_2CH_2CH_2CH_2NH_2$ ($pK_b = 3.47$). Draw the structure of lysine at pH 1, physiological pH (~7), and pH 13.
(d) The R group of glutamic acid is $-CH_2CH_2COOH$ ($pK_a = 4.07$). Of the forms of glutamic acid that are shown below, which predominates at (i) pH 1; (ii) physiological pH (~7); (iii) pH 13?

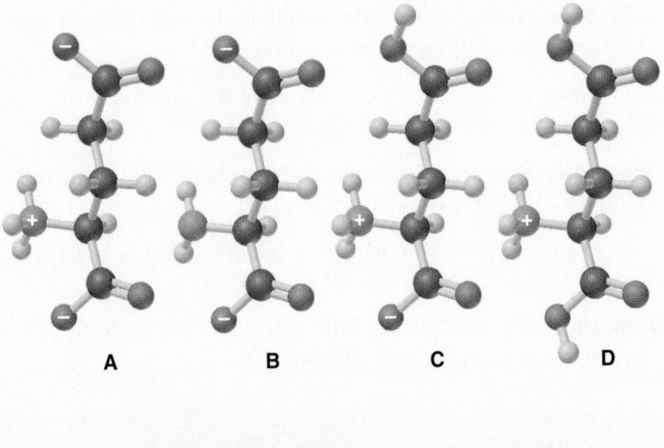

17.130 The scene at the right depicts a saturated solution of $MCl_2(s)$ in the presence of dilute aqueous NaCl. Each sphere represents 1.0×10^{-6} mol of ion, and the volume is 250.0 mL. (Solid MCl_2 is shown as *green* chunks; Na^+ ions and water molecules are not shown.)

(a) Calculate the K_{sp} of MCl_2.
(b) If $M(NO_3)_2(s)$ is added, is there an increase, decrease, or no change in (i) the number of Cl^- particles; (ii) K_{sp}; (iii) the mass of $MCl_2(s)$?

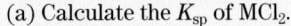

17.131 Tooth enamel consists of hydroxyapatite, $Ca_5(PO_4)_3OH$ ($K_{sp} = 6.8 \times 10^{-37}$). Fluoride ion added to drinking water reacts with $Ca_5(PO_4)_3OH$ to form the more tooth decay–resistant fluorapatite, $Ca_5(PO_4)_3F$ ($K_{sp} = 1.0 \times 10^{-60}$). Fluoridated water has dramatically decreased cavities among children. Calculate the solubility of $Ca_5(PO_4)_3OH$ and of $Ca_5(PO_4)_3F$ in water.

17.132 The acid-base indicator ethyl orange turns from red to yellow over the pH range 3.4 to 4.8. Estimate K_a for ethyl orange.

17.133 Use the values obtained in Problem 17.58 to sketch a curve of $[H_3O^+]$ versus volume (mL) of added titrant. Are there advantages or disadvantages to viewing the results in this form? Explain.

17.134 Instrumental acid-base titrations use a pH meter to monitor the changes in pH and volume. The equivalence point is found from the volume at which the curve has the steepest slope.
(a) Use the data in Figure 17.9 to calculate (i) the slope ($\Delta pH/\Delta V$) for all pairs of adjacent points; (ii) the average volume (V_{avg}) for each interval.
(b) Plot $\Delta pH/\Delta V$ versus V_{avg} to find the steepest slope, and thus the volume at the equivalence point. (For example, the first pair of points gives $\Delta pH = 0.22$, $\Delta V = 10.00$ mL; hence, $\Delta pH/\Delta V = 0.022$ mL^{-1}, and $V_{avg} = 5.00$ mL.)

17.135 What is the pH of a solution of 6.5×10^{-9} mol of $Ca(OH)_2$ in 10.0 L of water (K_{sp} of $Ca(OH)_2 = 6.5 \times 10^{-6}$)?

17.136 Muscle physiologists study the accumulation of lactic acid, $CH_3CH(OH)COOH$, during exercise. Food chemists study its occurrence in sour milk, beer, wine, and fruit. Industrial microbiologists study its formation by various bacterial species from carbohydrates. A biochemist prepares a lactic acid–lactate buffer by mixing 225 mL of 0.85 mol/L lactic acid ($K_a = 1.38 \times 10^{-4}$) with 435 mL of 0.68 mol/L sodium lactate. What is the buffer pH?

17.137 A student wants to dissolve the maximum amount of CaF_2 to make 1 L of aqueous solution (K_{sp} of $CaF_2 = 3.2 \times 10^{-11}$).
(a) Into which of the following solvents should she dissolve the salt?
(i) Pure water (ii) 0.01 mol/L HF
(iii) 0.01 mol/L NaOH (iv) 0.01 mol/L HCl
(v) 0.01 mol/L $Ca(OH)_2$
(b) Which solvent would dissolve the least amount of salt?

17.138 A 500. mL solution consists of 0.050 mol of solid NaOH and 0.13 mol of hypochlorous acid (HClO; $K_a = 3.0 \times 10^{-8}$) dissolved in water.
(a) Aside from water, what is the concentration of each species that is present?
(b) What is the pH of the solution?
(c) What is the pH after adding 0.0050 mol of HCl to the flask?

17.139 Calcium ion in water supplies is easily precipitated as calcite ($CaCO_3$):

$$Ca^{2+}(aq) + CO_3^{2-}(aq) \rightleftharpoons CaCO_3(s)$$

Because the K_{sp} decreases with temperature, heating hard water forms a calcite "scale," which clogs pipes and water heaters. Find the solubility of calcite in water at (a) 10°C ($K_{sp} = 4.4 \times 10^{-9}$); (b) 30°C ($K_{sp} = 3.1 \times 10^{-9}$).

17.140 Calculate the molar solubility of $Hg_2C_2O_4$ ($K_{sp} = 1.75 \times 10^{-13}$) in 0.13 mol/L $Hg_2(NO_3)_2$.

17.141 Environmental engineers use alkalinity as a measure of the capacity of carbonate buffering systems in water samples:

$$\text{Alkalinity(mol/L)} = [HCO_3^-] + 2[CO_3^{2-}] + [OH^-] - [H^-]$$

Find the alkalinity of a water sample that has a pH of 9.5, 26.0 mg/L CO_3^{2-}, and 65.0 mg/L HCO_3^-.

17.142 Human blood contains one buffer system based on phosphate species and another buffer system based on carbonate species. Assuming that blood has a normal pH of 7.4, what are the principal phosphate and carbonate species present? What is the ratio of the two phosphate species? (In the presence of the dissolved ions and other species in blood, K_{a1} of $H_3PO_4 = 1.3 \times 10^{-2}$, $K_{a2} = 2.3 \times 10^{-7}$, and $K_{a3} = 6 \times 10^{-12}$; K_{a1} of $H_2CO_3 = 8 \times 10^{-7}$ and $K_{a2} = 1.6 \times 10^{-10}$.)

17.143 A 0.050 mol/L H_2S solution contains 0.15 mol/L $NiCl_2$ and 0.35 mol/L $Hg(NO_3)_2$. What pH is required to precipitate the maximum amount of HgS but none of the NiS? (See Appendix C.)

17.144 Quantitative analysis of Cl^- ion is often performed by a titration with silver nitrate, using sodium chromate as an indicator. As standardized $AgNO_3$ is added, both white AgCl and red Ag_2CrO_4 precipitate, but so long as some Cl^- remains, the Ag_2CrO_4 redissolves as the mixture is stirred. When the red colour is permanent, the equivalence point has been reached.
(a) Calculate the equilibrium constant for the following reaction:

$$2AgCl(s) + CrO_4^{2-}(aq) \rightleftharpoons Ag_2CrO_4(s) + 2Cl^-(aq)$$

(b) Explain why the silver chromate redissolves.
(c) If 25.00 cm^3 of 0.1000 mol/L NaCl is mixed with 25.00 cm^3 of 0.1000 mol/L $AgNO_3$, what is the concentration of Ag^+ remaining in solution? Is this sufficient to precipitate any silver chromate?

17.145 An ecobotanist separates the components of a tropical bark extract by chromatography. He discovers a large proportion of quinidine, a dextrorotatory isomer of quinine used for control of arrhythmic heartbeat. Quinidine has two basic nitrogens ($K_{b1} = 4.0 \times 10^{-6}$ and $K_{b2} = 1.0 \times 10^{-10}$). To measure the concentration, the ecobotanist carries out a titration. Because of the low solubility of quinidine, he first protonates both nitrogens with excess HCl and then titrates the acidified solution with standardized base. A 33.85 mg sample of quinidine ($\mathcal{M} = 324.41$ g/mol) is acidified with 6.55 mL of 0.150 mol/L HCl.
(a) What volume of 0.0133 mol/L NaOH is needed to titrate the excess HCl?
(b) What additional volume of titrant is needed to reach the first equivalence point of quinidine dihydrochloride?
(c) What is the pH at the first equivalence point?

17.146 Some kidney stones form by the precipitation of calcium oxalate monohydrate ($CaC_2O_4 \cdot H_2O$; $K_{sp} = 2.3 \times 10^{-9}$). The pH of urine varies from 5.5 to 7.0, and the average $[Ca^{2+}]$ in urine is 2.6×10^{-3} mol/L.
(a) If the [oxalic acid] in urine is 3.0×10^{-13} mol/L, will kidney stones form at (i) pH = 5.5; (ii) pH = 7.0?
(b) Vegetarians have a urine pH above 7. Are they more or less likely to form kidney stones?

17.147 A biochemist needs a medium for acid-producing bacteria. The pH of the medium must not change by more than 0.05 for every 0.0010 mol of H_3O^+ generated by the organisms per litre of medium. A buffer consisting of 0.10 mol/L HA and 0.10 mol/L A^- is included in the medium to control its pH. What volume of this buffer must be included in 1.0 L of medium?

17.148 A 35.00 mL solution of 0.2500 mol/L HF is titrated with a standardized 0.1532 mol/L solution of NaOH at 25°C.
(a) What is the pH of the HF solution before the titrant is added?
(b) What volume of titrant is required to reach the equivalence point?
(c) What is the pH at 0.50 mL before the equivalence point?
(d) What is the pH at the equivalence point?
(e) What is the pH at 0.50 mL after the equivalence point?

17.149 Because of the toxicity of mercury compounds, mercury(I) chloride is used in antibacterial salves. The mercury(I) ion (Hg_2^{2+}) consists of two bound Hg^+ ions.
(a) What is the empirical formula for mercury(I) chloride?
(b) Calculate $[Hg_2^{2+}]$ in a saturated solution of mercury(I) chloride (K_{sp} of mercury(I) chloride $= 1.5 \times 10^{-18}$).
(c) A seawater sample contains 0.024 g of NaCl per cm^3. Find $[Hg_2^{2+}]$ if the seawater is saturated with mercury(I) chloride.
(d) What mass of mercury(I) chloride is needed to saturate 4900 km^3 of pure water (the volume of Lake Michigan)?
(e) What mass of mercury(I) chloride is needed to saturate 4900 km^3 of seawater?

17.150 A 35.0 mL solution of 0.075 mol/L $CaCl_2$ is mixed with 25.0 mL of 0.090 mol/L $BaCl_2$.
(a) If aqueous KF is added, which fluoride precipitates first?
(b) Describe how the metal ions can be separated using KF to form the fluorides.
(c) Calculate the fluoride ion concentration that will accomplish the separation.

17.151 Even before the industrial age, rainwater was slightly acidic due to dissolved CO_2. Use the following data to calculate the pH of unpolluted rainwater at 25°C: vol % of CO_2 in air = 0.033 vol %; solubility of CO_2 in pure water at 25°C and 1 bar = 88 mL CO_2/100 mL H_2O; K_{a1} of $H_2CO_3 = 4.5 \times 10^{-7}$.

17.152 Seawater at the surface has a pH of about 8.5.
(a) Which of the following species has the highest concentration at this pH? Explain.
(i) H_2CO_3 (ii) HCO_3^- (iii) CO_3^{2-}
(b) What are the concentration ratios (i) $[CO_3^{2-}]/[HCO_3^-]$ and (ii) $[HCO_3^-]/[H_2CO_3]$ at this pH?
(c) In the deep sea, light levels are low, and the pH is around 7.5. Suggest a reason for the lower pH at the greater ocean depth. (*Hint*: Consider the presence or absence of plant and animal life, and the effects on carbon dioxide concentrations.)

17.153 Ethylenediaminetetraethanoic acid (abbreviated H_4EDTA) is a tetraprotic acid. Its salts are used to treat toxic metal poisoning by forming soluble complex ions that are then excreted. Because $EDTA^{4-}$ also binds essential calcium ions, it is often administered as the calcium disodium salt. For example, when $Na_2Ca(EDTA)$ is given to a patient, the $[Ca(EDTA)]^{2-}$ ions react with circulating Pb^{2+} ions and the metal ions are exchanged:

$$[Ca(EDTA)]^{2}(aq) + Pb^{2+}(aq) \rightleftharpoons [Pb(EDTA)]^{2-}(aq) + Ca^{2+}(aq) \quad K = 2.5 \times 10^{7}$$

A child has a dangerous blood lead level of 120 μg/100 mL. If the child is administered 100. mL of 0.10 mol/L $Na_2Ca(EDTA)$, assuming the exchange reaction and excretion process are 100% efficient, what is the final concentration of Pb^{2+} in μg/100 mL blood? (Total blood volume is 1.5 L.)

17.154 Buffers that are based on 3-morpholinopropanesulfonic acid (MOPS) are often used in RNA analysis. The useful pH range of a MOPS buffer is 6.5 to 7.9. Estimate the K_a of MOPS.

17.155 NaCl is purified by adding HCl to a saturated solution of NaCl (317 g/L). When 28.5 mL of 8.65 mol/L HCl is added to 0.100 L of saturated solution, what mass (g) of pure NaCl precipitates?

17.156 Scenes A to D represent tiny portions of 0.10 mol/L aqueous solutions of a weak acid HA (*red and blue*; $K_a = 4.5 \times 10^{-5}$), its conjugate base A^- (*red*), or a mixture of the two. (Only these species are shown.)

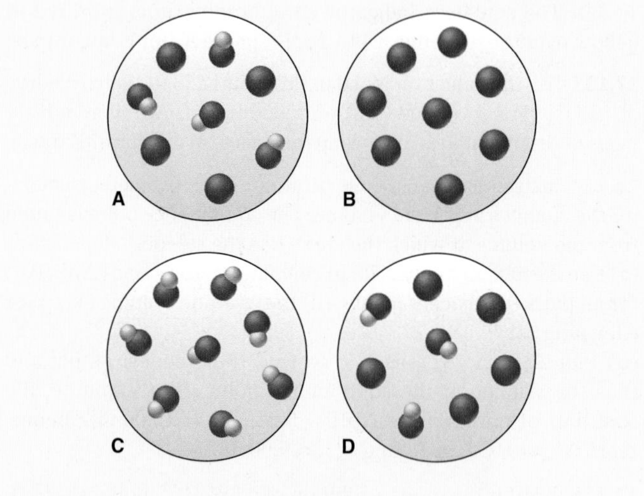

(a) Which scene(s) show(s) a buffer?
(b) What is the pH of each solution?
(c) Arrange the scenes in sequence, assuming that they represent stages in a weak acid–strong base titration.
(d) Which scene represents the titration at its equivalence point?

17.157 Scenes A to C represent aqueous solutions of the slightly soluble salt MZ. (Only the ions of this salt are shown.)

$$MZ(s) \rightleftharpoons M^{2+}(aq) + Z^{2-}(aq)$$

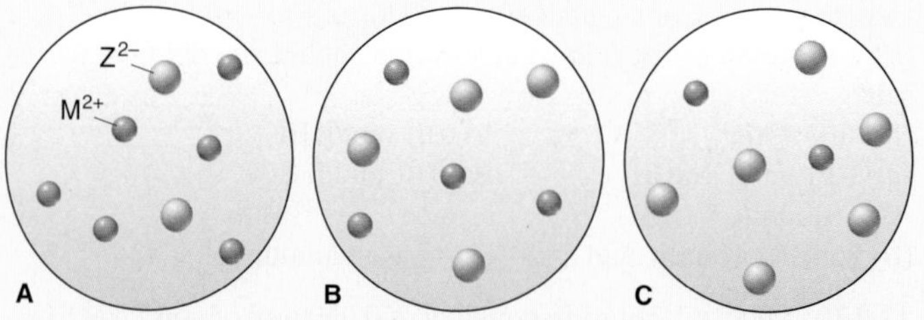

(a) Which scene represents the solution just after solid MZ is stirred thoroughly in distilled water?
(b) If each sphere represents 2.5×10^{-6} mol/L of ions, what is the K_{sp} of MZ?
(c) Which scene represents the solution after $Na_2Z(aq)$ has been added?
(d) If Z^{2-} is CO_3^{2-}, which scene represents the solution after the pH has been lowered?

17.158 The solubility of Ag(I) in aqueous solutions containing different concentrations of Cl^- is based on the following equilibria:

$$Ag^+(aq) + Cl^-(aq) \rightleftharpoons AgCl(s) \qquad K_{sp} = 1.8 \times 10^{-10}$$

$$Ag^+(aq) + 2\,Cl^-(aq) \rightleftharpoons [AgCl_2]^-(aq) \quad K_f = 1.8 \times 10^5$$

When solid AgCl is shaken with a solution containing Cl^-, Ag(I) is present as both Ag^+ and $[AgCl_2]^-$. The solubility of AgCl is the sum of the concentrations of Ag^+ and $[AgCl_2]^-$.
(a) Show that $[Ag^+]$ in solution is given by

$$[Ag^+] = \frac{1.8 \times 10^{-10}}{[Cl^-]}$$

and that $[[AgCl_2]^-]$ in solution is given by

$$[[AgCl_2]^-] = (3.2 \times 10^{-5})([Cl^-])$$

(b) Find the $[Cl^-]$ at which $[Ag^+] = [[AgCl_2]^-]$.
(c) Explain the shape of a plot of AgCl solubility versus $[Cl^-]$.
(d) Find the solubility of AgCl at the $[Cl^-]$ of part (b), which is the minimum solubility of AgCl in the presence of Cl^-.

17.159 EDTA binds metal ions to form complex ions (see Problem 17.153), so it is used to determine the concentrations of metal ions in solution:

$$M^{n+}(aq) + [EDTA]^{4-}(aq) \longrightarrow [MEDTA]^{n-4}(aq)$$

A 50.0 mL sample of 0.048 mol/L Co^{2+} is titrated with 0.050 mol/L $EDTA^{4-}$. Find $[Co^{2+}]$ and $[EDTA^{4-}]$ after (a) 25.0 mL and (b) 75.0 mL of $EDTA^{4-}$ are added ($-\log K_f$ of $CoEDTA^{2-} = 16.31$).

Thermodynamics: Entropy, Gibbs Energy, and the Direction of Chemical Reactions

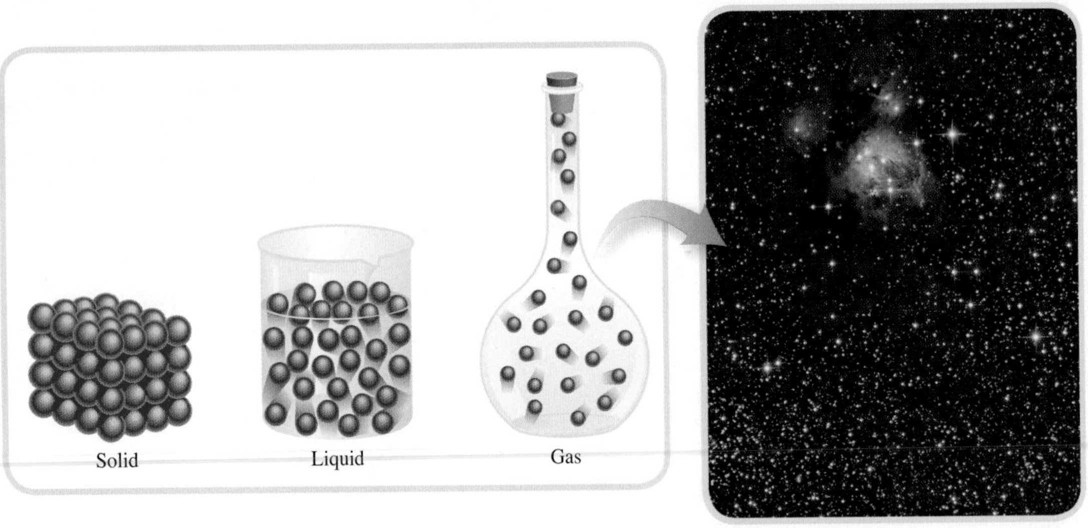

IN THIS CHAPTER . . . We discuss why changes occur and how we can use them, by focusing on entropy and Gibbs energy and their relation to the direction of a thermodynamically allowed (or spontaneous) change. By the end of this chapter, you should be able to

- Explain the need for a criterion to predict the direction of a thermodynamically allowed change
- Review the first law of thermodynamics and show that it accounts for the energy of a change but not the direction
- Show that the sign of the enthalpy change does not predict direction
- Determine the criterion for predicting the direction of a thermodynamically allowed change in the second law of thermodynamics and its relation to entropy (S), a state function based on the natural tendency of a system's energy to become dispersed
- Calculate entropy changes for exothermic and endothermic processes
- Develop the concept of Gibbs energy as a simplified criterion for a thermodynamically allowed change and show how it relates to the work that a system can do
- Describe and calculate the key relationship between the Gibbs energy change of a reaction and its equilibrium constant

18

Concepts and Skills to Review before Studying This Chapter

- Internal energy, heat, and work (Section 5.1)
- State functions (Section 5.1) and standard states (Section 5.6)
- Enthalpy, ΔH, and Hess's law (Sections 5.2 and 5.5)
- Entropy and solution formation (Section 12.2)
- Comparison of Q and K to find reaction direction (Section 15.4)

The universe we live in is continuously expanding. Despite our best efforts, we age. Although it takes heat energy for ice to melt, it does so spontaneously under certain conditions. What do all these examples have in common? In Chapter 5, we learned that reactions where heat is released (exothermic processes) have a tendency to occur naturally (be spontaneous). However, there are reactions that require heat (such as the melting of ice) that are nonetheless spontaneous. In this chapter, we will learn about entropy, another thermodynamic variable that is also a determinant for whether a reaction will occur or not. We see examples of entropy in our daily lives, as the examples above demonstrate, but what *is* entropy, and how do we determine its effects? Can enthalpy or entropy determine if a reaction is spontaneous? We will define a new variable called the Gibbs energy and learn why it alone can predict whether a reaction occurs spontaneously. We will connect these thermodynamic variables to the ideas we have established about equilibrium in the three previous chapters and see how these concepts are related.

18.1 The Second Law of Thermodynamics: Predicting Spontaneous Change

A thermodynamically allowed change, or **spontaneous change**, of a system is a change that occurs under specified conditions without a continuous input of energy from outside the system. The freezing of water, for example, is spontaneous at 1 bar and −5°C. A spontaneous process, such as burning or falling, may need a little "push" to get started—a spark to ignite the gasoline vapours in the engine of a car, a shove to knock a book off your desk—but once the process begins, it supplies the energy to keep going. In contrast, a *nonspontaneous* change occurs only if the surroundings *continuously* supply the system with an input of energy. Under given conditions, *if a change is spontaneous in one direction, it is* ***not*** *spontaneous in the other.*

The term *spontaneous* does not mean *instantaneous*, nor does it reveal anything about how long a process takes to occur; it means that, given enough time, the process will happen by itself. (As a reminder that the word *spontaneous* is used in a specific chemical context in this chapter, we will interchangeably use the terms *spontaneous* and *thermodynamically allowed*.) Many processes, such as ripening, rusting, and aging, are spontaneous but slow. *A chemical reaction proceeding toward equilibrium is a spontaneous change.* Recall that we can predict the net direction of the reaction—its spontaneous direction—by comparing the reaction quotient (Q) with the equilibrium constant (K). However, *why* is there a drive to attain equilibrium? What determines the value of the equilibrium constant? Most important, can we predict the direction of a spontaneous change in situations that are not as obvious as burning gasoline and falling books?

The First Law of Thermodynamics Does Not Predict Spontaneous Change

Let us see whether energy changes can clarify the criterion for spontaneity. Recall, from Chapter 5, that the first law of thermodynamics (the law of conservation of

Freezing of water

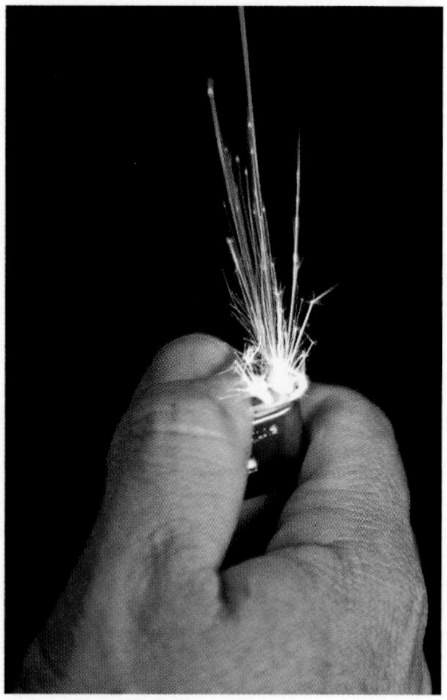

Spark needed for combustion

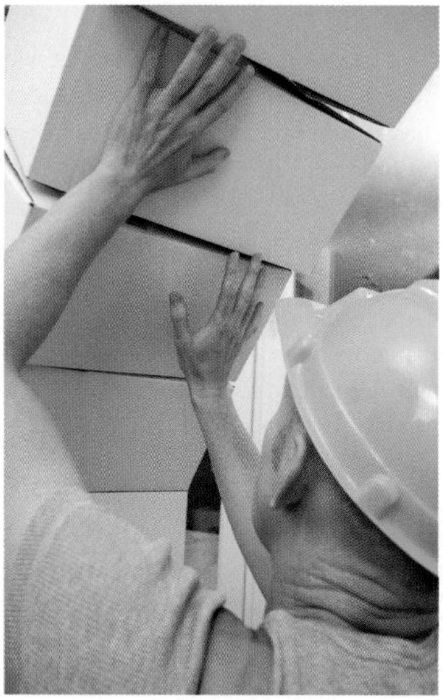

Spontaneous tipping of boxes unless held

Expansion of the universe and dissipation of the Sun's energy

energy) states that the internal energy (U) of a system, the sum of the kinetic and potential energies of its particles, changes when heat (q) and/or work (w) are absorbed or released:

$$\Delta U = q + w$$

Whatever is not part of the system (sys) is part of the surroundings (surr); thus, *the change in energy and, therefore, the heat and/or work absorbed by the system are released by the surroundings*, and vice versa:

$$\Delta_{sys}U = -\Delta_{surr}U \qquad \text{or} \qquad (q + w)_{sys} = -(q + w)_{surr}$$

Since the system plus the surroundings is the universe (univ), it follows that *the total energy of the universe is constant, so the change in the energy of the universe is zero:**

$$\Delta_{sys}U + \Delta_{surr}U = 0 = \Delta_{univ}U$$

The first law accounts for the energy, but not the direction, of a process. When gasoline burns in a car engine, the first law states that the potential energy difference between the bonds in the fuel mixture and the bonds in the exhaust gases is converted to the kinetic energy of the moving car and its parts plus the heat released

*Any modern statement of conservation of energy must take into account mass-energy equivalence and the processes in stars, which convert enormous amounts of matter into energy. Thus, the total *mass-energy* of the universe is constant.

to the environment. However, why does the heat released in the engine not convert exhaust fumes back into gasoline and oxygen? When an ice cube melts in your hand, the first law tells us that energy from your hand is converted to kinetic energy as the solid changes to a liquid. Why does the pool of water in your cupped hand not transfer the heat back to your hand and refreeze? Neither of these events violates the first law—if we measured the work and heat in each event, we would find that energy is conserved—but these reverse changes never happen. That is, *the first law by itself does not predict the* ***direction*** *of a thermodynamically allowed change.*

The Sign of ΔH Does Not Predict Spontaneous Change

Perhaps the sign of the enthalpy change (ΔH), the heat gained or lost at constant pressure (q_p), is the criterion for spontaneity. If so, we can expect exothermic processes ($\Delta H < 0$) to be spontaneous and endothermic processes ($\Delta H > 0$) to be nonspontaneous. Let us examine some examples to see if this is true.

1. *Spontaneous processes with* $\Delta H < 0$. All freezing and condensing processes are exothermic *and* spontaneous at certain conditions:

$$H_2O(l) \longrightarrow H_2O(s) \qquad \Delta_r H^\circ = -\Delta_{fus}H^\circ = -6.02\ \text{kJ/mol (1 bar; } T < 0^\circ\text{C)}$$

All combustion reactions are spontaneous and exothermic:

$$CH_4(g) + 2O_2(g) \longrightarrow CO_2(g) + 2H_2O(g) \qquad \Delta_r H^\circ = -802\ \text{kJ/mol}$$

The oxidation of iron and other metals occurs spontaneously and exothermically:

$$2Fe(s) + \frac{3}{2}O_2(g) \longrightarrow Fe_2O_3(s) \qquad \Delta_r H^\circ = -826\ \text{kJ/mol}$$

Ionic compounds form spontaneously and exothermically from their elements:

$$Na(s) + \frac{1}{2}Cl_2(g) \longrightarrow NaCl(s) \qquad \Delta_r H^\circ = -411\ \text{kJ/mol}$$

Often, however, an exothermic process occurs spontaneously under one set of conditions, whereas the opposite, endothermic, process occurs spontaneously under another set of conditions.

2. *Spontaneous processes with* $\Delta H > 0$. All melting and vaporizing processes are endothermic *and* spontaneous at certain conditions:

$$H_2O(s) \longrightarrow H_2O(l) \qquad \Delta_r H^\circ = \Delta_{fus}H^\circ = +6.02\ \text{kJ/mol (} p = 1\ \text{bar; } T = 0^\circ\text{C)}$$

At ordinary pressure, water vaporizes spontaneously:

$$H_2O(l) \longrightarrow H_2O(g) \qquad \Delta_r H^\circ = \Delta_{vap}H^\circ = +44.0\ \text{kJ/mol (} p = 1\ \text{bar; } T = 100^\circ\text{C)}$$

Most soluble salts dissolve endothermically *and* spontaneously:

$$NaCl(s) \xrightarrow{H_2O} Na^+(aq) + Cl^-(aq) \qquad \Delta_{soln}H^\circ = +3.9\ \text{kJ/mol}$$

$$RbClO_3(s) \xrightarrow{H_2O} Rb^+(aq) + ClO_3^-(aq) \qquad \Delta_{soln}H^\circ = +47.7\ \text{kJ/mol}$$

$$NH_4NO_3(s) \xrightarrow{H_2O} NH_4^+(aq) + NO_3^-(aq) \qquad \Delta_{soln}H^\circ = +25.7\ \text{kJ/mol}$$

Even some endothermic reactions are spontaneous:

$$N_2O_5(s) \longrightarrow 2NO_2(g) + \frac{1}{2}O_2(g) \qquad \Delta_r H^\circ = +109.5\ \text{kJ/mol}$$

$$Ba(OH)_2{\cdot}8H_2O(s) + 2NH_4NO_3(s) \longrightarrow Ba^{2+}(aq) + NO_3^-(aq) + 2NH_3(aq) + 10H_2O(l) \qquad \Delta_r H^\circ = +62.3\ \text{kJ/mol}$$

In the latter reaction, crystalline solids are mixed (Figure 18.1A) and the release of waters of hydration solvates the ions. The reaction mixture absorbs heat from the surroundings so quickly that the beaker freezes to a wet block of wood (Figure 18.1B).

Given just these few examples, we can see that *the sign of* ΔH *by itself does not predict the* ***direction*** *of a spontaneous change.* This is the same as what we saw for the first law.

FIGURE 18.1 A spontaneous endothermic reaction

Freedom of Particle Motion and Dispersal of Particle Energy

When we look closely at the previous examples of spontaneous *endothermic* processes, they have one major feature in common: in every process, the chemical entities—atoms, molecules, or ions—have more freedom of motion *after* the change. Put another way, after the change, the particles have a wider range of energy of motion (kinetic energy); we say that the energy has become more dispersed, distributed, or spread out:

- The phase changes convert a solid, in which motion is restricted, to a liquid, in which particles have more freedom to move around each other, and then to a gas, in which the particles have much greater freedom of motion. Thus, the energy of motion is more dispersed.
- Dissolving a salt changes a crystalline solid and a pure liquid into separate ions and solvent molecules moving and interacting, so their freedom of motion is greater and their energy of motion is more dispersed.
- In the chemical reactions, a *smaller* amount (mol) of crystalline solids produces a *larger* amount (mol) of gases and/or solvated ions, so, once again, the particles have greater freedom of motion and their energy of motion is more dispersed:

	less freedom of particle motion	$\longrightarrow$	more freedom of particle motion
	localized energy of motion	$\longrightarrow$	dispersed energy of motion
Phase change:	solid $\longrightarrow$ liquid	$\longrightarrow$	gas
Dissolving of salt:	crystalline solid + liquid	$\longrightarrow$	ions in solution
Chemical change:	crystalline solids	$\longrightarrow$	gases + ions in solution

In thermodynamic terms, a change in the freedom of motion of particles in a system—that is, in the dispersal of their energy of motion—is a key factor affecting the ***direction*** *of a spontaneous process.*

Entropy and the Number of Microstates

Earlier, we discussed the quantized *electronic* energy levels of an atom (Chapter 6) and a molecule (Chapter 10). In this section, we will see that the energy state of a whole system of particles is quantized, too. (In Chapter 22, we will examine the quantized *kinetic* energy levels—vibrational, rotational, and translational—of a molecule.)

Energy Dispersal and the Meaning of Entropy Let us see why freedom of motion and dispersal of energy relate to a thermodynamically allowed change:

- *Quantization of energy.* Picture a system of, say, 1 mol of N_2 gas, and focus on one molecule. At any instant, the molecule is moving through space (translating) at some speed and rotating at some frequency, and its atoms are vibrating at some frequency. In the next instant, the molecule collides with another molecule or with the container, and these motional (kinetic) energy states change to different values. The complete quantum state of the molecule at any instant consists of its electronic states and these translational, rotational, and vibrational states. In this discussion, we focus on the latter three states.
- *Number of microstates.* The energy of the other molecules is similarly quantized. Each quantized state of the system of molecules is called a *microstate.* At any instant, the total energy of the system is dispersed throughout one microstate. In the next instant, it is dispersed throughout a different microstate. The number of microstates possible for a system of 1 mol of molecules is staggering, on the order of $10^{10^{23}}$.
- *Dispersal of energy.* Under a given set of conditions, each microstate has the *same* total energy as any other microstate. Therefore, each microstate is equally possible for the system, and the laws of probability say that, over time, all microstates are equally likely. The number of microstates for a system is the number of ways that the system can disperse (distribute or spread) its kinetic energy among the various motions of all its particles.

In 1877, the Austrian mathematician and physicist Ludwig Boltzmann related the number of microstates (Ω) to the **entropy (S)** of a system:

$$S = k \ln \Omega \qquad (18.1)$$

In this relationship, k, the *Boltzmann constant*, is the universal gas *constant* (R) divided by Avogadro's number (N_A), or $\frac{R}{N_A}$, and equals 1.38×10^{-23} J/K. The term Ω is the number of microstates, so it has no units; therefore, S has units of J/(mol•K). Here the mole unit in the denominator signifies the value of S per mole of reactant or per mole of reaction, as we saw for enthalpy. Thus, the following apply:

- A system with fewer microstates (smaller Ω) has *lower entropy (lower S)*.
- A system with more microstates (larger Ω) has *higher entropy (higher S)*.

For our earlier examples of endothermic processes,

lower entropy (fewer microstates) $\longrightarrow$ higher entropy (more microstates)

Phase change:	solid $\longrightarrow$ liquid $\longrightarrow$ gas
Dissolving of salt:	crystalline solid + liquid $\longrightarrow$ ions in solution
Chemical change:	crystalline solids $\longrightarrow$ gases + ions in solution

(Recall, from Chapter 12, that entropy is a factor in the formation of solutions.)

Entropy as a State Function If a change results in a greater number of microstates, there are more ways to disperse the energy of the system and the entropy increases:

$$S_{\text{more microstates}} > S_{\text{fewer microstates}}$$

If a change results in a lower number of microstates, the entropy decreases. Like internal energy (U) and enthalpy (H), *entropy is a state function*, so it depends only on the present state of the system, not on how the system arrived at this state (see Section 5.1). Therefore, the change in entropy of the system ($\Delta_{\text{sys}}S$) depends only on the *difference* between the final and initial values:

$$\Delta_{\text{sys}}S = S_{\text{final}} - S_{\text{initial}}$$

Like any state function, when the entropy increases during a change, $\Delta_{\text{sys}}S > 0$. For example, the entropy increases during the phase change that occurs when dry ice sublimes:

$$CO_2(s) \longrightarrow CO_2(g) \qquad \Delta_{\text{sys}}S = S_{\text{final}} - S_{\text{initial}} = S_{\text{gaseous } CO_2} - S_{\text{solid } CO_2} > 0$$

When the entropy decreases during a change, as when water vapour condenses, $\Delta_{\text{sys}}S < 0$:

$$H_2O(g) \longrightarrow H_2O(l) \qquad \Delta_{\text{sys}}S = S_{\text{liquid } H_2O} - S_{\text{gaseous } H_2O} < 0$$

As an example of a reaction during which entropy increases, consider the decomposition of dinitrogen tetroxide (written as $O_2N{-}NO_2$):

$$O_2N{-}NO_2(g) \longrightarrow 2NO_2(g)$$

When the N—N bond in 1 mol of dinitrogen tetroxide breaks, the 2 mol of NO_2 molecules has many more possible motions; thus, at any instant, the energy of the system is dispersed into any one of a larger number of microstates. Therefore, the change in entropy of the system, which is the change in entropy of the reaction ($\Delta_r S$), goes up:

$$\Delta_{\text{sys}}S = \Delta_r S = S_{\text{final}} - S_{\text{initial}} = S_{\text{products}} - S_{\text{reactants}} = 2S_{NO_2} - S_{N_2O_4} > 0$$

Quantitative Meaning of an Entropy Change There are two approaches for quantifying an entropy change, but they give the same result. The first approach is based on the number of microstates that are possible for a system, and the second approach is based on the heat absorbed (or released) by the system. We will examine both approaches with a system of 1 mol of a gas expanding from 1 L to 2 L and behaving ideally, much as neon does at 298 K:

1 mol neon (initial: 1 L and 298 K) $\longrightarrow$ 1 mol neon (final: 2 L and 298 K)

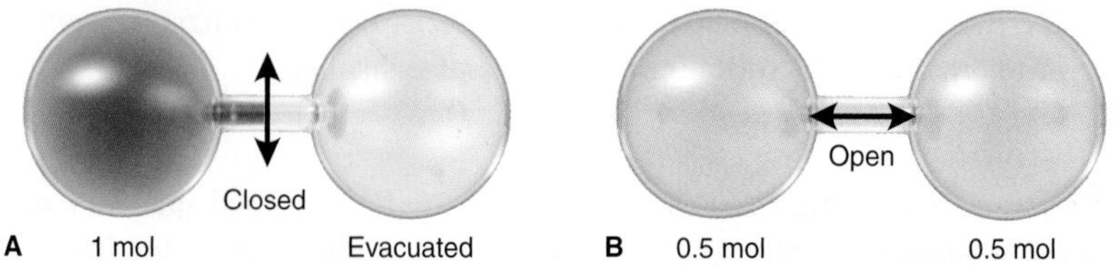

FIGURE 18.2 Spontaneous expansion of a gas. A. With the stopcock closed, the left flask contains 1 mol of Ne. **B.** With the stopcock open, the gas expands, and each flask contains 0.5 mol of Ne.

1. *Quantifying $\Delta_{sys}S$ from the number of microstates.* Figure 18.2 shows two flasks connected by a stopcock: the right flask is evacuated, and the left flask contains 1 mol of neon. When we open the stopcock, the gas expands until each flask contains 0.5 mol. But *why*? Opening the stopcock increases the volume, which increases the number of translational energy levels that the particles can occupy as they move to more locations. Thus, the number of microstates—and the entropy—increase.

Figure 18.3 presents this idea with particles on energy levels in a box of changeable volume. When the stopcock opens, there are more energy levels and, on average, they are closer together, so more distributions of particles are possible.

In Figure 18.4, the number of microstates is represented by the placement of particles in the left and/or right flasks:

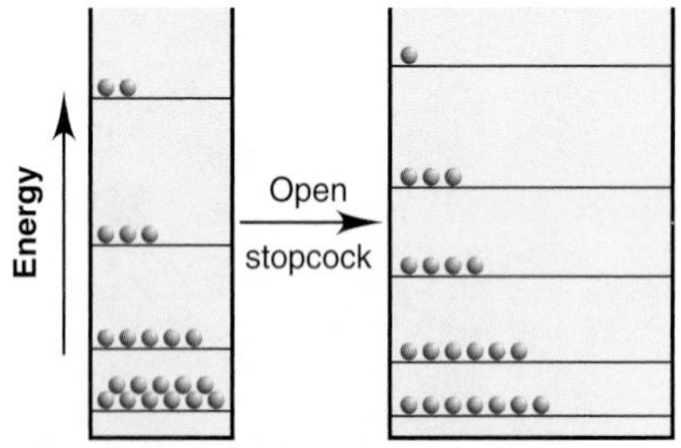

FIGURE 18.3 The entropy increase due to expansion of a gas. Energy levels are shown as lines in a box of narrow width (*left*). Each distribution of energies for the 21 particles is one microstate. When the stopcock is opened, the box is wider (volume increases, *right*), and the particles have more energy levels available.

- *One Ne atom.* At a given instant, an Ne atom in the left flask has its energy in one of some number (Ω) of microstates. Opening the stopcock increases the volume, which increases the number of possible locations and the number of translational energy levels. Thus, the system has 2^1, or 2, times as many microstates available when the atom moves through both flasks (final state, Ω_{final}) as when it is confined to the left flask (initial state, $\Omega_{initial}$).
- *Two Ne atoms.* When atoms A and B move through both flasks, there are 2^2, or 4, times as many microstates as there were when the atoms were initially in the left flask: one microstate with both A and B in the left flask, two microstates with

FIGURE 18.4 Expansion of a gas and the increase in the number of microstates. Each set of particle locations represents a different microstate. When the volume increases (stopcock opens), the number of microstates is 2^n, where *n* is the number of particles.

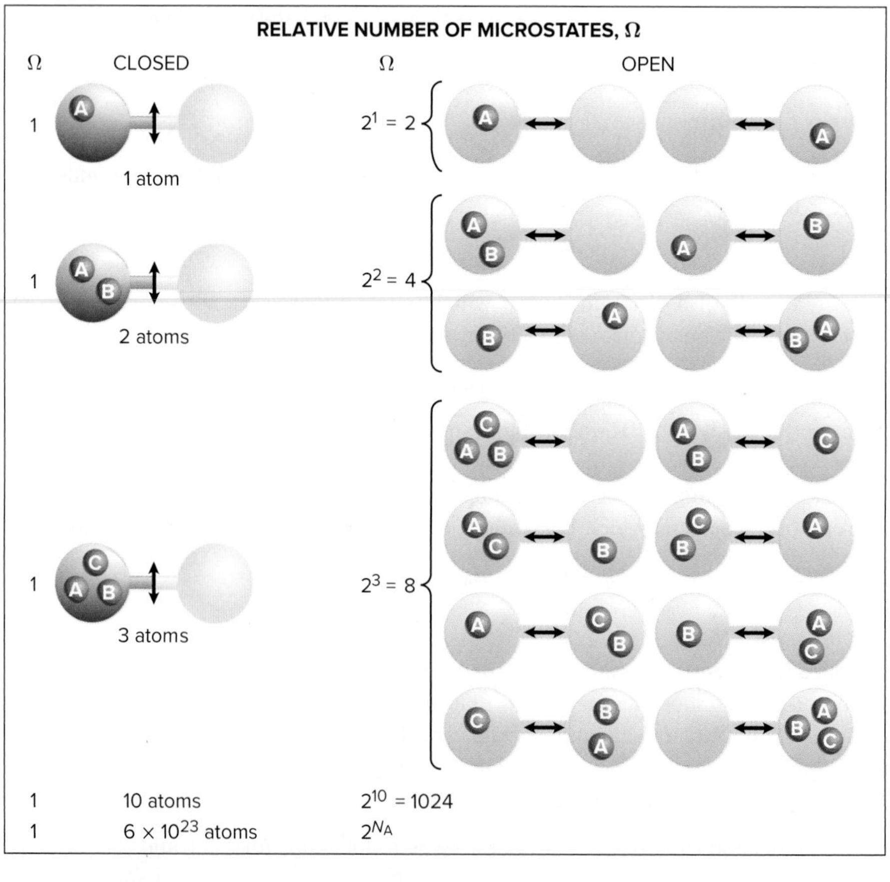

A in the left flask and B in the right flask or with B in the left flask and A in the right flask, and one microstate with both A and B in the right flask.

- *Three Ne atoms.* Add another atom, and there are 2^3, or 8, times as many microstates when the stopcock is open.
- *Ten Ne atoms.* With 10 Ne atoms, there are 2^{10}, or 1024, times as many microstates for the atoms in both flasks as there were for the 10 atoms in the left flask.
- *One mole of Ne atoms.* With 1 mol (N_A) of Ne, there are 2^{N_A} times as many microstates for the atoms in the larger volume (Ω_{final}) than there were in the smaller volume ($\Omega_{initial}$):

$$\frac{\Omega_{final}}{\Omega_{initial}} = 2^{N_A}$$

Now let us find $\Delta_{sys}S$ through the Boltzmann equation, $S = k \ln \Omega$. From the properties of logarithms (Appendix A), we know that $\ln A - \ln B = \ln (A/B)$. Thus,

$$\Delta_{sys}S = S_{final} - S_{initial} = k \ln \Omega_{final} - k \ln \Omega_{initial} = k \ln\left(\frac{\Omega_{final}}{\Omega_{initial}}\right) = k \ln 2^{N_A}$$

Also, from Appendix A, $\ln A^y = y \ln A$. Thus, with $k = \frac{R}{N_A}$, we have

$$\Delta_{sys}S = \left(\frac{R}{N_A}\right) \ln 2^{N_A} = \left(\frac{R}{N_A}\right) N_A \ln 2 = R \ln 2$$

$$= [8.314\ \text{J/(mol·K)}](0.693) = 5.76\ \text{J/(mol·K)}$$

2. *Quantifying $\Delta_{sys}S$ from the changes in heat.* Now let us compare the $\Delta_{sys}S$ we just found for a gas expanding into an evacuated flask with the $\Delta_{sys}S$ for a gas that is heated and does work on the surroundings. This approach uses the following equation, which is an integrated form of a more basic thermodynamic relationship:

$$\Delta_{sys}S = \frac{q_{rev}}{T} \qquad (18.2)$$

where T is the temperature at which the heat change occurs and q is the heat absorbed. The subscript "rev" refers to a *reversible process*, one that occurs in such tiny increments that the system remains at equilibrium and the direction of the change can be reversed by an *infinitesimal* reversal of conditions.

We can approximate a reversible expansion by placing Ne gas in a piston-cylinder assembly within a heat reservoir (to maintain a constant T of 298 K). We start by confining Ne to a volume of 1 L by the "pressure" of a beaker of sand on the piston (Figure 18.5). We remove one grain of sand (an "infinitesimal" decrease in pressure) with a pair of tweezers, and the gas expands a tiny amount, raising the piston and doing work, $-w$. Assuming that Ne behaves ideally, it absorbs an equivalent tiny increment of heat, q, from the heat reservoir. With each grain of sand removed, the expanding gas absorbs another tiny increment of heat. This process simulates a reversible expansion because we can reverse it by putting back a grain of sand, which causes the surroundings to do a tiny quantity of work compressing the gas and, thus, to release a tiny quantity of heat to the reservoir.

If we continue this nearly reversible expansion to 2 L and use calculus to integrate the tiny increments of heat together, q_{rev} is 1718 J. From Equation 18.2,

$$\Delta_{sys}S = \frac{q_{rev}}{T} = \frac{1718\ \text{J}}{298\ \text{K}} = 5.77\ \text{J/K}$$

This is the same value (within rounding) of $\Delta_{sys}S$ that we obtained based on the number of microstates.

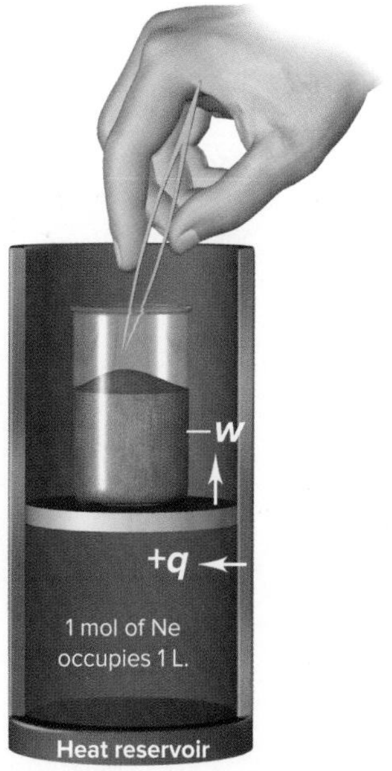

FIGURE 18.5 Simulating a reversible process

Entropy and the Second Law of Thermodynamics

While it is true that the change in entropy determines the direction of a thermodynamically allowed process, we must consider more than the entropy change of the system. After all, some processes, such as ice melting, occur spontaneously and $\Delta_{sys}S$ goes up, whereas others, such as water freezing, occur spontaneously and $\Delta_{sys}S$ goes down. When we consider *both* the system *and* its surroundings, however, we find that *all real processes occur spontaneously in the direction that increases the entropy of the universe (system plus surroundings).* This is one way to state the **second law of thermodynamics**.

According to the second law, *either* the entropy change of the system *or* the entropy change of the surroundings may be negative, but, for a process to be thermodynamically allowed, the *sum* of the two entropy changes must be positive. If the entropy of the system decreases, the entropy of the surroundings must increase even more to offset this decrease, so that the entropy of the universe (system *plus* surroundings) increases. A quantitative statement of the second law is, for any real spontaneous process,

$$\Delta_{univ}S = \Delta_{sys}S + \Delta_{surr}S > 0 \qquad (18.3)$$

Standard Molar Entropies and the Third Law

Entropy and enthalpy are state functions, but their values differ in a fundamental way:

- For *enthalpy*, there is no zero point, so we can measure only *changes* in the value of enthalpy.
- For *entropy*, there *is* a zero point, and we can determine *absolute* values of entropy by applying the **third law of thermodynamics**: *a perfect crystal has zero entropy at absolute zero*, or $S_{sys} = 0$ at 0 K.

A "perfect" crystal means that all the particles are aligned flawlessly. At absolute zero, the particles have minimum energy, so there is only one microstate. Thus, in Equation 18.1,

$$\Omega = 1, \quad \text{so} \quad S = k \ln 1 = 0$$

When we warm the crystal to any temperature above 0 K, the total energy increases, so it can be dispersed into more than one microstate. Thus,

$$\Omega > 1 \quad \text{and} \quad \ln \Omega > 0, \quad \text{so} \quad S > 0$$

In principle, to find S of a substance at a given temperature, we cool it as close to 0 K as possible. Then we heat it in small increments, dividing q by T to get the increase in S for each increment. We add up all the entropy increases to the temperature of interest, usually 298 K. Therefore, S of a substance at a given temperature is an *absolute* value. As for other thermodynamic variables, we must do the following:

- We must compare entropy values for substances at the temperature of interest in their *standard states: 1 bar for gases, 1 mol/L for solutions, and the pure substance in its most stable form for solids or liquids.*
- Because entropy is an *extensive* property—a property that depends on the amount of substance—we must specify the **standard molar entropy ($S°$)**, in units of J/(mol·K) (or J·mol^{-1}·K^{-1}). ($S°$ values at 298 K for many elements, compounds, and ions are given, with other thermodynamic variables, in Appendix B.)

Predicting the Relative S° of a System

Let us see how the standard molar entropy of a substance is affected by several parameters: temperature, physical state, dissolution, and atomic size or molecular complexity. (Unless stated otherwise, the $S°$ values refer to the system at 298 K.)

Temperature Changes Temperature has the most important effect. For any substance, $S°$ *increases as T rises*, as the following values for copper metal show:

T (K)	273	295	298
S° [J/(mol·K)]	31.0	32.9	33.2

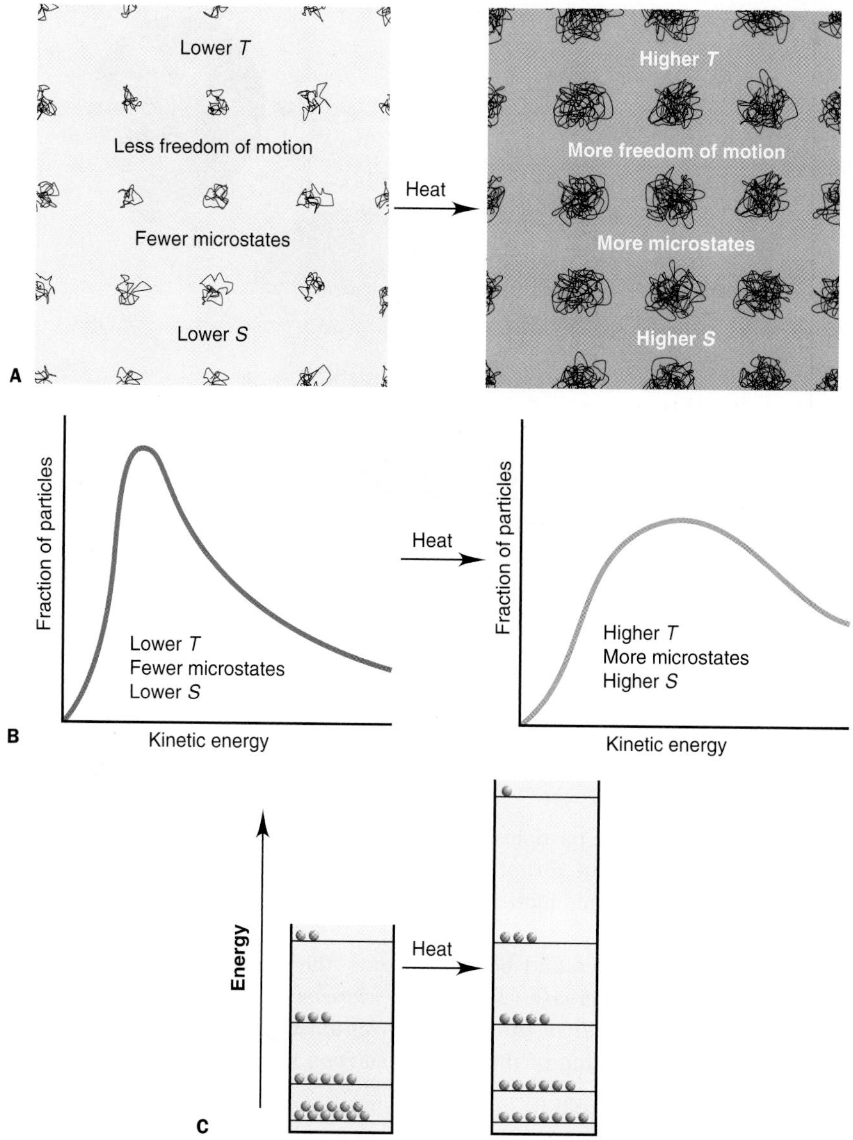

FIGURE 18.6 Visualizing the effect of temperature on entropy. A. Computer simulations show each particle in a crystal moving about its lattice position. Adding heat increases T and the total energy, so the particles have greater freedom of motion, and their energy is more dispersed. Thus, S increases. **B.** At any T, there is a range of occupied energy levels and, thus, a certain number of microstates. Adding heat increases the total energy (*area under curve*), so the range of occupied energy levels becomes greater, as does the number of microstates (higher S). **C.** A system of 21 particles occupies energy levels (*lines*) in a box whose height represents the total energy. When heat is added, the total energy increases (*box is higher*) *and* becomes more dispersed (*more lines*), so S increases.

As heat is absorbed ($q > 0$), temperature, which is a measure of the average kinetic energy of the particles, increases. Recall that the kinetic energies of gas particles are distributed over a range, which becomes wider as T rises (Figure 4.13); liquids and solids have the same behaviour. Thus, at any instant, more microstates are available in which the energy can be dispersed, so the entropy of the substance goes up. Figure 18.6 presents three ways to view the effect of temperature on entropy.

Physical States and Phase Changes In melting or vaporizing, heat is absorbed ($q > 0$). The particles have more freedom of motion, and their energy is more dispersed. Thus, $S°$ *increases as the physical state of a substance changes from solid to liquid to gas*:

	Na	H_2O	C(graphite)
$S°$(s or *l*) [J/(mol·K)]	51.4(s)	69.9(*l*)	5.7(s)
$S°$(*g*) [J/(mol·K)]	153.6	188.7	158.0

Figure 18.7 plots entropy versus T as solid O_2 is heated and changes to liquid and then to gas, with $S°$ values at various points; this behaviour is typical of many substances. At the molecular scale, several stages occur:

- Particles in the solid vibrate about their positions but, on average, remain fixed. The energy of the solid is dispersed least—that is, it has the fewest microstates—so the solid has the lowest entropy.

FIGURE 18.7 The increase in entropy of O_2 during phase changes from solid to liquid to gas

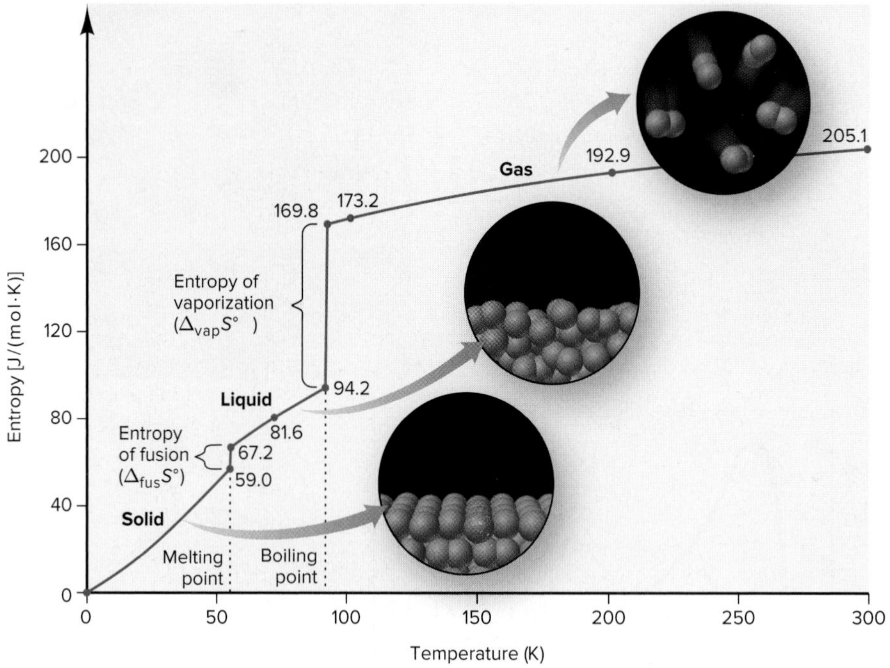

- As T rises, the entropy increases gradually as the kinetic energy of the particles increases.
- When the solid melts, the particles move much more freely between and around one another, so there is an abrupt increase in entropy ($\Delta_{fus}S°$).
- Further heating of the liquid increases the speed of the particles, and the entropy increases gradually.
- When the liquid vaporizes and becomes a gas, the particles undergo a much larger, abrupt entropy increase ($\Delta_{vap}S°$); *the increase in entropy from liquid to gas is much larger than it is from solid to liquid*: $\Delta_{vap}S° \gg \Delta_{fus}S°$.
- Finally, with further heating of the gas, the entropy increases gradually.

Dissolving a Solid or a Liquid in Water Recall, from Chapter 12, that, in general, entropy increases when a solute dissolves in a solvent: $S_{soln} > (S_{solute} + S_{solvent})$. When water is the solvent, however, the entropy may also depend on the nature of solute *and* solvent interactions and includes two opposing events:

	NaCl	$AlCl_3$	CH_3OH
S°(s or l) [J/(mol·K)]	72.1(s)	167(s)	127(l)
S°(aq) [J/(mol·K)]	115.1	−148	132

1. *For ionic solutes*, when the crystal dissolves in water, the ions have much more freedom of motion and their energy is dispersed into more microstates. That is, the entropy of the ions themselves is greater in the solution. However, some water molecules become arranged around the ions (Figure 18.8), which limits the freedom of motion of the molecules (see also Figure 12.2). In fact, around small, multiply charged ions, H_2O molecules become so organized that their energy of motion becomes *less* dispersed. This negative portion of the total entropy change can lead to *negative* $S°$ values for the ions in solution. For example, the $Al^{3+}(aq)$ ion in $AlCl_3$ has such a negative $S°$ value [−313 J/(mol·K)] that when $AlCl_3$ dissolves in water, even though $S°$ of $Cl^-(aq)$ is positive, the entropy of aqueous $AlCl_3$ is lower than the entropy of solid $AlCl_3$.*

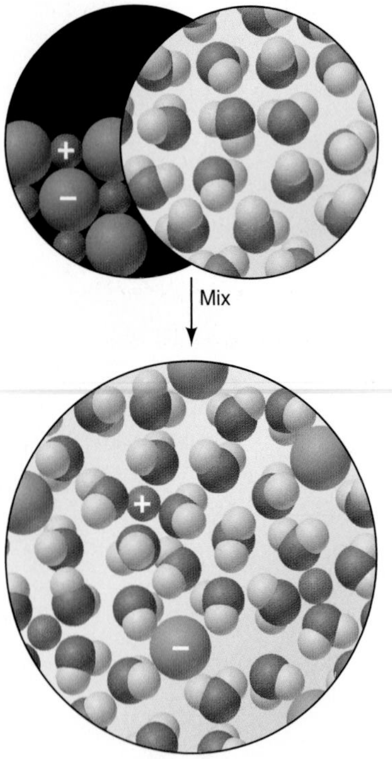

FIGURE 18.8 The entropy change accompanying the dissolution of a salt. The entropy of a salt solution is usually *greater* than the entropy of the solid and the entropy of water, but it is affected by water molecules becoming organized around each ion.

*An S° value for a hydrated ion can be negative because it is relative to the S° value for the hydrated proton, $H^+(aq)$, which is assigned a value of zero. In other words, $Al^{3+}(aq)$ has a lower entropy than $H^+(aq)$.

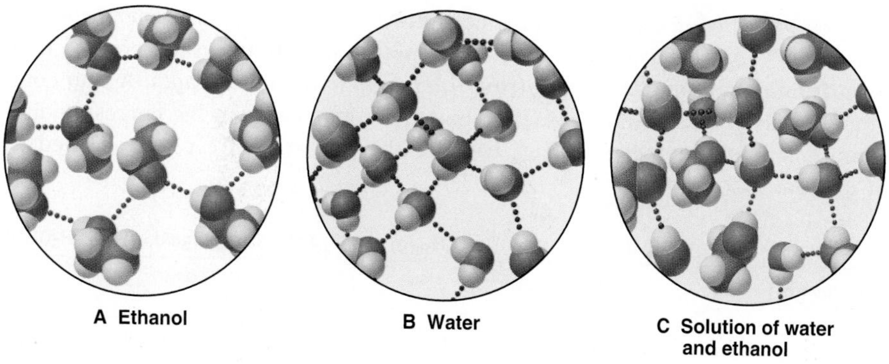

FIGURE 18.9 The small increase in entropy when ethanol dissolves in water. Both ethanol **(A)** and water **(B)** have many hydrogen bonds to themselves. In solution **(C)**, they form hydrogen bonds to each other, so their freedom of motion does not change significantly.

2. *For molecular solutes*, the increase in entropy upon dissolving is typically much smaller than it is for ionic solutes. After all, for a solid such as glucose, there is no separation into ions; for a liquid such as methanol or ethanol (Figure 18.9), there is no breakdown of a crystal structure. Furthermore, in these small alcohols and in pure water, the molecules form many hydrogen bonds, so there is relatively little change in their freedom of motion either before or after they are mixed.

Dissolving a Gas in a Liquid The particles in a gas already have so much freedom of motion and, thus, such highly dispersed energy that they lose some of this energy when they dissolve in a liquid or a solid. Therefore, the entropy of a solution of a gas in a liquid or a solid is always *less* than the entropy of the gas itself. For instance, when gaseous O_2 [$S°(g)$ = 205.0 J/(mol·K)] dissolves in water, its entropy decreases considerably [$S°(aq)$ = 110.9 J/(mol·K)] (Figure 18.10). When a gas dissolves in another gas, however, the entropy increases as a result of the separation and mixing of the molecules.

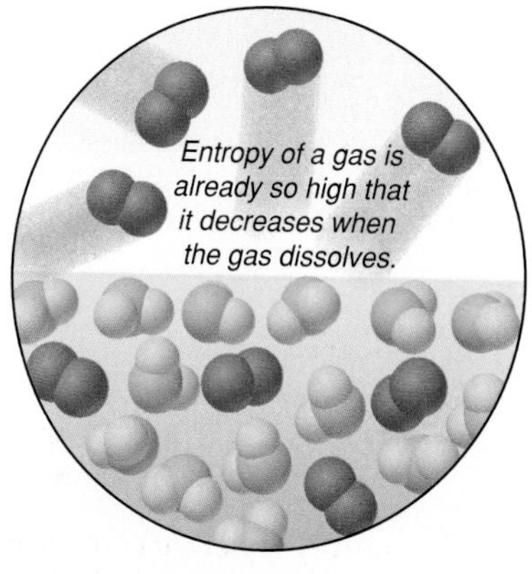

FIGURE 18.10 The entropy of a gas dissolved in a liquid

Atomic Size or Molecular Complexity Differences in $S°$ values for substances in the same phase are usually based on atomic size and molecular complexity.

1. *Within a periodic group*, energy levels become closer together for larger, heavier atoms, so the number of microstates and, thus, the molar entropy increase:

	Li	Na	K	Rb	Cs
Atomic radius (pm)	152	186	227	248	265
Molar mass (g/mol)	6.941	22.99	39.10	85.47	132.9
$S°(s)$ [J/(mol·K)]	29.1	51.4	64.7	69.5	85.2

The same trend of increasing entropy holds for similar compounds down a group:

	HF	HCl	HBr	HI
Molar mass (g/mol)	20.01	36.46	80.91	127.9
$S°(g)$ [J/(mol·K)]	173.7	186.8	198.6	206.3

2. *For different forms of an element (allotropes)*, the entropy is *higher* in the form that allows the atoms more freedom of motion. For example, $S°$ of graphite is 5.69 J/(mol·K), whereas $S°$ of diamond is 2.44 J/(mol·K). In graphite, covalent bonds extend within a two-dimensional sheet, and the sheets move past each other easily; in diamond, covalent bonds extend in three dimensions, allowing the atoms little movement (Table 11.6).

3. *For compounds*, entropy increases with chemical complexity, that is, with the number of atoms in the formula. This trend holds for both ionic and covalent substances:

	NaCl	$AlCl_3$	P_4O_{10}	NO	NO_2	N_2O_4
$S°(s)$ [J/(mol·K)]	72.1	167	229			
$S°(g)$ [J/(mol·K)]				211	240	304

This trend is also based on the different types of motion. For example, among the nitrogen oxides listed, the number of different vibrational motions increases with the number of atoms in the molecule (Figure 18.11; also see Section 22.5).

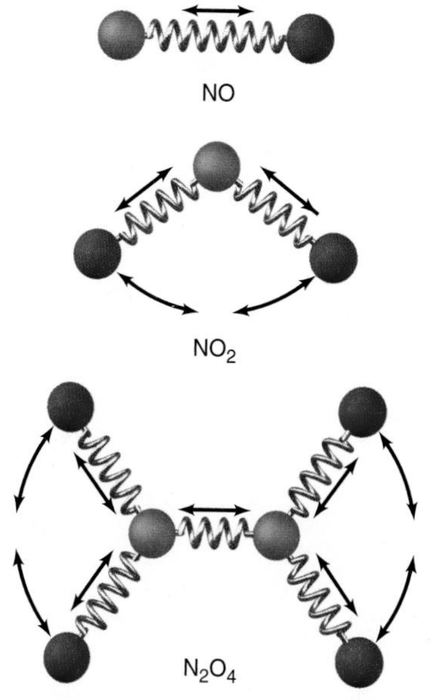

FIGURE 18.11 Entropy, vibrational motion, and molecular complexity

For larger compounds, we also consider the motion of different parts of a molecule. A long hydrocarbon chain can rotate and vibrate in more ways than a short hydrocarbon chain, so *entropy increases with chain length.* A ring compound, such as cyclopentane (C_5H_{10}), has lower entropy than the corresponding chain compound, pentene (C_5H_{10}), because the ring structure restricts freedom of motion:

	$CH_4(g)$	$C_2H_6(g)$	$C_3H_8(g)$	$C_4H_{10}(g)$	$C_5H_{10}(g)$	C_5H_{10}(cyclo, g)	$C_2H_5OH(l)$
S° [J/(mol·K)]	186	230	270	310	348	293	161

Remember, these trends hold only for *substances in the same physical state.* Gaseous methane (CH_4) has higher entropy than liquid ethanol (C_2H_5OH), even though ethanol molecules are more complex. When gases are compared with liquids, *the effect of physical state dominates the effect of molecular complexity.*

Sample Problem 18.1 Predicting Relative Entropy Values

Problem Choose the substance with the higher entropy in each pair, and justify your choice. [Assume constant temperature, except in part (e).]

(a) 1 mol of $SO_2(g)$ or 1 mol of $SO_3(g)$

(b) 1 mol of $CO_2(s)$ or 1 mol of $CO_2(g)$

(c) 3 mol of $O_2(g)$ or 2 mol of $O_3(g)$

(d) 1 mol of KBr(s) or 1 mol of KBr(aq)

(e) Seawater at 2°C or at 23°C

(f) 1 mol of $CF_4(g)$ or 1 mol of $CCl_4(g)$

Plan In general, particles with more freedom of motion have more microstates in which to disperse their kinetic energy, so they have higher entropy. We know that either raising the temperature or having *more* particles increases entropy. We apply the general criteria we have just examined to choose the member with the higher entropy.

Solution (a) 1 mol of $SO_3(g)$: For equal amounts (mol) of substances with the same types of atoms in the same physical state, the more atoms in the molecule, the more types of motion available, and, thus, the higher the entropy.

(b) 1 mol of $CO_2(g)$: For a given substance, entropy increases in the sequence $s < l < g$.

(c) 3 mol of $O_2(g)$: The two samples contain the same number of oxygen atoms, but different numbers of molecules. Despite the greater complexity of O_3, the greater number of molecules dominates because there are many more microstates possible for 3 mol of particles than there are for 2 mol of particles.

(d) 1 mol of KBr(aq): The two samples have the same number of ions, but their motion is more limited and their energy is less dispersed in the solid than in the solution.

(e) Seawater at 23°C: Entropy increases with rising temperature.

(f) 1 mol of $CCl_4(g)$: For similar compounds, entropy increases with molar mass.

Follow-Up Problem 18.1 Choose the substance with the higher entropy in each pair, and justify your choice. (Assume that there is 1 mol of each substance at the same temperature.)

(a) $PCl_3(g)$ or $PCl_5(g)$

(b) $CaF_2(s)$ or $BaCl_2(s)$

(c) $Br_2(g)$ or $Br_2(l)$

(See Brief Solutions to Follow-Up Problems at the end of the chapter.)

SUMMARY OF SECTION 18.1

- A change is thermodynamically allowed under specified conditions if it occurs without a continuous input of energy.
- Neither the first law of thermodynamics nor the sign of ΔH predicts the direction of a spontaneous change.
- Many thermodynamically allowed processes involve an increase in the freedom of motion of the particles in the system and, thus, in the dispersal of the energy of motion of the system.
- Entropy is a state function that measures the extent of energy dispersed into the number of microstates possible for a system. Each microstate consists of the quantized energy levels of the system at a given instant.
- The second law of thermodynamics states that, in a spontaneous process, the entropy of the universe (system plus surroundings) increases.
- Absolute entropy values can be found because perfect crystals have zero entropy at 0 K (third law of thermodynamics).
- Standard molar entropy, $S°$ [J/(mol·K)], is affected by temperature, phase changes, dissolution, and atomic size or molecular complexity.

18.2 Calculating the Entropy Change of a Reaction

Chemists are especially interested in learning how to predict the sign *and* calculate the value of the entropy change that occurs during a reaction.

Entropy Changes in the System: Standard Entropy of Reaction ($\Delta_r S$)

The **standard entropy of reaction ($\Delta_r S°$)** is the entropy change that occurs when all the reactants and products are in their standard states.

Predicting the Sign of $\Delta_r S°$ A deciding factor in predicting the sign of $\Delta_r S°$ is a change in the amount (mol) of gas, because gases have such great freedom of motion and, thus, high molar entropies: *if the amount (mol) of gas increases, $\Delta_r S°$ is positive; if the* amount (mol) *of gas decreases, $\Delta_r S°$ is negative.* Here are a few examples:

- *Increase in the amount of gas.* When gaseous H_2 reacts with solid I_2 to form gaseous HI, the total amount (mol) of *substance* stays the same. Nevertheless, the sign of $\Delta_r S°$ is positive (entropy increases) because the amount (mol) of *gas* increases:

$$H_2(g) + I_2(s) \longrightarrow 2HI(g) \qquad \Delta_r S° = S°_{products} - \Delta S°_{reactants} > 0$$

- *Decrease in the amount of gas.* When ammonia forms from its elements, 4 mol of gas produces 2 mol of gas, so $\Delta_r S°$ is negative (entropy decreases):

$$N_2(g) + 3H_2(g) \rightleftharpoons 2NH_3(g) \qquad \Delta_r S° = S°_{products} - S°_{reactants} < 0$$

- *No change in the amount of gas, but change in structure.* When the amount (mol) of gas does not change, we *cannot* use Δn_{gas} to predict the sign of $\Delta_r S°$. However, a change in one of the structures can make it easier to predict the sign of $\Delta_r S°$. For example, when cyclopropane is heated to 500°C, the ring opens and propene forms. The chain has more freedom of motion than the ring, so $\Delta_r S°$ is positive:

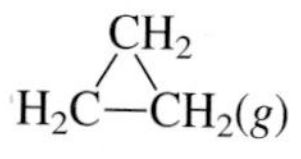

$$\xrightarrow{\Delta} CH_3{-}CH{=}CH_2(g) \qquad \Delta_r S° = S°_{products} - S°_{reactants} > 0$$

Keep in mind, however, that, in general, we cannot predict the sign of the entropy change unless the reaction involves a change in the amount (mol) of gas.

Calculating $\Delta_r S°$ from $S°$ Values By applying Hess's law (Chapter 5), we combined $\Delta_f H°$ values to find $\Delta_r H°$. Similarly, we combine $S°$ values to find the standard entropy of reaction, $\Delta_r S°$:

$$\Delta_r S° = \sum mS°_{\text{products}} - \sum nS°_{\text{reactants}} \quad (18.4)$$

where m and n are the amounts (mol) of products and reactants, respectively, given by the coefficients in the balanced equation. For the formation of ammonia, we have

$$\Delta_r S° = [2(S° \text{ of } NH_3)] - [1(S° \text{ of } N_2) + 3(S° \text{ of } H_2)]$$

From Appendix B, we find the $S°$ values:

$$\Delta_r S° = [2(193\ \text{J/[mol·K]})] - [1(191.5\ \text{J/[mol·K]}) + 3(130.6\ \text{J/[mol·K]})]$$
$$= -197\ \text{J/(mol·K)}$$

As we predicted from the decrease in amount (mol) of gas, $\Delta_r S° < 0$.

Sample Problem 18.2 Calculating the Standard Entropy of Reaction, $\Delta_r S°$

Problem Predict the sign of $\Delta_r S°$, if possible, and calculate its value for the combustion of 1 mol of propane at 25°C:

$$C_3H_8(g) + 5O_2(g) \longrightarrow 3CO_2(g) + 4H_2O(l)$$

Plan From the change in the amount (mol) of gas (6 mol yields 3 mol), the entropy should decrease ($\Delta_r S° < 0$). To find $\Delta_r S°$, we use Equation 18.4.

Solution Calculate $\Delta_r S°$ with values from Appendix B:

$$\Delta_r S° = [3(S° \text{ of } CO_2) + 4(S° \text{ of } H_2O)] - [1(S° \text{ of } C_3H_8) + 5(S° \text{ of } O_2)]$$
$$= [3(213.7\ \text{J/[mol·K]}) + 4(69.9\ \text{J/[mol·K]})] - [1(269.9\ \text{J/[mol·K]}) + 5(205.0\ \text{J/[mol·K]})]$$
$$= -374\ \text{J/(mol·K)}$$

Check $\Delta_r S° < 0$, so our prediction is correct. Rounding gives [3(200) + 4(70)] − [270 + 5(200)] = 880 − 1270 = −390, which is close to the calculated value.

Follow-Up Problem 18.2 Balance each equation, predict the sign of $\Delta_r S°$ if possible, and calculate the value of $\Delta_r S°$ at 25°C:

(a) $NaOH(s) + CO_2(g) \longrightarrow Na_2CO_3(s) + H_2O(l)$

(b) $Fe(s) + H_2O(g) \longrightarrow Fe_2O_3(s) + H_2(g)$

Entropy Changes in the Surroundings: The Other Part of the Total

In the synthesis of ammonia, the combustion of propane, and many other spontaneous reactions, the entropy of the system decreases ($\Delta_r S° < 0$). Remember that the second law dictates that, for a thermodynamically allowed process, a ***decrease*** *in the entropy of the system is outweighed by an* ***increase*** *in the entropy of the surroundings.* In this section, we examine the influence of the surroundings—in particular, the addition (or removal) of heat and the temperature at which this heat flow occurs—on the *total* entropy change.

The Role of the Surroundings In essence, the surroundings *add heat to or remove heat from the system.* That is, the surroundings function as an enormous heat source or heat sink, one so large that its temperature remains constant, even though its entropy changes. The surroundings participate in the two types of enthalpy changes as follows:

1. *In an exothermic change, heat released by the system is absorbed by the surroundings.* More heat increases the freedom of motion of the particles and makes the energy more dispersed, so the entropy of the surroundings increases.

For an exothermic change: $q_{sys} < 0$, $q_{surr} > 0$, and $\Delta_{surr}S > 0$

2. *In an endothermic change, heat absorbed by the system is released by the surroundings.* Less heat reduces the freedom of motion of the particles and makes the energy less dispersed, so the entropy of the surroundings decreases.

For an endothermic change: $q_{sys} > 0, \; q_{surr} < 0, \;$ and $\; \Delta_{surr}S < 0$

Temperature at Which Heat Is Transferred The *temperature* of the surroundings at the time when the heat is transferred also affects $\Delta_{surr}S$. Consider the effect of an exothermic reaction at a low temperature or a high temperature:

- At a low temperature, such as 20 K, there is little motion in the surroundings due to little energy. This means that there are few energy levels in each microstate and few microstates in which to disperse the energy. Transferring a given quantity of heat to these surroundings causes a relatively large change in how much energy is dispersed.
- At a high temperature, such as 298 K, the surroundings have a large quantity of energy dispersed. There are more energy levels in each microstate and a greater number of microstates. Transferring the same quantity of heat to these surroundings causes a relatively small change in how much energy is dispersed. ■

A Chequebook Analogy for Heating the Surroundings If you have $10 in your chequing account, a $10 deposit represents a 100% increase in your net worth; that is, a given change to a low initial state has a large impact. If, however, you have a $1000 balance, a $10 deposit represents only a 1% increase. Thus, the same absolute change to a high initial state has a smaller impact.

In other words, $\Delta_{surr}S$ is greater when heat is added at a lower T. Putting these ideas together, $\Delta_{surr}S$ *is directly related to an opposite change in the heat of the system* (q_{sys}) *and inversely related to the temperature at which the heat is transferred:*

$$\Delta_{surr}S = -\frac{q_{sys}}{T}$$

For a process at *constant pressure*, the heat (q_p) is ΔH (Section 5.2), so

$$\Delta_{surr}S = -\frac{\Delta_{sys}H}{T} \qquad (18.5)$$

Thus, we find $\Delta_{surr}S$ by measuring $\Delta_{sys}H$ and T at which the change takes place.

The *main point* can be summarized as follows: if a thermodynamically allowed reaction has a negative $\Delta_{sys}S$ (fewer microstates into which energy can be dispersed), $\Delta_{surr}S$ must be positive enough (even more microstates into which energy can be dispersed) for $\Delta_{univ}S$ to be positive (net increase in the number of microstates to disperse the energy).

Sample Problem 18.3 Determining Reaction Spontaneity

Problem At 298 K, the formation of ammonia has a negative $\Delta_{sys}S°$:

$$N_2(g) + 3H_2(g) \longrightarrow 2NH_3(g) \qquad \Delta_{sys}S° = -197 \text{ J/(mol·K)}$$

Calculate $\Delta_{univ}S$, and state whether the reaction occurs spontaneously at this temperature.

Plan For the reaction to occur spontaneously, $\Delta_{univ}S > 0$, so $\Delta_{surr}S$ must be greater than +197 J/K. To find $\Delta_{surr}S$, we need $\Delta_{sys}H°$, which is the same as $\Delta_rH°$. We use $\Delta_fH°$ values from Appendix B to find $\Delta_rH°$. Then we divide $\Delta_rH°$ by the given T (298 K) to find $\Delta_{surr}S$. To find $\Delta_{univ}S$, we add the calculated $\Delta_{surr}S$ to the given $\Delta_{sys}S°$ (−197 J/K).

Solution Calculate $\Delta_{sys}H°$:

$$\begin{aligned}\Delta_{sys}H° &= \Delta_rH° \\ &= 2(-45.9 \text{ kJ/mol}) - [1(0 \text{ kJ/mol}) + 3(0 \text{ kJ/mol})] \\ &= -91.8 \text{ kJ/mol}\end{aligned}$$

Calculate $\Delta_{surr}S$:

$$\Delta_{surr}S = -\frac{\Delta_{sys}H°}{T} = -\frac{-91.8\dfrac{\text{kJ}}{\text{mol}} \times \dfrac{1000 \text{ J}}{1 \text{ kJ}}}{298 \text{ K}} = 308 \text{ J/(mol·K)}$$

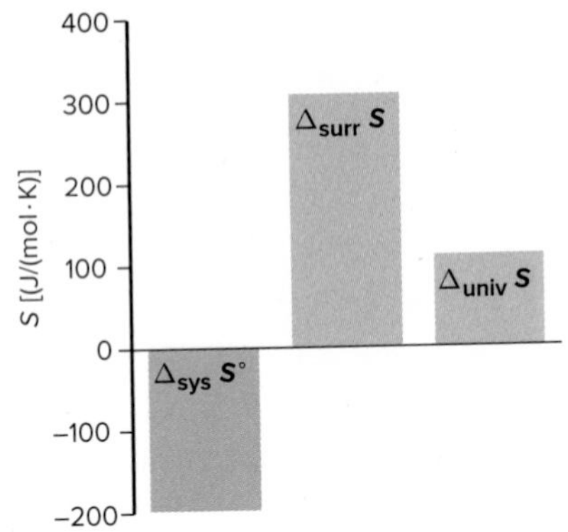

Determine $\Delta_{univ}S$:

$$\Delta_{univ}S = \Delta_{sys}S^\circ + \Delta_{surr}S = -197\text{ J/(mol·K)} + 308\text{ J/(mol·K)} = 111\text{ J/(mol·K)}$$

$\Delta_{univ}S > 0$, so the reaction occurs spontaneously at 298 K *(see margin)*.

Check Rounding to check the math, we have

$$\Delta_r H^\circ \approx 2(-45\text{kJ}) = -90\text{ kJ/mol}$$
$$\Delta_{surr}S \approx -\frac{(-90\,000\text{ J})}{300\text{ K}} = 300\text{ J/(mol·K)}$$
$$\Delta_{univ}S \approx -200\text{ J/(mol·K)} + 300\text{ J/(mol·K)} = 100\text{ J/(mol·K)}$$

Given the negative $\Delta_r H^\circ$, Le Châtelier's principle says that low temperature favours NH_3 formation, so the answer is reasonable (see Section 15.6).

Comment **1.** Because ΔH° has units of kJ/mol and ΔS has units of J/(mol·K), remember to convert kJ to J. Otherwise, you will introduce a large error.

2. This example highlights the distinction between thermodynamics and kinetics. NH_3 forms spontaneously, but so slowly that catalysts are required to achieve a practical rate.

Follow-Up Problem 18.3 Does the oxidation of FeO(*s*) to Fe_2O_3(*s*) occur spontaneously at 298 K?

Do Organisms Obey the Laws of Thermodynamics? Taking the surroundings into account is crucial, not only for determining reaction spontaneity, as in Sample Problem 18.3, but also for understanding the relevance of thermodynamics to biology. Let us examine the first and second laws to see if they apply to living systems.

1. *Do organisms comply with the first law?* The chemical bond energy in food and oxygen is converted into other forms of energy, such as the mechanical energy of jumping, flying, crawling, swimming, and countless other movements; the electrical energy of nerve conduction; and the thermal energy of warming the body. Many experiments have demonstrated that the total energy is conserved in these situations. Some of the earliest experiments were performed by Lavoisier, who showed that "animal heat" was produced by slow, continuous combustion. In experiments with guinea pigs, he invented a calorimeter to measure the heat released from the intake of food and O_2 and the output of CO_2 and water, and he included respiration in his new theory of combustion. Modern room-size calorimeters measure these and other variables to confirm the conservation of energy for an exercising human (Figure 18.12).

FIGURE 18.12 A whole-body calorimeter. In this room-size apparatus, a person exercises while respiratory gases, energy input and output, and other physiological variables are monitored.

2. *Do organisms comply with the second law?* Mature humans are far more complex than the egg and sperm cells from which they developed, and modern organisms are far more complex than the one-celled ancestral specks from which they evolved. Are the growth of an organism and the evolution of life exceptions to the spontaneous tendency of natural processes to disperse their energy? For an organism to grow or a species to evolve, a large amount (mol) of oxygen and nutrients—carbohydrates, proteins, and fats—undergo exothermic reactions to form a much larger amount (mol) of gaseous CO_2 and H_2O. Formation of these waste gases and the accompanying release of heat result in an enormous *increase* in the entropy of the surroundings. Thus, the localization of energy and the synthesis of macromolecular structures required for the growth and evolution of organisms cause a far greater dispersal of energy and freedom of motion of small molecules in the environment. When both a system and its surroundings are considered together, the entropy of the universe increases.

The Entropy Change and the Equilibrium State

For a process approaching equilibrium, $\Delta_{univ}S > 0$. When the process reaches equilibrium, there is no further *net* change, $\Delta_{univ}S = 0$, because any entropy change in the system is balanced by an opposite entropy change in the surroundings.

At equilibrium: $\Delta_{univ}S = \Delta_{sys}S + \Delta_{surr}S = 0$ so $\Delta_{sys}S = -\Delta_{surr}S$

As an example, let us calculate $\Delta_{univ}S$ for the vaporization-condensation of 1 mol of water at 100°C (373 K):

$$H_2O(l; 373\ K) \rightleftharpoons H_2O(g; 373\ K)$$

First, we find $\Delta_{sys}S°$ for the forward change (vaporization) of 1 mol of water:

$$\Delta_{sys}S° = \Sigma mS°_{products} - \Sigma nS°_{reactants} = S° \text{ of } H_2O(g; 373\ K) - S° \text{ of } H_2O(l; 373\ K)$$
$$= 195.9\ J/(mol{\cdot}K) - 86.8\ J/(mol{\cdot}K) = 109.1\ J/(mol{\cdot}K)$$

As we expected, the entropy of the system increases ($\Delta_{sys}S° > 0$) as the liquid absorbs heat and changes to a gas.

For $\Delta_{surr}S$ of the vaporization step, we have

$$\Delta_{surr}S = -\frac{\Delta_{sys}H°}{T}$$

where $\Delta_{sys}H° = \Delta_{vap}H°$ at 373 K = 40.7 kJ/mol = 40.7×10^3 J/mol. For 1 mol of water, we have

$$\Delta_{surr}S = -\frac{\Delta_{vap}H°}{T} = -\frac{40.7 \times 10^3\ J}{373\ K} = -109\ J/K$$

The surroundings lose heat, and the negative sign means that the entropy of the surroundings decreases. The two entropy changes have the same magnitude (within rounding) but opposite signs, so they cancel:

$$\Delta_{univ}S = 109\ J/K + (-109\ J/K) = 0$$

For the reverse change (condensation), $\Delta_{univ}S$ also equals zero, but the signs of $\Delta_{sys}S°$ and $\Delta_{surr}S$ are opposite the signs for vaporization.

A similar treatment of a chemical change shows the same result: the entropy change of the forward reaction is *equal in magnitude but opposite in sign* to the entropy change of the reverse reaction. Thus, *when a system reaches equilibrium, neither the forward reaction nor the reverse reaction is spontaneous,* so there is no net reaction in either direction.

Spontaneous Exothermic and Endothermic Changes

No matter what its *enthalpy* change, a reaction occurs because the total *entropy* of the reacting system *and* its surroundings increases. There are two ways that this can happen:

1. *In an exothermic reaction* ($\Delta_{sys}H < 0$), the heat released by the system increases the freedom of motion and the dispersal of energy in the surroundings; thus, $\Delta_{surr}S > 0$.

- If the entropy of the products is *more* than the entropy of the reactants ($\Delta_{sys}S > 0$), the total entropy change ($\Delta_{sys}S + \Delta_{surr}S$) is positive (Figure 18.13A). For example, consider the oxidation of glucose, an essential reaction for all higher organisms:

$$C_6H_{12}O_6(s) + 6O_2(g) \longrightarrow 6CO_2(g) + 6H_2O(g) + \textbf{heat}$$

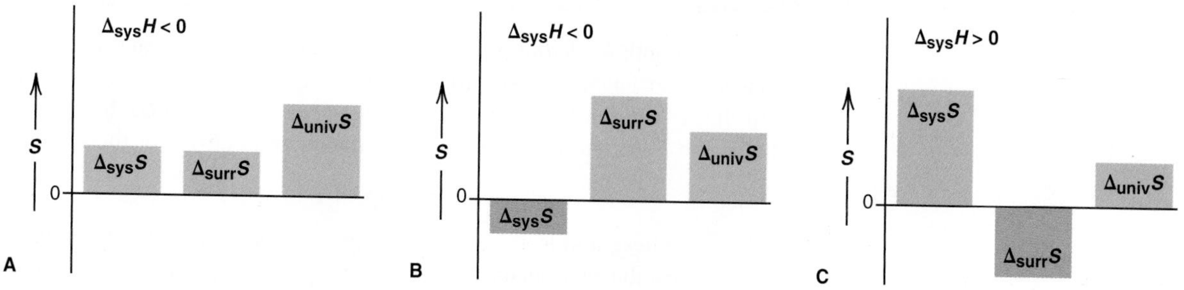

FIGURE 18.13 Components of $\Delta_{univ}S$ for thermodynamically allowed reactions. For a reaction to occur spontaneously, $\Delta_{univ}S$ must be positive. **A.** In an exothermic reaction in which $\Delta_{sys}S$ increases, the size of $\Delta_{surr}S$ is not important. **B.** In an exothermic reaction in which $\Delta_{sys}S$ decreases, $\Delta_{surr}S$ must be larger than $\Delta_{sys}S$. **C.** In an endothermic reaction in which $\Delta_{sys}S$ increases, $\Delta_{surr}S$ must be smaller than $\Delta_{sys}S$.

In this reaction, 6 mol of gas yields 12 mol of gas; thus, $\Delta_{sys}S > 0$, $\Delta_{surr}S > 0$, and $\Delta_{univ}S > 0$.

- If the entropy of the products is *less* than the entropy of the reactants ($\Delta_{sys}S < 0$), the entropy of the surroundings must increase even more ($\Delta_{surr}S \gg 0$) to make the total ΔS positive (Figure 18.13B). For example, when calcium oxide and carbon dioxide form calcium carbonate, the amount (mol) of gas decreases from 1 to 0:

$$CaO(s) + CO_2(g) \longrightarrow CaCO_3(s) + \textbf{\textit{heat}}$$

However, even though the entropy of the system goes down, the heat that is released increases the entropy of the surroundings even more; thus, $\Delta_{sys}S < 0$, but $\Delta_{surr}S \gg 0$, so $\Delta_{surr}S > 0$.

2. *In an endothermic reaction* ($\Delta_{sys}H > 0$), the heat absorbed by the system decreases the molecular freedom of motion and the dispersal of energy in the surroundings; thus, $\Delta_{surr}S < 0$. Therefore, an endothermic reaction can only occur spontaneously if $\Delta_{sys}S$ is positive and large enough to outweigh the negative $\Delta_{surr}S$ (Figure 18.13C).

- In the solution process for many ionic compounds, heat is absorbed to form the solution, so the entropy of the surroundings decreases ($\Delta_{surr}S < 0$). However, when the crystalline solid becomes freely moving ions, the entropy increase is so large ($\Delta_{sys}S \gg 0$) that it outweighs the negative $\Delta_{surr}S$; thus, $\Delta_{univ}S$ is positive.
- Spontaneous endothermic reactions are similar. Recall the reaction between barium hydroxide octahydrate and ammonium nitrate (see Figure 18.1):

$$\textbf{\textit{heat}} + Ba(OH)_2{\cdot}8H_2O(s) + 2NH_4NO_3(s) \longrightarrow Ba^{2+}(aq) + 2NO_3^-(aq) + 2NH_3(aq) + 10H_2O(l)$$

In this reaction, 3 mol of crystalline solids absorbs heat from the surroundings ($\Delta_{surr}S < 0$) and yields 15 mol of dissolved ions and molecules, which have much more freedom of motion and, therefore, much greater entropy ($\Delta_{sys}S \gg 0$).

SUMMARY OF SECTION 18.2

- The standard entropy of reaction, $\Delta_r S°$, is calculated from $S°$ values.
- When the amount (mol) of gas increases in a reaction, usually $\Delta_r S° > 0$.
- $\Delta_{surr}S$ is related directly to $\Delta_{sys}H°$ and inversely to the temperature at which a change occurs.
- In a spontaneous change, the entropy of the system can decrease only if the entropy of the surroundings increases even more, so that $\Delta_{univ}S > 0$.
- The second law is obeyed in living systems when we consider the system *plus* its surroundings.
- For a system at equilibrium, $\Delta_{univ}S = 0$, so $\Delta_{sys}S° = -\Delta_{surr}S$.
- Even if $\Delta_{sys}S° < 0$, an exothermic reaction ($\Delta_r H° < 0$) is spontaneous ($\Delta_{univ}S > 0$) if $\Delta_{surr}S \gg 0$. An endothermic reaction ($\Delta_r H° > 0$) is spontaneous only if $\Delta_{sys}S° > \Delta_{surr}S$.

18.3 Entropy, Gibbs Energy, and Work

By measuring both $\Delta_{sys}S$ *and* $\Delta_{surr}S$, we can predict whether a reaction will be spontaneous at a particular temperature. This section introduces *one* criterion for spontaneity that is determined by measuring the system only. The **Gibbs energy (*G*)**, or simply the free energy, combines the enthalpy and entropy of the system:

$$G = H - TS$$

One of the greatest, and least recognized, American scientists, Josiah Willard Gibbs (1839–1903), established chemical thermodynamics as well as major principles of equilibrium and electrochemistry. Although the great European scientists of his time, James Clerk Maxwell and Henri Le Châtelier, realized Gibbs's achievements, he was not recognized by his American colleagues until nearly 50 years after his death!

Gibbs Energy Change and Reaction Spontaneity

The Gibbs energy change (ΔG) is a measure of the spontaneity of a process and the useful energy that is available from it.

Deriving the Gibbs Equation Let us derive ΔG from the second law. By definition, the entropy change of the universe is the sum of the entropy changes of the system and its surroundings:

$$\Delta_{univ}S = \Delta_{sys}S + \Delta_{surr}S$$

At constant pressure,

$$\Delta_{surr}S = -\frac{\Delta_{sys}H}{T}$$

Substituting for $\Delta_{surr}S$ gives a relationship that relies solely on the system:

$$\Delta_{univ}S = \Delta_{sys}S - \frac{\Delta_{sys}H}{T}$$

Multiplying both sides by $-T$ gives

$$-T\Delta_{univ}S = \Delta_{sys}H - T\Delta_{sys}S$$

By setting $-T\Delta_{univ}S$ equal to the newly defined quantity $\Delta_{sys}G$, we obtain the *Gibbs equation* for the *change* in the Gibbs energy of the system ($\Delta_{sys}G$) at constant temperature and pressure:

$$\Delta_{sys}G = \Delta_{sys}H - T\Delta_{sys}S \qquad (18.6)$$

Significance of the Sign of ΔG Let us see how the *sign* of ΔG tells if a reaction is spontaneous. According to the second law,

- $\Delta_{univ}S > 0$ for a spontaneous process
- $\Delta_{univ}S < 0$ for a nonspontaneous process
- $\Delta_{univ}S = 0$ for a process at equilibrium

Since the absolute temperature is always positive, for a spontaneous process,

$$T\Delta_{univ}S > 0 \qquad \text{so} \qquad -T\Delta_{univ}S < 0$$

From our derivation above, $\Delta G = -T\Delta_{univ}S$, so we have

- $\Delta G < 0$ for a spontaneous process
- $\Delta G > 0$ for a nonspontaneous process
- $\Delta G = 0$ for a process at equilibrium

Calculating Standard Gibbs Energy Changes

The *sign* of ΔG reveals *whether* a reaction is thermodynamically allowed, but the *magnitude* of ΔG tells *how* spontaneous the reaction is. Because Gibbs energy (G) combines three state functions—H, S, and T—it is also a state function. As we do with enthalpy, we focus on the Gibbs energy *change* (ΔG). As we do with other thermodynamic variables, to compare the Gibbs energy changes of different reactions, we calculate the **standard Gibbs energy change ($\Delta G°$)**, which occurs when all the components of the system are in their standard states.

Using the Gibbs Equation to Find $\Delta G°$ One way to calculate $\Delta G°$ is by writing the Gibbs equation (18.6) at standard-state conditions and using Appendix B to find $\Delta_{sys}H°$ and $\Delta_{sys}S°$. Adapting the Gibbs equation, we have the following relationship:

$$\Delta_{sys}G° = \Delta_{sys}H° - T\Delta_{sys}S° \qquad (18.7)$$

This important relationship is used to find any one of the four variables, given the other three, as in Sample Problem 18.4.

Fireworks explode spontaneously over Niagara Falls.

Sample Problem 18.4 Calculating $\Delta_r G°$ from Enthalpy and Entropy Values

Problem Potassium chlorate, a common oxidizing agent in fireworks (*see photo*) and match heads, undergoes a solid-state disproportionation reaction when heated:

$$4\overset{+5}{KClO_3}(s) \xrightarrow{\Delta} 3\overset{+7}{KClO_4}(s) + \overset{-1}{KCl}(s)$$

Use $\Delta_f H°$ and $S°$ values to calculate $\Delta_{sys}G°$ (which is $\Delta_r G°$) for this reaction at 25°C.

Plan To solve for $\Delta G°$, we need to use values from Appendix B. We use $\Delta_f H°$ values to calculate $\Delta_r H°$ ($\Delta_{sys}H°$), use $S°$ values to calculate $\Delta_r S°$ ($\Delta_{sys}S°$), and then apply Equation 18.7.

Solution Calculate $\Delta_{sys}H°$ from $\Delta_f H°$ values (with Equation 5.9):

$$\begin{aligned}\Delta_{sys}H° &= \Delta_r H° = \Sigma m\Delta_f H°_{(products)} - \Sigma n\Delta_f H°_{(reactants)} \\ &= [3(\Delta_f H° \text{ of } KClO_4) + 1(\Delta_f H° \text{of } KCl)] - 4(\Delta_f H° \text{ of } KClO_3) \\ &= [3(-432.8 \text{ kJ/mol}) + 1\,(-436.7 \text{ kJ/mol})] - 4(-397.7 \text{ kJ/mol}) \\ &= -144.3 \text{ kJ/mol}\end{aligned}$$

Calculate $\Delta_{sys}S°$ from $S°$ values (with Equation 18.4):

$$\begin{aligned}\Delta_{sys}S° &= \Delta_r S° = \Sigma mS°_{products} - \Sigma nS°_{reactants} \\ &= [3(S° \text{ of } KClO_4) + 1(S° \text{ of } KCl)] - 4(S° \text{ of } KClO_3) \\ &= [3(151.0 \text{ J/[mol·K]}) + 1(82.6 \text{ J/[mol·K]})] - 4(143.1 \text{ J/[mol·K]}) \\ &= -36.8 \text{ J/(mol·K)}\end{aligned}$$

Calculate $\Delta_{sys}G°$ at 298 K:

$$\Delta_{sys}G° = \Delta_{sys}H° - T\Delta_{sys}S° = -144.3 \text{ kJ/mol} - \left[(298 \text{ K})(-36.8 \text{ J/[mol·K]})\left(\frac{1 \text{ kJ}}{1000 \text{ J}}\right)\right]$$

$$= -133 \text{ kJ/mol}$$

Check Round to check the math:

$$\begin{aligned}\Delta H° &\approx [3(-433 \text{ kJ/mol}) + (-440 \text{ kJ/mol})] - 4(-400 \text{ kJ/mol}) \\ &= -1740 \text{ kJ/mol} + 1600 \text{ kJ/mol} = -140 \text{ kJ/mol} \\ \Delta S° &\approx [3(150 \text{ J/[mol·K]}) + 85 \text{ J/(mol·K)}] - 4[145 \text{ J/(mol·K)}] \\ &= 535 \text{ J/(mol·K)} - 580 \text{ J/(mol·K)} = -45 \text{ J/(mol·K)} \\ \Delta G° &\approx -140 \text{ kJ/mol} - 300 \text{ K}(-0.04 \text{ kJ/[mol·K]}) \\ &= -140 \text{ kJ/mol} + 12 \text{ kJ/mol} = -128 \text{ kJ/mol}\end{aligned}$$

Comment Recall, from Section 18.1, that reaction spontaneity tells us nothing about rate. Even though this reaction is thermodynamically allowed, the rate is very low in the solid. When $KClO_3$ is heated slightly above its melting point, the ions can move and the reaction occurs readily.

Follow-Up Problem 18.4 Determine the standard Gibbs energy change at 298 K for the following reaction:

$$2NO(g) + O_2(g) \longrightarrow 2NO_2(g)$$

Using Standard Gibbs Energies of Formation to Find $\Delta_r G°$ Another way to calculate $\Delta_r G°$ is with values for the **standard Gibbs energy of formation ($\Delta_f G°$)** of the components. Similar to the standard enthalpy of formation, $\Delta_f H°$ (Section 5.6), $\Delta_f G°$ is the Gibbs energy change that occurs when 1 mol of compound is made *from its elements*, with all the components in their standard states. Because Gibbs energy is a state function, we can apply Hess's law and combine the $\Delta_f G°$ values of the reactants and products to calculate $\Delta_r G°$, no matter how the reaction takes place:

$$\Delta_r G° = \Sigma m\Delta_f G°_{(products)} - \Sigma n\Delta_f G°_{(reactants)} \qquad (18.8)$$

$\Delta_f G°$ values have properties similar to $\Delta_f H°$ values:

- $\Delta_f G°$ of an element in its standard state is zero.
- $\Delta_f G°$ is multiplied by a coefficient (m or n above) in the equation.
- Reversing a reaction changes the sign of $\Delta_f G°$.

Many $\Delta_f G°$ values are given, along with $\Delta_f H°$ and $S°$ values, in Appendix B.

Sample Problem 18.5 Calculating $\Delta_r G°$ from $\Delta_f G°$ Values

Problem Use $\Delta_f G°$ values to calculate $\Delta_r G°$ for the reaction in Sample Problem 18.4:

$$4KClO_3(s) \longrightarrow 3KClO_4(s) + KCl(s)$$

Plan Use Equation 18.8 to calculate $\Delta_r G°$.

Solution Use Equation 18.8 with values from Appendix B:

$$\begin{aligned}\Delta_r G° &= \Sigma m \Delta_f G°_{(products)} - \Sigma n \Delta_f G°_{(reactants)}\\ &= [3(\Delta_f G° \text{ of } KClO_4) + 1(\Delta_f G° \text{ of } KCl)] - 4(\Delta_f G° \text{ of } KClO_3)\\ &= [3(-303.2 \text{ kJ/mol}) + 1(-409.2 \text{ kJ/mol})] - 4(-296.3 \text{ kJ/mol})\\ &= -133.6 \text{ kJ/mol}\end{aligned}$$

Check Round to check the math:

$$\begin{aligned}\Delta_r G° &\approx [3(-300 \text{ kJ/mol}) + 1(-400 \text{ kJ/mol})] - 4(-300 \text{ kJ/mol})\\ &= -1300 \text{ kJ/mol} + 1200 \text{ kJ/mol} = -100 \text{ kJ/mol}\end{aligned}$$

Comment The slight discrepancy between this answer and the answer obtained in Sample Problem 18.4 is due to rounding. As you can see, when $\Delta_f G°$ values are available, this method is simpler arithmetically than the method used in Sample Problem 18.4.

Follow-Up Problem 18.5 Use $\Delta_f G°$ values to calculate the Gibbs energy change for each reaction at 25°C:

(a) $2NO(g) + O_2(g) \longrightarrow 2NO_2(g)$ (from Follow-Up Problem 18.4)

(b) $2C(\text{graphite}) + O_2(g) \longrightarrow 2CO(g)$

ΔG and the Work a System Can Do

Thermodynamics developed after the invention of the steam engine, a major advance that spawned a new generation of machines. Thus, some of the key ideas in the field applied the relationships between the Gibbs energy change and the work that a system can do:

- ΔG is the *maximum useful work* done ***by*** a system during a *spontaneous* process at constant temperature and pressure:

$$\Delta G = w_{max} \quad (18.9)$$

- ΔG is the *minimum work* done ***to*** a system to make a *nonspontaneous* process occur at constant temperature and pressure.

The Gibbs energy change is the maximum work that the system can *possibly* do. However, the work that it *actually* does is always less and depends on how the Gibbs energy is released. Let us consider the work done by an expanding gas, a car engine, and a battery.

1. *Work done by an expanding gas.* Suppose that a gas at $V_{initial}$ is confined in a piston-cylinder assembly attached to a 1 kg weight (Figure 18.14A). The gas expands to some final volume, V_{final}, lifting the weight in one step and doing a certain quantity of work (Figure 18.14B). The expanding gas can do more work by lifting a 2 kg weight to one-half V_{final} and then lifting a 1 kg weight to

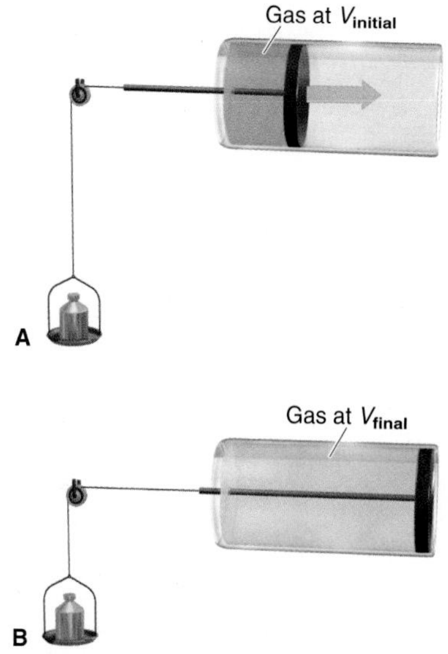

FIGURE 18.14 An expanding gas lifting a weight

V_{final}, that is, by doing the work in two steps. Lifting a 3 kg weight to one-third of V_{final}, a 2 kg weight to two-thirds V_{final}, and a 1 kg weight all the way to V_{final}—that is, in three steps—would do even more work. *As the number of steps increases, the quantity of work done by the gas increases.* Thus, the gas would do *close* to the maximum work if the weight were replaced by a container of sand that could be lifted by removing one grain at a time, that is, by using a *very* high number of steps. (We described a similar method for finding the entropy change in Section 18.1.)

However, the *maximum* work is done by a spontaneous process *only if the work is carried out in an infinite number of steps*, that is, *reversibly*. Of course, in any *real* process, work is performed in a *finite* number of steps, that is, *irreversibly*, so *the maximum work is never done.* Any Gibbs energy that is not used for work is lost to the surroundings as heat. This "unharnessed" energy is a consequence of any real process.

2. *"Useful" work done by a car engine.* Gasoline (represented by octane, C_8H_{18}) is burned in a car engine as follows:

$$C_8H_{18}(l) + \frac{25}{2}O_2(g) \longrightarrow 8CO_2(g) + 9H_2O(g)$$

A large amount of energy is released as heat ($\Delta_{sys}H < 0$), and, because the amount (mol) of gas increases, the entropy of the system increases ($\Delta_{sys}S > 0$). Therefore, the reaction is spontaneous ($\Delta_{sys}G < 0$). The Gibbs energy that is released turns the wheels, moves the belts, plays the radio, and so on—all examples of "useful" work. However, only if the Gibbs energy is released *reversibly*—that is, in an infinite number of steps—can this reaction do the *maximum* useful work. In reality, the reaction occurs *irreversibly*, and much of the total Gibbs energy just warms the engine and the outside air, which increases the freedom of motion of the particles in the universe, in accordance with the second law.

3. *"Useful" work done by a battery.* As you will see in Chapter 19, a battery is essentially a packaged spontaneous redox reaction that releases Gibbs energy to the surroundings (flashlight, radio, motor, and so on). If we connect the battery terminals to each other through a short piece of wire, the Gibbs energy change is released all at once but does no work—it just heats the wire and battery. If we connect the terminals to a motor, a significant portion of the Gibbs energy runs the motor, but some is still converted to heat. If we connect the battery to a more efficient device, a device that discharges the Gibbs energy still more slowly, more of the energy does work and less is converted to heat. However, as with all systems, only when the battery discharges infinitely slowly can it do the maximum amount of work.

Efficiency can be defined as the percentage of work output relative to the energy input. The range of efficiencies is very large: an incandescent light bulb converts <7% of the incoming electricity to light, the rest being given off as heat. At the other extreme, an electrical generator converts 95% of the incoming mechanical energy to electricity. Here are the efficiencies of some other devices: home oil furnace, 65%; hand-tool motor, 63%; liquid fuel rocket, 50%; car engine, <30%; compact fluorescent bulb, 18%; solar cell, ~15%. LED bulbs, which are becoming increasingly versatile in their applications, are extremely efficient. Even the least efficient LED bulb has an efficiency of about 67%, while many LED bulbs have efficiencies above 90%.

Therefore, all engineers must face the fact that *no **real** process uses all the available Gibbs energy to do work because some is always "wasted" as heat.* Let us summarize the relationship between the Gibbs energy change of a reaction and the work it can do:

- A spontaneous reaction ($\Delta_{sys}G < 0$) will do work on the surroundings ($-w$). In any real machine, the actual work done is *always less than the maximum* because some of the ΔG is released as heat.
- A nonspontaneous reaction ($\Delta_{sys}G > 0$) will occur only if the surroundings do work on the system ($+w$). In any real machine, the actual work done on the

system is *always more than the minimum* because some of the added Gibbs energy is wasted as heat.

- A reaction at equilibrium ($\Delta_{sys}G = 0$) can no longer do any work.

The Effect of Temperature on Reaction Spontaneity

In most cases, the enthalpy contribution (ΔH) to the Gibbs energy change (ΔG) is much *larger* than the entropy contribution ($T\Delta S$). In fact, the reason most exothermic reactions are spontaneous is that the large negative ΔH makes ΔG negative. However, the *temperature of a reaction influences the magnitude of the $T\Delta S$ term*, so, for many reactions, the overall spontaneity depends on the temperature. From the signs of ΔH and ΔS, we can predict how the temperature affects the sign of ΔG. (The values we will use below for the thermodynamic variables are standard-state values, but we show them without the degree sign to emphasize that the relationships among ΔG, ΔH, and ΔS are valid at any conditions. Also, we assume that ΔH and ΔS change little with temperature, which is true as long as no phase change occurs.)

Let us examine the four combinations of positive and negative ΔH and ΔS—two that are independent of temperature and two that are dependent on temperature.

- *Temperature-independent cases.* When ΔH and ΔS have *opposite* signs, the reaction occurs spontaneously either at all temperatures or at none (nonspontaneous).
  1. *The reaction is spontaneous at all temperatures: $\Delta H < 0$ and $\Delta S > 0$.* Since ΔS is positive, $-T\Delta S$ is negative; thus, both contributions favour a negative ΔG. Most combustion reactions are spontaneous at all temperatures. The decomposition of hydrogen peroxide, a common disinfectant, is also spontaneous at all temperatures:

$$2H_2O_2(l) \longrightarrow 2H_2O(l) + O_2(g)$$
$$\Delta H = -196 \text{ kJ/mol} \quad \text{and} \quad \Delta S = 125 \text{ J/(mol·K)}$$

  2. *The reaction is nonspontaneous at all temperatures: $\Delta H > 0$ and $\Delta S < 0$.* Both contributions oppose spontaneity: ΔH is positive and ΔS is negative, so $-T\Delta S$ is positive; thus, ΔG is always positive. The formation of ozone from oxygen requires a continuous energy input, so it is not spontaneous at any temperature:

$$3O_2(g) \longrightarrow 2O_3(g)$$
$$\Delta H = 286 \text{ kJ/mol} \quad \text{and} \quad \Delta S = -137 \text{ J/(mol·K)}$$

- *Temperature-dependent cases.* When ΔH and ΔS have the *same* sign, the relative magnitudes of $-T\Delta S$ and ΔH determine the sign of ΔG. In these cases, the *direction* of the change in temperature is crucial.
  3. *The reaction becomes spontaneous as the temperature increases: $\Delta H > 0$ and $\Delta S > 0$.* With a positive ΔH, the reaction occurs spontaneously only when $-T\Delta S$ becomes large enough to make ΔG negative, which happens as the temperature rises. For example, the oxidation of N_2O occurs spontaneously at any $T > 994$ K:

$$2N_2O(g) + O_2(g) \longrightarrow 4NO(g)$$
$$\Delta H = 197.1 \text{ kJ/mol} \quad \text{and} \quad \Delta S = 198.2 \text{ J/(mol·K)}$$

  4. *The reaction becomes spontaneous as the temperature decreases: $\Delta H < 0$ and $\Delta S < 0$.* Here, ΔH favours spontaneity, but ΔS does not ($-T\Delta S > 0$). The reaction occurs spontaneously only when $-T\Delta S$ becomes smaller than ΔH, and this happens as the temperature drops. For example, the production of iron(III) oxide occurs spontaneously at any $T < 3005$ K:

$$4Fe(s) + 3O_2(g) \longrightarrow 2Fe_2O_3(s)$$
$$\Delta H = -1651 \text{ kJ/mol} \quad \text{and} \quad \Delta S = -549.4 \text{ J/(mol·K)}$$

Table 18.1 summarizes these four possible combinations of ΔH and ΔS, and Sample Problem 18.6 applies them.

TABLE 18.1 Reaction Spontaneity and the Signs of ΔH, ΔS, and ΔG

ΔH	ΔS	$-T\Delta S$	ΔG	Description
−	+	−	−	Spontaneous at all *T*
+	−	+	+	Nonspontaneous at all *T*
+	+	−	+ or −	Spontaneous at higher *T*; nonspontaneous at lower *T*
−	−	+	+ or −	Spontaneous at lower *T*; nonspontaneous at higher *T*

Sample Problem 18.6 Using Molecular Scenes to Determine the Signs of ΔH, ΔS, and ΔG

Problem The following scenes represent a familiar phase change for water (*blue spheres*):

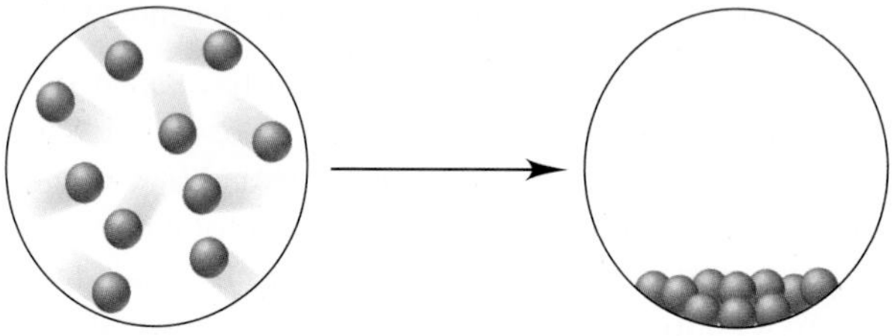

(a) What are the signs of ΔH and ΔS for this process? Explain.

(b) Is the process spontaneous at all *T*, no *T*, low *T*, or high *T*? Explain.

Plan (a) From the scenes, we determine any change in the amount of gas (which indicates the sign of ΔS) and any change in the freedom of motion of the particles (which indicates whether heat is absorbed or released) and, thus, the sign of ΔH. **(b)** The question refers to the sign of ΔG (+ or −) at the different temperature possibilities, so we apply Equation 18.6 and refer to the previous discussion and Table 18.1.

Solution (a) The scene represents the condensation of water vapour, so the amount of gas decreases dramatically and the separated molecules give up energy as they come closer together. Therefore, $\Delta S < 0$ and $\Delta H < 0$.
(b) With ΔS negative, the $-T\Delta S$ term is positive. For $\Delta G < 0$, the magnitude of *T* must be small. Therefore, the process is spontaneous at low *T*.

Check The answer to part (b) seems reasonable, based on our analysis in part (a). The answer makes sense because we know, from everyday experience, that water condenses spontaneously, and it does so at low temperatures.

Follow-Up Problem 18.6 The following molecular scenes represent the gas-phase decomposition of X_2Y_2 to X_2 (*red*) and Y_2 (*blue*):

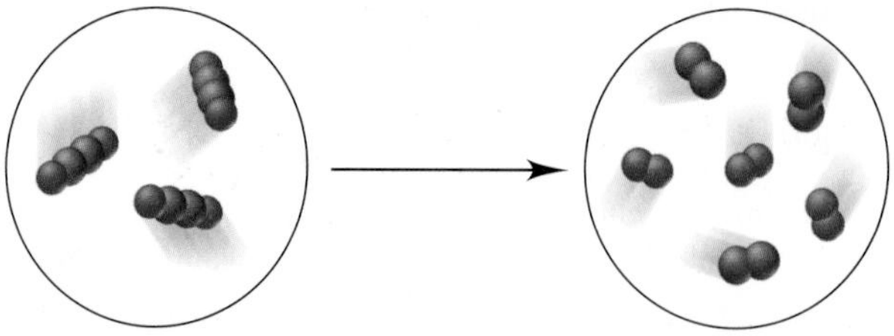

(a) What is the sign of ΔS for the reaction?

(b) If the reaction is spontaneous only above 325°C, what is the sign of ΔH? Explain.

As you saw in Sample Problem 18.4, one way to calculate ΔG is from enthalpy and entropy changes. As long as phase changes do not occur, ΔH and ΔS usually change little with temperature. In Sample Problem 18.7, we use the values of ΔH and ΔS at 298 K to examine the effect of *T* on ΔG and, thus, on reaction spontaneity.

Sample Problem 18.7 Determining the Effect of Temperature on ΔG

Problem A key step in the production of sulfuric acid is the oxidation of $SO_2(g)$ to $SO_3(g)$:

$$2SO_2(g) + O_2(g) \longrightarrow 2SO_3(g)$$

At 298 K, $\Delta G = -141.6$ kJ/mol, $\Delta H = -198.4$ kJ/mol, and $\Delta S = -187.9$ J/K.

(a) Use the data to decide if this reaction is spontaneous at 25°C, and predict how ΔG will change with increasing temperature.

(b) Assuming that ΔH and ΔS are constant with temperature (no phase change occurs), is the reaction spontaneous at 900.°C?

Plan (a) We note the sign of ΔG to see if the reaction is spontaneous and the signs of ΔH and ΔS to see the effect of T. **(b)** We use Equation 18.6 to calculate ΔG from the given ΔH and ΔS at the higher T (in K).

Solution (a) $\Delta G < 0$, so the reaction is spontaneous at 298 K. SO_2 and O_2 will form SO_3 spontaneously. With $\Delta S < 0$, the term $-T\Delta S > 0$, and this term will become more positive at higher T. Therefore, ΔG will become less negative, and the reaction will become less spontaneous, with increasing T.

(b) Calculate ΔG at 900.°C ($T = 273 + 900. = 1173$ K):

$$\Delta G = \Delta H - T\Delta S = -198.4\ \text{kJ/mol} - [(1173\ \text{K})(-187.9\ \text{J/[mol·K]})(1\ \text{kJ}/1000\ \text{J})]$$
$$= 22.0\ \text{kJ/mol}$$

$\Delta G > 0$, so the reaction is nonspontaneous at the higher T.

Check The answer in part (b) seems reasonable based on our prediction in part (a). The arithmetic seems correct, given considerable rounding:

$$\Delta G \approx -200\ \text{kJ/mol} - [(1200\ \text{K})(-200\ \text{J/[mol·K]})(1\ \text{kJ}/1000\ \text{J}] = +40\ \text{kJ/mol}$$

Follow-Up Problem 18.7 A reaction is nonspontaneous at room temperature but *is* spontaneous at −40°C. What can you say about the signs and relative magnitudes of ΔH, ΔS, and $-T\Delta S$?

The Temperature at Which a Reaction Becomes Spontaneous As we have just seen, when the signs are the same for ΔH and ΔS of a reaction, the reaction can be nonspontaneous at one temperature and spontaneous at another. The crossover temperature occurs when a positive ΔG switches to a negative ΔG because of the magnitude of the $-T\Delta S$ term. We find this temperature by setting ΔG equal to zero and solving for T:

$$\Delta G = \Delta H - T\Delta S = 0$$

Therefore,

$$\Delta H = T\Delta S \quad \text{and} \quad T = \frac{\Delta H}{\Delta S} \tag{18.10}$$

Consider the reaction of copper(I) oxide with carbon. It does *not* occur at a low temperature, but it does at a high temperature and is used to extract copper from a copper ore:

$$Cu_2O(s) + C(s) \xrightarrow{\Delta} 2Cu(s) + CO(g)$$

We predict that this reaction has a positive ΔS because the amount (mol) of gas increases; in fact, $\Delta S = 165$ J/(mol·K). Furthermore, because the reaction is *nonspontaneous* at lower temperatures, it must have a positive ΔH; the actual value is 58.1 kJ/mol. As the $-T\Delta S$ term becomes more negative with higher T, it eventually outweighs the positive ΔH term, so ΔG becomes negative and the reaction occurs spontaneously.

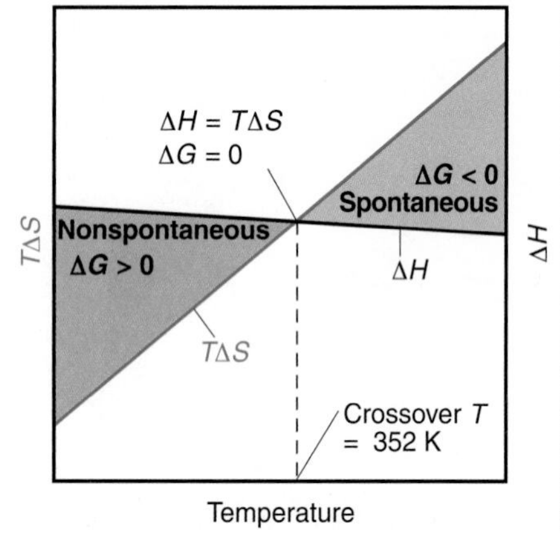

FIGURE 18.15 The effect of temperature on reaction spontaneity. At low T, $\Delta G > 0$ because ΔH dominates. At 352 K, $\Delta H = T\Delta S$, so $\Delta G = 0$. At any higher T, $\Delta G < 0$ because $-T\Delta S$ dominates.

Sample Problem 18.8 Finding the Temperature at Which a Reaction Becomes Spontaneous

Problem At 25°C (298 K), the reduction of copper(I) oxide to copper is nonspontaneous ($\Delta G = 8.9$ kJ/mol). Calculate the temperature at which the reaction becomes spontaneous.

Plan As just discussed, we want the temperature at which ΔG crosses over from a positive value to a negative value. We set ΔG equal to zero and use Equation 18.10 to solve for T, using the values for ΔH (58.1 kJ/mol) and ΔS [165 J/(mol·K)] from Appendix B.

Solution From $\Delta G = \Delta H - T\Delta S = 0$, we have

$$T = \frac{\Delta H}{\Delta S} = \frac{58.1 \frac{\text{kJ}}{\text{mol}} \times \frac{1000 \text{ J}}{1 \text{ kJ}}}{165 \frac{\text{J}}{\text{mol·K}}} = 352 \text{ K}$$

Thus, at any temperature above 352 K (79°C), which is a moderate temperature for extracting a metal from its ore, $\Delta G < 0$, so the reaction becomes spontaneous.

Check Round to check the math quickly:

$$T = \frac{60\,000 \text{ J/mol}}{150 \frac{\text{J}}{\text{mol·K}}} = 400 \text{ K}$$

This is close to the answer.

Comment Figure 18.15 shows that the line for $T\Delta S$ rises steadily (and thus the $-T\Delta S$ term becomes more negative) with increasing temperature. The line crosses the relatively constant ΔH line at 352 K. At any higher temperature, the $-T\Delta S$ term is greater than the ΔH term, so ΔG is negative.

Follow-Up Problem 18.8 Use values from Appendix B to find the temperature at which the following reaction becomes spontaneous. (Assume that ΔH and ΔS are constant with T.)

$$CaO(s) + CO_2(g) \longrightarrow CaCO_3(s)$$

Coupling of Reactions to Drive a Nonspontaneous Change

In a complex, multistep reaction, we often see a nonspontaneous step driven by a spontaneous step. In such a **coupling of reactions**, *one step supplies enough Gibbs energy for the other to occur*, just as burning gasoline supplies enough Gibbs energy to move a car.

Look again at the reduction of copper(I) oxide by carbon. Previously, we found that the *overall* reaction becomes spontaneous at any temperature above 352 K. Dividing the reaction into two steps, however, we find that even at a higher temperature, such as 375 K, copper(I) oxide does not spontaneously decompose to its elements:

$$Cu_2O(s) \longrightarrow 2Cu(s) + \frac{1}{2}O_2(g) \qquad \Delta_{375}G = 140.0 \text{ kJ/mol}$$

However, the oxidation of carbon to CO at 375 K is quite spontaneous:

$$C(s) + \frac{1}{2}O_2(g) \longrightarrow CO(g) \qquad \Delta_{375}G = -143.8 \text{ kJ/mol}$$

Coupling these reactions means having the carbon in contact with the Cu_2O, which allows the reaction with the larger negative ΔG to "drive" the reaction with the smaller positive ΔG. Adding the reactions together and cancelling the common substance ($\frac{1}{2}O_2$) gives an overall reaction with a negative ΔG:

$$Cu_2O(s) + C(s) \longrightarrow 2Cu(s) + CO(g) \qquad \Delta_{375}G = -3.8 \text{ kJ/mol}$$

Many biochemical reactions are also nonspontaneous, including key steps in the synthesis of proteins and nucleic acids, the formation of fatty acids, the maintenance of ion balance, and the breakdown of nutrients. Driving a nonspontaneous step by coupling it to a spontaneous step is a life-sustaining strategy that is common to all organisms, as you will see in the Chemical Connections section.

SUMMARY OF SECTION 18.3

- The sign of the Gibbs energy change, $\Delta G = \Delta H - T\Delta S$, is directly related to reaction spontaneity: a negative ΔG corresponds to a positive $\Delta_{univ}S$.
- We use the standard Gibbs energy of formation ($\Delta_f G°$) to calculate $\Delta_r G°$ at 298 K.
- The maximum work that a system can do is never obtained from a real (irreversible) process because some Gibbs energy is always converted to heat.
- The magnitude of T influences the spontaneity of a temperature-dependent reaction (same signs for ΔH and ΔS) by affecting the size of $T\Delta S$. For such a reaction, the temperature at which the reaction becomes spontaneous can be found by setting $\Delta G = 0$.
- A nonspontaneous reaction ($\Delta G > 0$) can be coupled to a more spontaneous reaction ($\Delta G \ll 0$) to make the nonspontaneous reaction occur. In organisms, for example, the hydrolysis of ATP drives many reactions with a positive ΔG.

18.4 Gibbs Energy, Equilibrium, and Reaction Direction

As you know from discussions in earlier chapters, the sign of ΔG is not the only way to predict the direction of a reaction. In Chapter 15, we did this by comparing the values of the reaction quotient (Q) and the equilibrium constant (K). Recall the following:

- If $Q < K$ ($\frac{Q}{K} < 1$), the reaction proceeds spontaneously to the right.
- If $Q > K$ ($\frac{Q}{K} > 1$), the reaction proceeds spontaneously to the left.
- If $Q = K$ ($\frac{Q}{K} = 1$), the reaction has attained equilibrium and no longer proceeds spontaneously in either direction.

It is easier to see the relationship between these two ways of predicting reaction spontaneity (that is, the sign of ΔG and the magnitude of $\frac{Q}{K}$) when we compare the sign of the natural logarithm of $\frac{Q}{K}$ ($\ln \frac{Q}{K}$) with the sign of ΔG (refer to Appendix A if necessary):

- If $\frac{Q}{K} < 1$, then $\ln \frac{Q}{K} < 0$ and the reaction proceeds spontaneously to the right ($\Delta G < 0$).
- If $\frac{Q}{K} > 1$, then $\ln \frac{Q}{K} > 0$ and the reaction proceeds spontaneously to the left ($\Delta G > 0$).
- If $\frac{Q}{K} = 1$, then $\ln \frac{Q}{K} = 0$ and the reaction is at equilibrium ($\Delta G = 0$).

Note that the signs of ΔG and $\ln \frac{Q}{K}$ are the same for a given direction; in fact, ΔG equals $\ln \frac{Q}{K}$ multiplied by the proportionality constant RT:

$$\Delta G = RT \ln \frac{Q}{K} = RT \ln Q - RT \ln K \tag{18.11}$$

Q represents the reaction quotient and is calculated using activities (which we normally approximate with concentrations or pressures) of the components of a system at any time during the reaction. K represents these quantities at equilibrium. Therefore, according to Equation 18.11, ΔG is a measure of how different the activities at any time, Q, are from the activities at equilibrium, K:

- If Q and K are very different, the reaction releases (absorbs) a *lot* of Gibbs energy.
- If Q and K are nearly the same, the reaction releases (absorbs) relatively *little* energy.

CHEMICAL CONNECTIONS TO BIOLOGICAL ENERGETICS

The Universal Role of ATP

Despite their incredible diversity, virtually all organisms use the same amino acids to make their proteins, the same nucleotides to make their nucleic acids, and the same carbohydrate (glucose) to provide energy. As well, *all organisms use the same spontaneous reaction to drive a variety of nonspontaneous reactions.* This spontaneous reaction is the hydrolysis of **adenosine triphosphate (ATP)** to adenosine diphosphate (ADP):*

$$ATP^{4+}(aq) + H_2O(l) \rightleftharpoons ADP^{3-}(aq) + HPO_4^{2-}(aq) + H^+(aq)$$
$$\Delta G^{\circ\prime} = -30.5 \text{ kJ/mol}$$

In the metabolic breakdown of glucose, for example, the first step, the addition of HPO_4^{2-} to glucose, is nonspontaneous:

$$\text{Glucose}(aq) + HPO_4^{2-}(aq) + H^+(aq) \rightleftharpoons$$
$$[\text{glucose phosphate}]^-(aq) + H_2O(l) \quad \Delta G^{\circ\prime} = 13.8 \text{ kJ/mol}$$

Coupling this nonspontaneous reaction to ATP hydrolysis makes the overall process spontaneous. If we add the two reactions, HPO_4^{2-}, H^+, and H_2O cancel:

$$\text{Glucose}(aq) + ATP^{4+}(aq) \rightleftharpoons$$
$$[\text{glucose phosphate}]^-(aq) + ADP^{3-}(aq)$$
$$\Delta G^{\circ\prime} = -16.7 \text{ kJ/mol}$$

Coupling cannot occur if reactions are physically separated, so these reactions take place on an enzyme (Section 14.7) that simultaneously binds glucose and ATP, and the phosphate group of ATP that will be transferred lies next to the —OH group of glucose that will bind it (Figure B18.1).

The ADP produced in energy-releasing reactions combines with phosphate to regenerate ATP in energy-absorbing reactions catalyzed by other enzymes. Thus, there is a continuous cycling of ATP to ADP and then back to ATP to supply energy to the cells (Figure B18.2).

*In biochemical systems, the standard-state concentration of H^+ is 10^{-7} mol/L, not the usual 1 mol/L, and the standard Gibbs energy change has the symbol $\Delta G^{\circ\prime}$.

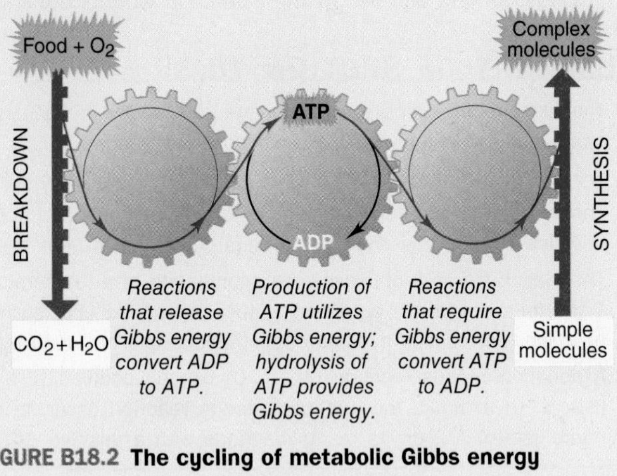

FIGURE B18.2 The cycling of metabolic Gibbs energy

By examining the phosphate portions of ATP, ADP, and HPO_4^{2-}, we can see two basic chemical reasons that ATP hydrolysis supplies so much Gibbs energy (Figure B18.3):

1. *Charge repulsion.* At physiological pH (~7), the triphosphate group of ATP has four negative charges close together. This *high charge repulsion* is reduced in ADP (Figure B18.3A).

2. *Electron delocalization.* Once HPO_4^{2-} is free, extensive delocalization and resonance stabilization of the π electrons occur (Figure B18.3B).

Thus, greater charge repulsion and less electron delocalization make ATP higher in energy (less stable) than the sum of the energies of ADP and HPO_4^{2-}. When ATP is hydrolyzed, some of this

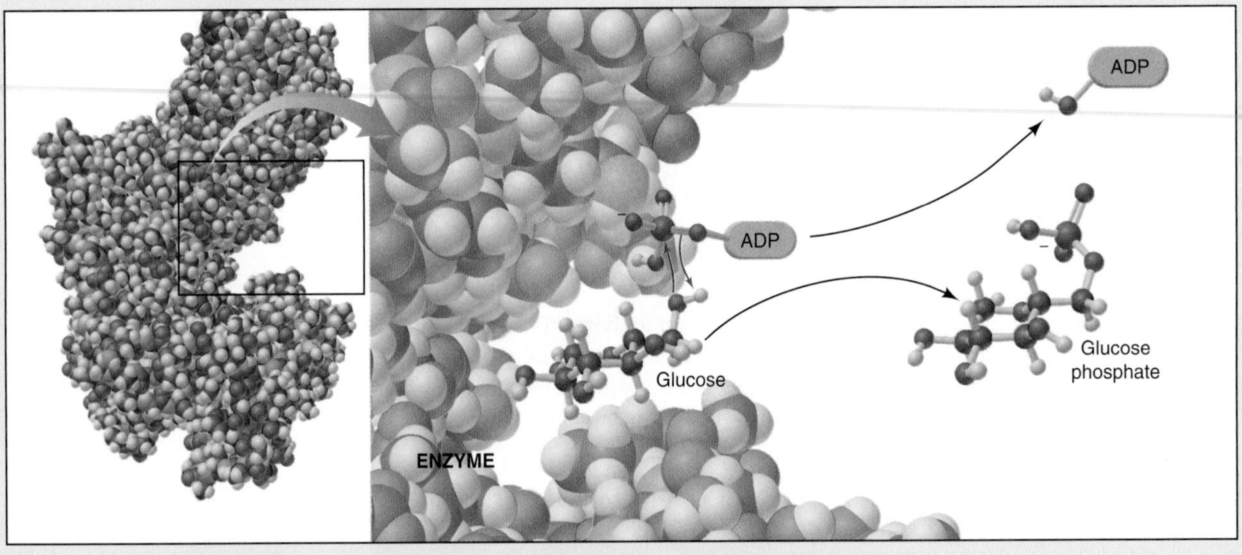

FIGURE B18.1 The coupling of a nonspontaneous reaction to the hydrolysis of ATP. Glucose lies next to ATP (shown as ADP—O—PO_3H) in the enzyme's active site. ADP (shown as ADP—OH) and glucose phosphate are released.

(Continued)

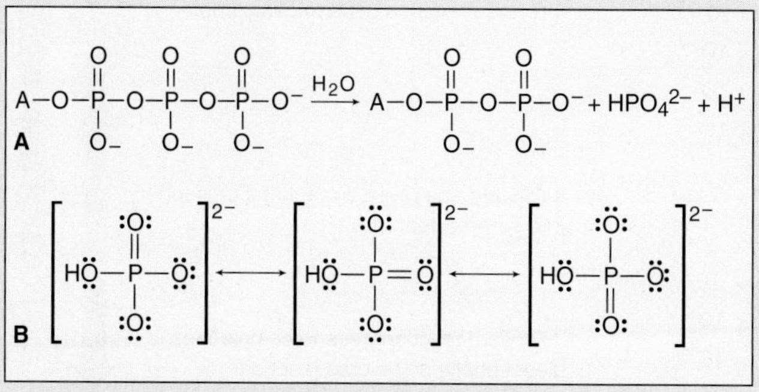

FIGURE B18.3 ATP is a high-energy molecule.

additional energy is released and harnessed by the organism to drive metabolic reactions that could not otherwise take place.

Problems

B18.1 The oxidation of 1 mol of glucose supplies enough metabolic energy to form 36 mol of ATP. Oxidation of 1 mol of a typical dietary fat, such as tristearin ($C_{57}H_{116}O_6$), yields enough energy to form 458 mol of ATP. How many molecules of ATP can form per gram of (a) glucose; (b) tristearin?

B18.2 Nonspontaneous processes, such as muscle contraction, protein synthesis, and nerve conduction, are coupled to the spontaneous hydrolysis of ATP to ADP. ATP is then regenerated by coupling its synthesis to energy-yielding reactions such as these:

$$\text{creatine phosphate} \longrightarrow \text{creatine} + \text{phosphate} \qquad \Delta G^{\circ\prime} = -43.1 \text{ kJ/mol}$$

$$\text{ADP} + \text{phosphate} \longrightarrow \text{ATP} \qquad \Delta G^{\circ\prime} = +30.5 \text{ kJ/mol}$$

Find $\Delta G^{\circ\prime}$ for the overall reaction that regenerates ATP.

A very important reminder at this time is that, to use K or Q in the equation with ΔG, the value *must* be the thermodynamic equilibrium constant or reaction quotient. Hence, if the reaction contains only gaseous terms, the pressures (bar) must be used in the approximation of K or Q; similarly, if the reaction contains only aqueous terms, the concentrations (mol/L) must be used to approximate the value of K or Q. If the reaction contains both aqueous and gaseous terms, then the value of K or Q must be calculated using pressures for the gaseous terms and concentrations for the aqueous terms.

The Standard Gibbs Energy Change and the Equilibrium Constant When we choose standard-state values for Q (activities of 1 for all states, and for substances in solution) in Equation 18.11, ΔG becomes, by definition, ΔG°, and Q equals 1:

$$\Delta G^\circ = RT \ln 1 - RT \ln K$$

Since $\ln 1 = 0$, the term $RT \ln Q$ drops out, which allows us to find the standard Gibbs energy change of a reaction ($\Delta_r G^\circ$) from its equilibrium constant, or vice versa:

$$\Delta G^\circ = -RT \ln K \qquad (18.12)$$

Table 18.2 shows that, because of their logarithmic relationship, a small change in ΔG° causes a large change in K. Notice the following:

- As ΔG° becomes more positive, K becomes smaller: for example, if $\Delta G^\circ = +10$ kJ/mol, K is 0.02, so the product terms are $\frac{1}{50}$ the size of the reactant terms.
- As ΔG° becomes more negative, K becomes larger: for example, if $\Delta G^\circ = -10$ kJ/mol, the product terms are 50 times the reactant terms.

Finding the Gibbs Energy Change under Any Conditions In reality, reactions rarely begin with all the components in their standard states. By substituting the relationship between ΔG° and K (Equation 18.12) into the expression for ΔG (Equation 18.11), we obtain a relationship that applies to *any starting concentrations*:

$$\Delta G = \Delta G^\circ + RT \ln Q \qquad (18.13)$$

Sample Problem 18.9 uses molecular scenes to explore these ideas, and Sample Problem 18.10 applies these ideas in an important industrial reaction.

TABLE 18.2 The Relationship between ΔG° and K at 298 K

ΔG° (kJ/mol)	K	Significance
200	9×10^{-36}	Essentially no forward reaction; reverse reaction goes to completion
100	3×10^{-18}	
50	2×10^{-9}	
10	2×10^{-2}	
1	7×10^{-1}	
0	1	
−1	1.5	Forward and reverse reactions proceed to same extent
−10	5×10^{1}	
−50	6×10^{8}	
−100	3×10^{17}	
−200	1×10^{35}	Forward reaction goes to completion; essentially no reverse reaction

FORWARD REACTION
REVERSE REACTION

Sample Problem 18.9 Using Molecular Scenes to Find ΔG for a Reaction at Nonstandard Conditions

Problem These molecular scenes represent three mixtures in which A_2 (*black*) and B_2 (*green*) are forming AB. Each molecule represents 0.10 bar.

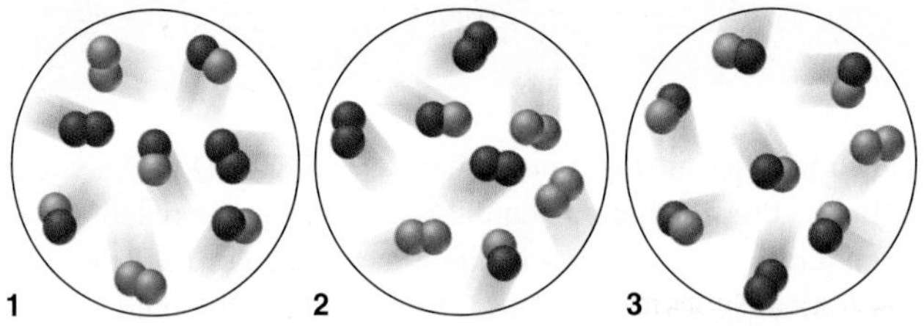

The equation is given below:

$$A_2(g) + B_2(g) \rightleftharpoons 2AB(g) \qquad \Delta G^\circ = -3.4 \text{ kJ/mol}$$

(a) If mixture 1 is at equilibrium, calculate K.

(b) Which mixture has the most negative ΔG, and which has the most positive ΔG?

(c) Is the reaction spontaneous at the standard state, that is, at $p_{A_2} = p_{B_2} = p_{AB} = 1.0$ bar?

Plan **(a)** Mixture 1 is at equilibrium, so we first write the expression for Q. We then find the partial pressure of each substance from the numbers of molecules and calculate K. **(b)** To find ΔG, we apply Equation 18.13. We are given ΔG° (−3.4 kJ/mol) and know R [8.314 J/(mol·K)], but we still need to find T. We calculate T from Equation 18.12, using K from part (a), and substitute the partial pressure of each substance (by counting particles) to get Q. **(c)** Once again, we use Equation 18.12. We are given ΔG°, and we know the partial pressures of all three gases. It may appear as though we are missing information, but, if we substitute into Equation 18.12, we can see that the pressures correspond to standard-state pressures. This gives us a Q value of 1, making $\ln Q = 0$. Thus, $G = \Delta G^\circ$.

Solution **(a)** Write the expression for Q and calculate K:

$$A_2(g) + B_2(g) \rightleftharpoons 2AB(g) \qquad Q = \frac{p_{AB}^2}{p_{A_2} \times p_{B_2}} \qquad K = \frac{(0.40)^2}{(0.20)(0.20)} = 4.0$$

(b) Calculate T from Equation 18.12 for use in Equation 18.13:

$$\Delta G^\circ = -RT \ln K = -3.4 \frac{\text{kJ}}{\text{mol}} = -\left(8.314 \frac{\text{J}}{\text{mol·K}}\right) T \ln 4.0$$

$$T = \frac{-3.4 \dfrac{\text{kJ}}{\text{mol}}\left(\dfrac{1000 \text{ J}}{1 \text{ kJ}}\right)}{-\left(8.314 \dfrac{\text{J}}{\text{mol·K}}\right) \ln 4.0} = 295 \text{ K}$$

Calculate ΔG from Equation 18.13 for each reaction mixture:

Mixture 1:

$$\Delta G = \Delta G^\circ + RT \ln Q = -3.4 \text{ kJ/mol} + RT \ln 4.0$$
$$= -3.4 \text{ kJ/mol}\left(\frac{1000 \text{ J}}{1 \text{ kJ}}\right) + \left(8.314 \frac{\text{J}}{\text{mol·K}}\right)(295 \text{ K}) \ln 4.0$$
$$= -3400 \text{ J/mol} + 3400 \text{ J/mol} = 0.0$$

Mixture 2:

$$\Delta G = -3.4 \text{ kJ/mol} + RT \ln \frac{(0.20)^2}{(0.30)(0.30)}$$
$$= -3.4 \text{ kJ/mol}\left(\frac{1000 \text{ J}}{1 \text{ kJ}}\right) + \left(8.314 \frac{\text{J}}{\text{mol·K}}\right)(295 \text{ K}) \ln 0.44$$
$$= -5.4 \times 10^3 \text{ J/mol}$$

Mixture 3:

$$\Delta G = -3.4 \text{ kJ/mol} + RT \ln \frac{(0.60)^2}{(0.10)(0.10)}$$
$$= -3.4 \text{ kJ/mol}\left(\frac{1000 \text{ J}}{1 \text{ kJ}}\right) + \left(8.314 \frac{\text{J}}{\text{mol·K}}\right)(295 \text{ K}) \ln 36$$
$$= 5.4 \times 10^3 \text{ J/mol}$$

Mixture 2 has the most negative ΔG, and mixture 3 has the most positive ΔG.

(c) Find ΔG when $p_{A_2} = p_{B_2} = p_{AB} = 1.0$ bar:

$$\Delta G = \Delta G^\circ + RT \ln Q = -3.4 \text{ kJ/mol} + RT \ln \frac{(1.0)^2}{(1.0)(1.0)}$$
$$= -3.4 \text{ kJ/mol} + RT \ln 1.0 = -3.4 \text{ kJ/mol}$$

Yes, the reaction is spontaneous when the components are in their standard states.

Check We can round to check the arithmetic in part (b); for example, for mixture 3, $\Delta G \approx -3000 \text{ J/mol} + [8 \text{ J/(mol·K)}](300 \text{ K})4 \approx 7000 \text{ J/mol}$, which is in the ballpark.

Comment **1.** By using the properties of logarithms, we did not have to calculate T and ΔG in (b). In mixture 2, $Q < 1$, so $\ln Q$ is negative, which makes ΔG more negative. Also, note that Q (0.44) $< K$ (4.0), so $\Delta G < 0$. In mixture 3, $Q > 1$ (and is greater than it is in mixture 1), so $\ln Q$ is positive, which makes ΔG positive. Also, Q (36) $> K$ (4.0), so $\Delta G > 0$.

2. In (b), the value of zero for ΔG of the equilibrium mixture (mixture 1) makes sense, because a system at equilibrium has released all of its Gibbs energy.

Follow-Up Problem 18.9 The following scenes depict mixtures in which X_2 (*tan*) and Y_2 (*blue*) are forming XY_2. Each molecule represents 0.10 mol, and the volume is 1 L.

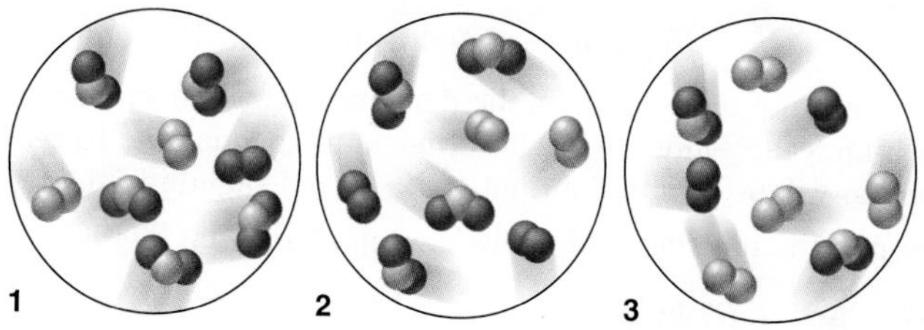

The equation is given below:

$$X_2(g) + 2Y_2(g) \rightleftharpoons 2XY_2(g) \qquad \Delta G^\circ = -1.3 \text{ kJ/mol}$$

(a) If, at the temperature at which the scenes are shown, $K_C = 20$, which mixture is at equilibrium?

(b) Rank the three mixtures from lowest (most negative) ΔG to highest (most positive) ΔG.

(c) What is the sign of ΔG for the change that occurs as each nonequilibrium mixture approaches equilibrium?

Sample Problem 18.10 Calculating ΔG at Nonstandard Conditions

Problem The oxidation of $SO_2(g)$ is too slow at 298 K to be useful in the manufacture of sulfuric acid, so the reaction is run at high T:

$$2SO_2(g) + O_2(g) \longrightarrow 2SO_3(g)$$

(a) Calculate K at 298 K and 973 K. ($\Delta_{298}G° = -141.6$ kJ/mol of reaction as written; using $\Delta H°$ and $\Delta S°$ values at 973 K, $\Delta_{973}G° = -12.12$ kJ/mol of reaction as written.)

(b) Two containers are filled with 0.500 bar of SO_2, 0.0100 bar of O_2, and 0.100 bar of SO_3; one is kept at 25°C, and the other is kept at 700.°C. In which direction, if any, will the reaction proceed to reach equilibrium at each temperature?

(c) Calculate ΔG for the system in part (b) at each temperature.

Plan **(a)** We know $\Delta G°$, T, and R, so we can calculate the K values from Equation 18.12. **(b)** To determine if a net reaction will occur, we find Q from the given partial pressures and compare it with each K from part (a). **(c)** These are *not* standard-state pressures, so we find ΔG at each T using Equation 18.13 with the values of $\Delta G°$ (given) and Q [from part (b)].

Solution **(a)** Calculate K at the two temperatures:

$$\Delta G° = -RT \ln K$$

so

$$K = e^{-[\Delta G°/(RT)]}$$

At 298 K, the exponent is

$$-[\Delta G°/(RT)] = -\left(\frac{-141.6 \text{ kJ/mol} \times \dfrac{1000 \text{ J}}{1 \text{ kJ}}}{8.314 \dfrac{\text{J}}{\text{mol·K}} \times 298 \text{ K}}\right) = 57.2$$

so

$$K = e^{-[\Delta G°/(RT)]} = e^{57.2} = 7 \times 10^{24}$$

At 973 K, the exponent is

$$-[\Delta G°/(RT)] = -\left(\frac{-12.12 \text{ kJ/mol} \times \dfrac{1000 \text{ J}}{1 \text{ kJ}}}{8.314 \dfrac{\text{J}}{\text{mol·K}} \times 973 \text{ K}}\right) = 1.50$$

so

$$K = e^{-[\Delta G°/(RT)]} = e^{1.50} = 4.5$$

(b) Calculate the value of Q:

$$Q = \frac{p^2_{SO_3}}{p^2_{SO_2} \times p_{O_2}} = \frac{0.100^2}{0.500^2 \times 0.0100} = 4.00$$

Because $Q < K$ at both temperatures, the denominator will decrease and the numerator will increase—more SO_3 will form—until Q equals K. To reach equilibrium, the reaction will go far to the right at 298 K and slightly to the right at 973 K.

(c) Calculate ΔG, the nonstandard Gibbs energy change, at 298 K:

$$\Delta_{298}G = \Delta G° + RT \ln Q$$

$$= -141.6 \text{ kJ/mol} + \left(8.314 \frac{\text{J}}{\text{mol·K}} \times \frac{1 \text{ kJ}}{1000 \text{ J}} \times 298 \text{ K} \times \ln 4.00\right)$$

$$= -138.2 \text{ kJ/mol}$$

Calculate ΔG at 973 K:

$$\Delta_{973}G = \Delta G° + RT \ln Q$$

$$= -12.12 \text{ kJ/mol} + \left(8.314 \frac{\text{J}}{\text{mol·K}} \times \frac{1 \text{ kJ}}{1000 \text{ J}} \times 973 \text{ K} \times \ln 4.00\right)$$

$$= -0.9 \text{ kJ/mol}$$

Check Note, in parts (a) and (c), that we made the Gibbs energy units (kJ/mol) consistent with the units in R (J). For significant figures in addition and subtraction, we retain one digit to the right of the decimal point in part (c).

Comment For these starting gas pressures at 973 K, the process is barely spontaneous ($\Delta G = -0.9$ kJ/mol), so why use a higher temperature? As in the synthesis of NH_3 (Section 15.6), where the *yield* is greater at a lower temperature, this process is carried out at a higher temperature *with a catalyst* to attain a higher *rate*. We will discuss the details of the industrial production of sulfuric acid in Chapter 23.

Follow-Up Problem 18.10 At 298 K, hypobromous acid (HBrO) dissociates in water with a K_a of 2.3×10^{-9}.

(a) Calculate ΔG° for the dissociation of HBrO.

(b) Calculate ΔG if $[H_3O^+] = 6.0 \times 10^{-4}$ mol/L, $[BrO^-] = 0.10$ mol/L, and $[HBrO] = 0.20$ mol/L.

Another Look at the Meaning of Spontaneity At this point, we introduce two terms related to *spontaneous* and *nonspontaneous*:

1. *Product-favoured reaction.* For the general reaction $A \rightleftharpoons B$, $K = \frac{[B]}{[A]} > 1$ and, therefore, the reaction proceeds largely from left to right (Figure 18.16A). From pure A to equilibrium, $Q < K$ and the curved *green* arrow in the figure indicates that the reaction is spontaneous ($\Delta G < 0$). From there on, the curved *red* arrow indicates that the reaction is nonspontaneous ($\Delta G > 0$). Similarly, from pure B to equilibrium, $Q > K$ and the reaction is spontaneous ($\Delta G < 0$), but not thereafter. In the graphs in Figure 18.16, the *change* in Gibbs energy, ΔG, can also be interpreted as the slope of the tangent to the curve at any point in the reaction. In either case, *Gibbs energy decreases until the reaction reaches a minimum at the equilibrium mixture*: $Q = K$ *and* $\Delta G = 0$. For the overall reaction $A \rightleftharpoons B$ (starting with all the components in their standard states), G°_B is smaller than G°_A, so ΔG° is negative, which corresponds to $K > 1$. We call this a *product-favoured* reaction because the system contains mostly product in its final state.

2. *Reactant-favoured reaction.* For the opposite reaction, $C \rightleftharpoons D$, $K = \frac{[D]}{[C]} < 1$ and the reaction proceeds slightly from left to right (Figure 18.16B). Here, too, whether we start with pure C or pure D, the reaction is spontaneous ($\Delta G < 0$) until equilibrium. In this reaction, however, the equilibrium mixture contains mostly C (the reactant), so we say the reaction is *reactant favoured.* Here, G°_D is *larger* than G°_C, so ΔG° is *positive*, which corresponds to $K < 1$.

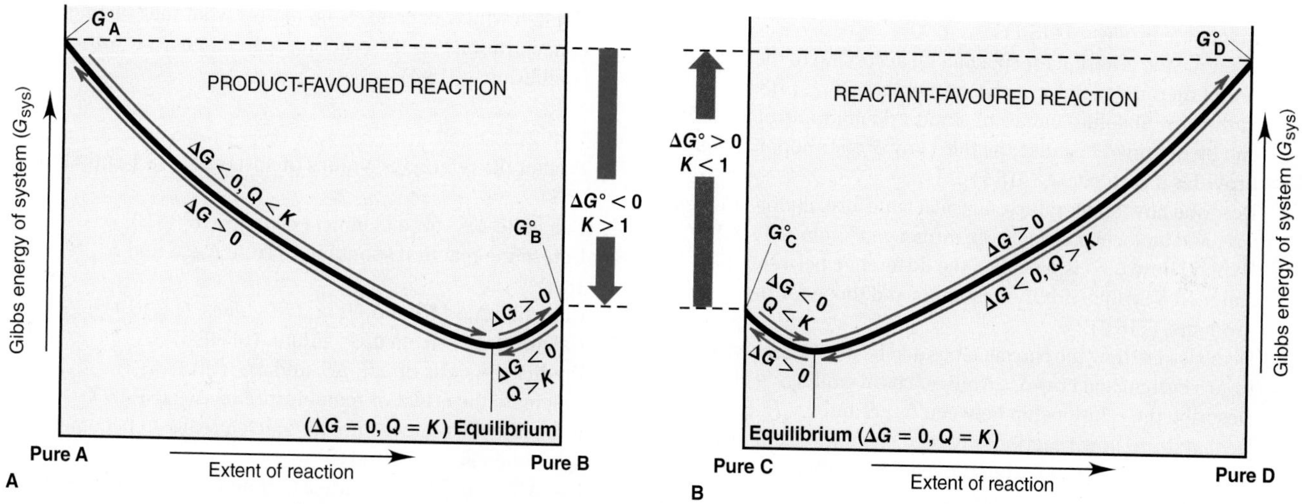

FIGURE 18.16 Gibbs energy and the extent of reaction. Each reaction proceeds spontaneously (*curved green arrows*) from reactants (A or C) or products (B or D) to the equilibrium mixture, at which point $\Delta G = 0$. After that, the reaction is nonspontaneous (*curved red arrows*). **A.** For the product-favoured reaction $A \rightleftharpoons B$, $G^\circ_A > G^\circ_B$, so $\Delta G^\circ < 0$ and $K > 1$. **B.** For the reactant-favoured reaction $C \rightleftharpoons D$, $G^\circ_D > G^\circ_C$, so $\Delta G^\circ > 0$ and $K < 1$.

Thus, *spontaneous* refers to the *portion* of a reaction in which the Gibbs energy decreases—from the starting mixture to the equilibrium mixture. A product-favoured reaction goes predominantly, but *not* completely, toward product, and a reactant-favoured reaction goes relatively little toward product (see Table 18.2).

SUMMARY OF SECTION 18.4

- Reaction spontaneity can be predicted from the sign of ΔG or from the value of $\frac{Q}{K}$. These variables are related to each other by $\Delta G = RT \ln \frac{Q}{K}$. When $Q = K$, $\frac{Q}{K} = 1$ and $\ln \frac{Q}{K} = 0$. Thus, the system is at equilibrium and can release no more Gibbs energy.
- Beginning with Q at the standard state, the Gibbs energy change is $\Delta G°$ and is related to the equilibrium constant: $\Delta G° = -RT \ln K$.
- Any nonequilibrium mixture of reactants and products moves spontaneously ($\Delta G < 0$) toward the equilibrium mixture.
- A product-favoured reaction goes predominantly toward product and, thus, has $K > 1$ and $\Delta G° < 0$. A reactant-favoured reaction has $K < 1$ and $\Delta G° > 0$.

CHAPTER REVIEW GUIDE

Learning Objectives

Relevant section (§) and/or sample problem (SP) numbers appear in parentheses.

Concepts

1. Distinguish between the tendency of a process to occur by itself and how long the process takes to occur. (Introduction)
2. Distinguish between a spontaneous change and a nonspontaneous change. (§18.1)
3. Explain why the first law of thermodynamics and the sign of $\Delta H°$ cannot predict the direction of a spontaneous process. (§18.1)
4. Describe how the entropy (S) of a system is defined by the number of microstates over which its energy is dispersed. (§18.1)
5. Describe how entropy is alternatively defined by the heat absorbed or released at a constant temperature in a reversible process. (§18.1)
6. Describe the criterion for spontaneity according to the second law of thermodynamics: a change increases S_{univ}. (§18.1)
7. Show how absolute values of standard molar entropies ($S°$) can be obtained because the third law of thermodynamics provides a *zero point*. (§18.1)
8. Describe how temperature, physical state, dissolution, atomic size, and molecular complexity influence $S°$ values. (§18.1)
9. Explain how $\Delta_r S°$ is based on the difference between the summed $S°$ values for the reactants and those for the products. (§18.2)
10. Demonstrate how the surroundings add heat to or remove heat from a system and how $\Delta_{surr}S$ influences overall $\Delta_r S°$. (§18.2)
11. Describe the relationship between $\Delta_{surr}S$ and $\Delta_{sys}H$. (§18.2)
12. Demonstrate how reactions proceed spontaneously toward equilibrium ($\Delta_{univ}S > 0$), but proceed no further at equilibrium ($\Delta_{univ}S = 0$). (§18.2)
13. Show how the Gibbs energy change (ΔG) combines a system's entropy and enthalpy changes. (§18.3)
14. Describe how the expression for the Gibbs energy change is derived from the second law. (§18.3)
15. Describe the relationship between ΔG and the maximum work that a system can perform, and explain why this quantity of work is never performed in a real process. (§18.3)
16. Demonstrate how temperature determines spontaneity for reactions in which ΔS and ΔH have the same sign. (§18.3)
17. Explain why the temperature at which a reaction becomes spontaneous occurs when $\Delta G = 0$. (§18.3)
18. Describe how a spontaneous change can be coupled to a nonspontaneous change to make it occur. (§18.3)
19. Demonstrate how ΔG is related to the ratio of Q to K. (§18.4)
20. Describe the meaning of $\Delta G°$ and its relation to K. (§18.4)
21. Describe the relationship of ΔG to $\Delta G°$ and Q. (§18.4)
22. Explain why G decreases, no matter what the starting concentrations, as the reacting system moves toward equilibrium. (§18.4)

Skills

1. Predict the relative $S°$ values of systems. (§18.1 and SP 18.1)
2. Calculate $\Delta_r S°$ for a chemical change. (SP 18.2)
3. Determine reaction spontaneity from $\Delta_{surr}S$ and $\Delta_{sys}H°$. (SP 18.3)
4. Calculate $\Delta_r G°$ from $\Delta_f H°$ and $S°$ values. (SP 18.4)
5. Calculate $\Delta_r G°$ from $\Delta_f G°$ values. (SP 18.5)
6. Predict the signs of ΔH, ΔS, and ΔG. (SP 18.6)
7. Calculate the effect of temperature on ΔG. (SP 18.7)
8. Calculate the temperature at which a reaction becomes spontaneous. (§18.3 and SP 18.8)
9. Use $\Delta G°$ and Q to calculate ΔG at any conditions. (SPs 18.9, 18.10)
10. Calculate K from $\Delta G°$. (§18.4 and SP 18.10)

Key Terms

Section 18.1
spontaneous change
entropy (S)
second law of thermodynamics
third law of thermodynamics
standard molar entropy (S°)

Section 18.2
standard entropy of reaction ($\Delta_r S^\circ$)

Section 18.3
Gibbs energy (G)
standard Gibbs energy change (ΔG°)
standard Gibbs energy of formation ($\Delta_f G^\circ$)
coupling of reactions

Section 18.4
adenosine triphosphate (ATP)

Key Equations and Relationships

18.1 Quantifying entropy in terms of the number of microstates (Ω) over which the energy of a system can be distributed:

$$S = k \ln \Omega$$

18.2 Quantifying the entropy change in terms of heat absorbed (or released) in a reversible process:

$$\Delta_{sys} S = \frac{q_{rev}}{T}$$

18.3 Stating the second law of thermodynamics for a spontaneous process:

$$\Delta_{univ} S = \Delta_{sys} S + \Delta_{surr} S > 0$$

18.4 Calculating the standard entropy of reaction from the standard molar entropies of reactants and products:

$$\Delta_r S^\circ = \Sigma m S^\circ_{products} - \Sigma n S^\circ_{reactants}$$

18.5 Relating the entropy change in the surroundings to the enthalpy change of the system and the temperature:

$$\Delta_{surr} S = -\frac{\Delta_{sys} H}{T}$$

18.6 Expressing the Gibbs energy change of a system in terms of its component enthalpy and entropy changes (Gibbs equation):

$$\Delta_{sys} G = \Delta_{sys} H - T\Delta_{sys} S$$

18.7 Calculating the standard Gibbs energy change from standard enthalpy and entropy changes:

$$\Delta_{sys} G^\circ = \Delta_{sys} H^\circ - T\Delta_{sys} S^\circ$$

18.8 Calculating the standard Gibbs energy change from the standard free energies of formation:

$$\Delta_r G^\circ = \Sigma m \Delta_f G^\circ_{(products)} - \Sigma n \Delta_f S^\circ_{(reactants)}$$

18.9 Relating the Gibbs energy change to the maximum work a process can perform:

$$\Delta G = w_{max}$$

18.10 Finding the temperature at which a reaction becomes spontaneous:

$$T = \frac{\Delta H}{\Delta S}$$

18.11 Expressing the Gibbs energy change in terms of Q and K:

$$\Delta G = RT \ln \frac{Q}{K} = RT \ln Q - RT \ln K$$

18.12 Expressing the Gibbs energy change when Q is at standard-state conditions:

$$\Delta G^\circ = -RT \ln K$$

18.13 Expressing the Gibbs energy change for nonstandard initial conditions:

$$\Delta G = \Delta G^\circ + RT \ln Q$$

Brief Solutions to Follow-Up Problems

18.1 (a) $PCl_5(g)$: It has a higher molar mass and is a more complex molecule.
(b) $BaCl_2(s)$: It has a higher molar mass.
(c) $Br_2(g)$: Gases have more freedom of motion and dispersal of energy than liquids.

18.2 (a) $2NaOH(s) + CO_2(g) \longrightarrow Na_2CO_3(s) + H_2O(l)$
$\Delta n_{gas} = -1$, so $\Delta_r S^\circ < 0$.

$$\begin{aligned}\Delta_r S^\circ &= [1(69.9\ \text{J/[mol·K]}) + 1(139\ \text{J/[mol·K]})] \\ &\quad - [2(64.5\ \text{J/[mol·K]}) + 1(213.7\ \text{J/[mol·K]})] \\ &= -134\ \text{J/(mol·K)}\end{aligned}$$

(b) $2Fe(s) + 3H_2O(g) \longrightarrow Fe_2O_3(s) + 3H_2(g)$
Since $\Delta n_{gas} = 0$, we cannot predict the sign of $\Delta_r S^\circ$.

$$\begin{aligned}\Delta_r S^\circ &= [1(87.4\ \text{J/[mol·K]}) + 3(130.6\ \text{J/[mol·K]})] \\ &\quad - [2(27.3\ \text{J/[mol·K]}) + 3(188.7\ \text{J/[mol·K]})] \\ &= -141.5\ \text{J/(mol·K)}\end{aligned}$$

18.3 $2FeO(s) + \frac{1}{2}O_2(g) \longrightarrow Fe_2O_3(s)$

$$\begin{aligned}\Delta_{sys} S^\circ &= 1[87.4\ \text{J/(mol·K)}] \\ &\quad - [2(60.75\ \text{J/[mol·K]}) + \tfrac{1}{2}(205.0\ \text{J/[mol·K]})] \\ &= -136.6\ \text{J/(mol·K)}\end{aligned}$$

$$\begin{aligned}\Delta_{sys} H^\circ &= 1(-825.5\ \text{kJ/mol}) \\ &\quad - [2(-272.0\ \text{kJ/mol}) + \tfrac{1}{2}(0\ \text{kJ/mol})] \\ &= -281.5\ \text{kJ/mol}\end{aligned}$$

$$\begin{aligned}\Delta_{surr} S &= -\frac{\Delta_{sys} H^\circ}{T} = -\frac{(-281.5\ \text{kJ/mol} \times 1000\ \text{J/kJ})}{298\ \text{K}} \\ &= +945\ \text{J/(mol·K)}\end{aligned}$$

$$\begin{aligned}\Delta_{univ} S &= \Delta_{sys} S^\circ + \Delta_{surr} S \\ &= -136.6\ \text{J/(mol·K)} + 945\ \text{J/(mol·K)} \\ &= 808\ \text{J/(mol·K)}\end{aligned}$$

The oxidation occurs spontaneously at 298 K.

Brief Solutions to Follow-Up Problems (continued)

18.4 Using $\Delta_f H^\circ$ and S° values from Appendix B:

$$\Delta_r H^\circ = -114.2 \text{ kJ/mol}$$
$$\Delta_r S^\circ = -146.5 \text{ J/(mol·K)}$$
$$\begin{aligned}\Delta_r G^\circ &= \Delta_r H^\circ - T\Delta_r S^\circ \\ &= -114.2 \text{ kJ/mol} \\ &\quad - [(298 \text{ K})(-146.5 \text{ J/[mol·K]})(1 \text{ kJ}/1000 \text{ J})] \\ &= -70.5 \text{ kJ/mol}\end{aligned}$$

18.5 (a) $\Delta_r G^\circ = 2(51 \text{ kJ/mol}) - [2(86.60 \text{ kJ/mol}) + 1(0 \text{ kJ/mol})] = -71 \text{ kJ/mol}$

(b) $\Delta_r G^\circ = 2(-137.2 \text{ kJ/mol}) - [2(0 \text{ kJ/mol}) + 1(0 \text{ kJ/mol})] = -274.4 \text{ kJ/mol}$

18.6 (a) A larger amount (mol) of gas is present after the reaction, so $\Delta S > 0$.
(b) The reaction is spontaneous ($\Delta G < 0$) only above 325°C, which implies high T. If $\Delta S > 0$, $-T\Delta S < 0$, so ΔG will become negative at higher T only if $\Delta H > 0$.

18.7 ΔG becomes negative at lower T, so $\Delta H < 0$, $\Delta S < 0$, and $-T\Delta S > 0$. At lower T, the negative ΔH value becomes larger than the positive $-T\Delta S$ value.

18.8

$$\begin{aligned}\Delta H &= \Delta_f H^\circ \text{ CaCO}_3 - (\Delta_f H^\circ \text{ CaO} + \Delta_f H^\circ \text{ CO}_2) \\ &= -1206.9 \text{ kJ/mol} - (-635.1 \text{ kJ/mol} - 393.5 \text{ kJ/mol}) \\ &= -178.3 \text{ kJ/mol}\end{aligned}$$

$$\begin{aligned}\Delta S &= S^\circ \text{ CaCO}_3 - (S^\circ \text{ CaO} + S^\circ \text{ CO}_2) \\ &= 92.9 \text{ J/(mol·K)} - [38.2 \text{ J/(mol·K)} + 213.7 \text{ J/(mol·K)}] \\ &= -159.0 \text{ J/(mol·K)}\end{aligned}$$

$$T = \frac{\Delta H}{\Delta S} = \frac{-178.3 \dfrac{\text{kJ}}{\text{mol}} \times \dfrac{1000 \text{ J}}{1 \text{ kJ}}}{-159.0 \dfrac{\text{J}}{\text{mol·K}}} = 1121 \text{ K}$$

The reaction becomes spontaneous ($\Delta G < 0$) at any $T < 1121$ K.

18.9 (a) Mixture 2 is at equilibrium.
(b) 3 (most negative) < 2 < 1 (most positive)
(c) Any reaction mixture moves spontaneously toward equilibrium, so both changes have a negative ΔG.

18.10 (a)

$$\begin{aligned}\Delta G^\circ = -RT \ln K &= -8.314 \frac{\text{J}}{\text{mol·K}} \times \frac{1 \text{ kJ}}{1000 \text{ J}} \times 298 \text{ K} \\ &\quad \times \ln(2.3 \times 10^{-9}) \\ &= 49.3 \text{ kJ/mol}\end{aligned}$$

(b) $$Q = \frac{[\text{H}_3\text{O}^+][\text{BrO}^-]}{[\text{HBrO}]} = \frac{(6.0 \times 10^{-4})(0.10)}{0.20} = 3.0 \times 10^{-4}$$

$$\begin{aligned}\Delta G &= \Delta G^\circ + RT \ln Q \\ &= 49.3 \text{ kJ/mol} \\ &\quad + \left[8.314 \frac{\text{J}}{\text{mol·K}} \times \frac{1 \text{ kJ}}{1000 \text{ J}} \times 298 \text{ K} \times \ln(3.0 \times 10^{-4})\right] \\ &= 29.2 \text{ kJ/mol}\end{aligned}$$

PROBLEMS

Problems with **red** numbers are answered in Appendix G and worked in detail in the Student Solutions Manual. Problem sections match those in this book and provide the numbers of relevant sample problems. Most offer Concept Review Questions, Skill-Building Exercises (grouped in pairs covering the same concept), and Problems in Context. The Comprehensive Problems are based on material from any section or previous chapter.

Note: Unless stated otherwise, all problems refer to systems at 298 K (25°C). Solving these problems may require values from Appendix B.

The Second Law of Thermodynamics: Predicting Spontaneous Change

(Sample Problem 18.1)

Concept Review Questions

18.1 (a) Distinguish between the terms *spontaneous* (or *thermodynamically allowed*) and *instantaneous*.
(b) Give an example of a process that is spontaneous but very slow, and a process that is very fast but not spontaneous.

18.2 (a) Distinguish between the terms *spontaneous* and *nonspontaneous*.
(b) Can a nonspontaneous process occur? Explain.

18.3 (a) State the first law of thermodynamics in terms of (i) the energy of the universe; (ii) the creation or destruction of energy; (iii) the energy change of a system and its surroundings.
(b) Does the first law reveal the direction of spontaneous change? Explain.

18.4 (a) State qualitatively the relationship between entropy and freedom of particle motion.
(b) Use this relationship to explain why you will probably never (i) be suffocated because all the air near you has moved to the other side of the room; (ii) see half the water in your cup of tea freeze while the other half boils.

18.5 Why is $\Delta_{vap}S$ of a substance always larger than $\Delta_{fus}S$?

18.6 (a) How does the entropy of the surroundings change during (i) an exothermic reaction; (ii) an endothermic reaction?
(b) Other than the examples in this book, describe a spontaneous endothermic reaction.

18.7 (a) What is the entropy of a perfect crystal at 0 K?
(b) Does entropy increase or decrease as the temperature rises?
(c) Why is $\Delta_f H^\circ = 0$ but $S^\circ > 0$ for an element?
(d) Why does Appendix B list $\Delta_f H^\circ$ values but not $\Delta_f S^\circ$ values?

Skill-Building Exercises (grouped in similar pairs)

18.8 Which processes are thermodynamically allowed?
(a) Water evaporates from a puddle. (b) A lion chases an antelope.
(c) An isotope undergoes radioactive disintegration.

18.9 Which processes are thermodynamically allowed?
(a) Earth moves around the Sun. (b) A boulder rolls up a hill.
(c) Sodium metal and chlorine gas form solid sodium chloride.

18.10 Which processes are spontaneous?
(a) Methane burns in air. (b) A teaspoonful of sugar dissolves in a cup of hot coffee. (c) A soft-boiled egg becomes raw.

18.11 Which processes are spontaneous?
(a) A satellite falls to Earth. (b) Water decomposes to H_2 and O_2 at 298 K and 1 bar. (c) Average car prices increase.

18.12 Predict the sign of $\Delta_{sys}S$ for each process:
(a) A piece of wax melts. (b) Silver chloride precipitates from solution. (c) Dew forms on a lawn in the morning.

18.13 Predict the sign of $\Delta_{sys}S$ for each process:
(a) Gasoline vapours mix with air in a car engine. (b) Hot air expands. (c) Humidity condenses in cold air.

18.14 Predict the sign of $\Delta_{sys}S$ for each process:
(a) Alcohol evaporates. (b) A solid explosive converts to a gas.
(c) Perfume vapours diffuse through a room.

18.15 Predict the sign of $\Delta_{sys}S$ for each process:
(a) A pond freezes in the winter. (b) Atmospheric CO_2 dissolves in the ocean. (c) An apple tree bears fruit.

18.16 Without using Appendix B, predict the sign of $\Delta S°$ for each reaction:
(a) $2K(s) + F_2(g) \longrightarrow 2KF(s)$
(b) $NH_3(g) + HBr(g) \longrightarrow NH_4Br(s)$
(c) $NaClO_3(s) \longrightarrow Na^+(aq) + ClO_3^-(aq)$

18.17 Without using Appendix B, predict the sign of $\Delta S°$ for each reaction:
(a) $H_2S(g) + \frac{1}{2}O_2(g) \longrightarrow \frac{1}{8}S_8(s) + H_2O(g)$
(b) $HCl(aq) + NaOH(aq) \longrightarrow NaCl(aq) + H_2O(l)$
(c) $2NO_2(g) \longrightarrow N_2O_4(g)$

18.18 Without using Appendix B, predict the sign of $\Delta S°$ for each reaction:
(a) $CaCO_3(s) + 2HCl(aq) \longrightarrow CaCl_2(aq) + H_2O(l) + CO_2(g)$
(b) $2NO(g) + O_2(g) \longrightarrow 2NO_2(g)$
(c) $2KClO_3(s) \longrightarrow 2KCl(s) + 3O_2(g)$

18.19 Without using Appendix B, predict the sign of $\Delta S°$ for each reaction:
(a) $Ag^+(aq) + Cl^-(aq) \longrightarrow AgCl(s)$
(b) $KBr(s) \longrightarrow KBr(aq)$
(c) $CH_3CH{=}CH_2(g) \longrightarrow$ 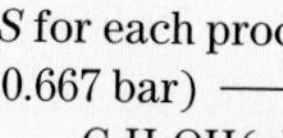

18.20 Predict the sign of ΔS for each process:
(a) $C_2H_5OH(g)$ (350 K and 0.667 bar) $\longrightarrow$ $C_2H_5OH(g)$ (350 K and 0.333 bar)
(b) $N_2(g)$ (298 K and 1 bar) $\longrightarrow$ $N_2(aq)$ (298 K and 1 bar)
(c) $O_2(aq)$ (303 K and 1 bar) $\longrightarrow$ $O_2(g)$ (303 K and 1 bar)

18.21 Predict the sign of ΔS for each process:
(a) $O_2(g)$ (1.0 L at 1 bar) $\longrightarrow$ $O_2(g)$ (0.10 L at 10 bar)
(b) $Cu(s)$ (350°C and 2.5 bar) $\longrightarrow$ $Cu(s)$ (450°C and 2.5 bar)
(c) $Cl_2(g)$ (100°C and 1 bar) $\longrightarrow$ $Cl_2(g)$ (10°C and 1 bar)

18.22 Predict which substance in each pair has greater molar entropy. Explain your prediction.
(a) Butane $[CH_3CH_2CH_2CH_3(g)]$ or 2-butene $[CH_3CH{—}CHCH_3(g)]$
(b) $Ne(g)$ or $Xe(g)$ (c) $CH_4(g)$ or $CCl_4(l)$

18.23 Predict which substance in each pair has greater molar entropy. Explain your prediction.
(a) $NO_2(g)$ or $N_2O_4(g)$ (b) $CH_3OCH_3(l)$ or $CH_3CH_2OH(l)$
(c) $HCl(g)$ or $HBr(g)$

18.24 Predict which substance in each pair has greater molar entropy. Explain your prediction.
(a) $CH_3OH(l)$ or $C_2H_5OH(l)$ (b) $KClO_3(s)$ or $KClO_3(aq)$
(c) $Na(s)$ or $K(s)$

18.25 Predict which substance in each pair has greater molar entropy. Explain your prediction.
(a) $P_4(g)$ or $P_2(g)$ (b) $HNO_3(aq)$ or $HNO_3(l)$
(c) $CuSO_4(s)$ or $CuSO_4{\cdot}5H_2O(s)$

18.26 Without consulting Appendix B, arrange the substances in each group in order of *increasing* standard molar entropy ($S°$). Explain your arrangement.
(a) Graphite, diamond, charcoal
(b) Ice (solid water), water vapour, liquid water
(c) O_2, O_3, O atoms

18.27 Without consulting Appendix B, arrange the substances in each group in order of *increasing* standard molar entropy ($S°$). Explain your arrangement.
(a) Glucose ($C_6H_{12}O_6$), sucrose ($C_{12}H_{22}O_{11}$), ribose ($C_5H_{10}O_5$)
(b) $CaCO_3$, $Ca + C + \frac{3}{2}O_2$, $CaO + CO_2$
(c) $SF_6(g)$, $SF_4(g)$, $S_2F_{10}(g)$

18.28 Without consulting Appendix B, arrange the substances in each group in order of *decreasing* standard molar entropy ($S°$). Explain your arrangement.
(a) $ClO_4^-(aq)$, $ClO_2^-(aq)$, $ClO_3^-(aq)$
(b) $NO_2(g)$, $NO(g)$, $N_2(g)$
(c) $Fe_2O_3(s)$, $Al_2O_3(s)$, $Fe_3O_4(s)$

18.29 Without consulting Appendix B, arrange the substances in each group in order of *decreasing* standard molar entropy ($S°$). Explain your arrangement.
(a) Mg metal, Ca metal, Ba metal
(b) Hexane (C_6H_{14}), benzene (C_6H_6), cyclohexane (C_6H_{12})
(c) $PF_2Cl_3(g)$, $PF_5(g)$, $PF_3(g)$

Calculating the Entropy Change of a Reaction

(Sample Problems 18.2 and 18.3)

Concept Review Questions

18.30 For the reaction depicted in the following molecular scenes, X is *red* and Y is *green*:

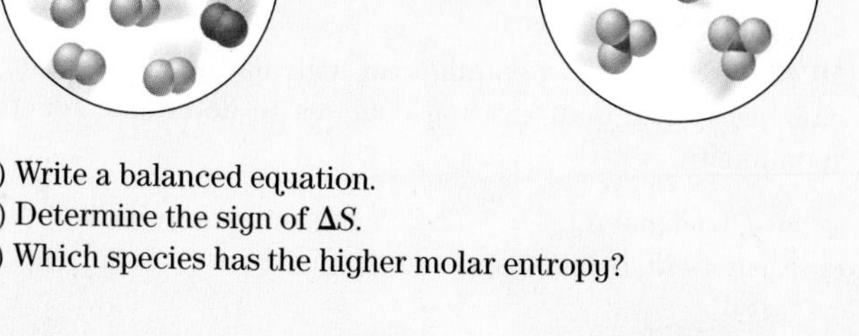

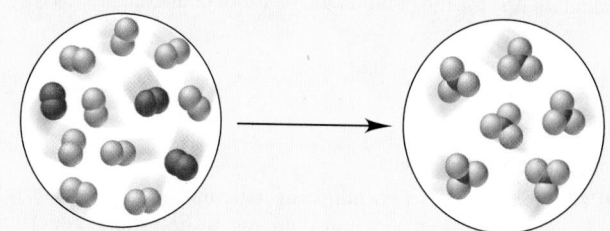

(a) Write a balanced equation.
(b) Determine the sign of ΔS.
(c) Which species has the higher molar entropy?

18.31 Describe the equilibrium condition in terms of the entropy changes of a system and its surroundings. What does this description mean about the entropy change of the universe?

18.32 Suppose that you know $\Delta_r S^\circ$ and S° of $HClO(g)$ and $H_2O(g)$ in the following reaction:

$$H_2O(g) + Cl_2O(g) \longrightarrow 2HClO(g)$$

Write an expression that you could use to determine S° of $Cl_2O(g)$.

Skill-Building Exercises (grouped in similar pairs)

18.33 For each reaction, predict the sign of ΔS° and find its value:
(a) $3NO(g) \longrightarrow N_2O(g) + NO_2(g)$
(b) $3H_2(g) + Fe_2O_3(s) \longrightarrow 2Fe(s) + 3H_2O(g)$
(c) $P_4(s) + 5O_2(g) \longrightarrow P_4O_{10}(s)$

18.34 For each reaction, predict the sign of ΔS° and find its value:
(a) $3NO_2(g) + H_2O(l) \longrightarrow 2HNO_3(l) + NO(g)$
(b) $N_2(g) + 3F_2(g) \longrightarrow 2NF_3(g)$
(c) $C_6H_{12}O_6(s) + 6O_2(g) \longrightarrow 6CO_2(g) + 6H_2O(g)$

18.35 Find ΔS° for the combustion of ethane (C_2H_6) to carbon dioxide and gaseous water. Is the sign of ΔS° what you expected?

18.36 Find ΔS° for the combustion of methane to carbon dioxide and liquid water. Is the sign of ΔS° what you expected?

18.37 Find ΔS° for the reaction of nitrogen monoxide with hydrogen to form ammonia and water vapour. Is the sign of ΔS° what you expected?

18.38 Find ΔS° for the combustion of ammonia to nitrogen dioxide and water vapour. Is the sign of ΔS° what you expected?

18.39 Find ΔS° for the formation of $Cu_2O(s)$ from its elements.

18.40 Find ΔS° for the formation of $HI(g)$ from its elements.

18.41 Find ΔS° for the formation of $CH_3OH(l)$ from its elements.

18.42 Find ΔS° for the formation of $PCl_5(g)$ from its elements.

Problems in Context

18.43 Sulfur dioxide is released in the combustion of coal. Scrubbers use aqueous slurries of calcium hydroxide to remove the SO_2 from the flue gases. Write a balanced equation for this reaction, and calculate ΔS° at 298 K [S° of $CaSO_3(s)$ = 101.4 J/(mol•K)].

18.44 Oxyacetylene welding is used to repair metal structures, including bridges, buildings, and even the Parliament Buildings. Calculate ΔS° for the combustion of 1 mol of acetylene (C_2H_2).

Entropy, Gibbs Energy, and Work
(Sample Problems 18.4 to 18.8)

Concept Review Questions

18.45 What is the advantage of calculating Gibbs energy changes rather than entropy changes to determine reaction spontaneity?

18.46 Given that $\Delta_{sys}G = -T\Delta_{univ}S$, explain how the sign of $\Delta_{sys}G$ correlates with reaction spontaneity.

18.47 (a) Is an endothermic reaction more likely to be spontaneous at higher temperatures or at lower temperatures? Explain.
(b) The change depicted below occurs at constant pressure:

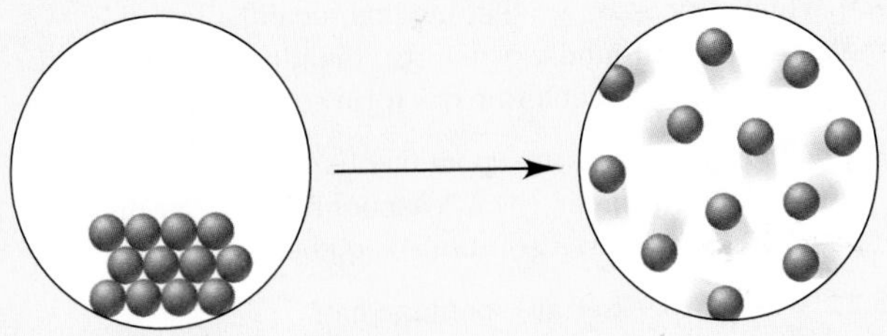

Answer the following questions, and explain your answers:
(i) What is the sign of ΔH?
(ii) What is the sign of ΔS?
(iii) What is the sign of $\Delta_{surr}S$?
(iv) How does the sign of ΔG vary with temperature?

18.48 With its components in their standard states, a certain reaction is spontaneous only at high temperatures. What do you know about the signs of ΔH° and ΔS°? Describe a process for which this is true.

18.49 How can ΔS° be relatively independent of temperature if S° of each reactant and product increases with temperature?

Skill-Building Exercises (grouped in similar pairs)

18.50 Calculate ΔG° for each reaction using $\Delta_f G^\circ$ values:
(a) $2Mg(s) + O_2(g) \longrightarrow 2MgO(s)$
(b) $2CH_3OH(g) + 3O_2(g) \longrightarrow 2CO_2(g) + 4H_2O(g)$
(c) $BaO(s) + CO_2(g) \longrightarrow BaCO_3(s)$

18.51 Calculate ΔG° for each reaction using $\Delta_f G^\circ$ values:
(a) $H_2(g) + I_2(s) \longrightarrow 2HI(g)$
(b) $MnO_2(s) + 2CO(g) \longrightarrow Mn(s) + 2CO_2(g)$
(c) $NH_4Cl(s) \longrightarrow NH_3(g) + HCl(g)$

18.52 Find ΔG° for the reactions in Problem 18.50 using $\Delta_f H^\circ$ and S° values.

18.53 Find ΔG° for the reactions in Problem 18.51 using $\Delta_f H^\circ$ and S° values.

18.54 Consider the oxidation of carbon monoxide:

$$CO(g) + \frac{1}{2}O_2(g) \longrightarrow CO_2(g)$$

(a) Predict the signs of ΔS° and ΔH°. Explain.
(b) Calculate ΔG° by two different methods.

18.55 Consider the combustion of butane gas:

$$C_4H_{10}(g) + \frac{13}{2}O_2(g) \longrightarrow 4CO_2(g) + 5H_2O(g)$$

(a) Predict the signs of ΔS° and ΔH°. Explain.
(b) Calculate ΔG° by two different methods.

18.56 Xenon hexafluoride is formed in the gaseous reaction of xenon and fluorine.
(a) Calculate ΔS° at 298 K (ΔH° = −402 kJ/mol; ΔG° = −280. kJ/mol).
(b) Assuming that ΔS° and ΔH° change little with temperature, calculate ΔG° at 500. K.

18.57 Phosgene ($COCl_2$) is formed in the gaseous reaction of carbon monoxide and chlorine.
(a) Calculate $\Delta S°$ at 298 K ($\Delta H° = -220.$ kJ/mol and $\Delta G° = -206$ kJ/mol).
(b) Assuming that $\Delta S°$ and $\Delta H°$ change little with temperature, calculate $\Delta G°$ at 450. K.

18.58 One reaction that is used to produce small quantities of pure H_2 is given below:

$$CH_3OH(g) \rightleftharpoons CO(g) + 2H_2(g)$$

(a) Determine $\Delta H°$ and $\Delta S°$ for the reaction at 298 K.
(b) Assuming that these values are relatively independent of temperature, calculate $\Delta G°$ at 28°C, 128°C, and 228°C.
(c) What is the significance of the different values of $\Delta G°$?

18.59 A reaction that occurs in an internal combustion engine is given below:

$$N_2(g) + O_2(g) \rightleftharpoons NO(g) \text{ (unbalanced)}$$

(a) Determine $\Delta H°$ and $\Delta S°$ for the reaction at 298 K.
(b) Assuming that these values are relatively independent of temperature, calculate $\Delta G°$ at 100.°C, 2560.°C, and 3540.°C.
(c) What is the significance of the different values of $\Delta G°$?

18.60 Use $\Delta H°$ and $\Delta S°$ values for the following process at 1 bar to find the normal boiling point of Br_2:

$$Br_2(l) \rightleftharpoons Br_2(g)$$

18.61 Use $\Delta H°$ and $\Delta S°$ values to find the temperature at which these sulfur allotropes reach equilibrium at 1 bar:

$$\text{S(rhombic)} \rightleftharpoons \text{S(monoclinic)}$$

Problems in Context

18.62 As a fuel, $H_2(g)$ produces only nonpolluting $H_2O(g)$ when it burns. Moreover, it combines with $O_2(g)$ in a fuel cell (Chapter 19) to provide electrical energy. (a) Calculate $\Delta H°$, $\Delta S°$, and $\Delta G°$ per mole of H_2 at 298 K. (b) Does the spontaneity of this reaction depend on temperature? Explain. (c) At what temperature does the reaction become spontaneous?

18.63 The Canadian government requires automobile fuels to contain a renewable component. Fermentation of glucose from corn yields ethanol, which, when added to gasoline, fulfills this requirement:

$$C_6H_{12}O_6(s) \longrightarrow 2C_2H_5OH(l) + 2CO_2(g)$$

(a) Calculate $\Delta H°$, $\Delta S°$, and $\Delta G°$ for the reaction at 25°C.
(b) Does the spontaneity of the reaction depend on temperature? Explain.

Gibbs Energy, Equilibrium, and Reaction Direction

(Sample Problems 18.9 and 18.10)

Concept Review Questions

18.64 (a) If $K \ll 1$ for a reaction, what do you know about the sign and magnitude of $\Delta G°$?
(b) If $\Delta G° \ll 0$ for a reaction, what do you know about the magnitude of K and the magnitude of Q?

18.65 How is the Gibbs energy change of a process related to the work that can be obtained from the process? Is this quantity of work obtainable in practice? Explain.

18.66 The following scenes and graph relate to the reaction of $X_2(g)$ (*black*) with $Y_2(g)$ (*orange*) to form $XY(g)$. (a) If the reactants and products are in their standard states, what quantity is represented by x on the graph? (b) Which scene represents point 1 on the graph? Explain. (c) Which scene represents point 2 on the graph? Explain.

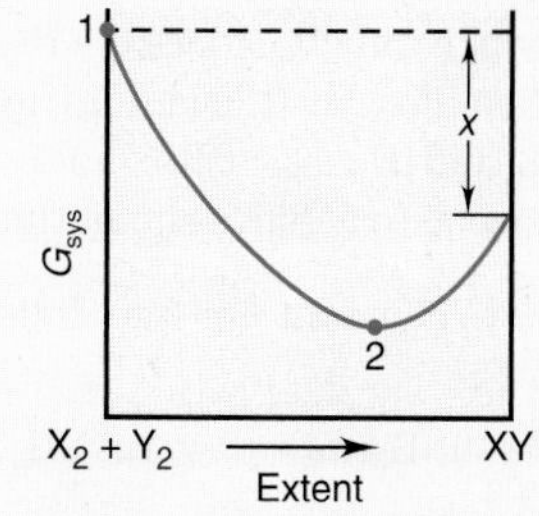

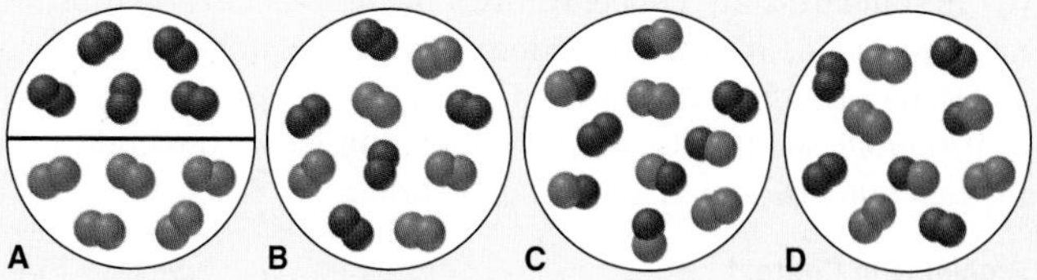

18.67 What is the difference between $\Delta G°$ and ΔG? Under what circumstances does $\Delta G = \Delta G°$?

Skill-Building Exercises (grouped in similar pairs)

18.68 Calculate K at 298 K for each reaction:
(a) $NO(g) + \frac{1}{2}O_2(g) \rightleftharpoons NO_2(g)$
(b) $2HCl(g) \rightleftharpoons H_2(g) + Cl_2(g)$
(c) $2C(\text{graphite}) + O_2(g) \rightleftharpoons 2CO(g)$

18.69 Calculate K at 298 K for each reaction:
(a) $MgCO_3(s) \rightleftharpoons Mg^{2+}(aq) + CO_3^{2-}(aq)$
(b) $2HCl(g) + Br_2(l) \rightleftharpoons 2HBr(g) + Cl_2(g)$
(c) $H_2(g) + O_2(g) \rightleftharpoons H_2O_2(l)$

18.70 Calculate K at 298 K for each reaction:
(a) $2H_2S(g) + 3O_2(g) \rightleftharpoons 2H_2O(g) + 2SO_2(g)$
(b) $H_2SO_4(l) \rightleftharpoons H_2O(l) + SO_3(g)$
(c) $HCN(aq) + NaOH(aq) \rightleftharpoons NaCN(aq) + H_2O(l)$

18.71 Calculate K at 298 K for each reaction:
(a) $SrSO_4(s) \rightleftharpoons Sr^{2+}(aq) + SO_4^{2-}(aq)$
(b) $2NO(g) + Cl_2(g) \rightleftharpoons 2NOCl(g)$
(c) $Cu_2S(s) + O_2(g) \rightleftharpoons 2Cu(s) + SO_2(g)$

18.72 Use Appendix B to determine the K_{sp} of Ag_2S.

18.73 Use Appendix B to determine the K_{sp} of CaF_2.

18.74 For the following reaction, calculate K at 25°C [$\Delta_f G°$ of $ICl(g) = -6.075$ kJ/mol]:

$$I_2(g) + Cl_2(g) \rightleftharpoons 2ICl(g)$$

18.75 For the following reaction, calculate the equilibrium p_{CO_2} at 25°C:

$$CaCO_3(s) \rightleftharpoons CaO(s) + CO_2(g)$$

18.76 The K_{sp} of $PbCl_2$ is 1.7×10^{-5} at 25°C. What is $\Delta G°$? Is it possible to prepare a solution that contains $Pb^{2+}(aq)$ and $Cl^-(aq)$ at their standard-state concentrations?

18.77 The K_{sp} of ZnF_2 is 3.0×10^{-2} at 25°C. What is $\Delta G°$? Is it possible to prepare a solution that contains $Zn^{2+}(aq)$ and $F^-(aq)$ at their standard-state concentrations?

18.78 The equilibrium constant for the following reaction is $K = 9.1 \times 10^6$ at 298 K:

$$2Fe^{3+}(aq) + Hg_2^{2+}(aq) \rightleftharpoons 2Fe^{2+}(aq) + 2Hg^{2+}(aq)$$

(a) What is $\Delta G°$ at this temperature?
(b) If standard-state concentrations of the reactants and products are mixed, in which direction does the reaction proceed?
(c) Calculate ΔG when $[Fe^{3+}] = 0.20$ mol/L, $[Hg_2^{2+}] = 0.010$ mol/L, $[Fe^{2+}] = 0.010$ mol/L, and $[Hg^{2+}] = 0.025$ mol/L. In which direction will the reaction proceed to achieve equilibrium?

18.79 The formation constant for the following reaction is $K_f = 5.6 \times 10^8$ at 25°C:

$$Ni^{2+}(aq) + 6NH_3(aq) \rightleftharpoons Ni(NH_3)_6^{2+}(aq)$$

(a) What is $\Delta G°$ at this temperature?
(b) If standard-state concentrations of the reactants and products are mixed, in which direction does the reaction proceed?
(c) Determine ΔG when $[Ni(NH_3)_6^{2+}] = 0.010$ mol/L, $[Ni^{2+}] = 0.0010$ mol/L, and $[NH_3] = 0.0050$ mol/L. In which direction will the reaction proceed to achieve equilibrium?

Problems in Context

18.80 High levels of ozone (O_3) cause rubber to deteriorate, green plants to turn brown, and many people to have difficulty breathing.
(a) Is the formation of O_3 from O_2 favoured at all T, no T, high T, or low T?
(b) Calculate $\Delta G°$ for this reaction at 298 K.
(c) Calculate ΔG at 298 K for this reaction in urban smog where $[O_2] = 0.21$ bar and $[O_3] = 5 \times 10^{-7}$ bar.

18.81 A $BaSO_4$ slurry is ingested before the gastrointestinal tract is X-rayed because it is opaque to X-rays and defines the contours of the tract:

$$BaSO_4(s) \rightleftharpoons Ba^{2+}(aq) + SO_4^{2-}(aq)$$

Ba^{2+} ion is toxic, but the compound is nearly insoluble. If $\Delta G°$ at 37°C (body temperature) is 59.1 kJ/mol for the process, what is $[Ba^{2+}]$ in the intestinal tract? (Assume that the only source of SO_4^{2-} is the ingested slurry.)

Comprehensive Problems

18.82 According to advertisements, "a diamond is forever."
(a) Calculate $\Delta H°$, $\Delta S°$, and $\Delta G°$ at 298 K for the following phase change:

$$\text{diamond} \longrightarrow \text{graphite}$$

(b) Given the conditions under which diamond jewellery is normally kept, argue for and against the phrase used in advertisements.
(c) Given your answer to part (a), what would need to be done to make synthetic diamonds from graphite?
(d) Assuming $\Delta H°$ and $\Delta S°$ do not change with temperature, can graphite be converted to diamond spontaneously at 1 bar?

18.83 Replace each question mark with the correct information:

	$\Delta_r S$	$\Delta_r H$	$\Delta_r G$	Comment
(a)	+	−	−	?
(b)	?	0	−	Spontaneous
(c)	−	+	?	Nonspontaneous
(d)	0	?	−	Spontaneous
(e)	?	0	+	?
(f)	+	+	?	$T\Delta S > \Delta H$

18.84 Among the many complex ions of cobalt are the following, where "en" stands for ethylenediamine, $H_2NCH_2CH_2NH_2$:

$$Co(NH_3)_6^{3+}(aq) + 3en(aq) \rightleftharpoons Co(en)_3^{3+}(aq) + 6NH_3(aq)$$

Six Co—N bonds are broken and six Co—N bonds are formed in this reaction, so $\Delta_r H° \approx 0$, yet $K > 1$. What are the signs of $\Delta S°$ and $\Delta G°$? What drives the reaction?

18.85 What is the change in entropy when 0.200 mol of potassium freezes at 63.7°C ($\Delta_{fus}H = 2.39$ kJ/mol)?

18.86 Is each statement true or false? If a statement is false, correct it.
(a) All spontaneous reactions occur quickly.
(b) The reverse of a spontaneous reaction is nonspontaneous.
(c) All spontaneous processes release heat.
(d) The boiling of water at 100°C and 1 bar is spontaneous.
(e) If a process increases the freedom of motion of the particles of a system, the entropy of the system decreases.
(f) The energy of the universe is constant; the entropy of the universe decreases toward a minimum.
(g) All systems disperse their energy spontaneously.
(h) Both $\Delta_{sys}S$ and $\Delta_{surr}S$ equal zero at equilibrium.

18.87 Hemoglobin carries O_2 from the lungs to tissue cells, where the O_2 is released. The protein is represented as Hb in its unoxygenated form and as $Hb{\cdot}O_2$ in its oxygenated form. One reason CO is toxic is that it competes with O_2 in binding to Hb:

$$Hb{\cdot}O_2(aq) + CO(g) \rightleftharpoons Hb{\cdot}CO(aq) + O_2(g)$$

(a) If $\Delta G° \approx -14$ kJ/mol at 37°C (body temperature), what is the ratio of $[Hb{\cdot}CO]$ to $[Hb{\cdot}O_2]$ at 37°C with $[O_2] = [CO]$?
(b) How is Le Châtelier's principle used to treat CO poisoning?

18.88 Magnesia (MgO) is used for fire brick, crucibles, and furnace linings because of its high melting point. It is produced by decomposing magnesite ($MgCO_3$) at around 1200°C.
(a) Write a balanced equation for magnesite decomposition.
(b) Use $\Delta H°$ and $S°$ values to find $\Delta G°$ at 298 K.
(c) Assuming that $\Delta H°$ and $S°$ do not change with temperature, find the minimum temperature at which the reaction is spontaneous.
(d) Calculate the equilibrium p_{CO_2} above $MgCO_3$ at 298 K.
(e) Calculate the equilibrium p_{CO_2} above $MgCO_3$ at 1200 K.

18.89 To prepare nuclear fuel, U_3O_8 ("yellow cake") is converted to $UO_2(NO_3)_2$, which is then converted to UO_3 and finally to UO_2. The fuel is enriched (the proportion of the ^{235}U is increased) by a two-step conversion of UO_2 into UF_6, a volatile solid, followed by a gaseous-diffusion separation of the ^{235}U and ^{238}U isotopes:

$$(1)\ UO_2(s) + 4HF(g) \longrightarrow UF_4(s) + 2H_2O(g)$$
$$(2)\ UF_4(s) + F_2(g) \longrightarrow UF_6(s)$$

Calculate $\Delta G°$ for the overall process at 85°C, given the following data:

Compound	$\Delta_f H°$ (kJ/mol)	$S°$ [J/(mol·K)]	$\Delta_f G°$ (kJ/mol)
$UO_2(s)$	−1085	77.0	−1032
$UF_4(s)$	−1921	152	−1830.
$UF_6(s)$	−2197	225	−2068

18.90 Methanol, a major industrial feedstock, is made by several catalyzed reactions, such as the equation below:

$$CO(g) + 2H_2(g) \longrightarrow CH_3OH(l)$$

(a) Show that this reaction is thermodynamically feasible.
(b) Is it favoured at a low temperature or a high temperature?
(c) One concern about using CH_3OH as an automobile fuel is its oxidation in air to yield formaldehyde, $CH_2O(g)$, which poses a health hazard. Calculate $\Delta G°$ at 100.°C for this oxidation.

18.91 (a) Write a balanced equation for the gaseous reaction between N_2O_5 and F_2 to form NF_3 and O_2.
(b) Determine $\Delta_r G°$.
(c) Find $\Delta_r G$ at 298 K if $p_{N_2O_5} = p_{F_2} = 0.20$ bar, $p_{NF_3} = 0.25$ bar, and $p_{O_2} = 0.50$ bar.

18.92 Consider the following reaction:

$$2NOBr(g) \rightleftharpoons 2NO(g) + Br_2(g) \qquad K = 0.42 \text{ at } 373 \text{ K}$$

$S°$ of NOBr(g) is 272.6 J/(mol•K), and $\Delta_r S°$ and $\Delta_r H°$ are constant with temperature. Find (a) $\Delta_r S°$ at 298 K; (b) $\Delta_r G°$ at 373 K; (c) $\Delta_r H°$ at 373 K; (d) $\Delta_f H°$ of NOBr at 298 K; (e) $\Delta_r G°$ at 298 K; (f) $\Delta_f G°$ of NOBr at 298 K.

18.93 Hydrogenation is the addition of H_2 to double (or triple) carbon-carbon bonds. Peanut butter (except "natural" kinds) and most commercial baked goods include hydrogenated oils. Find $\Delta H°$, $\Delta S°$, and $\Delta G°$ for the hydrogenation of ethene (C_2H_4) to ethane (C_2H_6) at 25°C.

18.94 Styrene is produced by the catalytic dehydrogenation of ethylbenzene at a high temperature in the presence of superheated steam.
(a) Find $\Delta_r H°$, $\Delta_r G°$, and $\Delta_r S°$ given the following data at 298 K:

Compound	$\Delta_f H°$ (kJ/mol)	$\Delta_f G°$ (kJ/mol)	$S°$ [J/(mol·K)]
Ethylbenzene, $C_6H_5—CH_2CH_3$	−12.5	119.7	255
Styrene, $C_6H_5—CH{=}CH_2$	103.8	202.5	238

(b) At what temperature is the reaction spontaneous?
(c) What are $\Delta_r G°$ and K at 600.°C?
(d) With 5.0 parts steam to 1.0 part ethylbenzene in the reactant mixture and the total pressure kept constant at 1.3 bar, what is ΔG at 50.% conversion (when 50.% of the ethylbenzene has reacted)?

18.95 Propene (propylene; $CH_3CH{=}CH_2$) is used to produce polypropylene and many other chemicals. Although most propene is obtained from the cracking of petroleum, about 2% is produced by the catalytic dehydrogenation of propane ($CH_3CH_2CH_3$):

$$CH_3CH_2CH_3 \xrightarrow{Pt/Al_2O_3} CH_3CH{=}CH_2 + H_2$$

Because this reaction is endothermic, heaters are placed between the reactor vessels to maintain the required temperature.
(a) If the molar entropy, $S°$, of propene is 267.1 J/(mol•K), find its entropy of formation, $S_f°$.
(b) Find $\Delta_f G°$ of propylene ($\Delta_f H°$ of propene = 20.4 kJ/mol).
(c) Calculate $\Delta_r H°$ and $\Delta_r G°$ for the dehydrogenation.
(d) What is the theoretical yield of propene at 580°C if the initial pressure of propane is 1.00 bar?
(e) Would the yield change if the reactor walls were permeable to H_2? Explain.
(f) At what temperature is the dehydrogenation spontaneous, with all substances in the standard state?

Note: Problems 18.96 and 18.97 relate to the thermodynamics of adenosine triphosphate (ATP). Refer to Chemical Connections in Section 18.4.

18.96 (a) Find K for (i) the hydrolysis of ATP; (ii) the dehydration-condensation to form glucose phosphate; (iii) the coupled reaction between ATP and glucose.
(b) How does each K change when the temperature changes from 25°C to 37°C?

18.97 Energy from ATP hydrolysis drives many nonspontaneous cell reactions:

$$ATP^{4-}(aq) + H_2O(l) \rightleftharpoons ADP^{3-}(aq) + HPO_4^{\;2-}(aq) + H^+(aq) \qquad \Delta G°' = -30.5 \text{ kJ/mol}$$

Energy for the reverse process comes ultimately from glucose metabolism:

$$C_6H_{12}O_6(s) + 6O_2(g) \longrightarrow 6CO_2(g) + 6H_2O(l)$$

(a) Find K for the hydrolysis of ATP at 37°C.
(b) Find $\Delta_r G°'$ for the metabolism of 1 mol of glucose.
(c) What amount (mol) of ATP can be produced by the metabolism of 1 mol of glucose?
(d) If 36 mol of ATP is formed, what is the actual yield?

18.98 From the following reaction and data, find (a) $S°$ of $SOCl_2$; (b) the temperature at which the reaction becomes nonspontaneous:

$$SO_3(g) + SCl_2(l) \longrightarrow SOCl_2(l) + SO_2(g) \qquad \Delta_r G° = -75.2 \text{ kJ/mol}$$

	$SO_3(g)$	$SCl_2(l)$	$SOCl_2(l)$	$SO_2(g)$
$\Delta_f H°$ (kJ/mol)	−396	−50.0	−245.6	−296.8
$S°$ [J/(mol·K)]	256.7	184	—	248.1

18.99 Write equations for the oxidation of Fe and the oxidation of Al. Use $\Delta_f G°$ to determine whether either process is spontaneous at 25°C.

18.100 The following molecular scene depicts a gaseous equilibrium mixture for the reaction of H_2 (*blue*) and I_2 (*purple*) to form HI at 460°C. Each molecule represents 0.010 mol, and the volume of the container is 1.0 L. (a) Is K_c greater than, equal to, or less than 1? (b) Is K greater than, equal to, or less than K_c? (c) Calculate $\Delta_r G°$. (d) How would the value of $\Delta_r G°$ change if the purple molecules represented H_2 and the blue molecules represented I_2? Explain.

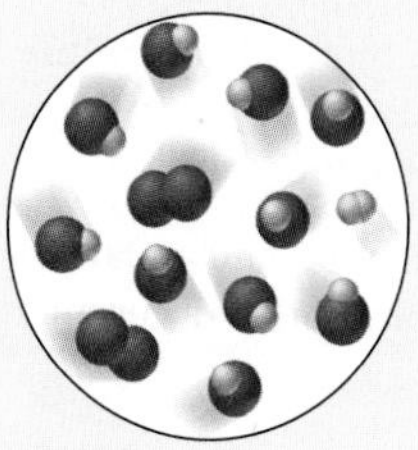

18.101 A key step in the metabolism of glucose for energy is the isomerization of glucose-6-phosphate (G6P) to fructose-6-phosphate (F6P):

$$\text{G6P} \rightleftharpoons \text{F6P} \quad K = 0.510 \text{ at } 298 \text{ K}$$

(a) Calculate $\Delta G°$ at 298 K.
(b) Calculate ΔG when Q, the [F6P]/[G6P] ratio, equals 10.0.
(c) Calculate ΔG when $Q = 0.100$.
(d) Calculate Q if $\Delta G = -2.50$ kJ/mol.

18.102 A chemical reaction, such as HI forming from its elements, can reach equilibrium at many temperatures. In contrast, a phase change, such as ice melting, is in equilibrium at a given pressure only at the melting point. Each graph depicts G_{sys} versus the extent of change.

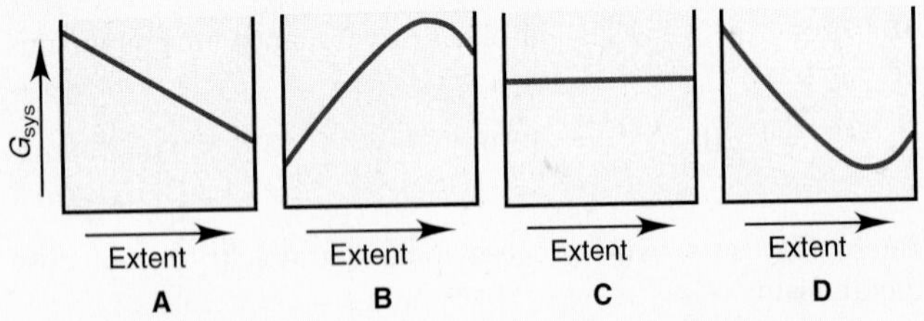

(a) Which graph depicts how G_{sys} changes for the formation of HI? Explain. (b) Which graph depicts how G_{sys} changes as ice melts at 1°C and 1 bar? Explain.

18.103 When heated, the DNA double helix separates into two random-coil single strands. When cooled, the random coils re-form the double helix:

$$\text{double helix} \rightleftharpoons 2 \text{ random coils}$$

(a) What is the sign of ΔS for the forward process? Why?
(b) Energy must be added to overcome hydrogen bonds and dispersion forces between the strands. What is the sign of ΔG for the forward process when $T\Delta S$ is smaller than ΔH?
(c) Write an expression that shows T in terms of ΔH and ΔS when the reaction is at equilibrium. (This temperature is called the *melting temperature* of the nucleic acid.)

18.104 In the process of respiration, glucose is oxidized completely. In fermentation, O_2 is absent and glucose is broken down to ethanol and CO_2. Ethanol is oxidized to CO_2 and H_2O.
(a) Balance each equation for these processes:

Respiration: $C_6H_{12}O_6(s) + O_2(g) \longrightarrow CO_2(g) + H_2O(l)$
Fermentation: $C_6H_{12}O_6(s) \longrightarrow C_2H_5OH(l) + CO_2(g)$
Ethanol oxidation: $C_2H_5OH(l) + O_2(g) \longrightarrow CO_2(g) + H_2O(l)$

(b) Calculate $\Delta_r G°$ for the respiration of 1.00 g of glucose.
(c) Calculate $\Delta_r G°$ for the fermentation of 1.00 g of glucose.
(d) Calculate $\Delta_r G°$ for the oxidation of the ethanol from part (c).

18.105 Consider the formation of ammonia:

$$N_2(g) + 3H_2(g) \rightleftharpoons 2NH_3(g)$$

(a) Assuming that $\Delta H°$ and $\Delta S°$ are constant with temperature, find the temperature at which $K = 1.00$.
(b) Find K at 400.°C, a typical temperature for NH_3 production.
(c) Given the lower K at the higher temperature, why are these conditions used industrially?

18.106 Kyanite, sillimanite, and andalusite all have the formula Al_2SiO_5. Each is stable under different conditions (*see graph below*). At the point where the three phases intersect, which mineral, if any, has (a) the lowest Gibbs energy; (b) the lowest enthalpy; (c) the highest entropy; (d) the lowest density?

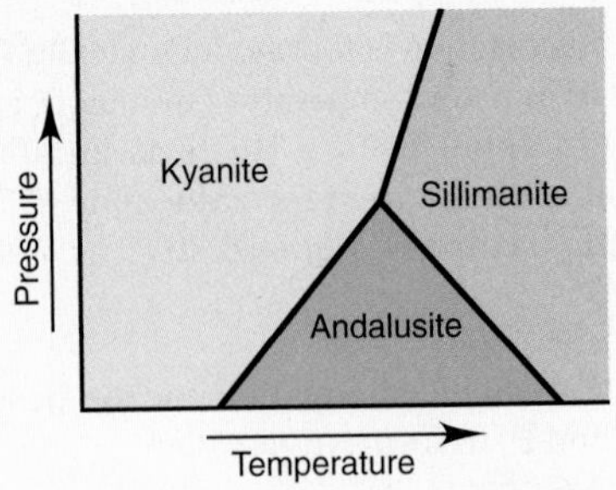

18.107 Ethyne (acetylene) is produced commercially by the partial oxidation of methane:

$$2CH_4 + \frac{1}{2}O_2 \longrightarrow C_2H_2 + 2H_2 + H_2O$$

At 1500°C and pressures of 1 bar to 10 bar, the yield of ethyne is about 20%. The major side product is carbon monoxide, and some soot and carbon dioxide also form.
(a) At what temperature is the reaction spontaneous?
(b) Ethyne can also be made by the reaction of its elements, carbon (graphite) and hydrogen. At what temperature is this formation reaction spontaneous?
(c) Why must the reaction mixture described in part (b) be immediately cooled?

18.108 Synthesis gas, a mixture that includes the fuels CO and H_2, is used to produce liquid hydrocarbons and methanol. It is made at pressures up to 100 bar by the oxidation of methane followed by the steam re-forming and water-gas shift reactions. Because the process is exothermic, temperatures reach 950°C to 1100°C, and the conditions are such that the amounts of H_2, CO, CO_2, CH_4, and H_2O leaving the reactor are close to the equilibrium amounts for the steam re-forming and water-gas shift reactions:

$$CH_4(g) + H_2O(g) \rightleftharpoons CO(g) + 3H_2(g) \quad \text{(steam re-forming)}$$
$$CO(g) + H_2O(g) \rightleftharpoons CO_2(g) + H_2(g) \quad \text{(water-gas shift)}$$

(a) At 1000.°C, what are $\Delta G°$ and $\Delta H°$ for (i) the steam re-forming reaction; (ii) the water-gas shift reaction?
(b) By doubling the steam re-forming reaction and adding it to the water-gas shift reaction, we obtain the following combined reaction:

$$2CH_4(g) + 3H_2O(g) \rightleftharpoons CO_2(g) + CO(g) + 7H_2(g)$$

(i) Is this reaction spontaneous at 1000.°C in the standard state?
(ii) Is it spontaneous at 98 bar and 50.% conversion (when 50.% of the starting materials have reacted)?
(iii) Is it spontaneous at 98 bar and 90.% conversion?

18.109 One industrial reaction of economic importance is the manufacture of sulfur trioxide gas by the equilibrium reaction of sulfur dioxide gas with oxygen gas.
(a) Write the balanced reaction for this process.
(b) Use the thermodynamic data in Appendix B to calculate the values of $\Delta_r H°$ and $\Delta_r S°$ for the reaction.

(c) If the reaction is carried out at 700. K, what is the value of $\Delta_r G°$?
(d) Calculate the equilibrium constant at this temperature for this reaction.
(e) Describe one way to maximize the yield of the product, and explain why this method would work.
(f) Calculate the value of $\Delta_r G°$ at 1000. K.

18.110 One potential side reaction in the production of phosphoric acid from fluorapatite is the equilibrium reaction between the concentrated liquid sulfuric acid and calcium carbonate in the rock to form calcium sulfate, liquid water, and carbon dioxide.
(a) Write the balanced reaction for this process.
(b) Use the thermodynamic data in Appendix B to calculate the values of $\Delta_r H°$ and $\Delta_r S°$ for the reaction.
(c) If the reaction is carried out at 350. K, what is the value of $\Delta_r G°$?
(d) At this temperature, calculate the value of the equilibrium constant.
(e) This is an undesired side reaction. What conditions would minimize the formation of products in this process? Justify your answer.
(f) At what temperature will the pressure of CO_2 be lower by a factor of 1×10^5? (*Hint*: Remember that $\Delta_r G°$ is also affected by a change in *T*.)

Electrochemistry: Chemical Change and Electrical Work

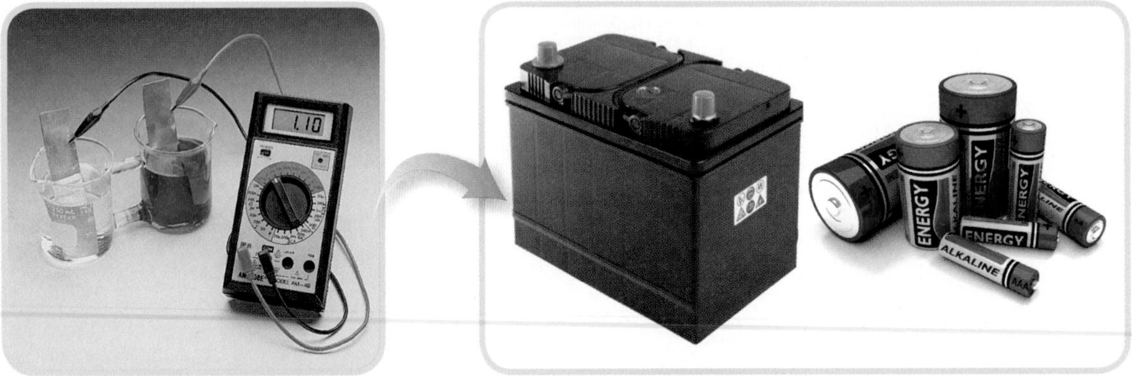

IN THIS CHAPTER . . . We examine the essential features of electrochemical cells and quantify the relationship between Gibbs energy and electrical work. By the end of this chapter, you should be able to

- Discuss oxidation-reduction reactions and describe methods for balancing redox equations that take place in electrochemical cells
- Differentiate between the two types of electrochemical cells
- Describe voltaic cells, which use chemical energy to do electrical work, and explain how they operate as a combination of two half-cells
- Calculate the electric potential of a cell, using the reduction potentials of the two half-reactions
- Determine the relative strengths of redox couples using half-cell potentials, and write spontaneous redox equations
- Describe and explain the relationship between the standard Gibbs energy change, the standard cell potential, and the equilibrium constant
- Discuss how differences in concentration of the half-cells can be harnessed to generate electricity
- Analyze the makeup and operation of the major types of batteries
- Analyze the basis for the problem of corrosion, which has key similarities to the operation of voltaic cells
- Describe electrolytic cells and how they absorb energy from an external source to do electrical work, predict the products that form under different conditions, and quantify the relationship between charge and amount of substance
- Discuss and describe the redox system that generates energy in living cells

19

With our increased understanding of how the combustion of fossil fuels affects our environment, there is a growing drive to find cleaner sources of energy. A huge amount of money is being invested in alternative ways to produce and store electrical energy from the wind (windmills), from the Sun (solar panels), and from chemicals (batteries and fuel cells). The energy generated by windmills and solar panels is either fed directly back into an energy grid or stored in a battery or fuel cell. What, then, is a battery, and how does it work? To understand the answer to this question, it is essential to understand how the movement of electrons is related to energy. In this chapter, we will describe the different ways electrons can move in a chemical reaction and how differences in energy can create a potential energy difference. We will see how electrochemical cells can be designed and constructed and how we can calculate the energy of a cell. We will see how some cells produce energy spontaneously and how other cells must be driven to produce energy. With this background, we can then appreciate what a battery is and how it works. Finally, we will look at the system that generates energy in living cells to give us a better understanding of how our bodies use food to provide us with the energy we need.

Concepts and Skills to Review before Studying This Chapter

- Trends in ionization energy (Section 7.3) and electronegativity (Section 8.5)
- Gibbs energy, work, and equilibrium (Sections 18.3 and 18.4)
- *Q* versus *K* (Section 15.4) and ΔG versus $\Delta G°$ (Section 18.4)

19.1 Oxidation-Reduction (Redox) Reactions

Oxidation-reduction (redox) reactions include the formation of a compound from its elements (and the reverse process), all combustion processes, the generation of electricity in batteries, the production of cellular energy, and many other reactions. In fact, redox reactions are so widespread that many do not occur in aqueous solution. In this section, we examine the key event in a redox reaction, discuss important terminology, see one way to balance redox equations, and learn how to quantify them.

The Key Event: Movement of Electrons between Reactants

The key chemical event in an **oxidation-reduction reaction (redox reaction)** is the *net movement of electrons from one reactant to another*. This movement occurs from the reactant (or atom in the reactant) with *less* attraction for electrons to the reactant (or atom) with *more* attraction for electrons.

This event occurs in the formation of both ionic and covalent compounds.

- *Ionic compounds: transfer of electrons.* Consider the reaction that forms MgO from its elements (see Figure 3.7):

$$2Mg(s) + O_2(g) \longrightarrow 2MgO(s)$$

Figure 19.1A shows that each Mg atom loses two electrons and each O atom gains these electrons. This loss and gain is a ***transfer*** *of electrons* away from each Mg atom to each O atom. The resulting Mg^{2+} and O^{2-} ions aggregate into an ionic solid.

- *Covalent compounds: shift of electrons.* During the formation of a covalent compound from its elements, there is more of a ***shift*** *of electrons* than a full transfer. Thus, *ions do not form*. Consider the formation of HCl gas:

$$H_2(g) + Cl_2(g) \longrightarrow 2HCl(g)$$

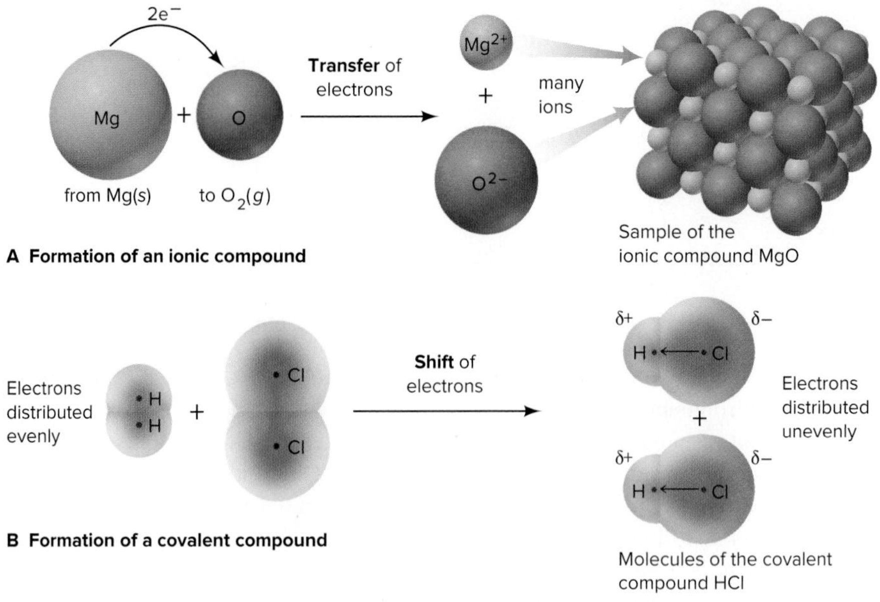

FIGURE 19.1 The redox process in the formation of ionic (A) and covalent (B) compounds from their elements

To see the electron movement, we will compare the electron distribution in the reactants and product. As Figure 19.1B shows, in H_2 and Cl_2, the electrons are shared equally between the atoms (*symmetrical shading*). Because the Cl atom attracts electrons more than the H atom does, the electrons in HCl are shared unequally (*asymmetrical shading*). The electrons shift away from H and toward Cl, so the Cl atom has more negative charge (δ–, *red* in Figure 19.1B) than it had in Cl_2, and the H atom has less negative charge (δ+, *blue* in Figure 19.1B) than it had in H_2.

Some Essential Redox Terminology

Certain key terms describe the redox process and explain its name:

- **Oxidation** is the *loss* of electrons.
- **Reduction** is the *gain* of electrons.

During the formation of MgO, Mg undergoes oxidation (loss of electrons), and O_2 undergoes reduction (gain of electrons). The loss and gain are simultaneous, but we can imagine them occurring separately:

Oxidation (electron loss by Mg): $Mg \longrightarrow Mg^{2+} + 2e^-$

Reduction (electron gain by O_2): $\frac{1}{2}O_2 + 2e^- \longrightarrow O^{2}$

(Throughout the chapter, blue type indicates oxidation, and red type indicates reduction.)

- The **oxidizing agent** helps the other species to be oxidized.
- The **reducing agent** helps the other species to be reduced.

One reactant acts on the other. During the reaction that forms MgO, *O_2 oxidizes Mg*, so O_2 is the oxidizing agent, and *Mg reduces O_2*, so Mg is the reducing agent.

Notice the give and take of electrons: O_2 takes the electrons that Mg gives up, or, put the other way around, Mg gives up the electrons that O_2 takes. This can be summarized as follows:

- *The oxidizing agent is reduced*: it takes electrons (and thus gains them).
- *The reducing agent is oxidized*: it gives up electrons (and thus loses them).

In the formation of HCl, *Cl_2 oxidizes H_2* (H loses some electron charge and Cl gains it), which is the same as saying that *H_2 reduces Cl_2*. The reducing agent, H_2, is oxidized, and the oxidizing agent, Cl_2, is reduced.

Using Oxidation Numbers to Monitor Electron Charge

Chemists have devised a "bookkeeping" system to monitor which atom loses electron charge and which atom gains it: each atom in a molecule (or formula unit) is assigned an **oxidation number (oxidation state)** (first seen in Section 8.5), which is the charge that the atom would have *if* the electrons were transferred completely, not shared.

Each element in a binary *ionic* compound has a full charge because the atom transferred its electron(s); thus, the oxidation number of the atom equals the ionic charge. However, each element in a *covalent* compound (or in a polyatomic ion) has a partial charge because the electrons shifted away from one atom and toward the other atom. Therefore, we use a set of rules to determine the oxidation number for an element in a covalent compound. These rules are outlined in Table 19.1. (You learned the atomic basis of these rules in Chapters 7 and 8.)

An oxidation number has the sign *before* the number (for example, +2), whereas an ionic charge has the sign *after* the number (for example, 2+). Also, unlike a unitary ionic charge, as in Na^+ or Cl^-, an oxidation number of +1 or −1 retains the numeral. For example, we do not write the sodium ion as Na^{1+}, but the oxidation number of the Na^+ ion is +1, not +.

TABLE 19.1 Rules for Assigning an Oxidation Number

General Rules

1. For an atom in its elemental form (such as Na, O_2, and Cl_2,), oxidation number = 0.
2. For a monatomic ion, oxidation number = ion charge (with the sign *before* the numeral).
3. The sum of the oxidation number values for the atoms in a molecule or formula unit of a compound equals zero. The sum of the oxidation number values for the atoms in a polyatomic ion equals the charge of the ion.

Rules for Specific Atoms or Periodic Table Groups

1. For group 1:	Oxidation number = +1 in all compounds
2. For group 2:	Oxidation number = +2 in all compounds
3. For hydrogen:	Oxidation number = +1 in combination with nonmetals
	Oxidation number = −1 in combination with metals and boron
4. For fluorine:	Oxidation number = −1 in all compounds
5. For oxygen:	Oxidation number = −2 in most cases, unless coupled to a more electronegative centre (such as F) or a group 1 or group 2 metal (in which case, it might be +2 or −1)
6. For group 17:	Oxidation number = −1 in combination with metals, nonmetals (except O), and other halogens lower in the group

Sample Problem 19.1 Determining the Oxidation Number of Each Element in a Compound (or Ion)

Problem Determine the oxidation number of each element in **(a)** zinc chloride; **(b)** sulfur trioxide; **(c)** nitric acid; **(d)** dichromate ion.

Plan We determine the formulas and consult Table 19.1, including the general rules that the oxidation numbers for a compound add up to zero and the oxidation numbers for a polyatomic ion add up to the ion's charge. We can express these rules as a formula, as shown below:

Net charge on species
$$= \Sigma(\text{oxidation number of atom X})(\text{number of X atoms in the species})$$

where the net charge of a neutral species (a molecule or an ionic compound) is zero and the net charge of an ion is the net charge on the ion.

Solution (a) $ZnCl_2$: The sum of the oxidation numbers must equal zero. From Table 19.1, the oxidation number of each Cl^- ion is −1. Using the formula above, and representing the oxidation number of Zn as y, we get

$$0 = (y)(1) + (-1)(2)$$

Therefore, $y = 2$ and the oxidation number of Zn is $+2$.

(b) SO_3: The oxidation number of each oxygen atom is -2. Representing the oxidation number of S as y,

$$0 = (y)(1) + (-2)(3)$$

Therefore, $y = 6$ and the oxidation number of S is $+6$.

(c) HNO_3: The oxidation number of H is $+1$, so the oxidation numbers of the atoms in NO_3^- must add up to -1 to equal the charge of the polyatomic ion and give zero for the compound. The oxidation number of each O is -2. Representing the oxidation number of N as y,

$$-1 = (y)(1) + (-2)(3)$$

Therefore, $y = 5$ and the oxidation number of N is $+5$.

(d) $Cr_2O_7^{2-}$. The oxidation number of each O is -2. Representing the oxidation number of Cr as y,

$$-2 = (y)(2) + (-2)(7) \text{ or } 2y = 12$$

Therefore, $y = 6$ and the oxidation number of Cr is $+6$.

Follow-Up Problem 19.1 Determine the oxidation number of each element in **(a)** scandium oxide (Sc_2O_3); **(b)** gallium chloride ($GaCl_3$); **(c)** hydrogen phosphate ion; **(d)** iodine trifluoride.

(See Brief Solutions to Follow-Up Problems at the end of the chapter.)

Using Oxidation Numbers to Identify Redox Reactions A redox reaction is defined as a reaction in which the oxidation numbers of the species change. Sample Problem 19.2 introduces the *tie-line*, a useful device for keeping track of these changes.

Sample Problem 19.2 Identifying Redox Reactions

Problem Use oxidation numbers to decide whether each equation represents a redox reaction:

(a) $CaO(s) + CO_2(g) \longrightarrow CaCO_3(s)$

(b) $4KNO_3(s) \longrightarrow 2K_2O(s) + 2N_2(g) + 5O_2(g)$

(c) $H_2SO_4(aq) + 2NaOH(aq) \longrightarrow Na_2SO_4(aq) + 2H_2O(l)$

Plan To determine whether an equation represents an oxidation-reduction reaction, we use Table 19.1 to assign an oxidation number to each atom. If an oxidation number changes during the reaction, we draw a tie-line from the atom on the left side to the atom on the right side and note the change.

Solution (a)

$$\overset{+2}{Ca}\overset{-2}{O}(s) + \overset{+4}{C}\overset{-2}{O_2}(g) \longrightarrow \overset{+2}{Ca}\overset{+4}{C}\overset{-2}{O_3}(s)$$

Because each atom in the product has the same oxidation number that it had in the reactants, we conclude that this is *not* a redox reaction.

(b)

Oxidation number decreased: *reduction*

$$4\overset{+1}{K}\overset{+5}{N}\overset{-2}{O_3}(s) \longrightarrow 2\overset{+1}{K_2}\overset{-2}{O}(s) + 2\overset{0}{N_2}(g) + 5\overset{0}{O_2}(g)$$

Oxidation number increased: *oxidation*

The oxidation number of N changes from $+5$ to 0, and the oxidation number of O changes from -2 to 0, so this *is* a redox reaction.

(c)

+6 / +1 / +6

+1 −2   +1 −2   +1 −2   +1 −2

$$H_2SO_4(aq) + 2NaOH(aq) \longrightarrow Na_2SO_4(aq) + 2H_2O(l)$$

The oxidation numbers do not change, so this is *not* a redox reaction.

Comment The reaction in (c) is an acid-base reaction in which H_2SO_4 transfers two H^+ ions to two OH^- ions to form two H_2O molecules. In the net ionic equation, we see that the oxidation numbers remain the same on both sides of the equation. Therefore, *an acid-base reaction is* ***not*** *a redox reaction.*

+1   −2 +1   +1 −2

$$H^+(aq) + OH^-(aq) \longrightarrow H_2O(l)$$

Follow-Up Problem 19.2 Use oxidation numbers to decide whether each of these equations represents a redox reaction:

(a) $NCl_3(l) + 3H_2O(l) \longrightarrow NH_3(aq) + 3HOCl(aq)$

(b) $AgNO_3(aq) + NH_4I(aq) \longrightarrow AgI(s) + NH_4NO_3(aq)$

(c) $2H_2S(g) + 3O_2(g) \longrightarrow 2SO_2(g) + 2H_2O(g)$

Using Oxidation Numbers to Identify Oxidizing and Reducing Agents By assigning an oxidation number to each atom, we can see which species was oxidized and which was reduced and, thus, which is the oxidizing agent and which is the reducing agent.

- If an atom has a higher (more positive or less negative) oxidation number in the product than it had in the reactant, the reactant that contains this atom was oxidized (lost electrons) and is the reducing agent. Thus, *oxidation is shown by an increase in oxidation number.*
- If an atom has a lower (more negative or less positive) oxidation number in the product than it had in the reactant, the reactant that contains this atom was reduced (gained electrons) and is the oxidizing agent. Thus, *reduction is shown by a decrease in oxidation number.*

Sample Problem 19.3 Identifying Oxidizing and Reducing Agents

Problem Identify the oxidizing agent and the reducing agent in each reaction:

(a) $2Al(s) + 3H_2SO_4(aq) \longrightarrow Al_2(SO_4)_3(aq) + 3H_2(g)$

(b) $PbO(s) + CO(g) \longrightarrow Pb(s) + CO_2(g)$

(c) $2H_2(g) + O_2(g) \longrightarrow 2H_2O(g)$

Plan We assign an oxidation number to each atom (or ion). The reducing agent contains an atom that is oxidized (oxidation number *increased* from left to right in the equation). The oxidizing agent contains an atom that is reduced (oxidation number *decreased*). We mark the changes with tie-lines.

Solution (a) Assign oxidation numbers:

oxidation

+6   +6

0   +1 −2   +3 −2   0

$$2Al(s) + 3H_2SO_4(aq) \longrightarrow Al_2(SO_4)_3(aq) + 3H_2(g)$$

reduction

The oxidation number of Al increased from 0 to +3 (Al lost electrons), so Al was oxidized; Al is the reducing agent because it allowed H^+ to be reduced.

The oxidation number of H decreased from +1 to 0 (H gained electrons), so H^+ was reduced; H_2SO_4 is the oxidizing agent because it allowed Al to be oxidized.

(b) Assign oxidation numbers:

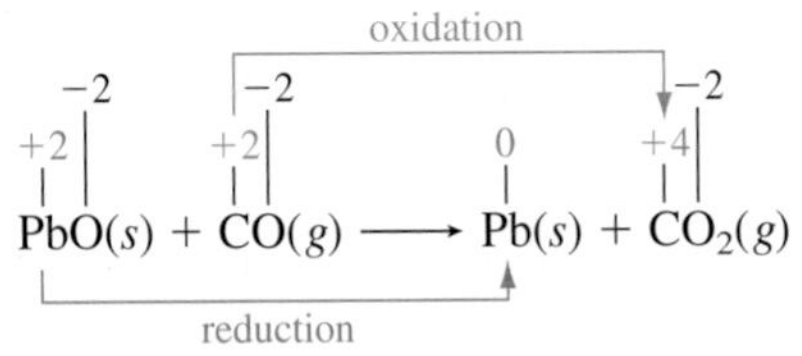

The oxidation number of Pb decreased from +2 to 0, so PbO was reduced; PbO is the oxidizing agent.

The oxidation number of C increased from +2 to +4, so CO was oxidized; CO is the reducing agent.

When a reactant (in this case, CO) becomes a product with more O atoms (CO_2), it is oxidized; when a reactant (PbO) becomes a product with fewer O atoms (Pb), it is reduced.

(c) Assign oxidation numbers:

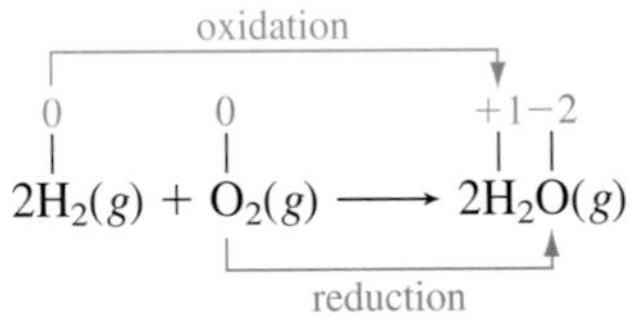

The oxidation number of O decreased from 0 to −2, so O_2 was reduced; O_2 is the oxidizing agent.

The oxidation number of H increased from 0 to +1, so H_2 was oxidized; H_2 is the reducing agent.

Oxygen is always the oxidizing agent in a combustion reaction.

Follow-Up Problem 19.3 Identify each oxidizing agent and each reducing agent:

(a) $2Fe(s) + 3Cl_2(g) \longrightarrow 2FeCl_3(s)$

(b) $2C_2H_6(g) + 7O_2(g) \longrightarrow 4CO_2(g) + 6H_2O(g)$

(c) $5CO(g) + I_2O_5(s) \longrightarrow I_2(s) + 5CO_2(g)$

Be sure to remember that *transferred electrons are never free because the reducing agent loses electrons and the oxidizing agent gains them* ***simultaneously***. In other words, a complete reaction *cannot* be "an oxidation" *or* "a reduction"; it must be an oxidation-reduction. Figure 19.2 summarizes redox terminology.

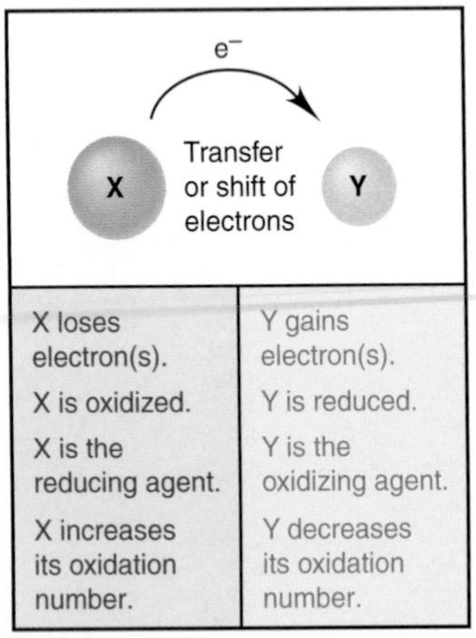

FIGURE 19.2 A summary of terminology for redox reactions

Balancing Redox Equations

We balance a redox equation by making sure that *the number of electrons lost by the reducing agent equals the number of electrons gained by the oxidizing agent.*

Two methods used to balance redox equations are the *oxidation number* method and the *half-reaction* method. Both methods are described here, since both have their advantages. As well, Sample Problems 19.4 and 19.5 show how these methods are used. You can choose which method you find easier to use.

The Oxidation Number Method The **oxidation number method** consists of five steps that use the changes in oxidation numbers to generate balancing coefficients. The first two steps are identical to those used in Sample Problem 19.3:

Step 1. Assign oxidation numbers to all atoms.

Step 2. From the changes in oxidation numbers, identify the oxidized and reduced reactants.

Step 3. Compute the number of electrons lost in the oxidation and gained in the reduction, and draw tie-lines from the reactant atom to the product atom to show the change.

Step 4. Multiply the number of electrons by factor(s) that make the electrons lost equal to the electrons gained, and use the factor(s) as balancing coefficients.

Step 5. Complete the balancing by inspection, verify that the atoms *and* the charges are balanced, and add the states of matter.

Sample Problem 19.4 Balancing Redox Equations by the Oxidation Number Method

Problem Use the oxidation number method to balance each equation:

(a) $Cu(s) + HNO_3(aq) \longrightarrow Cu(NO_3)_2(aq) + NO_2(g) + H_2O(l)$

(b) $PbS(s) + O_2(g) \longrightarrow PbO(s) + SO_2(g)$

Solution

(a) The equation represents the reaction of copper and nitric acid.

Step 1. Assign oxidation numbers to all atoms:

$$\overset{0}{Cu} + \overset{+1}{H}\overset{+5}{N}\overset{-2}{O_3} \longrightarrow \overset{+2}{Cu}(\overset{+5}{N}\overset{-2}{O_3})_2 + \overset{+4}{N}\overset{-2}{O_2} + \overset{+1}{H_2}\overset{-2}{O}$$

Copper in nitric acid

Step 2. Identify the oxidized and reduced reactants. The oxidation number of Cu increased from 0 (in Cu metal) to +2 (in Cu^{2+}), so Cu was oxidized. The oxidation number of N decreased from +5 (in HNO_3) to +4 (in NO_2), so HNO_3 was reduced. Notice that some NO_3^- also acts as a spectator ion, appearing unchanged in $Cu(NO_3)_2$.

Step 3. Compute e^- lost and e^- gained, and draw tie-lines. In the oxidation, Cu lost $2e^-$. In the reduction, N gained $1e^-$:

loses $2e^-$

$$Cu + HNO_3 \longrightarrow Cu(NO_3)_2 + NO_2 + H_2O$$

gains $1e^-$

Step 4. Multiply by factor(s) to make e^- lost equal e^- gained, and use the factor(s) as coefficients. Cu lost $2e^-$, so the $1e^-$ gained by N should be multiplied by 2. We put the coefficient 2 before *both* NO_2 and HNO_3:

$$Cu + 2HNO_3 \longrightarrow Cu(NO_3)_2 + 2NO_2 + H_2O$$

Step 5. Complete the balancing by inspection. Balancing N atoms requires a 4 in front of HNO_3 because two additional N atoms are in the NO_3^- ions in $Cu(NO_3)_2$:

$$Cu + 4HNO_3 \longrightarrow Cu(NO_3)_2 + 2NO_2 + H_2O$$

Then, balancing H atoms requires a 2 in front of H_2O, and we add the states of matter:

$$Cu(s) + 4HNO_3(aq) \longrightarrow Cu(NO_3)_2(aq) + 2NO_2(g) + 2H_2O(l)$$

(b) The equation represents the reaction of lead(II) sulfide and oxygen.

Step 1. Assign oxidation numbers to all atoms:

$$\overset{+2}{Pb}\overset{-2}{S} + \overset{0}{O_2} \longrightarrow \overset{+2}{Pb}\overset{-2}{O} + \overset{+4}{S}\overset{-2}{O_2}$$

Step 2. Identify the oxidized and reduced reactants. The oxidation number of S increased from −2 in PbS to +4 in SO_2, so PbS was oxidized. The oxidation number of O decreased from 0 in O_2 to −2 in both PbO and SO_2, so O_2 was reduced.

Step 3. Compute e^- lost and e^- gained, and draw tie-lines. In the oxidation, the S lost $6e^-$. In the reduction, each O gained $2e^-$:

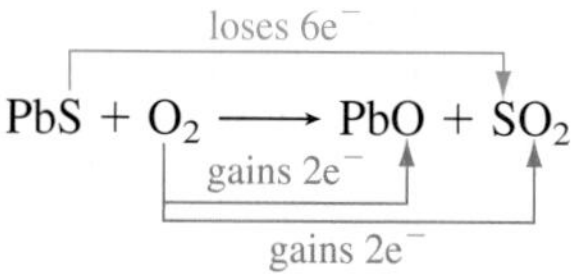

Step 4. Multiply by factor(s) to make e^- lost equal e^- gained. The S atom loses $6e^-$, and each O in O_2 gains $2e^-$, for a total gain of $4e^-$. Placing the coefficient $\frac{3}{2}$ before O_2 gives 3 O atoms that gain $2e^-$ each, for a total gain of $6e^-$:

$$PbS + \frac{3}{2}O_2 \longrightarrow PbO + SO_2$$

Step 5. Complete the balancing by inspection. The atoms are balanced, but we multiply all the coefficients by 2 to obtain integers. Then we add the states of matter:

$$2PbS(s) + 3O_2(g) \longrightarrow 2PbO(s) + 2SO_2(g)$$

Check Verify that the atoms are balanced on both sides of the arrow; then verify that the charges are balanced on both sides of the arrow.

(a) Verify that the atoms are balanced:

reactants (1 Cu, 4 H, 4 N, 12 O) $\longrightarrow$ products [1 Cu, 4 H, (2 + 2) N, (6 + 4 + 2) O]

Then verify that the charges are balanced. In this equation, both sides are 0.

(b) Verify that the atoms and charges are balanced:

reactants (2 Pb, 2 S, 6 O) $\longrightarrow$ products [2 Pb, 2 S, (2 + 4) O]

Comment We often find that some of a species reacts and the rest does not. In part (a), 2 mol of NO_3^- is reduced to NO_2, and 2 mol of NO_3^- is spectator ions.

Follow-Up Problem 19.4 Use the oxidation number method to balance the following equation:

$$K_2Cr_2O_7(aq) + HI(aq) \longrightarrow KI(aq) + CrI_3(aq) + I_2(s) + H_2O(l)$$

The Half-Reaction Method The other method, the **half-reaction method**, offers several advantages when studying electrochemistry:

- It *divides the overall redox reaction* into oxidation and reduction *half-reactions*, which reflect their actual physical separation in electrochemical cells.
- It is easier to apply to reactions in acidic or basic solutions, which are common in cells.
- It (usually) does *not* require assigning oxidation numbers. However, we often determine the oxidation numbers anyway. (In cases where the half-reactions are not obvious, we assign oxidation numbers to determine which atoms undergo a change and write half-reactions with the species that contain those atoms.)

The balancing process begins with a "skeleton" ionic equation, which consists only of species that are oxidized and reduced. The steps are as follows:

Step 1. Divide the skeleton equation into two half-reactions, each containing the oxidized and reduced forms of one of the species: *if the oxidized form of a species is on the left side, the reduced form of this species is on the right side, and vice versa.*

Step 2. Balance the atoms and charges in each half-reaction.

- Atoms are balanced *in the following order*: atoms other than O and H, then O, then H. Atoms other than O and H are balanced using stoichiometric coefficients. O is balanced by adding *only* H_2O (not O_2). H is balanced by adding H^+ (not H_2, for example).

- Charge is balanced by *adding* electrons (e^-) to the side that is more positive: *to the left side in the reduction half-reaction* because the reactant gains them and *to the right side in the oxidation half-reaction* because the reactant loses them.

Step 3. If necessary, multiply one or both half-reactions by an integer so that

$$\text{number of } e^- \text{ gained in reduction} = \text{number of } e^- \text{ lost in oxidation}$$

Step 4. Add the balanced half-reactions, and include the states of matter.
Step 5. Check that the atoms and charges are balanced.

We will go through the steps of balancing a redox reaction that occurs in acidic solution below. Then, in Sample Problem 19.5, we will balance a redox reaction that occurs in basic solution.

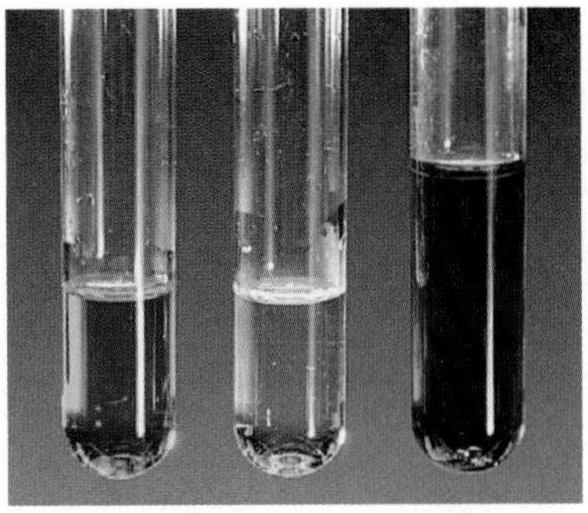

Dichromate ion (*left*) and iodide ion (*centre*) form chromium(III) ion and solid iodine (*right*).

Balancing Redox Reactions in Acidic Solution For a reaction in acidic solution, H_2O molecules and H^+ ions are present for the balancing. As mentioned in previous chapters, H^+ and H_3O^+ are equivalent. To illustrate this, we will balance the following example with both H^+ and H_3O^+ and show that the only difference is the number of water molecules.

The reaction between dichromate ion and iodide ion to form chromium(III) ion and solid iodine occurs in acidic solution (*see photo*). The skeleton ionic equation is as follows:

$$Cr_2O_7^{2-}(aq) + I^-(aq) \longrightarrow Cr^{3+}(aq) + I_2(s) \qquad \text{(acidic solution)}$$

Step 1. Divide the reaction into half-reactions. Each half-reaction contains the oxidized and reduced forms of one species:

$$Cr_2O_7^{2-} \longrightarrow Cr^{3+}$$
$$I^- \longrightarrow I_2$$

Step 2. Balance the atoms and charges in each half-reaction. We use H_2O to balance O, H^+ to balance H, and e^- to balance the charges.

- For the $Cr_2O_7^{2-}/Cr^{3+}$ half-reaction:
  (a) *Balance atoms other than O and H.* We balance the two Cr on the left with a coefficient of 2 on the right:

$$Cr_2O_7^{2-} \longrightarrow 2Cr^{3+}$$

  (b) *Balance the O atoms by adding H_2O molecules.* Each H_2O has one O atom, so we add seven H_2O on the right to balance the seven O in $Cr_2O_7^{2-}$:

$$Cr_2O_7^{2-} \longrightarrow 2Cr^{3+} + 7H_2O$$

  (c) *Balance the H atoms by adding H^+ ions.* Each H_2O contains 2 H, and we added 7 H_2O, so we add 14 H^+ ions on the left:

$$14H^+ + Cr_2O_7^{2-} \longrightarrow 2Cr^{3+} + 7H_2O$$

  (d) *Balance the charge by adding electrons.* Each H^+ ion has a 1+ charge, and 14 H^+ plus $Cr_2O_7^{2-}$ gives 12+ on the left. Two Cr^{3+} gives 6+ on the right. The left side is more positive, since there is an excess of 6+, so we add 6 e^- to the left:

$$6e^- + 14H^+ + Cr_2O_7^{2-} \longrightarrow 2Cr^{3+} + 7H_2O$$

This half-reaction is balanced. It is a *reduction* because there is a *gain* of electrons; thus, electrons appear on the *left, as reactants*: $Cr_2O_7^{2-}$ gains electrons (is reduced), so $Cr_2O_7^{2-}$ is the *oxidizing agent*. The oxidation number of Cr decreases from +6 on the left to +3 on the right.

- For the I^-/I_2 half-reaction:
  (a) *Balance atoms other than O and H.* Two I atoms on the right require a coefficient of 2 on the left:

$$2I^- \longrightarrow I_2$$

(b) *Balance the O atoms with H_2O.* Not needed; there are no O atoms.
(c) *Balance the H atoms with H^+.* Not needed; there are no H atoms.
(d) *Balance the charge with e^-.* To balance the 2^- on the left, we add two e^- to the more positive side, which is the right:

$$2I^- \longrightarrow I_2 + 2e^-$$

This half-reaction is balanced. It is an *oxidation* because there is a *loss* of electrons; thus, electrons appear on the *right, as products*: the reactant I^- loses electrons (is oxidized), so I^- is the *reducing agent.* The oxidation number of I increases from −1 to 0.

Step 3. *Multiply each half-reaction, if necessary, by an integer* so that the number of e^- lost in the oxidation equals the number of e^- gained in the reduction. Two e^- are lost in the oxidation and six e^- are gained in the reduction, so we multiply the oxidation by 3:

$$\times 3 \ (2I^- \longrightarrow I_2 + 2e^-)$$
$$6I^- \longrightarrow 3I_2 + 6e^-$$

Step 4. *Add the half-reactions,* cancelling substances that appear on both sides, and include the states of matter. In this example, only the electrons cancel:

$$6e^- + 14H^+ + Cr_2O_7^{2-} \longrightarrow 2Cr^{3+} + 7H_2O$$
$$6I^- \longrightarrow 3I_2 + 6e^-$$
$$6I^-(aq) + 14H^+(aq) + Cr_2O_7^{2-}(aq) \longrightarrow 3I_2(s) + 7H_2O(l) + 2Cr^{3+}(aq)$$

Step 5. *Check that the atoms and charges are balanced:*

reactants (6 I, 14 H, 2 Cr, 7 O; 6 +) ⟶ products (6 I, 14 H, 2 Cr, 7 O; 6 +)

Using H_3O^+, Instead of H^+, to Balance H Now let us see what happens if we use H_3O^+ to supply H atoms, instead of H^+. In step 2 for the reduction half-reaction, there is no difference in balancing Cr atoms and balancing O atoms with H_2O, so we have

$$Cr_2O_7^{2-} \longrightarrow 2Cr^{3+} + 7H_2O$$

Now we use H_3O^+ to balance the H atoms. Because each H_3O^+ is an H^+ bonded to an H_2O, we balance the 14 H on the right with 14 H_3O^+ on the left and immediately add 14 more H_2O on the right:

$$14H_3O^+ + Cr_2O_7^{2-} \longrightarrow 2Cr^{3+} + 7H_2O + 14H_2O$$

Taking the sum of the H_2O molecules on the right gives

$$14H_3O^+ + Cr_2O_7^{2-} \longrightarrow 2Cr^{3+} + 21H_2O$$

None of this affects the redox charge, so balancing the charge still requires six e^- on the left. We obtain the balanced reduction half-reaction:

$$6e^- + 14H_3O^+ + Cr_2O_7^{2-} \longrightarrow 2Cr^{3+} + 21H_2O$$

Adding the balanced oxidation half-reaction gives the balanced redox equation:

$$6I^-(aq) + 14H_3O^+(aq) + Cr_2O_7^{2-}(aq) \longrightarrow 3I_2(s) + 21H_2O(l) + 2Cr^{3+}(aq)$$

Notice that, as mentioned earlier, the only difference is the number of H_2O molecules: 21 H_2O instead of 7 H_2O. So, even though this approach uses H_3O^+, which is more accurate, only the number of water molecules changes. Because the balancing involves more steps, however, we will continue to use the form H^+ to balance H atoms.

Balancing Redox Reactions in Basic Solution In acidic solution, H_2O molecules and H^+ ions are available to balance a redox reaction; in basic solution, H_2O molecules and OH^- ions are available.

We need only one extra step to balance a redox equation that takes place in basic solution. This step is included after we balance the half-reactions *as if they occur in acidic solution* and are combined (step 4). At that point, *we add one OH^-* ***to both sides of the equation*** *for every H^+ present.* (This step is labelled *Step 4: Basic* in Sample Problem 19.5.) The OH^- ions added on the side with the H^+ ions combine with them to form H_2O, whereas the OH^- ions on the other side remain in the equation, and excess H_2O is cancelled.

Sample Problem 19.5 Balancing a Redox Reaction in Basic Solution

Problem Permanganate ion reacts in basic solution with oxalate ion to form carbonate ion and solid manganese dioxide. Balance the skeleton ionic equation for the reaction between $NaMnO_4$ and $Na_2C_2O_4$ in basic solution:

$$MnO_4^-(aq) + C_2O_4^{2-}(aq) \longrightarrow MnO_2(s) + CO_3^{2-}(aq) \qquad \text{(basic solution)}$$

Plan We follow the numbered steps, as described in the text, and proceed through step 4 as if this reaction occurs in acidic solution. Then we add the appropriate number of OH^- ions and cancel excess H_2O molecules (*Step 4: Basic*).

Solution

Step 1. Identify and separate the two half-reactions:

$$MnO_4^- \longrightarrow MnO_2 \qquad\qquad C_2O_4^{2-} \longrightarrow CO_3^{2-}$$

Step 2. Balance the atoms and charges.

(a) Atoms other than O and H:

Not needed

(a) Atoms other than O and H:

$$C_2O_4^{2-} \longrightarrow 2CO_3^{2-}$$

(b) O atoms with H_2O:

$$MnO_4^- \longrightarrow MnO_2 + 2H_2O$$

(b) O atoms with H_2O:

$$2H_2O + C_2O_4^{2-} \longrightarrow 2CO_3^{2-}$$

(c) H atoms with H^+:

$$4H^+ + MnO_4^- \longrightarrow MnO_2 + 2H_2O$$

(c) H atoms with H^+:

$$2H_2O + C_2O_4^{2-} \longrightarrow 2CO_3^{2-} + 4H^+$$

(d) Charge with e^-:

$$3e^- + 4H^+ + MnO_4^- \longrightarrow MnO_2 + 2H_2O$$

(reduction)

(d) Charge with e^-:

$$2H_2O + C_2O_4^{2-} \longrightarrow 2CO_3^{2-} + 4H^+ + 2e^-$$

(oxidation)

Step 3. Multiply each half-reaction, if necessary, by an integer to make e^- lost equal e^- gained.

$$2 \times (3e^- + 4H^- + MnO_4^- \longrightarrow MnO_2 + 2H_2O)$$

$$6e^- + 8H^+ + 2MnO_4^- \longrightarrow 2MnO_2 + 4H_2O$$

$$3 \times (2H_2O + C_2O_4^{2-} \longrightarrow 2CO_3^{2-} + 4H^+ + 2e^-)$$

$$6H_2O + 3C_2O_4^{2-} \longrightarrow 6CO_3^{2-} + 12H^+ + 6e^-$$

Step 4. Add the half-reactions, and cancel the substances that appear on both sides. The six e^- cancel, eight H^+ cancel to leave four H^+ on the right, and four H_2O cancel to leave two H_2O on the left:

$$\cancel{6e^-} + \cancel{8H^+} + 2MnO_4^- \longrightarrow 2MnO_2 + \cancel{4H_2O}$$

$$2\ \cancel{6}H_2O + 3C_2O_4^{2-} \longrightarrow 6CO_3^{2-} + 4\ \cancel{12H^+} + \cancel{6e^-}$$

$$2MnO_4^- + 2H_2O + 3C_2O_4^{2-} \longrightarrow 2MnO_2 + 6CO_3^{2-} + 4H^+$$

Step 4: Basic. Add the same amount of OH^- to both sides to neutralize the H^+. Then cancel the appropriate number of H_2O molecules by comparing the left and right sides of the reaction. *Remember*: If you add x OH^- ions to one side, you *must* add x OH^- ions to the other side as well!

Adding four OH^- to both sides forms four H_2O on the right. Two of these cancel the two H_2O on the left, leaving two H_2O on the right:

$$2MnO_4^- + 2H_2O + 3C_2O_4^{2-} + 4OH^- \longrightarrow 2MnO_2 + 6CO_3^{2-} + [4H^+ + 4OH^-]$$
$$2MnO_4^- + \cancel{2H_2O} + 3C_2O_4^{2-} + 4OH^- \longrightarrow 2MnO_2 + 6CO_3^{2-} + 2\ \cancel{4}H_2O$$

Adding the states of matter gives the final balanced equation:

$$2MnO_4^-(aq) + 3C_2O_4^{2-}(aq) + 4OH^-(aq) \longrightarrow 2MnO_2(s) + 6CO_3^{2-}(aq) + 2H_2O(l)$$

Step 5. Check that the atoms and charges balance.

$$(2\ Mn, 24\ O, 6\ C, 4\ H; 12\ -) \longrightarrow (2\ Mn, 24\ O, 6\ C, 4\ H; 12\ -)$$

Comment As a final step, let us see how to obtain the balanced *molecular* equation for this reaction. We note the amount (mol) of each anion in the balanced ionic equation and add the correct amount (mol) of spectator ions (Na^+, as given in the problem) to obtain neutral compounds. The balanced molecular equation is

$$2NaMnO_4(aq) + 3Na_2C_2O_4(aq) + 4NaOH(aq) \longrightarrow 2MnO_2(s) + 6Na_2CO_3(aq) + 2H_2O(l)$$

Follow-Up Problem 19.5 Write a balanced molecular equation for the reaction between $KMnO_4$ and KI in basic solution. The skeleton ionic equation is given below:

$$MnO_4^-(aq) + I^-(aq) \longrightarrow MnO_4^{2-}(aq) + IO_3^-(aq) \qquad \text{(basic solution)}$$

Quantifying Redox Reactions by Titration

In an acid-base titration, a known concentration of a base is used to find an unknown concentration of an acid (or vice versa). Similarly, in a redox titration, a known concentration of an oxidizing agent is used to find an unknown concentration of a reducing agent (or vice versa). This kind of titration has many applications, from measuring the iron content in drinking water to quantifying vitamin C in fruits and vegetables.

The permanganate ion, MnO_4^-, is a common oxidizing agent, and, because of its deep purple colour, also serves as an indicator. In Figure 19.3, MnO_4^- is used to determine the oxalate ion ($C_2O_4^{2-}$) concentration. As long as $C_2O_4^{2-}$ is present, it reduces the added MnO_4^- to faint pink (nearly colourless) Mn^{2+} (Figure 19.3, *left*). When all the $C_2O_4^{2-}$ has been oxidized, the next drop of MnO_4^- turns the solution light purple (Figure 19.3, *right*). This colour change indicates the *end point*, which we assume is the same as the *equivalence point*, the point at which the electrons lost by the oxidized species ($C_2O_4^{2-}$) equal the electrons gained by the reduced species (MnO_4^-). We can calculate the $C_2O_4^{2-}$ concentration from the known volume of the $Na_2C_2O_4$ solution and the known volume and concentration of the $KMnO_4$ solution.

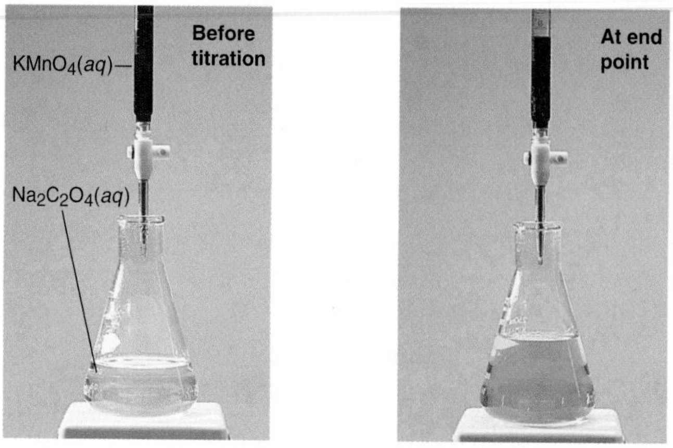

FIGURE 19.3 The redox titration of $C_2O_4^{2-}$ with MnO_4^-

Net ionic equation

$$\overset{+7}{2MnO_4^-(aq)} + \overset{+3}{5C_2O_4^{2-}(aq)} + 16H^+(aq) \longrightarrow \overset{+2}{2Mn^{2+}(aq)} + \overset{+4}{10CO_2(g)} + 8H_2O(l)$$

Sample Problem 19.6 Finding the Amount of Reducing Agent by Titration

Problem Calcium ion (Ca^{2+}) is necessary for blood clotting and many other physiological processes. To measure the Ca^{2+} concentration in 1.00 mL of human blood, $Na_2C_2O_4$ solution is added, causing Ca^{2+} to precipitate as solid CaC_2O_4. This solid is dissolved in dilute H_2SO_4 to release $C_2O_4^{2-}$, and 2.05 mL of 4.88×10^{-4} mol/L $KMnO_4$ is required to reach the end point. The balanced equation is

$$2KMnO_4(aq) + 5CaC_2O_4(s) + 8H_2SO_4(aq) \longrightarrow 2MnSO_4(aq) + K_2SO_4(aq) + 5CaSO_4(s) + 10CO_2(g) + 8H_2O(l)$$

Calculate the amount (mol) of Ca^{2+} in 1.00 mL of blood.

Plan We have to find the amount (mol) of Ca^{2+} from the volume (2.05 mL) and concentration (4.88×10^{-4} mol/L) of $KMnO_4$ used to oxidize the CaC_2O_4 precipitated from 1.00 mL of blood. We find the amount (mol) of $KMnO_4$ needed to reach the end point and use the molar ratio to find the amount (mol) of CaC_2O_4. Then we use the chemical formula, which shows 1 mol of Ca^{2+} for every mol of CaC_2O_4 (*see road map*).

Road Map

Volume (mL) of $KMnO_4$ solution
↓ 10^3 mL = 1 L
Volume (L) of $KMnO_4$ solution
↓ Multiply by concentration (mol/L).
Amount (mol) of $KMnO_4$
↓ Molar ratio
Amount (mol) of CaC_2O_4
↓ Ratio of elements in chemical formula
Amount (mol) of Ca^{2+}

Solution Convert volume (mL) and concentration (mol/L) to amount (mol) of $KMnO_4$:

$$n_{KMnO_4} = V \times c$$
$$= 2.05 \text{ mL soln} \times \frac{1 \text{ L}}{1000 \text{ mL}} \times \frac{4.88 \times 10^{-4} \text{ mol } KMnO_4}{1 \text{ L soln}}$$
$$= 1.00 \times 10^{-6} \text{ mol } KMnO_4$$

Use the molar ratio to convert amount (mol) of $KMnO_4$ to amount (mol) of CaC_2O_4:

$$n_{CaC_2O_4} = 1.00 \times 10^{-6} \text{ mol } KMnO_4 \times \frac{5 \text{ mol } CaC_2O_4}{2 \text{ mol } KMnO_4}$$
$$= 2.50 \times 10^{-6} \text{ mol } CaC_2O_4$$

Find the amount (mol) of Ca^{2+}:

$$n_{Ca^{2+}} = 2.50 \times 10^{-6} \text{ mol } CaC_2O_4 \times \frac{1 \text{ mol } Ca^{2+}}{1 \text{ mol } CaC_2O_4}$$
$$= 2.50 \times 10^{-6} \text{ mol } Ca^{2+}$$

Check A very small volume of dilute $KMnO_4$ is needed, so 10^{-6} mol of $KMnO_4$ seems reasonable. The molar ratio of CaC_2O_4 to $KMnO_4$ is 5/2, which gives 2.5×10^{-6} mol of CaC_2O_4 and, thus, 2.5×10^{-6} mol of Ca^{2+}.

Comment **1.** When blood is donated, the receiving bag contains $Na_2C_2O_4$ solution, which precipitates the Ca^{2+} ion to prevent clotting.
2. The normal Ca^{2+} concentration in human adult blood is 9.0 mg to 11.5 mg Ca^{2+}/100 mL. If we multiply the amount of Ca^{2+} in 1 mL by 100 and then multiply by the molar mass of Ca (40.08 g/mol), we get 1.0×10^{-2} g, or 10.0 mg Ca^{2+} in 100 mL:

$$m_{Ca^{2+}} = \frac{2.50 \times 10^{-6} \text{ mol } Ca^{2+}}{1 \text{ mL}} \times 100 \text{ mL} \times \frac{40.08 \text{ g } Ca^{2+}}{1 \text{ mol } Ca^{2+}} \times \frac{1000 \text{ mg}}{1 \text{ g}}$$
$$= 10.0 \text{ mg } Ca^{2+}$$

Follow-Up Problem 19.6 When 2.50 mL of low-fat milk was treated with sodium oxalate and the precipitate was dissolved in H_2SO_4, 6.53 mL of 4.56×10^{-3} mol/L $KMnO_4$ was required to reach the end point.

(a) Calculate the concentration (mol/L) of Ca^{2+} in the milk sample.

(b) What is the concentration of Ca^{2+} in g/L? Is this concentration consistent with the typical concentration for milk, which is about 1.2 g Ca^{2+}/L?

SUMMARY OF SECTION 19.1

- When one reactant has a greater attraction for electrons than another reactant does, there is a net movement of electrons, and a redox reaction takes place. Electron gain (reduction) and electron loss (oxidation) occur simultaneously.
- Assigning oxidation numbers to all the atoms in a reaction is a method for identifying a redox reaction. The species that is oxidized (contains an atom that increases in oxidation number) is the reducing agent; the species that is reduced (contains an atom that decreases in oxidation number) is the oxidizing agent.
- An oxidation-reduction (redox) reaction involves the transfer of electrons from a reducing agent to an oxidizing agent.
- Two methods that can be used to balance a redox reaction are explained: the oxidation number method and the half-reaction method:
  - The oxidation number method keeps track of the changes in oxidation number.
  - The half-reaction method divides the overall reaction into half-reactions that are balanced separately and then recombined.
- A redox titration determines the concentration of an oxidizing agent from the known concentration of the reducing agent (or vice versa).

19.2 Voltaic Cells: Electrochemical Cells Using Spontaneous Reactions to Generate Electrical Energy

An Overview of Electrochemical Cells

The fundamental difference between the two types of electrochemical cells is based on whether the overall redox reaction in the cell is spontaneous (Gibbs energy is released) or nonspontaneous (Gibbs energy is absorbed):

1. A **voltaic cell (galvanic cell)** uses a *spontaneous* reaction ($\Delta G < 0$) to generate electrical energy. In the cell reaction, the difference in Gibbs energy between higher energy reactants and lower energy products is converted into electrical energy, which operates the load: for example, flashlight, DVD player, or car starter motor. Thus, *the system does work on the surroundings*. All batteries contain voltaic cells.

2. An **electrolytic cell** uses electrical energy to drive a *nonspontaneous* reaction ($\Delta G > 0$). In the cell reaction, an external source supplies Gibbs energy to convert lower energy reactants into higher energy products. Thus, *the surroundings do work on the system*. Electroplating and the recovery of metals from ores utilize electrolytic cells.

These two types of cells have several features in common. Two **electrodes**, which conduct the electricity between the cell and its surroundings, are dipped into an **electrolyte**, a mixture of ions (often in aqueous solution) that are involved in the reaction or carry the charge (Figure 19.4). An electrode is identified as either an **anode** or a **cathode**, depending on the half-reaction that takes place there: ■

■ **Which Half-Reaction Occurs at Which Electrode?** Here are some memory aids to help you remember which half-reaction occurs at which electrode:

1. **LEO** the lion says **GER**: Loss of Electrons is Oxidation and Gain of Electrons is Reduction.
2. The words *anode* and *oxidation* start with vowels; the words *cathode* and *reduction* start with consonants.
3. Alphabetically, the *A* in anode comes before the *C* in cathode, and the *O* in oxidation comes before the *R* in reduction.
4. Look at the first syllables, and use your imagination:

ANode, OXidation; REDuction, CAThode ⇒ AN OX and a RED CAT

- *The oxidation half-reaction occurs at the anode.* Electrons lost by the substance being oxidized (reducing agent) *leave the oxidation half-cell* at the anode.
- *The reduction half-reaction occurs at the cathode.* Electrons gained by the substance being reduced (oxidizing agent) *enter the reduction half-cell* at the cathode.

Note that, for reasons we will discuss shortly, *the relative charges of the electrodes have* ***opposite signs*** *in the two types of cells.*

When you put a strip of zinc metal in a solution of Cu^{2+} ion, the blue colour of the solution fades and a brown-black crust of Cu metal forms on the Zn strip (Figure 19.5). During this spontaneous reaction, Cu^{2+} ion is reduced to Cu metal,

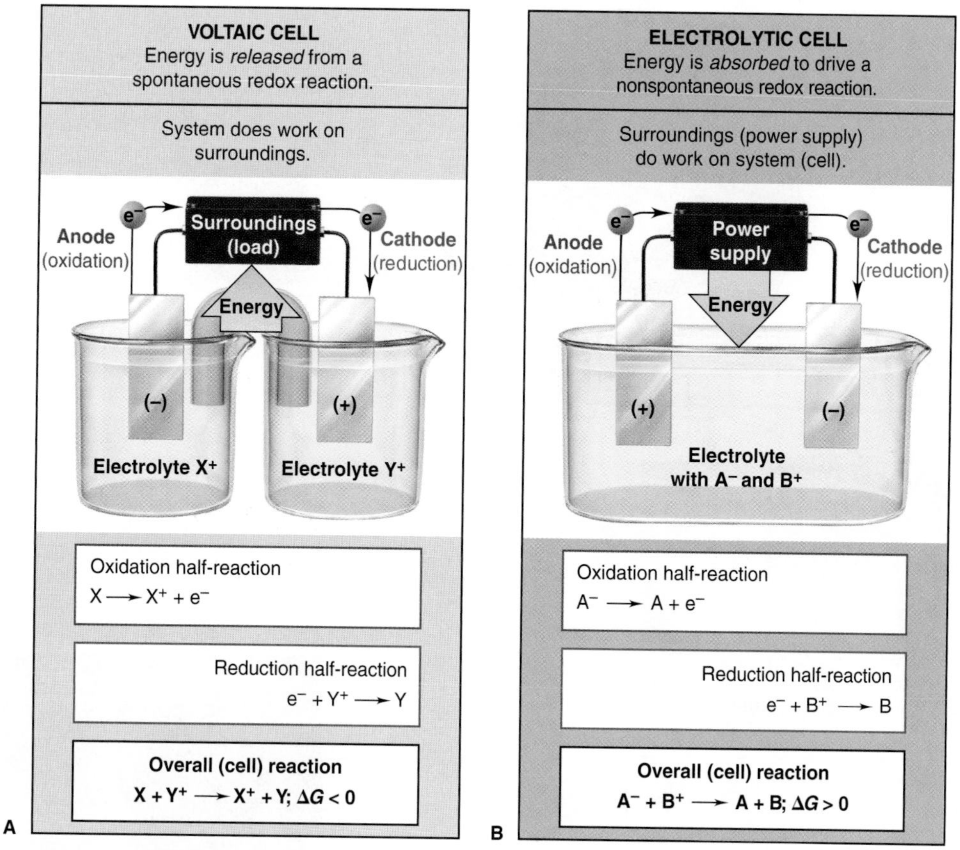

FIGURE 19.4 General characteristics of voltaic cells (A) and electrolytic cells (B)

while Zn metal is oxidized to Zn^{2+} ion. The overall reaction consists of two half-reactions:

$$\begin{array}{rcll} Cu^{2+}(aq) + 2e^- & \longrightarrow & Cu(s) & \text{(reduction)} \\ Zn(s) & \longrightarrow & Zn^{2+}(aq) + 2e^- & \text{(oxidation)} \\ \hline Zn(s) + Cu^{2+}(aq) & \longrightarrow & Zn^{2+}(aq) + Cu(s) & \text{(overall reaction)} \end{array}$$

Let us examine this spontaneous reaction as the basis of a voltaic cell.

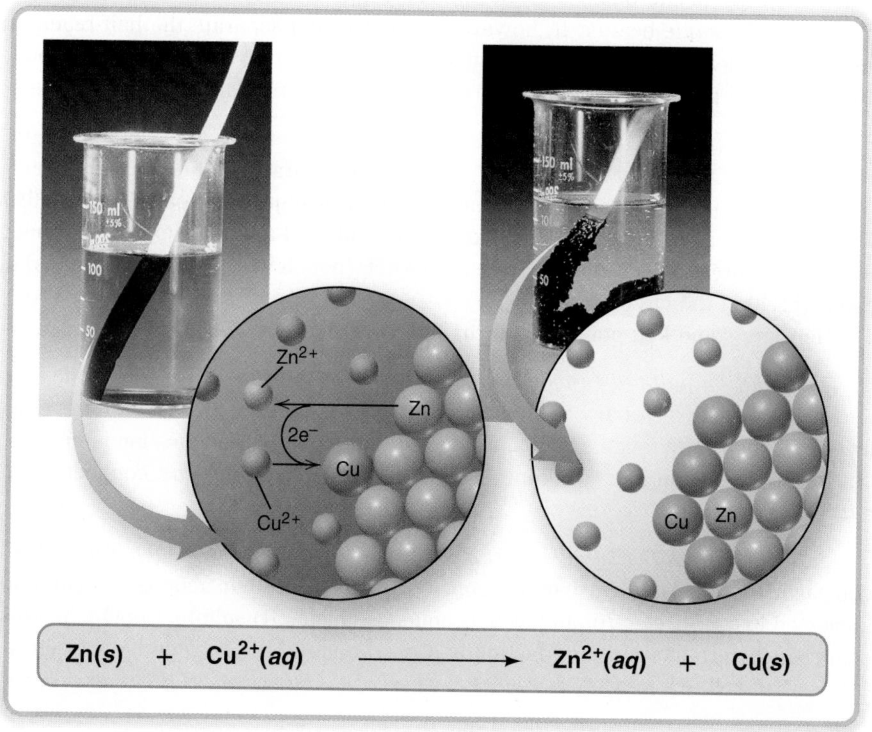

FIGURE 19.5 The spontaneous reaction between zinc and copper(II) ion. When zinc metal is placed in a solution of Cu^{2+} ion (*left*), zinc is oxidized to Zn^{2+}, and Cu^{2+} is reduced to copper metal (*right*). (The very finely divided Cu appears black.)

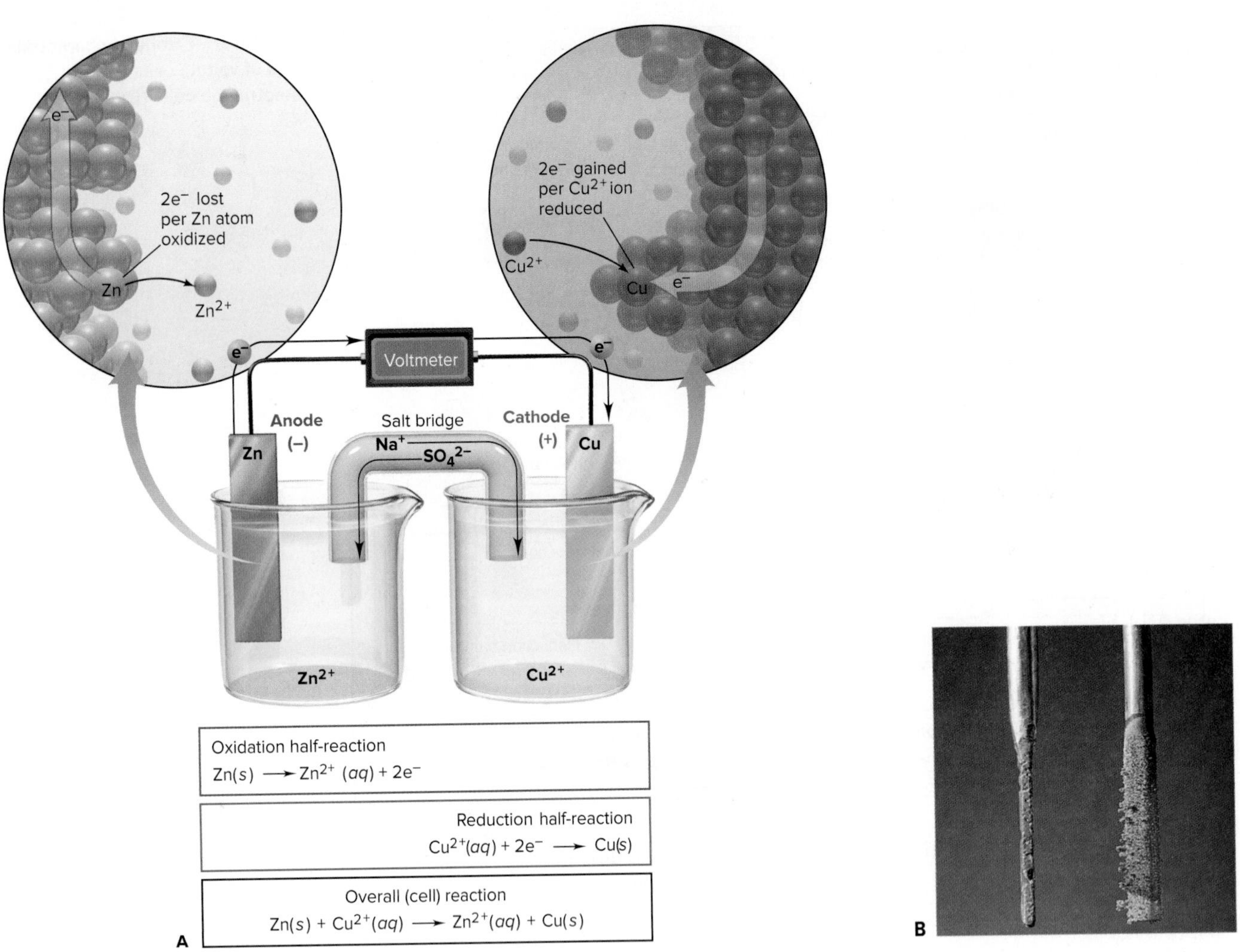

FIGURE 19.6 A voltaic cell based on the zinc-copper reaction. A. In the anode half-cell (oxidation; *left*), two electrons from a Zn atom move through the Zn bar as the Zn^{2+} ion enters the solution. In the cathode half-cell (reduction; *right*), the electrons reduce Cu^{2+} ions to Cu atoms. **B.** After several hours, the Zn anode weighs less and the Cu cathode weighs more.

Construction and Operation of a Voltaic Cell

Electrons are being transferred in the Zn/Cu^{2+} reaction, but the system does not generate electrical energy because the oxidizing agent (Cu^{2+}) and the reducing agent (Zn) are in the *same* beaker. If, however, we physically separate the half-reactions and connect them by an external circuit, the electrons lost by the zinc travel through the circuit and produce an electric current as the copper ions gain them.

Basis of the Voltaic Cell: Separation of Half-Reactions In any voltaic cell, the components of each half-reaction are placed in a separate container, or **half-cell**, which consists of one electrode dipping into an electrolyte solution (Figure 19.6A). The two half-cells are joined by the external circuit, and a digital voltmeter or multimeter measures the voltage generated. A switch (not shown) closes (completes) or opens (breaks) the circuit.

Here are some key points about the half-cells in the Zn/Cu^{2+} voltaic cell:

1. *Oxidation half-cell (anode compartment, shown on the left).* The anode compartment consists of a bar of zinc (the electrode that acts as the anode) immersed in a Zn^{2+} electrolyte (such as aqueous zinc sulfate, $ZnSO_4$). Metallic zinc is the reactant in the oxidation half-reaction, and the bar loses electrons *and* conducts them *out* of this half-cell.

2. *Reduction half-cell (cathode compartment, shown on the right).* The cathode compartment consists of a bar of copper (the electrode that acts as the cathode) immersed in a Cu^{2+} electrolyte [such as aqueous copper(II) sulfate, $CuSO_4$]. Metallic copper is the product in the reduction half-reaction, and the bar conducts electrons *into* its half-cell, where Cu^{2+} is reduced.

Charges of the Electrodes The charges of the electrodes are determined by the *source of the electrons* and the *direction of electron flow* through the circuit. In the Zn/Cu^{2+} voltaic cell, Zn atoms are oxidized at the anode to Zn^{2+} ions and electrons. The Zn^{2+} ions enter the half-cell electrolyte, while the electrons move through the bar and into the wire. *The electrons flow from anode to cathode* (in Figure 19.6A, *from left to right*) through the wire to the cathode, where the Cu^{2+} ions in this electrolyte are reduced to Cu atoms. As the cell operates, electrons are continuously generated at the anode and consumed at the cathode. Therefore, the anode has an excess of electrons and the cathode has a deficit of electrons: *in any **voltaic** cell, the anode has a negative charge and the cathode has a positive charge.*

Completing the Circuit with a Salt Bridge A cell cannot operate unless the circuit is complete. The oxidation half-cell originally contains a neutral solution of Zn^{2+} and SO_4^{2-} ions. As Zn atoms in the bar lose electrons, however, the Zn^{2+} ions that form enter the solution and give it a net positive charge. Similarly, in the reduction half-cell, the neutral solution of Cu^{2+} and SO_4^{2-} ions develops a net negative charge as Cu^{2+} ions leave the solution and form Cu atoms.

The existence of such a charge imbalance between the half-cells makes cell operation impossible. The only way for the cell to operate at all is by using a **salt bridge**. It joins the half-cells and acts like a "liquid wire," allowing ions to flow through both compartments and complete the circuit. The salt bridge is an inverted U tube that contains nonreacting ions, such as Na^+ and SO_4^{2-}, dissolved in a gel. The gel does not flow out of the tube but allows ions to diffuse into or out of the half-cells:

- *Maintaining a neutral reduction half-cell (cathode compartment; right side of Figure 19.6A).* As Cu^{2+} ions change to Cu atoms, Na^+ ions move from the salt bridge into the electrolyte solution (and some SO_4^{2-} ions move from the solution into the salt bridge).
- *Maintaining a neutral oxidation half-cell (anode compartment; left side of Figure 19.6A).* As Zn atoms change to Zn^{2+} ions, SO_4^{2-} ions move from the salt bridge into the electrolyte solution (and some Zn^{2+} ions move from the solution into the salt bridge).

Thus, the wire and the salt bridge complete the circuit:

- *Electrons move from anode to cathode* through the wire.
- *Anions move from the cathode compartment to the anode compartment* through the salt bridge.
- *Cations move from the anode compartment to the cathode compartment* through the salt bridge.

Active versus Inactive Electrodes The electrodes in the Zn/Cu^{2+} cell are *active* because the metals themselves are components of the half-reactions. During the operation of the cell, the mass of the zinc bar gradually decreases as the $[Zn^{2+}]$ in the anode half-cell increases. At the same time, the mass of the copper bar increases as the $[Cu^{2+}]$ in the cathode half-cell decreases and the ions form Cu atoms that deposit on the electrode (Figure 19.6B).

In many cases, however, no reaction components can be physically used as an electrode, so *inactive* electrodes are used. Most commonly, inactive electrodes are rods of *graphite* or *platinum*. In a voltaic cell based on the following half-reactions, for instance, the reacting species cannot be made into electrodes:

$$2I^-(aq) \longrightarrow I_2(s) + 2e^- \qquad \text{(anode; oxidation)}$$

$$MnO_4^-(aq) + 8H^+(aq) + 5e^- \longrightarrow Mn^{2+}(aq) + 4H_2O(l) \qquad \text{(cathode; reduction)}$$

Each half-cell consists of an inactive electrode immersed in an electrolyte that contains *all the reactant species involved in this half-reaction* (Figure 19.7):

- In the *anode half-cell*, I^- ions are oxidized to solid I_2, and the released electrons flow into the graphite electrode (C) and through the wire.
- From the wire, the electrons enter the graphite electrode in the *cathode half-cell* and reduce MnO_4^- ions to Mn^{2+} ions. (A KNO_3 salt bridge is used.)

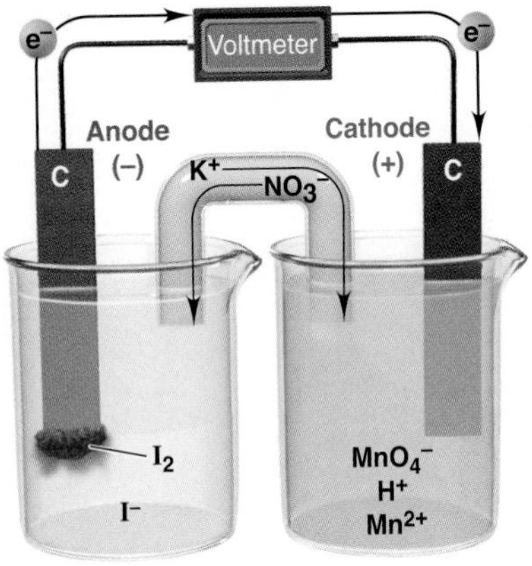

Oxidation half-reaction
$2I^-(aq) \longrightarrow I_2(s) + 2e^-$

Reduction half-reaction
$MnO_4^-(aq) + 8H^+(aq) + 5e^- \longrightarrow Mn^{2+}(aq) + 4H_2O(l)$

Overall (cell) reaction
$2MnO_4^-(aq) + 16H^+(aq) + 10I^-(aq) \longrightarrow 2Mn^{2+}(aq) + 5I_2(s) + 8H_2O(l)$

FIGURE 19.7 A voltaic cell using inactive electrodes

Diagram of a Voltaic Cell As shown in Figures 19.6A and 19.7, there are certain consistent features in the *diagram* of any voltaic cell:

- Components of the half-cells include electrode materials, electrolyte ions, and other species involved in the reaction.
- The name of the electrode (anode or cathode) and its charge are shown. By convention, the anode compartment appears *on the left*.
- Each half-reaction and the overall cell reaction are given.
- The direction of electron flow in the external circuit is from left to right.
- The nature of the ions and the direction of ion flow in the salt bridge are shown, with cations moving toward the cathode (*to the right* in a conventional cell diagram) and anions moving toward the anode (*to the left* in a conventional cell diagram).

Notation for a Voltaic Cell

A useful shorthand notation describes the components of a voltaic cell. The notation for the Zn/Cu^{2+} cell is

$$Zn(s)|\,Zn^{2+}(aq)||Cu^{2-}(aq)|Cu(s)$$

This notation includes the following key parts:

- The components of the anode compartment (oxidation half-cell) are written *to the left* of the components of the cathode compartment (reduction half-cell).
- A double vertical line indicates that the half-cells are physically separated.
- A single vertical line represents a phase boundary. For example, $Zn(s)|Zn^{2+}(aq)$ indicates that *solid* Zn is a *different* phase from *aqueous* Zn^{2+}.
- A comma separates the half-cell components that are in the *same* phase. For example, the notation for the voltaic cell shown in Figure 19.7 is

$$\text{graphite}|I^-(aq)|I_2(s)||MnO_4^-(aq),\ H^+(aq),\ Mn^{2+}(aq)|\text{graphite}$$

That is, in the cathode compartment, MnO_4^-, H^+, and Mn^{2+} ions are in an aqueous solution, with solid graphite immersed in it.

- If needed, the concentrations of dissolved components are given in parentheses; for example, if the concentrations of Zn^{2+} and Cu^{2+} are 1 mol/L, we write

$$Zn(s)|Zn^{2+}(1\ \text{mol/L})||Cu^{2+}(1\ \text{mol/L})|Cu(s)$$

- Half-cell components usually appear in the same order they occur in the half-reaction, and electrodes appear at the *far left* (anode) and *far right* (cathode) of the notation.
- Ions in the salt bridge are not part of the reaction, so they are not in the notation.

Sample Problem 19.7 Describing a Voltaic Cell with a Diagram and Notation

Problem Draw a diagram, show balanced equations, and write the notation for a voltaic cell that consists of one half-cell with a Cr bar in a $Cr(NO_3)_3$ solution, another half-cell with an Ag bar in an $AgNO_3$ solution, and a KNO_3 salt bridge. Measurement indicates that the Cr electrode is negative relative to the Ag electrode.

Plan From the given contents of the half-cells, we write the half-reactions. To determine which of the two half-reactions is the anode (oxidation) and which is the cathode (reduction), we note the relative electrode charges (which are based on the direction of the spontaneous redox reaction). Electrons are released into the anode during oxidation, so it has a negative charge. We are told that Cr is negative, so it is the anode and Ag is the cathode.

Solution The cell diagram is shown in the margin. We write the balanced half-reactions. The Ag half-reaction consumes e^-:

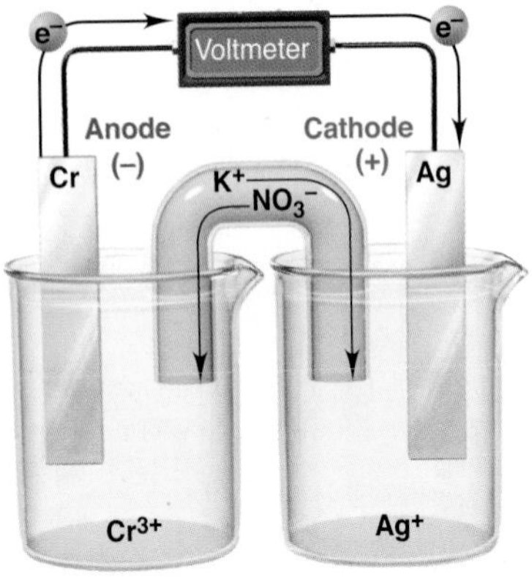

$$Ag^+(aq) + e^- \longrightarrow Ag(s) \quad \text{(reduction; cathode)}$$

The Cr half-reaction releases e^-:

$$Cr(s) \longrightarrow Cr^{3+}(aq) + 3e^- \quad \text{(oxidation; anode)}$$

To write the balanced overall cell equation, we triple the reduction half-reaction to balance e^- and combine the half-reactions:

$$Cr(s) + 3Ag^+(aq) \longrightarrow Cr^{3+}(aq) + 3Ag(s)$$

Finally, we determine the direction of electron and ion flow. The released electrons in the Cr electrode (negative) flow through the external circuit to the Ag electrode (positive). As Cr^{3+} ions enter the anode electrolyte, NO_3^- ions enter from the salt bridge to maintain neutrality. As Ag^+ ions leave the cathode electrolyte and plate out on the Ag electrode, K^+ ions enter from the salt bridge to maintain neutrality.

Write the cell notation:

$$Cr(s)|Cr^{3+}(aq)||Ag^+(aq)|Ag(s)$$

Check Always be sure that the half-reactions and the cell reaction are balanced, the half-cells contain *all* the components of the half-reactions, and the electron and ion flow are shown. Write the half-reactions from the cell notation as a check.

Comment The diagram of a voltaic cell relies on the *direction of the spontaneous reaction* to give the oxidation (anode; negative) and reduction (cathode; positive) half-reactions.

Follow-Up Problem 19.7 In one compartment of a voltaic cell, a graphite rod dips into an acidic solution of $K_2Cr_2O_7$ and $Cr(NO_3)_3$; in the other compartment, a tin bar dips into a $Sn(NO_3)_2$ solution. A KNO_3 salt bridge joins them. The tin electrode is negative relative to the graphite. Draw a diagram of the cell, and write the balanced equations and the cell notation.

Why Does a Voltaic Cell Work?

What principles explain *how* the Zn/Cu^{2+} cell reaction takes place and *why* electrons flow in the direction they do? Let us examine what is happening when the switch is open and no reaction is occurring.

In each half-cell, the metal electrode is in equilibrium, with the metal ions in the electrolyte and the electrons within the metal:

$$Zn(s) \rightleftharpoons Zn^{2+}(aq) + 2e^- \text{ (in Zn metal)}$$
$$Cu(s) \rightleftharpoons Cu^{2+}(aq) + 2e^- \text{ (in Cu metal)}$$

Given the direction of the overall spontaneous reaction, Zn loses its electrons more easily than Cu does; thus, Zn is a stronger reducing agent. Therefore, the equilibrium position of the Zn half-reaction lies farther to the right. In this reaction, the copper has a greater hold on its electrons than the zinc does. The copper ion also has a greater ability to pull electrons toward itself than the zinc ion does. When zinc metal comes into contact with a copper ion, the copper ion's ability to pull the electrons from the zinc is greater than the ability of the zinc to pull the electrons from the copper ion. As a result, zinc loses electrons to form the zinc ion and copper ions gain electrons to form copper metal. In the "tug of war" between copper ion and zinc metal, the copper ion wins and pulls the electrons toward itself. When viewed in terms of potential energy, the species with the more positive value of potential energy will be reduced and the species with the less positive value will be oxidized. As an analogy, any object with a higher potential energy will move in the direction that lowers its potential energy (water flows downhill, a piece of chalk on a slanted table will roll down), creating a potential difference (which is the difference between the water or chalk at its higher level and the water or chalk at its lower level). Thus, *the spontaneous reaction occurs as a result of the different abilities of these metals to pull electrons toward themselves.* ■

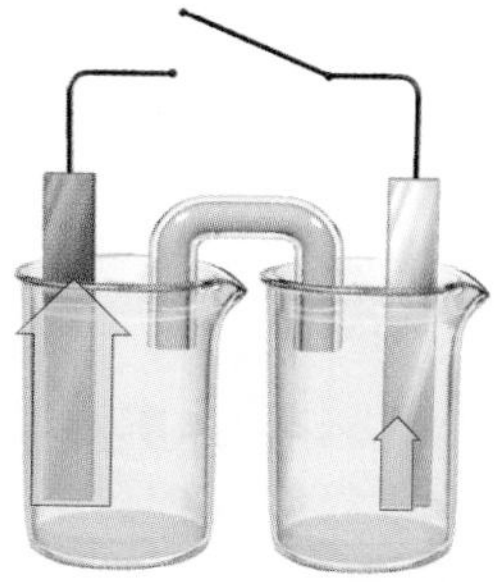

■ The two metals are on either side of the "rope" in the tug of war. The metal with the more positive reduction potential will win the tug of war and pull the electrons toward itself. The voltage we observe when the two metals come into contact is the potential difference we observe, and the flow of electrons can be likened to the movement of water down a hill (from a higher potential energy to a lower potential energy).

SUMMARY OF SECTION 19.2

- There are two types of electrochemical cells. In a voltaic cell, a spontaneous reaction generates electricity and does work on the surroundings. In an electrolytic cell, the surroundings supply electricity that does work to drive a nonspontaneous reaction.
- A voltaic cell consists of oxidation (anode) and reduction (cathode) half-cells, connected by a wire to conduct electrons and a salt bridge to maintain charge neutrality.
- Electrons move from the anode to the cathode (left to right in a standard cell diagram), while cations move from the salt bridge into the cathode half-cell and anions move from the salt bridge into the anode half-cell.
- The cell notation shows the species and their phases in each half-cell, as well as the direction of current flow.
- A voltaic cell operates because the species in the two half-cells differ in their tendency to pull electrons toward themselves (their strength as an oxidizing agent).

19.3 Cell Potential: Output of a Voltaic Cell

A voltaic cell converts the Gibbs energy change of a spontaneous reaction into electrical energy, the kinetic energy of electrons moving through an external circuit. This electrical energy is proportional to the *difference in electric potential between the two electrodes*, which is called the **cell potential (E_{cell})**; it is also commonly called the **voltage** of the cell.

Electrons flow spontaneously from the anode to the cathode, that is, toward the electrode with the more positive electrode potential (think of water flowing *spontaneously* downhill). Thus, when the cell operates *spontaneously*, there is a *positive* cell potential:

$$E_{cell} > 0 \text{ for a spontaneous process} \tag{19.1}$$

- A *positive* E_{cell} arises from a spontaneous reaction. The more positive it is, the more work the cell can do, and the farther the reaction proceeds to the right as written.
- A *negative* E_{cell}, on the other hand, is associated with a *nonspontaneous* cell reaction.
- If $E_{cell} = 0$, the reaction has reached equilibrium and the cell can do no more work. (We return to these ideas in Section 19.4.)

Standard Cell Potentials

Cell potential refers to the energy available to do the work of moving a charge between electrodes. The SI unit of electric potential is the **volt (V)**, and the SI unit of electric charge is the **coulomb (C)**. By definition, if the electric potential difference between two electrodes is 1 V (volt), 1 J (joule) of energy is released (that is, 1 J of work can be done) for each coulomb of charge that moves between the electrodes:

$$1 \text{ V} = 1 \text{ J/C} \tag{19.2}$$

Table 19.2 lists the voltages of some commercial and natural voltaic cells.

The *measured* cell potential is affected by changes in concentration as the reaction proceeds and by energy losses from heating the cell and external circuit. Therefore, as with other thermodynamic quantities, to compare the potentials of different cells, we obtain a **standard cell potential (E°_{cell})**. The standard cell potential is measured at a specified temperature (usually 298 K) with no current flowing,* *using the activities of all the components in their standard states.* In this chapter, we use the standard pressure of 1 bar for gases, the standard concentration of 1 mol/L for solutions, and an activity of 1 for the pure solid of an electrode. For example, the

*The tiny current that is required to operate modern digital voltmeters or multimeters makes a negligible difference in the value of E°_{cell}.

TABLE 19.2 Voltages of Some Voltaic Cells

Voltaic Cell	Potential (V)
Common alkaline flashlight battery	1.5
Lead-acid car battery (6 cells ≈ 12 V)	2.1
Calculator battery (mercury)	1.3
Lithium-ion laptop battery	3.7
Electric eel (~5000 cells in 1.8 m eel = 750 V)	0.15
Nerve of a giant squid (across cell membrane)	0.070

zinc-copper cell produces 1.10 V when it operates at 298 K with $[Zn^{2+}] = [Cu^{2+}] =$ 1 mol/L (Figure 19.8):

$$Zn(s) + Cu^{2+}(aq; 1\ mol/L) \longrightarrow Zn^{2+}(aq; 1\ mol/L) + Cu(s) \qquad E^\circ_{cell} = 1.10\ V$$

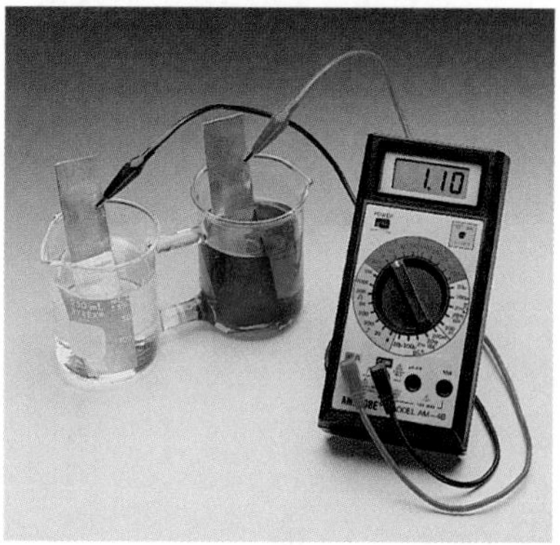

FIGURE 19.8 Measuring the standard cell potential of a zinc-copper cell

Standard Electrode (Half-Cell) Potentials Just as each half-reaction makes up part of the overall reaction, the potential of each half-cell makes up part of the overall cell potential. The **standard electrode potential ($E^\circ_{half\text{-}cell}$)** is the potential of a given half-reaction (electrode compartment) with all the components in their standard states relative to the standard hydrogen electrode.

By convention (as given in reference tables), *a standard electrode potential refers to the half-reaction written as a **reduction*** (as determined relative to a standard hydrogen electrode). For the zinc-copper reaction, both the zinc half-reaction (E°_{zinc}, anode compartment) and the copper half-reaction (E°_{copper}, cathode compartment) are written as reductions:

$$Zn^{2+}(aq) + 2e^- \longrightarrow Zn(s) \qquad E^\circ_{zinc}\ (E^\circ_{anode}) \qquad \text{(reduction)}$$
$$Cu^{2+}(aq) + 2e^- \longrightarrow Cu(s) \qquad E^\circ_{copper}\ (E^\circ_{cathode}) \qquad \text{(reduction)}$$

However, since the overall cell reaction involves the *oxidation* of zinc, not the *reduction* of Zn^{2+}, we reverse the zinc half-reaction:

$$Zn(s) \longrightarrow Zn^{2+}(aq) + 2e^- \qquad \text{(oxidation)}$$
$$Cu^{2+}(aq) + 2e^- \longrightarrow Cu(s) \qquad \text{(reduction)}$$

Calculating E°_{cell} from $E^\circ_{half\text{-}cell}$ The overall redox reaction for the zinc-copper cell is the sum of its half-reactions:

$$Zn(s) + Cu^{2+}(aq) \longrightarrow Zn^{2+}(aq) + Cu(s)$$

Because electrons flow spontaneously from the anode to the cathode, the copper electrode (cathode) has a more positive $E^\circ_{half\text{-}cell}$ than the zinc electrode (anode). Arithmetically, to obtain a positive E°_{cell} for this spontaneous redox reaction, we must subtract E°_{zinc} from E°_{copper}:

$$E^\circ_{cell} = E^\circ_{copper} - E^\circ_{zinc}$$

We can generalize this result for any voltaic cell: *the standard cell potential is the difference between the standard reduction potential of the cathode half-cell and the standard reduction potential of the anode half-cell:*

$$E^\circ_{cell} = E^\circ_{cathode} - E^\circ_{anode} \qquad (19.3)$$

For a *spontaneous* reaction at standard conditions, this calculation gives $E^\circ_{cell} > 0$.

Determining $E^\circ_{half\text{-}cell}$ with the Standard Hydrogen Electrode Suppose that we want to determine the portion of E°_{cell} contributed by the anode half-cell (oxidation of Zn) and the portion contributed by the cathode half-cell (reduction of Cu^{2+}). But how can we find potentials of the individual half-cells if we can measure only the potential of the overall cell?

Half-cell potentials, such as E°_{zinc} and E°_{copper}, are actually determined *relative* to a standard reference half-cell, *which has a standard electrode potential defined as*

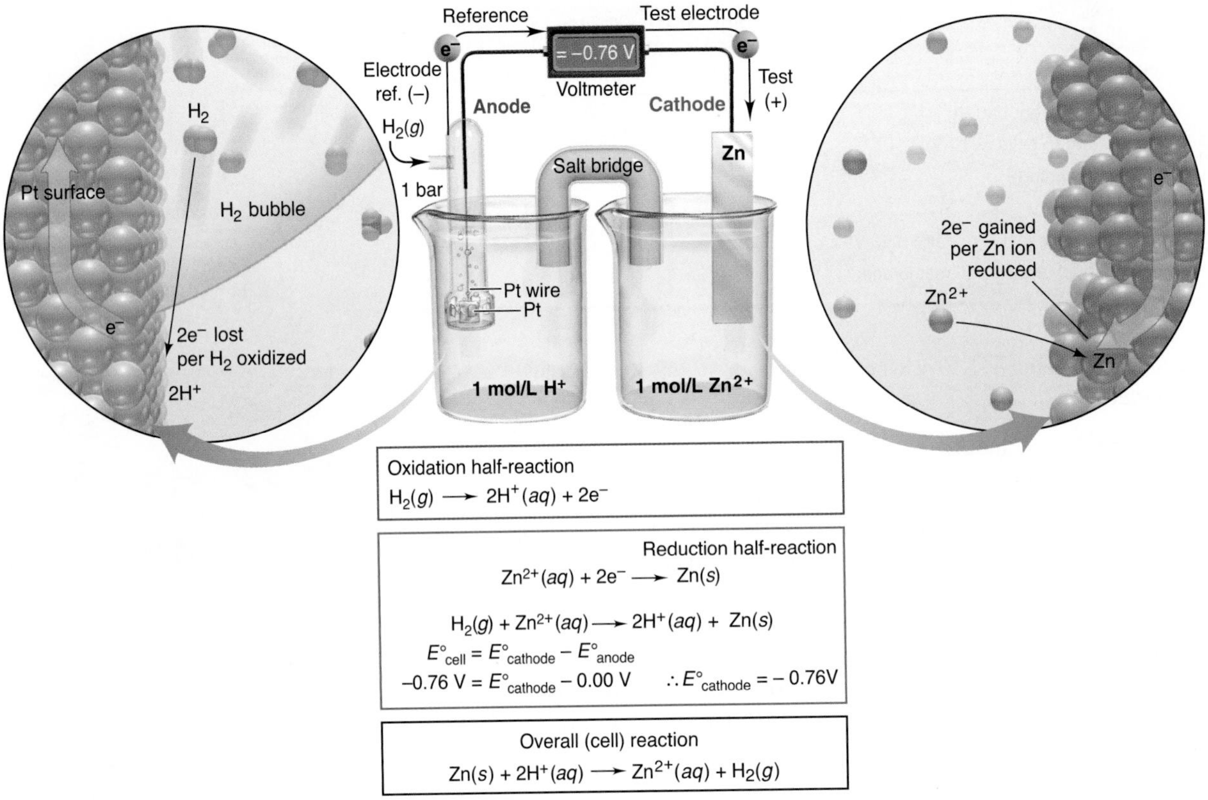

FIGURE 19.9 Determining an unknown $E^\circ_{half\text{-}cell}$ with the standard reference (hydrogen) electrode. The magnified view of the hydrogen half-reaction (*left*) shows two H^+ ions being reduced to an H_2 molecule, which enters the H_2 bubble.

zero ($E^\circ_{reference} \equiv 0.00$ V). The **standard reference half-cell (standard hydrogen electrode)** consists of a platinum electrode that has H_2 gas at 1 bar bubbling through it and is immersed in 1 mol/L strong acid, $H^+(aq)$ [or $H_3O^+(aq)$]. Thus, the reference half-reaction is

$$2H^+(aq;\ 1\text{ mol/L}) + 2e^- \rightleftharpoons H_2(g;\ 1\text{ bar}) \qquad E^\circ_{reference} = 0.00\text{ V}$$

To find an unknown standard electrode potential ($E^\circ_{unknown}$), we construct a voltaic cell consisting of this reference half-cell and the unknown half-cell. Since $E^\circ_{reference}$ is zero, the overall E°_{cell} gives $E^\circ_{unknown}$.

If we make the reference half-cell the anode, the overall cell potential as read on a digital voltmeter or multimeter is the standard reduction potential of the unknown cell:

- When the overall cell potential (which is $E^\circ_{cathode} - E^\circ_{anode}$) is positive, we can immediately see that $E^\circ_{cathode}$ is positive. This, in turn, tells us that the reaction at the cathode is a reduction and $H_2(g)$ is being oxidized to form H^+ ions:

$$E^\circ_{cell} = E^\circ_{cathode} - E^\circ_{anode} = E^\circ_{unknown} - E^\circ_{reference} = E^\circ_{unknown} - 0.00\text{ V} = E^\circ_{unknown}$$

- When the overall cell potential is negative, we can immediately see that $E^\circ_{cathode}$ is negative. This tells us that the spontaneous reaction occurring in the cell is actually the reverse of the reaction as written. The spontaneous reaction is the reaction in which the standard hydrogen electrode is the *cathode* and the unknown cell is the *anode*. In the spontaneous cell, H^+ ions are reduced to form $H_2(g)$.

Figure 19.9 shows a cell that has the Zn/Zn^{2+} half-reaction in one compartment and the H^+/H_2 half-reaction in the other compartment. The measured E°_{cell} is −0.76 V; this tells us that the reaction representing the reduction of the Zn^{2+} to $Zn(s)$ and the oxidation of $H_2(g)$ to $2H^+(aq)$ is *not* spontaneous. We can use the measured cell potential ($E^\circ_{cell} = -0.76$ V) to find the unknown standard electrode potential, E°_{zinc}:

$$\begin{array}{lll} H_2(g) \longrightarrow 2H^+(aq) + 2e^- & E^\circ_{reference} = 0.00\text{ V} & \text{(anode; oxidation)} \\ Zn^{2+}(aq) + 2e^- \longrightarrow Zn(s) & E^\circ_{zinc} = ?\text{ V} & \text{(cathode; reduction)} \\ \hline Zn^{2+}(aq) + H_2(g) \longrightarrow Zn(s) + 2H^+(aq) & E^\circ_{cell} = -0.76\text{ V} & \end{array}$$

$$E^\circ_{\text{cell}} = E^\circ_{\text{cathode}} - E^\circ_{\text{anode}} = E^\circ_{\text{zinc}} - E^\circ_{\text{reference}}$$
$$E^\circ_{\text{zinc}} = E^\circ_{\text{reference}} + E^\circ_{\text{cell}} = 0.00 \text{ V} + (-0.76 \text{ V}) = -0.76 \text{ V}$$

The spontaneous reaction between the standard hydrogen electrode and the Zn/Zn^{2+} electrode is the reaction in which the Zn(*s*) is oxidized to $Zn^{2+}(aq)$ and the $H^+(aq)$ is reduced to $H_2(g)$. This reaction has a standard cell potential of +0.76 V.

Now we can return to the zinc-copper cell and use the measured E°_{cell} (1.10 V) and the value we just found for E°_{zinc} to calculate E°_{copper}:

$$E^\circ_{\text{cell}} = E^\circ_{\text{cathode}} - E^\circ_{\text{anode}} = E^\circ_{\text{copper}} - E^\circ_{\text{zinc}}$$
$$E^\circ_{\text{copper}} = E^\circ_{\text{cell}} - E^\circ_{\text{zinc}} = 1.10 \text{ V} + (-0.76 \text{ V}) = 0.34 \text{ V}$$

By continuing this process of constructing cells with one known and one unknown electrode potential, we can find other standard electrode potentials.

Sample Problem 19.8 Calculating an Unknown $E^\circ_{\text{half-cell}}$ from E°_{cell}

Problem A voltaic cell houses the reaction between aqueous bromine and zinc metal:

$$Br_2(aq) + Zn(s) \longrightarrow Zn^{2+}(aq) + 2Br^-(aq) \qquad E^\circ_{\text{cell}} = 1.83 \text{ V}$$

Calculate E°_{bromine}, given $E^\circ_{\text{zinc}} = -0.76$ V.

Plan E°_{cell} is positive, so the reaction is spontaneous as written. By separating the reaction into half-reactions, we see that Br_2 is reduced and Zn is oxidized; thus, the zinc half-cell contains the anode. We use Equation 19.3 and the known E°_{zinc} to find E°_{unknown} (E°_{bromine}).

Solution Separate the reaction into half-reactions:

$$Br_2(aq) + 2e^- \longrightarrow 2Br^-(aq) \qquad E^\circ_{\text{unknown}} = E^\circ_{\text{bromine}} = ?\text{ V}$$
$$Zn(s) \longrightarrow Zn^{2+}(aq) + 2e^- \qquad E^\circ_{\text{zinc}} = -0.76 \text{ V}$$

Calculate E°_{bromine}:

$$E^\circ_{\text{cell}} = E^\circ_{\text{cathode}} - E^\circ_{\text{anode}} = E^\circ_{\text{bromine}} - E^\circ_{\text{zinc}}$$
$$E^\circ_{\text{bromine}} = E^\circ_{\text{cell}} - E^\circ_{\text{zinc}} = 1.83 \text{ V} + (-0.76 \text{ V}) = 1.07 \text{ V}$$

Check A good check is to make sure that calculating $E^\circ_{\text{bromine}} - E^\circ_{\text{zinc}}$ gives E°_{cell}: 1.07 V − (−0.76 V) = 1.83 V.

Comment Keep in mind that, regardless of which half-cell potential is the unknown, the reduction half-reaction occurs at the cathode and the oxidation half-reaction occurs at the anode. *Always* subtract the reduction potential E°_{anode} from the reduction potential E°_{cathode} to get E°_{cell}.

Follow-Up Problem 19.8 A voltaic cell based on the reaction between aqueous Br_2 and vanadium(III) ions has $E^\circ_{\text{cell}} = 1.39$ V:

$$Br_2(aq) + 2V^{3+}(aq) + 2H_2O(l) \longrightarrow 2VO^{2+}(aq) + 4H^+(aq) + 2Br^-(aq)$$

What is $E^\circ_{\text{vanadium}}$, the standard electrode potential for the reduction of VO^{2+} to V^{3+}?

Relative Strengths of Oxidizing and Reducing Agents

Chemists learn the relative strengths of oxidizing and reducing agents by measuring cell potentials. Three oxidizing agents just discussed are Cu^{2+}, H^+, and Zn^{2+}. We can rank their relative oxidizing strengths by writing each half-reaction as a reduction (gain of electrons), with its corresponding standard electrode potential:

$$Cu^{2}(aq) + 2e^- \longrightarrow Cu(s) \qquad E^\circ = 0.34 \text{ V}$$
$$2H^+(aq) + 2e^- \longrightarrow H_2(g) \qquad E^\circ = 0.00 \text{ V}$$
$$Zn^{2}(aq) + 2e^- \longrightarrow Zn(s) \qquad E^\circ = -0.76 \text{ V}$$

TABLE 19.3 Selected Standard Electrode Potentials (298 K)

Half-Reaction	$E^\circ_{half\text{-}cell}$ (V)
$F_2(g) + 2e^- \rightleftharpoons 2F^-(aq)$	+2.87
$Cl_2(g) + 2e^- \rightleftharpoons 2Cl^-(aq)$	+1.36
$MnO_2(s) + 4H^+(aq) + 2e^- \rightleftharpoons Mn^{2+}(aq) + 2H_2O(l)$	+1.23
$NO_3^-(aq) + 4H^+(aq) + 3e^- \rightleftharpoons NO(g) + 2H_2O(l)$	+0.96
$Ag^+(aq) + e^- \rightleftharpoons Ag(s)$	+0.80
$Fe^{3+}(aq) + e^- \rightleftharpoons Fe^{2+}(aq)$	+0.77
$O_2(g) + 2H_2O(l) + 4e^- \rightleftharpoons 4OH^-(aq)$	+0.40
$Cu^{2+}(aq) + 2e^- \rightleftharpoons Cu(s)$	+0.34
$2H^+(aq) + 2e^- \rightleftharpoons H_2(g)$	0.00
$N_2(g) + 5H^+(aq) + 4e^- \rightleftharpoons N_2H_5^+(aq)$	−0.23
$Fe^{2+}(aq) + 2e^- \rightleftharpoons Fe(s)$	−0.44
$Zn^{2+}(aq) + 2e^- \rightleftharpoons Zn(s)$	−0.76
$2H_2O(l) + 2e^- \rightleftharpoons H_2(g) + 2OH^-(aq)$	−0.83
$Na^+(aq) + e^- \rightleftharpoons Na(s)$	−2.71
$Li^+(aq) + e^- \rightleftharpoons Li(s)$	−3.05

Strength of oxidizing agent (increases upward)

Strength of reducing agent (increases downward)

The more positive the E° value, the more readily the reaction (written as a reduction) occurs; thus, Cu^{2+} gains two e^- more readily than H^+, which gains them more readily than Zn^{2+}:

- Strength as a reducing agent: $Zn > H_2 > Cu$
- Strength as an oxidizing agent: $Cu^{2+} > H^+ > Zn^{2+}$

By continuing this process with other half-cells, we can create a list of reduction half-reactions in *decreasing* order of standard electrode potential (from most positive to most negative). Such a list, called a *table of standard electrode potentials*, appears in Appendix D; a few examples are presented in Table 19.3.

There are several key points to keep in mind:

- All the values are relative to the standard hydrogen (reference) electrode.
- Since the half-reactions are written as *reductions, reactants are oxidizing agents* and *products are reducing agents.*
- The more positive the $E^\circ_{half\text{-}cell}$, the more readily the half-reaction occurs as written.
- Half-reactions are shown with an equilibrium arrow because each half-reaction can occur as a reduction (at the cathode) or an oxidation (at the anode), depending on the $E^\circ_{half\text{-}cell}$ of the half-reaction with which it is paired.
- As Appendix D (or Table 19.3) is arranged, the strength of the oxidizing agent (reactant) *increases going up (more positive standard reduction potential)*, and the strength of the reducing agent (product) *increases going down (less positive standard reduction potential).*

Thus, $F_2(g)$ is the strongest oxidizing agent (has the largest positive E°), which means that $F^-(aq)$ is the weakest reducing agent. Similarly, $Li^+(aq)$ is the weakest oxidizing agent (has the lowest value of E°), which means that $Li(s)$ is the strongest reducing agent. In other words, a *strong oxidizing agent forms a weak reducing agent*, and vice versa. Rely on your knowledge of the elements (Chapter 7) if you forget the rankings in the table:

- F_2 is very electronegative and occurs as F^-, so it is easily reduced (gains electrons) and must be a strong oxidizing agent (large and positive E°).
- Li metal has a low ionization energy and occurs as Li^+, so it is easily oxidized (loses electrons) and must be a strong reducing agent (large and negative E°).

Writing Spontaneous Redox Reactions

Every redox reaction is the sum of two half-reactions, so there is a reducing agent and an oxidizing agent on each side. In the zinc-copper reaction, for instance, Zn and Cu are the reducing agents, and Cu^{2+} and Zn^{2+} are the oxidizing agents. The stronger oxidizing and reducing agents react spontaneously to form the weaker oxidizing and reducing agents:

$$\underset{\substack{\text{stronger}\\\text{reducing agent}}}{Zn(s)} + \underset{\substack{\text{stronger}\\\text{oxidizing agent}}}{Cu^{2+}(aq)} \longrightarrow \underset{\substack{\text{weaker}\\\text{oxidizing agent}}}{Zn^{2+}(aq)} + \underset{\substack{\text{weaker}\\\text{reducing agent}}}{Cu(s)}$$

Based on the order of the $E°$ values in Appendix D, and as we just saw for the Cu^{2+}/Cu, H^+/H_2, and Zn^{2+}/Zn redox pairs (or redox *couples*), *the stronger oxidizing agent (species on the left) has a half-reaction with a larger (more positive or less negative) $E°$ value, and the stronger reducing agent (species on the right) has a half-reaction with a smaller (less positive or more negative) $E°$ value.* Therefore, we can use Appendix D to choose a redox reaction for constructing a voltaic cell.

Writing a Spontaneous Reaction with Appendix D A spontaneous reaction ($E°_{cell} > 0$) will occur between an oxidizing agent and any reducing agent that lies *below* it in the list of standard electrode potentials in Appendix D. In other words, *for a spontaneous reaction to occur, the half-reaction higher in the list proceeds at the cathode as written, and the half-reaction lower in the list proceeds at the anode in reverse.* This pairing ensures that the stronger oxidizing agent and stronger reducing agent will be the reactants. For example, two half-reactions in the order they appear in Appendix D are

$$Cl_2(g) + 2e^- \longrightarrow 2Cl^-(aq) \qquad E°_{chlorine} = 1.36\text{ V}$$
$$Ni^{2+}(aq) + 2e^- \longrightarrow Ni(s) \qquad E°_{nickel} = -0.25\text{ V}$$

We reverse the nickel half-reaction (lower in the list); notice, however, that we do *not* reverse the sign of $E°_{half\text{-}cell}$ because the minus sign in Equation 19.3 does that. Next, we have to be sure that electrons lost are equal to electrons gained. If they are not, we multiply one or both half-reactions by coefficients so that they are. In this case, they are, so we can skip this step. Adding the half-reactions and applying Equation 19.3 gives the balanced equation and $E°_{cell}$:

$$Cl_2(g) + Ni(g) \longrightarrow Ni^{2+}(aq) + 2Cl^-(aq)$$
$$E°_{cell} = E°_{cathode} - E°_{anode} = E°_{chlorine} - E°_{nickel} = 1.36\text{ V} - (-0.25\text{ V}) = 1.61\text{ V}$$

Writing a Spontaneous Reaction without Appendix D Even when a list like the one in Appendix D is not available, we can write a spontaneous redox reaction from a given pair of half-reactions. For example, here are two half-reactions:

$$Sn^{2+}(aq) + 2e^- \longrightarrow Sn(s) \qquad E°_{tin} = -0.14\text{ V}$$
$$Ag^+(aq) + e^- \longrightarrow Ag(s) \qquad E°_{silver} = 0.80\text{ V}$$

Two steps are required:

Step 1. Reverse one of the half-reactions into an oxidation step so the difference between the electrode potentials (cathode *minus* anode) gives a *positive* $E°_{cell}$. (Remember that, as above, when we reverse the half-reaction, we do *not* reverse the sign of $E°_{half\text{-}cell}$.)

Step 2. Multiply by coefficients to make electrons lost equal to electrons gained, add the rearranged half-reactions to get a balanced overall equation, and cancel the species that are common to both sides.

We want the reactants to be the stronger oxidizing and reducing agents:

- The larger (more positive) $E°$ value for the silver half-reaction means that Ag^+ is a stronger oxidizing agent (gains electrons more readily) than Sn^{2+}.
- The smaller (more negative) $E°$ value for the tin half-reaction means that Sn is a stronger reducing agent (loses electrons more readily) than Ag.

For step 1, we reverse the tin half-reaction (but *not* the sign of $E°_{tin}$):

$$Sn(s) \longrightarrow Sn^{2+}(aq) + 2e^- \qquad E°_{tin} = -0.14\text{ V}$$

Now, when we subtract $E^\circ_{\text{half-cell}}$ of the tin half-reaction (anode, oxidation) from $E^\circ_{\text{half-cell}}$ of the silver half-reaction (cathode, reduction), we get a positive E°_{cell}:

$$0.80\text{ V} - (-0.14\text{ V}) = 0.94\text{ V}$$

For step 2, the number of electrons lost in the oxidation must equal the number of electrons gained in the reduction, so we double the silver (reduction) half-reaction. Adding the half-reactions and applying Equation 19.3 gives the following:

$$\begin{array}{llr} 2\text{Ag}^+(aq) + 2\text{e}^- \longrightarrow 2\text{Ag}(s) & E^\circ_{\text{silver}} = 0.80\text{ V} & \text{(reduction)} \\ \text{Sn}(s) \longrightarrow \text{Sn}^{2}(aq) + 2\text{e}^- & E^\circ_{\text{tin}} = -0.14\text{ V} & \text{(oxidation)} \\ \hline \text{Sn}(s) + 2\text{Ag}^+(aq) \longrightarrow \text{Sn}^{2+}(aq) + 2\text{Ag}(s) & E^\circ_{\text{cell}} = E^\circ_{\text{silver}} - E^\circ_{\text{tin}} = 0.94\text{ V} & \end{array}$$

Very important: When we double the coefficients of the silver half-reaction, we do *not* double its $E^\circ_{\text{half-cell}}$. *Changing the coefficients of a half-reaction does* **not** *change its* $E^\circ_{\text{half-cell}}$ because standard electrode potential is an *intensive* property, a property that does *not* depend on the amount of substance. Let us see why. The potential is the *ratio* of energy to charge. When we change the coefficients to change the amounts, the energy *and* the charge change proportionately, so their ratio stays the same. (Similarly, density is an intensive property because the mass *and* the volume change proportionately; the density of a drop of pure water is the same as the density of a lake full of pure water.)

Sample Problem 19.9 Writing Spontaneous Redox Reactions and Ranking Oxidizing and Reducing Agents by Strength

Problem **(a)** Combine the following three half-reactions into three balanced equations for spontaneous reactions (A, B, and C), and calculate E°_{cell} for each:

(1) $\text{NO}_3^-(aq) + 4\text{H}^+(aq) + 3\text{e}^- \longrightarrow \text{NO}(g) + 2\text{H}_2\text{O}(l)$ $\quad E^\circ = 0.96\text{ V}$
(2) $\text{N}_2(g) + 5\text{H}^+(aq) + 4\text{e}^- \longrightarrow \text{N}_2\text{H}_5^+(aq)$ $\quad E^\circ = -0.23\text{ V}$
(3) $\text{MnO}_2(s) + 4\text{H}^+(aq) + 2\text{e}^- \longrightarrow \text{Mn}^{2+}(aq) + 2\text{H}_2\text{O}(l)$ $\quad E^\circ = 1.23\text{ V}$

(b) Rank the relative strengths of the oxidizing and reducing agents.

Plan **(a)** To write the redox equations, we combine the possible pairs of half-reactions: (1) and (2) give reaction A, (1) and (3) give reaction B, and (2) and (3) give reaction C. They are all written as reductions, so the oxidizing agents appear as reactants and the reducing agents appear as products. In each pair, we reverse the half-reaction that has the smaller (less positive or more negative) E° value to an oxidation in order to obtain a positive value for E°_{cell}. We make the number of electrons lost equal to the number of electrons gained, without changing the E° value; add the half-reactions together; and then use Equation 19.3 to find E°_{cell}. **(b)** Because each reaction is spontaneous as written, the stronger oxidizing agent of one reaction and the stronger reducing agent of the second reaction are the reactants. To obtain the overall ranking, we first rank the relative strengths within each equation and then compare them.

Solution **(a)** Combining half-reactions 1 and 2 gives equation A. The E° value for half-reaction 1 is larger (more positive) than the E° value for reaction 2, so we reverse reaction 2 to obtain a positive E°_{cell}:

$$\begin{array}{llr} (1)\ \text{NO}_3^-(aq) + 4\text{H}^+(aq) + 3\text{e}^- \longrightarrow \text{NO}(g) + 2\text{H}_2\text{O}(l) & & E^\circ = 0.96\text{ V} \\ (\text{rev. }2)\quad \text{N}_2\text{H}_5^+(aq) \longrightarrow \text{N}_2(g) + 5\text{H}^+(aq) + 4\text{e}^- & & E^\circ = -0.23\text{ V} \end{array}$$

To make the number of electrons lost equal to the number of electrons gained, we multiply equation 1 by 4 and the reversed equation 2 by 3; then we add the half-reactions and cancel appropriate numbers of common species (H^+ and e^-). *Note*: When multiplying the equations, we do *not* multiply the values of E°_{cell} by the same factor.

$$\begin{array}{lr} 4\text{NO}_3^-(aq) + \cancel{16}\text{H}^+(aq) + \cancel{12\text{e}^-} \longrightarrow 4\text{NO}(g) + 8\text{H}_2\text{O}(l) & E^\circ = 0.96\text{ V} \\ 3\text{N}_2\text{H}_5^+(aq) \longrightarrow 3\text{N}_2(g) + \cancel{15\text{H}^+(aq)} + \cancel{12\text{e}^-} & E^\circ = -0.23\text{ V} \\ \hline \text{(A) } 3\text{N}_2\text{H}_5^+(aq) + 4\text{NO}_3^-(aq) + \text{H}^+(aq) \longrightarrow 3\text{N}_2(g) + 4\text{NO}(g) + 8\text{H}_2\text{O}(l) & \\ E^\circ_{\text{cell}} = 0.96\text{ V} - (-0.23\text{ V}) = 1.19\text{ V} & \end{array}$$

Combining half-reactions 1 and 3 gives equation B. Half-reaction 1 has a smaller $E°$, so it is reversed:

$$\text{(rev. 1)}\quad NO(g) + 2H_2O(l) \longrightarrow NO_3^-(aq) + 4H^+(aq) + 3e^- \qquad E° = 0.96\text{ V}$$
$$\text{(3)}\quad MnO_2(s) + 4H^+(aq) + 2e^- \longrightarrow Mn^{2+}(aq) + 2H_2O(l) \qquad E° = 1.23\text{ V}$$

We multiply reversed equation 1 by 2 and equation 3 by 3, and then add and cancel:

$$2NO(g) + \cancel{4H_2O(l)} \longrightarrow 2NO_3^-(aq) + \cancel{8H^+(aq)} + \cancel{6e^-} \qquad E° = 0.96\text{ V}$$
$$3MnO_2(s) + 4\,\cancel{12}H^+(aq) + \cancel{6e^-} \longrightarrow 3Mn^{2+}(aq) + 2\,\cancel{6}H_2O(l) \qquad E° = 1.23\text{ V}$$

$$\text{(B)}\quad 3MnO_2(s) + 4H^+(aq) + 2NO(g) \longrightarrow 3Mn^{2+}(aq) + 2H_2O(l) + 2NO_3^-(aq)$$
$$E°_{cell} = 1.23\text{ V} - 0.96\text{ V} = 0.27\text{ V}$$

Combining half-reactions 2 and 3 gives equation C. Half-reaction 2 has a smaller $E°$, so it is reversed:

$$\text{(rev. 2)}\quad N_2H_5^+(aq) \longrightarrow N_2(g) + 5H^+(aq) + 4e^- \qquad E° = -0.23\text{ V}$$
$$\text{(3)}\quad MnO_2(s) + 4H^+(aq) + 2e^- \longrightarrow Mn^{2}(aq) + 2H_2O(l) \qquad E° = 1.23\text{ V}$$

We multiply reaction 3 by 2, add the half-reactions, and cancel:

$$N_2H_5^+(aq) \longrightarrow N_2(g) + \cancel{5H^+(aq)} + \cancel{4e^-} \qquad E° = -0.23\text{ V}$$
$$2MnO_2(s) + 3\,\cancel{8}H^+(aq) + \cancel{4e^-} \longrightarrow 2Mn^{2+}(aq) + 4H_2O(l) \qquad E° = 1.23\text{ V}$$

$$\text{(C)}\quad N_2H_5^+(aq) + 2MnO_2(s) + 3H^+(aq) \longrightarrow N_2(s) + 2Mn^{2+}(aq) + 4H_2O(l)$$
$$E°_{cell} = 1.23\text{ V} - (-0.23\text{ V}) = 1.46\text{ V}$$

(b) Rank the oxidizing and reducing agents within each equation:

Equation A	Oxidizing agents: $NO_3^- > N_2$	Reducing agents: $N_2H_5^+ > NO$
Equation B	Oxidizing agents: $MnO_2 > NO_3^-$	Reducing agents: $NO > Mn^{2+}$
Equation C	Oxidizing agents: $MnO_2 > N_2$	Reducing agents: $N_2H_5^+ > Mn^{2+}$

Comparing the relative strengths from the three balanced equations gives the following rankings:

Oxidizing agents: $MnO_2 > NO_3^- > N_2$
Reducing agents: $N_2H_5^+ > NO > Mn^{2+}$

Check As always, check that the atoms and charges balance on both sides of each equation. A good way to check the equations and rankings is to list the given half-reactions in order of decreasing $E°$ value:

$$MnO_2(s) + 4H^+(aq) + 2e^- \longrightarrow Mn^{2+}(aq) + 2H_2O(l) \qquad E° = 1.23\text{ V}$$
$$NO_3^-(aq) + 4H^+(aq) + 3e^- \longrightarrow NO(g) + 2H_2O(l) \qquad E° = 0.96\text{ V}$$
$$N_2(g) + 5H^+(aq) + 4e^- \longrightarrow N_2H_5^+(aq) \qquad E° = -0.23\text{ V}$$

Then the oxidizing agents (reactants) decrease in strength going down the list, so the reducing agents (products) decrease in strength going up. Moreover, each of the three spontaneous reactions (A, B, and C) should combine a reactant with a product that is lower down on the list.

Follow-Up Problem 19.9 Is the following reaction spontaneous as written?

$$3Fe^{2+}(aq) \longrightarrow Fe(s) + 2Fe^{3+}(aq)$$

If not, write the equation for the spontaneous reaction, calculate $E°_{cell}$, and rank the three species of iron in order of decreasing reducing strength.

Explaining the Activity Series of the Metals

Chemists often refer to the activity series of the metals (see Figure 19.10), which ranks the metals by their ability to "displace" one another from aqueous solution. We can use standard reduction potentials to see *why* this displacement occurs, as well as why many, but not all, metals react with acid to form H_2, and why a few metals form H_2 even in water.

1. *Metals that can displace H_2 from acid.* The standard hydrogen half-reaction represents the reduction of H^+ ions from an acid to H_2:

$$2H^+(aq) + 2e^- \longrightarrow H_2(g) \qquad E° = 0.00\text{ V}$$

FIGURE 19.10 The activity series of the metals. The most active metal (strongest reducing agent) is at the top, and the least active metal (weakest reducing agent) is at the bottom.

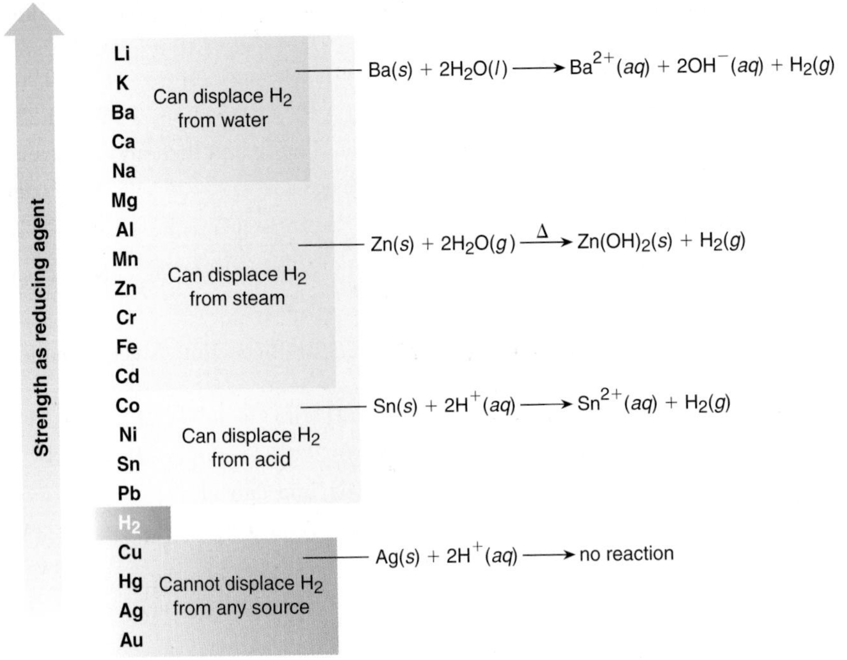

To see which metals reduce H^+ (or *displace* H_2) from acids, choose a metal, write its half-reaction as an oxidation, combine this half-reaction with the hydrogen half-reaction, and see if E°_{cell} is positive. We find that the metals Li through Pb, those that lie *below* the standard hydrogen (reference) half-reaction in Appendix D, give a positive E°_{cell} when reducing H^+. Iron, for example, reduces H^+ to H_2:

$$\begin{array}{lll} Fe(s) \longrightarrow Fe^{2+}(aq) + 2e^- & E^\circ = -0.44\text{ V} & \text{(anode; oxidation)} \\ 2H^+(aq) + 2e^- \longrightarrow H_2(g) & E^\circ = 0.00\text{ V} & \text{(cathode; reduction)} \\ \hline Fe(s) + 2H^+(aq) \longrightarrow H_2(g) + Fe^{2+}(aq) & E^\circ_{cell} = 0.00\text{ V} - (-0.44\text{ V}) = 0.44\text{ V} & \end{array}$$

The lower the metal in the list in Appendix D, the stronger it is as a reducing agent; therefore, *if E°_{cell} for the reduction of H^+ is more positive with metal A than with metal B, metal A is a stronger reducing agent than metal B and a more* ***active*** *metal.*

2. *Metals that cannot displace H_2 from acid.* For metals that are above the standard hydrogen (reference) half-reaction, E°_{cell} is negative when we reverse the metal half-reaction, so the reaction does not occur. For example, the coinage metals—copper, silver, and gold, which are in group 11—are not strong enough reducing agents to reduce H^+ from acids:

$$\begin{array}{lll} Ag(s) \longrightarrow Ag^+(aq) + e^- & E^\circ = 0.80\text{ V} & \text{(anode; oxidation)} \\ 2H^+(aq) + 2e^- \longrightarrow H_2(g) & E^\circ = 0.00\text{ V} & \text{(cathode; reduction)} \\ \hline 2Ag(s) + 2H^+(aq) \longrightarrow 2Ag^+(aq) + H_2(g) & E^\circ_{cell} = 0.00\text{ V} - 0.80\text{ V} = -0.80\text{ V} & \end{array}$$

The *higher* the metal in the list in Appendix D, the *more negative* its E°_{cell} for the reduction of H^+ to H_2, the *lower* its reducing strength, and the *less active* it is. Thus, gold is less active than silver, which is less active than copper.

3. *Metals that can displace H_2 from water.* Metals that are active enough to reduce H_2O *lie below the half-reaction*

$$2H_2O(l) + 2e^- \longrightarrow H_2(g) + 2OH^-(aq) \qquad E = -0.42\text{ V}$$

(The value of the potential shown here is the *nonstandard* electrode potential because, in pure water, $[OH^-]$ is 1.0×10^{-7} mol/L, not the standard-state value of 1 mol/L.) For example, consider the reaction of sodium in water (with the Na^+/Na half-reaction reversed and doubled):

$$\begin{array}{llll} 2Na(s) \longrightarrow 2Na^+(aq) + 2e^- & E° = -2.71\text{ V} & \text{(anode; oxidation)} \\ 2H_2O(l) + 2e^- \longrightarrow H_2(g) + 2OH^-(aq) & E° = -0.42\text{ V} & \text{(cathode; reduction)} \\ \hline 2Na(s) + 2H_2O(l) \longrightarrow 2Na^+(aq) + H_2(g) + 2OH^-(aq) & & \\ \qquad E°_{cell} = -0.42\text{ V} - (-2.71\text{ V}) = 2.29\text{ V} & & \end{array}$$

The alkali metals (group 1) and the larger alkaline earth metals (group 2) can reduce water (displace H_2 from H_2O). Calcium is shown doing this in Figure 19.11.

FIGURE 19.11 The reaction of calcium in water

4. *Metals that displace other metals from solution.* We can also predict whether one metal can reduce the aqueous ion of another metal. Any metal that is lower in the list in Appendix D can reduce the ion of a metal that is higher in the list and, thus, displace this metal from solution. For example, zinc can displace iron from solution:

$$\begin{array}{llll} Zn(s) \longrightarrow Zn^{2+}(aq) + 2e^- & E° = -0.76\text{ V} & \text{(anode oxidation)} \\ Fe^{2}(aq) + 2e^- \longrightarrow Fe(s) & E° = -0.44\text{ V} & \text{(cathode reduction)} \\ \hline Zn(s) + Fe^{2+}(aq) \longrightarrow Zn^{2+}(aq) + Fe(s) & E°_{cell} = -0.44\text{ V} - (-0.76\text{ V}) = 0.32\text{ V} & \end{array}$$

This particular reaction has tremendous economic importance in protecting iron from rusting, as we will discuss in Section 19.6.

A common incident involving the reducing power of metals occurs when we bite down, with a filled tooth, on a scrap of aluminum foil left on a piece of food. The foil acts as an active anode ($E°_{aluminum} = -1.66$ V), saliva acts as the electrolyte, and the filling (usually a silver/tin/mercury alloy) acts as an inactive cathode when O_2 is reduced to water. The circuit between the foil and the filling creates a current that is sensed as pain by the nerve of the tooth (Figure 19.12).

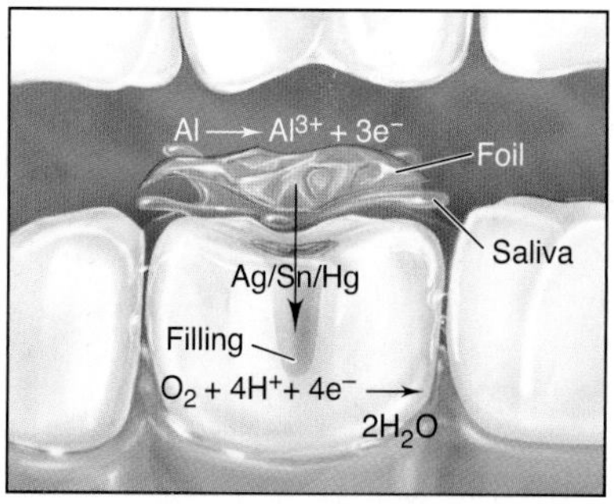

FIGURE 19.12 A dental "voltaic cell"

SUMMARY OF SECTION 19.3

- The output of a cell is the cell potential (E_{cell}), measured in volts (1 V = 1 J/C).
- With all substances in their standard states, the output is the standard cell potential ($E°_{cell}$).
- $E°_{cell} > 0$ for a spontaneous reaction at standard-state conditions.
- By convention, a standard electrode potential ($E°_{half\text{-}cell}$) refers to the *reduction* half-reaction.
- $E°_{cell}$ equals $E°_{half\text{-}cell}$ of the cathode *minus* $E°_{half\text{-}cell}$ of the anode.
- Using a standard hydrogen (reference) electrode ($E°_{reference} = 0$ V), $E°_{half\text{-}cell}$ values can be measured and used to rank oxidizing (or reducing) agents.
- Spontaneous redox reactions combine stronger oxidizing agents from one half-reaction with stronger reducing agents from another half-reaction to form weaker reducing and oxidizing agents, respectively.
- A metal can reduce another species (H^+, H_2O, or an ion of another metal) if $E°_{cell}$ for the overall reaction is positive.

19.4 Gibbs Energy and Electrical Work

In this section, we examine the relationship among useful work, Gibbs energy, and the equilibrium constant in the context of electrochemical cells and see the effect of concentration on cell potential.

Standard Cell Potential and the Equilibrium Constant

The signs of ΔG and E_{cell} are *opposite* for a spontaneous reaction: a *negative* Gibbs energy change ($\Delta G < 0$; Section 18.3) and a *positive* cell potential ($E_{cell} > 0$). These two indicators of spontaneity are proportional to each other:

$$\Delta G \propto -E_{cell}$$

Let us determine the proportionality constant. The electrical work done (w, in J) is the product of the potential (E_{cell}, in V) and the charge that flows (in C). Since E_{cell} is measured with no current flowing and, thus, no energy is lost as heat, it is the maximum voltage possible and, therefore, the maximum work possible (w_{max}).* Work done *by* the cell *on* the surroundings has a negative sign:

$$w_{max} = -E_{cell} \times \text{charge}$$

The maximum work done *on* the surroundings is equal to ΔG (Equation 18.9):

$$w_{max} = \Delta G = -E_{cell} \times \text{charge}$$

The charge, Q, that flows through the cell equals the amount (mol) of electrons transferred (z) multiplied by the charge of 1 mol of electrons (which has the symbol F):

$$\text{Charge} = \text{amount (mol) of } e^- \times \frac{\text{charge}}{\text{mol } e^-} \quad \text{or} \quad Q = zF$$

The charge of 1 mol of electrons is the **Faraday constant (*F*)**, named for Michael Faraday, the 19th-century British scientist who pioneered the study of electrochemistry:

$$F = \frac{96\,485 \text{ C}}{\text{mol } e^-}$$

Because 1 V = 1 J/C, we have 1 C = 1 J/V, and

$$F = 96\,485 \frac{\text{J}}{\text{V}\cdot\text{mol } e^-} = 96\,485 \frac{\text{C}}{\text{mol}} \qquad (19.4)$$

Substituting for charge, the proportionality constant is zF:

$$\Delta G = -zFE_{cell} \qquad (19.5)$$

And, when all components are in their standard states, we have

$$\Delta G^\circ = -zFE^\circ_{cell} \qquad (19.6)$$

Using this relationship, we can relate the standard cell potential to the equilibrium constant of the redox reaction. Recall, from Equation 18.12, that

$$\Delta G^\circ = -RT \ln K$$

Substituting for ΔG° from Equation 19.6 gives

$$-zFE^\circ_{cell} = -RT \ln K$$

Solving for E°_{cell} gives

$$E^\circ_{cell} = \frac{RT}{zF} \ln K \qquad (19.7)$$

Figure 19.13 summarizes the interconnections among the standard Gibbs energy change, the equilibrium constant, and the standard cell potential. In Chapter 18, we determined K from ΔG°, which we found either from ΔH° and ΔS° values or from

*Recall, from Chapter 18, that only a reversible process can do maximum work. For no current to flow and the process to be reversible, E_{cell} must be opposed by an equal potential in the measuring circuit: if the opposing potential is infinitesimally smaller, the cell reaction goes forward; if the opposing potential is infinitesimally larger, the reaction goes backward.

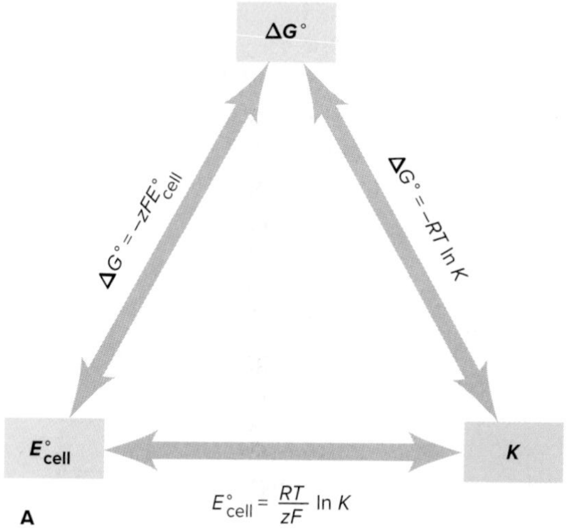

Reaction Parameters at the Standard State			
$\Delta G°$	K	$E°_{cell}$	Reaction at Standard-State Conditions
<0	>1	>0	Spontaneous
0	1	0	At equilibrium
>0	<1	<0	Nonspontaneous

B

FIGURE 19.13 The interrelationship among $\Delta G°$, $E°_{cell}$, and K. A. Any one parameter can be used to find the other two. **B.** The signs of $\Delta G°$ and $E°_{cell}$ determine the direction of the reaction.

$\Delta G°_f$ values. Now, for redox reactions, we have a direct experimental method for determining K *and* $\Delta G°$: measure $E°_{cell}$.

Sample Problem 19.10 Calculating K and $\Delta G°$ from $E°_{cell}$

Problem Lead can displace silver from solution, and silver occurs in trace amounts in some ores of lead:

$$Pb(s) + 2Ag^+(aq) \longrightarrow Pb^{2+}(aq) + 2Ag(s)$$

As a result, silver is a valuable by-product in the industrial extraction of lead from its ores. Calculate K and $\Delta G°$ at 298.15 K for this reaction.

Plan We divide the spontaneous redox equation into the half-reactions and use values from Appendix D to calculate $E°_{cell}$. Then we substitute the result into Equation 19.7 to find K and into Equation 19.6 to find $\Delta G°$.

Solution Write the half-reactions with their $E°$ values:

(1) $Ag^+(aq) + e^- \longrightarrow Ag(s) \qquad E° = 0.80\ V$

(2) $Pb^{2+}(aq) + 2e \longrightarrow Pb(s) \qquad E° = -0.13\ V$

To calculate $E°_{cell}$, we double equation 1, reverse equation 2, add the half-reactions, and subtract $E°_{lead}$ from $E°_{silver}$:

$$\begin{array}{lll} 2Ag^+(aq) + 2e^- \longrightarrow 2Ag(s) & & E° = 0.80\ V \\ Pb(s) \longrightarrow Pb^{2+}(aq) + 2e^- & & E° = -0.13\ V \\ \hline Pb(s) + 2Ag^+(aq) \longrightarrow Pb^{2+}(aq) + 2Ag(s) & & E°_{cell} = 0.80\ V - (-0.13\ V) = 0.93\ V \end{array}$$

Next, we need to calculate K with Equation 19.7. The adjusted half-reactions show that 2 mol of electrons is transferred per mole of reaction as written, so $z = 2$. Running the cell at 25°C (298 K), we have

$$E°_{cell} = \frac{RT}{zF} \ln K = 0.93\ V$$

So,

$$\ln K = \frac{(0.93\ V)(2\ mol\ e^-)(96\,485\ C/mol)}{[8.314\ J/(mol{\cdot}K)](298\ K)} = 72.43 \quad \text{and} \quad K = 2.9 \times 10^{31}$$

Calculate $\Delta G°$ (Equation 19.6):

$$\Delta G° = -zFE°_{cell} = -2\ mol\ e^- \times \frac{96\,485\ C}{mol\ e^-} \times 0.93\ V = -1.8 \times 10^5\ J/mol$$

$$= -1.8 \times 10^5\ J/mol = -1.8 \times 10^2\ kJ/mol$$

Check The three variables are consistent with the reaction being spontaneous at standard-state conditions: $E^\circ_{cell} > 0$, $\Delta G^\circ < 0$, and $K > 1$. Be sure to round and check the order of magnitude: to find ΔG°, for instance, $\Delta G^\circ \approx -2 \times 100 \times 1 = -200$, so the overall arithmetic seems right. Another check is to obtain ΔG° directly from its relation with K:

$$\Delta G^\circ = -RT \ln K = -8.314\ \text{J/(mol·K)} \times 298.15\ \text{K} \times \ln(2.6 \times 10^{31})$$
$$= -1.8 \times 10^5\ \text{J/mol} = -1.8 \times 10^2\ \text{kJ/mol}$$

Remember: All values of the Gibbs energy are per mole of the reaction as written. The value of z is also for the reaction as written.

Follow-Up Problem 19.10 When cadmium metal reduces Cu^{2+} in solution, Cd^{2+} forms in addition to copper metal. Given that $\Delta G^\circ = -143$ kJ/mol, calculate K at 25°C. What is E°_{cell} of a voltaic cell that uses this reaction?

The Effect of Concentration on Cell Potential

So far, we have considered cells at standard-state conditions. However, most cells do not start with standard-state concentrations, and, even if they did, concentrations change as a cell operates. Moreover, in all batteries, reactant concentrations are far from the standard state.

To determine E_{cell}, the cell potential under *nonstandard* conditions, we need to derive an expression for the relationship between E_{cell} and concentration based on the relationship between ΔG and concentration. Recall from Chapter 18, Equation 18.13, that

$$\Delta G = \Delta G^\circ + RT \ln Q$$

ΔG is related to E_{cell} and ΔG° to E°_{cell} (Equations 19.5 and 19.6), so we substitute for them and get

$$-zFE_{cell} = -zFE^\circ_{cell} + RT \ln Q$$

Dividing both sides by $-zF$, we obtain the equation developed by the great German chemist Walther Hermann Nernst when he was only 25 years old. (In his career, which culminated in the 1920 Nobel Prize, he also formulated the third law of thermodynamics and established the concept of the solubility product.) According to the **Nernst equation**, E_{cell} depends on E°_{cell} *and* a term for the potential at any ratio of activities:

$$E_{cell} = E^\circ_{cell} - \frac{RT}{zF} \ln Q \qquad (19.8)$$

How do changes in Q affect cell potential? From Equation 19.8, we see the following:

- When $Q < 1$ and thus [reactant] > [product], $\ln Q < 0$, so $E_{cell} > E^\circ_{cell}$.
- When $Q = 1$ and thus [reactant] = [product], $\ln Q = 0$, so $E_{cell} = E^\circ_{cell}$.
- When $Q > 1$ and thus [reactant] < [product], $\ln Q > 0$, so $E_{cell} < E^\circ_{cell}$.

Remember that the expression for Q *contains only species whose activities are other than 1 (that is, species with concentrations and/or pressures that can vary)*; thus, solids are not included, even when they are the electrodes. Also remember that, for aqueous terms, the concentration (mol/L) must be used as an approximation for the activity and, for gaseous terms, the pressure (bar) must be used as an approximation for the activity. For example, in the reaction between cadmium and silver ion, the Cd and Ag electrodes do not appear in the expression for Q:

$$Cd(s) + 2Ag^+(aq) \longrightarrow Cd^{2+}(aq) + 2Ag(s) \qquad Q = \frac{[Cd^{2+}]}{[Ag^+]^2}$$

Sample Problem 19.11 Using the Nernst Equation to Calculate E_{cell}

Problem In a test of a new reference electrode, a chemist constructs a voltaic cell consisting of a Zn/Zn^{2+} half-cell and an H_2/H^+ half-cell under the following conditions:

$$[Zn^{2+}] = 0.010 \text{ mol/L} \qquad [H^+] = 2.5 \text{ mol/L} \qquad p_{H_2} = 0.30 \text{ bar}$$

Calculate E_{cell} at 298 K.

Plan To apply the Nernst equation and determine E_{cell}, we must know E°_{cell} and Q. We write the spontaneous reaction and calculate E°_{cell} from the standard electrode potentials (Appendix D). Then we substitute into Equation 19.8.

Solution Determine the cell reaction and E°_{cell}:

$$2H^+(aq) + 2e^- \longrightarrow H_2(g) \qquad E^\circ = 0.00 \text{ V}$$
$$Zn(s) \longrightarrow Zn^{2+}(aq) + 2e^- \qquad E^\circ = -0.76 \text{ V}$$
$$2H^+(aq) + Zn(s) \longrightarrow H_2(g) + Zn^{2+}(aq) \qquad E^\circ_{cell} = 0.00 \text{ V} - (-0.76\text{V}) = 0.76 \text{ V}$$

In the expression for Q, we must use the activities of the reactants and products. We can approximate the activities of $Zn^{2+}(aq)$ and $H^+(aq)$ as being their concentrations (mol/L). We can approximate the activity of $H_2(g)$ as being its pressure (bar).

Calculate Q:

$$Q = \frac{p_{H_2}[Zn^{2+}]}{[H^+]^2} = \frac{(0.30)(0.010)}{(2.5)^2} = 4.8 \times 10^{-4}$$

Solve for E_{cell} at 298 K, with $z = 2$:

$$E_{cell} = E^\circ_{cell} - \frac{RT}{zF}\ln Q$$
$$= 0.76 \text{ V} - \left[\frac{[8.314 \text{ J/(mol·K)}](298 \text{ K})}{(2 \text{ mol})(96\,485 \text{ C/mol})} \ln (4.8 \times 10^{-4})\right]$$
$$= 0.76 \text{ V} - (-0.098 \text{ V}) = 0.86 \text{ V}$$

Check After you check the arithmetic, reason through the answer: $E_{cell} > E^\circ_{cell}$ ($0.86 > 0.76$) because the $\ln Q$ term was negative, which is consistent with $Q < 1$.

Follow-Up Problem 19.11 Consider a voltaic cell based on the following reaction:

$$Fe(g) + Cu^{2+}(aq) \longrightarrow Fe^{2+}(aq) + Cu(s)$$

If $[Cu^{2+}] = 0.30$ mol/L, what must $[Fe^{2+}]$ be to increase E_{cell} by 0.25 V above E°_{cell} at 25°C?

Changes in Potential during Cell Operation

As with any voltaic cell, the potential of the zinc-copper cell changes during cell operation because the concentrations of the components change. Since both of the other components are solids, the only variables are $[Cu^{2+}]$ and $[Zn^{2+}]$:

$$Zn(s) + Cu^{2+}(aq) \longrightarrow Zn^{2+}(aq) + Cu(s) \qquad Q = \frac{[Zn^{2+}]}{[Cu^{2+}]}$$

In this section, we follow the potential as the zinc-copper cell operates.

1. *Starting point of cell operation.* The positive E°_{cell} (1.10 V) means that the reaction proceeds *spontaneously* to the right at standard-state conditions, $[Zn^{2+}] = [Cu^{2+}] = 1$ mol/L ($Q = 1$). However, if we start the cell when $[Zn^{2+}] < [Cu^{2+}]$ ($Q < 1$)—for example, when $[Zn^{2+}] = 1.0 \times 10^{-4}$ mol/L and $[Cu^{2+}] = 2.0$ mol/L—the cell potential starts out *higher* than the standard cell potential:

$$E_{cell} = E^\circ_{cell} - \frac{RT}{zF}\ln\frac{[Zn^{2+}]}{[Cu^{2+}]} = 1.10\text{V} - \left[\frac{(8.314 \text{ J/[mol·K]})(298 \text{ K})}{(2 \text{ mol})(96\,485 \text{ C/mol})}\ln\left(\frac{1.0 \times 10^{-4}}{2.0}\right)\right]$$
$$= 1.10 \text{ V} - \left[\frac{(8.314 \text{ J/[mol·K]})(298 \text{ K})}{(2 \text{ mol})(96\,485 \text{ C/mol})}(-9.90)\right] = 1.10 \text{ V} + 0.127 \text{ V} = 1.23 \text{ V}$$

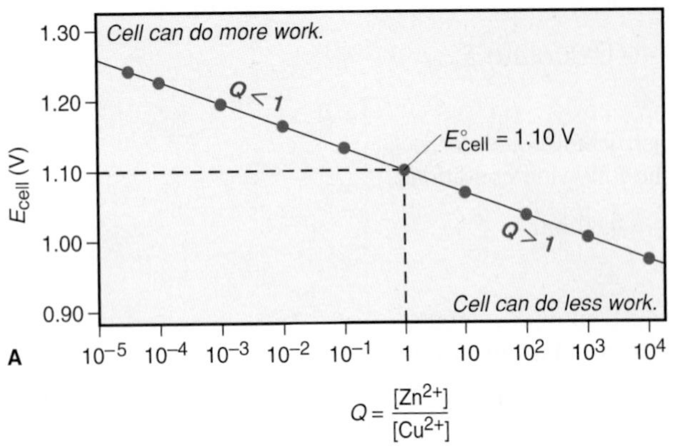

A

B

Changes in E_{cell} and Concentration			
Stage in Cell Operation	**Q**	**Relative [P] and [R]**	$\left(\frac{RT}{zF}\right)$ **ln Q**
1. $E > E°$	<1	[P] < [R]	<0
2. $E = E°$	=1	[P] = [R]	=0
3. $E < E°$	>1	[P] > [R]	>0
4. $E = 0$	=K	[P] ≫ [R]	=$E°$

FIGURE 19.14 The relation between E_{cell} and Q for the zinc-copper cell. **A.** A plot of E_{cell} versus Q (on a logarithmic scale) decreases linearly. When $Q < 1$ (*left*), the cell does relatively more work. When $Q = 1$, $E_{cell} = E^\circ_{cell}$. When $Q > 1$ (*right*), the cell does relatively less work. **B.** A summary of the changes in E_{cell} as this or any voltaic cell operates. ($[Zn^{2+}]$ is [P] for [product], and $[Cu^{2+}]$ is [R] for [reactant].)

2. *Key stages during cell operation.* Using Equation 19.8, we identify four key stages of operation. Figure 19.14A shows the first three stages. Notice that, *as the cell operates, its potential decreases.* As $[Zn^{2+}]$ increases and $[Cu^{2+}]$ decreases, Q becomes larger, the $\frac{RT}{zF}\ln Q$ term becomes less negative (more positive), and E_{cell} decreases:

Stage 1. $E_{cell} > E^\circ_{cell}$ when $Q < 1$: When the cell begins operation, $[Cu^{2+}] > [Zn^{2+}]$, so $\frac{RT}{zF}\ln Q < 0$ and $E_{cell} > E^\circ_{cell}$.

Stage 2. $E_{cell} = E^\circ_{cell}$ when $Q = 1$: At the point when $[Cu^{2+}] = [Zn^{2+}]$, $Q = 1$, so $\frac{RT}{zF}\ln Q = 0$ and $E_{cell} = E^\circ_{cell}$.

Stage 3. $E_{cell} < E^\circ_{cell}$ when $Q > 1$: As the $\frac{[Zn^{2+}]}{[Cu^{2+}]}$ ratio continues to increase, $\frac{RT}{zF}\ln Q > 0$, so $E_{cell} < E^\circ_{cell}$.

Stage 4. $E_{cell} = 0$ when $Q = K$: Eventually, $\frac{RT}{zF}\ln Q$ becomes so large that it equals E°_{cell}, which means that E_{cell} equals zero. This occurs *when the system reaches* ***equilibrium***: *no more Gibbs energy is released, so the cell can do no more work.* At this point, we say that a battery is "dead." (In real life, however, the battery is effectively dead long before the cell potential reaches 0 V, since most applications require a minimum operating voltage well above 0 V.)

Figure 19.14B summarizes these four key stages during the operation of a voltaic cell.

3. *Q/K and the work the cell can do.* At equilibrium, Equation 19.8 becomes

$$0 = E^\circ_{cell} - \frac{RT}{zF}\ln K$$

which rearranges to

$$E^\circ_{cell} = \frac{RT}{zF}\ln K$$

Notice that this result is identical to Equation 19.7, which we obtained from ΔG°. Solving for K of the zinc-copper cell ($E^\circ_{cell} = 1.10$ V),

$$\ln K = \frac{zFE^\circ_{cell}}{RT} \quad \text{so} \quad K = e^{[(2\text{ mol})(96\,485\text{ C/mol})(1.10\text{ V})]/[(8.314\text{ J/[mol·K]})(298\text{ K})]} = e^{85.68} = 1.6 \times 10^{37}$$

As you can see, this cell does work until $\frac{[Zn^{2+}]}{[Cu^{2+}]}$ is *very* high.

The three relations between initial $\frac{Q}{K}$ and E_{cell} are summarized below:

- If $\frac{Q}{K} < 1$, E_{cell} is positive for the reaction *as written*. The smaller $\frac{Q}{K}$ is, the greater the value of E_{cell}, and the more electrical work the cell can do.
- If $\frac{Q}{K} = 1$, $E_{cell} = 0$. The cell is at equilibrium and can no longer do work.
- If $\frac{Q}{K} > 1$, E_{cell} is negative for the reaction *as written*. The reverse reaction will take place, and the cell will do work until $\frac{Q}{K}$ equals 1 at equilibrium.

Concentration Cells

If you mix a concentrated solution of a substance with a dilute solution of the substance, the final solution has an intermediate concentration. A **concentration cell** employs this simple, spontaneous change to generate electrical energy. The two solutions are in separate half-cells, so they do not mix, but their concentrations become equal as the cell operates.

Finding E_{cell} for a Concentration Cell Suppose that a voltaic cell has the Cu/Cu^{2+} half-reaction in both compartments. The cell reaction is the sum of identical half-reactions, written in opposite directions. The *standard* cell potential, E°_{cell}, is zero because the *standard* electrode potentials are both based on 1 mol/L Cu^{2+}, so they cancel. *In a concentration cell, however, the concentrations are **different**.* Thus, even though E°_{cell} is still zero, the *nonstandard* cell potential, E_{cell}, depends on the *ratio of concentrations*, so it is *not* zero.

For the final concentration to be equal, a concentration cell must have the dilute solution in the anode compartment and the concentrated solution in the cathode compartment. For example, let us use 0.10 mol/L Cu^{2+} in the anode half-cell and 1.0 mol/L Cu^{2+}, a 10-fold higher concentration, in the cathode half-cell (Figure 19.15A):

$$\mathrm{Cu}(s) \longrightarrow \mathrm{Cu}^{2+}(aq;\ 0.10\ \mathrm{mol/L}) + 2e^- \quad \text{(anode; oxidation)}$$

$$\mathrm{Cu}^{2+}(aq;\ 1.0\ \mathrm{mol/L}) + 2e^- \longrightarrow \mathrm{Cu}(s) \quad \text{(cathode; reduction)}$$

The overall cell reaction is the sum of the half-reactions:

$$\mathrm{Cu}^{2+}(aq;\ 1.0\ \mathrm{mol/L}) \longrightarrow \mathrm{Cu}^{2+}(aq;\ 0.10\ \mathrm{mol/L}) \qquad E_{cell} = ?$$

The cell potential at the initial concentrations of 0.10 mol/L (dilute) and 1.0 mol/L (concentrated), with $z = 2$, is obtained from the Nernst equation:

$$E_{cell} = E^\circ_{cell} - \frac{RT}{zF}\ln\frac{[\mathrm{Cu}^{2+}]_{dil}}{[\mathrm{Cu}^{2+}]_{conc}} = 0\ \mathrm{V} - \left[\frac{(8.314\ \mathrm{J/[mol{\cdot}K]})(298\ \mathrm{K})}{(2\ \mathrm{mol})(96\,485\ \mathrm{C/mol})}\ln\left(\frac{0.10\ \mathrm{mol/L}}{1.0\ \mathrm{mol/L}}\right)\right]$$

$$= 0\ \mathrm{V} - \left[\frac{(8.314\ \mathrm{J/[mol{\cdot}K]})(298\ \mathrm{K})}{(2\ \mathrm{mol})(96\,485\ \mathrm{C/mol})}(-2.30)\right] = 0.0296\ \mathrm{V}$$

Since E°_{cell} is zero, E_{cell} depends entirely on the $\frac{RT}{zF}\ln Q$ term. Although this seems to be an insignificantly small voltage, it does correspond to a ΔG value of -5.71 kJ, which is reasonably large.

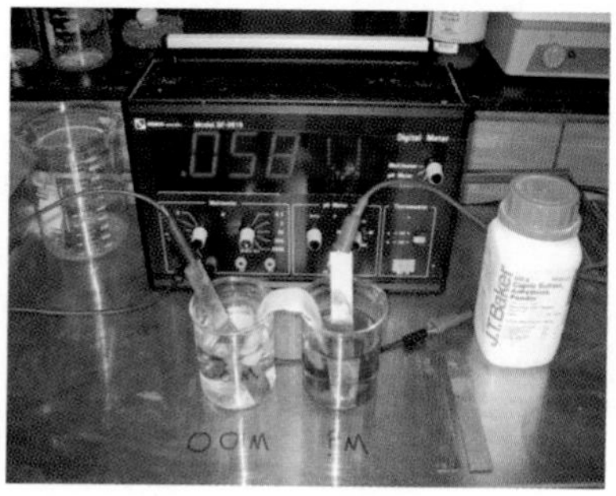

An example of a concentration cell with $[Cu^{2+}] = 0.01$ mol/L on the left and $[Cu^{2+}] = 1.0$ mol/L on the right

How a Concentration Cell Works Let us see what is happening as the voltaic cell with the Cu/Cu^{2+} half-reactions operates:

- *In the anode (dilute) half-cell,* Cu atoms in the electrode give up electrons. The resulting Cu^{2+} ions enter the solution and make it *more* concentrated.
- *In the cathode (concentrated) half-cell,* Cu^{2+} ions gain the electrons. The resulting Cu atoms plate out on the electrode, which makes this solution *less* concentrated.

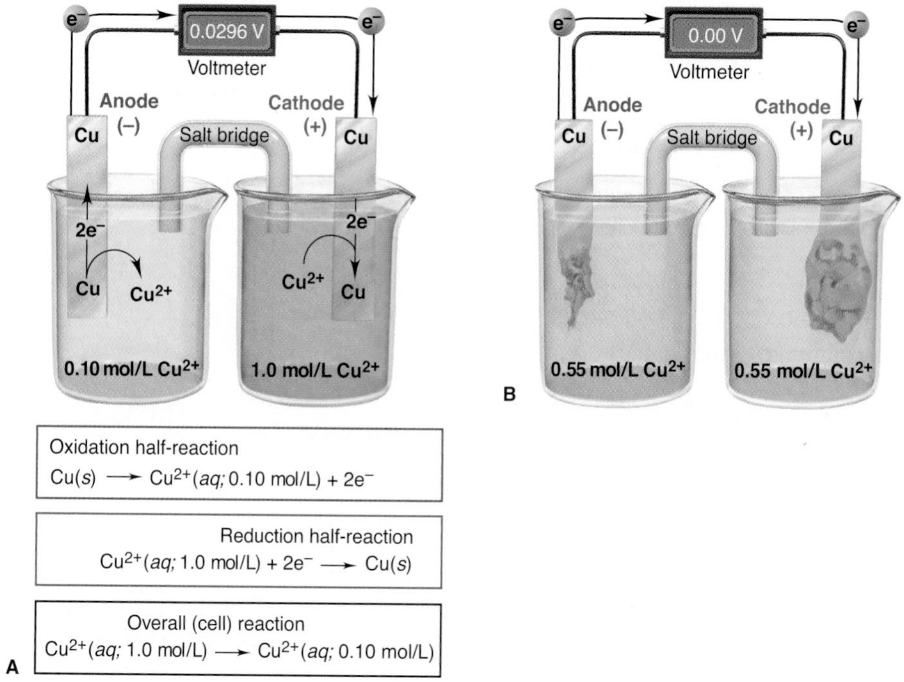

FIGURE 19.15 A concentration cell based on the Cu/Cu^{2+} half-reaction. **A.** $E_{cell} > 0$ as long as the half-cell concentrations are different. **B.** Half-cell concentrations are equal (same colour), and the sizes of the electrodes (exaggerated for clarity) are different.

As in any voltaic cell, E_{cell} decreases until equilibrium is attained, which happens when $[Cu^{2+}]$ is the same in both half-cells (Figure 19.15B). The same final concentration would result if we mixed the two solutions, but no electrical work would be done.

Sample Problem 19.12 Calculating the Potential of a Concentration Cell

Problem A concentration cell consists of two Ag/Ag^+ half-cells. In half-cell A, the electrolyte is 0.010 mol/L $AgNO_3$; in half-cell B, the electrolyte is 4.0×10^{-4} mol/L $AgNO_3$. What is the cell potential at 298 K?

Plan The standard half-cell reactions are identical, so E°_{cell} is zero, and we find E_{cell} from the Nernst equation. Half-cell A has a higher $[Ag^+]$, so Ag^+ ions are reduced and plate out on electrode A. In half-cell B, Ag atoms of the electrode are oxidized and Ag^+ ions enter the solution. As in all voltaic cells, reduction occurs at the cathode, so it is positive.

Solution First, we need to write the spontaneous reaction. The $[Ag^+]$ decreases in half-cell A and increases in half-cell B, so the spontaneous reaction is

$$Ag^+(aq;\ 0.010\ \text{mol/L})\ [\text{half-cell A}] \longrightarrow Ag^+(aq;\ 4.0\times10^{-4}\ \text{mol/L})\ [\text{half-cell B}]$$

Calculate E_{cell}, with $z = 1$:

$$E_{cell} = E^\circ_{cell} - \frac{RT}{zF}\ln\frac{[Ag^+]_{dil}}{[Ag^+]_{conc}} = 0\ \text{V} - \left[\frac{[8.314\ \text{J/(mol·K)}](298\ \text{K})}{(1\ \text{mol})(96\,485\ \text{C/mol})}\ln\left(\frac{4.0\times10^{-4}}{0.010}\right)\right]$$

$$= 0.0827\ \text{V}$$

Follow-Up Problem 19.12 A concentration cell is built using two Au/Au^{3+} half-cells. In half-cell A, $[Au^{3+}] = 7.0 \times 10^{-4}$ mol/L; in half-cell B, $[Au^{3+}] = 2.5 \times 10^{-2}$ mol/L. What is E_{cell}, and which electrode is negative?

Applications of Concentration Cells The principle of a concentration cell has many applications. We will discuss three of these applications here.

1. *The pH meter.* The most important laboratory application of this principle involves measuring $[H^+]$. If we construct a concentration cell in which the cathode compartment is the standard hydrogen electrode and the anode compartment has the same apparatus dipping into a solution of unknown $[H^+]$, the half-reactions and overall reaction are

$$\begin{array}{lll} H_2(g;\ 1\ \text{bar}) \longrightarrow 2H^+(aq;\ \text{unknown}) + 2e^- & & (\text{anode; oxidation}) \\ 2H^+(aq;\ 1\ \text{mol/L}) + 2e^- \longrightarrow H_2(g;\ 1\ \text{bar}) & & (\text{cathode; reduction}) \\ \hline 2H^+(aq;\ 1\ \text{mol/L}) \longrightarrow 2H^+(aq;\ \text{unknown}) & E_{cell} = ? & \end{array}$$

E°_{cell} is zero, but E_{cell} is *not*, because the half-cells differ in $[H^+]$. From the Nernst equation, with $z = 2$, we have

$$E_{cell} = E^\circ_{cell} - \frac{RT}{zF}\ln\frac{[H^+]^2_{unknown}}{[H^+]^2_{standard}}$$

Substituting 1 mol/L for $[H^+]_{standard}$ and 0 V for E°_{cell} gives

$$E_{cell} = 0\ \text{V} - \frac{RT}{zF}\ln\frac{[H^+]^2_{unknown}}{1^2} = -\frac{[8.314\ \text{J/(mol·K)}](298\ \text{K})}{(2\ \text{mol})(96\,485\ \text{C/mol})}\ln[H^+]^2_{unknown}$$

Because $\ln x^2 = 2\ln x$ (see Appendix A), we obtain

$$E_{cell} = -0.0257\ \text{V}\ln[H^+]$$

Thus, by measuring E_{cell}, we can find the pH.

For routine lab measurement of pH, a concentration cell made of two hydrogen electrodes is too bulky and difficult to maintain. Instead, a pH meter (Figure 19.16A)

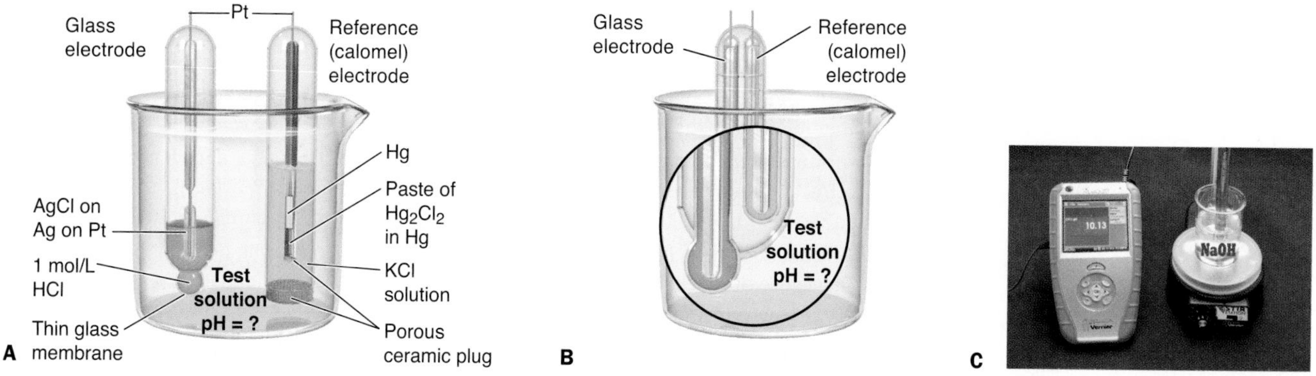

FIGURE 19.16 **Laboratory measurement of pH. A.** An older-style pH meter includes a glass electrode (*left*) and a reference calomel electrode (*right*). **B.** Modern pH meters use a combination electrode. **C.** A modern pH meter.

is used. In a common, but older, design, two separate electrodes dip into the solution being tested:

- The *glass electrode* consists of an Ag/AgCl half-reaction immersed in HCl solution (usually 1.000 mol/L) and enclosed in a thin (~0.05 mm) membrane made of a glass that is very sensitive to H^+ ions.
- The *reference electrode*, usually a *saturated calomel electrode*, consists of a platinum wire immersed in calomel (Hg_2Cl_2) paste, liquid Hg, and saturated KCl solution.

The glass electrode monitors the solution's $[H^+]$ relative to its own fixed internal $[H^+]$, and the meter converts the potential difference that develops at the glass membrane/solution interface into a measured pH. Both the Ag/AgCl electrode and the external reference electrode provide constant offset. In a modern pH meter, a *combination* electrode, which houses both electrodes in one tube, is used (Figure 19.16B and 19.16C).

TABLE 19.4 Some Ions Measured with Ion-Specific Electrodes

Species Detected	Typical Sample
NH_3/NH_4^+	Industrial wastewater, seawater
CO_2/HCO_3^-	Blood, groundwater
F^-	Drinking water, urine, soil, industrial stack gases
Br^-	Grain, plant tissue
I^-	Milk, pharmaceuticals
NO_3^-	Soil, fertilizer, drinking water
K^+	Blood serum, soil, wine
H^+	Laboratory solutions, soil, natural waters

2. *Measuring ions selectively.* The pH electrode is one type of *ion-selective (or ion-specific) electrode.* This type of electrode is designed with specialized membranes to measure specific ion concentrations in a mixture of many ions, as in natural waters and soils. Biologists implant a tiny ion-selective electrode in a single cell to study ion channels and receptors. Recent advances allow measurement in the femtomolar (10^{-15} mol/L) range. Table 19.4 shows a few of the ions that are studied.

3. *Concentration cells in nerves.* A nerve cell membrane is embedded with enzyme "gates" that use the energy of one-third of the body's ATP to create an ion gradient of low $[Na^+]$ and high $[K^+]$ inside, and high $[Na^+]$ and low $[K^+]$ outside. As a result of these differences, the outside of a nerve cell is more positive than the inside. (The 1997 Nobel Prize in Chemistry was shared by Jens C. Skou for elucidating this mechanism.) When the nerve membrane is stimulated, an electrical impulse is created as Na^+ ions rush in spontaneously, causing the inside to become more positive than the outside. This is followed by K^+ ions rushing out spontaneously, causing the outside to become more positive again. Together, the whole process takes only about 0.002 s! The large changes in charge in one membrane region stimulate the neighbouring region, and the electrical impulse moves along the cell (Figure 19.17).

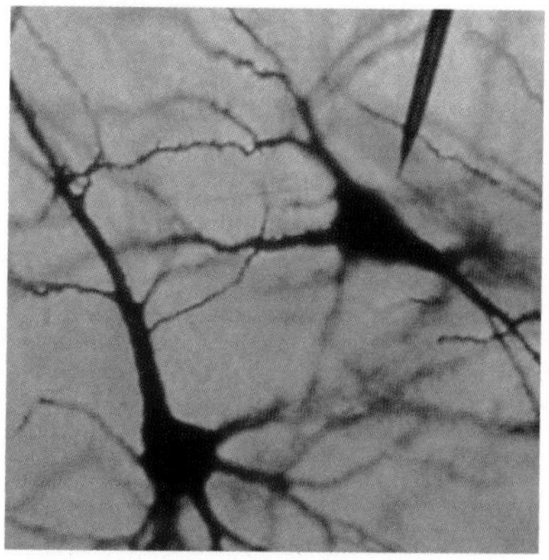

FIGURE 19.17 **Minimicroanalysis.** A microelectrode records electrical impulses of a single neuron in the visual cortex of a monkey.

SUMMARY OF SECTION 19.4

- A spontaneous process has a negative ΔG and a positive E_{cell}: $\Delta G = -zFE_{cell}$. The ΔG of the cell reaction represents the maximum electrical work the cell can do.
- The standard Gibbs energy change, $\Delta G°$, is related to $E°_{cell}$ and K.
- For nonstandard conditions, the Nernst equation shows that E_{cell} depends on $E°_{cell}$ and a correction term based on Q. E_{cell} is high when Q is small (high [reactant]), and it decreases as the cell operates. At equilibrium, ΔG and E_{cell} are zero, which means that $Q = K$.
- Concentration cells have identical half-reactions, but solutions with different concentrations. They generate electrical energy as the concentrations become equal.
- Ion-specific electrodes, such as the pH electrode, measure the concentration of one species.
- The principle of the concentration cell—spontaneous movement of ions across a concentration gradient—creates an electrical impulse in a nerve cell.

19.5 Electrochemical Processes in Batteries

Because of their compactness and mobility, batteries play a major role in everyday life. In our increasingly wireless world, this role is growing. In general, a **battery** consists of self-contained voltaic cells arranged in series (plus-to-minus-to-plus, and so on), so that their individual voltages are added. In this section, we examine the three categories of batteries: primary, secondary, and fuel cell.

Primary (Nonrechargeable) Batteries

A *primary battery* cannot be recharged, so it is discarded when the cell reaction has reached equilibrium, that is, when the battery is "dead." We will discuss the alkaline battery, mercury and silver "button" batteries, and the primary lithium battery.

Alkaline Battery Invented in the 1860s, the common dry cell—the precursor of today's ubiquitous alkaline battery—was a familiar item into the 1970s. The anode is a zinc can that houses a mixture of MnO_2 and a weakly acidic electrolyte paste, consisting of NH_4Cl, $ZnCl_2$, H_2O, and starch. Powdered graphite improves conductivity, and the cathode is an inactive graphite rod. Even today, the cathode half-reaction is not completely understood, but we know that it involves the reduction of $MnO_2(s)$ to $Mn_2O_3(s)$ and an acid-base reaction between NH_4^+ and OH^-. At a high current drain, this reaction yields NH_3 gas, which can cause a large voltage drop. Moreover, because the zinc anode reacts with the acidic NH_4^+, dry cells have a short shelf life.

The alkaline battery avoids these drawbacks. The same electrode materials, zinc and manganese dioxide, are used, but the electrolyte is a basic paste of KOH and water (Figure 19.18). The half-reactions are essentially the same, but KOH eliminates the buildup of NH_3 gas and maintains the Zn electrode:

$$Zn(s) + 2OH^-(aq) \longrightarrow ZnO(s) + H_2O(l) + 2e^- \quad \text{(anode; oxidation)}$$
$$MnO_2(s) + 2H_2O(l) + 2e^- \longrightarrow Mn(OH)_2(s) + 2OH^-(aq) \quad \text{(cathode; reduction)}$$
$$Zn(s) + MnO_2(s) + H_2O(l) \longrightarrow ZnO(s) + Mn(OH)_2(s) \qquad E_{cell} = 1.5\text{ V}$$

FIGURE 19.18 The alkaline battery

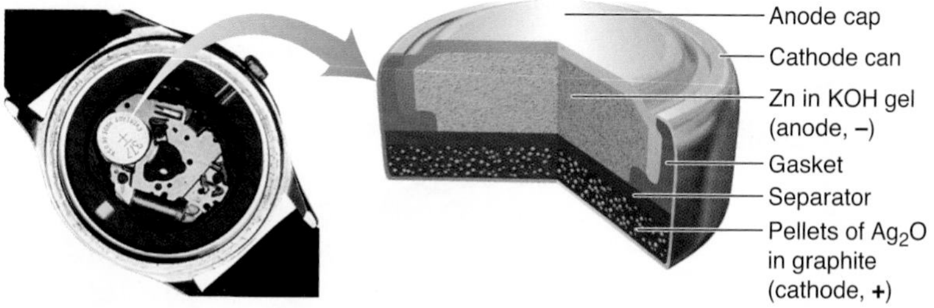

FIGURE 19.19 The silver button battery

The alkaline battery is used to power portable radios, toys, flashlights, and so on. It is safe and comes in many sizes. It has no voltage drop, a longer shelf life, and better performance than the dry cell in terms of power capability and stored energy.

Silver (Button) Battery The silver battery uses a zinc container in a basic medium as the anode (reducing agent) and Ag_2O inside a steel can as the cathode (oxidizing agent). The solid zinc, mixed with KOH, is separated from the Ag_2O by moist paper. The half-reactions are

$$\mathrm{Zn}(s) + 2\mathrm{OH}^-(aq) \longrightarrow \mathrm{ZnO}(s) + \mathrm{H_2O}(l) + 2e^- \quad \text{(anode; oxidation)}$$

$$\mathrm{Ag_2O}(s) + \mathrm{H_2O}(l) + 2e^- \longrightarrow 2\mathrm{Ag}(s) + 2\mathrm{OH}^-(aq) \quad \text{(cathode; reduction)(silver)}$$

$$\mathrm{Zn}(s) + \mathrm{Ag_2O}(s) \longrightarrow \mathrm{ZnO}(s) + 2\mathrm{Ag}(s) \qquad E_{\text{cell}} = 1.6\ \mathrm{V}$$

The silver battery is manufactured as a button-sized battery (Figure 19.19), which is used in watches, cameras, heart pacemakers, and hearing aids because of its very steady output. Its major disadvantage is the increasingly high cost of silver.

Primary Lithium Battery The primary lithium battery is also widely used in watches, implanted medical products, and remote-controlled devices. It offers an extremely high energy/mass ratio, producing 1 mol of electrons (1 F) from less than 7 g of metal ($\mathcal{M}$ of Li = 6.941 g/mol). The anode is lithium foil in a nonaqueous electrolyte. The cathode is one of several metal oxides in which lithium ions lie between oxide layers. Some pacemakers have a silver vanadium oxide (SVO; $AgV_2O_{5.5}$) cathode and can deliver up to 3.5 V for several years, but at a low rate (Figure 19.20). Lithium batteries often have high energy density, but, since they use nonaqueous lower conductivity electrolytes, they offer lower power capabilities. The half-reactions are

$$3.5\mathrm{Li}(s) \longrightarrow 3.5\mathrm{Li}^+ + 3.5e^- \quad \text{(anode; oxidation)}$$

$$\mathrm{AgV_2O_{5.5}} + 3.5\mathrm{Li}^+ + 3.5e^- \longrightarrow \mathrm{Li_{3.5}AgV_2O_{5.5}} \quad \text{(cathode; reduction)}$$

$$\mathrm{AgV_2O_{5.5}} + 3.5\mathrm{Li}(s) \longrightarrow \mathrm{Li_{3.5}AgV_2O_{5.5}}$$

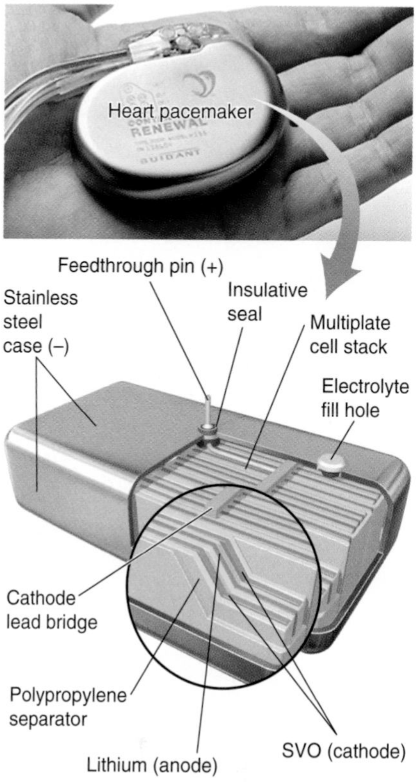

FIGURE 19.20 The lithium battery

Secondary (Rechargeable) Batteries

In contrast to a primary battery, a *secondary battery* is *rechargeable*; when it runs down, *electrical energy is supplied to reverse the cell reaction* and form more reactant. In other words, in a secondary battery, the voltaic cells are periodically converted to electrolytic cells to restore *nonequilibrium* concentrations of the cell components. We will discuss the common lead-acid car battery, the nickel–metal hydride battery, and the lithium-ion battery (a secondary lithium battery).

Lead-Acid Battery A typical lead-acid car battery has six cells connected in series, each delivering about 2.1 V for a total of about 12 V. Each cell contains two lead grids packed with high-surface-area (spongy) Pb in the anode and high-surface-area PbO_2 in the cathode. The grids are immersed in a solution of ~4.5 mol/L H_2SO_4. Fibreglass sheets between the grids prevent shorting due to physical contact (Figure 19.21).

1. *Discharging.* When the cell discharges as a voltaic cell, it generates electrical energy:

$$\mathrm{Pb}(s) + \mathrm{HSO_4}^-(aq) \longrightarrow \mathrm{PbSO_4}(s) + \mathrm{H}^- + 2e^- \quad \text{(anode; oxidation)}$$

$$\mathrm{PbO_2}(s) + 3\mathrm{H}^+(aq) + \mathrm{HSO_4}^-(aq) + 2e^- \longrightarrow \mathrm{PbSO_4}(s) + 2\mathrm{H_2O}(l) \quad \text{(cathode; reduction)}$$

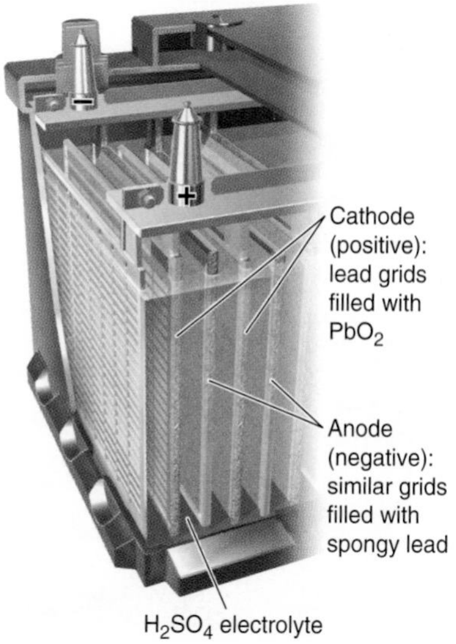

FIGURE 19.21 The lead-acid battery

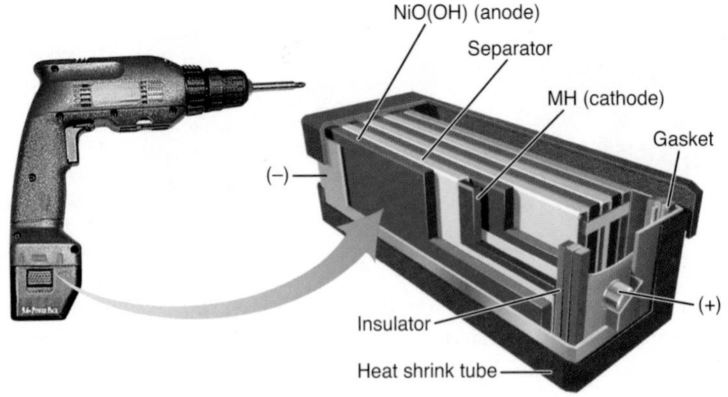

FIGURE 19.22 The nickel–metal hydride battery

Both half-reactions form Pb^{2+} ions, one through the oxidation of Pb, the other through the reduction of PbO_2. The Pb^{2+} forms $PbSO_4(s)$ at both electrodes by reacting with HSO_4^-.

Overall (cell) reaction (discharge):

$$PbO_2(s) + Pb(s) + 2H_2SO_4(aq) \longrightarrow 2PbSO_4(s) + 2H_2O(l) \quad E_{cell} = 2.1 \text{ V}$$

2. *Recharging.* When the cell recharges as an electrolytic cell, it uses electrical energy, and the half-cell and overall reactions are reversed.

Overall (cell) reaction (recharge):

$$2PbSO_4(s) + 2H_2O(l) \longrightarrow PbO_2(s) + Pb(s) + 2H_2SO_4(aq)$$

Car and truck owners have relied on the lead-acid battery for over a century to provide the large burst of current needed to start the engine—even in hot and cold weather. The main problems with the lead-acid battery are loss of capacity due to corrosion of the positive (Pb) grid, detachment of the active material due to normal mechanical bumping, and formation of large $PbSO_4$ crystals that hinder recharging. Increasingly, another problem posed by the lead-acid battery is lead toxicity, causing difficulty with disposing of or recycling the spent batteries.

Nickel–Metal Hydride (Ni-MH) Battery Concerns about the toxicity of cadmium in the nickel-cadmium (nicad) battery have led to its replacement by the nickel–metal hydride battery. The anode half-reaction oxidizes the hydrogen absorbed within a metal alloy (such as $LaNi_5$; designated M) in a basic (KOH) electrolyte, while nickel(III) in the form of NiO(OH) is reduced at the cathode (Figure 19.22):

$$MH(s) + OH^-(aq) \longrightarrow M(s) + H_2O(l) + e^- \quad \text{(anode; oxidation)}$$

$$NiO(OH)(s) + H_2O(l) + e^- \longrightarrow Ni(OH)_2(s) + OH^-(aq) \quad \text{(cathode; reduction)}$$

$$MH(s) + NiO(OH)(s) \longrightarrow M(s) + Ni(OH)_2(s) \quad E_{cell} = 1.4 \text{ V}$$

The cell reaction is reversed during recharging. The Ni-MH battery is common in cordless razors, camera flash units, and power tools. It is lightweight, has high power, and is nontoxic, but it discharges significantly during storage. Other issues of consideration are the cost of these batteries and their low-temperature performance.

Lithium-Ion Battery The secondary lithium-ion battery has an anode of Li atoms that lie between sheets of graphite (designated Li_xC_6). The cathode is a lithium metal oxide, such as $LiMn_2O_4$ or $LiCoO_2$, and a typical electrolyte is 1 mol/L $LiPF_6$ in an organic solvent, such as dimethyl carbonate mixed with methylethyl carbonate. Electrons flow through the circuit, while solvated Li^+ ions flow from anode to cathode within the cell (Figure 19.23). The cell reactions are

$$Li_xC_6 \longrightarrow xLi^+ + xe^- + C_6(s) \quad \text{(anode; oxidation)}$$

$$Li_{1-x}Mn_2O_4(s) + xLi^+ + xe^- \longrightarrow LiMn_2O_4(s) \quad \text{(cathode; reduction)}$$

$$Li_xC_6 + Li_{1-x}Mn_2O_4(s) \longrightarrow LiMn_2O_4(s) + C_6(s) \quad E_{cell} = 3.7 \text{ V}$$

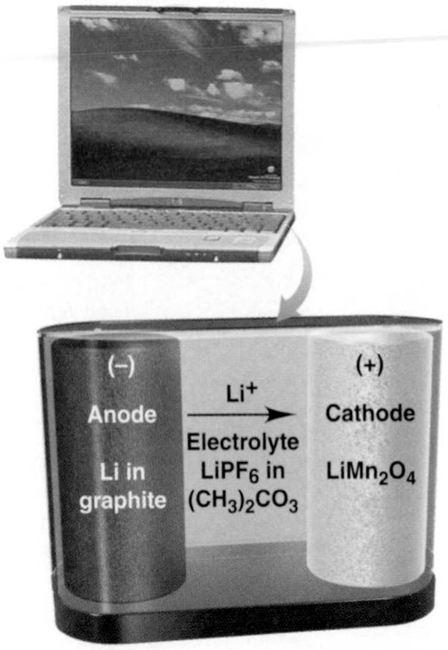

FIGURE 19.23 The lithium-ion battery

The cell reaction is reversed during recharging. The lithium-ion battery powers laptop computers, cellphones, and digital cameras. Its key drawbacks are cost and flammability of the organic solvent.

New Advances in Batteries In spite of all the amazing technology described above, research into new batteries continues with increasing intensity due to the number of devices needing batteries and the desire for shorter recharging times, longer life, and decreasing cost. Some of the incredible work being done involves new flexible batteries such as the ones being developed by ProLogium, a company based in Taiwan. This solid state cell is capable of being bent and twisted like a piece of gum, and it continues to function even if parts of it are cut, poked, or pierced, making it unique (normally, batteries may explode if they are damaged in any way). A nonflexible version of this battery has been incorporated into the Power Flip case for the HTC One Max, but the applications for the flexible battery are endless.

Another amazing development is the aluminum graphite battery developed at Stanford (Figure 19.24). It is potentially capable of recharging a cellphone in 60 s and, in tests, has undergone 7500 charge-discharge cycles without loss of capacity, as opposed to a typical lithium ion battery, which can only go through 1000 cycles. The aluminum graphite battery is also flexible—it can be bent and folded—and not only is aluminum cheaper than lithium, it has none of the flammability issues of lithium.

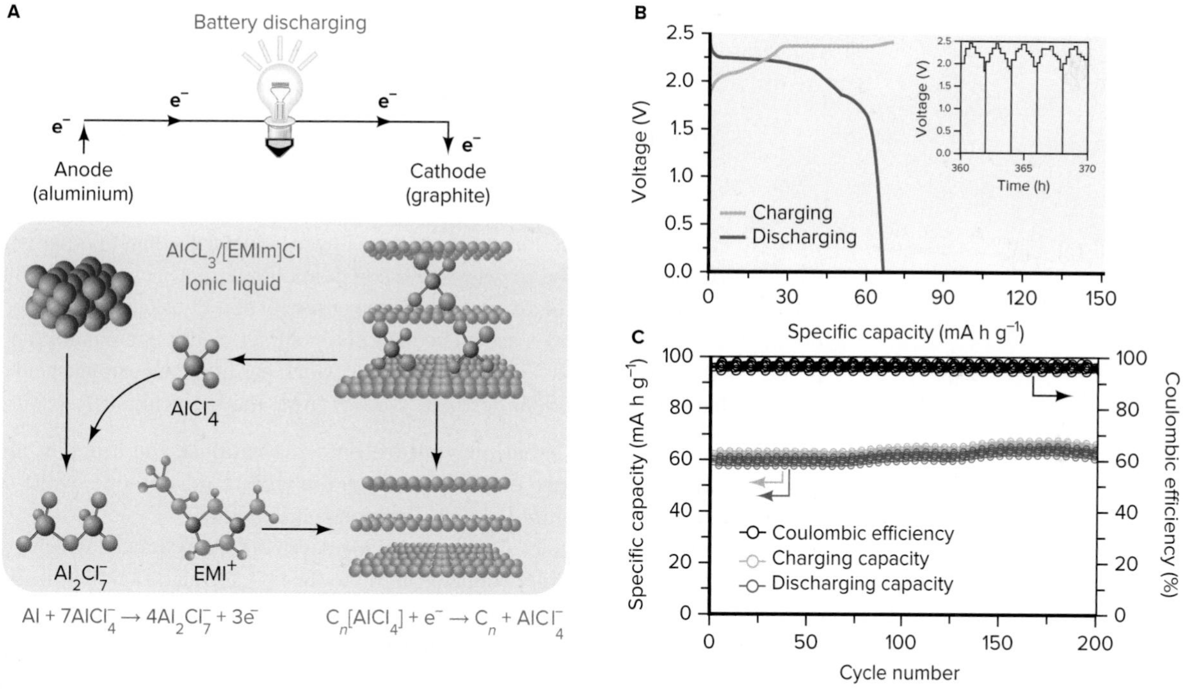

FIGURE 19.24 Some information about the Al-Graphite battery being developed at Stanford. A. The basic working of the battery and its composition. **B.** The voltage as a function of the specific capacity of the battery for charging and discharging. **C.** The battery maintains its efficiency and capacity over a number of recharges.

Source: M-C Lin et al., *Nature* 520, 324–328 (16 April 2015).

While there is still progress required before these batteries can be fully commercially applied, they are just a few examples of the new applications of electrochemistry in the field of battery development.

Fuel Cells

In contrast to primary and secondary batteries, a **fuel cell** is not self-contained. The reactants (usually a fuel and oxygen) enter the cell, and the products leave, *generating electricity through controlled combustion* of the fuel. The fuel does not burn because, as in other voltaic cells, the half-reactions are separated, and the electrons move through an external circuit.

FIGURE 19.25 The hydrogen fuel cell

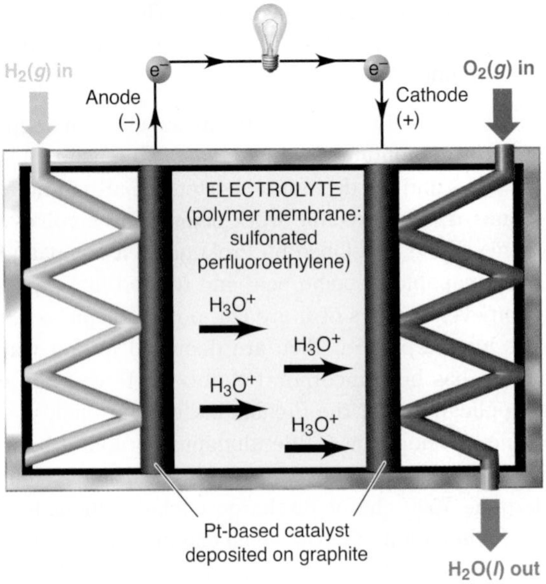

The most common fuel cell being developed for use in cars is the *proton exchange membrane (PEM) cell,* which uses H_2 as the fuel and has an operating temperature of around 80°C (Figure 19.25). The cell reactions are

$$\begin{array}{rcll} 2H_2(g) & \longrightarrow & 4H^+(aq) + 4e^- & \text{(anode; oxidation)} \\ O_2(g) + 4H^+(aq) + 4e^- & \longrightarrow & 2H_2O(g) & \text{(cathode; reduction)} \\ \hline 2H_2(g) + O_2(g) & \longrightarrow & 2H_2O(g) & E_{\text{cell}} = 1.2\ \text{V} \end{array}$$

How H_2 Fuel Cells Work Reaction rates are lower in fuel cells than in other batteries, so an *electrocatalyst* is used to decrease the activation energy (Section 14.7). The PEM cell's electrodes are made of a nanocomposite consisting of a Pt-based catalyst deposited on graphite. These electrodes are embedded in a polymer electrolyte membrane with a perfluoroethylene backbone ($—[F_2C—CF_2]_n—$), which has attached sulfonic acid groups (RSO_3^-) that play a key role in ferrying protons from anode to cathode.

- *At the anode,* two H_2 molecules adsorb onto the catalyst and are split and oxidized. From each H_2, two e^- travel through the wire to the cathode, while two H^+ are hydrated and migrate through the electrolyte as H_3O^+.
- *At the cathode,* an O_2 molecule is believed to adsorb onto the catalyst, which provides an e^- to form O_2^-. One H_3O^+ donates its H^+ to the O_2^-, forming HO_2 (that is, HO—O). The O—O bond stretches and breaks as another H_3O^+ gives its H^+ and the catalyst provides another e^-: the first H_2O has formed. In a similar manner, a third H^+ and e^- attach to the freed O atom to form OH, and a fourth H^+ and e^- are transferred to form the second H_2O. Both water molecules desorb and leave the cell.

Applications of Fuel Cells Hydrogen fuel cells have been used for years to provide electricity *and* water during space flights. Similar fuel cells have begun to supply electrical power for residential needs, and every major car manufacturer has a fuel-cell prototype. Fuel cells produce no pollutants and convert about 75% of a fuel's bond energy into power, compared with 40% for a coal-fired power plant and 25% for a gasoline engine.

SUMMARY OF SECTION 19.5

- Batteries are voltaic cells arranged in series. They are classified as primary (for example, alkaline, silver, and lithium), secondary (for example, lead-acid, nickel–metal hydride, and lithium-ion), or fuel cell.
- Supplying electricity to a rechargeable (secondary) battery reverses the redox reaction, re-forming the reactant.
- Fuel cells generate a current through the controlled oxidation of a fuel such as H_2.

19.6 Corrosion: An Environmental Voltaic Cell

If you think all spontaneous electrochemical processes are useful, consider the problem of **corrosion**, which causes tens of billions of dollars of damage to cars, ships, buildings, and bridges each year. This natural process, which oxidizes metals to their oxides and sulfides, has similarities to the operation of a voltaic cell. In this section, we focus on the corrosion of iron, but remember that many metals, such as copper and silver, also corrode.

The Corrosion of Iron

The most common and economically destructive form of corrosion is the rusting of iron. Amounts as high as 25% of the steel produced in North America each year are used to replace steel in which the iron has corroded. Rust is *not* a direct product of the reaction between iron and oxygen, but it arises through a complex electrochemical process. Let us look at the facts of iron corrosion and then use the features of a voltaic cell to explain them:

Fact 1. Iron does not rust in dry air; moisture must be present.
Fact 2. Iron does not rust in air-free water; oxygen must be present.
Fact 3. Iron loss and rust formation occur at *different* places on the *same* object.
Fact 4. Iron rusts more quickly at a low pH (a high $[H^+]$).
Fact 5. Iron rusts more quickly in ionic solutions.
Fact 6. Iron rusts more quickly in contact with a less active metal (such as Cu) and more slowly in contact with a more active metal (such as Zn).

Two separate redox processes occur during corrosion:

1. *The loss of iron.* Picture the surface of a piece of iron (Figure 19.26). A strain, ridge, or dent in contact with water is typically the site of iron loss (fact 1). This site is called an *anodic region* because the following half-reaction occurs there:

$$Fe(s) \longrightarrow Fe^{2+}(aq) + 2e^- \quad \text{(anodic region; oxidation)}$$

Once the iron atoms lose electrons, the damage to the object has been done, and a pit forms where the iron has been lost.

The freed electrons move through the external circuit—the piece of iron itself—and reach a region of relatively high $[O_2]$ (fact 2), usually the air near the edge of a water droplet that surrounds the newly formed pit. At this *cathodic region*, the electrons released from the iron atoms reduce O_2:

$$O_2(g) + 4H^+(aq) + 4e^- \longrightarrow 2H_2O(l) \quad \text{(cathodic region; reduction)}$$

This part of the corrosion process (the sum of these two half-reactions) occurs without any rust forming:

$$2Fe(s) + O_2(g) + 4H^+(aq) \longrightarrow 2Fe^{2+}(aq) + 2H_2O(l) \quad \text{(overall)}$$

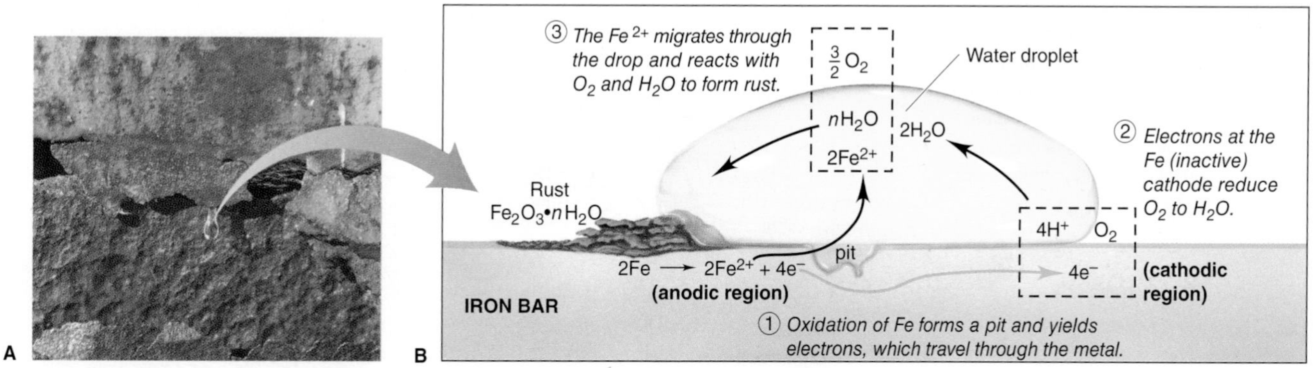

FIGURE 19.26 The corrosion of iron. **A.** False-colour close-up view of an iron surface. Corrosion usually occurs at a surface irregularity. **B.** A small area of the surface, showing the steps in the corrosion process.

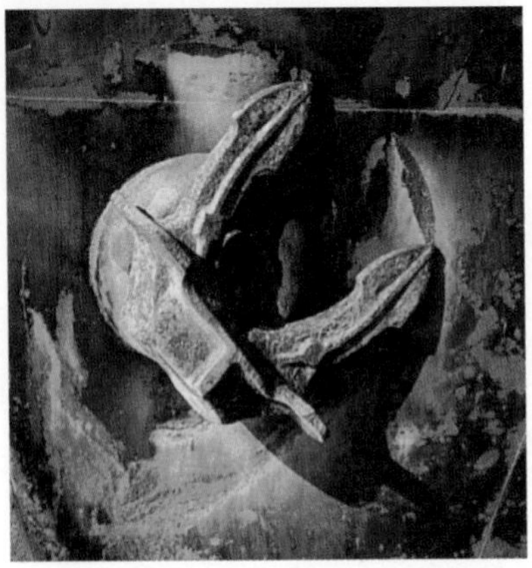

FIGURE 19.27 Enhanced corrosion at sea. The high ion concentration of seawater enhances the corrosion of iron in hulls and anchors.

2. *The rusting process.* Rust forms in another redox reaction. The Fe^{2+} ions that are formed at the anodic region disperse through the water and react with O_2, often away from the pit, to form the Fe^{3+} in rust (fact 3). The overall reaction for this step is

$$2Fe^{2+}(aq) + \frac{1}{2}O_2(g) + (2+n)H_2O(l) \longrightarrow Fe_2O_3{\cdot}nH_2O(s) + 4H^+(aq)$$

(H_2O has the coefficient n because rust, $Fe_2O_3{\cdot}nH_2O$, has a variable number of waters of hydration.) The rust deposit is incidental to the real damage, which is the loss of iron that weakens the strength of the object. Adding the two previous equations gives the overall equation for the loss and rusting of iron:

$$2Fe(s) + \frac{3}{2}O_2(g) + nH_2O(l) + \cancel{4H^+(aq)} \longrightarrow Fe_2O_3{\cdot}nH_2O(s) + \cancel{4H^+(aq)}$$

Other species ($2Fe^{2+}$ and $2H_2O$) also cancel, but we show the cancelled H^+ ions to emphasize that they act as a catalyst: they speed up the process as they are used in one step and created in another. For this reason, rusting is faster at a low pH (high $[H^+]$) (fact 4). Ionic solutions speed rusting by improving the conductivity of the aqueous medium near the anodic and cathodic regions (fact 5). The effect of ions is especially evident on ocean-going vessels (Figure 19.27) and on the underbodies and around the wheel wells of cars driven in cold climates, where salts are used to melt ice on slippery roads. (We will discuss fact 6 in the next subsection.)

Thus, in some key ways, corrosion resembles the operation of a voltaic cell:

- Anodic and cathodic regions are physically separated.
- The regions are connected via an external circuit through which the electrons travel.
- Iron behaves like an active electrode in the anodic region but is inactive in the cathodic region.
- The moisture surrounding the pit functions somewhat like an electrolyte and salt bridge, a solution of ions and a means for them to move and keep the solution neutral.

Protecting against the Corrosion of Iron

Corrosion is prevented by eliminating corrosive factors. Washing road salt off a car removes ions. Painting an object keeps out O_2 and moisture. Plating chromium on plumbing fixtures is a more permanent method, as is "blueing" of gun barrels and other steel objects, in which a coating of Fe_3O_4 (magnetite) is bonded to the surface.

The final fact about corrosion (fact 6) concerns the relative activity of other metals in contact with iron. The essential idea is that *iron functions as both the anode and the cathode in the rusting process, but it is lost only at the anode*:

1. *Corrosion increases when iron behaves more like the anode.* When iron is in contact with a *less* active metal (weaker reducing agent), such as copper, it loses electrons more readily (its anodic function is enhanced; Figure 19.28A). For example, when iron plumbing is connected directly to copper plumbing, the iron pipe corrodes rapidly. Nonconducting rubber or plastic spacers are placed between the metals to avoid this problem.

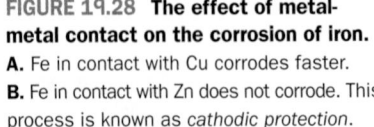

FIGURE 19.28 The effect of metal-metal contact on the corrosion of iron.
A. Fe in contact with Cu corrodes faster.
B. Fe in contact with Zn does not corrode. This process is known as *cathodic protection*.

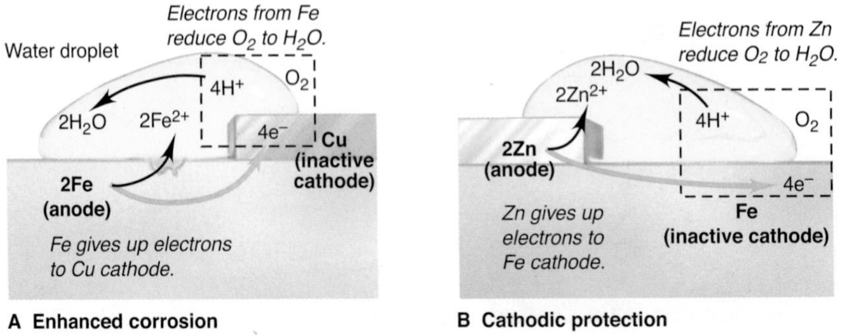

2. *Corrosion decreases when iron behaves more like the cathode.* In *cathodic protection*, the most effective way to prevent corrosion, iron makes contact with a *more* active metal (stronger reducing agent), such as zinc. The iron becomes the cathode and remains intact, while the zinc acts as the anode and loses electrons (Figure 19.28B). Coating steel with a "sacrificial" layer of zinc is called *galvanizing*. In addition to blocking physical contact with H_2O and O_2, the zinc (or another active metal) is "sacrificed" (oxidized) instead of the iron. Sacrificial anodes are used underwater and underground to protect iron and steel pipes, tanks, oil rigs, and so on. Magnesium and aluminum are often used because they are much more active than iron and, thus, act as the anode (Figure 19.29). Moreover, they form adherent oxide coatings, which slow their own corrosion.

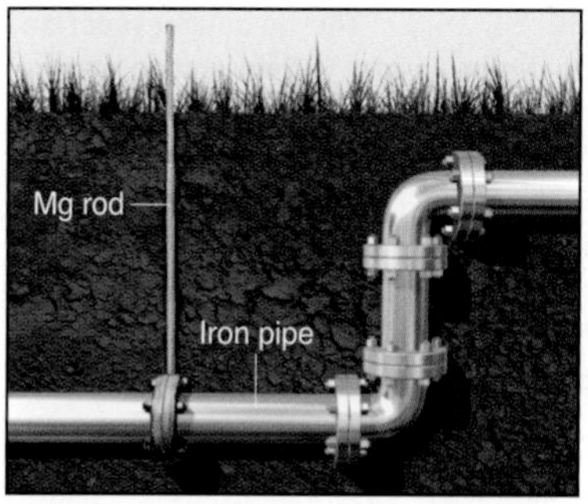

FIGURE 19.29 The use of a sacrificial anode to prevent iron corrosion. In cathodic protection, an active metal, such as zinc, magnesium, or aluminum, acts as the anode and is sacrificed instead of the iron.

SUMMARY OF SECTION 19.6

- Corrosion damages metal structures through a natural electrochemical process.
- Iron corrosion occurs in the presence of oxygen and moisture and is increased by high $[H^+]$; high ion concentrations; or contact with a less active metal, such as Cu.
- Fe is oxidized and O_2 is reduced in one redox reaction, while Fe^{2+} is oxidized and O_2 is reduced to form rust (hydrated form of Fe_2O_3) in another redox reaction, which often takes place at a different location.
- Because Fe functions as both the anode and the cathode in the corrosion process, an iron object can be protected by physically covering it or by joining it to a more active metal (such as Zn, Mg, or Al), which acts as the anode in place of the Fe.

19.7 Electrolytic Cells: Electrochemical Cells Using Electrical Energy to Drive Nonspontaneous Reactions

In contrast to a voltaic cell, which generates electrical energy from a spontaneous reaction, *an electrolytic cell requires electrical energy from an external source to drive a nonspontaneous redox reaction.* (We will discuss industrial electrolysis in Chapter 23.)

Construction and Operation of an Electrolytic Cell

Let us see how an electrolytic cell operates by constructing one from a voltaic cell. Consider a tin-copper voltaic cell (Figure 19.30A). The Sn anode will gradually be oxidized to Sn^{2+} ions, which enter the electrolyte, and the Cu^{2+} ions will gradually be reduced and plate out on the Cu cathode because the cell reaction is spontaneous in that direction:

For the standard-state voltaic cell,

$$\begin{array}{llll} \mathrm{Sn}(s) & \longrightarrow & \mathrm{Sn}^{2+}(aq) + 2e^- & \text{(anode; oxidation)} \\ \mathrm{Cu}^{2+}(aq) + 2e^- & \longrightarrow & \mathrm{Cu}(s) & \text{(cathode; reduction)} \\ \hline \mathrm{Sn}(s) + \mathrm{Cu}^{2+}(aq) & \longrightarrow & \mathrm{Sn}^{2+}(aq) + \mathrm{Cu}(s) & E^\circ_{\text{cell}} = 0.48\ \text{V and } \Delta G^\circ = -93\ \text{kJ/mol} \end{array}$$

Therefore, the *reverse* standard-state cell reaction is *non*spontaneous and never happens on its own. However, we can make it happen by supplying an electric potential *greater than* E°_{cell} from an external source. In effect, we convert the voltaic cell into an electrolytic cell—the anode becomes the cathode, and the cathode becomes the anode (Figure 19.30B):

For the electrolytic cell,

$$\begin{array}{llll} \mathrm{Cu}(s) & \longrightarrow & \mathrm{Cu}^{2+}(aq) + 2e^- & \text{(anode; oxidation)} \\ \mathrm{Sn}^{2+}(aq) + 2e^- & \longrightarrow & \mathrm{Sn}(s) & \text{(cathode; reduction)} \\ \hline \mathrm{Cu}(s) + \mathrm{Sn}^{2+}(aq) & \longrightarrow & \mathrm{Cu}^{2+}(aq) + \mathrm{Sn}(s) & E^\circ_{\text{cell}} = -0.48\ \text{V and } \Delta G^\circ = 93\ \text{kJ/mol} \end{array}$$

In an electrolytic cell, as in a voltaic cell, *oxidation takes place at the anode and reduction takes place at the cathode.* Notice, however, that *the signs of the electrodes are reversed, causing the electrons to flow in the opposite direction (but still from the anode to the cathode).*

FIGURE 19.30 The tin-copper reaction as the basis of a voltaic and an electrolytic cell. A. The spontaneous reaction between Sn and Cu^{2+} generates 0.48 V in a voltaic cell. **B.** If more than 0.48 V is supplied, the nonspontaneous (reverse) reaction between Cu and Sn^{2+} occurs. Notice the changes in electrode charges and the direction of electron flow.

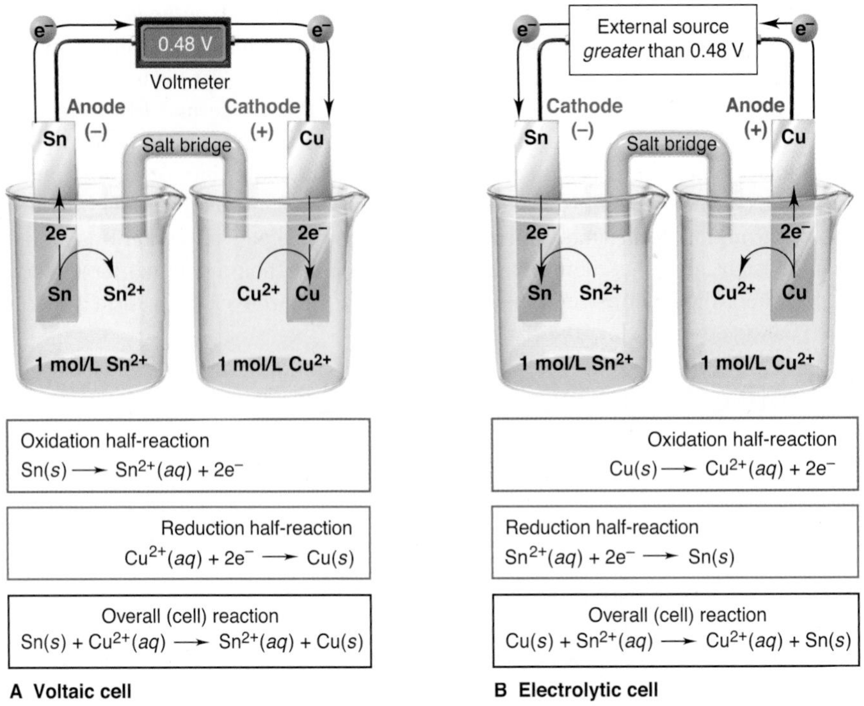

To understand these differences, we focus on the *cause* of the electron flow:

- In a voltaic cell, electrons are generated *in* the anode, so the anode is *negative*, and *removed from* the cathode, so the cathode is *positive*.
- In an electrolytic cell, an external power source supplies electrons *to* the cathode, so the cathode is *negative*, and removes them *from* the anode, so it is *positive*.

A rechargeable battery is a voltaic cell when it is discharging and an electrolytic cell when it is recharging. Let us compare these two functions in a lead-acid car battery (Figure 19.31; recall also Section 19.5):

- In the discharge mode (voltaic cell), oxidation occurs at electrode I, making the *negative* electrode the anode and the *positive* electrode (electrode II) the cathode.

FIGURE 19.31 The processes that occur during the discharge (I) and recharge (II) of a lead-acid battery

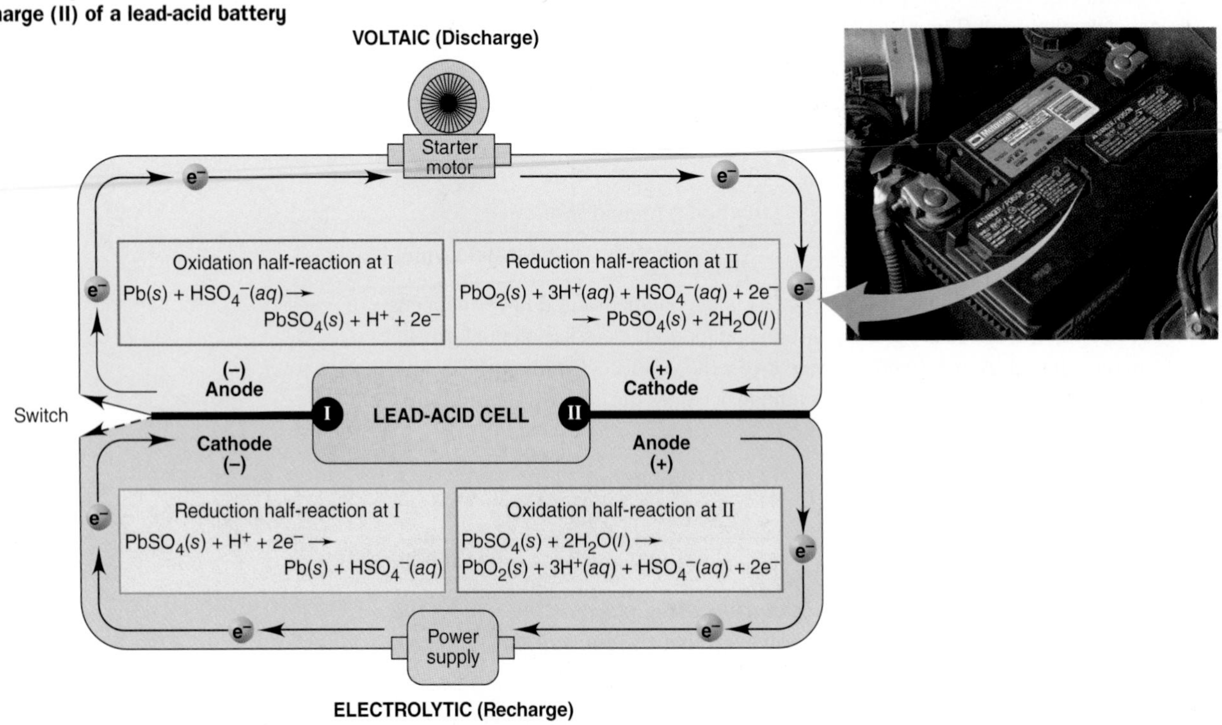

TABLE 19.5 Comparison of Voltaic and Electrolytic Cells

Cell Type	ΔG	E_{cell}	Electrode Name	Electrode Process	Electrode Sign
Voltaic	<0	>0	Anode	Oxidation	−
			Cathode	Reduction	+
Electrolytic	>0	<0	Anode	Oxidation	+
			Cathode	Reduction	−

- In the recharge mode (electrolytic cell), reduction occurs at electrode I, making the *negative* electrode the cathode and the *positive* electrode (electrode II) the anode.

Table 19.5 summarizes the processes and signs in the two types of cells.

Predicting the Products of Electrolysis

Electrolysis is the splitting (lysing) of a substance by the input of electrical energy. It is often used to decompose a compound into its elements, as you will see for chlorine, aluminum, copper, and several other elements in Chapter 23. Water was first electrolyzed to H_2 and O_2 in 1800, and the process is still used to produce these gases in ultrahigh purity. The electrolyte in an electrolytic cell can be a pure compound (such as H_2O or a molten salt), a mixture of molten salts, or an aqueous solution of a salt. As you will see, the products depend on atomic properties and several other factors.

Electrolysis of Pure Molten Salts If the salt is pure, predicting the products is straightforward: *the cation will be reduced and the anion will be oxidized.* The electrolyte is the molten salt itself, and the ions are attracted by the oppositely charged electrodes.

Consider the electrolysis of molten (fused) calcium chloride. The two species present are Ca^{2+} and Cl^-, so Ca^{2+} ion is reduced and Cl^- ion is oxidized:

$$2Cl^-(l) \longrightarrow Cl_2(g) + 2e^- \quad \text{(anode; oxidation)}$$
$$Ca^{2+}(l) + 2e^- \longrightarrow Ca(s) \quad \text{(cathode; reduction)}$$
$$Ca^{2+}(l) + 2Cl^-(l) \longrightarrow Ca(s) + Cl_2(g) \quad \text{(overall)}$$

Calcium is prepared industrially this way, as are several other active metals, such as Na and Mg, and the halogens Cl_2 and Br_2.

Electrolysis of Mixed Molten Salts More typically, the electrolyte is a mixture of molten salts being electrolyzed to obtain one of the metals. How can we tell which species will react at which electrode? The general rule for all electrolytic cells is that *the more easily oxidized species (stronger reducing agent) will react at the anode, and the more easily reduced species (stronger oxidizing agent) will react at the cathode.*

Unfortunately, for molten salts, we *cannot* use $E°$ values to tell the relative strength of the oxidizing and reducing agents, because these values refer to the reduction of the *aqueous ion to the free element* under standard-state conditions:

$$M^{n+}(aq) + ne^- \longrightarrow M(s)$$

Instead, we use periodic trends of atomic properties to predict which ion will gain or lose electrons more easily (Sections 7.3 and 8.5).

Sample Problem 19.13 Predicting the Electrolysis Products of a Molten Salt Mixture

Problem A chemical engineer melts a naturally occurring mixture of NaBr and $MgCl_2$ and then decomposes it in an electrolytic cell. Predict the substance formed at each electrode, and write balanced half-reactions and the overall cell reaction.

Plan We have to determine which metal and nonmetal will form more easily at the electrodes. We list the ions as oxidizing or reducing agents:

- If a metal holds its electrons *more* tightly than another, it has a higher ionization energy (IE). Thus, as a cation, it gains electrons more easily, so it is the stronger oxidizing agent and is reduced at the cathode.
- If a nonmetal holds its electrons *less* tightly than another, it has a lower electronegativity (χ). Thus, as an anion, it loses electrons more easily, so it is the stronger reducing agent and is oxidized at the anode.

Solution First, we list the ions as oxidizing or reducing agents:
- The possible oxidizing agents are Na^+ and Mg^{2+}.
- The possible reducing agents are Br^- and Cl^-.

Next, we determine the cathode product (more easily reduced cation). Mg is to the right of Na in period 3. IE increases from left to right, so it is harder to remove e^- from Mg. Thus, Mg^{2+} has a greater attraction for e^- and is more easily reduced (stronger oxidizing agent):

$$Mg^{2+}(l) + 2e^- \longrightarrow Mg(l) \quad \text{(cathode; reduction)}$$

Finally, we determine the anode product (more easily oxidized anion). Br is below Cl in group 17. χ decreases down the group, so Br accepts e^- less readily. Thus, Br^- loses its e^- more easily, so it is more easily oxidized (stronger reducing agent):

$$2Br^-(l) \longrightarrow Br_2(g) + 2e^- \quad \text{(anode; oxidation)}$$

Write the overall cell reaction:

$$Mg^{2+}(l) + 2Br^-(l) \longrightarrow Mg(l) + Br_2(g) \quad \text{(overall)}$$

Comment The cell temperature must be high enough to keep the salt mixture molten. In this cell, the temperature is greater than the melting point of Mg, so Mg appears as a liquid in the equation. The temperature is greater than the boiling point of Br_2, so Br_2 appears as a gas.

Follow-Up Problem 19.13 A sample of $AlBr_3$ contaminated with KF is melted and electrolyzed. Determine the electrode products, and write the overall cell reaction.

Electrolysis of Water and Nonstandard Half-Cell Potentials Before we analyze the electrolysis products of aqueous salt solutions, let us examine the electrolysis of water itself. Very pure water is difficult to electrolyze because few ions are present to conduct a current. However, if we add a small amount of a salt that cannot be electrolyzed (such as Na_2SO_4), electrolysis proceeds rapidly. An electrolytic cell with separate compartments for H_2 and O_2 is used (Figure 19.32).

At the anode, water is oxidized; notice that the oxidation number of O changes from −2 to 0:

$$2H_2O(l) \longrightarrow O_2(g) + 4H^+(aq) + 4e^- \quad E = 0.82\ V \quad \text{(anode; oxidation)}$$

At the cathode, water is reduced; notice that the oxidation number of H changes from +1 to 0:

$$2H_2O(l) + 2e^- \longrightarrow H_2(g) + 2OH^-(aq) \quad E = -0.42\ V \quad \text{(cathode; reduction)}$$

Doubling the cathode half-reaction to make e^- loss equal to e^- gain, adding the half-reactions (which involves combining the H^+ and OH^- into H_2O and cancelling e^- and excess H_2O), and calculating E_{cell} gives the overall reaction:

$$2H_2O(l) \longrightarrow 2H_2(g) + O_2(g) \quad E_{cell} = -0.42\ V - 0.82\ V = -1.24\ V \quad \text{(overall)}$$

Notice that these electrode potentials are *not* standard electrode potentials (and not designated with $E°$). The $[H^+]$ and $[OH^-]$ are 1.0×10^{-7} mol/L rather than the standard-state value of 1 mol/L. These E values are obtained by applying the Nernst equation. For example, the calculation for the reduction potential at the anode (with $z = 4$) is

$$E_{cell} = E°_{cell} - \frac{RT}{zF} \ln \frac{1}{(p_{O_2} \times [H^+]^4)}$$

Oxidation half-reaction
$2H_2O(l) \longrightarrow O_2(g) + 4H^+(aq) + 4e^-$

Reduction half-reaction
$2H_2O(l) + 2e^- \longrightarrow H_2(g) + 2OH^-(aq)$

Overall (cell) reaction
$2H_2O(l) \longrightarrow 2H_2(g) + O_2(g)$

FIGURE 19.32 In the electrolysis of water, twice as much H_2 forms as O_2.

The standard potential for the *reduction* of oxygen to water is 1.23 V (from Appendix D) and $p_{O_2} \approx 1$ bar in the half-cell, so we have

$$E_{cell} = 1.23\text{ V} - \left\{\frac{(8.314\text{ J/[mol·K]})(298\text{ K})}{(4\text{ mol})(96\,485\text{ C/mol})} \times -[\ln 1 + 4 \ln (1.0 \times 10^{-7})]\right\} = 0.82\text{ V}$$

In aqueous ionic solutions, $[H^+]$ and $[OH^-]$ are also approximately 10^{-7} mol/L, so we use these nonstandard E_{cell} values to predict the electrode products.

Electrolysis of Aqueous Salt Solutions and the Effect of Overvoltage Aqueous salt solutions are mixtures of ions *and* water, so we have to compare the various electrode potentials to predict the electrode products.

1. *Predicting the electrode products.* What happens when two half-reactions are possible at an electrode?

- *The reduction with the less negative (more positive) electrode potential occurs.*
- *The oxidation with the less positive (more negative) electrode potential occurs.*

For example, what happens when a solution of potassium iodide is electrolyzed?

- The possible *oxidizing agents* are K^+ and H_2O; their reduction half-reactions and the corresponding reduction potentials are

$$K^+(aq) + e^- \longrightarrow K(s) \qquad E^\circ = -2.93\text{ V}$$
$$2H_2O(l) + 2e^- \longrightarrow H_2(g) + 2OH^-(aq) \qquad E = -0.42\text{ V}$$

The less *negative* electrode potential for water means that it is much easier to reduce than K^+, so H_2 forms at the cathode.

- The possible *reducing agents* are I^- and H_2O; their oxidation half-reactions and the corresponding reduction potentials are

$$2I^-(aq) \longrightarrow I_2(s) + 2e^- \qquad E^\circ = 0.53\text{ V}$$
$$2H_2O(l) \longrightarrow O_2(g) + 4H^+(aq) + 4e^- \qquad E = 0.82\text{ V}$$

The less *positive* electrode potential for I^- means that it is easier to oxidize than H_2O, so I_2 forms at the anode.

2. *The effect of overpotential.* The products predicted from a comparison of electrode potentials are not always the actual products. For most reactions to occur, additional voltage is required. This increment above the expected voltage is the **overpotential**. It is 0.4 V to 0.6 V for $H_2(g)$ or $O_2(g)$, and it is due to the large activation energy (Section 14.5) needed to form gases at the electrode.

Overvoltage has major practical significance. A multibillion-dollar example is the industrial production of chlorine from concentrated NaCl solution. Water is easier to reduce than Na^+, so H_2 forms at the cathode, even *with* an overpotential of 0.6 V:

$$Na^+(aq) + e^- \longrightarrow Na(s) \qquad E^\circ = -2.71\text{ V}$$
$$2H_2O(l) + 2e^- \longrightarrow H_2(g) + 2OH^-(aq) \qquad E = -0.42\text{ V } (\sim -1\text{ V with overvoltage})$$

However, Cl_2 *does* form at the anode, even though the electrode potentials *themselves* would lead us to predict that O_2 should form:

$$2H_2O(l) \longrightarrow O_2(g) + 4H^+(aq) + 4e^- \qquad E = 0.82\text{ V } (\sim 1.4\text{ V with overvoltage})$$
$$2Cl^-(aq) \longrightarrow Cl_2(g) + 2e^- \qquad E^\circ = 1.36\text{ V}$$

An overpotential of ~0.6 V for O_2 makes Cl_2 the product that is easier to form. Keeping $[Cl^-]$ high also favours this step. Thus, Cl_2, one of the 10 most heavily produced chemicals, is formed from plentiful natural sources of aqueous sodium chloride (Chapter 23).

3. *A summary: Which product forms at which electrode?* Experiments have shown the elements that can be prepared electrolytically from aqueous solutions of their salts:

- Cations of less active metals, including gold, silver, copper, chromium, platinum, and cadmium, *are* reduced to the metal.
- Cations of more active metals, including those in groups 1 and 2, and Al in group 13, *are not* reduced. Water is reduced to H_2 and OH^- instead.

- Anions that *are* oxidized, because of an overpotential from O_2 formation, include the halides ([Cl^-] must be high), except for F^-.
- Anions that *are not* oxidized include F^- and common oxoanions, such as SO_4^{2-}, CO_3^{2-}, NO_3^-, and PO_4^{3-}, because the central nonmetal in these oxoanions is already in its highest oxidation state. Water is oxidized to O_2 and H^+ instead.

Sample Problem 19.14 Predicting the Electrolysis Products of Aqueous Salt Solutions

Problem What products form at which electrode during the electrolysis of aqueous salt solutions?

(a) KBr **(b)** $AgNO_3$ **(c)** $MgSO_4$

Plan We identify the reacting ions and compare their electrode potentials with the electrode potential of water, taking into account the 0.4 V to 0.6 V overpotential. The reduction half-reaction with the less negative electrode potential occurs at the cathode, and the oxidation half-reaction with the less positive electrode potential occurs at the anode.

Solution

(a) Reduction half-reaction:

$$K^+(aq) + e^- \longrightarrow K(s) \qquad E^\circ = -2.93\text{ V}$$
$$2H_2O(l) + 2e^- \longrightarrow H_2(g) + 2OH^-(aq) \qquad E = -0.42\text{ V}$$

Despite the overpotential, which makes E for the reduction of water between -0.8 V and -1.0 V, H_2O is still easier to reduce than K^+, so $H_2(g)$ forms at the cathode.

Oxidation half-reaction:

$$2Br^-(aq) \longrightarrow Br_2(l) + 2e^- \qquad E^\circ = 1.07\text{ V}$$
$$2H_2O(l) \longrightarrow O_2(g) + 4H^+(aq) + 4e^- \qquad E = 0.82\text{ V}$$

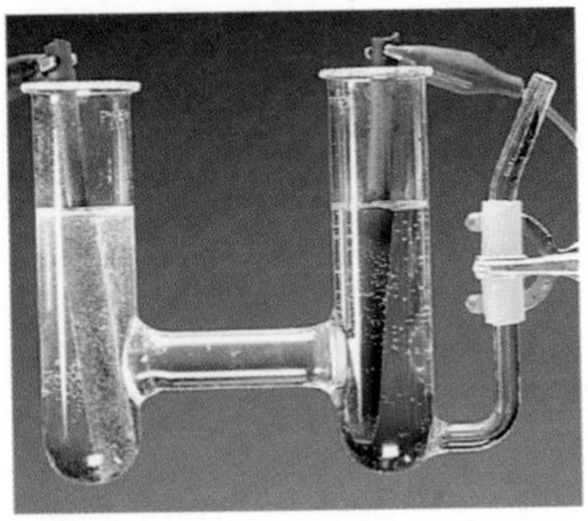
Electrolysis of aqueous KBr

Because of the overpotential, which makes E for the oxidation of water between 1.2 V and 1.4 V, Br^- is easier to oxidize than water, so $Br_2(l)$ forms at the anode (*see photo*).

(b)
$$Ag^+(aq) + e^- \longrightarrow Ag(s) \qquad E^\circ = 0.80\text{ V}$$
$$2H_2O(l) + 2e^- \longrightarrow H_2(g) + 2OH^-(aq) \qquad E = -0.42\text{ V}$$

As the cation of an inactive metal, Ag^+ is a better oxidizing agent than H_2O, so $Ag(s)$ forms at the cathode. NO_3^- cannot be oxidized, because N is already in its highest (+5) oxidation state. Thus, $O_2(g)$ forms at the anode:

$$2H_2O(l) \longrightarrow O_2(g) + 4H^+(aq) + 4e^-$$

(c)
$$Mg^{2+}(aq) + 2e^- \longrightarrow Mg(s) \qquad E^\circ = -2.37\text{ V}$$
$$2H_2O(l) + 2e^- \longrightarrow H_2(g) + 2OH^-(aq) \qquad E = -0.42\text{ V}$$

Like K^+ in part (a), Mg^{2+} cannot be reduced in the presence of water. Despite the overpotential, which makes E for the reduction of water between -0.8 V and -1.0 V, H_2O is still easier to reduce than K^+, so $H_2(g)$ forms at the cathode.

The SO_4^{2-} ion cannot be oxidized because S is in its highest (+6) oxidation state. Thus, H_2O is oxidized, and $O_2(g)$ forms at the anode:

$$2H_2O(l) \longrightarrow O_2(g) + 4H^+(aq) + 4e^-$$

Follow-Up Problem 19.14 Write half-reactions for the products you predict will form in the electrolysis of aqueous $AuBr_3$.

The Stoichiometry of Electrolysis: The Relation between Charge and the Amount of Products

Since charge flowing through an electrolytic cell yields products at the electrodes, it makes sense that the more charge that flows, the more product will form. In fact, this relationship was first determined experimentally by Michael Faraday and is referred to as *Faraday's law of electrolysis: the amount of substance produced at each electrode is directly proportional to the quantity of charge flowing through the cell.*

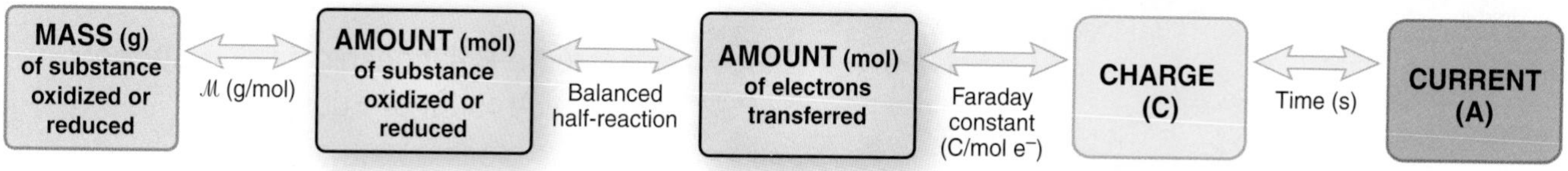

FIGURE 19.33 **A summary diagram for the stoichiometry of electrolysis**

Each balanced half-reaction shows the amounts (mol) of reactant, electrons, and product involved in the change, so it contains the information we need to answer such questions as "How much material will form from a given quantity of charge?" or, conversely, "How much charge is needed to produce a given amount of material?" To apply Faraday's law, follow these steps:

Step 1. Balance the half-reaction to find the amount (mol) of electrons needed per mole of product.

Step 2. Use the Faraday constant (F = 96 485 C/mol e^-) to find the quantity of charge.

Step 3. Use the molar mass to find the charge needed for a given mass of product.

Measuring Current to Find Charge We cannot measure charge directly, but we *can* measure *current*, the charge flowing per unit time. The SI unit of current is the **ampere (A)**, which is defined as a charge of 1 C (coulomb) flowing through a conductor in 1 s:

$$1 \text{ ampere} = 1 \text{ coulomb/second} \qquad \text{or} \qquad 1 \text{ A} = 1 \text{ C/s}$$

Thus, the current is defined as charge per unit time:

$$I = \frac{Q}{t}$$

$$\text{Current} = \frac{\text{charge}}{\text{time}} \qquad (19.9)$$

Therefore, by measuring the current *and* the time during which the current flows, we find the charge, which relates to the amount of product (Figure 19.33). It is important to note that this product of current and time *only* produces charge at a constant current.

Problems Involving Stoichiometry of Electrolysis Problems based on Faraday's law often ask us to calculate current, mass of material, or time. As we mentioned, the electrode half-reaction provides the key to solving these problems because it is related to the mass for a certain quantity of charge.

Here is a typical problem in practical electrolysis: How long does it take to produce 3.0 g of $Cl_2(g)$ by electrolysis of aqueous NaCl using a power supply with a current of 12 A? This problem asks for the time needed to produce a certain mass, so let us first relate mass to amount (mol) of electrons to find the charge needed. Then, we will relate the charge to the current to find the time.

The half-reaction tells us that 2 mol of electrons is lost to form 1 mol of Cl_2. We will use this relationship as a conversion factor:

$$2Cl^-(aq) \longrightarrow Cl_2(g) + 2e^-$$

We convert the given mass of Cl_2 to amount of Cl_2, use the conversion factor from the half-reaction, and multiply by the Faraday constant to find the total charge:

$$Q = 3.0 \text{ g } Cl_2 \times \frac{1 \text{ mol } Cl_2}{70.90 \text{ g } Cl_2} \times \frac{2 \text{ mol } e^-}{1 \text{ mol } Cl_2} \times \frac{96\,485 \text{ C}}{1 \text{ mol } e^-} = 8.2 \times 10^3 \text{ C}$$

Now we use the relationship between charge and current to find the time needed:

$$t = \frac{Q}{I} = 8.2 \times 10^3 \text{ C} \times \frac{1 \text{ s}}{12 \text{ C}} = 6.8 \times 10^2 \text{ s } (\sim 11 \text{ min})$$

Notice that the entire calculation follows Figure 19.33, including the last step, which asks for the time that the given current must flow:

$$m_{Cl_2} \Rightarrow n_{Cl_2} \Rightarrow z \Rightarrow Q \Rightarrow t$$

Sample Problem 19.15 applies these ideas in an important industrial setting.

Sample Problem 19.15 Applying the Relationship among Current, Time, and Amount of Substance

Road Map

Mass (g) of Cr needed

↓ Divide by $\mathcal{M}$ (g/mol).

Amount (mol) of Cr needed

↓ 3 mol e^- = 1 mol Cr

Amount (mol) of e^- transferred

↓ 1 mol e^- = 96 485 C

Charge (C)

↓ Divide by time (convert min to s).

Current (A)

Problem A technician plates a faucet with 0.86 g of Cr metal by electrolysis of aqueous $Cr_2(SO_4)_3$. If 12.5 min is allowed for the plating, what current is needed?

Plan To find the current, we divide the charge by the time, so we need to find the charge. We write the half-reaction for Cr^{3+} reduction to get the amount (mol) of e^- transferred per mole of Cr. To find the charge, we convert the mass of Cr needed (0.86 g) to the amount (mol) of Cr. Then we use the Faraday constant (96 485 C/mol e^-) to find the charge and divide by the time (12.5 min, converted to seconds) to obtain the current (*see road map*).

Solution Write the balanced half-reaction:

$$Cr^{3+}(aq) + 3e^- \longrightarrow Cr(s)$$

Combine steps to find the amount (mol) of e^- transferred for the mass of Cr needed:

$$z = 0.86 \text{ g Cr} \times \frac{1 \text{ mol Cr}}{52.00 \text{ g Cr}} \times \frac{3 \text{ mol e}^-}{1 \text{ mol Cr}} = 0.050 \text{ mol e}^-$$

Calculate the charge:

$$Q = 0.050 \text{ mol e}^- \times \frac{96\,485 \text{ C}}{1 \text{ mol e}^-} = 4.8 \times 10^3 \text{ C}$$

Calculate the current:

$$I = \frac{Q}{t} = \frac{4.8 \times 10^3 \text{ C}}{12.5 \text{ min}} \times \frac{1 \text{ min}}{60 \text{ s}} = 6.4 \text{ C/s} = 6.4 \text{ A}$$

Check Rounding gives

$$(0.9 \text{ g})(1 \text{ mol Cr}/50 \text{ g})(3 \text{ mol e}^-/1 \text{ mol Cr}) = 5 \times 10^{-2} \text{ mol e}^-$$

Then,

$$(5 \times 10^{-2} \text{ mol e}^-)(\sim 1 \times 10^5 \text{ C/mol e}^-) = 5 \times 10^3 \text{ C}$$

and

$$(5 \times 10^3 \text{ C}/12 \text{ min})(1 \text{ min}/60 \text{ s}) = 7 \text{ A}$$

Comment To introduce Faraday's law, we neglected some details about actual electroplating. In practice, electroplating chromium is only 30% to 40% efficient and must be run at a specific temperature range for the plate to appear bright. Nearly 10 000 t (tonne; 2×10^8 mol) of chromium is used annually for electroplating.

Follow-Up Problem 19.15 Using a current of 4.75 A, how much time does it take to plate a sculpture with 1.50 g of Cu from a $CuSO_4$ solution?

The following Chemical Connections section links several themes of this chapter to the generation of energy in a living cell.

SUMMARY OF SECTION 19.7

- An electrolytic cell uses electrical energy to drive a nonspontaneous reaction.
- Oxidation occurs at the anode and reduction occurs at the cathode, but the direction of electron flow and the charges of the electrodes are opposite those in voltaic cells.
- When two products can form at each electrode, the more easily oxidized substance reacts at the anode and the more easily reduced substance reacts at the cathode.
- The reduction or oxidation of water takes place at nonstandard conditions.
- Overpotential causes the actual voltage required to be unexpectedly high (especially for gases, such as H_2 and O_2) and can affect the electrode product that forms.
- The amount of product that forms depends on the quantity of charge flowing through the cell, which is related to the time that the charge flows and the current.
- Biological redox systems combine aspects of voltaic, concentration, and electrolytic cells to convert bond energy in food into electrochemical potential and then into the bond energy of ATP (see Chemical Connections).

Cellular Electrochemistry and the Production of ATP

Biological cells apply the principles of electrochemical cells to generate energy. The complex multistep process can be divided into two steps:

Step 1. Bond energy in food generates an electrochemical potential.

Step 2. The electrochemical potential creates the bond energy of ATP (see Chemical Connections in Section 18.4).

These steps are part of the *electron transport chain* (ETC), which lies on the inner membranes of *mitochondria*, the subcellular particles that produce energy (Figure B19.1).

The ETC is a series of large molecules (mostly proteins), each of which contains a *redox couple* (the oxidized and reduced forms of a species), such as Fe^{3+}/Fe^{2+}, that passes electrons along the chain. At three points, large potential differences supply enough Gibbs energy to convert adenosine diphosphate (ADP) into ATP.

Bond Energy to Electrochemical Potential Cells use the energy in food by releasing it in controlled steps rather than all at once. The reaction that ultimately powers the ETC is the oxidation of hydrogen to form water:

$$H_2 + \frac{1}{2}O_2 \longrightarrow H_2O$$

However, instead of H_2 gas, which does not occur in organisms, the hydrogen takes the form of two H^+ ions and two electrons. The biological oxidizing agent NAD^+ (*n*icotinamide *a*denine *d*inucleotide) acquires these protons and electrons in the process of oxidizing the molecules in food. To show this process, we use the following half-reaction (without cancelling an H^+ on both sides):

$$NAD^+(aq) + 2H^+(aq) + 2e^- \longrightarrow NADH(aq) + H^+(aq)$$

At the mitochondrial inner membrane, the NADH and H^+ transfer the two electrons to the first redox couple of the ETC and release the two H^+. The electrons are transported down the chain of redox couples, where they finally reduce O_2. The overall process, with standard electrode potentials,* is given below:

$$NADH(aq) + H^+(aq) \longrightarrow NAD^+(aq) + 2H^+(aq) + 2e^- \qquad E^{\circ\prime} = -0.315\text{ V}$$

$$\tfrac{1}{2}O_2(aq) + 2e^- + 2H^+(aq) \longrightarrow H_2O(l) \qquad E^{\circ\prime} = 0.815\text{ V}$$

$$NADH(aq) + H^+(aq) + \tfrac{1}{2}O_2(aq) \longrightarrow NAD^+(aq) + H_2O(l)$$

$$E^{\circ\prime}_{\text{overall}} = 0.815\text{ V} - (-0.315\text{ V}) = 1.130\text{ V}$$

Thus, for every 2 mol of e^-, 1 mol of NADH enters the ETC, and the Gibbs energy equivalent of 1.13 V is available:

$$\begin{aligned}\Delta G^{\circ\prime} &= -zFE^{\circ\prime}\\ &= -(2\text{ mol e}^-/\text{mol NADH})[96.5\text{ kJ/(V·mol e}^-)](1.130\text{ V})\\ &= -218\text{ kJ/mol NADH}\end{aligned}$$

Note that this aspect of the process functions like a voltaic cell: a spontaneous reaction, the reduction of O_2 to H_2O, is used to generate a potential. In contrast to the operation of a voltaic cell, however, this reaction occurs in many small steps. Figure B19.2 is a highly simplified diagram showing the three steps in the ETC that generate a high enough potential to form ATP from ADP.

Each step consists of several components, in which electrons are passed from one redox couple to the next. Since most of the ETC components are iron-containing proteins, the redox change consists of the oxidation of Fe^{2+} to Fe^{3+} in one component accompanied by the reduction of Fe^{3+} to Fe^{2+} in the other component:

$$Fe^{2+}\text{ (in A)} + Fe^{3+}\text{ (in B)} \longrightarrow Fe^{3+}\text{ (in A)} + Fe^{2+}\text{ (in B)}$$

In other words, metal ions bound within the ETC proteins are the actual species undergoing the redox reactions.

*In biological systems, standard potentials are designated $E^{\circ\prime}$ and the standard state includes a pH of 7.0 ($[H^+] = 1 \times 10^{-7}$ mol/L).

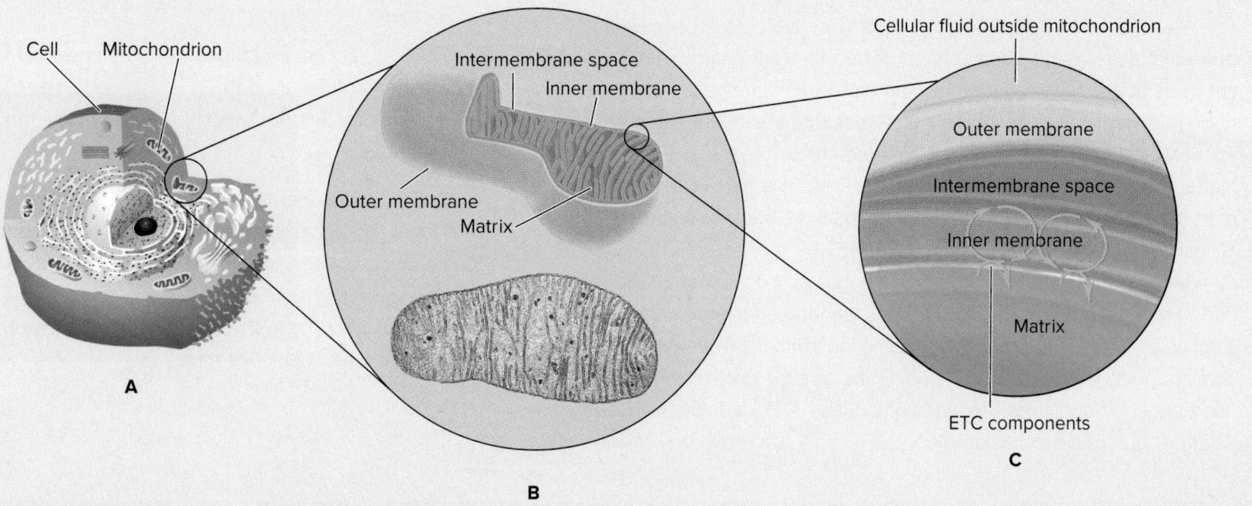

FIGURE B19.1 The mitochondrion. **A.** Mitochondria are subcellular particles outside the cell nucleus. **B.** A mitochondrion has a smooth outer membrane and a highly folded inner membrane, shown schematically and in an electron micrograph. **C.** The components of the ETC are attached to the inner membrane.

(Continued)

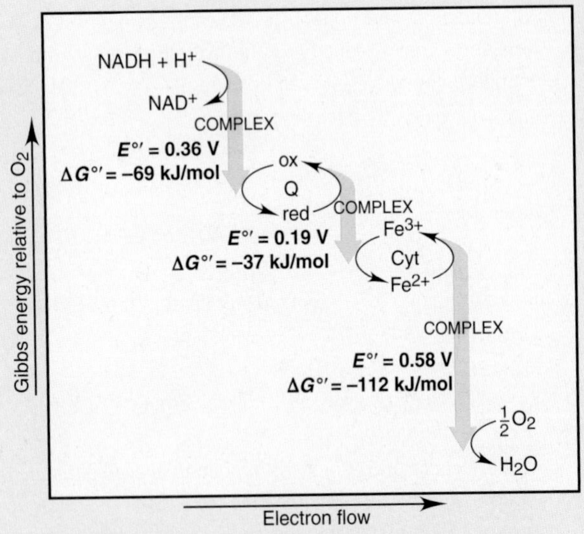

FIGURE B19.2 The main energy-yielding steps in the ETC. At the three points shown, $E^{\circ\prime}$ (or $\Delta G^{\circ\prime}$) is large enough to form ATP from ADP. (A complex consists of many components; Q is a large organic molecule; Cyt stands for a cytochrome, a protein that contains a metal-ion redox couple.)

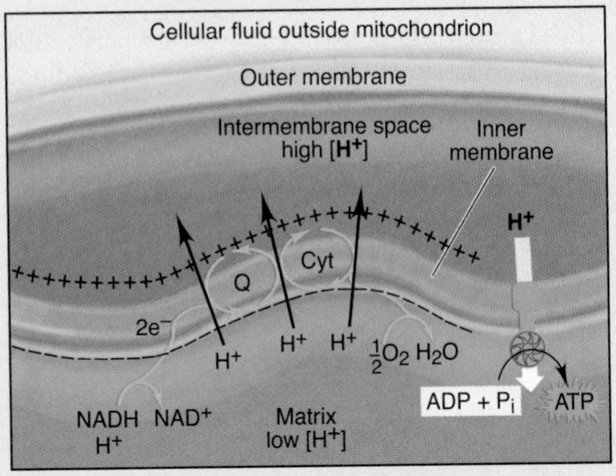

FIGURE B19.3 Coupling electron transport to proton transport to ATP synthesis. Electrons are transported along the ETC (*curved yellow arrows*). Protons pumped into the intermembrane space create an $[H^+]$ difference that generates a potential across the inner membrane. When this difference is large enough, H^+ flows back into the inner space, and the Gibbs energy drives the formation of ATP.

Electrochemical Potential to Bond Energy The formation of ATP requires a lot of energy:

$$ADP^{3-}(aq) + HPO_4^{2-}(aq) + H^+(aq) \longrightarrow ATP^{4-}(aq) + H_2O(l)$$
$$\Delta G^{\circ\prime} = 30.5 \text{ kJ/mol}$$

Notice that, at each of the three points in Figure B19.2, the Gibbs energy *released* exceeds the 30.5 kJ needed to form 1 mol of ATP. Thus, as in an electrolytic cell, an electrochemical potential is supplied to drive a nonspontaneous reaction.

We have followed the flow of *electrons* through the ETC. To see how electrochemical potential is converted to bond energy in ATP, we focus on the released *protons*. The Gibbs energy released at the three key steps is used to force the H^+ ions into the intermembrane space; as a result, $[H^+]$ of the intermembrane space becomes higher than $[H^+]$ of the matrix (Figure B19.3). This part of the process uses the Gibbs energy released by the three steps to *create an H^+ concentration cell across the membrane*.

When the $[H^+]$ difference across the membrane reaches a "trigger" point of about 2.5-fold, H^+ ions spontaneously flow back through the membrane (in effect, closing the switch and allowing the concentration cell to operate). The Gibbs energy released in this spontaneous flow drives the nonspontaneous ATP formation via an enzyme-catalyzed mechanism.

Thus, the mitochondrion uses the *electron-motive force* of redox couples on the membrane to generate a *proton-motive force* across the membrane, which converts a potential difference into bond energy.

Problems

B19.1 In the final steps of the ETC, there is a large protein complex that contains iron and copper ions, which interact with each other.

(a) Write balanced equations for the one-electron half-reactions of Fe^{3+} and Cu^+.

(b) Write a balanced overall equation for this redox reaction.

B19.2 As the ETC proceeds, a difference in $[H^+]$ develops across the inner mitochondrial membrane. The proton-motive force, which is used to form ATP, is based largely on this difference and is equivalent to an electrical potential of 0.224 V. What is the Gibbs energy change (kJ/mol) associated with the proton-motive force?

CHAPTER REVIEW GUIDE

Learning Objectives

Relevant section (§) and/or sample problem (SP) numbers appear in parentheses.

Concepts

1. Differentiate between the following pairs of terms: *oxidation* and *reduction*; *oxidizing agent* and *reducing agent*. (§19.1)
2. Explain why an oxidizing agent is reduced and a reducing agent is oxidized; describe what oxidation numbers are and how to determine them. (§19.1)
3. Explain how the oxidation number and half-reaction methods are used to balance redox reactions in acidic or basic solutions. (§19.1)
4. Describe how to use redox titrations to determine the concentration of an unknown solution. (§19.1)
5. Distinguish between voltaic and electrolytic cells in terms of the sign of ΔG. (§19.2)
6. Describe how voltaic cells use a spontaneous reaction to release electrical energy. (§19.2)
7. Demonstrate the physical makeup of a voltaic cell, giving importance to arrangement and composition of half-cells, relative charges of electrodes, and the purpose of a salt bridge. (§19.2)
8. Explain how the difference in reducing strength of the electrodes determines the direction of electron flow. (§19.2)
9. Describe the correspondence between a positive E_{cell} and a spontaneous cell reaction. (§19.3)
10. Explain the usefulness and significance of standard electrode potentials. ($E^\circ_{half\text{-}cell}$). (§19.3)
11. Describe how $E^\circ_{half\text{-}cell}$ values are combined to give E°_{cell}. (§19.3)
12. Demonstrate how a standard reference electrode is used to find an unknown $E^\circ_{half\text{-}cell}$. (§19.3)
13. Show how a list of standard electrode potentials (such as Table 19.3 or Appendix D) can be used to write spontaneous redox reactions. (§19.3)
14. Show how the relative reactivity of a metal is determined by its reducing power and is related to the negative of its $E^\circ_{half\text{-}cell}$. (§19.3)
15. Describe how E_{cell} (the nonstandard cell potential) is related to ΔG (maximum work) and the charge [amount (mol) of electrons times the Faraday constant] flowing through the cell. (§19.4)
16. Demonstrate the interrelationship among ΔG°, E°_{cell}, and K. (§19.4)
17. Describe how E_{cell} changes as the cell operates (Q changes). (§19.4)
18. Explain why a voltaic cell can do work until $Q = K$. (§19.4)
19. Describe how a concentration cell does work until the half-cell concentrations are equal. (§19.4)
20. Distinguish between primary (nonrechargeable) batteries and secondary (rechargeable) batteries; explain the nature of fuel cells. (§19.5)
21. Describe how corrosion occurs and is prevented; compare and contrast a corroding metal and a voltaic cell. (§19.6)
22. Describe how electrolytic cells use nonspontaneous redox reactions driven by an external source of electricity. (§19.7)
23. Explain how atomic properties (ionization energy and electronegativity) determine the products of the electrolysis of a molten salt mixture. (§19.7)
24. Demonstrate how the electrolysis of water influences the products of aqueous electrolysis; explain the importance of overpotential. (§19.7)
25. Describe the relationship between the quantity of charge flowing through a cell and the amount of product formed. (§19.7)

Skills

1. Determine oxidation numbers, and decide whether or not a reaction is a redox reaction. (§19.1 and SPs 19.1–19.3)
2. Balance redox reactions by the oxidation number method. (§19.1 and SP 19.4)
3. Balance redox reactions by the half-reaction method. (§19.1 and SP 19.5)
4. Determine the concentration of an unknown solution using a redox titration. (§19.1 and SP 19.6)
5. Draw and label a diagram of a voltaic cell. (§19.2 and SP 19.7)
6. Combine $E^\circ_{half\text{-}cell}$ values to obtain E°_{cell}. (§19.3)
7. Use E°_{cell} and a known $E^\circ_{half\text{-}cell}$ to find an unknown $E^\circ_{half\text{-}cell}$. (SP 19.8)
8. Manipulate half-reactions to write a spontaneous redox reaction and calculate its E°_{cell}. (SP 19.9)
9. Rank the relative strengths of oxidizing and reducing agents in a redox reaction. (SP 19.9)
10. Predict whether a metal can displace hydrogen or another metal from solution. (§19.3)
11. Use the interrelationship among ΔG°, E°_{cell}, and K to calculate one of the three given the other two. (§19.4 and SP 19.10)
12. Use the Nernst equation to calculate E_{cell}. (SP 19.11)
13. Calculate E_{cell} of a concentration cell. (SP 19.12)
14. Predict the products of the electrolysis of a molten salt mixture. (SP 19.13)
15. Predict the products of the electrolysis of aqueous salt solutions. (SP 19.14)
16. Calculate the current (or time) needed to produce a given amount of product by electrolysis. (SP 19.15)

Key Terms

Section 19.1

oxidation-reduction reaction (redox reaction)
oxidation
reduction
oxidizing agent
reducing agent
oxidation number (oxidation state)
oxidation number method
half-reaction method

Section 19.2

voltaic cell (galvanic cell)
electrolytic cell
electrodes
electrolyte
anode
cathode
half-cell
salt bridge

Section 19.3
cell potential (E_{cell})
voltage
volt (V)
coulomb (C)
standard cell potential (E°_{cell})
standard electrode potential ($E^\circ_{half\text{-}cell}$)
standard reference half-cell (standard hydrogen electrode)

Section 19.4
Faraday constant (F)
Nernst equation
concentration cell

Section 19.5
battery
fuel cell

Section 19.6
corrosion

Section 19.7
electrolysis
overpotential
ampere (A)

Key Equations and Relationships

19.1 Relating the spontaneity of a process to the sign of the cell potential:

$$E_{cell} > 0 \text{ for a spontaneous process}$$

19.2 Relating the electric potential to the energy and charge in SI units:

$$\text{Potential} = \frac{\text{energy}}{\text{charge}} \quad \text{or} \quad 1\text{ V} = 1\text{ J/C}$$

19.3 Relating the standard cell potential to the standard electrode potentials in a voltaic cell:

$$E^\circ_{cell} = E^\circ_{cathode\ (reduction)} - E^\circ_{anode\ (oxidation)}$$

19.4 Defining the Faraday constant:

$$F = 96\,485 \frac{\text{J}}{\text{V}\cdot\text{mol e}^-} = 96\,485 \frac{\text{C}}{\text{mol}}$$

19.5 Relating the Gibbs energy change to the cell potential:

$$\Delta G = -zFE_{cell}$$

19.6 Finding the standard Gibbs energy change from the standard cell potential:

$$\Delta G^\circ = -zFE^\circ_{cell}$$

19.7 Finding the equilibrium constant from the standard cell potential:

$$E^\circ_{cell} = \frac{RT}{zF}\ln K$$

19.8 Calculating the nonstandard cell potential (Nernst equation):

$$E_{cell} = E^\circ_{cell} - \frac{RT}{zF}\ln Q$$

19.9 Relating current to charge and time:

$$I = \frac{Q}{t} \quad \text{with units} \quad 1\text{ A} = 1\text{ C/s}$$

Brief Solutions to Follow-Up Problems

19.1 (a) Oxidation number of Sc = +3; oxidation number of O = −2
(b) Oxidation number of Ga = +3; oxidation number of Cl = −1
(c) Oxidation number of H = +1; oxidation number of P = +5; oxidation number of O = −2
(d) Oxidation number of I = +3; oxidation number of F = −1

19.2 (a) Oxidation number decreased: *reduction*

$$\overset{+3\ -1}{NCl_3}(l) + 3\overset{+1\ -2}{H_2O}(l) \longrightarrow \overset{-3\ +1}{NH_3}(aq) + 3\overset{+1\ -2\ +1}{HOCl}(aq)$$

Oxidation number increased: *oxidation*

Redox reaction; the oxidation number of N decreases, and the oxidation number of Cl increases.

(b)

$$\overset{+1\ +5\ -2}{AgNO_3}(aq) + \overset{-3\ +1\ -1}{NH_4I}(aq) \longrightarrow \overset{+1\ -1}{AgI}(s) + \overset{-3\ +1\ +5\ -2}{NH_4NO_3}(aq)$$

Not a redox reaction; there are no changes in the oxidation numbers.

(c) Oxidation number decreased: *reduction*

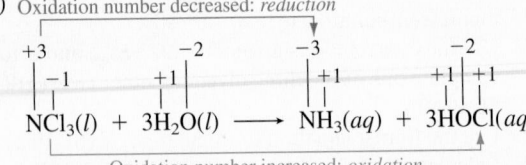

$$2\overset{+1\ -2}{H_2S}(g) + 3\overset{0}{O_2}(g) \longrightarrow 2\overset{+4\ -2}{SO_2}(g) + 2\overset{+1\ -2}{H_2O}(g)$$

Oxidation number increased: *oxidation*

Redox reaction; the oxidation number of S increases, and the oxidation number of O decreases.

19.3 (a) Fe is the reducing agent; Cl_2 is the oxidizing agent.
(b) C_2H_6 is the reducing agent; O_2 is the oxidizing agent.
(c) CO is the reducing agent; I_2O_5 is the oxidizing agent.

19.4 $K_2Cr_2O_7(aq) + 14HI(aq) \longrightarrow 2KI(aq) + 2CrI_3(aq) + 3I_2(s) + 7H_2O(l)$

19.5 $6KMnO_4(aq) + 6KOH(aq) + KI(aq) \longrightarrow 6K_2MnO_4(aq) + KIO_3(aq) + 3H_2O(l)$

19.6 (a) Amount (mol) of Ca^{2+}

$$= 6.53\text{ mL soln} \times \frac{1\text{ L}}{10^3\text{ mL}} \times \frac{4.56 \times 10^{-3}\text{ mol KMnO}_4}{1\text{ L soln}} \times \frac{5\text{ mol CaC}_2\text{O}_4}{2\text{ mol KMnO}_4} \times \frac{1\text{ mol Ca}^{2+}}{1\text{ mol CaC}_2\text{O}}$$

$$= 7.44 \times 10^{-5}\text{ mol Ca}^{2+}$$

$$\text{Concentration of Ca}^{2+} = \frac{7.44 \times 10^{-5}\text{ mol Ca}^{2+}}{2.50\text{ mL milk}} \times \frac{10^3\text{ mL}}{1\text{ L}} = 2.98 \times 10^{-2}\text{ mol/L Ca}^{2+}$$

(b) Concentration of Ca^{2+} (g/L)

$$= \frac{2.98 \times 10^{-2}\text{ mol Ca}^{2+}}{1\text{ L}} \times \frac{40.08\text{ g Ca}^{2+}}{1\text{ mol Ca}^{2+}} = \frac{1.19\text{ g Ca}^{2+}}{1\text{ L}}$$

Yes, this concentration is consistent with the typical concentration.

19.7

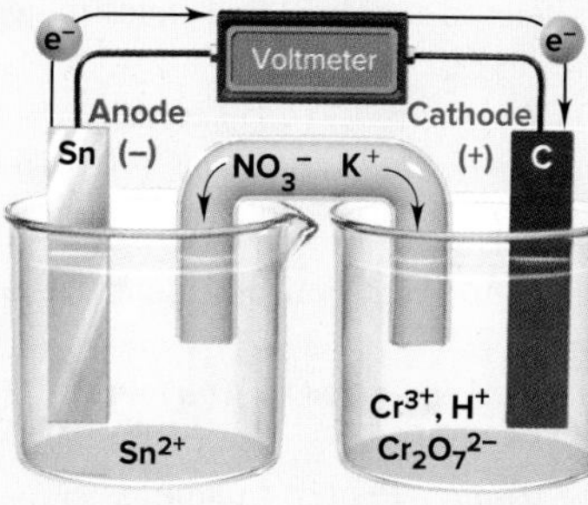

$Sn(s) \longrightarrow Sn^{2+}(aq) + 2e^-$ (anode; oxidation)

$6e^- + 14H^+(aq) + Cr_2O_7^{2-}(aq) \longrightarrow 2Cr^{3+}(aq) + 7H_2O(l)$ (cathode; reduction)

$3Sn(s) + Cr_2O_7^{2-}(aq) + 14H^+(aq) \longrightarrow 3Sn^{2+}(aq) + 2Cr^{3+}(aq) + 7H_2O(l)$ (overall)

Cell notation:

$Sn(s)\,|\,Sn^{2+}(aq)\,||\,H^+(aq), Cr_2O_7^{2-}(aq), Cr^{3+}(aq)\,|\,\text{graphite}$

19.8 $Br_2(aq) + 2e^- \longrightarrow 2Br^-(aq)$

$E^\circ_{bromine} = 1.07$ V (cathode)

$2V^{3+}(aq) + 2H_2O(l) \longrightarrow 2VO^{2+}(aq) + 4H^+(aq) + 2e^-$

$E^\circ_{vanadium} = ?$ (anode)

$E^\circ_{vanadium} = E^\circ_{bromine} - E^\circ_{cell} = 1.07\text{ V} - 1.39\text{ V} = -0.32\text{ V}$

19.9 $Fe^{2+}(aq) + 2e^- \longrightarrow Fe(s)$ $\quad E^\circ = -0.44$ V

$2[Fe^{2+}(aq) \longrightarrow Fe^{3+}(aq) + e^-]$ $\quad E^\circ = 0.77$ V

$3Fe^{2+}(aq) \longrightarrow 2Fe^{3+}(aq) + Fe(s)$

$E^\circ_{cell} = -0.44\text{ V} - 0.77\text{ V} = -1.21\text{ V}$

The reaction is nonspontaneous. The spontaneous reaction is

$2Fe^{3+}(aq) + Fe(s) \longrightarrow 3Fe^{2+}(aq)$ $\quad E^\circ_{cell} = 1.21$ V

Ranking: $Fe > Fe^{2+} > Fe^{3+}$

19.10 $Cd(s) + Cu^{2+}(aq) \longrightarrow Cd^{2+}(aq) + Cu(s)$

$\Delta G^\circ = -RT \ln K = -8.314\text{ J/(mol·K)} \times 298\text{ K} \times \ln K$

$= -143\text{ kJ/mol}; K = 1.2 \times 10^{25}$

$$E^\circ_{cell} = \frac{[(8.314\text{ J/(mol·K)}](298\text{ K})}{(2\text{ mol})(96\,485\text{ C/mol})} \ln(1.2 \times 10^{25}) = 0.74\text{ V}$$

19.11 $Fe(s) \longrightarrow Fe^{2+}(aq) + 2e^-$ $\quad E^\circ = -0.44$ V

$Cu^{2+}(aq) + 2e^- \longrightarrow Cu(s)$ $\quad E^\circ = 0.34$ V

$Fe(s) + Cu^{2+}(aq) \longrightarrow Fe^{2+}(aq) + Cu(s)$ $\quad E^\circ_{cell} = 0.78$ V

So, $E_{cell} = 0.78\text{ V} + 0.25\text{ V} = 1.03\text{ V}$.

$$1.03\text{ V} = 0.78\text{ V} - \frac{[8.314\text{ J/(mol·K)}](298\text{ K})}{(2\text{ mol})(96\,485\text{ C/mol})} \ln \frac{[Fe^{2+}]}{[Cu^{2+}]}$$

$$\frac{[Fe^{2+}]}{[Cu^{2+}]} = 3.5 \times 10^{-9}$$

$[Fe^{2+}] = 3.5 \times 10^{-9} \times 0.30\text{ mol/L} = 1.0 \times 10^{-9}\text{ mol/L}$

19.12 $Au^{3+}(aq; 2.5 \times 10^{-2}\text{ mol/L})$ [B] $\longrightarrow$ $Au^{3+}(aq; 7.0 \times 10^{-4}\text{ mol/L})$ [A]

$$E_{cell} = 0\text{ V} - \left[\frac{[8.314\text{ J/(mol·K)}](298\text{ K})}{(3\text{ mol})(96\,485\text{ C/mol})} \times \ln\left(\frac{7.0 \times 10^{-4}}{2.5 \times 10^{-2}}\right)\right]$$

$= 0.0306$ V

The electrode in half-cell A is negative, so it is the anode.

19.13 The oxidizing agents are K^+ and Al^{3+}. The reducing agents are F^- and Br^-.

Al is above and to the right of K in the periodic table, so it has a higher IE:

$Al^{3+}(l) + 3e^- \longrightarrow Al(s)$ (cathode; reduction)

Br is below F in group 17, so it has a lower χ:

$2Br^-(l) \longrightarrow Br_2(g) + 2e^-$ (anode; oxidation)

$2Al^{3+}(l) + 6Br^-(l) \longrightarrow 2Al(s) + 3Br_2(g)$ (overall)

19.14 The reduction with the more positive electrode potential is

$Au^{3+}(aq) + 3e^- \longrightarrow Au(s)$ $\quad E^\circ = 1.50$ V (cathode; reduction)

Because of overpotential, O_2 will not form at the anode, so Br_2 will form:

$2Br^-(aq) \longrightarrow Br_2(l) + 2e^-$ $\quad E^\circ = 1.07$ V (cathode; oxidation)

19.15 $Cu^{2+}(aq) + 2e^- \longrightarrow Cu(s)$

Therefore, 2 mol e^-/1 mol Cu = 2 mol e^-/63.55 g Cu.

$$\text{Time (min)} = 1.50\text{ g Cu} \times \frac{2\text{ mol e}^-}{63.55\text{ g Cu}} \times \frac{96\,485\text{ C}}{1\text{ mol e}^-} \times \frac{1\text{ s}}{4.75\text{ C}} \times \frac{1\text{ min}}{60\text{ s}} = 16.0\text{ min}$$

PROBLEMS

Problems with **red** numbers are answered in Appendix G and worked in detail in the Student Solutions Manual. Problem sections match those in this book and provide the numbers of relevant sample problems. Most offer Concept Review Questions, Skill-Building Exercises (grouped in pairs covering the same concept), and Problems in Context. The Comprehensive Problems are based on material from any section or previous chapter.

Note: Unless stated otherwise, all problems refer to systems at 298 K (25°C).

Oxidation-Reduction (Redox) Reactions

(Sample Problems 19.1 to 19.6)

Concept Review Questions

19.1 Define *oxidation* and *reduction* in terms of electron transfer and change in oxidation number.

19.2 Why must an electrochemical process involve a redox reaction?

19.3 Can one half-reaction in a redox reaction take place independently of the other? Explain.

19.4 Water is used to balance O atoms in the half-reaction method. Why can O^{2-} ions not be used instead?

19.5 During the redox balancing process, what step is taken to ensure that e^- loss is equal to e^- gain?

19.6 How are protons removed when balancing a redox reaction in basic solution?

19.7 (a) Are spectator ions used to balance the half-reactions in a redox reaction?
(b) At what stage might spectator ions enter the balancing process?

19.8 (a) Which type of electrochemical cell has $\Delta_{sys}G < 0$?
(b) Which type of electrochemical cell shows an increase in Gibbs energy?

19.9 Which statements are true? Correct any statements that are false.
(a) In a voltaic cell, the anode is negative relative to the cathode.
(b) Oxidation occurs at the anode of a voltaic or electrolytic cell.
(c) Electrons flow into the cathode of an electrolytic cell.
(d) In a voltaic cell, the surroundings do work on the system.
(e) A metal that plates out of an electrolytic cell appears on the cathode.
(f) The cell electrolyte provides a solution of mobile electrons.

Skill-Building Exercises (grouped in similar pairs)

19.10 Consider the following balanced redox reaction:

$$16H^+(aq) + 2MnO_4^-(aq) + 10Cl^-(aq) \longrightarrow 2Mn^{2+}(aq) + 5Cl_2(g) + 8H_2O(l)$$

(a) Which species is being oxidized?
(b) Which species is being reduced?
(c) Which species is the oxidizing agent?
(d) Which species is the reducing agent?
(e) From which species to which species does electron transfer occur?
(f) Write the balanced molecular equation, with K^+ and SO_4^{2-} as the spectator ions.

19.11 Consider the following balanced redox reaction:

$$2CrO_2^-(aq) + 2H_2O(l) + 6ClO^-(aq) \longrightarrow 2CrO_4^{2-}(aq) + 3Cl_2(g) + 4OH^-(aq)$$

(a) Which species is being oxidized?
(b) Which species is being reduced?
(c) Which species is the oxidizing agent?
(d) Which species is the reducing agent?
(e) From which species to which species does electron transfer occur?
(f) Write the balanced molecular equation, with Na^+ as the spectator ion.

19.12 Balance each skeleton equation, and identify the oxidizing and reducing agents:
(a) $ClO_3^-(aq) + I^-(aq) \longrightarrow I_2(s) + Cl^-(aq)$ (acidic)
(b) $MnO_4^-(aq) + SO_3^{2-}(aq) \longrightarrow MnO_2(s) + SO_4^{2-}(aq)$ (basic)
(c) $MnO_4^-(aq) + H_2O_2(aq) \longrightarrow Mn^{2+}(aq) + O_2(g)$ (acidic)

19.13 Balance each skeleton equation, and identify the oxidizing and reducing agents:
(a) $O_2(g) + NO(g) \longrightarrow NO_3^-(aq)$ (acidic)
(b) $CrO_4^{2-}(aq) + Cu(s) \longrightarrow Cr(OH)_3(s) + Cu(OH)_2(s)$ (basic)
(c) $AsO_4^{3-}(aq) + NO_2^-(aq) \longrightarrow AsO_2^-(aq) + NO_3^-(aq)$ (basic)

19.14 Balance each skeleton equation, and identify the oxidizing and reducing agents:
(a) $Cr_2O_7^{2-}(aq) + Zn(s) \longrightarrow Zn^{2+}(aq) + Cr^{3+}(aq)$ (acidic)
(b) $Fe(OH)_2(s) + MnO_4^-(aq) \longrightarrow MnO_2(s) + Fe(OH)_3(s)$ (basic)
(c) $Zn(s) + NO_3^-(aq) \longrightarrow Zn^{2+}(aq) + N_2(g)$ (acidic)

19.15 Balance each skeleton equation, and identify the oxidizing and reducing agents:
(a) $BH_4^-(aq) + ClO_3^-(aq) \longrightarrow H_2BO_3^-(aq) + Cl^-(aq)$ (basic)
(b) $CrO_4^{2-}(aq) + N_2O(g) \longrightarrow Cr^{3+}(aq) + NO(g)$ (acidic)
(c) $Br_2(l) \longrightarrow BrO_3^-(aq) + Br^-(aq)$ (basic)

19.16 Balance each skeleton equation, and identify the oxidizing and reducing agents:
(a) $Sb(s) + NO_3^-(aq) \longrightarrow Sb_4O_6(s) + NO(g)$ (acidic)
(b) $Mn^{2+}(aq) + BiO_3^-(aq) \longrightarrow MnO_4^-(aq) + Bi^{3+}(aq)$ (acidic)
(c) $Fe(OH)_2(s) + Pb(OH)_3^-(aq) \longrightarrow Fe(OH)_3(s) + Pb(s)$ (basic)

19.17 Balance each skeleton equation, and identify the oxidizing and reducing agents:
(a) $NO_2(g) \longrightarrow NO_3^-(aq) + NO_2^-(aq)$ (basic)
(b) $Zn(s) + NO_3^-(aq) \longrightarrow Zn(OH)_4^{2-}(aq) + NH_3(g)$ (basic)
(c) $H_2S(g) + NO_3^-(aq) \longrightarrow S_8(s) + NO(g)$ (acidic)

19.18 Balance each skeleton equation, and identify the oxidizing and reducing agents:
(a) $As_4O_6(s) + MnO_4^-(aq) \longrightarrow AsO_4^{3-}(aq) + Mn^{2+}(aq)$ (acidic)
(b) $P_4(s) \longrightarrow HPO_3^{2-}(aq) + PH_3(g)$ (acidic)
(c) $MnO_4^-(aq) + CN^-(aq) \longrightarrow MnO_2(s) + CNO^-(aq)$ (basic)

19.19 Balance each skeleton equation, and identify the oxidizing and reducing agents:
(a) $SO_3^{2-}(aq) + Cl_2(g) \longrightarrow SO_4^{2-}(aq) + Cl^-(aq)$ (basic)
(b) $Fe(CN)_6^{3-}(aq) + Re(s) \longrightarrow Fe(CN)_6^{4-}(aq) + ReO_4^-(aq)$ (basic)
(c) $MnO_4^-(aq) + HCOOH(aq) \longrightarrow Mn^{2+}(aq) + CO_2(g)$ (acidic)

Problems in Context

19.20 In many residential water systems, the aqueous Fe^{3+} concentration is high enough to stain sinks and turn drinking water light brown. The iron content is analyzed by first reducing the Fe^{3+} to Fe^{2+} and then titrating with MnO_4^- in acidic solution. Balance the following skeleton equation of the titration step:

$$Fe^{2+}(aq) + MnO_4^-(aq) \longrightarrow Mn^{2+}(aq) + Fe^{3+}(aq)$$

19.21 *Aqua regia*, a mixture of concentrated HNO_3 and HCl, was developed by alchemists as a means to "dissolve" gold. The process is a redox reaction, with this simplified skeleton equation:

$$Au(s) + NO_3^-(aq) + Cl^-(aq) \longrightarrow AuCl_4^-(aq) + NO_2(g)$$

(a) Balance the equation by the half-reaction method.
(b) What are the oxidizing and reducing agents?
(c) What is the function of HCl in *aqua regia*?

Voltaic Cells: Electrochemical Cells Using Spontaneous Reactions to Generate Electrical Energy

(Sample Problem 19.7)

Concept Review Questions

19.22 Consider the following general voltaic cell:

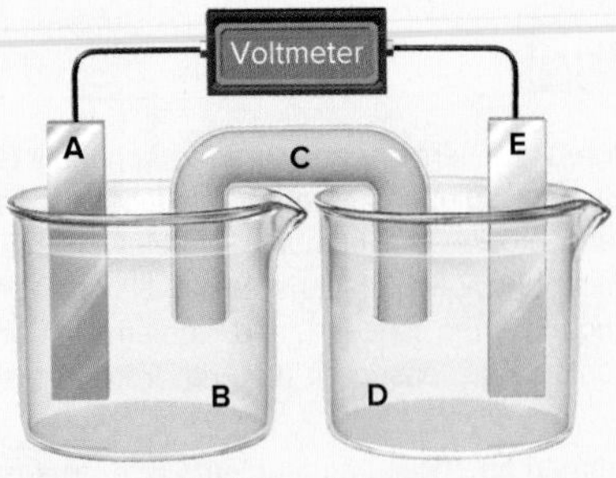

Identify (a) the anode; (b) the cathode; (c) the salt bridge; (d) the electrode at which e^- leaves the cell; (e) the electrode with a positive charge; (f) the electrode that gains mass as the cell operates (assuming that a metal plates out).

19.23 Why does a voltaic cell not operate unless the two compartments are connected through an external circuit?

19.24 What purpose does the salt bridge serve in a voltaic cell, and how does it accomplish this purpose?

19.25 (a) What is the difference between an active electrode and an inactive electrode?
(b) Why are inactive electrodes used?
(c) Name two substances that are commonly used for inactive electrodes.

19.26 When a piece of metal A is placed in a solution that contains ions of metal B, metal B plates out on the piece of A.
(a) Which metal is being oxidized?
(b) Which metal is being displaced?
(c) Which metal would you use as the anode in a voltaic cell incorporating these two metals?
(d) If bubbles of H_2 form when B is placed in an acid, will they form if A is placed in the acid? Explain.

Skill-Building Exercises (grouped in similar pairs)

19.27 A voltaic cell is constructed with an Sn/Sn^{2+} half-cell and a Zn/Zn^{2+} half-cell. The zinc electrode is negative.
(a) Write balanced half-reactions and the overall reaction.
(b) Draw a diagram of the cell, labelling the electrodes with their charges and showing the direction of electron flow in the circuit and the directions of cation and anion flow in the salt bridge.

19.28 A voltaic cell is constructed with an Ag/Ag^+ half-cell and a Pb/Pb^{2+} half-cell. The silver electrode is positive.
(a) Write balanced half-reactions and the overall reaction.
(b) Draw a diagram of the cell, labelling the electrodes with their charges and showing the direction of electron flow in the circuit and the directions of cation and anion flow in the salt bridge.

19.29 Consider the following voltaic cell:

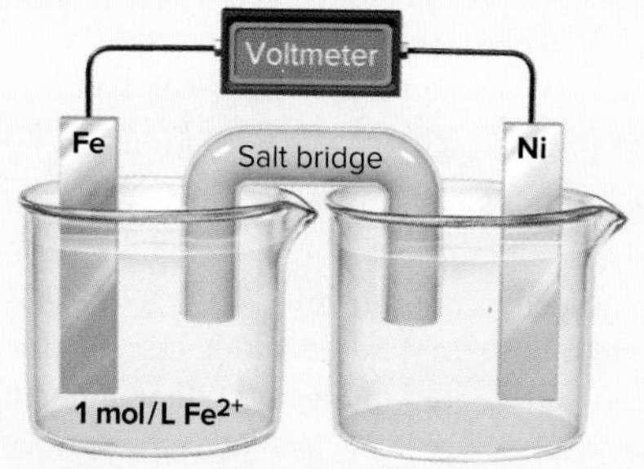

(a) In which direction do electrons flow in the external circuit?
(b) In which half-cell does oxidation occur?
(c) In which half-cell do electrons enter the cell?
(d) At which electrode are electrons consumed?
(e) Which electrode is negatively charged?
(f) Which electrode decreases in mass during cell operation?
(g) Suggest a solution for the cathode electrolyte.
(h) Suggest a pair of ions for the salt bridge.
(i) For which electrode could you use an inactive material?
(j) In which direction do anions within the salt bridge move to maintain charge neutrality?
(k) Write balanced half-reactions and the overall cell reaction.

19.30 Consider the following voltaic cell:

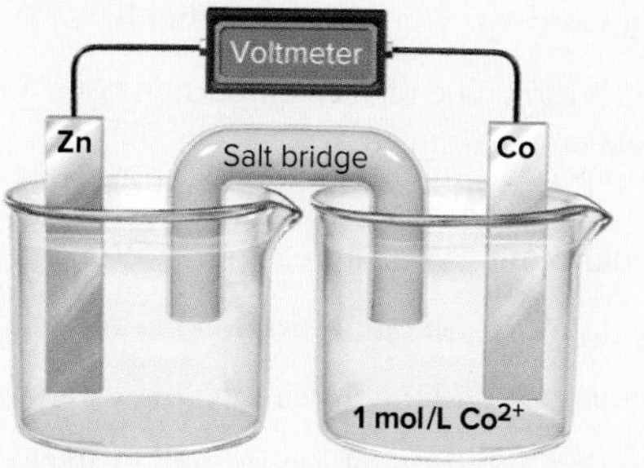

(a) In which direction do electrons flow in the external circuit?
(b) In which half-cell does reduction occur?
(c) In which half-cell do electrons leave the cell?
(d) At which electrode are electrons generated?
(e) Which electrode is positively charged?
(f) Which electrode increases in mass during cell operation?
(g) Suggest a solution for the anode electrolyte.
(h) Suggest a pair of ions for the salt bridge.
(i) For which electrode could you use an inactive material?
(j) In which direction do cations within the salt bridge move to maintain charge neutrality?
(k) Write balanced half-reactions and the overall cell reaction.

19.31 A voltaic cell is constructed with an Fe/Fe^{2+} half-cell and an Mn/Mn^{2+} half-cell. The iron electrode is positive.
(a) Write balanced half-reactions and the overall reaction.
(b) Draw a diagram of the cell, labelling the electrodes with their charges and showing the direction of electron flow in the circuit and the directions of cation and anion flow in the salt bridge.

19.32 A voltaic cell is constructed with a Cu/Cu^{2+} half-cell and an Ni/Ni^{2+} half-cell. The nickel electrode is negative.
(a) Write balanced half-reactions and the overall reaction.
(b) Draw a diagram of the cell, labelling the electrodes with their charges and showing the direction of electron flow in the circuit and the directions of cation and anion flow in the salt bridge.

19.33 Write the cell notation for the voltaic cell that incorporates each redox reaction:
(a) $Al(s) + Cr^{3+}(aq) \longrightarrow Al^{3+}(aq) + Cr(s)$
(b) $Cu^{2+}(aq) + SO_2(g) + 2H_2O(l) \longrightarrow Cu(s) + SO_4^{2-}(aq) + 4H^+(aq)$

19.34 Write a balanced equation from each cell notation:
(a) $Mn(s)|\,Mn^{2+}(aq)\,||\,Cd^{2+}(aq)|Cd(s)$
(b) $Fe(s)|Fe^{2+}(aq)||NO_3^-(aq)|\,NO(g)|\,Pt(s)$

Cell Potential: Output of a Voltaic Cell

(Sample Problems 19.8 and 19.9)

Concept Review Questions

19.35 How is a standard reference electrode used to determine unknown $E^\circ_{\text{half-cell}}$ values?

19.36 (a) What does a negative E°_{cell} indicate about a redox reaction?
(b) What does a negative E°_{cell} indicate about the reverse reaction?

19.37 The standard cell potential is a thermodynamic state function.
(a) How are E° values treated similarly to ΔH°, ΔG°, and S° values?
(b) How are E° values treated differently?

Skill-Building Exercises (grouped in similar pairs)

19.38 In basic solution, Se^{2-} and SO_3^{2-} ions react spontaneously:

$$2Se^{2-}(aq) + 2SO_3^{2-}(aq) + 3H_2O(l) \longrightarrow 2Se(s) + 6OH^-(aq) + S_2O_3^{2-}(aq) \quad E^\circ_{\text{cell}} = 0.35\text{ V}$$

(a) Write balanced half-reactions for the process.
(b) If E°_{sulfite} is -0.57 V, calculate $E^\circ_{\text{selenium}}$.

19.39 In acidic solution, O_3 and Mn^{2+} ion react spontaneously:

$$O_3(g) + Mn^{2+}(aq) + H_2O(l) \longrightarrow O_2(g) + MnO_2(s) + 2H^+(aq) \quad E^\circ_{\text{cell}} = 0.84\text{ V}$$

(a) Write the balanced half-reactions.
(b) Using Appendix D to find E°_{ozone}, calculate $E^\circ_{\text{manganese}}$.

19.40 Use the list of standard electrode potentials in Appendix D to arrange the given species as indicated:
(a) In order of *decreasing* strength as *oxidizing* agents: Fe^{3+}, Br_2, Cu^{2+}
(b) In order of *increasing* strength as *oxidizing* agents: Ca^{2+}, $Cr_2O_7^{2-}$, Ag^+

19.41 Use the list of standard electrode potentials in Appendix D to arrange the given species as indicated:
(a) In order of *decreasing* strength as *reducing* agents: SO_2, $PbSO_4$, MnO_2
(b) In order of *increasing* strength as *reducing* agents: Hg, Fe, Sn

19.42 Balance each skeleton equation, calculate E°_{cell}, and state whether the reaction is spontaneous:
(a) $Co(s) + H^+(aq) \longrightarrow Co^{2+}(aq) + H_2(g)$
(b) $Mn^{2+}(aq) + Br_2(l) \longrightarrow MnO_4^-(aq) + Br^-(aq)$ (acidic)
(c) $Hg_2^{2+}(aq) \longrightarrow Hg^{2+}(aq) + Hg(l)$

19.43 Balance each skeleton equation, calculate E°_{cell}, and state whether the reaction is spontaneous:
(a) $Cl_2(g) + Fe^{2+}(aq) \longrightarrow Cl^-(aq) + Fe^{3+}(aq)$
(b) $Mn^{2+}(aq) + Co^{3+}(aq) \longrightarrow MnO_2(s) + Co^{2+}(aq)$ (acidic)
(c) $AgCl(s) + NO(g) \longrightarrow Ag(s) + Cl^-(aq) + NO_3^-(aq)$ (acidic)

19.44 Balance each skeleton equation, calculate E°_{cell}, and state whether the reaction is spontaneous:
(a) $Ag(s) + Cu^{2+}(aq) \longrightarrow Ag^+(aq) + Cu(s)$
(b) $Cd(s) + Cr_2O_7^{2-}(aq) \longrightarrow Cd^{2+}(aq) + Cr^{3+}(aq)$
(c) $Ni^{2+}(aq) + Pb(s) \longrightarrow Ni(s) + Pb^{2+}(aq)$

19.45 Balance each skeleton equation, calculate E°_{cell}, and state whether the reaction is spontaneous:
(a) $Cu^+(aq) + PbO_2(s) + SO_4^{2-}(aq) \longrightarrow PbSO_4(s) + Cu^{2+}(aq)$ (acidic)
(b) $H_2O_2(aq) + Ni^{2+}(aq) \longrightarrow O_2(g) + Ni(s)$ (acidic)
(c) $MnO_2(s) + Ag^+(aq) \longrightarrow MnO_4^-(aq) + Ag(s)$ (basic)

19.46 Use the following half-reactions to write three spontaneous reactions. Then calculate E°_{cell} for each reaction, and rank the strengths of the oxidizing and reducing agents:
(1) $Al^{3+}(aq) + 3e^- \longrightarrow Al(s)$ $E^\circ = -1.66$ V
(2) $N_2O_4(g) + 2e^- \longrightarrow 2NO_2^-(aq)$ $E^\circ = 0.867$ V
(3) $SO_4^{2-}(aq) + H_2O(l) + 2e^- \longrightarrow SO_3^{2-}(aq) + 2OH^-(aq)$ $E^\circ = 0.93$ V

19.47 Use the following half-reactions to write three spontaneous reactions. Then calculate E°_{cell} for each reaction, and rank the strengths of the oxidizing and reducing agents:
(1) $Au^+(aq) + e^- \longrightarrow Au(s)$ $E^\circ = 1.69$ V
(2) $N_2O(g) + 2H^+(aq) + 2e^- \longrightarrow N_2(g) + H_2O(l)$ $E^\circ = 1.77$ V
(3) $Cr^{3+}(aq) + 3e^- \longrightarrow Cr(s)$ $E^\circ = -0.74$ V

19.48 Use the following half-reactions to write three spontaneous reactions. Then calculate E°_{cell} for each reaction, and rank the strengths of the oxidizing and reducing agents:
(1) $2HClO(aq) + 2H^+(aq) + 2e^- \longrightarrow Cl_2(g) + 2H_2O(l)$ $E^\circ = 1.63$ V
(2) $Pt^{2+}(aq) + 2e^- \longrightarrow Pt(s)$ $E^\circ = 1.20$ V
(3) $PbSO_4(s) + 2e^- \longrightarrow Pb(s) + SO_4^{2-}(aq)$ $E^\circ = -0.31$ V

19.49 Use the following half-reactions to write three spontaneous reactions. Then calculate E°_{cell} for each reaction, and rank the strengths of the oxidizing and reducing agents:
(1) $I_2(s) + 2e^- \longrightarrow 2I^-(aq)$ $E^\circ = 0.53$ V
(2) $S_2O_8^{2-}(aq) + 2e^- \longrightarrow 2SO_4^{2-}(aq)$ $E^\circ = 2.01$ V
(3) $Cr_2O_7^{2-}(aq) + 14H^+(aq) + 6e^- \longrightarrow 2Cr^{3+}(aq) + 7H_2O(l)$ $E^\circ = 1.33$ V

Problems in Context

19.50 When metal A is placed in a solution of a salt of metal B, the surface of metal A changes colour. When metal B is placed in an acid solution, gas bubbles form on the surface of the metal. When metal A is placed in a solution of a salt of metal C, no change is observed in the solution or on the surface of metal A. Will metal C cause the formation of H_2 when placed in an acid solution? Rank metals A, B, and C in order of *decreasing* reducing strength.

19.51 When a clean iron nail is placed in an aqueous solution of copper(II) sulfate, the nail becomes coated with a brownish black material.
(a) What is the material coating the iron?
(b) What are the oxidizing and reducing agents?
(c) Can this reaction be made into a voltaic cell?
(d) Write the balanced equation for the reaction.
(e) Calculate E°_{cell} for the process.

Gibbs Energy and Electrical Work

(Sample Problems 19.10 to 19.12)

Concept Review Questions

19.52 (a) How do the relative magnitudes of Q and K relate to the signs of ΔG and E_{cell}? Explain.
(b) Can a cell do work when $\frac{Q}{K} > 1$ or $\frac{Q}{K} < 1$? Explain.

19.53 A voltaic cell consists of A/A$^+$ and B/B$^+$ half-cells, where A and B are metals and the A electrode is negative. The initial $[A^+]/[B^+]$ is such that $E_{cell} > E^\circ_{cell}$.
(a) How do $[A^+]$ and $[B^+]$ change as the cell operates?
(b) How does E_{cell} change as the cell operates?
(c) What is $[A^+]/[B^+]$ when $E_{cell} = E^\circ_{cell}$? Explain.
(d) Is it possible for E_{cell} to be less than E°_{cell}? Explain.

19.54 Explain whether E_{cell} of a voltaic cell will increase or decrease with each change:
(a) Decrease in cell temperature
(b) Increase in [active ion] in the anode compartment
(c) Increase in [active ion] in the cathode compartment
(d) Increase in pressure of a gaseous reactant in the cathode compartment

19.55 In a concentration cell, is the more concentrated electrolyte in the cathode or the anode compartment? Explain.

Skill-Building Exercises (grouped in similar pairs)

19.56 What is the value of the equilibrium constant for the reaction between each pair at 25°C?
(a) Ni(*s*) and $Ag^+(aq)$ (b) Fe(*s*) and $Cr^{3+}(aq)$

19.57 What is the value of the equilibrium constant for the reaction between each pair at 25°C?
(a) Al(*s*) and $Cd^{2+}(aq)$ (b) $I_2(s)$ and $Br^-(aq)$

19.58 What is the value of the equilibrium constant for the reaction between each pair at 25°C?
(a) Ag(*s*) and $Mn^{2+}(aq)$ (b) $Cl_2(g)$ and $Br^-(aq)$

19.59 What is the value of the equilibrium constant for the reaction between each pair at 25°C?
(a) Cr(*s*) and $Cu^{2+}(aq)$ (b) Sn(*s*) and $Pb^{2+}(aq)$

19.60 Calculate ΔG° for each reaction in Problem 19.56.

19.61 Calculate ΔG° for each reaction in Problem 19.57.

19.62 Calculate ΔG° for each reaction in Problem 19.58.

19.63 Calculate ΔG° for each reaction in Problem 19.59.

19.64 What are E°_{cell} and ΔG° of a redox reaction at 25°C if $z = 1$ and $K = 5.0 \times 10^4$?

19.65 What are E°_{cell} and ΔG° of a redox reaction at 25°C if $z = 1$ and $K = 5.0 \times 10^{-6}$?

19.66 What are E°_{cell} and ΔG° of a redox reaction at 25°C if $z = 2$ and $K = 65$?

19.67 What are E°_{cell} and ΔG° of a redox reaction at 25°C if $z = 2$ and $K = 0.065$?

19.68 A voltaic cell consists of a standard reference half-cell and a Cu/Cu^{2+} half-cell. Calculate $[Cu^{2+}]$ when E_{cell} is 0.22 V.

19.69 A voltaic cell consists of a Mn/Mn^{2+} half-cell and a Pb/Pb^{2+} half-cell. Calculate $[Pb^{2+}]$ when $[Mn^{2+}]$ is 1.4 mol/L and E_{cell} is 0.44 V.

19.70 A voltaic cell with Ni/Ni^{2+} and Co/Co^{2+} half-cells has the following initial concentrations: $[Ni^{2+}] = 0.80$ mol/L and $[Co^{2+}] = 0.20$ mol/L.
(a) What is the initial E_{cell}?
(b) What is $[Ni^{2+}]$ when E_{cell} reaches 0.03 V?
(c) What are the equilibrium concentrations of the ions?

19.71 A voltaic cell with Mn/Mn^{2+} and Cd/Cd^{2+} half-cells has the following initial concentrations: $[Mn^{2+}] = 0.090$ mol/L and $[Cd^{2+}] = 0.060$ mol/L.
(a) What is the initial E_{cell}?
(b) What is E_{cell} when $[Cd^{2+}]$ reaches 0.050 mol/L?
(c) What is $[Mn^{2+}]$ when E_{cell} reaches 0.055 V?
(d) What are the equilibrium concentrations of the ions?

19.72 A concentration cell consists of two H_2/H^+ half-cells. Half-cell A has H_2 at 0.95 bar bubbling into 0.10 mol/L HCl. Half-cell B has H_2 at 0.60 bar bubbling into 2.0 mol/L HCl.
(a) Which half-cell houses the anode?
(b) What is the voltage of the cell?

19.73 A concentration cell consists of two Sn/Sn^{2+} half-cells, A and B. The electrolyte in A is 0.13 mol/L $Sn(NO_3)_2$. The electrolyte in B is 0.87 mol/L $Sn(NO_3)_2$.
(a) Which half-cell houses the cathode?
(b) What is the voltage of the cell?

Electrochemical Processes in Batteries

Concept Review Questions

19.74 What is the direction of electron flow with respect to the anode and the cathode in a battery? Explain.

19.75 In the everyday batteries used for items such as flashlights and toys, no salt bridge is evident. What is used in these cells to separate the anode and cathode compartments?

19.76 Both a D-size and a AAA-size alkaline battery have an output of 1.5 V.
(a) What property of the cell potential allows this to occur?
(b) What is different about these two batteries?

Problems in Context

19.77 Many common electrical devices require the use of more than one battery.
(a) How many alkaline batteries must be placed in series to light a flashlight with a 6.0 V bulb?
(b) What is the voltage requirement of a camera that uses six silver batteries?
(c) How many volts can a car battery deliver if two of its anode/cathode cells are shorted?

Corrosion: An Environmental Voltaic Cell

Concept Review Questions

19.78 Because of the extreme cold and heavy snowfall in many places in Canada, much of recent Canadian architecture employs PTFE (or Teflon) grease between joints connecting different types of metal, such as copper and iron. What purpose does the PTFE grease serve?

19.79 Why do steel bridge supports rust at the waterline but not above or below the waterline?

19.80 After the 1930s, chromium replaced nickel for corrosion resistance and appearance on car bumpers and trim. How does chromium protect steel from corrosion?

19.81 Which metals are suitable for use as sacrificial anodes to protect against corrosion of underground iron pipes? If any are not suitable, explain why.
(a) Aluminum; (b) magnesium; (c) sodium; (d) lead; (e) nickel; (f) zinc; (g) chromium

Electrolytic Cells: Electrochemical Cells Using Electrical Energy to Drive Nonspontaneous Reactions

(Sample Problems 19.13 to 19.15)

Concept Review Questions

Note: Unless stated otherwise, assume that the electrolytic cells in the following problems operate at 100% efficiency.

19.82 Consider the following general electrolytic cell:

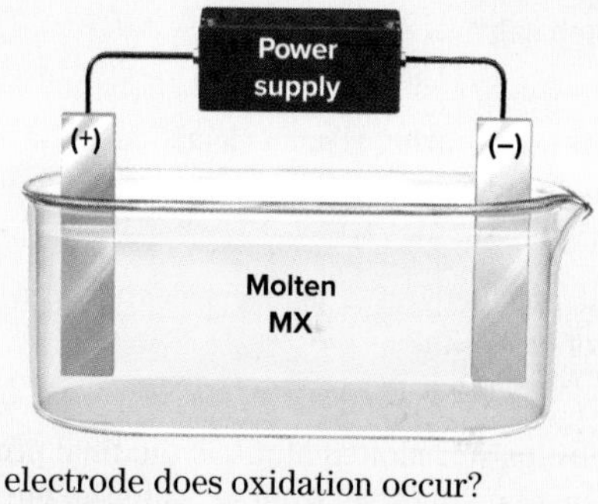

(a) At which electrode does oxidation occur?
(b) At which electrode does elemental M form?
(c) At which electrode are electrons being released by ions?
(d) At which electrode are electrons entering the cell?

19.83 A voltaic cell consists of Cr/Cr^{3+} and Cd/Cd^{2+} half-cells, with all the components in their standard states. After 10 min of operation, a thin coating of cadmium metal has plated out on the cathode. Describe what will happen if the negative terminal of a dry cell (1.5 V) is attached to the cell cathode and the positive terminal of the dry cell is attached to the cell anode.

19.84 Why are the $E_{half\text{-}cell}$ values for the oxidation and reduction of water different from the $E^\circ_{half\text{-}cell}$ values for the same processes?

19.85 In an aqueous electrolytic cell, nitrate ions never react at the anode, but nitrite ions do. Explain why.

19.86 How does overpotential influence the products in the electrolysis of aqueous salts?

Skill-Building Exercises (grouped in similar pairs)

19.87 Consider the electrolysis of molten NaBr.
(a) What product forms at the anode?
(b) What product forms at the cathode?

19.88 Consider the electrolysis of molten BaI_2.
(a) What product forms at the negative electrode?
(b) What product forms at the positive electrode?

19.89 In the electrolysis of a molten mixture of KI and MgF_2, identify the product that forms at the anode and the product that forms at the cathode.

19.90 In the electrolysis of a molten mixture of CsBr and $SrCl_2$, identify the product that forms at the negative electrode and the product that forms at the positive electrode.

19.91 In the electrolysis of a molten mixture of NaCl and $CaBr_2$, identify the product that forms at the anode and the product that forms at the cathode.

19.92 In the electrolysis of a molten mixture of RbF and $CaCl_2$, identify the product that forms at the negative electrode and the product that forms at the positive electrode.

19.93 Which of these elements can be prepared by electrolysis of their aqueous salts: copper, barium, aluminum, bromine?

19.94 Which of these elements can be prepared by electrolysis of their aqueous salts: strontium, gold, tin, chlorine?

19.95 Which of these elements can be prepared by electrolysis of their aqueous salts: lithium, iodine, zinc, silver?

19.96 Which of these elements can be prepared by electrolysis of their aqueous salts: fluorine, manganese, iron, cadmium?

19.97 What product forms at each electrode in the aqueous electrolysis of each salt?
(a) LiF (b) $SnSO_4$

19.98 What product forms at each electrode in the aqueous electrolysis of each salt?
(a) $ZnBr_2$ (b) $Cu(HCO_3)_2$

19.99 What product forms at each electrode in the aqueous electrolysis of each salt?
(a) $Cr(NO_3)_3$ (b) $MnCl_2$

19.100 What product forms at each electrode in the aqueous electrolysis of each salt?
(a) FeI_2 (b) K_3PO_4

19.101 Electrolysis of molten $MgCl_2$ is the final production step in the isolation of magnesium from seawater by the Dow process (Section 23.4). Answer the following questions, assuming that 45.6 g of Mg metal forms:
(a) What amount (mol) of electrons is required?
(b) What charge (C) is required?
(c) What current (A) will produce this amount in 3.50 h?

19.102 Electrolysis of molten NaCl in a Downs cell is the major isolation step in the production of sodium metal (Section 23.4). Answer the following questions, assuming that 215 g of Na metal forms:
(a) What amount (mol) of electrons is required?
(b) What charge (C) is required?
(c) What current (A) will produce this amount in 9.50 h?

19.103 What mass of radium can form by passing 235 C through an electrolytic cell containing a molten radium salt?

19.104 What mass of aluminum can form by passing 305 C through an electrolytic cell containing a molten aluminum salt?

19.105 How much time (s) does it take to deposit 65.5 g of Zn on a steel gate when 21.0 A is passed through a $ZnSO_4$ solution?

19.106 How much time (s) does it take to deposit 1.63 g of Ni on a decorative drawer handle when 13.7 A is passed through a $Ni(NO_3)_2$ solution?

Problems in Context

19.107 A professor adds Na_2SO_4 to water to facilitate its electrolysis in a lecture demonstration.
(a) What is the purpose of the Na_2SO_4?
(b) Why is the water electrolyzed instead of the salt?

19.108 Subterranean brines in certain parts of Canada are rich in iodides and bromides and serve as an industrial source of these elements. In one recovery method, the brines are evaporated to dryness and then melted and electrolyzed. Which halogen is more likely to form from this treatment? Why?

19.109 Zinc plating (galvanizing) is an important method of corrosion protection. Although galvanizing is usually done by dipping an object into molten zinc, the metal can also be electroplated from aqueous solutions. What mass of zinc can be deposited on a steel tank from a $ZnSO_4$ solution when a 0.855 A current flows for 2.50 days?

Comprehensive Problems

19.110 The MnO_2 that is used in alkaline batteries can be produced by an electrochemical process. One half-reaction in this process is given below:

$$Mn^{2+}(aq) + 2H_2O(l) \longrightarrow MnO_2(s) + 4H^+(aq) + 2e^-$$

If a current of 25.0 A is used, how much time is needed to produce 1.00 kg of MnO_2? At which electrode is the MnO_2 formed?

19.111 Car manufacturers are developing engines that use H_2 as fuel. In Iceland, Sweden, and other parts of Scandinavia, where hydroelectric plants produce inexpensive electrical power, the H_2 can be made industrially by the electrolysis of water.
(a) What charge is needed to produce 3.5×10^6 L of H_2 gas at 12.0 bar and 25°C? (Assume that the ideal gas law applies.)
(b) If the charge is supplied at 1.44 V, how much energy is produced?
(c) If the combustion of oil yields 4.0×10^4 kJ/kg, what mass of oil must be burned to yield the energy in part (b)?

19.112 The overall cell reaction occurring in an alkaline battery is given below:

$$Zn(s) + MnO_2(s) + H_2O(l) \longrightarrow ZnO(s) + Mn(OH)_2(s)$$

(a) What amount (mol) of electrons flows per mole of reaction?
(b) If 4.50 g of zinc is oxidized, what mass of manganese dioxide and what mass of water are consumed?
(c) What is the total mass of reactants consumed in part (b)?
(d) What charge is produced in part (b)?
(e) In practice, voltaic cells of a given capacity (C) are heavier than the calculation in part (c) indicates. Explain why.

19.113 An inexpensive and accurate method for measuring the quantity of electricity flowing through a circuit is to pass the current through a solution of a metal ion and weigh the metal deposited. A silver electrode immersed in an Ag^+ solution weighs 1.7854 g before and 1.8016 g after the current has passed through it. What charge has passed through the cell?

19.114 Brass, an alloy of copper and zinc, can be produced by simultaneously electroplating the two metals from a solution containing their 2^+ ions. If 65.0% of the total current is used to plate the copper, while 35.0% is used to plate the zinc, what is the mass percent of copper in the brass?

19.115 A thin circular-disk earring, 4.00 cm in diameter, is plated with a coating of gold, 0.25 mm thick, from an Au^{3+} bath.
(a) How long does it take to deposit the gold on one side of one earring if the current is 0.013 A (d of gold = 19.3 g/cm^3)?

(b) How long does it take to deposit the gold on both sides of the pair of earrings?
(c) If the price of gold is \$1615 per troy ounce (31.10 g), what is the total cost of the gold plating?

19.116 (a) How long does it take to form 10.0 L of O_2, measured at 99.8 kPa and 28°C, from water if a current of 1.3 A passes through the electrolytic cell?
(b) What mass of H_2 forms?

19.117 Trains that are powered by electricity, including subways, use direct current. The overhead wire (or "third rail" for subways) is one conductor, and the rails on which the wheels run are the other conductor. The rails are on supports in contact with the ground. To minimize corrosion, should the overhead wire or the rails be connected to the positive terminal? Explain.

19.118 A silver (button) battery used in a watch contains 0.75 g of zinc and can run until 80% of the zinc is consumed.
(a) How long can the battery run at a current of 0.85 μA (microampere; 10^{-6} A)?
(b) When the battery dies, 95% of the Ag_2O has been consumed. What mass of Ag was used to make the battery?
(c) If Ag costs \$27.48 per troy ounce (31.10 g), what is the cost of the Ag consumed each day that the watch runs?

19.119 Like any piece of apparatus, an electrolytic cell operates at less than 100% efficiency. A cell depositing Cu from a Cu^{2+} bath operates for 10 h with an average current of 5.8 A. If 53.4 g of copper is deposited, at what efficiency is the cell operating?

19.120 Commercial electrolysis is performed on both molten NaCl and aqueous NaCl solutions. Identify the anode product, cathode product, species reduced, and species oxidized for (a) the molten electrolysis; (b) the aqueous electrolysis.

19.121 To examine the effect of ion removal on cell voltage, a chemist constructs two voltaic cells, each with a standard hydrogen electrode in one compartment. One cell also contains a Pb/Pb^{2+} half-cell; the other contains a Cu/Cu^{2+} half-cell.
(a) What is $E°$ of each cell at 298 K?
(b) Which electrode in each cell is negative?
(c) When Na_2S solution is added to the Pb^{2+} electrolyte, solid PbS forms. What happens to the cell potential?
(d) When sufficient Na_2S is added to the Cu^{2+} electrolyte, CuS forms and $[Cu^{2+}]$ drops to 1×10^{-16} mol/L. Find the cell potential.

19.122 Electrodes used in electrocardiography are disposable, and many incorporate silver. The metal is deposited in a thin layer on a small plastic "button," and then some is converted to AgCl:

$$Ag(s) + Cl^-(aq) \rightleftharpoons AgCl(s) + e^-$$

(a) If the surface area of the button is 2.0 cm^2 and the thickness of the silver layer is 7.5×10^{-6} m, calculate the volume (cm^3) of Ag used in one electrode.
(b) The density of silver metal is 10.5 g/cm^3. What mass of silver is used per electrode?
(c) If Ag is plated on the button from an Ag^+ solution with a current of 12.0 mA, how long does the plating take?
(d) If bulk silver costs \$27.48 per troy ounce (31.10 g), what is the cost (in cents) of the silver in one disposable electrode?

19.123 Commercial aluminum production is done by electrolysis of a bath containing Al_2O_3 dissolved in molten Na_3AlF_6. Why is it not done by electrolysis of an aqueous $AlCl_3$ solution?

19.124 Comparing the standard electrode potentials ($E°$) of the group 1 metals Li, Na, and K with the negative of their first ionization energies reveals a discrepancy:

Ionization process reversed: $M^+(g) + e^- \rightleftharpoons M(g)\quad(-IE)$
Electrode reaction: $M^+(aq) + e^- \rightleftharpoons M(s)\quad(E°)$

Metal	–IE (kJ/mol)	E°(V)
Li	−520	−3.05
Na	−496	−2.71
K	−419	−2.93

The electrode potentials do not decrease smoothly down the group, as the ionization energies do. We might expect that if it is more difficult to remove an electron from an atom to form a gaseous ion (larger IE), then it would be less difficult to add an electron to an aqueous ion to form an atom (smaller $E°$), yet $Li^+(aq)$ is *more* difficult to reduce than $Na^+(aq)$. Applying Hess's law, use an approach similar to a Born-Haber cycle to break down the process occurring at the electrode into three steps and label the energy involved in each step. How can you account for the discrepancy?

19.125 In Appendix D, standard electrode potentials range from about +3 V to −3 V. Thus, it might seem possible to use a half-cell from each end of this range to construct a cell with a voltage of approximately 6 V. However, most commercial aqueous voltaic cells have $E°$ values of 1.5 V to 2 V. Why are there no aqueous cells with significantly higher potentials?

19.126 Tin is used to coat "tin" cans used for food storage. If the tin is scratched, exposing the iron of the can, will the iron corrode more or less rapidly than if the tin were not present? Inside the can, the tin itself is coated with a clear varnish. Explain.

19.127 Commercial electrolytic cells for producing aluminum operate at 5.0 V and 100 000 A.
(a) How long does it take to produce exactly 1 t (tonne; 1000 kg) of aluminum?
(b) How much electrical power, in kilowatt-hours (kW•h), is used (1 W = 1 J/s; 1 kW•h = 3.6×10^3 kJ)?
(c) If electricity costs \$0.117 per kW•h and cell efficiency is 90.%, what is the cost of electricity to produce exactly 454 g of aluminum?

19.128 Magnesium bars are connected electrically to underground iron pipes to serve as sacrificial anodes.
(a) Do electrons flow from the bar to the pipe, or the reverse?
(b) A 12 kg Mg bar is attached to an iron pipe, and it takes 8.5 years for the Mg to be consumed. What is the average current flowing between the Mg and the Fe during this period?

19.129 Bubbles of H_2 form when metal D is placed in hot H_2O. No reaction occurs when metal D is placed in a solution of a salt of metal E, but metal D is discoloured and coated immediately when placed in a solution of a salt of metal F.
(a) What happens if metal E is placed in a solution of a salt of metal F?
(b) Rank metals D, E, and F in order of *increasing* reducing strength.

19.130 In addition to reacting with gold (see Problem 19.21), *aqua regia* is used to bring other precious metals into solution. Balance the following skeleton equation for the reaction with Pt:

$$Pt(s) + NO_3^-(aq) + Cl^-(aq) \longrightarrow PtCl_6^{2-}(aq) + NO(g)$$

19.131 The following reactions are used in batteries:
(1) $2H_2(g) + O_2(g) \longrightarrow 2H_2O(l)\qquad E_{cell} = 1.23$ V
(2) $Pb(s) + PbO_2(s) + 2H_2SO_4(aq) \longrightarrow 2PbSO_4(s) + 2H_2O(l)\qquad E_{cell} = 2.04$ V
(3) $2Na(l) + FeCl_2(s) \longrightarrow 2NaCl(s) + Fe(s)\qquad E_{cell} = 2.35$ V

Reaction 1 is used in a fuel cell, reaction 2 is used in an automobile lead-acid battery, and reaction 3 is used in an experimental high-temperature battery for powering electric vehicles. The aim is to obtain as much work as possible from a cell, while keeping its weight to a minimum.
(a) In each cell, find the amount (mol) of electrons transferred and ΔG.
(b) Calculate the ratio (kJ/g) of w_{max} to mass of reactants for each cell. Which has the highest ratio, which the lowest ratio, and why? (*Note*: For simplicity, ignore the masses of cell components that do not appear in the cell as reactants, including electrode materials, electrolytes, separators, cell casing, and wiring.)

19.132 A current is applied to two electrolytic cells in series. In the first cell, silver is deposited; in the second cell, a zinc electrode is consumed. How much Ag is plated out if 1.2 g of Zn dissolves?

19.133 You are investigating a particular chemical reaction. State all the types of data available in standard tables that enable you to calculate the equilibrium constant for the reaction at 298 K.

19.134 In an electrical power plant, personnel monitor the O_2 content of boiler feed water to prevent corrosion of the boiler tubes. Why does Fe corrode faster in steam and hot water than it does in cold water?

19.135 A voltaic cell using Cu/Cu^{2+} and Sn/Sn^{2+} half-cells is set up at standard conditions, and each compartment has a volume of 345 mL. The cell delivers 0.17 A for 48.0 h.
(a) What mass of Cu(*s*) is deposited?
(b) What is the $[Cu^{2+}]$ remaining?

19.136 The $E_{half\text{-}cell}$ value for the reduction of water is very different from the $E^\circ_{half\text{-}cell}$ value. Calculate the $E_{half\text{-}cell}$ value when H_2 is in its standard state (pH = 7, T = 25°C).

19.137 From the skeleton equations below, create a list of balanced half-reactions in which the strongest oxidizing agent is on the top and the weakest oxidizing agent is on the bottom:

$$U^{3+}(aq) + Cr^{3+}(aq) \longrightarrow Cr^{+}(aq) + U^{4+}(aq)$$
$$Fe(s) + Sn^{2+}(aq) \longrightarrow Sn(s) + Fe^{2+}(aq)$$
$$Fe(s) + U^{4+}(aq) \longrightarrow \text{no reaction}$$
$$Cr^{3+}(aq) + Fe(s) \longrightarrow Cr^{2+}(aq) + Fe^{+}(aq)$$
$$Cr^{2+}(aq) + Sn^{2+}(aq) \longrightarrow Sn(s) + Cr^{3+}(aq)$$

19.138 Consider the following three half-reactions:
(1) $Fe^{3+}(aq) + e^- \rightleftharpoons Fe^{2+}(aq)$
(2) $Fe^{2+}(aq) + 2e^- \rightleftharpoons Fe(s)$
(3) $Fe^{3+}(aq) + 3e^- \rightleftharpoons Fe(s)$
(a) Use the $E^\circ_{half\text{-}cell}$ values for (1) and (2) to find the $E^\circ_{half\text{-}cell}$ value for (3).
(b) Calculate ΔG° for (1) and (2) from their $E^\circ_{half\text{-}cell}$ values.
(c) Calculate ΔG° for (3) from (1) and (2).
(d) Calculate $E^\circ_{half\text{-}cell}$ for (3) from its ΔG°.
(e) What is the relationship between the $E^\circ_{half\text{-}cell}$ values for (1) and (2) and the $E^\circ_{half\text{-}cell}$ value for (3)?

19.139 Use the half-reaction method to balance the following equation for the conversion of ethanol to ethanoic (acetic) acid in acid solution:

$$CH_3CH_2OH + Cr_2O_7^{2-} \longrightarrow CH_3COOH + Cr^{3+}$$

19.140 When zinc is refined by electrolysis, the desired half-reaction at the cathode is

$$Zn^{2+}(aq) + 2e^- \longrightarrow Zn(s)$$

A competing reaction, which lowers the yield, is the formation of hydrogen gas:

$$2H^{+}(aq) + 2e^- \longrightarrow H_2(g)$$

If 91.50% of the current flowing results in zinc being deposited, while 8.50% produces hydrogen gas, what volume of H_2, measured at STP, forms per kilogram of zinc?

19.141 A chemist designs an ion-specific probe for measuring $[Ag^+]$ in an NaCl solution saturated with AgCl. One half-cell has an Ag-wire electrode immersed in the unknown AgCl-saturated NaCl solution. It is connected through a salt bridge to the other half-cell, which has a calomel reference electrode (a platinum wire immersed in a paste of mercury and calomel, Hg_2Cl_2) in a saturated KCl solution. The measured E_{cell} is 0.060 V.
(a) Given the following standard half-reactions, calculate $[Ag^+]$. (*Hint*: Assume that $[Cl^-]$ is so high that it is essentially constant.)

Calomel: $Hg_2Cl_2(s) + 2e^- \longrightarrow 2Hg(l) + 2Cl^-(aq)$ $E^\circ = 0.24$ V

Silver: $Ag^+(aq) + e^- \longrightarrow Ag(s)$ $E^\circ = 0.80$ V

(b) A mining engineer wants an ore sample analyzed with the Ag^+-selective probe. After pretreating the ore sample, the chemist measures the cell voltage as 0.53 V. What is $[Ag^+]$?

19.142 Use Appendix D to calculate the K_{sp} of AgCl.

19.143 Black-and-white photographic film is coated with silver halides. Because silver is expensive, the manufacturer monitors the Ag^+ content of the waste stream, $[Ag^+]_{waste}$, from the plant with an Ag^+-selective electrode at 25°C. A stream of known Ag^+ concentration, $[Ag^+]_{standard}$, is passed over the electrode, alternating with the waste stream, and the data are recorded by a computer.
(a) Write the equations relating the nonstandard cell potential to the standard cell potential and $[Ag^+]$ for each solution.
(b) Combine your equations from part (a) into a single equation to find $[Ag^+]_{waste}$.
(c) Rewrite your equation from part (b) to find $[Ag^+]_{waste}$ in ng/L.
(d) If E_{waste} is 0.003 V higher than $E_{standard}$, and the standard solution contains 1000. ng/L, what is $[Ag^+]_{waste}$?
(e) Rewrite your equation from part (b) to find $[Ag^+]_{waste}$ for a system in which T changes and T_{waste} and $T_{standard}$ may be different.

19.144 Calculate the K_f of $Ag(NH_3)_2^+$ from the following half-reactions:

$Ag^+(aq) + e^- \rightleftharpoons Ag(s)$ $E^\circ = 0.80$ V
$Ag(NH_3)_2^+(aq) + e^- \rightleftharpoons Ag(s) + 2NH_3(aq)$ $E^\circ = 0.37$ V

19.145 Even though the toxicity of cadmium has become a concern, nickel-cadmium (nicad) batteries are still used in many devices. The overall cell reaction is given below:

$$Cd(s) + 2NiO(OH)(s) + 2H_2O(l) \longrightarrow 2Ni(OH)(s) + Cd(OH)_2(s)$$

A certain nicad battery weighs 18.3 g and has a capacity of 300. mA·h. (That is, the cell can store a charge equivalent to a current of 300. mA flowing for 1 h.)
(a) What is the capacity (C) of this cell?
(b) What mass of reactants is needed to deliver 300. mA·h?
(c) What percentage of the cell mass consists of reactants?

19.146 The zinc-air battery is a less expensive alternative to silver batteries for use in hearing aids. The cell reaction is given below:

$$2Zn(s) + O_2(g) \longrightarrow 2ZnO(s)$$

A new battery weighs 0.275 g. The zinc accounts for exactly $\frac{1}{10}$ of the mass, and the oxygen does not contribute to the mass because it is supplied by the air.
(a) How much electricity (C) can the battery deliver?
(b) How much Gibbs energy (J) is released if E_{cell} is 1.3 V?

19.147 Use Appendix D to create an activity series with Mn, Fe, Ag, Sn, Cr, Cu, Ba, Al, Na, Hg, Ni, Li, Au, Zn, and Pb. Rank these metals in order of decreasing reducing strength, and divide them into three groups: metals that displace H_2 from water, metals that displace H_2 from acid, and metals that cannot displace H_2.

19.148 Both Ti and V are reactive enough to displace H_2 from water. The difference in their $E^\circ_{\text{half-cell}}$ values is 0.43 V. Given the following reaction, use Appendix D to calculate the $E^\circ_{\text{half-cell}}$ values for V and Ti:

$$V(s) + Cu^{2+}(aq) \longrightarrow V^{2+}(aq) + Cu(s) \quad \Delta G^\circ = -298 \text{ kJ/mol}$$

19.149 Consider the following reaction:

$$S_4O_6^{2-}(aq) + 2I^-(aq) \longrightarrow I_2(s) + S_2O_3^{2-}(aq) \quad \Delta G^\circ = 87.8 \text{ kJ/mol}$$

(a) Identify the oxidizing and reducing agents.
(b) Calculate E°_{cell}.
(c) For the reduction half-reaction, write a balanced equation, give the oxidation number of each element, and calculate $E^\circ_{\text{half-cell}}$.

19.150 Two concentration cells are prepared, both with 90.0 mL of 0.0100 mol/L $Cu(NO_3)_2$ and a Cu bar in each half-cell.
(a) In the first concentration cell, 10.0 mL of 0.500 mol/L NH_3 is added to one half-cell. The complex ion $Cu(NH_3)_4^{2+}$ forms, and E_{cell} is 0.129 V. Calculate K_f for the formation of the complex ion.
(b) Calculate E_{cell} when an additional 10.0 mL of 0.500 mol/L NH_3 is added.
(c) In the second concentration cell, 10.0 mL of 0.500 mol/L NaOH is added to one half-cell. The precipitate $Cu(OH)_2$ forms (K_{sp} = 2.2×10^{-20}). Calculate E°_{cell}.
(d) What would the concentration of NaOH have to be for the addition of 10.0 mL to result in an E°_{cell} of 0.340 V?

19.151 Two voltaic cells are to be joined so that one will run the other as an electrolytic cell. In the first cell, one half-cell has Au foil in 1.00 mol/L $Au(NO_3)_3$, and the other half-cell has a Cr bar in 1.00 mol/L $Cr(NO_3)_3$. In the second cell, one half-cell has a Co bar in 1.00 mol/L $Co(NO_3)_2$, and the other half-cell has a Zn bar in 1.00 mol/L $Zn(NO_3)_2$.
(a) Calculate E°_{cell} for each cell.
(b) Calculate the total potential if the two cells are connected as voltaic cells in series.
(c) When the electrode wires are switched in one of the cells, which cell will run as the voltaic cell and which will run as the electrolytic cell?
(d) Which metal ion is being reduced in each cell?
(e) If 2.00 g of metal plates out in the voltaic cell, how much metal ion plates out in the electrolytic cell?

19.152 A voltaic cell has one half-cell with a Cu bar in a 1.00 mol/L Cu^{2+} salt, and the other half-cell with a Cd bar in the same volume of a 1.00 mol/L Cd^{2+} salt.
(a) Find E°_{cell}, ΔG°, and K.
(b) As the cell operates, $[Cd^{2+}]$ increases. Find E_{cell} and ΔG when $[Cd^{2+}]$ is 1.95 mol/L.
(c) Find E_{cell}, ΔG, and $[Cu^{2+}]$ at equilibrium.

19.153 Gasoline is a mixture of hydrocarbons, but the heat released when it burns is close to the heat released when octane, $C_8H_{18}(l)$, burns ($\Delta_f H^\circ$ of octane = −250.1 kJ/mol). As an alternative to gasoline, research is under way to use H_2 from the electrolysis of water in fuel cells to power cars.
(a) Calculate ΔH° when 4.00 L of gasoline burns to produce carbon dioxide gas and water vapour (d of gasoline = 0.7028 g/mL).
(b) What volume of H_2 at 25°C and 1.00 bar must burn to produce this quantity of energy?
(c) How long would it take to produce this amount of H_2 by electrolysis with a current of 1.00×10^3 A at 6.00 V?
(d) How much power, in kilowatt•hours (kW•h), is required to generate this amount of H_2 (1 W = 1 J/s; 1 J = 1 C·V; 1 kW•h = 3.6×10^6 J)?
(e) If the cell is 88.0% efficient and electricity costs $0.123 per kW•h, what is the cost of producing the amount of H_2 equivalent to 4.00 L of gasoline?

19.154 If the E_{cell} of the following cell is 0.915 V, what is the pH in the anode compartment?

$$Pt(s)|\ H_2(1.00 \text{ bar})|H^+(aq)\ ||\ Ag^+(0.100 \text{ mol/L})\ |Ag(s)$$

19.155 A concentration cell was constructed as follows in the lab at 298 K: Two 150.0 mL beakers (A and B) were placed on a bench and 50.0 mL of 0.10 mol/L $CuSO_4(aq)$ were added to both. A strip of Cu metal was placed in each beaker and then a salt bridge was used to connect the two beakers. The two strips of copper were connected to a voltmeter and the initial potential was measured as 0.000 V. 50.0 mL of distilled water was added to beaker A, and 50.0 mL of 1.0 mol/L $NH_3(aq)$ was added to beaker B. The contents of beaker B immediately turned a dark indigo due to the formation of $Cu(NH_3)_4^{2+}(aq)$, and the voltmeter displayed a potential of 0.289 V. (a) Write the equilibrium redox reaction occurring in the copper concentration cell. (b) Calculate the concentration of $Cu^{2+}(aq)$ in beaker A (remember dilution factors!). (c) Calculate the concentration of $Cu^{2+}(aq)$ in beaker B using the Nernst equation. (d) Calculate the value of K_f for the complex. (*Hint*: after writing the balanced equation for the formation of the complex, assume that all the copper ion reacts to form the complex for the purpose of completing the ICE table; remember to use the value from part (c) to calculate the K_f value, however!)

19.156 In the lab, a concentration cell was constructed as follows, at 298 K. Two 250.0 mL beakers (A and B) were placed on a bench. In beaker A, 100.0 mL of 0.100 mol/L $AgNO_3(aq)$ was placed. In beaker B, 100.0 mL of a saturated solution of $Ag_2S(aq)$ was placed. A strip of silver was placed in each beaker, and the strips were connected to a voltmeter. The two beakers were connected using a salt bridge. What was the reading on the voltmeter?

19.157 A decorative tray was plated with chromium using electrolysis. A current of 8.75 A was applied to a solution of $Cr^{3+}(aq)$ for 15.5 min. (a) What mass of chromium was used to plate the tray? (b) How long would it take to plate the same tray with the same mass of silver using a current of 9.25 A?

Organic Compounds and the Atomic Properties of Carbon

IN THIS CHAPTER . . . We apply the text's central theme to the enormous field of organic chemistry to see how the structures and reactivities of organic molecules emerge naturally from the properties of their component atoms. By the end of this chapter, you should be able to

- Discuss the atomic properties of carbon and determine how the properties lead to the complex structures and reactivity of organic molecules
- Write and name hydrocarbons, as a prelude to writing and naming other types of organic compounds

20

Carbon—this small element plays a huge role in our existence, our lives, and our economies. From the building blocks of our bodies to an entire industry based on petrochemicals, the complex chemistry of carbon continuously captures our curiosity. How can one tiny atom have such a huge influence on whether life exists or not? Especially today, almost everything we use is connected in some way with the chemistry of carbon. In this chapter, we will learn how we define organic chemistry and what its nomenclature is. We will study **organic compounds** and their different structures and determine what it is about this element that makes it capable of forming over nine million compounds.

Concepts and Skills to Review before Studying this Chapter

- Naming straight-chain alkanes (Section 2.8)
- Constitutional isomerism (Section 3.2)
- $\Delta\chi$ (difference in electronegativity) and bond polarity (Section 8.5)
- Resonance structures (Section 9.1)
- VSEPR theory (Section 9.1)
- Orbital hybridization (Section 10.1)
- σ and π bonding (Section 10.2)
- Intermolecular forces (Section 11.3)
- Properties of the period 2 elements (Section 13.2)
- Properties of the group 14 elements (Section 13.6)

20.1 The Special Nature of Carbon and the Characteristics of Organic Molecules

Although there is nothing mystical about organic molecules, their indispensable role in biology and industry leads us to ask if carbon has some extraordinary attributes that give it a special chemical "personality." Of course, each element has its own specific properties, and carbon is no more unique than sodium, hafnium, or any other element. But the atomic properties of carbon do give it bonding capabilities beyond those of any other element, which in turn lead to the two obvious characteristics of organic molecules—structural complexity and chemical diversity.

The Structural Complexity of Organic Molecules

In general, organic molecules have more complex structures than most inorganic molecules. A quick review of the atomic properties of the carbon atom and its behaviour when bonding shows why (Section 13.6).

Electron Configuration, Electronegativity, and Covalent Bonding Carbon forms covalent bonds in all its elemental forms and compounds, sharing electrons to attain a filled outer (valence) shell. Carbon's formation of covalent bonds, rather than ionic bonds, is the result of its electron configuration and its electronegativity value:

- The ground-state electron configuration of $[He]2s^22p^2$ for the carbon atom—four electrons more than He and four fewer than Ne—means that the formation of carbon ions is energetically impossible under ordinary conditions. The loss of four electrons to form the C^{4+} cation requires energy equal to the sum of IE_1 through IE_4; the gain of four electrons to form the C^{4-} anion requires the sum of EA_1 through EA_4, the last three steps of which are endothermic.
- Lying at the centre of period 2, carbon has an electronegativity ($\chi = 2.5$) that is midway between that of the most metallic element (Li, $\chi = 1.0$) and the most nonmetallic element (F, $\chi = 4.0$) of period 2 (Figure 20.1).

FIGURE 20.1 The position of carbon in the periodic table

Bond Properties, Catenation, and Molecular Shape The *number* and *strength* of carbon's bonds lead to its property of **catenation**, the ability to bond to itself, which in turn allows it to form a multitude of chemically and thermally stable chain, ring, and branched compounds. The small size of the carbon atom and its ability to form hybrid orbitals and multiple bonds affect molecular shape, thereby increasing the number of different molecules:

- Through the process of orbital hybridization (Section 10.1), carbon forms four bonds in virtually all its compounds, and they point in as many as four different directions.
- The small size of carbon allows for a close approach to another atom and thus greater orbital overlap, so carbon forms relatively short, strong bonds with other carbon atoms and with many other elements.
- The C—C bond is short enough to allow side-to-side overlap of half-filled, unhybridized *p* orbitals and the formation of multiple bonds, which restrict rotation of attached groups (see Figure 10.14). Carbon can form single, double, and triple bonds.

Molecular Stability Although silicon and several other elements also catenate, none can form chains as stable as those of carbon. Atomic and bonding properties confer three crucial differences between C and Si chains that explain why C chains are so stable and, therefore, so common:

- *Atomic size and bond strength.* As atomic size increases down group 14, bonds between identical atoms become longer and weaker. Thus, a C—C bond (347 kJ/mol) is much stronger than an Si—Si bond (226 kJ/mol).
- *Relative enthalpies of reaction.* A C—C bond (347 kJ/mol), a C—O bond (358 kJ/mol), and a C—Cl bond (339 kJ/mol) have nearly the same energy, so relatively little heat is released when a C chain reacts and one bond replaces the other. In contrast, an Si—O bond (368 kJ/mol) or an Si—Cl bond (381 kJ/mol) is much stronger than an Si—Si bond (226 kJ/mol), so a large quantity of heat is released when an Si chain reacts.
- *Orbitals available for reaction.* Unlike C, Si has low-energy *d* orbitals that can be attacked (occupied) by the lone pairs of incoming reactants. For example, ethane ($H_3C—CH_3$) is stable in water and does not react in air unless sparked, whereas disilane ($H_3Si—SiH_3$) breaks down in water and ignites spontaneously in air.

The Chemical Diversity of Organic Molecules

In addition to their complex geometries, organic compounds are noted for their sheer number and diverse chemical behaviour. Several million organic compounds are known, and thousands more are discovered or synthesized each year. This incredible diversity is also founded on atomic and bonding behaviour and is due to three interrelated factors, discussed below.

In this chapter and in Chapters 21 and 22, you will learn about the structure, bonding, and reactivity of many different types of organic molecules. While we have included many different drawings, photos, and three-dimensional structures to help you, you may find it helpful to approach these chapters with a molecular model kit in hand. You can form the structures as they are discussed, which will help you to clearly visualize the many properties that are explained. If you do not have a molecular model kit, coloured mini-marshmallows and toothpicks work well too!

Bonding to Heteroatoms Many organic compounds contain **heteroatoms**, that is, atoms other than C or H. The most common heteroatoms are N and O, but S, P, and the halogens often occur, and organic compounds with other elements are known as well. Figure 20.2 shows that 23 different molecular structures are possible from various arrangements of four C atoms bonded singly to each other, as well as to just one O atom (either singly or doubly bonded) and the necessary number of H atoms. Several other structures are possible from arrangements of four C atoms bonded to each other with multiple bonds.

Electron Density and Reactivity Most reactions start—that is, a new bond begins to form—*when a region of high electron density on one molecule meets a region* *of low electron density on another.* These regions may be due to the presence of a multiple bond or to the properties of carbon-heteroatom bonds. Consider the reactivities of four bonds commonly found in organic molecules:

- *The C—C bond.* When C is singly bonded to another C, as occurs in portions of nearly every organic molecule, the χ values are equal and the bond is nonpolar. Therefore, in general, *C—C bonds are unreactive.*
- *The C—H bond.* This bond, which also occurs in nearly every organic molecule, is short (109 pm) and very nearly nonpolar, with χ values of 2.1 for H and 2.5 for C. Thus, *C—H bonds are largely unreactive.*
- *The C—O bond.* This bond, which occurs in many types of organic molecules, is highly polar ($\Delta\chi = 1.0$), with the O end of the bond electron rich and the C end electron poor. As a result of this imbalance in electron density, *the C—O bond is reactive*, and, given appropriate conditions, a reaction will occur at this bond.
- *Bonds to other heteroatoms.* Even when a carbon-heteroatom bond has a small $\Delta\chi$, such as for C—Br ($\Delta\chi = 0.3$), or none at all, as for C—S ($\Delta\chi = 0$), the heteroatoms are generally large, so their bonds to carbon are long; weak; and, thus, *reactive.*

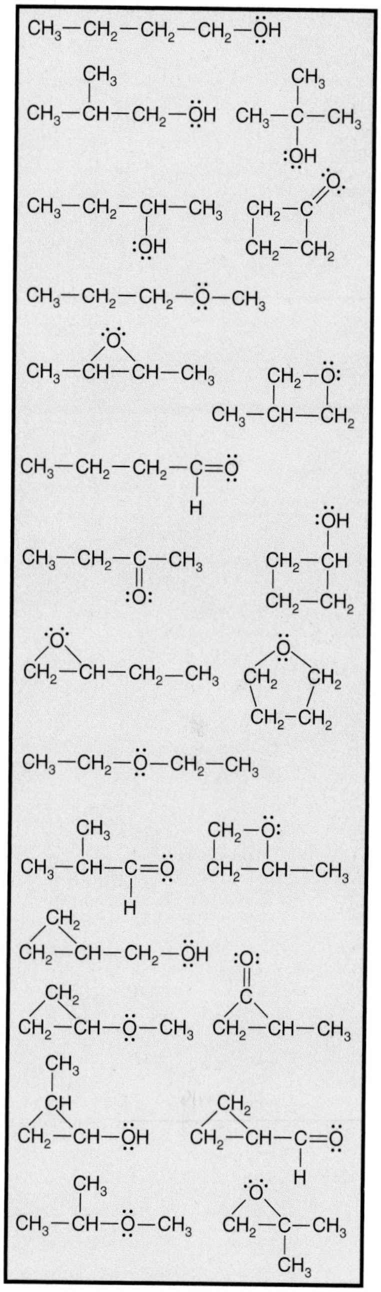

FIGURE 20.2 Heteroatoms (e.g., oxygen) and different bonding arrangements lead to great chemical diversity.

Importance of Functional Groups One of the most important concepts in organic chemistry is that of the **functional group**, a specific combination of bonded atoms that reacts in a *characteristic* way, no matter what molecule it occurs in. In nearly every case, *the reaction of an organic compound takes place at the functional group.* A functional group is reactive because it contains one or more of the following: π bond, polar bond, and lone pair of electrons. As you will see in Chapter 21, we often substitute a general symbol for the remainder of the organic molecule because it usually stays the same while the functional group reacts. Functional groups include carbon-carbon multiple bonds and several combinations of carbon-heteroatom bonds, and each has its own pattern of reactivity. A particular bond may be a functional group itself or *part* of one or more functional groups.

The chemistry of every organic molecule is determined by the functional group(s) it contains. The following are some examples of functional groups (*in blue*) that we will discuss later in greater detail.

- Carbon bonded by a double or a triple bond to another carbon:

Alkenes
Compounds that contain at least one carbon-to-carbon double bond (one π bond between carbon atoms)

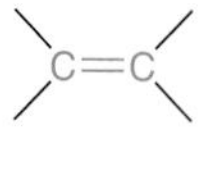

Alkynes
Compounds that contain at least one carbon-to-carbon triple bond (two π bonds between carbon atoms)

- Carbon singly bonded to a more *electronegative* atom:

Haloalkanes (also known as *alkyl halides*)
Compounds in which an H atom of an alkane has been replaced by a halide:
X = F, Cl, Br, or I

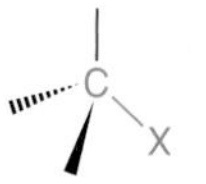

Ethers
Compounds in which an O atom is bonded to two C atoms

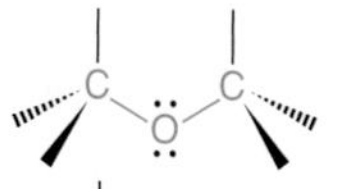

Alcohols
Compounds in which an H atom of an alkane has been replaced by an OH group

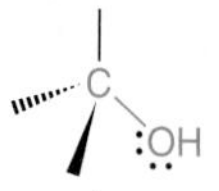

Amines
Compounds in which one or more H atoms of ammonia have been replaced by C atoms

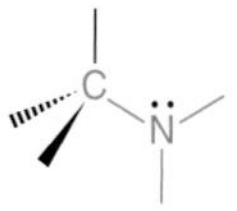

- Carbon-oxygen double bond:

  Aldehydes
  Compounds that contain a C═O bond to which there is also at least one H atom bonded

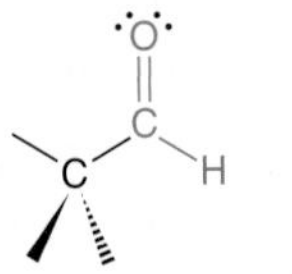

  Ketones
  Compounds that contain a C═O bond, the C atom of which is also bonded to adjacent C atoms

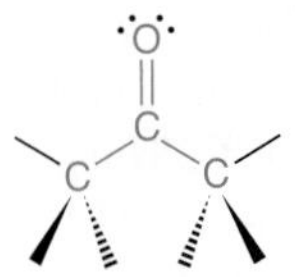

  Carboxylic acids
  Compounds that contain a C═O bond, the C atom of which is also bonded to an OH group

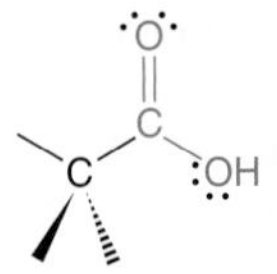

  Esters
  Compounds that contain a C═O bond, the C atom of which is also bonded to an O atom that is bonded to another C atom

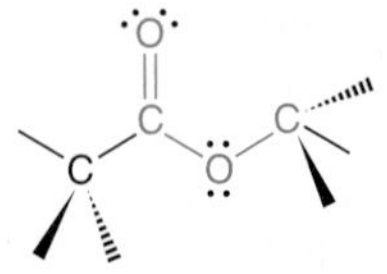

  Amide
  Compounds that contain an N—C═O bond

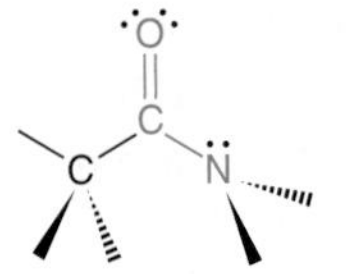

- Carbon-nitrogen triple bond:

  Nitriles
  Compounds that contain a C≡N bond

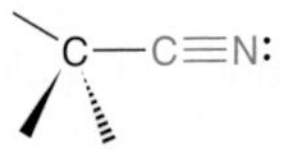

SUMMARY OF SECTION 20.1

- Carbon's small size, intermediate electronegativity, four valence electrons, and ability to form multiple bonds result in the structural complexity of organic compounds.
- These factors lead to the ability of the carbon atom to catenate, which creates chains, branches, and rings of C atoms. The small size of the carbon atom and the absence of *d* orbitals in its valence level lead to strong, chemically resistant bonds that point in as many as four directions from each C atom.
- The ability of the carbon atom to bond to many other elements, including O and N, creating polar bonds and resulting in greater reactivity, leads to the chemical diversity of organic compounds.
- Most organic compounds contain functional groups, that is, specific groupings of bonded atoms that react in characteristic ways.

20.2 The Structures of Hydrocarbons and Hydrocarbons with Single Bonds

A fanciful, anatomical analogy can be made between an organic molecule and an animal. The carbon-carbon bonds form the skeleton: the longest continuous chain is the backbone, and any branches are the limbs. Covering the skeleton is a skin of hydrogen atoms, with functional groups protruding at specific locations, like chemical fingers ready to grab an incoming reactant. In this section, we "dissect" one group of compounds down to their skeletons and see how to name and draw them.

Hydrocarbons, the simplest type of organic compound, are a large group of substances containing only H and C atoms. Some common fuels, such as natural gas and gasoline, are hydrocarbon mixtures. Some hydrocarbons, such as ethane (commonly known as ethylene), ethyne (better known as acetylene), and benzene, are important *feedstocks*, precursor reactants used to make other compounds.

Carbon Skeletons and Hydrogen Skins

Let us begin by examining the possible bonding arrangements of C atoms only (leaving off the H atoms for now) in simple skeletons without multiple bonds or rings.

To distinguish different skeletons, focus on the *arrangement* of C atoms (that is, the successive linkages of one to another) and keep in mind that *groups joined by a single (sigma) bond are relatively free to rotate* (Section 10.2). Notice that this will not be the final way of representing a hydrocarbon molecule. Since each carbon atom forms four bonds, once the skeletal structure of the carbons has been drawn, we need to complete the molecule by adding the correct number of hydrogen atoms.

Structures with one, two, or three carbon atoms can be arranged in only one way. Whether you draw three C atoms in a line or with a bend, the arrangement is the same. Four C atoms, however, have two possible arrangements—a four-carbon chain or a three-carbon chain with a one-carbon branch at the central C:

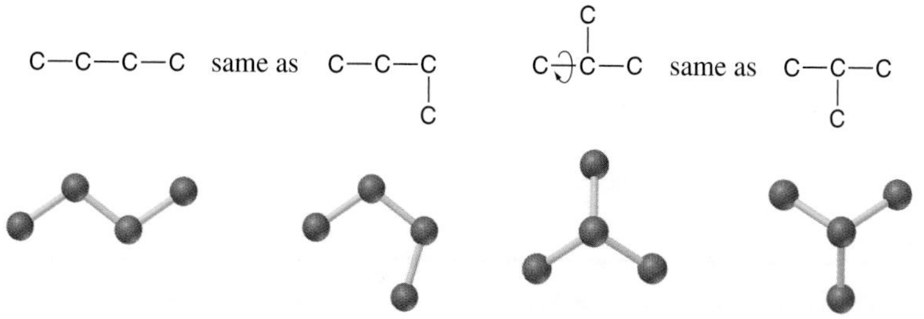

Even when we show the chains more realistically, with the bends due to the tetrahedral shape around the C atoms (as in the ball-and-stick models above), the situation is the same. Notice that if the branch is added to either end of the three-carbon chain, it is simply a bend in a four-carbon chain, *not* a different arrangement. Similarly, if the branch points down instead of up, it represents the same arrangement because groups joined by a single bond can rotate.

As the total number of C atoms increases, the number of different arrangements increases as well. Five C atoms have 3 possible arrangements (Figure 20.3A); 6 C atoms can be arranged in 5 ways, 7 C atoms in 9 ways, 10 C atoms in 75 ways, and 20 C atoms in more than 300 000 ways! If we include multiple bonds and rings, the number of arrangements increases further. For example, including one C=C bond in the five-carbon skeletons creates five more arrangements (Figure 20.3B), and including one ring creates five more (Figure 20.3C).

When determining the number of different skeletons for a given number of C atoms, remember the following:

- Each C atom can form a *maximum* of four single bonds, or two single and one double bond, or one single and one triple bond.
- The *arrangement* of C atoms determines the skeleton, so a straight chain and a bent chain represent the same skeleton.
- Groups joined by a single bond can *rotate*, so a branch pointing down is the same as one pointing up. (Recall that a double bond restricts rotation.)

If we put a hydrogen "skin" on a carbon skeleton, we obtain a hydrocarbon. Different carbon atoms can then be identified with respect to the number of carbon neighbours (*see margin figure*). Figure 20.4 shows that a skeleton has the correct number of H atoms when each C has four bonds. The carbon atom at the end of the

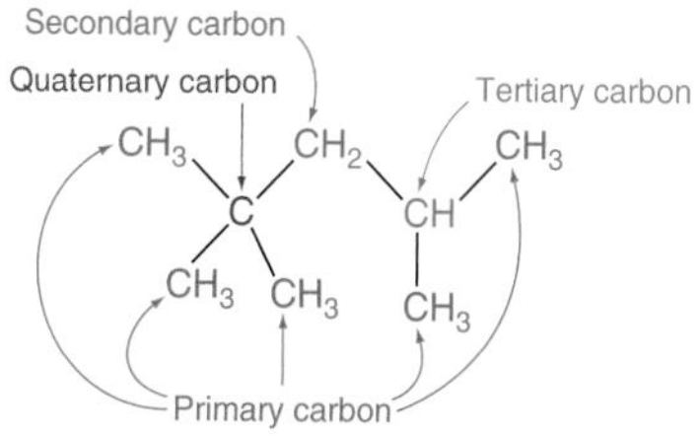

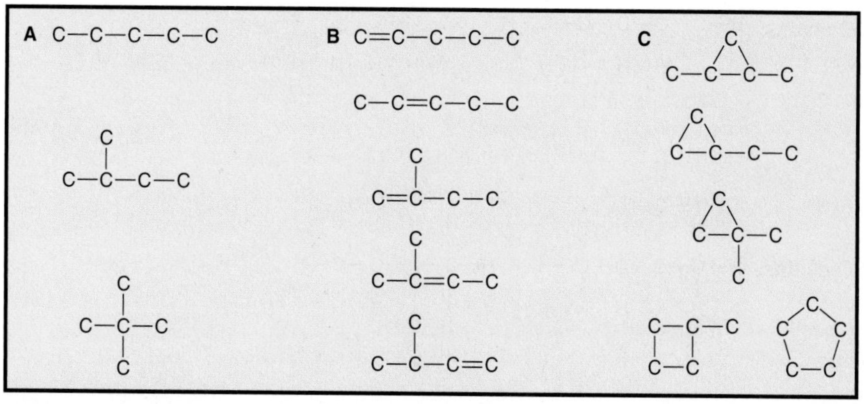

FIGURE 20.3 Some five-carbon skeletons. Keep in mind that this is not the final representation because the hydrogen atoms have not been included.

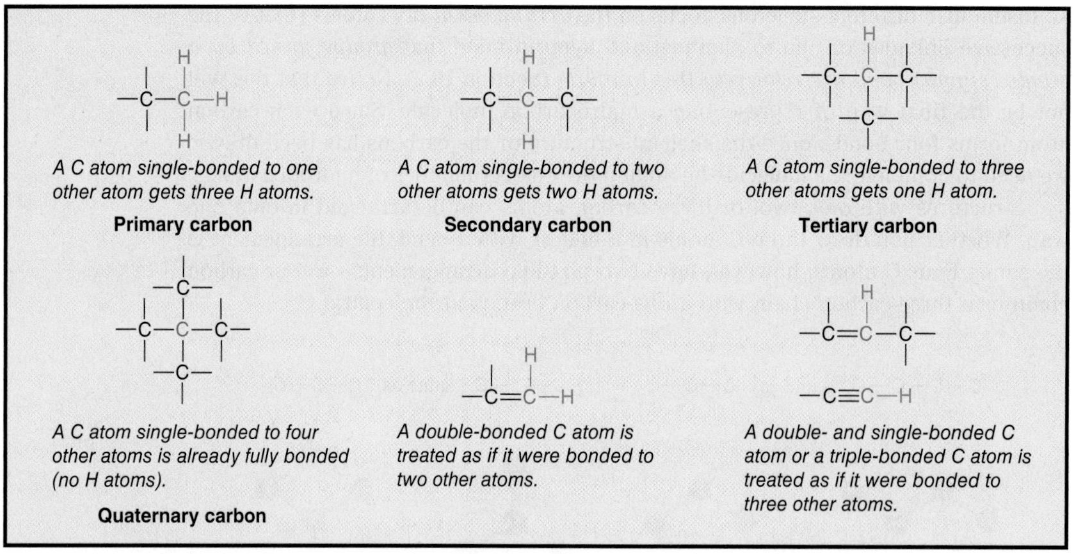
A C atom single-bonded to one other atom gets three H atoms.
Primary carbon
A C atom single-bonded to two other atoms gets two H atoms.
Secondary carbon
A C atom single-bonded to three other atoms gets one H atom.
Tertiary carbon
A C atom single-bonded to four other atoms is already fully bonded (no H atoms).
Quaternary carbon
A double-bonded C atom is treated as if it were bonded to two other atoms.
A double- and single-bonded C atom or a triple-bonded C atom is treated as if it were bonded to three other atoms.

FIGURE 20.4 Adding the H-atom skin to the C-atom skeleton. Each carbon atom should have four bonds.

chain is always a *primary carbon*, since it is bonded to only one other carbon atom. When the carbon atom is bonded to two other carbon atoms, it is a *secondary carbon*. One bonded to three carbon atoms is a *tertiary carbon*, and a *quaternary carbon* is the one bonded to four other carbon atoms.

Sample Problem 20.1 provides practice drawing hydrocarbons.

Sample Problem 20.1 Drawing Hydrocarbons

Problem Draw structures that have different atom arrangements for hydrocarbons as follows:

(a) Six C atoms, no multiple bonds, and no rings

(b) Four C atoms, one double bond, and no rings

(c) Four C atoms, no multiple bonds, and one ring

Plan In each case, we draw the longest carbon chain first and then work down to smaller chains with branches at different points along them. The process typically involves trial and error. Then, we add H atoms to give each C atom a total of four bonds.

Solution **(a)** Compounds with six C atoms:

6-C chain:

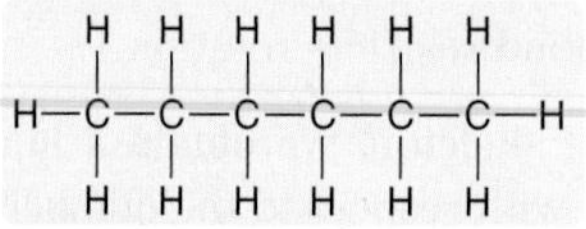

5-C chains:

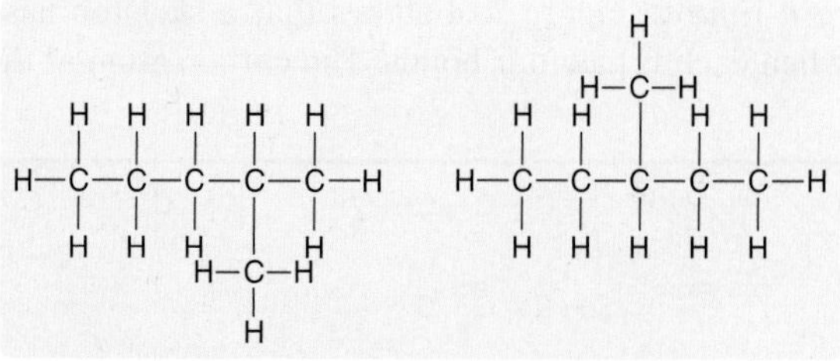

4-C chains:

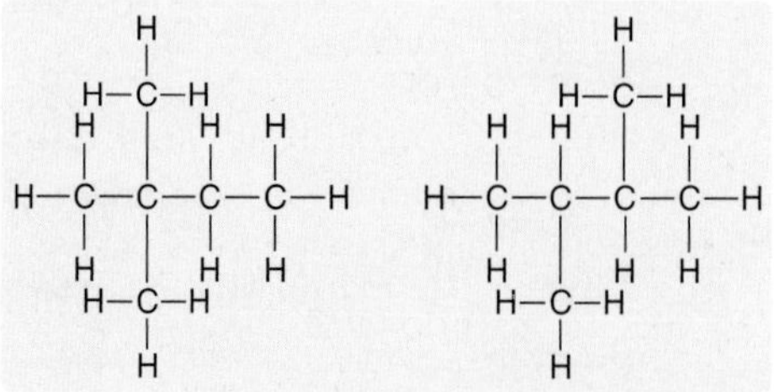

(b) Compounds with four C atoms and one double bond:

4-C chains: 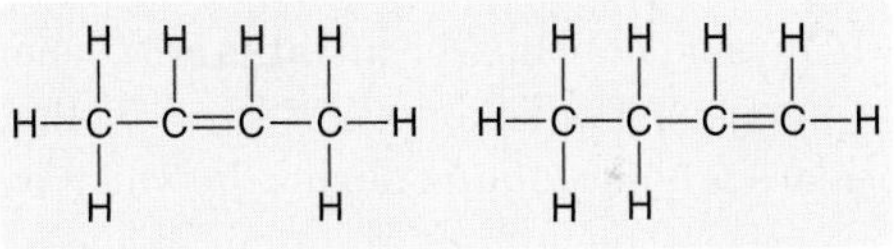 3-C chain: 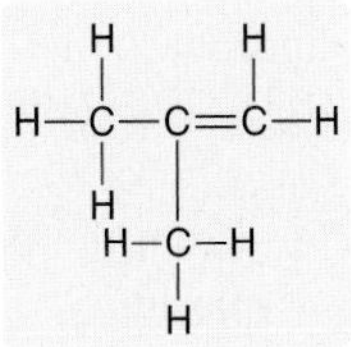

(c) Compounds with four C atoms and one ring:

4-C ring: 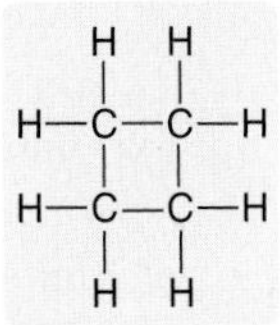 3-C ring: 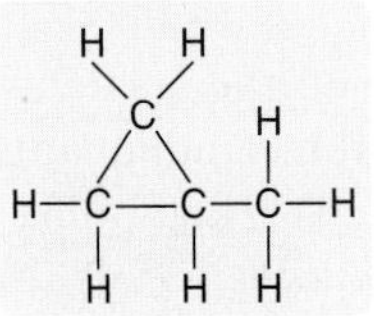

Check Be sure each skeleton has the correct number of C atoms, multiple bonds, and/or rings, and no arrangements are repeated or omitted; remember that a double bond counts as two bonds.

Comment Avoid some *common mistakes*:

In (a):

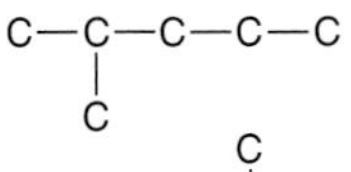 is the same skeleton as 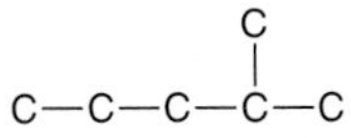

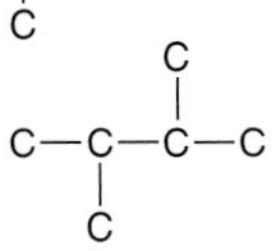 is the same skeleton as 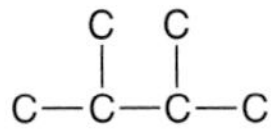

In (b): C—C—C=C is the same skeleton as C=C—C—C

The double bond restricts rotation; therefore, in addition to the *cis* form shown in part (b), another possibility is the *trans* form:

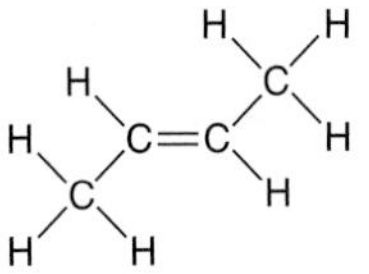

(We will discuss *cis* and *trans* forms more fully in Sections 20.4 and 20.5.)

Also, avoid drawing too many bonds to one C, as here: 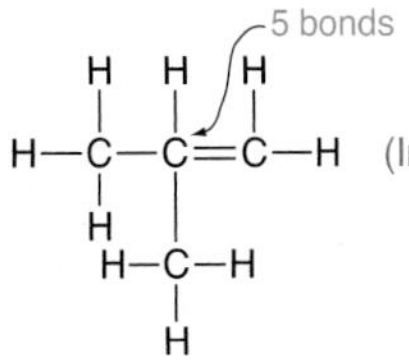 (Incorrect)

For (c): there are too many bonds to one C in 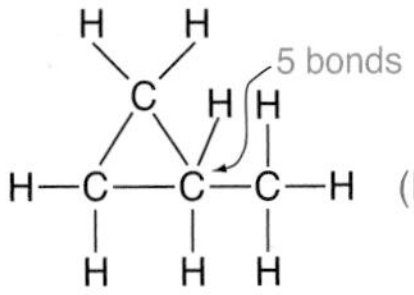 (Incorrect)

Follow-Up Problem 20.1 Draw all hydrocarbons that have different atom arrangements with **(a)** seven C atoms, no multiple bonds, and no rings (nine arrangements); **(b)** five C atoms, one triple bond, and no rings (three arrangements). (See Brief Solutions to Follow-Up Problems at the end of the chapter.)

Hydrocarbons can be classified into four main groups: alkanes, alkenes, alkynes, and aromatic. In the remainder of this section and in the following sections, we will discuss the naming of hydrocarbons, as well as some structural features and physical properties of each group. In Chapter 21, we will discuss the chemical behaviour of the hydrocarbons.

Alkanes: Hydrocarbons with Only Single Bonds

A hydrocarbon that contains only single bonds is an **alkane** (general formula C_nH_{2n+2}, where n is a positive integer). For example, if $n = 5$, the formula is $C_5H_{[(2\times5)+2]}$, or C_5H_{12}. The alkanes are a **homologous series**, in which each member differs from the next by a —CH_2— (methylene) group. In an alkane, each C atom is sp^3 hybridized. Because each C is bonded to the *maximum number of other atoms* (C or H), alkanes are referred to as **saturated hydrocarbons**.

Naming Alkanes You learned the straight-chain alkanes in Section 2.8. Here, we discuss general rules for naming any alkane and, by extension, other organic compounds as well. The key point is that *each chain, branch, or ring has a name based on the number of C atoms*. The name of a compound has three parts:

$$\text{PREFIX} + \text{ROOT} + \text{SUFFIX}$$

TABLE 20.1 Numerical Roots for Carbon Chains and Branches

Roots	Number of C Atoms
meth-	1
eth-	2
prop-	3
but-	4
pent-	5
hex-	6
hept-	7
oct-	8
non-	9
dec-	10

- *Root*: The root tells the number of C atoms in the longest *continuous* chain in the molecule. The roots for the 10 smallest alkanes are shown in Table 20.1. As you can see, there are special roots for compounds with chains of one to four C atoms; roots of longer chains are based on Greek numbers.
- *Suffix*: The suffix tells the *type of organic compound* that is being named; that is, it identifies the key functional group that the molecule possesses. The suffix is placed *after* the root.
- *Prefix*: Each prefix identifies a *group attached to the main chain* and the number of the carbon to which it is attached. Prefixes identifying hydrocarbon branches are the same as root names (Table 20.1) but have *-yl* as their ending. Each prefix is placed *before* the root.

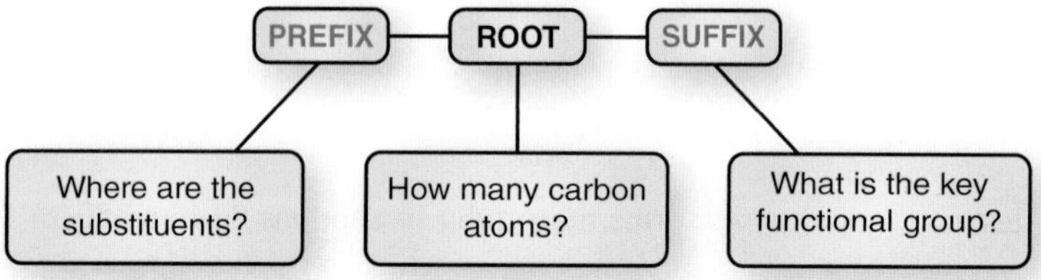

For example, in the name 2-methylbutane, *2-methyl-* is the prefix (a one-carbon branch is attached to C-2 of the main chain), *-but-* is the root (the main chain has four C atoms), and *-ane* is the suffix (the compound is an alkane). Notice that the numbers are separated from the letters by hyphens. If there are several numbers, they will be separated by commas. Do *not* leave spaces.

The systematic name of a compound is obtained as follows:

1. Name the longest chain (root).
2. Add the compound type (suffix).
3. Name any branches (prefix).

Table 20.2 presents the rules for naming any organic compound. Steps 1 to 3 apply these rules to an alkane component of gasoline. Other organic compounds are named with a variety of other prefixes and suffixes (see Table 20.5). In addition to these *systematic* names, some *common* names are still in use.

Depicting Alkanes with Formulas and Models Chemists have several ways to depict organic compounds. Expanded, condensed, and carbon-skeleton formulas are easy to draw; ball-and-stick and space-filling models show the actual shapes.

The **expanded formula** shows each atom and bond. One type of **condensed formula** groups each C atom with its H atoms. A **carbon-skeleton formula** shows only carbon-carbon bonds and appear as zig-zag lines, often with branches. Each

TABLE 20.2 Rules for Naming an Organic Compound

1. Naming the longest chain (root):
   (a) Find the longest *continuous* chain of C atoms.
   (b) Select the root that corresponds to the number of C atoms in this chain.

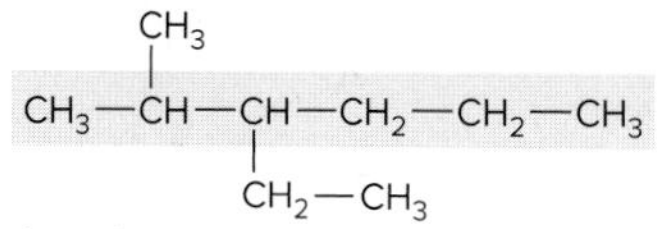

6 carbons ⟹ hex-

2. Naming the compound type (suffix):
   (a) For alkanes, add the suffix *-ane* to the chain root. (Other suffixes appear in Table 20.5 with their functional group and compound type.)
   (b) If the chain forms a ring, the name is preceded by *cyclo-*.

hex- + -ane ⟹ hexane

3. Naming the branches (prefixes) (if the compound has no branches, the name consists of the root and suffix):
   (a) Each branch name consists of a subroot (number of C atoms) and the ending *-yl* to signify that it is not part of the main chain.
   (b) Branch names precede the chain name. When two or more branches are present, their names appear in *alphabetical* order.

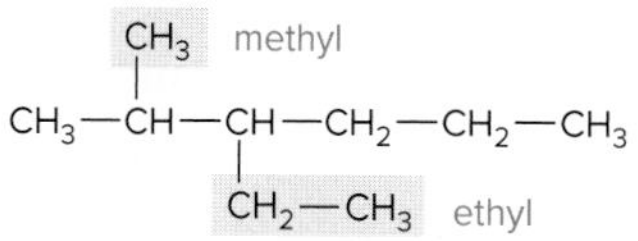
methyl
ethyl

ethylmethylhexane

   (c) To specify where the branch occurs along the chain, number the main-chain C atoms consecutively, starting at the end *closer* to a branch, to achieve the *lowest* numbers for the branches. Precede each branch name with the number of the main-chain C to which that branch is attached.

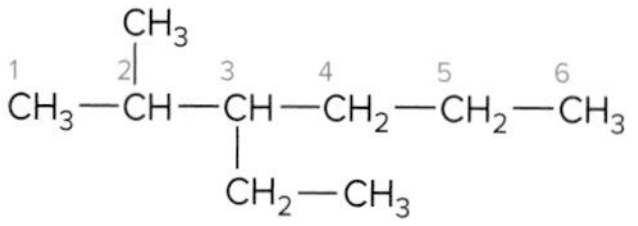

3-ethyl-2-methylhexane

4. For multiple branches (more than one of the same alkyl group), use the prefixes di-, tri-, tetra-. Arrange the branches alphabetically by their names, not by their prefixes.
5. If chains of equal length are competing for selection as the parent chain in a saturated branched acyclic hydrocarbon, then the choice goes to the chain that has the most side chains.

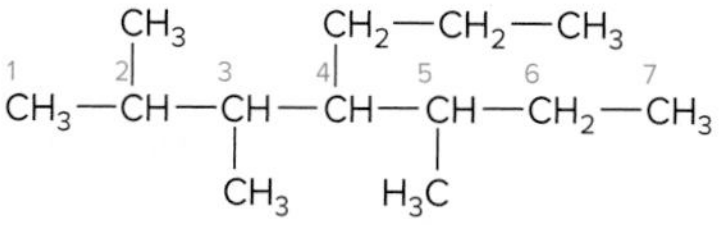

2,3,5-trimethyl-4-propylheptane

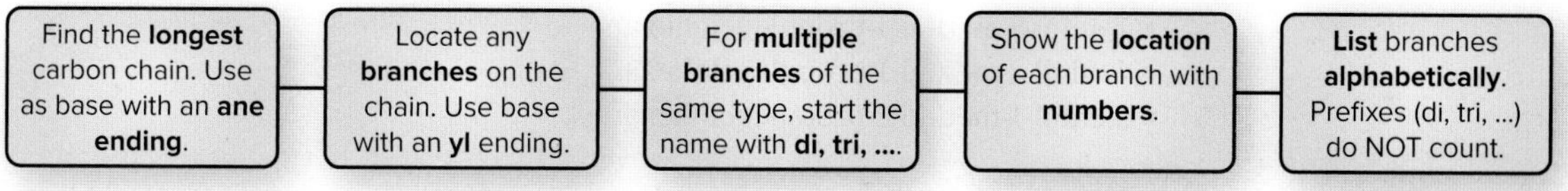
Find the **longest** carbon chain. Use as base with an **ane** ending.
Locate any **branches** on the chain. Use base with an **yl** ending.
For **multiple branches** of the same type, start the name with **di, tri, ...**.
Show the **location** of each branch with **numbers**.
List branches **alphabetically**. Prefixes (di, tri, ...) do NOT count.

end or bend of a zig-zag line or branch represents a C atom attached to the number of H atoms that gives it a total of four bonds.

Propane

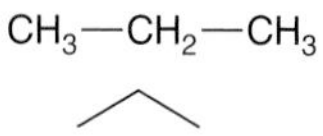

2,3-dimethylbutane

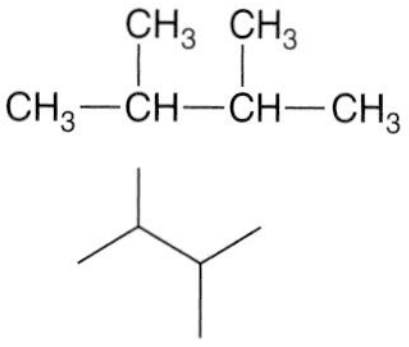

Figure 20.5 shows these formulas (and models) of 3-ethyl-2-methylhexane, one of the compounds named in Table 20.2.

Propane (C_3H_8) is normally a gas, but it can be compressed into a liquid. It is commonly used as a fuel for portable stoves, engines, and barbecues.

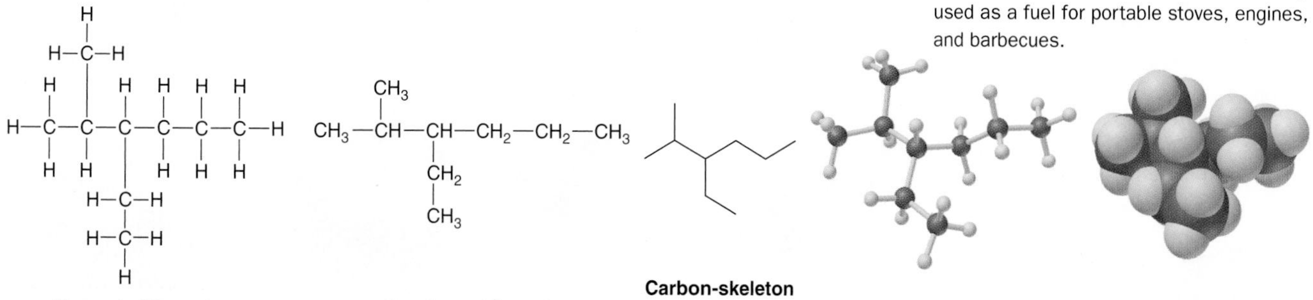

Expanded formula **Condensed formula** **Carbon-skeleton formula** **Ball-and-stick model** **Space-filling model**

FIGURE 20.5 Ways of depicting the alkane 3-ethyl-2-methylhexane

Cycloalkanes

A **cyclic hydrocarbon** contains one or more rings in its structure. When a straight-chain alkane (C_nH_{2n+2}) forms a ring, two H atoms are lost as the C—C bond forms to join the two ends of the chain. Thus, *cycloalkanes* have the general formula C_nH_{2n}. Cyclic hydrocarbons are often drawn with carbon-skeleton formulas (Figure 20.6, *top row*). Except for three-carbon rings, *cycloalkanes are nonplanar*, as the models show. This structural feature arises from the tetrahedral shape around each C atom and the need to minimize electron repulsions between adjacent H atoms. As a result, orbital overlap of adjacent C atoms is maximized. The most stable form of cyclohexane is called the *chair conformation* (see Figure 20.12). We discuss conformations of linear alkanes and cycloalkanes in Section 20.3.

Cycloalkanes are very important components of food, pharmaceutical drugs, and much more. They are named according to the rules previously described, but the name is based on the number of carbon atoms in the ring itself. The smallest cycloalkanes must have at least three carbon atoms in the ring. They are named by adding "cyclo" to the beginning of their root alkane name.

If the ring has only one alkyl substituent, there is no need to number the position; however, if there are two substituents, name them in alphabetical order, giving the number 1 position to whichever substituent results in a second substituent getting as low a number as possible. If there are more than two substituents, name them either clockwise or counterclockwise in the direction that gives them the lowest possible number. Essentially, we minimize the number at the "first point of difference." One way to assure the lowest number possible is to number the carbon atoms so that when the numbers corresponding to the substituents are added, their sum is the lowest possible.

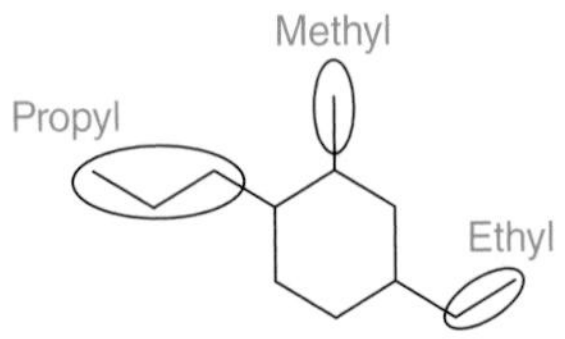

The molecule in the margin is a six-member ring (cyclohexane) with three alkyl substituents: propyl, methyl, and ethyl. Alphabetically, ethyl should be named first, then methyl, and then propyl. Based on the information above, how would we name this compound?

There are three naming options, but only one correct answer:

- 1-ethyl-3-methyl-4-propylcyclohexane
- 4-ethyl-2-methyl-1-propylcyclohexane
- 5-ethyl-1-methyl-2-propylcyclohexane

FIGURE 20.6 Depicting cycloalkanes

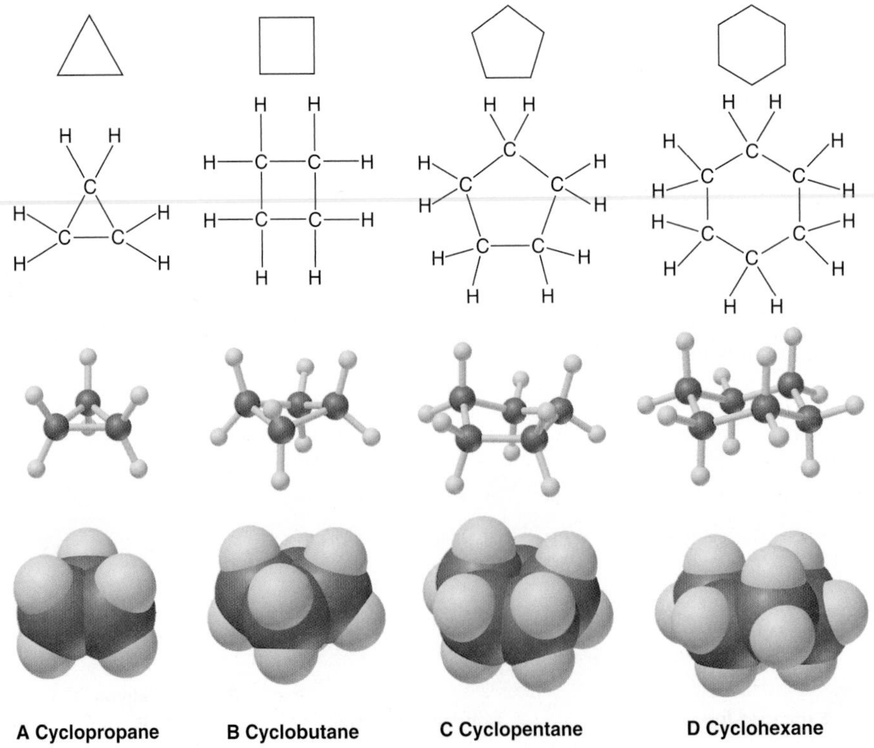

TABLE 20.3 Physical Properties of Linear Alkanes with Molecular Formulas C_4H_{10} and C_5H_{12}

Systematic Name (Common Name)	Expanded Formula	Condensed and Skeleton Formulas	Space-Filling Model	Density (g/mL)	Boiling Point (°C)
Butane	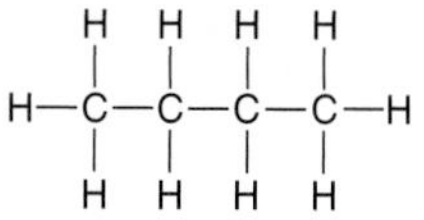	$CH_3{-}CH_2{-}CH_2{-}CH_3$ 	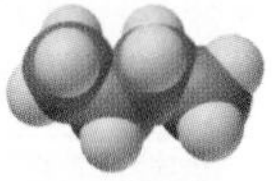	0.579	−0.5
2-methylpropane (isobutane)	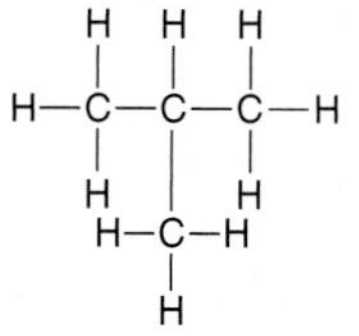	$CH_3{-}CH(CH_3){-}CH_3$ 	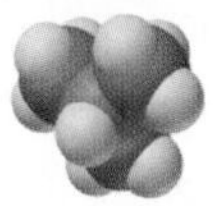	0.549	−11.6
Pentane	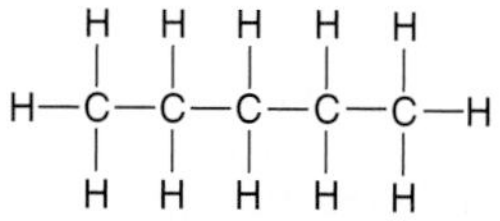	$CH_3{-}CH_2{-}CH_2{-}CH_2{-}CH_3$ 		0.626	36.1
2-methylbutane (isopentane)	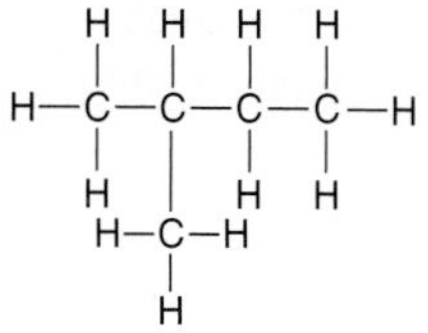	$CH_3{-}CH(CH_3){-}CH_2{-}CH_3$ 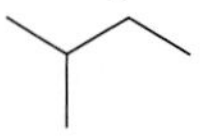		0.620	27.8
2,2-dimethylpropane (neopentane)	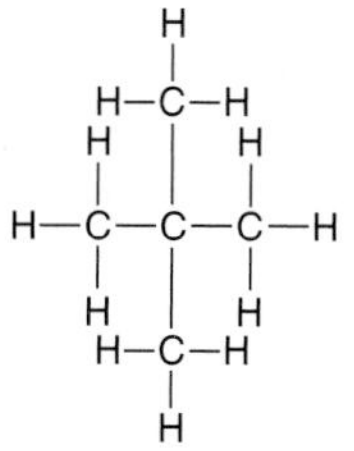	$CH_3{-}C(CH_3)_2{-}CH_3$ 		0.614	9.5

To satisfy all the naming rules, the correct answer is 4-ethyl-2-methyl-1-propylcyclohexane.

Physical Properties of Alkanes

Recall from Section 3.2 that two or more compounds with the same molecular formula but different properties are called *isomers*. Those with *different arrangements of bonded atoms* are *constitutional* (or *structural*) *isomers*; alkanes with the same number of C atoms but different skeletons are examples. (We will study isomers in more detail in Section 20.4).

Dispersion Forces and Boiling Points Because alkanes are nearly nonpolar, we expect their physical properties to be determined by dispersion forces, and the boiling points in Table 20.3 bear this out. The four-carbon alkanes boil at a lower temperature than the five-carbon compounds. Moreover, within each group of isomers (compounds with the same molecular formula but different chemical structure), the more spherical member (2,2-dimethylpropane, isobutane, or neopentane) boils at a lower temperature than the more elongated ones (butane and pentane). As you saw in Chapter 11, this trend occurs because a spherical shape leads to less intermolecular contact, and thus lower total dispersion forces, than does an elongated shape.

A particularly clear example of the effect of dispersion forces on physical properties occurs among the unbranched alkanes (*n*-alkanes). Among these compounds, boiling points increase steadily with chain length: the longer the chain, the greater the intermolecular contact, the stronger the dispersion forces, and the higher the boiling point (Figure 20.7). Pentane (five C atoms) is the smallest *n*-alkane that exists as a liquid at room temperature. The solubility of alkanes, and of all hydrocarbons, is easy

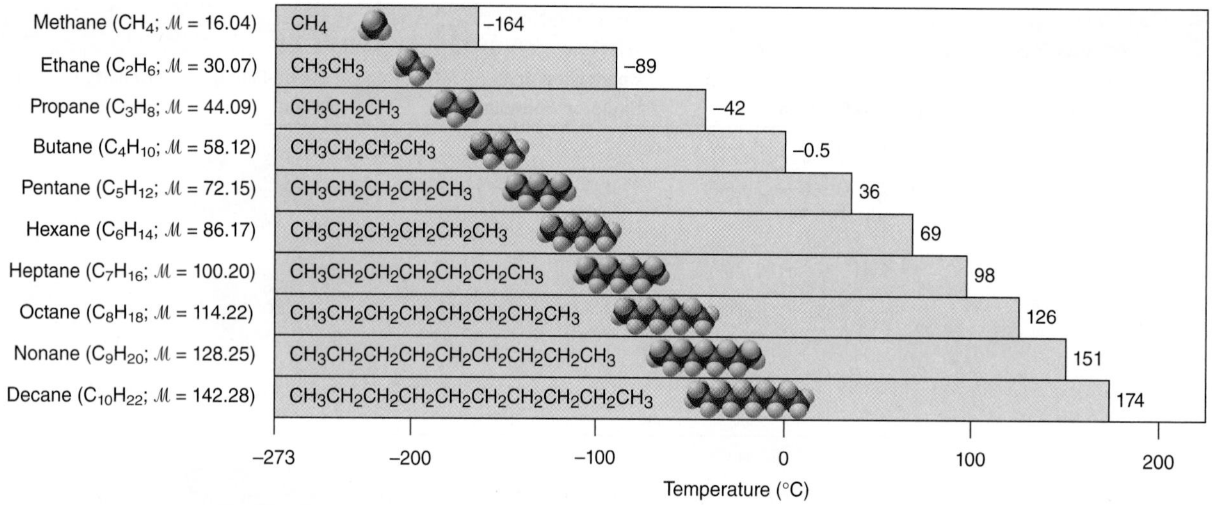

FIGURE 20.7 Formulas, molar masses (in g/mol), structures, and boiling points (at 1 bar pressure) of the first 10 unbranched alkanes

to predict from the like-dissolves-like rule (Section 12.1). Alkanes are miscible in each other and in other nonpolar solvents, such as benzene, but are nearly insoluble in water. The solubility of pentane in water, for example, is only 0.36 g/L at room temperature.

SUMMARY OF SECTION 20.2

- Hydrocarbons contain only C and H atoms, so their physical properties depend on the strength of their dispersion forces.
- Names of organic compounds have a root for the longest chain, a prefix for any attached group, and a suffix for the type of compound.
- Alkanes (C_nH_{2n+2}) have only single bonds. Cycloalkanes (C_nH_{2n}) have ring structures that are typically nonplanar.

20.3 Conformation of Alkanes

Conformation refers to spatial arrangements or three-dimensional structures that can be adopted by a molecule as a result of rotation of its atoms about single bonds, and conformational analysis studies the energy changes that occur during those rotations. This is important because the structure of a molecule can have a significant influence on its properties and reactions.

Conformation of Linear Alkanes

Alkanes of two or more carbon atoms can be twisted into a number of different three-dimensional arrangements by rotation about a carbon-carbon bond or bonds. Any three-dimensional arrangement of atoms that results by rotation about single bonds is called a conformation. A specific conformation is called a **conformer**. The different conformers have the same connections of atoms and they interconvert rapidly. The most stable conformer is the one with lower energy, resulting in lower interactions between the connected atoms. In the following sections, we will learn different ways of viewing the different conformers.

Sawhorse Projection The **sawhorse projection** views the carbon atoms with their substituents from an oblique angle. Sawhorse diagrams are similar to wedge-bond perspective drawings (introduced in Chapter 10) but without using "shading" to denote the perspective. Figure 20.8 shows some sawhorse projections for a butane

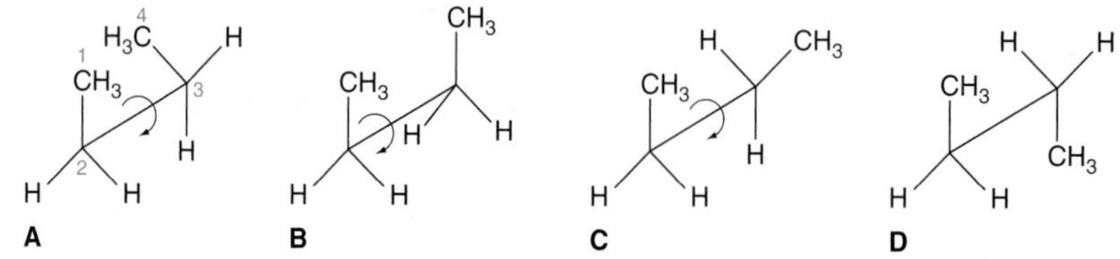

FIGURE 20.8 Butane sawhorse projections (a few of several possible ones). These projections make it easy to visualize the molecule in 3D.

molecule. The C-2—C-3 bond can rotate, creating several conformers, some of them more stable than others, depending on the interaction the substituents might have in each conformation. Let us start with conformer A. If the rear carbon (C-3) rotates 60°, we get conformer B, which has two methyl groups in the same region (not very stable). Rotating C-3 a further 60°, we get conformer C (with the same energy as conformer A), and if we rotate it 120° we get the most stable conformer, D, which has fewer interactions between the substituents.

Newman Projection The **Newman projection** is very useful when representing alkanes. It represents the spatial arrangement of bonds on two adjacent atoms in a molecule. According to IUPAC, the structure appears as viewed along the bond between these two atoms, and the bonds from them to other groups are drawn as projections in the plane of the paper. The bonds from the atom nearer to the observer are drawn to meet at the centre of a circle representing that atom. Those from the farther atom are drawn as if projecting from behind the circle.

One can observe the angle between a substituent on the front atom and a substituent on the back atom in the Newman projection, which is called the *dihedral angle* or *torsion angle* and is represented by the Greek letter theta (θ). For the propane molecule, for example, we could draw the following Newman projection:

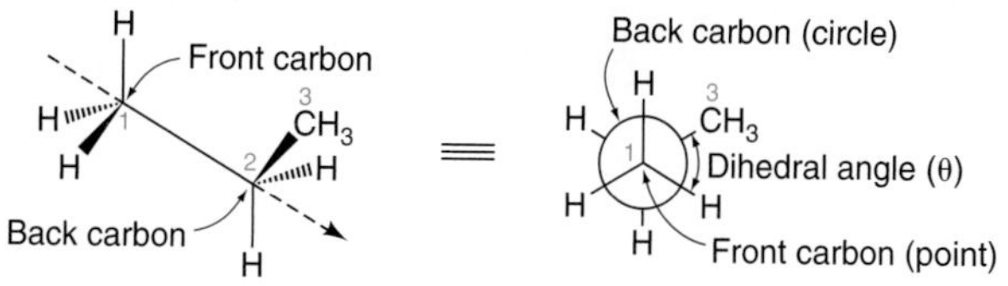

Now, try drawing the Newman projection, but looking from the right (the front carbon will now be C-2).

There is not a perfectly free rotation between the two carbon atoms. There is a barrier to rotation because some conformations are more stable than others. Although an infinite number of conformations are possible, the *staggered* and *eclipsed* conformations, which represent the most and least stable, respectively, are the two most important. In the *staggered* conformation, the substituents are at the maximum distance from each other. When the dihedral angle is a multiple of 60°, the *staggered* conformations are named *gauche*, and when the bulkiest substituents are opposite to each other, the *staggered* conformation is named *anti*. In the *eclipsed* conformation, the substituents on adjacent atoms are in closest proximity. The conformation with the bulkiest substituents *eclipsed* is named *syn*.

Figure 20.9 shows some Newman projections of butane conformers, considering C-2 the front carbon and C-3 the back carbon. Remember that butane is C_4H_{10}, or $CH_3CH_2CH_2CH_3$. Carbon atoms C-2 and C-3 are connected and each of them has two hydrogen atoms and a methyl group:

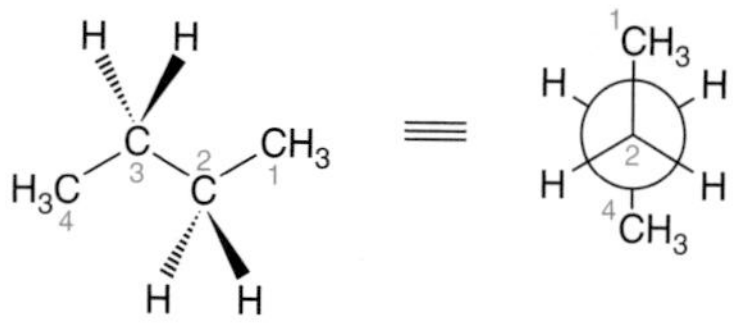

In the *anti* conformation (when the dihedral angle = 180°), the potential energy is at its lowest and, therefore, most stable. This is due to the methyl groups being the farthest away from each other. When the dihedral angle = 120° or 240° (*eclipsed* conformation), the methyl groups align themselves with hydrogen atoms on the same plane, resulting in the butane having a higher potential energy than the gauche conformations, although still having a lower potential energy than the *syn* conformation, where the two methyl groups are close to the molecule, and therefore the potential energy is at its highest. When the dihedral angle (θ) is 60° or 300° (*gauche* conformation), the methyl groups are farther apart and therefore the potential energy drops. The maximum energy occurs when the two methyl groups eclipse each other (θ = 0° and 360°) in a *syn* position (*eclipsed*). The conformer with the minimum energy has the larger groups (methyl groups) opposite to each other (θ = 180°) in an *anti* position (*staggered*).

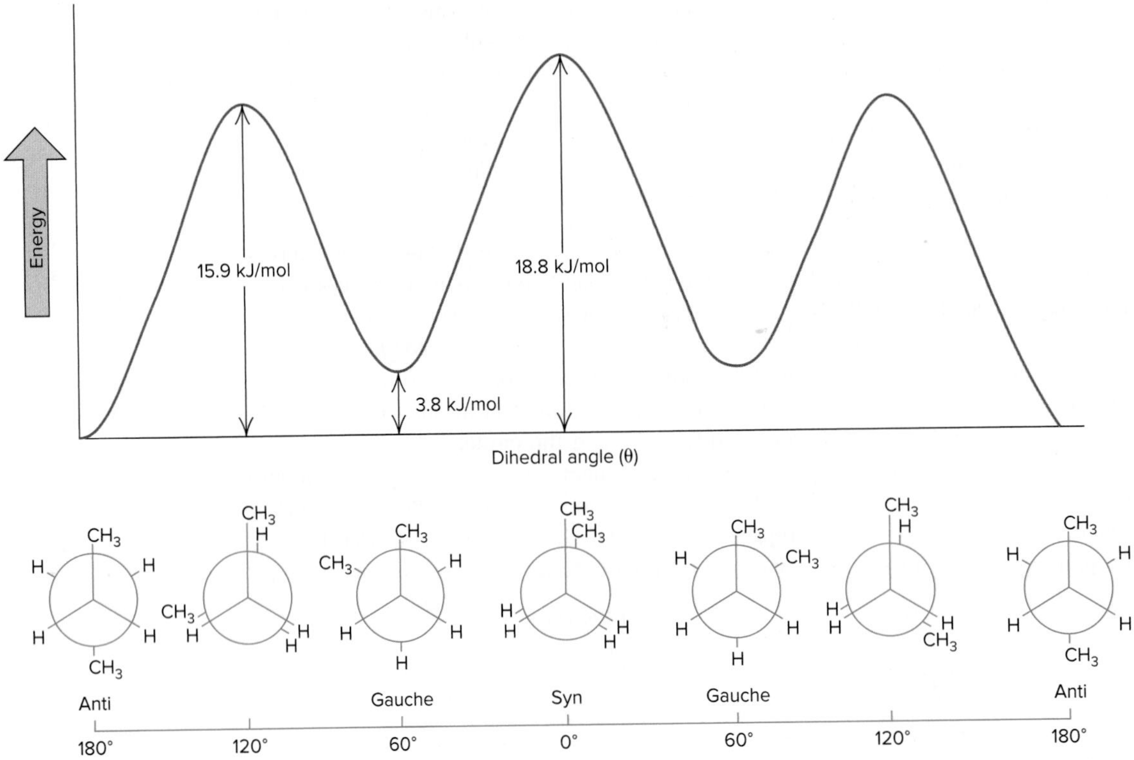

FIGURE 20.9 Diagram of potential energy versus rotation for butane

Fischer Projection The **Fischer projection** is a two-dimensional representation of a three-dimensional organic molecule. All bonds are depicted as horizontal or vertical lines. The carbon chain is depicted vertically, with the carbon atoms at the intersection with the horizontal lines. The chain is oriented so C-1 is at the top. These presentations are mostly used for molecules that contain *stereogenic centres* (also called *stereocentres*). Stereogenic centres are atoms that have different entities or groups bonded to them so that an interchanging of two groups leads to a stereoisomer, as we will study in Section 20.4. Figure 20.10 illustrates how to draw Fischer projections.

The following is the Fischer projection of D-glucose:

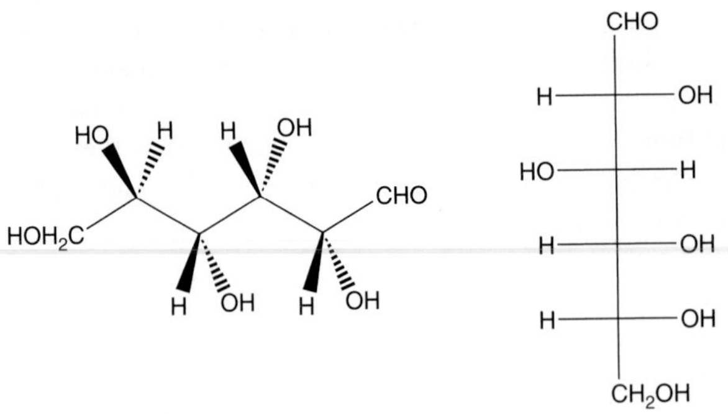

FIGURE 20.10 Fischer projections. A. A projection of a tetrahedral molecule onto a planar surface. **B.** The Fischer projection consists of both horizontal and vertical lines, where the horizontal lines represent the atoms that are pointed toward the viewer, and the vertical line represents atoms that are pointed away from the viewer. Each point of intersection of the horizontal and vertical lines represents a central carbon atom.

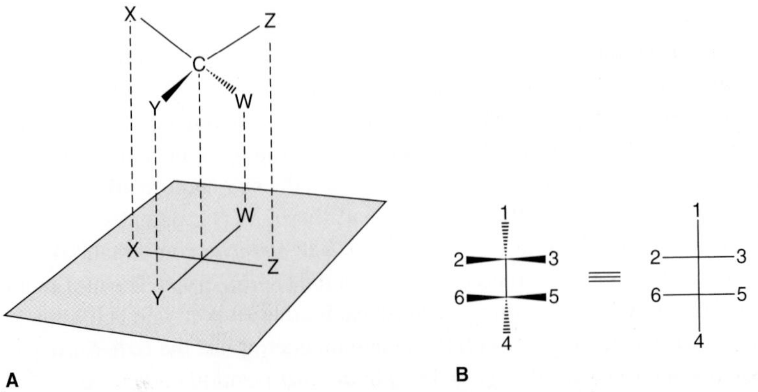

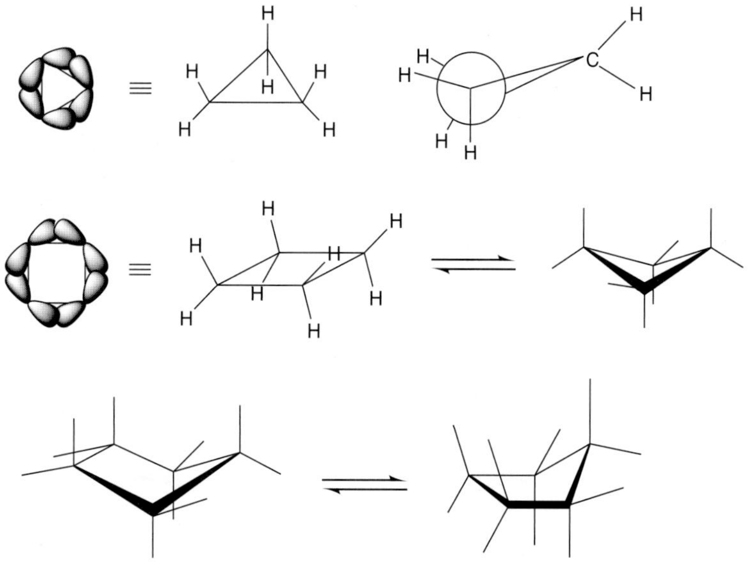

FIGURE 20.11 Cycloalkane conformations. A. A carbon frame of cyclopropane, which is a planar structure. The less than optimal orbital overlap due to bent bonds leads to high bond angle strain. All the hydrogen atoms are eclipsed, which leads to an additional torsional strain. **B.** Two conformers of cyclobutane. The planar one (*left*), with internal angles of 90°, exhibits a high bond angle strain as well as torsional strain due to the eclipsed hydrogen atoms. The puckered one (*right*), with angles of 88°, has slightly more angle strain but less eclipsing strain. **C.** Two puckered conformations of cyclopentane, the envelope (*left*) and the half chair (*right*). The planar conformation would have a 108° angle, but all of the hydrogen atoms eclipse. Adopting the envelope position reduces the angle (increasing the bond angle strain) but reduces the torsional strain and therefore the ring strain as a whole.

Conformation of Cycloalkanes

Cycloalkanes are less stable than their straight-chain counterparts. They try to adopt minimum energy conformations to avoid ring strain due to the following:

- **Angle strain**, induced in a molecule when the bond *deviates* from the ideal tetrahedral value
- **Steric strain**, due to *eclipsing* of bonds on adjacent atoms
- **Torsional strain**, due to *repulsive interactions* between their substituents

A tetrahedral carbon atom has a bond angle of 109.5°. However, to accommodate the geometry of cycloalkanes, these bond angles are forced into other angles, resulting in angle strain. Both cyclopropane (Figure 20.11A) and cyclobutane (Figure 20.11B) have large ring strains of 131.8 kJ/mol and 110.5 kJ/mol, respectively. The large amount of ring strain in cyclopropane and cyclobutane can be explained by (1) the large deviation of the respectively required 60° and 90° bond angles from 109.5°, and (2) torsional strain, because all the hydrogen atoms are eclipsed. Cyclopentane has much less ring strain at 27.2 kJ/mol (Figure 20.11C). Cyclohexane is the only cycloalkane that has no ring strain. Cycloheptane and higher cycloalkanes tend to have modest amounts of transannular ring strain, which is defined as the repulsive interaction between substituents attached to nonadjacent ring atoms.

Drawing Cyclohexanes The cyclohexane molecule has several conformations. The more stable ones are the two chair conformations. There are other less stable conformations, such as the boat and the twist boat. They can interconvert into each other by conformational rotation, also known as ring-flipping. In the chair conformation of cyclohexane, there are two types of positions: *axial* and *equatorial*. Six of the 12 carbon-hydrogen bonds end up almost perpendicular to the mean ring plane and almost parallel to the symmetry axis, with alternating directions, and are said to be *axial* (H^a). The other C-H bonds lie almost parallel to the mean plane and are said to be *equatorial* (H^e):

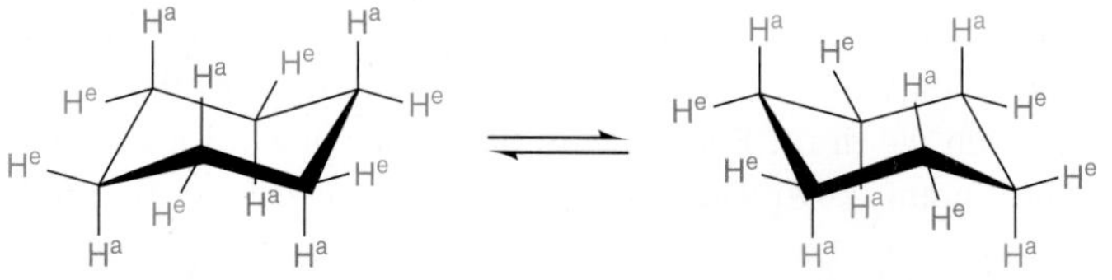

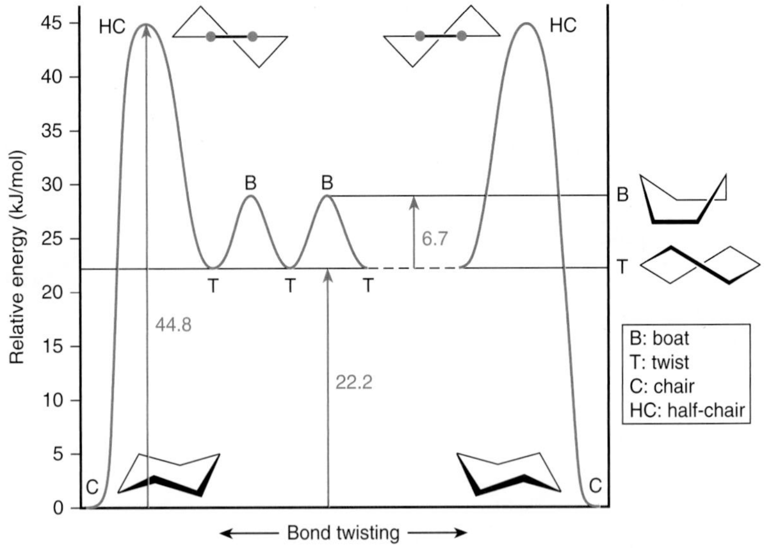

FIGURE 20.12 Schematic energy profile for the interconversion of the cyclohexane ring

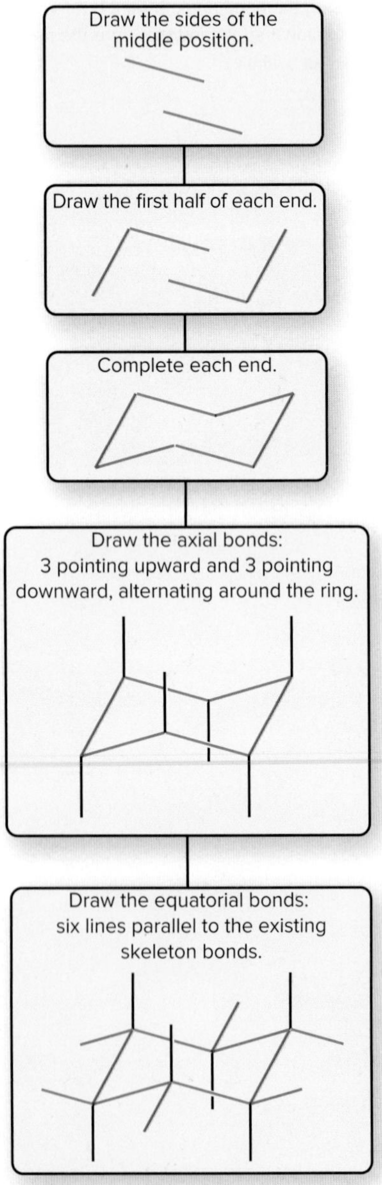

FIGURE 20.13 Guide to drawing cyclohexanes. The chair conformation can be obtained by drawing sets of parallel lines opposite to each other.

The transition from one chair to another is not a single-step process. It occurs through several energy minima, representing half-chair, boat, and twisted conformations (Figure 20.12). The boat and twist conformations have high energy and are thus only observed as the predominant form very infrequently, such as when the six-membered ring is fashioned with a large substituent or when it is included in a fused bicyclic structure.

It is important that you learn how to draw cyclohexane chair conformations. Figure 20.13 outlines the steps to follow to draw them.

The two chair conformations can interconvert into each other going through other conformers with higher energy, like the boat conformation, as was described in Figure 20.12. Now that we have shown how to draw cyclohexanes, we can demonstrate how the ring flip should be performed. Notice that an axial substituent in one chair becomes an equatorial substituent in the ring-flipped chair:

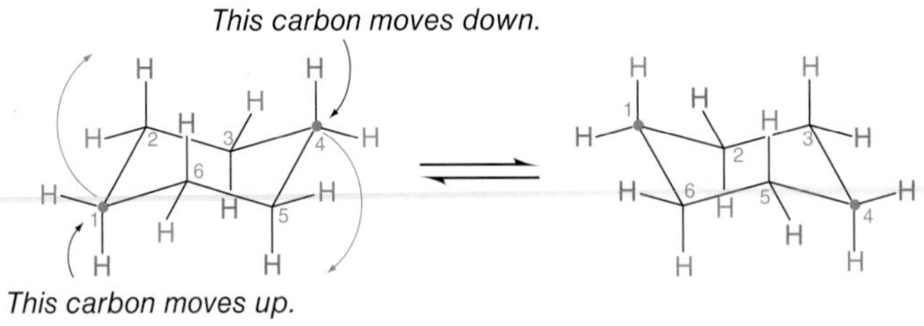

An easy way to determine the most stable cyclohexane conformation is by drawing Newman projections. Figure 20.14 shows two cyclohexane conformers and how to "look" at the molecules to draw their Newman projections. In the first conformer (A), C-1 and C-5 are in the front, connected through C-6, whereas C-2 and C-4 are in the back, connected by C-3. In conformer B, C-6 and C-4 are the front atoms connected by C-5, and C-1 and C-3 are in the back connected by C-2.

Once we know how to draw cyclohexane chairs, we can look at *substituted cyclohexane conformations*. The most stable conformation will be the one with the least steric strain. Keep in mind that whenever we draw the cyclohexane molecule, those substituents that are pointing upward will keep this position, no matter what chair conformation we draw. Substituents that point upward are above the plane, and those that point downward are below the plane. In the following flat

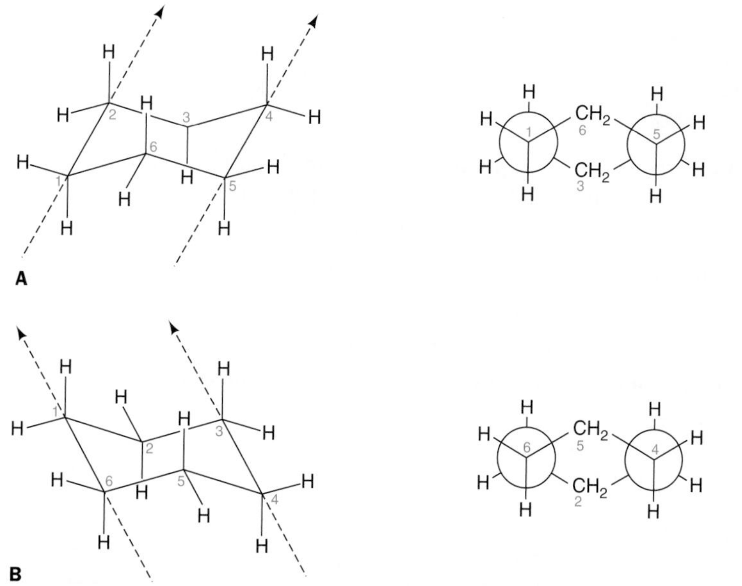

FIGURE 20.14 How to draw cyclohexane Newman projections. **A** and **B** are the two chair conformations. The bonds of the atoms nearer to the observer are drawn to meet at the centre of the circles representing those atoms.

cyclohexane conformer, the red lines point out of the plane, and the blue ones are pointing into the plane.

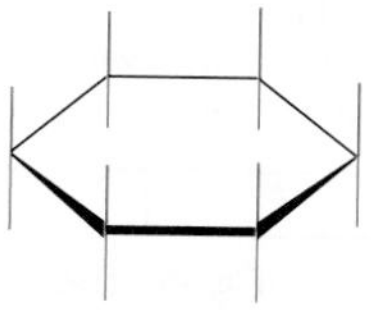

For example, when we draw the monosubstituted cyclohexane methylhexane molecule (Figure 20.15), the methyl substituent could be in an equatorial or an axial position. The conformer with the methyl group in equatorial position A is favoured over the one with the methyl in axial position B. In addition to all the repulsions between hydrogen atoms in axial positions in both conformers, in conformer B there will be a repulsion between the methyl group and the hydrogen atoms when they are in axial positions (*1,3-diaxial repulsions*). Since each CH_3-H repulsion is 3.77 kJ/mol, the energy of conformer B will be 7.54 kJ/mol higher. In disubstituted cyclohexanes, the substituents can be above, below, or both above and below the plane of the ring. If they are both on the same side of the plane of the ring, the name will start with *cis* (from the Latin for *same side*), and if they are on opposite sides of the plane, the name will start with *trans* (from the Latin for *across*). The *cis* and the *trans* isomers cannot interconvert into each other by rotation of bonds.

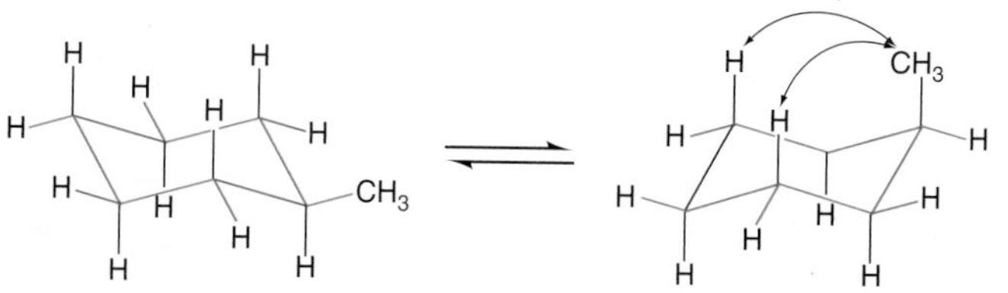

FIGURE 20.15 Methylhexane (a monosubstituted cyclohexane) chair conformations*

*Note that 1,3-diaxial interactions are really gauche butane interactions—each 1,3 interaction is approximately equivalent to one gauche butane interaction of 3.77 kJ/mol, for a total of 7.54 kJ/mol strain for an axial methyl cyclohexane.

Sample Problem 20.2 Predicting Which Conformer of a Polysubstituted Cycloalkane Is More Stable

Problem Calculate the energy and draw the more stable conformation of the following disubstituted cycloalkanes, knowing the following:

1,3-diaxial CH_3-H repulsions = 3.77 kJ/mol

1,3-diaxial CH_3-CH_3 repulsions = 15.1 kJ/mol

CH_3-CH_3 gauche interactions = 3.77 kJ/mol

(a) *trans*-1,4-dimethylcyclohexane **(b)** *cis*-1,4-dimethylcyclohexane

(c) *trans*-1,2-dimethylcyclohexane **(d)** *cis*-1,3-dimethylcyclohexane

Plan

Draw the cycloalkane ring using wedge-dash diagrams.
If both substituents are *cis*, draw both of them pointing into the plane or both of them pointing out of the plane.
If the substituents are *trans*, draw one pointing in and the other pointing out.

↓

Draw the planar conformer with the solid wedge substituent pointing upward and the dashed wedge substituent pointing downward.
You can skip this step, since this conformer is the one with highest energy, but it helps to visualize the substituents in 3D.

↓

Draw one cyclohexane chair conformation showing the axial and the equatorial bonds. Add the substituents to the ring: pointing upward if the bond was a solid wedge or downward if the bond was a dashed wedge.
The first substituent might be added at any carbon atom in the ring.

↓

Flip the chair.
The substituents that were equatorial become axial, and the ones in axial position become equatorial.
The substituents will be pointing upward or downward regardless of the chair conformation drawn.

↓

Look for 1,3-diaxial interactions and calculate the energy due to these repulsions.
If the substituents are in adjacent carbons, draw the Newman projection of each chair. Look for gauche interactions and calculate the energy.

↓

Calculate the total energy for each conformer.
The more stable conformation is the one with lower energy.

Solution

(a) *trans*-1,4-dimethylcyclohexane

Four 1,3-diaxial CH_3-H repulsions
4×3.77 kJ/mol = 15.1 kJ/mol

Methyl groups equatorial
No 1,3-diaxial repulsions

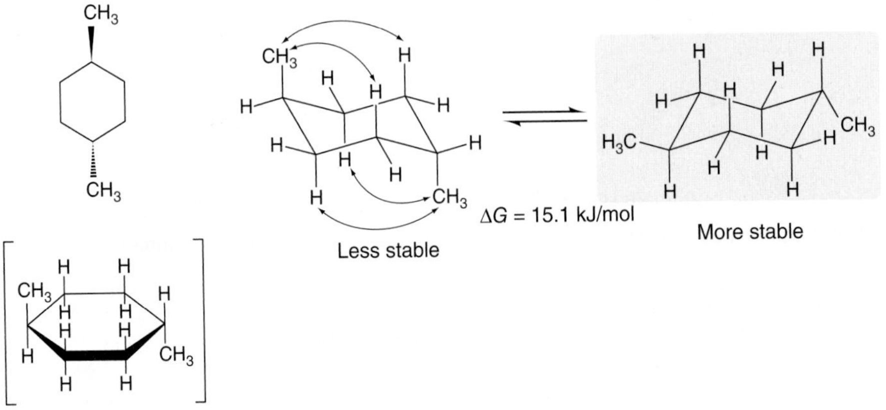

(b) *cis*-1,4-dimethylcyclohexane

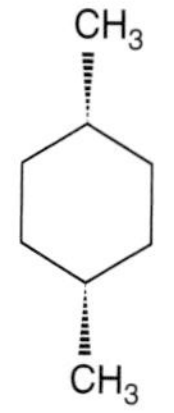

Two 1,3-diaxial CH_3-H repulsions
2×3.77 kJ/mol = 7.54 kJ/mol

Two 1,3-diaxial CH_3-H repulsions
2×3.77 kJ/mol = 7.54 kJ/mol

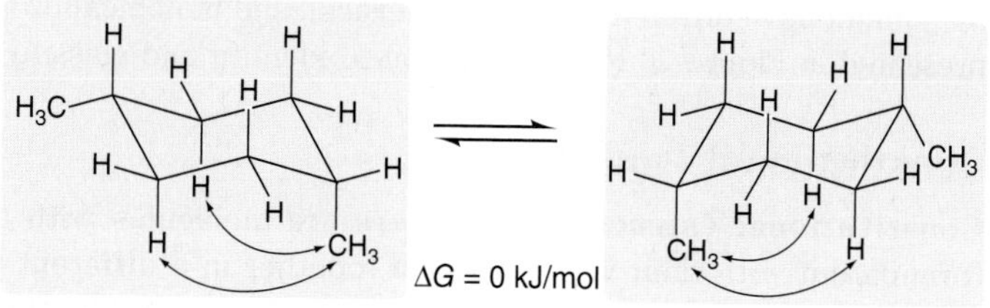

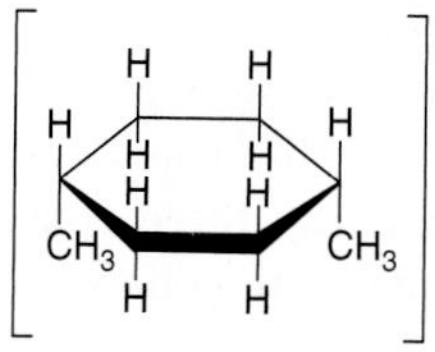

Both conformations are equally stable.

(c) *trans*-1,2-dimethylcyclohexane

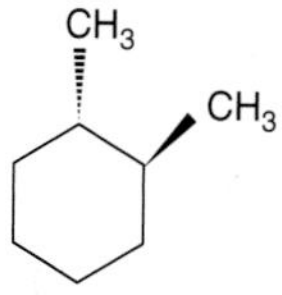

Four 1,3-diaxial CH_3-H repulsions
4×3.77 kJ/mol = 15.1 kJ/mol

Methyl groups equatorial
No 1,3-diaxial repulsions

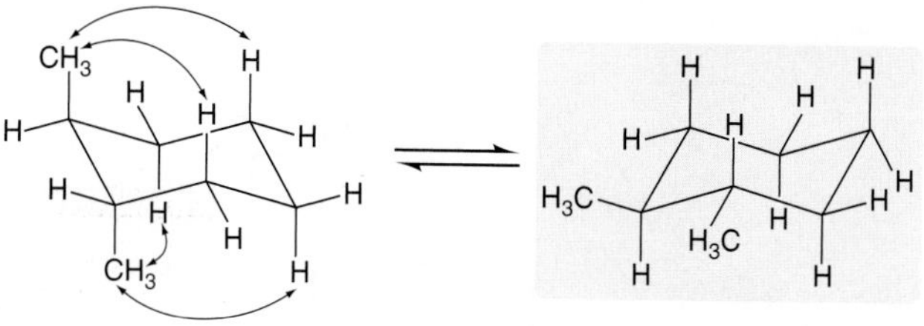

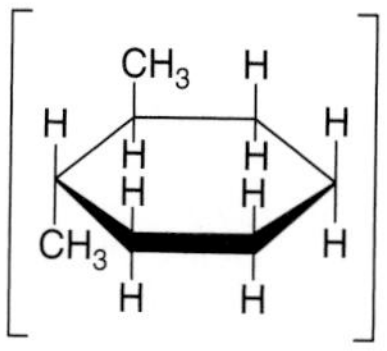

One gauche CH_3-CH_3 interaction = 3.77 kJ/mol

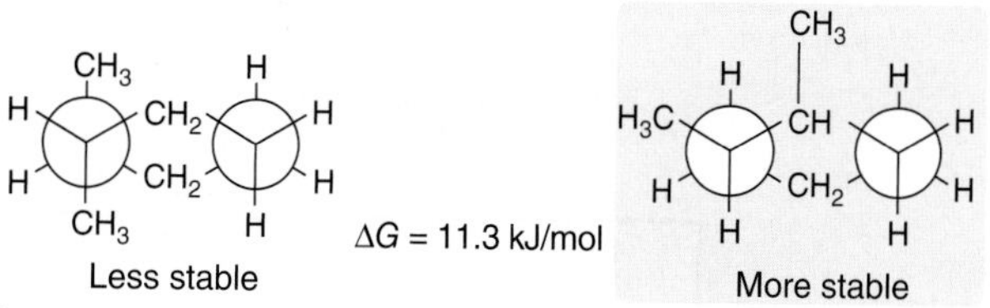

(d) *cis*-1,3-dimethylcyclohexane

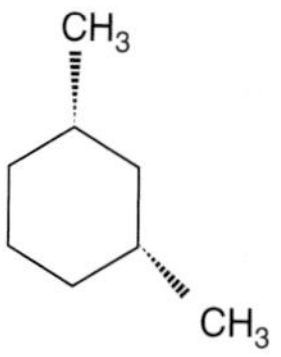

Methyl groups equatorial
No 1,3-diaxial CH_3-H repulsions

Two 1,3-diaxial CH_3-H repulsions
2×3.77 kJ/mol = 7.54 kJ/mol
One 1,3-diaxial CH_3-CH_3 repulsion
1×15.1 kJ/mol = 15.1 kJ/mol

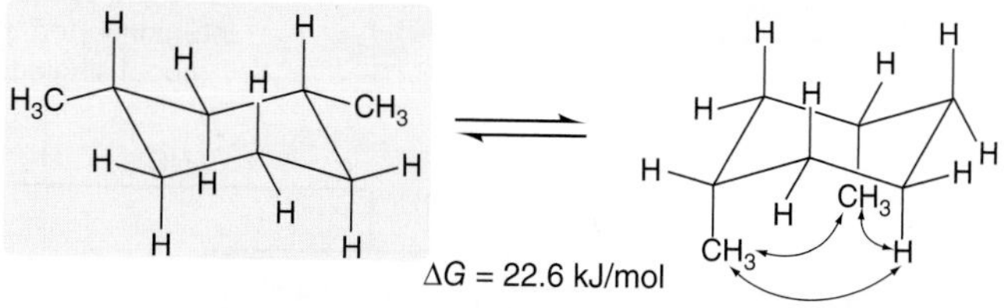

More stable

Less stable

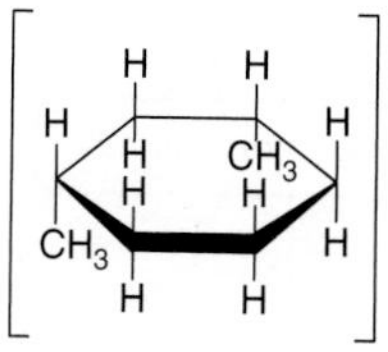

Follow-Up Problem 20.2 Draw the more stable conformation for
(a) *cis*-1,2-dimethylcyclohexane
(b) *trans*-1,2-dimethylcyclohexane

SUMMARY OF SECTION 20.3

- A conformation is any three-dimensional arrangement of atoms that is the result of rotation about single bonds.
- Sawhorse, Newman, and Fischer projections are ways of viewing the different conformers.
- The most stable conformer is the one with lower energy, resulting in lower interaction between the connected atoms.

20.4 Isomers and Isomerism

There are different types of isomers, and it is important to recognize them because they might have different chemical, physical, and biological properties. The diagram presented in Figure 20.16 will help you to identify and classify them.

Constitutional Isomers

Constitutional (structural) isomers are molecules with the same molecular formula but with their atoms bonded together in a different order. Alkanes with the same number of C atoms but different skeletons, like the ones in Table 20.3, are examples. The smallest alkane to exhibit constitutional isomerism has four C atoms: two different compounds have the formula C_4H_{10}. The unbranched one is butane, and the other is 2-methylpropane (isobutane). Similarly, three compounds have the formula C_5H_{12}. The unbranched isomer is pentane; the one with

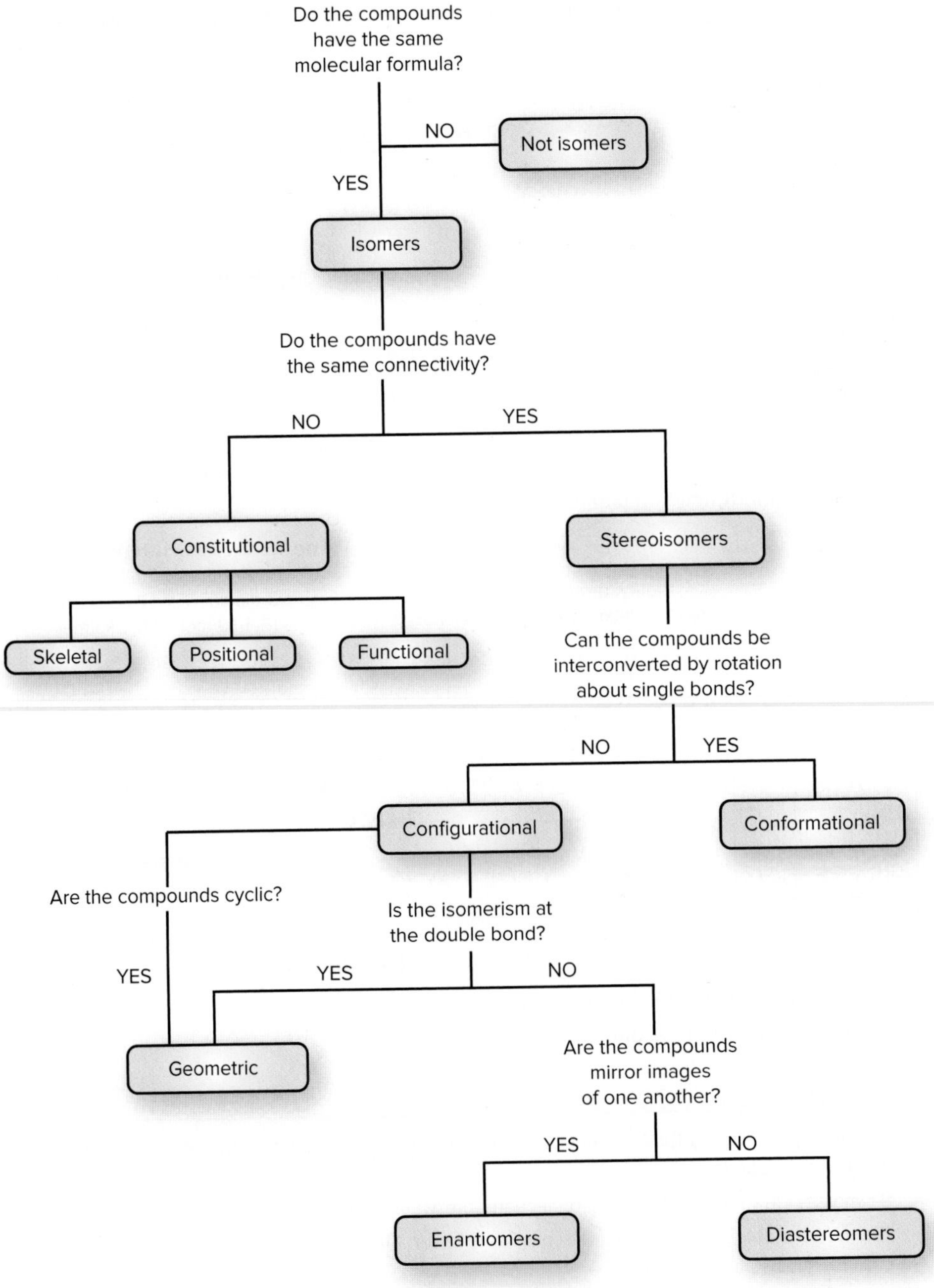

FIGURE 20.16 Methodology of identifying isomers

a methyl group at C-2 of a four-carbon chain is 2-methylbutane (isopentane). The third isomer has two methyl branches on C-2 of a three-carbon chain, so its name is 2,2-dimethylpropane (neopentane).

There are three categories of constitutional isomers: skeletal or chain isomers, positional isomers, and functional isomers. Following is a description of each category, with examples that will introduce the nomenclature of compounds with different functional groups.

Skeletal or Chain Isomers Skeletal or chain isomers are constitutional isomers with the atoms, usually carbon, arranged in a different way. For example, an alkane with five carbon atoms (molecular formula C_5H_{12}) could have the following skeletal isomers:

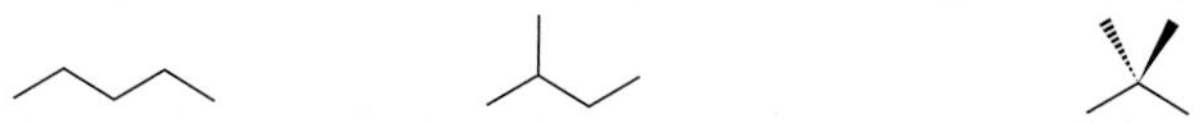

Pentane 2-methylbutane (isopentane) 2,2-dimethylpropane (neopentane)

Positional Isomers Positional isomers have a functional group or another substituent located in a different position on the parent structure. For example, the hydroxyl group could be attached to different carbon atoms on a pentane. Each compound will have different properties:

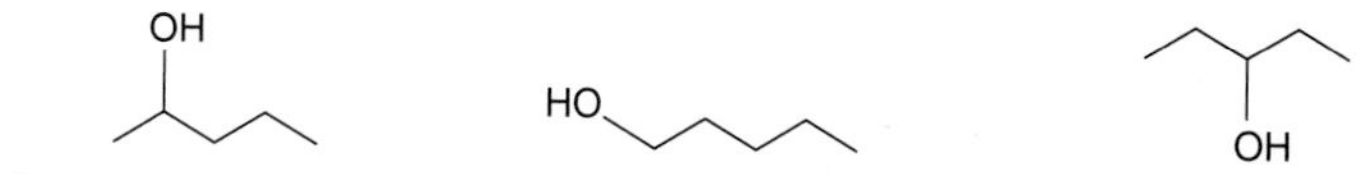

Pentan-2-ol (2-pentanol) Pentan-1-ol (1-pentanol) Pentan-3-ol (3-pentanol)

Functional Isomers Functional isomers are compounds with the same molecular formula that have the atoms connected in a different way, so that the functional groups are different. Examples of functional isomers of a compound with molecular formula C_3H_8O are as follows:

Methoxyethane (ethyl methyl ether) Propan-2-ol (2-propanol, isopropyl alcohol, or isopropanol)

Stereoisomers

Stereoisomers are isomeric molecules that have the same molecular formula and sequence of bonded atoms, differing only in the three-dimensional orientation of the atoms in space. Stereoisomers can be subdivided into various subfamilies, and the first question we need to ask is, "Do bonds need to be broken to interconvert by σ bond rotation?" If the answer is yes, then we continue comparing the molecules to determine what kind of stereoisomers they are. Remember that whenever we talk about isomers, we need to consider at least two molecules.

Conformational Isomers As we studied in the previous section, **conformational isomers**, also called conformers, rotational isomers, or rotamers, are molecules that can often interconvert rapidly into each other by rotation about a single bond at room temperature. Different conformations can have different energies and *can be isolated very rarely*. Those conformers that can be isolated (usually at very low temperature) are considered stereoisomers. Conformational isomers are thus distinct from the other classes of stereoisomers, for which the process of interconversion involves breaking and reforming chemical bonds. The activation energy required to interconvert conformers is the rotational barrier, or barrier to rotation. To determine if two compounds are conformational isomers, we could use the sawhorse projection, the Newman projection, or the Fischer projection discussed in Section 20.3. For example, the following conformers, when isolated, are stereoisomers. They interconvert to each other at room temperature by rotation about a single bond (the sequence of the substituents attached to the front and the back carbon is the same):

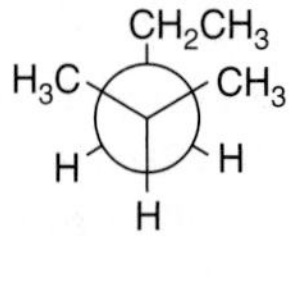

Configurational Isomers Configurational isomers are stereoisomers that cannot be interconverted at room temperature by rotation around a σ bond. If they have different spatial arrangements of the atoms/groups attached to a double bond, or if the cycloalkanes have two or more substituents attached to the ring on opposite sides, they are named *geometric stereoisomers*; if the arrangement is different around a single bond, they are named *enantiomers* when they are mirror images and nonsuperimposable on one another or *diastereomers* if not mirror images.

Geometric isomers differ in the geometric arrangement of the groups attached to the double bond or to the ring. Let us take a look at the but-2-ene (2-butene) molecule. We could draw it in two different ways. If the two larger groups (methyl groups) are on the same side of the double bond, the isomer receives the prefix *cis*, and if they are on opposite sides, the prefix is *trans*. The two isomers clearly have the same structural framework, but they differ in the arrangement of this framework in space. Therefore, they are stereoisomers.

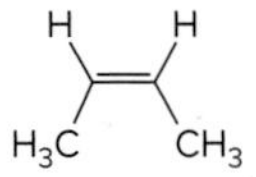

cis-but-2-ene

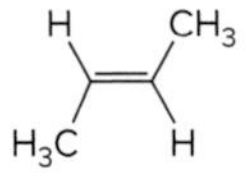

trans-but-2-ene

Since the double bond is rigid, there is no free rotation about this bond. Therefore, the geometric isomers do not interconvert without breaking the double bond, and they exist as different compounds, each with its own chemical and physical properties.

Note: There will *not* be *cis-trans* isomers of alkenes in which one end of the double bond carries identical groups; for example, but-1-ene (1-butene) does not have geometric isomers:

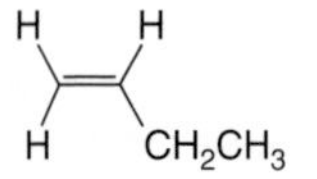

is identical to
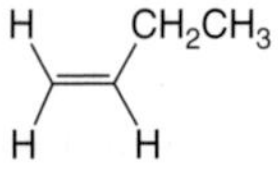

But-1-ene

Cis and *trans* cycloalkanes cannot be interconverted into each other without breaking a σ bond. Molecules like *cis*-1,2-dimethylcyclohexane and *trans*-1,2-di-methylcyclohexane are also geometric isomers:*

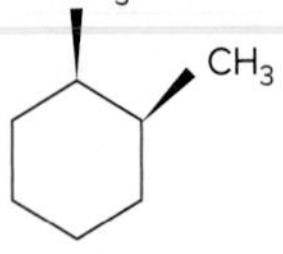

cis-1,2-dimethylcyclohexane

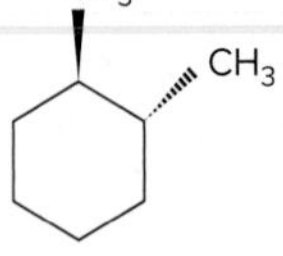

trans-1,2-dimethylcyclohexane

Enantiomers are stereoisomers that are mirror images of each other. Furthermore, the molecules are *nonsuperimposable* on one another. For example, let us consider the following two molecules of 2-bromobutane:

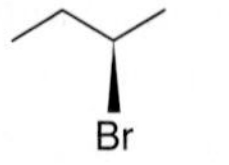

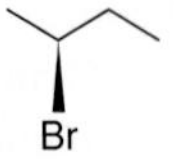

These molecules are mirror images of each other (enantiomers). When the molecule on the right is flipped 180°, the stereochemistry is different.

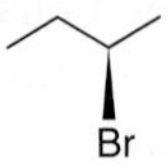

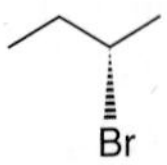

*Note that these and the following molecules do not show the H atoms.

In the molecule on the left, the bromine is oriented in front of the plane of the page, whereas in the molecule on the right, the bromine is oriented behind the plane. These molecules can also be viewed using Newman projections. Notice that these molecules are never identical at any point throughout the rotation around the central carbon bond. Therefore, the molecules are nonsuperimposable:

Diastereomers are pairs of stereoisomers that have opposite configurations at one or more of the stereogenic centres and are not mirror images of each other. For example, consider the following molecules:

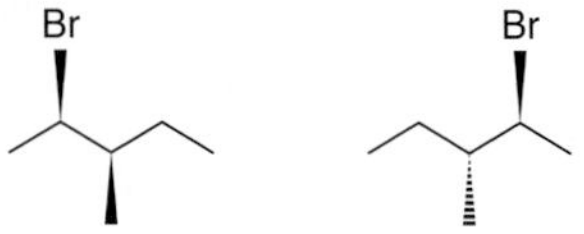

These molecules are not mirror images of each other. Additionally, they are nonsuperimposable: if the molecule on the right is flipped 180°, the stereochemistry is the same at one carbon (C-2) and different at another carbon (C-3).

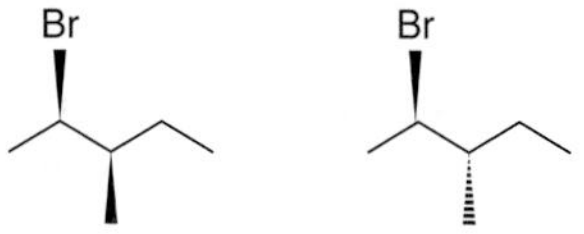

Enantiomers of a compound with more than one stereocentre are also diastereomers of the other stereoisomers that are not their mirror image. Pure samples of enantiomers have identical physical properties (for example, boiling point, density, freezing point), whereas diastereomers have different physical properties and different reactivity. Remember, enantiomers or diastereomers are not necessarily locked into their positions, but they cannot be converted into one another, even by a rotation around a single bond.

A **stereogenic centre (stereocentre)** is an atom where exchanging any two groups leads to a stereoisomer. It is important to recognize the presence of stereogenic centres in a molecule. The following molecules have their stereogenic centres identified with a blue dot. The only stereogenic centre in A is C-2 (it has four different attachments; exchanging any two of these attachments will produce the mirror image). B has only one stereogenic centre (C-3). C-2 is not a stereogenic centre. C has no stereogenic centres. D and E have two stereogenic centres each (where exchanging the two groups on the second stereogenic centre of D produces the *cis* isomer shown in E):

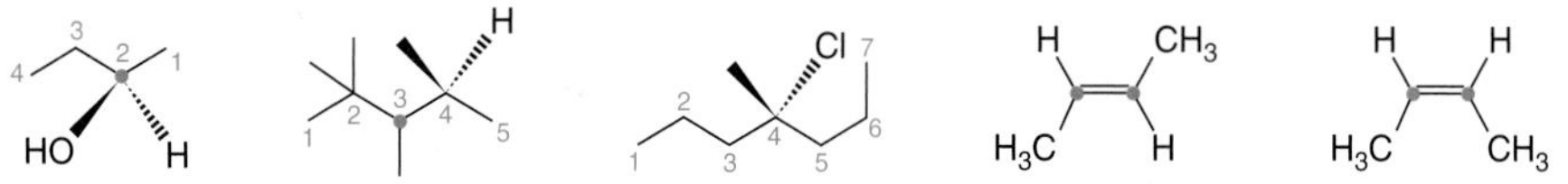

A B C D E

Chiral Molecules and Optical Isomerism

As we have just seen, when two molecules are mirror images of each other and cannot be superimposed, they are enantiomers (also called **optical isomers**). To use a familiar example, your right hand is an optical isomer of your left. Look at your right hand in a mirror: the *image* is identical to your left hand (Figure 20.17). No matter how you twist your arms around, however, your hands cannot lie on top of each other with your palms facing in the same direction and be superimposed. They are not superimposable because each is *asymmetric*: there is no plane of symmetry that divides your hand into two identical parts.

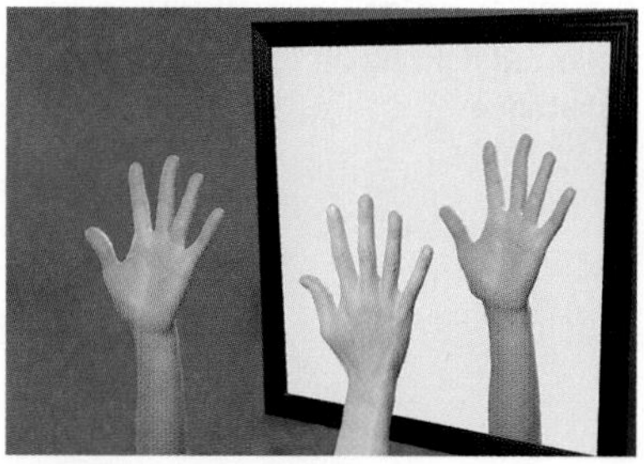

FIGURE 20.17 An analogy for optical isomers

Asymmetry and Chirality An asymmetric carbon is a tetrahedral C atom bonded to four different groups. A trigonal planar carbon, for example, is not an asymmetric C atom, because it is not bonded to four different groups; it is planar and

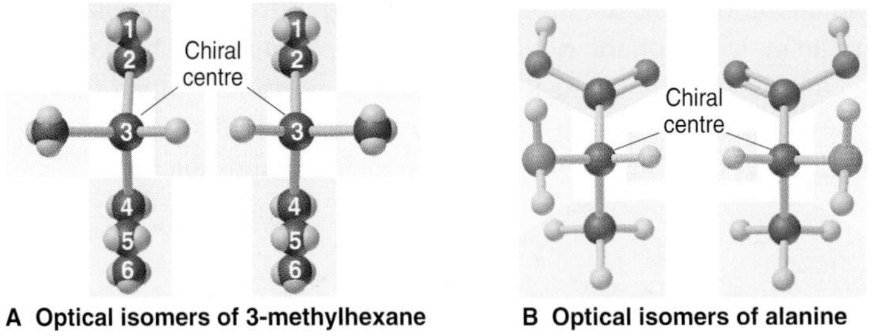

FIGURE 20.18 Two chiral molecules

therefore contains a plane of symmetry. A molecule must have an asymmetric carbon atom to be **chiral** (Greek *cheir*, meaning "hand"). Generally, an organic molecule is chiral if it contains a carbon atom that is bonded to four different groups and is nonsuperimposable on its mirror image. In 3-methylhexane, for example, C-3 is an asymmetric carbon or chiral centre because it is bonded to four different groups: —H, —CH_3, —CH_2—CH_3, and —CH_2—CH_2—CH_3 (Figure 20.18A). Like your two hands, the two forms are mirror images and cannot be superimposed on each other: when two of the groups are superimposed, the other two are opposite each other. Thus, the two forms are enantiomers. The central C atom in the amino acid alanine is also a chiral centre (Figure 20.18B).

Properties of Optical Isomers Unlike constitutional isomers, which have different physical properties such as boiling point, optical isomers are identical in all but two aspects:

1. In their physical properties, optical isomers differ only in the direction that each isomer rotates the plane of polarized light. A **polarimeter** is used to measure the angle that the plane is rotated (Figure 20.19). A beam of light consists of waves that oscillate in all planes. A polarizing filter blocks all waves except those in one plane, so the light emerging through the filter is *plane polarized.* An optical isomer is **optically active** because it rotates the plane of this polarized light. The *dextrorotatory* isomer (designated *d* or +) rotates the plane of light clockwise; the *levorotatory* isomer (designated *l* or −) is the mirror image of the *d* isomer and rotates the plane counterclockwise. An equimolar mixture of the two isomers (called a ***racemic** mixture*) does not rotate the plane of light because the dextrorotation cancels the levorotation. The *specific rotation* is a characteristic, measurable property of an optical isomer at a certain temperature, concentration, and wavelength of light.

2. In their chemical properties, optical isomers differ only in a chiral (asymmetric) chemical environment, one that distinguishes "right-handed" from "left-handed" molecules (chiral molecules and ions have different chemical properties only when they are in chiral environments). As an analogy, your right hand fits well in your right glove but not in your left glove. Typically, in chemical reactions, one isomer of an optically active reactant is added to a mixture of optical isomers of another compound. The products of the reaction have different properties and can be separated. As mentioned before, a mixture of equal parts of two optically active isomers is termed *racemic* and has a net rotation of plane-polarized light of zero.

FIGURE 20.19 The rotation of plane-polarized light by an optically active substance

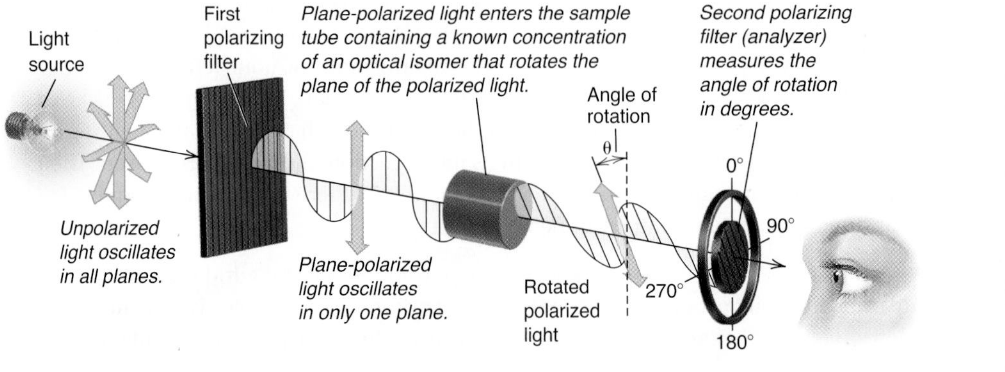

Naming Enantiomers Using the Cahn-Ingold-Prelog Rules

The **Cahn-Ingold-Prelog (CIP) rules** are a set of rules used in organic chemistry to name stereoisomers. They were developed by two British chemists (Robert Sidney Cahn and Sir Christopher Kelk Ingold) and the Croatian Nobel Prize winner Vladimir Prelog. Since each stereogenic centre and each double bond in a molecule gives rise to two possible configurations, the configuration of the entire molecule needs to be specified by including descriptors in its systematic name. The purpose of the CIP system is to assign an *R* or *S* descriptor to the stereogenic centres and an *E* or *Z* descriptor to the double bonds. They are assigned based on a system of ranking priorities:

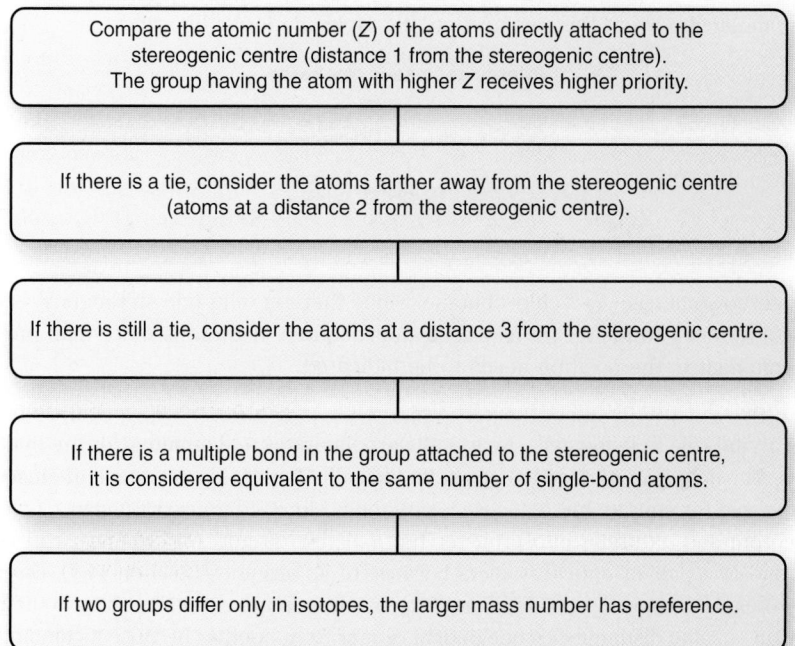

Compare the atomic number (*Z*) of the atoms directly attached to the stereogenic centre (distance 1 from the stereogenic centre). The group having the atom with higher *Z* receives higher priority.

If there is a tie, consider the atoms farther away from the stereogenic centre (atoms at a distance 2 from the stereogenic centre).

If there is still a tie, consider the atoms at a distance 3 from the stereogenic centre.

If there is a multiple bond in the group attached to the stereogenic centre, it is considered equivalent to the same number of single-bond atoms.

If two groups differ only in isotopes, the larger mass number has preference.

The *R*/*S* system is a very important nomenclature system for denoting enantiomers and diastereomers. It labels each stereogenic centre *R* or *S* according to the CIP rules. If the centre is oriented so that the lowest priority of the four substituents is pointed away from a viewer, the viewer will then see two possibilities: if the priority of the remaining three substituents decreases in a clockwise direction, it is labelled *R* (for *rectus*, Latin for "right"); if it decreases in a counterclockwise direction, it is *S* (for *sinister*, Latin for "left"). This system labels each stereogenic centre in a molecule. The *R*/*S* system has no fixed relation to the (+)/(−) system. An *R* isomer can be either dextrorotatory or levorotatory, depending on its exact substituents. The procedure for using the CIP rules for a given stereogenic centre is illustrated with the following enantiomer of 2-chlorobutane (you will want to use molecular models for this process):

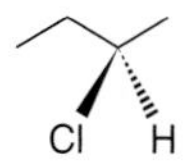

1. Look at the atoms directly attached to the stereogenic centre. Rank them according to decreasing atomic number:

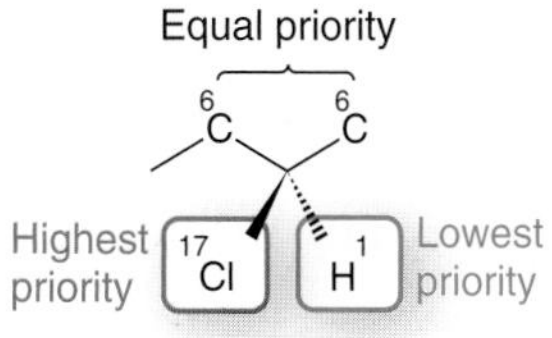

2. If the atoms have the same priority, then move one atom farther out from the stereogenic centre until you find a difference:

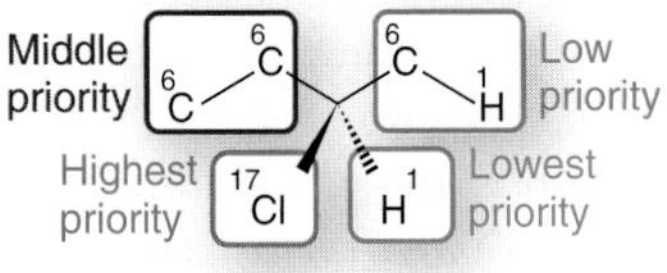

3. Change your view of the molecule so that you are looking along the bond between the stereogenic centre and the lowest priority group, with the lowest priority group facing away from you. If you are using molecular models, use the bond between the stereogenic centre and the lowest priority group as a handle to hold the molecule. From this perspective, the molecule looks like this:

Priority decreases clockwise

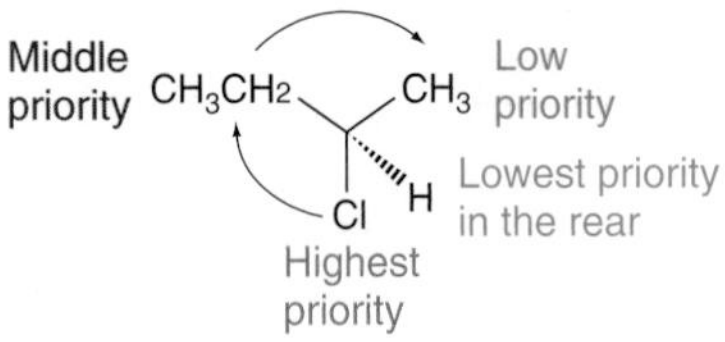

This stereoisomer is (*R*)-2-chlorobutane. Since there is only one stereogenic centre, there is no need to identify it as (2*R*)-2-chlorobutane. If there is more than one stereogenic centre, the position needs to be indicated.

The Role of Optical Isomerism in Organisms and Medicines Optical isomerism plays a vital role in living cells. Nearly all carbohydrates and amino acids are optically active, but only one of the isomers is biologically usable. For example, *d*-glucose is metabolized for energy, but *l*-glucose is not and is excreted unused. Similarly, *l*-alanine is incorporated naturally into proteins, but *d*-alanine is not. An organism can utilize only one of a pair of optical isomers because of its enzymes (Section 14.7). Enzymes are proteins that speed virtually every reaction in a living cell by binding to the reactants; an enzyme distinguishes one optical isomer from another because its binding site is chiral (asymmetric) (Figure 20.20). The shape of one optical isomer fits at the binding site, but the mirror image shape of the other isomer does not fit, so it cannot bind.

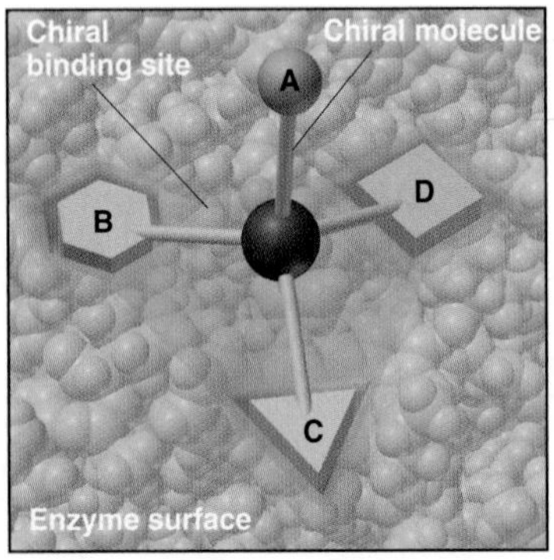

FIGURE 20.20 The binding site of an enzyme

Many drugs are chiral molecules. One optical isomer has certain biological activity, and the other has either a different type of activity or none at all. Naproxen (*see margin figure*), a pain reliever and anti-inflammatory agent, is an example: one isomer is active as an anti-arthritic agent, and the other is a potent liver toxin that must be removed from the mixture during synthesis. The notorious drug thalidomide is another example: one optical isomer is active against depression, whereas the other causes fetal mutations and deaths. This drug as the racemic mixture was available as a sample tablet in the late 1950s and sold in the early 1960s to mothers who had it prescribed to relieve "morning sickness" during pregnancy, tragically causing limb malformations in many newborns. It was also found that thalidomide's enantiomers can interconvert in vivo; that is, if a human is given a pure optical isomer, both isomers will later be found in the body.

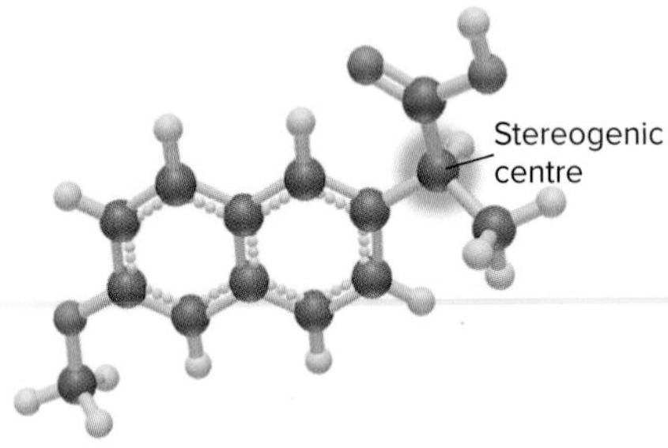

Naproxen (Aleve)

Geometric Isomers and the Chemistry of Vision The difference in properties between geometric isomers has profound effects in biological systems. For example, the first step in the sequence of events that allows us to see relies on the different shapes of a pair of geometric isomers. *Retinal*, a 20-carbon compound consisting of a 15-carbon chain with five 1-C branches and five C=C bonds, is part of the molecule responsible for receiving light energy that is converted to electrical signals that are transmitted to the brain. There are two biologically occurring isomers of retinal, which have very different shapes. The all-*trans* isomer has a *trans* orientation around all five double bonds and is elongated. The 11-*cis* isomer has a *cis* orientation around the bond between C-11 and C-12, which results in a bend in the structure at that double bond.

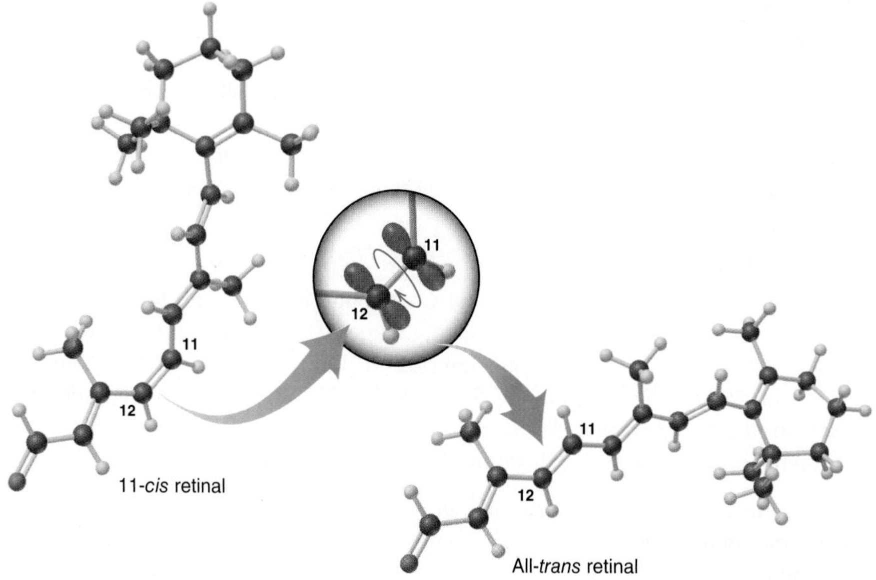

FIGURE 20.21 The initial chemical event in vision and the change in the shape of retinal

Certain cells of the retina are densely packed with *rhodopsin*, a large protein covalently bonded to 11-*cis* retinal. The initial chemical event in vision occurs when rhodopsin absorbs a photon of visible light. The energy range of visible photons (165–293 kJ/mol) encompasses the energy needed to break a C=C π bond (about 250 kJ/mol). Within a few millionths of a second after rhodopsin absorbs a photon, the 11-*cis* π bond of retinal breaks, the intact σ bond between C-11 and C-12 rotates, and the π bond re-forms to produce all-*trans* retinal (Figure 20.21).

The rapid and significant change in shape of retinal causes the attached protein to change its shape as well, triggering a flow of ions into the retina's cells and initiating electrical impulses, which the optic nerve conducts to the brain. Meanwhile, the free all-*trans* retinal diffuses away from the protein and is changed back to the *cis* form, which then binds to the protein portion again. Because of the speed and efficiency with which light causes such a large structural change in retinal, natural selection has made it the photon absorber in organisms as different as purple bacteria, molluscs, insects, and vertebrates.

SUMMARY OF SECTION 20.4

- Isomers are compounds with the same molecular formula but different properties.
- Constitutional (structural) isomers have different atom arrangements.
- Stereoisomers have the same arrangement of atoms, but their atoms are oriented differently in space.
- The stereogenic centre (also called stereocentre) is an atom bearing groups such that an interchanging of any two groups leads to a stereoisomer.
- An asymmetric carbon atom is a carbon atom bonded to four different identities.
- Conformational isomers can interconvert rapidly into each other at room temperature by rotation about a single bond and can only be isolated very rarely.
- Enantiomers are optical isomers that are mirror images of one another and cannot be superimposed on each other. They have identical physical and chemical properties, except in their rotation of plane-polarized light and their reaction with chiral reagents.
- Diastereomers are isomers that have opposite configurations at one or more of the stereogenic centres and are not mirror images of each other.
- Geometric (*cis-trans*) isomers have groups oriented differently around a C=C bond or to the ring, which restricts rotation.
- The Cahn-Ingold-Prelog (CIP) rules are used in organic chemistry to name stereoisomers.
- Light converts a *cis* isomer of retinal to the all-*trans* form, which initiates the process of vision.

20.5 Hydrocarbons with Multiple Bonds

Alkenes, alkynes, and aromatic compounds are hydrocarbons with one or more π bonds between adjacent carbon atoms. In this section, we will learn how to name and recognize these unsaturated hydrocarbons. A large number of compounds are possible, since there can be other substituents attached to these molecules instead of hydrogen atoms.

Alkenes: Hydrocarbons with Double Bonds

■ Alkenes have diverse applications in industry. They are used as starting materials in the production of plastics, detergents, fuels, and alcohols. Ethene, propene, and 1,3-butadiene are the most important alkenes in the chemical industry.

A hydrocarbon that contains at least one C=C bond is called an **alkene**. With two H atoms removed to make the double bond, alkenes have the general formula C_nH_{2n}. The double-bonded C atoms are sp^2 hybridized. Because their carbon atoms are bonded to less than the maximum of four atoms each, alkenes are considered to be **unsaturated hydrocarbons.** ■

Alkene names differ from those of alkanes in two respects:

1. The main chain (root) *must* contain both C atoms of the double bond, even if it is not the longest chain. The chain is numbered from the end *closer* to the C=C bond, and the position of the bond is indicated by the number of the *first* C atom in it.

2. When more than one double bond is present, the compound is named as a diene, a triene, or an equivalent prefix indicating the number of double bonds, and each double bond is assigned a locator number.

The suffix for alkenes is *-ene.*

For example, there are three four-carbon alkenes (C_4H_8), two unbranched and one branched (see Sample Problem 20.1b). The branched isomer is 2-methylpropene; the unbranched isomer with the C=C bond between C-1 and C-2 is but-1-ene; the unbranched isomer with the C=C bond between C-2 and C-3 is but-2-ene. As you will see next, there are two isomers of but-2-ene, but they are of a different sort (discussed briefly in Section 20.4).

Select the **longest** carbon chain containing the C=C bond. Replace the "**ane**" ending of alkane with "**ene**."

Number the alkene chain **beginning** at the end **nearer** the **C=C** bond.

The **position of the bond** is indicated by the number of the **first C** atom in it.

The C=C Bond and Geometric (*cis-trans* and *E/Z*) Isomerism There are two major structural differences between alkanes and alkenes.

- Alkanes have a *tetrahedral* geometry (bond angles of ~109.5°) around each C atom, whereas the double-bonded C atoms in alkenes are *trigonal planar* (~120°).
- The C—C bond *allows* rotation of bonded groups, so the atoms in an alkane continuously change their relative positions; in contrast, the π bond of the alkene C=C bond *restricts* rotation, so the relative positions of the atoms attached to the double bond are fixed.

This rotational restriction leads to the geometric isomers (also called *cis-trans* isomers), which have different orientations of groups around a double bond (or similar structural feature). Table 20.4 shows the two geometric isomers of but-2-ene (2-butene) (see also Comment, Sample Problem 20.1). One isomer, *cis*-but-2-ene, has the CH_3 groups on the *same* side of the C=C bond, whereas the other isomer, *trans*-but-2-ene, has them on *opposite* sides of the C=C bond. In general, the *cis* isomer has the *larger portions of the main chain* (in this case, two CH_3 groups) *on the same side* of the double bond, and the *trans* isomer has them on opposite sides. For a molecule to have geometric isomers, *each C atom in the C=C bond must be bonded to two different groups.*

There is also another method for naming alkenes, which is the *E* and *Z* nomenclature (*Z*, from German *zusammen* for "together," and *E*, *entgegen* for "opposite").

Whereas the *cis-trans* style is based on the longest chain, the *E*/*Z* style is based on a set of priority groups. If there is more than one double bond that can be *E*/*Z*, then the location needs to be included with the locant, for example, (2*E*, 4*Z*):

E *Z*

To assign the group priority, we use the CIP rules:

1. Look at the atoms directly attached to each carbon of the double bond. Rank them according to decreasing atomic number. Since I > Br > Cl > S > F > O > N > C > H, 2-chlorobut-2-ene, for example, has two geometric isomers:

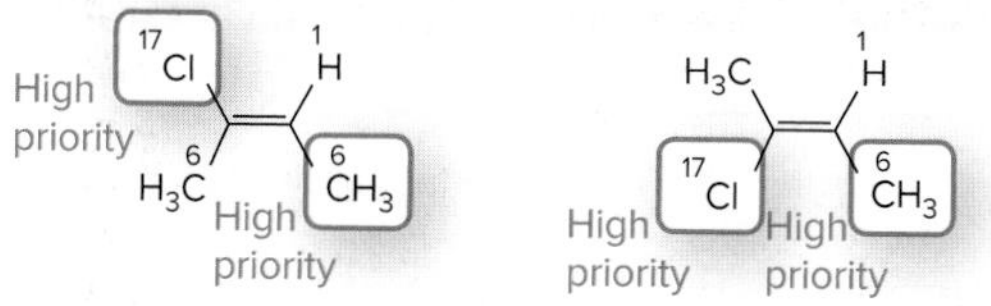

(*E*)-2-chlorobut-2-ene (*Z*)-2-chlorobut-2-ene

2. If the two atoms attached to the double bond are identical, look at all the atoms directly attached to the identical atoms in question. Assign priorities to all these atoms based on atomic numbers, and compare the highest priority atoms. If a difference cannot be found, move out to the next highest priority group and repeat the process:

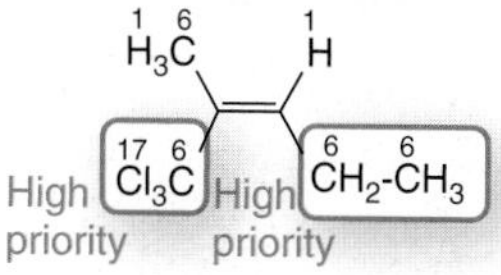

3. Multiple bonds are considered equivalent to the same number of single-bond atoms:

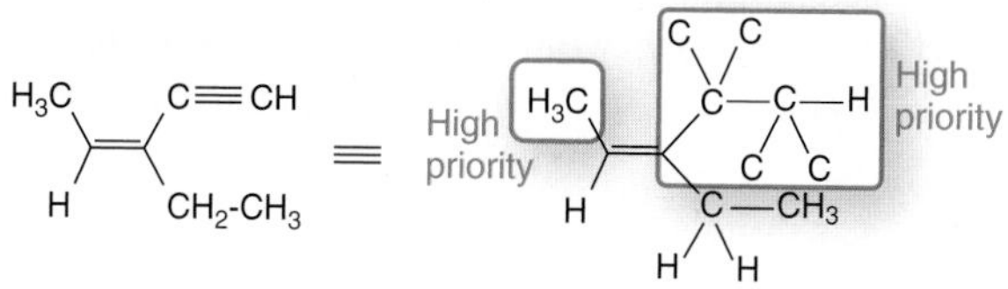

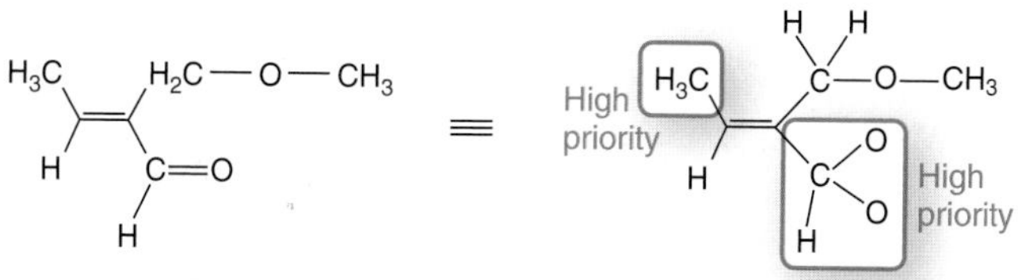

Like structural isomers, geometric isomers have different physical properties. Note in Table 20.4 that the two but-2-enes differ in molecular shape *and* physical properties. The *cis* isomer has a bend in the chain that the *trans* isomer lacks. In Chapters 9 and 11, you saw how such a difference affects molecular polarity and physical properties, which arise from differing strengths of intermolecular attractions.

TABLE 20.4 Geometric Isomers of But-2-ene

Systematic Name	Condensed and Skeleton Formulas	Space-Filling Model	Density (g/mL)	Boiling Point (°C)
cis-but-2-ene	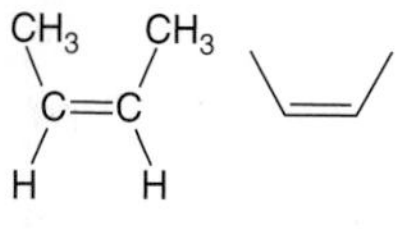	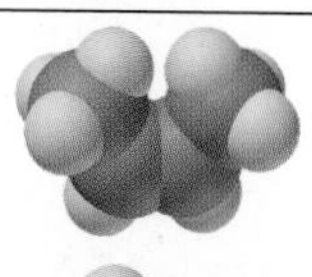	0.621	3.7
trans-but-2-ene	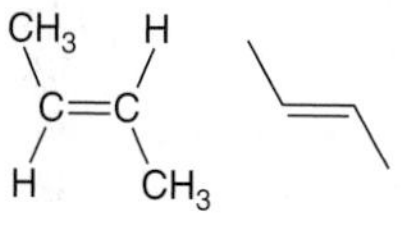	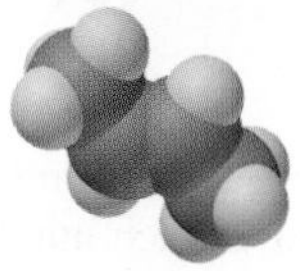	0.604	0.9

Alkynes: Hydrocarbons with Triple Bonds

A hydrocarbon that contains at least one C≡C bond is called an **alkyne**. Its general formula is C_nH_{2n-2} because it has two H atoms fewer than an alkene with the same number of carbon atoms. Because a carbon involved in a C≡C bond can bond to only one other atom, the geometry around each C atom is linear (180°): each C is *sp* hybridized. Alkynes are named in the same way as alkenes, except that the suffix is *-yne*. Because of their localized π electrons, C=C and C≡C bonds are electron rich and act as functional groups. Thus, alkenes and alkynes are much more reactive than alkanes.

Select the **longest** carbon chain containing the C≡C bond. Replace the "**ane**" ending of alkane with "**yne.**" — Number the alkyne chain **beginning** at the end **nearer** the C≡C bond. — The **position of the bond** is indicated by the number of the **first C** atom in it.

A propane/air flame burns at approximately 2000°C, a propane/oxygen flame burns at around 2500°C, and an ethyne/oxygen flame burns at around 3500°C.

For example,

$HC{\equiv}C{-}CH_2{-}CH_2{-}CH_3$ Pent-1-yne

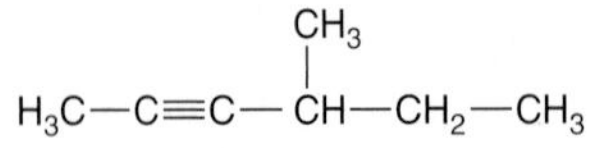

4-methylhex-2-yne

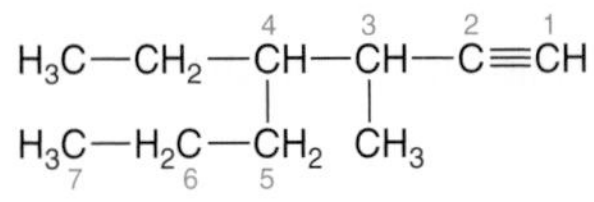

4-ethyl-3-methylhept-1-yne

Sample Problem 20.3 Naming Alkanes, Alkenes, and Alkynes

Problem Give the systematic name for each of the following, indicate the stereogenic centre in part (d), and draw two geometric isomers for part (e):

(a)

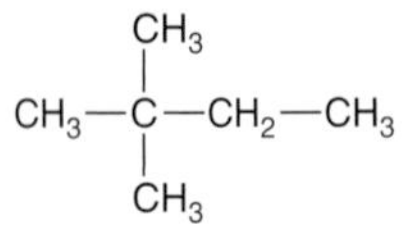

(b)

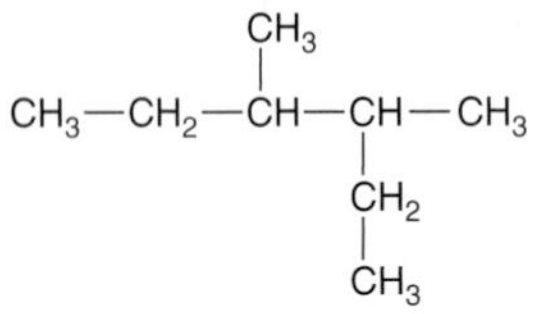

(c)

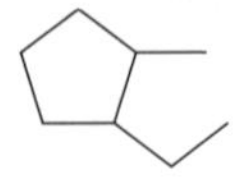

(d)

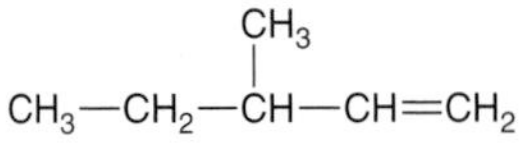

(e)

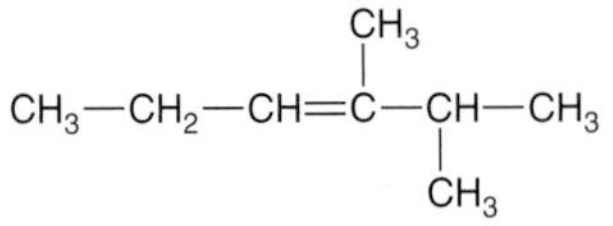

Plan For (a) to (c), we refer to Table 20.2. We first name the longest chain (*root-*) and add the suffix *-ane* because there are only single bonds. Then we find the *lowest* branch numbers by counting C atoms from the end *closer* to a branch. Finally, we name each branch (*root-* + *-yl*) and put the names alphabetically before the root name. For (d) and (e), the longest chain that *includes* the multiple bond is numbered from the end closer to it. For (d), the stereogenic centre is the C atom bonded to four different groups. In (e), the *cis* isomer has larger groups on the same side of the double bond, and the *trans* isomer has them on opposite sides.

Solution

(a)

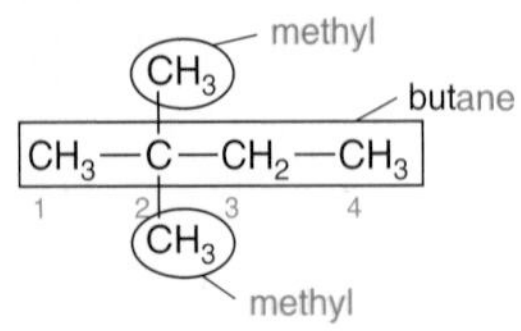

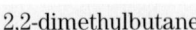

2,2-dimethylbutane

When a type of branch appears more than once, we group the branch numbers and indicate the number of branches with a prefix, as in 2,2-*di*methyl.

(b)

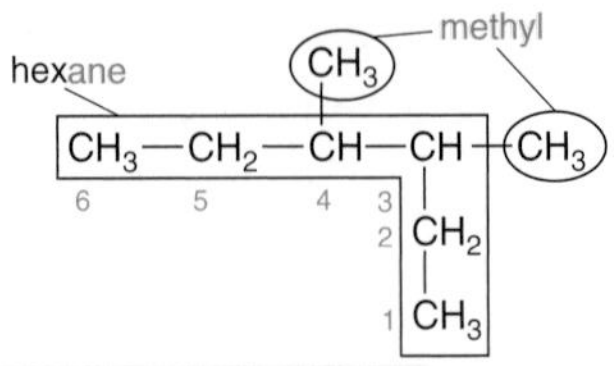

3,4-dimethylhexane

In this case, we can number the chain from either end because the branches are the same and are attached to the two central C atoms.

(c)

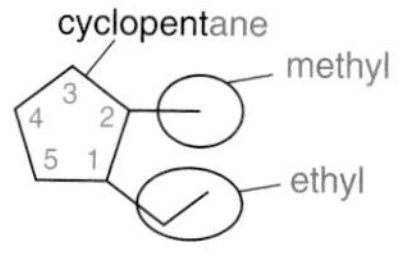

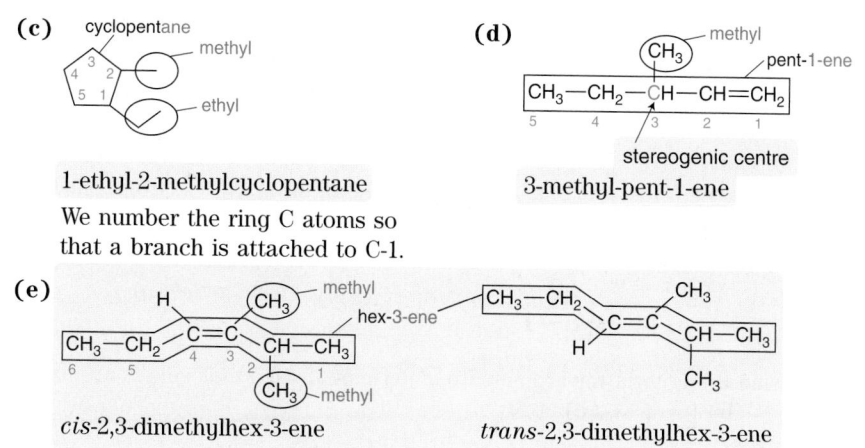

1-ethyl-2-methylcyclopentane

We number the ring C atoms so that a branch is attached to C-1.

(d)

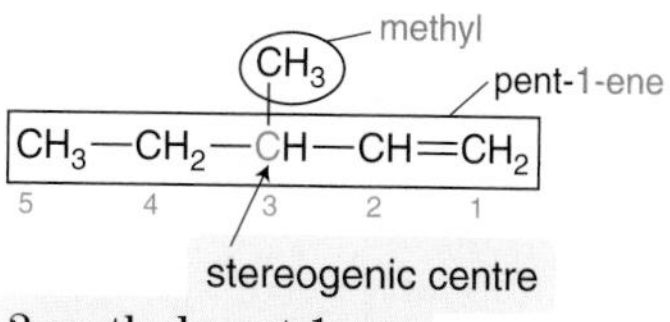

3-methyl-pent-1-ene

(e)

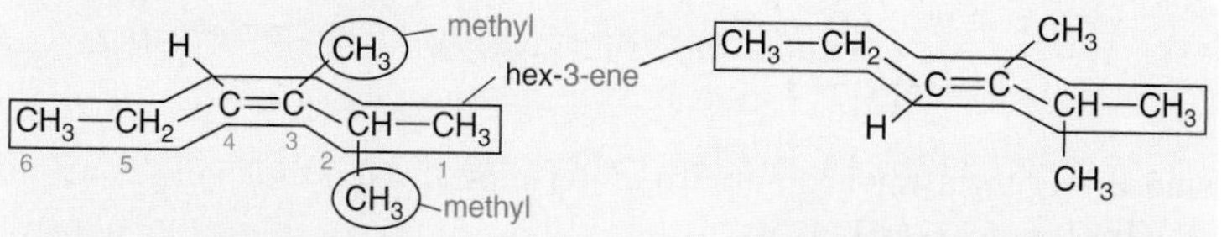

cis-2,3-dimethylhex-3-ene *trans*-2,3-dimethylhex-3-ene

Check A good check (and excellent practice) is to reverse the process by drawing structures for the names to see if you come up with the structures given in the problem.

Comment **1.** In part (b), C-3 and C-4 are asymmetric carbon atoms, as are C-1 and C-2 in part (c). However, in part (b), the molecule is not chiral: it has a plane of symmetry between C-3 and C-4, so each half of the molecule rotates light in opposite directions. **2.** Avoid these common mistakes: In part (b), 2-ethyl-3-methylpentane is wrong; the longest chain is *hexane*. In part (c), 1-methyl-2-ethylcyclopentane is wrong; the branch names should appear *alphabetically*.

Follow-Up Problem 20.3 Draw condensed formulas for each compound: **(a)** 3-ethyl-3-methyloctane; **(b)** 1-ethyl-3-propylcyclohexane (also draw a carbon-skeleton formula for this compound); **(c)** 3,3-diethylhex-1-yne; **(d)** *trans*-3-methylhept-3-ene.

Aromatic Hydrocarbons: Cyclic Molecules with Delocalized π Electrons

Unlike the cycloalkanes, **aromatic hydrocarbons** are planar molecules, usually with one or more rings of six C atoms, and are often drawn with alternating single and double bonds. As you learned for benzene (Section 8.6), however, all the ring bonds are identical, with values of length and strength *between* those of a C—C and a C=C bond. To indicate this, benzene is also shown as a resonance hybrid, with a circle (or dashed circle) representing the delocalized character of the π electrons (Figure 20.22A). An orbital picture shows the two lobes of the delocalized π cloud above and below the hexagonal plane of the π-bonded C atoms (Figure 20.22B). When deciding if a compound is aromatic, check if it follows the following criteria (if it does not meet one of them, it is likely not aromatic):

- The molecule is cyclic.
- The molecule is planar (all atoms lie in the same plane).
- The molecule is fully conjugated, having *p* orbitals at every atom in the ring.
- It follows **Hückel's rule**: the molecule has to have $4n + 2$ π electrons, where $n = 0$ or any positive integer.

To apply the $4n + 2$ rule, first count the number of π electrons in the molecule. Then set this number equal to $4n + 2$ and solve for n. If n turns out to be 0 or any positive integer (1, 2, 3, . . .), then the rule has been met. For example, benzene has six

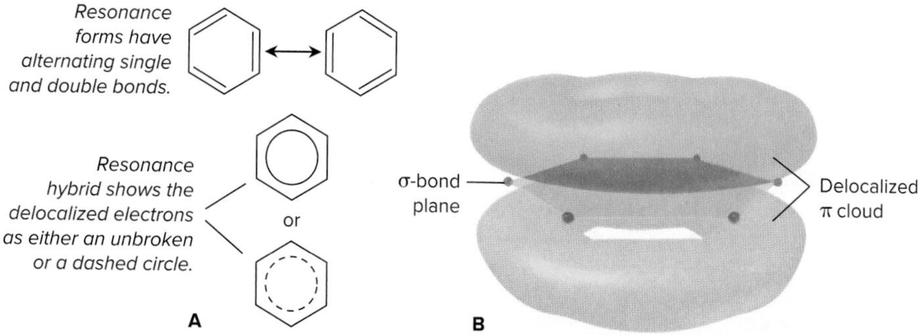

FIGURE 20.22 Representations of benzene

(delocalized) π electrons; therefore, $4n + 2 = 6$. Solving for n, we find $n = 1$. According to Hückel's molecular orbital (MO) theory, a compound is particularly stable if all of its bonding MOs are filled with paired electrons. With aromatic compounds, two electrons fill the lowest energy MO, and four electrons fill each subsequent energy level, leaving all bonding orbitals filled and no anti-bonding orbitals occupied, as we saw in the MO diagram for benzene (Figure 10.27). This gives a total of $4n + 2$ π electrons. Hückel's rule can only be theoretically justified for monocyclic systems.

Sample Problem 20.4 Predicting Which Molecule Is Aromatic

Problem Using the criteria for aromaticity, determine if each molecule is aromatic:

(a)  **(b)** 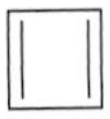 **(c)** 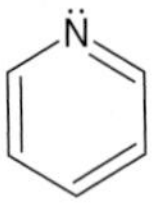

Plan For each molecule, we check that in addition to being a cyclic structure, it follows the other three criteria for being aromatic: Is the molecule planar? Do all the atoms in the ring have a p orbital? Does the molecule follow Hückel's rule?

Solution

(a) Not aromatic. The top C is sp^3 hybridized. Since the molecule is not fully conjugated, it is not aromatic.

(b) Not aromatic. The molecule is fully conjugated. Because it has two double bonds, we can tell that there are four π electrons. We apply Hückel's rule and solve for n:

$$4n + 2 = \text{number of } \pi \text{ electrons}$$
$$4n + 2 = 4$$
$$4n = (4 - 2)$$
$$4n = 2$$
$$n = \frac{1}{2}$$

Since n is not an integer, the molecule is not aromatic.

(c) Aromatic. The molecule is fully conjugated, and since the N atom is using its one p orbital for the electrons in the double bond, its lone pair of electrons are not π electrons; therefore, the total number of π electrons is 6. Applying Hückel's rule, we find that $n = 1$. Since the molecule follows all criteria for aromaticity, we conclude that it is aromatic.

Follow-Up Problem 20.4 Determine if each molecule is aromatic:

(a)  **(b)** 

(c) 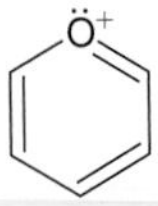

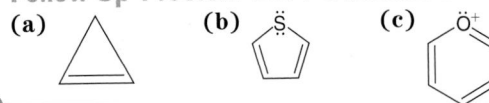

The systematic naming of simple aromatic compounds is quite straightforward. Usually, benzene is the parent compound, and attached groups, or *substituents*, are named as prefixes. However, many common names are still in use. For example, benzene with one methyl group attached is systematically named *methylbenzene* but is better known by its common name, *toluene*. With only one substituent present in toluene, we do not number the ring C atoms; when two or more groups are attached, however, we number so that one of the groups is attached to ring C-1:

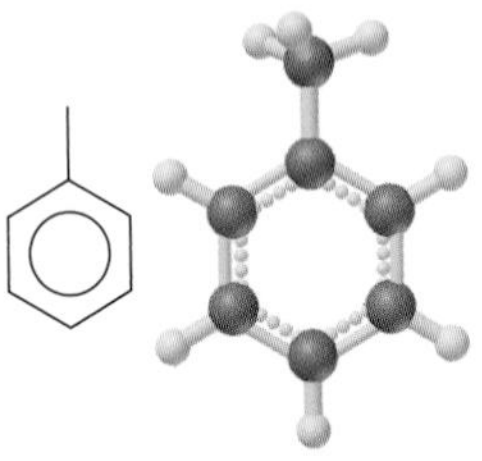

Methylbenzene
(toluene)
bp = 110.6°C

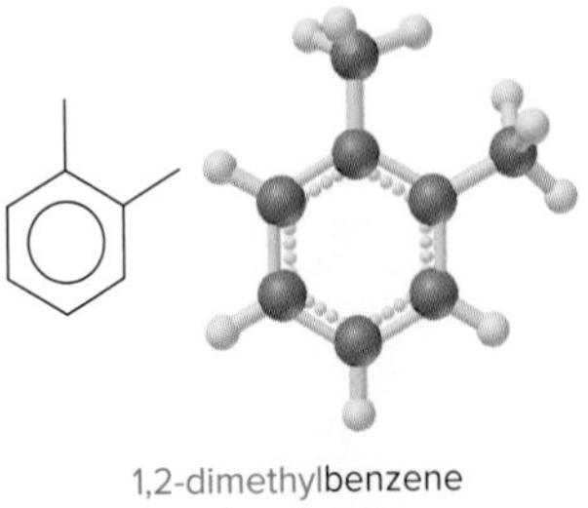

1,2-dimethylbenzene
(*o*-xylene)
bp = 144.4°C

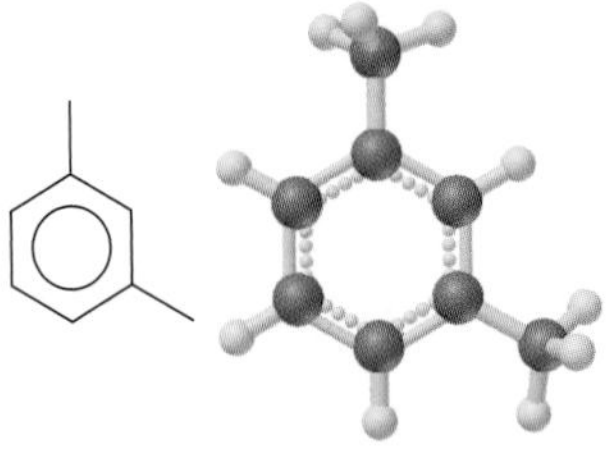

1,3-dimethylbenzene
(*m*-xylene)
bp = 139.1°C

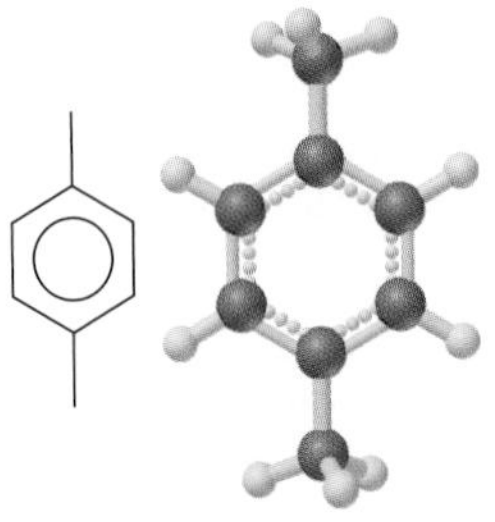

1,4-dimethylbenzene
(*p*-xylene)
bp = 138.3°C

In common names, the positions of two groups are indicated by *o*- (*ortho*) for groups on adjacent ring C atoms, *m*- (*meta*) for groups separated by one ring C atom, and *p*- (*para*) for groups on opposite ring C atoms. The following are isomers of dimethylbenzene, showing the substituents in different positions:

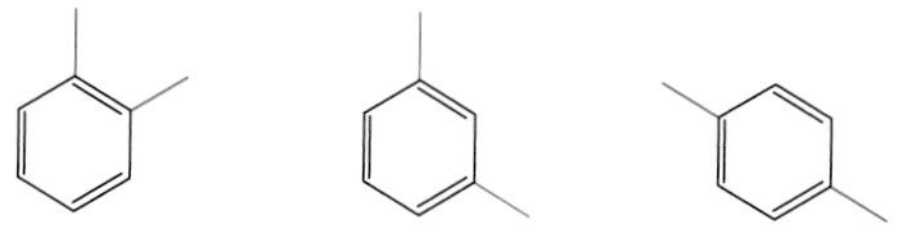

ortho (*o*-) *meta* (*m*-) *para* (*p*-)

The dimethylbenzenes above (commonly known as *xylenes*) are important solvents and feedstocks for polyester fibres and dyes. Benzene and many other aromatic hydrocarbons have been shown to have carcinogenic (cancer-causing) activity.

The number of isomers increases with more than two attached groups. For example, there are six isomers for a compound with one methyl and three nitro (—NO_2) groups attached to a benzene ring; the explosive TNT, shown below, is one of the isomers:

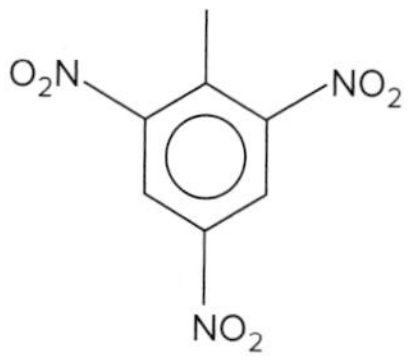

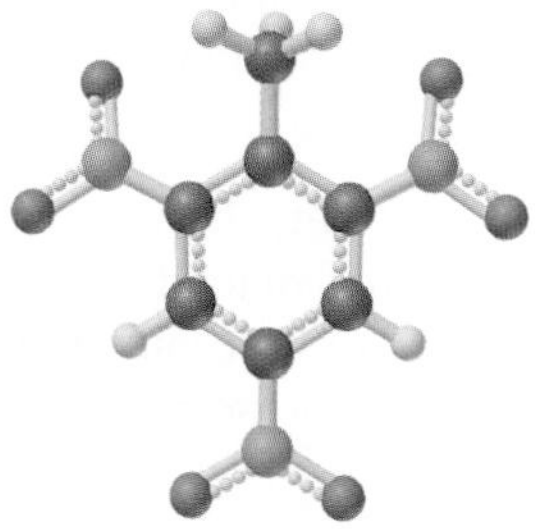

2,4,6-trinitromethylbenzene
(2,4,6-trinitrotoluene, TNT)

Methods for determining the structures of organic molecules are discussed in Section 22.5.

Variations on a Theme: Catenated Inorganic Hydrides

In brief discussions called Variations on a Theme in the next section, we examine similarities between organic and inorganic compounds. From this perspective, you will see that the behaviour of carbon is remarkable, but not unique, in the chemistry of the elements.

Although no other element has as many different hydrides as carbon, catenated hydrides occur for several other elements, and many ring, chain, and cage structures are known:

- *Boranes.* One such group of inorganic compounds is the boron hydrides, or boranes (Section 13.5). Although their varied shapes rival those of the hydrocarbons, the weakness of their unusual bridge bonds renders most of them thermally and chemically unstable.
- *Silanes.* An obvious structural similarity exists between alkanes and the silicon hydrides, or silanes. Silanes even have an analogous general formula ($Si_nH_{2n}+_2$). Branched silanes are known, but no cyclic or unsaturated (containing an Si═Si bond) compounds had been prepared until very recently. Unlike alkanes, silanes are thermally unstable and ignite spontaneously in air.
- *Polysulfanes.* Sulfur's ability to catenate is second only to carbon's, and many chains and rings occur among its allotropes (Section 13.8). A number of sulfur hydrides, or polysulfanes, are known. However, these molecules are unbranched chains with H atoms only at the ends (H—S_n—H). Like the silanes, the polysulfanes are oxidized easily and decompose readily, yielding sulfur's only stable hydride, H_2S, and its most stable allotrope, cyclo-S_8.

SUMMARY OF SECTION 20.5

- Alkenes (C_nH_{2n}) have at least one C═C bond.
- Alkynes (C_nH_{2n-2}) have at least one C≡C bond.
- Aromatic hydrocarbons have at least one planar ring with delocalized π electrons.
- Boron, silicon, and sulfur also form catenated hydrides, but these are unstable.

20.6 Common Functional Groups

The central organizing principle of organic chemistry is the *functional group*. The distribution of electron density in a functional group is a major factor in the reactivity of the compound, as we will study in Chapter 21. The electron density can be high, as in the C═C and C≡C bonds, or it can be low at one end of a bond and high at the other, as in the C—Cl and C—O bonds. Such electron-rich or polar bonds enhance the oppositely charged pole in the other reactant. As a result, the reactants attract each other and begin a sequence of bond-forming and bond-breaking steps that lead to a product. Thus, *the intermolecular forces that affect physical properties and solubility also affect reactivity*. Table 20.5 lists some of the important functional groups in organic compounds.

When we classify functional groups by bond order (single, double, and so forth), they tend to follow certain patterns of reactivity:

- Functional groups with only single bonds undergo elimination or substitution.
- Functional groups with double or triple bonds undergo addition.
- Functional groups with both single and double bonds undergo substitution and addition.

In this section, we will learn how to name compounds that contain each functional group and examine some of their common properties and uses. In Chapter 21, we will study substitution, elimination, and addition reactions, as well as other common reactions for organic compounds.

Peppermint oil is often used in aromatherapy. It is extracted from the leaves of the peppermint plant (*Mentha pipertita*). It has the following structure:

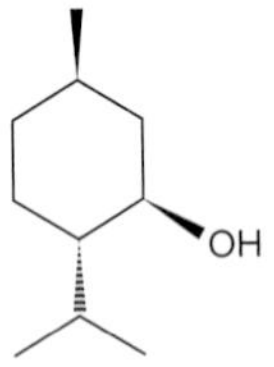

Functional Groups with Only Single Bonds

The most common functional groups with only single bonds are alcohols, ethers, haloalkanes, and amines.

Alcohols The **alcohol** functional group consists of a carbon bonded to a hydroxyl (—OH) group, —C—Ö—H, and the general formula for an alcohol is R–OH. Alcohols are named by dropping the final *-e* from the parent hydrocarbon name and adding the suffix *-ol*. Thus, the two-carbon alcohol is ethanol (ethan- + -ol). The common name is the hydrocarbon *root-* + *-yl*, followed by "alcohol"; thus, the common name of ethanol is ethyl alcohol. (This substance, obtained from fermented grain, has been consumed by people as an intoxicant in beverages since ancient times.)

Select the **longest** carbon chain containing the hydroxyl group and replace the "**e**" ending of alkane with "**ol**." → Number the alkane chain **beginning** at the end **nearer** the **OH** group. → Number the substituents according to their **positions** on the chain (listed in **alphabetical** order).

If the hydroxyl group is attached to a primary carbon (the one at the end of the chain), the alcohol is also named *primary* (1°). *Secondary* (2°) alcohols have the hydroxyl group attached to a secondary carbon, and *tertiary* (3°) alcohols have it attached to a tertiary carbon. There are no quaternary alcohols. When the hydroxyl group is attached to a benzene ring, the compound (hydroxybenzene or benzenol) is named *phenol*. The following are some examples of aliphatic and aromatic alcohols:

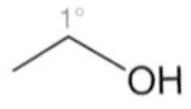

Ethanol (ethyl alcohol) is a primary alcohol.

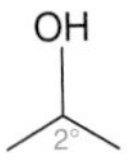

Propan-2-ol (isopropanol or rubbing alcohol) is a secondary alcohol.

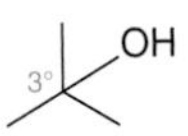

Methylpropan-2-ol (tertbutyl alcohol) is a tertiary alcohol.

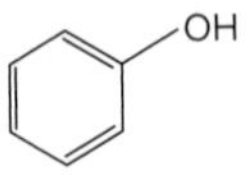

Phenol is an aromatic alcohol.

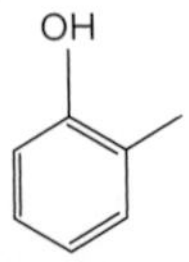

2-methylphenol (o-cresol) is an aromatic alcohol.

TABLE 20.5 Important Functional Groups in Organic Compounds

Functional Group	Compound Type	Prefix or Suffix of Name	Example: Lewis Structure	Example: Ball-and-Stick Model	Example: Systematic Name (Common Name)
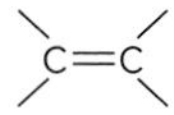	Alkene	-ene	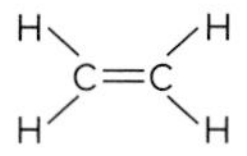	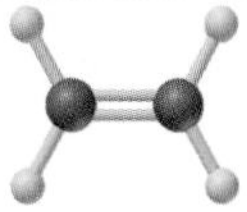	Ethene (ethylene)
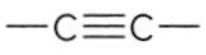	Alkyne	-yne			Ethyne (acetylene)
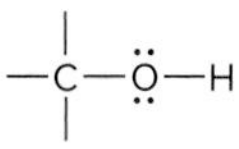	Alcohol	-ol	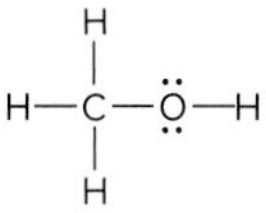	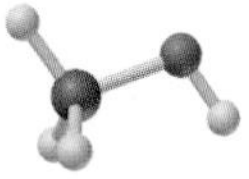	Methanol (methyl alcohol)
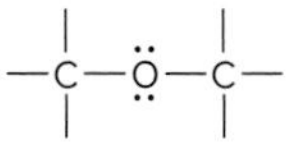	Ether	-oxy	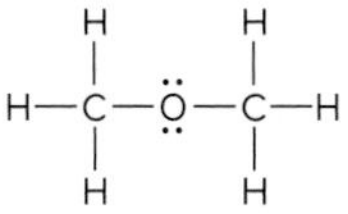	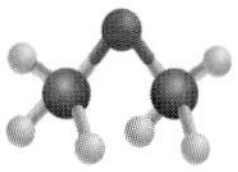	Methoxymethane (dimethyl ether)
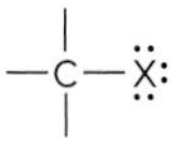 (X = halogen)	Haloalkane	halo-	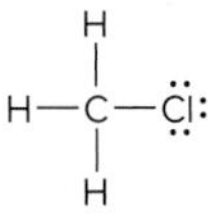	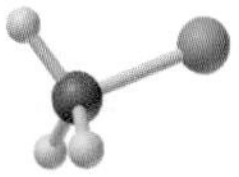	Chloromethane (methyl chloride)
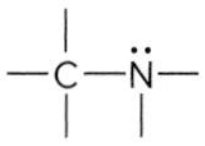	Amine	-amine	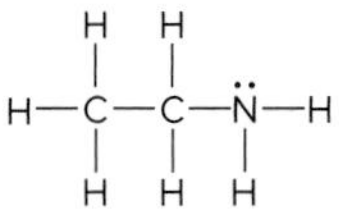	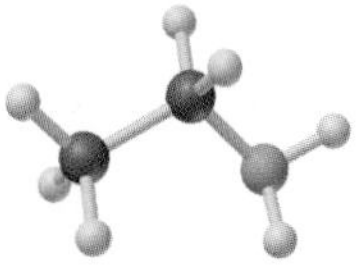	Ethanamine (ethylamine)
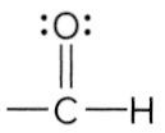	Aldehyde	-al	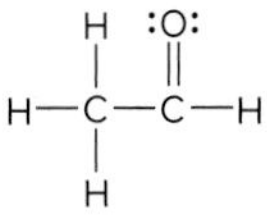	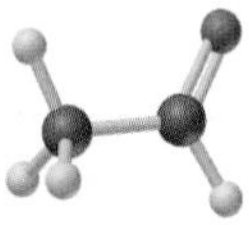	Ethanal (acetaldehyde)
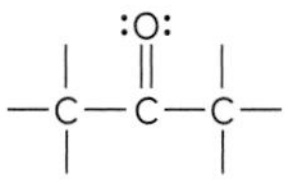	Ketone	-one	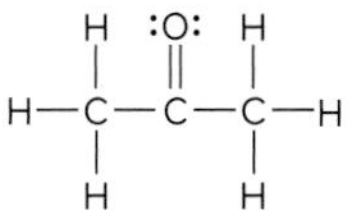	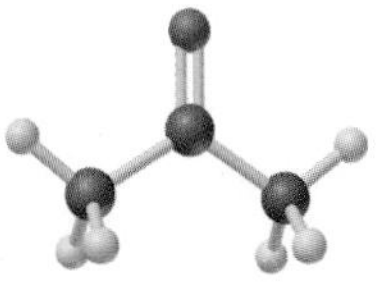	Propan-2-one (acetone)
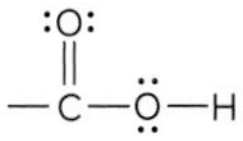	Carboxylic acid	-oic acid	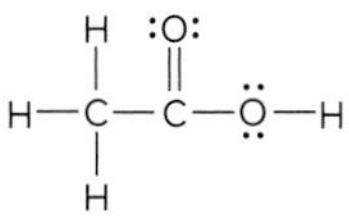	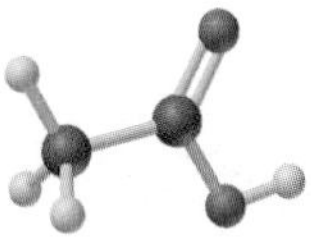	Ethanoic acid (acetic acid)
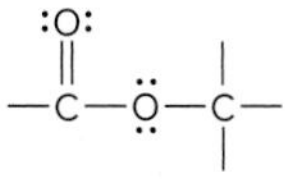	Ester	-oate	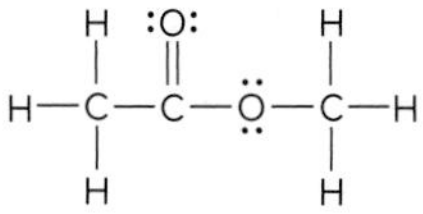	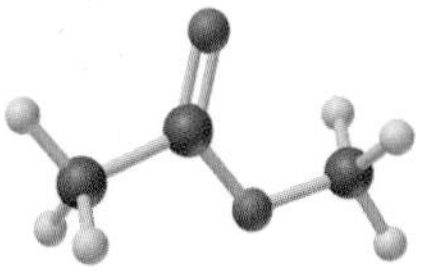	Methyl ethanoate (methyl acetate)
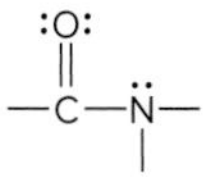	Amide	-amide	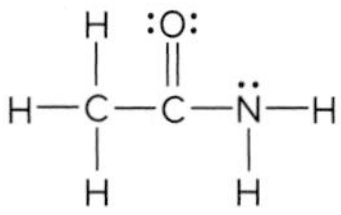	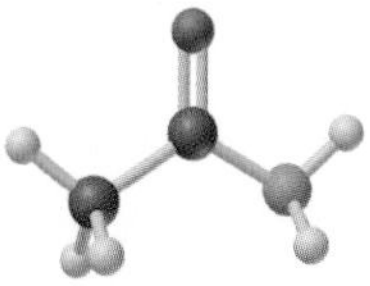	Ethanamide (acetamide)
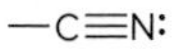	Nitrile	-nitrile	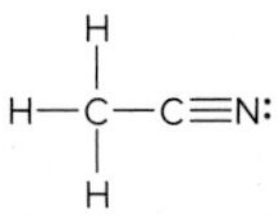	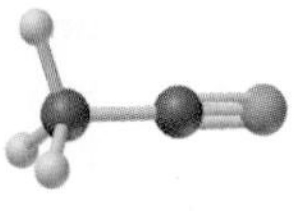	Ethanenitrile (acetonitrile, methyl cyanide)

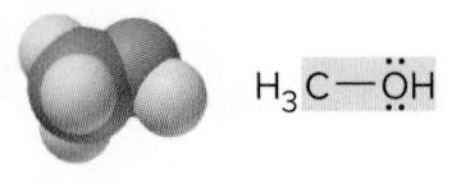

Methanol (methyl alcohol)
By-product in coal gasification; de-icing agent; gasoline substitute; precursor of organic compounds

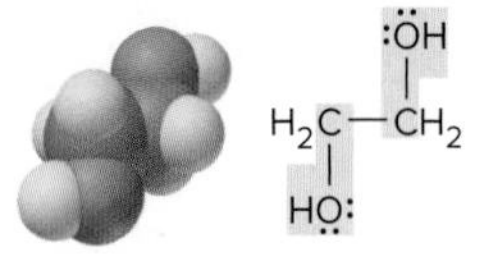

Ethane-1,2-diol (ethylene glycol)
Main component of auto antifreeze

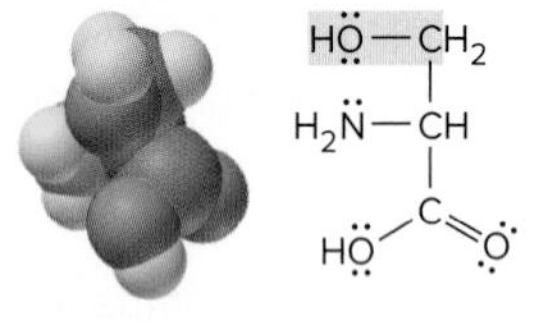

Serine
Amino acid found in most proteins

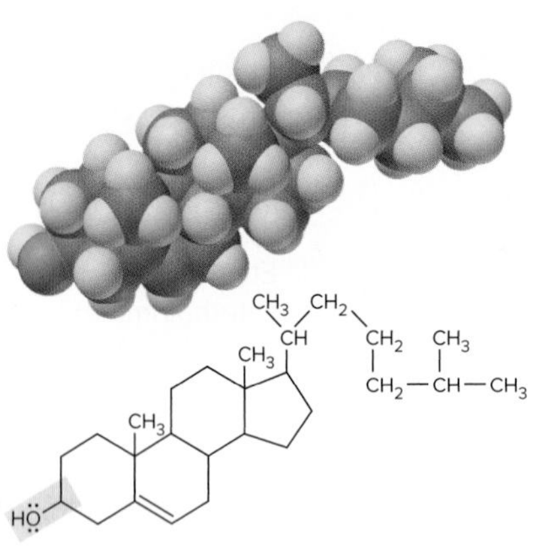

Cholesterol
Major sterol in animals; essential for cell membranes; precursor of steroid hormones

FIGURE 20.23 Some molecules with the alcohol functional group

If the molecule has more than one —OH group, name the parent followed by *-diol*, *-triol*, etc.:

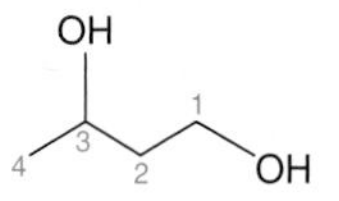

Butane-1,3-diol

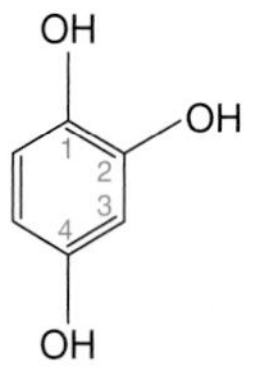

Benzene-1,2,4-triol

Alcohols are common organic reactants, and the functional group occurs in many biomolecules, including carbohydrates, sterols, and some amino acids. Figure 20.23 shows the names, structures, and uses of some important compounds that contain the alcohol group.

The *physical properties* of the smaller alcohols are similar to those of water. They have high melting and boiling points as a result of hydrogen bonding, and they dissolve polar molecules and some salts:

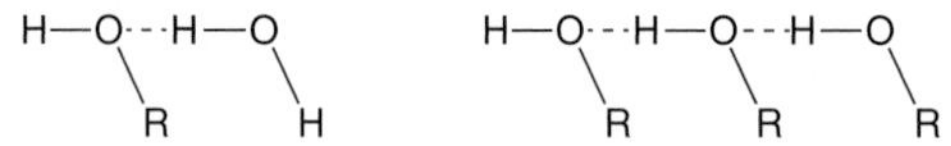

Alcohols form hydrogen bonds with water. They also form hydrogen bonds with other alcohol molecules. That is why their boiling points are much higher than those of corresponding alkanes. Isomeric alcohols show the following boiling point order:

$$\text{primary} > \text{secondary} > \text{tertiary}$$

Smaller alcohols are colourless, volatile, toxic, and flammable liquids that are soluble in water; however, their solubility decreases with increase in molecular weight, due to increasing of the hydrophobic alkyl chain. Above C_{12}, alcohols are solid. Alcohols are neutral and do not influence the pH of a solution, and thus alcohols are weaker acids than water. Alcohol's toxicity is as follows: ethyl alcohol (grain alcohol) < isopropyl alcohol (rubbing alcohol) < methyl alcohol (wood alcohol).

Ethers The **ether** functional group consists of an oxygen bonded to two carbon atoms (a 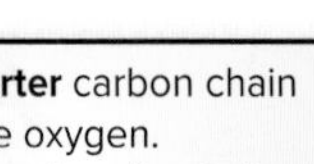 group), and the general formula is R—O—R. To name the ethers, the shorter of the two chains will become the first part of the name, changing the *-ane* suffix to *-oxy*. The longer chain will become the suffix of the name of the ether. If the oxygen is not attached to the end of the main alkane chain, then the whole shorter R—O group is treated as a side-chain and prefixed with its bonding position on the main chain.

Select the **shorter** carbon chain attached to the oxygen. Change the "**ane**" ending to "**oxy.**" It becomes the first part of the name.	The **longer** carbon chain becomes the suffix of the name of the ether.	Number the substituents according to their **positions** on the chain (listed in **alphabetical** order).

Some examples are given below:

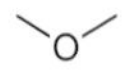

Methoxymethane
(dimethyl ether)

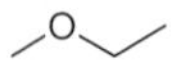

Methoxyethane
(ethyl methyl ether)

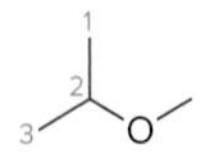

2-methoxypropane
(isopropyl methyl ether)

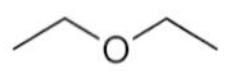

Ethoxyethane
(diethyl ether)

Physical properties. Ethers are colourless, highly volatile, usually sweet smelling, and very flammable liquids. They do not form hydrogen bonds between themselves; therefore, their boiling points are lower than the corresponding alcohols. Ethers are weakly polar. The C—O—C bond angle is approximately 110°, so the C—O dipole moment does not cancel out.

Uses. Diethyl ether is used as an anaesthetic agent since its vapours cause unconsciousness. Methoxymethane is used as an aerosol spray propeller and is a potential renewable fuel for diesel engines. Tetrahydrofuran (THF), a cyclic ether, is one of the most polar simple ethers used as a solvent. Linear polyethers, like polyethylene glycol, are used in cosmetics and pharmaceutical products, and crown ethers are used as phase-transfer catalysts, since they bind certain cations strongly, forming complexes (their oxygen atoms coordinate with cations located at the centre of the ring, whereas the exterior of the ring is hydrophobic). The 12-crown-4 (*left*) has high affinity for the lithium cation, whereas the 15-crown-5 (*right*) has a high affinity for the sodium ion.

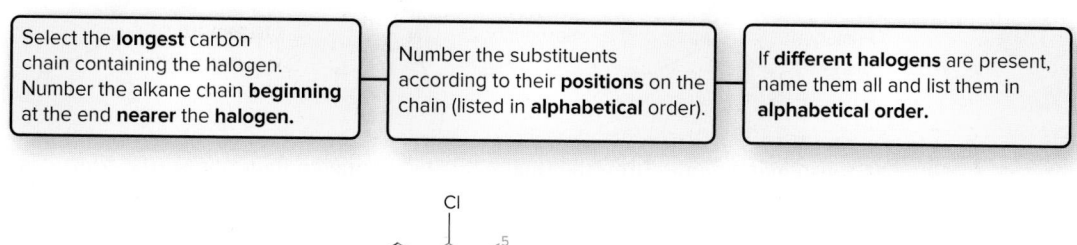

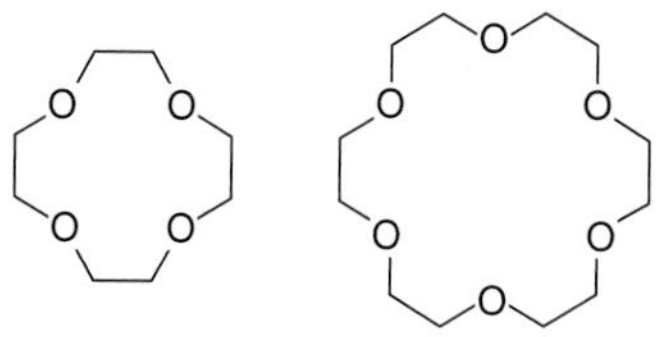

Crown ethers

Tetrahydrofuran

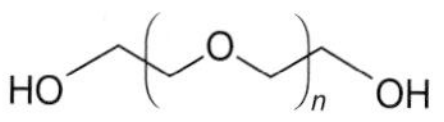

Polyethylene glycol

Haloalkanes A *halogen* atom (X) bonded to a carbon gives the **haloalkane (alkyl halide)** functional group, —C—X:, and compounds with the general formula R–X. Haloalkanes are named by identifying the halogen with a prefix on the hydrocarbon name and numbering the C atom to which the halogen is attached, as in bromomethane, 2-chloropropane, or 1,3-diiodohexane. If different halogens are present, number all and list them in alphabetical order.

Select the **longest** carbon chain containing the halogen. Number the alkane chain **beginning** at the end **nearer** the **halogen.**

Number the substituents according to their **positions** on the chain (listed in **alphabetical** order).

If **different halogens** are present, name them all and list them in **alphabetical order.**

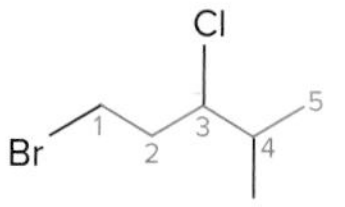

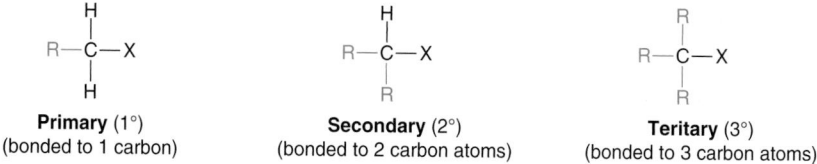

1-bromo-3-chloro-4-methylpentane

Three terms are used to refer to the carbon bearing the halogen in the classification of haloalkanes:

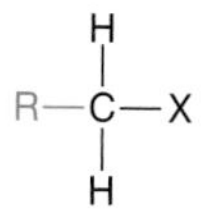

Primary (1°)
(bonded to 1 carbon)

Secondary (2°)
(bonded to 2 carbon atoms)

Teritary (3°)
(bonded to 3 carbon atoms)

where R is a generalized organic group and X is the halogen.

Physical properties. Haloalkanes are colourless when pure, but bromo and iodo alkanes develop colour when exposed to light. They are relatively odourless and hydrophobic, making them only very slightly soluble in water. Their boiling points are higher than those of the corresponding alkanes and increase with the atomic weight and with the number of halides present. The boiling point of isomeric haloalkanes decreases with an increase in branching, and the boiling points of dihalobenzenes are almost the same. However, the *para*-isomers have higher melting points because they are able to form stronger intermolecular interactions than *ortho*- and *meta*-isomers. Halo-alkanes are generally more dense than the alkanes they are derived from and are usually denser than water. Density increases with the number of carbon atoms and with the mass of the halogen atom.

Uses. Haloalkanes have many important uses. A large number of pharmaceuticals contain halogens, especially fluorine (present in about one-fifth of pharmaceuticals),

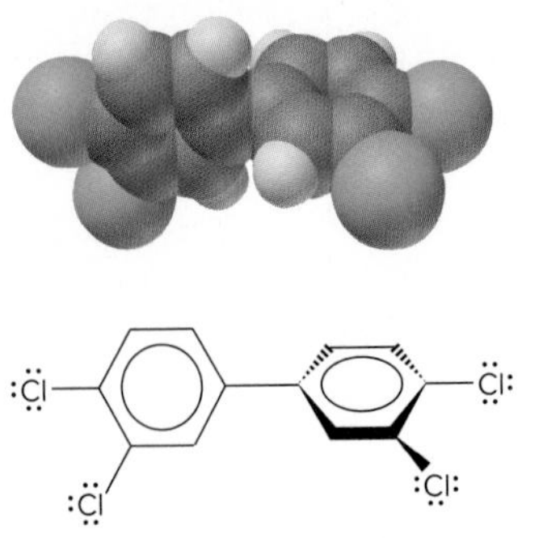

FIGURE 20.24 A tetrachlorobiphenyl, one of 209 polychlorinated biphenyls (PCBs)

such as fluoxetine (Prozac—antidepressant), paroxetine (Paxil—antidepressant), ciprofloxacin (Cipro—antibiotic), mefloquine (Lariam or Mefaquin—antimalarial), and fluconazole (Diflucan—antifungal). The nonflammable and not very toxic chlorofluorocarbons (CFCs), such as CCl_3F and CCl_2F_2, are used for general degreasing purposes, for dry cleaning, as refrigerants, as foamed plastics such as expanded polystyrene or polyurethane foam, and as propellants for aerosols. Unfortunately, they cause serious environmental problems (Chemical Connections, Chapter 23). Also, some halogenated aromatic hydrocarbons (aryl halides) are carcinogenic in mammals; have severe neurological effects in humans; and, to make matters worse, are very stable and accumulate in the environment. For example, polychlorinated biphenyls (PCBs) (Figure 20.24), used as insulating fluids in electrical transformers, have accumulated for decades in rivers and lakes and have been incorporated into the food chain. PCBs in natural waters pose health risks and present an enormous cleanup problem.

Amines The **amine** functional group is —C—N̈. Chemists classify amines as derivatives of ammonia, with R groups in place of one or more of the H atoms in NH_3. *Primary* (1°) amines are RNH_2, *secondary* (2°) amines are R_2NH, and *tertiary* (3°) amines are R_3N. When the NH_2 group is attached to a benzene ring, the compound is named *aniline*. Like ammonia, amines have trigonal pyramidal shapes and a lone pair of electrons on a partially negative N atom, which is the key to amine reactivity (Figure 20.25).

FIGURE 20.25 General structures of amines

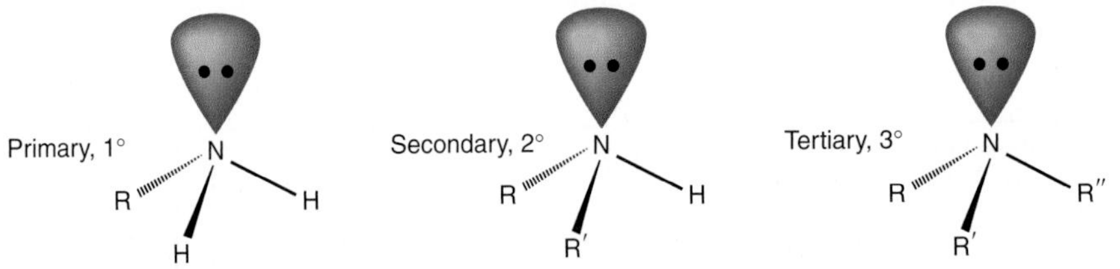

Systematic names drop the final *-e* of the alkane and add the suffix *-amine*, as in ethanamine. However, there is still wide usage of common names, in which the suffix *-amine* follows the name of the alkyl group; thus, methanamine (methylamine) has one methyl group attached to N, *N*-ethylethanamine (diethylamine) has two ethyl groups attached, and so forth. Figure 20.26 shows that the amine functional group occurs in many biomolecules.

Replace the "**ane**" ending of alkane with "**anamine.**"

Find and name the longest chain. List the alkyl groups in **alphabetical** order.

Use ***N*** to show the location of an alkyl group that is attached to the nitrogen atom.

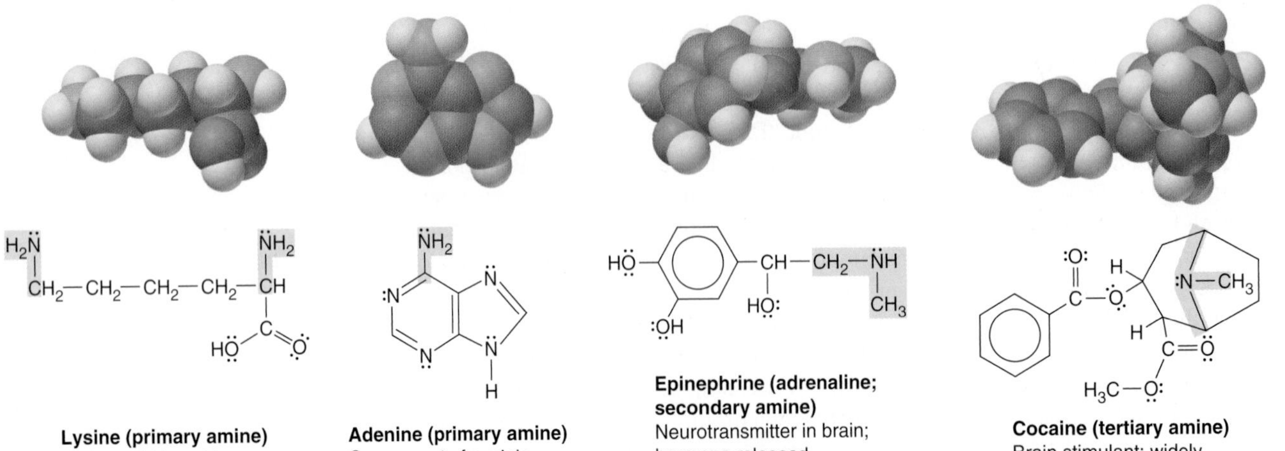

Lysine (primary amine)
Amino acid found in most proteins

Adenine (primary amine)
Component of nucleic acids

Epinephrine (adrenaline; secondary amine)
Neurotransmitter in brain; hormone released during stress

Cocaine (tertiary amine)
Brain stimulant; widely abused drug

FIGURE 20.26 Some biomolecules with the amine functional group

The following molecules are examples of primary, secondary, and tertiary amines:

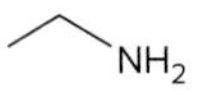

Ethanamine
(ethylamine)
1°

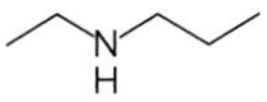

N-ethylpropan-1-amine
(ethylpropylamine)
2°

N,*N*-dimethylmethanamine
(trimethylamine)
3°

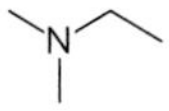

N,*N*-dimethylethanamine
(dimethylethylamine)
3°

Primary and secondary amines can form hydrogen bonds, so they have higher melting and boiling points than hydrocarbons and haloalkanes of similar molar mass ($\mathcal{M}$). For example, *N*-methylmethanamine ($\mathcal{M}$ = 45.09 g/mol) boils 45°C higher than ethylfluoride does ($\mathcal{M}$ = 48.06 g/mol). *N*,*N*-dimethylmethanamine has a greater molar mass than does *N*-methylmethanamine, but it melts more than 20°C *lower* because *N*,*N*-dimethylmethanamine molecules do not have an H atom attached to the N for hydrogen bonding.

Amines of low molar mass are fishy smelling, water soluble, and weakly basic because of the lone pair on N. The reaction with water proceeds only slightly to the right to reach equilibrium:

$$CH_3-\ddot{N}H_2 + H_2O \rightleftharpoons CH_3-\overset{+}{N}H_3 + OH^-$$

Uses. Amines and their salts are used as corrosion inhibitors in boilers and in lubricating oils (morpholine), as antioxidants for rubber and roofing asphalt (diarylamines), as stabilizers for cellulose, as nitrate explosives (diphenylamine), as protectants against damage from gamma radiation (diarylamines), as developers in photography (aromatic diamines), as flotation agents in mining, as anti-cling and waterproofing agents for textiles, as fabric softeners, in paper coating, and for solubilizing herbicides. Some polyfunctional amines are valuable pharmaceuticals, such as the two enantiomers (1*R*,2*S*)-2-(methylamino)-1-phenylpropan-1-ol and (1*S*,2*R*)-2-(methylamino)-1-phenylpropan-1-ol (*ephedrine*), as well as (*R*)-4-(1-hydroxy-2-(methylamino)ethyl)benzene-1,2-diol (*epinephrine* or *adrenaline*), and anaesthetics, such as 2-diethylaminoethyl-4-aminobenzoate (*novocaine*).

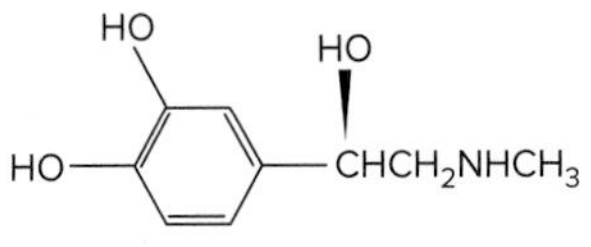

Adrenaline

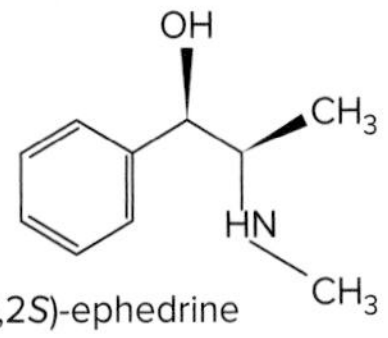

(1*R*,2*S*)-ephedrine

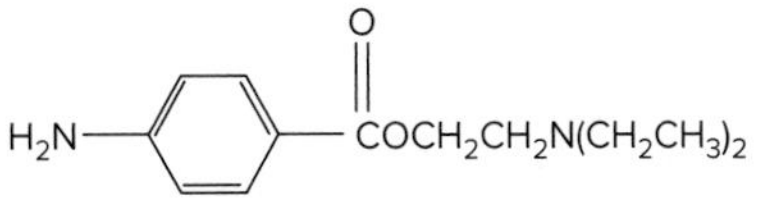

Novocaine

Variations on a Theme: Inorganic Compounds with Single Bonds to O, X, and N The —OH group occurs frequently in inorganic compounds. All oxoacids contain at least one —OH group, bonded to a nonmetal atom, which in most cases is bonded to other O atoms. Oxoacids are acidic in water because these additional O atoms pull electron density from the central nonmetal, which pulls electron density from the O—H bond, releasing an H^+ ion and stabilizing the conjugate base (oxoanion) through resonance. That is why carboxylic acids are more acidic than alcohols, which lack the additional O atom.

The bonds between nitrogen and larger nonmetals, such as Si, P, and S, have significant double-bond character, which affects structure and reactivity. For example, $(SiH_3)_3N$, trisilylamine, the Si analogue of *N*,*N*-dimethylmethanamine (trimethylamine), has a trigonal planar rather than a trigonal pyramidal structure because of the formation of a double bond between N and Si, where the lone pair on N is delocalized in this π bond. Trisilylamine is not basic.

Functional Groups with Double Bonds

The most important functional groups with double bonds are the C=C group of alkenes and the C=O group of aldehydes and ketones. Both appear in many organic and biological molecules. As we will see in Chapter 21, their most common reaction type is *addition*.

Alkenes The C=C bond is the essential portion of the *alkene* functional group, 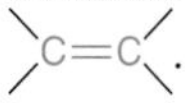. We learned how to name them and discussed their physical properties in Section 20.5.

Aldehydes and Ketones The C=O bond, or **carbonyl group**, is one of the most chemically versatile.

- In the **aldehyde** functional group, the carbonyl C is bonded to H (and often to another C), so it occurs *at the end of a chain,* 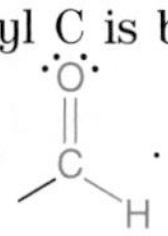. Aldehyde names drop the final *-e* from the alkane name and add *-al*; thus, the three-carbon aldehyde is propanal:

Propanal 2-methylbutanal

The main chain must contain the **carbonyl group**, where there is at least one H bonded to the C=O. → Replace the "**e**" ending of alkane with "**al.**" → The **CHO carbon** is numbered as **C-1.**

- In the **ketone** functional group, the carbonyl C is bonded to two other C atoms, 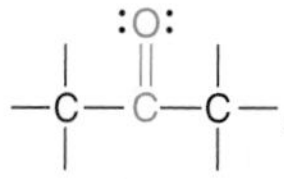, so it occurs *within the chain.* Ketones, 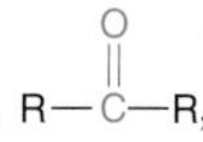, are named by numbering the carbonyl C, dropping the final *-e* from the alkane name, and adding *-one.* For example, the unbranched, five-carbon ketone with the carbonyl C as C-2 in the chain is named pentan-2-one.When more than one carbonyl group is present, use prefixes *di-*, *tri-*, etc. Figure 20.27 shows some common carbonyl compounds.

Pentan-2-one 1-phenylpentane-2,3-dione

The main chain must contain the **carbonyl group**, where there are two alkyl groups bonded to the C=O. → Replace the "**e**" ending of alkane with "**one.**" → **Numbering** begins at the end nearer the **C=O carbon**.

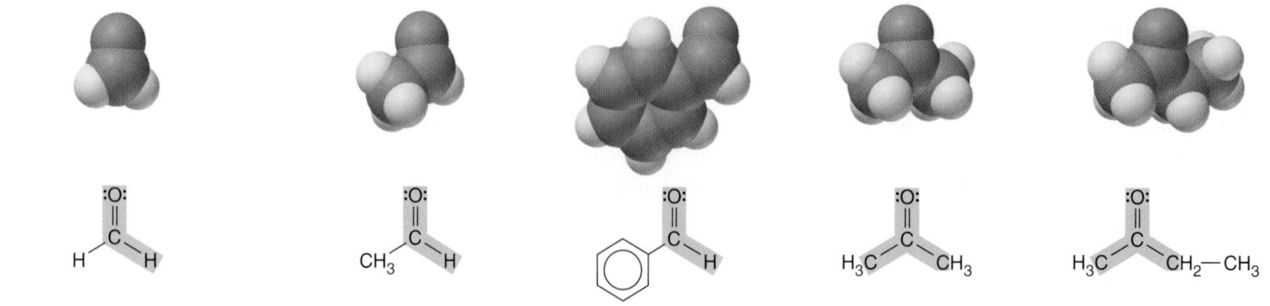

Methanal (formaldehyde) Used to make resins in plywood, dishware, countertops; biological preservative

Ethanal (acetaldehyde) Narcotic product of ethanol metabolism; used to make perfumes, flavours, plastics, other chemicals

Benzaldehyde Artificial almond flavouring

Propan-2-one (acetone) Solvent for fat, rubber, plastic, varnish, lacquer; chemical feedstock

Butan-2-one (methyl ethyl ketone) Important solvent

FIGURE 20.27 Some common aldehydes and ketones

If another functional group has the priority as the principal characteristic group, aldehydes and ketones are described by the prefix *oxo-*. Aldehydes and ketones are formed by the oxidation of alcohols:

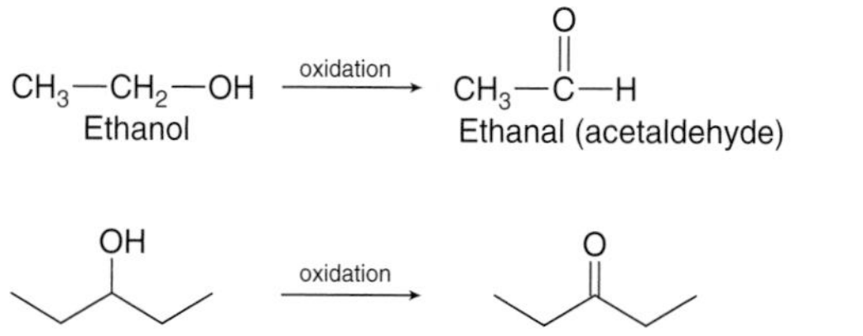

Ethanol → Ethanal (acetaldehyde)

Pentan-3-ol → Pentan-3-one (diethylketone)

Like the C═C bond, the C═O bond is *electron rich*; unlike the C═C bond, it is *highly polar* ($\Delta\chi = 1.0$). Figure 20.28 emphasizes this polarity with an orbital contour model (Figure 20.28A) and a charged resonance form (Figure 20.28B).

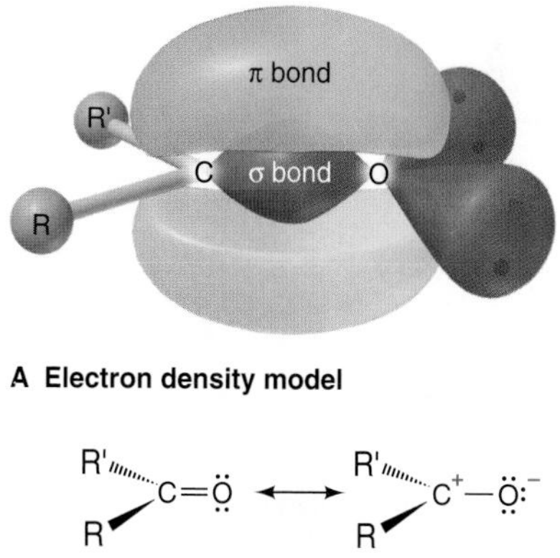

A Electron density model

B Charged resonance form

FIGURE 20.28 The polar carbonyl group

Variations on a Theme: Inorganic Compounds with Double Bonds Homonuclear (same kind of atom) double bonds are rare for atoms other than C. However, we have seen many double bonds between O and other nonmetals, for example, in the oxides of S, N, and the halogens.

Functional Groups with Both Single and Double Bonds

A family of three functional groups contains C double-bonded to O (a carbonyl group) *and* single-bonded to O or N. The parent of the family is the **carboxylic acid** group, —C(═O)—OH, also called the *carboxyl group* and written —COOH.

Carboxylic Acids Carboxylic acids, R—C(═O)—OH, are named by dropping the *-e* from the alkane name and adding *-oic acid*; however, many common names are used. For example, the four-carbon acid is butanoic acid (the carboxyl C is counted when choosing the root); its common name is butyric acid. Figure 20.29 shows some important carboxylic acids. The carboxyl C already has three bonds, so it forms only one other. In methanoic acid (formic acid), the carboxyl C is bonded to an H atom, but in all other carboxylic acids it is bonded to a chain or ring.

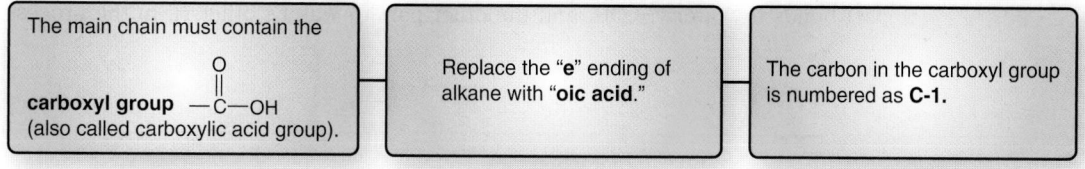

The main chain must contain the **carboxyl group** —C(═O)—OH (also called carboxylic acid group).	Replace the "**e**" ending of alkane with "**oic acid.**"	The carbon in the carboxyl group is numbered as **C-1.**

Carboxylic acids are weak acids in water:

$$CH_3{-}\overset{\overset{\large O}{\|}}{C}{-}OH(l) + H_2O(l) \rightleftharpoons CH_3{-}\overset{\overset{\large O}{\|}}{C}{-}O^-(aq) + H_3O^+(aq)$$

Ethanoic acid (acetic acid)

At equilibrium in a solution of typical concentration, more than 99% of the acid molecules are undissociated at any given moment. In a strong base, however, a carboxylic acid reacts completely to form a salt and water:

$$CH_3{-}\overset{\overset{\large O}{\|}}{C}{-}OH(l) + NaOH(aq) \longrightarrow CH_3{-}\overset{\overset{\large O}{\|}}{C}{-}O^-(aq) + Na^+(aq) + H_2O(l)$$

The anion of the salt is the *carboxylate ion*, named by dropping *-oic acid* and adding *-oate*; the sodium salt of butanoic acid, for instance, is sodium butanoate. The resonance stabilization of the conjugate base explains the acidity of carboxylic acids.

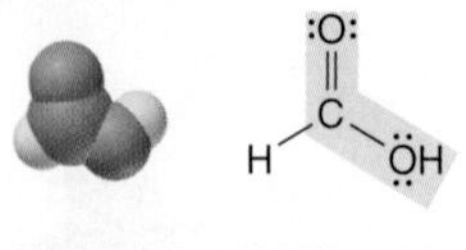

Methanoic acid (formic acid)
An irritant component of ant and bee stings

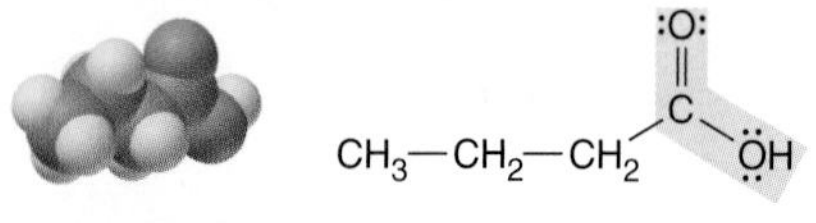

Butanoic acid (butyric acid)
Odour of rancid butter; suspected component of monkey sex attractant

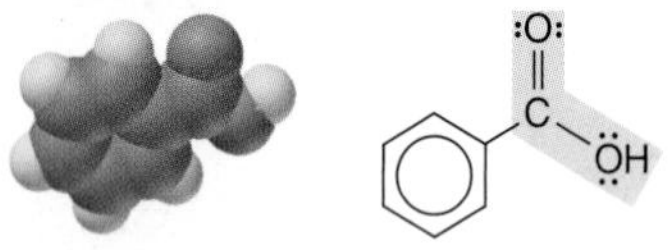

Benzoic acid
Calorimetric standard; used in preserving food, dyeing fabric, curing tobacco

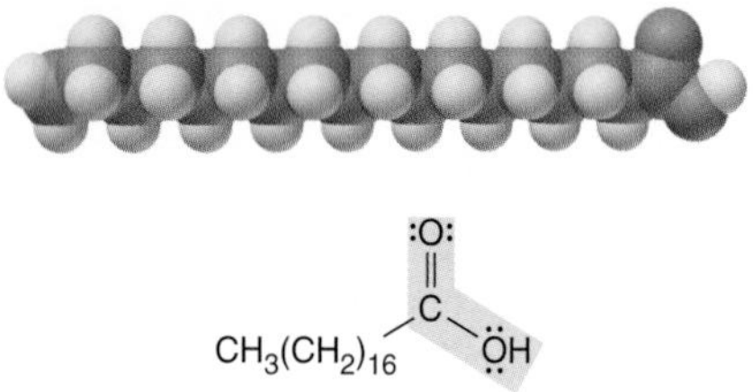

Octadecanoic acid (stearic acid)
Found in animal fats; used in making candles and soaps

FIGURE 20.29 Some molecules with the carboxylic acid functional group

Carboxylic acids with long hydrocarbon chains are *fatty acids*, an essential group of compounds found in all cells. Animal fatty acids have saturated chains, whereas many fatty acids from plants are unsaturated, usually with the C═C bonds in the *cis* configuration. The double bonds make them much easier to metabolize. Nearly all fatty acid skeletons have an even number of C atoms—16- and 18-carbon atoms are very common—because cells use two-carbon units in synthesizing them. Fatty acid salts, usually with a cation from group 1 or 2, are soaps (Section 22.4).

Esters The **ester** group, $-\overset{\overset{O}{\|}}{C}-O-R$, is formed from an alcohol and a carboxylic acid. The first part of an ester name designates the alcohol portion, and the second is the acid portion (named in the same way as the carboxylate ion). For example, the ester formed between methanol and ethanoic acid (acetic acid) is methyl ethanoate (methyl acetate), used as a solvent in fast-drying paints, glues, and nail polish removers, and for the manufacture of celluloid adhesives from waste film.

First name the alkyl group replacing the H in the carboxyl group. — Replace the "**oic**" acid ending from the parent carboxylic acid with "**oate**."

The ester group occurs commonly in *lipids*, a large group of fatty biological substances. Most dietary fats are *triglycerides*, esters that are composed of three fatty acids linked to the alcohol 1,2,3-trihydroxypropane (glycerol) and that function as energy stores. Some important lipids are shown in Figure 20.30; lecithin is one of several phospholipids that make up the lipid bilayer in all cell membranes (Section 22.4).

An ester, like an **acid anhydride** (see Figure 20.32), forms through a dehydration-condensation reaction; in this case, it is called an *esterification*:

$$R-\overset{\overset{O}{\|}}{C}-OH + H-O-R' \overset{H^+}{\rightleftharpoons} R-\overset{\overset{O}{\|}}{C}-O-R' + HOH$$

Note that the esterification reaction is reversible. The opposite of dehydration-condensation is **hydrolysis**, in which the O atom of water is attracted to the partially positive C atom of the ester, cleaving (lysing) the ester molecule into two parts. One part bonds to water's —OH, and the other part to water's other H. In the process

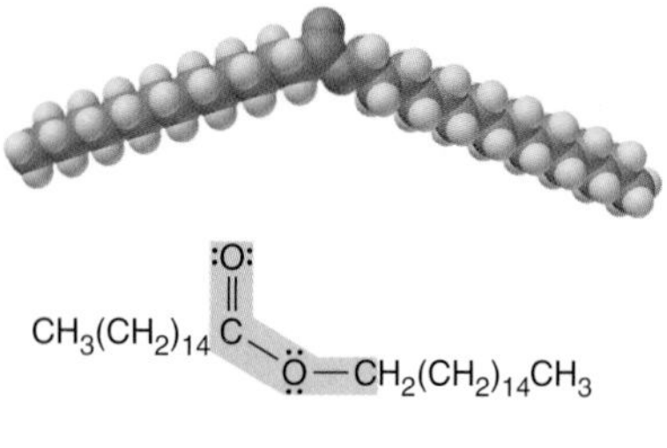

Cetyl palmitate The most common lipid in whale blubber

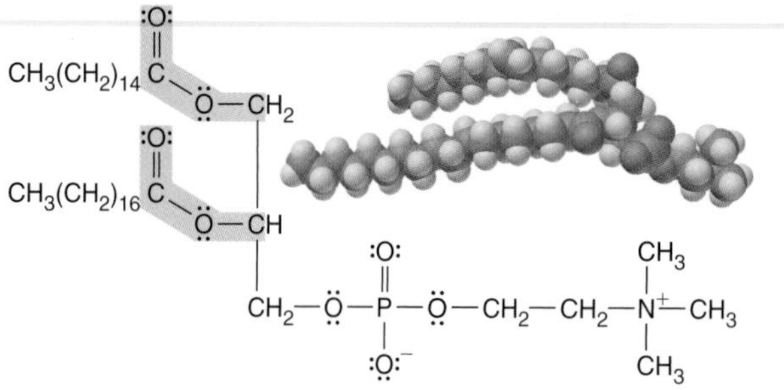

Lecithin Phospholipid found in all cell membranes

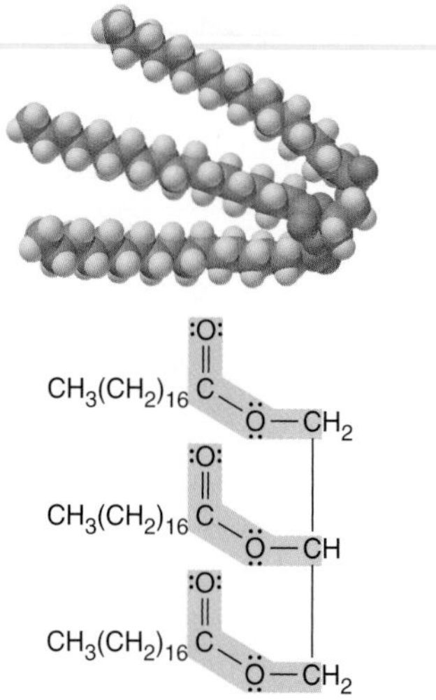

Tristearin Typical dietary fat used as an energy store in animals

FIGURE 20.30 Some lipid molecules with the ester functional group

of making soap, or *saponification* (Latin *sapon* for "soap"), begun in ancient times, the ester bonds in animal or vegetable fats are hydrolyzed with a strong base:

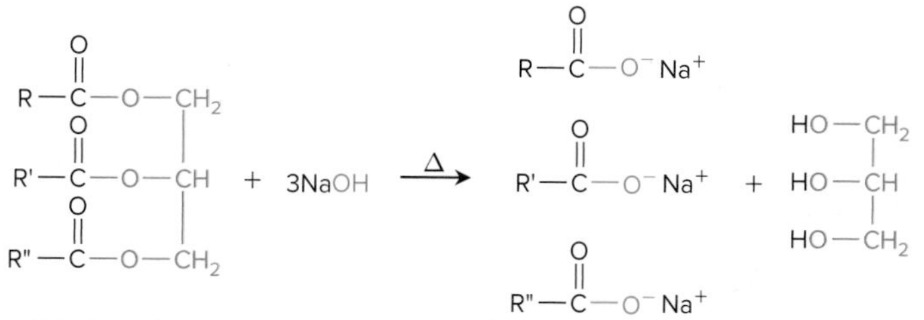

A triglyceride Three soap molecules Glycerol

Carboxylic acids and esters have some familiar and distinctive odours.

Whereas carboxylic acids have pungent, vinegary, and cheesy odours, esters have pleasant odours. For example, ethyl butanoate has the scent of pineapple, and pentyl pentanoate has the aroma of apples (*see photo*). Naturally occurring and synthetic esters are used to add fruity, floral, and herbal odours to foods, cosmetics, household deodorizers, and medicines. In Table 20.6 you will find examples of esters used for flavours and fragrances.

Amides The product of a substitution between an amine (or NH_3) and an ester is an **amide** with the functional group —C(=O)—N—. The partially negative N of the amine is attracted to the partially positive C of the ester, an alcohol (ROH) is lost, and an amide forms:

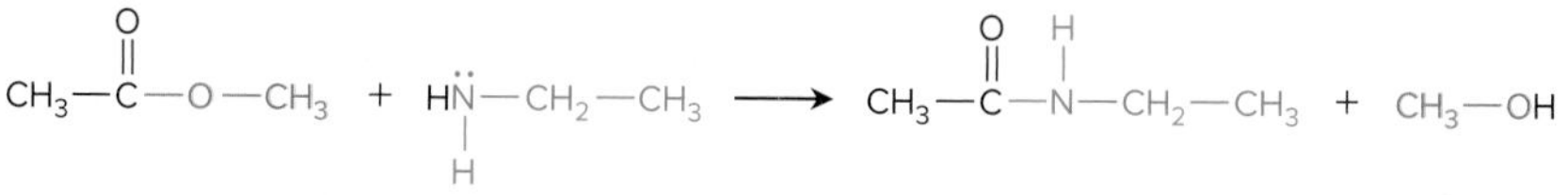

Methyl ethanoate (methyl acetate) Ethanamine (ethylamine) *N*-ethylethanamide (*N*-ethylacetamide) Methanol

Amides are named by denoting the amine portion with *N*- and replacing *-oic acid* from the parent carboxylic acid with *-amide*. In the amide produced in the preceding reaction, the ethyl group comes from the amine, and the acid portion comes from ethanoic acid (acetic acid).

Name the **alkyl** group(s) attached to the N atom in place of H atom(s), using ***N***- to show their location.	Replace the "**oic**" acid ending from the parent carboxylic acid with "**amide**."

Some amides are shown in Figure 20.31. The most important example of the amide group is the *peptide bond* (discussed in Sections 22.3 and 22.4), which links amino acids in a protein.

FIGURE 20.31 Some molecules with the amide functional group

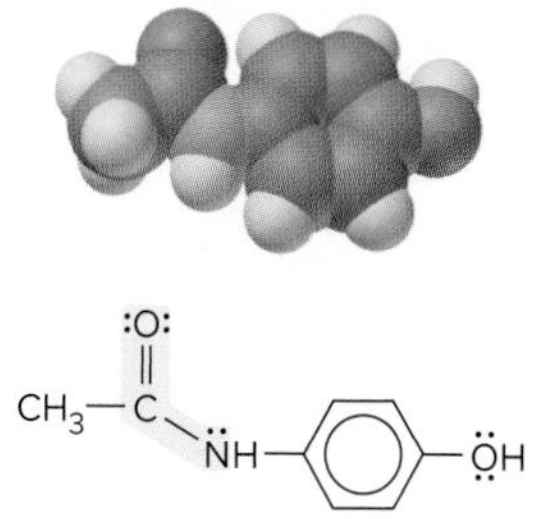

Acetaminophen
Active ingredient in some pain relievers; used to make dyes and photographic chemicals

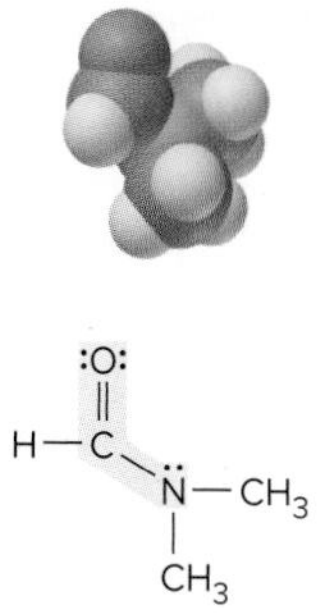

***N,N*-dimethylmethanamide (dimethylformamide)**
Major organic solvent; used in production of synthetic fibres

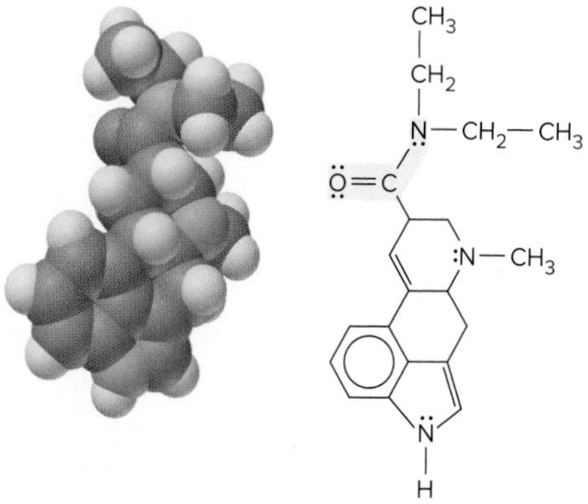

Lysergic acid diethylamide (LSD-25) A potent hallucinogen

TABLE 20.6 Common Esters Used for Fragrances and Flavours

		Prepared from	
Ester	**Smells like**	**Alcohol**	**Acid**
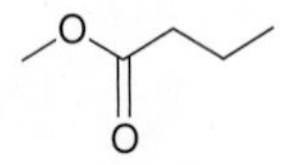 Methyl butanoate (methyl butyrate)	 Apples	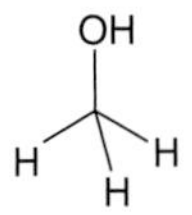 Methanol (methyl alcohol)	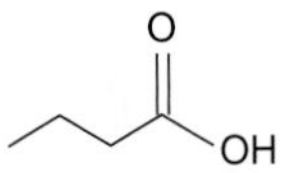 Butanoic acid (butyric acid)
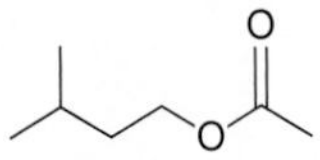 3-methylbut-1-yl ethanoate (isopentyl acetate or isoamyl acetate)	 Bananas	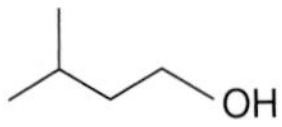 3-methylbutan-1-ol (isopentanol or isoamyl alcohol)	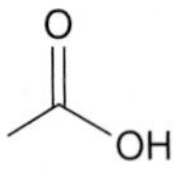 Ethanoic acid (acetic acid)
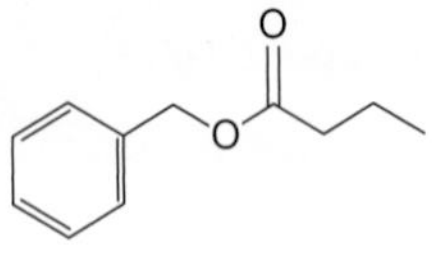 Benzyl butanoate (benzyl butyrate)	 Flowers	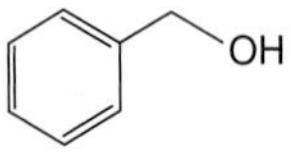 Phenylmethanol (benzyl alcohol)	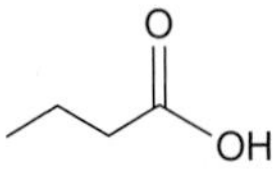 Butanoic acid (butyric acid)
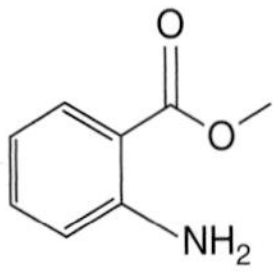 Methyl 2-aminobenzoate (methyl anthranilate)	 Grapes	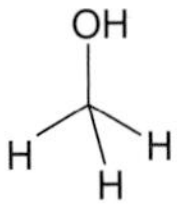 Methanol (methyl alcohol)	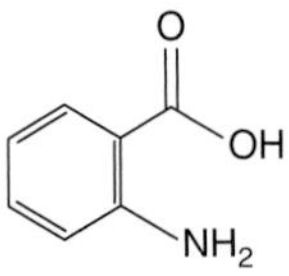 2-aminobenzoic acid (anthranilic acid)
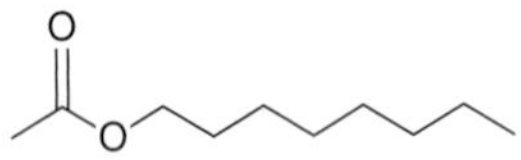 Octyl ethanoate (octyl acetate)	 Oranges	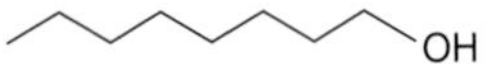 Octan-1-ol (1-octanol or capryl alcohol)	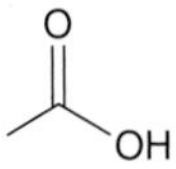 Ethanoic acid (acetic acid)
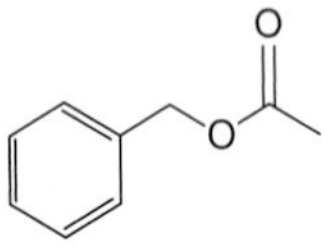 Phenylmethyl ethanoate (benzyl acetate)	 Peaches	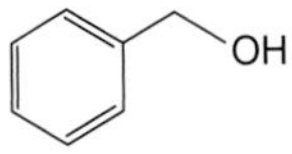 Phenylmethanol (benzyl alcohol)	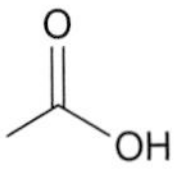 Ethanoic acid (acetic acid)
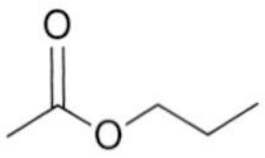 Propyl ethanoate (propyl acetate)	 Pears	 Propan-1-ol (n-propyl alcohol)	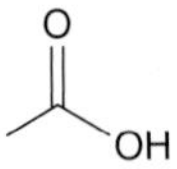 Ethanoic acid (acetic acid)
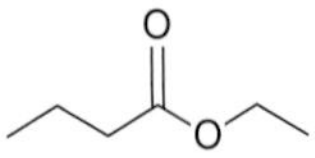 Ethyl butanoate (ethyl butyrate)	 Pineapple	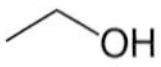 Ethanol (ethyl alcohol)	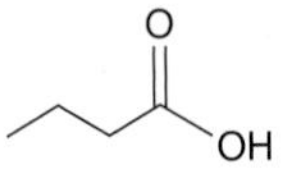 Butanoic acid (butyric acid)
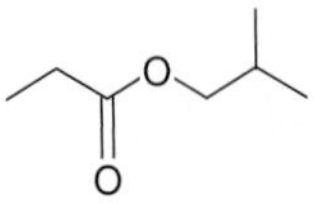 2-methylprop-1-yl propanoate (isobutyl propionate)	 Rum	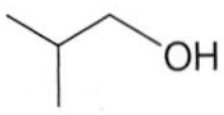 2-methylpropan-1-ol (isobutyl alcohol)	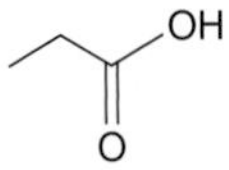 Propanoic acid (propionic acid)

An amide is hydrolyzed in hot water (or a base) to a carboxylic acid and an amine. Amides can be prepared from carboxylic acids, acid chlorides, and acid anhydrides; they can be viewed as the result of a reversible dehydration-condensation reaction:

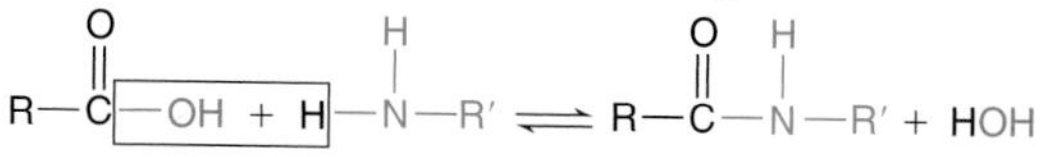

Variations on a Theme: Oxoacids, Esters, and Amides of Other Nonmetals

A nonmetal that has both double and single bonds to oxygen atoms occurs in most inorganic oxoacids, such as phosphoric, sulfuric, and chlorous acids. Those with additional O atoms are stronger acids than carboxylic acids.

Diphosphoric and disulfuric acids are acid anhydrides formed by dehydration-condensation reactions, like the one that yields a carboxylic acid anhydride (Figure 20.32). Inorganic oxoacids form esters and amides that are part of many biological molecules. We already saw that certain lipids are phosphate esters (see Figure 20.30). The first compound formed when glucose is digested is a phosphate ester (Figure 20.33A); a similar phosphate ester is a major structural feature of nucleic acids, as we will see shortly. Amides of organic sulfur-containing oxoacids, called *sulfonamides*, are potent antibiotics; the simplest of these is depicted in Figure 20.33B. More than 10 000 different sulfonamides have been synthesized.

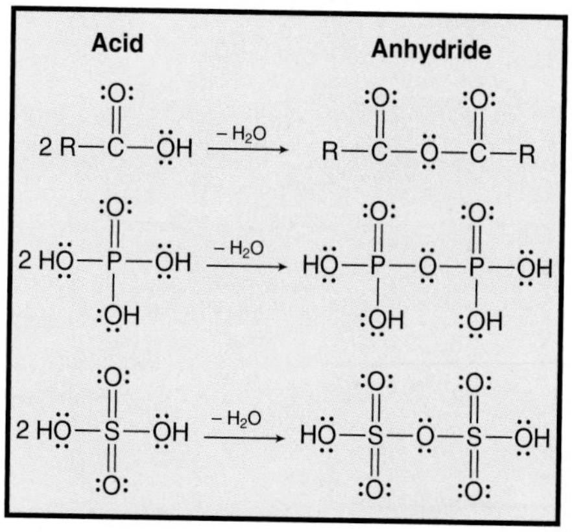

FIGURE 20.32 The formation of carboxylic, phosphoric, and sulfuric acid anhydrides

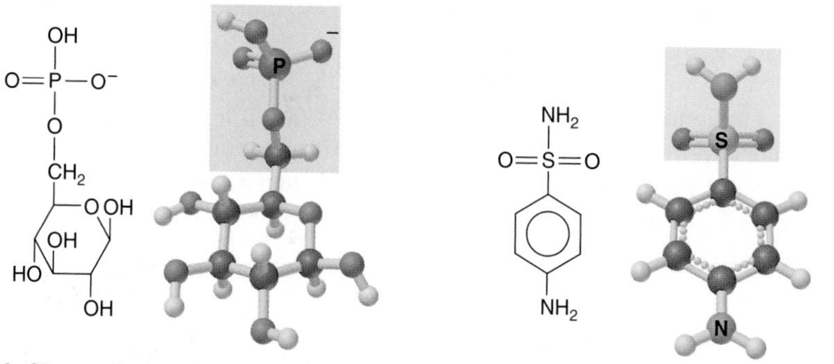

A Glucose-6-phosphate **B Sulfanilamide**

FIGURE 20.33 A phosphate ester and a sulfonamide

Functional Groups with Triple Bonds

Alkynes and nitriles are the only two important functional groups with triple bonds.

Alkynes Alkynes, with an electron-rich C≡C group, undergo addition (by H_2O, H_2, HX, X_2, etc.) to form double-bonded or saturated compounds:

$$\underset{\text{Propyne}}{CH_3{-}C{\equiv}CH} \xrightarrow{H_2} \underset{\text{Propene}}{CH_3{-}CH{=}CH_2} \xrightarrow{H_2} \underset{\text{Propane}}{CH_3{-}CH_2{-}CH_3}$$

Alkynes have physical properties similar to alkanes and alkenes. The boiling points and melting points increase with increase in the number of carbon atoms. Alkynes with up to four C atoms are gases, those containing five to thirteen C atoms are liquids, and those with more than thirteen C atoms are solids.

Alkynes are generally nonpolar molecules with little solubility in polar solvents, such as water, but are very soluble in nonpolar solvents, or solvents with low polarity, such as benzene, ether, and acetone.

Nitriles Nitriles, generically called cyanides (R—C≡N), contain the **nitrile** group (—C≡N:) and are made by substituting a CN^- (cyanide) ion for X^- in a reaction with an alkyl halide:

$$CH_3{-}CH_2{-}Cl + NaCN \longrightarrow CH_3{-}CH_2{-}C{\equiv}N + NaCl$$

This reaction is useful because it *increases the chain by one C atom.* Nitriles are versatile because once they are formed, they can be reduced to amines or hydrolyzed to carboxylic acids:

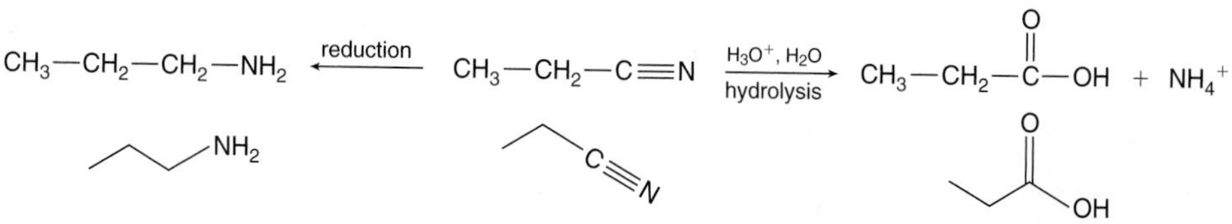

Nitriles in which —C≡N may be considered to have replaced the —COOH group of an acid are named by changing the *-oic acid* (or *-ic acid*) ending of the name of the acid to *-nitrile* (or *-onitrile*), or changing the *-carboxylic acid* suffix to *-carbonitrile.*

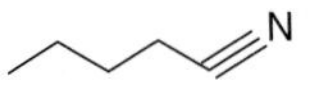

Pentanenitrile

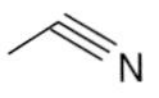

Ethanenitrile (acetonitrile)

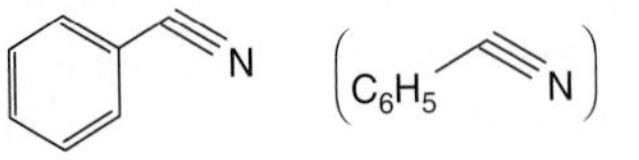

Benzonitrile

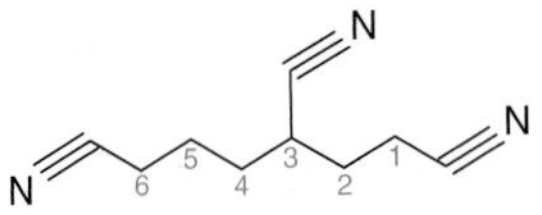

Hexane-1,3,6-tricarbonitrile

Physical properties. The small nitriles are liquids at room temperature. Their boiling points are very high for the size of the molecules, similar to what you would expect if they were capable of forming hydrogen bonds. However, they do *not* form hydrogen bonds because they do not have an H atom attached to the N atom. They are just very polar molecules. Ethanenitrile is very soluble in water, and the solubility decreases as the chain length increases.

Uses. Nitrile rubber is used in nonlatex gloves, hoses, automotive transmission belts, and synthetic leather. The nitrile functional group is found in several drugs, such as Vildagliptin (an antidiabetic drug), Anastrazole (for treating breast cancer), Pericyazine and Periciazine (antipsychotic drugs), and Citalopram (an antidepressant). The nitrile group is quite robust and, in most cases, is not readily metabolized but passes through the body unchanged.

Variations on a Theme: Inorganic Compounds with Triple Bonds Triple bonds are as scarce in the inorganic world as in the organic world. Carbon monoxide (:C≡O:), elemental nitrogen (:N≡N:), and the cyanide ion ($[:C{\equiv}N:]^-$) are the only common examples.

You have seen quite a few functional groups by this time, and it is especially important that you can recognize them in a complex organic molecule. Sample Problem 20.5 provides some practice.

Sample Problem 20.5 Recognizing Functional Groups

Problem Circle and name the functional groups in each molecule:

(a) 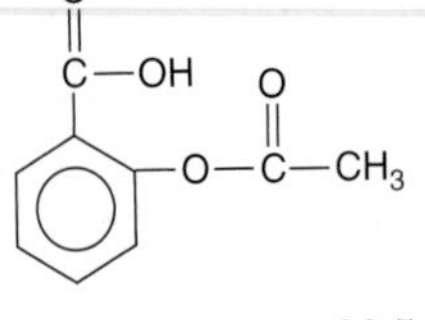

(b) 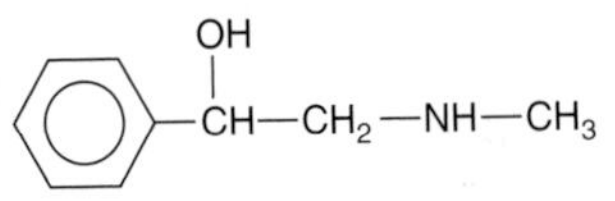

(c) 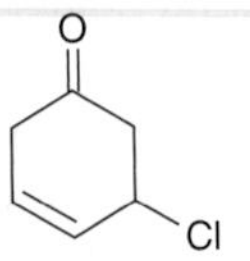

Plan We use Table 20.5 to identify the various functional groups.

Solution

(a) 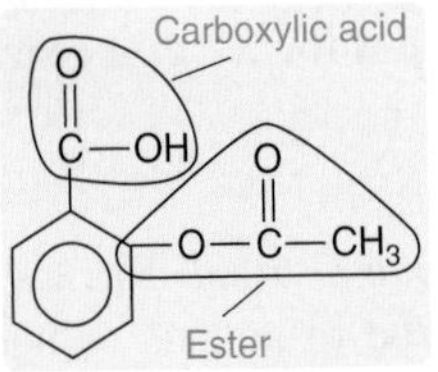

(b) 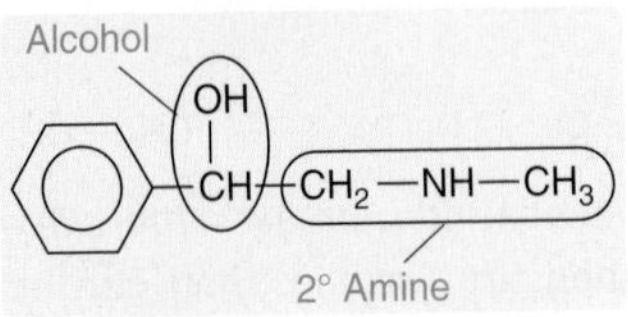

(c) 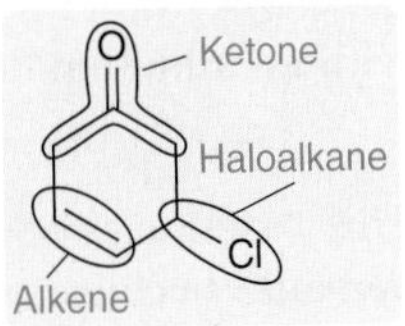

Follow-Up Problem 20.5 Circle and name the functional groups:

(a) 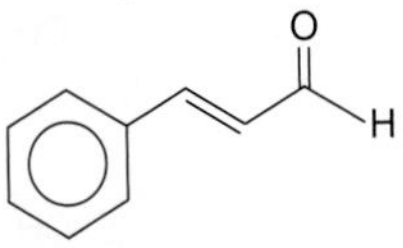

(b) 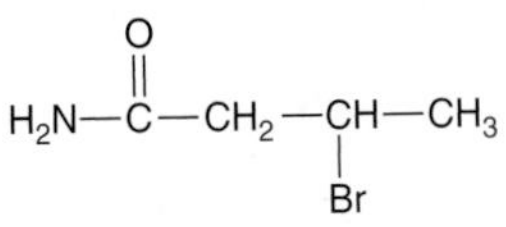

Nomenclature Priorities

When compounds contain more than one functional group, the group with higher priority takes the *suffix* form, whereas the others take the *prefix* form; however, double and triple bonds take only the *suffix* form (-*en* and –*yn*) and are used with other suffixes.

The substituents are ordered alphabetically (*di-*, *tri-*, etc., do not count); for example, the correct name is 4-ethyl-2,3-dimethylheptane, *not* 2,3-dimethyl-4-ethylheptane. If there are multiple functional groups of the same type, the position numbers are ordered numerically; an example is propane-1,2-diol, *not* propane-2,1-diol. Table 20.7 summarizes the nomenclature of the functional groups according to their priority (more examples of compounds with several functional groups can be found in Table 20.8). Thus, the name of the following compound is 2-bromo-6-ethyl-4-hydroxycyclohexanone, *not* 2-bromo-4-hydroxy-6-ethylcyclohexanone:

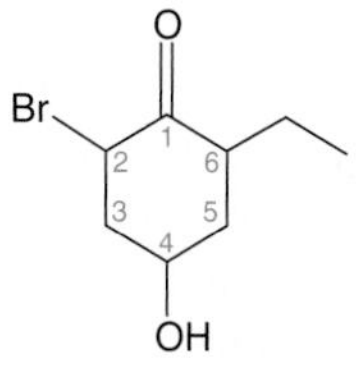

TABLE 20.7 Functional Group Nomenclature Organized by Priority

Increasing priority (↑)

Class	Prefix Name	Suffix Name
Carboxylic acids	carboxy-	-oic acid
Esters	R-oxycarbonyl-	-R-oate
Amides	carbamoyl-	-amide
Nitriles	cyano-	-nitrile
Aldehydes	oxo (formyl)	-al (carbaldehyde)
Ketones	oxo-	-one
Alcohols	hydroxy-	-ol
Amines	amino-	-amine
Alkenes	alkenyl-	-ene
Alkynes	alkynyl-	-yne
Alkane*	alkyl-	-ane
Ethers*	–	-oxy
Haloalkanes*	–	halo-

*Alkanes, ethers, and haloalkanes are all of equal priority (which is to say, of no priority).

TABLE 20.8 Examples of Compounds with More Than One Functional Group

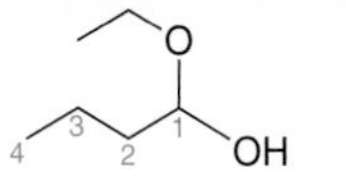	1-ethoxybutan-1-ol
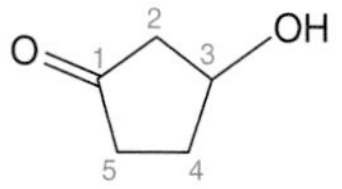	3-hydroxycyclopentanone
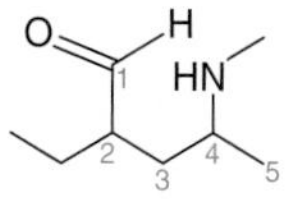	2-ethyl-4-(methylamino)pentanal
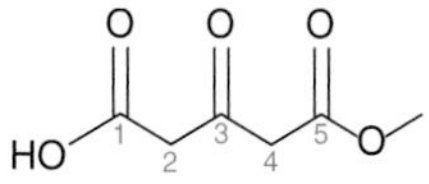	5-methoxy-3,5-dioxopentanoic acid
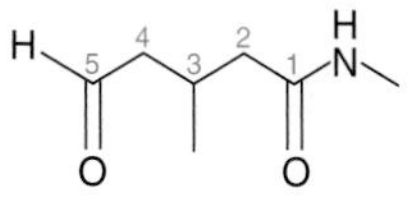	*N*,3-dimethyl-5-oxopentanamide

Naming Carbon Atoms Greek letters (α, β, γ, δ, etc.) are sometimes used to identify carbon atoms. The first carbon atom attached to the functional group is called the alpha carbon (C_α), the second is the beta carbon (C_β), the third is the gamma carbon (C_γ), and so on. If there is a functional group at another carbon, it may be named with the Greek letter; for example, the γ-amine in γ-aminobutanoic acid is on the third carbon of the carbon chain attached to the carboxylic acid group:

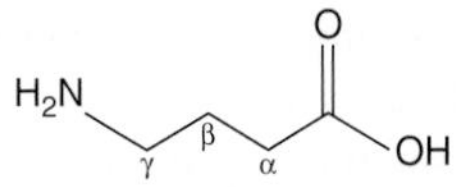

Degrees of Unsaturation: Molecular Formula and Molecular Structure

Frequently, students are asked to draw one or more legitimate structural formulas for a compound with a particular molecular formula. This process can be simplified considerably if you understand that a molecular formula dictates not only the number and type of atoms that must appear in the structural formula but also the number and types of bonds that must be present.

The **degrees of unsaturation (DU)**, also known as the index of hydrogen deficiency or IHD, is the number of π bonds (double and triple bonds, where a double bond contains one π bond and a triple bond contains two π bonds) plus the number of rings in each structure, compared with an alkane with the same number of carbon atoms. DU can be calculated in different ways. It does not give the exact number of rings or double or triple bonds, but rather the sum of the number of rings and double bonds plus twice the number of triple bonds present.

A saturated open-chain molecule contains only single bonds and no rings, which means that a saturated molecule has the maximum number of hydrogen atoms possible to be an acyclic alkane. Thus, the number of hydrogen atoms can be represented by 2C + 2, which is the general molecular representation of an alkane (C_nH_{2n+2}, where n is the number of carbon atoms or C). As an example, for the molecular formula C_3H_4, the number of actual hydrogen atoms needed for the compound to be saturated is 8 [$2C + 2 = (2 \times 3) + 2 = 8$]. The compound needs 4 more hydrogen atoms to be fully saturated (*expected number of hydrogen atoms minus observed number of hydrogen atoms* = $8 - 4 = 4$). Each π bond or ring removes two hydrogen atoms from a structure. This means that the molecule can have two double bonds, one double bond and a ring, or one triple bond. Each of these structures has 2 DU:

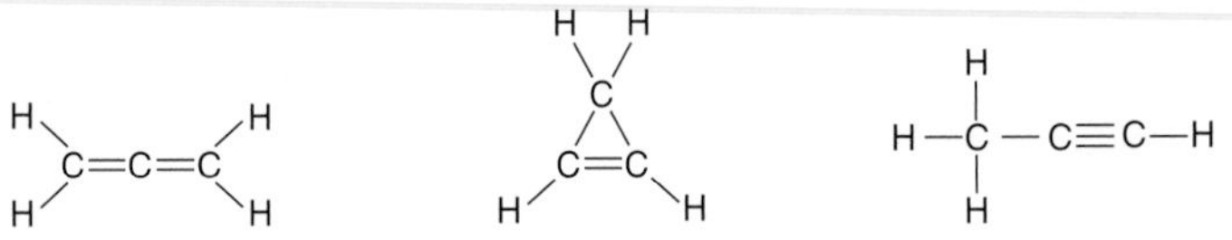

The halogen replaces hydrogen in a compound (it is like adding one hydrogen atom to the structural formula in place of each halogen present). For instance, in chloroethane (C_2H_5Cl), there is one less hydrogen compared with ethane (C_2H_6), but having a halogen makes DU zero. For each nitrogen (or phosphorus) atom present, we deduct one hydrogen atom from the structural formula. For example, there are four chemical structures (isomers) for a compound with molecular formula C_3H_9N, all of them saturated (DU = 0):

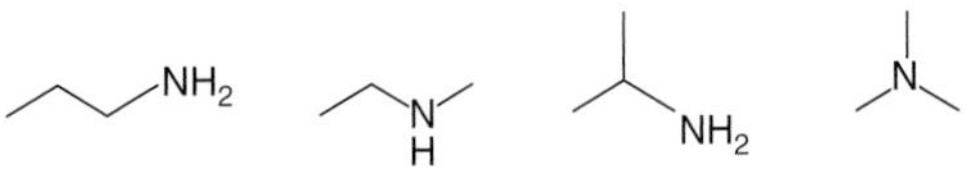

Oxygen and sulfur atoms do not affect DU. As seen in alcohols, ethanol (C_2H_5OH) has the same number of hydrogen atoms as ethane (C_2H_6).

Therefore, we can calculate DU using the following general formula:

$$DU = \frac{2(IV) + (III) - (I) + 2}{2}$$

where

IV is the number of tetravalent atoms (C, Si, Ge)
III is the number of trivalent atoms (N, P)
I is the number of monovalent atoms (H, halogens)

Since carbon, nitrogen, halogens, and hydrogen are the atoms most frequently present, we can write the expression

$$DU = \frac{2C + N - H - X + 2}{2}$$

where

C = number of carbon atoms
N = number of nitrogen atoms
X = number of halogen atoms (F, Cl, Br, I)
H = number of hydrogen atoms

DU only gives the sum of double bonds, triple bonds, and/or rings. For instance, a DU = 3 can contain 2 triple bonds + 1 ring, 1 double bond + 2 rings, 2 double bonds + 1 ring, 3 rings, 3 double bonds, and so on. The following chart illustrates the possible combinations of the number of double bond(s), triple bond(s), and/or ring(s) for a given degree of unsaturation. Any good Lewis structure with that formula will have the same DU. Each row corresponds to a different combination:

DU	Possible Combinations of Rings/Bonds		
	No. of Rings	No. of Double Bonds	No. of Triple Bonds
1	1	0	0
	0	1	0
2	2	0	0
	0	2	0
	0	0	1
	1	1	0

For example, the molecular formula for benzene is C_6H_6. Thus, DU = 4, where C = 6, N = 0, X = 0, and H = 6. This corresponds to benzene containing 1 ring and 3 double bonds:

However, when given the molecular formula C_6H_6, benzene is only one of many possible structures. The following structures all have DU of 4 and have the same molecular formula as benzene:

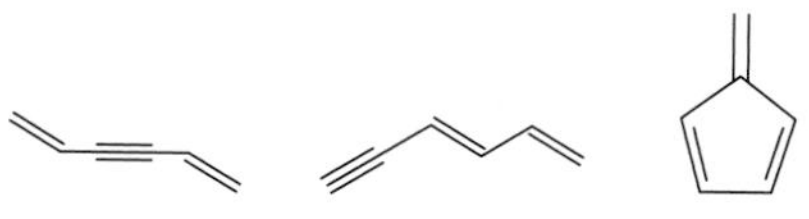

Sample Problem 20.6 Determining DU and Drawing Possible Structures for a Compound

Problem Determine DU and draw four structures that have different atom arrangements for the compound with molecular formula C_6H_6O.

Plan

Calculate Du.
Since the compound has C, H, and O atoms, we can use this formula:

$$DU = \frac{2C + N - H - X + 2}{2}$$

When drawing multiple bonds, one of them is a σ bond and the others are π bonds: each double bond has one π bond (1 unsaturation), each triple bond has two π bonds (2 unsaturations), and each ring has one unsaturation.

Recall functional groups with one oxygen atom: alcohols and ethers have no unsaturation in the functional group, whereas aldehydes and ketones have one unsaturation in the functional group.

Draw each structure following the previous steps.
Check the DU and the number of each atom present.

Solution First, calculate DU knowing that

$C = 6$

$H = 6$

$N = 0$

$X = 0$

$$DU = \frac{2C + N - H - X + 2}{2} = \frac{(2 \times 6) + 0 - 6 - 0 + 2}{2} = 4$$

How can we show 4 DU in a structure? One choice might be to have an oxygen atom with a double bond (aldehyde or ketone) and either three π bonds or two π bonds and a ring.

The following four structures have the molecular formula C_6H_6O and 4 DU:

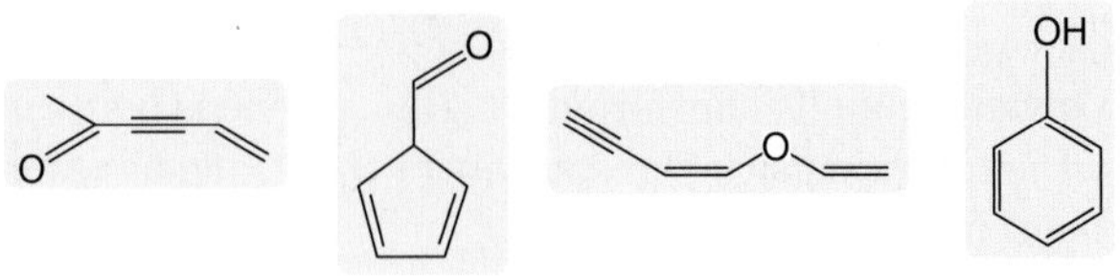

Follow-Up Problem 20.6 **(a)** Draw three more structures of compounds with molecular formula C_6H_6O. **(b)** Calculate DU and draw three possible structures for the compounds with molecular formulas C_5H_7Cl and C_4H_7N.

SUMMARY OF SECTION 20.6

- Common functional groups can be those with only single bonds (alcohols, ethers, haloalkanes, and amines), only double bonds (alkenes, aldehydes, and ketones), both single and double bonds (carboxylic acids, esters, and amides), or triple bonds (alkynes and nitriles).
- The functional group confers specific chemical and physical properties to the molecules.
- DU helps to relate the molecular formula to the molecular structure of a compound.

CHAPTER REVIEW GUIDE

Learning Objectives

Relevant section (§) and/or sample problem (SP) numbers appear in parentheses.

Concepts

1. Explain how the atomic properties of the carbon atom give rise to its ability to form four strong covalent bonds, multiple bonds, and chains, which result in the vast structural diversity of organic compounds. (§20.1)
2. Explain how the atomic properties of the carbon atom give rise to its ability to bond to various heteroatoms, which creates regions of charge imbalance that result in functional groups. (§20.1)
3. Describe common structures and explain how to assign names of alkanes, alkenes, and alkynes. (§20.2 and §20.5)
4. Predict which conformer of an alkane and of a polysubstituted cycloalkane is more stable. (§20.3)
5. Distinguish between the following: constitutional isomers and stereoisomers; constitutional and conformational stereoisomers; and geometric isomers, enantiomers, and diastereomers. (§20.4)
6. Explain the importance of optical isomerism in organisms. (§20.4)
7. Explain how to decide if a compound is aromatic. (§20.5)
8. Draw structures and assign names to compounds with functional groups; describe the properties of the various functional groups. (§20.6)

Skills

1. Draw hydrocarbon structures, given the number(s) of C atoms, multiple bonds, and rings. (SP 20.1)
2. Draw conformations of alkanes and cycloalkanes. (§20.3 and SP 20.2)
3. Name hydrocarbons and draw expanded, condensed, and carbon-skeleton formulas. (§20.2 and §20.5, SP 20.3)
4. Draw geometric isomers and identify stereogenic centres of molecules. (§20.4)
5. Recognize which molecule is aromatic. (SP 20.4)
6. Recognize and name the functional groups in organic molecules. (SP 20.5)
7. Determine the degrees of unsaturation and draw possible structures for a compound. (SP 20.6)

Key Terms

Introduction
organic compounds

Section 20.1
catenation
heteroatoms
functional group

Section 20.2
hydrocarbons
alkane
homologous series
saturated hydrocarbons
expanded formula
condensed formula
carbon-skeleton formula
cyclic hydrocarbon

Section 20.3
conformation
conformer
sawhorse projection
Newman projection
Fischer projection
angle strain
steric strain
torsional strain

Section 20.4
constitutional (structural) isomers
stereoisomers
conformational isomers
geometric isomers
enantiomers
diastereomers
stereogenic centre (stereocentre)
optical isomers
chiral
polarimeter
optically active
Cahn-Ingold-Prelog (CIP) rules

Section 20.5
alkene
unsaturated hydrocarbons
alkyne
aromatic hydrocarbons
Hückel's rule

Section 20.6
alcohol
ether
haloalkane (alkyl halide)
amine
carbonyl group
aldehyde
ketone
carboxylic acid
ester
acid anhydride
hydrolysis
amide
nitrile
degrees of unsaturation (DU)

Brief Solutions to Follow-Up Problems

20.1 (a)

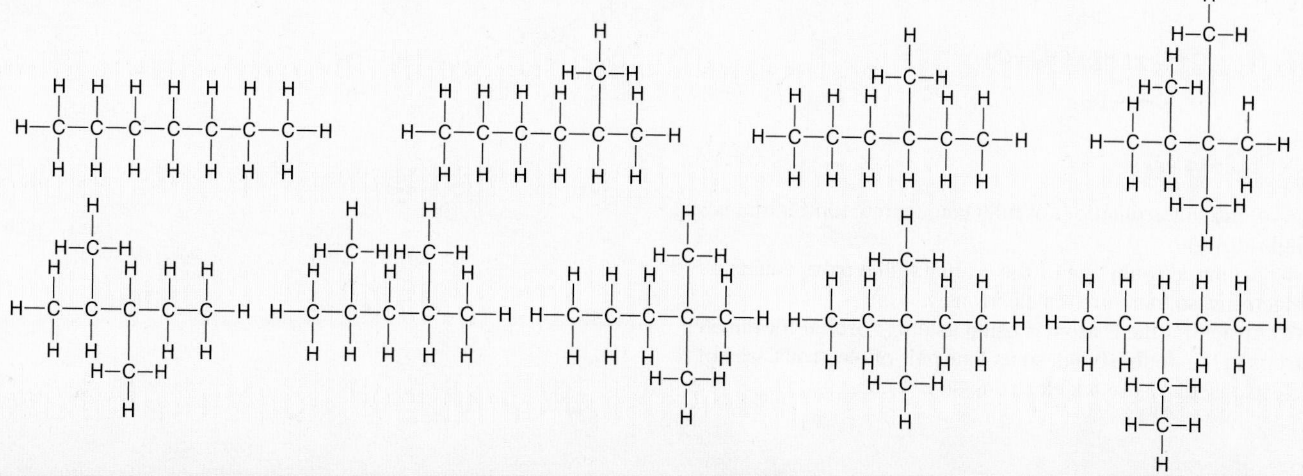

Brief Solutions to Follow-Up Problems (continued)

(b)

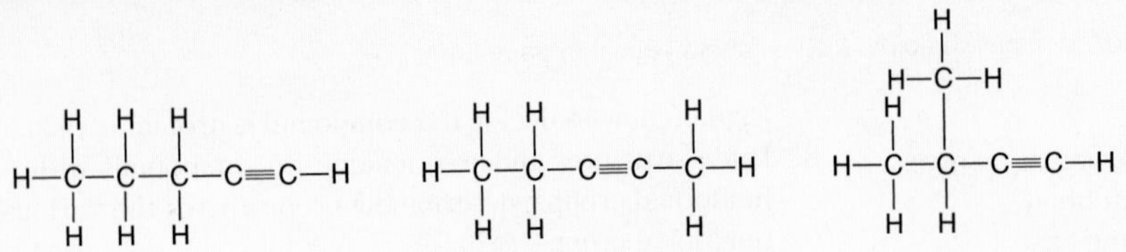

20.2 (a)

Two 1,3-diaxial CH_3-H repulsions
2×3.77 kJ/mol = 7.54 kJ/mol

Two 1,3-diaxial CH_3-H repulsions
2×3.77 kJ/mol = 7.54 kJ/mol

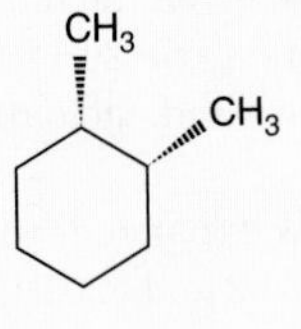

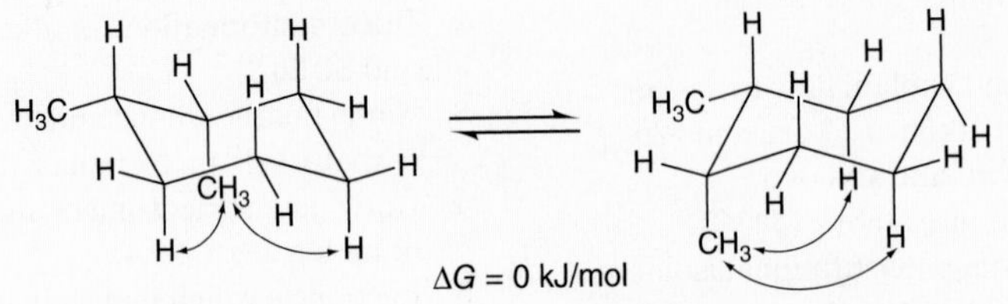

$\Delta G = 0$ kJ/mol

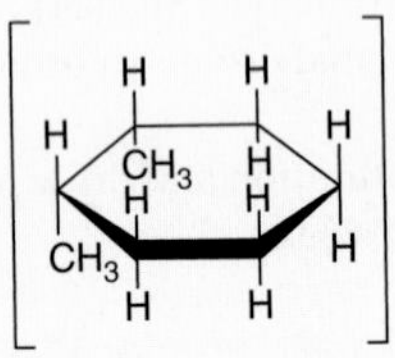

Both conformations are equally stable.

(b)

Four 1,3-diaxial CH_3-H repulsions
4×3.77 kJ/mol = 15.1 kJ/mol

Methyl groups equatorial
No 1,3-diaxial repulsions

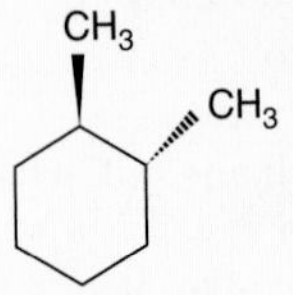

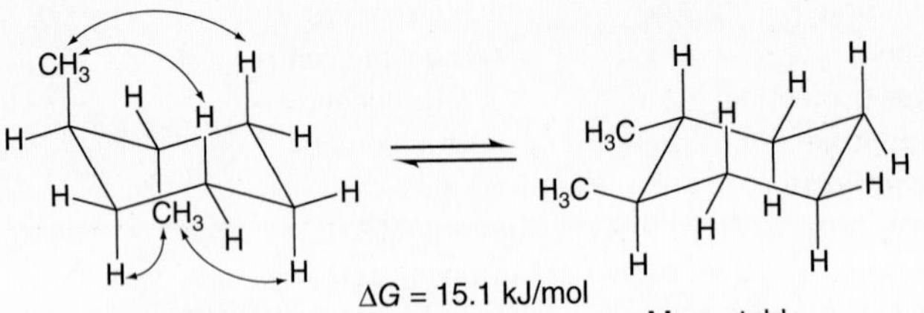

$\Delta G = 15.1$ kJ/mol

Less stable

More stable

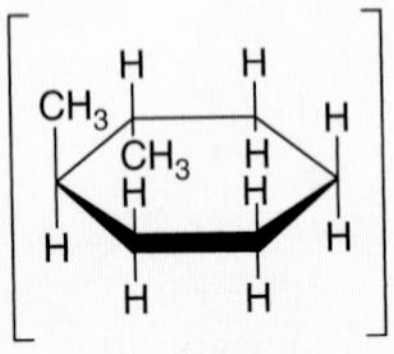

20.3 (a)

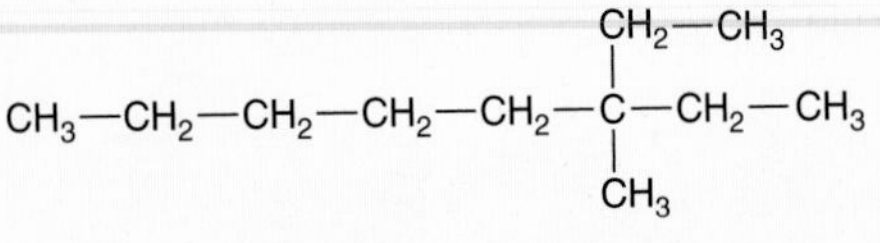

(b)

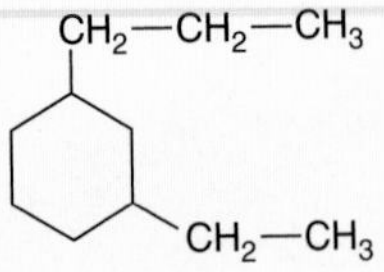

and

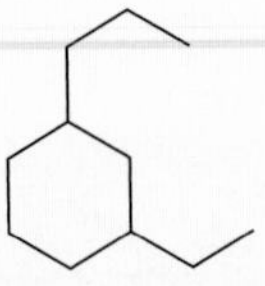

(c)

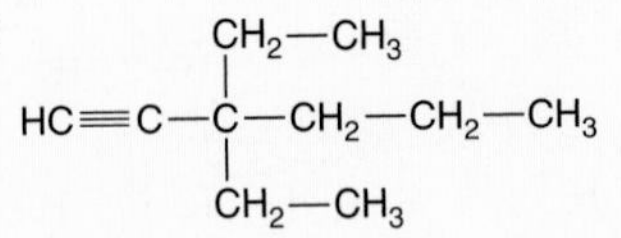

(d)

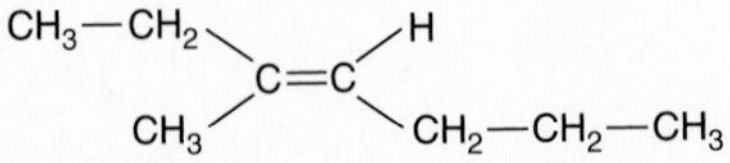

20.4 (a) Not aromatic (not fully conjugated, top C atom is sp^3 hybridized)
(b) Aromatic (only one of the S atom's lone pairs counts as π electrons, so there are 6 π electrons, $n = 1$)
(c) Aromatic (the O atom is using its one p orbital for the electrons in the double bond, so its lone pair of electrons are not π electrons; there are 6 π electrons, so $n = 1$)

20.5 (a)

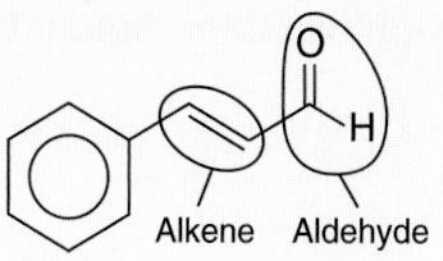

(b)

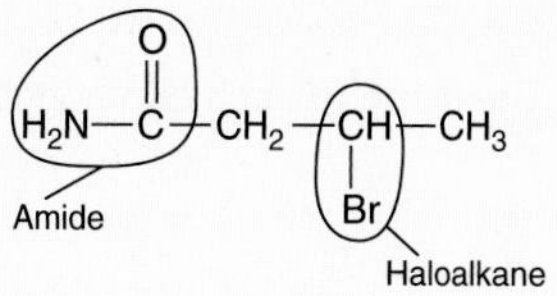

20.6 (a)

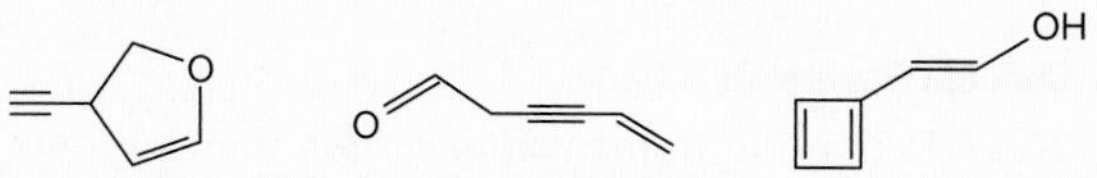

(b) C_5H_7Cl $\quad DU = \dfrac{2C + N - H - X + 2}{2} = \dfrac{(2 \times 5) + 0 - 7 - 1 + 2}{2} = 2$

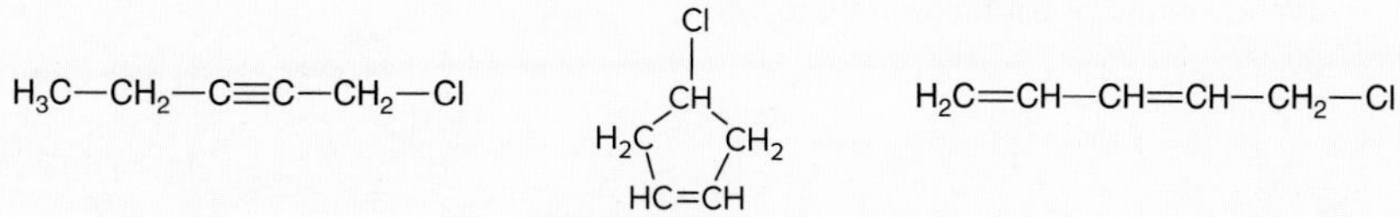

C_4H_7N $\quad DU = \dfrac{2C + N - H - X + 2}{2} = \dfrac{(2 \times 4) + 1 - 7 - 0 + 2}{2} = 2$

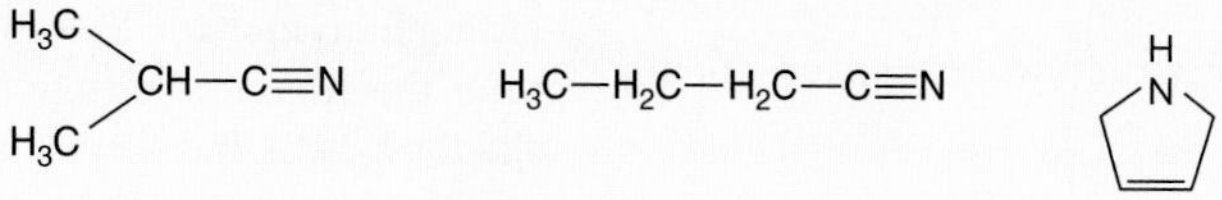

PROBLEMS

Problems with **red** numbers are answered in Appendix G and worked in detail in the Student Solutions Manual. Problem sections match those in the text and provide the numbers of relevant sample problems. Most offer Concept Review Questions, Skill-Building Exercises (usually grouped in pairs covering the same concept), and Problems in Context. The Comprehensive Problems are based on material from any section or previous chapter.

(*Note*: The following abbreviations might be used: Me = methyl, Et = ethyl, Pr = propyl, *t*-Bu = *tert*-butyl or 1,1-dimethylethyl, *i*-Pr = isopropyl or 2-methylpropyl, Ph = phenyl, AcOH = ethanoic acid or acetic acid, OAc = ethanoate or acetate.)

The Special Nature of Carbon and the Characteristics of Organic Molecules

Concept Review Questions

20.1 Give the names and formulas of two carbon compounds that are organic and two that are inorganic.

20.2 Explain each statement in terms of atomic properties:
(a) Carbon engages in covalent rather than ionic bonding.
(b) Carbon has four bonds in all its organic compounds.
(c) Carbon forms neither stable cations, like many metals, nor stable anions, like many nonmetals.
(d) Carbon bonds to itself more extensively than does any other element.
(e) Carbon forms stable multiple bonds.

20.3 Carbon bonds to many elements other than itself.
(a) Name six elements that commonly bond to carbon in organic compounds.
(b) Which of these elements are heteroatoms?
(c) Which of these elements are more electronegative than carbon? Less electronegative?
(d) How does bonding of carbon to heteroatoms increase the number of organic compounds?

20.4 Silicon lies just below carbon in group 14 and also forms four covalent bonds. Why are there not as many silicon compounds as carbon compounds?

20.5 What is the range of oxidation states for carbon? Name a compound in which carbon has its highest oxidation state and one in which it has its lowest.

20.6 Which of these bonds to carbon would you expect to be relatively reactive: C—H, C—C, C—I, C=O, C—Li? Explain.

The Structures of Hydrocarbons and Hydrocarbons with Single Bonds

(Sample Problem 20.1)

Concept Review Questions

20.7 (a) What structural feature is associated with alkanes and cycloalkanes?
(b) Give the general formula for each of them.

20.8 Draw correct structures, by making a single change, for any that are incorrect:

(a)

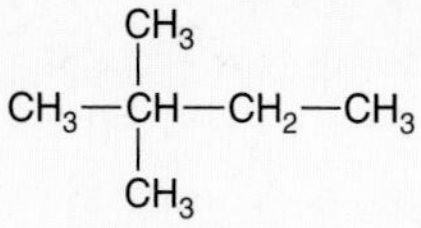

(b)

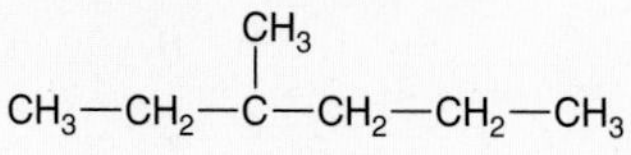

(c)

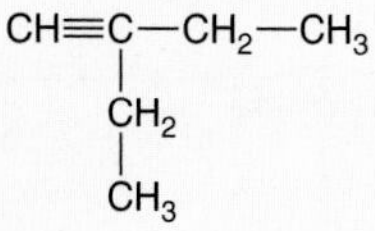

(c)

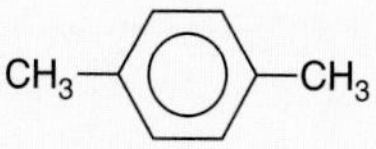

Skill-Building Exercises

20.9 Draw the structure or give the name of each compound:
(a) 2,3-dimethyloctane
(b) 1-ethyl-3-methylcyclohexane

(c) 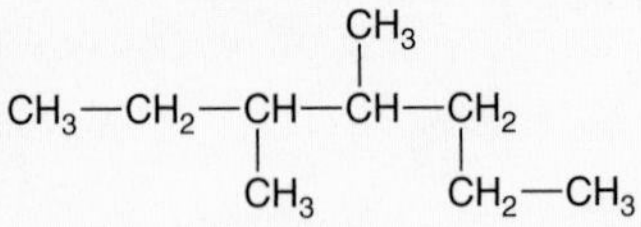

(d) 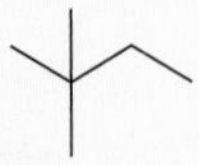

20.10 Draw the structure or give the name of each compound:

(a) 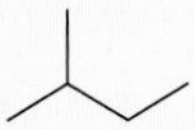 (b) 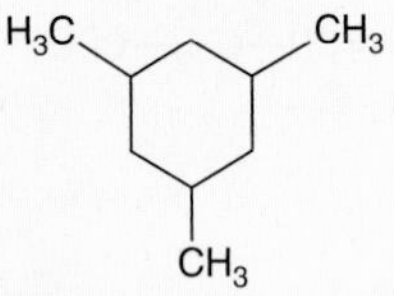

(c) 1,2-diethylcyclopentane
(d) 2,4,5-trimethylnonane

20.11 Each name is wrong. Draw structures based on each name, and correct the name:
(a) 4-methylhexane
(b) 2-ethylpentane
(c) 2-methylcyclohexane
(d) 3,3-methyl-4-ethyloctane

20.12 Each name is wrong. Draw structures based on each name, and correct the name:
(a) 3,3-dimethylbutane
(b) 1,1,1-trimethylheptane
(c) 1,4-diethylcyclopentane
(d) 1-propylcyclohexane

20.13 Identify which centres are primary, secondary, tertiary, or quaternary in the following structure:

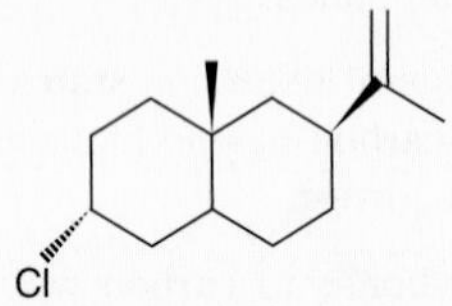

20.14 Convert the structure below into an "expanded" formula representation:

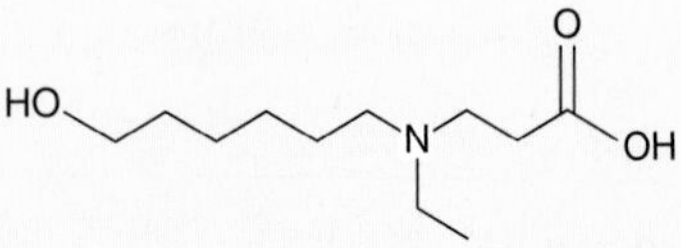

20.15 Convert the structure below into a carbon-skeleton formula:

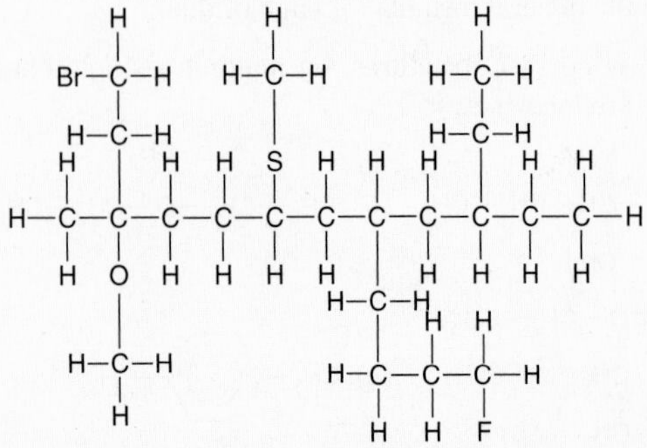

20.16 Represent 5-bromo-4-chloro-2-methylcyclohexanol in carbon-skeleton form.

Conformation of Alkanes

(Sample Problem 20.2)

Concept Review Questions

20.17 Define the terms *conformation* and *configuration*. Explain the difference between the two.

20.18 Below are three conformations of butane. Which is most stable? Why?

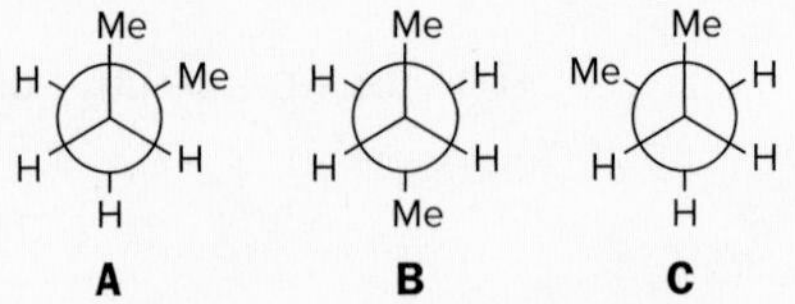

20.19 The proportion of **A** present in the following equilibrium when X = Me, X = Et, and X = *i*-Pr is approximately the same, but much greater for X = *t*-Bu. Given that larger groups should prefer the **A** conformation, explain the apparent anomalous results for X = Et and X = *i*-Pr.

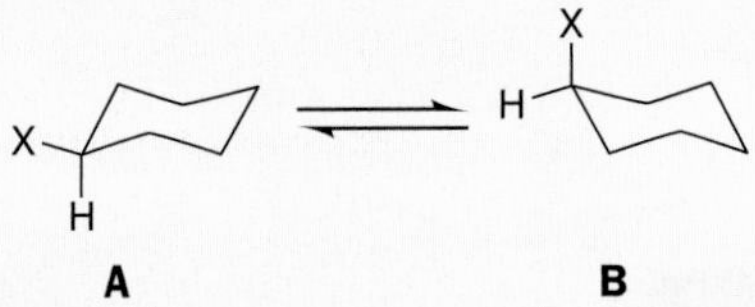

Skill-Building Exercises

20.20 Draw the three staggered conformations of the molecule given below. Which is least stable?

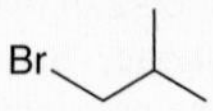

20.21 Convert the following Fischer projection into a carbon-skeleton representation:

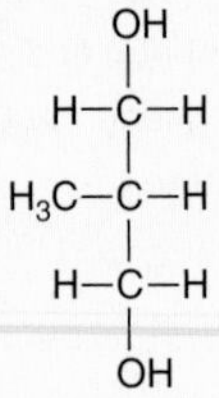

20.22 Convert the Newman projection given below into both a carbon-skeleton representation and a Fischer representation:

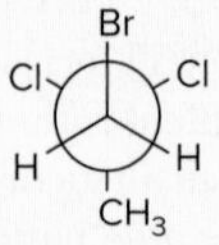

20.23 By considering the two possible chair conformations, predict the most stable conformation of the molecule depicted below:

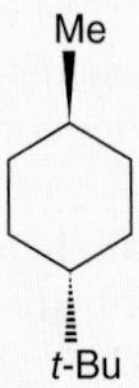

20.24 By considering the two possible conformations, and depicting them in chair representation and Newman projections, predict the most stable conformation of the molecule depicted below:

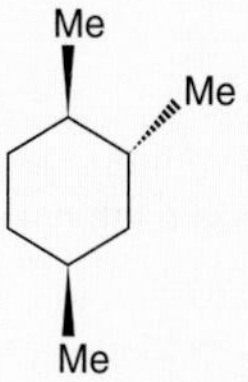

Isomers and Isomerism

Concept Review Questions

20.25 Define each type of isomer: (a) constitutional; (b) geometric; (c) optical. Which types of isomers are stereoisomers?

20.26 Among alkenes, alkynes, and aromatic hydrocarbons, only alkenes exhibit *cis-trans* isomerism. Why do the others not?

20.27 Treatment of alkenes with light can lead to the isomerization of the molecule to give the most stable isomer. Predict what the major product of treatment of each alkene with light would be:

(a) 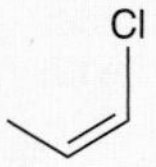

(b) 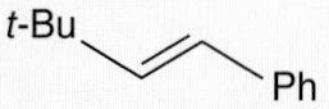

(c) 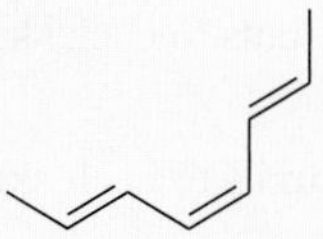

20.28 Define each compound as *E* or *Z* and state whether it is *cis* or *trans* (with reference to the longest carbon chain):

(a) 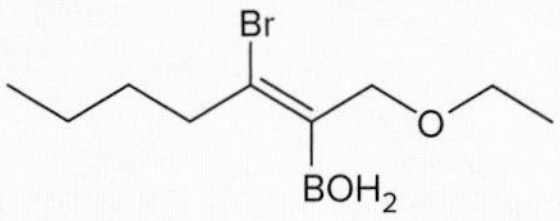

(b) 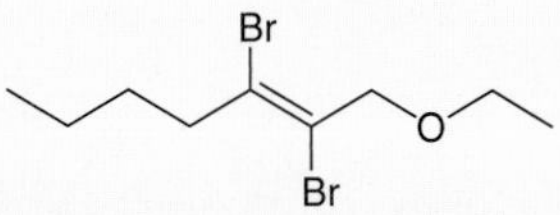

20.29 The following molecule can be isolated as a mixture of equal parts *cis* and *trans* isomers. When compared with a pure sample of the *cis* isomer, what differences would you expect to see in the optical activity of the two samples?

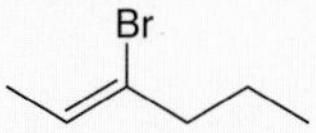

20.30 Many molecules contain chiral centres. Define the requirements of an asymmetric carbon.

20.31 Define the term *racemic*. How would you expect a racemic substance to affect polarized light?

20.32 List the following atoms in order of priority, according to the CIP rules, from lowest to highest: H, Br, D, O, P, I.

20.33 Which objects are asymmetric (have no plane of symmetry): (a) a circular clock face; (b) a football; (c) a dime; (d) a brick; (e) a hammer; (f) a spring?

20.34 Explain how a polarimeter works and what it measures.

20.35 How does an aromatic hydrocarbon differ from a cycloalkane in terms of its bonding? How does this difference affect structure?

20.36 There are two compounds with the name 2-methylhex-3-ene, but only one with the name 2-methylhex-2-ene. Explain with structures.

20.37 Any tetrahedral atom with four different groups attached can be a chiral centre. Which of these species are optically active?
(a) CHClBrF
(b) $NBrCl_2H^+$
(c) $PFClBrI^+$
(d) SeFClBrH

20.38 Give each molecule the appropriate stereochemical assignment:

(a) 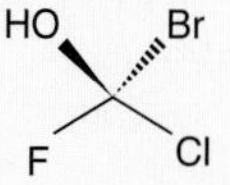

(b) 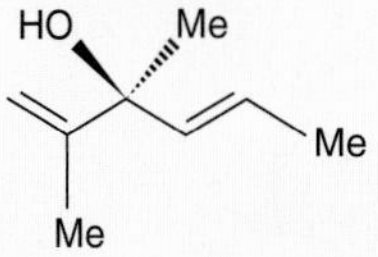

(c) 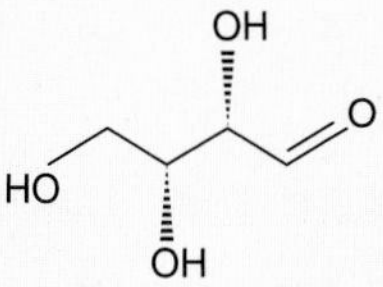

d-threose

20.39 Give the chiral centres in the following Fischer projection the appropriate stereochemical assignments:

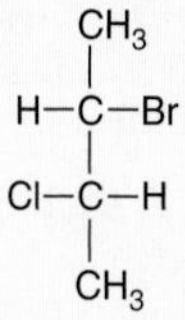

20.40 Deuterium labelling of compounds can be useful in understanding the mechanism of chemical processes. Give the correct stereochemical assignment of the following deuterium-labelled compound:

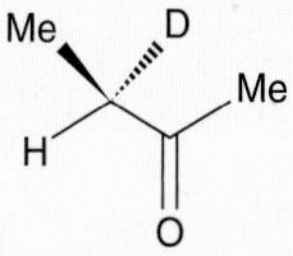

20.41 Draw all the isomers of methylphenol, indicating whether the substituents are *ortho*, *meta*, or *para* substituted.

20.42 Below are two isomers of a compound with the formula C_7H_8O. Which is more stable and why?

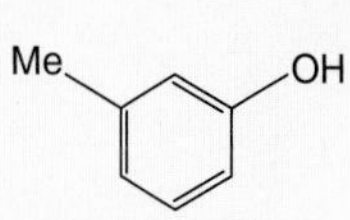

A

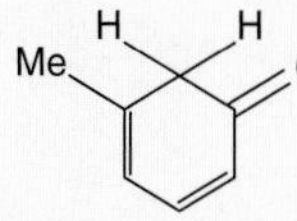

B

20.43 One isomer of trinitrotoluene is shown in Section 20.5. Draw the other five possible isomers.

Skill-Building Exercises (grouped in similar pairs)

20.44 Each compound can exhibit optical activity. Circle the chiral centre(s) in each:

(a) 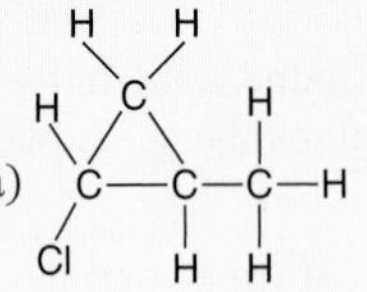

(b) 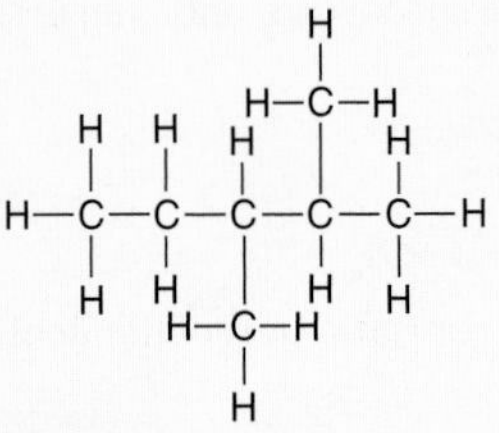

20.45 Each compound can exhibit optical activity. Circle the chiral centre(s) in each:

(a) 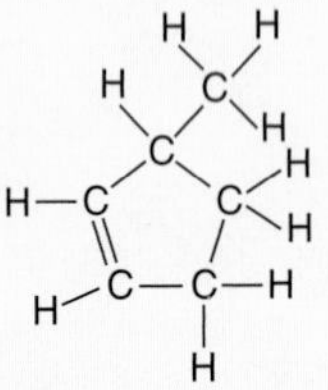

(b) 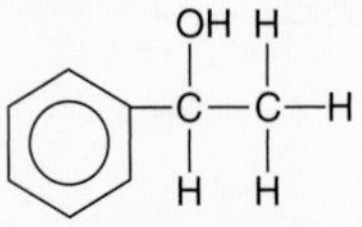

20.46 Draw a structure from each name, and determine which compounds are optically active:
(a) 3-bromohexane
(b) 3-chloro-3-methylpentane
(c) 1,2-dibromo-2-methylbutane

20.47 Draw a structure from each name, and determine which compounds are optically active:
(a) 1,3-dichloropentane
(b) 3-chloro-2,2,5-trimethylhexane
(c) 1-bromo-1-chlorobutane

20.48 Which structures exhibit geometric isomerism? Draw and name the two isomers in each case:

(a) $CH_3—CH_2—CH{=}CH—CH_3$

(b) 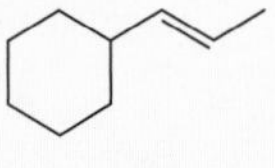

(c) 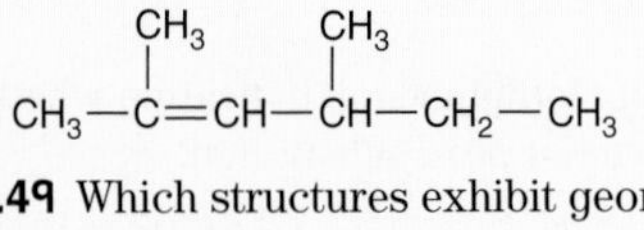

20.49 Which structures exhibit geometric isomerism? Draw and name the two isomers in each case:

(a) 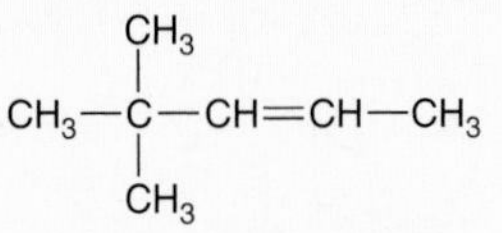

(b) 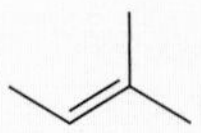

(c) 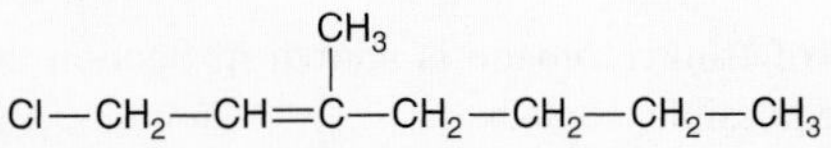

20.50 Which compounds exhibit geometric isomerism? Draw and name the two isomers in each case:
(a) propene
(b) hex-3-ene
(c) 1,1-dichloroethene
(d) 1,2-dichloroethene

20.51 Which compounds exhibit geometric isomerism? Draw and name the two isomers in each case:
(a) pent-1-ene
(b) pent-2-ene
(c) 1-chloropropene
(d) 2-chloropropene

Hydrocarbons with Multiple Bonds

(Sample Problems 20.3 and 20.4)

Concept Review Questions

20.52 (a) What structural feature is associated with alkenes and alkynes?
(b) Give the general formula for each of them.

20.53 Draw all possible skeletons for a 7-C compound with
(a) a 6-C chain and 1 double bond
(b) a 5-C chain and 1 double bond
(c) a 5-C ring and no double bonds

20.54 Draw all possible skeletons for a 6-C compound with
(a) a 5-C chain and 2 double bonds
(b) a 5-C chain and 1 triple bond
(c) a 4-C ring and no double bonds

20.55 Add the correct number of hydrogen atoms to each skeleton in Problem 20.53.

20.56 Add the correct number of hydrogen atoms to each skeleton in Problem 20.54.

20.57 Draw correct structures, by making a single change, for any of the following that are incorrect:

(a) $CH_3{=}CH—CH_2—CH_3$

(b) 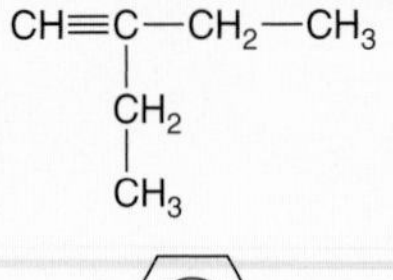

(c) 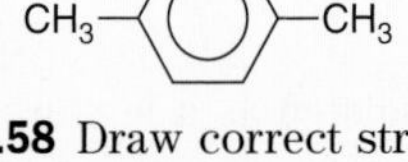

20.58 Draw correct structures, by making a single change, for any of the following that are incorrect:

(a) $CH_3—CH{=}CH—CH_2—CH_3$

(b) 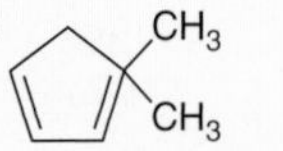

(c) $CH_3—C{\equiv}CH—CH_2—CH_3$

20.59 Explain why, although a C—C single bond is approximately 154 pm and a C=C double bond is approximately 133 pm, the bond lengths between all carbons in benzene are equal and approximately 140 pm.

20.60 By reviewing the chapter, give the common names of methylbenzene, benzenol, methylbenzenol, and dimethylbenzene.

20.61 Consider the structure of the drug epinephrine (Figure 20.26). Can you identify the aromatic component?

20.62 Taking the nitro group as your reference, label the following substituents as *ortho*, *meta*, or *para*:

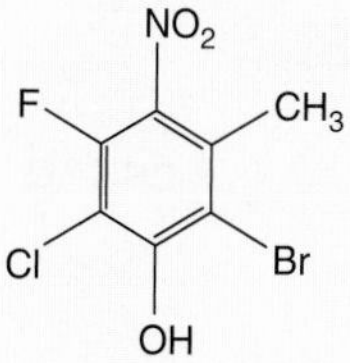

20.63 Why might molecules with a large amount of aromatic components be useful as dyes?

20.64 Name each molecule:

(a)
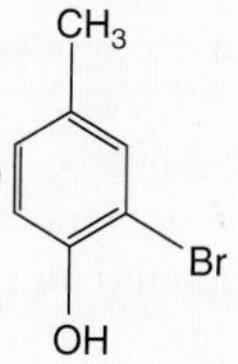

(b)
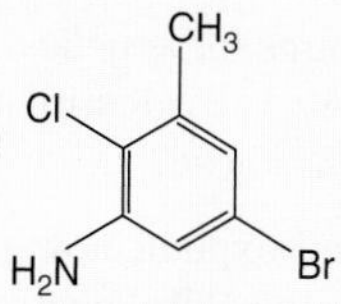

(c)
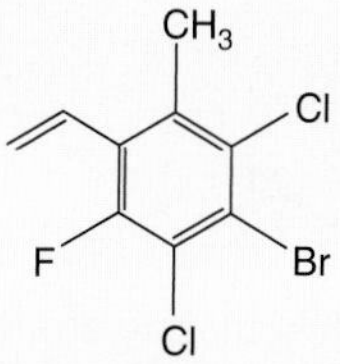

20.65 Which molecule is aromatic?

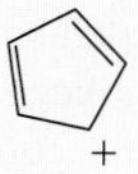
A
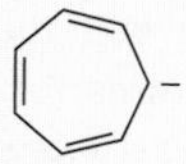
B
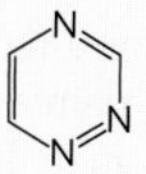
C
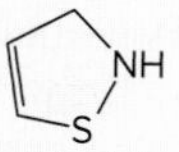
D

20.66 Which molecule is not aromatic?

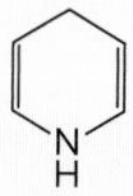
A
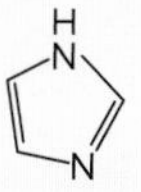
B
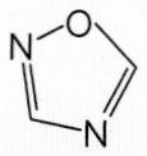
C
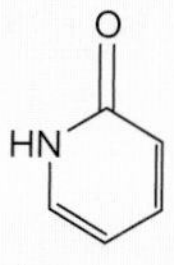
D
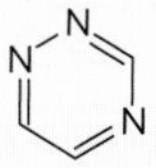
E

Skill-Building Exercises

20.67 Represent 3,5-dichloro-4-methylcyclohex-2-enol in a carbon-skeleton representation:

20.68 Name each compound:

(a)
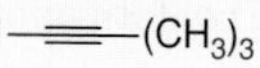

(b)
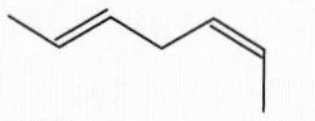

(c)
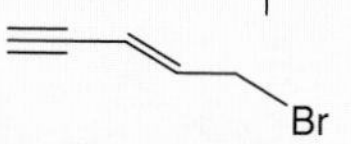

20.69 Draw and name all the constitutional isomers of dichlorobenzene.

20.70 Draw and name all the constitutional isomers of trimethylbenzene.

20.71 Butylated hydroxytoluene (BHT) is a common preservative added to cereals and other dry foods. Its systematic name is 1-hydroxy-2,6-di-*tert*-butyl-4-methylbenzene. Draw the structure of BHT.

Common Functional Groups

(Sample Problems 20.5 and 20.6)

Concept Review Questions

20.72 Arrange the following alcohols in order of increasing predicted boiling point. Why are they in this order?

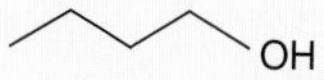
A
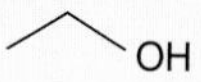
B
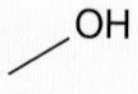
C
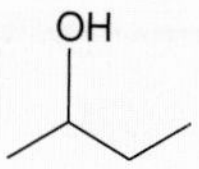
D

20.73 Ethoxyethane (diethyl ether) and butan-1-ol are isomers. Explain why ethoxyethane has a much lower boiling point than butan-1-ol.

20.74 Which molecule would you predict to have the highest melting point?

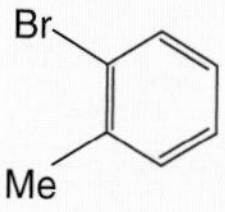
A
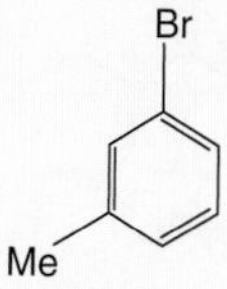
B
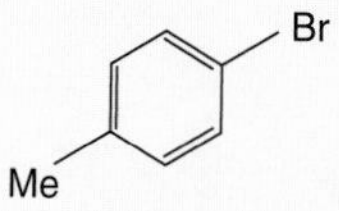
C

Skill-Building Exercises (grouped in similar pairs)

20.75 Name the type of organic compound from each description of the functional group: (a) a polar group that has only single bonds and does not include O or N; (b) a group that is polar and has a triple bond; (c) a group that has single and double bonds and is acidic in water; (d) a group that has a double bond and must be at the end of a C chain.

20.76 Name the type of organic compound from each description of the functional group: (a) an N-containing group with single and double bonds; (b) a group that is not polar and has a double bond; (c) a polar group that has a double bond and cannot be at the end of a C chain; (d) a group that has only single bonds and is basic in water.

20.77 Circle and name the functional group(s) in each compound:

(a) $CH_3—CH=CH—CH_2—OH$

(b)
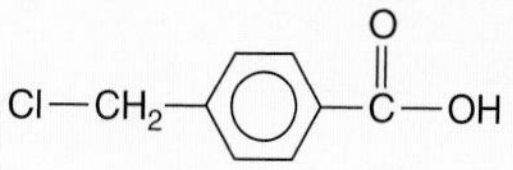

(c)
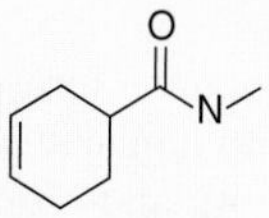

(d) $N\equiv C—CH_2—\overset{\displaystyle O}{\underset{}{C}}—CH_3$ (C=O)

(e)
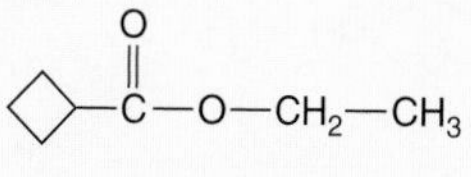

20.78 Circle and name the functional group(s) in each compound:

(a)

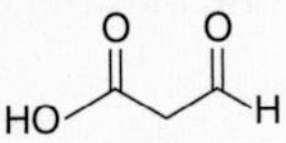

(b) $I{-}CH_2{-}CH_2{-}C{\equiv}CH$

(c) $CH_2{=}CH{-}CH_2{-}\overset{O}{\overset{\|}{C}}{-}O{-}CH_3$

(d) $CH_3{-}NH{-}\overset{O}{\overset{\|}{C}}{-}\overset{O}{\overset{\|}{C}}{-}O{-}CH_3$

(e) $CH_3{-}\underset{}{\overset{Br}{\overset{|}{C}H}}{-}CH{=}CH{-}CH_2{-}NH{-}CH_3$

20.79 Draw all alcohols with the formula $C_5H_{12}O$.

20.80 Draw all aldehydes and ketones with the formula $C_5H_{10}O$.

20.81 Draw all amines with the formula $C_4H_{11}N$.

20.82 Draw all carboxylic acids with the formula $C_5H_{10}O_2$.

20.83 Determine DU for the following compound:

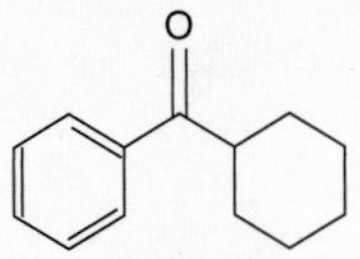

$C_{13}H_{16}O$

20.84 Determine DU for the following compound

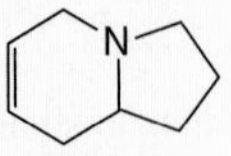

$C_8H_{13}N$

20.85 Determine DU for the following compound: $C_6H_6N_2O_2Cl_2$

20.86 Determine DU for the following compound: C_6H_7N

Problems in Context

20.87 A compound has the molecular formula $C_5H_8O_2$. Given that the molecule contains two alcohols, calculate DU and give a possible structure of the molecule.

20.88 A compound has the molecular formula $C_7H_6O_2$. Given that the molecule reacts with a base to form a salt, calculate DU and give a possible structure of the molecule.

20.89 A compound has the molecular formula $C_7H_{11}N$. Given that when the molecule is reduced it yields an amine but does not contain an alkene, calculate DU and give a possible structure of the molecule.

Comprehensive Problems

20.90 Structures A, B, and C show the three common forms of vitamin A. Match each structure with the correct vitamin A compound: retinal (an aldehyde), retinol (an alcohol), and retinoic acid (a carboxylic acid):

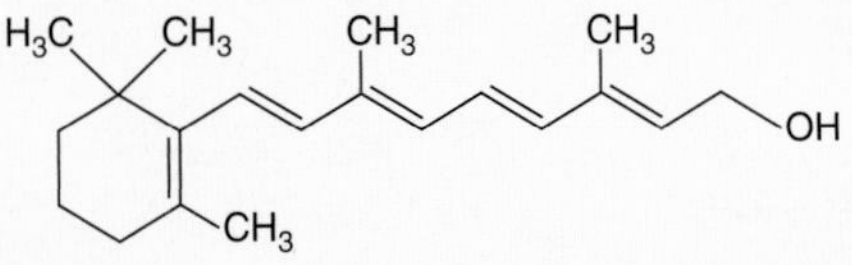

A

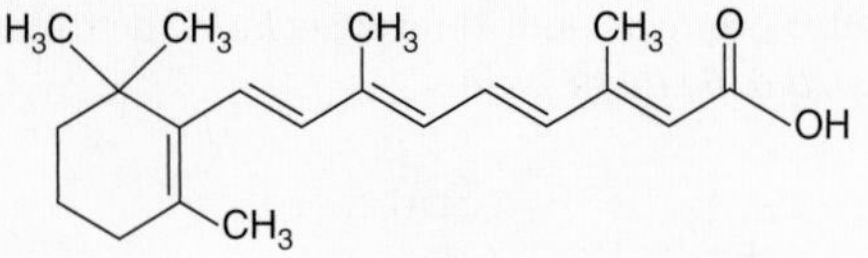

B

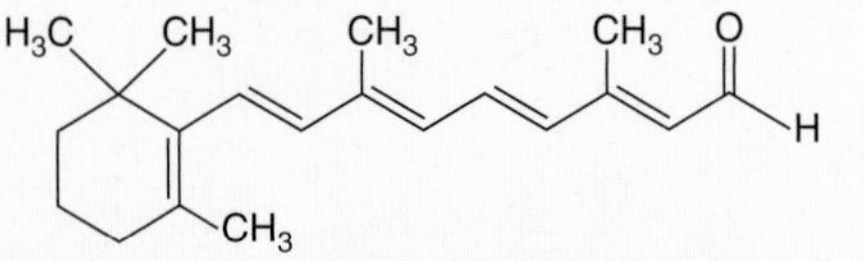

C

20.91 Draw and name all the possible structures for molecules with the general formula C_nH_{2n+2}, where $n = 5$. What type of molecules are these? Rank them in order of increasing boiling point.

20.92 A five-carbon linear alkane has had two bromine atoms substituted for two of its hydrogen atoms.
(a) Draw all possible isomers, ignoring the stereochemistry.
(b) Draw all possible stereoisomers of 2,3-dibromopentane and label the diastereomer pairs. Name the isomers following the CIP naming rules.

20.93 Draw and name all the possible structures for acyclic molecules with the general formula C_nH_{2n}, where $n = 5$. Identify any geometric isomers. What type of molecules are these?

20.94 (a) Name the following molecules and explain how you chose the priority for each group attached to the chiral centre:

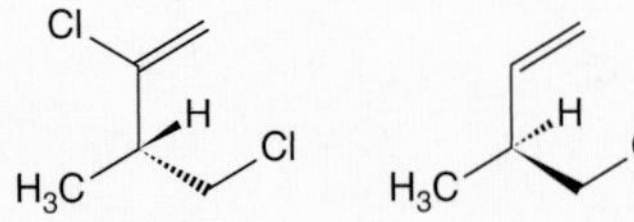

(b) How would the naming be affected if the chlorine atom attached to the alkene were moved to the terminal position? Provide the names of any new isomers generated.

20.95 Determine DU for a molecule with the molecular formula $C_7H_{10}O_2$. Draw four possible structures for this molecular formula, being certain to include at least *one* structure containing each of the following groups: aldehyde, alkyne, aromatic ring, ketone. Is it possible to draw a single structure with this formula that contains ALL of these groups?

20.96 Draw both chair conformations of the following molecules:

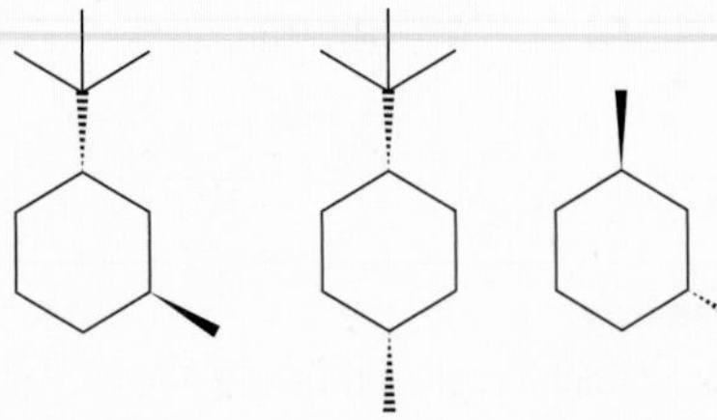

A B C

Indicate which chair conformation is the most stable and name each compound. Explain how you chose the most stable conformation.

20.97 Use Hückel's rules to determine which compounds are aromatic:

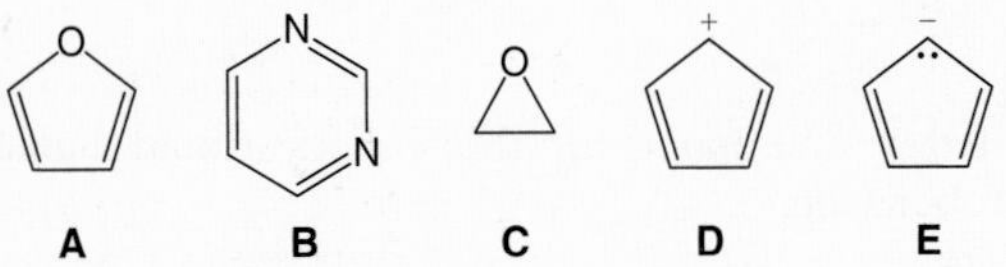

A B C D E

Indicate the chiral centre(s) on each molecule.

20.98 The amide (*R*)-*N*-((*S*)-2-bromopropyl)-3-chloropentanamide can be hydrolyzed to yield a carboxylic acid and a primary amine. Draw the structure of (*R*)-*N*-((*S*)-2-bromopropyl)-3-chloropentanamide as well as the structures of the two hydrolysis products. How would the name of the hydrolysis products change if the bromine atom were replaced with an —OH group?

20.99 Name each compound using both the common name (*ortho*, *meta*, *para*) and the standard numbering system:

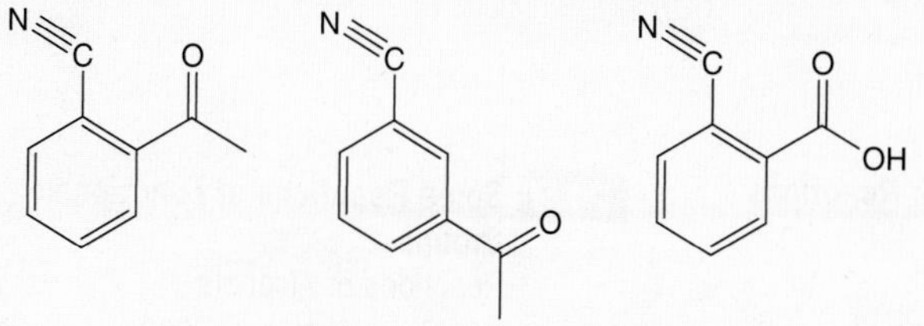

A B C

Why does the name change for the third molecule? Which compound would you expect to have the highest boiling point?

20.100 (a) Draw and name the molecule that has four carbon atoms in its linear backbone and a heteroatom functional group at the second carbon. The molecule is weakly polar and capable of accepting hydrogen bonds. (*Hint*: If you are not careful when working with this molecule, you may soon find yourself unconscious or with an explosive mixture.)
(b) Draw and name the molecule that has six carbons in its linear backbone and contains a functional group attached at the fifth carbon that tends to give compounds a fishy odour. The first carbon contains the functional group, which takes priority over all other groups.
(c) Do any of the molecules drawn have a chiral centre?

Organic Reaction Mechanisms

IN THIS CHAPTER . . . We study how organic compounds react. By the end of this chapter, you should be able to

- Describe and differentiate between the main types of organic reactions, including nucleophilic substitution, elimination, and electrophilic aromatic substitution, and discuss the factors that affect these reactions
- Describe reaction mechanisms, and explain what the processes of bond breaking and bond forming are and how they produce new products
- Classify organic reactions in terms of the key functional groups that characterize the different families of organic compounds

21

Concepts and Skills to Review before Studying This Chapter

- $\Delta\chi$ (difference in electronegativity) and bond polarity (Section 8.5)
- Resonance structures (Section 8.6)
- Orbital hybridization (Section 10.1)
- VSEPR theory (Section 9.1)
- σ and π bonding (Section 10.2)
- Intermolecular forces (Section 11.3)
- Transition state (Section 14.5)
- Reaction mechanisms (Section 14.6)
- Acid-base strength (Section 16.1)
- Brønsted–Lowry acids and bases (Section 16.1)
- Lewis acids and bases (Section 16.8)
- The structures of hydrocarbons and hydrocarbons with single bonds (Section 20.2)
- Conformation of alkanes (Section 20.3)
- Isomers and isomerism (Section 20.4)
- Hydrocarbons with multiple bonds (Section 20.5)
- Common functional groups (Section 20.6)

At this moment, if you look around you, without moving, you will likely see a computer; a phone; clothes; shoes; a table; possibly food; and, if so, something to hold the food. Unless it is metal, ceramic, or glass, the probability is very high that what you are looking at is organic. Computers contain huge amount of plastics, as do phones; our clothes are either natural fibre or synthetic; our shoes contain plastics and synthetics; and the containers for food, whether plastic and reusable or plastic wrap, are organic as well. Much of our food has been modified, as well, or contains organic additives. There are millions of organic compounds and yet, interestingly, they are made using just a few types of organic reactions. In this chapter, we will look at some of the key types of organic reactions that provide us with so many of our necessities of life. A sound understanding of these reactions will lead us to a better comprehension of the more advanced organic chemistry we will see in Chapter 22.

21.1 Some Important Classes of Organic Reactions

Thus far, we have classified chemical reactions based on the chemical process involved (precipitation, acid-base, or redox). We take a similar approach here with organic reactions, but in reverse order.

From here on, we use an uppercase R with a single bond, R—, to signify a general organic group attached to one of the atoms shown; you can usually picture R— as an **alkyl group**, a saturated hydrocarbon group with one bond available to link to another atom. Thus, R—CH_2—Br has an alkyl group attached to a CH_2 group bonded to a Br atom, R—CH=CH_2 is an alkene with an alkyl group attached to one of the carbon atoms in the double bond, and so forth. (Often, when more than one R group is present, we write R, R′, R″, and so forth, to indicate that these groups may be different.)

Types of Organic Reactions

Three important organic reaction types are *substitution*, *elimination*, and *addition* reactions, which can be identified by comparing the *number of bonds to C* in reactants and products.

Substitution Reactions A **substitution** reaction occurs when an atom (or group) from an added reactant substitutes for one attached to a carbon in the organic reactant:

Note that the C atom is bonded to the *same number* of atoms in the product as in the reactant. The C atom may be saturated or unsaturated, and X and Y can be many different atoms, but generally *not* C. The main flavour ingredient in banana oil (3-methylbutyl ethanoate or isopentylacetate), for instance, forms through a substitution reaction between ethanoyl chloride (acetyl chloride) and 3-methylbutan-1-ol. Note that the O substitutes for the Cl:

$$
\underset{\text{O}}{\overset{\text{O}}{CH_3-\overset{\|}{C}-Cl}} \;+\; HO-CH_2-CH_2-\underset{CH_3}{\overset{|}{CH}}-CH_3 \;\longrightarrow\; CH_3-\overset{\|}{\underset{\text{O}}{C}}-O-CH_2-CH_2-\underset{CH_3}{\overset{|}{CH}}-CH_3 \;+\; H-Cl
$$

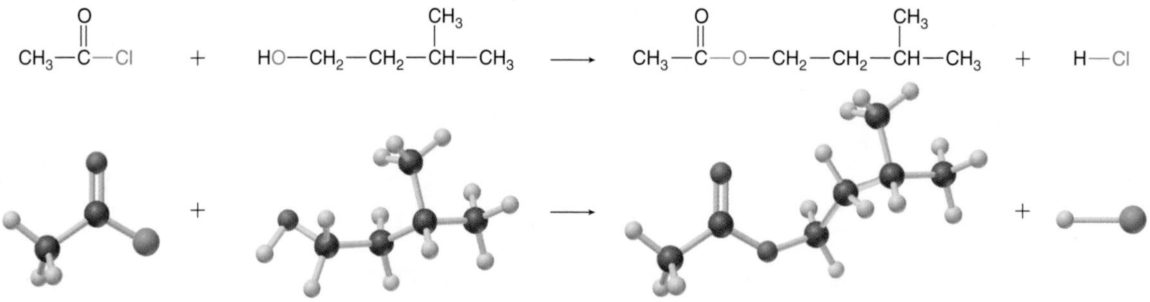

Elimination Reactions **Elimination** reactions occur when a saturated reactant becomes an unsaturated product:

$$
R-\underset{\overset{|}{X}}{\overset{\overset{Y}{|}}{CH}}-CH_2 \;\longrightarrow\; R-CH{=}CH_2 \;+\; X-Y
$$

$$
R-\overset{\overset{Y}{|}}{\underset{\underset{H}{|}}{C}}-\overset{\overset{X}{|}}{\underset{\underset{H}{|}}{C}}-R' \;\longrightarrow\; R-C{\equiv}C-R' \;+\; H-X \;+\; H-Y
$$

$$
R-\overset{\overset{H}{|}}{\underset{\underset{H}{|}}{C}}-\overset{\overset{X}{|}}{\underset{\underset{Y}{|}}{C}}-R' \;\longrightarrow\; R-C{\equiv}C-R' \;+\; H-X \;+\; H-Y
$$

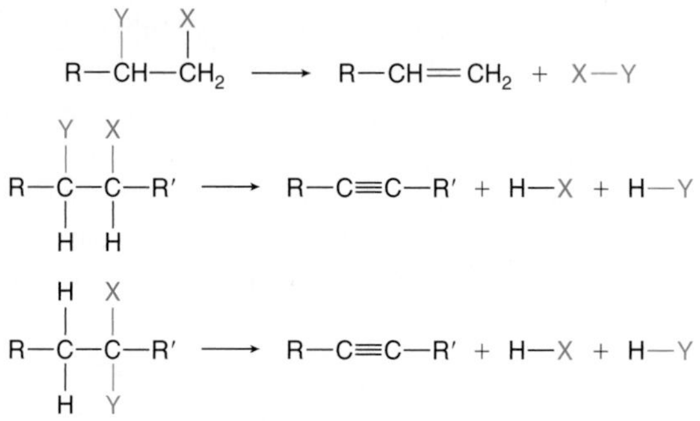

Note that for every two σ bonds that break, one π bond is formed. Therefore, the C atoms are bonded to *fewer* atoms in the product than in the reactant. A pair of halogen atoms, an H atom and a halogen atom, or an H atom and an —OH group are typically eliminated to form each π bond, but C atoms are not. Thus, the driving force for many elimination reactions is the formation of a small, stable molecule, such as HCl(*g*) or H₂O, that increases the entropy of the system (Section 12.2):

$$
CH_3-\overset{\overset{OH}{|}}{CH}-\overset{\overset{H}{|}}{CH_2} \;\xrightarrow{H_2SO_4}\; CH_3-CH{=}CH_2 \;+\; H-OH
$$

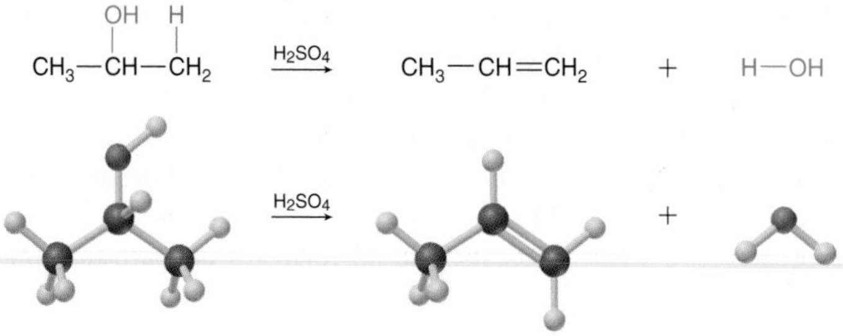

Addition Reactions **Addition** reactions are the opposite of elimination reactions. An addition reaction occurs when an unsaturated reactant becomes a saturated product:

$$
R-CH{=}CH-R' \;+\; X-Y \;\longrightarrow\; R-\overset{\overset{X}{|}}{CH}-\overset{\overset{Y}{|}}{CH}-R'
$$

$$
R-C{\equiv}C-R' \;+\; X-Y \;\longrightarrow\; R-\overset{\overset{Y}{|}}{C}{=}\overset{\overset{X}{|}}{C}-R'
$$

$$
R-\overset{\overset{O}{\|}}{C}-H \;+\; H-Y \;\longrightarrow\; R-\overset{\overset{OH}{|}}{\underset{\underset{H}{|}}{C}}-Y
$$

$$
R-\overset{\overset{O}{\|}}{C}-R' \;+\; H-Y \;\longrightarrow\; R-\overset{\overset{OH}{|}}{\underset{\underset{R'}{|}}{C}}-Y
$$

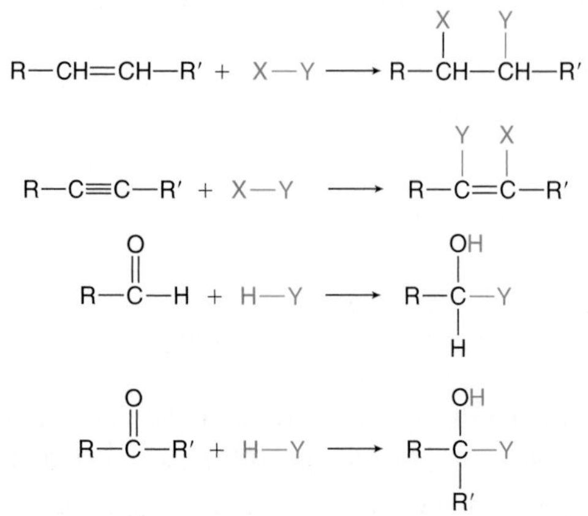

Note that for every π bond that breaks, two σ bonds are formed. Therefore, the C atoms are bonded to *more* atoms in the product than in the reactant.

The C=C and C≡C bonds and the C=O bond commonly undergo addition reactions. In each case, the π bond breaks, leaving the σ bond intact. In the product, the two C atoms (or C and O) form two additional σ bonds. Let us examine the standard enthalpy of reaction ($\Delta_r H°$) for a typical addition to see why they occur. (*Note*: This is only an approximation, and it is only valid in the gas phase.) Consider the reaction between ethene (common name, ethylene) and HCl:

$$CH_2{=}CH_2 \quad + \quad H{-}Cl \quad \longrightarrow \quad H{-}CH_2{-}CH_2{-}Cl$$

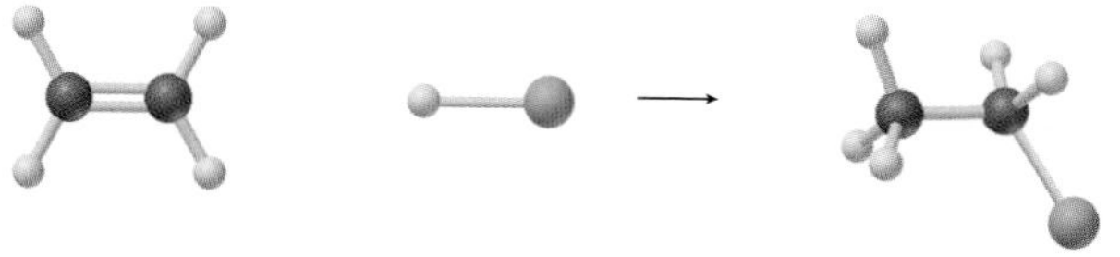

Reactants (bonds broken)	Product (bonds formed)
1 C=C = 614 kJ/mol	1 C—C = −347 kJ/mol
4 C—C = 1652 kJ/mol	5 C—H = −2065 kJ/mol
1 H—Cl = 427 kJ/mol	1 C—Cl = −339 kJ/mol
Total = 2693 kJ/mol	Total = −2751 kJ/mol

$$\Delta_r H° = \Sigma\Delta_{\text{bonds broken}}H° + \Sigma\Delta_{\text{bonds formed}}H° = 2693\text{ kJ/mol} + (-2751\text{ kJ/mol}) = -58\text{ kJ/mol}$$

The reaction is exothermic. By looking at the *net* change in bonds, we see that the driving force for many additions is the formation of two σ bonds (in this case, C—H and C—Cl) from one σ bond (in this case, H—Cl) and one relatively weak π bond. An addition reaction is the basis of a colour test for the presence of π bonds (in C=C and C≡C groups, and in C=O groups with alpha hydrogen atoms) (Figure 21.1).

A B

FIGURE 21.1 A colour test for π bonds (C=C and C≡C groups, and C=O groups with alpha hydrogen atoms). The *brown liquid* Br_2 (in pipette) is expected to react with a colourless compound that has a π bond (in beaker **A**), and its orange-brown colour disappears. For example,

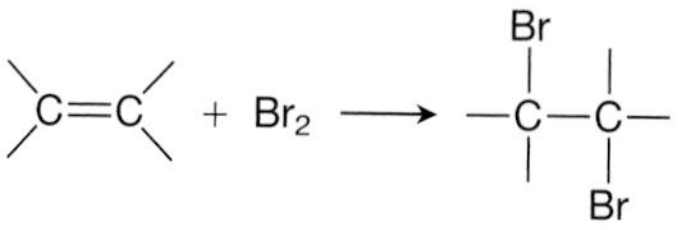

B This compound has no π bonds, so the Br_2 does not react and just colours the mixture.

Sample Problem 21.1 Recognizing the Type of Organic Reaction

Problem State whether each reaction is an addition, an elimination, or a substitution:

(a) $CH_3{-}CH_2{-}CH_2{-}Br \longrightarrow CH_3{-}CH{=}CH_2 + HBr$

(b) 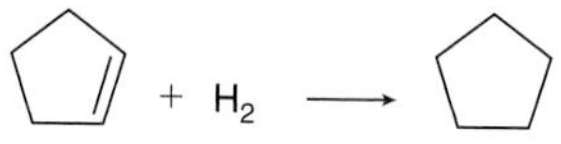

(c) $CH_3C({=}O){-}Br + CH_3CH_2OH \longrightarrow CH_3C({=}O){-}OCH_2CH_3 + HBr$

Plan We determine the type of reaction by looking for any change in the number of atoms bonded to C:

- More atoms bonded to C—an *addition*
- Fewer atoms bonded to C—an *elimination*
- Same number of atoms bonded to C—a *substitution*

Solution (a) Elimination: Two bonds in the reactant, C—H and C—Br, are absent in the product, so fewer atoms are bonded to C.

(b) Addition: Two more C—H bonds have formed in the product, so more atoms are bonded to C.

(c) Substitution: The reactant C—Br bond becomes a C—O bond in the product, so the same number of atoms are bonded to C.

Follow-Up Problem 21.1 Write a balanced equation for each reaction:

(a) An addition reaction between but-2-ene and Cl_2

(b) A substitution reaction between $CH_3{-}CH_2{-}CH_2{-}Br$ and OH^-

(c) The elimination of H_2O from $(CH_3)_3C{-}OH$

(See Brief Solutions to Follow-Up Problems at the end of the chapter.)

The Redox Process in Organic Reactions

An important process in many organic reactions is *oxidation-reduction*. Even though a **redox** reaction always involves both an oxidation and a reduction, organic chemists typically *focus on the organic reactant only.* Instead of monitoring the change in oxidation numbers of the various C atoms in a redox reaction, chemists usually note the movement of electron density around a C atom by counting the number of bonds to more electronegative atoms (usually O) or to less electronegative atoms (usually H). A more electronegative atom takes some electron density from the C, whereas a less electronegative atom gives some electron density to the C.

- When a C atom in the organic reactant forms more bonds to O or fewer bonds to H, thus losing some electron density, the reactant is oxidized, and the reaction is called an **oxidation**.
- When a C atom in the organic reactant forms fewer bonds to O or more bonds to H, thus gaining some electron density, the reactant is reduced, and the reaction is called a **reduction**.

The most dramatic redox reactions are combustion reactions. All organic compounds contain C and H atoms and burn in excess O_2 to form CO_2 and H_2O. For ethane, the reaction is

$$2CH_3{-}CH_3 + 7O_2 \longrightarrow 4CO_2 + 6H_2O$$

Obviously, when ethane is converted to CO_2 and H_2O, each of its C atoms has more bonds to O and fewer bonds to H. Thus, ethane is oxidized, and this reaction is referred to as an *oxidation*, even though O_2 is reduced as well.

Most oxidations do not involve a breakup of the entire molecule, however. When propan-2-ol reacts with potassium dichromate in acidic solution (a common oxidizing agent in organic reactions), it forms propan-2-one (acetone):

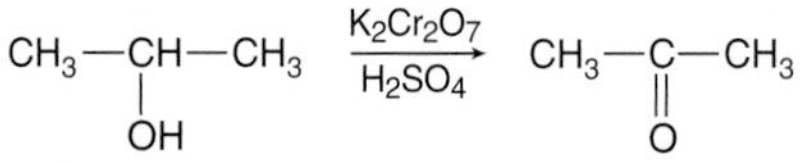

Propan-2-ol Propan-2-one

Note that C-2 has one less bond to H and one more bond to O in propan-2-one than it does in propan-2-ol. Thus, propan-2-ol is oxidized, so this is an *oxidation*. Do not forget, however, that the dichromate ion is reduced at the same time:

$$Cr_2O_7^{2-} + 14H^+ + 6e^- \longrightarrow 2Cr^{3+} + 7H_2O$$

The addition reaction between H_2 and an alkene is also a *reduction*:

$$CH_2{=}CH_2 + H_2 \xrightarrow{\text{Pd}} CH_3{-}CH_3$$

Note that each C has more bonds to H in ethane than it has in ethene, so the ethene is reduced. The H_2 is oxidized in the process, and the palladium shown over the arrow acts as a catalyst to speed up the reaction (Section 14.7). Therefore, it is worth pointing out that many redox reactions are also additions, eliminations, or substitutions.

SUMMARY OF SECTION 21.1

- In a substitution reaction, one atom bonded to C is replaced by another, but the total number of atoms bonded to the C does not change.
- In an elimination reaction, a π bond forms, with the two C atoms bonded to fewer atoms in the product.
- In an addition reaction, a π bond breaks, and the two C atoms (or one C and one O) are bonded to more atoms in the product.
- In an organic redox reaction, the organic reactant is oxidized if at least one of its C atoms forms more bonds to O atoms (or fewer bonds to H atoms), and it is reduced if a C atom forms more bonds to H (or fewer bonds to O).

21.2 Understanding Reactions

The key is to understand where the available electrons are and where they are going, since the reactions are interactions between electron-deficient and electron-rich molecules. Electron-rich molecules are more likely to share electrons by making a new bond and are known as **nucleophiles** (*nucleus-loving*). Electron-deficient molecules want to accept electrons by making a new bond and are known as **electrophiles** (*electron-loving*). Therefore, all molecules or ions with a free pair of electrons or at least a π bond can act as a nucleophile and donate electrons. So, by definition, nucleophiles are *Lewis bases*. On the other hand, electrophiles can be positively charged species or uncharged species, such as a *Lewis acid*, which are attracted to an electron-rich centre.

Reaction Mechanisms: Bond Breaking and Bond Making

A reaction mechanism is a step-by-step sequence of how bonds are broken and formed. It explains how the overall reaction proceeds from the reactant(s) or starting material to the product(s).

A reaction can occur in one step, in a *concerted reaction*, or in a series of steps, in a *stepwise reaction*. In both cases, two variables need to be considered: *kinetics*, which deals with the rate of the reaction (see Chapter 14), and *thermodynamics*, which tells us on which side of the equilibrium the reaction lies (see Chapter 18).

- In a **concerted reaction**, the reactants are converted directly into products:

$$\text{reactants} \longrightarrow \text{products}$$

The following is a general reaction energy diagram for a *concerted* mechanism:

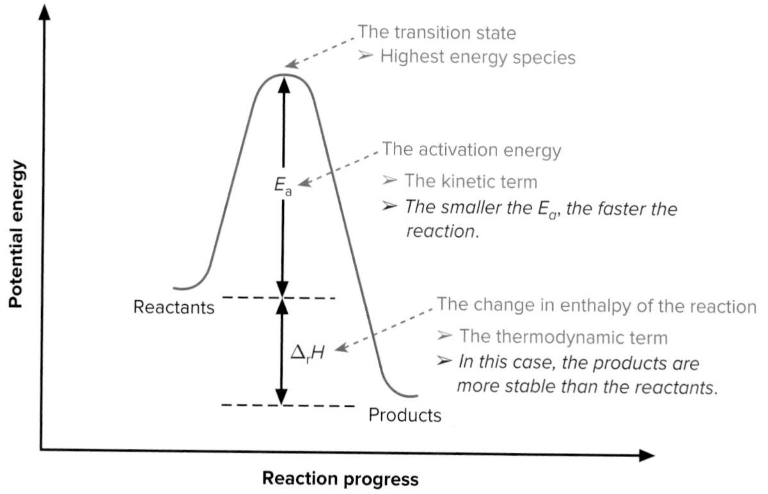

- In a **stepwise reaction**, the reactants form a less stable intermediate, which goes on to form the products:

$$\text{reactants} \longrightarrow \text{intermediate} \longrightarrow \text{products}$$

The following is a general reaction energy diagram for a *stepwise* mechanism. The *rate-determining step* (RDS) has the highest activation energy:

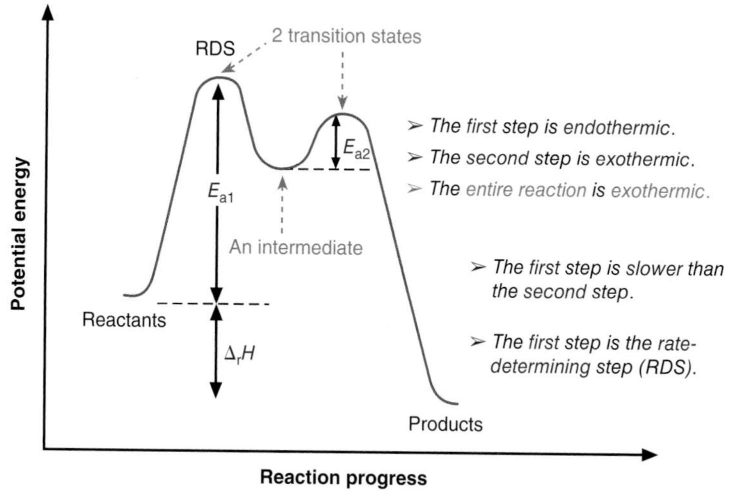

Bond Cleavage and Bond Formation We can indicate the flow of electrons in a reaction with an arrow. A full-headed curved arrow (↷) shows the movement of two electrons (an electron pair), whereas a half-headed curved arrow (⇀) shows the movement of a single electron.

- **Bond cleavage.** When the electrons in the bond are equally divided, we call it a **homolytic cleavage**, forming a pair of **radicals** (reactive intermediates with a single unpaired electron):

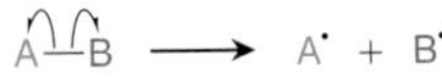

When the bond is unequally divided, the more electronegative atom gets the electrons, and we call it a **heterolytic cleavage (heterolysis)**:

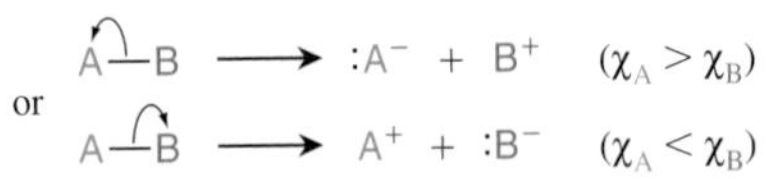

- **Bond formation.** Like bond cleavage, bond formation can occur in two ways. Two *radicals* (atoms, molecules, or ions with unpaired electrons) can get together, sharing their single electrons to form a σ bond:

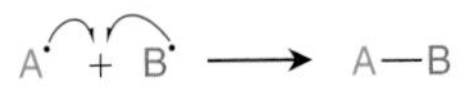

Alternatively, a **Lewis base** (nucleophile) donates a pair of electrons to a **Lewis acid** (electrophile) to form a σ bond:

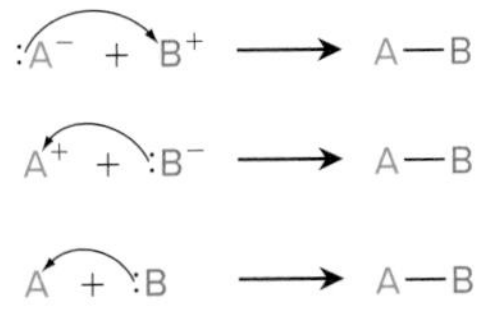

Sample Problem 21.2 Using Curved Arrows to Show the Flow of Electrons in a Reaction

Problem **1.** Write the product(s) for each reaction based on the curved arrows:

(a) 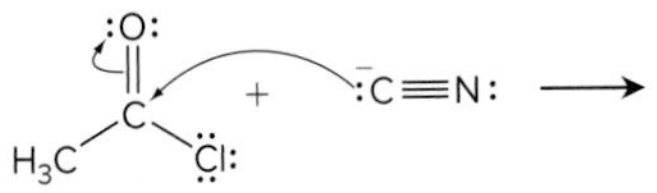

(b) 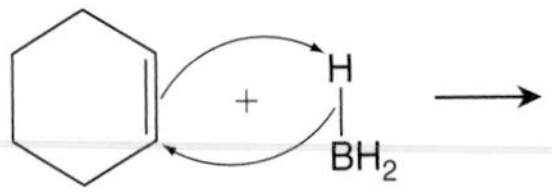

(c) 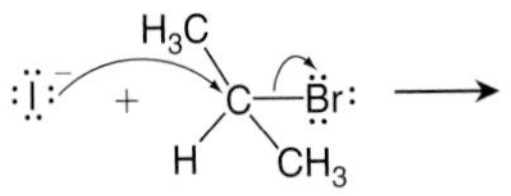

2. Draw curved arrows to show how each reaction occurs:

(a) 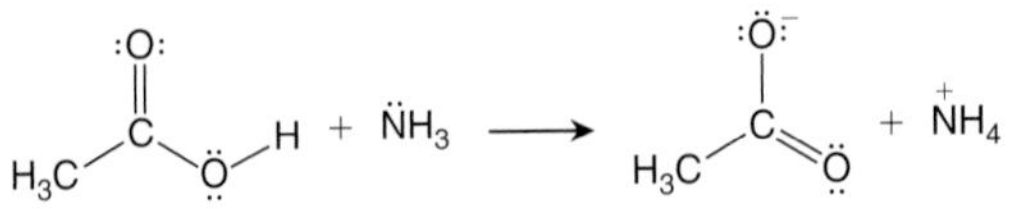

(b) 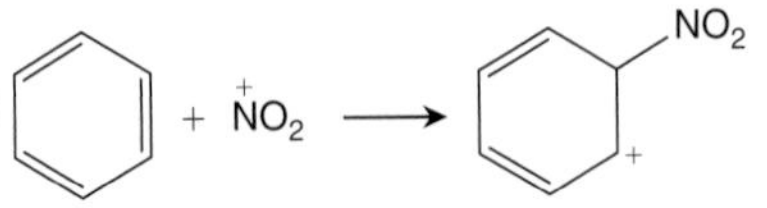

Plan The arrows show the movement of electrons. If the arrow starts at a bond between two atoms, then that bond is broken. If the arrow ends between two atoms, then a new bond is formed between those atoms.

Solution **1. (a)** 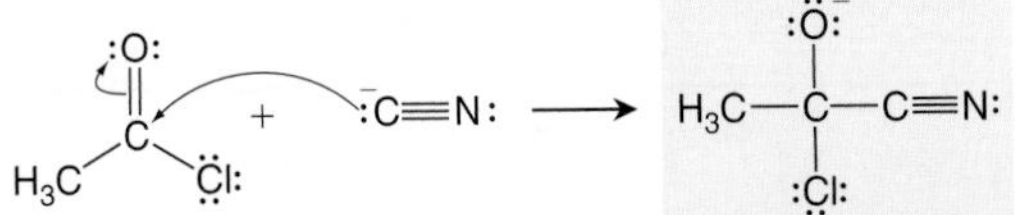

(b) 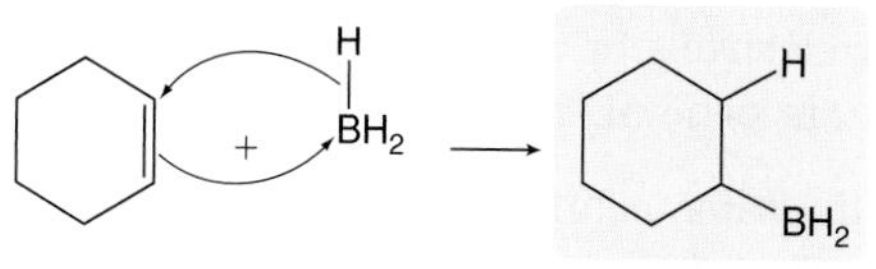

(c) 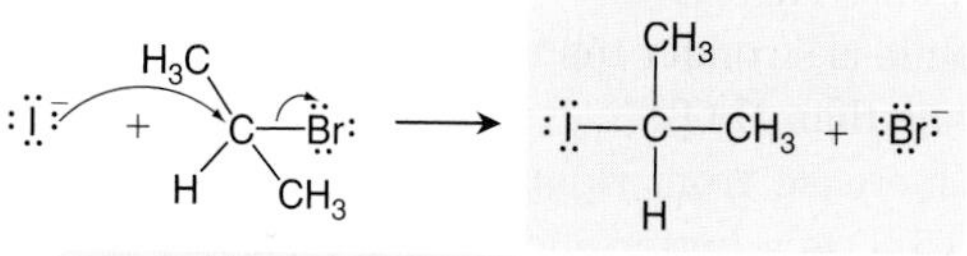

2. (a) 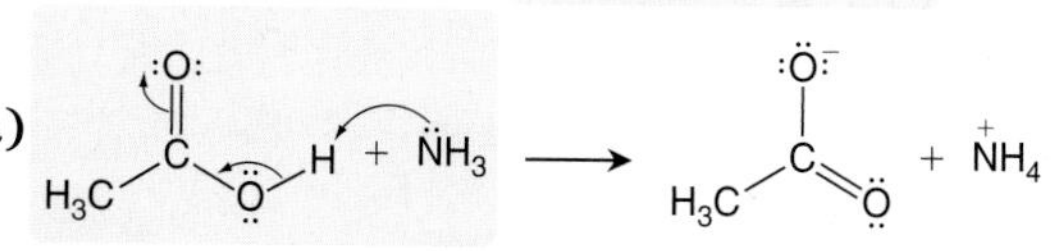

(b) 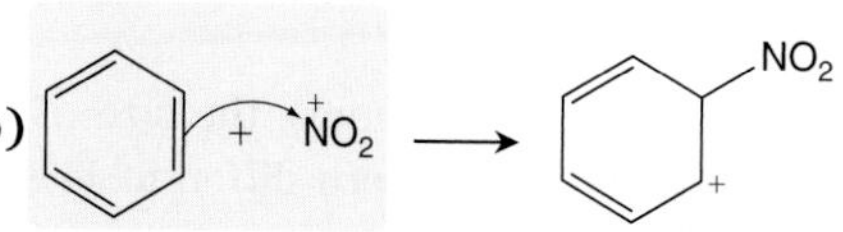

Follow-Up Problem 21.2 **1.** Write the product(s) for each reaction based on the curved arrows:

(a) 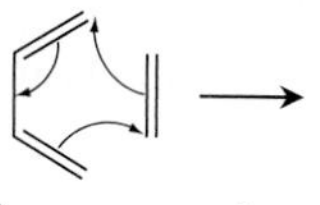 (b) 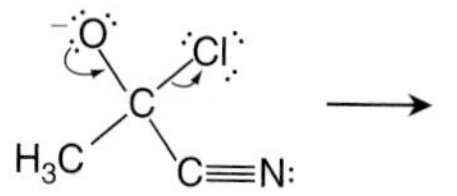 (c) 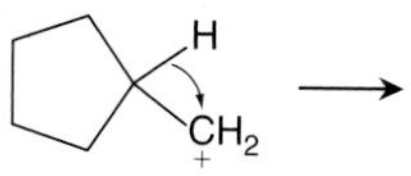

2. Draw curved arrows to show how each reaction might occur:

(a) 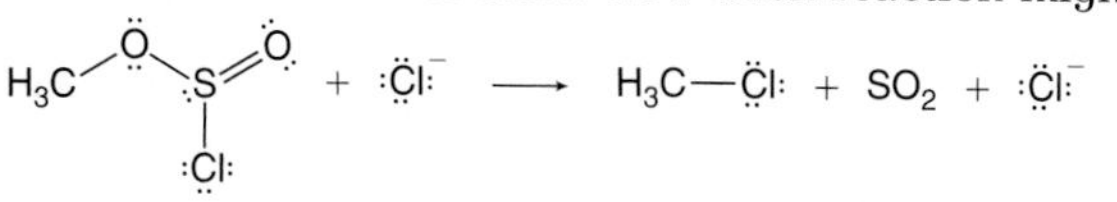

(b) 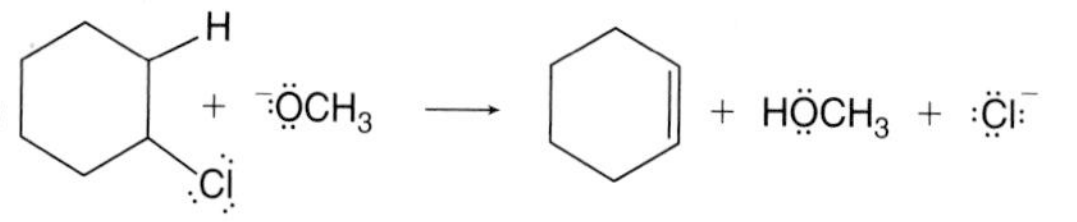

(c) 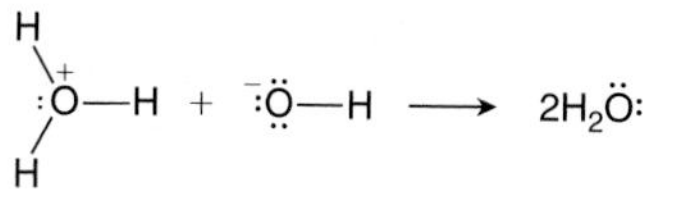

Nucleophiles and Electrophiles

The foundation of most reactions in organic chemistry (besides radical reactions) can be summarized in one statement: "*Nucleophiles attack electrophiles.*" The *nucleophile* (electron-rich species that may or may not be negatively charged or species with π bonding electrons) donates an electron pair to an *electrophile* (electron-poor species that may or may not be positively charged) to form a chemical bond in a reaction. As we learned when we studied molecular orbital (MO) theory (Section 10.3), the highest energy electrons are donated—often they are nonbonding (that is, lone pairs), but they can be π bonding electrons.

Recall that the driving force for polar reactions is an electrostatic attraction, so look for polar bonds in molecules:

- $^{\delta+}C—X^{\delta-}$ (the carbon atom is bonded to a more electronegative atom, such as F, Cl, Br, N, O)
- $^{\delta-}C—M^{\delta+}$ (the carbon atom is bonded to a less electronegative atom, such as a metal)

It is extremely important to bear in mind that we always *push the arrow from the nucleophile to the electrophile*, as described in the following general polar reaction, where curved arrows indicate the direction of *electron flow*, *not* atom flow:

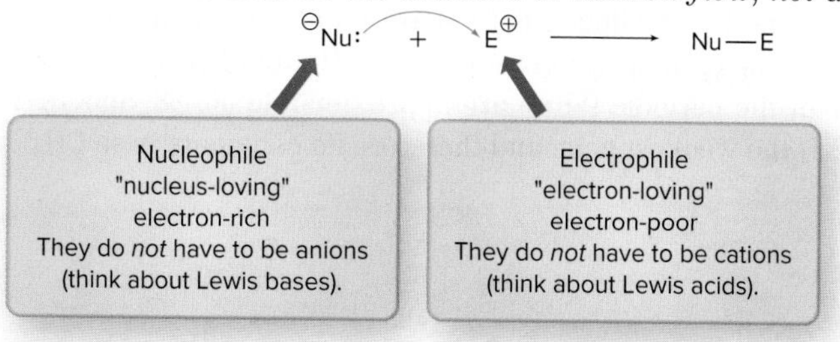

A *nucleophile* must have a lone pair of electrons or a π bond: the more accessible the lone pair, the better the nucleophile. In general, we have the following:

- When comparing nucleophiles with the same central atom, the more basic compound will be the better nucleophile. Therefore,
  - a negatively charged nucleophile is stronger than its conjugate acid (for example, OH^- is a better nucleophile than H_2O)
  - nucleophilicity and basicity increase from right to left across a row of the periodic table (for example, CH_3^- is a better nucleophile than F^-)

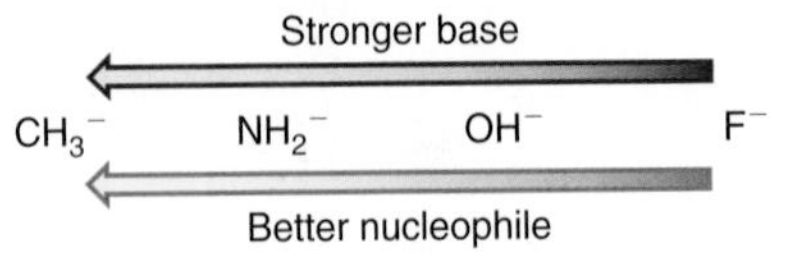

- When comparing nucleophiles with different nucleophilic atoms, the size of the atoms must be considered. If the nucleophilic atoms are very different in size, both *polarizability* and *solvent effects* must be considered.

Polarizability is the ease with which electrons can be delocalized over the surface of the atom. Larger ions, like I^-, which have more loosely held electrons, are more polarizable than smaller ions, like F^-. Therefore, larger ions are more likely to act as a nucleophile.

Solvent effects include polar protic solvents and polar aprotic solvents:

- **Polar protic solvents** (those that have O—H, N—H, or S—H bonds, such as water and methanol) can form hydrogen bonds and can solvate both cations and anions. In smaller ions, the solvent shell is held more tightly and is therefore more stabilizing, making the anion less nucleophilic. Because F^- is most tightly solvated and I^- is least, the nucleophilicity order in polar protic solvents is $I^- > Br^- > Cl^- > F^-$.
- **Polar aprotic solvents**, such as dimethyl sulfoxide (DMSO), *N,N*-dimethylformamide (DMF), and acetonitrile, do not have O—H, N—H, or S—H bonds. They do not form hydrogen bonds between themselves and can solvate cations better than anions. Therefore, the unsolvated anion (nucleophile) is not hindered by a solvent shell and is available to share electrons with an electrophile:

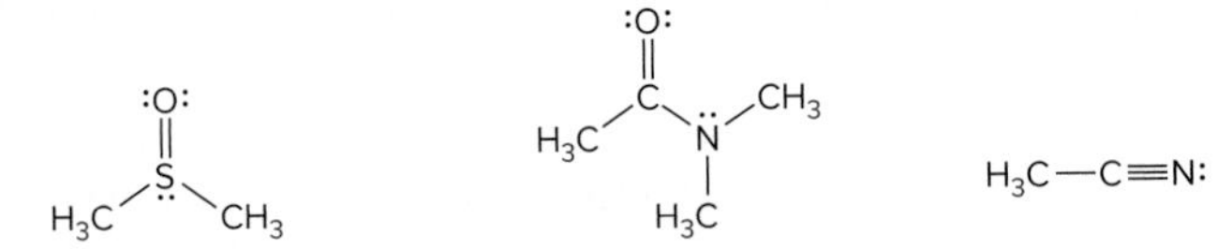

Dimethyl sulfoxide (DMSO) *N,N*-dimethylformamide (DMF) Acetonitrile

Polar aprotic solvents

Factors Influencing the Base Strength Recall from Chapter 16 that the greater the strength of the acid, the weaker its conjugate base, and vice versa (the higher the pK_a of the conjugate acid, the stronger the base). The *base strength* is the most important factor in determining the mechanism for elimination that we will study in Section 21.4. Therefore, we will review the factors that influence the base strength here.

- *Electronegativity.* Electronegativity is important to consider when comparing atoms in the same row of the periodic table. The greater the electronegativity of an atom, the less willing it is to share its electrons, and thus it is a weaker base. For example, the pK_a values of CH_4, NH_3, and H_2O (C, N, and O are in the same row in the periodic table) are approximately 50, 38, and 16, respectively, making CH_4 the weakest acid, and therefore its conjugate base CH_3^- is the strongest base:

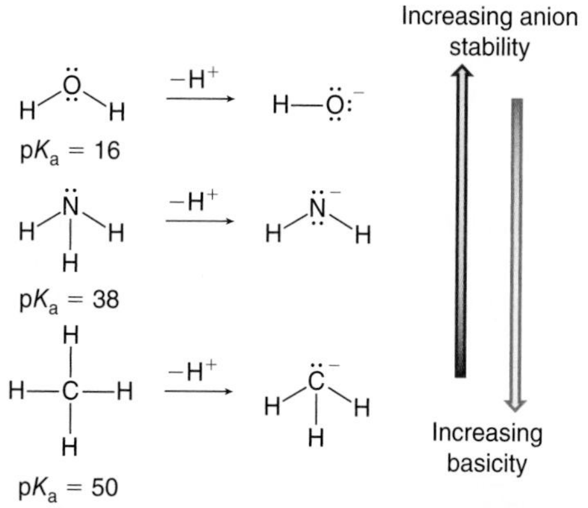

- *Atomic size.* When comparing atoms of equal formal charge, the negative charge is more dispersed in a larger atom, leading to a greater stabilization of the ion. Therefore, smaller atoms are stronger bases. This principle applies when considering atoms in the same column of the periodic table (because size changes are significant). Because size does not greatly change across a period, this principle does not apply. When comparing atoms across a period, we consider electronegativity:

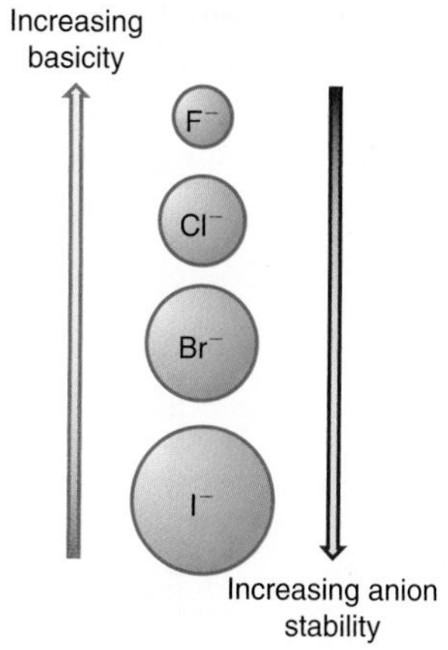

Remember that the weaker the conjugate acid (higher pK_a), the stronger the base. HF, for example, has a higher pK_a value than HCl; thus HF (pK_a = 3.17) is a weaker acid than HCl (pK_a = −7). Therefore, its conjugate base (F^-) is a stronger base.

- *Resonance.* In most cases, the presence of **resonance** leads to less basicity, since resonance delocalizes the electron density and thus leads to greater stabilization. For example, the ethanolate ion (ethoxide ion, conjugate base of ethanol) is a stronger base than the ethanoate (acetate) ion [conjugate base of ethanoic (acetic) acid]:

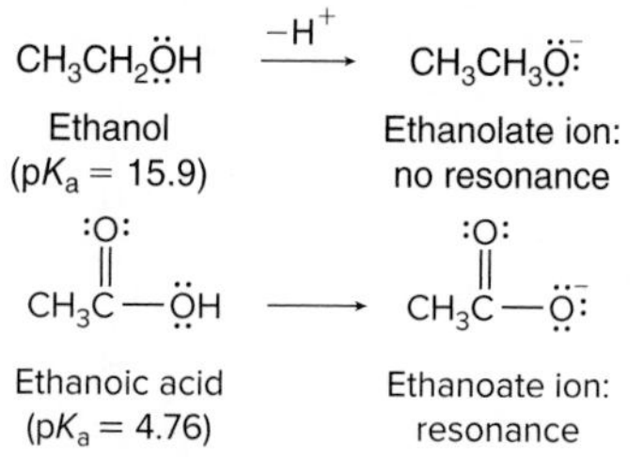

- *Inductive effect.* Nearby atoms may add to or detract from the electron density of the atom sharing electrons with the proton. This electrostatic **inductive effect** of the transmission of electron density through a chain of atoms in a molecule influences its basicity. In many cases, electronegative atoms pull away the electron density and make it less basic because its driving force to share electron density is decreased. This is the electron-withdrawing inductive effect, also known as the *−I effect.* Other groups, like alkyl groups, are less electron withdrawing than hydrogen and are considered electron releasing. This effect is also known as the +*I effect* and increases the basicity. The inductive effect often comes into play when the same functional groups (with similar resonance, atomic size, and electronegativity) are present on the molecules being compared. Relative inductive effects have been experimentally measured with reference to hydrogen. It has been found, for example, that NO_2 decreases the basicity more than F (NO_2 detracts more from the electron density of the atom sharing electrons with the proton than F); therefore, NO_2 is more electron-withdrawing (has a higher −I effect) than F:

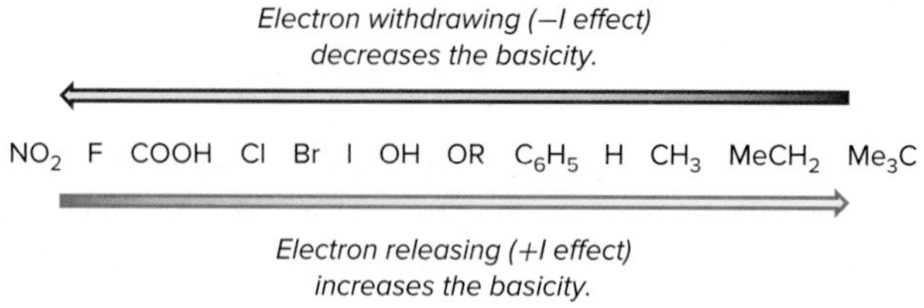

Relationship among Size, Basicity, and Nucleophilicity Remember that **basicity** measures the tendency of an electron pair donor to react with a proton (H^+) or hydronium ion (H_3O^+), whereas **nucleophilicity** measures the ability of the nucleophile to attack other electron-deficient atoms (usually carbon atoms). As we will see later, if an anion is reacting as a Lewis base, the resulting reaction will be a substitution, but if it reacts as a Brønsted-Lowry base, then the resulting reaction will be an elimination.

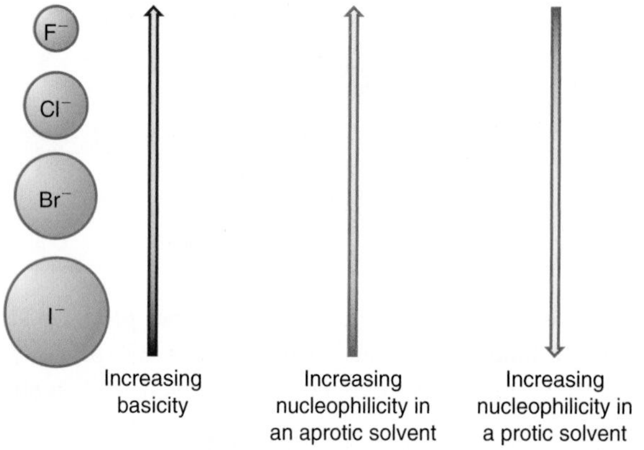

A charged nucleophile is stronger than the corresponding conjugate acid (uncharged nucleophile):

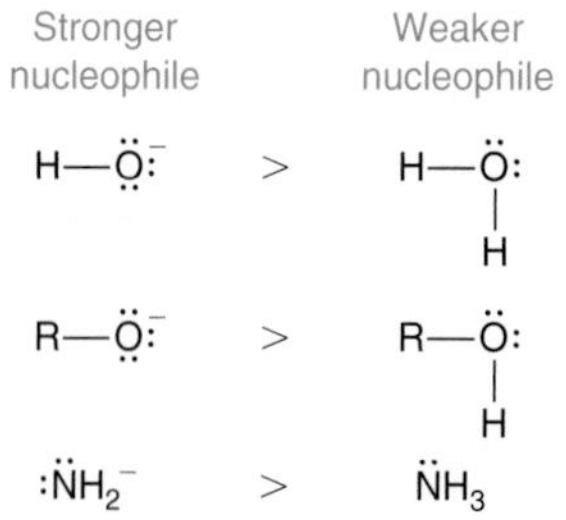

The following are some common nucleophiles:

- *Strong* nucleophiles:
  - Very good nucleophiles: I^-, HS^-, RS^-
  - Good nucleophiles: Br^-, OH^-, RO^-, CN^-, N_3^-
- *Fair* nucleophiles: NH_3, Cl^-, F^-, RCO_2^-
- *Poor* nucleophiles:
  - Weak nucleophiles: H_2O, ROH
  - Very weak nucleophiles: RCO_2H

In summary, we note the following:

- Nucleophilicity increases from right to left across a row in the periodic table (opposite order as for electronegativity): $C^- > N^- > O^- > F^-$.
- In protic solvents, nucleophilicity increases going down a column in the periodic table: $I^- > Br^- > Cl^- > F^-$.
- In aprotic solvents, nucleophilicity decreases going down a column in the periodic table. This is the same order as for basicity: $F^- > Cl^- > Br^- > I^-$.
- Polarizability and size increase going down a column in the periodic table.
- Anions are stronger nucleophiles than their neutral conjugate acids.

SUMMARY OF SECTION 21.2

- A reaction mechanism is a step-by-step sequence of how bonds are broken and formed.
- The flow of electrons in a reaction can be shown with an arrow. Curved arrows indicate the direction of electron flow, not atom flow.
- Nucleophiles are electron-rich species that use their electrons to form bonds to electrophiles (electron-deficient species).
- In a concerted reaction, the reactants are converted directly into products. In a stepwise reaction, the reactants form an unstable intermediate, which goes on to form the products.

21.3 Nucleophilic Substitution Reactions

A **nucleophilic substitution** is the reaction of an electron-pair donor (the nucleophile, Nu or Nu^-) with an electron pair acceptor (the electrophile). A tetrahedral (sp^3-hybridized) electrophile must have a leaving group (X) for the reaction to take place.

Leaving Group The **leaving group** is a molecular fragment that departs with a pair of electrons in heterolytic cleavage. The leaving group's ability is related to its basicity: the weaker the base, the better the leaving group, and the better the leaving group, the more likely it is to depart.

Leaving groups can be as follows:

- *Anions.* Anions of sulfonate esters, such as p-toluenesulfonate (OTs, tosylate) or methanesulfonic acid (OMs), and halides, such as I^-, Br^-, and Cl^-, are common ionic leaving groups:

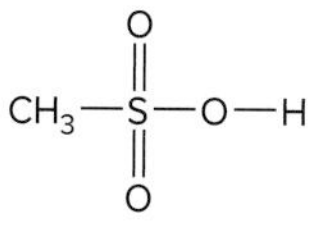
Methanesulfonic acid

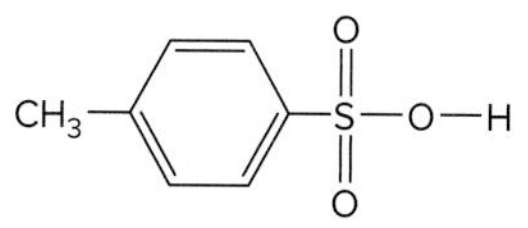
p-toluenesulfonic acid

If we look at the relative reactivity of halide leaving groups, we can establish the following order of reactivity of the haloalkanes: RF << RCl < RBr < RI. Recall that F^- is the strongest base and, therefore, the worst leaving group, whereas I^- is the weakest base (the best leaving group). Remember that a base is weak because it is stable. That is why a weak base is a good leaving group. To determine how strong or weak the base is, you can look at the pK_a value of its conjugate acid: the lower the pK_a of the conjugate acid, the weaker the base. Appendix C shows a list of K_a values of selected acids.

- *Neutral molecules,* like water (H_2O) and ammonia (NH_3).

The table below shows leaving groups ordered by their ability to leave:

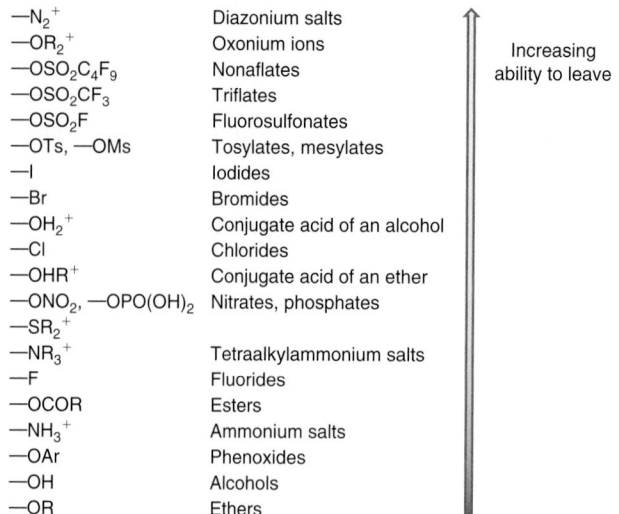

$—N_2^+$	Diazonium salts	↑ Increasing ability to leave
$—OR_2^+$	Oxonium ions	
$—OSO_2C_4F_9$	Nonaflates	
$—OSO_2CF_3$	Triflates	
$—OSO_2F$	Fluorosulfonates	
—OTs, —OMs	Tosylates, mesylates	
—I	Iodides	
—Br	Bromides	
$—OH_2^+$	Conjugate acid of an alcohol	
—Cl	Chlorides	
$—OHR^+$	Conjugate acid of an ether	
$—ONO_2$, $—OPO(OH)_2$	Nitrates, phosphates	
$—SR_2^+$		
$—NR_3^+$	Tetraalkylammonium salts	
—F	Fluorides	
—OCOR	Esters	
$—NH_3^+$	Ammonium salts	
—OAr	Phenoxides	
—OH	Alcohols	
—OR	Ethers	

Nucleophile The nucleophile can be any of the following species.

1. An anion:

$$R{-}X + Nu{:}^- \longrightarrow R{-}Nu + X^-$$

Different nucleophiles produce different products:

$$R{-}X + {:}\ddot{O}H^- \longrightarrow R{-}\ddot{O}H + X^-$$

Alcohol

$$R{-}X + {}^-{:}\ddot{O}{-}CH_3 \longrightarrow R{-}\ddot{O}{-}CH_3 + X^-$$

Ether

$$R{-}X + {}^-{:}\ddot{S}H \longrightarrow R{-}\ddot{S}H + X^-$$

Thiol

$$R{-}X + {}^-{:}\ddot{S}CH_3 \longrightarrow R{-}\ddot{S}CH_3 + X^-$$

Thioether

$$R{-}X + {}^-{:}\ddot{Br}{:} \longrightarrow R{-}\ddot{Br}{:} + X^-$$

Alkyl bromide

$$R{-}X + {}^-{:}\ddot{I}{:} \longrightarrow R{-}\ddot{I}{:} + X^-$$

Alkyl iodide

$$R{-}X + {}^-{:}C{\equiv}N{:} \longrightarrow R{-}C{\equiv}N{:} + X^-$$

Nitrile

$$R{-}X + {}^-{:}C{\equiv}C{-}CH_3 \longrightarrow R{-}C{\equiv}C{-}CH_3 + X^-$$

Alkyne

2. A neutral molecule with lone pairs of electrons:

$$Nu{:} + R{-}X \longrightarrow R{-}Nu^+ + X^-$$

Amines, for example, are used to form alkyl ammonium salts:

$$R{-}X + {:}NH_3 \longrightarrow R{-}NH_3^+ + X^-$$

$$R{-}X + H_3C{-}\ddot{N}H{-}H \longrightarrow H_3C{-}N^+H_2{-}R + X^-$$

The oxygen atom in the water molecule has two lone pairs of electrons. Therefore, water can act as a nucleophile, for example, in hydrolysis reactions of haloalkanes. Although OH^- is a better nucleophile to produce alcohols from haloalkanes, water can still form the same products:

$$R{-}X + H_2\ddot{O}{:} \longrightarrow R{-}\ddot{O}H + H^+ + X^-$$

Note that a neutral molecule with a lone pair of electrons will be a better nucleophile the less electronegative it is (for example, F_2 is not a good nucleophile despite having many lone pairs).

3. Molecules with π bonding electrons. The π electrons have a high electron density in the π electron cloud, which can be easily polarized (give up the electrons) and can act like a nucleophile. This behaviour is very similar to the behaviour that the lone-pair electrons have on a Lewis base:

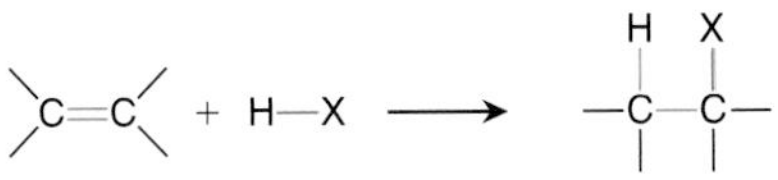

Sample Problem 21.3 Predicting the Product in Nucleophilic Substitution Reactions

Problem Predict the products in each nucleophilic substitution reaction:

(a) $CH_3CH_2CH_2CH_2Cl + {}^{-}OH \longrightarrow$

(b)

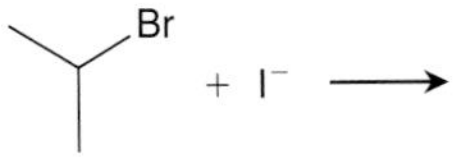

(c)

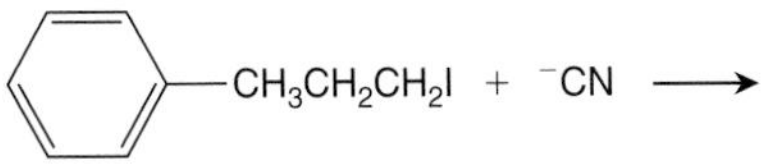

Plan We write the starting material without the leaving group and then connect the nucleophile to the same C atom where the leaving group was previously attached. Since the equation has to be balanced (mass and charge), we add the leaving group to the equation for mass balance. If the charges are not balanced at this point, we add a charge to the leaving group that balances the charges in the equation.

Solution (a) $CH_3CH_2CH_2CH_2Cl + {}^{-}OH \longrightarrow CH_3CH_2CH_2CH_2OH + Cl^-$

(b)

(c)

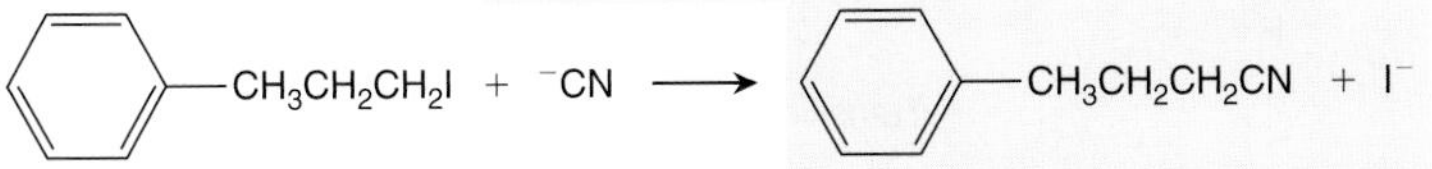

Follow-Up Problem 21.3 Predict the products in each nucleophilic substitution reaction:

(a)

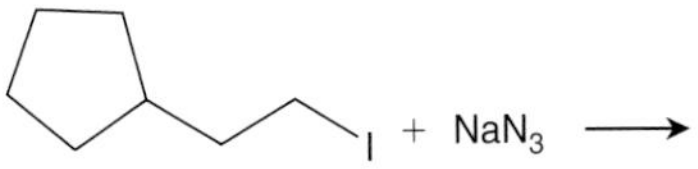

(b)

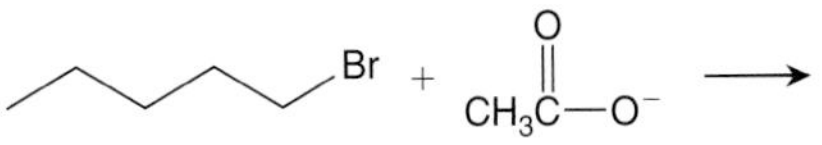

Mechanisms of Nucleophilic Substitution Reactions (S_N1 and S_N2)

Nucleophilic substitutions at a tetrahedral (sp^3-hybridized) carbon involve the breaking of the σ bond to the leaving group (X) and the formation of a new σ bond to the nucleophile.

1. If the bond-breaking and bond-forming steps occur simultaneously, the reaction goes through at least a **one-step (concerted) mechanism**, and it is known as **S_N2** (substitution, nucleophilic, bimolecular). A **bimolecular reaction** is one in which the reactants proceed to products without an intermediate, only through a transition state. The *2* in S_N2 tells you that there are two molecules involved in the **rate-determining step** (RDS). The RDS is the step in which the leaving group departs. Both the nucleophile and the electrophile (R-X) are involved in the RDS. The energy diagram will have at least one hump, and the rate of such a bimolecular

reaction depends on the concentration of both species (**second-order reaction**). S_N2 reactions can involve a protonation and/or deprotonation step as well:

$$\text{Rate} = k[\text{Nu}^-][\text{R-X}]$$

S_N2 mechanism:

2. If the R-X bond is broken first, a **carbocation** (C^+) is formed as an intermediate, and a new R-nucleophile bond is formed; the reaction goes through at least a **two-step mechanism** and is known as **S_N1** (substitution, nucleophilic, unimolecular). **Unimolecular** means that the rearrangement of a single molecule produces one or more molecules of product. Therefore, the *1* in S_N1 tells you that there is only one molecule involved in the RDS. Also, in S_N1 reactions, the RDS is the step in which the leaving group departs. The energy diagram has two or more humps, and the rate of the reaction depends only on the concentration of the R-X species (**first-order reaction**):

$$\text{Rate} = k[\text{R-X}]$$

S_N1 mechanism:

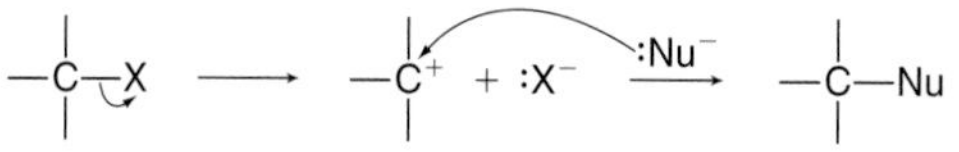

(Carbocation)

The S_N2 Mechanism In this concerted mechanism, the nucleophile attacks **back-side** of the carbon with the leaving group. The RDS is also the step in which the leaving group departs:

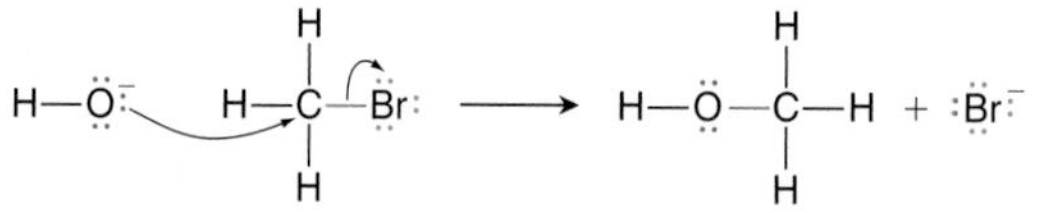

Notice that the leaving group and the nucleophile will be on opposite sides in the transition state:

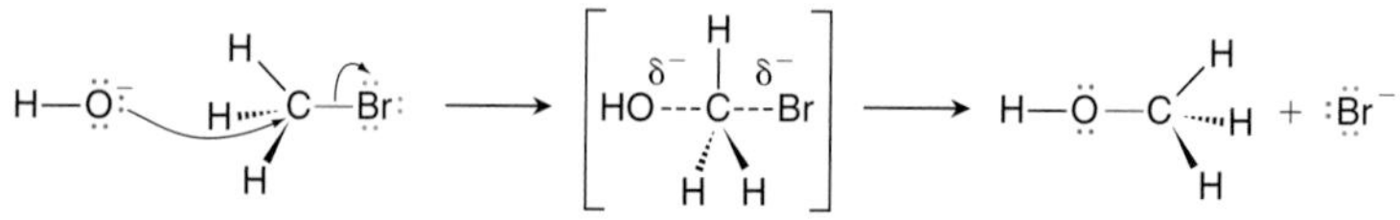

Transition state

If the nucleophile attacks a stereogenic centre, there will be an *inversion in the configuration*:

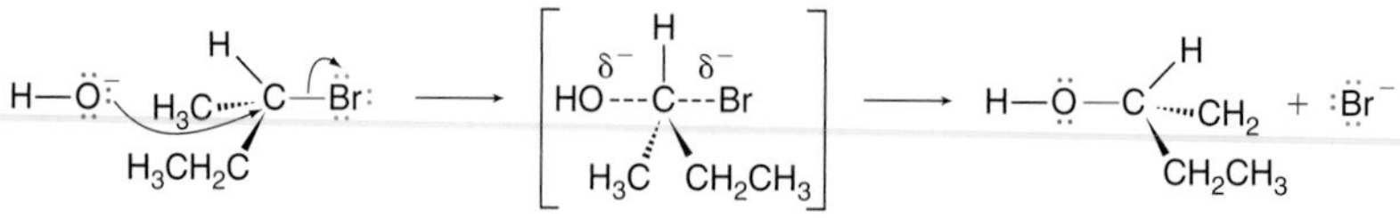

(*S*) Transition state (*R*)

Steric Effects in the S_N2 Mechanism Each atom within the molecule occupies a certain amount of space. Since there is a back-side attack by the nucleophile, it is easier for the nucleophile to get close to the carbon attached to the leaving group, making the reaction go faster. Thus, as the number of R groups on the carbon with the leaving group increases, the rate of an S_N2 reaction of haloalkanes decreases. The S_N2 reaction rate of haloalkanes has the following order:

methyl halide > primary halide > secondary halide >>>>> tertiary halide (almost no reaction)

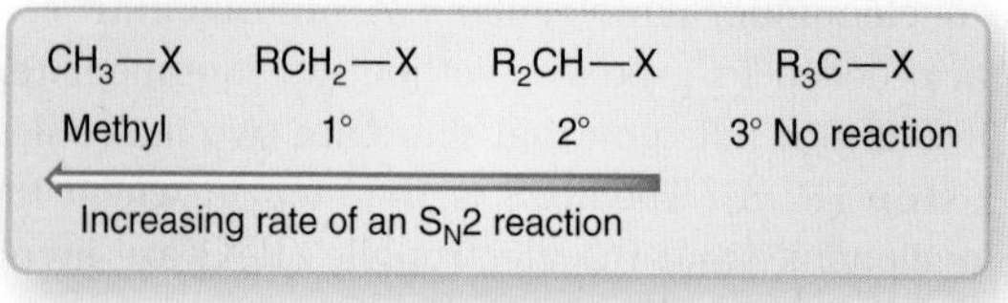

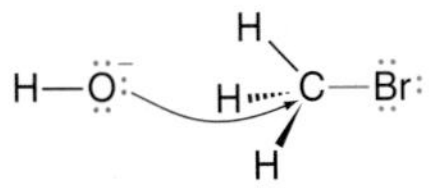

Methyl halide

The **steric effect** increases with the substitution, as we can see in the following molecules:

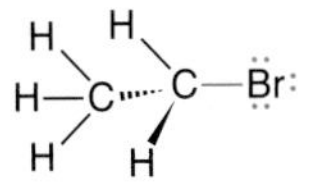

Primary halide

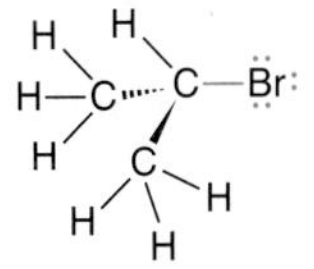

Secondary halide

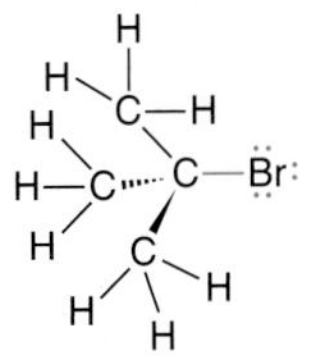

Tertiary halide

Effect of the Leaving Group in the S_N2 Mechanism As mentioned before, strong bases such as OH^- and RO^- are poor leaving groups. However, they can be converted into good leaving groups by protonating them or by turning them into the sulfonate ester.

- Nucleophilic substitutions of alcohols are performed in highly concentrated H_2SO_4 to turn the poor leaving group OH^- into the good leaving group H_2O:

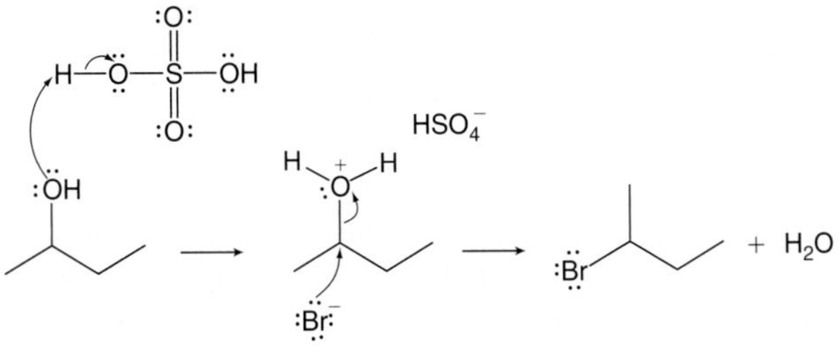

- We saw that alcohols are poor leaving groups in S_N2 reactions, but if they are converted to alkyl tosylates, the S_N2 reaction occurs in the presence of a good nucleophile:

Alcohols can be converted into tosylates by reacting them with p-tosyl chloride in the presence of a base such as pyridine to neutralize the hydrochloric acid by-product. The stereochemistry is retained, since there is no bond breaking at the stereogenic centre (the oxygen from the original alcohol is retained):

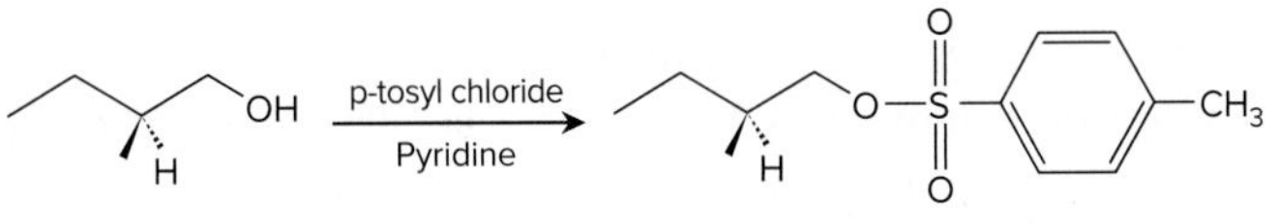

(S) (S)

It was found that the reaction rate increases 10^6 times when the leaving group is OTs^- instead of I^- in S_N2 reactions.

The S_N1 Mechanism The S_N1 reaction is at least a two-step reaction in which the leaving group departs, forming a planar *carbocation*, which then reacts further with the nucleophile. Since the nucleophile is free to attack from either side, this reaction is associated with *racemization*. **Racemization** occurs when a mixture containing only one enantiomer (also called an enantiomerically pure mixture) is converted into a mixture where more than one of the enantiomers are present or into a mixture of diastereomers (when dealing with molecules with more than one stereogenic centre).

The S_N1 mechanism explains substitution of compounds such as tertiary halides, which cannot undergo S_N2 reactions, as well as substitution reactions where the nucleophile is a neutral molecule.

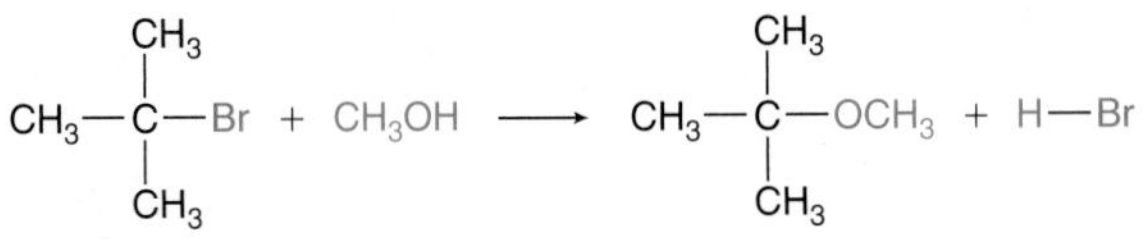

MTBE

The S_N1 mechanism follows at least two steps. In this example, a neutral product is generated in a third step; however, the substitution is actually accomplished in two.

Step 1. Loss of the leaving group to generate a carbocation (the slowest step in the reaction and, therefore, the RDS):

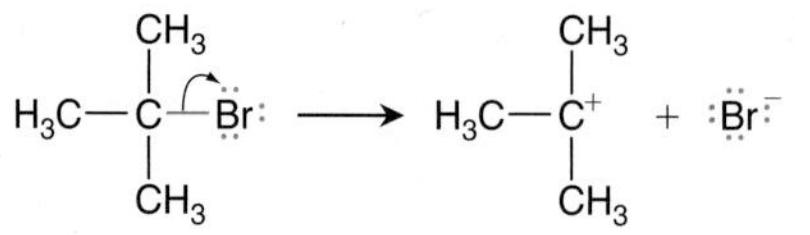

Step 2. Addition of the nucleophile to the carbocation:

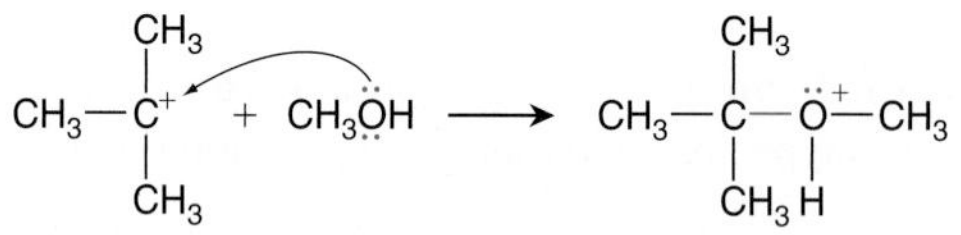

Step 3. Loss of proton to solvent to generate a neutral product:

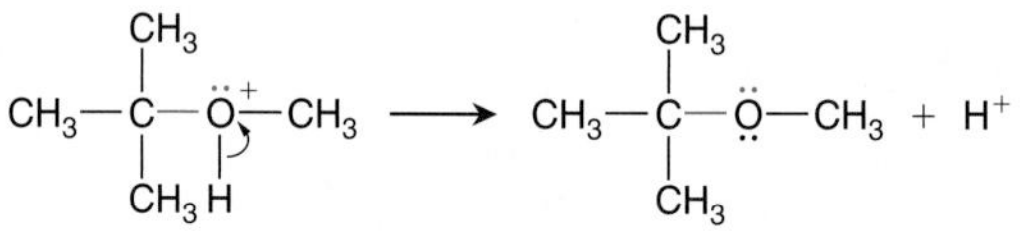

Since the intermediate is a carbocation, the more stable the carbocation (the lower its energy), the easier it is to form. The general stability order of alkyl carbocations is

(most stable) tertiary C^+ > secondary C^+ > primary C^+ > methyl C^+(least stable)

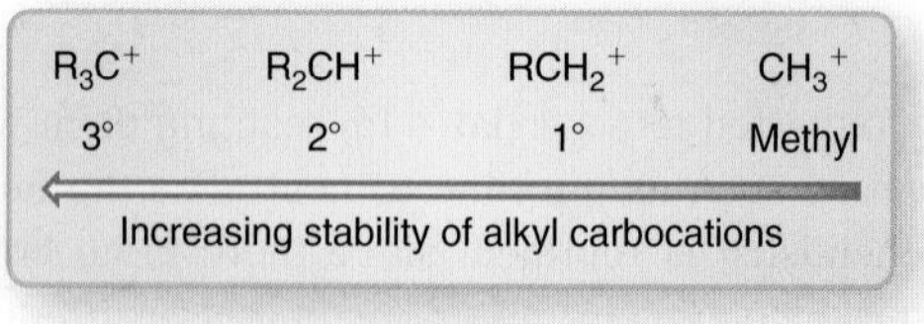

Resonance effects, when present, can further stabilize carbocations. Carbocations are prone to **rearrangement** via 1,2-hydride or 1,2-alkyl shifts, if this generates a more stable carbocation. For example,

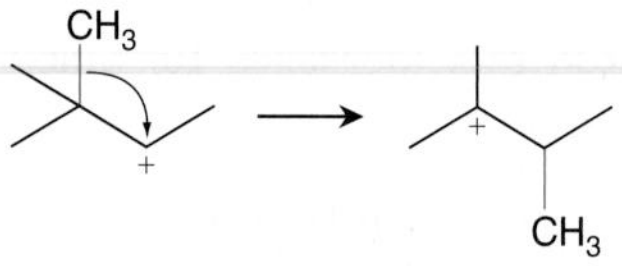

1,2-alkyl shift
2° carbocation to 3° carbocation

Carbocations are sp^2 hybridized, thus forming planar structures. As shown below, the nucleophile can attack back-side (a) or front-side (b), leading to a mixture of *S* (a) and *R* (b) configurations at the reacting stereogenic centre:

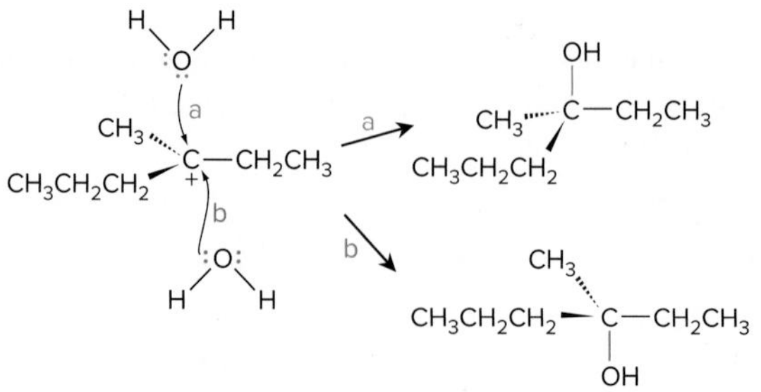

Substitution Reactions of Allylic and Benzylic Substrates Allylic and benzylic substrates can undergo an S_N2 reaction if there is very *little hindrance*, for example,

$$CH_3CH=CHCH_2Br + OH^- \longrightarrow CH_3CH=CHCH_2OH + Br^- \quad (S_N2)$$

$$\text{⬡}-CH_2Cl + CH_3O^- \longrightarrow \text{⬡}-CH_2OCH_3 + Cl^- \quad (S_N2)$$

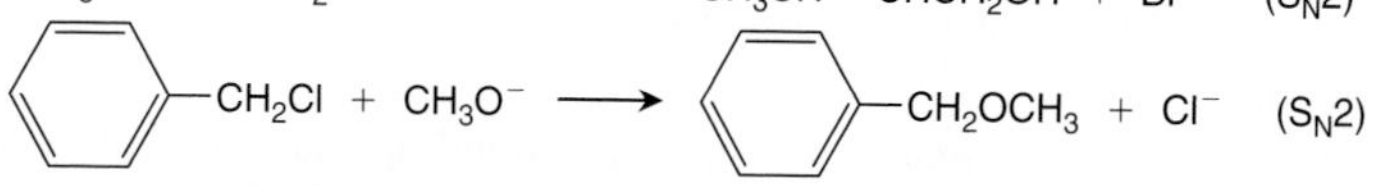

Since the allylic and benzylic substrates can form resonance *stabilized cations*,* they can also have S_N1 reactions:

$$CH_3CH=CHCH_2Br \rightleftharpoons CH_3CH=CH\overset{+}{C}H_2 + Br^- \xrightarrow{H_2O} CH_3CH=CHCH_2OH + H^+ \quad (S_N1)$$

Solvent Effect in S_N1 and S_N2 Reactions

The solvent can affect the rate of a substitution reaction.

- *The rate of an S_N1 reaction increases in polar solvents* (solvents with high dielectric constants). Recall that the higher the dielectric constant, the larger the ability to support separated positively and negatively charged species. Polar solvents with high dielectric constants help to stabilize the transition state, since it helps the charge separation.

  Polar protic solvents favour S_N1 reactions:

H_2O	CH_3OH	CH_3CH_2OH	$CH_3\overset{O}{\overset{\|}{C}}-OH$
Water	Methanol	Ethanol	Ethanoic acid (acetic acid)

- *Polar aprotic solvents favour S_N2 reactions.* The solvent must be polar enough to dissolve the nucleophile (especially if it is an ion). Protic solvents with an OH group will solvate the nucleophile, decreasing the rate of the S_N2 reaction. That is why aprotic solvents (lacking the OH group) increase the rate of an S_N2 reaction:

$CH_3\overset{O}{\overset{\|}{C}}CH_3$	$\overset{CH_3}{\underset{CH_3}{\diagdown}}N-\overset{O}{\overset{\|}{C}}-H$	$\overset{CH_3}{\underset{CH_3}{\diagdown}}S=O$
Propan-2-one (acetone)	*N,N*-dimethylformamide (DMF)	Dimethyl sulfoxide (DMSO)

Note: Keep in mind that the role of the solvent is to stabilize reactants, transition states, and/or intermediates and that the solvent itself is very much a secondary consideration. If the reactants are not reactive, choosing a reaction-friendly solvent will not make the reaction go.

Summary of S_N1 and S_N2 Reactions

- Effect of substrate (RX):
  - Tertiary halides only undergo S_N1 reactions.
  - Primary and methyl halides only undergo S_N2 reactions.
  - Secondary halides can undergo either S_N2 or S_N1 reactions.

---

*Cation resonance structures:

$$\text{⬡}-CH_2Cl \rightleftharpoons \text{⬡}-\overset{+}{C}H_2 + Cl^- \xrightarrow{CH_3OH} \text{⬡}-CH_2OCH_3 + H^+ \quad (S_N1)$$

$$\left[\text{⬡}-\overset{+}{C}H_2 \longleftrightarrow \text{⬡}=CH_2 \longleftrightarrow ^+\text{⬡}=CH_2 \longleftrightarrow \overset{+}{\text{⬡}}=CH_2 \longleftrightarrow \text{⬡}-\overset{+}{C}H_2 \right]$$

$$\left[H_3C-CH=CH-\overset{+}{C}H_2 \longleftrightarrow H_3C-\overset{+}{C}H-CH=CH_2 \right]$$

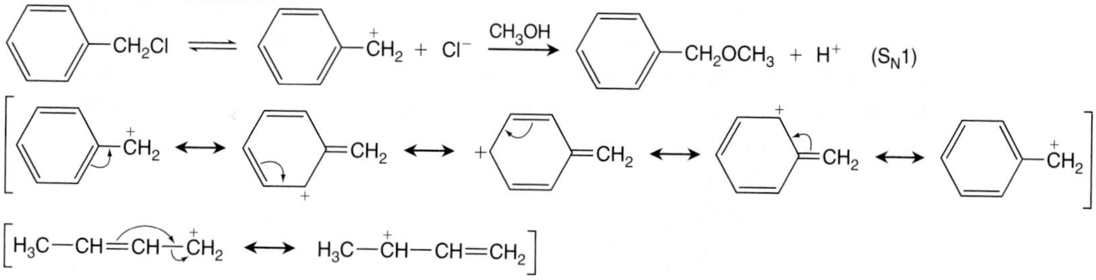

- Effect of nucleophile: Remember that nucleophilicity *only* plays a role in S_N2, but not in S_N1, reactions.
  - Good nucleophiles (anions) favour S_N2 reactions if they can.
  - Poor nucleophiles (neutral species) favour S_N1 reactions.
- Effect of solvent:
  - Polar protic solvents (capable of hydrogen bonding, having an OH, an NH, or an NH_2 group) favour S_N1 reactions.
  - Polar aprotic solvents (without OH, NH, or NH_2) favour S_N2 reactions.
- Increasing the concentration of the nucleophile increases the rate of an S_N2 reaction but has no effect on the rate of an S_N1 reaction.
- Effect of resonance:
  - Primary halides can undergo S_N2 reactions with enough resonance stabilization of the carbocation (such as benzyl chloride).
- Poor nucleophiles only favour S_N1 reactions because they cannot undergo S_N2 reactions.

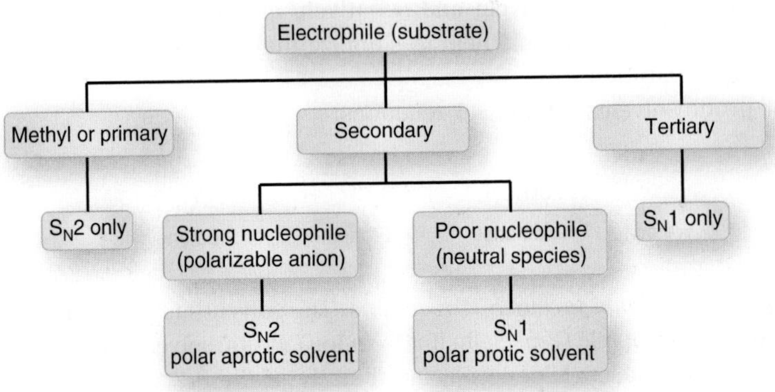

SUMMARY OF SECTION 21.3

- The leaving group is a molecular fragment that departs with a pair of electrons in heterolytic cleavage.
- An S_N2 reaction goes through a concerted mechanism. The nucleophile attacks back-side of the carbon with the leaving group, leading to an inversion in the configuration. The S_N2 reaction rate depends on the concentration of both the substrate (electrophile) and the nucleophile.
- An S_N1 reaction goes through a multistep mechanism, with the formation of a carbocation intermediate. Therefore, it generates a mixture of configurations at the reacting stereogenic centre. Its rate depends on the concentration of the substrate but not the nucleophile.
- To predict the mechanism of a substitution reaction, it is important to analyze the substrate, the nucleophile that will attack the substrate, and the solvent in which the reaction will occur.

21.4 Elimination Reactions

In an elimination reaction, two substituents are removed from a molecule in either a one- or a two-step mechanism to form a π bond, increasing the unsaturation of the molecule. In most organic elimination reactions, hydrogen atoms are lost to form the double bond. In this section, we will focus on elimination reactions involving haloalkanes, with good leaving groups, reacting with a nucleophile (*a strong Lewis base*) to form an alkene:

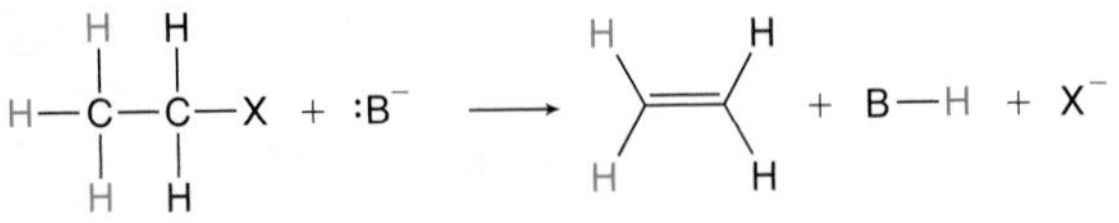

Mechanisms of Elimination Reactions (E1 and E2)

Elimination reactions involve the removal of a proton, the breaking of a σ bond to the leaving group (X), and the formation of a π bond.

1. If the bond-breaking and the bond-forming steps occur simultaneously, the reaction goes through at least a *one-step mechanism* (a *concerted mechanism*), and it is known as **E2** (elimination, bimolecular). Both the nucleophile and the electrophile (R-X) are involved in the rate-determining step. The energy diagram has one hump, and the rate of such a bimolecular reaction depends on the concentration of both species (*second-order reaction*):

$$\text{Rate} = k[\text{B}^-][\text{R-X}]$$

E2 mechanism:

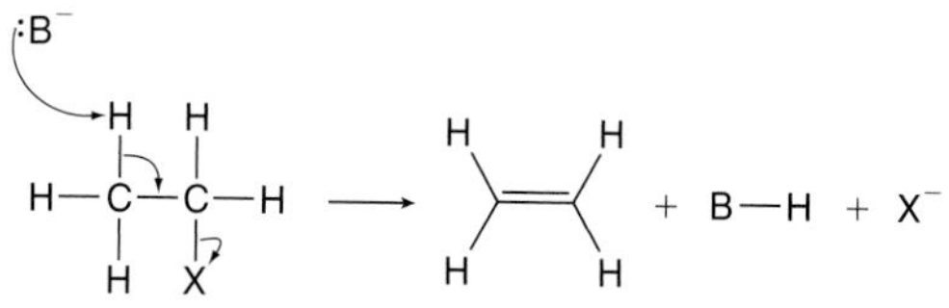

2. If the R-X bond is broken first, a carbocation (C^+) is formed as an intermediate, and a new C=C bond is formed; the reaction goes through at least a *two-step mechanism* and is known as **E1** (*elimination, unimolecular*). The first step is the rate-determining step. The energy diagram has at least two humps, and the rate of the reaction depends only on the concentration of the R-X species (*first-order reaction*):

$$\text{Rate} = [\text{R-X}]$$

E1 mechanism:

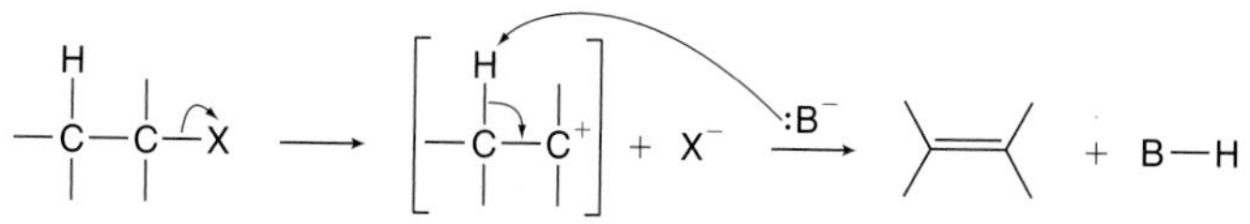

The E2 Mechanism In this concerted mechanism, the proton and the leaving group must be *anti* to each other (staggered conformation). The nucleophile, which is usually *a strong base*, attacks an H atom on the C atom adjacent to the one with the leaving group at the same time as the double bond starts to form and the leaving group starts to leave. E2 reactions occur faster for *antiperiplanar* conformations: when the H-C bond and the C-X bonds are at 180° with respect to each other. Therefore, the transition state has a staggered conformation, which has lower energy than an eclipsed conformation. The E2 reaction needs a chemical strong enough to remove a weakly acidic hydrogen. For the π bond to be created, the adjacent carbon atoms change from a tetrahedral to a trigonal planar geometry (the hybridization changes from sp^3 to sp^2). The E2 mechanism is very similar to the S_N2 reaction mechanism:

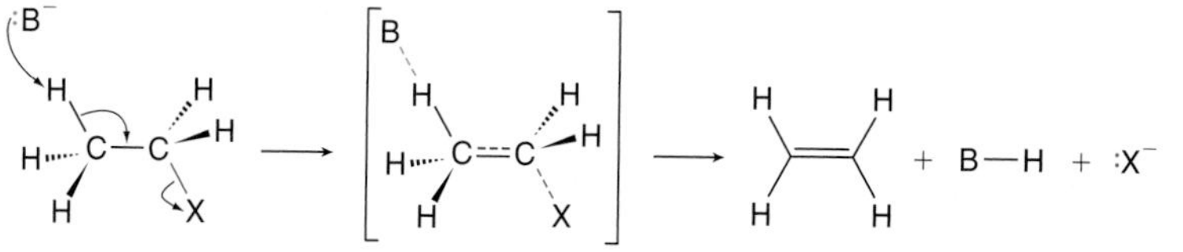

The E2 mechanism forms a C=C bond between the carbon containing the leaving group and an adjacent carbon with a proton. It will form the most stable alkene, as stated by **Zaitsev's rule** (Russian, Зайцев): *If more than one alkene can be formed by an elimination reaction, the more stable alkene is the major product. In general, the compound that has a more highly substituted C=C double bond is more stable owing to the electron-donating properties of the alkyl group.* For example, in the following reaction, the proton will be lost mostly from C-3 instead of C-1, since it will give a more substituted C=C double bond:

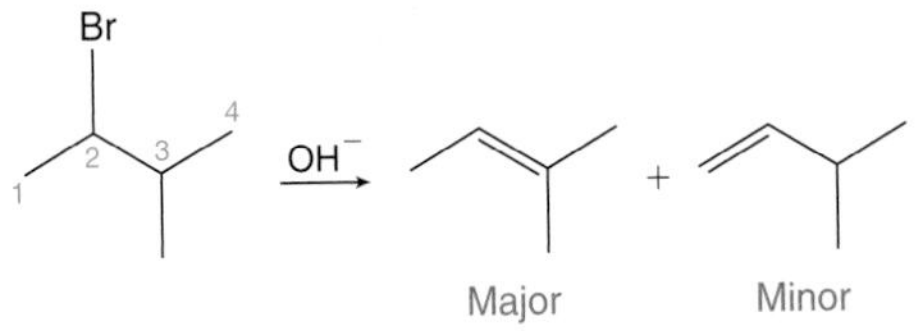

During the elimination reaction, the base that causes the double bond to form should be sterically unhindered for the reaction to follow Zaitsev's rule. If the base, for example, is $(CH_3)_3CONa$, the bulkiness prohibits the base from pulling the proton off the most substituted carbon, and the alkene with fewer substituents will predominate. On the other hand, if only one of the carbon atoms adjacent to the carbon with the leaving group has hydrogen atoms, the H will be removed from there. The following elimination reaction, for example, will give only one product, since only one of the carbon atoms adjacent to the carbon containing the leaving group can be deprotonated:

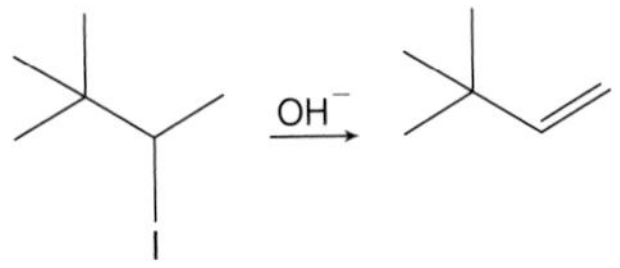

The E1 Mechanism The E1 mechanism is a multistep process of elimination:

- *Ionization*, where the carbon-leaving group (usually a halogen) breaks, forming a carbocation
- *Deprotonation* of the carbocation and formation of the π bond

E1 typically takes place with tertiary haloalkanes, but it is also possible with secondary haloalkanes, and with alcohols as long as they are first protonated to give water as a leaving group. The reaction usually occurs in the absence of a base, or with weak bases. The first step of the E1 and S_N1 mechanisms is identical (they have a common carbocation intermediate). This is one reason that these pathways compete. If S_N1 and E1 pathways are competing, the E1 pathway can be favoured by increasing the temperature:

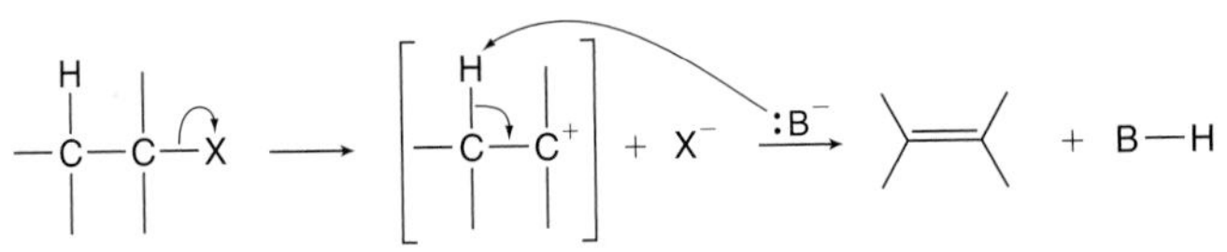

E1 reactions happen with highly substituted haloalkanes for two main reasons:

- Highly substituted haloalkanes are bulky, limiting the room for the E2 one-step mechanism; therefore, the two-step E1 mechanism is favoured.
- Highly substituted carbocations are more stable than methyl or primary substituted cations. Such stability favours the two-step E1 mechanism.

Summary of S_N1, S_N2, E1, and E2 Reactions

In general, there are competitions between E2 and S_N2 reactions and also between E1 and S_N1 reactions, but elimination is generally favoured over substitution when the nucleophile is poor or when the following variables are increased:

- Temperature
- Steric hindrance
- Basicity
- Steric bulk of the base

Table 21.1 compares the substitution and elimination reactions. Notice the following:

- Strong bases, such as OH^-, NH_2^-, CH_3O^-, or other alkoxide ions (conjugate base of an alcohol, R-O^-) favour elimination reactions (E2). Strong, hindered bases favour elimination over substitution.

- Weak bases that are strong nucleophiles (halide ions, NH_3, PH_3) favour substitution over elimination.
- With S_N1 reactions, we usually get some E1 as well (common intermediate).
- Species that are good nucleophiles but weak bases promote S_N2 reactions (I^-, Br^-, Cl^-, HS^-, NH_3, PH_3).
- Species that are good nucleophiles but strong bases, such as OH^-, RO^-, and H_2N^-, promote both S_N2 and E2 reactions (depends on substrate).
- Species that are poor nucleophiles and weak bases, such as H_2O and ROH, promote both S_N1 and E1 reactions.

TABLE 21.1 Comparison of Substitution and Elimination Reactions

Substrate (RX)	Good Nucleophile, Weak Base	Good Nucleophile, Strong Base	Poor Nucleophile, Weak Base (Protic Solvents)	Poor Nucleophile, Weak Base (Aprotic Solvents)	Strong, Bulky Base
	I^-, Br^-, HS^-, NH_3, PH_3	OH^-, RO^-, NH_2^-	H_2O, ROH, RNH_2	DMSO, propan-2-one (acetone), acetonitrile, DMF	$(CH_3)_3CO^-$
Methyl	S_N2	S_N2	No reaction	No reaction	S_N2
Primary	S_N2	Mostly S_N2, some E2	No reaction	No reaction	E2
Secondary	S_N2	E2	S_N1 and E1	S_N2 or no reaction	E2
Tertiary	No reaction	E2	S_N1 and E1 (increasing temperature favours E1)	No reaction	E2

Sample Problem 21.4 Predicting the Product in Reactions

Problem Identify the mechanism involved in each reaction and predict the products:

(a) 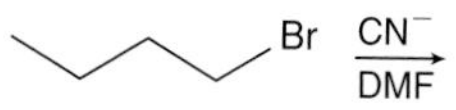

(b) 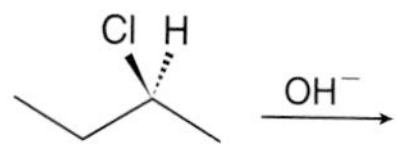

(c) 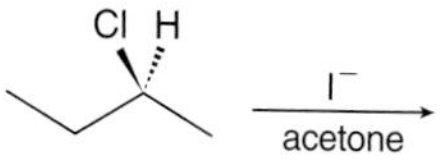

(d) 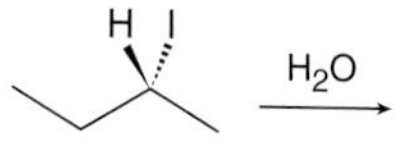

Plan To identify the mechanism that each reaction will follow, we need to pay attention not only to the location of the leaving group but also to the strength of the nucleophile or the base with which the starting material is reacting. We also need to consider the solvent in which the reaction is taking place. Based on all of these and on Table 21.1, we will then be able to identify the mechanism.

Solution **(a)** S_N2: 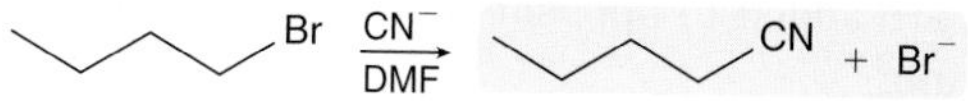

Since the leaving group is attached to a primary carbon, no stable carbocation will be formed. This means that the reaction will go through a concerted mechanism. CN^- is a strong nucleophile, as well as a strong base; however, DMF is a polar aprotic solvent, which will favour an S_N2 reaction over E2 reactions.

(b) E2: 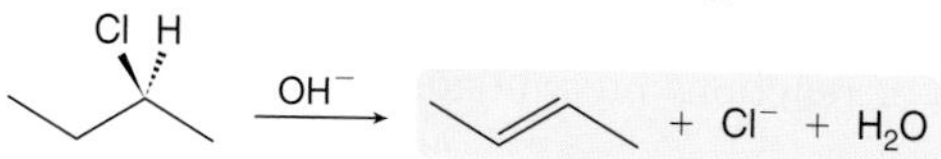

In this reaction, we observe that the leaving group is attached to a secondary carbon, and that the starting material is reacting with a strong base. Based on these observations, we can predict that elimination will take place. Since chlorine is not a good leaving group, a concerted mechanism will be favoured over a stepwise one.

(c) S_N2:

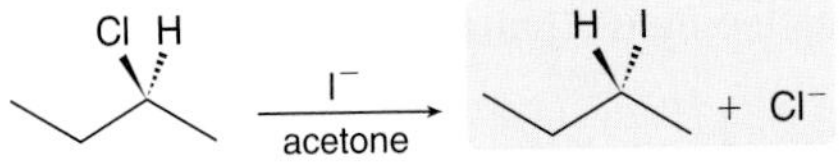

We notice that the leaving group is attached to a secondary carbon, so there is the option to go through a stepwise or a concerted mechanism. The latter will take place because chlorine is not a good leaving group. On the other hand, the fact that the reaction takes place in an aprotic solvent (acetone) and that I^- is a strong nucleophile but a weak base will favour an S_N2 over an E2 reaction. Therefore, we will observe an inversion in the configuration.

(d) E1 and S_N1:

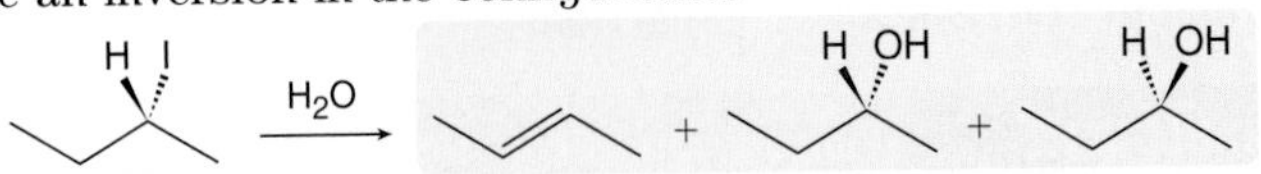

The iodine is a good leaving group attached to a secondary carbon, so we can expect the formation of a carbocation. The water is simultaneously the solvent (polar protic) and the reactant (weak base and poor nucleophile). Therefore, this stepwise reaction will undergo both elimination and nucleophilic substitution mechanisms.

Follow-Up Problem 21.4 Predict the products in each reaction:

(a) 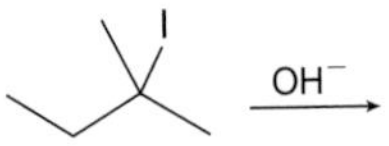

(b) 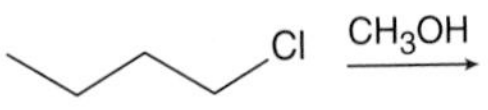

(c) 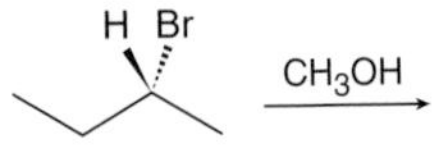

SUMMARY OF SECTION 21.4

- An E2 reaction goes through a concerted mechanism. The nucleophile, usually a strong base, attacks hydrogen on the back-side to the neighbouring carbon with the leaving group (antiparallel geometry). The E2 reaction rate depends on the concentration of both the substrate and the base.
- An E1 reaction goes through at least a two-step mechanism, with the formation of a carbocation intermediate.
- Elimination reactions will usually produce the most stable alkene, as stated in Zaitsev's rule.
- To predict the mechanism of an elimination reaction, it is important to analyze the leaving group ability, the substrate, the strength of the base, and the solvent in which the reaction will occur.
- There is competition between S_N1 and E1 reactions, as well as between S_N2 and E2 reactions.

21.5 Addition Reactions: Reactions of Alkenes

An **addition** reaction may be considered the reverse of an elimination reaction and can be summarized as

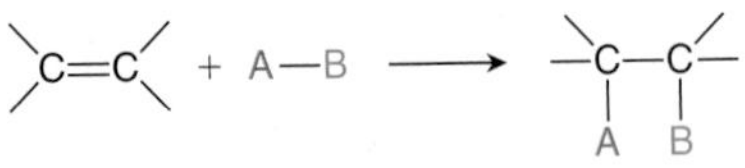

where A and B are the atoms added to the double bond. When the substituents are added to the same side or face of the double or triple bond, it is called a **syn addition**; when they are added to opposite sides or faces of a double or triple bond, it is an **anti addition**:

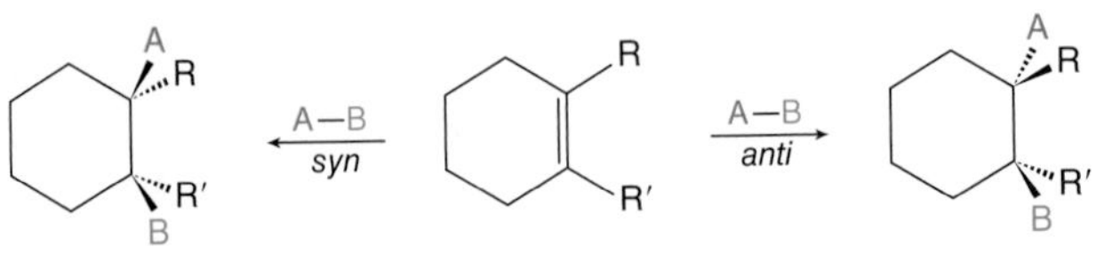

The following are examples of reactions on π bonds (that is, double or triple):

- Hydrogenation
- Electrophilic additions of haloalkanes and sulfuric acid
- Acid-catalyzed hydration

- Halogens
- Hydroboration-oxidation
- Epoxidation
- Ozonolysis
- Polymerization

In each case, a π bond breaks, two new σ bonds are created, the bonds formed are stronger than the bonds broken, and the reaction is exothermic.

Hydrogenation

A **hydrogenation** reaction involves the addition of molecular hydrogen (H_2), usually in the presence of a very finely divided metal catalyst (Ni, Pd, Rh Pt), although non-catalytic hydrogenation can be done at very high temperatures. A pair of H atoms is added to a molecule with a double bond to produce saturated organic compounds.

Product Formation by Hydrogenation Both hydrogen atoms add from the *same side* of the double bond (*syn* addition):

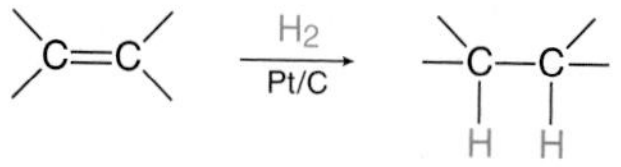

The heat evolved in this exothermic reaction is termed the **heat of hydrogenation**. In the previous example, where two new C-H bonds are formed, the $\Delta_r H°$ of hydrogenation is −136 kJ/mol. The heat of hydrogenation measures the stability of the alkene. The greater the heat evolved, the less stable the alkene. In the following, alkenes are ordered by decreasing stability:

- tetrasubstituted > trisubstituted > disubstituted > monosubstituted
- *trans* > *cis* and *E* > *Z*

Stereochemistry of the Hydrogenation

- The addition of the hydrogen atoms occurs at the same face of the double bond, as we can see in the following two examples:

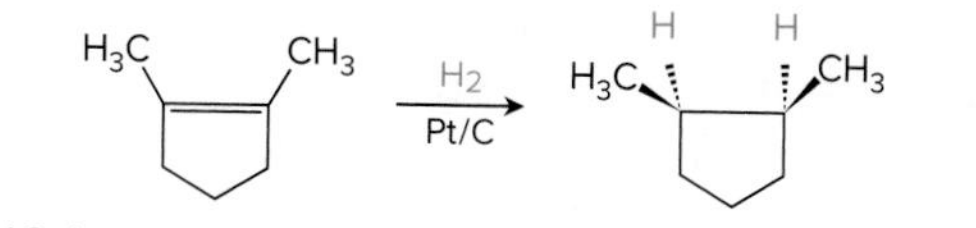

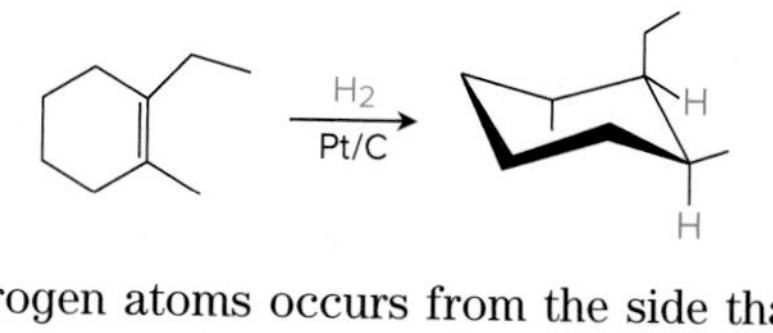

- The addition of hydrogen atoms occurs from the side that is less hindered. This is particularly important when dealing with cyclic structures, since this stereoselective reaction forms one stereoisomer in greater quantity:

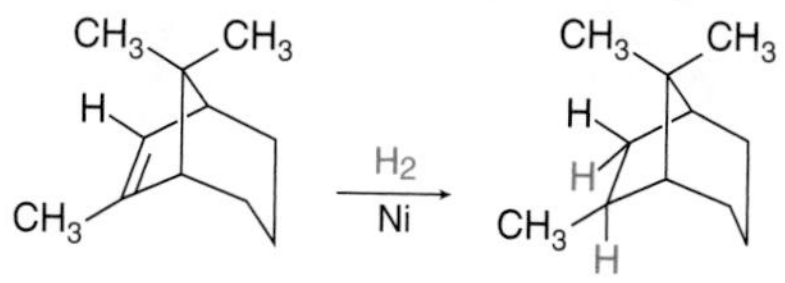

Electrophilic Additions

Addition of a Haloalkane (HX = HI, HBr, HCl) For *symmetrical alkenes*, it does not matter where the haloalkane adds, since just one product is formed:

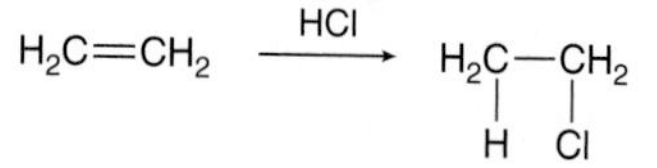

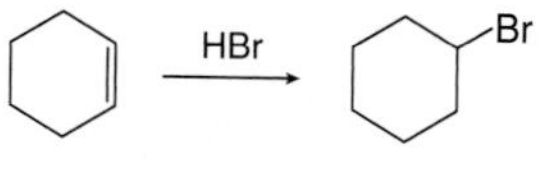

However, *asymmetrical alkenes*, which could theoretically form two products, only form one, in which the hydrogen adds to the carbon in the C=C bond that has more hydrogen atoms, and the halogen adds to the carbon with fewer hydrogen atoms (**Markovnikov's rule**):

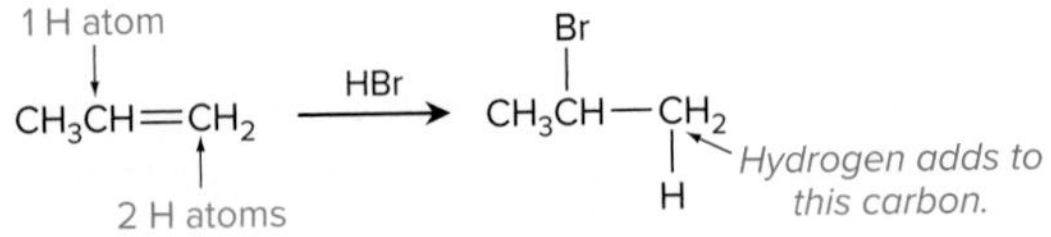

The carbon with more hydrogen atoms gets the hydrogen.

Mechanism of Addition of Haloalkanes The mechanism of addition of haloalkanes can be described in the following steps:

Step 1. H^+ adds to the double bond, forming an intermediate as a carbocation (recall that the π bonding electrons are basic/nucleophilic):

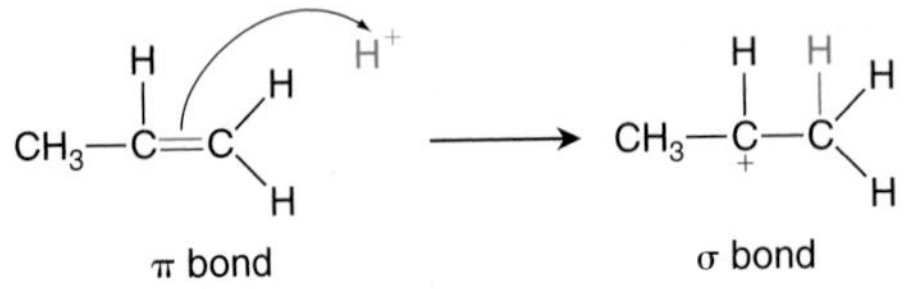

Step 2. The halogen ion adds to the carbocation:

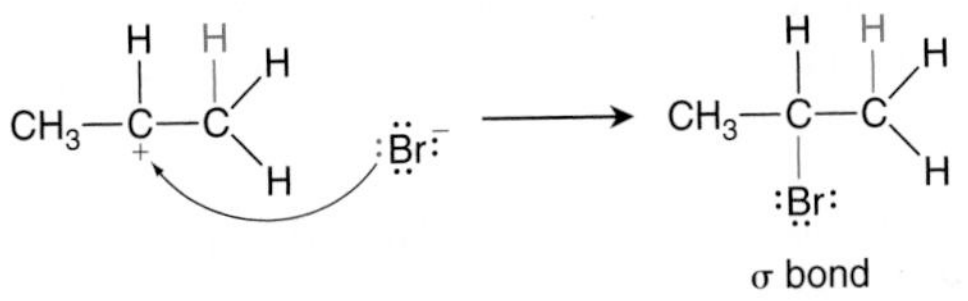

The overall process of the addition of an electrophile to an alkene or alkyne can be summarized in the following sequence, and the reaction energy diagram is shown in Figure 21.2:

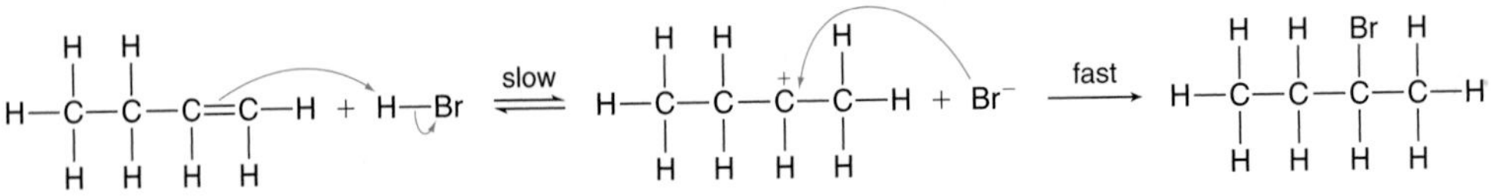

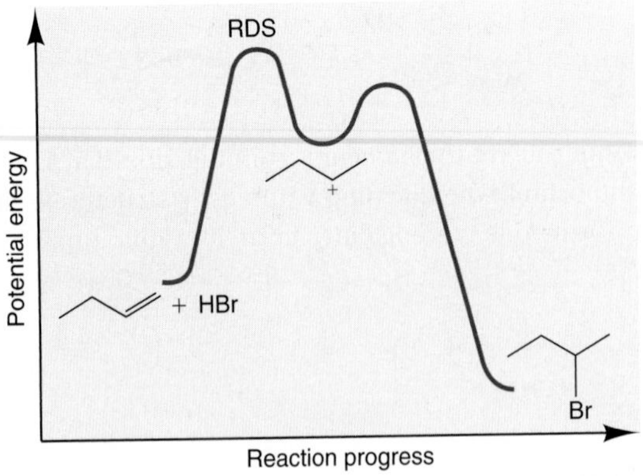

FIGURE 21.2 Reaction energy diagram of an electrophilic addition, showing the rate-determining step (RDS)

As discussed before, in regular haloalkane additions to a π bond, the halogen gets attached to the carbon with fewer hydrogen atoms. There is, however, a way to get the halogen attached to the carbon with more hydrogen atoms, and it is through a free-radical mechanism. This regiospecific reaction might be initiated by peroxides (H_2O_2, ROOR) or by photochemical means and is considered an *anti-Markovnikov* reaction. For example,

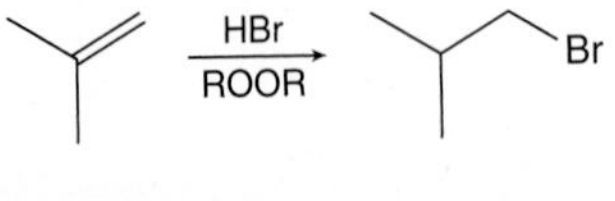

which can also be shown as follows:

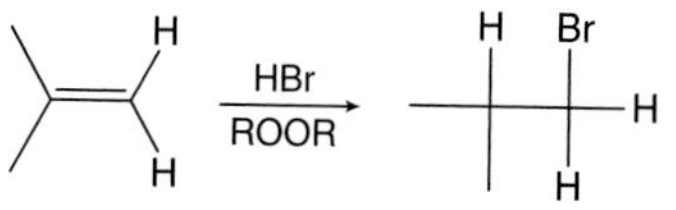

Sample Problem 21.5 Predicting the Product and Showing the Mechanism of Electrophilic Addition of Haloalkanes to a π Bond

Problem Show the mechanism involved in each electrophilic addition reaction, and predict the products.

Plan First we identify which carbon (if any) is the most substituted carbon, since that will be the one that will form the carbocation. Then we follow the two steps of the mechanism of addition of haloalkanes to the π bond: (1) the electrophilic addition of the hydrogen atom to the starting material and the formation of a carbocation on the most substituted carbon atom, and (2) the addition of the halogen ion to the carbocation.

Solution

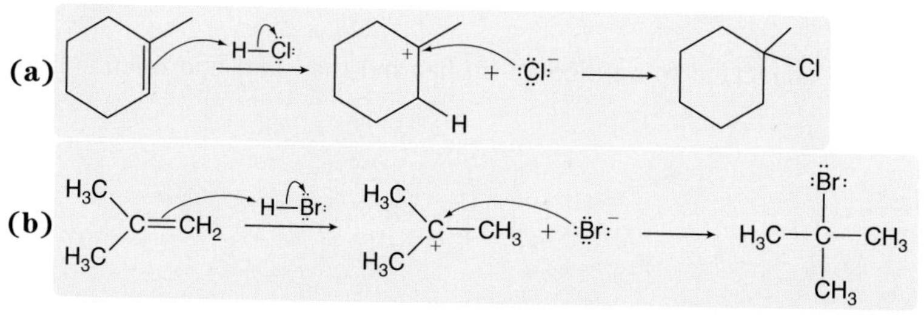

Check The addition of haloalkanes to the double bond follows Markovnikov's rule. Therefore, we will check that the hydrogen atom was added to the carbon in the C=C bond that has more hydrogen atoms and that the halogen atom was added to the carbon with fewer hydrogen atoms.

Follow-Up Problem 21.5 Predict the products in each reaction:

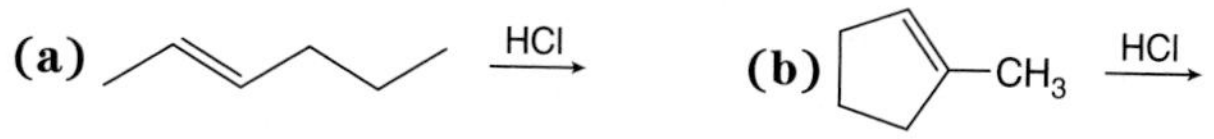

Sample Problem 21.6 Predicting the Major Product in Hydrogenation and in Electrophilic Addition of Haloalkanes to the π Bond

Problem Predict the major product in each addition reaction:

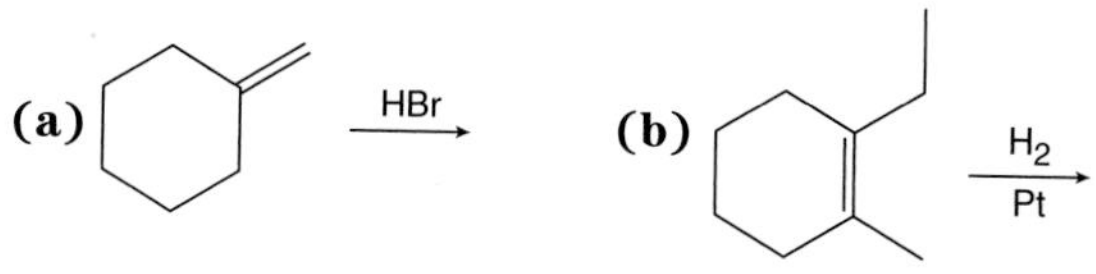

Plan In the addition of a haloalkane to the double bond, the hydrogen atom will add to the less-substituted carbon atom, whereas in the hydrogenation reaction, the hydrogen atoms add from the same side of the double bond.

Solution

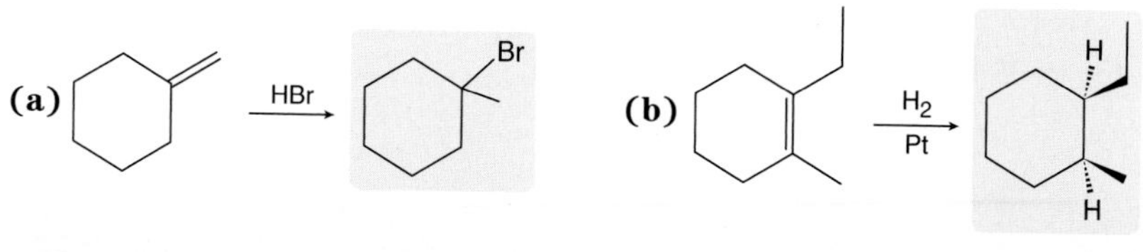

Follow-Up Problem 21.6 Predict the products in each reaction:

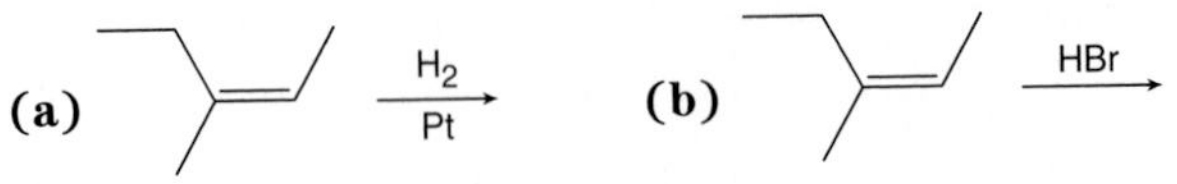

Addition of Sulfuric Acid The general equation when H_2SO_4 ($HOSO_2OH$) is added to an alkene is

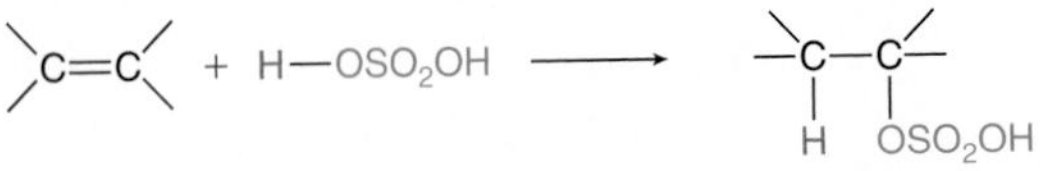

Mechanism of Addition of Sulfuric Acid The mechanism of addition of sulfuric acid can be described in the following steps:

Step 1. H^+ adds to the double bond, forming an intermediate as a carbocation:

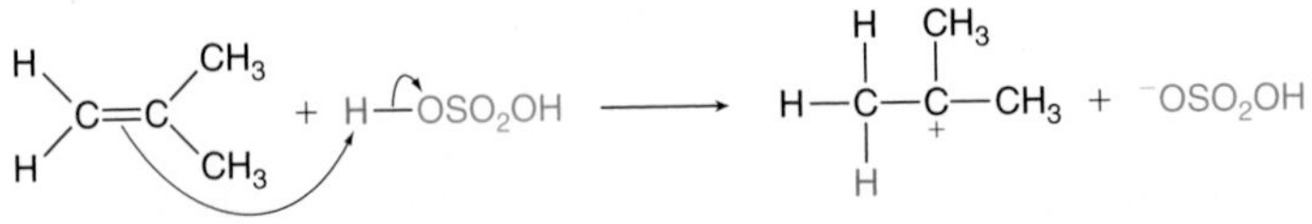

Step 2. The carbocation reacts with the hydrogen sulfate ion. It follows a Markovnikov orientation:

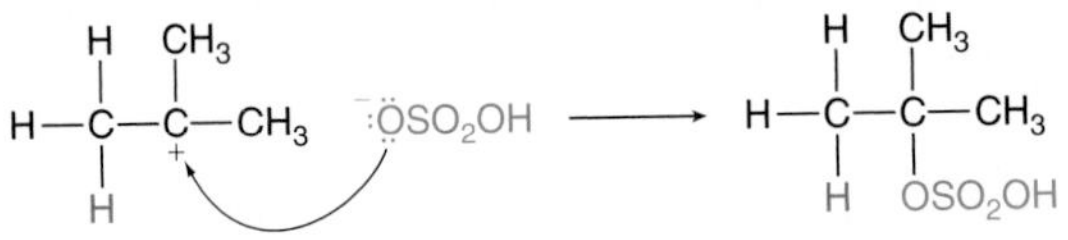

The product can be converted to an alcohol by heating the alkyl hydrogen sulfate with water:

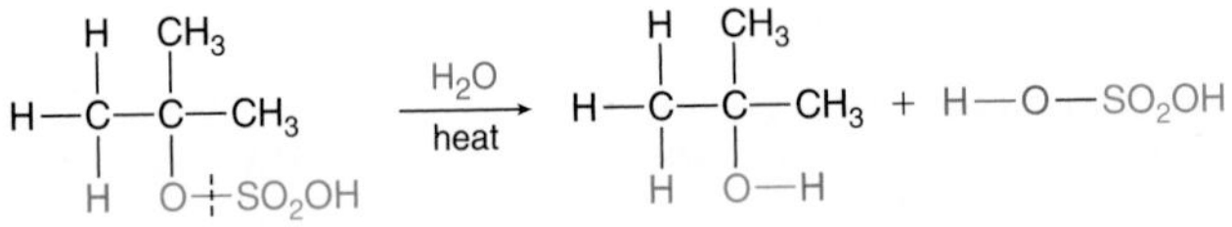

Acid-Catalyzed Hydration

The addition of water to alkenes in an acidic medium to form alcohols follows Markovnikov's rule (the hydrogen goes to the carbon in the π bond that has more hydrogen atoms, and the OH adds to the carbon with fewer hydrogen atoms):

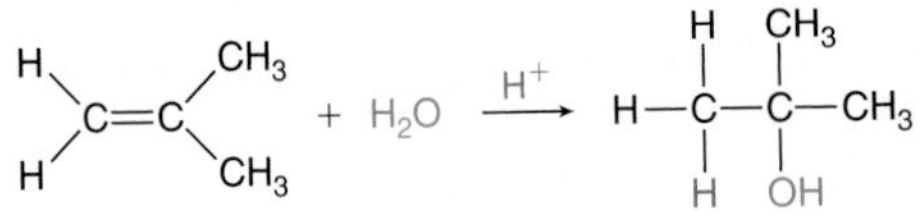

Mechanism of Acid-Catalyzed Hydration The mechanism of acid-catalyzed hydration can be described in the following steps:

Step 1. Protonation to form the carbocation:

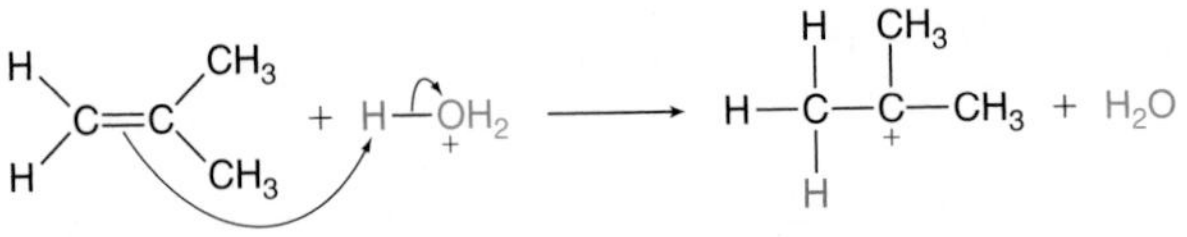

Step 2. Addition of water:

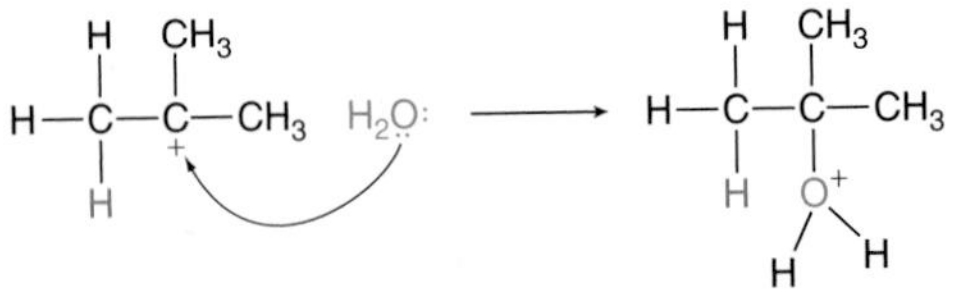

Step 3. Deprotonation:

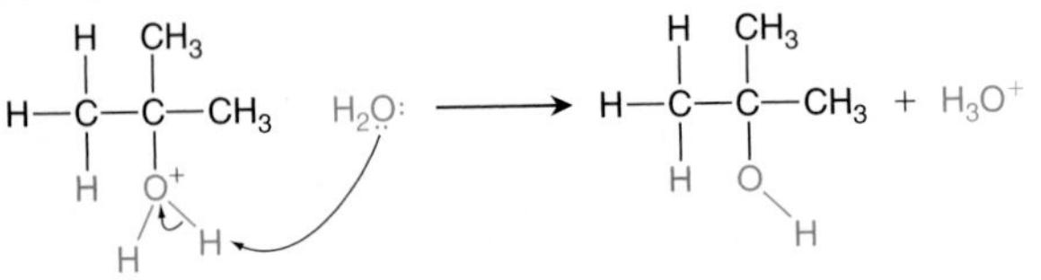

Sample Problem 21.7 Predicting the Product in Acid-Catalyzed Hydration Reactions

Problem Predict the major product in this addition reaction (Me = methyl):

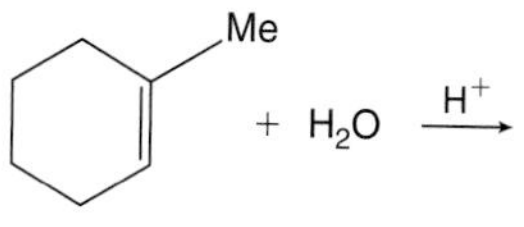

Plan We check if one of the carbon atoms forming the π bond is more substituted than the other, since that is the one that will form the carbocation where the water will add and then get deprotonated.

Solution

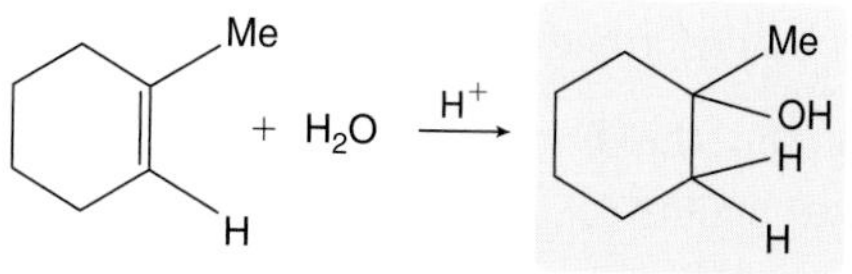

Check Following Markovnikov's rule, when the π bond is broken, the more-substituted carbon atom will form a σ bond with the hydroxyl group, whereas the less-substituted carbon (the one with more hydrogen atoms) will form a σ bond with a hydrogen atom.

Follow-Up Problem 21.7 Predict the products in each reaction:

(a)

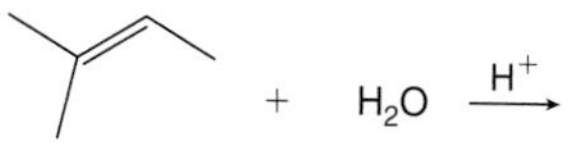

(b)

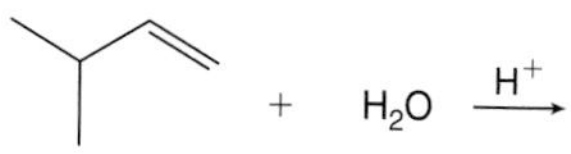

Addition of Halogens

The halogen atoms (X) add *trans* to each other in an *anti* addition. The halogen addition forms a vicinal dihalide, and the reactivity increases with carbon substitution. It works with I_2, Br_2, and Cl_2:

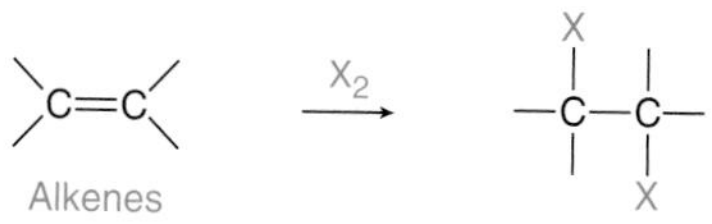

Mechanism of Halogen Addition The mechanism of *halogen addition* (such as Br_2) can be described in the following steps:

Step 1. Addition of the electrophilic bromine with formation of a cyclic bromonium ion:

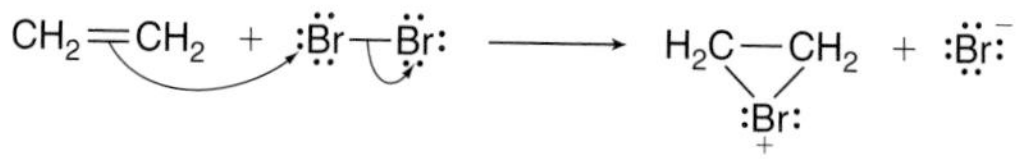

Step 2. Nucleophilic attack on the cyclic bromonium ion, where the nucleophile attacks the side opposite to the bromonium ion:

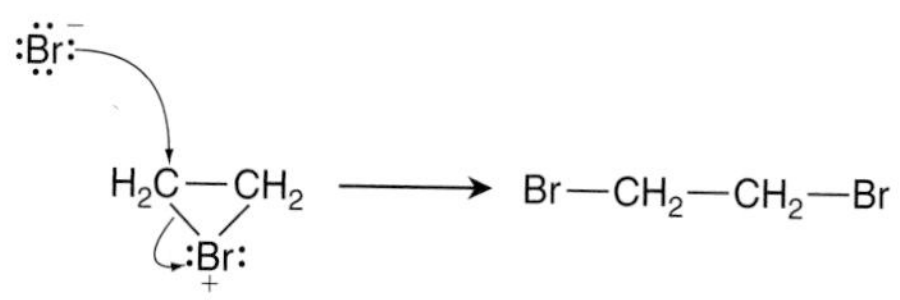

For example, the mechanism of the electrophilic addition of bromine to cyclohexane is

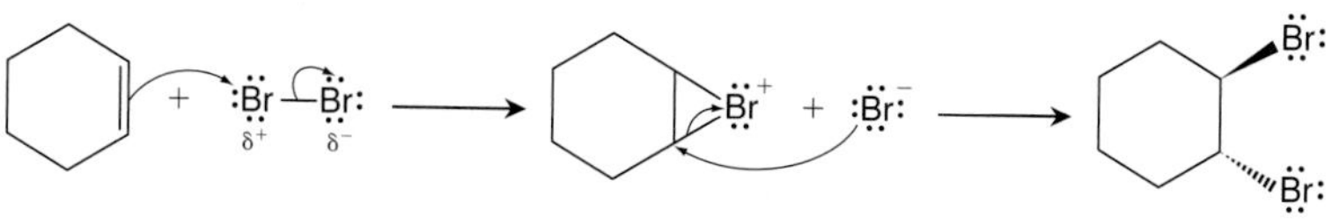

Hydroboration-Oxidation

In this regioselective* (*anti-Markovnikov*) reaction, the borane goes to the less-hindered side. It is a *syn* addition (H and OH get attached on the same side of the product) that forms an alcohol:

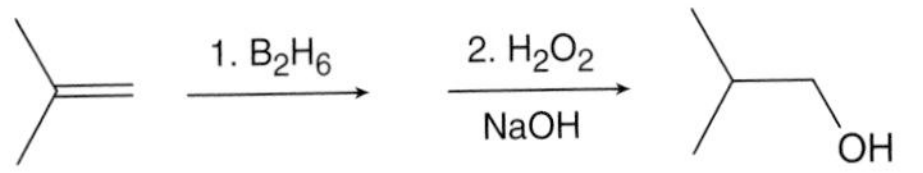

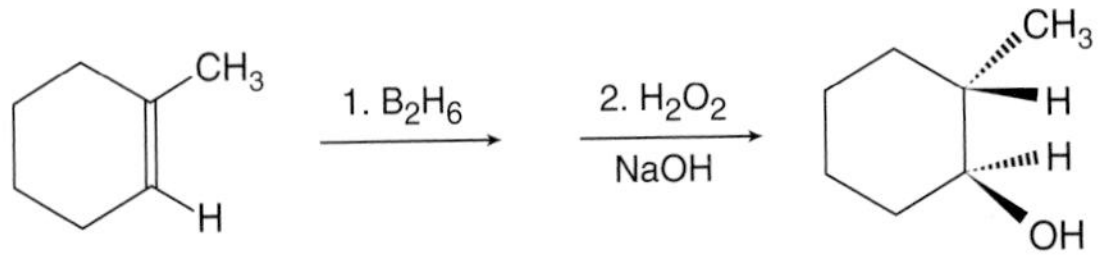

Mechanism of Hydroboration-Oxidation The mechanism of the hydroboration-oxidation reaction can be described in the following steps:

Step 1. Hydroboration by the borane (BH_3, B_2H_6), which is electron deficient (strongly electronegative):

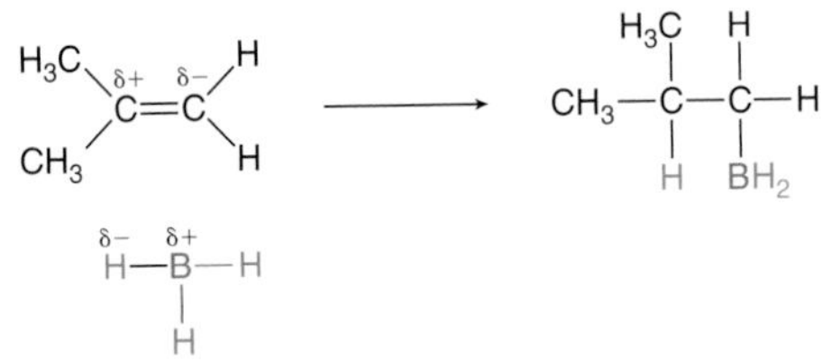

It goes through a four-centred transition state. The boron goes to the carbon with more hydrogen atoms, and the hydride goes to the carbon with fewer hydrogen atoms (more substituents):

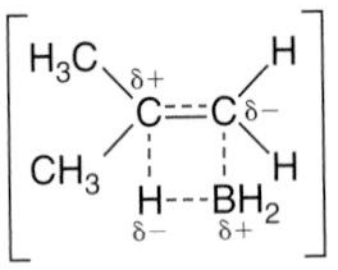

Step 2. Oxidation of the organoborane, where alkaline hydroperoxide replaces the boron atom with OH, with the oxygen coming from the peroxide:

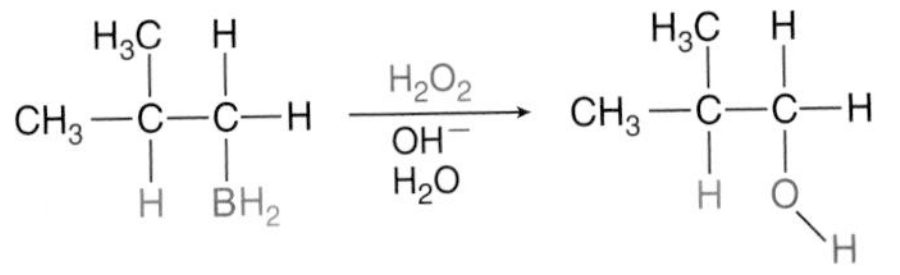

Sample Problem 21.8 Predicting the Product in Hydroboration-Oxidation Reactions

Problem Predict the major product in each addition reaction:

(a) 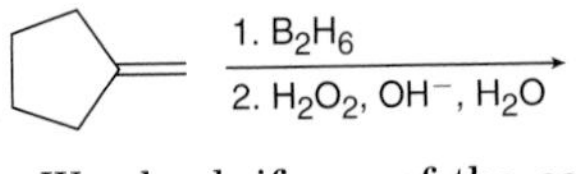

(b) 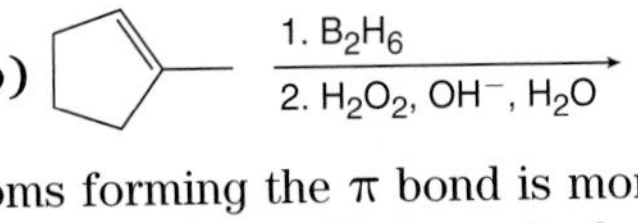

Plan We check if one of the carbon atoms forming the π bond is more substituted than the other, since the hydride goes to the carbon with fewer hydrogen atoms, and the boron from the borane goes to the carbon with more hydrogen atoms to form a four-centred transition state. The oxidation of the organoborane replaces the boron atom with OH. Recall that this is a *syn* addition.

*A regioselective reaction is one in which one direction of bond making or bond breaking occurs preferentially over all other possible directions.

Solution

(a)
(b)

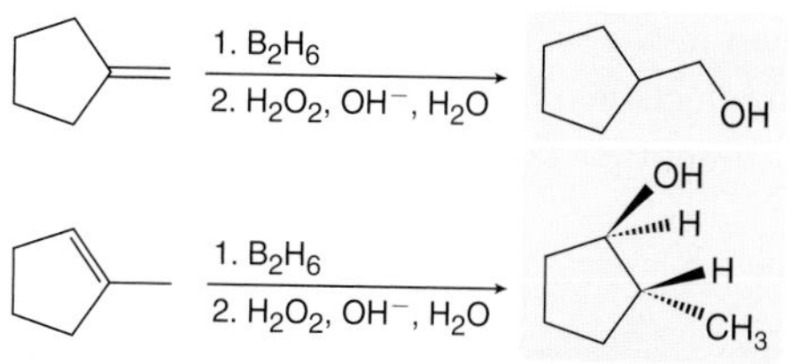

Check The hydroboration-oxidation reaction is a *syn* addition that does not follow Markovnikov's rule. When the π bond is broken, the more substituted carbon atom will form a σ bond with a hydrogen atom, whereas the less substituted carbon (the one with more hydrogen atoms) will form a σ bond with the hydroxyl group. Both bonds should be on the same side of the starting material.

Follow-Up Problem 21.8 Predict the products in each reaction:

(a)

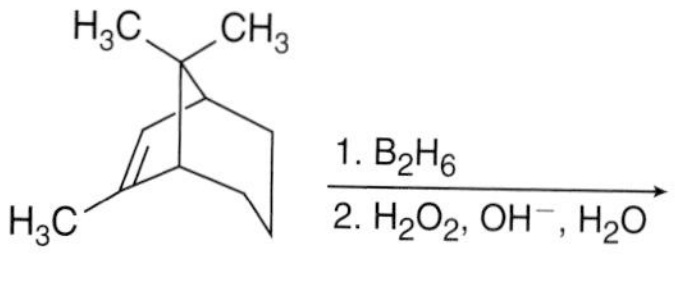

(b)

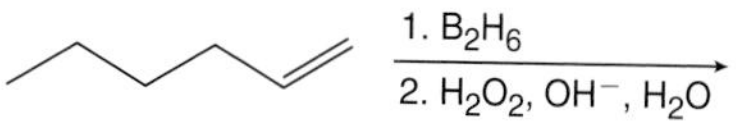

Epoxidation

Epoxidation is a syn addition of an oxygen atom from a peroxyacid that forms a three-membered ring:

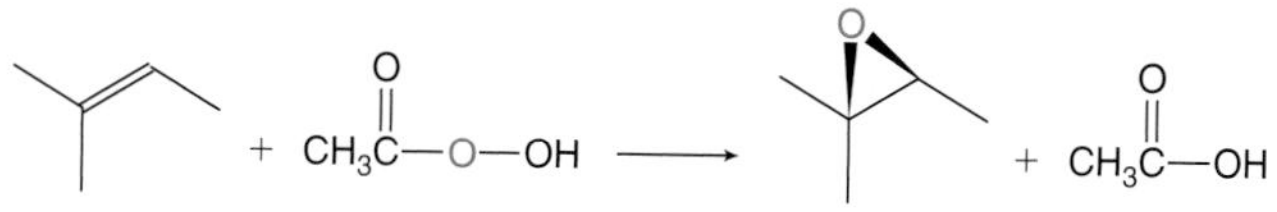

Mechanism of Epoxidation The epoxidation reaction is considered to be concerted:

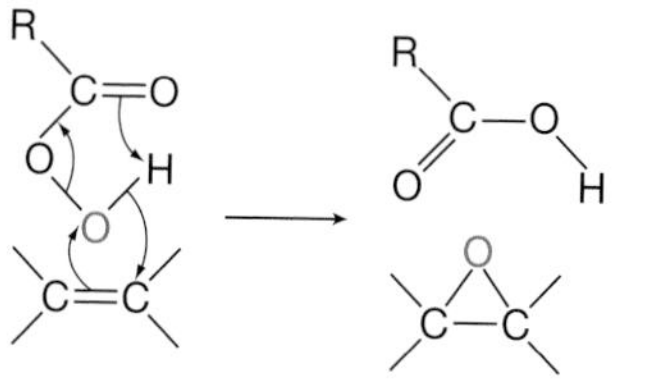

Ozonolysis

Ozone is a strong oxidizing agent that breaks double and triple bonds and replaces them with oxygen. **Ozonolysis** can be done on alkenes to obtain ketones and/or aldehydes, and on terminal alkynes to form carboxylic acids and carbon dioxide. Internal alkynes give two carboxylic acids:

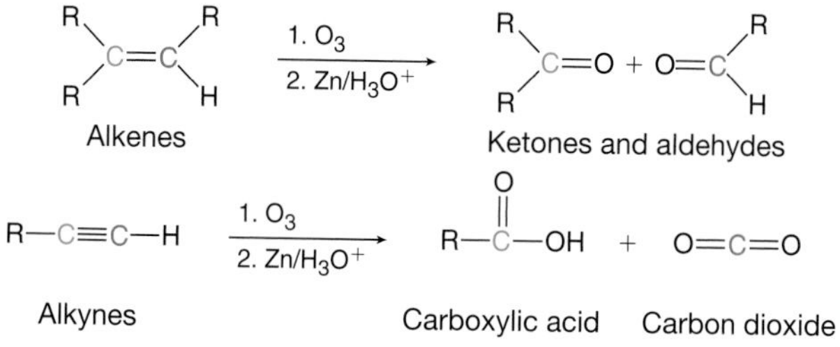

Mechanism of Ozonolysis The reaction of an alkene with ozone produces ozonide, which hydrolyzes to give carbonyl compounds:

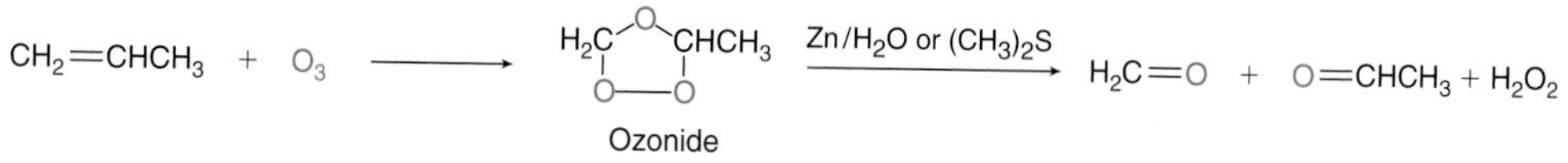

The purpose of the Zn is to reduce the H_2O_2 that is generated in the hydrolysis of the ozonide intermediate.

The ozonolysis mechanism is a good exercise in arrow pushing. Notice that both σ and π bonds are broken to produce two new C=O double bonds. Remember that each arrow indicates the motion of two electrons, from a bond to a lone pair, from a lone pair to a bond, or from a bond to another bond:

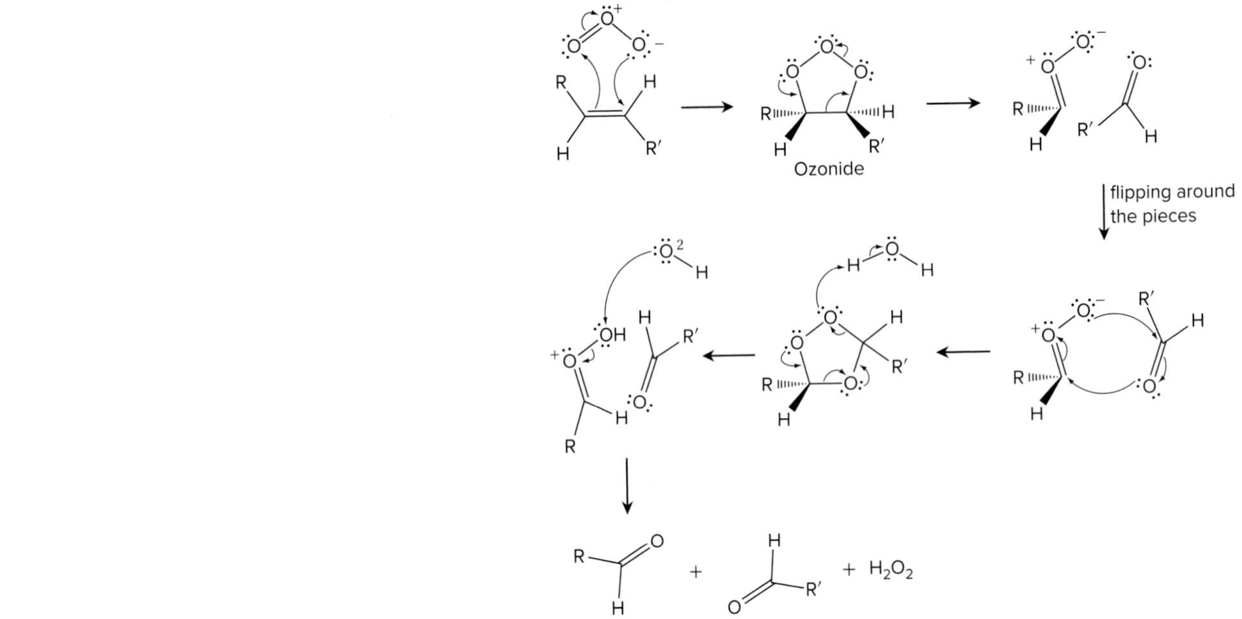

The following reaction summarizes the products expected in the ozonolysis:

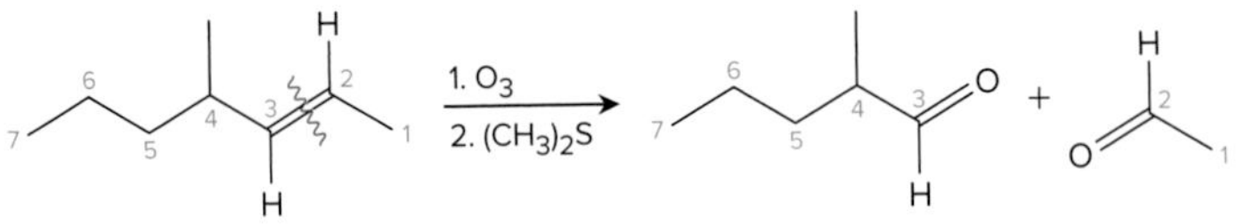

Sample Problem 21.9 Predicting the Products and the Starting Material of Ozonolysis

Problem

(a) Predict the products of this reaction:

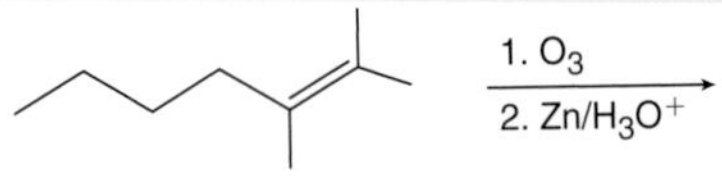

(b) What is the structure of the starting material in this reaction?

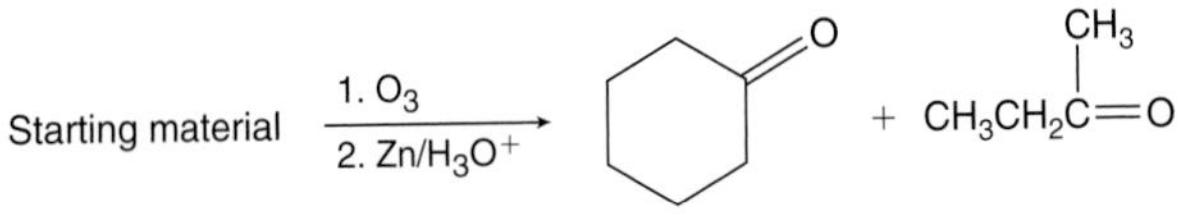

Plan In the ozonolysis of an alkene, both σ and π bonds are broken to produce two new C=O double bonds.

Solution

(a)

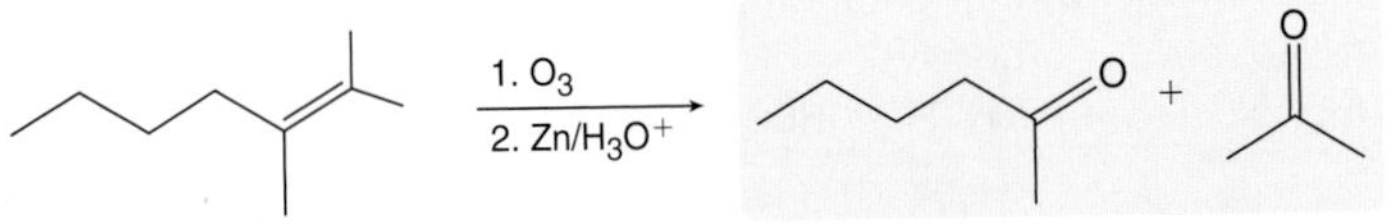

(b)

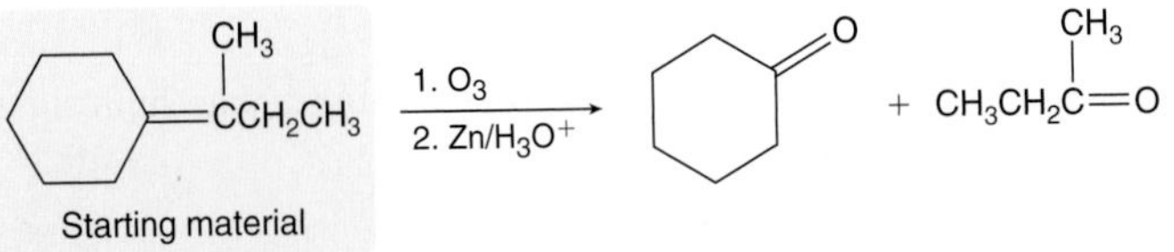

Check The carbon atoms forming double bonds are the same before and after the reaction.

Follow-Up Problem 21.9

What is the structure of the starting material in this reaction?

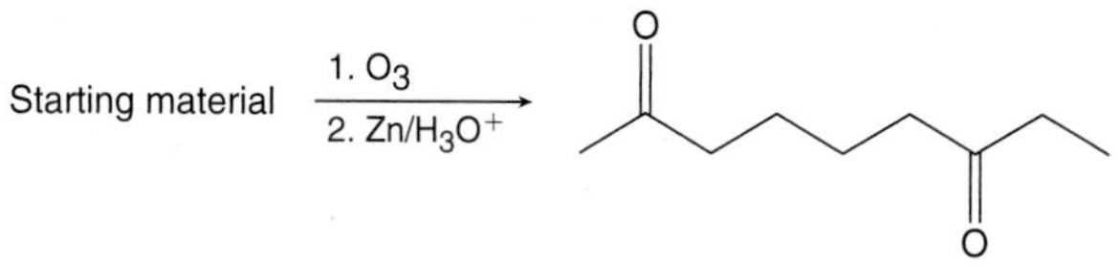

Polymerization

Polymerization is a chemical reaction in which monomers, which may or may not be all the same units, form a polymer. A polymer is a high-molecular-weight molecule formed by joining many thousands of single units (monomers). In industry, the reaction is usually done with a catalyst, often under high pressure or heat:

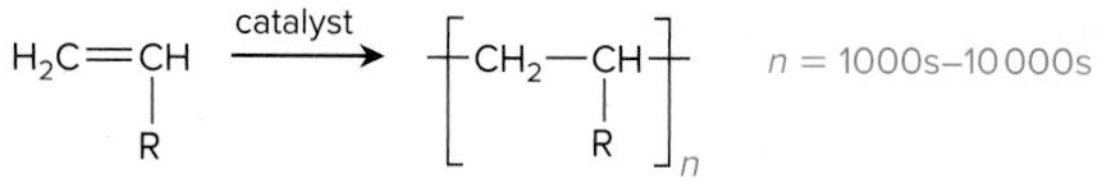

The following are examples of synthetic polymers. These and other polymers will be described in more detail in Chapter 22:

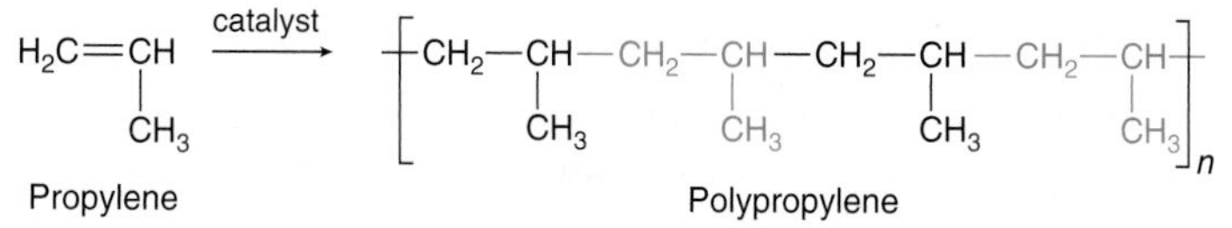

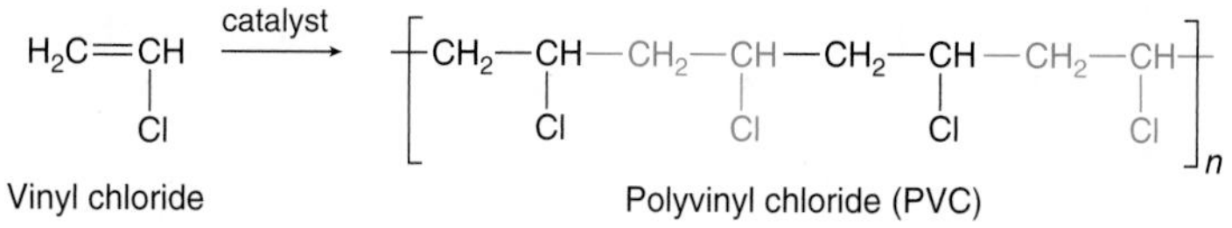

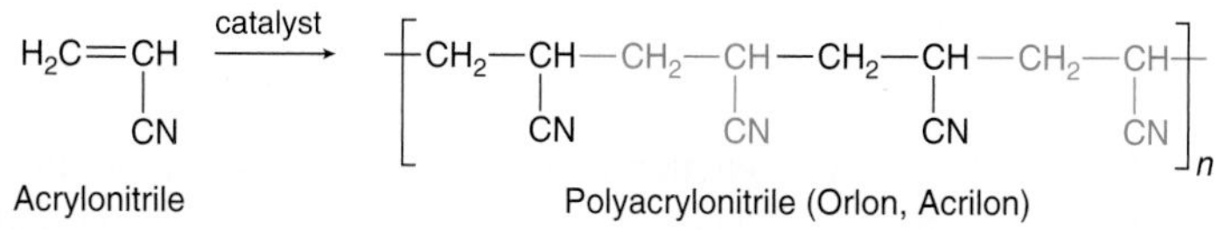

SUMMARY OF SECTION 21.5

- In an addition reaction, the carbon atoms change from tetrahedral to trigonal planar (their hybridization changes from sp^2 to sp^3).
- There are several addition reactions to the double bond, each of them with its own mechanism.
- If there is an addition of a hydrogen to the carbon in the C=C bond that has more hydrogen atoms, it is said to follow Markovnikov's rule. If the hydrogen adds to the carbon with fewer hydrogen atoms, it is named anti-Markovnikov addition.

21.6 Electrophilic Aromatic Substitution: Reactions of the Benzene Ring

The benzene ring can undergo electrophilic substitution, nucleophilic substitution, elimination, and addition reactions. Benzylic derivatives (an alkyl-substituted benzene ring) can undergo S_N1, S_N2, and E1 reactions; however, the benzene ring itself can also react in what is called electrophilic aromatic substitution (EArS).

Comparing the Reactivities of Alkenes and Aromatic Compounds

The *delocalized* unsaturation of aromatic compounds is very different from the *localized* unsaturation of alkenes. That is, despite the way we depict its resonance forms, benzene does *not* have double bonds. Thus, benzene does not decolourize Br_2 by an

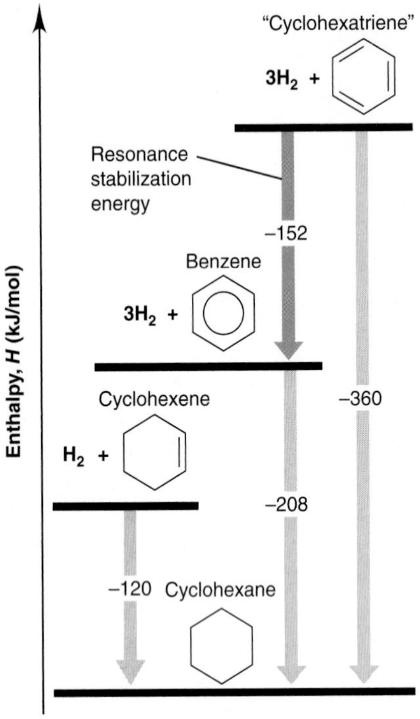

FIGURE 21.3 The stability of benzene

addition reaction (see Figure 21.1B). As we saw previously, alkenes and alkynes usually react in exactly the same ways because the π electrons react. Remember from MO theory (Section 10.3) that σ bonding electrons are lower in energy than π bonding electrons.

In general, aromatic rings are much *less* reactive than alkenes because of their delocalized π electrons. For example, let us compare the enthalpies of reaction for the addition of H_2. Hydrogenation of cyclohexene, with one C═C bond, has a $\Delta_r H°$ of −120 kJ/mol. Then, for the imaginary molecule "cyclohexatriene" (that is, the structure with three C═C bonds), the hypothetical $\Delta_r H°$ for hydrogenation is three times as much, or −360 kJ/mol. Hydrogenation of benzene has a $\Delta_r H°$ of −208 kJ/mol. Thus, the $\Delta_r H°$ for hydrogenation of benzene is 152 kJ/mol *less* than that for "cyclohexatriene." This energy difference is attributed to the aromatic resonance stabilization of benzene (Figure 21.3).

This extra energy needed to break up the delocalized π system means *addition* reactions with benzene rarely occur. However, benzene does undergo many *substitution* reactions, in which the delocalization is retained when an H atom attached to a ring C is replaced by another group:

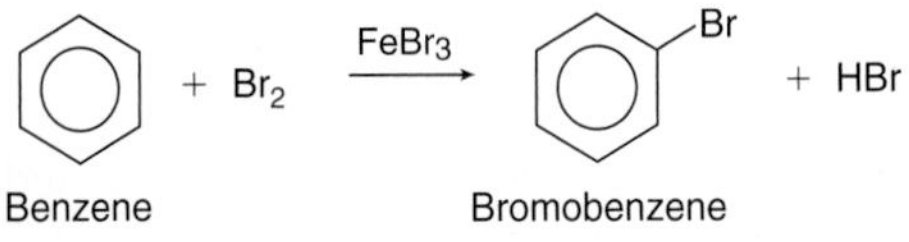

Electrophilic Aromatic Substitution

Electrophilic substitution happens in many of the reactions of compounds containing benzene rings. The benzene ring is highly attractive to electrophiles because of the delocalized electrons exposed above and below the plane of the rest of the molecule (Figure 20.22). The electrophile can be either a positive ion or the slightly positive end of a polar molecule. An **electrophilic aromatic substitution** (EArS) can be represented as follows:

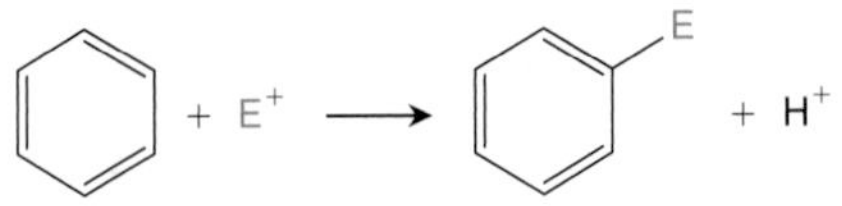

The mechanism of the EArS reaction involves the formation of an intermediate carbocation. The reaction of the electrophile E^+ with the arene (aromatic hydrocarbon) is the slow step since it results in the loss of aromaticity. Remember that it is impossible to get a positive ion on its own, so the anion associated with the cation E^+ will attract the leaving hydrogen atom in the intermediate step.

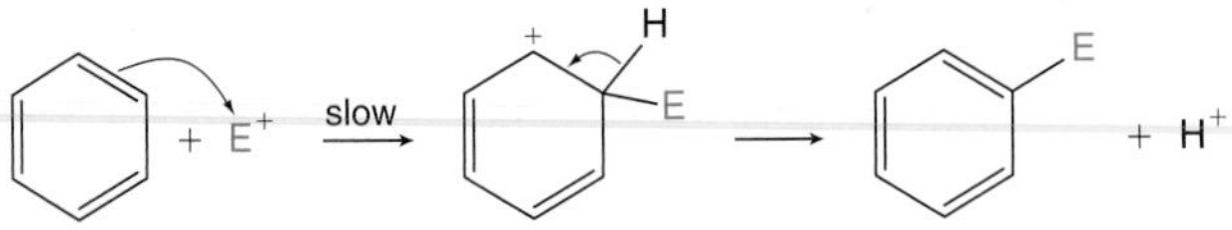

The positive charge on the intermediate carbocation that is formed, also described as the **cyclohexdienyl cation (arenium ion, sigma-complex)**, is stabilized by resonance throughout the molecule:

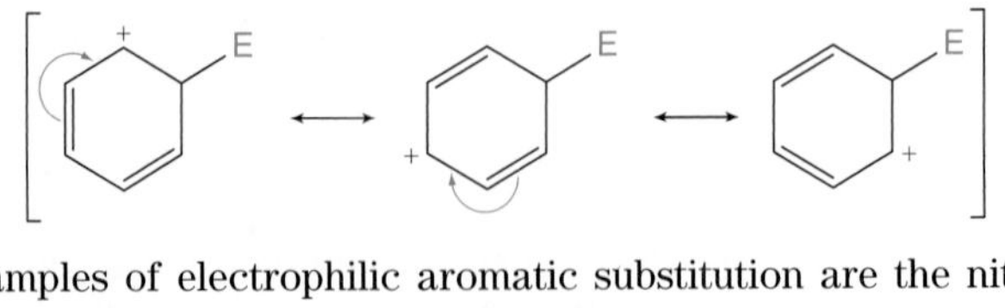

Some examples of electrophilic aromatic substitution are the nitration of benzene, the Friedel-Crafts reaction, and the halogenation of benzene.

Nitration of Benzene The nitration of benzene occurs by the action of the nitronium ion (NO_2^+), which is formed by the loss of water from HNO_3, as the electrophile:

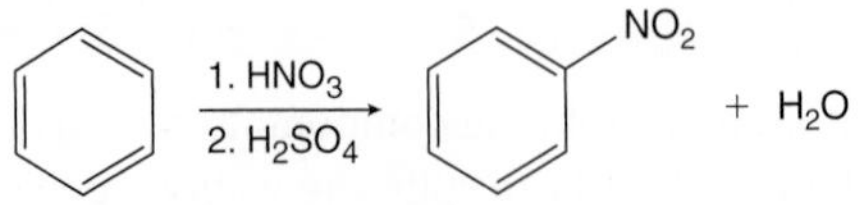

The mechanism for nitration of benzene is as follows. The rate-determining step is when the electrophilic nitronium ion reacts with the arene as it destroys the aromaticity of the arene:

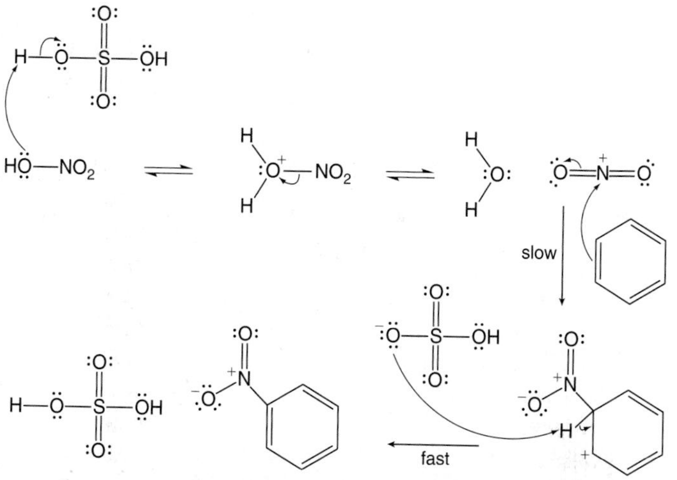

A simplified mechanism showing the aromatic ring resonance structures is as follows:

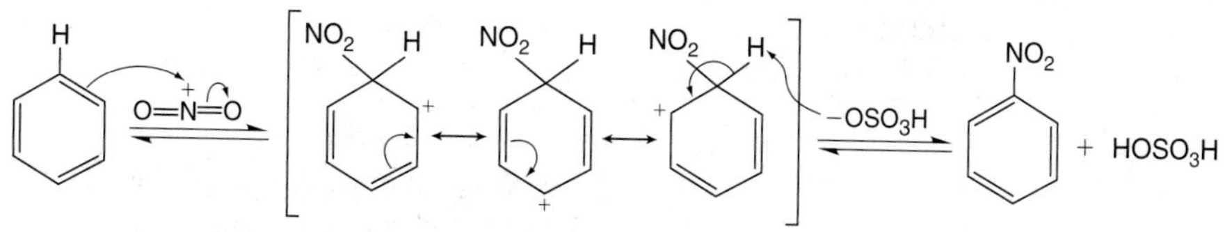

Friedel-Crafts Reaction The **Friedel-Crafts reaction** can be performed either as an acylation or as an alkylation:

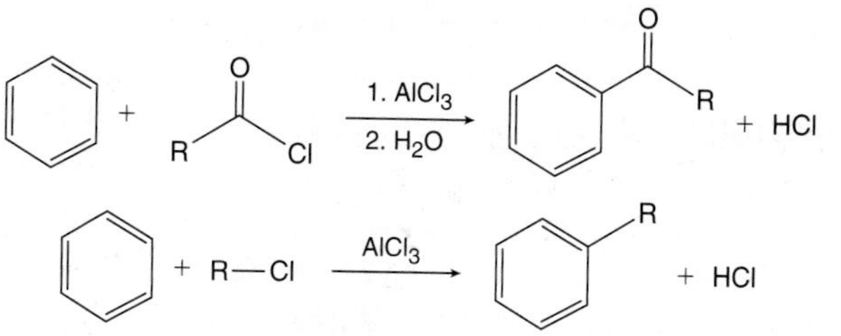

A general form of the Friedel-Crafts **alkylation** mechanism is as follows:

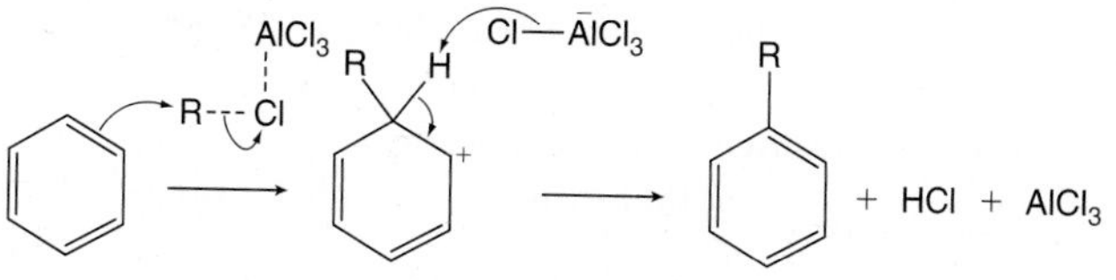

The Friedel-Crafts **acylation** of benzene follows a similar mechanism:

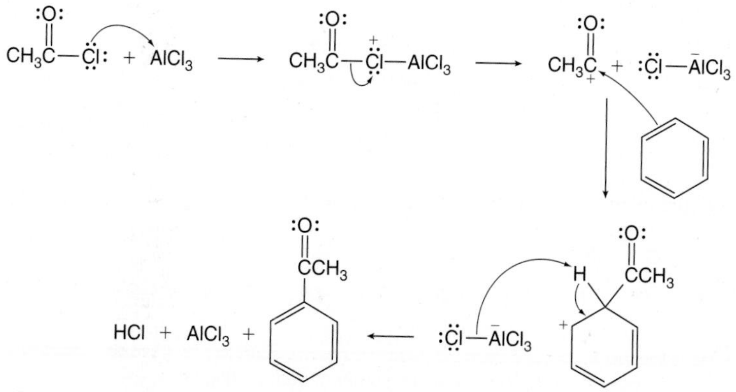

Halogenation of Benzene The aromatic **halogenation of benzene** with bromine, chlorine, or iodine is catalyzed by an iron salt or aluminum trihalide:

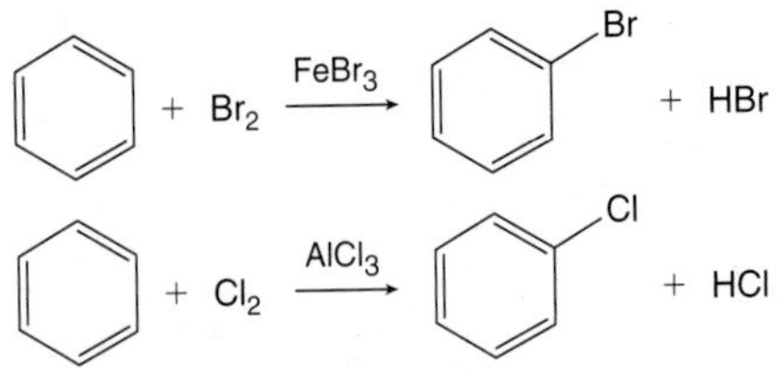

The mechanism for halogenation of benzene is as follows:

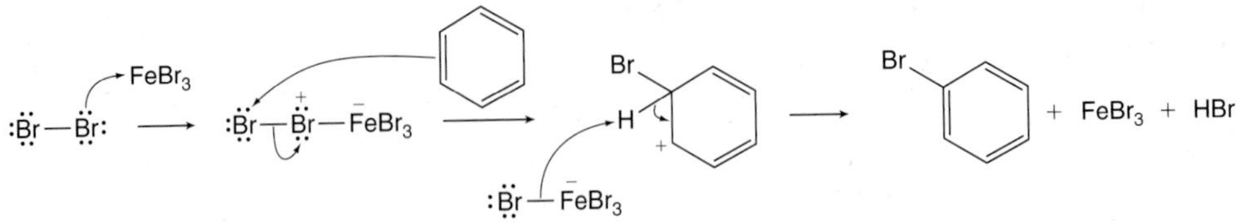

Ortho/Para- and *Meta*-Directing Substituents

EArS of already substituted benzene can be directed to *ortho*, *meta*, or *para* positions. The question that arises is, *where does E_2 substitute relative to E_1 when there are three possibilities*?

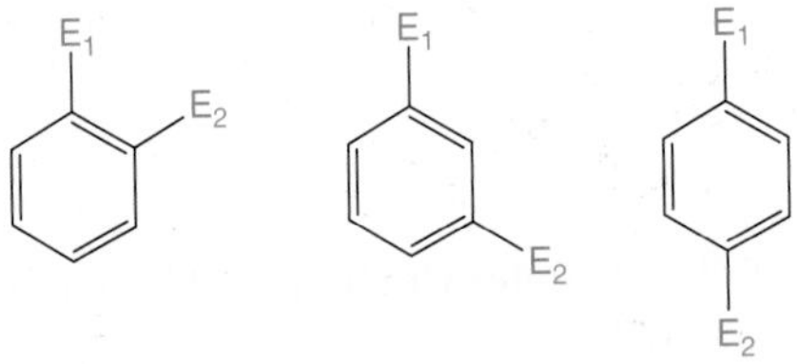

We can actually predict where the second substituent will attach to the benzene ring. Some groups are *ortho/para* directors and others are only *meta* directors:

- **Electron-donating groups** (EDGs) add electron density to the π system, making it *more nucleophilic.* They can be recognized by lone pairs on the atom adjacent to the π system, for example, $—OCH_3$. EDGs are *activating groups* that direct *ortho/para* substitution.
- **Electron-withdrawing groups** (EWGs) remove electron density from the π system, making it *less nucleophilic.* They can be recognized *either* by the atom adjacent to the π system having several bonds to more electronegative atoms *or* from having a formal positive or negative charge, for example, $—CO_2R$, $—NO_2$. EWGs are *deactivating groups* that direct *meta* substitution.

Having H as a reference, substituents in a benzene ring can be ordered as follows:

1. *Ortho/para* directors (from strongest to weakest):

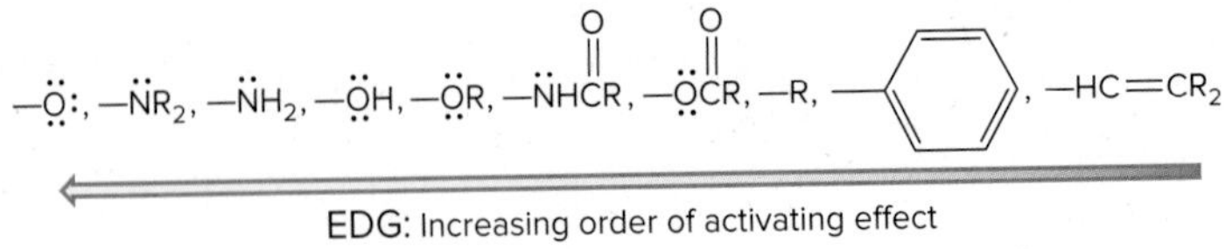

2. *Meta* directors (from weakest to strongest):

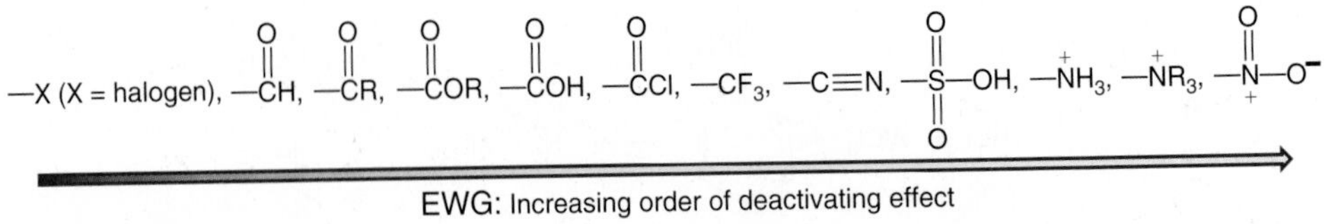

SUMMARY OF SECTION 21.6

- Aromatic rings are much less reactive than alkenes because of their delocalized π electrons.
- Electrophilic aromatic substitution (EArS) is one of the reactions the benzene ring undergoes. Nitration, Friedel-Crafts (formation of alkyl and acyl derivatives) reactions, and halogenation are examples of EArS.
- A ring substituent can direct the substitution of a second electrophile to *ortho/para* positions or to a *meta* position.

21.7 Some Reactions of Functional Groups

Functional groups are responsible for the chemical reactions of the molecules. Each functional group undergoes a similar chemical reaction regardless of the size of the molecule it is part of; however, other functional groups that are present can modify its reactivity. Several reactions are given the name of the scientist who discovered that particular reaction; for example, the Grignard reaction, a method for generating carbon-carbon bonds using magnesium to couple ketones and haloalkanes, honours French chemist and Nobel Prize recipient François Auguste Victor Grignard. Table 21.2 summarizes some of the most common reactions of some functional groups. This section will examine a few of these reactions.

Reactions of Alcohols

Alcohols undergo elimination and substitution reactions.

- The elimination of H and OH, called **dehydration**, requires acid and forms an alkene:

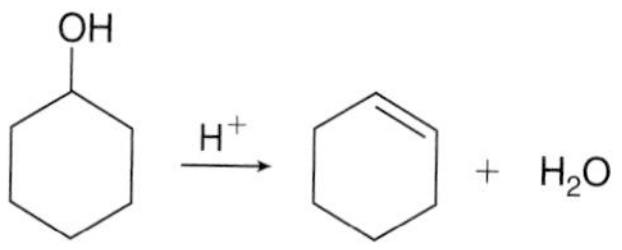

Cyclohexanol Cyclohexene

The elimination of two H atoms is an *oxidation* reaction and requires an inorganic oxidizing agent, such as potassium dichromate ($K_2Cr_2O_7$) in aqueous H_2SO_4. The product has a C═O group:

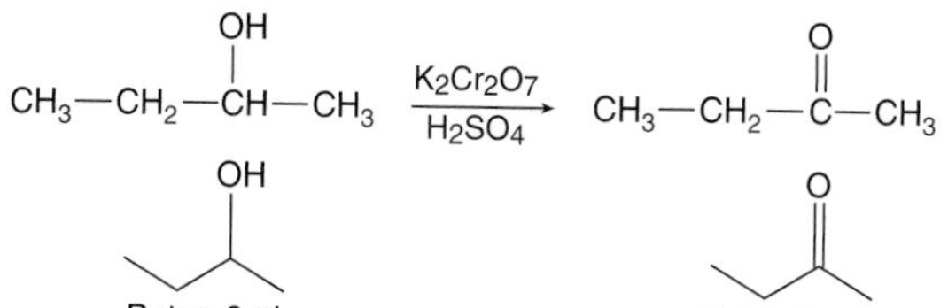

Butan-2-ol Butan-2-one

Alcohols with an OH group at the end of the chain (R—CH_2—OH) can be oxidized further to acids. (This is *not* an elimination reaction.) Wine turns sour, for example, when the ethanol in contact with air is oxidized to ethanoic acid (acetic acid):

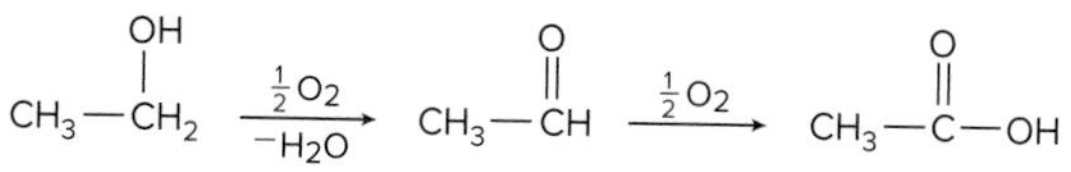

- Substitution yields products with other single-bonded functional groups. Reactions of hydrohalic acids with many alcohols give haloalkanes:

$$R_2CH\text{—}OH + HBr \longrightarrow R_2CH\text{—}Br + HOH$$

As we saw before, *the C atom undergoing the change in a substitution is bonded to a more electronegative element*, which makes it partially positive and, thus, a target for a negatively charged or an electron-rich group of an incoming reactant.

TABLE 21.2 Characteristic Reactions of Some Functional Groups

Functional Group	Formula	Characteristic Reactions
Alkanes	C—C, C—H	Combustion Radical substitution (of H by halogen)
Alkenes	C═C	Electrophilic additions Radical additions and addition reactions to dienes Allylic substitutions
Alkynes	C≡C—H	Addition reactions Nucleophilic addition and reduction reactions Substitution of H
Haloalkanes	H—C—C—X	Substitution of X Elimination of H, X, and dihalides
Alcohols	H—C—C—O—H	Substitution of the hydroxyl H and of the hydroxyl group Elimination of water Oxidation of aliphatic alcohols and phenols Electrophilic substitution of the phenol aromatic ring
Ethers	R—O—R	Acid cleavage Peroxide formation
Benzene ring	C_6H_6	Electrophilic substitution Nucleophilic substitution, elimination, and addition reactions
Amines	C—NRH	Electrophilic substitution at nitrogen Reaction with nitrous acid Elimination
Aldehydes, ketones	(α)C—C═O	Reversible and irreversible addition reactions Hydration and hemiacetal formation Acetal, imine, enamine, and cyanohydrin formation Complex metal hydrides Organometallic reactants Carbonyl group modifications Reduction and oxidation Electrophilic α-substitutions Reactions at the α-carbon Aldol reaction Alkylation of enolate anions
Carboxylic acids	(α)C—CO_2H	Salt formation Substitution (of the hydroxyl H and of the hydroxyl group) Reduction and oxidation
Carboxylic derivatives	(α)C—CZ═O (Z = OR, Cl, NHR, etc.)	Acyl (alkanoyl) group substitution Reduction Reaction with organometallic reactants Reactions to the α-carbon Claisen condensation

Reactions of Haloalkanes

Like alcohols, haloalkanes undergo substitution and elimination reactions:

- In the same way as many alcohols undergo substitution to form haloalkanes when treated with halide ions in acid, many halides undergo substitution to form alcohols in a base. For example, OH^- attacks the C end of the C—X bond and displaces X^-:

$$CH_3—CH_2—CH_2—CH_2—Br + OH^- \longrightarrow CH_3—CH_2—CH_2—CH_2—OH + Br^-$$

1-bromobutane Butan-1-ol

Substitutions by groups such as —CN, —SH, —OR, and $—NH_2$ allow chemists to convert haloalkanes to a host of other families of compounds.

- Just as addition of HX *to* an alkene produces haloalkanes, elimination of HX *from* a haloalkane by reaction with a strong base, such as potassium ethanolate (potassium ethoxide), produces an alkene:

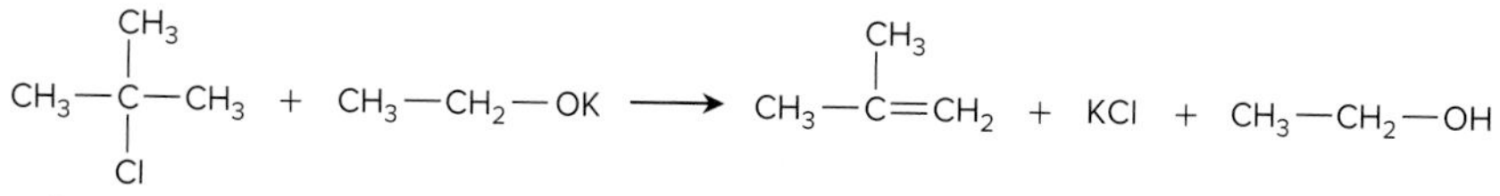

2-chloro-2-methylpropane Potassium ethanolate (potassium ethoxide) 2-methylpropene

Halides of nearly every nonmetal are known, and many undergo substitution reactions in a base. As in the case of a haloalkane, the process involves an attack on the partially positive central atom by OH^-:

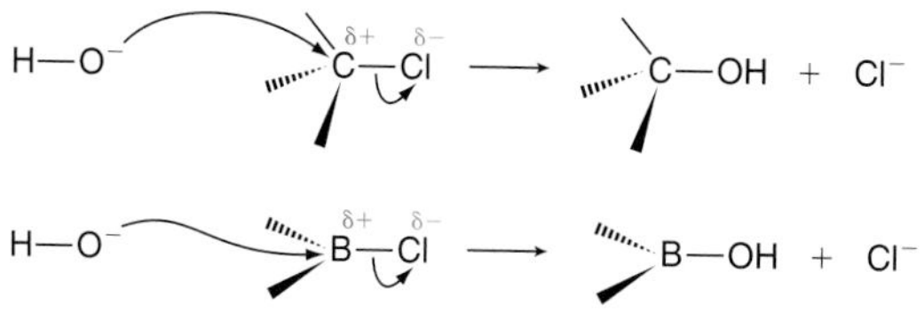

Thus, haloalkanes undergo the same general reaction as other nonmetal halides, such as BCl_3, SiF_4, and PCl_5.

Reactions of Amines

The lone pair is the reason amines undergo nucleophilic substitution reactions: the lone pair attacks the partially positive C in haloalkanes to displace X^- and form a larger, more substituted amine:

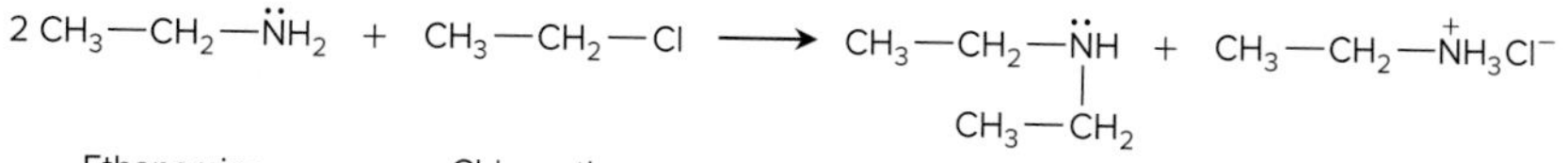

Ethanamine (ethylamine) Chloroethane *N*-ethylethanamine (diethylamine) Ethylammonium chloride

Two molecules of ethanamine are needed: one attacks the chloroethane, and the other binds the released H^+ to remove it from the *N*-ethylethanamine product.

Nitrous acid (HNO_2 or HONO) reacts with aliphatic amines, thereby providing a useful test for distinguishing primary, secondary, and tertiary amines:

1° amines + HONO (cold acidic solution) ⟶ Nitrogen gas evolution from a clear solution

2° amines + HONO (cold acidic solution) ⟶ An insoluble oil (*N*-nitrosamine)

3° amines + HONO (cold acidic solution) ⟶ A clear solution without the evolution of nitrogen gas bubbles by formation of an ammonium salt

Sample Problem 21.10 Predicting the Reactions of Alcohols, Haloalkanes, and Amines

Problem Determine the reaction type and predict the product(s) for each reaction:

(a) $CH_3—CH_2—CH_2—I + NaOH \longrightarrow$

(b) $CH_3—CH_2—Br + 2\,CH_3—CH_2—CH_2—NH_2 \longrightarrow$

(c) $CH_3—CH(OH)—CH_3 \xrightarrow[H_2SO_4]{Cr_2O_7^{2-}}$

Plan We first determine the functional group(s) of the reactant(s) and then examine any inorganic reactant(s) to decide on the reaction type, keeping in mind that, in general, these functional groups undergo substitution or elimination. In (a), the reactant is a haloalkane, so the OH^- of the inorganic reactant substitutes for I^-. In (b), the reactants are an amine and a haloalkane, so the N of the amine substitutes for the Br. In (c), the reactant is an alcohol, the inorganic reactants form a strong oxidizing agent, and the alcohol group undergoes oxidation to form a $C=O$.

Solution **(a)** Substitution: The products are $CH_3-CH_2-CH_2-OH$ + NaI.

(b) Substitution: The products are $CH_3-CH_2-CH_2-NH$ + $CH_3-CH_2-CH_2-\overset{+}{N}H_3Br^-$. 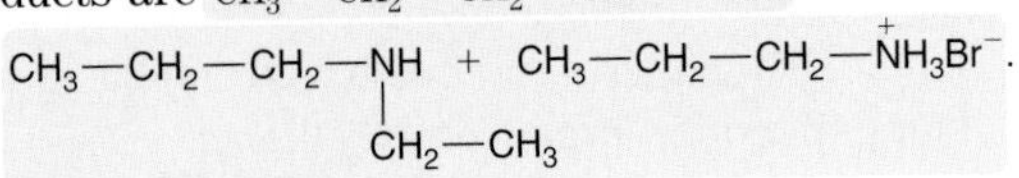
$$\underset{CH_2-CH_3}{|}$$

(c) Elimination (oxidation): The product is $CH_3-\underset{\underset{O}{\|}}{C}-CH_3$. 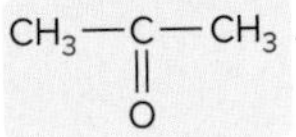

Check The only changes should be at the functional group.

Follow-Up Problem 21.10 Fill in the blank in each reaction. (*Hint*: Examine any inorganic compounds and the organic product to determine the organic reactant.)

(a) _____ + CH_3-ONa ⟶ $CH_3-CH=\overset{\overset{CH_3}{|}}{C}-CH_3$ + NaCl + CH_3-OH

(b) _____ $\xrightarrow[H_2SO_4]{Cr_2O_7^{2-}}$ $CH_3-CH_2-\overset{\overset{O}{\|}}{C}-OH$

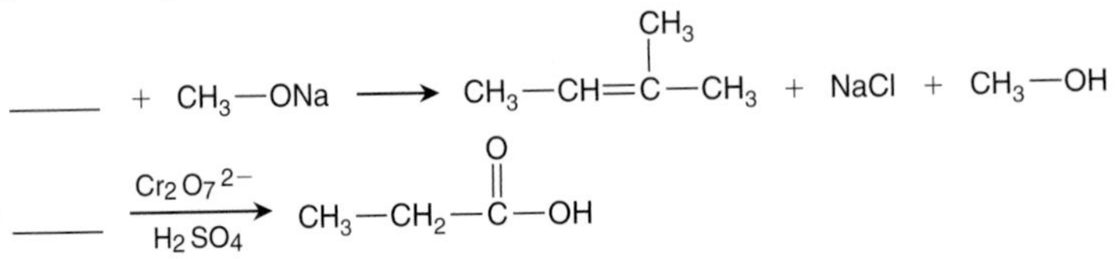

Sample Problem 21.11   Predicting the Steps in a Reaction Sequence

Problem Fill in the blanks in this reaction sequence:

$$CH_3-CH_2-\overset{\overset{Br}{|}}{CH}-CH_3 \xrightarrow{OH^-} \underline{\quad\quad} \xrightarrow[H_2SO_4]{Cr_2O_7^{2-}} \underline{\quad\quad} \xrightarrow{CH_3-Li} \xrightarrow{H_2O} \underline{\quad\quad}$$

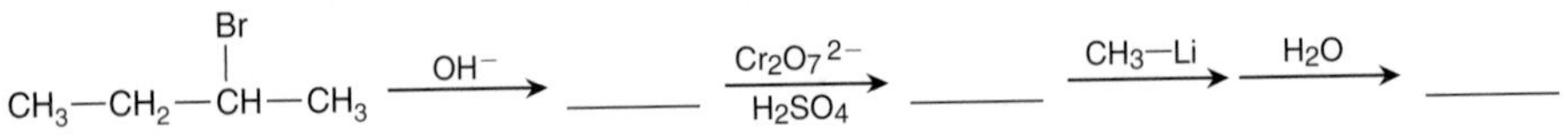

Plan For each step, we examine the functional group of the reactant and the reactant above the yield arrow to decide on the most likely product.

Solution The sequence starts with a haloalkane reacting with OH^-. Substitution gives an alcohol. Oxidation of this alcohol with acidic dichromate gives a ketone. Finally, a two-step reaction of a ketone with CH_3-Li and then water forms an alcohol with a carbon skeleton that has the $-CH_3$ group attached to the carbonyl C:

$$CH_3-CH_2-\overset{\overset{Br}{|}}{CH}-CH_3 \xrightarrow{OH^-} CH_3-CH_2-\overset{\overset{OH}{|}}{CH}-CH_3 \xrightarrow[H_2SO_4]{Cr_2O_7^{2-}}$$
          2-bromobutane   Butan-2-ol

$$CH_3-CH_2-\overset{\overset{O}{\|}}{C}-CH_3 \xrightarrow{CH_3-Li} \xrightarrow{H_2O} CH_3-CH_2-\overset{\overset{OH}{|}}{\underset{\underset{CH_3}{|}}{C}}-CH_3$$
          Butan-2-one   2-methylbutan-2-ol

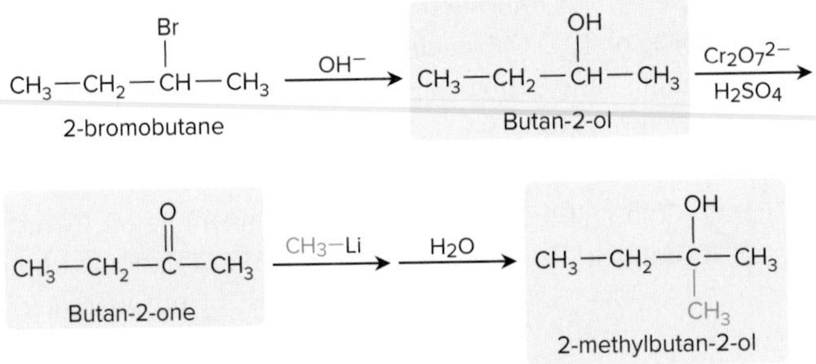

Check In this case, make sure that the first two reactions alter the functional group only and that the final steps change the C skeleton.

Follow-Up Problem 21.11 Choose reactants to obtain each product:

(a) _____ $\xrightarrow[H_2SO_4]{Cr_2O_7^{2-}}$ 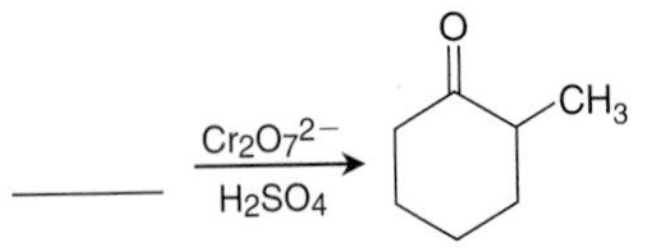

(b) _____ $\xrightarrow{CH_3-CH_2-Li} \xrightarrow{H_2O}$ 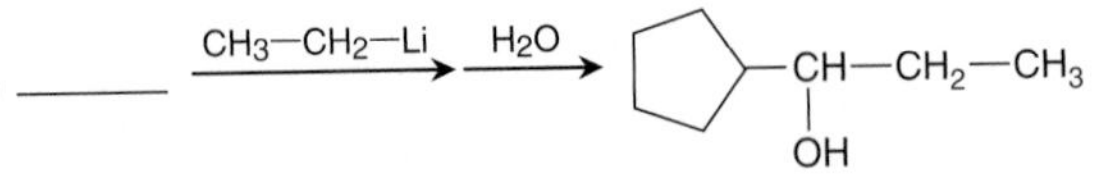

Reactions of Ethers

Ethers are less reactive than other functional groups. They do not react with active metals, strong bases, or reducing and oxidizing agents. Ethers are highly flammable and form an explosive mixture with air, giving CO_2 and water:

$$C_2H_5OC_2H_5 + 6O_2 \longrightarrow 4CO_2 + 5H_2O$$

Ethers usually show good solvent properties for many nonpolar organic compounds. This property and their low reactivity make ethers good solvents in which to run reactions. It is important to know that ethers can become dangerous with storage because they tend to form explosive peroxides with age. Exposure to light and air enhance the formation of the peroxides. ■ Ethers are cleaved by the strong acids HBr and HI. Primary and secondary alkyl ethers react by an S_N2 mechanism, whereas tertiary, benzylic, and allylic ethers cleave by an S_N1 mechanism:

$$R-O-R' \xrightarrow{\text{HX (X=I, Br)}} R-O-H + R'-X$$

Because of the presence of two lone pairs of electrons on the oxygen atom, ethers behave as Lewis bases and form a salt with strong acids like HCl:

$$\underset{R'}{\overset{R}{\diagdown}}\ddot{O}: + HCl \longrightarrow \left[\underset{R'}{\overset{R}{\diagdown}}\ddot{O}-H\right]^{+} Cl^{-}$$

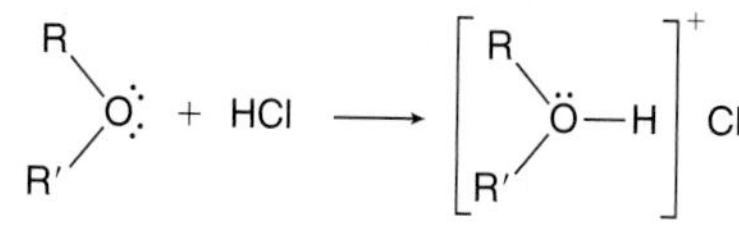

Dialkyl oxonium chloride

The alkyl group of ethers might undergo substitution reactions with Cl_2 and Br_2.

Reactions of Aldehydes and Ketones

Aldehydes and ketones undergo a variety of reactions that lead to many different products. The most common reactions are *nucleophilic addition* reactions to the C=O double bond, which lead to the formation of alcohols, alkenes, diols, cyanohydrins (RCH(OH)C≡N, and imines (R_2C=NR), to mention just a few representative examples. The following are some addition reactions of the carbonyl group.

- Addition of water:

$$CH_3-\overset{\overset{\displaystyle O}{\|}}{C}-H \xrightarrow[H^+]{H_2O} CH_3-\underset{\underset{\displaystyle OH}{|}}{\overset{\overset{\displaystyle OH}{|}}{C}}-H$$

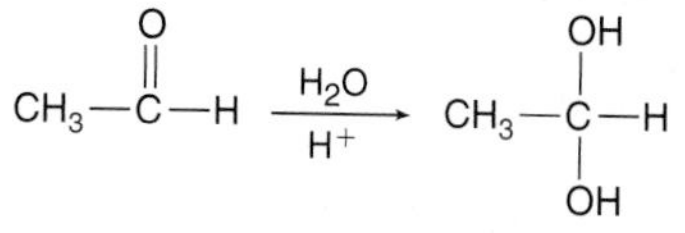

A hydrate

- Addition of alcohol:

$$CH_3-\overset{\overset{\displaystyle O}{\|}}{C}-H \xrightarrow[CH_3OH]{\Delta} CH_3-\underset{\underset{\displaystyle OCH_3}{|}}{\overset{\overset{\displaystyle OH}{|}}{C}}-H$$

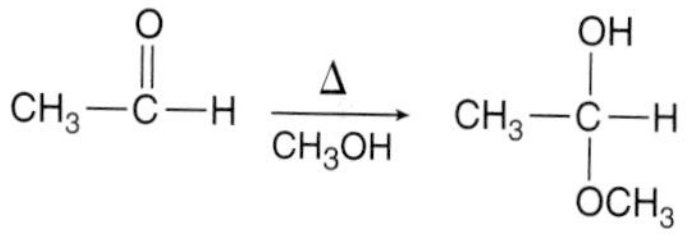

A hemiacetal

$$CH_3-\overset{\overset{\displaystyle O}{\|}}{C}-H \xrightarrow[\substack{CH_3OH \\ HCl}]{\Delta} CH_3-\underset{\underset{\displaystyle OCH_3}{|}}{\overset{\overset{\displaystyle OCH_3}{|}}{C}}-H$$

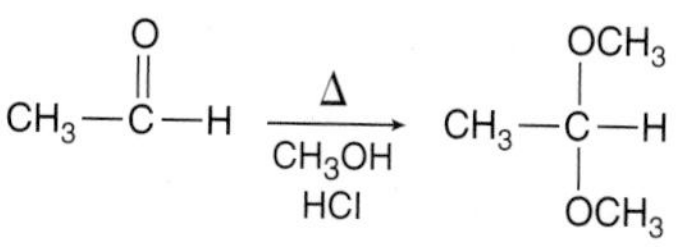

An acetal

- Addition of hydrogen cyanide:

$$CH_3-CH_2-\overset{\overset{\displaystyle O}{\|}}{C}-H + HCN \rightleftharpoons CH_3-CH_2-\underset{\underset{\displaystyle CN}{|}}{\overset{\overset{\displaystyle OH}{|}}{C}}-H$$

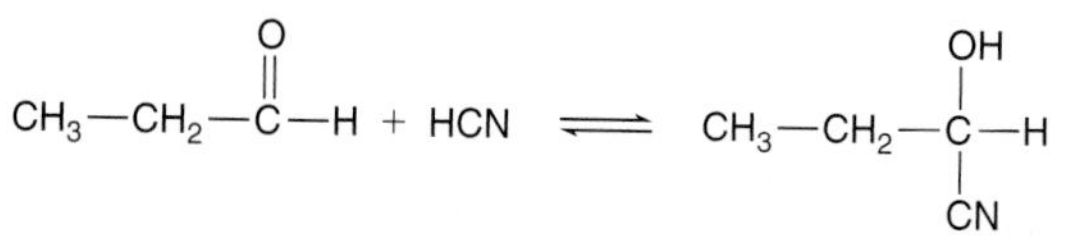

Propanal   A cyanohydrin

$$CH_3-\overset{\overset{\displaystyle O}{\|}}{C}-CH_3 + HCN \rightleftharpoons CH_3-\underset{\underset{\displaystyle CN}{|}}{\overset{\overset{\displaystyle OH}{|}}{C}}-CH_3$$

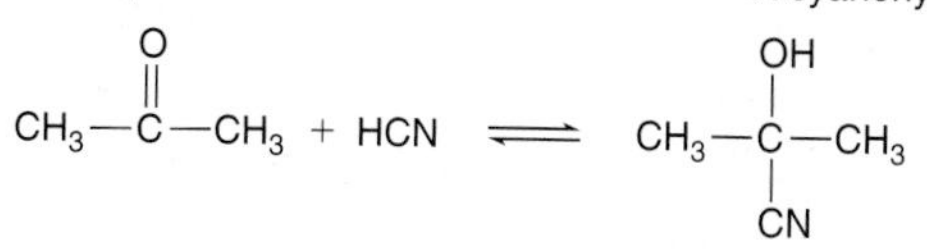

Propan-2-one (acetone)   A cyanohydrin

■ Peroxides can form in freshly distilled and unstabilized ethers within two weeks, in ethoxyethane within eight days, and in oxolane (tetrahydrofuran, THF) within three days. Peroxide crystals tend to form on the inner surfaces of the container. The crystals may cause an explosion if subjected to impact or friction.

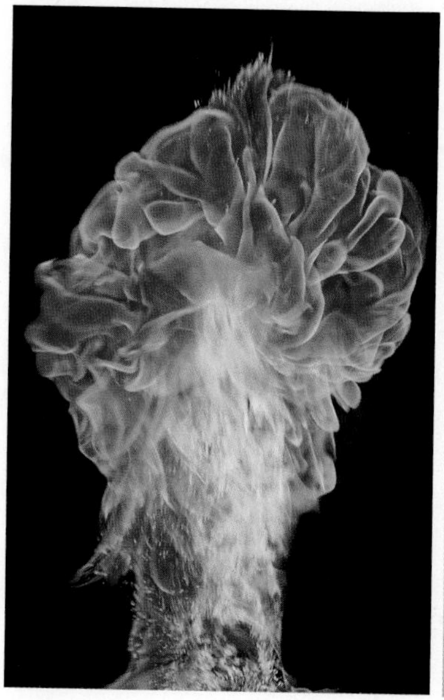

- Addition of organometallic reagents (like **Grignard reagents**):

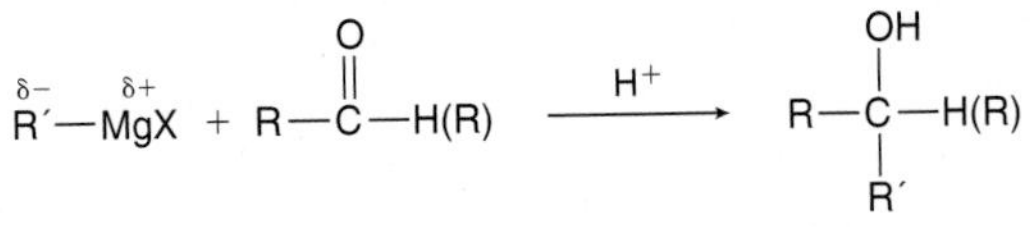

Aldehydes and ketones undergo several other reactions, including *oxidation* and *reduction*. For example, aldehydes can be oxidized to carboxylic acids with both mild and strong oxidizing agents. However, ketones can be oxidized to various types of compounds only by using extremely strong oxidizing agents. In the presence of a base, ketones with α-hydrogen atoms react to form α-haloketones. Recall that, as a result of their unsaturation, carbonyl compounds can undergo *addition* reactions and be reduced to alcohols. In general terms, reduction of an aldehyde leads to a *primary alcohol*, whereas reduction of ketones leads to a *secondary alcohol*. All of these reactions are covered in more advanced organic chemistry courses.

Reactions of Carboxylic Acids, Esters, and Amides

Substitution of carboxylic acids and other members of this family occurs through a two-step sequence: *addition then elimination equals substitution*. Addition to the trigonal planar C atom gives an unstable tetrahedral intermediate, which immediately undergoes elimination to revert to a trigonal planar product (in this case, X is OH):

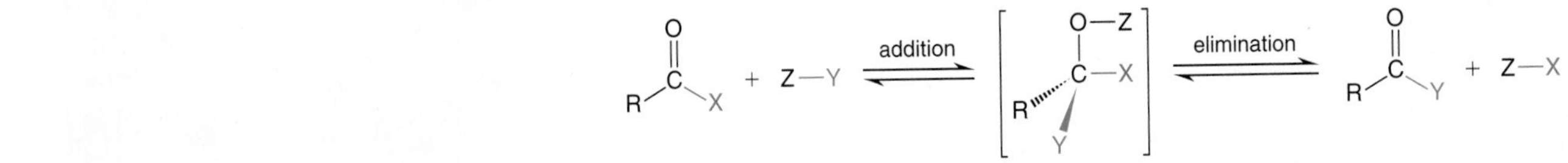

Carboxylic acids undergo reactions to produce derivatives of the acid. The most common derivatives formed are esters, acid halides, acid anhydrides, and amides:

- *Esters* can be formed by the reaction of carboxylic acids with alcohols in an acidic medium:

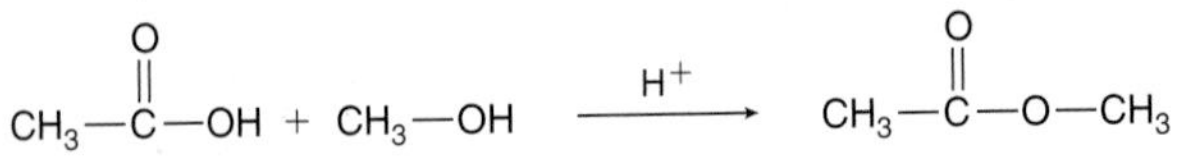

Ethanoic acid (acetic acid) Methanol Methyl ethanoate (methyl acetate)

- *Acid halides* are products of the reaction of carboxylic acids and phosphorous trichloride (PCl_3), phosphorous pentachloride (PCl_5), thionyl chloride (SOC_{12}), or phosphorous tribromide (PBr_3):

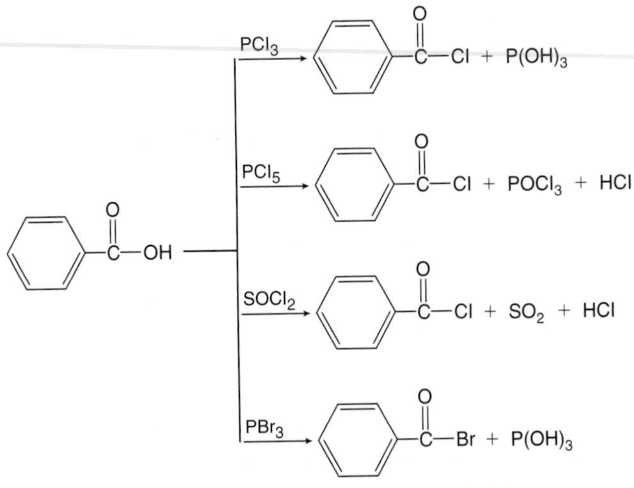

- *Acid anhydrides* form by reaction of the salt of a carboxylic acid with an acyl halide:

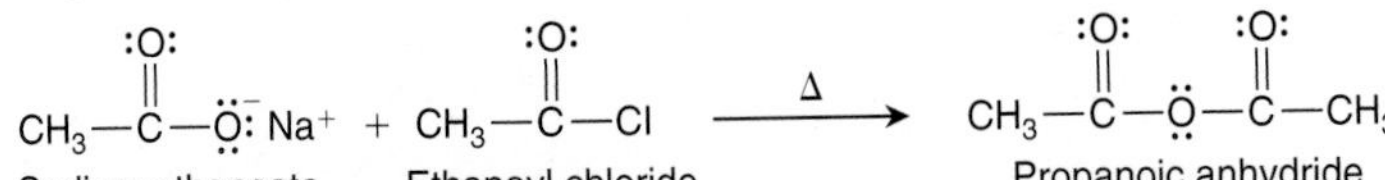

Sodium ethanoate (sodium acetate) Ethanoyl chloride (acetyl chloride) Propanoic anhydride (acetic anhydride)

• *Amides* are generally prepared by a reaction of acid chlorides with ammonia or amines:

$$CH_3-CH_2-\overset{\overset{\displaystyle O}{\|}}{C}-Cl \xrightarrow[\Delta]{NHR_2} CH_3-CH_2-\overset{\overset{\displaystyle O}{\|}}{C}-NR_2 + NH_4Cl$$

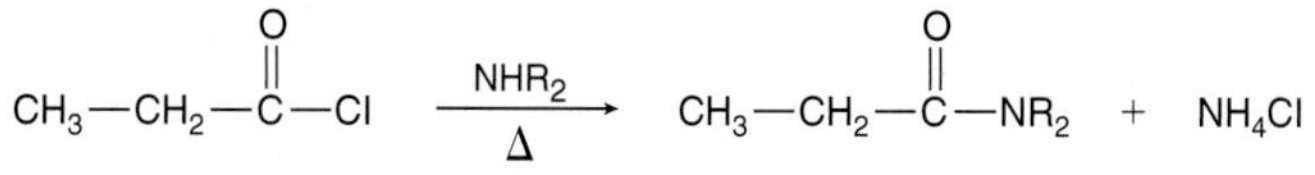

where R can be hydrogen atoms or alkyl groups, forming amides or substituted amides.

Compounds in the carboxylic acid family also undergo reduction to form other functional groups. For example, certain inorganic *reducing agents* convert acids or esters to alcohols and convert amides to amines:

$$R-\overset{\overset{\displaystyle O}{\|}}{C}-OH \text{ (or } R-\overset{\overset{\displaystyle O}{\|}}{C}-O-R') \xrightarrow{reduction} R-CH_2-OH + HOH \text{ (or } R'-OH)$$

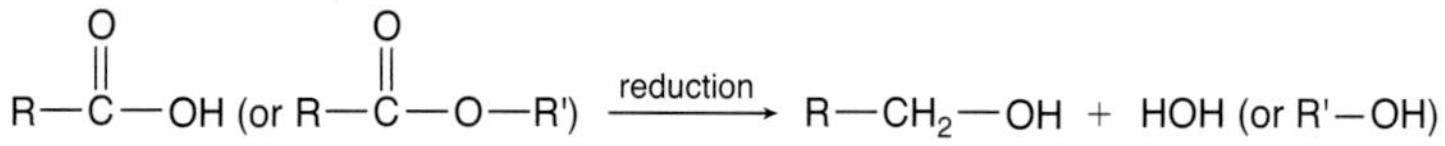

$$R-\overset{\overset{\displaystyle O}{\|}}{C}-NH-R' \xrightarrow{reduction} R-CH_2-NH-R' + H_2O$$

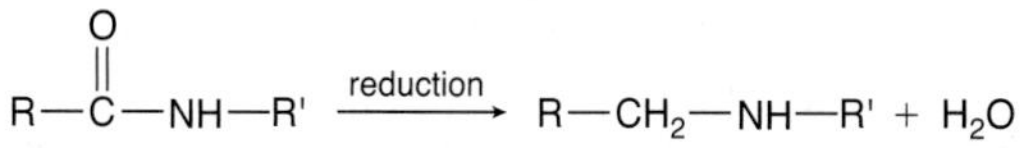

Hydrolysis of esters in a basic or an acidic medium is a way to produce carboxylic acids and alcohols:

$$R-\overset{\overset{\displaystyle O}{\|}}{C}\boxed{-OR' + H}OH \xrightarrow{H^+ \text{ or } OH^-} RCOOH + R'OH$$

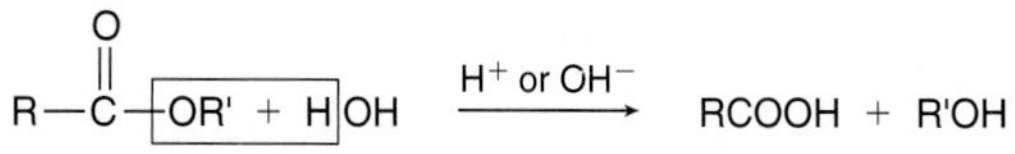

Sample Problem 21.12	Predicting Reactions of the Carboxylic Acid Family

Problem Predict the product(s) of each reaction:

(a) $CH_3-CH_2-CH_2-\overset{\overset{\displaystyle O}{\|}}{C}-OH + CH_3-\overset{\overset{\displaystyle OH}{|}}{CH}-CH_3 \underset{}{\overset{H^+}{\rightleftharpoons}}$ 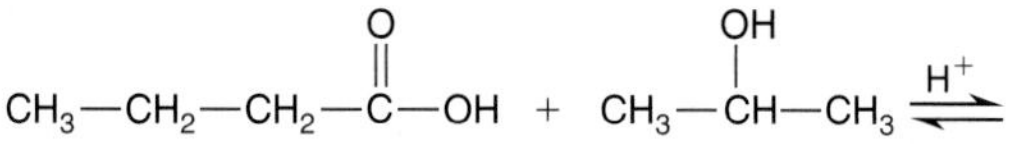

(b) $CH_3-\overset{\overset{\displaystyle CH_3}{|}}{CH}-CH_2-CH_2-\overset{\overset{\displaystyle O}{\|}}{C}-NH-CH_2-CH_3 \xrightarrow[H_2O]{NaOH}$ 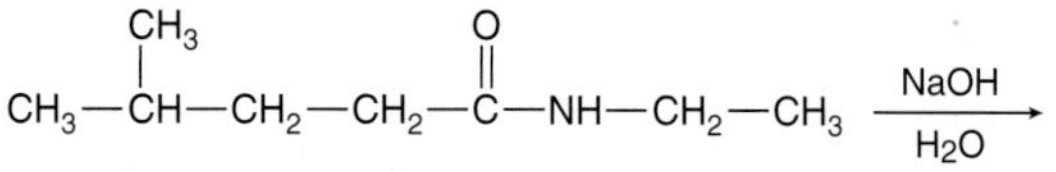

Plan We discussed substitution reactions (including addition-elimination and dehydration-condensation) and hydrolysis. **(a)** A carboxylic acid and an alcohol react, so the reaction must be a substitution to form an ester and water. **(b)** An amide reacts with OH⁻, so it is hydrolyzed to an amine and a sodium carboxylate.

Solution **(a)** Formation of an ester:

$$CH_3-CH_2-CH_2-\overset{\overset{\displaystyle O}{\|}}{C}-O-\overset{\overset{\displaystyle CH_3}{|}}{CH}-CH_3 + H_2O$$

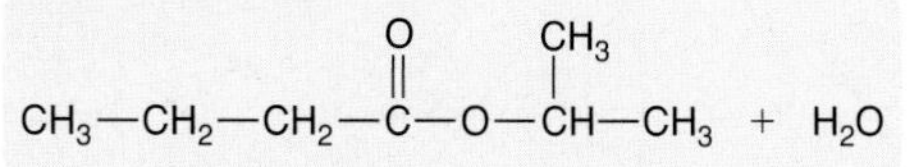

(b) Basic hydrolysis of an amide:

$$CH_3-\overset{\overset{\displaystyle CH_3}{|}}{CH}-CH_2-CH_2-\overset{\overset{\displaystyle O}{\|}}{C}-O^- + Na^+ + H_2N-CH_2-CH_3$$

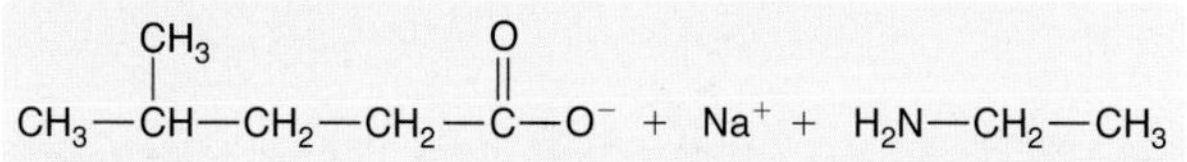

Check Note that in part (b), the carboxylate ion forms, rather than the acid, because the aqueous NaOH that is present reacts with the carboxylic acid.

Follow-Up Problem 21.12 Fill in the blanks in each reaction:

(a) _____ + $CH_3-OH \rightleftharpoons^{H^+}$ ⬡$-CH_2-\overset{\overset{\displaystyle O}{\|}}{C}-O-CH_3 + H_2O$ 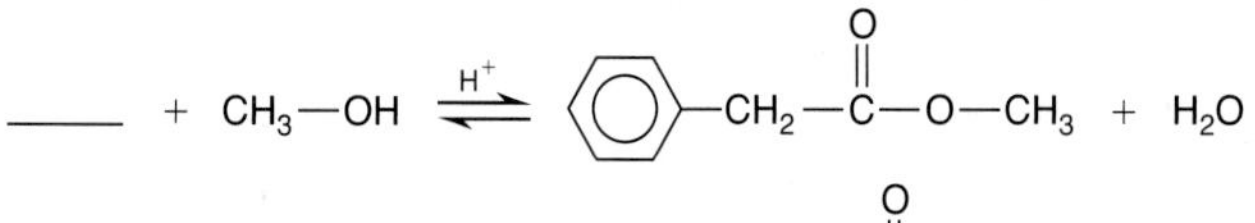

(b) _____ + _____ $\longrightarrow CH_3-CH_2-CH_2-\overset{\overset{\displaystyle O}{\|}}{C}-NH-CH_2-CH_3 + CH_3-OH$ 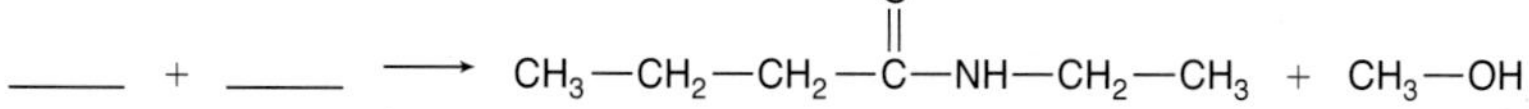

Reactions of Inorganic Compounds with Double Bonds

Like carbonyl compounds, these substances undergo addition reactions. For example, the partially negative O of water attacks the partially positive S of SO_3 to form sulfuric acid:

$$\underset{\text{S bonded to 3 atoms}}{H-\overset{\delta-}{\underset{\overset{\displaystyle |}{H}}{O}}: \quad \overset{\overset{\displaystyle O}{\|}}{\underset{O}{S}}\overset{\delta+}{\diagdown}O} \longrightarrow \longrightarrow \underset{\text{S bonded to 4 atoms}}{HO-\overset{\overset{\displaystyle O}{\|}}{\underset{\overset{\displaystyle \|}{O}}{S}}-OH}$$

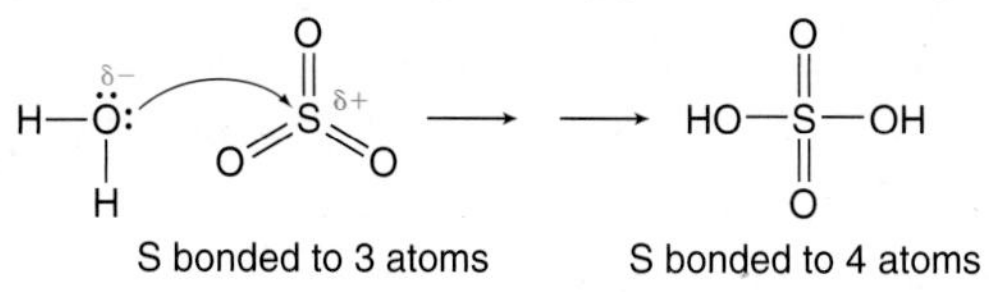

CHEMICAL CONNECTIONS
TO ORGANIC SYNTHESIS

The synthesis of organic compounds has developed into one of the most important branches of organic chemistry. Compounds that were previously isolated from natural sources, as well as new organic compounds, can now be synthesized in the laboratory. We learned the different types of organic reactions in this chapter (substitution, elimination, addition, reduction, oxidation) and, therefore, some of the basic functional group transformations and the use of reactants and different reaction conditions. The key in organic synthesis is to determine the route to take to obtain the target molecule. In several cases, the desired molecule is considered first, and then different steps have to be proposed, one at a time, to lead back to the appropriate starting materials. This process is known as **retrosynthesis**.

Remember that a reaction is a relationship among three components: the starting material (SM), the product (P), and the reactant (R):

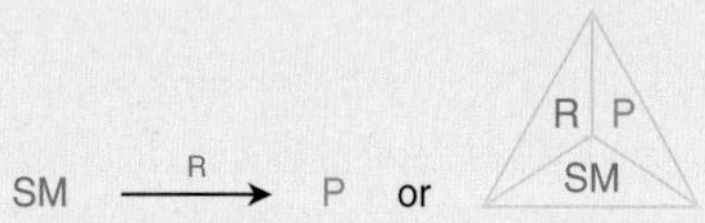

If we know two of the reaction components, we should be able to figure out the other one:

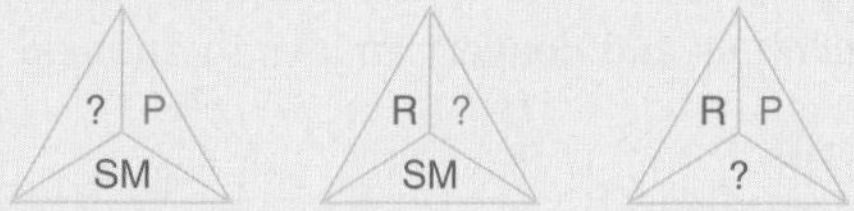

The flow chart diagram (Figure B21.1) shows several common organic group transformations. It is a helpful guide to complete a chemical reaction when we do not know one or more of the previously mentioned organic reaction components. As you can see, we can apply it to several of the examples of the previous sections.

Table B21.1 summarizes some of the organic reactions that we studied in this chapter, and which functional groups undergo each of them.

You now have enough information to (a) predict the product of an organic chemical reaction, and the mechanism involved in the reaction; (b) determine the reactant needed to obtain the desired product with a given starting material; and (c) determine, by retrosynthesis, the reactant you need to start with to get the target molecule.

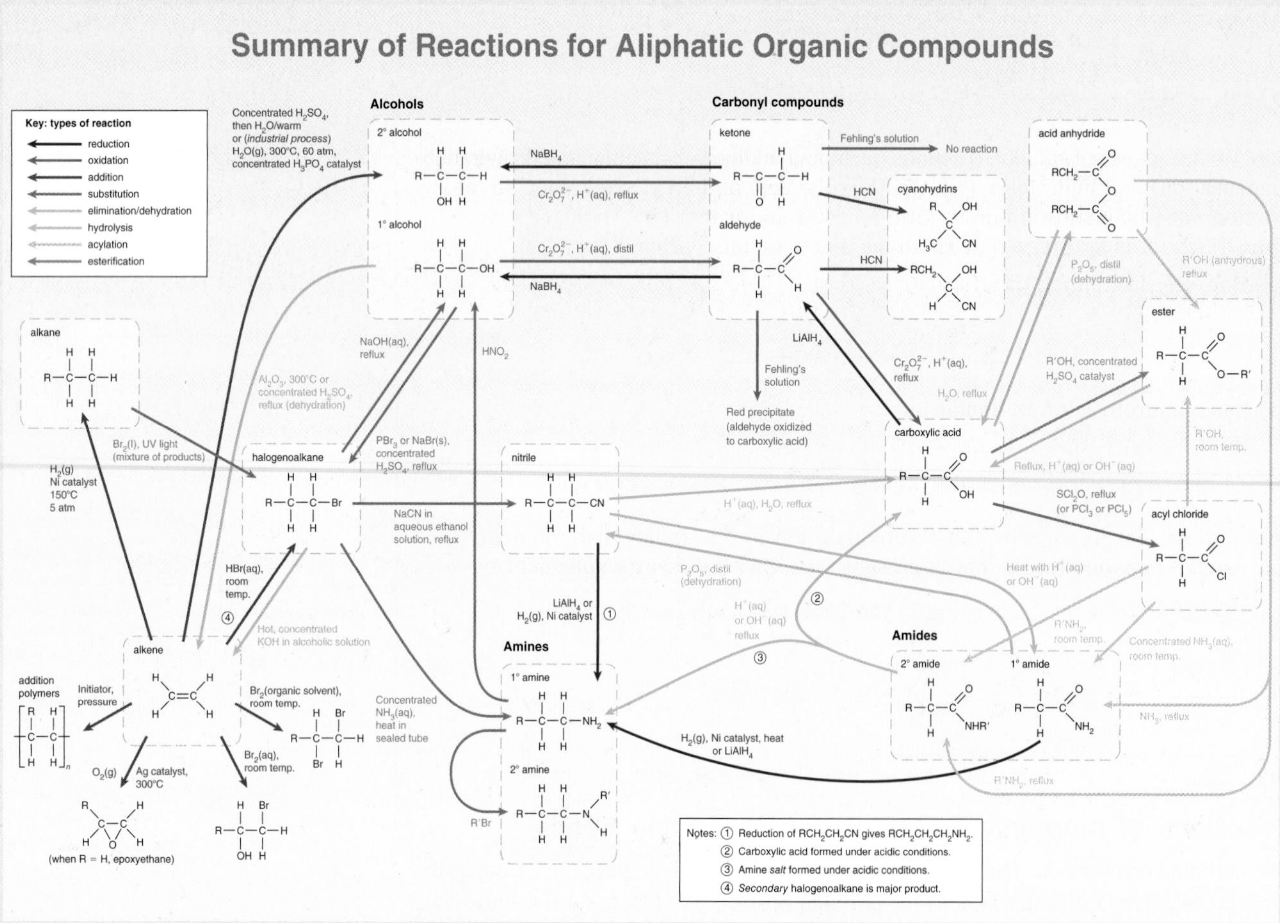

FIGURE B21.1 Summary of reactions for aliphatic organic compounds [Source: http://malleshchemist.blogspot.ca/2011/01/summary-of-reactions-for-aliphatic.html. First published *Chemistry Review*, Vol. 15, No. 1, pp. 26–27, available from Philip Allan Updates.]

(*Continued*)

TABLE B21.1 Summary of Some of the Reactions That Common Functional Groups Undergo

Reaction	Example	Functional Group
Nucleophilic substitution		Haloalkanes Alcohols and tosylates Epoxides Ethers
Elimination	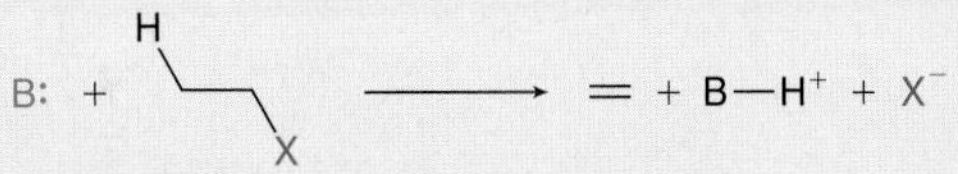	Haloalkanes Alcohols and tosylates Amines
Electrophilic addition	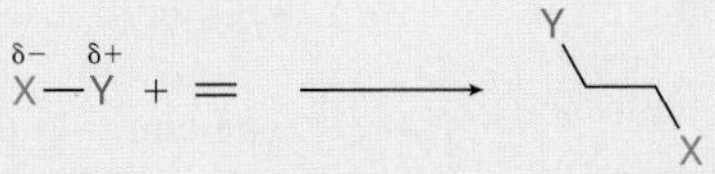	Alkenes Alkynes
Electrophilic aromatic substitution (EArS)	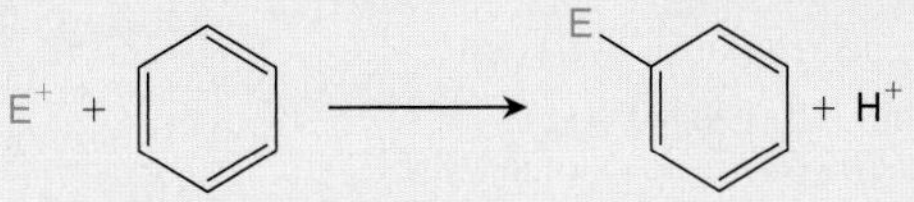	Arenes
Nucleophilic addition	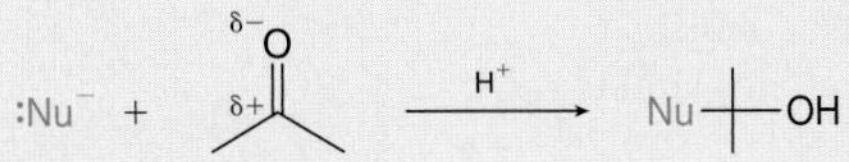	Aldehydes and ketones
Nucleophilic acyl substitution	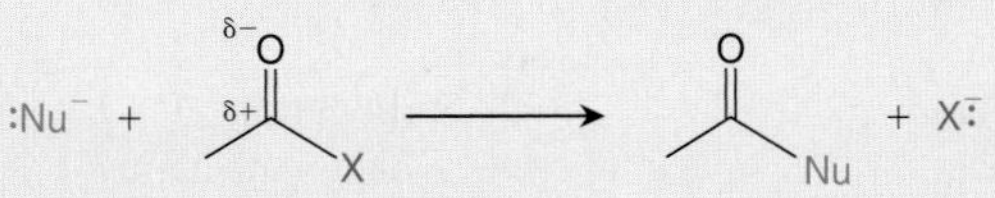	Carboxylic acids Acyl halides Anhydrides Esters Amides

SUMMARY OF SECTION 21.7

- Table 21.2 summarizes some characteristic reactions of some functional groups. (Their mechanism and more reactions are covered in more advanced organic chemistry courses.)
- Alcohols and haloalkanes undergo elimination and substitution reactions.
- Amines are nucleophiles and undergo addition to the nitrogen atom.
- Ethers are less reactive than other functional groups. They are cleaved by HI and HBr and form a salt with HCl.
- Aldehydes and ketones undergo addition reactions at the carbonyl group, as well as reduction and oxidation.
- Carboxylic acids and derivatives undergo substitution reactions through a two-step sequence: addition to the carbonyl group and then elimination.
- Inorganic compounds with double bonds undergo addition reactions.

CHAPTER REVIEW GUIDE

Learning Objectives

Relevant section (§) and/or sample problem (SP) numbers appear in parentheses.

Concepts

1. Recognize the different types of organic reactions. (§21.1)
2. Explain what a reaction mechanism is and how to draw arrows to show the movement of electrons. (§21.2)
3. Differentiate between a nucleophile and an electrophile, and the factors that influence the strength of nucleophiles and bases. (§21.2)
4. Differentiate between S_N2 and S_N1 mechanisms, and the factors that affect them. (§21.3)
5. Differentiate between E1 and E2 mechanisms, and the factors that affect them. (§21.4)
6. Describe the difference between anti and syn addition mechanisms and the steps that lead to the final product(s) in addition reactions. (§21.5)
7. Explain the mechanism of the electrophilic aromatic substitution (EArS) and how a ring substituent can direct substitution of a second electrophile to *ortho/para* positions or to a *meta* position. (§21.6)
8. Predict the products of some reactions of alcohols, haloalkanes, amines, ethers, aldehydes, ketones, and the carboxylic acid family. (§21.7)
9. Explain what retrosynthesis is and how it can be used to obtain target molecules (Chemical Connections to Organic Synthesis).

Skills

1. Recognize the different types of organic reactions. (SP 21.1)
2. Draw arrows to indicate the flow of electrons. (§21.2 and SP 21.2)
3. Predict the product and show the mechanisms of nucleophilic substitution reactions. (§21.3 and SP 21.3 and 21.4)
4. Predict the product and show the mechanism of elimination reactions. (§21.4 and SP 21.4)
5. Identify the mechanism that a reaction will follow and predict the products. (SP 21.4)
6. Predict the product and show the mechanism for electrophilic addition reactions. (SP. 21.5, 21.6, 21.7, 21.8, and 21.9)
7. Predict the reaction of alcohols, haloalkanes, and amines. (SP 21.10)
8. Predict the steps in a reaction sequence. (SP. 21.11)
9. Predict the products of reactions of the carboxylic acid family. (SP 21.12)

Key Terms

Section 21.1
alkyl group
substitution
elimination
addition
redox
oxidation
reduction

Section 21.2
concerted reaction
stepwise reaction
bond cleavage
homolytic cleavage
radicals
heterolytic cleavage (heterolysis)
bond formation
nucleophiles
electrophiles
Lewis acid
Lewis base
polarizability
solvent effects
polar protic solvents
polar aprotic solvents
resonance
inductive effect
basicity
nucleophilicity

Section 21.3
nucleophilic substitution
leaving group
one-step (concerted) mechanism
S_N2
bimolecular reaction
rate-determining step
second-order reaction
carbocation
two-step mechanism
unimolecular
S_N1
first-order reaction
back-side
steric effect
racemization
rearrangement

Section 21.4
E2
E1
Zaitsev's rule

Section 21.5
addition
syn addition
anti addition
hydrogenation
heat of hydrogenation
Markovnikov's rule
epoxidation
ozonolysis
polymerization

Section 21.6
electrophilic aromatic substitution
cyclohexdienyl cation (arenium ion, sigma-complex)
Friedel-Crafts reaction
alkylation
acylation
halogenation of benzene
electron-donating groups
electron-withdrawing groups

Section 21.7
dehydration
Grignard reagents
hydrolysis
retrosynthesis

Brief Solutions to Follow-Up Problems

21.1 (a)

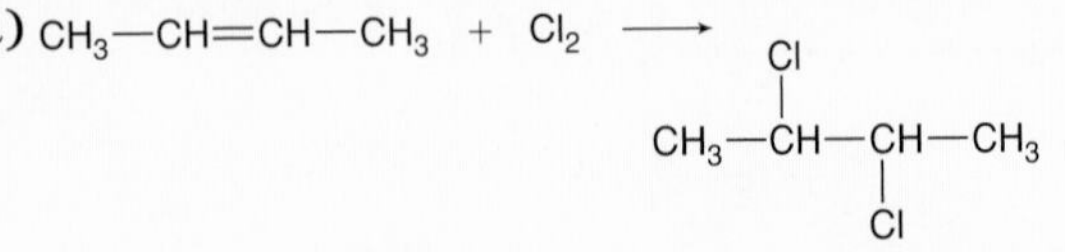

(b) $CH_3—CH_2—CH_2—Br + OH^- \longrightarrow CH_3—CH_2—CH_2—OH + Br^-$

(c)

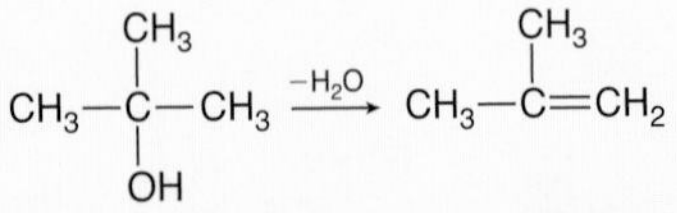

21.2

1. (a) 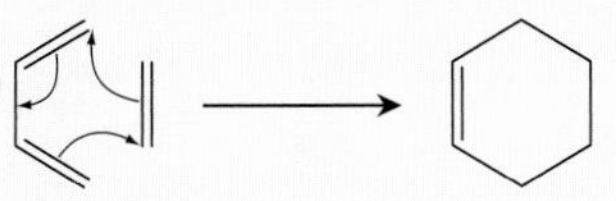 **(b)** 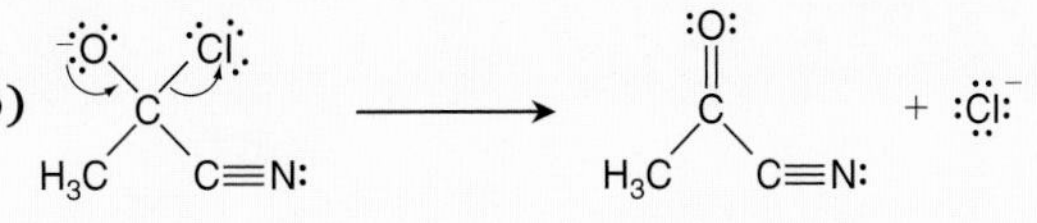

(c) 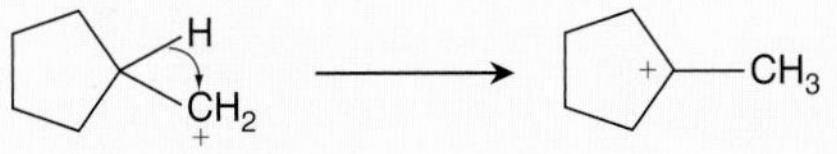

2. (a) 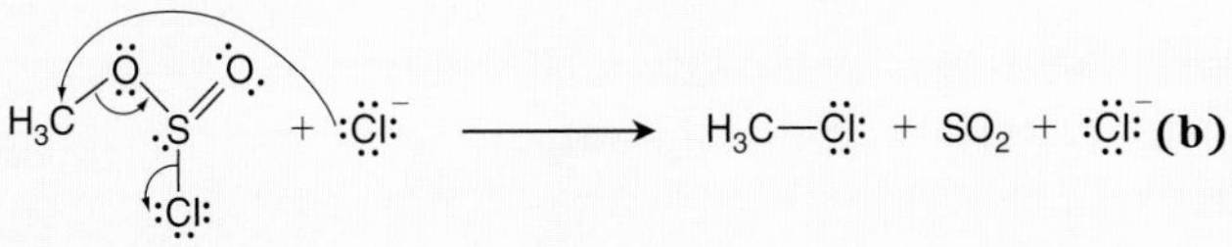 **(b)** 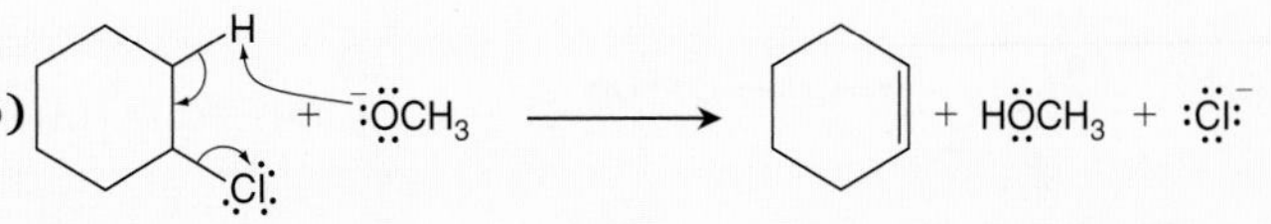

(c) 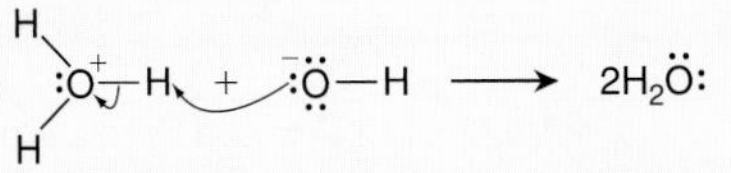

21.3 (a) 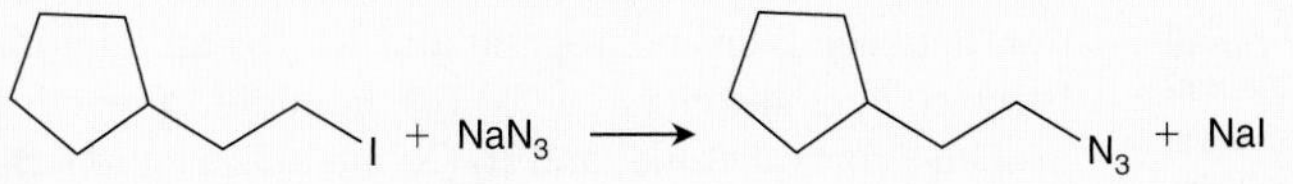

(b) 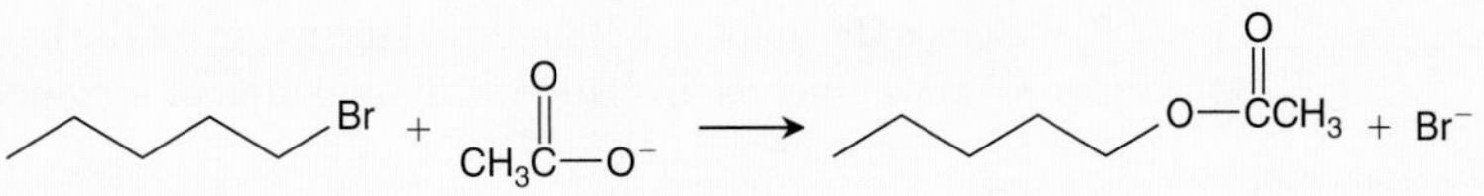

21.4 (a) E2: 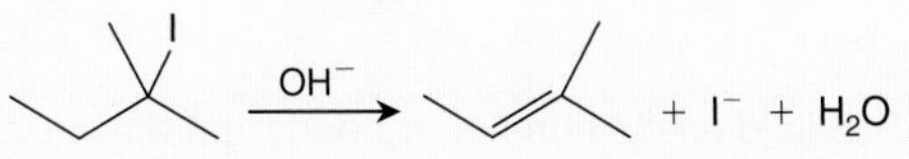

(b) No reaction

(c) S_N1 and E1: 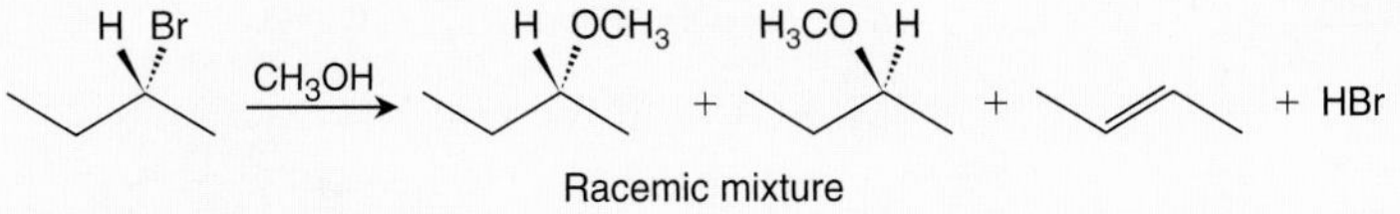

21.5 (a) The reaction of HCl with pent-2-ene yields a mixture of two addition products, since each of the secondary carbon atoms will form a carbocation with equal stability:

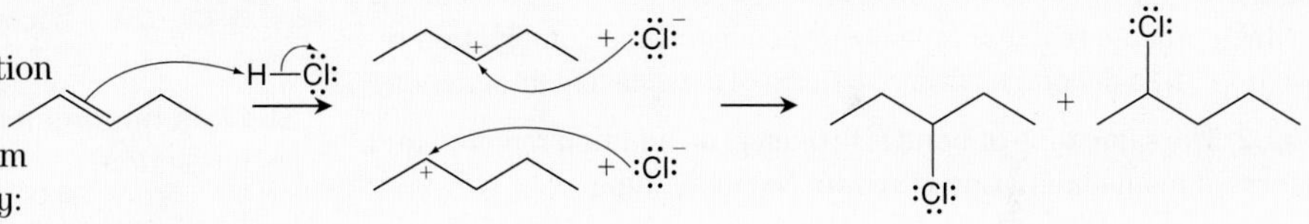

(b) 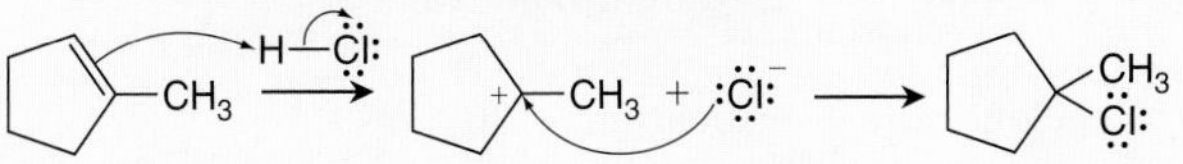

21.6 (a) 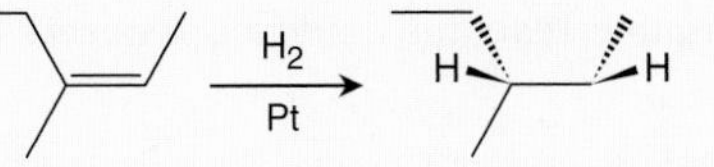 **(b)** 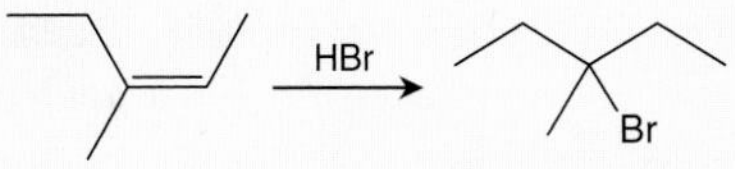

21.7 (a) 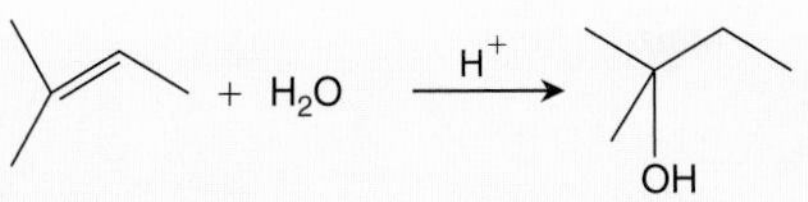 **(b)** 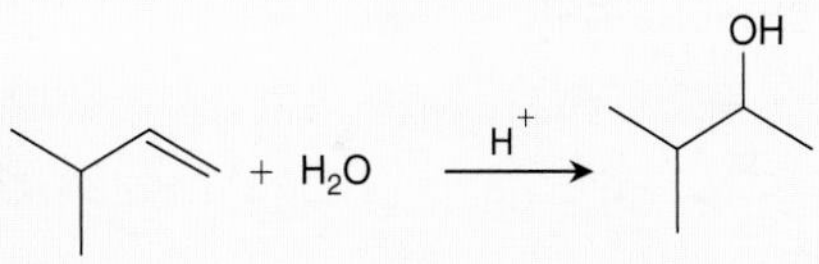

21.8 (a) 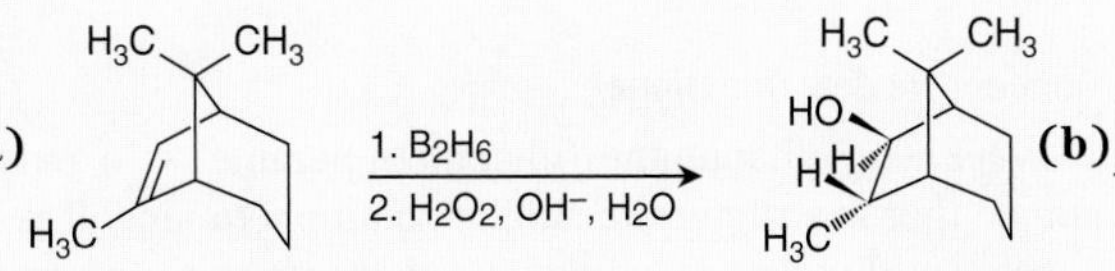 **(b)** 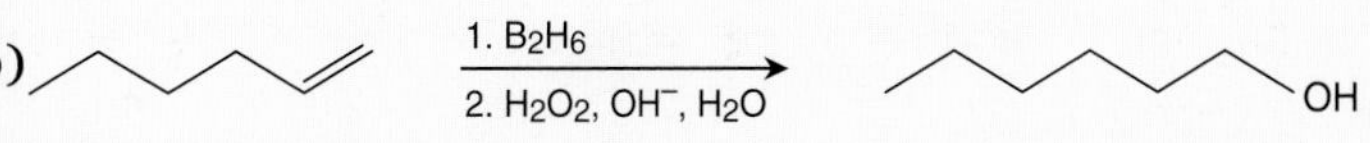

21.9 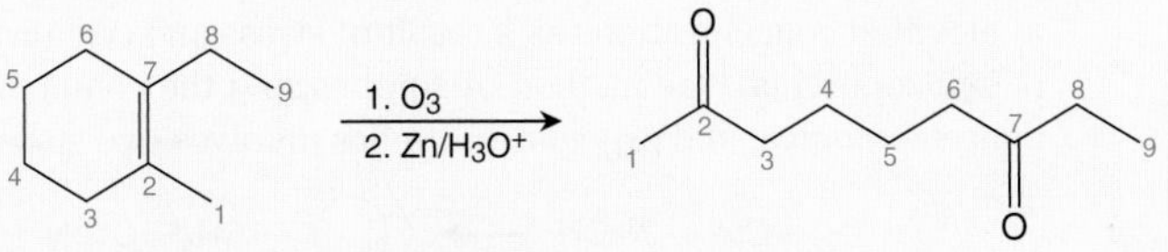

Starting material

21.10 (a) 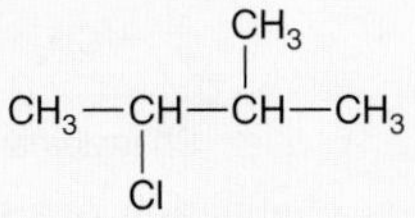 or 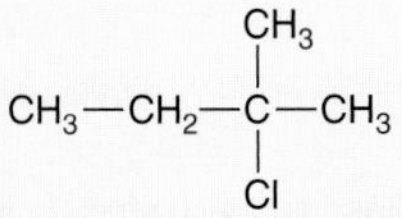 **(b)** $CH_3—CH_2—CH_2—OH$

Brief Solutions to Follow-Up Problems (continued)

21.11 (a) 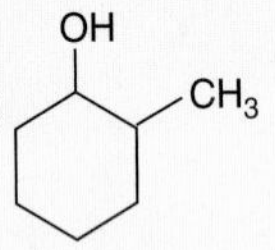
(b) 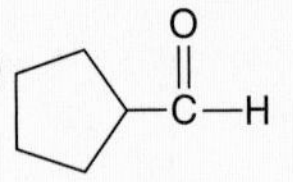

21.12 (a) 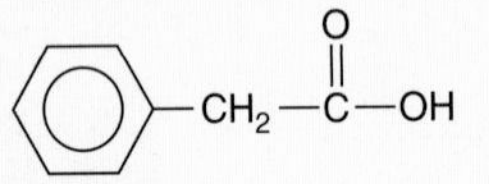
(b) $CH_3—CH_2—CH_2—\overset{O}{\overset{\|}{C}}—O—CH_3 + CH_3—CH_2—NH_2$

PROBLEMS

Problems with **red** numbers are answered in Appendix G and worked in detail in the Student Solutions Manual. Problem sections match those in the text and provide the numbers of relevant sample problems. Most offer Concept Review Questions, Skill-Building Exercises (usually grouped in pairs covering the same concept), and Problems in Context. The Comprehensive Problems are based on material from any section or previous chapter.

(*Note*: The following abbreviations might be used: Me = methyl, Et = ethyl, Pr = propyl, *t*-Bu = *tert*-butyl or 1,1-dimethylethyl, *i*-Pr = isopropyl or 2-methylpropyl, Ph = phenyl, AcOH = ethanoic acid or acetic acid, OAc = ethanoate or acetate, *m*-CPBA = *meta*-chloroperoxybenzoic acid.)
(*Note:* Reflux is a distillation technique that involves the condensation of vapours and the return of the concentrate to where it originated.)

Some Important Classes of Organic Reactions
(Sample Problem 21.1)

Concept Review Questions

21.1 In terms of numbers of reactant and product substances, which organic reaction type corresponds to (a) a combination reaction; (b) a decomposition reaction; (c) a displacement reaction?

21.2 The same type of bond is broken in an addition reaction and formed in an elimination reaction. Name the type.

21.3 Can a redox reaction also be an addition, elimination, or substitution reaction? Explain with examples.

21.4 Determine the type of each reaction:

(a) $CH_3—CH_2—\overset{Br}{\overset{|}{C}H}—CH_3 \xrightarrow[\Delta]{NaOH} CH_3—CH{=}CH—CH_3 + NaBr + H_2O$

(b) $CH_3—CH{=}CH—CH_2—CH_3 + H_2 \xrightarrow{Pt} CH_3—CH_2—CH_2—CH_2—CH_3$

21.5 Determine the type of each reaction:

(a) $CH_3—\overset{O}{\overset{\|}{C}}H + HCN \longrightarrow CH_3—\overset{OH}{\overset{|}{C}H}—CN$

(b) $CH_3—\overset{O}{\overset{\|}{C}}—O—CH_3 + CH_3—NH_2 \xrightarrow{H^+} CH_3—\overset{O}{\overset{\|}{C}}—NH—CH_3 + CH_3—OH$

Skill-Building Exercises (grouped in similar pairs)

21.6 Write an equation for each reaction: (a) an addition reaction between H_2O and hex-3-ene (H^+ is a catalyst); (b) an elimination reaction between 2-bromopropane and hot potassium ethanolate (potassium ethoxide), $CH_3—CH_2—OK$ (KBr and ethanol are also products); (c) a light-induced substitution reaction between Cl_2 and ethane to form 1,1-dichloroethane.

21.7 Write an equation for each reaction: (a) a substitution reaction between 2-bromopropane and KI; (b) an addition reaction between cyclohexene and Cl_2; (c) an addition reaction between propan-2-one (acetone) and H_2 (Ni metal is a catalyst).

21.8 Based on the number of bonds and the nature of the bonded atoms, state whether each change is an oxidation or a reduction:
(a) $=CH_2$ becomes $—CH_2—OH$
(b) $=CH—$ becomes $—CH_2—$
(c) $=C—$ becomes $—CH_2—$

21.9 Based on the number of bonds and the nature of the bonded atoms, state whether each change is an oxidation or a reduction:
(a) $—\overset{|}{\underset{|}{C}}—OH$ becomes $—\underset{|}{C}{=}O$
(b) $—CH_2—OH$ becomes $=CH_2$
(c) $—\overset{O}{\overset{\|}{C}}—\underset{|}{C}—$ becomes $—\overset{O}{\overset{\|}{C}}—O—$

21.10 Is the organic reactant oxidized, reduced, or neither in each reaction?
(a) hex-2-ene $\xrightarrow[\text{cold } OH^-]{KM_nO_4}$ 2,3-dihydroxyhexane
(b) cyclohexane $\xrightarrow[\text{catalyst}]{\Delta}$ benzene + $3H_2$

21.11 Is the organic reactant oxidized, reduced, or neither in each reaction?
(a) but-1-yne + H_2 $\xrightarrow{Pt}$ but-1-ene
(b) toluene $\xrightarrow[H_3O^+, \Delta]{KM_nO_4}$ benzoic acid

Understanding Reactions
(Sample Problem 21.2)

Concept Review Questions

21.12 Phenylethanamine (phenylethylamine) is a natural substance that is structurally similar to amphetamine. It is found in sources as diverse as almond oil and human urine, where it occurs at elevated concentrations as a result of stress and certain forms of schizophrenia. One method of synthesizing the compound for pharmacological and psychiatric studies involves two steps:

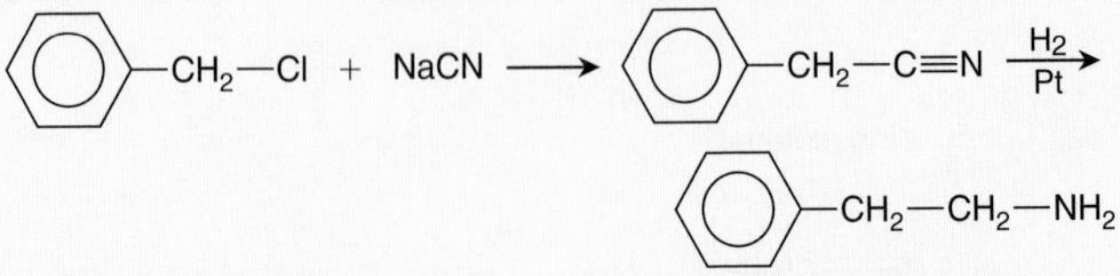

Phenylethanamine

Classify each step as an addition, an elimination, or a substitution.

21.13 Which reaction energy diagram best describes a concerted exothermic process?

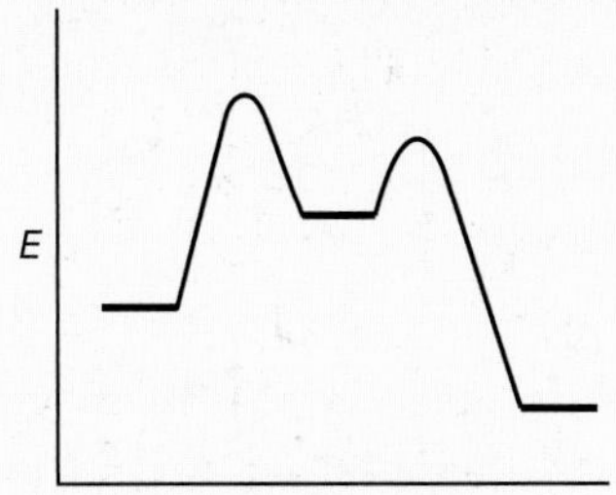

(a)

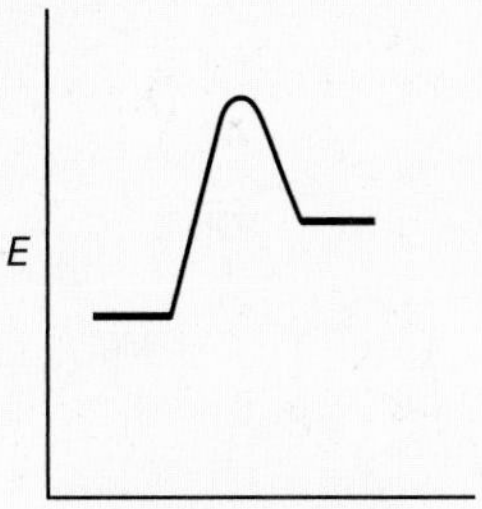

(b)

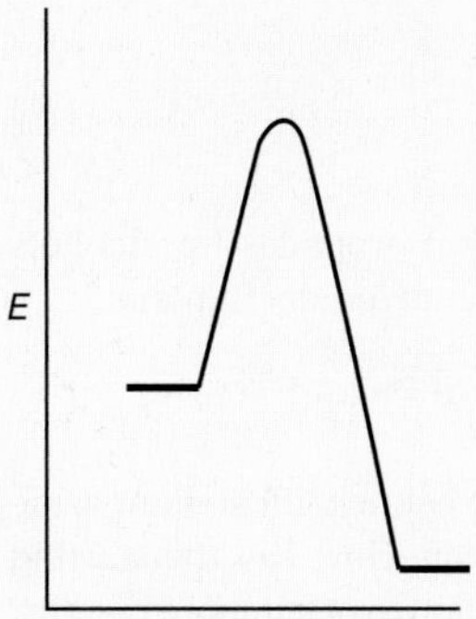

(c)

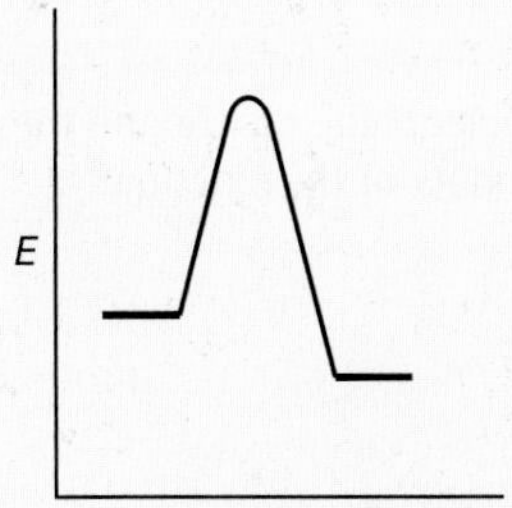

(c)

21.14 Below is a reaction energy diagram showing a general reaction:

Reactants ⟶ [Intermediate] ⟶ Products

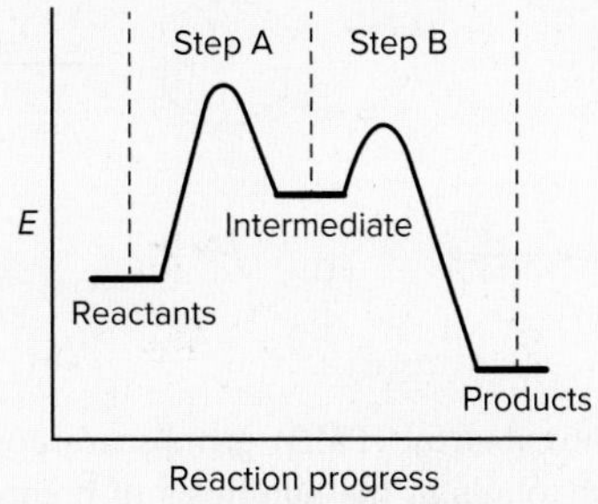

(a) Is the reaction concerted or stepwise?
(b) Is the reaction endothermic or exothermic overall?
(c) Which step is rate-determining: step A or step B?

21.15 Which reaction energy diagram describes the fastest reaction?

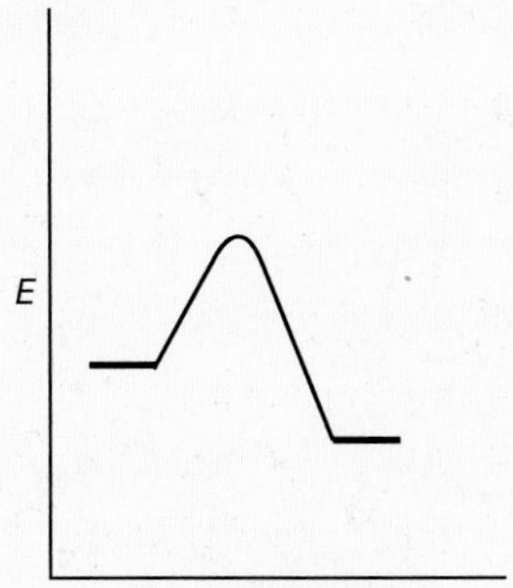

(a)

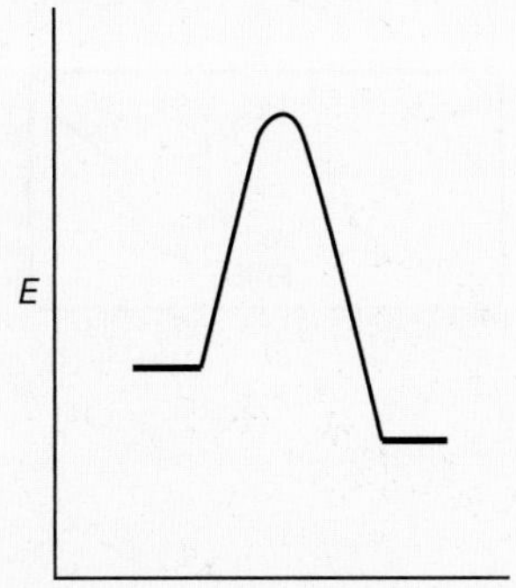

(b)

21.16 Classify each bond-breaking event as homolytic or heterolytic cleavage:

(a) 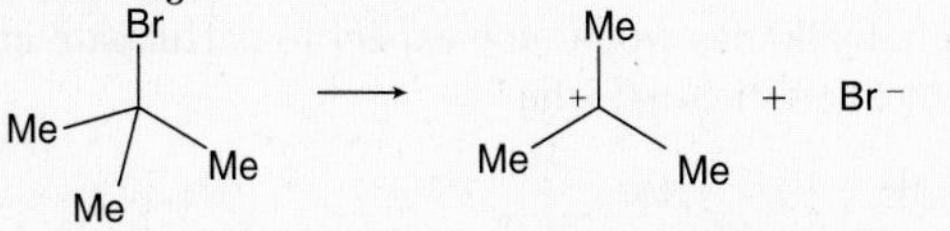

(b) I—I ⟶ I· + I·

(c) 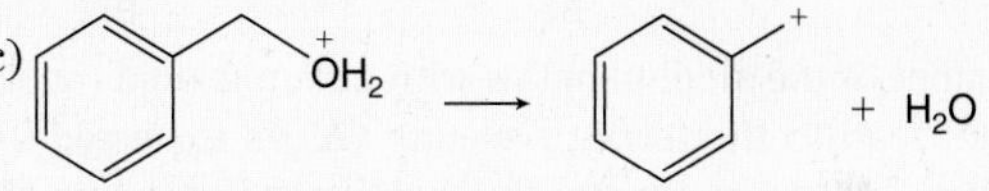

21.17 Define the term *electrophile*. Which carbon atom would you expect to be the most electrophilic? Why?

(a) 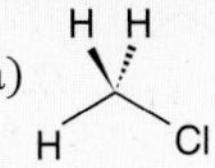

(b) 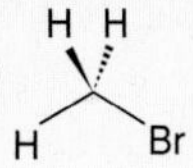

(c) 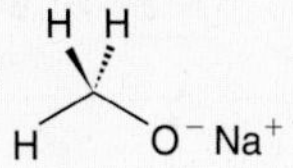

21.18 Which oxygen-containing molecule is the most nucleophilic? Why?

(a) 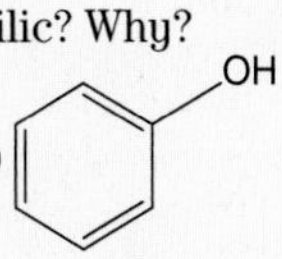

(b) 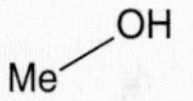

(c) 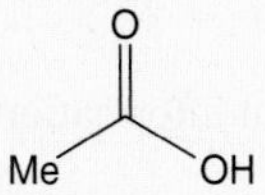

(d) 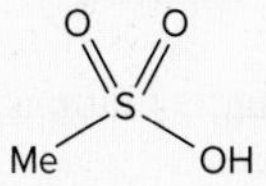

21.19 Which species is the most nucleophilic? Why?
(a) NaCl (b) $NaNH_2$ (c) NaOH (d) $NaCH_3$

21.20 Hexamethylphosphoramide, sometimes referred to as HMPA (shown below), can be used to make the reactions of nucleophiles (such as NaF) faster. How can HMPA achieve this?

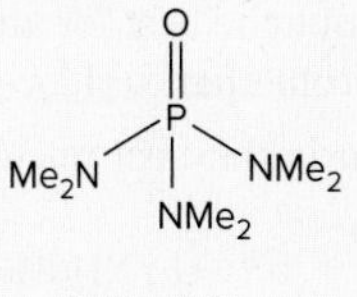

21.21 Thiophenol (shown below) is more nucleophilic than phenol. Explain.

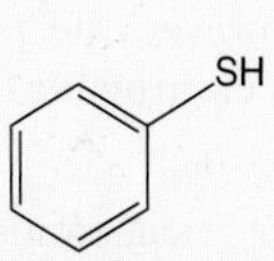

Thiophenol

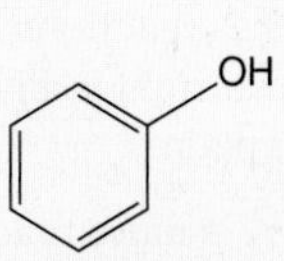

Phenol

Nucleophilic Substitution Reactions

(Sample Problem 21.3)

Concept Review Questions

21.22 The substitution reaction of alkyl bromides by alcohols can be catalyzed by the addition of sodium iodide. Explain.

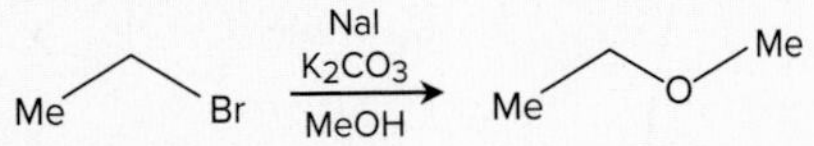

21.23 Treatment of the alcohol shown below with sodium azide does not lead to an efficient S_N2 reaction. How can the starting material be derivatized to allow the reaction to occur more readily?

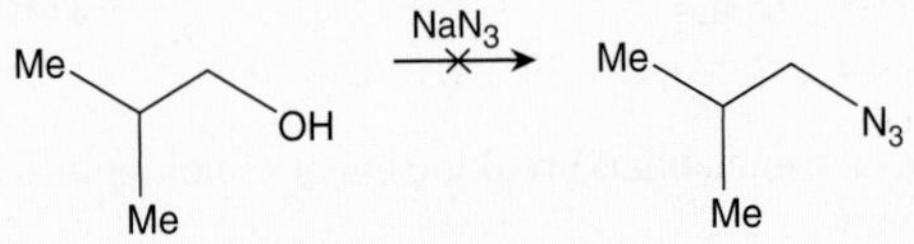

21.24 Which haloalkanes would you expect to participate in an S_N1 reaction most efficiently? Why?

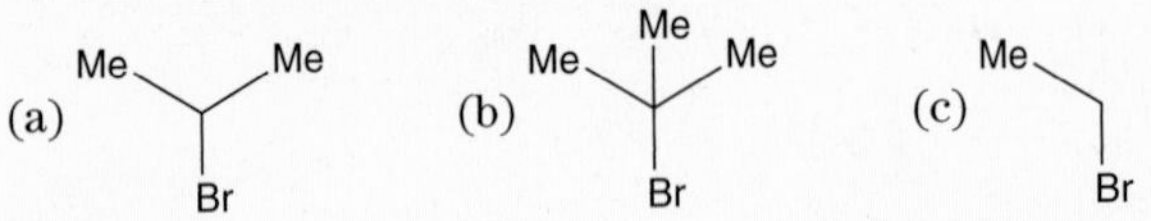

21.25 Treatment of the alcohol below with ethanoic acid (acetic acid) does not lead to the tertiary acetate (**A**) as expected but gives a mixture of two products, the major product being the primary acetate (**B**) shown below. Explain.

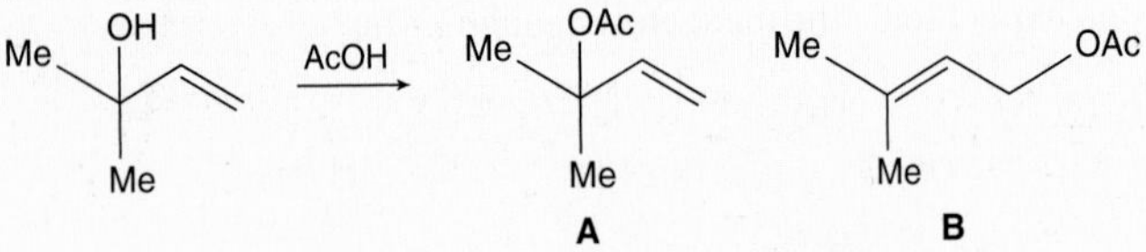

21.26 Explain the formation of the product reaction:

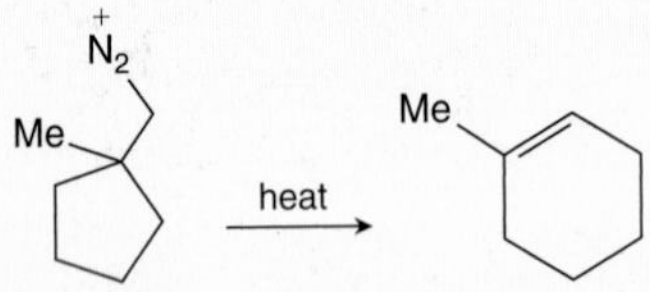

21.27 Under the following conditions, the chiral information in molecule **A** is lost. Explain.

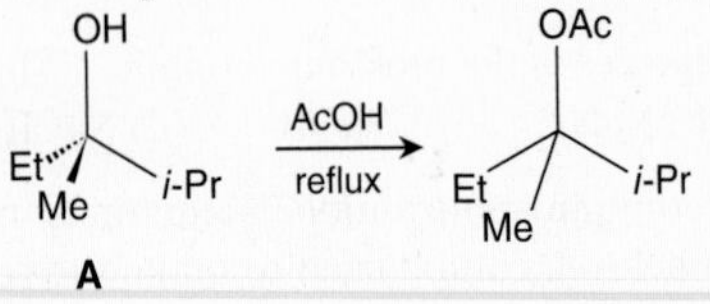

21.28 What type of solvents should be used to favour S_N2 reactions? Which solvent would be best for an S_N2 reaction?
(a) hexane (b) trichloromethane (chloroform) (c) acetonitrile

21.29 Which substance is the best nucleophile in an S_N2 reaction? Why?
(a) NEt_3 (b) NaN_3 (c) NH*t*-Bu$_2$

21.30 Primary haloalkanes are generally good substrates for S_N2 reactions; however, the primary haloalkane at right is a poor electrophile for S_N2 reactions. Explain.

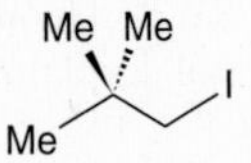

21.31 Molecule **A** is more reactive in substitution reactions than molecule **B**. Explain.

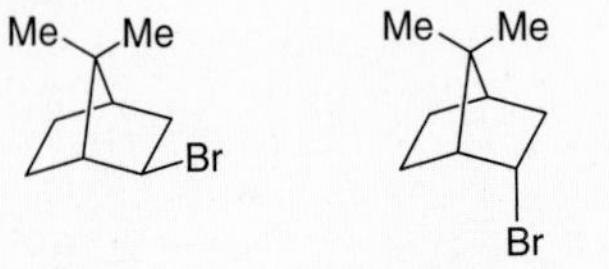

A **B**

21.32 The following alcohol is available as a single enantiomer. Give conditions for its conversion to both the ethers shown below:

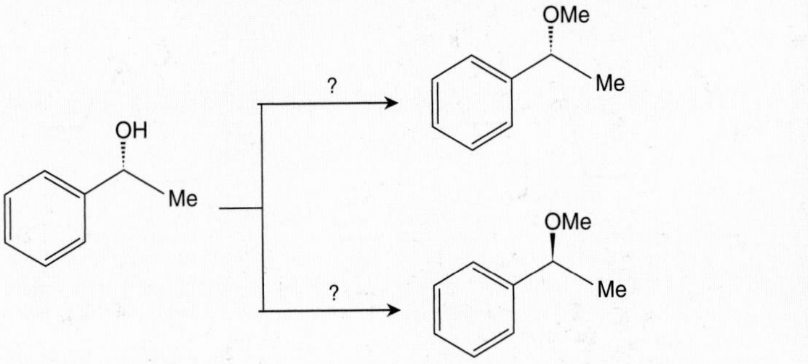

Skill-Building Exercises

21.33 In each reaction, identify the nucleophilic centre and the leaving group. Can you propose the product of the reaction?

(a) 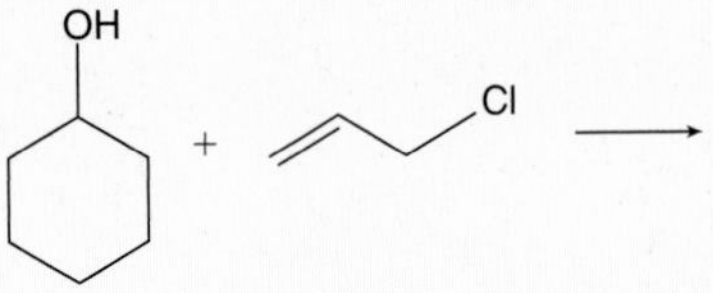

(b) NaN_3 + 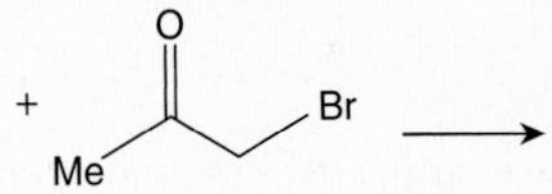

(c) 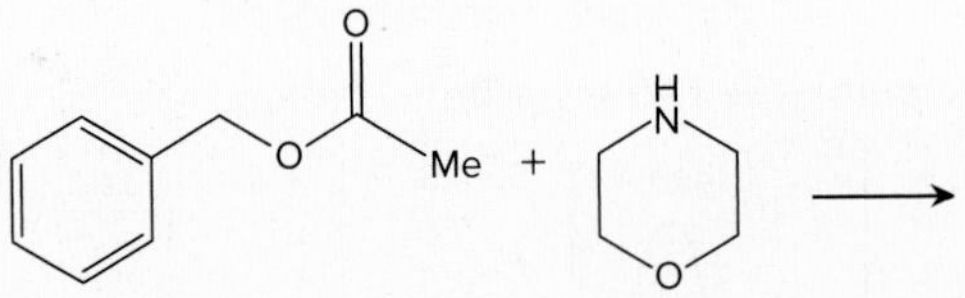

(d) 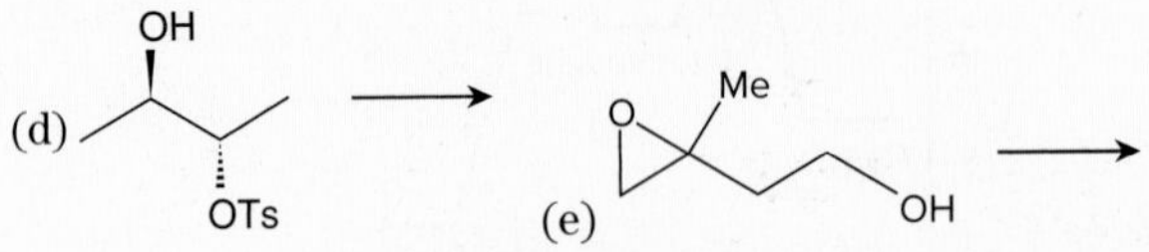 (e)

21.34 Paramethoxybenzyl (PMB) groups are common protecting groups for alcohols in the synthesis of complex molecules. They can be removed under acidic conditions, generating the free alcohol. Using your knowledge of S_N1 reactions, can you propose a mechanism for the following reaction?

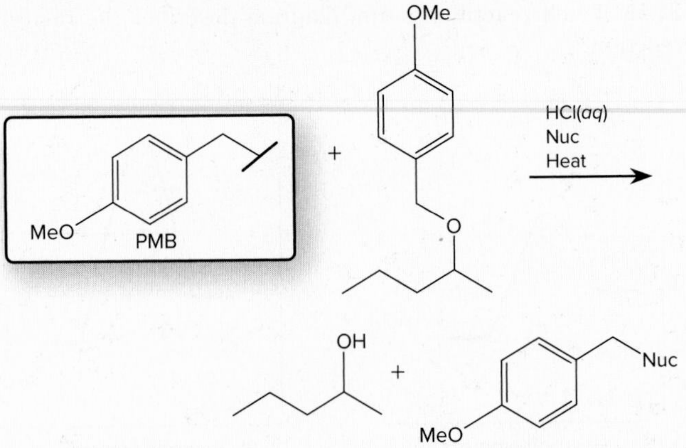

21.35 Predict the product of treating the alcohol below with ethanoic acid (acetic acid).

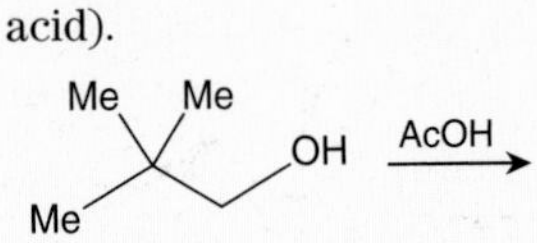

21.36 Draw a reaction mechanism explaining the formation of the following aldehyde under the reaction conditions:

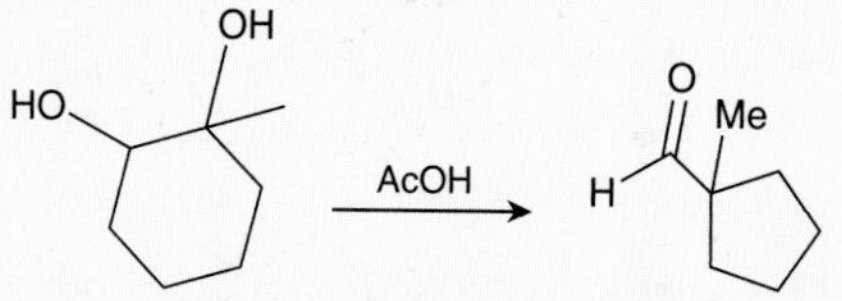

21.37 Provide starting materials for the synthesis of each compound.

(a) 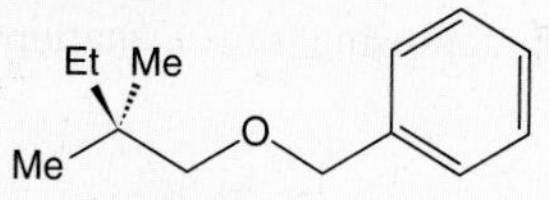

(b) 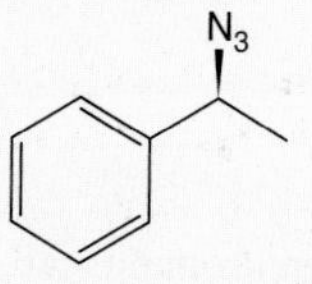 (c) 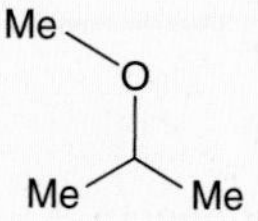

Elimination Reactions

(Sample Problem 21.4)

Concept Review Questions

21.38 Which substrate would you expect to undergo an E1 reaction most readily?

(a) 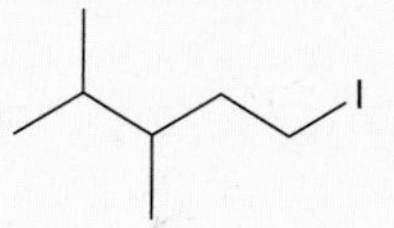 (b) 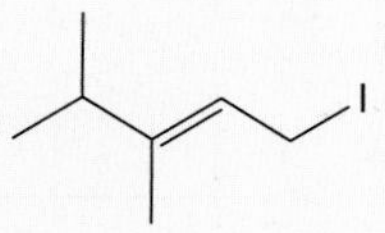

(c) 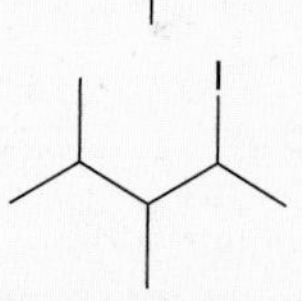

21.39 Which substrate would you expect to undergo an E1 reaction most readily?

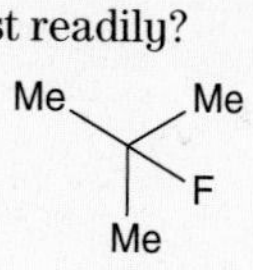
(a)

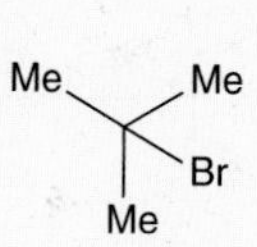
(b)

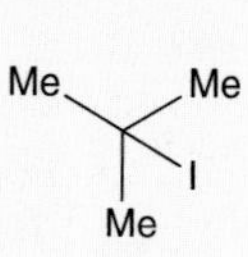
(c)

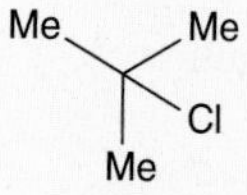
(d)

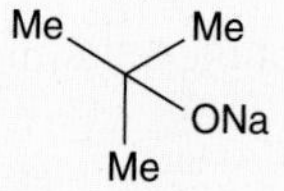
(e)

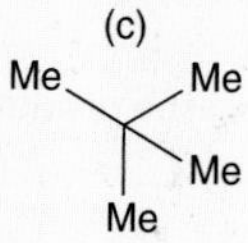
(f)

21.40 Predict the major product of the following reaction. Is this the kinetic or thermodynamic product?

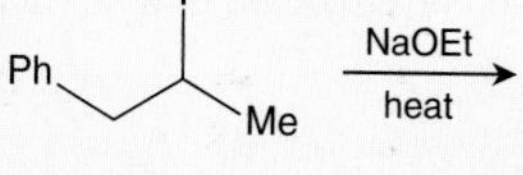

21.41 Which substrate would you expect to undergo an E2 reaction most readily?

(a) 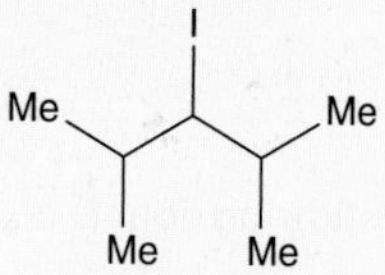 (b) 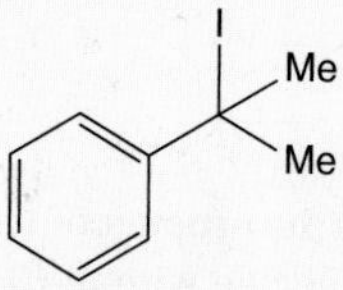

(c) 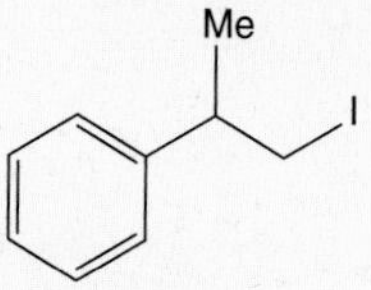

21.42 Which E2 reaction would you expect to progress most readily?

(a) 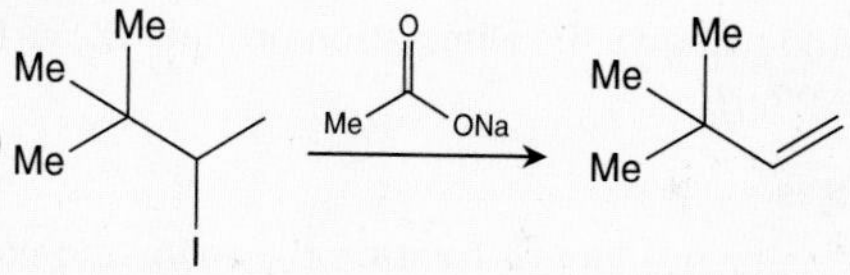

(b) 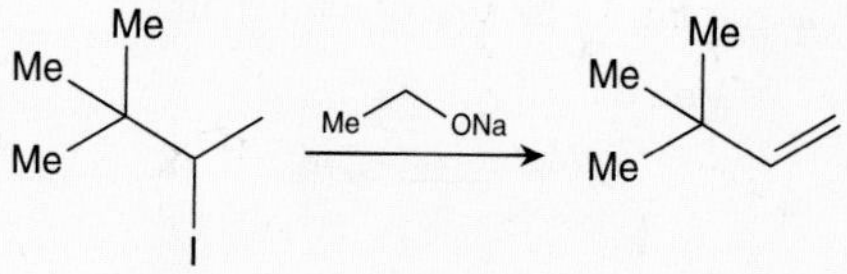

(c) 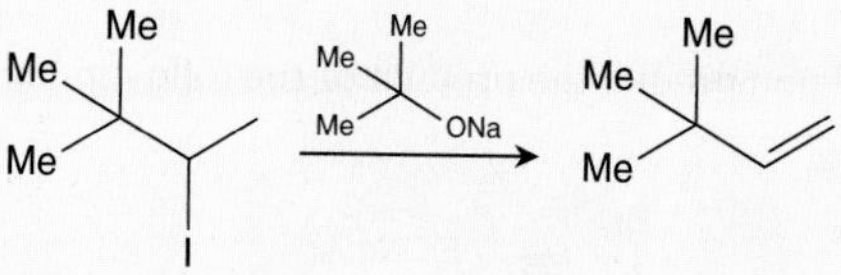

21.43 Which molecule would you expect to be the most basic? Which is the least basic? Why?

(a)  (b) 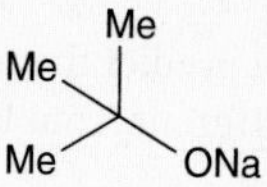

(c) 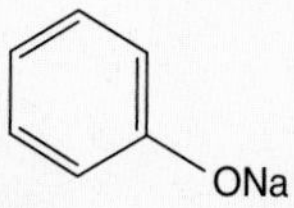 (d) 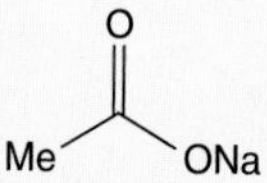

21.44 Which molecule would you expect to be the most basic? Why?
(a) NaOH (b) $NaNH_2$ (c) NaI

21.45 Molecule **A** rapidly undergoes elimination under basic conditions (NaOEt), whereas molecule B eliminates much more slowly. Explain.

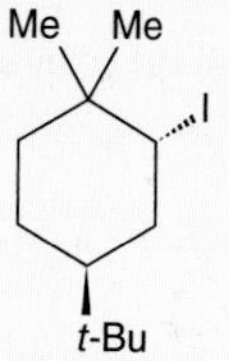
A

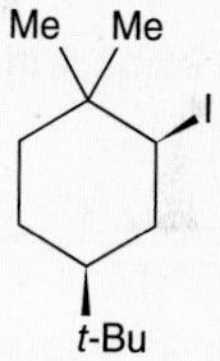
B

21.46 What is the rate-determining step of an E1 reaction? Can you draw a reaction energy diagram for a general E1 reaction?

Consider the following reaction schemes for Problems 21.47 and 21.48 (r.t. stands for room temperature).

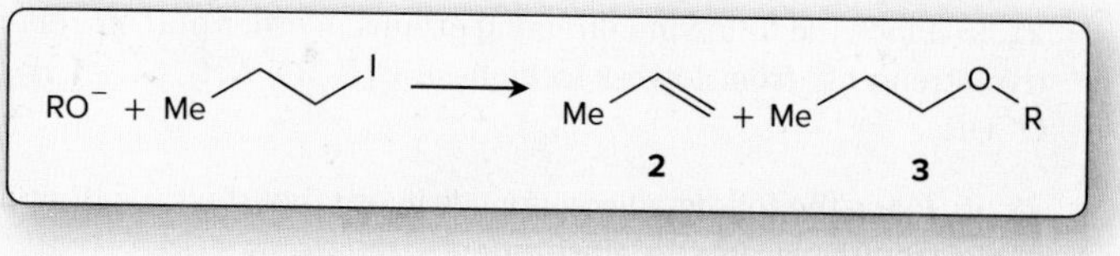

(a)

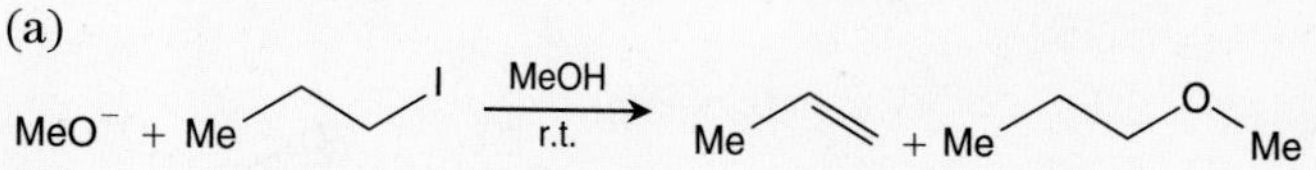

(b)

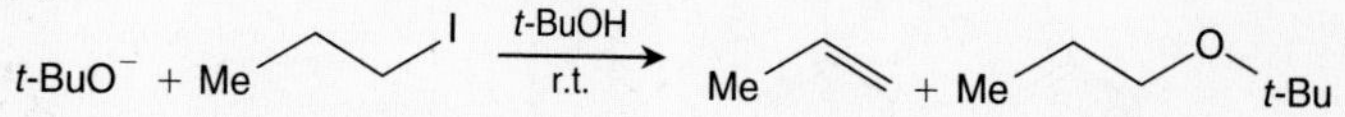

(c)

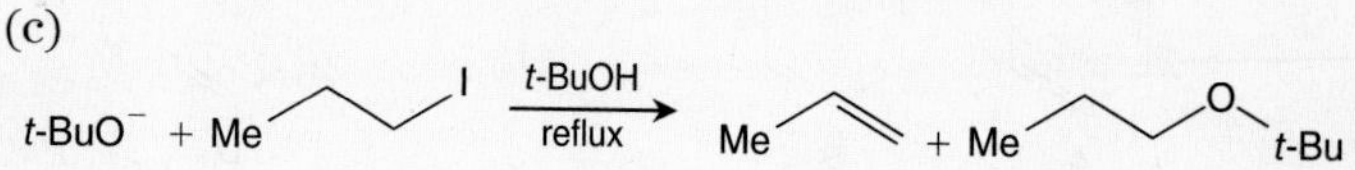

21.47 Which reaction will give the substitution product (**3**) as the dominant product? Why?

21.48 Which reaction will give the elimination product (**2**) as the dominant product? Why?

Skill-Building Exercises

21.49 Predict the major product of treating the following compound with NaOEt, assuming an E2 reaction mechanism.

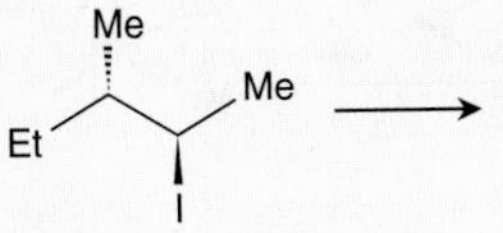

21.50 What would you predict the product of the following reaction to be?

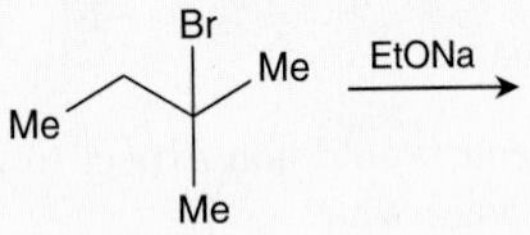

21.51 What would you predict the product of the following reaction to be? Why is it different from Problem 21.50?

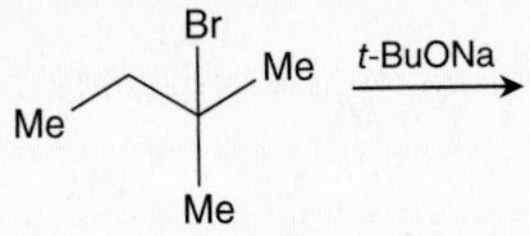

Addition Reactions: Reactions of Alkenes

(Sample Problems 21.5 to 21.9)

Concept Review Questions

21.52 Explain Markovnikov's rule. Use the addition of HCl to but-1-ene to illustrate your answer.

21.53 Fill in each blank with a general formula for the type of compound formed:

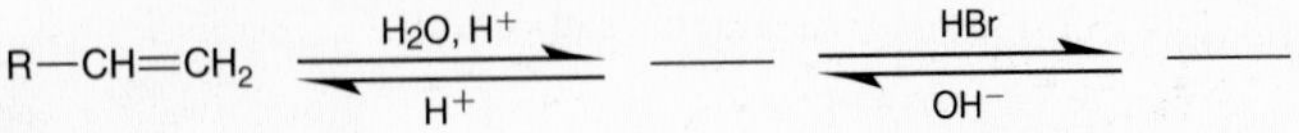

21.54 Explain the selectivity observed in the following hydrogenation:

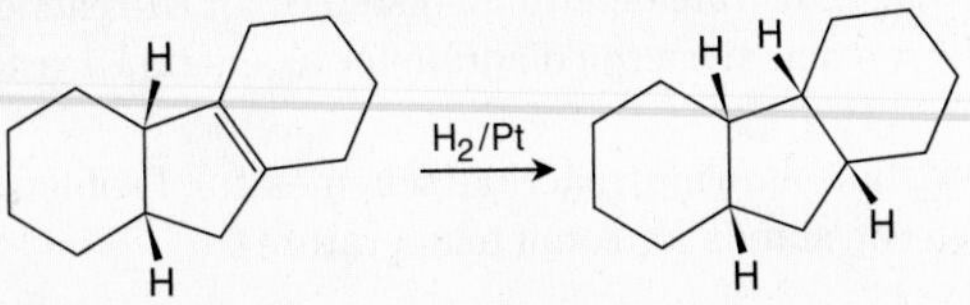

21.55 Place the following directing groups in order of their relative strengths, from lowest to highest: OH, Me, NH_2, $CH{=}CH_2$, OCOR

21.56 Place the following compounds in order of their reactivity in electrophilic aromatic substitution.

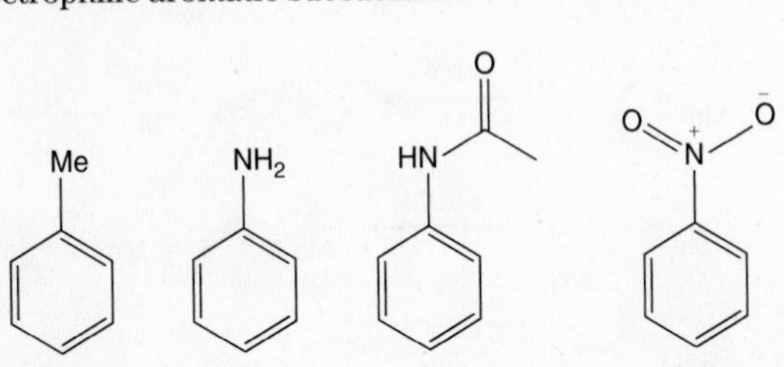

Skill-Building Exercises

21.57 Give the product of each reaction:

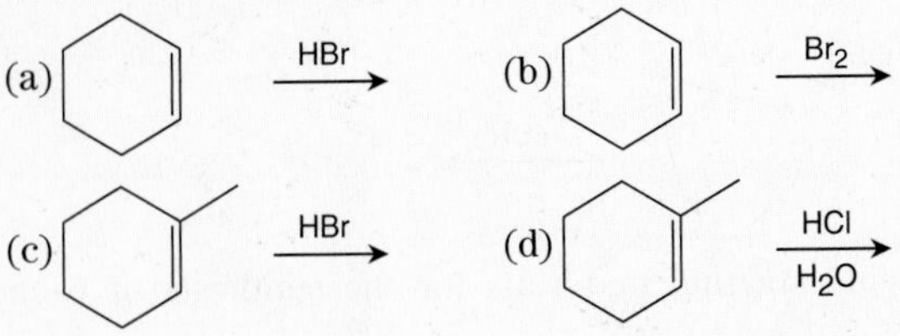

21.58 Give reactants to achieve the following transformation:

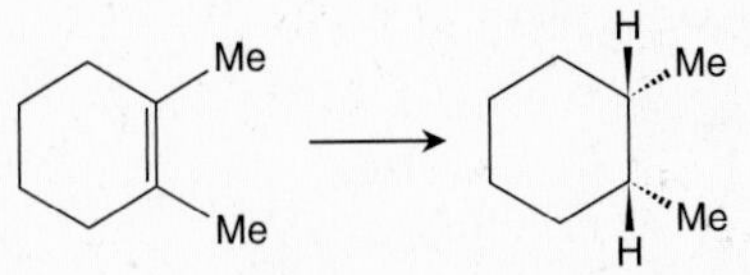

21.59 Given that treatment of **A** with hydrogen bromide leads to product **B**, what conditions are required to synthesize **C**?

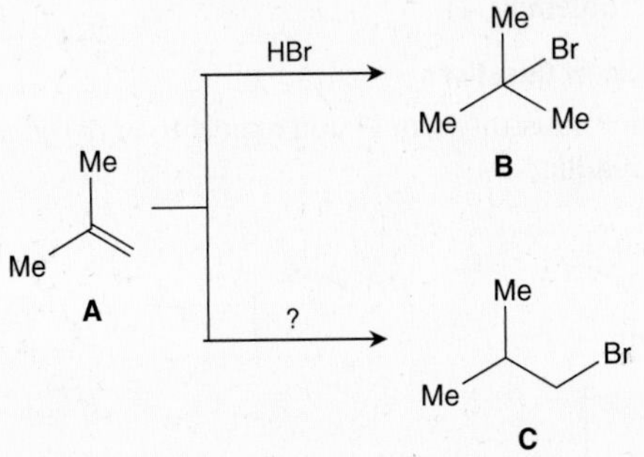

21.60 Give the product(s) of the following reaction:

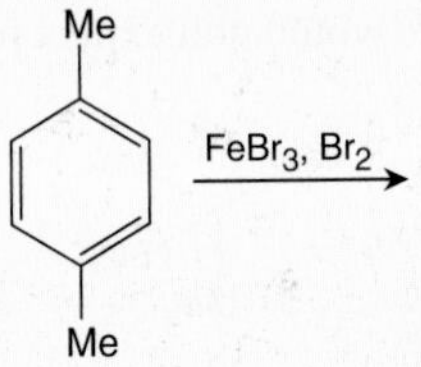

21.61 Give the product of the following reaction:

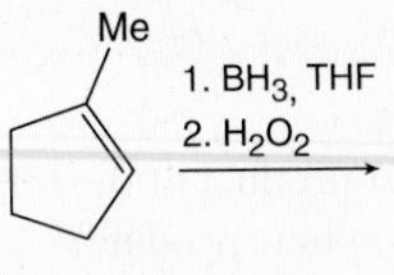

21.62 Explain the selectivity seen in the reaction below:

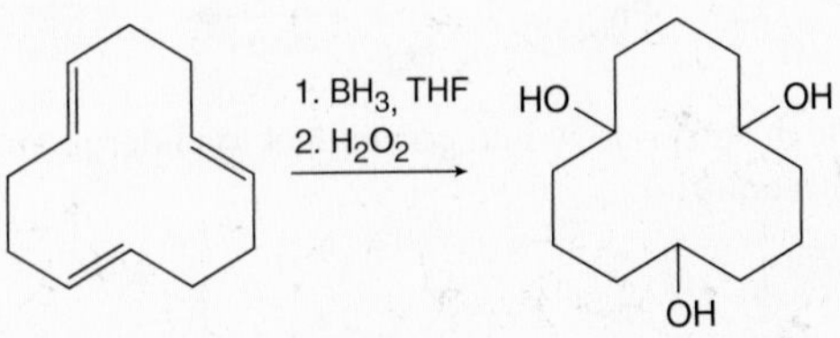

21.63 What is the product of treating the following double bond–containing molecule with *m*-CPBA?

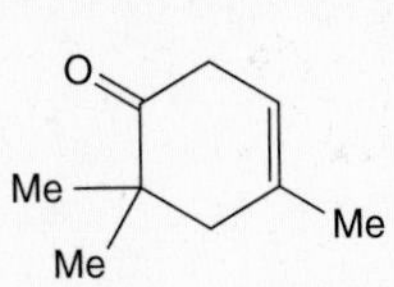

21.64 Given that epoxidation is stereospecific, draw the double bond–containing molecule you would need to synthesize the following epoxide with *m*-CPBA:

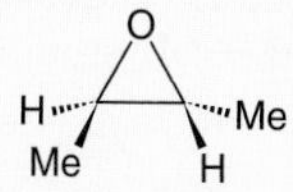

21.65 By considering the mechanism of epoxidation, which double bonds will be epoxidized most rapidly by *m*-CPBA?

(a) 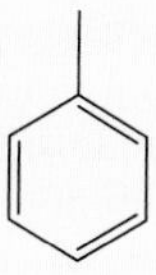

(b) 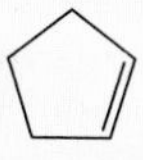

(c) 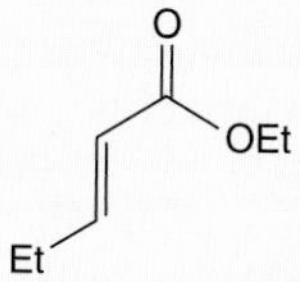

21.66 Give the products of ozonolysis of the following molecule:

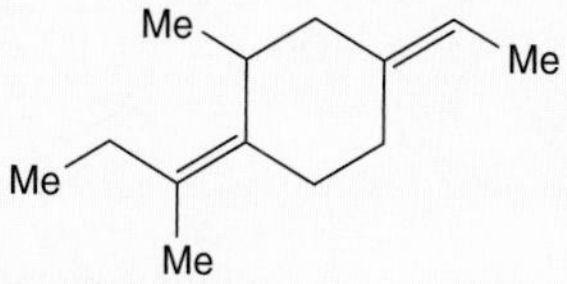

21.67 Complete ozonolysis of a compound with the formula C_8H_{10} gave the following molecule. What is the structure of the starting material?

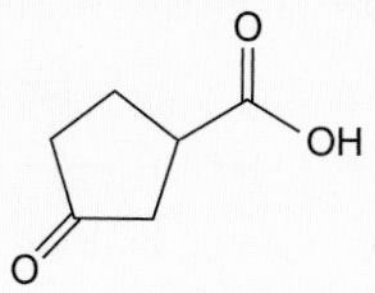

Electrophilic Aromatic Substitution: Reactions of the Benzene Ring

Concept Review Questions

21.68 What is the main difference in reactivity between isolated C=C double bonds and aromatic C=C double bonds? (That is, what types of reaction does each type of bond undergo?)

21.69 Of the three major types of organic reactions, which do *not* occur readily with benzene? Why?

21.70 Which statement is most true for the following general reaction scheme?

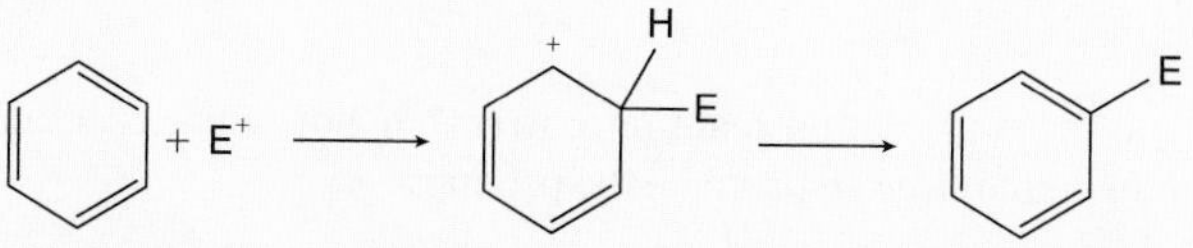

(a) The rate-determining step of this reaction is activation of the electrophile.
(b) The rate-determining step of this reaction is re-aromatization of the ring.
(c) The rate-determining step of this reaction is breaking the aromaticity of the ring.
(d) This reaction does not have an rate-determining step.

Skill-Building Exercises

21.71 Give the product(s) of the following reaction:

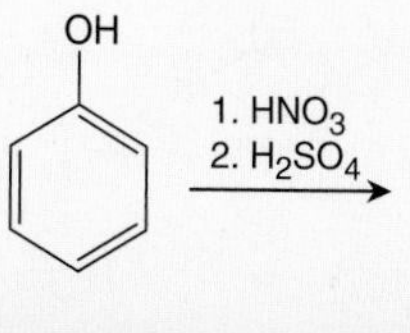

21.72 Give the major product(s) of the following reaction:

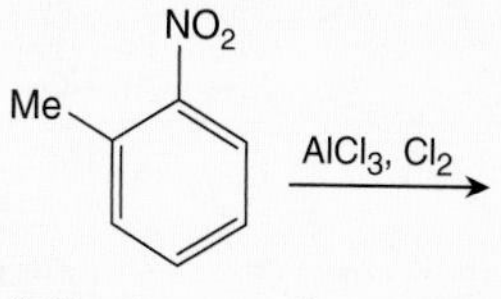

21.73 Explain the following results in terms of the directing groups and reaction conditions:

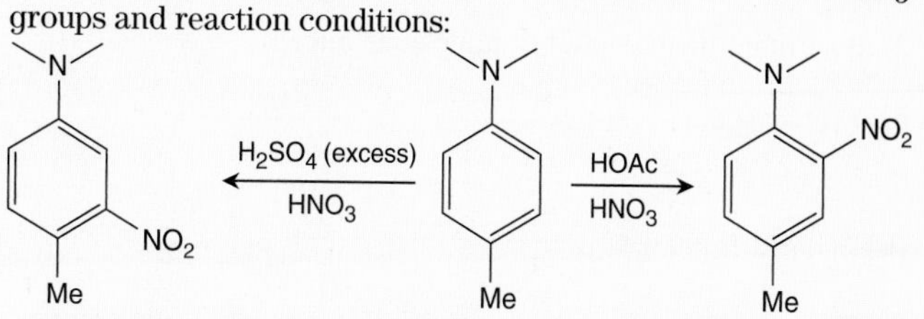

Some Reactions of Functional Groups

(Sample Problems 21.10 to 21.12)

Concept Review Questions

21.74 Why does the C=O group react differently from the C=C group? Show an example of the difference.

21.75 Many substitution reactions are initiated by electrostatic attraction between reactants. Show where this attraction arises in the formation of an amide from an amine and an ester.

21.76 Although both carboxylic acids and alcohols contain an —OH group, one is more acidic in water than the other. Explain.

21.77 What reaction type is common to the formation of esters and acid anhydrides? What is the other product?

21.78 Both alcohols and carboxylic acids undergo substitution, but the processes are very different. Explain.

Skill-Building Exercises (grouped in similar pairs)

21.79 Draw the product resulting from mild oxidation of (a) butan-2-ol; (b) 2-methylpropanal; (c) cyclopentanol.

21.80 Draw the alcohol you would oxidize to produce (a) 2-methy-l-propanal; (b) pentan-2-one; (c) 3-methylbutanoic acid.

21.81 Draw the organic product formed when the following compounds undergo a substitution reaction: (a) ethanoic acid (acetic acid) and methanamine (methylamine); (b) butanoic acid and propan-2-ol; (c) methanoic acid (formic acid) and 2-methylpropan-1-ol.

21.82 Draw the organic product formed when the following compounds undergo a substitution reaction: (a) ethanoic acid (acetic acid) and hexan-1-ol; (b) propanoic acid and *N*-methylmethanamine (dimethylamine); (c) ethanoic acid (acetic acid) and *N*-ethylethanamine (diethylamine).

21.83 Draw condensed formulas for the carboxylic acid and alcohol that form each ester:

(a) 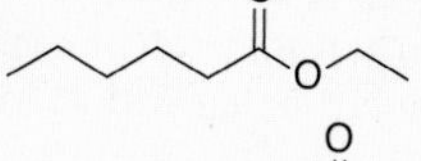

(b) 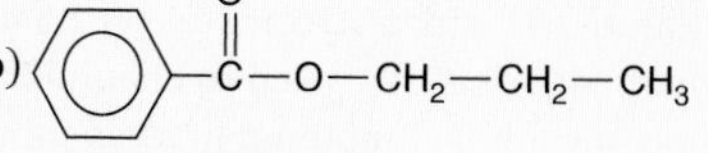

(c) 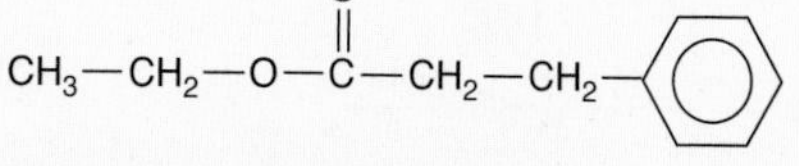

21.84 Draw condensed formulas for the carboxylic acid and amine that form each amide:

(a) 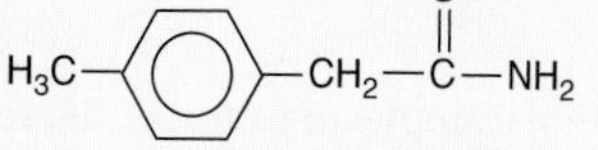

(b) 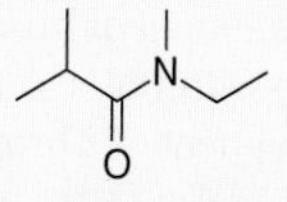

(c) 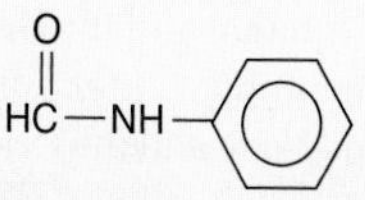

21.85 Fill in the expected organic substances:

(a) 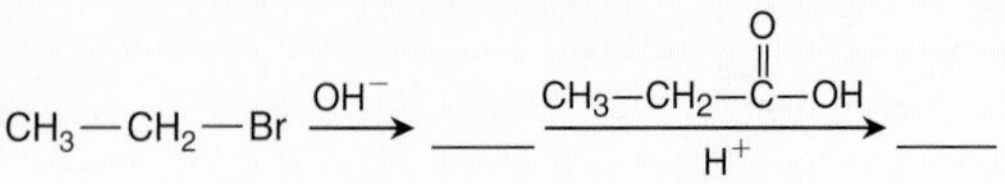

(b) 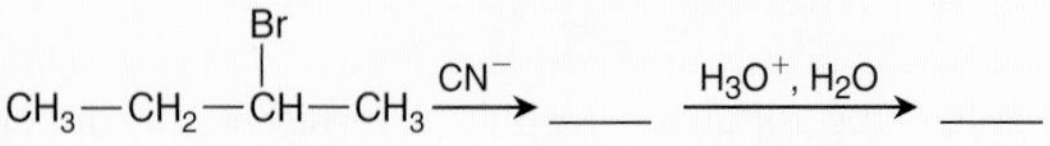

21.86 Fill in the expected organic substances:

(a) $CH_3—CH_2—CH{=}CH_2 \xrightarrow{H^+,\ H_2O}$ ____ $\xrightarrow{Cr_2O_7^{2-},\ H^+}$ ____

(b) 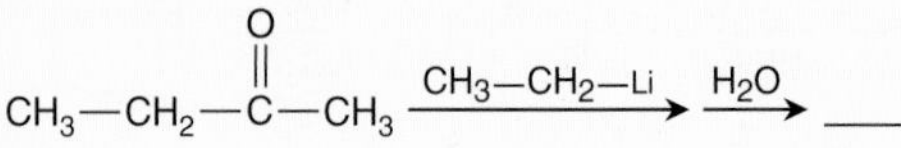

21.87 Supply the missing organic and/or inorganic substances:

(a) 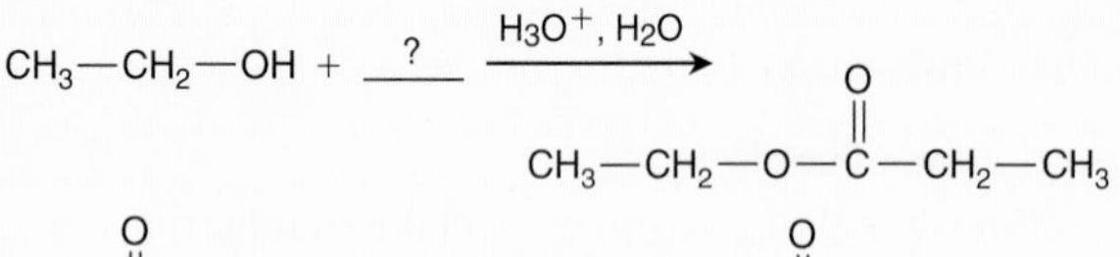

(b) 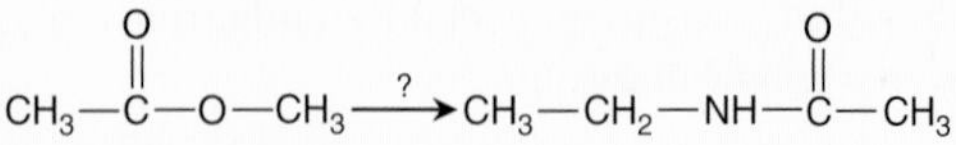

21.88 Supply the missing organic and/or inorganic substances:

(a) 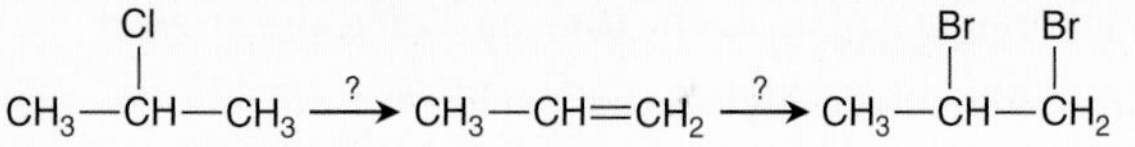

(b)

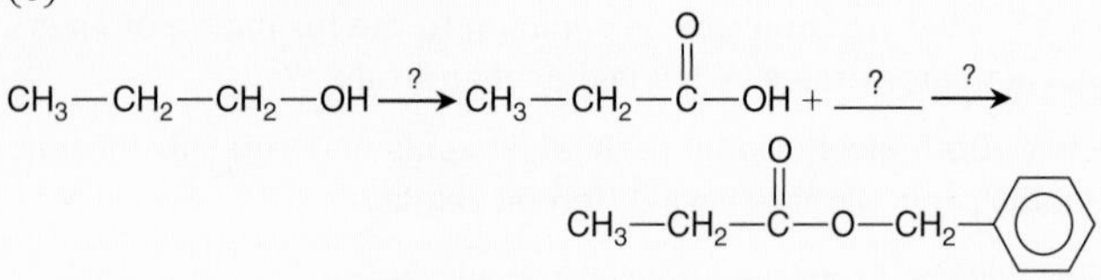

Comprehensive Problems

21.89 (a) Draw the four isomers of $C_5H_{12}O$ that can be oxidized to an aldehyde. (b) Draw the three isomers of $C_5H_{12}O$ that can be oxidized to a ketone. (c) Draw the isomers of $C_5H_{12}O$ that cannot be easily oxidized to an aldehyde or ketone. (d) Name any isomer that is an alcohol.

21.90 Ethyl methanoate (ethyl formate, $HC(=O)—O—CH_2—CH_3$) is added to foods to give them the flavour of rum. How would you synthesize ethyl methanoate from ethanol, methanol, and any inorganic reactants?

21.91 An alcohol is oxidized to a carboxylic acid, and 0.2003 g of the acid is titrated with 45.25 mL of 0.038 11 mol/L NaOH. (a) What is the molar mass of the acid? (b) What is the molar mass of the alcohol?

21.92 Some of the most useful compounds for organic synthesis are Grignard reagents (general formula R—MgXs, where X is a halogen), which are made by combining a haloalkane, R—X, with Mg. They are used to change the carbon skeleton of a starting carbonyl compound in a reaction similar to that with R—Li:

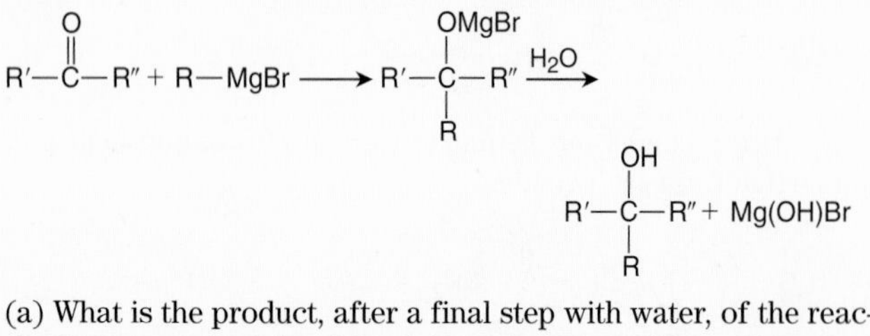

(a) What is the product, after a final step with water, of the reaction between ethanal and the Grignard reagent of bromobenzene? (b) What is the product, after a final step with water, of the reaction between butan-2-one and the Grignard reagent of 2-bromopropane? (c) There are often two (or more) combinations of Grignard reagent and carbonyl compound that will give the same product. Choose another pair of reagants to give the product in (a). (d) What carbonyl compound must react with a Grignard reagent to yield a product with the —OH group at the *end* of the carbon chain? (e) What Grignard and carbonyl compound would you use to prepare 2-methylbutan-2-ol?

21.93 Ibuprofen is one of the most common anti-inflammatory drugs. (a) Identify the functional group(s) and chiral centre(s) in ibuprofen. (b) Write a four-step synthesis of a racemic mixture of ibuprofen from 4-(2-methylpropyl)benzaldehyde, also known as 4-(2-methylpropyl)benzaldehyde, using inorganic reactants and one organometallic reactant (see Problem 21.92).

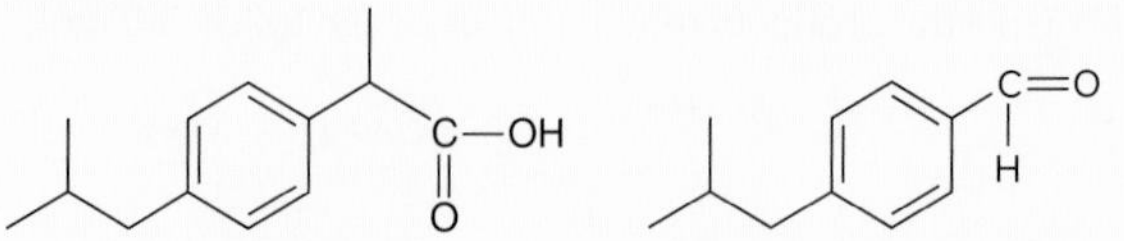

Ibuprofen 4-*iso*butylbenzaldehyde

21.94 Starting with the given organic reactant and any necessary inorganic reactants, explain how you would perform each synthesis:

(a) 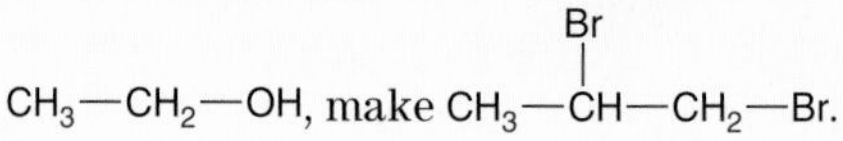

(b) 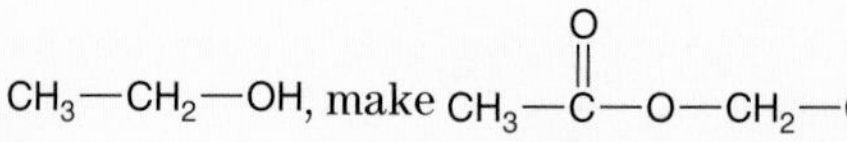

21.95 Butan-2-one is reduced by hydride ion donors, such as sodium borohydride ($NaBH_4$), to butan-2-ol. Even though the alcohol has a chiral centre, the product isolated from the redox reaction is not optically active. Explain.

21.96 Classify each reaction as an addition, an elimination, or a substitution reaction:

(a) 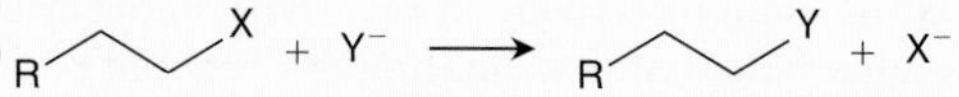

If X = Br and Y = I, would this reaction still proceed? Explain your answer.

(b) 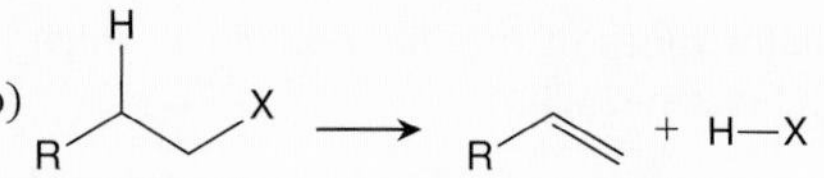

(c) 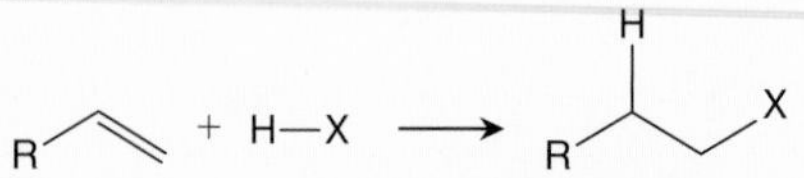

Is the product of this reaction correct? If not, what reactants would be required to produce the product?

21.97 Examine the reactions below, especially the major and minor products. Draw the mechanisms for each reaction depicted. Are the major and minor products correct in each case? If not, explain what is wrong and show the correct major product.

(a) 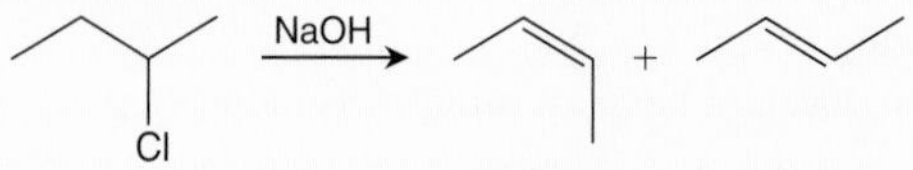

major minor

(b) 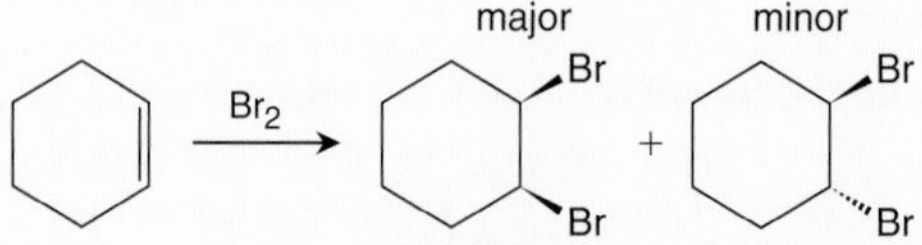

major minor

(c)

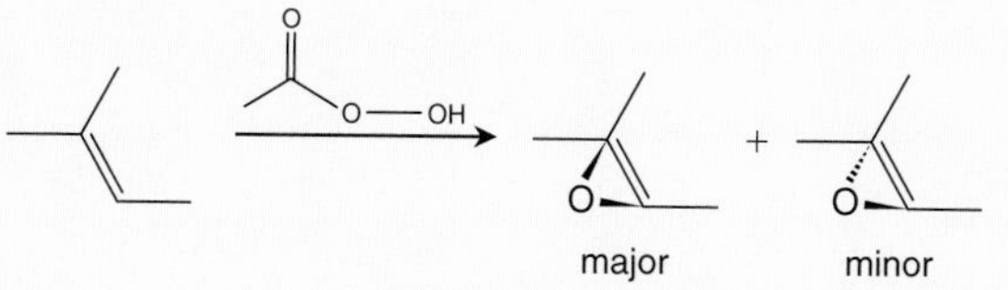

21.98 Draw a phenol molecule and perform a Friedel-Crafts alkylation with chloroethane, showing all the reactants required. Show your mechanism for the reaction and include any relevant resonance structures. Name all possible final compounds and identify the major and minor products, and give a reason for your assignments.

21.99 Identify each reaction, and provide the missing molecules or reactants:

(a)

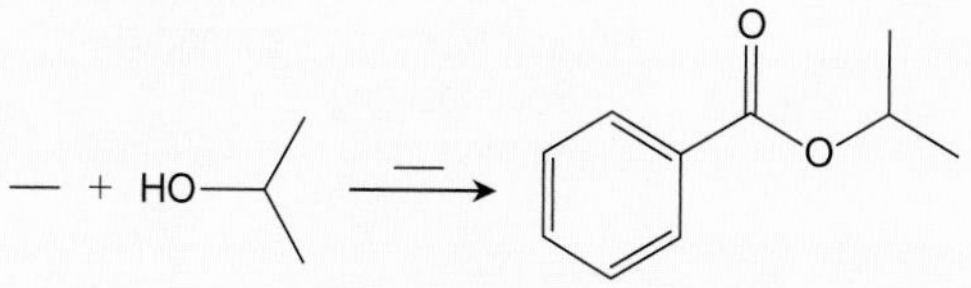

(b)

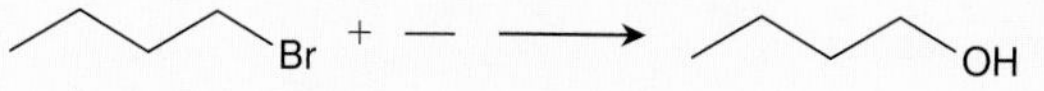

(c)

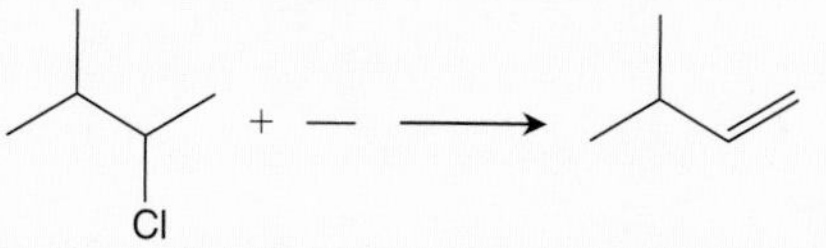

21.100 (a) Starting from 2-methylbut-1-ene, what reactants would you need to produce 2-bromo-2-methylbutane? If you wanted to convert the 2-bromo-2-methylbut-1-ene to 2-methoxy-2-methylbutane, what additional reactants might be required?

(b) If you wanted to use the 2-bromo-2-methylbutane to form the starting material, 2-methylbut-1-ene, would E1 conditions be desirable? If so, why? If not, why not? Identify any minor products formed in the final step.

21.101 Starting from a three-carbon linear haloalkane, identify each of the compounds in the blanks. Draw the structure of the final compound after the reduction step.

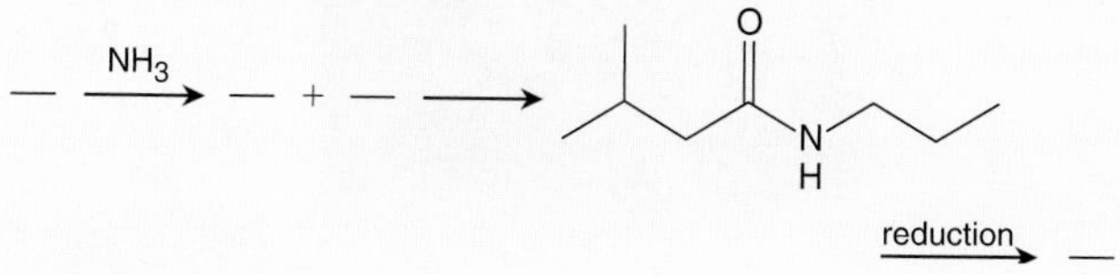

21.102 (a) Fill in the blanks in the following reaction sequence (Ph = phenyl):

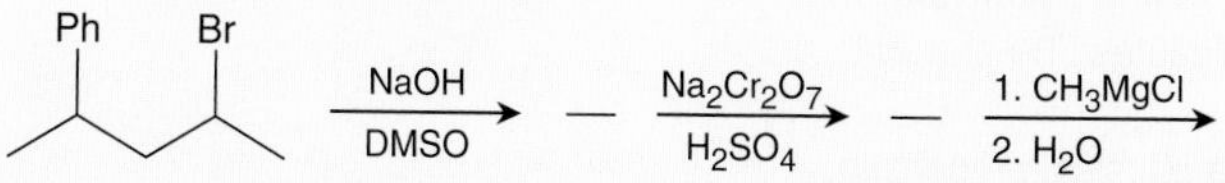

(b) What would be the final product if the solvent in the first step were changed to CH_3OH?

21.103 Provide all missing reactants and structures for the following reaction scheme:

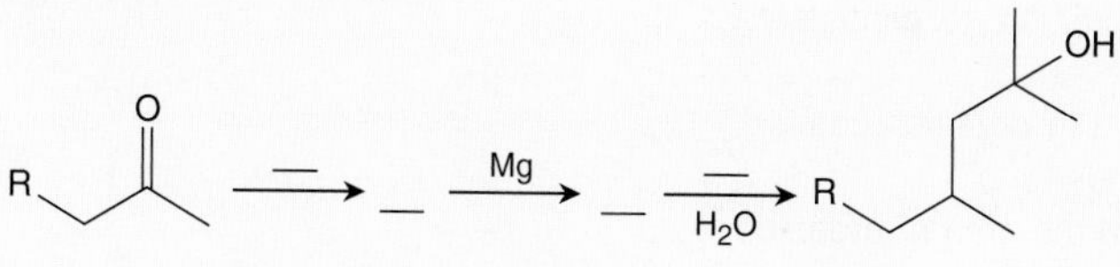

21.104 (a) Provide the identities of the missing compounds for the following reaction:

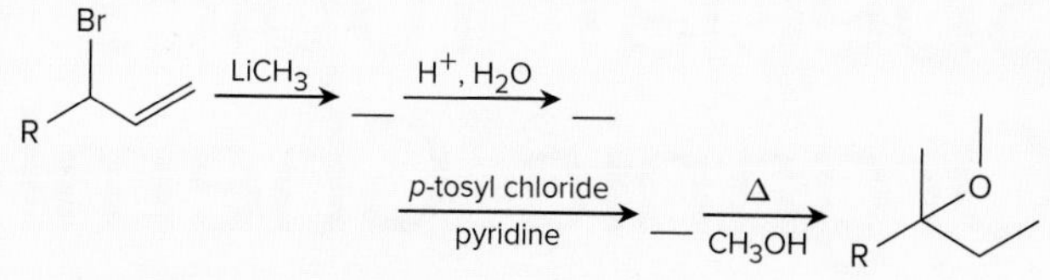

(b) What would happen to the reaction rate if you doubled the concentration of $LiCH_3$ in the first reaction?
(c) What would be the effect on the reaction rate if you halved the concentration of CH_3OH in the final step?

21.105 (a) Why do the hydrogenation and hydroboration-oxidation reactions of alkenes always produce syn- products, whereas the halogenation of alkenes does not?
(b) If you wish to be able to control the stereochemistry of a chiral centre, which of the types of reactions are not suitable (S_N2, S_N1, E_2, and E_1)?

21.106 By considering the following reactions, provide a possible structure of A:

$$A \xrightarrow{NaH} C_4H_7NaO_2 \xrightarrow{MeI} C_5H_{10}O_2$$

21.107 When A is exposed to the following reaction conditions, it gives two molecules with the formulas $C_8H_8O_2$ and C_6H_6N. Can you propose a structure for A and the products of reaction?

$$A \xrightarrow[\text{Heat}]{\text{6 mol/L NaOH, } H_2O} C_8H_8O_2 + C_6H_6N$$

21.108 When A is exposed to the following reaction conditions, it gives a molecule with the formula $C_6H_{12}O$. Can you propose a structure for the product?

$$A \xrightarrow{MeMgBr} C_6H_{12}O$$

21.109 Cadaverine (1,5-diaminopentane) and putrescine (1,4-diaminobutane) are two compounds that are formed by bacterial action and are responsible for the odour of rotting flesh. Draw their structures. Suggest a series of reactions to synthesize putrescine from 1,2-dibromoethane and any inorganic reactants.

21.110 Pyrethrins, such as jasmolin II (*below*), are a group of natural compounds synthesized by flowers of the genus *Chrysanthemum* (known as pyrethrum flowers) to act as insecticides.
(a) Circle and name the functional groups in jasmolin II.
(b) What is the hybridization of the numbered carbons?
(c) Which, if any, of the numbered carbons are chiral centres?

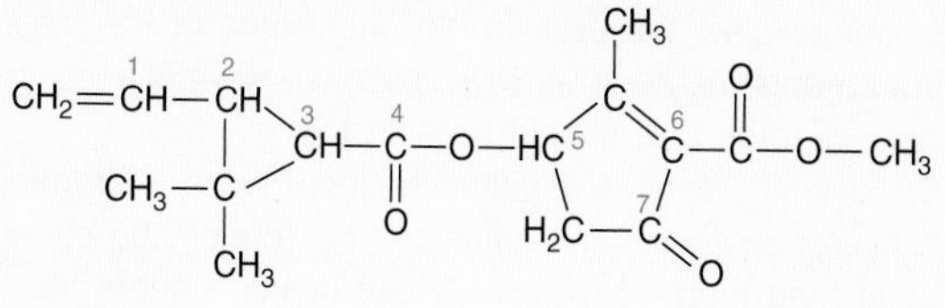

Special Topics in Organic Chemistry

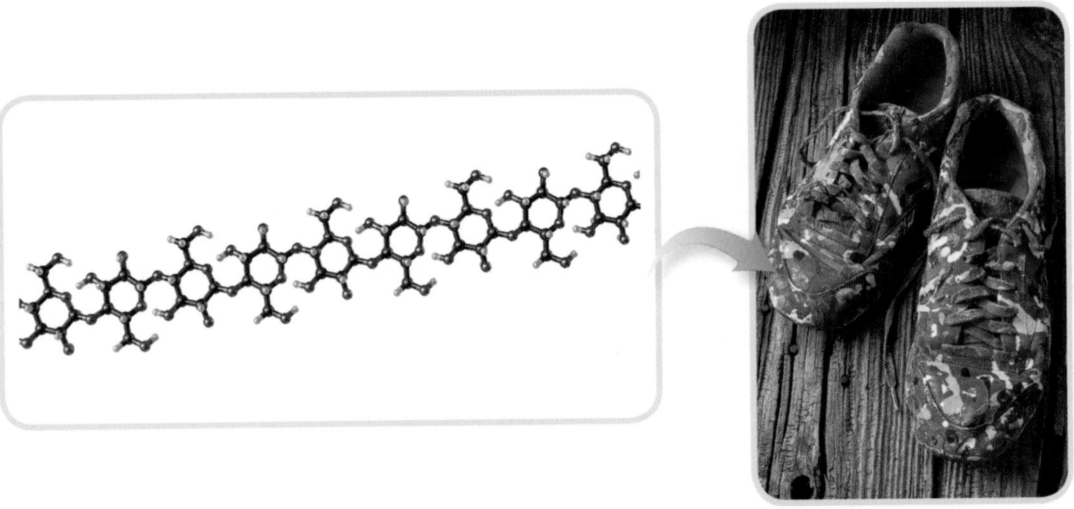

IN THIS CHAPTER . . . We focus on a very interesting type of carbon compound: macromolecules. We present not only some of the macromolecules that we can find in living organisms, such as DNA, but also several synthetic macromolecules. By the end of this chapter, you should be able to

- Discuss key features of polymers, focusing on their mass, shape, and physical properties
- Explain how these molecules can be formed, by either addition or condensation reactions, and then examine their use in our modern life
- Describe macromolecules that can be found in living organisms, including polymers such as sugars and polysaccharides, amino acids and proteins, and nucleotides and nucleic acids
- Apply previous knowledge about intermolecular forces from Chapter 11 to biological macromolecules
- Identify and critically examine techniques that are commonly used in organic chemistry laboratories to separate, purify, and identify organic compounds
- Describe some of the fascinating research being done in Canada in the area of organic chemistry

22

Macromolecules, whether natural or synthetic, form the backbone of society. We live in a world where new and exciting applications for polymers are being discovered daily. These discoveries are used across various industries, such as biomedical, pharmaceutical, food, and environmental. For example, we use cellulose as a biofuel, where previously it was mostly used to produce paper products and textiles; as building materials; or in the pharmaceutical and chemical industries. Additionally, we use dendrimers and chitin for drug delivery. Chitin and its derivatives are also used as antioxidants and antimicrobials, are immunostimulating, and are known for their anti-inflammatory and anticancer effects. From our DNA to our clothes, we are surrounded by macromolecules. In this chapter, we will learn how certain key polymers are formed and the reactions they undergo. We will also examine characterization techniques of many organic compounds and the tools used to characterize them. We will end this chapter by looking at some biological macromolecules and how they affect our lives.

Concepts and Skills to Review before Studying This Chapter

- Naming of straight-chain alkanes (Sections 2.8 and 20.2)
- Formulas and models used to represent molecules (Section 2.8)
- Classification and separation of mixtures (Section 2.9)
- Intermolecular forces (Section 11.3)
- Structure, properties, and bonding of solids (Section 11.6)
- Advanced materials (Section 11.7)
- Common functional groups (Section 20.6)
- Reaction of alkenes: addition to the double bond (Section 21.5)
- Some reactions of functional groups (Section 21.7)

22.1 Polymeric Materials

A **polymer** (Greek *poly*, meaning "many") is an extremely large molecule, or macromolecule, consisting of a covalently linked chain of smaller molecules called **monomers** (Greek *mono*, meaning "one"). A monomer is a *repeat unit* of a polymer, and a polymer may have hundreds to hundreds of thousands of repeat units. The many types of monomers give polymers the complete repertoire of intermolecular forces. There are two types of polymers:

- *Synthetic* polymers, which are created in a laboratory
- *Natural* polymers (or *biopolymers*), which are created within organisms

In this section, we examine the physical nature of synthetic polymers and explore the role of intermolecular forces in their properties and uses. In the following sections, we will examine the preparation of synthetic polymers and the monomers, structures, and functions of some biopolymers.

Dimensions of a Polymer Chain

Polymers differ from smaller molecules in terms of molar mass, chain length, shape, and size. As we discuss polymers, we will use polyethene as an example. Polyethene is also known as polyethylene, and it is, by far, the most common synthetic polymer.

Polymer Mass The molar mass of a polymer chain ($\mathcal{M}_{\text{polymer}}$, in g/mol, often referred to as the *molecular weight*) depends on the molar mass of the repeat unit ($\mathcal{M}_{\text{repeat}}$) and the **degree of polymerization (n)**, the number of repeat units in the chain:

$$\mathcal{M}_{\text{polymer}} = \mathcal{M}_{\text{repeat}} \times n$$

For example, the molar mass of the ethene (ethylene) repeat unit (—CH_2—CH_2—) is 28 g/mol. If an individual polyethene chain in a plastic grocery bag has a degree of polymerization of 7100, the molar mass of this chain is

$$\mathcal{M}_{\text{polymer}} = \mathcal{M}_{\text{repeat}} \times n = (28\text{ g/mol})(7.1 \times 10^3) = 2.0 \times 10^5\text{ g/mol}$$

TABLE 22.1 Molar Masses of Some Common Polymers

Name	$\mathcal{M}_{polymer}$ (g/mol)	n	Uses
Poly(prop-2-enoic acid) (acrylates)	2×10^5	2×10^3	Rugs, carpets
Polyamide (nylons)	1.5×10^4	1.2×10^2	Tires, fishing line
Polycarbonate	1×10^5	4×10^2	Compact discs
Polyethene (polyethylene)	3×10^5	1×10^4	Grocery bags
Polyethene (ultra-high molecular weight)	5×10^6	2×10^5	Hip joints
Poly(dimethyl benzene-1,4-dicarboxylate; ethane-1,2-diol) [poly(ethylene terephthalate) or PET]	2×10^4	1×10^2	Soft drink bottles
Polyphenylethene (polystyrene, PS)	3×10^5	3×10^3	Packing, coffee cups
Polychloroethene [poly(vinyl chloride) or PVC]	1×10^5	1.5×10^3	Plumbing

Table 22.1 shows some other examples.

However, even though the monomer repeat unit in any given chain has the same molar mass, the degree of polymerization often varies considerably from chain to chain. Thus, *a sample of a synthetic polymer has a distribution of chain lengths and molar masses*. Polymer chemists use various definitions to find the *average* molar mass $\mathcal{M}_w$; a common definition is the *number-average molar mass*, $\mathcal{M}_n$:

$$\mathcal{M}_n = \frac{\text{total mass of all chains}}{\text{amount (mol) of chains}}$$

If the number-average molar mass of the polyethene in a grocery bag is, for example, 1.6×10^5 g/mol, the chains may vary in molar mass from about 7.0×10^4 g/mol to 3.0×10^5 g/mol. The ratio of $\mathcal{M}_w$ to $\mathcal{M}_n$ is known as the *polydispersity index* (PDI) and provides a rough indication of the breadth of the distribution.

Polymer Chain Length The long axis of a polymer chain is called its *backbone*. The length of an *extended* backbone is simply the number of repeat units (degree of polymerization, n) times the length of each repeat unit (l_0). For instance, the length of an ethene repeat unit is about 250 pm, so the extended length of our particular grocery-bag polyethene chain is

$$\text{length of extended chain} = n \times l_0 = (7.1 \times 10^3)(2.5 \times 10^2 \text{ pm}) = 1.8 \times 10^6 \text{ pm}$$

Comparing this length with the thickness of the chain, which is about 40 pm, shows the threadlike nature of the extended chain (with a length that is about 50 000 times the width).

Coil Shape and Size A polymer molecule, however, whether pure or in solution, does not exist as an extended chain. In principle, the shape of the molecule arises from free rotation around all the single bonds. Thus, as each repeat unit rotates randomly, the chain continuously changes direction, turning back on itself many times and eventually arriving at the random **coil shape** that most polymers adopt (Figure 22.1). In reality, rotation is not completely free because intermolecular forces between chain portions, between different chains, and/or between the chain and solvent have significant effects on the actual shape of a polymer chain.

The size of the coiled chain is expressed by its **radius of gyration (R_g)**, the average distance from the centre of mass of the molecule to the outer edge of the coil (Figure 22.1). Even though R_g is reported as a single value for a given polymer, it represents an average value for many chains. The mathematical expression for the radius of gyration includes the length of each repeat unit and the degree of polymerization:*

$$R_g = \sqrt{\frac{nl_0^2}{6}}$$

As expected, the radius of gyration increases with the degree of polymerization and, thus, with the molar mass as well. Light-scattering measurements correlate with

*The mathematical derivation of R_g is beyond the scope of this book, but it is analogous to the two-dimensional "walk of the drunken sailor." With each step, the sailor stumbles in random directions and, given enough time, ends up very close to the starting position. The radius of gyration quantifies how far the end of the polymer chain (the sailor) has gone from the origin.

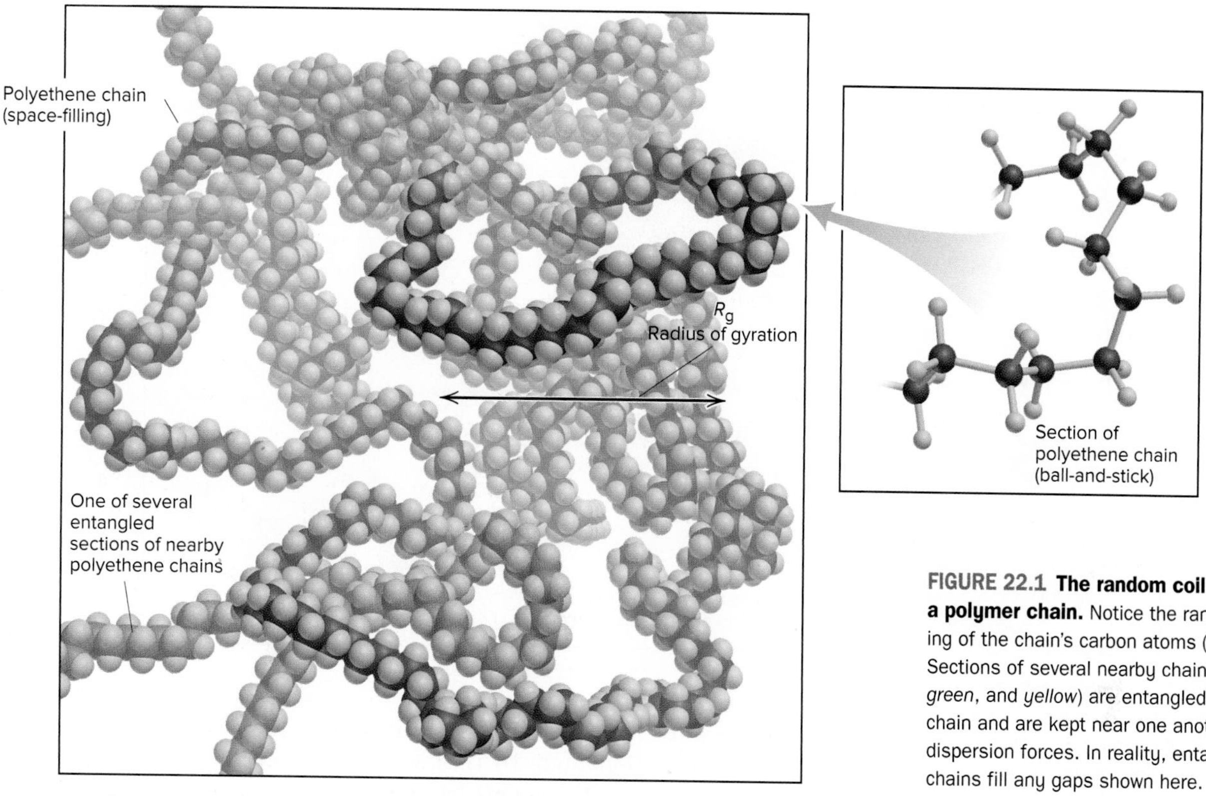

FIGURE 22.1 The random coil shape of a polymer chain. Notice the random coiling of the chain's carbon atoms (*black*). Sections of several nearby chains (*red*, *green*, and *yellow*) are entangled with this chain and are kept near one another by dispersion forces. In reality, entangling chains fill any gaps shown here.

these calculated results, so the radius of gyration of many polymers can be determined experimentally.

For our grocery-bag polyethene chain, the radius of gyration can be calculated as follows:

$$R_g = \sqrt{\frac{nl_0^{\,2}}{6}} = \sqrt{\frac{(7.1 \times 10^3)(2.5 \times 10^2\text{pm})^2}{6}} = 8.6 \times 10^3\,\text{pm}$$

Doubling the radius gives a diameter of 1.7×10^4 pm for the coiled chain, less than one-hundredth the length of the extended chain!

Polymer Crystallinity

A sample of a given polymer is not just a disorderly jumble of chains. If the molecular structure allows neighbouring chains to pack together, and if the chemical groups lead to favourable dipole-dipole, hydrogen-bonding, and/or dispersion forces, portions of the chains can align regularly and exhibit **crystallinity**.

However, the crystallinity of a polymer is very different from the crystal structures we discussed in Chapter 11. In those structures, the orderly array extends over many atoms or molecules, and the unit cell includes at least one atom or molecule. In contrast, the orderly regions of a polymer rarely involve even one whole molecule (Figure 22.2). At best, a polymer is **semicrystalline**, because parts of the molecule align with parts of neighbouring molecules (or with other parts of the same chain), while most of the chain remains as a random coil.

Viscosity of Pure and Dissolved Polymers

Some of the most important uses of polymers arise from their ability to change the **viscosity** of a solvent in which they are dissolved and to undergo temperature-dependent changes in their own viscosity.

Dissolved Polymers When an appreciable amount of polymer (about 5 to 15 mass %) dissolves, the viscosity of the solution is much higher than the viscosity of the pure solvent. In fact, polymers are added to increase the viscosity of many common materials, such as motor oil, paint, and salad dressing. As the random coil of a

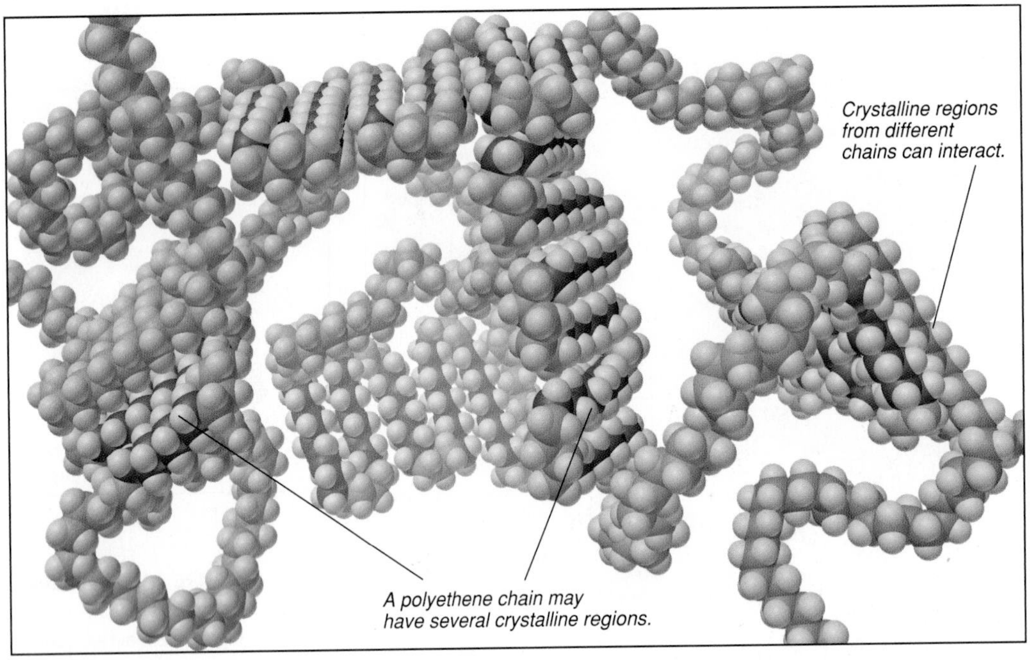
Crystalline regions
from different
chains can interact.
A polyethene chain may
have several crystalline regions.

FIGURE 22.2 The semicrystallinity of a polymer chain. Several crystalline regions (*darker colour*) have randomly coiled regions between them. Crystalline regions from nearby chains (*red* and *yellow*) align with crystalline regions in the main chain.

polymer moves through a solution, solvent molecules are attracted to it through intermolecular forces (Figure 22.3). Thus, as the solution flows, the polymer coil drags along solvent molecules that are interacting with other solvent molecules and other coils, and flow is reduced. Increasing the polymer concentration increases the viscosity because coils are more likely to become entangled.

Viscosity is a property of a particular polymer-solvent pair at a given temperature. The size (radius of gyration) of a random coil in solution increases with molar mass, and so does the viscosity. To improve polymer manufacture, chemists have developed equations to predict the viscosity of polymers of different molar masses in a variety of solvents.

Pure Polymers Intermolecular forces also play a major role in the flow of a pure polymer sample. At temperatures high enough to melt them, many polymers exist as viscous liquids, flowing more like honey than water. The forces between the chains, as well as the entangling of chains, hinder the flow. As the temperature decreases, the intermolecular attractions exert a greater effect, and the eventual result is a rigid solid. If the chains do not crystallize, the resulting material is called a *polymer glass*. The transition from a liquid to a glass occurs over a narrow temperature range (10°C to 20°C) for a given polymer; the temperature at the midpoint of the range is called the **glass transition temperature (T_g)**. Like window glass, many polymer glasses are transparent, such as the polyphenylethene, also known as polystyrene, used in drinking cups and the polycarbonate used in eyeglass lenses.

Plasticity of Polymers

The flow-related properties of polymers give rise to their familiar plastic mechanical behaviour, or **plasticity**. The word *plastic* refers to a material that, when deformed, retains its new shape; in contrast, when an *elastic* object is deformed, it returns to its original shape. Many polymers can be deformed (stretched, bent, and/or twisted) when warm and retain their deformed shape when cooled. This allows them to be made into milk bottles, car parts, and countless other everyday objects.

Molecular Architecture of Polymers

A polymer's architecture—its overall spatial layout and molecular structure—is crucial to its properties. In addition to linear chains, more complex architectures arise through branching and crosslinking.

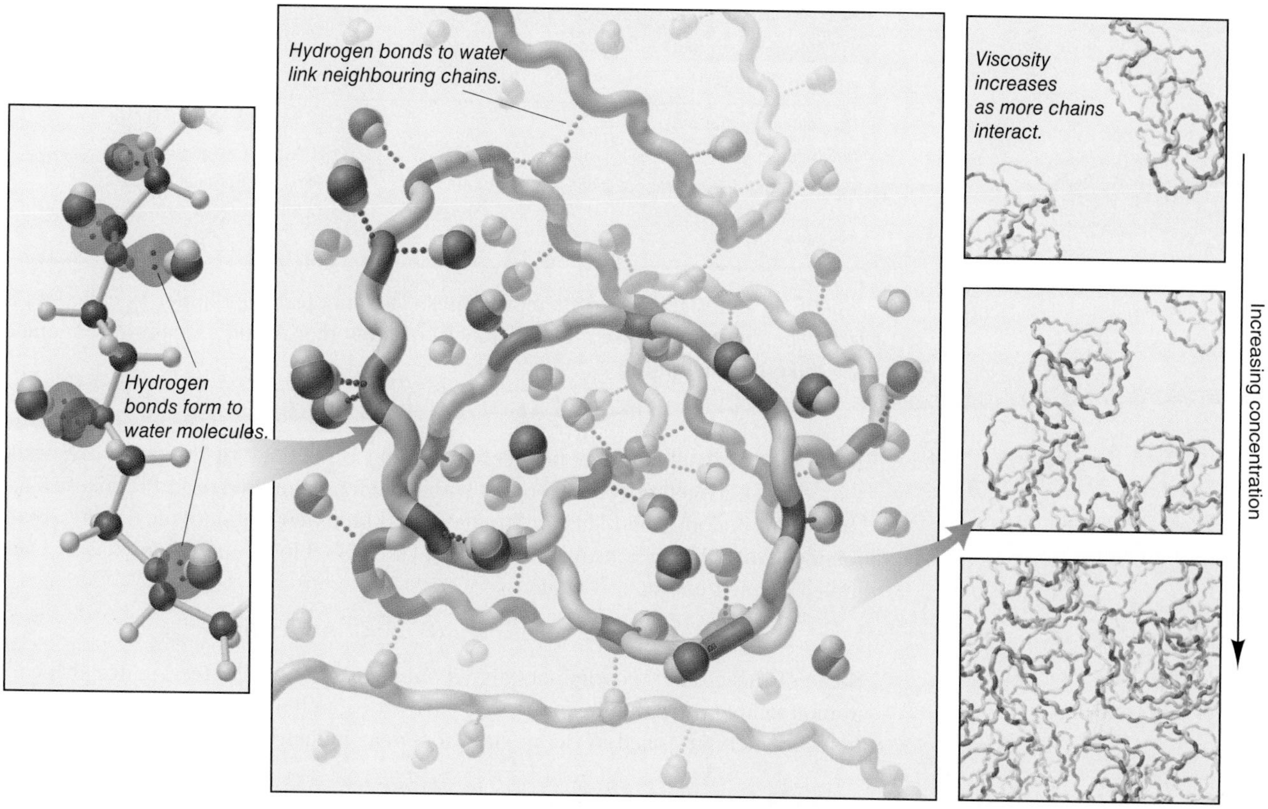

FIGURE 22.3 The viscosity of a polymer in aqueous solution. A section of a polyethane-1,2-diol chain (*left*) forms hydrogen bonds between the lone pairs of the chain's O atoms and the H atoms of water molecules. A polymer chain (*blue and red coiled rod, centre*) forms many hydrogen bonds with water molecules that allow the chain to interact with other chains. As the concentration of polymer increases (*right*), the viscosity of the solution increases.

Branches Smaller chains appended to a polymer backbone are called **branches**. As the number of branches increases, the chains cannot pack as well, so the degree of crystallinity decreases and the polymer is less rigid. A small amount of branching occurs as a side reaction in the preparation of high-density polyethene (HDPE), but HDPE is still largely linear and rigid enough for use in milk containers. In contrast, much more branching is intentionally induced to prepare low-density polyethene (LDPE). The chains cannot pack well, so crystallinity is low. LDPE is used in flexible, transparent food storage bags.

Dendrimers are prepared from monomers with three or more attachment points, so each monomer forms branches. In essence, then, dendrimers have no backbone and consist only of branches. A dendrimer has a constantly increasing number of branches and an incredibly large number of end groups at its outer edge. Chemists use dendrimers to bind one polymer to another to create films and fibres and to deliver drugs in medical applications.

Crosslinks Branches that link one chain to another are called **crosslinks**. The extent of crosslinking leads to remarkable differences in properties. A small degree of crosslinking often yields a *thermoplastic* polymer, one that still flows at high temperatures. As the extent of crosslinking increases, however, a thermoplastic polymer is transformed into a *thermoset* polymer, one that can no longer flow because it has become a single network. Below their glass transition temperatures, some thermosets are extremely rigid and strong, making them ideal as matrix materials in high-strength composites (*see photo*).

Bicycle helmet containing a thermoset polymer

Above their glass transition temperatures, many thermosets become **elastomers**, polymers that can be stretched and immediately spring back to their initial shapes when released, like a trampoline or an elastic band. When an elastic band is stretched, individual polymer chains flow for only a short distance before the connectivity of the network returns them to their original positions. Table 22.2 lists some elastomers.

Impact of Monomer Sequence

Differences in monomer sequences influence polymer properties as well. A *homopolymer* consists of one type of monomer (A—A—A—A—A— ···), whereas a

TABLE 22.2 Some Common Elastomers

Name	T_g (°C)	Uses
Poly(dimethyl siloxane), PDMS	−123	Breast implants
Polybutadiene	−106	Elastic bands
Poly(2-methyl-1,3-butadiene) (polyisoprene rubber, IR)	−65	Surgical gloves
Poly(2-chlorobuta-1,3-diene) (neoprene, chloroprene rubber, CR)	−43	Footwear; medical tubing

copolymer consists of two or more types. The simplest copolymer is called an *AB block copolymer*, because a chain (block) of monomer A and a chain of monomer B are linked at one point:

$$\cdots\text{—A—A—A—A—A—B—B—B—B—B—}\cdots$$

If the intermolecular forces between the A and B portions of the chain are weaker than those between different regions within each portion, the A and B portions form their own random coils. This ability makes AB block copolymers ideal *adhesives* for joining two polymer surfaces covalently. An ABA block copolymer has A chains linked at each end of a B chain:

$$\cdots\text{—A—A—A—B—(B)}_n\text{—B—A—A—A—}\cdots$$

Some of these block copolymers act as *thermoplastic elastomers*, materials shaped at high temperature that become elastomers at room temperature. Many thermoplastic elastomers are used in the modern footwear industry.

SUMMARY OF SECTION 22.1

- Polymers are extremely large molecules that are made of repeat units called monomers. They can adopt the shape of a random coil as a result of intermolecular forces.
- A polymer sample has an average molar mass because it consists of chains with a range of lengths.
- The high viscosity of a polymer arises from attractions between chains or, in a dissolved polymer, between chains and solvent.
- By varying the degrees of branching, crosslinking, and ordering (crystallinity), chemists tailor polymers with specific properties.

22.2 Synthetic Macromolecules

In the previous section, we saw that polymers are extremely large molecules that consist of many monomeric repeat units. In this section, we see how polymers are named and discuss the two types of organic reactions that link monomers covalently into a chain. To name a polymer, just add the prefix *poly-* to the monomer name, as in *polyethene* or *polyphenylethene*. When the monomer has a two-word name or numbers, parentheses are used, as in *poly(ethenyl ethanoate)* or *poly(1,1-dichloroethene)*

The two major types of reaction processes that form synthetic polymers lend their names to the resulting classes of polymer: addition and condensation.

Addition Polymers

Addition polymers form when monomers undergo an addition reaction with one another. These substances are also called *chain-reaction* (or *chain-growth*) *polymers* because, as each monomer adds to the chain, it forms a new reactive site to continue the process. The monomers of most addition polymers have the following grouping:

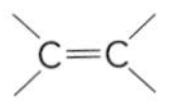

As you can see from Table 22.3, the essential chemical differences between a plastic bag, an acrylic sweater, and a bowling ball are due to the different groups that are attached to the double-bonded C atoms of the monomer.

The *free-radical polymerization* of ethene ($CH_2{=}CH_2$) to polyethene, also known as polyethylene, is a simple example of the addition process (Figure 22.4). The monomer reacts to form a *free radical*, a species with an unpaired electron, which seeks an electron from another monomer to form a covalent bond:

Step 1. The process begins when an *initiator*, usually a peroxide, generates a free radical.

Step 2. The free radical attacks the π bond of an ethene molecule, forming a σ bond with one of the p electrons and leaving the other unpaired, as the new free radical.

Step 3. This new free radical then attacks the π bond of another ethene molecule, joining it to the chain end and allowing the backbone of the polymer to grow. As each ethene adds, it leaves an unpaired electron on the growing end to find an electron "mate" and make the chain one repeat unit longer.

Step 4. This process stops when two free radicals form a covalent bond or when a very stable free radical is formed by the addition of an *inhibitor* molecule.

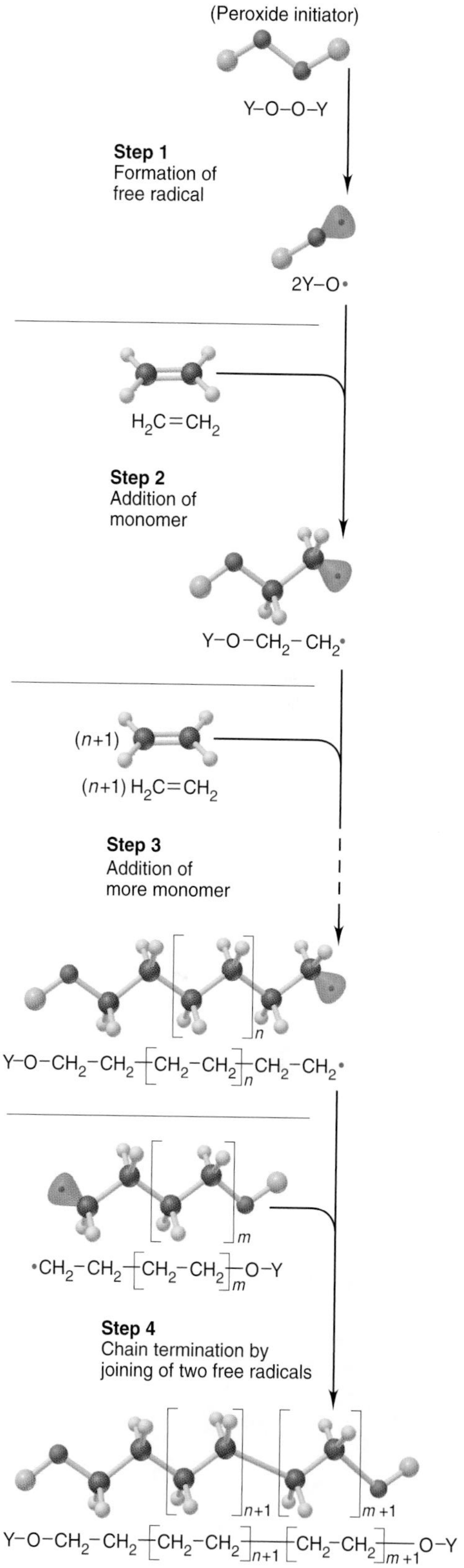

FIGURE 22.4 Steps in the free-radical polymerization of ethene (also known as ethylene)

Recent progress in controlling the high reactivity of free-radical species promises an even wider range of polymers. In one method, polymerization is initiated by the formation of a cation (or anion) instead of a free radical. The cationic (or anionic) reactive end of the chain attacks the π bond of another monomer to form a new cationic (or anionic) end, and the process continues.

The most important polymerization reactions take place under relatively mild conditions through the use of catalysts that incorporate transition metals. In 1963, Karl Ziegler and Giulio Natta received the Nobel Prize in Chemistry for developing *Ziegler-Natta catalysts*, which employ an organoaluminum compound, such as $Al(C_2H_5)_3$, and the tetrachloride of titanium or vanadium. Today, chemists use organometallic catalysts that are *stereoselective* to create polymers whose repeat units have groups that are spatially oriented in particular ways. Through the use of catalysts, such as the Ziegler-Natta or the *Phillips catalyst* [prepared by depositing chromium(VI) oxide on silica], polyethene chains with molar masses of 10^4 g/mol to 10^5 g/mol are made by varying conditions and reagents.

Similar methods are used to make polypropenes that have all the CH_3 groups of the repeat units oriented either on one side of the chain or on alternating sides:

$$\left[CH_2-\underset{\displaystyle CH_3}{\underset{|}{CH}} \right]_n$$

The different orientations lead to different packing efficiencies of the chains and, thus, different degrees of crystallinity, which, in turn, lead to variations in physical properties such as density, rigidity, and elasticity.

Condensation Polymers

The monomer of a **condensation polymer** must have *two functional groups*; we can designate such a monomer as A—**R**—B (where A and B may or may not be the same, and **R** is the rest of the molecule). Most commonly, the monomers link when an A group on one undergoes a *dehydration-condensation reaction* with a B group on another:

$$\frac{1}{2}n\text{H—A—}\mathbf{R}\text{—B—OH} + \frac{1}{2}n\text{H—A—}\mathbf{R}\text{—B—OH} \xrightarrow{-(n-1)\text{HOH}} \text{H}\!\left[\text{A—}\mathbf{R}\text{—B}\right]_n\!\text{OH}$$

TABLE 22.3 Some Major Addition Polymers

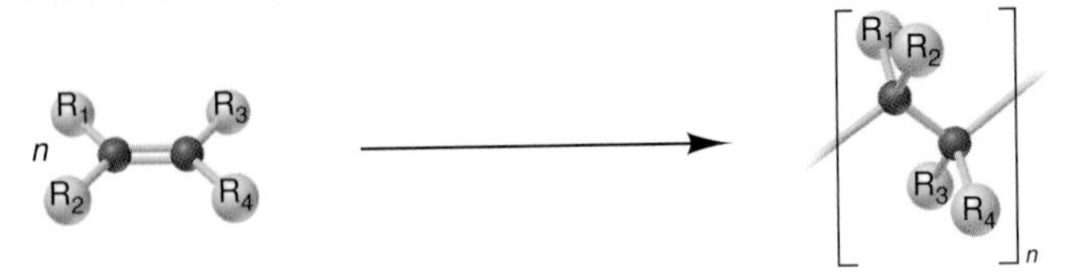

Monomer		Polymer	Applications
$H_2C{=}CH_2$		Polyethene (polyethylene)	Plastic bags, bottles, toys
$F_2C{=}CF_2$	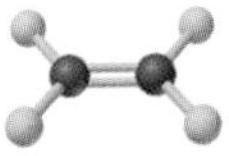	Polytetrafluoroethene (polytetrafluoroethylene)	Cooking utensils (e.g., Teflon)
$H_2C{=}CH{-}CH_3$	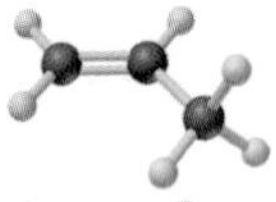	Polypropene (polypropylene)	Carpeting (indoor-outdoor), bottles
$H_2C{=}CH{-}Cl$		Polychloroethene [poly(vinyl chloride)]	Plastic wrap, garden hose, indoor plumbing
$H_2C{=}CH{-}C_6H_5$	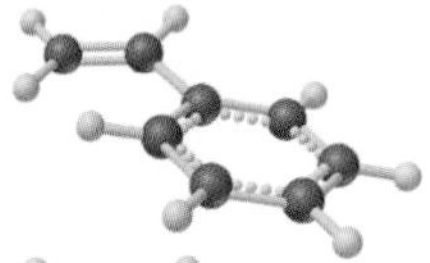	Polyphenylethene (polystyrene)	Insulation, furniture, packing materials
$H_2C{=}CH{-}CN$	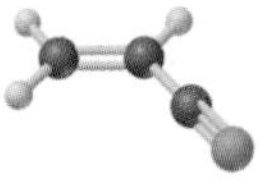	Poly(prop-2-enenitrile) (polyacrylonitrite)	Yarns (e.g., Orlon, Acrilan), fabrics, wigs
$H_2C{=}CH{-}O{-}C({=}O){-}CH_3$	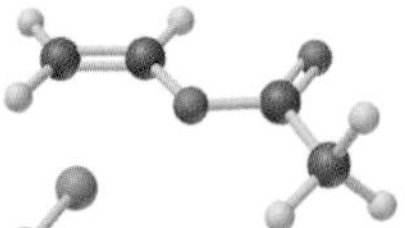	Poly(ethenyl ethanoate) [poly(vinyl acetate)]	Adhesives, paints, textile coatings, computer disks
$H_2C{=}CCl_2$	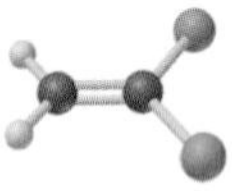	Poly(1,1-dichloroethene) [poly(vinylidene chloride)]	Food wrap (e.g., Saran)
$H_2C{=}C(CH_3){-}C({=}O){-}O{-}CH_3$	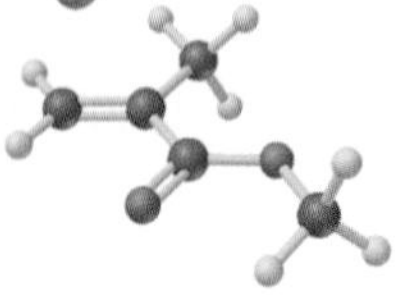	Poly(methyl 2-methylpropenoate) [poly(methyl methacrylate)]	Glass substitute (e.g., Lucite, Plexiglas), bowling balls, paint

FIGURE 22.5 The formation of nylon 6-6. In the laboratory, the six-carbon diacid chloride, which is more reactive than the diacid, is used as one monomer; the polyamide forms between the two liquid phases.

Many condensation polymers are *copolymers*, those consisting of two or more different repeat units (Section 22.1). Two major types are polyamides and polyesters.

1. *Polyamides.* Condensation of carboxylic acid and amine monomers forms *polyamides (nylons)*. One of the most common is *nylon 6-6* (also referred to as nylon 66), which is manufactured by mixing equimolar amounts of a six-carbon diamine (1,6-diamino-hexane) and a six-carbon diacid (1,6-hexanedioic acid). The basic amine reacts with the acid to form a "nylon salt." Heating drives off water and forms the amide bonds:

$$n\,HO{-}\overset{O}{\overset{\|}{C}}{-}(CH_2)_4{-}\overset{O}{\overset{\|}{C}}{-}OH \;+\; n\,H_2N{-}(CH_2)_6{-}NH_2$$

$$\xrightarrow[-(2n-1)H_2O]{\Delta} HO{-}\left[\overset{O}{\overset{\|}{C}}{-}(CH_2)_4{-}\overset{O}{\overset{\|}{C}}{-}NH{-}(CH_2)_6{-}NH\right]_n{-}H$$

In the laboratory, this nylon is made without heating by using a more reactive acid component (Figure 22.5). Covalent bonds within the chains give nylons great strength, and hydrogen bonds between the chains give them great flexibility (see Table 2.8). About half of all nylons are made to reinforce automobile tires; the other nylons are used for rugs, clothing, and fishing line, among other things.

2. *Polyesters.* Condensation of carboxylic acid and alcohol monomers forms *polyesters*. Dacron, a popular polyester fibre, is woven from polymer strands that are formed when equimolar amounts of 1,4-benzenedicarboxylic acid and ethane-1,2-diol react. Blending Dacron with various amounts of cotton gives fabrics that are durable, easily dyed, and crease resistant. Extremely thin Mylar films, used for recording tape and food packaging, are also made from this polymer.

Inorganic Polymers We already know that some synthetic polymers have inorganic backbones. In Chapter 13, we discussed the silicones, which are polymers with the repeat unit —$(R_2)Si$—O—. Depending on the chain crosslinks and the R groups, silicones range from oily liquids to elastic sheets to rigid solids and have applications that include artificial limbs and spacesuits. Polyphosphazenes exist as flexible chains, even at low temperatures, and have the repeat unit —$(R_2)P{=}N$—.

SUMMARY OF SECTION 22.2

- Polymers are extremely large molecules made of repeat units called monomers.
- Addition polymers are formed from unsaturated monomers that commonly link through free-radical reactions.
- Most condensation polymers are formed by linking monomers, each with two functional groups, through a dehydration-condensation reaction.
- Reaction conditions, catalysts, and monomers can be varied to produce polymers with different properties.

22.3 Biological Macromolecules

With about 3200 companies employing over 90 000 workers, Canada's $26.3-billion plastics industry is a sophisticated, multifaceted sector that encompasses the manufacturing of plastic products, as well as machinery, moulds, and resins.

[Source: Statistics Canada, 2011]

The monomer-polymer theme was being played out in nature eons before humans employed it to such great advantage. Biological macromolecules are condensation polymers created by nature's reaction chemistry and improved through evolution. These remarkable molecules are the best demonstration of the versatility of carbon and its handful of atomic partners.

Natural polymers, such as polysaccharides, proteins, and nucleic acids, are the "building blocks of life." Some have structures that make wood strong, hair curly, fingernails hard, and wool flexible. Others speed up the myriad reactions that occur in every cell or defend the body against infection. Still others possess the genetic information that organisms need to forge other biomolecules. Remarkable as these giant molecules are, the functional groups of their monomers and the reactions that link them are identical to those of other, smaller organic molecules. Moreover, as you will see in Section 22.4, the same intermolecular forces that dissolve smaller molecules stabilize these giant molecules in the aqueous medium of the cell.

Sugars and Polysaccharides

Raymond Lemieux (Figure 22.6) is considered the pioneer of modern carbohydrate chemistry. He and George Huber were the first to synthesize sucrose (ordinary sugar). Lemieux perfected techniques that allowed him to synthesize simple carbohydrates in their correct three-dimensional configurations. In addition to being a pioneer in configurational determination by nuclear magnetic resonance spectroscopy, he started several research companies, including Chembiomed Ltd. with the University of Alberta. His discoveries led to many important medical applications, including improved treatments for leukemia and hemophilia, as well as the development of new antibiotics, blood reagents, and antirejection drugs for organ transplants. As well, he synthesized the carbohydrate sequences for six different blood group determinants—a great achievement that was very important to the chemistry of immunology. He also developed the concept of the anomeric effect, which is the

FIGURE 22.6 Raymond Urgel Lemieux (1920–2000) was a world-famous Canadian carbohydrate chemist. He and George Huber were the first to synthesize sucrose.

thermodynamic preference for polar groups, shown below as the blue X bonded to C-1 (shown in *red* to indicate the axial position; OAc = ethanoate, also known as acetate):

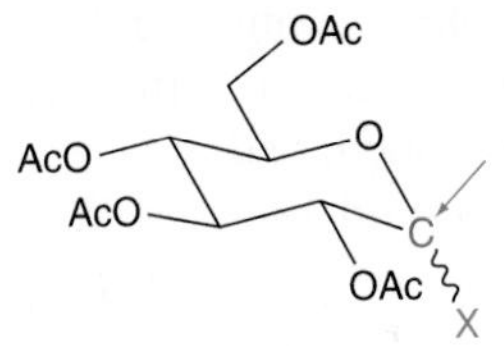

Monomer Structure and Linkage Glucose and other simple sugars, from the three-carbon *trioses* to the seven-carbon *heptoses*, are called **monosaccharides** and consist of carbon chains with attached hydroxyl and carbonyl groups. In addition to their roles as individual molecules engaged in energy metabolism, they serve as the monomer units of **polysaccharides**. Most natural polysaccharides are formed from five-carbon and six-carbon units. In aqueous solution, *an alcohol group and the aldehyde (or ketone) group of the same monosaccharide react with each other to form a cyclic molecule with either a five-membered or a six-membered ring.* For example, a molecule of glucose undergoes an internal addition reaction between the aldehyde group of C-1 and the alcohol group of C-5 (Figure 22.7A).

When two monosaccharides undergo a dehydration-condensation reaction, a **disaccharide** forms. For example, sucrose (table sugar) is a disaccharide of glucose (linked at C-1) and fructose (linked at C-2) (Figure 22.7B); lactose (milk sugar) is a disaccharide of glucose (C-1) and galactose (C-4); and maltose, used in brewing and as a sweetener, is a disaccharide of two glucose units (C-1 to C-4).

Types of Polysaccharides A polysaccharide consists of *many* monosaccharide units linked together. The three major natural polysaccharides—cellulose, starch, and glycogen—consist entirely of glucose units, but they differ in the ring positions of the links, in the orientation of certain bonds, and in the extent of cross-linking. Some other polysaccharides contain nitrogen in their attached groups. One of these nitrogen-containing polysaccharides is *chitin* (pronounced "KY-tin"), the main component of the tough, brittle, external skeletons of insects and crustaceans (*see photo*).

Lobster shells contain a polysaccharide called *chitin*.

1. *Cellulose* is the most abundant organic chemical on Earth. More than 50% of the carbon in plants occurs in the cellulose of stems and leaves; wood is largely cellulose, and cotton is more than 90% cellulose. Cellulose consists of long chains of glucose. The great strength of wood is due largely to the countless hydrogen bonds between cellulose chains. The monomers are linked in a

FIGURE 22.7 The structure of glucose in aqueous solution and the formation of a disaccharide

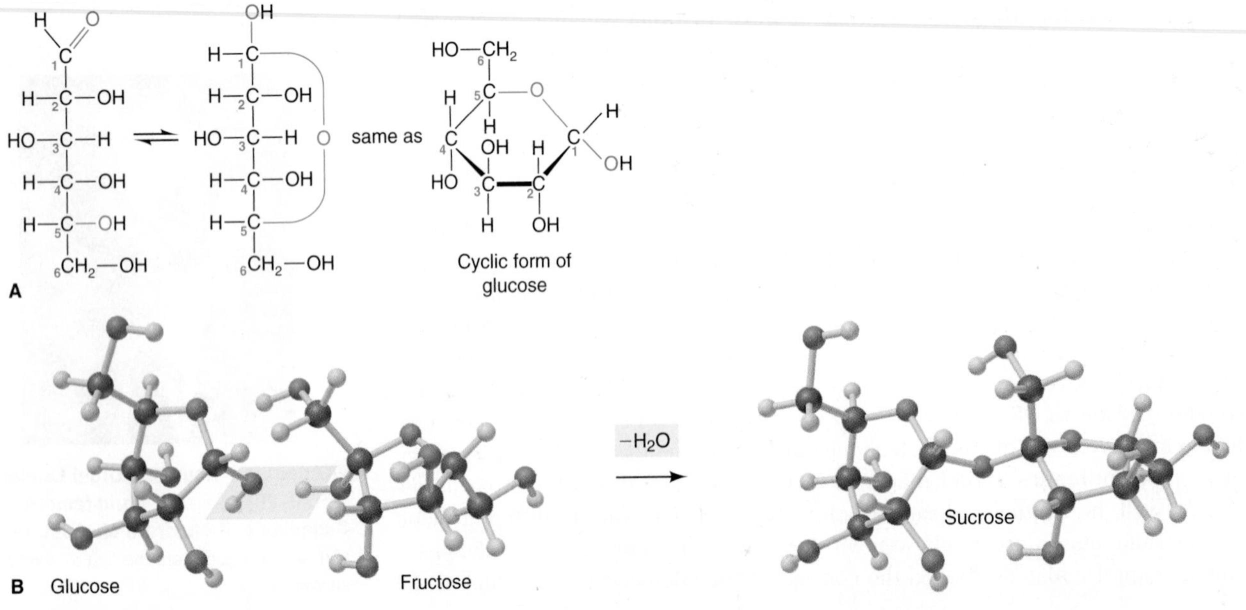

particular way, from C-1 in one unit to C-4 in the next unit. Humans lack the enzymes to break this link, so we cannot digest cellulose (unfortunately!); however, micro-organisms in the digestive tracts of some animals, such as cows, sheep, and termites, can.

2. *Starch* is a mixture of polysaccharides of glucose and serves as an *energy store* in plants. When a plant needs energy, some starch is broken down by hydrolysis of the bonds between units, and the released glucose is oxidized through a multistep metabolic pathway. Starch occurs in plant cells as insoluble granules of amylose, a helical molecule of several thousand glucose units, and amylopectin, a highly branched, bushlike molecule of up to a million glucose units. Most of the glucose units are linked by C-1 to C-4 bonds, as in cellulose, but a different orientation around the chiral C-1 allows our digestive enzymes to break starch down into monomers. A C-6 to C-1 crosslink joins chains every 24 to 30 units:

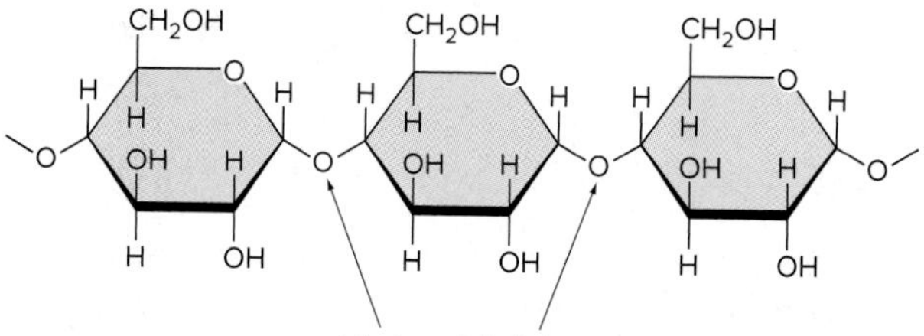

3. *Glycogen* functions as the energy storage molecule in animals. It occurs in liver and muscle cells as large, insoluble granules, which consist of glycogen molecules made from 1000 to more than 500 000 glucose units. In glycogen, these units are linked by C-1 to C-4 bonds, as in starch; however, glycogen is more highly crosslinked than starch, with C-6 to C-1 crosslinks every 8 to 12 units:

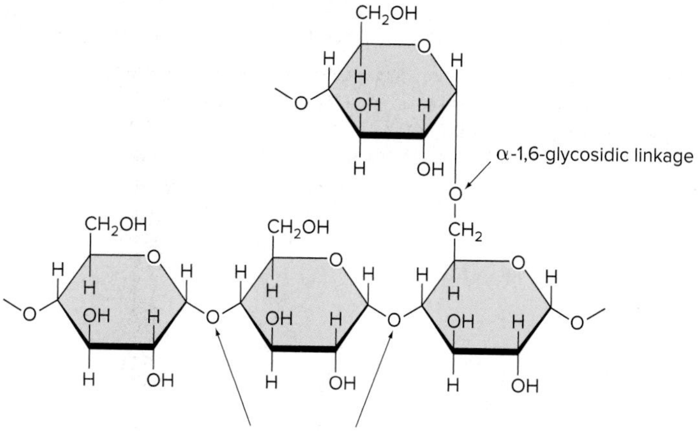

Amino Acids and Proteins

As you saw in Section 22.2, synthetic polyamides (such as nylon 6-6) are formed from two monomers, one with a carboxylic acid or carboxyl group at each end and the other with an amine group at each end. **Proteins**, the polyamides of nature, are unbranched polymers formed from monomers called **amino acids**. *Each amino acid molecule has both a carboxyl group and an amine group.*

Monomer Structure and Linkage An amino acid has both its carboxyl group and its amine group attached to the *α-carbon*, the second C atom in the chain. Proteins are made up of about 20 different types of amino acids, each with its own particular R group, ranging from an H atom to a polycyclic N-containing aromatic structure (Figure 22.8). The R groups play a major role in the shape and function of a protein.

In the aqueous cell fluid, the NH_2 and COOH groups of amino acids are charged because each carboxyl group transfers an H^+ ion to H_2O to form H_3O^+. The H_3O^+

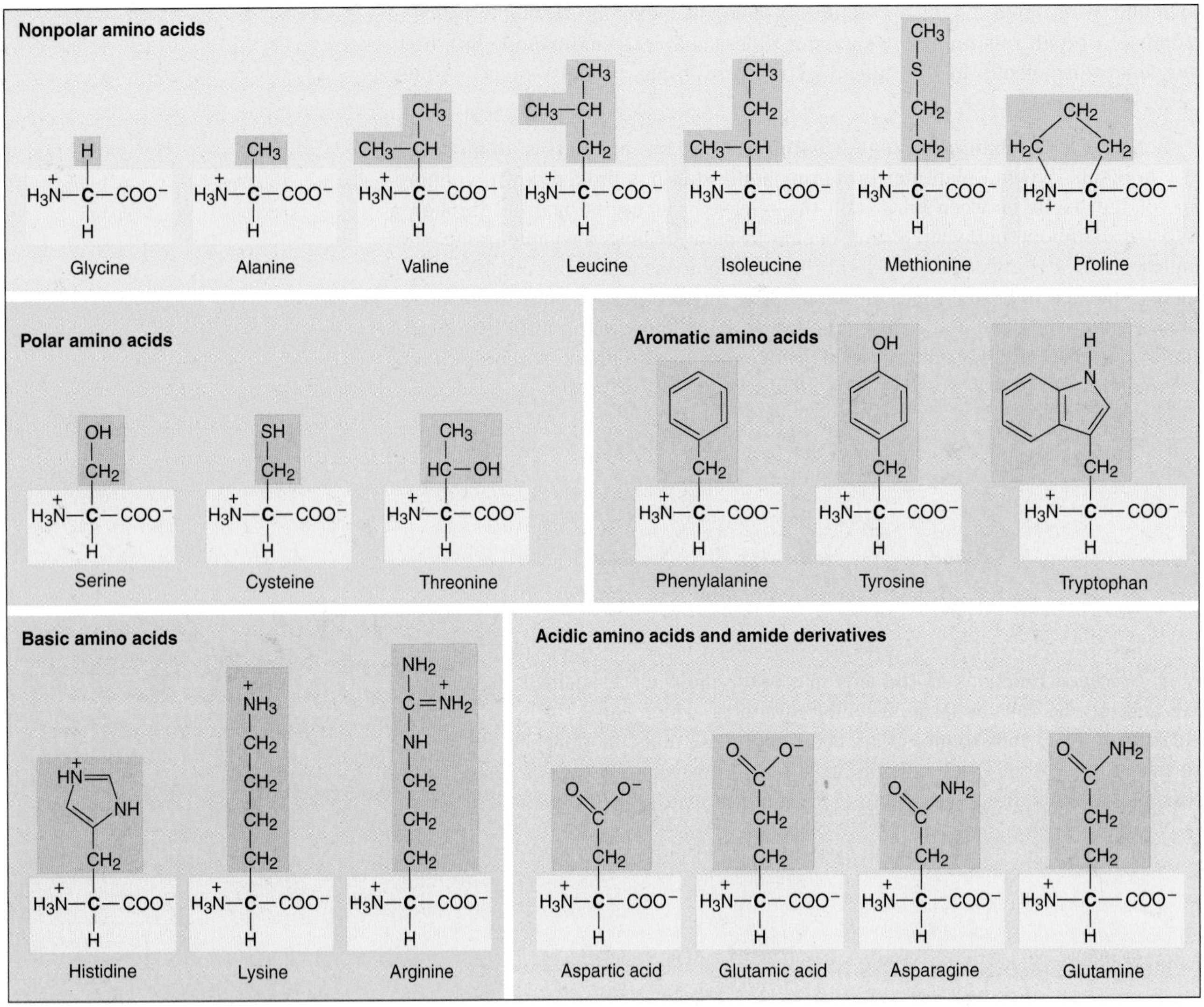

FIGURE 22.8 The common amino acids. The R groups are indicated by *grey screens*, and the α-carbons (*boldface*), with carboxyl and amino groups, are indicated by *yellow screens*. Here the amino acids are shown with the charges they have under physiological conditions.

ions transfer the H^+ to the amine groups. The overall process can be viewed as an intramolecular acid-base reaction:

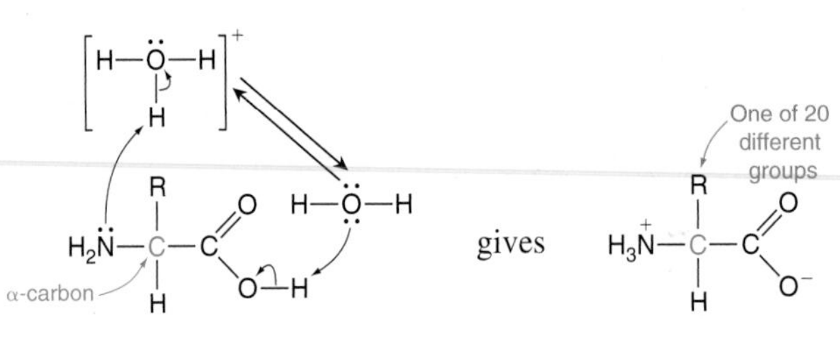

An H atom is the third group bonded to the α-carbon, and the R group (also called the *side chain*) is the fourth group.

Each amino acid is linked to the next through a *peptide (amide) bond.* This bond is formed by a dehydration-condensation reaction in which the carboxyl group of one monomer reacts with the amine group of the next:

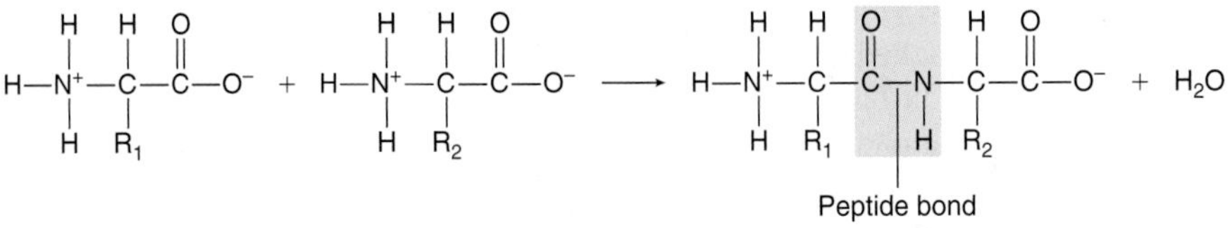

Therefore, as noted previously, the polypeptide chain—the backbone of the protein—has a repeating structure that consists of an *α-carbon bonded to an amide*

group bonded to the next α-carbon bonded to the next amide group, in a continuing pattern (see Figure 22.16). The various R groups dangle from the α-carbons on alternate sides of the chain.

The Hierarchy of Protein Structure *Each type of protein has its own amino acid composition*, specific numbers and proportions of the various amino acids. However, this composition is not what defines a protein's role in the cell; rather, *the sequence of amino acids determines the protein's shape and function*. Proteins contain from about 50 to several thousand amino acids, yet, from a purely mathematical point of view, even a small protein of 100 amino acids has a virtually limitless number of possible sequences of the 20 types of amino acids ($20^{100} \approx 10^{130}$). In reality, though, only a tiny fraction of these possibilities occur in actual proteins. For example, even in an organism as complex as a human, there are only about 10^5 different types of proteins.

A protein folds into its native shape as it is being synthesized in a cell. Some shapes are simple: long helical tubes or undulating sheets. Other shapes are far more complex: baskets, Y shapes, spheroid blobs, and countless other globular forms. Biochemists define a hierarchy for the overall structure of a protein (Figure 22.9):

1. *Primary (1°) structure*, the most basic level, refers to the sequence of covalently bonded amino acids in the polypeptide chain.
2. *Secondary (2°) structure* refers to sections of the chain that, as a result of intermolecular forces (including hydrogen bonding) between nearby peptide groupings, adopt shapes called α-helices and β-pleated sheets. Disulfide bonds (also called SS bonds or disulfide bridges), which are usually derived by the coupling of two thiol groups, are covalent bonds that also generate a secondary structure.
3. *Tertiary (3°) structure* refers to the three-dimensional folding of the whole polypeptide chain. Research shows that, in many proteins, certain folding patterns form characteristic regions that play a role in the protein's function, for example, binding a hormone, attaching to a membrane, or forming a membrane channel.
4. *Quaternary (4°) structure*, the most complex level, occurs in proteins made up of several polypeptide chains (subunits) and refers to the way that the chains assemble into the overall multi-subunit protein. Hemoglobin, for example, consists of four subunits arranged as shown in Figure 22.9.

Note that *only the 1° and some 2° structures (like disulfide or SS bonds) involve covalent bonds; most 2°, 3°, and 4° structures rely primarily on intermolecular forces*, which we will discuss in Section 22.4.

FIGURE 22.9 The structural hierarchy of proteins

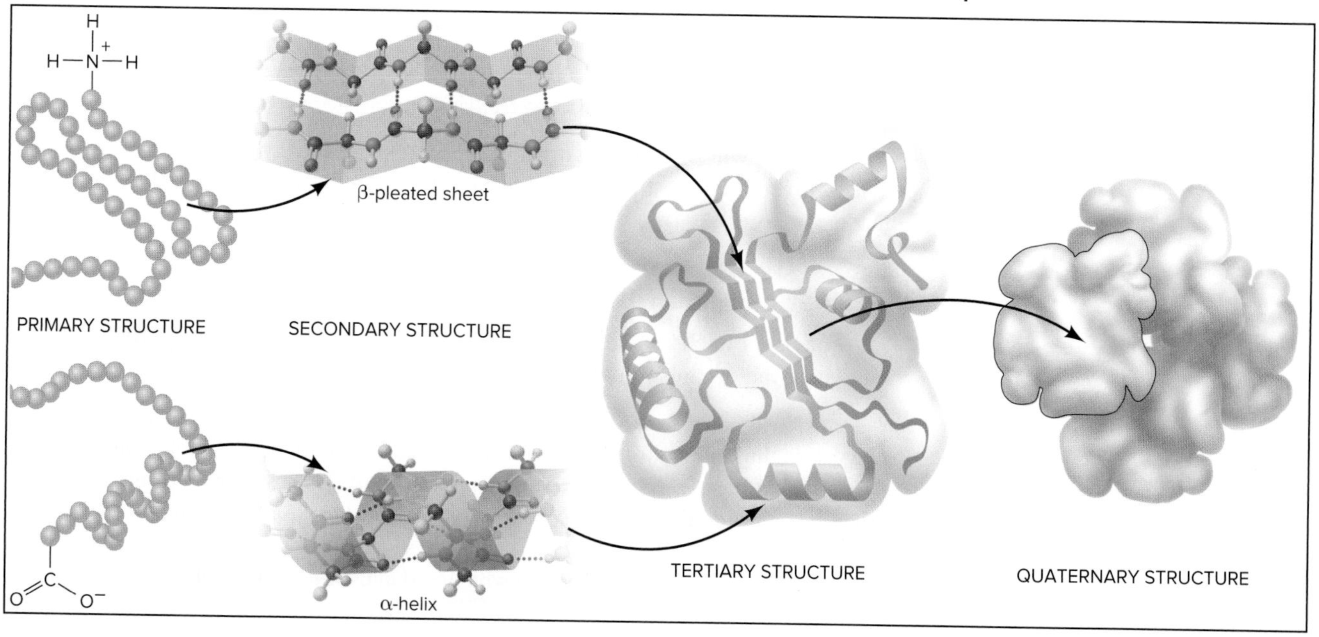

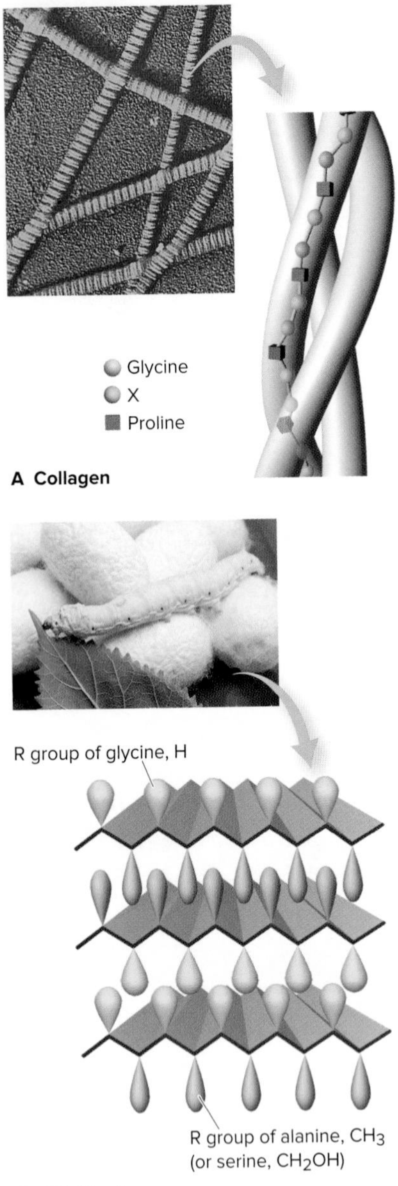

FIGURE 22.10 The shapes of fibrous proteins

The Relation between Structure and Function Two broad classes of proteins differ in the complexity of their amino acid compositions and sequences and, therefore, in their structures and functions:

1. *Fibrous proteins* have relatively simple amino acid compositions and correspondingly simple structures. They are key components of hair, wool, skin, and connective tissue—materials that require strength and flexibility. Like synthetic polymers, these proteins have a small number of different amino acids in a repeating sequence.

- *Collagen*, the most common animal protein, makes up as much as 40% of human body weight. More than 30% of its amino acids are glycine (G), and another 20% are proline (P). Collagen is formed of three chains, each an extended helix, that wind tightly around each other. The peptide C═O groups in one chain form hydrogen bonds to the peptide N—H groups in another chain. The result is a long, triple-helical cable with the sequence —G—X—P—G—X—P— and so on (where X is another amino acid), as shown in Figure 22.10A. As the main protein component of tendons, skin, and blood vessel walls, collagen has a high tensile strength; in fact, a 1-mm-thick strand can support a 10 kg weight!
- In *silk fibroin*, secreted by the silk moth caterpillar, more than 85% of the amino acids are glycine, alanine, and serine (Figure 22.10B). Fibroin chain segments bend back and forth, running alongside each other, and form interchain hydrogen bonds to create a *pleated sheet*. Stacks of sheets interact through dispersion forces, which make fibroin strong and flexible but not very extendable—perfect for a silkworm's cocoon.

2. *Globular proteins* have much more complex compositions, often containing all 20 amino acids in varying proportions. As the name implies, globular proteins are typically more rounded and compact, with a wide variety of shapes and a correspondingly wide range of functions, for example, defenders against bacterial invasion, messengers that trigger cell actions, catalysts of chemical change, and membrane gatekeepers that maintain cellular concentrations.

The locations of particular amino acid R groups are crucial to a protein's function. For example, in catalytic proteins, a few R groups form a crevice that closely matches the shape of reactant molecules. These groups typically hold the reactants through intermolecular forces and speed their reaction to products by bringing them together and twisting and stretching their bonds (Section 14.7). Experiment has established that a slight change in one of these critical R groups decreases the catalytic function dramatically. This supports the essential idea that *a protein's amino acid sequence determines its structure, which, in turn, determines its function*:

$$\text{SEQUENCE} \Rightarrow \text{STRUCTURE} \Rightarrow \text{FUNCTION}$$

Next, we will see how the amino acid sequence of every protein in every organism is prescribed by the genetic information that is held within the organism's nucleic acids.

Nucleotides and Nucleic Acids

An organism's nucleic acids construct its proteins. Given that these proteins determine how the organism looks and behaves, no job could be more essential.

Monomer Structure and Linkage **Nucleic acids** are unbranched polymers that consist of linked monomer units called **mononucleotides**, which consist of an N-containing base, a sugar, and a phosphate group. The two types of nucleic acids, ***ribonucleic acid*** (RNA) and ***deoxyribonucleic acid*** (DNA), differ most obviously in the sugar portions of their mononucleotides. RNA contains *ribose*, a five-carbon sugar. DNA contains *deoxyribose*, in which —H substitutes for —OH on the 2′ position of ribose. (Carbon atoms in the sugar portion are given numbers with a prime to distinguish them from carbon atoms in the base.)

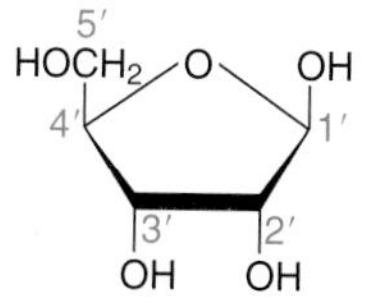

Ribose in RNA

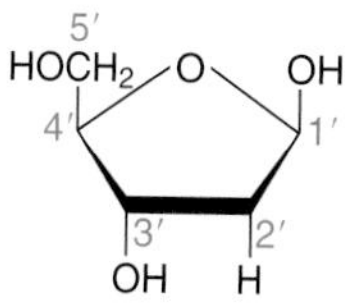

Deoxyribose in RNA

Attached to the sugar molecule at the 1′ carbon is one of four N-containing bases, either a pyrimidine (six-membered ring) or a purine (six- and five-membered rings sharing a side). The pyrimidines are cytosine (C), thymine (T), and uracil (U); thymine occurs only in DNA, and uracil occurs only in RNA. The purines are guanine (G) and adenine (A):

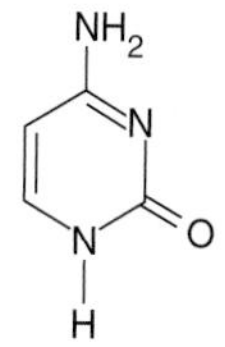

Cytosine (C)
DNA, RNA

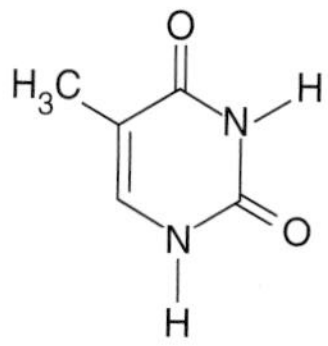

Thymine (T)
DNA

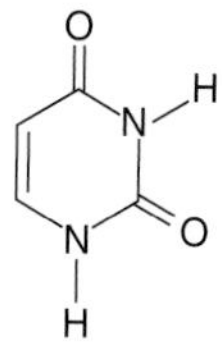

Uracil (U)
RNA

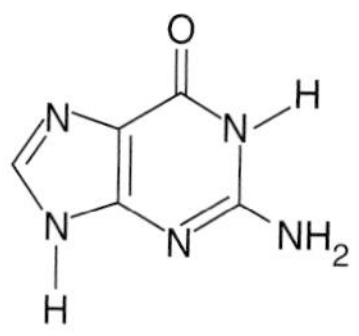

Guanine (G)
DNA, RNA

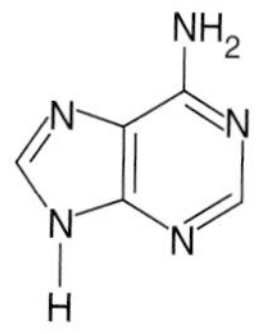

Adenine (A)
DNA, RNA

A phosphate group is attached to the 5′ carbon of the sugar. The cellular precursors that form a nucleic acid are *nucleoside triphosphates* (Figure 22.11A). Dehydration-condensation reactions between nucleoside triphosphates release inorganic diphosphate ($H_2P_2O_7^{2-}$) and create a phosphodiester bond between the phosphate group on the 5′ carbon of one sugar and the —OH group on the 3′ carbon of a second sugar to form a polynucleotide chain. Therefore, *the repeating pattern of the nucleic acid backbone is sugar-phosphate-sugar-phosphate*, and so on (Figure 22.11B), with the bases dangling off the chain, much as R groups dangle off the polypeptide chain of a protein.

The Central Importance of Base Pairing In the nucleus of a cell, DNA exists as two chains wrapped around each other in a **double helix** (Figure 22.12). The polar sugar-phosphate backbone of each DNA chain faces the watery outside, and the *bases on each chain form hydrogen bonds to each other* in the DNA core, which holds the two chains together. A double-helical DNA molecule may contain many millions of hydrogen-bonded bases in specified pairs. Two features of these **base pairs** are crucial to the structure and function of DNA:

- A pyrimidine and a purine are always paired, which gives the double helix a constant diameter.
- Each base is always paired with the same partner: A with T, and G with C. Thus, *the base sequence on one chain is the complement of the base sequence on the other*. For example, the sequence A—C—T on one chain is *always* paired with T—G—A on the other: A with T, C with G, and T with A. The pairing is due to these bases having just the right geometry to hydrogen bond with each other.

Each DNA molecule is folded into a tangled mass that forms one of the cell's *chromosomes*. The DNA molecule is amazingly long and thin. If the largest human chromosome were stretched out, it would be 4 cm long; in the cell nucleus, however, it is wound into a structure that is only 5 nm in diameter—one 8-millionth the length! Segments of the DNA chains are the *genes* that contain the chemical information for synthesizing the organism's proteins.

Protein Synthesis *The information content of a gene resides in its base sequence.* In the **genetic code**, each base acts as a "letter," each three-base sequence acts as a "word," and *each word codes for a specific amino acid*. For example, the sequence

FIGURE 22.11 Nucleic acid precursors and their linkage

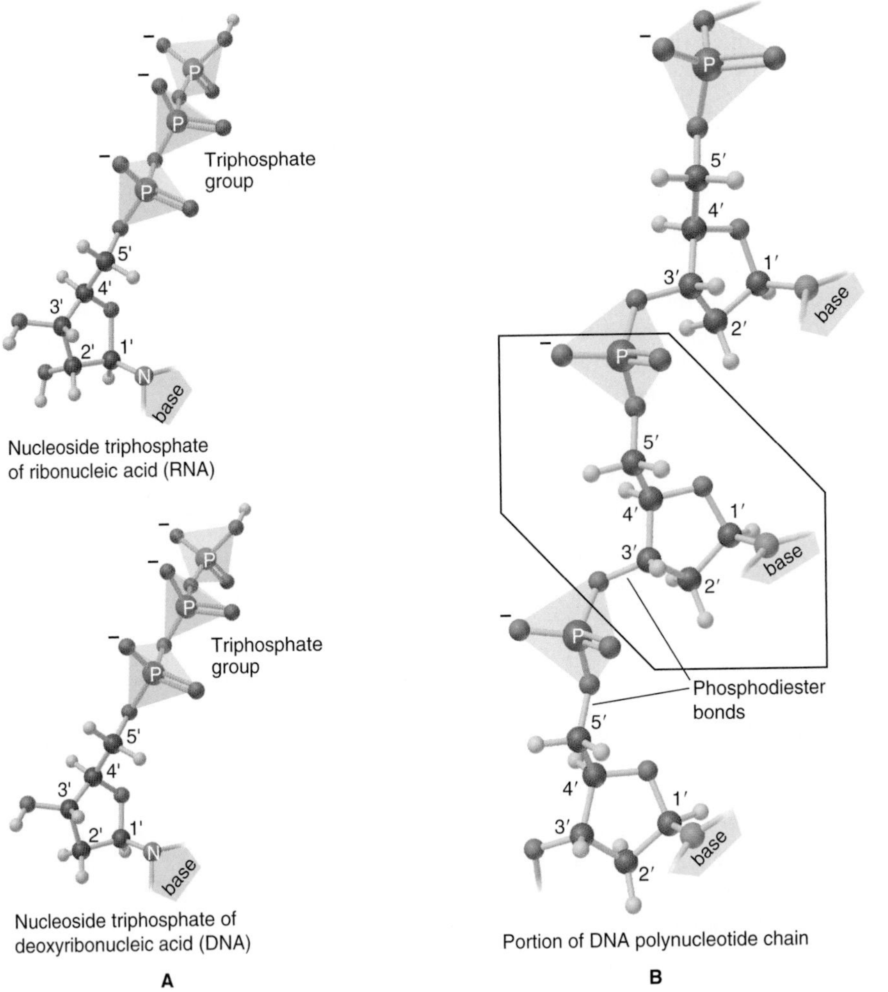

FIGURE 22.12 The double helix of DNA and a section showing the base pairs

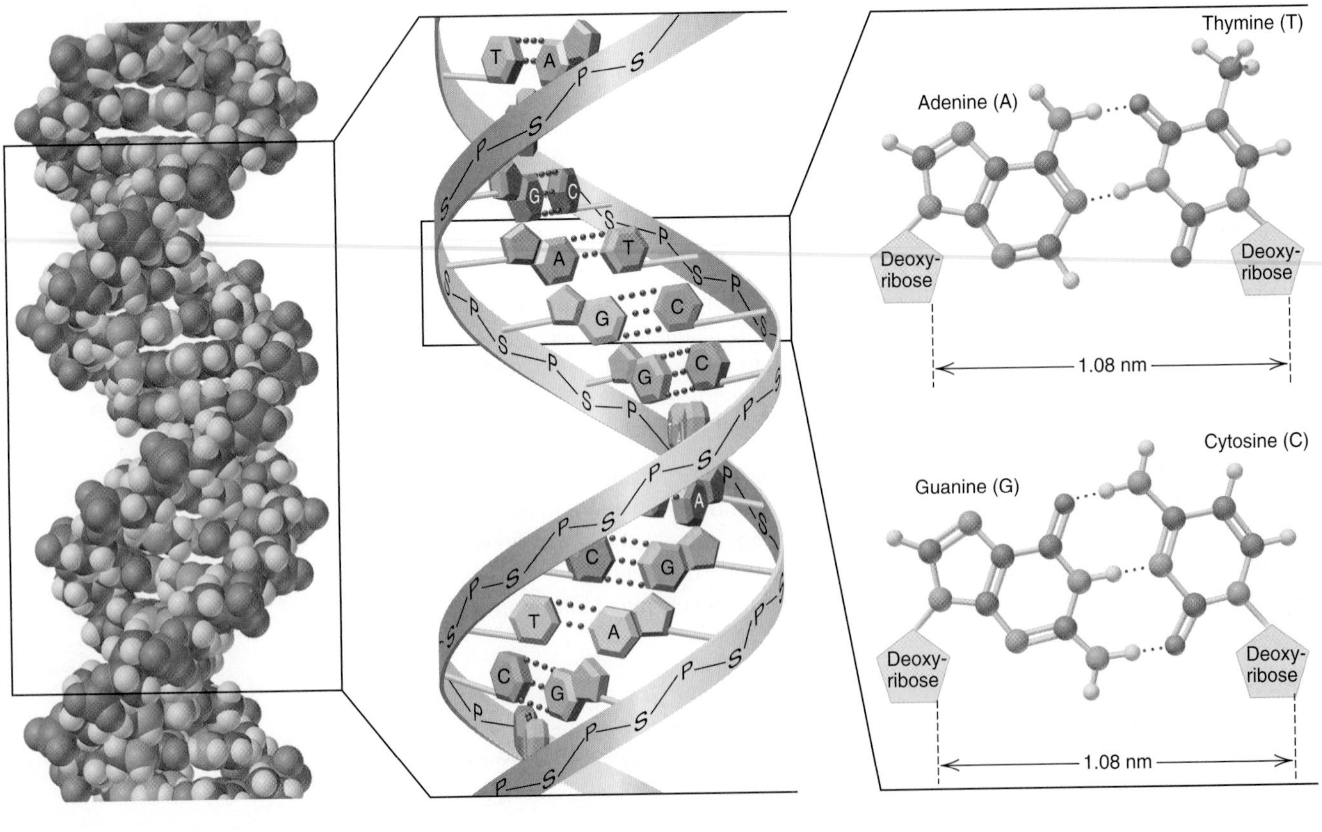

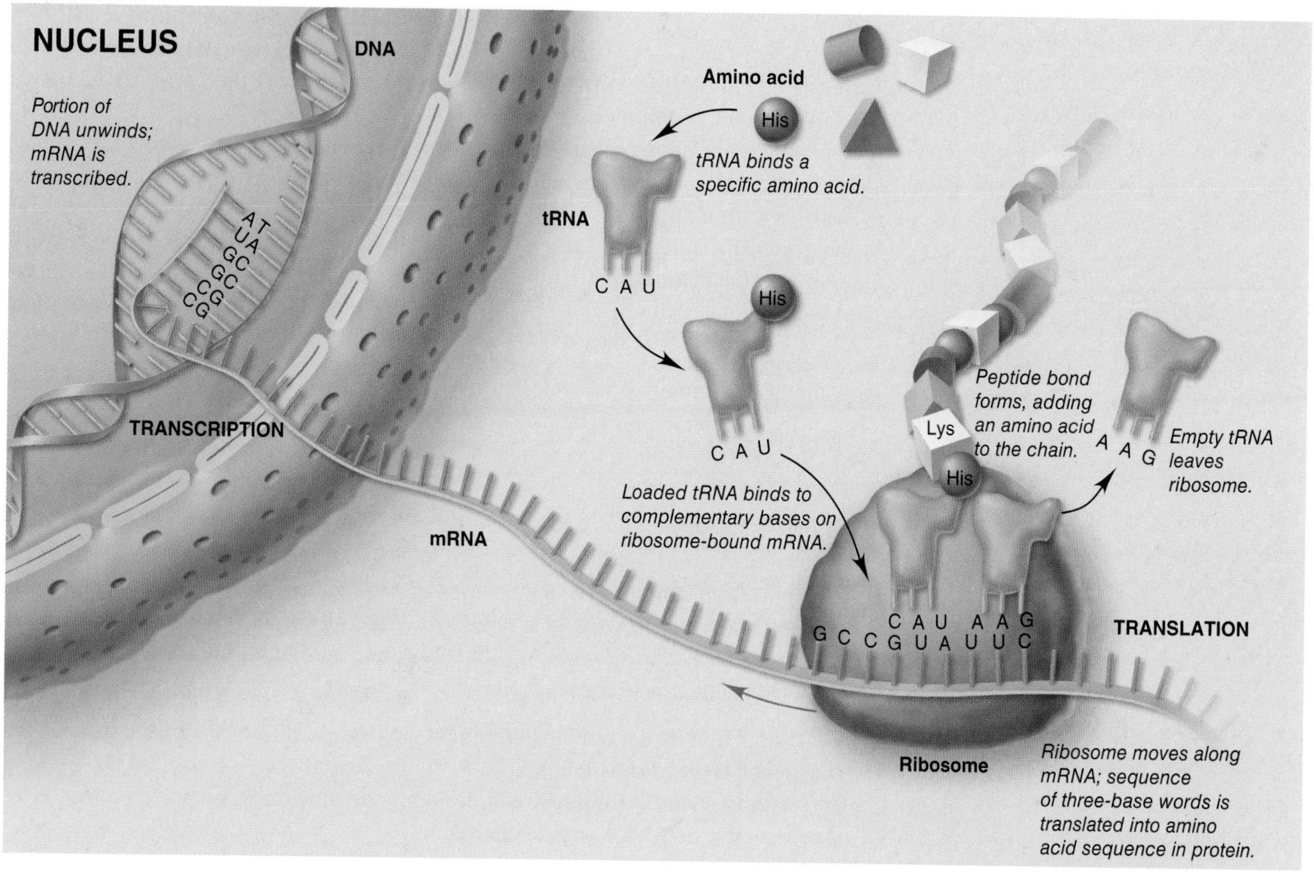
NUCLEUS
DNA
Portion of DNA unwinds; mRNA is transcribed.
Amino acid
His
tRNA binds a specific amino acid.
tRNA
C A U
His
TRANSCRIPTION
C A U
Loaded tRNA binds to complementary bases on ribosome-bound mRNA.
mRNA
Peptide bond forms, adding an amino acid to the chain.
Lys
His
A A G
Empty tRNA leaves ribosome.
C A U A A G
G C C G U A U U C
TRANSLATION
Ribosome
Ribosome moves along mRNA; sequence of three-base words is translated into amino acid sequence in protein.

FIGURE 22.13 Key stages in protein synthesis

C—A—C codes for the amino acid histidine, the sequence A—A—G codes for lysine, and so on. Through a complex series of interactions (greatly simplified in the overview in Figure 22.13), one amino acid at a time is positioned and linked to the next amino acid to gradually synthesize a protein. To have a full appreciation of this amazingly complex process, keep in mind that *it occurs largely through hydrogen bonding between base pairs.*

There are two main processes in protein synthesis: transcription and translation. DNA occurs in the cell nucleus, but the genetic message is decoded outside the nucleus in the cytoplasm of the cell. Therefore, the information must be sent from the DNA to the synthesis site. RNA serves this messenger role.

Transcription is the process of producing a *messenger RNA* (mRNA) copy of the DNA information:

1. A portion of the DNA is temporarily unwound. One chain segment acts as a template for the formation of a complementary chain of mRNA, which is made by linking individual mononucleoside triphosphates.
2. The DNA code words are transcribed into RNA code words through base pairing. Cytosine (C) and guanine (G) are complementary bases. Adenine (A) is the complementary base for thymine (T) in DNA and for uracil (U) in RNA.
3. The DNA rewinds, and mRNA moves out of the nucleus through pores in its membrane.

Translation is the process that uses the mRNA produced through transcription to synthesize the protein:

1. The mRNA binds, again through base pairing, to an RNA-rich particle in the cell. This particle is called a *ribosome,* and it moves along the mRNA strand.
2. *Transfer RNA* (tRNA) is used to bring amino acid molecules to the mRNA that is bound to the ribosome. The tRNA molecule has two key portions on opposite

ends. On one end is a three-base sequence that is *the complement of a word on the mRNA*. On the opposite end is a *covalent binding site for the amino acid* that is coded by this word. So, tRNA is used to "shuttle" the correct amino acid to the correct position on the mRNA.

3. The ribosome moves along the bound mRNA, one three-letter word at a time, while the tRNA molecules hydrogen-bond to the mRNA to position their amino acids near one another.
4. When a tRNA molecule attaches to the mRNA at the ribosome, it releases its amino acid, and an enzyme (not shown) catalyzes the formation of a peptide bond to the growing chain of amino acids in the protein. Once empty of its amino acid, the tRNA molecule leaves the site.

To summarize, protein synthesis involves *transcribing* the DNA three-base words into an RNA message of three-base words and then *translating* the words, via three-base RNA carriers, into the sequence of linked amino acids that make up a protein:

DNA base sequence ⇒ RNA base sequence ⇒ Protein amino acid sequence

DNA Replication Another complex series of interactions leads to *DNA replication*, the process by which DNA copies itself. When a cell divides, its chromosomes are replicated, or reproduced, ensuring that the new cells have the same number and types of genes. In this process (Figure 22.14), the following stages occur:

1. A small portion of the double helix is "unzipped," so that each DNA chain can act as a template for a new chain.
2. The bases in each DNA chain are hydrogen-bonded with the complementary bases of free mononucleoside triphosphate units.
3. The new units are linked through phosphodiester bonds into a new chain, in a process catalyzed by an enzyme called *DNA polymerase*.
4. Gradually, each of the unzipped chains forms the complementary half of a double helix, which results in *two* double helices.

Because each base always pairs with its complement, the original double helix is copied and the genetic makeup of the cell is preserved. (Base pairing is central to methods for learning the sequence of nucleotides in genes, and for analyzing evidence in criminal investigations, as the upcoming Chemical Connections section will show.)

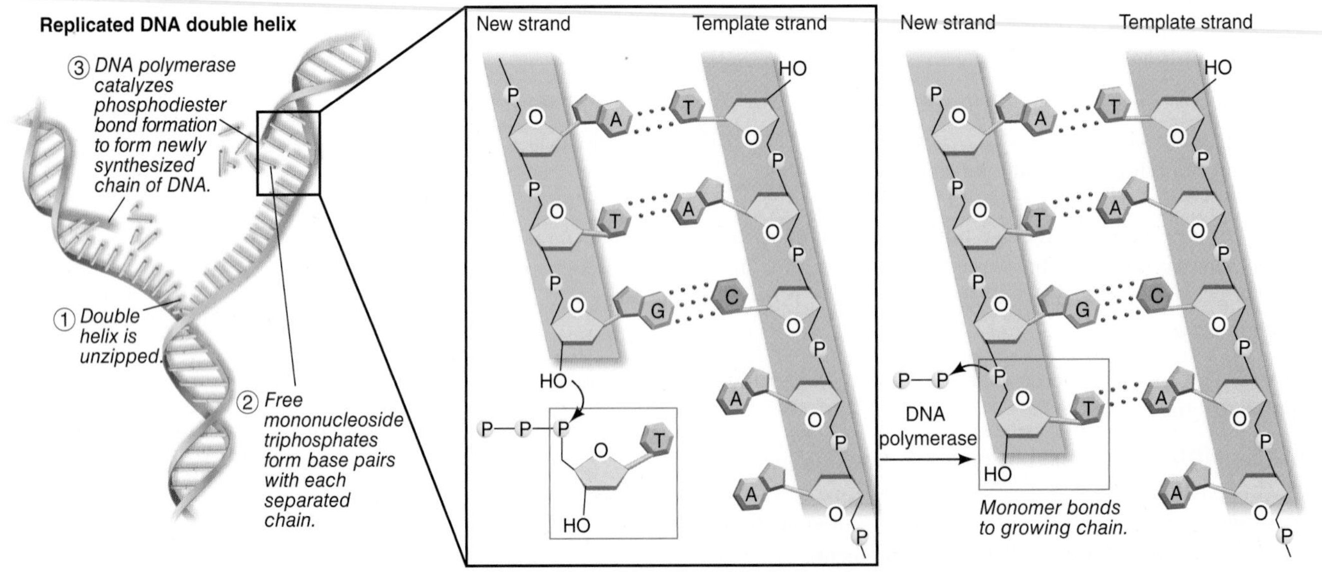

FIGURE 22.14 Key stages in DNA replication

DNA Sequencing and Fingerprinting

CHEMICAL CONNECTIONS TO GENETICS AND FORENSICS

Thanks to *DNA sequencing*, the process of determining the identity and order of DNA bases, we now know the sequence of the 3 billion nucleotide base pairs in the DNA of the entire human genome! Sequencing has become indispensable to molecular biology and biochemical genetics.

A different process that is just as indispensable to forensic science is *DNA fingerprinting* (or *DNA profiling*). Since the discovery that portions of an individual's DNA are as unique as fingerprints, this technique has been applied to countless situations in forensic science (such as identifying victims, apprehending criminals, and overturning incorrect convictions) and medical and nutritional research.

DNA Sequencing A given chromosome in the nucleus may have 100 million nucleotide bases, but the sequencing process can only deal with DNA fragments that are about 2000 bases long. Therefore, the chromosome is first broken into pieces by enzymes that cleave at specific sites. Then, to obtain a large enough sample for analysis, the DNA is replicated through a variety of *amplification* methods, which make many copies of the individual DNA *target fragments*.

The most popular sequencing method is the *Sanger chain-termination method*, which uses chemically altered bases to stop the growth of a complementary DNA chain at specific locations. As you have learned, the chain consists of linked 2′-deoxyribonucleoside monophosphate units (dNMPs, where N represents A, T, G, or C). The link is a phosphodiester bond from the 3′-OH of one unit to the 5′-OH of the next unit. The free monomers that are used to construct the chain are 2′-deoxyribonucleoside triphosphates (dNTPs) (Figure B22.1A). The Sanger method uses a modified monomer, called a *di*deoxyribonucleoside triphosphate (ddNTP), in which the 3′-OH group is also missing from the ribose unit (Figure B22.1B). As soon as the ddNTP is incorporated into the growing chain, polymerization stops because there is no —OH group on the 3′ position to form a phosphodiester bond to the next dNTP unit.

The Sanger method is shown in Figure B22.2. After several preparation steps, the sample to be sequenced consists of single-stranded DNA target fragments, each of which is attached to one strand of a double-stranded segment of DNA (Figure B22.2A). This

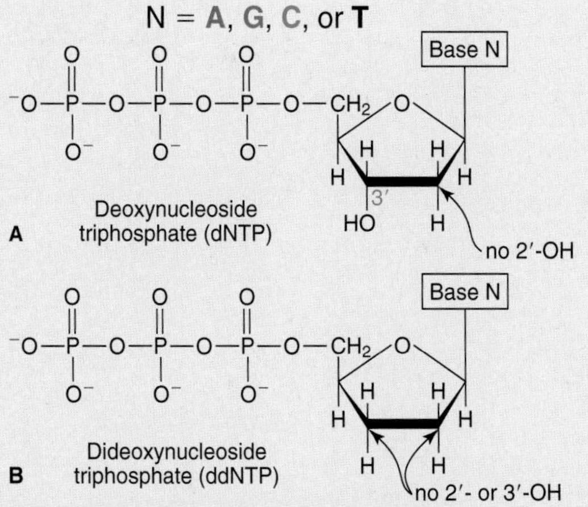

FIGURE B22.1 Nucleoside triphosphate monomers

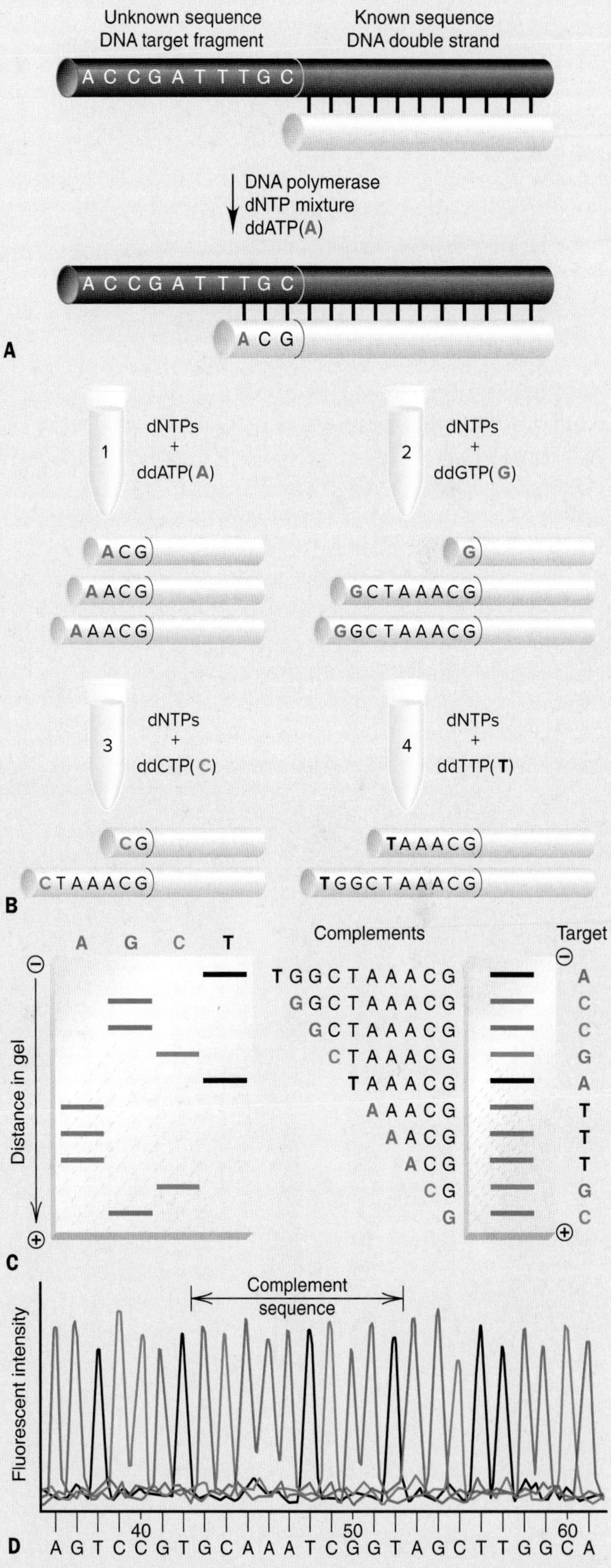

FIGURE B22.2 Steps in the Sanger method of DNA sequencing

(*Continued*)

sample is divided into four tubes, and to each tube is added a mixture of DNA polymerase, large amounts of all four dNTPs, and a small amount of one of the four ddNTPs. Thus, tube 1 contains polymerase, dATP, dGTP, dCTP, and dTTP, and, say, ddATP; tube 2 contains the same mixture, except with ddGTP instead of ddATP. The same is true for the other components. After the polymerization reaction is complete, each tube contains the original target fragments paired to complementary chains of varying lengths (Figure B22.2B). The chain lengths vary because each complementary chain ends in ddA (designated **A** in the figure) in tube 1, ddG (**G**) in tube 2, ddC (**C**) in tube 3, and ddT (**T**) in tube 4.

Each double-stranded product is divided into single strands, and then the complementary chains are separated using *high-resolution poly(2-propenamide)* (or *polyacrylamide) gel electrophoresis*. In this technique, an electric field is applied to separate charged species through differences in their rate of migration through pores in a gel: the smaller the species, the faster it moves. Polynucleotide fragments are commonly separated by electrophoresis because they have charged phosphate groups. High-resolution gels have pores that vary so slightly in size that they can separate fragments differing by only a single nucleotide.

During electrophoresis, each sample is applied to its own "lane" on a gel. After electrophoresis, the gel is scanned to locate the chains, which appear in bands. Because all four ddNTPs were used, all possible chain fragments are formed, so the sequence of the original DNA fragment can be determined (Figure B22.2C).

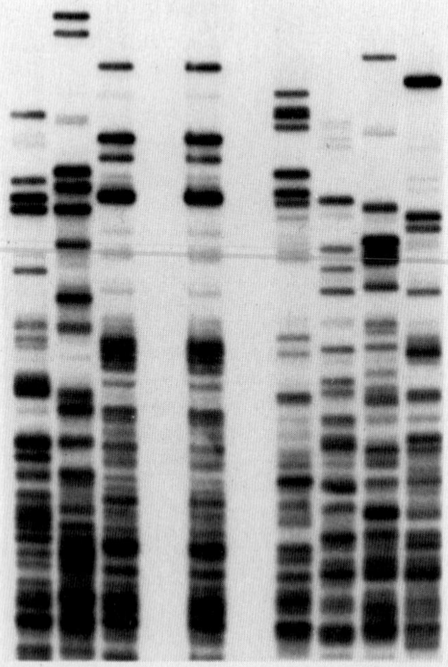

FIGURE B22.3 **STR analysis of DNA in the blood of seven suspects and DNA in the blood found at a crime scene *(centre)***

An automated approach begins with each ddNTP tagged using a different fluorescent dye, which emits light of a distinct colour. The entire mixture of complementary chains is introduced onto the gel and separated. The gel is scanned by a laser, which activates the dyes. The fluorescent intensity versus the distance along the chain is detected and plotted by a computer (Figure B22.2D).

DNA Fingerprinting Modern techniques of DNA fingerprinting require less than 1 ng of DNA. Thus, in addition to blood and semen, common items, such as licked stamps and envelopes, chewed gum, and clothing containing dead skin cells, can also be sources of DNA.

Currently, the most successful and widely used DNA profiling procedure is called *short tandem repeat (STR) analysis*. STRs are specific areas on a chromosome that contain short sequences (3 to 7 bases) that repeat themselves within the DNA molecule. There are hundreds of different types of STRs, but each person has a unique number and assortment of types. Therefore, the more STRs that are characterized, the more discriminating the analysis can be.

Once the STRs have been amplified, they are separated by electrophoresis, and an analyst determines the number of base repeats within each STR. In Figure B22.3, stained gels containing the STRs of DNA in blood samples from several suspects are compared with those in blood found at a crime scene. Suspect 3 has a pattern of STRs identical to the pattern in the crime scene sample (*centre*).

Problems

B22.1 If you were analyzing a DNA fragment with the base sequence TACAGGTTCAGT, how many complementary chain pieces would you obtain in the tube containing ddATP? How many would you obtain in the tube containing ddCTP?

B22.2 A DNA fragment is sequenced, and the following complementary chain pieces are obtained from the four tubes:
(1) The ddATP tube contains ATG and ATATG.
(2) The ddTTP tube contains TG and TATG.
(3) The ddCTP tube contains CATATG.
(4) The ddGPT tube contains G.

What is the sequence of bases in the DNA fragment? Make a sketch of the electrophoresis gel for this fragment, similar to the diagram shown in Figure B22.2C.

SUMMARY OF SECTION 22.3

- The three types of natural polymers—polysaccharides, proteins, and nucleic acids—are formed by dehydration-condensation reactions.
- Polysaccharides are formed from cyclic monosaccharides, such as glucose. Cellulose, starch, and glycogen have structural or energy storage roles.
- Proteins are polyamides formed from as many as 20 different types of amino acids. Fibrous proteins have extended shapes and play structural roles. Globular proteins have compact shapes and play metabolic, immunologic, and hormonal roles. The amino acid sequence of a protein determines its shape and function.
- Nucleic acids (DNA and RNA) are polynucleotides that consist of four different mononucleotides. The base sequence of the DNA chain determines the sequence of amino acids in an organism's proteins. Hydrogen bonding between specific base pairs is the key to protein synthesis and DNA replication.
- DNA sequencing is used to determine the identity and order of nucleotides in a fragment of a DNA chain. DNA fingerprinting is used to identify a person from the unique pattern of repeating base sequences in his or her DNA.

22.4 Intermolecular Forces and Biological Macromolecules

The three-dimensional shapes of proteins, nucleic acids, and cell membranes and the functions of soaps and antibiotics depend on intermolecular forces and are explained by the following two ideas:

- Polar and ionic groups attract water, but nonpolar groups do not.
- Just as separate molecules attract each other, so do distant groups on the same molecule.

The Structures of Proteins

As discussed in Section 22.3, proteins are unbranched polymers formed from about 20 different monomers called amino acids. They range in size from about 50 amino acids ($\mathcal{M} \approx 5 \times 10^3$ g/mol) to several thousand ($\mathcal{M} \approx 5 \times 10^5$ g/mol). Proteins with a few types of amino acids in repeating patterns have extended helical or sheetlike shapes and give structure to many parts of the body, such as hair and skin. Proteins with many types of amino acids have complex, globular shapes and function as antibodies, enzymes, and so on. Let us see how intermolecular forces among amino acids influence the shapes of globular proteins.

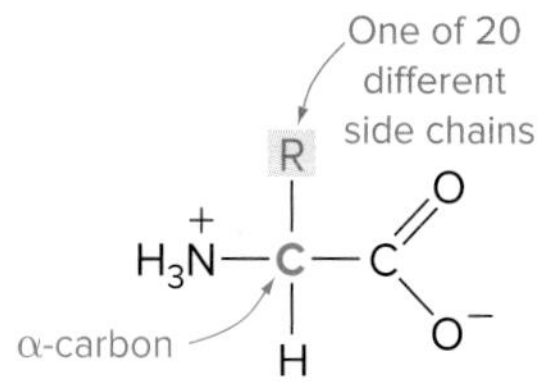

FIGURE 22.15 The charged form of an amino acid under physiological conditions

The Polarity of Amino Acid Side Chains In the cell, a free amino acid has four groups bonded to its α-carbon (Figure 22.15): charged carboxyl (—COO^-) and amine (—NH_3^+) groups; an H atom; and a side chain, which ranges from an H atom to a two-ringed C_9H_8N group. Amino acids can be classified by the polarity or charge of their side chains: nonpolar, polar, and ionic. A few examples are shown below:

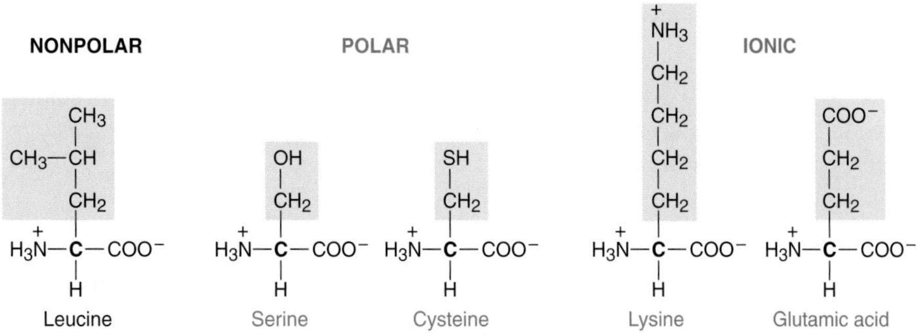

FIGURE 22.16 A portion of a polypeptide chain. Three peptide bonds (*tan screens*) join four amino acids in this portion.

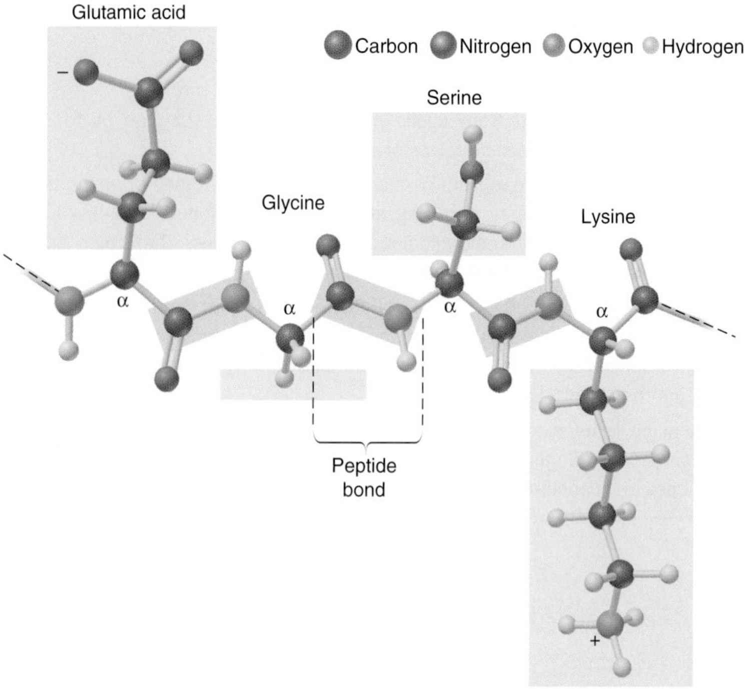

In a protein, the carboxyl group of one amino acid is linked covalently to the amine group of the next amino acid by a *peptide* bond. Thus, as Figure 22.16 shows, the *backbone* of a protein is a polypeptide chain: an *α-carbon connected through a peptide bond (tan screen) to the next α-carbon*, and so on. The various side chains (*grey screens*) dangle from the α-carbons on alternate sides of the chain.

Intermolecular Forces and Protein Shape The same forces that act between separate molecules are responsible for a protein's shape, because *distant groups on the protein chain end up near each other as the chain bends.* Figure 22.17 depicts the forces within a protein and the forces between the protein and the aqueous medium of the cell. These forces are listed below in their order of importance:

- Covalent peptide bonds create the backbone (polypeptide chain).
- Helical and sheetlike segments arise from *hydrogen bonds between the C═O of one peptide bond and the N—H of another.*
- Polar and ionic side chains protrude into the surrounding cell fluid, interacting with water through ion-dipole forces and hydrogen bonds.
- Nonpolar side chains interact through dispersion forces within the nonaqueous protein interior.
- The —SH ends of two cysteine side chains form a covalent —S—S— bond, or *disulfide bridge*, between distant parts of the chain. The disulfide bridge fixes a bend in the protein chain.
- Oppositely charged ends of ionic side chains ($—COO^-$ and $—NH_3^+$ groups) that lie near each other form an electrostatic *salt link* (or *ion pair*) that creates a bend in the protein chain.
- Other hydrogen bonds between side chains keep distant chain portions near each other.

Thus, *soluble proteins have polar-ionic exteriors and nonpolar interiors.* As emphasized in the previous section, *the amino acid sequence of a protein determines its shape, which determines its function.*

Dual Polarity in Soaps, Membranes, and Antibiotics

Dual polarity, which we discussed in relation to the solubility of alcohols, also helps to explain how soaps, cell membranes, and antibiotics function.

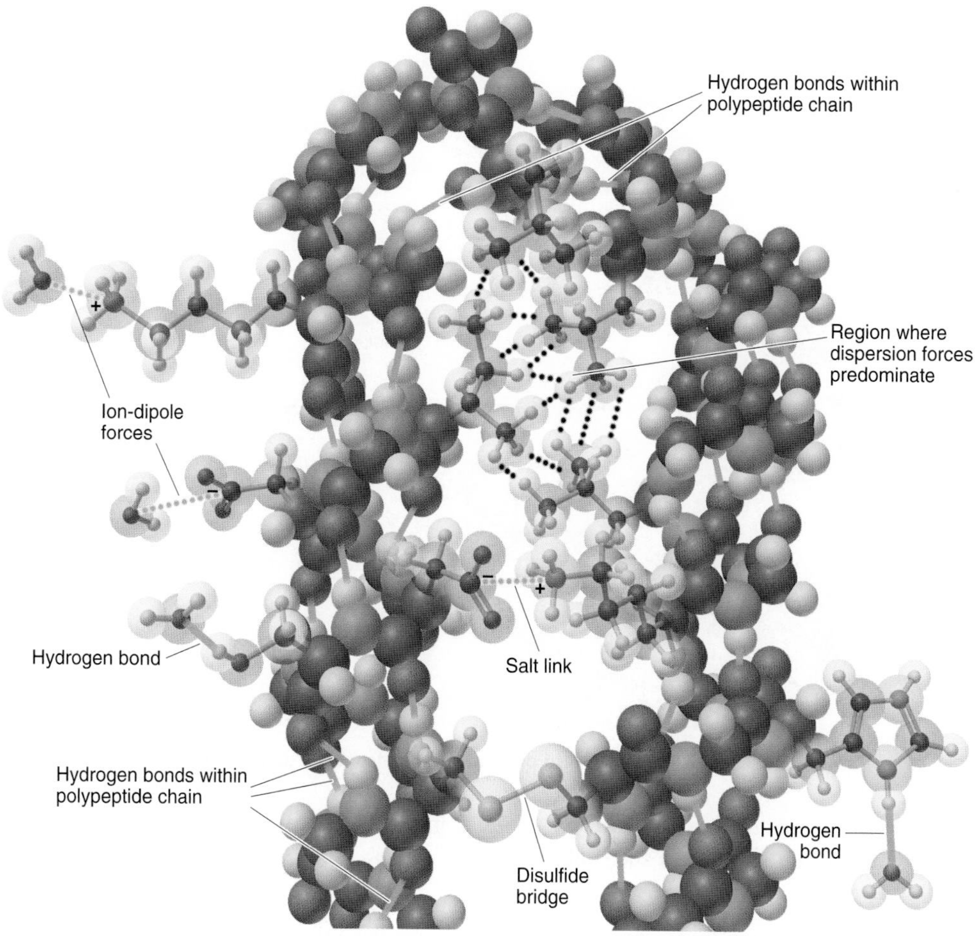

FIGURE 22.17 The forces that maintain protein structure. Covalent, ionic, and intermolecular forces act between parts of a protein and between the protein and surrounding H_2O molecules to determine the protein's shape. (Water molecules and some amino acid side chains are shown as ball-and-stick models within space-filling contours.)

Action of Soaps A **soap** is the salt that is formed when a strong base (metal hydroxide) reacts with a *fatty acid*, a carboxylic acid with a long hydrocarbon chain. A typical soap molecule is made up of a nonpolar "tail," 15 to 19 carbons long, and a polar-ionic "head" that consists of a $—COO^-$ group and the cation of the strong base. The cation greatly influences the properties of a soap. Lithium soaps are hard and high melting and used in car lubricants. Potassium soaps are low melting and used in their liquid form. Softer, more water-soluble sodium soaps, including sodium stearate, $CH_3(CH_2)_{16}COONa$, are components of common bar soaps:

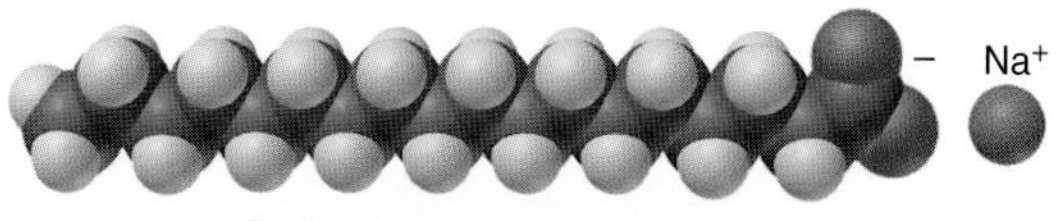

Sodium stearate ($C_{17}H_{35}COONa$)

When grease on your hands or clothes is immersed in soapy water, the soap molecule's nonpolar tails interact with the nonpolar grease molecules through dispersion forces, while the polar-ionic heads attract water molecules through ion-dipole forces and hydrogen bonds. Tiny aggregates of grease molecules, embedded with soap molecules whose polar-ionic heads stick into the water, can be flushed away by added water (Figure 22.18).

Lipid Bilayers and the Structure of the Cell Membrane The most abundant molecules in cell membranes are *phospholipids*. Like soap, a phospholipid has dual polarity: a nonpolar tail that consists of two fatty acid chains, and an organophosphate group that is the polar-ionic head (Figure 22.19). Remarkably, phospholipids self-assemble in water into a sheetlike double layer called a **lipid bilayer**, with the tails of the two layers touching and the heads in the water. In a laboratory, bilayers

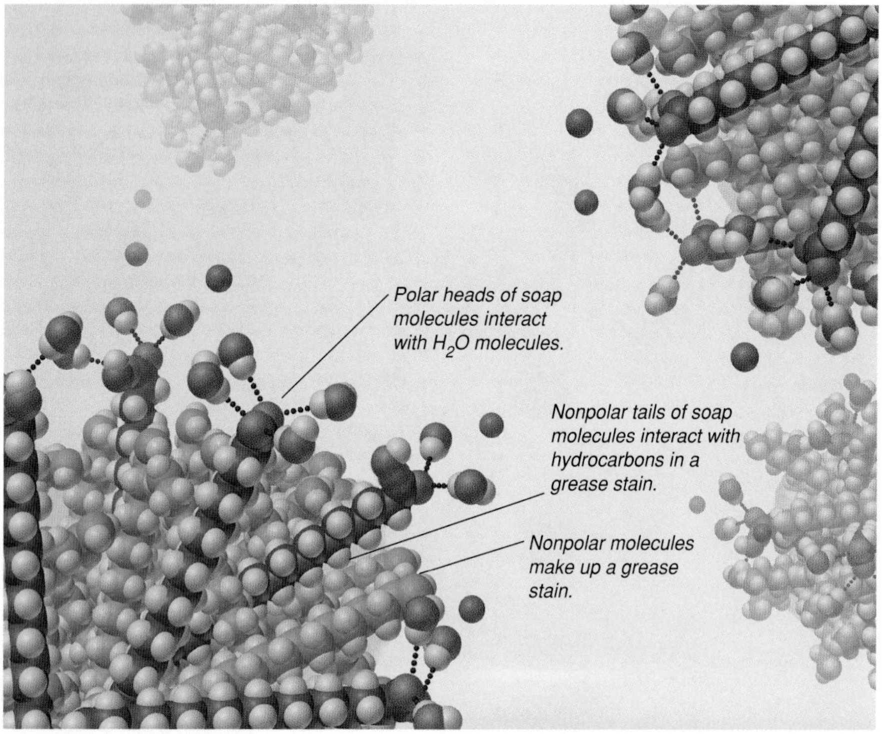

FIGURE 22.18 The cleaning ability of a soap depends on the dual polarity of its molecules.

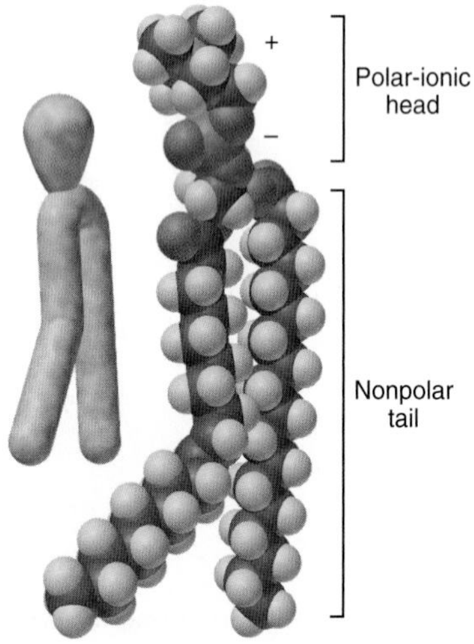

FIGURE 22.19 A membrane phospholipid. Lecithin (phosphatidylcholine), a phospholipid, is shown as a space-filling model and as a simplified purple-and-grey shape.

form spherical vesicles that trap water inside. These structures are favoured energetically because of their intermolecular forces:

- Ion-dipole forces occur between the polar heads and the water inside and outside.
- Dispersion forces occur between the nonpolar tails within the bilayer interior.
- Minimal contact exists between the nonpolar tails and water.

A typical animal cell membrane consists of a phospholipid bilayer with proteins partially embedded in it (Figure 22.20). Membrane proteins, which play countless essential roles, differ fundamentally from soluble proteins in terms of their dual polarity:

- Soluble proteins have polar exteriors and nonpolar interiors. They form ion-dipole and hydrogen-bonding forces between water and *polar groups on the exterior* and dispersion forces between *nonpolar groups in the interior* (see Figure 22.17).

FIGURE 22.20 Intermolecular forces and cell membrane structure

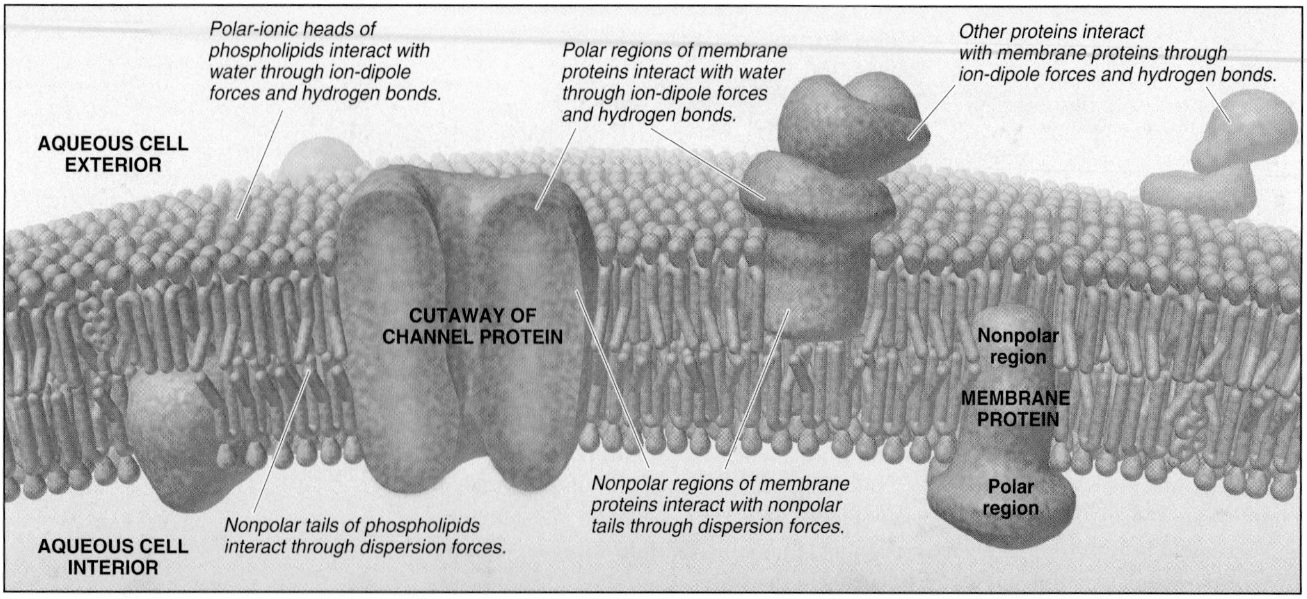

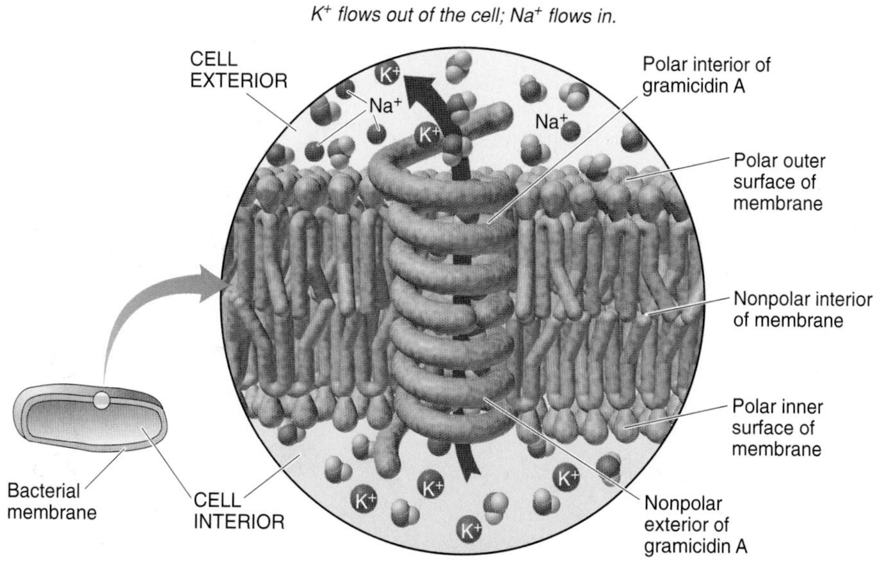

FIGURE 22.21 The mode of action of the antibiotic gramicidin A. The K^+ ions are shown leaving the cell. At the same time (not shown), Na^+ ions are entering the cell.

- Membrane proteins have exteriors that are partially polar and partially nonpolar. They have polar groups on the *exterior portion that juts into the aqueous surroundings* and nonpolar groups on *the exterior portion embedded in the membrane.* The nonpolar groups can, thus, form dispersion forces with the phospholipid tails of the bilayer. Channel proteins also have polar groups lining the aqueous channel.

Action of Antibiotics A key function of the membrane is to balance internal and external ion concentrations: Na^+ is excluded from the cell, and K^+ is kept inside. Gramicidin A and similar antibiotics act by forming channels in the bacterial membrane through which ions flow (Figure 22.21). Two helical gramicidin A molecules, with nonpolar groups outside and polar groups inside, lie end to end to form a channel through the membrane. The nonpolar exterior stabilizes the molecules in the membrane through dispersion forces, and the polar interior passes the ions along using ion-dipole forces, like a "bucket brigade" (a method for transporting items where they are passed from one stationary person to the next). Over 10^7 ions pass in and out through each of these channels per second, and the micro-organism soon dies.

The Structure of DNA

The chemical information that guides the design and construction, and therefore the function, of all proteins is contained in nucleic acids. As discussed in Section 22.3, nucleic acids are unbranched polymers made up of monomers called mononucleotides. Each mononucleotide consists of an N-containing base, a sugar, and a phosphate group (Figure 22.22). In DNA, the sugar is *2-deoxyribose*, a sugar in which —H substitutes for —OH on the second C atom of the five-carbon sugar ribose.

The repeating pattern of the DNA chain is *sugar linked to phosphate linked to sugar linked to phosphate*, and so on. Attached to each sugar is one of four N-containing bases, flat ring structures that dangle off the polynucleotide chain, similar to the way that amino acid side chains dangle off the polypeptide chain.

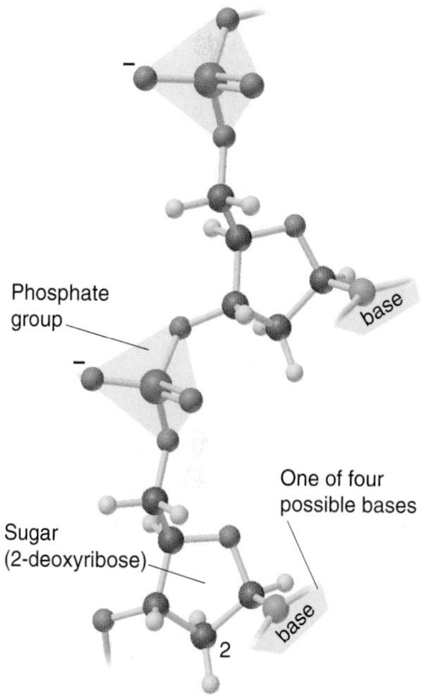

FIGURE 22.22 A short portion of the polynucleotide chain of DNA

Intermolecular Forces and the Double Helix DNA exists as two chains wrapped around each other in a double helix stabilized by intermolecular forces (Figure 22.23):

- *On the more polar exterior*, negatively charged sugar-phosphate groups interact with the aqueous surroundings via ion-dipole forces and hydrogen bonds.
- *In the less polar interior*, flat, N-containing bases stack above each other and interact by dispersion forces.
- *Bases form specific interchain hydrogen bonds*; that is, each base in one chain is always hydrogen-bonded with its complementary base in the other chain. Thus, *the base sequence of one chain is the hydrogen-bonded complement of the base sequence of the other chain.*

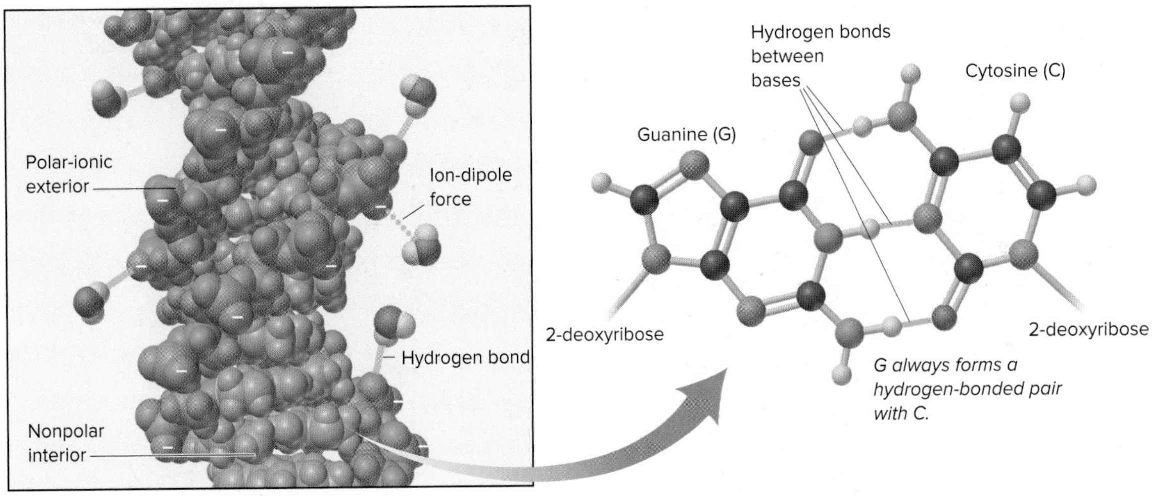

FIGURE 22.23 The double helix of DNA. A segment of DNA (*left*) has its polar-ionic sugar-phosphate portion (*pink*) facing the water and the nonpolar bases (*grey*) stacking in the interior. The expanded portion (*right*) shows a hydrogen-bonded pair of the bases guanine and cytosine.

A DNA molecule contains millions of hydrogen bonds linking bases in these prescribed pairs. The *total* energy of the hydrogen bonds keeps the chains together, but each hydrogen bond is weak enough (around 5% of a typical covalent single bond) that a few at a time can break as the chains separate during crucial cellular processes.

SUMMARY OF SECTION 22.4

- In soluble proteins, polar and ionic amino acid side chains on the exterior interact with surrounding water, and nonpolar side chains in the interior interact with each other.
- With a polar-ionic head *and* a nonpolar tail, soap dissolves grease *and* interacts with water.
- Like soaps, phospholipids also have dual polarity. They assemble into a water-impermeable lipid bilayer. In a cell membrane, the embedded portion of membrane proteins has exterior nonpolar side chains that face the nonpolar tails in the lipid bilayer. Some antibiotics form channels with nonpolar exteriors and polar interiors that shuttle ions through the cell membrane.
- DNA forms a double helix with a sugar-phosphate, polar-ionic exterior. In the interior, N-containing bases hydrogen-bond in specific pairs and stack through dispersion forces.

22.5 Instrumental Analysis

Infrared Spectroscopy

Infrared (IR) spectroscopy is an instrumental technique that is most often used to study covalently bonded molecules. The key components of an IR spectrometer are the same as the components of other types of spectrometers (see Figure B6.3). The source emits radiation of many wavelengths, but only wavelengths in the infrared (IR) region are selected. The sample is typically a pure liquid or solid that absorbs varying amounts of certain IR wavelengths. An IR spectrum consists of peaks that indicate these absorptions.

Molecular Vibrations and IR Absorption An IR spectrum indicates the types of bonds in a molecule based on their **molecular vibrations**. All molecules undergo movement through space, rotation around several axes, and vibrations of bonds. Consider a sample of ethane gas. The $H_3C—CH_3$ molecules zoom throughout the container, but let us disregard motion through space and focus on other motions. The whole molecule rotates, and its two CH_3 groups rotate about the C—C bond. More important to IR spectroscopy, each pair of atoms is vibrating as though the bonds were springs. Figure 22.24 depicts the vibrations of diatomic and triatomic molecules; larger molecules vibrate in many more ways.

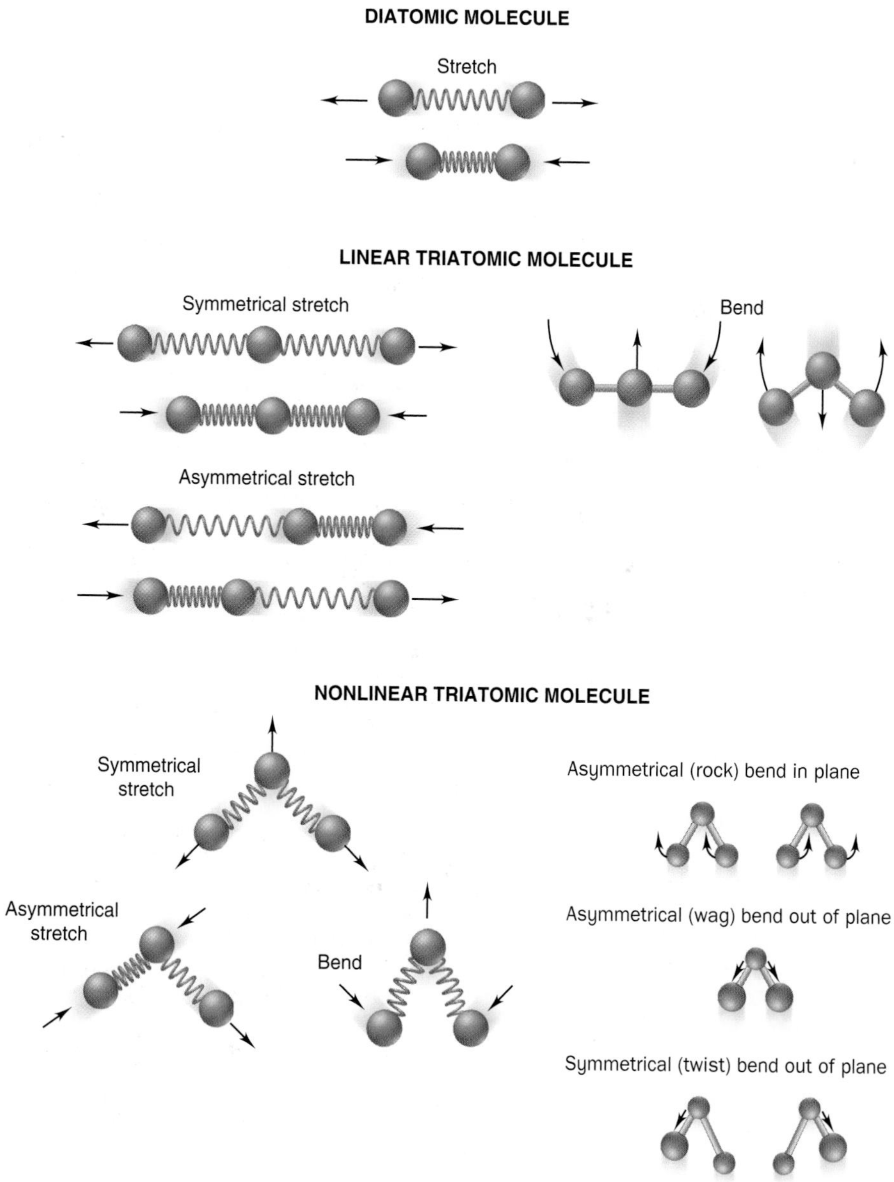

FIGURE 22.24 Vibrational motions in general diatomic and triatomic molecules

The energies of IR photons are in the range of these vibrational energies. *Each vibration has a frequency based on the type of motion, the masses of the atoms, and the strength of the bond.* The frequencies correspond to IR wavelengths between 2.5 μm and 25 μm. *The energy of these vibrations is quantized.* Just as an atom can absorb a photon whose energy equals the difference between the *electron* energy levels, a molecule can absorb an IR photon whose energy equals the difference between the *vibrational* energy levels.

IR Radiation and Global Warming Carbon dioxide, O=C=O, is a linear molecule that bends and stretches symmetrically and asymmetrically when it absorbs IR radiation. Sunlight is absorbed by Earth's surface and re-emitted as heat, much of which is IR radiation. Atmospheric CO_2 absorbs this radiation and re-emits it, thus warming the atmosphere (see Chemical Connections in Chapter 5).

Compound Identification An IR spectrum can be used to identify a compound for three related reasons:

1. *Each kind of bond absorbs a specific range of wavelengths.* That is, a C—C bond absorbs a different range than other bonds, such as a C=C bond, a C—H bond, and a C=O bond, do.

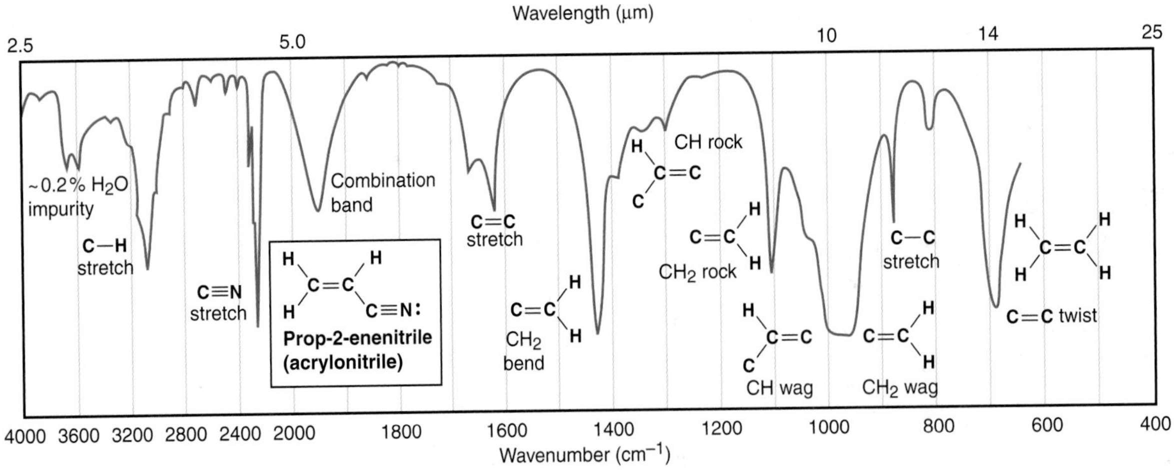

FIGURE 22.25 The infrared (IR) spectrum of prop-2-enenitrile (acrylonitrile). In this typical IR spectrum, there are many absorption bands (peaks) of differing depths and sharpness. Most correspond to a particular type of vibration (stretch, bend, rock, wag, or twist). Some broad peaks (e.g., "combination band") represent several overlapping vibrations. Notice that the bottom axis is labelled "Wavenumber," which is the inverse of wavelength and has units of cm^{-1}. (The scale expands to the right of 2000 cm^{-1}.)

2. *Different types of organic compounds have characteristic spectra.* The different groupings of atoms that define an alcohol, a carboxylic acid, an ether, and other specific functional groups (see Chapter 20) absorb differently. (See also Appendix E.)

3. *Each compound has a unique spectrum.* The IR spectrum acts like a fingerprint to identify the compound, because the quantity of each wavelength absorbed depends on the detailed molecular structure. For example, no other compound has the IR spectrum of prop-2-enenitrile (acrylonitrile), a compound used to make plastics (Figure 22.25).

Recall, from Chapter 20, that constitutional (structural) isomers have the same molecular formula but different structures. We should see very different IR spectra for the isomers ethoxyethane (diethyl ether) and butan-2-ol because their molecular structures are so dissimilar (Figure 22.26). However, even the very similar 1,3-dimethylbenzene and 1,4-dimethylbenzene have different spectra (Figure 22.27).

Nuclear Magnetic Resonance Spectroscopy

In addition to mass spectrometry (Chapter 2) and infrared (IR) spectroscopy, one of the most useful tools for analyzing organic and biochemical structures is **nuclear magnetic resonance (NMR) spectroscopy**, which measures the molecular environments of certain nuclei in a molecule.

Like electrons, several types of nuclei, such as ^{13}C, ^{19}F, ^{31}P, and ^{1}H, can spin in either of two directions, each of which creates a tiny magnetic field. In this discussion,

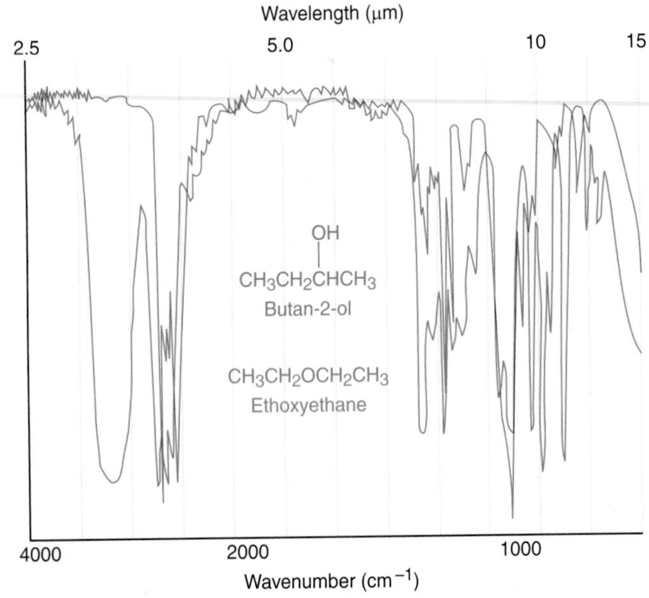

FIGURE 22.26 The infrared spectra of butan-2-ol (*green*) and ethoxyethane (diethyl ether; *red*)

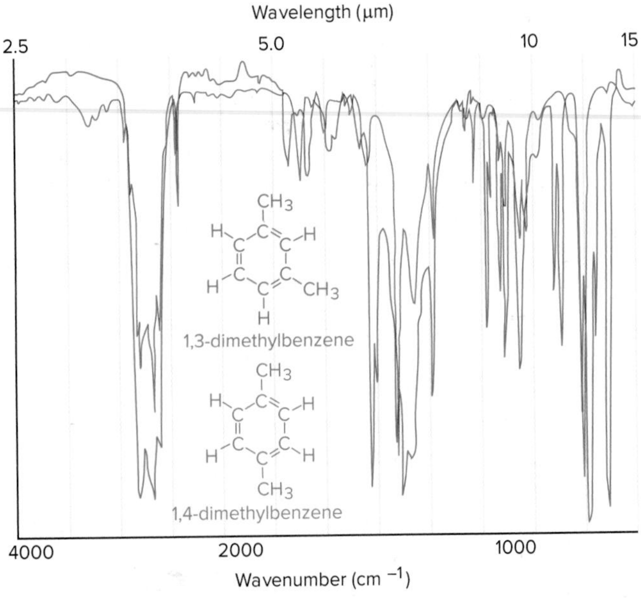

FIGURE 22.27 The infrared spectra of 1,3-dimethylbenzene (*green*) and 1,4-dimethylbenzene (*red*)

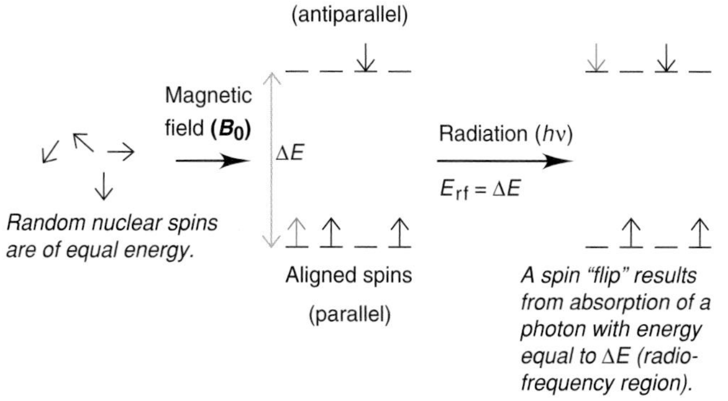

FIGURE 22.28 The basis of proton spin resonance

we focus primarily on ¹H NMR spectroscopy, which examines the proton, ¹H, the nucleus of the most common isotope of hydrogen. Generally oriented randomly, the magnetic fields of all the protons in a sample of a compound, when placed in a strong external magnetic field (B_0), become aligned either with the external field (parallel) or against it (antiparallel). Most nuclei adopt the parallel orientation, which is slightly lower in energy. The energy difference (ΔE) between the two energy states (spin states) lies in the radio-frequency (rf) region of the electromagnetic spectrum (Figure 22.28).

When a proton in the lower energy (parallel) spin state absorbs a photon in the rf region with energy equal to ΔE, the proton "flips," in a process called *resonance*, to the higher energy (antiparallel) spin state. The system then re-emits the energy, which is detected by the rf receiver of the ¹H NMR spectrometer. The ΔE between the two states depends on the *actual* magnetic field that is felt by each proton, which is affected by the tiny magnetic fields of the *electrons* of atoms adjacent to this proton. Thus, the ΔE required for the resonance of each proton depends on the specific molecular environment of the proton—the C atoms, electronegative atoms, multiple bonds, and aromatic rings around it. The protons in different molecular environments produce different peaks in the ¹H NMR spectrum.

A ¹H NMR spectrum, which is unique for each compound, is a series of peaks that represent the resonance of the protons as a function of the changing magnetic field. The *chemical shift* of the protons in a given environment is where a peak appears (see Appendix F). The chemical shifts are shown relative to the chemical shift of an added standard, tetramethylsilane [$(CH_3)_4Si$, or TMS]. TMS has 12 protons bonded to four C atoms that are bonded to one Si atom in a tetrahedral arrangement, so all 12 protons are in identical environments and produce only one peak. The *chemical shift* of a given proton is affected by the properties of the other groups that are attached to the carbon to which it is attached. For example, if an electron-withdrawing group is attached to the carbon, it will pull the electron density away from the carbon. This will *deshield* the proton of interest, leading to a **downfield** signal (or a signal with a higher chemical shift). Also, if the proton is attached to a carbon with π electrons, the proton will be **deshielded** (shifted downfield) by the magnetic field created by the circulating π electrons. The splitting pattern of a given nucleus can be predicted by the $n + 1$ rule. In the spectra of propan-2-one (Figure 22.29) and dimethoxymethane (Figure 22.30), the peaks are all singlets (one peak), since n (the neighbouring spin-coupled nuclei with the same, or a very similar, coupling constant*) is zero in all cases; however, doublets, triplets, quartets, and so on, can be observed if there are one, two, three, and so on, respectively, neighbouring spin-coupled nuclei in the molecule (Figure 22.31C).

In the ¹H NMR spectrum of propan-2-one, also known as acetone (Figure 22.29), the six protons of acetone have identical environments: all six are bonded to two C atoms, each of which is bonded to the C atom in a C=O bond. So, one peak is produced, but at a

FIGURE 22.29 The ¹H NMR spectrum of propan-2-one, also known as acetone

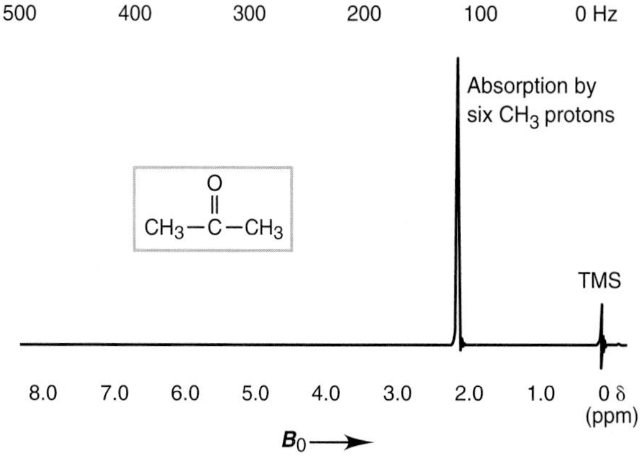

*The coupling constant (J) is the distance between peaks in a multiplet. It tells us how strong the interaction is between two neighbouring hydrogen nuclei.

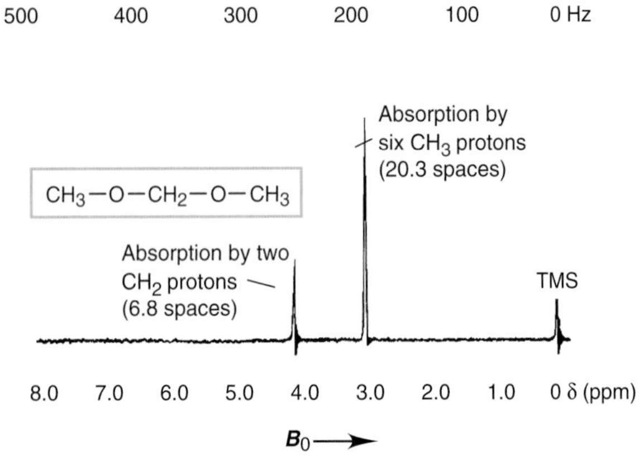

FIGURE 22.30 The ^{1}H NMR spectrum of dimethoxymethane

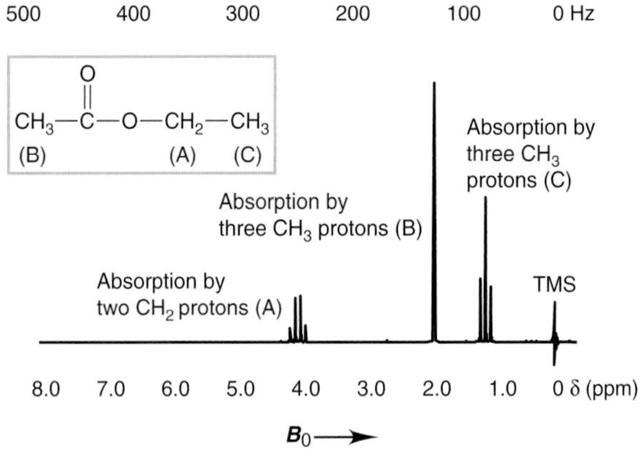

FIGURE 22.31 The ^{1}H NMR spectrum of ethyl ethanoate (also known as ethyl acetate)

different position from the TMS peak. The spectrum of dimethoxymethane (Figure 22.30) shows *two* peaks in addition to the TMS peak, since there are protons in two different environments. The taller peak is due to the six CH_3 protons, and the shorter peak is due to the two CH_2 protons. The difference in chemical shift of the two kinds of protons can be explained by the number of electronegative (electron-withdrawing) atoms attached to each C. The area under each peak (given here in units of chart-paper spaces) is proportional to *the number of protons in a given environment.* Note that the area ratio is $20.3 / 6.8 \approx 3 / 1$, the same as the ratio of six CH_3 protons to two CH_2 protons. Thus, by analyzing the chemical shifts and peak areas, a chemist can learn the type and number of hydrogen atoms in the compound.

The ^{1}H NMR spectrum of ethyl ethanoate, also known as ethyl acetate (Figure 22.31), shows three sets of peaks in addition to the TMS peak (the protons are in three different environments). The quartet at a higher chemical shift (shifted downfield) is due to the two CH_2 protons. These protons have three neighbouring spin-coupled nuclei ($n = 3$; therefore $n + 1 = 4$), and the carbon atom they are bonded to is directly bonded to an electron-withdrawing atom (oxygen). The singlet is due to the three CH_3 protons bonded to a C atom bonded to another C atom with π electrons (since $n = 0$, $n + 1 = 1$). The triplet at higher field (lower chemical shift) is due to the three CH_3 protons, which have two neighbouring spin-coupled nuclei ($n = 2$, so $n + 1 = 3$). These CH_2 protons are the less *deshielded* protons.

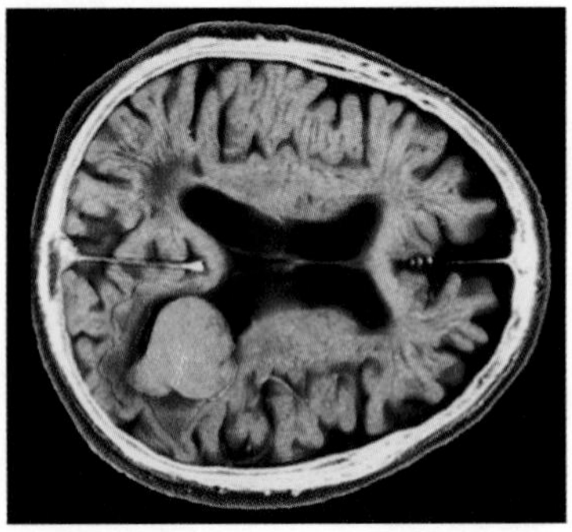

FIGURE 22.32 An MRI scan showing a brain tumour

NMR has many applications in biochemistry and medicine. Applying the principle to the imaging of organs and other body parts is known as computer-aided magnetic resonance imaging, or MRI. For example, an MRI scan of the head (Figure 22.32) can reveal a brain tumour by mapping levels of metabolic activity in different regions of the brain.

Chromatography

As we saw in Chapter 2, chromatography is a physical method of separation that is commonly used in organic chemistry. The separation of the components of a mixture is based on the relative attraction (intermolecular forces) of the components of the mobile phase (liquid or gas) for a stationary phase (solid or liquid) as it passes over the stationary phase.

Chromatography may be preparative or analytical. Preparative chromatography is a form of purification that separates the components of a mixture for more advanced use. Analytical chromatography, on the other hand, is normally done with smaller amounts of material and is used to measure the relative proportions of *analytes* in a mixture. The two types of chromatography are not mutually exclusive.

The four most common chromatography methods are gas chromatography, liquid chromatography, size-exclusion chromatography, and thin-layer chromatography.

Gas Chromatography The gas chromatography (GC) method is used to separate compounds that are volatile. The mobile phase is a gas, and the stationary phase is usually a liquid on a solid support or an adsorbent solid. The organic compounds

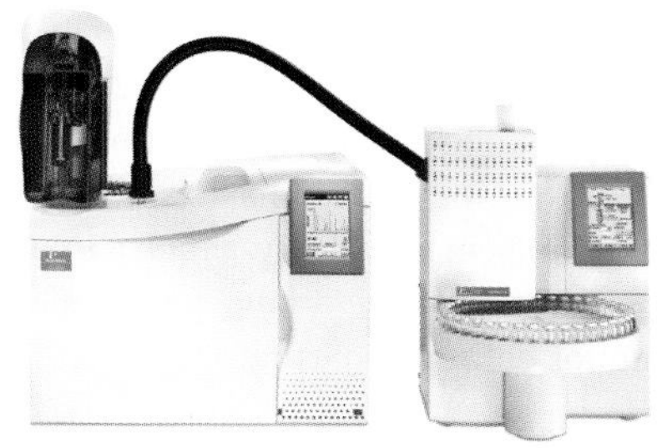

A

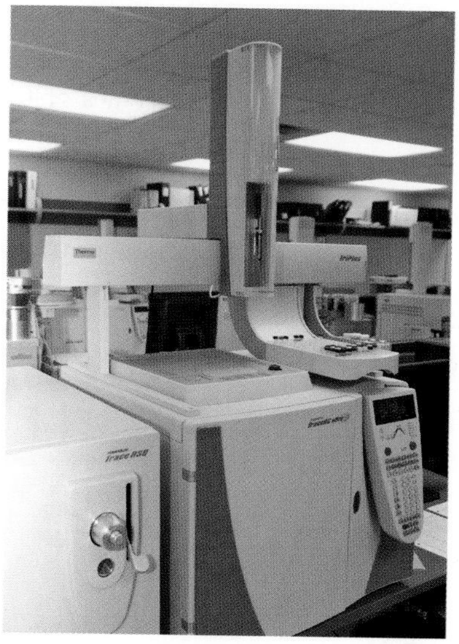

B

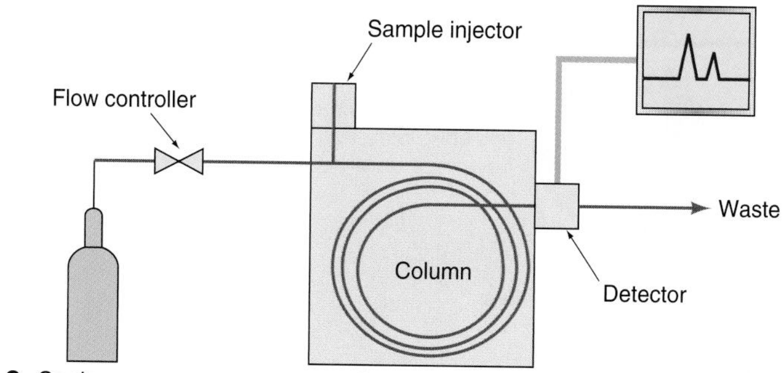

C

FIGURE 22.33 Gas chromatographs. **A.** A Perkin-Elmer Autosystem gas chromatograph. **B.** A gas chromatograph with a headspace sampler. **C.** A diagram of a gas chromatograph.

are separated based on differences in their partitioning behaviour between both phases. Figures 22.33A and 22.33B are photos of gas chromatographs. The diagram in Figure 22.33C shows the basic parts of a gas chromatograph.

Liquid Chromatography Liquid chromatography (LC) is used to separate ions or molecules dissolved in a solvent. The mobile phase is a solvent, and the stationary phase is a liquid on a solid support, a solid, or an ion-exchange resin. The different solutes are separated from each other by interacting with the stationary phase in differing degrees, due to differences in intermolecular forces: the molecules that are held more loosely move through the column more quickly than the molecules that are more strongly attracted to the stationary phase. The different transit times of the solutes through a column allow their separation. In conventional LC, the *eluent* (the solvent that carries the analyte) moves down the column by gravity. It is most commonly used in preparative-scale work to purify and isolate some components of a mixture. A more sophisticated version of this LC is HPLC (high-performance liquid chromatography), which uses a pump to move the mobile phase and sample components through a densely packed column, which provides a better separation with a shorter length. Figure 22.34 shows a schematic of a simple liquid chromatographic separation.

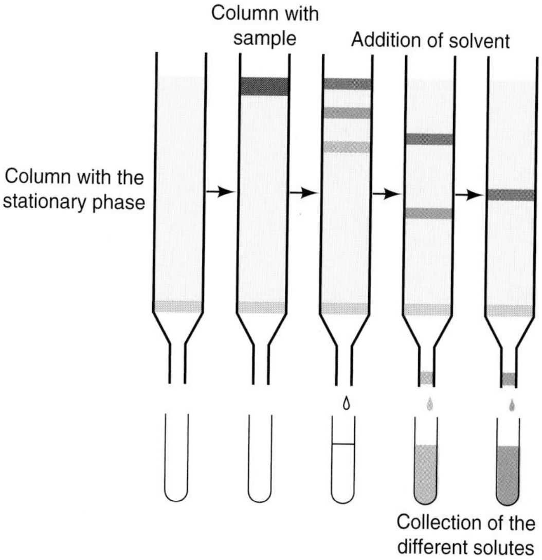

FIGURE 22.34 Schematic of a simple liquid chromatographic separation

Size-Exclusion Chromatography Size-exclusion chromatography [SEC; also called gel-permeation chromatography (GPC)] is mostly used to separate biological molecules or to determine the molecular weights or molecular weight distributions of polymers. The mobile phase of SEC is a solvent, and the stationary phase is a packing of porous particles. As shown in Figure 22.35, molecules that are smaller than the pore size can enter the porous particles and, therefore, have a longer path and longer transit time than larger molecules that cannot enter the particles.

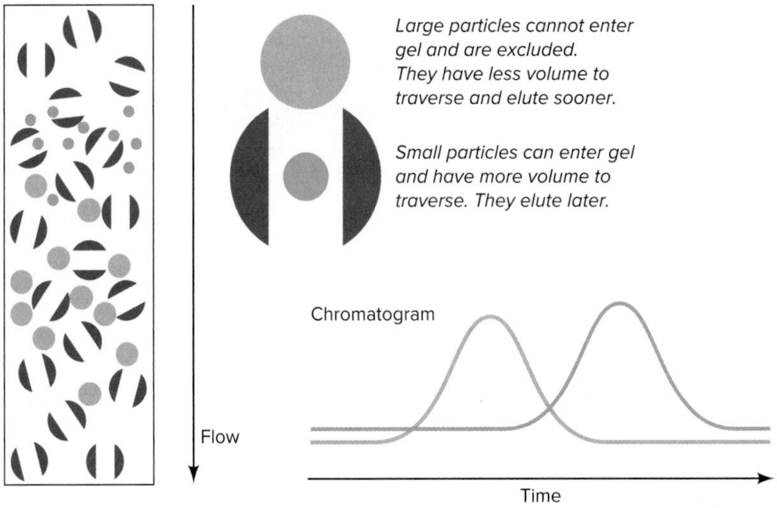
Large particles cannot enter gel and are excluded.
They have less volume to traverse and elute sooner.
Small particles can enter gel and have more volume to traverse. They elute later.
Chromatogram
Flow
Time

FIGURE 22.35 Schematic of SEC

Thin-Layer Chromatography Thin-layer chromatography (TLC) is a simple and rapid method that is used to monitor the extent of a reaction or to check the purity of organic compounds. The mobile phase is a solvent, and the stationary phase is a solid adsorbent on a flat support like a glass or a plastic plate. The sample, either a liquid or a solid dissolved in a volatile solvent, is deposited as a spot in the stationary phase (Figure 22.36A). The constituents of a sample can be identified by simultaneously running standards with the unknown. The bottom edge of the plate is placed in a solvent tank, and the solvent moves up the plate by capillary action. The different components in the mixture move up the plate at different rates as a result of differences in their partitioning behaviour between the mobile liquid phase and the stationary phase (Figure 22.36B). When the solvent front is close to the top of the stationary phase, the plate is removed from the solvent tank, and the separated spots can be seen by using ultraviolet light or placing the plate in iodine vapour.

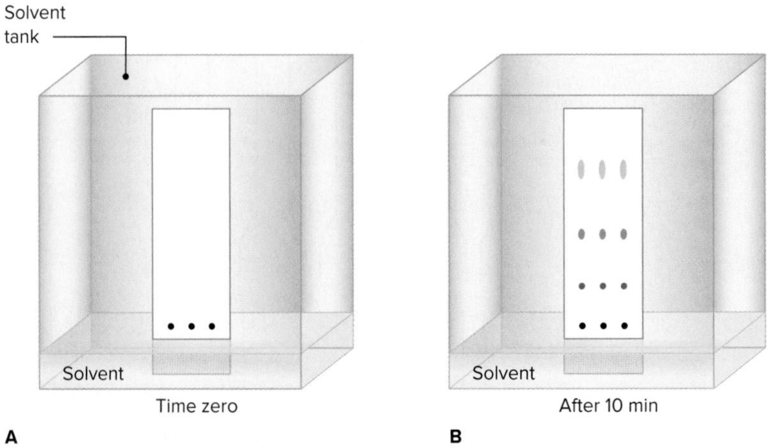
Solvent tank
Solvent
Time zero
A
Solvent
After 10 min
B

FIGURE 22.36 Schematic of TLC. A. A sample spotted on a TLC plate. **B.** Running the TLC plate in a solvent.

Chemical Glycobiology How complex carbohydrates contribute, at a molecular level, to the functioning of the human body is one of the frontiers of modern molecular biology. Different types of monosaccharides are found linked together to form glycans. These glycans decorate proteins that circulate within the body, as well as proteins found at the surface of all cells, and modulate how cells respond to their environment. Other glycans are found attached to lipids that are embedded in the cell membrane. Still others comprise long and highly elaborate polymers found outside cells that encode information read by proteins and, thus, influence how nearby cells function. These complex carbohydrate structures are regulated by many different enzymes that act to put them together and break them down. Chemists are attempting to decode the roles played by glycans in health and disease by studying these complicated structures and generating chemical tools with which to study and regulate these enzymes.

Dr. David Vocadlo (Figure B22.4) and his group in the Department of Chemistry at Simon Fraser University have taken a great interest in one type of monosaccharide that is linked to proteins inside cells. This sugar is known as O-GlcNAc, and its functions are slowly being uncovered. Dr. Vocadlo and his group have synthesized substrates of the enzyme, known as O-GlcNAcase, that removes O-GlcNAc from proteins (Figure B22.5A). Using these substrates, in combination with traditional chemical kinetic studies, they have been able to determine, in great detail, how the enzyme works, including how it facilitates the removal of O-GlcNAc from proteins. Using their findings, they have chemically synthesized molecules that mimic the transition state of the O-GlcNAcase-catalyzed reaction. These molecules, including Thiamet-G, are potent inhibitors of O-GlcNAcase and penetrate into cells and even the brains of animals. Their findings have prompted them to investigate the role of this sugar modification in Alzheimer's disease.

Alzheimer's disease is the most common neurodegenerative disease, affecting over 5 million people in North America. There is no cure and no known medicine or therapy that can slow its progression. One of the key features of Alzheimer's is the clumping together of a protein called *tau*. Tau is known to have O-GlcNAc on it, but the levels of this sugar modification are much lower in patients suffering from Alzheimer's. Speculating that increasing the levels of this sugar modification could prevent tau from clumping together, Dr. Vocadlo and his group found that Thiamet-G (Figure B22.5B) can block the clumping in mice and prevent the associated neurodegeneration. Their findings highlight how basic chemical research into how enzymes work and the chemical synthesis of new enzyme inhibitors can lead to new knowledge and chemical tools that can have translational potential. Their knowledge of O-GlcNAc, as well as the new tools they have developed, led them to establish a biotechnology spin-off company from Simon Fraser University. This company, called Alectos Therapeutics, is focusing on therapeutics for medical needs that have not yet been met, including the slowing or elimination of Alzheimer's disease.*

*Source: Text and illustrations by Dr. David Vocadlo

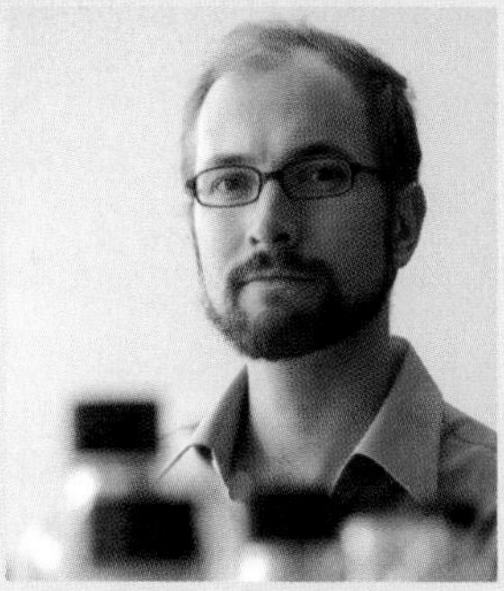

FIGURE B22.4 David Vocadlo (Simon Fraser University, Vancouver)

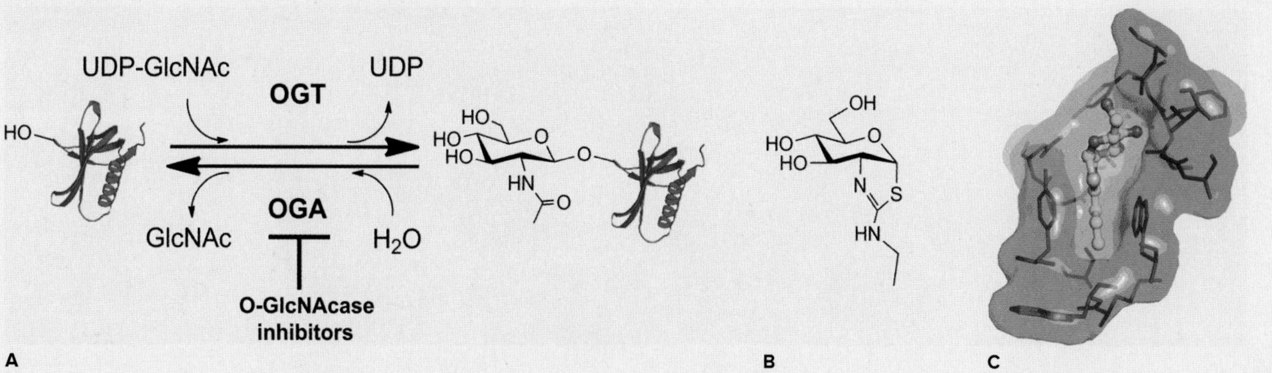

A B C

FIGURE B22.5 **A.** OGT puts O-GlcNAc onto proteins, while OGA acts to remove O-GlcNAc from proteins. **B.** The chemical structure of Thiamet-G. **C.** A related OGA inhibitor bound to the active site of a homologue of OGA, showing how the inhibitor fits closely into the pocket of the enzyme.

(Continued)

Polymer Chemistry Nanoparticle drug delivery systems present exciting opportunities for safer and more effective anticancer drug therapy. Dr. Molly Shoichet (Figure B22.6) and her group at the University of Toronto* are engineering intelligent biomaterials for these applications, giving rise to a platform technology that can be used to target and destroy more cancer cells, with greater specificity. Toxic chemotherapeutics, when given in their free form, distribute broadly throughout the body. By redirecting more of the drug dose toward tumour sites, however, Dr. Shoichet and her group aim to reduce the systemic side effects (suppression of the immune system, irritation in the digestive tract, and accelerated hair loss) to make anticancer treatment safer.

Nanomedicine uses a completely different approach to cancer targeting. As tumours grow, they develop new blood vessels to provide nutrients to rapidly dividing cancer cells. These blood vessels, however, are poorly formed and have abnormal gaps in their walls. Consequently, large molecules that would ordinarily be contained by healthy blood vessels can cross the gaps. The abnormal tumour blood vessels, also called leaky vasculature, are targeted using nanoparticles. Although free drugs are small enough to pass through healthy blood vessels, packaging the same drugs in nanoparticles channels a greater portion of the drug dose through leaky vasculature into tumour tissue. This size-based method is called passive targeting (Figure B22.7).

FIGURE B22.6 Molly Shoichet

*Text and illustration provided by Karyn Ho and Molly Shoichet at the University of Toronto (source: www.ecf.utoronto.ca/~molly/mini_team/Targeted%20Delivery%20Project%20Overview.pdf).

To take advantage of leaky tumour vasculature, Meng Shi and Karyn Ho from the Shoichet group synthesized a novel biodegradable polymer with hydrophobic and hydrophilic segments to form drug-loaded nanoparticles. The dual behaviour of this polymer allows the polymer strands to come together and form nanoparticles through self-assembly in water. The core-forming hydrophobic segments entrap hydrophobic anticancer drugs, and the outer hydrophilic shell facilitates the long circulation in the bloodstream that is needed to achieve passive targeting. As a result, Shi and Ho observed enhanced retention of the drug at tumour sites over equal doses of the same free drug. To target cancer cells actively, they also designed the polymer to be easily modified with antibodies to bind known markers that are present on the surface of cancer cells. Their innovative use of Diels-Alder chemistry allowed them to use mild reaction and processing conditions, better preserving the binding activity of the antibodies during nanoparticle attachment, and further enhancing their platform's ability to target cancer cells. Shi and Ho also improved the uniformity of the nanoparticle size by reducing variation in the lengths of the polymer strands through advanced synthesis

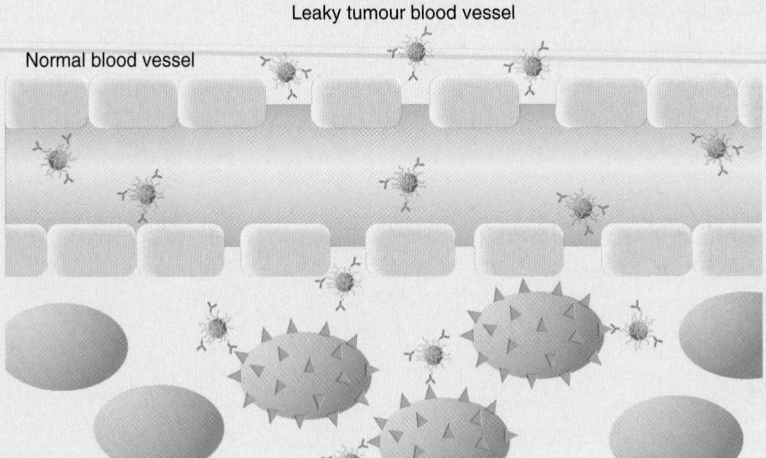

FIGURE B22.7 Passive and active targeting (antibody binding to antigen-expressing cells)

(Continued)

methods, and they have shown that these nanoparticles are stable in biologically relevant media.

Bio-organic Chemistry Dr. Kelvin Kenneth Ogilvie (Figure B22.8A) served as president of Acadia University, in Wolfville, Nova Scotia, from 1993 to 2003, and led the development and implementation of the Acadia Advantage Program, which ensured that all students at Acadia had a personal laptop to aid their studies. Dr. Ogilvie is a leading expert on biotechnology, bio-organic chemistry, and genetic engineering. In 1980, he developed the *Gene Machine*, an automated process for the manufacture of DNA. The Gene Machine allowed DNA sequences to be built in a matter of hours, rather than months. Dr. Ogilvie has many patents, including Ganciclovir (Figure 22.8B), a drug used around the world to fight infections that occur when the immune system is weakened. Both of these achievements were recognized in 2000 as "Milestones of Canadian Chemistry in the 20th Century" by the Canadian Society of Chemistry.

Dr. Ogilvie also developed a general method for the chemical synthesis of large RNA molecules, demonstrated by the first total chemical synthesis of a functional transfer RNA (tRNA) molecule, which is still the basis for RNA synthesis worldwide. In medical research, synthetic RNA is making possible the development of new drugs that were previously beyond the reach of science. The drugs will consist of RNA sequences tailor-made to attach to certain types of viruses and interfere with their ability to replicate themselves. Ganciclovir is one such drug. It kills CMV (cytomegalovirus), a type of herpes that attacks people with weakened immune systems, such as people with AIDS, organ transplant recipients, and cancer patients.

Dr. Ogilvie preferred chemistry over physics and mathematics courses at university and gravitated to chemistry, and more specifically to organic chemistry synthesis. "The excitement over DNA and RNA led me to want to try to be able to chemically construct these exciting molecules to (a) better understand life at the molecular level and (b) to perhaps be able to use that knowledge to improve human health." He was appointed to the Senate of the government of Canada in 2009 and (as of the writing of this book) serves as Chair of the Senate Standing Committee on Social Affairs, Science and Technology. He is a member of the Health Minister's and the Environment Minister's Advisory Caucus Committees and the Chair of Research Canada's Health Research Caucus.

Dr. Ogilvie has written and spoken extensively on the challenges facing Canada as a nation, the role of the "knowledge" economy, postsecondary education, and entrepreneurship. In 2013, Simon Fraser University conferred the degree of Doctor of Science, honoris causa, on Dr. Ogilvie.*

*Source: science.ca and parl.gc.ca/SenatorsBio/senator_biography.aspx?senator_id=2855.

Synthesis of Natural Products, Asymmetric Synthesis, and Biocatalysis A natural product is a compound that is produced by a living organism. Many natural products have a pharmacological or biological activity that can be used in pharmaceutical drug discovery and design. Natural products are a major source of inspiration for drug discovery.

Dr. Tomas Hudlicky (Figure B22.9) and his group at Brock University are engaged in a program in organic synthesis and biocatalysis. The use of enzymatic methods in combination with traditional organic synthesis leads to increased levels of efficiency as well as to more environmentally benign manufacturing. Recently, the Advanced Biomanufacturing Centre was established at Brock University to promote interdisciplinary collaborations between chemists and biologists.

A

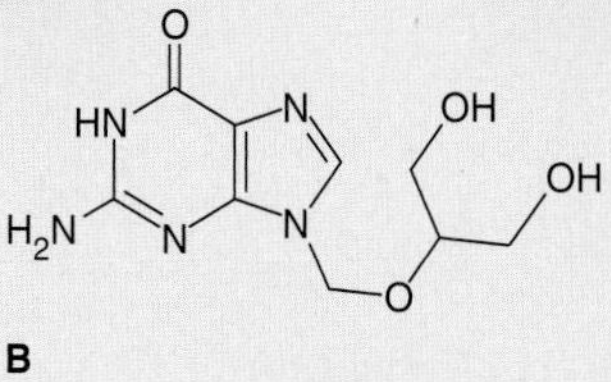

B

FIGURE B22.8 **A.** Kelvin Kenneth Ogilvie. **B.** Ganciclovir (2-amino-9-{[(1,3-dihydroxypropan-2-yl)oxy]methyl}-6,9-dihydro-3H-purin-6-one), which is used to treat or prevent cytomegalovirus infections, was invented by Dr. Ogilvie.

FIGURE B22.9 Tomas Hudlicky (Brock University, Saint Catharines, Ontario)

(*Continued*)

Current research interests in the Hudlicky group include the development of enantioselective synthetic methods, bacterial dioxygenase-mediated degradation of aromatics, design and synthesis of fluorinated inhalation anaesthetic agents, synthesis of amaryllidaceae and morphine alkaloids, and design of unnatural oligosaccharide conjugates with new molecular properties. Some of the targets currently being pursued are shown in Figure B22.10A. The enzymatic dihydroxylation of aromatic compounds, shown in Figure B22.10B, represents a key contribution to the effective design of oxygenated natural products. The program in the Hudlicky group is further focused on the following areas:

Amaryllidaceae alkaloids and their unnatural derivatives. Several total syntheses of these alkaloids have been accomplished by multigenerational design. Unnatural derivatives with nanomolar activities against several cancer cell lines have also been attained.

Morphine alkaloids and opiate-derived pharmaceutical agents. Codeine, hydrocodone, morphine, and hydromorphone, as well as their enantiomers, have been synthesized recently. Current research in the morphinan area is focused on further improvements in efficiency of the total syntheses as well as in process development of analgesic and/or antagonist derivatives such as buprenorphine, naltrexone, naloxone, and others. Included in the process development agendas are new chemical and biological methods for N- and O-demethylation of morphine alkaloids.

Cyclitol and aminocyclitol polymerizations directed at chiral materials. Cyclitol compounds derived from the homochiral diene diols are used in various polymerization schemes to produce chiral polymers and oligomers that can be used as templates for asymmetric synthesis of materials for separation and/or resolution protocols.

Investigations directed at new metabolites for asymmetric synthesis. Toluene and naphthalene dioxygenase-mediated production of new metabolites is an ongoing endeavour in the biocatalytic program to expand the pool of available homochiral intermediates.

Electrochemical methods of oxidation and reduction. These methods are being investigated as potential replacement technology for the use of metal-based reagents.*

Agricultural Chemistry Despite the extensive use of fungicides, insecticides, and herbicides, plant diseases and pests continue to cause enormous crop yield losses worldwide. However, the increasing demand for food requires substantial increases in the production of staple crops. At the same time, environmental issues stress that sustainable agricultural practices cannot rely

*Source: Text and illustrations by Dr. Tomas Hudlicky.

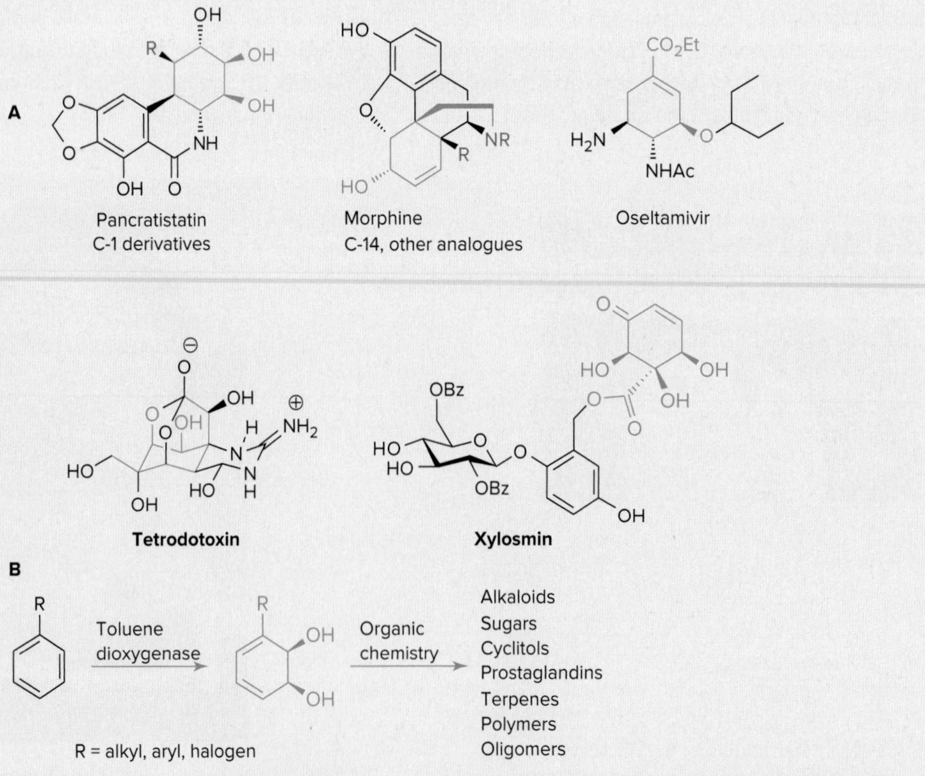

FIGURE B22.10 A. Current targets for total synthesis. **B.** Enzymatic dihydroxylation of aromatic substrates in asymmetric synthesis.

(Continued)

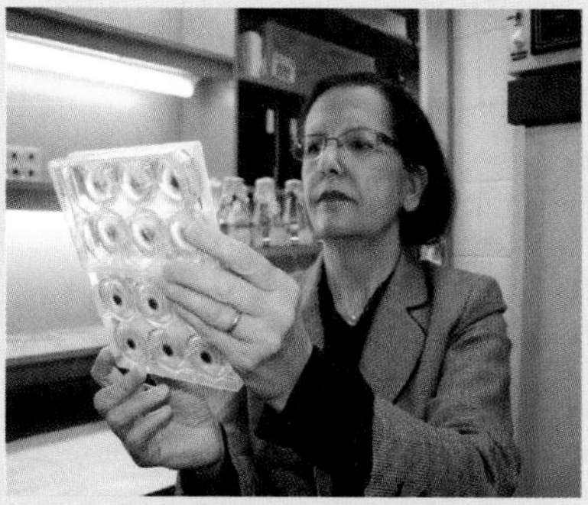

FIGURE B22.11 **Soledade Pedras (University of Saskatchewan, Saskatoon)**

on the continuous application of pesticides. Dr. Soledade Pedras (Figure B22.11) and her group at the University of Saskatchewan are using bio-organic, biochemical, and biological techniques to understand economically important diseases that affect crucifer oilseeds such as canola, rapeseed, and mustard; vegetables such as rutabaga, broccoli, and turnip; and condiments such as mustard and wasabi. In particular, the interaction of crucifers with blackleg, blackspot, root rot, stem rot, and white rust fungi is being investigated (Figure B22.12). The experimental work of Dr. Pedras and her group combines a wide variety of chemical and biological studies, including the following:

- Determination of the chemical structure of bioactive metabolites synthesized by pathogens (such as phytotoxins and elicitors) and plants (such as phytoalexins and phytoanticipins) with the development of bioassays
- Biosynthesis and metabolism of these bioactive compounds in pathogens and plants
- Chemical synthesis of bioactive compounds and intermediates/products of detoxification pathways
- Isolation and characterization of detoxifying enzymes
- Design of paldoxins (inhibitors of phytoalexin detoxifying enzymes)*

*Source: Text and illustrations by Dr. Soledade Pedras.

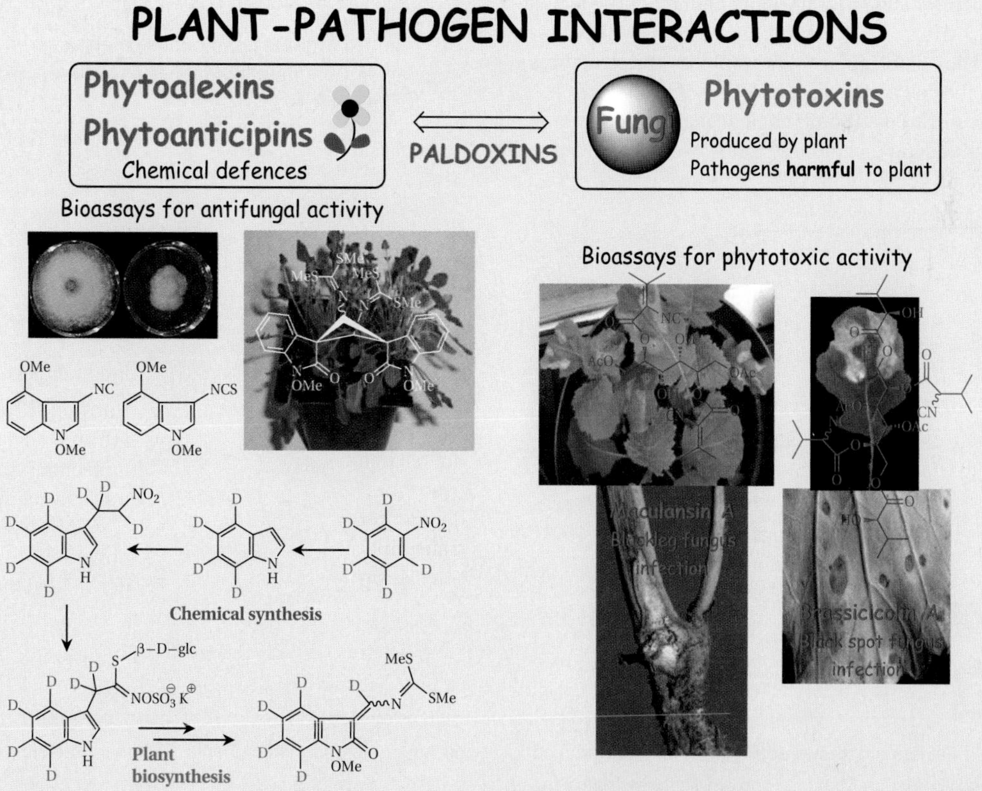

FIGURE B22.12 **Overview of the current research being done by Dr. Pedras's group to understand the chemistry of bio-organic and natural products.** This research includes the chemical synthesis of bioactive compounds and intermediates/products of biosynthetic metabolism and detoxification pathways.

SUMMARY OF SECTION 22.5

- Infrared (IR) spectroscopy is an instrumental technique used mostly to study covalently bonded molecules. The IR spectrum indicates the type of bonds in a molecule based on their molecular vibrations.
- Nuclear magnetic resonance (NMR) gives information about the molecular environment of certain nuclei in a molecule. When the nuclei spin, they create a small magnetic field that results in a magnetic moment proportional to the spin and produces an NMR signal.
- Chromatography is a physical method of separating the components in a mixture, based on their relative attraction for a stationary phase.

CHAPTER REVIEW GUIDE

Learning Objectives

Relevant section (§) numbers appear in parentheses.

Concepts

1. Describe the physical properties of polymers. (§22.1)
2. Explain how addition and condensation polymers form. (§22.2)
3. Describe the three types of biopolymers and their monomers. (§22.3)
4. Explain how an amino acid sequence determines the shape of a protein, which determines its function. (§22.3)
5. Explain how complementary base pairing controls the processes of protein synthesis and DNA replication. (§22.3)
6. Explain how a DNA base sequence determines the RNA base sequence, which determines the amino acid sequence. (§22.3)
7. Explain how intermolecular forces stabilize the structures of proteins, the cell membrane, and DNA. (§22.4)
8. Explain how infrared (IR) and nuclear magnetic resonance (NMR) spectroscopy techniques can be used to analyze and identify organic compounds. (§22.5)
9. Describe the different chromatography methods; explain how they can be used as physical methods of separating and purifying organic compounds. (§22.5)

Skills

1. Draw an abbreviated synthetic polymer structure based on monomer structures. (§22.2)
2. Draw small peptides from amino acid structures. (§22.3)
3. Use the base-paired sequence of one DNA strand to predict the sequence of the other. (§22.3)
4. Identify the peptide bonds and the types of intermolecular forces that are present in biological macromolecules. (§22.4)
5. Identify organic compounds using infrared (IR) and nuclear magnetic resonance (NMR) spectroscopy techniques. (§22.5)

Key Terms

Section 22.1
polymer
monomers
degree of polymerization (n)
coil shape
radius of gyration (R_g)
crystallinity
semicrystalline
viscosity
glass transition temperature (T_g)
plasticity
branches
crosslinks
elastomers

Section 22.2
addition polymers
condensation polymer

Section 22.3
monosaccharides
polysaccharides
disaccharide
proteins
amino acids
nucleic acids
mononucleotides
double helix
base pairs
genetic code
transcription
translation

Section 22.4
soap
lipid bilayer

Section 22.5
infrared (IR) spectroscopy
molecular vibrations
nuclear magnetic resonance (NMR) spectroscopy
downfield
deshielded

PROBLEMS

Problems with red numbers are answered in Appendix G and worked in detail in the Student Solutions Manual. Problem sections match those in this book. Most offer Concept Review Questions, Skill-Building Exercises, and Problems in Context. The Comprehensive Problems are based on material from any section or previous chapter.

(*Note*: The following abbreviations might be used: Me = methyl, Et = ethyl, Pr = propyl, *t*-Bu = *tert*-butyl or 1,1-dimethylethyl, *i*-Pr = isopropyl or 2-methylpropyl, Ph = phenyl, AcOH = ethanoic (acetic) acid, OAc = ethanoate (acetate).)

Polymeric Materials

Concept Review Questions

22.1 If an individual polyethene chain has a degree of polymerization of 6600, calculate the molar mass of the polymer chain (or polymer mass).

22.2 If an individual polychloroethene (polyvinyl chloride) chain has a degree of polymerization of 8000, calculate the molar mass of the polymer chain (or polymer mass).

22.3 If an individual polyphenylethene (polystyrene) chain has a degree of polymerization of 10 000, calculate the molar mass of the polymer chain (or polymer mass).

22.4 If a sample of polyphenylethene (polystyrene) has a number-average molar mass of 80 000 g/mol, calculate the total mass of 0.23 mol of polyphenylethene (polystyrene).

22.5 If a polyethene (polyethylene) repeat unit is approximately 250 pm, calculate the polymer chain length of an individual chain in Problem 22.1.

22.6 If a polychloroethene (polyvinyl chloride) repeat unit is approximately 250 pm, calculate the polymer chain length of an individual chain in Problem 22.2.

22.7 If a polyphenylethene (polystyrene) repeat unit is approximately 250 pm, calculate the polymer chain length of an individual chain in Problem 22.3.

22.8 Calculate the radius of gyration of the polyethene (polyethylene) chain in Problem 22.5. Give the answer to four significant figures.

22.9 Calculate the radius of gyration of the polychloroethene (polyvinyl chloride) chain in Problem 22.6. Give the answer to four significant figures.

22.10 Calculate the radius of gyration of the polyphenylethene (polystyrene) chain in Problem 22.7. Give the answer to four significant figures.

22.11 Define the term *glass transition temperature*. By way of an example, give the glass transition temperature for a polymer that begins to form a glass at 12°C and is complete at 0°C.

22.12 How would you expect the viscosity of a solution to change if the amount of polymer dissolved in the solution was increased?

22.13 Explain the difference between the terms *branching* and *crosslinking*. How do these affect the physical properties of polymers?

22.14 Define the terms *thermoplastic* and *thermosetting plastics*. What are the differences between the molecular structures of these polymer types?

22.15 Define the term *block polymer*. Include a diagram of an ABA block copolymer to illustrate your answer.

Synthetic Macromolecules

Concept Review Questions

22.16 Name the reaction processes that lead to the two types of synthetic polymers.

22.17 Which functional group occurs in the monomers of addition polymers? How are these polymers different from one another?

22.18 What is a free radical? How is it involved in polymer formation?

22.19 Which intermolecular force is primarily responsible for the different types of polyethene? Explain.

22.20 Which of the two types of synthetic polymer is more similar chemically to biopolymers? Explain.

22.21 Which functional groups react to form nylons? Polyesters?

22.22 In condensation polymerization, two monomers are joined together and liberate an additional molecule. What is this molecule?

22.23 Draw an abbreviated structure for each polymer, with brackets around the repeat unit:

(a) polychloroethene or poly (vinyl chloride) (PVC) from

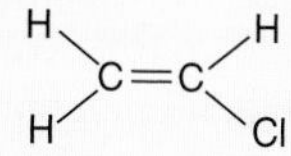

(b) polypropene from

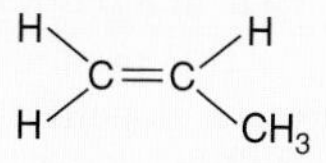

22.24 Draw an abbreviated structure for each polymer, with brackets around the repeat unit:

(a) Teflon from

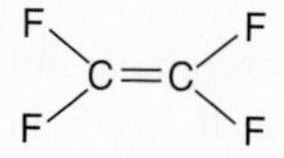

(b) polyphenylethene from

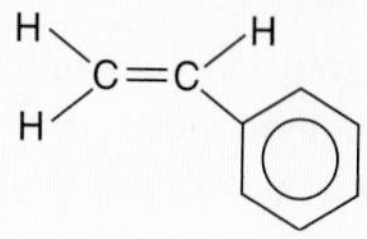

22.25 Polyethene terephthalate (PET) is used to make synthetic fibres such as Dacron, thin films such as Mylar, and bottles for carbonated beverages:

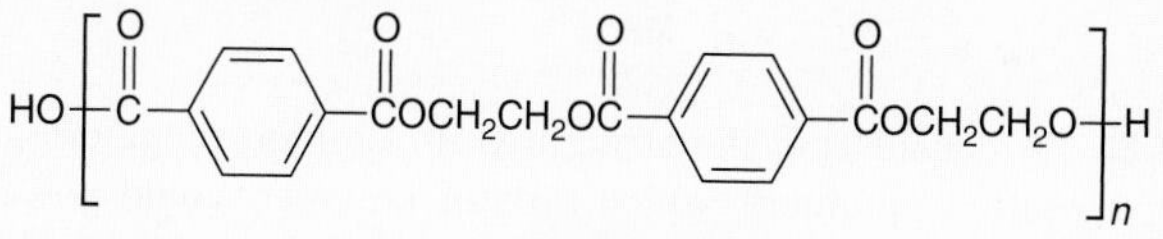

PET

PET is produced from ethane-1,2-diol (ethylene glycol) and either of two monomers, depending on whether the reaction proceeds by dehydration-condensation or by displacement. Write equations for the two syntheses. (*Hint*: The displacement is reversed by adding methanol to PET at high *T* and *p*.)

22.26 Write a balanced equation for the reaction of dihydroxydimethylsilane (below) to form the condensation polymer known as Silly Putty:

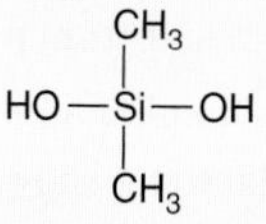

Skill-Building Exercises

22.27 Ethers (general formula R—O—R′) have many important uses. Until recently, 2-methoxy-2-methylpropane (methyl *tert*-butyl ether, MTBE, *below*) was used as an octane booster and fuel additive for gasoline. It increases the oxygen content of the fuel, which reduces CO emissions. MTBE is synthesized by the catalyzed reaction of 2-methylpropene with methanol:

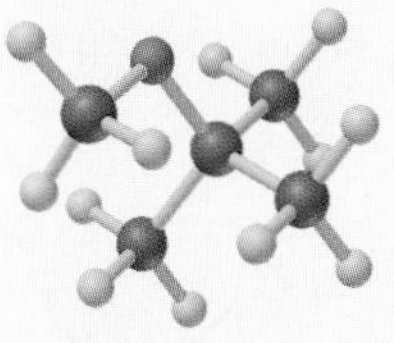

(a) Write a balanced equation for the synthesis of MTBE. (*Hint*: Alcohols add to alkenes similarly to the way water does.)

(b) If the government requires that auto fuel mixtures contain 2.7% oxygen by mass to reduce CO emissions, what amount of MTBE (grams) must be added to each 100 g of gasoline?
(c) What volume (litres) of MTBE would be in each litre of fuel mixture? (The density of both gasoline and MTBE is 0.740 g/mL.)
(d) What volume (litres) of air (21% O_2 by volume) is needed at 24°C and 1.00 bar to fully combust 1.00 L of MTBE?

22.28 In addition to their use in water treatment, ion-exchange resins are used to extract Au, Ag, and Pt ions from solution. One of the most common resins consists of a polymer with a benzene-containing backbone to which sulfonic acid groups (—SO_3H) are added. (a) What monomer can be used to prepare the polymer backbone? (b) This polymer typically contains 4% to 16% cross-linking. Draw the structure of the benzene-containing monomer used to crosslink the polymer.

22.29 Draw a mechanism for the polymerization of chloroethene (vinyl chloride) making sure you highlight the key steps in the reaction.

22.30 AIBN (azobisisobutyronitrile) is a common radical initiator used in polymerization. Draw a mechanism to show how AIBN can thermally decompose, generating nitrogen gas to produce a radical:

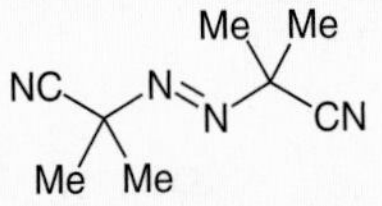

22.31 By considering the structures of addition polymers and condensation polymers, which type of polymer would you expect to be more biodegradable?

22.32 Below are diagrams showing different possible structures of polypropene. Explain which type of polypropene you would expect to have the highest melting point.

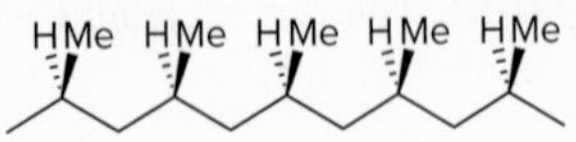

Isotatic

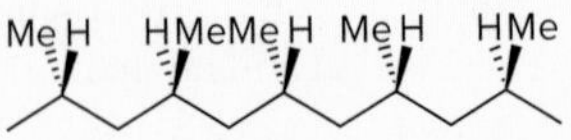

Atatic

22.33 By considering the mechanism of polymerization, explain how chain branching can result from polymerization.

22.34 Comparing the polyamide polymer (caprolactam) and the polyester polymer (caprolactone) shown below, which would you predict to have the higher glass transition temperature (T_g)? Why?

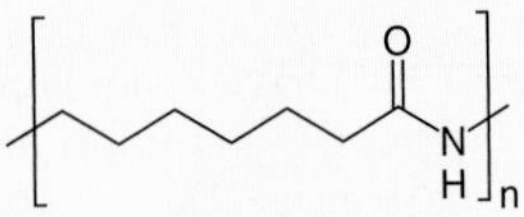

Polycaprolactam

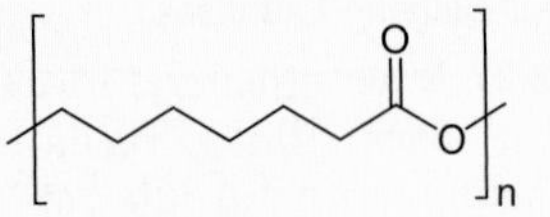

Polycaprolactone

Biological Macromolecules

Concept Review Questions

22.35 In the structure below, identify the anomeric position on the sugar:

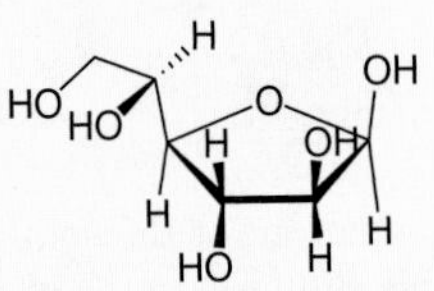

22.36 Which type of polymer is formed from each monomer: (a) amino acid; (b) alkene; (c) simple sugar; (d) mononucleotide?

22.37 What linkage joins the monomers in each strand of DNA?

22.38 What is base pairing? How does it pertain to DNA structure?

22.39 RNA base sequence, protein amino acid sequence, and DNA base sequence are interrelated. Which determines which in the process of protein synthesis?

22.40 Complete hydrolysis of a 100.00 g sample of a peptide gave the following amounts of individual amino acids (molar masses, in g/mol, appear in parentheses):

3.00 g of glycine (75.07) 0.90 g of alanine (89.10)
3.70 g of valine (117.15) 6.90 g of proline (115.13)
7.30 g of serine (105.10) 86.00 g of arginine (174.21)

(a) Why does the total mass of amino acids exceed the mass of the peptide? (b) What are the relative numbers of amino acids in the peptide? (c) What is the minimum molar mass of the peptide?

22.41 Describe which structural features of glycogen make it ideal for energy storage compared with cellulose.

22.42 Using the terms *monosaccharide* and *disaccharide*, explain the differences between glucose, fructose, and sucrose.

22.43 Using the terms *ketose* (a sugar containing a ketone) and *aldose* (a sugar containing an aldehyde), classify glucose and fructose appropriately.

22.44 *Nucleotides*, *nucleosides*, and *nucleobase* are often confused. By illustrating your answer with uracil, explain the differences between these terms.

Skill-Building Exercises

22.45 Draw the R group of (a) alanine; (b) histidine; (c) methionine.

22.46 Draw the R group of (a) glycine; (b) isoleucine; (c) tyrosine.

22.47 Draw the structure of each tripeptide:
(a) Aspartic acid-histidine-tryptophan
(b) Glycine-cysteine-tyrosine with the charges that exist in cell fluid

22.48 Draw the structure of each tripeptide:
(a) Lysine-phenylalanine-threonine
(b) Alanine-leucine-valine with the charges that exist in cell fluid

22.49 Write the sequence of the complementary DNA strand that pairs with each DNA base sequence:
(a) TTAGCC (b) AGACAT

22.50 Write the sequence of the complementary DNA strand that pairs with each DNA base sequence:
(a) GGTTAC (b) CCCGAA

22.51 Write the base sequence of the DNA template from which this RNA sequence was derived: UGUUACGGA. How many amino acids are coded for in this sequence?

22.52 Write the base sequence of the DNA template from which this RNA sequence was derived: GUAUCAAUGAACUUG. How many amino acids are coded for in this sequence?

22.53 DNA chains are more stable than RNA to degradation. By considering the structural differences between these two molecules, explain why this may be the case.

Intermolecular Forces and Biological Macromolecules

Concept Review Questions

22.54 Name three intermolecular forces that stabilize the shape of a soluble globular protein, and explain how they act.

22.55 Name three intermolecular forces that stabilize the structure of DNA, and explain how they act.

22.56 How can relatively weak hydrogen bonds hold the double helix together and yet allow DNA to function?

22.57 Is the sodium salt of propanoic acid as effective a soap as sodium stearate? Explain.

22.58 What intermolecular forces stabilize a lipid bilayer?

22.59 In what way do proteins embedded in a membrane differ structurally from soluble proteins?

22.60 Histones are proteins that control gene function by attaching through salt links to exterior regions of DNA. Name an amino acid whose side chain is often found on the exterior of histones.

22.61 β-pleated sheets are an important structural feature of proteins. What type of structural class are they part of: primary, secondary, tertiary, or quaternary?

22.62 What is the key structural difference between fibrous and globular proteins? How is it related, in general, to the proteins' amino acid composition?

22.63 Protein shape, function, and amino acid sequence are interrelated. Which determines which?

22.64 Protein shapes are maintained by a variety of forces that arise from interactions between the amino acid R groups. Name the amino acid that possesses each R group and the force that could arise in each interaction:

(a) $-CH_2-SH$ with $HS-CH_2-$

(b) $-(CH_2)_4-NH_3^+$ with $^-O-C(=O)-CH_2-$

(c) $-CH_2-C(=O)-NH_2$ with $HO-CH_2-$

(d) 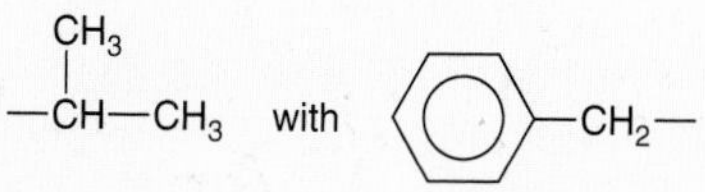

22.65 Amino acids have an average molar mass of 100 g/mol. How many bases on a single strand of DNA are needed to code for a protein with a molar mass of 5×10^5 g/mol?

Skill-Building Exercises

22.66 The polypeptide chain in proteins does not exhibit free rotation because of the partial double-bond character of the peptide bond. Explain this fact with resonance structures.

22.67 Draw the base pairing of guanine and cytosine, and highlight the key intermolecular forces responsible for recognition.

22.68 Identify the key inter- and intramolecular forces responsible for maintaining the tertiary structure of proteins.

Instrumental Analysis

Concept Review Questions

22.69 IR is an important tool for the analysis of compounds. What part of the molecule does it provide information about?

22.70 IR signals can be viewed as a measure of bond strength. The IR signal for an OH stretch is characteristically broad. Can you explain why?

22.71 As stated in Problem 22.70, IR signals can be viewed as a measure of bond strength, with larger wavenumbers corresponding to stronger bonds. Given this, explain why the carbonyl of propan-2-one (acetone) absorbs at 1715 cm^{-1} and ethanoyl chloride (acetyl chloride) resonates at 1799 cm^{-1}.

22.72 As stated in Problem 22.70, IR signals can be viewed as a measure of bond strength, with larger wavenumbers corresponding to stronger bonds. Given this, explain why the carbonyl of propan-2-one (acetone) absorbs at 1715 cm^{-1}, whereas amides resonate at approximately 1650 cm^{-1}.

22.73 NMR is an important tool for the analysis of molecules. What part of a molecule does ^1H NMR examine?

22.74 How is the ^1H NMR scale referenced? Why is this a good choice of reference?

22.75 What is the main purpose of chromatography? Explain in general terms how it works.

Skill-Building Exercises

22.76 Ignoring coupling, how many peaks would you expect to see in the ^1H NMR spectrum of propan-2-one (acetone)?

22.77 Ignoring coupling, how many peaks would you expect to see in the ^1H NMR spectrum of benzaldehyde?

22.78 Ignoring coupling, how many peaks would you expect to see in the ^1H NMR spectrum of tetrachloromethane?

22.79 The ^1H NMR spectrum of DMF (*N,N*-dimethylformamide) shows three signals. Explain why.

22.80 Which protons in the following molecule would you expect to have the highest ^1H NMR chemical shift? Explain why.

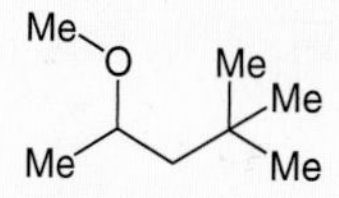

22.81 Alcohol (OH) resonances have a wide range of ^1H NMR chemical shifts. A common way to confirm the presence of an alcohol is with a technique called a "D_2O shake." In this process, a solution of the compound in $CDCl_3$ (where D is deuterium, ^2H), where the OH signal is visible, is mixed with a small quantity of D_2O. When the ^1H NMR is recorded again, the OH shift will have disappeared. Explain.

22.82 A compound with the formula $C_4H_{10}O$ has one large peak in the ^1H NMR spectrum at approximately 1 ppm and a broad peak at 3 ppm, which disappears following a D_2O shake (see Problem 22.81). Provide a possible structure of the molecule.

22.83 ^{13}C NMR works similarly to ^1H NMR, except it detects the number of carbon environments. How many peaks would you expect to see in the ^{13}C NMR spectrum of propan-2-one (acetone)?

22.84 How many peaks would you expect to see in the ^{13}C NMR spectrum of toluene (see Problem 22.83)?

22.85 Considering coupling, describe the multiplicity and number of peaks in a ^{1}H NMR spectrum of ethyl ethanoate (ethyl acetate) in $CDCl_3$.

22.86 Considering coupling, describe the multiplicity and number of peaks in a ^{1}H NMR spectrum of *N,N*-diethylethanamine (triethylamine) in $CDCl_3$.

22.87 Considering coupling, describe the multiplicity and number of peaks in a ^{1}H NMR spectrum of 1,2-dichloroethane.

Comprehensive Problems

22.88 In a preparatory synthetic chemistry lab, column chromatography is usually conducted on the achiral material silica gel (or silicon dioxide). If you were provided a mixture of the following three compounds, which would you be able to separate and why?

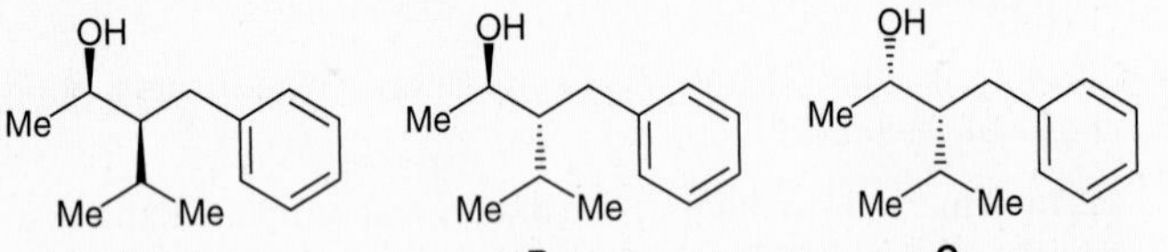

A B C

22.89 Compound A, composed of C, H, and O, is heated at 1.01 bar in a 1.00 L flask to 160.°C until all of the A has vaporized and displaced the air. The flask is then cooled, and 2.48 g of A remains. When 0.500 g of A burns in O_2, 0.409 g of H_2O and 1.00 g of CO_2 are produced. Compound A is not acidic, but it can be oxidized to compound B, which is weakly acidic: 1.000 g of B is neutralized with 33.9 mL of 0.5 mol/L sodium hydroxide. When B is heated to 260°C, it gives off water and forms compound C, whose solution in $CDCl_3$ (where D is deuterium, ^{2}H) has one peak in its ^{1}H NMR spectrum. (a) What are the structures of A, B, and C? (b) Compound A is a controlled substance because it is metabolized to the weakly acidic "date rape" drug GHB ($C_4H_8O_3$). What are the structure and name of GHB?

22.90 The structure of Kevlar is shown below. Which monomer units are required to generate this polymer?

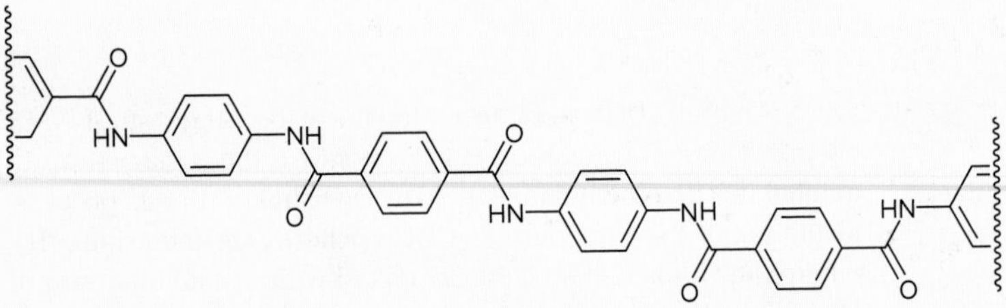

22.91 The structure of Nomex is shown below. Which monomer units are required to generate this polymer?

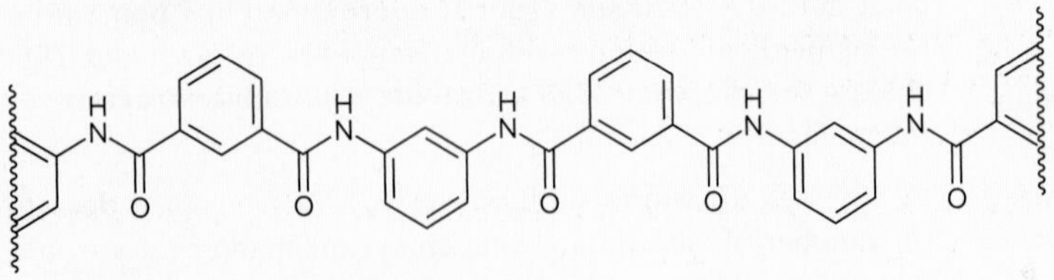

22.92 By comparing the structural features of each polymer, suggest why Kevlar is more rigid and dense than Nomex (see Problems 22.90 and 22.91).

22.93 A polymer has an IR spectrum with a strong sharp peak at 2280 cm^{-1} and a ^{1}H NMR spectrum with two signals: a quintuplet at 2.5 ppm and a triplet at 1.9 ppm. If the polymer chain has a molar mass ($\mathcal{M}_{polymer}$) of 77 096 g/mol and a degree of polymerization of 1453, what is the monomer unit? Label your structure to show which groups are responsible for the IR and NMR signals. What type of polymerization is this?

22.94 (a) Describe in broad terms the order of events that must occur to generate a protein, starting from DNA in the nucleus.
(b) If a large amount of proteins were generated in the first step and then dissolved in ethanol, what would be the effect on the solution as the concentration of the protein in solution increased?
(c) If the secondary structure in a single protein unit contains covalent crosslinks, which amino acid must be present in this protein?
(d) Can this protein attain a quaternary structure?

22.95 (a) A short DNA nucleotide contains the following sequence: GATCGACTA. What is the complementary DNA sequence? Is it different from the complementary RNA sequence? If so, explain why and provide the new sequence.
(b) How many amino acids are coded for in this sequence?
(c) Could gas chromatography be used to separate this nucleotide sequence from a mixture of other nucleotides? If not, propose a better separation method.

22.96 SIBS (**S**tyrene-**I**so**B**utene-**S**tyrene) is an ABA block copolymer composed of styrene and 2-methylpropene (isobutene). If a sample of SIBS composed of 60 mol% *iso*butene and 40 mol% styrene were analyzed by size-exclusion chromatography (SEC) and found to have a particle size (diameter) of 5.5 nm, what would be the molar mass of the polymer ($\mathcal{M}_{polymer}$)? The repeat unit length for styrene is 260 pm and for *iso*butene is 240 pm. This material is a *thermoplastic elastomer*; define what that means. How many distinct peaks would you expect in the ^{1}H NMR spectrum (ignore splitting)?

22.97 (a) Nylon 6-10 can be synthesized in the laboratory by combining 1,6-diaminohexane with sebacoyl dichloride (shown below). Draw the reaction mechanism and indicate what type of polymerization this is:

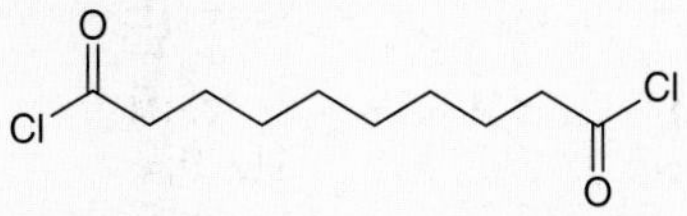

(b) If the glass transition temperature (T_g) of nylon 6-10 is 50°C, would this polymer be suitable for making rubber gloves?
(c) What would be the best chromatography technique to analyze this polymer?

22.98 Two polyethene polymers were separated using size-exclusion chromatography (SEC), with one polymer eluting at 6 min and the second polymer eluting at 9 min. Dynamic light-scattering measurements recorded the diameter of one polymer to be 16.2 nm and the other polymer to be 5.3 nm. Unfortunately, a power outage occurred during the analysis and the data were garbled before the molar mass could be determined. The instrument will now be in the repair shop for 3 months and you need the molar mass values for an important presentation in 3 h. Using what you know about SEC, determine which light-scattering measurement corresponded to which elution time, and also calculate the molar mass of each polymer. The length of the repeat unit for the polymer you were analyzing is 250 pm.

The Elements in Nature and Industry

IN THIS CHAPTER . . . We examine the distribution of elements on Earth, the intertwining natural cycles in which certain elements take part, and the methods we have developed to extract, purify, and use them. By the end of this chapter, you should be able to

- Identify the abundances and sources of elements in the various regions of Earth
- Describe how three essential elements—carbon, nitrogen, and phosphorus—cycle through the environment and how humans influence these cycles
- Explain the isolation and purification of metals, discussing general redox and metallurgical procedures for extracting an element from its ore
- Describe in detail the isolation and uses of seven key elements: sodium, potassium, iron, copper, aluminum, magnesium, and hydrogen
- Discuss the relevance and importance of two key processes in chemical manufacturing: the production of sulfuric acid and the isolation of chlorine

23

While we cannot function without our toiletry items, cosmetics, medications, or electronic items, we rarely put much thought into how these products come to us in their final form. Elements, with the exception of a very few, are not found in their pure form on Earth, and chemists and engineers put a lot of work into purifying and refining the raw materials so they can be converted into useful products. There are also many industrially important chemicals whose value we never hear about. In this chapter, we will describe how the elements are distributed on Earth and examine important natural and industrial processes that affect our lives. We will also learn the importance of topics studied in the previous chapters and their effects on the production of the many items we consume on a daily basis.

Concepts and Skills to Review before Studying This Chapter

- Trends in atomic properties (Sections 7.3, 7.4, 8.2, and 8.5)
- Catalysts and reaction rate (Section 14.7)
- Le Châtelier's principle (Section 15.6)
- Acid-base equilibria (Section 16.1)
- Solubility and complex-ion equilibria (Sections 17.3 and 17.4)
- Temperature and reaction spontaneity (Section 18.3)
- Gibbs energy and equilibrium (Section 18.4)
- Standard electrode potentials (Section 19.3)
- Electrolysis of molten salts and aqueous salt solutions (Section 19.7)

23.1 How the Elements Occur in Nature

To begin our examination of how we use the elements, let us take inventory of our elemental stock—the distribution and relative amounts of the elements on Earth, especially the thin outer portion of Earth that we can reach.

Earth's Structure and the Abundance of the Elements

Any attempt to isolate an element must begin with knowledge of its **abundance**, the amount of the element in a particular region of the natural world. The abundances of the elements result from details of our planet's evolution.

Formation and Layering of Earth About 4.5 billion years ago, vast clouds of cold gases and interstellar debris from exploded older stars gradually coalesced into the Sun and planets. At first, Earth was a cold, solid sphere of uniformly distributed elements and simple compounds. Over the next billion years or so, heat from radioactive decay and continuous meteor impacts raised the planet's temperature to around 10^4 K, sufficient to form an enormous molten mass. Any remaining gaseous elements, such as the cosmically abundant hydrogen and helium, were ejected into space.

As Earth cooled, chemical and physical processes resulted in its **differentiation**, the formation of regions of different composition and density. Differentiation gave Earth an internal structure that consists of three layers (Figure 23.1):

- The dense **core**, which has a density of 10 g/cm^3 to 15 g/cm^3, is composed of a molten outer core and a solid Moon-size inner core. Remarkably, the inner core is nearly as hot as the surface of the Sun and spins within the molten outer core slightly faster than Earth itself does!
- The thick, homogeneous **mantle**, which has an overall average density of 4 g/cm^3 to 6 g/cm^3, lies around the core.
- A thin, homogeneous **crust** lies on top of the mantle. The average density of the crust, on which all life takes place, is 2.8 g/cm^3.

Table 23.1 compares the abundance of some key elements in the universe, on Earth as a whole (actually the core plus the mantle, which account for more than 99% of

FIGURE 23.1 The layered internal structure of Earth

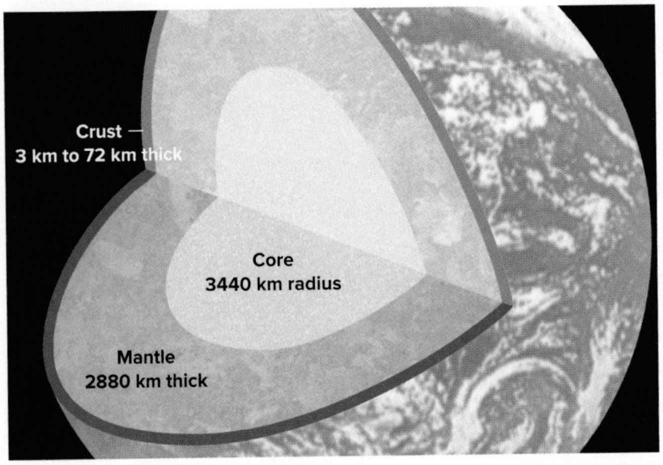

Earth's mass), and in Earth's three regions. (Because the deepest terrestrial sampling can penetrate only a few kilometres into the crust, some of these data represent extrapolations from meteor samples and from seismic studies of earthquakes.) Several points stand out:

- Cosmic and whole-Earth abundances are very different, particularly for H and He. These two elements account for the majority of the mass of the universe, but are not abundant in elemental form on Earth.
- The elements O, Si, Fe, and Mg are abundant both in the universe and on Earth. Together, they account for more than 90% of Earth's mass.
- The core is particularly rich in the dense groups 8 through 10 metals: Co, Ni, and especially Fe, the most abundant element in the whole Earth.
- Crustal abundances are very different from whole-Earth abundances. The crust makes up only 0.4% of Earth's mass but has the largest share of nonmetals; metalloids; and light, active metals: Al, Ca, Na, and K. The mantle contains much smaller proportions of these metals, and the core has none. Oxygen is the most abundant element in the crust and mantle, but it is absent from the core.

TABLE 23.1 Cosmic and Terrestrial Abundances (Mass %) of Selected Elements*

Element	Universe	Earth	Crust	Mantle	Core
O	1.07	29.5	49.5	43.7	—
Si	0.06	15.2	25.7	21.6	—
Al	—	1.1	7.5	1.8	—
Fe	0.19	34.6	4.7	13.3	88.6
Ca	0.007	1.1	3.4	2.1	—
Mg	0.06	12.7	2.8	16.6	—
Na	—	0.6	2.6	0.8	—
K	—	0.07	2.4	0.2	—
H	73.9	—	0.87	—	—
Ti	—	0.05	0.58	0.18	0.01
Cl	—	—	0.19	—	—
P	—	—	0.12	—	—
Mn	—	—	0.09	—	—
C	0.46	—	0.08	—	—
S	0.04	1.9	0.06	>2	—
Ni	0.006	2.4	0.008	0.3	8.5
Co	—	0.13	0.003	0.04	0.6
He	24.0	—	—	—	—

*A missing abundance value indicates that either reliable data are not available or the value is less than 0.001% by mass.

These differences in Earth's major layers arose from the effects of thermal energy. When Earth was molten, gravity and convection caused more-dense materials to sink and less-dense materials to rise, yielding several compositional *phases*:

- Most of the Fe sank to form the core, or *iron phase.*
- In the light outer phase, oxygen combined with Si, Al, Mg, and some Fe to form silicates, the material of rocks. This *silicate phase* later separated into the mantle and the crust.
- The *sulfide phase,* intermediate in density and insoluble in the other two phases, consisted mostly of iron sulfide and mixed with parts of the silicate phase above and the iron phase below.
- The thin, primitive atmosphere, probably a mixture of water vapour (which gave rise to the oceans), carbon monoxide, and nitrogen (or ammonia), was produced by *outgassing* (the expulsion of trapped gases).

The distribution of the remaining elements (discussed here using only the new periodic table group numbers) was controlled by their chemical affinity for one of the three phases. In general terms, as Figure 23.2 shows, they are as follows:

- Elements with low or high electronegativity—active metals (those in groups 1 through 5, Cr, and Mn) and nonmetals (O, lighter members of groups 13 to 15, and all of group 17)—tended to congregate in the silicate phase as ionic compounds.
- Metals with intermediate electronegativities (many from groups 6 to 10) dissolved in the iron phase.
- Lower-melting transition metals and many metals and metalloids in groups 11 to 16 became concentrated in the sulfide phase.

The Impact of Life on Crustal Abundances At present, the crust is the only physically accessible portion of Earth, so only crustal abundances have practical significance. The crust is divided into the **lithosphere** (the solid region of the crust), the **hydrosphere** (the liquid region), and the **atmosphere** (the gaseous region). Over billions of years, weathering and volcanic upsurges have dramatically altered the composition of the crust. The **biosphere**, which consists of the living systems that have inhabited the planet, has been another *major influence on crustal element composition and distribution.*

When the earliest rocks were forming and the ocean basins were filling with water, binary inorganic molecules in the atmosphere were reacting to form first simple, and then more complex, organic molecules. The energy for these (mostly) endothermic changes was supplied by lightning, solar radiation, geologic heating, and meteoric impact. In an amazingly short period of time, probably no more than

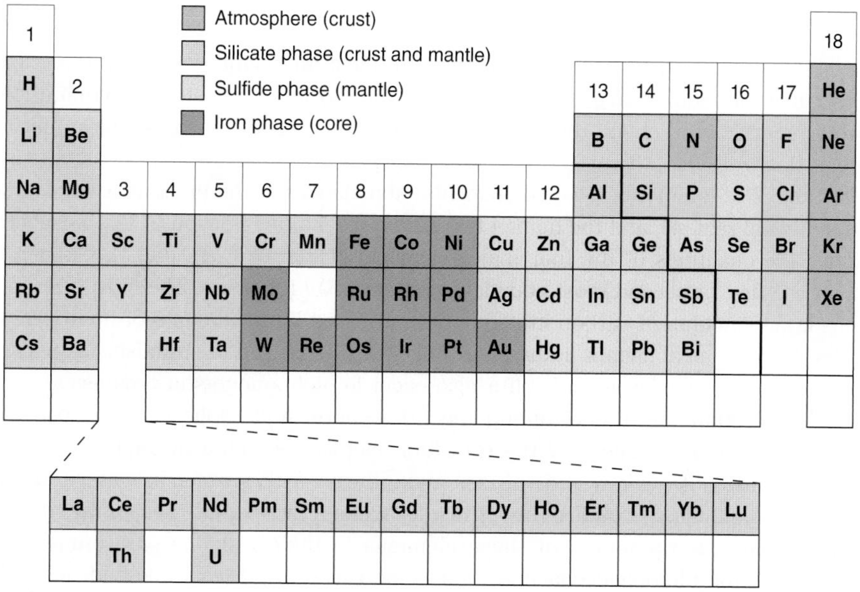

FIGURE 23.2 Geochemical differentiation of the elements

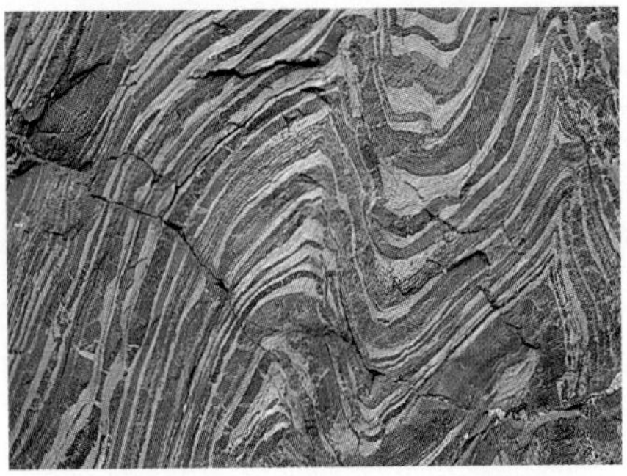

A

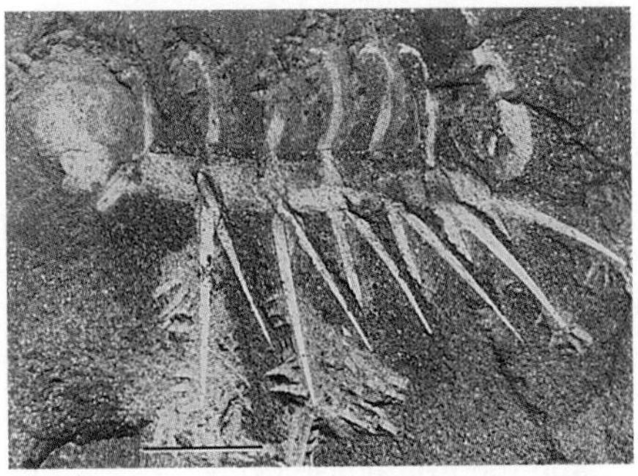

B

FIGURE 23.3 Ancient effects of an O_2-rich atmosphere. A. Banded-iron formations containing Fe_2O_3. **B.** A fossil of an early multicellular organism.

500 million years, the first organisms appeared. In less than another billion years, these organisms evolved into simple algae that could derive metabolic energy from photosynthesis, converting CO_2 and H_2O into organic molecules and releasing O_2 as a by-product.

Let us look at some of the essential ways that the evolution of organisms affected the chemistry of the crust.

1. Owing to photosynthesis, the atmosphere gradually became richer in O_2 over the 300 million years following the emergence of algae. Thus, *oxidation became the major source of free energy in the crust and biosphere.* As a result of this oxidizing environment, the predominating Fe(II)-containing minerals were oxidized to Fe(III)-containing minerals, such as hematite (Fe_2O_3), shown in Figure 23.3A in ancient banded-iron formations (red beds). The increase in oxygen also led to an explosion of O_2-using life forms, a few of which evolved into the organisms of today. Figure 23.3B shows a fossil of one of the multicellular organisms whose appearance coincided with the increase in O_2.

2. The K^+ concentration in the oceans is much lower than the Na^+ concentration. Forming over eons from outgassed water vapour condensing into rain and streaming over the land, the oceans became complex ionic solutions of dissolved minerals with 30 times as much Na^+ as K^+. One reason for the low K^+ concentration is that plants, which require K^+ for growth, absorbed dissolved K^+ that would otherwise have washed down to the oceans.

Dolomite Mountains, Italy

3. There are enormous subterranean deposits of organic carbon. These deposits formed when ancient plants, buried deep and decomposing under high pressure and temperature in the absence of free O_2, gradually turned into coal, or when animals, buried under similar conditions, turned into petroleum. Crustal deposits of organic carbon still provide the fuels that move our cars, heat our homes, and electrify our cities.

4. Carbon, oxygen, and calcium (the fifth most abundant element in the crust) occur in vast sedimentary deposits all over the world as limestone, dolomite (*see photo*), marble, and chalk, the fossilized skeletal remains of countless early marine organisms.

Table 23.2 compares selected elemental abundances in the crust as a whole; the three crustal regions; and the human body, a representative portion of the biosphere. Notice the quantities of the four major elements of life: O, C, H, and N. Oxygen is either the first or second most abundant element in all the areas. The biosphere contains large amounts of carbon in its biomolecules and large amounts of hydrogen as water. Nitrogen is abundant in the atmosphere as free N_2 and in organisms as part of their proteins. Phosphorus and sulfur also occur in high amounts in organisms.

One striking difference among the lithosphere, hydrosphere, and biosphere (human) is the abundances of the transition metals vanadium through zinc. Their relatively insoluble oxides and sulfides make them much scarcer in water than on land. Yet organisms, which evolved in the oceans, developed the ability to concentrate the trace amounts of these elements in their aqueous environment. In every area, the biological concentration is increased by at least 100-fold, with Mn

TABLE 23.2 Abundances (Mass %) of Selected Elements in the Crust, Its Regions, and the Human Body as Representative of the Biosphere

		Crustal Regions			
Element	**Crust**	**Lithosphere**	**Hydrosphere**	**Atmosphere**	**Human Body**
O	49.5	45.5	85.8	23.0	65.0
C	0.08	0.018	—	0.01	18.0
H	0.87	0.15	10.7	0.02	10.0
N	0.03	0.002	—	75.5	3.0
P	0.12	0.11	—	—	1.0
Mg	1.9	2.76	0.13	—	0.50
K	2.4	1.84	0.04	—	0.34
Ca	3.4	4.66	0.05	—	2.4
S	0.06	0.034	—	—	0.26
Na	2.6	2.27	1.1	—	0.14
Cl	0.19	0.013	2.1	—	0.15
Fe	4.7	6.2	—	—	0.005
Zn	0.013	0.008	—	—	0.003
Cr	0.02	0.012	—	—	3×10^{-6}
Co	0.003	0.003	—	—	3×10^{-6}
Cu	0.007	0.007	—	—	4×10^{-4}
Mn	0.09	0.11	—	—	1×10^{-4}
Ni	0.008	0.010	—	—	3×10^{-6}
V	0.015	0.014	—	—	3×10^{-6}

increasing about 1000-fold and Cu, Zn, and Fe increasing even more. Each of these elements performs an essential role in living systems.

Sources of the Elements

Given an element's abundance in a region of the crust, we must determine its **occurrence (source)**, the form(s) in which the element exists. Practical considerations often determine the commercial source. Oxygen, for example, is abundant in all three crustal regions, but the atmosphere is its primary industrial source because it occurs there as the free element. Nitrogen and the noble gases (except helium) are also obtained from the atmosphere. Several other elements occur uncombined, formed in large deposits by prehistoric biological action, such as sulfur in caprock salt domes and nearly pure carbon in coal. Gold and platinum, two relatively unreactive elements, also occur in an uncombined (*native*) state.

The overwhelming majority of elements, however, occur in **ores**, natural compounds or mixtures from which an element can be extracted economically (*see photo*): *the financial costs of mining, isolating, and purifying must be considered when choosing a process to obtain an element.*

Deposit of borate (tufa), the ore of boron

Figure 23.4 shows the most useful sources of the elements. Notice that elements with the same types of ores tend to be grouped together in the periodic table:

- Alkali metal halides are ores for both of their component groups of elements, group 1 and group 17.
- Group 2 metals occur as carbonates in the marble and limestone of mountain ranges, although magnesium is very abundant in seawater (making seawater the preferred source, as we will discuss later).
- Even though many elements occur as silicates, most of these compounds are very stable thermodynamically. Thus, the cost that would be required to process them prohibits their use. Aside from silicon, only lithium and beryllium are obtained from their silicates.
- The ores of most industrially important metals are either oxides, which dominate the left half of the transition series, or sulfides, which dominate the right half of the transition series, and a few of the main groups beyond the right-hand transition elements.

FIGURE 23.4 Sources of the elements

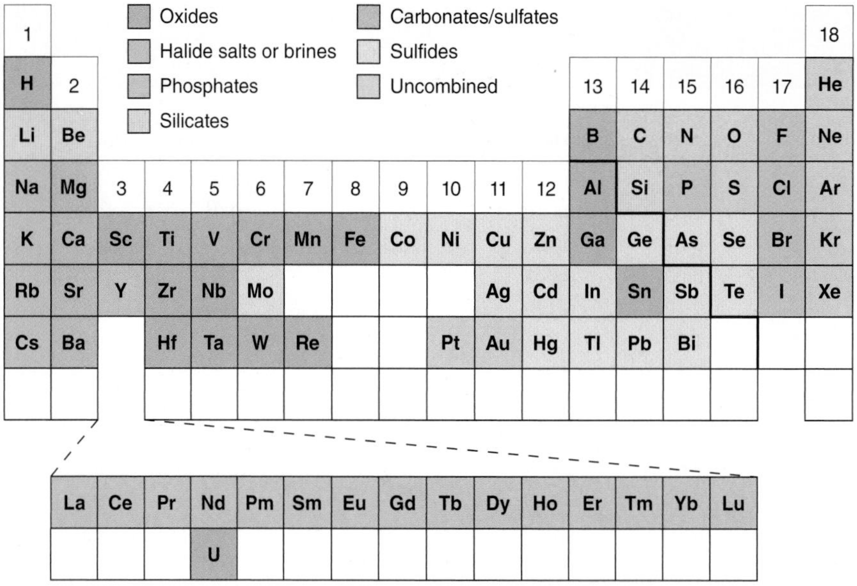

The reasons for the prominence of oxide and sulfide ores of metallic elements are complex and include weathering processes, selective precipitation, and relative solubilities. Certain atomic properties are relevant as well. Metals toward the left side of the periodic table have lower ionization energies and electronegativities, so they tend to give up electrons or hold electrons loosely in bonds. The O^{2-} ion is small enough to approach a metal cation closely, which results in a high lattice energy for the oxide. In contrast, metals toward the right side have higher ionization energies and electronegativities. Thus, they tend to form bonds that are more covalent, which suits the larger, more polarizable S^{2-} ion.

SUMMARY OF SECTION 23.1

- As the young Earth cooled, the elements became differentiated into a dense metallic core, a silicate-rich mantle, and a low-density crust.
- High abundances of light metals, metalloids, and nonmetals are concentrated in the crust, which consists of three regions: lithosphere (solid), hydrosphere (liquid), and atmosphere (gaseous).
- The biosphere (living systems) profoundly affected crustal chemistry by producing free O_2 and, thus, an oxidizing environment.
- Some elements occur in their native state, but most are combined in ores. The ores of most important metallic elements are oxides or sulfides.

23.2 The Cycling of Elements through the Environment

The distributions of many elements change at widely differing rates. The physical, chemical, and biological paths that atoms of an element take through regions of the crust constitute the element's **environmental cycle**. In this section, we consider the cycles for carbon, nitrogen, and phosphorus, highlighting the effects of humans on these cycles.

The Carbon Cycle

Carbon is one of a handful of elements that appear in all three regions of Earth's crust. In the lithosphere, it occurs in elemental form as graphite and diamond; in fully oxidized form in carbonate minerals; in fully reduced form in petroleum hydrocarbons; and in complex mixtures, such as coal and living matter. In the hydrosphere, it occurs in living matter, in carbonate minerals formed by the action of coral reef

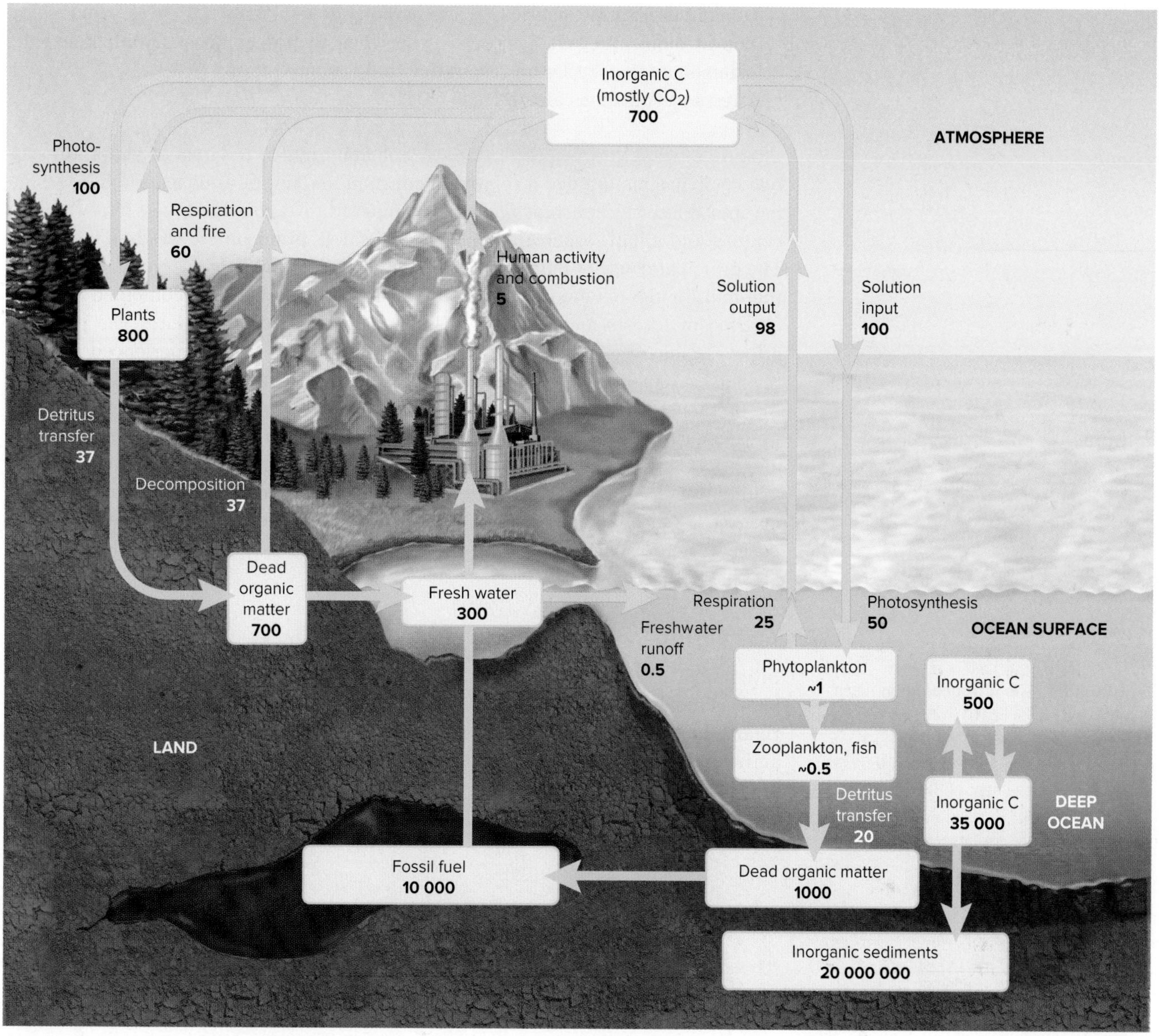

FIGURE 23.5 The carbon cycle. Each number in a box refers to the size of the source; each number along an arrow refers to the annual movement of the element from source to source. Values are in 10^9 t of C (1 t = 1000 kg).

organisms, and in dissolved CO_2 and related oxoanions. In the atmosphere, it occurs principally in gaseous CO_2, a minor but essential component that exists in equilibrium with the aqueous fraction.

Key Interactions within the Carbon Cycle Figure 23.5 depicts the complex interplay of sources and the effect of the biosphere on carbon's environmental cycle. It provides an estimate of the size of each source and, where possible, the amount of carbon moving annually between sources. Notice these key points:

- The parts of the cycle are linked by the atmosphere—one link between oceans and air, the other link between land and air. The 2.6×10^{12} t (tonne) of CO_2 in air cycles through the oceans and atmosphere about once every 300 years but spends a much longer time in carbonate minerals.
- Atmospheric CO_2 accounts for only 0.003% of crustal carbon but moves the element through the other regions. A CO_2 molecule spends, on average, 3.5 years in the atmosphere.
- The land and oceans are in contact with the largest sources of carbon, immobilized as carbonates, coal, and oil in rocky sediment beneath the soil.
- *Photosynthesis*, *respiration*, and *decomposition* are major factors in the carbon cycle. **Fixation** is the process of converting a gaseous substance into a condensed form. Via photosynthesis, marine plankton and terrestrial plants use

sunlight to fix atmospheric CO_2 into carbohydrates. Plants release CO_2 by respiration at night and, much more slowly, when they decompose. Animals eat the plants and release CO_2 by respiration and decomposition.
- Fires and volcanoes also release CO_2 into the air.

Buildup of CO_2 and Global Warming For hundreds of millions of years, the carbon cycle has maintained a relatively constant amount of atmospheric CO_2, resulting in a relatively constant planetary temperature range. However, over the past century and a half, especially since World War II, atmospheric CO_2 has increased. The principal cause is human activities: the combustion of coal, wood, and oil for fuel and for the decomposition of limestone to make cement, combined with the clearing of forests and jungles for lumber, paper, and agriculture. The principal effect is climate change. The decades since the 1990s have experienced some of the highest temperatures recorded, and there is increasingly abundant evidence that the previously mentioned human activities have led to global warming through the greenhouse effect (see Chemical Connections in Chapter 5). Higher temperatures, alterations in dry and rainy seasons, melting of polar ice, and increasing ocean acidity, with associated changes in carbonate equilibria, are already being observed. Nearly all science policy experts are pressing for programs that combine conservation of carbon-based fuels, an end to deforestation, and development of alternative energy sources.

The Nitrogen Cycle

In contrast to the carbon cycle, the nitrogen cycle includes a *direct* interaction between land and sea (Figure 23.6). All nitrites and nitrates are soluble, so rain and runoff contribute huge amounts of nitrogen to lakes, rivers, and oceans. Human activities, through the use of fertilizer, play a major role in this cycle.

FIGURE 23.6 The nitrogen cycle. Each number in a box is in 10^9 t of N and refers to the size of the source; each number along an arrow is in 10^6 t of N and refers to the annual movement of the element between sources.

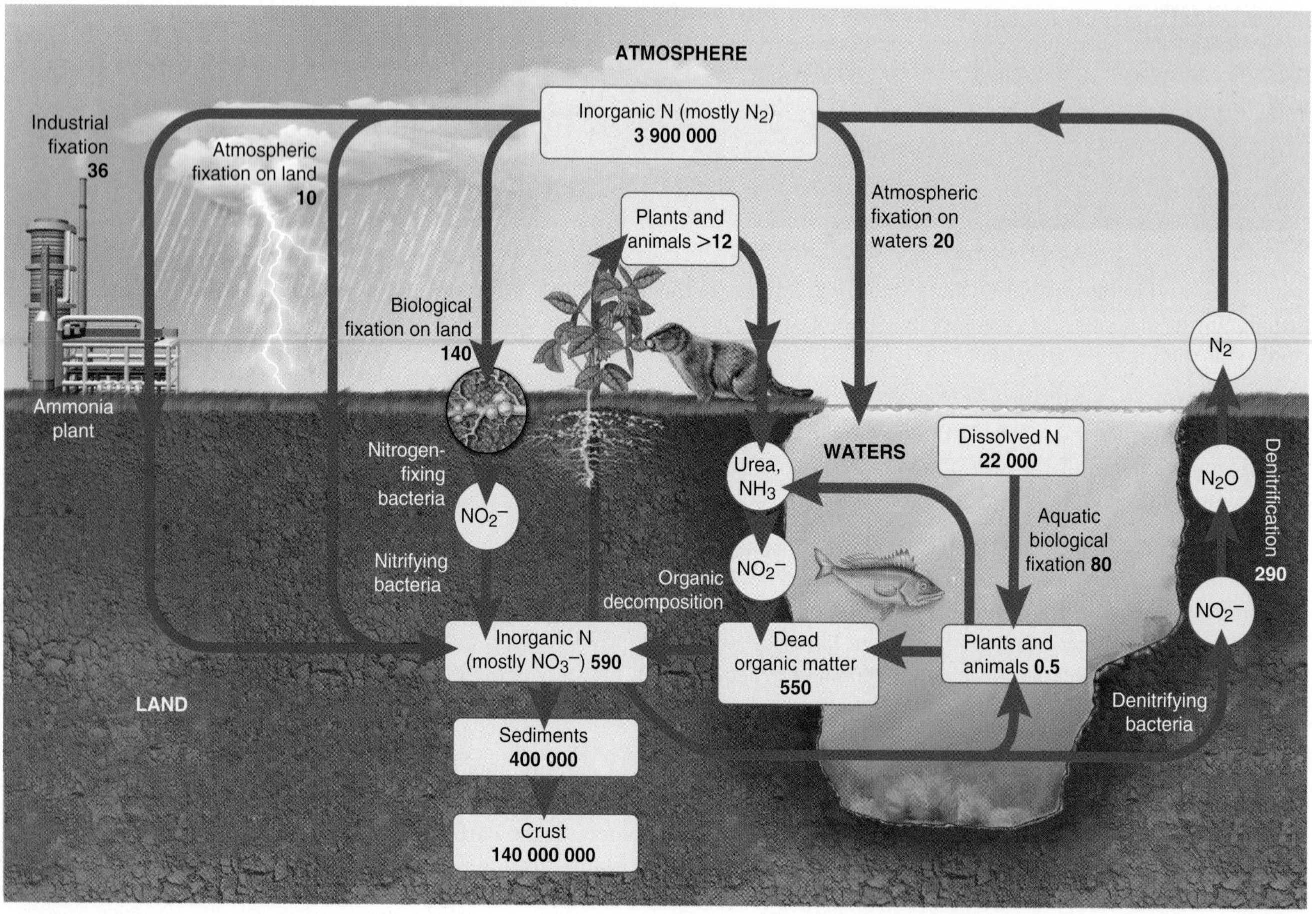

Like carbon dioxide, atmospheric nitrogen must be fixed. However, whereas CO_2 can be incorporated by plants in either its gaseous or its aqueous form, the great stability of N_2 prevents plants from using it directly. Fixation of N_2 requires a great deal of energy and occurs through atmospheric, industrial, and biological processes.

Atmospheric Fixation Lightning causes the high-temperature endothermic reaction of N_2 and O_2 to form NO, which is then oxidized by ozone exothermically to NO_2:

$$N_2(g) + O_2(g) \longrightarrow 2NO(g) \qquad \Delta H^\circ = 180.6\ \text{kJ}$$
$$NO(g) + O_3(g) \longrightarrow NO_2(g) + O_2(g) \qquad \Delta H^\circ = -198.9\ \text{kJ}$$

During the day, NO_2 reacts with the hydroxyl radical to form HNO_3:

$$NO_2(g) + HO\cdot(g) \longrightarrow HNO_3(g)$$

At night, a multistep reaction between NO_2 and ozone is involved. When rain dissociates the nitric acid, $NO_3^-(aq)$ enters both oceans and land to be used by plants.

Industrial Fixation Most human-caused fixation occurs industrially during ammonia synthesis via the Haber process (Chapter 15):

$$N_2(g) + 3H_2(g) \rightleftharpoons 2NH_3(g)$$

The Haber process takes place on an enormous scale, and NH_3 ranks first, on a mole basis, among compounds produced industrially. Some of this NH_3 is converted to HNO_3 in the Ostwald process (Chapter 13), but most is used as fertilizer, either directly or in the form of urea and ammonium salts (sulfate, phosphate, and nitrate), which enter the biosphere when taken up by plants (see Figure 23.6).

In recent decades, high-temperature combustion in electrical power plants and in car, truck, and airplane engines has become an important contributor to total fixed nitrogen. High operating temperatures in the engines mimic lightning to form NO from the air taken in to burn the hydrocarbon fuel. The NO in exhaust gases reacts to form nitric acid in the atmosphere, which adds to the nitrate load that reaches the ground.

The overuse of fertilizers *and* automobiles presents an increasingly serious water pollution problem in many areas. Leaching of the land by rain causes nitrate from fertilizer use and vehicle operation to enter lakes, rivers, and coastal estuaries and causes *eutrophication*, the depletion of O_2 and the death of aquatic animal life from excessive algal and plant growth and decomposition. Excess nitrate also spoils nearby drinkable water.

Biological Fixation The biological fixation of atmospheric N_2 occurs in blue-green algae and in nitrogen-fixing bacteria that live on the roots of leguminous plants (such as peas, alfalfa, and clover). These microbial processes dwarf the previous two, fixing more than seven times as much nitrogen as the atmosphere and six times as much as industry. Root bacteria fix N_2 by enzymatically reducing it to NH_3 and NH_4^+. Enzymes in other soil bacteria catalyze the multistep oxidation of NH_4^+ to NO_2^- and finally NO_3^-, which the plants reduce again to make proteins. When the plants die, still other soil bacteria oxidize the proteins to NO_3^-.

Animals eat the plants, break down the plant proteins to make their own proteins, and excrete nitrogenous wastes, such as urea, $(H_2N)_2C{=}O$. The nitrogen in the proteins is released when the animals die and decompose and is converted by soil bacteria to NO_2^- and NO_3^- again. This central pool of inorganic nitrate has three main fates: some enters marine and terrestrial plants; some enters the enormous sediment store of mineral nitrates; and some is reduced by denitrifying bacteria to NO_2^- and then reduced further to N_2O and N_2, which reenter the atmosphere to complete the cycle.

The Phosphorus Cycle

Virtually all the mineral sources of phosphorus contain the phosphate group, PO_4^{3-}. The most commercially important ores are **apatites**, compounds with the

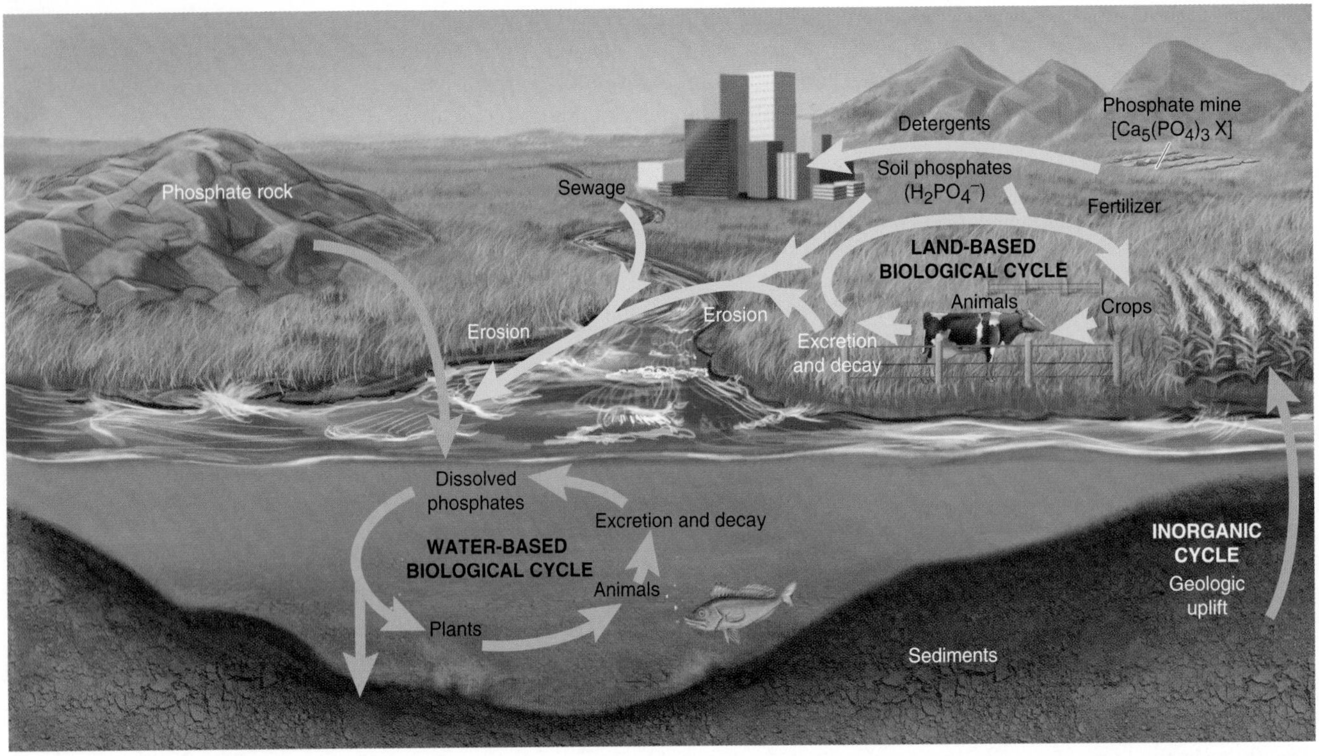

FIGURE 23.7 The phosphorus cycle

general formula $Ca_5(PO_4)_3X$, where X is usually F, Cl, or OH. The cycling of phosphorus through the environment involves three interlocking subcycles (Figure 23.7). Two rapid biological cycles—a land-based cycle (*yellow arrows*), completed over several years, and a water-based cycle (*blue arrows*), completed in weeks to years—are superimposed on an inorganic cycle (*purple arrows*) that takes millions of years to complete. Unlike the carbon and nitrogen cycles, the phosphorus cycle has no gaseous component and, thus, does *not* involve the atmosphere. Since phosphorus is a nonrenewable resource, it is important to note that the world's supply of readily available ore containing phosphorus has already peaked. It has been predicted that global commercial phosphate reserves will be depleted in 50 to 100 years at the current rates of extraction, leaving reserves that are either lower quality or that require greater expense to extract the phosphorus that remains.

The Inorganic Cycle Although most phosphorus occurs in phosphate rock formed when Earth's lithosphere solidified, a sizable amount has come from outer space in meteorites. About 100 t (tonne) of meteorites enters Earth's atmosphere every day. With an average P content of 0.1% by mass, these meteorites contribute about 10^{11} t of phosphorus to the crust over Earth's lifetime. The most important components of phosphate rock are nearly insoluble phosphate salts:

$$K_{sp} \text{ of } Ca_3(PO_4)_2 \approx 10^{-29}$$
$$K_{sp} \text{ of } Ca_5(PO_4)_3OH \approx 10^{-51}$$
$$K_{sp} \text{ of } Ca_5(PO_4)_3F \approx 10^{-60}$$

Weathering slowly leaches phosphates from the soil and carries the ions through rivers to the oceans. Plants in the land-based biological cycle speed this process. In the oceans, some phosphate is absorbed by organisms in the water-based biological cycle, but the majority is precipitated again by Ca^{2+} ion and deposited on the continental shelves. Geologic activity lifts the continental shelves, returning the phosphate to the land.

The Land-Based Biological Cycle The biological cycles involve the incorporation of phosphate into organisms (biomolecules, bones, teeth, and so on) and the release of phosphate through their excretion and decomposition. In the land-based cycle, plants continuously remove phosphate from the inorganic cycle. Recall that there are three phosphate oxoanions, which exist in equilibrium in aqueous solution:

$$H_2PO_4^- \underset{+\,H^+}{\overset{-\,H^+}{\rightleftharpoons}} HPO_4^{2-} \underset{+\,H^+}{\overset{-\,H^+}{\rightleftharpoons}} PO_4^{3-}$$

In topsoil, phosphates occur as insoluble compounds of Ca^{2+}, Fe^{3+}, and Al^{3+}. Because plants can absorb only the water-soluble dihydrogen phosphate, they have evolved the ability to secrete acids near their roots to convert the insoluble salts gradually into the soluble ion:

$$Ca_3(PO_4)_2(s;\ \text{soil}) + 4H^+(aq;\ \text{secreted by plants}) \longrightarrow 3Ca^{2+}(aq) + 2H_2PO_4^-(aq;\ \text{absorbed by plants})$$

Animals that eat the plants excrete soluble phosphate, which is used by newly growing plants. As plants and animals excrete, die, and decompose, some phosphate is washed into rivers and oceans. Most of the 2 million t of phosphate that washes from land to oceans each year comes from this biological source. Thus, the biosphere greatly increases the movement of phosphate from the lithosphere to the hydrosphere.

The Water-Based Biological Cycle Various phosphate oxoanions continuously enter the aquatic environment. Studies that incorporate trace amounts of radioactive phosphorus into $H_2PO_4^-$ and monitor its uptake show that, within 1 min, 50% of the ion is taken up by photosynthetic algae, which use it to synthesize their biomolecules (denoted by the top curved arrow in the following equation, which shows the overall process):

$$106CO_2(g) + 16NO_3^-(aq) + H_2PO_4^-(aq) + 122H_2O(l) + 17H^+(aq) \underset{\text{decomposition}}{\overset{\text{synthesis}}{\rightleftharpoons}} C_{106}H_{263}O_{110}N_{16}P(aq;\ \text{in algal cell fluid}) + 138O_2(g)$$

(The complex formula on the second line represents the total composition of algal biomolecules, not a particular compound.)

As on land, animals eat the plants and are eaten by other animals, and they all excrete phosphate, die, and decompose (denoted by the bottom curved arrow in the equation above). Some of the released phosphate is used by other aquatic organisms, and some returns to the land when fish-eating animals (mostly humans and birds) excrete phosphate, die, and decompose. In addition, some aquatic phosphate precipitates with Ca^{2+}, sinks to the seabed, and returns to the long-term mineral deposits of the inorganic cycle.

Human Effects on Phosphorus Movement From prehistoric through preindustrial times, the three interlocking phosphorus subcycles were balanced. Modern human activities, however, have altered the movement of phosphorus considerably. Annually, we use more than 97 million t of phosphate rock. Figure 23.8 shows the major end products, with the box height proportional to the mass percent of phosphate rock used annually. Removal of phosphate rock has minimal influence on the inorganic cycle, but the end products have unbalanced the two biological cycles.

The imbalance arises from our overuse of *soluble phosphate fertilizers*, such as $Ca(H_2PO_4)_2$ and $NH_4H_2PO_4$, the end products of about 85% of the phosphate rock mined. Some fertilizers find their way into rivers, lakes, and oceans to enter the water-based cycle. Much larger pollution sources, however, are the crops grown with the fertilizers and the detergents made from the phosphate rock. The great majority of the phosphate pollution arrives eventually in cities as crops; as animals that were fed crops; and as consumer products, such as the tripolyphosphates in detergents. Human garbage, excrement, wash water with detergents,

FIGURE 23.8 Industrial uses of phosphorus

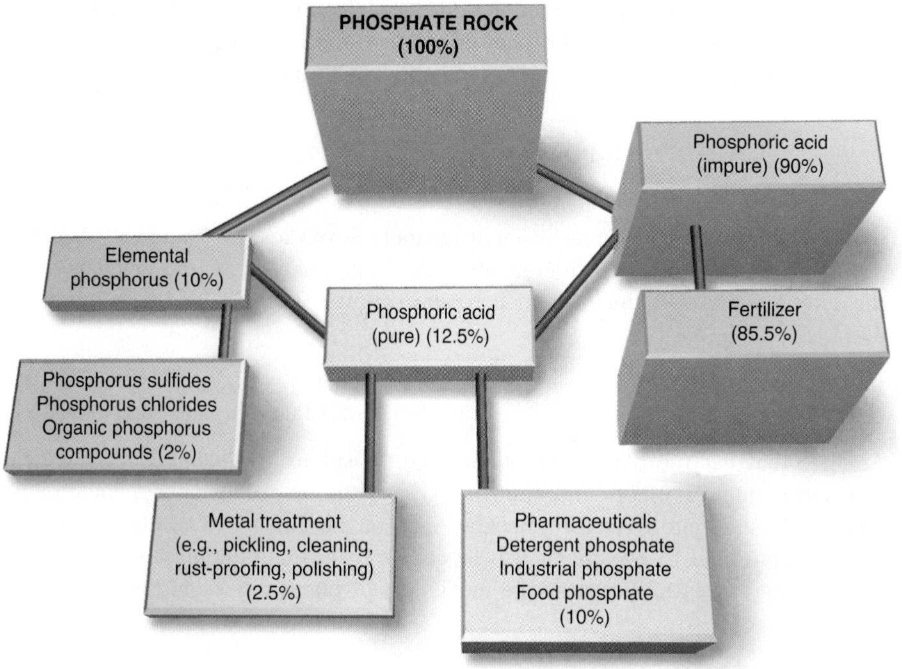

and industrial wastewater containing phosphate are carried through sewers to the aquatic system. This human contribution equals the natural contribution—another 2 million t of phosphate per year. The increased concentration in rivers and lakes causes eutrophication, which robs the water of O_2 so that it cannot support life. Such "dead" rivers and lakes are no longer usable for fishing, drinking, or recreation.

A century ago, in 1912, Lake Zurich in Switzerland was choked with algae and became devoid of fish. Treatment with $FeCl_3$ precipitated enough phosphate to gradually return the lake to its natural state. Soluble aluminum compounds have the same effect; however, these reactions do not occur at neutral pH:

$$FeCl_3(aq) + PO_4^{3-}(aq) \longrightarrow FePO_4(s) + 3Cl^-(aq)$$

$$KAl(SO_4)_2\cdot 12H_2O(aq) + PO_4^{3-}(aq) \longrightarrow AlPO_4(s) + K^+(aq) + 2SO_4^{2-}(aq) + 12H_2O(l)$$

New research proposes the removal of PO_4^{3-} by surface complexation, using hydrous ferric oxide (HFO, $Fe_2O_3\cdot nH_2O$).*

SUMMARY OF SECTION 23.2

- The environmental distribution of many elements changes cyclically with time and is affected in major ways by organisms.
- Carbon occurs in all three regions of Earth's crust, with atmospheric CO_2 linking the other two regions. Photosynthesis and decomposition of organisms alter the amount of carbon in the land and oceans. Human activities have increased atmospheric CO_2, leading to climate change.
- Nitrogen is fixed by lightning, by industries, and primarily by micro-organisms. When plants or animals decompose, bacteria convert proteins to nitrites and nitrates, and eventually to N_2, which returns to the atmosphere. Through extensive use of fertilizers, humans have added excess nitrogen to freshwater rivers and lakes.
- The phosphorus cycle has no gaseous component. Inorganic phosphates leach slowly into land and water, where the biological cycles interact. When plants and animals decompose, they release phosphate to natural bodies of water, where it is then absorbed by other plants and animals. Through overuse of phosphate fertilizers and detergents, humans double the amount of phosphorus that enters the aqueous environment.

*D. S. Smith, I. Takács, S. Murthy, G. T. Daigger, and A. Szabó, "Phosphate complexation model and its implications for chemical phosphorus removal," *Water Environment Research* 80, 5 (2008), pp. 428–438.

Depletion of Earth's Ozone Layer

The stratospheric ozone layer absorbs UV radiation with wavelengths between 280 nm and 320 nm. If it reaches Earth's surface, this radiation has enough energy to damage genes by breaking bonds in DNA. Thus, depletion of stratospheric ozone increases human health risks, particularly the risks of skin cancer and cataracts (clouding of the eye's lens). The radiation emitted by the Sun may also damage plant life, especially forms at the bottom of the food chain. Both homogeneous and heterogeneous catalysts play roles in the depletion of ozone.

Before 1976, stratospheric ozone concentration varied seasonally but remained nearly constant from year to year as a result of a series of atmospheric reactions:

$$O_2 \xrightarrow{UV} 2O \quad \text{(dissociation of } O_2 \text{ by UV)}$$
$$O + O_2 \longrightarrow O_3 \quad \text{(ozone formation)}$$
$$O_3 + O \longrightarrow 2O_2 \quad \text{(ozone breakdown)}$$

Then, in 1985, British scientists reported that a severe reduction of ozone, an *ozone hole*, had appeared in the stratosphere over the Antarctic when the Sun ended the long winter darkness. Subsequent research by Paul J. Crutzen, Mario J. Molina, and F. Sherwood Rowland, for which they received the Nobel Prize in Chemistry in 1995, revealed that industrial chlorofluorocarbons (CFCs) were lowering $[O_3]$ in the stratosphere by catalyzing the breakdown reaction.

Widely used as aerosol propellants, foaming agents, and air-conditioning coolants, large quantities of CFCs were released for years. Unreactive in the troposphere, CFC molecules gradually reach the stratosphere, where UV radiation splits them:

$$CF_2Cl_2 \xrightarrow{UV} CF_2Cl\cdot + Cl\cdot$$

(The dots are lone electrons resulting from C—Cl bond cleavage.)

Like many species with a lone electron (free radicals), atomic Cl is very reactive. It reacts with stratospheric O_3 to produce the intermediate chlorine monoxide ($ClO\cdot$), which then reacts with a free O atom to regenerate a Cl atom:

$$O_3 + Cl\cdot \longrightarrow ClO\cdot + O_2$$
$$ClO\cdot + O \longrightarrow Cl\cdot + O_2$$

The sum of these steps is the ozone breakdown reaction:

$$O_3 + \cancel{Cl\cdot} + \cancel{ClO\cdot} + O \longrightarrow \cancel{ClO\cdot} + O_2 + \cancel{Cl\cdot} + O_2$$

or

$$O_3 + O \longrightarrow 2O_2$$

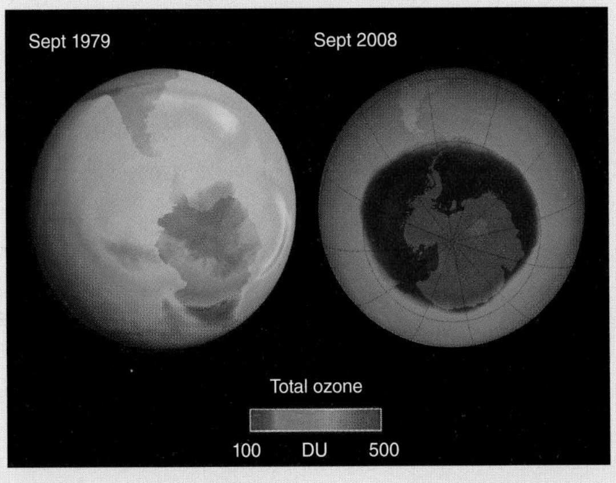

FIGURE B23.1 Satellite images of the increasing size of the Antarctic ozone hole (*purple*)

Studies finding high $[ClO\cdot]$ over Antarctica supported this mechanism. Figure B23.1 shows the ozone hole expanding over the years in which CFCs accumulated.

Figure B23.2 shows the reaction energy diagram for the process. Note that *Cl acts as a homogeneous catalyst*: it exists in the same phase as the reactants, lowers the total activation energy via a different mechanism, and is regenerated. Each Cl atom has a stratospheric half-life of about 2 years, during which it speeds the breakdown of about 100 000 ozone molecules.

Later studies showed that the ozone hole enlarges by heterogeneous catalysis. Stratospheric clouds provide a surface for reactions that convert inactive chlorine compounds, such as HCl and chlorine nitrate ($ClONO_2$), to substances, such as Cl_2, that are cleaved by UV radiation to Cl atoms. Fine particles in the stratosphere act in the same way: dust from the 1991 eruption of Mount Pinatubo reduced stratospheric ozone for two years. The Montreal Protocol of 1987 and later amendments curtailed CFC production and proposed substituting hydrocarbon propellants, such as methylpropane [isobutane, $(CH_3)_3CH$]. Nevertheless, because of the long lifetimes of CFCs, complete recovery of the ozone layer may take another century! The good news is that tropospheric halogen levels have begun to fall.

However, the *National Post* reported in 2011 that a massive Arctic ozone hole similar in size to the Antarctic ozone hole (about 2 million km^2), had appeared over the Northern Hemisphere. The hole allowed high levels of UV radiation to penetrate to large parts of northern Canada, Europe, and Russia.

According to Environment Canada, a rising trend of the average outdoor ambient concentration of ground-level ozone was detected from 1990 to 2010 (Figure B23.3). The increase is mainly due to warmer and drier years and to transboundary pollution from the United States.

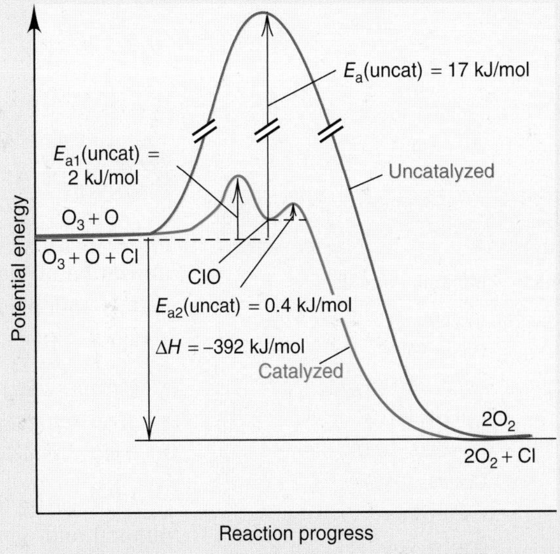

Figure B23.2 Reaction energy diagram for breakdown of O_3 by Cl atoms. (Not drawn to scale.)

(Continued)

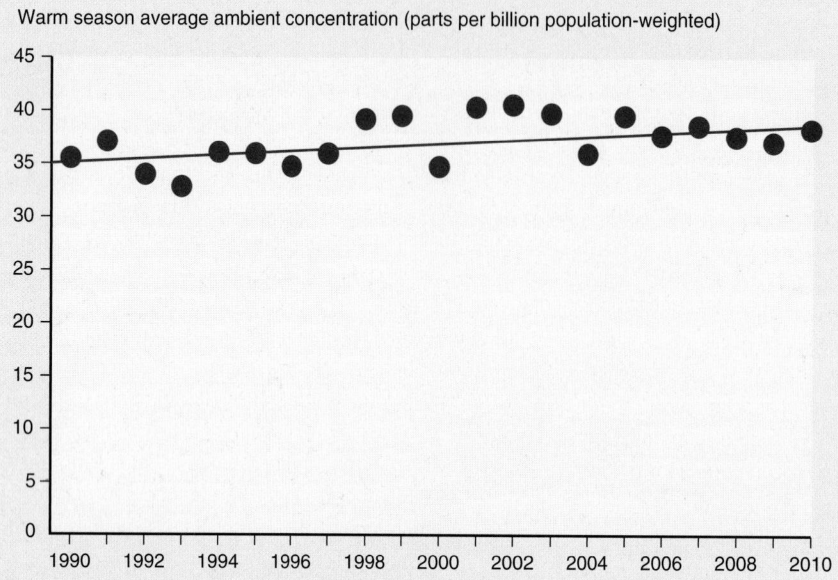

FIGURE B23.3 Ground-level ozone concentrations, Canada, 1990 to 2010

Problems

B23.1 The catalytic destruction of ozone occurs via a two-step mechanism, where X can be one of several species:

(1) $X + O_3 \longrightarrow XO + O_2$ (slow)

(2) $XO + O \longrightarrow X + O_2$ (fast)

(a) Write the overall reaction and the rate law for each step.

(b) X acts as _____, and XO acts as _____.

B23.2 Aircraft in the stratosphere release NO, which catalyzes ozone breakdown by the mechanism shown in Problem B23.1.

(a) Write the mechanism.

(b) When $[O_3]$ is 5×10^{12} molecules/cm^3 and [NO] is 1.0×10^9 molecules/cm^3, what is the rate of O_3 depletion [k for the rate-determining step is 6×10^{-15} cm^3/(molecule·s)]?

23.3 Metallurgy: Extracting a Metal from Its Ore

TABLE 23.3 Common Mineral Sources of Some Elements

Element	Mineral, Formula
Al	Gibbsite (in bauxite), $Al(OH)_3$
Ba	Barite, $BaSO_4$
Be	Beryl, $Be_3Al_2Si_6O_{18}$
Ca	Limestone, $CaCO_3$
Fe	Hematite, Fe_2O_3
Hg	Cinnabar, HgS
Na	Halite, NaCl
Pb	Galena, PbS
Sn	Cassiterite, SnO_2
Zn	Sphalerite, ZnS

Metallurgy is the branch of materials science that focuses on the extraction and utilization of metals. In this section, we discuss the general procedures for isolating metals and, occasionally, the nonmetals that are found in their ores. The extraction process applies one or more of these three types of metallurgy:

- *Pyrometallurgy,* which uses heat to obtain the metal
- *Electrometallurgy,* which employs an electrochemical step
- *Hydrometallurgy,* which relies on the metal's aqueous solution chemistry

The extraction of an element begins with *mining the ore.* Most ores consist of mineral and gangue. The **mineral** contains the element; it is defined as a naturally occurring, homogeneous, crystalline inorganic solid, with a well-defined composition. The **gangue** is the portion of the ore that has no commercial value, such as sand, rock, or clay. Table 23.3 lists the common mineral sources of some metals.

Canada is one of the top mining nations in the world, producing more than 60 minerals and metals. There are approximately 1262 mining operations in Canada, which contribute nearly 7.2% of our gross domestic product and account for 19.6% of our total exports. The products of mining provide many of the essential consumer goods we use, as well as much of the infrastructure we rely on daily, such as highways, electrical and communications networks, and housing. It is difficult to imagine a world without mining products! Check out the following examples of mining products in action:

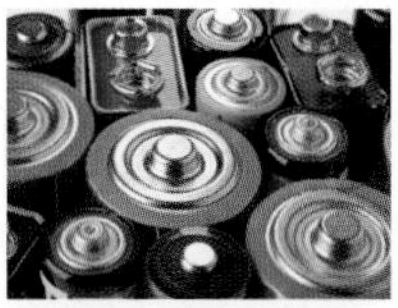

Batteries: cadmium, lithium, nickel, and cobalt

Circuitry: gold, copper, aluminum, steel, silver, lead, and zinc

Computer and television screens: silicon, boron, lead, phosphorus, and indium

Cosmetics and jewellery: gold, diamonds, silver, platinum, iron oxide, zinc, and titanium dioxide

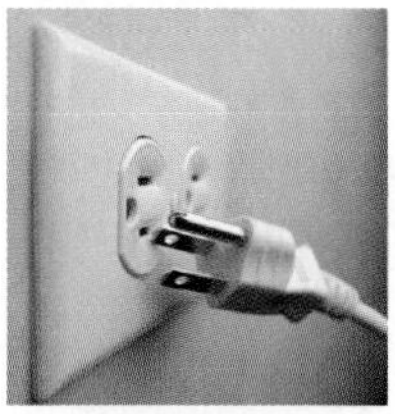

Electricity: coal and uranium

Eyeglass lenses: limestone, feldspar, and soda ash

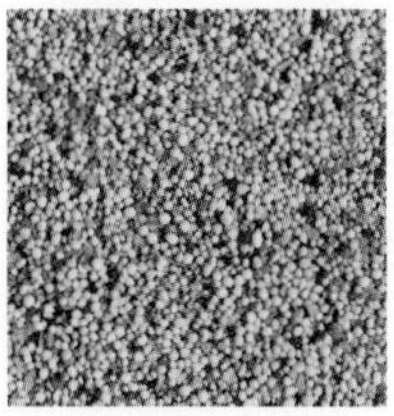

Fertilizers: phosphate, nitrogen, sulfur, and potash

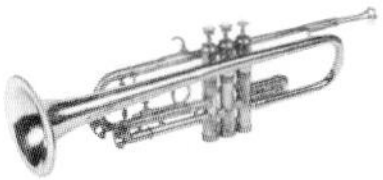

Musical instruments: copper, silver, steel, nickel, brass, cobalt, iron, and aluminum

Sports equipment: graphite, aluminum, and titanium

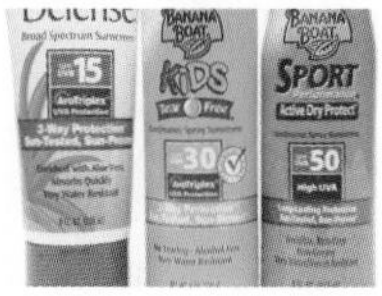

Sun protection and medical ointments: zinc

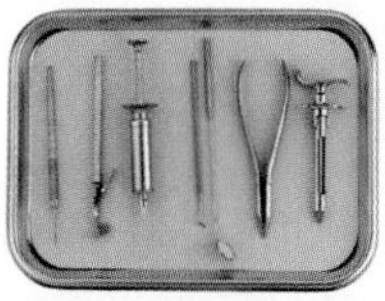

Surgical instruments: stainless steel

Vehicles and tires: steel, copper, zinc, barium, graphite, sulfur, and iodine

Housing construction: gypsum, clay, limestone, sand, and gravel

Hybrid car components: rare earth elements such as dysprosium, lanthanum, neodymium, and samarium

TABLE 23.4 Top Canadian Mining Companies

Canadian Mining Company	Gross Revenues in Billions of Canadian Dollars ($), 2013
Agrium Inc.	16.2
Barrick Gold	12.9
Suncor Energy	12.3
Syncrude Canada	11.4
Teck Resources	9.4
PotashCorp	7.5
Canadian Oil Sands	4.2
Kinross Gold	3.9
Goldcorp	3.8
First Quantum	3.7
Cameco	2.4
Walter Energy	1.9
Yamana Gold	1.9
Agnico Eagle	1.7
IamGold	1.2
Eldorado Gold	1.2
KGHM International	1.1
Centerra Gold	1.0
Pan American Silver	0.8
New Gold	0.8
Lundin Mining	0.7
Silver Wheaton	0.7
Osisko	0.7
B2Gold	0.6
Hudbay Minerals	0.5

The top Canadian mining companies are given in Table 23.4.

Humans have been removing metals for thousands of years, so most of the concentrated sources are gone, and many ores contain very low mass percents of the metals. The general procedure for extracting most metals (and many nonmetals) involves a few basic steps, which are described in the upcoming subsections (Figure 23.9).

Pretreating the Ore

Following a crushing, grinding, or pulverizing step, which can be very expensive, pretreatment uses a physical or chemical difference to separate the mineral from the gangue. The following are common pretreatment techniques:

- *Magnetic attraction.* For magnetic minerals, such as magnetite (Fe_3O_4), a magnet can remove the mineral and leave the gangue behind.
- *Density separation.* When large density differences exist, a *cyclone separator* blows high-pressure air through the pulverized mixture to separate the particles.

FIGURE 23.9 Steps in metallurgy

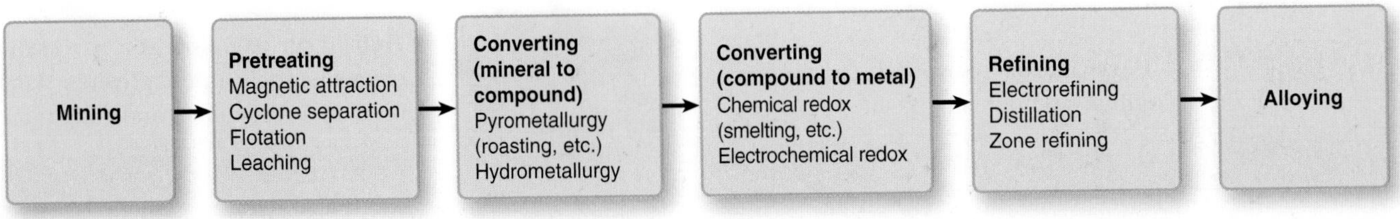

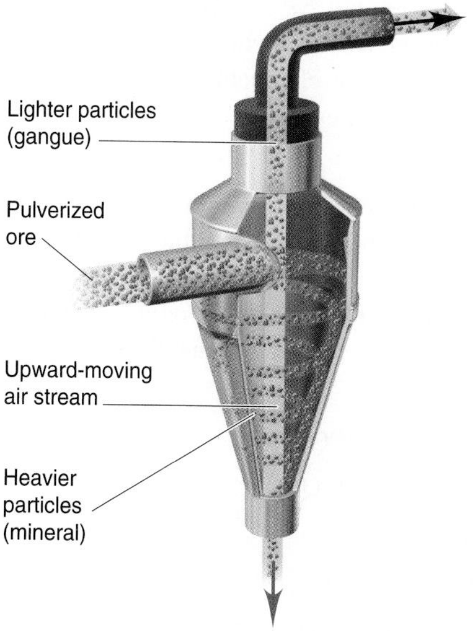

FIGURE 23.10 The cyclone separator

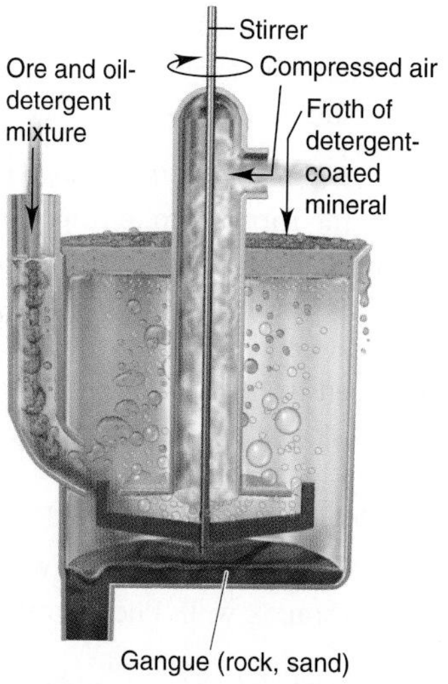

FIGURE 23.11 The flotation process

The lighter silicate-rich gangue is blown away, while the denser mineral-rich particles hit the walls of the separator and fall through the open bottom (Figure 23.10).

- *Flotation.* In **flotation**, an oil-detergent mixture is stirred with the pulverized ore in water to form a slurry (Figure 23.11). The surfaces of the mineral and the gangue become wet, to different extents, with water and detergent. Rapid mixing with air produces an oily, mineral-rich froth that floats, while the silicate particles sink. Skimming, followed by solvent removal, isolates the concentrated mineral fraction.
- *Leaching.* Some metals are extracted by a hydrometallurgical process known as **leaching**, usually accomplished via the formation of a complex ion. The modern extraction of gold is a good example of this process. In the past, prospectors panned for gold (*see photo*), a process that relied on the very large density difference between gold and sand. In the leaching process, the crushed ore, which often contains as little as 25 ppm of gold, is contained in a plastic-lined pool, treated with a cyanide ion solution, and aerated. In the presence of CN^-, the O_2 in air oxidizes the gold metal to gold(I) ion, Au^+, which forms the soluble complex ion, $Au(CN)_2^-$:

$$4Au(s) + O_2(g) + 8CN^-(aq) + 2H_2O(l) \longrightarrow 4Au(CN)_2^-(aq) + 4OH^-(aq)$$

This process is controversial, however, because cyanide enters streams and lakes, where it poisons fish and birds.

Panning for gold in Alberta. Canada is one of the leading mining nations in the world, producing natural gas, copper, zinc, nickel, aluminum, sulfur, uranium, potash, and gold.

Converting Mineral to Element

After the mineral has been freed of debris and concentrated, it may undergo several chemical steps during its conversion to the element.

Converting the Mineral to Another Compound First, the mineral is often converted to another compound, one that has different solubility properties, is easier to reduce, or is free of an impurity. Conversion to an oxide is common because oxides can be reduced easily. Carbonates are heated to convert them to oxides:

$$CaCO_3(s) \xrightarrow{\Delta} CaO(s) + CO_2(g)$$

Metal sulfides, such as ZnS, can be converted to oxides by **roasting** in air:

$$2ZnS(s) + 3O_2(g) \xrightarrow{\Delta} 2ZnO(s) + 2SO_2(g)$$

In most countries, hydrometallurgical methods are now used to avoid the atmospheric release of SO_2 from roasting. One method for processing copper, for example, involves bubbling air through an acidic slurry of insoluble Cu_2S, which completes both the pretreatment and conversion steps:

$$2Cu_2S(s) + 5O_2(g) + 4H^+(aq) \longrightarrow 4Cu^{2+}(aq) + 2SO_4^{2-}(aq) + 2H_2O(l)$$

Converting Compound to Element through Chemical Redox The next step converts the new mineral form (usually an oxide) to the free element by either chemical or electrochemical redox methods. Let us consider the chemical methods first. In these methods, a reducing agent reacts directly with the compound. The most common reducing agents are carbon, hydrogen, and a more active metal.

1. *Reduction with carbon.* Because of its low cost and ready availability, carbon, in the form of coke (a porous residue from the incomplete combustion of coal) or charcoal, is a common reducing agent. Heating an oxide with a reducing agent such as coke to obtain the metal is called **smelting**. Many metal oxides, such as zinc oxide and tin(IV) oxide, are smelted with carbon to free the metal, which may then need to be condensed and solidified:

$$ZnO(s) + C(s) \longrightarrow Zn(g) + CO(g)$$
$$SnO_2(s) + 2C(s) \longrightarrow Sn(l) + 2CO(g)$$

Several nonmetals that occur with positive oxidation states in minerals can be reduced with carbon as well. Phosphorus, for example, is produced from calcium phosphate:

$$2Ca_3(PO_4)_2(s) + 10C(s) + 6SiO_2(s) \longrightarrow 6CaSiO_3(s) + 10CO(g) + P_4(s)$$

(Metallic calcium is a much stronger reducing agent than carbon, so it is not formed.)

- *Thermodynamic considerations.* Thermodynamic principles explain why carbon is often such an effective reducing agent. Consider the standard free energy change ($\Delta G°$) for the reduction of tin(IV) oxide:

$$SnO_2(s) + 2C(s) \longrightarrow Sn(s) + 2CO(g) \quad \Delta G° = 245 \text{ kJ at } 25°C \text{ (298 K)}$$

The magnitude and sign of $\Delta G°$ indicate a highly *non*spontaneous reaction at standard-state conditions and 25°C (298 K). However, because solid C becomes gaseous CO, the standard molar entropy change is very positive ($\Delta S° \approx 380$ J/K). Therefore, *the* $-T\Delta S°$ *term of* $\Delta G°$ *becomes more negative with higher temperature* (Section 18.3). As the temperature increases, $\Delta G°$ decreases; at a high enough temperature, the reaction becomes spontaneous ($\Delta G° < 0$). At 1000°C (1273 K), for example, $\Delta G° = -62.8$ kJ/mol. The actual process is carried out at a temperature of around 1250°C (1523 K), so $\Delta G°$ is even more negative. In effect, the nonspontaneous reduction of SnO_2 is coupled with the spontaneous oxidation of C.

- *Multistep nature of the process.* As you know, overall reactions often mask several intermediate steps. For example, the reduction of tin(IV) oxide involves the formation of tin(II) oxide:

$$SnO_2(s) + C(s) \longrightarrow \cancel{SnO(s)} + CO(g)$$
$$\cancel{SnO(s)} + C(s) \longrightarrow Sn(l) + CO(g)$$

[Molten tin (mp = 232 K) is obtained at this temperature.] The second step may even be composed of others, in which CO, not C, is the actual reducing agent:

$$SnO(s) + \cancel{CO(g)} \longrightarrow Sn(l) + \cancel{CO_2(g)}$$
$$\cancel{CO_2(g)} + C(s) \rightleftharpoons 2CO(g)$$

Other pyrometallurgical processes are similarly complex. For instance, you will see that the overall reaction for smelting iron is

$$2Fe_2O_3(s) + 3C(s) \longrightarrow 4Fe(l) + 3CO_2(g)$$

However, the actual process is a multistep process, with CO as the reducing agent. The advantage of CO over C is the far greater contact that the gaseous reducing agent can make with the other reactant, which speeds up the process.

2. *Reduction with hydrogen.* For oxides of some metals, especially some members of groups 6 and 7, reduction with carbon forms metal carbides that are difficult to convert further. Hydrogen is used as the reducing agent instead, especially for less active metals such as tungsten:

$$WO_3(s) + 3H_2(g) \longrightarrow W(s) + 3H_2O(g)$$

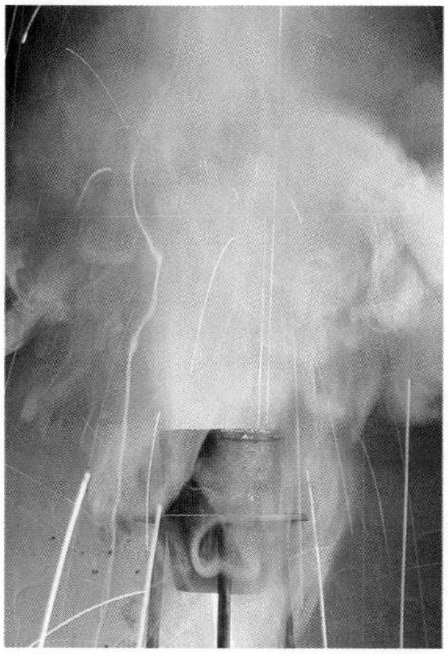

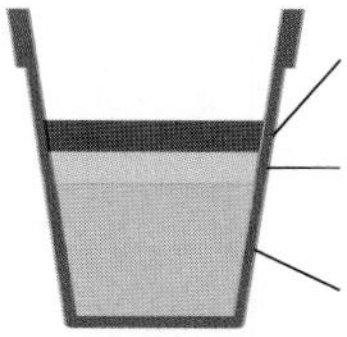

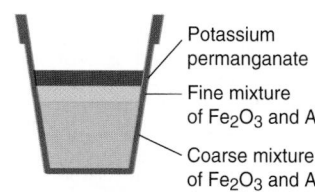

FIGURE 23.12 The thermite reaction

One step in the purification of the metalloid germanium also uses hydrogen:

$$GeO_2(s) + 2H_2(g) \longrightarrow Ge(s) + 2H_2O(g)$$

3. *Reduction with a more active metal.* If a metal forms an undesirable hydride as well, its oxide is reduced by a more active metal. In the *thermite reaction,* aluminum powder reduces the metal oxide in a spectacular exothermic reaction to give the molten metal (Figure 23.12). The reaction for chromium is

$$Cr_2O_3(s) + 2Al(s) \longrightarrow 2Cr(l) + Al_2O_3(s) \quad \Delta H^\circ \ll 0$$

Reduction by an active metal is also used when the new mineral is not an oxide. In the extraction of gold, after the pretreatment by leaching, the gold(I) complex ion is reduced with zinc:

$$2Au(CN)_2^-(aq) + Zn(s) \longrightarrow 2Au(s) + Zn(CN)_4^{2-}(aq)$$

Sometimes, an active metal, such as calcium, is used to recover an even more active metal, such as rubidium, from its molten salt:

$$Ca(l) + 2RbCl(l) \longrightarrow CaCl_2(l) + 2Rb(g)$$

A similar process (described in Section 23.4) is used to recover sodium.

4. A variation uses the opposite process because, just as *reduction* of a mineral is used to obtain the metal, *oxidation* of a mineral can be used to obtain the nonmetal. A stronger oxidizing agent removes electrons from the nonmetal anion to give the free nonmetal, as in the production of iodine from concentrated brines:

$$2I^-(aq) + Cl_2(g) \longrightarrow 2Cl^-(aq) + I_2(s)$$

Converting Compound to Element through Electrochemical Redox In electrochemical redox processes, the minerals are converted to the elements in an electrolytic cell (Section 19.7). Sometimes, the *pure* mineral, in the form of the molten halide or oxide, is used to prevent unwanted side reactions. The cation is reduced to the metal at the cathode, and the anion is oxidized to the nonmetal at the anode:

$$BeCl_2(l) \longrightarrow Be(s) + Cl_2(g)$$

High-purity hydrogen gas is prepared by electrochemical reduction:

$$2H_2O(l) \longrightarrow 2H_2(g) + O_2(g)$$

Specially designed cells separate the products to prevent recombination. An inexpensive source of electricity is essential for large-scale methods. The current and voltage requirements depend on the electrochemical potential and any overpotential (overvoltage) (Section 19.7).

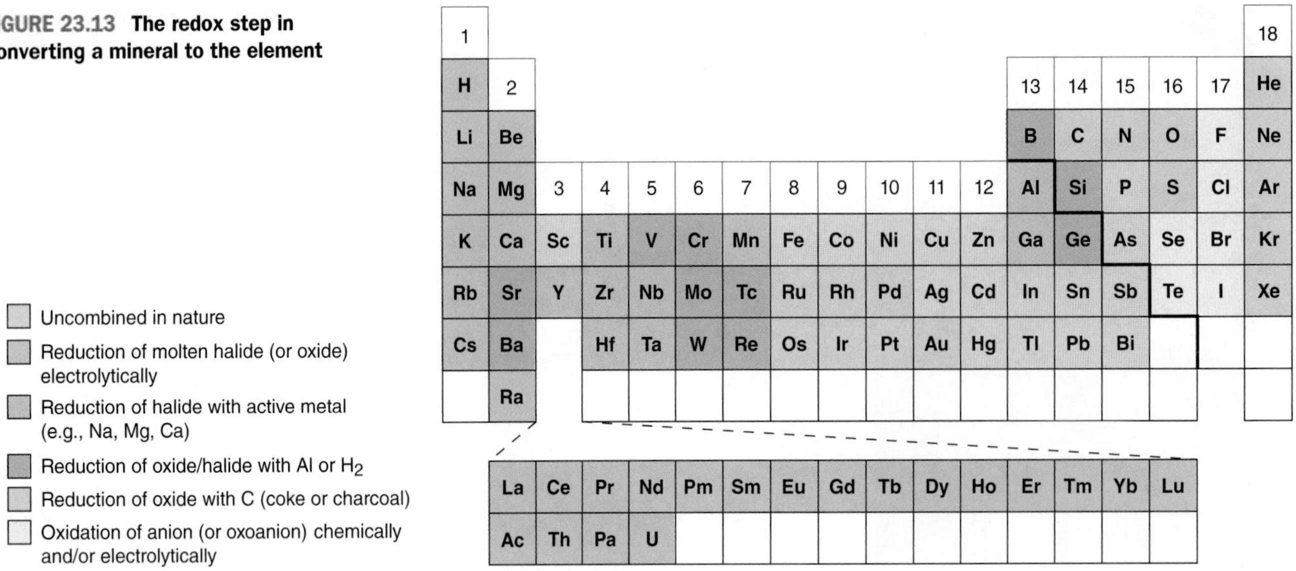

FIGURE 23.13 The redox step in converting a mineral to the element

Figure 23.13 shows the most common redox step used in obtaining each free element from its mineral.

Refining and Alloying the Element

Usually, after being isolated, the element still contains impurities, so it must be refined to purify it further. Then, metallic elements are often alloyed to improve their properties.

Refining (Purifying) the Element Refining is often carried out by one of three common methods:

1. *Electrorefining.* An electrolytic cell is employed in **electrorefining**, in which the *impure metal acts as the anode and the pure metal acts as the cathode.* As the reaction proceeds, the anode disintegrates, and the metal ions are reduced to deposit on the cathode. In some cases, impurities that fall beneath the anode are the source of several valuable and less abundant elements.

2. *Distillation.* Metals with relatively low boiling points, such as zinc and mercury, are refined by *distillation.*

3. *Zone refining.* In the process of **zone refining**, a rod of an impure metal or metalloid (such as silicon) is passed slowly through a heating coil in an inert atmosphere. The first narrow zone of the impure solid melts, and, as the next zone melts, the dissolved impurities from the first zone lower the freezing point, so the solvent (purer solid of first zone) refreezes. The process continues, zone by zone, for the entire rod. After several passes through the coil, in which each zone's impurities move into the adjacent zone, the refrozen solid at the end of the rod becomes extremely pure (Figure 23.14). Metalloids used in electronic semiconductors, such as silicon and germanium, are zone-refined to greater than 99.999999% purity.

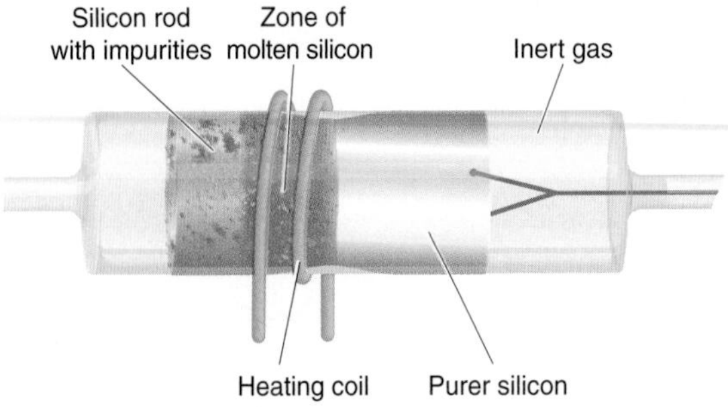

FIGURE 23.14 Zone refining of silicon

In 1969, Canadian scientist Willard S. Boyle (Figure 23.15) and American applied physicist George E. Smith invented the first charge-coupled device (CCD), an integrated circuit etched onto a *silicon* surface that transforms light into electric signals, which produce a large number of image points, or pixels, in a short time (Figure 23.16). This invention made possible the development of the digital camera, and the technology is also commonly used in medical imaging devices, bar code readers, camcorders, and photocopiers. Both Boyle and Smith were avid sailors, and they took many trips together. They jointly received the Franklin Institute's Stuart Ballantine Medal in 1973, the 1974 IEEE Morris N. Liebmann Memorial Award, the 2006 Charles Stark Draper Prize, and the 2009 Nobel Prize in Physics.

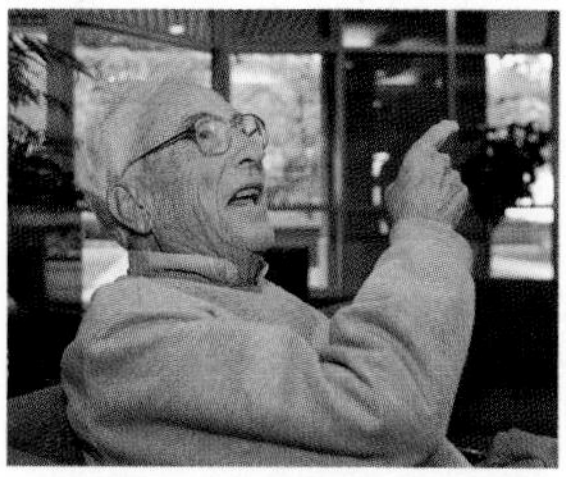

FIGURE 23.15 Canadian scientist Willard Sterling Boyle (1924–2011) was born in Amherst, Nova Scotia. He shared the 2009 Nobel Prize in Physics with George E. Smith for the invention of an imaging semiconductor circuit, the CCD (charge-coupled device) sensor, which is used as a digital camera's electronic eye. Dr. Boyle also invented (with Don Nelson) the first continuously operating ruby laser, a device used in CD players. The technology that Dr. Boyle helped to invent is commonly used in medical imaging and astronomy, as well.

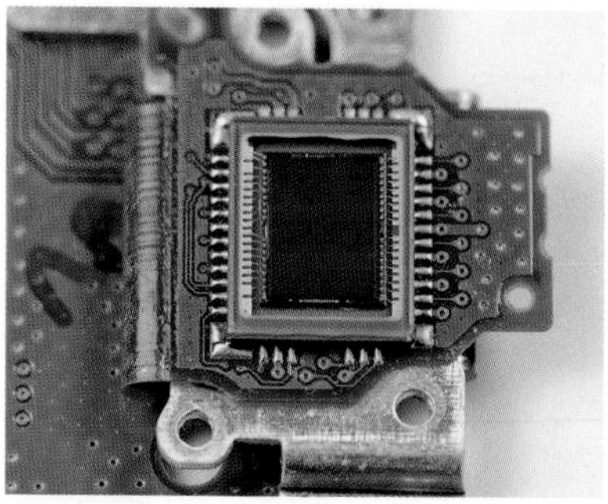

FIGURE 23.16 A charge-coupled device (CCD) used for UV imaging in a wire-bonded package. In a CCD sensor, every pixel's charge is transferred through a very limited number of output nodes to be converted to voltage, buffered, and sent off-chip as an analogue signal. All the pixels can be devoted to light capture, so the output's uniformity, a key factor in image quality, is high.

Alloying the Purified Element **Alloys** are metal-like mixtures that consist of solid phases of two or more pure elements (a solid solution) or distinct intermediate phases (these alloys are sometimes referred to as *intermetallic compounds*) that are so finely divided they can only be distinguished microscopically. Alloying a metal with other metals (and sometimes nonmetals) is done to alter the metal's melting point and to enhance properties such as lustre, conductivity, malleability, ductility, and strength.

Iron is used only when alloyed. In pure form, it is soft and corrodes easily. However, when alloyed with carbon and other metals, such as Mo for hardness and Cr and Ni for corrosion resistance, it forms various steels. Copper stiffens when zinc is added to make brass. Mercury solidifies when it is alloyed with sodium, and vanadium becomes extremely tough when carbon is added. Table 23.5 shows the composition and uses of some common alloys.

The simplest alloys, called *binary alloys*, contain only two elements, which can combine in two ways:

- *The added element enters interstices*, spaces between the parent metal atoms in the crystal structure (Section 11.6). In vanadium carbide, for instance, carbon atoms occupy holes of a face-centred cubic vanadium unit cell (Figure 23.17A).
- The *added element substitutes for atoms of the parent*. In many types of brass, for example, relatively few zinc atoms substitute randomly for copper atoms in the face-centred cubic copper unit cell (see Figure 11.38A). On the other hand,

TABLE 23.5 Some Important Alloys and Their Composition

Name	Composition (Mass %)	Uses
Mostly Fe		
Stainless steel	73–79 Fe, 14–18 Cr, 7–9 Ni	Cutlery, instruments
Nickel steel	96–98 Fe, 2–4 Ni	Cables, gears
High-speed steels	80–94 Fe, 14–20 W (or 6–12 Mo)	Cutting tools
Mostly Cu		
Bronzes	70–95 Cu, 1–25 Zn, 1–18 Sn	Statues, castings
Brasses	50–80 Cu, 20–50 Zn	Plating, ornamental objects
Mostly precious metals		
Sterling silver	92.5 Ag, 7.5 Cu	Jewellery, tableware
14-karat gold	58 Au, 4–28 Ag, 14–28 Cu	Jewellery
18-karat white gold	75 Au, 12.5 Ag, 12.5 Cu	Jewellery
Dental amalgam	69 Ag, 18 Sn, 12 Cu, 1 Zn (dissolved in Hg)	Dental fillings
Other		
Permalloy	78 Ni, 22 Fe	Ocean cables
Typical tin solder	67 Pb, 33 Sn	Electrical connections

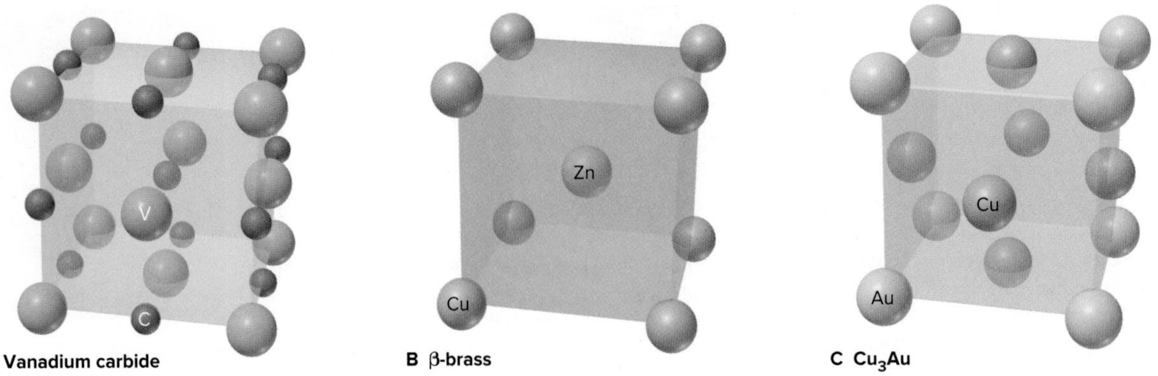

FIGURE 23.17 Three binary alloys

β-brass is an intermediate phase that crystallizes as the Zn/Cu ratio approaches 1/1 in a structure with a central Zn atom surrounded by eight Cu atoms (Figure 23.17B). In one alloy of copper and gold, Cu atoms occupy the faces of a face-centred cube, and Au atoms lie at the corners (Figure 23.17C).

Atomic properties, including size, electron configuration, and number of valence electrons, determine which metals form stable alloys. For example, transition metals from the left half of the *d* block (groups 3 to 5) often form alloys with metals from the right half (groups 9 to 11): electron-poor Zr, Nb, and Ta form strong alloys with electron-rich Ir, Pt, or Au; an example is the very stable $ZrPt_3$.

SUMMARY OF SECTION 23.3

- Metallurgy involves mining an ore, separating it from debris, pretreating it to concentrate the mineral source, converting the mineral to another compound that is easier to process further, reducing this compound to the metal, purifying the metal, and often alloying the metal to obtain a more useful material.

23.4 Tapping the Crust: Isolation and Uses of Selected Elements

The isolation process depends on the physical and chemical properties of the source; thus, we isolate an element that occurs uncombined in the air differently from an element that is dissolved in the oceans or found in a rocky ore. In this section, we detail methods for recovering some important elements.

Producing the Alkali Metals: Sodium and Potassium

The alkali metals are among the most reactive elements and, thus, are always found as ions in nature, either in solid minerals or in aqueous solution. In fact, they are so reactive that, in a laboratory, the free elements must be protected from contact with air (usually by being stored under mineral oil) to prevent their immediate oxidation and to minimize the possibility of fires. The two most important alkali metals are sodium and potassium. Their abundant water-soluble compounds are used by industry and in research, and the Na^+ and K^+ ions are essential to organisms.

Industrial Production of Sodium and Potassium Different methods are used in the industrial production of these two metals:

1. *Production of sodium*. The sodium ore is *halite* (largely NaCl), which is obtained either by evaporating concentrated salt solutions called *brines* or by mining the vast salt deposits formed from the evaporation of prehistoric seas. The Cheshire salt field in Britain, for example, is 60 km by 24 km by 400 m thick and contains more than 90% NaCl. Other large deposits are found in Ontario, Nova Scotia, Saskatchewan, New Mexico, Michigan, New York, and Kansas.

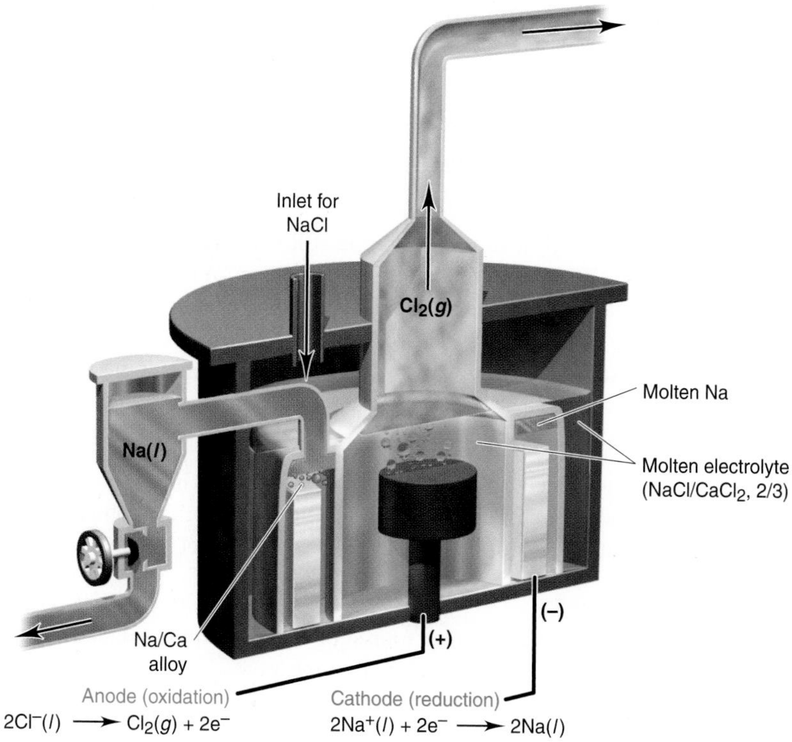

FIGURE 23.18 The Downs cell, used for the production of sodium

The brine is evaporated and the solid is crushed and fused (melted) for use in an electrolytic apparatus called the **Downs cell** (Figure 23.18). To reduce heating costs, the NaCl (mp = 801°C) is mixed with $1\frac{1}{2}$ parts $CaCl_2$ to form a mixture that melts at only 580°C. Reduction of the metal ions to Na and Ca takes place at a cylindrical steel cathode, with the molten metals floating on the denser molten salt mixture. As the molten metals rise through a short collecting pipe, the liquid Na is siphoned off, while a higher melting Na/Ca alloy solidifies and falls back into the molten electrolyte. Chloride ions are oxidized to Cl_2 gas at a large anode within an inverted cone-shaped chamber, which separates the metals from the Cl_2 to prevent their explosive recombination. The Cl_2 is purified and sold as a valuable by-product.

As mentioned before, Canada has salt deposits in Ontario, the western provinces, and the Atlantic provinces. The Goderich rock salt mine in Ontario, for example, produces 6.6 million t of salt annually (*see photo*). Evaporation plants around Canada produce another 426 000 t annually.

Inside the Goderich, Ontario, salt mine—the largest rock salt mine in the world

2. *Production of potassium.* Sylvite (mostly KCl) is the major ore of potassium. The metal is too soluble in molten KCl to be obtained by a method similar to that used for sodium. Instead, chemical reduction of K^+ ions by Na is based on atomic properties and the nature of equilibrium systems. An Na atom is smaller than a K atom, so it holds its outer electron more tightly: IE_1 of Na = 496 kJ/mol; IE_1 of K = 419 kJ/mol. Thus, based on this atomic property, Na would not be effective at reducing K^+. However, the reduction is carried out at 850°C (1123 K), which is above the boiling point of K, so the equilibrium mixture contains *gaseous* K:

$$Na(l) + K^+(l) \rightleftharpoons Na^+(l) + K(g)$$

As the K gas is removed, Le Châtelier's principle predicts that the system will shift to produce more K. The gas is then condensed and purified by fractional distillation. The same general method (with Ca as the reducing agent) is used to produce rubidium and cesium.

Uses of Sodium and Potassium The compounds of Na and K (and, in fact, of all the alkali metals) have many more uses than the elements themselves. Nevertheless, there are some interesting uses of the metals that take advantage of their strong reducing power:

1. *Uses of sodium.* Large amounts of Na were used as an alloy with lead to make gasoline antiknock additives, such as tetraethyllead:

$$4C_2H_5Cl(g) + 4Na(s) + Pb(s) \longrightarrow (C_2H_5)_4Pb(l) + 4NaCl(s)$$

The toxic effects of environmental lead have limited this use to aviation fuel in Canada. Moreover, although leaded gasoline is still used in some developing countries, the global market for tetraethyllead is declining more than 15% annually.

If certain types of nuclear reactors, called *breeder reactors* (Chapter 25), become a practical way to generate energy, Na production will increase enormously. Its low melting point, low viscosity, and low absorption of neutrons, combined with its high thermal conductivity and heat capacity, make it perfect for cooling the reactors and exchanging heat with the steam generators.

Firefighter with KO_2 breathing mask

2. *Uses of potassium.* Today, the major use of potassium is in an alloy with sodium for use as a heat exchanger in chemical and nuclear reactors. Another application relies on production of its superoxide, which it forms by direct contact with O_2:

$$K(s) + O_2(g) \longrightarrow KO_2(s)$$

This material is used as an emergency source of O_2 in breathing masks for miners, divers, submarine crews, and firefighters (*see photo*):

$$4KO_2(s) + 4CO_2(g) + 2H_2O(g) \longrightarrow 4KHCO_3(s) + 3O_2(g)$$

The Indispensable Three: Iron, Copper, and Aluminum

The innumerable applications of iron (in the form of steel), copper, and aluminum make them indispensable materials in our industrial society.

TABLE 23.6 Important Minerals of Iron

Mineral Type	Mineral, Formula
Oxide	Hematite, Fe_2O_3
	Magnetite, Fe_3O_4
	Ilmenite, $FeTiO_3$
Carbonate	Siderite, $FeCO_3$
Sulfide	Pyrite, FeS_2
	Pyrrhotite, FeS

Metallurgy of Iron Although humans have practised iron smelting for more than 3000 years, it was only about 240 years ago that iron assumed its current dominant role. In 1773, an inexpensive process to convert coal to carbon, in the form of coke, was discovered. The coke process made iron smelting cheap and efficient and led to large-scale iron production, which ushered in the Industrial Revolution. Modern society rests, quite literally, on the various alloys of iron known as **steel**. Although steel production has grown enormously since the 18th century, the process of recovering iron from its ores still uses *reduction by carbon in a blast furnace.* The most abundant minerals of iron are listed in Table 23.6. Only the first four minerals are used for steelmaking, because traces of sulfur from the sulfide minerals make the steel brittle.

Canadians have made notable contributions to the advancement of the iron and steel industry. In the early 1960s, Canadian Liquid Air designed an injector that allowed pure oxygen to be introduced through the bottom of the basic-oxygen process vessels (se Figure 23.20). The first successful continuous casting machine for steel in North America was developed by Atlas Steels in Welland, Ontario, in 1954. Stelco, now part of the United States Steel Corporation, developed the Stelmor rod-cooling process and the Coilbox, a major energy-saving device used in hot-strip rolling mills, as well as the Ardox spiral nail. Ipsco, in Regina, Saskatchewan, was the first company to install a spiral-weld pipe mill.

Canada is the world's tenth-largest producer and fourth-largest exporter of iron ore. The conversion from ore to metal involves a series of redox and acid-base reactions that appear simple, even though their detailed chemistry is not completely understood even today. A modern **blast furnace** (Figure 23.19) is a tower about 14 m wide by 40 m high, made of brick. The *charge,* which usually consists of hematite, coke, and limestone, is fed through the top. More coke is burned in air at the bottom. The charge falls and meets a *blast* of rapidly rising hot air created by the burning coke:

$$2C(s) + O_2(g) \longrightarrow 2CO(g) + heat$$

At the bottom of the furnace, the temperature exceeds 2000°C (2273 K), while at the top, it reaches only 200°C (473 K). Because the blast of hot air passes through the entire furnace in only 10 s, the various gas-solid reactions do *not* reach equilibrium,

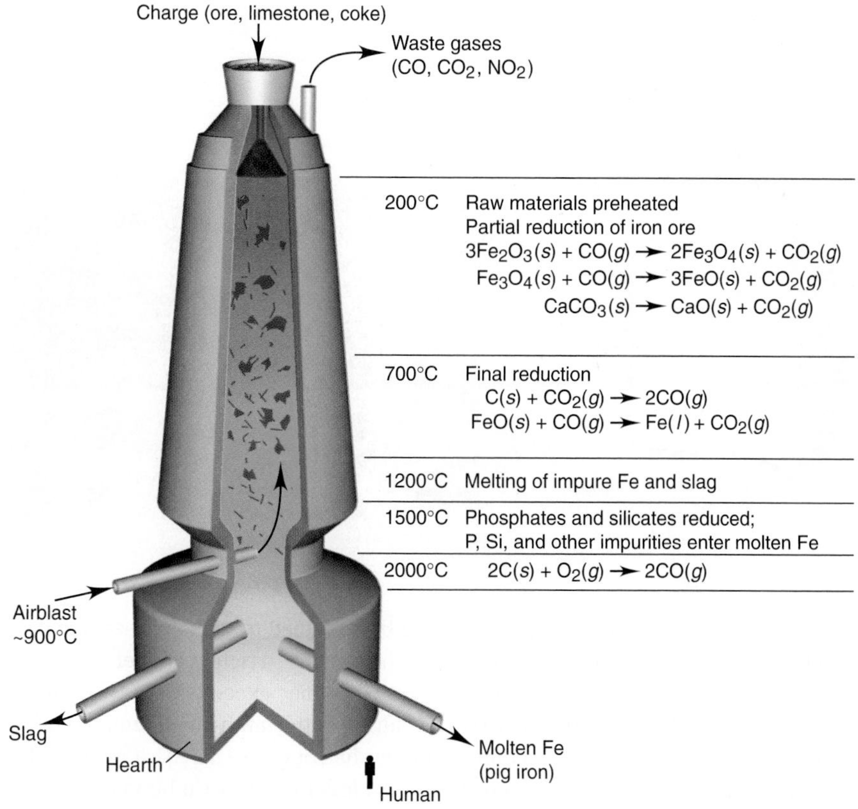

FIGURE 23.19 The major reactions in a blast furnace

and many intermediate products form. As a result, different stages of the overall reaction process occur at different heights:

1. In the upper part of the furnace, at 200°C (473 K) to 700°C (973 K), the charge is preheated and a *partial reduction* occurs. The hematite (Fe_2O_3) is reduced to magnetite (Fe_3O_4) and then to iron(II) oxide (FeO) by CO, which is the actual reducing agent. Carbon dioxide is also formed, and limestone ($CaCO_3$) decomposes to form CO_2 and the basic oxide CaO.

2. Lower down, at 700°C (973 K) to 1200°C (1473 K), a *final reduction* step occurs, as some of the coke reduces CO_2 to form more CO, which reduces the FeO to Fe.

3. At temperatures between 1200°C (1473 K) and 1500°C (1773 K), the iron melts and drips to the bottom of the furnace. Acidic silica particles from the gangue react with the CaO in a Lewis acid-base reaction to form a molten waste product called **slag**:

$$CaO(s) + SiO_2(s) \longrightarrow CaSiO_3(l)$$

The slag drips down and floats on the denser iron.

4. Some unwanted reactions occur between 1500°C (1773 K) and 2000°C (2273 K). Any remaining phosphates and silicates are reduced to P and Si, and some Mn and traces of S dissolve into the molten iron along with carbon. The resulting impure product is called *pig iron* and contains about 3% to 4% C. A small amount of pig iron is used to make *cast iron*, but most is purified and alloyed to make various kinds of steel.

From Pig Iron to Steel Pig iron is converted to steel in a separate furnace by means of the **basic-oxygen process** (Figure 23.20). High-pressure O_2 is blown over and through the molten iron, so that impurities (C, Si, P, Mn, and S) are oxidized rapidly. The highly negative standard enthalpies of formation of their oxides ($\Delta_f H°$ of $CO_2 = -394$ kJ/mol and $\Delta_f H°$ of $SiO_2 = -911$ kJ/mol) raise the temperature, which speeds the reaction. A lime (CaO) flux is added, which converts the oxides to a molten slag [primarily $CaSiO_3$ and $Ca_3(PO_4)_2$] that is decanted from the molten steel. The product is **carbon steel**, which contains 1% to 1.5% C and other impurities. It is alloyed with metals that prevent corrosion and increase its strength or flexibility.

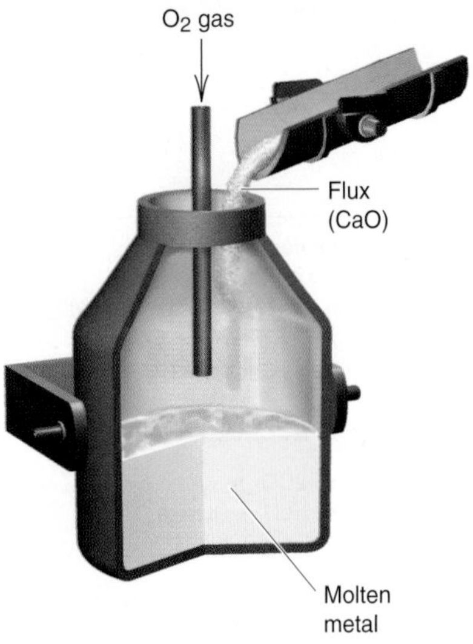

A

B

FIGURE 23.20 A. The basic-oxygen process for making steel. **B.** The typical size of an actual steel furnace.

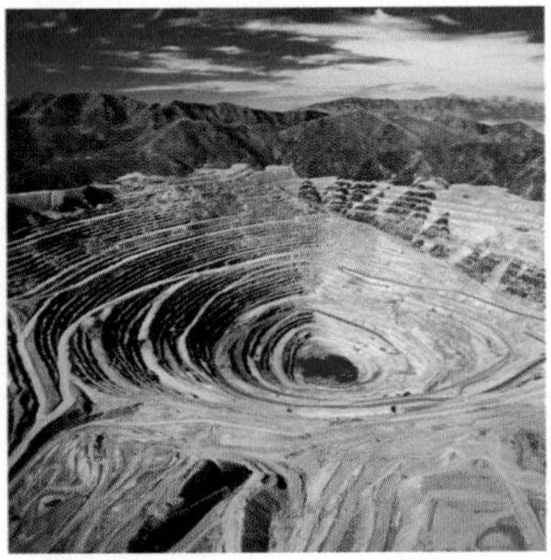

Mining chalcopyrite

Isolation of Copper After many centuries of being mined to make bronze and brass, copper ores have become much less plentiful and much less rich in copper (*see photo*). Despite this, Canada's annual mine production of copper exceeds 600 000 t, and our refined copper production is over 500 000 t. (Canada is the ninth-largest mine producer of copper in the world.) Our most common ore is chalcopyrite, $CuFeS_2$, a mixed sulfide of FeS and CuS. Most remaining deposits contain less than 0.5% Cu by mass. To "win" this small amount of copper requires several metallurgical steps:

1. *Flotation.* The low copper content in chalcopyrite must be enriched by removing the iron. The first step is pretreatment by flotation (see Figure 23.11), which concentrates the ore to around 15% Cu by mass.

2. *Roasting.* The next step in many processing plants is a controlled roasting step, which oxidizes the FeS but not the CuS:

$$2FeCuS_2(s) + 3O_2(g) \longrightarrow 2CuS(s) + 2FeO(s) + 2SO_2(s)$$

3. *Heating with sand.* To remove the FeO and convert the CuS to a more convenient form, the mixture is heated to 1100°C (1373 K) with sand and more of the concentrated ore. Several reactions occur in this step. The FeO reacts with sand to form a molten slag:

$$FeO(s) + SiO_2(s) \longrightarrow FeSiO_3(l)$$

The CuS is thermodynamically unstable at the elevated temperature and decomposes to yield Cu_2S, which is drawn off as a liquid.

4. *Smelting.* In the final smelting step, the Cu_2S is roasted in air, which converts some of it to Cu_2O:

$$2Cu_2S(s) + 3O_2(g) \longrightarrow 2Cu_2O(s) + 2SO_2(g)$$

The two copper(I) compounds then react, with sulfide ion acting as the reducing agent:

$$Cu_2S(s) + 2Cu_2O(s) \longrightarrow 6Cu(l) + SO_2(g)$$

The copper obtained at this stage is usable for plumbing, but it must be further purified for electrical applications by removing the unwanted Fe and Ni, as well as the valuable by-products Ag, Au, and Pt.

Electrorefining of Copper Achieving the desired 99.99% purity needed for wiring (copper's most important use) is accomplished by *electrorefining*, which involves the oxidation of Cu and the formation of Cu^{2+} ions in solution, followed by the reduction of these ions and the plating out of Cu metal (Figure 23.21). The impure copper that is obtained from smelting is cast into plates to be used as anodes, and

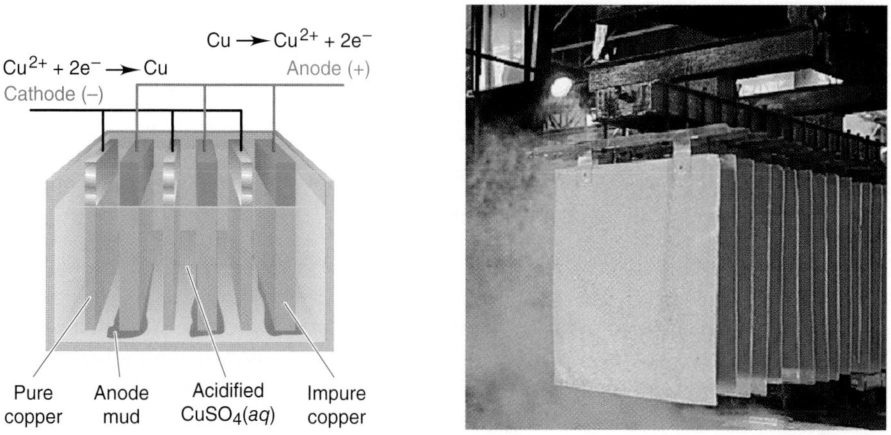

FIGURE 23.21 The electrorefining of copper

cathodes are made from already purified copper. The electrodes are immersed in acidified $CuSO_4$ solution, and a voltage is applied to accomplish the following two tasks simultaneously:

- Copper and the more active impurities (Fe and Ni) are oxidized to their cations, while the less active impurities (Ag, Au, and Pt) are not. As the anode slabs react, these unoxidized metals fall off as valuable "anode mud" and are purified separately. Sale of the precious metals in the anode mud nearly offsets the cost of electricity to operate the cell, making Cu wire inexpensive.
- Because Cu is much less active than the Fe and Ni impurities, Cu^{2+} ions are reduced at the cathode, but Fe^{2+} and Ni^{2+} ions remain in solution:

$$Cu^{2+}(aq) + 2e^- \longrightarrow Cu(s) \qquad E° = 0.34\text{ V}$$
$$Ni^{2+}(aq) + 2e^- \longrightarrow Ni(s) \qquad E° = -0.25\text{ V}$$
$$Fe^{2+}(aq) + 2e^- \longrightarrow Fe(s) \qquad E° = -0.44\text{ V}$$

Isolation of Aluminum Aluminum is the most abundant metal, by mass, in Earth's crust and the third most abundant element (after O and Si). It is found in numerous aluminosilicate minerals, such as feldspars, micas, and clays, and in the rare gems garnet, beryl, spinel, and turquoise. Corundum, pure aluminum oxide (Al_2O_3), is extremely hard; mixed with traces of transition metals, it exists as ruby and sapphire. Impure Al_2O_3 is used in sandpaper and other abrasives.

Through eons of weathering, certain clays became *bauxite,* the major ore of aluminum. This mixed oxide-hydroxide occurs in enormous surface deposits in Mediterranean and tropical regions (*see photo*). However, with world aluminum production near 49 million t annually, it may someday be scarce. In addition to hydrated Al_2O_3 (about 75%), industrial-grade bauxite also contains Fe_2O_3, SiO_2, and TiO_2, which are removed during the extraction. The overall two-step process combines hydrometallurgical and electrometallurgical techniques:

Mining bauxite

1. *Isolating Al_2O_3 from bauxite.* After mining, bauxite is pretreated by boiling in 30% NaOH in the *Bayer process*, which involves acid-base, solubility, and complex-ion equilibria. The acidic SiO_2 and the amphoteric Al_2O_3 dissolve in the base, but the basic Fe_2O_3 and TiO_2 do not:

$$SiO_2(s) + 2NaOH(aq) + 2H_2O(l) \longrightarrow Na_2Si(OH)_6(aq)$$
$$Al_2O_3(s) + 2NaOH(aq) + 3H_2O(l) \longrightarrow 2NaAl(OH)_4(aq)$$
$$Fe_2O_3(s) + NaOH(aq) \longrightarrow \text{no reaction}$$
$$TiO_2(s) + NaOH(aq) \longrightarrow \text{no reaction}$$

Further heating precipitates the $Na_2Si(OH)_6$ as an aluminosilicate, which is filtered out with the insoluble Fe_2O_3 and TiO_2 ("red mud").

Acidifying the filtrate precipitates Al^{3+} as $Al(OH)_3$. Recall that the aluminate ion, $Al(OH)_4^-(aq)$, is actually the complex ion $Al(H_2O)_2(OH)_4^-$ (see Figure 17.19). Weakly acidic CO_2 is added to produce a small amount of H^+ ion, which reacts with this complex ion to form $Al(H_2O)_3(OH)_3$. Cooling supersaturates the solution, and the solid that forms is filtered out:

$$CO_2(g) + H_2O(l) \rightleftharpoons H^+(aq) + HCO_3^-(aq)$$
$$Al(H_2O)_2(OH)_4^-(aq) + H^+(aq) \longrightarrow Al(H_2O)_3(OH)_3(s)$$

[Recall that we usually write $Al(H_2O)_3(OH)_3$ more simply as $Al(OH)_3$.] Drying at high temperature converts the hydroxide to the oxide:

$$2Al(H_2O)_3(OH)_2(s) \xrightarrow{\Delta} Al_2O_3(s) + 9H_2O(g)$$

2. *Converting Al_2O_3 to the free metal: the Hall-Heroult process.* Aluminum is an active metal, much too strong a reducing agent to be formed at the cathode from aqueous solution (Section 19.7), so the oxide itself is electrolyzed. Since Al_2O_3 has a very high melting point (2030°C), it is dissolved in molten *cryolite* (Na_3AlF_6) to give a mixture that electrolyzes at ~1000°C. Using cryolite provides major energy savings (and, thus, cost savings as well), but the only sizable cryolite mines cannot supply enough of the mineral, so production of synthetic cryolite has become a major subsidiary industry in aluminum manufacture.

The electrolytic step, called the *Hall-Heroult process*, takes place in a graphite-lined furnace, with the lining itself acting as the cathode. Anodes of graphite dip into the molten Al_2O_3-Na_3AlF_6 mixture (Figure 23.22). The cell typically operates at a moderate voltage of 4.5 V, but with an enormous current flow of 1.0×10^5 A to 2.5×10^5 A.

The Hall-Heroult process is complex, and its details are still not entirely known. The specific reactions shown below are among several possibilities. Molten cryolite contains several ions (including AlF_6^{3-}, AlF_4^-, and F^-), which react with Al_2O_3 to form fluoro-oxy ions (including $AlOF_3^{2-}$, $Al_2OF_6^{2-}$, and $Al_2O_2F_4^{2-}$) that dissolve in the mixture, for example,

$$2Al_2O_3(s) + 2AlF_6^{3-}(l) \longrightarrow 3Al_2O_2F_4^{2-}(l)$$

Al forms at the cathode (reduction), shown here with AlF_6^{3-} as reactant:

$$AlF_6^{3-}(l) + 3e^- \longrightarrow Al(l) + 6F^-(l) \quad \text{(cathode; reduction)}$$

The graphite anodes are oxidized and form carbon dioxide gas. Using one of the fluoro-oxy species as an example, the anode reaction is

$$Al_2O_2F_4^-(l) + 8F^-(l) + C(\text{graphite}) \longrightarrow 2AlF_6^{3-}(l) + CO_2(g) + 4e^- \quad \text{(anode; oxidation)}$$

FIGURE 23.22 The electrolytic cell in the manufacture of aluminum

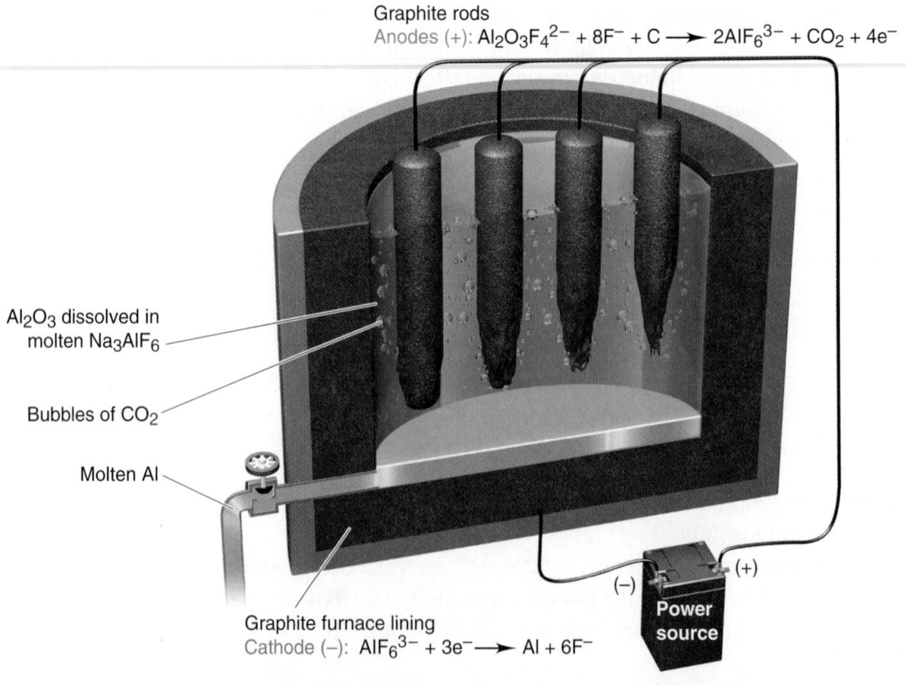

Thus, the *graphite anodes themselves are consumed* in this half-reaction and must be replaced frequently.

Combining the three previous equations and making sure that e^- gained at the cathode are equal to e^- lost at the anode gives the overall reaction:

$$2Al_2O_3(\text{in } Na_3AlF_6) + 3C(\text{graphite}) \longrightarrow 4Al(l) + 3CO_2(g) \quad \text{(overall/cell reaction)}$$

3. *Energy considerations.* The Hall-Heroult process uses an enormous quantity of energy: aluminum production accounts for more than 5% of the total U.S. electrical usage. The most basic reason for such high energy usage is the electron configuration of Al ([Ne]$3s^23p^1$). Each Al^{3+} ion needs $3e^-$ to form an Al atom, and the atomic mass of Al is so low (~27 g/mol) that 1 mol of e^- (96 500 C) produces only 9 g of Al. An aluminum-air battery can turn this disadvantage around. Once produced, Al represents a concentrated form of electrical energy that can deliver a high output per gram: 1 mol of e^- for every 9 g of Al consumed in the battery.

Uses of and Recycling of Aluminum Aluminum is a superb decorative, functional, and structural metal. It is lightweight, attractive, and easy to work with, and it forms strong alloys. Although Al is very active, it does not corrode readily because of an adherent oxide layer that forms rapidly in air and prevents more O_2 from penetrating. Nevertheless, when it is in contact with less active metals, such as Fe, Cu, and Pb, aluminum becomes the anode in a voltaic-like arrangement and deteriorates rapidly (Section 19.6). To prevent this, aluminum objects are often *anodized*, that is, made to act as the anode in an electrolytic cell that coats them with an oxide layer. The object is immersed in a 20% H_2SO_4 bath and connected to a graphite cathode:

$$6H^+(aq) + 6e^- \longrightarrow 3H_2(g) \quad \text{(cathode; reduction)}$$
$$2Al(s) + 3H_2O(l) \longrightarrow Al_2O_3(s) + 6H^+(aq) + 6e^- \quad \text{(anode; oxidation)}$$
$$2Al(s) + 3H_2O(l) \longrightarrow Al_2O_3(s) + 3H_2(g) \quad \text{(overall/cell reaction)}$$

The Al_2O_3 layer deposited is typically from 10 μ to 100 μ thick, depending on the object's intended use. Figure 23.23 presents a breakdown of the many uses of aluminum.

More than 1.8 million t of aluminum cans and packaging are discarded each year—a waste of one of the most useful materials in the world *and* the energy needed to make it. A quick calculation of the energy needed to prepare 1 mol of Al from *purified* Al_2O_3, compared with the energy needed to recycle aluminum, conveys a clear message. For the overall cell reaction in the Hall-Heroult process, $\Delta H° = 2272$ kJ and $\Delta S° = 635.4$ J/K. If we consider *only* the standard free energy change of the reaction at 1000.°C, for 1 mol of Al, we obtain

$$\Delta G° = \Delta H° - T\Delta S° = \frac{2272 \text{ kJ}}{4 \text{ mol Al}} - \left(1273 \text{ K} \times \frac{0.6354 \text{ kJ/K}}{4 \text{ mol Al}}\right) = 365.8 \text{ kJ/mol Al}$$

The molar mass of Al is nearly 2.5 times the mass of a soft-drink or beer can, so the electrolysis step requires nearly 150 kJ of energy for each can!

When aluminum is recycled, the major energy input (which is needed to melt the cans and foil) has been calculated as ~26 kJ/mol Al. The ratio of these energy inputs is

$$\frac{\text{energy to recycle 1 mol Al}}{\text{energy for electrolysis of 1 mol Al}} = \frac{26 \text{ kJ}}{365.8 \text{ kJ}} = 0.071$$

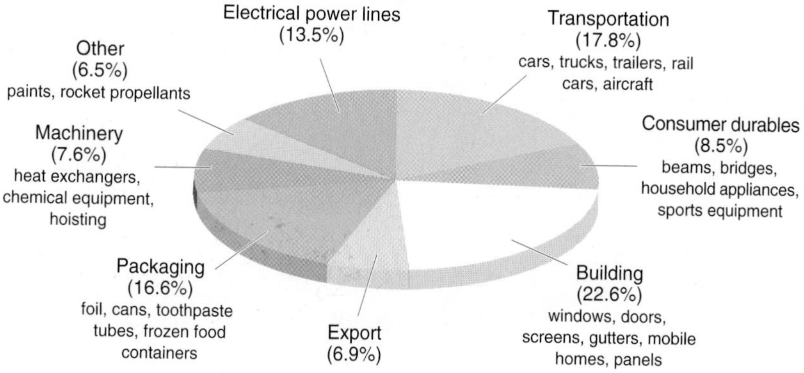

FIGURE 23.23 The many familiar and essential uses of aluminum

Crushed aluminum cans, ready for recycling

Based on just these parts of the process, recycling uses about 7% as much energy as electrolysis. Recent energy estimates for the entire manufacturing process (including mining, pretreating, maintaining operating conditions, electrolyzing, and so on) are about 6000 kJ/mol Al, which means that recycling requires less than 1% as much energy as manufacturing. The economic advantages, not to mention the environmental advantages, are obvious, and the recycling of aluminum has become common in Canada (*see photo*).

Mining the Sea for Magnesium

As terrestrial sources of certain elements become scarce or too costly to mine, the oceans will become an important source. In fact, despite the abundant distribution of magnesium on land, it is already obtained from the oceans.

Isolation of Magnesium The *Dow process* for isolating magnesium from the oceans involves steps that are similar to those used for rocky ores (Figure 23.24):

1. *Mining.* Intake of seawater and straining of the debris are the "mining" steps. No pretreatment is needed.

2. *Converting to mineral.* The dissolved Mg^{2+} ion is converted to the mineral $Mg(OH)_2$ with $Ca(OH)_2$, which is generated on-site (at the plant). Seashells ($CaCO_3$) are crushed, decomposed with heat to CaO, and mixed with water to make slaked lime, $Ca(OH)_2$. The slaked lime is pumped into the intake tank to precipitate the Mg^{2+} as the hydroxide ($K_{sp} \approx 10^{-9}$):

$$Ca(OH)_2(aq) + Mg^{2+}(aq) \longrightarrow Mg(OH)_2(s) + Ca^{2+}(aq)$$

3. *Converting to compound.* The solid $Mg(OH)_2$ is filtered and mixed with excess HCl, which is also made on-site, to form aqueous $MgCl_2$:

$$Mg(OH)_2(s) + 2HCl(aq) \longrightarrow MgCl_2(aq) + 2H_2O(l)$$

The water is evaporated in stages to give solid, hydrated $MgCl_2{\cdot}nH_2O$.

4. *Electrochemical redox.* Heating above 700°C drives off the water of hydration and melts the $MgCl_2$. Electrolysis gives chlorine gas and the molten metal, which floats on the denser molten salt:

$$MgCl_2(l) \longrightarrow Mg(l) + Cl_2(g)$$

The Cl_2 that forms is recycled to make the HCl used in step 3.

Uses of Magnesium Magnesium is the lightest structural metal available (about one-half the density of Al and one-fifth the density of steel). Although Mg is quite

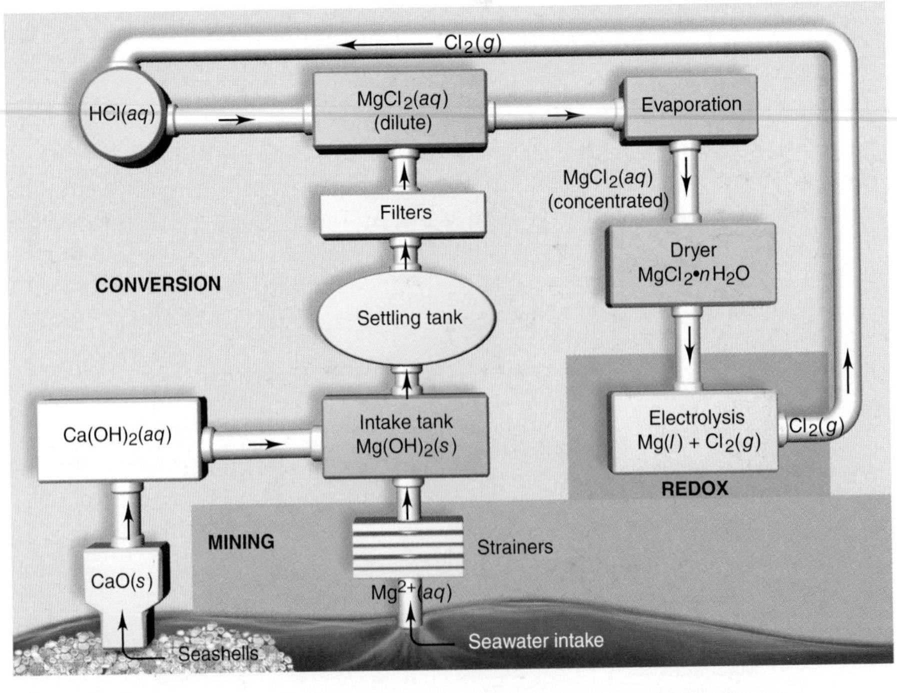

FIGURE 23.24 The Dow process for the isolation of elemental Mg from seawater

reactive, it forms an extremely adherent, high-melting oxide layer (MgO); thus, it finds many uses in metal alloys and can be machined into any form. Magnesium alloys occur in everything from aircraft bodies to camera bodies and from luggage to auto engine blocks. The pure metal is a strong reducing agent, which makes it useful for sacrificial anodes (Section 19.6) and for the metallurgical extraction of other metals, such as Be, Ti, Zr, Hf, and U. For example, titanium, an essential component of jet engines (*see photo*), is made from its major ore, ilmenite, in two steps:

Jet engine assembly

(1) $2FeTiO_3(s) + 7Cl_2(g) + 6C(s) \longrightarrow 2TiCl_4(l) + 2FeCl_3(s) + 6CO(g)$

(2) $TiCl_4(l) + 2Mg(l) \longrightarrow Ti(s) + 2MgCl_2(l)$

The Sources and Uses of Hydrogen

Although hydrogen accounts for 90% of the atoms in the universe, it makes up only 15% of the atoms in Earth's crust. Moreover, hydrogen exists elsewhere in the universe mostly as H_2 molecules and free H atoms, but virtually all the hydrogen in the crust is combined with other elements—either oxygen in natural water or carbon in biomass, petroleum, and coal.

Industrial Production of Hydrogen Hydrogen gas (H_2) is produced on an industrial scale worldwide. Canada is one of the world's top producers and users of hydrogen, primarily for the upgrade of heavy oil and for feedstock in the chemical industry. All production methods are energy intensive, so the choice of method is determined by energy costs. In Scandinavia, where hydroelectric power is plentiful, electrolysis is used; in the United States and Great Britain, where natural gas from oil refineries is plentiful, thermal methods are used.

1. *Electrolysis.* Very pure H_2 is prepared through electrolysis of water with Pt (or Ni) electrodes:

$$2H_2O(l) + 2e^- \longrightarrow H_2(g) + 2OH^-(aq) \qquad E = -0.42\text{ V} \quad \text{(cathode; reduction)}$$
$$H_2O(l) \longrightarrow \tfrac{1}{2}O_2(g) + 2H^+(aq) + 2e^- \qquad E = 0.82\text{ V} \quad \text{(anode; oxidation)}$$
$$H_2O(l) \longrightarrow H_2(g) + \tfrac{1}{2}O_2(g) \qquad E_{cell} = -0.42\text{ V} - 0.82\text{ V} = -1.24\text{ V}$$

Overpotential (overvoltage) makes the cell potential about −2 V (Section 19.7). Therefore, under typical operating conditions, it takes about 400 kJ of energy to produce 1 mol of H_2:

$$\Delta G = -nFE = (-2\text{ mol e}^-)\left(\frac{96.5\text{ kJ}}{\text{V}\cdot\text{mol e}^-}\right)(-2\text{ V}) = 4 \times 10^2\text{ kJ/mol H}_2$$

High-purity O_2 is a valuable by-product of this process, and it offsets some of the costs of isolating H_2.

2. *Thermal methods.* The most common thermal methods use water and a simple alkane such as methane, which has the highest H/C ratio of any hydrocarbon. The reactants are heated to around 1000°C over a nickel-based catalyst in an endothermic *steam-reforming process*:

$$CH_4(g) + H_2O(g) \longrightarrow CO(g) + 3H_2(g) \qquad \Delta H^\circ = 206\text{ kJ/mol}$$

Heat is supplied by burning methane. To generate more H_2, the product mixture (called *water gas*) is heated with steam at 400°C over an iron or cobalt oxide catalyst in the exothermic *water-gas shift reaction*, and the CO reacts as follows (see Chemical Connections in Chapter 5):

$$H_2O(g) + CO(g) \rightleftharpoons CO_2(g) + H_2(g) \qquad \Delta H^\circ = -41\text{ kJ/mol}$$

The reaction mixture is recycled several times, which decreases the CO to around 0.2% by volume. Passing the mixture through liquid water removes the more soluble CO_2 (solubility = 0.034 mol/L) from the H_2 (solubility < 0.001 mol/L). Calcium oxide can also be used to remove CO_2 by the formation of $CaCO_3$. By removing CO_2, these steps shift the equilibrium position to the right and produce H_2 that is about 98% pure. To filter out nearly all the molecules that are larger than H_2 and, thus, attain greater purity (~99.9%), the gas mixture is passed through a *synthetic zeolite.* This material is an aluminosilicate whose framework of polyhedra contains small cavities of various sizes, which trap some molecules and allow others to pass through.

Industrial Uses of Hydrogen Typically, a plant produces H_2 to make another product. More than 95% of H_2 produced industrially is consumed in ammonia or petrochemical plants.

1. *Ammonia synthesis.* In a plant that synthesizes NH_3, the reactant gases, N_2 and H_2, are formed through a series of reactions that involve methane. Here is a typical reaction series in an ammonia plant:

- The steam-reforming reaction is performed with excess CH_4, which depletes the reaction mixture of H_2O:

$$CH_4(g;\text{ excess}) + H_2O(g) \longrightarrow CO(g) + 3H_2(g)$$

- An excess of the product mixture (CH_4, CO, and H_2) is burned in an amount of air ($N_2 + O_2$) that is insufficient for complete combustion but is enough to consume the O_2, heat the mixture to 1100°C, and form additional H_2O:

$$4CH_4(g;\text{ excess}) + 7O_2(g) \longrightarrow 2CO_2(g) + 2CO(g) + 8H_2O(g)$$
$$2H_2(g;\text{ excess}) + O_2(g) \longrightarrow 2H_2O(g)$$
$$2CO(g;\text{ excess}) + O_2(g) \longrightarrow 2CO_2(g)$$

- Any remaining CH_4 reacts by the steam-reforming reaction, and the remaining CO reacts by the water-gas shift reaction to form more H_2:

$$H_2O + CO \rightleftharpoons CO_2 + H_2$$

  The CO_2 is then removed with CaO.

- The amounts are carefully adjusted to produce a final mixture that contains a 1/3 ratio of N_2 (from the added air) to H_2 (with traces of CH_4, Ar, and CO). The mixture is used directly in the synthesis of ammonia (Section 15.6).

2. *Hydrogenation.* A second major use of H_2 is *hydrogenation* of C═C bonds in liquid oils to form C—C bonds in solid fats and margarine. This process uses H_2 in contact with transition metal catalysts, such as powdered nickel (see Figure 14.25). Solid fats are also used in commercial baked goods. Look at the list of ingredients on most packages of bread, cake, and cookies, and you will see the "partially hydrogenated vegetable oils" made by this process.

3. *Bulk chemicals and energy production.* Hydrogen is also essential in the manufacture of numerous "bulk" chemicals, those produced in large amounts because they have many further uses. One application that has been gaining great attention is the production of methanol. In this process, carbon monoxide reacts with hydrogen over a copper–zinc oxide catalyst:

$$CO(g) + 2H_2(g) \xrightarrow{\text{Cu-ZnO catalyst}} CH_3OH(l)$$

Methanol is already being used as a gasoline additive. As less expensive sources become available, hydrogen will be used increasingly in fuel cells (Section 19.5).

Production and Uses of Deuterium; Kinetic Isotope Effect In addition to ordinary hydrogen (^1H), or protium, there are two other naturally occurring isotopes of significantly lower abundance. Deuterium (^2H or D) has one neutron, and rare radioactive tritium (^3H or T) has two neutrons. Like H_2, both occur as diatomic molecules: D_2 (or $^2H_2$) and T_2 (or $^3H_2$). Table 23.7 compares some of their molecular and physical properties. As you can see, the heavier the isotope and, thus, the higher the molar mass of the molecule, the higher the melting point, boiling point, and heats of phase change.

Because hydrogen is so light, the relative difference in mass of its isotopes is enormous compared with the difference in the masses of isotopes of other common elements. For example, ^2H has 2 times the mass of ^1H, whereas ^{13}C has only 1.08 times the mass of ^{12}C. *The mass difference leads to different bond energies, which affect reactivity.* As a result, an H atom bonded to a given atom vibrates at a higher frequency than a D atom does, so the bond to H is higher in energy. Thus, *the bond*

TABLE 23.7 Some Molecular and Physical Properties of Diatomic Protium, Deuterium, and Tritium

Property	H_2	D_2	T_2
Molar mass (g/mol)	2.016	4.028	6.032
Bond length (pm)	74.14	74.14	74.14
Melting point (K)	13.96	18.73	20.62
Boiling point (K)	20.39	23.67	25.04
$\Delta_{fus}H°$ (kJ/mol)	0.117	0.197	0.250
$\Delta_{vap}H°$ (kJ/mol)	0.904	1.226	1.393
Bond energy (kJ/mol at 298 K)	432	443	447

is weaker and easier to break. Therefore, any reaction that includes breaking a bond to H or D in the rate-determining step occurs *faster with H than with D*. This phenomenon, called a *kinetic isotope effect*, is also used to isolate deuterium.

Deuterium and its compounds are produced from D_2O (heavy water), which is present as a minor component (0.016 mol %) in normal water and is isolated on the multitonne scale by *electrolytic enrichment*. Owing to the kinetic isotope effect, there is a *higher rate of bond breaking for O—H bonds than for O—D bonds* and, thus, a higher rate of electrolysis of H_2O compared with D_2O.

For example, with Pt electrodes, H_2O is electrolyzed about 14 times as fast as D_2O. As some of the liquid decomposes to the elemental gases, the remainder becomes enriched in D_2O. Thus, by the time the volume of water has been reduced to 1/20 000 its original volume, the remaining water is around 99% D_2O. By combining samples and repeating the electrolysis, more than 99.9% D_2O is obtained.

Deuterium gas is produced by electrolysis of D_2O or by any of the chemical reactions that produce hydrogen gas from water, such as the following reaction:

$$2\text{Na}(s) + 2\text{D}_2\text{O}(l) \longrightarrow 2\text{Na}^+(aq) + 2\text{OD}^-(aq) + \text{D}_2(g)$$

Compounds that contain deuterium (or tritium) are produced from reactions that give rise to the corresponding hydrogen-containing compound, for example,

$$\text{SiCl}_4(l) + 2\text{D}_2\text{O}(l) \longrightarrow \text{SiO}_2(s) + 4\text{DCl}(g)$$

Compounds with acidic protons undergo hydrogen/deuterium exchange:

$$\text{CH}_3\text{COOH}(l) + \text{D}_2\text{O}(l;\ \text{excess}) \longrightarrow \text{CH}_3\text{COOD}(l) + \text{DHO}(l;\ \text{small amount})$$

We will discuss the natural and synthetic formation of tritium in Chapter 25.

SUMMARY OF SECTION 23.4

- Na is isolated by the electrolysis of molten NaCl in the Downs process; Cl_2 is a by-product.
- K is produced by reduction with Na in a thermal process.
- Fe is produced through a multistep high-temperature process in a blast furnace. The crude pig iron is converted to carbon steel in the basic-oxygen process and then alloyed with other metals to make different steels.
- Cu is produced by concentration of the ore through flotation, reduction to the metal by smelting, and purification by electrorefining. The metal has extensive electrical and plumbing uses.
- Al is extracted from bauxite by pretreatment of the ore with a concentrated base, followed by electrolysis of the product Al_2O_3 mixed with molten cryolite. Al alloys are widely used in homes and industry. The total energy needed to extract Al from its ore is over 100 times the total energy needed to recycle it.
- The Mg^{2+} in seawater is converted to $Mg(OH)_2$ and then to $MgCl_2$, which is electrolyzed to obtain the metal; Mg forms strong, lightweight alloys.
- H_2 is produced by the electrolysis of water or the formation of gaseous fuels from hydrocarbons. It is used in NH_3 production, hydrogenation of vegetable oils, and energy production. The isotopes of hydrogen differ significantly in atomic mass and, thus, in the rate at which their bonds to other atoms break (kinetic isotope effect). This difference is used to obtain D_2O from water.

23.5 Chemical Manufacturing: Two Case Studies

In this final section of Chapter 23, we examine the interplay of theory and practice in two of the most important processes in the inorganic chemical industry: the contact process for the production of sulfuric acid, and the chlor-alkali process for the production of chlorine.

Sulfuric Acid, the Most Important Chemical

Considering its countless uses, sulfuric acid is produced throughout the world on a gigantic scale—more than 200 million t in 2011. The modern **contact process** is based on the *catalyzed oxidation of SO_2*. Here are the key steps:

1. *Production of elemental sulfur.* In most countries today, the production of sulfuric acid starts with the production of elemental sulfur, often by the *Claus process*, in which the H_2S in "sour" natural gas is chemically separated and then oxidized:

$$2H_2S(g) + 2O_2(g) \xrightarrow{\text{low temperature}} \frac{1}{8}S_8(g) + SO_2(g) + 2H_2O(g)$$

$$2H_2S(g) + SO_2(g) \xrightarrow{Fe_2O_3\text{ catalyst}} \frac{3}{8}S_8(g) + 2H_2O(g)$$

The *Claus process* (see below) is conducted on a large scale in oil refineries. This is how sulfur is produced in Canada, by the very important Alberta sour gas.

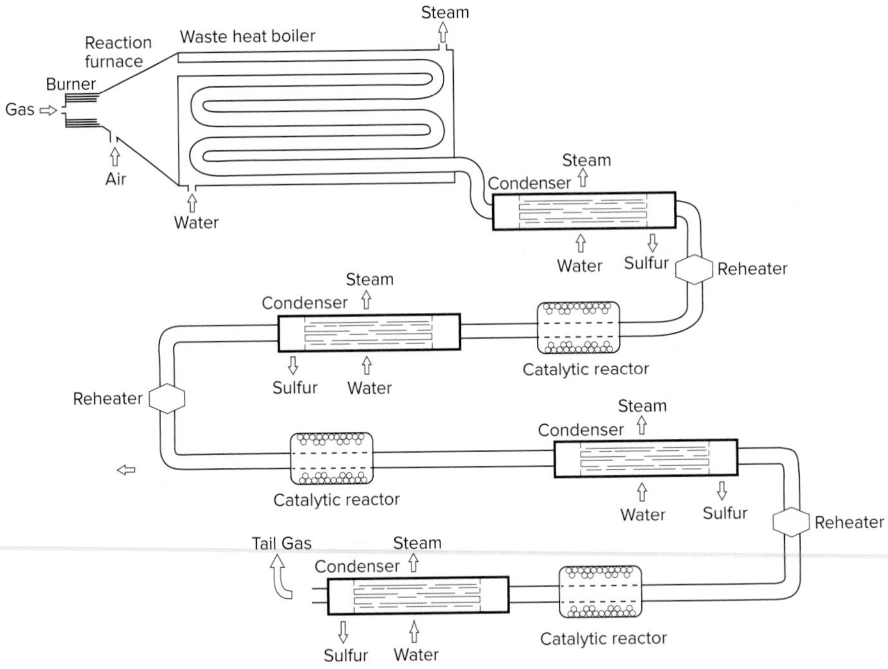

Where natural deposits of the element are found, sulfur is obtained by a nonchemical method called the *Frasch process*. A hole is drilled into the deposit, and superheated water (about 160°C) is pumped down two outer concentric pipes to melt the sulfur (Figure 23.25A). A combination of the hydrostatic pressure in an outer pipe and the pressure of compressed air sent through a narrow inner pipe then forces the sulfur to the surface (Figures 23.25B and C). The costs of drilling, pumping, and supplying water (2×10^7 L per day) are balanced somewhat by the purity of the product (~99.7% sulfur).

2. *From sulfur to sulfur dioxide.* The sulfur is burned in air to form SO_2:

$$\frac{1}{8}S_8(s) + O_2(g) \longrightarrow SO_2(g) \qquad \Delta H^\circ = -297 \text{ kJ/mol}$$

Some SO_2 is also obtained by roasting metal sulfide ores and by employing the Claus process. About 90% of processed sulfur is used to make sulfur dioxide for the production of sulfuric acid. Indeed, this end product is so important that a nation's level

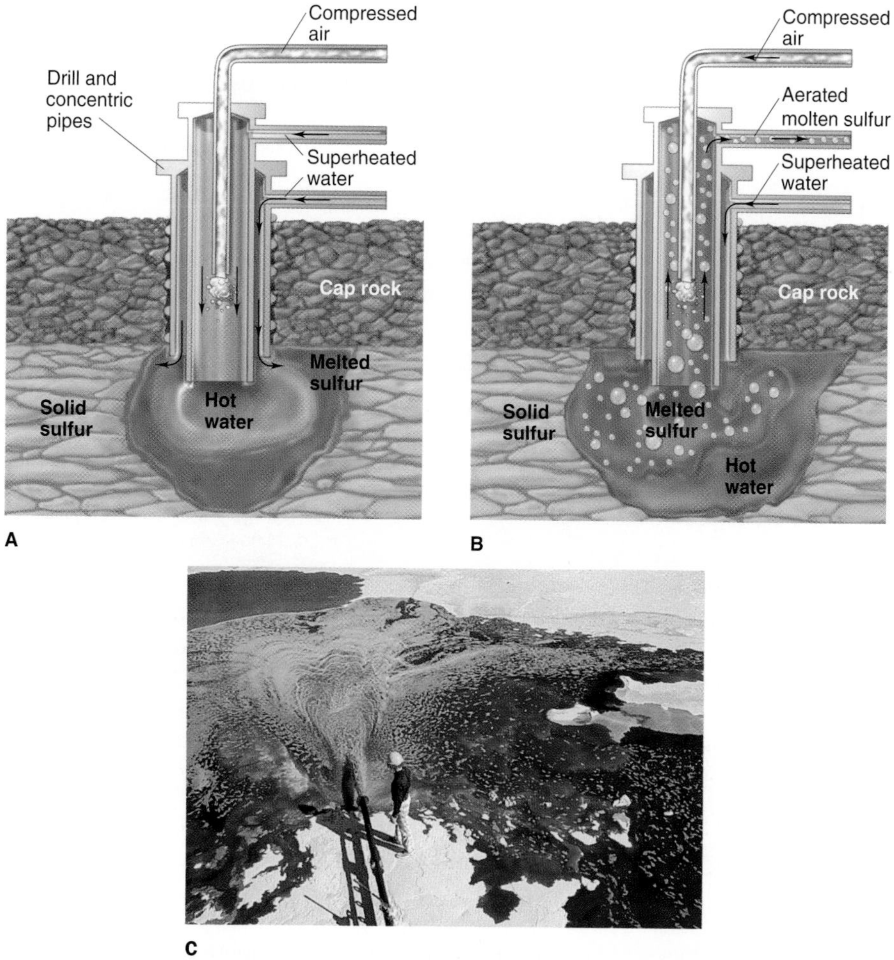

FIGURE 23.25 The Frasch process for mining elemental sulfur

of sulfur production is a reliable indicator of its overall industrial capacity: China, the United States, Russia, Canada, Saudi Arabia, and Germany were the top six sulfur producers in 2012.

3. *From sulfur dioxide to trioxide.* The *contact process* oxidizes SO_2 to SO_3:

$$SO_2(g) + \frac{1}{2}O_2(g) \rightleftharpoons SO_3(g) \qquad \Delta H^\circ = -99 \text{ kJ/mol}$$

The reaction is *exothermic* and very *slow* at room temperature. From Le Châtelier's principle (Section 15.6), the yield of SO_3 can be increased by (1) changing the temperature, (2) increasing the pressure [larger amount (mol) of gas on the left than on the right], or (3) adjusting the concentrations (adding excess O_2 and removing SO_3). In this case, the pressure effect is small and economically not worth exploiting. The other two effects are more useful:

- *Effect of changing the temperature.* Adding heat (raising the temperature) increases the frequency and energy of SO_2-O_2 collisions and, thus, increases the *rate* of SO_3 formation. However, because SO_3 formation is exothermic, removing heat (lowering the temperature) shifts the equilibrium position to the right and increases the *yield* of SO_3. This is a classic situation that calls for the use of a catalyst. By lowering the activation energy, *a catalyst allows equilibrium to be reached more quickly and at a lower temperature*; therefore, rate *and* yield are optimized (Section 14.7). The catalyst in the contact process is V_2O_5 on inert silica, which is active between 400°C and 600°C.
- *Effect of changing the concentration.* Providing an excess of O_2 in the form of a 5/1 mixture of air to SO_2, which is about 1/1 O_2 to SO_2, supplies about twice as much O_2 as in the balanced equation. The mixture is passed over catalyst beds in four stages, and the SO_3 is removed to favour more SO_3 formation. The overall yield of SO_3 is 99.5%.

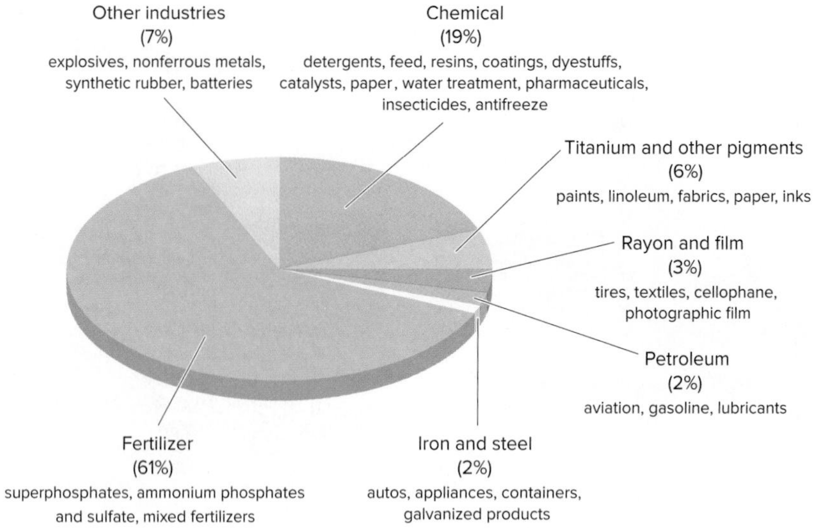

FIGURE 23.26 The many indispensable applications of sulfuric acid

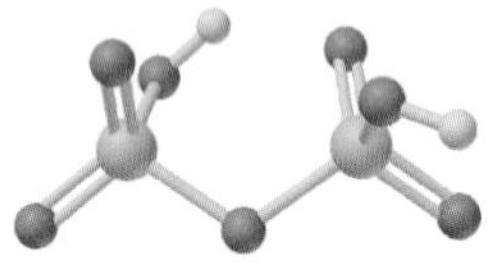

Disulfuric acid

4. *From sulfur trioxide to acid.* Sulfur trioxide is the anhydride of sulfuric acid. However, SO_3 cannot be added to water because, at the operating temperature, it would first meet water vapour, which catalyzes its polymerization to $(SO_3)_x$ and results in a smoke of solid particles that yields little acid (reaction 1). To avoid this, previously formed H_2SO_4 absorbs the SO_3 (reaction 2) and forms pyrosulfuric acid (or disulfuric acid, $H_2S_2O_7$; *see margin*), which is then hydrolyzed with sufficient water (reaction 3):

$$\text{(1)} \quad SO_3(g) + H_2O(l) \longrightarrow H_2SO_4(l) \quad \text{(low yield)}$$
$$\text{(2)} \quad SO_3(g) + H_2SO_4(l) \longrightarrow H_2S_2O_7(l)$$
$$\text{(3)} \quad H_2S_2O_7(l) + H_2O(l) \longrightarrow 2H_2SO_4(l)$$

The industrial uses of sulfuric acid are numerous, as Figure 23.26 indicates.

Sulfuric acid is remarkably inexpensive (about \$150/t), largely because each step in the process is exothermic—burning S ($\Delta H° = -297$ kJ/mol), oxidizing SO_2 ($\Delta H° = -99$ kJ/mol), hydrating SO_3 ($\Delta H° = -132$ kJ/mol)—and the heat is a valuable by-product. Three-quarters of the heat is sold as steam, and the rest is used to pump gases through the plant. A typical plant making 748 t of H_2SO_4 per day produces enough steam to generate 7×10^6 W of electric power.

The Chlor-Alkali Process

Chlorine is produced and used in amounts that are many times greater than all the other halogens combined, ranking among the top five inorganic chemicals produced in Canada. The **chlor-alkali process**, which forms the basis of one of the largest inorganic chemical industries, *electrolyzes concentrated aqueous NaCl to produce* Cl_2, as well as several other important chemicals. There are three versions of this process.

Diaphragm-Cell Method As you learned in Section 19.7, the electrolysis of aqueous NaCl does not yield both of the component elements. Because of the overpotential, Cl^- ions rather than H_2O molecules are oxidized at the anode. However, Na^+ ions are not reduced at the cathode because the half-cell potential (−2.71 V) is much more negative than the half-cell potential for the reduction of H_2O (−0.42 V), even with the normal overpotential (around −0.6 V). Therefore, the half-reactions for electrolysis of aqueous NaCl are

$$2Cl^-(aq) \longrightarrow Cl_2(g) + 2e^- \qquad E° = 1.36\ \text{V} \quad \text{(anode; oxidation)}$$
$$2H_2O(l) + 2e^- \longrightarrow 2OH^-(aq) + H_2(g) \qquad E \approx -1.0\ \text{V} \quad \text{(cathode; reduction)}$$
$$2Cl^-(aq) + 2H_2O(l) \longrightarrow 2OH^-(aq) + H_2(g) + Cl_2(g) \qquad E_{cell} = -1.0\ \text{V} - 1.36\ \text{V} = -2.4\ \text{V}$$

To obtain commercially meaningful amounts of Cl_2, however, a voltage almost twice this value and a current in excess of 3×10^4 A are used.

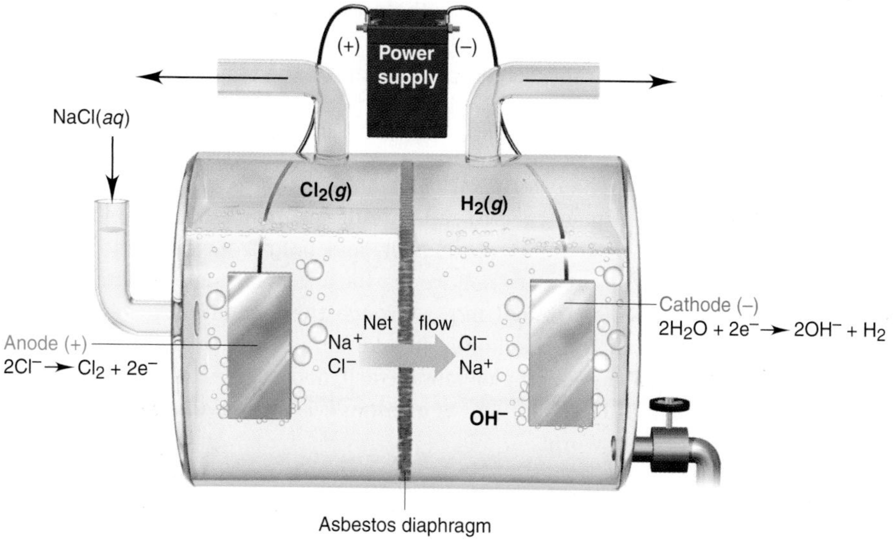

FIGURE 23.27 A diaphragm cell for the chlor-alkali process

When we include the spectator ion Na^+, the total ionic equation shows another important product made by the process:

$$2Na^+(aq) + 2Cl^-(aq) + 2H_2O(l) \longrightarrow 2Na^+(aq) + 2OH^-(aq) + H_2(g) + Cl_2(g)$$

As Figure 23.27 shows, the sodium salts in the cathode compartment exist as an aqueous mixture of NaCl and NaOH; the NaCl is removed by fractional crystallization. Thus, in this version of the chlor-alkali process, which uses an *asbestos diaphragm* to separate the anode and cathode compartments, electrolysis of NaCl brines yields Cl_2, H_2, and industrial-grade NaOH, an important base. Like other reactive products, H_2 and Cl_2 are kept apart to prevent explosive recombination. Notice the higher liquid level in the anode (*left*) compartment. This slight hydrostatic pressure difference minimizes backflow of NaOH, which prevents the disproportionation reactions of Cl_2 that occur in the presence of OH^- (Section 13.9), such as the reaction below:

$$Cl_2(g) + 2OH^-(aq) \longrightarrow Cl^-(aq) + ClO^-(aq) + H_2O(l)$$

Mercury-Cell Method If high-purity NaOH is desired, a slightly different version, called the *chlor-alkali mercury-cell process*, is employed. Mercury is used as the cathode, which creates such a large overpotential for reduction of H_2O to H_2 that the process *does* favour the reduction of Na^+. The sodium dissolves in the mercury to form sodium amalgam, Na(Hg). In the mercury-cell version, the half-reactions are

$$2Cl^-(aq) \longrightarrow Cl_2(g) + 2e^- \quad \text{(anode; oxidation)}$$
$$2Na^+(aq) + 2e^- \xrightarrow{\text{Hg}} 2Na(Hg) \quad \text{(cathode; reduction)}$$

To obtain sodium hydroxide, the sodium amalgam is pumped out of the system and treated with H_2O, which is reduced by the Na:

$$2Na(Hg) + 2H_2O(l) \xrightarrow{-\text{Hg}} 2Na^+(aq) + 2OH^-(aq) + H_2(g)$$

The mercury released in this step is recycled back to the electrolysis bath. Therefore, *the products are the same in both versions*, but the purity of the NaOH from the mercury-cell method is much higher.

The mercury-cell method is being steadily phased out because, as the mercury is recycled, small amounts are lost in the industrial wastewater. On average, 220 g of Hg is lost per tonne of Cl_2 produced. In the 1980s, annual U.S. production via the mercury-cell method was 2.5 million t of Cl_2; thus, during that decade, 550 000 kg of mercury, a toxic heavy metal, was flowing into U.S. waterways each year! In 1970, the Ontario Minamata disease (a neurological syndrome caused by severe mercury poisoning) severely affected two First Nations communities in northwestern Ontario following the consumption of local fish contaminated with mercury, as well as one First Nations community in southern Ontario due to the illegal disposal of industrial chemical waste. The disease was named after the infamous case of severe mercury poisoning in the fishing community of Minamata, Japan. It became known as Minamata disease because it devastated only the residents of this community.

Membrane-Cell Method The *chlor-alkali membrane-cell process* replaces the diaphragm with a polymeric membrane to separate the cell compartments. The membrane allows only cations to move through it, and only from anode to cathode compartments. Thus, as Cl^- ions are removed at the anode through oxidation to Cl_2, Na^+ ions in the anode compartment move through the membrane to the cathode compartment and form an NaOH solution. In addition to forming purer NaOH than the older diaphragm-cell method, the membrane-cell method uses less electricity and eliminates the problem of Hg pollution. As a result, it has been adopted throughout much of the industrialized world.

SUMMARY OF SECTION 23.5

- Sulfuric acid production starts with the extraction of sulfur, either by the oxidation of H_2S or the mining of sulfur deposits. The sulfur is roasted to SO_2 and oxidized to SO_3 by the catalyzed contact process, which optimizes the yield at lower temperatures. Sulfuric acid is formed by absorption of the SO_3 into H_2SO_4, followed by hydration.
- In the diaphragm-cell chlor-alkali process, aqueous NaCl is electrolyzed to form Cl_2, H_2, and low-purity NaOH. The mercury-cell method produces high-purity NaOH but has been almost completely phased out because of mercury pollution. The membrane-cell method requires less electricity and does not use Hg.

CHAPTER REVIEW GUIDE

Learning Objectives

Relevant section (§) numbers appear in parentheses.

Concepts

1. Explain how gravity, thermal convection, and elemental properties led to the silicate, sulfide, and iron phases and the predominance of certain elements in Earth's crust, mantle, and core. (§23.1)
2. Explain how organisms affect the crustal abundances of the elements, especially oxygen, carbon, calcium, and some transition metals, and how they influenced the onset of oxidation as an energy source. (§23.1)
3. Discuss how atomic properties influence which elements have oxide ores and which have sulfide ores. (§23.1)
4. Explain the central role of CO_2, and discuss the importance of photosynthesis, respiration, and decomposition in the carbon cycle. (§23.2)
5. Explain the central role of N_2, and discuss the importance of atmospheric, industrial, and biological fixation in the nitrogen cycle. (§23.2)
6. Discuss the absence of a gaseous component, the interactions of the inorganic and biological cycles, and the impact of humans on the phosphorus cycle. (§23.2)
7. Explain how the pyrometallurgical, electrometallurgical, and hydrometallurgical processes are used to extract a metal from its ore; describe the importance of the reduction step from compound to metal; summarize the refining and alloying processes. (§23.3)
8. Describe the functioning of the Downs cell for Na production and the application of Le Châtelier's principle for K production. (§23.4)
9. Explain how iron ore is reduced in a blast furnace and how pig iron is purified by the basic-oxygen process. (§23.4)
10. Explain how Fe is removed from copper ore and impure Cu is electrorefined. (§23.4)
11. Discuss the importance of amphoterism in the Bayer process for isolating Al_2O_3 from bauxite; explain the significance of cryolite in the electrolytic step; discuss the energy advantage of Al recycling. (§23.4)
12. Discuss the steps in the Dow process for extracting Mg from seawater. (§23.4)
13. Explain how H_2 production, whether by chemical or electrolytic means, is tied to NH_3 production. (§23.4)
14. Explain how the kinetic isotope effect is applied to produce deuterium. (§23.4)
15. Explain how the Frasch process is used to obtain sulfur from natural deposits. (§23.5)
16. Discuss the importance of equilibrium and kinetic factors in H_2SO_4 production. (§23.5)
17. Explain how overpotential allows the electrolysis of aqueous NaCl to form Cl_2 gas in the chlor-alkali process; analyze the coproduction of NaOH; compare the diaphragm-cell, mercury-cell, and membrane-cell methods. (§23.5)

Key Terms

Section 23.1
abundance
differentiation
core
mantle
crust
lithosphere
hydrosphere
atmosphere
biosphere
occurrence (source)
ores

Section 23.2
environmental cycle
fixation
apatites

Section 23.3
metallurgy
mineral
gangue
flotation
leaching
roasting
smelting
electrorefining
zone refining
alloys

Section 23.4
Downs cell
steel
blast furnace
slag
basic-oxygen process
carbon steel

Section 23.5
contact process
chlor-alkali process

PROBLEMS

Problems with **red** numbers are answered in Appendix G and worked in detail in the Student Solutions Manual. Problem sections match those in this book and offer Concept Review Questions and Problems in Context. The Comprehensive Problems are based on material from any section or previous chapter.

How the Elements Occur in Nature

Concept Review Questions

23.1 Hydrogen is, by far, the most abundant element cosmically. In interstellar space, it exists mainly as H_2. On Earth, however, it exists very rarely as H_2 and is ninth in abundance in the crust. Why is hydrogen so abundant in the universe? Why is hydrogen so rare as a diatomic gas in Earth's atmosphere?

23.2 Metallic elements can be recovered from ores that are oxides, carbonates, halides, or sulfides. Give an example of each type of ore.

23.3 The location of elements in different regions of Earth has enormous practical importance.
(a) Define *differentiation*, and explain which physical property of a substance is primarily responsible for this process.
(b) What are the four most abundant elements in the crust?
(c) Which element is abundant in the crust and the mantle, but not in the core?

23.4 How does the position of a metal in the periodic table relate to whether it occurs primarily as an oxide or as a sulfide?

Problems in Context

23.5 What material is the source for commercial production of (a) aluminum; (b) nitrogen; (c) chlorine; (d) calcium; (e) sodium?

23.6 Aluminum is widely distributed throughout the world in the form of aluminosilicates. What property of these minerals prevents them from being a source of aluminum?

23.7 Describe two ways in which the biosphere has influenced the composition of Earth's crust.

The Cycling of Elements through the Environment

Concept Review Questions

23.8 Use atomic and molecular properties to explain why life is based on carbon rather than some other element, such as silicon.

23.9 Define *fixation*. Name two elements that undergo environmental fixation. What natural forms of these elements are fixed?

23.10 Carbon dioxide enters the atmosphere by natural processes and from human activities. Why is the latter a cause of concern?

23.11 Diagrams of environmental cycles are simplified to omit minor contributors. For example, the production of lime from limestone is not shown in the carbon cycle diagram in Section 23.2 (Figure 23.5). Which labelled category in the diagram includes the production of lime? Name two other processes that contribute to this category.

23.12 Describe three pathways for the use of atmospheric nitrogen. Are human activities a significant factor? Explain.

23.13 Why do the N-containing species in Figure 23.6 not include rings or long chains with N—N bonds?

23.14 (a) Which region of Earth's crust is not involved in the phosphorus cycle?
(b) Name two roles of organisms in the phosphorus cycle.

Problems in Context

23.15 Nitrogen fixation requires a great deal of energy because the N_2 bond is strong.
(a) How do the processes of atmospheric fixation and industrial fixation reflect this energy requirement?
(b) How do the thermodynamics of the two processes differ? (*Hint*: Examine their heats of formation.)
(c) In view of the mild conditions for biological fixation, what must be the source of the "great deal of energy"?
(d) What would be the most obvious environmental result of a low activation energy for N_2 fixation?

23.16 The following steps are *unbalanced* half-reactions involved in the nitrogen cycle. Balance each half-reaction to show the number of electrons lost or gained, and state whether it is an oxidation or a reduction (all occur in acidic conditions):
(a) $N_2(g) \longrightarrow NO(g)$
(b) $N_2O(g) \longrightarrow NO_2(g)$
(c) $NH_3(aq) \longrightarrow NO_2^-(aq)$
(d) $NO_3^-(aq) \longrightarrow NO_2^-(aq)$
(e) $N_2(g) \longrightarrow NO_3^-(aq)$

23.17 The use of silica to form slag in the production of phosphorus from phosphate rock was introduced by Robert Boyle more than 300 years ago. When fluorapatite, $Ca_5(PO_4)_3F$, is used in phosphorus production, most of the fluorine atoms appear in the slag, but some end up in the toxic and corrosive gas SiF_4.
(a) If 15% by mass of the fluorine in 100. kg of $Ca_5(PO_4)_3F$ forms SiF_4, what volume of this gas is collected at 101.3 kPa and the industrial furnace temperature of 1450.°C?
(b) In some facilities, the SiF_4 is used to produce sodium hexafluorosilicate (Na_2SiF_6), which is sold for water fluoridation:

$$2SiF_4(g) + Na_2CO_3(s) + H_2O(l) \longrightarrow Na_2SiF_6(aq) + SiO_2(s) + CO_2(g) + 2HF(aq)$$

What volume (m^3) of drinking water can be fluoridated to a level of 1.0 ppm of F^- using the volume of SiF_4 produced in part (a)?

23.18 An impurity that is sometimes found in $Ca_3(PO_4)_2$ is Fe_2O_3, which is removed during the production of phosphorus as *ferrophosphorus* (Fe_2P).
(a) Why is this impurity troubling from an economic standpoint?
(b) If 50. t of crude $Ca_3(PO_4)_2$ contains 2.0% Fe_2O_3 by mass, and the overall yield of phosphorus is 90.%, what mass (t) of P_4 can be isolated?

Metallurgy: Extracting a Metal from Its Ore

Concept Review Questions

23.19 Define (a) *ore*; (b) *mineral*; (c) *gangue*; (d) *brine*.

23.20 Define (a) *roasting*; (b) *smelting*; (c) *flotation*; (d) *refining*.

23.21 What factors determine which reducing agent is selected to produce a specific metal?

23.22 Use atomic properties to explain the reduction of a less active metal by a more active metal (a) in aqueous solution; (b) in the molten state. Give a specific example of each process.

23.23 What class of element is obtained by oxidation of a mineral? What class of element is obtained by reduction of a mineral?

Problems in Context

23.24 Which set of elements gives (a) brass; (b) stainless steel; (c) bronze; (d) sterling silver?

(1) Cu, Ag	(2) Cu, Sn, Zn	(3) Ag, Au
(4) Fe, Cr, Ni	(5) Fe, V	(6) Cu, Zn

Tapping the Crust: Isolation and Uses of Selected Elements

Concept Review Questions

23.25 How is each of the following involved in iron metallurgy?
(a) Slag (b) Pig iron
(c) Steel (d) Basic-oxygen process

23.26 (a) What are the distinguishing features of (i) pyrometallurgy; (ii) electrometallurgy; (iii) hydrometallurgy?
(b) Explain briefly how the three types of metallurgy in part (a) are used in the production of (i) Fe; (ii) Na; (iii) Au; (iv) Al.

23.27 What property allows copper to be purified in the presence of iron and nickel impurities? Explain.

23.28 Why is cryolite used in the electrolysis of aluminum oxide?

23.29 (a) What is a kinetic isotope effect?
(b) Do compounds of hydrogen exhibit a relatively large or small kinetic isotope effect? Explain.
(c) Carbon compounds also exhibit a kinetic isotope effect. How do you expect their kinetic isotope effect to compare in magnitude with the kinetic isotope effect of hydrogen compounds? Why?

23.30 How is Le Châtelier's principle involved in the production of elemental potassium?

Problems in Context

23.31 Elemental Li and Na are prepared by electrolysis of a molten salt, whereas K, Rb, and Cs are prepared by chemical reduction.
(a) In general terms, explain why the alkali metals cannot be prepared by electrolysis of their aqueous salt solutions.
(b) Use ionization energies from the Family Portraits in Sections 13.3 and 13.4 to explain why calcium should *not* be able to isolate Rb from molten RbX (X = halide).
(c) Use physical properties to explain why calcium *is* used to isolate Rb from molten RbX.
(d) Can Ca be used to isolate Cs from molten CsX? Explain.

23.32 A Downs cell operating at 77.0 A produced 31.0 kg of Na.
(a) What volume of $Cl_2(g)$ was produced at 101.3 kPa and 540.°C?
(b) What amount of electric charge (coulomb, C) was passed through the cell?
(c) How long did the cell operate?

23.33 (a) In the industrial production of iron, what reducing substance is loaded into the blast furnace?
(b) In addition to furnishing the reducing power, what other function does this substance serve?
(c) What is the formula for the active reducing agent in the process?
(d) Write equations for the step-by-step reduction of Fe_2O_3 to iron in the furnace.

23.34 One of the substances loaded into a blast furnace is limestone, which produces lime in the furnace.
(a) Give the chemical equation for the reaction that forms lime.
(b) Explain the function of lime in the furnace. A substance that acts like lime is often called *flux*. What is the derivation of this term, and how does it relate to the function of lime?
(c) Write a chemical equation describing the action of lime flux.

23.35 The last step in the Dow process for the production of magnesium metal involves the electrolysis of molten $MgCl_2$.
(a) Why is the electrolysis not carried out with aqueous $MgCl_2$? What are the products of this aqueous electrolysis?
(b) Do the high temperatures required to melt $MgCl_2$ favour products or reactants? (*Hint*: Consider the $\Delta_f H°$ of $MgCl_2$.)

23.36 Iodine is the only halogen that occurs in a positive oxidation state. It occurs in $NaIO_3$ impurities within Chile saltpeter, $NaNO_3$.
(a) Is this mode of occurrence consistent with iodine's location in the periodic table? Explain.
(b) In the production of I_2, IO_3^- reacts with HSO_3^-:

$$IO_3^-(aq) + HSO_3^-(aq) \longrightarrow HSO_4^-(aq) + SO_4^{2-}(aq) + H_2O(l) + I_2(s) \quad \text{(unbalanced)}$$

Identify the oxidizing and reducing agents.
(c) If 0.78 mol % of an $NaNO_3$ deposit is $NaIO_3$, what amount of $I_2(g)$ can be obtained from 1000. t of the deposit?

23.37 Selenium is prepared by the reaction of H_2SeO_3 with gaseous SO_2.
(a) What redox process does the sulfur dioxide undergo? What is the oxidation state of sulfur in the product?
(b) Given that the reaction occurs in acidic aqueous solution, what is the formula for the sulfur-containing species?
(c) Write the balanced redox equation for the process.

23.38 F_2 and Cl_2 are produced by electrolytic oxidation, whereas Br_2 and I_2 are produced by chemical oxidation of the halide ions in a concentrated aqueous solution (brine) by a more electronegative halogen. Give two reasons that Cl_2 is not prepared this way.

23.39 Silicon is prepared by the reduction of K_2SiF_6 with Al. Write the equation for this reaction. (*Hint*: Can F^- be oxidized in this reaction? Can K^+ be reduced?)

23.40 What is the mass percent of iron in each iron ore?
(a) Fe_2O_3 (b) Fe_3O_4 (c) FeS_2

23.41 Phosphorus is one of the impurities in pig iron that is removed in the basic-oxygen process. Assuming that phosphorus is present as P atoms, write equations for its oxidation and subsequent reaction in the basic slag.

23.42 The final step in the smelting of $FeCuS_2$ is

$$Cu_2S(s) + 2Cu_2O(s) \longrightarrow 6Cu(l) + SO_2(g)$$

(a) Give the oxidation states of copper in (i) Cu_2S; (ii) Cu_2O; (iii) Cu.
(b) What are the oxidizing and reducing agents in this reaction?

23.43 Use equations to show how acid-base properties are used to separate Fe_2O_3 and TiO_2 from Al_2O_3 in the Bayer process.

23.44 A piece of Al, with a surface area of 2.5 m^2, is anodized to produce a film of Al_2O_3 that is 23 μm (23×10^{-6} m) thick.
(a) What amount of electrical charge (coulomb, C) flows through the cell in this process? (Assume that the density of the Al_2O_3 layer is 3.97 g/cm^3.)
(b) If it takes 18 min to produce this film, what current must flow through the cell?

23.45 The production of H_2 gas by the electrolysis of water typically requires about 400 kJ of energy per mole.
(a) Use the relationship between work and cell potential (Section 19.4) to calculate the minimum work needed to form 1.0 mol of H_2 gas at a cell potential of 1.24 V.
(b) What is the energy efficiency of the cell operation?
(c) Find the cost of producing 500. mol of H_2 if electricity costs \$0.06 per kilowatt•hour (1 W•s = 1 J).

23.46 (a) What are the components of the reaction mixture following the water-gas shift reaction?
(b) Explain how zeolites are used to purify the H_2 formed.

23.47 Metal sulfides are often first converted to oxides by roasting in air and then reduced with carbon to produce the metal. Why are the metal sulfides not reduced directly by carbon to yield CS_2? Give a thermodynamic analysis of both processes for ZnS.

Chemical Manufacturing: Two Case Studies

Concept Review Questions

23.48 Explain, in detail, why a catalyst is used to produce SO_3.

23.49 Among the exothermic steps in the manufacture of sulfuric acid is the process of hydrating SO_3.
(a) Write two chemical reactions that show this process.
(b) Why is the direct reaction of SO_3 with water not feasible?

23.50 Why is commercial H_2SO_4 so inexpensive?

23.51 (a) What are the three commercial products formed in the chlor-alkali process?
(b) State an advantage and a disadvantage of the mercury-cell method for this process.

Problems in Context

23.52 Consider the reaction of SO_2 to form SO_3 at standard conditions.
(a) Calculate $\Delta G°$ at 25°C. Is the reaction spontaneous?
(b) Why is the reaction not performed at 25°C?
(c) Is the reaction spontaneous at 500°C? (Assume that $\Delta H°$ and $\Delta S°$ are constant with changing temperature.)
(d) Compare K at 500.°C with K at 25°C.
(e) What is the highest temperature at which the reaction is spontaneous?

23.53 If a chlor-alkali cell used a current of 3×10^4 A, what mass (kg) of Cl_2 would be produced in a typical 8 h operating day?

23.54 The products of the chlor-alkali process, Cl_2 and NaOH, are kept separated.
(a) Why is this separation necessary when Cl_2 is the desired product?
(b) ClO^- or ClO_3^- may form by disproportionation of Cl_2 in basic solution. What determines which product forms?
(c) What mole ratio of Cl_2 to OH^- will produce (i) ClO^-; (ii) ClO_3^-?

Comprehensive Problems

23.55 The key step in the manufacture of sulfuric acid is the oxidation of sulfur dioxide in the presence of a catalyst, such as V_2O_5. At 727°C, 0.010 mol of SO_2 is injected into an empty 2.00 L container ($K = 0.0314\ kPa^{-1}$).
(a) What equilibrium pressure of O_2 is needed to maintain a 1/1 mole ratio of SO_3 to SO_2?
(b) What equilibrium pressure of O_2 is needed to maintain a 95/5 mole ratio of SO_3 to SO_2?

23.56 Tetraphosphorus decoxide (P_4O_{10}) is made from phosphate rock and used as a drying agent in the laboratory.
(a) Write a balanced equation for its reaction with water.
(b) What is the pH of a solution formed from the addition of 8.5 g of P_4O_{10} in sufficient water to form 0.750 L? (See Table 16.7 for additional information.)

23.57 Heavy water (D_2O) is used to make deuterated chemicals.
(a) What major species, aside from the starting compounds, do you expect to find in a solution of CH_3OH and D_2O?
(b) Write equations to explain how these major species arise. (*Hint*: Consider the autoionization of both components.)

23.58 A blast furnace uses Fe_2O_3 to produce 7620. t of Fe per day.
(a) What mass of CO_2 is produced each day?
(b) Compare this amount of CO_2 with the amount produced by 1.0 million automobiles, each burning 19 L of gasoline a day. Assume that gasoline has the formula C_8H_{18} and a density of 0.74 g/mL, and that it burns completely. (Note that Canadian gasoline consumption is over 1×10^8 L/day.)

23.59 A major use of Cl_2 is in the manufacture of chloroethene (vinyl chloride), the monomer of poly(vinyl chloride). The two-step sequence for formation of vinyl chloride is depicted below.

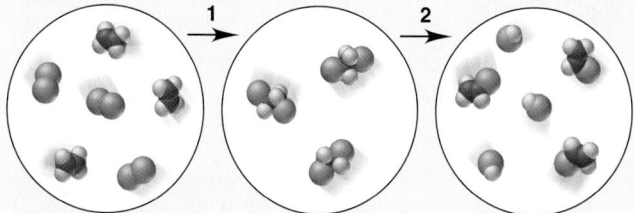

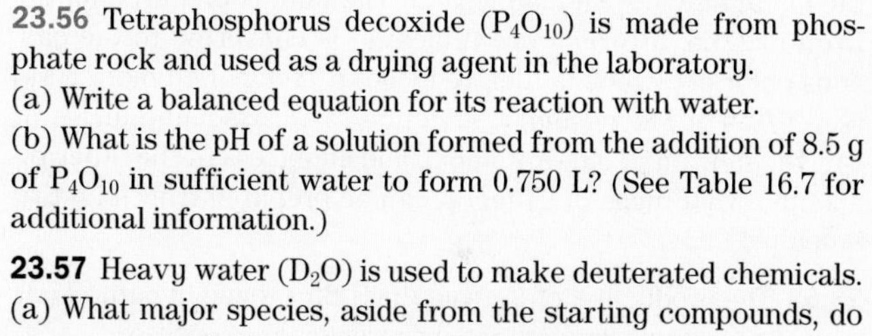

(a) Write a balanced equation for each step.
(b) Write the overall equation.
(c) What type of organic reaction is shown in step 1?
(d) What type of organic reaction is shown in step 2?
(e) If each molecule depicted in the initial reaction mixture represents 0.15 mol of substance, what mass (g) of vinyl chloride forms?

23.60 In the production of magnesium, $Mg(OH)_2$ is precipitated by using $Ca(OH)_2$, which itself is insoluble.
(a) Use K_{sp} values to show that $Mg(OH)_2$ can be precipitated from seawater in which $[Mg^{2+}]$ is initially 0.052 mol/L. (b) If the seawater is saturated with $Ca(OH)_2$, what fraction of the Mg^{2+} is precipitated?

23.61 Step 1 of the Ostwald process for nitric acid production is

$$4NH_3(g) + 5O_2(g) \xrightarrow{\text{Pt/Rh catalyst}} 4NO(g) + 6H_2O(g)$$

An unwanted side reaction for this step is

$$4NH_3(g) + 3O_2(g) \longrightarrow 2N_2(g) + 6H_2O(g)$$

(a) Calculate K for these two NH_3 oxidations at 25°C.
(b) Calculate K for these two NH_3 oxidations at 900.°C.
(c) The Pt/Rh catalyst is one of the most efficient in the chemical industry, achieving 96% yield in 1 millisecond (ms) of contact with the reactants. However, at normal operating conditions (506.6 kPa and 850°C), about 175 mg of Pt is lost per tonne (t) of HNO_3 produced. If the annual Canadian production of HNO_3 is 6.6×10^6 t and the market price of Pt is \$1557/oz t (troy oz), what is the annual cost of the lost Pt (1 kg = 32.15 oz t)?
(d) Because of the high price of Pt, a filtering unit composed of ceramic fibre is often installed, which recovers as much as 75% of the lost Pt. What is the value of the Pt captured by a recovery unit with 72% efficiency?

23.62 Several transition metals are prepared by reduction of the metal halide with magnesium. Titanium is prepared by the Kroll method, in which ore (ilmenite) is converted to the gaseous chloride, which is then reduced to Ti metal by molten Mg (see "Uses of Magnesium" in Section 23.4). Assuming yields of 84% for step 1 and 93% for step 2, and an excess of the other reactants, what mass of Ti metal can be prepared from 21.5 t of ilmenite?

23.63 The production of S_8 from the $H_2S(g)$ found in natural gas deposits occurs through the Claus process (Section 23.5):
(a) Use these two unbalanced steps to write an overall balanced equation for the Claus process:
(1) $H_2S(g) + O_2(g) \longrightarrow S_8(g) + SO_2(g) + H_2O(g)$
(2) $H_2S(g) + SO_2(g) \longrightarrow S_8(g) + H_2O(g)$
(b) Write the overall reaction with Cl_2 as the oxidizing agent, instead of O_2. Use thermodynamic data to show whether $Cl_2(g)$ can be used to oxidize $H_2S(g)$.
(c) Why is oxidation by O_2 preferred to oxidation by Cl_2?

23.64 Acid mine drainage (AMD) occurs when geologic deposits containing pyrite (FeS_2) are exposed to oxygen and moisture. AMD is generated in a multistep process catalyzed by acidophilic (acid-loving) bacteria. Balance each step, and identify the steps that increase acidity:
(1) $FeS_2(s) + O_2(g) \longrightarrow Fe^{2+}(aq) + SO_4^{2-}(aq)$
(2) $Fe^{2+}(aq) + O_2(g) \longrightarrow Fe^{3+}(aq) + H_2O(l)$
(3) $Fe^{3+}(aq) + H_2O(l) \longrightarrow Fe(OH)_3(s) + H^+(aq)$
(4) $FeS_2(s) + Fe^{3+}(aq) \longrightarrow Fe^{2+}(aq) + SO_4^{2-}(aq)$

23.65 Below 912°C, pure iron crystallizes in a body-centred cubic structure (ferrite), with d = 7.86 g/cm^3. From 912°C to 1394°C, it adopts a face-centred cubic structure (austenite), with d = 7.40 g/cm^3. Both types of iron form interstitial alloys with carbon. The maximum amount of carbon is 0.0218 mass % in ferrite and 2.08 mass % in austenite. Calculate the density of each alloy.

23.66 Why is nitric acid not produced by oxidizing N_2 as follows?

$$\begin{array}{lrcl}
(1) & N_2(g) + 2O_2(g) & \longrightarrow & 2NO_2(g) \\
(2) & 3NO_2(g) + H_2O(l) & \longrightarrow & 2HNO_3(aq) + NO(g) \\
(3) & 2NO(g) + O_2(g) & \longrightarrow & 2NO_2(g) \\
\hline
 & 3N_2(g) + 6O_2(g) + 2H_2O(l) & \longrightarrow & 4HNO_3(aq) + 2NO(g)
\end{array}$$

(*Hint*: Evaluate the thermodynamics of each step.)

23.67 Before the development of the Downs cell, the Castner cell was used for the industrial production of Na metal. The Castner cell was based on the electrolysis of molten NaOH.
(a) Write balanced cathode and anode half-reactions for this cell.
(b) A major problem with this cell was that the water produced at one electrode diffused to the other electrode and reacted with the Na. If all the water that was produced reacted with Na, what would be the maximum efficiency of the Castner cell, expressed as amount (mol) of Na produced per mole of electrons flowing through the cell?

23.68 When gold ores are leached with CN^- solutions, gold forms a complex ion, $Au(CN)_2^-$.
(a) Find E_{cell} for the oxidation in air (p_{O_2} = 0.21) of Au to Au^+ in basic solution (pH = 13.55), with $[Au^+]$ = 0.50 mol/L. Is the following reaction spontaneous?

$$Au^+(aq) + e^- \longrightarrow Au(s) \qquad E° = 1.68 \text{ V}$$

(b) How does formation of the complex ion change $E°$ so the oxidation can be accomplished?

23.69 Nitric oxide occurs in the tropospheric nitrogen cycle, but it destroys ozone in the stratosphere.
(a) Write equations for its reaction with ozone and for the reverse reaction.
(b) If the forward and reverse steps are first order in each component, write general rate laws for them.
(c) Calculate $\Delta G°$ for this reaction at 280. K, the average temperature in the stratosphere. (Assume that the $\Delta H°$ and $S°$ values in Appendix B do not change with temperature.)
(d) What ratio of rate constants is consistent with K at this temperature?

23.70 A key part of the carbon cycle is the fixation of CO_2 by photosynthesis to produce carbohydrates and oxygen gas.
(a) Using the formula $(CH_2O)_n$ to represent a carbohydrate, write a balanced equation for the photosynthetic reaction.
(b) If a tree fixes 48 g of CO_2 per day, what volume of O_2 gas, measured at 101.3 kPa and 25.6°C, does the tree produce per day?
(c) What volume of air (0.035 mol % CO_2), at the same conditions, contains this amount of CO_2?

23.71 Farmers use ammonium sulfate as a fertilizer. In the soil, nitrifying bacteria oxidize NH_4^+ to NO_3^-, a groundwater contaminant that causes methemoglobinemia ("blue baby" syndrome). The World Health Organization standard for maximum $[NO_3^-]$ in groundwater is 45 mg/L. A farmer adds 210. kg of $(NH_4)_2SO_4$ to a field, and 37% is oxidized to NO_3^-. What is the groundwater $[NO_3^-]$ (mg/L) if 1000. m^3 of the water is contaminated?

23.72 The key reaction (unbalanced) in the manufacture of synthetic cryolite for aluminum electrolysis is given below:

$$HF(g) + Al(OH)_3(s) + NaOH(aq) \longrightarrow Na_3AlF_6(aq) + H_2O(l)$$

Assuming a 95.6% yield of dried, crystallized product, what mass (kg) of cryolite can be obtained from the reaction of 365 kg of $Al(OH)_3$, 1.20 m^3 of 50.0% by mass aqueous NaOH (d = 1.53 g/mL), and 265 m^3 of gaseous HF at 305 kPa and 91.5°C? (Assume that the ideal gas law holds.)

23.73 Because of their different molar masses, H_2 and D_2 effuse at different rates (Section 4.5).
(a) If it takes 16.5 min for 0.10 mol of H_2 to effuse, how long does it take for 0.10 mol of D_2 to effuse in the same apparatus at the same temperature and pressure?
(b) How many effusion steps does it take to separate an equimolar mixture of D_2 and H_2 to 99 mol % purity?

23.74 The disproportionation of CO to graphite and CO_2 is thermodynamically favoured, but slow.
(a) What does this mean in terms of the magnitudes of the equilibrium constant (K), rate constant (k), and activation energy (E_a)?
(b) Write a balanced equation for the disproportionation of CO.
(c) Calculate K at 298 K.
(d) Calculate K_p at 298 K.

23.75 The overall cell reaction for aluminum production is

$$2Al_2O_3(\text{in } Na_3AlF_6) + 3C(\text{graphite}) \longrightarrow 4Al(l) + 3CO_2(g)$$

(a) Assuming 100% efficiency, what mass (t) of Al_2O_3 is consumed per tonne of Al produced?
(b) Assuming 100% efficiency, what mass (t) of the graphite anode is consumed per tonne of Al produced?
(c) Actual conditions in an aluminum plant require 1.72 t of Al_2O_3 and 0.45 t of graphite per tonne of Al. What is the percent yield of Al with respect to Al_2O_3?
(d) What is the percent yield of Al with respect to graphite?
(e) What volume of CO_2 (m^3) is produced per tonne of Al at operating conditions of 960.°C and exactly 101.3 kPa?

23.76 World production of chromite ($FeCr_2O_4$), the main ore of chromium, was 1.97×10^7 t in 2006. To isolate chromium, a mixture of chromite and sodium carbonate is heated in air to form sodium chromate, iron(III) oxide, and carbon dioxide. The sodium chromate is dissolved in water, and this solution is acidified with sulfuric acid to produce the less-soluble sodium dichromate. The sodium dichromate is filtered out and reduced with carbon to produce chromium(III) oxide, sodium carbonate, and carbon monoxide. The chromium(III) oxide is then reduced to chromium with aluminum metal.
(a) Write a balanced equation for each step.
(b) What mass of chromium (kg) could be prepared if the world production of chromite is 1.97×10^7 t?

23.77 Like heavy water (D_2O), so-called semi-heavy water (HDO) undergoes H/D exchange. The following molecular scenes depict an initial mixture of HDO and H_2 reaching equilibrium:

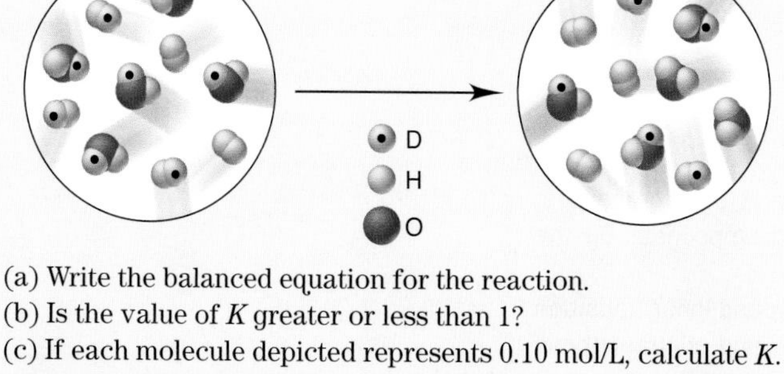

(a) Write the balanced equation for the reaction.
(b) Is the value of K greater or less than 1?
(c) If each molecule depicted represents 0.10 mol/L, calculate K.

23.78 Even though most metal sulfides are sparingly soluble in water, their solubilities differ by several orders of magnitude. This difference is sometimes used to separate the metals by controlling the pH. Use the following data to find the pH at which you can separate 0.10 mol/L Cu^{2+} and 0.10 mol/L Ni^{2+}:
Saturated $[H_2S] = 0.10$ mol/L
K_{a1} of $H_2S = 9 \times 10^{-8}$ K_{a2} of $H_2S = 1 \times 10^{-17}$
K_{sp} of $NiS = 1.1 \times 10^{-18}$ K_{sp} of $CuS = 8 \times 10^{-34}$

23.79 Ores with as little as 0.25% by mass of copper are used as sources of the metal.
(a) What mass (kg) of such an ore would be needed for another Statue of Liberty, which contains 9.1×10^4 kg of copper?
(b) If the mineral in the ore is chalcopyrite ($FeCuS_2$), what is the mass percent of chalcopyrite in the ore?

23.80 How does acid rain affect the leaching of phosphate into groundwater from terrestrial phosphate rock? Calculate the solubility of $Ca_3(PO_4)_2$ in each of the following:
(a) Pure water, pH = 7.0 (Assume that PO_4^{3-} does not react with water.)
(b) Moderately acidic rainwater, pH = 4.5 (Assume that all the phosphate exists in the form that predominates at this pH.)

23.81 The lead(IV) oxide used in a car battery is prepared by coating the electrode plate with PbO and then oxidizing it to lead dioxide (PbO_2). Despite its name, PbO_2 has a nonstoichiometric ratio of lead to oxygen of about 1/1.98. In fact, the holes in the PbO_2 crystal structure, due to missing O atoms, are responsible for the oxide's conductivity.
(a) What is the mole percent of O missing from the PbO_2 structure?
(b) What is the molar mass of the nonstoichiometric compound?

23.82 Chemosynthetic bacteria reduce CO_2 by "splitting" $H_2S(g)$ instead of the $H_2O(g)$ used by photosynthetic organisms. Compare the free energy change for splitting H_2S with the free energy change for splitting H_2O. Is there an advantage to using H_2S instead of H_2O?

23.83 Silver has a face-centred cubic structure with a unit cell edge length of 408.6 pm. Sterling silver is a substitutional alloy that contains 7.5% copper atoms. Assuming that the unit cell remains the same, find the density of silver and the density of sterling silver.

23.84 Earth's mass is estimated to be 5.98×10^{24} kg, and titanium represents 0.05% by mass of this total.
(a) What amount (mol) of Ti is present?
(b) If half of the Ti is found as ilmenite ($FeTiO_3$), what mass of ilmenite is present?
(c) If the airline and auto industries use 9.07×10^4 t of Ti per year, how long (years) will it take to use up all the Ti?

23.85 In 1790, Nicolas Leblanc found a way to form Na_2CO_3 from NaCl. His process, now obsolete, consisted of three steps:
(1) $2NaCl(s) + H_2SO_4(aq) \longrightarrow Na_2SO_4(aq) + 2HCl(g)$
(2) $Na_2SO_4(s) + 2C(s) \longrightarrow Na_2S(s) + 2CO_2(g)$
(3) $Na_2S(s) + CaCO_3(s) \longrightarrow Na_2CO_3(s) + CaS(s)$
(a) Write a balanced overall equation for Leblanc's process.
(b) Calculate the $\Delta_f H°$ of CaS if $\Delta_r H°$ is 351.8 kJ/mol.
(c) Is the overall process spontaneous at standard-state conditions and 298 K?
(d) What mass (g) of Na_2CO_3 forms from 250. g of NaCl if the process is 73% efficient?

23.86 Limestone ($CaCO_3$) is the second most abundant mineral on Earth after SiO_2. For many uses, it is first decomposed thermally to quicklime (CaO). MgO is prepared similarly from $MgCO_3$.
(a) At what temperature is each decomposition spontaneous?
(b) Quicklime reacts with SiO_2 to form a slag ($CaSiO_3$), a byproduct of steelmaking. In 2012, Canadian steel production was 940 000 t per month. If 45. kg of slag is produced per tonne of steel, what mass (kg) of limestone was used to make slag in 2012?

Transition Elements and Their Coordination Compounds

Unrefined ilmenite, a source of Ti

Refined Ti

IN THIS CHAPTER . . . We focus on the transition metals and their properties, and especially on the bonding and structure of their coordination compounds. By the end of this chapter, you should be able to

- Discuss key atomic, physical, and chemical properties of transition and inner transition elements—with attention to the effects of filling inner sublevels—and contrast these properties with the properties of the main-group elements
- Describe the nomenclature, structure, and isomerism of transition elements' coordination compounds, species that contain complex ions
- Apply valence bond theory to explain the bonding and molecular shape of complex ions
- Apply crystal field theory to explain the colours and magnetic properties of coordination compounds, which relate directly to their structure
- Discuss some examples of transition metal ions in biochemistry

24

The vibrant colour in our universe comes largely from the transition metals around us. What is it about **transition metals** that makes them so colourful compared to the other metals in the periodic table? This is just one of the many questions that will be answered in this chapter. We use rare earth and actinide elements, those that make up the *f* block of the periodic table, daily, without even realizing it. In fact, their prevalence in our electronic gadgets is a source of worldwide economic concern, causing a race around the globe to find new sources. Many Canadian universities and researchers are involved in the study of applications of the transition and inner transition elements, and this chapter will highlight some of the incredible work being done in Canada.

24.1 Properties of the Transition Elements

The transition elements differ considerably in physical and chemical behaviour from the main-group elements:

- *All transition elements are metals,* whereas main-group elements in each period change from metal to nonmetal.
- *Many transition metal compounds are coloured and paramagnetic,* whereas most main-group ionic compounds are colourless and diamagnetic.

We first discuss electron configurations of the atoms and ions, and then we examine key properties of transition elements to see how they contrast with the same properties of the main-group elements.

Electron Configurations of the Transition Metals and Their Ions

Like the properties of all elements, the properties of the transition elements and their compounds arise largely from the electron configurations of their atoms (Section 7.2) and ions (Section 7.4). As Figure 24.1 shows, *d*-block elements occur in four series that lie within periods 4 through 7 between the last *ns*-block element (group 2) and the first *np*-block element (group 13). Each of the four series contains 10 elements, based on filling five *d* orbitals, for a total of 40 transition elements. Lying between the first and second members of the series in periods 6 and 7 are the inner transition elements, which have filled *f* orbitals. The following three features are important to review:

1. *Electron configurations of the atoms.* Despite several exceptions, the atoms in each *d*-block series, in general, have the following *valence shell* electron configuration:

$$[\text{noble gas}](n-1)d^x ns^2\text{, with } n = 4 \text{ to } 7 \text{ and } x = 1 \text{ to } 10$$

The valence shell electron configuration for the elements in periods 6 and 7 includes the *f* sublevel:

$$[\text{noble gas}](n-2)f^{14}(n-1)d^x ns^2\text{, with } n = 6 \text{ or } 7 \text{ and } x = 1 \text{ to } 10$$

The *partial* (valence-level) electron configuration for the *d*-block elements excludes the noble gas core and the filled inner *f* sublevel:

$$(n-1)d^x ns^2$$

Concepts and Skills to Review before Studying This Chapter

- Properties of light (Section 6.1)
- Electron shielding of nuclear charge (Section 7.1)
- Electron configuration, ionic size, and magnetic behaviour (Sections 7.2 and 7.4)
- Valence bond theory (Section 10.1)
- Constitutional isomerism, geometric isomerism, and enantiomerism (Section 20.4)
- Lewis acid-base concepts (Section 16.8)
- Complex ion formation (Section 17.4)
- Redox behaviour and standard electrode potentials (Section 19.3)

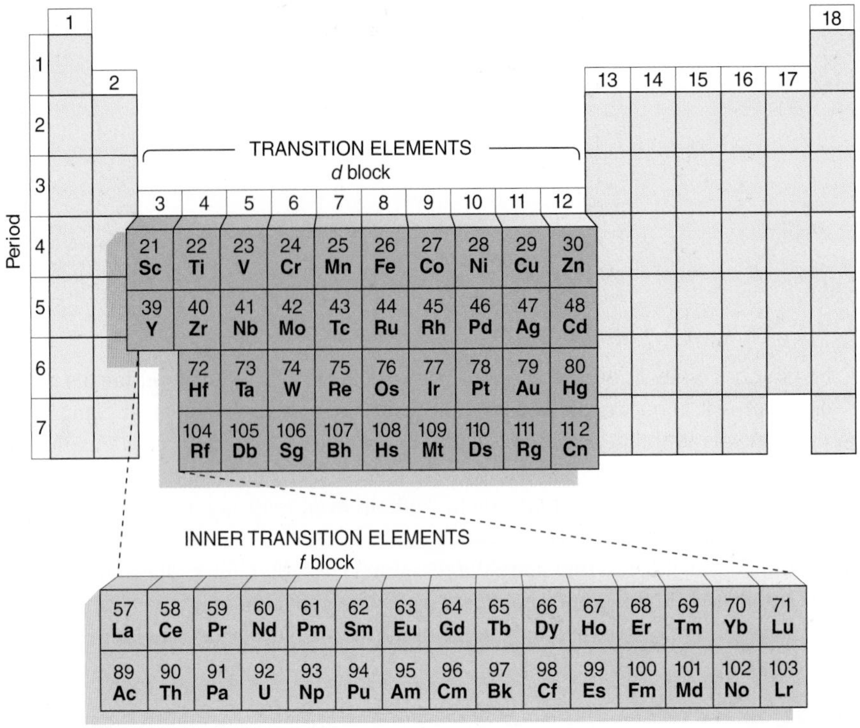

FIGURE 24.1 The transition elements (*d* block) and inner transition elements (*f* block) in the periodic table

2. *Filling pattern in period 4.* The first (period 4) transition series consists of scandium (Sc) through zinc (Zn) (Figure 24.2 and Table 24.1). Scandium has the electron configuration $[Ar]3d^14s^2$, and the addition of one electron at a time first half-fills, and then fills, the 3*d* orbitals to zinc. Recall that chromium and copper are exceptions to this pattern: the 4*s* and 3*d* orbitals in Cr are half-filled to give $[Ar]3d^54s^1$, and the 4*s* orbital in Cu is half-filled to give $[Ar]3d^{10}4s^1$. These exceptions are due to the relative energies of the 4*s* and 3*d* sublevels as electrons are added (see Figure 7.23), as well as the unusual stability of half-filled and filled sublevels.

3. *Electron configurations of the ions.* Transition metal ions form through *the loss of ns electrons before (n − 1)d electrons*. Thus, the electron configuration of Ti^{2+} is $[Ar]3d^2$, *not* $[Ar]4s^2$, and Ti^{2+} is referred to as a d^2 ion. Ions of different metals with the same configuration often have similar properties. For example, Mn^{2+} and Fe^{3+} are d^5 ions: both have pale colours in solution and form complex ions with similar magnetic properties.

Table 24.1 shows partial orbital box diagrams for the period 4 elements. In general, for all periods, the number of *unpaired* electrons (or half-filled orbitals)

FIGURE 24.2 The period 4 transition metals. The 10 elements are shown in periodic-table order.

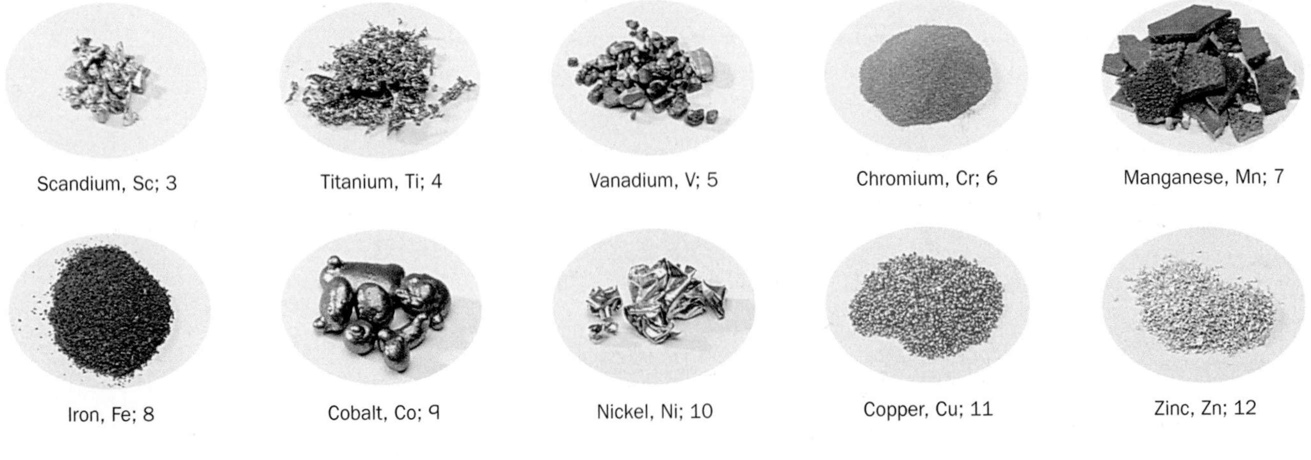

TABLE 24.1 Orbital Occupancy of the Period 4 Transition Metals

Element	4s	3d					4p			Unpaired Electrons
Sc	↑↓	↑								1
Ti	↑↓	↑	↑							2
V	↑↓	↑	↑	↑						3
Cr	↑	↑	↑	↑	↑	↑				6
Mn	↑↓	↑	↑	↑	↑	↑				5
Fe	↑↓	↑↓	↑	↑	↑	↑				4
Co	↑↓	↑↓	↑↓	↑	↑	↑				3
Ni	↑↓	↑↓	↑↓	↑↓	↑	↑				2
Cu	↑	↑↓	↑↓	↑↓	↑↓	↑↓				1
Zn	↑↓	↑↓	↑↓	↑↓	↑↓	↑↓				0

increases in the first half of the series and, when pairing begins, *decreases in the second half*. A key point to note, in upcoming discussions, is that the electron configuration of the *atom* (oxidation number = 0) correlates with the physical properties of the *element*, whereas the electron configuration of the *ion* (and other species of higher oxidation number) correlates with the chemical properties of the *compounds*.

Sample Problem 24.1 Writing Electron Configurations of Transition Metal Atoms and Ions

Problem Write a *condensed* electron configuration for **(a)** Zr; **(b)** V^{3+}; **(c)** Mo^{3+}. (Assume that elements in higher periods behave like elements in period 4.)

Plan We locate each element in the periodic table and count its position in the respective transition series. These elements are in periods 4 and 5, so the general configuration is [noble gas]$(n-1)d^x ns^2$. For the ions, we recall that ns electrons are lost first.

Solution **(a)** Zr is the second element in the $4d$ series: [Kr]$4d^2 5s^2$.

(b) V is the third element in the $3d$ series: [Ar]$3d^3 4s^2$. In forming V^{3+}, three electrons are lost (two $4s$ and one $3d$), so V^{3+} is a d^2 ion: [Ar]$3d^2$.

(c) Mo lies below Cr in group 6, so it is an exception to the filling pattern, like Cr is. Thus, Mo is [Kr]$4d^5 5s^1$. To form the ion, Mo loses the one $5s$ and two of the $4d$ electrons, so Mo^{3+} is a d^3 ion: [Kr]$4d^3$.

Check Figure 7.10 shows that we are correct for the atoms. Be sure that charge plus number of d electrons in the ion equals the sum of outer s and d electrons in the atom.

Follow-Up Problem 24.1 Write a *partial* electron configuration (no noble gas core or filled inner sublevels) for **(a)** Ag^+; **(b)** Cd^{2+}; **(c)** Ir^{3+}. (See Brief Solutions to Follow-Up Problems at the end of the chapter.)

Atomic and Physical Properties of the Transition Elements

The properties of the transition elements contrast with the properties of the main-group elements in several ways.

Trends across a Period Consider the variations in atomic size, electronegativity, and ionization energy across period 4 (Figure 24.3):

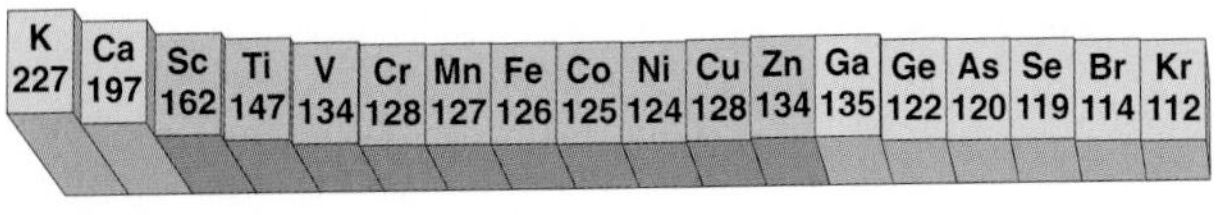

A Atomic radius (pm)

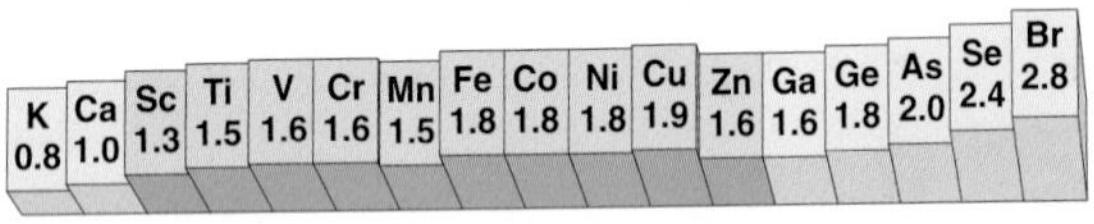

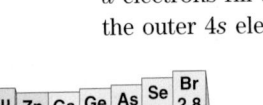

B Electronegativity

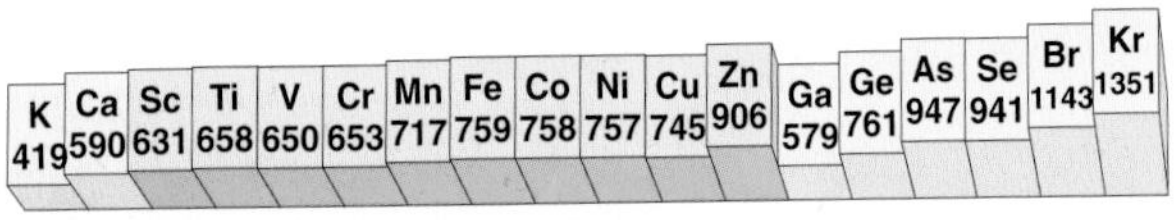

C First ionization energy (kJ/mol)

FIGURE 24.3 Trends in the key atomic properties of period 4 elements. Atomic radius **(A)**, electronegativity **(B)**, and first ionization energy **(C)** of all the period 4 elements are shown as the heights of the posts, with darker shading for the transition series.

- *Atomic size.* Atomic size decreases overall across the period (Figure 24.3A). There is a smooth, steady decrease across the main groups because the electrons are added to *outer* orbitals, which shield the increasing nuclear charge poorly. This decrease is not steady in the transition series, where *atomic size decreases at first but then remains relatively constant.* This is because the *d* electrons fill *inner* orbitals, so they shield outer electrons very efficiently, and the outer 4*s* electrons are *not* pulled closer.
- *Electronegativity* (χ). Electronegativity generally increases across the period, but, once again, the main groups show a steady, steep increase between the metal potassium (0.8) and the nonmetal bromine (2.8), whereas the transition elements have *relatively constant electronegativity* (Figure 24.3B), consistent with their relatively constant size. The transition elements all have intermediate electronegativity values, much like the large, metallic members of groups 13 to 15.
- *Ionization energy* (IE_1). The ionization energies of the period 4 main-group elements rise steeply from left to right, more than tripling from potassium (419 kJ/mol) to krypton (1351 kJ/mol), as electrons become more difficult to remove from the poorly shielded, increasing nuclear charge. In the transition metals, *IE_1 values increase relatively little* because the inner 3*d* electrons shield more effectively (Figure 24.3C). (Recall, from Section 7.3, that the drop at group 13 occurs because the first electron is relatively easy to remove from the outer *np* orbital.)

Trends within a Group Vertical trends for transition elements are also different from the main-group trends.

1. *Atomic size.* As expected, atomic size increases from period 4 to period 5, as it does for the main-group elements, but there is virtually *no size increase from period 5 to period 6* (Figure 24.4A). Since the lanthanides, with their buried 4*f* sublevel, appear between the 4*d* (period 5) and 5*d* (period 6) series, an element in period 6 is separated from the element above it by 32 elements (ten 4*d*, six 5*p*, two 6*s*, and fourteen 4*f* orbitals) instead of just 18. The extra shrinkage from the increase in nuclear charge, due to the 14 additional protons, and relativistic effects is called the **lanthanide contraction.*** It is estimated that roughly 10% of the lanthanide contraction is due to relativistic effects. By coincidence, this size *decrease* is about equal to the normal *increase* between periods, so the transition elements in periods 5 and 6 have about the same atomic sizes. As a result, while this contraction makes it easier to separate elements in period 5, it makes it harder to separate elements in periods 5 and 6.

2. *Electronegativity.* The vertical trend in electronegativity is opposite to the decreasing trend in the main groups. Electronegativity *increases* from period 4 to period 5, but there is no further increase in period 6 (Figure 24.4B). The heavier elements, especially gold (χ = 2.4), become quite electronegative, with values

*For the transition metals, one other factor that becomes important is that the radius of an atom depends on the mass of the electron. When Schrödinger first developed quantum theory, he did not take into consideration Einstein's theory of relativity. For atoms with low molar mass, this has very little effect, but as the molar mass increases, the relativistic effects become more pronounced. Thus, electrons in high molar mass atoms are subject to relativistic effects that result in their apparent masses being significantly larger than the mass of an electron in a hydrogen atom, for example. This, in turn, results in the atomic radius being much smaller than expected. In the case of gold, for example, its radius shrinks by approximately 22%. [Source: I. B. Bersuker, *Electronic Structure and Properties of Transition Metal Compounds* (Hoboken, NJ: Wiley, 2010).]

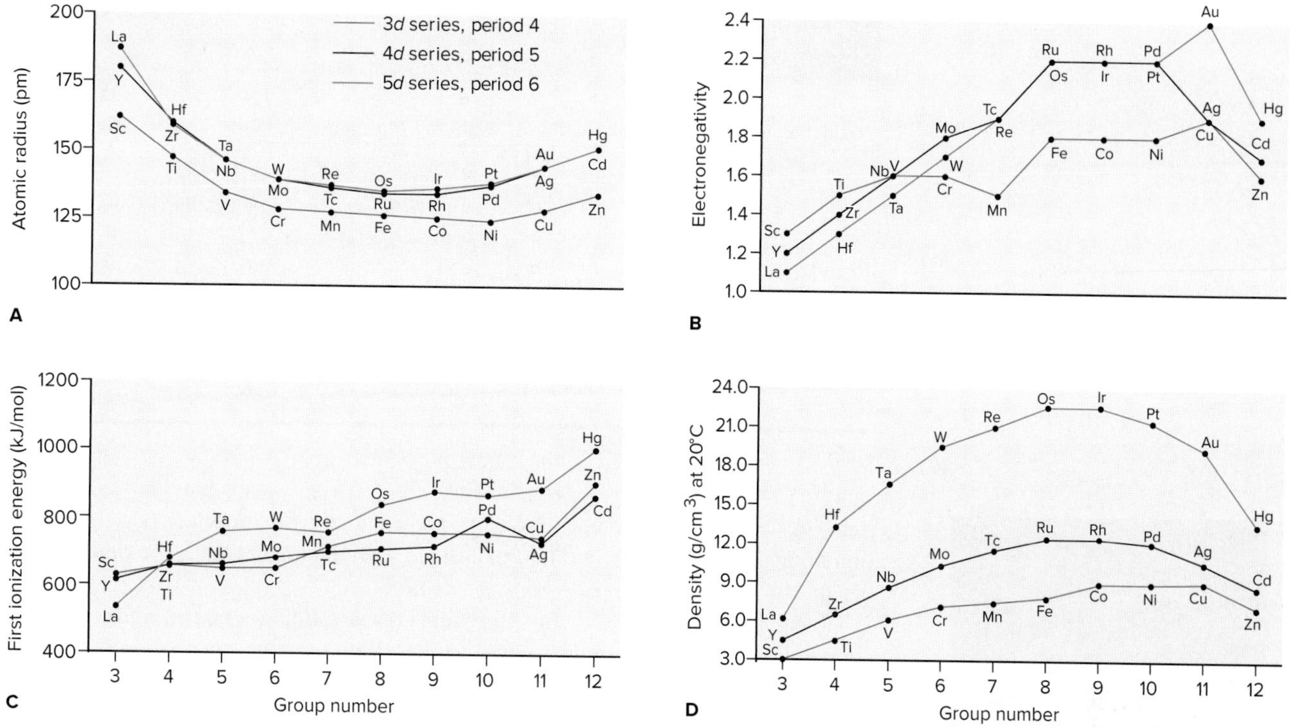

FIGURE 24.4 Vertical trends in key properties within the transition elements. The trends are unlike those for the main-group elements in several ways: **A.** The second and third members of a transition metal group are nearly the same size. **B.** Electronegativity increases down a transition group. **C.** First ionization energies are highest at the bottom of a transition group. **D.** Density increases down a transition group because mass increases faster than volume.

higher than the values of most metalloids and even some nonmetals. (For example, the electronegativity of both Te and P equals 2.1.) In fact, with very electropositive cesium ($\chi = 0.7$), gold forms the saltlike CsAu. Although atomic size increases slightly from top to bottom in a group, the nuclear charge increases much more. Therefore, the heavier transition metals form bonds with more covalent character because they attract electrons more strongly than main-group metals do.

3. *Ionization energy.* The small increase in size combined with the large increase in nuclear charge also explains why *IE_1 values generally increase* down a transition group (Figure 24.4C). This trend also runs counter to the main-group trend, where heavier members are so much larger that the outer electron is easier to remove.

4. *Density.* Atomic size and, therefore, volume are inversely related to density. Across a period, density increases, then levels off, and finally dips a bit at the end of a series (Figure 24.4D). Down a transition group, density increases dramatically because atomic volume changes little from period 5 to period 6, but atomic mass increases significantly. As a result, the period 6 series contains some of the densest elements known: tungsten, rhenium, osmium, iridium, platinum, and gold have densities about 20 times the density of water and twice the density of lead.

Chemical Properties of the Transition Elements

Like their atomic and physical properties, the chemical properties of transition elements are very different from the properties of main-group elements. In this section, we examine key properties in the period 4 series and then see how behaviour changes within a group.

Multiple Oxidation States One of the most characteristic chemical properties of the transition metals is the occurrence of *multiple oxidation states*; main-group metals display one or, at most, two states. However, vanadium, for example, has two common oxidation states, chromium and manganese have three (Figure 24.5A), and many others have states that are seen less often (Table 24.2). Since the *ns* and $(n-1)d$ electrons are so close in energy, transition elements can use all or most of these electrons in bonding.

The highest oxidation state of elements in groups 3 through 7 equals the group number. These states are seen when the elements combine with highly electronegative oxygen or fluorine. For instance, Figure 24.5B shows vanadium as the vanadate

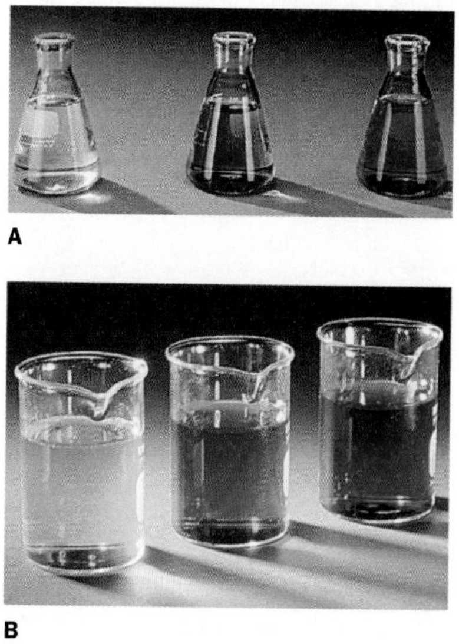

FIGURE 24.5 Aqueous oxoanions of transition elements. A. Mn can have the +2 (Mn^{2+}, *left*), +6 (MnO_4^{2-}, *middle*), or +7 (MnO_4^-, *right*) oxidation state. **B.** The +5 state in VO_4^{3-} (*left*), the +6 state in $Cr_2O_7^{2-}$ (*middle*), and the +7 state in MnO_4^- (*right*).

TABLE 24.2 Oxidation States and *d*-Orbital Occupancy of the Period 4 Transition Metals*

Oxidation State	3 Sc	4 Ti	5 V	6 Cr	7 Mn	8 Fe	9 Co	10 Ni	11 Cu	12 Zn
0	d^1	d^2	d^3	d^5	d^5	d^6	d^7	d^8	d^{10}	d^{10}
+1			d^3	d^5	d^5	d^6	d^7	d^8	d^{10}	
+2		d^2	d^3	d^4	d^5	d^6	d^7	d^8	d^9	d^{10}
+3	d^0	d^1	d^2	d^3	d^4	d^5	d^6	d^7	d^8	
+4		d^0	d^1	d^2	d^3	d^4	d^5	d^6		
+5			d^0	d^1	d^2		d^4			
+6				d^0	d^1	d^4				
+7					d^0					

*The most important orbital occupancies are in red.

Dr. Sandro Gambarotta and his group at the University of Ottawa are working on the chemistry of vanadium, with a view to its application in catalysis. Their paper on vanadium provided valuable information about the direct involvement of the ligand system in the organometallic chemistry of the metal centre. Some of their most recent work involves a breakthrough in making families of single-component catalysts with high polymerization activity. Dr. Gambarotta's group has also provided unique insight into dinitrogen cleavage and the promotion of hydrogenation by an *f*-block element.

ion (VO_4^{3-}; oxidation number of V = +5), chromium as dichromate ($Cr_2O_7^{2-}$; oxidation number of Cr = +6), and manganese as permanganate (MnO_4^-; oxidation number of Mn = +7).

Elements in groups 8, 9, and 10 exhibit fewer oxidation states, and the highest state is less common and never equal to the group number. For example, we never encounter iron in the +8 state and only rarely in the +6 state. The +2 and +3 states are the most common states for iron* and cobalt, and the +2 state is the most common state for nickel, copper, and zinc. *The +2 oxidation state is common because ns^2 electrons are readily lost.*

Metallic Behaviour, Oxide Acidity, and Valence-State Electronegativity Atomic size and oxidation state have a major effect on the nature of bonding in transition metal compounds. Like the metals in groups 13, 14, and 15, the transition elements in their *lower* oxidation states behave chemically more like metals. That is, *ionic bonding is more prevalent for the lower oxidation states, and covalent bonding is more prevalent for the higher oxidation states.* For example, at room temperature, $TiCl_2$ (oxidation number = +2) is an ionic solid, whereas $TiCl_4$ (oxidation number = +4) is a molecular liquid (see also Figure 13.11). In the higher oxidation states, the atoms have higher charge densities, so they polarize the electron clouds of the nonmetal ions more strongly and the bonding becomes more covalent. For the same reason, the oxides become less basic (more acidic) as the oxidation state increases: TiO is weakly basic in water, whereas TiO_2 is amphoteric (reacts with both acids and bases).

Why does oxide acidity increase with oxidation state? In addition to forming oxides, how can a metal like chromium or manganese form an oxoanion? The answers involve a type of "effective" electronegativity called *valence-state electronegativity*, which also has numerical values. A metal atom in a positive oxidation state has a greater attraction to the bonded electrons—that is, a higher electronegativity—than it does in its zero oxidation state. Thus, for example, the electronegativity of elemental chromium is 1.6, close to the electronegativity of aluminum (1.5), another active metal. For chromium(III), the value increases to 1.7, still characteristic of a metal. However, for chromium(VI), the value is 2.3, which is close to the values for some nonmetals, such as phosphorus (2.1) and sulfur (2.5). Thus, like P in PO_4^{3-} and S in SO_4^{2-}, Cr in CrO_4^{2-} is covalently bonded at the centre of an oxoanion of a relatively strong acid, H_2CrO_4, and manganese(VII) in MnO_4^- behaves similarly in the strong acid $HMnO_4$.

TABLE 24.3 Standard Electrode Potentials of Period 4 M^{2+} Ions

Half-Reaction	E° (V)
$Ti^{2+}(aq) + 2e^- \rightleftharpoons Ti(s)$	−1.63
$V^{2+}(aq) + 2e^- \rightleftharpoons V(s)$	−1.19
$Cr^{2+}(aq) + 2e^- \rightleftharpoons Cr(s)$	−0.91
$Mn^{2+}(aq) + 2e^- \rightleftharpoons Mn(s)$	−1.18
$Fe^{2+}(aq) + 2e^- \rightleftharpoons Fe(s)$	−0.44
$Co^{2+}(aq) + 2e^- \rightleftharpoons Co(s)$	−0.28
$Ni^{2+}(aq) + 2e^- \rightleftharpoons Ni(s)$	−0.25
$Cu^{2+}(aq) + 2e^- \rightleftharpoons Cu(s)$	0.34
$Zn^{2+}(aq) + 2e^- \rightleftharpoons Zn(s)$	−0.76

Reducing Strength Table 24.3 shows the standard electrode potentials of the period 4 transition metals in their +2 oxidation state in acid solution. Note that, in general,

*Iron seems to have unusual oxidation states in magnetite (Fe_3O_4) and pyrite (FeS_2), but it does not. In magnetite, one-third of the metal ions are Fe^{2+} and two-thirds are Fe^{3+}, which gives a 1/1 ratio of FeO/Fe_2O_3 and a formula of Fe_3O_4. Pyrite contains Fe^{2+} combined with the disulfide ion, S_2^{2-}.

FIGURE 24.6 Colours of representative compounds of the period 4 transition metals. Staggered from left to right, the compounds are scandium oxide (*white*), titanium(IV) oxide (*white*), vanadyl sulfate dihydrate (*light blue*), sodium chromate (*yellow*), manganese(II) chloride tetrahydrate (*light pink*), potassium ferricyanide (*red-orange*), cobalt(II) chloride hexahydrate (*violet*), nickel(II) nitrate hexahydrate (*green*), copper(II) sulfate pentahydrate (*blue*), and zinc sulfate heptahydrate (*white*).

reducing strength decreases across the series. All the period 4 transition metals, except copper, are active enough to reduce H^+ from aqueous acid to form hydrogen gas. In contrast to the rapid reaction of the group 1 and 2 metals with water at room temperature, however, most transition metals have an oxide coating that allows rapid reaction only with hot water or steam.

Colour and Magnetism of Compounds *Most main-group ionic compounds are colourless* because their metal ions have a filled outer level (noble gas electron configuration). With only much higher energy orbitals available to receive an excited electron, their ions do not absorb visible light. In contrast, electrons in a partially filled *d* sublevel can absorb visible wavelengths and move to slightly higher energy orbitals. As a result, *many transition metal compounds have striking colours.* Exceptions are the compounds of scandium, titanium(IV), and zinc, which are colourless because their metal ions have either an empty *d* sublevel (Sc^{3+} or Ti^{4+}: $[Ar]3d^0$) or a filled *d* sublevel (Zn^{2+}: $[Ar]3d^{10}$) (Figure 24.6).

Magnetic properties are also related to sublevel occupancy (Section 7.4). *Most main-group metal ions are diamagnetic* for the same reason they are colourless: all their electrons are paired. In contrast, *many transition metal compounds are paramagnetic because of their unpaired d electrons.* Transition metal ions with a d^0 or d^{10} configuration are also colourless and diamagnetic.

Chemical Behaviour within a Group The *increase* in reactivity going down a group of main-group metals due to the *decrease* in IE_1 does *not* occur going down a group of transition metals. The chromium (Cr) group (group 6) shows a typical pattern (Table 24.4). IE_1 *increases* down the group, which makes the two heavier members *less* reactive than the lightest member. Chromium is also a much stronger reducing agent than molybdenum (Mo) or tungsten (W), as shown by the standard electrode potentials.

The similarity in atomic size of elements in periods 5 and 6 leads to similar chemical behaviour, which has some important consequences. Because Mo and W compounds behave similarly, their ores often occur together in nature, which makes these elements very difficult to separate from each other. The same situation occurs with zirconium and hafnium in group 4 and with niobium and tantalum in group 5.

TABLE 24.4 Some Properties of Group 6 Elements

Element	Atomic Radius (pm)	IE_1 (kJ/mol)	$E°$ (V) for $M^{3+}(aq)\|M(s)$
Cr	128	653	−0.74
Mo	139	685	−0.20
W	139	770	−0.11

HUMAN HEMOGLOBIN

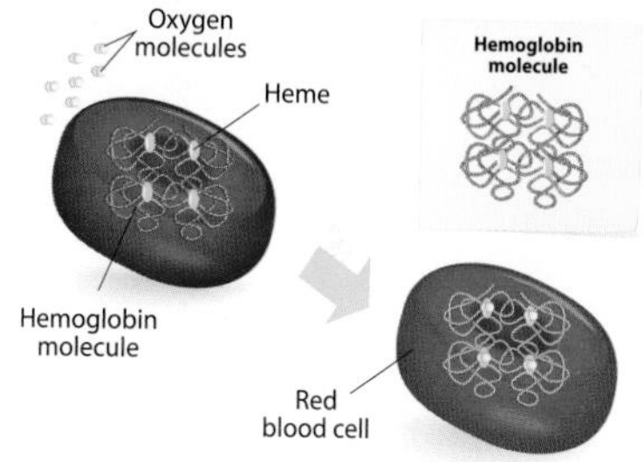

In mammals, hemoglobin is a protein containing iron found in red blood cells. It has the important job of transporting oxygen from the lungs to various tissues in the body, including, most important, the brain. Each molecule of hemoglobin can transport four oxygen molecules, which bind reversibly, thus allowing the oxygen to be adsorbed and desorbed readily. Hemoglobin is also capable of transporting nitric oxide, which it releases at the same time as oxygen, and carbon dioxide, which we exhale. If we are exposed to high levels of carbon dioxide, it will bind in place of oxygen, which can be very dangerous, but if we are re-exposed to oxygen, the oxygen will replace the carbon dioxide and our cells will recover. On the other hand, if we are exposed to high levels of carbon monoxide, it will bind preferentially and more strongly than oxygen, causing carbon monoxide poisoning (two symptoms of which are cherry red cheeks and light-headedness). Not only does the carbon monoxide occupy sites that would normally be occupied by oxygen, but it modifies the chemistry of hemoglobin such that the bound oxygen is less likely to be released. If removal from the carbon monoxide atmosphere into a normal atmosphere is insufficient to release the carbon monoxide, sometimes placement in a hyperbaric chamber (where greater than normal pressures of oxygen are administered) forces the carbon monoxide to be released and shifts the equilibrium in favour of adsorbing oxygen.

SUMMARY OF SECTION 24.1

- All transition elements are metals.
- Atoms of the *d*-block elements have $(n - 1)d$ orbitals filled, and their ions have an empty *ns* orbital.
- Unlike the trends for the main-group elements, atomic size, electronegativity, and first ionization energy change relatively little across a transition series. Because of the lanthanide contraction, atomic size changes little from period 5 to period 6 in a transition metal group; thus, electronegativity, first ionization energy, and density *increase* down a transition metal group.
- Transition metals typically have several oxidation states, with the +2 state being the most common. The elements exhibit less metallic behaviour in their higher states, since they have higher valence state electronegativity.
- Most period 4 transition metals are active enough to reduce hydrogen ions from acid solution.
- Many transition metal compounds are coloured and paramagnetic because their metal ions have unpaired *d* electrons.
- In contrast to main-group metals, reactivity decreases down a transition group.

24.2 The Inner Transition Elements

The 14 **lanthanides**—cerium (Ce; $Z = 58$) through lutetium (Lu; $Z = 71$)—and the 14 **actinides** below them—thorium (Th; $Z = 90$) through lawrencium (Lr; $Z = 103$)—are called **inner transition elements** because their seven inner $4f$ or $5f$ orbitals are filled. In the more recent periodic tables, lanthanum (La) and actinium (Ac) are also included as part of the inner transition elements because their properties more closely coincide with the properties of the inner transition elements.

The Lanthanides

The lanthanides are also known as the *rare earth elements*, but many are not rare: cerium (Ce), for instance, ranks 26th in abundance (by mass percent), or five times as abundant as lead. All the lanthanides are silvery, high-melting metals.

As with other transition elements, atomic properties vary little across the period. As a result, chemical properties are so similar that the two ores of lanthanides, ceria and yttria, are mixtures of compounds of all 14 of them. This natural co-occurrence arises because the elements exist as M^{3+} ions of very similar radii. Most lanthanides have the ground-state electron configuration $[Xe]6s^24f^x5d^0$, where x varies across the series. The three exceptions (Ce, Gd, and Lu) have a single electron in one of their $5d$ orbitals: cerium ($[Xe]6s^24f^15d^1$) forms a stable 4+ ion with an empty (f^0) sublevel, and the Gd^{3+} and Lu^{3+} ions have a stable half-filled (f^7) or filled (f^{14}) sublevel.

Dr. Daniel Leznoff and his group at Simon Fraser University, in Vancouver, are working on making polymeric materials containing transition metals and lanthanides for many applications. For example, by using lanthanides with gold or silver cyanides, they are preparing highly emissive materials that could also act as sensors for vapours of environmental pollutants. They can tune the colour of the emitting light by mixing different fractions of lanthanide metals, so these materials may also have applications in the electronic display industry. Dr. Leznoff's group is also studying the chemistry of actinides.

Lanthanide compounds are used in tinted glass, electronic devices (magnets, batteries, and lasers), and high-quality camera lenses, but two industrial applications account for over 60% of their total usage. In gasoline refining, catalysts that are used to "crack" hydrocarbons into smaller molecules are 5% by mass rare earth oxides. In steelmaking, a mixture of lanthanides, called *misch metal*, removes carbon impurities from molten iron and steel. Recently, shortages of some lanthanides have become a concern for many manufacturers. The shortages have been caused because China, a major producer and exporter of rare earth elements, has drastically cut back the supply of these elements, citing increased local need and environmental concerns as the reasons. Although there are other large deposits of these elements around the world, notably in Africa, the United States, and Canada, large investments of time and resources are required to begin production. Canada has several companies that will begin mining and production in 2015/2016 (e.g., Avalon Rare Metals Inc. and Great Western Minerals Group Ltd.) and other companies that are in the process of completing the research and exploration phase (Midland Exploration Inc., Pele Mountain Resources, Matamec Explorations Inc., and Quest) (*The Canadian Business Journal*, 2015, Volume 8, Issue 9). It is estimated that Canada has almost 40% of the world's supply of rare earth metals and, most important, the rare earth metals that are projected as being in short supply in the future, namely europium, terbium, dysprosium, yttrium, and neodymium. Canada is viewed as being a stable country with the scientific and engineering expertise required to produce the rare earth metals and will likely play an important role in the world export of these elements (*Ottawa Citizen*, 2014; House of Commons Standing Committee on Natural Resources, "The Rare Earth Elements Industry in Canada—Summary of Evidence," June 2014).

Sample Problem 24.2 Finding the Number of Unpaired Electrons

Problem The alloy $SmCo_5$ forms a permanent magnet because both samarium and cobalt have unpaired electrons. How many unpaired electrons does Sm ($Z = 62$) have?

Plan We write the condensed electron configuration of Sm and then, using Hund's rule and the Aufbau principle, place the electrons in a partial orbital diagram and count the unpaired electrons.

Solution Samarium is the eighth element after Xe. Two electrons go into the 6*s* sublevel. In general, the 4*f* sublevel fills before the 5*d* sublevel (among the lanthanides, only Ce, Gd, and Lu have 5*d* electrons), so the remaining electrons go into the 4*f* sublevel. Thus, the valence shell electron configuration of Sm is $[Xe]4f^66s^2$. There are seven *f* orbitals, so each of the six *f* electrons enters a separate orbital:

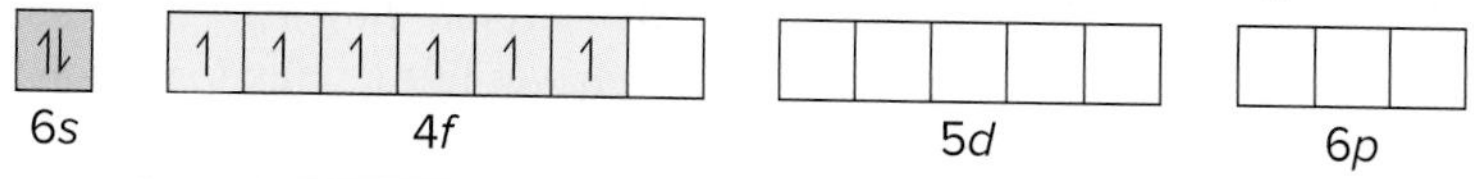

Therefore, Sm has six unpaired electrons.

Check Six 4*f* e^- plus two 6*s* e^- plus the 54 e^- in Xe gives 62, the atomic number of Sm.

Follow-Up Problem 24.2 How many unpaired electrons does the Er^{3+} ion have?

The Actinides

All actinides are radioactive. Like the lanthanides, they have very similar physical and chemical properties. Thorium and uranium occur in nature, but the transuranium elements, those with *Z* greater than 92, have been synthesized in particle accelerators (Section 25.3). The actinides are silvery and chemically reactive and, like the lanthanides, form highly coloured compounds. The actinides and lanthanides have similar outer-electron configurations. Although the +3 oxidation state is characteristic of the actinides, as it is for the lanthanides, other states also occur. For example, uranium exhibits +3 through +6 states, with the +6 state being the most prevalent; thus, the most common oxide of uranium is UO_3.

SUMMARY OF SECTION 24.2

- There are two series of inner transition elements: the lanthanides and the actinides.
- The lanthanides (4*f* series) have a common +3 oxidation state and exhibit very similar properties.
- The actinides (5*f* series) are radioactive. All actinides have a +3 oxidation state; several, including uranium, have higher states as well.

24.3 Coordination Compounds

The most distinctive feature of transition metal chemistry is the common occurrence of **coordination compounds** (or *complexes*). These species contain at least one **complex ion**, which consists of *a central metal cation bonded to molecules and/or anions called* **ligands** *(see margin)*. To maintain charge neutrality, the complex ion is associated with **counterions**. In the coordination compound $[Co(NH_3)_6]Cl_3$ (Figure 24.7A), the complex ion (in square brackets) is $[Co(NH_3)_6]^{3+}$, the six NH_3 molecules bonded to the Co^{3+} ion are ligands, and the three Cl^- ions are counterions.

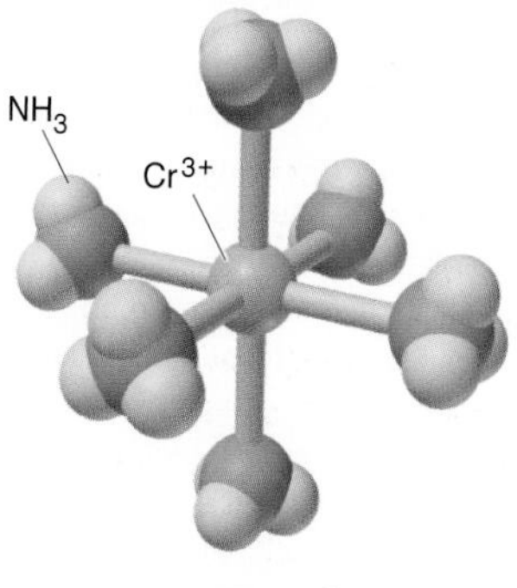

The complex ion $[Cr(NH_3)_6]^{3+}$ has a central Cr^{3+} ion bonded to six NH_3 ligands.

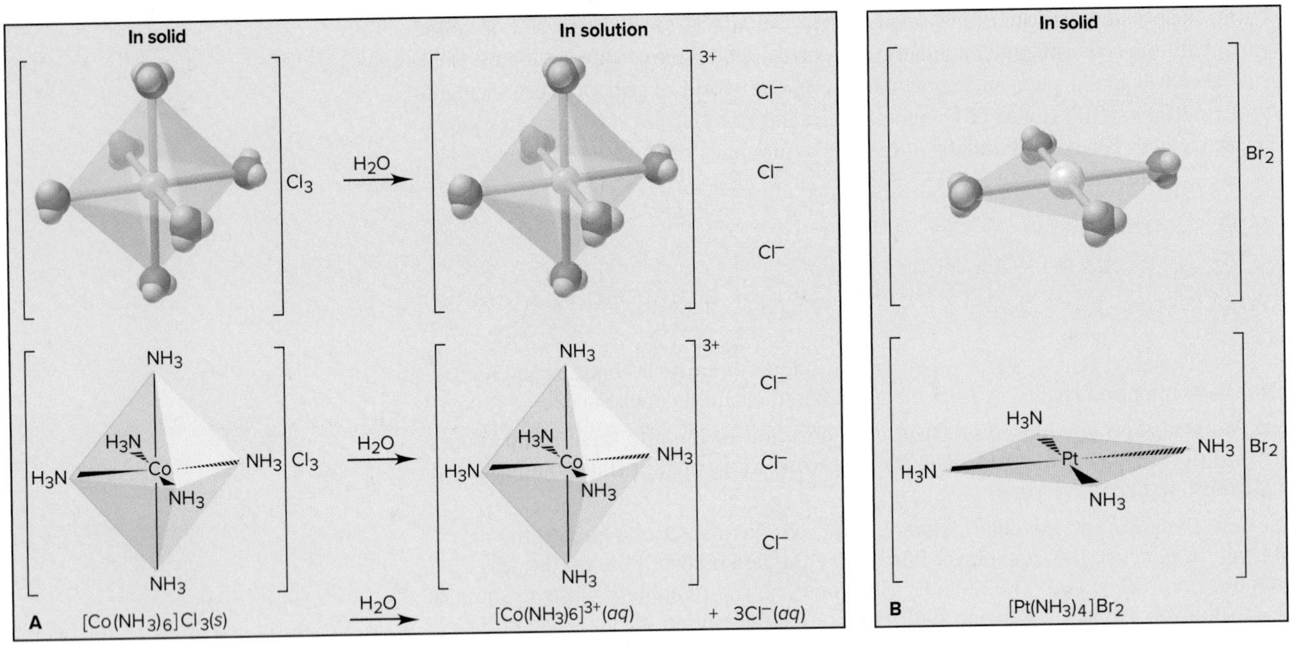

FIGURE 24.7 Components of a coordination compound. Coordination compounds are shown as models (*top*), wedge-bond perspective drawings (*middle*), and formulas (*bottom*). **A.** $[Co(NH_3)_6]Cl_3$ dissolves in water. The six ligands remain bound in the complex ion. **B.** $[Pt(NH_3)_4]Br_2$ has four NH_3 ligands and two Br^- counterions.

A coordination compound behaves like an electrolyte in water: the complex ion and counterions separate, but the complex ion behaves like a polyatomic ion because *the ligands and central metal ion remain attached*. Thus, as Figure 24.7A shows, 1 mol of $[Co(NH_3)_6]Cl_3$ yields 1 mol of $[Co(NH_3)_6]^{3+}$ ions and 3 mol of Cl^- ions. This section covers the structure, naming, and properties of complex ions.

Complex Ions: Coordination Numbers, Geometries, and Ligands

A complex ion is described by the metal ion and the number and types of ligands attached to it. The structure of a complex ion is related to its coordination number and geometry.

Coordination Numbers The **coordination number** is the *number of ligand atoms* bonded directly to the central metal ion. It is *specific* for a given metal ion in a particular oxidation state and compound. The coordination number of the Co^{3+} ion in $[Co(NH_3)_6]^{3+}$ is 6 because six ligand atoms (N of NH_3) are bonded to it. The coordination number of the Pt^{2+} ion in many of its complexes is 4, whereas the coordination number of the Pt^{4+} ion in its complexes is 6. Copper(II) may have a coordination number of 2, 4, or 6 in different complex ions. In general, *the most common coordination number in complex ions is 6*, but 2 and 4 are often seen, and some higher coordination numbers are known.

Geometries The geometry (shape) depends on *the coordination number and the metal ion*. Table 24.5 shows the geometries for coordination numbers 2, 4, and 6. A complex ion whose metal ion has a coordination number of 2, such as $[Ag(NH_3)_2]^+$, is *linear*. The coordination number 4 gives rise to either of two geometries. Most d^8 metal ions form *square planar* complex ions (Figure 24.7B). The d^{10} ions form *tetrahedral* complex ions. A coordination number of 6 results in an *octahedral* geometry, as shown by $[Co(NH_3)_6]^{3+}$ (Figure 24.7A). Notice the similarity with some of the molecular shapes in VSEPR theory (Section 9.1).

Ligands The ligands of complex ions are *molecules or anions* with one or more **donor atoms**. Each *donates a lone pair of electrons* to the metal ion, thus forming a covalent bond. Because donor atoms must have at least one lone pair, they often come from group 15, 16, or 17.

Ligands are classified in terms of their number of donor atoms, or "teeth":

- *Monodentate* (Latin, meaning "one-toothed") ligands bond through a single donor atom.
- *Bidentate* ligands have two donor atoms, each of which bonds to the metal ion.
- *Polydentate* ligands have more than two donor atoms.

Dr. Jennifer Love and her group at the University of British Columbia, in Vancouver, are working to create new catalysts that use transition metals to prepare more structurally and functionally diverse synthetic compounds. They are focusing on late transition metals, since these metals are easier to work with and are compatible with a broad range of functional groups. Dr. Love's group is also studying the reactivity of transition metal bonds with elements such as sulfur, oxygen, fluorine, and phosphorus.

TABLE 24.5 Coordination Numbers and Shapes of Some Complex Ions

Coordination Number	Shape		Examples
2	Linear		$[CuCl_2]^-$, $[Ag(NH_3)_2]^+$, $[AuCl_2]^-$
4	Square planar	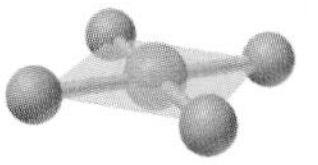	$[Ni(CN)_4]^{2-}$, $[PdCl_4]^{2-}$, $[Pt(NH_3)_4]^{2+}$, $[Cu(NH_3)_4]^{2+}$
4	Tetrahedral	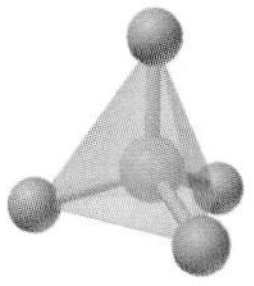	$[Cu(CN)_4]^{3-}$, $[Zn(NH_3)_4]^{2+}$, $[CdCl_4]^{2-}$, $[MnCl_4]^{2-}$
6	Octahedral	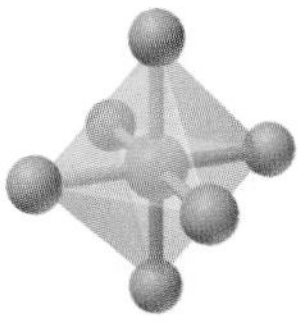	$[Ti(H_2O)_6]^{3+}$, $[V(CN)_6]^{4-}$, $[Cr(NH_3)_4Cl_2]^+$, $[Mn(H_2O)_6]^{2+}$, $[FeCl_6]^{3-}$, $[Co(en)_3]^{3+}$

Bidentate and polydentate ligands give rise to *rings* in the complex ion. For example, ethylenediamine (abbreviated *en*) has a chain of four atoms (:N—C—C—N:), so it forms a five-membered ring, with the two electron-donating N atoms bonding to the metal ion. Such ligands seem to grab the metal ion like claws, so a complex ion that contains them is also called a **chelate** (pronounced "KEY-late"; Greek *chela*, meaning "crab's claw"). With its six donor atoms, the ethylenediaminetetraacetate ($EDTA^{4-}$) ion forms very stable complexes with metal ions. This property makes EDTA useful for treating *heavy-metal poisoning.* When ingested by a patient, the $EDTA^{4-}$ ion removes lead (*see margin*) and other heavy-metal ions from the blood and other body fluids.

Table 24.6 shows some common ligands in coordination compounds. Notice that each ligand has one or more donor atoms (*blue*), each with a lone pair of electrons that it can donate to form a covalent bond to the metal ion.

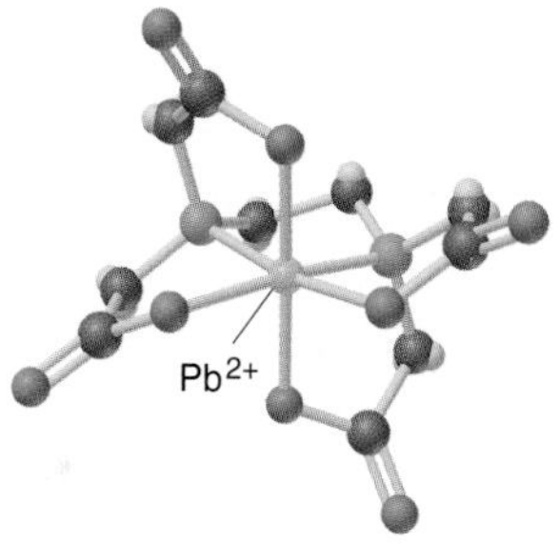

$[Pb(EDTA)]^{2-}$

Formulas and Names of Coordination Compounds

The combination of ions in a coordination compound is the key to writing its formula and name. A coordination compound can consist of a complex cation with simple anionic counterions, a complex anion with simple cationic counterions, or even a complex cation with a complex anion as a counterion.

TABLE 24.6 Some Common Ligands in Coordination Compounds

Ligand Type	Examples			
Monodentate	$H_2\ddot{O}$: Water	:F:$^-$ Fluoride ion	$[:C{\equiv}N:]^-$ Cyanide ion	$[:\ddot{O}—H]^-$ Hydroxide ion
	$:NH_3$ Ammonia	:Cl:$^-$ Chloride ion	$[:\ddot{S}{=}C{=}\ddot{N}:]^-$ Thiocyanate ion	$[:\ddot{O}—\ddot{N}{=}\ddot{O}:]^-$ Nitrite ion
Bidentate	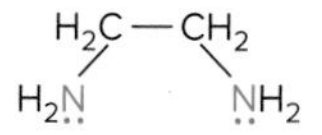 Ethylenediamine (en)	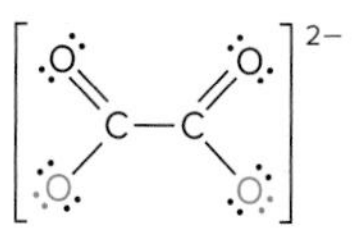 Oxalate ion		
Polydentate	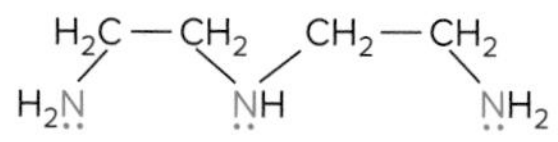 Diethylenetriamine	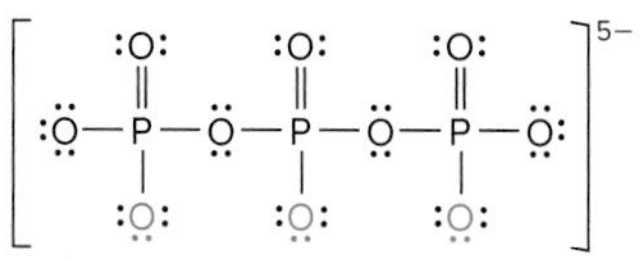 Triphosphate ion	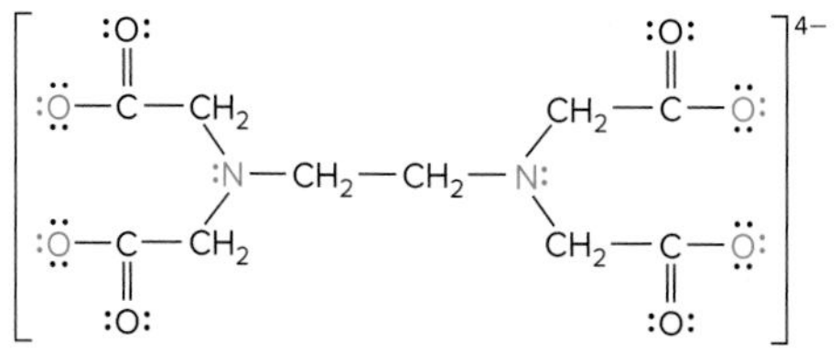 Ethylenediaminetetraacetate ion ($EDTA^{4-}$)	

Determining the Charge of the Metal Ion To know the number of cations and anions, we must know the ion charges, and we use basic arithmetic to find them:

- For a compound with a *complex anion*, such as $K_2[Co(NH_3)_2Cl_4]$, we know that the K^+ ions balance the charge of the complex anion, which contains two NH_3 molecules and four Cl^- ions as ligands. The two NH_3 are neutral, the four Cl^- have a total charge of 4^-, and the entire complex ion must have a charge of 2− to balance the two K^+. The charge of the central metal ion can be calculated as follows:

$$\text{Charge of complex ion} = \text{charge of metal ion} + \text{total charge of ligands}$$
$$2- = \text{charge of metal ion} + [(2 \times 0) + (4 \times 1-)]$$

So,

$$\text{Charge of metal ion} = (2-) - (4-) = 2+, \text{ that is, } Co^{2+}$$

- For a compound with a *complex cation*, such as $[Co(NH_3)_4Cl_2]Cl$, the complex ion is $[Co(NH_3)_4Cl_2]^+$, and one Cl^- is the counterion. The four NH_3 ligands are neutral, the two Cl^- ligands have a total charge of 2−, and the complex cation has a charge of 1+, so the central metal ion must be Co^{3+}:

$$\text{Charge of metal ion} = \text{charge of complex ion} - \text{total charge of ligands}$$
$$= 1+ \quad \{-[(4 \times 0) + (2 \times 1-)]\} = 3+$$

Rules for Writing Formulas There are three rules for writing the formulas for coordination compounds. (The first two are the same as those for writing the formula for any ionic compound.)

1. *The cation is written before the anion.*
2. *The charge of the cation(s) is balanced by the charge of the anion(s).*
3. *For the complex ion, neutral ligands are written before anionic ligands, and the formula for the whole ion is placed in brackets.*

Rules for Naming Coordination Compounds Originally named after their discoverer or colour, coordination compounds are named systematically with a set of rules. Let us see how to name $[Co(NH_3)_4Cl_2]Cl$. As we go through the naming steps, refer to Tables 24.7 and 24.8:

1. *The cation is named before the anion.* Thus, we name the $[Co(NH_3)_4Cl_2]^+$ ion before the Cl^- ion.
2. *Within the complex ion, the ligands are named in alphabetical order* ***before*** *the metal ion.* In the $[Co(NH_3)_4Cl_2]^+$ ion, four NH_3 and two Cl^- are named before Co^{3+}.
3. *Neutral ligands generally have the molecule name*, but there are a few exceptions (Table 24.7). *Anionic ligands drop the -ide and add -o after the root name*; thus, the anion name *fluoride* for an F^- ion becomes the ligand name *fluoro*. The two ligands in $[Co(NH_3)_4Cl_2]^+$ are *ammine* (NH_3) and *chloro* (Cl^-), with *ammine* coming before *chloro* alphabetically.
4. *A numerical prefix indicates the number of ligands of a particular type.* For example, *tetra*ammine denotes *four* NH_3, and *di*chloro denotes *two* Cl^-. Other prefixes are *tri-*, *penta-*, and *hexa-*. However, prefixes do *not* affect the alphabetical order: *tetraammine* comes before *dichloro*. For ligand names that include a numerical prefix (such as ethylene*di*amine), we use *bis* (2), *tris* (3), or *tetrakis* (4) to indicate the

TABLE 24.7 Names of Some Neutral and Anionic Ligands

Neutral		Anionic	
Name	**Formula**	**Name**	**Formula**
Aqua	H_2O	Fluoro	F^-
Ammine	NH_3	Chloro	Cl^-
Carbonyl	CO	Bromo	Br^-
Nitrosyl	NO	Iodo	I^-
		Hydroxo	OH^-
		Cyano	CN^-

number of ligands, followed by the ligand name in parentheses. For example, a complex ion that has two ethylenediamine ligands has *bis(ethylenediamine)* in its name.

5. *The oxidation state of the metal ion has a roman numeral (in parentheses) only if the metal ion can have more than one state.* Since cobalt can have +2 and +3 states, we add a III to name the complex ion. Thus, the compound is tetraamminedichlorocobalt(III) chloride. The only typographical space in the name comes between the cation and the anion.
6. *If the complex ion is an anion, we drop the ending of the metal name and add -ate.* Thus, the name for $K[Pt(NH_3)Cl_5]$ is potassium amminepentachloroplatinate(IV). There is one K^+ counterion, so the complex anion has a charge of 1−. The five Cl^- ligands have a total charge of 5−, so Pt must be in the +4 oxidation state. For some metals, we use the Latin root with the *-ate* ending (Table 24.8). For example, the name for $Na_4[FeBr_6]$ is sodium hexabromoferrate(II).

TABLE 24.8 Names of Some Metal Ions in Complex Anions

Metal	Name in Anion
Iron	Ferrate
Copper	Cuprate
Lead	Plumbate
Silver	Argentate
Gold	Aurate
Tin	Stannate

Sample Problem 24.3 Writing Names and Formulas for Coordination Compounds

Problem **(a)** What is the systematic name of $Na_3[AlF_6]$?

(b) What is the systematic name of $[Co(en)_2Cl_2]NO_3$?

(c) What is the formula for tetraamminebromochloroplatinum(IV) chloride?

(d) What is the formula for hexaamminecobalt(III) tetrachloroferrate(III)?

Plan We use the rules that were just presented, as well as Tables 24.7 and 24.8.

Solution **(a)** The complex ion is $[AlF_6]^{3-}$. There are six (*hexa-*) F^- ions (*fluoro*) as ligands, so we have *hexafluoro*. The complex ion is an anion, so the ending of the metal ion (aluminum) must be changed to *-ate*: hexafluoroaluminate. Aluminum has only the +3 oxidation state, so we do *not* use a roman numeral. The positive counterion is named first and separated from the anion by a space: sodium hexafluoroaluminate.

(b) Listed alphabetically, there are two Cl^- (*dichloro*) and two en [*bis(ethylenediamine)*] as ligands. The complex ion is a cation, so the metal name is unchanged, but we specify its oxidation state because cobalt can have several. One NO_3^- balances the 1+ cation charge. With 2− for two Cl^- and 0 for two en, the metal must be *cobalt(III)*. The word *nitrate* follows a space: dichlorobis(ethylenediamine)cobalt(III) nitrate.

(c) The central metal ion is written first, followed by the neutral ligands and then (in alphabetical order) by the negative ligands. *Tetraammine* is four NH_3, *bromo* is one Br^-, *chloro* is one Cl^-, and *platinate(IV)* is Pt^{4+}, so the complex ion is $[Pt(NH_3)_4BrCl]^{2+}$. Its 2+ charge is the sum of 4+ for Pt^{4+}, 0 for four NH_3, 1− for one Br^-, and 1− for one Cl^-. To balance the 2+ charge, we need two Cl^- counterions: $[Pt(NH_3)_4BrCl]Cl_2$.

(d) This compound consists of two different complex ions. In the cation, *hexaammine* is six NH_3 and *cobalt(III)* is Co^{3+}, so the cation is $[Co(NH_3)_6]^{3+}$. The 3+ charge is the sum of 3+ for Co^{3+} and 0 for six NH_3. In the anion, *tetrachloro* is four Cl^-, and *ferrate(III)* is Fe^{3+}, so the anion is $[FeCl_4]^-$. The 1− charge is the sum of 3+ for Fe^{3+} and 4− for four Cl^-. In the neutral compound, one 3+ cation must be balanced by three 1− anions: $[Co(NH_3)_6][FeCl_4]_3$.

Check Reverse the process to be sure that you obtain the name or formula in the problem.

Follow-Up Problem 24.3 **(a)** What is the name of $[Cr(H_2O)_5Br]Cl_2$?

(b) What is the formula for barium hexacyanocobaltate(III)?

Isomerism in Coordination Compounds

Isomers are compounds that have the same chemical formula but different properties. Recall the discussion of isomerism in organic compounds (Section 20.4); coordination compounds have the same two broad categories: constitutional isomers and stereoisomers.

FIGURE 24.8 A pair of linkage (constitutional) isomers

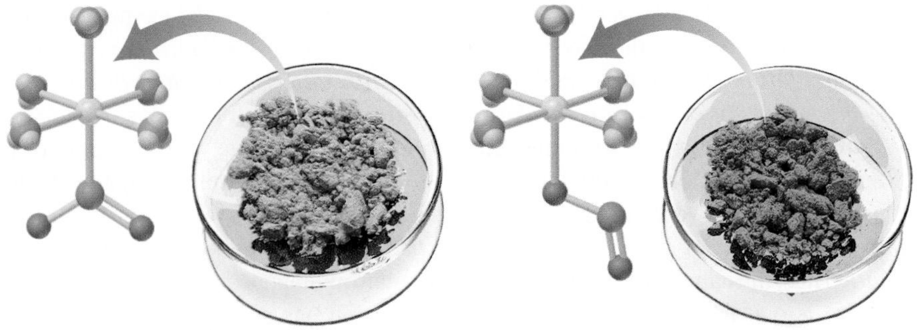

Nitro isomer, $[Co(NH_3)_5(NO_2)]Cl_2$ **Nitrito** isomer, $[Co(NH_3)_5(ONO)]Cl_2$

Constitutional Isomers: Atoms Connected Differently Compounds that have the same formula, but the atoms connected differently, are **constitutional (structural) isomers**. There are two types of constitutional isomers: coordination isomers and linkage isomers.

1. **Coordination isomers** occur when the composition of the complex ion, but not the compound, is different. This type of isomerism occurs in two ways:

- *Exchange of ligand and counterion.* For example, in $[Pt(NH_3)_4Cl_2](NO_2)_2$, the Cl^- ions are the ligands, and the NO_2^- ions are the counterions; in $[Pt(NH_3)_4(NO_2)_2]Cl_2$, the roles are reversed. A common test to see whether Cl^- is a ligand or a counterion is to treat a solution of the compound with $AgNO_3$; the Cl^- counterion will form a white precipitate of AgCl, but the Cl^- ligand will not. Thus, a $[Pt(NH_3)_4Cl_2](NO_2)_2$ solution does not form AgCl because Cl^- is bound to the metal ion, but a $[Pt(NH_3)_4(NO_2)_2]Cl_2$ solution forms 2 mol of AgCl per mole of compound.
- *Exchange of ligands.* This type of isomerism occurs in compounds that consist of two complex ions. For example, in $[Cr(NH_3)_6][Co(CN)_6]$ and $[Co(NH_3)_6][Cr(CN)_6]$, NH_3 is a ligand of Cr^{3+} in one compound and a ligand of Co^{3+} in the other.

2. **Linkage isomers** occur when the composition of the complex ion is the same but the ligand donor atom is different. Some ligands can bind to the metal ion through *either of two donor atoms.* For example, the nitrite ion can bind through the N atom (*nitro*, O_2N:) or through either of the O atoms (*nitrito*, ONO:) to give linkage isomers, such as pentaammine*nitro*cobalt(III) chloride $[Co(NH_3)_5(NO_2)]Cl_2$ and pentaammine*nitrito*cobalt(III) chloride $[Co(NH_3)_5(ONO)]Cl_2$ (Figure 24.8).

Other examples of ligands that have more than one donor atom are the cyanate ion, which can attach via the O atom (*cyanato*, NCO:) or the N atom (*isocyanato*, OCN:), and the thiocyanate ion, which can attach via the S atom or the N atom:

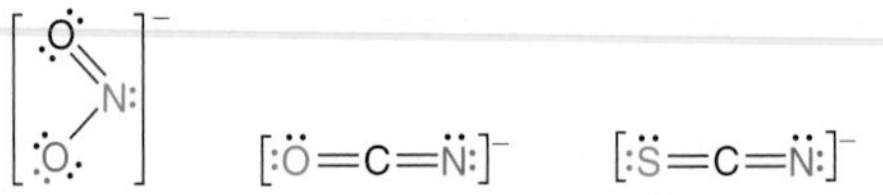

nitrite cyanate thiocyanate

Stereoisomers: Atoms Arranged Differently in Space **Stereoisomers** are compounds that have the same atomic connections but different spatial arrangements of the atoms. *Geometric* isomers and enantiomers, which we discussed for organic compounds, occur with coordination compounds as well:

1. **Geometric (*cis-trans*) isomers** occur when atoms or groups of atoms are arranged differently in space relative to the central metal ion. For example, the square planar $[Pt(NH_3)_2Cl_2]$ has two arrangements, which give rise to two different compounds (Figure 24.9A). The isomer with identical ligands *next* to each other is *cis*-diamminedichloroplatinum(II), and the isomer with identical ligands *across* from each other is *trans*-diamminedichloroplatinum(II). The *cis-trans* isomers can be considered to be *diastereomers* when they are not mirror images of one another and when they are nonsuperimposable (Section 20.4).

These isomers have remarkably different biological behaviours. In the mid-1960s, Barnett Rosenberg and his colleagues found that *cis*-$[Pt(NH_3)_2Cl_2]$ (cisplatin)

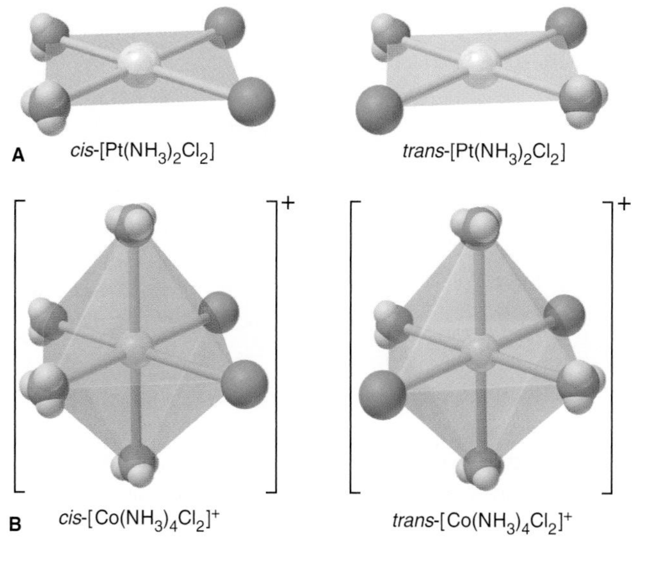

FIGURE 24.9 Geometric (*cis-trans*) isomerism. A. The *cis* and *trans* isomers of $[Pt(NH_3)_2Cl_2]$. **B.** The *cis* and *trans* isomers of $[Co(NH_3)_4Cl_2]^+$. The coloured shapes indicate the actual colours of the species.

was an antitumour agent. Cisplatin and several closely related platinum(II) complexes are still used to treat certain types of cancer. In contrast, *trans*-$[Pt(NH_3)_2Cl_2]$ has no antitumour effect. Cisplatin may work by becoming oriented within a cancer cell such that a donor atom on each DNA strand can replace a Cl^- ligand. Cisplatin binds the platinum(II) strongly, preventing DNA replication (Section 22.3).

Octahedral complexes also exhibit *cis-trans* isomerism (Figure 24.9B). The *cis* isomer of the $[Co(NH_3)_4Cl_2]^+$ ion has the two Cl^- ligands at any two adjacent positions of the ion's octahedral shape, whereas the *trans* isomer has these ligands across from each other.

2. **Enantiomers** occur when a molecule and its mirror image cannot be superimposed (see Figures 20.17 to 20.19). Unlike other types of isomers, which have distinct physical properties, enantiomers are physically identical except for *the direction in which they rotate the plane of polarized light.* Many octahedral complex ions show enantiomerism, which we can determine by rotating one isomer and seeing if it is superimposable on the other isomer (its mirror image). For example, in Figure 24.10A, the two structures (I and II) of *cis*-$[Co(en)_2Cl_2]^+$, the *cis*-dichlorobis(ethylenediamine)cobalt(III) ion, are mirror images of each other. Rotate structure I 180° around a vertical axis, and you obtain structure III. The Cl^- ligands

FIGURE 24.10 Enantiomerism in an octahedral complex ion. A. Structure I and its mirror image, structure II, are enantiomers of *cis*-$[Co(en)_2Cl_2]$. (The curved wedges represent the bidentate ligand ethylenediamine, $H_2N—CH_2—CH_2—NH_2$.) **B.** The *trans* isomer does *not* have enantiomers.

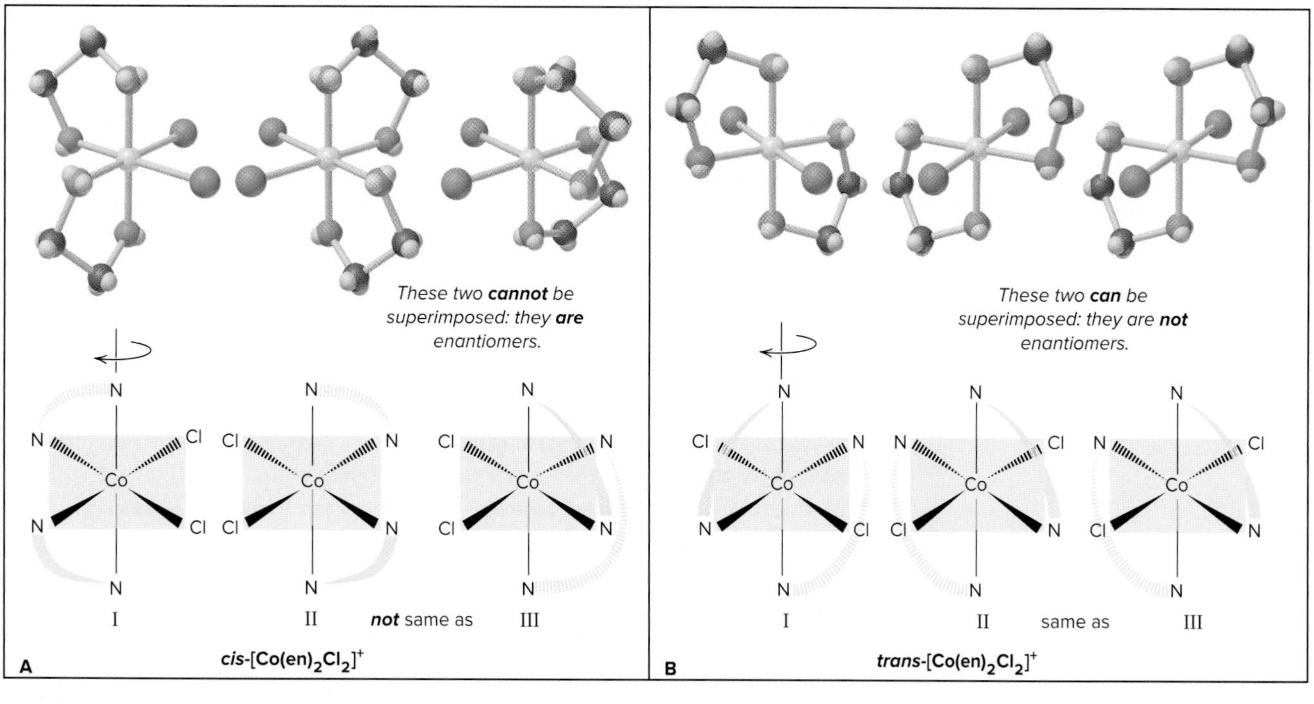

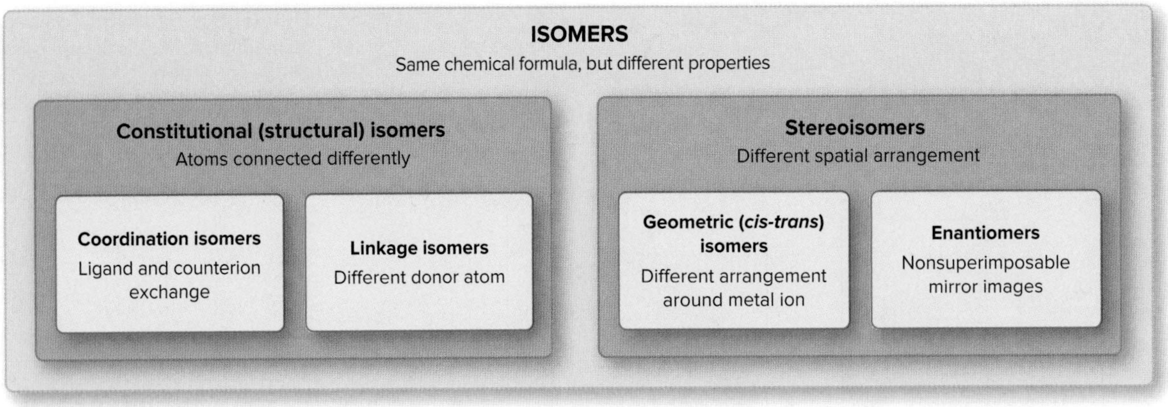

FIGURE 24.11 Important types of isomerism in coordination compounds

of structure III match those of structure II, but the en ligands do not: structures II and III (which is structure I rotated) are not superimposable: they are enantiomers. One isomer is designated *d*-$[Co(en)_2Cl_2]^+$ and the other is designated *l*-$[Co(en)_2Cl_2]^+$, depending on whether it rotates polarized light to the right (*d*- for *dextro*-) or the left (*l*- for *levo*-). (The *d*- or *l*- designation can only be determined by experiment.)

In contrast, as shown in Figure 24.10B, the two structures of the *trans*-dichlorobis(ethylenediamine)cobalt(III) ion are *not* enantiomers. Rotate structure I 90° around a vertical axis and you obtain structure III, which *is* superimposable on structure II.

Figure 24.11 is an overview of the most common types of isomerism in coordination compounds.

Sample Problem 24.4 Determining the Type of Stereoisomerism

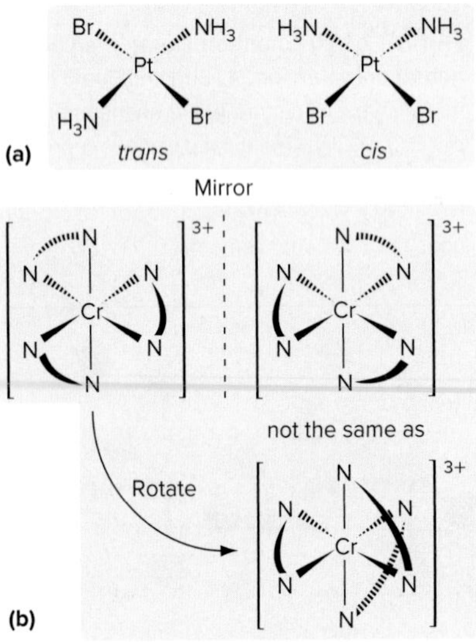

Problem Draw stereoisomers for each complex, and state the type of isomerism:
(a) $[Pt(NH_3)_2Br_2]$ (square planar) **(b)** $[Cr(en)_3]^{3+}$ (en = $H_2\ddot{N}CH_2CH_2\ddot{N}H_2$)

Plan We determine the geometry around each metal ion and the nature of the ligands. If there are different ligands that can be placed in different positions relative to each other, geometric (*cis-trans*) isomerism occurs. Then we see whether the mirror image of an isomer is superimposable on the original. If it is *not*, they are enantiomers.

Solution (a) The square planar Pt(II) complex has two different monodentate ligands. Each pair of ligands can lie next to or across from each other (*see margin*). Thus, geometric isomerism occurs. Each isomer *is* superimposable on a mirror image of itself, so there are no enantiomers.

(b) Ethylenediamine (en) is a bidentate ligand. The Cr^{3+} has a coordination number of 6 and an octahedral geometry, like Co^{3+}. The three bidentate ligands are identical, so there is no geometric isomerism. However, the complex ion has a nonsuperimposable mirror image (*see margin*). Thus, it has enantiomers.

Follow-Up Problem 24.4 What stereoisomers, if any, are possible for the $[Co(NH_3)_2(en)Cl_2]^+$ ion?

SUMMARY OF SECTION 24.3

- Coordination compounds consist of a complex ion and charge-balancing counterions. The complex ion has a central metal ion bonded to neutral and/or anionic ligands, which have one or more donor atoms, each with a lone pair of electrons.
- The most common complex-ion geometry is octahedral (six ligand atoms bonding).
- Formulas and names of coordination compounds follow systematic rules.
- Coordination compounds can exhibit constitutional isomerism (coordination and linkage) and stereoisomerism (geometric and optical).

24.4 Theoretical Basis for the Bonding and Properties of Complexes

In this section, we consider valence bond theory and crystal field theory, which address different features of complexes: how metal-ligand bonds form, why certain geometries are preferred, and why complexes are brightly coloured and often paramagnetic. Finally, we will conclude with a brief consideration of molecular orbital (MO) theory and why it may provide a more comprehensive explanation for experimental observations in metal ligand bonding.

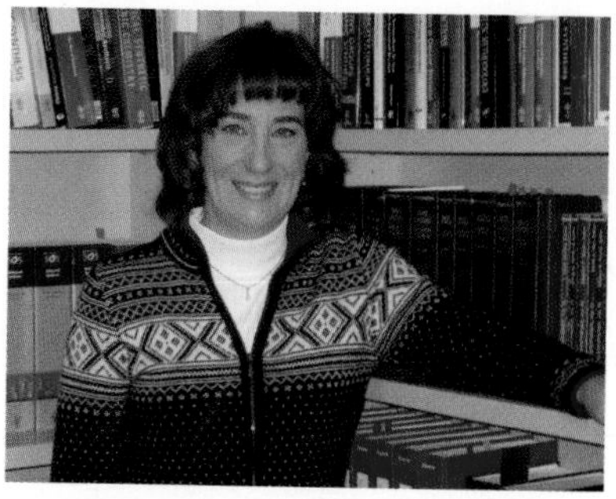

Dr. Hélène Lebel and her group at the University of Montréal are studying new synthetic organic methodologies using transition metal catalyzed processes. They have developed a series of one-pot procedures using a combination of transition metal catalysts, with the goal of obtaining specific chemical reactions. They are studying the reaction of nitrogen with transition metal complexes from the viewpoint of organic reactions, rather than the viewpoint of inorganic chemistry. As well, they are developing new chiral ligands for transition metals, with a view to creating efficient chiral catalysts for the stereocontrol of organic reactions.

Applying Valence Bond Theory to Complex Ions

Valence bond (VB) theory, which helped to explain bonding and structure in main-group compounds (Section 10.1), is also used to describe bonding in complex ions. In the formation of a complex ion, the *filled* ligand orbital overlaps an *empty* metal-ion orbital: *the ligand (Lewis base) donates an electron pair, and the metal ion (Lewis acid) accepts it to form a covalent bond in the complex ion (Lewis adduct)* (Section 16.8). A bond in which one atom contributes both electrons is called a **coordinate covalent bond**; once formed, it is identical to any covalent single bond.

Recall that the VB concept of hybridization proposes mixing particular combinations of *s*, *p*, and *d* orbitals to obtain sets of hybrid orbitals, which have specific geometries. For coordination compounds, the model proposes that *the type of metal-ion orbital hybridization determines the geometry of the complex ion.* Let us discuss orbital combinations that lead to octahedral, square planar, and tetrahedral geometries.

Octahedral Complexes The hexaamminechromium(III) ion, $[Cr(NH_3)_6]^{3+}$, is an *octahedral complex* (Figure 24.12). The six lowest energy, *empty* orbitals of the Cr^{3+} ion—two 3*d*, one 4*s*, and three 4*p*—mix and become six equivalent d^2sp^3 hybrid orbitals that point toward the corners of an octahedron.* Six NH_3 molecules donate lone pairs from their N atoms to form six metal-ligand bonds. Three unpaired 3*d* electrons of the central Cr^{3+} ion ($[Ar]3d^3$) remain in unhybridized orbitals and make the complex ion paramagnetic.

Square Planar Complexes Metal ions with a d^8 configuration usually form *square planar complexes* (Figure 24.13). In the $[Ni(CN)_4]^{2-}$ ion, for example, the model proposes that one 3*d*, one 4*s*, and two 4*p* orbitals of Ni^{2+} mix and form four dsp^2 hybrid orbitals, which point to the corners of a square and accept one electron pair from each of four CN^- ligands.

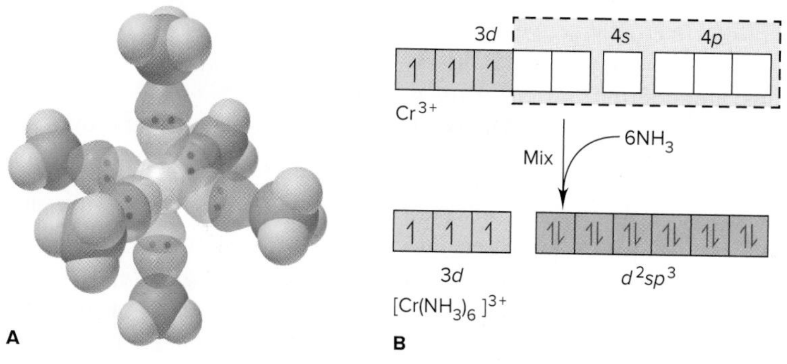

FIGURE 24.12 Hybrid orbitals and bonding in the octahedral $[Cr(NH_3)_6]^{3+}$ ion. **A.** Orbital contour depiction of $[Cr(NH_3)_6]^{3+}$. **B.** Partial orbital diagrams depict the formation of six d^2sp^3 hybrid orbitals, which are filled with six NH_3 lone pairs (*red*).

*Note the distinction between hybrid-orbital designations here and in Chapter 10. Both designations give the orbitals in energy order. In $[Cr(NH_3)_6]^{3+}$, the 3*d* orbitals have a *lower n* value than the 4*s* and 4*p* orbitals, so the hybrid orbitals are d^2sp^3. In SF_6, however, the 3*d* orbitals have the *same n* value as the 3*s* and 3*p* orbitals of S, so the designation is sp^3d^2.

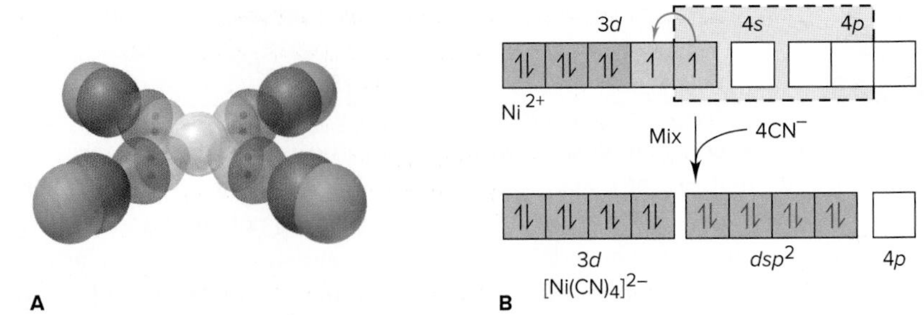

FIGURE 24.13 Hybrid orbitals and bonding in the square planar $[Ni(CN)_4]^{2-}$ ion. **A.** Contour depiction of $[Ni(CN)_4]^{2-}$. **B.** Two lone 3*d* electrons pair up, which frees one 3*d* orbital for hybridization. Four dsp^2 orbitals are filled with lone pairs (*red*) from four CN^- ligands.

A look at the ground-state electron configuration of the Ni^{2+} ion, however, raises a key question: How can the Ni^{2+} ion ($[Ar]3d^8$) offer an empty 3*d* orbital for accepting a lone pair if its eight 3*d* electrons lie in three filled and two half-filled orbitals? Apparently, in the d^8 configuration of Ni^{2+}, electrons in the half-filled orbitals *pair up* and leave one 3*d* orbital empty. This explanation is consistent with the fact that the complex is diamagnetic (no unpaired electrons). Moreover, it means that the energy *gained* by using a 3*d* orbital for bonding in the hybrid orbital is greater than the energy *required* to overcome repulsions from pairing the 3*d* electrons.

Tetrahedral Complexes Metal ions that have a filled *d* sublevel, such as Zn^{2+} ($[Ar]3d^{10}$), often form diamagnetic *tetrahedral complexes* (Figure 24.14). In the $[Zn(OH)_4]^{2-}$ ion, for example, VB theory proposes that the lowest *empty* Zn^{2+} orbitals—one 4*s* and three 4*p*—mix to become four sp^3 hybrid orbitals that point to the corners of a tetrahedron and are occupied by lone pairs, one from each of the four OH^- ligands.

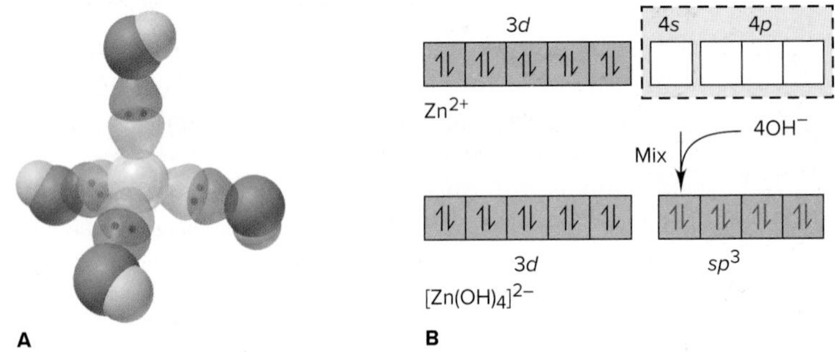

FIGURE 24.14 Hybrid orbitals and bonding in the tetrahedral $[Zn(OH)_4]^{2-}$ ion. **A.** Contour depiction of $[Zn(OH)_4]^{2-}$. **B.** Formation and filling of four sp^3 hybrid orbitals.

Crystal Field Theory

The VB model is easy to picture, and it rationalizes bonding and shape. However, it treats the orbitals as little more than empty "slots" for accepting electron pairs. Moreover, it gives no insight into the colours of these compounds and sometimes predicts their magnetic properties incorrectly. In contrast, **crystal field theory** provides little insight into metal-ligand bonding, but it explains colour and magnetism by highlighting the *effect on d-orbital energies of the metal ion as the ligands approach.* Before we discuss this theory, let us consider why a substance is coloured.

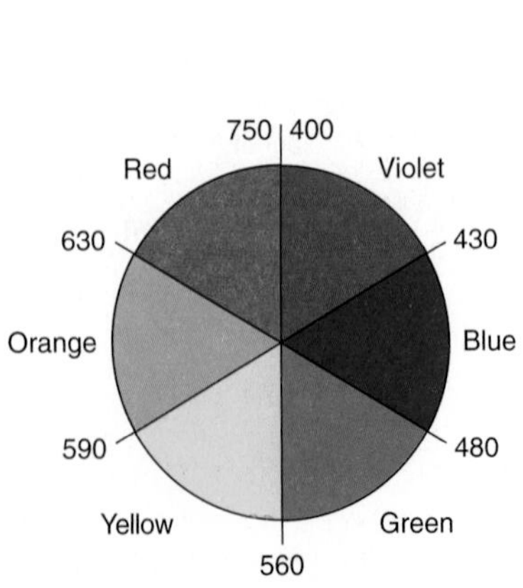

FIGURE 24.15 An artist's wheel. The colours, with approximate wavelength ranges (nm), are shown as wedges.

What Is Colour? White light consists of all wavelengths (λ) in the visible range (Section 6.1) and can be dispersed into colours of a narrower wavelength range. Objects appear coloured in white light because they absorb only certain wavelengths: an opaque object *reflects* the other wavelengths, and a clear object *transmits* them. If an object absorbs all visible wavelengths, it appears black; if it reflects all visible wavelengths, it appears white.

Each colour has a *complementary* colour; for example, green and red are complementary colours. Figure 24.15 presents these relationships on an artist's colour wheel, with complementary colours shown as wedges opposite each other. A mixture of complementary colours absorbs all visible wavelengths and appears black.

TABLE 24.9 Relation between Absorbed and Observed Colours

Absorbed Colour	λ (nm)	Observed Colour	λ (nm)
Violet	400	Green-yellow	560
Blue	450	Yellow	600
Blue-green	490	Red	620
Yellow-green	570	Violet	410
Yellow	580	Dark blue	430
Orange	600	Blue	450
Red	650	Green	520

An object has a particular colour for one of two reasons:

- It reflects (or transmits) light of *this* colour. Thus, if an object absorbs all wavelengths *except* green, the reflected (or transmitted) light is seen as green.
- It absorbs light of the *complementary* colour. Thus, if the object absorbs only red, the *complement* of green, the remaining mixture of reflected (or transmitted) wavelengths is also seen as green.

Table 24.9 lists the colours absorbed and the resulting colours observed. The relationship between absorbed and observed colours is demonstrated by the seasonal colours of deciduous trees. In the spring and summer, leaves contain high concentrations of the photosynthetic pigment *chlorophyll* and lower concentrations of other pigments called *xanthophylls.* Chlorophyll absorbs blue and red strongly, reflecting mostly green. In the fall, photosynthesis slows, so the leaves no longer make chlorophyll. The green fades as the chlorophyll decomposes, revealing the xanthophylls that were present but masked by the chlorophyll. Xanthophylls absorb green and blue strongly, reflecting bright yellow and red (*see photo*).

Photograph of Cape Breton fall colours

Splitting of *d* Orbitals in an Octahedral Field of Ligands The crystal field model explains that the properties of complexes result from the splitting of *d*-orbital energies, which is caused by *electrostatic attractions between the metal cation and the negative charge of the ligands.* The negative charge of the ligands is either partial, as in a polar covalent ligand like NH_3, or full, as in an anionic ligand like Cl^-. Picture six ligands approaching a metal ion along the mutually perpendicular x, y, and z axes, thus forming an octahedral arrangement (Figure 24.16A). Let us follow the orientation of ligand and orbital, and see how the approach affects orbital energies.

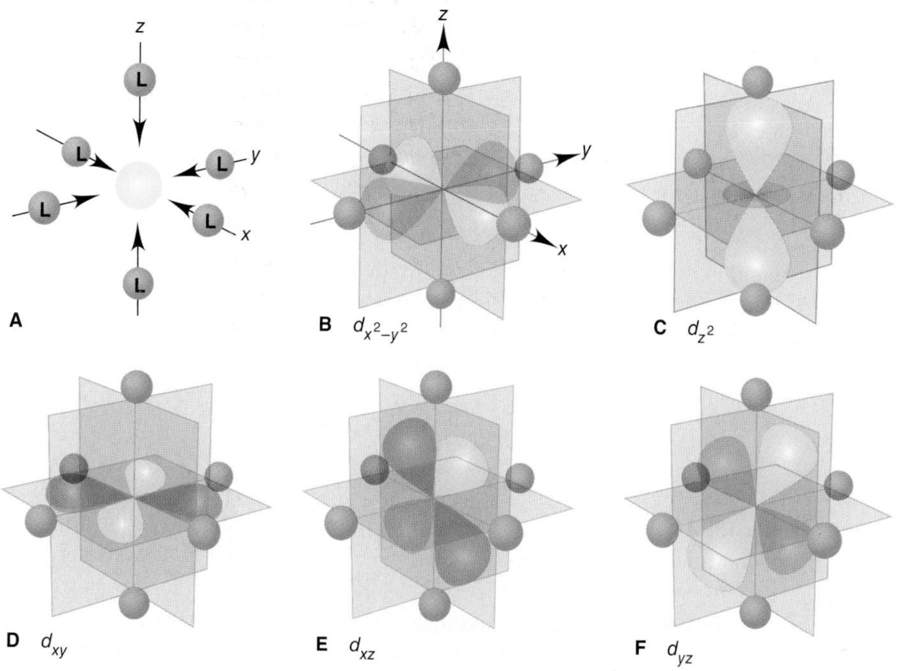

FIGURE 24.16 The five *d* orbitals in an octahedral field of ligands. A. Ligands approach along the x, y, and z axes. **B** and **C.** The $d_{x^2-y^2}$ and d_{z^2} orbitals point *directly* at some of the ligands. **D** to **F.** The d_{xy}, d_{xz}, and d_{yz} orbitals point *between* the ligands.

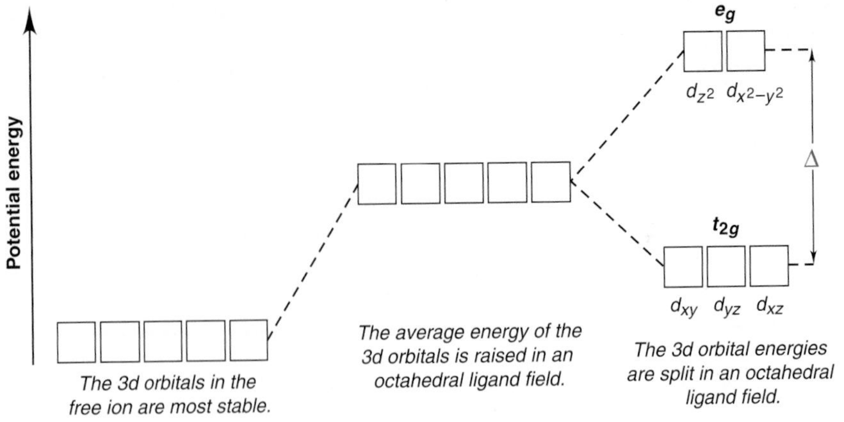

Potential energy
e_g
d_{z^2} $d_{x^2-y^2}$
Δ
t_{2g}
d_{xy} d_{yz} d_{xz}
The 3d orbitals in the free ion are most stable.
The average energy of the 3d orbitals is raised in an octahedral ligand field.
The 3d orbital energies are split in an octahedral ligand field.

FIGURE 24.17 Splitting of d-orbital energies in an octahedral field of ligands

Dr. Mark Stradiotto and his group at Dalhousie University, in Halifax, Nova Scotia, are directing their research efforts toward the development of new classes of transition metal complexes that exhibit interesting and unusual reactivity, with a view to incorporating this reactivity into useful substrate transformations. As well, they are studying the development of new late metal catalyst complexes for use in cross-coupling reactions and for the hydroamination of unsaturated substrates. Dr. Stradiotto's group is also studying zwitterionic relatives of traditional cationic late metal complexes, hoping that they may prove useful in a range of catalytic substrate transformations.

1. *Orientation of ligand and metal ion orbitals.* As ligands approach, their electron pairs repel electrons in the five d orbitals of the metal ion. In the isolated ion, the d orbitals have different orientations but *equal* energies. In the negative field of ligands, however, the d electrons are *repelled unequally because of their different orbital orientations.* The ligands moving *along* the x, y, and z axes approach as follows:

- *Directly toward* the lobes of the $d_{x^2-y^2}$ and d_{z^2} orbitals (Figure 24.16B and 24.16C)
- *Between* the lobes of the d_{xy}, d_{xz}, and d_{yz} orbitals (Figure 24.16D to 24.16F)

2. *Effect on d-orbital energies.* As a result of these different orientations, electrons in the $d_{x^2-y^2}$ and d_{z^2} orbitals experience *stronger* repulsions than electrons in the d_{xy}, d_{xz}, and d_{yz} orbitals do. An energy diagram shows that the five d orbitals are most stable in the free ion and their *average* energy is higher in the ligand field. However, *the orbital energies split, with two d orbitals higher in energy and three d orbitals lower in energy* (Figure 24.17):

- The two higher energy orbitals are $\boldsymbol{e_g}$ **orbitals** and arise from the $d_{x^2-y^2}$ and d_{z^2} orbitals.
- The three lower energy orbitals are $\boldsymbol{t_{2g}}$ **orbitals** and arise from the d_{xy}, d_{xz}, and d_{yz} orbitals.

The names of orbitals e_g and t_{2g} are assigned on the basis of the theory of molecular symmetry.

3. *The crystal field effect.* This splitting of orbital energies is called the *crystal field effect*, and the energy difference between the e_g and t_{2g} orbitals is the **crystal field splitting energy (Δ)**. Different ligands create crystal fields of different strength:

- **Strong-field ligands** lead to a *larger* splitting energy (larger Δ).
- **Weak-field ligands** lead to a *smaller* splitting energy (smaller Δ).

For instance, H_2O is a weak-field ligand, and CN^- is a strong-field ligand (Figure 24.18). Notice the different orbital occupancies; we will discuss the reason for these differences below.

Explaining the Colours of Transition Metal Complexes The colour of a coordination compound is determined by the crystal field splitting energy of its complex ion. When the ion absorbs radiant energy, electrons can move from the lower energy t_{2g} level to the higher energy e_g level. Recall, from Chapter 6, that the *difference* between two atomic energy levels is equal to the energy (and inversely related to the wavelength) of the absorbed photon:

$$\Delta E_{\text{electron}} = E_{\text{photon}} = h\nu = \frac{hc}{\lambda}$$

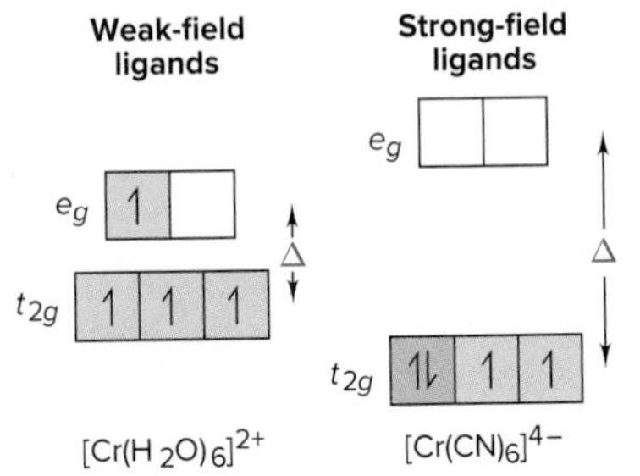

Weak-field ligands
Strong-field ligands
e_g
t_{2g}
Δ
$[Cr(H_2O)_6]^{2+}$
$[Cr(CN)_6]^{4-}$

FIGURE 24.18 The effect of ligands and splitting energy on orbital occupancy

Consider the $[Ti(H_2O)_6]^{3+}$ ion, which appears purple in aqueous solution (Figure 24.19). Hydrated Ti^{3+} has its one d electron in one of the three lower energy t_{2g} orbitals. The energy difference (Δ) between the t_{2g} and e_g orbitals in this ion corresponds to photons between the green and yellow range. When white light shines on the solution, these colours of light are absorbed, and the electron jumps to one of the e_g orbitals. Red, blue, and violet light is transmitted, so the solution appears purple.

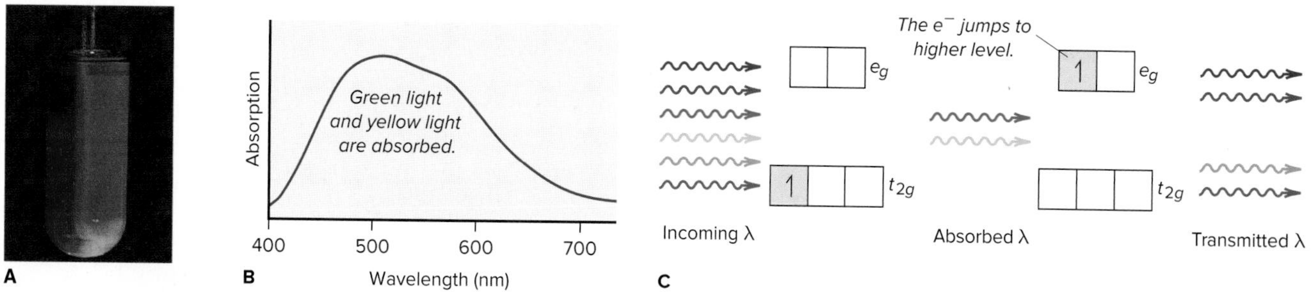

FIGURE 24.19 The colour of $[Ti(H_2O)_6]^{3+}$. **A.** The hydrated Ti^{3+} ion is purple. **B.** An absorption spectrum shows that green light and yellow light are absorbed and other wavelengths are transmitted. **C.** An orbital diagram depicts the colours absorbed when the *d* electron jumps to the higher level.

Absorption spectra can show the wavelengths absorbed by (1) a metal ion with different ligands and (2) different metal ions with the same ligand. Such data allow us to relate the energy of the absorbed light to Δ and make two key observations:

- *For a given ligand,* colour depends on the *oxidation state of the metal ion.* In Figure 24.20A, aqueous $[V(H_2O)_6]^{2+}$ (*left*) is violet and $[V(H_2O)_6]^{3+}$ (*right*) is yellow.
- *For a given metal ion,* colour depends on the *ligand.* A single ligand substitution can affect the wavelengths absorbed and, thus, the colour (Figure 24.20B).

FIGURE 24.20 Effects of oxidation state and ligand on colour. **A.** Solutions of two hydrated vanadium ions: the oxidation number of V is +2 (*left*); the oxidation number of V is +3 (*right*). **B.** A change in one ligand can influence the colour. $[Cr(NH_3)_6]^{3+}$ is yellow (*left*), and $[Cr(NH_3)_5Cl]^{2+}$ is purple (*right*).

The Spectrochemical Series The fact that colour depends on the ligand allows us to create a **spectrochemical series**, which ranks the ability of a ligand to split *d*-orbital energies (Figure 24.21). Using this series, we can predict the *relative* magnitude of Δ for a series of octahedral complexes of a *given* metal ion. Although we cannot predict the actual colour of a given complex, we can determine whether a complex will absorb longer or shorter wavelengths than other complexes in the series do.

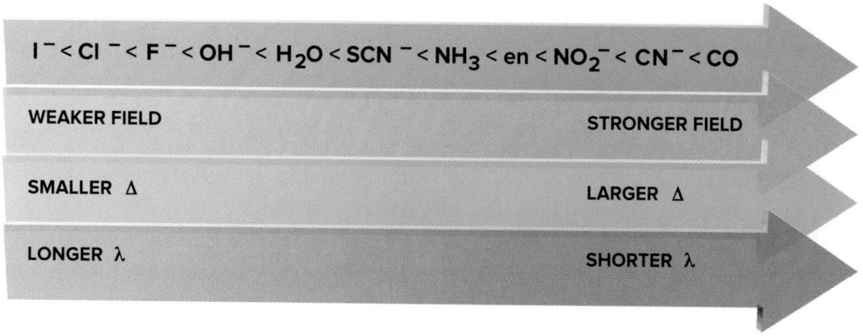

FIGURE 24.21 The spectrochemical series. As Δ increases, shorter wavelengths (higher energies) of light must be absorbed to excite electrons. For reference, water is a weak-field ligand.

Sample Problem 24.5 Ranking Crystal Field Splitting Energies (Δ) for Complex Ions of a Metal

Problem Rank $[Ti(H_2O)_6]^{3+}$, $[Ti(CN)_6]^{3-}$, and $[Ti(NH_3)_6]^{3+}$ in terms of D and the energy of visible light absorbed.

Plan The formulas show that Ti has an oxidation state of +3 in the three ions. From Figure 24.21, we rank the ligands by crystal field strength: the stronger the ligand, the greater the splitting and the higher the energy of light absorbed.

Solution The ligand field strength is in the order $CN^- > NH_3 > H_2O$, so the relative size of Δ and the energy of light absorbed is

$$Ti(CN)_6^{3-} > Ti(NH_3)_6^{3+} > Ti(H_2O)_6^{3+}$$

Follow-Up Problem 24.5 Which complex ion absorbs visible light of higher energy: $[V(H_2O)_6]^{3+}$ or $[V(NH_3)_6]^{3+}$?

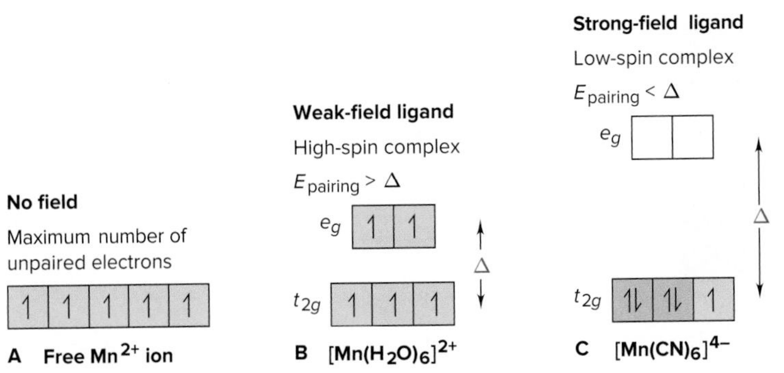

FIGURE 24.22 High-spin and low-spin octahedral complex ions of Mn^{2+}. **A.** Free Mn^{2+} has five unpaired electrons. **B.** Bonded to a weak-field ligand, Mn^{2+} still has five unpaired electrons (high-spin complex). **C.** Bonded to a strong-field ligand, Mn^{2+} has one unpaired electron (low-spin complex).

Explaining the Magnetic Properties of Transition Metal Complexes The splitting of energy levels gives rise to magnetic properties based on the number of *unpaired* electrons in the metal ion's d orbitals. Based on Hund's rule, electrons occupy orbitals of equal energy one at a time. When all lower energy orbitals are half-filled, the next electron can do the following:

- Enter a half-filled orbital and pair up by overcoming a repulsive *pairing energy* ($E_{pairing}$)
- Enter an empty, higher energy orbital by overcoming Δ

Thus, *the relative sizes of $E_{pairing}$ and Δ determine the occupancy of the d orbitals*, which determines the number of unpaired electrons and, thus, the magnetic behaviour of the ion.

As an example, the isolated Mn^{2+} ion ($[Ar]3d^5$) has five unpaired electrons of equal energy (Figure 24.22A). In an octahedral field of ligands, orbital occupancy is affected by the ligand in one of two ways:

- *Weak-field ligands and high-spin complexes.* Weak-field ligands, such as H_2O in $[Mn(H_2O)_6]^{2+}$, cause a *small* splitting energy, so *less* energy is needed for d electrons to jump to the e_g set and stay unpaired than to pair up in the t_{2g} set (Figure 24.22B). Thus, for weak-field ligands, $E_{pairing} > \Delta$. Therefore, *the number of unpaired electrons in a complex ion is the* ***same*** *as the number in the free ion*: a weak-field ligand creates a **high-spin complex**, which has the *maximum* number of unpaired electrons.
- *Strong-field ligands and low-spin complexes.* In contrast, strong-field ligands, such as CN^- in $[Mn(CN)_6]^{4-}$, cause a *large* splitting energy, so *more* energy is needed for electrons to jump to the e_g set than to pair up in the t_{2g} set (Figure 24.22C). Thus, for strong-field ligands, $E_{pairing} < \Delta$. Therefore, *the number of unpaired electrons in a complex ion is* ***less*** *than the number in the free ion*: a strong-field ligand creates a **low-spin complex**, which has *fewer* unpaired electrons.

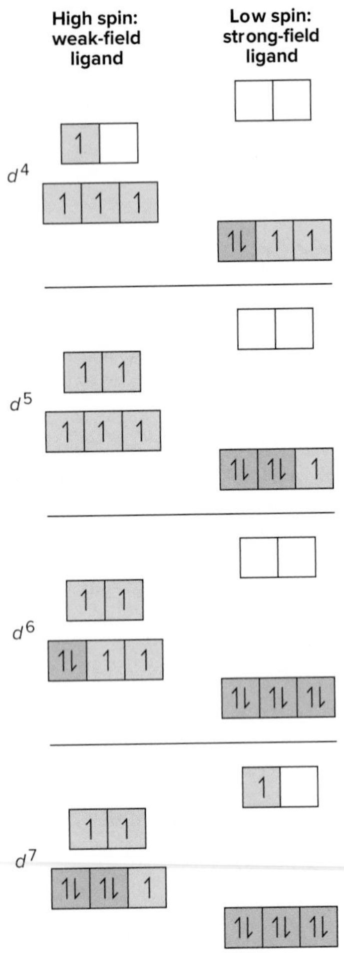

FIGURE 24.23 Orbital occupancy for high-spin and low-spin octahedral complexes of d^4 through d^7 metal ions

Orbital diagrams for d^1 through d^9 ions in octahedral complexes show that high-spin *and* low-spin options are possible only for d^4, d^5, d^6, and d^7 ions (Figure 24.23). With three t_{2g} orbitals available, d^1, d^2, and d^3 ions always form high-spin complexes because there is no need to pair up. Similarly, d^8 and d^9 ions always form high-spin complexes because the t_{2g} set is filled with six electrons, so the e_g orbitals *must* have either two (d^8) unpaired electrons or one (d^9) unpaired electron.

Sample Problem 24.6 Identifying High-Spin and Low-Spin Complex Ions

Problem Iron(II) forms a complex in hemoglobin. For each of the two octahedral complex ions $[Fe(H_2O)_6]^{2+}$ and $[Fe(CN)_6]^{4-}$, draw an energy diagram showing orbital splitting, predict the number of unpaired electrons, and identify the ion as low spin or high spin.

Plan The Fe^{2+} electron configuration shows the number of d electrons, and the spectrochemical series (Figure 24.21) shows the relative ligand strengths. We draw energy diagrams and separate the t_{2g} and e_g orbital sets more for the strong-field ligand. Then we add electrons, noting that a weak-field ligand gives the *maximum* number of unpaired electrons and a high-spin complex, whereas a strong-field ligand gives the *minimum* number of unpaired electrons and a low-spin complex.

Solution Fe^{2+} has the [Ar]$3d^6$ configuration. H_2O produces smaller splitting than CN^-. The energy diagrams are shown in the margin. The $[Fe(H_2O)_6]^{2+}$ ion has four unpaired electrons (high spin), and the $[Fe(CN)_6]^{4-}$ ion has no unpaired electrons (low spin).

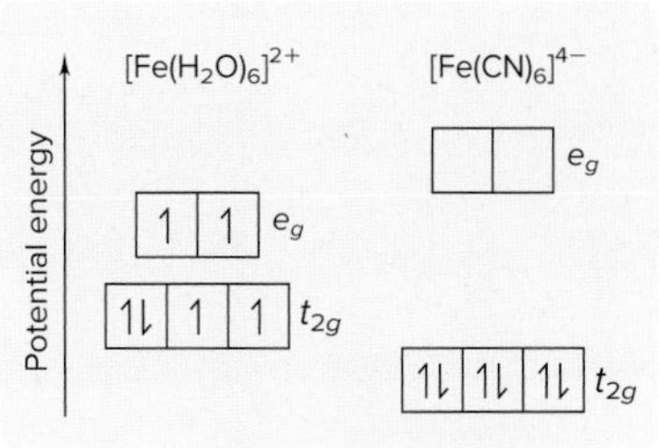

Comment **1.** H_2O is a weak-field ligand, so it forms high-spin complexes.
2. We cannot confidently predict the spin of a complex without having actual values for Δ and $E_{pairing}$.
3. Cyanide ions and carbon monoxide are toxic because they bind to the iron complexes in the proteins involved in cellular energy (Chemical Connections, Chapter 19).

Follow-Up Problem 24.6 How many unpaired electrons do you expect for $[Mn(CN)_6]^{3-}$? Is this a high-spin or a low-spin complex ion?

Dr. Deryn Fogg and her group at the University of Ottawa have focused on defining the parameters that limit the lifetimes of the dominant ruthenium catalysts used in olefin metathesis and using this insight to develop methodologies that will aid industrial uptake. They have recently discovered that electron-rich ruthenium complexes strongly bind and activate N_2, a finding that has major implications for worldwide energy use. As well, Dr. Fogg's group is studying Canadian "waste" biomass as a strategic resource that can be tapped via forefront methodologies in olefin metathesis.

Crystal Field Splitting in Tetrahedral and Square Planar Complexes Four ligands around a metal ion also cause *d*-orbital splitting, but the magnitude and pattern of the splitting depend on whether the ligands approach from a tetrahedral orientation or a square planar orientation.

1. *Tetrahedral complexes.* When the ligands approach from the corners of a tetrahedron, none of the five *d* orbitals are directly in their paths (Figure 24.24A). Thus, the overall attraction of ligand and metal ion is weaker. As a result, the splitting of *d*-orbital energies is *less* in a tetrahedral complex than it is in an octahedral complex with the same ligands:

$$\Delta_{tetrahedral} < \Delta_{octahedral}$$

Repulsions are minimized when the ligands approach the d_{xy}, d_{xz}, and d_{yz} orbitals closer than they approach the $d_{x^2-y^2}$ and d_{z^2} orbitals. This situation is the *opposite of the octahedral situation*, so the relative *d*-orbital energies are reversed: the d_{xy}, d_{xz}, and d_{yz} orbitals become *higher* in energy than the $d_{x^2-y^2}$ and d_{z^2} orbitals. *Only high-spin tetrahedral complexes are known* because the magnitude of Δ is so small.

2. *Square planar complexes.* The effects of the ligand field in the square planar orientation are easier to picture if we imagine starting with an octahedral geometry and then removing the two ligands along the *z* axis (Figure 24.24B). With no *z* axis interactions present, any *d* orbital with a *z* axis component has lower energy, with the d_{z^2} orbital decreasing the most, and the d_{xz} and d_{yz} orbitals also decreasing. In contrast, the two *d* orbitals in the *xy* plane interact strongly with the ligands, and because

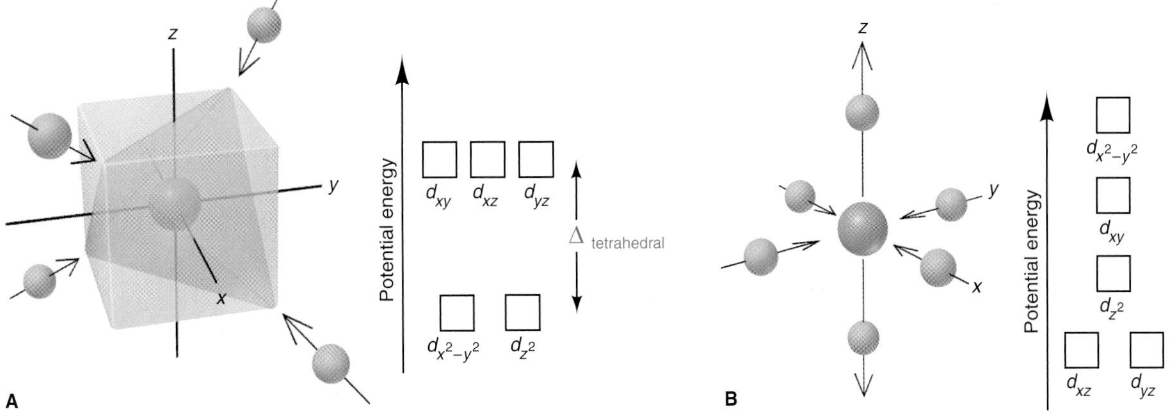

FIGURE 24.24 Splitting of *d*-orbital energies. A. The pattern of splitting by a tetrahedral field of ligands is the opposite of the octahedral pattern. **B.** Splitting by a square planar field of ligands decreases the energies of d_{xz}, d_{yz}, and especially d_{z^2} orbitals, relative to the octahedral pattern.

the $d_{x^2-y^2}$ orbital has its lobes *on* the axes, its energy is the highest. The d^8 metal ions, such as $[PdCl_4]^{2-}$, form square planar complexes. They are *low spin* and usually *diamagnetic* because the four pairs of *d* electrons fill the four lowest energy orbitals.

Molecular Orbital Theory

While VB theory provides an excellent explanation for structures of metal ligand complexes, it has not been able to satisfactorily explain the electronic transitions giving rise to the ultraviolet, visible, and infrared spectra observed for these complexes. The inability to account for the experimentally observed spectra was a shortcoming addressed by crystal field theory. Crystal field theory provided a theoretical explanation and allowed for calculation of values that matched either exactly or very closely the spectral lines observed experimentally in the complexes. The criticism that crystal field theory has faced is that it uses the approximation of point charges or point charges and point dipoles to obtain the spectral values. If a more realistic ionic model (in terms of the behaviour of the particular ion) is used, the model often provides less accurate values or fails entirely. Further, the spectrochemical series is almost impossible to understand in terms of an ionic model. Crystal field theory does not allow for covalency; however, there is an overwhelming amount of experimental evidence that suggests that covalent bonding is actually present within metal complexes.

Molecular orbital (MO) theory (first seen in Section 10.3) provides the ability to explain in better terms the bonding that is experimentally observed in metal ligand complexes as well as the spectra that are obtained experimentally. In addition, it explains the experimentally observed spectrochemical series, the magnitudes of Δ in the different complexes, the effects of different complex geometries (square planar > octahedral > tetrahedral), the effect of charge on the central metal atom, the effect of increasing *n* value in the *d* valence orbitals, and the structure and magnetic properties of the complexes.

As an example, if we look at the two complexes already studied using crystal field theory, since water is a weak-field ligand and CN^- is a strong-field ligand, we expect $Mn(H_2O)_6^{2+}$ to be a high-spin complex and $Mn(CN^-)_6^{4-}$ to be a low-spin complex. If we look at these two complexes using MO theory, a very simplistic view using only the sigma-bonding orbitals would give us energy level diagrams such as those in Figure 24.25.

Since water is a weak-field ligand, the electrons of Mn^{2+} remain unpaired as the lower energies of the MOs of water combine only to a limited extent with the MOs of the metal ion (Figure 24.25A). The MOs of the cyanide ion, however, are higher

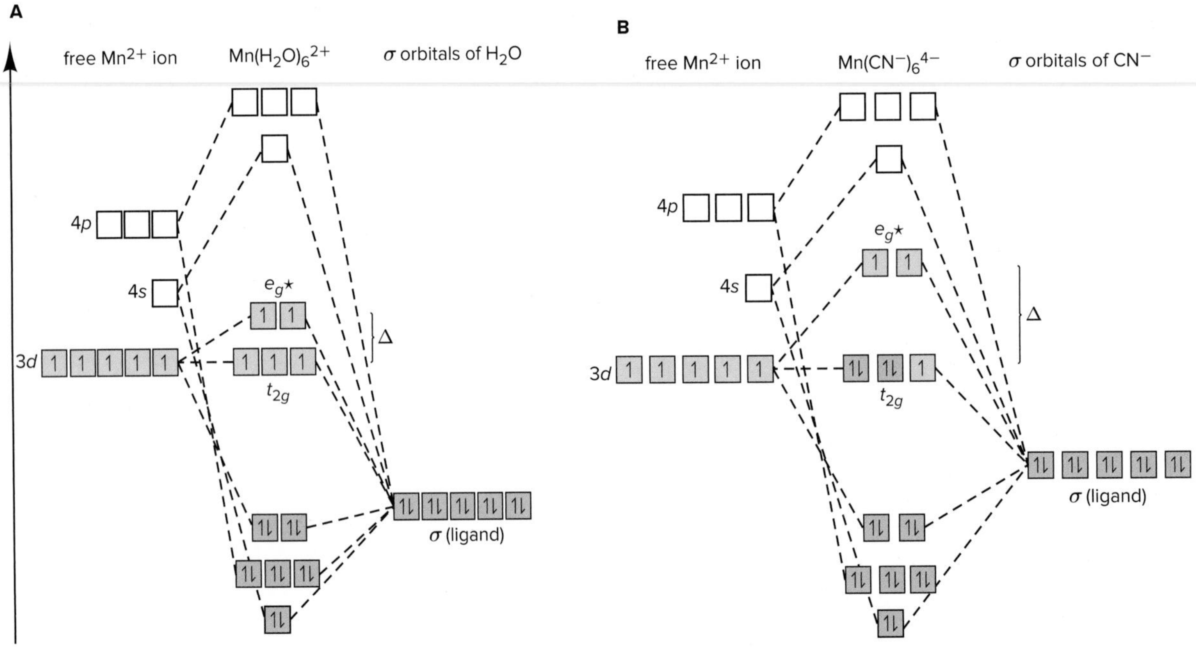

FIGURE 24.25 **A.** MO energy levels for $Mn(H_2O)_6^{2+}$. **B.** MO energy levels for $Mn(CN^-)_6^{4-}$.

in energy and can thus combine to a greater extent with the MOs of the metal ion, causing the splitting to be larger (Figure 24.25B).

The mathematics involved is more complicated due to the number of bonding and antibonding orbitals both on the central metal ion and on the ligands. For more information on this subject, consult the excellent paper by H. B. Gray, "Molecular orbital theory for transition metal complexes," *Journal of Chemical Education*, 41(1) (1964), pp. 2–11.

SUMMARY OF SECTION 24.4

- According to VB theory, complex ions have coordinate covalent bonds between ligands (Lewis bases) and metal ions (Lewis acids).
- Ligand lone pairs occupy hybridized metal-ion orbitals, leading to the characteristic shapes of complex ions.
- In crystal field theory, the surrounding field of ligands splits the metal ion's *d*-orbital energies. The crystal field splitting energy (Δ) depends on the charge of the metal ion and the crystal field strength of the ligands.
- Δ influences the colour (energy of the photons absorbed) and paramagnetism (number of unpaired *d* electrons). Strong-field ligands create a large Δ and produce low-spin complexes that absorb light of higher energy (shorter wavelength); the reverse is true of weak-field ligands.
- MO theory provides a more complete approach to the understanding of metal ligand complexes as it explains both the coordination and the spectra, in addition to other properties of these complexes.

CHAPTER REVIEW GUIDE

Learning Objectives

Relevant section (§) and/or sample problem (SP) numbers appear in parentheses.

Concepts

1. Describe the positions of the *d*-block and *f*-block elements and the general forms of their atomic and ionic electron configurations. (§24.1)
2. Explain how atomic size, ionization energy, and electronegativity vary across a period and down a group of transition elements and how these trends differ from those of the main-group elements; explain why the densities of period 6 transition elements are so high. (§24.1)
3. Describe why the transition elements often have multiple oxidation states and why the +2 state is common. (§24.1)
4. Demonstrate why metallic behaviour (the prevalence of ionic bonding and basic oxides) decreases as the oxidation state increases; show how valence-state electronegativity explains metal atoms in oxoanions. (§24.1)
5. Explain why many transition metal compounds are coloured and paramagnetic. (§24.1)
6. Explain the common +3 oxidation state of lanthanides and the similarity in their M^{3+} radii; explain the radioactivity of actinides. (§24.2)
7. Describe the coordination numbers, geometries, and ligand structures of complex ions. (§24.3)
8. Explain how coordination compounds are named and how their formulas are written. (§24.3)
9. Differentiate between the types of constitutional isomerism (coordination and linkage) and stereoisomerism (geometric isomers and enantiomers) of coordination compounds. (§24.3)
10. Describe how valence bond (VB) theory uses hybridization to account for the shapes of octahedral, square planar, and tetrahedral complexes. (§24.4)
11. Describe how crystal field theory explains that approaching ligands cause *d*-orbital energies to split. (§24.4)
12. Show how the relative crystal field strength of ligands (spectrochemical series) affects the *d*-orbital splitting energy (Δ). (§24.4)
13. Explain how the magnitude of Δ accounts for the energy of light absorbed and, thus, the colour of a complex. (§24.4)
14. Describe how the relative sizes of pairing energy and Δ determine the occupancy of *d* orbitals and, thus, the magnetic properties of complexes. (§24.4)
15. Show how *d*-orbital splitting in tetrahedral and square planar complexes differs from *d*-orbital splitting in octahedral complexes. (§24.4)
16. Explain why molecular orbital (MO) theory is better at explaining the electronic structure of metal ligand complexes than hybridization or crystal field theory.

Skills

1. Write electron configurations of transition metal atoms and ions. (SP 24.1)
2. Use a partial orbital diagram to determine the number of unpaired electrons in a transition metal atom or ion. (SP 24.2)
3. Recognize the structural components of complex ions. (§24.3)
4. Name and write the formulas for coordination compounds. (SP 24.3)

5. Determine the type of stereoisomerism in complexes. (SP 24.4)
6. Correlate the shape of a complex ion with the number and type of hybrid orbitals of the central metal ion. (§24.4)
7. Use the spectrochemical series to rank complex ions in terms of Δ and the energy of light absorbed. (SP 24.5)
8. Use the spectrochemical series to determine if a complex ion is high-spin or low-spin. (SP 24.6)

Key Terms

Introduction
transition metals

Section 24.1
lanthanide contraction

Section 24.2
lanthanides
actinides
inner transition elements

Section 24.3
coordination compounds
complex ion
ligands
counterions
coordination number
donor atoms
chelate
isomers
constitutional (structural) isomers
coordination isomers
linkage isomers
stereoisomers
geometric (*cis-trans*) isomers
enantiomers

Section 24.4
coordinate covalent bond
crystal field theory
e_g orbitals
t_{2g} orbitals
crystal field splitting energy (Δ)
strong-field ligands
weak-field ligands
spectrochemical series
high-spin complex
low-spin complex

Brief Solutions to Follow-Up Problems

24.1 (a) Ag^+: $4d^{10}$
(b) Cd^{2+}: $4d^{10}$
(c) Ir^{3+}: $5d^6$

24.2 The Er^{3+} ion has three unpaired electrons. Er^{3+} is $[Xe]4f^{11}$:

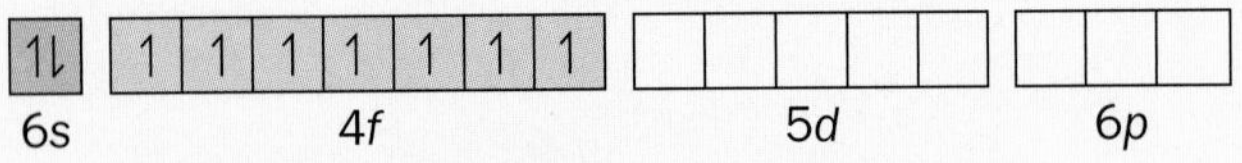

24.3 (a) Pentaaquabromochromium(III) chloride
(b) $Ba_3[Co(CN)_6]_2$

24.4 Two sets of *cis-trans* isomers are possible, and the two *cis* isomers are enantiomers:

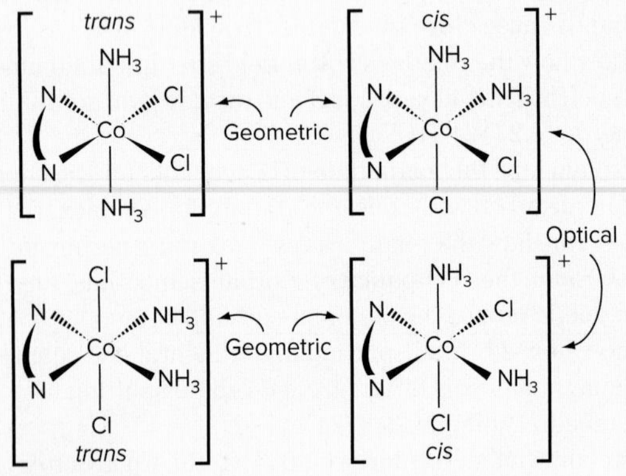

24.5 Both metal ions are V^{3+}; in terms of ligand field energy, $NH_3 > H_2O$, so $[V(NH_3)_6]^{3+}$ absorbs visible light of higher energy.

24.6 The metal ion is Mn^{3+}: $[Ar]3d^4$.

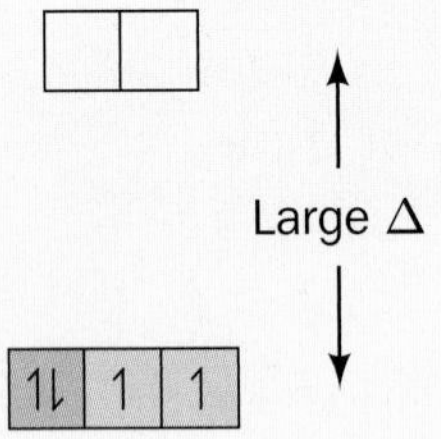

There are two unpaired *d* electrons, so this is a low-spin complex ion.

PROBLEMS

Problems with **red** numbers are answered in Appendix G and worked in detail in the Student Solutions Manual. Problem sections match those in this book and provide the numbers of relevant sample problems. Most offer Concept Review Questions, Skill-Building Exercises (grouped in pairs covering the same concept), and Problems in Context. The Comprehensive Problems are based on material from any section or previous chapter. **Note**: In these problems, the term *electron configuration* refers to the valence shell ground-state electron configuration.

Properties of the Transition Elements
(Sample Problem 24.1)

Concept Review Questions

24.1 How is the *n* value of the *d* sublevel of a transition element related to the period number of the element?

24.2 Write the general electron configuration of a transition element in (a) period 5; (b) period 6.

24.3 (a) What is the general rule for the order in which electrons are removed from a transition metal atom to form an ion? Give an example from group 5.
(b) Name two types of measurements that are used to study the electron configurations of ions.

24.4 (a) What is the maximum number of unpaired *d* electrons that an atom or ion can possess? (b) Give an example of an atom and an ion that have this number of unpaired *d* electrons.

24.5 How does the variation in atomic size across a transition series contrast with the variation across the main-group elements of the same period? Why?

24.6 (a) What is the lanthanide contraction?
(b) How does the lanthanide contraction affect atomic size down a group of transition elements?
(c) How does the lanthanide contraction influence the densities of the period 6 transition elements?

24.7 (a) What is the range in electronegativity across the first (3*d*) transition series?
(b) What is the range in electronegativity across period 4 of main-group elements?
(c) Explain the difference in these ranges of electronegativity.

24.8 (a) Explain the major difference between the number of oxidation states of most transition elements and the number of oxidation states of most main-group elements. (b) Why is the +2 oxidation state so common among transition elements? (c) What is valence-state electronegativity? Is the electronegativity of Cr different in CrO, Cr_2O_3, and CrO_3? Explain.

24.9 (a) What behaviour distinguishes paramagnetic and diamagnetic substances? (b) Why are paramagnetic ions common among transition elements, but not among main-group elements? (c) Why are coloured solutions of metal ions common among transition elements, but not among main-group elements?

Skill-Building Exercises (grouped in similar pairs)

24.10 Using the periodic table to locate each element, write the electron configuration of (a) V; (b) Y; (c) Hg.

24.11 Using the periodic table to locate each element, write the electron configuration of (a) Ru; (b) Cu; (c) Ni.

24.12 Using the periodic table to locate each element, write the electron configuration of (a) Os; (b) Co; (c) Ag.

24.13 Using the periodic table to locate each element, write the electron configuration of (a) Zn; (b) Mn; (c) Re.

24.14 Give the electron configuration and the number of unpaired electrons for (a) Sc^{3+}; (b) Cu^{2+}; (c) Fe^{3+}; (d) Nb^{3+}.

24.15 Give the electron configuration and the number of unpaired electrons for (a) Cr^{3+}; (b) Ti^{4+}; (c) Co^{3+}; (d) Ta^{2+}.

24.16 What is the highest oxidation state for (a) Ta; (b) Zr; (c) Mn?

24.17 What is the highest oxidation state for (a) Nb; (b) Y; (c) Tc?

24.18 Which transition metals have a maximum oxidation number of +6?

24.19 Which transition metals have a maximum oxidation number of +4?

24.20 In which compound does Cr exhibit greater metallic behaviour: CrF_2 or CrF_6? Explain.

24.21 VF_5 is a liquid that boils at 48°C, whereas VF_3 is a solid that melts above 800°C. Explain this difference in properties.

24.22 Is it more difficult to oxidize Cr or Mo? Explain.

24.23 Is MnO_4^- or ReO_4^- a stronger oxidizing agent? Explain.

24.24 Which oxide, CrO_3 or CrO, is more acidic in water? Why?

24.25 Which oxide, Mn_2O_3 or Mn_2O_7, is more basic in water? Why?

Problem in Context

24.26 The green patina of Cu-alloy roofs results from corrosion in the presence of O_2, H_2O, CO_2, and sulfur compounds. The other members of group 11, Ag and Au, do not form a patina. Corrosion of Cu and Ag in the presence of sulfur compounds leads to a black tarnish, but Au does not tarnish. This pattern is different from the pattern in group 1, where ease of oxidation *increases* down the group. Explain these different group patterns.

The Inner Transition Elements

(Sample Problem 24.2)

Concept Review Questions

24.27 What atomic property of the lanthanides leads to their remarkably similar chemical properties?

24.28 (a) What is the maximum number of unpaired electrons in a lanthanide ion?
(b) How does this number relate to occupancy of the 4*f* sublevel?

24.29 Which of the actinides are radioactive?

Skill-Building Exercises (grouped in similar pairs)

24.30 Give the valence shell electron configuration of (a) La; (b) Ce^{3+}; (c) Es; (d) U^{4+}.

24.31 Give the valence shell electron configuration of (a) Pm; (b) Lu^{3+}; (c) Th; (d) Fm^{3+}.

Problems in Context

24.32 Only a few lanthanides show an oxidation state other than +3. Two of these, europium (Eu) and terbium (Tb), are found near the middle of the series, and their unusual oxidation states can be associated with a half-filled *f* sublevel.
(a) Write the valence shell electron configurations of Eu^{2+}, Eu^{3+}, and Eu^{4+}. Why is Eu^{2+} a common ion, whereas Eu^{4+} is unknown?
(b) Write the valence shell electron configurations of Tb^{2+}, Tb^{3+}, and Tb^{4+}. Would you expect Tb to show a +2 or a +4 oxidation state? Explain.

24.33 Cerium (Ce) and ytterbium (Yb) exhibit some oxidation states in addition to +3.
(a) Write the valence shell electron configurations of Ce^{2+}, Ce^{3+}, and Ce^{4+}.
(b) Write the valence shell electron configurations of Yb^{2+}, Yb^{3+}, and Yb^{4+}.
(c) In addition to the +3 ions, the ions Ce^{4+} and Yb^{2+} are stable. Suggest a reason for this stability.

24.34 Which lanthanide has the maximum number of unpaired electrons in both its atom and its +3 ion? Give the number of unpaired electrons in the atom and ion.

Coordination Compounds

(Sample Problems 24.3 and 24.4)

Concept Review Questions

24.35 Describe the makeup of a complex ion, including the nature of the ligands and their interaction with the central metal ion. Explain how a complex ion can be positive or negative and how it occurs as part of a neutral coordination compound.

24.36 What electronic feature must a donor atom of a ligand have?

24.37 What is the coordination number of a metal ion in a complex ion? How does it differ from the oxidation number?

24.38 What structural feature is characteristic of a chelate?

24.39 What geometries are associated with the coordination numbers 2, 4, and 6?

24.40 What are the coordination numbers of (a) cobalt(III), (b) platinum(II), and (c) platinum(IV) in complexes?

24.41 How is a complex ion a Lewis adduct?

24.42 What does the ending *-ate* signify in the name of a complex ion?

24.43 In what order are the metal ion and ligands given in the name of a complex ion?

24.44 Is a linkage isomer a type of constitutional isomer or stereoisomer? Explain.

Skill-Building Exercises (grouped in similar pairs)

24.45 Give the systematic name of (a) $[Ni(H_2O)_6]Cl_2$; (b) $[Cr(en)_3](ClO_4)_3$; (c) $K_4[Mn(CN)_6]$.

24.46 Give the systematic name of (a) $[Co(NH_3)_4(NO_2)_2]Cl$; (b) $[Cr(NH_3)_6][Cr(CN)_6]$; (c) $K_2[CuCl_4]$.

24.47 What are the charge and coordination number of the central metal ion(s) in each compound in Problem 24.45?

24.48 What are the charge and coordination number of the central metal ion(s) in each compound in Problem 24.46?

24.49 Give the systematic name of (a) $K[Ag(CN)_2]$; (b) $Na_2[CdCl_4]$; (c) $[Co(NH_3)_4(H_2O)Br]Br_2$.

24.50 Give the systematic name of (a) $K[Pt(NH_3)Cl_5]$; (b) $[Cu(en)(NH_3)_2][Co(en)Cl_4]$; (c) $[Pt(en)_2Br_2](ClO_4)_2$.

24.51 What are the charge and coordination number of the central metal ion(s) in each compound in Problem 24.49?

24.52 What are the charge and coordination number of the central metal ion(s) in each compound in Problem 24.50?

24.53 Give the formula for (a) tetraamminezinc sulfate; (b) pentaamminechlorochromium(III) chloride; (c) sodium bis(thiosulfato)argentate(I).

24.54 Give the formula for (a) dibromobis(ethylenediamine) cobalt(III) sulfate; (b) hexaamminechromium(III) tetrachlorocuprate(II); (c) potassium hexacyanoferrate(II).

24.55 What is the coordination number of the metal ion and the number of individual ions per formula unit in each compound in Problem 24.53?

24.56 What is the coordination number of the metal ion and the number of individual ions per formula unit in each compound in Problem 24.54?

24.57 Give the formula for (a) hexaaquachromium(III) sulfate; (b) barium tetrabromoferrate(III); (c) bis(ethylenediamine) platinum(II) carbonate.

24.58 Give the formula for (a) potassium tris(oxalato)chromate(III); (b) tris(ethylenediamine)cobalt(III) pentacyanoiodomanganate(II); (c) diamminediaquabromochloroaluminum nitrate.

24.59 Give the coordination number of the metal ion and the number of ions per formula unit in each compound in Problem 24.57.

24.60 Give the coordination number of the metal ion and the number of ions per formula unit in each compound in Problem 24.58.

24.61 Which of the following ligands can participate in linkage isomerism? Explain with Lewis structures.
(a) NO_2^- (b) SO_2 (c) NO_3^-

24.62 Which of the following ligands can participate in linkage isomerism? Explain with Lewis structures.
(a) SCN^- (b) $S_2O_3^{2-}$ (thiosulfate) (c) HS^-

24.63 If any of the following complexes can exist as isomers, state the type of isomerism and draw the structures:
(a) $[Pt(CH_3NH_2)_2Br_2]$ (b) $[Pt(NH_3)_2FCl]$ (c) $[Pt(H_2O)(NH_3)FCl]$

24.64 If any of the following complexes can exist as isomers, state the type of isomerism and draw the structures:
(a) $[Zn(en)F_2]$ (b) $[Zn(H_2O)(NH_3)FCl]$ (c) $[Pd(CN)_2(OH)_2]^{2-}$

24.65 If any of the following complexes can exist as isomers, state the type of isomerism and draw the structures:
(a) $[PtCl_2Br_2]^{2-}$ (b) $[Cr(NH_3)_5(NO_2)]^{2+}$ (c) $[Pt(NH_3)_4I_2]^{2+}$

24.66 If any of the following complexes can exist as isomers, state the type of isomerism and draw the structures:
(a) $[Co(NH_3)_5Cl]Br_2$ (b) $[Pt(CH_3NH_2)_3Cl]Br$
(c) $[Fe(H_2O)_4(NH_3)_2]^{2+}$

Problems in Context

24.67 Chromium(III), like cobalt(III), has a coordination number of 6 in many of its complex ions. Before Alfred Werner, in the 1890s, established the idea of a complex ion, coordination compounds had traditional formulas. Compounds with the traditional formula $CrCl_3 \cdot nNH_3$, where $n = 3$ to 6, are known. Which of these compounds has an electrical conductivity in aqueous solution similar to that of an equimolar NaCl solution?

24.68 When $MCl_4(NH_3)_2$ is dissolved in water and treated with $AgNO_3$, 2 mol of AgCl precipitates immediately for each mole of $MCl_4(NH_3)_2$. Give the coordination number of M in the complex.

24.69 Palladium, like its group neighbour platinum, forms four-coordinate Pd(II) and six-coordinate Pd(IV) complexes. Write formulas for the complexes with these compositions:
(a) $PdK(NH_3)Cl_3$ (b) $PdCl_2(NH_3)_2$
(c) PdK_2Cl_6 (d) $Pd(NH_3)_4Cl_4$

Theoretical Basis for the Bonding and Properties of Complexes
(Sample Problems 24.5 and 24.6)

Concept Review Questions

24.70 (a) What is a coordinate covalent bond?
(b) Is it involved when $FeCl_3$ dissolves in water? Explain.
(c) Is it involved when HCl gas dissolves in water? Explain.

24.71 According to VB theory, what set of orbitals is used by a period 4 metal ion when forming (a) a square planar complex; (b) a tetrahedral complex?

24.72 A metal ion uses d^2sp^3 orbitals when forming a complex. What is its coordination number and the shape of the complex?

24.73 A complex in solution absorbs green light. What is the colour of the solution?

24.74 In terms of the theory of colour absorption, explain two ways that a solution can be blue.

24.75 (a) What is the crystal field splitting energy (Δ)?
(b) How does Δ arise for an octahedral field of ligands?
(c) How is Δ different for a tetrahedral field of ligands?

24.76 What is the distinction between a weak-field ligand and a strong-field ligand? Give an example of each.

24.77 Is a complex with the same number of unpaired electrons as the free gaseous metal ion a high-spin or a low-spin complex?

24.78 How do the relative magnitudes of $E_{pairing}$ and Δ affect the paramagnetism of a complex?

24.79 Why are there both high-spin and low-spin octahedral complexes, but only high-spin tetrahedral complexes?

Skill-Building Exercises (grouped in similar pairs)

24.80 Give the number of d electrons (n of d^n) for the central metal ion in (a) $[TiCl_6]^{2-}$; (b) $K[AuCl_4]$; (c) $[RhCl_6]^{3-}$.

24.81 Give the number of d electrons (n of d^n) for the central metal ion in (a) $[Cr(H_2O)_6](ClO_3)_2$; (b) $[Mn(CN)_6]^{2-}$; (c) $[Ru(NO)(en)_2Cl]Br$.

24.82 How many d electrons (n of d^n) are in the central metal ion in (a) $Ca[IrF_6]$; (b) $[HgI_4]^{2-}$; (c) $[Co(EDTA)]^{2-}$?

24.83 How many d electrons (n of d^n) are in the central metal ion in (a) $[Ru(NH_3)_5Cl]SO_4$; (b) $Na_2[Os(CN)_6]$; (c) $[Co(NH_3)_4CO_3I]$?

24.84 Which of these ions *cannot* form both high-spin and low-spin octahedral complexes: Ti^{3+}, Co^{2+}, Fe^{2+}, Cu^{2+}?

24.85 Which of these ions *cannot* form both high-spin and low-spin octahedral complexes: Mn^{3+}, Nb^{3+}, Ru^{3+}, Ni^{2+}?

24.86 Draw orbital-energy splitting diagrams and use the spectrochemical series to show the orbital occupancy for each complex (assuming that H_2O is a weak-field ligand):
(a) $[Cr(H_2O)_6]^{3+}$ (b) $[Cu(H_2O)_4]^{2+}$ (c) $[FeF_6]^{3-}$

24.87 Draw orbital-energy splitting diagrams and use the spectrochemical series to show the orbital occupancy for each complex (assuming that H_2O is a weak-field ligand):
(a) $[Cr(CN)_6]^{3-}$ (b) $[Rh(CO)_6]^{3+}$ (c) $[Co(OH)_6]^{4-}$

24.88 Draw orbital-energy splitting diagrams and use the spectrochemical series to show the orbital occupancy for each complex (assuming that H_2O is a weak-field ligand):
(a) $[MoCl_6]^{3-}$ (b) $[Ni(H_2O)_6]^{2+}$ (c) $[Ni(CN)_4]^{2-}$

24.89 Draw orbital-energy splitting diagrams and use the spectrochemical series to show the orbital occupancy for each complex (assuming that H_2O is a weak-field ligand):
(a) $[Fe(C_2O_4)_3]^{3-}$ ($C_2O_4^{2-}$ creates a weaker field than H_2O does.)
(b) $[Co(CN)_6]^{4-}$
(c) $[MnCl_6]^{4-}$

24.90 Rank the following complexes in order of *increasing* Δ and the energy of light absorbed: $[Cr(NH_3)_6]^{3+}$, $[Cr(H_2O)_6]^{3+}$, $[Cr(NO_2)_6]^{3-}$.

24.91 Rank the following complexes in order of *decreasing* Δ and the energy of light absorbed: $[Cr(en)_3]^{3+}$, $[Cr(CN)_6]^{3-}$, $[CrCl_6]^{3-}$.

Problems in Context

24.92 Explain the splitting and the relative energies of the d_{xy} and the $d_{x^2-y^2}$ orbitals by sketching the orientation of the orbitals relative to the ligands in an octahedral complex.

24.93 The two e_g orbitals are identical in energy in an octahedral complex but have different energies in a square planar complex, with the d_{z^2} orbital being much lower in energy than the $d_{x^2-y^2}$ orbital. Explain, using orbital sketches.

24.94 A complex, ML_6^{2+}, is violet. The same metal forms a complex with another ligand, Q, that creates a weaker field. What colour might MQ_6^{2+} be expected to show? Explain.

24.95 $[Cr(H_2O)_6]^{2+}$ is violet. Another CrL_6 complex is green. Can ligand L be CN^-? Can it be Cl^-? Explain.

24.96 Octahedral $[Ni(NH_3)_6]^{2+}$ is paramagnetic, whereas planar $[Pt(NH_3)_4]^{2+}$ is diamagnetic, even though both metal ions are d^8 species. Explain.

24.97 The hexaaqua complex $[Ni(H_2O)_6]^{2+}$ is green, whereas the hexaammonia complex $[Ni(NH_3)_6]^{2+}$ is violet. Explain.

24.98 Three of the complex ions that are formed by Co^{3+} are $[Co(H_2O)_6]^{3+}$, $[Co(NH_3)_6]^{3+}$, and $[CoF_6]^{3-}$. These ions have the observed colours (listed in random order) yellow-orange, green, and blue. Match each complex with its colour. Explain.

Comprehensive Problems

24.99 When neptunium (Np) and plutonium (Pu) were discovered, the periodic table did not include the actinides, so these elements were placed in groups 7 and 8. When americium (Am) and curium (Cm) were synthesized, they were placed in groups 9 and 10. However, during chemical isolation procedures, Glenn Seaborg and his colleagues, who had synthesized these elements, could not find their compounds among other compounds of members of the same groups, which led Seaborg to suggest that they were part of a new inner transition series.
(a) How do the electron configurations of these elements support Seaborg's suggestion?
(b) The highest fluorides of Np and Pu are hexafluorides, as is the highest fluoride of uranium. How does this chemical evidence support the placement of Np and Pu as *inner* transition elements, rather than transition elements?

24.100 How many different formulas are there for octahedral complexes with a metal, M, and four ligands, A, B, C, and D? Give the number of isomers for each formula, and describe the isomers.

24.101 At one time, it was common to write the formula for copper(I) chloride as Cu_2Cl_2, instead of CuCl, similar to Hg_2Cl_2 for mercury(I) chloride. Use electron configurations to explain why Hg_2Cl_2 and CuCl are both correct.

24.102 For the compound $[Co(en)_2Cl_2]Cl$, give the following:
(a) The coordination number of the metal ion
(b) The oxidation number of the central metal ion
(c) The number of individual ions per formula unit
(d) The amount (mol) of AgCl that precipitates when 1 mol of compound is dissolved in water and treated with $AgNO_3$

24.103 Hexafluorocobaltate(III) ion is a high-spin complex. Draw the orbital-energy splitting diagram for its d orbitals.

24.104 A salt of each of the ions in Table 24.3 is dissolved in water. A Pt electrode is immersed in each solution and connected to a 0.38 V battery. All of the electrolytic cells are run for the same amount of time with the same current.
(a) In which cell(s) will a metal plate out? Explain.
(b) Which cell will plate out the least mass of metal? Explain.

24.105 Criticize and correct the following statement: Strong-field ligands always give rise to low-spin complexes.

24.106 Two major bidentate ligands that are used in analytical chemistry are bipyridyl (bipy) and *ortho*-phenanthroline (*o*-phen):

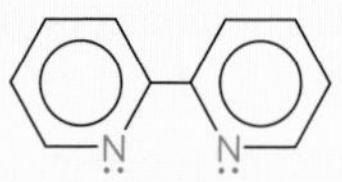

Bipyridyl

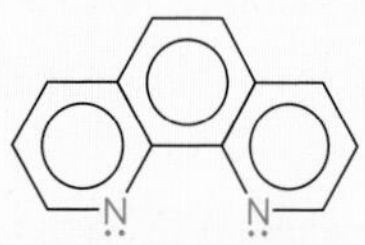

o-phenanthroline

Draw structures and discuss the possibility of isomers for (a) $[Pt(bipy)Cl_2]$; (b) $[Fe(o\text{-phen})_3]^{3+}$; (c) $[Co(bipy)_2F_2]^+$; (d) $[Co(o\text{-phen})(NH_3)_3Cl]^{2+}$.

24.107 In many species, a transition metal has an unusually high or low oxidation state. Write a balanced equation for each reaction, and find the oxidation state of the transition metal product:
(a) Iron(III) ion reacts with hypochlorite ion in basic solution to form ferrate ion (FeO_4^{2-}), Cl^-, and water.
(b) Potassium hexacyanomanganate(II) reacts with K metal to form $K_6[Mn(CN)_6]$.
(c) Heating sodium superoxide (NaO_2) with Co_3O_4 produces Na_4CoO_4 and O_2 gas.
(d) Vanadium(III) chloride reacts with Na metal under a CO atmosphere to produce $Na[V(CO)_6]$ and NaCl.
(e) Barium peroxide reacts with nickel(II) ions in basic solution to produce $BaNiO_3$.
(f) Bubbling CO through a basic solution of cobalt(II) ion produces $[Co(CO)_4]^-$, CO_3^{2-}, and water.
(g) Heating cesium tetrafluorocuprate(II) with F_2 gas under pressure gives Cs_2CuF_6.
(h) Heating tantalum(V) chloride with Na metal produces NaCl and Ta_6Cl_{15}, in which half of the Ta is in the +2 state.
(i) Potassium tetracyanonickelate(II) reacts with hydrazine (N_2H_4) in a basic solution to form $K_4[Ni_2(CN)_6]$ and N_2 gas.

24.108 Draw a Lewis structure with lowest formal charges for MnO_4^-.

24.109 The coordination compound $[Pt(NH_3)_2(SCN)_2]$ displays two types of isomerism. Name the types, and give the names and structures for the six possible isomers.

24.110 An octahedral complex with three different ligands (A, B, and C) can have formulas with three different ratios of the ligands. For each example, give the name, state the type(s) of isomerism present, and draw all the isomers:
(a) $[MA_4BC]^{n+}$, such as $[Co(NH_3)_4(H_2O)Cl]^{2+}$
(b) $[MA_3B_2C]^{n+}$, such as $[Cr(H_2O)_3Br_2Cl]$
(c) $[MA_2B_2C_2]^{n+}$, such as $[Cr(NH_3)_2(H_2O)_2Br_2]^+$

24.111 In $[Cr(NH_3)_6]Cl_3$, the $[Cr(NH_3)_6]^{3+}$ ion absorbs visible light in the blue-violet range, and the compound is yellow-orange. In $[Cr(H_2O)_6]Br_3$, the $[Cr(H_2O)_6]^{3+}$ ion absorbs visible light in the red range, and the compound is blue-grey. Explain these differences in light absorbed and the colour of the compound.

24.112 The actinides Pa, U, and Np form a series of complex ions, such as the anion in the compound $Na_3[UF_8]$, in which the central metal ion has an unusual geometry and oxidation state. In the crystal structure, the complex ion can be pictured as resulting from the interpenetration of simple cubic arrays of uranium and fluoride ions.
(a) What is the coordination number of the metal ion in the complex ion?
(b) What is the oxidation state of uranium in the compound?
(c) Sketch the complex ion.

24.113 Consider the square planar complex shown at the right. Which of structures A to F are geometric isomers of it?

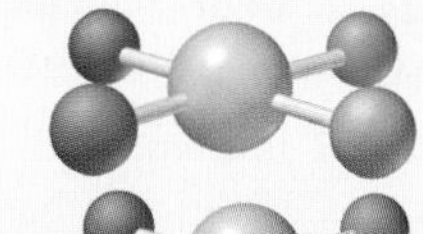

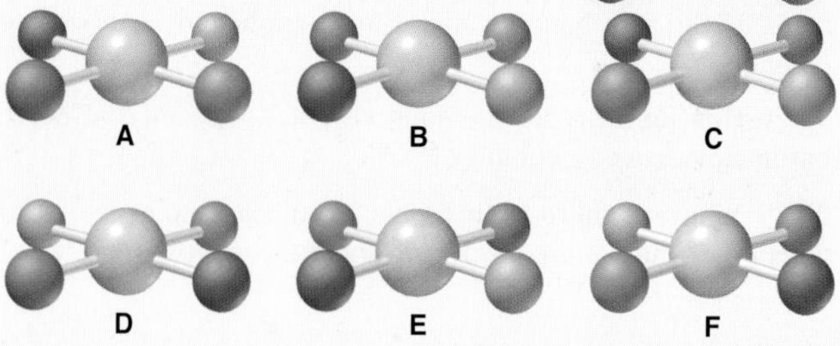

24.114 Several coordination isomers, with both Co and Cr as +3 ions, have the molecular formula $CoCrC_6H_{18}N_{12}$.
(a) Give the name and formula for the isomer in which the Co complex ion has six NH_3 groups.
(b) Give the name and formula for the isomer in which the Co complex ion has one CN group and five NH_3 groups.

24.115 A shortcut to finding enantiomers is to see if the complex has a *plane of symmetry*: a plane passing through the metal atom such that every atom on one side of the plane is matched by an identical atom at the same distance from the plane on the other side. Any planar complex has a plane of symmetry, since all atoms lie in one plane. Use this approach to determine whether the following complexes exist as enantiomers:
(a) $[Zn(NH_3)_2Cl_2]$ (tetrahedral) (b) $[Pt(en)_2]^{2+}$
(c) *trans*-$[PtBr_4Cl_2]^{2-}$ (d) *trans*-$[Co(en)_2F_2]^+$
(e) *cis*-$[Co(en)_2F_2]^+$

24.116 Alfred Werner (see Problem 24.67) prepared two compounds by heating a solution of $PtCl_2$ with triethyl phosphine, $P(C_2H_5)_3$, which is a ligand for Pt. Both compounds have, by mass, the following elements: Pt, 38.8%; Cl, 14.1%; C, 28.7%; P, 12.4%; and H, 6.02%. Write formulas, structures, and systematic names for the two isomers.

24.117 Some octahedral complexes have distorted shapes. In some, two metal-ligand bonds that are 180° apart are shorter than the other four bonds. In $[Cu(NH_3)_6]^{2+}$, for example, two Cu—N bonds are 207 pm long, and the other four bonds are 262 pm long.
(a) Calculate the longest distance between two N atoms in this complex.
(b) Calculate the shortest distance between two N atoms.

24.118 The effect of entropy on reactions appears in the stabilities of certain complexes.
(a) In terms of number of product particles, predict which of the following will be favoured in terms of $\Delta_r S°$:

$$[Cu(NH_3)_4]^{2+}(aq) + 4H_2O(l) \longrightarrow [Cu(H_2O)_4]^{2+}(aq) + 4NH_3(aq)$$

$$[Cu(H_2NCH_2CH_2NH_2)_2]^{2+}(aq) + 4H_2O(l) \longrightarrow [Cu(H_2O)_4]^{2+}(aq) + 2en(aq)$$

(b) Given that the Cu—N bond strength is approximately the same in both complexes, which complex will be more stable with respect to ligand exchange in water? Explain.

24.119 You know the following information about a coordination compound:
(1) The partial empirical formula is $KM(CrO_4)Cl_2(NH_3)_4$.
(2) The compound has A (red) and B (blue) crystal forms.
(3) When 1.0 mol of A or B reacts with 1.0 mol of $AgNO_3$, 0.50 mol of a red precipitate forms immediately.

(4) After the reaction in (3), 1.0 mol of A reacts very slowly with 1.0 mol of silver oxalate ($Ag_2C_2O_4$) to form 2.0 mol of a white precipitate. (Oxalate can displace other ligands.)
(5) After the reaction in (3), 1.0 mol of B does not react further with 1.0 mol of $AgNO_3$.
From this information, determine the following:
(a) The coordination number of M
(b) The group(s) bonded to M ionically and covalently
(c) The stereochemistry of the red and blue forms

24.120 The extent of crystal field splitting is often determined from spectra.
(a) Given the wavelength (λ) of maximum absorption, find the crystal field splitting energy (Δ), in kJ/mol, for each complex ion:

Ion	λ (nm)	Ion	λ (nm)
(i) $[Cr(H_2O)_6]^{3+}$	562	(vi) $[Fe(H_2O)_6]^{2+}$	966
(ii) $[Cr(CN)_6]^{3-}$	381	(vii) $[Fe(H_2O)_6]^{3+}$	730
(iii) $[CrCl_6]^{3-}$	735	(viii) $[Co(NH_3)_6]^{3+}$	405
(iv) $[Cr(NH_3)_6]^{3+}$	462	(ix) $[Rh(NH_3)_6]^{3+}$	295
(v) $[Ir(NH_3)_6]^{3+}$	244		

(b) Write a spectrochemical series for the ligands in Cr complexes.

(c) Use the Fe data to describe how the oxidation state affects Δ.

(d) Use the Co, Rh, and Ir data to state how the period number affects Δ.

24.121 Ionic liquids have many applications in engineering and materials science. The dissolution of the metavanadate ion in chloroaluminate ionic liquids has been studied:

$$VO_3^- + AlCl_4^- \longrightarrow VO_2Cl_2^- + AlOCl_2^-$$

(a) What is the oxidation number of V and Al in each ion?
(b) In reactions of V_2O_5 with HCl, the acid concentration affects the product. At low acid concentration, $VO_2Cl_2^-$ and VO_3^- form:

$$V_2O_5 + HCl \longrightarrow VO_2Cl_2^- + VO_3^- + H^+$$

At high acid concentration, $VOCl_3$ forms:

$$V_2O_5 + HCl \longrightarrow VOCl_3 + H_2O$$

Balance each equation, and state which, if either, is a redox process.
(c) What mass of $VO_2Cl_2^-$ or $VOCl_3$ can form from 12.5 g of V_2O_5 and the appropriate concentration of acid?

24.122 The orbital occupancies for the *d* orbitals of several complex ions are shown below.

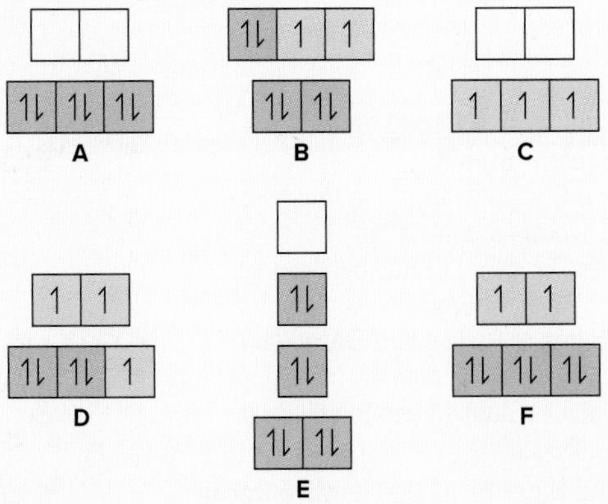

(a) Which diagram corresponds to the orbital occupancy of the cobalt ion in $[Co(CN)_6]^{3-}$?
(b) If diagram D depicts the orbital occupancy of the cobalt ion in $[CoF_6]^n$, what is the value of n?
(c) $[NiCl_4]^{2-}$ is paramagnetic, and $[Ni(CN)_4]^{2-}$ is diamagnetic. Which diagrams correspond to the orbital occupancies of the nickel ions in these species?
(d) Diagram C shows the orbital occupancy of V^{2+} in the octahedral complex VL_6. Can you determine whether L is a strong-field or a weak-field ligand? Explain.

Nuclear Reactions and Their Applications

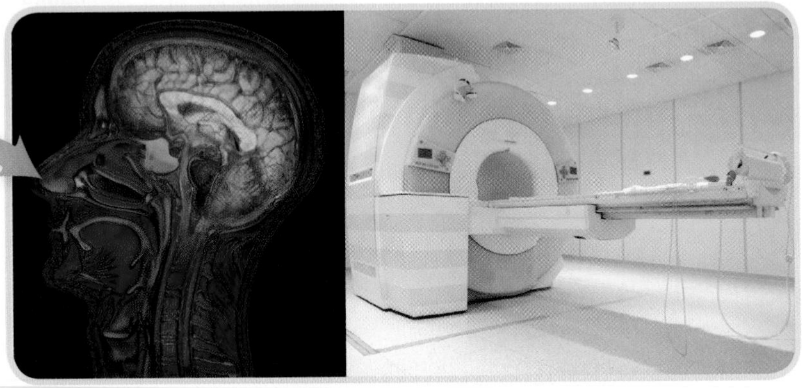

IN THIS CHAPTER . . . We survey the field of nuclear chemistry, examining radioactive nuclei and their decay processes, especially the effects and applications of radioactivity and the interconversion of matter and energy. By the end of this chapter, you should be able to

- Discuss nuclear stability and determine why some nuclei are stable, whereas other nuclei are not and undergo radioactive decay
- Explain how radioactivity is detected and how the kinetics of decay is applied
- Discuss how nuclei synthesized in particle accelerators have extended the periodic table
- Determine the effects of radioactive emissions on matter, especially living tissue, and correlate them with some major applications of radioisotopes in science, technology, and medicine
- Determine the mass difference that arises when a nucleus forms from its subatomic particles and the energy that is equivalent to this difference
- Debate current and future attempts to harness the energy that is released when heavy nuclei split and lighter nuclei fuse
- Correlate the nuclear processes in stars to the processes used to create chemical elements

25

Concepts and Skills to Review before Studying This Chapter

- Discovery of the atomic nucleus (Section 2.4)
- Protons, neutrons, mass number, and the $^{A}_{Z}X$ notation (Section 2.5)
- Half-life and first-order reaction rate (Section 14.4)

So far, we have discussed or explained many different types of reactions, why reactions absorb or release heat energy, why some reactions happen quickly while others take so long, why some reactions go to completion while others do not, and the importance to society of many reactions. These reactions all deal with interactions between atoms and movement of electrons. In this chapter, we will explore the final frontier: the nucleus of the atom and the tremendous amount of energy contained within it. What are the components of the nucleus, and, when it breaks down, what are the products? Why is the energy released by the nucleus so powerful and potentially so harmful to us? How can seemingly small changes in the nucleus have such a large impact? From radioactive isotopes to nuclear fusion, this chapter will delve into the mysteries of the atomic nucleus.

The changes that occur in atomic nuclei differ strikingly from chemical changes. In chemical reactions, electrons are shared or transferred to form *compounds,* while nuclei remain unchanged. In nuclear reactions, the roles are reversed: electrons take part much less often, while nuclei undergo changes that, in nearly every case, form different *elements.* Nuclear reactions are often accompanied by energy changes a million times greater than those for chemical reactions—energy changes so large that changes in mass *are* detectable. Moreover, the yields and rates of nuclear reactions are *not* subject to the effects of pressure, temperature, and catalysis that the yields and rates of chemical reactions are. Table 25.1 summarizes these general differences.

25.1 Radioactive Decay and Nuclear Stability

A stable nucleus remains intact indefinitely, but *the great majority of possible nuclei contain excess neutrons or protons and are therefore β-unstable.* An unstable nucleus exhibits radioactivity, spontaneously decaying by emitting radiation. In Section 25.2, you will see that each type of unstable nucleus has its own characteristic *rate* of radioactive decay. In this section, we cover important terms and notation for nuclei, discuss some key events in the discovery of **radioactivity**, define the types of emission, and describe various modes of radioactive decay and how to predict which occurs for a given nucleus.

The Components of the Nucleus: Terms and Notation

Recall, from Chapter 2, that the nucleus contains essentially all the atom's mass but is only about 10^{-5} times its radius (or 10^{-15} times its volume), making the nucleus incredibly dense: about 10^{14} g/mL. ■ *Protons* and *neutrons,* the elementary particles that make up the nucleus, are called **nucleons**. A **nuclide** is a nucleus with a particular composition, that is, with specific numbers of the two types of nucleons. Most elements occur in nature as a mixture of **isotopes**, atoms with the characteristic number of protons of the element but different numbers of neutrons. Thus, each isotope of an element has a different nuclide. For example, oxygen has three naturally occurring isotopes; the most abundant contains eight protons and eight neutrons, whereas the least abundant contains eight protons and nine neutrons.

■ **Big Atom, Massive Core** If you could strip the electrons from the atoms in an object and compress the nuclei together, the object would lose only a fraction of a percent of its mass, but it would shrink to 0.000 000 000 000 1% (10^{-13}%) of its volume!

TABLE 25.1 Comparison of Chemical and Nuclear Reactions

Chemical Reactions	Nuclear Reactions
1. One substance is converted into another, but the atoms never change identity.	1. Atoms of one element are usually converted into atoms of another element.
2. Electrons in orbitals are involved as bonds break and form; nuclear particles do not take part.	2. Protons, neutrons, and other nuclear particles are involved; electrons in orbitals take part much less often.
3. Reactions are accompanied by relatively small changes in energy and no measurable changes in mass.	3. Reactions are accompanied by relatively large changes in energy and measurable changes in mass.
4. Reaction rates are influenced by temperature, concentration, catalysts, and the compound in which an element occurs.	4. Reaction rates depend on the number of nuclei but are not affected by temperature; catalysts; or, except on rare occasions, the compound in which an element occurs.

The relative mass and charge of a particle—nucleon, elementary particle, or nuclide—is described by the notation $^{A}_{Z}X$, where X is the *symbol* for the particle; A is the *mass number*, or the total number of nucleons; and Z is the *charge* of the particle; for a nucleus, A is the *sum of the protons and neutrons* and Z is the *number of protons* (atomic number). In this notation, the three subatomic elementary particles can be written as follows:

$$^{0}_{-1}e \text{ (electron)}, \ ^{1}_{1}p \text{ (proton), and } ^{1}_{0}n \text{ (neutron)}$$

(Sometimes, a proton is also represented as $^{1}_{1}H^{+}$.) The number of neutrons (N) in a nucleus is the mass number (A) minus the atomic number (Z):

$$N = A - Z$$

The two naturally occurring stable isotopes of chlorine, for example, have 17 protons ($Z = 17$), but one has 18 neutrons ($^{35}_{17}Cl$, also written ^{35}Cl) and the other has 20 ($^{37}_{17}Cl$, or ^{37}Cl). Alternatively, nuclides can be designated with the element name followed by the mass number: for example, chlorine-35 and chlorine-37. In naturally occurring samples of an element or its compounds, *the isotopes of the element are present in specific proportions* that vary only very slightly. Thus, in a sample of sodium chloride (or any Cl-containing substance), 75.77% of the Cl atoms are chlorine-35 and the remaining 24.23% are chlorine-37.

To understand this chapter, it is very important that you be comfortable with nuclear notations. Please take a moment to review Sample Problem 2.4 and Problems 2.39 to 2.46.

The Discovery of Radioactivity and the Types of Emissions

In 1896, the French physicist Antoine-Henri Becquerel accidentally discovered that uranium minerals, even when wrapped in paper and stored in the dark, emit a penetrating radiation that can expose a photographic plate. Becquerel also found that the radiation creates an electric discharge in air, thus providing a means for measuring its intensity.

Two years later, a young doctoral student named Marie Sklodowska Curie began a search for other minerals that behaved the same way. She found that thorium minerals also emit radiation and, most important, that *the intensity of the radiation is directly proportional to the concentration of the element in the mineral, not to the formula for the mineral or compound.* Curie named the emissions *radioactivity* and showed that they are *unaffected by temperature, pressure, or other physical and chemical conditions.*

After months of painstaking chemical work, Curie and her husband, physicist Pierre Curie, isolated two new elements from pitchblende, the principal ore of uranium: polonium (Po; $Z = 84$), the most metallic member of group 16, and radium (Ra; $Z = 88$), the heaviest alkaline earth metal.

Then, starting with several tonnes of pitchblende residues from which the uranium had been extracted, and working arduously for four years, Curie isolated 0.1 g of radium chloride, which she melted and electrolyzed to obtain pure metallic radium. Curie is the only person to have received two Nobel Prizes in Science, one in physics for her research into radioactivity, and the other in chemistry for her discoveries of polonium and radium.

During the next few years, Henri Becquerel, the Curies, and P. Villard in France, as well as Ernest Rutherford and his coworkers in England, studied the nature of radioactive emissions. A key finding was the observation of Rutherford and his colleague Frederick Soddy that elements other than radium were formed when radium decayed. They were at McGill University, in Montreal, in the early 1900s and, in 1902, proposed that radioactive emission results in the change of one element into another. This explanation seemed like a return to alchemy and was met with disbelief and ridicule. We now know it to be true: *when a nuclide of one element decays, it emits radiation and usually changes into a nuclide of a different element.*

Harriet Brooks (Figure 25.1A) was the first female Canadian nuclear physicist. She was the first graduate student of Ernest Rutherford (then professor at McGill University), under whom she worked immediately after graduating in 1901. She was the first woman to receive a Master's degree at McGill, and she was one of the group of researchers to discover radon and try to determine its atomic mass. For a brief period, she also worked under the supervision of Marie Curie (Figure 25.1B). She is most famous for her research on nuclear transmutations and radioactivity. In 1904, Brooks was appointed to the faculty of Barnard College. She worked there until 1907, when she married Frank Pitcher and resigned. (At that time, universities required women to resign after getting married.) Harriet Brooks was regarded by her contemporaries as, "next to Marie Curie, the most outstanding woman in the field of radioactivity."

There are three natural types of radioactive emissions:

- **Alpha particles (α, ${}^4_2\alpha$, or ${}^4_2He^{2+}$)** are identical to helium-4 nuclei.
- **Beta particles (β, β^-, or sometimes ${}^{\ 0}_{-1}\beta$)** are high-speed electrons. (The emission of electrons from the nucleus may seem strange, but, as you will see shortly, they result from a nuclear reaction.)
- **Gamma rays (γ, or sometimes ${}^0_0\gamma$)** are photons with very high energy.

A 

B 

FIGURE 25.1 **A.** Harriet Brooks (1876–1933). **B.** Marie Sklodowska Curie (1867–1934).

Figure 25.2 illustrates the behaviour of these emissions in an electric field: the positively charged α particles curve to a small extent toward the negative plate, the negatively charged β particles curve to a greater extent toward the positive plate (because they have lower mass), and the uncharged γ rays are not affected by the electric field.

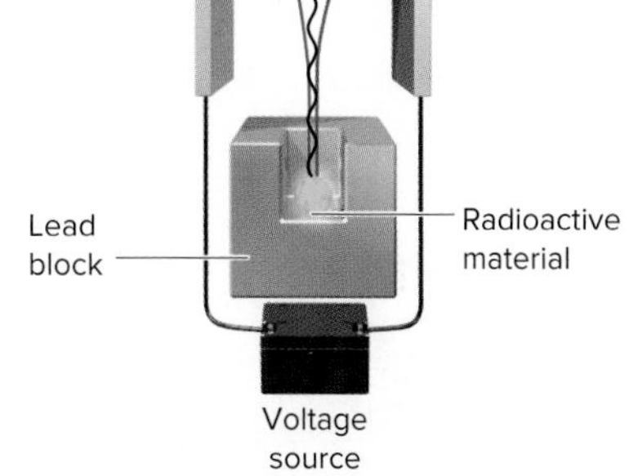

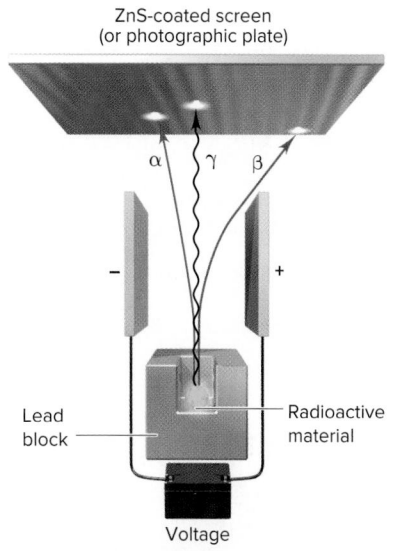

FIGURE 25.2 How the three types of radioactive emissions behave in an electric field

Modes of Radioactive Decay and Nuclear Equations

When a nuclide decays, it forms a nuclide of lower energy, and the excess energy is carried off by the emitted radiation and the recoiling nucleus. The decaying, or reactant, nuclide is called the *parent*; the product nuclide is called the *daughter*. Nuclides can decay in several ways. As each of the major modes of decay is introduced (summarized in Table 25.2), we will show examples of the mode and apply the key principle used to balance nuclear reactions: *the total Z (charge, or number of protons) and the total A (sum of protons and neutrons) of the reactants equal those of the products:*

$$^{\text{Total } A}_{\text{Total } Z}\text{Reactants} = ^{\text{Total } A}_{\text{Total } Z}\text{Products} \tag{25.1}$$

TABLE 25.2 Modes of Radioactive Decay

Mode	Emission	Decay Process	Change in A	Change in Z	Change in N
α decay	α^+ ($^4_2He^{2+}$)	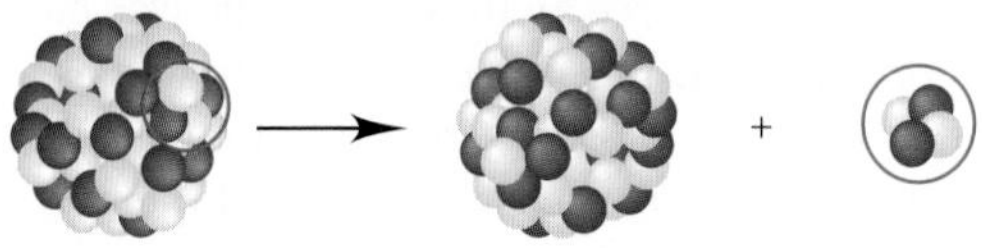 α expelled	−4	−2	−2
β^- decay	β^- ($^{\ 0}_{-1}\beta$)	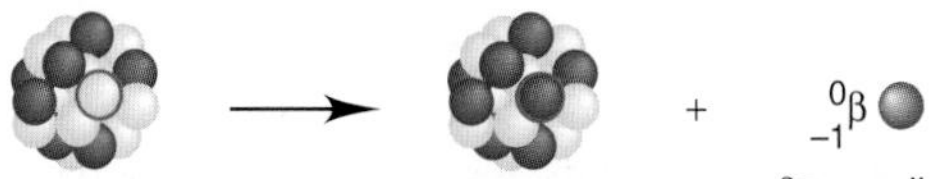 nucleus with xp^+ and yn^0 → nucleus with $(x+1)p^+$ and $(y-1)n^0$ + $^{\ 0}_{-1}\beta$ β^- expelled	0	+1	−1
	Net	^1_0n in nucleus → ^1_1p in nucleus + $^{\ 0}_{-1}\beta$ β^- expelled			
Positron (β^+) emission	β^+ ($^0_1\beta$)	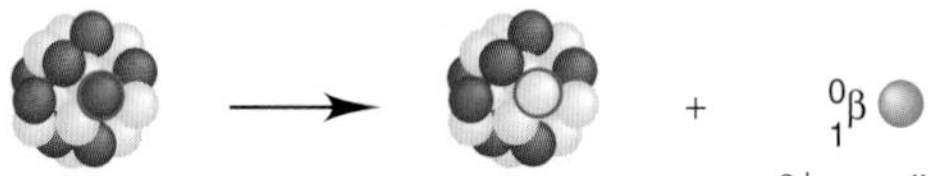 nucleus with xp^+ and yn^0 → nucleus with $(x-1)p^+$ and $(y+1)n^0$ + $^0_1\beta$ β^+ expelled	0	−1	+1
	Net	^1_1p in nucleus → ^1_0n in nucleus + $^0_1\beta$ β^+ expelled			
Electron (e^-) capture (EC)	X-ray	low-energy orbital 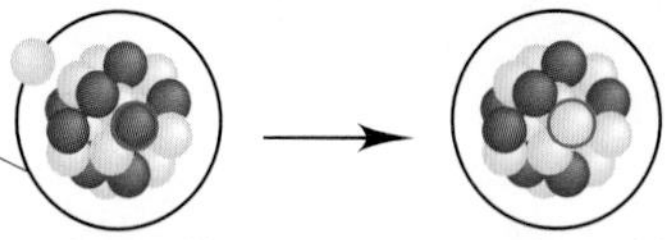 nucleus with xp^+ and yn^0 → nucleus with $(x-1)p^+$ and $(y+1)n^0$	0	−1	+1
	Net	$^{\ 0}_{-1}e$ absorbed from low-energy orbital + ^1_1p in nucleus → ^1_0n in nucleus			
Gamma (γ) emission	γ	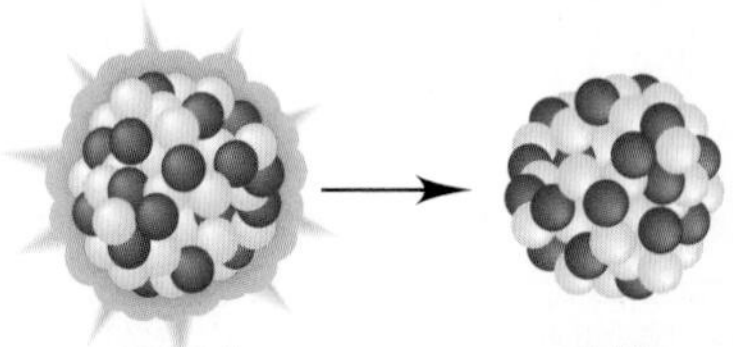 excited nucleus → stable nucleus + γ γ photon radiated	0	0	0

1. **Alpha (α) decay** involves the loss of an α particle from a nucleus. For each α particle emitted by the parent, *A decreases by 4 and Z decreases by 2* in the daughter. Every element beyond bismuth (Bi; $Z = 83$) is radioactive and exhibits α decay, which is *the most common means for a heavy unstable nucleus to become more stable.* For example, radium undergoes α decay to yield radon (Rn; $Z = 86$):

$$^{226}_{88}\text{Ra} \longrightarrow {}^{222}_{86}\text{Rn} + {}^{4}_{2}\alpha$$

Note that the A value for Ra equals the sum of the A values for Rn and α ($226 = 222 + 4$) and that the Z value for Ra equals the sum of the Z values for Rn and α ($88 = 86 + 2$).

2. **Beta (β) decay** is a more general class of radioactive decay. It includes three modes: β^- decay, β^+ emission (positron emission), and electron capture. Neutrinos (ν) or antineutrinos ($\bar{\nu}$) are also formed during all three modes of beta decay. Neutrinos are similar to electrons, with one crucial difference: neutrinos do not carry electric charge. Because neutrinos are electrically neutral, they are not affected by the electromagnetic forces that act on electrons; thus, neutrinos can carry a large amount of the energy from many nuclear reactions. The antineutrino is a lepton, or an antimatter particle, the counterpart to the neutrino. Although we will not include antineutrinos in other equations in this chapter, keep in mind that they are always expelled during β^- decay, and neutrinos are expelled during β^+ emission and e^- capture.

- **β^- decay** (or *negatron emission*) occurs through the ejection of a β^- particle from the nucleus. This change does not involve expulsion of a β^- particle that was in the nucleus; rather, *a neutron is converted into a proton, which remains in the nucleus, and a β^- particle, which is expelled immediately*:

$$^{1}_{0}\text{n} \longrightarrow {}^{1}_{1}\text{p} + {}^{0}_{-1}\beta$$

  As always, the totals of the A and Z values for the reactant and products are equal. Radioactive nickel-63 becomes stable copper-63 through β^- decay:

$$^{63}_{28}\text{Ni} \longrightarrow {}^{63}_{29}\text{Cu} + {}^{0}_{-1}\beta$$

  Another example is the β^- decay of carbon-14, used in radiocarbon dating:

$$^{14}_{6}\text{C} \longrightarrow {}^{14}_{7}\text{N} + {}^{0}_{-1}\beta$$

  Note that *β^- decay results in a product nuclide with the same A, but with Z one higher (one more proton) than in the reactant nuclide.* In other words, an atom of the element with the next *higher* atomic number is formed. Although equations in this chapter do not include it, an antineutrino (a neutral particle) is also emitted during β^- decay. The mass of the antineutrino is thought to be much less than 10^{-4} times the mass of an electron.

- **Positron (β^+) emission** is the emission of a β^+ particle from the nucleus. A key idea of modern physics is that most fundamental particles have corresponding *antiparticles* with the same mass but opposite charge. The **positron** is the antiparticle of the electron. Positron emission occurs through a process in which *a proton in the nucleus is converted into a neutron, and a positron is expelled:*

$$^{1}_{1}p \longrightarrow {}^{1}_{0}n + {}^{0}_{1}\beta$$

  Also emitted during this process is a *neutrino*, the antiparticle of the antineutrino. In terms of the effect on A and Z, *positron emission has the opposite effect of β^- decay: the daughter has the same A, but Z is one lower (one less proton) than the parent.* Thus, an atom of the element with the next *lower* atomic number forms. Carbon-11, a synthetic radioisotope, decays to a stable boron isotope through β^+ emission:

$$^{11}_{6}\text{C} \longrightarrow {}^{11}_{5}\text{B} + {}^{0}_{1}\beta$$

- **Electron (e^-) capture (EC)** occurs when the nucleus interacts with an electron in a low atomic energy level. The net effect is that *a proton is transformed into a neutron*:

$$^{1}_{1}\text{p} + {}^{0}_{-1}\text{e} \longrightarrow {}^{1}_{0}\text{n}$$

(We use the symbol "e" to distinguish an orbital electron from a beta particle, β.) The orbital vacancy is quickly filled by an electron that moves down from a higher energy level, and this process continues through still higher energy levels, with X-ray photons and neutrinos carrying off the energy difference in each step. Radioactive iron forms stable manganese through electron capture:

$$^{55}_{26}\text{Fe} + {}^{0}_{-1}\text{e} \longrightarrow {}^{55}_{25}\text{Mn} + h\nu \quad \text{(X-rays and neutrinos)}$$

Even though the processes are different, *electron capture has the same net effect as positron emission*: Z is lower by 1, and A is unchanged.

3. **Gamma (γ) emission** involves the radiation of high-energy γ photons (also called γ rays) from an excited nucleus. Just as an atom in an excited *electronic* state reduces its energy by emitting photons, usually in the UV and visible ranges (Section 6.2), a nucleus in an excited state lowers its energy by emitting γ photons, which have much higher energy (much shorter wavelength) than UV photons. Many nuclear processes leave the nucleus in an excited state, so *γ emission accompanies many other (mostly β) types of decay.* Several γ photons of different energies can be emitted from an excited nucleus as it returns to the ground state. Some of Marie Curie's experiments involved the release of γ rays, for example,

$$^{215}_{84}\text{Po} \longrightarrow {}^{211}_{82}\text{Pb} + {}^{4}_{2}\alpha \quad \text{(several } \gamma \text{ emitted)}$$

Gamma emission often accompanies β^- decay:

$$^{99}_{43}\text{Tc} \longrightarrow {}^{99}_{44}\text{Ru} + {}^{0}_{-1}\beta \quad \text{(several } \gamma \text{ emitted)}$$

Because γ rays have no mass or charge, *γ emission does not change A or Z.* Two gamma rays are emitted when a particle and an antiparticle annihilate each other. In the medical technique known as *positron-emission tomography* (which we will discuss in Section 25.5), a positron and an electron annihilate each other (with all the A and Z values shown):

$$^{0}_{1}\beta + {}^{0}_{-1}\text{e} \longrightarrow 2\,{}^{0}_{0}\gamma$$

Sample Problem 25.1 Writing Equations for Nuclear Reactions

Problem Write a balanced equation for each nuclear reaction:

(a) Naturally occurring thorium-232 undergoes α decay.

(b) Zirconium-86 undergoes electron capture.

Plan First, we write a skeleton equation that includes the mass numbers, atomic numbers, and symbols of all the particles on the correct sides of the equation, showing the unknown product particle as ${}^{A}_{Z}\text{X}$. Then, because the total of the mass numbers and the total of the charges on the left side and the right side must be equal, we solve for A and Z and use Z to determine X from the periodic table.

Solution (a) Write the skeleton equation, with the α particle as a product:

$$^{232}_{90}\text{Th} \longrightarrow {}^{A}_{Z}\text{X} + {}^{4}_{2}\alpha$$

We need to solve for A and Z and balance the equation. For A, $232 = A + 4$, so $A = 228$. For Z, $90 = Z + 2$, so $Z = 88$.

From the periodic table, we see that the element with $Z = 88$ is radium (Ra). Thus, the balanced equation is

$$^{232}_{90}\text{Th} \longrightarrow {}^{228}_{88}\text{Ra} + {}^{4}_{2}\alpha$$

(b) Write the skeleton equation, with the captured electron as a reactant:

$$^{86}_{40}\text{Zr} + {}^{0}_{-1}\text{e} \longrightarrow {}^{A}_{Z}\text{X}$$

We need to solve for A and Z and balance the equation. For A, $86 + 0 = A$, so $A = 86$. For Z, $40 + (-1) = Z$, so $Z = 39$.

The element with $Z = 39$ is yttrium (Y), so we have

$$^{86}_{40}\text{Zr} + ^{0}_{-1}\text{e} \longrightarrow ^{86}_{39}\text{Y}$$

Check Always read across superscripts and then across subscripts, with the yield arrow as an equal sign, to check your arithmetic. In part (a), for example, $232 = 228 + 4$, and $90 = 88 + 2$.

Follow-Up Problem 25.1 Write a balanced equation for the reaction in which a nuclide undergoes β^- decay and changes to cesium-133. (See Brief Solutions to Follow-Up Problems at the end of the chapter.)

Nuclear Stability and the Mode of Decay

Can we predict whether, and how, an unstable nuclide will decay? Our knowledge of the nucleus is much less complete than our knowledge of the whole atom, but some patterns emerge by observing naturally occurring nuclides.

The Band of Stability Two key factors determine the stability of a nuclide:

- The number of neutrons (N), the number of protons (Z), and their ratio (N/Z), which we calculate from $\frac{A-Z}{Z}$; this factor relates primarily to nuclides that undergo one of the three modes of β decay
- The total mass of the nuclide, which mostly relates to nuclides that undergo α decay

Figure 25.3A is a plot of number of neutrons versus number of protons for all stable nuclides. Note the following:

- The points form a narrow **band of stability** that gradually curves above the line for $N = Z$ ($N/Z = 1$).
- The only stable nuclides with $N/Z < 1$ are $^{1}_{1}\text{H}$ and $^{3}_{2}\text{He}$.
- Many lighter nuclides with $N = Z$ are stable, for example, $^{4}_{2}\text{He}$, $^{12}_{6}\text{C}$, $^{16}_{8}\text{O}$, and $^{20}_{10}\text{Ne}$. The heaviest of these is $^{40}_{20}\text{Ca}$. Thus, for lighter nuclides, one neutron for each proton ($N = Z$) is enough to provide stability.
- The N/Z ratio of stable nuclides gradually increases as Z increases. For example, for $^{56}_{26}\text{Fe}$, $N/Z = 1.15$; for $^{107}_{47}\text{Ag}$, $N/Z = 1.28$; for $^{184}_{74}\text{W}$, $N/Z = 1.49$; and, finally, for $^{209}_{83}\text{Bi}$, $N/Z = 1.52$. Thus, for heavier stable nuclides, $N > Z$ ($N/Z > 1$), and N increases faster than Z. As we discuss below, and show in Figure 25.3B, if N/Z of a nuclide is either too high (above the band) or not high enough (below the band), the nuclide is unstable and undergoes one of the three modes of beta decay.
- All nuclides with $Z > 83$ are unstable.* This means that all nuclides with a higher Z—the largest members of the main groups, actinium and the actinides ($Z = 89$ to 103), and the other elements of the fourth (6*d*) transition series ($Z = 104$ to 112)—are radioactive, and, as we point out below, undergo α decay.

Stability and Nuclear Structure Given that protons are positively charged and neutrons are uncharged, the first question to ask is, what holds the nucleus together? Nuclear scientists answer this question and explain the importance of N and Z values in terms of two opposing forces: electrostatic repulsive forces and an attractive force called the **strong force**. The electrostatic repulsive forces between protons would break the nucleus apart if not for the strong force, which exists between all nucleons (protons and neutrons). This force is 137 times as strong as the repulsive force but *operates only over the short distances within the nucleus*. Thus, the *attractive* strong force overwhelms the much weaker *repulsive* electrostatic force and keeps the nucleus together.

*In 2003, French nuclear physicists discovered that $^{209}_{83}\text{Bi}$ actually undergoes α decay, but the half-life is 1.9×10^{19} years, about a billion times longer than the estimated age of the universe. Bismuth ($Z = 83$) can, thus, be considered stable.

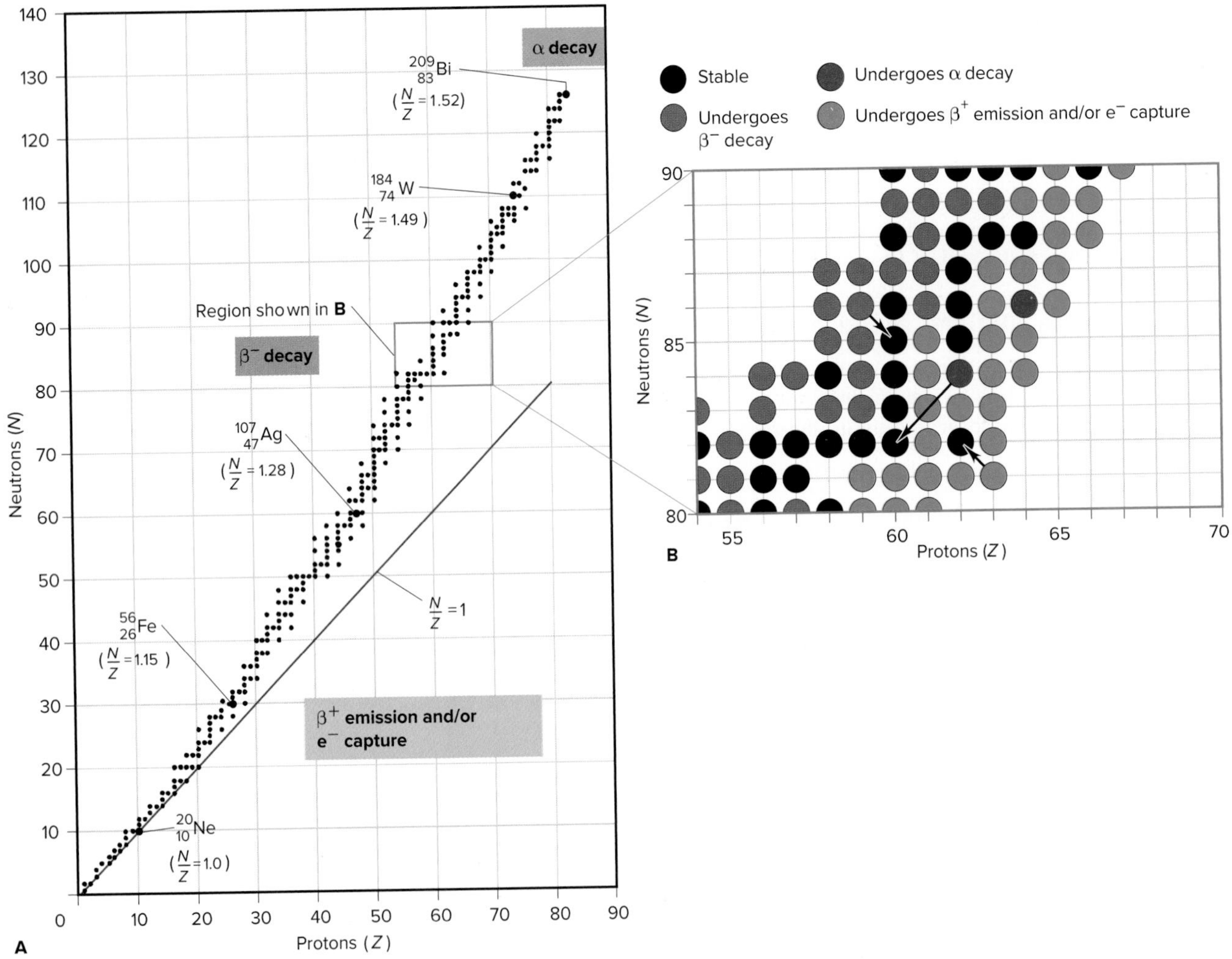

FIGURE 25.3 **A plot of number of neutrons versus number of protons for the stable nuclides**

TABLE 25.3 Number of Stable Nuclides for Elements 48 to 54*

Element	Atomic No. (Z)	No. of Nuclides
Cd	**48**	**8**
In	49	2
Sn	**50**	**10**
Sb	51	2
Te	**52**	**8**
I	53	1
Xe	**54**	**9**

*Elements with even values of Z are shown in boldface.

TABLE 25.4 An Even-Odd Breakdown of the Stable Nuclides

Z	N	No. of Nuclides
Even	Even	157
Even	Odd	53
Odd	Even	50
Odd	Odd	4
	Total	264

The oddness or evenness of N and Z values is related to some important patterns of nuclear stability. Two interesting points become apparent when we classify the known stable nuclides:

1. Elements with an even Z (number of protons) usually have a larger number of stable nuclides than elements with an odd Z. Table 25.3 demonstrates this point for cadmium ($Z = 48$) through xenon ($Z = 54$).

2. Well over half the stable nuclides have *both* even N and even Z (Table 25.4). Only four nuclides with odd N and odd Z are stable: $^{2}_{1}H$, $^{6}_{3}Li$, $^{10}_{5}B$, and $^{14}_{7}N$.

One model of nuclear structure that attempts to explain the stability of even values of N and Z postulates that protons and neutrons lie in *nucleon energy levels* and that greater stability results from the *pairing of spins* of like nucleons. (Note the analogy to electron energy levels and the stability from pairing of electron spins.)

Just as noble gases—with 2, 10, 18, 36, 54, and 86 electrons—are exceptionally stable because they have filled *electron* levels, nuclides with N or Z values of 2, 8, 20, 28, 50, and 82 (as well as $N = 126$) are exceptionally stable. These so-called magic numbers are thought to correspond to the numbers of protons or neutrons in filled *nucleon* levels. A few examples are $^{50}_{22}Ti$ ($N = 28$), $^{88}_{38}Sr$ ($N = 50$), and the 10 stable nuclides of tin ($Z = 50$). Some extremely stable nuclides have two magic numbers: $^{4}_{2}He$, $^{16}_{8}O$, $^{40}_{20}Ca$, and $^{208}_{82}Pb$ ($N = 126$).

Sample Problem 25.2 Predicting Nuclear Stability

Problem Which nuclides would you predict to be stable, and which would you predict to be radioactive? Explain.

(a) $^{18}_{10}Ne$ **(b)** $^{32}_{16}S$ **(c)** $^{236}_{90}Th$ **(d)** $^{123}_{56}Ba$

Plan To evaluate the stability of each nuclide, we determine the N and Z values, the N/Z ratio from $\frac{A-Z}{Z}$, the value of Z, stable N/Z ratios (from Figure 25.3A), and whether Z and N are even or odd.

Solution (a) This nuclide is radioactive. It has $N = 8$ (18 − 10) and $Z = 10$, so $N/Z = \frac{18-10}{10} = 0.8$. Except for hydrogen-1 and helium-3, no nuclides with $N < Z$ are stable.

Despite its even N and Z, this nuclide has too few neutrons to be stable.

(b) This nuclide is stable. It has $N = Z = 16$, so $N/Z = 1.0$. With $Z < 20$ and even N and Z, this nuclide is most likely stable.

(c) This nuclide is radioactive. It has $Z = 90$, and every nuclide with $Z > 83$ is radioactive.

(d) This nuclide is radioactive. It has $N = 67$ and $Z = 56$, so $N/Z = 1.20$. For Z values of 55 to 60, Figure 25.3A shows that $N/Z \geq 1.3$, so this nuclide has too few neutrons to be stable.

Check By consulting a table of isotopes, such as the one in the *CRC Handbook of Chemistry and Physics*, we find that our predictions are correct.

Follow-Up Problem 25.2 Why is $^{31}_{15}P$ stable, but $^{30}_{15}P$ unstable?

Predicting the Mode of Decay An unstable nuclide generally decays in a mode that shifts its N/Z ratio toward the band of stability. This is illustrated in Figure 25.3B, which expands the small region of $Z = 54$ to 70 in Figure 25.3A to show the stable *and* many of the unstable nuclides in that region, as well as their modes of decay. Note the following points:

1. *Neutron-rich nuclides*, nuclides with too many neutrons for stability (a high N/Z), lie above the band of stability. They undergo *β^- decay*, which converts a neutron into a proton, thus reducing the value of N/Z.

2. *Proton-rich nuclides*, nuclides with too many protons for stability (a low N/Z), lie below the band of stability. They undergo *β^+ emission* and/or *e^- capture*, both of which convert a proton into a neutron, thus increasing the value of N/Z. (The rate of e^- capture increases with Z, so β^+ emission is more common among lighter elements and e^- capture is more common among heavier elements.)

3. *Heavy nuclides*, nuclides with $Z > 83$ (shown in Figure 25.3A), are too heavy to be stable and undergo *α decay*. This reduces their Z and N values by two units per emission.

With the information in Figure 25.3, predicting the mode of decay of an unstable nuclide simply involves comparing its N/Z ratio with the N/Z ratios in the nearby region of the band of stability. However, even without Figure 25.3, we can often make an educated guess about the mode of decay. The atomic mass of an element is the weighted average of its naturally occurring isotopes. Therefore, we know the following:

- The mass number, A, of a *stable* nuclide is relatively close to the atomic mass.
- If an *unstable* nuclide of an element (given Z) has an A value much higher than the atomic mass, it is neutron rich and will probably decay by β^- emission.
- If, on the other hand, the unstable nuclide has an A value much lower than the atomic mass, it is proton rich and will probably decay by β^+ emission and/or e^- capture.

In the next sample problem, we predict the mode of decay of some unstable nuclides.

Sample Problem 25.3 Predicting the Mode of Nuclear Decay

Problem Use the atomic mass of the element to predict the mode(s) of decay of **(a)** $^{12}_{5}B$; **(b)** $^{234}_{92}U$; **(c)** $^{81}_{33}As$; **(d)** $^{127}_{57}La$.

Plan If the nuclide is too heavy to be stable ($Z > 83$), it undergoes α decay. For other nuclides, we can use the Z value to obtain its atomic mass from the periodic table. If the mass number of the nuclide is higher than the atomic mass, the nuclide has too many neutrons: N too high ⇒ β^- decay. If the mass number is lower than the atomic mass, the nuclide has too many protons: Z too high ⇒ β^+ emission and/or e^- capture.

Solution (a) This nuclide has $Z = 5$, which is boron (B). Its atomic mass is 10.81. The nuclide's A value of 12 is significantly higher than its atomic mass, so this nuclide is neutron rich. It will probably undergo β^- decay.

(b) This nuclide has $Z = 92$, which is neptunium (Np). It will undergo α decay and decrease its total mass.

(c) This nuclide has $Z = 33$, which is arsenic (As). Its atomic mass is 74.92. The A value of 81 is much higher, so this nuclide is neutron rich and will probably undergo β^- decay.

(d) This nuclide has $Z = 57$, which is lanthanum (La). Its atomic mass is 138.9. The A value of 127 is much lower, so this nuclide is proton rich and will probably undergo β^+emission and/or e^- capture.

Check To confirm our predictions in (a), (c), and (d), let us compare each nuclide's N/Z ratio to the N/Z ratios in the band of stability. **(a)** This nuclide has $N = 7$ and $Z = 5$, so $N/Z = 1.40$, which is too high for the band of stability, so it will undergo β^- decay. **(c)** This nuclide has $N = 48$ and $Z = 33$, so $N/Z = 1.45$, which is too high for the band of stability, so it will undergo β^- decay. **(d)** This nuclide has $N = 70$ and $Z = 57$, so $N/Z = 1.23$, which is too low for the band of stability, so it will undergo β^+ emission and/or e^- capture. Our predictions based on N/Z values are the same as our predictions based on atomic mass.

Comment Both possible modes of decay are observed for the nuclide in part (d).

Follow-Up Problem 25.3 Use the A value for the nuclide and the atomic mass in the periodic table to predict the mode of decay of **(a)** $^{61}_{26}Fe$; **(b)** $^{241}_{95}Am$.

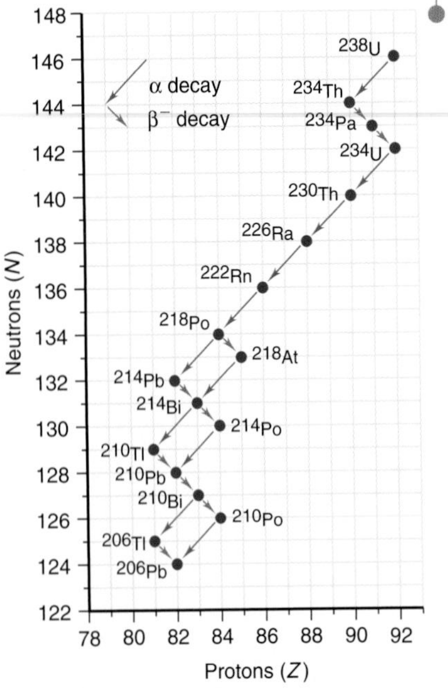

FIGURE 25.4 The ^{238}U decay series

Decay Series A parent nuclide may undergo a series of decay steps before a stable daughter nuclide forms. This succession of steps is called a **decay series (disintegration series)** and is typically depicted on a gridlike display. Figure 25.4 shows the decay series from uranium-238 to lead-206. Numbers of neutrons (N) are plotted against numbers of protons (Z) to form the grid, which displays a series of α and β^- decays. The typical zig-zag pattern arises because $N > Z$, which means that α decay, which reduces both N and Z by two units, decreases Z slightly more than it decreases N. Therefore, α decays result in neutron-rich daughters, which undergo β^- decay to gain more stability. Note that a given nuclide can undergo both modes of decay. (Gamma emission accompanies many of these steps, but it does not affect the type of nuclide.)

The series in Figure 25.4 is one of three that occur in nature. All end with isotopes of lead whose nuclides have one ($Z = 82$) or two ($N = 126$, $Z = 82$) magic numbers. A second series begins with uranium-235 and ends with lead-207, and a third series begins with thorium-232 and ends with lead-208. (Neptunium-237 began a fourth series, but its half-life is so much less than the age of Earth that only traces of it remain today.)

SUMMARY OF SECTION 25.1

- To become more stable, a radioactive nuclide may emit α particles ($^{4}_{2}He^{2+}$; helium-4 nuclei), β particles (β^- or $^{0}_{-1}\beta$; high-speed electrons), positrons (β^+ or $^{0}_{1}\beta$), or γ rays (high-energy photons), or it may capture an orbital electron ($^{0}_{-1}e$).
- A narrow band of neutron-to-proton ratios (N/Z) includes the ratios of all the stable nuclides.
- Even values of N and Z are associated with stable nuclides, as are certain "magic numbers" of neutrons and protons.
- By comparing a nuclide's mass number with its atomic mass, and its N/Z ratio with the ratios in the band of stability, we can predict that, in general, neutron-rich nuclides undergo β^- decay and proton-rich nuclides undergo β^+ emission and/or e^- capture. Heavy nuclides ($Z > 83$) undergo α decay.
- Three naturally occurring decay series all end in isotopes of lead.

25.2 The Kinetics of Radioactive Decay

Both chemical and nuclear systems tend toward maximum stability. Just as the concentrations in a chemical system change in a predictable direction to give a stable equilibrium ratio, the type and number of nucleons in an unstable nucleus change in a predictable direction to give a stable N/Z ratio. As you know, however, the tendency of a chemical system to become more stable tells nothing about how long the process will take, and the same holds true for a nuclear system. In this section, we first see how radioactivity is detected and measured and then examine the kinetics of nuclear change; later, we will examine the energetics of nuclear change.

Detection and Measurement of Radioactivity

Radioactive emissions interact with surrounding atoms. To determine the rate of nuclear decay, we measure radioactivity by observing the effects of these interactions over time. Because the effects can be electrically amplified billions of times, it is possible to detect the decay of a single nucleus. Ionization counters and scintillation counters are two devices that are used to measure radioactive emissions.

Ionization Counters An *ionization counter* detects radioactive emissions as they ionize a gas. Ionization produces free electrons and gaseous cations, which are attracted to electrodes that conduct a current to a recording device. The most common type of ionization counter is a **Geiger-Müller counter** (Figure 25.5). It consists of a tube filled with an argon-methane mixture; the tube housing acts as the cathode, and a thin wire in the centre of the tube acts as the anode. Emissions from the sample enter the tube through a thin window and strike argon atoms, producing Ar^+ ions that migrate toward the cathode and free electrons that are then accelerated toward the anode. These electrons collide with other argon atoms and free more electrons in an *avalanche effect.* The current created is amplified and appears as a meter reading and/or an audible click. An initial release of one electron can release 10^{10} electrons in a microsecond, giving the Geiger-Müller counter great sensitivity.

Scintillation Counters In a **scintillation counter**, radioactive emissions are detected by their ability to excite atoms and cause them to emit light. The light-emitting substance in the counter, called a *phosphor*, is coated onto part of a *photomultiplier tube*, a device that increases the original electrical signal. Incoming radioactive particles strike the phosphor, which emits photons. Each photon, in turn, strikes a cathode, releasing an electron through the photoelectric effect (Section 6.1). The electron hits other parts of the tube, which release increasing numbers of electrons, and the resulting current is recorded. Liquid scintillation counters use an organic mixture that contains a phosphor and a solvent (Figure 25.6). This "cocktail" dissolves the radioactive sample *and* emits pulses of light when excited by the emission. The number of pulses is proportional to the concentration of the radioactive substance. Scintillation counters are often used to measure β^- emissions from dissolved biological samples, particularly those containing compounds of ^{3}H and ^{14}C.

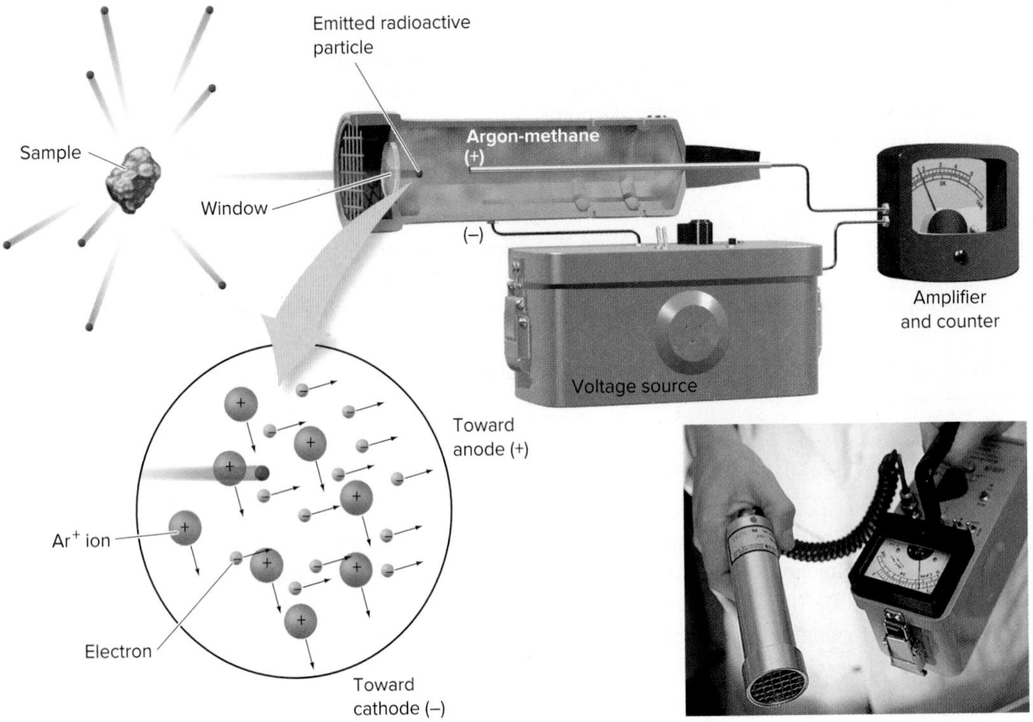

FIGURE 25.5 Detection of radioactivity by an ionization counter

FIGURE 25.6 A scintillation "cocktail" in tubes, which will be placed in the counter

The Rate of Radioactive Decay

Radioactive nuclei decay at a characteristic rate, regardless of the chemical substance in which they occur. The *decay rate*, or **activity ($\mathcal{A}$)**, of a radioactive sample is the change in the number of nuclei ($\mathcal{N}$) divided by the change in time (t). As with chemical reaction rates (Section 14.2), because the number of nuclei is *decreasing*, a minus sign precedes the expression to obtain a positive decay rate:

$$\text{Decay rate } (\mathcal{A}) = -\frac{\Delta \mathcal{N}}{\Delta t}$$

The SI unit of radioactivity is the **becquerel (Bq)**, defined as one disintegration per second (d/s): 1 Bq = 1 d/s. A much larger and more common unit of radioactivity is the **curie (Ci)**, which was originally defined as the number of disintegrations per second in 1 g of radium-226, but is now a fixed quantity:

$$1 \text{ Ci} = 3.70 \times 10^{10} \text{ d/s} \qquad (25.2)$$

Therefore, 1 Bq = 2.703×10^{-11} Ci. Because the curie is so large, the millicurie (mCi) and microcurie (μCi) are commonly seen. The radioactivity of a sample is often given as a *specific activity*, the decay rate per gram.

An activity is meaningful only when we consider the large number of nuclei in a macroscopic sample. Suppose there are 1×10^{15} radioactive nuclei of a particular type in a sample and they decay at a rate of 10% per hour. Although any particular nucleus in the sample might decay in a microsecond or in a million hours, the *average* of all decays results in 10% of the entire collection of nuclei disintegrating each hour. During the first hour, 10% of the *original* number, or 1×10^{14} nuclei, will decay. During the next hour, 10% of the remaining 9×10^{14} nuclei, or 9×10^{13} nuclei, will decay. During the next hour, 10% of those remaining will decay, and so forth. Thus, for a large collection of radioactive nuclei, *the number decaying per unit time is proportional to the number present*:

$$\text{Decay rate } (\mathcal{A}) \propto \mathcal{N} \qquad \text{or} \qquad \mathcal{A} = k\mathcal{N}$$

where k is called the **decay constant** and is characteristic of each type of nuclide. The larger the value of k, the higher the decay rate: larger $k \Rightarrow$ higher $\mathcal{A}$.

Combining the two rate expressions just given, we obtain

$$\mathcal{A} = -\frac{\Delta \mathcal{N}}{\Delta t} = k\mathcal{N} \qquad (25.3)$$

Note that the activity depends only on $\mathcal{N}$ raised to the first power (and on the constant value of k). Therefore, *radioactive decay is a first-order process* (Section 14.4), but with respect to the *number* of nuclei rather than their concentration.

Half-Life of Radioactive Decay Decay rates are also commonly expressed in terms of the fraction of nuclei that decay over a given time interval. The **half-life ($t_{1/2}$)** of a nuclide has the same meaning as for a chemical change (Section 14.4) and can be expressed in terms of number of nuclei, mass of radioactive substance, and activity:

- *Number of nuclei.* Half-life is the time it takes for half the nuclei in a sample to decay—*the number of nuclei remaining is halved after each half-life.* Figure 25.7 shows the decay of carbon-14, which has a half-life of 5730 years, in terms of number of ^{14}C nuclei remaining:

$$^{14}_{6}C \longrightarrow {}^{14}_{7}N + {}^{0}_{-1}\beta$$

- *Mass.* As ^{14}C decays, the mass of ^{14}C decreases while the mass of ^{14}N increases. If we start with 1.0 g of ^{14}C, half that mass of ^{14}C (0.50 g) will be left after 5730 years, half of that mass (0.25 g) after another 5730 years, and so on.
- *Activity.* The activity depends on the number of nuclei, so the activity is halved after each succeeding half-life.

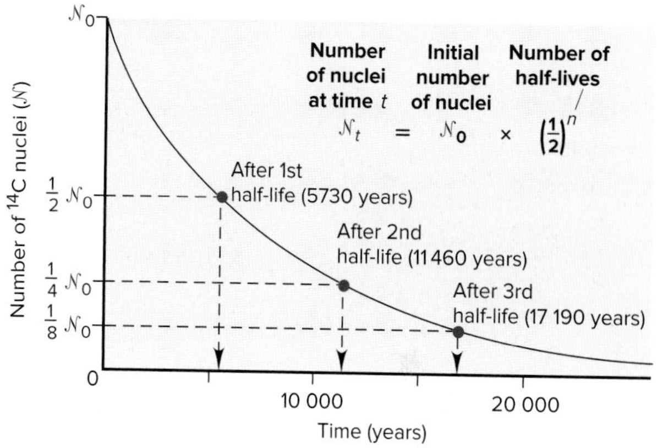

FIGURE 25.7 Decrease in number of ^{14}C nuclei over time

We determine the half-life of a nuclear reaction from its rate constant. Rearranging Equation 25.3 and integrating over time gives an expression for finding the number of nuclei remaining, $\mathcal{N}_t$, after a given time t:

$$\ln\frac{\mathcal{N}_t}{\mathcal{N}_0} = -kt \quad \text{or} \quad \mathcal{N}_t = \mathcal{N}_0 e^{-kt} \quad \text{and} \quad \ln\frac{\mathcal{N}_0}{\mathcal{N}_t} = kt \qquad (25.4)$$

where $\mathcal{N}_0$ is the number of nuclei at $t = 0$. (Note the similarity to Equation 14.4.) To calculate the half-life ($t_{1/2}$), we set $\mathcal{N}_t$ equal to $\frac{1}{2}\mathcal{N}_0$ and solve for $t_{1/2}$:

$$\ln\frac{\mathcal{N}_0}{\frac{1}{2}\mathcal{N}_0} = kt_{1/2} \quad \text{so} \quad t_{1/2} = \frac{\ln 2}{k} \qquad (25.5)$$

Exactly analogous to the half-life of a first-order chemical change, *this half-life does* ***not*** *depend on the number of nuclei and is inversely related to the decay constant:*

$$\text{large } k \Rightarrow \text{short } t_{1/2} \qquad \text{and} \qquad \text{small } k \Rightarrow \text{long } t_{1/2}$$

The decay constants and half-lives of radioactive nuclides vary over a very wide range, even for the nuclides of a given element (Table 25.5). Table 25.6 shows representative half-lives of common radioactive isotopes.

TABLE 25.5 Decay Constants (k) and Half-Lives ($t_{1/2}$) of Beryllium Isotopes

Nuclide	k	$t_{1/2}$
$^{7}_{4}Be$	1.30×10^{-2}/day	53.3 days
$^{8}_{4}B$	1.0×10^{16}/s	6.7×10^{-17} s
$^{9}_{4}Be$	Stable	
$^{10}_{4}Be$	4.3×10^{-7}/year	1.6×10^{6} years
$^{11}_{4}B$	5.02×10^{-2}/s	13.8 s

TABLE 25.6 Half-Lives of Common Radioactive Isotopes

Isotope	Emits	Half-Life	Used in/Comes from/Affects
Uranium-238	Alpha	4.5 billion years	Used in new depleted uranium weapons and tank armour
Uranium-235	Alpha	700 million years	Used in atomic weapons and poisoning fabrication factories
Uranium-234	Alpha and gamma	245 000 years	Left from uranium ore milling and enrichment
Plutonium-239	Alpha	24 300 years	Used in hydrogen bombs; seeks liver, lungs, and bone
Cesium-137	Beta and gamma	30.2 years	Left in large quantities from bomb production and in reactor wastes; contaminates whole body and muscle
Strontium-90	Beta	28 years	Spewed by accidents at Chernobyl and Fukushima; vented in routine "allowable" releases by all operating nuclear power reactors; seeks bone
Cobalt-60	Beta and gamma	5 years	Left from hydrogen bomb production and used in food irradiation; contaminates whole body
Iodine-125 and iodine-131	Beta and gamma	8.1 days	Spewed in large quantities during reactor accidents and in fallout from above-ground bomb testing; contaminates the thyroid gland

Sample Problem 25.4 Finding the Number of Radioactive Nuclei

Problem Strontium-90 is a radioactive by-product of nuclear reactors that behaves biologically like calcium, the element above it in group 2. When ^{90}Sr is ingested by mammals, it is found in their milk and eventually in the bones of those drinking the milk. If a sample of ^{90}Sr has an activity of 1.2×10^{12} d/s, what are the activity and the fraction of nuclei that have decayed after 59 years ($t_{1/2}$ of ^{90}Sr = 29 years)?

Plan The fraction of nuclei that have decayed is the change in number of nuclei, expressed as a fraction of the starting number. The activity of the sample ($\mathscr{A}$) is proportional to the number of nuclei ($\mathscr{N}$), so we know that

$$\text{Fraction decayed} = \frac{\mathscr{N}_0 - \mathscr{N}_t}{\mathscr{N}_0} = \frac{\mathscr{A}_0 - \mathscr{A}_t}{\mathscr{A}_0}$$

We are given $\mathscr{A}_0$ (1.2×10^{12} d/s), so we find $\mathscr{A}_t$ from the integrated form of the first-order rate equation (Equation 25.4), in which t is 59 years. To solve that equation, we first need k, which we can calculate from the given $t_{1/2}$ (29 years).

Solution Calculate the decay constant, k:

$$t_{1/2} = \frac{\ln 2}{k} \quad \text{so} \quad k = \frac{\ln 2}{t_{1/2}} = \frac{0.693}{29 \text{ years}} = 0.024 \text{ year}^{-1}$$

Applying Equation 25.4 to calculate $\mathscr{A}_t$, the activity remaining at time t is

$$\ln \frac{\mathscr{N}_0}{\mathscr{N}_t} = \ln \frac{\mathscr{A}_0}{\mathscr{A}_t} = kt \quad \text{or} \quad \ln \mathscr{A}_0 - \ln \mathscr{A}_t = kt$$

So,

$$\ln \mathscr{A}_t = -kt + \ln \mathscr{A}_0 = (-0.024 \text{ year}^{-1} \times 59 \text{ years}) + \ln (1.2 \times 10^{12} \text{ d/s})$$
$$= -1.4 + 27.81 = 26.4$$
$$\mathscr{A}_t = 2.9 \times 10^{11} \text{ d/s}$$

All the data contain two significant figures, so we retained two in the answer. Calculate the fraction decayed:

$$\text{Fraction decayed} = \frac{\mathscr{A}_0 - \mathscr{A}_t}{\mathscr{A}_0} = \frac{1.2 \times 10^{12} \text{ d/s} - 2.9 \times 10^{11} \text{ d/s}}{1.2 \times 10^{12} \text{ d/s}} = 0.76$$

Check The answer is reasonable: t is about 2 half-lives, so $\mathscr{A}$ should be about $\frac{1}{4}\mathscr{A}_0$, or about 0.3×10^{12}; therefore, the activity should have decreased by about $\frac{3}{4}$.

Comment 1. A useful substitution of Equation 25.4 for finding $\mathscr{A}_t$, the activity at time t, is $\mathscr{A}_t = \mathscr{A}_0 e^{-kt}$.

2. Another way to find the fraction of activity (or nuclei) remaining incorporates the number of half-lives ($\frac{t}{t_{1/2}}$). By combining Equations 25.4 and 25.5 and substituting $\frac{\ln 2}{t_{1/2}}$ for k, we obtain

$$\ln\left(\frac{\mathscr{N}_0}{\mathscr{N}_t}\right) = \left(\frac{\ln 2}{t_{1/2}}\right)t = \frac{t}{t_{1/2}} \ln 2 = \ln 2^{t/t_{1/2}}$$

Inverting the ratio gives

$$\ln \frac{\mathscr{N}_t}{\mathscr{N}_0} = \ln \left(\frac{1}{2}\right)^{t/t_{1/2}}$$

Taking the antilog gives

$$\text{Fraction remaining} = \frac{\mathscr{N}_t}{\mathscr{N}_0} = \left(\frac{1}{2}\right)^{t/t_{1/2}} = \left(\frac{1}{2}\right)^{59/29} = 0.24$$

So,

$$\text{Fraction decayed} = 1.00 - 0.24 = 0.76$$

Follow-Up Problem 25.4 Sodium-24 ($t_{1/2}$ = 15 h) is used to study blood circulation. A patient is injected with an aqueous solution of $^{24}NaCl$, whose activity is 2.5×10^9 d/s. How much of the activity is present in the patient's body and excreted fluids after 4.0 days?

Radioisotopic Dating

The historical record fades rapidly with time and virtually disappears for events of more than a few thousand years ago. Much knowledge of prehistory comes from **radioisotopic dating**, which uses **radioisotopes** to determine the ages of objects. This technique supplies data for fields such as art history, archaeology, geology, and paleontology.

The technique of *radiocarbon dating*, for which the American chemist Willard F. Libby won the Nobel Prize in Chemistry in 1960, is based on measuring the amounts of ^{14}C and ^{12}C in materials of biological origin. The accuracy of the method falls off after about six half-lives of ^{14}C ($t_{1/2}$ = 5730 years), so it is used to date objects up to about 36 000 years old.

Here is how radiocarbon dating works:

1. High-energy cosmic rays, consisting mainly of protons, enter the atmosphere from outer space and initiate a cascade of nuclear reactions, some of which produce neutrons that bombard ordinary ^{14}N atoms to form ^{14}C:

$$^{14}_{7}N + ^{1}_{0}n \longrightarrow ^{14}_{6}C + ^{1}_{1}p$$

Through the competing processes of formation and radioactive decay, the amount of ^{14}C in the atmosphere has remained nearly constant.*

2. The ^{14}C atoms combine with O_2, diffuse throughout the lower atmosphere, and enter the total carbon pool as gaseous $^{14}CO_2$ and aqueous $H^{14}CO_3^-$. They mix with ordinary $^{12}CO_2$ and $H^{12}CO_3^-$, reaching a constant $^{12}C/^{14}C$ ratio of about $10^{12}/1$.

3. The CO_2 is taken up by plants during photosynthesis and then taken up and excreted by animals that eat the plants. Thus, the $^{12}C/^{14}C$ ratio of a living organism is the same as the ratio in the environment.

4. When an organism dies, it no longer absorbs or releases CO_2, so the $^{12}C/^{14}C$ ratio steadily increases because *the amount of ^{14}C decreases as it decays*:

$$^{14}_{6}C \longrightarrow ^{14}_{7}N + ^{0}_{-1}\beta$$

The difference between the $^{12}C/^{14}C$ ratio in a dead organism and the ratio in living organisms reflects the time elapsed since the organism died.

As you saw in Sample Problem 25.4, the first-order rate equation can be expressed in terms of a ratio of activities:

$$\ln \frac{\mathcal{N}_0}{\mathcal{N}_t} = \ln \frac{\mathcal{A}_0}{\mathcal{A}_t} = kt$$

We use this expression in radiocarbon dating, where $\mathcal{A}_0$ is the activity in a living organism and $\mathcal{A}_t$ is the activity in the object whose age is unknown. Solving for t gives the age of the object:

$$t = \frac{1}{k} \ln \frac{\mathcal{A}_0}{\mathcal{A}_t} \qquad (25.6)$$

A useful graphical method in radioisotopic dating shows a plot of the natural logarithm of the specific activity versus time, which gives a straight line with a slope of $-k$, the negative of the decay constant. Using such a plot and measuring the ^{14}C specific activity of an object, we can determine its age; several examples appear in Figure 25.8.

One well-known example of the use of radiocarbon dating was the determination of the age of the Shroud of Turin (*see photo*). One of the holiest Christian relics, the shroud is a piece of linen that bears a faint image of a man's body and was thought to be the burial cloth used to wrap the body of Jesus Christ. Three labs in Europe and

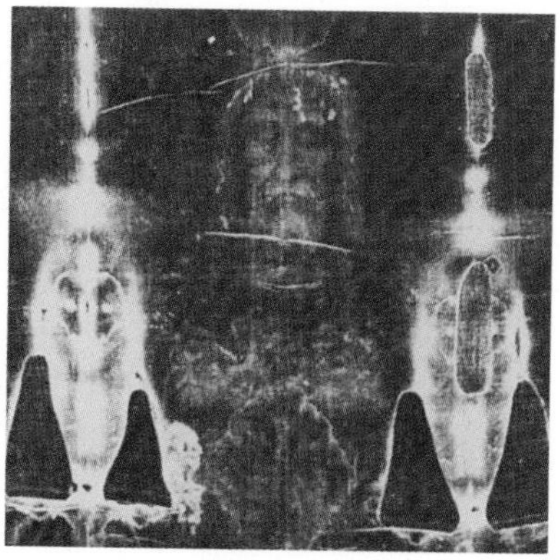
The Shroud of Turin

*Cosmic ray intensity does vary slightly with time, which affects the amount of atmospheric ^{14}C. From ^{14}C activity in ancient trees, we know that the amount fell slightly about 3000 years ago to current levels. Recently, nuclear testing and fossil fuel combustion have also altered the fraction of ^{14}C slightly. Taking these factors into account improves the accuracy of the dating method.

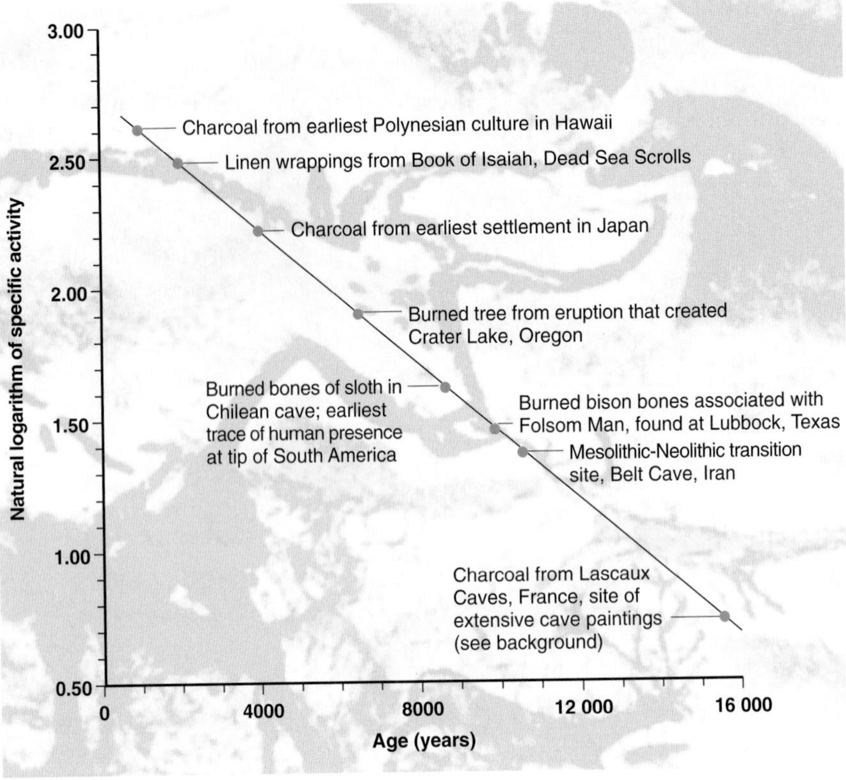

FIGURE 25.8 Ages of several objects, as determined by radiocarbon dating

the United States independently measured the $^{12}C/^{14}C$ ratio of a 50 mg piece of linen from the shroud and determined that the flax from which the linen had been made had been grown between 1260 C.E. and 1390 C.E. However, the age of the shroud determined by the radiocarbon dating was flawed because the piece of linen that was tested was a newer piece, which had been incorporated into the shroud later, as a repair.

To determine the ages of more ancient objects or the ages of objects that do not contain carbon, different radioisotopes are measured. For example, comparing the ratio of ^{238}U ($t_{1/2} = 4.5 \times 10^9$ years) to its final decay product, ^{206}Pb, in meteorites gives 4.7 billion years for the age of the Solar System, and therefore Earth. From this and other isotope ratios, such as $^{40}K/^{40}Ar$ ($t_{1/2}$ of $^{40}K = 1.3 \times 10^9$ years) and $^{87}Rb/^{87}Sr$ ($t_{1/2}$ of $^{87}Rb = 4.9 \times 10^{10}$ years), Moon rocks collected by the Apollo astronauts have been dated at 4.2 billion years old.

Sample Problem 25.5 Applying Radiocarbon Dating

Problem The charred bones of a sloth in a cave in Chile represent the earliest evidence of human presence at the southern tip of South America. A sample of the bone has a specific activity of 5.22 disintegrations per minute per gram of carbon [d/(min·g)]. If the $^{12}C/^{14}C$ ratio for living organisms results in a specific activity of 15.3 d/(min·g), how old are the bones ($t_{1/2}$ of ^{14}C = 5730 years)?

Plan We calculate k from the given $t_{1/2}$ (5730 years). Then we apply Equation 25.6 to find the age (t) of the bones, using the given activities of the bones ($\mathscr{A}_t$ = 5.22 d/min·g) and a living organism ($\mathscr{A}_0$ = 15.3 d/(min·g)).

Solution Calculate k for ^{14}C decay:

$$k = \frac{\ln 2}{t_{1/2}} = \frac{0.693}{5730 \text{ years}} = 1.21 \times 10^{-4} \text{ year}^{-1}$$

Calculate the age (t) of the bones:

$$t = \frac{1}{k} \ln \frac{\mathscr{A}_0}{\mathscr{A}_t} = \frac{1}{1.21 \times 10^{-4} \text{ year}^{-1}} \ln \left[\frac{15.3 \text{ d/(min·g)}}{5.22 \text{ d/(min·g)}}\right] = 8.89 \times 10^3 \text{ years}$$

The bones are about 8900 years old.

Check The activity of the bones is between $\frac{1}{2}$ and $\frac{1}{4}$ the activity of a living organism, so the age should be between one and two half-lives (5730 and 11 460 years).

Follow-Up Problem 25.5 A sample of wood from an Egyptian mummy case has a specific activity of 9.41 d/(min•g). How old is the case?

SUMMARY OF SECTION 25.2

- Ionization and scintillation counters measure the number of emissions from a radioactive sample.
- The decay rate (activity) of a sample is proportional to the number of radioactive nuclei. Nuclear decay is a first-order process, so the half-life is constant.
- Radioisotopic methods, such as ^{14}C dating, determine the age of an object by measuring the ratio of specific isotopes in the object.

25.3 Nuclear Transmutation: Induced Changes in Nuclei

The alchemists' dream of changing base metals into gold was never realized, but, in the early 20th century, nuclear physicists *did* change one element into another. Research into **nuclear transmutation**, the *induced* conversion of the nucleus of one element into the nucleus of another, was closely linked to research into atomic structure and led to the discovery of the neutron and the production of artificial radioisotopes. Later, high-energy bombardment of nuclei in particle accelerators began the ongoing effort to create new nuclides and new elements and, most recently, to answer fundamental questions about matter and energy.

Early Transmutation Experiments and Nuclear Shorthand Notation

The first recognized transmutation occurred in 1919, when Ernest Rutherford showed that α particles emitted from radium bombarded atmospheric nitrogen to form a proton and oxygen-17:

$$^{14}_{7}N + ^{4}_{2}\alpha \longrightarrow ^{1}_{1}p + ^{17}_{8}O$$

By 1926, experimenters had found that α bombardment transmuted most elements with low atomic numbers to the next higher element, with the ejection of a proton.

A shorthand notation used specifically for nuclear transmutation reactions shows the reactant (target) nucleus to the left and the product nucleus to the right of a set of parentheses, within which a comma separates the projectile particle from the ejected particle(s):

$$\text{reactant nucleus (particle in, particle(s) out) product nucleus} \qquad (25.7)$$

Using this notation, we can write the previous reaction as $^{14}N\ (\alpha, p)\ ^{17}O$.

An unexpected finding in a transmutation experiment led to *the discovery of the neutron*. When lithium, beryllium, and boron were bombarded with α particles, they emitted highly penetrating radiation that could not be deflected by a magnetic field or an electric field. Unlike γ radiation, these emissions were massive enough to eject protons from the substances they penetrated. In 1932, James Chadwick, a student of Rutherford's, proposed that the emissions consisted of neutral particles with a mass similar to the mass of a proton. He named the neutral particles *neutrons*. He received the Nobel Prize in Physics in 1935 for his discovery.

In 1933, Irene and Frederic Joliot-Curie (*see photo*), the daughter and son-in-law of Marie and Pierre Curie, created the first artificial radioisotope, phosphorus-30. When they bombarded aluminum foil with α particles, phosphorus-30 and neutrons were formed:

$$^{27}_{13}Al + ^{4}_{2}\alpha \longrightarrow ^{1}_{0}n + ^{30}_{15}P \qquad \text{or} \qquad ^{27}Al\ (\alpha, n)\ ^{30}P$$

Since then, other techniques for producing artificial radioisotopes have been developed. In fact, the majority of the nearly 1000 known radionuclides have been produced artificially.

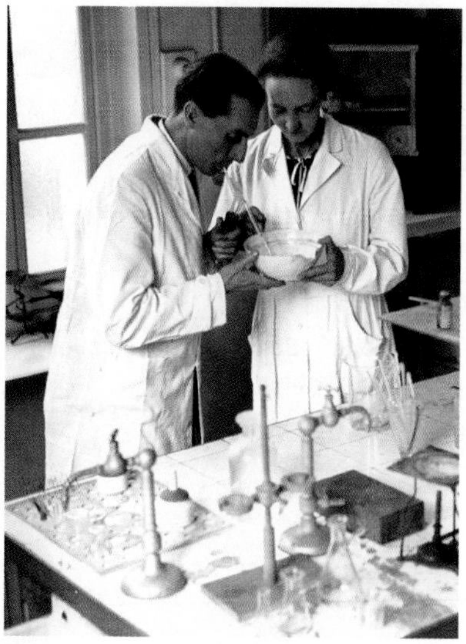
Frederic and Irene Joliot-Curie in their laboratory

Particle Accelerators and the Transuranium Elements

During the 1930s and 1940s, researchers probing the nucleus bombarded elements with neutrons, α particles, protons, and **deuterons** (nuclei of the stable hydrogen isotope deuterium, ^{2}H). Neutrons are especially useful as projectiles because they have no charge and, thus, are not repelled as they approach a target nucleus. The other particles are all positive, so early researchers found it difficult to give them enough energy to overcome their repulsion by the target nuclei. Beginning in the 1930s, however, **particle accelerators** were invented to impart high kinetic energies to particles by placing them in an electric field, usually in combination with a magnetic field. In the simplest and earliest design, protons are introduced at one end of a tube and attracted to the other end of the tube by a potential difference.

Linear Accelerator A major advance was the *linear accelerator*, a series of separated tubes of increasing length that, through a source of alternating voltage, change their charge from positive to negative in synchrony with the movement of the particle through them (Figure 25.9).

A proton, for example, exits the first tube just when this tube becomes positive and the next tube becomes negative. Repelled by the first tube and attracted by the second, the proton accelerates across the gap between them. The process is performed in stages to achieve high particle energies without having to apply a single high voltage. A 12 m linear accelerator with 46 tubes, built in California after World War II, accelerated protons to speeds several million times faster than earlier accelerators.

Other Early Particle Accelerators The *cyclotron* (Figure 25.10), invented by E. O. Lawrence in 1930, applies the principle of the linear accelerator but uses electromagnets to give the particle a spiral path to save space. The magnets lie within an evacuated chamber above and below two "dees," open D-shaped electrodes that act like the tubes in the linear design. The particle is accelerated as it passes from one dee, which is momentarily positive, to the other dee, which is momentarily negative. Its speed and path radius increase until it is deflected toward the target nucleus.

The *synchrotron* uses a synchronously increasing magnetic field to make the particle's path circular rather than spiral.

Powerful Modern Accelerators Some very powerful accelerators, which are used to study the physics of high-energy particle collisions, include both a linear section and a synchrotron section. The *tevatron* at the Fermi Lab near Chicago has a circumference of 6.3 km and can accelerate particles to an energy slightly less than 1×10^{12} eV (electron volts), or 1 TeV, before colliding them together.

A small section of the Large Hadron Collider

The world's most powerful accelerator is the Large Hadron Collider (LHC) near Geneva, Switzerland (*see photo*). The first successful collision of protons in the LHC was achieved in early 2010. When operating at peak capacity, the LHC can accelerate protons to an energy of 7 TeV as they travel around the main ring at a speed of 11 000 revolutions per second, attaining a final speed about 99.99% of the speed of light before colliding with other protons. Physicists are eager to study the results of these high-energy subatomic collisions for information on the nature of matter and the early universe.

Synthesis of Transuranium Elements Scientists use accelerators for many applications, from producing the radioisotopes used in medical applications to studying the fundamental nature of matter. Perhaps the most specific application for chemists is the synthesis of **transuranium elements**, elements with atomic numbers higher than

FIGURE 25.9 Schematic diagram of a linear accelerator

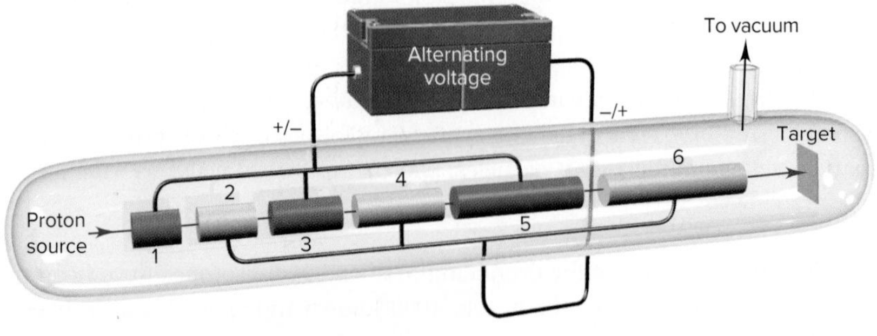

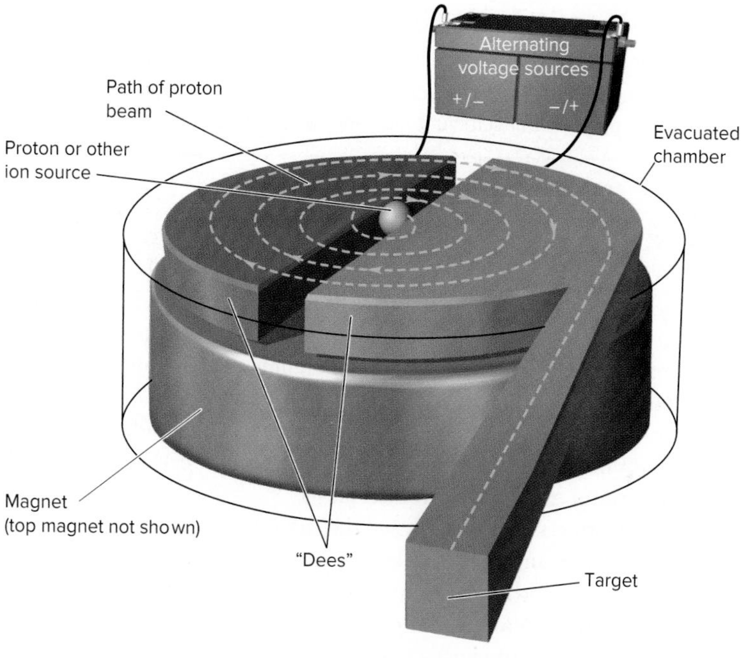

FIGURE 25.10 Schematic diagram of a cyclotron

uranium, the heaviest naturally occurring element. Some of the reactions that were used to form several transuranium elements appear in Table 25.7. The transuranium elements include the remaining actinides (Z = 93 to 103), in which the $5f$ sublevel is filled; the elements in the fourth transition series (Z = 104 to 112), in which the $6d$ sublevel is filled; and the remaining elements in the $7p$ sublevel (Z = 113 to 118).

Conflicting claims of discovery by scientists in the United States and the former Soviet Union led to controversies about names for some of the more recently synthesized elements. To provide interim names until the disputes could be settled, the International Union of Pure and Applied Chemistry (IUPAC) adopted a system that uses the atomic number as the basis for a Latin name. Thus, for example, element 104 was named unnilquadium (un = 1, nil = 0, quad = 4, ium = metal suffix), with the symbol Unq. After much compromise, IUPAC suggested the following names: 104, rutherfordium (Rf); 105, dubnium (Db); 106, seaborgium (Sg); 107, bohrium (Bh); 108, hassium (Hs); 109, meitnerium (Mt); 110, darmstadtium (Ds); 111, roentgenium (Rg); and 112, copernicium (Cn).

In 2012, IUPAC announced the official names of elements 114 and 116. Element 114 now has the official name flerovium and the chemical symbol Fl. It is named after Soviet physicist Georgy Flyorov, the founder of the Joint Institute for Nuclear Research (JINR) in Dubna, Moscow Oblast, Russia, where the element was discovered. Element 116 now has the official name livermorium and the chemical symbol Lv. This element is named after the Lawrence Livermore National Laboratory, within the city of Livermore, California, which collaborated with JINR on the discovery of the element in 2000. Other elements with higher atomic numbers have not yet been named.

TABLE 25.7 Formation of Some Transuranium Nuclides*

Reaction							Half-Life of Product
$^{239}_{94}Pu$	+	$2^{1}_{0}n$	$\longrightarrow$	$^{241}_{95}Am$	+	$^{0}_{-1}\beta$	432 years
$^{239}_{94}Pu$	+	$^{4}_{2}\alpha$	$\longrightarrow$	$^{242}_{96}Cm$	+	$^{1}_{0}n$	163 days
$^{241}_{95}Am$	+	$^{4}_{2}\alpha$	$\longrightarrow$	$^{243}_{97}Bk$	+	$2^{1}_{0}n$	4.5 h
$^{242}_{96}Cm$	+	$^{4}_{2}\alpha$	$\longrightarrow$	$^{245}_{98}Cf$	+	$^{1}_{0}n$	45 min
$^{253}_{99}Es$	+	$^{4}_{2}\alpha$	$\longrightarrow$	$^{256}_{101}Md$	+	$^{1}_{0}n$	76 min
$^{243}_{95}Am$	+	$^{18}_{8}O$	$\longrightarrow$	$^{256}_{103}Lr$	+	$5^{1}_{0}n$	28 s

*Like chemical reactions, nuclear reactions may occur in several steps. For example, the first reaction in this table is actually an overall process that occurs in three steps:

(1) $^{239}_{94}Pu + ^{1}_{0}n \longrightarrow ^{240}_{94}Pu$ (2) $^{240}_{94}Pu + ^{1}_{0}n \longrightarrow ^{241}_{94}Pu$ (3) $^{241}_{94}Pu \longrightarrow ^{241}_{95}Am + ^{0}_{-1}\beta$

SUMMARY OF SECTION 25.3

- One nucleus can be transmuted to another through bombardment with high-energy particles.
- Accelerators increase the kinetic energy of particles in nuclear bombardment experiments and are used to produce transuranium elements and radioisotopes for medical use.

25.4 Effects of Nuclear Radiation on Matter

In 1986, an accident at the Chernobyl nuclear facility in Ukraine released radioactivity that, according to the World Health Organization, has caused and will cause thousands of cancer deaths (31 people died as an immediate result, and an estimated 15 000 more people died in the surrounding area after exposure to the radiation). In the same year, isotopes used in medical treatment emitted radioactivity that prevented thousands of cancer deaths. In this section and Section 25.5, we examine radioactivity's harmful and beneficial effects.

Both types of effects occur because *nuclear changes cause chemical changes in surrounding matter.* In other words, even though the nucleus of an atom may undergo a reaction with little or no involvement of the atom's electrons, the emissions from that reaction *do* affect the electrons of nearby atoms.

Virtually all radioactivity causes **ionization** in surrounding matter as the emissions collide with atoms and dislodge electrons:

$$\text{atom} \xrightarrow{\text{ionizing radiation}} \text{ion}^+ + e^-$$

From each ionization event, a cation and a free electron result, and the number of such *cation-electron pairs* produced is directly related to the energy of the incoming **ionizing radiation**.

Many applications of ionizing radiation depend not only on the ionizing event itself, but also on secondary processes. For example, in a Geiger-Müller counter, the free electron of the cation-electron pair often collides with another atom, which ejects a second electron. In a scintillation counter, the initial ionization eventually results in the emission of light. In nuclear power plants (Section 25.7), the initial process, as well as several secondary processes, causes the release of heat that makes steam to create electricity.

Effects of Ionizing Radiation on Living Tissue

Ionizing radiation has a destructive effect on living tissue. If the ionized atom is part of a key biological macromolecule or cell membrane, the results can be devastating to the cell and perhaps the organism.

The danger from a radionuclide depends on three factors: the type of radiation; its half-life; and, most important, its biological behaviour. For example, both ^{235}U and ^{239}Pu emit α particles and have long half-lives. However, uranium is rapidly excreted by the body, whereas plutonium behaves like calcium and, thus, is incorporated into bones and teeth. One of the most dangerous radionuclides is $^{90}Sr^{2+}$ (formed during a nuclear explosion) because it behaves very similarly to Ca^{2+} and is absorbed rapidly by bones. Let us first see how to measure radiation dose; then we will look in more detail at the damaging power of ionizing radiation.

Units of Radiation Dose and Its Effects To measure the effects of ionizing radiation, we need a unit for radiation dose. Units of radioactive decay, such as the becquerel and the curie, measure the number of decay events in a given time, but not their energy or absorption by matter. However, the number of cation-electron pairs produced in a given amount of living tissue *is* a measure of the energy absorbed by the tissue. The SI unit for such energy absorption is the **gray (Gy)**; it is equal to one joule of energy absorbed per kilogram of body tissue: 1 Gy = 1 J/kg. A more widely used unit is the **rad (radiation-absorbed dose)**, which is one-hundredth as much:

$$1 \text{ rad} = 0.01 \text{ J/kg} = 0.01 \text{ Gy}$$

To measure actual tissue damage, we must account for differences in the strength of the radiation, the exposure time, and the type of tissue. To do this, we multiply the number of rad by a relative biological effectiveness (RBE) factor, which depends on the effect of a given type of radiation on a given tissue or body part. The product is the **rem (roentgen equivalent for man)**, the unit of radiation dosage equivalent to a given amount of tissue damage in a human:

$$\text{No. of rem} = \text{no. of rad} \times \text{RBE}$$

Doses are often expressed in millisievert (10^{-3} Sv). The **sievert (Sv)**, which is the SI unit for dosage equivalent, is defined in the same way as the rem, but with the absorbed dose in gray (Gy); thus, 1 Sv = 100 rem.

Penetrating Power of Emissions The effect of a radiation dose on living tissue depends on the penetrating power *and* ionizing ability of the radiation. Since water is the main component of living tissue, the penetrating power is often measured in terms of the depth of water that stops 50% of the incoming radiation. In Figure 25.11, the average values of the penetrating distances are shown, in actual sizes, for the three common types of emissions. Note, in general, that *penetrating power is inversely related to the mass, charge, and energy of the emission.* In other words, if a particle interacts strongly with matter, it penetrates only slightly, and vice versa.

1. *Alpha particles.* Alpha particles are massive and highly charged, which means that they interact with matter the most strongly of the three common emissions. As a result, they penetrate so little that a piece of paper, light clothing, or the outer layer of skin can stop α radiation from an external source. However, if ingested, an α emitter can cause grave localized internal damage through extensive ionization. For example, in the early 20th century, wristwatch and clock dials were painted by hand with paint containing radium so they would glow in the dark. To write the numbers clearly, the women who applied the paint "tipped" the fine brushes repeatedly between their lips. Small amounts of ingested $^{226}Ra^{2+}$ were incorporated into their bones along with normal Ca^{2+}, which led to numerous cases of bone fracture and jaw cancer.

2. *Beta particles and positrons.* Beta particles (β^-) and positrons (β^+) have less charge and much less mass than α particles, so they interact less strongly with matter. Even though a given particle has less chance of causing ionization, a β^- (or β^+) emitter is a more destructive external source because the particles penetrate deeper. Specialized heavy clothing or a thick (0.5 cm) piece of metal is required to stop these particles. The relative danger from α and β^- emissions is shown by the poisoning of two former Russian intelligence agents. In 1957, an agent was poisoned by thallium-204, a β^- emitter, but he survived because β^- particles do relatively little radiation damage. In contrast, in 2006, another agent was poisoned with polonium-210, an α emitter, and died in three weeks because of the much greater radiation damage.

3. *Gamma rays.* Neutral, massless γ rays interact least with matter and, thus, penetrate the most. A block of lead 40 cm thick is needed to stop them. Therefore, an external γ ray source is the most dangerous because the energy can ionize many layers of living tissue.

The extent of interaction with matter is variable for emitted *neutrons*, which are neutral and massive. Outside a nucleus, neutrons decay with a half-life of 10.6 min into a proton and an electron (and an antineutrino). With energies ranging from very low to very high, they may be scattered or captured by interaction with nuclei.

Molecular Interactions How does the damage take place on the molecular level? When ionizing radiation interacts with molecules, it causes the *loss of an electron from a bond or a lone pair.* The resulting charged species go on to form **free radicals**, molecular or atomic species with one or more unpaired electrons, which, as you know, are very reactive (Section 8.6). They form electron pairs by attacking bonds in other molecules, often forming more free radicals.

When γ radiation strikes living tissue, for instance, the most likely molecule to absorb it is water, yielding an electron and a water ion radical:

$$H_2O + \gamma \longrightarrow H_2O\cdot^+ + e^-$$

The products collide with other water molecules to form more free radicals:

$$H_2O\cdot^+ + H_2O \longrightarrow H_3O^+ + \cdot OH \qquad \text{and} \qquad e^- + H_2O \longrightarrow H\cdot + OH^-$$

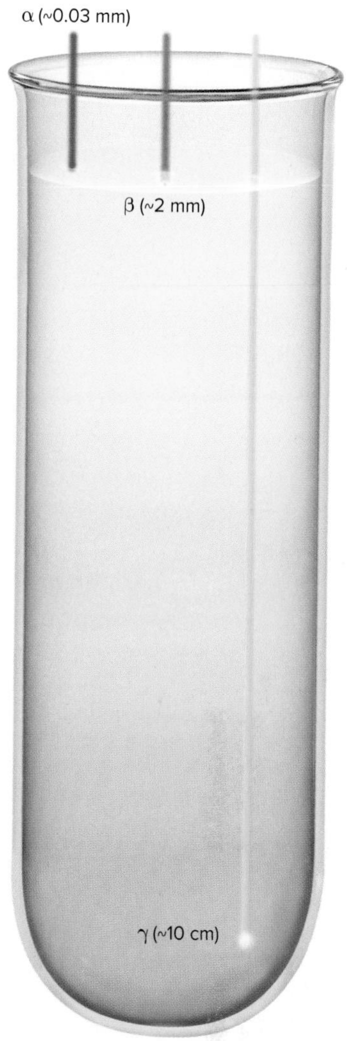

FIGURE 25.11 Penetrating power of radioactive emissions

These free radicals attack more H_2O and surrounding biomolecules, whose bonding and structure are intimately connected with their function (Section 22.3).

The double bonds in membrane lipids are highly susceptible to free-radical attack:

$$H\cdot + RCH{=}CHR' \longrightarrow RCH_2{-}\dot{C}HR'$$

In this reaction, one electron of the π bond forms a C—H bond between one of the double-bonded carbons and the H, and the other electron resides on the other carbon to form a free radical. Changes to lipid structure cause damage that results in leakage through cell membranes and destruction of the protective fatty tissue around organs. Changes to critical bonds in enzymes lead to their malfunction as catalysts of metabolic reactions. Changes in the nucleic acids and proteins that govern the rate of cell division can cause cancer. Genetic damage and mutations may occur when bonds in the DNA of sperm and egg cells are altered by free radicals.

Sources of Ionizing Radiation

We are continuously exposed to ionizing radiation from natural and artificial sources (Table 25.8). Indeed, life evolved in the presence of natural ionizing radiation, called **background radiation**. The same radiation that alters bonds in DNA and causes harmful mutations also causes beneficial mutations that allow species to evolve.

Background radiation has several natural sources:

- *Cosmic radiation* increases with altitude because of decreased absorption by the atmosphere. Thus, people in Lake Louise, Alberta (1661 m above sea level), absorb twice as much cosmic radiation as people in Vancouver (between 100 m and 150 m above sea level); even a jet flight involves measurable absorption.
- *Thorium and uranium minerals* are present in rocks and soil. Radon, the heaviest noble gas in group 18, is a radioactive product of uranium and thorium decay. Its concentration in the air we breathe is related to the presence of trace minerals in building materials and to the uranium content of local soil and rocks (Figure 25.12). Once it enters the body, radon poses a serious potential hazard as

TABLE 25.8 Typical Radiation Doses from Natural and Artificial Sources

Source of Radiation	Average Adult Exposure
Natural	
Cosmic radiation	0.30–0.50 mSv/year
Radiation from the ground	
From clay soil and rocks	~0.25–1.70 mSv/year
In wooden houses	0.10–0.20 mSv/year
In brick houses	0.60–0.70 mSv/year
In concrete (cinder block) houses	0.60–1.60 mSv/year
Radiation from the air (mainly radon)	
Outdoors, average value	0.20 mSv/year
In wooden houses	0.70 mSv/year
In brick houses	1.30 mSv/year
In concrete (cinder block) houses	2.60 mSv/year
Internal radiation from minerals in tap water and daily intake of food (^{40}K, ^{14}C, Ra)	~0.40 mSv/year
Artificial	
Diagnostic X-ray methods	
Lung (local)	0.04–0.2 rad/film
Kidney (local)	1.5–3 rad/film
Dental (dose to the skin)	≤ 1 rad/film
Therapeutic radiation treatment	Locally ≤10 000 rad
Other sources	
Jet flight (4 h)	~0.01 mSv
Nuclear testing	<0.04 mSv/year
Nuclear power industry	<0.01 mSv/year
Total average value	1.00–2.00 mSv/year

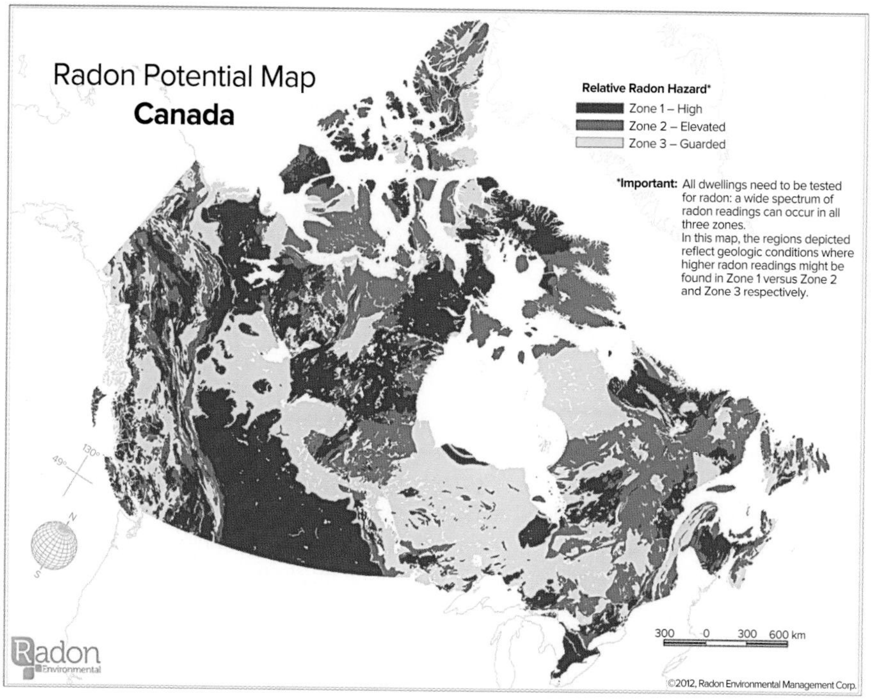

FIGURE 25.12 Radon distribution in Canada*

*Courtesy Radon Management Corp. www.radoncorp.com.

it decays to radioactive nuclides of Po, Pb, and Bi, through α, β^-, and γ emissions. The emissions damage lung tissue, and the heavy-metal atoms that are formed aggravate the problem. In 2009, the World Health Organization said that radon contributes to 15% of annual lung cancer deaths worldwide.

- About 150 g of K^+ *ions* is dissolved in the water in the tissues of an average adult, and 0.0118% of these ions are radioactive ^{40}K. The presence of these ions, as well as atmospheric $^{14}CO_2$, makes *all* food (including bananas and Brazil nuts), water, clothing, and building materials slightly radioactive.

The largest artificial source of radiation, and the source that is easiest to control, is from medical diagnostic techniques, especially X-rays. The radiation dosages from nuclear testing and radioactive waste disposal are miniscule for most people, but exposures for those living near test sites or disposal areas may be much higher.

Assessing the Risk from Ionizing Radiation

How much radiation is too much? To answer this question, we must ask several others: How strong is the exposure? How long is the exposure? Which tissue is exposed? Are children affected? One reason we lack clear data to answer these questions is that scientific ethical standards forbid the intentional exposure of humans in an experimental setting. However, accidentally exposed radiation workers and Japanese survivors of atomic bombs have been studied extensively.

Table 25.9 summarizes the immediate effects of an acute single dose of ionizing radiation to the whole human body. The severity of the effects increases with dose; a dose of 5 Sv will kill about 50% of the exposed population within a month.

Most data come from laboratory animals, whose biological systems may differ greatly from ours. Nevertheless, studies with mice and dogs show that lesions and cancers appear after massive whole-body exposure, with rapidly dividing cells affected first. In an adult animal, these are cells of the bone marrow, organ linings, and reproductive organs. However, many other tissues are affected in an immature animal or fetus. Studies in both animals and humans show an increase in the incidence of cancer from either a high single exposure or a low chronic exposure.

Two current models of radiation risk are as follows:

- The *linear response model* proposes that radiation effects, such as cancer risks, accumulate over time regardless of dose, and that populations should not be exposed to any radiation above background levels.
- The *S-shaped response model* implies that there is a threshold above which the effects are more significant: very low risk at low dose and high risk at high dose.

TABLE 25.9 Acute Effects of a Single Dose of Whole-Body Irradiation

Dose (mSv)	Effect	Lethal Dose Population (%)	Lethal Dose No. of Days
0.05–0.20	Possible late effect; possible chromosomal aberrations	—	—
0.20–1.00	Temporary reduction in white blood cells	—	—
0.50+	Temporary sterility in men (1.00+ Sv = 1-year duration)	—	—
1.00–2.00	"Mild radiation sickness": vomiting, diarrhea, tiredness in a few hours	—	—
	Reduction in infection resistance		
	Possible bone growth retardation in children		
3.00+	Permanent sterility in women	—	—
5.00	"Serious radiation sickness": marrow/intestine destruction	50–70	30
4.00–10.00	Acute illness; early death	60–95	30
30.00+	Acute illness; death in hours to days	100	2

Reliable data on the genetic effects of radiation are scarce. Studies on fruit flies show a linear increase in genetic defects with both dose and exposure time. However, in a mouse, whose genetic system is obviously much more similar to ours than the fruit fly's genetic system, a total dose given over a long period created one-third as many genetic defects as the same dose given over a short period. Therefore, the rate of exposure is a key factor. The children of atomic bomb survivors show higher-than-normal childhood cancer rates, implying that their parents' reproductive systems were affected.

SUMMARY OF SECTION 25.4

- All radioactive emissions cause ionization.
- The effect of ionizing radiation on living matter depends on the quantity of energy absorbed and the extent of ionization in a given type of tissue. The radiation dose for the human body is measured in sievert (Sv).
- Ionization forms free radicals, some of which proliferate and destroy biomolecular function.
- All organisms are exposed to varying quantities of natural ionizing radiation.
- Studies show that either a large acute dose or a chronic small dose is harmful.

25.5 Applications of Radioisotopes

Radioisotopes are powerful tools for studying processes in biochemistry, medicine, materials science, environmental studies, and many other scientific and industrial fields. Such uses depend on the fact that *isotopes of an element exhibit* ***very*** *similar chemical and physical behaviour.* In other words, except for having a less stable nucleus, a radioisotope of an element has nearly the same properties as a nonradioactive isotope of the element.* For example, the fact that $^{14}CO_2$ is used by a plant in the same way as $^{12}CO_2$ is used forms the basis of radiocarbon dating.

Radioactive Tracers

Just think how useful it could be to follow a substance through a complex process or from one region of a system to another. A tiny amount of a radioisotope mixed with a large amount of the stable isotope can act as a **tracer**, a chemical "beacon" emitting radiation that signals the presence of the substance.

Reaction Pathways Tracers are used to identify simple and complex reaction pathways.

1. *Inorganic systems: the periodate-iodide reaction.* One well-studied example is the reaction between periodate and iodide ions:

$$IO_4^-(aq) + 2I^-(aq) + H_2O(l) \longrightarrow I_2(s) + IO_3^-(aq) + 2OH^-(aq)$$

*Although this statement is generally correct, differences in isotopic mass *can* influence bond strengths and therefore reaction rates. Such behaviour is called a *kinetic isotope effect* and is particularly important for isotopes of hydrogen—^1H, ^2H, and ^3H—because their masses differ by such large proportions. Section 23.4 discussed how the kinetic isotope effect is employed in the industrial production of heavy water, D_2O.

Is IO_3^- the result of IO_4^- reduction or I^- oxidation? When we add "cold" (nonradioactive) IO_4^- to a solution of I^- that contains some "hot" (radioactive, indicated in red) $^{131}I^-$, we find that the I_2 is radioactive, not the IO_3^-:

$$IO_4^-(aq) + 2\,^{131}I^-(aq) + H_2O(l) \longrightarrow {}^{131}I_2(s) + IO_3^-(aq) + 2OH^-(aq)$$

These results show that IO_3^- forms through the reduction of IO_4^-, and that I_2 forms through the oxidation of I^-. To confirm this pathway, we add IO_4^- containing some $^{131}IO_4^-$ to a solution of I^-. As we expected, the IO_3^- is radioactive, not the I_2:

$$^{131}IO_4^-(aq) + 2I^-(aq) + H_2O(l) \longrightarrow I_2(s) + {}^{131}IO_3^-(aq) + 2OH^-(aq)$$

Thus, tracers act like "handles" we can "hold" to follow the changing reactants.

2. *Biochemical pathways: photosynthesis.* Far more complex pathways can also be followed with tracers. The photosynthetic pathway, the most essential and widespread metabolic process on Earth, in which energy from sunlight is used to form the chemical bonds of glucose, has an overall reaction that looks quite simple:

$$6CO_2(g) + 6H_2O(l) \xrightarrow[\text{chlorophyll}]{\text{light}} C_6H_{12}O_6(s) + 6O_2(g)$$

However, the actual process is extremely complex: 13 enzyme-catalyzed steps are required to incorporate each C atom from CO_2, so the six CO_2 molecules incorporated to form a molecule of $C_6H_{12}O_6$ require six repetitions of the pathway. Melvin Calvin and his coworkers took seven years to determine the pathway, using ^{14}C in CO_2 as the tracer and chromatography as the method for separating the products formed after different times of light exposure. Calvin won the Nobel Prize in Chemistry in 1961 for this remarkable achievement.

Physiological Studies Tracers are used in many studies of physiological function. Some recent studies examine the challenges of living in outer space. In an animal study of red blood cell loss during extended space flight, blood plasma volume was measured using albumin (a blood protein) labelled with ^{125}I, and the survival of blood cells was assessed using red blood cells labelled with ^{51}Cr as a trace substitute for Fe. In another study, blood flow in skin under long periods of microgravity was monitored using injected ^{133}Xe.

Material Flow Tracers are used in studies of solid surfaces and the flow of materials. Metal atoms, hundreds of layers deep within a solid, have been shown to exchange with metal ions from the surrounding solution within a matter of minutes. Chemists and engineers use tracers to study material movement in semiconductor chips, paint, and metal plating; in detergent action; and in the process of corrosion, to mention just a few of the many applications.

Hydrologic engineers use tracers to study the volume and flow of large bodies of water. By following radionuclides that formed during atmospheric nuclear bomb tests (^3H in H_2O, $^{90}Sr^{2+}$, and $^{137}Cs^+$), scientists have mapped the flow of water from land to lakes and streams to oceans. They also use tracers to study the surface and deep ocean currents that circulate around the globe, the mechanisms of hurricane formation, and the mixing of the troposphere and stratosphere. Industries use tracers to study material flow during manufacturing processes, such as the flow of ore pellets in smelting kilns, the paths of wood chips and bleach in paper mills, and the diffusion of fungicide into lumber. Tracers are also used in a particularly important application, the porosity and leakage of oil and gas wells in geologic formations.

Activation Analysis Another use of tracers is in *neutron activation analysis* (NAA). In NAA, neutrons bombard a nonradioactive sample, converting a small fraction of its atoms to radioisotopes. These radioisotopes exhibit characteristic decay patterns, such as γ-ray spectra, that reveal the elements present. Unlike chemical analysis, NAA leaves the sample virtually intact, so the method can be used to determine the composition of a valuable object or a very small sample. For example, a painting thought to be a 16th-century Dutch masterpiece was shown through NAA to be a 20th-century forgery, because a microgram-size sample of its pigment contained much less silver and antimony than the pigments used by the Dutch masters. Forensic chemists use NAA to detect traces of ammunition on a suspect's hand or traces of arsenic in the hair of a victim of poisoning.

Automotive engineers employ NAA and γ-ray detectors to measure the friction and wear of moving parts without having to take an engine apart. For example, when a steel surface that has been neutron-activated to form radioactive ^{59}Fe moves against a second steel surface, the amount of radioactivity on the second surface indicates the amount of material rubbing off. The radioactivity appearing in a lubricant placed between the surfaces can demonstrate the lubricant's ability to reduce the wear.

Medical Diagnosis The largest use of radioisotopes is in medical science. Over 10 000 hospitals worldwide use radioisotopes in medicine, and about 90% of the procedures are for diagnosis. Most medical radioisotopes that are made in nuclear reactors are sourced from relatively few research reactors, including NRU (National Research Universal) at Chalk River near Ottawa. ■

■ Canada's NRU reactor at Chalk River and the Netherlands' High Flux Reactor (HFR) at Petten account together for nearly two-thirds of the world's medical isotope supply. The Government of Canada is "looking to transform the way medical isotopes are produced in Canada, and in particular Tc-99m, so that Canadian production is: on a sound commercial footing without government subsidization; scaled to the needs of Canadians; sustainable in terms of environmental impacts, health, safety and security, and so that Canada remains a global technological leader. We believe that this transformation will best serve the needs of Canadians for a secure supply of medical isotopes in the medium and longer term." (Source: http://www.parl.gc.ca/HousePublications/Publication.aspx?DocId=5063422&Language=E&Mode=1&Parl=40&Ses=3).

Tracers with half-lives of a few minutes to a few days are used to observe specific organs and body parts. They can be given by injection or inhalation, or orally. For example, a healthy thyroid gland incorporates dietary I^- into iodine-containing hormones at a known rate. To assess thyroid function, the patient drinks a solution containing a trace amount of $Na^{131}I$; the thyroid gland absorbs $^{131}I^-$ ions, which undergo β^- decay, and the emissions produce an image of the gland (Figure 25.13A). Technetium-99 ($Z = 43$) is also used for imaging the thyroid (Figure 25.13B), as well as the heart, lungs, and liver. Technetium does not occur naturally, so the radioisotope (actually a metastable form, ^{99m}Tc) is prepared from radioactive molybdenum just before use:

$$^{90}_{42}\text{Mo} \longrightarrow {}^{99m}_{43}\text{Tc} + {}^{0}_{-1}\beta$$

TABLE 25.10	Some Radioisotopes Used as Medical Tracers
Isotope	**Body Part or Process**
^{11}C, ^{18}F, ^{13}N, ^{15}O	PET studies of brain and heart
^{60}Co, ^{192}Ir	Cancer therapy
^{64}Cu	Metabolism of copper
^{59}Fe	Blood flow, spleen
^{67}Ga	Tumour imaging
^{123}I, ^{131}I	Thyroid
^{111}In	Brain, colon
^{42}K	Blood flow
^{81m}Kr	Lungs
^{99m}Tc	Heart, thyroid, liver, lung, bone
^{201}Tl	Heart muscle
^{90}Y	Cancer, arthritis

Tracers are used to measure physiological processes, such as blood flow, as well. The rate at which the heart pumps blood, for example, can be observed by injecting ^{59}Fe, which becomes incorporated into the hemoglobin of blood cells. Several radioisotopes that are used in medical diagnosis are listed in Table 25.10.

Positron-emission tomography (PET) is a powerful imaging method for observing brain structure and function. A biological substance is synthesized, with one of its atoms replaced by an isotope that emits positrons. The substance is injected into a patient's bloodstream, from which it is taken up into the brain. The isotope emits positrons, each of which annihilates a nearby electron. In this process, two γ photons are emitted simultaneously, 180° from each other:

$$^{1}_{0}\beta + {}^{0}_{-1}\text{e} \longrightarrow 2\,^{0}_{0}\gamma$$

An array of detectors around the patient's head pinpoint the sites of γ emission, and the image is analyzed by a computer. Two of the isotopes used are ^{15}O, injected as

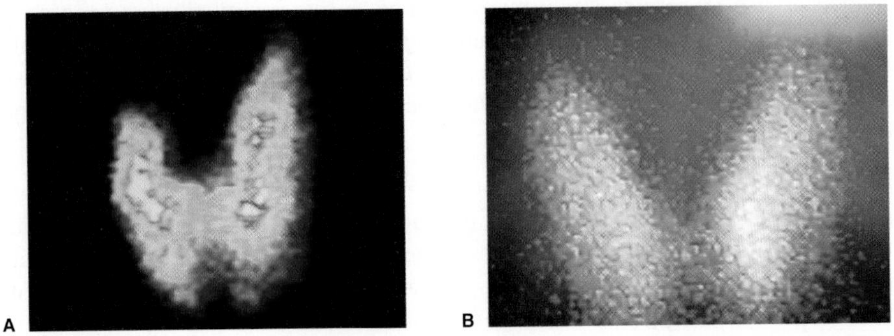

FIGURE 25.13 The use of radioisotopes to image the thyroid gland. A. This ^{131}I scan shows an asymmetric image that is indicative of disease. **B.** A ^{99}Tc scan of a healthy thyroid.

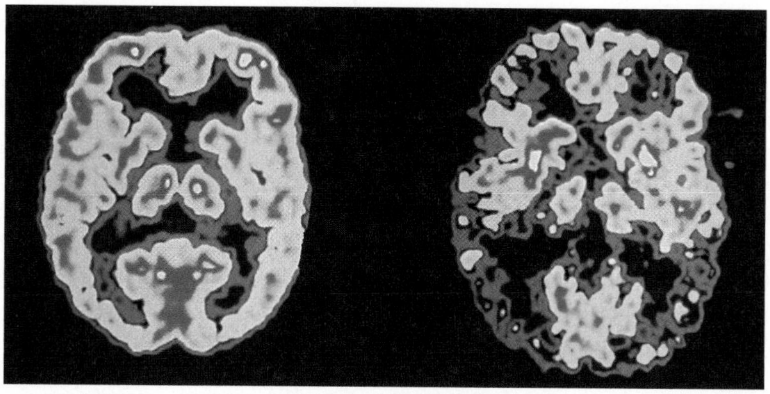

FIGURE 25.14 PET and brain activity. These PET scans show brain activity in a normal person (*left*) and in a patient with Alzheimer's disease (*right*). Red and yellow indicate relatively high activity within a region.

$H_2{}^{15}O$ to measure blood flow, and ^{18}F, bonded to a glucose analogue to measure glucose uptake, which is an indicator of energy metabolism. Among many fascinating PET findings are those that show how changes in blood flow and glucose uptake accompany normal or abnormal brain activity (Figure 25.14). Also, substances incorporating ^{11}C and ^{15}O are being investigated using PET to learn how molecules interact with and move along the surface of a catalyst.

Additional Applications of Ionizing Radiation

Many other uses of radioisotopes involve higher energy ionizing radiation:

- *Radiation therapy.* Cancer cells divide more rapidly than normal cells, so radioisotopes that interfere with cell division kill more cancer cells than normal cells. Implants of ^{198}Au or ^{90}Sr isotopes, which decay to the γ-emitting ^{90}Y, have been used to destroy pituitary and breast tumour cells, and γ rays from ^{60}Co have been used to destroy tumours of the brain and other body parts.
- *Destruction of microbes.* Irradiation of food increases shelf life by killing microorganisms that cause rotting or spoilage (Figure 25.15), but the practice is controversial. Advocates point to the benefits of preserving fresh foods, grains, and seeds for long periods, whereas opponents suggest that irradiation might lower the food's nutritional content or produce harmful by-products. Irradiation also provides a way to destroy newer, more resistant bacterial strains that survive the increasing use of the more common antibiotics in animal feed. The United Nations has approved irradiation for potatoes, wheat, chicken, and strawberries, and, in 2002, Health Canada approved it for select food products as well.
- *Insect control.* Ionizing radiation has been used to control harmful insects. Captured males are sterilized by radiation and released to mate, thereby reducing the number of offspring. This method has been used to control the Mediterranean fruit fly in California and disease-causing insects, such as the tsetse fly and malarial mosquito, in other parts of the world.
- *Power for spacecraft instruments.* A nonharmful use of ionizing radiation relies on its secondary processes. Most spacecraft use solar energy to provide power for instruments. However, in deep-space missions, when solar energy is too weak, a radioisotope heater unit (RHU) has been used. It consists of a $^{238}PuO_2$ fuel pellet, the size of a pencil eraser, that is clad within a multilayered graphite and metal shell a little smaller than a flashlight battery. The whole RHU weighs only 40 g.

FIGURE 25.15 The increased shelf life of irradiated food

Canadian Applications

Canadian Light Source Synchrotron The Canadian Light Source (CLS) synchrotron is a third-generation 2.9 GeV synchrotron located in Saskatoon, Saskatchewan (*see photo*). It opened

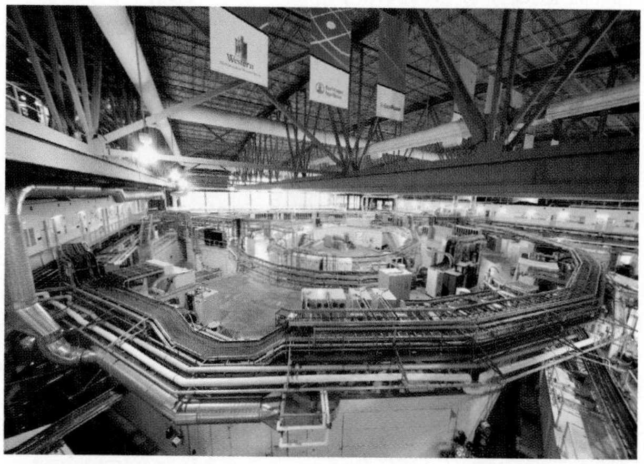

The Canadian Light Source synchrotron in Saskatoon, Saskatchewan

on October 22, 2004, after three years of construction, and cost $173.5 million. One of 42 such facilities in the world, and the only one in Canada, it occupies a footprint the size of a football field on the grounds of the University of Saskatchewan. Using powerful magnets and radio-frequency waves, the synchrotron accelerates electrons to nearly the speed of light, producing intense light beams for probing matter with unprecedented precision. The CLS synchrotron can be used to probe the structure of matter and analyze a host of physical, chemical, geologic, and biological processes. Information obtained by scientists can be used to design new drugs and medical treatments (such as medical imaging of tumours and other biological tissues), develop new materials for safer medical implants, examine the structure of surfaces to develop more effective motor oils, build more powerful microchips, improve food products, and discover ways to clean up the environment and eliminate mining wastes. More than 2500 academic and industrial researchers from across Canada, and from other countries, have used the facility so far.

TIGRESS

TRIUMF TRIUMF (Tri-University Meson Facility) is Canada's national laboratory for particle and nuclear physics. It is owned and operated as a joint venture by a consortium of 17 Canadian universities via a contribution through the National Research Council, with building capital funds provided by the Government of British Columbia.

TRIUMF has the world's largest cyclotron, which is used to generate exotic atoms for unlocking the secrets of the stars and discovering the medical isotopes of the future. New methods for disease detection and treatment are also being developed at TRIUMF. At its Centre for Molecular and Materials Science, scientists measure and test high-temperature superconductors; physical, green, and materials design chemistry; exotic magnetism, strongly correlated systems, and quantum phase transitions; hydrogen-materials interactions; and industrial basic research for automobile technology. As well, TRIUMF operates a proton and neutron irradiation facility.

The TRIUMF-ISAC Gamma-Ray Escape Suppressed Spectrometer, or TIGRESS (*see photo*), is a versatile γ-ray spectrometer that is used at TRIUMF's Isotope Separator and Accelerator (ISAC) radioactive beam facility. In a typical experiment, accelerated isotopes are delivered down the beam line. The isotopes hit a target inside the ball, and a nuclear reaction takes place. Gamma rays created in the reaction are detected by TIGRESS, and this, in conjunction with the response from the detectors in the ball, allows scientists to identify the products of the reaction, providing a greater understanding of the structure and behaviour of the nucleus.

■ Arthur B. McDonald is a Canadian astrophysicist; Professor Emeritus at Queen's University in Kingston, Ontario; and was the Director of the Sudbury Neutrino Observatory. He shared the 2015 Nobel Prize in Physics with Takaaki Kajita "for the discovery of neutrino oscillations, which shows that neutrinos have mass."

Sudbury Neutrino Observatory The Sudbury Neutrino Observatory (SNO) was constructed to study the fundamental properties of neutrinos, in particular the mass and mixing parameters. It is located 2039 m below the surface in Vale Inco's Creighton Mine in Sudbury, Ontario. The detector at SNO was designed to detect solar neutrinos through their interactions with a large tank of heavy water. The long distance to the Sun makes the search for neutrino mass sensitive to much smaller mass splittings than can be studied with terrestrial sources. The detector was turned on in May 1999 and was turned off on November 28, 2006. Although new data are no longer being collected, the SNO collaboration will continue for the next several years to analyze the data that were collected. The underground laboratory has been enlarged, and other experiments continue to operate at SNOLAB. ■

SUMMARY OF SECTION 25.5

- Radioisotopic tracers have been used to study reaction mechanisms, material flow, elemental composition, and medical conditions.
- Ionizing radiation has been used to destroy cancer tissue, kill organisms that spoil food, control insect populations, and power spacecraft instruments.

The Sudbury Neutrino Detector

25.6 The Interconversion of Mass and Energy

Most of the nuclear processes we have considered so far involve radioactive decay, in which a nucleus emits one or a few small particles or photons to become a more stable, slightly lighter nucleus. Two other nuclear processes cause much greater mass changes. In nuclear **fission**, a heavy nucleus splits into two much lighter nuclei, emitting several small particles at the same time. In nuclear **fusion**, the opposite process occurs: two lighter nuclei combine to form a heavier nucleus. Both fission and fusion release enormous quantities of energy. Let us take a look at the origins of this energy by first examining the change in mass that accompanies the breakup of a nucleus into its nucleons and then considering the energy that is equivalent to this mass change.

The Mass Difference between a Nucleus and Its Nucleons

Since the early 20th century, we have known that mass and energy are interconvertible. The separate mass and energy conservation laws are combined to state that *the total quantity of mass-energy in the universe is constant.* Therefore, when *any* reacting system releases or absorbs energy, it also loses or gains mass.

Mass Difference and Chemical Reactions The interconversion of mass and energy is not important for chemical reactions because the energy changes involved in breaking or forming chemical bonds are so small that the mass changes are negligible. For example, when 1 mol of water breaks up into its atoms, heat is absorbed:

$$H_2O(g) \longrightarrow 2H(g) + O(g) \qquad \Delta_r H^\circ = 2 \times \text{BE of O—H} = 934\ \text{kJ/mol}$$

We find the mass that is equivalent to this energy from *Einstein's equation*:

$$E = mc^2 \quad \text{or} \quad \Delta E = \Delta mc^2 \quad \text{so} \quad \Delta m = \frac{\Delta E}{c^2} \qquad (25.8)$$

where Δm is the mass difference between the reactants and products:

$$\Delta m = m_{\text{products}} - m_{\text{reactants}}$$

Substituting the enthalpy of reaction (in J/mol) for ΔE and the numerical value for c (2.9979×10^8 m/s), we obtain

$$\Delta m = \frac{9.34 \times 10^5\ \text{J/mol}}{(2.9979 \times 10^8\ \text{m/s})^2} = 1.04 \times 10^{-11}\ \text{kg/mol} = 1.04 \times 10^{-8}\ \text{g/mol}$$

(Units of kg/mol are obtained because the joule includes the kilogram: $1\ \text{J} = 1\ \text{kg·m}^2/\text{s}^2$.) The mass of 1 mol of H_2O molecules (reactant) is about 10 ng *less* than the combined masses of 2 mol of H atoms and 1 mol of O atoms (products), a change that is difficult to measure with even the most sophisticated balance. Such minute mass changes when bonds break or form allow us to assume that, for all practical purposes, mass is conserved in *chemical* reactions.

Mass Difference and Nuclear Reactions The much larger mass change that accompanies a *nuclear* process is related to the enormous energy required to bind the nucleus together from its parts. In an analogy with the calculation above, involving the water molecule, consider the change in mass that occurs when one ^{12}C nucleus breaks apart into its nucleons—six protons and six neutrons:

$$^{12}C \longrightarrow 6\,^1_1p + 6\,^1_0n$$

We calculate this mass difference in a special way. By combining the mass of six H *atoms* and six neutrons and then subtracting the mass of one ^{12}C *atom*, the masses of the electrons cancel: six e^- (in six ^1H atoms) cancel six e^- (in one

^{12}C atom). The mass of one ^{1}H atom is 1.007 825 u, and the mass of one neutron is 1.008 665 u, so

$$\text{Mass of six } {}^{1}\text{H atoms} = 6 \times 1.007\,825 \text{ u} = 6.046\,950 \text{ u}$$
$$\text{Mass of six neutrons} = 6 \times 1.008\,665 \text{ u} = 6.051\,990 \text{ u}$$
$$\text{Total mass} = 12.098\,940 \text{ u}$$

The mass of the reactant, one ^{12}C atom, is 12 u (exactly). The mass difference (Δm) we obtain is the total mass of the nucleons minus the mass of the nucleus:

$$\begin{aligned}\Delta m &= 12.098\,940 \text{ u} - 12.000\,000 \text{ u} \\ &= 0.098\,940 \text{ u}/{}^{12}\text{C} = 0.098\,940 \text{ g/mol } {}^{12}\text{C}\end{aligned}$$

Two key points emerge from these calculations:

- *The mass of the nucleus is* **less** *than the combined masses of its nucleons*: there is *always* a mass decrease when nucleons form a nucleus.
- The mass change of this nuclear process (9.89×10^{-2} g/mol) is nearly 10 million times the mass change of the chemical process (10.4×10^{-9} g/mol) we examined earlier and *can* be observed on any laboratory balance.

Nuclear Binding Energy and the Binding Energy per Nucleon

Einstein's equation for the relation between mass and energy allows us to find the energy equivalent of any mass change. For 1 mol of ^{12}C, after converting grams to kilograms, we have

$$\begin{aligned}\Delta E = \Delta mc^2 &= (9.8940 \times 10^{-5} \text{ kg/mol})(2.9979 \times 10^{8} \text{ m/s})^2 \\ &= 8.8921 \times 10^{12} \text{ J/mol} = 8.8921 \times 10^{9} \text{ kJ/mol}\end{aligned}$$

This quantity of energy is called the **nuclear binding energy** for carbon-12, and the positive value means that *energy is absorbed.* This is the energy required to break 1 mol of ^{12}C atoms into neutrons and hydrogen atoms. Thus, the *nuclear binding energy* is the energy to *break 1 mol of nuclei into individual nucleons*:

$$\text{nucleus} + \text{nuclear binding energy} \longrightarrow \text{nucleons}$$

Thus, the nuclear binding energy is qualitatively analogous to the sum of the bond energies of a covalent compound or the lattice energy of an ionic compound. Quantitatively, however, nuclear binding energies are typically several million times greater.

The Electron Volt as the Unit of Binding Energy We use joules to express the binding energy per mole of nuclei, but the joule is much too large a unit to express the binding energy of a single nucleus. Instead, nuclear scientists use the **electron volt (eV)**, the energy that an electron acquires when it moves through a potential difference of 1 V (volt):

$$1 \text{ eV} = 1.602 \times 10^{-19} \text{ J}$$

Binding energies are commonly expressed in millions of electron volts, that is, in *mega–electron volts* (MeV):

$$1 \text{ MeV} = 10^{6} \text{ eV} = 1.602 \times 10^{-13} \text{ J}$$

A particularly useful factor converts the atomic mass unit to its energy equivalent in electron volts:

$$1 \text{ u} = 931.5 \times 10^{6} \text{ eV} = 931.5 \text{ MeV} \qquad (25.9)$$

Earlier, we found the mass change when ^{12}C breaks apart into its nucleons to be 0.098 940 u. The binding energy per ^{12}C nucleus, expressed in MeV, is

$$\frac{\text{binding energy}}{{}^{12}\text{C nucleus}} = 0.098\,940 \text{ u} \times \frac{931.5 \text{ MeV}}{1 \text{ u}} = 92.16 \text{ MeV}$$

We can compare the stability of nuclides of different elements by determining the *binding energy per nucleon*. For ^{12}C, we have

$$\text{Binding energy per nucleon} = \frac{\text{binding energy}}{\text{no. of nucleons}} = \frac{92.16\ \text{MeV}}{12\ \text{nucleons}} = 7.680\ \text{MeV/nucleon}$$

Sample Problem 25.6 Calculating the Binding Energy per Nucleon

Problem Iron-56 is an extremely stable nuclide. Calculate the binding energy per nucleon for ^{56}Fe, and compare it with the binding energy for ^{12}C. (Masses: ^{56}Fe atom = 55.934 939 u; ^{1}H atom = 1.007 825 u; neutron = 1.008 665 u.)

Plan Iron-56 has 26 protons and 30 neutrons. We calculate the mass difference, Δm, when the nucleus forms, by subtracting the given mass of one ^{56}Fe atom from the sum of the masses of 26 ^{1}H atoms and 30 neutrons. To find the binding energy per nucleon, we multiply Δm by the equivalent in MeV (931.5 MeV/u) and divide by the number of nucleons (56).

Solution Calculate the mass difference, Δm:

$$\begin{aligned}\text{Mass difference} &= [(26 \times \text{mass } ^{1}\text{H atom}) + (30 \times \text{mass neutron})] - \text{mass } ^{56}\text{Fe atom} \\ &= [(26)(1.007\,825\ \text{u}) + (30)(1.008\,665\ \text{u})] - 55.934\,939\ \text{u} \\ &= 0.528\,46\ \text{u}\end{aligned}$$

Calculate the binding energy per nucleon:

$$\text{Binding energy per nucleon} = \frac{0.528\,46\ \text{u} \times 931.5\ \text{MeV/u}}{56\ \text{nucleons}} = 8.790\ \text{MeV/nucleon}$$

An ^{56}Fe nucleus would require more energy per nucleon to break up into its nucleons than ^{12}C would (7.680 MeV/nucleon), so ^{56}Fe is more stable than ^{12}C.

Check The answer is consistent with the great stability of ^{56}Fe. Given the number of decimal places in the values, rounding to check the math is useful only to find a *major* error. The number of nucleons (56) is an exact number, so we retain four significant figures.

Follow-Up Problem 25.6 Uranium-235 is an essential component of the fuel in nuclear power plants. Calculate the binding energy per nucleon for ^{235}U. Is this nuclide more or less stable than ^{12}C? (Mass of ^{235}U atom = 235.043 924 u.)

Fission or Fusion: Means of Increasing the Binding Energy per Nucleon

Calculations for other nuclides, similar to those in Sample Problem 25.6, show that the binding energy per nucleon varies considerably. The essential point is that *the greater the binding energy per nucleon, the more stable the nuclide.*

Figure 25.16 shows a plot of the binding energy per nucleon versus mass number. It provides information about nuclide stability and the two possible processes that nuclides can undergo to form more stable nuclides. Most nuclides with fewer than 10 nucleons have a relatively small binding energy per nucleon. The ^{4}He nucleus is an exception: it is stable enough to be emitted intact as an α particle. Above A = 12, the binding energy per nucleon varies from about 7.6 MeV to 8.8 MeV.

The most important observation is that *the binding energy per nucleon peaks at elements with $A \approx 60$.* In other words, nuclides become more stable with increasing mass number up to around 60 nucleons and then become less stable with higher numbers of nucleons. The existence of a peak of stability suggests that there are two ways nuclides can increase their binding energy per nucleon:

- *Fission.* A heavier nucleus can *split into lighter nuclei (closer to $A \approx 60$)* by undergoing fission. The product nuclei have greater binding energy per nucleon (are more stable) than the reactant nucleus, and the difference in *energy is*

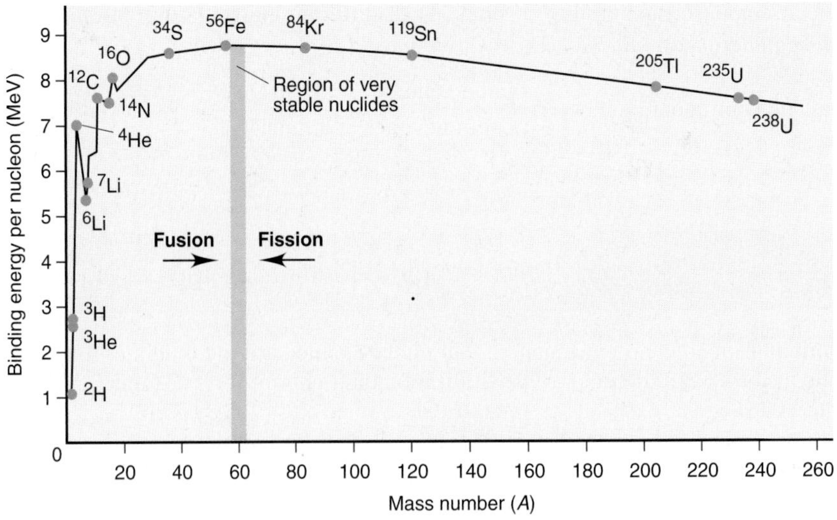

FIGURE 25.16 The variation in binding energy per nucleon

released. Nuclear power plants generate energy through fission, as do atomic bombs (Section 25.7).

- *Fusion.* Lighter nuclei, on the other hand, can *combine to form a heavier nucleus (closer to $A \approx 60$)* by undergoing fusion. Once again, the product is more stable than the reactants, and *energy is released.* The Sun and other stars generate energy through fusion, as do thermonuclear (hydrogen) bombs. In these examples, and in all current research efforts for developing fusion as a useful energy source, hydrogen nuclei fuse to form the very stable helium-4 nucleus.

In Section 25.7, we will examine fission and fusion and their applications.

SUMMARY OF SECTION 25.6

- The mass of a nucleus is less than the sum of the masses of its nucleons. The energy equivalent to this mass difference is the nuclear binding energy, often expressed in units of mega–electron volts (MeV).
- The binding energy per nucleon is a measure of nuclide stability and varies with the number of nucleons. Nuclides with $A \approx 60$ are the most stable.
- Lighter nuclides join (fusion) or heavier nuclides split (fission) to create more stable products.

25.7 Applications of Fission and Fusion

Of the many beneficial applications of nuclear reactions, the greatest is the potential for abundant quantities of energy, which is based on the multimillion-fold increase in energy yield of nuclear reactions over chemical reactions. Our experience with nuclear energy from power plants in the recent past, however, has shown that we must improve how we tap this energy source safely and economically and deal with the waste generated. In this section, we discuss how fission and fusion occur and how we are applying them.

The Process of Nuclear Fission

During the mid-1930s, Enrico Fermi and coworkers bombarded uranium ($Z = 92$) with neutrons in an attempt to synthesize transuranium elements. Many of the unstable nuclides produced were tentatively identified as having $Z > 92$, but other

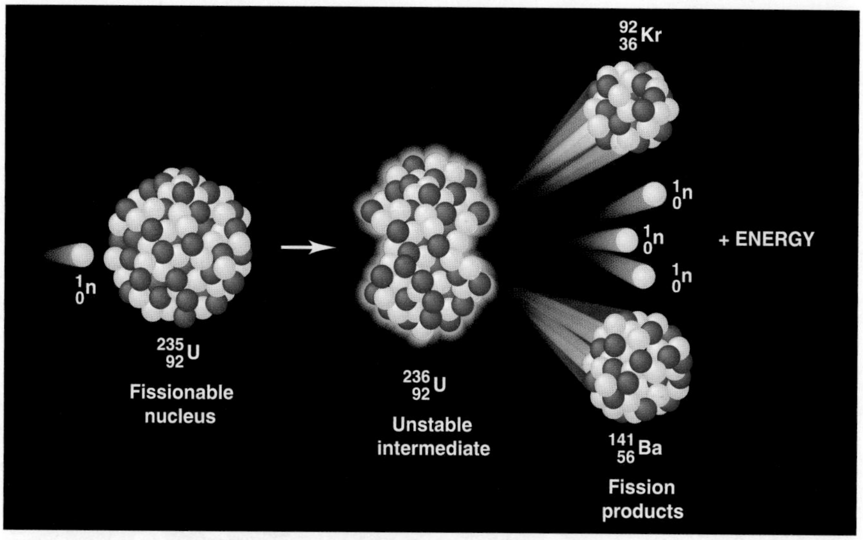

FIGURE 25.17 Fission of ^{235}U caused by neutron bombardment

scientists were skeptical. Four years later, the German chemist Otto Hahn and his associate F. Strassmann showed that one of these unstable nuclides was an isotope of barium ($Z = 56$). The Austrian physicist Lise Meitner, a coworker of Hahn, and her nephew Otto Frisch proposed that barium resulted from the *splitting* of the uranium nucleus into *smaller* nuclei, a process that they called *fission* as an analogy to cell division in biology. Element 109 was named *meitnerium* in honor of this extraordinary physicist, who proposed the correct explanation of nuclear fission and discovered the element protactinium (Pa; $Z = 91$), as well as numerous radioisotopes.

The ^{235}U nucleus can split in many different ways, giving rise to various daughter nuclei, but all routes have the same general features. Figure 25.17 depicts one of these fission patterns. Neutron bombardment results in a highly excited ^{236}U nucleus, which splits apart in 10^{-14} s. The products are two nuclei of unequal masses, two to four neutrons (average of 2.4), and a large quantity of energy. A single ^{235}U nucleus releases 3.5×10^{-11} J when it splits; 1 mol of ^{235}U (about 0.25 kg) releases 2.1×10^{13} J—a billion times as much energy as burning 0.25 kg of coal (about 2×10^{4} J)!

Chain Reaction and Critical Mass We harness the energy of nuclear fission, much of which eventually appears as heat, by means of a **chain reaction** (Figure 25.18): the few neutrons that are released by the fission of one nucleus collide with other fissionable nuclei and cause them to split, releasing more neutrons, and so on, in a self-sustaining process (with each step shown as a vertical dashed line). In this way, the energy released increases rapidly because each fission event in a chain reaction releases about two and a half times as much energy as the preceding fission event.

Whether a chain reaction occurs depends on the mass (and thus the volume) of the fissionable sample. If the piece of uranium is large enough, the product neutrons strike another fissionable nucleus *before* flying out of the sample, and a chain reaction takes place. The mass required to achieve a chain reaction is called the **critical mass**. If the sample has less than the critical mass (a *subcritical mass*), too many product neutrons leave the sample before they collide with and cause the fission of another ^{235}U nucleus, and a chain reaction does not occur.

Uncontrolled Fission: The Atomic Bomb An uncontrolled chain reaction can be adapted to make an extremely powerful explosive, as several of the world's leading atomic physicists hypothesized just prior to World War II. In August 1939, Albert Einstein wrote to the president of the United States, Franklin Delano Roosevelt, warning him about the danger of allowing the Nazi government to develop this

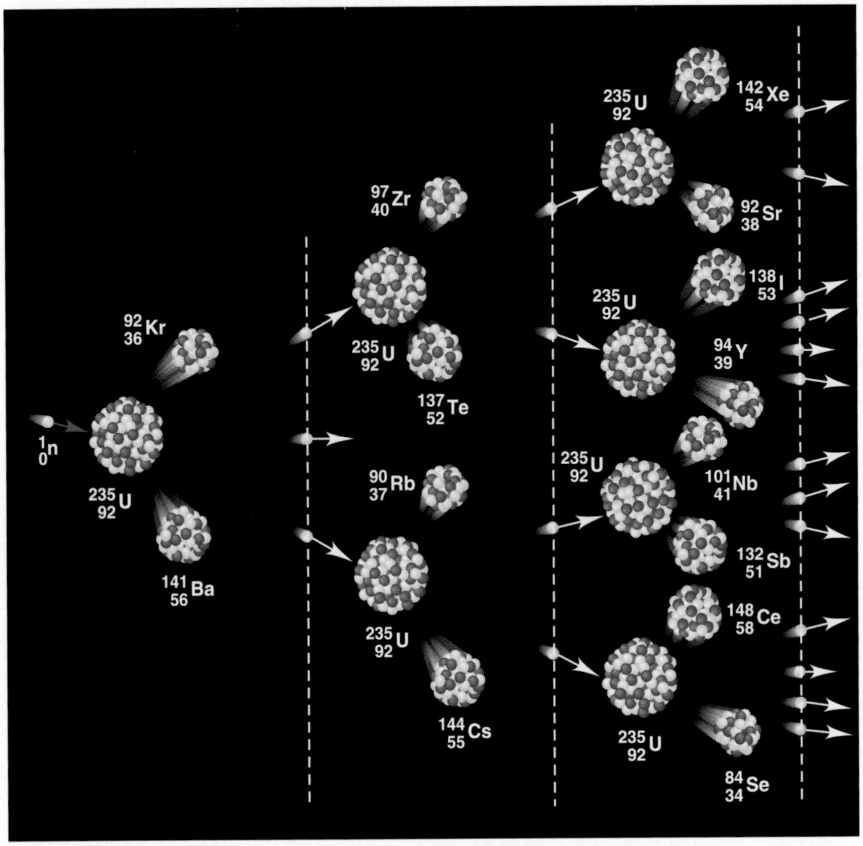

FIGURE 25.18 A chain reaction involving the fission of ^{235}U

FIGURE 25.19 Louis Slotin (1910–1946) was a Canadian physicist and chemist who took part in the Manhattan Project, the secret U.S. project during World War II that developed the atomic bomb. As part of the Manhattan Project, Slotin performed experiments with uranium and plutonium cores to determine their critical mass values. After World War II, Slotin continued his research at Los Alamos National Laboratory. On May 21, 1946, Slotin accidentally began a fission reaction, which released a burst of hard radiation. He was rushed to hospital, but died of radiation sickness nine days later, on May 30.

power first. It was his concern that led to the Manhattan Project, an enormous scientific effort to develop a bomb based on nuclear fission, which was initiated in 1941.* Louis Alexander Slotin (Figure 25.19), a Canadian physicist and chemist, was invited to participate in the project. In August 1945, the United States detonated two atomic bombs over Japan. The horrible destructive power of these bombs was a major factor in the surrender of the Japanese a few days later.

In an atomic bomb, small explosions of trinitrotoluene (TNT) bring subcritical masses of fissionable material together to exceed the critical mass, and the ensuing chain reaction brings about the explosion (Figure 25.20). This system was used in the first atomic bombs. Today, atomic bombs are based on plutonium-239, not uranium-235. Although any combination of plutonium isotopes can be used to make a nuclear weapon, not all combinations are equally convenient or efficient. The most common isotope, ^{239}Pu, is produced when the most common isotope of uranium, ^{238}U, absorbs a neutron and then quickly decays to plutonium. It is this plutonium isotope that is most useful for making nuclear weapons, and it is produced in varying quantities in virtually all operating nuclear reactors. The proliferation of nuclear power plants, which use fissionable materials to generate energy for electricity, has increased concern that more countries (and unscrupulous individuals) may have access to these materials for making bombs.

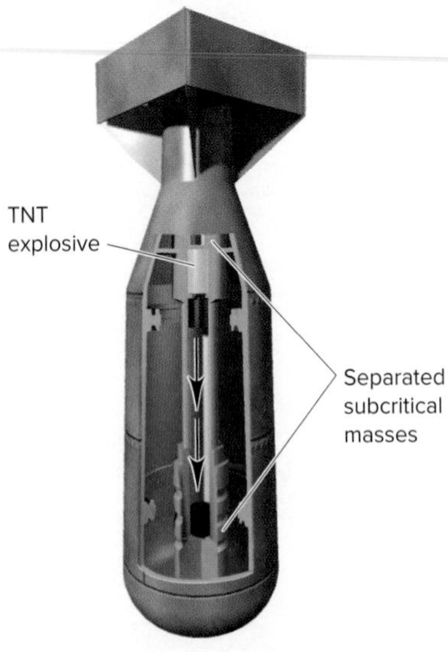

FIGURE 25.20 An atomic bomb based on ^{235}U

Controlled Fission: Nuclear Energy Reactors Like a coal-fired power plant, *a nuclear power plant generates heat to produce steam, which turns a turbine attached to an electrical generator.* However, a nuclear power plant has the potential to produce electrical power much more cleanly than that produced by the combustion of coal.

1. *Operation of a nuclear power plant.* Heat generation takes place in the **reactor core** of a nuclear plant (Figure 25.21). The core contains the *fuel rods,* which

*For an excellent scientific and historical account of the development of the atomic bomb, see R. Rhodes, *The Making of the Atomic Bomb* (New York: Simon and Schuster, 1986).

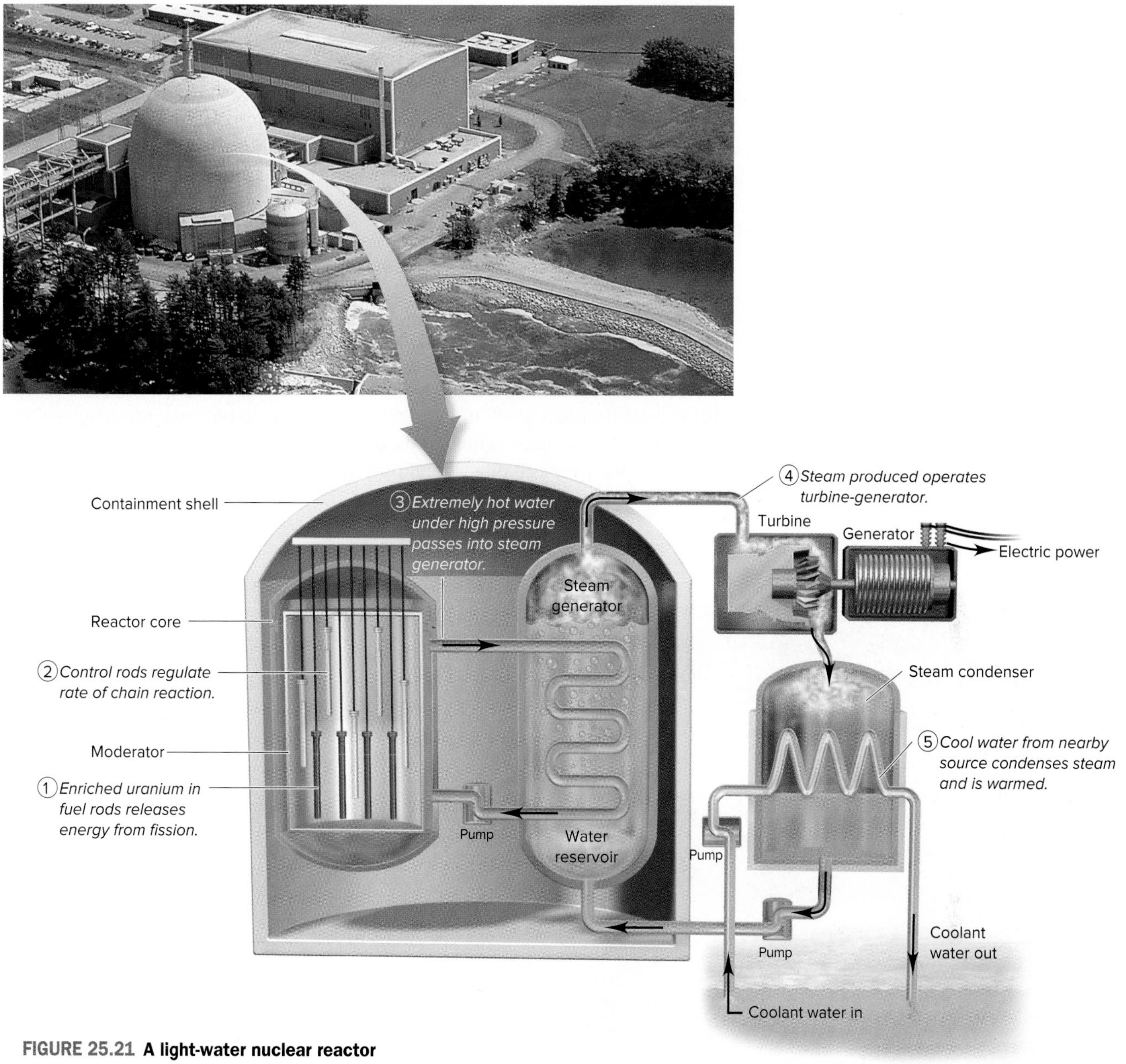

FIGURE 25.21 A light-water nuclear reactor

consist of fuel enclosed in tubes of a corrosion-resistant zirconium alloy. The fuel is uranium(IV) oxide (UO_2) that has been *enriched* from 0.7% ^{235}U, the natural abundance of this fissionable isotope, to the 3% to 4% ^{235}U required to sustain a chain reaction in a practical volume. (Enrichment of nuclear fuel is the most important application of Graham's law; see Section 4.5.) Sandwiched between the fuel rods are movable *control rods* made of cadmium or boron (or, in nuclear submarines, hafnium), substances that absorb neutrons very efficiently. When the control rods are lowered between the fuel rods, the chain reaction slows because fewer neutrons are available to bombard uranium atoms; when they are raised, the chain reaction speeds up. Neutrons that leave the fuel-rod assembly collide with a *reflector*, usually made from a beryllium alloy, which absorbs very few neutrons. The chain reaction can be speeded up by reflecting the neutrons back to the fuel rods.

Flowing around the fuel and control rods in the reactor core is the *moderator*, a substance that slows the neutrons, making them much better at causing fission than the fast neutrons that are emerging directly from the fission event. In most modern reactors, the moderator also acts as the *coolant*, the fluid that transfers the released heat to the steam-producing region. *Light-water reactors* use H_2O as the moderator because ^1H absorbs neutrons to some extent; in heavy-water reactors, D_2O is used. The advantage of D_2O is that it absorbs very few neutrons, leaving more available for fission, so heavy-water reactors can use uranium that has been *less enriched.* As the coolant flows around the encased fuel, pumps circulate it through coils that transfer its heat to the water reservoir. Steam formed in the reservoir turns the turbine that runs the generator. The steam then enters a condenser, where water from a lake or river is used to cool it back to liquid, which is returned to the water reservoir. Often (although not shown here), large cooling towers aid in this step (see Figure 12.11).

2. *Canada Deuterium Uranium (CANDU) reactors.* The CANDU reactor is a Canadian-invented pressurized heavy-water reactor (see Figure 25.22). It is similar to most light-water reactors in principle; however, the design has considerable differences. Some of the unique features of the CANDU design are described below:

- *Use of online refuelling.* CANDU plants use robotic machines to fuel the reactor with natural uranium while it is in operation. They do not undergo batch refuelling, since two machines simply hook up to the reactor faces, open the end caps (located on the pressure tubes), and push in the new fuel, while spent fuel comes out the other end.

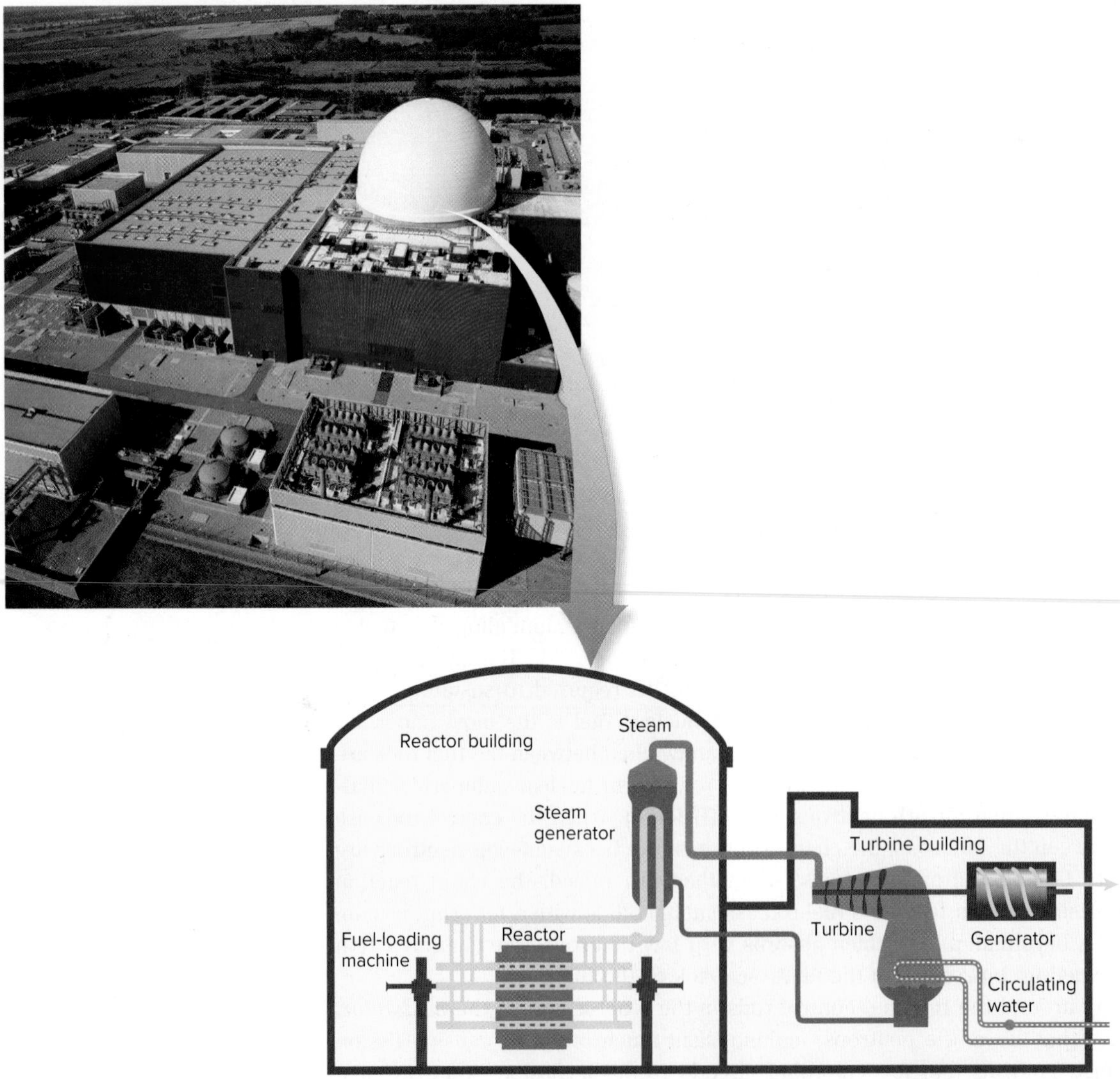

FIGURE 25.22 A four-unit CANDU (CANada Deuterium Uranium) reactor facility (Darlington Nuclear Generating Station). It supplies about 20% of Ontario's electricity needs.

- *Use of natural uranium.* By using a heavy-water moderator and heavy-water coolant, the CANDU reactor maintains a very high neutron economy. Therefore, the subsequent neutrons resulting from fission are used more effectively (with less loss), allowing the use of natural uranium as the fuel source and saving the cost of enrichment.
- *Pressure-tube design.* Each pressure tube is inside calendria tubes, and there are normally 380 to 480 such tubes assembled in a reactor. This design enables the use of online refuelling, as well as many other unique features of CANDU reactors.

All the nuclear power reactors built in Canada are CANDU reactors. CANDU reactors are also marketed abroad, and there are CANDU-type units operating in India, Pakistan, Argentina, South Korea, Romania, and China.

3. *Breeder reactors.* Because ^{235}U is not an abundant isotope, the *breeder reactor* was designed to consume one type of nuclear fuel as it produces another. Outside the moderator, fuel rods are surrounded by natural U_3O_8, which contains 99.3% *nonfissionable* ^{238}U atoms. As neutrons formed during ^{235}U fission escape the fuel rod, they collide with ^{238}U, transmuting it into ^{239}Pu, another fissionable nucleus:

$$^{238}_{92}U + ^{1}_{0}n \longrightarrow ^{239}_{92}U \qquad (t_{1/2} \text{ of } ^{239}_{92}U = 23.5 \text{ min})$$
$$^{239}_{92}U \longrightarrow ^{239}_{93}Np + ^{0}_{-1}\beta \qquad (t_{1/2} \text{ of } ^{239}_{93}Np = 2.35 \text{ days})$$
$$^{239}_{93}Np \longrightarrow ^{239}_{93}Pu + ^{0}_{-1}\beta \qquad (t_{1/2} \text{ of } ^{239}_{93}Pu = 2.4 \times 10^4 \text{ years})$$

Although breeder reactors can make fuel as they operate, they are difficult and expensive to build, and $^{239}_{94}Pu$ is extremely toxic and long lived. Breeder reactors are not used in the United States, although several are operating in Europe and Japan.

4. *Power plant accidents and other concerns.* Some major accidents at nuclear plants have caused decidedly negative public reactions. In 1979, malfunctions of coolant pumps and valves at the Three Mile Island facility in Pennsylvania led to the melting of some of the fuel and damage to the reactor core but, fortunately, the release of only a very small amount (about 1 Ci) of radioactive gases into the atmosphere. In 1986, a million times as much radioactivity (1 MCi) was released when a cooling system failure at the Chernobyl plant in Ukraine caused a much greater melting of fuel and an uncontrolled reaction. High-pressure steam and ignited graphite moderator rods caused the reactor building to explode and expel radioactive debris. Carried by prevailing winds, the radioactive particles contaminated vegetables and milk in much of Europe. As mentioned earlier, health officials have evidence that thousands of people living near the accident have already developed or may eventually develop cancer from radiation exposure. The design of the Chernobyl plant was particularly unsafe because, unlike reactors in North America and western Europe, the reactor was not contained in a massive concrete building.*

The largest nuclear disaster since the Chernobyl disaster happened in Fukushima, Japan, following the Tōhoku earthquake and tsunami on March 11, 2011. The Fukushima I Nuclear Power Plant consisted of six light-water boiling-water reactors designed by General Electric. These reactors drove electrical generators with a combined power of 4.7 GW (gigawatts), making Fukushima I one of the 25 largest nuclear power stations in the world. The tsunami broke the reactors' connection to the power grid, causing the reactors to begin to overheat. The Japanese government has estimated that the total amount of radioactivity released into the atmosphere was approximately one-tenth as much as was released during the Chernobyl disaster. Significant amounts of radioactive material have also been released into the groundwater and oceans.

5. *Current use of nuclear power.* Despite potential safety problems, nuclear power remains an important source of electricity. Worldwide, there are over 435

*An even greater release of radioactivity (1.1 MCi), which involved several radionuclides, including the extremely dangerous ^{90}Sr, occurred in 1957 in Kyshtym, a town in the South Ural Mountains of Russia. As a result, radioactivity spread globally, but the accident was kept a secret by the former Soviet government until 1980.

commercial nuclear power reactors in over 31 countries producing over 375 000 MWe of electricity, with at least another 70 reactors still under construction. They are expected to provide 11% of the world's clean electricity requirements. As our need for energy grows and climate change from fossil-fuel consumption worsens, nuclear power is likely to be seen as an increasingly clean and safe alternative.

6. *Thermal pollution and waste disposal.* Even a smoothly operating plant has certain inherent problems. The problem of *thermal pollution* is common to all power plants. The water that is used to condense the steam is several degrees warmer when it is returned to its source, which can harm aquatic organisms (Section 12.3). A more serious problem is *nuclear waste disposal.* Many of the fission products formed in nuclear reactors have long half-lives, and no satisfactory plan for their permanent disposal has yet been devised. Proposals to place the waste in containers and bury them in deep bedrock cannot possibly be field-tested for the thousands of years the material will remain harmful. Leakage of radioactive material into groundwater is a danger, and earthquakes can occur even in geologically stable regions. It remains to be seen whether we can operate fission reactors *and* dispose of their waste safely and economically.

The Promise of Nuclear Fusion

Nuclear fusion in the Sun is the ultimate source of nearly all the energy—and chemical elements—on Earth. In fact, *all the elements heavier than hydrogen were formed in fusion and decay processes within stars*, as described in the following Chemical Connections section.

Much research is being devoted to making nuclear fusion a practical, direct source of energy on Earth. To understand the advantages of fusion, let us consider one of the most discussed fusion reactions, in which deuterium and tritium react:

$$ {}^{2}_{1}\text{H} + {}^{3}_{1}\text{H} \longrightarrow {}^{4}_{2}\text{He} + {}^{1}_{0}\text{n} $$

This reaction produces 1.7×10^9 kJ/mol of energy, an enormous quantity of energy with no radioactive by-products. Moreover, the reactant nuclei are relatively easy to come by. We obtain deuterium from the electrolysis of water (Section 23.4). In nature, tritium forms through the cosmic (neutron) irradiation of ^{14}N:

$$ {}^{14}_{7}\text{N} + {}^{1}_{0}\text{n} \longrightarrow {}^{3}_{1}\text{H} + {}^{12}_{6}\text{C} $$

However, this process results in a natural abundance of only $10^{-7}\%$ ^{3}H. More practically, tritium is produced in nuclear accelerators by bombarding lithium-6 or by surrounding the fusion reactor itself with material containing lithium-6:

$$ {}^{6}_{3}\text{Li} + {}^{1}_{0}\text{n} \longrightarrow {}^{3}_{1}\text{H} + {}^{4}_{2}\text{He} $$

Thus, in principle, fusion *seems* promising and may represent an ideal source of power. Unfortunately, some extremely difficult technical problems remain. Fusion requires enormous energy in the form of heat to give the positively charged nuclei enough kinetic energy to force themselves together. The fusion of deuterium and tritium, for example, occurs at about 10^8 K, a temperature that is hotter than the Sun's core! How can such conditions be achieved? Two current research approaches have promise. In one approach, atoms are stripped of their electrons at high temperatures, which results in a gaseous *plasma*, a neutral mixture of positive nuclei and electrons. Because of the extreme temperatures needed for fusion, however, no *material* can contain the plasma. The most successful solution to this problem, to date, involves enclosing the plasma within a magnetic field. The *tokamak* design has a doughnut-shaped container in which a helical magnetic field confines the plasma and prevents it from contacting the walls (Figure 25.23). Scientists at the Princeton University Plasma Physics facility have achieved some success in generating energy from fusion this way. In another approach, the high temperature is reached by using many focused lasers to compress and heat the fusion reactants. With either approach, one or more major breakthroughs are needed before fusion will be a practical, everyday source of energy. Researchers all over the world are working on this. Some examples are the ITER project (which is a tokamak

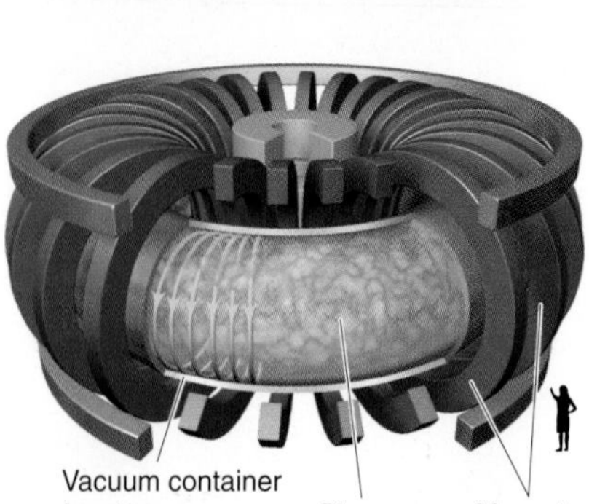

FIGURE 25.23 The tokamak design for the magnetic containment of a fusion plasma

design to be used as a pre-commercial reactor) and the stellarator (a fusion device developed in Germany that has many attributes that could make them much better prospects for a commercial fusion power plant than the tokamak). Here in Canada, General Fusion in Burnaby, British Columbia, is working on developing a fusion generator (see the TED talk given by Dr. Michel Laberge at http://www.ted.com/talks/michel_laberge_how_synchronized_hammer_strikes_could_generate_nuclear_fusion).

Advantages and Disadvantages of Nuclear Fission Energy

There are some pros and cons of using nuclear power generation.

Advantages 1. *Little pollution.* Nuclear power generation emits relatively low amounts of carbon dioxide (CO_2), whereas coal, gas, and oil-burning power plants increase the level of air pollution by emitting not only CO_2 but also gaseous by-products that contain sulfur. When the sulfur is absorbed in clouds, precipitation becomes sulfuric acid. Coal-fired plants emit more radiation into the air than a nuclear power plant does.

2. *Safety.* Although the results of a compromised reactor core can be disastrous, nuclear power is one the safest methods of producing energy. The reactors' many safety mechanisms make the chances of accidents very low. Many more people die from respiratory diseases due to coal burning or from mining and transportation accidents than from reactor accidents or from exposure to nuclear power plants. Although uranium mining leaves residue that leads to radon exposure, coal burning leaves ashes that will increase future radon exposure more significantly.

3. *Reliability.* Besides the fact that it is possible to generate a high amount of electrical power in one single plant, nuclear power plants need little fuel, so they are less vulnerable to shortages because of strikes or natural disasters. Kazakhstan produces the largest share of uranium from mines (33% of the world supply from mines), followed by Canada (18%) and Australia (11%). However, since uranium is evenly deposited around the globe, international relations will have little effect on the supply of fuel to the reactors.

4. *Technology.* One of the advantages of nuclear power generation is that the technology is readily available; it does not have to be developed first.

Disadvantages 1. *Meltdown.* The rods that contain the uranium fuel pellets overheat and dissolve if there is a loss of coolant water in a fission reactor, leaving the fuel exposed. Emergency water reservoirs are designed to flood the core immediately if there is a sudden loss of coolant, but they do not always work in extreme circumstances. Some known nuclear disasters are the 1979 accidental partial core meltdown at Three Mile Island, Pennsylvania (although the core was completely destroyed, the radioactive mass never penetrated the steel outlining the containment structure); the fire at the Chernobyl, Ukraine, facility in 1986, which released radioactive isotopes into the atmosphere; and the Fukushima Daiichi nuclear disaster after the tsunami that hit Japan in 2011. There have been several minor nuclear accidents in Chalk River, Ontario, that prompted temporary shutdowns. It is worth mentioning that the 2009 shutdown occurred at a time when only one of the other four regular medical isotope sourcing reactors in the world was producing, resulting in a worldwide shortage.

2. *Radioactive waste.* This problem is still unsolved. The by-products of the fissioning of ^{235}U remain radioactive for thousands of years, requiring safe disposal. Storage facilities are not sufficient to store the world's nuclear waste, which limits the amount of nuclear fuel that can be used per year. Transportation of the waste has its own risks, as well.

3. *Radiation.* The background radiation that each person receives per year is an average of 2.00 mSv. If all our power came from nuclear plants, we would receive an extra 0.002 mSv of radiation per year. Although the three major effects of radiation (cancer, radiation sickness, and genetic mutation) are nearly untraceable at levels below about 0.50 mSv, there is certainly a concern associated with it.

CHEMICAL CONNECTIONS
TO COSMOLOGY

Origin of the Elements in the Stars

How did the universe begin? Where did matter come from? How were the elements formed? The most accepted model proposes that a sphere of unimaginable properties—a diameter of 10^{-28} cm, a density of 10^{96} g/mL (density of a nucleus $\approx 10^{14}$ g/mL), and a temperature of 10^{32} K—exploded in a "big bang," for reasons not yet known, and distributed its contents throughout the void of space. Cosmologists consider this moment the beginning of time.

One second later, the universe was an expanding mixture of neutrons, protons, and electrons, denser than rock and hotter than an exploding hydrogen bomb (about 10^{10} K). In the next few minutes, it became a gigantic fusion reactor, creating the first atomic nuclei other than ^{1}H: ^{2}H, ^{3}He, and ^{4}He. After 10 min, more than 25% of the mass of the universe existed as ^{4}He, and only about 0.025% as ^{2}H. About 100 million years later, or about 15 billion years ago, gravitational forces pulled this cosmic mixture into primitive, contracting stars.

This account of the origin of the universe is based on the observation of spectra from the Sun, other stars, nearby galaxies, and cosmic (interstellar) dust. Spectral analysis of planets and chemical analysis of Earth and Moon rocks, meteorites, and cosmic-ray particles furnish data about isotope abundance. From these, a model has been developed for **stellar nucleosynthesis**, the creation of the elements in stars. The overall process occurs in several stages during a star's evolution, and the entire sequence of steps occurs only in very massive stars with 10 to 100 times the mass of the Sun. Each step involves a contraction of the volume of the star that results in higher temperatures and yields heavier nuclei. Such events are forming elements in stars today. The key stages in the process are shown in Figure B25.1 and described below:

1. *Hydrogen burning produces He*. The initial contraction of a star heats its core to about 10^{7} K, at which point a fusion process called *hydrogen burning* begins. Through three possible reactions (one is shown below), helium nuclei are produced from the abundant protons:

$$4\,^{1}_{1}\text{H} \longrightarrow \,^{4}_{2}\text{He} + 2\,^{0}_{1}\beta + 2\nu + \text{energy}$$

2. *Helium burning produces C, O, Ne, and Mg*. After several billion years of hydrogen burning, about 10% of the ^{1}H is consumed, and the star contracts further. The ^{4}He forms a dense core, hot enough (2×10^{8} K) to fuse ^{4}He nuclei. The energy released during *helium burning* expands the remaining ^{1}H into a vast envelope: the star becomes a *red giant*, more than 100 times its original diameter. Within its core, pairs of ^{4}He nuclei (α particles) fuse into unstable ^{8}Be nuclei ($t_{1/2} = 7 \times 10^{-17}$ s), which play the role of an activated complex in chemical reactions. An ^{8}Be nucleus collides with another ^{4}He to form ^{12}C. Then further fusion with ^{4}He creates nuclei up to ^{24}Mg:

$$^{12}\text{C} \xrightarrow{\alpha} \,^{16}\text{O} \xrightarrow{\alpha} \,^{20}\text{Ne} \xrightarrow{\alpha} \,^{24}\text{Mg}$$

3. *Elements through Fe and Ni form*. For another 10 million years, ^{4}He is consumed, and the heavier nuclei that are created form a core. This core contracts and heats, expanding the star into a *supergiant*. Within the hot core (7×10^{8} K), *carbon and oxygen burning occurs*:

$$^{12}\text{C} + \,^{12}\text{C} \longrightarrow \,^{23}\text{Na} + \,^{1}\text{H}$$
$$^{12}\text{C} + \,^{16}\text{O} \longrightarrow \,^{28}\text{Si} + \gamma$$

Absorption of α particles forms nuclei up to ^{40}Ca:

$$^{12}\text{C} \xrightarrow{\alpha} \,^{16}\text{O} \xrightarrow{\alpha} \,^{20}\text{Ne} \xrightarrow{\alpha} \,^{24}\text{Mg} \xrightarrow{\alpha}$$
$$^{28}\text{Si} \xrightarrow{\alpha} \,^{32}\text{S} \xrightarrow{\alpha} \,^{36}\text{Ar} \xrightarrow{\alpha} \,^{40}\text{Ca}$$

Further contraction and heating to a temperature of 3×10^{9} K allow reactions in which nuclei release neutrons, protons, and α particles and then recapture them. As a result, nuclei with lower binding energies supply nucleons to create nuclei with higher binding energies. In stars of moderate mass, less than 10 times the mass of the Sun, this process stops at iron-56 and nickel-58, the nuclei with the highest binding energies.

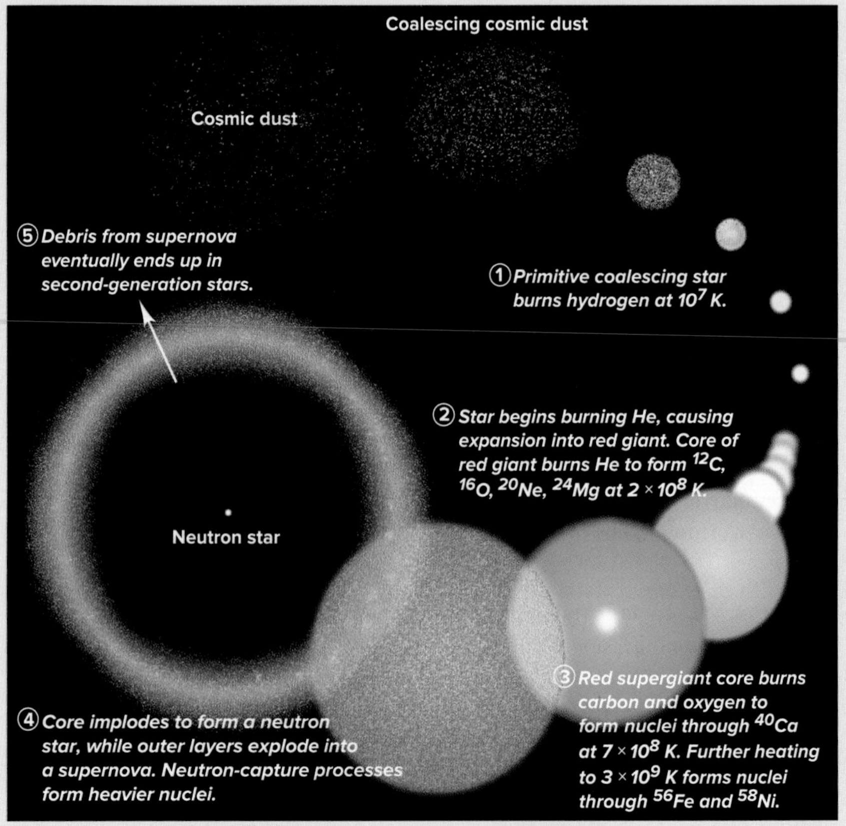

Coalescing cosmic dust
Cosmic dust
(5) Debris from supernova eventually ends up in second-generation stars.
(1) Primitive coalescing star burns hydrogen at 10^{7} K.
(2) Star begins burning He, causing expansion into red giant. Core of red giant burns He to form ^{12}C, ^{16}O, ^{20}Ne, ^{24}Mg at 2×10^{8} K.
Neutron star
(3) Red supergiant core burns carbon and oxygen to form nuclei through ^{40}Ca at 7×10^{8} K. Further heating to 3×10^{9} K forms nuclei through ^{56}Fe and ^{58}Ni.
(4) Core implodes to form a neutron star, while outer layers explode into a supernova. Neutron-capture processes form heavier nuclei.

FIGURE B25.1 Element synthesis in the life cycle of a star

(Continued)

4. *Heavier elements form*. In very massive stars, the next stage is the most spectacular. With all the fuel consumed, the core collapses within a second. Many Fe and Ni nuclei break down into neutrons and protons. Protons capture electrons to form neutrons, and the entire core forms an incredibly dense *neutron star*. (An Earth-sized star that became a neutron star would fit in Montreal's Olympic Stadium!) As the core implodes, the outer layers explode in a *supernova*, which expels material throughout space. A supernova occurs an average of every few hundred years in each galaxy, or once every second somewhere in the universe. The heavier elements are formed during supernova events and are found in *second-generation stars*, those that coalesce from interstellar ^{1}H and ^{4}He and the debris of exploded first-generation stars.

Heavier elements form through *neutron-capture* processes. In the *s-process* (*slow* neutron-capture process), a nucleus captures a neutron, sometime over a period of 10 to 1000 years. If the resulting nucleus is unstable, it undergoes β^- decay to form the next element, as in the following conversion of ^{68}Zn to ^{70}Ge:

$$^{68}Zn \xrightarrow{n} {}^{69}Zn \xrightarrow{\beta^-} {}^{69}Ga \xrightarrow{n} {}^{70}Ga \xrightarrow{\beta^-} {}^{70}Ge$$

The stable isotopes of most heavy elements, up to ^{209}Bi, form by the s-process.

Less stable isotopes, and those with *A* greater than 230, cannot form by the s-process because their half-lives are too short. These isotopes form by the *r-process* (*rapid* neutron-capture process) during the fury of the supernova. Many neutron captures, followed by many β^- decays, occur in a second, as when ^{56}Fe is converted to ^{79}Br:

$$^{56}_{26}Fe + 23\,^{1}_{0}n \longrightarrow {}^{79}_{26}Fe \longrightarrow {}^{79}_{35}Br + 9\,^{0}_{-1}\beta$$

We know, from the heavy elements present in the Sun, that the Sun is at least a second-generation star, which is currently undergoing hydrogen burning. Together with its planets, it was formed from the dust of exploded stars about 4.6×10^9 years ago. This means that many of the atoms on Earth, including some within you, came from exploded stars and are older than the Solar System itself!

Any theory of element formation must be consistent with the element abundances we observe (Section 23.1). Although local compositions, such as those of Earth and the Sun, differ, and mineral grains in meteorites may have abnormal isotopic compositions, large regions of the universe have, on average, similar compositions. Scientists believe that element forming reaches a dynamic equilibrium, resulting in *relatively constant amounts of the isotopes*.

Problems

B25.1 Compare the s-process and r-process in stellar nucleosynthesis in terms of rate, number of neutrons absorbed, and types of isotopes formed.

B25.2 The overall reaction that takes place during hydrogen burning in a young star is given below:

$$4\,^{1}_{1}H \longrightarrow {}^{4}_{2}He + 2\,^{0}_{1}\beta + 2\nu + \text{energy}$$

How much energy (MeV) is released **(a)** per He nucleus formed; **(b)** per mole of He formed? (Masses: $^{1}_{1}H$ atom = 1.007 825 u; $^{4}_{2}He$ atom = 4.002 60 u; positron = 5.485 80 $\times 10^{-4}$ u.)

B25.3 Cosmologists who have modelled the origin of the elements postulate nuclides with very short half-lives. One of these short-lived nuclides, ^{8}Be ($t_{1/2} = 7 \times 10^{-17}$ s), is thought to play a key role in stellar nucleosynthesis by fusing with ^{4}He to form ^{12}C. Another model proposes the simultaneous fusion of three ^{4}He nuclei to form ^{12}C. Comment on the validity of this alternative mechanism.

B25.4 The s-process produces isotopes of elements up to ^{209}Bi. This process is terminated when ^{209}Bi captures a neutron to produce ^{210}Bi, which subsequently undergoes β^- decay to produce nuclide A, which then decays by emitting an α particle to form nuclide B. This nuclide captures three neutrons to produce nuclide C, which undergoes β^- decay to produce nuclide D, starting the cycle again. Write balanced nuclear equations for the formation of nuclides A through D.

SUMMARY OF SECTION 25.7

- In nuclear fission, neutron bombardment causes a nucleus to split into two smaller nuclei and release neutrons that split other nuclei, giving rise to a chain reaction.
- A nuclear power plant controls the rate of the chain reaction to produce heat that creates steam, which is used to generate electricity.
- Potential hazards, such as radiation leaks, thermal pollution, and disposal of nuclear waste, remain current concerns.
- Nuclear fusion holds great promise as a source of clean abundant energy, but it requires extremely high temperatures and is not yet practical.
- The elements were formed through a complex series of nuclear reactions in evolving stars.
- There are both pros and cons to using nuclear energy.

CHAPTER REVIEW GUIDE

Learning Objectives

Relevant section (§) and/or sample problem (SP) numbers appear in parentheses.

Concepts

1. Explain how nuclear changes differ, in general, from chemical changes. (Introduction)
2. Discuss the meanings of *radioactivity, nucleon, nuclide,* and *isotope.* (Introduction and §25.1)
3. Describe the characteristics of three types of radioactive emissions: α, β, and γ. (§25.1)
4. Describe the various modes of radioactive decay, and explain how each mode changes the values of *A* and *Z*. (§25.1)
5. Explain how the *N*/*Z* ratio, the even-odd nature of *N* and *Z*, and magic numbers correlate with nuclear stability. (§25.1)
6. Explain how an unstable nuclide's mass number or *N*/*Z* ratio correlates with its mode of decay. (§25.1)
7. Explain how a decay series combines numerous decay steps and ends with a stable nuclide. (§25.1)
8. Explain how ionization and scintillation counters detect and measure radioactivity. (§25.2)
9. Explain why radioactive decay is a first-order process, and define *decay constant* and *activity*. (§25.2)
10. Explain the meaning of *half-life* in the context of radioactive decay. (§25.2)
11. Explain how the specific activity of an isotope in an object is used to determine the object's age. (§25.2)
12. Explain how particle accelerators and reactors are used to synthesize new nuclides. (§25.3)
13. Discuss the units of radiation dose, the effects on living tissue of various dosage levels, and the inverse relationship between the mass and charge of an emission and its penetrating power. (§25.4)
14. Explain how ionizing radiation creates free radicals that damage tissue, and describe the sources and risks of ionizing radiation. (§25.4)
15. Explain how radioisotopes are used in research, analysis, and medical diagnosis. (§25.5)
16. Explain why the mass of a nucleus is less than the total mass of its nucleons and how this mass difference is related to the nuclear binding energy. (§25.6)
17. Explain how nuclear stability is related to binding energy per nucleon. (§25.6)
18. Explain how heavy nuclides undergo fission and lighter nuclides undergo fusion to increase the binding energy per nucleon. (§25.6)
19. Describe the current application of fission and the potential application of fusion to produce energy. (§25.7)

Skills

1. Express the mass and charge of a particle with the $^{A}_{Z}X$ notation. (§25.1; see also §2.5)
2. Use changes in the values of *A* and *Z* to write and balance nuclear equations. (SP 25.1)
3. Use the *N*/*Z* ratio, the even-odd nature of *N* and *Z*, and the presence of magic numbers to predict nuclear stability. (SP 25.2)
4. Use the atomic mass of the element or the *N*/*Z* ratio to predict the mode of decay of an unstable nuclide. (SP 25.3)
5. Convert units of radioactivity. (§25.2)
6. Calculate specific activity, decay constant, half-life, and number of radioactive nuclei. (§25.2 and SP 25.4)
7. Estimate the age of an object from the specific activity and half-life of carbon-14. (SP 25.5)
8. Write notations and balancing equations for nuclear transmutation. (§25.3)
9. Calculate radiation dose, and convert the units. (§25.4)
10. Calculate the mass difference between a nucleus and its nucleons, as well as the energy equivalent. (§25.6)
11. Calculate the binding energy per nucleon, and use it to compare the stabilities of nuclides. (SP 25.6)

Key Terms

Section 25.1
radioactivity
nucleons
nuclide
isotopes
alpha particles (α, $^{4}_{2}\alpha$, or $^{4}_{2}He^{2+}$)
beta particles (β, β^-, or sometimes $^{0}_{-1}\beta$)
gamma rays (γ, or sometimes $^{0}_{0}\gamma$)
alpha (α) decay
beta (β) decay
β^- decay
positron (β^+) emission
positron
electron (e^-) capture (EC)
gamma (γ) emission
band of stability
strong force
decay series (disintegration series)

Section 25.2
Geiger-Müller counter
scintillation counter
activity ($\mathscr{A}$)
becquerel (Bq)
curie (Ci)
decay constant
half-life ($t_{1/2}$)
radioisotopic dating
radioisotopes

Section 25.3
nuclear transmutation
deuterons
particle accelerators
transuranium elements

Section 25.4
ionization
ionizing radiation
gray (Gy)
rad (radiation-absorbed dose)
rem (roentgen equivalent for man)
sievert (Sv)
free radicals
background radiation

Section 25.5
tracer

Section 25.6
fission
fusion
nuclear binding energy
electron volt (eV)

Section 25.7
chain reaction
critical mass
reactor core
stellar nucleosynthesis

Key Equations and Relationships

25.1 Balancing a nuclear equation:

$$^{\text{Total } A}_{\text{Total } Z}\text{Reactants} = {}^{\text{Total } A}_{\text{Total } Z}\text{Products}$$

25.2 Defining the SI derived unit of radioactivity (becquerel, Bq):

$$1 \text{ Bq} = 2.703 \times 10^{-11} \text{ Ci}$$

where Ci (curie) is the activity of 1 g of the radium isotope ^{226}Ra:

$$1 \text{ Ci} = 3.70 \times 10^{10} \text{ d/s} \quad \text{(disintegrations per second)}$$

25.3 Expressing the decay rate (activity) for radioactive nuclei:

$$\text{Decay rate } (\mathcal{A}) = -\frac{\Delta\mathcal{N}}{\Delta t} = k\mathcal{N}$$

25.4 Finding the number of nuclei remaining after a given time, $\mathcal{N}_t$:

$$\ln\frac{\mathcal{N}_t}{\mathcal{N}_0} = -kt \quad \text{or} \quad \mathcal{N}_t = \mathcal{N}_0 e^{-kt} \quad \text{and} \quad \ln\frac{\mathcal{N}_0}{\mathcal{N}_t} = kt$$

25.5 Finding the half-life of a radioactive nuclide:

$$t_{1/2} = \frac{\ln 2}{k}$$

25.6 Calculating the time to reach a given specific activity (age of an object in radioisotopic dating):

$$t = \frac{1}{k}\ln\frac{\mathcal{A}_0}{\mathcal{A}_t}$$

25.7 Writing a transmutation in shorthand notation:

reactant nucleus (particle in, particle(s) out) product nucleus

25.8 Adapting Einstein's equation to calculate mass difference and/or nuclear binding energy:

$$\Delta m = \frac{\Delta E}{c^2} \quad \text{or} \quad \Delta E = \Delta mc^2$$

25.9 Relating the atomic mass unit to its energy equivalent in mega–electron volts (MeV):

$$1 \text{ u} = 931.5 \times 10^6 \text{ eV} = 931.5 \text{ MeV}$$

Brief Solutions to Follow-Up Problems

25.1 $^{133}_{54}\text{Xe} \longrightarrow {}^{133}_{55}\text{Cs} + {}^{0}_{-1}\beta$

25.2 ^{31}P has an even N (16), but ^{30}P has both N and Z odd. ^{31}P also has a slightly higher N/Z ratio, which is closer to the band of stability.

25.3 (a) $Z = 26$ is iron (Fe), and the atomic mass is 55.85. The A value of 61 is higher: β^- decay

(b) $Z > 83$, which is too high for stability: α decay

25.4 $\ln \mathcal{A}_t = -kt + \ln \mathcal{A}_0$

$$= -\left(\frac{\ln 2}{15 \text{ h}} \times 4.0 \text{ days} \times \frac{24 \text{ h}}{1 \text{ day}}\right) + \ln(2.5 \times 10^9)$$

$$= 17.20$$

$$\mathcal{A} = 3.0 \times 10^7 \text{ d/s}$$

25.5 $t = \dfrac{1}{k}\ln\dfrac{\mathcal{A}_0}{\mathcal{A}_t} = \dfrac{5730 \text{ years}}{\ln 2}\ln\left[\dfrac{15.3 \text{ d/(min·g)}}{9.41 \text{ d/(min·g)}}\right]$

$$= 4.02 \times 10^3 \text{ years}$$

The mummy case is about 4000 years old.

25.6 ^{235}U has 92 $^1_1$p and 143 $^1_0$n:

$$\Delta m = [(92 \times 1.007\,825 \text{ u}) + (143 \times 1.008\,665 \text{ u})] - 235.043\,924 \text{ u}$$
$$= 1.9151 \text{ u}$$

$$\frac{\text{binding energy}}{\text{nucleon}} = \frac{1.9151 \text{ u} \times \dfrac{931.5 \text{ MeV}}{1 \text{ u}}}{235 \text{ nucleons}}$$

$$= 7.591 \text{ MeV/nucleon}$$

Therefore, ^{235}U is less stable than ^{12}C.

PROBLEMS

Problems with **red** numbers are answered in Appendix G and worked in detail in the Student Solutions Manual. Problem sections match those in this book and provide the numbers of relevant sample problems. Most offer Concept Review Questions, Skill-Building Exercises (grouped in pairs covering the same concept), and Problems in Context. The Comprehensive Problems are based on material from any section or previous chapter.

Radioactive Decay and Nuclear Stability

(Sample Problems 25.1 to 25.3)

Concept Review Questions

25.1 How do chemical and nuclear reactions differ in (a) magnitude of the energy change; (b) effect on the rate of increasing temperature; (c) effect on the rate of higher reactant concentration; (d) effect on the yield of higher reactant concentration?

25.2 Sulfur has four naturally occurring stable isotopes. The isotope with the lowest mass number is sulfur-32, which is also the most abundant (95.02%).
(a) What percentage of the S atoms in a match head are ^{32}S?
(b) The isotopic mass of ^{32}S is 31.972 070 u. Is the atomic mass of S larger than, smaller than, or equal to this mass? Explain.

25.3 What led Marie Curie to draw the following conclusions?
(a) Radioactivity is a property of the element, not the compound in which it is found.

(b) A highly radioactive element, aside from uranium, occurs in pitchblende.

25.4 Which processes produce an atom of a *different* element? Show how Z and N change, if at all, in each process.
(a) α decay (b) β^- decay (c) γ emission
(d) β^+ emission (e) e^- capture

25.5 Why is ^3_2He stable, but ^2_2He has never been detected?

25.6 How do the modes of decay differ for a neutron-rich nuclide and a proton-rich nuclide?

25.7 Why might it be difficult to use only a nuclide's N/Z ratio to predict whether it will decay by β^+ emission or by e^- capture? What other factor is important?

Skill-Building Exercises (grouped in similar pairs)

25.8 Write a balanced nuclear equation for (a) alpha decay of $^{234}_{92}U$; (b) electron capture by neptunium-232; (c) positron emission by $^{12}_{7}N$.

25.9 Write a balanced nuclear equation for (a) β^- decay of sodium-26; (b) β^- decay of francium-223; (c) alpha decay of $^{212}_{83}Bi$.

25.10 Write a balanced nuclear equation for (a) β^- emission by magnesium-27; (b) β^+ emission by $^{23}_{12}Mg$; (c) electron capture by $^{103}_{46}Pd$.

25.11 Write a balanced nuclear equation for (a) β^- decay of silicon-32; (b) alpha decay of polonium-218; (c) electron capture by $^{110}_{49}In$.

25.12 Write a balanced nuclear equation for the formation of (a) $^{48}_{22}Ti$ through positron emission; (b) silver-107 through electron capture; (c) polonium-206 through α decay.

25.13 Write a balanced nuclear equation for the formation of (a) $^{241}_{95}Am$ through β^- decay; (b) $^{228}_{89}Ac$ through β^- decay; (c) $^{203}_{83}Bi$ through α decay.

25.14 Write a balanced nuclear equation for the formation of (a) ^{186}Ir through electron capture; (b) francium-221 through α decay; (c) iodine-129 through β^- decay.

25.15 Write a balanced nuclear equation for the formation of (a) ^{52}Mn through positron emission; (b) polonium-215 through α decay; (c) ^{81}Kr through electron capture.

25.16 Which nuclide(s) would you predict to be stable? Why?
(a) $^{20}_{8}O$ (b) $^{59}_{27}Co$ (c) $^{9}_{3}Li$

25.17 Which nuclide(s) would you predict to be stable? Why?
(a) $^{146}_{60}Nd$ (b) $^{114}_{48}Cd$ (c) $^{88}_{42}Mo$

25.18 Which nuclide(s) would you predict to be stable? Why?
(a) ^{127}I (b) tin-106 (c) ^{68}As

25.19 Which nuclide(s) would you predict to be stable? Why?
(a) ^{48}K (b) ^{79}Br (c) argon-32

25.20 What is the most likely mode of decay for (a) $^{238}_{92}U$; (b) $^{48}_{24}Cr$; (c) $^{50}_{25}Mn$?

25.21 What is the most likely mode of decay for (a) $^{111}_{47}Ag$; (b) $^{41}_{17}Cl$; (c) $^{110}_{44}Ru$?

25.22 What is the most likely mode of decay for (a) ^{15}C; (b) ^{120}Xe; (c) ^{224}Th?

25.23 What is the most likely mode of decay for (a) ^{106}In; (b) ^{141}Eu; (c) ^{241}Am?

25.24 Why is $^{52}_{24}Cr$ the most stable isotope of chromium?

25.25 Why is $^{40}_{20}Ca$ the most stable isotope of calcium?

Problems in Context

25.26 ^{237}Np is the parent nuclide of a decay series that starts with α emission, followed by β^- emission, and then two more α emissions. Write a balanced nuclear equation for each step.

25.27 Why is helium found in deposits of uranium and thorium ores? What kind of radioactive emission produces it?

25.28 In a natural decay series, how many α and β^- emissions per atom of uranium-235 result in an atom of lead-207?

The Kinetics of Radioactive Decay

(Sample Problems 25.4 and 25.5)

Concept Review Questions

25.29 What electronic process is the basis for detecting radioactivity in (a) a scintillation counter; (b) a Geiger-Müller counter?

25.30 What is the reaction order of radioactive decay? Explain.

25.31 After 1 min, three radioactive nuclei remain from an original sample of six. Is it valid to conclude that $t_{1/2}$ equals 1 min? Is this conclusion valid if the original sample contained 6×10^{12} nuclei, and 3×10^{12} remain after 1 min? Explain.

25.32 Radioisotopic dating depends on the constant rate of decay and the formation of various nuclides in a sample. How is the proportion of ^{14}C kept relatively constant in living organisms?

Skill-Building Exercises (grouped in similar pairs)

25.33 What is the specific activity (Ci/g) if 1.65 mg of an isotope emits 1.56×10^6 α particles per second?

25.34 What is the specific activity (Ci/g) if 2.6 g of an isotope emits 4.13×10^8 β^- particles per hour?

25.35 What is the specific activity (Bq/g) if 8.58 μg of an isotope emits 7.4×10^4 α particles per minute?

25.36 What is the specific activity (Bq/g) if 1.07 kg of an isotope emits 3.77×10^7 β^- particles per minute?

25.37 If one-trillionth ($1/10^{12}$) of the atoms of a radioactive isotope disintegrate each day, what is the decay constant of the process?

25.38 If 2.8×10^{-10}% of the atoms of a radioactive isotope disintegrate in 1.0 year, what is the decay constant of the process?

25.39 If 1.00×10^{-12} mol of ^{135}Cs emits 1.39×10^5 β^- particles in 1.00 year, what is the decay constant?

25.40 If 6.40×10^{-9} mol of ^{176}W emits 1.07×10^{15} β^+ particles in 1.00 h, what is the decay constant?

25.41 The isotope $^{212}_{83}Bi$ has a half-life of 1.01 years. What mass (mg) of a 2.00 mg sample will remain after 3.75×10^3 h?

25.42 The half-life of radium-226 is 1.60×10^3 years. How much time (h) will it take for a 2.50 g sample to decay to the point where 0.185 g of the isotope remains?

25.43 A rock contains 270 μmol of ^{238}U ($t_{1/2} = 4.5 \times 10^9$ years) and 110 μmol of ^{206}Pb. Assuming that all the ^{206}Pb comes from decay of the ^{238}U, estimate the age of the rock.

25.44 A fabric remnant from a burial site has a $^{14}C/^{12}C$ ratio of 0.735 of the original value. How old is the fabric?

Problems in Context

25.45 Due to decay of ^{40}K, cow's milk has a specific activity of about 6×10^{-11} mCi/mL. How many disintegrations of ^{40}K nuclei occur per minute in a 0.2 L glass of milk?

25.46 Plutonium-239 ($t_{1/2} = 2.41 \times 10^4$ years) represents a serious nuclear waste hazard. If seven half-lives are required to reach a tolerable level of radioactivity, how long must ^{239}Pu be stored?

25.47 A rock that contains 3.1×10^{-15} mol of ^{232}Th ($t_{1/2} = 1.4 \times 10^{10}$ years) has 9.5×10^4 fission tracks, each track representing the fission of one atom of ^{232}Th. How old is the rock?

25.48 A volcanic eruption melts a large chunk of rock, and all the gases are expelled. After cooling, $^{40}_{18}Ar$ accumulates from the ongoing decay of $^{40}_{19}K$ in the rock ($t_{1/2} = 1.25 \times 10^9$ years). When a piece of rock is analyzed, it is found to contain 1.38 mmol of ^{40}K and 1.14 mmol of ^{40}Ar. How long ago did the rock cool?

Nuclear Transmutation: Induced Changes in Nuclei

Concept Review Questions

25.49 Irene and Frederic Joliot-Curie converted $^{27}_{13}Al$ to $^{30}_{15}P$ in 1933. Why was this transmutation significant?

25.50 Early workers mistakenly thought that neutron beams were γ radiation. Why? What evidence led to the correct conclusion?

25.51 Why must the electrical polarity of the tubes in a linear accelerator be reversed at very short time intervals?

25.52 Why does bombardment with protons usually require higher energies than bombardment with neutrons?

Skill-Building Exercises (grouped in similar pairs)

25.53 Determine the missing species in each transmutation, and write a full nuclear equation from the shorthand notation:
(a) ^{10}B (α, n) __
(b) ^{28}Si (d, __) ^{29}P (where d is a deuteron, ^2H)
(c) __ (α, 2n) ^{244}Cf

25.54 Name the unidentified species in each description, and write each transmutation process in shorthand notation:
(a) Gamma irradiation of a nuclide yields a proton, a neutron, and ^{29}Si.
(b) Bombardment of ^{252}Cf with ^{10}B yields five neutrons and a nuclide.
(c) Bombardment of ^{238}U with a particle yields three neutrons and ^{239}Pu.

Problem in Context

25.55 Elements 104, 105, and 106 have been named rutherfordium (Rf), dubnium (Db), and seaborgium (Sg), respectively. These elements are synthesized from californium-249 by bombarding it with carbon-12, nitrogen-15, and oxygen-18 nuclei, respectively. Four neutrons are formed in each reaction as well.
(a) Write balanced nuclear equations for the formation of these elements.
(b) Write the equations in shorthand notation.

Effects of Nuclear Radiation on Matter

Concept Review Questions

25.56 The effects of γ rays and α particles on matter are different. Explain.

25.57 What is a cation-electron pair, and how does it form?

25.58 Why is ionizing radiation more harmful to children than adults?

25.59 Why is $\cdot OH$ more dangerous than OH^- in an organism?

Skill-Building Exercises (grouped in similar pairs)

25.60 A 61.2 kg person absorbs 3.3×10^{-7} J of energy from radioactive emissions.
(a) What amount (rad) does she receive?
(b) What amount (Gy) does she receive?

25.61 A 3.6 kg laboratory animal receives a single dose of 8.92×10^{-4} Gy.
(a) What amount (rad) does the animal receive?
(b) What energy (J) does the animal absorb?

25.62 A 70. kg person who is exposed to ^{90}Sr absorbs 6.0×10^5 β^- particles, each with an energy of 8.74×10^{-14} J.
(a) What dose (Gy) does he receive?
(b) If the RBE (relative biological effectiveness) is 1.0, what is the equivalent dose in millirem (mrem)?
(c) What is the equivalent dose in sievert (Sv)?

25.63 A laboratory rat weighs 265 g and absorbs 1.77×10^{10} β^- particles, each with an energy of 2.20×10^{-13} J.
(a) What amount (rad) does the animal receive?
(b) What is this dose in gray (Gy)?
(c) If the RBE (relative biological effectiveness) is 0.75, what is the equivalent dose in sievert (Sv)?

Problems in Context

25.64 If 2.50 pCi of radioactivity from ^{239}Pu is emitted in a 95 kg human for 65 h, and each disintegration has an energy of 8.25×10^{-13} J, what dose (Gy) does the person receive [1 pCi (picocurie) = 1×10^{-12} Ci]?

25.65 A small region of a cancer patient's brain is exposed for 24.0 min to 475 Bq of radioactivity from ^{60}Co to treat a tumour. If the brain mass exposed is 1.858 g and each β^- particle emitted has an energy of 5.05×10^{-14} J, what is the dose in rad?

Applications of Radioisotopes

Concept Review Questions

25.66 In what two ways are radioactive tracers used in organisms?

25.67 Why is neutron activation analysis (NAA) useful to art historians and criminologists?

25.68 Positrons cannot penetrate matter more than a few atomic diameters, but positron emission of radiotracers can be monitored in medical diagnosis. Explain.

25.69 A steel part is treated to form iron-59. Oil used to lubricate the part emits 298 β^- particles (with the energy characteristic of ^{59}Fe) per minute per millilitre of oil. What other information would you need to calculate the rate of removal of the steel from the part during use?

Problems in Context

25.70 The oxidation of methanol to methanal (formaldehyde) can be accomplished by reaction with chromic acid:

$$6H^+(aq) + 3CH_3OH(aq) + 2H_2CrO_4(aq) \longrightarrow 3CH_2O(aq) + 2Cr^{3+}(aq) + 8H_2O(l)$$

This reaction can be studied with the stable isotope tracer ^{18}O and mass spectrometry. When a small amount of $CH_3{}^{18}OH$ is present in the alcohol reactant, $CH_2{}^{18}O$ forms. When a small amount of $H_2Cr^{18}O_4$ is present, $H_2{}^{18}O$ forms. Does chromic acid or methanol supply the O atom to the aldehyde? Explain.

The Interconversion of Mass and Energy

(Sample Problem 25.6)

Note: Use the following data for the problems in this section: mass of ^1H atom = 1.007 825 u; mass of neutron = 1.008 665 u.

Concept Review Questions

25.71 Many scientists at first reacted skeptically to Einstein's equation, $E = mc^2$. Why?

25.72 How does a change in mass arise when a nuclide forms from nucleons?

25.73 When a nucleus forms from nucleons, is energy absorbed or released? Why?

25.74 What is the binding energy per nucleon? Why is the binding energy per nucleon, rather than per nuclide, used to compare nuclide stability?

Skill-Building Exercises (grouped in similar pairs)

25.75 A ^{3}H nucleus decays with an energy of 0.018 61 MeV. Convert this energy into (a) eV; (b) J.

25.76 Arsenic-84 decays with an energy of 1.57×10^{-15} kJ per nucleus. Convert this energy into (a) eV; (b) MeV.

25.77 How much energy (J) is released when 1.5 mol of ^{239}Pu decays, if each nucleus releases 5.243 MeV?

25.78 How much energy (MeV) is released per nucleus when 3.2×10^{-3} mol of chromium-49 releases 8.11×10^{5} kJ?

25.79 Oxygen-16 is one of the most stable nuclides. The mass of a ^{16}O atom is 15.994 915 u. Calculate the binding energy (a) per nucleon in MeV; (b) per atom in MeV; (c) per mole in kJ.

25.80 Lead-206 is the end product of ^{238}U decay. One ^{206}Pb atom has a mass of 205.974 440 u. Calculate the binding energy (a) per nucleon in MeV; (b) per atom in MeV; (c) per mole in kJ.

25.81 Cobalt-59 is the only stable isotope of this transition metal. One ^{59}Co atom has a mass of 58.933 198 u. Calculate the binding energy (a) per nucleon in MeV; (b) per atom in MeV; (c) per mole in kJ.

25.82 Iodine-131 is one of the most important isotopes used in the diagnosis of thyroid cancer. One atom has a mass of 130.906 114 u. Calculate the binding energy (a) per nucleon in MeV; (b) per atom in MeV; (c) per mole in kJ.

Problem in Context

25.83 The ^{80}Br nuclide decays either by β^{-} decay or by electron capture.
(a) What is the product of each process?
(b) Which process releases more energy? (Masses of atoms: ^{80}Br = 79.918 528 u; ^{80}Kr = 79.916 380 u; ^{80}Se = 79.916 520 u; neglect the mass of the electrons involved because these are atomic, not nuclear, masses.)

Applications of Fission and Fusion

Concept Review Questions

25.84 What is the minimum number of neutrons from each fission event that must be absorbed by other nuclei for a chain reaction to be sustained?

25.85 In what main way is fission different from radioactive decay? Are all fission events in a chain reaction identical? Explain.

25.86 What is the purpose of enrichment in the preparation of fuel rods? How is it accomplished?

25.87 Describe the nature and purpose of (a) the control rods, (b) the moderator, and (c) the reflector in a nuclear reactor.

25.88 State an advantage and a disadvantage of heavy-water reactors, compared with light-water reactors.

25.89 What are the expected advantages of fusion reactors over fission reactors?

25.90 Why is iron the most abundant element in Earth's core?

Problem in Context

25.91 The reaction that will probably power the first commercial fusion reactor is

$$^{3}_{1}H + ^{2}_{1}H \longrightarrow ^{4}_{2}He + ^{1}_{0}n$$

How much energy would be produced per mole of reaction (masses: $^{3}_{1}H$ = 3.016 05 u; $^{2}_{1}H$ = 2.0140 u; $^{4}_{2}He$ = 4.002 60 u; $^{1}_{0}n$ = 1.008 665 u)?

Comprehensive Problems

25.92 Some $^{243}_{94}Am$ was present when Earth formed, but it all decayed in the next billion years. The first three steps in this decay series are emissions of an α particle, a β^{-} particle, and another α particle. What other isotopes were present on the young Earth in a rock that contained some $^{243}_{95}Am$?

25.93 Curium-243 undergoes α decay to plutonium-239:

$$^{243}Cm \longrightarrow ^{239}Pu + \alpha$$

(a) Find the change in mass, Δm (kg). (Masses: ^{243}Cm = 243.0614 u; ^{239}Pu = 239.0522 u; ^{4}He = 4.0026 u; 1 u = 1.661×10^{-24} g.)

(b) Find the energy released in joules.

(c) Find the energy released in kJ/mol of reaction, and comment on the difference between this value and a typical heat of reaction for a chemical change, which is a few hundred kJ/mol.

25.94 Plutonium "triggers" for nuclear weapons were manufactured at the Rocky Flats plant in Colorado. An 85 kg worker inhaled a dust particle containing 1.00 μg of $^{239}_{94}Pu$, which resided in his body for 16 h. What dose did the worker receive (a) in rad; (b) in Gy? ($t_{1/2}$ of ^{239}Pu = 2.41×10^{4} years; each disintegration released 5.15 MeV.)

25.95 Archaeologists removed some charcoal from a First Nations campfire, burned it in O_2, and bubbled the CO_2 formed into $Ca(OH)_2$ solution (limewater). The $CaCO_3$ that precipitated was filtered and dried. If 4.58 g of the $CaCO_3$ had a radioactivity of 3.2 d/min, how long ago was the campfire lit?

25.96 A 5.4 μg sample of $^{226}RaCl_2$ has a radioactivity of 1.5×10^{5} Bq. Calculate $t_{1/2}$ of ^{226}Ra.

25.97 What dose (rad) does a 65 kg human receive each year from the approximately 10^{-8} g of $^{14}_{6}C$ naturally present in her body? ($t_{1/2}$ = 5730 years; each disintegration releases 0.156 MeV.)

25.98 A sample of AgCl emits 175 nCi/g. A saturated solution prepared from the solid emits 1.25×10^{-2} Bq/mL due to radioactive Ag^{+} ions. What is the molar solubility of AgCl?

25.99 The scene below depicts a neutron bombarding ^{235}U:

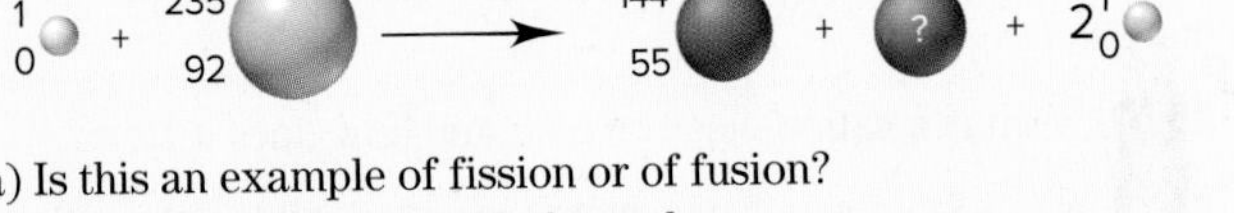

(a) Is this an example of fission or of fusion?
(b) Identify the other nuclide formed.
(c) What is the most likely mode of decay of the nuclide with Z = 55?

25.100 What fraction of the ^{235}U ($t_{1/2}$ = 7.0×10^{8} years) created when Earth was formed would remain after 2.8×10^{9} years?

25.101 ^{238}U ($t_{1/2}$ = 4.5×10^{9} years) begins a decay series that ultimately forms ^{206}Pb. The scene below depicts the relative number of ^{238}U atoms (red) and ^{206}Pb atoms (green) in a mineral. If all the Pb comes from ^{238}U, calculate the age of the sample.

25.102 Technetium-99m is a metastable nuclide that is used in numerous cancer diagnostic and treatment programs. It is prepared just before use because it decays rapidly through γ emission:

$$^{99m}Tc \longrightarrow {}^{99}Tc + \gamma$$

Use the data below to determine (a) the half-life of ^{99m}Tc; (b) the percentage of the isotope that is lost if 2.0 h is needed to prepare and administer the dose

Time (h)	γ Emission (photons/s)
0	5000.
4	3150.
8	2000.
12	1250.
16	788
20	495

25.103 What amount of radioactivity is produced by 1.0 mol of ^{40}K ($t_{1/2}$ = 1.25 × 10^9 years)? Express your answer (a) in Ci (curie); (b) in Bq (becquerel).

25.104 The fraction of a radioactive isotope remaining at time t is $(\frac{1}{2})^{t/t_{1/2}}$, where $t_{1/2}$ is the half-life.
(a) The half-life of carbon-14 is 5730 years. What fraction of carbon-14 in a piece of charcoal remains after (i) 10.0 years; (ii) 10.0 × 10^3 years; (iii) 10.0 × 10^4 years?
(b) In part (a), why is radiocarbon dating more reliable for the fraction remaining in (ii) than the fraction remaining in (i) or in (iii)?

25.105 The isotopic mass of $^{210}_{86}Rn$ is 209.989 669 u. When this nuclide decays by electron capture, it emits 2.368 MeV. What is the isotopic mass of the resulting nuclide (Mass e^- = 0.000 549 u)?

25.106 Exactly 0.1 of the radioactive nuclei in a sample decays per hour. Thus, after n hours, the fraction of nuclei remaining is $(0.900)^n$. Find the value of n equal to one half-life.

25.107 In neutron activation analysis (NAA), stable isotopes are bombarded with neutrons. Depending on the isotope and the energy of the neutron, various emissions are observed. What are the products when the following neutron-activated species decay? Write an overall equation for the reaction in shorthand notation, starting with the stable isotope before neutron activation.
(a) $^{52}_{23}V^* \longrightarrow$ [β^- emission]
(b) $^{64}_{29}Cu^* \longrightarrow$ [β^+ emission]
(c) $^{28}_{13}Al^* \longrightarrow$ [β^- emission]

25.108 In the 1950s, radioactive material was spread over the land where above-ground nuclear tests were conducted. A man drinks some contaminated milk and ingests 0.0500 g of ^{90}Sr, which is taken up by bones and teeth and not eliminated.
(a) How much ^{90}Sr ($t_{1/2}$ = 29 years) is present in his body after 10 years?
(b) How long will it take for 99.9% of the ^{90}Sr to decay?

25.109 The scene below represents a reaction that occurs during the lifetime of a star. (The neutrons are grey, and the protons are purple.)

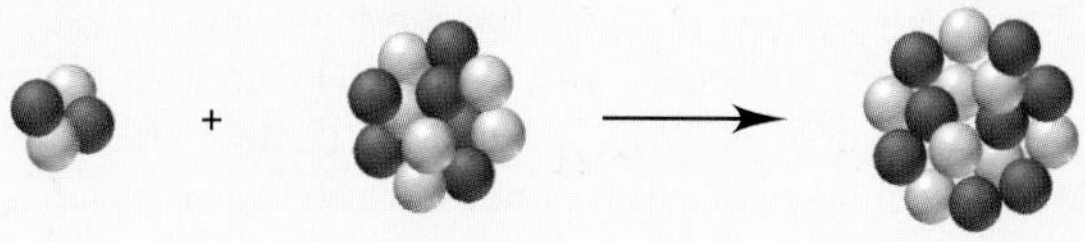

(a) Write a balanced nuclear equation for the reaction.
(b) If the mass difference is 7.7 × 10^{-2} u, find the energy (kJ) released.

25.110 What volume of radon will be produced per hour at STP from 1.000 g of ^{226}Ra? ($t_{1/2}$ = 1599 years; 1 year = 8766 h; mass of one ^{226}Ra atom = 226.025 402 u.)

25.111 ^{90}Kr ($t_{1/2}$ = 32 s) is used to study respiration. How soon after being made must a sample be administered to the patient if the activity must be at least 90% of the original activity?

25.112 Which isotope in each pair is more stable? Why?
(a) $^{140}_{55}Cs$ or $^{133}_{55}Cs$ (b) $^{79}_{35}Br$ or $^{78}_{35}Br$
(c) $^{28}_{12}Mg$ or $^{24}_{12}Mg$ (d) $^{14}_{7}N$ or $^{18}_{7}N$

25.113 A bone sample containing strontium-90 ($t_{1/2}$ = 29 years) emits 7.0 × 10^4 β^- particles per month. How long will it take for the emission to decrease to 1.0 × 10^4 particles per month?

25.114 The 23rd-century starship *Enterprise* uses a substance called "dilithium crystals" as its fuel.
(a) Assuming this material is the result of fusion, what is the product of the fusion of two ^6Li nuclei?
(b) How much energy is released per kilogram of dilithium formed? (Mass of one ^6Li atom = 6.015 121 u.)
(c) When four ^1H atoms fuse to form ^4He, how many positrons are released?
(d) To determine the energy potential of the fusion processes in parts (b) and (c), compare the changes in mass per kilogram of dilithium and ^4He.
(e) Compare the change in mass per kilogram in part (b) with the change in mass for the formation of ^4He by the method used in current fusion reactors. (For masses, see Problem 25.91.)
(f) Using early 21st-century fusion technology, how much tritium can be produced per kilogram of ^6Li in the following reaction?

$$^6_3Li + {}^1_0n \longrightarrow {}^4_2He + {}^3_1H$$

When this amount of tritium is fused with deuterium, what is the change in mass? How does the change in mass compare with the use of dilithium in part (b)?

25.115 Uranium and radium are found in many rocky soils throughout the world. Both undergo radioactive decay, and one of the products is radon-222, the heaviest noble gas ($t_{1/2}$ = 3.82 days). Inhalation of this gas contributes to many lung cancers. According to Health Canada recommendations, the level of radioactivity from radon in homes should not exceed 4.0 pCi/L of air.
(a) What is the safe level of radon in Bq/L of air?

(b) A home has a radon measurement of 41.5 pCi/L. The owner vents the basement so that no more radon enters the living area. What is the activity of the radon remaining in the room air (Bq/L) after 9.5 days?
(c) How long (days) does it take to reach the Health Canada recommended level?

25.116 Nuclear disarmament could be accomplished if weapons were not "replenished." The tritium in warheads decays to helium with a half-life of 12.26 years. If the tritium is not replaced, the weapon is useless. What fraction of the tritium is lost in 5.50 years?

25.117 A decay series starts with the synthetic isotope ${}^{239}_{92}U$. The first four steps are emissions of a β^- particle, another β^-, an α particle, and another α. Write a balanced nuclear equation for each step. Which natural series could start by this sequence?

25.118 For how long can a 24.5 kg child be exposed to 1.0 mCi of radiation from ${}^{222}Rn$ before accumulating 1.0 mrad, if the energy of each disintegration is 5.59 MeV?

25.119 An earthquake in the area of present-day San Francisco is to be dated by measuring the ${}^{14}C$ activity ($t_{1/2}$ = 5730 years) of parts of a tree uprooted during the event. The tree parts have an activity of 12.9 d/(min•g C), and a living tree has an activity of 15.3 d/(min•g C). How long ago did the earthquake occur?

25.120 Were organisms a billion years ago exposed to more or less ionizing radiation than similar organisms today? Explain.

25.121 Tritium (${}^{3}H$; $t_{1/2}$ = 12.26 years) is continuously formed in the upper troposphere by the interaction of solar particles with nitrogen. As a result, natural waters contain a small amount of tritium. Two samples of wine are analyzed, one known to be made in 1941 and the other made earlier. The water in the 1941 wine has 2.23 times as much tritium as the water in the other wine. When was the other wine produced?

25.122 Even though plutonium-239 ($t_{1/2}$ = 2.41 × 10^4 years) is one of the main fission fuels, it is still a radiation hazard in spent uranium fuel from nuclear power plants. How long (years) does it take for 99% of the plutonium-239 in spent fuel to decay?

25.123 Carbon from the remains of an extinct Australian marsupial, called *Diprotodon*, has a specific activity of 0.61 pCi/g. Modern carbon has a specific activity of 6.89 pCi/g. How long ago did the *Diprotodon* apparently become extinct?

25.124 The reaction that allows for radiocarbon dating is the continuous formation of carbon-14 in the upper atmosphere:

$${}^{14}_{7}N + {}^{1}_{0}n \longrightarrow {}^{14}_{6}C + {}^{1}_{1}H$$

What is the energy change associated with this process in eV/reaction and in kJ/mol reaction? (Masses of atoms: ${}^{14}_{7}N$ = 14.003 074 u; ${}^{14}_{6}C$ = 14.003 241 u; ${}^{1}_{1}H$ = 1.007 825 u; ${}^{1}_{0}n$ = 1.008 665 u.)

25.125 What is the nuclear binding energy of a lithium-7 nucleus in units of kJ/mol and eV/nucleus? (Mass of a lithium-7 atom = 7.016 003 u.)

25.126 Gadolinium-146 undergoes electron capture (EC). Identify the product, and use Figure 25.3 to find the modes of decay and the other two nuclides in the series shown here:

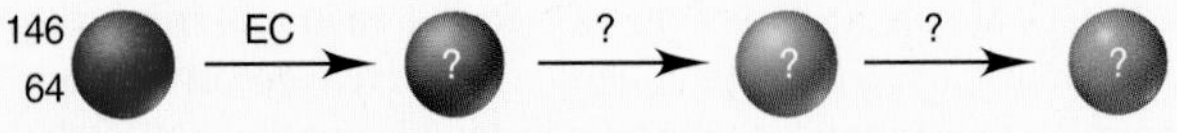

25.127 Using 21st-century technology, hydrogen fusion requires temperatures around 10^8 K. However, lower initial temperatures can be used if the hydrogen is compressed. In the late 24th century, the starship *Leinad* uses such methods to fuse hydrogen at 10^6 K.
(a) What is the kinetic energy of an H atom at 1.00 × 10^6 K?
(b) How many H atoms are heated to 1.00 × 10^6 K from the energy of one H and one anti-H atom annihilating each other?
(c) If these H atoms fuse into ${}^{4}He$ atoms (with the loss of two positrons per ${}^{4}He$ formed), how much energy (J) is generated?
(d) How much more energy is generated by the fusion in part (c) than by the hydrogen-antihydrogen collision in part (b)?
(e) Should the captain of the *Leinad* change the technology and produce ${}^{3}He$ (mass = 3.016 03 u) instead of ${}^{4}He$?

25.128 A metastable (excited) form of ${}^{50}Sc$ changes to its stable form by emitting γ radiation with a wavelength of 8.73 pm. What is the change in the mass of 1 mol of the isotope when it undergoes this change?

25.129 A sample of cobalt-60 ($t_{1/2}$ = 5.27 years), a powerful γ emitter that is used to treat cancer, was purchased by a hospital on March 1, 2016. The sample must be replaced when its activity decreases to 70% of the original value. On what date must the sample be replaced?

25.130 Uranium-233 decays to thorium-229 by α decay, but the emissions have different energies and products: 83% emit an α particle with energy of 4.816 MeV and give ${}^{229}Th$ in its ground state; 15% emit an α particle of 4.773 MeV and give ${}^{229}Th$ in excited state I; 2% emit a lower energy α particle and give ${}^{229}Th$ in the higher excited state II. Excited state II emits a γ ray of 0.060 MeV to reach excited state I.
(a) Find the γ-ray energy and wavelength that would convert excited state I to the ground state.
(b) Find the energy of the α particle that would convert ${}^{233}U$ to excited state II.

25.131 Uranium-238 undergoes a slow decay step ($t_{1/2}$ = 4.5 × 10^9 years) followed by a series of fast steps to form the stable isotope ${}^{206}Pb$. Thus, on a time scale of billions of years, ${}^{238}U$ effectively decays "directly" to ${}^{206}Pb$, and the relative amounts of these isotopes can be used to find the age of a rock. Two students derive equations relating the number of half-lives (n) since the rock formed to the amounts of the isotopes:

Student 1: $$\left(\tfrac{1}{2}\right)^n = \frac{{}^{238}_{92}U}{{}^{206}_{82}Pb}$$

Student 2: $$\left(\tfrac{1}{2}\right)^n = \frac{{}^{238}_{92}U}{{}^{238}_{92}U + {}^{206}_{82}Pb}$$

(a) Which equation is correct, and why?
(b) If a rock contains exactly twice as much ${}^{238}U$ as ${}^{206}Pb$, what is its age in years?

25.132 In the naturally occurring thorium-232 decay series, the steps emit this sequence of particles: α, β^-, β^-, α, α, α, α, β^-, β^-, and α. Write a balanced equation for each step.

25.133 At death, a nobleman in ancient Egypt was mummified, at which time his body contained 1.4 × 10^{-3} g of ${}^{40}K$ ($t_{1/2}$ = 1.25 × 10^9 years), 1.2 × 10^{-8} g of ${}^{14}C$ ($t_{1/2}$ = 5730 years), and 4.8 × 10^{-14} g of ${}^{3}H$ ($t_{1/2}$ = 12.26 years). Which nuclide would give the most accurate estimate of the mummy's age? Explain.

25.134 Assuming that many radioactive nuclides can be considered safe after 20 half-lives, how long will it take for each nuclide to be safe?
(a) ${}^{242}Cm$ ($t_{1/2}$ = 163 days)
(b) ${}^{214}Po$ ($t_{1/2}$ = 1.6 × 10^{-4} s)
(c) ${}^{232}Th$ ($t_{1/2}$ = 1.39 × 10^{10} years)

25.135 An ancient sword has a blade from the early Roman Empire, around 100 C.E., but the wooden handle, inlaid wooden decorations, leather ribbon, and leather sheath have different styles. Given the following activities, estimate the age of each part. Which part was made near the time of the blade [$t_{1/2}$ of ^{14}C = 5730 years; $\mathcal{A}_0$ = 15.3 d/(min·g)]?

Part	$\mathcal{A}_t$ [d/(min·g)]
Handle	10.1
Inlaid wood	13.8
Ribbon	12.1
Sheath	15.0

25.136 The starship *Voyager*, like many other vessels of a newly designed 24th-century fleet, uses antimatter as fuel.
(a) How much energy is released when 1.00 kg of antimatter and 1.00 kg of matter annihilate each other?
(b) When the antimatter is atomic antihydrogen, a small amount of it is mixed with excess atomic hydrogen, which was gathered from interstellar space during a flight. The annihilation releases so much heat that the remaining hydrogen nuclei fuse to form ^{4}He. If each hydrogen-antihydrogen collision releases enough heat to fuse 1.00×10^5 hydrogen atoms, how much energy (kJ) is released per kilogram of antihydrogen?
(c) Which produces more energy per kilogram of antihydrogen: the procedure in part (a) or the procedure in part (b)?

25.137 Use Einstein's equation, the mass (g) of 1 u, and the relation between eV and J to find the energy equivalent (MeV) of a mass difference of 1 u.

25.138 Determine the age of a rock containing 0.065 g of uranium-238 ($t_{1/2} = 4.5 \times 10^9$ years) and 0.023 g of lead-206. (Assume that all the lead-206 came from ^{238}U decay.)

25.139 Plutonium-242 decays to uranium-238 by the emission of an α particle with an energy of 4.853 MeV. The ^{238}U that forms is unstable and emits a γ ray (λ = 0.02757 nm).
(a) Write balanced equations for these reactions.
(b) What would be the energy of the α particle if ^{242}Pu decayed directly to the more stable ^{238}U?

25.140 Seaborgium-263 (Sg) is the first isotope of element 106 that was synthesized. It was synthesized, together with four neutrons, by bombarding californium-249 with oxygen-18. It then underwent a series of decays starting with three α emissions. Write balanced equations for the synthesis and the three α emissions of ^{263}Sg.

25.141 Some nuclear power plants use plutonium-239, which is produced in breeder reactors. The rate-determining step is the second β^- emission. How long does it take to make 1.00 kg of ^{239}Pu if the reaction is complete when the product is 90.% ^{239}Pu?

25.142 A random-number generator can be used to simulate the probability of a given atom decaying over a given time. For example, the formula "=RAND()" in a Microsoft Excel spreadsheet returns a random number between 0 and 1; thus, for one radioactive atom and a time of one half-life, a number less than 0.5 means that the atom decays, and a number greater than or equal to 0.5 means that the atom does not decay.
(a) Place the "=RAND()" formula in cells A1 to A10 of a Microsoft Excel spreadsheet. In cell B1, place "=IF(A1<0.5, 0, 1)." This formula returns 0 if A1 is less than 0.5 (the atom decays) and 1 if A1 is greater than or equal to 0.5 (the atom does not decay). Place analogous formulas in cells B2 to B10 (using the "Fill Down" procedure in Microsoft Excel). To determine the number of atoms remaining after one half-life, sum cells B1 to B10 by placing "=SUM(B1:B10)" in cell B12. To create a new set of random numbers, click on an empty cell (such as B13) and hit "Delete." Perform 10 simulations, each time recording the total number of atoms remaining. Do half of the atoms remain after each half-life? If not, why not?
(b) Increase the number of atoms to 100 by placing suitable formulas in cells A1 to A100, B1 to B100, and B102. Perform 10 simulations, and record the number of atoms remaining each time. Are these results more realistic for radioactive decay? Explain.

25.143 In this Microsoft Excel–based simulation, the fate of 256 atoms is followed over five half-lives. Set up formulas in columns A and B, as in Problem 25.142, and simulate the fate of the sample of 256 atoms over one half-life. Cells B1 to B256 should contain 1 or 0. In cell C1, enter "=IF(B1=0, 0, RAND())." This returns 0 if the original atom decayed in the previous half-life or a random number between 0 and 1 if it did not. Fill the formula in C1 down to cell C256. Column D should have formulas similar to those in B, but with modified references, as should columns F, H, and J. Columns E, G, and I should have formulas similar to those in C, but with modified references. In cell B258, enter "=SUM(B1:B256)" to record the number of atoms remaining after the first half-life. Put formulas in cells D258, F258, H258, and J258 to record the number of atoms remaining after subsequent half-lives.
(a) Ideally, how many atoms should remain after each half-life?
(b) Make a table to record the atoms remaining after each half-life in four separate simulations. Compare these outcomes with the ideal outcome. How would you make the results more realistic?

25.144 Representations of three nuclei are shown below. (The neutrons are grey, and the protons are purple.) Nucleus 1 is stable, but nuclei 2 and 3 are not.

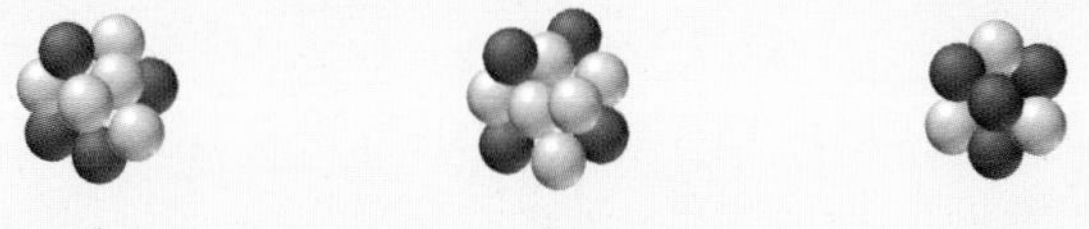

1 2 3

(a) Write the symbol for each isotope.
(b) What is (are) the most likely mode(s) of decay for (i) nucleus 2; (ii) nucleus 3?

Common Mathematical Operations in Chemistry

In addition to basic arithmetic and algebra, four mathematical techniques are used frequently in general chemistry: manipulating logarithms, using exponential (scientific) notation, solving quadratic equations, and graphing data. Each is discussed briefly below.

Manipulating Logarithms

Meaning and Properties of Logarithms

A *logarithm* is an exponent. Specifically, if $x^n = A$, we can say that the logarithm to the base x of the number A is n, and we can denote it as

$$\log_x A = n$$

Because logarithms are exponents, they have the following properties:

$$\log_x 1 = 0$$

$$\log_x (A \times B) = \log_x A + \log_x B$$

$$\log_x \frac{A}{B} = \log_x A - \log_x B$$

$$\log_x A^y = y \log_x A$$

Types of Logarithms

Common and natural logarithms are used in chemistry and the other sciences.

1. *Common* logarithms have base (x in the examples above) 10, but they are written without specifying the base; that is, $\log_{10} A$ is written more simply as $\log A$; thus, the notation *log* means the logarithm with base 10. The common logarithm of 1000 is 3; in other words, you must raise 10 to the 3rd power to obtain 1000:

$$\log 1000 = 3 \qquad \text{or} \qquad 10^3 = 1000$$

Similarly, we have

$$\log 10 = 1 \qquad \text{or} \qquad 10^1 = 10$$

$$\log 1\,000\,000 = 6 \qquad \text{or} \qquad 10^6 = 1\,000\,000$$

$$\log 0.001 = -3 \qquad \text{or} \qquad 10^{-3} = 0.001$$

$$\log 853 = 2.931 \qquad \text{or} \qquad 10^{2.931} = 853$$

The last example illustrates an important point about significant figures with all logarithms: *the number of significant figures in the number equals the number of digits to the right of the decimal point in the logarithm.* That is, the number 853 has three significant figures, and the logarithm 2.931 has three digits to the right of the decimal point.

2. For *natural* logarithms, the base is the number e, which is 2.71828 . . . , and $\log_e A$ is written $\ln A$; thus, the notation *ln* means base e. The relationship between the common and natural logarithms is*

$$\ln A = 2.303 \log A$$

Antilogarithms

The *antilogarithm* or inverse logarithm is the base raised to the logarithm:

$$\text{antilogarithm (antilog) of } n \text{ is } 10^n$$

Using two of the earlier examples, the antilog of 3 is 1000, and the antilog of 2.931 is 853.

Using Exponential (Scientific) Notation

Many quantities in chemistry are very large or very small. For example, in the conventional way of writing numbers, the number of gold atoms in 1 g of gold is

59 060 000 000 000 000 000 000 atoms (to four significant figures)

As another example, the mass (g) of one gold atom is

0.000 000 000 000 000 000 000 3272 g (to four significant figures)

Exponential (scientific) notation provides a much more practical way of writing such numbers. In exponential notation, we express numbers in the form

$$A \times 10^n$$

where A (the coefficient) is greater than or equal to 1 and less than 10 (that is, $1 \le A < 10$), and n (the exponent) is a nonzero integer.

If the number we want to express in exponential notation is larger than 1, the exponent is positive ($n > 0$); if the number is greater than 0 but less than 1, the exponent is negative ($n < 0$). The size of n tells the number of places the decimal point (in conventional notation) must be moved to obtain a coefficient A greater than or equal to 1 and less than 10 (in exponential notation). In exponential notation, 1 g of gold contains 5.906×10^{22} atoms, and each gold atom has a mass of 3.272×10^{-22} g.

Changing between Conventional and Exponential Notation

To use exponential notation, you must be able to convert to it from conventional notation, and vice versa.

1. To change a number from conventional to exponential notation, move the decimal point to the left for numbers equal to or greater than 10 and to the right for numbers between 0 and 1:

   75 000 000 changes to 7.5×10^7 (decimal point 7 places to the left)

   0.006 042 changes to 6.042×10^{-3} (decimal point 3 places to the right)

2. To change a number from exponential to conventional notation, move the decimal point the number of places indicated by the exponent to the right for numbers with positive exponents and to the left for numbers with negative exponents:

   1.38×10^5 changes to 138 000 (decimal point 5 places to the right)

   8.41×10^{-6} changes to 0.000 008 41 (decimal point 6 places to the left)

*For those who are interested, the actual proof is as follows:

$x = \ln A \Rightarrow A = e^x$

$y = \log A \Rightarrow A = 10^y$

$e^x = 10^y$

Take the natural logarithm of both sides:

$x \ln e = y \ln 10$

$x = 2.303y$

$\ln A = 2.303 \log A$

3. An exponential number with a coefficient greater than 10 or less than 1 can be changed to the standard exponential form by converting the coefficient to the standard form and adding the exponents:

$$582.3 \times 10^6 \text{ changes to } 5.823 \times 10^2 \times 10^6 = 5.823 \times 10^{(2+6)} = 5.823 \times 10^8$$

$$0.0043 \times 10^{-4} \text{ changes to } 4.3 \times 10^{-3} \times 10^{-4} = 4.3 \times 10^{[(-3)+(-4)]} = 4.3 \times 10^{-7}$$

Using Exponential Notation in Calculations

In calculations, you can treat the coefficient and exponents separately and apply the properties of exponents (see earlier section on logarithms).

1. To multiply exponential numbers, multiply the coefficients, add the exponents, and reconstruct the number in standard exponential notation:

$$(5.5 \times 10^3)(3.1 \times 10^5) = (5.5 \times 3.1) \times 10^{(3+5)} = 17 \times 10^8 = 1.7 \times 10^9$$

$$(9.7 \times 10^{14})(4.3 \times 10^{-20}) = (9.7 \times 4.3) \times 10^{[14+(-20)]} = 42 \times 10^{-6} = 4.2 \times 10^{-5}$$

2. To divide exponential numbers, divide the coefficients, subtract the exponents, and reconstruct the number in standard exponential notation:

$$\frac{2.6 \times 10^6}{5.8 \times 10^2} = \frac{2.6}{5.8} \times 10^{(6-2)} = 0.45 \times 10^4 = 4.5 \times 10^3$$

$$\frac{1.7 \times 10^{-5}}{8.2 \times 10^{-8}} = \frac{1.7}{8.2} \times 10^{[(-5)-(-8)]} = 0.21 \times 10^3 = 2.1 \times 10^2$$

3. To add or subtract exponential numbers, change all numbers so that they have the same exponent; then add or subtract the coefficients:

$$(1.45 \times 10^4) + (3.2 \times 10^3) = (1.45 \times 10^4) + (0.32 \times 10^4) = 1.77 \times 10^4$$

$$(3.22 \times 10^5) - (9.02 \times 10^4) = (3.22 \times 10^5) - (0.902 \times 10^5) = 2.32 \times 10^5$$

Solving Quadratic Equations

A *quadratic equation* is one in which the highest power of x is 2. The general form of a quadratic equation is

$$ax^2 + bx + c = 0$$

where a, b, and c are numbers. For given values of a, b, and c, the values of x that satisfy the equation are called *solutions* of the equation. We calculate x with the quadratic formula:

$$x = \frac{-b \pm \sqrt{b^2 - 4ac}}{2a}$$

We commonly require the quadratic formula when solving for concentration in an equilibrium problem. For example, we might have an expression that is rearranged into the quadratic equation

$$\underset{a}{4.3x^2} + \underset{b}{0.65x} - \underset{c}{8.7} = 0$$

Applying the quadratic formula, with $a = 4.3$, $b = 0.65$, and $c = -8.7$, gives

$$x = \frac{-0.65 \pm \sqrt{(0.65)^2 - 4(4.3)(-8.7)}}{2(4.3)}$$

The "plus or minus" sign ($\pm$) indicates that there are always two possible values for x (if $b^2 - 4ac > 0$). In this case, they are

$$x = 1.3 \text{ and } x = -1.5$$

In any real physical system, however, only one of the values will have any meaning. For example, if x were $[H_3O^+]$, the negative value would give a negative concentration, which has no physical meaning.

Graphing Data in the Form of a Straight Line

Visualizing changes in variables by means of a graph is used throughout science. In many cases, it is most useful if the data can be graphed in the form of a straight line. Any equation will appear as a straight line if it has, or can be rearranged to have, the following general form:

$$y = mx + b$$

where y is the dependent variable (typically plotted along the vertical axis), x is the independent variable (typically plotted along the horizontal axis), m is the slope of the line, and b is the intercept of the line on the y axis. The intercept is the value of y when $x = 0$:

$$y = m(0) + b = b$$

The slope of the line is the change in y for a given change in x:

$$\text{Slope } (m) = \frac{y_2 - y_1}{x_2 - x_1} = \frac{\Delta y}{\Delta x}$$

The *sign* of the slope tells the *direction* of the line. If y increases as x increases, m is positive, and the line slopes upward with higher values of x; if y decreases as x increases, m is negative, and the line slopes downward with higher values of x. The *magnitude* of the slope indicates the *steepness* of the line. A line with $m = 3$ is three times as steep (y changes three times as much for a given change in x) as a line with $m = 1$.

Consider the linear equation $y = 2x + 1$. A graph of this equation is shown in Figure A.1. In practice, you can find the slope by drawing a right triangle to the line, using the line as the hypotenuse. Then, one leg gives Δy, and the other gives Δx. In the figure, $\Delta y = 8$ and $\Delta x = 4$.

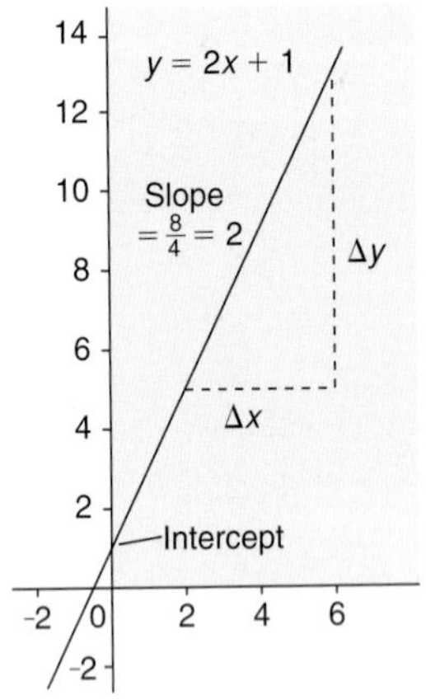

FIGURE A.1

In several places in the text, an equation is rearranged into the form of a straight line to determine information from the slope and/or the intercept. For example, in Chapter 14, we obtained the following expression:

$$\ln\left(\frac{[\text{A}]_0}{[\text{A}]_t}\right) = kt$$

Based on the properties of logarithms, we have

$$\ln [\text{A}]_0 - \ln [\text{A}]_t = kt$$

Rearranging into the form of an equation for a straight line gives

$$\ln [\text{A}]_t = -kt + \ln [\text{A}]_0$$
$$y \quad = mx + \quad b$$

Thus, a plot of $\ln [\text{A}]_t$ versus t is a straight line, from which you can see that the slope is $-k$ (the negative of the rate constant) and the intercept is $\ln [\text{A}]_0$ (the natural logarithm of the initial concentration of A).

Standard Thermodynamic Values for Selected Substances*

Substance or Ion	$\Delta_f H^\circ$ (kJ/mol)	$\Delta_f G^\circ$ (kJ/mol)	S° [J/(mol·K)]
$e^-(g)$	0	0	20.87
Aluminum			
$Al(s)$	0	0	28.3
$Al^{3+}(aq)$	−524.7	−481.2	−313
$AlCl_3(s)$	−704.2	−628.9	110.7
$Al_2O_3(s)$	−1676	−1582	50.94
Barium			
$Ba(s)$	0	0	62.5
$Ba(g)$	175.6	144.8	170.28
$Ba^{2+}(g)$	1649.9	—	—
$Ba^{2+}(aq)$	−538.36	−560.7	13
$BaCl_2(s)$	−806.06	−810.9	126
$BaCO_3(s)$	−1219	−1139	112
$BaO(s)$	−548.1	−520.4	72.07
$BaSO_4(s)$	−1465	−1353	132
Boron			
B(β-rhombohedral)	0	0	5.87
$BF_3(g)$	−1137.0	−1120.3	254.0
$BCl_3(g)$	−403.8	−388.7	290.0
$B_2H_6(g)$	35	86.6	232.0
$B_2O_3(s)$	−1272	−1193	53.8
$H_3BO_3(s)$	−1094.3	−969.01	88.83
Bromine			
$Br_2(l)$	0	0	152.23
$Br_2(g)$	30.91	3.13	245.38
$Br(g)$	111.9	82.40	174.90
$Br^-(g)$	−218.9	—	—
$Br^-(aq)$	−120.9	−102.82	80.71
$HBr(g)$	−36.3	−53.5	198.59
Cadmium			
$Cd(s)$	0	0	51.5
$Cd(g)$	112.8	78.20	167.64
$Cd^{2+}(aq)$	−72.38	−77.74	−61.1
$CdS(s)$	−144	−141	71
Calcium			
$Ca(s)$	0	0	41.6
$Ca(g)$	192.6	158.9	154.78
$Ca^{2+}(g)$	1934.1	—	—
$Ca^{2+}(aq)$	−542.96	−553.04	−55.2
$CaF_2(s)$	−1215	−1162	68.87
$CaCl_2(s)$	−795.0	−750.2	114
$CaCO_3(s)$	−1206.9	−1128.8	92.9
$CaO(s)$	−635.1	−603.5	38.2
$Ca(OH)_2(s)$	−986.09	−898.56	83.39
$Ca_3(PO_4)_2(s)$	−4138	−3899	263
$CaSO_4(s)$	−1432.7	−1320.3	107
Carbon			
C(graphite)	0	0	5.686
C(diamond)	1.896	2.866	2.439
$C(g)$	715.0	669.6	158.0
$CO(g)$	−110.5	−137.2	197.5
$CO_2(g)$	−393.5	−394.4	213.7
$CO_2(aq)$	−412.9	−386.2	121
$CO_3^{2-}(aq)$	−676.26	−528.10	−53.1
$HCO_3^-(aq)$	−691.11	587.06	95.0
$H_2CO_3(aq)$	−698.7	−623.42	191
$CH_4(g)$	−74.87	−50.81	186.1
$C_2H_2(g)$	227	209	200.85
$C_2H_4(g)$	52.47	68.36	219.22
$C_2H_6(g)$	−84.667	−32.89	229.5
$C_3H_8(g)$	−105	−24.5	269.9
$C_4H_{10}(g)$	−126	−16.7	310
$C_6H_6(l)$	49.0	124.5	172.8
$CH_3OH(g)$	−201.2	−161.9	238
$CH_3OH(l)$	−238.6	−166.2	127
$HCHO(g)$	−116	−110	219
$HCOO^-(aq)$	−410	−335	91.6
$HCOOH(l)$	−409	−346	129.0
$HCOOH(aq)$	−410	−356	164
$C_2H_5OH(g)$	−235.1	−168.6	282.6
$C_2H_5OH(l)$	−277.63	−174.8	161
$CH_3CHO(g)$	−166	−133.7	266
$CH_3COOH(l)$	−487.0	−392	160
$C_6H_{12}O_6(s)$	−1273.3	−910.56	212.1
$C_{12}H_{22}O_{11}(s)$	−2221.7	−1544.3	360.24
$CN^-(aq)$	151	166	118
$HCN(g)$	135	125	201.7
$HCN(l)$	105	121	112.8
$HCN(aq)$	105	112	129
$CS_2(g)$	117	66.9	237.79
$CS_2(l)$	87.9	63.6	151.0
$CH_3Cl(g)$	−83.7	−60.2	234
$CH_2Cl_2(l)$	−117	−63.2	179

*All values at 298 K.

(Continued)

Substance or Ion	$\Delta_f H^\circ$ (kJ/mol)	$\Delta_f G^\circ$ (kJ/mol)	S° [J/(mol·K)]
$CHCl_3(l)$	−132	−71.5	203
$CCl_4(g)$	−96.0	−53.7	309.7
$CCl_4(l)$	−139	−68.6	214.4
$COCl_2(g)$	−220	−206	283.74
Cesium			
$Cs(s)$	0	0	85.15
$Cs(g)$	76.7	49.7	175.5
$Cs^+(g)$	458.5	427.1	169.72
$Cs^+(aq)$	−248	−282.0	133
$CsF(s)$	−554.7	−525.4	88
$CsCl(s)$	−442.8	−414	101.18
$CsBr(s)$	−395	−383	121
$CsI(s)$	−337	−333	130
Chlorine			
$Cl_2(g)$	0	0	223.0
$Cl(g)$	121.0	105.0	165.1
$Cl^-(g)$	−234	−240	153.25
$Cl^-(aq)$	−167.46	−131.17	55.10
$HCl(g)$	−92.31	−95.30	186.79
$HCl(aq)$	−167.46	−131.17	55.06
$ClO_2(g)$	102	120	256.7
$Cl_2O(g)$	80.3	97.9	266.1
Chromium			
$Cr(s)$	0	0	23.8
$Cr^{3+}(aq)$	−1971	—	—
$CrO_4^{2-}(aq)$	−863.2	−706.3	38
$Cr_2O_7^{2-}(aq)$	−1461	−1257	214
Copper			
$Cu(s)$	0	0	33.1
$Cu(g)$	341.1	301.4	166.29
$Cu^+(aq)$	51.9	50.2	−26
$Cu^{2+}(aq)$	64.39	64.98	−98.7
$Cu_2O(s)$	−168.6	−146.0	93.1
$CuO(s)$	−157.3	−130	42.63
$Cu_2S(s)$	−79.5	−86.2	120.9
$CuS(s)$	−53.1	−53.6	66.5
Fluorine			
$F_2(g)$	0	0	202.7
$F(g)$	78.9	61.8	158.64
$F^-(g)$	−255.6	−262.5	145.47
$F^-(aq)$	−329.1	−276.5	−9.6
$HF(g)$	−273	−275	173.67
Hydrogen			
$H_2(g)$	0	0	130.6
$H(g)$	218.0	203.30	114.60
$H^+(aq)$	0	0	0
$H^+(g)$	1536.3	1517.1	108.83
Iodine			
$I_2(s)$	0	0	116.14
$I_2(g)$	62.442	19.38	260.58
$I(g)$	106.8	70.21	180.67
$I^-(g)$	−194.7	—	—
$I^-(aq)$	−55.94	−51.67	109.4
$HI(g)$	25.9	1.3	206.33
Iron			
$Fe(s)$	0	0	27.3
$Fe^{3+}(aq)$	−47.7	−10.5	−293
$Fe^{2+}(aq)$	−87.9	−84.94	113
$FeCl_2(s)$	−341.8	−302.3	117.9
$FeCl_3(s)$	−399.5	−334.1	142
$FeO(s)$	−272.0	−251.4	60.75
$Fe_2O_3(s)$	−825.5	−743.6	87.400
$Fe_3O_4(s)$	−1121	−1018	145.3
Lead			
$Pb(s)$	0	0	64.785
$Pb^{2+}(aq)$	1.6	−24.3	21
$PbCl_2(s)$	−359	−314	136
$PbO(s)$	−218	−198	68.70
$PbO_2(s)$	−276.6	−219.0	76.6
$PbS(s)$	−98.3	−96.7	91.3
$PbSO_4(s)$	−918.39	−811.24	147
Lithium			
$Li(s)$	0	0	29.10
$Li(g)$	161	128	138.67
$Li^+(g)$	687.163	649.989	132.91
$Li^+(aq)$	−278.46	−293.8	14
$LiF(s)$	−616.9	−588.7	35.66
$LiCl(s)$	−408	−384	59.30
$LiBr(s)$	−351	−342	74.1
$LiI(s)$	−270	−270	85.8
Magnesium			
$Mg(s)$	0	0	32.69
$Mg(g)$	150	115	148.55
$Mg^{2+}(g)$	2351	—	—
$Mg^{2+}(aq)$	−461.96	−456.01	118
$MgCl_2(s)$	−641.6	−592.1	89.630
$MgCO_3(s)$	−1112	−1028	65.86
$MgO(s)$	−601.2	−569.0	26.9
$Mg_3N_2(s)$	−461	−401	88
Manganese			
$Mn(s, \alpha)$	0	0	31.8
$Mn^{2+}(aq)$	−219	−223	−84
$MnO_2(s)$	−520.9	−466.1	53.1
$MnO_4^-(aq)$	−518.4	−425.1	190
Mercury			
$Hg(l)$	0	0	76.027
$Hg(g)$	61.30	31.8	174.87
$Hg^{2+}(aq)$	171	164.4	−32
$Hg_2^{2+}(aq)$	172	153.6	84.5
$HgCl_2(s)$	−230	−184	144
$Hg_2Cl_2(s)$	−264.9	−210.66	196
$HgO(s)$	−90.79	−58.50	70.27
Nitrogen			
$N_2(g)$	0	0	191.5
$N(g)$	473	456	153.2
$N_2O(g)$	82.05	104.2	219.7
$NO(g)$	90.29	86.60	210.65
$NO_2(g)$	33.2	51	239.9
$N_2O_4(g)$	9.16	97.7	304.3
$N_2O_5(g)$	11	118	346
$N_2O_5(s)$	−43.1	114	178
$NH_3(g)$	−45.9	−16	193
$NH_3(aq)$	−80.83	26.7	110
$N_2H_4(l)$	50.63	149.2	121.2

(Continued)

Substance or Ion	$\Delta_f H^\circ$ (kJ/mol)	$\Delta_f G^\circ$ (kJ/mol)	S° [J/(mol·K)]
$NO_3^-(aq)$	−206.57	−110.5	146
$HNO_3(l)$	−173.23	−79.914	155.6
$HNO_3(aq)$	−206.57	−110.5	146
$NF_3(g)$	−125	−83.3	260.6
$NOCl(g)$	51.71	66.07	261.6
$NH_4Cl(s)$	−314.4	−203.0	94.6
Oxygen			
$O_2(g)$	0	0	205.0
$O(g)$	249.2	231.7	160.95
$O_3(g)$	143	163	238.82
$OH^-(aq)$	−229.94	−157.30	−10.54
$H_2O(g)$	−241.826	−228.60	188.72
$H_2O(l)$	−285.840	−237.192	69.940
$H_2O_2(l)$	−187.8	−120.4	110
$H_2O_2(aq)$	−191.2	−134.1	144
Phosphorus			
$P(s$, white)	0	0	41.1
$P(g)$	314.6	278.3	163.1
$P(s$, red)	−17.6	−12.1	22.8
$P_2(g)$	144	104	218
$P_4(g)$	58.9	24.5	280
$PCl_3(g)$	−287	−268	312
$PCl_3(l)$	−320	−272	217
$PCl_5(g)$	−402	−323	353
$PCl_5(s)$	−443.5	—	—
$P_4O_{10}(s)$	−2984	−2698	229
$PO_4^{3-}(aq)$	−1266	−1013	−218
$HPO_4^{2-}(aq)$	−1281	−1082	−36
$H_2PO_4^-(aq)$	−1285	−1135	89.1
$H_3PO_4(aq)$	−1277	−1019	228
Potassium			
$K(s)$	0	0	64.672
$K(g)$	89.2	60.7	160.23
$K^+(g)$	514.197	481.202	154.47
$K^+(aq)$	−251.2	−282.28	103
$KF(s)$	−568.6	−538.9	66.55
$KCl(s)$	−436.7	−409.2	82.59
$KBr(s)$	−394	−380	95.94
$KI(s)$	−328	−323	106.39
$KOH(s)$	−424.8	−379.1	78.87
$KClO_3(s)$	−397.7	−296.3	143.1
$KClO_4(s)$	−432.75	−303.2	151.0
Rubidium			
$Rb(s)$	0	0	69.5
$Rb(g)$	85.81	55.86	169.99
$Rb^+(g)$	495.04	—	—
$Rb^+(aq)$	−246	−282.2	124
$RbF(s)$	−549.28	—	—
$RbCl(s)$	−435.35	−407.8	95.90
$RbBr(s)$	−389.2	−378.1	108.3
$RbI(s)$	−328	−326	118.0
Silicon			
$Si(s)$	0	0	18.0
$SiF_4(g)$	−1614.9	−1572.7	282.4
$SiO_2(s)$	−910.9	−856.5	41.5
Silver			
$Ag(s)$	0	0	42.702
$Ag(g)$	289.2	250.4	172.892
$Ag^+(aq)$	105.9	77.111	73.93
$AgF(s)$	−203	−185	84
$AgCl(s)$	−127.03	−109.72	96.11
$AgBr(s)$	−99.51	−95.939	107.1
$AgI(s)$	−62.38	−66.32	114
$AgNO_3(s)$	−45.06	19.1	128.2
$Ag_2S(s)$	−31.8	−40.3	146
Sodium			
$Na(s)$	0	0	51.446
$Na(g)$	107.76	77.299	153.61
$Na^+(g)$	609.839	574.877	147.85
$Na^+(aq)$	−239.66	−261.87	60.2
$NaF(s)$	−575.4	−545.1	51.21
$NaCl(s)$	−411.1	−384.0	72.12
$NaBr(s)$	−361	−349	86.82
$NaOH(s)$	−425.609	−379.53	64.454
$Na_2CO_3(s)$	−1130.8	−1048.1	139
$NaHCO_3(s)$	−947.7	−851.9	102
$NaI(s)$	−288	−285	98.5
Strontium			
$Sr(s)$	0	0	54.4
$Sr(g)$	164	110	164.54
$Sr^{2+}(g)$	1784	—	—
$Sr^{2+}(aq)$	−545.51	−557.3	−39
$SrCl_2(s)$	−828.4	−781.2	117
$SrCO_3(s)$	−1218	−1138	97.1
$SrO(s)$	−592.0	−562.4	55.5
$SrSO_4(s)$	−1445	−1334	122
Sulfur			
S(rhombic)	0	0	31.9
S(monoclinic)	0.3	0.096	32.6
$S(g)$	279	239	168
$S_2(g)$	129	80.1	228.1
$S_8(g)$	101	49.1	430.211
$S^{2-}(aq)$	41.8	83.7	22
$HS^-(aq)$	−17.7	12.6	61.1
$H_2S(g)$	−20.2	−33	205.6
$H_2S(aq)$	−39	−27.4	122
$SO_2(g)$	−296.8	−300.2	248.1
$SO_3(g)$	−396	−371	256.66
$SO_4^{2-}(aq)$	−907.51	−741.99	17
$HSO_4^-(aq)$	−885.75	−752.87	126.9
$H_2SO_4(l)$	−813.989	−690.059	156.90
$H_2SO_4(aq)$	−907.51	−741.99	17
Tin			
Sn(white)	0	0	51.5
Sn(gray)	3	4.6	44.8
$SnCl_4(l)$	−545.2	−474.0	259
$SnO_2(s)$	−580.7	−519.7	52.3
Titanium			
$Ti(s)$	0	0	30.7
$TiCl_4(l)$	−804.2	−737.2	252.3
$TiO_2(s)$	−944.0	−888.8	50.6
Zinc			
$Zn(s)$	0	0	41.6
$Zn(g)$	130.5	94.93	160.9
$Zn^{2+}(aq)$	−152.4	−147.21	−106.5
$ZnO(s)$	−348.0	−318.2	43.9
$ZnS(s$, zinc blende)	−203	−198	57.7

APPENDIX C

Equilibrium Constants for Selected Substances*†

Dissociation (Ionization) Constants (K_a) in Water of Selected Acids

Name and Formula	Lewis Structure‡	K_{a1}	K_{a2}	K_{a3}
2-acetyloxybenzoic acid (Acetylsalicylic acid) $CH_3COOC_6H_4COOH$	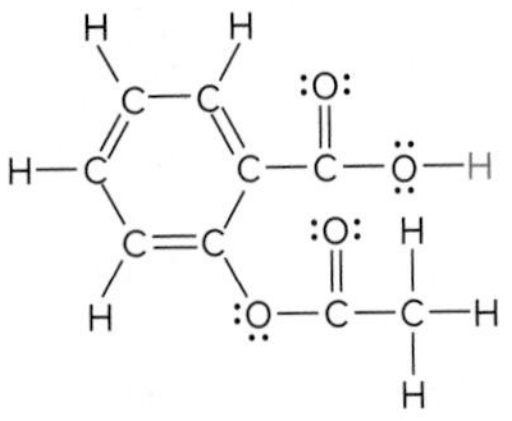	3.6×10^{-4}		
Arsenic acid H_3AsO_4	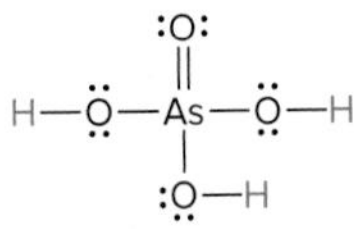	6×10^{-3}	1.1×10^{-7}	3×10^{-12}
Benzoic acid C_6H_5COOH	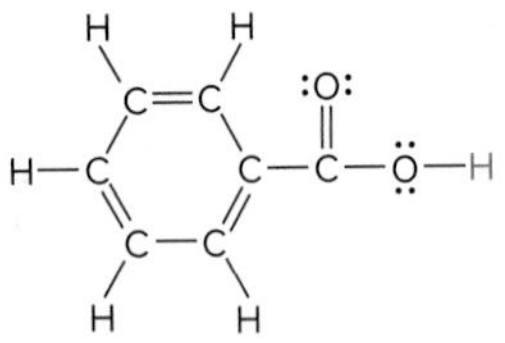	6.3×10^{-5}		
Butanedioic acid (succinic acid) $HOOCCH_2CH_2COOH$	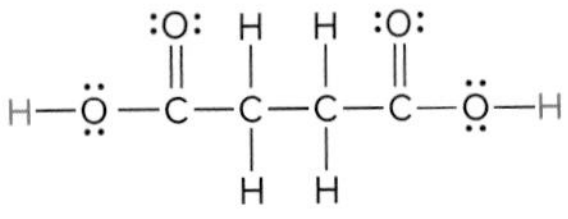	6.2×10^{-5}	2.3×10^{-6}	
Carbonic acid H_2CO_3	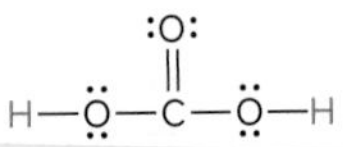	4.5×10^{-7}	4.7×10^{-11}	
Chloroacetic acid $ClCH_2COOH$	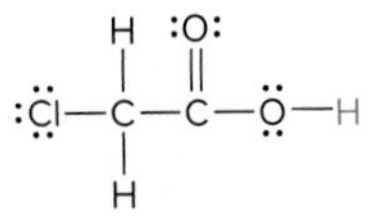	1.4×10^{-3}		
Chlorous acid $HClO_2$	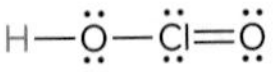	1.1×10^{-2}		

(Continued)

*Listed according to their Preferred IUPAC Name (PIN).
†All values at 298 K, except for 2-acetyloxybenzoic acid (acetylsalicylic acid), which is at 37°C (310 K) in 0.15 mol/L NaCl.
‡Acidic (ionizable) proton(s) shown in red. Structures have lowest formal charges. Benzene rings show one resonance form.

Dissociation (Ionization) Constants (K_a) in Water of Selected Acids (*continued*)

Name and Formula	Lewis Structure	K_{a1}	K_{a2}	K_{a3}
2,3-dihydroxypropanoic acid (glyceric acid) $HOCH_2CH(OH)COOH$	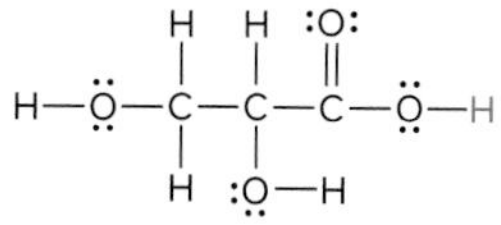	2.9×10^{-4}		
Ethanoic acid (acetic acid) CH_3COOH	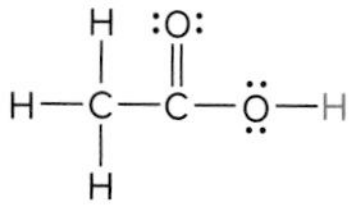	1.8×10^{-5}		
Formonitrile (hydrocyanic acid) HCN	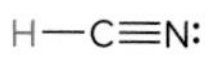	6.2×10^{-10}		
Hexanedioic acid (adipic acid) $HOOC(CH_2)_4COOH$	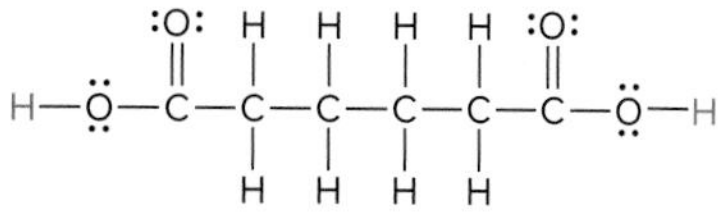	3.8×10^{-5}	3.8×10^{-6}	
Hydroxyacetic acid (glycolic acid) $HOCH_2COOH$	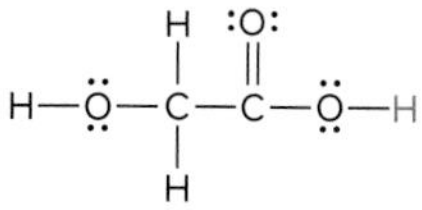	1.5×10^{-4}		
Hydrofluoric acid HF	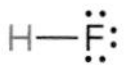	6.8×10^{-4}		
Hydrosulfuric acid H_2S	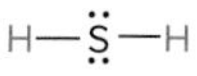	9×10^{-8}	1×10^{-17}	
Hypobromous acid HBrO	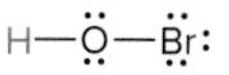	2.3×10^{-9}		
Hypochlorous acid HClO	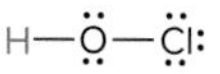	2.9×10^{-8}		
Hypoiodous acid HIO	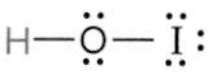	2.3×10^{-11}		
2-hydroxypropanoic acid (lactic acid) $CH_3CH(OH)COOH$	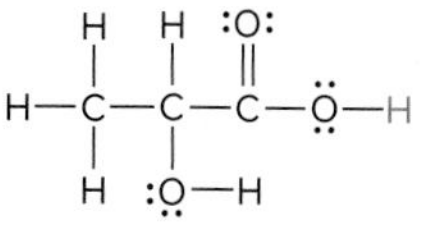	1.4×10^{-4}		
2-hydroxypropane-1,2,3-tricarboxylic acid (citric acid) $HOOCCH_2C(OH)(COOH)CH_2COOH$	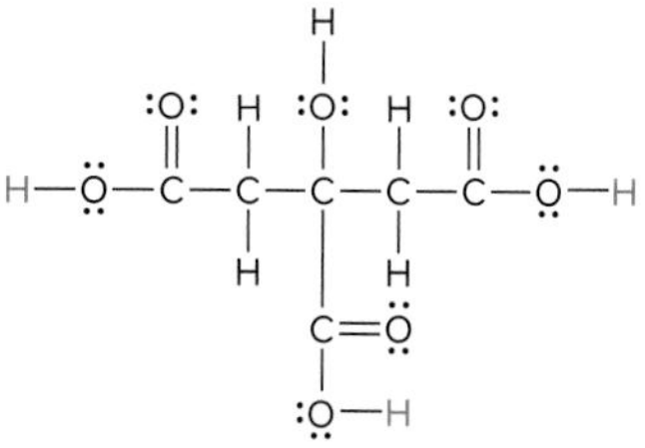	7.4×10^{-4}	1.7×10^{-5}	4.0×10^{-7}
Iodic acid HIO_3	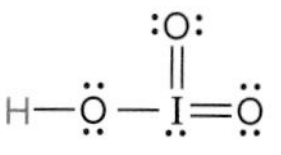	1.6×10^{-1}		
Methanoic acid (formic acid) HCOOH	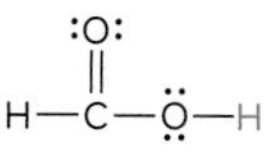	1.8×10^{-4}		
Nitrous acid HNO_2	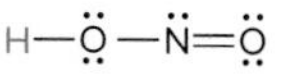	7.1×10^{-4}		
Oxaldehydic acid (glyoxylic acid) HC(O)COOH	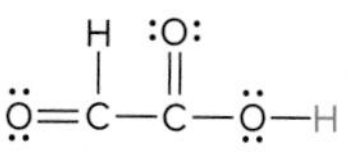	3.5×10^{-4}		

(*Continued*)

Dissociation (Ionization) Constants (K_a) in Water of Selected Acids (*continued*)

Name and Formula	Lewis Structure	K_{a1}	K_{a2}	K_{a3}
Oxalic acid HOOCCOOH	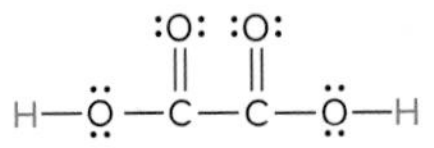	5.6×10^{-2}	5.4×10^{-5}	
2-oxopropanoic acid (pyruvic acid) $CH_3C(O)COOH$	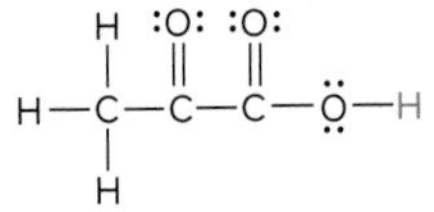	2.8×10^{-3}		
Phenol C_6H_5OH	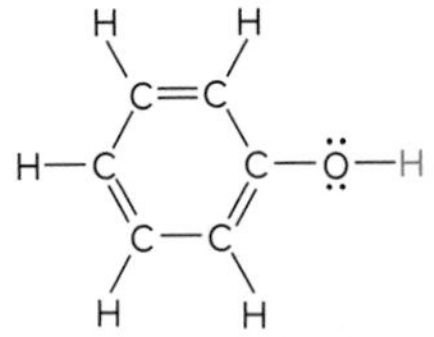	1.0×10^{-10}		
Phenylacetic acid $C_6H_5CH_2COOH$	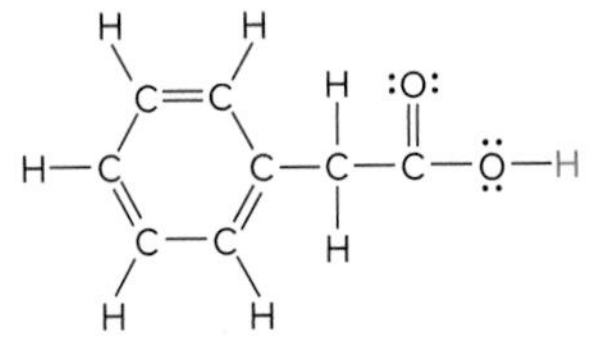	4.9×10^{-5}		
Phosphoric acid H_3PO_4	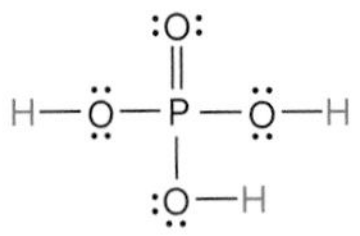	7.2×10^{-3}	6.3×10^{-8}	4.2×10^{-13}
Phosphorous acid $HPO(OH)_2$	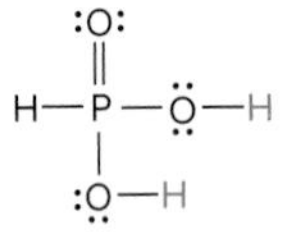	3×10^{-2}	1.7×10^{-7}	
Propanedioic acid (malonic acid) $HOOCCH_2COOH$	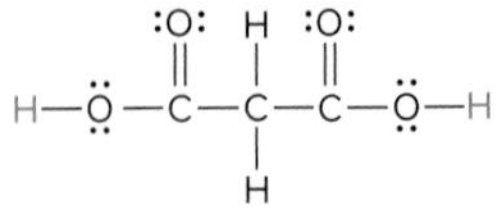	1.4×10^{-3}	2.0×10^{-6}	
Propanoic acid CH_3CH_2COOH	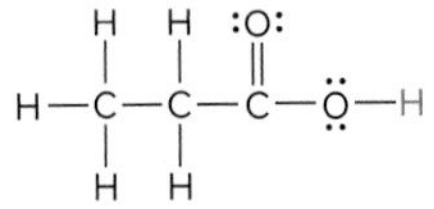	1.3×10^{-5}		
(R)-5-((S)-1,2-dihydroxyethyl)-3,4-dihydroxyfuran-2(5H)-one (ascorbic acid) $H_2C_6H_6O_6$	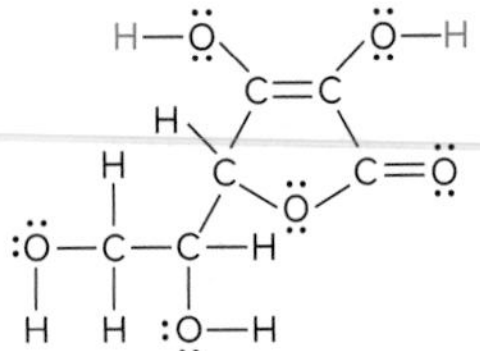	1.0×10^{-5}	5×10^{-12}	
Sulfuric acid H_2SO_4	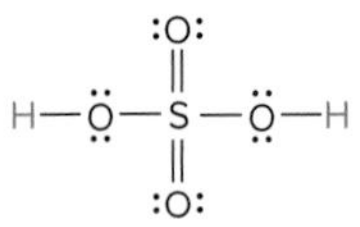	Very large	1.0×10^{-2}	
Sulfurous acid H_2SO_3	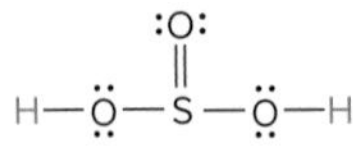	1.4×10^{-2}	6.5×10^{-8}	
(*Z*)-but-2-endoic acid (maleic acid) HOOCCH=CHCOOH	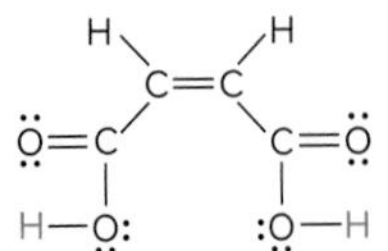	1.2×10^{-2}	4.7×10^{-7}	

(*Continued*)

Dissociation (Ionization) Constants (K_b) in Water of Selected Amine Bases

Name and Formula	Lewis Structure*	K_{b1}	K_{b2}
Ammonia NH_3	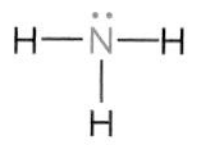	1.76×10^{-5}	
2-aminoethanol (ethanolamine) $HOCH_2CH_2NH_2$	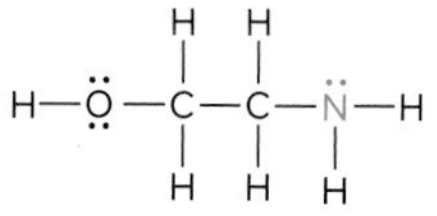	3.2×10^{-5}	
Aniline $C_6H_5NH_2$	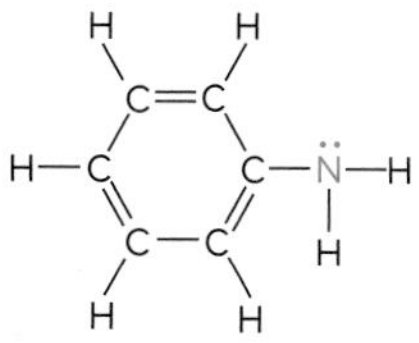	4.0×10^{-10}	
Ethanamine (ethylamine) $CH_3CH_2NH_2$	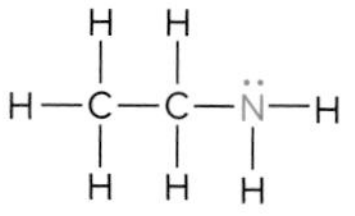	4.3×10^{-4}	
Ethane-1,2-diamine (ethylenediamine) $H_2NCH_2CH_2NH_2$	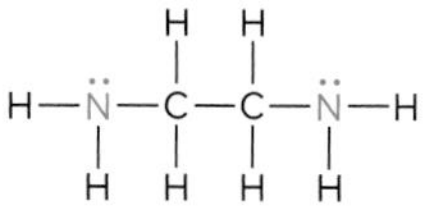	8.5×10^{-5}	7.1×10^{-8}
Methanamine (methylamine) CH_3NH_2	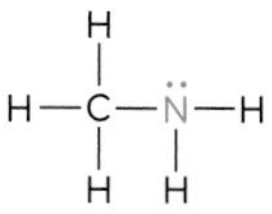	4.4×10^{-4}	
2-methylpropan-2-amine (*tert*-butylamine) $(CH_3)_3CNH_2$	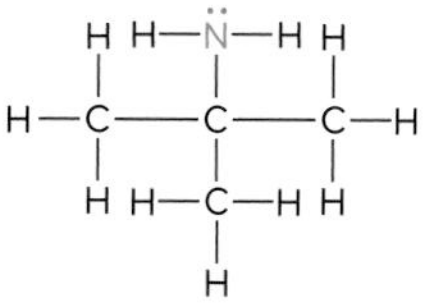	4.8×10^{-4}	
N-ethylethanamine (diethylamine) $(CH_3CH_2)_2NH$	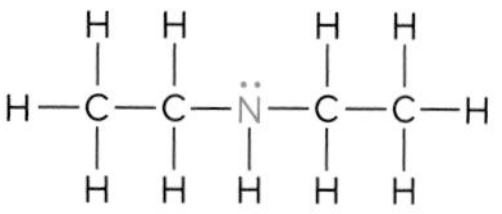	8.6×10^{-4}	
N-methylmethanamine (dimethylamine) $(CH_3)_2NH$	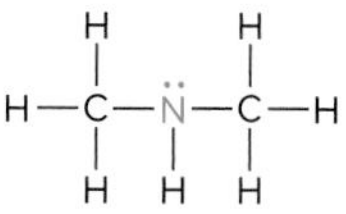	5.9×10^{-4}	
N,N-diethylethanamine (triethylamine) $(CH_3CH_2)_3N$	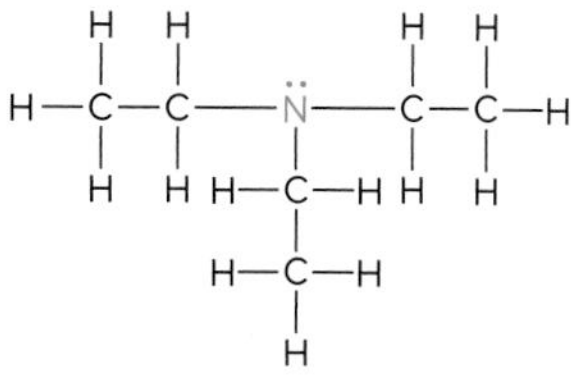	5.2×10^{-4}	
N,N-dimethylmethanamine (trimethylamine) $(CH_3)_3N$	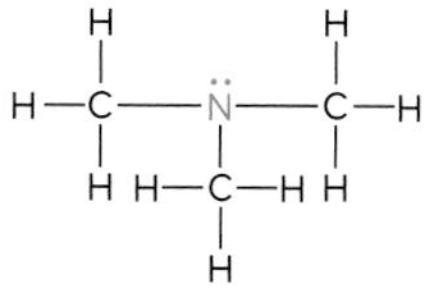	6.3×10^{-5}	
Piperidine $C_5H_{10}NH$	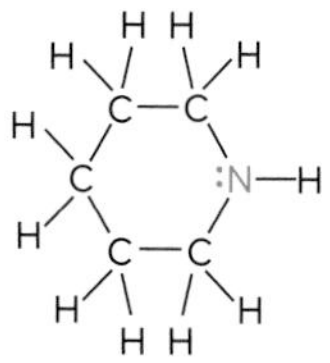	1.3×10^{-3}	
Propanamine (*n*-propylamine) $CH_3CH_2CH_2NH_2$	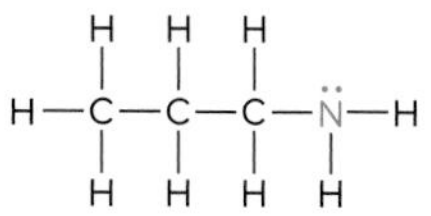	3.5×10^{-4}	

*Blue type indicates the basic nitrogen and its lone pair.

(*Continued*)

Dissociation (Ionization) Constants (K_b) in Water of Selected Amine Bases (*continued*)

Name and Formula	Lewis Structure	K_{b1}	K_{b2}
Propane-1,3-diamine (1,3-propylenediamine) $H_2NCH_2CH_2CH_2NH_2$	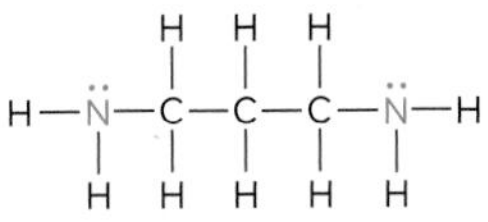	3.1×10^{-4}	3.0×10^{-6}
Propan-2-amine (isopropylamine) $(CH_3)_2CHNH_2$	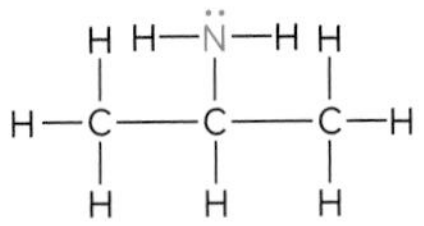	4.7×10^{-4}	
Pyridine C_5H_5N	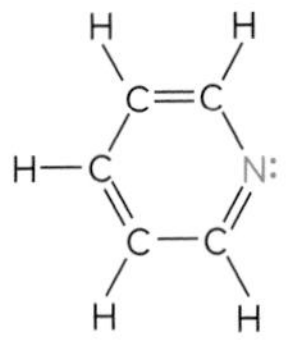	1.7×10^{-9}	

Dissociation (Ionization) Constants (K_a) of Some Hydrated Metal Ions

Free Ion	Hydrated Ion	K_a
Fe^{3+}	$Fe(H_2O)_6^{3+}(aq)$	6×10^{-3}
Sn^{2+}	$Sn(H_2O)_6^{2+}(aq)$	4×10^{-4}
Cr^{3+}	$Cr(H_2O)_6^{3+}(aq)$	1×10^{-4}
Al^{3+}	$Al(H_2O)_6^{3+}(aq)$	1×10^{-5}
Cu^{2+}	$Cu(H_2O)_6^{2+}(aq)$	3×10^{-8}
Pb^{2+}	$Pb(H_2O)_6^{2+}(aq)$	3×10^{-8}
Zn^{2+}	$Zn(H_2O)_6^{2+}(aq)$	1×10^{-9}
Co^{2+}	$Co(H_2O)_6^{2+}(aq)$	2×10^{-10}
Ni^{2+}	$Ni(H_2O)_6^{2+}(aq)$	1×10^{-10}

Formation Constants (K_f) of Some Complex Ions

Complex Ion	K_f
$Ag(CN)_2^-$	3.0×10^{20}
$Ag(NH_3)_2^+$	1.7×10^{7}
$Ag(S_2O_3)_2^{3-}$	4.7×10^{13}
AlF_6^{3-}	4×10^{19}
$Al(OH)_4^-$	3×10^{33}
$Be(OH)_4^{2-}$	4×10^{18}
CdI_4^{2-}	1×10^{6}
$Co(OH)_4^{2-}$	5×10^{9}
$Cr(OH)_4^-$	8.0×10^{29}
$Cu(NH_3)_4^{2+}$	5.6×10^{11}
$Fe(CN)_6^{4-}$	3×10^{35}
$Fe(CN)_6^{3-}$	4.0×10^{43}
$Hg(CN)_4^{2-}$	9.3×10^{38}
$Ni(NH_3)_6^{2+}$	2.0×10^{8}
$Pb(OH)_3^-$	8×10^{13}
$Sn(OH)_3^-$	3×10^{25}
$Zn(CN)_4^{2-}$	4.2×10^{19}
$Zn(NH_3)_4^{2+}$	7.8×10^{8}
$Zn(OH)_4^{2-}$	3×10^{15}

Solubility-Product Constants (K_{sp}) of Slightly Soluble Ionic Compounds

Name, Formula	K_{sp}
Carbonates	
Barium carbonate, $BaCO_3$	2.0×10^{-9}
Cadmium carbonate, $CdCO_3$	1.8×10^{-14}
Calcium carbonate, $CaCO_3$	3.3×10^{-9}
Cobalt(II) carbonate, $CoCO_3$	1.0×10^{-10}
Copper(II) carbonate, $CuCO_3$	3×10^{-12}
Lead(II) carbonate, $PbCO_3$	7.4×10^{-14}
Magnesium carbonate, $MgCO_3$	3.5×10^{-8}
Mercury(I) carbonate, Hg_2CO_3	8.9×10^{-17}
Nickel(II) carbonate, $NiCO_3$	1.3×10^{-7}
Strontium carbonate, $SrCO_3$	5.4×10^{-10}
Zinc carbonate, $ZnCO_3$	1.0×10^{-10}
Chromates	
Barium chromate, $BaCrO_4$	2.1×10^{-10}
Calcium chromate, $CaCrO_4$	1×10^{-8}
Lead(II) chromate, $PbCrO_4$	2.3×10^{-13}
Silver chromate, Ag_2CrO_4	2.6×10^{-12}
Cyanides	
Mercury(I) cyanide, $Hg_2(CN)_2$	5×10^{-40}
Silver cyanide, AgCN	2.2×10^{-16}
Halides	
Fluorides	
Barium fluoride, BaF_2	1.5×10^{-6}
Calcium fluoride, CaF_2	3.2×10^{-11}
Lead(II) fluoride, PbF_2	3.6×10^{-8}
Magnesium fluoride, MgF_2	7.4×10^{-9}
Strontium fluoride, SrF_2	2.6×10^{-9}
Chlorides	
Copper(I) chloride, CuCl	1.9×10^{-7}
Lead(II) chloride, $PbCl_2$	1.7×10^{-5}
Silver chloride, AgCl	1.8×10^{-10}
Bromides	
Copper(I) bromide, CuBr	5×10^{-9}
Silver bromide, AgBr	5.0×10^{-13}
Iodides	
Copper(I) iodide, CuI	1×10^{-12}
Lead(II) iodide, PbI_2	7.9×10^{-9}
Mercury(I) iodide, Hg_2I_2	4.7×10^{-29}
Silver iodide, AgI	8.3×10^{-17}
Hydroxides	
Aluminum hydroxide, $Al(OH)_3$	3×10^{-34}
Cadmium hydroxide, $Cd(OH)_2$	7.2×10^{-15}
Calcium hydroxide, $Ca(OH)_2$	6.5×10^{-6}
Cobalt(II) hydroxide, $Co(OH)_2$	1.3×10^{-15}
Copper(II) hydroxide, $Cu(OH)_2$	2.2×10^{-20}
Iron(II) hydroxide, $Fe(OH)_2$	4.1×10^{-15}
Iron(III) hydroxide, $Fe(OH)_3$	1.6×10^{-39}
Magnesium hydroxide, $Mg(OH)_2$	6.3×10^{-10}
Manganese(II) hydroxide, $Mn(OH)_2$	1.6×10^{-13}
Nickel(II) hydroxide, $Ni(OH)_2$	6×10^{-16}
Zinc hydroxide, $Zn(OH)_2$	3×10^{-16}
Iodates	
Barium iodate, $Ba(IO_3)_2$	1.5×10^{-9}
Calcium iodate, $Ca(IO_3)_2$	7.1×10^{-7}
Lead(II) iodate, $Pb(IO_3)_2$	2.5×10^{-13}
Silver iodate, $AgIO_3$	3.1×10^{-8}
Strontium iodate, $Sr(IO_3)_2$	3.3×10^{-7}
Zinc iodate, $Zn(IO_3)_2$	3.9×10^{-6}
Oxalates	
Barium oxalate dihydrate, $BaC_2O_4{\cdot}2H_2O$	1.1×10^{-7}
Calcium oxalate monohydrate, $CaC_2O_4{\cdot}H_2O$	2.3×10^{-9}
Strontium oxalate monohydrate, $SrC_2O_4{\cdot}H_2O$	5.6×10^{-8}
Phosphates	
Calcium phosphate, $Ca_3(PO_4)_2$	1.2×10^{-29}
Magnesium phosphate, $Mg_3(PO_4)_2$	5.2×10^{-24}
Silver phosphate, Ag_3PO_4	2.6×10^{-18}
Sulfates	
Barium sulfate, $BaSO_4$	1.1×10^{-10}
Calcium sulfate, $CaSO_4$	2.4×10^{-5}
Lead(II) sulfate, $PbSO_4$	1.6×10^{-8}
Radium sulfate, $RaSO_4$	2×10^{-11}
Silver sulfate, Ag_2SO_4	1.5×10^{-5}
Strontium sulfate, $SrSO_4$	3.2×10^{-7}
Sulfides	
Cadmium sulfide, CdS	1.0×10^{-24}
Copper(II) sulfide, CuS	8×10^{-34}
Iron(II) sulfide, FeS	8×10^{-16}
Lead(II) sulfide, PbS	3×10^{-25}
Manganese(II) sulfide, MnS	3×10^{-11}
Mercury(II) sulfide, HgS	2×10^{-50}
Nickel(II) sulfide, NiS	3×10^{-16}
Silver sulfide, Ag_2S	8×10^{-48}
Tin(II) sulfide, SnS	1.3×10^{-23}
Zinc sulfide, ZnS	2.0×10^{-22}

APPENDIX D

Standard Electrode (Half-Cell) Potentials*

Half-Reaction	$E°$ (V)
$F_2(g) + 2e^- \rightleftharpoons 2F^-(aq)$	+2.87
$O_3(g) + 2H^+(aq) + 2e^- \rightleftharpoons O_2(g) + H_2O(l)$	+2.07
$Co^{3+}(aq) + e^- \rightleftharpoons Co^{2+}(aq)$	+1.82
$H_2O_2(aq) + 2H^+(aq) + 2e^- \rightleftharpoons 2H_2O(l)$	+1.77
$PbO_2(s) + 3H^+(aq) + HSO_4^-(aq) + 2e^- \rightleftharpoons PbSO_4(s) + 2H_2O(l)$	+1.70
$Ce^{4+}(aq) + e^- \rightleftharpoons Ce^{3+}(aq)$	+1.61
$MnO_4^-(aq) + 8H^+(aq) + 5e^- \rightleftharpoons Mn^{2+}(aq) + 4H_2O(l)$	+1.51
$Au^{3+}(aq) + 3e^- \rightleftharpoons Au(s)$	+1.50
$Cl_2(g) + 2e^- \rightleftharpoons 2Cl^-(aq)$	+1.36
$Cr_2O_7^{2-}(aq) + 14H^+(aq) + 6e^- \rightleftharpoons 2Cr^{3+}(aq) + 7H_2O(l)$	+1.33
$MnO_2(s) + 4H^+(aq) + 2e^- \rightleftharpoons Mn^{2+}(aq) + 2H_2O(l)$	+1.23
$O_2(g) + 4H^+(aq) + 4e^- \rightleftharpoons 2H_2O(l)$	+1.23
$Br_2(l) + 2e^- \rightleftharpoons 2Br^-(aq)$	+1.07
$NO_3^-(aq) + 4H^+(aq) + 3e^- \rightleftharpoons NO(g) + 2H_2O(l)$	+0.96
$2Hg^{2+}(aq) + 2e^- \rightleftharpoons Hg_2^{2+}(aq)$	+0.92
$Hg_2^{2+}(aq) + 2e^- \rightleftharpoons 2Hg(l)$	+0.85
$Ag^+(aq) + e^- \rightleftharpoons Ag(s)$	+0.80
$Fe^{3+}(aq) + e^- \rightleftharpoons Fe^{2+}(aq)$	+0.77
$O_2(g) + 2H^+(aq) + 2e^- \rightleftharpoons H_2O_2(aq)$	+0.68
$MnO_4^-(aq) + 2H_2O(l) + 3e^- \rightleftharpoons MnO_2(s) + 4OH^-(aq)$	+0.59
$I_2(s) + 2e^- \rightleftharpoons 2I^-(aq)$	+0.53
$O_2(g) + 2H_2O(l) + 4e^- \rightleftharpoons 4OH^-(aq)$	+0.40
$Cu^{2+}(aq) + 2e^- \rightleftharpoons Cu(s)$	+0.34
$AgCl(s) + e^- \rightleftharpoons Ag(s) + Cl^-(aq)$	+0.22
$SO_4^{2-}(aq) + 4H^+(aq) + 2e^- \rightleftharpoons SO_2(g) + 2H_2O(l)$	+0.20
$Cu^{2+}(aq) + e^- \rightleftharpoons Cu^+(aq)$	+0.15
$Sn^{4+}(aq) + 2e^- \rightleftharpoons Sn^{2+}(aq)$	+0.13
$2H^+(aq) + 2e^- \rightleftharpoons H_2(g)$	0.00
$Pb^{2+}(aq) + 2e^- \rightleftharpoons Pb(s)$	−0.13
$Sn^{2+}(aq) + 2e^- \rightleftharpoons Sn(s)$	−0.14
$N_2(g) + 5H^+(aq) + 4e^- \rightleftharpoons N_2H_5^+(aq)$	−0.23
$Ni^{2+}(aq) + 2e^- \rightleftharpoons Ni(s)$	−0.25
$Co^{2+}(aq) + 2e^- \rightleftharpoons Co(s)$	−0.28
$PbSO_4(s) + H^+(aq) + 2e^- \rightleftharpoons Pb(s) + HSO_4^-(aq)$	−0.31
$Cd^{2+}(aq) + 2e^- \rightleftharpoons Cd(s)$	−0.40
$Fe^{2+}(aq) + 2e^- \rightleftharpoons Fe(s)$	−0.44
$Cr^{3+}(aq) + 3e^- \rightleftharpoons Cr(s)$	−0.74
$Zn^{2+}(aq) + 2e^- \rightleftharpoons Zn(s)$	−0.76
$2H_2O(l) + 2e^- \rightleftharpoons H_2(g) + 2OH^-(aq)$	−0.83
$Mn^{2+}(aq) + 2e^- \rightleftharpoons Mn(s)$	−1.18
$Al^{3+}(aq) + 3e^- \rightleftharpoons Al(s)$	−1.66
$Mg^{2+}(aq) + 2e^- \rightleftharpoons Mg(s)$	−2.37
$Na^+(aq) + e^- \rightleftharpoons Na(s)$	−2.71
$Ca^{2+}(aq) + 2e^- \rightleftharpoons Ca(s)$	−2.87
$Sr^{2+}(aq) + 2e^- \rightleftharpoons Sr(s)$	−2.89
$Ba^{2+}(aq) + 2e^- \rightleftharpoons Ba(s)$	−2.90
$K^+(aq) + e^- \rightleftharpoons K(s)$	−2.93
$Li^+(aq) + e^- \rightleftharpoons Li(s)$	−3.05

*All values at 298 K. Written as reductions; $E°$ value refers to all components in their standard states: 1 mol/L for dissolved species; 1 bar pressure for the gas behaving ideally; the pure substance for solids and liquids.

APPENDIX E

Infrared Absorptions for Representative Functional Groups

Functional Group	Molecular Motion	Wavenumber (cm^{-1})
Alkanes	C—H stretch	2960–2850
	CH_2 bend	1465
	CH_3 bend	1375
	CH_2 bend (4 or more)	720
Alkenes	=CH stretch	3100–3020
	C=C stretch (isolated)	1650
	C=C stretch (conjugated)	1640–1610
	C—H in-plane bend	1430–1290
	C—H bend (monosubstituted)	990 & 910
	C—H bend (disubstituted, E)	970
	C—H bend (disubstituted, 1,1)	890
	C—H bend (disubstituted, Z)	700
	C—H bend (trisubstituted)	815
Alkynes	C—H stretch (acetylenic)	3300
	C—H bend (acetylenic)	650–600
	C≡C stretch	2260–2100
Aromatics	C—H stretch	3020–3000
	C=C stretch	1600–1500
	C=C bend	1700–1500
	C—H bend (mono)	770–730 & 715–685
	C—H bend (ortho)	770–735
	C—H bend (meta)	880 & 780 & 690
	C—H bend (para)	850–800
Alcohols	O—H stretch	3650–3400
	C—O stretch	1150–1050
Ethers	C—O—C stretch (dialkyl)	1300–1000
	C—O—C stretch (diaryl)	1250 & 1120
Amines	N—H stretch (1 per N—H bond)	3500–3300
	N—H bend	1640–1500
	C—N stretch (alkyl)	1180–1025
	C—N stretch (aryl)	1360–1250
	N—H bend [out-of-plane (oop)]	800
Aldehydes	C—H aldehyde stretch	2850 & 2750
	C=O stretch	1725
Ketones	C=O stretch	1720–1700
	C—C stretch	1300–1100

(Continued)

Functional Group	Molecular Motion	Wavenumber (cm^{-1})
Carboxylic acids	O—H stretch	3000–2500
	C═O stretch	1730–1700
	C—O stretch	1320–1210
	O—H bend	1440–1395
Esters	C═O stretch	1750–1735
	C—C(O)—C stretch (ethanoates or acetates)	1240
	C—C(O)—C stretch (all others)	1210–1160
Amides	N—H stretch	3500–3300
	C═O stretch	1680–1640
	N—H bend	1640–1550
	N—H bend (1°)	1570–1515
Alkyl halides	C—F stretch	1400–1000
	C—Cl stretch	800–600
	C—Br stretch	650–500
	C—I stretch	500
Nitriles	C≡N stretch	2250

^{1}H Nuclear Magnetic Resonance (NMR) Shifts

Type of Proton	Type of Compound	Chemical Shift Range (ppm)
RCH_3	alkane, 1° aliphatic	0.9
R_2CH_2	alkane, 2° aliphatic	1.3
R_3CH	alkane, 3° aliphatic	1.7
C═C—H	vinylic (ethenylic)	4.6–5.9
C═C—H	vinylic (ethenylic), conjugated	5.5–7.5
C≡C—H	acetylenic (ethynylic)	2–3
Ar—H	aromatic	6.5–8
Ar—C—H	benzylic	2.2–3
C═C—CH_3	allylic (2-propenylic)	1.5–2.5
HC—F	fluorides	4–4.5
HC—Cl	chlorides	3–4
HC—Br	bromides	2.7–4
HC—I	iodides	2.2–4
HC—OH	alcohols	3.4–4
HC—OR	ethers	3.4–4.0
RCOO—CH	esters	3.7–4.1
HC—COOR	esters	2–2.2
HC—COOH	acids	2–2.6
HC—C═O	carbonyl compounds	2–2.5
RCHO	aldehydic	9–10
ROH	hydroxylic	2–4
ArOH	phenolic	4–12
C═C—OH	enolic	15–17
RCOOH	carboxylic	10–12
RNH_2	amino	1–5

APPENDIX G

BRIEF ANSWERS TO SELECTED PROBLEMS

Chapter 1

1.2 Gas molecules fill the entire container; the volume of a gas is the volume of the container. Solids and liquids have a definite volume. The volume of the container does not affect the volume of a solid or liquid. (a) gas (b) liquid (c) liquid **1.4** Physical property: a characteristic shown by a substance itself, without any interaction with or change into other substances. Chemical property: a characteristic of a substance that appears as it interacts with, or transforms into, other substances. (a) Colour (yellow-green and silvery to white) and physical state (gas and metal to crystals) are physical properties. The interaction between chlorine gas and sodium metal is a chemical property. (b) Colour and magnetism are physical properties. No chemical changes. **1.6** (a) Physical change; there is only a temperature change. (b) Chemical change; the change in appearance indicates an irreversible chemical change. (c) Physical change; there is only a change in size, not composition. (d) Chemical change; the wood (and air) become different substances with different compositions. **1.8** (a) fuel (b) wood **1.13** Lavoisier measured the total mass of the reactants and products, not just the mass of the solids. The total mass of the reactants and products remained constant. His measurements showed that a gas was involved in the reaction. He called this gas *oxygen* (one of his key discoveries). **1.16** A well-designed experiment must have the following essential features: (1) at least two variables that are expected to be related; (2) a way to control all the variables, so that only one at a time may be changed; (3) reproducible results. **1.19** (a) $(1\ m)^2/(100\ cm)^2$ (b) $(1000\ m)^2/(1\ km)^2$ and $(100\ cm)^2/(1\ m)^2$ (c) (1000 m/1 km) and (1 h/3600 s) (d) (1000 g/1 kg) and $(1\ m)^3/(100\ cm)^3$ **1.21** An extensive property depends on the amount of material present. An intensive property is the same regardless of how much material is present. (a) extensive property (b) intensive property (c) extensive property (d) intensive property **1.23** (a) increases (b) remains the same (c) decreases (d) increases (e) remains the same **1.26** 1.43 nm **1.28** 1×10^{11} nm **1.30** (a) 2.07×10^{-9} km^2 (b) \$$6.73 \times 10^3$ **1.34** (a) 5.52×10^3 kg/m^3 (b) 5.52 mg/mm^3 **1.36** (a) 2.56×10^{-9} mm^3/cell (b) 10^{-10} L **1.38** (a) 9.626 cm^3 (b) 64.92 g **1.40** 2.70 g/cm^3 **1.42** (a) 291 K (b) 109 K (c) −273°C **1.45** (a) 2.47×10^{-7} m (b) 6.76 nm **1.52** (a) none (b) none (c) 0.0410 (d) 4.0100×10^4 **1.54** (a) 0.00036 (b) 35.83 (c) 22.5 **1.56** 6×10^2 **1.58** (a) 134 m (b) 21 621 mm^3 (c) 443 cm **1.60** (a) $1.310\ 000 \times 10^5$ (b) 4.7×10^{-4} (c) $2.100\ 06 \times 10^5$ (d) 2.1605×10^3 **1.62** (a) 5550 (b) 10 070 (c) 0.000 000 885 (d) 0.003 004 **1.64** (a) 8.025×10^4 (b) 1.0098×10^{-3} (c) 7.7×10^{-11} **1.66** (a) 4.06×10^{-19} J (b) 1.61×10^{24} molecules (c) 1.82×10^5 J/mol **1.68** (a) height measured, not exact (b) planets counted, exact (c) Number of grams in a pound is not a unit definition; not exact. (d) definition of "millimetre," exact **1.70** 7.50 ± 0.05 cm **1.72** (a) $I_{avg} = 8.72$ g; $II_{avg} = 8.72$ g; $III_{avg} = 8.50$ g; $IV_{avg} = 8.56$ g; sets I and II are most accurate. (b) Set III is the most precise but is the least accurate. (c) Set I has the best combination of high accuracy and high precision. (d) Set IV has both low accuracy and low precision.

1.74 (a)

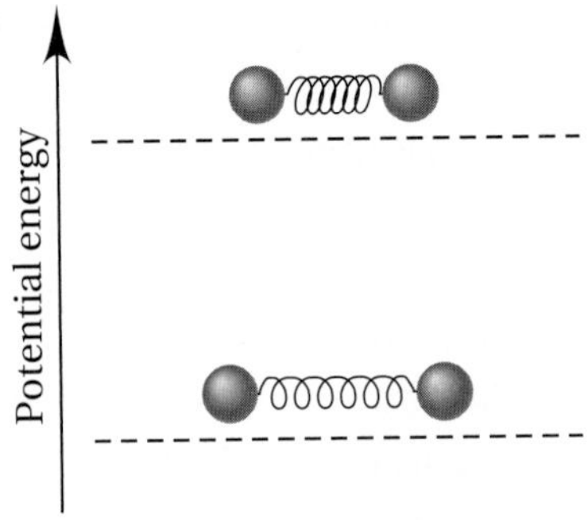

Compressed spring less stable—energy stored in spring

(b)

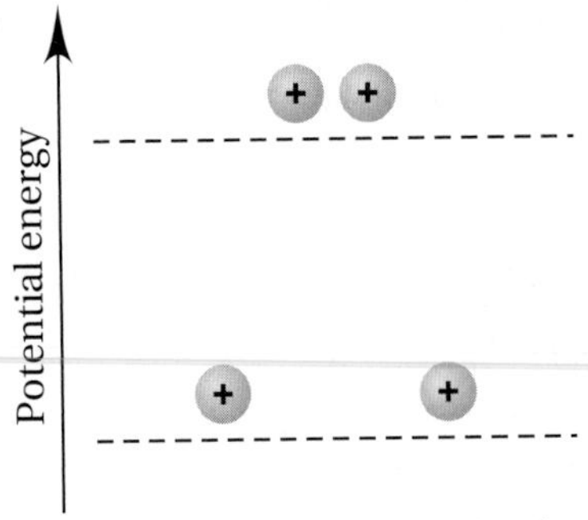

Two charges near each other less stable—repulsion of like charges

1.76 7.7/1 **1.78** (a) density = 0.21 g/L, will float (b) CO_2 is denser than air, will sink (c) density = 0.30 g/L, will float (d) O_2 is denser than air, will sink (e) density = 1.38 g/L, will sink (f) 0.55 g for empty ball; 0.50 g for ball filled with hydrogen **1.80** (a) 8.0×10^{12} g (b) 4.1×10^5 m^3 (c) \$$4.1 \times 10^{14}$ **1.82** (a) −195.79°C (b) 5.05 L **1.83** (a) 2.6 m/s (b) 15 km (c) 12:45 pm **1.85** freezing point = −3.7°X; boiling point = 63.3°X **1.86** 2.3×10^{25} g oxygen; 1.4×10^{25} g silicon; 5×10^{15} g each of ruthenium and rhodium

Chapter 2

Answers to Boxed Reading Problems: B2.1 (a) 5 peaks (b) *m/e* ratio of heaviest particle = 74; *m/e* ratio of lightest particle = 35 **B2.2** 13 **B2.3** (a) Since salt dissolves in water and pepper does not, add water to mixture and filter to remove solid pepper. Evaporate water to recover solid salt. (b) The water/soot mixture can be filtered; the water will flow through the filter paper, leaving the soot collected on the filter paper. (c) Allow the mixture to warm up, and then pour off the melted ice (water), or add water, and the glass will sink and the ice will float. (d) Heat the mixture; the ethanol will boil off (distill), while the sugar will remain behind. (e) The spinach leaves can be extracted with a solvent that dissolves the pigments. Chromatography can be used to separate one pigment from the other.

2.1 Compounds contain different types of atoms; there is only one type of atom in an element. **2.4** (a) The presence of more than one element makes pure calcium chloride a compound. (b) There is only one kind of atom, so sulfur is an element. (c) The presence of more than one compound makes baking powder a mixture. (d) The presence of more than one type of atom means cytosine cannot be an element. The specific, not variable, arrangement means it is a compound. **2.12** (a) elements, compounds, and mixtures (b) compounds (c) compounds **2.14** (a) Law of definite composition: the composition is the same regardless of its source. (b) Law of mass conservation: the total quantity of matter does not change. (c) Law of multiple proportions: two elements can combine to form two different compounds that have different proportions of those elements. **2.16** (a) No; the percent by mass of each element in a compound is fixed. (b) Yes; the *mass* of each element in a compound depends on the amount of compound. **2.18** The two experiments demonstrate the law of definite composition. The unknown compound decomposes the same way both times. The experiments also demonstrate the law of conservation of mass since the total mass before reaction equals the total mass after reaction. **2.20** (a) 1.34 g F (b) 0.514 Ca; 0.486 F (c) 51.4 mass % Ca; 48.6 mass % F **2.22** (a) 0.603 (b) 322 g Mg **2.24** 3.498×10^6 g Cu; 1.766×10^6 g S **2.26** compound 1: 0.905 S/Cl; compound 2: 0.451 S/Cl; ratio: 2.00/1.00 **2.29** Coal A **2.31** Dalton postulated that atoms of an element are identical and that compounds result from the chemical combination of specific ratios of different elements. **2.32** If you know the ratio of any two quantities and the value of one of them, you can always calculate the other; in this case, you know the charge and the mass/charge ratio. **2.36** The atomic number is the number of protons in an atom's nucleus. When the atomic number changes, the identity of the element changes. The mass number is the total number of protons and neutrons in the nucleus. The identity of an element is based on the number of protons, not the number of neutrons. The mass number can vary (by a change in number of neutrons) without changing the identity of the element. **2.39** All three isotopes have 18 protons and 18 electrons. Their respective mass numbers are 36, 38, and 40, with the respective numbers of neutrons being 18, 20, and 22. **2.41** (a) These have the same number of protons and electrons, but different numbers of neutrons; same *Z*. (b) These have the same number of neutrons, but different numbers of protons and electrons; same *N*. (c) These have different numbers of protons, neutrons, and electrons; same *A*. **2.43** (a) $^{38}_{18}Ar$ (b) $^{55}_{25}Mn$ (c) $^{109}_{47}Ag$

2.45 (a) $^{48}_{22}Ti$ (b) $^{79}_{34}Se$ (c) $^{11}_{5}B$

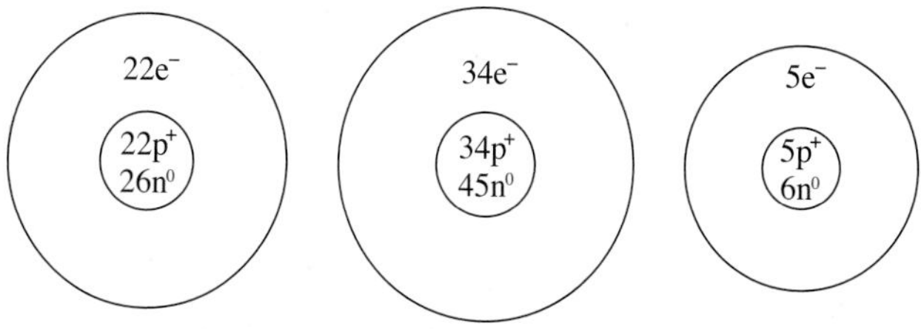

2.47 69.72 u **2.49** % abundance (^{35}Cl) = 75.774%, % abundance (^{37}Cl) = 24.226% **2.52** (a) In the modern periodic table, the elements are arranged in order of increasing atomic number. (b) Elements in a group (or family) have similar chemical properties. (c) Elements can be classified as metals, metalloids, or nonmetals. **2.55** The alkali metals [group 1] are metals and readily lose one electron to form cations; the halogens [group 17] are nonmetals and readily gain one electron to form anions. **2.56** (a) germanium; Ge; 14; metalloid (b) phosphorus; P; 15; nonmetal (c) helium; He; 18; nonmetal (d) lithium; Li; 1; metal (e) molybdenum; Mo; 6; metal **2.58** (a) Ra; 88 (b) Si; 14 (c) Cu; 63.55 u (d) Br; 79.90 u **2.60** Atoms of these two kinds of substances will form ionic bonds, in which one or more electrons are transferred from the metal atom to the nonmetal atom to form a cation and an anion, respectively. **2.63** Coulomb's law states that the energy of attraction in an ionic bond is directly proportional to the product of charges and inversely proportional to the distance between charges. The absolute value of the product of charges in MgO [(+2)(−2)] is greater than that in LiF [(+1)(−1)]. Thus, MgO has stronger ionic bonding. **2.66** The group 1 elements form cations, and the group 17 elements form anions. **2.68** Each potassium atom loses one electron to form an ion with a +1 charge. Each sulfur atom gains two electrons to form an ion with a −2 charge. Two potassiums, losing one electron each, are required for each sulfur, which gains two electrons. The oppositely charged ions attract each other to form an ionic solid, K_2S. **2.70** K^+; I^- **2.72** (a) oxygen; 17; 16; 2 (b) fluorine; 19; 17; 2 (c) calcium; 40; 2; 4 **2.74** Lithium forms the Li^+ ion; oxygen forms the O^{2-} ion. Number of O^{2-} ions = 4.2×10^{21}. **2.76** NaCl **2.78** The subscripts in a formula give the numbers of ions in a formula unit of the compound. The subscripts indicate that there are two F^- ions for each Mg^{-1} ion. Using this information and the mass of each element, we can calculate the mass percent of each element in the compound. **2.80** The two samples are similar in that both contain 20 billion oxygen atoms and 20 billion hydrogen atoms. They differ in that they contain different types of molecules: H_2O_2 molecules in the hydrogen peroxide sample, and H_2 and O_2 molecules in the mixture. In addition, the mixture contains 20 billion molecules (10 billion

H_2 and 10 billion O_2), while the hydrogen peroxide sample contains 10 billion molecules. **2.84** (a) Na_3N, sodium nitride (b) SrO, strontium oxide (c) $AlCl_3$, aluminum chloride **2.86** (a) MgF_2, magnesium fluoride (b) ZnS, zinc sulfide (c) $SrCl_2$, strontium chloride **2.88** (a) $SnCl_4$ (b) iron(III) bromide (c) CuBr (d) manganese(III) oxide **2.90** (a) cobalt(II) oxide (b) Hg_2Cl_2 (c) lead(II) acetate trihydrate (d) Cr_2O_3 **2.92** (a) BaO (b) $Fe(NO_3)_2$ (c) MgS **2.94** (a) H_2SO_4; sulfuric acid (b) HIO_3; iodic acid (c) HCN; hydrocyanic acid (d) H_2S; hydrosulfuric acid **2.96** (a) ammonium ion, NH_4^+; ammonia, NH_3 (b) magnesium sulfide, MgS; magnesium sulfite, $MgSO_3$; magnesium sulfate, $MgSO_4$ (c) hydrochloric acid, HCl; chloric acid, $HClO_3$; chlorous acid, $HClO_2$ (d) cuprous bromide, CuBr; cupric bromide, $CuBr_2$ **2.98** disulfur tetrafluoride, S_2F_4 **2.100** (a) calcium chloride (b) copper(I) oxide (c) stannic fluoride (d) hydrochloric acid **2.102** (a) 12 oxygen atoms; 342.2 u (b) 9 hydrogen atoms; 132.06 u (c) 8 oxygen atoms; 344.6 u **2.104** (a) $(NH_4)_2SO_4$; 132.15 u (b) NaH_2PO_4; 119.98 u (c) $KHCO_3$; 100.12 u **2.106** (a) 108.02 u (b) 331.2 u (c) 72.08 u **2.108** (a) SO_3; sulfur trioxide; 80.07 u (b) C_3H_8; propane; 44.09 u **2.112** Separating the components of a mixture requires physical methods only; that is, no chemical changes (no changes in composition) take place, and the components maintain their chemical identities and properties throughout. Separating the components of a compound requires a chemical change (change in composition). **2.115** (a) compound (b) homogeneous mixture (c) heterogeneous mixture (d) homogeneous mixture (e) homogeneous mixture **2.117** (a) filtration (b) extraction or chromatography **2.119** (a) fraction of volume = 5.2×10^{-13} (b) mass of nucleus = $6.644\,66 \times 10^{-24}$ g; fraction of mass = 0.999 726 **2.120** strongest ionic bonding: MgO; weakest ionic bonding: RbI **2.124** (a) I = NO; II = N_2O_3; III = N_2O_5 (b) I has 1.14 g O per 1.00 g N; II, 1.71 g O; III, 2.86 g O **2.128** (a) Cl^-, 1.898 mass %; Na^+, 1.056 mass %; SO_4^{2-}, 0.265 mass %; Mg^{2+}, 0.127 mass %; Ca^{2+}, 0.04 mass %; K^+, 0.038 mass %; HCO_3^-, 0.014 mass % (b) 30.72% (c) alkaline earth metal ions, total mass % = 0.167%; alkali metal ions, total mass % = 1.094% (d) Anions (2.177 mass %) make up a larger mass fraction than cations (1.26 mass %). **2.130** molecular formula, $C_4H_6O_4$; molecular mass, 118.09 u; 40.68% by mass C; 5.122% by mass H; 54.20% by mass O **2.133** (a) formulas and masses in u: $^{15}N_2{}^{18}O$, 48; $^{15}N_2{}^{16}O$, 46; $^{14}N_2{}^{18}O$, 46; $^{14}N_2{}^{16}O$, 44; $^{15}N^{14}N^{18}O$, 47; $^{15}N^{14}N^{16}O$, 45 (b) $^{15}N_2{}^{18}O$, least common; $^{14}N_2{}^{16}O$, most common **2.136** 58.091 u **2.138** nitroglycerin, 39.64 mass % NO; isoamyl nitrate, 22.54 mass % NO **2.139** 0.370 kg C; 0.0222 kg H; 0.423 kg O; 0.185 kg N **2.143** (1) chemical change (2) physical change (3) chemical change (4) chemical change (5) physical change

Chapter 3

3.2 (a) 12 mol C atoms (b) 1.445×10^{25} C atoms **3.7** (a) left (b) left (c) left (d) neither **3.8** (a) 121.64 g/mol (b) 76.02 g/mol (c) 106.44 g/mol (d) 152.00 g/mol **3.10** (a) 134.7 g/mol (b) 175.3 g/mol (c) 342.17 g/mol (d) 125.84 g/mol **3.12** (a) 1.1×10^2 g $KMnO_4$ (b) 0.188 mol O atoms (c) 1.5×10^{20} O atoms **3.14** (a) 9.73 g $MnSO_4$ (b) 44.6 mol $Fe(ClO_4)_3$ (c) 1.74×10^{21} N atoms **3.16** (a) 1.56×10^3 g Cu_2CO_3 (b) 0.0725 g N_2O_5 (c) 0.644 mol $NaClO_4$; 3.88×10^{23} formula units $NaClO_4$ (d) 3.88×10^{23} Na^+ ions; 3.88×10^{23} ClO_4^- ions; 3.88×10^{23} Cl atoms; 1.55×10^{24} O atoms **3.18** (a) 6.375 mass % H (b) 71.52 mass % O **3.20** (a) 0.1252 mass fraction C (b) 0.3428 mass fraction O **3.23** (a) 0.9507 mol cisplatin (b) 3.5×10^{24} H atoms **3.25** (a) 195 mol rust (b) 195 mol Fe_2O_3 (c) 2.18×10^4 g Fe **3.27** $CO(NH_2)_2 > NH_4NO_3 > (NH_4)_2SO_4 > KNO_3$ **3.28** (a) 3.12×10^4 mol PbS (b) 1.88×10^{25} Pb atoms **3.32** (b) From the mass percent, determine the empirical formula. Add up the total number of atoms in the empirical formula, and divide that number into the total number of atoms in the molecule. The result is the multiplier that converts the empirical formula into the molecular formula.

Road Map

Mass (g) of each element (express mass percent directly as grams)

↓ Divide by $\mathcal{M}$ (g/mol).

Amount (mol) of each element

↓ Use amount (mol) as subscripts.

Preliminary empirical formula

↓ Change to integer subscripts.

Empirical formula

↓ Divide total number of atoms in molecule by the number of atoms in the empirical formula and multiply the empirical formula by that factor.

Molecular formula

(c) Find the empirical formula from the mass percents. Compare the number of atoms given for the one element to the number in the empirical formula. Multiply the empirical formula by the factor that is needed to obtain the given number of atoms for that element.

Road Map

(Same first three steps as in part (b).)

Empirical formula

↓ Divide the number of atoms of the one element in the molecule by the number of atoms of that element in the empirical formula and multiply the empirical formula by that factor.

Molecular formula

(e) Count the numbers of the various types of atoms in the structural formula and put these into a molecular formula.

Road Map

Structural formula

↓ Count the number of atoms of each element and use these numbers as subscripts.

Molecular formula

3.34 (a) CH_2; 14.03 g/mol (b) CH_3O; 31.03 g/mol (c) N_2O_5; 108.02 g/mol (d) $Ba_3(PO_4)_2$; 601.8 g/mol (e) TeI_4; 635.2 g/mol **3.36** disulfur dichloride; SCl; 135.04 g/mol **3.38** (a) C_3H_6 (b) N_2H_4 (c) N_2O_4 (d) $C_5H_5N_5$ **3.40** (a) Cl_2O_7 (b) $SiCl_4$

(c) CO_2 **3.42** (a) NO_2 (b) N_2O_4 **3.44** (a) 1.20 mol F (b) 24.0 g M (c) calcium **3.47** $C_{21}H_{30}O_5$ **3.49** $C_{10}H_{20}O$ **3.50** A balanced equation provides information on the identities of reactants and products, the physical states of reactants and products, and the molar ratios by which reactants form products. **3.53** b

3.54 (a) $16Cu(s) + S_8(s) \longrightarrow 8Cu_2S(s)$
(b) $P_4O_{10}(s) + 6H_2O(l) \longrightarrow 4H_3PO_4(l)$
(c) $B_2O_3(s) + 6NaOH(aq) \longrightarrow 2Na_3BO_3(aq) + 3H_2O(l)$
(d) $4CH_3NH_2(g) + 9O_2(g) \longrightarrow 4CO_2(g) + 10H_2O(g) + 2N_2(g)$

3.56 (a) $2SO_2(g) + O_2(g) \longrightarrow 2SO_3(g)$
(b) $Sc_2O_3(s) + 3H_2O(l) \longrightarrow 2Sc(OH)_3(s)$
(c) $H_3PO_4(aq) + 2NaOH(aq) \longrightarrow Na_2HPO_4(aq) + 2H_2O(l)$
(d) $C_6H_{10}O_5(s) + 6O_2(g) \longrightarrow 6CO_2(g) + 5H_2O(g)$

3.58 (a) $4Ga(s) + 3O_2(g) \longrightarrow 2Ga_2O_3(s)$
(b) $2C_6H_{14}(l) + 19O_2(g) \longrightarrow 12CO_2(g) + 14H_2O(g)$
(c) $3CaCl_2(aq) + 2Na_3PO_4(aq) \longrightarrow Ca_3(PO_4)_2(s) + 6NaCl(aq)$

3.64 Balance the equation for the reaction: $a\text{A} + b\text{B} \longrightarrow c\text{C}$. Since A is the limiting reactant, A is used to determine the amount of C. Divide the mass of A by its molar mass to obtain the amount (mol) of A. Use the molar ratio from the balanced equation to find the amount (mol) of C. Multiply the amount (mol) of C by its molar mass to obtain the mass of C.

Road Map

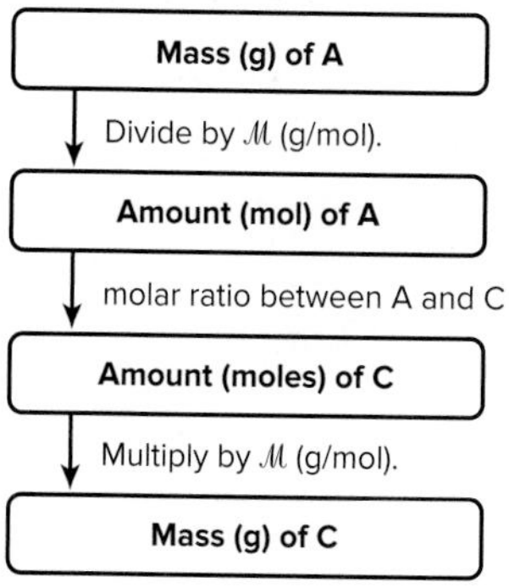

3.66 (a) 0.455 mol Cl_2 (b) 32.3 g Cl_2 **3.68** (a) 1.42×10^3 mol KNO_3 (b) 1.43×10^5 g KNO_3 **3.70** 195.8 g H_3BO_3; 19.16 g H_2 **3.72** 2.60×10^3 g Cl_2

3.74 (a) $I_2(s) + Cl_2(g) \longrightarrow 2ICl(s)$;
$ICl(s) + Cl_2(g) \longrightarrow ICl_3(s)$
(b) $I_2(s) + 3Cl_2(g) \longrightarrow 2ICl_3(s)$
(c) 1.33×10^3 gI_2

3.76 (a) 0.105 mol CaO (b) 0.175 mol CaO (c) calcium (d) 5.88 g CaO **3.78** 1.36 mol HIO_3, 239 g HIO_3; 44.9 g H_2O in excess **3.80** 4.40 g CO_2; 4.80 g O_2 in excess **3.82** 12.2 g $Al(NO_2)_3$, no NH_4Cl, 48.7 g $AlCl_3$, 30.7 g N_2, 39.5 g H_2O **3.84** 50.% **3.86** 90.5% **3.88** 24.0 g CH_3Cl **3.90** 39.7 g CF_4 **3.91** A **3.95** (a) C (b) B (c) C (d) B **3.97** No; instructions should read: "Take 100.0 mL of the 10.0 mol/L solution and, with stirring, add water until the total volume is 1000. mL." **3.99** (a) 7.85 g $Ca(C_2H_3O_2)_2$ (b) 0.254 mol/L KI (c) 124 mol NaCN **3.101** (a) 4.65 g K_2SO_4 (b) 0.0653 mol/L $CaCl_2$ (c) 1.11×10^{20} Mg^{2+} ions **3.103** (a) 0.0617 mol/L KCl (b) 0.00363 mol/L $(NH_4)_2SO_4$ (c) 0.138 mol/L Na^+ **3.105** (a) 987 g HNO_3/L (b) 15.7 mol/L HNO_3 **3.107** 845 mL **3.109** 0.88 g $BaSO_4$ **3.112** (a) Instructions: Be sure to wear goggles to protect your eyes! Pour approximately 2.0 L of water into the container. Add to the water, slowly and with mixing, 0.90 L of concentrated HCl. Dilute to 3.0 L with more water. (b) 22.6 mL **3.115** ionic or polar covalent compounds **3.116** Ions must be present, and they come from ionic compounds or from electrolytes such as acids and bases. **3.119** B **3.123** (a) Benzene is likely to be insoluble in water because it is nonpolar and water is polar. (b) Sodium hydroxide, an ionic compound, is likely to be very soluble in water. (c) Ethanol (CH_3CH_2OH) is likely to be soluble in water because the alcohol group (—OH) is very polar, like the water molecule. (d) Potassium acetate, an ionic compound, is likely to be very soluble in water. **3.125** (a) Yes; CsBr is a soluble salt. (b) Yes; HI is a strong acid. **3.127** (a) 0.64 mol (b) 0.242 mol (c) 1.18×10^{-4} mol **3.129** (a) 3.0 mol (b) 7.57×10^{-5} mol (c) 0.148 mol **3.131** (a) 0.058 mol Al^{3+}; 3.5×10^{22} Al^{3+} ions; 0.18 mol Cl^-; 1.1×10^{23} Cl^- ions (b) 4.62×10^{-4} mol Li^+; 2.78×10^{20} Li^+ ions; 2.31×10^{-4} mol SO_4^{2-}; 1.39×10^{20} SO_4^{2-} ions (c) 1.50×10^{-2} mol K^+; 9.02×10^{21} K^+ ions; 1.50×10^{-2} mol Br^-; 9.02×10^{21} Br^- ions **3.133** (a) 0.35 mol H^+ (b) 6.3×10^{-3} mol H^+ (c) 0.22 mol H^+ **3.137** Spectator ions do not appear in a net ionic equation because they are not involved in the reaction and serve only to balance charges. **3.142** $x = 3$ **3.143** ethane > propane > cetyl palmitate > ethanol > benzene **3.148** (a) $Fe_2O_3(s) + 3CO(g) \longrightarrow 2Fe(s) + 3CO_2(g)$ (b) 3.39×10^7 g CO **3.150** 89.8% **3.152** (a) $2AB_2 + B_2 \longrightarrow 2AB_3$ (b) AB_2 (c) 5.0 mol AB_3 (d) 0.5 mol B_2 **3.154** B, C, and D have the same empirical formula, C_2H_4O; 44.05 g/mol **3.155** 44.3% **3.158** (a) C (b) B (c) D **3.164** (a) 586 g CO_2 (b) 10.5% CH_4 by mass **3.167** 10/0.66/1.0 **3.172** (a) 192.12 g/mol; $C_6H_8O_7$ (b) 0.580 mol

3.173 (a) $N_2(g) + O_2(g) \longrightarrow 2NO(g)$
$2NO(g) + O_2(g) \longrightarrow 2NO_2(g)$
$3NO_2(g) + H_2O(g) \longrightarrow 2HNO_3(aq) + NO(g)$
(b) $2N_2(g) + 5O_2(g) + 2H_2O(g) \longrightarrow 4HNO_3(aq)$
(c) 6.07×10^3 t HNO_3

3.174 A **3.176** (a) 0.039 g heme (b) 6.3×10^{-5} mol heme (c) 3.5×10^{-3} g Fe (d) 4.1×10^{-2} g hemin **3.178** (a) 46.65 mass % N in urea; 31.98 mass % N in arginine; 21.04 mass % N in ornithine (b) 28.45 g N **3.180** 29.54% **3.182** (a) 89.3% (b) 1.47 g ethylene **3.184** (a) 125 g salt (b) 65.6 L H_2O

Chapter 4

Answers to Boxed Reading Problems: B4.1 The density of the atmosphere decreases with increasing altitude. High density causes more drag on the aircraft. At high altitudes, low density means that there are relatively few gas particles present to collide with the aircraft. **B4.2** Saturn **B4.3** 0.934%, 946 kPa **B4.4** (a) 1.78×10^{20} mol (b) 4.4×10^{21} L

4.1 (a) The volume of the liquid remains constant, but the volume of the gas increases to the volume of the larger container. (b) The volume of the container holding the gas sample increases when heated, but the volume of the container holding the liquid sample remains essentially constant when heated. (c) The volume of the liquid remains essentially constant, but the volume of the gas is reduced.

4.6 990 cm H_2O **4.8** (a) 75.5 kPa (b) 1.32 bar (c) 3.65×10^4 Pa (d) 107 kPa **4.10** 0.953 bar **4.12** 0.979 bar **4.18** At constant temperature and volume, the pressure of a gas is directly proportional to the amount (mol) of the gas. **4.20** (a) Volume decreases to one-third of the original volume. (b) Volume increases by a factor of 3.0. (c) Volume increases by a factor of 4. **4.22** (a) Volume decreases by a factor of 2. (b) Volume increases by a factor of 1.48. (c) Volume decreases by a factor of 3. **4.24** −144°C **4.26** 35.8 L **4.28** 0.063 mol Cl_2 **4.30** 0.873 g ClF_3 **4.33** yes **4.35** Beaker is inverted for H_2 and upright for CO_2. The molar mass of CO_2 is greater than the molar mass of air, which, in turn, has a greater molar mass than H_2. **4.39** 5.78 g/L **4.41** 1.76×10^{-3} mol AsH_3; 3.43 g/L **4.43** 51.1 g/mol **4.45** 1.35 bar **4.47** 38.7 g P_4 **4.49** 41.2 g PH_3 **4.51** 0.0249 g Al **4.55** C_5H_{12} **4.57** (a) 0.90 mol (b) 0.00898 bar **4.58** 286 mL SO_2 **4.60** 10.1 kPa SiF_4 **4.65** At STP, the volume occupied by 1 mol of any gas is the same. At the same temperature, all gases have the same average kinetic energy, resulting in the same pressure. **4.68** (a) $P_A > P_B > P_C$ (b) $E_A = E_B = E_C$ (c) $\text{rate}_A > \text{rate}_B > \text{rate}_C$ (d) total $E_A >$ total $E_B >$ total E_C (e) $d_A = d_B = d_C$ (f) collision frequency in A > collision frequency in B > collision frequency in C **4.69** 13.21 **4.71** (a) curve 1 (b) curve 1 (c) curve 1; fluorine and argon have about the same molar mass **4.73** 14.9 min **4.75** 4 atoms per molecule **4.78** negative deviations; $N_2 < Kr < CO_2$ **4.80** At 1 bar; at lower pressures, the gas molecules are farther apart and intermolecular forces are less important. **4.83** 6.89×10^4 g/mol **4.86** (a) 2.24×10^3 kPa (b) 2124 kPa **4.90** (a) 79.4 kPa N_2; 21.1 kPa O_2; 0.04 kPa CO_2; 0.46 kPa H_2O (b) 74.2 mol % N_2; 13.6 mol % O_2; 5.2 mol % CO_2; 6.1 mol % H_2O (c) 1.6×10^{21} molecules O_2 **4.92** (a) 4×10^2 mL (b) 0.013 mol N_2 **4.93** 36.7 L NO_2 **4.98** Al_2Cl_6 **4.100** 1.52×10^{-2} mol SO_3 **4.104** (a) 1.95×10^3 g Ni (b) 3.5×10^4 g Ni (c) 62 m^3 CO **4.106** (a) 9 volumes of O_2 (g) (b) CH_5N **4.109** The lungs would expand by a factor of 4.86; the diver can safely ascend 15.99 m to a depth of 22 m. **4.111** 6.00 g H_2O_2 **4.116** 6.53×10^{-3} g N_2 **4.120** (a) xenon (b) water vapour (c) mercury vapour (d) water vapour **4.124** 17.0 g CO_2; 18.0 g Kr **4.130** Ne, 676 m/s; Ar, 481 m/s; He, 1.52×10^3 m/s **4.132** (a) 0.052 g (b) 1.1 mL **4.139** (a) 16.5 L CO_2 (b) $P_{H_2O} = 6.51$ kPa; $P_{O_2} = P_{CO_2} = 49$ kPa **4.144** 332 steps **4.146** 1.4 **4.150** $P_{total} = 0.327$ bar; $P_{I_2} = 33.4 \times 10^{-3}$ bar

Chapter 5

Answers to Boxed Reading Problems:

B5.1 -5.49×10^5 kJ **B5.2** (a) 2C(s, coal) + $2H_2O(g) \longrightarrow CH_4(g) + CO_2(g)$ (b) 12 kJ/mol (c) -3.30×10^4 kJ

5.4 Increase: eating food, lying in the sun, taking a hot bath. Decrease: exercising, taking a cold bath, going outside on a cold day. **5.6** The amount of the change in internal energy is the same for heater and air conditioner. Since both devices consume the same amount of electrical energy, the change in energy of the heater equals that of the air conditioner. **5.8** 0 J **5.10** 1.54×10^3 J/mol **5.12** (a) 6.6×10^7 kJ (b) 1.6×10^7 kcal **5.15** 8.8 h **5.17** Measuring the heat transfer at constant pressure is more convenient than measuring at constant volume. **5.19** (a) exothermic (b) endothermic (c) exothermic (d) exothermic (e) endothermic (f) endothermic (g) exothermic

5.22

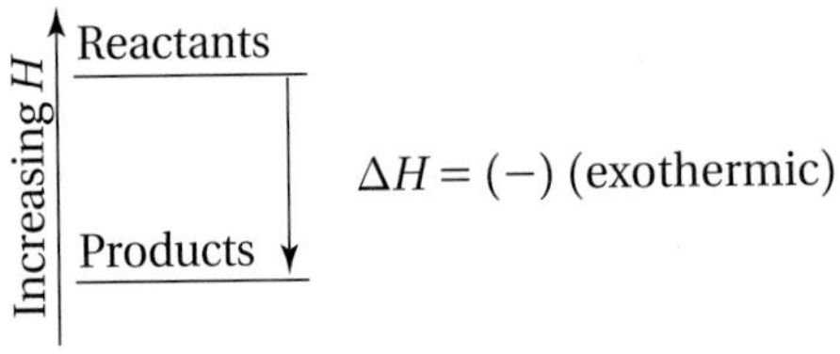

5.24 (a) Combustion of ethane: $2C_2H_6(g) + 7O_2(g) \longrightarrow 4CO_2(g) + 6H_2O(g) + \text{heat}$

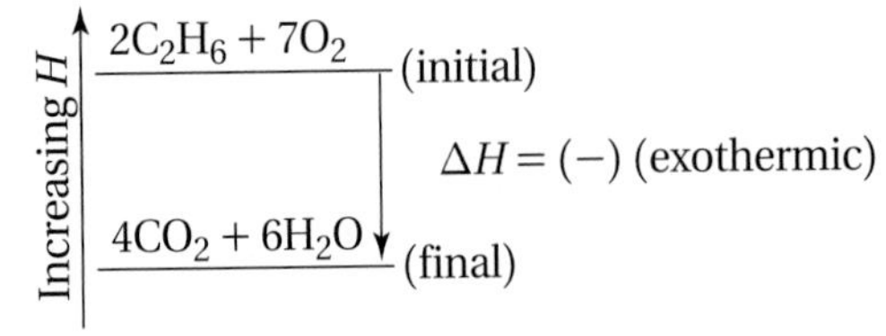

(b) Freezing of water: $H_2O(l) \longrightarrow H_2O(s) + \text{heat}$

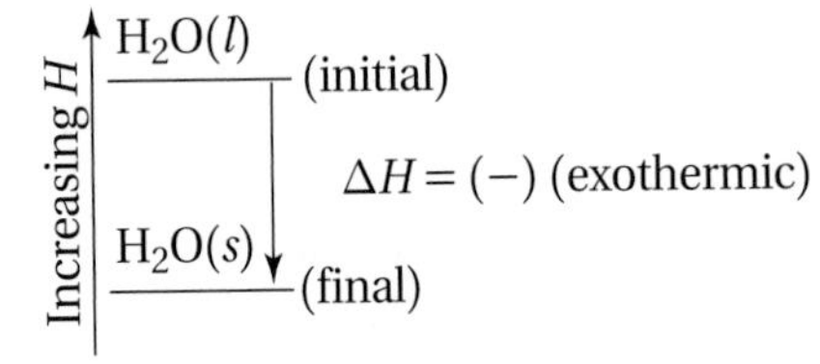

5.26 (a) $2CH_3OH(l) + 3O_2(g) \longrightarrow 2CO_2(g) + 4H_2O(g) + \text{heat}$

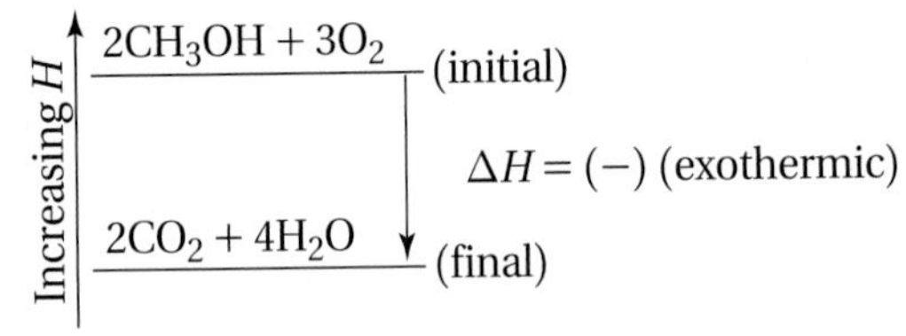

(b) $\frac{1}{2}N_2(g) + O_2(g) + \text{heat} \longrightarrow NO_2(g)$

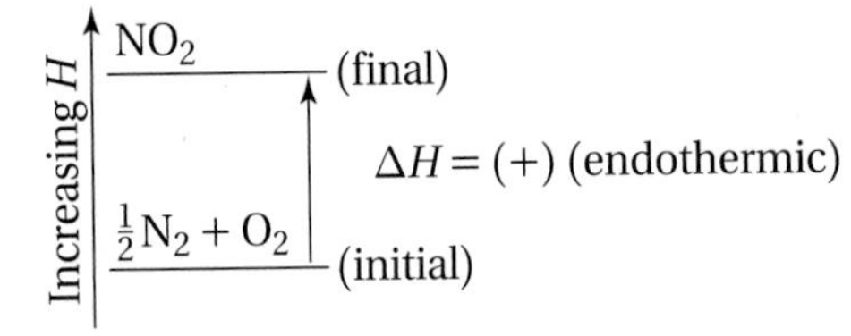

5.28 (a) This is a phase change from the solid phase to the gas phase. Heat is absorbed by the system so q_{sys} is positive (+). (b) The volume of the system is expanding because a greater amount (mol) of gas is present after the phase change than before the phase change. So the system has done work of expansion, and w is negative. Since $\Delta U_{sys} = q + w$, q is positive, and w is negative, the sign of ΔU_{sys} cannot be predicted. It will be positive if $q > w$ and negative if $q > w$. (c) $\Delta U_{univ} = 0$. If the system loses energy, the surroundings gain an equal amount of energy. The sum of the energy of the system and the energy of the surroundings remains constant. **5.31** To determine the specific heat capacity of a substance, you need its mass, the heat added (or lost), and the change in temperature. **5.33** Heat capacity is the quantity of heat required to raise the temperature of an object 1 K. Specific heat capacity is the quantity of heat required to raise the temperature of 1 g of a material by 1 K. Molar heat capacity is the quantity of heat required to raise the temperature of 1 mol of a substance by 1 K.

5.35 6.9×10^3 J **5.37** 295°C **5.39** 77.5°C **5.41** 45°C **5.43** 36.6°C **5.50** The reaction has a positive ΔH because it requires the input of energy to break the oxygen-oxygen bond. **5.51** ΔH is negative; it is opposite in sign and half of the value for the vaporization of 2 mol of H_2O.
5.52 (a) exothermic (b) 20.2 kJ per $\frac{1}{8}$ mol of S_8 produced (c) $q = -4.2 \times 10^2$ kJ (d) $q = -15.7$ kJ
5.54 (a) $\frac{1}{2}N_2(g) + \frac{1}{2}O_2(g) \longrightarrow NO(g)$; $\Delta H = 90.29$ kJ/mol (b) $q = -10.5$ kJ
5.56 $q = -1.88 \times 10^6$ kJ
5.60 (a) $C_2H_4(g) + 3O_2(g) \longrightarrow 2CO_2(g) + 2H_2O(g)$; $\Delta_r H = -1411$ kJ/mol
(b) 1.39 g C_2H_4
5.64 −110.5 kJ/mol **5.65** −813.4 kJ/mol **5.67** $N_2(g) + 2O_2(g) \longrightarrow 2NO_2(g)$; $\Delta_r H = 66.4$ kJ/mol; equation 1 is A, equation 2 is B, and equation 3 is C. **5.69** 44.0 kJ/mol
5.72 The standard enthalpy of reaction, $\Delta_r H°$, is the enthalpy change for a reaction where all substances are in their standard states. The standard enthalpy of formation, $\Delta_f H°$, is the enthalpy change that accompanies the formation of 1 mol of a compound in its standard state from elements in their standard states.
5.74 (a) $\frac{1}{2}Cl_2(g) + Na(s) \longrightarrow NaCl(s)$
(b) $H_2(g) + \frac{1}{2}O_2(g) \longrightarrow H_2O(g)$
(c) no changes
5.75 (a) $Ca(s) + Cl_2(g) \longrightarrow CaCl_2(s)$
(b) $Na(s) + \frac{1}{2}H_2(g) + C(s, \text{graphite}) + \frac{3}{2}O_2(g) \longrightarrow NaHCO_3(s)$
(c) $C(s, \text{graphite}) + 2Cl_2(g) \longrightarrow CCl_4(l)$
(d) $\frac{1}{2}H_2(g) + \frac{1}{2}N_2(g) + \frac{3}{2}O_2(g) \longrightarrow HNO_3(l)$
5.77 (a) −1036.9 kJ/mol (b) −433 kJ/mol **5.79** −157.3 kJ/mol
5.81 (a) 503.9 kJ/mol (b) $-\Delta_1 H + 2\Delta_2 H = 504$ kJ/mol
5.82 (a) $C_{18}H_{36}O_2(s) + 26O_2(g) \longrightarrow 18CO_2(g) + 18H_2O(g)$ (b) −10 488 kJ/mol (c) −36.9 kJ; −8.81 kcal (d) 8.81 kcal/g × 11.0 g = 96.9 kcal of energy released **5.84** (a) initial = 23.7 L/mol; final = 24.9 L/mol (b) 187 J (c) -1.2×10^2 J (d) 3.1×10^2 J (e) 310 J (f) $\Delta H = \Delta U + P\Delta V = \Delta U - w = (q + w) - w = q_P$ **5.93** (a) 1.2×10^2 mol CH_4 (b) \$0.0053/mol (c) \$0.90 **5.98** (a) $\Delta_{r1}H° = -657.0$ kJ/mol; $\Delta_{r2}H° = 32.9$ kJ/mol (b) −106.6 kJ/mol **5.99** (a) -6.81×10^3 J (b) +243°C
5.100 −22.2 kJ/mol **5.101** (a) 34 kJ/mol (b) −757 kJ
5.103 (a) -1.25×10^3 kJ (b) 2.24×10^3°C

Chapter 6

Answers to Boxed Reading Problems: B6.1 (a) slope = 1.3×10^4 L/mol; y-intercept = 0.00 (b) diluted solution = 1.8×10^{-5} mol/L; original solution = 1.4×10^{-4} mol/L **B6.2** (a) red: $\nu = 4.47 \times 10^{14}$ s^{-1} (b) blue: $\nu = 6.62 \times 10^{14}$ s^{-1} (c) yellow-orange: $\nu = 5.09 \times 10^{14}$ s^{-1}

6.2 (a) X-ray < ultraviolet < visible < infrared < microwave < radio waves (b) radio < microwave < infrared < visible < ultraviolet < X-ray (c) radio < microwave < infrared < visible < ultraviolet < X-ray **6.7** 316 m; 3.16×10^{11} nm
6.9 2.5×10^{-23} J **6.11** red < yellow < blue **6.13** 1.3483×10^7 nm
6.16 (a) 1.24×10^{15} s^{-1}; 8.21×10^{-19} J (b) 1.4×10^{15} s^{-1}; 9.0×10^{-19} J **6.18** Bohr's key assumption was that the electron in an atom does not radiate energy while in a stationary state, and it can move to a different orbit only by absorbing or emitting a photon whose energy is equal to the difference in energy between two states. These differences in energy correspond to the wavelengths in the known line spectra for the hydrogen atom. A solar system model does not allow for the movement of electrons between levels. **6.20** (a) absorption (b) emission (c) emission (d) absorption **6.22** Yes; the predicted line spectra are accurate. The energies could be predicted from $E_n = \dfrac{-(Z^2)(2.18 \times 10^{-18}\text{ J})}{n^2}$, where Z is the atomic number for the atom or ion. The energy levels for Be^{3+} will be greater by a factor of 16 ($Z = 4$) than those for the hydrogen atom. This means that the pattern of lines will be similar, but the lines will be at different wavelengths. **6.23** 434.17 nm **6.25** 1875.6 nm **6.27** -2.76×10^5 J/mol **6.29** d < a < c < b **6.31** $n = 4$ **6.37** Macroscopic objects do exhibit a wavelike motion, but the wavelength is too small for humans to perceive. **6.39** (a) 1.15×10^{-36} m (b) 2×10^{-35} m **6.41** 2.2×10^{-26} m/s **6.43** 3.75×10^{-36} kg/photon **6.47** The total probability of finding an electron at 52.9 pm is much greater for the 1s orbital than for the 2s orbital. **6.48** (a) principal determinant of the electron's energy or distance from the nucleus (b) determines the shape of the orbital (c) determines the orientation of the orbital in three-dimensional space **6.49** (a) one (b) five (c) three (d) nine **6.51** (a) $m_l = -2, -1, 0, +1, +2$ (b) $m_l = 0$ (if $n = 1$, then $l = 0$) (c) $m_l = -3, -2, -1, 0, +1, +2, +3$
6.53 (a)

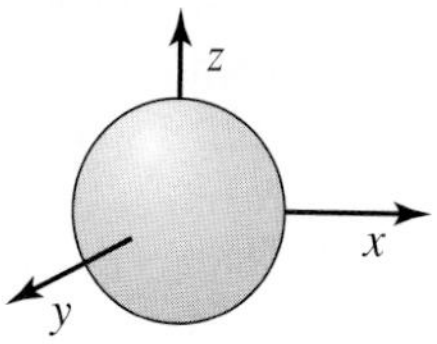

(b)

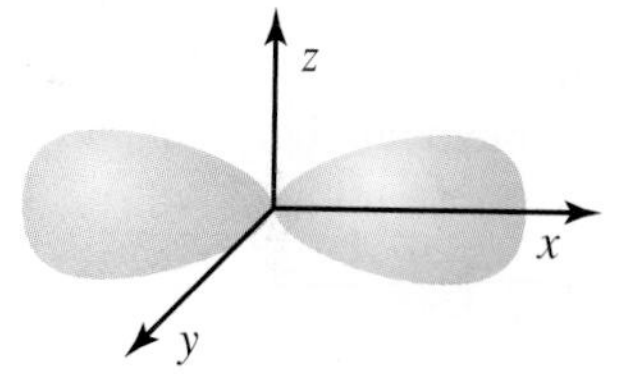

6.55

Subshell	Allowable m_l Values	# of Orbitals
(a) d (l = 2)	−2, −1, 0, +1, +2	5
(b) p (l = 1)	−1, 0, +1	3
(c) f (l = 3)	−3, −2, −1, 0, +1, +2, +3	7

6.57 (a) $n = 5$ and $l = 0$; one orbital (b) $n = 3$ and $l = 1$; three orbitals (c) $n = 4$ and $l = 3$; seven orbitals **6.59** (a) no; correct: $n = 2, l = 1, m_l = -1$ or $n = 2, l = 0, m_l = 0$ (b) allowed (c) allowed (d) no; correct: $n = 5, l = 3, m_l = +3$ or $n = 5, l = 2, m_l = 0$ **6.62** (a) $E = -(2.180 \times 10^{-18}\text{ J})(1/n^2)$. This is identical to the expression from Bohr's theory. (b) 3.028×10^{-19} J (c) 656.1 nm **6.63** (a) The attraction of the nucleus for the electrons must be overcome. (b) The electrons in silver are more tightly held by the nucleus. (c) silver (d) Once the electron is freed from the atom, its energy increases in proportion to the frequency of the light. **6.66** Li^{2+} **6.68** (a) 2 ⟶ 1 (b) 5 ⟶ 2 (c) 4 ⟶ 2 (d) 3 ⟶ 2 (e) 6 ⟶ 3 **6.72** (a) l = 1 or 2 (b) l = 1 or 2 (c) l = 3, 4, 5, or 6 (d) l = 2 or 3
6.74 (a)

$$\Delta E = (-2.18 \times 10^{-18}\text{ J})\left(\frac{1}{\infty^2} - \frac{1}{n^2_{\text{initial}}}\right)Z^2\left(\frac{6.022 \times 10^{23}}{1\text{ mol}}\right)$$

(b) 3.28×10^7 J/mol (c) 205 nm (d) 22.8 nm **6.76** (a) 5.293 $\times 10^{-11}$ m (b) 5.293×10^{-9} m **6.78** 6.4×10^{27} photons **6.80** (a) no overlap (b) overlap (c) two lines (d) At longer wavelengths, the hydrogen spectrum starts becoming a continuous band. **6.82** (a) 7.56×10^{-18} J; 2.63×10^{-8} m (b) 5.122×10^{-17} J; 3.881×10^{-9} m (c) 1.2×10^{-18} J; 1.66×10^{-7} m **6.84** (a) 1.87×10^{-19} J (b) 3.58×10^{-19} J **6.86** (a) red (Sr); green (Ba) (b) 5.89 kJ (Sr); 5.83 kJ (Ba) **6.88** (a) This is the wavelength of maximum absorbance, so it gives the highest sensitivity. (b) ultraviolet region (c) 1.93×10^{-2} g vitamin A/g oil **6.92** 1.0×10^{18} photons/s **6.95** $3s \longrightarrow 2p$; $3d \longrightarrow 2p$; $4s \longrightarrow 2p$; $3p \longrightarrow 2s$

Chapter 7

7.1 Elements are listed in the periodic table in an ordered, systematic way that correlates with the periodicity of their chemical and physical properties. The theoretical basis for the table in terms of atomic number and electron configuration does not allow for a "new element" between Sn and Sb. **7.3** (a) predicted atomic mass = 54.23 u (b) predicted melting point = 6.3°C **7.6** The quantum number m_s relates to just the electron; all the others describe the orbital. **7.9** Shielding occurs when electrons protect or shield other electrons from the full nuclear attraction. The effective nuclear charge is the nuclear charge an electron actually experiences. As the number of electrons, especially core electrons, increases, the effective nuclear charge decreases. **7.11** (a) 6 (b) 10 (c) 2 **7.13** (a) 6 (b) 2 (c) 14 **7.16** Hund's rule states that electrons will occupy empty orbitals in a given subshell (with parallel spins) before filling half-filled orbitals. The lowest energy arrangement has the maximum number of unpaired electrons with parallel spins.

A correct **B** incorrect

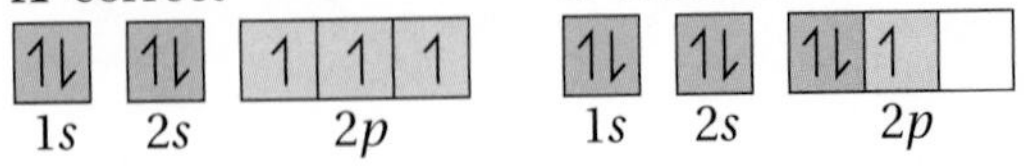

7.18 Main-group elements from the same group have similar valence electron configurations. Valence electron configurations in a period (row) vary, with each succeeding element having an additional electron. **7.20** The maximum number of electrons in any energy level, n, is $2n^2$, so the $n = 4$ energy level holds a maximum of $2(4^2) = 32$ electrons. **7.21** (a) $n = 5, l = 0, m_l = 0$, and $m_s = +\frac{1}{2}$ or $-\frac{1}{2}$ (b) $n = 3; l = 1; m_l = +1, 0$, or -1; and $m_s = +\frac{1}{2}$ or $-\frac{1}{2}$ (c) $n = 5, l = 0, m_l = 0$, and $m_s = +\frac{1}{2}$ or $-\frac{1}{2}$ (d) $n = 2$, $l = 1, m_l = +1$, and $m_s = +\frac{1}{2}$ or $-\frac{1}{2}$

7.23 (a) Rb: $1s^22s^22p^63s^23p^64s^23d^{10}4p^65s^1$
(b) Ge: $1s^22s^22p^63s^23p^64s^23d^{10}4p^2$
(c) Ar: $1s^22s^22p^63s^23p^6$

7.25 (a) Cl: $1s^22s^22p^63s^23p^5$ (b) Si: $1s^22s^22p^63s^23p^2$
(c) Sr: $1s^22s^22p^63s^23p^64s^23d^{10}4p^65s^2$

7.27 (a) Ti: [Ar] $4s^23d^2$

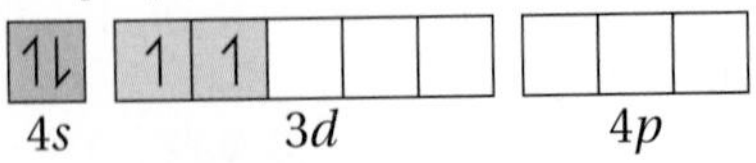

(b) Cl: [Ne] $3s^23p^5$

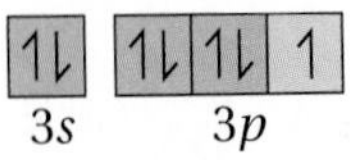

(c) V: [Ar] $4s^23d^3$

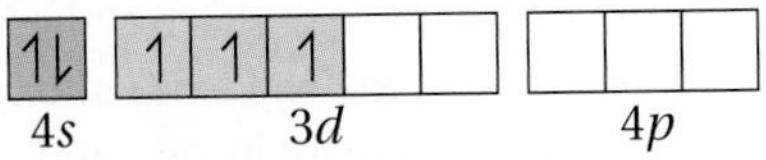

7.29 (a) Mn: [Ar] $4s^23d^5$

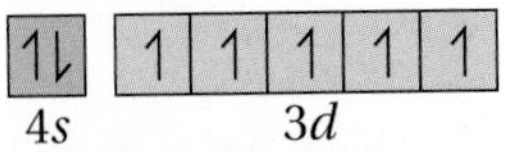

(b) P: [Ne] $3s^23p^3$

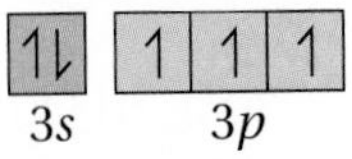

(c) Fe: [Ar] $4s^23d^6$

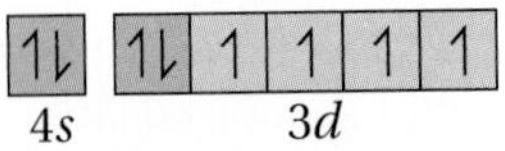

7.31 (a) O; group 16; period 2

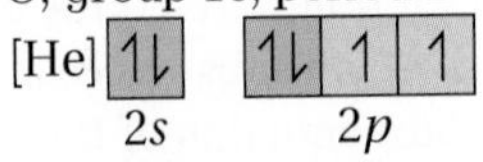

(b) P; group 15; period 3

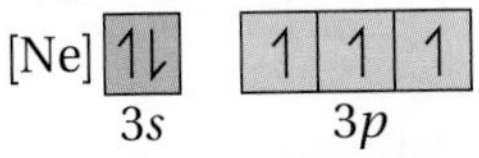

7.33 (a) Cl; group 17; period 3

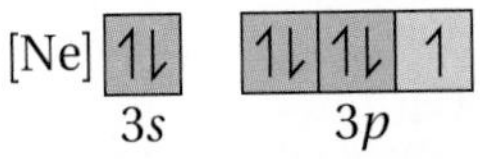

(b) As; group 15; period 4

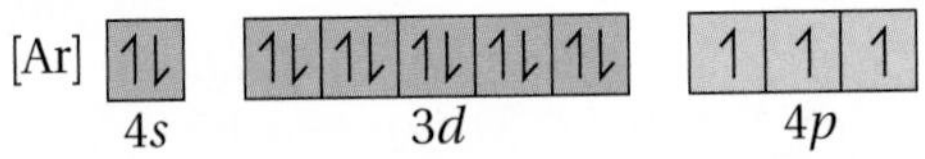

7.35 (a) [Ar] $4s^23d^{10}4p^1$; group 13 (b) [He] $2s^22p^6$; group 18

7.37

	Core Electrons	Valence Electrons
(a) O	2	6
(b) Sn	46	4
(c) Ca	18	2
(d) Fe	18	8
(e) Se	28	6

7.39 (a) B; Al, Ga, In, and Tl (b) S; O, Se, Te, and Po (c) La; Sc, Y, and Ac **7.41** (a) C; Si, Ge, Sn, and Pb (b) V; Nb, Ta, and Db (c) P; N, As, Sb, and Bi

7.43 Na (first excited state): $1s^22s^22p^63p^1$

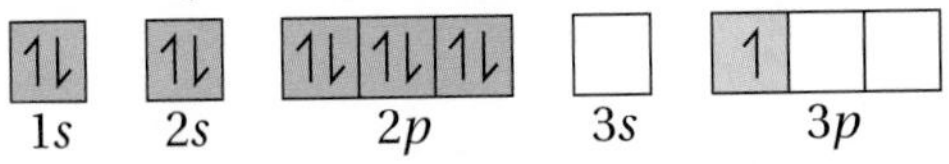

7.50 A high IE_1 and a very negative EA_1 suggest that the elements are halogens, in group 17, which form −1 ions. **7.53** (a) K < Rb < Cs (b) O < C < Be (c) Cl < S < K (d) Mg < Ca < K **7.55** (a) Ba < Sr < Ca (b) B < N < Ne (c) Rb < Se < Br (d) Sn < Sb < As **7.57** $1s^22s^22p^1$ (boron, B) **7.59** (a) Na (b) Na (c) Be **7.61** (1) Metals conduct electricity; nonmetals do not. (2) Metal ions have a positive charge; nonmetal ions have a negative charge. (3) Metal oxides are mostly ionic and act as bases; nonmetal oxides are mostly covalent and act as acids. **7.62** Metallic character increases down a group and decreases to the right across a period. These trends are the same as those for atomic size and opposite those for ionization energy.

7.63 Possible ions are +2 and +4. **7.67** (a) Rb (b) Ra (c) I **7.69** (a) As (b) P (c) Be **7.71** (a) Cl^-: $1s^22s^22p^63s^23p^6$ (b) Na^+: $1s^22s^22p^6$ (c) Ca^{2+}: $1s^22s^22p^63s^23p^6$ **7.73** (a) Al^{3+}: $1s^22s^22p^6$ (b) S^{2-}: $1s^22s^22p^63s^23p^6$ (c) Sr^{2+}: $1s^22s^22p^63s^23p^64s^23d^{10}4p^6$ **7.75** (a) 0 (b) 3 (c) 0 (d) 1 **7.77** a, b, and d are paramagnetic **7.79** (a) V^{3+}: [Ar] $3d^2$, paramagnetic (b) Cd^{2+}: [Kr] $4d^{10}$, diamagnetic (c) Co^{3+}: [Ar] $3d^6$, paramagnetic (d) Ag^+: [Kr] $4d^{10}$, diamagnetic **7.81** For palladium to be diamagnetic, all of its electrons must be paired. (a) You might first write the condensed electron configuration for Pd as [Kr] $5s^24d^8$. However, the partial orbital diagram is not consistent with diamagnetism.

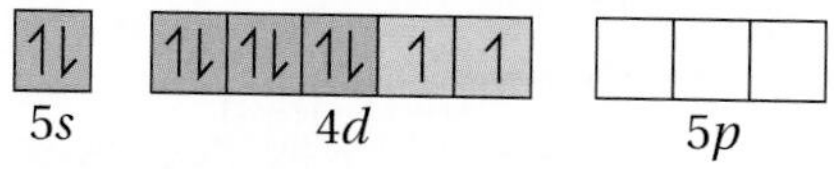

(b) This is the only configuration that supports diamagnetism, [Kr] $4d^{10}$.

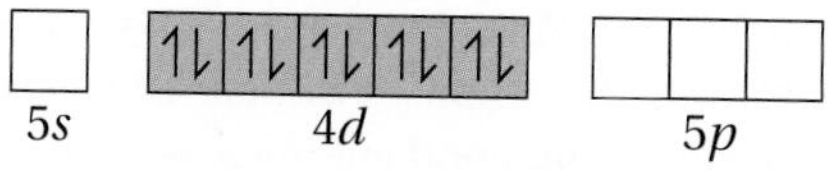

(c) Promoting an *s* electron into the *d* subshell still leaves two electrons unpaired.

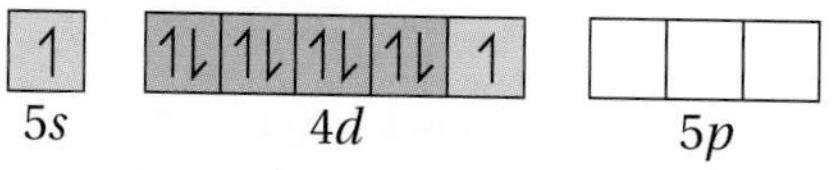

7.83 (a) $Li^+ < Na^+ < K^+$ (b) $Rb^+ < Br^- < Se^{2-}$ (c) $F^- < O^{2-} < N^{3-}$ **7.86** Ce: [Xe] $6s^24f^15d^1$; Ce^{4+}: [Xe]; Eu: [Xe] $6s^24f^7$; Eu^{2+}: [Xe] $4f^7$. Ce^{4+} has a noble-gas configuration; Eu^{2+} has a half-filled *f* subshell. **7.89** (a) $SrBr_2$, strontium bromide (b) CaS, calcium sulfide (c) ZnF_2, zinc fluoride (d) LiF, lithium fluoride **7.90** (a) 2009 kJ/mol (b) −549 kJ/mol **7.91** All ions except Fe^{8+} and Fe^{14+} are paramagnetic; Fe^+ and Fe^{3+} would be most attracted to a magnetic field.

Chapter 8

8.1 (a) Greater ionization energy decreases metallic character. (b) Larger atomic radius increases metallic character. (c) Higher number of outer electrons decreases metallic character. (d) Larger effective nuclear charge decreases metallic character. **8.4** (a) Cs (b) Rb (c) As **8.6** (a) ionic (b) covalent (c) metallic **8.8** (a) covalent (b) ionic (c) covalent **8.10** (a) Rb· (b) ·Ṡi· (c) :Ï· **8.12** (a) ·Sr· (b) :Ṗ· (c) :Ṡ· **8.14** (a) 16; [noble gas] ns^2np^4 (b) 13; [noble gas] ns^2np^1 **8.20** (a) Ba^{2+}, [Xe]; Cl^-, [Ne]$3s^23p^6$, :C̈l:$^-$; $BaCl_2$ (b) Sr^{2+}, [Kr]; O^{2-}, [He]$2s^22p^6$, :Ö:$^{2-}$; SrO (c) Al^{3+}, [Ne]; F^-, [He]$2s^22p^6$, :F̈:$^-$; AlF_3 (d) Rb^+, [Kr]; O^{2-}, [He]$2s^22p^6$, :Ö:$^{2-}$; Rb_2O **8.22** (a) 2 (b) 16 (c) 1 **8.24** (a) 13 (b) 2 (c) 16 **8.26** (a) BaS; the charge on each ion is twice the charge on the ions in CsCl. (b) LiCl; Li^+ is smaller than Cs^+. **8.28** (a) BaS; Ba^{2+} is larger than Ca^{2+}. (b) NaF; the charge on each ion is half the charge on the ions in MgO. **8.30** 788 kJ/mol; the lattice energy for NaCl is less than that for LiF, because the Na^+ and Cl^- ions are larger than Li^+ and F^- ions. **8.33** −336 kJ/mol **8.34** When two chlorine atoms are far apart, there is no interaction between them. As the atoms move closer together, the nucleus of each atom attracts the electrons of the other atom. The closer the atoms, the greater this attraction; however, the repulsions of the two nuclei and the electrons also increase at the same time. The final internuclear distance is the distance at which maximum attraction is achieved in spite of the repulsion. **8.35** The bond energy is the energy required to break the bond between H atoms and Cl atoms in 1 mol of HCl molecules in the gaseous state. Energy is needed to break bonds, so bond breaking is always endothermic and $\Delta_{\text{bond breaking}}H°$ is positive. The quantity of energy needed to break the bond is released upon its formation, so $\Delta_{\text{bond forming}}H°$ has the same magnitude as $\Delta_{\text{bond breaking}}H°$ but is opposite in sign (always exothermic and negative). **8.39** (a) I—I < Br—Br < Cl—Cl (b) S—Br < S—Cl < S—H (c) C—N < C═N < C≡N **8.41** (a) C—O < C═O; the C═O bond (bond order = 2) is stronger than the C—O bond (bond order = 1). (b) C—H < O—H; O is smaller than C so the O—H bond is shorter and stronger than the C—H bond. **8.43** Less energy is required to break weak bonds. **8.45** Both are one-carbon molecules. Since methane contains fewer carbon-oxygen bonds, it will have the greater heat of reaction per mole for combustion. **8.47** −168 kJ/mol **8.49** −22 kJ/mol **8.50** −59 kJ/mol **8.51** Electronegativity increases from left to right and increases from bottom to top within a group. Fluorine and oxygen are the two most electronegative elements. Cesium and francium are the two least electronegative elements. **8.53** The H—O bond in water is polar covalent. A nonpolar covalent bond occurs between two atoms with identical electronegativities. A polar covalent bond occurs when the atoms have differing electronegativities. Ionic bonds result from electron transfer between atoms. **8.56** (a) Si < S < O (b) Mg < As < P **8.58** (a) N > P > Si (b) As > Ga > Ca
8.60

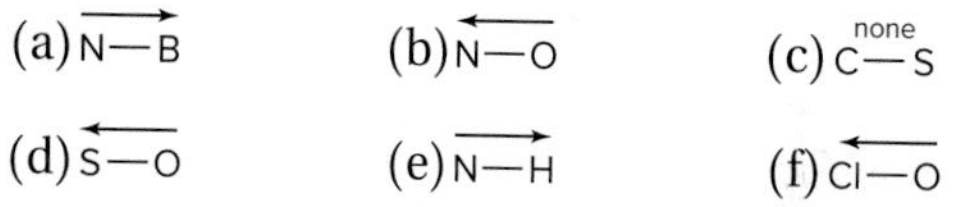

8.62 a, d, and e **8.64** (a) nonpolar covalent (b) ionic (c) polar covalent (d) polar covalent (e) nonpolar covalent (f) polar covalent; $SCl_2 < SF_2 < PF_3$
8.66

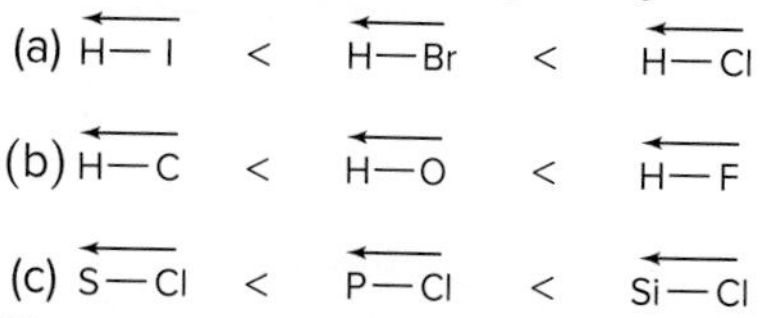

8.69 He cannot serve as a central atom because it does not bond. H cannot because it forms only one bond. Fluorine cannot because it needs only one electron to complete its valence level, and it does not have *d* orbitals available to expand its valence level. Thus, it can bond to only one other atom. **8.71** All the structures obey the octet rule except c and g.
8.73

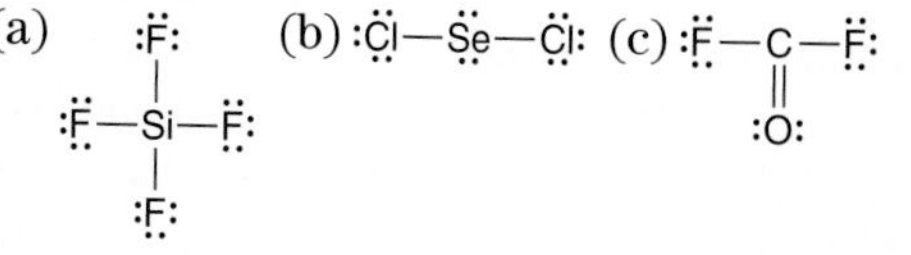

8.75

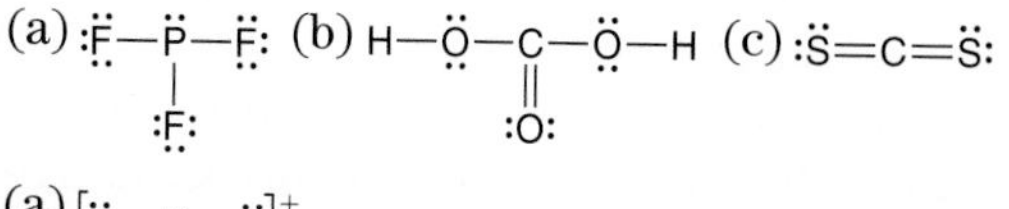

8.77

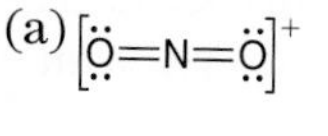

(b)
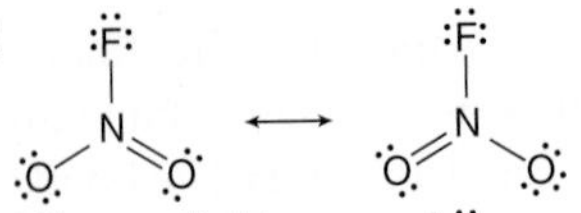

8.79 (a)
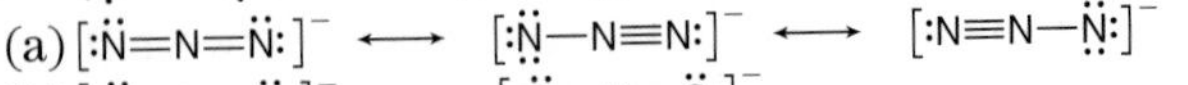
(b)

8.81 (a)
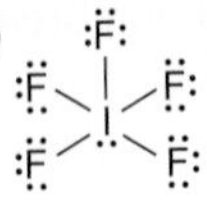
formal charges: $FC_I = 0$, $FC_F = 0$

(b)
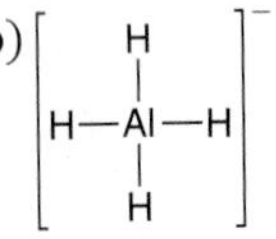
formal charges: $FC_H = 0$, $FC_{Al} = -1$

8.83 (a)
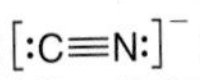
formal charges: $FC_C = -1$, $FC_N = 0$
(b)
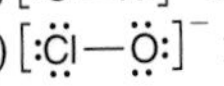
formal charges: $FC_{Cl} = 0$, $FC_O = -1$

8.85 (a)
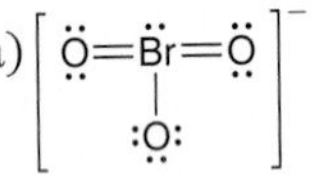
formal charges: $FC_{Br} = 0$, doubly bonded $FC_O = 0$, singly bonded $FC_O = -1$; oxidation numbers: oxidation number$_{Br} = +5$; oxidation number$_O = -2$

(b)
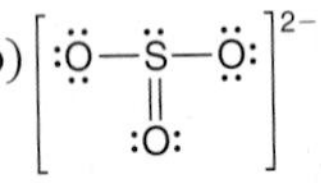
formal charges: $FS_S = 0$, singly bonded $FS_O = -1$, doubly bonded $FS_O = 0$; oxidation numbers: oxidation number$_S = +4$; oxidation number$_O = -2$

8.87 (a) B has 6 valence electrons in BH_3, so the molecule is electron deficient. (b) As has an expanded valence level with 10 electrons. (c) Se has an expanded valence level with 10 electrons.

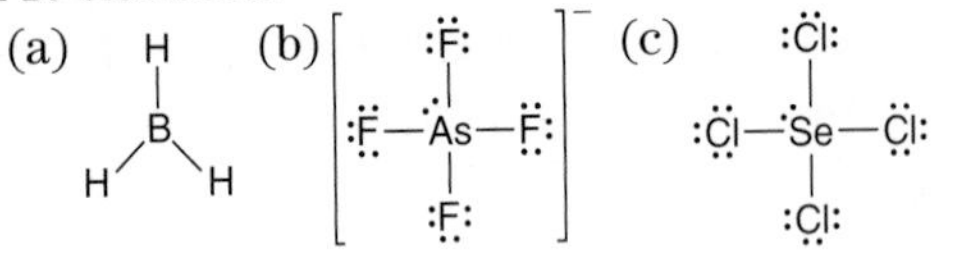

8.89 (a) Br expands its valence level to 10 electrons. (b) I has an expanded valence level of 10 electrons. (c) Be has only 4 valence electrons in BeF_2, so the molecule is electron deficient.

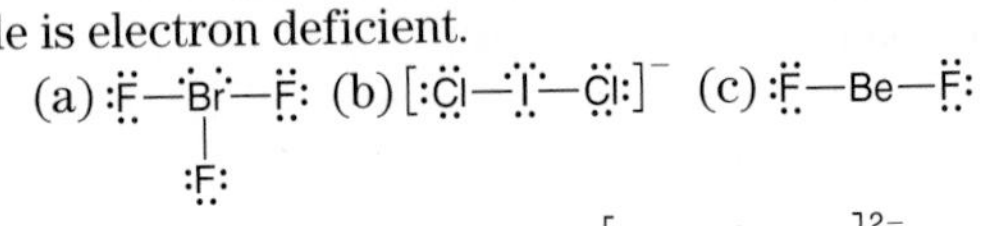

8.91
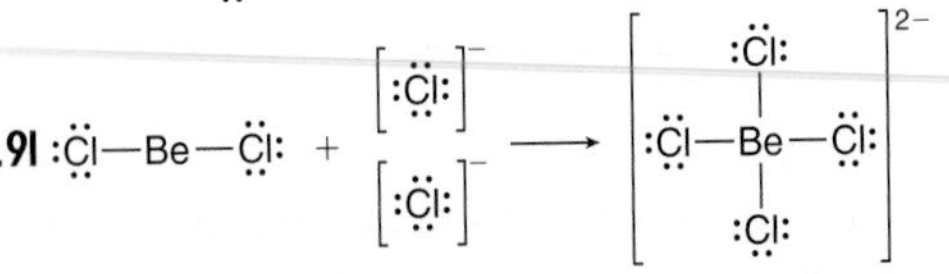

8.94 structure A **8.95** (a) shiny, conducts heat, conducts electricity, and is malleable (b) Metals lose electrons to form positive ions **8.99** (a) 800. kJ/mol, which is lower than the value in Table 8.2 (b) -2.417×10^4 kJ (c) 1690. g CO_2 (d) 65.2 L O_2 **8.101** (a) −125 kJ/mol (b) yes, since $\Delta_f H°$ is negative (c) −392 kJ/mol (d) No; $\Delta_f H°$ for $MgCl_2$ is much more negative than that for MgCl. **8.103** (a) 406 nm (b) 2.93×10^{-19} J (c) 1.87×10^4 m/s **8.106** C—Cl: 3.53×10^{-7} m; bond in O_2: 2.40×10^{-7} m **8.107** XeF_2: 132 kJ/mol; XeF_4: 150. kJ/mol; XeF_6: 146 kJ/mol **8.109** (a) The presence of the very electronegative fluorine atoms bonded to one of the carbons makes the C—C bond polar. This polar bond will tend to undergo heterolytic rather than homolytic cleavage. More energy is required to achieve heterolytic cleavage. (b) 1420 kJ/mol

8.112 13 286 kJ/mol **8.114** 8.70×10^{14} s^{-1}; 3.45×10^{-7} m; the ultraviolet region of the electromagnetic spectrum
8.116 (a) $CH_3OCH_3(g)$: −326 kJ/mol; $CH_3CH_2OH(g)$: −369 kJ/mol (b) The formation of gaseous ethanol is more exothermic. (c) 43 kJ/mol
8.118 (a)
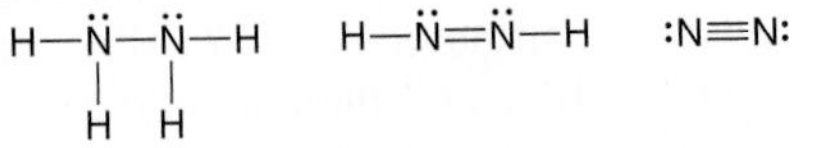
Hydrazine Diazene Nitrogen

The single N—N bond (bond order = 1) is weaker and longer than the others. The triple bond (bond order = 3) is stronger and shorter than the others. The double bond (bond order = 2) has an intermediate strength and length.
(b)
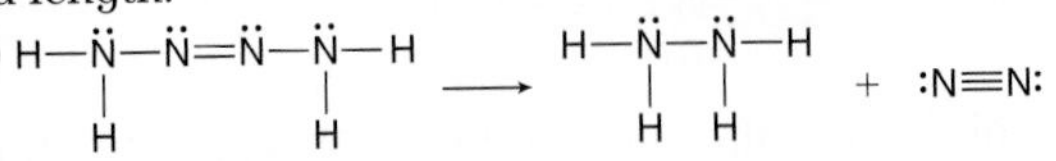
$\Delta_r H° = -367$ kJ/mol
8.122 (a) −1267 kJ/mol (b) −1226 kJ/mol (c) −1234.8 kJ/mol The two answers differ by less than 10 kJ/mol. This is very good agreement since average bond energies were used to calculate answers a and b. (d) −37 kJ/mol
8.124
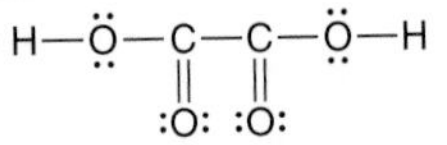

8.127 CH_4: −409 kJ/mol O_2; H_2S: −398 kJ/mol O_2 **8.129** (a) The O in the OH species has only seven valence electrons, which is less than an octet, and one electron is unpaired. (b) 426 kJ/mol (c) 508 kJ/mol
8.132 $H_2C_2O_4$:
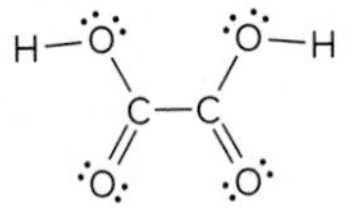

$HC_2O_4^-$:
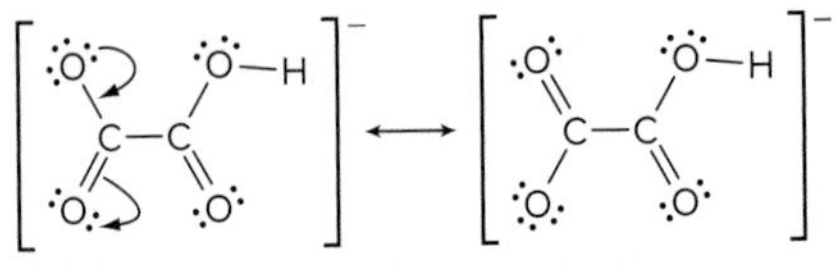

$C_2O_4^{2-}$:
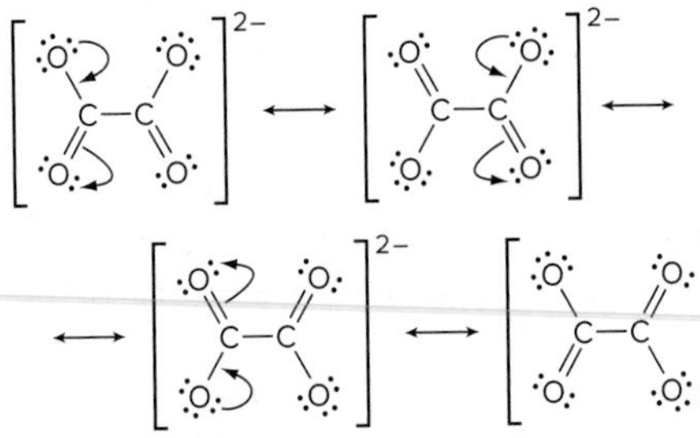

In $H_2C_2O_4$, there are two shorter and stronger C=O bonds and two longer and weaker C—O bonds. In $HC_2O_4^-$, the carbon-oxygen bonds on the side retaining the H remain as one long, weak C—O and one short, strong C=O. The carbon-oxygen bonds on the other side of the molecule have resonance forms with a bond order of 1.5, so they are intermediate in length and strength. In $C_2O_4^{2-}$, all the carbon-oxygen bonds have a bond order of 1.5. **8.135** 22 kJ/mol

Chapter 9

Answers to Boxed Reading Problems: B9.1 resonance form on the left: trigonal planar around C, trigonal pyramidal around N; resonance form on the right: trigonal planar around both C and N. **B9.2** The top portions of both molecules are similar, so the top portions will interact with biomolecules in a similar manner. The mescaline molecule may fit into the same nerve receptors as dopamine due to the similar molecular shape.

9.2 The molecular shape and the electron-group arrangement are the same when no lone pairs are present on the central atom. **9.4** tetrahedral, AX_4; trigonal pyramidal, AX_3E_1; bent or V shaped, AX_2E_2
9.6

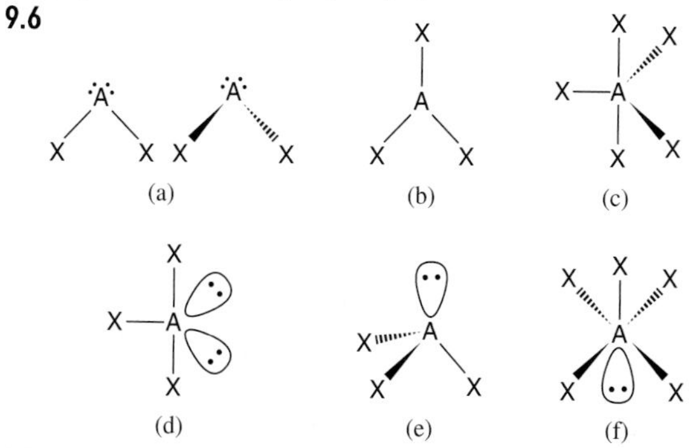

9.8 (a) trigonal planar, bent, 120° (b) tetrahedral, trigonal pyramidal, 109.5° (c) tetrahedral, trigonal pyramidal, 109.5° **9.10** (a) trigonal planar, trigonal planar, 120° (b) trigonal planar, bent, 120° (c) tetrahedral, tetrahedral, 109.5° **9.12** (a) trigonal planar, AX_3, 120° (b) trigonal pyramidal, AX_3E_1, 109.5° (c) trigonal bipyramidal, AX_5, 90° and 120° **9.14** (a) bent, 109.5°, less than 109.5° (b) trigonal bipyramidal, 90° and 120°, angles are ideal (c) seesaw, 90° and 120°, less than ideal (d) linear, 180°, angle is ideal **9.16** (a) C: tetrahedral, 109.5°; O: bent, < 109.5° (b) N: trigonal planar, 120° **9.18** (a) C in CH_3: tetrahedral, 109.5°; C with C=O: trigonal planar, 120°; O with H: bent, < 109.5° (b) O: bent, < 109.5° **9.20** $OF_2 < NF_3 < CF_4 < BF_3 < BeF_2$ **9.22** (a) The C and N each have three groups, so the ideal angles are 120°; the O has four groups, so the ideal angle is 109.5°. The N and O have lone pairs, so the angles are less than ideal. (b) All central atoms have four pairs, so the ideal angles are 109.5°. The lone pairs on the O reduce this value. (c) The B has three groups and an ideal bond angle of 120°. All the oxygen atoms have four groups (ideal bond angles of 109.5°), two of which are lone pairs that reduce the angle.

9.25

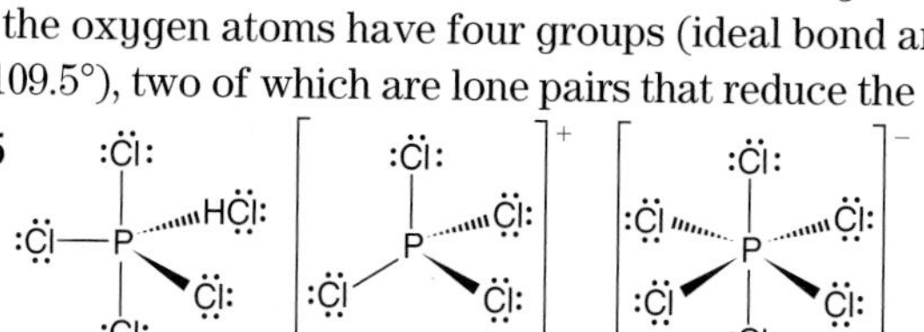

In the gas phase, PCl_5 is AX_5, so the shape is trigonal bipyramidal, and the bond angles are 120° and 90°. The PCl_4^+ ion is AX_4, so the shape is tetrahedral, and the bond angles are 109.5°. The PCl_6^- ion is AX_6, so the shape is octahedral, and the bond angles are 90°. **9.26** Molecules are polar if they have polar bonds that are not arranged to cancel each other. A polar bond is present any time there is a bond between elements with differing electronegativities. **9.29** (a) CF_4 (b) BrCl and SCl_2 **9.31** (a) SO_2, because it is polar and SO_3 is not. (b) IF has a greater electronegativity difference between its atoms. (c) SF_4, because it is polar and SiF_4 is not (d) H_2O has a greater electronegativity difference between its atoms.

9.33

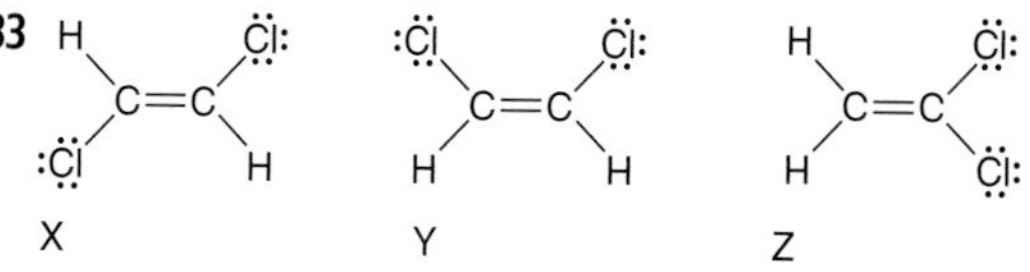

Yes; compound Y has a dipole moment.
9.37 (a) formal charges for Al_2Cl_6: $FC_{Al} = -1$, $FC_{end\ Cl} = 0$, $FC_{bridging\ Cl} = +1$; formal charges for I_2Cl_6: $FC_I = -1$, $FC_{end\ Cl} = 0$, $FC_{bridging\ Cl} = +1$ (b) The iodine atoms are each AX_4E_2, and the shape around each is square planar. These square planar portions are adjacent, giving a planar molecule.
9.42

(a) In epoxypropane, the shape around each C is tetrahedral, with ideal angles of 109.5°. (b) The C that is not part of the three-membered ring should have close to the ideal angle. The atoms in the ring form an equilateral triangle, so the angles around the two carbon atoms in the ring are reduced from the ideal 109.5° to 60°. **9.45** (a) The F atoms will substitute at the axial positions first. (b) PF_5 and PCl_3F_2 **9.47** Trigonal planar molecules are nonpolar, so AY_3 cannot be that shape. Trigonal pyramidal molecules and T-shaped molecules are polar, so AY_3 could have either of these shapes. **9.50** (a) 339 pm (b) 316 pm and 223 pm (c) 270 pm

Chapter 10

10.1 (a) sp^2 (b) sp^3d^2 (c) sp (d) sp^3 (e) sp^3d **10.3** C has only 2*s* and 2*p* atomic orbitals, allowing for a maximum of four hybrid orbitals. Si has 3*s*, 3*p*, and 3*d* atomic orbitals, allowing it to form more than four hybrid orbitals. **10.5** (a) six, sp^3d^2 (b) four, sp^3 **10.7** (a) sp^2 (b) sp^2 (c) sp^2 **10.9** (a) sp^3 (b) sp^3 (c) sp^3 **10.11** (a) Si: one *s* and three *p* atomic orbitals form four sp^3 hybrid orbitals. (b) C: one *s* and one *p* atomic orbitals form two *sp* hybrid orbitals. (c) S: one *s*, three *p*, and one *d* atomic orbitals mix to form five sp^3d hybrid orbitals. (d) N: one *s* and three *p* atomic orbitals mix to form four sp^3 hybrid orbitals. **10.13** (a) B ($sp^3 \longrightarrow sp^3$) (b) A ($sp^2 \longrightarrow sp^3$)

10.15

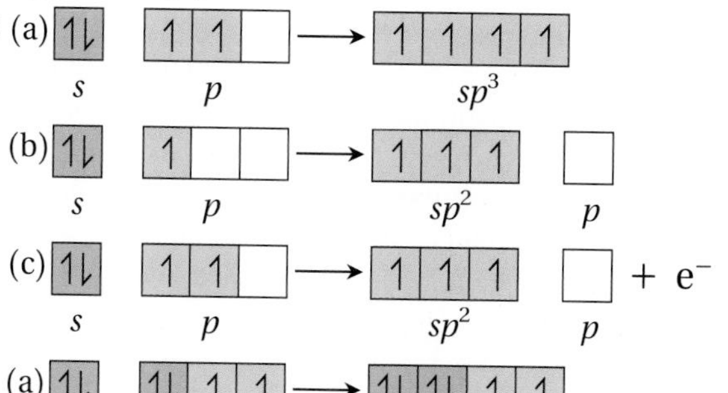

10.17

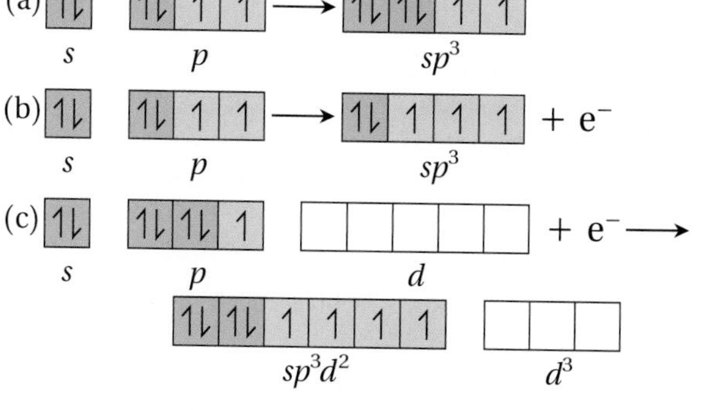

10.20 (a) False. A double bond is one σ and one π bond. (b) False. A triple bond consists of one σ and two π bonds. (c) true (d) true (e) False. A π bond consists of a second pair of electrons after a σ bond has been previously formed. (f) False. End-to-end overlap results in a bond with electron density along the bond axis. **10.21** (a) Nitrogen uses sp^2 to form three σ bonds and one π bond. (b) Carbon uses sp to form two σ bonds and two π bonds. (c) Carbon uses sp^2 to form three σ bonds and one π bond.
10.23 (a) N: sp^2, forming 2 σ bonds and 1 π bond

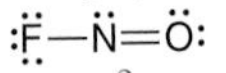

(b) C: sp^2, forming 3 σ bonds and 1 π bond

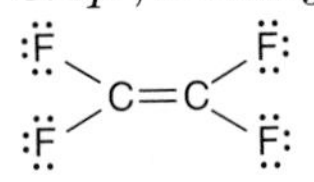

(c) C: sp, forming 2 σ bonds and 2 π bonds

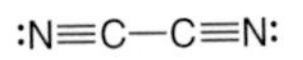

10.25

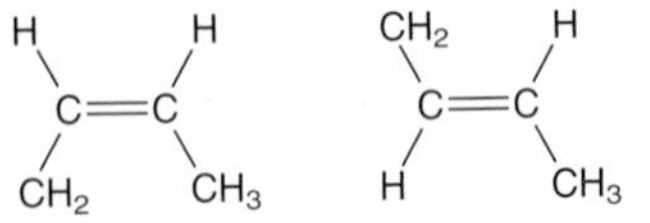

cis *trans*

The single bonds are all σ bonds. The double bond is one σ bond and one π bond. **10.26** Four MOs form from the four p atomic orbitals. The total number of MOs must equal the number of atomic orbitals. **10.28** (a) Bonding MOs have lower energy than antibonding MOs. Lower energy means more stable. (b) Bonding MOs do not have a nodal plane perpendicular to the bond. (c) Bonding MOs have higher electron density between nuclei than antibonding MOs. **10.30** (a) two (b) two (c) four **10.32** (a) A is π^*_{2p}, B is σ_{2p}, C is π_{2p}, and D is σ^*_{2p}. (b) π^*_{2p} (A), σ_{2p} (B), and π_{2p} (C) have at least one electron. (c) π^*_{2p} (A) has only one electron. **10.34** (a) stable (b) paramagnetic (c) $(\sigma_{2s})^2(\sigma^*_{2s})^1$ **10.36** (a) $C_2^- < C_2 < C_2^+$ (b) $C_2^- < C_2 < C_2^+$ **10.40** (a) C (ring): sp^2; C (all others): sp^3; O (all): sp^3; N: sp^3 (b) 26 σ bonds (c) 6 π electrons **10.42** (a) 17 σ bonds (b) All carbons are sp^2; the ring N is sp^2; the other N atoms are sp^3. **10.44** (a) B changes from sp^2 to sp^3. (b) P changes from sp^3 to sp^3d. (c) C changes from sp to sp^2. Two electron groups surround C in C_2H_2, and three electron groups surround C in C_2H_4. (d) Si changes from sp^3 to sp^3d^2. (e) no change for S **10.46** P: tetrahedral, sp^3; N: trigonal pyramidal, sp^3; C_1 and C_2: tetrahedral, sp^3; C_3: trigonal planar, sp^2 **10.51** (a) B and D are present. (b) yes; sp hybrid orbitals (c) two sets of sp orbitals, four sets of sp^2 orbitals, and three sets of sp^3 orbitals **10.52** Through resonance, the C—N bond gains some double-bond character, which hinders rotation about that bond.

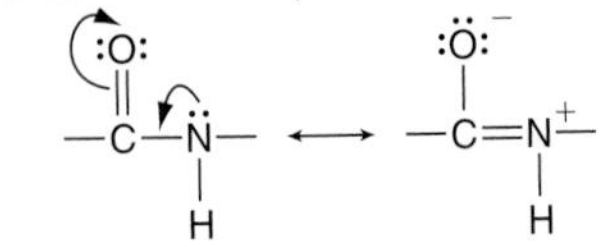

10.55 (a) C in —CH_3: sp^3; all other C atoms: sp^2; O in two C—O bonds: sp^3; O in two C=O bonds: sp^2 (b) two (c) eight; one **10.56** (a) four (b) eight

Chapter 11

Answers to Boxed Reading Problems: B11.1 1.76×10^{-10} m **B11.2** (a) 155 pm (b) 33.2°

11.1 In a solid, the energy of attraction of the particles is greater than their energy of motion; in a gas, it is less. Gases have high compressibility and the ability to flow, while solids have neither. **11.4** (a) because the intermolecular forces are only partially overcome when fusion occurs but need to be totally overcome in vaporization (b) because solids have greater intermolecular forces than liquids do (c) $\Delta_{vap}H = -\Delta_{cond}H$ **11.5** (a) intermolecular (b) intermolecular (c) intramolecular (d) intramolecular **11.7** (a) condensation (b) fusion (c) evaporation **11.9** The gas molecules slow down as the gas is compressed. Therefore, much of the kinetic energy lost by the propane molecules is released to the surroundings. **11.13** At first, the vaporization of liquid molecules from the surface predominates, which increases the number of gas molecules and hence the vapour pressure. As more molecules enter the gas phase, more gas molecules hit the surface of the liquid and "stick" more frequently, so the condensation rate increases. When the vaporization and condensation rates become equal, the vapour pressure becomes constant. **11.14** As intermolecular forces increase, (a) critical temperature increases, (b) boiling point increases, (c) vapour pressure decreases, and (d) heat of vaporization increases. **11.18** because the condensation of the vapour supplies an additional 41 kJ/mol **11.19** 7.67×10^3 J **11.21** 78.7 kPa **11.23** 21.3 kJ/mol
11.25

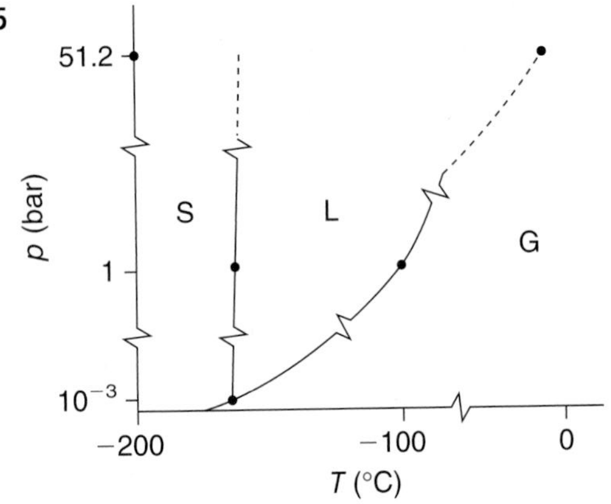

Solid ethene is denser than liquid ethene. **11.28** 3280 kPa **11.32** O is smaller and more electronegative than Se, so the electron density on O is greater, which attracts H more strongly. **11.34** All particles (atoms and molecules) exhibit dispersion forces, but the total force is weak for small molecules. Dipole-dipole forces between small polar molecules dominate the dispersion forces. **11.37** (a) hydrogen bonding (b) dispersion forces (c) dispersion forces **11.39** (a) dipole-dipole forces (b) dispersion forces (c) hydrogen bonding
11.41 (a)

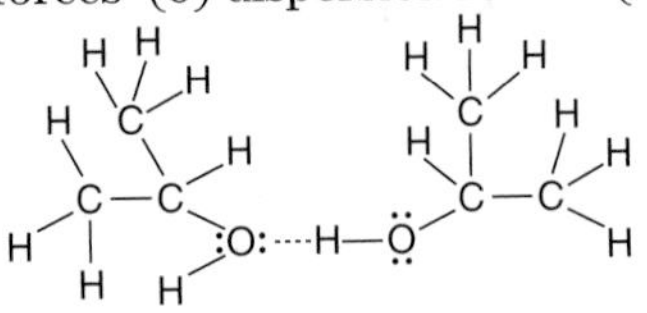

(b)

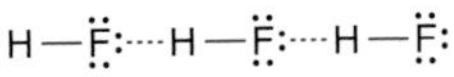

11.43 (a) dispersion forces (b) hydrogen bonding (c) dispersion forces **11.45** (a) I^- (b) $CH_2{=}CH_2$ (c) H_2Se. In (a) and (c) the larger particle has the higher polarizability. In (b), the less tightly held π electron clouds are more easily distorted. **11.47** (a) C_2H_6; it is a smaller molecule exhibiting weaker dispersion forces than C_4H_{10}. (b) CH_3CH_2F; it has no H—F bonds, so it exhibits only dipole-dipole forces, which are weaker than the hydrogen bonds of CH_3CH_2OH. (c) PH_3; it has weaker intermolecular forces (dipole-dipole) than NH_3 (hydrogen bonding). **11.49** (a) HCl; it has dipole-dipole forces, and there are stronger ionic bonds in LiCl. (b) PH_3; it has dipole-dipole forces, and there is stronger hydrogen bonding in NH_3. (c) Xe; it exhibits weaker dispersion forces since its smaller size results in lower polarizability than the larger I_2 molecules. **11.51** (a) C_4H_8 (cyclobutane); it is more compact than C_4H_{10}. (b) PBr_3; the dipole-dipole forces in PBr_3 are weaker than the ionic bonds in NaBr. (c) HBr; the dipole-dipole forces in HBr are weaker than the hydrogen bonds in water. **11.53** As atomic size decreases and electronegativity increases, the electron density of an atom increases. Thus, the attraction to an H atom on another molecule increases while its bonded H atom becomes more positive. Fluorine is the smallest of the three and the most electronegative, so the hydrogen bonds in hydrogen fluoride are the strongest. Oxygen is smaller and more electronegative than nitrogen, so hydrogen bonds in water are stronger than hydrogen bonds in ammonia. **11.57** The cohesive forces in water and mercury are stronger than the adhesive forces to the nonpolar wax on the floor. Weak adhesive forces result in spherical drops. The adhesive forces overcome the even weaker cohesive forces in the oil, so the oil drop spreads out. **11.59** Surface tension is defined as the energy needed to increase the surface area by a given amount, so units of energy per area are appropriate. **11.61** $CH_3CH_2CH_2OH < HOCH_2CH_2OH < HOCH_2CH(OH)CH_2OH$ More hydrogen bonding means more attraction between molecules, so more energy is needed to increase surface area. **11.63** $HOCH_2CH(OH)CH_2OH > HOCH_2CH_2OH > CH_3CH_2CH_2OH$ More hydrogen bonding means more attraction between neighbouring molecules, so they flow less easily. **11.68** Water is a good solvent for polar and ionic substances and a poor solvent for nonpolar substances. Water is a polar molecule and dissolves polar substances because their intermolecular forces are of similar strength. **11.69** A single water molecule can form four hydrogen bonds. The two hydrogen atoms each form one hydrogen bond to oxygen atoms on neighbouring water molecules. The two lone pairs on the oxygen atom form hydrogen bonds with hydrogen atoms on two neighbouring molecules. **11.72** Water exhibits strong capillary action, which allows it to be easily absorbed by the plant's roots and transported upward to the leaves. **11.78** simple cubic **11.81** The energy gap is the energy difference between the highest filled energy level (valence band) and the lowest unfilled energy level (conduction band). In conductors and superconductors, the energy gap is zero because the valence band overlaps the conduction band. In semiconductors, the energy gap is small. In insulators, the gap is large. **11.83** atomic mass and atomic radius **11.84** (a) face-centred cubic (b) body-centred cubic (c) face-centred cubic **11.86** (a) The change in unit cell is from a sodium chloride structure in CdO to a zinc blende structure in CdSe. (b) Yes; the coordination number of Cd changes from 6 in CdO to 4 in CdSe. **11.88** (a) Nickel forms a metallic solid since it is a metal whose atoms are held together by metallic bonds. (b) Fluorine forms a molecular solid since the F_2 molecules are held together by dispersion forces. (c) Methanol forms a molecular solid since the CH_3OH molecules are held together by hydrogen bonds. (d) Tin forms a metallic solid since it is a metal whose atoms are held together by metallic bonds. (e) Silicon is in the same group as carbon, so it exhibits similar bonding properties. Since diamond and graphite are both network covalent solids, it makes sense that Si forms a network covalent solid as well. (f) Xe is an atomic solid since its individual atoms are held together by dispersion forces. **11.90** four **11.92** (a) four Se^{2-} ions, four Zn^{2+} ions (b) 577.48 u (c) 1.77×10^{-22} cm^3 (d) 5.61×10^{-8} cm **11.94** (a) insulator (b) conductor (c) semiconductor **11.96** (a) Conductivity increases. (b) Conductivity increases. (c) Conductivity decreases. **11.98** 1.68×10^{-8} cm **11.105** A substance whose properties are the same in all directions is isotropic; otherwise, the substance is anisotropic. Liquid crystals have a high degree of order only in certain directions, so they are anisotropic. **11.107** (a) n-type semiconductor (b) p-type semiconductor **11.110** (a) 2.6453 kPa (b) 0.0486 g **11.115** 259 K **11.116** (a) 2.26 kPa (b) 6.24 L

11.119 (a) 2-furoic acid

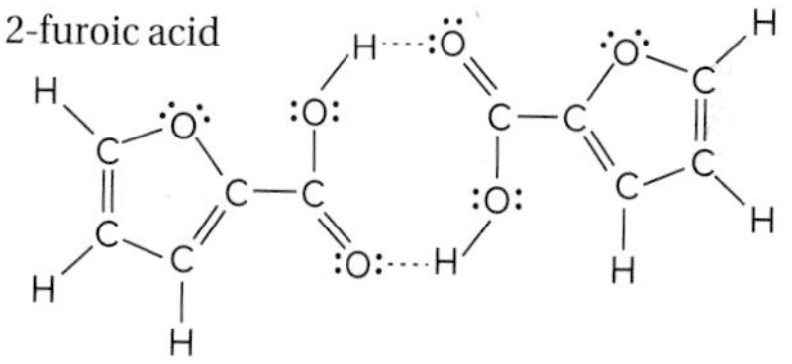

Furfuryl alcohol

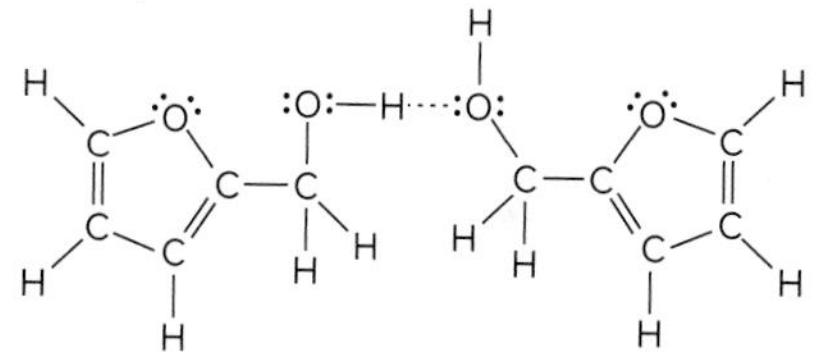

(b) 2-furoic acid Furfuryl alcohol

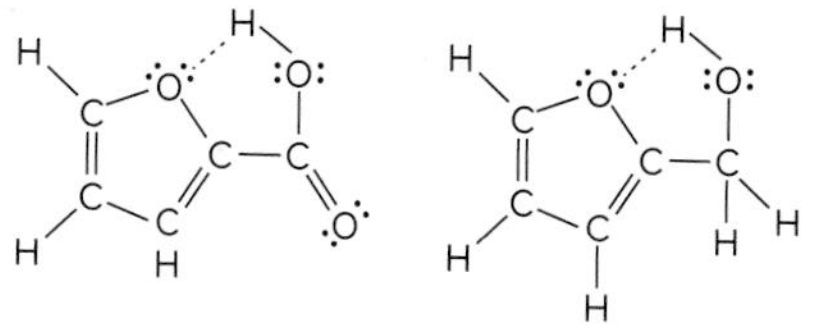

11.120 (a) 50.1 t H_2O (b) -1.12×10^8 kJ **11.121** 2.9 g/m^3 **11.126** (a) 1.1 min (b) 10. min

(c)

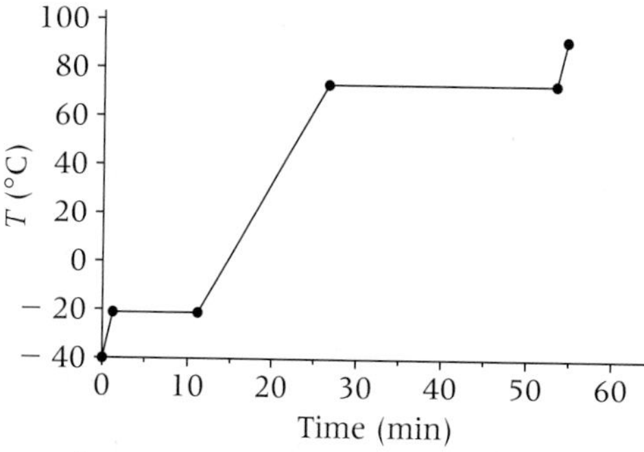

11.127 2.98×10^5 g BN **11.130** 45.98 u

Chapter 12

Answers to Boxed Reading Problems: B12.1 (a) The colloidal particles in water generally have negatively charged surfaces and so repel each other, slowing the settling process. Cake alum, $Al_2(SO_4)_3$, is added to coagulate the colloids. The Al^{3+} ions neutralize the negative surface charges and allow the particles to aggregate and settle. (b) Water that contains large amounts of divalent cations (such as Ca^{2+} and Mg^{2+}) is called hard water. During cleaning, these ions combine with the fatty-acid anions in soaps to produce insoluble deposits. (c) In reverse osmosis, a pressure greater than the osmotic pressure is applied to the solution, forcing the water back through the membrane, leaving the ions behind. (d) Chlorine may give the water an unpleasant odor and can form carcinogenic chlorinated compounds. (e) The high concentration of NaCl displaces the divalent and polyvalent ions from the ion-exchange resin. **B12.2** 25.3 kPa

12.2 When a salt such as NaCl dissolves, ion-dipole forces cause the ions to separate, and many water molecules cluster around each ion in hydration shells. Ion-dipole forces bind the outermost shell to an ion. The water molecules in that shell form hydrogen bonds to others to create the next shell, and so on. **12.4** Sodium stearate is a more effective soap because the hydrocarbon chain in the stearate ion is longer than that in the ethanoate (acetate) ion. The longer chain allows for more dispersion forces with grease molecules. **12.7** KNO_3 is an ionic compound and is therefore more soluble in water. **12.9** (a) ion-dipole forces (b) hydrogen bonding (c) dipole–induced dipole forces **12.11** (a) hydrogen bonding (b) dipole–induced dipole forces (c) dispersion forces **12.13** (a) $HCl(g)$, because the molecular interactions (dipole-dipole forces) in ethoxyethane are like those in HCl but not like the ionic bonding in NaCl (b) $CH_3CHO(l)$, because the molecular interactions with ethoxyethane (dipole-dipole) are like those between CH_3CHO, but not like the hydrogen bonds in water (c) $CH_3CH_2MgBr(s)$, because the molecular interactions (dipole-dipole and dispersion forces) are greater than between ethoxyethane and the ions in $MgBr_2$ **12.16** Gluconic acid is soluble in water due to extensive hydrogen bonding by the —OH groups attached to five of its carbons. The dispersion forces in the nonpolar tail of caproic acid are more similar to the dispersion forces in hexane; thus, caproic acid is soluble in hexane. **12.18** The energy changes needed to separate solvent into particles ($\Delta_{solvent}H$), and that needed to mix the solvent and solute particles ($\Delta_{mix}H$), would be combined to obtain $\Delta_{solution}H$. **12.22** very soluble, because a decrease in enthalpy and an increase in entropy both favour the formation of a solution

12.23

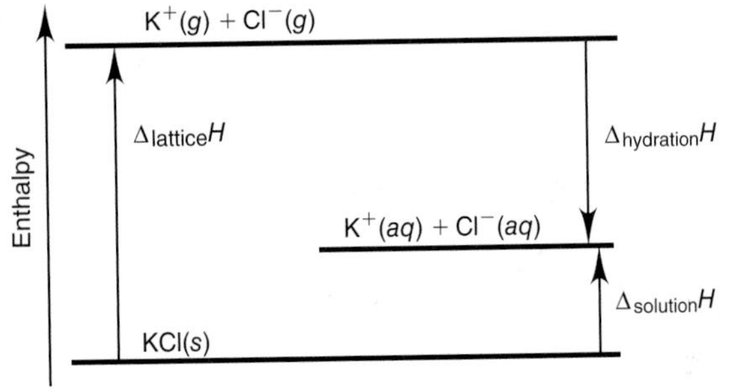

12.25 (a) The volume of Na^+ is smaller. (b) Sr^{2+} has a larger ionic charge and a smaller volume. (c) Na^+ is smaller than Cl^-. (d) O^{2-} has a larger ionic charge with a similar ion volume. (e) OH^- has a smaller volume than SH^-. (f) Mg^{2+} has a smaller volume. (g) Mg^{2+} has both a smaller volume and a larger ionic charge. (h) CO_3^{2-} has both a smaller volume and a larger ionic charge. **12.27** (a) Na^+ (b) Sr^{2+} (c) Na^+ (d) O^{2-} (e) OH^- (f) Mg^{2+} (g) Mg^{2+} (h) CO_3^{2-} **12.29** (a) -704 kJ/mol (b) The K^+ ion contributes more because it is smaller and, therefore, has a greater charge density. **12.31** (a) increases (b) decreases (c) increases **12.34** Add a pinch of X to each solution. Addition of a "seed" crystal of solute to a supersaturated solution causes the excess solute to crystallize immediately, leaving behind a saturated solution. The solution in which the added X dissolves is the unsaturated solution. The solution in which the added X remains undissolved is the saturated solution. **12.37** (a) increase (b) decrease **12.39** (a) 0.102 g O_2 (b) 0.0214 g O_2 **12.42** 0.20 mol/L **12.45** (a) concentration (mol/L) and parts-by-volume (% w/v or % v/v) (b) parts-by-mass (% w/w) (c) molality **12.47** With just this information, you can convert between molality and concentration (mol/L), but you need to know the molar mass of the solvent to convert to mole fraction. **12.49** (a) 0.944 mol/L $C_{12}H_{22}O_{11}$ (b) 0.167 mol/L $LiNO_3$ **12.51** (a) 0.0749 mol/L NaOH (b) 0.36 mol/L HNO_3 **12.53** (a) Add enough distilled water to 4.25 g KH_2PO_4 to make 365 mL of aqueous solution. (b) Add enough distilled water to 125 mL of 1.25 mol/L NaOH to make 465 mL of solution. **12.55** (a) Weigh out 48.0 g KBr, dissolve it in about 1 L distilled water, and then dilute to 1.40 L with distilled water. (b) Measure 82.7 mL of the 0.264 mol/L $LiNO_3$ solution and add distilled water to make a total of 255 mL. **12.57** (a) 0.896 mol/kg glycine (b) 1.21 mol/kg glycerol **12.59** 4.48 mol/kg C_6H_6 **12.61** (a) Add 2.39 g $C_2H_6O_2$ to 308 g H_2O. (b) Add 0.0508 kg of 52.0 mass % HNO_3 to 1.15 kg H_2O to make 1.20 kg of 2.20 mass % HNO_3. **12.63** (a) 0.29 (b) 58 mass % (c) 23 mol/kg C_3H_7OH **12.65** 42.6 g CsBr; mole fraction $= 7.16 \times 10^{-3}$; 7.84 mass % **12.67** 5.11 mol/kg NH_3; 4.53 mol/L NH_3; mole fraction $= 0.0843$ **12.69** 2.5 ppm Ca^{2+}; 0.56 ppm Mg^{2+} **12.73** It conducts a large current. A strong electrolyte dissociates completely into ions in solution. **12.75** The boiling point is higher and the freezing point is lower for the solution than for the solvent. **12.78** A dilute solution of an electrolyte behaves more ideally than a concentrated one. With increasing concentration, the effective concentration deviates from the molar concentration because of ionic attractions. Thus, 0.050 mol/kg NaF has a boiling point closer to its predicted value. **12.81** (a) strong electrolyte (b) strong electrolyte (c) nonelectrolyte (d) weak electrolyte **12.83** (a) 0.6 mol of solute particles (b) 0.13 mol (c) 2×10^{-4} mol (d) 0.06 mol **12.85** (a) CH_3OH in H_2O (b) H_2O in CH_3OH solution **12.87** (a) $\Pi_{II} < \Pi_{I} < \Pi_{III}$ (b) $bp_{II} < bp_{I} < bp_{III}$ (c) $fp_{III} < fp_{I} < fp_{II}$ (d) $vp_{III} < vp_{I} < vp_{II}$ **12.89** 3.13 kPa **12.91** -0.467°C **12.93** 79.5°C **12.95** 1.18×10^4 g $C_2H_6O_2$ **12.97** (a) NaCl: 0.173 mol/kg and $i = 1.84$ (b) CH_3COOH: 0.0837 mol/kg and $i = 1.02$ **12.100** 27.8 kPa for CH_2Cl_2; 6.40 kPa for CCl_4 **12.101** The fluid inside a bacterial cell is both a solution and a colloid. It is a solution of ions and small molecules and a colloid of large molecules (proteins and nucleic acids).

12.105 Soap micelles have nonpolar tails pointed inward and anionic heads pointed outward. The like charges on the heads of one micelle repel those on the heads of a neighbouring micelle. This repulsion between micelles keeps them from coagulating. Soap is more effective in fresh water than in seawater because the divalent cations in seawater combine with the anionic heads to form a precipitate. **12.109** 3.4×10^9 L **12.113** 0.0°C: 4.53×10^{-4} mol/L O_2; 20.0°C: 2.83×10^{-4} mol/L O_2; 40.0°C: 2.00×10^{-4} mol/L O_2 **12.115** (a) 89.9 g/mol (b) C_2H_5O; $C_4H_{10}O_2$ (c) Forms H bonds:

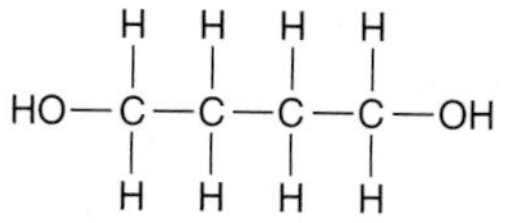

Does not form H bonds:

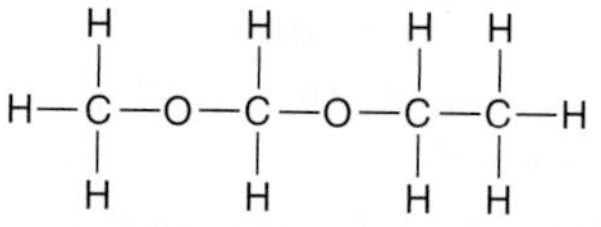

12.119 (a) 9.45 g NaF (b) 0.0017 g F^- **12.122** (a) 68 g/mol (b) 2.1×10^2 g/mol (c) The molar mass of CaN_2O_6 is 164.10 g/mol. This value is less than the 2.1×10^2 g/mol calculated when the compound is assumed to be a strong electrolyte and is greater than the 68 g/mol calculated when the compound is assumed to be a nonelectrolyte. Thus, the compound does not dissociate completely in solution. (d) $i = 2.4$ **12.126** (a) 1.82×10^4 g/mol (b) 3.41×10^{-5}°C **12.127** 8.2×10^5 ng/L **12.131** (a) 0.02 (b) 5×10^{-1} kPa·L/mol (c) yes **12.133** Weigh 3.11 g of $NaHCO_3$ and dissolve in 247 g of water. **12.137** c (mol solute/L solution) = m (kg solvent/L solution) = $m \times d_{solution}$. Thus, for very dilute solutions, molality × density = concentration (mol/L). For an aqueous solution, the volume (L) of solution is approximately the same as the mass (kg) of solvent because the density of water is close to 1 kg/L, so $m = c$. **12.139** (a) 2.5×10^{-3} mol/L SO_2 (b) The base reacts with the sulfur dioxide to produce calcium sulfite. The reaction of sulfur dioxide makes "room" for more sulfur dioxide to dissolve. **12.144** (a) 7.74×10^{-3} mol/(L·kPa) (b) 4×10^{-5} mol/L (c) 3×10^{-6} (d) 1 ppm **12.148** (a) 2.09×10^{-4} mol/(L·kPa) (b) 8.84 ppm (c) k_H: $C_6F_{14} > C_6H_{14} >$ ethanol > water. To dissolve oxygen in a solvent, the solvent molecules must be moved apart to make room for the gas. The stronger the intermolecular forces in the solvent, the more difficult it is to separate solvent particles and the lower the solubility of the gas. Both C_6F_{14} and C_6H_{14} have weak dispersion forces, with C_6F_{14} having the weaker forces due to the electronegative fluorine atoms repelling each other. In both ethanol and water, the molecules are held together by strong hydrogen bonds with those bonds being stronger in water, as the boiling point indicates. **12.150** (a) Yes; the phases of water can still coexist at some temperature and can therefore establish equilibrium. (b) The triple point would occur at a lower pressure and lower temperature because the dissolved air would lower the vapour pressure of the solvent. (c) Yes; this is possible because the gas-solid phase boundary exists below the new triple point. (d) No; at both temperatures, the presence of the solute lowers the vapour pressure of the liquid. **12.152** (a) 2.8×10^{-4} g/mL (b) 81 mL

Chapter 13

13.1 The outermost electron is attracted by a smaller effective nuclear charge in Li because of shielding by the inner electrons, and it is farther from the nucleus in Li than in H. Both of these factors lead to a lower ionization energy for Li. **13.3** (a) NH_3 will form hydrogen bonds.

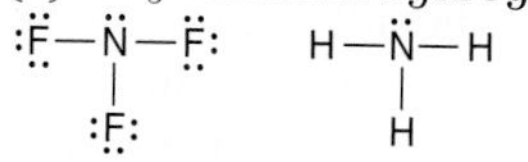

(b) CH_3CH_2OH will form hydrogen bonds.

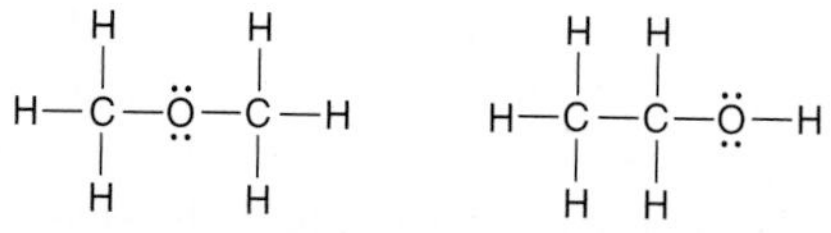

13.5 (a) $2Al(s) + 6HCl(aq) \longrightarrow 2AlCl_3(aq) + 3H_2(g)$
(b) $LiH(s) + H_2O(l) \longrightarrow LiOH(aq) + H_2(g)$
13.7 (a) $NaBH_4$: +1 for Na, +3 for B, −1 for H
$Al(BH_4)_3$: +3 for Al, +3 for B, −1 for H
$LiAlH_4$: +1 for Li, +3 for Al, −1 for H

(b) tetrahedral 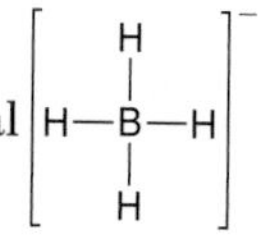

13.12 (a) group 13 or 3 (b) If E is in group 3, the oxide will be more basic and the fluoride will be more ionic; if E is in group 13, the oxide will be more acidic and the fluoride will be more covalent. **13.15** (a) reducing agent (b) Alkali metals have relatively low ionization energies, which means they easily lose the outermost electron. (c) $2Na(s) + 2H_2O(l) \longrightarrow 2Na^+(aq) + 2OH^2(aq) + H_2(g)$ and $2Na(s) + Cl_2(g) \longrightarrow 2NaCl(s)$ **13.17** Density and ionic size increase down a group; the other three properties decrease down a group. **13.19** $2Na(s) + O_2(g) \longrightarrow Na_2O_2(s)$ **13.21** $K_2CO_3(s) + 2HI(aq) \longrightarrow 2KI(aq) + H_2O(l) + CO_2(g)$ **13.25** Group 2 metals have an additional valence electron. The greater number of shared electrons increases the strength of metallic bonding, which leads to higher melting points, higher boiling points, greater hardness, and greater density. **13.26** (a) $CaO(s) + H_2O(l) \longrightarrow Ca(OH)_2(s)$ (b) $2Ca(s) + O_2(g) \longrightarrow 2CaO(s)$ **13.29** (a) $BeO(s) + H_2O(l) \longrightarrow$ no reaction (b) $BeCl_2(l) + 2Cl^-(\text{solvated}) \longrightarrow BeCl_4^{2-}(\text{solvated})$; Be behaves like other group 2 elements in reaction (b). **13.32** The electron removed from group 2 atoms occupies the outer *s* orbital, whereas in group 13 atoms, the electron occupies the outer *p* orbital. For example, the electron configuration for Be is $1s^22s^2$ and for B it is $1s^22s^22p^1$. It is easier to remove the *p* electron of B than an *s* electron of Be, because the energy of a *p* orbital is higher than that of the *s* orbital of the same level. Even though atomic size decreases because of increasing Z_{eff}, IE decreases from 2 to 13. **13.33** (a) Most atoms form stable compounds when they complete their valence shell (octet). Some compounds of group 13 elements, like boron, have only six electrons around the central atom. Having fewer than eight electrons is called *electron deficiency*.

(b) $BF_3(g) + NH_3(g) \longrightarrow F_3B^-NH_3(g)$
$B(OH)_3(aq) + OH^-(aq) \longrightarrow B(OH)_4^-(aq)$
13.35 $In_2O_3 < Ga_2O_3 < Al_2O_3$ **13.37** Apparent O.N., +3; actual O.N., +1; $[:\ddot{I}—\ddot{I}—\ddot{I}:]^-$ The anion I_3^- has the general formula AX_2E_3 and bond angles of 180°. $(Tl^{3+})(I^-)_3$ does not exist because of the low strength of the Tl—I bond. **13.42** (a) $B_2O_3(s) + 2NH_3(g) \longrightarrow 2BN(s) + 3H_2O(g)$ (b) 1.30×10^2 kJ/mol (c) 5.3 kg borax **13.43** Basicity in water is greater for the oxide of a metal. Tin(IV) oxide is more basic in water than carbon dioxide because tin has more metallic character than carbon. **13.45** (a) Ionization energy generally decreases down a group. (b) The deviations (increases) from the expected trend are due to the presence of the first transition series between Si and Ge and of the lanthanides between Sn and Pb. (c) group 13 **13.48** Atomic size increases down a group. As atomic size increases, ionization energy decreases and so it is easier to form a positive ion. An atom that is easier to ionize exhibits greater metallic character.

13.50 (a) 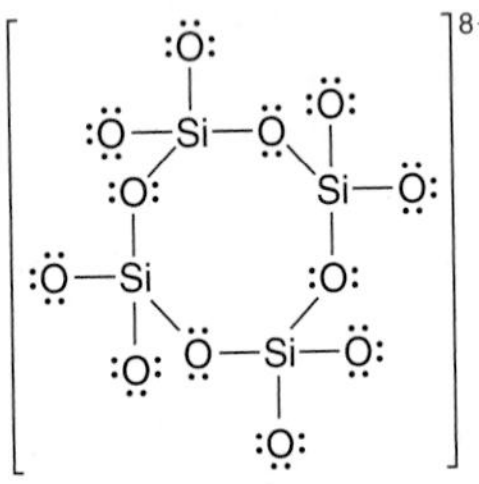 (b) 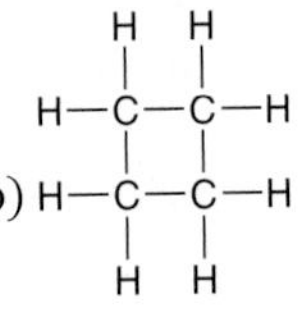

13.53 (a) diamond, C (b) calcium carbonate, $CaCO_3$ (c) carbon dioxide, CO_2 (d) carbon monoxide, CO (e) silicon, Si **13.57** (a) −3 to +5 (b) For a group of nonmetals, the oxidation states range from the lowest, group number − 18, or 15 − 18 = −3 for group 15, to the highest, group number − 10, or 15 − 10 = +5 for group 15. **13.60** $H_3AsO_4 < H_3PO_4 < HNO_3$ **13.62** (a) $4As(s) + 5O_2(g) \longrightarrow 2As_2O_5(s)$ (b) $2Bi(s) + 3F_2(g) \longrightarrow 2BiF_3(s)$ (c) $Ca_3As_2(s) + 6H_2O(l) \longrightarrow 3Ca(OH)_2(s) + 2AsH_3(g)$ **13.64** (a) $N_2(g) + 2Al(s) \xrightarrow{\Delta} 2AlN(s)$ (b) $PF_5(g) + 4H_2O(l) \longrightarrow H_3PO_4(aq) + 5HF(g)$ **13.66** Trigonal bipyramidal, with axial F atoms and equatorial Cl atoms;

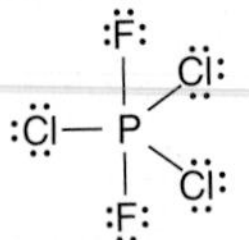

13.70 (a) $\ddot{O}=\ddot{N}—\ddot{N}=\ddot{O}$ (b) $\ddot{O}=\ddot{N}—\ddot{O}=\ddot{N}:$

(c) 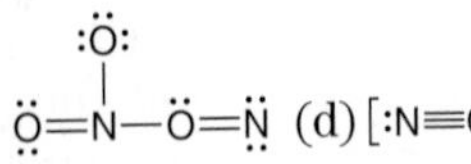 (d) $[:N\equiv O:]^+$ and 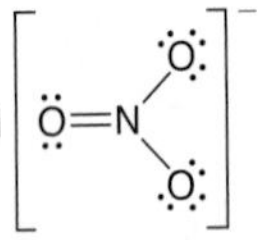

13.72 (a) $2KNO_3(s) \xrightarrow{\Delta} 2KNO_2(s) + O_2(g)$ (b) $4KNO_3(s) \longrightarrow 2K_2O(s) + 2N_2(g) + 5O_2(g)$ **13.74** (a) Boiling point and conductivity vary in similar ways down both groups. (b) Degree of metallic character and types of bonding vary in similar ways down both groups. (c) Both P and S have allotropes, and both bond covalently with almost every other nonmetal. (d) Both nitrogen and oxygen are diatomic gases at normal temperature and pressure.

(e) O_2 is a reactive gas, whereas N_2 is not. Nitrogen can have any of six oxidation states, whereas oxygen has two. **13.76** (a) $NaHSO_4(aq) + NaOH(aq) \longrightarrow Na_2SO_4(aq) + H_2O(l)$ (b) $S_8(s) + 24F_2(g) \longrightarrow 8SF_6(g)$ (c) $FeS(s) + 2HCl(aq) \longrightarrow H_2S(g) + FeCl_2(aq)$ (d) $Te(s) + 2I_2(s) \longrightarrow TeI_4(s)$ **13.78** (a) acidic (b) acidic (c) basic (d) amphoteric (e) basic **13.80** $H_2O < H_2S < H_2Te$ **13.83** (a) O_3, ozone (b) SO_3, sulfur trioxide (c) SO_2, sulfur dioxide (d) H_2SO_4, sulfuric acid (e) $Na_2S_2O_3.5H_2O$, sodium thiosulfate pentahydrate **13.85** $S_2F_{10}(g) \longrightarrow SF_4(g) + SF_6(g)$; O.N. of S in S_2F_{10} is +5; O.N. of S in SF_4 is +4; O.N. of S in SF_6 is +6. **13.87** (a) −1, +1, +3, +5, +7 (b) The electron configuration for Cl is [Ne] $3s^23p^5$. By gaining one electron, Cl achieves an octet. By forming covalent bonds, Cl completes or expands its valence level by maintaining electron pairs in bonds or as lone pairs. (c) Fluorine has only the −1 oxidation state because its small size and absence of *d* orbitals prevent it from forming more than one covalent bond. **13.89** (a) Cl—Cl bond is stronger than Br—Br bond. (b) Br—Br bond is stronger than I—I bond. (c) Cl—Cl bond is stronger than F—F bond. The fluorine atoms are so small that electron-electron repulsion of the lone pairs decreases the strength of the bond. **13.91** (a) $3Br_2(l) + 6OH^-(aq) \longrightarrow 5Br^-(aq) + BrO_3^-(aq) + 3H_2O(l)$ (b) $ClF_5(l) + 6OH^-(aq) \longrightarrow 5F^-(aq) + ClO_3^-(aq) + 3H_2O(l)$ **13.92** (a) $2Rb(s) + Br_2(l) \longrightarrow 2RbBr(s)$ (b) $I_2(s) + H_2O(l) \longrightarrow HI(aq) + HIO(aq)$ (c) $Br_2(l) + 2I^-(aq) \longrightarrow I_2(s) + 2Br^-(aq)$ (d) $CaF_2(s) + H_2SO_4(l) \longrightarrow CaSO_4(s) + 2HF(g)$ **13.94** HIO < HBrO < HClO < $HClO_2$ **13.98** $I_2 < Br_2 < Cl_2$, because Cl_2 is able to oxidize Re to the +6 oxidation state, Br_2 only to +5, and I_2 only to +4. **13.101** Only dispersion forces hold atoms of noble gases together. **13.105** $\Delta_{soln}H = -411$ kJ/mol **13.109** (a) Second ionization energies for alkali metals are so high because the electron being removed is from the next lower energy level and these are very tightly held by the nucleus. Also, the alkali metal would lose its noble-gas electron configuration. (b) $2CsF_2(s) \longrightarrow 2CsF(s) + F_2(g)$; −405 kJ/mol **13.111** (a) hyponitrous acid, $H_2N_2O_2$; nitroxyl, HNO
(b) $H—\ddot{O}—\ddot{N}=\ddot{N}—\ddot{O}—H$ $\quad H—\ddot{N}=\ddot{O}:$
(c) In both species the shape is bent about the N atoms.
(d) 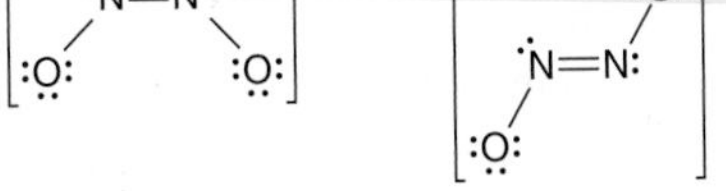

cis *trans*

13.115 13 t **13.117** In a disproportionation reaction, a substance acts as both a reducing agent and an oxidizing agent because atoms of an element within the substance attain both higher and lower oxidation states in the products. The disproportionation reactions are b, c, d, e, and f. **13.119** (a) group 15 (b) group 17 (c) group 16 (d) group 1 (e) group 13 (f) group 18 **13.121** 117.2 kJ/mol

13.126 $[\ddot{O}=\ddot{N}—\ddot{O}:]^- \longleftrightarrow [:\ddot{O}—\ddot{N}=\ddot{O}]^-$
$\ddot{O}=\dot{N}—\ddot{O}: \longleftrightarrow :\ddot{O}—\dot{N}=\ddot{O}$
$[\ddot{O}=N=\ddot{O}]^+$

The nitronium ion (NO_2^+) has a linear shape because the central N atom has two surrounding electron groups, which achieve maximum repulsion at 180°. The nitrite ion (NO_2^-) bond angle is more compressed than the nitrogen dioxide (NO_2) bond angle because the lone pair of electrons takes up more space than the lone electron. **13.128** 2.29×10^4 g UF_6 **13.132** O^+, O^-, and O^{2+} **13.133** (a) 39.96 mass % of As in $CuHAsO_3$; 62.42 mass % of As in $(CH_3)_3As$ (b) 0.35 g $CuHAsO_3$

Chapter 14

Answers to Boxed Reading Problems: B14.1 (a) $k = 2.5 \times 10^{-48}\ s^{-1}$ (b) $k = 5.2 \times 10^8\ s^{-1}$ (c) $E_a = 4 \times 10^5$ J/mol **B14.2** (a) (i) unimolecular (ii) bimolecular (iii) unimolecular (iv) bimolecular (v) bimolecular (b) CH_3, C_2H_5, H

14.2 Reaction rate is proportional to concentration. An increase in pressure will increase the concentration, resulting in an increased rate. **14.3** The addition of water will lower the concentrations of all dissolved solutes, and the rate will decrease. **14.5** An increase in temperature increases the rate by increasing the number of collisions between particles, but, more important, it increases the energy of collisions. **14.8** (a) The instantaneous rate is the rate at one point in time during the reaction. The average rate is the average over a period of time. On a graph of reactant concentration versus time, the instantaneous rate is the slope of the tangent to the curve at any point. The average rate is the slope of the line connecting two points on the curve. The closer together the two points (the shorter the time interval), the closer the average rate is to the instantaneous rate. (b) The initial rate is the instantaneous rate at the point on the graph where $t = 0$, that is, when reactants are mixed.

14.10

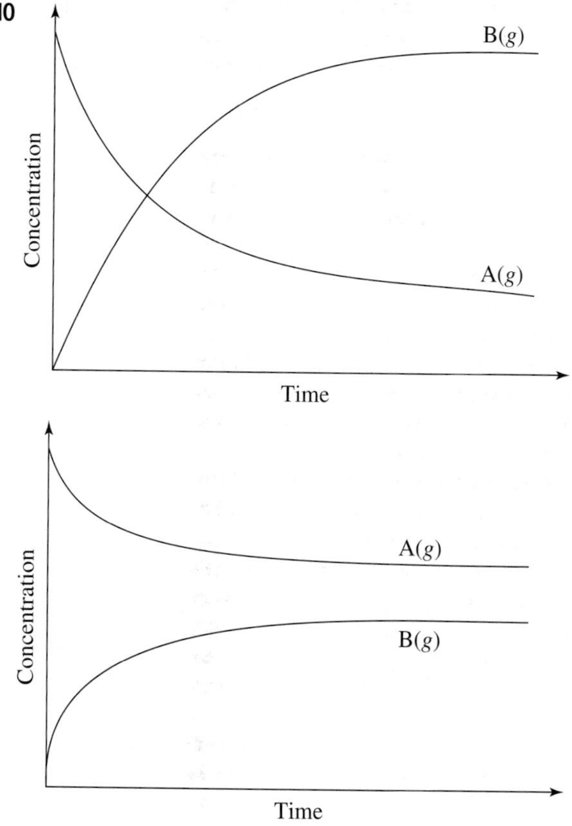

14.12 (a) $\text{Rate} = -\dfrac{1}{2}\dfrac{\Delta[AX_2]}{\Delta t}$

$$= -\frac{1}{2}\frac{(0.0088\ \text{mol/L} - 0.0500\ \text{mol/L})}{(20.0\ \text{s} - 0\ \text{s})}$$

$$= 0.001\,03\ \text{mol/(L·s)}$$

$$= 0.0010\ \text{mol/(L·s)}$$

(b)

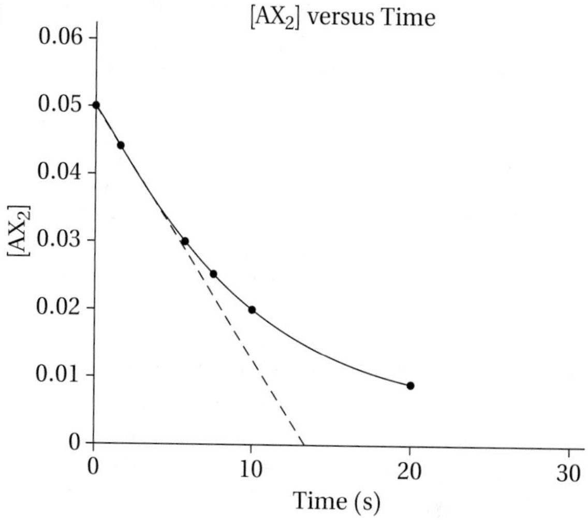

The initial rate is higher than the average rate because the rate decreases as reactant concentration decreases.

14.14 $\text{rate} = -\dfrac{1}{2}\dfrac{\Delta[A]}{\Delta t} = \dfrac{\Delta[B]}{\Delta t} = \dfrac{\Delta[C]}{\Delta t} = -4\ \text{mol/(L·s)}$

14.16 $\text{rate} = -\dfrac{\Delta[A]}{\Delta t} = -\dfrac{1}{2}\dfrac{\Delta[B]}{\Delta t} = \dfrac{\Delta[C]}{\Delta t} = -0.2\ \text{mol/(L·s)}$

14.18 $2N_2O_5(g) \longrightarrow 4NO_2(g) + O_2(g)$

14.21 $\text{rate} = -\dfrac{\Delta[N_2]}{\Delta t} = -\dfrac{1}{3}\dfrac{\Delta[H_2]}{\Delta t} = \dfrac{1}{2}\dfrac{\Delta[NH_3]}{\Delta t}$

14.22 (a) $\text{rate} = -\dfrac{1}{3}\dfrac{\Delta[O_2]}{\Delta t} = \dfrac{1}{2}\dfrac{\Delta[O_3]}{\Delta t}$

(b) 1.45×10^{-5} mol/(L·s)

14.23 (a) k is the rate constant, the proportionality constant in the rate law; it is reaction and temperature specific. (b) m represents the order of the reaction with respect to [A], and n represents the order of the reaction with respect to [B]. The order of a reactant does not necessarily equal its stoichiometric coefficient in the balanced equation. (c) $L^2/(mol^2 \cdot min)$ **14.25** (a) Rate doubles. (b) Rate decreases by a factor of 4. (c) Rate increases by a factor of 9. **14.26** first order in BrO_3^-; first order in Br^-; second order in H^+; fourth order overall **14.28** (a) Rate doubles. (b) Rate is halved. (c) The rate increases by a factor of 16. **14.30** second order in NO_2; first order in Cl_2; third order overall **14.32** (a) Rate increases by a factor of 9. (b) Rate increases by a factor of 8. (c) Rate is halved. **14.34** (a) second order in A; first order in B (b) rate $= k[A]^2[B]$ (c) $5.00 \times 10^3\ L^2/(mol^2 \cdot min)$ **14.36** (a) $time^{-1}$ (b) L/(mol·time) (c) $L^2/(mol^2 \cdot time)$ (d) $L^{3/2}/(mol^{3/2} \cdot time)$ **14.39** (a) first order (b) second

order (c) zero order **14.41** 7 s **14.43** (a) $k = 0.0660\ \text{min}^{-1}$ (b) 21.0 min
14.45 (a)

x-axis (time, s)	$[NH_3]$	*y*-axis (ln $[NH_3]$)
0	4.000 mol/L	1.38629
1.000	3.986 mol/L	1.38279
2.000	3.974 mol/L	1.37977

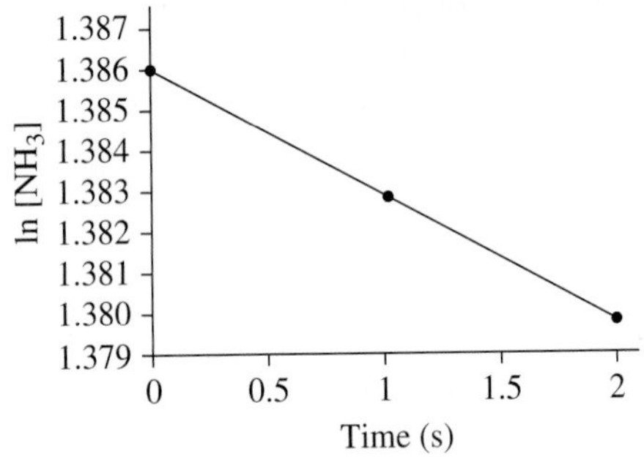

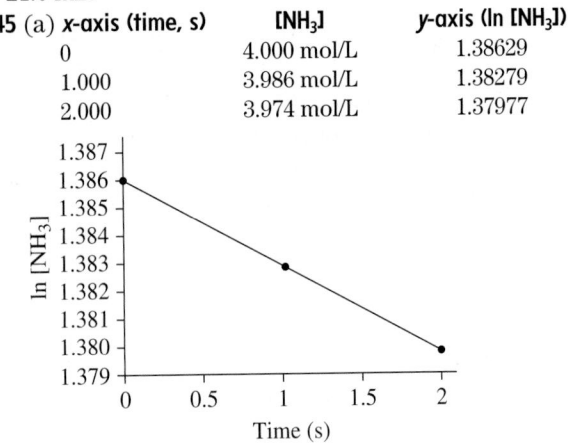

$k = 3 \times 10^{-3}\ \text{s}^{-1}$ (b) $t_{1/2} = 2 \times 10^2$ s
14.47 No; other factors that affect the rate are the energy and orientation of the collisions. **14.50** Measure the rate constant at a series of temperatures and plot ln k versus $1/T$. The slope of the line equals $-E_a/R$. **14.53** No; reaction is reversible and will eventually reach a state where the forward and reverse rates are equal. When this occurs, there are no concentrations equal to zero. Since some reactants are reformed from EF, the amount of EF will be less than 4×10^{-5} mol. **14.54** At the same temperature, both reaction mixtures have the same average kinetic energy, but the reactant molecules do not have the same average velocity. The trimethylamine molecule has greater mass than the ammonia molecule, so trimethylamine molecules will collide less often with HCl. Moreover, the bulky groups bonded to nitrogen in trimethylamine mean that collisions with HCl having the correct orientation occur less frequently. Therefore, the rate of the reaction between NH_3 and HCl is higher. **14.55** 12 unique collisions **14.57** 2.96×10^{-18} **14.59** $0.033\ \text{s}^{-1}$
14.61 (a)

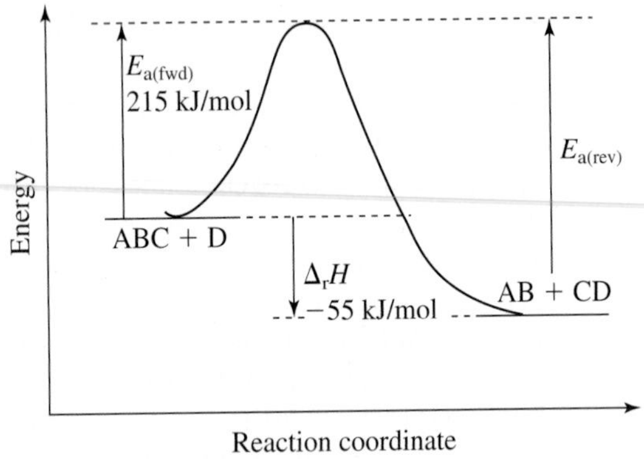

(b) 2.70×10^2 kJ/mol
(c)

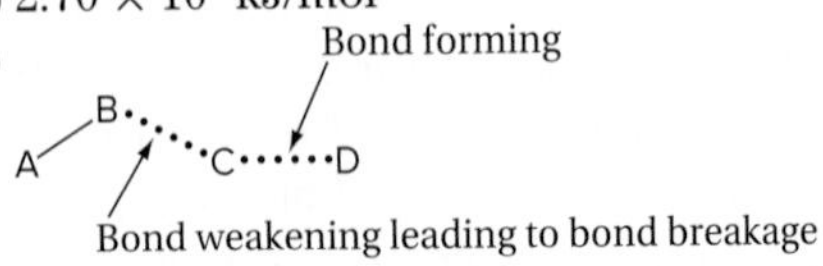

14.64 (a) Because the enthalpy change is positive, the reaction is endothermic.

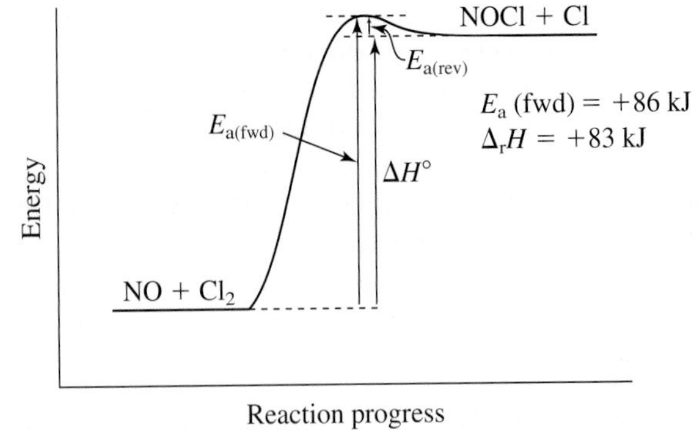

(b) 3 kJ (c)

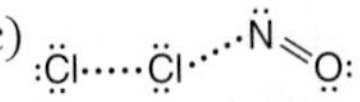

14.65 The rate of an overall reaction depends on the rate of the slowest step. The rate of the overall reaction will be lower than the average of the individual rates because the average includes higher rates as well. **14.69** The probability of three particles colliding with one another with the proper energy and orientation is much less than the probability for two particles. **14.70** No; the overall rate law must contain only reactants (no intermediates), and the overall rate is determined by the slow step.
14.72 (a) $A(g) + B(g) + C(g) \longrightarrow D(g)$
(b) X and Y are intermediates.
(c)

Step	Molecularity	Rate Law
$A(g) + B(g) \longrightarrow X(g)$	bimolecular	$\text{rate}_1 = k_1[A][B]$
$X(g) + C(g) \longrightarrow Y(g)$	bimolecular	$\text{rate}_2 = k_2[X][C]$
$Y(g) \longrightarrow D(g)$	unimolecular	$\text{rate}_3 = k_3[Y]$

(d) yes (e) yes **14.74** The proposed mechanism is valid because the individual steps are chemically reasonable, they add to give the overall equation, and the rate law for the mechanism matches the observed rate law. **14.77** No. A catalyst changes the mechanism of a reaction to one with lower activation energy. Lower activation energy means a faster reaction. An increase in temperature does not influence the activation energy, but increases the fraction of collisions with sufficient energy to equal or exceed the activation energy. **14.78** (a) No. The spark provides energy that is absorbed by the H_2 and O_2 molecules to achieve the activation energy. (b) Yes. The powdered metal acts as a heterogeneous catalyst, providing a surface on which the oxygen-hydrogen reaction can proceed with a lower activation energy.
14.83 (a) Water does not appear as a reactant in the rate-determining step.
(b) Step 1: $\text{rate}_1 = k_1[(CH_3)_3CBr]$
Step 2: $\text{rate}_2 = k_2[(CH_3)_3C^+]$
Step 3: $\text{rate}_3 = k_3[(CH_3)_3COH_2^{\ +}]$
(c) $(CH_3)_3C^+$ and $(CH_3)_3COH_2^{\ +}$
(d) The rate-determining step is step 1. The rate law for this step agrees with the rate law observed with $k = k_1$.
14.85 4.61×10^4 J/mol **14.89** (a) Rate increases 2.5 times. (b) Rate is halved. (c) Rate decreases by a factor of 0.01. (d) Rate does not change. **14.90** second order **14.93** 57 years **14.96** (a) $0.21\ \text{h}^{-1}$; 3.3 h (b) 6.6 h (c) If the concentration of sucrose is relatively low, the concentration of water remains nearly constant even with small changes in the

amount of water. This gives an apparent zero-order reaction with respect to water. Thus, the reaction is first order overall because the rate does not change with changes in the amount of water. **14.99** (a) 0.68 mol/L (b) 0.57 **14.102** 71 kPa **14.106** 7.3×10^3 J/mol **14.108** (a) 2.4×10^{-15} mol/L (b) 2.4×10^{-11} mol/(L•s) **14.111** (a) 2.8 days (b) 7.4 days (c) 4.5 mol/m^3 **14.114** (a) $\text{rate}_1 = 1.7 \times 10^{-5}$ mol/(L•s); $\text{rate}_2 = 3.4 \times 10^{-5}$ mol/(L•s); $\text{rate}_3 = 3.4 \times 10^{-5}$ mol/(L•s) (b) zero order with respect to $S_2O_8^{2-}$; first order with respect to I^- (c) 4.3×10^{-4} s^{-1} (d) rate = $(4.3 \times 10^{-4}\ s^{-1})[I^-]$ **14.117** (a) 7×10^4 cells/L (b) 2.0×10^1 min **14.120** (a) Use the Monod equation.

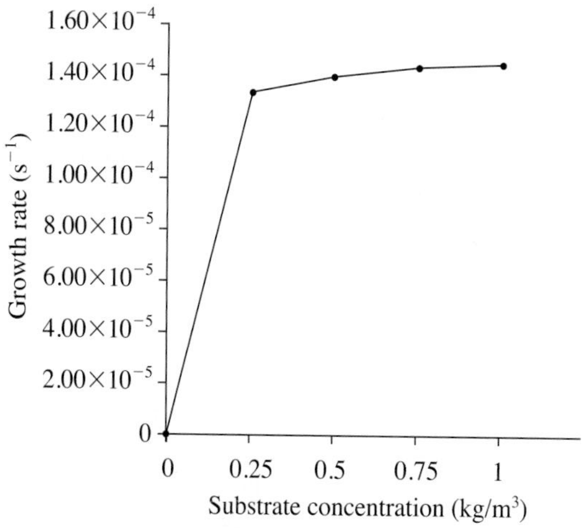

(b) 8.2×10^3 cells/m^3 (c) 8.4×10^3 cells/m^3 **14.125** (a) 8.0×10^1 s (b) 59 s (c) 2.6×10^2 s

Chapter 15

Answers to Boxed Reading Problem: B15.1 (a) enzyme 3 (b) enzyme 6 (c) If F inhibited enzyme 1, then neither branch of the reaction would take place when enough F was produced. (d) If F inhibited enzyme 6, then the second branch would not take place when enough F was made.

15.1 If the change is one of concentrations, it results temporarily in more products and less reactants. After equilibrium is re-established, K_c remains unchanged because the ratio of products and reactants remains the same. **15.7** The equilibrium constant expression is $K = p_{O_2}$. If the temperature remains constant, K remains constant. If the initial amount of Li_2O_2 present is sufficient to reach equilibrium, the amount of O_2 obtained will be constant.

15.8 (a) $Q = \dfrac{p_{HI}^2}{p_{H_2} p_{I_2}}$

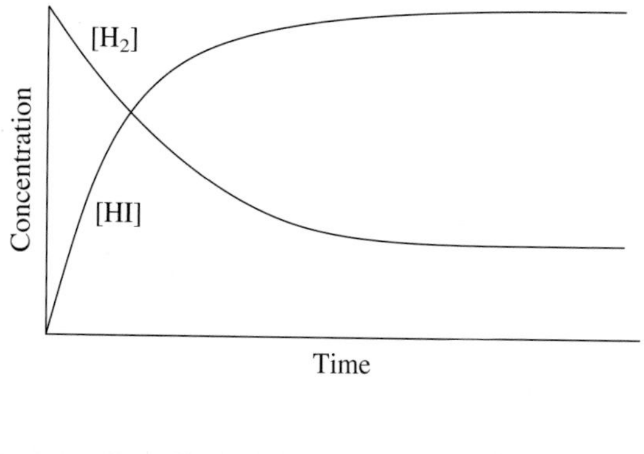

The value of Q increases as a function of time until it reaches the value of K. (b) no **15.11** Yes. If Q_1 is for the formation of 1 mol NH_3 from H_2 and N_2, and Q_2 is for the formation of NH_3 from H_2 and 1 mol of N_2, then $Q_2 = Q_1^2$. **15.12** (a) $4NO(g) + O_2(g) \rightleftharpoons 2N_2O_3(g)$;

$$Q = \frac{p_{N_2O_3}^2}{p_{NO}^4 p_{O_2}}$$

(b) $SF_6(g) + 2SO_3(g) \rightleftharpoons 3SO_2F_2(g)$;

$$Q = \frac{p_{SO_2F_2}^3}{p_{SF_6} p_{SO_3}^2}$$

(c) $2SClF_5(g) + H_2(g) \rightleftharpoons S_2F_{10}(g) + 2HCl(g)$;

$$Q = \frac{p_{S_2F_{10}} p_{HCl}^2}{p_{SClF_5}^2 p_{H_2}}$$

15.14 (a) $2NO_2Cl(g) \rightleftharpoons 2NO_2(g) + Cl_2(g)$;

$$Q = \frac{p_{NO_2}^2 p_{Cl_2}}{p_{NO_2Cl}^2}$$

(b) $2POCl_3(g) \rightleftharpoons 2PCl_3(g) + O_2(g)$;

$$Q = \frac{p_{PCl_3}^2 p_{O_2}}{p_{POCl_3}^2}$$

(c) $4NH_3(g) + 3O_2(g) \rightleftharpoons 2N_2(g) + 6H_2O(g)$;

$$Q = \frac{p_{N_2}^2 p_{H_2O}^6}{p_{NH_3}^4 p_{O_2}^3}$$

15.16 (a) 7.9 (b) 3.2×10^{-5} **15.18** (a) $2Na_2O_2(s) + 2CO_2(g) \rightleftharpoons 2Na_2CO_3(s) + O_2(g)$;

$$Q = \frac{p_{O_2}}{p_{CO_2}^2}$$

(b) $H_2O(l) \rightleftharpoons H_2O(g)$; $Q = p_{H_2O}$
(c) $NH_4Cl(s) \rightleftharpoons NH_3(g) + HCl(g)$; $Q = p_{NH_3} p_{HCl}$
15.20 (a) $2NaHCO_3(s) \rightleftharpoons Na_2CO_3(s) + CO_2(g) + H_2O(g)$;
$Q = p_{CO_2} p_{H_2O}$
(b) $SnO_2(s) + 2H_2(g) \rightleftharpoons Sn(s) + 2H_2O(g)$;

$$Q = \frac{p_{H_2O}^2}{p_{H_2}^2}$$

(c) $H_2SO_4(l) + SO_3(g) \rightleftharpoons H_2S_2O_7(l)$;

$$Q = \frac{1}{p_{SO_3}}$$

15.23 (a)

(1)	$Cl_2(g) + F_2(g) \rightleftharpoons 2ClF(g)$
(2)	$2ClF(g) + 2F_2(g) \rightleftharpoons 2ClF_3(g)$
Overall:	$Cl_2(g) + 3F_2(g) \rightleftharpoons 2ClF_3(g)$

$$Q_1 = \frac{p_{ClF}^2}{p_{Cl_2} p_{F_2}}, \quad Q_2 = \frac{p_{ClF_3}^2}{p_{ClF}^2 p_{F_2}^2}, \quad Q_{overall} = \frac{p_{ClF_3}^2}{p_{Cl_2} p_{F_2}^3}$$

(b) $Q_{overall} = \dfrac{p_{ClF_3}^2}{p_{Cl_2} p_{F_2}^3}$

15.25 K_c and K are equal when $\Delta n_{gas} = 0$. **15.26** (a) smaller (b) Assuming that $RT > 1$ ($T > 12.2$ K), $K > K_c$ if the amount (mol) of gaseous products exceeds the amount (mol) of reactant at equilibrium, and $K < K_c$ if there is a larger amount (mol) of gaseous reactants than gaseous products. **15.27** (a) 3 (b) −1 (c) 3 **15.29** (a) 3.2 (b) 28.5 **15.31** (a) 0.15 (b) 3.5×10^{-7} **15.33** The reaction quotient (Q) and equilibrium constant (K) are determined by the ratio [products]/

[reactants]. When $Q < K$, the reaction proceeds to the right to form more products. **15.35** no; to the left **15.38** At equilibrium, equal concentrations of $CFCl_3$ and HCl exist, regardless of starting reactant concentrations, because the product coefficients are equal. **15.40** (a) The approximation applies when the change in concentration from initial concentration to equilibrium concentration is so small that it is insignificant; this occurs when K is small and initial concentration is large. (b) This approximation should not be used when the change in concentration is greater than 5%. This can occur when $[\text{reactant}]_{\text{initial}}$ is very small or when change in [reactant] is relatively large due to a large K.
15.41 50.8
15.43

Concentration (*M*)	$PCl_5(g)$	$\rightleftharpoons$	$PCl_3(g)$ +	$Cl_2(g)$
Initial	0.075		0	0
Change	$-x$		$+x$	$+x$
Equilibrium	$0.075 - x$		x	x

15.45 28 bar **15.47** 0.33 bar **15.49** 3.5×10^{-3} mol/L **15.51** [HI] = 0.0152 mol/L; $[I_2]$ = 0.00328 mol/L **15.53** $[I_2]_{eq} = [Cl_2]_{eq}$ = 0.0200 mol/L; $[ICl]_{eq}$ = 0.060 mol/L **15.55** 6.01×10^{-6}
15.58 Equilibrium position refers to the specific concentrations or pressures of reactants and products that exist at equilibrium, whereas equilibrium constant is the overall ratio of equilibrium concentrations or pressures. Equilibrium position changes as a result of a change in reactant and product concentrations. **15.59** (a) B, because the amount of product increases with temperature (b) A, because the lowest temperature will give the least product **15.63** (a) shifts toward products (b) shifts toward products (c) does not shift (d) shifts toward reactants **15.65** (a) more F and less F_2 (b) more C_2H_2 and H_2 and less CH_4 **15.67** (a) no effect (b) less H_2 and O_2 and more H_2O **15.69** (a) no change (b) increase volume **15.71** (a) amount decreases (b) amounts increase (c) amounts increase (d) amount decreases
15.73 2.0 **15.76** (a) lower temperature, higher pressure (b) Q decreases; no change in K (c) Reaction rates are lower at lower temperatures, so a catalyst is used to speed up the reaction. **15.78** (a) p_{N_2} = 31 bar; p_{H_2} = 93 bar; p_{total} = 174 bar (b) p_{N_2} = 18 bar; p_{H_2} = 111 bar; p_{total} = 179 bar; not a valid argument **15.81** 0.206 bar **15.84** (a) 3×10^{-3} bar (b) high pressure; low temperature (c) 2×10^5 (d) no, because water condenses at a higher temperature
15.87 (a) 0.016 bar (b) $K_c = 5.6 \times 10^2$; p_{O_2} = 0.17 bar
15.89 12.5 g $CaCO_3$ **15.93** Both concentrations increased by a factor of 2.2. **15.95** (a) 3.0×10^{-14} bar (b) 0.013 pg CO/L
15.97 (a) 98.0% (b) 99.0% (c) 2.60×10^5 J/mol
15.99 (a) $2CH_4(g) + O_2(g) + 2H_2O(g) \rightleftharpoons 2CO_2(g) + 6H_2(g)$ (b) 1.76×10^{29} (c) 3.02×10^{23} (d) 48 bar
15.100 (a) 4.0×10^{-21} bar (b) 5.5×10^{-8} bar (c) 29 N atoms/L; 4.0×10^{14} H atoms/L (d) The more reasonable step is $N_2(g) + H(g) \longrightarrow NH(g) + N(g)$. With only 29 N atoms in 1.0 L, the first reaction would produce virtually no $NH(g)$ molecules. There are orders of magnitude more N_2 molecules than N atoms, so the second reaction is the more reasonable step. **15.103** (a) p_{N_2} = 0.780 bar; p_{O_2} = 0.210 bar; $p_{NO} = 2.67 \times 10^{-16}$ bar (b) 0.990 bar (c) $K_c = K = 4.35 \times 10^{-31}$
15.105 (a) 1.26×10^{-3} (b) 794 (c) 10.6 kJ/mol (d) 1.2×10^4 J/mol **15.109** (a) 1.52 (b) 0.975 bar (c) 0.2000 mol CO (d) 0.010 93 mol/L.

Chapter 16

16.2 (a) All Arrhenius acids have H in their formula and produce hydronium ion (H_3O^+) in aqueous solution. All Arrhenius bases have OH in their formula and produce hydroxide ion (OH^-) in aqueous solution. (b) Neutralization occurs when each H_3O^+ ion combines with an OH^- ion to form two molecules of H_2O. Chemists found the reaction of any strong base with any strong acid always had a ΔH of −56 kJ/mol H_2O produced. **16.4** (a) The Brønsted-Lowry theory defines acids as proton donors and bases as proton acceptors, while the Arrhenius definition looks at acids as containing ionizable hydrogen atoms and at bases as containing hydroxide ions. In both definitions, an acid produces H^+ (H_3O^+) ions and a base produces OH^- ions when added to water. (b) Ammonia and carbonate ion are two Brønsted-Lowry bases that are not Arrhenius bases because they do not contain OH^- ions. Brønsted-Lowry acids must contain an ionizable hydrogen atom in order to be proton donors, so a Brønsted-Lowry acid is also an Arrhenius acid. **16.7** An amphiprotic species is one that can lose a proton to act as an acid or gain a proton to act as a base. The dihydrogen phosphate ion, $H_2PO_4^-$, is an example: $H_2PO_4^-(aq) + OH^-(aq) \longrightarrow H_2O(aq) + HPO_4^{2-}(aq)$. In the presence of a strong acid (HCl), the dihydrogen phosphate ion acts like a base by accepting hydrogen: $H_2PO_4^-(aq) + HCl(aq) \longrightarrow H_3PO_4(aq) + Cl^-(aq)$.
16.8 (a) Strong acids and bases dissociate completely into ions when dissolved in water. Weak acids and bases dissociate only partially. (b) The characteristic property of all weak acids and bases is that a great majority of the molecules are undissociated. **16.9** a, c, and d **16.11** b and d

16.13 (a) $K_a = \dfrac{[H_3O^+][CN^-]}{[HCN]}$

(b) $K_a = \dfrac{[H_3O^+][CO_3^{2-}]}{[HCO_3^-]}$

(c) $K_a = \dfrac{[H_3O^+][HCOO^-]}{[HCOOH]}$

16.15 (a) $K_a = \dfrac{[H_3O^+][NO_2^-]}{[HNO_2]}$

(b) $K_a = \dfrac{[H_3O^+][CH_3COO^-]}{[CH_3COOH]}$

(c) $K_a = \dfrac{[H_3O^+][BrO_2^-]}{[HBrO_2]}$

16.17 (a) $H_3PO_4(aq) + H_2O(l) \rightleftharpoons H_2PO_4^-(aq) + H_3O^+(aq)$; $K_a = \dfrac{[H_3O^+][H_2PO_4^-]}{[H_3PO_4]}$

(b) $C_6H_5COOH(aq) + H_2O(l) \rightleftharpoons C_6H_5COO^-(aq) + H_3O^+(aq)$; $K_a = \dfrac{[H_3O^+][C_6H_5COO^-]}{[C_6H_5COOH]}$

(c) $HSO_4^-(aq) + H_2O(l) \rightleftharpoons SO_4^{2-}(aq) + H_3O^+(aq)$; $K_a = \dfrac{[H_3O^+][SO_4^{2-}]}{[HSO_4^-]}$

16.19 (a) Cl^- (b) HCO_3^- (c) OH^- **16.21** (a) NH_4^+ (b) NH_3 (c) $C_{10}H_{14}N_2H^+$

16.23 (a) $HCl + H_2O \rightleftharpoons Cl^- + H_3O^+$
acid base conjugate base conjugate acid
Conjugate acid-base pairs: HCl/Cl^- and H_3O^+/H_2O
(b) $HClO_4 + H_2SO_4 \rightleftharpoons ClO_4^- + H_3SO_4^+$
acid base conjugate base conjugate acid
Conjugate acid-base pairs: $HClO_4/ClO_4^-$ and $H_3SO_4^+/H_2SO_4$
(c) $HPO_4^{2-} + H_2SO_4 \rightleftharpoons H_2PO_4^- + HSO_4^-$
base acid conjugate acid conjugate base
Conjugate acid-base pairs: H_2SO_4/HSO_4^- and $H_2PO_4^-/HPO_4^{2-}$
16.25 (a) $NH_3 + H_3PO_4 \rightleftharpoons NH_4^+ + H_2PO_4^-$
base acid conjugate acid conjugate base
Conjugate acid-base pairs: $H_3PO_4/H_2PO_4^-$; NH_4^+/NH_3
(b) $CH_3O^- + NH_3 \rightleftharpoons CH_3OH + NH_2^-$
base acid conjugate acid conjugate base
Conjugate acid-base pairs: NH_3/NH_2^-; CH_3OH/CH_3O^-
(c) $HPO_4^{2-} + HSO_4^- \rightleftharpoons H_2PO_4^- + SO_4^{2-}$
base acid conjugate acid conjugate base
Conjugate acid-base pairs: HSO_4^-/SO_4^{2-}; $H_2PO_4^-/HPO_4^{2-}$
16.27 (a) $OH^-(aq) + H_2PO_4^-(aq) \rightleftharpoons H_2O(l) + HPO_4^{2-}(aq)$
Conjugate acid-base pairs: $H_2PO_4^-/HPO_4^{2-}$ and H_2O/OH^-
(b) $HSO_4^-(aq) + CO_3^{2-}(aq) \rightleftharpoons SO_4^{2-}(aq) + HCO_3^-(aq)$
Conjugate acid-base pairs: HSO_4^-/SO_4^{2-}; HCO_3^-/CO_3^{2-}
16.29 $K > 1$; $HS^- + HCl \rightleftharpoons H_2S + Cl^-$
$K > 1$; $H_2S + Cl^- \rightleftharpoons HS^- + HCl$
16.31 $K > 1$ for both a and b. **16.33** $K_c < 1$ for both a and b. **16.35** $CH_3COOH < HF < HIO_3 < HI$ **16.37** (a) weak acid (b) strong base (c) weak acid (d) strong acid **16.39** (a) strong base (b) strong acid (c) weak acid (d) weak acid **16.44** (a) The acid with the smaller K_a (4×10^{-5}) has the higher pH, because less dissociation yields fewer hydronium ions. (b) The acid with the larger pK_a (3.5) has the higher pH, because it has a smaller K_a and, thus, lower $[H_3O^+]$. (c) Lower concentration (0.01 mol/L) contains fewer hydrogen (hydronium) ions. (d) A 0.1 mol/L weak acid solution contains fewer hydronium ions. (e) The 0.01 mol/L base solution contains more hydroxide ions, so fewer hydronium ions. (f) The solution that has pOH = 6.0 has the higher pH: pH = 14.0 − 6.0 = 8.0. **16.45** (a) 12.05; basic (b) 11.13; acidic **16.47** (a) pH = 2.212; acidic (b) pOH = 0.708; basic
16.49 (a) $[H^+] = 1.4 \times 10^{-10}$ mol/L, $[OH^-] = 7.1 \times 10^{-5}$ mol/L, pOH = 4.15
(b) $[H^+] = 2.7 \times 10^{-5}$ mol/L, $[OH^-] = 3.7 \times 10^{-10}$ mol/L, pH = 4.57
16.51 (a) $[H^+] = 1.7 \times 10^{-5}$ mol/L, $[OH^-] = 5.9 \times 10^{-10}$ mol/L, pOH = 9.23
(b) $[H^+] = 4.5 \times 10^{-9}$ mol/L, $[OH^-] = 2.2 \times 10^{-6}$ mol/L; pH = 8.35
16.53 4.8×10^{-4} mol OH^-/L **16.55** 1.4×10^{-4} mol OH^-
16.58 (a) Rising temperature increases the value of K_w. (b) $K_w = 2.5 \times 10^{-14}$; pH = pOH = 6.80; $[H^+] = [OH^-] = 1.6 \times 10^{-7}$mol/ **16.70** (a) A strong acid is 100% dissociated, so the acid concentration will be very different after dissociation. (b) A weak acid dissociates to a very small extent, so the acid concentration before and after dissociation is nearly the same. (c) same as b, but with the extent of dissociation greater (d) same as a **16.71** No. HCl is a strong acid and dissociates to a greater extent than the weak acid CH_3COOH. The K_a of the acid, not the concentration of H_3O^+, determines the strength of the acid. **16.74** 1.5×10^{-5} **16.76** 11.79 **16.78** 11.45 **16.80** $[H^+] = [NO_2^-] = 2.1 \times 10^{-2}$ mol/L; $[OH^-] = 4.8 \times 10^{-13}$ mol/L **16.82** $[H^+] = [ClCH_2COO^-] = 0.041$ mol/L; $[ClCH_2COOH] = 1.21$ mol/L; pH = 1.39 **16.84** (a) 11.19 (b) 5.56 **16.86** (a) 8.78 (b) 4.66 **16.88** (a) $[H^+] = 6.0 \times 10^{-3}$ mol/L; pH = 2.22; $[OH^-] = 1.7 \times 10^{-12}$ mol/L; pOH = 11.78 (b) 1.9×10^{-4} **16.90** 2.2×10^{-7} **16.92** (a) 2.37 (b) 11.53 **16.94** (a) 2.290 (b) 12.699 **16.96** 1.1% **16.98** $[H^+] = [HS^-] = 9 \times 10^{-5}$ mol/L; pH = 4.0; $[OH^-] = 1 \times 10^{-10}$ mol/L; pOH = 10.0; $[H_2S] = 0.10$ mol/L; $[S^{2-}] = 1 \times 10^{-17}$ mol/L **16.101** 1.5% **16.102** $[OH^-] = 5.5 \times 10^{-4}$ mol/L; pH = 10.74 **16.104** As a nonmetal becomes more electronegative, the acidity of its binary hydride increases. The electronegative nonmetal attracts the electrons more strongly in the polar bond, shifting the electron density away from H^+, thus making H^+ more easily transferred to a water molecule to form H_3O^+. **16.107** Chlorine is more electronegative than iodine, and $HClO_4$ has more oxygen atoms than HIO. **16.108** (a) H_2SeO_4 (b) H_3PO_4 (c) H_2Te **16.110** (a) H_2Se (b) $B(OH)_3$ (c) $HBrO_2$ **16.112** (a) 0.5 mol/L $AlBr_3$ (b) 0.3 mol/L $SnCl_2$ **16.114** (a) 0.2 mol/L $Ni(NO_3)_2$ (b) 0.35 mol/L $Al(NO_3)_3$ **16.117** NaF contains the anion of the weak acid HF, so F^- acts as a base. NaCl contains the anion of the strong acid HCl.
16.119 (a) $KBr(s) \xrightarrow{H_2O} K^+(aq) + Br^-(aq)$; neutral
(b) $NH_4I(s) \xrightarrow{H_2O} NH_4^+(aq) + I^-(aq)$; acidic
(c) $KCN(s) \xrightarrow{H_2O} K^+(aq) + CN^-(aq)$; basic
16.121 (a) $Na_2CO_3(s) \xrightarrow{H_2O} 2Na^+(aq) + CO_3^{2-}(aq)$; basic
(b) $CaCl_2(s) \xrightarrow{H_2O} Ca^{2+}(aq) + 2Cl^-(aq)$; neutral
(c) $Cu(NO_3)_2(s) \xrightarrow{H_2O} Cu^{2+}(aq) + 2NO_3^-(aq)$; acidic
16.123 (a) A solution of strontium bromide is neutral because Sr^{2+} is the conjugate acid of a strong base, $Sr(OH)_2$, and Br^- is the conjugate base of a strong acid, HBr, so neither changes the pH of the solution. (b) A solution of barium acetate is basic because CH_3COO^- is the conjugate base of a weak acid and therefore forms OH^- in solution, whereas Ba^{2+} is the conjugate acid of a strong base, $Ba(OH)_2$, and does not influence solution pH. The base-dissociation reaction of ethanoate (acetate) ion is $CH_3COO^-(aq) + H_2O(l) \rightleftharpoons CH_3COOH(aq) + OH^-(aq)$. (c) A solution of dimethylammonium bromide is acidic because $(CH_3)_2NH_2^+$ is the conjugate acid of a weak base and therefore forms H_3O^+ in solution, whereas Br^- is the conjugate base of a strong acid and does not influence the pH of the solution. The acid-dissociation reaction for methylammonium ion is $(CH_3)_2NH_2^+(aq) + H_2O(l) \rightleftharpoons (CH_3)_2NH(aq) + H_3O^+(aq)$.
16.125 (a) $NH_4^+(aq) + H_2O(l) \rightleftharpoons NH_3(aq) + H_3O^+(aq)$
$PO_4^{3-}(aq) + H_2O(l) \rightleftharpoons HPO_4^{2-}(aq) + OH^-(aq)$
$K_b > K_a$; basic
(b) $SO_4^{2-}(aq) + H_2O(l) \rightleftharpoons HSO_4^-(aq) + OH^-(aq)$
Na^+ gives no reaction; basic
(c) $ClO^-(aq) + H_2O(l) \rightleftharpoons HClO(aq) + OH^-(aq)$
Li^+ gives no reaction; basic

16.127 (a) $Fe(NO_3)_2 < KNO_3 < K_2SO_3 < K_2S$
(b) $NaHSO_4 < NH_4NO_3 < NaHCO_3 < Na_2CO_3$
16.129 Since both bases produce OH^- ions in water, both bases appear equally strong. $CH_3O^-(aq) + H_2O(l) \longrightarrow OH^-(aq) + CH_3OH(aq)$ and $NH_2^-(aq) + H_2O(l) \longrightarrow OH^-(aq) + NH_3(aq)$ **16.131** Ammonia, NH_3, is a more basic solvent than H_2O. In a more basic solvent, weak acids such as HF act like strong acids and are 100% dissociated. **16.133** A Lewis acid is an electron-pair acceptor, while a Brønsted-Lowry acid is a proton donor. The proton of a Brønsted-Lowry acid is a Lewis acid because it accepts an electron pair when it bonds with a base. All Lewis acids are not Brønsted-Lowry acids. A Lewis base is an electron-pair donor and a Brønsted-Lowry base is a proton acceptor. All Brønsted-Lowry bases are Lewis bases, and vice versa. **16.134** (a) No; for example: $Ni^{2+}(aq) + 4H_2O(l) \rightleftharpoons Ni(H_2O)_4^{2+}(aq)$; water is a very weak Brønsted-Lowry base but forms the Zn complex fairly well and is a reasonably strong Lewis base. (b) cyanide ion and water (c) cyanide ion
16.137 (a) Lewis acid (b) Lewis base (c) Lewis acid (d) Lewis base **16.139** (a) Lewis acid (b) Lewis base (c) Lewis base (d) Lewis acid **16.141** (a) Lewis acid: Na^+; Lewis base: H_2O (b) Lewis acid: CO_2; Lewis base: H_2O (c) Lewis acid: BF_3; Lewis base: F^- **16.143** (a) Lewis (b) Brønsted-Lowry and Lewis (c) none (d) Lewis
16.146 3.5×10^{-8} to 4.5×10^{-8} mol/L H^+; 5.2×10^{-7} to 6.6×10^{-7} mol/L OH^- **16.147** (a) Acids vary in the extent of dissociation depending on the acid-base character of the solvent. (b) Methanol is a weaker base than water since phenol dissociates less in methanol than in water.
(c) $C_6H_5OH(solvated) + CH_3OH(l) \rightleftharpoons CH_3OH_2^+(solvated) + C_6H_5O^-(solvated)$ (d) $CH_3OH(l) + CH_3OH(l) \rightleftharpoons CH_3O^-(solvated) + CH_3OH_2^+(solvated)$ **16.150** (a) $SnCl_4$ is the Lewis acid; $(CH_3)_3N$ is the Lewis base (b) 5*d* **16.151** pH = 5.00, 6.00, 6.79, 6.98, 7.00 **16.157** 3×10^{18} **16.160** 10.48
16.162 2.41 **16.164** amylase, 2×10^{-7} mol/L; pepsin, 1×10^{-2} mol/L; trypsin, 3×10^{-10} mol/L **16.168** 1.47×10^{-3} **16.170** (a) Ca^{2+} does not react with water; $CH_3CH_2COO^-(aq) + H_2O(l) \rightleftharpoons CH_3CH_2COOH(aq) + OH^-(aq)$; basic (b) 9.08
16.176 4.5×10^{-5} **16.179** (a) The concentration of oxygen is higher in the lungs, so the equilibrium shifts to the right. (b) In an oxygen-deficient environment, the equilibrium shifts to the left to release oxygen. (c) A decrease in $[H_3O^+]$ shifts the equilibrium to the right. More oxygen is absorbed, but it will be more difficult to remove the O_2. (d) An increase in $[H_3O^+]$ shifts the equilibrium to the left. Less oxygen is bound to Hb, but it will be easier to remove it. **16.181** (a) 1.012 mol/L (b) 0.004 mol/L (c) 0.4%
16.182 (a) 10.0 (b) The pK_b for the 3° amine group is much smaller than that for the aromatic ring, so the K_b is significantly larger (yielding a much greater amount of OH^-). (c) 4.7 (d) 5.1

Chapter 17

Answers to Boxed Reading Problems: B17.1 3.5 to 4 **B17.2** (a) 64 mol (b) 6.28 (c) 3.9×10^3 g HCO_3^-

17.2 The acid component neutralizes added base, and the base component neutralizes added acid, so the pH of the buffer solution remains relatively constant. The components of a buffer do not neutralize one another because they are a conjugate acid-base pair. **17.7** The buffer range, the pH over which the buffer acts effectively, is greatest when the buffer-component concentration ratio is 1; the range decreases as this ratio deviates from 1. **17.9** (a) Ratio and pH increase; added OH^- reacts with HA. (b) Ratio and pH decrease; added H^+ reacts with A^-. (c) Ratio and pH increase; added A^- increases $[A^-]$. (d) Ratio and pH decrease; added HA increases [HA]. **17.11** $[H_3O^+] = 5.6 \times 10^{-6}$ mol/L; pH = 5.25
17.13 $[H_3O^+] = 5.2 \times 10^{-4}$ mol/L; pH = 3.28 **17.15** 3.89
17.17 10.03 **17.19** 9.47 **17.21** (a) K_{a_2} (b) 10.55 **17.23** 3.6
17.25 0.20 **17.27** 3.37 **17.29** 8.82 **17.31** (a) 4.81 (b) 0.66 g KOH
17.33 (a) $HOOC(CH_2)_4COOH/HOOC(CH_2)_4COO^-$ or $C_6H_5NH_3^+/C_6H_5NH_2$ (b) $H_2PO_4^-/HPO_4^{2-}$ or $H_2AsO_4^-/HAsO_4^{2-}$
17.35 (a) $HOCH_2CH(OH)COOH/HOCH_2CH(OH)COO^-$ or $CH_3COOC_6H_4COOH/CH_3COOC_6H_4COO^-$ (b) $C_5H_5NH^+/C_5H_5N$ **17.38** 1.6 **17.42** To see a distinct colour in a mixture of two colours, you need one to have about 10 times the intensity of the other. For this to be the case, the concentration ratio $[HIn]/[In^-]$ has to be greater than 10/1 or less than 1/10. This occurs when $pH = pK_a - 1$ or $pH = pK_a + 1$, respectively, giving a pH range of about 2 units. **17.44** The equivalence point in a titration is the point at which the amount (mol) of base is stoichiometrically equivalent to the amount (mol) of acid. The endpoint is the point at which the added indicator changes colour. If an appropriate indicator is selected, the endpoint is close to the equivalence point, but they are not usually the same. The pH at the endpoint, or colour change, may precede or follow the pH at the equivalence point, depending on the indicator chosen. **17.46** (a) initial pH: *strong acid*–strong base < *weak acid*–strong base < *weak base*–strong acid (b) pH at equivalence point: *weak base*–strong acid < *strong acid*–strong base < *weak acid*–strong base **17.48** At the centre of the buffer region, the concentrations of weak acid and conjugate base are equal, so the $pH = pK_a$ of the acid. **17.52** pH range from 7.5 to 9.5 **17.54** (a) bromthymol blue (b) thymol blue or phenolphthalein **17.56** (a) methyl red (b) bromthymol blue **17.58** (a) 1.0000 (b) 1.6368 (c) 2.898 (d) 3.903 (e) 7.00 (f) 10.10 (g) 12.05 **17.60** (a) 2.91 (b) 4.81 (c) 5.29 (d) 6.09 (e) 7.41 (f) 8.76 (g) 10.10 (h) 12.05
17.62 (a) 59.0 mL and 8.54 (b) 66.0 mL and 7.13; total 132.1 mL and 9.69 **17.64** (a) 123 mL and 5.17 (b) 194 mL and 5.80 **17.72** Fluoride ion is the conjugate base of a weak acid and reacts with H_2O: $F^-(aq) + H_2O(l) \rightleftharpoons HF(aq) + OH^-(aq)$. As the pH increases, the equilibrium shifts to the left and $[F^-]$ increases. As the pH decreases, the equilibrium shifts to the right and $[F^-]$ decreases. The changes in $[F^-]$ influence the solubility of BaF_2. Chloride ion is the conjugate base of a strong acid, so it does not react with water and its concentration is not influenced by pH.
17.74 The compound precipitates. **17.75** (a) $K_{sp} = [Ag^+]^2[CO_3^{2-}]$ (b) $K_{sp} = [Ba^{2+}][F^-]^2$ (c) $K_{sp} = [Cu^{2+}][HS^-][OH^-]$
17.77 (a) $K_{sp} = [Ca^{2+}][CrO_4^{2-}]$ (b) $K_{sp} = [Ag^+][CN^-]$ (c) $K_{sp} = [Ni^{2+}][HS^-][OH^-]$ **17.79** 1.3×10^{-4} **17.81** 2.8×10^{-11}
17.83 (a) 2.3×10^{-5} mol/L (b) 4.2×10^{-9} mol/L **17.85** (a) 1.7×10^{-3} mol/L (b) 2.0×10^{-4} mol/L **17.87** (a) $Mg(OH)_2$ (b) PbS (c) Ag_2SO_4 **17.89** (a) $CaSO_4$ (b) $Mg_3(PO_4)_2$ (c) $PbSO_4$

17.91 (a) $AgCl(s) \rightleftharpoons Ag^+(aq) + Cl^-(aq)$; the chloride ion is the anion of a strong acid, so it does not react with H_3O^+. No change with pH. (b) $SrCO_3(s) \rightleftharpoons Sr^{2+}(aq) + CO_3^{2-}$. The strontium ion is the cation of a strong base, so pH will not affect its solubility. The carbonate ion acts as a base: $CO_3^{2-}(aq) + H_2O(l) \rightleftharpoons HCO_3^-(aq) + OH^-(aq)$; also $CO_2(g)$ forms and escapes: $CO_3^{2-}(aq) + 2H_3O^+(aq) \longrightarrow CO_2(g) + 3H_2O(l)$. Therefore, the solubility of $SrCO_3$ will increase with addition of H_3O^+ (decreasing pH). **17.93** (a) $Fe(OH)_2(s) \rightleftharpoons Fe^2(aq) + 2OH^-(aq)$. The OH^- ion reacts with added H_3O^+: $OH^-(aq) + H_3O^+(aq) \longrightarrow 2H_2O(l)$. The added H_3O^+ consumes the OH^-, driving the equilibrium toward the right to dissolve more $Fe(OH)_2$. Solubility increases with addition of H_3O^+ (decreasing pH). (b) $CuS(s) + H_2O(l) \rightleftharpoons Cu^{2+}(aq) + HS^-(aq) + OH^-(aq)$. Both HS^- and OH^- are anions of weak acids, so both ions react with added H_3O^+. Solubility increases with addition of H_3O^+ (decreasing pH). **17.95** yes **17.97** yes **17.100** (a) $Fe(OH)_3$ (b) The two metal ions are separated by adding just enough NaOH to precipitate iron(III) hydroxide. (c) 2.0×10^{-7} mol/L **17.103** No, because it indicates that a complex ion forms between the lead ion and hydroxide ions:

$Pb^{2+}(aq) + nOH^-(aq) \rightleftharpoons Pb(OH)_n^{2-n}(aq)$

17.104 $Hg(H_2O)_4^{2+}(aq) + 4CN^-(aq) \rightleftharpoons Hg(CN)_4^{2-}(aq) + 4H_2O(l)$

17.106 $Ag(H_2O)_2^+(aq) + 2S_2O_3^{2-}(aq) \rightleftharpoons Ag(S_2O_3)_2^{3-}(aq) + 2H_2O(l)$

17.108 8.1×10^{-15} mol/L **17.110** 0.05 mol/L **17.112** 1.0×10^{-16} mol/L Zn^{2+}; 0.025 mol/L $Zn(CN)_4^{2-}$; 0.049 mol/L CN^- **17.114** 0.035 L of 2.00 mol/L NaOH and 0.465 L of 0.200 mol/L HCOOH **17.116** (a) 0.99 (b) assuming the volumes are additive: 0.468 L of 1.0 mol/L HCOOH and 0.232 L of 1.0 mol/L NaOH (c) 0.34 mol/L **17.119** 1.3×10^{-4} mol/L **17.122** (a) 14 (b) 1 g from the second beaker **17.125** (a) 0.088 (b) 0.14 **17.126** 0.260 mol/L TRIS; pH = 8.53 **17.128** Lower the pH below 6.6. **17.132** 8×10^{-5} **17.134** (a)

V (mL)	pH	ΔpH/ΔV	V_{avg} (mL)
0.00	1.00		
10.00	1.22	0.022	5.00
20.00	1.48	0.026	15.00
30.00	1.85	0.037	25.00
35.00	2.18	0.066	32.50
39.00	2.89	0.18	37.00
39.50	3.20	0.62	39.25
39.75	3.50	1.2	39.63
39.90	3.90	2.67	39.83
39.95	4.20	6	39.93
39.99	4.90	18	39.97
40.00	7.00	200	40.00
40.01	9.40	200	40.01
40.05	9.80	10	40.03
40.10	10.40	10	40.08
40.25	10.50	0.67	40.18
40.50	10.79	1.2	40.38
41.00	11.09	0.60	40.75
45.00	11.76	0.17	43.00
50.00	12.05	0.058	47.50
60.00	12.30	0.025	55.00
70.00	12.43	0.013	65.00
80.00	12.52	0.009	75.00

(b)

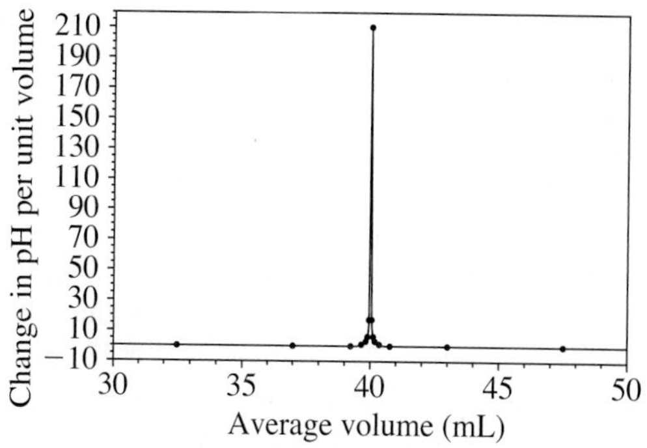

Maximum slope (equivalence point) is at $V_{avg} = 40.00$ mL. **17.136** 4.05 **17.142** H_2CO_3/HCO_3^- and $H_2PO_4^-/HPO_4^{2-}$; $[HPO_4^{2-}]/[H_2PO_4^-] = 5.8$ **17.143** 3.8 **17.145** (a) 58.2 mL (b) 7.85 mL (c) 6.30 **17.147** 170 mL **17.151** 5.68 **17.153** 3.9×10^{-9} μg Pb^{2+}/100 mL blood **17.155** No NaCl will precipitate. **17.156** (a) A and D (b) $pH_A = 4.35$; $pH_B = 8.67$; $pH_C = 2.67$; $pH_D = 4.57$ (c) C, A, D, B (d) B

Chapter 18

Answer to Boxed Reading Problems: B18.1 (a) 1.203×10^{23} molecules ATP/g glucose (b) 3.073×10^{23} molecules ATP/g tristearin **B18.2** −12.6 kJ/mol

18.2 (a) A spontaneous process occurs by itself, whereas a nonspontaneous process requires a continuous input of energy to make it happen. (b) It is possible to cause a nonspontaneous process to occur, but the process stops once the energy source is removed. A reaction that is nonspontaneous under one set of conditions may be spontaneous under a different set of conditions. **18.5** The transition from liquid to gas involves a greater increase in dispersal of energy and freedom of motion than does the transition from solid to liquid. **18.6** (a) (i) In an exothermic reaction, $\Delta_{surr}S > 0$. (ii) In an endothermic reaction, $\Delta_{surr}S < 0$. (b) A chemical cold pack for injuries is an example of an application using a spontaneous endothermic process. **18.8** a, b, and c **18.10** a and b **18.12** (a) positive (b) negative (c) negative **18.14** (a) positive (b) positive (c) positive **18.16** (a) negative (b) negative (c) positive **18.18** (a) positive (b) negative (c) positive **18.20** (a) positive (b) negative (c) positive **18.22** (a) Butane; the double bond in 2-butene restricts freedom of rotation. (b) $Xe(g)$—it has the greater molar mass. (c) $CH_4(g)$—gases have greater entropy than liquids. **18.24** (a) $C_2H_5OH(l)$—it is a more complex molecule. (b) $KClO_3(aq)$—ions in solution have their energy more dispersed than those in a solid. (c) $K(s)$—it has a greater molar mass. **18.26** (a) Diamond < graphite < charcoal; freedom of motion is least in the network solid; more freedom between graphite sheets; most freedom in amorphous solid. (b) Ice < liquid water < water vapour. Entropy increases as a substance changes from solid to liquid to gas. (c) O atoms < O_2 < O_3. Entropy increases with molecular complexity. **18.28** (a) $ClO_4^-(aq) > ClO_3^-(aq) > ClO_2^-(aq)$; decreasing molecular complexity (b) $NO_2(g) > NO(g) > N_2(g)$. N_2 has lower standard molar entropy because it consists of two of the same atoms; the other species have two different types of atoms. NO_2 is more complex than NO.

(c) $Fe_3O_4(s) > Fe_2O_3(s) > Al_2O_3(s)$. Fe_3O_4 is more complex and more massive. Fe_2O_3 is more massive than Al_2O_3. **18.31** For a system at equilibrium, $\Delta_{univ}S = \Delta_{sys}S + \Delta_{surr}S = 0$. For a system moving to equilibrium, $\Delta_{univ}S > 0$. **18.32** $S°$ of $Cl_2O = 2(S°$ of $HClO) - S°$ of $H_2O - \Delta_r S°$ **18.33** (a) negative; $\Delta S° = -172.4$ J/(mol·K) (b) positive; $\Delta S° = 141.6$ J/(mol·K) (c) negative; $\Delta S° = -837$ J/(mol·K) **18.35** $\Delta S° = 93.1$ J/(mol·K); yes; the positive sign of ΔS is expected because there is a net increase in the number of gas molecules. **18.37** $\Delta S° = -311$ J/(mol·K); yes; the negative entropy change matches the decrease in amount (mol) of gas. **18.39** -75.6 J/(mol·K) **18.41** -242 J/(mol·K) **18.44** -97.2 J/(mol·K) **18.46** A spontaneous process has $\Delta_{univ}S > 0$. Since the absolute temperature is always positive, $\Delta_{sys}G$ must be negative ($\Delta_{sys}G < 0$) for a spontaneous process. **18.48** $\Delta H°$ is positive and $\Delta S°$ is positive. Melting is an example. **18.49** The entropy changes little within a phase. As long as the substance does not change phase, the value of $\Delta S°$ is relatively unaffected by temperature. **18.50** (a) -1138.0 kJ/mol (b) -1379.4 kJ/mol (c) -224 kJ/mol **18.52** (a) -1138 kJ/mol (b) -1379 kJ/mol (c) -226 kJ/mol **18.54** (a) Entropy decreases ($\Delta S°$ is negative) because the amount (mol) of gas decreases. The combustion of CO releases energy ($\Delta H°$ is negative). (b) -257.2 kJ/mol or -257.3 kJ/mol, depending on the method **18.56** (a) -0.409 kJ/(mol·K) (b) -197 kJ/mol **18.58** (a) $\Delta_r H° = 90.7$ kJ/mol; $\Delta_r S° = 221$ J/(mol·K) (b) At 28°C, $\Delta G° = 24.3$ kJ/mol; at 128°C, $\Delta G° = 2.2$ kJ/mol; at 228°C, $\Delta G° = -19.9$ kJ/mol (c) For the substances in their standard states, the reaction is nonspontaneous at 28°C, near equilibrium at 128°C, and spontaneous at 228°C. **18.60** $\Delta_r H° = 30.91$ kJ/mol, $\Delta_r S° = 93.15$ J/(mol·K), $T = 331.8$ K **18.62** (a) $\Delta_r H° = -241.826$ kJ/mol, $\Delta_r S° = -44.4$ J/(mol·K), $\Delta_r G° = -228.60$ kJ/mol (b) Yes. The reaction will become nonspontaneous at higher temperatures. (c) The reaction is spontaneous below 5.45×10^3 K. **18.64** (a) $\Delta G°$ is a relatively large, positive value. (b) $K > 1$. Q depends on initial conditions, not equilibrium conditions. **18.67** The standard Gibbs energy change, $\Delta G°$, applies when all components of the system are in their standard states; $\Delta G° = \Delta G$ when all concentrations equal 1 mol/L and all partial pressures equal 1 bar. **18.68** (a) 1.7×10^6 (b) 3.89×10^{-34} (c) 1.26×10^{48} **18.70** (a) 6.57×10^{173} (b) 4.46×10^{-15} (c) 3.46×10^4 **18.72** 4.89×10^{-51} **18.74** 3.36×10^5 **18.76** 2.7×10^4 J/mol; no **18.78** (a) 2.9×10^4 J/mol (b) The reverse direction, formation of reactants, is spontaneous, so the reaction proceeds to the left. (c) 7.0×10^3 J/mol; the reaction proceeds to the left to reach equilibrium. **18.80** (a) no T (b) 163 kJ/mol (c) 1×10^2 kJ/mol **18.83** (a) spontaneous (b) + (c) + (d) – (e) –; not spontaneous (f) – **18.87** (a) 2.3×10^2 (b) The treatment is to administer oxygen-rich air; increasing the concentration of oxygen shifts the equilibrium to the left, in the direction of $Hb \cdot O_2$. **18.89** $-370.$ kJ/mol **18.91** (a) $2N_2O_5(g) + 6F_2(g) \longrightarrow 4NF_3(g) + 5O_2(g)$ (b) $\Delta_r G° = -569$ kJ/mol (c) $\Delta_r G = -5.60 \times 10^2$ kJ/mol **18.93** $\Delta_r H° = -137.14$ kJ/mol; $\Delta_r S° = -120.3$ J/(mol·K); $\Delta_r G° = -101.25$ kJ/mol **18.101** (a) 1.67×10^3 J/mol (b) 7.37×10^3 J/mol (c) -4.04×10^3 J/mol (d) 0.19 **18.105** (a) 465 K (b) 6.59×10^{-4} (c) The reaction rate is higher at the higher temperature. The shorter time required (kinetics) overshadows the lower yield (thermodynamics).

Chapter 19

Answers to Boxed Reading Problems: B19.1 (a) reduction: $Fe^{3+} + e^- \longrightarrow Fe^{2+}$; oxidation: $Cu^+ \longrightarrow Cu^{2+} + e^-$ (b) Overall: $Fe^{3+} + Cu^+ \longrightarrow Fe^{2+} + Cu^{2+}$ **B19.2** $\Delta G° = -43.2$ kJ/mol

19.1 Oxidation is the loss of electrons and results in a higher oxidation number; reduction is the gain of electrons and results in a lower oxidation number. **19.3** No; one half-reaction cannot take place independently because there is a transfer of electrons from one substance to another. If one substance loses electrons, another substance must gain them. **19.6** To remove H^+ ions from an equation, add an equal number of OH^- ions to both sides to neutralize the H^+ ions and produce water. **19.8** (a) Spontaneous reactions, for which $\Delta_{sys}G < 0$, take place in voltaic cells (also called galvanic cells). (b) Nonspontaneous reactions, for which $\Delta_{sys}G > 0$, take place in electrolytic cells. **19.10** (a) Cl^- (b) MnO_4^- (c) MnO_4^- (d) Cl^- (e) from Cl^- to MnO_4^- (f) $8H_2SO_4(aq) + 2KMnO_4(aq) + 10KCl(aq) \longrightarrow 2MnSO_4(aq) + 5Cl_2(g) + 8H_2O(l) + 6K_2SO_4(aq)$

19.12 (a) $ClO_3^-(aq) + 6H^+(aq) + 6I^-(aq) \longrightarrow Cl^-(aq) + 3H_2O(l) + 3I_2(s)$; oxidizing agent is ClO_3^- and reducing agent is I^-.
(b) $2MnO_4^-(aq) + H_2O(l) + 3SO_3^{2-}(aq) \longrightarrow 2MnO_2(s) + 3SO_4^{2-}(aq) + 2OH^-(aq)$; oxidizing agent is MnO_4^- and reducing agent is SO_3^{2-}.
(c) $2MnO_4^-(aq) + 6H^+(aq) + 5H_2O_2(aq) \longrightarrow 2Mn^{2+}(aq) + 8H_2O(l) + 5O_2(g)$; oxidizing agent is MnO_4^- and reducing agent is H_2O_2.

19.14 (a) $Cr_2O_7^{2-}(aq) + 14H^+(aq) + 3Zn(s) \longrightarrow 2Cr^{3+}(aq) + 7H_2O(l) + 3Zn^{2+}(aq)$; oxidizing agent is $Cr_2O_7^{2-}$ and reducing agent is Zn.
(b) $MnO_4^-(aq) + 3Fe(OH)_2(s) + 2H_2O(l) \longrightarrow MnO_2(s) + 3Fe(OH)_3(s) + OH^-(aq)$; oxidizing agent is MnO_4^- and reducing agent is $Fe(OH)_2$.
(c) $2NO_3^-(aq) + 12H^+(aq) + 5Zn(s) \longrightarrow N_2(g) + 6H_2O(l) + 5Zn^{2+}(aq)$; oxidizing agent is NO_3^- and reducing agent is Zn.

19.16 (a) $4NO_3^-(aq) + 4H^+(aq) + 4Sb(s) \longrightarrow 4NO(g) + 2H_2O(l) + Sb_4O_6(s)$; oxidizing agent is NO_3^- and reducing agent is Sb.
(b) $5BiO_3^-(aq) + 14H^+(aq) + 2Mn^{2+}(aq) \longrightarrow 5Bi^{3+}(aq) + 7H_2O(l) + 2MnO_4^-(aq)$; oxidizing agent is BiO_3^- and reducing agent is Mn^{2+}.
(c) $Pb(OH)_3^-(aq) + 2Fe(OH)_2(s) \longrightarrow Pb(s) + 2Fe(OH)_3(s) + OH^-(aq)$; oxidizing agent is $Pb(OH)_3^-$ and reducing agent is $Fe(OH)_2$.

19.18 (a) $5As_4O_6(s) + 8MnO_4^-(aq) + 18H_2O(l) \longrightarrow 20AsO_4^{3-}(aq) + 8Mn^{2+}(aq) + 36H^+(aq)$; oxidizing agent is MnO_4^- and reducing agent is As_4O_6.
(b) $P_4(s) + 6H_2O(l) \longrightarrow 2HPO_3^{2-}(aq) + 2PH_3(g) + 4H^+(aq)$; P_4 is both the oxidizing agent and the reducing agent.
(c) $2MnO_4^-(aq) + 3CN^-(aq) + H_2O(l) \longrightarrow 2MnO_2(s) + 3CNO^-(aq) + 2OH^-(aq)$; oxidizing agent is MnO_4^- and reducing agent is CN^-.

19.21 (a) $Au(s) + 3NO_3^-(aq) + 4Cl^-(aq) + 6H^+(aq) \longrightarrow AuCl_4^-(aq) + 3NO_2(g) + 3H_2O(l)$ (b) Oxidizing agent is NO_3^- and reducing agent is Au. (c) HCl provides chloride ions that combine with the gold(III) ion to form the stable $AuCl_4^-$ ion. **19.22** (a) A (b) E (c) C (d) A (e) E (f) E
19.25 (a) An active electrode is a reactant or product in the cell reaction. (b) An inactive electrode does not take part in the reaction and is present only to conduct a current. (c) Platinum and graphite are commonly used as inactive electrodes. **19.26** (a) A (b) B (c) A (d) Hydrogen bubbles will form when metal A is placed in acid. Metal A is a better reducing agent than metal B, so if metal B reduces H^+ in acid, then metal A will also.
19.27 (a) Oxidation: $Zn(s) \longrightarrow Zn^{2+}(aq) + 2e^-$
Reduction: $Sn^{2+}(aq) + 2e^- \longrightarrow Sn(s)$
Overall: $Zn(s) + Sn^{2+}(aq) \longrightarrow Zn^{2+}(aq) + Sn(s)$
(b)

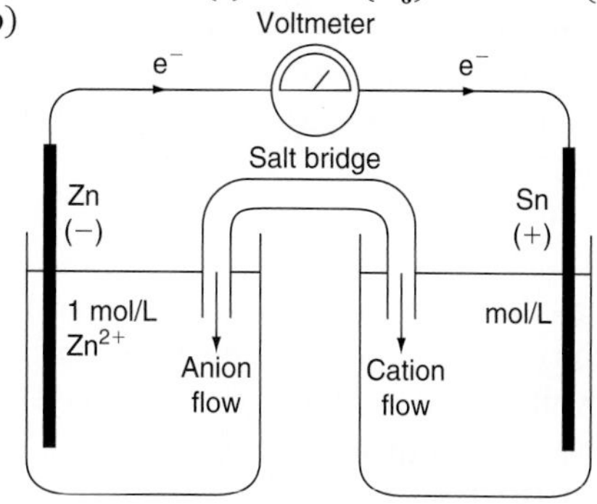

19.29 (a) left to right (b) left (c) right (d) Ni (e) Fe (f) Fe(g) 1 mol/L $NiSO_4$ (h) K^+ and NO_3^- (i) neither (j) from right to left
(k) Oxidation: $Fe(s) \longrightarrow Fe^{2+}(aq) + 2e^-$
Reduction: $Ni^{2+}(aq) + 2e^- \longrightarrow Ni(s)$
Overall: $Fe(s) + Ni^{2+}(aq) \longrightarrow Fe^{2+}(aq) + Ni(s)$
19.31 (a) Reduction: $Fe^{2+}(aq) + 2e^- \longrightarrow Fe(s)$
Oxidation: $Mn(s) \longrightarrow Mn^{2+}(aq) + 2e^-$
Overall: $Fe^{2+}(aq) + Mn(s) \longrightarrow Fe(s) + Mn^{2+}(aq)$
(b)

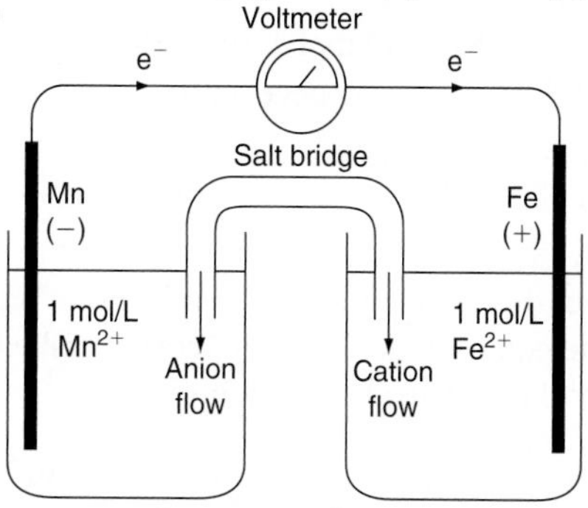

19.33 (a) $Al(s) \mid Al^{3+}(aq) \parallel Cr^{3+}(aq) \mid Cr(s)$
(b) $Pt \mid SO_2(g) \mid SO_4^{2-}(aq), H^+(aq) \parallel Cu^{2+}(aq) \mid Cu(s)$
19.36 (a) A negative E°_{cell} indicates that the redox reaction is not spontaneous at the standard state, that is, $\Delta G^\circ > 0$. (b) The reverse reaction is spontaneous with $E^\circ_{cell} > 0$.
19.37 (a) Similar to other state functions, E° changes sign when a reaction is reversed. (b) Unlike ΔG°, ΔH°, and S°, E° (the ratio of energy to charge) is an intensive property. When the coefficients in a reaction are multiplied by a factor, the values of ΔG°, ΔH°, and S° are multiplied by that factor. However, E° does not change because both the energy and charge are multiplied by the factor and thus their ratio remains unchanged.

19.38 (a) Oxidation: $Se^{2-}(aq) \longrightarrow Se(s) + 2e^-$
Reduction: $2SO_3^{2-}(aq) + 3H_2O(l) + 4e^- \longrightarrow S_2O_3^{2-}(aq) + 6OH^-(aq)$
(b) $E^\circ_{anode} = E^\circ_{cathode} - E^\circ_{cell} - 0.57\ V - 0.35\ V = -0.92\ V$
19.40 (a) $Br_2 > Fe^{3+} > Cu^{2+}$ (b) $Ca^{2+} < Ag^+ < Cr_2O_7^{2-}$
19.42 (a) $Co(s) + 2H^+(aq) \longrightarrow Co^{2+}(aq) + H_2(g)$; $E^\circ_{cell} = 0.28$ V; spontaneous
(b) $2Mn^{2+}(aq) + 5Br_2(l) + 8H_2O(l) \longrightarrow 2MnO_4^-(aq) + 10Br^-(aq) + 16H^+(aq)$; $E^\circ_{cell} = -0.44$V; not spontaneous
(c) $Hg_2^{2+}(aq) \longrightarrow Hg^{2+}(aq) + Hg(l)$; $E^\circ_{cell} = -0.07$ V; not spontaneous
19.44 (a) $2Ag(s) + Cu^{2+}(aq) \longrightarrow 2Ag^+(aq) + Cu(s)$; $E^\circ_{cell} = -0.46$ V; not spontaneous
(b) $Cr_2O_7^{2-}(aq) + 3Cd(s) + 14H^+(aq) \longrightarrow 2Cr^{3+}(aq) + 3Cd^{2+}(aq) + 7H_2O(l)$; $E^\circ_{cell} = 1.73$ V; spontaneous
(c) $Pb(s) + Ni^{2+}(aq) \longrightarrow Pb^{2+}(aq) + Ni(s)$; $E^\circ_{cell} = -0.12$ V; not spontaneous
19.46 $3N_2O_4(g) + 2Al(s) \longrightarrow 6NO_2^-(aq) + 2Al^{3+}(aq)$; $E^\circ_{cell} = 0.867\ V - (-1.66\ V) = 2.53$ V;
$2Al(s) + 3SO_4^{2-}(aq) + 3H_2O(l) \longrightarrow 2Al^{3+}(aq) + 3SO_3^{2-}(aq) + 6OH^-(aq)$; $E^\circ_{cell} = 2.59$ V;
$SO_4^{2-}(aq) + 2NO_2^-(aq) + H_2O(l) \longrightarrow SO_3^{2-}(aq) + N_2O_4(g) + 2OH^-(aq)$; $E^\circ_{cell} = 0.06$ V;
oxidizing agents: $Al^{3+} < N_2O_4 < SO_4^{2-}$;
reducing agents: $SO_3^{2-} < NO_2^- < Al$
19.48 $2HClO(aq) + Pt(s) + 2H^+(aq) \longrightarrow Cl_2(g) + Pt^{2+}(aq) + 2H_2O(l)$; $E^\circ_{cell} = 0.43$ V;
$2HClO(aq) + Pb(s) + SO_4^{2-}(aq) + 2H^+(aq) \longrightarrow Cl_2(g) + PbSO_4(s) + 2H_2O(l)$; $E^\circ_{cell} = 1.94$ V;
$Pt^{2+}(aq) + Pb(s) + SO_4^{2-}(aq) \longrightarrow Pt(s) + PbSO_4(s)$; $E^\circ_{cell} = 1.51$ V;
oxidizing agents: $PbSO_4 < Pt^{2+} < HClO$;
reducing agents: $Cl_2 < Pt < Pb$
19.50 yes; C > A > B **19.53** $A(s) + B^+(aq) \longrightarrow A^+(aq) + B(s)$ with $Q = [A^+]/[B^+]$ (a) $[A^+]$ increases and $[B^+]$ decreases. (b) E_{cell} decreases. (c) $E_{cell} = E^\circ_{cell} - (RT/nF) \ln ([A^+]/[B^+])$; $E_{cell} = E^\circ_{cell}$ when $(RT/nF) \ln ([A^+]/[B^+]) = 0$. This occurs when $\ln ([A^+]/[B^+]) = 0$, that is, $[A^+]$ equals $[B^+]$. (d) yes, when $[A^+] > [B^+]$ **19.55** In a concentration cell, the overall reaction decreases the concentration of the more concentrated electrolyte because it is being reduced. Reduction occurs in the cathode compartment. **19.56** (a) 3×10^{35} (b) 4×10^{-31} **19.58** (a) 1×10^{-67} (b) 6×10^9
19.60 (a) -2.03×10^5 J/mol (b) 1.7×10^5 J/mol
19.62 (a) 3.82×10^5 J/mol (b) -5.6×10^4 J/mol **19.64** $E^\circ = 0.28$ V; $\Delta G^\circ = -2.7 \times 10^4$ J/mol **19.66** $E^\circ_{cell} = 0.054$V; $\Delta G^\circ = -1.0 \times 10^4$ J/mol **19.68** 8.7×10^{-5} mol/L **19.70** (a) 0.05 V (b) 0.50 mol/L (c) $[Co^{2+}] = 0.91$ mol/L; $[Ni^{2+}] = 0.09$ mol/L
19.72 A; 0.083 V **19.74** Electrons flow from the anode, where oxidation occurs, to the cathode, where reduction occurs. The electrons always flow from the anode to the cathode

no matter what type of battery. **19.76** A D-sized alkaline battery is larger than an AAA-sized one, so it contains greater amounts of the cell components. (a) The cell potential is an intensive property and does not depend on the amounts of the cell components. (b) The total charge, however, does depend on the amount of cell components, so the D-sized battery produces more charge. **19.78** The Teflon spacers keep the two metals separated so that the copper cannot conduct electrons that would promote the corrosion (rusting) of the iron skeleton. **19.81** Sacrificial anodes are made of metals with $E°$ less than that of iron, -0.44 V, so they are more easily oxidized than iron. Only b, f, and g will work for iron: a will form an oxide coating that prevents further oxidation; c will react with groundwater quickly; d and e are less easily oxidized than iron.
19.83 To reverse the reaction requires 0.34 V with the cell in its standard state. A 1.5 V cell supplies more than enough potential, so the Cd metal is oxidized to Cd^{2+} and Cr metal plates out. **19.85** The oxidation number of N in NO_3^- is +5, the maximum O.N. for N. In the nitrite ion, NO_2^-, the O.N. of N is +3, so nitrogen can be further oxidized.
19.87 (a) Br_2 (b) Na **19.89** I_2 gas forms at the anode; magnesium (liquid) forms at the cathode. **19.91** Bromine gas forms at the anode; calcium metal forms at the cathode.
19.93 copper and bromine **19.95** iodine, zinc, and silver
19.97 (a) Anode: $2H_2O(l) \longrightarrow O_2(g) + 4H^+(aq) + 4e^-$
Cathode: $2H_2O(l) + 2e^- \longrightarrow H_2(g) + 2OH^-(aq)$
(b) Anode: $2H_2O(l) \longrightarrow O_2(g) + 4H^+(aq) + 4e^-$
Cathode: $Sn^{2+}(aq) + 2e^- \longrightarrow Sn(s)$
19.99 (a) Anode: $2H_2O(l) \longrightarrow O_2(g) + 4H^+(aq) + 4e^-$
Cathode: $NO_3^-(aq) + 4H^+(aq) + 3e^- \longrightarrow NO(g) + 2H_2O(l)$
(b) Anode: $2Cl^-(aq) \longrightarrow Cl_2(g) + 2e^-$
Cathode: $2H_2O(l) + 2e^- \longrightarrow H_2(g) + 2OH^-(aq)$
19.101 (a) 3.75 mol e^- (b) 3.62×10^5 C (c) 28.7 A
19.103 0.275 g Ra **19.105** 9.20×10^3 s **19.107** (a) The sodium and sulfate ions conduct a current, facilitating electrolysis. Pure water, which contains very low (10^{-7}mol/L) concentrations of H^+ and OH^-, conducts electricity very poorly. (b) The reduction of H_2O has a more positive half-potential than does the reduction of Na^+; the oxidation of H_2O is the only reaction possible because SO_4^{2-} cannot be oxidized. Thus, it is easier to reduce H_2O than Na^+ and easier to oxidize H_2O than SO_4^{2-}. **19.109** 62.6 g Zn **19.111** (a) 3.3×10^{11} C (b) 4.7×10^{11} J (c) 1.2×10^4 kg **19.114** 64.3 mass % Cu
19.115 (a) 8 days (b) 32 days (c) \$1300 **19.118** (a) 2.4×10^4 days (b) 2.1 g (c) \$CAD 7.3×10^{-5} **19.121** (a) Pb/Pb^{2+}: $E°_{cell} = 0.13$ V; Cu/Cu^{2+}: $E°_{cell} = 0.34$ V (b) The anode (negative electrode) is Pb. The anode in the other cell is platinum in the standard hydrogen electrode. (c) The precipitation of PbS decreases $[Pb^{2+}]$, which increases the potential. (d) -0.13 V
19.124 The three steps equivalent to the overall reaction $M^+(aq) + e^- \longrightarrow M(s)$ are
(1) $M^+(aq) \longrightarrow M^+(g)$ $\quad \Delta H$ is $-\Delta_{hydration}H$
(2) $M^+(g) + e^- \longrightarrow M(g)$ $\quad \Delta H$ is $-$IE
(3) $M(g) \longrightarrow M(s)$ $\quad \Delta H$ is $-\Delta_{atomization}H$
The energy for step 3 is similar for all three elements, so the difference in energy for the overall reaction depends on the values for $\Delta_{hydration}H$ and IE. The Li^+ ion has a much greater hydration energy than Na^+ and K^+ because it is smaller, with a larger charge density that holds the water molecules more tightly. The energy required to remove the waters surrounding Li^+ offsets the lower ionization energy, making the overall energy for the reduction of lithium larger than expected. **19.125** The very high and very low standard electrode potentials involve extremely reactive substances, such as F_2 (a powerful oxidizer) and Li (a powerful reducer). These substances react directly with water because any aqueous cell with a voltage of more than 1.23 V has the ability to electrolyze water into hydrogen and oxygen.
19.127 (a) 1×10^5 s (b) 1.5×10^4 kW·h (c) \$1.3
19.129 If metal E and a salt of metal F are mixed, the salt is reduced, producing metal F because E has the greatest reducing strength of the three metals; F < D < E.
19.131 (a) Cell I: 4 mol electrons; $\Delta G = -4.75 \times 10^5$ J/mol
Cell II: 2 mol electrons; $\Delta G = -3.94 \times 10^5$ J/mol
Cell III: 2 mol electrons; $\Delta G = -4.53 \times 10^5$ J/mol
(b) Cell I: -13.2 kJ/g
Cell II: -0.613 kJ/g
Cell III: -2.63 kJ/g
Cell I has the highest ratio (most energy released per gram) because the reactants have very low mass, while Cell II has the lowest ratio because the reactants have large masses.
19.135 (a) 9.7 g Cu (b) 0.56 mol/L Cu^{2+}
19.137 $Sn^{2+}(aq) + 2e^- \longrightarrow Sn(s)$
$Cr^{3+}(aq) + e^- \longrightarrow Cr^{2+}(aq)$
$Fe^{2+}(aq) + 2e^- \longrightarrow Fe(s)$
$U^{4+}(aq) + e^- \longrightarrow U^{3+}(aq)$
19.141 (a) 3.5×10^{-9}mol/L (b) 0.3 mol/L
19.143 (a) Nonstandard cell:

$$E_{waste} = E°_{cell} - \frac{[8.314\ J/(mol\cdot K)](298\ K)}{(1)(96\ 485\ C/mol)} \ln [Ag^+]_{waste}$$

Standard cell:

$$E_{standard} = E°_{cell} - \frac{[8.314\ J/(mol\cdot K)](298\ K)}{(1)(96\ 485\ C/mol)} \ln [Ag^+]_{standard}$$

(b) $$[Ag^+]_{waste} = \left[e^{\left(\frac{E_{standard} - E_{waste}}{0.025\ 678\ 3}\right)} \right] ([Ag^+]_{standard})$$

(c) If both silver ion concentrations are in the same units, in this case ng/L, the "conversions" cancel and the equation derived in part (b) applies if the standard concentration is in ng/L.

$$Conc.(Ag^+)_{waste} = \left[e^{\left(\frac{E_{standard} - E_{waste}}{0.025\ 678\ 3}\right)} \right] ([Ag^+]_{standard})$$

where C is concentration in ng/L.
(d) 900 ng/L
(e) $$[Ag^+]_{waste} = e^{\left(\frac{(E_{standard} - E_{waste})(zF/R) + T_{standard} \ln [Ag^+]_{standard}}{T_{waste}}\right)}$$

19.145 (a) 1.08×10^3 C (b) 0.629 g Cd, 1.03 g NiO(OH), 0.202 g H_2O; total mass of reactants = 1.86 g (c) 10.1%
19.147 Li > Ba > Na > Al > Mn > Zn > Cr > Fe > Ni > Sn > Pb > Cu > Ag > Hg > Au. Metals with potentials lower than that of water (-0.83 V) can displace H_2 from water: Li, Ba, Na, Al, and Mn. Metals with potentials lower than that of hydrogen (0.00 V) can displace H_2 from acid:

Li, Ba, Na, Al, Mn, Zn, Cr, Fe, Ni, Sn, and Pb. Metals with potentials greater than that of hydrogen (0.00 V) cannot displace H_2: Cu, Ag, Hg, and Au. **19.150** (a) 5.4×10^{-11} (b) 0.20 V (c) 0.43 V (d) 8.3×10^{-4} mol/L NaOH **19.153** (a) -1.25×10^5 kJ (b) 1.28×10^4 L (c) 9.97×10^4 s (d) 166 kW•h (e) $23.2 **19.154** 2.94

Chapter 20

20.2 (a) Carbon's electronegativity is midway between the most metallic and most nonmetallic elements of period 2. To attain a filled outer shell, carbon forms covalent bonds to other atoms in molecules, network covalent solids, and polyatomic ions. (b) Since carbon has four valence electrons, it forms four covalent bonds to attain an octet. (c) To reach the He electron configuration, a carbon atom must lose four electrons, requiring too much energy to form the C^{4+} cation. To reach the Ne electron configuration, the carbon atom must gain four electrons, also requiring too much energy to form the C^{4-} anion. (d) Carbon is able to bond to itself extensively because its small size allows for close approach and great orbital overlap. The extensive orbital overlap results in a strong, stable bond. (e) The C—C σ bond is short enough to allow sideways overlap of unhybridized *p* orbitals of neighbouring C atoms. The sideways overlap of *p* orbitals results in the π bonds that are part of double and triple bonds. **20.3** (a) H, O, N, P, S, and halogens (F, Cl, Br, I) (b) Heteroatoms are atoms of any element other than carbon and hydrogen. (c) More electronegative than C: N, O, F, Cl, and Br; less electronegative than C: H and P. Sulfur and iodine have the same electronegativity as carbon. (d) Since carbon can bond to a wide variety of heteroatoms and to carbon atoms, it can form many different compounds. **20.6** The C—H and C—C bonds are unreactive because the electronegativities are close and the bonds are short. The C—I bond is reactive because it is long and weak. The C═O bond is reactive because oxygen is more electronegative than carbon and the electron-rich π bond makes it attract electron-poor atoms. The C—Li bond is also reactive because the bond polarity results in an electron-rich region around carbon and an electron-poor region around lithium. **20.7** (a) An alkane is an organic compound consisting of carbon and hydrogen in which there are no multiple bonds between carbon atoms, only single bonds. A cycloalkane is an alkane in which the carbon chain is arranged in a ring. (b) The general formula for an alkane is C_nH_{2n+2}. The general formula for a cycloalkane is C_nH_{2n} (elimination of two hydrogen atoms is required to form the additional bond between carbon atoms in the ring).

20.9 (a)

(b)

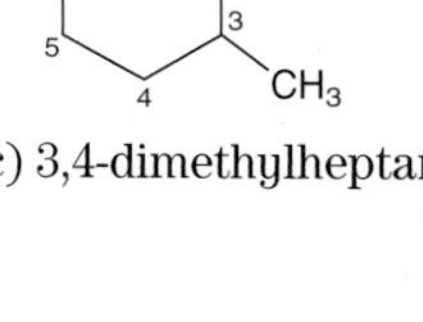

(c) 3,4-dimethylheptane (d) 2,2-dimethylbutane

20.11 (a) 4-methylhexane means a six-carbon chain with a methyl group on the fourth carbon:

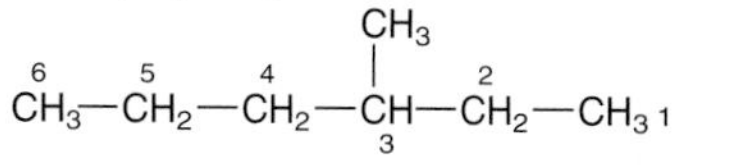

Numbering from the end carbon to give the lowest value for the methyl group gives the correct name of 3-methylhexane. (b) 2-ethylpentane means a five-carbon chain with an ethyl group on the second carbon. Numbering the longest chain gives the correct name, 3-methylhexane.

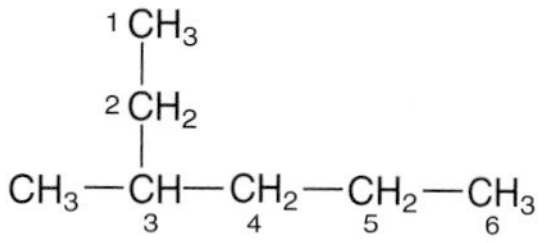

(c) 2-methylcyclohexane means a six-carbon ring with a methyl group on the second carbon:

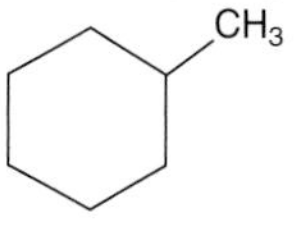

The correct name is methylcyclohexane.
(d) 3,3-methyl-4-ethyloctane means an eight-carbon chain with two methyl groups attached to the third carbon and one ethyl group to the fourth carbon.

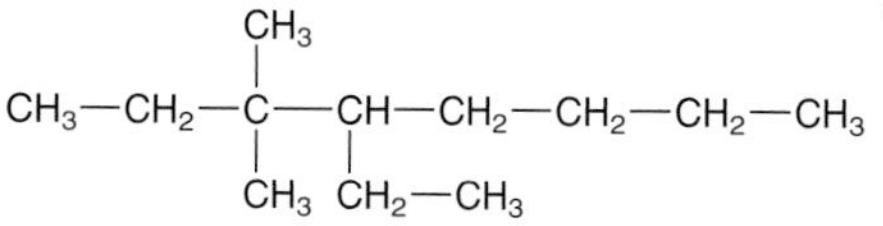

The correct name is 4-ethyl-3,3-dimethyloctane.

20.13

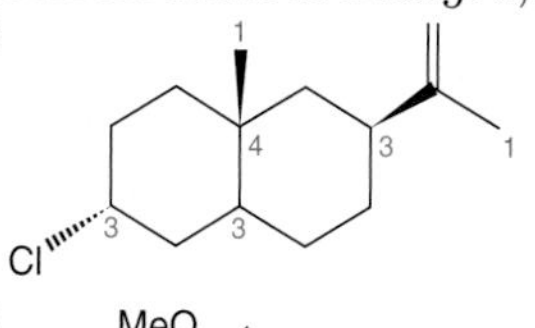

20.15

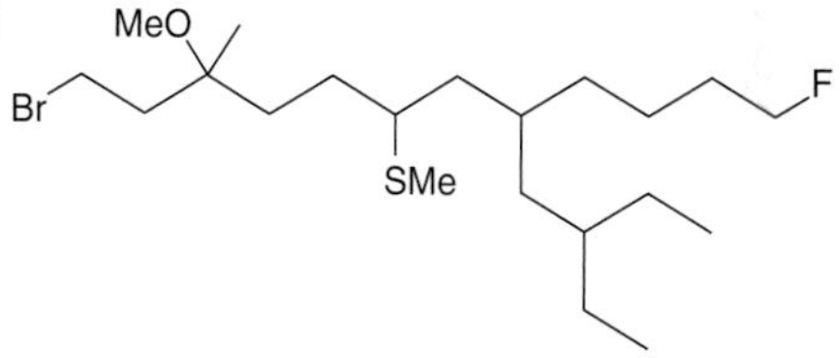

20.18 B is the most stable because it minimizes the interaction between the largest substituents (methyl groups). **20.20** B is the least stable. A and C are equally stable.

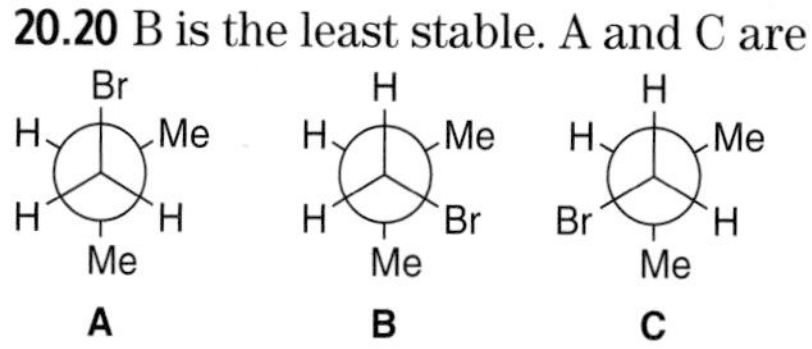

20.22 Carbon skeleton: 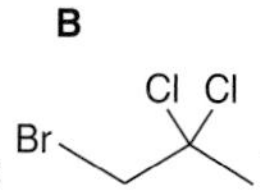 Fischer projection: 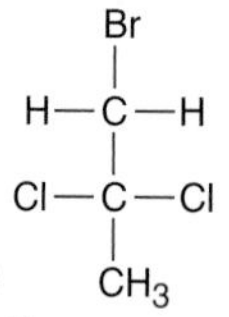

20.23 Structure B is more stable because the bulky groups (t-butyl and methyl) are in the equatorial position, which reduces the 1,3 diaxial interactions.

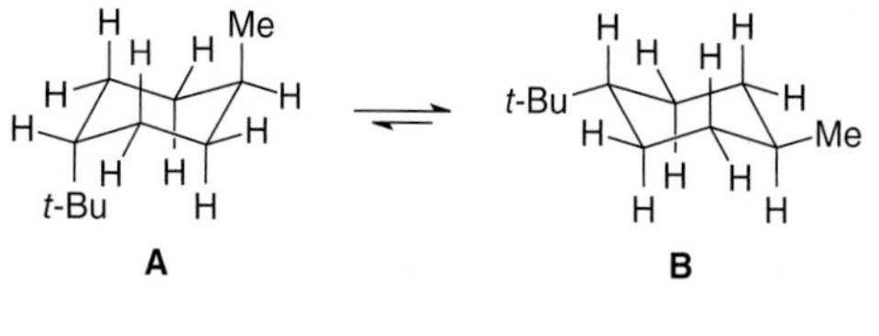

20.25 (a) Constitutional isomers are those with different sequences of bonded atoms. They are not stereoisomers. (b) Geometric isomers are a type of stereoisomers where there is a different orientation of groups around a double bond or a cyclic structure. (c) Optical isomers are a type of stereoisomers where a molecule and its mirror image cannot be superimposed on each other. They rotate the plane of polarized light in the opposite direction. **20.28** (a) *trans*, labelled Z (b) *trans*, labelled E **20.32** H, D, O, P, Br, I **20.33** (a) asymmetric (b) symmetric (c) asymmetric (d) symmetric (e) symmetric (f) asymmetric **20.36** The compound 2-methylhex-2-ene does not have *cis-trans* isomers because the second carbon atom is attached to two identical methyl ($-CH_3$) groups.

20.38 (a)

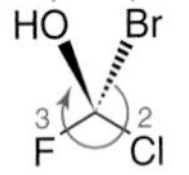

S

(b)

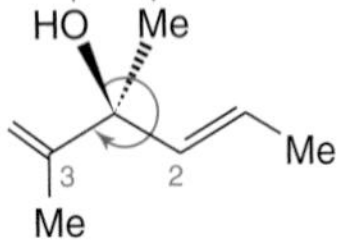

R and E

(c)

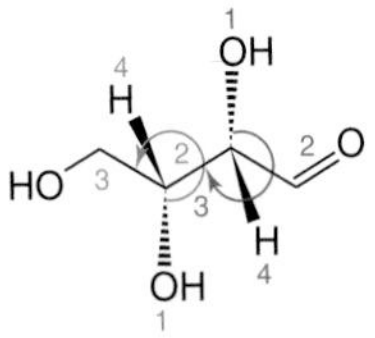

2*S* 3*R*

20.41

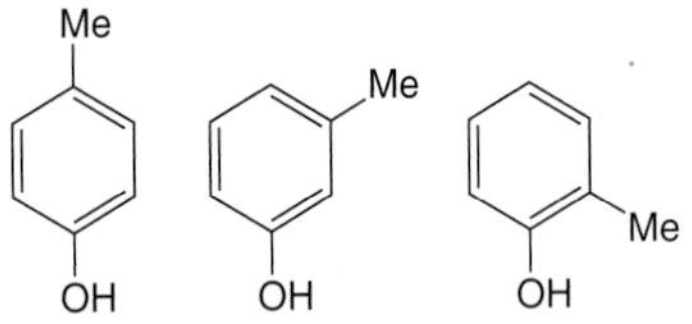

para *meta* *ortho*

20.44 (a)

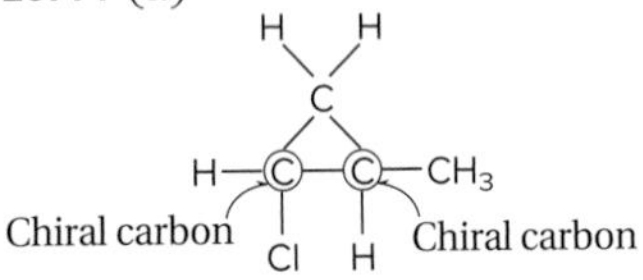

(b)

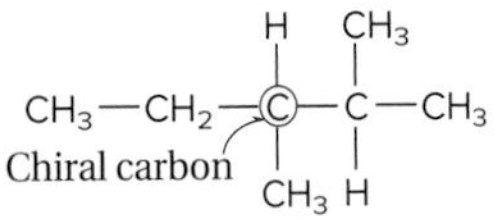

20.46 (a) 3-bromohexane is optically active.

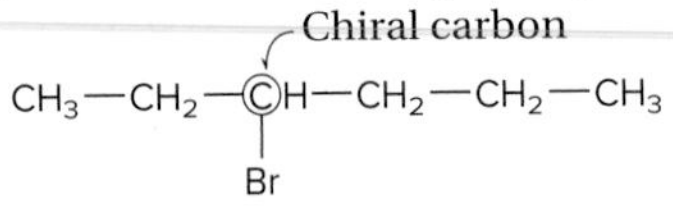

(b) 3-chloro-3-methylpentane is not optically active.

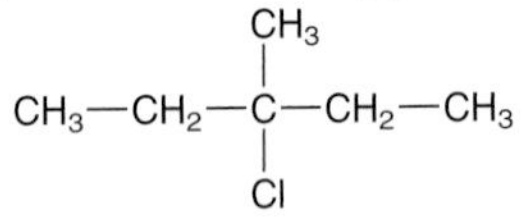

(c) 1,2-dibromo-2-methylbutane is optically active.

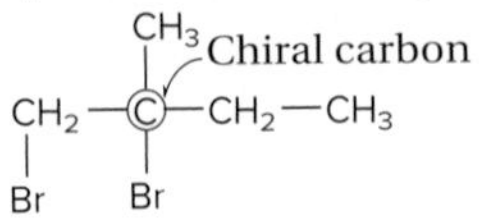

20.48 (a)

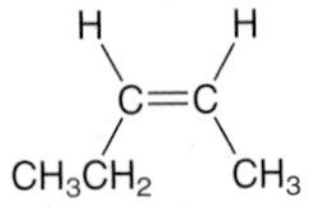

cis-pent-2-ene

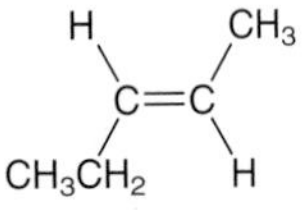

trans-pent-2-ene

(b)

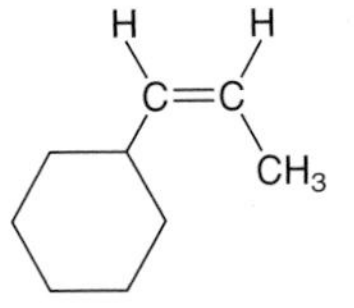

cis-cyclohexylprop-1-ene

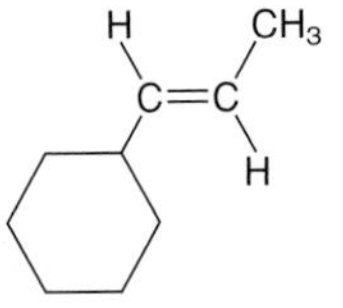

trans-cyclohexylprop-1-ene

(c) no geometric isomers

20.50 (a) no geometric isomers

(b)

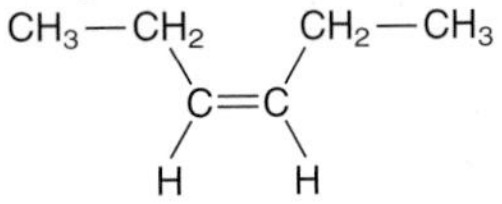

cis-hex-3-ene

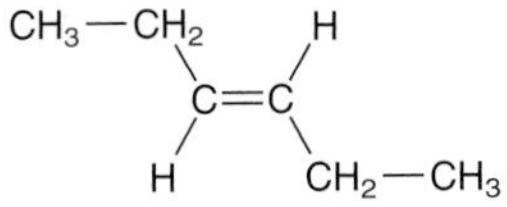

trans-hex-3-ene

(c) no geometric isomers

(d)

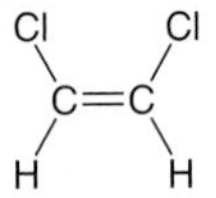

cis-1,2-dichloroethene

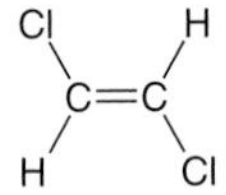

trans-1,2-dichloroethene

20.52 (a) An alkene is a hydrocarbon with at least one double bond between two carbon atoms. An alkyne is a hydrocarbon with at least one triple bond between two carbon atoms. (b) For an alkene, assuming only one double bond, the general formula is C_nH_{2n}. For an alkyne, assuming only one triple bond, the general formula is C_nH_{2n-2}.

20.53 (a)

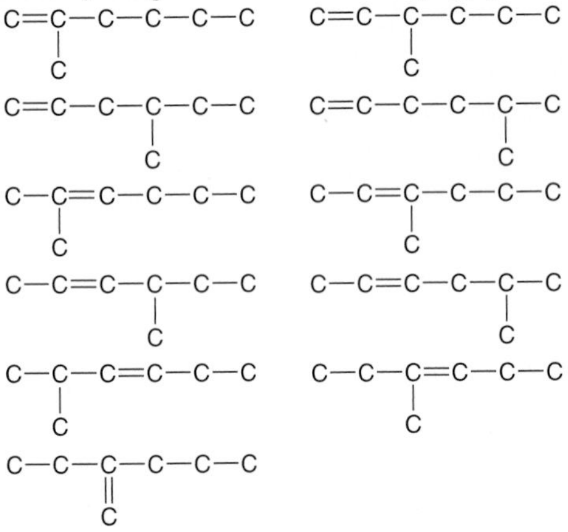

(b)

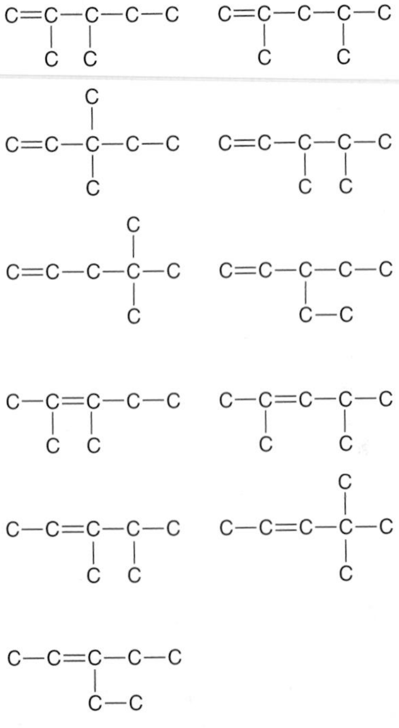

(c)

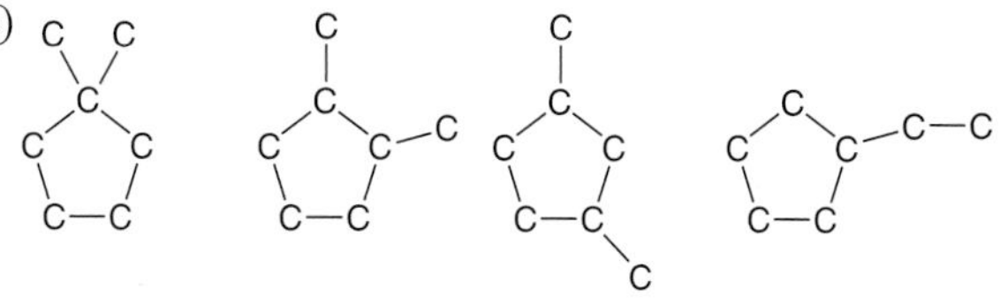

20.55 (a)

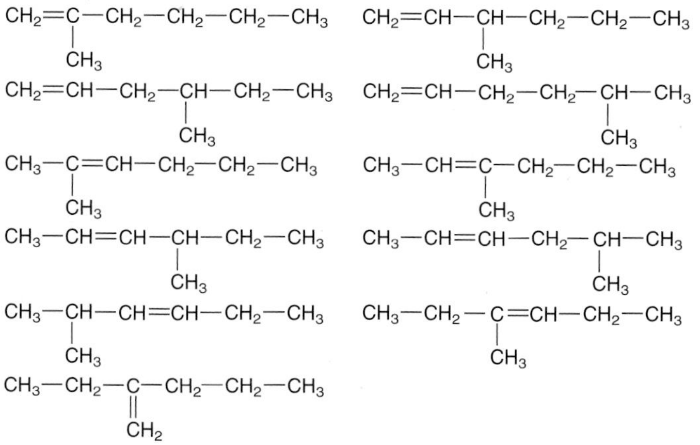

(b)

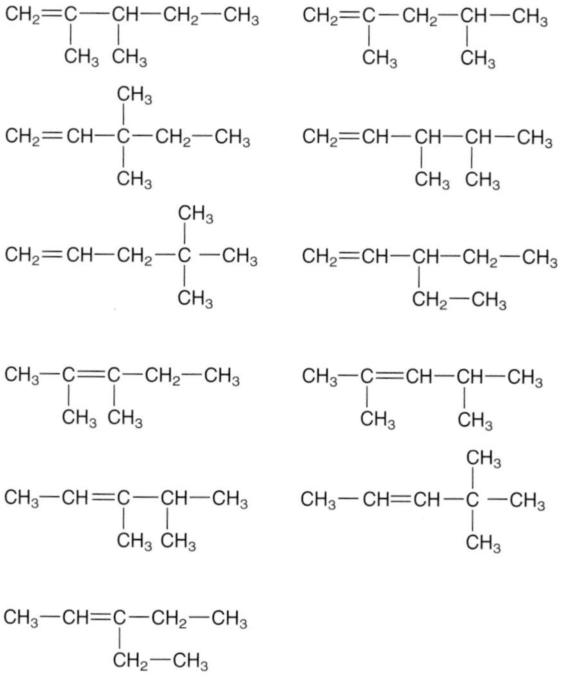

(c)

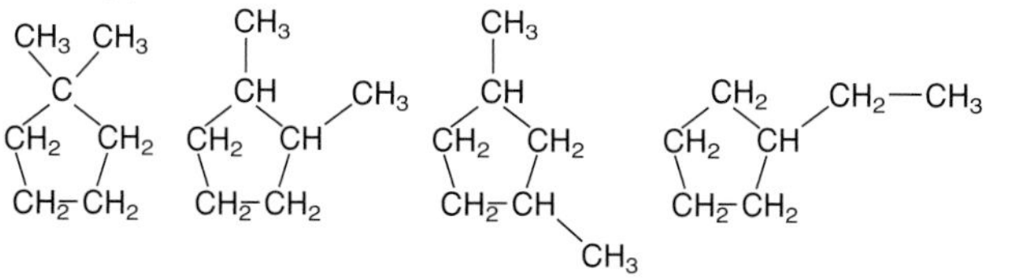

20.57 (a) $H_2C=CH-CH_2-CH_3$

(b)

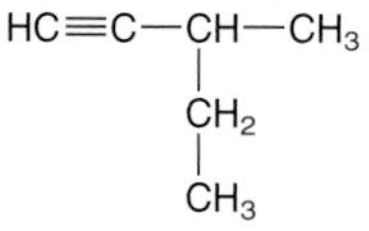

(c) Structure is correct.

20.59 Due to resonance structures, all the bonds in benzene are the same, having partial double bond and partial single bond characteristics. **20.62** methyl: *ortho*; hydroxy: *para*; bromo: *meta*; chloro: *meta*; fluoro: *ortho* **20.64** (a) 2-bromo-4-methylphenol (b) 5-bromo-2-chloro-3-methylaniline (c) 1-bromo-2,6-dichloro-3-fluoro-5-methyl-4-ethenylbenzene **20.65** A is not aromatic, B is not aromatic, C is aromatic, and D is not aromatic. **20.68** (a) 4,4-dimethylpent-2-yne (b) (2Z,5E)-hepta-2,5-diene (if counted from right to left) or (2E,5Z)-hepta-2,5-diene (if counted from left to right) (c) (E)-5-bromopent-3-en-1-yne

20.69

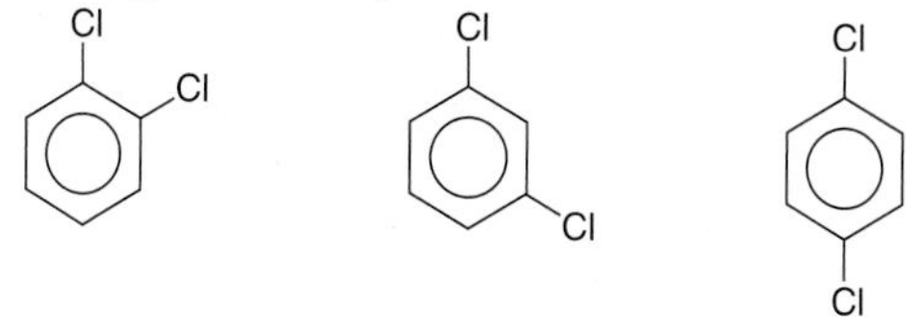

1,2-dichlorobenzene (*o*-dichlorobenzene) 1,3-dichlorobenzene (*m*-dichlorobenzene) 1,4-dichlorobenzene (*p*-dichlorobenzene)

20.71

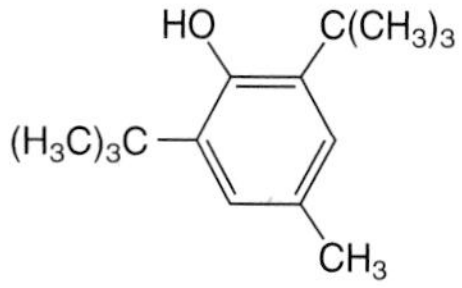

20.72 C < B < D < A **20.75** (a) haloalkane (b) nitrile (c) carboxylic acid (d) aldehyde

20.77 (a)

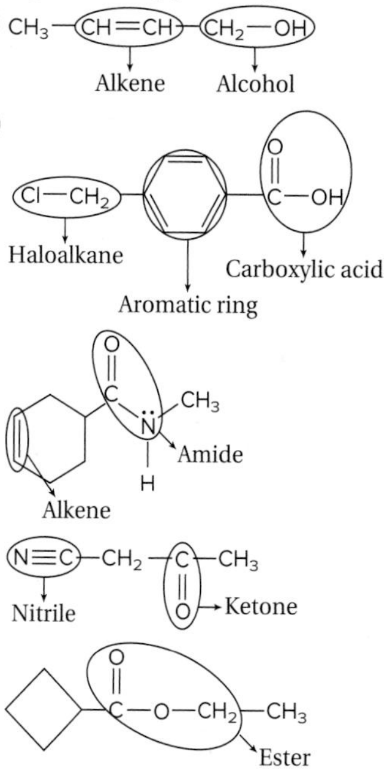

20.80

Aldehydes:

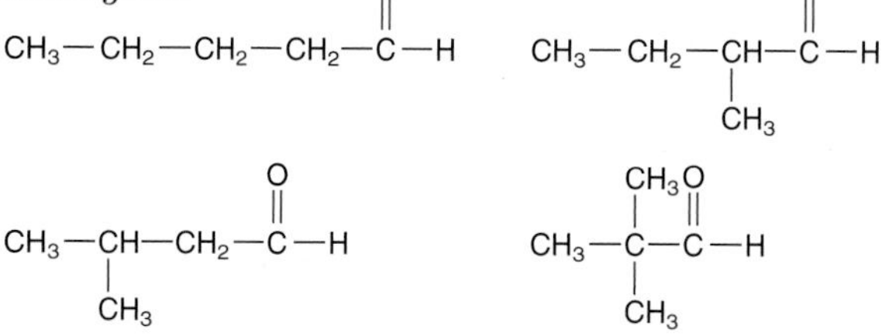

Ketones:

20.81

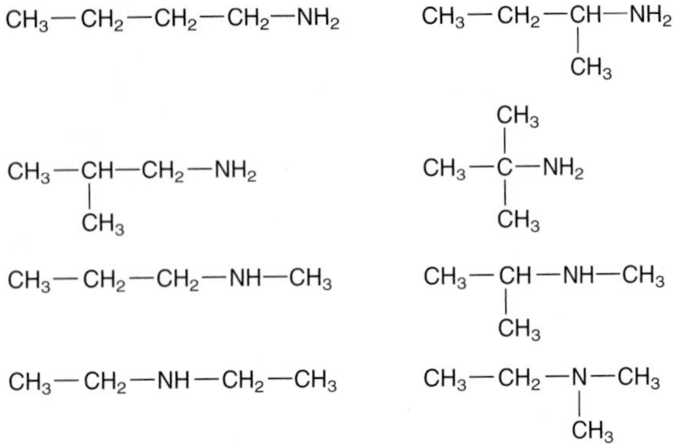

20.83 DU = 6 **20.85** DU = 4
20.88 DU = 5

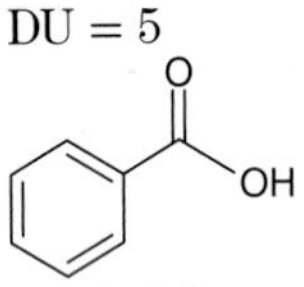

$C_7H_6O_2$

20.93

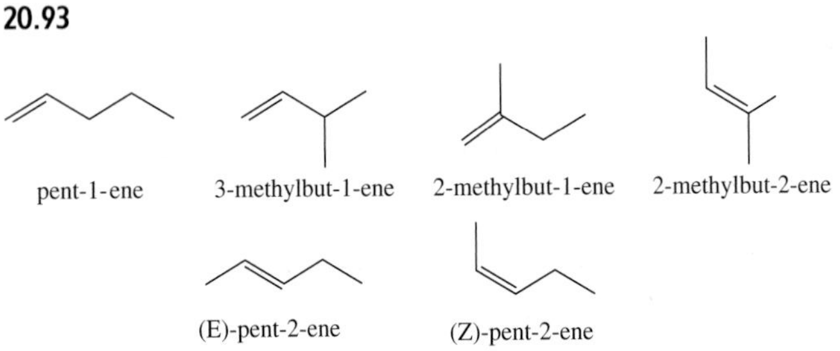

20.95 DU = 2

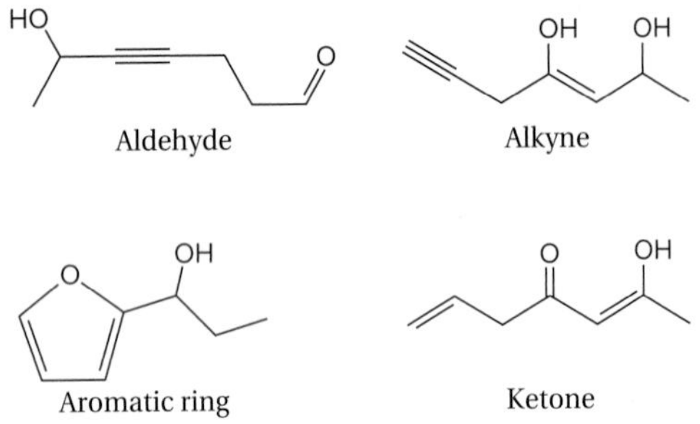

20.96

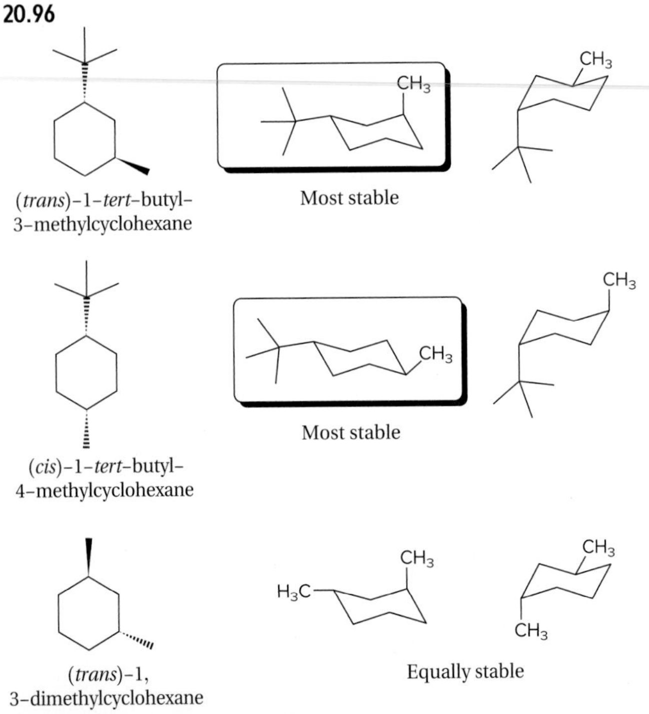

Chapter 21

21.4 (a) elimination (b) addition
21.6 (a) $CH_3CH_2CH{=}CHCH_2CH_3 + H_2O \xrightarrow{H^+} CH_3CH_2CH_2\underset{OH}{CH}CH_2CH_3$

(b) $CH_3CHBrCH_3 + CH_3CH_2OK \longrightarrow CH_3CH{=}CH_2 + CH_3CH_2OH + KBr$

(c) $CH_3CH_3 + 2Cl_2 \xrightarrow{h\nu} CHCl_2CH_3 + 2HCl$

21.8 (a) oxidation (b) reduction (c) reduction **21.10** (a) oxidized (b) oxidized **21.12** *Step 1* substitution *Step 2* addition **21.13** C **21.14** (a) stepwise (b) exothermic (c) step A **21.16** (a) heterolytic (b) hemolytic (c) heterolytic **21.18** b is the most nucleophilic **21.24** b; it forms a more stable carbocation. **21.27** Because the reaction goes by an S_N1 mechanism, there is a planar carbocation intermediate that can be attacked from either side, giving both enantiomers and resulting in a racemic mixture. **21.29** b; the azide (N_3^-) is negatively charged (making it highly nucleophilic).

21.32

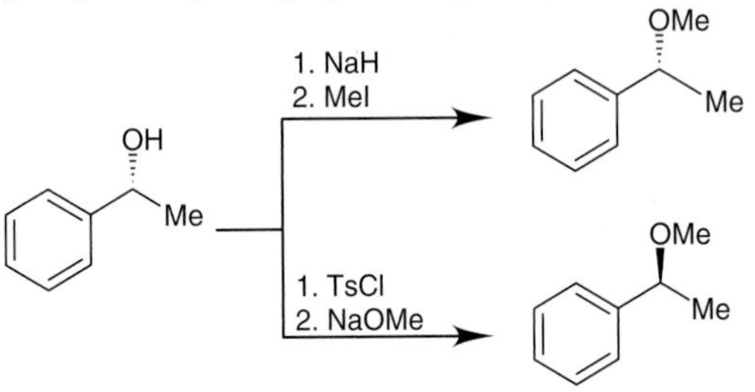

21.33 (a)

Nucleophile Leaving group

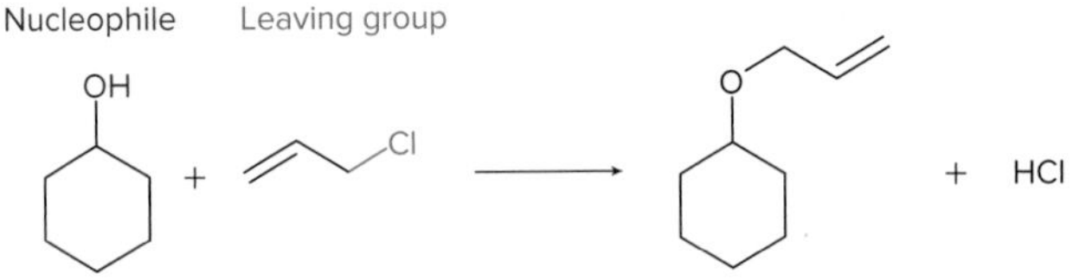

(b)

Nucleophile Leaving group

(c)

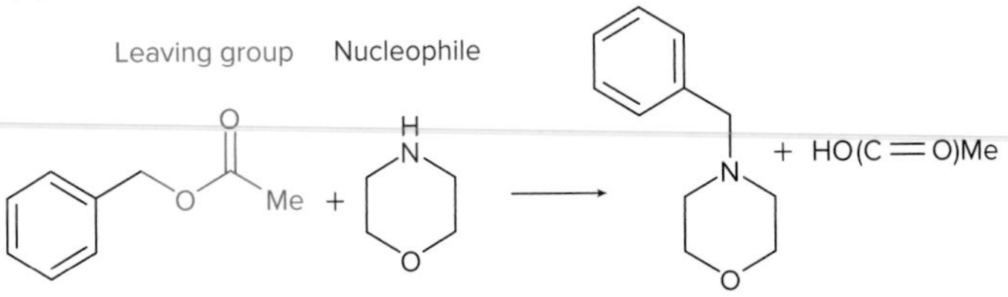

(d)

Nucleophile

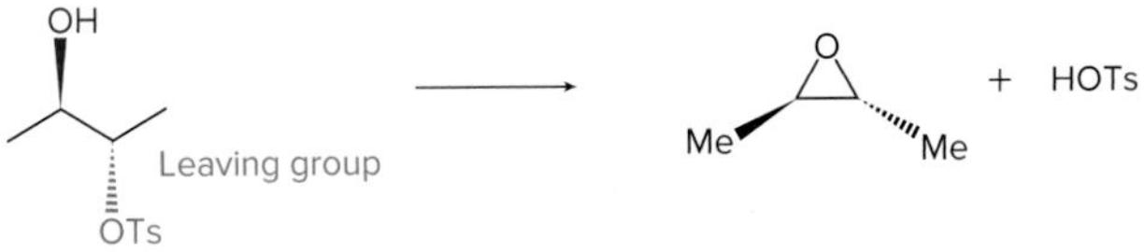

(e)

Leaving group

21.37 (a)

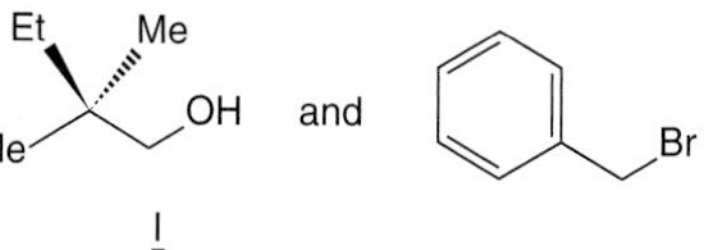

(b)

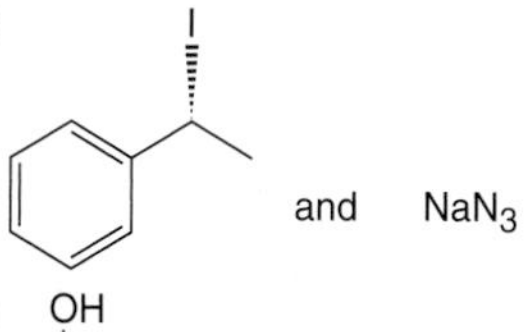

(c)

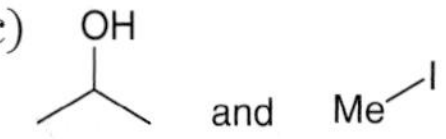

21.38 b **21.41** b **21.43** The most basic is b; the least basic is d. The more stable the anion, the less basic it is. **21.47** a; in the reaction with the unhindered base (MeO^-), the S_N2 reaction will dominate.

21.49

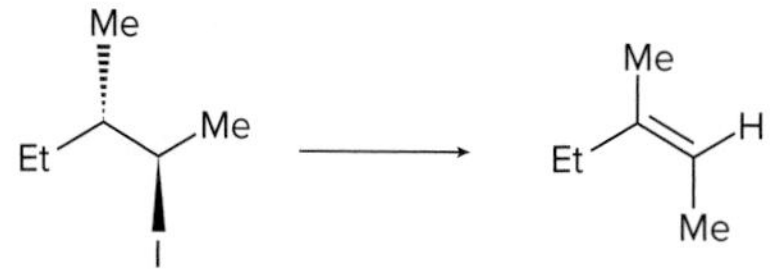

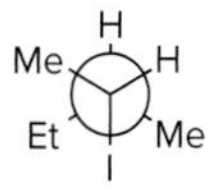

Anti-periplanar conformation

21.51 The sterically hindered base will abstract the more accessible proton.

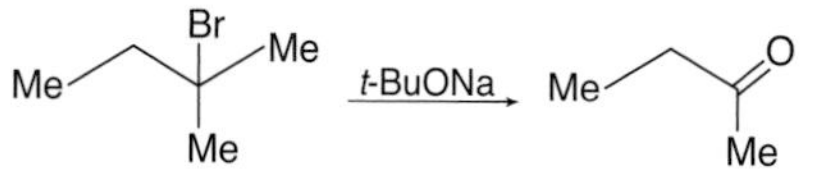

21.53

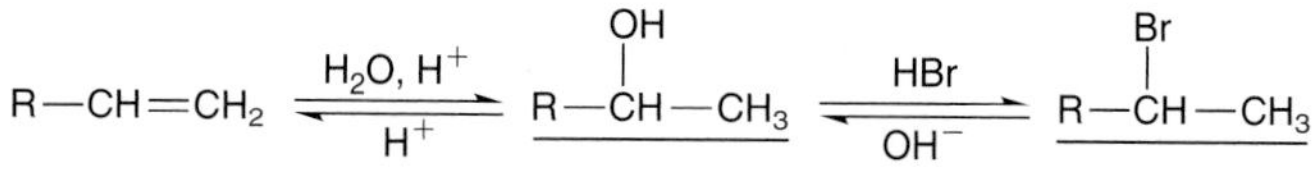

21.56 B > C > A > D

21.57 (a)

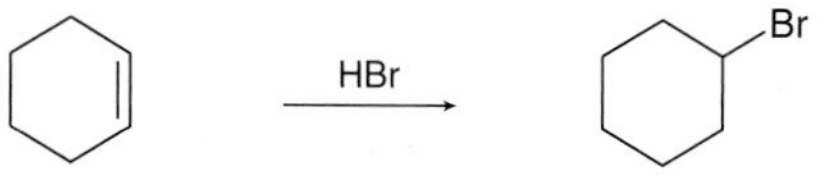

(b)

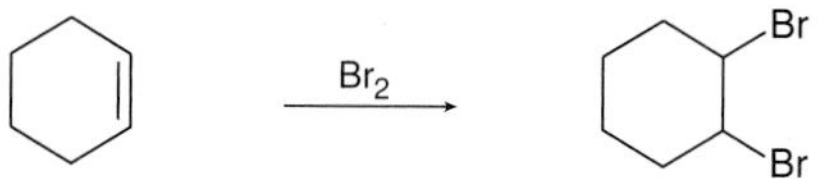

(c)

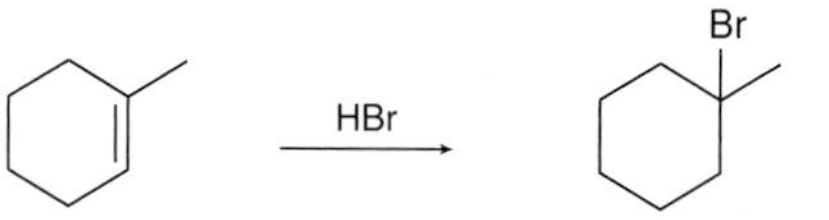

(d)

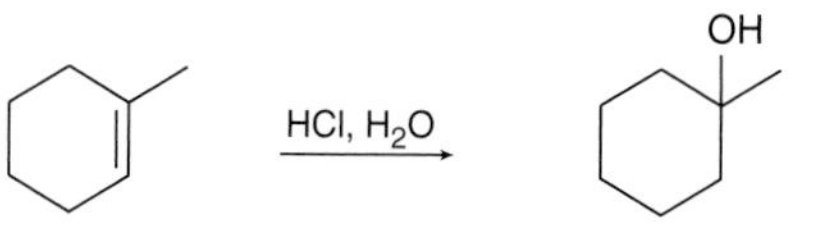

21.60

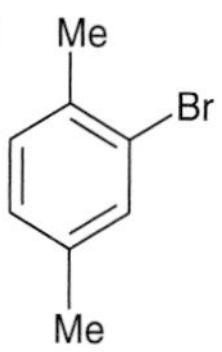

21.61

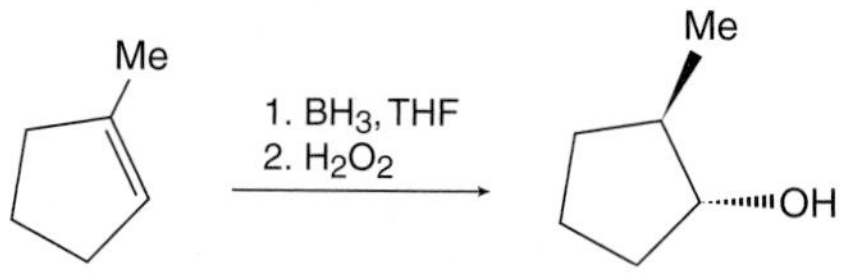

21.63

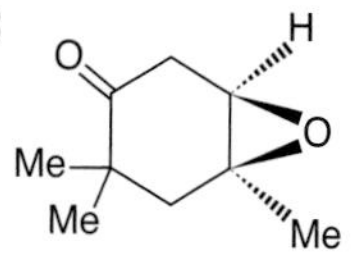

21.66

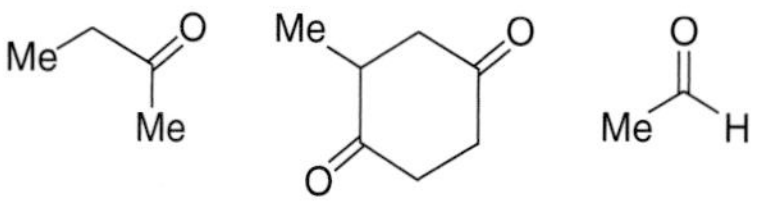

21.68 In general, aromatic rings are much less reactive (more stable) than compounds with isolated C=C double bonds because of their delocalized π electrons. The benzene ring can undergo electrophilic substitution, nucleophilic substitution, elimination, and addition reactions, while the alkenes mainly undergo addition reactions. **21.73** The aniline is a good *ortho/para* director and under weak acidic conditions (right-hand arrow) dominates the course of the reaction. Under strongly acidic conditions, the aniline is protonated, making it a weak *meta* director, and the methyl group (an *ortho/para* directing group) and the protonated aniline give the alternative product (left-hand arrow). **21.74** The C=C bond is nonpolar, while the C=O bond is polar, since oxygen is more electronegative than carbon. Therefore, in the case of addition to a C=O bond, the electron-rich group will bond to the carbon and the electron-poor group will bond to the oxygen, resulting in one product, while the addition to the alkene depends on the structure of the alkene.

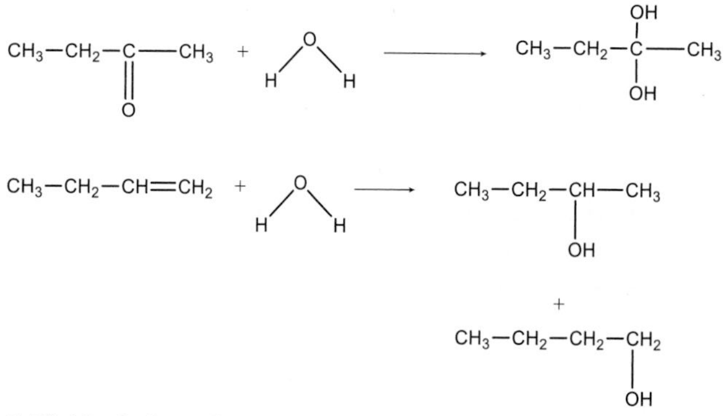

21.78 Alcohols undergo substitution at a saturated carbon, while acids undergo substitution at the carboxyl carbon.

21.81 (a)

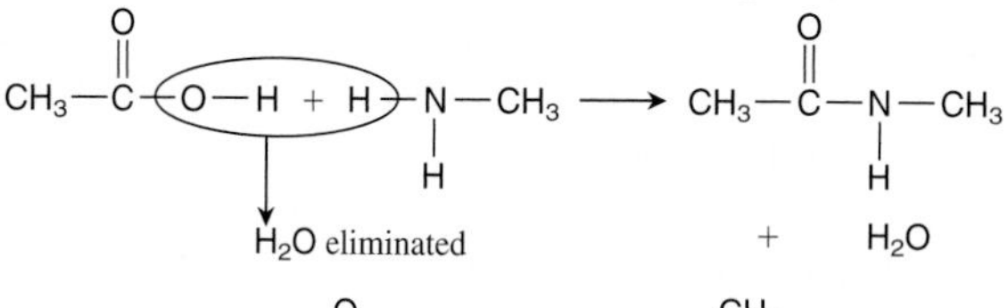

(b)

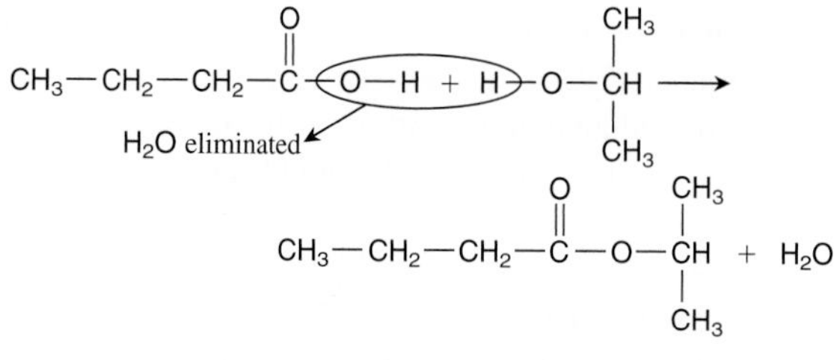

(c)

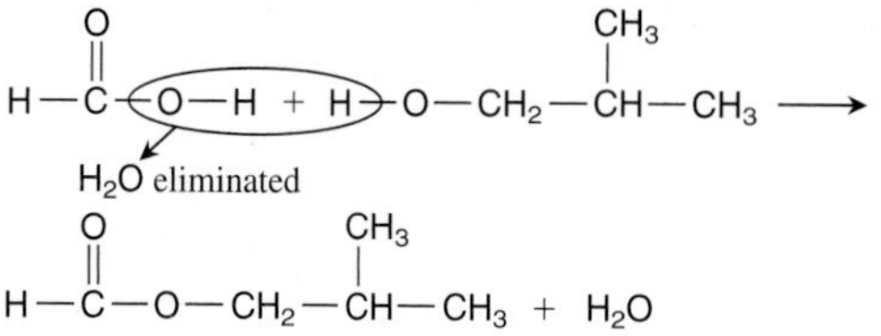

21.84 (a)

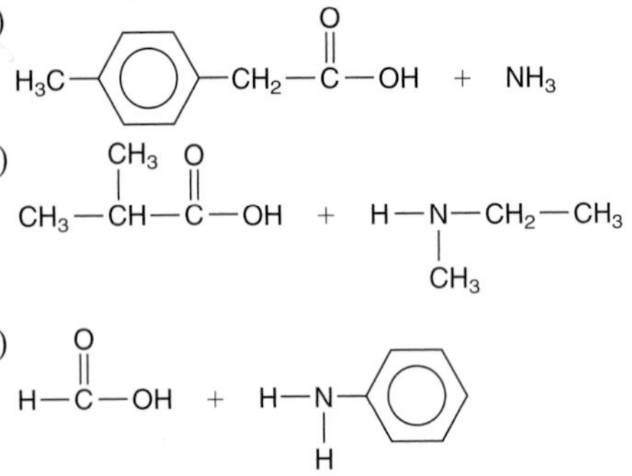

21.85 (a)

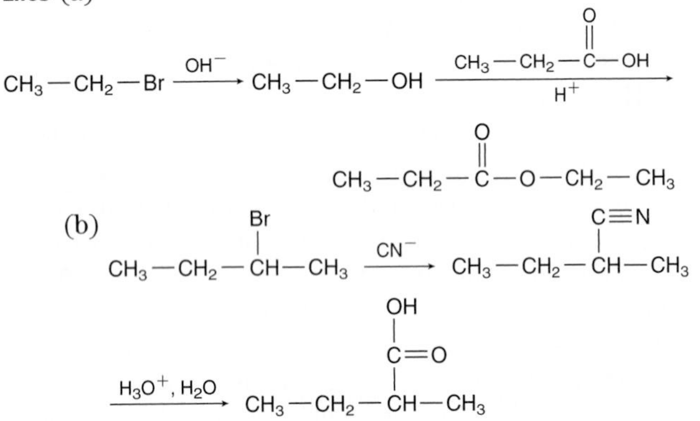

21.91 (a) 116.2 g/mol (b) 102.2 g/mol

21.92 (a) $CH_3CHO + C_6H_5—MgBr \longrightarrow C_6H_5CH(OH)CH_3$

(b)

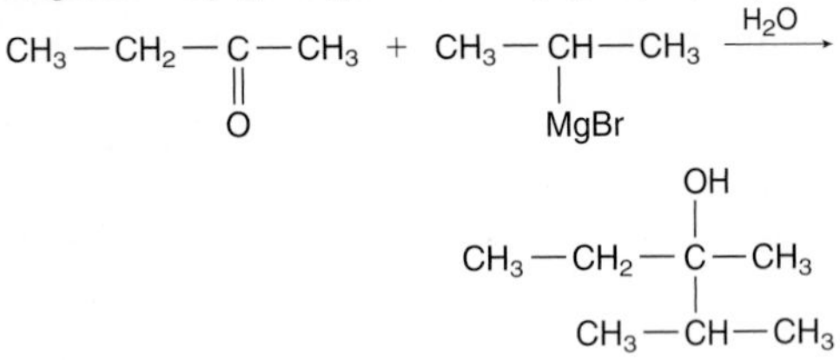

(c) CH_3MgBr and C_6H_5CHO (d) HCHO (methanal, also known as formaldehyde)

(e)

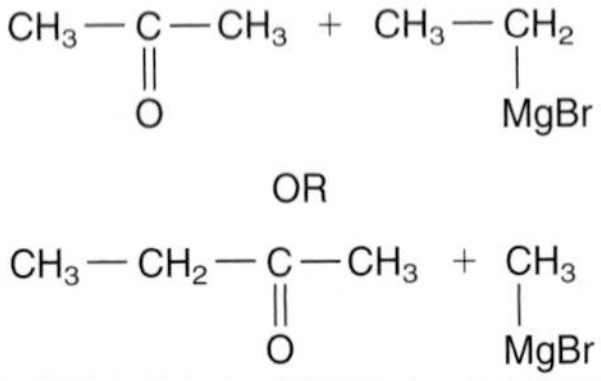

21.93 (a) The functional group in ibuprofen is the carboxylic acid group COOH. The chiral centre is $RC^{*}H(CH_3)COOH$. (b) React the aldehyde with methyl Grignard reagent, CH_3MgBr, to get the alcohol. React the alcohol with HBr to get the brominated product. React the bromide with cyanide ion to produce the nitrile. Then hydrolyze the nitrile with aqueous HCl to get the carboxylic acid. **21.94** (a) Perform an acid-catalyzed dehydration of the alcohol (elimination), followed by bromination of the double bond (addition of Br_2). (b) The product is an ester, so a carboxylic acid is needed to prepare the ester. First, oxidize 1 mol of ethanol to ethanoic acid (acetic acid). Then, react 1 mol of ethanoic acid with 1 mol of ethanol to form the ester.

21.97 (a)

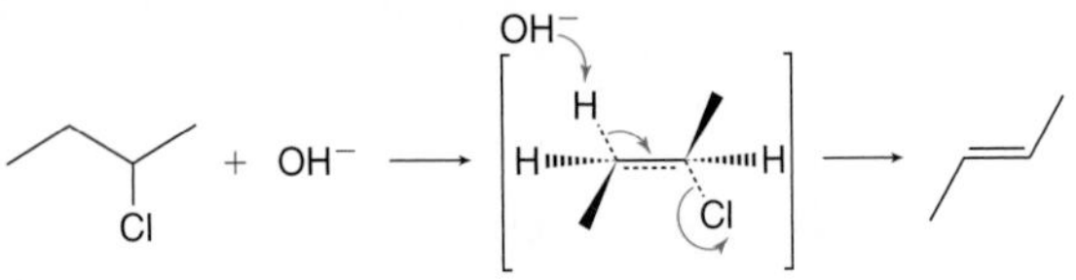

(b)

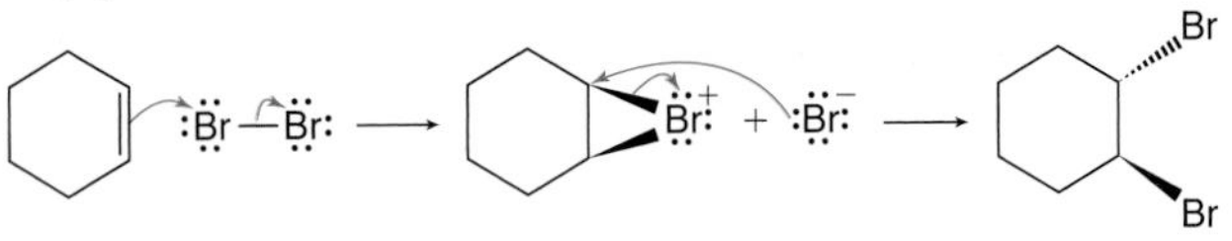

(c)

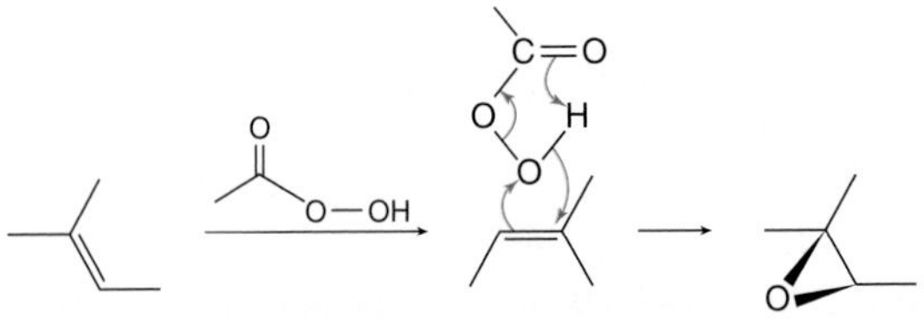

21.99 (a)

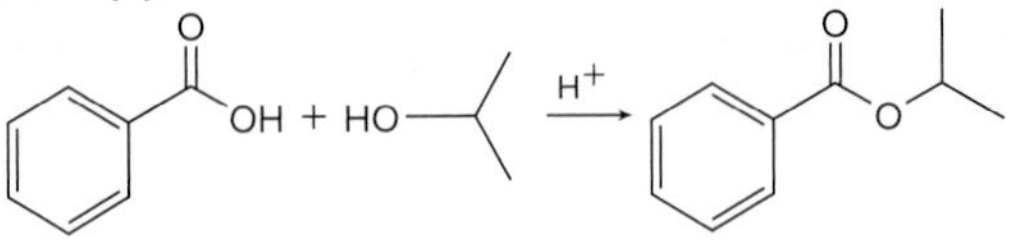

Esterification

(b)

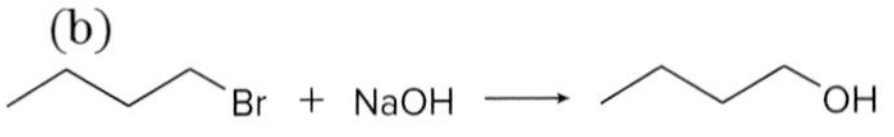

Substitution (S_N2)

(c)

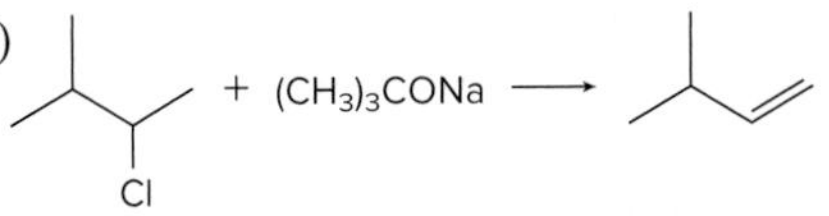

Elimination (E2)

21.101

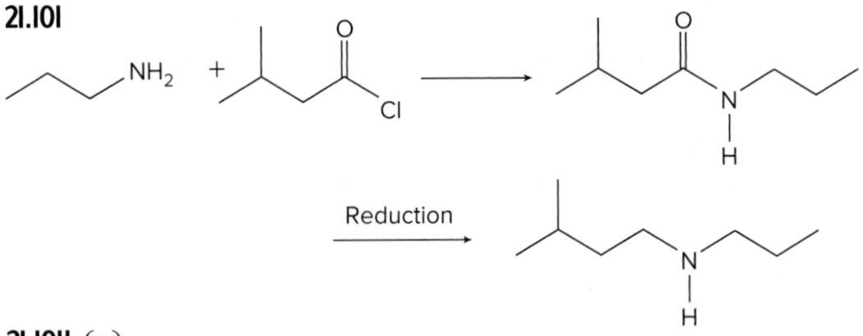

21.104 (a)

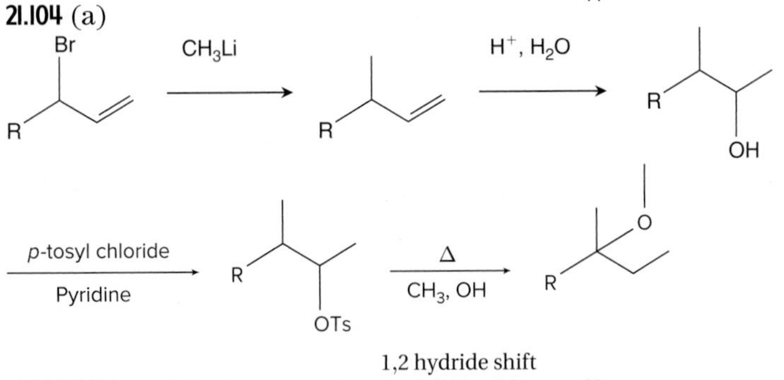

(b) The reaction rate would be doubled. (c) no effect

21.106

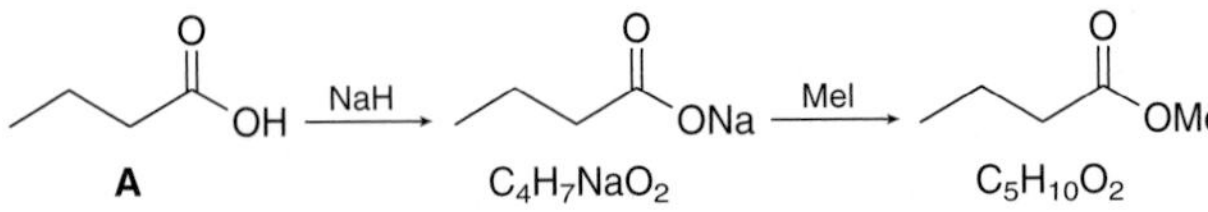

21.108

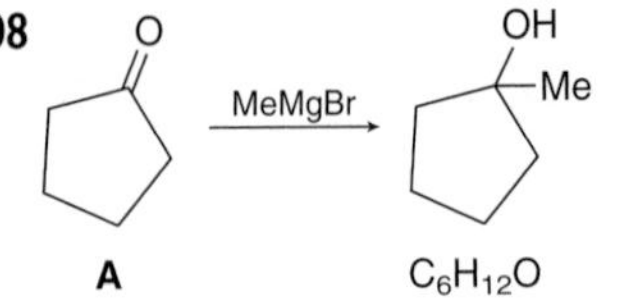

21.110 (a)

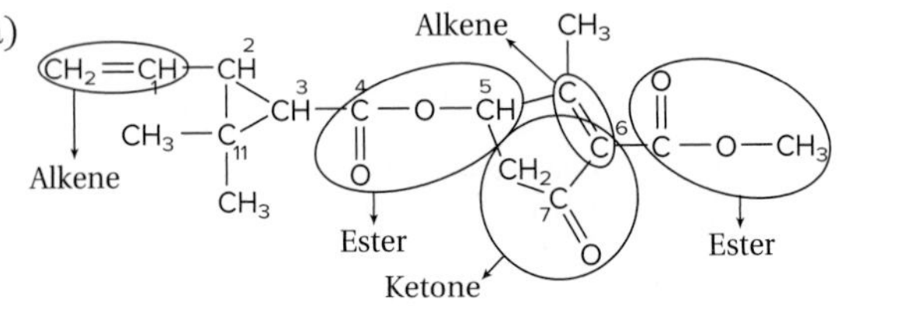

(b) Carbon 1 is sp^2 hybridized. Carbon 2 is sp^3 hybridized. Carbon 3 is sp^3 hybridized. Carbon 4 is sp^2 hybridized. Carbon 5 is sp^3 hybridized. Carbons 6 and 7 are sp^2 hybridized.
(c) Carbon atoms 2, 3, and 5 are chiral centres because they are each bonded to four different groups.

Chapter 22

Answers to Boxed Reading Problems: B22.1 ddATP: four complementary chain pieces; ddCTP: three complementary chain pieces **B22.2** GTATAC

22.1 1.8×10^5g/mol **22.5** 1.7×10^6 pm **22.8** $R_g = 8292$ pm **22.11** The glass transition temperature is the midpoint of the range of temperatures when a semicrystalline substance, like a polymer, changes from a molten or rubber-like substance into a hard brittle solid. The glass transition temperature for the polymer given is 6°C. **22.13** Both branching and cross-linking are bifurcations in a linear polymer, where instead of extending in only one linear direction, now two or more chains exist. In cross-linked polymers, these branches connect two chains together, whereas in a normal branched polymer they do not connect chains together. **22.15** a larger polymer made up from blocks of polymerized monomer units **22.21** Nylon: an amine and a carboxylic acid; polyester: a carboxylic acid and an alcohol **22.23** (a) 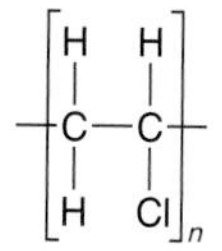 (b) 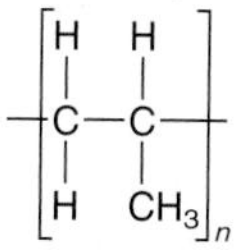

22.26 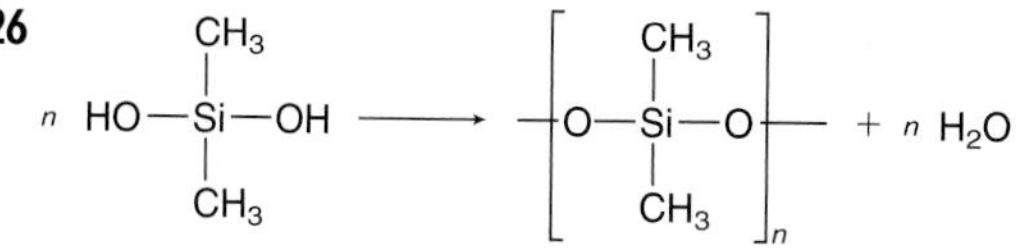

22.28 (a) $C_6H_5—CH═CH_2$ can be used to produce polyphenylethene or polystyrene, —(—CH(C_6H_5) —)—n.
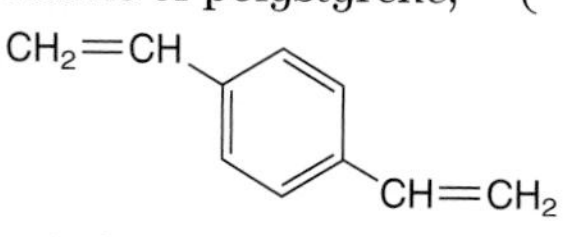

22.32 Isotatic has a regular structure, making its solid form more regular and semicrystalline. As this makes the solid more stable, it will have a higher melting point than the atatic variant. **22.34** Polyamide polymers; they can form hydrogen bonds that stabilize the solid form and raise the glass transition temperature. **22.36** (a) condensation (b) addition (c) condensation (d) condensation **22.40** (a) Water is eliminated when the peptide is formed. (b) glycine: 4, alanine: 1, valine: 3, proline: 6, serine: 7, arginine: 49 (c) 10 700 g/mol **22.44** nucleoside = sugar + nucleobase; nucleotide = sugar + nucleobase + phosphate

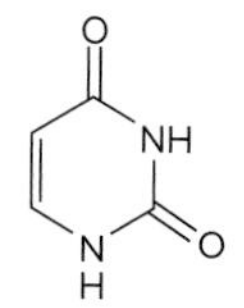
Nucleobase

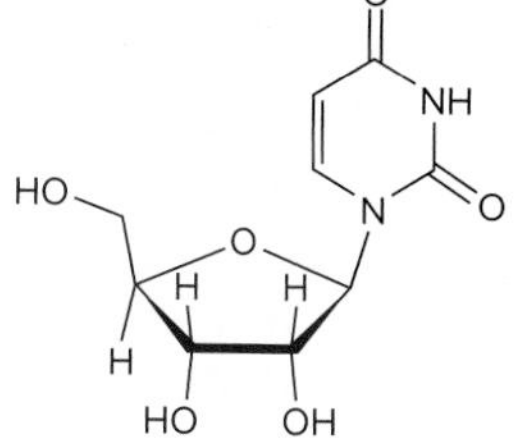
Nucleoside

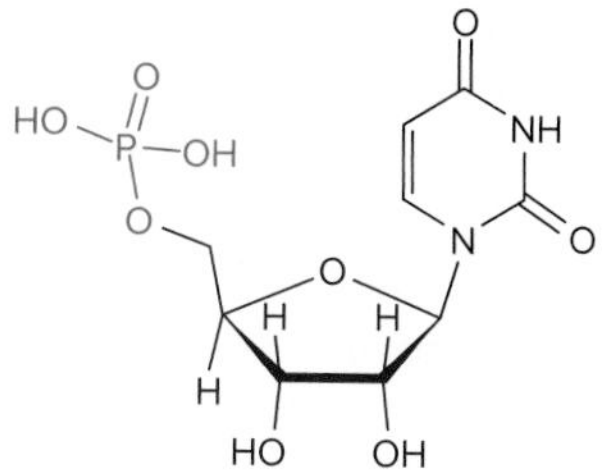
Nucleotide

22.47 (a)
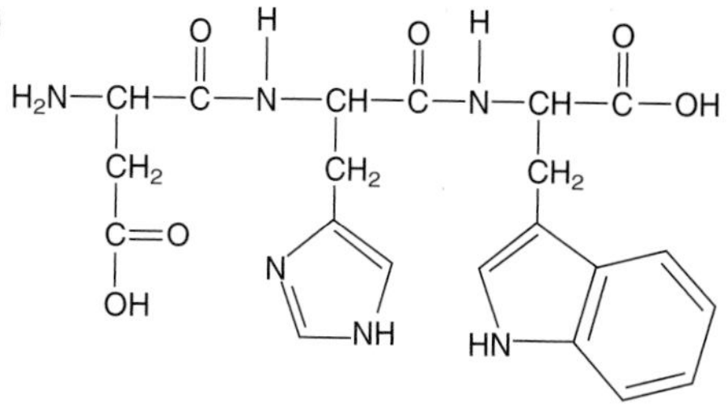

(b)
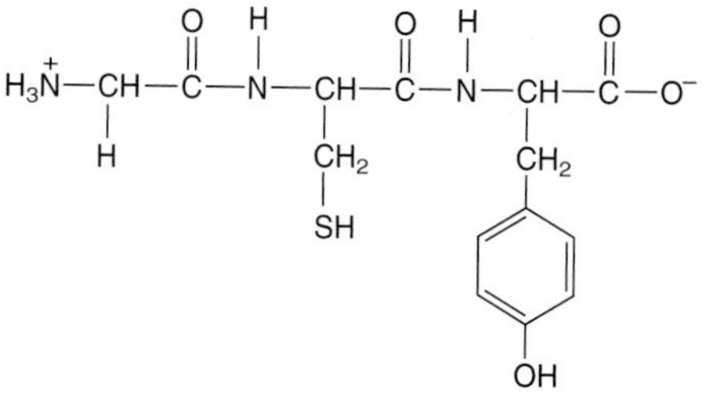

22.49 (a) AATCGG (b) TCTGTA **22.55** The nitrogen-containing bases form hydrogen bonds to their complementary bases. The flat, N-containing bases stack above each other, which allows extensive dispersion forces. The exterior negatively charged sugar-phosphate chains form ion-dipole and hydrogen bonds with water molecules in the aqueous surroundings, which also stabilizes the structure. **22.58** Dispersion forces are present between the nonpolar tails of the lipid molecules within the bilayer. The polar heads interact with the aqueous surroundings through hydrogen bonds and ion-dipole forces. **22.61** secondary **22.64** (a) Both R groups are from cysteine, which can form a disulfide bond (covalent bond). (b) Lysine and aspartic acid give a salt link. (c) Asparagine and serine will hydrogen-bond. (d) Valine and phenylalanine interact through dispersion forces.

22.66 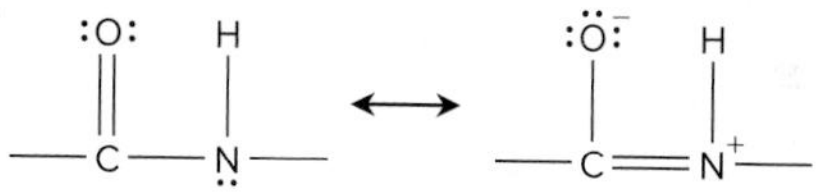

22.71 Because chlorine is a highly electronegative atom, it increases the positive charge on the carbon of the carbonyl compared to propan-2-one (acetone). This in turn makes the carbonyl more polarized and increases the bond strength. As stronger bonds vibrate at larger wavenumbers, it increases the frequency of the absorption. **22.77** four

22.80 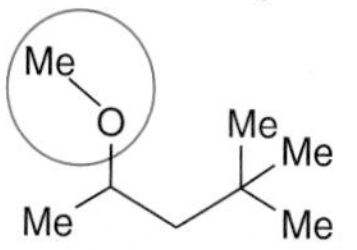

22.81 O—H bonds are readily exchangeable in $CDCl_3$. The protons attached to the hydroxyl groups are clearly seen. When the sample is shaken with D_2O, the deuterium in the D_2O can exchange with OH groups of the sample, converting them to OD groups. Because deuterium atoms cannot be seen in ^{1}H NMR, the OD groups can no longer been seen. **22.83** two **22.86** two; one high medium shifted peak (approx. 2.5 ppm), where multiplicity is a quartet (the CH_2); one high low shifted peak (approx. 1.0 ppm), where multiplicity is a triplet (the CH_3)

22.90 Kevlar is formed from 1,4-phenylenediamine with terephthaloyl chloride.

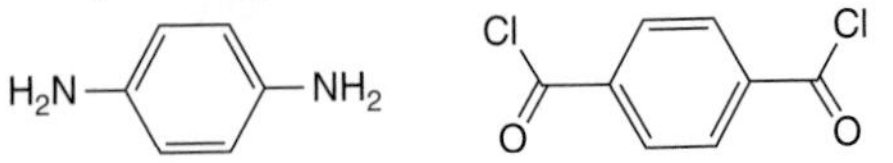

22.95 (a) The complementary sequence to GATCGACTA is CTAGCTGAT. The RNA sequence is different because RNA contains U instead of T, so the sequence is CUAGCUGAU. (b) Three bases are required to code for one amino acid, so this sequence could code for three amino acids. (c) GC is not a good technique to separate this nucleotide sequence from a mixture of other nucleotides because the molecular weight of the sequence and polarity mean that the nucleotide is not volatile enough. Liquid chromatography or better yet gel electrophoresis are better separation techniques.
22.97 (a) a condensation polymerization

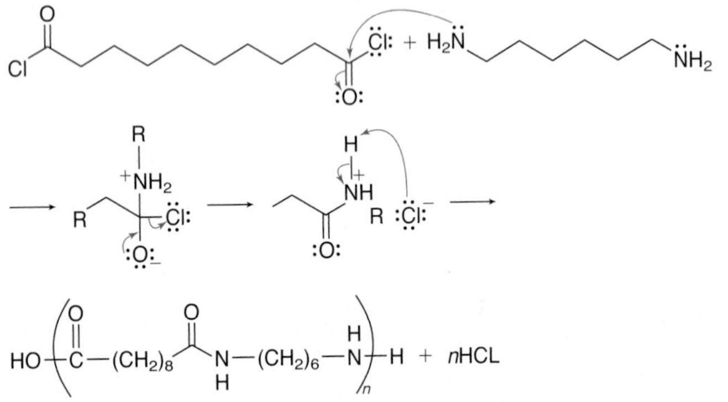

(b) Rubber gloves need to be flexible at room temperature, not hard and brittle, like glass, which is how this polymer becomes when it is cooled below its T_g (50°C). Therefore, it is not suitable for making rubber gloves. (c) Since this is a polymer, size exclusion chromatography (SEC) is likely the best technique to analyze it, though it might also be possible to use liquid chromatography.

Chapter 23

Answers to Boxed Reading Problems: B23.1 (a) overall reaction: $O_3 + O \longrightarrow 2O_2$; step 1: rate $= k[X][O_3]$; step 2: rate $= k[X][O]$ (b) catalyst; intermediate **B23.2** (a) (1) $NO + O_3 \longrightarrow XO + O_2$ [slow], (2) $XO + O \longrightarrow X + O_2$ [fast]
The rate-determining step is the slow step.
(b) rate $= 3 \times 10^7$ molecule/(cm$^3 \cdot$ s)

23.2 Fe from Fe_2O_3; Ca from $CaCO_3$; Na from NaCl; Zn from ZnS **23.3** (a) Differentiation refers to the processes involved in the formation of Earth into regions (core, mantle, and crust) of differing composition. Substances separated according to their densities, with the more dense material in the core and the less dense in the crust. (b) O, Si, Al, and Fe (c) O **23.7** Plants produced O_2, slowly increasing the oxygen concentration in the atmosphere and creating an environment for oxidizing metals. The oxygen-free decay of plant and animal material created large fossil fuel deposits. **23.9** Fixation refers to the process of converting a substance in the atmosphere into a form more readily usable by organisms. Carbon and nitrogen; fixation of carbon dioxide gas by plants and fixation of nitrogen gas by nitrogen-fixing bacteria. **23.12** Atmospheric nitrogen is fixed by three pathways: atmospheric, industrial, and biological. Atmospheric fixation requires high-temperature reactions (e.g., initiated by lightning) to convert N_2 into NO and other oxidized species. Industrial fixation involves mainly the formation of ammonia, NH_3, from N_2 and H_2. Biological fixation occurs in nitrogen-fixing bacteria that live on the roots of legumes. Human activity is an example of industrial fixation. It contributes about 17% of the fixed nitrogen. **23.14** (a) the atmosphere (b) Plants excrete acid from their roots to convert PO_4^{3-} ions into more soluble $H_2PO_4^-$ ions, which the plant can absorb. Through excretion and decay, organisms return soluble phosphate compounds to the cycle. **23.17** (a) 1.1×10^3 L (b) 4.2×10^2 m^3 **23.18** (a) The iron(II) ions form an insoluble salt, $Fe_3(PO_4)_2$, that decreases the yield of phosphorus. (b) 8.8 t **23.20** (a) Roasting consists of heating the mineral in air at high temperatures to convert the mineral to the oxide. (b) Smelting is the reduction of the metal oxide to the free metal using heat and a reducing agent such as coke. (c) Flotation is a separation process in which the ore is removed from the gangue by exploiting the difference in density in the presence of detergent. The gangue sinks to the bottom and the lighter ore-detergent mix is skimmed off the top. (d) Refining is the final step in the purification process to yield the pure metal. **23.25** (a) Slag is a by-product of steel-making and contains the impurity SiO_2. (b) Pig iron is the impure product of iron metallurgy (containing 3%–4% C and other impurities). (c) Steel refers to iron alloyed with other elements to attain desirable properties. (d) The basic-oxygen process is used to purify pig iron and obtain carbon steel. **23.27** Iron and nickel are more easily oxidized and less easily reduced than copper. They are separated from copper in the roasting step and converted to slag. In the electrorefining process, all three metals are in solution, but only Cu^{2+} ions are reduced at the cathode to form Cu(s). **23.30** According to Le Châtelier's principle, the system shifts toward formation of K when the potassium gas is removed as it is produced. **23.31** (a) $E^\circ_{\text{half-cell}} = -3.05$ V, -2.93 V, and -2.71 V for Li^+, K^+, and Na^+, respectively. In all of these cases, it is energetically more favourable to reduce H_2O to H_2 than to reduce M^+ to M. (b) $2RbX + Ca \longrightarrow CaX_2 + 2Rb$, where $\Delta H = IE_1(Ca) + IE_2(Ca) - 2IE_1(Rb) = 929$ kJ/mol. Based on the IEs and positive ΔH for the forward reaction, it seems more reasonable that Rb metal will reduce Ca^{2+} than the reverse. (c) If the reaction is carried out at a temperature greater than the boiling point of Rb, the product mixture will contain gaseous Rb, which can be removed from the reaction vessel; this would cause a shift in the equilibrium to form more Rb as product. (d) $2CsX + Ca \longrightarrow CaX_2 + 2Cs$, where $\Delta H = IE_1(Ca) + IE_2(Ca) - 2IE_1(Cs) = 983$ kJ/mol. This reaction is more unfavourable than for Rb, but Cs has a lower boiling point. Ca can be used to separate gaseous Cs from molten CsX if the reaction is carried out at a temperature between the boiling points of Cs and Ca. **23.32** (a) 4.5×10^4 L (b) 1.30×10^8 C (c) 1.69×10^6 s **23.35** (a) Mg^{2+} is more difficult to reduce

than H_2O, so $H_2(g)$ would be produced instead of Mg metal. $Cl_2(g)$ forms at the anode due to overvoltage. (b) The $\Delta_f H°$ of $MgCl_2(s)$ is −641.6 kJ/mol. High temperature favours the reverse (endothermic) reaction, the formation of magnesium metal and chlorine gas. **23.37** (a) Sulfur dioxide is the reducing agent and is oxidized to the +6 state (SO_4^{2-}). (b) $HSO_4^-(aq)$ (c) $H_2SeO_3(aq) + 2SO_2(g) + H_2O(l) \longrightarrow Se(s) + 2HSO_4^-(aq) + 2H^+(aq)$ **23.42** (a) O.N. for Cu: in Cu_2S, +1; in Cu_2O, +1; in Cu, 0 (b) Cu_2S is the reducing agent, and Cu_2O is the oxidizing agent. **23.44** (a) 1.3×10^6 C (b) 1.2×10^3 A **23.47** $2ZnS(s) + C(graphite) \longrightarrow 2Zn(s) + CS_2(g)$; $\Delta_r G° = 463$ kJ/mol. Since $\Delta_r G°$ is positive, this reaction is not spontaneous at standard-state conditions. $2ZnO(s) + C(s) \longrightarrow 2Zn(s) + CO_2(g)$; $\Delta_r G° = 242.0$ kJ/mol. This reaction is also not spontaneous, but it is less unfavourable. **23.48** The formation of sulfur trioxide is very slow at ordinary temperatures. Increasing the temperature can speed up the reaction, but because the reaction is exothermic, increasing the temperature decreases the yield. Adding a catalyst increases the rate of the reaction, allowing a lower temperature to be used to enhance the yield. **23.51** (a) Cl_2, H_2, and NaOH (b) The mercury-cell method yields higher purity NaOH, but releases some Hg, which is released into the environment. **23.52** (a) $\Delta G° = -142$ kJ/mol; yes (b) The rate of the reaction is very low at 25°C. (c) $\Delta_{500°C} G° = -53$ kJ/mol, so the reaction is spontaneous. (d) $K_{25°C} = 7.8 \times 10^{24}$; $K_{500°C} = 3.8 \times 10^3$ (e) 1.05×10^3 K **23.53** 3×10^2 kg Cl_2 **23.56** (a) $P_4O_{10}(s) + 6H_2O(l) \longrightarrow 4H_3PO_4(l)$ (b) 1.52 **23.58** (a) 9.007×10^9 g CO_2 (b) The 4.3×10^{10} g CO_2 produced by automobiles is much greater than that from the blast furnace. **23.60** (a) If $[OH^-] > 1.1 \times 10^{-4}$ mol/L (i.e., if pH > 10.04), $Mg(OH)_2$ will precipitate. (b) 1 (To the correct number of significant figures, all the magnesium precipitates.) **23.61** (a) $K_{25°C}$ (step 1) $= 1 \times 10^{168}$; $K_{25°C}$ (side rxn) $= 7 \times 10^{228}$; (b) $K_{900°C}$ (step 1) $= 4.5 \times 10^{49}$; $K_{900°C}$ (side rxn) $= 1.4 \times 10^{63}$ (c) \$5.8 × 10^7 (d) \$4.2 × 10^7
23.64 (1) $2H_2O(l) + 2FeS_2(s) + 7O_2(g) \longrightarrow 2Fe^{2+}(aq) + 4SO_4^{2-}(aq) + 4H^+(aq)$ increases acidity
(2) $4H^+(aq) + 4Fe^{2+}(aq) + O_2(g) \longrightarrow 4Fe^{3+}(aq) + 2H_2O(l)$
(3) $Fe^{3+}(aq) + 3H_2O(l) \longrightarrow Fe(OH)_3(s) + 3H^+(aq)$ increases acidity
(4) $8H_2O(l) + FeS_2(s) + 14Fe^{3+}(aq) \longrightarrow 15Fe^{2+}(aq) + 2SO_4^{2-}(aq) + 16H^+(aq)$ increases acidity
23.65 density of ferrite: 7.86 g/cm^3; density of austenite: 7.55 g/cm^3
23.67 (a) Cathode: $Na^+(l) + e^- \longrightarrow Na(l)$
Anode: $4OH^-(l) \longrightarrow O_2(g) + 2H_2O(g) + 4e^-$
(b) 50%
23.70 (a) $nCO_2(g) + nH_2O(l) \longrightarrow (CH_2O)_n(s) + nO_2(g)$ (b) 27 L (c) 7.6×10^4 L **23.71** 73 mg/L **23.72** 892 kg Na_3AlF_6 **23.73** (a) 23.2 min (b) 13 effusion steps **23.75** (a) 1.890 t Al_2O_3 (b) 0.3339 t C (c) 100% (d) 74% (e) 2.813×10^3 m^3 **23.80** Acid rain increases the leaching of phosphate into the groundwater, due to the protonation of PO_4^{3-} to form HPO_4^{2-} and $H_2PO_4^-$. (a) 6.4×10^{-7}mol/L (b) 1.1×10^{-2} mol/L **23.81** (a) 1.00 mol % (b) 238.9 g/mol **23.83** density of silver = 10.51 g/cm^3; density of sterling silver = 10.2 g/cm^3

Chapter 24

24.2 (a) $1s^22s^22p^63s^23p^64s^23d^{10}4p^65s^24d^x$ (b) $1s^22s^22p^63s^23p^64s^23d^{10}4p^65s^24d^{10}5p^66s^24f^{14}5d^x$ **24.4** (a) five; (b) Examples are Mn, [Ar]$4s^23d^5$, and Fe^{3+}, [Ar]$3d^5$. **24.6** (a) The elements should increase in size as they increase in mass from period 5 to period 6. Because there are 14 inner transition elements in period 6, the effective nuclear charge increases significantly, so the atomic size decreases, or "contracts." This effect is significant enough that Zr^{4+} and Hf^{4+} are almost the same size but differ greatly in atomic mass. (b) The atomic size increases from period 4 to period 5, but stays fairly constant from period 5 to period 6. (c) Atomic mass increases significantly from period 5 to period 6, but atomic radius (and thus volume) increase slightly, so period 6 elements are very dense. **24.9** (a) A paramagnetic substance is attracted to a magnetic field, while a diamagnetic substance is slightly repelled by one. (b) Ions of transition elements often have partially filled *d* orbitals whose unpaired electrons make the ions paramagnetic. Ions of main-group elements usually have a noble-gas configuration with no partially filled levels. (c) Some *d* orbitals in the transition element ions are empty, which allows an electron from one *d* orbital to move to a slightly higher energy one. The energy required for this transition is small and falls in the visible wavelength range. All orbitals are filled in ions of main-group elements, so enough energy would have to be added to move an electron to the next principal energy level, not just another orbital within the same energy level. This amount of energy is very large and much greater than the visible range of wavelengths. **24.10** (a) $1s^22s^22p^63s^23p^64s^23d^3$ (b) $1s^22s^22p^63s^23p^64s^23d^{10}4p^65s^24d^1$ (c) [Xe]$4f^{14}5d^{10}6s^2$ **24.12** (a) [Xe]$4f^{14}5d^66s^2$ (b) [Ar]$3d^74s^2$ (c) [Kr]$4d^{10}5s^1$ **24.14** (a) [Ar], no unpaired electrons (b) [Ar]$3d^9$, one unpaired electron (c) [Ar]$3d^5$, five unpaired electrons (d) [Kr]$4d^2$, two unpaired electrons **24.16** (a) +5 (b) +4 (c) +7 **24.18** Cr, Mo, and W **24.20** in CrF_2, because the chromium is in a lower oxidation state **24.22** Atomic size increases slightly down a group of transition elements, but nuclear charge increases much more, so the first ionization energy generally increases. The reduction potential for Mo is lower, so it is more difficult to oxidize Mo than Cr. In addition, the ionization energy of Mo is higher than that of Cr, so it is more difficult to remove electrons from Mo.
24.24 CrO_3, with Cr in a higher oxidation state, yields a more acidic aqueous solution. **24.28** (a) seven (b) This corresponds to a half-filled *f* subshell. **24.30** (a) [Xe]$5d^16s^2$ (b) [Xe]$4f^1$ (c) [Rn]$5f^{11}7s^2$ (d) [Rn]$5f^2$ **24.32** (a) Eu^{2+}: [Xe]$4f^7$; Eu^{3+}: [Xe]$4f^6$; Eu^{4+}: [Xe]$4f^5$. The stability of the half-filled *f* subshell makes Eu^{2+} most stable. (b) Tb^{2+}: [Xe]$4f^9$; Tb^{3+}: [Xe]$4f^8$; Tb^{4+}: [Xe]$4f^7$. Tb should show a +4 oxidation state because that gives a half-filled subshell. **24.34** Gd has the electron configuration [Xe]$4f^75d^16s^2$ with eight unpaired electrons. Gd^{3+} has seven unpaired

electrons: [Xe]$4f^7$. **24.37** The coordination number indicates the number of ligand atoms bonded to the metal ion. The oxidation number represents the number of electrons lost to form the ion. The coordination number is unrelated to the oxidation number. **24.39** 2, linear; 4, tetrahedral or square planar; 6, octahedral **24.42** The complex ion has a negative charge. **24.45** (a) hexaaquanickel(II) chloride (b) tris(ethylenediamine)chromium(III) perchlorate (c) potassium hexacyanomanganate(II) **24.47** (a) +2, 6 (b) +3, 6 (c) +2, 6 **24.49** (a) potassium dicyanoargentate(I) (b) sodium tetrachlorocadmate(II) (c) tetraammineaquabromocobalt(III) bromide **24.51** (a) +1, 2 (b) +2, 4 (c) +3, 6 **24.53** (a) $[Zn(NH_3)_4]SO_4$ (b) $[Cr(NH_3)_5Cl]Cl_2$ (c) $Na_3[Ag(S_2O_3)_2]$ **24.55** (a) 4, two ions (b) 6, three ions (c) 2, four ions **24.57** (a) $[Cr(H_2O)_6]_2(SO_4)_3$ (b) $Ba[FeBr_4]_2$ (c) $[Pt(en)_2]CO_3$ **24.59** (a) 6, five ions (b) 4, three ions (c) 4, two ions **24.61** (a) The nitrite ion forms linkage isomers because it can bind to the metal ion through the lone pair on the N atom or any lone pair on either O atom.

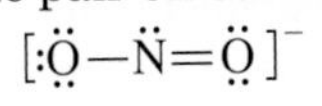

(b) Sulfur dioxide molecules form linkage isomers because the lone pair on the S atom or any lone pair on either O atom can bind the central metal ion.

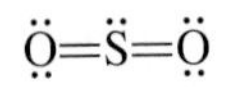

(c) Nitrate ions have an N atom, with no lone pair, and three O atoms, all with lone pairs that can bind to the metal ion. But all of the O atoms are equivalent, so these ions do not form linkage isomers.

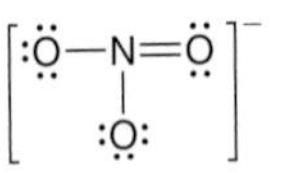

24.63 (a) geometric isomerism

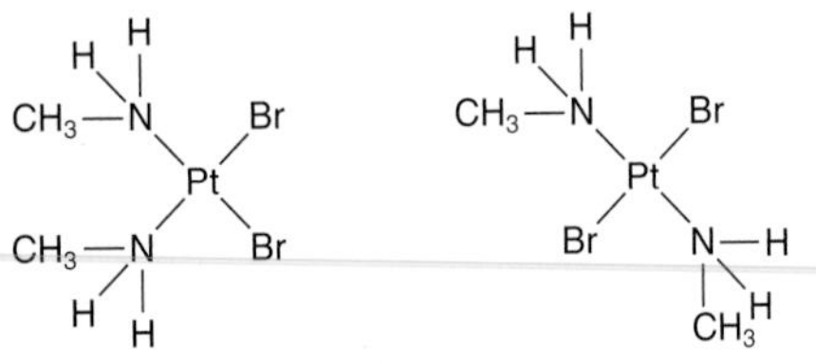

(b) geometric isomerism

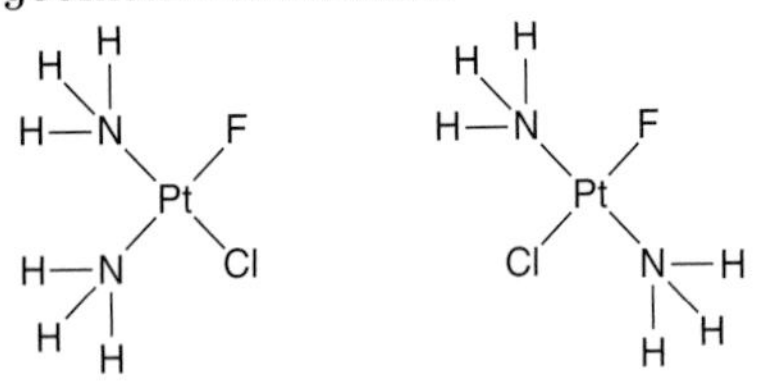

(c) geometric isomerism

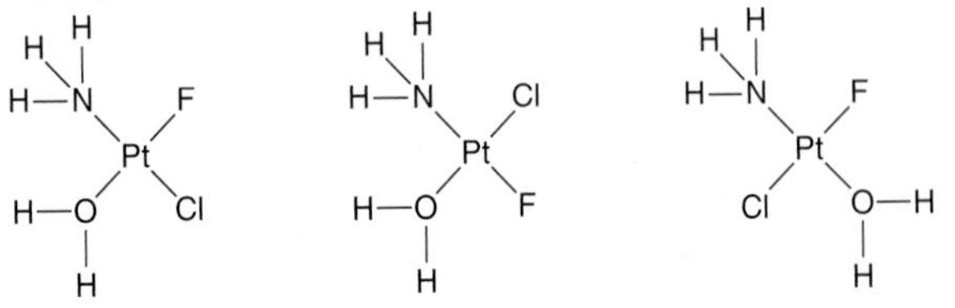

24.65 (a) geometric isomerism

(b) linkage isomerism

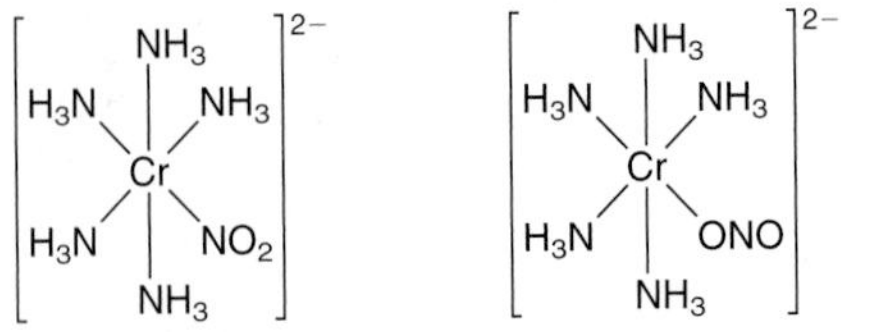

(c) geometric isomerism

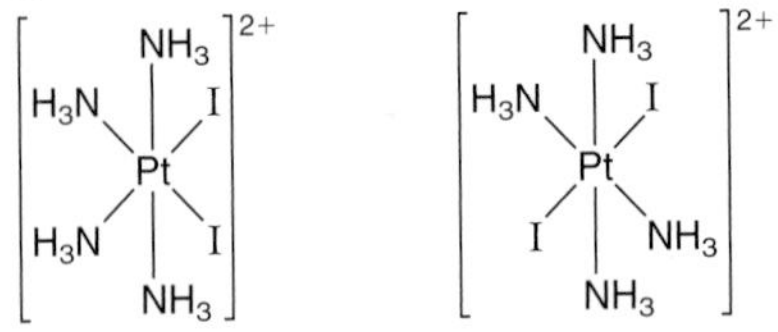

24.67 The compound with the traditional formula is $CrCl_3 \cdot 4NH_3$; the actual formula is $[Cr(NH_3)_4Cl_2]Cl$. **24.69** (a) $K[Pd(NH_3)Cl_3]$ (b) $[PdCl_2(NH_3)_2]$ (c) $K_2[PdCl_6]$ (d) $[Pd(NH_3)_4Cl_2]Cl_2$ **24.71** (a) dsp^2 (b) sp^3 **24.74** absorption of orange or yellow light **24.75** (a) The crystal field splitting energy (Δ) is the energy difference between the two sets of d orbitals that result from electrostatic effects of ligands on a central transition metal atom. (b) In an octahedral field of ligands, the ligands approach along the x, y, and z axes. The $d_{x^2-y^2}$ and d_{z^2} orbitals are located *along* the x, y, and z axes, so ligand interaction is higher in energy. The other orbital-ligand interactions are lower in energy because the d_{xy}, d_{yz}, and d_{xz} orbitals are located *between* the x, y, and z axes. (c) In a tetrahedral field of ligands, the ligands do not approach along the x, y, and z axes. The ligand interaction is greater for the d_{xy}, d_{yz}, and d_{xz} orbitals and lesser for the $d_{x^2-y^2}$ and d_{z^2} orbitals. Therefore, the crystal field splitting is reversed, and the d_{xy}, d_{yz}, and d_{xz} orbitals are higher in energy than the $d_{x^2-y^2}$ and d_{z^2} orbitals. **24.78** If Δ is greater than $E_{pairing}$, electrons will pair their spins in the lower energy set of d orbitals before entering the higher energy set of d orbitals as unpaired electrons. If Δ is less than $E_{pairing}$, electrons will occupy the higher energy set of d orbitals as unpaired electrons before pairing in the lower energy set of d orbitals. **24.80** (a) no d electrons (b) eight d electrons (c) six d electrons **24.82** (a) five (b) ten (c) seven **24.84** a and d cannot form high- and low-spin complexes

24.86 (a) 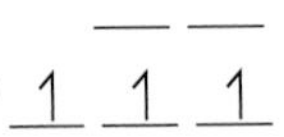 (b) 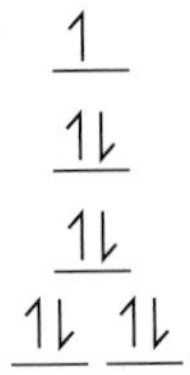 (c) 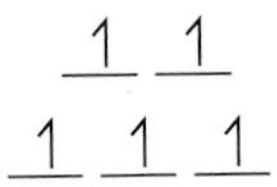

24.88 (a) 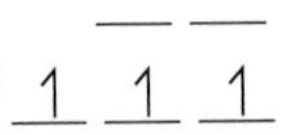 (b) 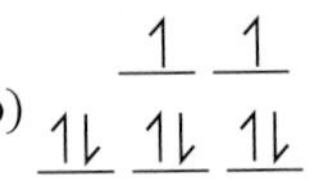 (c) 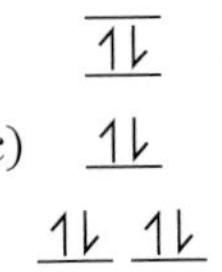

24.90 $[Cr(H_2O)_6]^{3+} < [Cr(NH_3)_6]^{3+} < [Cr(NO_2)_6]^{3-}$

24.92

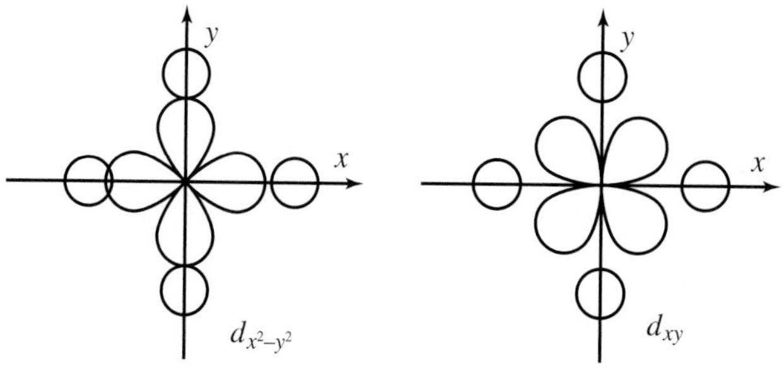

In an octahedral field of ligands, the ligands approach along the x, y, and z axes. The $d_{x^2-y^2}$ orbital is located along the x and y axes, so ligand interaction is greater. The d_{xy} orbital is offset from the x and y axes by 90°, so ligand interaction is less. The greater interaction of the $d_{x^2-y^2}$ orbital results in its higher energy. **24.94** A violet complex absorbs yellow-green light. The light absorbed by a complex with a weaker field ligand would be at a lower energy and higher wavelength. Light of lower energy than yellow-green light is yellow, orange, or red. The colour observed would be blue or green. **24.97** The H_2O ligand is weaker than the NH_3 ligand. The weaker field ligand results in a lower splitting energy and absorbs visible light of lower energy. The hexaaqua complex appears green because it absorbs red light. The hexaammine complex appears violet because it absorbs yellow light, which has higher energy (shorter λ) than red light. **24.101** Hg^+ is $[Xe]4f^{14}5d^{10}6s^1$ and Cu^+ is $[Ar]3d^{10}$. The mercury(I) ion has one electron in the 6s orbital that can form a covalent bond with the electron in the 6s orbital of another mercury(I) ion. In the copper(I) ion there are no electrons in the s orbital, so these ions cannot bond with one another. **24.102** (a) 6 (b) +3 (c) two (d) 1 mol
24.109 geometric (*cis-trans*) and linkage isomerism

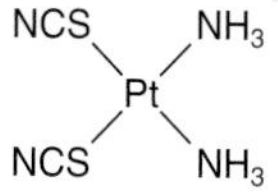

cis-diamminedithiocyanatoplatinum(II)

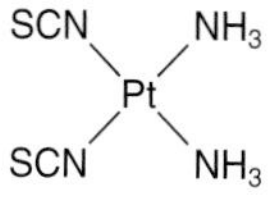

cis-diamminediisothiocyanatoplatinum(II)

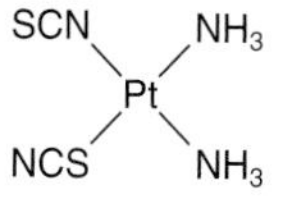

cis-diamminethiocyanatoisothiocyanatoplatinum(II)

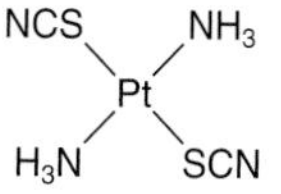

trans-diamminedithiocyanatoplatinum(II)

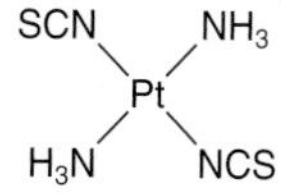

trans-diamminediisothiocyanatoplatinum(II)

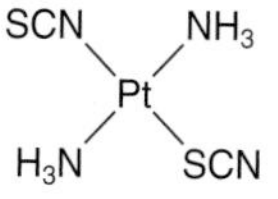

trans-diamminethiocyanatoisothiocyanatoplatinum(II)

24.110

(a) $[Co(NH_3)_4(H_2O)Cl]^{2+}$ tetraammineaquachlorocobalt(III) ion
2 geometric isomers

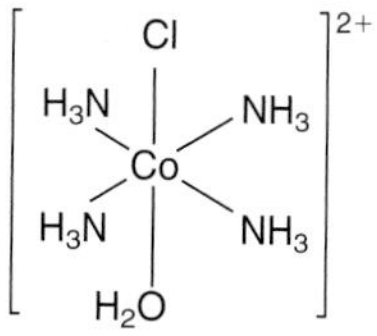

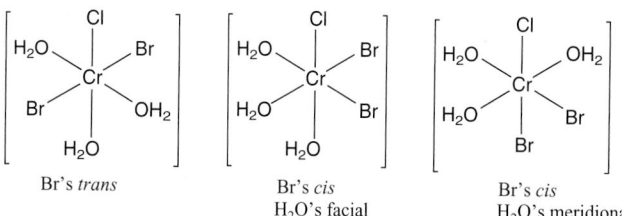

trans Cl and H_2O

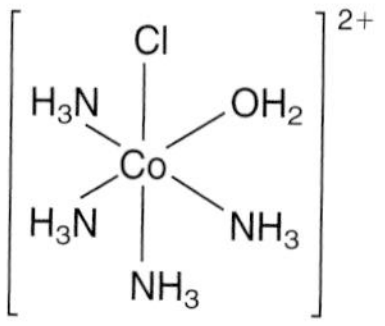

cis Cl and NH_3

(b) $[Cr(H_2O)_3Br_2Cl]$ triaquadibromochlorochromium(III)
3 geometric isomers

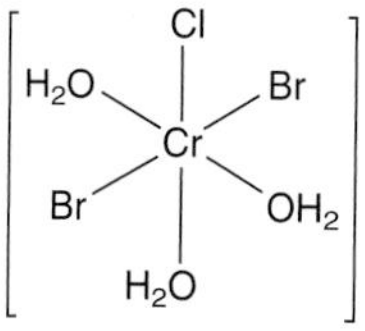

Br's *trans*

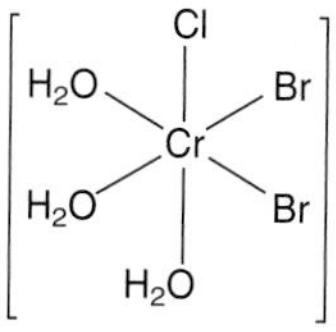

Br's *cis*
H_2O's facial

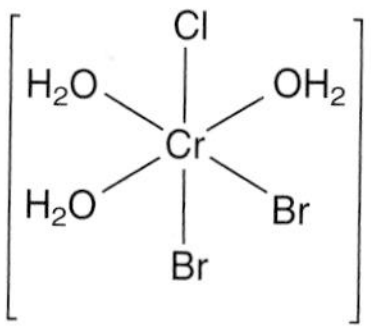

Br's *cis*
H_2O's meridional

(c) $[Cr(NH_3)_2(H_2O)_2Br_2]^+$ diamminediaquadibromochromium(III) ion
6 isomers (5 geometric)

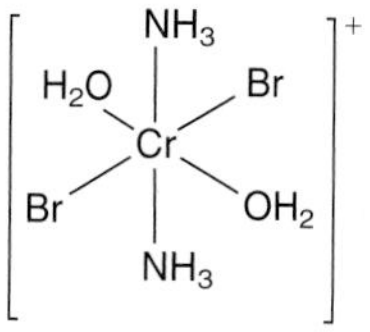

All pairs are *trans*

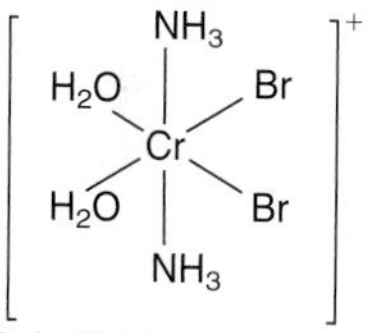

Only NH_3's are *trans*

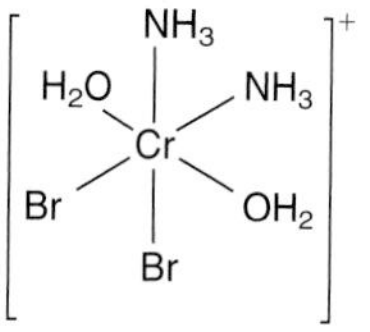

Only H_2O's are *trans*

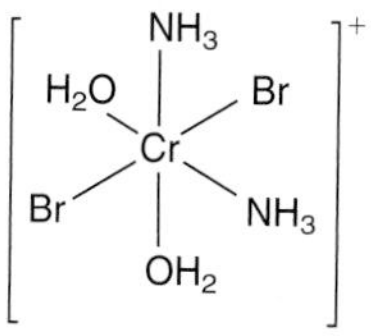

Only Br's are *trans*

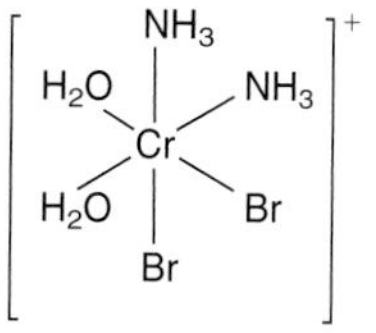

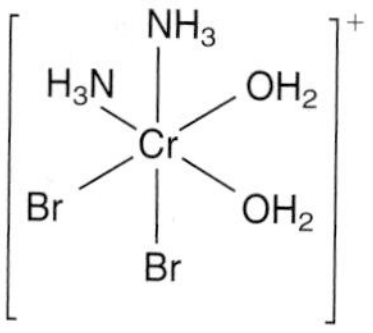

All pairs are *cis*. These are optical isomers of each other.

24.115 (a) no optical isomers (b) no optical isomers (c) no optical isomers (d) no optical isomers (e) optical isomers
24.116 $Pt[P(C_2H_5)_3]_2Cl_2$

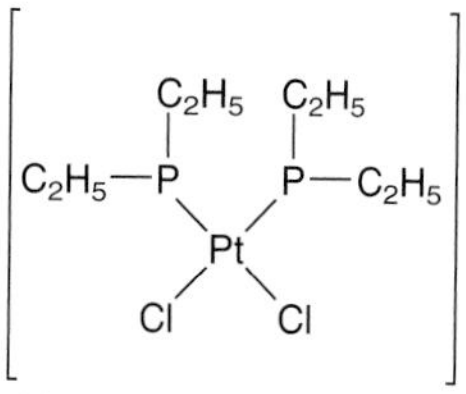

cis-dichlorobis(triethylphosphine)platinum(II)

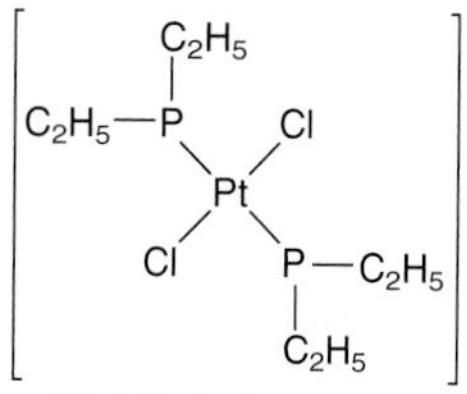

trans-dichlorobis(triethylphosphine)platinum(II)

24.118 (a) The first reaction shows no change in the number of particles. In the second reaction, the number of reactant particles is greater than the number of product particles. A decrease in the number of particles means a decrease in entropy. Based on entropy change only, the first reaction is favoured. (b) The ethylenediamine complex will be more stable with respect to ligand exchange in water because the entropy change for that exchange is unfavourable (negative).

Chapter 25

Answers to Boxed Reading Problems: B25.1 In the *s*-process, a nucleus captures a neutron sometime over a long period of time. Then the nucleus emits a beta particle to form another element. The stable isotopes of most heavy elements up to ^{209}Bi form by the s-process. The r-process very quickly forms less stable isotopes and those with *A* greater than 230 by multiple neutron captures, followed by multiple beta decays. **B25.2** Energy = 1.548×10^{25} MeV/mol **B25.3** The simultaneous fusion of three nuclei is a termolecular process. Termolecular processes have a very low probability of occurring. The bimolecular fusion of ^{8}Be with ^{4}He is more likely.

B25.4 $^{210}_{83}\text{Bi} \longrightarrow {}^{210}_{84}\text{Po} + {}^{0}_{-1}\beta$ (nuclide A);
$^{210}_{84}\text{Po} \longrightarrow {}^{206}_{82}\text{Pb} + {}^{4}_{2}\alpha$(nuclide B);
$^{206}_{82}\text{Pb} + 3{}^{1}_{0}\text{n} \longrightarrow {}^{209}_{82}\text{Pb}$(nuclide C);
$^{209}_{82}\text{Pb} \longrightarrow {}^{210}_{83}\text{Bi} + {}^{0}_{-1}\beta$(nuclide D)

25.1 (a) Chemical reactions are accompanied by relatively small changes in energy; nuclear reactions are accompanied by relatively large changes in energy. (b) Increasing temperature increases the rate of a chemical reaction but has no effect on a nuclear reaction. (c) Both chemical and nuclear reaction rates increase with higher reactant concentrations. (d) If the reactant is limiting in a chemical reaction, then more reactant produces more product and the yield increases. The presence of more radioactive reactant results in more decay product, so a higher reactant concentration increases the yield. **25.2** (a) 95.02% (b) The atomic mass is larger than the isotopic mass of ^{32}S. Sulfur-32 is the lightest isotope. **25.4** (a) *Z* down by 2, *N* down by 2 (b) *Z* up by 1, *N* down by 1 (c) no change in *Z* or *N* (d) *Z* down by 1, *N* up by 1 (e) *Z* down by 1, *N* up by 1. A different element is produced in all cases except c. **25.6** A neutron-rich nuclide decays by β^- decay. A neutron-poor nuclide undergoes positron decay or electron capture.

25.8 (a) $^{234}_{92}\text{U} \longrightarrow {}^{4}_{2}\alpha + {}^{230}_{90}\text{Th}$ (b) $^{232}_{93}\text{Np} + {}^{0}_{-1}\text{e} \longrightarrow {}^{232}_{92}\text{U}$
(c) $^{12}_{7}\text{N} \longrightarrow {}^{0}_{1}\beta + {}^{12}_{6}\text{C}$
25.10 (a) $^{27}_{12}\text{Mg} \longrightarrow {}^{0}_{-1}\beta + {}^{27}_{13}\text{Al}$ (b) $^{23}_{12}\text{Mg} \longrightarrow {}^{0}_{1}\beta + {}^{23}_{11}\text{Na}$
(c) $^{103}_{46}\text{Pd} + {}^{0}_{-1}\text{e} \longrightarrow {}^{103}_{45}\text{Rh}$
25.12 (a) $^{48}_{23}\text{V} \longrightarrow {}^{48}_{22}\text{Ti} + {}^{0}_{1}\beta$ (b) $^{107}_{48}\text{Cd} + {}^{0}_{-1}\text{e} \longrightarrow {}^{107}_{47}\text{Ag}$
(c) $^{210}_{86}\text{Rn} \longrightarrow {}^{206}_{84}\text{Po} + {}^{4}_{2}\alpha$
25.14 (a) $^{186}_{78}\text{Pt} + {}^{0}_{-1}\text{e} \longrightarrow {}^{186}_{77}\text{Ir}$ (b) $^{225}_{89}\text{Ac} \longrightarrow {}^{221}_{87}\text{Fr} + {}^{4}_{2}\alpha$
(c) $^{129}_{52}\text{Te} \longrightarrow {}^{129}_{53}\text{I} + {}^{0}_{-1}\beta$

25.16 (a) It appears stable because its *N* and *Z* values are both magic numbers, but its *N/Z* ratio (= 1.50) is too high; it is unstable. (b) It appears unstable because its *Z* value is an odd number, but its *N/Z* ratio (= 1.19) is in the band of stability, so it is stable. (c) unstable because its *N/Z* ratio is too high **25.18** (a) The *N/Z* ratio for ^{127}I is 1.4; it is stable. (b) The *N/Z* ratio for ^{106}Sn is 1.1; it is unstable because this ratio is too low. (c) The *N/Z* ratio is 1.1 for ^{68}As. The ratio is within the range of stability, but the nuclide is most likely unstable because there are odd numbers of both protons and neutrons. **25.20** (a) alpha decay (b) positron decay or electron capture (c) positron decay or electron capture **25.22** (a) β^- decay (b) positron decay or electron capture (c) alpha decay **25.24** Stability results from a favourable *N/Z* ratio, even-numbered *N* and/or *Z*, and the occurrence of magic numbers. The *N/Z* ratio of ^{52}Cr is 1.17, which is within the band of stability. The fact that *Z* is even does not account for the variation in stability because all isotopes of chromium have the same *Z*. However, ^{52}Cr has 28 neutrons, so *N* is both an even number and a magic number for this isotope only. **25.28** seven α emissions and four β^- emissions **25.31** No; it is not valid to conclude that $t_{1/2}$ equals 1 min because the number of nuclei is so small. Decay rate is an average rate and is only meaningful when the sample is macroscopic and contains a large number of nuclei. For the sample containing 6×10^{12} nuclei, the conclusion is valid. **25.33** 2.56×10^{-2} Ci/g **25.35** 1.4×10^{8} Bq/g **25.37** 1×10^{-12} day^{-1} **25.39** 2.31×10^{-7} year^{-1} **25.41** 1.49 mg **25.43** 2.2×10^{9} years **25.45** 27 dpm **25.47** 1.0×10^{6} years **25.50** Neither γ radiation nor neutron beams have charge, so neither is deflected by a magnetic or electric field. Neutron beams differ from γ radiation in that a neutron has a mass approximately equal to that of a proton. It was observed that a neutron beam induces the emission of protons from a substance; γ radiation does not cause such emission. **25.52** Protons are repelled from the target nuclei due to interaction with like (positive) charges. Higher energy is required to overcome the repulsion.

25.53 (a) $^{10}_{5}\text{B} + {}^{4}_{2}\alpha \longrightarrow {}^{1}_{0}\text{n} + {}^{13}_{7}\text{N}$
(b) $^{28}_{14}\text{Si} + {}^{2}_{1}\text{H} \longrightarrow {}^{1}_{0}\text{n} + {}^{29}_{15}\text{P}$
(c) $^{242}_{96}\text{Cm} + {}^{4}_{2}\alpha \longrightarrow 2{}^{1}_{0}\text{n} + {}^{244}_{98}\text{Cf}$

25.58 Ionizing radiation is more dangerous to children because their rapidly dividing cells are more susceptible to radiation than adults' slowly dividing cells. **25.60** (a) 5.4×10^{-7} rad (b) 5.4×10^{-9} Gy **25.62** (a) 7.5×10^{-10} Gy (b) 7.5×10^{-5} mrem (c) 7.5×10^{-10} Sv **25.65** 1.86×10^{-3} rad **25.67** NAA does not destroy the sample, while chemical analyses do. Neutrons bombard a nonradioactive sample, inducing some atoms within the sample to be radioactive. The radioisotopes decay by emitting radiation characteristic of each isotope. **25.73** Energy is released when a nucleus forms from nucleons. The nuclear binding energy is the quantity of energy holding 1 mol of nuclei together. This energy must be absorbed to break up the nucleus into nucleons and is released when nucleons come together. **25.75** (a) 1.861×10^{4} eV (b) 2.981×10^{-15} J **25.77** 7.6×10^{11} J **25.79** (a) 7.976 MeV/nucleon (b) 127.6 MeV/atom (c) 1.23134×10^{10} kJ/mol **25.81** (a) 8.768 MeV/nucleon (b) 517.3 MeV/atom (c) 4.99128×10^{10} kJ/mol **25.85** Radioactive decay is a spontaneous process in which unstable nuclei emit radioactive particles and energy. Fission occurs as a result of high-energy bombardment of nuclei with small particles that cause the nuclei to break into smaller nuclides, radioactive particles, and energy. All fission events are not the same. The nuclei split in a number of ways to produce several different products. **25.88** The water serves to slow the neutrons so that they are better able to cause a fission reaction. Heavy water is a better moderator because it does not absorb neutrons as well as light water does; thus, more neutrons are available to initiate the fission process. However, D_2O does not occur naturally in great abundance, so its production adds to the cost of a heavy-water reactor. **25.93** (a) 1.1×10^{-29} kg (b) 9.9×10^{-13} J (c) 5.9×10^8 kJ/mol; this is approximately 1 million times larger than a typical enthalpy of reaction. **25.95** 8.0×10^3 years **25.98** 1.35×10^{-5}mol/L **25.100** 6.2×10^{-2} **25.102** (a) 5.99 h (b) 21% **25.104** (a) 0.999 (b) 0.298 (c) 5.58×10^{-6} (d) Radiocarbon dating is more reliable for the fraction in part (ii) because a significant amount of ^{14}C has decayed and a significant amount remains. Therefore, a change in the amount of ^{14}C will be noticeable. For the fraction in part (i), very little ^{14}C has decayed, and for (iii) very little ^{14}C remains. In either case, it will be more difficult to measure the change, so the error will be relatively large. **25.106** 6.579 h **25.110** 4.969×10^{-9} L/h **25.113** 81 years **25.115** (a) 0.15 Bq/L (b) 0.27 Bq/L (c) 3.3 d; a total of 12.8 d **25.118** 7.4 s **25.121** 1926 **25.124** 6.27×10^5 eV, 6.05×10^7 kJ/mol **25.127** (a) 2.07×10^{-17} J/atom (b) 1.45×10^7 H atoms (c) 1.4960×10^{-5} J (d) 1.4959×10^{-5} J (e) No; the captain should continue using the current technology. **25.130** (a) 0.043 MeV, 2.9×10^{-11} m (b) 4.713 MeV **25.134** (a) 3.26×10^3 d (b) 3.2×10^{-3} s (c) 2.78×10^{11} years **25.136** (a) 1.80×10^{17} J (b) 6.15×10^{16} kJ (c) The procedure in part (b) produces more energy per kilogram of antihydrogen. **25.137** 9.316×10^2 MeV **25.141** 7.81 d

GLOSSARY

The numbers in parentheses refer to the sections in which the terms are introduced.

A

absorption spectrum The spectrum produced when atoms absorb specific wavelengths of incoming light and become excited from lower to higher energy levels (6.2)

abundance The amount of an element in a particular region of the natural world (23.1)

accuracy The closeness of a measurement to the actual value (1.6)

acid (See *Arrhenius acid-base definition, Brønsted-Lowry acid-base definition, Lewis acid*, and *Lewis acid-base definition.*)

acid anhydride A compound that consists of two alkanoyl (acyl) groups bonded to the same oxygen atom (acyl-O-acyl); it yields two molecules of acid when it reacts with water (20.6)

acid-base buffer A solution that resists changes in pH when a small amount of either strong acid or strong base is added; it consists of a mixture of a weak acid and its conjugate base (or a weak base and its conjugate acid) (17.1)

acid-base indicators Organic molecules that change colour with pH and are used to monitor the equivalence point of a titration or the pH of a solution (16.2)

acid-base reaction Any reaction between an acid and a base (17.2) (See also *neutralization reaction.*)

acid-base titration curve A plot of the pH of a solution of acid (or base) versus the volume of base (or acid) added to the solution (17.2)

acid-dissociation constant (acid-ionization constant), K_a An equilibrium constant for the dissociation of an acid (HA) in H_2O to yield the conjugate base (A^-) and H^+ or H_3O^+ (16.1):

$$K_a = \frac{[H_3O^+][A^-]}{[HA]}$$

actinides The period 7 elements that constitute the second inner transition series (5*f* block), which includes actinium (Ac; $Z = 89$) through lawrencium (Lr; $Z = 103$) (7.2, 24.2)

activated complex (See *transition state.*)

activation energy (E_a) The minimum energy with which molecules must collide to react (14.5)

active site The region of an enzyme formed by specific amino acid side chains, where substrates bind and undergo a catalytic reaction (14.7)

activity The "effective concentration" of a species in a solution; for a gas, the activity is calculated by multiplying the activity coefficient (γ) by the ratio of the pressure of the species over the standard pressure (taken as 1 bar); for an aqueous solution, the activity is calculated by multiplying the activity coefficient by the ratio of the concentration of the species over the standard concentration (taken as 1 mol/L); for a solution that is reasonably ideal, the activity can be approximated by the pressure (gas) or concentration (aqueous solution) of the species (15.2); ($\mathscr{A}$; also *decay rate*) the change in the number of nuclei (*n*) of a radioactive sample divided by the change in time (*t*) (25.2)

actual yield The amount of product actually obtained in a chemical reaction (3.4)

acylation The process in which an alkanoyl (acyl) group is added to a compound; an acyl group is a functional group derived by the removal of one or more hydroxyl groups from an oxoacid (an acid that contains oxygen) (21.6)

addition A type of organic reaction in which atoms linked by a multiple bond become bonded to more atoms; in an addition reaction, two or more molecules combine to form a larger one (21.1, 21.5)

addition polymers Polymers formed when monomers (usually containing C═C) combine through an addition reaction (22.2)

adduct The product of a Lewis acid-base reaction characterized by the formation of a new covalent bond (16.8)

adenosine triphosphate (ATP) A high-energy molecule that serves most commonly as a store and source of energy in organisms (18.4)

alcohol An organic compound (ending *-ol*) that contains a 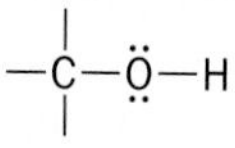 functional group (20.6)

aldehyde An organic compound (ending *-al*) that contains the carbonyl functional group 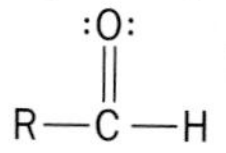 bonded to an H atom (20.6)

alkane A hydrocarbon that contains only single bonds and no rings; general formula C_nH_{2n+2} (20.2)

alkene A hydrocarbon that contains at least one C═C bond; general formula C_nH_{2n} (20.5)

alkyl group A saturated hydrocarbon chain with one bond available (21.1)

alkyl halide (See *haloalkane.*)

alkylation A process in which an alkyl group is added or substituted in a compound (21.6)

alkyne A hydrocarbon that contains at least one C≡C bond; general formula C_nH_{2n-2} (20.5)

allotropes Two or more crystalline or molecular forms of an element; in general, one allotrope is more stable than another at a particular pressure and temperature (13.6)

alloys Mixtures that have metallic properties and consist of solid phases of two or more pure elements, a solid-solid solution, or distinct intermediate phases (8.7, 12.1, 23.3)

alpha (α) decay A radioactive process in which an alpha particle is emitted from a nucleus (25.1)

alpha particles (α, ${}^4_2\alpha$, or ${}^4_2He^{2+}$) Positively charged particles that are identical to a helium-4 nucleus, and one of the common types of radioactive emissions (25.1)

amide An organic compound that contains the 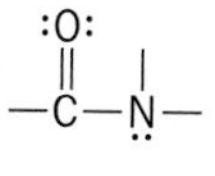 functional group (20.6)

amine An organic compound derived structurally by replacing one or more H atoms of ammonia with organic groups; a weak organic base; general formula 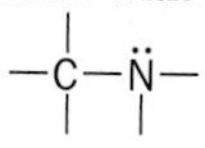 (20.6)

amino acids Organic compounds that are monomer units of a protein and have at least one carboxyl and one amine group on the same molecule; general formula H_2N—CH(R)—COOH (22.3)

amorphous solids Solids that have a poorly defined shape because they lack extensive molecular-level ordering of their particles (11.6)

ampere (A) The SI unit of electric current; 1 A of current results when 1 C (coulomb) flows through a conductor in 1 s (19.7)

amplitude The height of the crest (or the depth of the trough) of a wave; related to the intensity of the energy (brightness of the light) (6.1)

angle strain The distortion of bond angles from their ideal values in a cyclic compound, hence the resistance associated with bond angle compression or bond angle expansion (20.3)

angular momentum quantum number (*l*) An integer from 0 to $n - 1$ that is related to the shape of an atomic orbital (6.4)

anion A negatively charged ion (2.7)

anode The electrode at which oxidation occurs in an electrochemical cell; electrons are given up by the reducing agent and leave the cell at the anode (19.2)

anti addition The addition of two substituents to opposite sides (or faces) of a double (or triple) bond (21.5)

antibonding molecular orbital A molecular orbital formed when wave functions are subtracted from each other, decreasing the electron density between the nuclei and leaving a node; electrons in an antibonding molecular orbital destabilize the molecule (10.3)

apatites Compounds with the general formula $Ca_5(PO_4)_3X$, where X is generally F, Cl, or OH; a source of phosphorus (23.2)

aqueous solutions Solutions in which water is the solvent (2.9)

aromatic hydrocarbon A compound of C and H with one or more rings of C atoms (often drawn with alternating C—C and C═C bonds) and extensive delocalization of π electrons (20.5)

Arrhenius acid-base definition A model of acid-base behaviour in which an acid is a substance that has H in its formula and dissociates in water to yield H^+ or H_3O^+, and a base is a substance that has OH in its formula and produces OH^- in water (16.1)

Arrhenius equation An equation that expresses the exponential relationship between temperature and the rate constant: $k = Ae^{-\frac{E_a}{RT}}$ (14.5)

atmosphere A mixture of gases that extends from the surface of a planet and eventually merges with outer space; the gaseous portion of Earth's crust (4.5, 23.1) (For the unit, see *standard atmosphere.*)

atomic mass The average of the masses of the naturally occurring isotopes of an element weighted according to their abundances (2.5)

atomic mass unit (See *dalton [Da]* and *unified atomic mass unit [u].*)

atomic number (*Z*) The unique number of protons in the nucleus of each atom of an element (equal to the number of electrons in the neutral atom); an integer that expresses the positive charge of a nucleus or subatomic particle in multiples of the electronic charge (2.5)

atomic orbital (See *wave function.*)

atomic solid A solid that consists of individual atoms held together by dispersion forces; the frozen noble gases are the only examples of atomic solids (11.6)

atomic symbol A one- or two-letter abbreviation for the English, Latin, or Greek name of an element (2.5)

atoms The smallest particles of an element that retain the chemical nature of the element; neutral, spherical entities composed of a positively charged central nucleus surrounded by one or more negatively charged electrons (2.3)

Aufbau principle The conceptual approach for building up atoms by adding one proton at a time to the nucleus and one electron around the nucleus to obtain the ground-state electron configurations of the elements (7.2)

autoionization A reaction in which two molecules of a substance react to give ions; the most important example is for water (16.2):

$$2H_2O(l) \rightleftharpoons H_3O^+(aq) + OH^-(aq)$$

average rate The change in concentration of reactants (or products) divided by a finite time period (14.2)

Avogadro's law The gas law stating that, at fixed temperature and pressure, equal volumes of any ideal gas contain equal numbers of particles, and, therefore, the volume of a gas is directly proportional to its amount (mol): $V \propto n$ (4.3)

Avogadro's number, N_A A number (6.022×10^{23} to four significant figures) equal to the number of atoms in exactly 12 g of carbon-12; the number of atoms, molecules, or formula units in 1 mol of an element or a compound (3.1)

axial groups Atoms (or groups) that lie above or below the trigonal plane of a trigonal bipyramidal molecule or a similar structural feature in a molecule (9.1)

azeotrope A mixture of two or more liquids that cannot be separated by simple distillation, since its composition does not change with boiling; an azeotrope has a constant boiling point because the vapour has the same composition as the liquid mixture (12.5)

B

β^- decay A radioactive process in which a beta particle is emitted from a nucleus (25.1)

back-side How the nucleophile from the opposite side of a leaving group attacks, causing inversion of the configuration at the carbon that is undergoing substitution (21.3)

background radiation Natural ionizing radiation; the most important form of background radiation is cosmic radiation (25.4)

balancing (stoichiometric) coefficient The numerical multiplier of all the atoms in the formula that immediately follows it in a balanced chemical equation (3.3)

band of stability The band of stable nuclides that appears on a plot of number of neutrons versus number of protons for all nuclides (25.1)

band theory An extension of molecular orbital (MO) theory that explains many properties of metals and other solids, especially the differences in the electrical conductivity of conductors, semiconductors, and insulators (11.6)

bar A unit of pressure currently accepted as the SI standard; 1 bar is equal to 100 kPa (4.2)

barometer A device used to measure atmospheric pressure; most commonly, a tube that is open at one end, filled with mercury, and inverted into a dish of mercury (4.2)

base (See *Arrhenius acid-base definition, Brønsted-Lowry acid-base definition, Lewis acid-base definition,* and *Lewis base.*)

base-dissociation constant (base-ionization constant), K_b An equilibrium constant for the reaction of a base (B) with H_2O to yield the conjugate acid (BH^+) and OH^- (16.3):

$$K_b = \frac{[BH^+][OH^-]}{[B]}$$

base pairs Complementary mononucleotide bases that are hydrogen-bonded to each other; guanine (G) always pairs with cytosine (C), and adenine (A) always pairs with thymine (T) or uracil (U) (22.3)

base units (fundamental units) Units that define the standards for the seven physical quantities in the International System of Units (SI) (1.5)

basic-oxygen process The method used to convert pig iron to steel, in which O_2 is blown over and through molten pig iron to oxidize impurities and decrease the content of carbon (23.4)

basicity A measure of the strength of a base or, according to Brønsted, the tendency of a compound to act as an H^+ acceptor; in a more general way, basicity is based on electron pair availability: the greater the availability of the electrons to be donated to form a new bond, the stronger the base (21.2)

battery A group of voltaic cells arranged in series; primary and secondary batteries are self-contained, but flow batteries are not (19.5)

becquerel (Bq) The SI unit of radioactivity; 1 Bq = 1 d/s (disintegration per second) (25.2)

bent shape (V shape) A molecular shape that arises when a central atom is bonded to two other atoms and has one or two lone pairs or a single electron; occurs as the AX_2E_1 shape class (bond angle < 120°) in the trigonal planar arrangement and as the AX_2E_2 shape class (bond angle < 109.5°) in the tetrahedral arrangement (9.1)
beta (β) decay A radioactive change that encompasses any of three specific processes: β^- decay, β^+ emission, and e^- capture (25.1)
beta particles (β, β^-, or sometimes $^{0}_{-1}\beta$) Negatively charged particles that have been identified as high-speed electrons and are a common type of radioactive emission (25.1)
bimolecular reaction An elementary reaction that involves the collision of two reactant species (14.6, 21.3)
binary covalent compounds Compounds that consist of atoms of two elements, in which bonding occurs primarily through electron sharing (2.8)
binary ionic compound A compound that consists of the oppositely charged ions of two elements (2.7)
biomass conversion The process of applying chemical and biological methods to convert plant or animal matter into fuels (5.6)
biosphere The living systems that inhabit Earth (23.1)
blast furnace A tower-shaped furnace, made of brick material, in which intense heat and blasts of air are used to convert iron ore and coke to iron metal and carbon dioxide (23.4)
body-centred cubic unit cell A unit cell in which a particle lies at each corner and in the centre of a cube (11.6)
boiling point The temperature at which the vapour pressure of a gas equals the external (atmospheric) pressure (11.2)
boiling point elevation ($\Delta_b T$) The increase in the boiling point of a solvent caused by the presence of a dissolved solute (12.5)
bond angle The angle formed by the nuclei of two surrounding atoms, with the nucleus of the central atom at the vertex (9.1)
bond cleavage The breaking of a chemical bond (21.2)
bond energy (BE) The enthalpy change (always > 0) that accompanies the breakage of a given bond in 1 mol of gaseous molecules (8.3)
bond formation The construction of a chemical bond (21.2)
bond length The distance between the nuclei of two bonded atoms (8.3)
bond order The number of electron pairs that are shared by two bonded atoms (8.3)
bonding molecular orbital A molecular orbital formed when wave functions are added to each other, increasing electron density between the nuclei; electrons in a bonding molecular orbital stabilize the molecule (10.3)
bonding pair (shared pair) An electron pair that is shared by two nuclei; the mutual attraction between the nuclei and the electron pair forms a covalent bond (8.3)
Born-Haber cycle A series of hypothetical steps and their enthalpy changes, which convert elements to an ionic compound; the Born-Haber cycle is used to calculate the lattice energy (8.2)
Bose-Einstein condensate (BEC) A state of matter formed when temperatures of the matter are very close to absolute zero; the low energy of the particles in the state causes the sample to "clump," forming something akin to a "superatom"; the BEC exhibits interesting properties, such as superfluidity, and allows scientists to study quantum mechanics in an experimental setting (1.1)
Boyle's law The gas law stating that, at constant temperature and amount of gas, the volume occupied by a gas is inversely proportional to the applied (external) pressure: $V \propto \frac{1}{P}$ (4.3)
branches Side chains appended to a polymer backbone or longest sequence of atoms in an organic compound (22.1)
bridge bond A covalent bond in which three atoms are held together by two electrons (13.5)
Brønsted-Lowry acid-base definition A model of acid-base behaviour based on proton transfer, in which an acid and a base are defined, respectively, as species that donate and accept a proton (16.1)
buffer capacity A measure of the ability of a buffer to resist a change in pH; related to the total concentrations and relative proportions of the buffer components (17.1)
buffer range The pH range over which a buffer acts effectively; buffering action extends about 1 pH unit on either side of the buffer's pK_a (17.1)

C

Cahn-Ingold-Prelog (CIP) rules A set of rules used in organic chemistry to name the stereoisomers of a molecule (20.4)
calibration The process of comparing a measuring device with a known standard to account for systematic error (1.6)
calorie (cal) A unit of energy defined as exactly 4.184 J; originally defined as the heat needed to raise the temperature of 1 g of water by 1°C (from 14.5°C to 15.5°C) (5.1)
calorimeter A device used to measure the heat released or absorbed by a physical or chemical process taking place within it (5.3)
capillarity A property that results in a liquid rising through a narrow space against the pull of gravity (11.4)
carbocation An ion in which the positive charge is formally located on one or more carbon atoms (21.3)
carbon-skeleton formula a representation of a molecule with a chain of carbon atoms bonded together (leaving off the H atoms), forming the essential structure of the compound; can also be drawn with carbon atoms represented by line ends and intersections, while atoms other than hydrogen and carbon are represented by their elemental symbols
carbon steel The steel produced by the basic-oxygen process; contains about 1% to 1.5% C and other impurities and is alloyed with metals that prevent corrosion and increase its strength (23.4)
carbonyl group The C=O grouping of atoms (20.6)
carboxylic acid An organic compound (ending *-oic acid*) that contains the 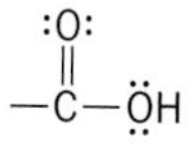 group (20.6)
catalyst A substance or mixture that increases the rate of a reaction; although the catalyst may participate in the reaction, it is regenerated and thus does not appear in the net reaction (14.7)
catenation The process by which atoms of an element bond to each other in chains; most common with carbon in organic compounds, but also occurs with boron, silicon, sulfur, and several other elements (20.1)
cathode The electrode at which reduction occurs in an electrochemical cell; electrons enter the cell and are acquired by the oxidizing agent at the cathode (19.2)
cathode rays Rays of light emitted by the cathode (negative electrode) in a gas discharge tube; travel in straight lines, unless deflected by magnetic or electric fields (2.4)
cation A positively charged ion (2.7)
cell potential (E_{cell}) The potential difference between the electrodes of an electrochemical cell when no current flows (19.3)
Celsius scale A temperature scale in which the freezing and boiling points of water are defined as 0°C and 100°C, respectively (1.5)
ceramics Nonmetallic, nonpolymeric solids that are hardened by heating them to high temperatures; most ceramics consist of silicate microcrystals suspended in a glassy cementing medium (11.7)
chain reaction In nuclear fission, a self-sustaining process in which neutrons released by the splitting of one nucleus cause other nuclei to split, releasing more neutrons, and so on (25.7)

change in enthalpy (ΔH) The change in internal energy plus the product of the constant pressure and the change in volume: $\Delta H = \Delta E + p\Delta V$; the heat lost or gained at constant pressure: $\Delta H = q_P$ (5.2)
charge density The ratio of the charge of an ion to its volume (12.2)
Charles's law The gas law stating that, at constant pressure, the volume occupied by a fixed amount of gas is directly proportional to the absolute temperature of the gas: $V \propto T$ (4.3)
chelate A complex ion in which a metal ion is bonded to a bidentate or polydentate ligand (24.3)
chemical bonds The forces that hold atoms together in a molecule (or formula unit) (2.7)
chemical change (chemical reaction) A change in which one or more substances are converted into one or more substances with different composition and properties (1.1)
chemical equation A statement that uses chemical formulas to express the identities and quantities of the substances involved in a chemical or physical change (3.3)
chemical formula A notation of atomic symbols and numerical subscripts that shows the type and number of each atom in a molecule or formula unit of a substance (2.8)
chemical kinetics The study of the rates and mechanisms of reactions (14.1)
chemical properties Characteristics of a substance that appear as the substance interacts with, or transforms into, other substances (1.1)
chemical reaction (See *chemical change.*)
chemistry The scientific study of matter and its properties, the changes it undergoes, and the energy associated with those changes (1.1)
chiral Describes a molecule that is not superimposable on its mirror image; usually an optically active molecule; in organic compounds, a chiral molecule typically contains a C atom bonded to four different groups (asymmetric C) (20.4)
chlor-alkali process An industrial method that electrolyzes concentrated aqueous NaCl and produces Cl_2, H_2, and NaOH (23.5)
chromatography A separation technique in which a mixture is dissolved in a fluid (gas or liquid) and the components are separated through differences in adsorption to (or solubility in) a solid surface (or viscous liquid) (2.9)
***cis-trans* isomers** (See *geometric isomers.*)
Clausius-Clapeyron equation An equation that expresses the linear relationship between vapour pressure, p, of a liquid and temperature, T; in two-point form, it is written as follows (11.2):

$$ln\frac{p_2}{p_1} = \frac{-\Delta_{vap}H}{R}\left(\frac{1}{T_2} - \frac{1}{T_1}\right)$$

closed system A system that does not allow transfer of mass between it and its surroundings but does allow heat transfer (5.1)
coal gasification An industrial process used to alter the large molecules in coal to sulfur-free gaseous fuels (5.6)
coil shape A spiral shape; shaped like a helix (22.1)
colligative properties Properties of a solution that depend on the number, not the identity, of solute particles (12.5) (See also *boiling point elevation, freezing point depression, osmotic pressure,* and *vapour pressure lowering.*)
collision frequency The average number of collisions per second that a particle undergoes (4.5)
collision theory A model that explains reaction rate as based on the number, energy, and orientation of colliding particles (14.5)
colloids Heterogeneous mixtures in which a solute-like phase is dispersed throughout a solvent-like phase (12.6)
combustion The process of burning in air, often with the release of heat and light (1.2)
combustion analysis A method for determining the formula of a compound from the amounts of its combustion products; commonly used for organic compounds (3.2)
common-ion effect The shift in the position of an ionic equilibrium away from an ion involved in the process; caused by the addition or presence of this ion (17.1)
complex (See *coordination compounds.*)
complex ion An ion that consists of a central metal ion bonded covalently to molecules and/or anions called ligands (17.4, 24.3)
composition The types and amounts of simpler substances that make up a sample of matter (1.1)
compound A substance that consists of two or more elements chemically combined in fixed proportions (2.1)
concentration, c A measure of the quantity of solute that is dissolved in a given quantity of solution (or solvent) (3.5)
concentration cell A voltaic cell in which both compartments contain the same components, but at different concentrations (19.4)
concerted reaction A chemical reaction in which all bond breakage and bond formation occurs in a single step (21.2)
condensation The process of a gas changing into a liquid (11.1)
condensation polymer A polymer formed by monomers with two functional groups that are linked together in a dehydration-condensation reaction (22.2)
condensed formula The formula of a molecule where all symbols of atoms are listed as they appear in the molecule's structure; bond dashes might be omitted or limited; one type groups each C atom with its H atoms, and uses dashes between groups; while another type uses parentheses to indicate multiple identical groups (20.2)
conduction band In band theory, the empty, higher energy portion of the band of molecular orbitals into which electrons move when conducting heat and electricity (11.6)
conductor A substance (usually a metal) that conducts an electric current well (11.6)
conformation Structural or spatial arrangements of the atoms in a molecule; a three-dimensional arrangement of atoms that results by rotation about single bonds (20.3)
conformational isomers Molecules that can often interconvert rapidly into each other by rotation about a single bond at room temperature (20.4)
conformer A specific conformation adopted by a molecule (20.3)
conjugate acid-base pair Two species that are related to each other through the gain or loss of a proton; the acid has one more proton than its conjugate base (16.1)
constitutional (structural) isomers Compounds with the same molecular formula but different arrangements of atoms (20.4, 24.3)
contact process An industrial process used to manufacture sulfuric acid, based on the catalyzed oxidation of SO_2 (23.5)
controlled experiment An experiment that measures the effect of one variable at a time by keeping the other variables constant (1.3)
conversion factor A ratio of equivalent quantities that is equal to 1 and used to convert the units of a quantity (1.4)
coordinate bond (dative covalent bond) A covalent bond in which both electrons come from the same atom (8.3)
coordinate covalent bond A covalent bond formed when one atom donates both electrons to provide the shared pair; once formed, this bond is identical to any covalent single bond (24.4)
coordination compounds Substances that contain at least one complex ion and counterion (24.3)
coordination isomers Two or more coordination compounds with the same composition in which the complex ions have different ligand arrangements (24.3)

coordination number In a crystal, the number of nearest neighbours surrounding a particle (11.6); in a complex, the number of ligand atoms bonded to the central metal ion (24.3)
core The dense, innermost region of Earth (23.1)
core electrons Electrons that fill all the energy levels of an atom except the valence level; also present in atoms of the previous noble gas and any completed transition series (7.2)
corrosion The natural redox process that results in the unwanted oxidation of a metal (19.6)
coulomb (C) The SI unit of electric charge; 1 C is the charge of 6.242×10^{18} electrons, and one electron possesses a charge of 1.602×10^{-19} C (19.3)
Coulomb's law A law stating that the electrostatic energy between particles A and B is directly proportional to the product of their charges and inversely proportional to the distance between them (8.2):

$$\text{Electrostatic energy} \propto \frac{\text{cation charge} \times \text{anion charge}}{\text{cation charge} + \text{anion charge}}$$
$$\propto \Delta_{lattice}H^\circ$$

counterions Simple ions associated with a complex ion in a coordination compound (24.3)
coupling of reactions The pairing of reactions in which one reaction releases enough free energy for the other reaction to occur (18.3)
covalent bond A type of bond in which atoms are bonded through the sharing of electrons; the mutual attraction of the nuclei and an electron pair that holds atoms together in a molecule (2.7, 8.3)
covalent bonding An idealized type of bonding based on localized electron-pair sharing between two atoms with little difference in their tendencies to lose or gain electrons (most commonly nonmetals) (8.1)
covalent compounds Compounds that consist of atoms bonded together by shared electron pairs (2.7)
covalent radius One-half the shortest distance between the nuclei of identical covalently bonded atoms (7.3)
critical mass The minimum mass that is needed to achieve a chain reaction (25.7)
critical point The point on a phase diagram above which the vapour cannot be condensed to a liquid; the end of the liquid-gas curve (11.2)
crosslinks Branches that covalently join one polymer chain to another (22.1)
crust The thin, light, heterogeneous outer layer of Earth; consists of gaseous, liquid, and solid regions (23.1)
crystal defects Any of a variety of disruptions in the regularity of a crystal structure (11.7)
crystal field splitting energy (Δ) The difference in energy between two sets of metal-ion *d* orbitals; results from electrostatic interactions with the surrounding ligands (24.4)
crystal field theory A model that explains the colour and magnetism of coordination compounds based on the effects of ligands on metal-ion *d*-orbital energies (24.4)
crystalline solids Solids with a well-defined shape because of the orderly arrangement of the atoms, molecules, or ions (11.6)
crystallinity The degree of structural order in a solid (22.1)
crystallization A technique used to separate and purify the components of a mixture through differences in solubility, causing a component to come out of solution as crystals (2.9, 12.5)
cubic closest packing A crystal structure based on the face-centred cubic unit cell in which the layers have an *abcabc* . . . pattern (11.6)
cubic metre (m^3) The SI-derived unit of volume (1.5)
curie (Ci) The most common unit of radioactivity, defined as the number of nuclei disintegrating each second in 1 g of radium-226; 1 Ci = 3.70×10^{10} d/s (disintegrations per second) (25.2)
cyclic hydrocarbon A hydrocarbon with one or more rings in its structure (20.2)
cyclohexadienyl cation (arenium ion, sigma-complex) The cation that forms when a benzene molecule reacts with H^+; a reactive intermediate in electrophilic aromatic substitution (21.6)

D

***d* orbital** An atomic orbital with $l = 2$ (6.4)
dalton (Da) A unit of mass that is identical to the *unified atomic mass unit* (2.5)
Dalton's law of partial pressures A gas law stating that, in a mixture of unreacting gases, the total pressure is the sum of the partial pressures of the individual gases: $p_{total} = p_1 + p_2 + \cdots$ (4.4)
data Quantitative information obtained by observation during the course of an experiment (1.3)
de Broglie wavelength The wavelength of a moving particle obtained from the de Broglie equation: $\lambda = \frac{h}{mu}$ (6.3)
decay constant The rate constant, *k*, for radioactive decay (25.2)
decay series (disintegration series) The succession of steps that a parent nucleus undergoes as it decays into a stable daughter nucleus (25.1)
degree of polymerization (*n*) The number of monomer repeat units in a polymer chain (22.1)
degrees of unsaturation (DU) The number of π bonds (double and triple bonds, where a double bond contains one π bond and a triple bond contains two π bonds) plus the number of rings in each structure, compared to alkane with the same number of carbon atoms (20.6)
dehydration A chemical reaction that involves the loss of a water molecule from the reacting molecule (21.7)
dehydration-condensation Describes a reaction in which an H_2O molecule is lost for every pair of OH groups that join (13.7)
density (*d*) An intensive physical property of a substance at a given temperature and pressure, defined as the ratio of the mass to the volume: $d = \frac{m}{v}$ (1.5)
deposition The process of changing directly from gas to solid without going through the liquid phase (11.1)
derived units Any of the various combinations of the seven SI base units (1.5)
desalination A process used to remove large amounts of ions from seawater, usually by reverse osmosis (12.6)
deshielded Refers to a nucleus whose chemical shift has been increased as a result of the removal of electron density, magnetic induction, or other effects (22.5)
deuterons Nuclei of the stable hydrogen isotope deuterium, ^2H (25.3)
diagonal relationships Physical and chemical similarities between a period 2 element and an element located diagonally down and to the right in period 3 (13.4)
diamagnetism The tendency of a species not to be attracted (or to be slightly repelled) by a magnetic field as a result of its electrons being paired (7.4)
diastereomers A pair of stereoisomers that are not mirror images and cannot be superimposed on each other (20.4) (See also *geometric isomers.*)
differentiation The geochemical process of forming regions in Earth based on differences in composition and density (23.1)
diffraction The phenomenon in which a wave striking the edge of an object bends around it; when a wave passes through a slit as wide as its wavelength, it forms a circular wave (6.1)

diffusion The movement of one fluid through another (4.5)
dimensional analysis A calculation method in which arithmetic steps are accompanied by the cancellation of units that represent physical dimensions (1.4)
dipole-dipole forces Intermolecular attractions between oppositely charged poles of nearby polar molecules (11.3)
dipole–induced dipole forces Intermolecular attractions that occur between a polar molecule and the oppositely charged pole it induces in a nearby molecule (12.1)
dipole moment (μ) A measure of molecular polarity; the magnitude of the partial charges on the ends of a molecule (C) times the distance between them (m) (9.2)
disaccharide An organic compound formed by a dehydration-condensation reaction between two simple sugars (monosaccharides) (22.3)
dispersion force (London force) The intermolecular attraction between all particles as a result of instantaneous polarizations of their electron clouds; the intermolecular force that is primarily responsible for the condensed states of nonpolar substances (11.3)
disproportionation reaction A reaction in which a given substance is both oxidized and reduced (13.7)
distillation A separation technique in which a more volatile component of a mixture vaporizes and condenses separately from the less volatile components (2.9)
donor atoms Atoms that donate a lone pair of electrons to form a covalent bond, usually from ligand to metal ion in a complex (24.3)
doping The process of adding small amounts of other elements into the crystal structure of a semiconductor to enhance a specific property, usually conductivity (11.7)
double bond A covalent bond that consists of two bonding pairs; two atoms sharing four electrons in the form of one *s* bond and one π bond (8.3)
double helix The two intertwined polynucleotide strands held together by hydrogen bonds that form the structure of DNA (deoxyribonucleic acid) (22.3)
downfield Refers to larger number in ppm units in an NMR spectrum (22.5)
Downs cell An industrial apparatus that electrolyzes molten NaCl to produce sodium and chlorine (23.4)

E

E1 Stands for *unimolecular elimination reaction*; an E1 reaction goes through at least a two-step mechanism (21.4)
E2 Stands for *bimolecular elimination reaction*; an E2 reaction goes through at least a one-step mechanism (21.4)
effective collisions Collisions in which the particles meet with sufficient energy and an orientation that allows them to react (14.5)
effective nuclear charge (Z_{eff}) The nuclear charge that an electron actually experiences as a result of shielding effects due to the presence of other electrons (7.1)
effusion The process by which a gas escapes from its container through a tiny hole into an evacuated space (4.5)
e_g orbitals The set of orbitals (composed of $d_{x^2-y^2}$ and d_{z^2}) that results when the energies of the metal-ion *d* orbitals are split by a ligand field; the e_g set is higher in energy than the t_{2g} set in an octahedral field of ligands and lower in energy in a tetrahedral field (24.4)
elastomers Polymeric materials that can be stretched and will spring back to their original shape when released (22.1)
electrodes Parts of an electrochemical cell that conduct the electricity between the cell and the surroundings (19.2)
electrolysis The nonspontaneous lysing (splitting) of a substance, often into its component elements, by supplying electrical energy (19.7)
electrolyte A substance that conducts a current when it dissolves in water (3.6, 12.5); a mixture of ions in which the electrodes of an electrochemical cell are immersed, thus conducting a current (19.2)
electrolytic cell An electrochemical system that uses electrical energy to drive a nonspontaneous chemical reaction ($\Delta G > 0$) (19.2)
electromagnetic radiation (also *electromagnetic energy* or *radiant energy*) Oscillating, perpendicular electric and magnetic fields that move simultaneously through space as waves and are manifested as visible light, X-rays, microwaves, radio waves, and so on (6.1)
electromagnetic spectrum The continuum of wavelengths of radiant energy (6.1)
electron (e^-) A subatomic particle that possesses a unit negative charge ($-1.602\ 18 \times 10^{-19}$ C) and occupies the space around the atomic nucleus (2.5)
electron affinity (EA) The energy change (kJ) accompanying the addition of 1 mol of electrons to 1 mol of gaseous atoms or ions (7.3)
electron cloud depiction An imaginary representation of an electron's rapidly changing position around the nucleus over time (6.4)
electron configuration The distribution of electrons within the orbitals of the atoms of an element; also the notation for such a distribution (Chapter 7)
electron deficient Having fewer than eight valence electrons; used to describe a bonded atom, such as Be or B, that has fewer than eight electrons around the central atom (8.6)
electron density diagram The pictorial representation for a given energy subshell of the quantity ψ^2 (the probability density of the electron lying within a particular tiny volume) as a function of *r* (distance from the nucleus) (6.4)
electron-donating groups (also *activating groups*) Atoms or groups of atoms that are attached to a benzene molecule and add electron density to the ring (21.6)
electron (e^-) capture (EC) A type of radioactive decay in which a nucleus draws in an orbital electron, usually from the lowest energy level, and releases energy (25.1)
electron-pair delocalization The process by which electron density is spread over several atoms rather than remaining between two atoms (8.6)
electron-sea model A qualitative description of metallic bonding, proposing that metal atoms pool their valence electrons into a delocalized "sea" of electrons in which the metal cores (metal ions) are submerged in an orderly array (8.7)
electron volt (eV) The energy (J) that an electron acquires when it moves through a potential difference of 1 V (volt); 1 eV = 1.602×10^{-19} J (25.6)
electron-withdrawing groups (also *deactivating groups*) Atoms or groups of atoms that are attached to a benzene molecule and remove electron density from the ring (21.6)
electronegativity (χ) The relative ability of a bonded atom to attract shared electrons (7.3)
electronegativity difference ($\Delta\chi$) The difference in electronegativities between the atoms in a bond (8.5)
electrophiles Chemical species that are electron-deficient and, therefore, attracted to an electron-rich centre (21.2)
electrophilic aromatic substitution An organic reaction in which an atom or group of atoms attached to an aromatic molecule is replaced by a strong electrophile (21.6)
electrorefining An industrial, electrolytic purification process in which a sample of impure metal acts as the anode and a sample of the pure metal acts as the cathode (23.3)
electrostatic potential maps Three-dimensional diagrams of molecules that enable us to visualize the sizes and shapes of the molecules, as well as their charge distribution and charge-related properties (9.2)

element The simplest type of substance with unique physical and chemical properties; an element consists of only one kind of atom, so it cannot be broken down into simpler substances (2.1)
element symbol (See *atomic symbol.*)
elementary reactions (elementary steps) Simple reactions that describe a single molecular event in a proposed reaction mechanism (14.6)
elimination A type of organic reaction in which carbon atoms are bonded to fewer atoms in the product than they are in the reactant, which leads to multiple bonding (21.1)
emission spectrum The line spectrum produced when excited atoms return to lower energy levels and emit photons that are characteristic of the element (6.2) (See also *line spectrum.*)
empirical formula A chemical formula that shows the smallest whole-number ratio of atoms of each element in a compound (3.2)
enantiomers A pair of stereoisomers that are mirror images but cannot be superimposed on each other (20.4, 24.3) (See also *geometric isomers* and *optical isomers.*)
end point The point in a titration at which the indicator changes colour (17.2)
endothermic process A process that occurs with an absorption of heat from the surroundings and, therefore, an increase in the enthalpy of the system: $\Delta H > 0$ (5.2)
energy The capacity to do work, that is, the capacity to move matter (1.1) (See also *kinetic energy* and *potential energy.*)
enthalpy (*H*) A thermodynamic quantity that is the sum of the internal energy and the product of the pressure and volume (5.2)
enthalpy diagram A graphical depiction of the enthalpy change of a system (5.2)
enthalpy of formation (See *standard enthalpy of formation.*)
enthalpy of fusion (See *heat [enthalpy] of fusion.*)
enthalpy of hydration (See *heat of hydration.*)
enthalpy of solution (See *heat of solution.*)
enthalpy of sublimation (See *heat [enthalpy] of sublimation.*)
enthalpy of vaporization (See *heat [enthalpy] of vaporization.*)
entropy (S) A thermodynamic quantity related to the number of ways that the energy of a system can be dispersed through the motions of its particles (12.2, 18.1)
environmental cycle The physical, chemical, and biological paths through which the atoms of an element move within Earth's crust (23.2)
enzyme A biological macromolecule (usually a protein) that acts as a catalyst (14.7)
enzyme-substrate complex (ES) The intermediate in an enzyme-catalyzed reaction, which consists of an enzyme and substrate(s); the concentration of the ES determines the rate of product formation (14.7)
epoxydation A chemical reaction in which an oxygen atom is joined to a π bond in an unsaturated molecule to form a cyclic, three-member ring (21.5)
equatorial groups Atoms (or groups) that lie in the trigonal plane of a trigonal bipyramidal molecule, or a similar structural feature in a molecule (9.1)
equilibrium constant (*K*) The value obtained when equilibrium concentrations are substituted into the reaction quotient (15.1)
equilibrium vapour pressure (See *vapour pressure.*)
equivalence point The point in a titration when the amount (mol) of the added species is stoichiometrically equivalent to the original amount (mol) of the other species (17.2)
ester An organic compound that contains the 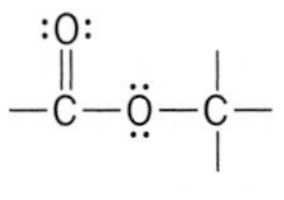 group (20.6)
ether An organic compound in which two hydrocarbon groups are linked by an oxygen atom (R-O-R′) (20.6)
exact number A quantity, usually obtained by counting or based on a unit definition, that has no uncertainty associated with it and, therefore, contains as many significant figures as a calculation requires (1.6)
excited state Any electron configuration of an atom or molecule other than the lowest energy (ground) state (6.2)
exclusion principle A principle, developed by Wolfgang Pauli, stating that no two electrons in an atom can have the same set of four quantum numbers; the exclusion principle arises from the fact that an orbital has a maximum occupancy of two electrons and their spins are paired (7.1)
exothermic process A process that occurs with a release of heat to the surroundings and, therefore, a decrease in the enthalpy of the system: $\Delta H < 0$ (5.2)
expanded valence shells Valence shells that can accommodate more than eight electrons by using available *d* orbitals; occur only with central nonmetal atoms from period 3 or higher (8.6)
experiment A set of procedural steps used to test a hypothesis (1.3)
extensive properties Properties, such as mass, that depend on the quantity of substance present (1.5)
extent of dissociation A measure of the amount of a species that has dissociated in a solution; for the dissociation $AB \longrightarrow A^+ + B^-$, the extent of dissociation is expressed as $\frac{[A^+]}{[AB]}$ and is a number less than or equal to 1 ($[B^-]$ can be used instead of $[A^+]$ in this calculation) (16.4)
extraction A separation technique used to isolate the components of a mixture based on differences in solubility (2.9)
Eyring equation An equation that allows for the calculation of the rate constant using transition state theory; while it closely resembles the Arrhenius equation, the latter is empirically derived while the former is derived entirely from theory (14.5)

F

face-centred cubic unit cell A unit cell in which a particle occurs at each corner and in the centre of each face of a cube (11.6)
Faraday constant (*F*) The physical constant that represents the charge of 1 mol of electrons: $F = 96\,485$ C/mol e^- (19.4)
filtration A method of separating the components of a mixture on the basis of differences in particle size (2.9)
first law of thermodynamics (See *law of conservation of energy.*)
first-order reaction A chemical reaction in which the rate of the reaction is directly proportional to the concentration of one of the reactants (21.3)
Fischer projection A two-dimensional representation of a three-dimensional organic molecule; all the bonds are depicted as horizontal or vertical lines (20.3)
fission The process by which a heavier nucleus splits into lighter nuclei with the release of energy (25.6)
fixation A chemical/biochemical process that converts a gaseous substance in the environment into a condensed form, which can be used by organisms (23.2)
flame test A procedure for identifying the presence of metal ions; in a flame test, a granule of a compound or a drop of its solution is placed in a flame to observe a characteristic colour (6.2)
flotation A metallurgical process in which oil and detergent are mixed with pulverized ore in water to create a slurry that separates the mineral from the gangue (23.3)
formal charge The hypothetical charge on an atom in a molecule or ion, equal to the number of valence electrons minus the sum of all the unshared and half the shared valence electrons (8.6)
formation constant (K_f) An equilibrium constant for the formation of a complex ion from the hydrated metal ion and ligands (17.4)

formation equation An equation in which 1 mol of a compound forms from its elements (5.6)
formula mass The sum (in *unified atomic mass units*) of the atomic masses of a formula unit of a (usually ionic) compound (2.8)
formula unit The chemical unit of a compound that contains the relative numbers of the types of atoms or ions expressed in the chemical formula for the compound (2.8)
fossil fuels Fuels, such as coal, petroleum, and natural gas, that are derived from the products of the decay of dead organisms (5.6)
fraction by mass (mass fraction) The portion of a compound's mass that is contributed by an element; the mass of an element in a compound divided by the mass of the compound (2.2)
fractional distillation A physical process, involving numerous vaporization-condensation steps, that is used to separate two or more volatile components with different boiling points (12.5)
free radicals Molecular or atomic species with one or more unpaired electrons, which typically make the species very reactive (8.6, 25.4)
freezing The process of cooling a liquid until it solidifies (11.1)
freezing point depression ($\Delta_f T$) The lowering of the freezing point of a solvent, caused by the presence of dissolved solute particles (12.5)
frequency (ν) The number of cycles that a wave undergoes per second, expressed in units of $\frac{1}{s}$, or s^{-1} (also called hertz, Hz); inversely related to wavelength (6.1)
Friedel-Crafts reaction An electrophilic aromatic substitution reaction, developed by Charles Friedel and James Crafts, to attach an alkyl or an alkanoyl (acyl) substituent to an aromatic ring (21.6)
fuel cell A battery, not self-contained, in which electricity is generated by the controlled oxidation of a fuel (19.5)
functional group A specific combination of atoms, usually containing a carbon-carbon multiple bond and/or a carbon-heteroatom bond; reacts in a characteristic way, no matter what molecule it occurs in (20.1)
fundamental units (See *base units.*)
fusion The process by which light nuclei combine to form a heavier nucleus, with the release of energy (25.6) (See also *melting.*)

G

galvanic cell (See *voltaic cell.*)
gamma (γ) emission The type of radioactive decay in which gamma rays are emitted from an excited nucleus (25.1)
gamma rays (γ, or sometimes ${}^0_0\gamma$) Very high energy photons (25.1)
gangue Debris, such as sand, rock, and clay, attached to the mineral in an ore (23.3)
gas One of the three common states of matter; a gas fills its container regardless of the shape because its particles are far apart (1.1)
Geiger-Müller counter An ionization counter that detects radioactive emissions through their ionization of gas atoms within the counter (25.2)
genetic code The set of three-base sequences that is translated into specific amino acids during the process of protein synthesis (22.3)
geometric (*cis-trans*) isomers Stereoisomers in which the molecules have the same connections between atoms in alkenes but differ in the spatial arrangements of the atoms; the *cis* isomer has the two larger groups on the same side of a double bond, and the *trans* isomer has the two larger groups on opposite sides; disubstituted cyclohexanes are *cis* when the substituents are on the same side of the ring, and *trans* if they are on the opposite sides (24.3)
Gibbs energy (G) A thermodynamic quantity that is the difference between the enthalpy and the product of the absolute temperature and the entropy: $G = H - TS$ (18.3)
glass transition temperature (T_g) The critical temperature of a noncrystalline material, such as a polymer, at which the material changes its behaviour; when cooled below this temperature, the material becomes hard and brittle, like glass (22.1)
Graham's law of effusion A gas law stating that the rate of effusion of a gas is inversely proportional to the square root of its density (or molar mass) (4.5):

$$\text{Rate of effusion} \propto \frac{1}{\sqrt{\mathcal{M}}}$$

gray (Gy) The SI unit of absorbed radiation dose; 1 Gy = 1 J/kg tissue (25.4)
green chemistry A field of chemistry that is focused on developing methods to synthesize compounds efficiently and reduce or prevent the release of harmful products into the environment (3.4)
Grignard reagents The reagents formed via the reaction of a haloalkane or a haloarene (an alkyl or aryl halide) with magnesium metal; used in organometallic chemical reactions (21.7)
ground state The electron configuration of an atom or ion that is lowest in energy (6.2)
groups The vertical columns in the periodic table; elements in a group usually have the same outer electron configuration and, thus, similar chemical behaviour (2.6)

H

Haber process An industrial process used to form ammonia from its elements (15.6)
half-cell The portion of an electrochemical cell in which a half-reaction takes place (19.2)
half-life ($t_{1/2}$) In chemical processes, the time required for half of the initial reactant concentration to be consumed (14.4); in nuclear processes, the time required for half of the initial number of nuclei in a sample to decay (25.2)
half-reaction method A method used to balance redox reactions by treating the oxidation and reduction half-reactions separately (19.1)
haloalkane (alkyl halide) A hydrocarbon with one or more halogen atoms (X) in place of H; contains a —C—X: group (20.6)
halogenation of benzene An electrophilic substitution reaction between benzene and chlorine or bromine (21.6)
heat capacity The quantity of heat required to change the temperature of an object by 1 K (5.3)
heat (enthalpy) of fusion ($\Delta_{fus}H^\circ$) The enthalpy change that occurs when 1 mol of a solid substance melts; $\Delta_{fus}H^\circ$ occurs at the standard state (11.1)
heat (enthalpy) of sublimation ($\Delta_{subl}H^\circ$) The enthalpy change that occurs when 1 mol of a solid substance changes directly to a gas; the sum of the heats of fusion and vaporization; $\Delta_{subl}H^\circ$ occurs at the standard state (11.1)
heat (enthalpy) of vaporization ($\Delta_{vap}H^\circ$) The enthalpy change that occurs when 1 mol of a liquid substance vaporizes; $\Delta_{vap}H^\circ$ occurs at the standard state (11.1)
heat of formation (See *standard enthalpy of formation.*)
heat of hydration ($\Delta_{hydr}H$) The enthalpy change that occurs when 1 mol of a gaseous species (often an ion) is hydrated; the sum of the enthalpies from separating water molecules and mixing the gaseous species with them; designated $\Delta_{hydr}H^\circ$ at the standard state (12.2)
heat of hydrogenation The heat released when 1 mol of an unsaturated compound, such as an alkene or alkyne, is completely hydrogenated to the corresponding alkane (21.5)

heat of solution ($\Delta_{soln}H$) The enthalpy change that occurs when a solution forms from solute and solvent; the sum of the enthalpies from separating the solute and solvent substances and mixing them; designated $\Delta_{soln}H^{\circ}$ at the standard state (12.2)

heat (q) The energy transferred between objects because of a difference in only their temperatures (1.5, 5.1)

heating-cooling curve A plot of temperature versus time for a substance when heat is absorbed or released by the system at a constant rate (11.2)

Henderson-Hasselbalch equation An equation for calculating the pH of a buffer system (17.1):

$$pH = pK_a + \log\left(\frac{[\text{conj. base}]}{[\text{acid}]}\right)$$

Henry's law A law stating that the solubility of a gas in a liquid is directly proportional to the partial pressure of the gas above the liquid: $S_{gas} = k_H \times p_{gas}$ (12.3)

Hess's law A law stating that the enthalpy change of an overall process is the sum of the enthalpy changes of the individual steps (5.5)

heteroatoms Any atoms in an organic compound, other than C or H (20.1)

heterogeneous catalyst A catalyst that occurs in a different phase from the reactants, usually a solid interacting with gaseous or liquid reactants (14.7)

heterogeneous mixture A mixture that has one or more visible boundaries among its components (2.9)

heterolytic cleavage (heterolysis) The breaking of a single covalent bond in which both electrons remain on one of the atoms (21.2)

hexagonal closest packing A crystal structure based on the hexagonal unit cell in which the layers have an *abab* . . . pattern (11.6)

high-spin complex A complex ion that has the same number of unpaired electrons as the isolated metal ion does; contains weak-field ligands (24.4)

homogeneous catalyst A catalyst (gas, liquid, or soluble solid) that exists in the same phase as the reactants (14.7)

homogeneous mixture (solution) A mixture that has no visible boundaries among its components (2.9)

homologous series A series of organic compounds in which each member differs from the next by a —CH_2— group (20.2)

homolytic cleavage The breaking of a single covalent bond in which one electron remains on each atom (21.2)

homonuclear diatomic molecules Molecules composed of two identical atoms (10.3)

Hückel's rule A quantum-mechanical approach stating that a planar ring molecule has to have $4n + 2$ π electrons, where $n =$ 0 or any positive integer, to be aromatic; used to predict whether the molecule has aromatic properties; not valid for many compounds that contain more than three fused aromatic nuclei (20.5)

Hund's rule A principle stating that when orbitals of equal energy are available, the electron configuration of lowest energy has the maximum number of unpaired electrons with parallel spins (7.2)

hybrid orbitals Atomic orbitals postulated to form during bonding by the mathematical mixing of specific combinations of nonequivalent orbitals in a given atom (10.1)

hybridization A postulated process of orbital mixing to form hybrid orbitals (10.1)

hydrates Compounds in which a specific number of water molecules are associated with each formula unit (2.8)

hydration Solvation in water (12.2)

hydration shells Oriented clusters of water molecules that surround an ion in aqueous solution (12.1)

hydrocarbons Organic compounds that contain only H and C atoms (20.2)

hydrogen bond A type of dipole-dipole force that arises between molecules that have an H atom bonded to a small, highly electronegative atom with lone pairs, usually N, O, or F; the hydrogen bonded to the electronegative atom of one molecule is attracted to the electronegative atom of a different molecule (11.3)

hydrogenation The addition of hydrogen to a carbon-carbon multiple bond, forming a carbon-carbon single bond (14.7, 21.5)

hydrolysis Cleaving a molecule by a reaction with water in which one part of the molecule bonds to the —OH in water and the other part bonds to the H (20.6, 21.7)

hydronium ion (H_3O^+) A proton that is covalently bonded to a water molecule; used to represent a solvated proton (16.1)

hydrosphere The liquid portion of Earth's crust (23.1)

hypothesis A testable proposal that is made to explain an observation; if a hypothesis is inconsistent with experimental results, it is revised or discarded (1.3)

I

ideal gas A hypothetical gas that exhibits linear relationships among volume, pressure, temperature, and amount (mol) at all conditions; approximated by simple gases at ordinary conditions (4.3)

ideal gas law (also *ideal gas equation*) An equation that expresses the relationships among volume, pressure, temperature, and amount (mol) of an ideal gas: $pV = nRT$ (4.3)

ideal solution A solution whose vapour pressure equals the mole fraction of the solvent times the vapour pressure of the pure solvent; approximated only by very dilute solutions (12.5) (See also *Raoult's law.*)

indicator (See *acid-base indicators.*)

induced-fit model A model of enzyme action that pictures the binding of the substrate as inducing the active site to change its shape and become catalytically active (14.7)

inductive effect An experimentally observable effect of the transmission of charge through a chain of atoms in a molecule by electrostatic attraction (21.2)

infrared (IR) The region of the electromagnetic spectrum between the microwave and visible regions (6.1)

infrared (IR) spectroscopy An instrumental technique for determining the types of bonds in a covalent molecule by measuring the absorption of IR radiation (22.5)

initial rate The instantaneous rate that occurs as soon as the reactants are mixed, at $t = 0$ (14.2)

inner transition elements The elements of the periodic table in which f orbitals are being filled; the lanthanides and actinides (7.2, 24.2)

instantaneous rate The reaction rate at a particular time, given by the slope of the tangent to a plot of reactant concentration versus time (14.2)

insulator A substance (usually a nonmetal) that does not conduct an electric current (11.6)

integrated rate laws Mathematical expressions for reactant concentration as a function of time (14.4)

intensive properties Characteristics, such as density, that do not depend on the quantity of substance that is present (1.5)

interhalogen compounds Compounds that consist entirely of halogens (13.9)

intermolecular (interparticle) forces The attractive and repulsive forces among the particles (molecules, atoms, or ions) in a sample of matter (11.1)

internal energy (U) The sum of the kinetic and potential energies of all the particles in a system (5.1)

ion-dipole force The intermolecular attractive force between an ion and a polar molecule (dipole) (11.3)

ion exchange A process of softening water by exchanging one type of ion (usually Ca^{2+}) for another type of ion (usually Na^+) by binding the ions on a specially designed resin (12.6)
ion–induced dipole forces The attractive forces that occur between an ion and the dipole it induces in the electron cloud of a nearby nonpolar molecule (12.1)
ion pairs Pairs of ions that form gaseous ionic molecules; sometimes formed when a salt boils (8.2)
ion-product constant for water (K_w) The equilibrium constant for the autoionization of water; equal to 1.0×10^{-14} at 298 K (16.2):

$$K_w = [H_3O^+][OH^-]$$

ionic atmosphere A cluster of ions of net opposite charge, surrounding a given ion in solution (12.5)
ionic bonding An idealized type of bonding based on the attraction of oppositely charged ions that arise through electron transfer between atoms with large differences in their tendencies to lose or gain electrons (typically metals and nonmetals) (8.1)
ionic compounds Compounds that consist of oppositely charged ions (2.7)
ionic radius The size of an ion as measured by the distance between the nuclei of adjacent ions in a crystalline ionic compound (7.4)
ionic solid A solid whose unit cell contains cations and anions (11.6)
ionization In nuclear chemistry, the process by which a substance absorbs energy from high-energy radioactive particles and loses an electron to become ionized (25.4)
ionization energy (IE) The energy (kJ) that is required to remove completely 1 mol of electrons from 1 mol of gaseous atoms or ions (7.3)
ionizing radiation The high-energy radiation, from natural and artificial sources, that forms ions in a substance by causing electron loss (25.4)
ions Charged particles that form from an atom (or covalently bonded group of atoms) when it gains or loses one or more electrons (2.7)
isoelectronic Having the same number and configuration of electrons as another species (7.4)
isolated system A system that allows neither mass nor heat to be transferred between it and its surroundings (5.1)
isomers Two or more compounds with the same molecular formula but different structural formulas, as a result of different arrangements of atoms (3.2, 24.3)
isotopes Atoms of a given atomic number (that is, of a specific element) that have different numbers of neutrons and, therefore, different mass numbers (2.5, 25.1)
isotopic mass The mass (in unified atomic mass units, u) of an isotope relative to the mass of carbon-12 (2.5)

J

joule (J) The SI unit of energy; $1\ J = 1\ kg{\cdot}m^2/s^2$ (5.1)

K

Kelvin (absolute) scale The preferred temperature scale in scientific work; has absolute zero (0 K or −273.15°C) as the lowest temperature (1.5)
kelvin (K) The SI base unit of temperature; the kelvin is the same size as the Celsius degree (1.5)
ketone An organic compound (ending *-one*) that contains a carbonyl group bonded to two other C atoms: 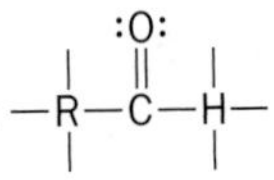 (20.6)
kilogram (kg) The SI base unit of mass (1.5)
kinetic energy The energy that an object has because of its motion (1.1)
kinetic-molecular theory The model that explains gas behaviour in terms of particles in random motion whose volumes and interactions are negligible (4.5)

L

lanthanide contraction The additional decrease in atomic and ionic size, beyond the expected trend, caused by poor shielding of the increasing nuclear charge by *f* electrons in the elements following the lanthanides (24.1)
lanthanides (also *rare earths*) The period 6 (4*f*) series of inner transition elements, which includes lanthanum (La; $Z = 57$) through lutetium (Lu; $Z = 71$) (7.2, 24.2)
lattice A three-dimensional arrangement of points that is created by choosing each point to be at the same location within each particle of a crystal; thus, a lattice consists of all points with identical surroundings (11.6)
lattice energy ($D_{\text{lattice}}H^\circ$) The enthalpy change (always positive) that accompanies the separation of 1 mol of a solid ionic compound into gaseous ions (8.2)
law (See *natural law.*)
law of chemical equilibrium (law of mass action) The law stating that when a system reaches equilibrium at a given temperature, the ratio of the quantities that make up the reaction quotient has a constant numerical value (15.2)
law of conservation of energy (first law of thermodynamics) A basic observation that the total energy of the universe is constant; thus, $\Delta_{\text{universe}}U = \Delta_{\text{system}}U + \Delta_{\text{surroundings}}U = 0$ (5.1)
law of definite (or constant) composition A mass law stating that, no matter what its source, a particular compound is composed of the same elements in the same parts (fractions) by mass (2.2)
law of mass action (See *law of chemical equilibrium.*)
law of mass conservation A mass law stating that the total mass of the substances in a chemical reaction does not change during the reaction (2.2)
law of multiple proportions A mass law stating that if elements A and B react to form two or more compounds, the different masses of B that combine with a fixed mass of A can be expressed as a ratio of small whole numbers (2.2)
Le Châtelier's principle A principle stating that if a system in a state of equilibrium is disturbed, it will undergo a change that shifts its equilibrium position in a direction that reduces the effect of the disturbance (15.6)
leaching A hydrometallurgical process that extracts a metal selectively, usually through the formation of a complex ion (23.3)
leaving group An atom or group of atoms that departs with a pair of electrons in a heterolytic cleavage (21.3)
levelling effect The inability of a solvent to distinguish the strength of an acid (or base) that is stronger than the conjugate acid (or conjugate base) of the solvent (16.7)
Lewis acid (electrophile) A substance (a compound or an ionic species) that can accept an electron pair from another substance (21.2)
Lewis acid-base definition A model of acid-base behaviour in which acids and bases are defined, respectively, as species that accept and donate an electron pair (16.8)
Lewis base (nucleophile) A substance (a compound or an ionic species) that can donate an electron pair to another substance (21.2)
Lewis electron-dot symbol A notation in which the element symbol represents the nucleus and inner electrons, and surrounding dots represent the valence electrons (8.1)

Lewis structure (Lewis formula) A structural formula that consists of electron-dot symbols, with lines as bonding pairs and dots as lone pairs (8.6)
ligands Molecules or anions that are bonded to a central metal ion in a complex ion (17.4, 24.3)
like-dissolves-like rule An empirical observation stating that substances with similar kinds of intermolecular forces dissolve in each other (12.1)
limiting reactant The reactant that is completely consumed when a reaction occurs and, therefore, is the reactant that determines the maximum amount of product that can form (3.4)
line spectrum A series of separated lines of different colours that represent photons whose wavelengths are characteristic of an element (6.2) (See also *emission spectrum.*)
linear arrangement The geometric arrangement obtained when two electron groups maximize their separation around a central atom (9.1)
linear shape A molecular shape formed by three atoms lying in a straight line, with a bond angle of 180° (shape class AX_2 or AX_2E_3) (9.1)
linkage isomers Coordination compounds with the same composition but with different ligand donor atoms linked to the central metal ion (24.3)
lipid bilayer An extended sheetlike double layer of phospholipid molecules that forms in water and has the charged heads of the molecules on the surfaces of the bilayer and the nonpolar tails within the interior (22.4)
liquid One of the three common states of matter; a liquid fills a container to the extent of its own volume and, thus, forms a surface (1.1)
liquid crystals Substances that flow like a liquid but pack like a crystalline solid at the molecular level (11.7)
lithosphere The solid portion of Earth's crust (23.1)
litre (L) A non-SI unit of volume equivalent to one cubic decimetre (0.001 m^3) (1.5)
lock-and-key model A model of enzyme function that pictures the enzyme's active site and the substrate as rigid shapes that fit together like a lock and key, respectively (14.7)
London force (See *dispersion force.*)
lone pair (unshared pair) An electron pair that is part of an atom's valence level but not involved in covalent bonding (8.3)
low-spin complex A complex ion that has fewer unpaired electrons than the free metal ion does, because of the presence of strong-field ligands (24.4)

M

magnetic quantum number (m_l) An integer from $-l$ through 0 to $+l$ that specifies the orientation of an atomic orbital in the three-dimensional space about the nucleus (6.4)
manometer A device used to measure the pressure of a gas in a container (4.2)
mantle A thick homogeneous layer of Earth's internal structure that lies between the core and the crust (23.1)
Markovnikov's rule A rule stating that, in an electrophilic addition reaction with asymmetrical alkenes, the hydrogen (for example, the hydrogen in a haloakane) will be added to the carbon atom with more hydrogen atoms (21.5)
mass The quantity of matter that an object contains; balances are designed to measure mass (1.5)
mass fraction (See *fraction by mass.*)
mass number (*A*) The total number of protons and neutrons in the nucleus of an atom (2.5)
mass percent [% (w/w)] (also *mass %* or *percent by mass*) A concentration term expressed as the mass of solute dissolved in 100. parts by mass of solution (12.4)
mass spectrometry An instrumental method for measuring the relative masses of particles in a sample by creating charged particles and separating the particles according to their mass-charge ratio (2.5)
matter Anything that possesses mass and occupies volume (1.1)
mean free path The average distance that a molecule travels between collisions at a given temperature and pressure (4.5)
melting (fusion) The change of a substance from a solid to a liquid (11.1)
melting point The temperature at which the solid and liquid forms of a substance are at equilibrium (11.2)
metabolic pathways Biochemical reaction sequences that flow in one direction and have each reaction enzyme catalyzed (15.6)
metallic bonding An idealized type of bonding based on the attraction between metal ions and their delocalized valence electrons (8.1) (See also *electron-sea model.*)
metallic radius One-half the shortest distance between the nuclei of adjacent individual atoms in a crystal of an element (7.3)
metallic solids Solids whose individual atoms are held together by metallic bonding (11.6)
metalloids (semimetals) Elements with properties between those of metals and nonmetals (2.6)
metallurgy The branch of materials science concerned with the extraction and use of metals (23.3)
metals Substances or mixtures that are relatively shiny and malleable and are good conductors of heat and electricity; in reactions, metals tend to transfer electrons to nonmetals and form ionic compounds (2.6)
methanogenesis The process of producing methane by the anaerobic biodegradation of plant and animal waste (5.6)
metre (m) The SI base unit of length; the distance that light travels in a vacuum in 1/299 792 458 s (1.5)
Michaelis constant A constant that represents the substrate concentration at which the reaction rate equals half of V_{max} (14.7)
Michaelis-Menten equation An equation that demonstrates the relationship between the rate of an enzyme-catalyzed reaction and the amount of enzyme-substrate complex (14.7)
millilitre (mL) A volume (0.001 L) that is equivalent to 1 cm^3 (1.5)
millimetre of mercury (mmHg) A unit of pressure based on the difference in the heights of mercury in a barometer or manometer; renamed the *Torr* in honor of Torricelli (4.2)
mineral The portion of an ore that contains the element of interest; a mineral is a naturally occurring, homogeneous, crystalline inorganic solid, with a well-defined composition (23.3)
miscible Soluble in any proportion (12.1)
mixture A group of two or more elements and/or compounds that are physically intermingled (2.1)
models (theories) Simplified conceptual pictures, based on experiments, that explain how natural phenomena occur (1.3)
molality (*m*) A concentration term expressed as the amount (mol) of solute dissolved in 1000 g (1 kg) of solvent (12.4)
molar heat capacity (C_m) The quantity of heat required to change the temperature of 1 mol of a substance by 1 K (5.3)
molar mass ($\mathcal{M}$) The mass of 1 mol of entities (atoms, molecules, or formula units) of a substance, in units of grams per mole (g/mol) (3.1)
molarity (*M*) One way to express the concentration of a solution; the amount (mol) of a solute per total volume (L) of the solution (3.5)
mole fraction (X) A concentration term expressed as the ratio of moles of one component in a mixture to the total moles of mixture (4.4, 12.4)

mole (mol) The SI base unit for amount of substance; the amount that contains a number of chemical entities equal to the number of atoms in exactly 12 g of carbon-12 (which is 6.022×10^{23}) (3.1)
molecular equation A chemical equation that shows a reaction in which the reactants and products appear as intact, undissociated compounds (3.7)
molecular formula A formula that shows the actual number of atoms of each element in a molecule of a compound; it is always a multiple of the empirical formula (2.8, 3.2)
molecular mass The sum (in unified atomic mass units, u) of the atomic masses of the elements in a molecule (or formula unit) of a compound (2.8)
molecular orbital bond order One-half the difference between the numbers of electrons in bonding and antibonding molecular orbitals (10.3)
molecular orbital diagram A depiction of the relative energy and number of electrons in each molecular orbital, as well as the atomic orbitals from which the molecular orbitals form (10.3)
molecular orbital (MO) theory A model that describes a molecule as a collection of nuclei and electrons in which the electrons occupy orbitals extending over the entire molecule (10.3)
molecular orbitals Orbitals of given energy and shape that extend over a molecule and can be occupied by no more than two paired electrons (10.3)
molecular polarity The overall distribution of electronic charge in a molecule, determined by its shape and bond polarities (9.2)
molecular shape The three-dimensional structure defined by the relative positions of the atomic nuclei in a molecule (9.1)
molecular solids Solids held together by intermolecular forces between individual molecules (11.6)
molecular vibrations The periodic motion of atoms in a molecule in alternately opposite directions (22.5)
molecularity The number of reactant particles that are involved in an elementary step (14.6)
molecule A structure that consists of two or more atoms bound chemically, behaving as an independent unit (2.1)
monatomic ion An ion derived from a single atom (2.7)
monomers Small molecules, linked covalently to others of the same or similar type to form a polymer; the repeat units of a polymer (22.1)
mononucleotides Monomer units of a nucleic acid; consist of an N-containing base, a sugar, and a phosphate group (22.3)
monosaccharides Simple sugars; polyhydroxy ketones or aldehydes with three to nine C atoms (22.3)

N

nanomachines Mechanical or electromechanical devices built from individual atoms, which range in size from 1 μm (one micrometre; 1×10^{-6} m) to 1 nm (one nanometre; 1×10^{-9} m) (11.7)
nanotechnology The science and engineering of nanoscale (1 nm to 100 nm) systems (11.7)
natural law A summary, often in mathematical form, of a universal observation (1.3)
Nernst equation An equation stating that the voltage of an electrochemical cell under any conditions depends on the standard cell voltage and the concentrations of the cell components (19.4):

$$E_{\text{cell}} = E^{\circ}_{\text{cell}} - \frac{RT}{zF}\ln Q$$

net ionic equation A chemical equation of a reaction in solution in which spectator ions have been eliminated to show the actual chemical change (3.7)
network covalent solids Solids in which all the atoms are bonded covalently so that individual molecules are not present (8.3, 11.6)
neutralization The process that occurs when a certain amount of acid reacts with a stoichiometrically equivalent amount of base (16.1)
neutralization reaction An acid-base reaction that yields water and a solution of a salt; when the H^+ ions of a strong acid react with an equivalent amount of the OH^- ions of a strong base, the solution is neutral (17.2) (See also *acid-base reaction.*)
neutron (n^0) An uncharged subatomic particle found in the nucleus, with a mass slightly greater than the mass of a proton (2.5)
Newman projection The representation of the conformation of a molecule in which the structure appears as viewed along the bond between two carbon atoms; the front carbon is represented as a point, and the back carbon is represented as a circle (20.3)
nitrile An organic compound that contains the —C≡N: group (20.6)
node A region of an orbital where the probability of finding an electron is zero (6.4)
nonbonding molecular orbitals Molecular orbitals that are not involved in bonding (10.3)
nonelectrolytes Substances whose aqueous solution does not conduct an electric current (3.6, 12.5)
nonmetals Elements that lack metallic properties; in reactions, nonmetals tend to share electrons with each other to form covalent compounds or accept electrons from metals to form ionic compounds (2.6)
nonpolar covalent bond A covalent bond between identical atoms that has the bonding pair shared equally (8.5)
nuclear binding energy The energy required to break 1 mol of nuclei of an element into individual nucleons (25.6)
nuclear magnetic resonance (NMR) spectroscopy An instrumental technique used to determine the molecular environment of a given type of nucleus, most often ^1H, from its absorption of electromagnetic radiation of a specific frequency in a strong magnetic field (22.5)
nuclear transmutation The induced conversion of one nucleus into another nucleus by bombardment with a particle (25.3)
nucleic acids Unbranched polymers that consist of mononucleotides; occur as two types, DNA and RNA (deoxyribonucleic and ribonucleic acids), which differ chemically in the nature of the sugar portion of the mononucleotides (22.3)
nucleons Subatomic particles found in the nucleus; protons or neutrons (25.18)
nucleophiles Atoms or molecules that donate an electron pair to an electrophile to form a chemical bond in a reaction (21.2)
nucleophilic substitution The reaction of an electron-pair donor (a nucleophile) with an electron-pair acceptor (an electrophile); the electrophile must have a leaving group for the reaction to take place (21.3)
nucleophilicity Refers to the nucleophilic character of a substance; measures the ability of the nucleophile to attack other electron-deficient atoms (usually carbon atoms) (21.1)
nucleus The central region of the atom that, although very small with respect to the volume of the atom, is positively charged and contains essentially all the mass of the atom (2.4)
nuclide A nuclear species with specified numbers of protons and neutrons (25.1)

O

observations Facts obtained with the senses, often with the aid of instruments; quantitative observations provide data that can be compared (1.3)
occurrence (source) The form(s) in which an element exists in nature (23.1)

octahedral arrangement The geometric arrangement obtained when six electron groups maximize their space around a central atom; when all six groups are bonding groups, the molecular shape is octahedral (AX_6; ideal bond angle = 90°) (9.1)
octahedral hole A hole surrounded by six atoms that are packed together in a closest-packing arrangement (the coordination number is six); the size of the hole is smaller than the size of the atoms surrounding it (11.6)
octet rule The observation that when atoms bond, they often lose, gain, or share electrons to attain a filled outer level of eight electrons (or two for H and Li) (8.1)
one-step (concerted) mechanism The step-by-step sequence of a reaction in which there is a simultaneous occurrence of bond making and bond breaking; written using a technique called *arrow pushing* to depict the flow or movement of electrons (21.3, 21.4)
open system A system that allows both mass and heat to be transferred between it and its surroundings (5.1)
optical isomers The name given in old literature to enantiomers, which are molecules that are mirror images of each other and cannot be superimposed; they were given this name because they were distinguished by how they rotated plane-polarized light (20.4) (See also *enantiomers*.)
optically active Able to rotate the plane of polarized light (20.4)
orbital diagram A depiction of orbital occupancy in terms of electron number and spin, shown using arrows in a series of small boxes, lines, or circles (7.2)
ores Naturally occurring compounds or mixtures of compounds from which an element can be profitably extracted (23.1)
organic compounds Compounds in which carbon is nearly always bonded to itself and to hydrogen, and often to other elements (Chapter 20)
osmosis The process by which solvent flows through a semipermeable membrane from a dilute solution to a concentrated solution (12.5)
osmotic pressure (Π) The pressure that results from the ability of solvent particles, but not solute particles, to cross a semipermeable membrane; the pressure required to prevent the net movement of solvent across the membrane (12.5)
outer electrons Electrons that occupy the highest energy level (highest n value) and are, on average, farthest from the nucleus (7.2)
overall (net) equation A chemical equation that is the sum of two or more balanced sequential equations, with a product of one equation becoming a reactant for the next (3.4)
overpotential The difference between the equilibrium potential (when the current is zero) and the potential at a given current (I) (19.7)
oxidation The loss of electrons by a species, accompanied by an increase in oxidation number (19.1, 21.1)
oxidation number method A method for balancing redox reactions in which the change in oxidation numbers is used to determine balancing coefficients (19.1)
oxidation number (oxidation state) A number equal to the magnitude of the charge that an atom would have if its shared electrons were held completely by the atom that attracts them more strongly (19.1)
oxidation-reduction reaction (redox reaction) A process in which there is a net movement of electrons from one reactant (reducing agent) to another (oxidizing agent) (19.1)
oxidation state (See *oxidation number*.)
oxidizing agent The substance that accepts electrons in a reaction and undergoes a decrease in oxidation number (19.1)
oxoanions Anions in which an element is bonded to one or more oxygen atoms (2.8)
ozonolysis The cleavage of an alkene or alkyne with ozone to form organic compounds in which the multiple carbon-carbon bond has been replaced by a double bond to oxygen (21.5)

P

***p* orbital** An atomic orbital with $l = 1$ (6.4)
packing efficiency The percentage of the available volume occupied by atoms, ions, or molecules in a unit cell (11.6)
paramagnetism The tendency of a species with unpaired electrons to be attracted by an external magnetic field (7.4)
partial ionic character An estimate of the actual charge separation in a bond (caused by the electronegativity difference of the bonded atoms) relative to complete separation (8.5)
partial pressure The portion of the total pressure contributed by a gas in a mixture of gases; hypothetical pressure of a gas if it were alone in the same volume and at the same temperature (4.4)
particle accelerators Devices used to impart high kinetic energies to nuclear particles (25.3)
pascal (Pa) The SI unit of pressure; 1 Pa = 1 N/m^2 (4.2)
penetration The process by which an outer electron moves through the region occupied by the core electrons to spend part of its time closer to the nucleus; penetration increases the average effective nuclear charge for this electron (7.1)
percent by mass (mass percent, mass %) The fraction by mass expressed as a percentage (2.2)
percent dissociation The extent of dissociation multiplied by 100% (16.4)
percent yield (% yield) The actual yield of a reaction expressed as a percentage of the theoretical yield (3.4)
periodic law A law stating that when the elements are arranged by atomic number, they exhibit a periodic recurrence of properties (Chapter 7)
periodic table of the elements A table in which the elements are arranged by atomic number into columns (groups) and rows (periods) (2.6)
periods The horizontal rows in the periodic table (2.6)
pH The negative common logarithm of $[H^+]$ (16.2)
phase A physically distinct and homogeneous part of a system (11.1)
phase changes Physical changes from one phase to another; usually refer to changes in physical state (11.1)
phase diagram A diagram used to describe the stable phases and phase changes of a substance as a function of temperature and pressure (11.2)
photoelectric effect The observation that when monochromatic light of sufficient frequency shines on a metal, electrons are ejected from the metal surface (6.1)
photons Quanta of electromagnetic radiation (6.1)
photovoltaic cell A device that is capable of converting light directly into electricity (5.6)
physical change A change in which the physical form (or state) of a substance, but not its composition, is altered (1.1)
physical properties Characteristics shown by a substance itself, without interacting with or changing into other substances (1.1)
pi (π) bond A covalent bond formed by the sideways overlap of two parallel atomic orbitals; has two regions of electron density, one above and one below the internuclear axis (10.2)
pi (π) molecular orbitals Molecular orbitals formed by the combination of two parallel atomic (usually p) orbitals, whose orientations are perpendicular to the internuclear axis (10.3)
Planck's constant (*h*) A proportionality constant that relates the energy and frequency of a photon; equal to 6.626×10^{-34} J·s (6.1)
plasma A state of matter formed at extremely high temperatures; it behaves like a gas in that the plasma occupies the entire volume

of the container, does not have a fixed shape, and is fluid; unlike a gas, which contains only neutral particles, a plasma contains atoms, ions, and electrons and is strongly affected by electrical and magnetic fields; plasma is considered to be the most common state of matter in the Universe, although it is not commonly found on Earth (1.1)

plasticity A property of a material that describes its ability to undergo a nonreversible change of shape in response to an applied force (22.1)

polar aprotic solvents Polar solvents, such as propanone, that cannot donate hydrogen (21.2)

polar covalent bond A covalent bond in which the electron pair is shared unequally, so the bond has partially negative and partially positive poles (8.5)

polar molecule A molecule with an unequal distribution of charge as a result of its polar bonds and shape (3.6)

polar protic solvents Polar solvents that have a dissociable H^+ (an acidic hydrogen) (21.2)

polarimeter A device used to measure the rotation of plane-polarized light by an optically active compound (20.4)

polarizability The ease with which a particle's electron cloud can be distorted (11.3, 21.2)

polyatomic ions Ions in which two or more atoms are bonded covalently (2.7)

polymer (also *macromolecule*) An extremely large molecule that results from the covalent linking of many simpler molecular units (monomers) (22.1)

polymerization A process that creates polymers by linking smaller molecules to make larger molecules (21.5)

polyprotic acid An acid with more than one ionizable proton (16.4)

polysaccharides Macromolecules that consist of many simple sugars linked covalently (22.3)

positron The antiparticle of an electron (25.1)

positron (β^+) emission A type of radioactive decay in which a positron is emitted from a nucleus (25.1)

potential energy The energy that an object has because of its position relative to other objects or its composition (1.1)

precision The closeness of a measurement to other measurements of the same phenomenon in a series of experiments (1.6)

pre-exponential factor The term that precedes the exponential term in the equation used to find the rate constant (k) at different temperatures with the activation energy; believed to be a representation of the number of collisions that occur (14.5)

pressure (p) The force exerted per unit of surface area (4.2)

pressure-volume work (pV work) A type of work in which a volume change occurs against an external pressure (5.2)

principal quantum number (n) A positive integer that specifies the energy and relative size of an atomic orbital; a number that specifies an energy level in an atom (6.4)

probability contour A shape that defines the volume around an atomic nucleus within which an electron spends a given percentage of its time (6.4)

products Substances formed in a chemical reaction (3.3)

properties Characteristics that give a substance its unique identity (1.1)

proteins Natural linear polymers composed of any of about 20 types of amino acid monomers linked together by peptide bonds (22.3)

proton acceptor A substance that accepts an H^+ ion; a Brønsted-Lowry base (16.1)

proton donor A substance that donates an H^+ ion; a Brønsted-Lowry acid (16.1)

proton (p^+) A subatomic particle that is found in the nucleus and has a unit positive charge ($1.602\ 18 \times 10^{-19}$ C) (2.5)

pseudo–noble gas configuration The $(n-1)d^{10}$ configuration of a p-block metal ion that has an empty outer energy level (7.4)

Q

qualitative When used to describe data, indicates that the data are nonnumerical; qualitative data are also called *observations* and can be used to describe anything that is seen, smelled, or heard during an experiment; some textures may also be described, but chemicals should never be tasted in a laboratory (1.3)

quantitative When used to describe data, indicates that the data are numerical; quantitative data are measured during an experiment and should always be noted with the correct units (1.3)

quantum A packet of energy that is equal to $h\nu$; the smallest quantity of energy that can be emitted or absorbed (6.1)

quantum mechanics The branch of physics that examines the wave motion of objects on the atomic scale (6.4)

quantum number A number that specifies a property of an orbital or an electron (6.1)

R

racemization A process by which a mixture that contains one enantiomer is converted into an equimolar mixture of two enantiomers (21.3)

rad (radiation-absorbed dose) The quantity of radiation that results in 0.01 J of energy being absorbed per kilogram of tissue; 1 rad = 0.01 J/kg tissue = 10^{-2} Gy (25.4)

radial probability distribution plot The graphic depiction of the total probability distribution (sum of ψ^2) of an electron in the region near the nucleus (6.4)

radicals Atoms or groups of atoms that contain one or more unpaired electrons; unpaired electrons in the outer shells do not affect the charge on the resultant molecule; free radicals can be negatively charged, positively charged, or electrically neutral, and are unstable and highly reactive (21.2)

radioactivity The emission of radiation by an atom undergoing spontaneous nuclear transformations (25.1)

radioisotopes Isotopes that have an unstable nucleus, which decays through radioactive emissions (25.2)

radioisotopic dating A method for determining the age of an object based on the rate of decay of a particular radioactive nuclide relative to a stable nuclide (25.2)

radius of gyration (R_g) A measure of the size of a coiled polymer chain, expressed as the average distance from the centre of mass of the chain to its outside edge (22.1)

random error The error that occurs in all measurements (with its size depending on the measurer's skill and the instrument's precision) and results in values *both* higher and lower than the actual value (1.6)

Raoult's law A law stating that the vapour pressure of a solvent above a solution equals the mole fraction of the solvent times the vapour pressure of the pure solvent: $p_{solvent} = X_{solvent} \times p^{\circ}_{solvent}$ (12.5)

rare earths (See *lanthanides.*)

rate constant The proportionality constant that relates the rate of a reaction to the reactant (and product) concentrations (14.3)

rate-determining step (also *rate-limiting step*) The slowest step in a reaction mechanism and, therefore, the step that limits the overall rate (14.6, 21.3)

rate law (rate equation) An equation that expresses the rate of a reaction as a function of the reactant (and product) concentrations (14.3)

reactants The starting substances in a chemical reaction (3.3)
reaction energy diagram A graph that shows the potential energy of a reacting system as it progresses from reactants to products (14.5)
reaction intermediate A substance that is formed and used up during an overall reaction and, therefore, does not appear in the overall equation (14.6)
reaction mechanism A series of elementary steps that sum to the overall reaction and is consistent with the rate law (14.6)
reaction orders The exponents of the reactant concentrations in a rate law, which show how the rate is affected by changes in these concentrations (14.3)
reaction quotient (Q) A ratio of terms for a given reaction, consisting of product concentrations multiplied together and divided by reactant concentrations multiplied together, each raised to the power of its balancing coefficient; the value of Q changes until the system reaches equilibrium, at which point it equals K (15.2)
reaction rate The change in the concentrations of reactants (or products) with time (14.1)
reactor core The part of a nuclear reactor that contains the fuel rods and generates heat from fission (25.7)
rearrangement A type of reaction in which the atoms of a molecule are rearranged to form a new isomer of the original molecule; involves a change in the connectivity of an atom or group of atoms (21.3)
redox A reaction that involves the transfer of electrons between two chemical species (See also *oxidation-reduction reaction.*) (21.1)
redox reaction (See *oxidation-reduction reaction.*)
reducing agent The substance that donates electrons in a redox reaction and undergoes an increase in oxidation number (19.1)
reduction The gain of electrons by a species, accompanied by a decrease in oxidation number (19.1, 21.1)
refraction A phenomenon in which a wave changes its speed and, therefore, its direction as it passes through a phase boundary into a different medium (6.1)
rem (*r*oentgen *e*quivalent for *m*an) The unit of radiation dosage for a human based on the product of the number of rads and a factor related to the biological tissue; 1 rem = 10^{-2} Sv (25.4)
resonance A method used to describe the delocalized electrons in molecules where the bonding cannot be expressed by a single Lewis structure; the connectivity is the same, but the electrons are distributed differently around the structure (21.2)
resonance hybrid The weighted average of the resonance structures of a molecule (8.6)
resonance structures (resonance forms) Two or more Lewis structures for a molecule that cannot be adequately depicted by a single structure; resonance structures differ only in the position of bonding and lone electron pairs (8.6)
retrosynthesis Working backwards to determine the route to obtain the target molecule; the desired molecule is considered first, and then different steps have to be proposed, one at a time, to lead back to the appropriate starting materials (21.7)
reverse osmosis A process for preparing drinkable water that uses an applied pressure greater than the osmotic pressure to remove ions from an aqueous solution, typically seawater (12.6)
rms speed (u_{rms}) The speed of a molecule with average kinetic energy; very close to the most probable speed (4.5)
roasting A pyrometallurgical process in which metal sulfides are converted to oxides (23.3)
round To remove digits based on a series of rules to obtain an answer with the proper number of significant figures (or decimal places) (1.6)

S

s orbital An atomic orbital with $l = 0$ (6.4)
salt bridge An inverted U tube containing a solution of a nonreacting electrolyte that connects the compartments of a voltaic cell and maintains neutrality by allowing ions to flow between compartments (19.2)
saturated hydrocarbons Hydrocarbons in which each carbon atom is bonded to four other atoms (20.2)
saturated solution A solution that contains the maximum amount of dissolved solute at a given temperature (prepared with undissolved solute present) (12.3)
sawhorse projection A view of a molecule along the axis of a model from an oblique angle; the groups connected to both the front and back carbon atoms are drawn using sticks at 120° angles (20.3)
scanning tunnelling microscopy An instrumental technique that uses electrons moving across a minute gap to observe the topography of a surface on the atomic scale (11.6)
Schrödinger equation An equation that describes how the electron matter-wave changes in space around the nucleus; solutions of the equation provide allowable energy levels of the atom (6.4)
scientific method A process of creative thinking and testing aimed at objective, verifiable discoveries of the causes of natural events (1.3)
scintillation counter A device used to measure radioactivity through its excitation of atoms and their subsequent emission of light (25.2)
second law of thermodynamics A law stating that a process occurs spontaneously in the direction that increases the entropy of the universe (18.1)
second-order reaction A chemical reaction in which the rate is proportional to the concentration of each of two reacting molecules or the square of the concentration of a single reactant; the sum of the exponents in the rate law is equal to 2 (21.3)
second (s) The SI base unit of time (1.5)
seesaw shape A molecular shape caused by the presence of one equatorial lone pair in a trigonal bipyramidal arrangement (AX_4E_1) (9.1)
selective precipitation The process of separating ions through differences in the solubility of their compounds with a given precipitating ion (17.3)
semiconductor A substance whose electrical conductivity is poor at room temperature but increases significantly with rising temperature (11.6)
semicrystalline When used to describe a polymer, indicates that the polymer is partially organized in orderly crystalline structures; therefore, some fraction of the polymer remains uncrystallized, or amorphous, when the polymer is cooled to room temperature (22.1)
semimetals (See *metalloids.*)
semipermeable membrane A membrane that allows solvent, but not solute, to pass through (12.5)
shared pair (See *bonding pair.*)
shells Specific energy states of an atom, given by the principal quantum number, n (6.4)
shielding The ability of other electrons, especially those occupying inner orbitals, to lessen the nuclear attraction for an outer electron (7.1)
SI unit A unit composed of one or more of the base units of the Système International d'Unités, a revised metric system (1.5)
side reactions Undesired chemical reactions that consume some of the reactant and reduce the overall yield of the desired product (3.4)

sievert (Sv) The SI unit of human radiation dosage; 1 Sv = 100 rem (25.4)
sigma (σ) bond A type of covalent bond that arises through end-to-end orbital overlap and has most of its electron density along an imaginary line joining the nuclei (10.2)
sigma (σ) molecular orbitals Molecular orbitals that are cylindrically symmetrical about an imaginary line that runs through the nuclei of the component atoms (10.3)
significant figures The digits obtained in a measurement; the greater the number of significant figures, the greater the certainty of the measurement (1.6)
silicates Compounds that consist of repeating —Si—O groupings and, in most cases, metal cations; found throughout rocks and soil (13.6)
silicones Synthetic polymers that contain —Si—O repeat units, with organic groups and crosslinks (13.6)
simple cubic unit cell A unit cell in which a particle occupies each corner of a cube (11.6)
single bond A bond that consists of one electron pair (8.3)
slag A molten waste product formed in a blast furnace by the reaction of acidic silica with a basic metal oxide (23.4)
smelting The process of heating a mineral with a reducing agent, such as coke, to obtain a metal (23.3)
S_N1 In organic chemistry, a substitution reaction that goes through at least a two-step mechanism; S_N stands for *nucleophilic substitution*, and the 1 represents the fact that the rate-determining step is unimolecular (21.3)
S_N2 In organic chemistry, a substitution reaction that goes through at least a one-step mechanism; S_N stands for *nucleophilic substitution*, and the 2 represents the fact that the rate-determining step is bimolecular (21.3)
soap The salt formed in a reaction between a fatty acid and a strong base, usually a group 1 or 2 hydroxide (22.4)
solid One of the three common states of matter; a solid has a fixed shape that does not conform to the shape of the container (1.1)
solubility (s) The maximum amount of solute that dissolves in a fixed quantity of a particular solvent at a specified temperature (2.9, 12.1)
solubility-product constant (K_{sp}) An equilibrium constant for a slightly soluble ionic compound dissolving in water (17.3)
solute The substance that dissolves in a solvent (3.5, 12.1)
solution (See *homogeneous mixture.*)
solvated Surrounded closely by solvent molecules (3.6)
solvation The process of surrounding a solute particle with solvent particles (12.2)
solvent The substance in which a solute dissolves (3.5, 12.1)
solvent effects The effects of solvents on chemical reactivity (21.2)
source (See *occurrence.*)
***sp* hybrid orbitals** Orbitals formed by the mixing of one *s* orbital and one *p* orbital of a central atom (10.1)
sp^2 hybrid orbitals Orbitals formed by the mixing of one *s* orbital and two *p* orbitals of a central atom (10.1)
sp^3 hybrid orbitals Orbitals formed by the mixing of one *s* orbital and three *p* orbitals of a central atom (10.1)
sp^3d hybrid orbitals Orbitals formed by the mixing of one *s* orbital, three *p* orbitals, and one *d* orbital of a central atom (10.1)
sp^3d^2 hybrid orbitals Orbitals formed by the mixing of one *s* orbital, three *p* orbitals, and two *d* orbitals of a central atom (10.1)
specific heat capacity (c) The quantity of heat required to change the temperature of 1 g of a material by 1 K (5.3)
spectator ions Ions that are present as part of a reactant but are not involved in the chemical change (3.7)
spectrochemical series A ranking of ligands in terms of their ability to split *d*-orbital energies (24.4)
spectrometry Any instrumental technique that uses a portion of the electromagnetic spectrum to measure the atomic and molecular energy levels of a substance (6.2)
speed The distance a wave moves per unit time (6.1)
speed of light (c) A fundamental constant that gives the speed at which electromagnetic radiation travels in a vacuum: $c = 2.997\,924\,58 \times 10^8$ m/s (6.1)
spin quantum number (m_s) A number, either $+\frac{1}{2}$ or $-\frac{1}{2}$, that indicates the direction of electron spin (7.1)
spontaneous change A change that occurs by itself, without an ongoing input of external energy (18.1)
square planar shape A molecular shape caused by the presence of two lone pairs at opposite vertices in an octahedral arrangement (AX_4E_2) (9.1)
square pyramidal shape A molecular shape caused by the presence of one lone pair in an octahedral arrangement (AX_5E_1) (9.1)
standard atmosphere (atm) The average atmospheric pressure measured at sea level and 0°C, defined as $1.013\,25 \times 10^5$ Pa (4.2)
standard cell potential (E°_{cell}) The potential of a cell measured with all the components in their standard states and no current flowing (19.3)
standard electrode potential ($E^\circ_{half\text{-}cell}$) The standard potential of a half-cell, with the half-reaction written as a reduction (19.3)
standard enthalpy of formation ($\Delta_f H^\circ$) The enthalpy change that occurs when 1 mol of a compound forms from its elements with all the components in their standard states (5.6)
standard enthalpy of reaction ($\Delta_r H^\circ$) (also *standard heat of reaction*) The enthalpy change that occurs during a reaction when all the components are in their standard states (5.6)
standard entropy of reaction ($\Delta_r S^\circ$) The entropy change that occurs during a reaction when all the components are in their standard states (18.2)
standard Gibbs energy change (ΔG°) The Gibbs energy change that occurs when all the components are in their standard states (18.3)
standard Gibbs energy of formation ($\Delta_f G^\circ$) The standard Gibbs energy change that occurs when 1 mol of a compound is made from its elements, with all the components in their standard states (18.3)
standard hydrogen electrode (See *standard reference half-cell.*)
standard molar entropy (S°) The entropy of 1 mol of a substance in its standard state (18.1)
standard molar volume The volume of 1 mol of an ideal gas at standard temperature and pressure: 22.7117 L (4.3)
standard pressure The IUPAC-defined pressure of 1 bar (10^5 Pa) that, along with standard temperature of 273.15 K, defines the standard conditions for a gas system (4.2)
standard reference half-cell (standard hydrogen electrode) A specially prepared platinum electrode immersed in 1 mol/L $H^+(aq)$ through which H_2 gas at 1 bar is bubbled; $E^\circ_{half\text{-}cell}$ is defined as 0 V (19.3)
standard states Specifications used to compare thermodynamic data: 1 bar for gases behaving ideally, 1 mol/L for dissolved species, or the pure substance for liquids and solids (5.6)
standard temperature and pressure (STP) The reference conditions for a gas: 0°C (273.15 K) and 1 bar (4.3)
state function A property of a system determined by its current state, regardless of how it arrived at this state (5.1)
states The three common physical forms of matter: solid, liquid, and gas (1.1)
stationary states In the Bohr model, the allowable energy levels of the atom in which it does not release or absorb energy (6.2)

steady state approximation A method used in chemical kinetics to help simplify complex reaction kinetics; assumes that the rate of change of the concentration of an intermediate in a reaction is zero or, in other words, that the rate at which an intermediate is produced is equal to the rate at which it is consumed (14.6)

steel An alloy of iron with small amounts of carbon and usually other metals (23.4)

stellar nucleosynthesis The process by which elements are formed in the stars through nuclear fusion (25.7)

stepwise reaction A chemical reaction that has one or more reaction intermediates and involves at least two consecutive elementary reactions (21.2)

stereogenic centre (stereocentre, chiral centre) A tetrahedral carbon atom with four different groups, or a trigonal planar carbon atom with three different groups, in which the exchanging of any two groups leads to a stereoisomer (20.4)

stereoisomers Molecules with the same connections of atoms but different orientations of groups in space (20.4, 24.3) (See also *geometric isomers* and *optical isomers*.)

steric effect The influence of the spatial configuration of reacting substances upon the rate, nature, and extent of a reaction (21.3)

steric strain The destabilization resulting from van der Waals repulsion between groups that are too close to each other (20.3)

stoichiometry The study of the mass-amount-number relationships of chemical formulas and reactions (Chapter 3)

strong-field ligands Ligands that cause larger crystal field splitting energy and, therefore, are part of a low-spin complex (24.4)

strong force An attractive force that exists between all nucleons and is many times stronger than the electrostatic repulsive force (25.1)

structural formula A formula that shows the actual number of atoms, their relative placement, and the bonds between them (2.8)

structural isomers (See *constitutional isomers*.)

sublimation The process by which a solid changes directly into a gas (11.1)

subshells Energy substates of atoms within levels; given by the n and l values, a subshell designates the size and shape of the atomic orbital (6.4)

substance A type of matter, either an element or a compound, that has a fixed composition (2.1)

substitution A reaction in which an atom or group of atoms is replaced by another atom or group of atoms (21.1)

substrates Reactants that bind to the active site in an enzyme-catalyzed reaction (14.7)

superconductivity The ability to conduct a current with no loss of energy to resistive heating (11.6)

supersaturated solution An unstable solution in which more solute is dissolved than in a saturated solution (12.3)

surface tension The energy required to increase the surface area of a liquid by a given amount (11.4)

surroundings All parts of the universe other than the system being considered (5.1)

suspension A heterogeneous mixture that contains particles that are distinct from the surrounding medium (12.6)

syn addition The addition of two substituents to the same side (or face) of a double (or triple) bond (21.5)

synthetic natural gas (SNG) A gaseous fuel mixture, mostly methane, that is formed from coal (5.6)

system The defined part of the universe under study (5.1)

systematic error A type of error that produces values that are all either higher or lower than the actual value, often caused by faulty equipment or a consistent flaw in technique (1.6)

T

t_{2g} orbitals The set of orbitals (composed of d_{xy}, d_{yz}, and d_{xz}) that results when the energies of the metal-ion d orbitals are split by a ligand field; this set of orbitals is lower in energy than the other (e_g) set in an octahedral field and higher in energy in a tetrahedral field (24.4)

temperature (T) A measure of how hot or cold a substance is, relative to another substance (1.5)

tetrahedral arrangement The geometric arrangement formed when four electron groups maximize their separation around a central atom; when all four groups are bonding groups, the molecular shape is tetrahedral (AX_4; ideal bond angle = 109.5°) (9.1)

tetrahedral hole A hole surrounded by four atoms that are packed together in a closest-packing arrangement (the coordination number is four); the size of the hole is smaller than the size of the atoms surrounding it (11.6)

theoretical yield The amount of product predicted by a stoichiometrically equivalent molar ratio in a balanced equation (3.4)

theories (See *models*.)

thermochemical equation A chemical equation that shows the heat involved for the amounts of substances specified (5.4)

thermometer A device for measuring temperature; contains a fluid that expands or contracts within a graduated tube (1.5)

third law of thermodynamics A law stating that the entropy of a perfect crystal is zero at 0 K (18.1)

titration A method for determining the concentration of a solution by monitoring relative amounts during a reaction between the solution and another solution of known concentration (17.2)

Torr A unit of pressure that is identical to 1 mmHg (4.2)

torsional strain The resistance to rotation through eclipsed and staggered conformations (20.3)

total ionic equation An equation for an aqueous reaction that shows all the soluble ionic substances dissociated into ions (3.7)

tracer A radioisotope that signals the presence of the species of interest by radioactive emissions (25.5)

transcription The process of producing messenger RNA from DNA (22.3)

transition elements Elements that occupy the d block of the periodic table; elements whose d orbitals are being filled (7.2)

transition metals Any elements that fall between the s block and the p block in the main periodic table; they are the elements that form the d block (Chapter 24)

transition state (activated complex) An unstable species formed in an effective collision of reactants; exists momentarily when the system is highest in energy and can either form products or re-form reactants (14.5)

transition state theory A model that explains how the energy of reactant collisions is used to form a high-energy transitional species that can change to a reactant or product (14.5)

translation The process by which messenger RNA specifies the sequence of amino acids and synthesizes proteins (22.3)

transuranium elements Elements with atomic numbers higher than the atomic number of uranium ($Z = 92$) (25.3)

trigonal bipyramidal arrangement The geometric arrangement formed when five electron groups maximize their separation around a central atom; when all five groups are bonding groups, the molecular shape is trigonal bipyramidal (AX_5; ideal bond angles: axial-centre-equatorial = 90° and equatorial-centre-equatorial = 120°) (9.1)

trigonal planar arrangement The geometric arrangement formed when three electron groups maximize their separation around a central atom; when all three groups are bonding groups, the molecular shape is trigonal planar (AX_3; ideal bond angle = 120°) (9.1)

trigonal pyramidal shape A molecular shape caused by the presence of one lone pair in a tetrahedral arrangement (AX_3E_1) (9.1)
triple bond A covalent bond that consists of three bonding pairs: two atoms sharing six electrons; one *s* bond and two π bonds (8.3)
triple point The pressure and temperature at which three phases of a substance are in equilibrium in a phase diagram (11.2)
T shape A molecular shape caused by the presence of two equatorial lone pairs in a trigonal bipyramidal arrangement (AX_3E_2) (9.1)
two-step mechanism The step-by-step sequence of a reaction that takes place in two steps; written using a technique called *arrow pushing* to depict the flow or movement of electrons (21.3)
Tyndall effect The scattering of light by a colloid (12.6)

U

ultraviolet (UV) Radiation in the region of the electromagnetic spectrum between the visible and the X-ray regions (6.1)
uncertainty A characteristic of every measurement that results from the inexactness of the measuring device and the need to estimate when taking a reading (1.6)
uncertainty principle A principle proposed by Werner Heisenberg, stating that it is impossible to know simultaneously the exact position and velocity of a particle; the uncertainty principle becomes important only for particles with very small masses (6.3)
unified atomic mass unit (u) A unit used to specify mass on an atomic scale; defined as $\frac{1}{12}$ the mass of a carbon-12 atom (2.5)
unimolecular Means that the rearrangement of a single molecule produces one or more molecules of product (21.3)
unimolecular reaction An elementary reaction involving the decomposition or rearrangement of a single species (molecule, ion) (14.6)
unit cell The smallest portion of a crystal that, if repeated in all three directions, gives the crystal (11.6)
universal gas constant (*R*) A proportionality constant that relates the energy, amount of substance, and temperature of a system; $R = 8.314\,472$ kPa·L/(mol·K) = 0.083 147 bar·L/(mol·K) (4.3)
unsaturated hydrocarbons Hydrocarbons with at least one carbon-carbon multiple bond; contain at least two C atoms bonded to fewer than four atoms each (20.5)
unsaturated solution A solution in which more solute can be dissolved at a given temperature (12.3)
unshared pair (See *lone pair.*)

V

V shape (See *bent shape.*)
valence band In band theory, the lower energy portion of the band of molecular orbitals, which is filled with valence electrons (11.6)
valence bond (VB) theory A model that attempts to reconcile the shapes of molecules with the shapes of atomic orbitals through the concepts of orbital overlap and hybridization (10.1)
valence electrons The electrons involved in compound formation; in main-group elements, the electrons in the valence (outer) shell (7.2)
valence-shell electron-pair repulsion (VSEPR) theory A model explaining that the shapes of molecules and ions result from minimizing electron-pair repulsions around a central atom (9.1)
van der Waals constants Experimentally determined positive numbers used in the van der Waals equation to account for the intermolecular attractions and molecular volumes of real gases (4.6)
van der Waals equation An equation that accounts for the behaviour of real gases (4.6)
van der Waals radius One-half of the shortest distance between the nuclei of identical nonbonded atoms (11.3)
vaporization The process of changing from a liquid to a gas (11.1)
vapour pressure The pressure exerted by a vapour at equilibrium with its liquid in a closed system (11.2)
vapour pressure lowering (Δp) The lowering of the vapour pressure of a solvent, caused by the presence of dissolved solute particles (12.5)
variables Quantities that can have more than a single value (1.3) (See also *controlled experiment.*)
viscosity A measure of the resistance of a liquid to flow (11.4, 22.1)
volatility The tendency of a substance to become a gas (2.9)
volt (V) The SI unit of electric potential: 1 V = 1 J/C (19.3)
voltage The current multiplied by the resistance in a circuit; in a chemical cell it can also represent the cell potential (19.3)
voltaic cell (galvanic cell) An electrochemical cell that uses a spontaneous redox reaction to generate electrical energy (19.2)
volume percent [% (v/v)] A concentration term defined as the volume of solute in 100. volumes of solution (12.4)
volume (*V*) The space occupied by a sample of matter (1.5)

W

wastewater Used water that is treated before being returned to the environment; usually contains industrial and/or residential waste (12.6)
water softening The process of replacing the hard-water ions Ca^{2+} and Mg^{2+} with Na^+ ions (12.6)
wave function (atomic orbital) A mathematical expression that describes the motion of an electron's matter-wave in terms of time and position in the region of the nucleus; this term is used qualitatively to mean the region of space in which there is a high probability of finding an electron (6.4)
wave-particle duality The principle stating that both matter and energy have wavelike and particle-like properties (6.3)
wavelength (λ) The distance between any point on a wave and the corresponding point on the next wave; the distance that a wave travels during one cycle (6.1)
weak-field ligands Ligands that cause smaller crystal field splitting energy and, therefore, are part of a high-spin complex (24.4)
weight The force exerted by a gravitational field on an object; the weight of an object is directly proportional to its mass (1.5)
work (*w*) The energy transferred when an object is moved by a force (5.1)

X

X-ray diffraction analysis An instrumental technique used to determine the dimensions of a crystal structure by measuring the diffraction patterns caused by X-rays impinging on the crystal (11.6)

Z

Zaitsev's rule An organic chemistry rule stating that if more than one alkene can be formed by elimination reactions, the major product will be the more stable product, which will generally be the alkene with more substituents on the carbon of the double bond (21.4)
zone refining A process used to purify metals and metalloids; impurities are removed from a bar of the element by concentrating them in a thin molten zone (23.3)

CREDITS

Chapter 1

Opener (left): Fototeca Storica Nazionale/Getty Images; Opener (right): © Klaus Tiedge/Blend Images LLC; Figure 1.1A: © Paul Morrell/Stone/Getty Images; Figure 1.1B: © McGraw-Hill Education. Stephen Frisch, photographer; Table 1.1 A&B: © McGraw-Hill Education. Stephen Frisch, photographer; Table 1.1C: © Ruth Melnick; Table 1.1 D–F: © McGraw-Hill Education. Stephen Frisch, photographer; Figure 1.4: © SSPL/The Image Works; Figure 1.5A: © George Haling/Photo Researchers, Inc.; Figure 1.5B: © Lukyslukys/GetStock.com; Figure 1.5C: © Herminutomo/GetStock.com; Figure 1.5D: © Kiselev Andrey Velerevich/Shutterstock.com; Figure 1.8A: © McGraw-Hill Education. Stephen Frisch, photographer; Figure 1.8B: BrandTech Scientific, Inc.; Figure 1.13A–B: © McGraw-Hill Education. Stephen Frisch, photographer.

Chapter 2

Opener (left): © Boris15/Dreamstime.com; Opener (right): © Steve Hamblin/Alamy; Figure 2.1: © McGraw-Hill Education. Stephen Frisch, photographer; Figure 2.2: © McGraw-Hill Education. Stephen Frisch, photographer; Figure 2.3 (top): © Punchstock RF; Figure 2.3 (bottom): © Alexander Cherednichenko/Shutterstock.com; Page 42: © Sprokop/GetStock.com; Page 49: © The Print Collector/GetStock.com; Figure B2.1: Pictorial Press Ltd./Alamy; Page 53 A–E: Courtesy of Rashmi Venkateswaran; Figure 2.10: © McGraw-Hill Education. Stephen Frisch, photographer; Figure 2.11: © McGraw-Hill Education. Stephen Frisch, photographer; Figure 2.16: Matteo Chinellato—ChinellatoPhoto/Getty Images; Figure 2.20: © McGraw-Hill Education. Stephen Frisch, photographer; Figure B2.4 (left): Courtesy of the Food and Drug Administration; Figure B2.4 (right): Datamax/Wikipedia.

Chapter 3

Opener (left): ollaweila/iStock/Thinkstock; Opener (right): © Tanya Constantine/Blend Images/Corbis; Figure 3.1: © McGraw-Hill Education. Charles Winters, photographer; Figure 3.7: © McGraw-Hill Education. Charles Winters, photographer; Page 107: Sampete/GetStock.com; Page 108: efesantik/iStock/Thinkstock; Figure 3.14: © McGraw-Hill Education. Stephen Frisch, photographer; Figure 3.15: © McGraw-Hill Education. Stephen Frisch, photographer; Figure 3.19: © McGraw-Hill Education. Stephen Frisch, photographer; Figure 3.20: © McGraw-Hill Education. Stephen Frisch, photographer.

Chapter 4

Opener (left): CHARLES D. WINTERS/SCIENCE PHOTO LIBRARY; Opener (right): Stefan Holm/Shutterstock.com; Figure 4.1: © McGraw-Hill Education. Stephen Frisch, photographer; Figure 4.2: © McGraw-Hill Education. Stephen Frisch, photographer;Page 166: © Charles F. McCarthy/Shutterstock.com; Page 171: © McGraw-Hill Education. Stephen Frisch, photographer.

Chapter 5

Opener (left): © McGraw-Hill Education; Opener (right): © Brendan Montgomery/Demotix/Corbis; Figure 5.1: © McGraw-Hill Education. Stephen Frisch, photographer.

Chapter 6

Opener (left): Courtesy of the Advanced Light Source, Lawrence Berkeley National Laboratory. Used with permission; Opener (right): STEVE MARCUS/Reuters/Landov; Figure 6.6 A: © Ravi/Shutterstock.com; Figure 6.6 B: © McGraw-Hill Education. Charles Winters, photographer; Figures 6.6 C: © Feng Yu/Shutterstock.com; Page 240: © Library of Congress/LC-DIG-ggbain-06493; Page 241: Ingram Publishing; Page 244: EMILIO SEGRE VISUAL ARCHIVES/AMERICAN INSTITUTE OF PHYSICS/SCIENCE PHOTO LIBRARY; Page 246: Library of Congress/LC-DIG-ggbain-35303; Page 248: University of Saskatchewan Archives, Photograph Collection, A-3234; Figure B6-1 A: © McGraw-Hill Education. Stephen Frisch, photographer; Figure B6-1 B: Washington DC Convention & Visitors Association; Page 251: © Corbis; Figure 6-14: PSSC Physics © 1965, Education Development Center, Inc.; DC Health & Company/Education Development Center, Inc.; Figure 6-15: © Dennis Kunkel/Phototake.com; Page 254: © INTERFOTO/Alamy; Page 257: © INTERFOTO/Alamy; Page 259: Courtesy of Tom Woo.

Chapter 7

Opener (left): Vera Kalyuzhnaya/Shutterstock; Opener (right): Ventin/GetStock.com; Page 273: SuperStock/SuperStock; Figure 7.9: © McGraw-Hill Education. Stephen Frisch, photographer; Figure 7.21: © McGraw-Hill Education. Stephen Frisch, photographer.

Chapter 8

Opener (left): KarenUpNorth/iStock/Thinkstock; Opener (right): Scoutts/iStock Editorial/Thinkstock; Figure 8.1: Used by permission of McMaster University; Page 307: LAWRENCE BERKELEY NATIONAL LABORATORY/SCIENCE PHOTO LIBRARY; Figure 8.7: © McGraw-Hill Education. Stephen Frisch, photographer; Figure 8.10: © McGraw-Hill Education. Stephen Frisch, photographer; Figure 8.11: © McGraw-Hill Education. Stephen Frisch, photographer; Figure 8.15 A: Courtesy of Paul G. Mezey; Figure 8.15 B: Used with permission of Dalhousie Photography Services; Figure 8.17: © McGraw-Hill Education. Stephen Frisch, photographer;Page 321: Danylo Fomin/iStock/Thinkstock; Figure 8.29: © McGraw-Hill Education. Stephen Frisch, photographer; Figure 8.33: © McGraw-Hill Education. Stephen Frisch, photographer; Page 346 (top): © Creative Control/Alamy; Page 346 (bottom): © John Lander/Alamy.

Chapter 9

Opener (left): ogwen/Shutterstock; Opener (right): lisadfindlay/iStock Editorial/Thinkstock; Figure 9.1 A: Used with permission from McMaster University; Figure 9.1 B: NPG x 165693, Sir Ronald Sydney Nyholm, by Godfrey Argent, bromide print, 9 October 1970, © National Portrait Gallery, London; Figure 9.2: © McGraw-Hill Education. Stephen Frisch, photographer;Figure 9.6: © McGraw-Hill Education. Stephen Frisch, photographer; Figure B9.3 A: © Jim West/Alamy; Figure B9.3 B: © jack thomas/Alamy; Figure B9.4: Courtesy of Simon Fraser University.

Chapter 10

Opener (left): © McGraw-Hill Education. Stephen Frisch, photographer; Opener (right): molekuul.be/Shutterstock; Figure 10.15 A: Courtesy of University of Calgary; Figure 10.15 B: Photo courtesy of University of Lethbridge; Figure 10.15 C: Photo by Mike Sturk. Courtesy of ISEEE; Figure 10.24: © McGraw-Hill Education. Charles Winters/Timeframe Photography, Inc.

Chapter 11

Opener (left): © IMEC; Opener (right): © billyfoto/iStock; Page 404: © Christopher Meder Photography/Shutterstock.com; Figure 11.1 A: © dinadesign/Shutterstock.com; Figure 11.1 B: © Thinkstock/Alamy; Figure 11.9: © McGraw-Hill Education. Richard Megna, photographer; Figure 11.12: Courtesy of Carleton University; Page 423: Nuridsany & Perennou/Photo Researchers, Inc.; Figure 11.25 B: © Scott Camazine/Photo Researchers, Inc.; Figure 11.27 A, C&D: © Jeffrey A. Scovil Photography (Wayne Thompson Minerals); Figure 11.27 B: Albert Russ/Shutterstock; Page 428: © Bryan Busovicki/Shutterstock.com; Figure B11.3: Courtesy National Institute of Standards and Technology; Figure 11.39: Courtesy of Arthur Mar; Figure 11.47 A: © J. R. Factor/Photo Researchers, Inc.; Figure 11.47 B: James Dennis/Phototake.com; Figure 11.50: Courtesy of Benoit Dubertret; Figure 11.51: © and by permission from Ferrotec (USA) Corporation. May not be reproduced without written permission; Figure 11.52: © Rob Fox/iStockphoto.com; Figure 11.53: © David Mack/Science Photo

Library; Figure B11.4 A: Courtesy of Geoffrey Ozin; Figure B11.4 B: Courtesy of Opalux Inc. www.opalux.com; Figure B11.5: DR. P. MARAZZI/SCIENCE PHOTO LIBRARY; Figure B11.6 A&B: Courtesy of Neil Branda; Figure B11.7B: Regularly faceted nanoprisms (nanoflowers) were synthesized by Nicole Cathcart. Electron microscopy images were made and arranged by Vladimir Kitaev. The research results have been published in Nanoscale. Cathcart, N.; Kitaev, V. Multifaceted prismatic silver nanoparticles: synthesis by chloride-directed selective growth from thiolate-protected clusters and SERS properties. Nanoscale, 2012, 4, 6981-6989.

Chapter 12

Opener (left): Mariyana M/Shutterstock; Opener (right): © DigitalVues/Alamy; Figure 12.4 A: © Ruth Melnick; Figure 12.4 B: © Ingram Publishing/AGE Fotostock RF; Figure 12.9: © McGraw-Hill Education. Stephen Frisch, photographer; Figure 12.11: © Photodisc/Getty RF; Page 471 (top): © Peter Scoones/Photo Researchers, Inc.; Page 471 (bottom): ROYAL INSTITUTION OF GREAT BRITAIN/SCIENCE PHOTO LIBRARY; Figure 12.13: © McGraw-Hill Education. Stephen Frisch, photographer; Page 478: *Journal of Physical Chemistry*, #27, 1898 © American Chemical Society; Page 482 (top): © KMI Images/Alamy; Page 482 (middle): Photodisc/Alamy; Page 482 (bottom): © Comstock/PunchStock; Page 488 (top): © Toronto Star Staff/GetStock.com; Page 488 (bottom): © Andrew J. Martinez/Science Photo Library; Figure 12.22: © David M. Phillips/Photo Researchers, Inc.; Figure 12.23 A: © McGraw-Hill Education. Charles Winters/Timeframe Photography, Inc.; Figure 12.23 B: © Corbis/RF; Figure 12.24: Pete Ryan/National Geographic/Getty Images; Figure B12.3: © Robert Essel/Corbis.

Chapter 13

Opener (left): Eyematrix/Getty Images; Opener (right): Neniia Lanti/iStock/Thinkstock; Table 13.1: (17 Fluorine) © Jumk.de Webprojects/http://images-of-elements.com/fluorine.php and (all others): © McGraw-Hill Education. Stephen Frisch, photographer; Page 511 (both): © McGraw-Hill Education. Stephen Frisch, photographer; Page 512: © McGraw-Hill Education. Stephen Frisch, photographer; Page 514: © McGraw-Hill Education. Stephen Frisch, photographer; Page 517: © McGraw-Hill Education. Stephen Frisch, photographer; Page 518: © McGraw-Hill Education. Stephen Frisch, photographer; Page 520: © McGraw-Hill Education. Stephen Frisch, photographer; Figure 13.11: © McGraw-Hill Education. Stephen Frisch, photographer; Page 524: © John Hryniuk; Page 524: Source Text by Dr. Doug Stephan; Figure 13.15: © Gontar/Shutterstock.com; Page 526: © McGraw-Hill Education. Stephen Frisch, photographer; Figure 13.16 A: Courtesy of Neil Burford; Figure 13.16 B: Courtesy of Michael D. Fryzuk; Page 530: THE CANADIAN PRESS/Adrian Wyld; Page 534: © McGraw-Hill Education. Stephen Frisch, photographer; Page 535: © Davo Blair/Alamy; Figure 13.22: Photo courtesy of Tris Chivers; Page 538: © Steve Cole/Getty RF; Figure 13.23: © McGraw-Hill Education. Stephen Frisch, photographer; Page 539: © McGraw-Hill Education. Stephen Frisch, photographer; Figure 13.25: © McGraw-Hill Education. Stephen Frisch, photographer; Figure 13.28: Courtesy of University of British Columbia Archives, [UBC 41.1/1944-2]; Figure 13.29: Argonne National Laboratory; Page 545: © McGraw-Hill Education. Stephen Frisch, photographer; Figure 13.30: Courtesy of Gary J. Schrobilgen.

Chapter 14

Opener (left): © 1990 Richard Megna, Fundamental Photographs, NYC; Opener (right): © Digital Vision/PunchStock; Figure 14.2 (left to right): © Til_de/Dreamstime.com/GetStock.com; © Hpphoto/Dreamstime.com/GetStock.com; © Joingate/Dreamstime.com/GetStock.com; © Fallsview/Dreamstime.com/GetStock.com; Figure 14.3: © McGraw-Hill Education. Stephen Frisch, photographer; Page 557: bikeriderlondon/Shutterstock; Page 580: THE CANADIAN PRESS/Sean Kilpatrick; Page 585: © Steve Russell/GetStock.com; Figure 14.24: © McGraw-Hill Education. Charles Winters, photographer; Page 596: Smithsonian Institution Archives, Acc. 90-105-Science Service; Figure B14.1: © Kevin J. Smith.

Chapter 15

Opener: Courtesy of the author; Figure 15.1: © McGraw-Hill Education. Stephen Frisch, photographer; Page 636: © AfriPics.com/Alamy.

Chapter 16

Opener (left): SCIENCE PHOTO LIBRARY; Opener (right): © All Canada Photos/Alamy; Page 659 (both): © McGraw-Hill Education. Stephen Frisch, photographer; Figure 16.5: © McGraw-Hill Education. Stephen Frisch, photographer; Figure 16.9: © McGraw-Hill Education. Stephen Frisch, photographer.

Chapter 17

Opener (left): phototake/iStock/Thinkstock; Opener (right): © Skypixel/Dreamstime.com—Citrus Fruit Photo; Figure 17.1: © McGraw-Hill Education. Stephen Frisch, photographer; Figure 17.2: © McGraw-Hill Education. Stephen Frisch, photographer; Figure 17.7: © McGraw-Hill Education. Stephen Frisch, photographer; Figure 17.8: © McGraw-Hill Education. Stephen Frisch, photographer; Figure 17.13: © Jackie Lewin, Royalty Free/Photo Researchers, Inc.; Page 734: © Joel Arem/Science Source; Figure 17.14: © McGraw-Hill Education. Stephen Frisch, photographer; Page 737: © CNRI/Science Source/Photo Researchers, Inc.; Figure 17.15: © McGraw-Hill Education. Stephen Frisch, photographer; Figure 17.16: © Ruth Melnick; Page 739: © McGraw-Hill Education. Stephen Frisch, photographer; Page 748: © McGraw-Hill Education. Mark Dierker, photographer; Figure 17.19: McGraw-Hill Education. Stephen Frisch, photographer.

Chapter 18

Opener (left): © McGraw-Hill Education; Opener (right): Courtesy of Dr. Howard Trottier; Page 764 (top left): Tatsuo Nakamura/Shutterstock; Page 764 (top right): Alta Oosthuizen/Shutterstock; Page 764 (bottom left): KatarzynaBialasiewicz/iStock/Thinkstock; Page 764 (bottom right): Bruce Hugo and Leslie Gaul/Adam Block/NOAO/AURA/NSF; Figure 18.1: © McGraw-Hill Education. Stephen Frisch, photographer; Figure 18.12: © Vanderbilt University—Clinical Nutrition Research; Page 782: © Kdrgreen/Dreamstime.com/GetStock.com.

Chapter 19

Opener (left): Hywit Dimyadi/Shutterstock; Opener (right): © kirstypargeter/iStock; Page 813: © McGraw-Hill Education. Stephen Frisch, photographer; Page 815: © McGraw-Hill Education. Stephen Frisch, photographer; Figure 19.3: © McGraw-Hill Education. Stephen Frisch, photographer; Figure 19.4: © McGraw-Hill Education. Stephen Frisch, photographer; Figure 19.5: © McGraw-Hill Education. Stephen Frisch, photographer; Figure 19.8: © Richard Megna/Fundamental Photographs, NYC; Figure 19.11: © McGraw-Hill Education. Stephen Frisch, photographer; Page 841: Courtesy John Straub, Boston University; Figure 19.16: © McGraw-Hill Education. Charles D. Winters/Timeframe Photography, Inc.; Figure 19.17: Photo © F.W. Goro; Figure 19.18: © Jill Braaten; Figure 19.19: © McGraw-Hill Education. Pat Watson, photographer; Figure 19.20: Courtesy of the Guidant Corporation; Figure 19.23: © AP/Wide World Photos; Figure 19.24: "An ultrafast rechargeable aluminum-ion battery," Lin et al. *Nature* 520, 324–328 (16 April 2015), doi:10.1038/nature14340; Figure 19.26: © Vincent Roy mastweb/Alamy; Figure 19.27: © David Weintraud/Photo Researchers, Inc.; Figure 19.31: © Jill Braaten; Figure 19.32: © McGraw-Hill Education. Stephen Frisch, photographer; Page 856: © McGraw-Hill Education. Stephen Frisch, photographer.

Chapter 20

Opener (left): everythingpossible/iStock/Thinkstock; Opener (right): McGraw-Hill Education; Page 881: Courtesy of Monika Schurmann; Figure 20.17: © Martin Bough/Fundamental Photographs, NYC; Page 902: © WorldFoto/Alamy; Page 906: Volosina/Shutterstock; Page 915: © Jill Braaten; Table 20.6 A: Purestock/SuperStock; Table 20.6 B: © Burke Triolo Productions/Getty Images; Table 20.6 C: Pixtal/AGE Fotostock; Table 20.6 D: © Ingram Publishing/AGE Fotostock; Table 20.6 E: © Creatas/PunchStock; Table 20.6 F: Ingram Publishing/SuperStock; Table 20.6 G: © Image Source/PunchStock; Table 20.6 H: © Ingram Publishing/AGE Fotostock; Table 20.6 I: © Vevesoran/Dreamstime.com/GetStock.com.

Chapter 21

Opener (left): © RGB Ventures/SuperStock/Alamy; Opener (right): Denis Kovin/Shutterstock; Page 971: Leigh Prather/Shutterstock.

Chapter 22

Opener (left): © Martin McCarthy/iStock; Opener (right): Garry Gay/Getty Images; Page 991: Used with permission from Seven Star Sports; Figure 22.5: © McGraw-Hill Education. Charles Winters, photographer; Figure 22.6: Photo courtesy of University of Alberta Archives Accession # 2008-05-289 (R. Lemieux fonds) and Dr. Laura Frost; Page 996: © McGraw-Hill Education. Pat Watson, photographer; Figure 22.10 A: © J. Gross/SPL/Science Source; Figure 22.10 B: fotohunter/Shutterstock; Figure B22.3: © Courtesy of Orchid Cellmark; Figure 22.32: © Zephyr/Photo Researchers, Inc.; Figure 22.33 A: © PerkinElmer, Inc.; Figure 22.33 B: © Mikael Karlsson/Alamy; Figure B22.6: Courtesy of Molly Shoichet; Figure B22.7: Courtesy of Molly Shoichet; Figure B22.8: Courtesy of Kelvin Kenneth Ogilvie; Figure B22.9: Courtesy of Tomas Hudlicky; Figure B22.10: Courtesy of Tomas Hudlicky; Figure B22.11: © Colleen MacPherson; Figure B22.12: Used by permission of Dr. Soledade Pedras.

Chapter 23

Opener (left): assistantua/iStock/Thinkstock; Opener (right): © Chris Cheadle/Alamy; Figure 23.3 A: © Doug Sherman/Geofile RF; Figure 23.3 B: © University of Cambridge, Department of Earth Sciences; Page 1034: © Ruth Melnick; Page 1035: © Robert Holmes/Corbis; Figure B23.1: NASA; Page 1045 A: Reed Richards/Alamy; Page 1045 B: Ingram Publishing/SuperStock; Page 1045 C: Stockbyte/Getty Images; Page 1045 D: Design Pics/Darren Greenwood; Page 1045 E: Don Farrall/Getty Images; Page 1045 F: © Martin Shields/Alamy; Page 1045 G: Ingram Publishing; Page 1045 H: Brand X Pictures/Punchstock; Page 1045 I: Adam Crowley/Getty Images; Page 1045 J: Design Pics/Darren Greenwood; Page 1045 K: Ingram Publishing; Page 1045 L: © Mrmarshall/Dreamstime.com/GetStock.com; Page 1045 M: © Luckydoor/Dreamstime.com/GetStock.com; Page 1047: © Patrick W. Grace/Science Photo Library; Figure 23.12: © Richard Megna/Fundamental Photographs, NYC; Figure 23.15: THE CANADIAN PRESS/Andrew Vaughan; Figure 23.16: asharkyu/Shutterstock; Page 1053: Richard Lautens/GetStock.com; Page 1054: © Bruce Ando/Photo Library/Getty Images; Figure 23.30: © Scanrail/Dreamstime.com/GetStock.com; Page 1056: © Lee Prince/Shutterstock.com; Figure 23.21: © Tom Hollyman/Photo Researchers, Inc.; Page 1057: John Carnemolla/Shutterstock; Page 1060: © Monty Rakusen/Cultura/Getty Images RF; Page 1061: © Ken Whitmore/Getty Images; Figure 23.25: © Nathan Benn/Woodfin Camp & Associates.

Chapter 24

Opener (left): Karol Kozlowski/Shutterstock; Opener (right): science photo/Shutterstock; Figure 24.2: © McGraw-Hill Education. Stephen Frisch, photographer; Figure 24.5: © McGraw-Hill Education. Stephen Frisch, photographer; Page 1080: Photo courtesy of Dr. Sandro Gambarotta; Figure 24.6: © McGraw-Hill Education. Stephen Frisch, photographer; Page 1081: Designua/Shutterstock; Page 1082: Courtesy of Daniel Leznoff; Page 1084: Photo courtesy of Dr. Jennifer Love; Figure 24.8: © McGraw-Hill Education. Charles Winters, Timeframe Photography, Inc.; Page 1091: Photo courtesy of Hélène Lebel; Page 1093: © Creatas/PunchStock; Page 1094: Courtesy of Mark Stradiotto; Figure 24.19: © McGraw-Hill Education. Pat Watson, photographer; Figure 24.20: © McGraw-Hill Education. Stephen Frisch, photographer; Page 1097: Courtesy of Dr. Deryn Fogg.

Chapter 25

Opener (left): © Sam Barnes/Alamy; Opener (middle): Don Farrall/Getty Images; Opener (right): parisvas/iStock/Thinkstock; Figure 25.1 A: SCIENCE PHOTO LIBRARY; Figure 25.1 B: Library of Congress Prints and Photographs Division; Figure 25.5: © Hank Morgan/Rainbow; Figure 25.6: © Meridan Biotechnologies Ltd.; Page 1121: © Copyright 1978 by Vernon Miller. For use by permission for Business Wire via Getty Images; Page 1123: © Hulton Archive/Getty Images; Page 1124: © Maximilien Brice/CERN; Figure 25.12: Courtesy Radon Environmental Management Corp. www.radoncorp.com; Figure 25.13 A: © Scott Camazine/Photo Researchers, Inc.; Figure 25.13 B: © T. Youssef/Custom Medical Stock Photo; Figure 25.14: © Dr. Robert Friedland/SPL/Photo Researchers, Inc.; Figure 24.15: © Dr. Dennis Olson/Met Lab, Iowa State University, Ames, IA; Page 1134 (top): © Canadian Light Source Inc.; Page 1134 (middle): © TRIUMF, Canada's national laboratory for particle and nuclear physics; Page 1134 (bottom): THE CANADIAN PRESS/Fred Chartrand; Page 1135: LAWRENCE BERKELEY NATIONAL LABORATORY/SCIENCE PHOTO LIBRARY; Figure 25.19: © Atomic/Alamy; Figure 25.21: © Skyscan Photo Library/Alamy; Figure 25.22: © David Cooper/GetStock.com; Figure 25.23: © Dietmar Krause/Princeton Plasma Physics Lab.

INDEX

E

G

H

J

K

T